EARTH

Portrait of a Planet

FIFTH EDITION

EARTH
Portrait of a Planet

FIFTH EDITION

Stephen Marshak
UNIVERSITY OF ILLINOIS

W. W. NORTON & COMPANY
NEW YORK · LONDON

W. W. Norton & Company has been independent since its founding in 1923, when William Warder Norton and Mary D. Herter Norton first published lectures delivered at the People's Institute, the adult education division of New York City's Cooper Union. The firm soon expanded their program beyond the Institute, publishing books by celebrated academics from America and abroad. By mid-century, the two major pillars of Norton's publishing program—trade books and college texts—were firmly established. In the 1950s, the Norton family transferred control of the company to its employees, and today—with a staff of four hundred and a comparable number of trade, college, and professional titles published each year—W. W. Norton & Company stands as the largest and oldest publishing house owned wholly by its employees.

Editor: Eric Svendsen
Senior Project Editor: Thomas Foley
Associate Production Director: Benjamin Reynolds
Copy Editor: Jude Grant
Managing Editor, College: Marian Johnson
Managing Editor, College Digital Media: Kim Yi
Media Editors: Robin Kimball and Rob Bellinger
Associate Media Editor: Cailin Barrett-Bressack
Media Project Editor: Danielle Belfiore
Media Editorial Assistant: Victoria Reuter
Marketing Manager, Geology: Meredith Leo
Design Director: Rubina Yeh
Designer: Alexandra Charitan
Photography Editor: Stephanie Romeo
Photo Researcher: Fay Torresyap
Permissions Manager: Megan Jackson
Editorial Assistants: Rachel Goodman and Lindsey Thomas
Developmental Editor for elements of the Fifth Edition: Sunny Hwang
Developmental Editor for the First Edition: Susan Gaustad
Composition and page layout by Precision Graphics / Lachina
Illustrations by Precision Graphics / Lachina
Senior Artist at Precision Graphics / Lachina: Stan Maddock
Project Manager at Precision Graphics / Lachina: Kristina Seymour
Manufacturing by Courier—Kendallville, IN

Permission to use copyrighted material is included in the backmatter of this book.

Library of Congress Cataloging-in-Publication Data

Marshak, Stephen, 1955-
   Earth : portrait of a planet / Stephen Marshak, University of Illinois.
-- Fifth edition.
      pages cm
   Includes bibliographical references and index.
   ISBN 978-0-393-93750-3 (pbk. : alk. paper)
1. Geology--Textbooks.   I. Title.
   QE26.3.M36 2015
   550--dc23

                                    2014047609

W. W. Norton & Company, Inc., 500 Fifth Avenue, New York, NY 10110
wwnorton.com
W. W. Norton & Company Ltd., Castle House, 75/76 Wells Street, London W1T 3QT

1 2 3 4 5 6 7 8 9 0

Dedication

To Kathy, David, Emma, and Michelle

Brief Contents

Preface • xxi

See for Yourself: Using *Google Earth*™ • xxiv

PRELUDE And Just What Is Geology? • 1

PART I ## Our Island in Space

CHAPTER 1 Cosmology and the Birth of Earth • 12
CHAPTER 2 Journey to the Center of the Earth • 36
CHAPTER 3 Drifting Continents and Spreading Seas • 61
CHAPTER 4 The Way the Earth Works: Plate Tectonics • 86

PART II ## Earth Materials

CHAPTER 5 Patterns in Nature: Minerals • 116
INTERLUDE A Introducing Rocks • 141
CHAPTER 6 Up from the Inferno: Magma and Igneous Rocks • 152
INTERLUDE B A Surface Veneer: Sediments and Soils • 183
CHAPTER 7 Pages of Earth's Past: Sedimentary Rocks • 202
CHAPTER 8 Metamorphism: A Process of Change • 233
INTERLUDE C The Rock Cycle in the Earth System • 261

PART III ## Tectonic Activity of a Dynamic Planet

CHAPTER 9 The Wrath of Vulcan: Volcanic Eruptions • 272
CHAPTER 10 A Violent Pulse: Earthquakes • 312
INTERLUDE D The Earth's Interior, Revisited: Seismic Layering, Gravity, and the Magnetic Field • 359
CHAPTER 11 Crags, Cracks, and Crumples: Crustal Deformation and Mountain Building • 379

PART IV ## History before History

INTERLUDE E Memories of Past Life: Fossils and Evolution • 418
CHAPTER 12 Deep Time: How Old Is Old? • 434
CHAPTER 13 A Biography of Earth • 467

PART V ## Earth Resources

CHAPTER 14 Squeezing Power from a Stone: Energy Resources • 504
CHAPTER 15 Riches in Rock: Mineral Resources • 545

PART VI ## Processes and Problems at the Earth's Surface

INTERLUDE F Ever-Changing Landscapes and the Hydrologic Cycle • 572
CHAPTER 16 Unsafe Ground: Landslides and Other Mass Movements • 586
CHAPTER 17 Streams and Floods: The Geology of Running Water • 614
CHAPTER 18 Restless Realm: Oceans and Coasts • 655
CHAPTER 19 A Hidden Reserve: Groundwater • 694
CHAPTER 20 An Envelope of Gas: Earth's Atmosphere and Climate • 728
CHAPTER 21 Dry Regions: The Geology of Deserts • 768
CHAPTER 22 Amazing Ice: Glaciers and Ice Ages • 795
CHAPTER 23 Global Change in the Earth System • 838

Appendix: Additional Maps and Charts • A-1
Glossary • G-1
Credits • C-1
Index • I-1

Special Features

WHAT A GEOLOGIST SEES

The Concept of Transform Faulting, Fig. 4.13a • 99

Hot-Spot Volcano Track, Fig. 4.17d • 103

Rifting, Fig. 4.18d • 104

Basalt Sill in Antarctica, Fig. 6.12c • 165

Dike near Shiprock, NM, Fig. 6.13a • 166

New York Palisades, Ft6.1 • 182

Grand Canyon, Fig. 7.2c • 205

Crossbeds, Fig. 7.15d • 220

Deposits of an Ancient River Channel, Fig. 7.18e • 225

Displacement on the San Andreas Fault, Fig. 10.4a • 317

Displacement and Fault Zone, Fig. 11.10a • 392

Slip on a Thrust Fault, Fig. 11.10b • 392

The San Andreas Fault, Fig. 11.10c • 392

Horsts and Grabens, Fig. 11.13e • 394

Train of Folds, Fig. 11.15d • 396

Plunging Anticline, Fig. 11.15e • 396

Flexural-Slip Fold, Fig. 11.16a • 397

Passive Fold, Fig. 11.16b • 397

Ramp Anticline, Fig. 11.17d • 398

Slaty Cleavage, Fig. 11.18b • 399

Horizontal Sandstone Beds, Fig. 12.4c • 439

Chilled Margin, Fig. 12.4g • 440

Unconformity in Scotland, Fig. 12.8a • 443

Unconformity in a Roadcut, Fig. 12.8b • 443

New York Outcrop, Ft. 12.1 • 466

Missouri Outcrop, Ft. 12.2 • 466

Topographic Profile, Fig. BxF.1e • 575

The Oso, Washington Mudslide, Fig. 16.5b • 593

Drainage Basins on a Ridge, Fig. 17.5b • 619

Floodplain in Utah, Fig. 17.17c • 630

Desert Pavement, Arizona, Fig. 21.20b • 787

GEOLOGY AT A GLANCE

Forming the Planets and the Earth-Moon System, Chapter 1 • 30–31

The Earth from Surface to Center, Chapter 2 • 56–57

Magnetic Reversals and Marine Magnetic Anomalies, Chapter 3 • 80–81

The Theory of Plate Tectonics, Chapter 4 • 108–109

Formation of Igneous Rocks, Chapter 6 • 171

Weathering, Sediment, and Soil Production, Interlude B • 192–193

The Formation of Sedimentary Rocks, Chapter 7 • 222–223

Environments of Metamorphism, Chapter 8 • 254–255

Rock-Forming Environments and the Rock Cycle, Interlude C • 266–267

Volcanoes, Chapter 9 • 286–287

Faulting in the Crust, Chapter 10 • 320–321

The Collision of India with Asia, Chapter 11 • 402–403

The Record in Rocks: Reconstructing Geologic History, Chapter 12 • 454–455

The Earth has a History, Chapter 13 • 498–499

Power from the Earth, Chapter 14 • 536–537

Forming and Processing Earth's Mineral Resources, Chapter 15 • 562–563

The Hydrologic Cycle, Interlude F • 580–581

Mass Movement, Chapter 16 • 602–603

River Systems, Chapter 17 • 642–643

Oceans and Coasts, Chapter 18 • 684–685

Caves and Karst Landscapes, Chapter 19 • 724–725

The Desert Realm, Chapter 21 • 784–785

Glaciers and Glacial Landforms, Chapter 22 • 820–821

The Earth System, Chapter 23 • 840–841

Contents

Preface · xxi

See for Yourself: Using *Google Earth™* · xxiv

PRELUDE
And Just What Is Geology? · 1

P.1   In Search of Ideas · 2
P.2   The Nature of Geology · 3
P.3   Themes of This Book · 5

BOX P.1   Consider This
The Scientific Method · 8

PART I
Our Island in Space

CHAPTER 1
Cosmology and the Birth of Earth · 12

1.1   Introduction · 13
1.2   An Image of Our Universe · 13

BOX 1.1   Science Toolbox
Force and Energy · 16

BOX 1.2   Consider This
How Do We Know That the Earth Rotates? · 20

1.3   Forming the Universe · 21

BOX 1.3   Science Toolbox
Atoms, Molecules, and the Energy They Contain · 24

1.4   We Are All Made of Stardust · 26

Geology at a Glance
Forming the Planets and the Earth-Moon System · 30–31

End-of-chapter material · 33

CHAPTER 2

Journey to the Center of the Earth • 36

2.1 Introduction • 37
2.2 Welcome to the Neighborhood • 37

BOX 2.1 Consider This
Comets and Asteroids—The Other Stuff of the Solar System • 39

2.3 Basic Characteristics of the Earth • 43
2.4 How Do We Know That the Earth Has Layers? • 47
2.5 What Are the Layers Made of? • 49

BOX 2.2 Consider This
Meteorites: Clues to What's Inside • 50

2.6 The Lithosphere and Asthenosphere • 53

BOX 2.3 Science Toolbox
Heat and Heat Transfer • 54

Geology at a Glance
The Earth from Surface to Center • 56–57

End-of-chapter material • 58

CHAPTER 3

Drifting Continents and Spreading Seas • 61

3.1 Introduction • 62
3.2 Wegener's Evidence for Continental Drift • 63
3.3 Paleomagnetism—Proving Continents Move • 67

BOX 3.1 Consider This
Finding Paleopoles • 71

3.4 The Discovery of Seafloor Spreading • 72
3.5 Evidence for Seafloor Spreading • 76

Geology at a Glance
Magnetic Reversals and Marine Magnetic Anomalies • 80–81

End-of-chapter material • 83

CHAPTER 4

The Way the Earth Works: Plate Tectonics • 86

4.1 Introduction • 87
4.2 What Do We Mean by Plate Tectonics? • 87

BOX 4.1 Consider This
Archimedes' Principle of Buoyancy • 90

4.3 Divergent-Plate Boundaries and Seafloor Spreading • 92
4.4 Convergent-Plate Boundaries and Subduction • 96
4.5 Transform-Plate Boundaries • 98
4.6 Special Locations in the Plate Mosaic • 100
4.7 How Do Plate Boundaries Form, and How Do They Die? • 102
4.8 Moving Plates • 106

Geology at a Glance
The Theory of Plate Tectonics • 108–109

End-of-chapter material • 112

PART II
Earth Materials

CHAPTER 5

Patterns in Nature: Minerals • 116

5.1 Introduction • 117
5.2 What Is a Mineral? • 118

BOX 5.1 Science Toolbox
Some Basic Concepts from Chemistry—A Quick Review • 120

5.3 Beauty in Patterns: Crystals and Their Structure • 122
5.4 How Can You Tell One Mineral from Another? • 127
5.5 Organizing Knowledge: Mineral Classification • 129

BOX 5.2 Consider This
Asbestos and Health: When Crystal Habit Matters! • 132

5.6 Something Precious—Gems! • 134

BOX 5.3 Consider This
Where Do Diamonds Come From? • 135

End-of-chapter material • 138

INTERLUDE A

Introducing Rocks • 141

A.1 Introduction • 141
A.2 What Is Rock? • 142
A.3 The Basis of Rock Classification • 144
A.4 Studying Rock • 147

CHAPTER 6

Up from the Inferno: Magma and Igneous Rocks • 152

6.1 Introduction • 153
6.2 Why Do Melts Form? • 153
6.3 What Is Molten Rock Made of? • 158
6.4 Movement and Solidification of Molten Rock • 159
6.5 Comparing Extrusive and Intrusive Environments • 162

BOX 6.1 Consider This
Bowen's Reaction Series • 164

6.6 How Do You Describe an Igneous Rock? • 166

Geology at a Glance
Formation of Igneous Rocks • 171

6.7   Plate Tectonic Context of Igneous Activity · 174

End-of-chapter material · 180

INTERLUDE B
A Surface Veneer: Sediments and Soils • 183

B.1   Introduction · 183
B.2   Weathering: Forming Sediment · 185

Geology at a Glance
Weathering, Sediment, and Soil Production · 192–193

B.3   Soil · 195

CHAPTER 7
Pages of Earth's Past: Sedimentary Rocks • 202

7.1   Introduction · 203
7.2   Classes of Sedimentary Rocks · 203
7.3   Sedimentary Structures · 215
7.4   How Do We Recognize Depositional Environments? · 220

Geology at a Glance
The Formation of Sedimentary Rocks · 222–223

7.5   Sedimentary Basins · 228

End-of-chapter material 230

CHAPTER 8
Metamorphism: A Process of Change • 233

8.1   Introduction · 234
8.2   Consequences and Causes of Metamorphism · 235
8.3   Types of Metamorphic Rocks · 241
8.4   Defining Metamorphic Intensity · 245

BOX 8.1   Consider This
Metamorphic Facies · 248

8.5   Where Does Metamorphism Occur? · 249

BOX 8.2   Consider This
Pottery Making—An Analog for Thermal Metamorphism · 252

Geology at a Glance
Environments of Metamorphism · 254–255

End-of-chapter material · 258

INTERLUDE C
The Rock Cycle in the Earth System • 261

C.1   Introduction · 262
C.2   Pathways through the Rock Cycle · 262

C.3   A Case Study of the Rock Cycle • 263
C.4   Cycles of the Earth System • 265

Geology at a Glance
Rock-Forming Environments and the Rock Cycle • 266-267

PART III
Tectonic Activity of a Dynamic Planet

CHAPTER 9

The Wrath of Vulcan: Volcanic Eruptions • 272

9.1   Introduction • 273
9.2   The Products of Volcanic Eruptions • 275
9.3   Structure and Eruptive Style • 282

Geology at a Glance
Volcanoes • 286–287

BOX 9.1   Consider This
Volcanic Explosions to Remember • 290

9.4   Geologic Settings of Volcanism • 292
9.5   Beware: Volcanoes Are Hazards! • 298
9.6   Protection from Vulcan's Wrath • 302
9.7   Effect of Volcanoes on Climate and Civilization • 305
9.8   Volcanoes on Other Planets • 309

End-of-chapter material • 309

CHAPTER 10

A Violent Pulse: Earthquakes • 312

10.1   Introduction • 313
10.2   What Causes Earthquakes? • 315

Geology at a Glance
Faulting in the Crust • 320–321

10.3   Seismic Waves and Their Measurement • 323
10.4   Defining the "Size" of Earthquakes • 328
10.5   Where and Why Do Earthquakes Occur? • 332
10.6   How Do Earthquakes Cause Damage? • 338

BOX 10.1   Consider This
The 2010 Haiti Catastrophe • 348

10.7   Can We Predict the "Big One"? • 350
10.8   Earthquake Engineering and Zoning • 354

BOX 10.2   Consider This
When Earthquake Waves Resonate—Beware! • 355

End-of-chapter material • 356

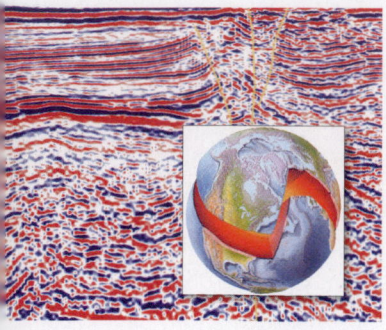

INTERLUDE D

The Earth's Interior, Revisited:
Seismic Layering, Gravity, and the Magnetic Field • 359

D.1   Introduction • 360
D.2   The Basis for Seismic Study of the Interior • 360
D.3   Results from Seismic Study of Earth's Interior • 362

BOX D.1   Consider This
Resolving the Details of Earth's Interior with EarthScope • 370

D.4   Earth's Gravity • 372
D.5   Earth's Magnetic Field, Revisited • 375

CHAPTER 11

Crags, Cracks, and Crumples:
Crustal Deformation and Mountain Building • 379

11.1   Introduction • 380
11.2   Rock Deformation in the Earth's Crust • 382
11.3   Brittle Structures • 387

BOX 11.1   Consider This
Describing the Orientation of Geologic Structures • 388

11.4   Folds and Foliations • 393
11.5   Causes of Mountain Building • 400

Geology at a Glance
The Collision of India with Asia • 402–403

11.6   Mountain Topography • 405
11.7   Basins and Domes in Cratons • 409
11.8   Life Story of a Mountain Range: A Case Study • 412

End-of-chapter material • 413

PART IV

History before History

INTERLUDE E

Memories of Past Life: Fossils and Evolution • 418

E.1   The Discovery of Fossils • 418
E.2   Fossilization • 420
E.3   Taxonomy and Identification • 425
E.4   The Fossil Record • 428
E.5   Evolution and Extinction • 430

CHAPTER 12

Deep Time: How Old Is Old? • 434

12.1   Introduction • 435

BOX 12.1   Consider This
Time: A Human Obsession • 436

12.2   The Concept of Geologic Time • 436
12.3   Geologic Principles Used for Defining Relative Age • 438
12.4   Unconformities: Gaps in the Record • 442
12.5   Stratigraphic Formations and Their Correlation • 445
12.6   The Geologic Column • 449
12.7   How Do We Determine Numerical Ages? • 453

Geology at a Glance
The Record in Rocks: Reconstructing Geologic History • 454–455

BOX 12.1   Consider This
Carbon-14 Dating • 457

12.8   Numerical Ages and Geologic Time • 460

End-of-chapter material • 464

CHAPTER 13

A Biography of Earth • 467

13.1   Introduction • 468
13.2   Methods for Studying the Past • 468
13.3   The Hadean and Before • 470
13.4   The Archean Eon: Birth of Continents and Life • 472
13.5   The Proterozoic Eon: The Earth in Transition • 476

BOX 13.1   Consider This
Where Was the Cradle of Life? • 477

BOX 13.2   Consider This
The Evolution of Atmospheric Oxygen • 481

13.6   The Paleozoic Era: Continents Reassemble and Life Gets Complex • 482

BOX 13.3   Consider This
Stratigraphic Sequences and Sea-Level Change • 486

13.7   The Mesozoic Era: When Dinosaurs Ruled • 487
13.8   The Cenozoic Era: The Modern World Comes to Be • 495

Geology at a Glance
The Earth has a History • 498–499

End-of-chapter material • 500

PART V
Earth Resources

CHAPTER 14

Squeezing Power from a Stone: Energy Resources • 504

14.1 Introduction • 505
14.2 Sources of Energy in the Earth System • 507
14.3 Introducing Hydrocarbon Resources • 508
14.4 Conventional Hydrocarbon Systems • 510

BOX 14.1 Consider This
Types of Oil and Gas Traps • 514

14.5 Unconventional Hydrocarbon Reserves • 517

BOX 14.2 Consider This
Hydrofracturing (*Fracking*) • 522

14.6 Coal: Energy from the Swamps of the Past • 524
14.7 Nuclear Power • 529
14.8 Other Energy Sources • 531
14.9 Energy Choices, Energy Problems • 535

Geology at a Glance
Power from the Earth • 536–537

BOX 14.3 Consider This
Offshore Drilling and the Deepwater Horizon Disaster • 540

End-of-chapter material • 542

CHAPTER 15

Riches in Rock: Mineral Resources • 545

15.1 Introduction • 546
15.2 Metals and Their Discovery • 547
15.3 Ores, Ore Minerals, and Ore Deposits • 549
15.4 Ore-Mineral Exploration and Production • 555
15.5 Nonmetallic Mineral Resources • 557

BOX 15.1 Consider This
The Amazing Chilean Mine Rescue of 2010 • 558

BOX 15.2 Consider This
The Sidewalks of New York • 560

Geology at a Glance
Forming and Processing Earth's Mineral Resources • 562–563

15.6 Global Mineral Needs • 564

End-of-chapter material • 567

PART VI
Processes and Problems at the Earth's Surface

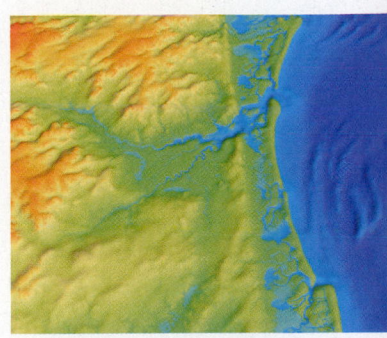

INTERLUDE F
Ever-Changing Landscapes and the Hydrologic Cycle • 572

F.1   Introduction • 572
F.2   Shaping the Earth's Surface • 574

BOX F.1   Consider This
Topographic Maps and Profiles • 575

F.3   Factors Controlling Landscape Development • 577
F.4   The Hydrologic Cycle • 579

Geology at a Glance
The Hydrologic Cycle • 580–581

F.5   Landscapes of Other Planets • 582

BOX F.2   Consider This
Water on Mars? • 584

CHAPTER 16
Unsafe Ground: Landslides and Other Mass Movements • 586

16.1   Introduction • 587
16.2   Types of Mass Movement • 588

BOX 16.1   Consider This
What Goes Up Must Come Down • 592

16.3   Why Do Mass Movements Occur? • 598

BOX 16.2   Consider This
The Storegga Slide and North Sea Tsunamis • 599

Geology at a Glance
Mass Movement • 602–603

16.4   Where Do Mass Movements Occur? • 606
16.5   How Can We Protect against Mass-Movement Disasters? • 608

End-of-chapter material • 612

CHAPTER 17
Streams and Floods: The Geology of Running Water • 614

17.1   Introduction • 615
17.2   Draining the Land • 615
17.3   Describing Flow in Streams: Discharge and Turbulence 621
17.4   The Work of Running Water • 623

17.5 How Do Streams Change along Their Length? • 626
17.6 Streams and Their Deposits in the Landscape • 628
17.7 The Evolution of Drainage • 636
17.8 Raging Waters • 640

Geology at a Glance
River Systems • 642–643

BOX 17.1 Consider This
The Johnstown Flood of 1889 • 645

17.9 Vanishing Rivers • 650

BOX 17.2 Consider This
Calculating the Threat Posed by Flooding • 651

End-of-chapter material • 652

CHAPTER 18

Restless Realm: Oceans and Coasts • 655

18.1 Introduction • 656
18.2 Landscapes beneath the Sea • 657
18.3 Ocean Water and Currents • 662

BOX 18.1 Consider This
The Coriolis Effect • 666

18.4 Tides • 667

BOX 18.2 Consider This
The Forces Causing Tides • 670

18.5 Wave Action • 672
18.6 Where Land Meets Sea: Coastal Landforms • 675
18.7 Causes of Coastal Variability • 683

Geology at a Glance
Oceans and Coasts • 684–685

18.8 Coastal Problems and Solutions • 688

End-of-chapter material • 692

CHAPTER 19

A Hidden Reserve: Groundwater • 694

19.1 Introduction • 695
19.2 Where Does Groundwater Reside? • 696
19.3 Characteristics of the Water Table • 701
19.4 Groundwater Flow • 703
19.5 Tapping Groundwater Supplies • 705

BOX 19.1 Consider This
Darcy's Law for Groundwater Flow • 706

BOX 19.2 Consider This
Oases • 709

19.6 Hot Springs and Geysers • 710
19.7 Groundwater Problems • 713
19.8 Caves and Karst • 719

Geology at a Glance
Caves and Karst Landscapes • 724–725

End-of-chapter material • 726

CHAPTER 20

An Envelope of Gas:
Earth's Atmosphere and Climate • 728

20.1 Introduction • 729
20.2 The Formation of the Atmosphere • 730
20.3 General Atmospheric Characteristics • 732

BOX 20.1 Consider This
Air Pollution • 733

BOX 20.2 Consider This
Why Is the Sky Blue • 734

20.4 Atmospheric Layers • 736
20.5 Wind and Global Circulation in the Atmosphere • 738

BOX 20.3 Consider This
The Earth's Tilt: The Cause of Seasons • 742

20.6 Weather and Its Causes • 744
20.7 Storms: Nature's Fury • 750
20.8 Global Climate • 761

End-of-chapter material • 766

CHAPTER 21

Dry Regions: The Geology of Deserts • 768

21.1 Introduction • 769
21.2 The Nature and Location of Deserts • 769
21.3 Producing Desert Landscapes • 773
21.4 Deposition in Deserts • 778
21.5 Desert Landforms and Life • 779

Geology at a Glance
The Desert Realm • 784–785

BOX 21.1 Consider This
Uluru (Ayers Rock) • 786

21.6 Desert Problems • 789

End-of-chapter material • 793

CHAPTER 22

Amazing Ice: Glaciers and Ice Ages • 795

22.1   Introduction • 796
22.2   Ice and the Nature of Glaciers • 797

BOX 22.1   Consider This
Polar Ice Caps on Mars • 802

22.3   Carving and Carrying by Ice • 808
22.4   Deposition Associated with Glaciation • 813
22.5   Other Consequences of Continental Glaciation • 819

Geology at a Glance
Glaciers and Glacial Landforms • 820–821

22.6   The Pleistocene Ice Age • 826

BOX 22.2   Consider This
So You Want to See Glaciation? • 827

22.7   The Causes of Ice Ages • 831

End-of-chapter material • 836

CHAPTER 23

Global Change in the Earth System • 838

23.1   Introduction • 839

Geology at a Glance
The Earth System • 840–841

23.2   Unidirectional Changes • 842
23.3   Cyclic Changes • 844
23.4   Global Climate Change • 847

BOX 23.1   Consider This
Global Climate Change and the Birth of Legends • 852

BOX 23.2   Consider This
Goldilocks and the Faint Young Sun • 854

23.5   Human Impact on Land and Life • 858
23.6   Recent Climate Change • 862
23.7   The Future of the Earth • 873

End-of-chapter material • 874

Appendix: Additional Maps and Charts • A-1

Glossary • G-1

Credits • C-1

Index • I-1

Preface

Narrative Themes

Why do earthquakes, volcanoes, floods, and landslides happen? What causes mountains to rise? How do beautiful landscapes develop? How have climate and life changed through time? When did the Earth form, and by what process? Where do we dig to find valuable metals, and where do we drill to find oil? Does sea level change? Do continents move? The study of geology addresses these important questions and many more. But from the birth of the discipline, in the late 18th century, until the mid-20th century, geologists considered each question largely in isolation, without pondering its relation to the others. This approach changed, beginning in the 1960s, in response to the formulation of two paradigm-shifting ideas that have unified thinking about the Earth and its features. The first idea, called the *theory of plate tectonics*, states that the Earth's outer shell, rather than being static, consists of discrete plates that slowly move, relative to each other, so that the map of our planet continuously changes. Plate interactions cause earthquakes and volcanoes, build mountains, provide gases that make up the atmosphere, and affect the distribution of life on Earth. The second idea, the *Earth System perspective*, emphasizes that our planet's water, land, atmosphere, and living inhabitants are dynamically interconnected, so that materials constantly cycle among various living and nonliving reservoirs on, above, and within the planet. In the context of this idea, we have come to realize that the history of life is intimately linked to the history of the physical Earth, and vice versa.

Earth: Portrait of a Planet, Fifth Edition, is an introduction to the study of our planet that uses the theory of plate tectonics as well as the Earth System perspective throughout, to weave together a number of narrative themes, including:

1. The solid Earth, the oceans, the atmosphere, and life interact in complex ways, yielding a planet that is unique in the Solar System.

2. Most geologic processes involve the interactions of plates, pieces of the outer, relatively rigid shell of the Earth.

3. The Earth is a planet formed, like other planets, from dust and gas. But, in contrast to other planets, the Earth is a dynamic place where new geologic features continue to form and old ones continue to be destroyed.

4. The Earth is very old—indeed, about 4.54 billion years have passed since its birth. During this time, the map of the planet and its surface features have changed, and life has evolved.

5. Internal processes (driven by Earth's internal heat) and external processes (driven by heat from the Sun) interact at the Earth's surface to produce complex landscapes.

6. Geologic knowledge can help society understand, and perhaps avoid or reduce, the danger of natural hazards, such as earthquakes, volcanoes, landslides, and floods.

7. Energy and mineral resources come from the Earth and are formed by geologic phenomena. Geologic study can help locate these resources and mitigate the consequences of their use.

8. Geology is a science, and the ideas of science come from observation, calculation, and experiment. Thus, people make scientific discoveries, and scientific understanding advances over time.

9. Geology utilizes ideas from physics, chemistry, and biology, so the study of geology provides an excellent means to improve science literacy overall.

These narrative themes serve as the take-home message of the book, a message that students hopefully will remember long after they finish their introductory geology course. In effect, they provide a mental framework on which students can organize and connect ideas, and develop a modern, coherent image of our planet.

Pedagogical Approach

Educational research demonstrates that students learn best when they actively engage with a combination of narrative text and narrative art. Some students respond more to the words of a textbook, which help to organize information, provide answers to questions, fill in the essential steps that link ideas together, and help a student develop a context for understanding ideas. Some students respond more to narrative art—art designed to tell a story—for visual images help students comprehend and remember processes. And some respond to question-and-answer-based active learning, an approach where students can, in effect, "practice" their knowledge. *Earth: Portrait of a Planet*, Fifth Edition, provides all three of these learning tools. The text has been crafted to be engaging, the art has been configured to tell a story, the chapters are laid out to help students internalize key principles, and the on-line activities have been designed to both engage students and provide active feedback. As before, the book's narrative doesn't just provide a dry statement of facts, but rather, it provides the story behind the story—meaning the reasoning and observation that led to our current understanding, as well as an explanation of the processes that cause a particular geological phenomenon.

Each chapter starts with a list of *Learning Objectives* that frames the most important pedagogical goals for each chapter. *Take-Home Message* panels, which include both a brief summary and a key question, appear at the end of each section to help students solidify key themes before proceeding to the next section. Throughout the chapter, *Did You Ever Wonder?* questions prompt students with real-life questions they may have already thought about—answers to these questions occur in the nearby text. *See for Yourself* panels guide students to key examples of spectacular geologic features, using the power of *Google Earth*™. They allow students to apply their newly acquired knowledge to the interpretation of real-world examples. Each chapter then concludes with a chapter summary that reinforces understanding and provides a concise study tool at the same time. *Review Questions* at the end of each chapter include two parts: the first addresses basic concepts, as defined by Bloom's Taxonomy; and the second, labeled *On Further Thought*, stimulates critical thinking opportunities that invite students to think beyond the basics.

To enhance active-learning opportunities, *SmartWork Online Homework* has been specifically developed for *Earth: Portrait of a Planet*, Fifth Edition. In addition to word questions, *SmartWork* also offers students visual drag and drop questions and figure-labeling exercises, all of which come with detailed feedback. *SmartWork* also boasts strong visual features with questions based on videos and vivid animations that display geologic processes.

Organization

The topics covered in this book have been arranged so that students can build their knowledge of geology on a foundation of overarching principles. Thus, the book starts with cosmology and the formation of the Earth, and then introduces the architecture of our planet, from surface to center. With this basic background, students are prepared to delve into plate tectonics theory. Plate tectonics appears early in the book, so that students can relate the content of subsequent chapters to the theory. Knowledge of plate tectonics, for example, helps students understand the suite of chapters on minerals, rocks, and the rock cycle. Knowledge of plate tectonics and rocks together, in turn, provides a basis for studying volcanoes, earthquakes, and mountains. And with this background, students are prepared to see how the map of the Earth has changed through the vast expanse of geologic time, and how energy and mineral resources have developed. The book's final chapters address processes and problems occurring at or near the Earth's surface, from the unstable slopes of hills, down the course of rivers, to the shores of the sea and beyond. This part concludes with a topic of growing concern in society—global change, particularly climate change.

In addition to numbered chapters, the book contains several *Interludes*. These are, in effect, "mini-chapters" in that they focus on topics that are self-contained but are not broad enough to require an entire chapter. By placing selected topics in interludes, we can keep chapters reasonable in length, and can provide additional flexibility in sequencing topics within a course.

Although the sequence of chapters and interludes was chosen for a reason, this book is designed to be flexible enough for instructors to choose their own strategies for teaching geology. The individual topics are so interrelated that there is not always a single best way to order them. Thus, each chapter is self-contained, reiterating relevant material where necessary. For example, if instructors prefer to introduce minerals and rocks before plate tectonics, they simply need to reorder the reading assignments. A low-cost, loose-leaf version of the book allows instructors to have students purchase only the chapters that they need.

We have used a different approach in highlighting terminology in *Earth: Portrait of a Planet*, Fifth Edition. Terminology, the basic vocabulary of a subject, serves an important purpose in simplifying the discussion of topics. For example, once students understand the formal definition of a mineral, the term can be used again in subsequent discussion without further explanation or redundancy. Too much new vocabulary, however, can be overwhelming. So we have tried to keep the book's key terms (set in boldface and referenced at the end of each chapter for studying purposes) to a minimum. But, since the field of geology has many important terms, we have also set other, less significant but still useful, terms in italic when first presented, to provide additional visual guidance for students. As in previous editions, we take care not to use vocabulary until it has been completely introduced and defined.

Special Features of this Edition

Earth: Portrait of a Planet, Fifth Edition, contains a number of new or revised features that distinguish it from all competing texts.

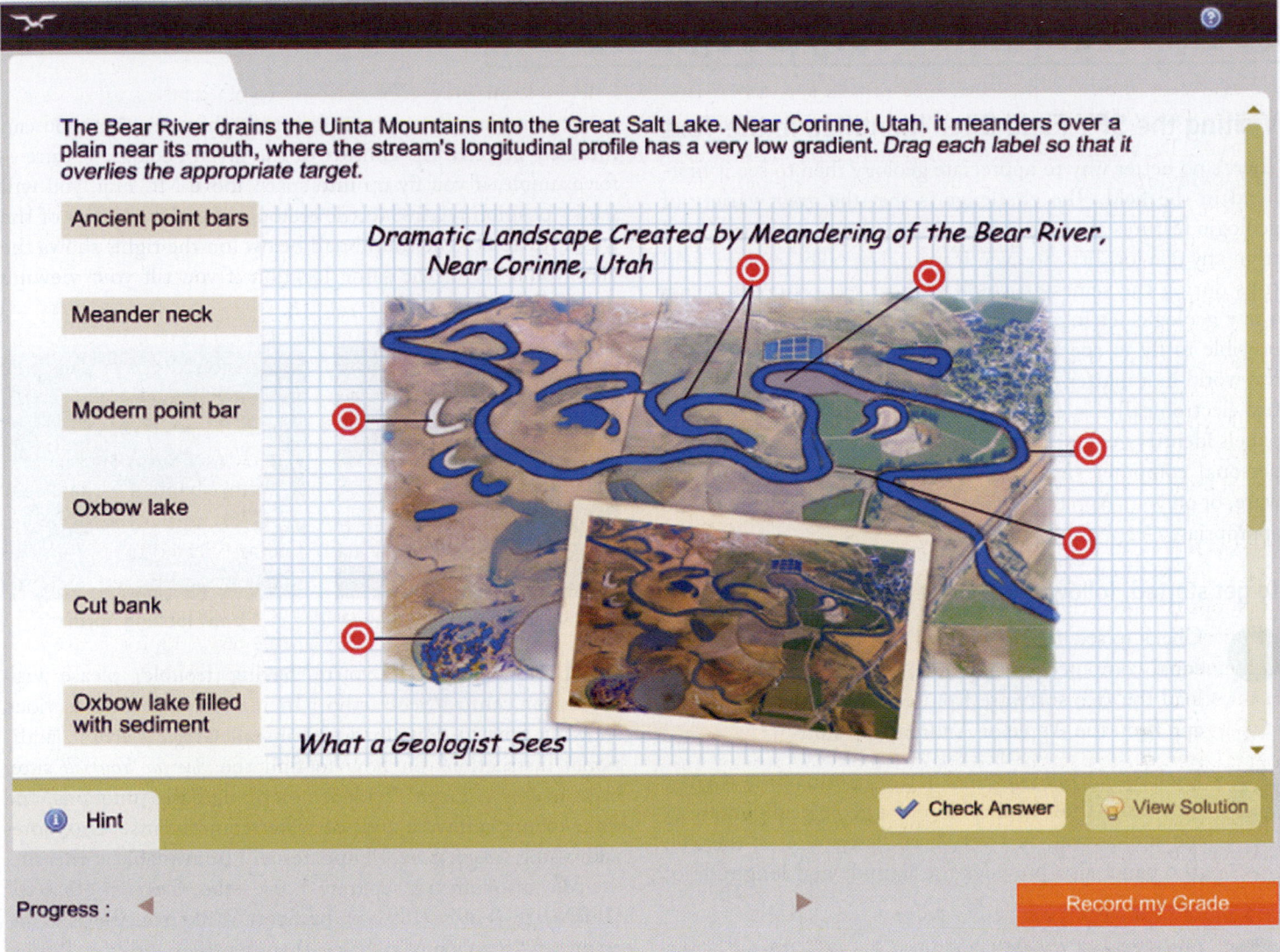

WHAT A GEOLOGIST SEES figures created just for *SmartWork*.

Narrative Art, *What a Geologist Sees*, and *See for Yourself*

It's difficult to understand many features of the Earth System without being able to see them. To help students visualize these and other features, this book is lavishly illustrated with figures that try to give a realistic context for the particular feature, without overwhelming students with too much extraneous detail. The talented artists who worked on the book have used the latest computer graphics software, resulting in the most sophisticated pedagogical art ever provided by a geoscience text. Many figures have been updated with an eye toward improving the 3-D visualization skills of students. They have also been reconfigured to make them more friendly and intuitive. In addition to the art, the book also boasts over 1,000 stunning photographs from all around the world. Many of the photographs were taken by the author, specifically to illustrate the exact concept under discussion. Where appropriate, photographs are accompanied by annotated sketches named *What a Geologist Sees*. These figures allow students to see how geologists perceive the world around them and to encourage students to start thinking like geologists.

Throughout the book, drawings and photographs have been integrated into *narrative art*, which has been laid out, labeled, and annotated to tell a story—the figures are drawn to teach! Subcaptions are positioned adjacent to relevant parts of each figure, labels point out key features, and balloons provide important annotation. Subparts are arranged to convey time progression, where relevant. The color schemes of drawings have been tied to those of relevant photos, so that students can easily relate features in the drawings to those in the photos. Further, all the art in this edition has been reworked to achieve a consistent style, using standard colors and textures for similar features across the book.

Google Earth™ provides an amazing opportunity for students to visit and tour important geologic sites wherever they occur. Throughout the book, we provide *See for Yourself* panels, which provide coordinates and descriptions of geologic

See for Yourself: Using *Google Earth*™

Visiting the SFY Field Sites Identified in the Text

There's no better way to appreciate geology then to see it first-hand in the field. The challenge is that the great variety of geologic features that we discuss in this book can't be visited from any one locality. So even if your class takes geology field trips during the semester, you'll at most see examples of just a few geologic settings. Fortunately, *Google Earth*™ makes it possible to fly to spectacular geologic field sites anywhere in the world in a matter of seconds—you can take a virtual field trip electronically. In each chapter in this book, *See for Yourself* panels identify geologic sites that you can explore on your own personal computer (Mac or PC) using *Google Earth*™ software, or on your Apple/Android smartphone or tablet with the appropriate *Google Earth*™ app.

To get started, follow these three simple steps:

1. Check to see if *Google Earth*™ is installed on your personal computer, smartphone, or tablet. If not, download the free software from **earth.google.com** or the app from the Apple or Android app store.

2. Each *See for Yourself* panel in the margin of the chapter provides a thumbnail photo of a geologically interesting site, as well as a very brief description of the site. The panel also provides the latitude and longitude of the site.

3. Open *Google Earth*™ and enter the coordinates of the site in the search window. As an example, let's find Mt. Fuji, a beautiful volcano in Japan. We note that the coordinates in the *See for Yourself* panel are as follows:

Latitude	35°21'41.78"N
Longitude	138°43'50.74"E

Type these coordinates into the search window of *Google Earth*™ as:

35 21 41.78N, 138 43 50.74E

with the degree, minute, and second symbols left blank. When you click enter or return, your device will bring you to the viewpoint right above Mt. Fuji, as illustrated by the following thumbnails.

Google Earth™ contains many built-in and easy-to-use tools that allow you to vary the elevation, tilt, orientation, and position of your viewpoint, so that you can tour around the feature, see it from many different perspectives, and thus develop a three-dimensional sense of the feature. In the case of Mt. Fuji, you'll be able to see its cone-like shape and the crater at its top. By zooming out to higher elevation, you can instantly perceive the context of the given geologic feature—for example, if you fly up into space above Mt. Fuji, you will see its position relative to the tectonic plate boundaries of the western Pacific. The thumbnail below (on the right) shows the view you'll see of the same location if you tilt your viewing direction and look north.

View looking down.   View looking north.

Need More Help? If you're having trouble, please visit wwnorton.com/rd/SeeEarth5. There, you will find a video showing how to download and install *Google Earth*™, additional instructions on how to find the *See for Yourself* sites, links to *Google Earth*™ videos describing basic functions, and links to any hardware and software requirements. Also, notes addressing *Google Earth*™ updates will be available at this site.

We also offer a separate book—the *Geotours Workbook* (ISBN 978-0-393-91891-5), by Scott Wilkerson, Beth Wilkerson, and Stephen Marshak—that identifies additional interesting geologic sites to visit, provides active-learning exercises linked to the sites, and explains how you can create your own virtual field trips.

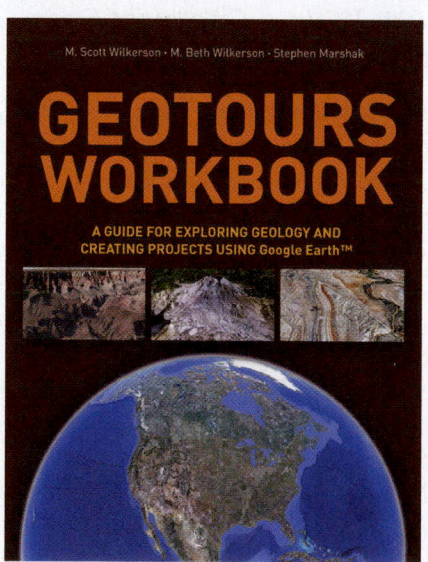

M. Scott Wilkerson · M. Beth Wilkerson · Stephen Marshak

GEOTOURS WORKBOOK

A GUIDE FOR EXPLORING GEOLOGY AND CREATING PROJECTS USING *Google Earth*™

features that students can visit at the touch of a finger, or the click of a mouse. The adjacent box provides a quick guide for using these panels.

Featured Paintings—*Geology at a Glance*

In addition to individual figures, British artist Gary Hincks has created spectacular two-page annotated paintings for each chapter. These paintings, called *Geology at a Glance*, integrate key concepts introduced in the chapters, visually emphasize the relationships between components of the Earth System, and allow students a way to review a subject . . . at a glance. The Fifth Edition includes a brand-new painting, illustrating the Earth's history, in Chapter 13.

Enhanced Coverage of Current Topics

To ensure that *Earth: Portrait of a Planet*, Fifth Edition, reflects the latest research discoveries and helps students understand geologic events that have been featured in current news, we have updated many topics throughout the book. For example: the energy chapter has been substantially revised to clarify the impact of the switch to unconventional gas and oil reserves; new geophysical data from the EarthScope project have been incorporated into discussions of the Earth's interior; discussions of Earth history incorporate the latest revisions to the geologic time scale; and data from the latest IPCC report contribute to the book's treatment of climate change. *Earth: Portrait of a Planet*, Fifth Edition, also covers the lessons learned from recent natural disasters such as Hurricane Sandy, Typhoon Haiyan, the Washington landslide, and the Tōhoku tsunami.

This book addresses geology's practical applications in several chapters. Students will learn about such topics as energy resources, mineral resources, global change, and mass wasting. Further, chapters on earthquakes, volcanoes, and landscapes highlight geologic hazards. And students are encouraged to apply their geologic understanding to environmental issues, where relevant. *Science and Society* features, updated by Geoffrey Cook, University of California, San Diego and Heather M. Cook, California State University, San Marcos are available through Norton Coursepacks to instructors. These features further challenge students to use material learned in class to interpret news articles and publicly available geologic data.

Supplementary Materials

SmartWork

The student experience of reading *Earth: Portrait of a Planet*, Fifth Edition, can be enhanced significantly through the use of *SmartWork Online Homework* for Geology. The *SmartWork* system features visual assignments that provide students with answer-specific feedback. Students get the coaching they need to work through the assignments, while instructors get real-time assessment of student progress with automatic grading and item analysis. Image-based drag-and-drop and labeling questions make use of carefully developed images. Also available are additional *What a Geologist Sees* figures created exclusively for *SmartWork*; additional video, animation, and conceptual questions that challenge students to apply their understanding of important concepts; reading quizzes for each chapter; and *Geotour*-guided inquiry activities using *Google Earth*™. Designed to be intuitive and easy to use (for both students and instructors), *SmartWork* makes it a snap to assign, assess, and report on student performance, and to keep the class on track.

Tablet- and Mobile-ready E-book

Earth Portrait of a Planet, Fifth Edition is available in a new format perfect for tablets and other mobile devices. Within the ebook, art expands for a closer look, links send you to geologic locations in *Google Maps*™, animations and videos link out from each chapter, and pop-up key terms provide a quick review. It's also easy to highlight, take notes, and search the text

The *Geotours Workbook*

Created by Scott Wilkerson, Beth Wilkerson, and Stephen Marshak, the *Geotours Workbook* provides active-learning opportunities that take students on virtual field trips to see outstanding examples of geology at localities around the world, using *Google Earth*™. Arranged by topic, questions in the *Geotours Workbook* have been designed for auto-grading, and are available as worksheets both in print format (these come free with the book and include complete user instructions and advanced instruction), or electronically with automatic grading through *SmartWork* or your campus LMS. The *Geotours Workbook* also provides instructions that will allow instructors or students to make their own geotours. Request a sample copy to preview each worksheet.

Art Files and PowerPoints

The publisher provides a variety of electronic presentations of art and photographs in the book to enhance the classroom experience. These include:

- *Enhanced Art PowerPoints*—Designed for instant classroom use, these slides utilize photographs and line art from the book in a form that has been optimized for use in the PowerPoint environment. The art has been relabeled and resized for projection formats. Enhanced Art PowerPoints also include supplemental photographs. For the Fifth Edition, both enhanced and lecture bullet PowerPoints were revised by Jennifer Sliko of Penn State University, Harrisburg.

- *Lecture Bullet PowerPoints*—These slides include both art and bulleted text for direct use either in lectures or as student handouts.

- *Labeled and Unlabeled Art PowerPoints*—These include all art from the book, formatted as JPEGs, pre-pasted into PowerPoints. We offer one set in which all labeling has been stripped, and one set in which the labeling has been retained.

- *Labeled and Unlabeled Art JPEGs*—We provide a complete file of individual JPEGs for all art and photographs presented in the book.

- *Monthly Update PowerPoints*—W. W. Norton & Company, Inc., offers a monthly update service that provides new PowerPoint slides, with instructor support, covering recent geologic events. Monthly updates are authored by Rich Oches of Bentley University.

New Animations and Videos

Earth: Portrait of a Planet, Fifth Edition, provides a rich collection of animations to illustrate geologic processes all created in a consistent style and with a 3-D perspective. The set includes additional new animations, developed by Alex Glass of Duke University, that allow you to control variables. Further, the new edition comes with a new set of Narrated Figure Videos, in which the author explains core concepts and figures from each chapter. Videos are free, require no special hardware, and are available in coursepacks, with links you can stream to your

ANIMATIONS illustrate geologic processes.

class, or for students to use. In addition, Melissa Hudley of the University of North Carolina–Chapel Hill, Heather Lehto of Angelo State University, and Meghan Lindsey of the University of South Florida developed over 50 videos of geologic processes and topics for our coursepacks and instructor support website. All animations and videos are ready to go and perfect for streaming in the classroom or for online use.

Instructor's Manual and Test Bank

The *Instructor's Manual*, prepared by John Werner of Seminole State College of Florida, is designed to help instructors prepare lectures and exams. It contains detailed *Learning Objectives*, *Chapter Summaries*, and complete answers to the end-of-chapter *Review* and *On Further Thought* questions for every chapter and interlude.

The *Test Bank*, written by Scott Marshall of Appalachian State University, Meredith Denton-Hedrick of Austin Community College, and Heather Lehto of University of South Florida, has been revised not only to correlate to this new Edition, but to provide greater, more rounded assessment than ever before. Expert accuracy checkers Geoffrey Cook of UC San Diego, and Mark Feigenson of Rutgers University have ensured that every question we've included in the *Test Bank* is scientifically reliable and truly tests students' understanding of the most important topics in each chapter, so that the questions can be assigned with confidence.

Instructor's Website—wwnorton.com/instructors

The Instructor's Website provides online access to a rich array of resources: the *Test Bank*, the *Instructor's Manual*, Power-Points, JPEGs, *Google Earth*™ file of sites from the book, art from the text, animations, new Narrated Figure and additional videos, and LMS-ready content.

Norton Media Library Instructor's DVD-ROM

Supporting the instructor's website, the instructor's DVD offers many of the Fifth Edition's multimedia resources, all structured around the text in a convenient package. Contact your Norton representative to obtain a copy.

Coursepacks

Available at no cost to professors or students, *Norton Course-packs* bring high-quality Norton digital media into a new or existing online course. *Coursepacks* contain ready-made content

for your campus LMS. For *Earth: Portrait of a Planet*, Fifth Edition, content includes new Narrated Figure Videos keyed to core figures in each chapter, the *Test Bank*, reading quizzes, new visual questions for each chapter, completely revised quiz questions by Cynthia Liutkus-Pierce of Appalachian State University, Geotour questions, animations, streaming video, *vocabulary flashcards*, *Science and Society* features, and links to the ebook.

Student Site—wwnorton.com/rd/SeeEarth5

Free and open for students, the *Student Site* provides Marshak's new Narrated Figure Videos, *vocabulary flashcards*, information on how to best utilize the *Google Earth*™ materials, as well as five new features that were developed by Rick Oches of Bentley University for this edition. The *Student Site* also includes a video designed to help with start-up, as well as a downloadable file of all the *See for Yourself* sites.

Acknowledgments

Many people contributed to the long and complex process of bringing this book from the concept stage to the shelf in the first place, and now to the continuous effort of improving the book to keep it current. Textbooks are, by definition, a work in progress.

First and foremost, I wish to thank my family, who allowed "the book" to become a member of our household, and have also tolerated the overabundance of photo stops on family trips. Production of this book is a partnership with my wife, Kathy. She has carried out the immense task of merging changes completed for *Essentials of Geology*, Fourth Edition, along with changes suggested by reviewers, to produce the initial manuscript for *Earth: Portrait of a Planet*, Fifth Edition. Kathy also edited new text, cross-checked many sets of proofs, and managed the never-ending inflow and outflow of proofs that perpetually occupy our dining-room table. Without her efforts, the updating of *Earth: Portrait of a Planet* through the years would not be possible. Our daughter, Emma, helped develop the concept of narrative art used in the book, provided several photos, served as scale in other photos, and provided invaluable feedback about how the book works. Our son, David, highlighted places where the writing could be improved, helped me to keep the project in perspective, and also served as scale in photographs.

I am very grateful to all of the staff of W. W. Norton & Company for their incredible efforts during the development of my books over the past two decades. It has been a privilege to work with an employee-owned company that is willing to collaborate so closely with its authors. In particular, I would like to thank Eric Svendsen, the geology editor at Norton, who continues to inject new enthusiasm and ideas into the project. His skill in editing, ability to "oversee many moving parts," and his friendly reminders of deadlines, have led this book to completion. Thom Foley, the book's senior project editor, continues to do an amazing job of guiding the book through production. He somehow keeps track of all the drafts, all the changes, and all the figures for a lengthy and complicated manuscript, while remaining incredibly calm. It's thanks to Thom that everything somehow manages to get done, and that mistakes are few and far between. Meredith Leo has done wonders as the marketing manager for the book, by helping to determine how to meet the needs of adopters worldwide. And as always, I would like to thank Jack Repcheck, who served as the geology editor for the first three editions of the book, before passing the baton to Eric for the Fourth and Fifth Editions. Jack suggested many of the original innovations that strengthened the book, and his instincts about what works in textbook publishing brought the book to the attention of a wider geological community than I ever thought possible. He remains an understanding friend and a fountain of sage advice.

Moving *Earth: Portrait of a Planet* from concept to completion involves a large team of professionals. Stan Maddock, Megan Stewart, Kristina Seymour, Stacy McDade, Eric Bramer, and the other artists and production staff at Precision Graphics in Champaign, Illinois, as always have created beauty and enhanced pedagogy with the line art that they have rendered and the page layout that they've created. I also wish to thank Jeff Mellander, founder and recently retired President of Precision Graphics, for his support in our interactions with his company over these many years. Stephanie Romeo and Jane Miller at Norton did a fantastic job with the Herculean task of finding, organizing, and crediting photographs, and Leah Scott creatively developed a clean and friendly page design. Trish Marx's efforts on the previous edition brought the management of the photo collection into the 21st century and has greatly streamlined the selection process. I am also grateful to Robin Kimball, Rob Bellinger, Cailin Barrett Bressack, Victoria Reuter, Kim Yi, Danielle Belfiore, Leah Clark, and Kristin Sheerin for their innovative approach to ancillary and e-media development and for overseeing the development of *Smart-Work*. Thanks also go to Kristian Sanford and Mateus Teixeira for their work on the Tablet and Mobile e-book, and to Ben Reynolds, who coordinated the back-and-forth between the publisher and various vendors and suppliers. Susan Gaustad, the outstanding developmental editor of the First Edition, helped to set the tone of the book and to weed out errors that otherwise might burden a new edition and Sunny Hwang now provides expert help crafting the text. Chris Thillen has kept that tradition going with her skillful copy editing of this Fifth Edition. And Editorial Assistants Lindsey Thomas and Rachel Goodman provided consistent editorial support and trouble-shooting throughout the process of making this book.

The five editions of this book and its cousin, *Essentials of Geology*, have benefited greatly from input by expert reviewers for specific chapters, by general reviewers of the entire book, and by comments from faculty and students who have used the book and were kind enough to contact me or the publisher with suggestions and corrections. Additional accuracy checking for the Fifth Edition was supplied by Andy Bobyarchick of UNC Charlotte, and Heather Lehto of Angelo State University. We gratefully acknowledge the contributions of the reviewers listed below, who have provided invaluable input into this and past editions. I apologize if I've inadvertently left anyone off the list.

- Jack C. Allen, Bucknell University
- David W. Anderson, San Jose State University
- Martin Appold, University of Missouri-Columbia
- Philip Astwood, University of South Carolina
- Eric Baer, Highline University
- Victor Baker, University of Arizona
- Julie Baldwin, University of Montana
- Miriam Barquero-Molina, University of Missouri
- Sandra Barr, Acadia University
- Keith Bell, Carleton University
- Mary Lou Bevier, University of British Columbia
- Jim Black, Tarrant County College
- Daniel Blake, University of Illinois
- Andy Bobyarchick, University of North Carolina—Charlotte
- Ted Bornhorst, Michigan Technological University
- Michael Bradley, Eastern Michigan University
- Mike Branney, University of Leicester, UK
- Sam Browning, Massachusetts Institute of Technology
- Bill Buhay, University of Winnipeg
- Rachel Burks, Towson University
- Peter Burns, University of Notre Dame
- Katherine Cashman, University of Oregon
- Cinzia Cervato, Iowa State University
- George S. Clark, University of Manitoba
- Kevin Cole, Grand Valley State University
- Patrick M. Colgan, Northeastern University
- Peter Copeland, University of Houston
- John W. Creasy, Bates College
- Norbert Cygan, Chevron Oil, retired
- Michael Dalman, Blinn College

- Peter DeCelles, University of Arizona
- Carlos Dengo, ExxonMobil Exploration Company
- Meredith Denton-Hedrick, Austin Community College—Cypress Creek
- John Dewey, University of California, Davis
- Charles Dimmick, Central Connecticut State University
- Robert T. Dodd, Stony Brook University
- Missy Eppes, University of North Carolina, Charlotte
- Eric Essene, University of Michigan
- James E. Evans, Bowling Green State University
- Susan Everett, University of Michigan, Dearborn
- Dori Farthing, State University of New York, Geneseo
- Mark Feigenson, Rutgers University
- Grant Ferguson, St. Francis Xavier University
- Eric Ferré, Southern Illinois University
- Leon Follmer, Illinois Geological Survey
- Nels Forman, University of North Dakota
- Bruce Fouke, University of Illinois
- David Furbish, Vanderbilt University
- Steve Gao, University of Missouri
- Grant Garvin, John Hopkins University
- Christopher Geiss, Trinity College, Connecticut
- Bryan Gibbs, Richland Community College
- Gayle Gleason, State University of New York, Cortland
- Cyrena Goodrich, Kingsborough Community College
- William D. Gosnold, University of North Dakota
- Lisa Greer, William & Mary College
- Steve Guggenheim, University of Illinois, Chicago
- Henry Halls, University of Toronto, Mississuaga
- Bryce M. Hand, Syracuse University
- Anders Hellstrom, Stockholm University
- Tom Henyey, University of South Carolina
- Bruce Herbert, Texas A & M University
- James Hinthorne, University of Texas, Pan American
- Paul Hoffman, Harvard University
- Curtis Hollabaugh, University of West Georgia
- Bernie Housen, Western Washington University
- Mary Hubbard, Kansas State University
- Paul Hudak, University of North Texas
- Melissa Hudley, University of North Carolina, Chapel Hill
- Warren Huff, University of Cincinnati
- Neal Iverson, Iowa State University
- Charles Jones, University of Pittsburgh
- Donna M. Jurdy, Northwestern University
- Thomas Juster, University of Southern Florida
- H. Karlsson, Texas Tech
- Daniel Karner, Sonoma State University
- Dennis Kent, Lamont Doherty/Rutgers

- Charles Kerton, Iowa State University
- Susan Kieffer, University of Illinois
- Jeffrey Knott, California State University, Fullerton
- Ulrich Kruse, University of Illinois
- Robert S. Kuhlman, Montgomery County Community College
- Lee Kump, Pennsylvania State University
- David R. Lageson, Montana State University
- Robert Lawrence, Oregon State University
- Heather Lehto, Angelo State University
- Scott Lockert, Bluefield Holdings
- Leland Timothy Long, Georgia Tech
- Craig Lundstrom, University of Illinois
- John A. Madsen, University of Delaware
- Jerry Magloughlin, Colorado State University
- Scott Marshall, Appalachian State University
- Jennifer McGuire, Texas A&M University
- Judy McIlrath, University of South Florida
- Paul Meijer, Utrecht University, Netherlands
- Jamie Dustin Mitchem, California University of Pennsylvania
- Alan Mix, Oregon State University
- Otto Muller, Alfred University
- Kristen Myshrall, University of Connecticut
- Kathy Nagy, University of Illinois, Chicago
- Pamela Nelson, Glendale Community College
- Robert Nowack, Purdue University
- Charlie Onasch, Bowling Green State University
- David Osleger, University of California, Davis
- Bill Patterson, University of Saskatchewan
- Eric Peterson, Illinois State University
- Ginny Peterson, Grand Valley State University
- Stephen Piercey, Laurentian University
- Adrian Pittari, University of Waikato, New Zealand
- Lisa M. Pratt, Indiana University
- Eriks Puris, Portland Community College
- Mark Ragan, University of Iowa
- Robert Rauber, University of Illinois
- Bob Reynolds, Central Oregon Community College
- Joshua J. Roering, University of Oregon
- Eric Sandvol, University of Missouri
- William E. Sanford, Colorado State University
- Jeffrey Schaffer, Napa Valley Community College
- Roy Schlische, Rutgers University
- Sahlemedhin Sertsu, Bowie State University
- Anne Sheehan, University of Colorado
- Roger D. Shew, University of North Carolina, Wilmington
- Doug Shakel, Pima Community College
- Norma Small-Warren, Howard University

- Donny Smoak, University of South Florida
- David Sparks, Texas A&M University
- Angela Speck, University of Missouri
- Larry Standlee, University of Texas—Arlington
- Tim Stark, University of Illinois
- Seth Stein, Northwestern University
- David Stetty, Jacksonville State University
- Kevin G. Stewart, University of North Carolina, Chapel Hill
- Michael Stewart, University of Illinois
- Don Stierman, University of Toledo
- Gina Marie Seegers Szablewski, University of Wisconsin, Milwaukee
- Barbara Tewksbury, Hamilton College
- Thomas M. Tharp, Purdue University
- Kathryn Thornbjarnarson, San Diego State University
- Basil Tikoff, University of Wisconsin
- Spencer Titley, University of Arizona
- Robert T. Todd, Stony Brook University
- Torbjörn Törnqvist, University of Illinois, Chicago
- Jon Tso, Radford University
- Stacey Verardo, George Mason University
- Barry Weaver, University of Oklahoma
- John Werner, Seminole State College of Florida
- Alan Whittington, University of Missouri
- John Wickham, University of Texas, Arlington
- Lorraine Wolf, Auburn University
- Christopher J. Woltemade, Shippensburg University
- Jackie Wood, Delgado Community College—City Park
- Kerry Workman-Ford, California State University-Fresno

Thanks!

I am very grateful to the faculty who have selected *Earth: Portrait of a Planet* for their classes, and to the students who engage so energetically with the book. I particularly appreciate the questions and corrections from readers that help to improve the book and keep it as accurate as possible. I continue to welcome comments and can be reached at smarshak@illinois.edu.

Stephen Marshak

It was during my enchanted days of travel that the idea came to me which, through the years, has come into my thoughts again and again and always happily—the idea that geology is the music of the Earth.

—Hans Cloos (German geologist, 1885–1951)

About the Author

Stephen Marshak is Professor of Geology at the University of Illinois, Urbana-Champaign, where he also serves as the Director of the School of Earth, Society, and Environment. He holds an A.B. from Cornell University, an M.S. from the University of Arizona, and a Ph.D. from Columbia University. Steve's research interests lie in structural geology and tectonics, and he has participated in field projects on a number of continents. Steve loves teaching and has won his college's and university's highest teaching awards. He also received the 2012 Neil Miner Award from the National Association of Geoscience Teachers (NAGT), for "exceptional contributions to the stimulation of interest in the Earth sciences." In addition to research papers and *Earth: Portrait of a Planet*, Steve has authored *Essentials of Geology*, and has co-authored *Laboratory Manual for Introductory Geology*; *Earth Structure: An Introduction to Structural Geology and Tectonics*, and *Basic Methods of Structural Geology*.

Another View Geology students enjoying the view from outcrops in northern Scotland.

EARTH
Portrait of a Planet

FIFTH EDITION

Students exploring a small canyon in Illinois. Geology is everywhere!

And Just What Is Geology?

Civilization exists by geological consent, subject to change without notice.

—*Will Durant (1885–1981)*

P.1   In Search of Ideas

We arrived in the late-night darkness, at a campsite in western Arizona. Here in the desert, so little rain falls over the course of a year that hardly any plants can survive, and rocks crop out on many hills. Under the dry sky, there's no need for tents, so we could sleep under the stars with our sleeping bags on a bed of sand. At dawn, the red rays of the first sunlight made the slope of the steep-sided hill near our campsite start to glow, and we could see our target, a prominent ledge of rusty-brown rock that formed a shelf at the top of the hill. To reach it, though, we'd have to climb a steep slope littered with jagged boulders.

After a quick breakfast, we loaded our day packs with water bottles and granola bars, slathered on a layer of sunscreen, and set off toward the slope. It was the breezeless morning of what was going to be a truly hot day, and we wanted to gain elevation before the sun rose too high in the sky. After a tiring hour finding our way through the boulder obstacle course, we reached the base of the ledge and decided to take a rest before ascending the final cliff. But just as we leaned back to rest our backs against a rock, we heard an unnerving vibration. Somewhere nearby, too close for comfort, a rattlesnake shook an urgent warning with its tail. Rest would have to wait, and we scrambled up the ledge. It was the right choice, for the view from the top of the surrounding landscape was amazing (**Fig. P.1a**). But the rocks beneath our feet were even more amazing. Close up, we could see curving ribbons of light and dark layers, cut by stripes of white quartz. The ledge preserved the story of a distant age in our planet's past when the rock we now stood on was kilometers below ground level and was able to flow like soft plastic, but ever so slowly (**Fig. P.1b**). We now set to the task of figuring out what it all meant.

Geologists—scientists who study the Earth—do indeed explore remote deserts, high mountains, damp rainforests, frigid glaciers, and deep canyons (**Fig. P.2**). Such efforts can strike people in other professions as a strange way to make a living. This sentiment underlies the Scottish poet Walter Scott's (1771–1832) description of geologists at work: "Some rin uphill and down dale, knappin' the chucky stones to pieces like sa' many roadmakers run daft. They say it is to see how the warld was made!" But Scott had it right—to see how the world was made, to see how it continues to evolve, to find its resources, to protect against its natural hazards, and to predict what its future may bring. These are the questions that have driven geologists to

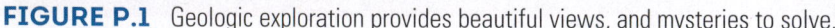

FIGURE P.1   Geologic exploration provides beautiful views, and mysteries to solve.

(a) A view of the western Arizona desert is not just beautiful—it holds clues to the Earth's past and to the changes taking place today.

(b) The contortions of the rock layers speak of a time when the rock flowed, like soft plastic.

then continue with "an earthquake occurred today off Japan," or "landslides will threaten the city," or "chemicals from a toxic waste dump will ruin the town's water supply," or "there's only a limited supply of oil left," or "the floods of the last few days are the worst on record," the scientists that the reports refer to are geologists. In fact, ask yourself the following questions, and you'll realize that geologic processes, phenomena, and materials play major roles in daily life:

- Do you live in a region threatened by landslides, volcanoes, earthquakes, or floods (**Fig. P.4a**)?
- Are you worried about the price of energy or about whether there will be a war in an oil-supplying country (**Fig. P.4b**)?
- Do you ever wonder where the copper in your home's wires comes from? Or the lithium in the battery of your cell phone?
- Have you seen fields of green crops surrounded by desert and wondered where the irrigation water comes from?
- Would you like to buy a dream house near a beach or a river?
- Are you following news stories about how toxic waste can migrate underground into your town's well water?

Addressing these questions requires a basic understanding of geology. This knowledge may help you avoid building your home on a hazardous floodplain or fault zone, on an unstable slope, or along a rapidly eroding coast. With an understanding of groundwater, you may be able to find a good site for a well. With knowledge of the geologic controls on resource distribution, you may be able to invest more wisely in the resource industry or understand the context of political choices regarding energy policy.

Second, the study of geology gives you an awareness of the planet that no other field can. As you will see, the Earth is a complicated world, where living organisms, oceans, atmosphere, and solid rock interact with one another in a great variety of ways. Geologic study reveals Earth's antiquity (it's about 4.54 billion years old) and demonstrates how the planet has changed profoundly during its existence. What our ancestors considered to be the center of the Universe has become, with the development of geologic perspective, our "island in space" today. And what was believed to be an unchanging orb originating at the same time as humanity has become a dynamic planet that existed long before people did—and continues to evolve.

Third, the study of geology puts the accomplishments and consequences of human civilization in a broader context. View the aftermath of a large earthquake, flood, or hurricane, and it's clear that the might of natural geologic phenomena greatly exceeds the strength of human-made structures. But watch a bulldozer clear a swath of forest, a dynamite explosion remove the top of a hill, or a prairie field evolve into a housing development, and it's clear that people can change the face of the Earth at rates often exceeding those of natural geologic processes.

Finally, when you finish reading this book, your view of the world may be forever colored by geologic curiosity. If you walk in the mountains, you may remember that mountains rise and fall over time in response to forces that shape and reshape the Earth's surface. If you hear about a natural disaster, you may think about the various phenomena that trigger disasters. And as you drive past rock exposures along a highway, you won't just see featureless masses of gray but will pick out layers and structures providing a record of the Earth's very long history.

P.3 Themes of this Book

A number of narrative themes appear—and reappear—throughout this text. These themes, listed below, can be viewed as the book's overall take-home message.

- *Geology is a synthesis of many sciences:* The study of geology can help you understand physical science in general, for

FIGURE P.4 Geology provides insight into natural hazards and resource exploration.

(a) Collapsed buildings in the aftermath of an earthquake.

(b) Coal is one of many resources that comes from the Earth.

geology applies many of the basic concepts of physics and chemistry to the interpretation of visible phenomena. As you learn about the Earth, you'll also be learning about the behavior of matter and energy, and about the nature of chemical reactions.

- *The Earth has an internal structure:* The Earth is not a homogeneous ball, but rather it consists of concentric layers. From center to surface, Earth has a core, mantle, and crust. We live on the surface of the crust, where it meets the atmosphere and the oceans.

- *The outer layer of the Earth consists of moving plates:* In the 1960s, geologists recognized that the crust, together with the uppermost part of the underlying mantle, forms a 100- to 150-km-thick semi-rigid shell called the **lithosphere**. Large cracks separate this shell into discrete pieces, called **plates**, which move very slowly relative to one another (**Fig. P.5**). The theory that describes this movement and its consequences is called the **theory of plate tectonics**, and it is the foundation for understanding most geologic phenomena. Plate movements yield earthquakes, volcanoes, and mountain ranges, and cause the map of Earth's surface to change very slowly over time.

- *We can picture the Earth as a complex system:* The Earth is not static, but rather it is a dynamic entity whose components can move and change over time. Our planet's interior, solid surface, oceans, atmosphere, and life all interact with one another in many ways to yield the land, oceans, and air in which we and other species of organisms can live. Geologists refer to this interconnected web of interacting materials and processes as the **Earth System**. Within the Earth System, certain materials cycle among different types of rock, among rock, sea, and air, and among all of these entities and life (**Fig. P.6**). Over time, the distribution of these materials among various components of the Earth System can change, as can the characteristics of the components.

- *The Earth is a planet:* Despite the uniqueness of the Earth System, we can think of Earth as a planet, formed like the other planets of the Solar System. But because of the way the Earth System operates, our planet differs from other planets by having plate tectonics, an oxygen-rich atmosphere and liquid-water ocean, and abundant life.

- *The Earth is very old:* Geologic data indicate that the Earth formed 4.54 billion years ago—plenty of time for geologic processes to generate and destroy landscapes, for life forms to evolve, and for the map of the planet to change. Plate movement at rates of only a few centimeters per year can move a continent thousands of kilometers if those movements continue for hundreds of millions of years. There is time enough to build mountains and time enough to grind them down, many times over. The Earth has a history, and it extends far into the past, long before human ancestors appeared.

- *The geologic time scale divides Earth's history into intervals:* To refer to specific portions of geologic time, geologists developed the **geologic time scale** (**Fig. P.7**). The last 541 million years comprise the *Phanerozoic Eon*, and all time before that falls in the *Precambrian*. The Precambrian can be further divided into three main intervals named, from oldest to youngest: the *Hadean*, the *Archean*, and the *Proterozoic Eons*. The Phanerozoic Eon, in turn, can be divided into three

FIGURE P.5 A simplified map of the Earth's plates. The arrows indicate the direction each plate is moving, and the length of the arrow indicates plate velocity (the longer the arrow, the faster the motion).

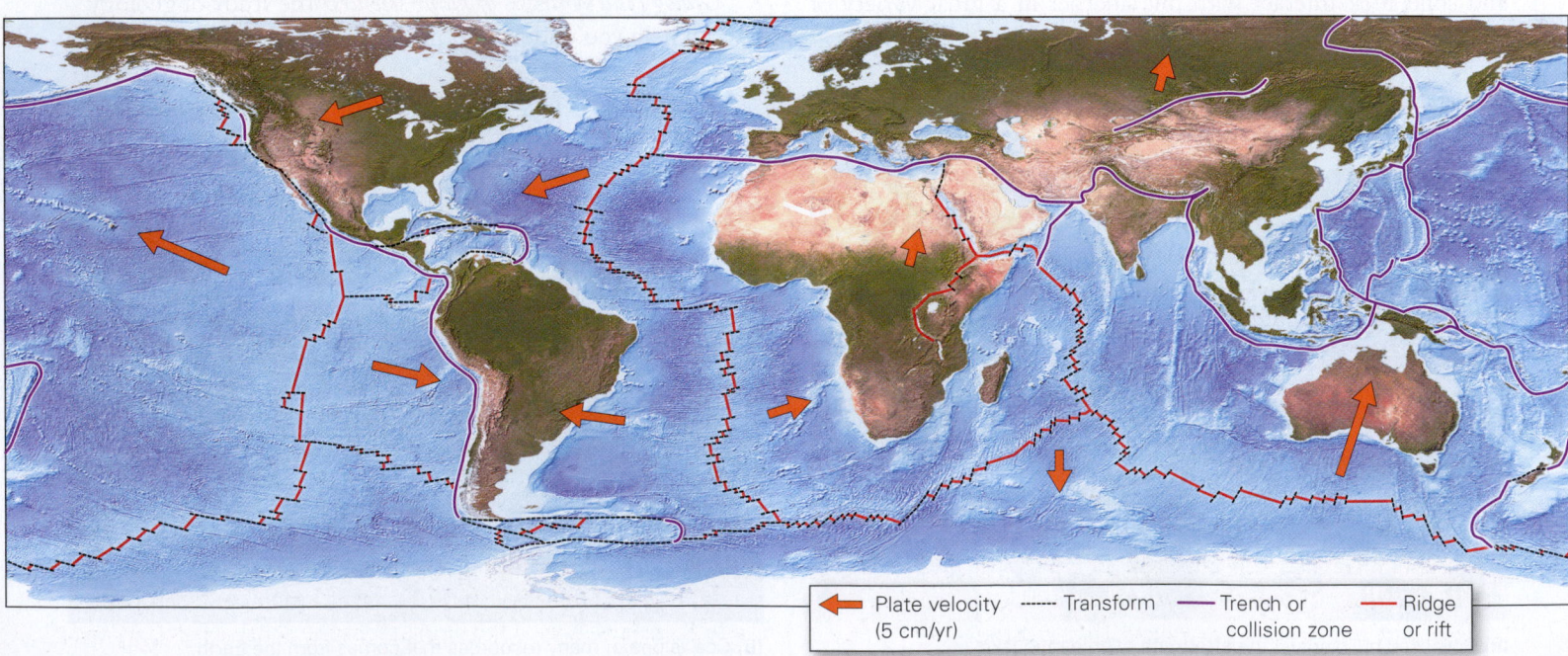

Plate velocity (5 cm/yr) ┈┈ Transform ─── Trench or collision zone ─── Ridge or rift

FIGURE P.6 In this scenic view in Switzerland, we see many aspects of the Earth System—air, water, ice, rock, life, and human activity.

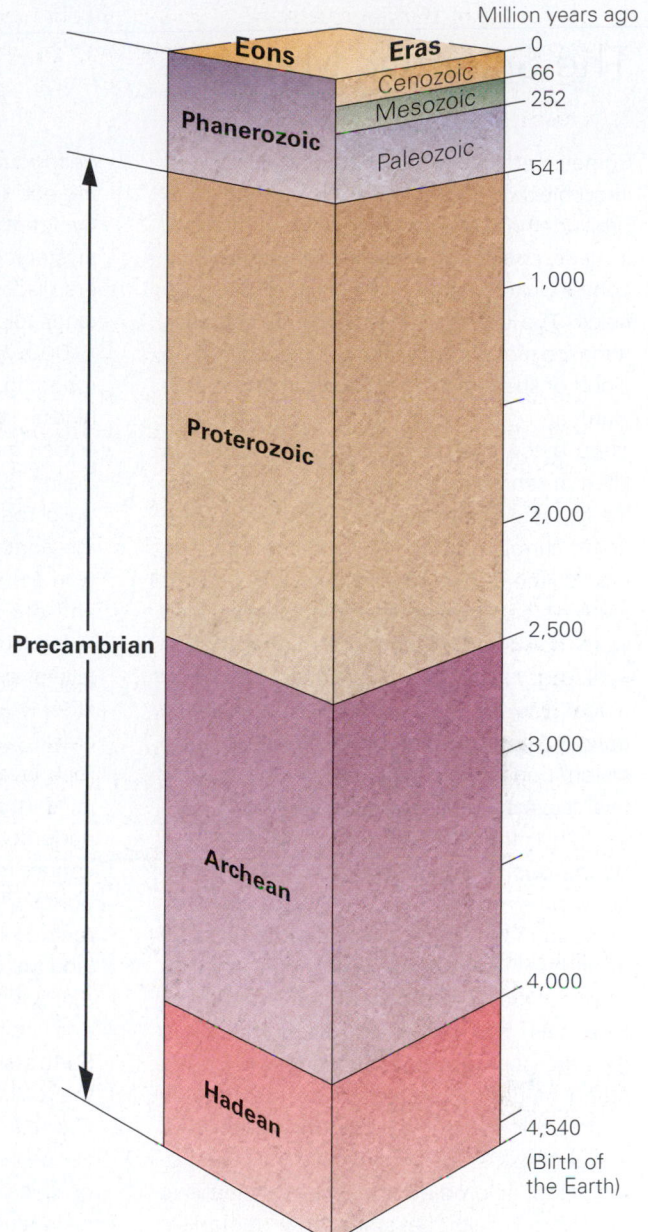

FIGURE P.7 The geologic time scale.

Million years ago

Eons	Eras	
		0
Phanerozoic	Cenozoic	66
	Mesozoic	252
	Paleozoic	541
Proterozoic		1,000
		2,000
		2,500
Archean		3,000
		4,000
Hadean		4,540 (Birth of the Earth)

Precambrian

(a) The scale has been divided into eons and eras.

One thousand years ago = 1 **Ka**
(Ka stands for kilo-annum)

One million years ago = 1 **Ma**
(Ma stands for mega-annum)

One billion years ago = 1 **Ga**
(Ga stands for giga-annum)

(b) Abbreviations for time units.

main intervals named, from oldest to youngest: the *Paleozoic*, the *Mesozoic*, and the *Cenozoic Eras*.

- *Internal and external processes drive geologic phenomena:* *Internal processes* are driven by heat from inside the Earth. Plate movement is an example. Because plate movements cause mountain building, earthquakes, and volcanoes, we consider all of these phenomena internal processes. *External processes* are driven by energy coming to the Earth from the Sun. The heat produced by this energy drives the movement of air and water, which grinds and sculpts the Earth's surface and transports the debris to new locations, where it accumulates. The interaction between internal and external processes forms the mountains, canyons, beaches, and plains of our planet. As we'll see, **gravity**—the pull that one mass exerts on another—plays an important role in both internal and external processes.

- *Geologic phenomena affect society:* Volcanoes, earthquakes, landslides, floods, groundwater, energy sources, and mineral reserves are of vital interest to every inhabitant of this planet. Therefore, throughout this book we emphasize the linkages among geology, the environment, and society.

- *Physical aspects of the Earth System are linked to life processes:* All life on this planet depends on such physical features as the minerals in soil; the temperature, humidity, and composition of the atmosphere; and the flow of surface and subsurface water. And life in turn affects and alters physical features. For example, the oxygen in Earth's atmosphere comes from photosynthesis, a life activity in plants. This oxygen in turn permits complex animals to survive and affects chemical reactions among air, water, and rock. Without the physical Earth, life could not exist; but without life, this planet's surface might have become a frozen wasteland, like that of Mars, or a cloud-enshrouded oven, like that of Venus.

- *The Earth has changed dramatically in many ways over geologic time and continues to change:* The landscape that you see

outside your window today is not what you would have seen a thousand, a million, or a billion years ago. Over Earth history, the planet's surface, composition of the atmosphere, and sea level have all changed. Change continues today, and

The Scientific Method

Sometime during the past 200 million years, a large block of rock or metal, which had been orbiting the Sun, slammed into our planet. It made contact at a site in what is now the central United States, a landscape of flat cornfields. The impact of this block, a meteorite, released more energy than a nuclear bomb—a cloud of shattered rock and dust blasted skyward, and once-horizontal layers of rock from deep below the ground sprang upward and tilted on end beneath the gaping hole left by the impact. When the dust had settled, a huge crater surrounded by debris marked the surface of the Earth at the impact site. Later in Earth history, running water and blowing wind wore down this jagged scar. Some 15,000 years ago, sand, gravel, and mud carried by a vast glacier buried what remained, hiding it entirely from view (**Fig. BxP.1**). Wow! So much history beneath a cornfield. How do we know this? It takes scientific investigation.

The movies often portray science as a dangerous tool, capable of creating Frankenstein's monster, and scientists as nerdy characters with thick glasses and poor taste in clothes. In reality, **science** is simply the use of observation, experiment, and calculation to explain how nature operates, and scientists are people who study and try to understand natural phenomena. Scientists guide their work using the **scientific method**, a sequence of steps for systematically analyzing scientific problems in a way that leads to verifiable results. Let's see how geologists employed the steps of the scientific method to come up with the meteorite-impact story.

1. *Recognizing the problem.* Any scientific project, like any detective story, begins by identifying a mystery. The cornfield mystery came to light when water drillers discovered limestone, a rock typically made of shell fragments, just below the 15,000-year-old glacial sediment. In surrounding regions, the rock beneath the glacial sediment consists of sandstone, a rock made of cemented-together sand grains. Since limestone can be used to build roads, make cement, and produce the agricultural lime used in treating soil, workers stripped off the glacial sediment and dug a quarry to excavate the limestone. They were amazed to find that rock layers exposed in the quarry were tilted steeply and had been shattered by large cracks. In the surrounding regions, all rock layers are horizontal like the layers in a birthday cake, the limestone layer lies underneath the sandstone, and the rocks contain relatively few cracks. Curious geologists came to investigate, and they soon realized that the geologic features of the land just beneath the cornfield presented a problem to be explained: what phenomena had brought limestone up close to the Earth's surface, had tilted the layering in the rocks, and had shattered the rocks?

2. *Collecting data.* The scientific method proceeds with the collection of observations or clues that point to an answer. Geologists studied the quarry and determined the age of its rocks, measured the orientation of the rock layers, and documented (made a written or photographic record of) the fractures that broke up the rocks.

3. *Proposing hypotheses.* A scientific **hypothesis** is merely a possible explanation, involving only natural processes, that can explain a set of observations. Scientists propose hypotheses during or after their initial data collection. In this example, the geologists working in the quarry came up with two alternative hypotheses: either the features in this region resulted from a volcanic explosion, or they were caused by a meteorite impact.

4. *Testing hypotheses.* Since a hypothesis is just an idea that can be either right or wrong, scientists must put hypotheses through a series of tests to see if they work. The geologists at the quarry compared their field observations with published observations made at other sites of volcanic explosions and meteorite impacts, and they studied the results of experiments designed to simulate such events. If the geologic features visible in the quarry were the result of volcanism, the quarry should contain rocks formed by the freezing of molten rock erupted by a volcano. But no such rocks were found. If, however, the features were produced by an impact, the rocks should contain **shatter cones**, tiny cracks that fan out from a point. Shatter cones can be overlooked, so the geologists returned to the quarry specifically to search for them and found them in abundance. The impact hypothesis passed the test!

---

aspects of the Earth System are changing faster than ever before because of human activity.

- *Most of the resources that we use come from geological materials:* Modern society uses vast quantities of oil, gas, coal, metal, concrete, and other materials. All of these come from the Earth.

- *Science comes from observation, and people make scientific discoveries:* Science does not consist of subjective guesses or arbitrary dogmas but rather of a consistent set of objective statements resulting from the application of the scientific method (**Box P.1**). Every scientific idea must be tested

thoroughly and should be used only when supported by documented observations. Further, scientific ideas do not appear out of nowhere; they are the result of human efforts. Wherever possible, this book shows where geologic ideas came from, and tries to answer the question, "How do we know that?"

As you read this book, please keep these themes in mind. Don't view geology as a list of words to memorize but rather as an interconnected set of concepts to digest. Most of all, enjoy yourself as you learn about the most fascinating planet in the Universe.

FIGURE BxP.1 An ancient meteorite impact excavates a crater and permanently changes rock beneath the surface.

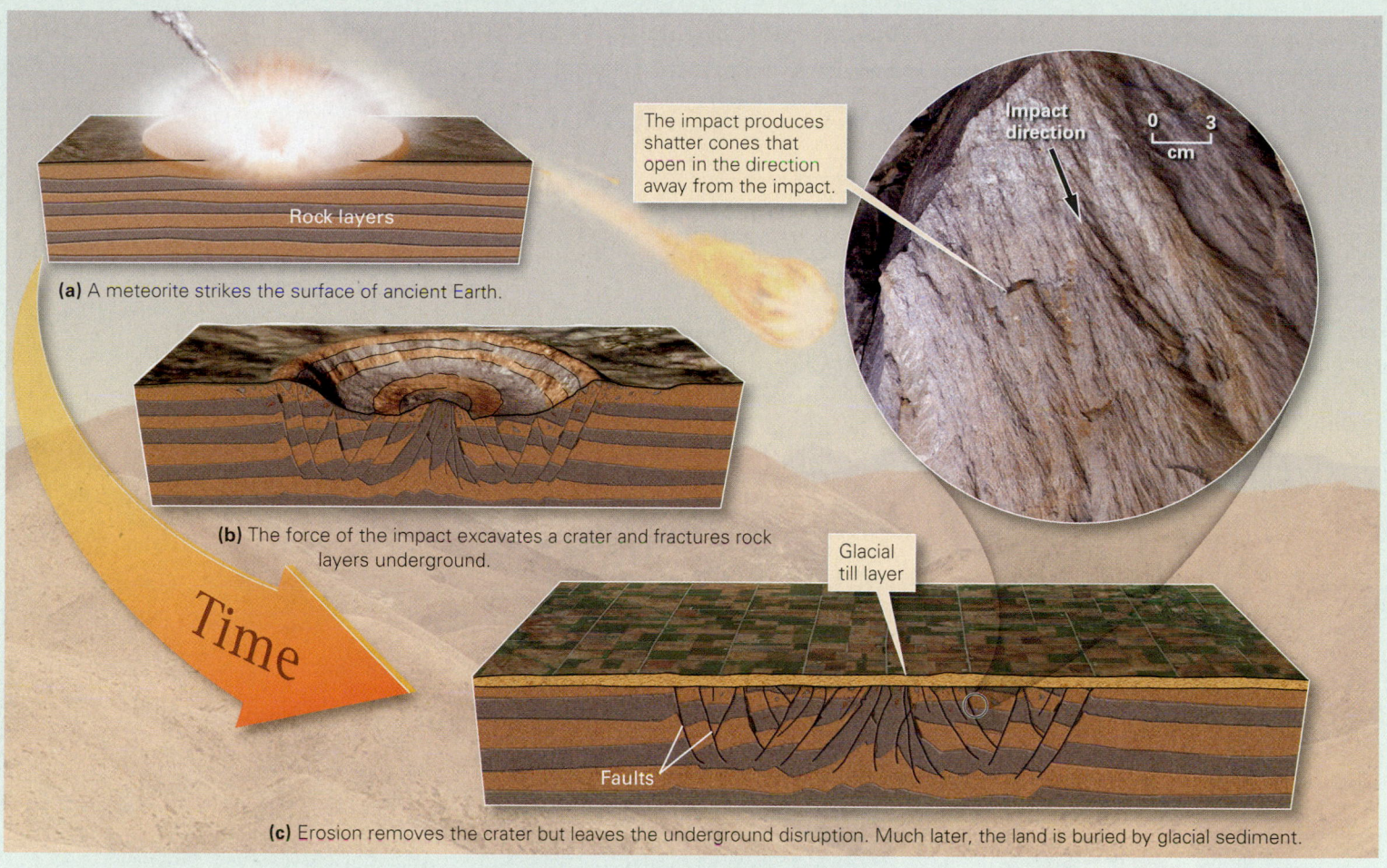

Rock layers

(a) A meteorite strikes the surface of ancient Earth.

The impact produces shatter cones that open in the direction away from the impact.

Impact direction

0   3 cm

(b) The force of the impact excavates a crater and fractures rock layers underground.

Time

Glacial till layer

Faults

(c) Erosion removes the crater but leaves the underground disruption. Much later, the land is buried by glacial sediment.

Theories are scientific ideas supported by an abundance of evidence; they have passed many tests and have failed none. Scientists are much more confident in the correctness of a theory than of a hypothesis. Continued study in the quarry eventually yielded so much evidence for impact that the impact hypothesis came to be viewed as a theory. Scientists continue to test theories over a long time. Successful theories withstand these tests and are supported by so many observations that they become part of a discipline's foundation. (As you will discover in Chapter 3, geologists consider the idea that continents have moved around the surface of the Earth to be a theory because so much evidence supports it.) However, some theories may eventually be disproven and replaced by better ones.

In a few cases, scientists have been able to devise concise statements that completely describe a specific relationship or phenomenon. Such statements, called **scientific laws**, apply without exception for a given range of conditions. Newton's law of gravitation serves as an example—it is a simple mathematical expression that always defines the invisible pull exerted by one mass upon another. Note that scientific laws do not in themselves explain a phenomenon, and in this way they differ from theories. For example, the law of gravity does not explain why gravity exists, but the theory of evolution does explain why evolution occurs.

GUIDE TERMS

Earth System (p. 6)

geologic time scale (p. 6)

geologist (p. 2)

geology (p. 3)

gravity (p. 7)

hypothesis (p. 8)

lithosphere (p. 6)

plate (p. 6)

science (p. 8)

scientific laws (p. 9)

scientific method (p. 8)

shatter cones (p. 8)

theory (p. 9)

theory of plate tectonics (p. 6)

PART 1

OUR ISLAND IN SPACE

When you gaze out toward the horizon from a mountaintop, the Earth seems endless, and before the modern era, many people thought it was. But to astronauts flying to the Moon, the Earth looks like a small, shining globe—they can see half the planet in a single glance. From the astronauts' perspective, we are living on an island in space.

Earth may not be endless, but it is a very special planet: its temperature and composition, unlike those of the other planets in the Solar System, make it habitable. In Part I of this book, we first learn in Chapter 1 scientific ideas about how the Earth, and the Universe around it, came to be. Then, in Chapter 2, we take a quick tour of the planet to get a sense of its composition and its various layers. With this background, we're ready to move on to Chapter 3, where we'll encounter the twentieth-century revolution in geology that yielded the set of ideas we now call the theory of plate tectonics. In Chapter 4, we'll delve into the physics behind this theory (which proposes that the outer layer of the Earth consists of plates that move with respect to each other) and its principles. We'll see that plate tectonics provides a rational explanation for a great variety of geologic features—from the formation of continents to the distribution of fossils—and is truly the unifying concept for all of geology.

1 **Cosmology and the Birth of Earth**
2 **Journey to the Center of the Earth**
3 **Drifting Continents and Spreading Seas**
4 **The Way the Earth Works: Plate Tectonics**

An astronaut taking a space walk outside of the International Space Station could not survive without a space suit. The hospitable Earth System—land, air, and water—lies 380 km (235 mi) below.

This Hubble Space Telescope image shows the Orion Nebula, a cloud of gas and dust 24 light years across, in which new stars are forming.

CHAPTER 1

Cosmology and the Birth of Earth

> I believe everyone should have a broad picture of how the Universe operates and our place in it. It is a basic human desire. And it also puts our worries in perspective.
>
> —Stephen Hawking (British cosmologist, born 1942)

1.1 Introduction

Sometime in the distant past, humans developed the capacity for complex, conscious thought. This amazing ability, which distinguishes our species from all others, brought with it the gift of curiosity, an innate desire to understand and explain the workings and origin of the **Universe**, meaning space and everything it contains, including our home, the Earth. For most of human history, such musings have spawned legends in which heroes, gods, and goddesses used supernatural powers to ignite the Sun, to make the Moon glow, to mold the Earth from nothingness and sculpt its surface into dramatic shapes, and to speckle the night with points of light. Only recently have people applied scientific principles (see Box P.1) to the systematic study of the overall structure and history of the Universe, thereby establishing the modern discipline of scientific **cosmology**. In the context of scientific cosmology, the Universe contains of two basic entities—**matter**, the substance that makes up objects, and **energy**, the inherent ability of a region of space and the matter within it to do "work" (to change itself and/or its surroundings). We can refer to the amount of matter in an object as its **mass**, so an object with greater mass contains more matter.

Scientific cosmology provides a foundation from which we can begin to explore the composition, structure, and evolution of the Earth. So we begin this book on geology, the study of the Earth, with a chapter that outlines key principles of scientific cosmology. In effect, we must look outward in order to be able understand what we'll see when we look inward at the Earth (**Fig. 1.1**). We start by characterizing the architecture of the Universe overall and of our Solar System in particular. Then we introduce the Big Bang theory, which researchers use to explain the formation of the Universe, and the production of the elements that eventually came together to form the Sun, Earth, and other **celestial objects** (naturally occurring bodies in the Universe). We conclude by discussing the nebular theory for the birth of the Earth itself.

1.2 An Image of Our Universe

What Is the Structure of the Universe?

Think about the mysterious spectacle of a clear night sky (**Fig. 1.2a**). What objects are up there? How big are they? How far away are they? How do they move? How are they arranged? Ancient cultures around the world thought long and hard about such questions, and by 3,000 years ago, keen observers—the

FIGURE 1.1 This iconic image of "Earthrise," taken by astronauts orbiting the Moon, profoundly influenced humanity's perception of our home planet. Since we live on an island in space, we need to understand space in order to understand the Earth.

FIGURE 1.2 The sky at night, showing a variety of celestial objects.

(a) Imagine what it would be like *not* to know what celestial objects, such as the Moon, are. Until the past few hundred years, we didn't.

(b) When viewed for the whole night, stars in the northern hemisphere appear to revolve around the North Star.

first **astronomers** (people who study celestial bodies)—had realized that what they could see above had a recognizable order. Specifically, most of the thousands of points of light visible to the naked eye move slowly across the sky nightly, as if revolving around a fixed point (**Fig. 1.2b**), and the positions of these points relative to each other remains fixed. These points became known as the *stars*. In contrast, a handful of the lights in the night sky etch seemingly complex paths across the night sky, moving relative to one another and relative to the backdrop of stars (**Fig. 1.2c**). These lights came to be known as the *planets* (from the Greek *planēs*, which means "wanderer").

Early observers did not know what the stars and planets were, and they didn't understand their own relationship with the Earth, Moon, and Sun. It wasn't at all obvious, in fact, that the Earth is a planet and the Sun is a star. In the days of the Greek philosopher Homer (ca. 850 B.C.E.), for example, people in the Mediterranean region held the belief that the Earth was a flat disk, with land toward the center and water around the margins, and that this disk lay at the center of a celestial sphere, a dome to which the stars were attached. Philosophers of Homer's day argued about the nature of the Sun and why it produced heat and light: to some, the Sun was a burning bowl of oil, while to others it was a ball of red-hot iron. Many favored the notion that movements of celestial bodies represented the activities of gods and goddesses, and they named *constellations*—distinctive arrangements of stars—after gods and goddesses. Other societies developed quite different mythologies, full of symbolism, in which to interpret the heavens.

In the Western world, two distinct schools of thought developed concerning the arrangement of stars and planets and their relationship to the Earth, Sun, and Moon. The first school advocated a **geocentric model** (**Fig. 1.3a**), in which the

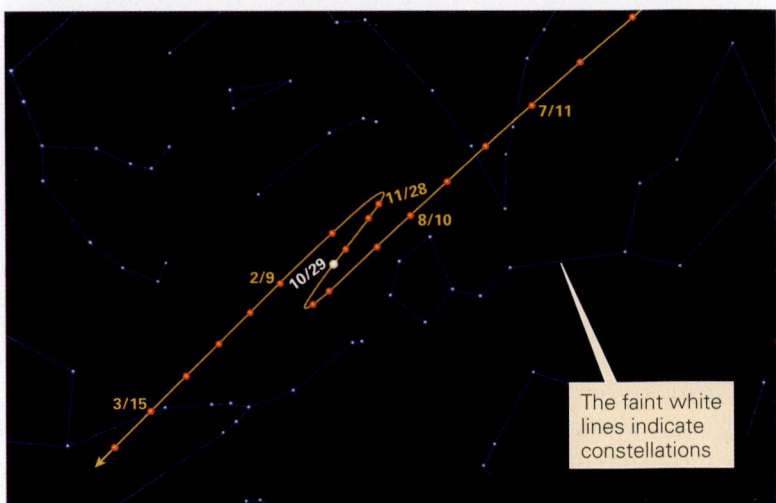

(c) If you track certain objects in the night sky, they appear to move relative to the backdrop of stars. These "wanderers" are the planets. The numbers are dates (month/day), indicating when the planet shown was at a given location.

The faint white lines indicate constellations

Earth sits without moving at the center of the Universe while the Moon and the planets whirl around it in circular orbits, all within a shell of stars. The second school advocated a **heliocentric model** (**Fig. 1.3b**), in which the Sun lies at the center of the Universe, with the Earth and other planets orbiting around it. The geocentric image gained a widespread following due to the influence of an Egyptian mathematician, Ptolemy (100–170 C.E.), who developed equations that seemed to predict the wanderings of the planets, in the context of the geocentric model, with remarkable accuracy. During the Middle Ages (ca. 476–1400 C.E.), church leaders in Europe adopted Ptolemy's geocentric image as dogma because it seemed to jus-

FIGURE 1.3 Contrasting views of the Universe, as drawn by artists hundreds of years ago.

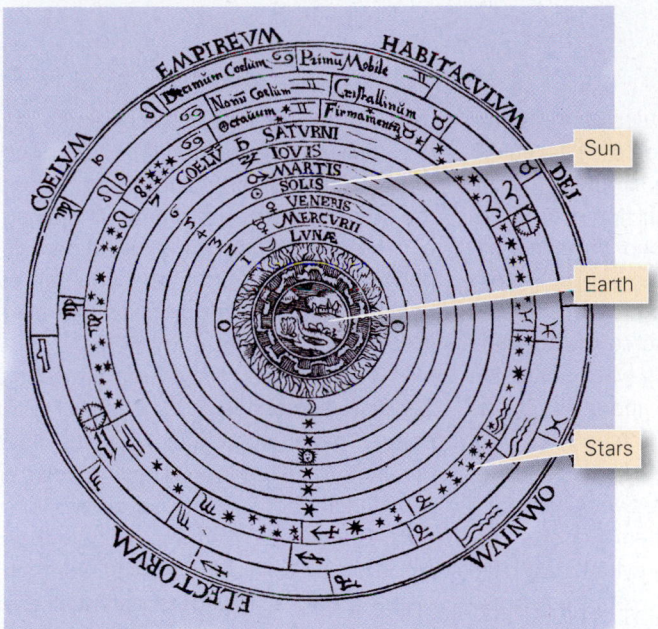

(a) The geocentric image of the Universe shows the Earth at the center, surrounded by air, fire, and the other planets, all contained within the globe of the stars.

(b) The heliocentric image of the Universe shows the Sun at the center, as envisioned by Copernicus.

tify the comforting thought that humanity's home occupies the most important place in the Universe. Eventually, anyone who disagreed with this view risked charges of heresy.

Then came the Renaissance. In 15th-century Europe, bold thinkers spawned a new age of exploration and scientific discovery. Thanks to the efforts of Nicolaus Copernicus (1473–1543) and Galileo Galilei (1564–1642), people eventually came to realize that the Earth and planets did indeed orbit the Sun, so

the Earth could not possibly be at the center of the Universe. The idea solidified when Johannes Kepler (1571–1630) showed that Ptolemy was wrong and that planets follow elliptical orbits. And when Isaac Newton (1643–1727) explained **gravity**, the attractive force that one mass exerts on another (**Box 1.1**), it finally became possible to understand why celestial objects display the motions that they do.

Stars and Galaxies

After Galileo popularized the telescope, astronomers gained the ability to see and measure features progressively farther into space and gradually refined our understanding of the Universe's structure. We now realize that, although it looks like a point of light, a **star** is actually an immense sphere of incandescent gas that emits intense energy—stars are just like our Sun but lie farther away. Furthermore, stars are not scattered randomly through the Universe; rather, gravity holds them together in immense groups called **galaxies**. Our Sun joins with over 300 billion other stars to form the *Milky Way galaxy*. From our vantage point on Earth, the Milky Way looks like a hazy band (**Fig. 1.4a**), but if we could view the Milky Way from a great distance, it would look like a flattened spiral with great curving arms slowly swirling around a glowing, disk-like center (**Fig. 1.4b**). Presently, our Sun lies near the outer edge of one of these arms. Astronomers estimate that more than 100 billion galaxies constitute the visible Universe (**Fig. 1.4c**).

Clearly, human understanding of Earth's place in the Universe evolved radically over the past few centuries. Neither the Earth nor the Sun, nor even the Milky Way, occupies the center of the Universe. While the Earth is quite special to us because we live here, there's nothing particularly special about its position in the Universe.

The Nature of Our Solar System

Our Sun's gravitational pull holds on to many objects which, together with the Sun, comprise the **Solar System** (**Fig. 1.5a**). Most of the mass of the Solar System—99.8%, to be exact—resides in the Sun itself. The remaining 0.2% includes a great variety of objects, the largest of which are planets. Astronomers define a **planet** as an object that orbits a star, is roughly spherical, and has "cleared its neighborhood of other objects." The last phrase in this definition sounds a bit strange at first, but it merely implies that a planet's gravity has pulled in all particles of matter in its orbit. According to this definition, formalized in 2005, our Solar System includes eight planets—Mercury, Venus, Earth, Mars, Jupiter, Saturn, Uranus, and Neptune. Until 2005, astronomers considered one more object, Pluto, to be a planet. But Pluto has not cleared its orbit, so it does not fit the modern definition of a planet and has been dropped from the roster. The

BOX 1.1 SCIENCE TOOLBOX . . .

Force and Energy

Isaac Newton defined a *force* as simply a push or pull that causes the velocity of an object to change in magnitude (speed) and/or direction. In your everyday experience, you constantly see and/or feel the effect of forces. Forces can speed objects up or slow them down. Forces can also tear, stretch, squash, spin, and twist objects and can make them float or sink. Any system (a defined volume of space and everything in it) contains *energy*, meaning that it has an inherent ability or capacity to do work. In this context, *work* refers to the product of the force and the distance over which the force is acting—you do work when you lift this book 10 cm, and you do more work when you lift it 20 cm. When work is done, energy transfers from one part of a system to another, but the total amount of energy in the system cannot be created or destroyed.

Physicists (scientists who study matter, energy, and the interactions between them) distinguish between two general types of force. The first type, a *contact force*, or *mechanical force*, results when one *mass* moves and comes in contact with another. You apply a mechanical force to a boulder when you push on it (**Fig. Bx1.1a**), and the wind applies a mechanical force to a sail when it blows. The second type, a non-contact force, or *field force*, applies across a distance; gravity and magnetism serve as examples. *Gravity* is the force of attraction between two masses—it is what holds you to the surface of the Earth and pulls objects from higher elevations to lower ones (**Fig. Bx1.1b**). The strength of gravity depends on the quantity of matter in the two masses and on the distance between them. For example, you feel a much stronger gravitational pull to the huge Earth than you do to a small baseball. Weight is the force that an object exerts due to gravitational pull, so the stronger the gravitational pull, the greater the weight of an object. For example, on the Moon, you weigh much less than on the Earth because the Moon is smaller so it exerts less gravitational pull. **Magnetism**, simplistically, is the force generated by electricity flowing in a wire, or by special materials called magnets.

Unlike gravity, magnetic force can be attractive (pulling objects together) or repulsive (pushing them apart). Over short distances, the magnetic force of even a small magnet can be larger than the gravitational force produced by the Earth; thus, you can overcome gravity and lift objects with a magnet (**Fig. Bx1.1c**).

Physicists also distinguish between two general types of energy. *Kinetic energy* is the energy that an object has when it's moving. For example, a boulder bounding down a hill has kinetic energy. *Potential energy*, in contrast, is the energy held in an object being acted on by a field force. A boulder sitting at the top of a hill stores potential energy because it is being pulled on by the Earth's gravity but isn't moving. Energy in one form can be transformed into energy in another. When the boulder starts rolling, its potential energy transforms into kinetic energy. Similarly, light energy (a form of electromagnetic energy) can be absorbed by a solar panel, which changes it into electricity that can cause a car to move.

FIGURE Bx1.1 Examples of the forces of nature in everyday life. Mechanical and field forces are very familiar.

(a) Pushing a boulder is a mechanical force.

(b) Gravity, a field force, pulls a person down a zip line.

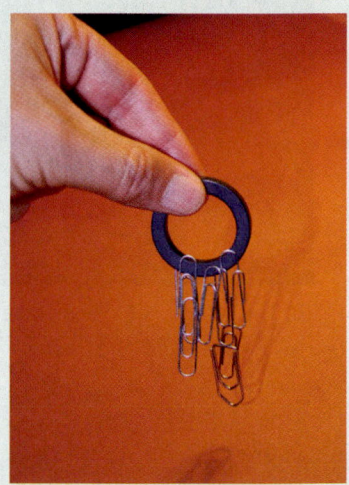

(c) A magnet produces a field force sufficient to hold onto these clips.

FIGURE 1.4 A galaxy may contain about 300 billion stars.

(a) The Milky Way on a clear night. The "haze" actually consists of millions of faraway stars.

(b) A spiral galaxy that looks like the Milky Way, as viewed from the top.

(c) A Hubble Space Telescope view of deep space showing some of the billions of galaxies in the Universe.

FIGURE 1.5 The relative sizes and positions of planets in the Solar System.

(a) Relative sizes of the planets. All are much smaller than the Sun, but the gas-giant planets are much larger than the terrestrial planets. Jupiter's diameter is about 11.2 times greater than that of Earth.

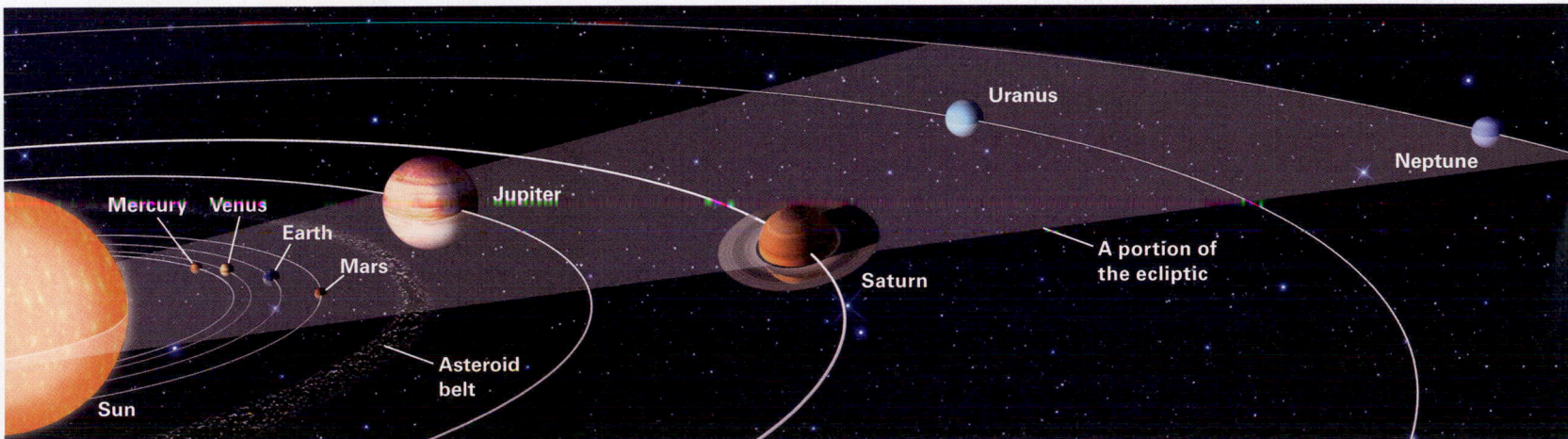

(b) Relative positions of the planets. This figure is not to scale. If the Sun in this figure was the size of a large orange, the Earth would be the size of a sesame seed 15 meters (49 feet) away. Note that all planetary orbits lie roughly in the same plane, called the ecliptic.

eight planets orbit the Sun in the same direction and more or less in the same plane, called the **ecliptic** (**Fig. 1.5b**).

Our Solar System is not alone in hosting planets. Land-based instruments, as well as those of the Kepler Space Telescope (launched into orbit in 2009), have allowed astronomers to locate about 2,000 *exoplanets* (planets that orbit stars other than our Sun) as of 2015. Some of these are similar in size to the Earth. In fact, astronomers now estimate that the Milky Way hosts about 14 billion Earth-sized planets.

Planets in our Solar System differ radically from one another both in size and composition. The inner planets (Mercury, Venus, Earth, and Mars), the ones closer to the Sun, are relatively small. These are the **terrestrial planets** because, like Earth, they consist of a shell of rock surrounding a ball of metal. The outer planets (Jupiter, Saturn, Uranus, and Neptune) are known as the **giant planets**, or Jovian planets. These planets are indeed huge—Jupiter, for example, contains 318 times as much mass as the Earth, and accounts for about 71% of the non-solar mass in the Solar System. The overall composition of giant planets differs markedly from that of terrestrial planets. Specifically, most of the mass of Neptune and Uranus consists of solid water, ammonia, and methane, so these planets are known as the *ice giants*. Most of the mass of Jupiter and Saturn consists of hydrogen and helium, in gas or liquid form—these planets are known as the *gas giants*.

In addition to the planets, the Solar System contains a great many smaller objects. A **moon** is a sizable body locked in orbit around a planet. All but two planets (Mercury and Venus) have moons in varying numbers—Earth has one, Mars has two, and Jupiter has at least 63. Some moons, such as Earth's Moon, are large and spherical, but many are small and have irregular shapes. **Asteroids** are rocky and/or metallic objects, with diameters ranging from less than 1 cm to about 930 km. Millions of asteroids occupy a belt between the orbits of Mars and Jupiter. About a trillion bodies of ice lie outside the orbit of Neptune. (We're using the word *ice* to mean not only frozen water but also other solid materials that could exist in gas form under conditions found at the Earth's surface; examples include carbon dioxide, methane, and ammonia.) Most of these icy bodies are tiny, but a few (including Pluto) are spheres with diameters of over 2,000 km and are known as *dwarf planets*. The largest dwarf planet, Eris, found in 2003, is about 20% larger than Pluto. The gravitational pull of planets has sent some of the icy objects on paths that take them into the inner part of the Solar System, where they release long tails of gas—these objects are **comets** (see Chapter 2).

Developing a Sense of Shape and Scale

The concept that "space is vast" has become ingrained in modern culture—indeed, we use the term *astronomical distance* to mean really, really far away. How did our sense of the scale of the Universe, and the objects in it, come to be?

Let's start by considering the dimensions of the Earth itself. The Greek astronomer Eratosthenes (ca. 276–194 B.C.E.), came up with the first good estimate. Eratosthenes served as chief of the library in Alexandria, Egypt, one of the great ancient centers of learning in the ancient Mediterranean region. One day, he came across a report noting that in the southern Egyptian city of Syene (modern-day Aswan), the Sun lit the base of a deep vertical well precisely at noon on the first day of summer. Eratosthenes deduced that the Sun's rays at noon on this day must be exactly perpendicular to the Earth's surface at Syene, and that if the Earth was spherical, then the Sun's rays could not simultaneously be perpendicular to the Earth's surface at Alexandria, 800 km to the north. So on the first day of summer, Eratosthenes measured the shadow cast by a tower in Alexandria at noon. The angle between the tower and the Sun's rays, as indicated by the shadow's length, proved to be 7.2° (**Fig. 1.6**). He then commanded a servant to pace out a

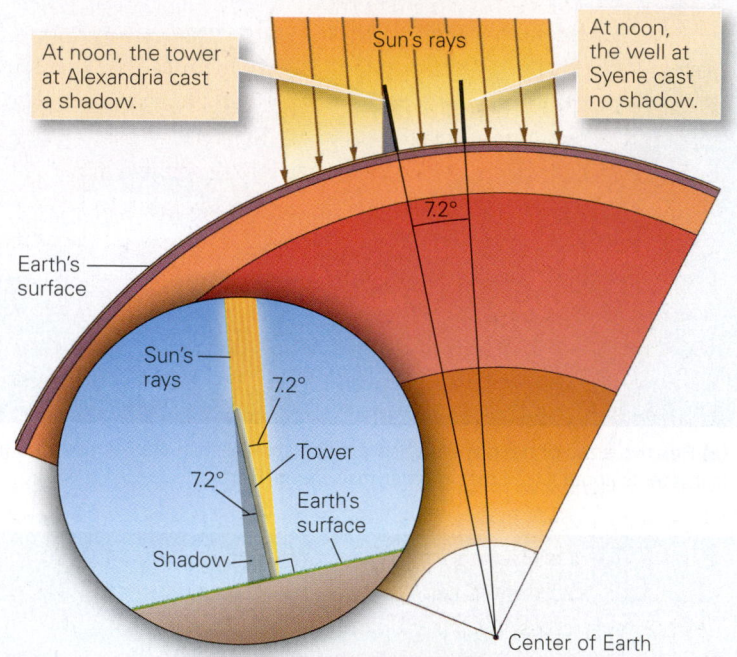

FIGURE 1.6 Over 2,000 years ago, Eratosthenes calculated the circumference of the Earth using simple geometry.

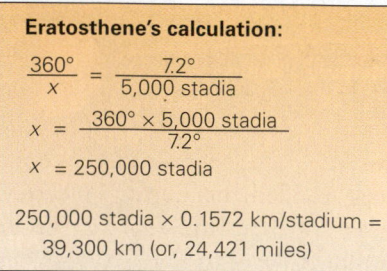

Eratosthene's calculation:

$$\frac{360°}{x} = \frac{7.2°}{5{,}000 \text{ stadia}}$$

$$x = \frac{360° \times 5{,}000 \text{ stadia}}{7.2°}$$

$$x = 250{,}000 \text{ stadia}$$

250,000 stadia × 0.1572 km/stadium = 39,300 km (or, 24,421 miles) straight line from Alexandria to Syene. The sore-footed servant found the distance to be 5,000 stadia (1 stadium = 0.1572 km). Knowing that a circle contains 360°, Eratosthenes then calculated the Earth's circumference using a simple equation, and his answer came within 2% of today's accepted value of 40,008 km (24,865 miles).

Not long after Eratosthenes' discovery, Greek mathematicians used ingenious geometric calculations to estimate that the distance to the Moon is about 30 times the Earth's diameter. This number comes close to the true distance, which on average is 381,555 km (about 237,100 miles). But it wasn't until the 17th century that astronomers figured out that the mean distance between the Earth and the Sun is 149,600,000 km (about 93,000,000 miles). As for the stars, the ancient Greeks realized that they must be much farther away than the Sun in order for them to appear as a fixed backdrop behind the Moon and planets, but the Greeks had no way of calculating the actual distance. Our modern documentation of the vastness of the Universe began in 1838, when astronomers determined that the nearest star to Earth, Alpha Centauri, lies 40.85 trillion km away.

Did you ever wonder . . .

how far away the stars are?

When researchers discovered that light travels at a constant speed of about 300,000 km (about 186,000 miles) per second in space, astronomers realized that they had a convenient way to describe the huge distances between objects in space. They defined a large distance by stating how long it takes for light to traverse that distance. For example, it takes light about 1.3 seconds to travel from the Earth to the Moon, so we can say that the Moon is about 1.3 light seconds away. Similarly, we can say that the Sun is 8.3 light minutes away. A **light year**, the distance that light travels in one Earth year, is about 9.5 trillion km (about 6 trillion miles). When you look up at Alpha Centauri, 4.37 light years distant, you see light that started on its journey to Earth 4.37 light years ago.

Astronomers didn't develop techniques for measuring the distance to very distant stars and galaxies until the 20th century. With these techniques (described in astronomy books), they determined that the Milky Way itself is about 100,000 light years across. Other galaxies are so far away that, to the naked eye, they look like stars in the night sky. The nearest spiral galaxy to ours, Andromeda, lies 2.2 million light years away. The farthest celestial objects that can be seen with the naked eye are less than 3 million light years away. Light from these objects that we see today started on its path to Earth a million years before the first hominids (direct human ancestors) walked the Earth. Powerful telescopes allow us to see much farther. The edge of the visible Universe lies over 13 billion light years away. When such numbers became known,

people came to the realization that the dimensions of the Universe are truly staggering!

It's hard to fathom the distances to planets and stars without visualizing more familiar objects. Imagine a *scale model* (a model in which dimensions are shrunk in proper proportion) in which the Sun is the size of an orange. At this scale, the Earth would be a sesame seed at a distance of 15 m (49 ft) from the orange, and Alpha Centauri would lie 2,000 km (about 1,243 miles) from the orange.

Motions of the Heavens

As you sit in your chair reading this book, you may think that you are motionless, but you aren't. It just seems that way because everything in the room around you is moving at exactly the same velocity as you are. Relative to an observer in intergalactic space, in fact, you're moving quite fast!

Let's consider the components of this motion. First of all, the Earth, like all planets, rotates or "spins" on its *axis*, the imaginary line that passes through the center of the Earth and pierces the planet's surface at its two poles. Earth's axis currently tilts at about 23.4° relative to the ecliptic. Because of this rotation, a person sitting on the **equator** (the circumference on the surface of the Earth, at a distance halfway between the poles) is moving at about 1,674 km per hour (1,040 mph)—faster than the speed of sound! It is due to this spin that the Sun and stars appear to cross the sky daily (**Box 1.2**).

The Earth also orbits the Sun, traveling counterclockwise along a 150 million km-long elliptical path that takes a full year to complete—it moves along this orbit at about 30 km per second (108,000 km per hour). And finally, the whole Solar System revolves around the center of the Milky Way about once every 250 million years, so we hurtle through space, relative to an observer standing outside the Milky Way, at about 200 km per second (720,000 km per hour).

Did you ever wonder . . .

how fast you are traveling through space?

Take-Home Message

People once thought the Earth lay at the center of the Universe. Now it's clear that it is one of eight spinning planets orbiting our Sun, one of 300 billion stars of the revolving, spiral-shaped Milky Way galaxy. Hundreds of billions of galaxies speckle the immense visible Universe.

QUICK QUESTION: Imagine that the distance between the Sun and Alpha Centauri is represented by the length of a city bus. At this scale, what is the diameter of the Milky Way?

BOX 1.2 CONSIDER THIS . . .

How Do We Know That the Earth Rotates?

How do we know that the Earth spins around its axis? The answer comes from observing the apparent motion of the stars (**Fig. Bx1.2a**). If you're in the northern hemisphere (i.e., north of the equator) and gaze at the night sky for many hours, you'll see the stars move in a circular path around the "North Star," which is currently Polaris.

Curiously, it wasn't until the middle of the 19th century that Léon Foucault (1819–68), a French physicist, actually proved that the Earth spins on its axis. He made

this discovery by setting a heavy pendulum, attached to a long cable, in motion (**Fig. Bx1.2b**). As the pendulum continued to swing, Foucault noted that the plane in which it oscillated was perpendicular to the Earth's surface and that this plane rotated around a vertical axis. If Newton's first law of motion—objects in motion remain in motion and objects at rest remain at rest—was correct, this phenomenon meant that the Earth was rotating under the pendulum while the pendulum continued to swing

in the same plane (**Fig. Bx1.2c**). Foucault displayed his discovery beneath the great dome of the Pantheon in Paris, to much acclaim.

We now know that the Earth's spin axis is not perfectly fixed in orientation; rather, it wobbles. This wobble, known as precession, is like the wobble of a toy top as it spins. As a result, Polaris isn't always the North Star. We'll see later in this book that the precession of the Earth's axis, which takes 23,000 years, may affect the planet's climate.

FIGURE Bx1.2 Proving the Earth rotates on its axis.

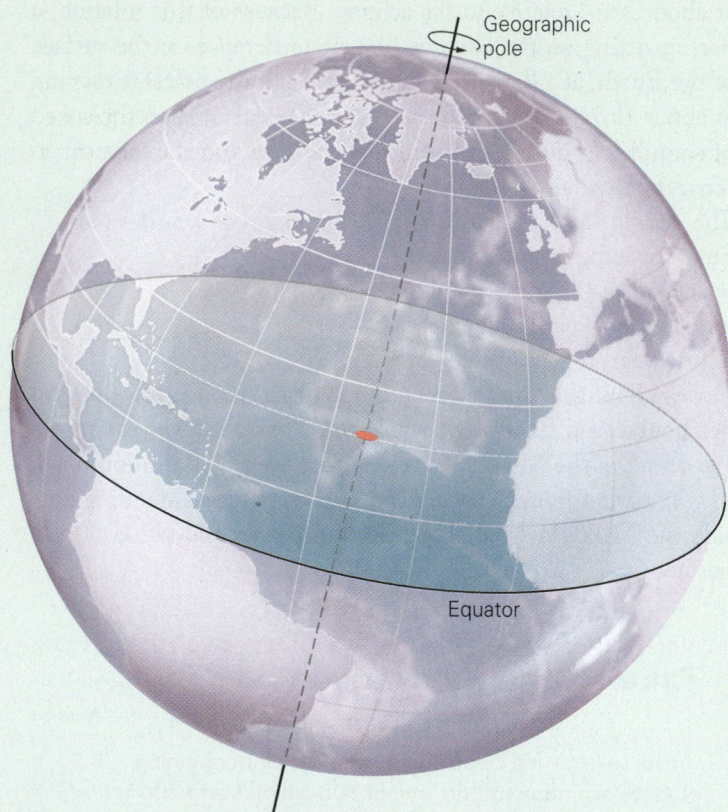

(a) The Earth is a sphere rotating around an axis. The axis, which pierces the Earth at the poles and goes through the center of the planet, is tilted relative to the ecliptic of the Earth's orbit.

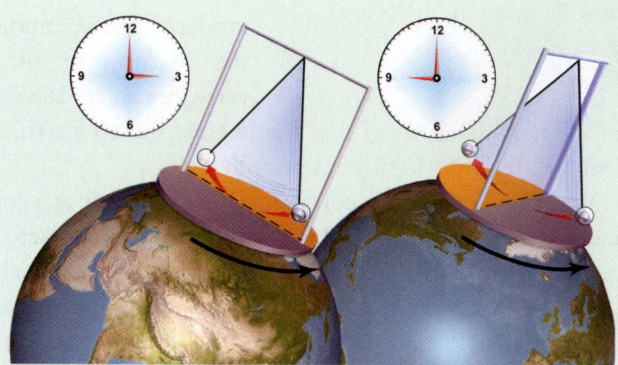

(b) Foucault's experiment. At Time 1, the plane in which the pendulum swings is the same as the plane of the frame. At Time 2, six hours later, the plane in which the pendulum swings is perpendicular to the plane of its frame.

(c) An exact replica of Foucault's original pendulum on display in the Panthéon, Paris.

1.3 Forming the Universe

We stand on a planet, in orbit around a star, speeding through space on the arm of a galaxy. Beyond our galaxy lie hundreds of billions of other galaxies. Where did all this "stuff"—the matter of the Universe—come from, and when did it first form? For most of human history, a scientific solution to these questions seemed intractable. But in the 1920s, unexpected observations about the nature of light from distant galaxies set astronomers on a path of discovery that ultimately led to a scientific model of Universe formation known as the *Big Bang theory*, and this idea has become the foundation of scientific cosmology. To explain these observations, we must first introduce an important phenomenon called the Doppler effect. Thus, we begin this section by developing an understanding of how the Doppler effect modifies the light seen in telescopes. We then show how this understanding leads to the idea that the Universe expands, and then to the conclusion that this expansion began during the Big Bang.

Waves and the Doppler Effect

When a train whistle screams, the sound you hear moved through the air from the whistle to your ear in the form of sound waves. *Waves* are disturbances that transmit energy from one point to another in the form of periodic motions. You're probably most familiar with water waves. As such waves pass, the surface of the water goes up and down in a direction perpendicular to the direction the wave moves. The ups form *crests*, and the downs form *troughs*. Sound waves are different. As a sound wave passes, air moves back and forth, alternately compressing and expanding. The whistle sound is not just one wave but rather is a succession of many waves. We refer to the distance between successive waves as the *wavelength* and the number of waves that pass a point in a given time interval as the **frequency**. If the wavelength decreases, more waves pass a point in a given time interval, so the frequency increases. The "pitch" of a sound, its note on the musical scale, depends on the frequency of the sound waves.

Now imagine that you are standing on a station platform, and a train moves toward you. The train whistle's sound gets louder as the train approaches, but its pitch remains the same. Then, the instant the train passes, the pitch abruptly changes—it sounds like a lower note in the musical scale. Why? When the train moves toward you, the sound has a higher frequency (the waves are closer together, so the wavelength is smaller) because the sound source, the whistle, has moved slightly closer to you between the instant that it emits one wave and the instant that it emits the next (**Fig. 1.7a, b**). When the train moves away from

you, the sound has a lower frequency (the waves are farther apart) because the whistle has moved slightly farther from you between the instant it emits one wave and the instant it emits the next. An Austrian physicist, C. J. Doppler (1803–53), first explained this phenomenon, and thus the change in frequency that happens when a wave source moves is now known as the **Doppler effect**.

Light energy also moves in the form of waves. Physicists consider light to be a form of **electromagnetic radiation**, energy that can be released by hot or glowing objects and can be transmitted through a vacuum. We can represent light waves symbolically by a periodic succession of crests and troughs, which shape-wise resemble water waves but are otherwise very different in character. Visible light comes in many colors—the colors of the rainbow. The color you see depends on the frequency of the light waves, just as a sound you hear depends on the pitch of the frequency of sound waves. Specifically, red light has a longer wavelength (lower frequency) than does blue light.

The Doppler effect also applies to light but can be noticed only if the light source moves very fast, at least a few percent of the speed of light. If a light source moves away from you, the light you see becomes redder, as the light shifts to longer wavelength or lower frequency. If the source moves toward you, the light you see becomes bluer, as the light shifts to higher frequency. We call these changes the **red shift** and the **blue shift**, respectively (**Fig. 1.7c**).

Does the Size of the Universe Change?

In the 1920s, astronomers such as Edwin Hubble, after whom the Hubble Space Telescope was named, braved many a frosty night beneath the open dome of a mountaintop observatory in order to aim telescopes into deep space. These researchers were searching for distant galaxies. At first, they only wanted to document the location and shape of newly discovered galaxies. Eventually, however, they also began to study the wavelength of light produced by the distant galaxies. The results yielded a surprise that would forever change humanity's perception of the Universe. To their amazement, astronomers found that the light coming to the Earth from distant galaxies displayed a red shift, relative to the light coming from nearby stars (**Fig. 1.7d**).

Around 1929, Hubble concluded that the red shift must be a consequence of the Doppler effect, and so galaxies exhibiting a red shift must be moving away from Earth at an immense velocity. At the time of Hubble's discovery, astronomers thought the Universe had a fixed size,

Did you ever wonder...

if galaxies move?

FIGURE 1.7 Manifestations of the Doppler effect for sound and for light.

The Doppler effect for sound

(a) The wavelength of sound waves emitted by a stationary train is the same in all directions.

(b) Waves behind a moving train have a longer wavelength than those in front.

The Doppler effect for light

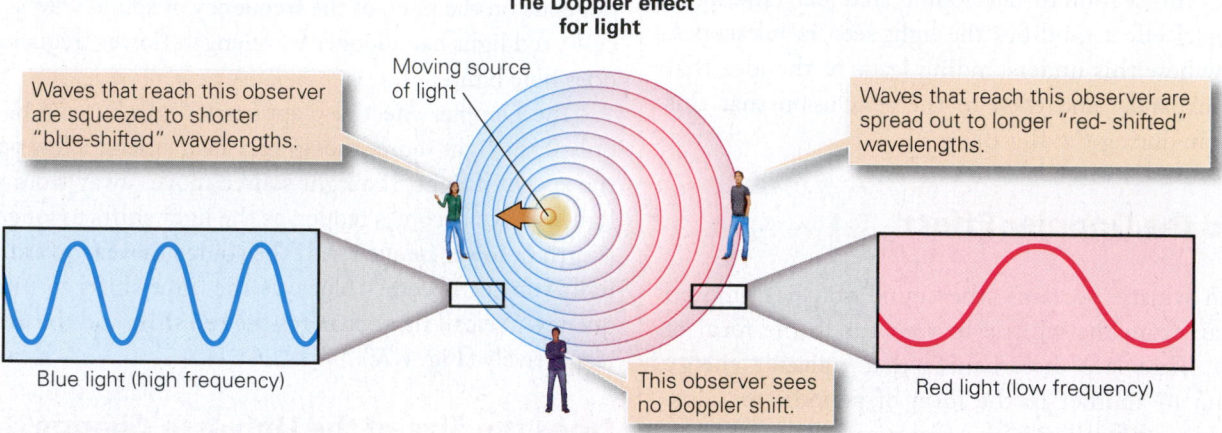

Moving source of light

Waves that reach this observer are squeezed to shorter "blue-shifted" wavelengths.

Waves that reach this observer are spread out to longer "red- shifted" wavelengths.

Blue light (high frequency)

This observer sees no Doppler shift.

Red light (low frequency)

(c) The wavelength of blue light is less than that of red light. If a light source moves very fast, the Doppler effect results in a shifting of the wavelengths. The observed shift depends on the position of the observer.

Sun

Distant galaxy

(d) The atoms in a star absorb certain specific wavelengths of light. We see these wavelengths as dark lines on a light spectrum. Note that the lines from a galaxy a billion light years away are shifted toward the red end of the spectrum (i.e., to the right), in comparison to the lines from our own Sun.

so Hubble initially assumed that if some galaxies were moving away from Earth, others must be moving toward Earth. But none could be found. The light from all distant galaxies, regardless of their direction from Earth, exhibits a red shift. In other words, all distant galaxies are moving rapidly away from us!

How can all galaxies be moving away from us, regardless of which direction we look? Hubble puzzled over this question and finally recognized the solution: the whole Universe must be expanding! This idea came to be known as the **expanding Universe theory**. To picture the expanding Universe, imagine a ball of bread dough with raisins scattered throughout. As the dough bakes and expands into a loaf, each raisin moves away from its neighbors, in every direction (**Fig. 1.8a**). Note that if two raisins were originally 1 cm apart, after a given time interval they spread to 2 cm apart and that during the same time interval raisins that were originally 4 cm apart become 8 cm apart—so the farther apart the raisins are to start with, the faster they move apart. By this analogy, galaxies that are farther away from the Earth are moving away from us faster than are galaxies closer to us, so farther galaxies exhibit a greater red shift than do nearer ones.

FIGURE 1.8 The concept of the expanding Universe and the Big Bang.

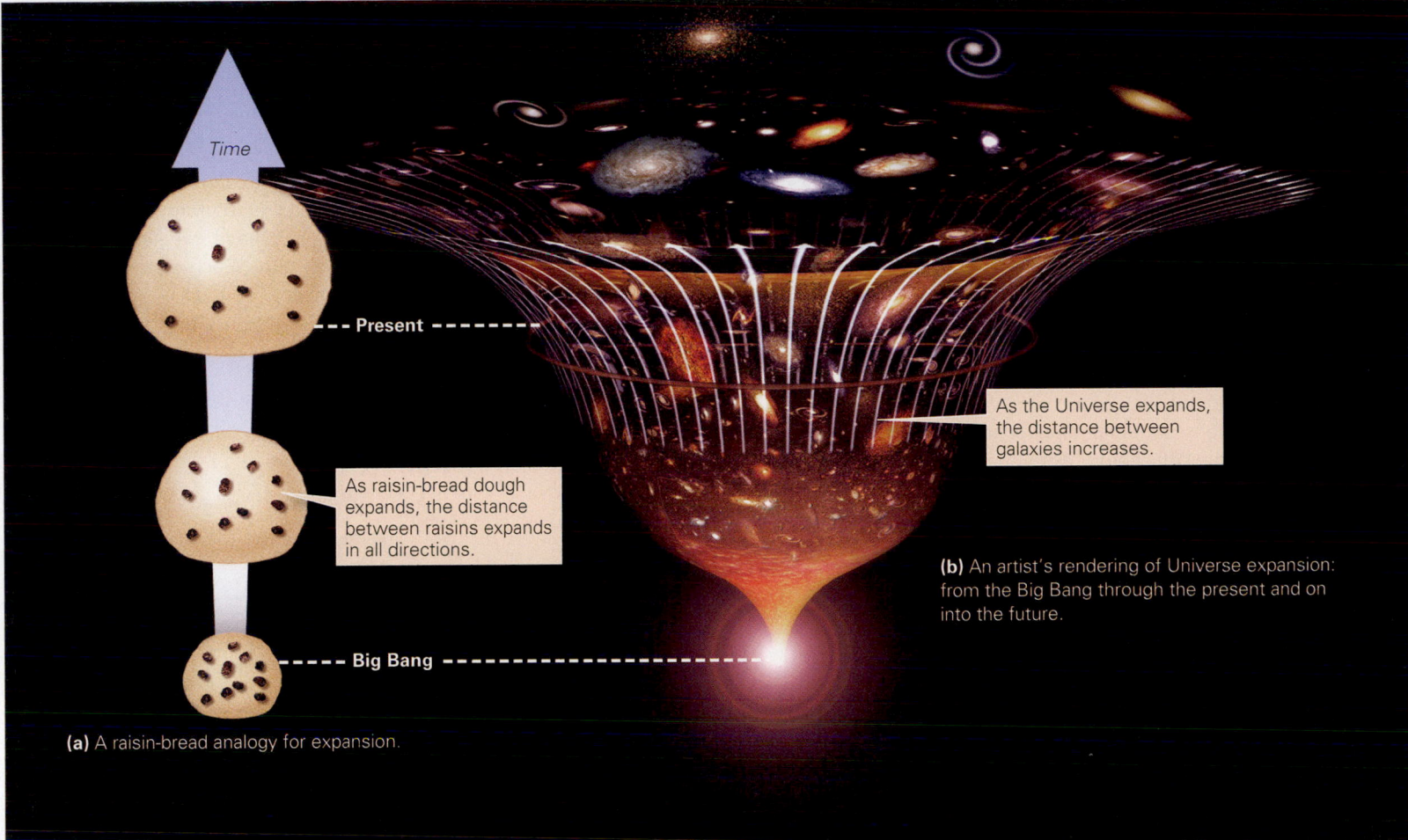

Time

Present

As raisin-bread dough
expands, the distance
between raisins expands
in all directions.

Big Bang

(a) A raisin-bread analogy for expansion.

As the Universe expands,
the distance between
galaxies increases.

(b) An artist's rendering of Universe expansion:
from the Big Bang through the present and on
into the future.

The Big Bang

Hubble's ideas started a revolution in cosmological thinking. Now we picture the Universe as an expanding bubble, in which galaxies race away from each other at incredible speeds. This image immediately triggers a key question of cosmology: Did the expansion begin at some specific time in the past? If it did, then that instant would mark the physical beginning of the Universe, the beginning of our space and time.

Most astronomers have concluded that expansion did indeed begin at a specific time, with a cataclysmic explosion called the Big Bang. According to the **Big Bang theory**, all matter and energy—everything that now constitutes the Universe was initially packed into an infinitesimally small point, called a "singularity." The point exploded and the Universe began, according to current estimates, 13.7 (±1%) Ga (billion years ago).

Of course, no one was present at the instant of the Big Bang, so no one actually saw it happen. But by combining clever calculations with careful observations, researchers have developed a consistent model of how the Universe evolved, beginning an instant after the explosion (**Fig. 1.8b**). According to this model, the Universe was so small, so dense, and so hot during the first instants of its existence that it consisted entirely of energy—atoms, or even the smallest subatomic particles that make up atoms, could not even exist. (See **Box 1.3** for a review of atoms and molecules.) Within a few seconds of cooling, however, hydrogen atoms could begin to form. And by the time the Universe reached an age of 3 minutes, when its temperature had fallen below 1 billion degrees and its diameter had grown to about 53 million km (35 million miles), hydrogen atoms could fuse together to form helium atoms. Formation of new nuclei in the first few minutes of time is called **Big Bang nucleosynthesis** because it happened before any stars existed. This process could produce only small atoms, meaning ones containing a small number of protons (an atomic number less than 5), and it happened very rapidly. In fact, virtually all of the new atomic nuclei consisted of hydrogen and helium, and all of these atoms existed by the end of the first 5 minutes.

BOX 1.3   SCIENCE TOOLBOX . . .

Atoms, Molecules, and the Energy They Contain

What does matter consist of? A Greek philosopher named Democritus (ca. 460–370 B.C.E.) thought that if you could keep dividing a volume of matter into progressively smaller pieces, you would eventually end up with nothing, but that since it's not possible to make something out of nothing, there must be a smallest piece of matter that can't be subdivided further. He proposed the name *atom* for these smallest pieces, from the Greek word *atomos*, which means indivisible.

Our modern understanding of matter developed in the 17th century, when *chemists* (scientists who study the properties, composition, and behavior of matter) recognized that certain substances, such as hydrogen and oxygen, cannot break down into other substances, whereas some substances, such as water and salt, can break down. The former came to be known as **elements**, and the latter came to be known as **compounds**. John Dalton (1766–1844) adopted the word *atom* for the smallest piece of an element that has the property of the element; the smallest piece of a compound that has the properties of the compound is a *molecule*. In 1869, Dmitri Mendeleev (1834–1907) recognized that groups of elements share similar characteristics, and he organized the elements into a chart that we now call the *periodic table* of the elements (see Appendix). The bonds holding atoms together in a molecule are called **chemical bonds**, which we discuss further in Chapter 5 As an example, chemical bonds hold two hydrogen atoms to form an H_2 molecule, and two hydrogen atoms plus an oxygen atom together form an H_2O (water) molecule (**Fig. Bx1.3a**).

During a **chemical reaction**, chemical bonds break and/or form, so atoms can separate from one another and can recombine in new molecules. *Chemical energy* is the potential energy stored in chemical bonds that can be released during a chemical reaction.

The vibration and movement of atoms and molecules produces thermal energy. Put another way, *thermal energy*, or *heat*, is the total kinetic energy contained in a material due to the motion of its particles. *Temperature* represents the average velocity particle movements and thus is a measure of hot and cold. The faster the particles are moving, the higher the temperature. We use a variety of scales—the Celsius scale in the metric system, and the Fahrenheit in the English system—to represent temperature. Water freezes at 0°C, or 32°F, and boils at 100°C, or 212°F.

Chemists in the 17th and 18th centuries identified 92 naturally occurring elements on Earth; physicists have created more than a dozen new ones. Each element has a name and a symbol (e.g., N = nitrogen; H = hydrogen; Fe = iron; Ag = silver). In

FIGURE Bx1.3   The nature of atoms.

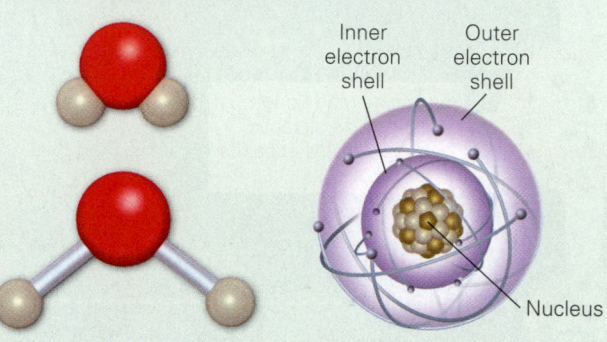

Inner electron shell   Outer electron shell

Nucleus

(a) Two ways of portraying a water molecule. The red ball is oxygen; the small balls are hydrogen.

(b) An image of an atom with a nucleus orbited by electrons.

Nuclear Reactions

Fission

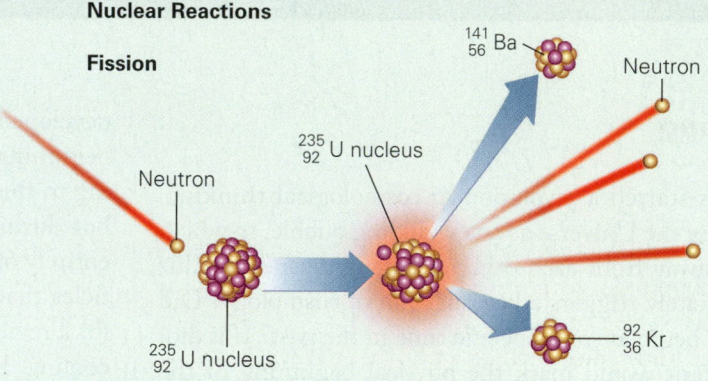

$^{141}_{56}$ Ba

Neutron

$^{235}_{92}$ U nucleus

Neutron

$^{92}_{36}$ Kr

$^{235}_{92}$ U nucleus

(c) A uranium atom splits during nuclear fission.

Fusion

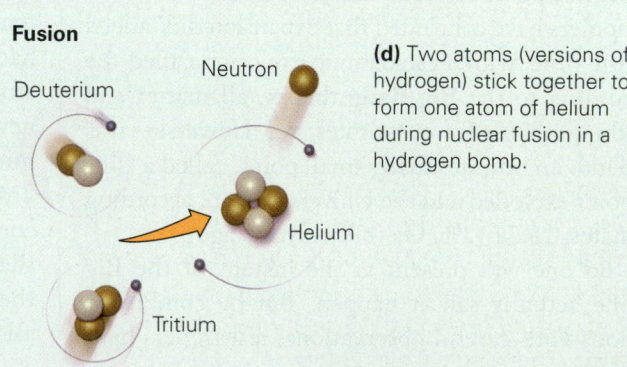

Neutron

Deuterium

Helium

Tritium

(d) Two atoms (versions of hydrogen) stick together to form one atom of helium during nuclear fusion in a hydrogen bomb.

1910, Ernest Rutherford, a British physicist, proved that, contrary to the view of Democritus, atoms actually can be divided into smaller pieces. Most of the mass in an atom clusters in a dense ball, called the *nucleus*, at the atom's center. The nucleus contains two types of subatomic particles: *neutrons*, which have a neutral electrical charge, and *protons*, which have a positive charge. A cloud of electrons surrounds the nucleus (**Fig. Bx1.3b**); an *electron* has a negative charge and contains only 1/1,836 as much mass as a proton. (*Charge*, simplistically, refers to the way in which a particle responds to a magnet or an electric current; unlike charges attract, while like charges repel).

Electron clouds have a complex internal structure—electrons are grouped into intervals called orbitals, energy levels, or *electron shells*. Electrons in inner shells concentrate closer to the nucleus, and those in outer shells are farther away. Roughly speaking, the diameter of an electron cloud is 10,000 times greater than that of the nucleus, yet the cloud contains only 0.05% of an atom's mass—thus, atoms are mostly empty space! Electrons in an orbital store potential energy, so when they move from one orbital to another, they release or absorb energy. Electromagnetic energy can drive this movement.

We distinguish atoms of different elements from one another by their *atomic number*, the number of protons in their nucleus. Smaller atoms have smaller atomic numbers, and larger ones have larger atomic numbers. The smallest atom, hydrogen, has an atomic number of 1, and the largest naturally occurring atom, uranium, has an atomic number of 92. Except for hydrogen nuclei, all nuclei also contain neutrons. In smaller atoms, the number of neutrons roughly equals the number of protons, but in larger atoms the number of neutrons exceeds the number of protons. The *atomic mass* of an atom is roughly the sum of the number of neutrons and the number of protons. For example, an oxygen nucleus contains 8 protons and 8 neutrons and thus has an atomic mass of 16.

The bonds holding subatomic particles in the nucleus, called *nuclear bonds*, store potential energy. Atoms can change only during *nuclear reactions*, when nuclear bonds break or form. Physicists recognize several types of nuclear reactions. For example, during *fission*, a large nucleus breaks apart to form two smaller atoms. Fission generates energy in nuclear power plants and in the explosion of an atomic bomb. During *fusion*, two smaller atoms come together to form a larger atom. Fusion reactions power the Sun and occur during the explosion of a hydrogen bomb (**Fig. Bx1.3c, d**).

Matter can exist in different forms depending on the degree to which the atoms are bonded together. These are called *states of matter* (discussed further in Chapter 5). A material in which all the atoms or molecules are bonded tightly together is a *solid*; solids can hold their shape and can't flow easily. A material in which the atoms are not held together so well and occur in disorganized clumps or chains that can flow past each other is a *liquid*; though a liquid can flow, it stays together in a confined volume. In a *gas*, atoms or molecules are not bonded to each other at all but rather move around rapidly in all directions; thus, a gas expands to fill any container it's placed in, completely. A substance can change state. For example, during *condensation*, a gas becomes a liquid and during *freezing*, a liquid becomes a solid.

Now that we've reviewed the basic structure of matter, we can quickly return to the discussion of energy that we started in Box 1.1. There are many types of energy that can be understood in the context of the behavior of matter, such as *thermal energy* or *heat* and *chemical energy*, as defined above. Also, nuclear reactions yield *nuclear energy* because when nuclear bonds break, a small amount of matter converts into energy following the famous equation $E = mc^2$, introduced by Albert Einstein (1879–1955); in this equation, E stands for energy, m stands for mass, and c stands for the speed of light.

The matter and energy that we've just discussed, according to calculations by physicists, accounts for only a small portion (4.6%) of the whole universe. The remainder consists of mysterious *dark matter* and *dark energy*. They are called dark because we cannot see them.

Eventually, the Universe cooled enough for chemical bonds to bind atoms together in molecules. Most notably, hydrogen atoms bonded to form molecular hydrogen (H_2). As the Universe expanded and cooled further, atoms and molecules slowed down and accumulated into patchy clouds called **nebulae.** The earliest nebulae of the Universe consisted almost entirely of hydrogen and helium gas.

Birth of the First Stars

When the Universe reached its 200 millionth birthday, it contained immense, slowly swirling, dark nebulae separated by vast voids of empty space (**Fig. 1.9**). The Universe could not remain this way forever, though, because of the invisible but persistent pull of gravity. Eventually, gravity began to remold the Universe pervasively and permanently.

All matter exerts gravitational pull—a type of force—on its surroundings, and as Isaac Newton first pointed out, the amount of pull depends on the amount of mass. Somewhere in the young Universe, the gravitational pull of an initially more massive region of a nebula began to suck in surrounding gas and, in a grand example of the rich getting richer, grew in mass and therefore **density** (mass per unit volume). As this denser region attracted progressively more gas, the gas compacted into a smaller region, and the initial swirling movement of gas transformed into a rotation around an axis. As gas continued to move inward, cramming into a progressively smaller volume, the rotation rate became faster and faster. (A similar phenomenon happens when a spinning ice skater pulls her arms inward.) Because of its increased rotation, the nebula evolved into a disk shape. As more and more matter rained down onto the disk, it continued to grow, until eventually, gravity collapsed the inner

FIGURE 1.9 A representation of a starless nebula in the early Universe. This rendition is modified from a Hubble Space Telescope photograph.

can be seen from Earth, where they appear as "new" stars in the sky. Thus, not long after the first generation of stars formed, the Universe began to be peppered with the first generation of supernovae.

Take-Home Message

Study of the red shift shows that all distant galaxies are moving away from us, an observation that requires the Universe to be expanding. According to the Big Bang theory, this expansion, and thus the beginning of our Universe, began with a cataclysmic explosion at 13.7 Ga. Atoms formed during the Big Bang collected into nebulae which, due to gravity, collapsed into dense balls that began to produce energy by nuclear fusion reactions. These objects were the first stars.

QUICK QUESTION: The farther a galaxy lies from the Earth, the greater the red shift it exhibits. Why?

1.4 We Are All Made of Stardust

Where Do Elements Come From?

Nebulae from which the first-generation stars formed consisted entirely of small atoms (atoms with atomic numbers less than 5) because only these small atoms were generated by Big Bang nucleosynthesis. In contrast, the Universe of today contains 92 naturally occurring elements. Where did the other 87 elements come from? In other words, how did such as carbon, sulfur, silicon, iron, gold, and uranium form? These elements, which are common on Earth, have atomic numbers that are greater than 5. For example, carbon has an atomic number of 6, and iron has an atomic number of 26. Physicists have shown that these elements form during the life cycle of stars, by the process of **stellar nucleosynthesis**. Because of stellar nucleosynthesis we can consider stars to be "element factories," constantly fashioning larger atoms out of smaller atoms.

What happens to the atoms produced by stellar nucleosynthesis? Some escape into space during the star's lifetime, simply by moving fast enough to overcome the star's gravitational pull. The stream of atoms emitted from a star during its lifetime is a **stellar wind** (**Fig. 1.10a**). Most escape only when a star dies. A low-mass star (like our Sun) releases a large shell of gas as it dies, ballooning into a *red giant* during the process, whereas a high-mass star blasts matter into space during a supernova explosion (**Fig. 1.10b**). Most very large atoms (those with atomic numbers greater than that of iron) require even more violent circumstances to form than generally occurs

portion of the disk into a dense ball. As the gas was compressed into a smaller and smaller space, its temperature increased dramatically. Eventually, the central ball of the disk became hot enough to glow, and at this point it became a **protostar**. The remaining mass of the disk, as we will see, eventually clumped into smaller spheres, the planets.

A protostar continues to grow, by pulling in successively more mass, until its core becomes extremely dense and its temperature reaches about 10,000,000°C. Under such conditions, hydrogen nuclei slam together so forcefully that they undergo a series of fusion reactions, ultimately forming helium nuclei. Such fusion reactions produce huge amounts of energy, and the mass becomes a fearsome furnace. When the first nuclear fusion reactions began in the first protostar, the body "ignited" and the first true star formed. When this happened, perhaps 800 million years after the Big Bang, the first starlight illuminated the newborn Universe. This process would soon happen again and again, and many *first-generation stars* came into existence.

First-generation stars tended to be very massive, perhaps 100 times the mass of the Sun. Astronomers have shown that the larger the star, the hotter it burns and the faster it runs out of fuel and dies. A huge star may survive only a few million years to a few tens of millions of years before it explodes. The explosion of a giant star produces a *supernova* (from the Latin word *nova*, meaning new), so named because closer examples

FIGURE 1.10 Element factories in space.

(a) A photo of solar (stellar) wind streaming into space.

(b) Very heavy elements form during supernova explosions. Here we see the rapidly expanding shell of gas ejected into space from an explosion whose light reached the Earth in 1054 C.E. This shell is called the Crab Nebula.

within a star. In fact, most very large atoms form by *supernova nucleogenesis*, which takes place in the inconceivable heat of a supernova explosion.

When the first generation of stars died, it left a legacy of new, heavier elements that then mixed with residual gas from the Big Bang. A second generation of stars and associ-

ated planets formed out of these compositionally more diverse nebulae. Second-generation stars lived and died and contributed heavier elements to third-generation stars. Succeeding generations contain a greater proportion of heavier elements. (Nevertheless, most of the visible mass of the Universe still consists of 74% hydrogen and 24% helium.) Because not all stars live for the same duration of time, at any given moment the Universe contains many different generations of stars. Our Sun may be a third-, fourth-, or fifth-generation star. Thus, the mix of elements we find on Earth includes relicts of primordial gas from the Big Bang as well as the disgorged guts of dead stars. Think of it—the elements that make up your body once resided inside a star! The proportions of elements reflect the processes of element formation and the degree to which stellar gases and supernova gases mix in (**Table 1.1**).

> **Did you ever wonder . . .**
>
> where the atoms in your body first formed?

The Nebular Theory for Forming the Solar System

Earlier in this chapter, we introduced scientific concepts of how stars form from nebulae that consisted mostly of small atoms formed by Big Bang nucleosynthesis. But we delayed addressing the question of how the planets and other objects in our Solar System originated until we had discussed the production of heavier elements, such as oxygen, silicon, and iron, because planets such as the Earth consist predominantly of these elements. Now that we've shown how stars and supernovae can

TABLE 1.1 Top 10 Elements in the Milky Way (out of 1 Million Atoms)

Hydrogen	739,000
Helium	240,000
Oxygen	10,400
Carbon	4,600
Neon	1,340
Iron	1,090
Nitrogen	960
Silicon	650
Magnesium	580
Sulfur	440
Other (approx.)	940

serve as element factories for larger atoms, we return to the early history of the Solar System and add an explanation for the origin of planets, moons, asteroids, and comets.

In the context of scientific cosmology, the Sun and all other objects in the Solar System formed from material that had been swirling about in a nebula, an idea now known as the **nebular theory** (**Fig. 1.11**). According to the theory, the process of Solar System formation involved several stages (**Geology at a Glance**, pp. 30–31). First, as we've seen, a swirling nebula coalesces into a spinning disk with a bulbous center. The central "bulb" of this disk became the Sun, whereas the remainder of the Solar System formed from the material in the flattened outer part of the disk, a region now known as the **protoplanetary disk**.

The protoplanetary disk contained all 92 elements, some as isolated atoms and some bonded to others in molecules, a mixture of gases produced during the Big Bang as well as gases expelled from earlier generations of stars and from supernova explosions. Geologists divide the materials formed from these atoms and molecules into two classes. **Volatile materials**— such as hydrogen, helium, methane, ammonia, water, and carbon dioxide—can exist as gas at the Earth's surface. In the pressure and temperature conditions of space, all volatile materials remain in a gaseous state closer to the Sun. But beyond a distance called the *frost line*, volatiles can freeze to form into ice. (Note again that we don't limit use of the word *ice* for frozen water alone but instead use it for any frozen volatile material.) **Refractory materials** are those that melt only at high temperatures, and they solidify to form solid soot-sized particles of "dust" in the coldness of space. Some of this dust consisted mostly of metal, but a large portion of the dust formed in this way contains molecules of silicon and oxygen bonded to various metal atoms. We'll see in Chapter 5 that such materials are known as *silicate minerals* and that they form most of the rock of the Earth, so clumps of silicate-mineral dust were rock-like in character.

Initially, the protoplanetary disk may have been fairly homogeneous, meaning that it had much the same composition throughout, and most volatiles were frozen. But as the proto-Sun began to form, the inner part of the disk became hotter and the frost line moved outward. The volatile materials closer to the Sun evaporated, and when the Sun became a nuclear inferno, the *solar wind* (the stellar wind produced by our Sun) blew these materials to the outer portions of the disk. Thus, the inner part of the protoplanetary disk ended up with relatively higher concentrations of dust, whereas the outer portions ended up with relatively higher concentrations of ice. As this was happening, gravity caused the gas, dust, and ice of the disk to separate into a series of concentric rings in which density exceeded that of the space between the rings.

How did the dusty, icy, and gassy rings transform into planets? Even before the proto-Sun ignited, the material of the surrounding rings began to clump and bind together due to gravity (**Fig 1.12**). First, soot-sized particles merged into sand-sized grains. Then these grains stuck together to form grainy basketball-sized blocks, which in turn collided. If the collision was slow, blocks stuck together or simply bounced apart. If the collision was fast, one or both of the blocks shattered, producing smaller fragments that recombined later. Eventually, enough blocks coalesced to form **planetesimals**, solid

FIGURE 1.11 Nebular theory for planet formation.

Time

Nebula

Protoplanetary disk

Rings of planetesimals

The 8 planets

FIGURE 1.12 Forming solids in the Solar System.

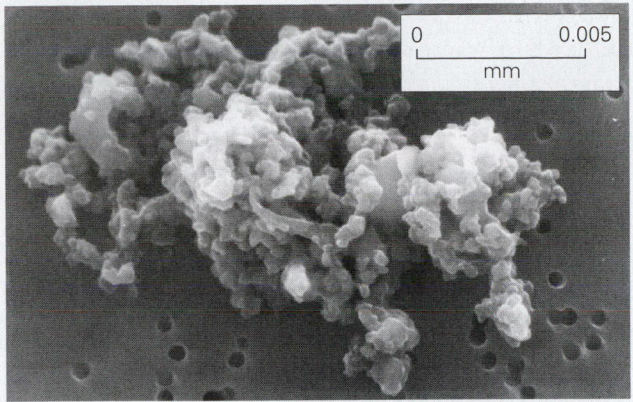

(a) At first, solids collected into dust-sized aggregates.

(c) The disk then evolved into a revolving ground of debris.

(b) Eventually, the dust collected into grainy masses, similar to this meteorite fragment.

bodies whose diameter exceeded about 1 km. Because of their mass, planetesimals exerted enough gravity to attract and pull in other objects that were nearby. Figuratively, planetesimals acted like vacuum cleaners, sucking in small pieces of dust and ice as well as smaller planetesimals that lay in their orbit, and in the process they grew progressively larger—astronomers refer to this process of planetary growth as *accretion*. Eventually, victors in the competition to attract mass grew into **protoplanets**, bodies approaching the size of today's inner planets. Once a protoplanet succeeded in incorporating virtually all the debris within its orbit, it became a full-fledged planet.

Early stages in the planet-forming process probably occurred very quickly—some computer models suggest that it may have taken less than 1 million years to go from the dust-and-gas stage to the large planetesimal stage. Planets may have grown from planetesimals in 10 million to 200 million years. In the inner orbits, where rings of the protoplanetary disk consisted mostly of dust (refractory materials), small terrestrial planets composed mostly of rock and metal formed. The outer

rings of the protoplanetary disk, in contrast, contained huge volumes of volatile materials. Protoplanets formed in the outer rings gathered up the volatile material to form the giant planets. Refractory materials that occur in these planets sank to the center, so these planets consist of a small refractory (metal and rock) ball, surrounding by very thick shells of volatile materials (in the form of gas, liquid, or ice).

When did the planets form? As we'll discuss later in the book, rocks now exposed on the surface of the Earth are much younger than the Solar System. But certain *meteorites*, objects that fell to Earth from space (see Chapter 2), appear to be leftover planetesimals. Using dating techniques introduced in Chapter 12, geologists have determined that the materials comprising these meteorites formed at about 4.54 Ga and thus consider that date to be the birth date of the Solar System. If this date is correct, it means that the Solar System formed about 9 billion years after the Big Bang and thus is only about a third as old as the Universe.

Differentiation of the Earth and Formation of the Moon

When planetesimals first formed, they had a fairly homogeneous distribution of material throughout because the smaller pieces from which they formed all had much the same composition and collected together in no particular order. But large planetesimals did not stay homogeneous for long because they

Forming the Planets and
the Earth-Moon System

1. Forming the Solar System, according to the nebular theory: A nebula forms from hydrogen and helium left over from the Big Bang, as well as from heavier elements that were produced by fusion reactions in stars or during explosions of stars.

2. Gravity pulls gas and dust inward to form an accretionary disk. Eventually a glowing ball—the proto-Sun—forms at the center of the disk.

5. Forming the planets from planetesimals: Planetesimals grow by continuous collisions. Gradually, an irregularly shaped proto-Earth develops. The interior heats up and becomes soft.

6. Gravity reshapes the proto-Earth into a sphere. The interior of the Earth differentiates into a core and mantle.

7. Soon after the Earth forms, a protoplanet collides with it, blasting debris that forms a ring around the Earth.

8. The Moon forms from the ring of debris.

3. "Dust" (particles of refractory materials) concentrates in the inner rings, while "ice" (particles of volatile materials) concentrates in the outer rings. Eventually, the dense ball of gas at the center of the disk becomes hot enough for fusion reactions to begin. When it ignites, it becomes the Sun.

4. Dust and ice particles collide and stick together, forming planetesimals.

9. Eventually, the atmosphere develops from volcanic gases. When the Earth becomes cool enough, moisture condenses and rains to create the oceans. Some gases may be added by passing comets.

began to heat up. The heat came primarily from three sources: the heat produced during collisions (similar to the phenomenon that happens when you bang on a nail with a hammer and they both get warm); the heat produced when matter is squeezed into a smaller volume; and the heat produced from the decay of radioactive elements (similar to the heat produced by a nuclear reactor). In bodies whose temperature rose sufficiently to cause internal melting, denser metals (mostly iron) separated out and sank to the center of the body, whereas lighter rocky materials (mostly silicate minerals) remained in a shell surrounding the center. By this process, called **differentiation**, protoplanets and large planetesimals developed internal layering early in their history (**Fig. 1.13**). As we will see later, the central ball of metal constitutes the body's *core* and the outer, rocky shell constitutes its *mantle*.

In the early days of the Solar System, planets continued to be bombarded by meteorites even after the Sun had ignited and differentiation had occurred. Heavy bombardment in the early days of the Solar System pulverized the surfaces of planets and eventually left huge numbers of craters. Bombardment also contributed to heating the planets. Most geologists favor a model in which a particularly large collision between the Earth and a large planetesimal or protoplanet at about 4.53 Ga produced our Moon. Moon formation happened because the collision was so cataclysmic that much of the colliding body disintegrated and evaporated, along with a large part of the Earth's mantle. A ring of debris formed around the remaining, now-molten Earth. This ring quickly coalesced by accretion to form the Moon. When first formed,

> **Did you ever wonder . . .**
>
> if the Moon is as old as the Earth?

the Moon was much closer to the Earth than it is today. Of note, not all moons in the Solar System necessarily formed in this manner. Some may have been independent protoplanets or comets that were captured by a larger planet's gravity.

Making the Earth Round

Small planetesimals were jagged or irregular in shape, and asteroids today have irregular shapes. Planets, on the other hand, are more or less spherical. Why? Simply put, when a protoplanet gets big enough, gravity can change its shape. To picture how, imagine a block of cheese warming in an oven. As the cheese gets softer and softer, gravity causes it to spread out in a pancake-like blob. This model shows that gravitational force alone can cause material to change shape if the material is soft enough. Now let's apply this model to planetary growth.

The rock composing a small planetesimal is cool and strong enough so that the force of gravity is not sufficient to cause the rock to flow. But once a planetesimal grows beyond a certain critical size (about 1,000 km in diameter), its interior becomes warm and soft enough to flow in response to gravity. As a consequence, protrusions are pulled inward toward the center, and indentations rise, so that the

FIGURE 1.13 Differentiation of the Earth's interior.

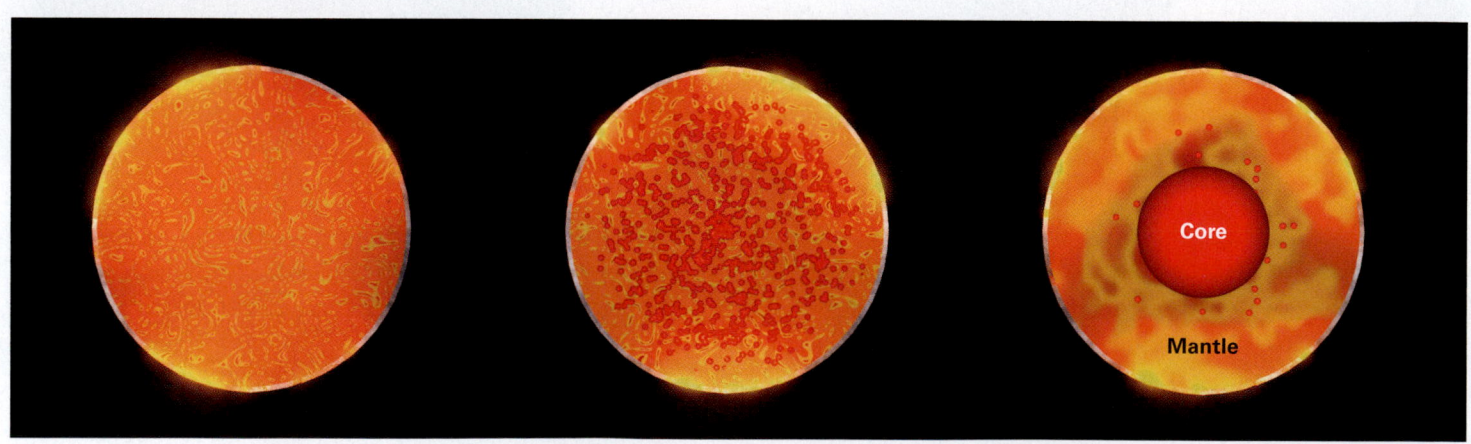

Time

(a) Early on, the Earth was fairly homogeneous inside.

(b) When the temperature got hot enough, iron began to melt.

(c) The iron accumulated at the center of the planet to form a metallic core.

Manicouagan Crater

LATITUDE
51°19'50.07"N

LONGITUDE
68°40'59.96"W

Zoom to an elevation of 175 km (110 mi) and look straight down.

Manicouagan Crater of Quebec, Canada is 65 km across. The rim of the crater has filled with a lake.

planetesimal re-forms into a special shape—a sphere—that permits the force of gravity to be nearly the same at all points on its surface, for in a sphere mass is evenly distributed around the center.

Forming the Ocean and Atmosphere

Unlike the other terrestrial planets, the Earth today has an ocean of liquid water and an atmosphere that consists primarily of molecular nitrogen (N_2) and molecular oxygen (O_2). Where did the ocean and atmosphere come from? Let's take a brief look at how these entities, without which life as we know it could not exist, came to be.

When the Earth first formed, its atmosphere probably consisted mostly of molecular hydrogen and helium. But as the planet warmed, the temperature of these gases increased and the atoms were able to move so fast that, like rockets, they escaped the pull of gravity and were blown away from the Earth by the solar wind. Gradually, they were replaced by gases released from erupting volcanoes. These gases (volatile materials) were originally bonded to solid material of the Earth's mantle. But when portions of the mantle melted to produce molten rock that rose to the surface, the volatiles separated from the solids and bubbled

out. Comets bombarding the Earth may have brought in additional gases. The atmosphere evolved into one consisting mostly of water, carbon dioxide (CO_2), ammonia, and methane—all gases released by "outgassing" of the interior. You would suffocate instantly if you were to breathe this early atmosphere.

When the Earth cooled sufficiently for water to condense (changing the state from gas to liquid), rain fell and the oceans accumulated. The concentration of water in the atmosphere decreased substantially as a consequence. The CO_2 from the atmosphere then dissolved in the water and precipitated into solids, which eventually became trapped in the crust as rock. Thus, the concentration of CO_2 diminished substantially. Because it does not react with other earth materials, N_2 remained in the atmosphere and eventually became the dominant component of the atmosphere. Only later, after the appearance of organisms capable of carrying out photosynthesis, an O_2-generating reaction, did molecular oxygen appear in the atmosphere. As discussed in Chapter 13, the concentration of O_2 did not become significant until 600 million years ago.

Take-Home Message

Heavier elements formed in stars and supernovae added to gases in nebulae from which new generations of stars formed. Planets formed from rings of dust and ice orbiting the stars, so we are all formed of stardust. As they formed, planets differentiated, with denser materials sinking to the center.

QUICK QUESTION: Why are the gas-giant planets further from the Sun than are the terrestrial planets?

CHAPTER SUMMARY

- The geocentric model of the Universe placed the Earth at the center of the Universe, with the planets and Sun orbiting around the Earth within a celestial sphere speckled with stars. The heliocentric model, which gained acceptance during the Renaissance, placed the Sun at the center.

- Eratosthenes was able to measure the size of the Earth in ancient times, but it was not until fairly recently that astronomers accurately determined the distances to the Sun, planets, and stars. Distances in the Universe are so large that they must be measured in light years.

- The Earth is one of eight planets orbiting the Sun, and this Solar System lies on the outer edge of a slowly revolving galaxy, the Milky Way, which is composed of about 300 billion stars. The Universe contains at least hundreds of billions of galaxies.

- The red shift of light from distant galaxies, a manifestation of the Doppler effect, indicates that all distant galaxies are moving away from the Earth. This observation leads to the expanding Universe theory. Most astronomers agree that this expansion began after the Big Bang, a cataclysmic explosion that occurred about 13.7 Ga.

- The first atoms (hydrogen and helium) of the Universe developed within minutes of the Big Bang. These atoms formed vast gas clouds called nebulae.
- Gravity caused clumps of gas in the nebulae to coalesce into revolving balls. As these balls of gas collapsed inward, they evolved into flattened disks with bulbous centers. The protostars at the center of these disks eventually became dense and hot enough that fusion reactions began in them. When this happened, they became true stars, emitting heat and light.
- Heavier elements form during fusion reactions in stars; the heaviest are mostly made during supernova explosions. The Earth and the life forms on it contain elements that could only have been produced during the life cycle of stars. Thus, we are all made of stardust.
- According to the nebular theory of planet formation, planets developed from the rings of gas and dust surrounding protostars. The gas and dust condensed into planetesimals, which then clumped together to form protoplanets and finally true planets. Inner rings became the terrestrial planets. Outer rings grew into giant planets.
- The Moon formed from debris ejected when a protoplanet collided with the Earth in the young Solar System.
- A planet assumes a near-spherical shape when it becomes so soft that gravity can smooth out irregularities.

GUIDE TERMS

asteroid (p. 18)

astronomer (p. 14)

Big Bang nucleosynthesis (p. 23)

Big Bang theory (p. 23)

blue shift (p. 21)

chemical bond (p. 24)

chemical reaction (p. 24)

celestial object (p. 13)

comet (p. 18)

compound (p. 24)

cosmology (p. 13)

density (p. 25)

differentiation (p. 32)

Doppler effect (p. 21)

ecliptic (p. 18)

electromagnetic radiation (p. 21)

element (p. 24)

energy (p. 13)

equator (p. 19)

expanding Universe theory (p. 22)

frequency (p. 21)

galaxy (p. 15)

giant planets (p. 18)

geocentric model (p. 14)

gravity (p. 15)

heliocentric model (p. 14)

light year (p. 19)

magnetism (p. 16)

mass (p. 13)

matter (p. 13)

moon (p. 18)

nebula (p. 25)

nebular theory (p. 28)

planet (p. 15)

planetesimal (p. 28)

protoplanetary disk (p. 28)

protoplanet (p. 29)

protostar (p. 26)

red shift (p. 21)

refractory materials (p. 28)

Solar System (p. 15)

star (p. 15)

stellar nucleosynthesis (p. 26)

stellar wind (p. 26)

terrestrial planets (p. 18)

Universe (p. 13)

volatile materials (p. 28)

REVIEW QUESTIONS

1. Why do planets appear to move with respect to stars?
2. Contrast the geocentric and heliocentric Universe concepts.
3. Describe how Foucault's pendulum demonstrates that the Earth is rotating on its axis.
4. How did Eratosthenes calculate the Earth's circumference?
5. Imagine you hear the main character in a low-budget science-fiction movie say he will "return ten light years from now." What's wrong with his usage of the term?
6. Describe how the Doppler effect works.
7. What does the red shift of the galaxies tell us about their motion with respect to the Earth?
8. What is the Big Bang, and when did it occur?
9. When did hydrogen and helium atoms form?
10. Where did heavier elements form?
11. Describe the steps in the formation of the Solar System according to the nebular theory.
12. Describe how the Moon was formed.
13. Why is the Earth round?

ON FURTHER THOUGHT

14. Look again at Figure 1.2b. The North Star, a particularly bright star, lies just about at the center of the circles of light tracked out by other stars. (a) What does this mean about the position of the North Star relative to Earth's spin axis? Why is it called the North Star? (b) Consider the wobble of Earth's axis. Will the North Star be in the same position in a photograph taken from the same location in the future? Why?

15. The horizon is the line separating sky from the Earth's surface. Consider the shape of the Earth. How does the distance from your eyes to the horizon change as your elevation above the ground increases? To answer this question, draw a semicircle to represent part of the Earth's surface; then draw a vertical tower up from the surface. With your ruler, draw a line from various elevations on the tower to where the line is tangent to the surface of the Earth. (A tangent is a line that touches a circle at one point and is perpendicular to a radius.)

16. Astronomers discovered that more-distant galaxies move away from the Earth more rapidly than do nearer ones. Why? To answer this question, make a model of the problem by drawing three equally spaced dots along a line; the dot at one end represents the Earth, and the other two represent galaxies. "Stretch" the line by drawing the line and dots again, but this time make the line twice as long. This stretching represents Universe expansion. Notice that the dots are now farther apart. Using the following equation for velocity (Velocity = Distance/Time)—if you pretend that it took 1 second to stretch the line (so Time = 1 second), measurement of the distance that each galaxy moved relative to the Earth allows you to calculate velocity.

17. Consider that the deaths of stars eject quantities of heavier elements into space and that these elements then become incorporated in nebulae from which the next generation of stars forms. Do you think that the ratio of heavier to lighter elements in, say, a sixth-generation star is larger or smaller than the ratio in a second-generation star? Why?

 smartwork.wwnorton.com

This chapter's Smartwork features:

- Labeling exercises on the correct relative positions of bodies in the Solar System.
- A visual exercise identifying the structure of an atom.
- In-depth questions on how the Doppler effect works.

GEOTOURS

This chapter's GeoTour exercise (A) features:

- Nebular supernova
- Spiral galaxy
- Meteorite impacts

Another View

(a) The Cat's Eye Nebula is about 3,300 light years away. It formed when a dying star shed a shell of gas.

(b) The Andromeda Galaxy, 2.2 million light years away. It's about the same size as our own Milky Way.

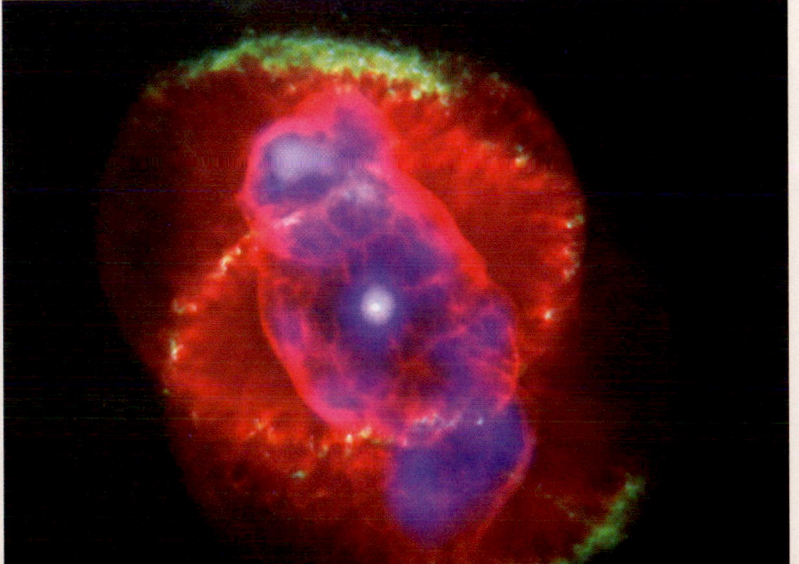

Even the height of these towering cliffs in the Andes Mountains of Peru represent only the top 0.01% of the Earth's radius.

CHAPTER 2

Journey to the Center of the Earth

> The Earth is not a mere fragment of dead history, stratum upon stratum like the leaves of a book . . .
> but living poetry like the leaves of a tree.
>
> —Henry David Thoreau (1817–1862)

LEARNING OBJECTIVES

By the end of this chapter, you should understand . . .

- that many objects besides the Sun and planets comprise the Solar System.
- that a magnetic field and an atmosphere surround our planet.
- that the Earth System includes many distinct, interacting realms.
- that the Earth has distinct internal layers (crust, mantle, and core).
- that the rigid lithosphere, Earth's outer shell, overlies a plastic asthenosphere.

2.1 Introduction

In 1961 a Russian cosmonaut, Yuri Gagarin, became the first human to orbit the Earth, and by the end of the decade two Americans, Neil Armstrong and Buzz Aldrin, became the first to walk on the Moon. These exploits were truly amazing: for the first time, humans saw their home planet from a distance and gained a true appreciation of its uniqueness and limits. Though people have not yet traveled farther than the Moon, we have sent spacecraft to other planets and beyond. In fact, in 2013 a spacecraft named *Voyager 1*, launched in 1977 to fly by Jupiter and Saturn, passed out of our Solar System. *Voyager 1* carries a message of greeting, inscribed on a copper disk, from humanity to whomever—or whatever—it might encounter tens of thousands of years in the future. Should *Voyager 1* ever reach one of the many exoplanets astronomers have found, the craft's instruments could provide interesting details about the body.

Let's turn the tables and speculate about what an imaginary spacecraft sent from exoplanet toward Earth would detect. In this chapter, we follow this "exoprobe" as it enters and traverses the Solar System and then explores the nature of Earth's immediate surroundings and finally, our planet's surface. Then we turn our attention downward, to characterize the interior of the Earth, from its surface to its center, as detected by measurements of the Earth's shape and gravitational pull and confirmed by instruments sent down to the surface. This high-speed fantasy tour provides a foundation from which we can develop geologic themes introduced through the remainder of this book.

2.2 Welcome to the Neighborhood

A Journey through the Solar System

For most of its journey toward our Solar System, the exoprobe travels through **interstellar space**, the region between stars. This region is a **vacuum** (an absence of matter) so profound that it contains less than one atom per liter—by comparison, air at sea level contains 27,000,000,000,000,000,000,000,000 (or, in scientific notation, 2.7×10^{22}) atoms per liter. The atoms in interstellar space are either parts of nebulae or are **cosmic rays**, high-energy atomic nuclei ejected into space at extreme velocity by supernova explosions. Eventually, the exoprobe begins to feel the ever-so-weak pull of the Sun's gravity—this happens at a distance of about 50,000 AU from the Sun. (An **AU**, or **astronomical unit**, is the distance between the Earth and the Sun—it equals about 150 million kilometers, or 93 million miles.) As it continues to move toward the Sun, the exoprobe detects specks, flakes, and balls of "ice" (frozen volatile materials such as water, carbon dioxide, ammonia, and methane), either leftovers of the nebulae from which our Solar System formed or fragments scattered into space soon after Solar System formation. Together, these objects make up the **Oort Cloud**, whose inner edge lies at a distance of about 3,500 AU from the Sun.

> **Did you ever wonder . . .**
>
> what defines the edge of our Solar System?

At a distance of about 200 AU, our spaceship crosses another invisible boundary and enters the bubble-like **heliosphere**, the inner edge of interstellar space. The region within the heliosphere contains predominantly solar-wind particles, electrons and protons ejected into space from our Sun, whereas the region outside contains predominantly cosmic rays ejected from other objects in space toward our Sun. It is the heliosphere, which technically defines the "edge" of the Solar System, that *Voyager 1* crossed. Then, at a distance of 30 to 55 AU, our spaceship traverses the **Kuiper Belt**, a diffuse ring of icy objects left over from the protoplanetary disk. About 100,000 of the objects in the Kuiper Belt have diameters over 100 km (**Fig. 2.1**), and some, such as Pluto and Eris, have diameters of over 1,200 km and are known as dwarf planets. *Comets* originate from the Kuiper Belt, and to a lesser extent, from the Oort Cloud (**Box 2.1**). All told, the Kuiper Belt and Oort Cloud could contain a trillion objects with a combined mass that may approach that of Jupiter.

The orbit of Neptune, the outermost true planet, defines the inner edge of the Kuiper Belt—once we've passed this orbit, we're traversing **interplanetary space**. In interplanetary space, the concentration of atoms increases to between 5,000 and 100,000 per liter. Thus, while this region is still a profound vacuum, it is much denser than interstellar space. As our probe zooms across the *ecliptic*, the plane containing the orbits of all the planets, we pass Uranus, the other ice-giant planet), the two gas-giant planets (Saturn and Jupiter), and then the *asteroid belt*, a diffuse band about 2.5 AU across that contains about 10 million small solid objects (see Box 2.1). In the inner part of the Solar System we find the four terrestrial planets—reddish Mars, bluish Earth (with its large Moon), greenish cloud-enshrouded Venus, and heavily cratered Mercury (**Fig. 2.2**). Even from the distance of space, Earth looks special, so the exoprobe's computer picks it out to approach and study more closely.

FIGURE 2.1 What a spacecraft would see if it traversed the Solar System and its surroundings.

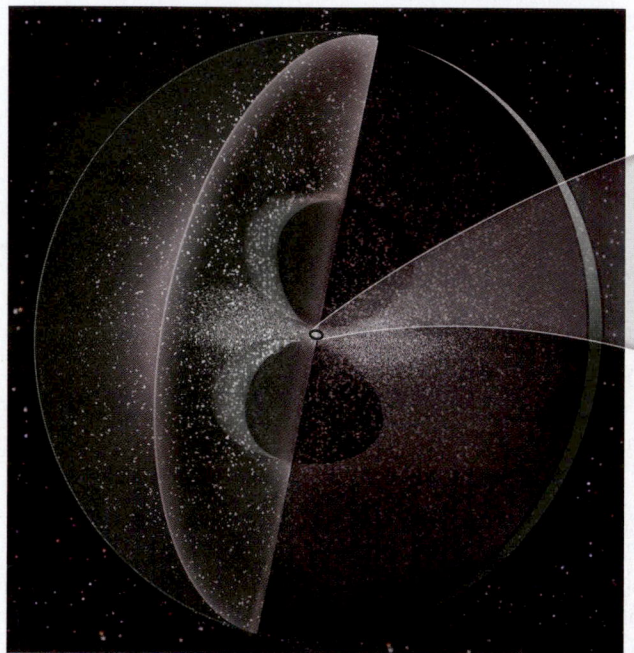

(a) The Oort Cloud is a diffuse cloud of icy particles held in by the Sun's gravity.

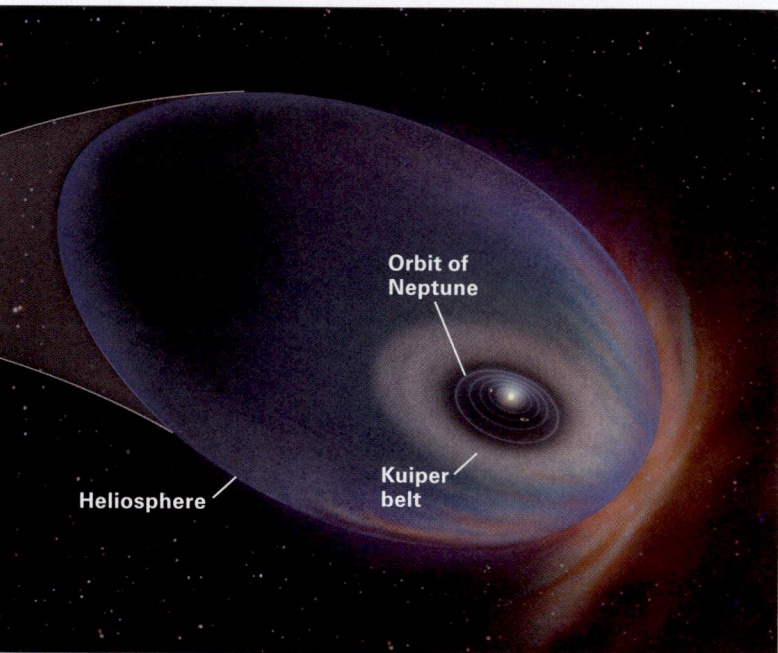

(b) The heliosphere, which is very tiny compared to the Oort Cloud, technically delimits the edge of the Solar System. The Kuiper Belt and planets lie within it.

(c) The asteroid belt lies outside of the orbit of Mars. Some asteroids (known as the Trojans and the Greeks) have been locked into Jupiter's orbit.

BOX 2.1   CONSIDER THIS . . .

Comets and Asteroids—The Other Stuff of the Solar System

Comets and asteroids are of growing concern to humanity because some of them have orbits that may cross the orbit of the Earth and thus could collide with our planet. What are comets and asteroids, and how do they differ from each other?

A **comet** is an icy planetesimal whose highly elliptical orbit brings it so close to the Sun that during part of its journey it evaporates and releases glowing gas and dust that forms a tail pointing away from the Sun (**Fig. Bx2.1a, b**). Comets that take less than 200 years to orbit the Sun originate from the Kuiper Belt, whereas those with longer orbits originate from the Oort Cloud. Objects become comets when gravity from planets tugs on them and sends them on a trajectory to the inner Solar System.

In recent decades, researchers have sent spacecraft to observe comets close-up. During an approach to Halley's Comet in 1986, *Giotto* photographed jets of gas and dust spurting from the comet's surface. *Stardust* visited a comet in 2004 and returned to Earth with samples, and in 2005 *Deep Impact* dropped a copper ball on a comet to analyze the debris ejected by the impact. Such studies confirm that comets consist primarily of frozen water (H_2O), carbon dioxide (CO_2), methane (CH_4), ammonia (NH_3), and other volatile compounds, along with a variety of organic chemicals and dust (tiny rocky or metallic particles). Considering these components, astronomers often refer to comets as "dirty snowballs." Some geologists have speculated that abundant impacts of comets with the Earth, early in Solar System history, might have brought water and even molecules from which life evolved.

An **asteroid** is a small body of solid rock and/or metal that orbits the Sun. Most reside in the asteroid belt between the orbits of Mars and Jupiter. Some asteroids are small rocky planetesimals that were never incorporated into larger bodies, whereas others are fragments of once-large planetesimals that had differentiated into a metal core surrounded by rocky mantle before being shattered into fragments by collisions early in the history of the Solar System. While most asteroids are very small—from dust to basketball sized—astronomers have found 1,000 asteroids with diameters greater than 30 km and estimate that there may be about 2 million more with diameters greater than 1 km.

Though asteroids are numerous, their combined mass equals only about 4% that of the Earth's Moon. Notably, half of the mass of the asteroid belt resides in the four largest asteroids. Of these, the most massive, Ceres, is spherical and has a diameter of 950 km; its rocky/metallic center is surrounded by water and ice. The second most massive asteroid, Vesta, differentiated into a mantle and core and is somewhat spherical. Other asteroids do not have differentiated interiors and are too small for their own gravity to reshape them into spheres, so they are irregular, pockmarked masses (**Fig. Bx2.1c**). The material in the asteroid belt never merged to form a planet because it is constantly churned by Jupiter's gravitational pull.

In recent decades, several space probes have visited asteroids. Some have flown by asteroids, and one, *Hayabusa* from Japan, landed on an asteroid, collected samples, and then returned to Earth. The *Dawn* spacecraft from the United States is currently mapping Ceres and Vesta in detail.

FIGURE Bx2.1   Images of asteroids and comets.

(a) Photograph of Comet Hale-Bopp, which approached the Earth in 1997. The head of this comet is about 40 km across.

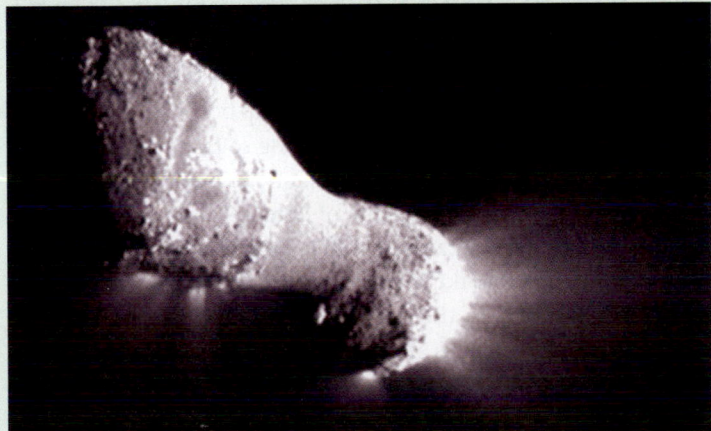

(b) Hartley 2, a comet viewed by the EPOXI spacecraft in 2010. The comet, which is about 1.5 km long, looks like a dirty snowball and emits jets of gas.

(c) Photograph of the asteroid Ida, a body that is about 56 km long.

FIGURE 2.2 An explorer from outer space would quickly realize that the four terrestrial planets look quite different from one another.

Mercury has a crater-pocked surface and icy poles.

Dense clouds hide the surface of Venus.

The surface of Mars varies in elevation, and is host to craters and polar ice caps.

Earth's surface has both land and sea. The atmosphere is largely transparent.

Earth's Magnetic Field

As the exoprobe nears the Earth, its instruments begin to detect the Earth's magnetic field, like a signpost shouting, "Approaching Earth!" A **magnetic field** is the region measurably affected by the force emanating from a magnet. This force, which grows progressively stronger as you approach the magnet, can attract or repel another magnet and can cause charged particles to move. Earth's magnetic field, like the familiar magnetic field around a bar magnet, is a **dipole**, meaning that it has two ends—a north pole and a south pole (**Fig. 2.3a, b**). If you bring two magnets close to each other, the unlike poles attract (pull toward each other), and the like poles repel (push away from each other). By convention, physicists represent the orientation of a magnetic dipole by an arrow that points from the south pole to the north pole, and they depict the magnetic field of a magnet by a set of invisible **magnetic field lines** that curve through the space around the magnet. Arrowheads along these lines point in a direction to complete a loop. Magnetized needles, such as iron filings or compass needles, when placed in a field, align with the magnetic field lines.

Because it is dipolar, we can simplistically represent the Earth's magnetic field as emanating from an imaginary bar magnet in the planet's interior. The north pole of this bar, as defined by physicists, lies near the south geographic pole of the Earth, whereas the south pole of the bar lies near the north geographic pole. (The **geographic poles** are the places where the spin axis of the Earth intersects the planet's surface. Presently, the spin axis tilts at angle of 23.4° relative to the ecliptic.) Nevertheless, geologists and geographers, by convention, refer to the magnetic pole closer to the north geographic pole as the

north magnetic pole, and the magnetic pole closer to the south geographic pole as the *south magnetic pole*. This way, the north-seeking end of a compass points toward the north geographic pole. The magnetic poles move at an observable rate (currently 50 to 60 km per year), so at any given time the position of the magnetic poles is not exactly the same as that of the geographic poles, but they generally have not been more than 15° of latitude apart over the course of geologic time.

The solar wind interacts with Earth's magnetic field, distorting it into a huge teardrop pointing away from the Sun. Fortunately, the magnetic field deflects most (but not all) solar-wind particles, so that they do not reach Earth's surface. In other words, the magnetic field serves as a shield against solar-wind particles, which is important for life on Earth because the particles can be dangerous to life forms. Physicists refer to the region inside this magnetic shield as the **magnetosphere** (**Fig. 2.3c**).

Though it protects the Earth from most of the solar wind, the magnetic field does not stop the exoprobe. At distances of 3,000 km to 10,500 km out from the Earth, the spacecraft encounters the *Van Allen Radiation Belts*, named for the physicist who first recognized them in 1959. The Van Allen Belts, a region in which the magnetic field starts to strengthen, trap both cosmic rays and the solar-wind particles that were moving so fast they could penetrate the weaker outer part of the magnetic field. Thus, the Van Allen Belts serve as a second line of defense that protects life on Earth from dangerous radiation. But they don't trap all incoming particles. Some make it past the Van Allen Belts and follow magnetic field lines to the polar regions of Earth. When these particles interact with gas atoms nearer the Earth, they cause the gases to glow, like the gases in neon signs, creating spectacular

FIGURE 2.3 A magnetic field permeates the space around the Earth. It can be symbolized by a bar magnet.

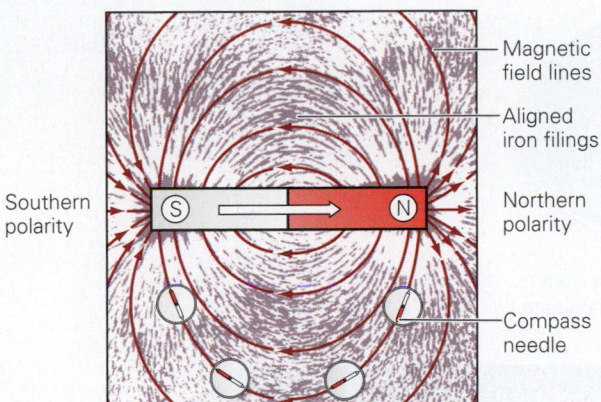

Magnetic field lines

Aligned iron filings

Southern polarity

Northern polarity

Compass needle

(a) A bar magnet produces a magnetic field. Magnetic field lines point into the "south pole" and out from the "north pole."

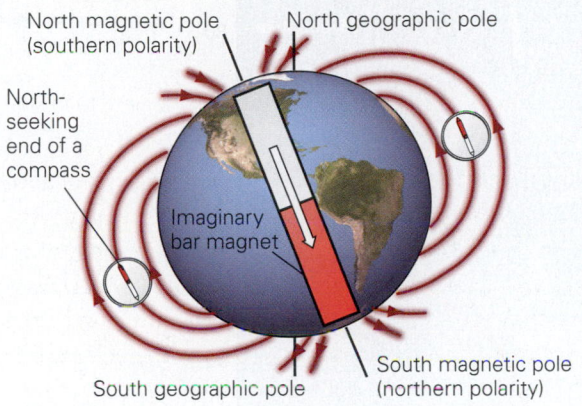

North magnetic pole (southern polarity)

North geographic pole

North-seeking end of a compass

Imaginary bar magnet

South geographic pole

South magnetic pole (northern polarity)

(b) We can represent the Earth's field by an imaginary bar magnet inside.

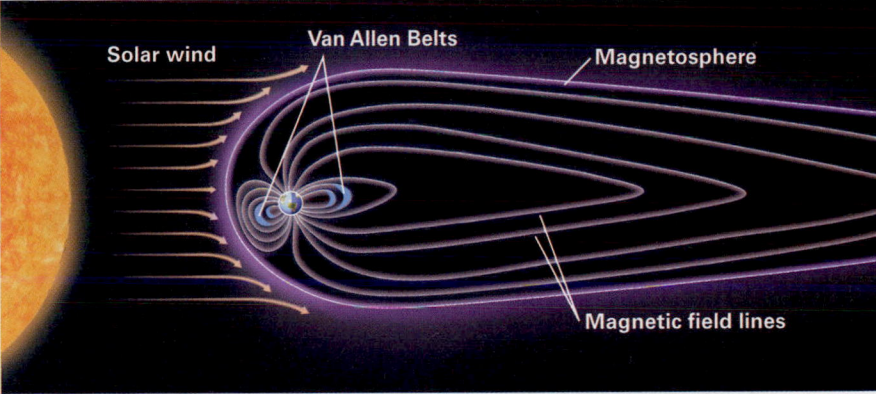

Solar wind

Van Allen Belts

Magnetosphere

Magnetic field lines

(c) Earth behaves like a magnetic dipole, but the field lines are distorted by the solar wind. The Van Allen radiation belts trap charged particles.

(d) Charged particles flow toward Earth's magnetic poles and cause gases in the atmosphere to glow, forming colorful aurorae in polar skies.

aurorae (**Fig. 2.3d**). The *aurora borealis* occurs in high latitudes of the northern hemisphere, while the *aurora australis* occurs in high latitudes of the southern hemisphere.

A Plunge through the Atmosphere

The exoprobe descends to an elevation of 600 km (370 miles) and goes into orbit around the planet, circling the globe once every 96 minutes. At this elevation, the spacecraft skims through the outer reaches of the Earth's **atmosphere**, the gaseous cloak that envelops the planet. This atmosphere contains a mixture of gases, which we refer to as *air*. The spacecraft's instruments determine that air consists of 78% nitrogen (N_2) and 21% oxygen (O_2), along with minor amounts (1% total) of other gases, including argon, carbon dioxide, neon, methane, ozone, carbon monoxide, and sulfur dioxide (**Fig. 2.4a, b**). Air also contains variable amounts of water (H_2O) gas, which, at lower elevations, locally condenses into whitish, translucent to opaque clouds that at any given time hide about 70% of the planet's surface; air without clouds is nearly transparent. Other terrestrial planets have atmospheres, but they are not like the Earth's—the atmospheres of Venus and Mars consists mostly of CO_2 gas, while that of Mercury consists of hydrogen and helium.

The density of the atmosphere progressively increases closer to the Earth, for the weight of overlying air squeezes on the air below, pushing gas molecules in the air closer together. At the Earth's surface, molecules are close enough together

Did you ever wonder...

how thick our atmosphere is?

FIGURE 2.4 Characteristics of the atmosphere that envelops the Earth.

(a) An orbiting astronaut's photograph shows the haze of the atmosphere fading up into the blackness of space.

(b) Composition of atmosphere. Nitrogen and oxygen dominate.

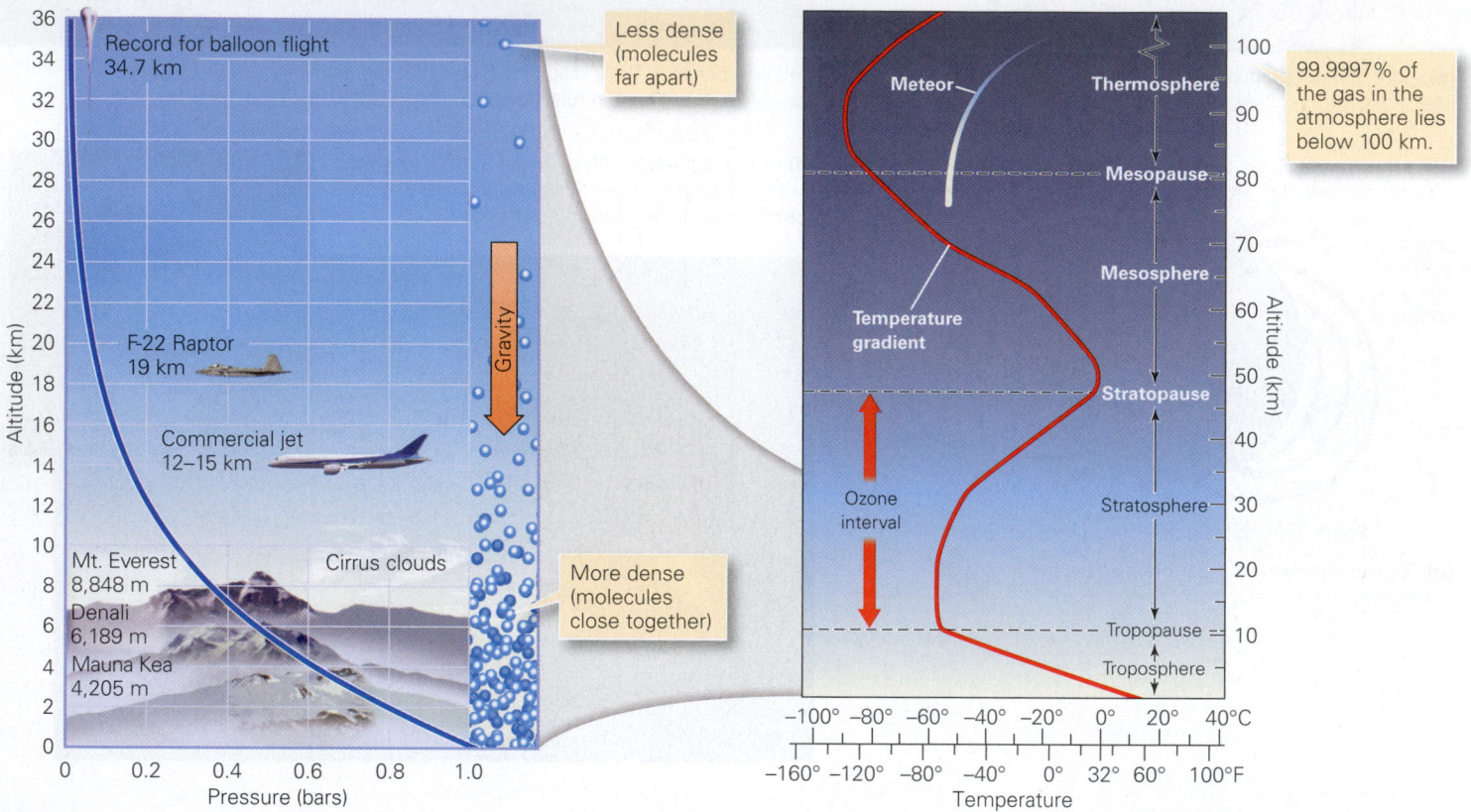

(c) Molecules pack together more tightly at the base of the atmosphere, so atmospheric pressure changes with elevation.

(d) The atmosphere can be divided into several distinct layers. We live in the troposphere.

that the atmosphere has a density of 1.2 g/L. While vastly denser than interplanetary space, this is only 12% that of water. (Notably, the density of the Earth's atmosphere lies between that of Venus, which is 6.5 g/L, and Mars, which is 0.02 g/L.) *Air pressure*, the amount of push that the air exerts on material surrounding it, also increases closer to the surface because of the weight of the overlying atmosphere (**Fig. 2.4c**). Technically, we specify pressure in units of force (push) per unit area. For example, we could specify atmospheric pressure in pounds (a force) per square inch (an area), or kilograms per square centimeter. Air pressure at sea level averages 14.7 lb/in^2, or 1.04 kg/cm^2. Researchers generally use other units, such

as atmospheres (abbreviated atm) and bars. In these units, air pressure at sea level is 1 atm. An atmosphere and a bar are almost the same: 1 atm = 1.01 bars.

Since air pressure decreases with increasing elevation, the air pressure at the peak of Mt. Everest, 8.85 km above sea level, is only 0.3 atm. People can't survive for long at elevations above about 5.5 km, where the air pressure is about 0.5 atm, 50% that of sea level, so since jet planes fly at altitudes above 5.5 km, their cabins must be pressurized by pumping air in.

The decrease in density with elevation means that 99% of atmospheric gas lies at elevations below 50 km, and the atmosphere is barely detectable at elevations above 120 km.

While molecules are still attracted to the Earth by gravity for half the distance to the Moon, there are so few molecules at elevations above about 600 km that the molecules no longer collide and interact like those of a gas. Thus, generally speaking, researchers consider the top of the atmosphere to lie at about 600 km.

The character of the atmosphere changes with increasing distance from the Earth's surface. Because of these changes, atmospheric scientists divide the atmosphere into layers. Most winds and clouds develop only in the lowest layer, the *troposphere*. The layers of the atmosphere that lie above the troposphere are named, in sequence from base to top: the stratosphere, the mesosphere, and the thermosphere (**Fig. 2.4d**). As described further in Chapter 20, the boundaries between layers are defined as elevations where temperature stops decreasing and starts increasing, or vice versa. Boundaries are named for the underlying layer. For example, the boundary between the troposphere and the overlying stratosphere is called the *tropopause*.

Take-Home Message

The Earth is one of eight planets and countless other, smaller objects that orbit the Sun. The Earth produces a magnetic field that deflects solar wind. An atmosphere composed mostly of N_2 and O_2 gas surrounds the planet; 99% of this atmosphere lies below an elevation of 50 km, so it is a thin envelope indeed.

QUICK QUESTION: What causes the aurorae?

2.3 Basic Characteristics of the Earth

The Earth System

Once in orbit, the exoprobe surveys the Earth's surface and detects several distinct components. Beneath the *atmosphere* (the gaseous envelope that we've just discussed) lie the *hydrosphere* (surface and near-surface liquid water), the *cryosphere* (surface and near-surface ice and snow), the *biosphere* (the great variety of living organisms), and the *solid Earth*. Geologists refer to the combination of these components, and the complex interactions among them, as the **Earth System**. Of all the planets in the Solar System, only Earth currently has liquid water. Earth lies within the **habitable zone**, the distance from the Sun in which temperatures are in the range that liquid water can exist (**Fig. 2.5**); on planets closer to the Sun than the habitable zone, all water will be hot enough to evaporate, and on planets farther away water can exist only as solid ice. Astronomers estimate that the habitable zone in our Solar System extends between about 0.8 and 2.5 AU. This region includes the orbits of Venus, Earth, and Mars and some of the asteroid belt. The fact that surface temperatures on Venus are too hot for life (460°C) and on Mars are too cold (−55°C) has to do with the respective abilities of their atmospheres to trap heat. Venus has a dense atmosphere that traps a lot of heat, and Mars has a very thin atmosphere that doesn't.

As we'll see throughout this book, the Earth System is a dynamic place. Its surface and the objects on it move, and its

---

FIGURE 2.5 In our Solar System, only Earth lies within the relatively narrow habitable zone. The scale is in astronomical units (AU).

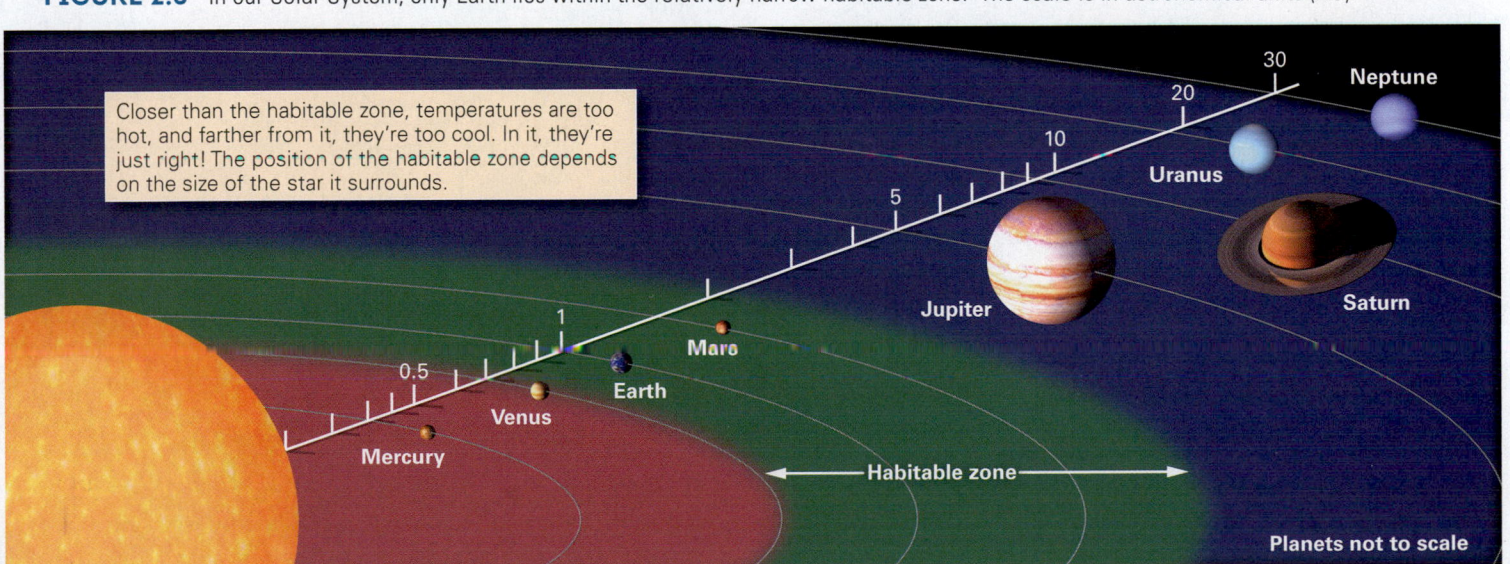

Closer than the habitable zone, temperatures are too hot, and farther from it, they're too cool. In it, they're just right! The position of the habitable zone depends on the size of the star it surrounds.

Neptune
30
20
Uranus
10
5
Saturn
Jupiter
1
Mars
0.5
Earth
Venus
Mercury
Habitable zone
Planets not to scale

SEE FOR YOURSELF . . .

The Whole Earth

LATITUDE
49°18'51.35"N

LONGITUDE
36°48'46.13"W

Zoom to an elevation of 20,000 km (12,400 mi) and look straight down.

A view of the whole Earth, showing land, sea, and ice. Green areas are vegetated, tan areas are not.

atmosphere and oceans circulate; materials from its interior spill out on the surface, and materials from the surface sink into the interior. The energy driving all this activity ultimately comes from heat inside the Earth, from gravity, and from the Sun's heat and light.

Land and Sea

As the exoprobe continues in orbit, it makes a basic map of the surface by collecting information about spatial variations in composition of the Earth's surface and determines that *dry land* (continents and islands) covers about 30% of the surface, whereas **surface water** covers the remaining 70% of the Earth (**Fig. 2.6**). Most surface water is salty and makes up the oceans (or sea), but some is fresh and fills lakes and rivers. Our instruments also detect **groundwater**, the water that fills cracks and holes (pores) beneath the

land surface. Finally, we find that ice covers significant areas of land and sea in polar regions and at high elevations, and that living organisms populate the land, sea, air, and even the upper few kilometers of the subsurface.

To finish off its map of the Earth's surface, the spacecraft detects that the planet's land surface has **topography**, variations in the elevation of the land surface, and distinguishes plains, mountains, and valleys. Notably, the highest point on land, the peak of Mt. Everest, lies about 8.9 km above sea level, while the lowest point, the Dead Sea, lies about 0.4 km below sea level. The exoprobe's instruments also measure **bathymetry**, the variation in elevation of the ocean floor. The sea's surface looks much the same everywhere, because by definition it lies on average at sea level (changing temporarily by centimeters to tens of meters as waves and/or tides pass by). However, the sea floor hidden below displays distinct bathymetric realms. Most of the sea floor comprises broad **abyssal plains**, where the flat seafloor

SEE FOR YOURSELF . . .

The Southern Alps

LATITUDE
44° 9'31.68"S

LONGITUDE
169°46'9.60"E

Zoom to an elevation of 25 km (15 mi) and tilt to look north.

Rugged topography of the Southern Alps of New Zealand.

FIGURE 2.6 This map of the Earth shows variations in elevation on both the land surface and the sea floor. Darker blues are deeper water in the ocean. Greens are lower elevation on land.

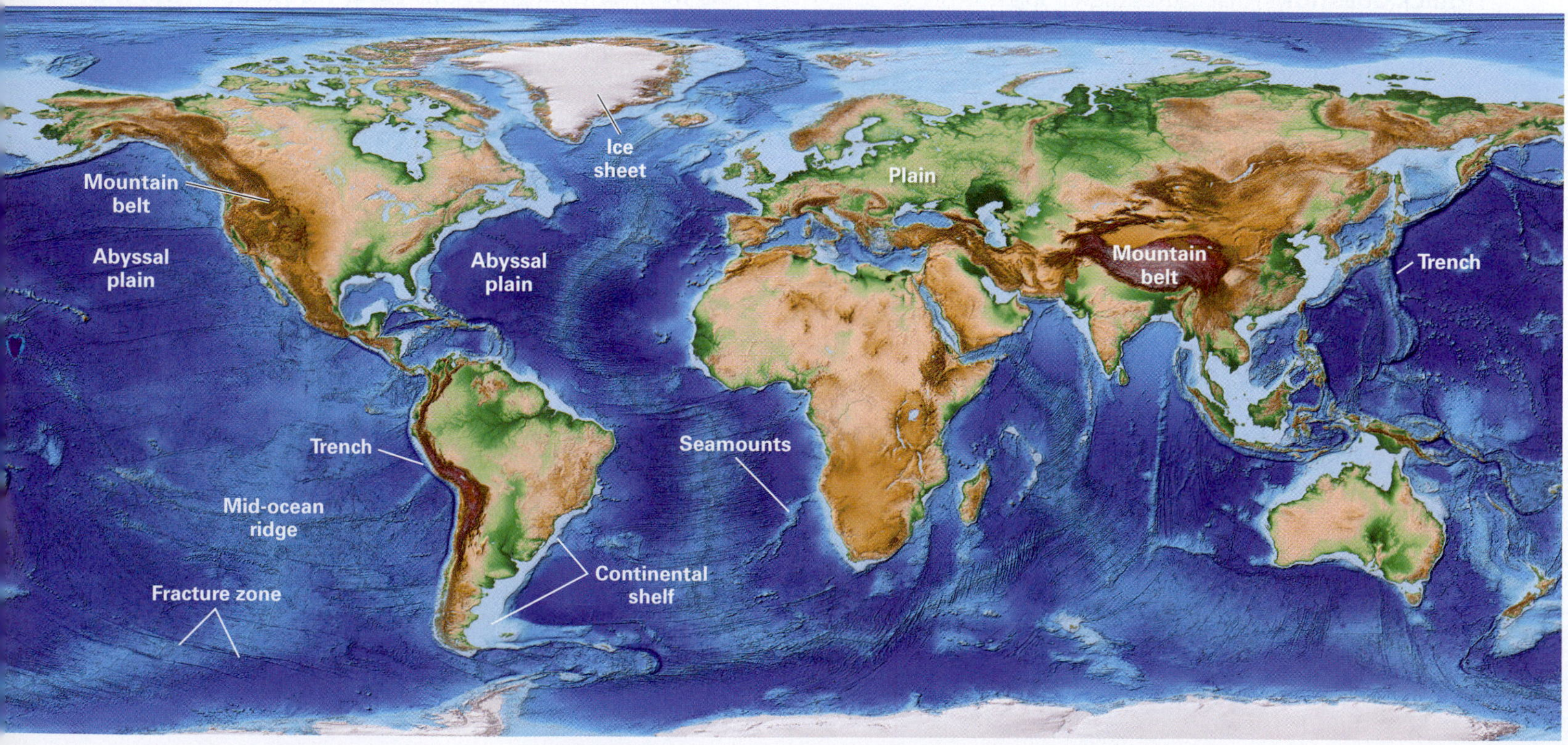

Did you ever wonder . . .

what the average elevation of the land is?

lies at a depth of 4 to 5 km below sea level. The sea floor rises to shallower depths along **mid-ocean ridges**, elongate submarine mountains that rise as much as 2.5 km above the abyssal plains, and descends to greater depths in the **deep-ocean trenches**, elongate troughs in which the sea floor reaches a depth of as much as 10.9 km below sea level. Note that Earth's total relief, from the floor of the deepest trench to the peak of the highest mountain, is 19.8 km, only 0.3% of Earth's radius (6,371 km). In fact, if the Earth were the size of a billiard ball, it would actually be smoother than a billiard ball.

A graph, called a **hypsometric curve**, plotting surface elevation on the vertical axis and the percentage of the Earth's surface on the horizontal axis, shows that a relatively small proportion of the Earth's surface occurs at very high elevations (mountains) or at great depths (deep trenches). In fact, most of the land surface lies just within a kilometer of sea level, and most of the sea floor lies between 4 and 5 km deep (**Fig. 2.7**). Thus, changes in sea level of tens to a couple of hundred meters would dramatically change the amount of dry land.

What Is the Solid Earth Made Of?

To complete its initial exploration of the Earth, the exoprobe sends robot landers down to the surface to sample and analyze the **Earth materials** that make up the solid planet. Of the 92 naturally occurring elements (produced by fusion reactions in stars and supernova explosions) that make up the Earth, 91.2% of the Earth's mass consists of only 4—iron, oxygen, silicon, and magnesium (**Fig. 2.8**). The elements of the Earth bond together to form a great variety of materials that can be classified into several basic categories, which we introduce here. (All will be discussed in more depth later in this book.)

- *Organic chemicals:* Carbon-containing compounds that either occur in living organisms or have characteristics that resemble compounds in living organisms are called organic chemicals.
- *Minerals:* A solid, natural substance in which atoms are arranged in an orderly pattern is a mineral. A coherent sample of a mineral that grew to its present shape is a **crystal**. Most minerals are *inorganic* chemicals.
- *Glasses:* A solid in which atoms are not arranged in an orderly pattern is called glass.
- *Rocks:* An aggregate of mineral crystals or grains, or a mass of natural glass, is called a rock. Geologists recognize three main groups of rocks: *Igneous rocks* develop when hot molten (liquid) rock cools and freezes solid. *Sedimentary rocks* form either from fragments that break off pre-existing rock and become cemented together or from minerals that precipitate out of a water solution at or near the Earth's surface. *Metamorphic rocks* form when

FIGURE 2.7 This graph shows a hypsometric curve, indicating the proportions of the Earth's solid surface at different elevations. Two principal zones—the continents and adjacent continental shelf areas (the submerged margins of continents), and the ocean floor—account for most of Earth's area. Mountains and deep trenches cover relatively little area.

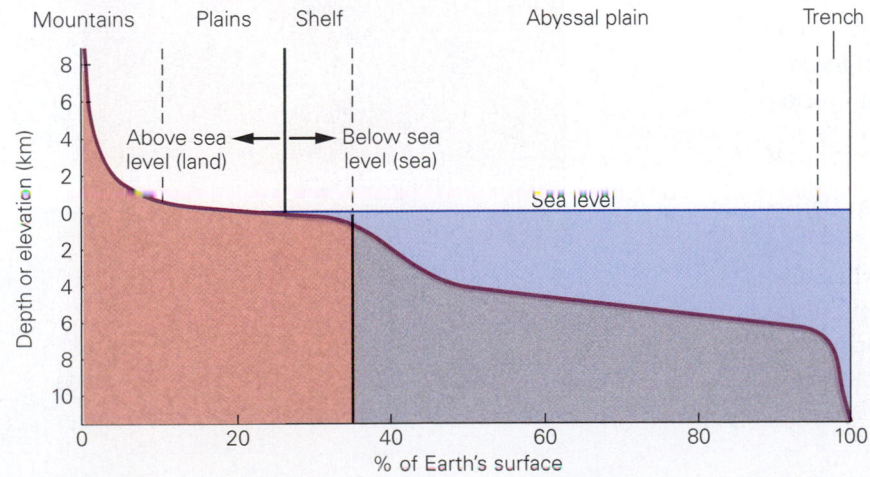

FIGURE 2.8 The proportions of major elements making up the mass of the whole Earth. Note that iron and oxygen account for most of the mass.

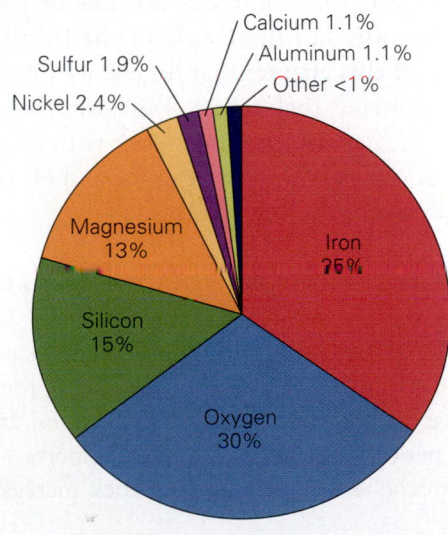

pre-existing rocks undergo changes in response to heat and pressure.

- *Grain:* The term *grain* is used either for individual crystals embedded within an igneous or metamorphic rock or for an individual fragment derived from a once-larger mineral sample or rock body. Grains derived from the fragmentation of rock can be composed of a single mineral, many minerals, and/or glass.
- *Sediment:* An accumulation of loose (unconsolidated) grains, meaning grains that have not been cemented together, is a *sediment*. Gravel and sand are types of sediment.
- *Metals:* Solids composed entirely of metal atoms (such as iron, aluminum, copper, and tin) are called *metals*. Metals can be stretched into wires or flattened into sheets; they tend to be shiny and can conduct electricity. An *alloy* is a mixture containing more than one type of metal.
- *Melts:* Melts form when solid materials become hot and transform into liquid. *Molten rock* is a type of melt. Geologists distinguish between *magma*, molten rock beneath the Earth's surface, and *lava*, molten rock that has flowed out onto the Earth's surface.
- *Volatiles:* As noted in Chapter 1, materials that can transform into gas at the relatively low temperatures found at the Earth's surface are called *volatiles*.

Most of the Earth consists of *silicate minerals*, which are minerals containing silica (SiO_2), either alone or bonded to other elements. Not surprisingly, rocks composed of silicate minerals are known as **silicate rocks**. Geologists distinguish among four classes of igneous silicate rocks based on a characteristic of their chemical composition, specifically, the proportion of silica to iron and magnesium that they contain. In order, from greatest to least proportion of silica to iron and magnesium, the names of these classes are: *felsic* (or *silicic*), *intermediate*, *mafic*, and *ultramafic*. (Chapter 6 provides further discussion of these classes.) Significantly, as the proportion of silica in a rock increases, the density decreases, so felsic rocks are less dense than mafic rocks. To simplify our discussion of the Earth's layers in the sections of this chapter that follow, we introduce the four common igneous rock types: (1) **granite**, a felsic rock with large grains; (2) **gabbro**, a mafic rock with large grains; (3) **basalt**, a mafic rock with small grains; and (4) **peridotite**, an ultramafic rock with large grains.

Pressure and Temperature inside the Earth

To keep underground tunnels from collapsing under the pressure created by the weight of overlying rock, mining engineers must design sturdy support structures. It is no surprise that deeper tunnels require stronger supports—the downward push from the weight of overlying rock increases with depth, sim-

ply because the mass of the overlying rock layer increases with depth. At the Earth's center, pressure probably reaches about 3,600,000 atm.

Temperature also increases with depth in the Earth. Even on a cool winter's day, miners who chisel away at gold veins exposed in tunnels 3.5 km below the surface swelter in temperatures of about 53°C (127°F). We refer to the rate of change in temperature with depth as the **geothermal gradient**. In the upper part of the crust, the geothermal gradient averages between 15° and 30°C per km. At greater depths, the rate decreases to 10°C per km or less. Thus, 35 km below the surface of a continent, the temperature reaches 400° to 700°C and the mantle-core boundary is about 3,500°C. No one has ever directly measured the temperature at the Earth's center, but calculations suggest it may exceed 4,700°C, close to the Sun's surface temperature of 5,500°C (**Fig. 2.9**).

Did you ever wonder...

how hot it gets at the center of this planet?

FIGURE 2.9 The Earth's geotherm shows how temperature increases with depth.

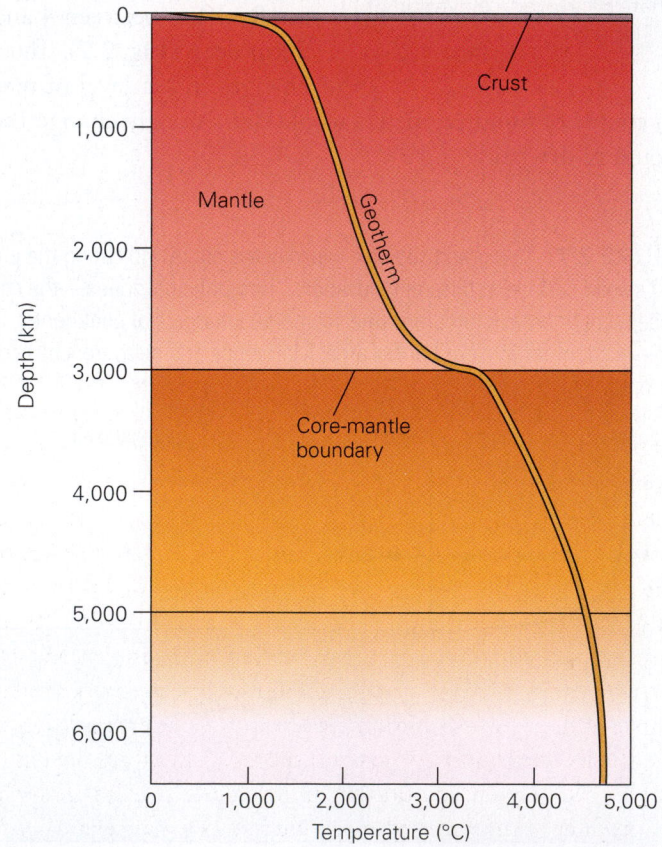

Take-Home Message

Geologists use the term *Earth System* in reference to the variety of interacting realms in, on, and around the planet. Thirty percent of the surface is land, while 70% is sea. Both the land surface and seafloor display notable variations in elevation and depth, respectively, but the difference between the highest point and the lowest is only about 0.3% of Earth's radius. The solid Earth consists of many materials, the most common of which is silicate rock.

QUICK QUESTION: How do geologists distinguish among classes of silicate rocks?

2.4 How Do We Know that the Earth Has Layers?

The world's deepest mine shaft penetrates gold-bearing rock that lies about 3.5 km (2 miles) beneath South Africa. Though miners seeking this gold begin their workday by plummeting straight down a vertical shaft for almost ten minutes aboard the world's fastest elevator, the shaft represents little more than a pinprick on Earth's surface when compared with the planet's average radius of 6,371 km. In fact, even the deepest well ever drilled, a 12.3-km-deep hole, penetrates only the upper 0.2% of the Earth. We literally live on the thin skin of our planet, its interior forever inaccessible to our direct observation.

Lacking the ability to observe the Earth's interior first-hand, pre-20th-century writers and artists dreamed up fanciful images of it. For example, to the ancient Greeks the subsurface was Hades, the "underworld," which they pictured to be a realm of gloom populated by the dead. In Chinese lore, the underworld was a complex maze, and in Buddhist mythology it was a succession of caverns. The English poet John Milton (1608–74) described the subsurface realm as "a dungeon horrible, on all sides round, as one great furnace flamed; yet from those flames, no light" (**Fig. 2.10**). In the 18th and 19th centuries, some European writers thought the Earth's interior resembled a sponge, containing open caverns variously filled with molten rock, water, or air, an image that seemed to explain the source of volcanoes and water springs. The French science fiction writer, Jules Verne, used this image as the basis of his popular 1864 novel *Journey to the Center of the Earth*, in which three explorers wander through interconnected caverns to reach the Earth's center. Modern scientific studies give a very different picture of the Earth's interior. Except in the upper few kilometers, the interior contains no open spaces. Rather, it consists of distinct rock shells surrounding an iron ball. This image is the end product of interpreting many clues, as we now see.

FIGURE 2.10 An image of the Earth's interior, following a description from *Paradise Lost,* an epic poem published by John Milton in 1667, painted by John Martin (1789–1854).

Early Clues to Characterize the Interior

The first key to understanding the Earth's interior came from studies that provided an estimate of this planet's average density (mass/volume). The volume of the Earth had been known since Eratosthenes first measured its dimensions, so all that was needed to determine the density was a measure of our planet's mass. In 1776, the British Royal Astronomer, Nevil Maskelyne, came up with the first realistic estimate of the mass. He obtained this estimate by observing the deflection, due to the gravitational attraction of a nearby mountain, of a lead ball (called a plumb bob) suspended from a wire. Maskelyne reasoned that if the mountain weren't there, the lead ball would be pulled straight down by the Earth's gravity. The mass of the mountain pulled the ball sideways, toward it, so the wire made a slight angle relative to vertical (**Fig. 2.11**). Since Newton's law of gravity states that the strength of gravitational pull depends on the amount of mass present, the amount of deflection represents the mass of the mountain relative to that of the whole Earth. From his measurements, Maskelyne calculated that the Earth had a density of 4.5 gm/cm³. In 1798, another British scientist used a different method for measurement and came up with a value of about 5.5 gm/cm³, the number we use today. Since average rocks at the Earth's surface have a density of only about 3.0 gm/cm³, researchers immediately realized that the interior of the Earth must contain denser material than its outermost layer and couldn't possibly be full of caverns or other open spaces.

FIGURE 2.11 A surveyor noticed that the plumb line was deflected by an angle β, owing to the gravitational attraction of the mountain. The angle represents the ratio between the mass of the mountain and the mass of the whole Earth.

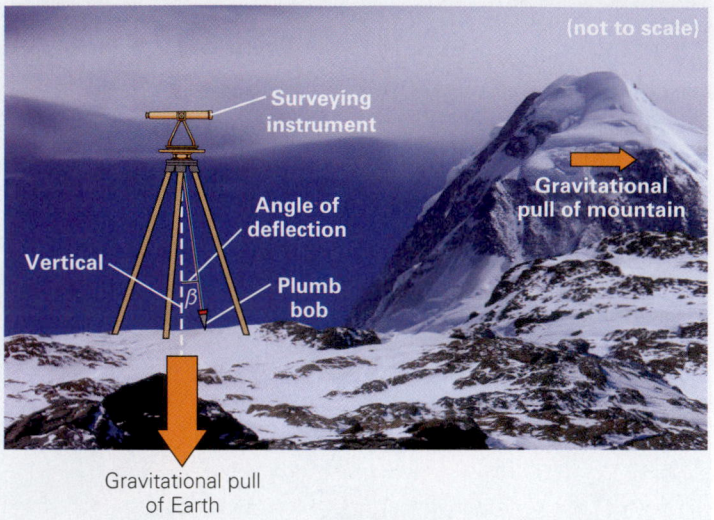

Gravitational pull
of Earth

What could the denser material of the Earth's interior be? Researchers eventually concluded that this material must be a metal, for only metals can have sufficiently high density. (At the Earth's surface, the densest nonmetal, iodine, has a density of only 4.9 gm/cm^3, while iron at the Earth's surface has a density of 7.9 gm/cm^3). With this idea in mind, they then addressed the question of where the metal of the Earth's interior resides. They realized that, since the Earth is nearly a sphere, the metal must be concentrated near the center—otherwise, centrifugal force due to the spin of the Earth on its axis would pull the equator out, and the planet would become somewhat disk-shaped. (To picture why, consider that when you swing a hammer in a circle, your hand feels more force if you hold the end of the light wooden shaft rather than the heavy metal head.) Laboratory studies show that at the immense pressures occurring in the core, the metal comprising the core is almost twice as dense as it would be on the Earth's surface because the pressure squeezes molecules closer together.

Researchers then wondered about the state of the materials in the Earth—are they solid or liquid (molten)? Eventually, they concluded that even though molten rock occasionally oozes out of the interior of volcanoes, and thus must exist somewhere inside the Earth, most of the interior must be solid, because if it weren't, the land surface would rise and fall due to daily tides much more than it actually does. (The forces that produce tides, as discussed in Chapter 18, cause the liquid sea surface to rise and fall by as much as 14 m per day, but the land rises and falls by less than 0.5 m per day.)

Taking into account all the above observations on the density, shape, and tidal behavior of the Earth, researchers real-

ized by the end of the 19th century that the Earth resembled a hard-boiled egg in that it had three principal layers: a not-so-dense *crust* (like an eggshell) composed of rocks such as granite, basalt, and gabbro; a denser solid *mantle* in the middle (like an egg white), composed of a then-unknown material; and a very dense *core* (like an egg yolk), composed of an unknown metal (**Fig. 2.12**). Many questions remained: How thick are the layers? Are the boundaries between layers sharp or gradational? And what exactly are the layers composed of? Data available at the time could not provide answers, and it would take 20th-century studies of earthquakes to give us the detailed image of the interior that we have today.

Notably, studies by spacecraft sent to other terrestrial planets in the Solar System reveal, based on the same kind of evidence presented above, that all terrestrial planets share the same basic internal structure—with a crust, mantle, and core—though the relative thicknesses of the different layers on other planets are not the same as those of the Earth. Also, since the other terrestrial planets are smaller and have cooled more, their interiors are not as soft and plastic as that of the Earth and thus do not appear to flow (and convect) like the Earth's mantle does.

Clues from the Study of Earthquakes: Refining the Image of the Interior

To understand how earthquakes can provide information about the Earth's interior, we first need to understand what an earthquake is. When rock within the outer portion of the Earth suddenly breaks along a *fault* (a fracture on which slip occurs), it generates energy that travels through the surrounding rock outward from the break. You can simulate this process, on

FIGURE 2.12 An early image of Earth's internal layers.

(a) The hard-boiled egg analogy for the Earth's interior.

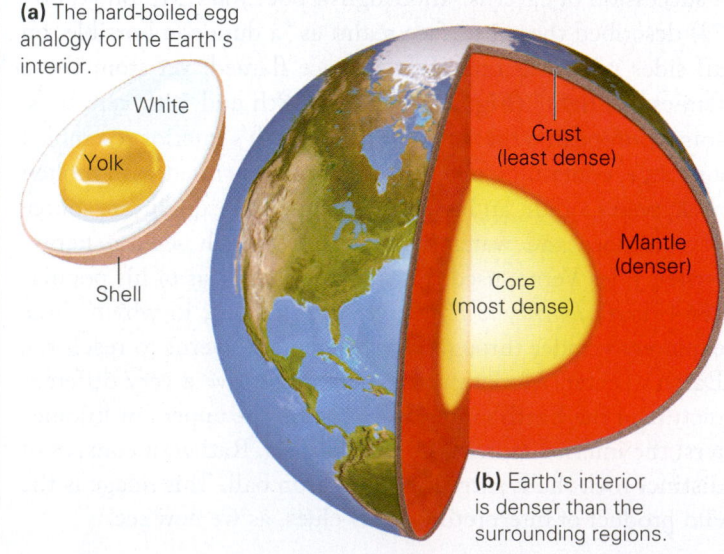

(b) Earth's interior is denser than the surrounding regions.

a small scale, by snapping a stick between your hands—the "shock" that you feel is energy that propagated along the stick from the break to your hands (**Fig. 2.13**). When energy reaches the Earth's surface and causes it to vibrate (move up and down or back and forth), it is called an **earthquake**, an episode of ground shaking.

In 1889, a physicist in Germany noticed that the pendulum in his lab appeared to begin moving back and forth without having been touched. He reasoned that the pendulum was actually standing still while the Earth vibrated under it. A few days later, he read in a newspaper that a large earthquake had taken place in Japan minutes before the movement of his pendulum began. The physicist deduced that the energy generated by the earthquake had traveled all the way through the Earth from Japan and had shaken his laboratory in Germany; the shaking was so gentle that humans couldn't feel it, but it was enough to make the pendulum appear to sway relative to the room. Earthquake energy moves through rock or along the Earth's surface in the form of waves, called either *seismic waves* or *earthquake waves*. You can get a sense of what such waves look like by pushing suddenly on a spring or by jerking the end of a rope. (Chapter 10 provides further detail about earthquakes and seismic waves.)

Geoscientists immediately realized that the study of seismic waves traveling through the Earth might provide a tool for exploring the Earth's interior, much as ultrasound today helps doctors study a patient's insides. Specifically, laboratory measurements demonstrated that seismic waves travel at different velocities (speeds) through different materials. Thus, by detecting depths at which seismic-wave velocities suddenly change, geoscientists pinpointed the boundaries between layers and even recognized subtler boundaries within the main layers, as we now see. (Interlude D shows how the study of earthquake waves defines the Earth's layers.)

Take-Home Message

Measurements of the Earth's mass, shape, and tidal response led to the conclusion that the Earth has three principal internal layers—the crust, the mantle, and the core. Study of earthquake waves passing through the interior has refined this image.

QUICK QUESTION: What observation led 19th-century researchers to conclude that the Earth has a metal core?

2.5 What Are the Layers Made of?

We've concluded that the material composing the Earth's insides must be much denser than familiar surface rocks such as granite and basalt. How can we determine what it consists of? Geoscientists have used several approaches to provide insight into the composition of the Earth's interior, including (1) examination of meteorite composition, for some meteorites are fragments that came from the interiors of planetesimals and thus may resemble materials deep inside the Earth (**Box 2.2**); (2) studies to characterize materials that could be the sources of the igneous rocks found in the crust; (3) analysis of mantle fragments that were carried up into the crust with igneous melts; and (4) measurement of the densities of known materials under high pressures and temperatures to see how these compare with estimated densities of the planet's interior.

As a result of this work, we now have a pretty clear sense of what the layers inside the Earth are made of, though this picture is constantly being adjusted as new findings become available. Let's now look at the properties of individual layers, starting with the outermost layer, the crust (**Fig. 2.14**).

The Crust

When you stand on the surface of the Earth, you are standing on top of its outermost layer, the **crust**. The crust is our home and the source of all our resources. Chemically, it is distinctly

FIGURE 2.13 Faulting and earthquakes.

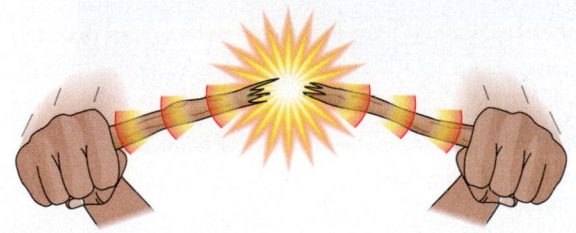

(a) Snapping a stick generates vibrations that pass through the stick to your hands.

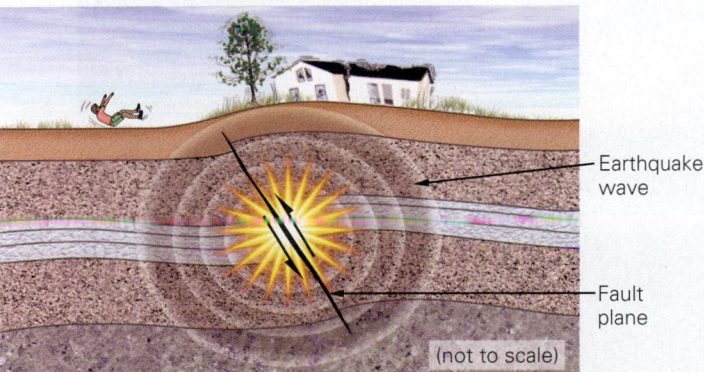

Earthquake wave

Fault plane

(not to scale)

(b) Similarly, when the rock inside the Earth suddenly breaks and slips, forming a fracture called a fault, it generates shock waves that pass through the Earth and shake the surface.

BOX 2.2 CONSIDER THIS . . .

Meteorites: Clues to What's Inside

During the early days of the Solar System, the Earth collided with and incorporated countless planetesimals and smaller fragments of solid material lying in its path. Intense bombardment ceased about 3.9 Ga (billion years ago), but even today collisions with space objects continue, and over 1,000 tons of material (rock, metal, dust, and ice) fall to Earth, on average, every year. The vast majority of this material consists of

fragments derived from comets and asteroids sent careening into the path of the Earth after billiard-ball-like collisions with each other out in space or because of the gravitational pull of a passing planet. Some of the material, however, consists of chips of the Moon or Mars, ejected into space when large objects collided with those bodies.

Astronomers refer to any object from space that enters the Earth's atmosphere as a *meteoroid*. Meteoroids move at speeds of 20 to 75 km/s (over 45,000 mph), so fast that when they reach an altitude of about 150 km, friction with the atmosphere causes them to heat up and evaporate, leaving a streak of bright, glowing gas. The glowing streak, an atmospheric phenomenon, is a *meteor*,

also known colloquially, though incorrectly, as a falling star (**Fig. Bx2.2a**). Most visible meteors completely evaporate by an altitude of about 30 km. But dust-sized ones may slow down sufficiently to float to Earth, and larger ones (fist-sized or bigger) can survive the heat of entry to reach the surface of the planet. In some cases, meteoroids explode as brilliant fireballs in midair.

Objects that strike the Earth are called **meteorites**. Almost all meteorites that have struck the Earth are small and have not caused notable damage on Earth. In fact, during human history, only a few have smashed through houses, dented cars, or bruised people. But two huge fireballs have caused significant damage. Specifically, a small comet exploded above

FIGURE Bx2.2 Meteors and meteorites.

(a) A shower of meteors over Hong Kong in 2001.

(c) Examples of stony meteorites (left) and iron meteorites (right).

The meteorite forming the crater was 50 m across.

(b) The Barringer meteor crater in Arizona. It formed about 50,000 years ago and is 1.1 km in diameter.

(d) In February 2013, a meteor exploded in the atmosphere above the Russian city of Chelyabinsk. The shock wave blew out windows and knocked people over.

Tunguska, Siberia, in 1908, flattening trees over an area of 2,150 square kilometers. More recently, in 2013, a small asteroid blew up 23 km above Chelyabinsk, Russia (**Fig. Bx2.2b**). This explosion was about 25 times larger than that of the atomic bomb over Hiroshima, and its shock waves blasted out windows and knocked down walls, injuring about 1,500 people. Huge impacts have been observed elsewhere in the Solar System. For example, in 1994, astronomers observed four huge impacts when a fragmented comet struck Jupiter. One of the impacts resulted in a 6-million-megaton explosion—this would be equivalent to blowing up 600 times the entire nuclear arsenal on Earth all at once! Even larger, catastrophic impacts in the geologic past may have been responsible for disrupting life on Earth, as we discuss later in this book. Some of these events have left huge craters that we can see today (**Fig. Bx2.2c**).

Most meteorites are asteroidal or planetary fragments, for the icy material of small cometary bodies is too fragile to survive the fall. Researchers recognize three basic classes of meteorites: *iron* (made of iron-nickel alloy), *stony* (made of rock), and *stony iron* (rock embedded in a matrix of metal). Of all known meteorites, about 93% are stony and 6% are iron (**Fig. Bx2.2d**). From their composition, researchers have concluded that some meteors (a special subcategory of stony meteorites called *chondrites*, because they contain small spherical nodules called chondrules) are asteroids derived from planetesimals that never underwent differentiation into a core and mantle. All other stony meteorites and all iron meteorites are asteroids derived from planetesimals that had differentiated into a metallic core and a rocky mantle early in Solar System history but later shattered into fragments during collisions with other planetesimals. Most meteorites appear to be about 4.54 Ga, but some chondrites as old as 4.57 Ga, are the oldest Solar System materials ever measured.

Since meteorites represent fragments of undifferentiated and differentiated planetesimals, geologists consider the average composition of meteorites to be representative of the average composition of the whole Earth. In other words, the estimates that geologists use for the proportions of different elements in the Earth are based largely on studying meteorites. Stony meteorites are probably similar in composition to the mantle, and iron meteorites are probably similar in composition to the core.

different from the whole Earth (**Fig. 2.15**). How thick is this all-important layer? Or, in other words, what is the depth to the crust-mantle boundary? An answer came from the work of Andrija Mohorovičić, a researcher working in Zagreb, Croatia. In 1909, Mohorovičić discovered that the velocity of seismic waves suddenly increased at a depth of a few tens of kilometers beneath the surface of continents, and he suggested that this increase was caused by an abrupt change in the properties of rock. He proposed that this change in seismic velocity represents the crust-mantle boundary. Today we refer to this boundary as the **Moho** in Mohorovičić's honor.

Studies since Mohorovičić's time show that the thickness of the crust, defined as the depth to the Moho, varies between 7 and 70 km, depending on location. Compared to the average radius of the Earth (6,371 km), the crust is only about 0.1% to 1.0% of the Earth's radius, so if the Earth were the size of a balloon, the crust would be about the thickness of the balloon's skin. The crust is not simply cooled mantle, like the skin on cooled chocolate pudding. Rather, it consists of a variety of rocks that differ in composition (chemical makeup) from underlying mantle rock. Geologists distinguish between two fundamentally different types of crust—**oceanic crust**, which underlies the seafloor, and **continental crust**, which underlies continents.

Oceanic crust is only 7 to 10 km thick. At highway speeds (100 km per hour), you could drive a distance equal to the thickness of the oceanic crust in only 5 minutes. The top portion of oceanic crust is a blanket of sediment, generally less than 1 km thick, that consists of clay and tiny shells that settled like snow out of sea water. Beneath this blanket, the oceanic crust consists of a layer of basalt and, below that, a layer of gabbro.

Continental crust, in contrast to oceanic crust, varies in thickness from 25 to 70 km. The thinnest crust lies beneath regions called rifts, where the crust is being stretched and pulled apart and therefore has been thinned. Very thick continental crust occurs beneath mountain belts forming where two continents are squeezing together, causing the crust to shorten horizontally and thicken vertically. The broad plains found in the interior of continents are generally 35 to 50 km thick, about four to six times the thickness of oceanic crust. Also, in contrast to oceanic crust, continental crust contains a great variety of rock types, ranging from mafic to felsic in composition. On average, upper continental crust is less mafic than oceanic crust—it has a felsic (granite-like) to intermediate composition—so continental crust overall is less dense than oceanic crust.

The Mantle

The **mantle** is a 2,885-km-thick shell that surrounds the core. In contrast to the crust, the mantle consists entirely of peridotite, a dark and dense ultramafic rock that's quite rare at the Earth's surface. Notably, the mantle accounts for most of the Earth's volume, and thus—perhaps surprisingly—peridotite, a rock that most people have never seen, is actually the most abundant rock in our planet!

Geoscientists have found that the velocity of seismic waves changes markedly, in a step-like manner, in the portion of the mantle that lies between 410 and 660 km deep. Based on this

> **Did you ever wonder . . .**
>
> what the most abundant rock of the Earth is?

observation, they divide the mantle into two sublayers—the **upper mantle**, down to a depth of 660 km, and the **lower mantle**, from 660 km down to 2,890 km. The portion of the upper mantle between 410 and 660 km, in which the steps in seismic velocity occur, is known as the **transition zone**. (Interlude D will explain why these steps exist.)

Almost all of the mantle is solid rock. But even though it's solid, mantle rock below a depth of about 100 km beneath the ocean floor, and of about 150 km beneath continents, is so hot that it's soft enough to flow. This flow, however, takes place extremely slowly—at a rate of less than 15 cm a year. Thus, "soft" in the context of the mantle does not mean liquid, and it does not mean that you could indent the mantle by pushing on with your finger—it simply means that over long periods of time (thousands to millions of years) mantle rock can change shape significantly without breaking. Note that we stated earlier that almost all of the mantle is solid. We said this because at a depth of 100 to 300 km beneath most ocean floor (and at a few other localities), up to a few percent of the mantle has melted. This melt occurs in thin films or tiny drops between the solid grains in mantle rock.

Although, overall, the temperature of the mantle increases with depth, temperature can also vary significantly with location even at the same depth. Warmer regions of mantle are less dense than adjacent cooler regions, so warmer regions are buoyant relative to cooler regions. As a result, warmer regions tend to flow upward, or "upwell," while cooler regions flow downward, or "downwell." In other words, the mantle undergoes very slow **convection**. The process resembles the flow of water in a simmering pot on a stove (**Box 2.3**).

The Core

Early calculations suggested that the **core** had the same density as gold, so people once dreamed that vast riches lay at the heart of our planet. Alas, geologists eventually concluded that the core consists of a far less glamorous material, iron alloy (>80% iron mixed with nickel and lesser amounts of sulfur, oxygen, and other elements). Studies of seismic waves led geoscientists to divide the core into two parts, the *outer core* (between 2,900 and 5,155 km deep) and the *inner core* (from 5,155 km down to the Earth's center at 6,371 km). The outer core consists of liquid iron alloy. It can exist as a liquid because the temperature in the outer core is so high that even the great pressures squeezing the region cannot keep atoms locked into a solid framework. The iron alloy of the outer core can flow, and this flow generates the

FIGURE 2.14 A modern view of Earth's interior layers.

Crustal stretching can thin the crust.

Mountain building can thicken the crust.

Oceanic crust

Continental crust

Moho

Mantle

Sea level

Crust

Upper mantle

Transition zone

Lower mantle

Outer core

Inner core

150 km
410 km
660 km

2,900 km

5,155 km

6,371 km

km
0
20
40
60

(a) There are two basic types of crust. Oceanic crust is thinner and consists of basalt and gabbro. Continental crust varies in thickness and rock type. By studying earthquake waves, geologists produced a refined image of Earth's interior, in which the mantle and core are subdivided.

Different earth materials have different densities.

Increasing density

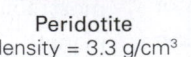

Granite	Basalt	Gabbro	Peridotite	Iron-nickel alloy
density = 2.7 g/cm³	density = 2.9 g/cm³	density = 3.0 g/cm³	density = 3.3 g/cm³	density = 7.5 g/cm³

(b) These are examples of materials that characterize different layers inside the Earth. Continental crust has an average composition similar to granite, whereas oceanic crust consists of basalt and gabbro. The mantle consists of peridotite and the core of iron-nickel alloy.

FIGURE 2.15 A table and a graph illustrating the abundance of elements in the Earth's crust.

Element	Symbol	% by weight	% by volume	% by atoms
Oxygen	O	46.3	93.8	60.5
Silicon	Si	28.0	0.9	20.5
Aluminum	Al	8.1	0.8	6.2
Iron	Fe	5.5	0.5	1.9
Calcium	Ca	3.4	1.0	1.9
Magnesium	Mg	2.8	0.3	1.4
Sodium	Na	2.4	1.2	2.5
Potassium	K	2.3	1.5	1.8
All others	—	1.2	>0.1	3.3

(a) A chart of relative abundances. Oxygen (O) and silicon (Si) account for almost three-quarters of the weight. By far, oxygen is the most abundant type of atom.

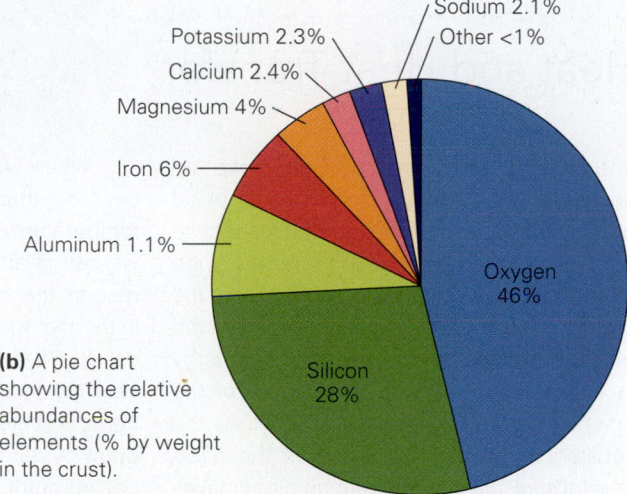

(b) A pie chart showing the relative abundances of elements (% by weight in the crust).

Earth's magnetic field (**Geology at a Glance**, pp. 56–57). Computer models suggest that because the Earth spins on its axis the flow in the outer core follows spiral paths aligned with the axis.

The inner core, with a radius of about 1,220 km, is a solid iron alloy that may reach a temperature of over 4,700°C. Even though it is hotter than the outer core, the inner core is a solid because it is deeper and is subjected to even greater pressure. The pressure keeps atoms locked together tightly in very dense crystals. As the Earth slowly cools, the base of the outer core solidifies and becomes, by definition, part of the inner core. Researchers estimate that the diameter of the inner core is growing, as a result, by about 1 mm per year.

2.6 The Lithosphere and the Asthenosphere

So far, we have identified three major layers (crust, mantle, and core) inside the Earth that differ compositionally from each other. Because of these differences, abrupt changes in seismic-wave velocity pinpoint the depths of these boundaries. An alternative way of thinking about Earth layers comes from studying the degree to which the material making up a layer can flow when subjected to pushes or pulls over a relatively short time period. In this context we distinguish between *rigid materials*, which can bend or break but do not flow, and *plastic materials*, which are relatively soft and can flow without breaking.

Let's apply this concept to the outer portion of the Earth. Geologists have determined that the outer 100 to 150 km of the Earth is relatively rigid. In other words, the Earth has an outer shell composed of rock that does not flow, overall, on a time scale of years to decades. This outer layer is called the **lithosphere**, and it consists of the crust plus the uppermost, and therefore cooler, part of the mantle. We refer to the portion of the mantle within the lithosphere as the **lithospheric mantle**. Note that the terms lithosphere and crust are not synonymous—the crust is just the upper part of the lithosphere; most of the lithosphere actually consists of mantle rock, specifically mantle rock that is cool enough to be rigid. (Cooler rock tends to be more rigid, while hotter rock tends to be more plastic. To picture this contrast, compare the behavior of a wax candle that you've just taken out of a freezer, to one that has been warmed by the summer Sun on a hot day.)

Geologists distinguish between two types of lithosphere (**Fig. 2.16**). Most *oceanic lithosphere*, meaning lithosphere topped by oceanic crust, has a thickness of about 100 km. (As we'll see later, oceanic lithosphere is thinner along mid-ocean ridges.) In contrast, *continental lithosphere*, topped by continental crust, generally has a thickness of 150 to 200 km.

The lithosphere overlies a region called the **asthenosphere**, the portion of the mantle that can flow and undergoes convection. The boundary between the lithosphere and asthenosphere occurs where the temperature reaches about 1,280°C. It is at this temperature that mantle rock (peridotite) becomes soft enough to flow, for rock gets softer as it gets hotter because

BOX 2.3 SCIENCE TOOLBOX...

Heat and Heat Transfer

The atoms and molecules that make up an object do not stay rigidly fixed in place but rather jiggle and jostle with respect to one another. This vibration creates *thermal energy*—the faster the atoms move, the greater the thermal energy and the hotter the object. Put another way, the thermal energy in a substance represents the sum of the kinetic energy (energy of motion) of all the substance's atoms. This includes the back-and-forth displacements that an atom makes as it vibrates, as well as the movement of an atom from one place to another.

When we say that one object is hotter or colder than another, we are describing its temperature. *Temperature* is a measure of warmth relative to some standard and represents the average kinetic energy of atoms in the material. In everyday life, we generally use the freezing or boiling point of water at sea level as the standard. In the Celsius (centigrade) scale, we arbitrarily set the freezing point of water (at sea level) as 0°C and the boiling point as 100°C, whereas in the Fahrenheit scale, we set the freezing point as 32°F and the boiling point as 212°F.

The coldest a substance can be is the temperature at which its atoms or molecules stand still. We call this temperature absolute zero, or 0K (pronounced "zero kay"), where K stands for Kelvin (after Lord Kelvin, 1824–1907, a British physicist), another unit of temperature; degrees in the Kelvin scale are the same increment as degrees in the Celsius scale. You simply can't get colder than absolute zero, meaning that you can't extract any thermal energy from a substance at 0K (−273.15°C).

Heat is the thermal energy transferred from one object to another. Heat can be measured in calories, defined so that 1,000 calories can heat 1 kilogram of water by 1°C. Heat can be transferred from one place or material to another. When heat is added to a substance, the substance warms in that its molecules start to vibrate or move more rapidly. When a substance cools, the motion of its molecules slows. There are four ways in which heat transfer takes place in the Earth System: radiation, conduction, convection, and advection.

Radiation is the process by which electromagnetic waves transmit heat into a body or out of a body (**Fig. Bx2.3a**). For example, when the Sun heats the ground during the day, radiative heating takes place. Similarly, when heat rises from the ground at night, radiative heating is occurring—in the opposite direction.

Conduction takes place when you stick the end of an iron bar in a fire (**Fig. Bx2.3b**). The iron atoms at the fire-licked end of the bar start to vibrate more energetically; they gradually incite atoms farther up the bar to start jiggling, and these atoms in turn set atoms even farther along in motion. In this way, heat slowly flows along the bar until you feel it with your hand. Conduction does not involve actual movement of atoms from one place to another.

FIGURE Bx2.3 The four processes of heat transfer.

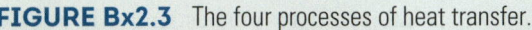

(a) Radiation from sunlight warms the Earth.

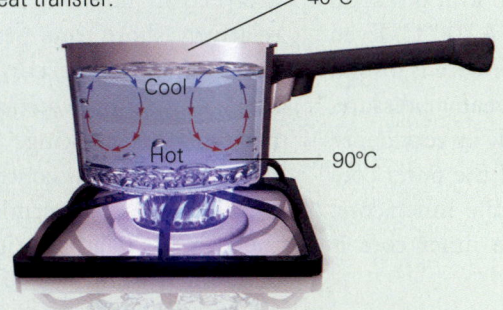

(c) Convection takes place when moving fluid carries heat with it. Hot fluid rises while cool fluid sinks, setting up a convective cell.

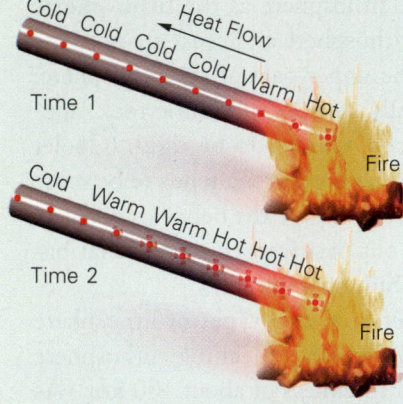

(b) Conduction occurs when you heat the end of an iron bar in a flame. Heat flows from the hot region toward the cold region as vibrating atoms cause their neighbors to vibrate.

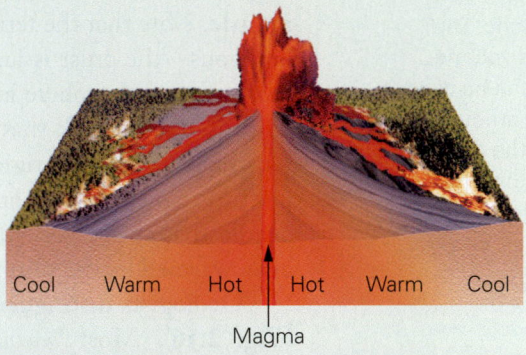

(d) During advection, a hot liquid (such as molten rock) rises into cooler material, and heat then conducts from the hot liquid into the cooler material.

Convection takes place when you set a pot of water on a stove (**Fig. Bx2.3c**). The heat from the stove warms the water at the base of the pot by making the molecules of water vibrate faster and move around more. As a consequence, the density of the water at the base of the pot decreases, for as you heat a liquid, the atoms move away from each other and the liquid expands. For a time, cold water remains at the top of the pot; but eventually the warm, less-dense water becomes buoyant relative to the cold, dense water. In a gravitational field, a buoyant material rises (like a Styrofoam ball in a pool of water) if the material above it is weak enough to flow out of the way. Since liquid water can flow easily, hot water rises. When this happens, cold water sinks to take its place. The new volume of cold water then heats up and rises itself. Thus, during convection, the actual flow of the material itself carries heat. The trajectory of flow defines *convective cells*.

Advection, a less-familiar process, happens when heat is carried by a fluid flowing through cracks and pores within a solid material (**Fig. Bx2.3d**). The heat brought by the fluid conductively heats up the adjacent solid that the fluid passes through. Advection takes place, for example, if you pass hot water through a metal pipe and the pipe itself gets hot. In the Earth, advection occurs where molten rock rises through the crust beneath a volcano and heats up the crust in the process.

FIGURE 2.16 A block diagram of the lithosphere emphasizing the difference between continental and oceanic lithosphere.

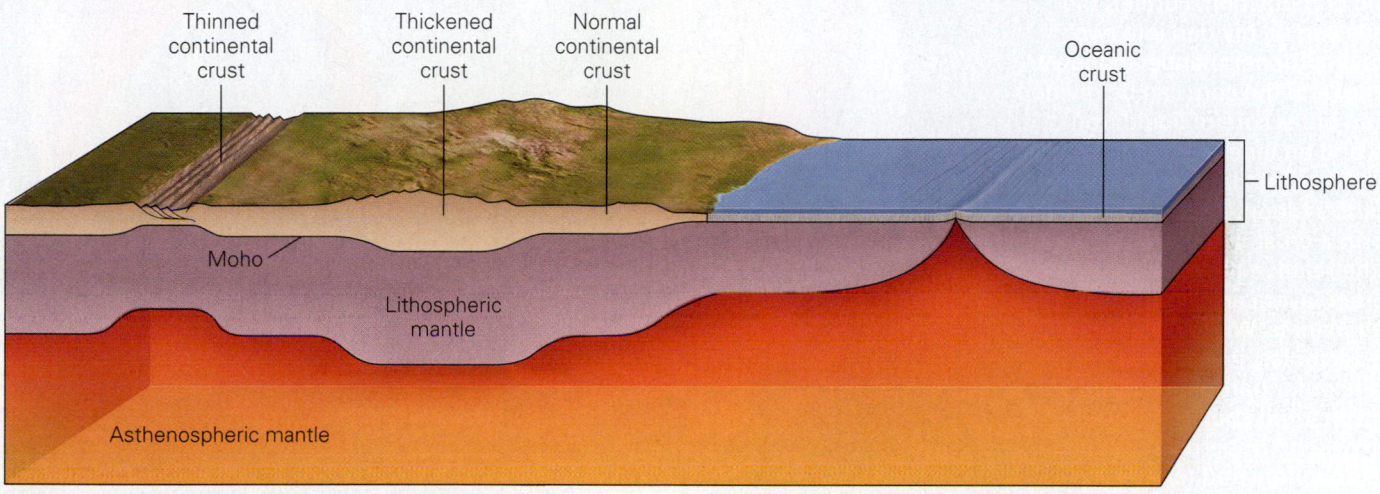

Thinned continental crust

Thickened continental crust

Normal continental crust

Oceanic crust

Lithosphere

Moho

Lithospheric mantle

Asthenospheric mantle

thermal energy causes bonds to break. Keep in mind that even though the asthenosphere flows, it is not molten overall. As we noted earlier, the mantle contains only tiny amounts of melt in films and drops between solid grains, and this zone of "partial melt" occurs only in the upper part of the asthenosphere (between depths of 100 and 300 km) beneath the ocean floor. In this interval, seismic waves travel more slowly, so this portion of the mantle has been called the *low-velocity zone*. While we can define the top of the asthenosphere as the base of the lithosphere (i.e., at a depth of about 100 km below oceanic abyssal plains and 150 to 200 km below the surface of continents), we can't really assign a specific depth to the base of the asthenosphere because all of the mantle has the ability flow. For purposes of discussion, however, some geologists place the base of the asthenosphere at the top of the lower mantle.

Take-Home Message

The crust and the outermost mantle together comprise the rigid lithosphere, a layer which overlies the softer, flowable asthenosphere. The behavior (rigid vs. plastic) of the mantle depends on the temperature.

QUICK QUESTION: Is the asthenosphere entirely a liquid?

The Earth from Surface to Center

If we could remove all the clouds and water that hide much of the solid surface from view, we would see that both the land areas and the seafloor have plains and mountains. And if we could break open the Earth, we would see that its interior consists of a series of concentric layers (crust, mantle, and core) that differ from one another in terms of their composition and seismic velocity. Notably, oceanic crust differs from continental crust, both in thickness and in composition. Further, the mantle can be divided into two sublayers (upper mantle and lower mantle) and the core can be divided into an outer core of liquid iron alloy and an inner core of solid iron alloy. All terrestrial planets, as well as Earth's Moon, have differentiated to form a crust, mantle, and core. But the relative thicknesses of the different layers are not the same for all planets. When discussing plate tectonics, it is convenient to call the outer part of the Earth, a relatively rigid shell composed of the crust and uppermost mantle, the lithosphere and to refer to the underlying warmer, more plastic portion of the mantle as the asthenosphere.

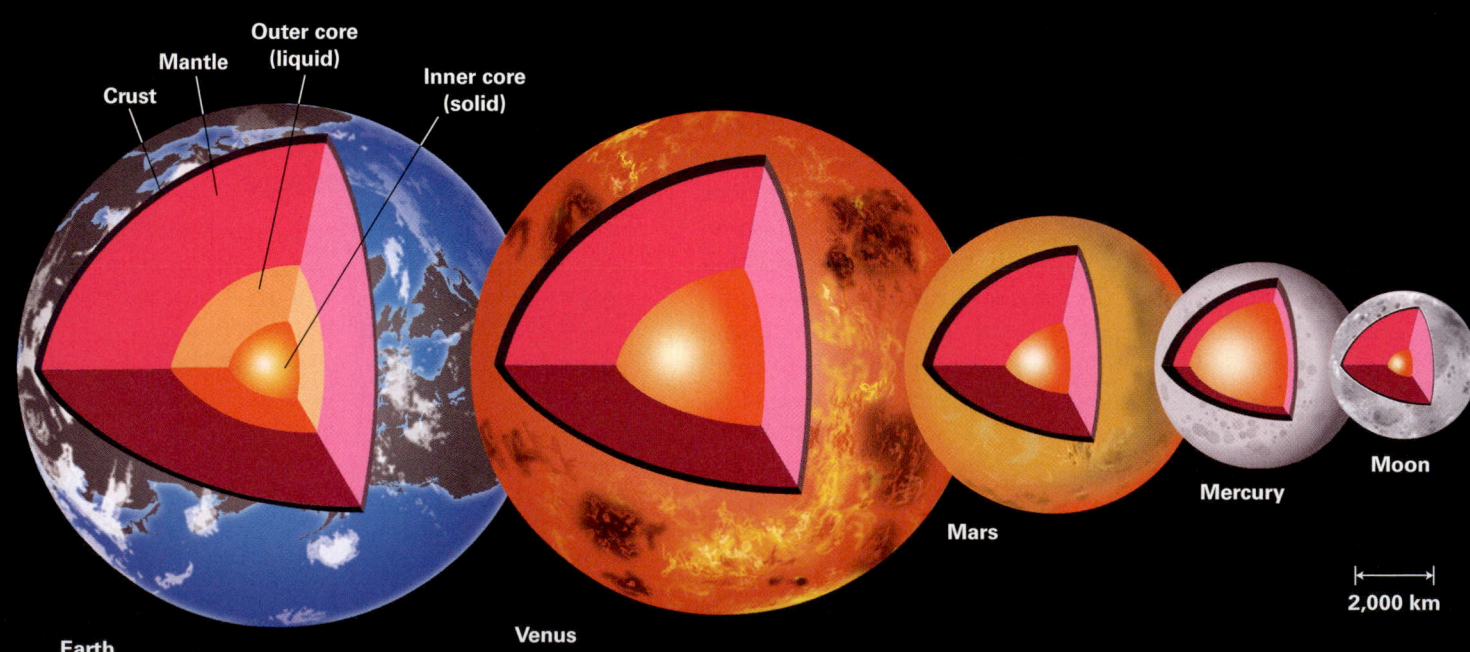

2,000 km

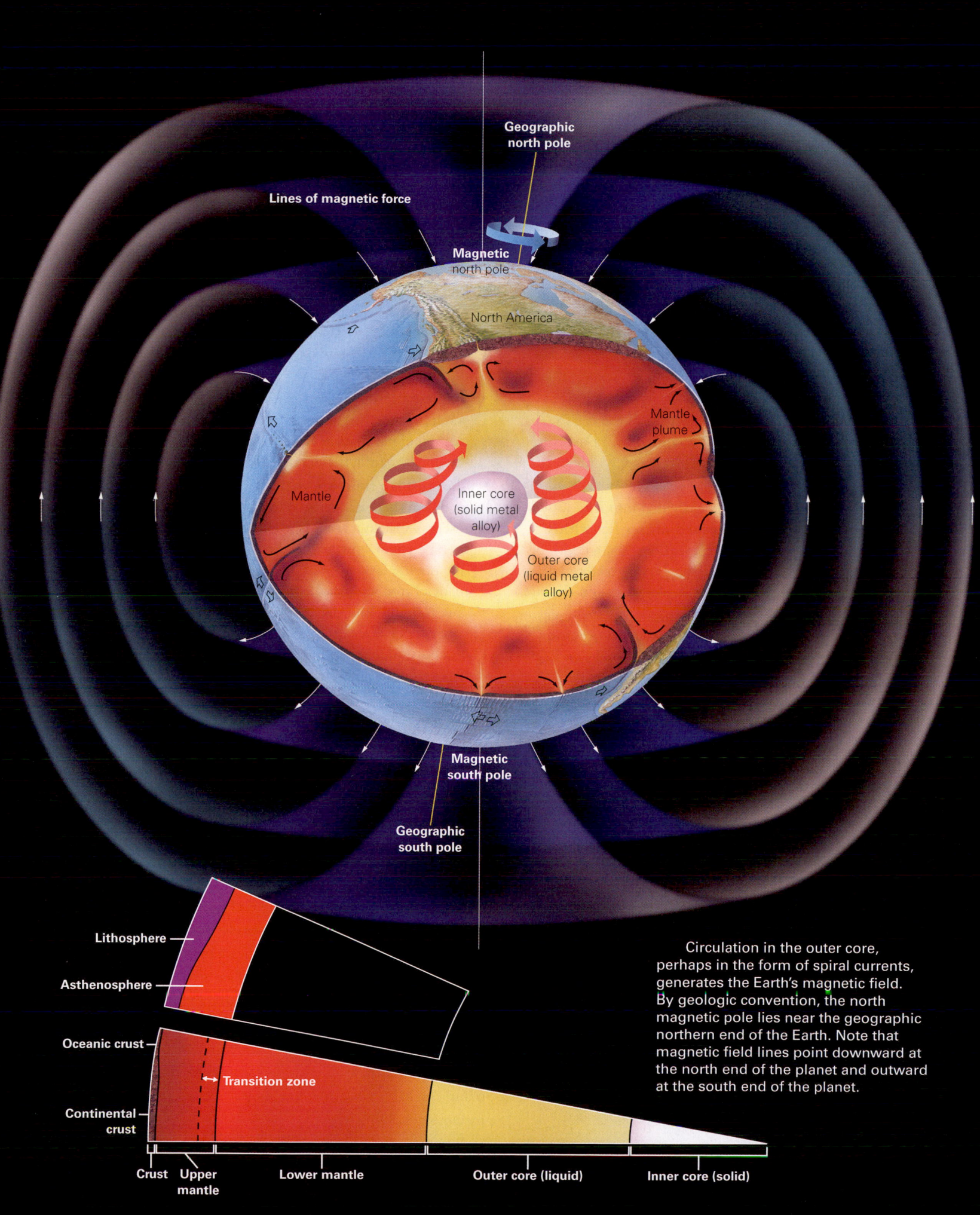

Geographic north pole

Lines of magnetic force

Magnetic north pole

North America

Mantle plume

Mantle

Inner core (solid metal alloy)

Outer core (liquid metal alloy)

Magnetic south pole

Geographic south pole

Lithosphere

Asthenosphere

Oceanic crust

Transition zone

Continental crust

Crust | Upper mantle | Lower mantle | Outer core (liquid) | Inner core (solid)

Circulation in the outer core, perhaps in the form of spiral currents, generates the Earth's magnetic field. By geologic convention, the north magnetic pole lies near the geographic northern end of the Earth. Note that magnetic field lines point downward at the north end of the planet and outward at the south end of the planet.

CHAPTER SUMMARY

- A traverse through the Solar System crosses many features. The Solar System is surrounded by the Oort Cloud of icy particles, attracted by the Sun's gravitational field. The edge of the Solar System itself, a bubble-like surface called the heliosphere, marks the limit at which the pressure of solar wind is countered by that of cosmic rays. Inboard lie the Kuiper Belt of icy objects, the outer planets, the asteroid belt of rocky and metallic material, and the inner planets.

- A magnetic field surrounds the Earth. The field shields it from solar wind. Closer to Earth, the field creates the Van Allen Belts, which also trap cosmic rays.

- A layer of gas, the atmosphere, surrounds the Earth. Air in the atmosphere consists of 78% nitrogen, 21% oxygen, and 1% other gases. Air pressure decreases with elevation, so 99% of the gas in the atmosphere resides below 50 km.

- The surface of the Earth can be divided into land (30%) and ocean (70%). Most of the land surface lies within 1 km of sea level, and most of the seafloor is at a depth of 4 to 5 km. Earth's land surface displays a great variety of landscapes because of variations in elevation and climate.

- Earth materials include organic chemicals, minerals, glasses, rocks (igneous, metamorphic, and sedimentary), grains, sediment, metals, melts, and volatiles. Most rocks on Earth contain silica (SiO_2) and thus are called silicate rocks. We distinguish among felsic, intermediate, mafic, and ultramafic igneous rocks based on the proportion of silica.

- The Earth's interior can be divided into three compositionally distinct layers, named from surface to center: the crust, the mantle, and the core. The first recognition of this division came from studying the density and shape of the Earth. The image has been refined by studying how the speed of seismic waves changes with depth.

- Pressure and temperature both increase with depth in the Earth. At the center, pressure is 3.6 million times greater than at the surface, and the temperature reaches over 4,700°C. The increase in temperature as depth increases is the geothermal gradient.

- Studies of seismic waves reveal the existence of sublayers in the core (liquid outer core and solid inner core) and mantle (upper mantle and lower mantle). The lower part of the upper mantle is the transition zone.

- The crust is a thin skin that varies in thickness from 7 to 10 km (beneath oceans) to 25 to 70 km (beneath continents). Oceanic crust is mafic in composition, whereas average upper continental crust is felsic to intermediate. The mantle is composed of ultramafic rock. The core is made of iron alloy and consists of two parts—the outer core is liquid, and the inner core is solid. Flow in the outer core generates the magnetic field.

- The crust plus the upper part of the mantle constitutes the lithosphere, a relatively rigid shell up to 150 km thick. The lithosphere lies over the asthenosphere, mantle that is capable of flowing and, therefore, convecting.

GUIDE TERMS

abyssal plain (p. 45)
asteroid (p. 39)
asthenosphere (p. 53)
astronomical unit (AU) (p. 37)
atmosphere (p. 41)
aurorae (p. 41)
basalt (p. 46)
bathymetry (p. 44)
comet (p. 39)
continental crust (p. 51)
convection (p. 52)
core (p. 52)
cosmic rays (p. 37)

crust (p. 49)
crystal (p. 45)
deep-ocean trench (p. 45)
dipole (p. 40)
Earth materials (p. 45)
earthquake (p. 49)
Earth System (p. 43)
gabbro (p. 46)
geographic pole (p. 40)
geothermal gradient (p. 46)
granite (p. 46)
groundwater (p. 44)
habitable zone (p. 43)

heliosphere (p. 37)
hypsometric curve (p. 45)
interplanetary space (p. 38)
interstellar space (p. 37)
Kuiper Belt (p. 37)
lithosphere (p. 53)
lithospheric mantle (p. 53)
lower mantle (p. 52)
magnetic field (p. 40)
magnetic field lines (p. 40)
magnetosphere (p. 40)
mantle (p. 51)
meteorite (p. 50)

mid-ocean ridge (p. 45)
Moho (p. 51)
oceanic crust (p. 51)
Oort Cloud (p. 37)
peridotite (p. 46)
silicate rock (p. 46)
surface water (p. 44)
topography (p. 44)
transition zone (p. 52)
upper mantle (p. 52)
vacuum (p. 37)

REVIEW QUESTIONS

1. Why do astronomers consider the space between planets to be a vacuum in comparison with the atmosphere near sea level?

2. Name the features that a spacecraft traversing the Solar System and its surroundings would encounter.

3. What is the Earth's magnetic field? Draw a representation of the field on a piece of paper. Where are the magnetic poles in relation to the geographic poles?

4. How does the magnetic field interact with solar wind and cosmic rays? Be sure to consider the magnetosphere, the Van Allen radiation belts, and the aurorae.

5. What is Earth's atmosphere composed of? Why would you die of suffocation if you were to parachute from an airplane at an elevation of 12 km without taking an oxygen tank with you?

6. What is the proportion of land area to sea area on Earth? From studies of the hypsometric curve, approximately what proportion of the Earth's surface lies at elevations above 2 km?

7. What are the two most abundant elements in the Earth? Describe the major categories of materials constituting the Earth. Does the crust have the same composition as the whole Earth?

8. What are silicate rocks? Give four examples of such rocks, and explain how they differ from one another in terms of their chemical composition.

9. How did researchers first obtain a realistic estimate of Earth's average density? What observations led to the realization that the Earth is largely solid and that a particularly dense core lies at the center?

10. What are seismic waves? Does the velocity at which an earthquake wave travels change or stay constant as the wave passes through the Earth?

11. What are the principal layers of the Earth? What happens to earthquake waves when they reach the boundary between layers?

12. How do temperature and pressure change with increasing depth in the Earth? Be sure to explain the geothermal gradient.

13. What is the Moho? How was it first recognized? Describe the differences between continental crust and oceanic crust.

14. What is the mantle composed of? What are the three sublayers within the mantle? Is there any melt within the mantle?

15. What is the core composed of? How do the inner core and outer core differ from each other? We can't sample the core directly, but geologists have studied samples of materials that are probably very similar in composition to the core. Where do these samples come from?

16. What is the difference between a meteor and a meteorite? Are all meteorites composed of the same material? Explain your answer.

17. What is the difference between lithosphere and asthenosphere? Which layer is softer and flows easily? At what depth does the lithosphere-asthenosphere boundary occur? Is this above or below the Moho?

ON FURTHER THOUGHT

18. (a) Recent observations suggest that the Moon has a very small, solid core that is less than 3% of its mass. In comparison, Earth's core is about 33% of its mass. Explain why this difference might exist. (Hint: Recall the model for Moon formation that we presented in Chapter 1.) (b) The Moon has virtually no magnetosphere. Why? (Hint: Remember what causes Earth's magnetic field.)

19. Popular media sometimes imply that the crust floats on a "sea of magma." Is this a correct image of the mantle just below the Moho? Explain your answer.

20. The measured temperature at the bottom of the deepest drill hole is about 180°C (356°F). What is the geothermal gradient at the location of this hole?

smartwork smartwork.wwnorton.com

This chapter's Smartwork features:

- Map labeling exercises covering geologic features.
- Visual questions that test knowledge of Earth's layers.
- Visual exercises on the parts of the lithosphere.

GEOTOURS

This chapter's GeoTour exercise (A) features:

- Meteorite impacts
- Transition from continental crust to oceanic crust

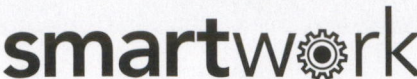

Another View Astronomers are discovering many planetary systems around other nearby stars in our galaxy. This is an artist's conception of what the nearest planetary system, Epsilon Eridani, might look like. This example has an asteroid belt and comets. Their tails follow the currents of the stellar wind.

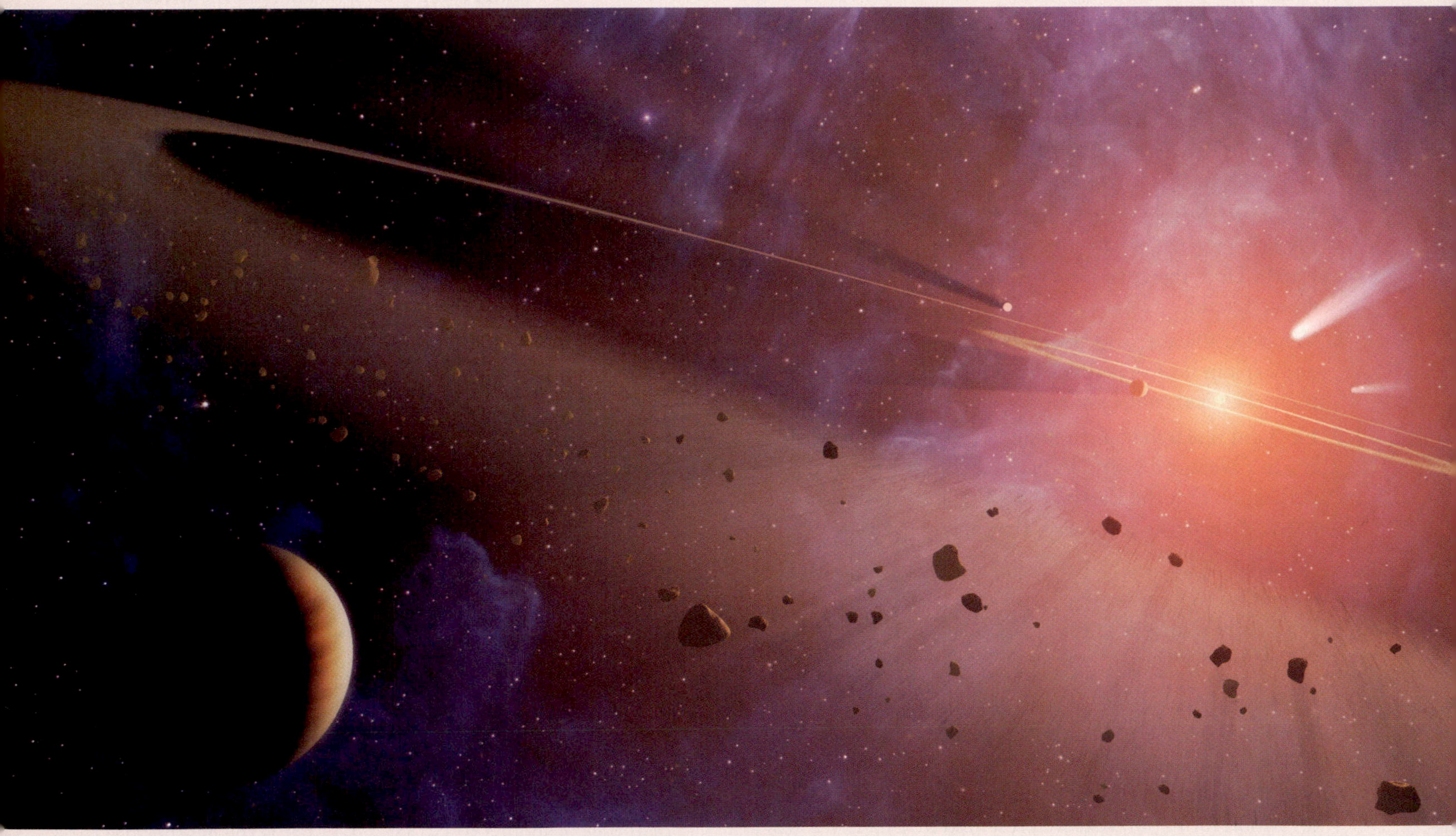

Fossilized remains of *Cynognathus*, an ancestor of mammals, occur on continents now separated by oceans. The organism couldn't swim across an ocean, so how could it have traveled so far? Mysteries such as this led Alfred Wegener to propose that the continents were once together and only later drifted apart.

CHAPTER 3

Drifting Continents and Spreading Seas

> It is only by combing the information furnished by all the earth sciences that we can hope to determine "truth" here.
>
> —*Alfred Wegener (1880–1930)*

3.1 Introduction

In September 1930, 15 explorers led by a German meteorologist, Alfred Wegener, set out across the endless snowfields of Greenland to resupply two weather observers stranded at a remote camp. The observers were planning to spend the long polar night recording wind speeds and temperatures on Greenland's polar plateau. Wegener was a scientist well known not only to researchers studying climate but also to geologists. Some 15 years earlier, he had published a small book, *The Origin of the Continents and Oceans*, in which he had dared to challenge geologists' long-held assumption that the continents had remained fixed in position through all of Earth history. Wegener proposed instead that the continents had once fit together like pieces of a giant jigsaw puzzle, making one vast **supercontinent**. This supercontinent, which he named **Pangaea** (pronounced pan-jee-ah) from the Greek *pan* (all) plus *gaia* (earth), later fragmented into separate continents that drifted apart, moving slowly to their present positions (**Fig. 3.1**). The phenomenon that Wegener proposed came to be known as *continental drift*.

Wegener presented many observations that he believed proved that continental drift had occurred, but he met strong resistance from his peers. At a widely publicized 1926 geology conference in New York City, a crowd of celebrated American professors challenged, "What force could possibly be great enough to move the immense mass of a continent?" Wegener's writings didn't provide a good answer, so most of the meeting's participants rejected continental drift. Four years later, Wegener faced an even greater challenge—survival itself. Sadly, he lost. On October 30, 1930, Wegener and a companion reached the observers, dropped off enough supplies to last the winter, and set out on the return trip the next day, but they never made it home.

Had Wegener survived to old age, he would have seen his hypothesis become the foundation of a scientific revolution. Today geologists accept Wegener's basic conclusion and take for granted the concept that the map of the Earth changes as continents seemingly waltz around this planet's surface, variously combining and breaking apart, through geologic time. In fact, Pangaea wasn't the only supercontinent in Earth history—others formed and broke into pieces that later combined again several times in the past few billion years. The scientific revolution began in 1960, when an American geologist, Harry Hess proposed that as continents move apart new ocean floor forms between them by a process that his contemporary, Robert Dietz, named *seafloor spreading*. Hess suggested that continents can move toward each other when the old ocean floor between them sinks back down into the Earth's interior, a process now called *subduction*. During the 1960s, geologists came to realize that continental movement, seafloor spreading, and subduction, along with a wide range of other geologic phenomena, were manifestations of the fact that the Earth's outer, relatively rigid shell is not a continuum but rather consists of about 20 distinct pieces—now called *plates*—that slowly move relative to each other. Because we can empirically confirm this idea, it has gained the status of a theory, which we now call the theory of *plate tectonics*, from the Greek word *tekton*, which means builder—plate movements effectively "build" regional geologic features. Geologists view plate tectonics as the grand unifying theory of geology, because it so successfully explains a great many geologic processes and features, as we will see.

In this chapter, we introduce the observations that led Wegener to propose his continental-drift hypothesis. Then we look at paleomagnetism, the record of Earth's magnetic field in the past, which provides a key proof of continental drift. Next we learn how observations about the seafloor, made by geologists during the mid-20th century, led to the proposal of seafloor spreading and how the idea was tested and shown to be correct. In Chapter 4 we will build on these concepts and describe the many facets of modern plate tectonics theory.

FIGURE 3.1 Alfred Wegener and his model of continental drift.

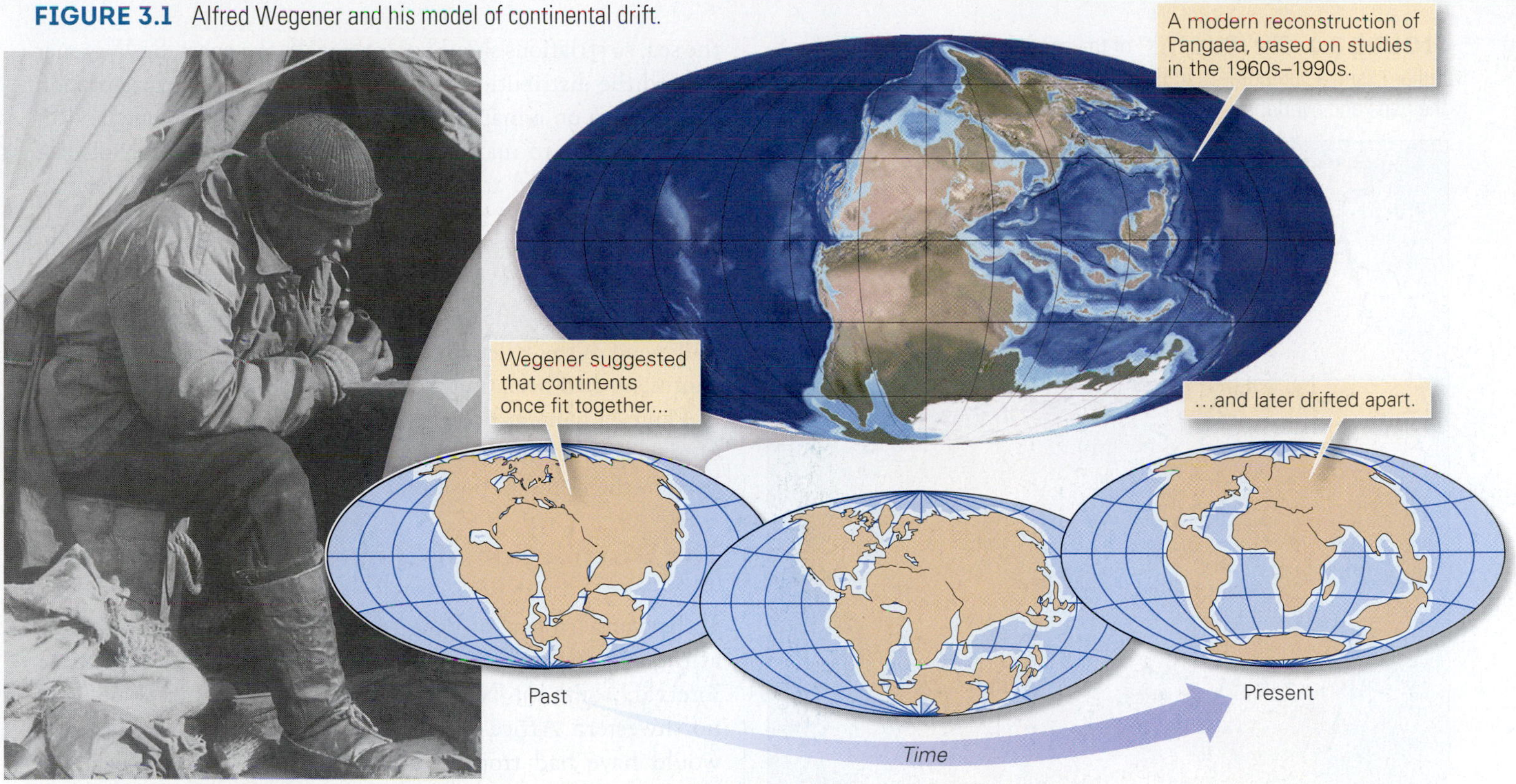

A modern reconstruction of Pangaea, based on studies in the 1960s–1990s.
Wegener suggested that continents once fit together...
...and later drifted apart.
Past
Present
Time

(a) Wegener in Greenland. **(b)** Wegener's maps illustrating continental drift.

3.2 Wegener's Evidence for Continental Drift

Before Wegener, geologists viewed the continents and oceans as "immobile"—fixed in position throughout geologic time. According to Wegener, however, the positions of continents change through time. Specifically, Wegener suggested that a vast supercontinent, Pangaea, existed until the end of the Paleozoic Era and that it broke apart during the Mesozoic (see Fig. P.7 for a simplified geologic time scale). The resulting smaller continents, the ones that exist today, then moved away from each other. Let's look at some of Wegener's arguments and see what led him to formulate this hypothesis of **continental drift**.

The Fit of the Continents

Almost as soon as maps of the Atlantic coastlines became available in the 1500s, scholars noticed the fit of the continents. Specifically, the northwestern coast of Africa looks like it could tuck in against the eastern coast of North America, and the bulge of eastern South America could nestle cozily into the indentation of southwestern Africa. Australia, Antarctica, and India could all connect to the southeast of Africa, while Greenland, Europe, and Asia could pack against the northeastern margin of North America (**Fig. 3.2**). In fact, all the continents could be joined, with remarkably few overlaps or gaps, to form a single supercontinent, Pangaea. Wegener concluded that the fit was too good to be coincidence and thus that the continents once did fit together.

Did you ever wonder . . .

why coasts of the Atlantic look like they fit together?

Locations of Past Glaciations

Glaciers are rivers or sheets of ice that flow across the land surface. As a glacier flows, it carries sediment grains of all sizes (clay, silt, sand, pebbles, and boulders; see Chapter 2 for a definition of *sediment*). Grains protruding from the base of the moving

SEE FOR YOURSELF . . .

The Fit of Continents

LATITUDE
33°31′25.77″N

LONGITUDE
39°26′23.32″W

Zoom to an elevation of 14,800 km (9,200 mi) and look straight down.

Note how northwest Africa could fit snuggly along eastern North America. Wegner used this fit as evidence for Pangaea.

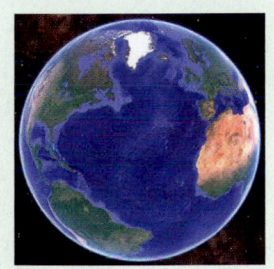

FIGURE 3.2 The "Bullard fit" of the continents. In 1965, Edward Bullard used a computer to fit the continents and demonstrate how minor the gaps and overlaps are, although the match still isn't perfect.

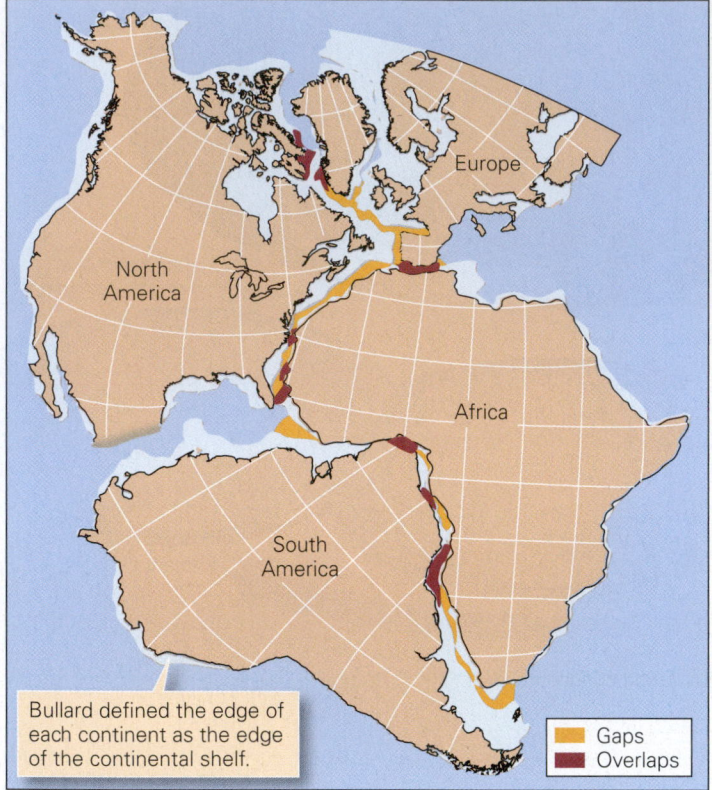

Bullard defined the edge of each continent as the edge of the continental shelf.

Gaps
Overlaps

ice carve scratches, called striations, into the substrate. In some cases, it's possible to tell the direction of slip from striations. When the ice melts, it leaves the sediment in a deposit called till, which may bury the striations. Thus, the occurrence of till and striations at a location serve as evidence that the location was covered by a glacier in the past. By studying the age of glacial till deposits, geologists have determined that large areas of land were covered by glaciers during discrete time intervals of Earth history called *ice ages*. One of these ice ages occurred from 280 to 260 Ma (million years ago), near the end of the Paleozoic Era.

Wegener was a climate scientist by training, and he studied the Arctic, so it's no surprise that he had a strong interest in glaciers. He knew that glaciers form at high (polar) latitudes today, so he was bothered by the observation that sediments indicative of late Paleozoic glaciation occurred in southern South America, southern Africa, southern India, Antarctica, and southern Australia. With the exception of Antarctica, these continents do not currently lie in high latitudes (**Fig. 3.3a–c**). Wegener also noted that most striations associated with these deposits seemed to point from the sea into the continents—this was puzzling because glaciers today form on land and flow to

the sea, so striations should point toward the coast. So Wegener plotted the distribution of glacial deposits and the orientation of striations on a map and then cut out the continents and fit them together to make Pangaea. To his amazement, all late Paleozoic glaciated areas lie adjacent to each other on his map of Pangaea, forming a single coherent ice sheet. Furthermore, when he determined the diretion of movement, he found that it was roughly outward from the center of this ice sheet. In other words, Wegener concluded that the distribution of glaciations at the end of the Paleozoic Era could easily be explained if the continents had been united in Pangaea, with the southern part of Pangaea lying at polar latitudes. The observed distribution of glaciation could not be explained if continents had always been in their present positions.

The Distribution of Climatic Belts

If the southern part of Pangaea had straddled the South Pole at the end of the Paleozoic Era, then during this same time interval, southern North America, southern Europe, and northwestern Africa would have straddled the equator and would have had tropical or subtropical climates. Wegener searched for evidence that this was so by studying characteristics of late Paleozoic sedimentary rocks from these regions, for the material making up sedimentary rocks can reveal clues to the climate at the time the sediment formed. For example, in the swamps and jungles of tropical regions, thick deposits of plant material accumulate, and when deeply buried, this material transforms into coal. And, in the clear, shallow seas of tropical regions, large reefs made from the shells of marine organisms develop. Finally, in subtropical regions, on either side of the tropical belt, where desert climates exist, salt deposits (from evaporating seawater or salt lakes) accumulate, and sand dunes grow. Wegener speculated that the distribution of late Paleozoic coal, reef, sand-dune, and salt deposits could define climate belts on Pangaea.

Sure enough, in the belt of Pangaea that Wegener expected to be equatorial, late Paleozoic sedimentary rock layers include abundant coal and the relicts of reefs. And in the belts of Pangaea that Wegener predicted

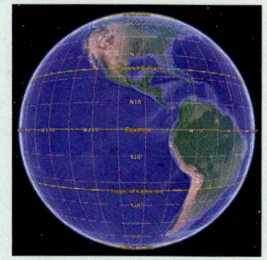

FIGURE 3.3 The distribution of Late Paleozoic glacial deposits, climate belts, and fossils.

(a) Glacial striations of Late Paleozoic age on the surface of bedrock along the south coast of Australia.

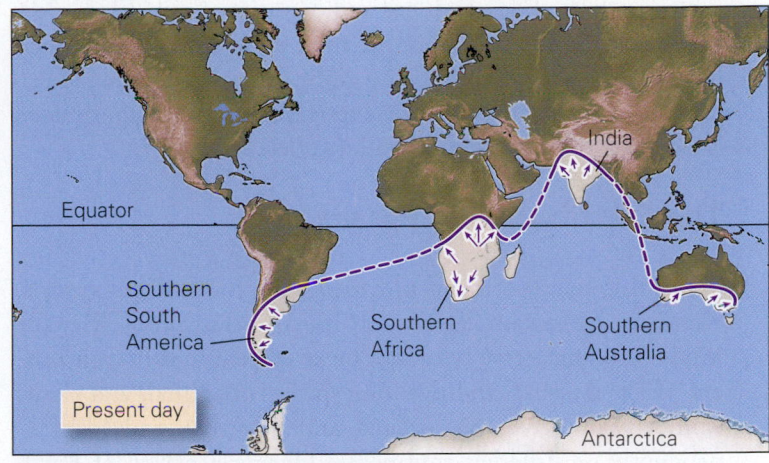

(b) A map showing the distribution of Late Paleozoic glacial deposits and the orientation of associated striations.

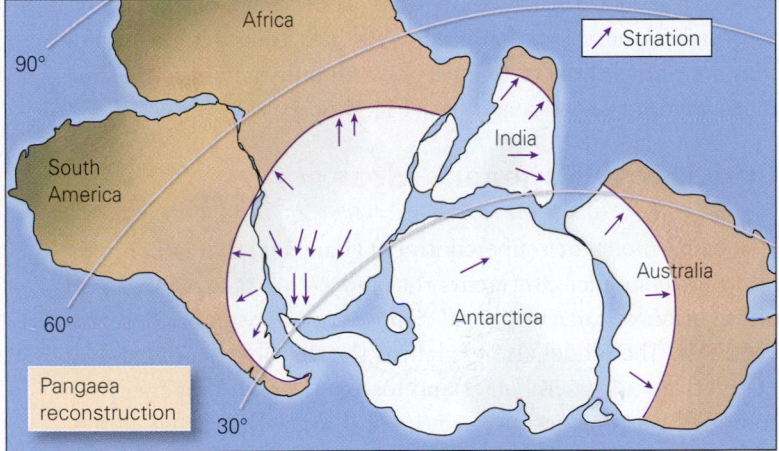

(c) On Wegener's reconstruction of Pangaea, the glaciated areas connect to outline a region of Late Paleozoic south polar ice caps.

Fossil leaves of *Glossopteris* from Australia.

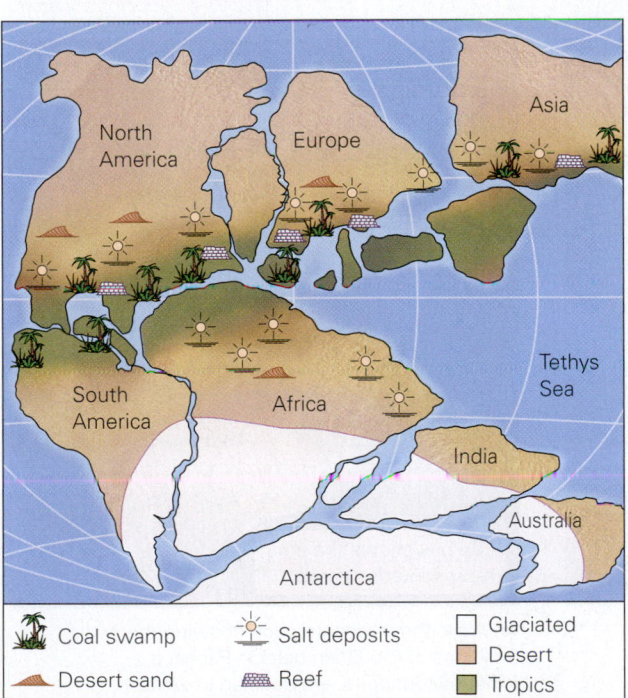

(d) Climate belts, as indicated by distinct rock types, make sense on a map of Pangaea.

Coal swamp   Salt deposits   Glaciated
Desert sand   Reef   Desert
Tropics

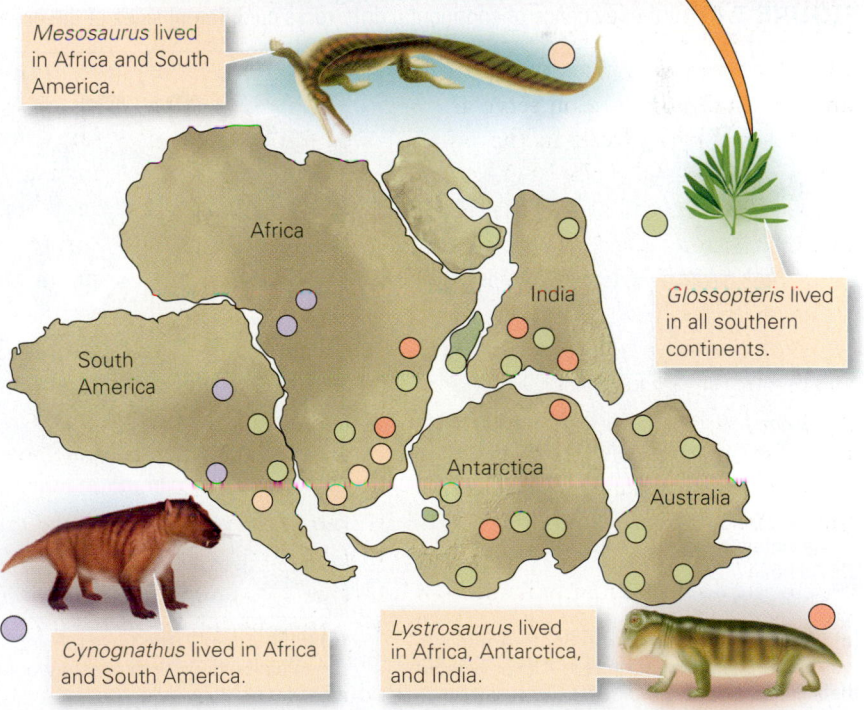

Mesosaurus lived in Africa and South America.

Glossopteris lived in all southern continents.

Cynognathus lived in Africa and South America.

Lystrosaurus lived in Africa, Antarctica, and India.

(e) Fossil localities shows that Mesozoic land-dwelling organisms occur on more than one continent. This would be hard to explain if oceans lay between these continents.

would be subtropical, late Paleozoic sedimentary rock layers include relics of desert dunes and deposits of salt (**Fig. 3.3d**). On a present-day map of our planet, exposures of these rock layers are scattered around the globe at a variety of latitudes. On Wegener's Pangaea, the exposures align in continuous bands that occupy appropriate latitudes.

The Distribution of Fossils

Today different continents provide homes for different species. Kangaroos, for example, live only in Australia. Similarly, many kinds of plants grow only on one continent and not on others. Why? Because land-dwelling species of animals and plants cannot swim across vast oceans, and thus they evolved independently on different continents. During a period of Earth history when all continents were in contact, however, land animals and plants could have migrated relatively easily among many continents.

Did you ever wonder...

if you could have once walked from New York to Paris?

With this concept in mind, Wegener plotted fossil occurrences of land-dwelling species that existed during the late Paleozoic and early Mesozoic Eras (between 300 and 210 Ma) and found that these species had indeed existed on several continents (**Fig. 3.3e**). Wegener argued that the distribution of these fossils required the continents to have been adjacent to one another in the late Paleozoic and early Mesozoic Eras.

Matching Geologic Units

Art historians can recognize a Picasso painting, and architects know what makes a building's style Victorian. Similarly, geologists can identify distinctive assemblages of rocks. Wegener found that the same distinctive Precambrian (before 541 Ma) rock assemblages occurred on the eastern coast of South America and the western coast of Africa, regions now separated by an ocean (**Fig. 3.4a**). If the continents had been joined to create Pangaea in the past, then these matching rock groups would have been adjacent to each other and thus could have composed continuous blocks or belts. Wegener also noted that features of the Appalachian mountain belt of the United States and Canada closely resemble those of mountain belts in southern Greenland, Great Britain, Scandinavia, and northwestern Africa (**Fig. 3.4b**), regions that would have lain adjacent to North America in Pangaea. Wegener thus demonstrated that not only did the coastlines of continents match, so did the rocks adjacent to the coastlines (**Fig. 3.4c**).

Criticism of Wegener's Ideas

Wegener's model of a supercontinent (Pangaea) that later broke up to form smaller continents that moved apart explained the distribution of ancient glacial deposits, coal swamps, deserts, and reefs. The model also explained the distribution of certain distinctive rock assemblages and fossils. Clearly, Wegener had compiled a strong case for continental drift. But, as we noted earlier, he could not adequately explain how or why continents

FIGURE 3.4 Further evidence of continental drift: rocks on different sides of the ocean match.

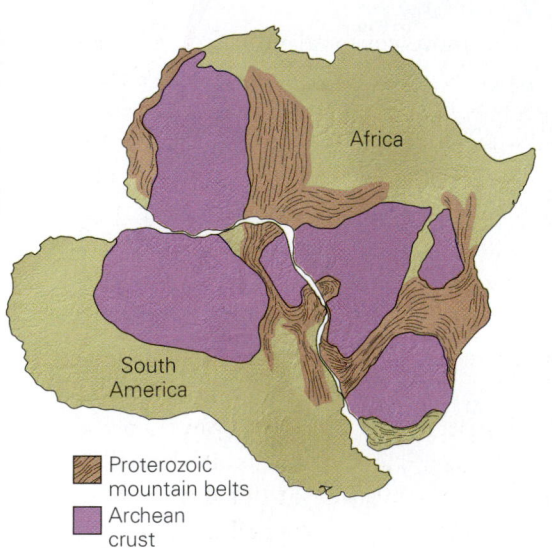

Proterozoic mountain belts
Archean crust

(a) Distinctive belts of rock in South America would align with similar ones in Africa, without the Atlantic.

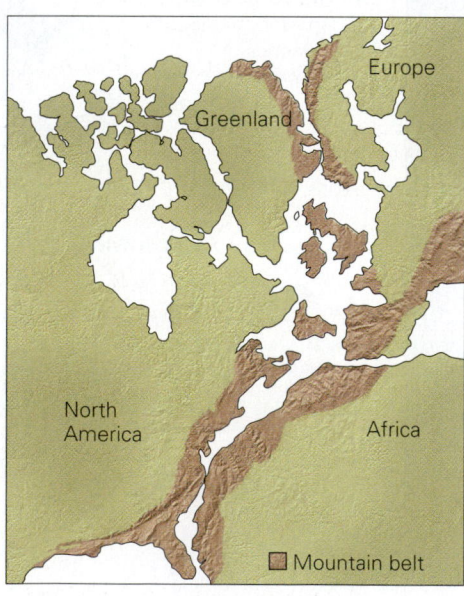

Mountain belt

(b) If the Atlantic didn't exist, Paleozoic mountain belts on both coasts would be adjacent.

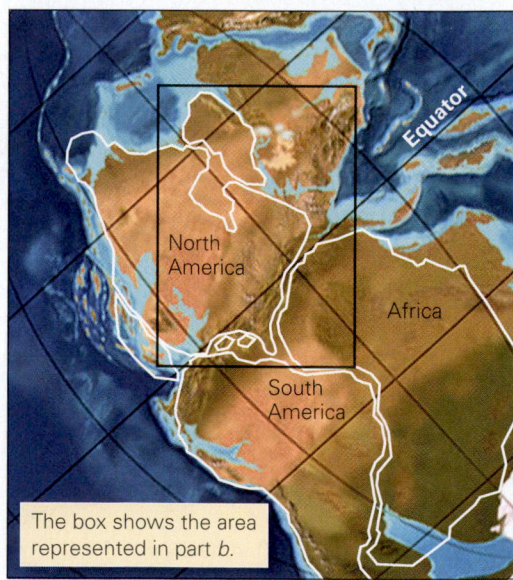

The box shows the area represented in part *b.*

(c) A modern reconstruction showing the positions of mountain belts in Pangaea. Modern continents are outlined in white.

moved. Wegener's writings gave the impression that continents somehow "plowed" through the ocean floor like the keel of a ship plows through water, but that's not possible because ocean floor rock is too strong. Wegener also suggested that centrifugal force, due to the Earth's spin, drove continental movement, but that's not possible because the force isn't strong enough. He left on his final expedition to Greenland having failed to convince his peers, and he died without knowing that his ideas, after lying dormant for decades, would be reborn as the basis of the broader theory of plate tectonics.

In effect, Wegener was ahead of his time. In the three decades that followed his death, a handful of iconoclasts continued to champion his notions. Among them was Arthur Holmes, a British geophysicist who argued that huge convection cells existed inside the Earth, slowly transporting hot rock from the deep mantle up to the base of the crust. Holmes speculated that continents might be forced apart in response to convective flow in the mantle and that the continents rode like rafts on the top of convective cells. But most geologists remained unconvinced and didn't realize that Wegener's bold idea would one day grow into a theory that would change the whole discipline of geology forever. The door to the discovery of plate tectonics opened in the mid-20th century, when technologies became available to provide such data. Between 1930 and 1960, geologists learned how to determine the age of rocks, how to analyze *paleomagnetism* (discussed below), and how to "see" the ocean floor. In the next two sections, we describe some of the key results of this work.

Take-Home Message

In the early 20th century, Alfred Wegener argued that the continents had once been connected in a supercontinent, Pangaea, that later broke up to produce smaller continents that "drifted" apart. He showed that the matching shapes of coastlines, as well as the distribution of ancient glaciers, climate belts, fossils, and rock units, make better sense if Pangaea existed. But Wegener couldn't convince his peers.

QUICK QUESTION: Why were Wegener's peers skeptical of continental drift?

3.3 Paleomagnetism— Proving Continents Move

More than 1,500 years ago, Chinese sailors discovered that a piece of lodestone, when suspended from a thread, points in a northerly direction and can help guide a voyage. Lodestone exhibits this behavior because it consists of magnetite, an iron-rich mineral that, like a compass needle, aligns with Earth's magnetic field lines. While not as magnetic as lodestone, several other rock types contain trace amounts of magnetite, or other magnetic minerals, and thus behave overall like weak magnets. In this section, we explain how the study of such magnetic behavior led to the realization that rocks preserve **paleomagnetism**, a record of Earth's magnetic field in the past. An understanding of paleomagnetism provided proof of continental drift and, as we'll see later in this chapter, contributed to the development of plate tectonics theory. As a foundation for introducing paleomagnetism, we first provide further detail on the basic nature of the Earth's magnetic field.

Did you ever wonder . . .

why compasses always point to the north?

Earth's Magnetic Field

Circulation of liquid iron alloy in the outer core of the Earth generates a magnetic field. (A similar phenomenon happens in an electrical dynamo—we'll discuss this concept further in Interlude D.) Earth's magnetic field resembles the field produced by a bar magnet in that it has two ends of opposite polarity, as introduced in Chapter 2. Thus, we can represent Earth's field by a **magnetic dipole**, an imaginary arrow (**Fig. 3.5a**). Earth's dipole intersects the surface of the planet at two points, known as the **magnetic poles**. By convention, the north magnetic pole lies at the end of the Earth nearest the north geographic pole (the point where the northern end of the spin axis intersects the surface), so that the north-seeking (red) end of a compass needle points to the north magnetic pole.

Earth's magnetic poles move constantly—in fact, at present the north magnetic pole is moving across the Arctic Ocean toward Russia at 50 to 60 km per year. Significantly, poles don't seem to stray farther than about 2,000 km (about 20° of latitude) from the geographic poles (**Fig. 3.5b**). Because of their overall fairly random movements, geologists assume that, averaged over thousands of years, the locations of the magnetic poles roughly coincide with Earth's geographic poles. This relationship presumably reflects the rotation of the Earth, for the spin may cause the flow to organize into patterns resembling spring-like spirals that align with the axis. At present, the magnetic poles lie hundreds of kilometers away from the geographic poles, so the magnetic dipole tilts at about 10° relative to the Earth's spin axis. Because of this difference, a compass today does not point exactly to geographic north. The angle between the direction that a compass needle points and a line of longitude at a given location is the **magnetic declination** (**Fig. 3.5c**).

FIGURE 3.5 Features of Earth's magnetic field.

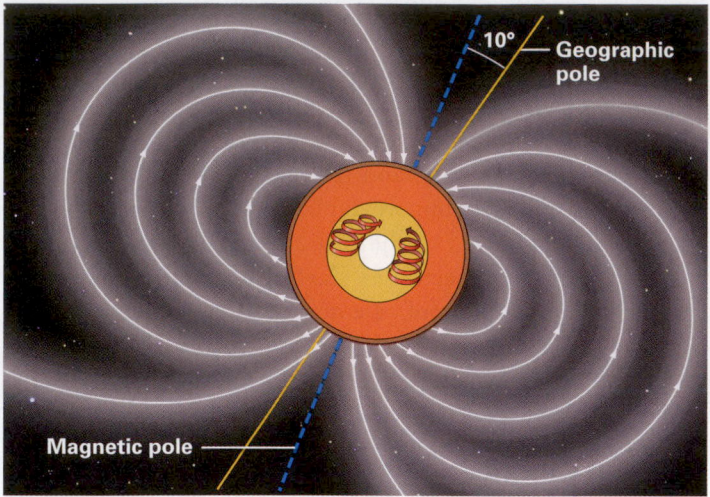

(a) The magnetic axis is not parallel to the spin axis. The field is due to flow in the outer core.

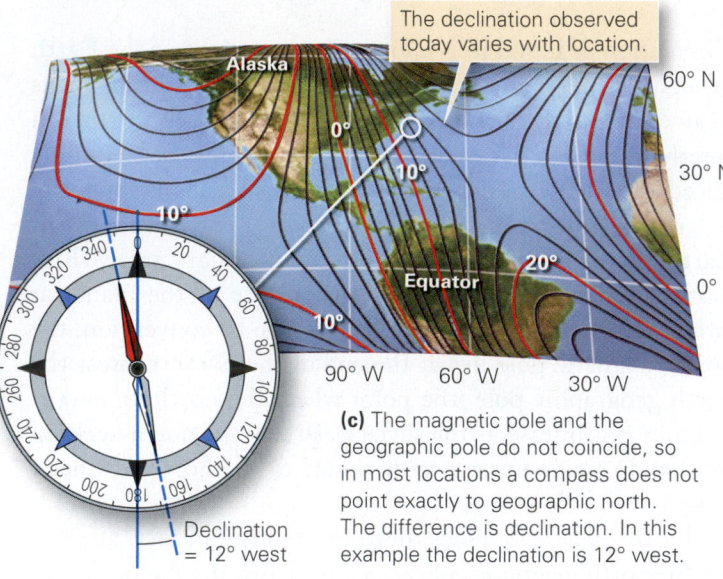

(c) The magnetic pole and the geographic pole do not coincide, so in most locations a compass does not point exactly to geographic north. The difference is declination. In this example the declination is 12° west.

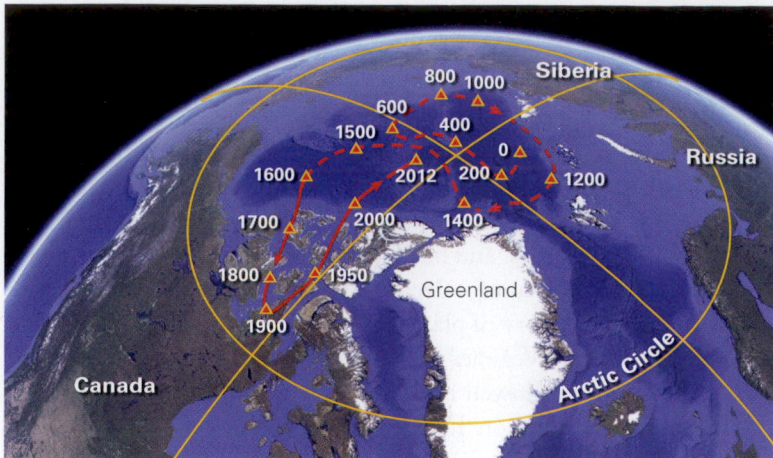

(b) A simplified map showing the changing position of the north magnetic pole over the past 2000 years. Before about 1600, the position was not as well constrained, so the path is dashed.

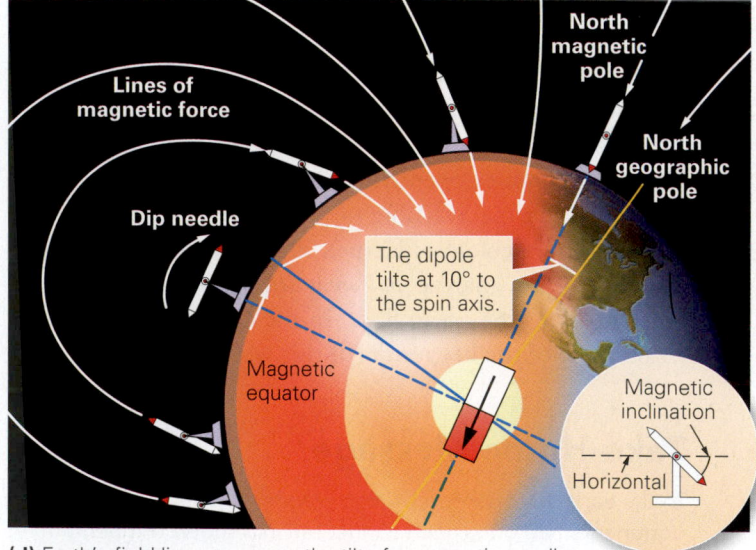

(d) Earth's field lines curve, so the tilt of a magnetic needle changes with latitude. This tilt is the magnetic inclination.

Invisible field lines curve through space between the magnetic poles (**Fig. 3.5d**). In a cross-sectional view, these lines lie parallel to the surface of the Earth (i.e., are horizontal) at the equator, tilt at an angle to the surface in midlatitudes, and plunge perpendicular to the surface (i.e., are vertical) at the magnetic poles. The angle between a magnetic field line and the surface of the Earth, at a given location, is called the **magnetic inclination**. If you place a magnetic needle on a horizontal axis so it can pivot up and down and then carry it from the magnetic equator to the magnetic pole, you'll see that the inclination varies with latitude—it is 0° at the magnetic equator and 90° at the magnetic poles. (Note that the compass you may carry with you on a hike does not show inclination because it has been balanced to remain horizontal and pivots on a vertical axis.)

What Is Paleomagnetism?

In the early 20th century, researchers developed instruments that could measure the very weak magnetic field produced by rocks and made a surprising discovery. In a rock that formed millions of years ago, the orientation of the dipole representing the magnetic field of the rock is not the same as that of present-day Earth (**Fig. 3.6a**). To understand this statement, imagine that you go to a locality and measure the very weak magnetic field emanating from a 90-million-year-old rock. (In reality, the signal is so weak that you would have to take a sample of the rock back to a laboratory and measure it with specialized instruments.) If you represent the rock's magnetic field by a bar magnet, you'll likely find that this magnet's declination differs from the declination that a compass at the location would

FIGURE 3.6 Paleomagnetism and how it can form.

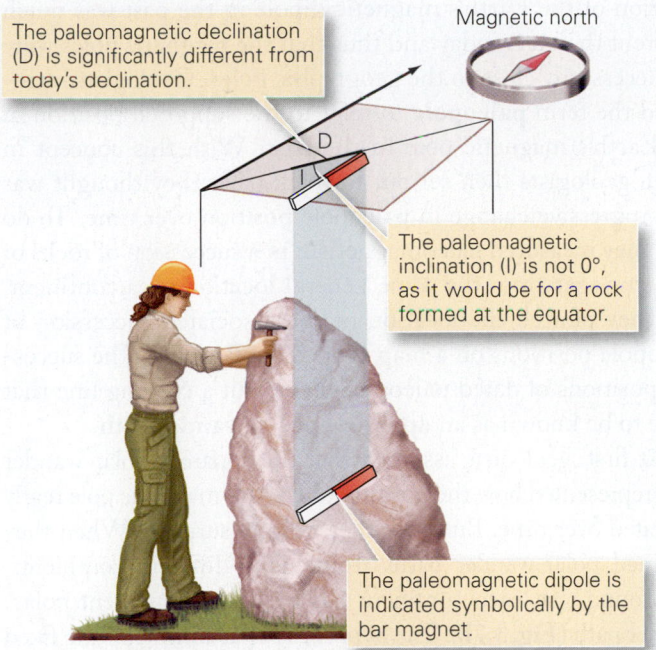

The paleomagnetic declination (D) is significantly different from today's declination.

Magnetic north

The paleomagnetic inclination (I) is not 0°, as it would be for a rock formed at the equator.

The paleomagnetic dipole is indicated symbolically by the bar magnet.

(a) A geologist finds an ancient rock sample at a location on the equator, where declination today is 0°. The orientation of the rock's paleomagnetism is different from that of today's field.

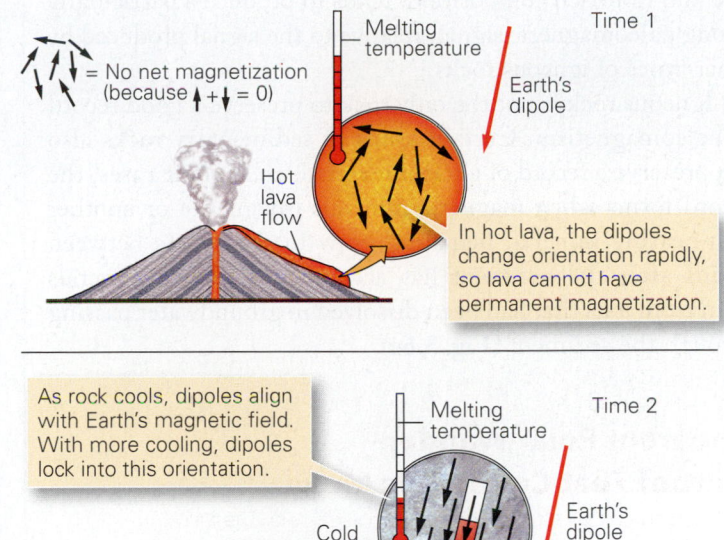

Time 1

Melting temperature

Earth's dipole

= No net magnetization (because ↑ + ↓ = 0)

Hot lava flow

In hot lava, the dipoles change orientation rapidly, so lava cannot have permanent magnetization.

As rock cools, dipoles align with Earth's magnetic field. With more cooling, dipoles lock into this orientation.

Melting temperature

Time 2

Cold basalt

Earth's dipole

(b) Paleomagnetism can form when lava cools and becomes solid rock.

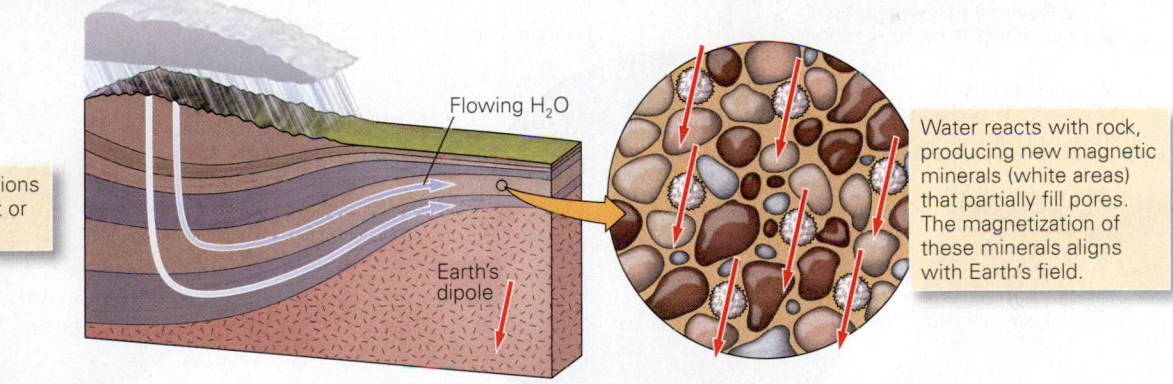

Flowing H₂O

Water carrying dissolved ions passes through sediment or sedimentary rocks.

Water reacts with rock, producing new magnetic minerals (white areas) that partially fill pores. The magnetization of these minerals aligns with Earth's field.

Earth's dipole

(c) Paleomagnetism can can also form when iron-bearing minerals precipitate out of groundwater passing through sediment.

display today. Also, you would likely find that this magnet's inclination is not the same as the inclination appropriate for the latitude of your sample. These differences arise because the rock is preserving a record of the orientation of the location of the magnetic pole, relative to the rock, at the time the rock formed. In geologic jargon, the rock is preserving paleomagnetism.

Paleomagnetism can develop in many ways. For example, the paleomagnetism of basalt forms when the rock cools from a melt (**Fig. 3.6b**). Let's follow the stages of this process. Imagine a flow of lava so hot that it contains no solid crystals. As the lava starts to cool and solidify into rock, tiny magnetite crystals begin to grow along with several other types of minerals. Each

magnetite crystal produces a tiny magnetic dipole. At first, thermal energy causes the magnetic dipole associated with each crystal to wobble and tumble chaotically. Thus, at any given instant, the dipoles of the crystals are randomly oriented and the magnetic forces they produce cancel each other out. As the rock cools, however, thermal energy decreases so the dipoles slow down and, like tiny compass needles, align with the Earth's magnetic field. Eventually, the rock cools down so much, that the dipoles can no longer move and lock into permanent parallelism with the Earth's magnetic field at the time this cooling takes place. Since the magnetic dipoles of all the grains point in the same direction, they can add together and produce a measurable field. Basalt, because of its small grain-size and iron-rich composition, tends to produce a particularly strong paleomagnetic signal, relative to the signal produced by other types of igneous rocks.

Igneous rock is not the only rock to preserve a good record of paleomagnetism. Certain kinds of sedimentary rocks also can preserve a record of ancient magnetism. In some cases, the record forms when magnetic minerals (magnetite or another iron-bearing mineral, hematite) grow in the spaces between grains after the sediment has accumulated. These minerals form from ions that had been dissolved in groundwater passing through the sediment (**Fig. 3.6c**).

Apparent Polar Wander–
A Proof That Continents Move

Why doesn't the paleomagnetic dipole in an ancient rock point to the present-day magnetic field? When geologists first attempted to answer this question, they assumed that conti-

nents were fixed in position; so they concluded that the orientation of the Earth's magnetic dipole in the past was much different than it is today and thus that the magnetic poles were not necessarily close to the geographic poles. Geologists introduced the term **paleopole** to refer to the supposed position of the Earth's magnetic pole in the past. With this concept in mind, geologists then set out to track what they thought was the progressive change in paleopole position over time. To do this, they measured paleomagnetism in a succession of rocks of different ages from the same general location on a continent, and they plotted the location of the associated succession of paleopole positions on a map (**Fig. 3.7a; Box 3.1**). The successive positions of dated paleopoles trace out a curving line that came to be known as an **apparent polar-wander path**.

At first, geologists assumed that the apparent polar-wander path represented how the position of Earth's magnetic pole really migrated over time. But were they in for a surprise! When they obtained polar-wander paths from many different continents, they found that each continent has a different apparent polar-wander path (**Fig. 3.7b**). The hypothesis that continents are fixed

FIGURE 3.7 Apparent polar-wander paths and their interpretation.

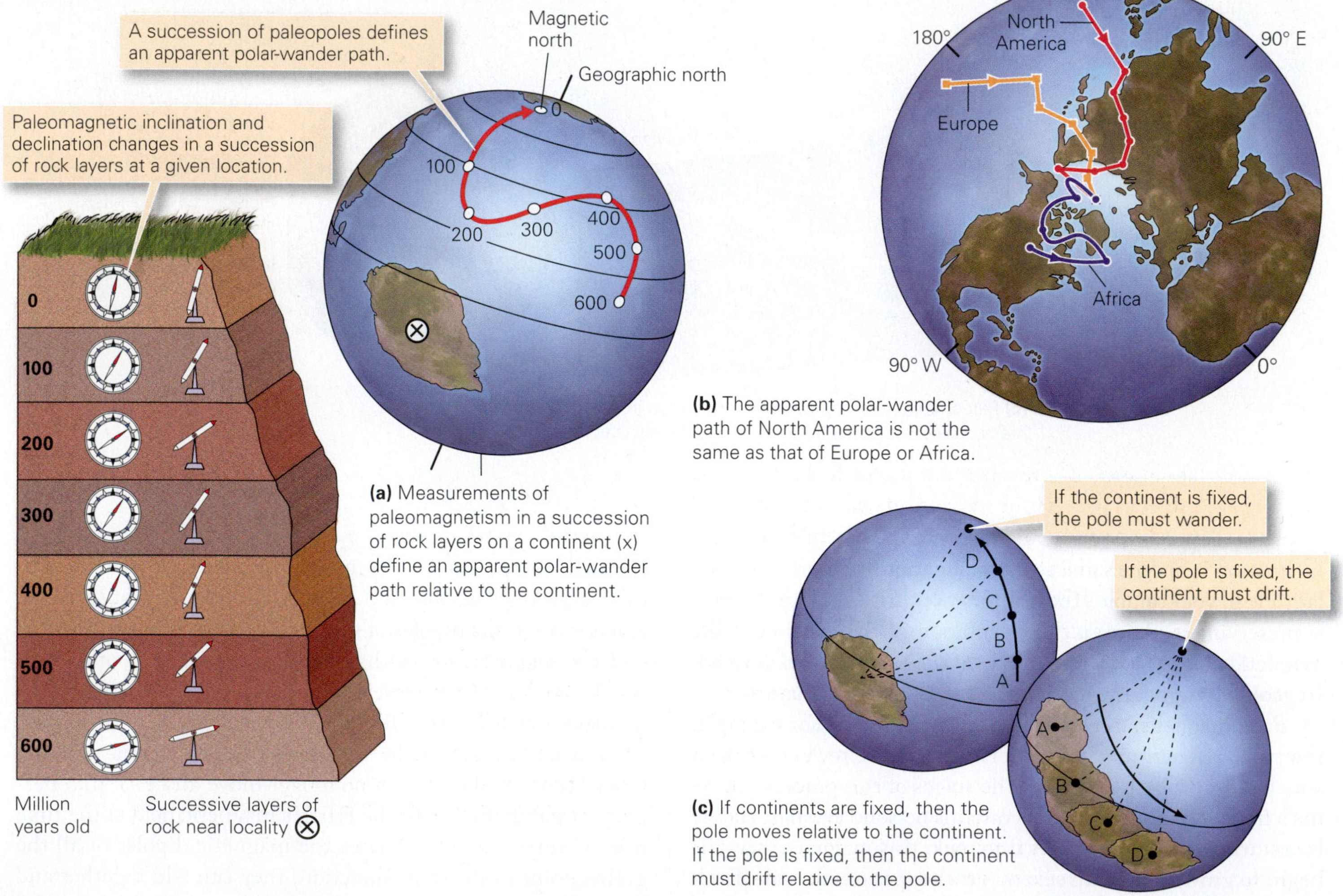

(a) Measurements of paleomagnetism in a succession of rock layers on a continent (x) define an apparent polar-wander path relative to the continent.

(b) The apparent polar-wander path of North America is not the same as that of Europe or Africa.

(c) If continents are fixed, then the pole moves relative to the continent. If the pole is fixed, then the continent must drift relative to the pole.

BOX 3.1   CONSIDER THIS . . .

Finding Paleopoles

How do you find a paleopole position from the orientation of a paleomagnetic dipole in a rock sample? The horizontal projection of the dipole arrow on the Earth's surface (picture the projection as the shadow cast on the Earth's surface by the arrow if the Sun were directly overhead) is like a compass needle that points to the paleopole; that is, the projection defines an imaginary great circle around the Earth that passes through the paleopole and the sample (**Fig. Bx3.1**). This great circle is like an imaginary "paleolongitude" line. Note that when drawing the circle, we assume that the declination at the time the sample was magnetized equals 0, because we assume that, averaged over time, the magnetic pole coincides with the geographic pole.

To find the specific position of the paleopole on this great circle, we must look at the inclination of the paleomagnetic dipole in the rock. Recall that inclination depends on latitude (see Fig. 3.5d). Thus, the inclination of the paleomagnetic dipole defines the paleo-latitude of the sample with respect to the paleopole, and paleolatitude simply represents the distance (measured in degrees) from the pole along the great circle to where the sample formed.

FIGURE Bx3.1 The basic concept of how to find a paleopole from a paleomagnetic measurement.

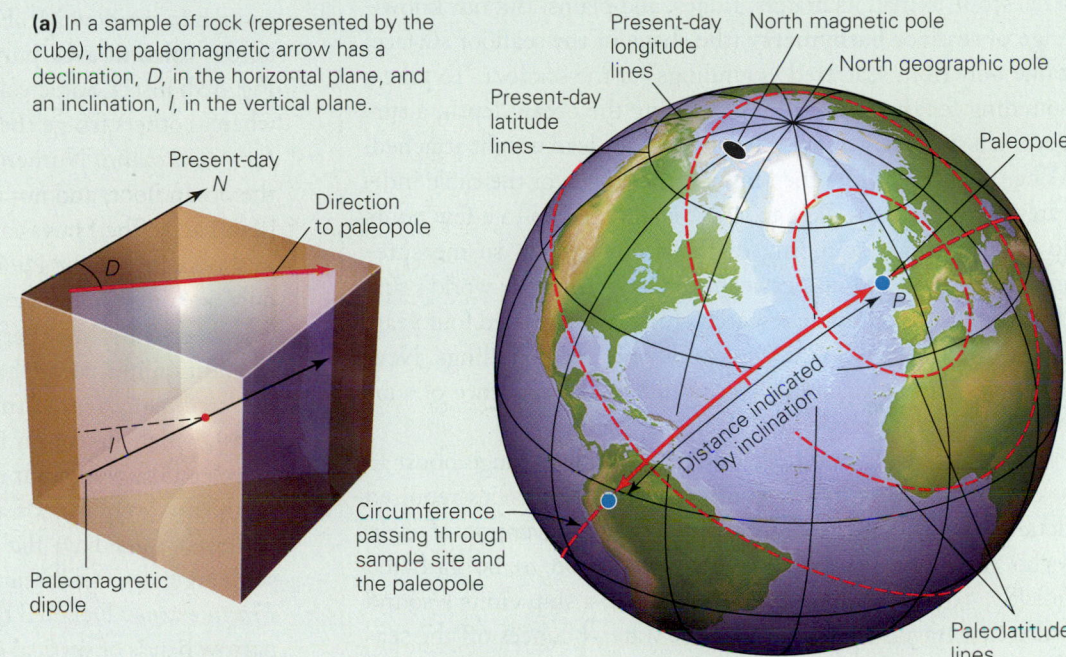

(a) In a sample of rock (represented by the cube), the paleomagnetic arrow has a declination, D, in the horizontal plane, and an inclination, I, in the vertical plane.

(b) The D of the sample points to the paleopole, P. Thus, the location of the rock outcrop and P lie on a circumference that represents a line of paleolongitude. The inclination indicates the paleolatitude of the sample and thus is a distance measured along the paleolongitude.

in position cannot explain this observation, for if the magnetic pole moved while all the continents stayed fixed, measurements from all continents should produce the same apparent polar-wander paths. Geologists suddenly realized that they were looking at apparent polar-wander paths in the wrong way. It's not the pole that moves relative to fixed continents but rather the continents that move relative to a fixed pole (**Fig. 3.7c**). And since each continent has its own unique polar-wander path, the continents must also be moving relative to one another. In effect, the interpretation of apparent polar-wander paths proved that Wegener was right all along—continents do move! Even so, much of the geologic community remained skeptical of continental movement because no one had yet been able to describe the mechanism that caused continents to move. But that was soon to come.

Take-Home Message

A rock can contain a record of the position of the Earth's magnetic poles, relative to the rock, at the time the rock formed. Study of such paleomagnetism indicates that the continents have moved relative to the Earth's magnetic poles. Each continent has a different apparent polar-wander path, which is possible only if the continents move relative to one another.

QUICK QUESTION: How does comparison of apparent polar-wander paths for different continents prove that continents move relative to each other?

3.4 The Discovery of Seafloor Spreading

New Images of Seafloor Bathymetry

Before World War II, we knew less about the shape of the ocean floor than we did about the shape of the Moon's surface. After all, we could at least see the surface of the Moon and could use a telescope to map its craters, ridges, and plains. But our knowledge of seafloor **bathymetry** (the shape of the seafloor surface) came only from scattered "soundings" of the seafloor. To take a sounding for the purpose of measuring the ocean depth, a surveyor lets out a length of cable with a heavy lead weight attached. When the weight hits the seafloor, the length of the cable indicates the depth. Needless to say, it could take up to a few hours to make a single sounding of the deep seafloor, so measurements were few and far between. In fact, during the world's first dedicated oceanographic research cruise, which lasted four years (1872–76), the HMS *Challenger* took only 360 soundings. Nevertheless, these measurements did hint at the existence of submarine mountain ranges and deep-sea troughs.

Military needs during World War II gave a huge boost to seafloor exploration; as submarine fleets grew, navies required detailed information about bathymetry. The invention of *sonar* (echo sounding) permitted such information to be gathered quickly. To make a sounding using sonar, a ship emits a sound pulse that travels down through the water, bounces off the seafloor, and returns up as an echo through the water to a receiver. Since sound waves travel at a known velocity, the time between the sound's emission and the echo's detection indicates the distance between the ship and the seafloor (Velocity = Distance ÷ Time; therefore: Distance = Velocity × Time). Because sound waves travel much faster than ships—a ship moves only about 2 m in the time it takes for a sound wave to travel to the deep seafloor and back—observers can obtain a continuous record of the depth to the seafloor and can produce a *bathymetric profile*, a graph showing how depth varies with location along a line. By cruising back and forth across the ocean many times at different locations, investigators eventually obtained enough bathymetric profiles to construct a *bathymetric map* of the seafloor. (Researchers now produce such maps much more rapidly and accurately using satellite data; **Fig. 3.8a**). Bathymetric maps reveal several important features (see Fig. 2.6).

- *Mid-ocean ridges:* The floor beneath oceans includes **abyssal plains**, broad flat regions of the ocean that lie at a depth of 4 to 5 km below sea level, and **mid-ocean ridges**, elongate submarine mountain ranges whose peaks lie about 2 to 2.5 km below sea level (**Fig. 3.8b, c**). Geologists call the crest of the mid-ocean ridge the **ridge axis**.

Mid-ocean ridges are roughly symmetrical, in that the bathymetry on one side of the ridge axis is more or less a mirror image of bathymetry on the other side.

- *Deep-ocean trenches:* Along much of the perimeter of the Pacific Ocean, and at several other localities as well, the ocean floor reaches depths greater than 5 km. These deep areas define elongate troughs that are now referred to as **trenches** (**Fig. 3.9**). Some trenches have depths in the range of 8 to 10 km, more than twice the depth of the abyssal plains. In fact, the deepest trench, the Mariana Trench of the western Pacific, reaches a depth of 10.9 km, deep enough to swallow Mt. Everest without a trace. All trenches border **volcanic arcs**, curving chains of active volcanoes (see Fig. 3.9b inset). Some volcanic arcs form a chain of islands, whereas others fringe the edge of continents.

- *Seamount chains:* Numerous volcanic islands poke up from the ocean floor, and not all of these are along volcanic island arcs. The Hawaiian Island chain, for example, lies in the middle of the Pacific. In contrast to an island arc, only one of the islands of the chain has active (erupting) volcanoes—all the other islands ceased erupting long ago. In addition to islands that rise above sea level, sonar has detected many **seamounts** (isolated submarine mountains), which also occur in chains. Seamounts originated by volcanic activity, but most are no longer active. Many seamounts were islands at one time but later sank beneath sea level. Some have flat tops due to reef growth before submergence—such seamounts are called *guyots*.

- *Fracture zones:* Detailed bathymetric surveys reveal that narrow bands of vertical cracks and broken-up rock locally dice up the seafloor of mid-ocean ridges. Notably, these bands, or **fracture zones**, trend at a high angle to the associated ridge axis and separate the ridge into small segments that do not align with one another (see Fig. 3.9a inset). Fracture zones become less distinct away from the ridge axis and are not visible on the surface of abyssal plains.

New Observations on the Nature of Oceanic Crust

By the mid-20th century, geologists had discovered many important characteristics of the seafloor crust and filled in huge blanks on the map of the Earth. These discoveries led them to realize that oceanic crust is quite different from continental crust and, further, that bathymetric features of the ocean floor provide clues to the origin of the crust. Of particular note:

1. Researchers found that a layer of sediment, composed of clay and the tiny shells of dead plankton, covers much of the ocean floor, but even at its thickest, given observed rates of sediment accumulation, the sediment layer is far too thin to have been accumulating for the

FIGURE 3.8 Bathymetry of the whole ocean, and of mid-ocean ridges in particular.

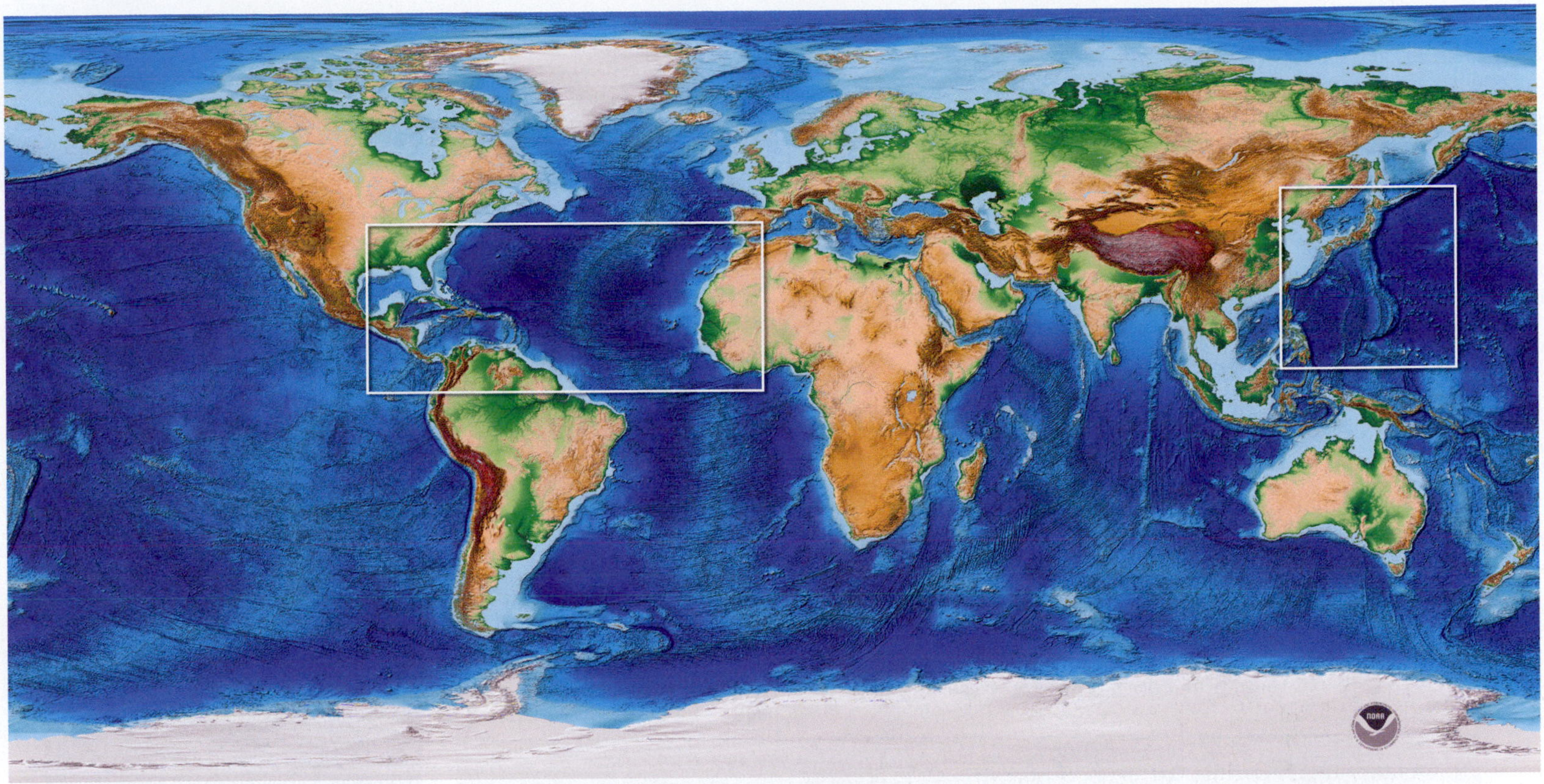

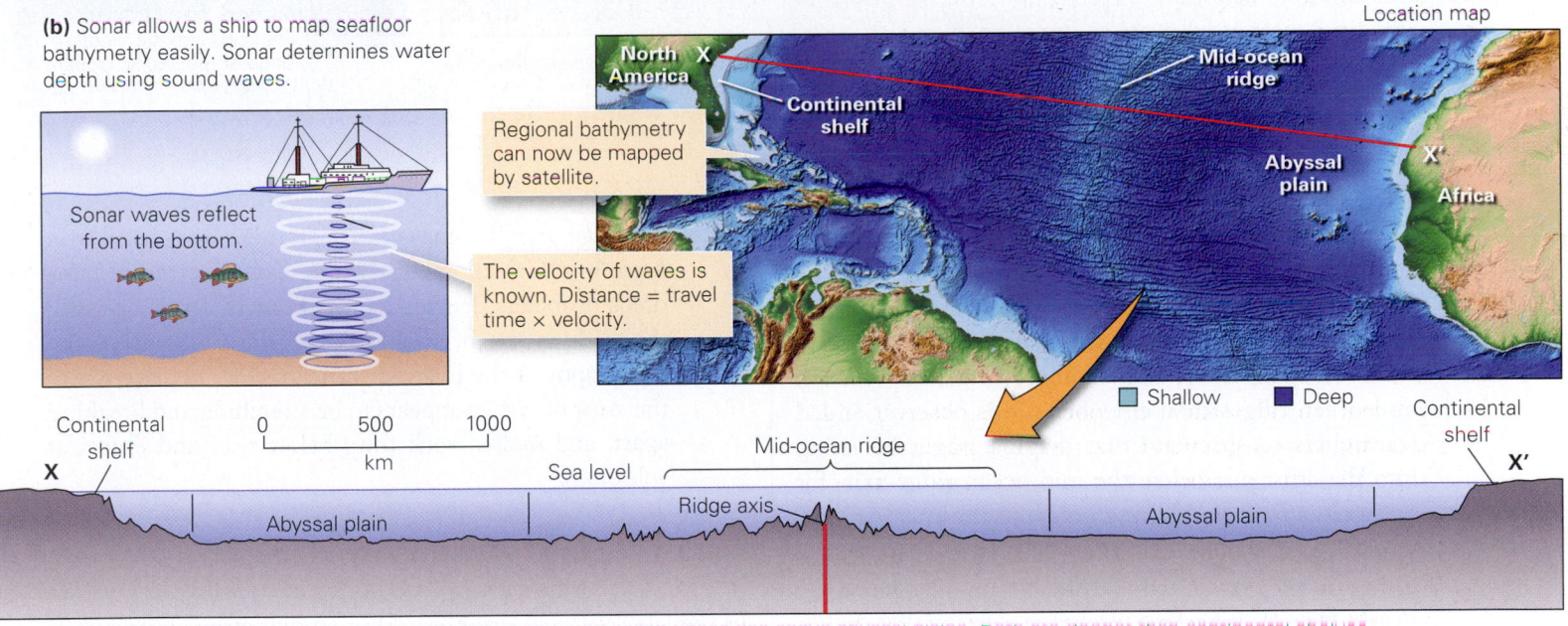

(a) A modern bathymetric map of the ocean floor. The squares indicate the locations of the seafloor in Figure 3.8b and the inset of Figure 3.9a.

(b) Sonar allows a ship to map seafloor bathymetry easily. Sonar determines water depth using sound waves.

Sonar waves reflect from the bottom.

Regional bathymetry can now be mapped by satellite.

The velocity of waves is known. Distance = travel time × velocity.

Location map

North America X

Continental shelf

Mid-ocean ridge

Abyssal plain

X'

Africa

☐ Shallow ☐ Deep

Continental shelf
X

0   500   1000
km

Sea level

Ridge axis

Mid-ocean ridge

Abyssal plain

Abyssal plain

Continental shelf
X'

(c) A bathymetric profile along line X–X' illustrates how mid-ocean ridges rise above abyssal plains. Both are deeper than continental shelves.

entirety of Earth history. Also of note, the sediment layer becomes progressively thicker away from the mid-ocean ridge axis—in fact, there's almost no sediment at all near the ridge axis.

2. By dredging up samples from the seafloor, geologists learned that oceanic crust is fundamentally different in composition from continental crust. Beneath its sediment cover, oceanic crust bedrock consists primarily

FIGURE 3.9 Other bathymetric features of the ocean floor.

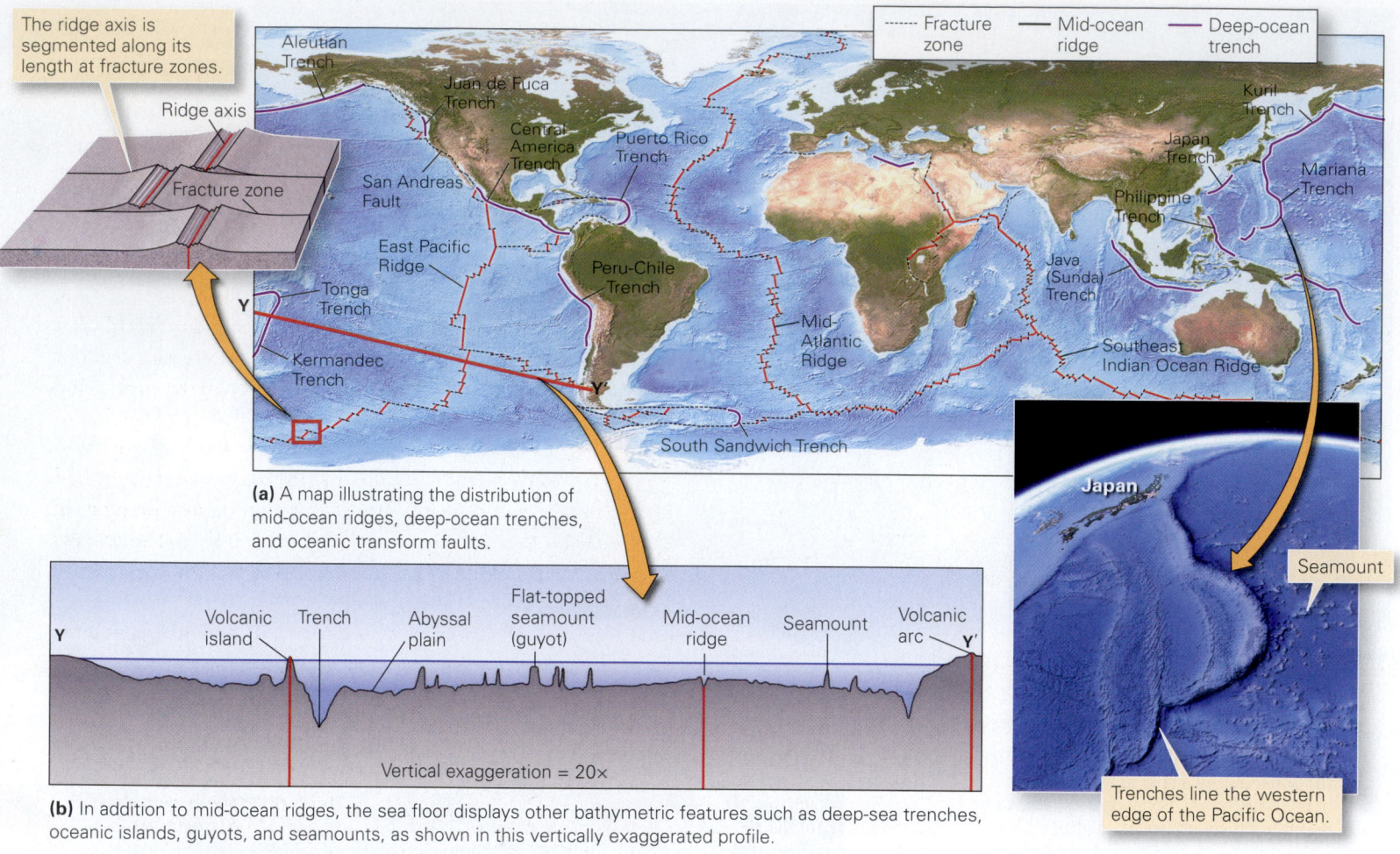

The ridge axis is segmented along its length at fracture zones.

Fracture zone ····· | Mid-ocean ridge —— | Deep-ocean trench ——

Ridge axis

Fracture zone

Aleutian Trench
Juan de Fuca Trench
Central America Trench
Puerto Rico Trench
San Andreas Fault
East Pacific Ridge
Tonga Trench
Peru-Chile Trench
Kermandec Trench
South Sandwich Trench
Mid-Atlantic Ridge
Kuril Trench
Japan Trench
Philippine Trench
Mariana Trench
Java (Sunda) Trench
Southeast Indian Ocean Ridge

Y — Y'

(a) A map illustrating the distribution of mid-ocean ridges, deep-ocean trenches, and oceanic transform faults.

Volcanic island | Trench | Abyssal plain | Flat-topped seamount (guyot) | Mid-ocean ridge | Seamount | Volcanic arc

Y — Y'

Vertical exaggeration = 20×

(b) In addition to mid-ocean ridges, the sea floor displays other bathymetric features such as deep-sea trenches, oceanic islands, guyots, and seamounts, as shown in this vertically exaggerated profile.

Japan

Seamount

Trenches line the western edge of the Pacific Ocean.

---

of basalt—it does not display the great variety of rock types found on continents.

3. **Heat flow**, the rate at which heat rises from the Earth's interior up through the crust, is not the same everywhere in the oceans. Rather, more heat rises beneath mid-ocean ridges than elsewhere. This observation led researchers to speculate that magma might be rising into the crust just below the mid-ocean ridge axis, for hot molten rock could bring heat into the crust.

4. When maps showing the distribution of earthquakes in oceanic regions became available in the years after World War II, it became clear that earthquakes do not occur randomly but rather occur in distinct zones called **seismic belts** (**Fig. 3.10**). Some belts follow trenches, some follow mid-ocean ridge axes, and others lie along portions of fracture zones. Since earthquakes define locations where rocks break and move, geologists concluded that these bathymetric features are places where movements of the crust are taking place.

5. The *ridge axis* of some mid-ocean ridges is marked by a narrow (a few kilometers wide), elongate trough hundreds of meters deeper than its borders. In this regard, the bathymetry of a mid-ocean ridge resembles the topography of the East African rift valley, a place where the crust of Africa appears to be stretching and breaking apart, and molten rock from below rises and erupts at volcanoes.

Hess's "Essay in Geopoetry"

In the late 1950s, Harry Hess, after studying the observations described above, concluded that because the sediment layer on the ocean floor was so thin overall, the ocean floor must be much younger than the continents, and because the sediment thickened progressively away from mid-ocean ridges, the ridges themselves likely were younger than the deeper parts of the ocean floor. If this was so, then somehow new ocean floor must be forming at the ridges, and thus an ocean could be get-

FIGURE 3.10 A 1953 map showing the distribution of earthquake locations in the ocean basins. Note that earthquakes occur in belts.

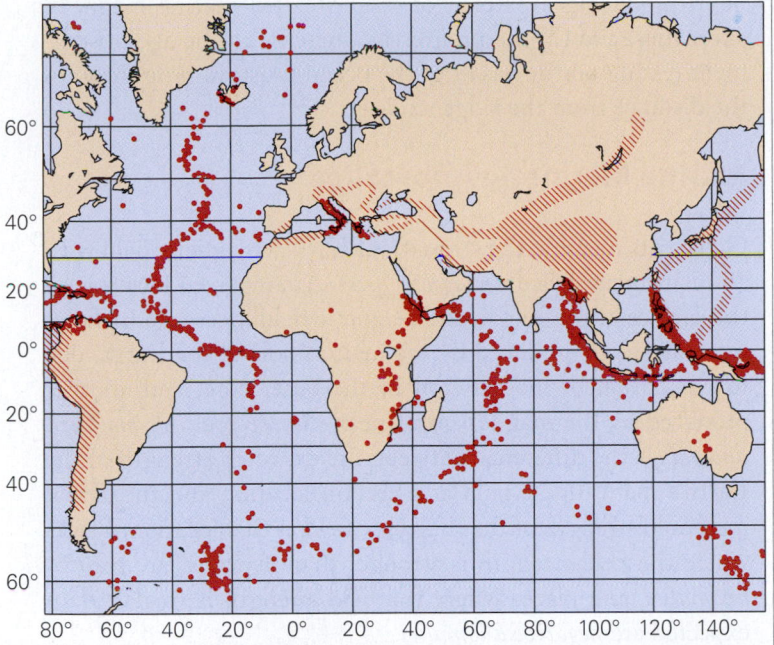

oceanic crust; and that, once formed, the new crust cracked, split apart, and moved away from the ridge (**Fig. 3.11**). As each increment of seafloor formed and moved away from the ridge axis, more melt rose from the mantle, filled the space, and became the next increment of seafloor. Robert Dietz, as we've noted, named the process **seafloor spreading**.

The concept that seafloor spreading takes place, allowing ocean basins to grow wider with time, led to a dilemma. Geologists realized that if new ocean floor formed, old ocean floor must be consumed or destroyed somewhere, or the Earth's circumference would have to increase, meaning the Earth would have to be expanding significantly, which the vast majority of studies had concluded wasn't possible. Hess suggested that deep-ocean trenches might be the places where the seafloor sank back into the mantle and that the earthquakes occurring at trenches were evidence of this movement. Part of the inspiration for this idea came from Hess's earlier study of guyots—he had found that guyots on the margins of trenches had tilted over, as if the seafloor that they had grown on was being bent and pulled into the trench. Geologists now refer to the process by which ocean floor bends and sinks back into the Earth's interior at trenches as **subduction**.

Hess and his contemporaries realized that the seafloor-spreading hypothesis instantly provided the long-sought explanation of how continental "drift" occurs. Continents passively move apart as the seafloor between them spreads at mid-ocean ridges, and they passively move together as the seafloor between them sinks back into the mantle at

ting wider with time. But how? The association of earthquakes with mid-ocean ridges suggested to him that the seafloor was breaking at the ridge. Furthermore, the discovery of high heat flow along mid-ocean ridge axes and the similarity of ridges to the East African rift indicated that molten rock was rising up beneath ridges and that the seafloor crust was stretching. In 1960, Hess finally saw how these observations fit together and wrote a manuscript in which he proposed that this material from the mantle rose beneath mid-ocean ridges; that at the ridge axis melt derived from the mantle solidified to form

Did you ever wonder...

if the distance between New York and Paris changes?

trenches. Researchers soon realized that the entity that was moving did not consist of crust alone but rather of the whole lithosphere (the crust plus the underlying cooler, and rigid,

FIGURE 3.11 Harry Hess's basic concept of seafloor spreading (1962). Hess implied, incorrectly, that only the crust moved. We will see that this sketch is an oversimplification and contains errors.

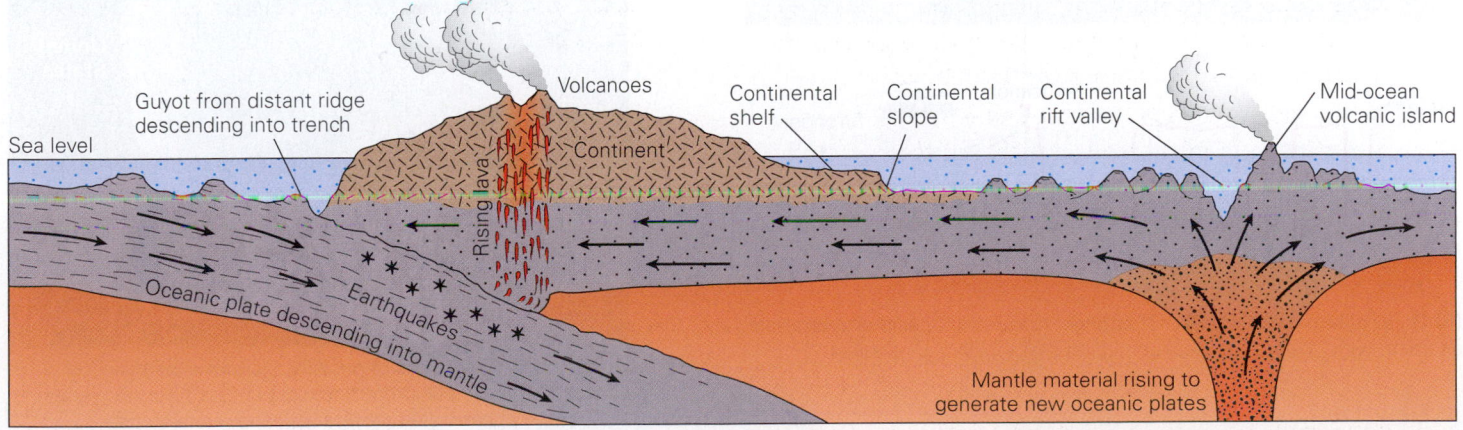

portion of the upper mantle). The idea that the outer shell of the earth was in motion, with ocean floor formed at ridges and consumed at trenches, seemed to be so good that Hess referred to his description of it as "an essay in geopoetry."

Take-Home Message

New observations about seafloor bathymetry, sediment cover, heat flow, and seismicity led to Hess's proposal of seafloor spreading–new seafloor forms at mid-ocean ridges and then moves away from the ridge axis, so ocean basins can get wider with time. As this happens, old ocean floor sinks back into the mantle by subduction.

QUICK QUESTION: How does seafloor spreading and subduction provide an explanation for how continents move?

3.5 Evidence for Seafloor Spreading

For a hypothesis to become a theory (see Box P.1), it must be tested—scientists must demonstrate that the idea really works. During the 1960s, geologists found that the seafloor-spreading hypothesis successfully explained several previously baffling observations. Here we discuss two: (1) the existence of orderly variations in the strength of the magnetic field over the seafloor, producing a pattern of stripes called marine magnetic anomalies, and (2) the progressive increase in the age of sediment resting on the basalt of the ocean crust, in proportion to the distance from the ridge axis.

Marine Magnetic Anomalies

Geologists measure the strength of Earth's magnetic field with an instrument called a *magnetometer*. At any given location on the surface of the Earth, the magnetic field measured includes two parts: one produced by the main dipole of the Earth, due to circulation of molten iron in the outer core, and another produced by the magnetism of near-surface rock. A *magnetic anomaly* is the difference between the expected strength of the Earth's main dipole field at a certain location and the actual measured strength of the magnetic field at that location. Places where the field strength is stronger than expected are *positive anomalies*, and places where the field strength is weaker than expected are *negative anomalies*.

Year after year, in the course of doing other oceanographic and/or seafloor studies, researchers towed magnetometers back and forth across the ocean to map variations in magnetic field strength (**Fig. 3.12a**). As a ship cruised along its course, they found that the magnetometer's gauge would first detect an interval of strong signal (a positive anomaly) and then suddenly an interval of weak signal (a negative anomaly). A graph

FIGURE 3.12 The discovery of marine magnetic anomalies.

(a) A ship towing a magnetometer detects changes in the strength of the magnetic field.

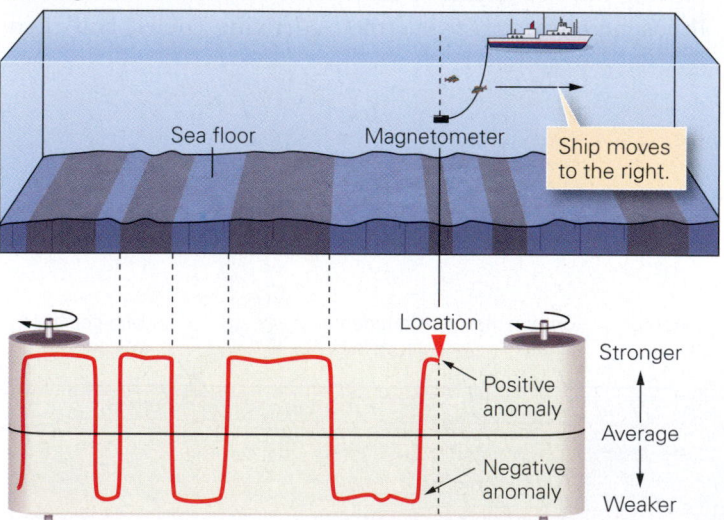

(b) On a paper record, intervals of stronger magnetism (positive anomalies) alternate with intervals of weaker magnetism (negative anomalies).

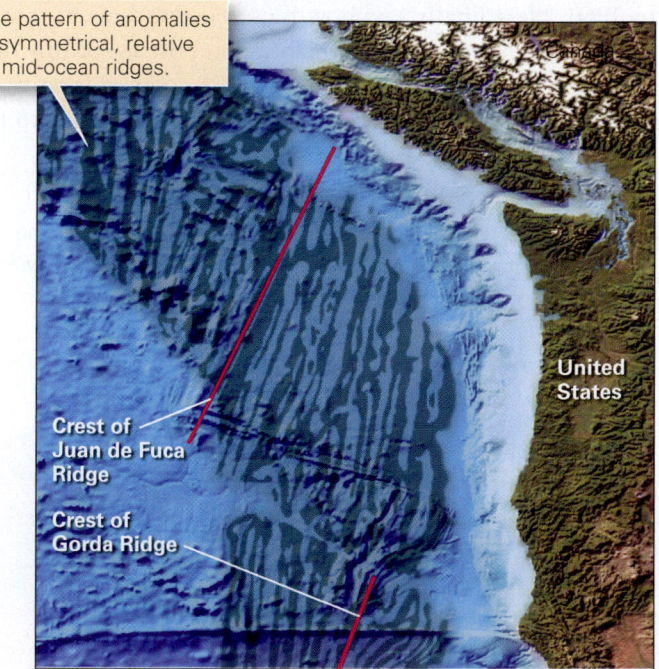

(c) A map showing areas of positive anomalies (dark) and negative anomalies (light) off the west coast of North America. The pattern of anomalies resembles candy-cane strips.

of signal strength versus distance along the traverse, therefore, has a sawtooth shape (**Fig. 3.12b**). When researchers compiled data from many parallel cruise lines on a map, they found that the sawtooth patterns lined up to define distinctive, alternating bands dubbed **marine magnetic anomalies**. If we color positive anomalies dark and negative anomalies light, the pattern made by the anomalies resembles the stripes on a candy cane (**Fig. 3.12c**). The mystery of this marine magnetic anomaly pattern, however, remained unsolved until geologists had discovered magnetic reversals.

Magnetic Reversals Recall that Earth's magnetic field can be depicted by an arrow, representing the dipole that presently points from the north magnetic pole to the south magnetic pole. When researchers measured the paleomagnetism of a succession of rock layers that had accumulated over a long period of time, they found that the polarity (which end of a magnet points north and which end points south) of the paleo-

magnetic field preserved in some layers was the same as that of Earth's present magnetic field, whereas in other layers it was the opposite.

At first, reversed polarity was thought to be the result of lightning strikes or of local chemical reactions between rock and water. But when repeated measurements from around the world revealed a systematic pattern of alternating normal and reversed polarity in rock layers, geologists realized that reversals were a worldwide, not a local, phenomenon. They reached the unavoidable conclusion that at various times during Earth history the polarity of Earth's magnetic field has suddenly flipped! In other words, sometimes the Earth has *normal polarity*, as it does today, and sometimes it has *reversed polarity* (**Fig. 3.13a**). When the Earth has reversed polarity, the south magnetic pole lies near the north geographic pole, and the north magnetic pole lies near the south geographic pole. Thus, if you were to use a compass during periods when the Earth's magnetic field was reversed, the north-seeking end of the needle would point

FIGURE 3.13 Magnetic polarity reversals and the chronology of reversals.

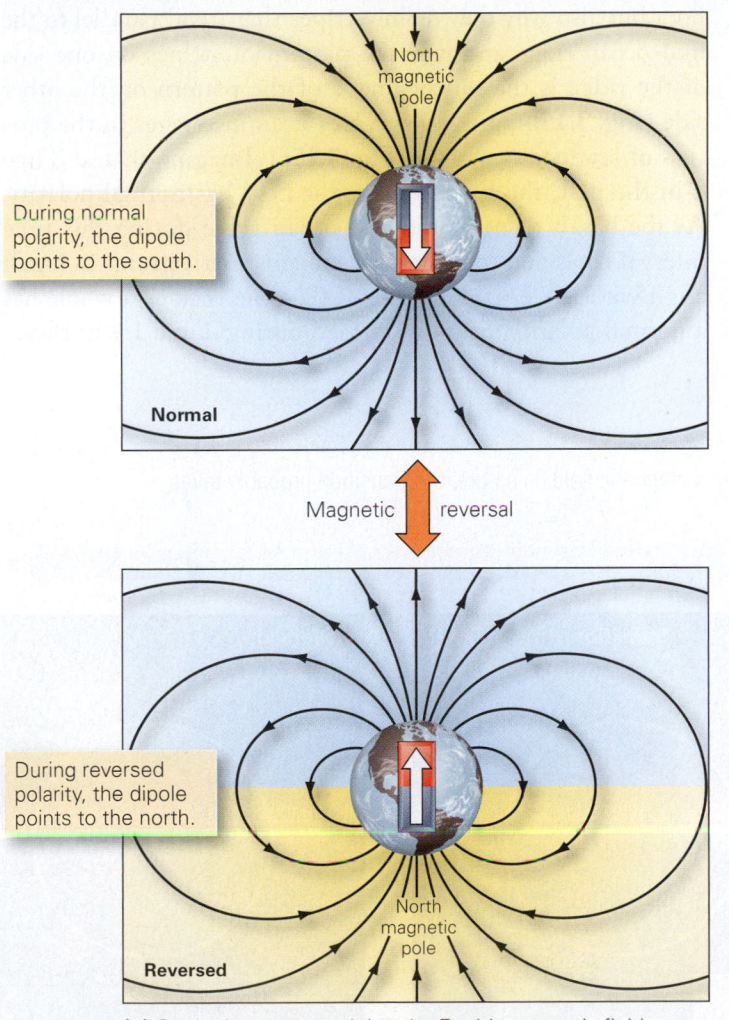

(a) Geologists proposed that the Earth's magnetic field reverses polarity every now and then.

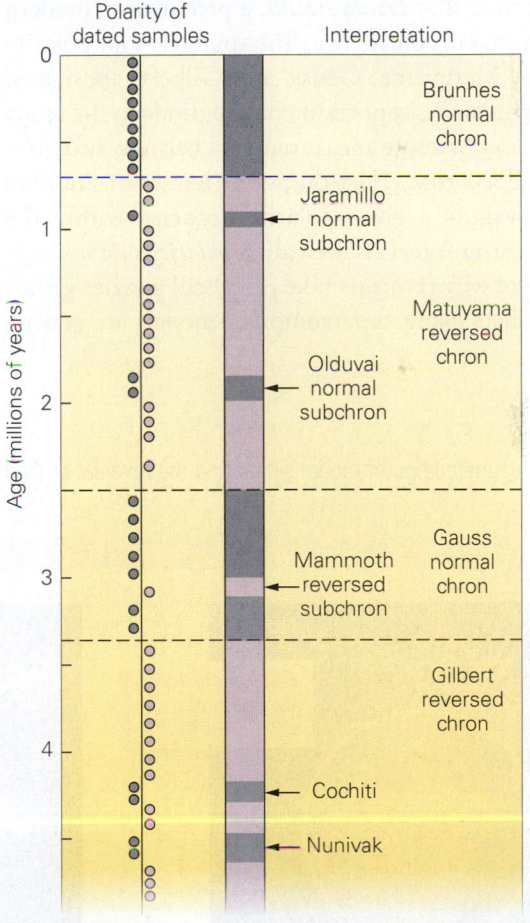

(b) Observations led to the production of a reversal chronology, with named polarity intervals.

to the south geographic pole. A time when the Earth's field flips from normal to reversed polarity, or vice versa, is called a **magnetic reversal**. Note that the Earth itself doesn't turn upside down—it is just the magnetic field that reverses.

In the 1950s, about the same time that polarity reversals were discovered, researchers developed a technique that permitted them to define the *numerical age* of a rock, meaning the age of the rock in years. (The technique, called isotopic dating, will be discussed in detail in Chapter 12.) Geologists applied the technique to determine the ages of the rock layers for which they obtained their paleomagnetic measurements and thus determined when the magnetic field of the Earth reversed. With this information, they constructed a history of magnetic reversals for the past 4.5 million years; this history is now called a **magnetic-reversal chronology**.

A diagram representing the Earth's magnetic-reversal chronology (**Fig. 3.13b**) shows that reversals do not occur periodically, so the lengths of different **polarity chrons**, the time intervals between reversals, are different. A polarity reversal from reverse (before) to normal (after) happened about 700,000 years ago. Thus, we are living in a normal polarity chron, which began about the time that *Homo erectus*, a precursor of modern humans, first learned to control fire. The youngest four polarity chrons (Brunhes, Matuyama, Gauss, and Gilbert) are named after scientists who made important contributions to the study of rock magnetism. As more measurements became available, investigators realized that short-duration (less than 200,000 years long) intervals of a given polarity occurred within the chrons—these shorter intervals are called *polarity subchrons*.

The question of why reversals take place still puzzles geologists, but researchers using supercomputer models are getting closer to an answer. The models show that changes in the fluid motion of the outer core can trigger reversals and that during reversals the magnetic field first weakens and becomes complicated (chaotic) before reconfiguring with a different polarity (**Fig. 3.14**).

Interpreting Marine Magnetic Anomalies Why do marine magnetic anomalies exist? In 1963, researchers in Britain and Canada proposed a solution to this riddle. Simply put, a positive anomaly occurs over areas of the seafloor where underlying basalt has normal polarity. In these areas, the weak magnetic force produced by the magnetite grains in basalt adds to the force produced by the Earth's dipole—the sum of these forces yields a stronger magnetic signal than expected due to the dipole alone (**Fig. 3.15a**). A negative anomaly occurs over regions of the seafloor where the underlying basalt has a reversed polarity. In these regions, the magnetic force of the basalt subtracts from the force produced by the Earth's dipole, so the measured magnetic signal is weaker than expected.

The seafloor-spreading model easily explains not only why positive and negative magnetic anomalies exist over the seafloor but also why they define stripes that trend parallel to the mid-ocean ridge and why the pattern of stripes on one side of the ridge is the mirror image of the pattern on the other side (**Fig. 3.15b**). To see why, let's examine stages in the process of seafloor spreading (**Fig. 3.15c**). Imagine that at Time 1 in the past, the Earth's magnetic field has normal polarity. As the basalt rising at the mid-ocean ridge during this time interval cools and solidifies, the tiny magnetic grains in basalt align with the Earth's field, and thus the rock as a whole has a normal polarity. Seafloor formed during Time 1 will there-

FIGURE 3.14 A supercomputer model simulating the reversal of the Earth's magnetic field. In nature, the transition probably takes thousands of years.

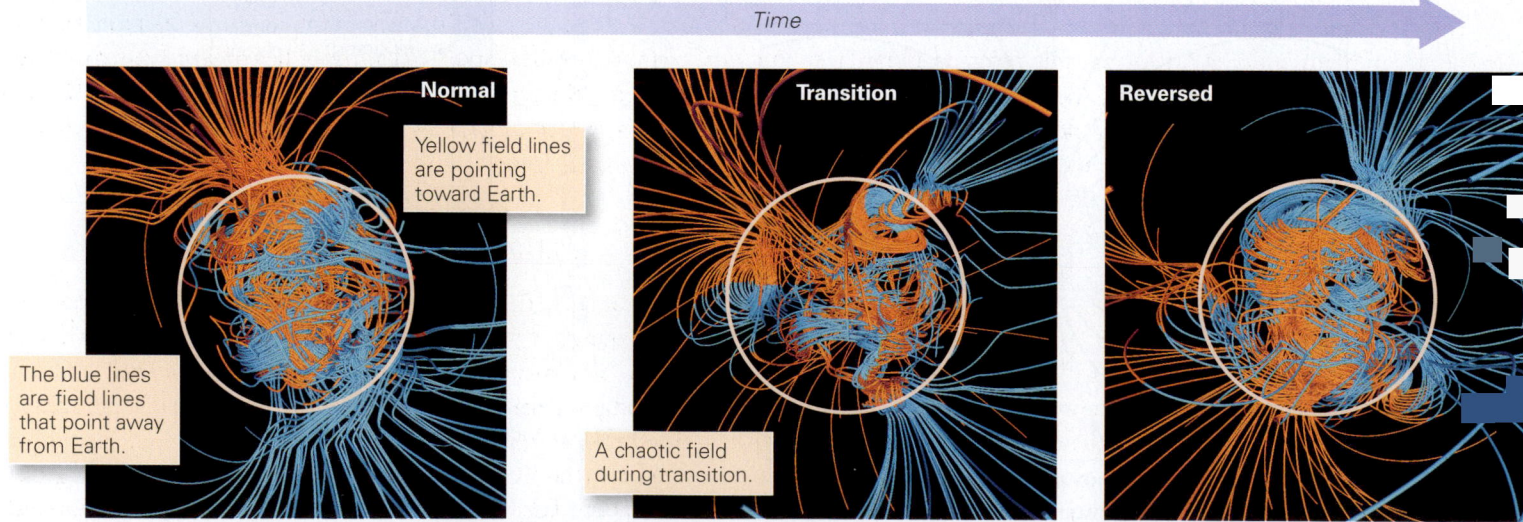

Time

Normal

Yellow field lines are pointing toward Earth.

The blue lines are field lines that point away from Earth.

Transition

A chaotic field during transition.

Reversed

FIGURE 3.15 The progressive development of magnetic anomalies and the long-term reversal chronology.

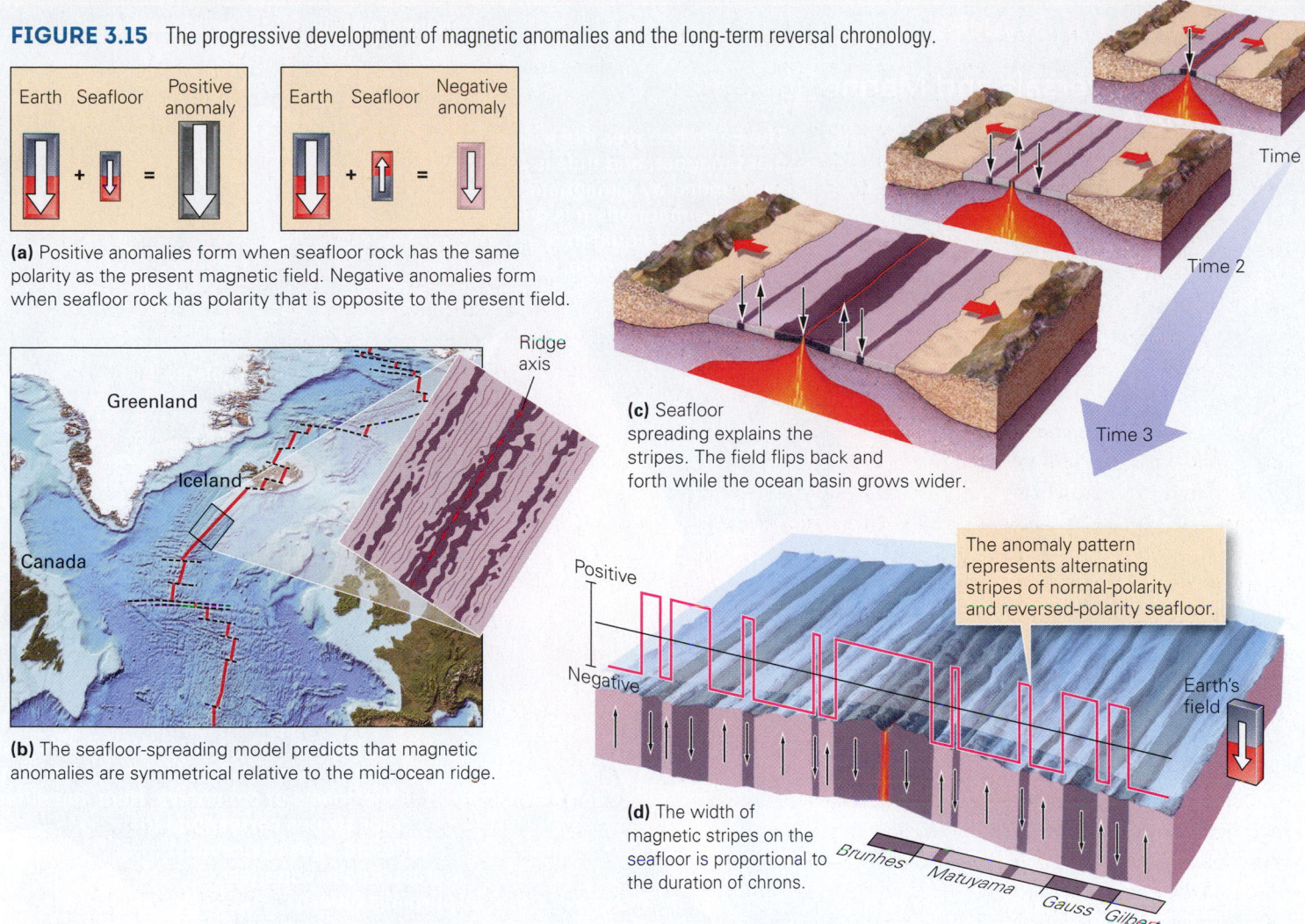

(a) Positive anomalies form when seafloor rock has the same polarity as the present magnetic field. Negative anomalies form when seafloor rock has polarity that is opposite to the present field.

(b) The seafloor-spreading model predicts that magnetic anomalies are symmetrical relative to the mid-ocean ridge.

(c) Seafloor spreading explains the stripes. The field flips back and forth while the ocean basin grows wider.

The anomaly pattern represents alternating stripes of normal-polarity and reversed-polarity seafloor.

(d) The width of magnetic stripes on the seafloor is proportional to the duration of chrons.

fore generate a positive anomaly and appear as a dark stripe on an anomaly map. As it forms, the rock of this stripe moves away from the ridge axis, so half goes to the right and half to the left. Now imagine that later, at Time 2, Earth's field has reversed polarity. Seafloor basalt formed during Time 2 therefore has reversed polarity and will appear as a light stripe on an anomaly map. As it forms, this reversed-polarity stripe moves away from the ridge axis, and even younger crust forms along the axis. The basalt in each new stripe of crust preserves the polarity that was present at the time it formed, so as the Earth's magnetic field flips back and forth, alternating positive and negative anomaly stripes form. A positive anomaly exists over the ridge axis today because seafloor is forming during the present chron of normal polarity.

Closer examination of a seafloor magnetic anomaly map reveals that anomalies are not all the same width. Geologists found that the relative widths of anomaly stripes near the Mid-Atlantic Ridge are the same as the relative durations of paleomagnetic chrons (**Fig. 3.15d**; **Geology at a Glance**, pp. 80–81). This relationship between anomaly-stripe width and polarity-

chron duration indicates that the rate of seafloor spreading has been fairly constant along the Mid-Atlantic Ridge for at least the last 4.5 million years.

If the spreading rate at a given mid-ocean ridge is constant over a long time, then we can use simple arithmetic to determine the rate of spreading. For example, in the North Atlantic Ocean, 4.5-million-year-old seafloor lies 45 km away from the ridge axis. Keeping in mind that Velocity = Distance/Time, the velocity (v) at which the seafloor moves away from the Mid-Atlantic Ridge axis can be calculated as follows:

$$v = \frac{45 \text{ km}}{4,500,000 \text{ years}} = \frac{4,500,000 \text{ cm}}{4,500,000 \text{ years}} = 1 \text{ cm/y}$$

This means that a point on one side of the ridge moves away from a point on the other side by 2 cm per year. We call this number the **spreading rate**. In the Pacific Ocean, seafloor spreading occurs at the East Pacific Rise. (Geographers named this a "rise" because it is not as rough and jagged as the Mid-Atlantic Ridge.) The anomaly stripes bordering the East Pacific Rise are much wider, and 4.5-million-year-old

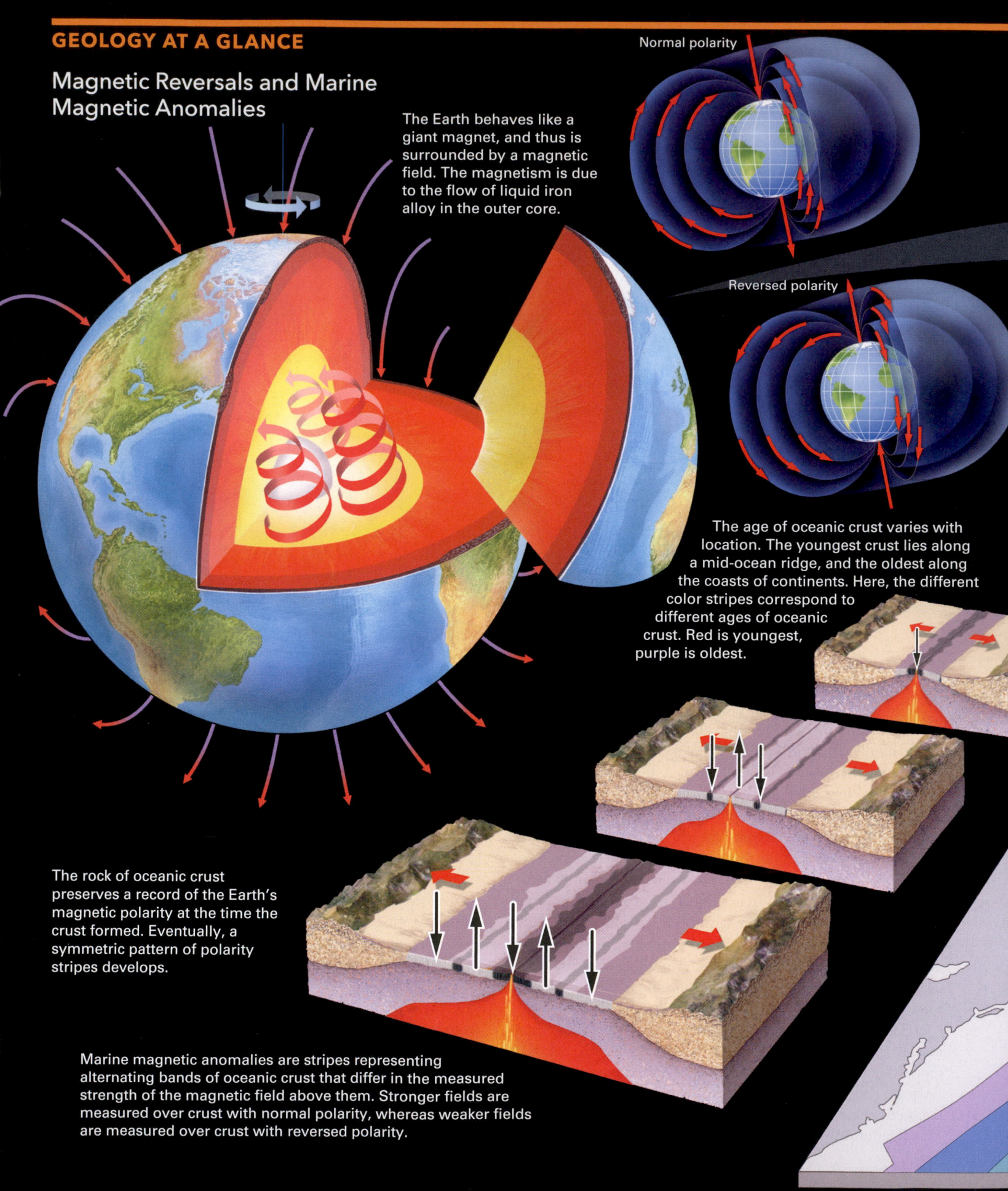

Magnetic Reversals and Marine Magnetic Anomalies

The Earth behaves like a giant magnet, and thus is surrounded by a magnetic field. The magnetism is due to the flow of liquid iron alloy in the outer core.

Normal polarity

Reversed polarity

The age of oceanic crust varies with location. The youngest crust lies along a mid-ocean ridge, and the oldest along the coasts of continents. Here, the different color stripes correspond to different ages of oceanic crust. Red is youngest, purple is oldest.

The rock of oceanic crust preserves a record of the Earth's magnetic polarity at the time the crust formed. Eventually, a symmetric pattern of polarity stripes develops.

Marine magnetic anomalies are stripes representing alternating bands of oceanic crust that differ in the measured strength of the magnetic field above them. Stronger fields are measured over crust with normal polarity, whereas weaker fields are measured over crust with reversed polarity.

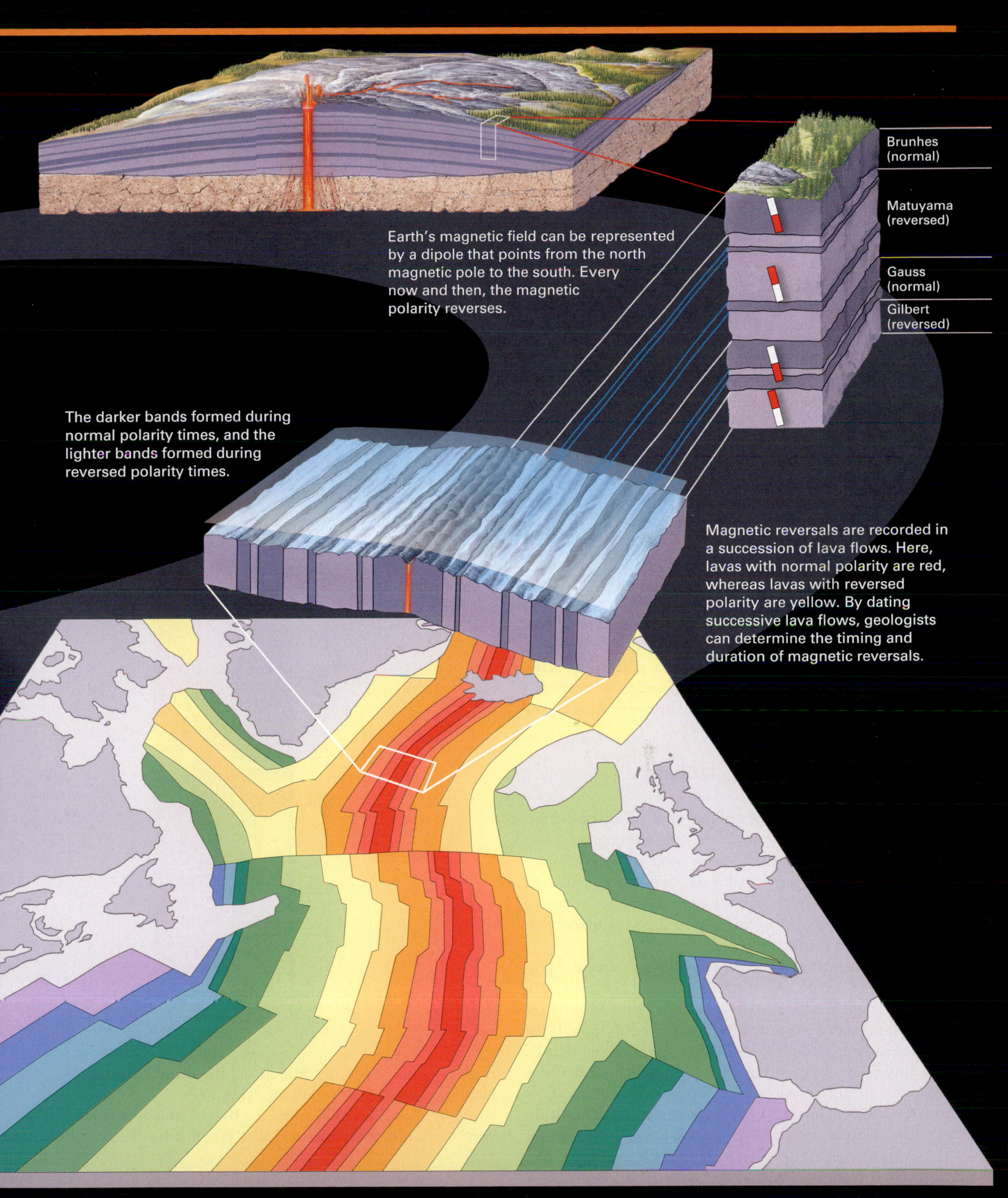

Earth's magnetic field can be represented by a dipole that points from the north magnetic pole to the south. Every now and then, the magnetic polarity reverses.

Brunhes (normal)

Matuyama (reversed)

Gauss (normal)

Gilbert (reversed)

The darker bands formed during normal polarity times, and the lighter bands formed during reversed polarity times.

Magnetic reversals are recorded in a succession of lava flows. Here, lavas with normal polarity are red, whereas lavas with reversed polarity are yellow. By dating successive lava flows, geologists can determine the timing and duration of magnetic reversals.

FIGURE 3.16 Magnetic anomalies across the width of the ocean permit determination of reversal chronology back further in time.

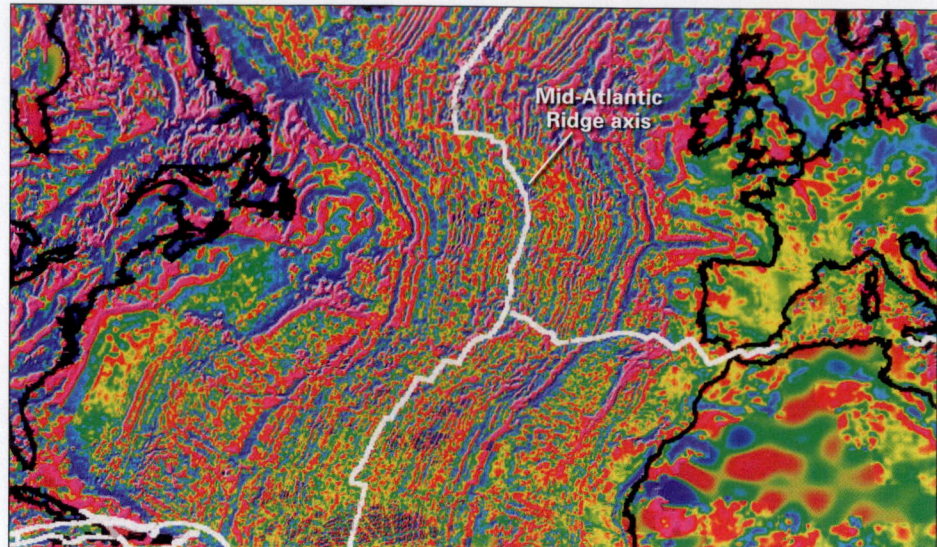

(a) A digital map showing the magnetic anomalies of the North Atlantic, and of adjacent continents. Note the striped pattern of the seafloor. (Anomalies on land don't show this pattern because they are controlled by the distribution of different rock types.)
Source: Korhonen, et al., 2007, © CCGM-CGMW.

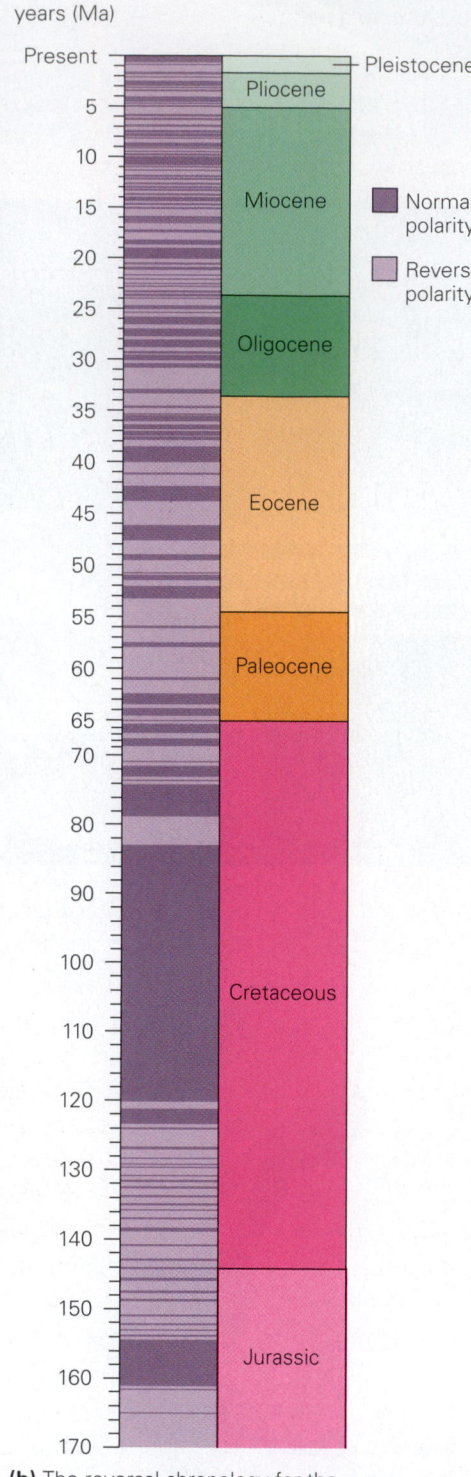

(b) The reversal chronology for the past 170 million years.

seafloor lies about 225 km from the rise axis. This requires the seafloor to move away from the rise at a rate of about 5 cm per year, so the spreading rate for the East Pacific Rise is about 10 cm per year.

If you assume that the spreading rate was constant for tens to hundreds of millions of years, then it is possible to estimate the age of stripes right up to the edge of the ocean (**Fig. 3.16**). Using this approach, the anomalies on the edges of the North Atlantic are about 175 Ma, so the oldest ocean floor of the North Atlantic formed about 175 Ma.

Evidence from Deep-Sea Drilling

In the late 1960s, a research drilling ship called the *Glomar Challenger* set out to sail around the ocean drilling holes into the seafloor. This amazing ship could lower enough drill pipe to drill in 5-km-deep water and could continue to drill until the hole reached a depth of about 1.7 km (1.1 miles) below the seafloor. Drillers brought up cores of rock and sediment that geoscientists then studied on board.

To test the seafloor-spreading hypothesis, researchers proposed that the *Glomar Challenger* drill a series of holes, spaced at progressively greater distances from the axis of the Mid-Atlantic Ridge, through seafloor sediment to the basalt layer. If the model of seafloor spreading was correct, then not only should the sediment layer be progressively thicker away from the ridge axis, as was already known based on earlier work, but the age of the oldest sediment just above the basalt (as well as the basalt just below the sediment) should be progressively older away from the

ridge axis, too. When the drilling and the analyses were complete, the prediction was confirmed (**Fig 3.17**).

So by the early 1960s, when the Beatles were topping the pop charts, it had become clear that Wegener had been right all along—continents do move. But, though the case for such movement had been greatly strengthened by the discovery of apparent polar-wander paths, it really took the proposal and proof of seafloor spreading to make believers of most geologists. Very quickly, as we will see in the next chapter, these ideas became the basis of the theory of plate tectonics.

FIGURE 3.17 Drilling into the sediment layer of the ocean floor confirmed that the age of the oldest sediment in contact with ocean-crust basalt gets older the farther away it is from the ridge. For example, Point A is older than Point B.

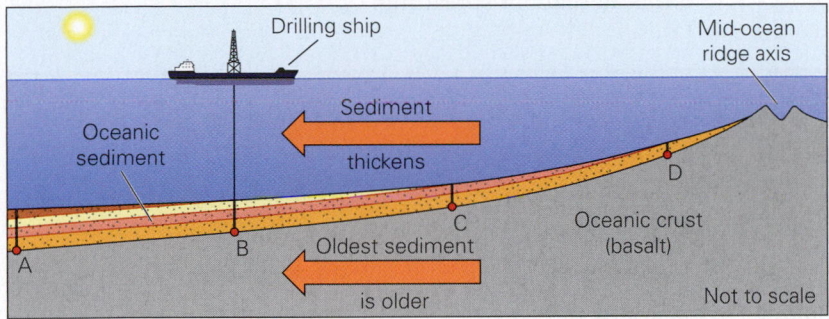

Take-Home Message

The Earth's magnetic polarity flips every now and then. As a result, different stripes of ocean floor formed at mid-ocean ridges preserve different polarities. These cause marine magnetic anomalies. The discovery of these anomalies, as well as documentation that the seafloor gets older away from the ridge axis, proved that the seafloor-spreading hypothesis is correct.

QUICK QUESTION: What determines the widths of marine magnetic anomalies?

CHAPTER SUMMARY

- Alfred Wegener proposed that continents had once been joined together to form a single huge supercontinent (Pangaea) and had subsequently drifted apart. This idea is the continental-drift hypothesis.

- Wegener drew from several different sources of data to support his hypothesis: (1) coastlines on opposite sides of the ocean match up; (2) the distribution of late Paleozoic glaciers can be explained if the glaciers made up a polar ice cap over the southern end of Pangaea; (3) the distribution of late Paleozoic equatorial climatic belts is compatible with the concept of Pangaea; (4) the distribution of fossil species suggests the existence of a supercontinent; (5) distinctive rock assemblages that are now on opposite sides of the ocean were adjacent on Pangaea.

- Despite all the observations that supported continental drift, most geologists did not initially accept the idea because no one could explain how continents could move. It took decades of new data collection before the idea could be reconsidered.

- Rocks retain a record of the Earth's magnetic field that existed at the time the rocks formed. This record is called paleomagnetism. By measuring paleomagnetism in successively older rocks, geologists found that the apparent position of the Earth's magnetic pole relative to the rocks changes through time. Successive positions of the pole define an apparent polar-wander path.

- Apparent polar-wander paths are different for different continents. This observation can be explained if continents move with respect to one another while the Earth's magnetic poles remain roughly fixed.

- The invention of sonar permitted explorers to make detailed maps of the seafloor. These maps revealed the existence of mid-ocean ridges, deep-ocean trenches, seamount chains, and fracture zones. Other measurement showed that heat flow is generally greater near the axis of a mid-ocean ridge.

- Hess's hypothesis of seafloor spreading states that new seafloor forms at mid-ocean ridges, above a band of upwelling mantle, then spreads symmetrically away from the ridge axis. As a consequence, an ocean basin gets progressively wider with time, and the continents on either side of the ocean basin drift apart. Eventually, the ocean floor sinks back into the mantle at deep-ocean trenches.

- Magnetometer surveys of the seafloor revealed marine magnetic anomalies. Positive anomalies (magnetic field strength is greater than expected) and negative anomalies (magnetic field strength is less than expected) are arranged in alternating stripes.

- During the 1950s, geologists documented that the Earth's magnetic field reverses polarity every now and then. The record of reversals, dated by isotopic techniques, is called the magnetic-reversal chronology.
- A proof of seafloor spreading came from the interpretation of marine magnetic anomalies. Seafloor that forms when the Earth has normal polarity results in positive anomalies, whereas seafloor that forms when the Earth has

reversed polarity results in negative anomalies. Anomalies are symmetric with respect to a mid-ocean ridge axis, and their widths are proportional to the duration of polarity chrons. Study of anomalies allows us to calculate the rate of spreading.
- Drilling of the seafloor confirmed that its age increases away from the mid-ocean ridge axis and served as another proof of seafloor spreading.

GUIDE TERMS

abyssal plain (p. 72)

apparent polar-wander path (p. 70)

bathymetry (p. 72)

continental drift (p. 63)

fracture zone (p. 72)

heat flow (p. 74)

magnetic declination (p. 67)

magnetic dipole (p. 67)

magnetic inclination (p. 68)

magnetic poles (p. 67)

magnetic-reversal chronology (p. 78)

magnetic reversal (p. 78)

marine magnetic anomaly (p. 77)

mid-ocean ridge (p. 72)

paleomagnetism (p. 67)

paleopole (p. 70)

Pangaea (p. 62)

polarity chron (p. 78)

ridge axis (p. 72)

seafloor spreading (p. 75)

seamount (p. 72)

seismic belt (p. 74)

spreading rate (p. 79)

subduction (p. 75)

supercontinent (p. 62)

trench (p. 72)

volcanic arc (p. 72)

REVIEW QUESTIONS

1. What was Wegener's continental-drift hypothesis?
2. How does the fit of the coastlines around the Atlantic support continental drift?
3. Explain the distribution of late Paleozoic glaciation.
4. How does the distribution of climatic belts support continental drift?
5. Was it possible for a dinosaur to walk from New York to Paris when Pangaea existed? Explain your answer.
6. Why were geologists initially skeptical of Wegener's continental-drift hypothesis?
7. What is paleomagnetism and how does it form?
8. Describe how the angle of inclination of the Earth's magnetic field varies with latitude. How can paleomagnetic inclination be used to determine the ancient latitude of a continent?
9. Describe the basic bathymetric characteristics of mid-ocean ridges, deep-ocean trenches, and seamount chains.
10. Describe the hypothesis of seafloor spreading.
11. How did the observations of heat flow and seismicity support the hypothesis of seafloor spreading?
12. What is a magnetic reversal?
13. What is a marine magnetic anomaly? How is it detected?
14. Describe the pattern of marine magnetic anomalies across a mid-ocean ridge. How do geologists explain the pattern?
15. How do geologists calculate rates of seafloor spreading?
16. Did drilling into the seafloor contribute further proof of seafloor spreading? If so, how?

ON FURTHER THOUGHT

The following questions will be answered, in large part, by Chapter 4. But by thinking about them now, you can get a feel for the excitement of discovery that geologists enjoyed in the wake of the proposal of seafloor spreading.

17. Alfred Wegener's writings implied that all continents had been linked to form Pangaea from the formation of the Earth until Pangaea's breakup in the Mesozoic. Modern geologists do not agree. Geologic evidence suggests that

Pangaea itself was formed by the late Paleozoic collision of continents that had been separate during most of the Paleozoic and that other supercontinents had formed and broken up prior to the Paleozoic. What geologic evidence led geologists to this conclusion? (Hint: Keep in mind that modern geologists, unlike Wegener, understand that mountain belts such as the Appalachians form when two continents collide and that modern geologists, unlike Wegener, are able to determine the age of rocks using isotopic dating.)

18. Dating methods indicate that the oldest rocks on continents are almost 4 billion years old, whereas the oldest ocean floor is only 200 million years old. Why?

19. The geologic record suggests that when supercontinents break up, a pulse of rapid evolution, with many new species appearing and many existing species becoming extinct, takes place. Why might this be? (Hint: Consider how the environment, both global and local, might change as a result of breakup, and keep in mind the widely held idea that competition for resources drives evolution.)

20. Why are the marine magnetic anomalies bordering the East Pacific Rise in the southeastern Pacific Ocean wider than those bordering the Mid-Atlantic Ridge in the South Atlantic Ocean?

smartwork

smartwork.wwnorton.com

This chapter's Smartwork features:

- Interactive labeling problems on Earth's magnetic field.
- Visual exercises on the movements of Pangaea.
- Detailed questions on the chronology of magnetic polarity reversals.

GEOTOURS

This chapter's GeoTour exercises (A, B) feature:

- Topography of the ocean floor
- Continental drift

Another View This image was produced by Christoph Hormann, using computer rendering techniques. It shows the Caucasus Mountains between the Black Sea and the Caspian Sea. This range is forming due to the collision between two moving continental masses.

Measurements with GPS indicate that Turkey is moving west relative to Asia. Modern measurements allow us to "see" plate movements. Each yellow arrow indicates the velocity of the point at the end of the arrow. Red arrows give overall plate-movement direction.

CHAPTER 4

The Way the Earth Works: Plate Tectonics

> If you start any large theory, such as quantum mechanics, plate tectonics, evolution, it takes about 40 years for mainstream science to come around.
>
> —James Lovelock (British scientist, born 1919)

4.1 Introduction

Thomas Kuhn, an influential historian of science, argued that scientific thinking advances in distinct steps. According to Kuhn, scientists mold their interpretations of the natural world to an established line of reasoning—a *paradigm*—that can remain unchanged for a long time until an inspired thinker proposes a radically new theory that works better and forms the basis of a new paradigm. Almost immediately, the broader scientific community scraps old ideas and formulates new ones consistent with the new theory. Kuhn called such abrupt changes in thinking a **scientific revolution**. Such a revolution happened in biology when Darwin proposed the theory of evolution by natural selection and in physics, when Einstein proposed the theory of relatively. In geology, a scientific revolution took place following the proposal of the theory of plate tectonics, or simply **plate tectonics**, which states that the outer layer of the Earth, the lithosphere, consists of separate pieces, or plates, that move with respect to one another. This theory required geoscientists to cast aside interpretations rooted in the older paradigm of fixed continents and led to a complete restructuring of how geologists think about Earth history.

Compare this book with a geology textbook from the 1950s, and you will instantly see the difference.

Alfred Wegener planted the seed of plate tectonics theory with his proposal of continental drift in 1915. But this seed lay dormant for 45 years, and geoscientists focused on collecting new data about the Earth. Discoveries about the ocean floor and about apparent polar wander set the stage for germination of the seed in 1960, when Harry Hess proposed seafloor spreading. The realization a few years later that the pattern of marine magnetic anomalies proved seafloor spreading led to a feeding frenzy as many investigators dropped what they'd been doing and turned their attention to examining the broader implications of seafloor spreading. By 1968, thanks primarily to the work of more than two dozen different investigators, the seafloor-spreading hypothesis had bloomed into plate tectonics theory. Researchers clarified the concept of a plate, described the types of plate boundaries, calculated plate motions, related plate tectonics to earthquakes and volcanoes, showed how plate motions could generate mountain belts and seamount chains, and defined the history of past plate motions. After these researchers presented their ideas to standing-room-only audiences at conferences between 1968 and 1970, the geoscience community, with few exceptions, embraced plate tectonics theory as the basis of a new paradigm and has been building on it ever since. In fact, plate tectonics has become geology's grand unifying theory.

To begin our explanation of the key elements of plate tectonics theory, we focus on lithosphere plates and distinguish among the three types of plate boundaries. Next, we characterize the nature of geologic activity that occurs at each boundary. We then look at hot spots and other special locations on plates and examine the breakup and collision of continents. Finally, we consider models proposed to explain why plates move and measurements that define rates of movement.

4.2 What Do We Mean By Plate Tectonics?

The Concept of a Lithosphere Plate

The **lithosphere**, which consists of the crust plus the uppermost part of the upper mantle, behaves as a relatively hard layer, meaning that when a force pushes or pulls on it, it does not flow but rather bends or breaks (**Fig. 4.1**). The lithosphere lies over a relatively soft layer called the **asthenosphere**, composed of mantle that can flow when acted on by force. As a result, the asthenosphere convects like water in a pot, though much more slowly (see Box 2.3). In geology, we say that the lithosphere is "rigid" whereas the asthenosphere is "plastic."

The base of the lithosphere lies in the upper mantle. In fact, both the mantle part of the lithosphere (the **lithospheric mantle**) and the asthenosphere consist of the same very dense ultramafic rock. The boundary between the two is defined by temperature. Lithospheric mantle consists of mantle rock that is cooler than 1,280°C, for such rock behaves rigidly, while asthenosphere consists of mantle rock that is warmer than 1,280°C, for such rock behaves plastically. (To picture why, compare the behavior of a wax candle that you pull out of a freezer to one that's been sitting in the hot sun.)

Continental lithosphere and oceanic lithosphere differ significantly from each other in two ways. First, the total thickness of continental lithosphere is greater than that of oceanic lithosphere—the former ranges between 150 to 200 km, whereas the latter ranges from <10 km beneath mid-ocean ridges to about 100 km beneath abyssal plains (**Fig. 4.2**). Second, the crust of continental lithosphere is thicker and less dense than

that of oceanic lithosphere. Continental crust ranges from 25 to 70 km thick and consists of relatively low-density felsic and intermediate rock. In contrast, the crust of oceanic lithosphere is only 7 to 10 km thick and consists of dense mafic rock.

In effect, we can picture the rigid lithosphere as "floating" on the soft asthenosphere. Because continental lithosphere is thicker and has a less dense crust, its surface sits higher than does the surface of the seafloor—that's why continents protrude above sea level and the ocean floor lies below sea level (**Fig. 4.3a, b**). To understand this concept, let's construct a model made of wood. In this model, we use oak (a dense wood) to represent lithospheric mantle, pine (a medium-density wood) to represent oceanic crust, and cork (a low-density wood) to represent continental crust. If we glue the different woods together in proportion to their thickness in the real world and place them in a tub of water, the composite block representing continental lithosphere floats higher than does the composite block representing oceanic lithosphere (**Fig. 4.3c**). The total mass of the cork-covered block exceeds the total mass of the pine-covered block, so the base of the cork-covered block sinks deeper into the water. But because the cork-covered block is thicker and has a lower overall density, it floats higher, as required by Archimedes' principle (**Box 4.1**).

FIGURE 4.1 The lithosphere is fairly rigid, but when a heavy load, such as a glacier or volcano, builds on its surface, the surface bends down. This can happen because underlying "plastic" asthenosphere can flow out of the way.

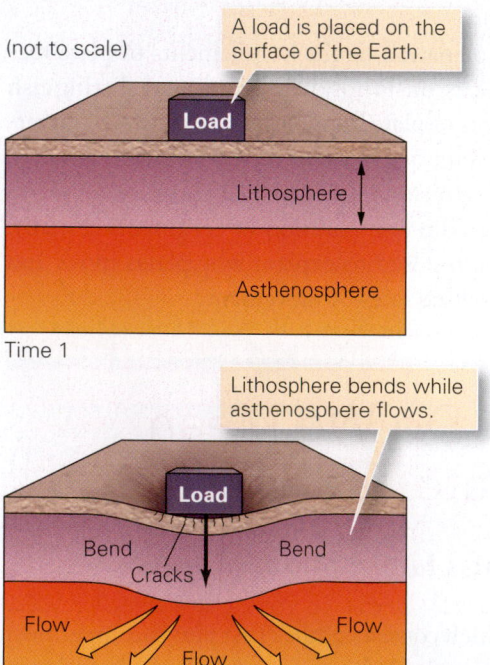

FIGURE 4.2 The lithosphere consists of the crust plus the uppermost mantle. It is thicker beneath continents than beneath oceans.

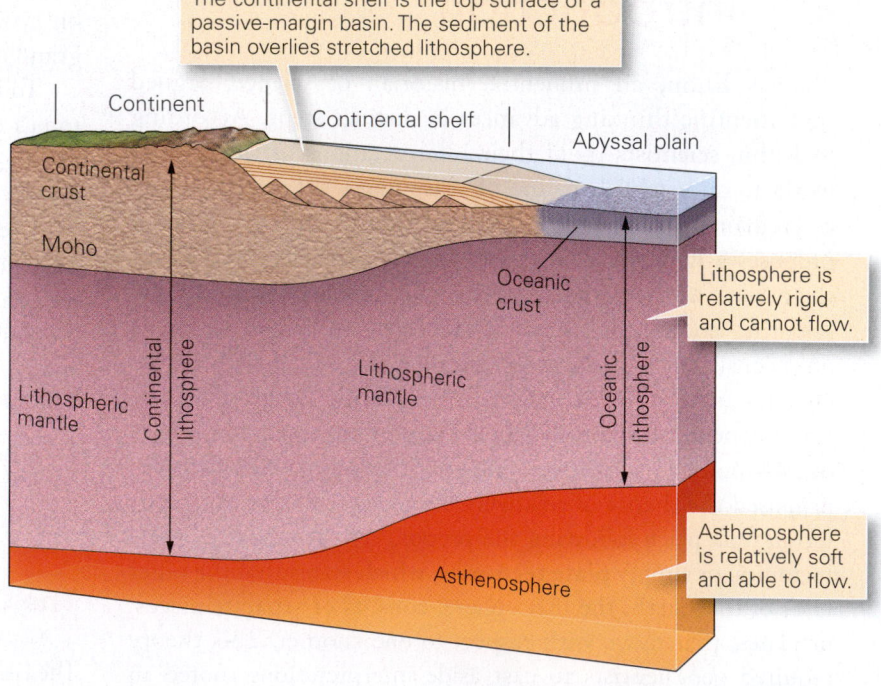

FIGURE 4.3 Why is the land surface higher than the seafloor surface?

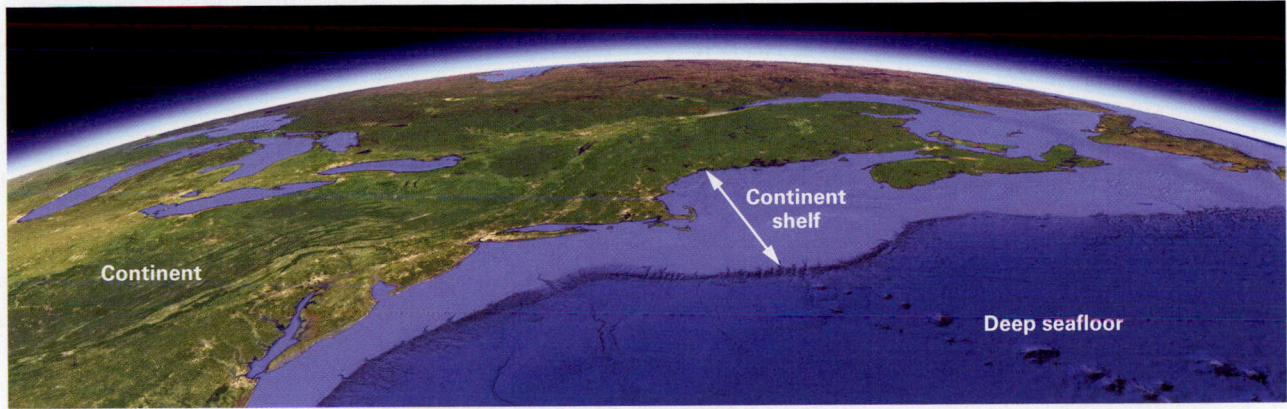

(a) An oblique view of the eastern coast of North America. Note the elevation difference between the land surface and the deep seafloor. This difference is due to contrasts between continental and oceanic lithosphere.

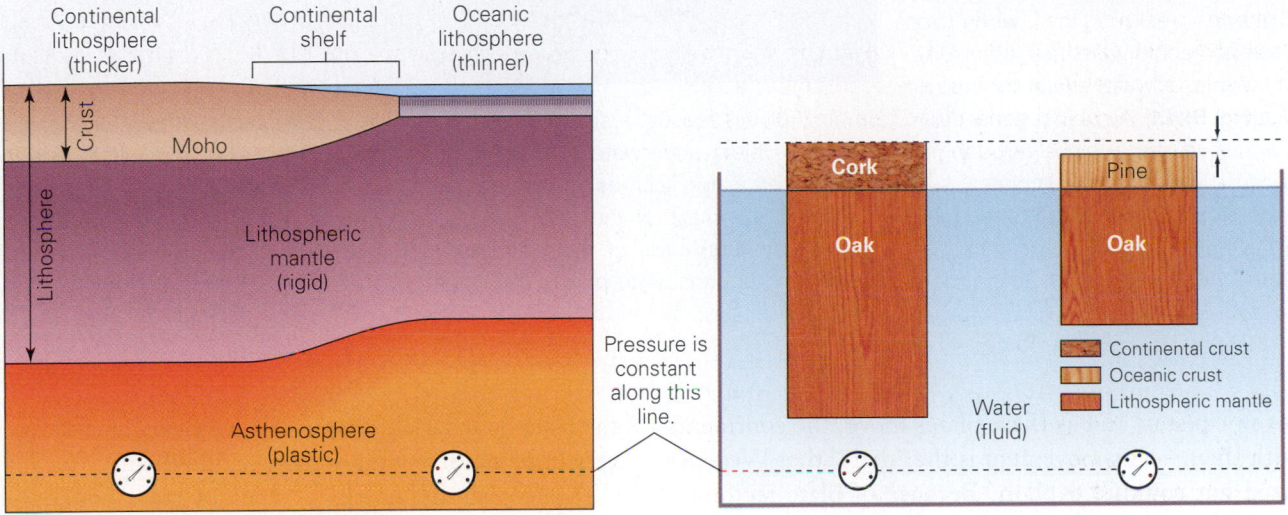

(b) The surface of continental lithosphere is higher than the surface of oceanic lithosphere because they have different crusts and, overall, different thicknesses.

(c) A model provides insight. The model is not ideal, for water is less dense than oak, but asthenosphere is less dense than lithospheric mantle.

Unlike the shell of a hen's egg, the lithosphere is discontinuous—it's broken into about 20 pieces. As noted earlier, we call the pieces **lithosphere plates**, or simply **plates**, and the contacts between them are **plate boundaries** (**Fig. 4.4a, b**). Of the 20 or so plates, geoscientists consider 12 of them to be *major plates* with large areas, and the rest to be *microplates*. Some plates consist entirely of oceanic lithosphere, whereas others consist of both oceanic and continental lithosphere. Some plates have familiar names (the North American Plate, the African Plate); some are more obscure (the Cocos Plate, the Juan de Fuca Plate, the Nazca Plate).

While several plate boundaries coincide with a *continental margin*, the boundary between a continent and an ocean, many do not. For this reason, we distinguish between **active margins**, which are plate boundaries, and **passive margins**, which are not plate boundaries. Along a passive margin, older continental crust is relatively thin and lies buried beneath an accumulation of sediment 10 to 15 km thick. Geologists refer to this accumulation as a *passive-margin basin*. The surface of this sediment pile is a broad area of shallow (less than 500 m deep) seafloor called the *continental shelf*, home to the major fisheries of the world.

The Basic Principles of Plate Tectonics Restated

With the background provided above, we can restate plate tectonics theory concisely as follows: The Earth's lithosphere is divided into plates that move relative to one another. As a plate moves, its internal area remains mostly, but not perfectly, rigid and intact; the motion of one plate relative to its neighbor takes place by slip along plate boundaries. Continents are

BOX 4.1 CONSIDER THIS . . .

Archimedes' Principle of Buoyancy

Archimedes (ca. 287–212 B.C.E.), a Greek mathematician and inventor, left an amazing legacy of discoveries. He described the geometry of spheres, cylinders, and spirals; introduced the concept of a center of gravity; and was the first to understand buoyancy. *Buoyancy* is the upward force acting on an object immersed or floating in a fluid. According to legend, Archimedes recognized this concept suddenly, while floating in a public bath, and was so inspired that he jumped out of the bath and ran home naked, shouting, "Eureka!"

Archimedes realized that when you place a solid object in water, the object displaces a volume of water equal in mass to the object (**Fig. Bx4.1**). An object denser than water, such as a stone, sinks through the water because even when completely submerged the stone's mass exceeds the mass of the water it displaces. When submerged, however, the stone weighs less than it does in air. (For this reason, a scuba diver can lift a heavy object underwater.) An object less dense than water, such as an iceberg, sinks only until the mass of the water displaced equals the total mass of the iceberg. This condition happens while part of the iceberg still protrudes up into the air. Put another way, an object placed in a fluid feels a "buoyancy force" that tends to push it up. If the object's weight is less than the buoyancy force, the object floats; but if its weight is greater than the buoyancy force, the object sinks.

FIGURE Bx4.1 According to Archimedes' principle of buoyancy, an iceberg sinks until the total mass of the water displaced equals the total mass of the whole iceberg. Since water is denser, the volume of the water displaced is less than the volume of the iceberg, so only 20% of the iceberg protrudes above the water.

parts of some plates, and as these plates move, the continents move with them—this movement is the "drift" that Wegener recognized but couldn't explain. Because of plate tectonics, the map of Earth's surface constantly changes.

Identifying Plate Boundaries

How do we recognize the location of a plate boundary? The answer becomes clear from looking at a map showing the locations of earthquakes (**Fig. 4.4c**). Recall from Chapter 2 that earthquakes are vibrations caused by shock waves generated where rock breaks and suddenly shears (slides) along a fault (a fracture on which sliding occurs). The focus of the earthquake is the spot where the fault slips, and the epicenter marks the point on the surface of the Earth directly above the focus. Earthquake epicenters do not speckle the Earth's surface randomly, like buckshot on a target. Rather, the majority occur in relatively narrow, distinct belts. These **seismic belts** define the position of plate boundaries, because the fracturing and slipping that takes place along these boundaries as plates move generate earthquakes. *Plate interiors*, regions away from the plate boundaries, remain relatively earthquake-free because very little movement takes place within them.

Geologists define three types of plate boundaries based simply on the *relative motions* of the plates on either side of the boundary (**Fig. 4.5**). A boundary at which two plates move apart from each other is a *divergent boundary*. A boundary at which two plates move toward each other so that one plate sinks beneath the other is a *convergent boundary*. And a boundary at which two plates slide sideways past each other is a *transform boundary*. Each type of boundary looks and behaves differently from the others, as we will now see.

> **Did you ever wonder . . .**
>
> why earthquakes don't occur everywhere?

Take-Home Message

Earth's rigid shell, its lithosphere, is divided into about 20 plates that move relative to one another. The motion of one plate relative to its neighbor is accommodated by sliding along plate boundaries. Seismic belts, therefore, define plate boundaries.

QUICK QUESTION: Can a plate include both continental and oceanic lithosphere? Are all continental margins seismically active?

FIGURE 4.4 The locations of plate boundaries and the distribution of earthquakes.

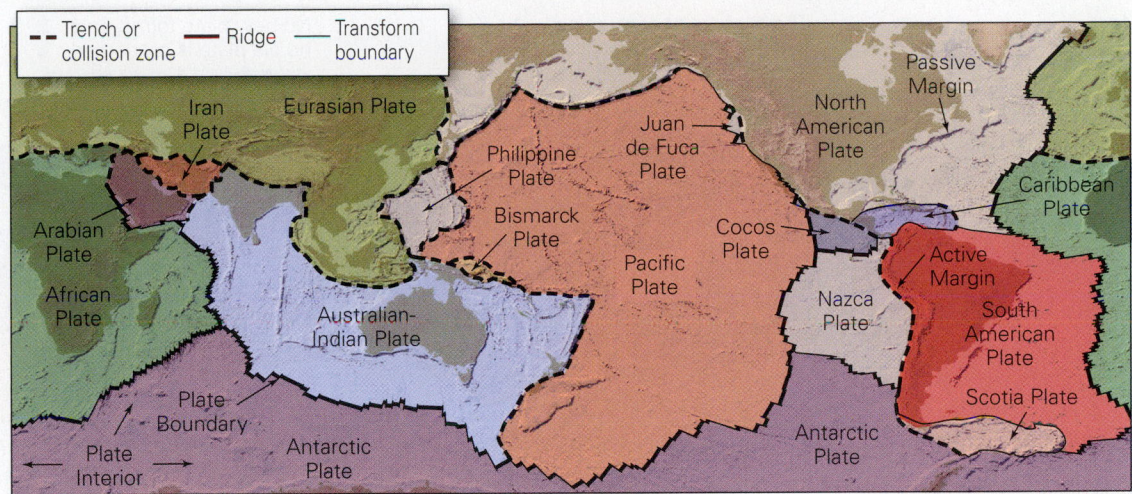

(a) A map of major plates shows that some consist entirely of oceanic lithosphere, whereas others consist of both continental and oceanic lithosphere. Active continental margins lie along plate boundaries; passive margins do not.

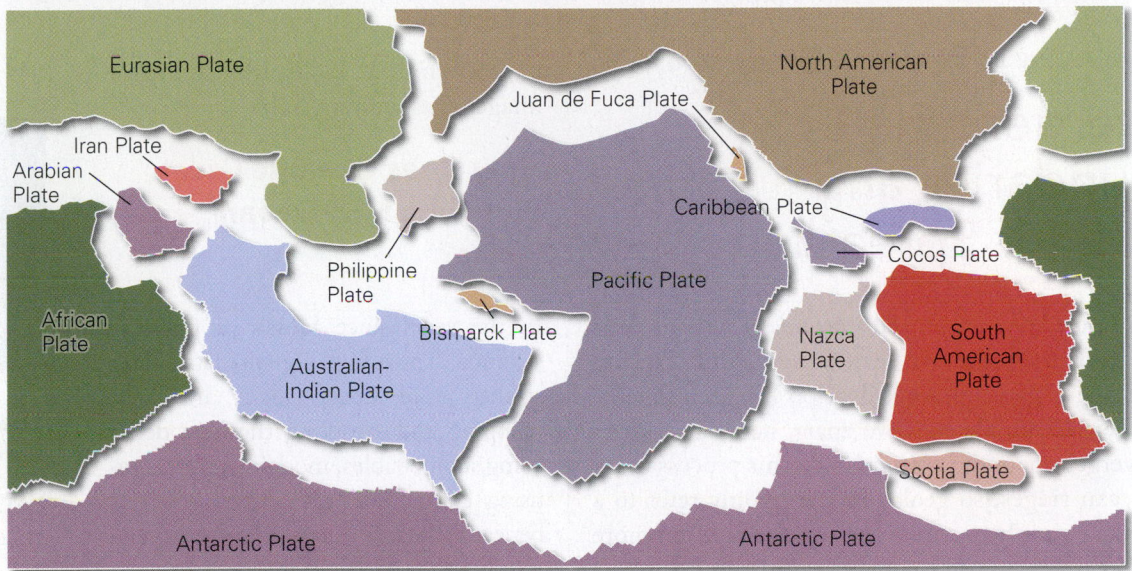

(b) An exploded view of the plates emphasizes the variation in shape and size of the plates.

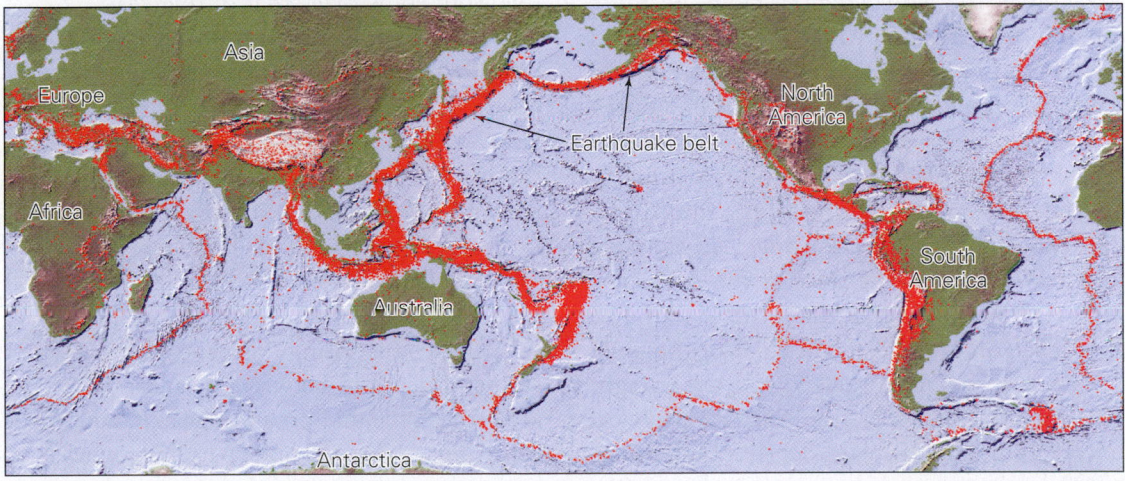

(c) The locations of earthquakes (red dots) mostly fall in distinct bands that correspond to plate boundaries. Relatively few earthquakes occur in the stabler plate interiors.

FIGURE 4.5 The three types of plate boundaries differ based on the nature of relative movement.

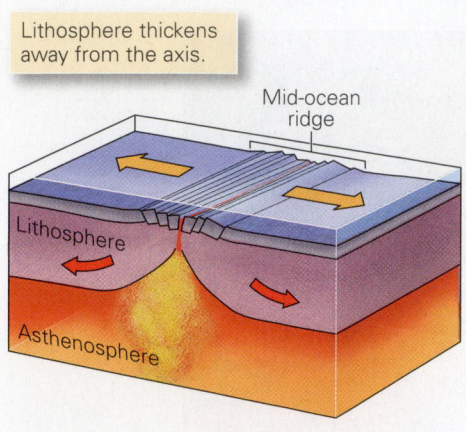

Lithosphere thickens away from the axis.

Mid-ocean ridge

Lithosphere

Asthenosphere

(a) At a divergent boundary, two plates move away from the axis of a mid-ocean ridge. New oceanic lithosphere forms.

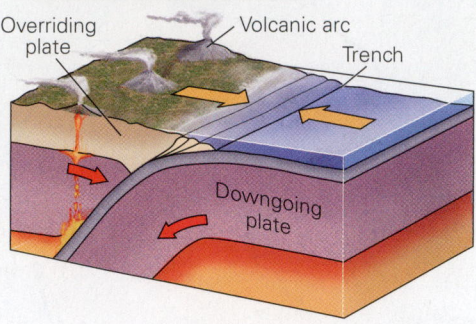

The process of consuming a plate is called subduction.

Overriding plate

Volcanic arc

Trench

Downgoing plate

(b) At a convergent boundary, two plates move toward each other; the downgoing plate sinks beneath the overriding plate.

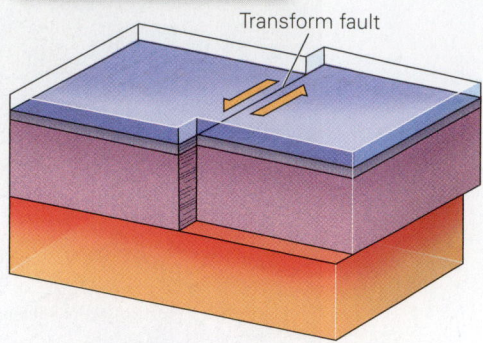

No new plate forms, and no old plate is consumed.

Transform fault

(c) At a transform boundary, two plates slide past each other on a vertical fault surface.

4.3 Divergent-Plate Boundaries and Seafloor Spreading

At a **divergent boundary**, or *spreading boundary*, two oceanic plates move apart by the process of seafloor spreading. During the process, an open space does not develop between diverging plates. Rather, as the plates move apart, new ocean floor forms at the divergent boundary (**Fig. 4.6a**). This process takes place at mid-ocean ridges, so geologists commonly refer to a divergent boundary simply as a mid-ocean ridge or, even more simply, a ridge.

To characterize a divergent boundary more completely, let's look at one example, the Mid-Atlantic Ridge, in more detail (**Fig. 4.6b**). The Mid-Atlantic Ridge extends from the waters between northern Greenland and northern Scandinavia southward across the equator to the latitude of the southern tip of South America. It rises by about 2 km above the depth of the Atlantic abyssal plains, and thus its crest lies at water depths of 2 to 2.5 km. The centerline of the mid-ocean ridge is called the *ridge axis*. Typically, a 1 km-deep elongate trough, the *median valley*, follows the trace of the ridge axis (**Fig. 4.6c**). Geologists have found that the formation of new oceanic crust takes place on or beneath the floor of the axial valley. A series of steep scarps form the walls of the trough. Outside of the trough, seafloor slopes from the ridge toward the abyssal plain, reaching abyssal plain depth at a distance of about 500 to 800 km from the ridge axis (see Fig. 3.8c). Roughly speaking, the Mid-Atlantic Ridge is symmetrical—

its eastern half looks like a mirror image of its western half.

How Does Oceanic Crust Form at a Mid-Ocean Ridge?

In the years since Hess's proposal of seafloor spreading, geologists have studied mid-ocean ridges intensely, mapping and sampling them in detail using submersibles, modeling the ridges using computers, and imaging what's beneath them by studying how seismic waves pass through the crust and mantle beneath them. The result of all this work emphasizes that during seafloor spreading hot asthenosphere rises beneath the ridge. But this asthenosphere does not reach the Earth's surface—thus, the oceanic crust is *not* just the frozen top of the mantle. Rather, the crust is shaped from melt that forms in the rising asthenosphere when it reaches shallower depths (**Fig. 4.7**). Let's look at this process more closely.

When the rising asthenosphere reaches a depth of between 60 and 30 km beneath the mid-ocean ridge axis, about 15% of it melts and turns into magma. (Chapter 6 explains why.) Not all the minerals in the ultramafic rock of the asthenosphere melt equally during this process, so the magma that forms has a different composition

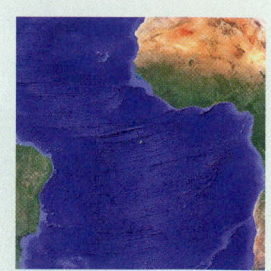

FIGURE 4.6 Divergent-plate boundaries are delineated by mid-ocean ridges, where seafloor spreading occurs.

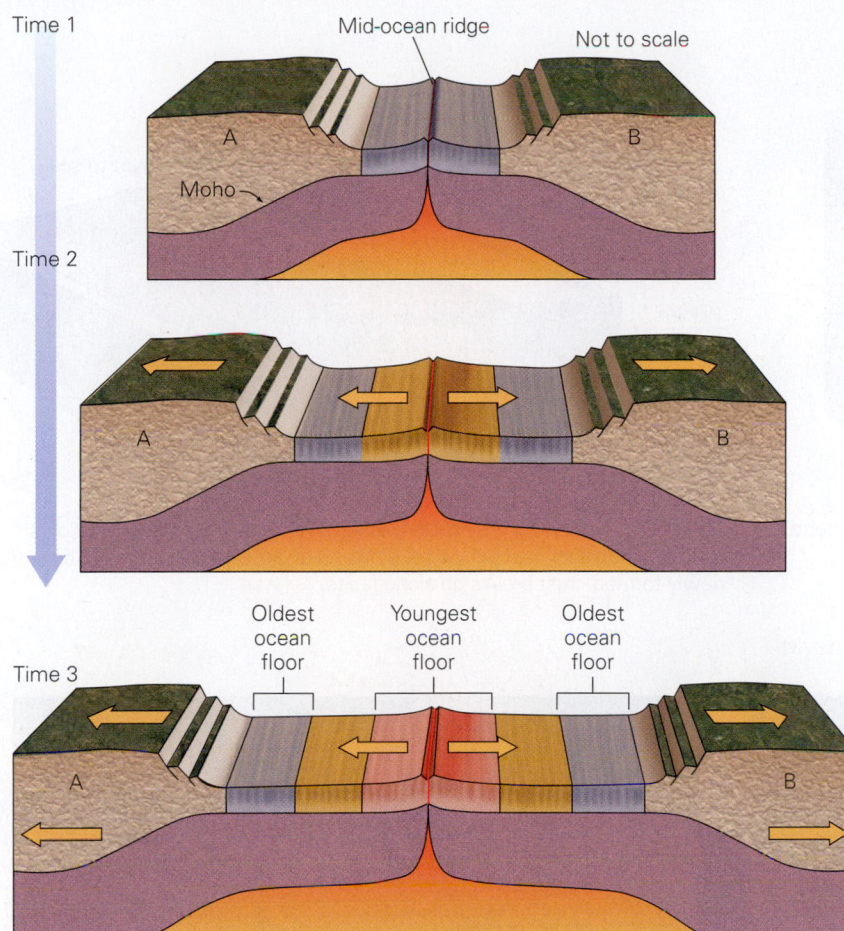

(a) During seafloor spreading, the ocean floor gets wider, and continents on either side move apart. New oceanic crust forms at the ridge axis. Rising asthenosphere melts beneath the axis.

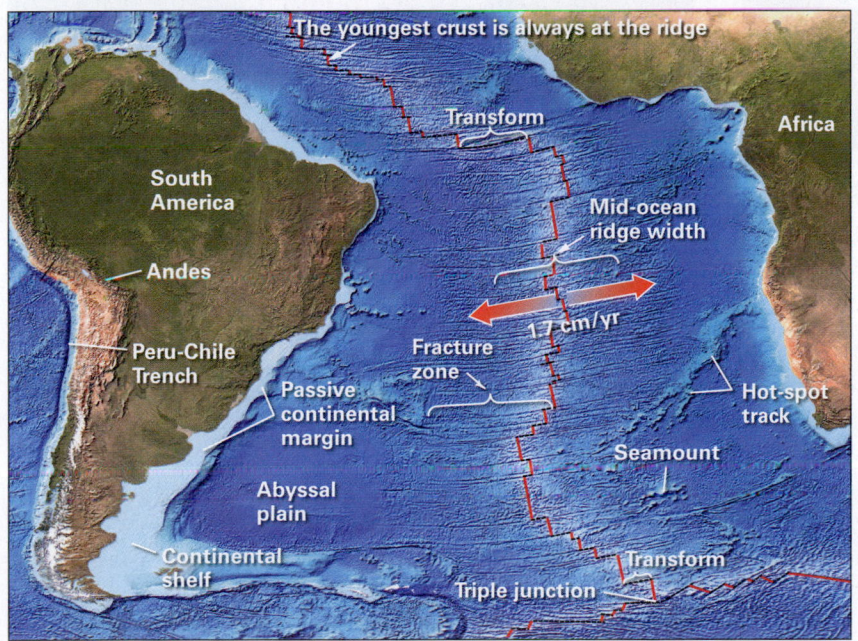

(b) The Mid-Atlantic Ridge in the South Atlantic Ocean. The lighter shades of blue are shallower water depths.

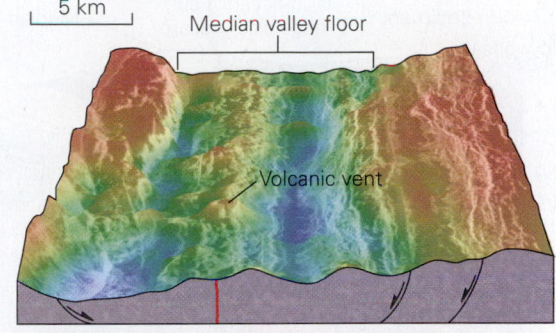

(c) A detailed digital map showing the surface of the median valley.

than its source rock. Specifically, the magma has a mafic composition—it contains relatively more silica than does the ultramafic rock from which it formed. The density of the magma is less than that of surrounding asthenosphere rock, so the magma behaves buoyantly and rises. Some of the magma accumulates in a region called a *magma chamber* that lies at a depth of 2 to 7 km beneath the floor of the median valley. At any given time, the magma chamber contains a "mush" of solid mineral crystals and liquid melt. Some of this material cools along the sides of the magma chamber and solidifies to become a coarse-grained, mafic igneous rock called *gabbro*. Some rises still higher into vertical cracks, in which it solidifies to form wall-like sheets of basalt—these sheets are called *basalt dikes*. Magma that makes it all the way to the surface of the sea-floor spills out of small submarine volcanoes as lava. This lava cools to form a layer of *pillow basalt*, consisting of piles of meter-wide basalt blobs, or "pillows" (**Fig. 4.8a**). Thus, the newly formed ocean crust consists of three distinct layers: a layer of pillow basalt overlies a layer of basalt dikes, which in turn overlies a layer of gabbro.

We can't easily see the submarine volcanoes of a mid-ocean ridge, because they occur beneath 2 to 3 km of water, but they have been detected by observers in research submersibles. These observers have also found chimneys spewing hot, mineralized water. These chimneys are known as **black smokers** because the water they emit looks like a cloud of dark smoke—the color comes from a suspension of tiny mineral grains that precipitate in cold seawater the instant that the rising mineralized water cools (**Fig. 4.8b**). Black smokers form because, over time, seawater percolates down into the oceanic crust through a network of cracks.

FIGURE 4.7 Formation of new oceanic crust occurs on and below the median valley.

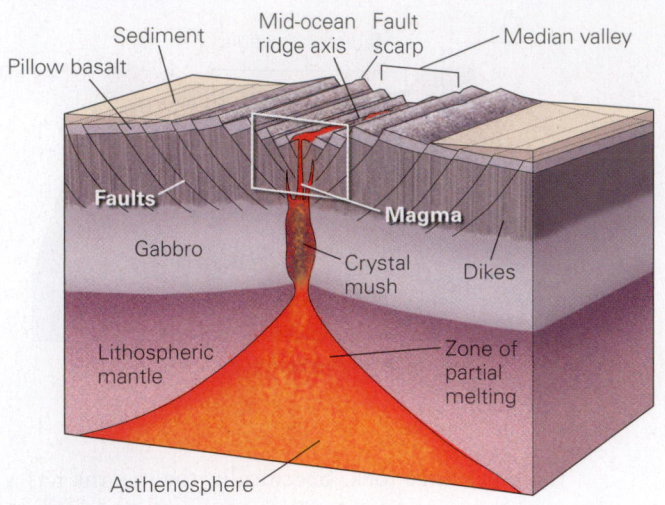

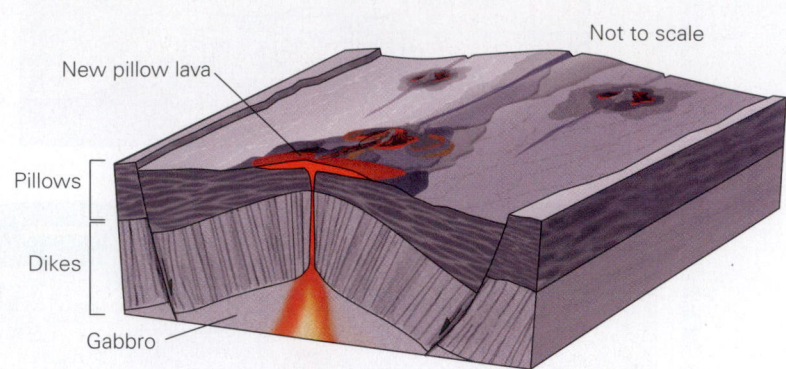

(a) Beneath a mid-ocean ridge, there is a magma chamber. Gabbro forms on the side of the magma chamber. Basalt dikes protrude upward.

(b) An enlargement of the median valley. illustrates that lenses of pillows spill out of distinct fissures where a dike gets close to the surface. The newly-formed crust breaks up along faults.

FIGURE 4.8 Eruption on mid-ocean ridges.

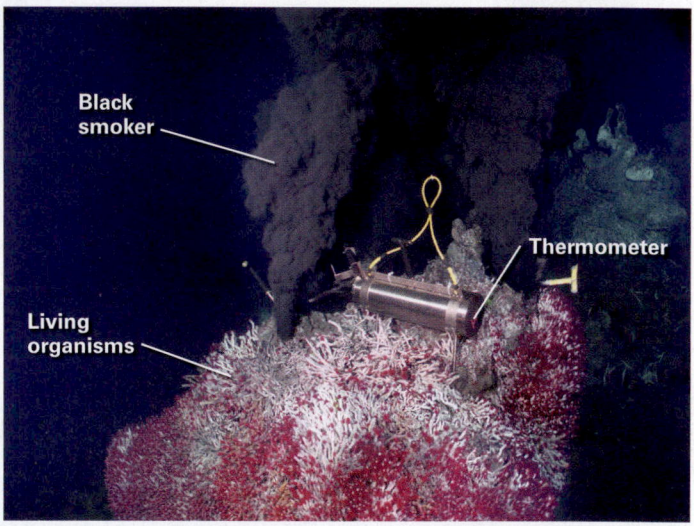

(a) Recently erupted pillow basalt from the Juan de Fuca Ridge.

(b) A column of superhot water gushing from a vent (known as a black smoker) along the mid-ocean ridge. The cloud of "smoke" actually consists of tiny mineral grains. A local ecosystem of bacteria, shrimp, and worms live around the vent.

At depth, heat from rising magma warms up the water, which then starts to rise. The hot water dissolves minerals and carries them along, in solution, as it spews back into the sea.

As soon as it forms, new oceanic crust moves away from the ridge axis, and when this happens, more magma rises from below, so still more crust forms. In other words, like a vast, continuously moving conveyor belt, magma from the mantle rises to the Earth's surface at the ridge, solidifies to form oceanic crust, and then moves laterally away from the ridge. Because all seafloor forms at mid-ocean ridges, the youngest seafloor occurs at the ridge axis, and seafloor becomes progressively older away from the axis. In the Atlantic Ocean, the oldest seafloor lies adjacent to the passive continental margins on either side of the ocean (**Fig. 4.9**). The oldest ocean floor on our planet underlies the western Pacific Ocean; this crust formed 200 Ma (million years ago).

The tension (stretching force) applied to newly formed solid crust as spreading takes place breaks this new crust, resulting in the formation of faults. Movement (slip) on these faults gen-

Did you ever wonder...

whether all the floors of the ocean are the same age?

FIGURE 4.9 This map of the world shows the age of the seafloor. Note how the seafloor grows older with increasing distance from the ridge axis.

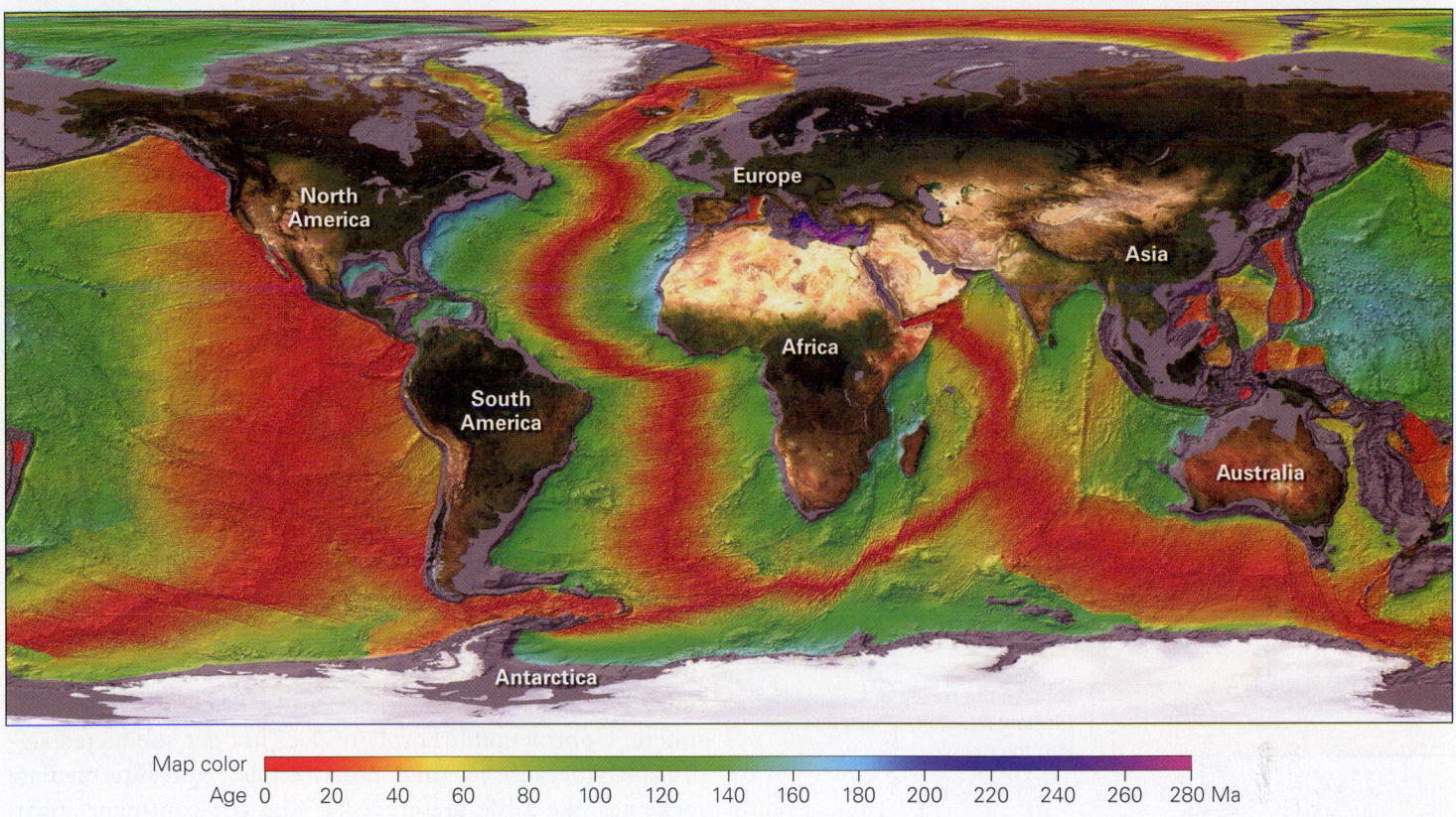

Map color

Age 0 20 40 60 80 100 120 140 160 180 200 220 240 260 280 Ma

erates earthquakes and produces scarps that border the median valley of the mid-ocean ridge.

How Does the Lithospheric Mantle Form at a Mid-Ocean Ridge?

We've seen that oceanic crust forms from magma formed beneath at mid-ocean ridges. What about the formation of the lithospheric mantle of ocean lithosphere? Recall that this part consists of the cooler uppermost layer of the mantle, in which temperatures are less than about 1,280°C. At the ridge axis, such temperatures occur almost at the base of the crust because of the presence of rising hot asthenosphere and hot magma, so lithospheric mantle beneath the ridge axis effectively doesn't exist. But as the newly formed oceanic crust moves away from the ridge axis, the crust and the uppermost mantle directly beneath it gradually cool by losing heat to the ocean above. As soon as mantle rock cools below 1,280°C, it becomes, by definition, part of the lithosphere because at this temperature it begins to behave rigidly. As oceanic lithosphere continues to move away from the ridge axis, it continues to cool, so the lithospheric mantle, and therefore the oceanic lithosphere as a whole, grows progressively thicker (**Fig. 4.10**).

Note that the process of forming the lithospheric mantle doesn't change the thickness of the overlying oceanic crust, for the crust formed entirely at the ridge axis. The rate at which cooling and lithospheric-mantle thickening occur decreases with distance from the ridge axis. In fact, by the time the lithosphere is about 80 million years old, it has just about reached maximum thickness. As lithosphere thickens and gets cooler and denser, it sinks down into the asthenosphere, like a ship taking on ballast. Thus, the ocean is deeper over older ocean floor than over younger ocean floor. That's why abyssal plains are deeper than mid-ocean ridges.

Take-Home Message

Seafloor spreading occurs at divergent-plate boundaries, defined by mid-ocean ridges. Hot asthenosphere rises beneath the mid-ocean ridge and starts to melt when it reaches shallower depths. The resulting magma rises still further. Some solidifies underground, as gabbro, some injects at a shallower level to produce vertical sheets (dikes) of basalt, and some erupts from small volcanoes along the ridge axis to form pillow basalt. The gabbro and basalt produced by these processes form new ocean crust. As plates move away from the axis, they cool, and the lithospheric mantle forms and thickens underneath this crust.

QUICK QUESTION: What are black smokers, and why do they form?

FIGURE 4.10 Changes accompanying the aging of lithosphere.

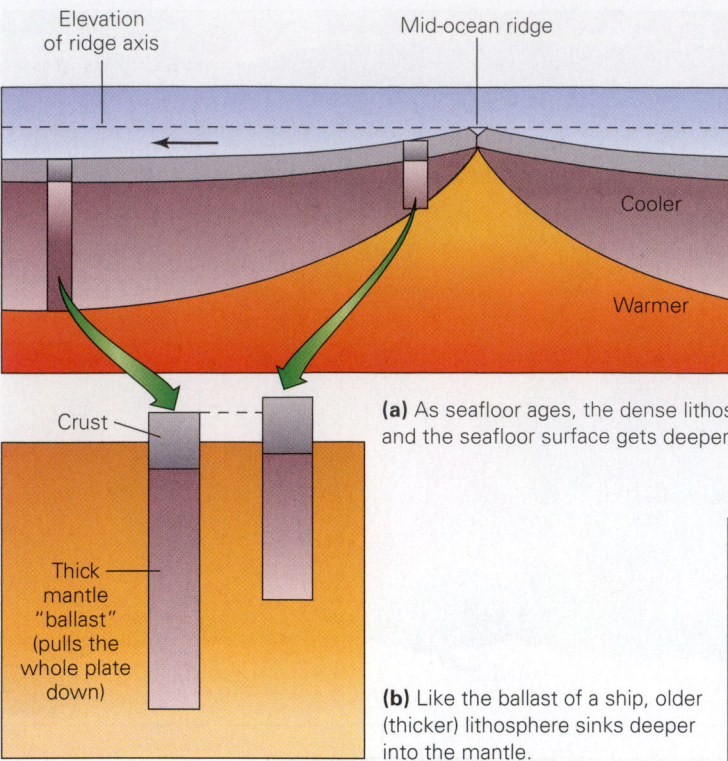

(a) As seafloor ages, the dense lithospheric mantle thickens and the seafloor surface gets deeper.

(b) Like the ballast of a ship, older (thicker) lithosphere sinks deeper into the mantle.

4.4 Convergent-Plate Boundaries and Subduction

At convergent-plate boundaries, two plates, at least one of which is oceanic, move toward each other. But rather than butting against each other like angry rams, one oceanic plate bends and sinks down into the asthenosphere beneath the other plate. Geologists refer to the sinking process as **subduction**, so convergent boundaries are also known as *subduction zones*. Because subduction at a **convergent boundary** consumes old ocean lithosphere and thus closes, or "consumes," oceanic basins, geologists also refer to convergent boundaries as *consuming boundaries*, and because they are delineated by deep-ocean trenches, they are sometimes simply called **trenches** (**Fig. 4.11a**). The amount of oceanic-plate consumption worldwide, averaged over time, equals the amount of seafloor spreading worldwide, so the surface area of the Earth remains constant through time.

Subduction occurs for a simple reason: the overall density of an oceanic plate, once the plate has aged at least 10 million years, exceeds that of the underlying asthenosphere. Therefore, where a plate bends down and starts to slip into the mantle, it will continue to sink like an anchor falling to the bottom of a lake

(**Fig. 4.11b**). As the lithosphere sinks, asthenosphere flows out of its way, just as water flows out of the way of an anchor. But unlike water, the *viscosity* of asthenosphere, meaning its resistance to flow, is large and thus it can flow only very slowly. Therefore, oceanic lithosphere can sink only very slowly—at a rate of less than about 15 cm per year. (To visualize the difference, imagine how much faster a coin can sink through water than it can through honey.) Because of the asthenosphere's high viscosity, lithosphere can sink only when it enters the asthenosphere at an angle, as occurs in a subduction zone—the broad horizontal interior of an oceanic plate can't just sink straight down because there's too much resistance to flow in the underlying asthenosphere.

Note that the "downgoing plate," the plate that has been subducted, must be composed of oceanic lithosphere. The "overriding plate," the one that is not sinking at subduction zone, can consist of either oceanic or continental lithosphere. Continental lithosphere does not get subducted significantly because its thick crust of felsic and intermediate rocks act like a life preserver keeping the continent afloat. If continental crust moves into a convergent margin, subduction eventually stops before crust gets taken down to a depth of 100 km and a "collision" occurs, as we discuss later. Because continental crust cannot be completely subducted, some continental crust has persisted for over 3.8 billion years; in contrast, the oldest oceanic lithosphere, as we have noted, is less than 200 million years old because all oceanic lithosphere older than that has been subducted.

Earthquakes and the Fate of Subducted Plates

At convergent-plate boundaries, the downgoing plate grinds along the base of the overriding plate, a process that generates large earthquakes. These earthquakes occur fairly close to the Earth's surface, so some of them cause massive destruction in coastal cities. But earthquakes also happen in downgoing plates at greater depths. In fact, geologists can detect earthquakes within downgoing plates to a depth of 660 km. The band of earthquakes in a downgoing plate is called a **Wadati-Benioff zone**, after its discoverers (**Fig. 4.11c**). At depths greater than 660 km, conditions leading to earthquakes in subducted lithosphere do not occur. Recent observations, however, indicate that downgoing plates do continue to sink below a depth of 660 km—they just do so without generating earthquakes. The lower mantle may be a graveyard for old subducted plates.

FIGURE 4.11 During the process of subduction, oceanic lithosphere sinks back into the deeper mantle.

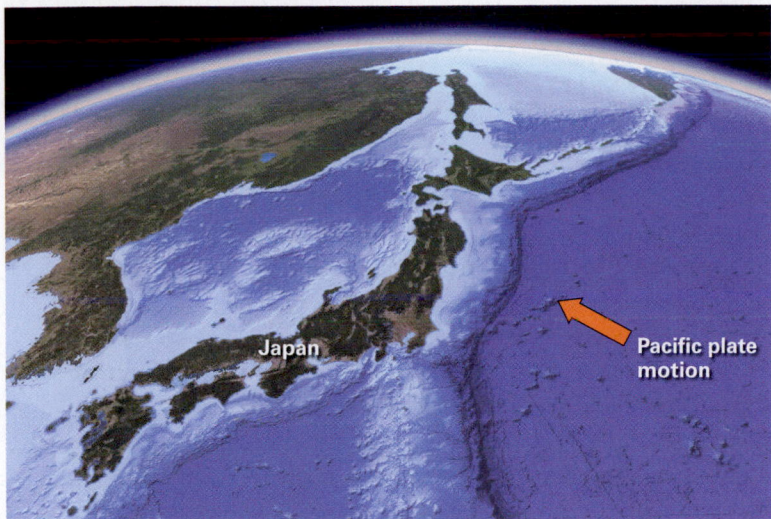

(a) The Pacific plate subducts underneath Japan.

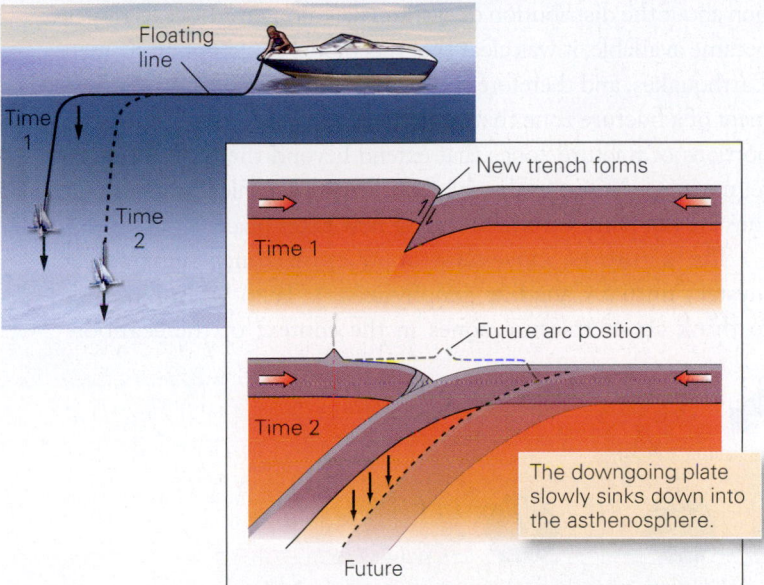

(b) Sinking of the downgoing plate resembles sinking of an anchor attached to a rope.

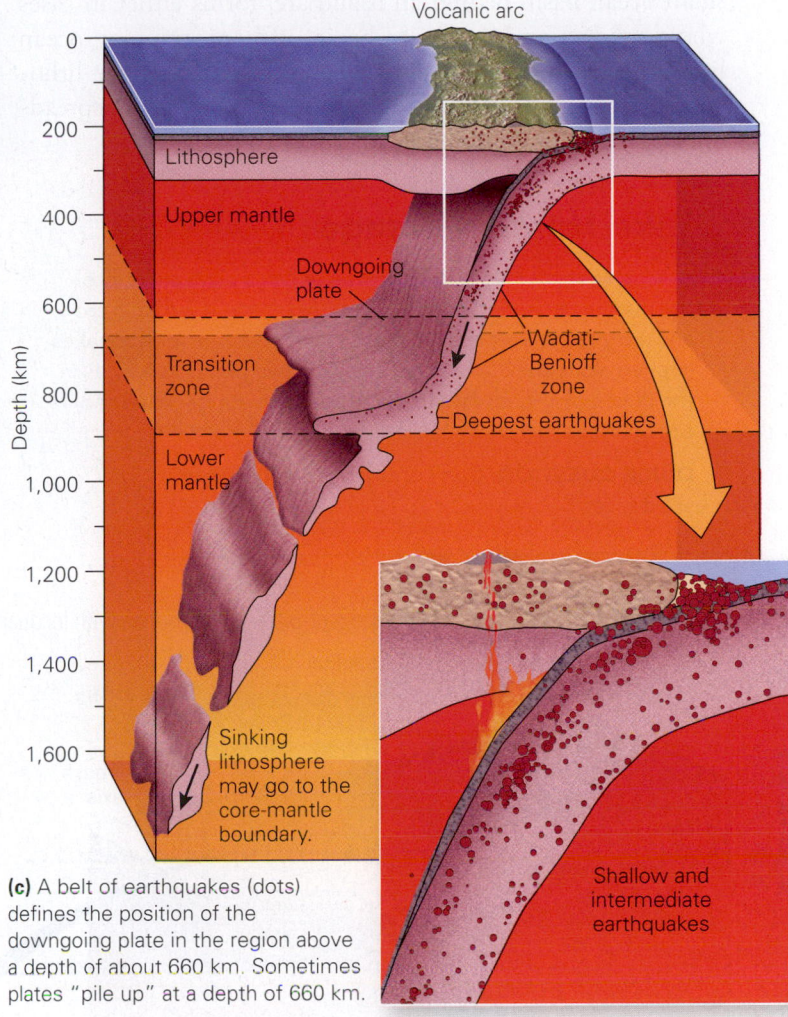

(c) A belt of earthquakes (dots) defines the position of the downgoing plate in the region above a depth of about 660 km. Sometimes plates "pile up" at a depth of 660 km.

Geologic Features of a Convergent Boundary

To become familiar with the various geologic features that occur along a convergent-plate boundary, let's look at an example, the boundary between the eastern coast of Japan and the western edge of the Pacific Plate. A deep-ocean trench, the Japan Trench, delineates this boundary (see Fig. 4.11a).

Trenches form where a plate bends as it starts to sink into the asthenosphere. In the Japan Trench, as the downgoing plate slides under the overriding plate, sediment (clay and plankton) that had settled on the surface of the downgoing plate, as well as sand that fell into the trench from the shores of South America, gets scraped up and incorporated in a wedge-shaped mass known as an **accretionary prism** (**Fig. 4.12a, b**). An accretionary prism forms in basically the same way as a pile of snow in front of a plow, and like the snow, the sediment becomes squashed and contorted during the process.

A chain of volcanoes known as a **volcanic arc** develops behind the accretionary prism. As we will see in Chapter 6, the magma that feeds these volcanoes forms above the surface of the downgoing plate where the plate reaches a depth of about 150 km below the Earth's surface. If the volcanic arc forms where an oceanic plate subducts beneath continental lithosphere, the resulting chain of volcanoes grows on the continent and forms a *continental volcanic arc*. In some cases, compression between the converging plates produces a belt of faults in the region behind the arc—this has happened, for example, on the east side of the Andes. If the volcanic arc forms where one oceanic plate subducts beneath another oceanic plate, the resulting volcanoes form a chain of islands known as a **volcanic**

island arc (**Fig. 4.12c**). A marginal sea, or *back-arc basin*, a small ocean basin behind an island arc, forms either in cases where subduction happens to begin offshore, trapping ocean lithosphere behind the arc, or where stretching of the lithosphere behind the arc leads to the formation of a small, spreading ridge behind the arc (**Fig. 4.12d**).

Take-Home Message

At a convergent-plate boundary, an oceanic plate sinks into the mantle beneath the edge of another plate. A volcanic arc and a trench delineate such plate boundaries, and earthquakes happen along the contact between the two plates as well as in the downgoing slab. Volcanic arcs can form on the edge of a continent or as a chain of islands in the sea.

QUICK QUESTION: Can continents be completely subducted?

FIGURE 4.12 The nature of a convergent-plate margin varies with location.

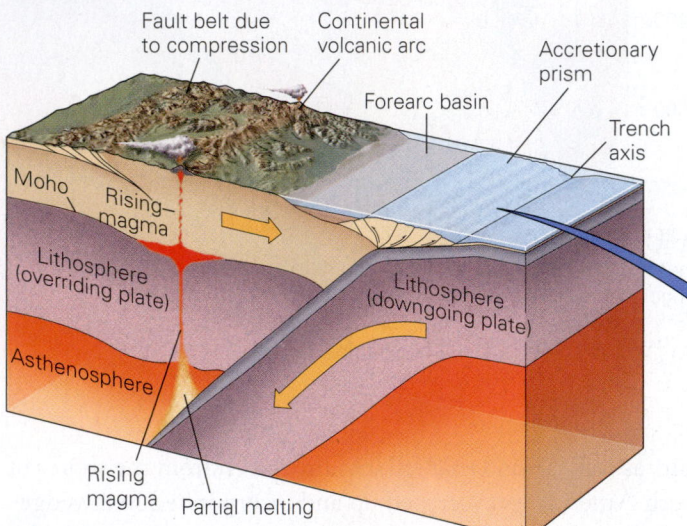

(a) A subduction zone along the edge of a continent. Here compression caused faulting behind the arc.

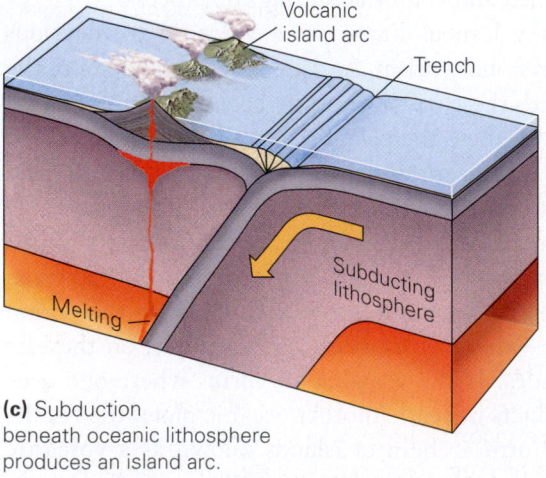

(c) Subduction beneath oceanic lithosphere produces an island arc.

4.5 Transform-Plate Boundaries

When researchers began to explore the bathymetry of mid-ocean ridges in detail, they discovered that mid-ocean ridges are not long, uninterrupted lines but rather consist of short segments that appear to be offset laterally from each other (**Fig. 4.13a**). Narrow belts of broken and irregular seafloor lie roughly at right angles to the ridge segments, intersect the ends of the segments, and extend beyond the ends of the segments. These belts are called **fracture zones**. Originally, researchers incorrectly assumed that the entire length of each fracture zone was a fault and that slip on a fracture zone had displaced segments of the mid-ocean ridge sideways, relative to one another. In other words, they imagined that a mid-ocean ridge initiated as a continuous, fence-like line that only later was broken up by faulting in fracture zones. But when information about the distribution of earthquakes along mid-ocean ridges became available, it was clear that this model could not be correct. Earthquakes, and therefore active fault slip, occur only on the segment of a fracture zone that lies between two ridge segments. The portions of fracture zones that extend beyond the edges of ridge segments, out into the abyssal plain, are not seismically active and thus are not faults on which sliding now takes place.

The nature of movement on fracture zones remained a mystery until a Canadian geophysicist, J. Tuzo Wilson, began to think about fracture zones in the context of the seafloor-

(b) The overriding plate acts like a bulldozer, scraping up sediment off the downgoing plate to build an accretionary prism.

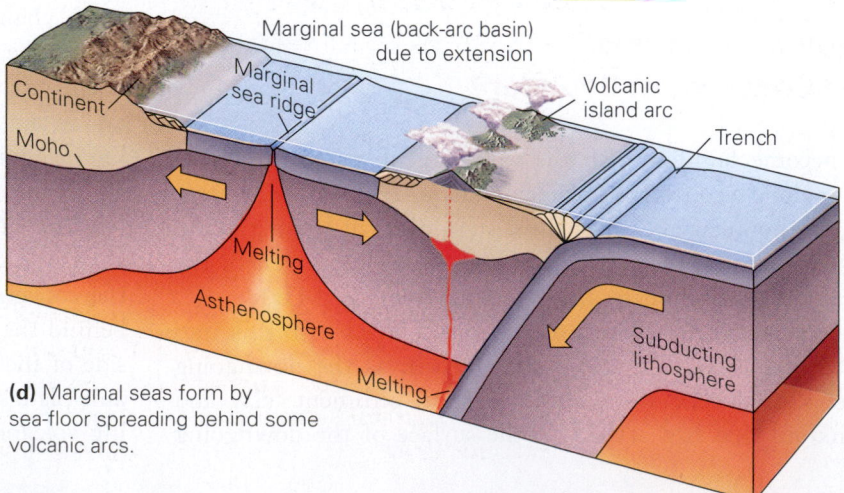

(d) Marginal seas form by sea-floor spreading behind some volcanic arcs.

FIGURE 4.13 The concept of transform faulting.

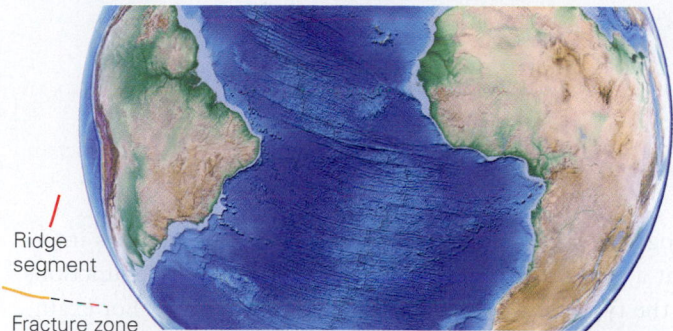

Ridge segment

Fracture zone (active transform where solid)

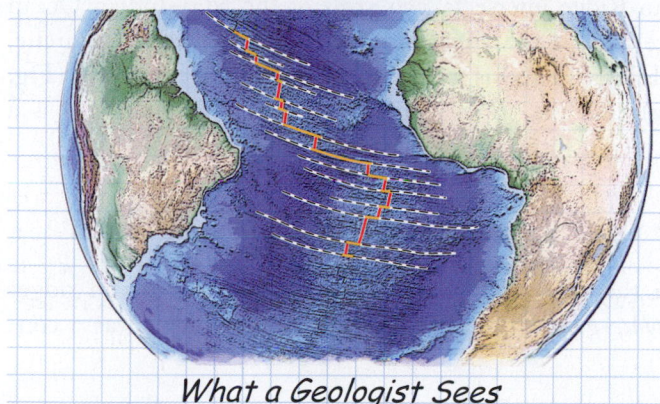

What a Geologist Sees

(a) Numerous transform faults segment the Mid-Atlantic Ridge.

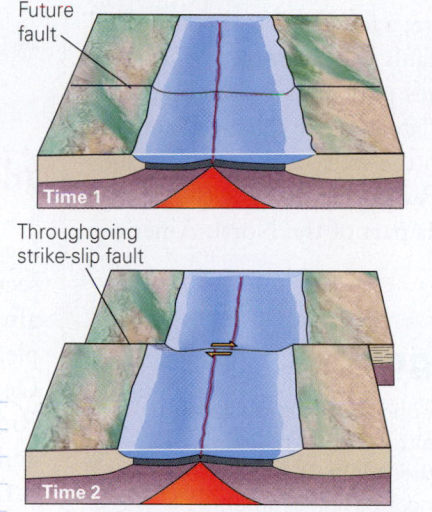

Incorrect

Future fault

Time 1

Throughgoing strike-slip fault

Time 2

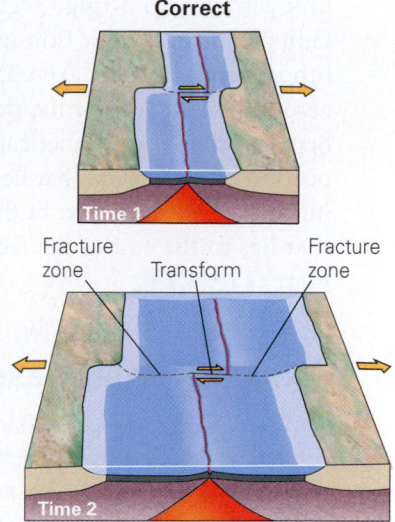

Correct

Time 1

Fracture zone   Transform   Fracture zone

Time 2

(b) A comparison of the old model of transform faults with the new model required by the seafloor-spreading hypothesis.

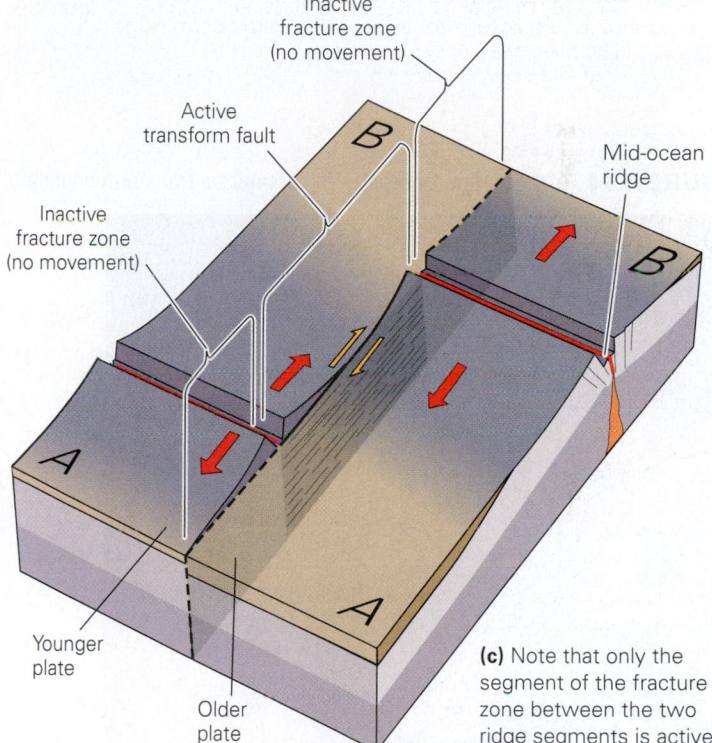

Inactive fracture zone (no movement)

Active transform fault

Inactive fracture zone (no movement)

Mid-ocean ridge

Younger plate

Older plate

(c) Note that only the segment of the fracture zone between the two ridge segments is active.

spreading concept. Wilson proposed that fracture zones formed *at the same time* as the ridge axis itself, and thus the ridge consisted of separate segments to start with. These segments were linked—not offset—by fracture zones. With this idea in mind, he drew a sketch map showing two ridge-axis segments linked by a fracture zone, and he drew arrows to indicate the direction that ocean crust was moving, relative to the ridge axis, as a result of seafloor spreading (**Fig. 4.13b**). Look at Wilson's arrows. Clearly, the movement direction on the active portion of the fracture zone must be compatible with the spreading direction and thus is opposite to the movement direction that researchers originally thought took place. Further, in Wilson's model, slip occurs only along the segment of the fracture zone between the two ridge segments. Plates on opposite sides of inactive fracture zones move together as one plate.

Wilson introduced the term **transform boundary** (or *transform fault*) for the actively slipping segment of a fracture zone between two ridge segments, and he pointed out that these are a third type of plate boundary. At a transform boundary, one plate slides sideways, relative to its neighbor, on a vertical fault. Thus, the slip direction on the fault is horizontal (parallel to the Earth's surface), and no new plate forms and no old plate is consumed, at a transform boundary (**Fig. 4.13c**).

So far we've discussed only transforms along mid-ocean ridges. Not all transforms link ridge segments. Some, such as the Alpine fault of New Zealand, link trenches, while others link a trench to a ridge segment. Further, not all transform faults occur in oceanic lithosphere; a few cut across continental lithosphere. The San Andreas fault, for example, which cuts across western California, defines part of the plate boundary between the North American Plate and the Pacific Plate. The portion of California that lies to the west of the fault (including Los Angeles) is part of the Pacific Plate, while the portion that lies to the east of the fault is part of the North American Plate (**Fig. 4.14**).

Take-Home Message

At transform-plate boundaries, one plate slips sideways past another along a vertical fault. Thus, there is no production or destruction of lithosphere at a transform boundary. Most transform boundaries link segments of mid-ocean ridges, but some, such as the San Andreas fault, cut across continental crust.

QUICK QUESTION: Why do earthquakes only occur on the portion of a fracture zone that links mid-ocean ridge segments?

4.6 Special Locations in the Plate Mosaic

Triple Junctions

Geologists refer to a place where three plate boundaries intersect at a point as a **triple junction**. We name triple junctions after the types of boundaries that intersect at them. For example, the triple junction formed where the Southwest Indian Ocean Ridge intersects two arms of the Mid–Indian Ocean Ridge (this is the triple junction of the African, Antarctic, and Indian Plates) is a ridge-ridge-ridge triple junction (**Fig. 4.15a**), and the triple junction north of San Francisco is a trench-transform-transform triple junction (**Fig. 4.15b**).

Hot Spots

Both the volcanoes of volcanic arcs and mid-ocean ridges are *plate-boundary volcanoes*, in that they formed as a con-

FIGURE 4.14 The San Andreas fault—a continental transform boundary.

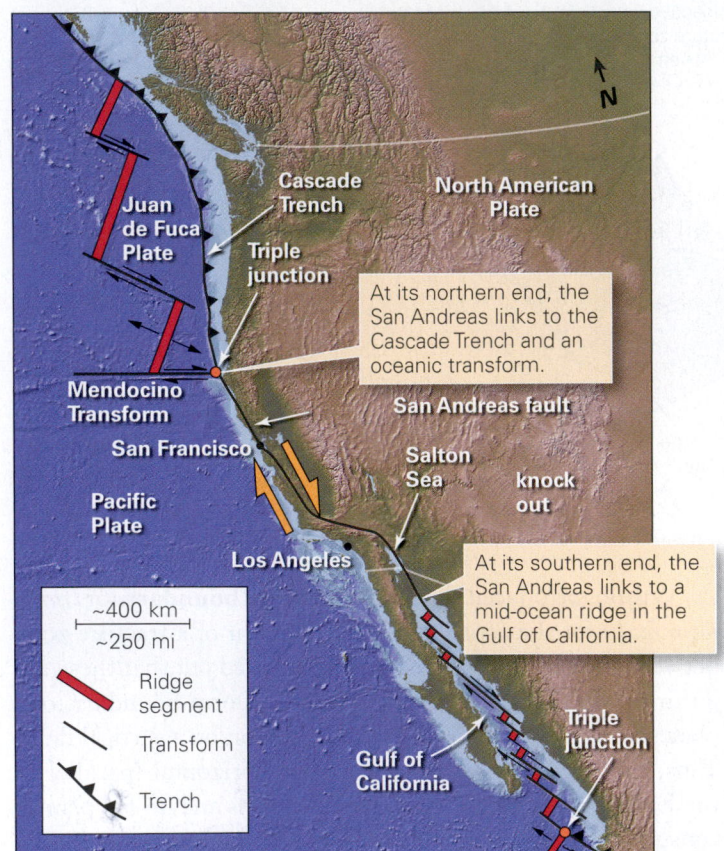

(a) The San Andreas fault is a transform plate boundary between the North American and Pacific Plates. The Pacific is moving northwest, relative to North America.

(b) In southern California, the San Andreas fault cuts a dry landscape. The fault trace is in the narrow valley. The land has been pushed up slightly along the fault.

FIGURE 4.15 Examples of triple junctions. The triple junctions are marked by dots.

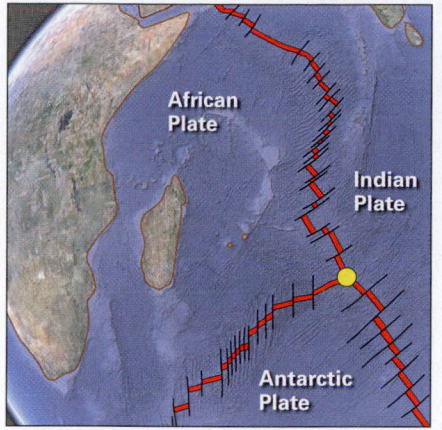

(a) A ridge-ridge-ridge triple junction occurs in the Indian Ocean.

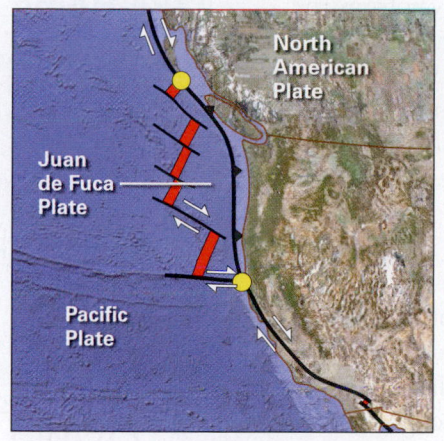

(b) A trench-transform-transform triple junction occurs at the north end of the San Andreas fault.

🔴 Ridge

⫽ Transform

🟡 Triple junction

➤ Subduction zone

╲ Fracture zone

100 volcanoes that exist as isolated points and are not a consequence of movement at a plate boundary. These are called *hot-spot volcanoes*, or simply **hot spots** (**Fig. 4.16**). Many hot spots are located in the interiors of plates, away from the boundaries, but some grew on mid-ocean ridges.

What causes hot-spot volcanoes? In the early 1960s, J. Tuzo Wilson noted that active hot-spot volcanoes (ones that are erupting or may erupt in the future) occur only at the end of a chain of inactive volcanic islands and seamounts (ones that will never erupt again). This configuration is different from that of volcanic arcs along convergent-plate boundaries—at volcanic arcs, all of the volcanoes are potentially active, because subduction is happening along the length of the arc. Wilson proposed that hot-spot volcanoes formed above a localized magma source whose position in the mantle was fixed relative to the moving plate above. In Wilson's model, the active volcano represents the present-day location of the magma source, whereas the chain of dead volcanic islands represents locations on the plate that were

sequence of movement along the boundary. Not all volcanoes on Earth are plate-boundary volcanoes, however. For example, the currently erupting volcanoes of the big island of Hawaii are over 4,000 kilometers from the nearest ridge or trench. And Yellowstone National Park, which occupies the remnants of a huge volcano that erupted during the past million years, lies well into the interior of the North American Plate. Worldwide, geoscientists have identified about

FIGURE 4.16 The dots represent the locations of selected hot-spot volcanoes. The red lines represent hot-spot tracks. The most recent volcano (dot) is at one end of this track. Some of these volcanoes are extinct. Some hot spots are fairly recent and do not have tracks. Dashed tracks were broken by seafloor spreading.

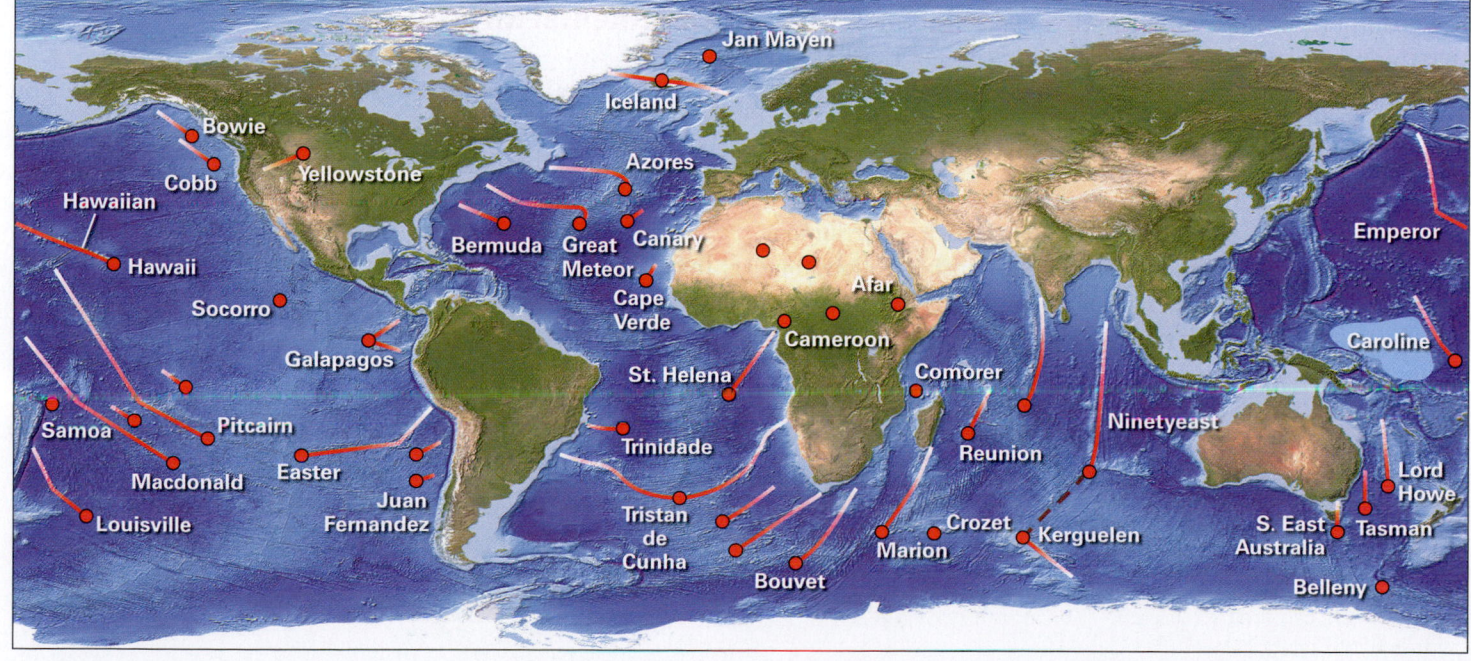

once over the magma source but progressively moved off as the plate moved. Once a volcano was carried away from the magma source, the volcano died, but as long as the plume exists, melt continues to be produced, so a new, younger volcano grows. Eventually, the younger volcano moves off the hot spot and dies, and another still younger one forms. The chain of extinct volcanoes that formed in succession is now known as a **hot-spot track**.

There isn't complete agreement why magma forms at hot spots, and the issue remains a topic of active research. Most geologists favor a model in which magma forms at the top of a **mantle plume**, a column of very hot rock rising up from deep in the mantle, up through the asthenosphere, to the base of the lithosphere (**Fig. 4.17a,b**). Rock in the plume, though solid, is soft enough to flow and rises buoyantly because it is less dense than the somewhat cooler rock of the surrounding mantle. According to the plume model, when the hot rock of the plume reaches the base of the lithosphere, it starts to melt and produces magma that seeps up through the lithosphere (see Chapter 6)—some of this magma reaches the Earth's surface and erupts to produce a hot-spot volcano.

We can apply Wilson's model of hot-spot track formation to the Hawaiian island chain and to its continuation, the Hawaiian-Emperor seamount chain (**Fig. 4.17c**). Volcanic eruptions occur today on the island of Hawaii, at the southeast end of the chain. All of the other islands and seamounts to the northwest are remnants of dead volcanoes that have sunk below sea level (**Fig. 4.17d**), and these get progressively older to the northwest. About 1,750 km northwest of Midway Island, the Hawaiian-Emperor hot-spot track bends abruptly to a more northerly trend. Geologists suggest that this bend reflects a change in the direction of Pacific Plate motion at about 47 Ma. The date comes from measuring the age of rocks obtained from the seamount at the bend.

> **Did you ever wonder...**
>
> why Hawaii rises above the middle of the ocean?

Some hot spots lie within continents. For example, several have been active in the interior of Africa and, as we've noted, one now underlies Yellowstone National Park. The famous geysers (natural steam and hot-water fountains) of Yellowstone exist because hot magma, formed above the Yellowstone hot spot, lies not far below the surface of the park. A few hot spots lie on mid-ocean ridges. Where this happens, a volcanic island may protrude above sea level, because the hot spot produces more magma than does a normal mid-ocean ridge. Iceland, for example, formed where a hot spot underlies the Mid-Atlantic Ridge. The extra volcanism of the hot spot built up the island of Iceland so that it rises almost 3 km above other places on the Mid-Atlantic Ridge.

Take-Home Message

A triple junction marks the point where three plate boundaries join. A hot spot is a point where volcanism occurs independently of plate-boundary movement. Hot spots may be due to melting at the top of a mantle plume. As a plate moves over a plume, a hot-spot track develops.

QUICK QUESTION: Why is the volcano at the end of a hot-spot track the only one to be active?

4.7 How Do Plate Boundaries Form, and How Do They Die?

The configuration of plates and plate boundaries visible on our planet today has not existed for all of geologic history and will not exist indefinitely into the future. Because of plate motion, oceanic plates form and are later consumed, while continents merge into supercontinents, which later split apart. How does a new divergent boundary come into existence, and how does an existing convergent boundary eventually cease to exist? Most new divergent boundaries form when a continent splits and separates into two continents. We call this process *rifting*. A convergent boundary ceases to exist when a piece of relatively buoyant lithosphere, such as a continent or an island arc, moves into the subduction zone and, in effect, jams up the system. We call this process *collision*.

Continental Rifting

A **continental rift** is a linear belt along which continental lithosphere undergoes horizontal stretching in a direction perpendicular to the trend of the rift, and thinning in the vertical direction (**Fig. 4.18a**). The overall process of stretching and thinning is called **rifting**. In the upper 15 km or so of the continent, where the continental crust is relatively cold and brittle, stretching causes rock to break and faults to develop. During this faulting, blocks of the upper crust slip down on fault surfaces, leading to the formation of a low area, a *rift valley*, that gradually becomes buried by sediment. At greater depth, continental rock is warmer so stretching can take place without breaking the rock.

> **Did you ever wonder...**
>
> if continents really split apart, and if so, how?

FIGURE 4.17 The deep mantle plume hypothesis for the formation of hot-spot tracks.

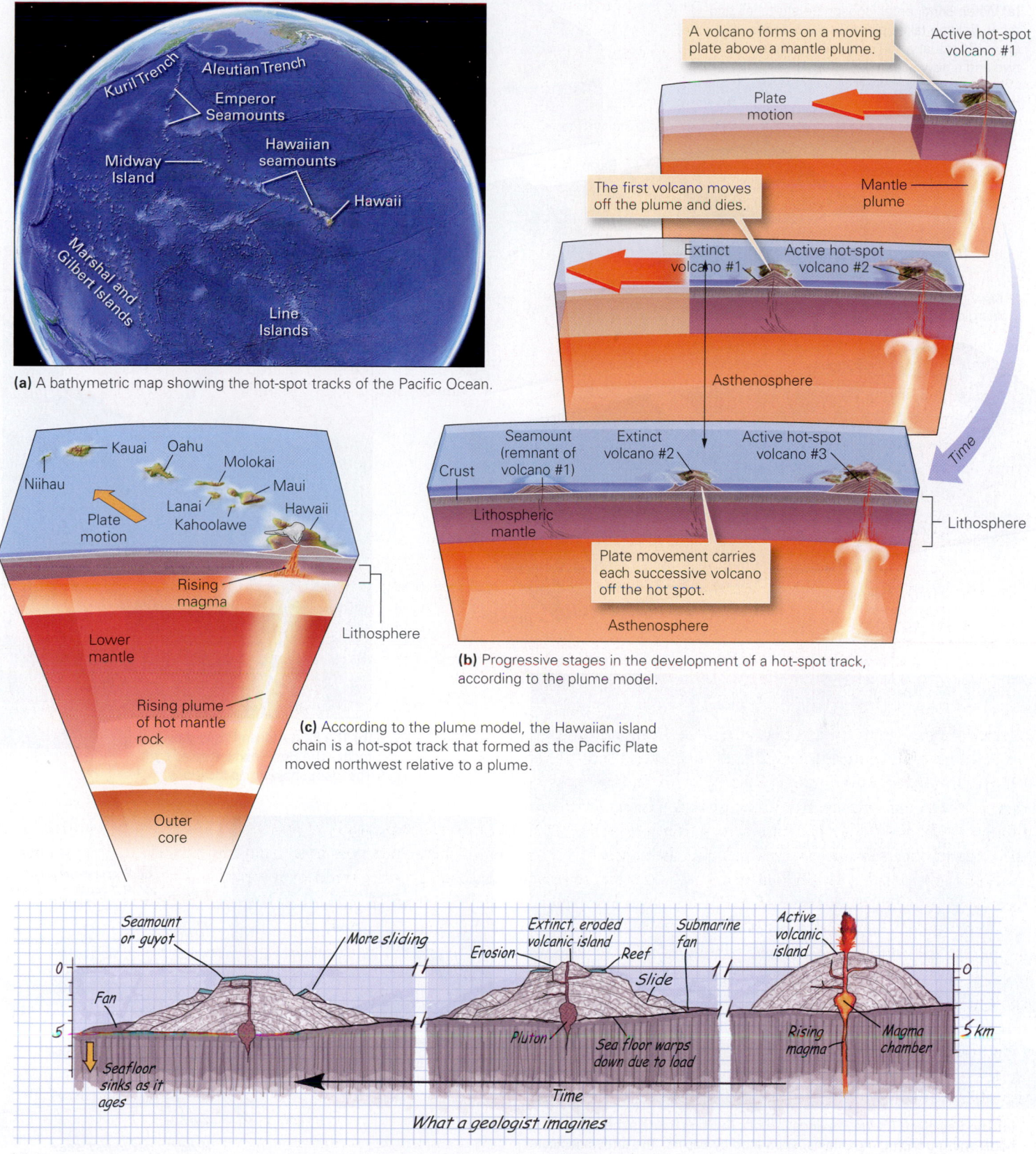

(a) A bathymetric map showing the hot-spot tracks of the Pacific Ocean.

A volcano forms on a moving plate above a mantle plume.

Active hot-spot volcano #1

Plate motion

Mantle plume

The first volcano moves off the plume and dies.

Extinct volcano #1

Active hot-spot volcano #2

Asthenosphere

Seamount (remnant of volcano #1)

Crust

Extinct volcano #2

Active hot-spot volcano #3

Lithospheric mantle

Plate movement carries each successive volcano off the hot spot.

Asthenosphere

Lithosphere

Time

(b) Progressive stages in the development of a hot-spot track, according to the plume model.

Kauai Oahu Molokai Maui Hawaii Lanai Kahoolawe

Niihau

Plate motion

Rising magma

Lower mantle

Rising plume of hot mantle rock

Outer core

Lithosphere

(c) According to the plume model, the Hawaiian island chain is a hot-spot track that formed as the Pacific Plate moved northwest relative to a plume.

Seamount or guyot

More sliding

Extinct, eroded volcanic island

Submarine fan

Active volcanic island

Erosion

Reef

Slide

Fan

Pluton

Sea floor warps down due to load

Rising magma

Magma chamber

Seafloor sinks as it ages

Time

What a geologist imagines

(d) As a volcano moves off the hot spot, it gradually sinks below sea level due to sinking of the plate, erosion, and submarine landslides.

FIGURE 4.18 During the process of rifting, lithosphere stretches.

(a) When continental lithosphere stretches and thins, faulting takes place, and volcanoes erupt. Eventually, the continent splits in two and a new ocean basin forms.

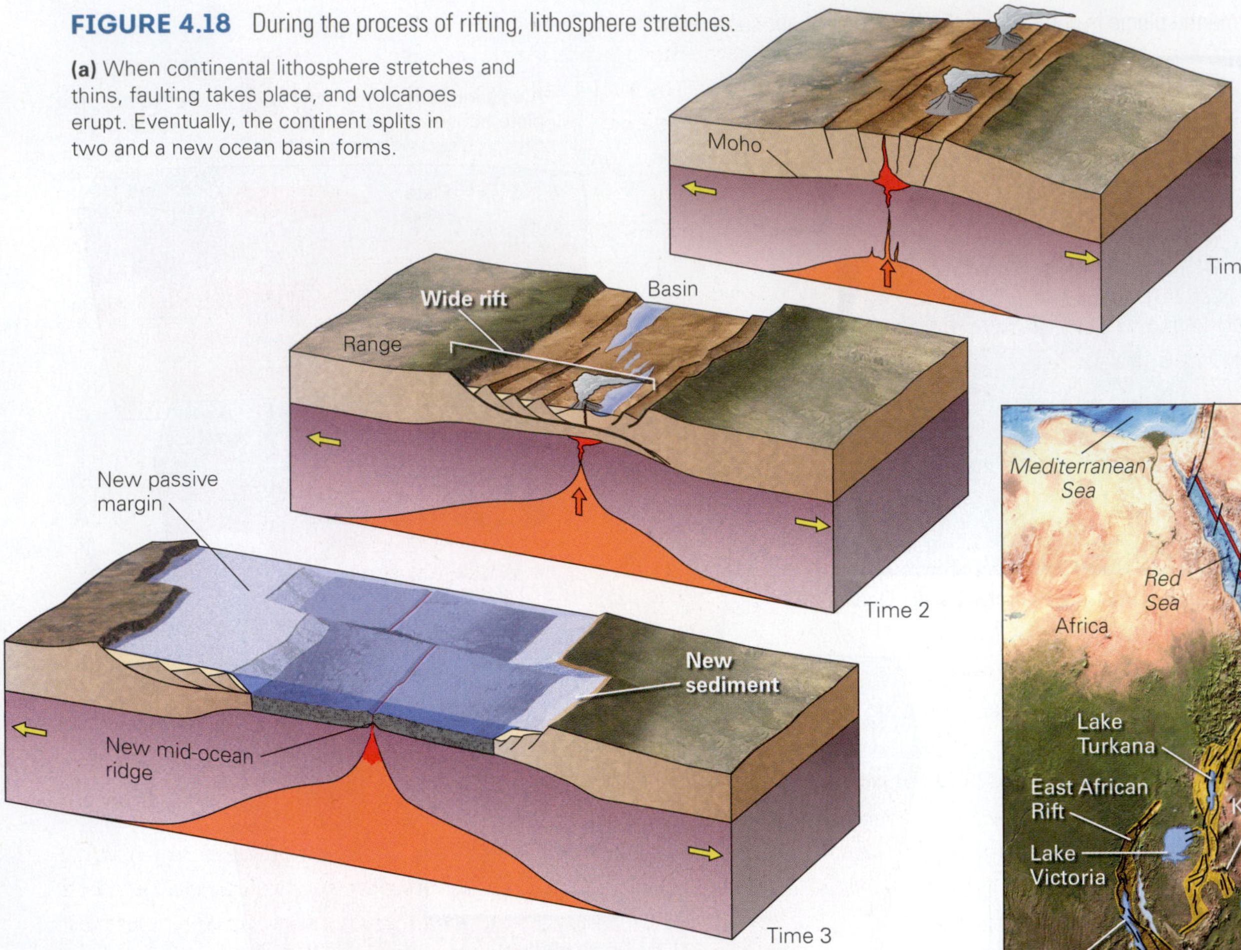

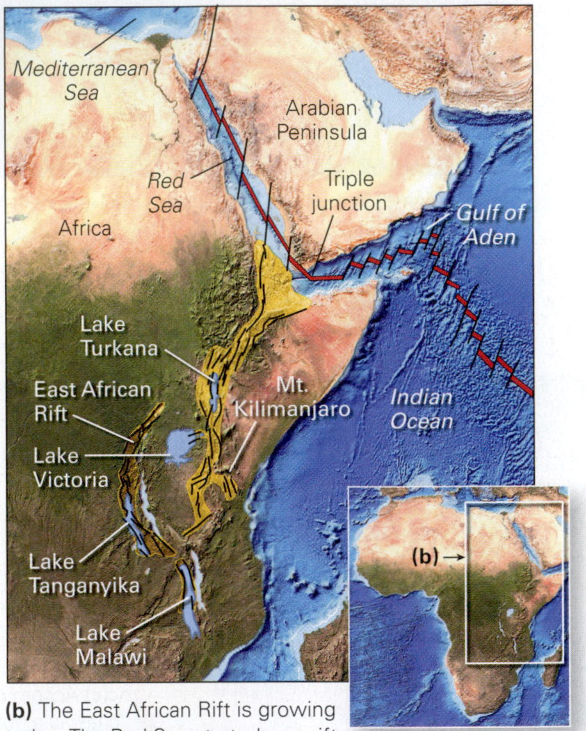

Moho

Time 1

Wide rift

Range

Basin

New passive margin

Time 2

New mid-ocean ridge

New sediment

Time 3

(b) The East African Rift is growing today. The Red Sea started as a rift. The inset shows map locations.

Mediterranean Sea

Arabian Peninsula

Red Sea

Triple junction

Africa

Gulf of Aden

Lake Turkana

East African Rift

Mt. Kilimanjaro

Indian Ocean

Lake Victoria

Lake Tanganyika

Lake Malawi

(b) →

(c) The Basin and Range Province is a rift. Faulting bounds the narrow north-south-trending mountains, separated by basins. The arrows indicate the direction of stretching.

Snake River Plain

Reno

Salt Lake City

Sierra Nevada

Basin and Range

Colorado Plateau

Rio Grande Rift

San Andreas fault

N

250 km

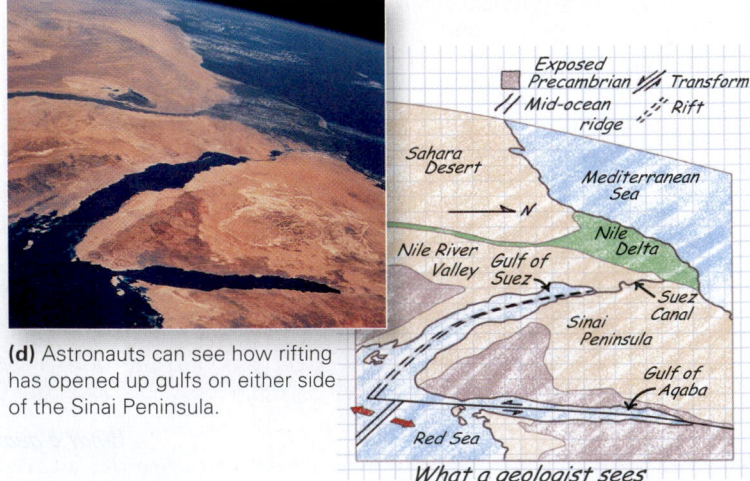

(d) Astronauts can see how rifting has opened up gulfs on either side of the Sinai Peninsula.

Exposed Precambrian

Transform

Mid-ocean ridge

Rift

Sahara Desert

Mediterranean Sea

N

Nile River Valley

Gulf of Suez

Nile Delta

Suez Canal

Sinai Peninsula

Gulf of Aqaba

Red Sea

What a geologist sees

As continental lithosphere thins during rifting, underlying hot asthenosphere rises, and as was the case beneath mid-ocean ridges, the asthenosphere starts to melt producing magma that rises up into the crust. Some of the molten rock reaches the surface and erupts at volcanoes along the rift. If rifting continues for a long enough time, the continent breaks in two, a new mid-ocean ridge forms, and seafloor spreading begins. The relict of the rift, where the continental lithosphere had been stretched and thinned, evolves into a passive margin—the triangular wedges shown in Figure 4.2 are blocks that were displaced by faulting during rifting. In some cases, however, rifting stops before the continent splits in two, and the low-lying rift valley eventually fills with sediment. Then the rift remains as a permanent scar in the crust, defined by a belt of faults, volcanic rocks, and a thick layer of sediment.

Geologists refer to places where rifting is happening today as *active rifts*. Active rifting produced the *Basin and Range Province*, a region in the western United States in which numerous narrow mountain ridges are separated by flat valleys (**Fig. 4.18b**). The mountain ridges are blocks of crust that slipped down faults, whereas the valleys are sediment-filled basins. The Basin and Range is the widest rift on this planet. Another active rift, the East African Rift extends in a north-south direction for over 3,500 km (**Fig. 4.18c**). To astronauts in orbit, the rift looks like a giant gash in the crust. On the ground, it consists of a deep rift valley bordered on both sides by high cliffs formed by faulting. Along the length of the rift, several major volcanoes smoke and fume—these include the snow-crested Mt. Kilimanjaro, towering over 6 km above the savannah. The East African rift links to the Red Sea and the Gulf of Afar, both of which are narrow oceans in which rifting has gone to completion and new mid-ocean ridges have formed. At its north end, the Red Sea rift dies out in the Gulf of Suez (**Fig. 14.8d**).

Collision

After the breakup of Pangaea, India was a small, isolated continent that lay far to the south of Asia. But subduction consumed the ocean between India and Asia, and India moved northward, finally slamming into the southern margin of Asia 40 to 50 Ma. Continental crust is too buoyant to subduct completely—it may seem strange to think of rock as being buoyant, but continental crustal rock has a density of about 2.8 g/cm³, which is much less than that of mantle rock, which has a density of about 3.4 g/cm³. So when India collided with Asia, the attached oceanic plate broke off and sank down into the deep mantle, but the crust of India slid only partly under Asia before it couldn't go

farther, so its continued northward movement pushed it into Asia, squeezing the rocks and sediment that once lay between the two continents into the 8-km-high welt that we now know as the Himalayan Mountains. During this process, not only did the surface of the Earth rise, but the crust became thicker. In fact, the crust beneath the Himalayas has attained a thickness of 60 to 70 km, about twice the thickness of normal continental crust. The boundary between what were once two separate continents is called a **suture**; slivers of ocean crust trapped between colliding continents locally occur along a suture.

Geoscientists refer to the process during which two buoyant pieces of lithosphere converge and squeeze together as collision (**Fig. 4.19**). Not all collisions involve two full-sized continents—some involve island arcs, continental slivers, hot-spot volcanoes, or broad areas of unusually thick ocean crust called oceanic plateaus. Regardless of what collides, when a collision is complete, the convergent-plate boundary that once existed between the two colliding pieces ceases to exist.

Collisions yield some of the highest mountain ranges on the planet, such as the Himalayas and the Alps. They also produced major mountain ranges in the geologic past. These old ranges eventually eroded away so that today we see only their relicts. For example, the Appalachian Mountains in the eastern United States formed as a consequence of three collisions. After the last one, a collision between Africa and North America at around 280 Ma, North America became part of the Pangaea supercontinent. We'll add more detail to this description in Chapter 11.

Take-Home Message

Rifting can split a continent in two and can lead to the formation of a new divergent-plate boundary. When two buoyant crustal blocks, such as continents and island arcs, collide, a mountain belt forms and subduction ceases.

QUICK QUESTION: Do rifting and collision affect crustal thickness? If so, how?

FIGURE 4.19 Continental collision (not to scale).

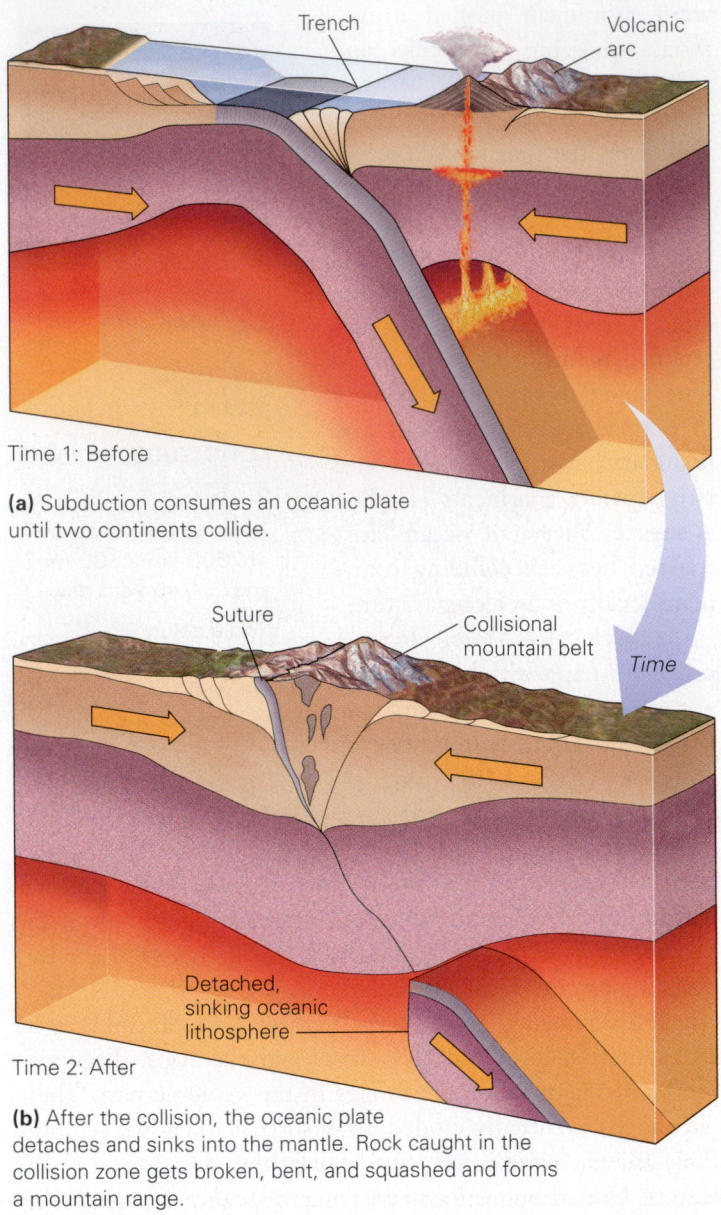

Time 1: Before

(a) Subduction consumes an oceanic plate until two continents collide.

Time 2: After

(b) After the collision, the oceanic plate detaches and sinks into the mantle. Rock caught in the collision zone gets broken, bent, and squashed and forms a mountain range.

4.8 Moving Plates

Forces Acting on Plates

We've now discussed the many facets of plate tectonics theory (**Geology at a Glance**, pp. 108–109). But to complete the story, we need to address the major question Wegener couldn't solve: what drives plate motion? When geoscientists first proposed plate tectonics, they speculated that the process happened simply because convective flow in the asthenosphere actively dragged plates along, as if the plates were rafts on a flowing river. Thus, early images depicting plate motion showed simple *convection cells*—elliptical flow paths—in the asthenosphere. In these images, the cells were positioned such that asthenosphere flowed up beneath mid-ocean ridges and sank down at subduction zones (**Fig. 4.20a**). At first glance, this model looked pretty good, but on closer examination, it failed. Among other reasons, it is impossible to draw a 3-D global configuration of convection cells that can really explain the complex geometry of real plate boundaries on Earth. Gradually, geoscientists came to the conclusion that while convective flow within the asthenosphere does occur, it does not drive motion alone. In other words, hot asthenosphere does rise in some places and sink in others, because of temperature contrasts, and this convection does influence plate motion. But the local directions of convective flow do not necessarily control the local directions of plate motion. Researchers now prefer a model in which convection plays a role in plate motion but is not the whole story—two other forces, ridge push and slab pull, contribute significantly to driving plates. Let's look at all three of these *plate-driving mechanisms* in turn.

Convection Recall that lithosphere forms at a mid-ocean ridge, and then it moves away from the ridge until eventually it sinks back into the mantle at a trench. Since the material forming the plate starts out hot, cools, and then sinks, we can view the plate itself as the top of a convection cell and plate motion as a form of convection. But in this view, convection is effectively a consequence of plate motion, not the cause. Can mantle convection beneath plates actually cause plates to move? The answer may come from studies that demonstrate that the mantle contains zones of **upwelling**, where warmer (buoyant) asthenosphere rises from depth, and zones of **downwelling**, where cooler asthenosphere sinks (**Fig. 4.20b**). As it moves from a zone of upwelling to a zone of downwelling, asthenospheric flow probably does exert a force or "shear" on the base of overlying plates. Conceivably, asthenosphere flow may either speed up or slow down plates depending on the orientation of the flow direction relative to the movement direction of the overlying plate. New research suggests that asthenosphere flow pushes on downgoing plates and affects the angle at which plates sink.

Ridge-push force Ridge-push force is an outward-directed force that contributes to moving plates away from a mid-ocean ridge axis. It develops because the lithosphere of mid-ocean ridges lies at a higher elevation than that of the adjacent abyssal plains (**Fig. 4.20c**). To understand ridge-push force, imagine you have a glass containing a layer of water over

FIGURE 4.20 Plate-driving forces.

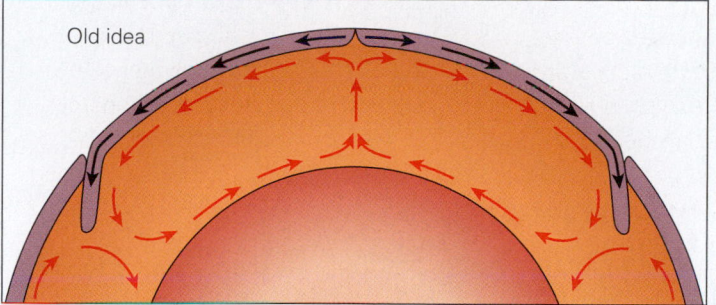

Old idea

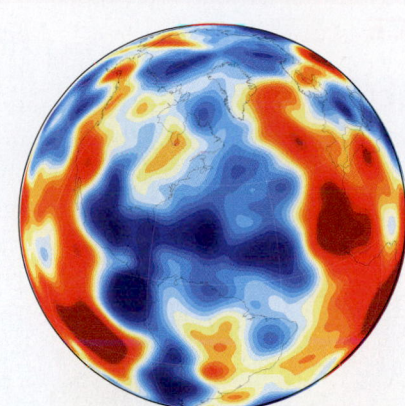

(a) The old, incorrect, image of simple convection cells.

(b) The colors represent the difference between the observed velocity of seismic waves and the expected velocity, at a depth of 2,800 km. Geologists interpret red areas to be warmer, upwelling regions and blue areas to be cooler, downwelling regions. Thus, this map gives an image of convection.

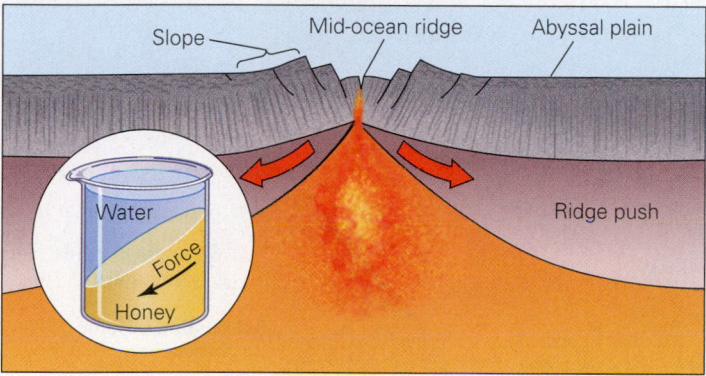

Slope   Mid-ocean ridge   Abyssal plain

Water

Force

Honey

Ridge push

(c) Ridge push develops because the region of a rift is elevated. Like a wedge of honey with a sloping surface, the mass of the ridge pushes sideways.

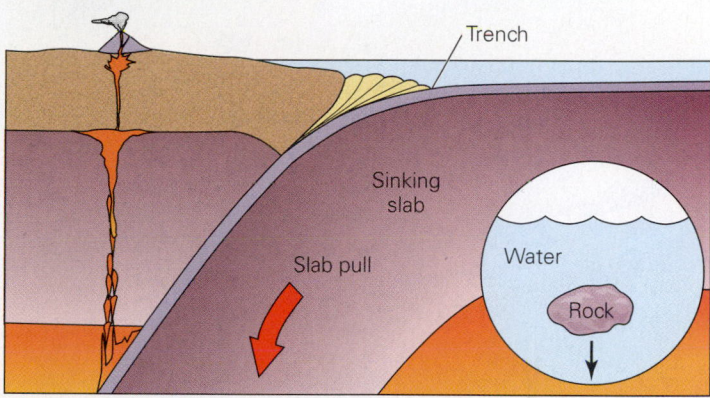

Trench

Sinking slab

Slab pull

Water

Rock

(d) Slab pull develops because lithosphere is denser than the underlying asthenosphere and sinks like a stone in water (though much more slowly).

a layer of honey. By tilting the glass momentarily and then returning it to its upright position, you can create a temporary slope in the boundary between these substances. While the boundary has this slope, gravity causes the elevated honey to push against the glass adjacent to the side where the honey surface lies at lower elevation. The geometry of a mid-ocean ridge resembles this situation, for the surface of the seafloor is higher along a mid-ocean ridge axis than in adjacent abyssal plains. Gravity causes the elevated lithosphere at the ridge axis to push on the lithosphere that lies farther from the axis, making it move away. As lithosphere moves away from the ridge axis, new hot asthenosphere continuously rises, providing material from which new crust and new lithospheric mantle forms. Note that the local upward movement of asthenosphere beneath a mid-ocean ridge is a *consequence* of seafloor spreading, not the cause.

Slab-pull force Slab-pull force arises simply because lithosphere that was formed more than 10 Ma is denser than asthenosphere, so it can sink into the asthenosphere (**Fig. 4.20d**). Thus, once an oceanic plate starts to sink, it gradually

pulls the rest of the plate along behind it, like an anchor pulling down the anchor line.

The Velocity of Plate Motions

How fast do plates move? It depends on your frame of reference. The movement of one plate with respect to another is called **relative plate velocity**, whereas the movement of a plate with respect to a fixed location inside the Earth is called **absolute plate velocity** (**Fig. 4.21**). To illustrate this concept, imagine two cars (A and B) speeding in the same direction down the highway. From the viewpoint of a tree along the side of the road, A zips by at 100 km an hour, while B moves at 80 km an hour. But relative to B, A moves at only 20 km an hour. The motion of a car relative to the tree is it's absolute motion, where as the motion of one car with respect to another is it's relative motion.

One way to determine relative plate motions comes from the study of marine magnetic anomalies. If you measure the distance between a mid-ocean ridge axis and a magnetic anomaly of known age on seafloor formed at the ridge, then

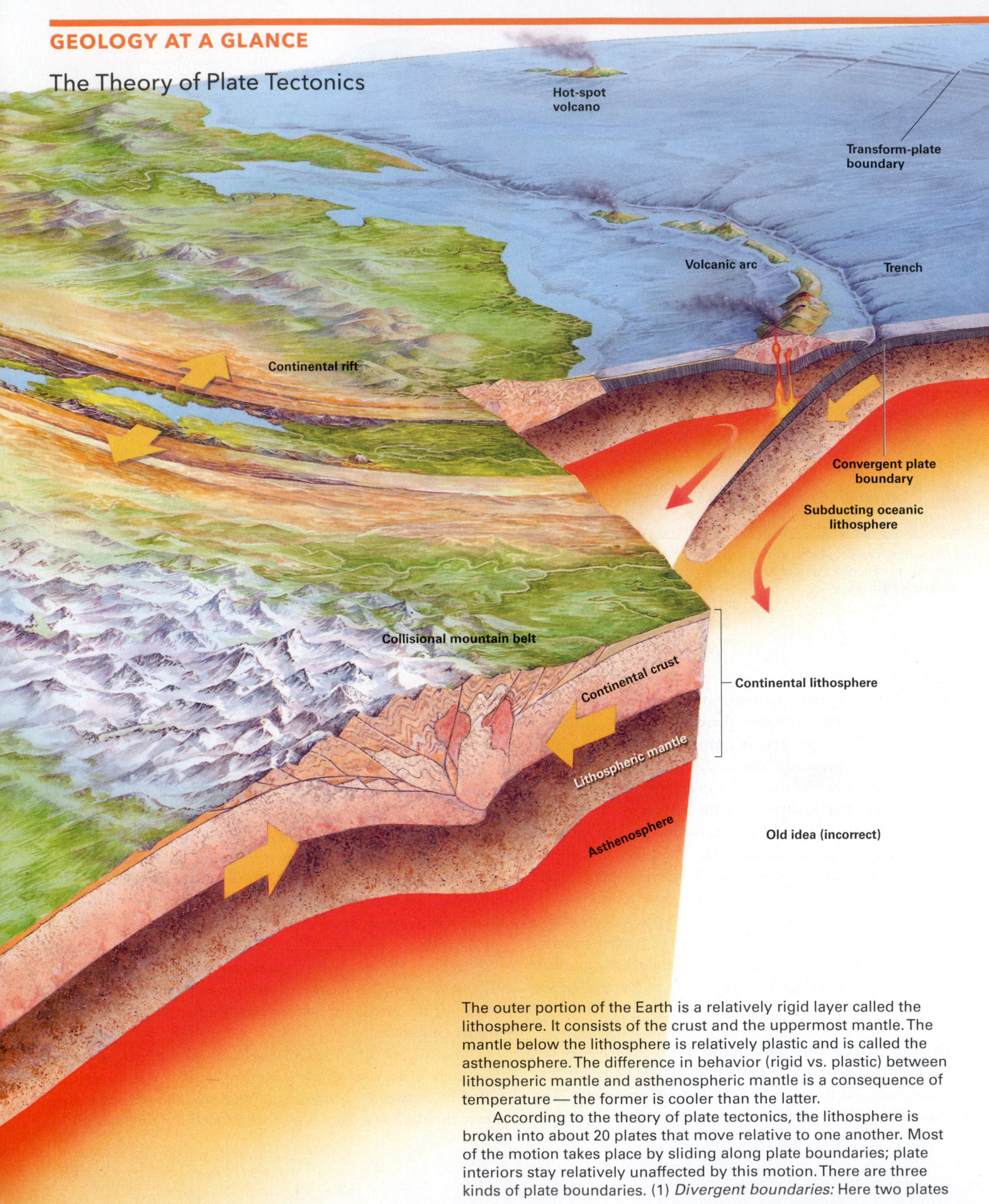

Hot-spot volcano

Transform-plate boundary

Volcanic arc

Trench

Continental rift

Convergent plate boundary

Subducting oceanic lithosphere

Collisional mountain belt

Continental crust

Continental lithosphere

Lithospheric mantle

Asthenosphere

Old idea (incorrect)

The outer portion of the Earth is a relatively rigid layer called the lithosphere. It consists of the crust and the uppermost mantle. The mantle below the lithosphere is relatively plastic and is called the asthenosphere. The difference in behavior (rigid vs. plastic) between lithospheric mantle and asthenospheric mantle is a consequence of temperature — the former is cooler than the latter.

According to the theory of plate tectonics, the lithosphere is broken into about 20 plates that move relative to one another. Most of the motion takes place by sliding along plate boundaries; plate interiors stay relatively unaffected by this motion. There are three kinds of plate boundaries. (1) *Divergent boundaries:* Here two plates

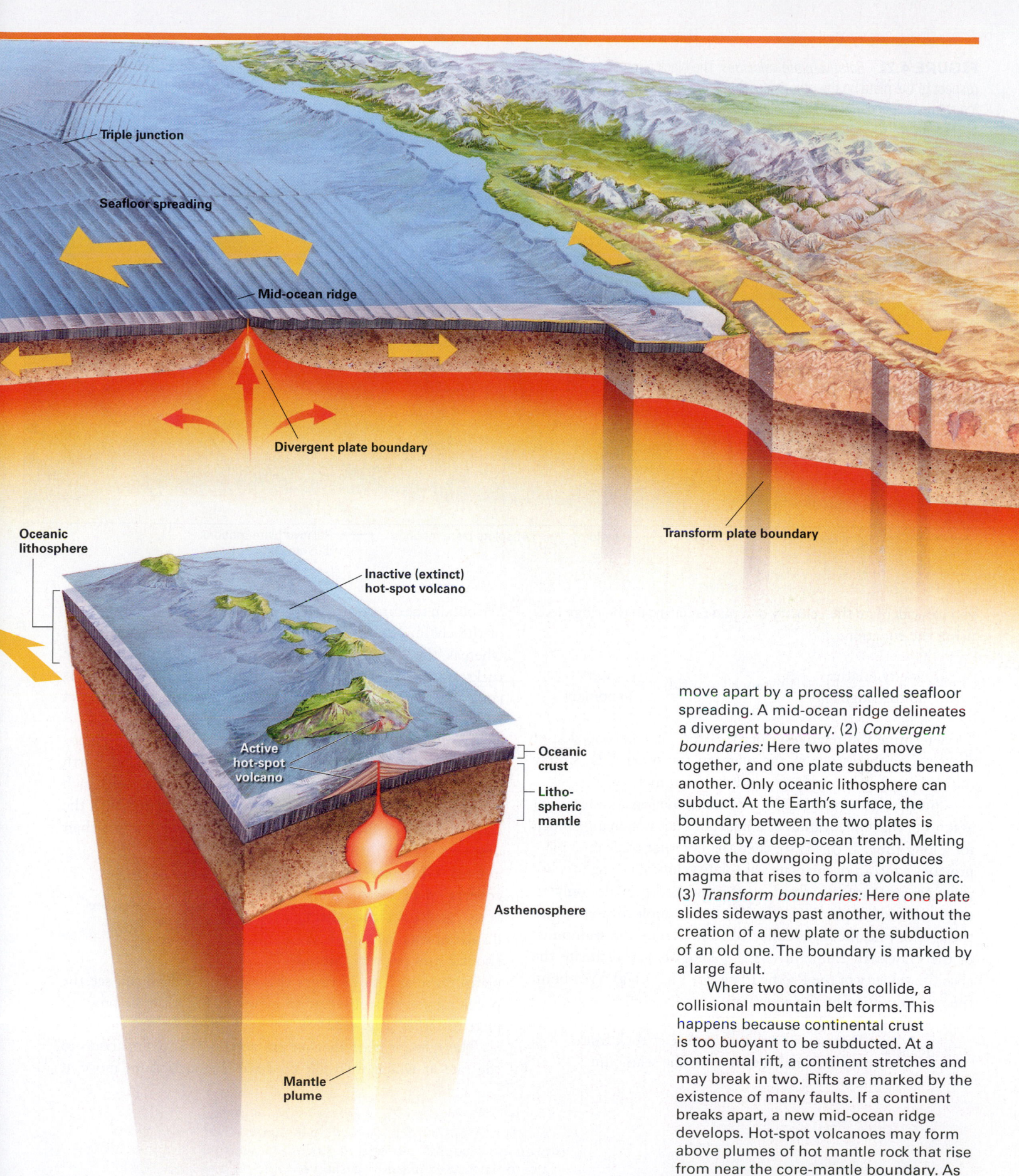

Triple junction

Seafloor spreading

Mid-ocean ridge

Divergent plate boundary

Oceanic lithosphere

Inactive (extinct) hot-spot volcano

Transform plate boundary

Active hot-spot volcano

Oceanic crust

Litho-spheric mantle

Asthenosphere

Mantle plume

move apart by a process called seafloor spreading. A mid-ocean ridge delineates a divergent boundary. (2) *Convergent boundaries:* Here two plates move together, and one plate subducts beneath another. Only oceanic lithosphere can subduct. At the Earth's surface, the boundary between the two plates is marked by a deep-ocean trench. Melting above the downgoing plate produces magma that rises to form a volcanic arc. (3) *Transform boundaries:* Here one plate slides sideways past another, without the creation of a new plate or the subduction of an old one. The boundary is marked by a large fault.

Where two continents collide, a collisional mountain belt forms. This happens because continental crust is too buoyant to be subducted. At a continental rift, a continent stretches and may break in two. Rifts are marked by the existence of many faults. If a continent breaks apart, a new mid-ocean ridge develops. Hot-spot volcanoes may form above plumes of hot mantle rock that rise from near the core-mantle boundary. As a plate drifts over a hot spot, it leaves a chain of extinct volcanoes.

FIGURE 4.21 *Relative plate velocities:* The black arrows show the rate and direction at which the plate on one side of the boundary is moving with respect to the plate on the other side. Outward-pointing arrows indicate spreading (divergent boundaries), inward-pointing arrows indicate subduction (convergent boundaries), and parallel arrows show transform motion. The length of an arrow represents the velocity. *Absolute plate velocities:* The red arrows show the velocity of the plates with respect to a fixed point in the mantle.

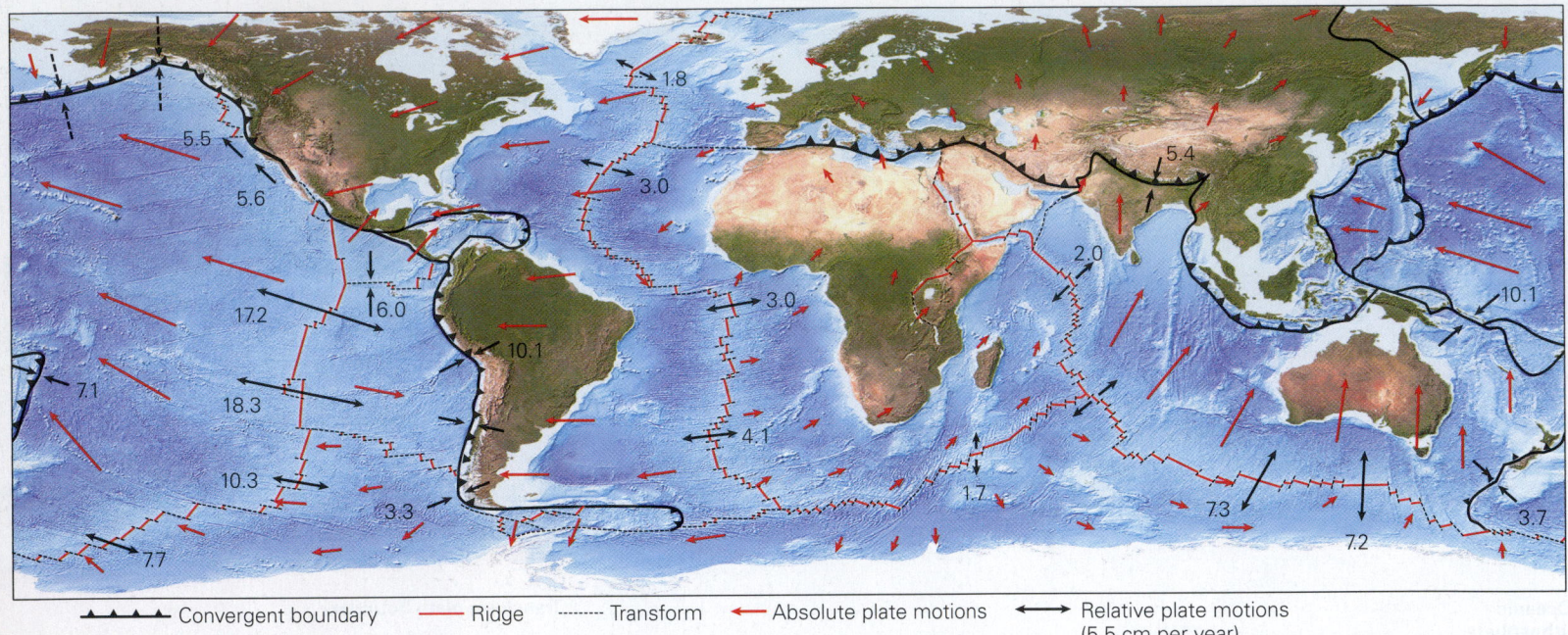

▲▲▲▲ Convergent boundary   ── Ridge   -------- Transform   ← Absolute plate motions   ←→ Relative plate motions (5.5 cm per year)

you can calculate the velocity of a plate relative to the ridge axis using the equation:

$$\text{Velocity relative to the ridge} = \text{Distance from ridge} \div \text{Age of anomaly}$$

To determine the relative velocity of the plate on one side of the ridge with respect to the plate on the other side, simply multiply by 2, for spreading at ridges is symmetrical.

One way to estimate absolute plate motions comes from assuming that the location of a hot spot does not change much for a long time. If this is so, then the hot-spot track on a plate provides a record of the plate's absolute velocity. (In reality, hot spots are not perfectly fixed, so this method provides only an approximation of absolute velocity.) For example, by measuring the ages of volcanic rocks collected from islands and seamounts of the Hawaiian-Emperor chain, geologists can estimate the absolute velocity of the Pacific Plate (see Fig. 4.17a). We obtain the rate of motion by the equation:

$$\text{Rate} = \frac{\text{Distance of a sea-}}{\text{mount from Hawaii}} \div \frac{\text{Age of rock from}}{\text{the seamount}}$$

We obtain the direction of motion by measuring the orientation of the chain. Note that the Hawaiian chain runs northwest, whereas the Emperor chain trends north-northwest. Since rocks from the seamount at the bend are 47 Ma, geologists conclude that the direction of Pacific Plate motion changed at 47 Ma.

Working from the calculations described above, geologists have determined that relative plate motions on Earth today occur at rates of 1 to 15 cm per year. Can we detect such slow rates? Yes, by using the *global positioning system* (*GPS*), the same technology that automobile drivers can use to find their destinations. By setting up a fixed GPS receiver that collects data over many years, geologists can detect displacements as small as about 2 mm per year. Since plates move at 5 to 75 times this rate, we indeed can see the plates move—this observation serves as the ultimate proof of plate tectonics (**Fig. 4.22**).

The rates of plate motion are very slow in comparison to the rate at which we walk or drive. In fact, plates move at

> **Did you ever wonder . . .**
>
> whether we can really "see" continents drift?

about the rate that your fingernails grow. But even at these slow rates, plate motions can yield large displacements given the immensity of geologic time. For example, at a rate of 2.5 cm per year, a plate can move 25 km in a million years, and 2,500 km in 100 million years! At this rate, the Atlantic Ocean opened to its present width, 4,500 km, in the 180 million years since the breakup of Pangaea. Taking into account many data sources that define the motion of plates, geologists have greatly refined the image of continental drift that Wegener tried so hard to prove nearly a century ago. We can now see how the map of our planet's surface has evolved radically during the past 400 million years and even before (**Fig. 4.23**).

FIGURE 4.22 GPS is used to measure plate motions at many locations on Earth.

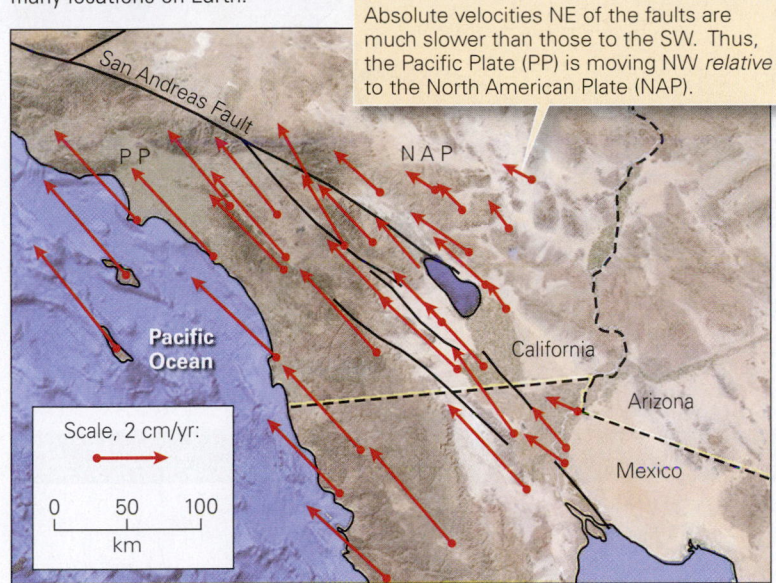

Absolute velocities NE of the faults are much slower than those to the SW. Thus, the Pacific Plate (PP) is moving NW *relative* to the North American Plate (NAP).

(a) GPS measurements in southern California show the region west of the San Andreas fault system, a plate boundary, is moving northwest up to 6 cm per year. The length of the arrows indicates the magnitude of velocity.

Take-Home Message

Plate motion takes place because plates are acted on by ridge push, slab pull, and convective shear. This motion takes place at rates of 1 to 15 cm per year. Relative motion specifies the rate that a plate moves relative to its neighbor, whereas absolute motion specifies the rate that a plate moves relative to a fixed point beneath the plate. GPS measurements can now detect relative plate motions directly.

QUICK QUESTION: What causes the map-view bend in the Hawaiian-Emperor seamount chain?

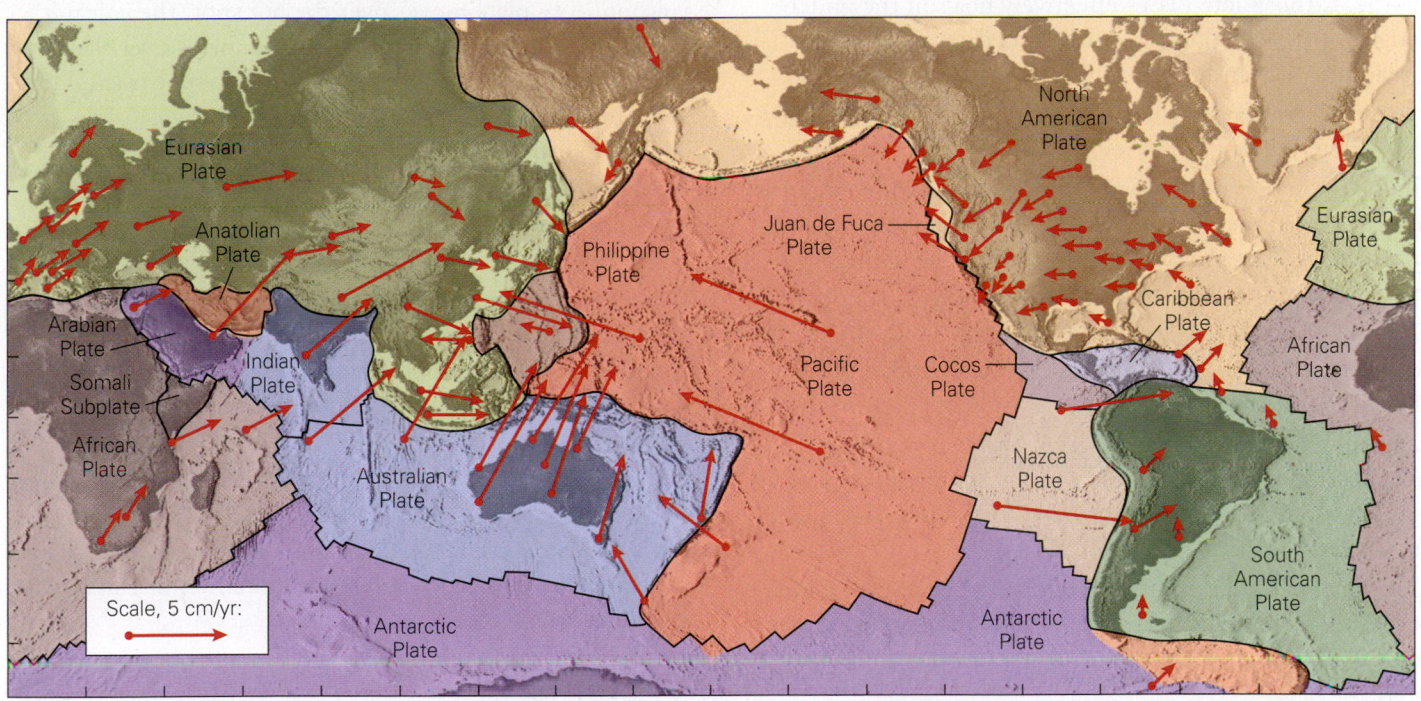

(b) A global map of plate velocities, determined by GPS.

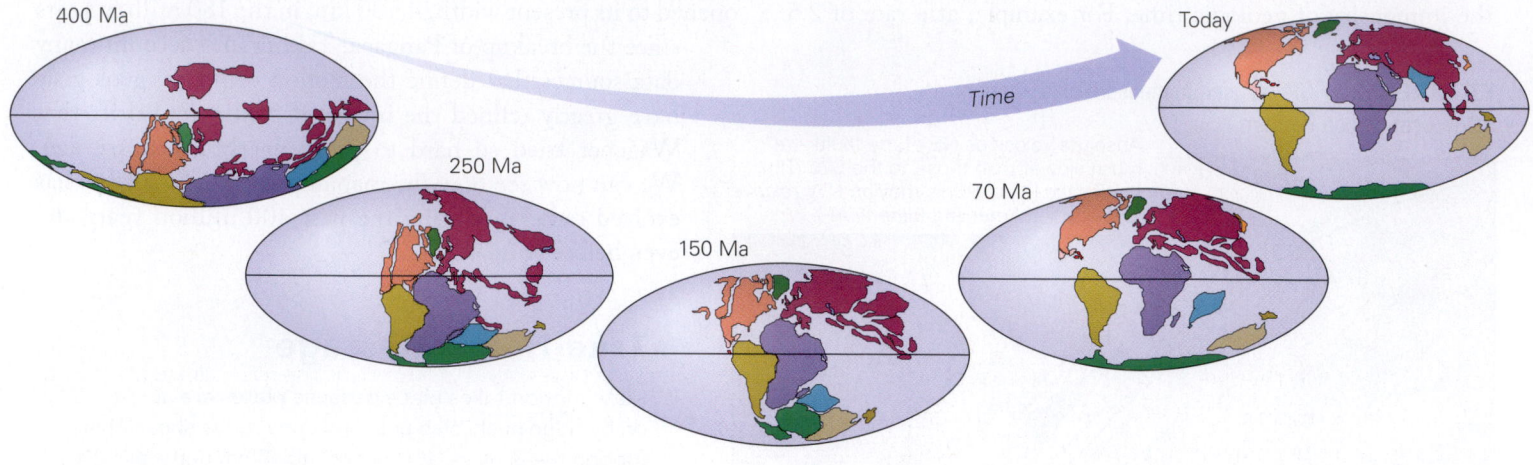

FIGURE 4.23 Due to plate tectonics, the map of Earth's surface slowly changes. Here we see the assembly, and later the breakup, of Pangaea during the past 400 million years.

400 Ma

250 Ma

150 Ma

70 Ma

Today

Time

CHAPTER SUMMARY

- The lithosphere, the rigid outer layer of the Earth, is broken into discrete plates that move relative to one another. Plates consist of the crust and the uppermost (cooler) mantle. Lithosphere plates effectively float on the underlying soft asthenosphere. Continental drift and seafloor spreading are manifestations of plate movement.

- Most earthquakes and volcanoes occur along plate boundaries; the interiors of plates remain relatively rigid and intact.

- There are three types of plate boundaries—divergent, convergent, and transform—distinguished from one another by the movement the plate on one side of the boundary makes relative to the plate on the other side.

- Divergent boundaries are marked by mid-ocean ridges. At divergent boundaries, seafloor spreading takes place, a process that forms new oceanic lithosphere.

- Convergent boundaries are marked by deep-ocean trenches and volcanic arcs. At convergent boundaries, oceanic lithosphere of the downgoing plate subducts beneath an overriding plate.

- Transform boundaries are marked by large faults at which one plate slides sideways past another. No new plate forms and no old plate is consumed at a transform boundary.

- Triple junctions are points where three plate boundaries intersect.

- Hot spots are places where volcanism occurs at an isolated volcano. As a plate moves over the hot spot, the volcano moves off and dies, and a new volcano forms over the hot spot; the chain of volcanoes defines a hot-spot track. Hot spots may be caused by mantle plumes.

- A large continent can split into two smaller ones by the process of rifting. During rifting, continental lithosphere stretches and thins. If it finally breaks apart, a new mid-ocean ridge forms and seafloor spreading begins.

- Convergent-plate boundaries cease to exist when a buoyant piece of crust (a continent or an island arc) moves into the subduction zone. When that happens, collision occurs. Collision can thicken crust and build large mountain belts.

- Ridge-push force and slab-pull force contribute to driving plate motions. Plates move at rates of about 1 to 15 cm per year. We can describe plate motions relative to each other or relative to a fixed point. Modern satellite measurements can detect these motions.

GUIDE TERMS

absolute plate velocity (p. 107)

accretionary prism (p. 97)

active margin (p. 89)

asthenosphere (p. 88)

black smoker (p. 93)

continental rift (p. 102)

convergent (consuming) boundary (p. 96)

divergent boundary (p. 92) lithospheric mantle (p. 88) ridge-push force (p. 106) transform boundary (p. 99)
downwelling (p. 106) mantle plume (p. 102) rifting (p. 102) trench (p. 96)
fracture zone (p. 98) passive margin (p. 89) scientific revolution (p. 87) triple junction (p. 100)
hot spot (p. 101) plate boundary (p. 89) seismic belt (p. 90) upwelling (p. 106)
hot-spot track (p. 102) plate (p. 89) slab-pull force (p. 107) volcanic arc (p. 97)
lithosphere (p. 87) plate tectonics (p. 87) subduction (p. 96) volcanic island arc (p. 97–98)
lithosphere plates (p. 89) relative plate velocity (p. 107) suture (p. 105) Wadati-Benioff zone (p. 96)

REVIEW QUESTIONS

1. What are the characteristics of a lithosphere plate?
2. How does oceanic lithosphere differ from continental lithosphere in thickness, composition, and density?
3. What are the basic premises of plate tectonics?
4. How do we identify a plate boundary?
5. Describe the three types of plate boundaries.
6. How does crust form along a mid-ocean ridge?
7. Why is subduction necessary on a nonexpanding Earth with spreading ridges?
8. Describe the major features of a convergent boundary.
9. Describe the motion that takes place on a transform boundary.
10. What is a triple junction?
11. How is a hot-spot track produced, and how can hot-spot tracks be used to track the past motions of a plate?
12. Describe the characteristics of a continental rift, and give examples of where this process is occurring today.
13. Describe the process of continental collision, and give examples of where this process has occurred.
14. Discuss the major forces that move lithosphere plates.
15. Explain the difference between relative plate velocity and absolute plate velocity.

ON FURTHER THOUGHT

16. Why are the marine magnetic anomalies bordering the East Pacific Rise in the southeastern Pacific Ocean wider than those bordering the Mid-Atlantic Ridge in the South Atlantic Ocean?
17. The Pacific Plate moves north relative to the North American Plate at a rate of 6 cm per year. How long will it take Los Angeles (a city on the Pacific Plate) to move northward by 480 km, the present distance between Los Angeles and San Francisco?
18. Look at a map of the western Pacific Ocean, and examine the position of Japan with respect to mainland Asia. Japan's older crust contains rocks similar to those of eastern Asia. Presently, there are many active volcanoes along the length of Japan. With these facts in mind, explain how the Japan Sea (the region between Japan and the mainland) formed.

smartwork smartwork.wwnorton.com

This chapter's Smartwork features:

- Video exercises on the life of a hotspot.
- What A Geologist Sees exercise on Mt. Saint Helens.
- Art questions testing knowledge of types of plate boundaries.

GEOTOURS

This chapter's GeoTour exercise (B) features:

- Divergent boundaries
- Convergent boundaries
- Transform boundaries
- Hot spots

PART II

EARTH MATERIALS

Imagine that you are standing in the prairie of the midwestern United States. The grass beneath your feet roots in a rich black soil, which grades downward into sediment (clay, sand, and gravel). The sediment covers countless layers of sedimentary rock (sandstone, shale, coal, and limestone), forming a veneer that in turn covers a "basement" of granite and other very ancient rocks. Descending still deeper takes you through 40 km of continental crust down to Moho. Below the Moho? Even more rock—almost 3,000 km of rock—down to the molten iron alloy of the outer core. How did these materials form? How do we describe and classify this material? Are Earth materials permanent, or can they change over time?

In this part of the book, we learn about the great variety of materials that make up the crust and mantle of the solid Earth. We begin, in Chapter 5, with minerals, the building blocks of rock. Then, in Interlude A, we see how geologists distinguish among three categories of rock–igneous, sedimentary, and metamorphic—based on how they form. In each succeeding chapter (6, 7, and 8), we examine one rock category in greater detail. Interlude B helps explain the origin of the sediments that develop into sedimentary rocks and also introduces soil. Finally, Interlude C shows us how materials in the dynamic Earth System can change over time as they pass through the rock cycle.

5 Patterns in Nature: Minerals

A Introducing Rocks

6 Up from the Inferno: Magma and Igneous Rocks

B A Surface Veneer: Sediments and Soils

7 Pages of Earth's Past: Sedimentary Rocks

8 Metamorphism: A Process of Change

C The Rock Cycle in the Earth System

A cluster of amethyst crystals . . . Nature's abstract art. Amethyst is one of about 4,000 known minerals, the building blocks of the Earth.

CHAPTER 5

Patterns in Nature: Minerals

> I died a mineral, and became a plant. I died as plant and rose to animal,
> I died as animal and I was Man. Why should I fear?
>
> —Jalal-Uddin Rumi (Persian mystic and poet, 1207–1273)

5.1 Introduction

In Greek legend, the god Dionysus, in a drunken rage, vowed to kill the next mortal that he saw. Just then, a beautiful young woman named Amethyst walked by, and Dionysus ordered two fearsome tigers to attack her. The goddess Artemis prevented a tragedy by changing Amethyst into a pure white statue made of quartz that was much harder than the tigers' teeth. The statue was so beautiful that Dionysus sobered up and regretted his rashness. In remorse, he spilled his wine onto the statue as an offering, and the wine stained the quartz, turning it into purple amethyst (see chapter opening photo). The word comes from the Greek *amethustos*, meaning not intoxicated. For centuries afterward, amethyst was thought of as an antidote for drunkenness. In fact, revelers would wear amulets of amethyst, or drink from goblets made of it, or put fragments of it into their drinks, thinking that it somehow neutralized the alcohol in their wine. It doesn't—amethyst has no effect on alcohol and can't prevent inebriation. But legend aside, amethyst is beautiful, one of many minerals that have been used in jewelry making for millennia (**Fig. 5.1**).

Amethyst, the maroon version of a common mineral, quartz, is one of about 4,000 minerals that **mineralogists**, people who specialize in the study of minerals, have identified so far. The list of known minerals continues to grow, for mineralogists discover about 50 new ones every year. Each mineral has a name. Some names come from Latin, Greek, German, or English words describing a certain characteristic of the mineral (for instance, albite comes from the Latin word for white, orthoclase comes from the German words meaning splits at right angles, and *olivine* is olive colored); some honor a person (sillimanite was named for Benjamin Silliman, a 19th-century mineralogist); some indicate the place where the mineral was first recognized (illite came from rocks in Illinois); and some refer to a particular element in the mineral (chromite contains chromium). The vast majority of mineral types are rare, forming only under special conditions. Fewer than 50 are considered common rock-forming minerals of the Earth's crust; in fact, most rocks that you pick up consist almost entirely of only 1 to 6 of these common minerals.

Though ancient Greek philosophers pondered minerals, and medieval alchemists puttered with them, true scientific study of minerals did not begin until 1556, when Georgius Agricola, a German physician, published *De re metallica* (On the nature of metals), in which he gave basic descriptions of minerals. In 1669, more than a century after Agricola's work, Nicholas Steno, a Danish scientist, discovered important geometric characteristics of minerals. Steno's work became the basis for the systematic description of minerals, a task that occupied many researchers during the next two centuries.

FIGURE 5.1 A royal crown containing a variety of valuable jewels.

The study of minerals with an optical microscope began in 1828, but though such studies helped in mineral identification, they could not reveal the arrangement of atoms inside minerals. That understanding came in the early 20th century, when researchers developed methods for using X-rays to study mineral structure. By the middle of the century, new tools, such as electron microscopes and microprobes, made it possible to see details of mineral structure and to analyze grain composition in a matter of seconds.

Why study minerals? Without exaggeration, we can say that minerals are the building blocks of our planet, for minerals make up most rocks and sediments of the solid Earth. Minerals are also important from a practical standpoint (see Chapter 15). For example, *industrial minerals* serve as the raw materials for manufacturing chemicals, concrete, and wallboard; *ore minerals* provide valuable metals, like copper and gold (**Fig. 5.2**), and provide energy resources, such as uranium; and *gems*—beautiful forms of certain minerals—delight the eye as jewelry. In recent decades, it's also become clear that certain minerals pose health and environmental hazards. No wonder **mineralogy**, the study of minerals, fascinates professionals and amateurs alike.

This chapter, an introduction to mineralogy, begins with the geologic definition of a mineral. We then look at how minerals form and at the main characteristics that enable us to identify minerals. Finally, we describe the basic scheme that mineralogists use to classify minerals. This chapter assumes that you understand fundamental concepts of matter and energy, especially the nature of atoms, molecules, and chemical bonds. If you are rusty on these topics, please study **Box 5.1** before going further.

5.2 What Is a Mineral?

In everyday English, the word *mineral* has many uses. When you play the game 20 Questions, a mineral is anything that's not animal or vegetable, and if you read food ingredients, the word refers to certain nutrients that people need in order to be healthy. Geologists use the term in a very specific way. To a geologist, a **mineral** is a naturally occurring solid, formed by geologic processes, that has a crystalline structure and a definable chemical composition. Almost all minerals are "inorganic." Let's pull apart this mouthful of a definition and examine its meaning in detail.

- *Naturally occurring:* True minerals are formed in nature, not in factories. In recent decades, chemists have learned how to manufacture materials that have characteristics virtually identical to those of real minerals. Such materials can be referred to as "synthetic minerals."
- *Formed by geologic processes:* Traditionally, this phrase implied that minerals were the result only of solidification of molten rock or direct precipitation from a water solution, processes that did not involve living organisms. Increasingly, however, geologists recognize that life is an integral part of the Earth System. So geologists now consider solid, crystalline materials produced by organisms to be minerals, too. To avoid confusion, the term **biogenic mineral** may be used when discussing such materials.
- *Solid:* A solid is a state of matter that can maintain its shape indefinitely, and thus will not conform to the shape

FIGURE 5.2 Copper ore is a useful mineral that serves as a source of copper metal.

Malachite grows by precipitation, in a succession of layers.

(a) Malachite is a type of copper ore ($Cu_2[CO_3][OH]_2$); it contains copper plus other chemicals.

(b) The copper for pots is produced by processing ore minerals.

of its container. Liquids (such as oil or water) and gases (such as air) are not minerals.

- *Crystalline structure:* The atoms that make up a mineral are fixed in a specific, orderly pattern. A material in which atoms are fixed in an orderly pattern is called a **crystalline solid** (**Fig. 5.3a**). Mineralogists refer to the pattern itself (the imaginary framework representing the arrangement of atoms) as a **crystal lattice**.

Did you ever wonder . . .

how a "cut crystal" glass differs from a clear quartz crystal?

- *Definable chemical composition:* This phrase simply means that it is possible to write a chemical formula for a mineral. Some minerals contain only one element, but most are compounds of two or more elements. For example, diamond and graphite have the formula C because they consist entirely of carbon. Quartz has the formula SiO_2— it contains the elements silicon and oxygen in the proportion of one silicon atom for every two oxygen atoms. Some mineral formulas are more complicated: for example, the formula for biotite is $K(Mg,Fe)_3(AlSi_3O_{10})(OH)_2$. According to this formula, the proportion of magnesium to iron can vary in biotite.

- *Inorganic:* To understand the meaning of *inorganic*, we must first understand what's meant by *organic*. Organic chemicals consist of molecules that include carbon-carbon and/or carbon-hydrogen bonds, and either form in living organisms or have structures similar to those that formed in living organisms. Sugar ($C_{12}H_{22}O_{11}$), fat, plastic, propane, and protein, for example, are organic chemicals. Some organic chemicals contain only carbon and hydrogen, while others include other elements, such as oxygen, nitrogen, and/or phosphorous. Almost all minerals are inorganic, in that they are not organic chemicals. But we have to add the qualifier "almost all" because mineralogists now consider a few dozen organic substances formed by "the action of geologic processes on organic materials" to be minerals. Examples include the crystals that grow in ancient deposits of bat guano.

With the geologic definition of a mineral in mind, we can distinguish between a mineral and a **glass**. Both minerals and glasses are solids, in that they can retain their shape indefinitely, but a mineral is crystalline, while glass is not. This means that the atoms, ions, or molecules in a mineral are ordered into a crystal lattice, like soldiers standing in formation, but those in a glass are arranged in a semi-chaotic way, in small clusters or chains that are neither oriented in the same way nor spaced at regular intervals, like guests at a party (**Fig. 5.3b**).

FIGURE 5.3 The nature of crystalline and noncrystalline materials.

A crystal face

Quartz crystal

A crystal lattice resembles intersection points in scaffolding.

Washington Monument

(a) Crystalline substances, like quartz, contain an orderly arrangement of atoms. The geometry of the arrangement defines the crystal lattice.

(b) A "cut crystal" bowl is actually made of glass. The atoms within do not have an orderly arrangement.

BOX 5.1 SCIENCE TOOLBOX . . .

Some Basic Concepts from Chemistry—A Quick Review

To describe minerals, we will be using the basic vocabulary of chemistry. We introduce this vocabulary below in an order that permits each successive term to build on previous terms.

- *Element:* A pure substance that cannot be separated into other materials is an **element**. There are 92 naturally occurring elements. Each has a name (e.g., hydrogen, carbon, silicon, oxygen, uranium) and a corresponding symbol (e.g., respectively, H, C, Si, O, U).
- *Atoms and their components:* The smallest piece of an element retaining the characteristics of the element is an **atom**. Atoms are so small that over 5 trillion (5,000,000,000,000) could fit on the head of a pin. An atom consists of a nucleus surrounded by a cloud of orbiting electrons. A nucleus is a compact ball of protons and neutrons. (The only exception is the element hydrogen—the nucleus of a hydrogen atom contains only one proton and no neutrons.) The mass of an electron is only about 1/1,836 that of a proton.

- *Electron cloud:* The electron cloud consists of distinct orbitals or electron shells, each of which contains a specific number of electrons (**Fig. Bx5.1a**). The outermost shell defines the "surface" of an atom. The diameter of an electron cloud is about 100,000 times greater than that of a nucleus, so an atom consists mostly of empty space.
- *Charge:* The charge of a particle characterizes the way the particle responds to an electrical current or to a magnet. A neutral particle has no charge. Electrons have a negative charge, protons have a positive charge, and neutrons have a neutral charge. Like charges repel (push away from each other) whereas unlike charges attract (pull toward each other).
- *Atomic mass number:* The number of protons in an atom of an element is the **atomic mass number** of the element. Hydrogen, the smallest atom, has an atomic number of 1, helium has an atomic number of 2, and

FIGURE Bx5.1 Examples of states of matter and chemical bonds.

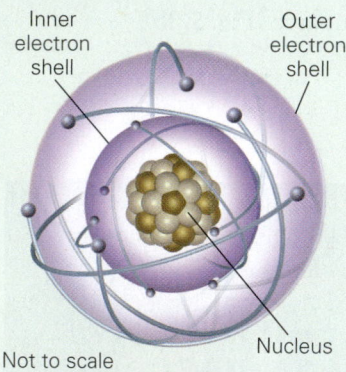

Not to scale

(a) A drawing of an atom.

Inner electron shell · Outer electron shell · Nucleus

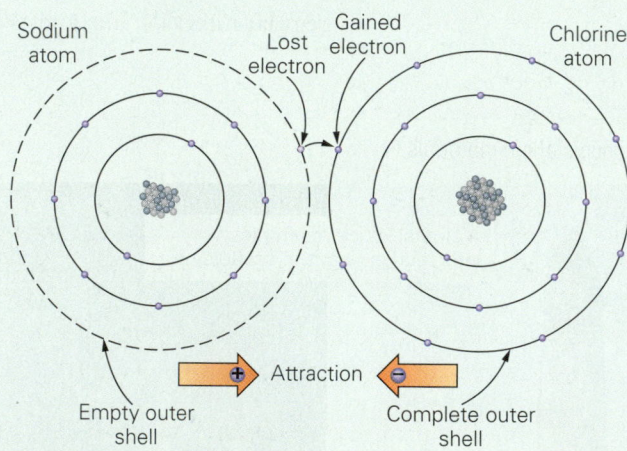

Sodium atom · Lost electron · Gained electron · Chlorine atom

+ Attraction −

Empty outer shell · Complete outer shell

(b) An ionic bond forms between a positive ion of sodium (Na⁺) and chloride (Cl⁻), a negative ion of chlorine. When sodium gives up one electron to chlorine, so that both have filled shells, halite (NaCl) is produced.

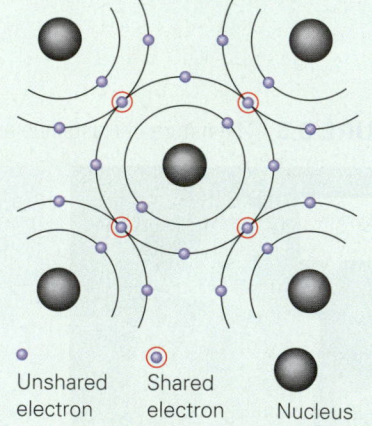

Unshared electron · Shared electron · Nucleus

(c) Covalent bonds form when carbon atoms share electrons so that all have filled electron shells.

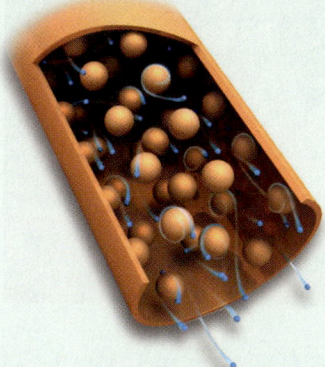

(d) In metallically bonded material, nuclei and their inner shells of electrons float in a "sea" of free electrons. The electrons stream through the metal if there is an electrical current.

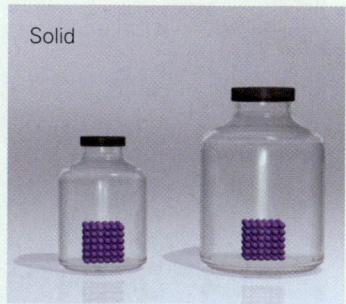

Solid

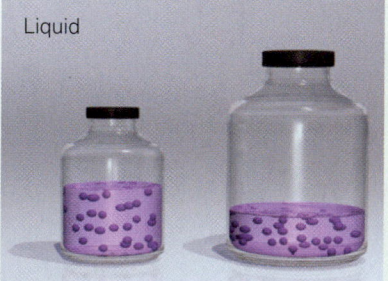

Liquid

Gas

(e) The three common states of matter. Solids keep their shape, liquids conform to the shape of their container without changing density, and gases expand to fill their container.

uranium, the largest naturally occurring atom, has an atomic number of 92. The atomic number determines the elemental identity of an atom.

- *Atomic mass:* The **atomic mass** is approximately equal to the number of protons plus neutrons in an atom of an element. (Technically, this sum is the "atomic mass number.") Hydrogen has an atomic mass of 1, helium has an atomic weight of 4, and the most common form of uranium has an atomic mass of 238.

- *Isotope:* Atoms that have the same atomic number, but a different atomic mass, are called *isotopes* of an element. For example, ^{238}U and ^{235}U are isotopes of uranium—both have atomic numbers of 92, but ^{238}U has 146 neutrons while ^{235}U has 143.

- *Ion:* An atom that has the same number of electrons as protons is neutral; an atom that is not neutral is an **ion**. An ion with excess negative charge (because it has more electrons than protons) is an *anion*, whereas an ion with excess positive charge (because it has more protons than electrons) is a *cation*. We indicate the charge with a superscript. For example, Cl^- has a single excess electron; Fe^{2+} is missing two electrons.

- *Chemical bond:* An attractive force that holds two or more atoms together is a **chemical bond** (**Fig. Bx5.1b–d**). There are several different types of chemical bonds. *Covalent bonds* form when atoms share electrons. *Ionic bonds* form when a cation and anion (ions with opposite charges) get close together and attract each other. In materials with *metallic bonds*, electrons from the outer shell can move freely.

- *Molecule:* Two or more atoms held together by chemical bonds comprise a **molecule**. The atoms may be of the same element or of different elements.

- *Ionic molecule:* If a molecule gains or loses electrons, it becomes an ionic molecule.

As an example, the carbonate ion (CO_3^{-2}) is an ionic molecule.

- *Compound:* A pure substance that can be subdivided into two or more elements is called a compound. The smallest piece of a compound that retains the characteristics of the compound is a molecule.

- *State of matter:* The form of a substance, which reflects the degree to which the atoms or molecules comprising the matter are bonded together, is called the state of matter. The most common states at the Earth's surface are solid, liquid, and gas (**Fig. Bx5.1e**). A solid can maintain its shape for a long time, a liquid will flow and conform to its container's shape, and a gas expands outward (when not confined). A fourth state, plasma, exists only at very high temperatures.

- *Changing of state:* Materials can change from one state of matter to another, usually when the pressure and/or temperature changes. Condensation is a change from gas to liquid, evaporation is a change from liquid to gas, freezing is a change from liquid to solid, melting is a change from solid to liquid, and sublimation is a change from solid to gas.

- *Chemical:* We can use the term **chemical** as a general name used for a pure substance (either an element or a compound).

- *Chemical formula:* A shorthand recipe that itemizes the various elements in a chemical and specifies their relative proportions is called a **chemical formula**. For example, the formula for water, H_2O, indicates that water consists of molecules in which two hydrogens bond to one oxygen.

- *Chemical reaction:* A **chemical reaction** is a process that involves the breaking or forming of chemical bonds and thus can cause molecules of a compound to break apart, and/or can lead to the formation of new molecules of a different compound.

Chemists represent chemical reactions in the form of an equation. Reactants are the starting elements or compounds that appear on the left side of the equation. Products, the new atoms or molecules formed during the reaction, appear on the right side of the equation. For example, the combustion of methane gas can be represented by the formula $CH_4 + O_2 \rightarrow CO_2 + H_2O$. This equation states that methane reacts with oxygen to form carbon dioxide and water.

- *Mixture:* A combination of two or more elements or compounds that can be separated without a chemical reaction is a mixture. For example, a cereal composed of bran flakes and raisins is a mixture—you can separate the raisins from the flakes without destroying either.

- *Solution:* A type of material in which one chemical (the solute) has dissolved in another (the solvent) is a solution. In solutions, a solute may separate into ions. For example, when salt (NaCl) dissolves in water, it separates into sodium (Na^+) and chloride (Cl^-) ions. In a solution, solute molecules fit between solvent molecules.

- *Concentration:* The amount of solute per unit volume of the solution is the **concentration** of the solute. For example, the concentration of salt in sea water is 3.5%. This means that every 100 g of seawater contains 3.5 grams of salt.

- *Precipitate:* When a solution becomes *oversaturated*, meaning that it contains more solute than it has room for, the excess solute forms solid particles that settle out of the solution. These particles are a **precipitate**. The word is also used as a verb in reference to the action of forming solid grains that settle from a solution. For example, we can say that when salt water evaporates, it becomes overconcentrated, and solid salt crystals start to precipitate.

If you ever need to figure out whether or not a substance is a mineral, just check it against the definition. Is motor oil a mineral? No—it's an organic liquid. Is table salt a mineral? Yes—it's a solid crystalline compound with the formula NaCl. Is the hard material making up the shell of an oyster considered to be a mineral? Yes, it's a biogenic mineral—it has the same composition and structure as an inorganic mineral. Is rock candy a mineral? No. Even though it is solid and crystalline, it's not made by geologic processes and it consists of sugar (an organic chemical).

Take-Home Message

Minerals are naturally formed solids with a crystalline structure (an orderly arrangement of atoms inside) and a definable chemical formula.

QUICK QUESTION: Is Styrofoam a mineral? Why or why not?

5.3 Beauty in Patterns: Crystals and Their Structure

What Is a Crystal?

The word *crystal* brings to mind sparkling chandeliers, elegant wine goblets, and shiny jewels. But, as is the case with the word *mineral*, geologists use a narrower definition in scientific discussion. A **crystal** is a single, continuous (i.e., uninterrupted) piece of a crystalline solid, typically bounded by flat surfaces called **crystal faces** that grow naturally as the mineral forms. The word comes from the Greek word *krystallos*, meaning ice. Many crystals have beautiful shapes that look like they belong in the pages of a geometry book. The angle between two adjacent crystal faces of one specimen is identical to the angle between the corresponding faces of another specimen. For example, a perfectly formed quartz crystal looks like an obelisk (**Fig. 5.4a**). The angle between the faces of the columnar part of a quartz crystal is exactly 120°. This rule, discovered by Nicolas Steno, holds regardless of whether the whole crystal is big or small and regardless of whether all of the faces are the same size. Crystals come in a great variety of shapes, including cubes, trapezoids, pyramids, octahedrons, hexagonal columns, blades, needles, columns, and obelisks (**Fig. 5.4b**).

Because crystals have a regular geometric form, people have always considered them to be special, perhaps even a source of "magical powers." For example, shamans of some cultures relied on talismans or amulets made of crystals, which supposedly brought power to their wearer or warded off evil spirits. Scientists have demonstrated, however, that crystals have no physical effect on health or mood. For millennia, crystals have inspired awe because of the way they sparkle; but such behavior is simply a consequence of how crystal structures interact with light.

Looking inside a Mineral

What do the insides of a mineral actually look like? This problem was the focus of study for centuries. An answer finally came from the work of a German physicist, Max von Laue, in 1912. He showed that an X-ray beam passing through a crystal breaks up into many tiny beams to create a pattern of dots on a screen. Physicists refer to this phenomenon as **diffraction**; it occurs when waves interact with regularly spaced objects whose spacing is close to the wavelength of the waves—you can see diffraction of ocean waves when they pass through gaps in a seawall (**Fig. 5.5a**). Von Laue concluded that for a crystal to cause diffraction, atoms within it must be regularly spaced and the spacing must

FIGURE 5.4 Some characteristics of crystals.

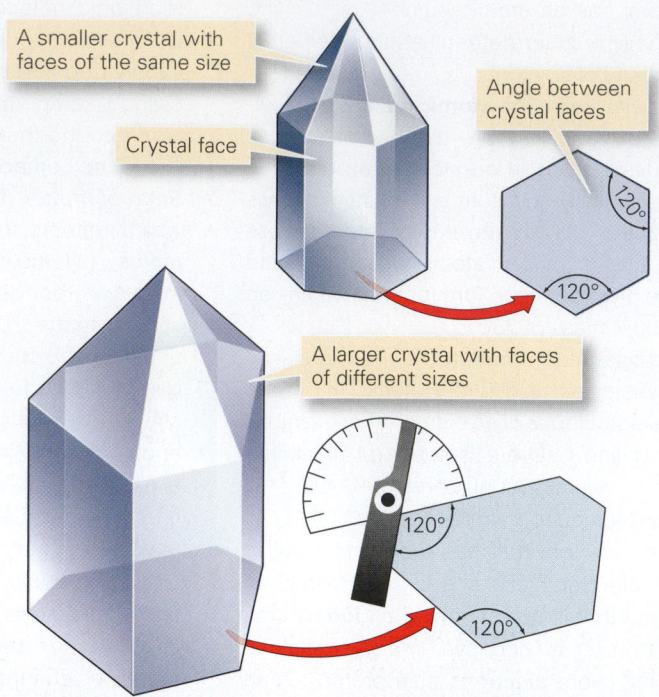

A smaller crystal with faces of the same size

Crystal face

Angle between crystal faces

A larger crystal with faces of different sizes

(a) Regardless of specimen size, the angle between two adjacent crystal faces is consistent in a particular mineral.

Halite — Diamond — Staurolite — Quartz

Garnet — Stibnite — Calcite — Kyanite

(b) Crystals come in a variety of shapes, including cubes, prisms, blades, and pyramids. Some terminate at a point and some terminate with flat surfaces.

be comparable to the wavelength of X-rays. Eventually, Von Laue and others learned how to use *X-ray diffraction patterns* as a basis for defining the specific arrangement of atoms in crystals (**Fig. 5.5b**). This arrangement defines the **crystal structure** of a mineral. Modern studies of minerals, using extremely powerful electron microscopes, now allow us to see the regular geometric arrangement of atoms in a crystal (**Fig. 5.5c**).

If you've ever looked at wallpaper, you've seen an example of a pattern (**Fig. 5.6a**). Crystal structures contain one of nature's most spectacular examples of such a pattern. In crys-

FIGURE 5.5 Using X-rays and electron beams to characterize the internal structure of minerals.

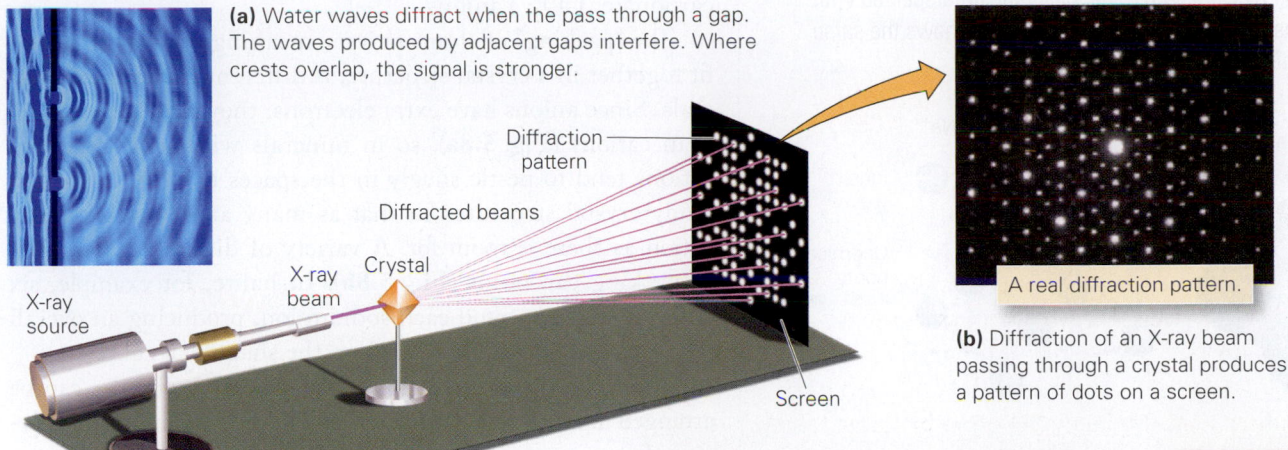

(a) Water waves diffract when the pass through a gap. The waves produced by adjacent gaps interfere. Where crests overlap, the signal is stronger.

Diffraction pattern

Diffracted beams

X-ray beam

Crystal

X-ray source

Screen

A real diffraction pattern.

(b) Diffraction of an X-ray beam passing through a crystal produces a pattern of dots on a screen.

Scattered electrons

Rows of atoms

Dark spot (shadow)

TEM image

Detector

4 nm

(c) A transmission electron microscope (TEM) shoots beams of electrons at a material. Some electrons scatter off atoms, but some pass between gaps and make a dark spot on a recorder. The result is an image (right) showing the pattern of atoms in the material.

tals, the pattern is defined by the regular spacing of atoms and, if the crystal contains more than one element, by the regular alternation of atoms (**Fig. 5.6b**). (Mineralogists refer to a 3-D geometry of points representing this pattern as a *crystal lattice*.) The pattern of atoms in a crystal may control the shape of a crystal. For example, if atoms in a crystal pack into the shape of a cube, the crystal may have faces that intersect at 90° angles—galena (PbS) and halite (NaCl) have such a cubic shape. Because of the pattern of atoms in a crystal structure, the structure has *symmetry*, meaning that the shape of one part of the structure is the mirror image of the shape of a neighboring part. For example, if you were to cut a halite crystal or a water crystal (snowflake) in half and place the half against a mirror, it would look whole again (**Fig. 5.6c**).

We can construct a model of a crystal's structure using a cluster of balls that are packed together (**Fig. 5.7**). In such models, different colors and/or sizes represent different atoms,

ions, or ionic molecules, and in some models sticks represent chemical bonds. For example, in diamond, which consists entirely of covalently bonded carbon atoms, all of the balls represent carbon atoms, whereas in halite (rock salt), composed of ionically bonded sodium and chloride ions, balls representing the cations of sodium (Na^+) are interspersed with balls representing anions of chloride (Cl^-). And in calcite ($CaCO_3$),

FIGURE 5.6 The concept of patterns and symmetry in minerals.

(a) The repetition of a flower motif on wallpaper.

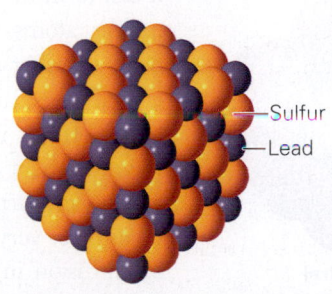

Sulfur

Lead

(b) The repetition of alternating sulfur and lead atoms in the mineral galena (PbS).

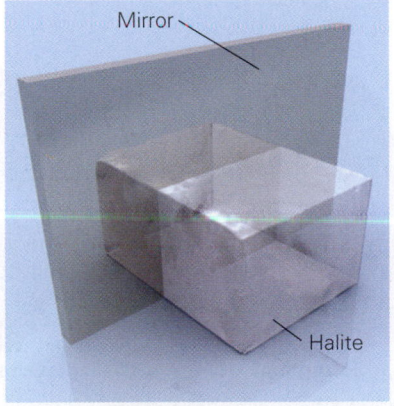

Mirror

Halite

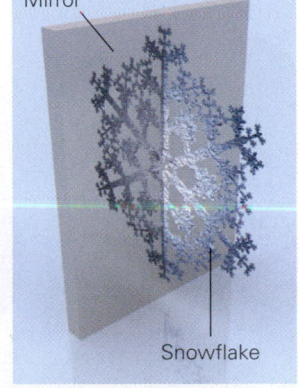

Mirror

Snowflake

(c) Minerals display symmetry. One-half of a halite crystal is a mirror image of the other.

FIGURE 5.7 A model of the crystal structure of halite (NaCl). The model on the left shows larger chloride ions (blue) interspersed with smaller sodium ions (white). The model on the right shows the same relation but uses sticks to represent chemical bonds.

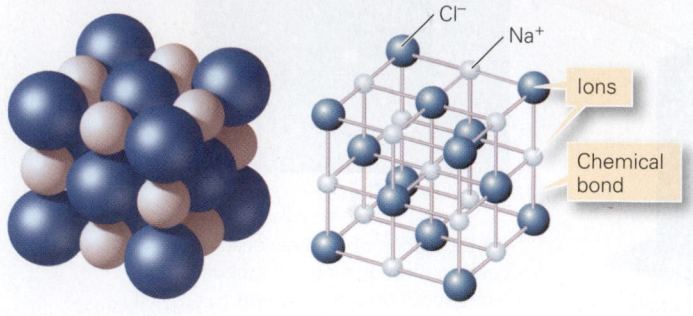

balls representing calcium (Ca^+) cations are interspersed with carbonate (CO^{3-}) anions.

The "packing" of atoms or ions, meaning the way that they fit together in a crystal structure, is different in different minerals. Since anions have extra electrons, they tend to be bigger than cations (**Fig. 5.8a**), so in minerals with ionic bonding, cations tend to nestle snugly in the spaces between anions in many crystal structures so that as many anions fit around a cation as there is room for. A variety of different geometries of packing can occur (**Fig. 5.8b**). In halite, for example, six chloride ions surround each sodium ion, producing an overall arrangement of atoms that defines the shape of a cube.

Some groups of atoms, ions, and/or ionic molecules can be arranged in more than one way, yielding different minerals. **Polymorphs** are two or more different minerals that have the same chemical composition but different crystal structures. As an example, let's consider diamond and graphite, polymorphs of carbon. In diamond, each atom packs together with four neighbors to form of a tetrahedron. As a result, some naturally formed diamond crystals have the shape of a double tetrahedron (**Fig. 5.9a**). Graphite, another mineral composed entirely of carbon, behaves very differently from diamond. In contrast to diamond, graphite is so soft that we use it as the "lead" in a pencil, for when a pencil moves across paper, tiny flakes of graphite peel off the pencil point and adhere to the paper. This behavior occurs because the carbon atoms in graphite are not arranged in tetrahedra, but rather they occur in sheets (**Fig. 5.9b**). The sheets are bonded to each other by weak bonds and thus can separate from each other easily.

The Formation and Destruction of Minerals

A new mineral crystal can form in one of five ways (**Fig. 5.10**). First, it can form by the solidification of a melt, meaning the *freezing* of a liquid. For example, ice crystals grow when water freezes. Second, it can form by *precipitation* from a solution, meaning that atoms, molecules, or ions dissolved in water bond together and separate out of the water. Salt crystals, for example, develop when salt water evaporates. Third, it can form by solid-state *diffusion*, the movement of atoms or ions through a solid to arrange into a new crystal structure, a process that takes place very slowly. For example, garnets grow by diffusion in solid rock, when the rock is subjected to high heat and pressure. Fourth, minerals can form at interfaces between the physical and bio-

FIGURE 5.8 The various sizes of ions and the ways they pack together in minerals.

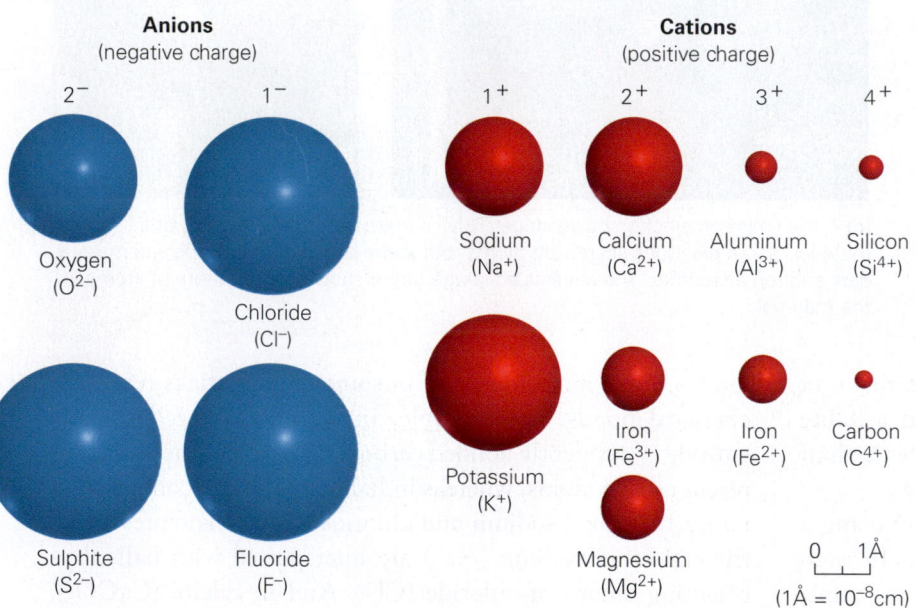

(a) Ions come in a wide range of sizes. The difference in size depends, in part, on the number of electrons they contain.

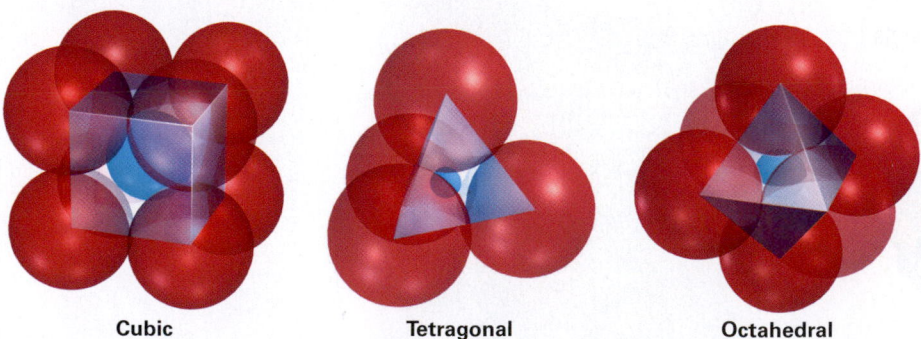

(b) Ions can pack together in different ways. Each configuration can be described by a geometric shape.

FIGURE 5.9 The nature of crystalline structure in minerals.

Carbon atoms   Strong bonds

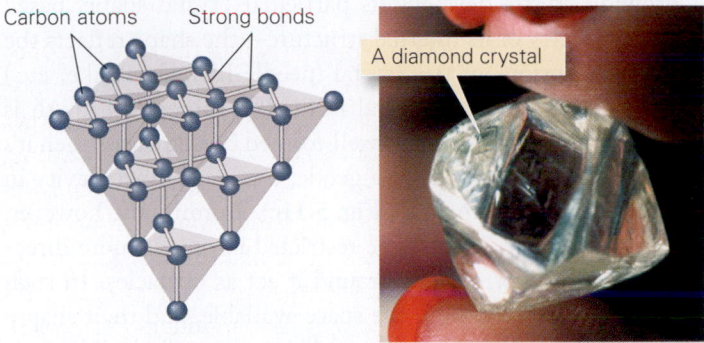

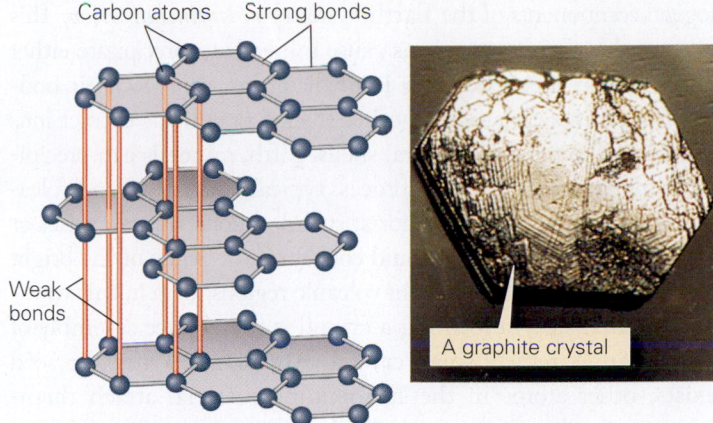

Carbon atoms   Strong bonds

A diamond crystal

Weak bonds

A graphite crystal

(a) In a diamond, carbon atoms are arranged in tetrahedra. All of the bonds are strong.

(b) Graphite consists of carbon atoms arranged in hexagonal sheets. The sheets are connected by weak bonds.

FIGURE 5.10 Processes that form new minerals.

~50 cm

(a) As molten rock cools, mineral crystals grow.

(b) Evaporation from a desert lake yields salt crystals.

10 cm

(c) These large garnets grew by diffusion within solid rock. (Coin for scale.)

5 cm

(d) Shells on a beach formed by biomineralization (Lens cap for scale.)

1 cm

(e) Sulfur crystals collect around a vent.

logical components of the Earth System by *biomineralization*. This occurs when living organisms cause minerals to precipitate either within or on their bodies or immediately adjacent to their bodies. For example, clams and other shelled organisms extract ions from water to produce mineral shells. Fifth, minerals can precipitate directly from a gas. This process typically occurs around volcanic vents or around geysers, for at such locations volcanic gases or steam enter the atmosphere and cool abruptly. Some of the bright yellow sulfur deposits found in volcanic regions form in this way.

The first step in forming a crystal is the chance assembly of a seed, an extremely small crystal (**Fig. 5.11a**). Once the seed exists, other atoms in the surrounding material attach themselves to the face of the seed. As the crystal grows, crystal faces move outward but maintain the same orientation (**Fig. 5.11b**),

so the youngest part of the new crystal occurs at its outer edge. A growing crystal develops its particular crystal shape, based on the geometry of its internal structure—the shape reflects the relative dimensions of the crystal (needle-like, sheet-like, etc.) and the angles between crystal faces. If a mineral's growth is uninhibited so that it displays well-formed crystal faces, then it's a *euhedral crystal*. Crystals in a geode, a mineral-lined cavity in rock, are typically euhedral (**Fig. 5.11c**). Commonly, however, the growth of a crystal may be restricted in one or more directions because other crystals around it act as obstacles. In such cases, minerals grow to fill the space available, and their shapes will be controlled by the shape of their surroundings. Minerals without well-formed crystal faces are known as *anhedral grains*.

Crystals formed by precipitation from a solution develop when the solution becomes oversaturated; that is, the number of dissolved ions per unit volume of solution becomes so great that they can get close enough to one another to bond together. If a solution is not saturated, dissolved ions are surrounded by solvent molecules, which shield the ions from the attractive forces of their neighbors. In the case of crystals formed by the solidification of a melt, atoms begin to attach to the seed when the melt becomes sufficiently cool that thermal vibrations can no longer break apart the attraction between the seed and the atoms in the melt.

A mineral can be destroyed by melting, dissolving, or some other chemical reaction. Melting involves heating a mineral to a temperature at which thermal vibration of the atoms or ions in the lattice break the chemical bonds holding them to the lattice. The atoms or ions then separate, either individually or in small groups, to move around again freely. Dissolution occurs when you immerse a mineral in a solvent, such as water. Atoms or ions then separate from the crystal face and are surrounded by solvent molecules. Chemical reactions can destroy a mineral when it comes in contact with reactive materials. For example, iron-bearing miner-

FIGURE 5.11 The growth of crystals.

Ions attach to the crystal face.

(a) New crystals nucleate and begin to precipitate out of a water solution. As time progresses, they grow into the open space.

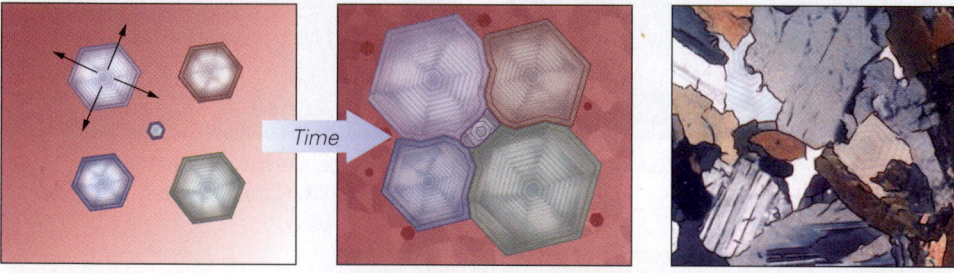

(b) New crystals grow outward from the central seed. As time passes, they maintain their shape until they interfere with each other. A crystal growing in a confined space will be anhedral.

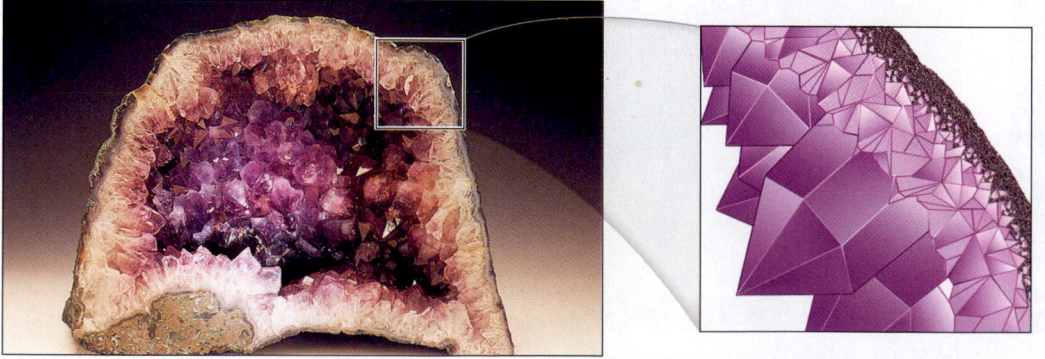

(c) A geode from Brazil consists of purple quartz crystals (amethyst) that grew from the wall into the center. The enlargement sketch indicates that the crystals are euhedral.

als react with air and water to form rust (iron oxide). Reactions that take place by diffusion in solid rock can effectively "digest" an assemblage of minerals and replace them with others. The action of microbes in the environment can also destroy minerals. In effect, some microbes can "eat" certain minerals; the microbes use the energy stored in the chemical bonds that hold the atoms of the mineral together as their source of energy for metabolism.

> ## Take-Home Message
>
> The crystal structure of minerals is defined by a regular geometric arrangement of atoms that has symmetry. Minerals can form by solidification of a melt, by precipitation from a water solution or a gas, or by rearrangement of atoms in a solid.
>
> **QUICK QUESTION:** How are minerals "destroyed" in nature?

5.4 How Can You Tell One Mineral from Another?

Amateur and professional mineralogists get a kick out of recognizing minerals. They might hover around a display case in a museum and name specimens without bothering to look at the labels. How do they do it? The trick lies in learning to recognize the basic physical properties (visual and material characteristics) that distinguish one mineral from another. Some physical properties, such as shape and color, can be seen from a distance. Others, such as hardness and magnetization, can be determined only by handling the specimen or by performing an identification test on it. Identification tests include scratching the mineral against another object, placing it near a magnet, weighing it, tasting it, or placing a drop of acid on it. Let's examine some of the physical properties most commonly used in mineral identification.

- *Color:* **Color** results from the way a mineral interacts with light. Sunlight contains the whole spectrum of colors, each with a different wavelength. A mineral absorbs certain wavelengths, so the color you see when looking at a specimen represents the wavelengths the mineral does not absorb. Certain minerals always have the same color, but many occur in a range of colors (**Fig. 5.12a**). Color variations in a mineral commonly reflect the presence of impurities. For example, trace amounts of iron may give quartz a reddish color.
- *Streak:* The **streak** of a mineral refers to the color of a powder produced by pulverizing the mineral. You can obtain a streak by scraping the mineral against an unglazed ceramic plate (**Fig. 5.12b**). The color of a mineral powder tends to be less variable than the color of a whole crystal and thus provides a fairly reliable clue to a mineral's

identity. Calcite, for example, always yields a white streak even though pieces of calcite may be white, pink, or clear.

- *Luster:* **Luster** refers to the way a mineral surface scatters light. Geoscientists describe luster simply by comparing the appearance of the mineral with the appearance of a familiar substance. For example, minerals that look like metal have a *metallic luster*, whereas those that do not have a *nonmetallic luster*—the adjectives are self-explanatory (**Fig. 5.12c, d**). Terms used for different versions of nonmetallic luster include: silky, glassy, satiny, resinous, pearly, or earthy.
- *Hardness:* **Hardness** is a measure of the relative ability of a mineral to resist scratching, and it therefore represents the ability of bonds in the crystal structure to resist being broken. Hard minerals can scratch soft minerals, but soft minerals cannot scratch hard ones because the atoms or ions in crystals of a hard mineral are more strongly bonded together than are those in a soft mineral. Diamond, the hardest mineral known, can scratch anything, which is why it is used to cut glass. In the early 1800s, a mineralogist named Friedrich Mohs listed some minerals in sequence of relative hardness. This list, the **Mohs hardness scale**, helps in mineral identification; a mineral with a Mohs hardness of 5 can scratch all minerals with a hardness of 5 or less. When you use the scale (**Table 5.1**), it helps to compare the

TABLE 5.1 Mohs Hardness Scale

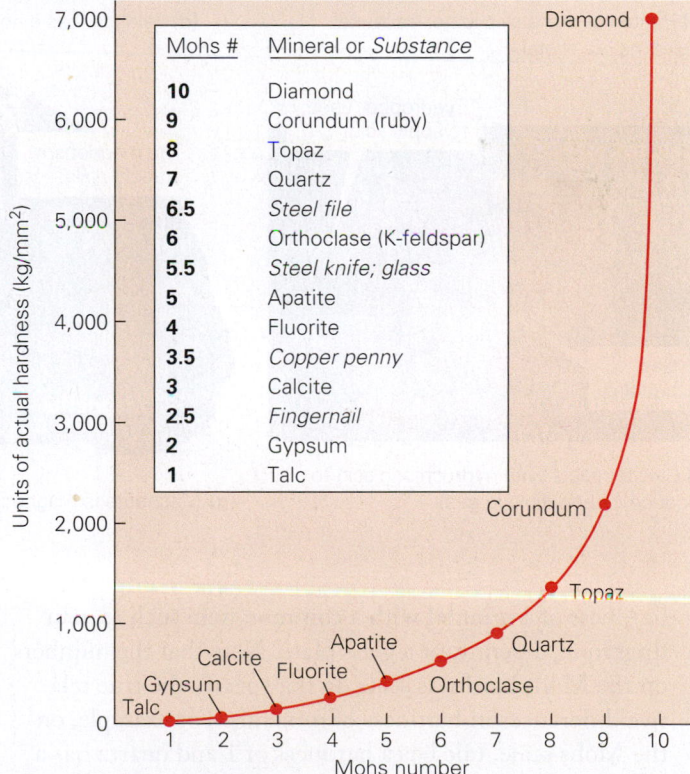

Mohs #	Mineral or *Substance*
10	Diamond
9	Corundum (ruby)
8	Topaz
7	Quartz
6.5	*Steel file*
6	Orthoclase (K-feldspar)
5.5	*Steel knife; glass*
5	Apatite
4	Fluorite
3.5	*Copper penny*
3	Calcite
2.5	*Fingernail*
2	Gypsum
1	Talc

Mohs's numbers are relative—in reality, diamond is 3.5 times harder than corundum, as the graph shows.

FIGURE 5.12 Physical characteristics of minerals.

(a) Color is diagnostic of some minerals, but not all. For example, quartz can come in many colors.

(b) To obtain the streak of a mineral, rub it against a porcelain plate. The streak consists of mineral powder.

(c) Pyrite has a metallic luster because it gleams like metal.

(d) Feldspar has a nonmetallic luster.

(e) Crystal habit refers to the shape or character of the crystal. The blue kyanite crystals on the right are bladed, and the chrysotile above is fibrous.

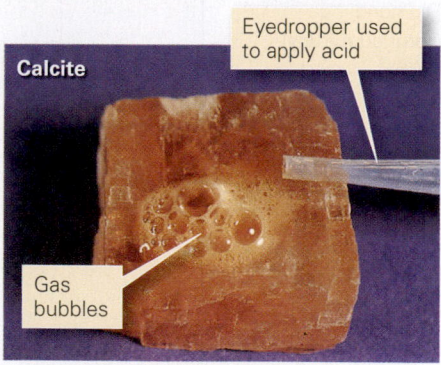

(f) Calcite reacts with hydrochloric acid to produce carbon dioxide gas.

(g) Magnetite is magnetic.

hardness of a mineral with a common item such as your fingernail, a penny, or a glass plate. Note that the numbers on the Mohs hardness scale do not specify the true relative differences in hardness of minerals. For example, on the Mohs scale, talc has a hardness of 1 and quartz has a hardness of 7. But this does not mean that quartz is 7 times harder than talc. Careful tests show that quartz is actually about 100 times harder than talc, as indicated by how difficult it is to make an indentation in the mineral.

- *Specific gravity:* **Specific gravity** represents the density of a mineral, as defined by the ratio between the weight of a volume of the mineral and the weight of an equal volume of water at 4°C. For example, one cubic centimeter of quartz has a weight of 2.65 grams, whereas one cubic centimeter of water has a weight of 1.00 gram. Thus, the specific gravity of quartz is 2.65. In practice, you can develop a sense of specific gravity by hefting minerals in your hands. A piece of galena (lead ore) "feels" heavier than a similar-sized piece of quartz.

- *Crystal habit:* The **crystal habit** of a mineral refers to the shape of a single crystal with well-formed crystal faces or to the character of an aggregate of many well-formed crystals that grew together as a group (**Fig. 5.12e**). The habit depends on the internal arrangement of atoms in the crystal. When describing habit, mineralogists commonly compare the mineral to a common geometric shape, by using adjectives such as cubic, prismatic, bladed, platy, or fibrous. The relative dimensions depend on relative rates of crystal growth in different directions. For example, crystals that grow rapidly in one direction but slowly in the other two directions are needle-like (**Box 5.2**).

- *Special properties:* Some minerals have distinctive properties that readily distinguish them from other minerals. For example, calcite ($CaCO_3$) reacts with dilute hydrochloric acid (HCl) to produce carbon dioxide (CO_2) gas (**Fig. 5.12f**). Dolomite ($CaMg[CO_3]_2$) also reacts with acid but not as strongly. Graphite makes a gray mark on paper, magnetite attracts a magnet (**Fig. 5.12g**), halite tastes salty, and plagioclase has striations (thin parallel corrugations or stripes) on its crystal faces.

- *Fracture and cleavage:* Different minerals fracture (break) in different ways, depending on the internal arrangement of atoms. If a mineral breaks to form distinct planar surfaces that have a specific orientation in relation to the crystal structure, then we say that the mineral has **cleavage** and we refer to each surface as a cleavage plane (**Fig. 5.13a–e**). Cleavage forms in directions where the bonds holding atoms together in the crystal are weaker. Some minerals have one direction of cleavage. For example, mica has very weak bonds in one direction but strong bonds in the other two directions, so it easily splits into parallel sheets; the surface of each sheet is a cleavage plane. Other minerals have two or three directions of cleavage that intersect at a specific angle. For example, halite has three sets of cleavage planes that intersect at right angles, so halite crystals break into little cubes. Cleavage planes are sometimes hard to distinguish from crystal faces (**Fig. 5.13f, g**). Materials that have no cleavage at all (because bonding is equally strong in all directions) break either by forming irregular fractures or by forming conchoidal fractures (**Fig. 5.13h**). **Conchoidal fractures** are smoothly curving, clamshell-shaped surfaces; they typically form in glass.

Take-Home Message

The properties of minerals (such as color, streak, luster, crystal habit, hardness, specific gravity, cleavage, magnetism, and reaction with acid) are a manifestation of the crystal structure and chemical composition of minerals and can be used for mineral identification.

QUICK QUESTION: Which minerals react with acid to produce CO_2 bubbles?

5.5 Organizing Knowledge: Mineral Classification

Just about every object you come across in daily life has been classified in some way, because classification schemes help organize information and streamline discussion. Biologists classify animals based on how they reproduce, whether they have skeletons or not, and whether they breathe oxygen in air or oxygen in water. Botanists classify plants according to the way they reproduce and by the shape of their leaves. In the case of minerals, a good means of classification eluded researchers until it became possible to determine the chemical makeup of minerals. A Swedish chemist, Baron Jöns Jacob Berzelius (1779–1848), analyzed minerals and noted chemical similarities among many of them. Berzelius, along with his students, established that most minerals can be classified by specifying the principal anion or anionic group within the mineral (see Box 5.1). Using this approach, it's possible to divide the 4,000 known minerals into a small number of groups, or **mineral classes**. We now take a look at principal mineral classes, focusing especially on silicates, the class that constitutes most of the rock in the Earth.

The Mineral Classes

Mineralogists distinguish several principal classes of minerals. Here are some of the major ones.

- *Silicates:* The fundamental component of **silicate minerals** is the SiO_4^{4-} anionic group. Examples include quartz (SiO_2; see Fig. 5.12a) and feldspar (e.g., $KAlSi_3O_8$). We will learn more about silicates in the next section.
- *Sulfides:* Sulfides consist of a metal cation bonded to a sulfide anion (S^{2-}). Examples include galena (PbS) and pyrite (FeS_2; see Fig. 5.12c).
- *Oxides:* Oxides consist of metal cations bonded to oxygen anions. Typical oxide minerals include hematite (Fe_2O_3; see Fig. 5.12b) and magnetite (Fe_3O_4; see Fig. 5.12g).
- *Halides:* The anion in a halide is a halogen ion (such as chloride, Cl^-, or fluoride, F^-). Halite, or rock salt (NaCl;

FIGURE 5.13 The nature of mineral cleavage and fracture.

(a) Mica has one strong plane of cleavage and splits into sheets.

(b) Pyroxene has two planes of cleavage that intersect at 90°.

(c) Amphibole has two planes of cleavage that intersect at 60°.

Halite breaks into cubes.

(d) Halite has three mutually perpendicular planes of cleavage.

Calcite breaks into rhombs.

(e) Calcite has three planes of cleavage, one of which is inclined.

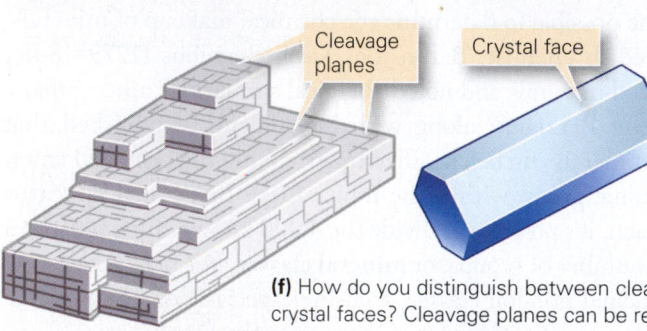

Cleavage planes

Crystal face

(f) How do you distinguish between cleavage places and crystal faces? Cleavage planes can be repeated, whereas a crystal face is a single surface.

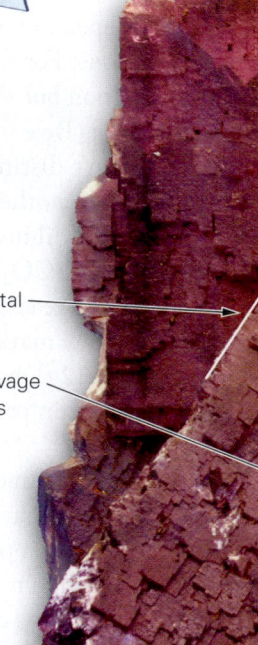

Crystal face

Cleavage steps

(g) Cleavage steps and crystal faces in a sample of fluorite (CaF_2).

Irregular fracture

Crystal face

Garnet

Conchoidal fracture

Quartz

(h) Minerals without cleavage can develop irregular or conchoidal fractures.

see Fig. 5.13d), and fluorite (CaF_2), a source of fluoride, are common examples.

- *Carbonates:* In **carbonate minerals**, the molecule CO_3^{2-} serves as the anionic group. Examples include calcite ($CaCO_3$; see Fig. 5.13e) and dolomite ($CaMg[CO_3]^2$).
- *Native metals:* Native metals consist of pure masses of a single metal. The metal atoms are bonded by metallic bonds. Copper and gold, for example, may occur as native metals. A gold nugget is a mass of native gold that has been broken out of a rock.
- *Sulfates:* Sulfates consist of metal cations bonded to SO_4^{2-} anionic groups. Many sulfates form by precipitation out of water at or near the Earth's surface (**Fig. 5.14**). An example is gypsum ($CaSO_4 \cdot 2H_2O$).

Silicates: The Major Rock-Forming Minerals

Silicate minerals, or simply **silicates**, make up over 95% of the continental crust. Rocks of the oceanic crust and of the Earth's mantle consist almost entirely of silicates. Thus, silicates are the most common minerals on Earth. As noted earlier, silicates contain the SiO_4^{4-} anionic group. In this group, four oxygen atoms surround a single silicon atom, thereby defining the corners of a tetrahedron, a pyramid-like shape with four triangular faces (**Fig. 5.15a**). We refer to this anionic group as the **silicon-oxygen tetrahedron** (or, informally, as the *silica tetrahedron*), and it acts, in effect, as the building block of silicate minerals.

Mineralogists distinguish among several groups of silicate minerals, detailed below, based on the way in which silica tetrahedra are arranged (**Fig. 5.15b**). The arrangement, in turn, determines the degree to which tetrahedra share oxygen atoms. Note that the number of shared oxygens determines the ratio of silicon (Si) to oxygen (O) in the mineral.

- *Independent tetrahedra:* In this group, tetrahedra are independent and do not share any oxygen atoms. The attraction between the tetrahedra and positive ions holds crystals together. This group includes olivine, a glassy green mineral (**Fig. 5.16a**), and garnet (see Fig. 5.10c and 5.13h).
- *Single chains:* In a single-chain silicate, tetrahedra link to form a chain by sharing two oxygen atoms. The most common of the many different types of single-chain silicates are pyroxenes (see Fig. 5.13b).
- *Double chains:* In a double-chain silicate, tetrahedra link to form a double chain by sharing two or three oxygen atoms. Amphiboles are the most common minerals in this group (see Fig. 5.13c).
- *Sheet silicates:* Tetrahedra in this group share three oxygen atoms and therefore link to form two-dimensional sheets. Other ions fit between the sheets in sheet silicates. Because of their structure, sheet silicates have

FIGURE 5.14 This photo is real, not a computer collage! We're seeing the world's largest-known mineral crystals jutting from the walls of a cave in Chihuahua, Mexico. The crystals are of the mineral gypsum ($CaSO_4 \cdot 2H_2O$), formed by precipitation from a water solution.

BOX 5.2 CONSIDER THIS . . .

Asbestos and Health: When Crystal Habit Matters!

Asbestos minerals have been in the news for decades because epidemiology studies show an association between asbestos inhalation and human health. Understanding the crystal habit of the minerals provides some insight into the origin of asbestos hazard.

Asbestos doesn't refer to a single mineral but rather is a general term used for six different minerals. The minerals share a key characteristic—all are fibrous in that they consist of clusters of needle-like crystals that are about 20 times longer than they are wide (**Fig. Bx5.2a**). "White asbestos," which accounts for more than 90% of the asbestos used, is the mineral chrysotile ($Mg_3[Si_2O_5]$

$[OH]_4$). It forms in serpentinite, a green rock that is produced when hot-water solutions interact with the olivine-rich rock that makes up oceanic lithosphere. "Brown asbestos" includes various types of amphibole minerals, such as grunerite ($Fe_7Si_8O_{22}[OH]_2$).

Between the mid-1800s and the 1990s, asbestos was used in a great variety of applications because its fibrous character allowed it to be woven into other materials and because it's strong, fire resistant, and chemical resistant. Thus, you can find asbestos in floor and ceiling tiles, insulation, roof shingles, fire-resistant clothing, gaskets, and brakes (**Fig. Bx5.2b**). When intact and

incorporated into other materials, asbestos is not dangerous because it is not poisonous and does not emit fumes. The problem arises when people inhale asbestos dust, as can happen during the mining of asbestos, the manufacturing of asbestos-containing materials, or the remodeling or demolition of rooms or buildings that contain asbestos-containing materials. Because asbestos has a fibrous habit, when it's handled it breaks into tiny (700 times smaller than human hair) needle-like shards that can be inhaled (**Fig. Bx5.2c**). These shards get embedded in the lungs where they cause irritation and clogging and inhibit exchange of oxygen, resulting in asbestosis. For reasons that are less well understood, the fibers can also interact chemically and/or mechanically with DNA in cells, causing genetic mutations that trigger various kinds of lung cancer. There is still debate about the relative dangers of different kinds of asbestos, but it appears that

FIGURE Bx5.2 Asbestos has distinctive characteristics that make it useful.

(a) Asbestos is a fibrous mineral.

(b) Asbestos, mixed in with vinyl, makes stronger floor tiles. But when old tiles break up, the resulting dust can contain asbestos.

(c) At high magnification, asbestos fibers are like tiny needles.

(d) To remove asbestos requires protective gear and for the area to be sealed off.

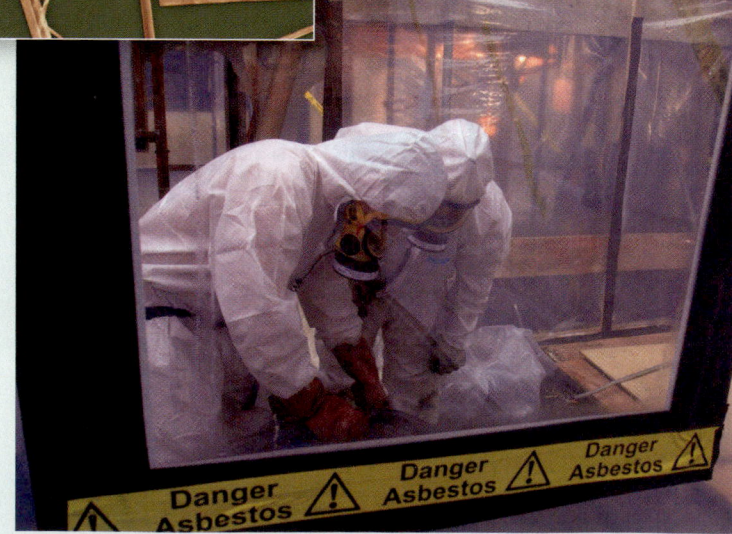

white asbestos is somewhat less dangerous than brown asbestos, perhaps because of its fiber shape.

When health concerns associated with asbestos became publicized in the 1970s, years of litigation followed, eventually result-ing in bans on the use of asbestos in the mid-1980s. As a consequence, renovations of rooms containing asbestos tiles or insu-lation must begin with a very expensive asbestos abatement, during which asbestos is removed by workers in protective cloth-ing, in an area sealed off from ventilation (**Fig. Bx5.2d**). In many cases, if asbestos is left alone and can be covered by a safer material, the hazard can be mitigated. It's only when asbestos materials are sanded or broken up that the fibers can enter the air.

FIGURE 5.15 The structure of silicate minerals.

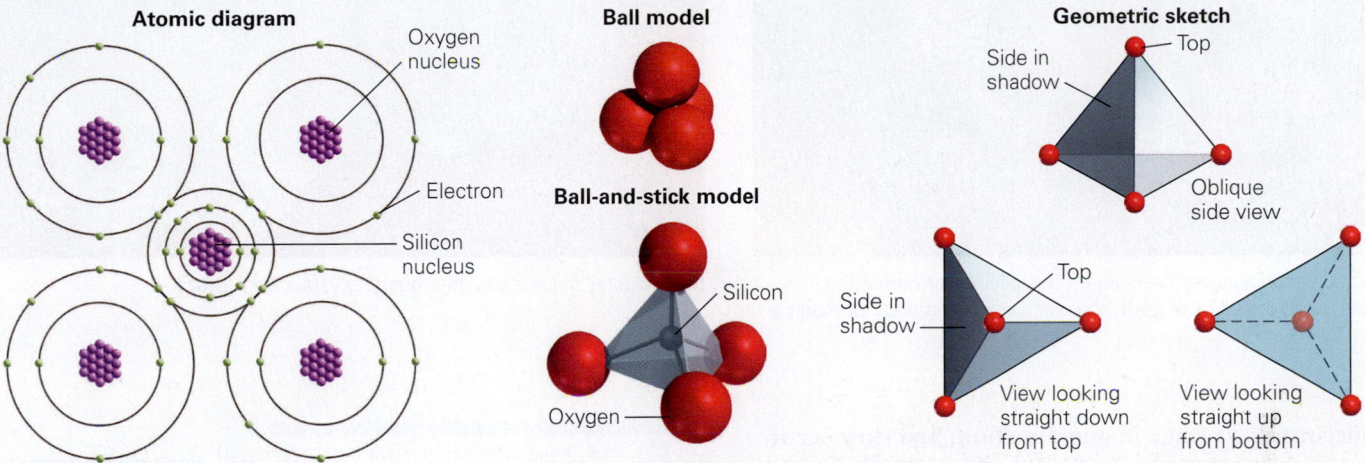

(a) The fundamental building block of a silicate mineral is the silicon-oxygen tetrahedron. Oxygens occupy the corners of the tetrahedron, and silicon lies at the center. Geologists portray the tetrahedron in a number of different ways.

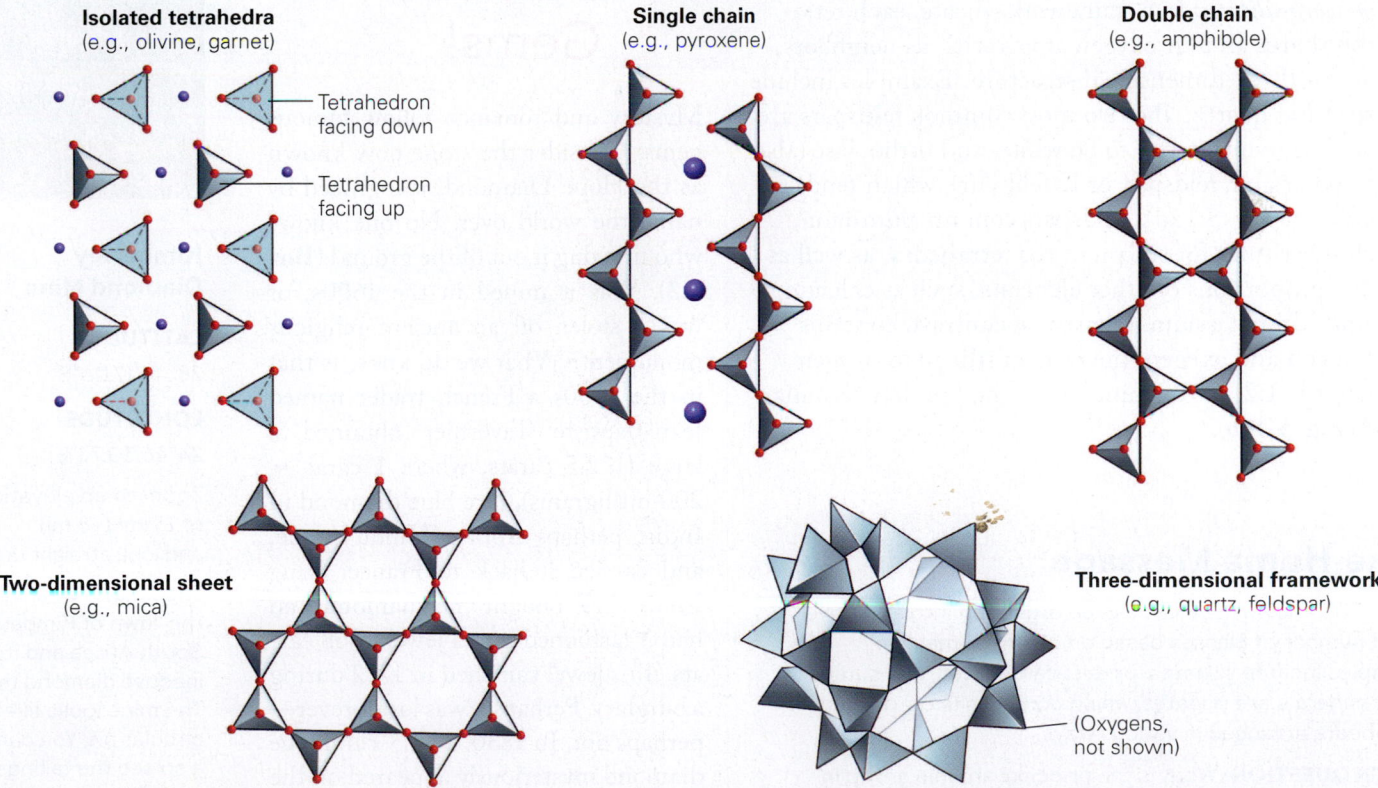

(b) The classes of silicate minerals differ from one another by the way in which the silicon-oxygen tetrahedra are linked. Where the tetrahedra link, they share an oxygen atom. Oxygen atoms are shown in red. Positive ions (not shown) occupy spaces between tetrahedra.

FIGURE 5.16 Examples of silicate materials.

(a) Olivine crystals in a sugar-like cluster. This olivine formed in the Earth's mantle and rode to the Earth's surface in a rising body of magma.

(b) A cluster of clear to white quartz crystals from Brazil.

a single strong cleavage in one direction, and they occur in books of very thin sheets. In this group, we find micas (see Fig. 5.13a) and clays. Clays occur only in extremely tiny flakes.

- *Framework silicates:* In a framework silicate, each tetrahedron shares all four oxygen atoms with its neighbors, forming a three-dimensional structure. Examples include feldspar and quartz. The two most common feldspars are plagioclase, which tends to be white, and orthoclase (also called potassium feldspar, or K-feldspar), which tends to be pink (see Fig. 5.12d). Feldspars contain aluminum, which substitutes for silicon in the tetrahedra, as well as varying proportions of other elements, such as calcium, sodium, and potassium. Quartz, in contrast, contains only silicon and oxygen; the ratio of silicon to oxygen in quartz is 1:2, so the mineral has the familiar formula SiO_2 (**Fig. 5.16b**).

Take-Home Message

The 4,000 known minerals can be organized into a relatively small number of classes based on chemical makeup. Examples include silicates, oxides, carbonates, and sulfides. Most minerals are silicates, which contain silicon-oxygen tetrahedra arranged in various ways.

QUICK QUESTION: What is the principal anionic group in carbonate minerals?

5.6 Something Precious— Gems!

Mystery and romance follow famous gems. Consider the stone now known as the Hope Diamond, recognized by name the world over. No one knows who first dug it out of the ground (**Box 5.3**). Was it mined in the 1600s, or was it stolen off an ancient religious monument? What we do know is that in the 1600s a French trader named Jean-Baptiste Tavernier obtained a large (112.5 carats, where 1 carat = 200 milligrams), rare blue diamond in India, perhaps from a Hindu statue, and carried it back to France. King Louis XIV bought the diamond and had it fashioned into a jewel of 68 carats. This jewel vanished in 1762 during a burglary. Perhaps it was lost forever— perhaps not. In 1830, a 44.5-carat blue diamond mysteriously appeared on the jewel market for sale. Henry Hope, a British banker, purchased the stone,

BOX 5.3 CONSIDER THIS . . .

Where Do Diamonds Come From?

Diamonds consist of carbon, which typically accumulates only at or near Earth's surface. Experiments demonstrate that the temperatures and pressures needed to form diamond are so extreme that in nature they generally occur only at depths of around 150 km below the Earth's surface. Under these conditions, the carbon atoms that were configured in hexagonal sheets in graphite rearrange to form the much stronger and more compact structure of diamond. (Because engineers can duplicate these conditions in the laboratory, corporations manufacture several tons of synthetic diamonds a year.)

How does carbon get down into the mantle, where it transforms into diamond? Geologists speculate that subduction or collision provides the means of carrying carbon-containing rocks and sediments from the Earth's surface down to depth. This carbon transforms into diamond, some of which becomes trapped beneath continents. But if diamonds form at great depth, then how do they return to the surface? One possibility is that the process of rifting cracks the continental crust and causes a small part of the underlying lithospheric mantle to melt. Magma generated during this process rises to the surface, bringing the diamonds up with it. Near the surface, the magma cools and solidifies to form a special kind of igneous rock called *kimberlite* (named for Kimberley, South Africa, where it was first found). Diamonds brought up with the magma are embedded in the kimberlite (**Fig. Bx5.3**). Kimberlite magma contains a lot of dissolved gas and thus froths to the surface very rapidly. Kimberlite rock commonly occurs in carrot-shaped bodies called kimberlite pipes that are 50 to 200 m across and at least 1 km deep.

Controversial measurements suggest that many of the diamonds that sparkle on engagement rings today were formed 3.2 billion years ago. The diamonds sat at depths of 150 km in the Earth until two rifting events, one of which took place in the late Precambrian and the other during the late Mesozoic, released them to the surface, like genies out of a bottle. The Mesozoic rifting event led to the breakup of Pangaea.

In places where diamonds occur in solid kimberlite, they can be obtained only by digging up the kimberlite and crushing it, to separate out the diamonds. But nature can also break diamonds free from the Earth on its own. In places where kimberlite has been exposed at the ground surface for a long time, the rock chemically reacts with water and air (a process called weathering; see Interlude B). These reactions cause most minerals in kimberlite to disintegrate, creating sediment that washes away in rivers. Diamonds are so hard that they remain as solid grains in river gravel. Thus, many diamonds have been obtained simply by separating them from recent or ancient river gravel.

Diamond-bearing kimberlite pipes occur in many places around the world, particularly where very old continental lithosphere exists. Southern and central Africa, Siberia, northwestern Canada, India, Brazil, Borneo, Australia, and the U.S. Rocky Mountains all have pipes. Rivers and glaciers, however, have transported diamond-bearing sediments great distances from their original sources. In fact, diamonds have even been found in farm fields of the midwestern United States.

Not all natural diamonds are valuable; value depends on color and clarity. Diamonds that contain imperfections (cracks or specks of other material) or are dark gray in color are not used for jewelry. These stones, called *industrial diamonds*, are used instead as abrasives, for diamond powder is so hard (10 on the Mohs hardness scale) that it can be used to grind away any other substance. Gem-quality diamonds come in a range of sizes. Jewelers measure diamond size in carats, where one *carat* equals 200 milligrams (0.2 grams). In English units of measurement, one ounce equals 142 carats. (Note that a carat measures gemstone weight, whereas a *karat* specifies the purity of gold.) The largest diamond ever found, a stone called the Cullinan Diamond, was discovered in South Africa in 1905. It weighed 3,106 carats (621 grams) before being cut. By comparison, the diamond on a typical engagement ring weighs less than one carat. Diamonds are rare but not as rare as their price perhaps suggests. A worldwide consortium of diamond producers stockpile the stones so as not to flood the market and drive the price down.

FIGURE Bx5.3 Diamond occurrences.

A diamond mine pit.

A diamond embedded in solid kimberlite.

which then became known as the Hope Diamond (**Fig. 5.17**). It changed hands several times until 1958, when the famous New York jeweler Harry Winston donated it to the Smithsonian Institution in Washington, D.C., where it now sits behind bulletproof glass in a heavily guarded display.

What makes stones such as the Hope Diamond so special that people risk life and fortune to obtain them? What is the difference among gemstones, gems, and other minerals? A gemstone is a mineral that has special value because it is rare and people consider it beautiful. A **gem** is a cut and finished gemstone ready to be set in jewelry. Jewelers distinguish between precious stones (such as diamond, ruby, sapphire, and emerald), which are particularly rare and expensive, and semiprecious stones (such as topaz, tourmaline, aquamarine, and garnet), which are not as rare or expensive. All the stones mentioned so far are transparent crystals, though most have some color (**Table 5.2**). The category of semiprecious stones also includes opaque or translucent minerals such as lapis, malachite (see Fig. 5.2a), and opal.

In everyday language, pearls and amber may also be considered gemstones. Unlike diamonds and garnets, which form inorganically in rocks, pearls form in living oysters when the oyster extracts calcium and carbonate ions from water and precipitates them around an impurity, such as a sand grain, embedded in its body. Thus, pearls are a result of biomineralization. Most pearls used in jewelry today are "cultured" pearls, made by artificially introducing round sand grains into oysters in order to stimulate pearl production. Amber is also formed by organic processes—it consists of fossilized tree sap. But because amber consists of organic compounds that are not arranged in a crystal structure, it does not meet the definition of a mineral.

"Rare" means hard to find, and some gemstones are indeed hard to find. Many diamond localities, for example, occur in isolated regions of Congo, South Africa, Brazil, Canada, Rus-

FIGURE 5.17 The Hope Diamond, now on display at the Smithsonian Institution in Washington, D.C.

sia, India, and Borneo (see Box 5.3). In some cases, it is not the mineral itself but rather the "gem-quality" versions of the mineral that are rare. For example, garnets are found in many rocks in such abundance that people use them as industrial abrasives. But most garnets are quite small and contain inclusions (specks of other minerals and/or bubbles) or fractures, so they are not particularly beautiful. Gem-quality garnets—clean, clear, large, unfractured crystals—are unusual. In some cases, gemstones are merely pretty and rare versions of more common minerals. For example, ruby is a special version of the common mineral corundum (Al_2O_3), emerald is a special version of the common mineral beryl (**Fig. 5.18a**), and peridot is the gem version of the common mineral olivine. As for the beauty of a gemstone, this quality lies basically in its color and, in the case of transparent gems, its "fire"—the way the mineral bends and internally reflects the light passing through it and disperses the light into a spectrum. Fire makes a diamond sparkle more than a similarly cut piece of glass.

Gemstones form in many ways. Some solidify from a melt, some form by diffusion, some precipitate out of a water solution in cracks, and some are a consequence of the chemical interaction of rock with water near the Earth's surface. Many gems come from pegmatites, particularly coarse-grained rocks formed by the solidification of steamy melt.

Most gems used in jewelry are "cut" stones, meaning that they are not raw crystals right from the ground but rather have been faceted. The smooth **facets** on a gem are ground and polished surfaces made with a faceting machine (**Fig. 5.18b**). Facets are not the natural crystal faces of the mineral, nor are they cleavage planes, though gem cutters sometimes make the facets parallel to cleavage directions and will try to break a

Did you ever wonder . . .

how jewelers make the facets on a jewel?

TABLE 5.2 Precious and semiprecious materials used in jewelry

Gem Name	Material/Formula	Comments
Amber	Fossilized tree sap	Composed of organic chemicals; amber is not strictly a mineral.
Amethyst	Quartz/SiO_2	The best examples precipitate from water in openings in igneous rocks; a deep purple version of quartz.
Aquamarine	Beryl/$Be_3Al_2Si_6O_{18}$	A bluish version of emerald.
Diamond	Diamond/C	Brought to the surface from the mantle in igneous bodies called diamond pipes; may later be mixed in deposits of sediment.
Emerald	Beryl/$Be_3Al_2Si_6O_{18}$	Occurs in coarse igneous rocks (pegmatites; see Chapter 6).
Garnet	Garnet/(e.g., $Mg_3Al_2[SiO_4]_3$)	A variety of types differ in composition (Ca, Fe, Mg, and Mn versions); occurs in metamorphic rocks (see Chapter 8).
Jade	Jadeite/$NaAlSi_2O_6$ Nephrite/$Ca_2(Mg,Fe)_5Si_8O_{22}(OH)_2$	Jade can be one of two minerals, jadeite (a pyroxene) or nephrite (an amphibole); both occur in metamorphic rocks.
Opal	Composed of microscopic spheres of hydrated silica packed together	Most opal comes from a single mining district in central Australia; occurs in bedrock that has reacted with water near the surface.
Pearl	Aragonite/$CaCO_3$	Formed by oysters, which secrete coatings around sand grains that are accidentally embedded in the soft parts of the organism. Cultured pearls are formed the same way, but the impurity is a spherical bead that is intentionally introduced.
Ruby	Corundum/Al_2O_3	The red color is due to chromium impurities; found in coarse igneous rocks called pegmatites and as a result of contact metamorphism (see Chapters 6 and 8).
Sapphire	Corundum/Al_2O_3	A blue version of ruby.
Topaz	$Al_2SiO_4(F,OH)_2$	Found in igneous rocks, a result of the reaction of rock with hot water.
Tourmaline	$Na(Mg,Fe)_3Al_6(BO_3)_3(Si_6O_{18})(OH,F)_4$	Forms in igneous and metamorphic rocks.
Turquoise	$CuAl_6(PO_4)_4(OH)_8 \cdot 4H_2O$	Found in copper-bearing rocks; a popular jewelry gem in the American Southwest.

large gemstone into smaller pieces by splitting it on a cleavage plane. A faceting machine consists of a doping arm, a device that holds a stone in a specific orientation, and a lap, a rotating disk covered with a wet paste of grinding powder and water. The gem cutter fixes a gemstone to the end of the doping arm and positions the arm so that it holds the stone against the moving lap. The movement of the lap grinds a facet. When the facet is complete, the gem cutter rotates the arm by a specific angle, lowers the stone, and grinds another facet. The geometry of the facets defines the cut of the stone. Different cuts have names, such as "brilliant," "French," "star," and "pear." Grinding facets is a lot of work—a typical engagement-ring diamond with a brilliant cut has 57 facets (**Fig. 5.18c**)!

Some mineral specimens have special value simply because their geometry and color before cutting are beautiful. Prize specimens exhibit shapes and colors reminiscent of fine art and may sell for tens of thousands of dollars or more (see **Fig. 5.18d** and **Another View**). It's no wonder that mineral "hounds" risk their necks looking for a cluster of crystals protruding from the dripping roof of a collapsing mine or hidden in a crack near the smoking summit of a volcano.

Take-Home Message

Gemstones are particularly rare and beautiful minerals. The gems or jewels found in jewelry have been faceted using a lap. The facets are not natural crystal faces or cleavage surfaces. The fire of a jewel comes from the way it reflects light internally.

QUICK QUESTION: What's the difference between a facet on a gem and a crystal face?

FIGURE 5.18 Cutting gemstones.

Non-gem-quality beryl

Rough emerald

Cut emerald

(a) Emerald is a green, transparent variety of the mineral beryl.

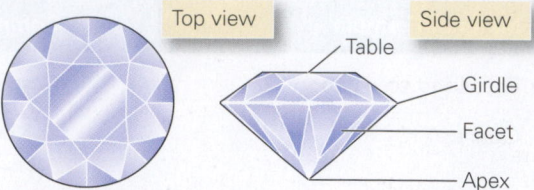

Top view

Side view

Table

Girdle

Facet

Apex

(c) There are many different "cuts" for a gem. Here we see the top and side views of a brilliant-cut diamond.

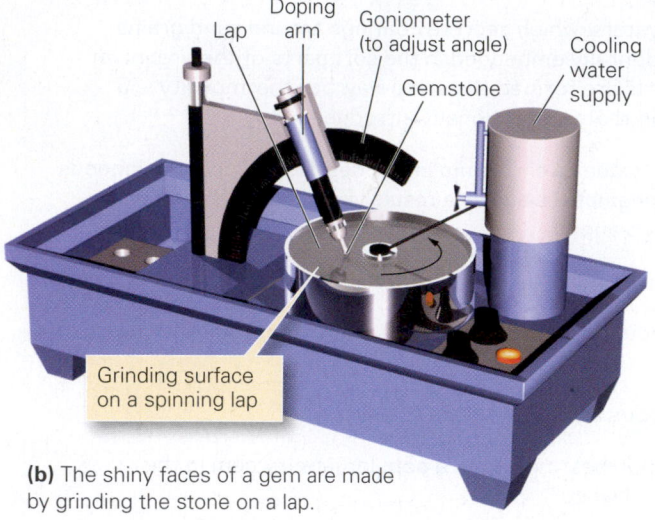

Lap

Doping arm

Goniometer (to adjust angle)

Gemstone

Cooling water supply

Grinding surface on a spinning lap

(b) The shiny faces of a gem are made by grinding the stone on a lap.

(d) Corundrum (Al_2O_3) comes in many colors. Gem-quality versions including ruby and sapphire can be "cut" into many shapes.

CHAPTER SUMMARY

- Minerals are naturally occurring, solid substances, formed by geologic processes, with a definable chemical composition and an internal structure characterized by an orderly arrangement of atoms, ions, or molecules in a crystalline lattice. Most minerals are inorganic.
- In the crystalline lattice of minerals, atoms occur in a specific pattern—one of nature's finest examples of ordering.

- Minerals can form by the solidification of a melt, precipitation from a water solution, diffusion through a solid, the metabolism of organisms, and precipitation from a gas.
- About 4,000 different types of minerals are known, each with a name and distinctive physical properties (such as color, streak, luster, hardness, specific gravity, crystal habit, and cleavage).

- The unique physical properties of a mineral reflect its chemical composition and crystal structure. By observing these physical properties, you can identify minerals.
- The most convenient way for classifying minerals is to group them according to their chemical composition. Mineral classes include silicates, oxides, sulfides, sulfates, halides, carbonates, and native metals.
- The silicate minerals are the most common on Earth. The silicon-oxygen tetrahedron, a silicon atom surrounded by four oxygen atoms, provides the fundamental building block of silicate minerals.
- Groups of silicate minerals are distinguished from each other by the ways in which the silicon-oxygen tetrahedra that constitute them are linked.
- Gemstones are minerals known for their beauty and rarity. The facets on cut gems used in jewelry are made by grinding and polishing the stones with a faceting machine.

GUIDE TERMS

atom (p. 120)
atomic number (p. 120)
atomic mass (p. 121)
biogenic mineral (p. 118)
carbonate mineral (p. 131)
chemical (p. 121)
chemical bond (p. 121)
chemical formula (p. 121)
chemical reaction (p. 121)
cleavage (p. 129)
color (p. 127)

concentration (p. 121)
conchoidal fracture (p. 129)
crystal (p. 122)
crystal face (p. 122)
crystal habit (p. 129)
crystal lattice (p. 119)
crystalline solid (p. 119)
crystal structure (p. 122)
diffraction (p. 122)
element (p. 120)
facet (p. 136)

gem (p. 136)
glass (p. 119)
hardness (p. 127)
ion (p. 121)
luster (p. 127)
mineral (p. 118)
mineral classes (p. 129)
mineralogist (p. 117)
mineralogy (p. 118)
Mohs hardness scale (p. 127)
molecule (p. 121)

polymorph (p. 124)
precipitate (p. 121)
silicate (p. 131)
silicate minerals (p. 129)
silicon-oxygen tetrahedron (p. 131)
specific gravity (p. 128)
streak (p. 127)

REVIEW QUESTIONS

1. What is a mineral, as geologists understand the term? How is this definition different from the everyday usage of the word?
2. Why is glass not a mineral?
3. Salt is a mineral, but the plastic making up an inexpensive pen is not. Why not?
4. Describe the several ways that mineral crystals can form.
5. Why do some minerals occur as euhedral crystals, whereas others occur as anhedral grains?
6. List and define the principal physical properties used to identify a mineral.
7. How can you determine the hardness of a mineral? What is the Mohs hardness scale?

8. How do you distinguish cleavage surfaces from crystal faces on a mineral? How does each type of surface form?
9. What is the prime characteristic that geologists use to separate minerals into classes?
10. On what basis do mineralogists organize silicate minerals into distinct groups?
11. What is the relationship between the way in which silicon-oxygen tetrahedra bond in micas and the characteristic cleavage of micas?
12. Why are some minerals considered gemstones? How do you make the facets on a gem?

ON FURTHER THOUGHT

13. Compare the chemical formula of magnetite with that of biotite. Why is magnetite mined to obtain iron supplies, but biotite is not?

14. Imagine that you are given two milky white crystals, each about 2 cm across. You are told that one of the crystals is composed of plagioclase and the other of quartz. How can you determine which is which?

15. Could you use crushed calcite to grind and form facets on a diamond? Why or why not?

 **smartwork**   smartwork.wwnorton.com

This chapter's Smartwork features:

- A "What a Geologist Sees" exercise on quartz crystal faces.
- Interactive problems covering the Mohs hardness scale.
- Video exercises on volcano deformation, gases, and seismic monitoring.

GEOTOURS

This chapter's GeoTour exercise (C) features:

- Diamond mines
- Mineral reactions after coal mining

Another View   A spectacular museum specimen of watermelon tourmaline. The beautiful arrangement of colors and shapes makes such specimens very valuable.

The first railroad through the Sierra Nevada followed the Truckee River. Countless tons of rock had to be blasted and moved by laborers.

INTERLUDE A

Introducing Rocks

LEARNING OBJECTIVES

By the end of this interlude, you should understand . . .

- the geologic definition of *rock*.
- the basis that geologists use to classify rock into three classes.
- the tools that can be used to study rocks.

A.1 Introduction

During the 1849 gold rush in the Sierra Nevada of California, only a few lucky individuals actually became rich. The rest of the "forty-niners" either slunk home in debt or took up less glamorous jobs in boom towns such as San Francisco. These towns grew rapidly, and soon the American west coast was demanding large quantities of manufactured goods from east-coast factories. Making the goods was no problem, but getting them to California meant either a stormy ocean voyage or a trek with stubborn mule teams through the deserts of Nevada and Utah. The time was ripe to build a railroad linking the east and west coasts of North America, so with much fanfare, a consortium of railroads set to work in 1863. The Union Pacific surveyed a route that crossed the Sierras, and as the Civil War raged, the company transported thousands of Chinese laborers across the Pacific in the squalor of unventilated cargo holds and set them to work chipping ledges around, and blasting tunnels through, the range's towering peaks. Sadly, untold numbers of laborers died of frostbite and exhaustion, and from mistimed blasts, landslides, and avalanches.

Through their efforts, transcontinental railroad builders certainly gained an intimate knowledge of how rock feels and behaves—it's solid, heavy, and mighty hard! They also found that some rocks break easily into layers but others do not, and some rocks are dark colored while others are light colored. Like anyone who looks closely at rock exposures, they realized that rocks are not just gray, featureless masses but rather come in a great variety of colors, textures, and configurations.

Why are there so many distinct types of rocks? The answer is simple: many different processes can produce rocks, and many different materials can form rocks. Because of the relationship between rock type and the process of formation, rocks provide a record of geologic events over Earth's history and give insight into interactions among components of the Earth System. The next few chapters are devoted to a discussion of rocks and a description of how rocks form. To provide a general introduction to these chapters, this Interlude provides the geological definition of the term *rock*, describes the basic components of rock, and characterizes the three principal classes of rocks. We also describe a few of the methods that geologists use to study rocks.

A.2 What Is Rock?

To a geologist, **rock** is a coherent, naturally occurring solid that consists of an aggregate of minerals or, less common, a body of glass. To understand this definition, let's look at its components more closely.

- *Coherent:* A rock holds together and thus must be broken in order to be separated into smaller pieces. As a result of its coherence, rock can form cliffs or can be carved into construction blocks or sculptures (**Fig. A.1**). A pile of unattached mineral grains does not constitute a rock.

- *Naturally occurring:* Geologists consider only materials formed by natural processes to be rocks. Manufactured products (concrete, bricks, or cinder blocks) are not rocks.

- *An aggregate of minerals or a body of glass:* The vast majority of rocks consist of an aggregate (a collection) of many mineral grains and/or crystals. Note that the terms *grain* and *crystal* are often used interchangeably. But to be picky, a **grain** is a more general term that can refer to a piece of mineral, to a fragment broken off a once larger piece of a mineral or off a pre-existing rock, or to a fragment of glass. The term **crystal** should be used only for a continuous (uninterrupted) piece of a single mineral that grew to its present shape and may display crystal faces. (A large crystal, or aggregate of crystals, of a single mineral may be called a *mineral specimen*, even if it is meters long.) Some rocks contain only one kind of mineral, whereas others contain several different kinds. A few rock types that form at volcanoes consist of natural glass, which may occur either as a homogeneous body or as an accumulation of tiny shards.

What holds rock together? Grains in rock connect to form a coherent mass either because they are stuck together by a natural **cement** composed of minerals that precipitated from water in the space between grains (**Fig. A.2a**) or because they interlock with one another and thus fit together like pieces in a jigsaw puzzle (**Fig. A.2b**). Rocks whose grains are held in place by cement are called **clastic rocks**, whereas rocks whose crystals interlock with one another are called **crystalline rocks**. A glassy rock can be coherent either because it originated as a continuous mass (that is, it does not contain separate grains) or because it formed when separate glass grains welded together while still hot.

FIGURE A.1 Rock is coherent, and thus can be very durable.

(a) The strength of rocks can hold up tall cliffs.

(b) Rock, because of its durability, was used to make this castle in Portugal.

FIGURE A.2 Rocks, aggregates of mineral grains and/or crystals, can be clastic or crystalline.

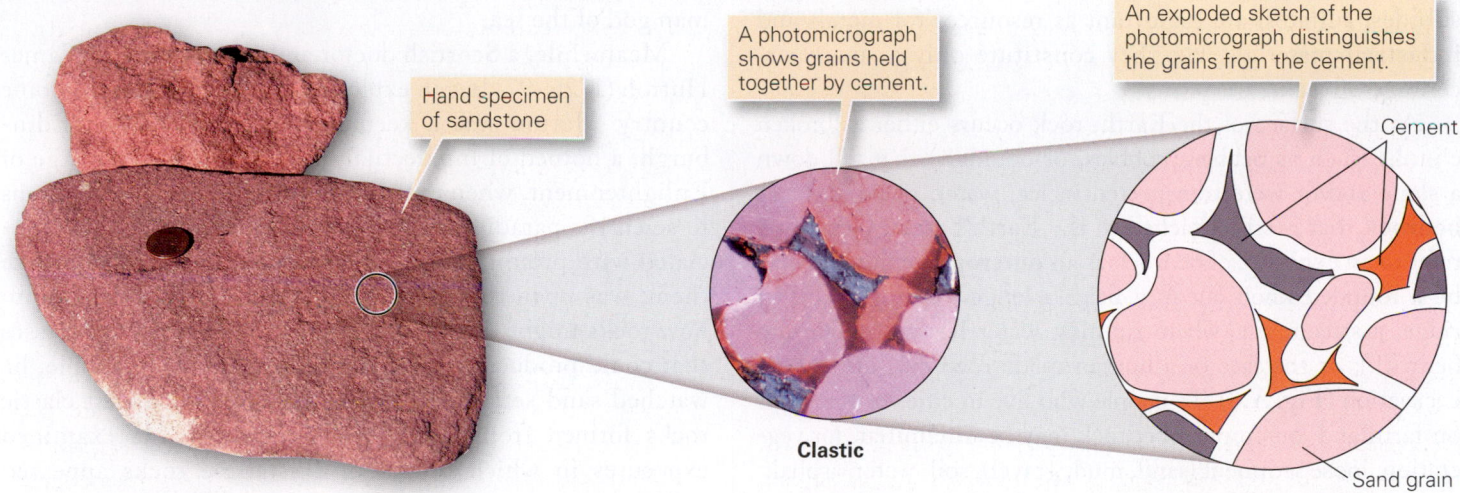

Hand specimen of sandstone

A photomicrograph shows grains held together by cement.

An exploded sketch of the photomicrograph distinguishes the grains from the cement.

Cement

Clastic

Sand grain

(a) Clastic texture is illustrated by the grains and cement in a sandstone.

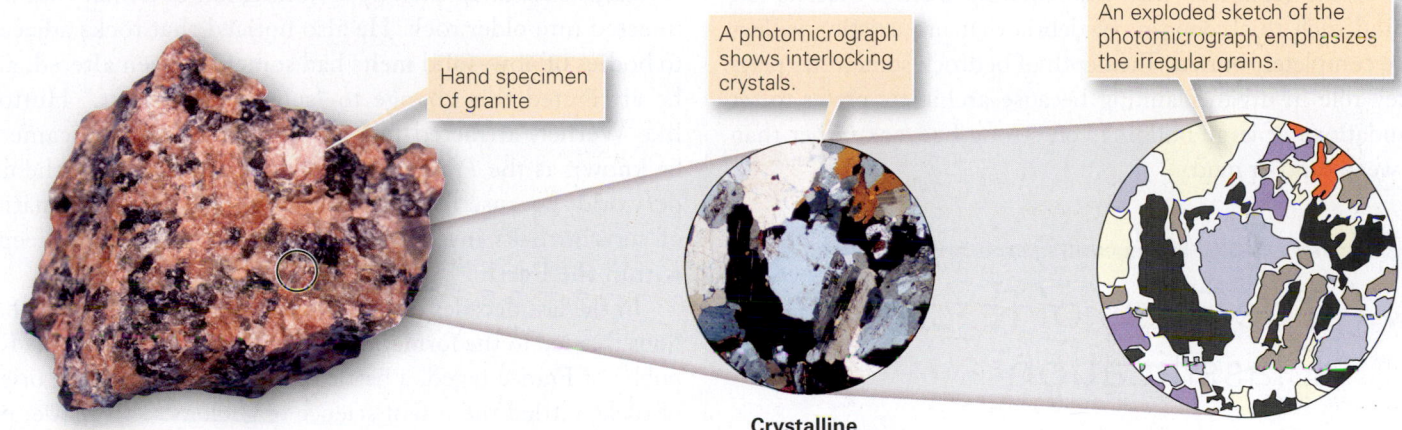

Hand specimen of granite

A photomicrograph shows interlocking crystals.

An exploded sketch of the photomicrograph emphasizes the irregular grains.

Crystalline

(b) Crystalline texture is illustrated by the interlocking crystals in a granite.

Regardless of whether a rock consists of glass or minerals, we can think of it, fundamentally, as a volume of chemicals. Significantly, not all rocks contain the same chemicals. For example, *granite*—a rock commonly used for gravestones, building facades, and kitchen counters—contains oxygen, silicon, aluminum, potassium, sodium, calcium, iron, and magnesium, whereas *marble*—a rock favored by sculptors—contains oxygen, carbon, and calcium. In both granite and marble, the elements are incorporated in minerals. Granite contains quartz, feldspar, mica, pyroxene, and amphibole, whereas marble contains mainly calcite.

By averaging the chemical compositions of rocks, geologists have determined that oxygen and silicon are by far the most common elements in rocks of the Earth's crust and mantle (**Table A.1**). That's because most rock consists of silicate minerals. At or near the Earth's surface, however, living organisms play a role in the formation of some rock types, so a significant proportion of rock in the upper crust of continents consists of carbonate minerals extracted

TABLE A.1 Chemical Composition of Rocks in the Earth (Weight %)

Element	Mantle	Crust (continental)
O	44.8	46.6
Si	21.5	27.7
Mg	22.8	1.5
Fe	5.8	5.0
Al	2.2	8.1
Ca	2.3	3.6
Na	0.3	2.8
K	0.03	2.6
Other	0.3	2.1

from water to form shells. Other minerals (such as oxides, sulfides, sulfates) are important as resources for metals and industrial materials, but they constitute only a small percentage of rocks.

At the surface of the Earth, rock occurs either as broken chunks (such as pebbles, cobbles, or boulders) that fell down a slope and/or were transported in ice, water, or wind, or as **bedrock** that is still attached to the Earth's crust. Geologists refer to an exposure of bedrock as an **outcrop**. An outcrop may be a rounded knob out in a field; a ledge forming a cliff or ridge; a stream cut (where running water has cut down into bedrock); or the face of a human-made road cut, rail cut, or excavation (**Fig. A.3**). To people who live in cities or forests or on farmland, outcrops of bedrock may be unfamiliar, for vegetation, loose sediment (sand, mud, gravel), soil, water, asphalt, concrete, or buildings cover bedrock. In fact, in the northern part of the American Midwest, melting ice-age glaciers left behind such thick deposits of debris that pre-existing valleys were completely buried. The depth of bedrock sometimes plays a key role in urban planning because architects prefer to set foundations of large buildings on strong bedrock rather than on weak sand or mud.

A.3 The Basis of Rock Classification

When the systematic study of geology began in the late 18th century, geologists began to grapple with the challenge of developing a meaningful classification scheme for rocks, for classification helps organize information and helps to emphasize similarities and differences. One of the earliest classification schemes, popular in the late 18th century, distinguished three groups of earth materials—so-called primary, secondary, and tertiary—based on the incorrect perception that the groups had formed in a time succession. According to this scheme, first formulated by Abraham Werner (1749–1817), a German mineralogist, a "universal ocean" containing dissolved minerals covered the Earth when the planet first formed. Precipitation of minerals from this solution produced early "primary materials," which included hard crystalline rocks, like granite. Then later, as sea level dropped, the action of rivers, waves, and wind wore down exposed primary rocks and produced debris that settled in the ocean to form "secondary materials," which occurred in layers over the primary material. Loose gravels, sands, and muds—materials that had not turned to rock—made up the most recent "tertiary materials." Because the classification scheme viewed most rock formation as a process that only took place in water, Werner's

followers came to be known as the *Neptunists*, after the Roman god of the sea.

Meanwhile, a Scottish doctor and scientist named James Hutton (1726–97) began exploring the outcrops in his home country. Hutton was a keen thinker who lived in Edinburgh, a hotbed of intellectual discourse during the Age of Enlightenment, when everything from political institutions to scientific paradigms became fodder for debate. He associated with prominent philosophers and scientists and, like them, was open to new ideas. Hutton sought insight into how rocks might form by looking for evidence of processes that could produce characteristics of rock. For example, he watched sand settle on a beach and concluded that clastic rocks formed from cementation of grains. He examined exposures in which bodies of crystalline rocks appeared to have pushed into other rocks and concluded that these crystalline rocks formed by solidification of a melt that had injected into older rock. He also noticed that rocks adjacent to bodies of now-solid melts had somehow been altered, and he attributed this change to "subterranean heat." Hutton, like Werner, attracted followers—Hutton's group came to be known as the *Plutonists*, after the Roman god of the underworld, because they favored the idea that the formation of certain rocks involved melts that had risen from deeper within the Earth.

In the last decades of the 18th century, as the armed rebellions that led to the formation of the United States and the Republic of France raged, a battle of ideas concerning the origin of rocks rattled the infant science of geology. This battle, pitting the Neptunists against the Plutonists, lasted for years. In the end, the Plutonists won, for they eventually demonstrated beyond a shadow of a doubt that certain crystalline rocks must have been in molten form when emplaced. Geologists came to understand that different rocks formed in different ways and concluded that rocks can best be classified on the basis of how they formed. This *genetic scheme* for classifying rocks—a scheme based on the origin (genesis) of rocks—became a foundation for the modern classification of rock. Because Hutton fostered this key idea, as well as many other ideas that we will describe later in the book, modern geologists revere Hutton as the "father of geology."

In the modern genetic scheme of rock classification, geologists recognize three basic rock classes: (1) **igneous rocks**, which form by the freezing (solidification) of molten rock (**Fig. A.4a**); (2) **sedimentary rocks**, which form either by the cementing together of grains broken off pre-existing rocks or by the precipitation of mineral crystals out of water solutions at or near the Earth's surface (**Fig. A.4b**); and (3) **metamorphic rocks**, which form when pre-existing rocks change character in response to a change in pressure and temperature conditions, and/or as a result of squashing,

FIGURE A.3 Types of rock exposures; the left column illustrates natural outcrops, the right column were produced during construction.

Outcrop in the woods, Illinois

Road cut, Maryland

Stream cut, New York

Quarry, Illinois

Mountain cliffs, Colorado

Railroad cut, Illinois

stretching, or shearing under conditions such that rock does not crack and break (**Fig. A.4c**). Metamorphic change occurs in the solid state, which means that it does not involve melting. In the context of modern plate tectonics theory, different rock types form in different geologic settings (**Fig. A.5**). In succeeding chapters, we will explore these settings in more detail.

Each of the three classes contains many different individual rock types, distinguished from one another by physical characteristics such as the following:

- *Grain size and shape:* The grains in rock come in a wide range of sizes. Some grains are so small that they can't be seen without a microscope, whereas others are as big as a car or larger. Rocks also differ in terms of the range

FIGURE A.4 Examples of the three major classes of rocks.

Igneous

(a) Lava (molten rock) freezes to form igneous rock. Here, the molten tip of a brand-new flow still glows red. Older flows are already solid.

Sedimentary

(b) Sand, formed from grains eroded off the rock cliffs, collects on the beach. If buried and turned to rock, it becomes layers of sandstone, such as those making up cliffs.

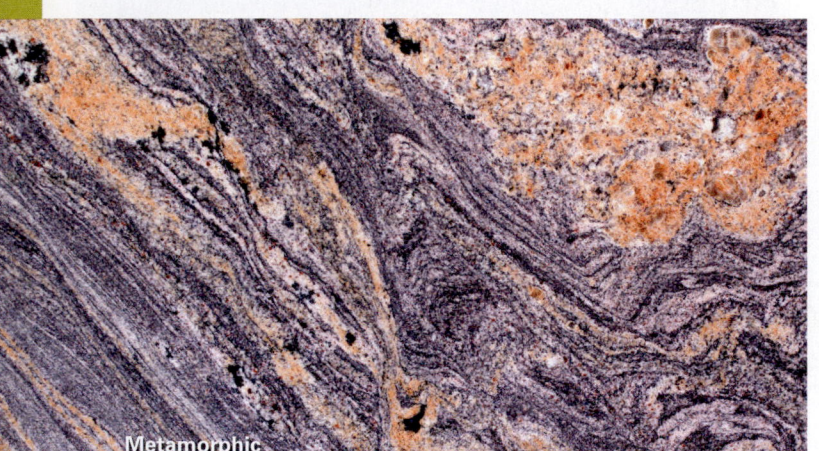

Metamorphic

(c) Metamorphic rock forms when preexisting rocks endure changes in temperature and pressure and/or are subjected to shearing, stretching, or shortening. New minerals and textures form.

of grain sizes that the rock contains and also in terms of grain shape (**Fig. A.6**). Specifically, in some rocks, all grains are the same size, whereas other rocks contain grains of many different sizes; and in some rocks, all grains are **equant**, meaning that they have the same dimensions in all direction, whereas in others the grains are **inequant**, meaning that the dimensions are not the same in all directions.

• *Composition:* The term **rock composition** refers to the proportions of different chemicals that make up the rock. The proportion of chemicals, in turn, affects the proportion of different minerals constituting the rock. As you will see, however, chemical composition alone does not completely control the assemblage of minerals present in a rock. For example, two rocks with exactly the same chemical

FIGURE A.5 A cross section illustrating various geologic settings in which rocks form.

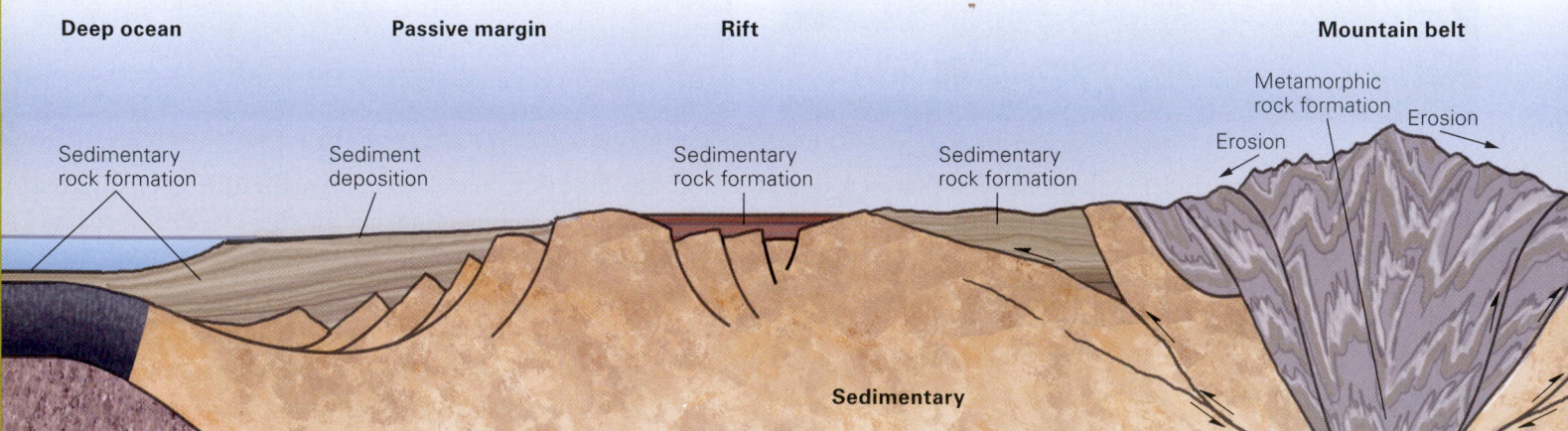

Deep ocean   Passive margin   Rift   Mountain belt

Metamorphic rock formation

Erosion

Erosion

Sedimentary rock formation

Sediment deposition

Sedimentary rock formation

Sedimentary rock formation

Sedimentary

composition can have totally different assemblages of minerals if each rock formed under different pressure and temperature conditions. That's because mineral formation is affected by environmental factors such as pressure and temperature.

- *Texture:* This term refers to the configuration of grains in a rock, that is, the way grains connect to one another and whether or not inequant grains are aligned parallel to each other. The concept of rock texture will become easier to grasp as we look at different examples of rocks in the following chapters.

- *Layering:* Some rock bodies appear to contain distinct layering, defined either by bands of different compositions, grain sizes, or textures or by the alignment of inequant grains so that they trend parallel to each other. Different types of layering occur in different kinds of rocks. For example, the layering in sedimentary rocks is called **bedding**, whereas the layering in metamorphic rocks is called **metamorphic foliation** (**Fig. A.7**).

Each distinct rock type has a name, which comes from a variety of sources. Some names reflect the dominant mineral making up the rock, some are derived from the name of the region where the rock was first discovered or is particularly abundant, some from ancient legends, some from a root word of Latin or Greek origin, and some from a traditional name used by people in an area where the rock is found. Many rock names date back to antiquity, but some were assigned only in recent decades. All told, there are hundreds of different rock names, though in this book we introduce only about 30.

FIGURE A.6 Describing grains in rock.

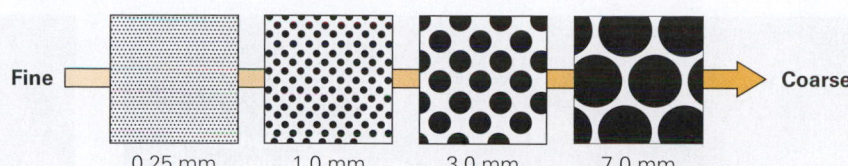

(a) Geologists define grain size by using this comparison chart.

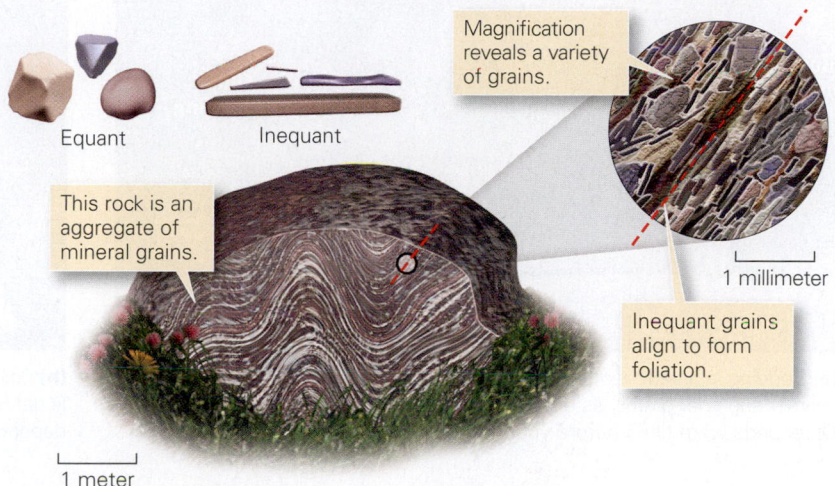

This rock is an aggregate of mineral grains.

Magnification reveals a variety of grains.

Inequant grains align to form foliation.

(b) Grains in rock come in a variety of shapes. Some are equant, whereas some are inequant. In this example of metamorphic rock, inequant grains align to define a foliation.

A.4 Studying Rock

Outcrop Observations

The study of rocks begins by examining a rock in an outcrop. If the outcrop is big enough, such an examination will reveal relationships between the rock you're interested in and the

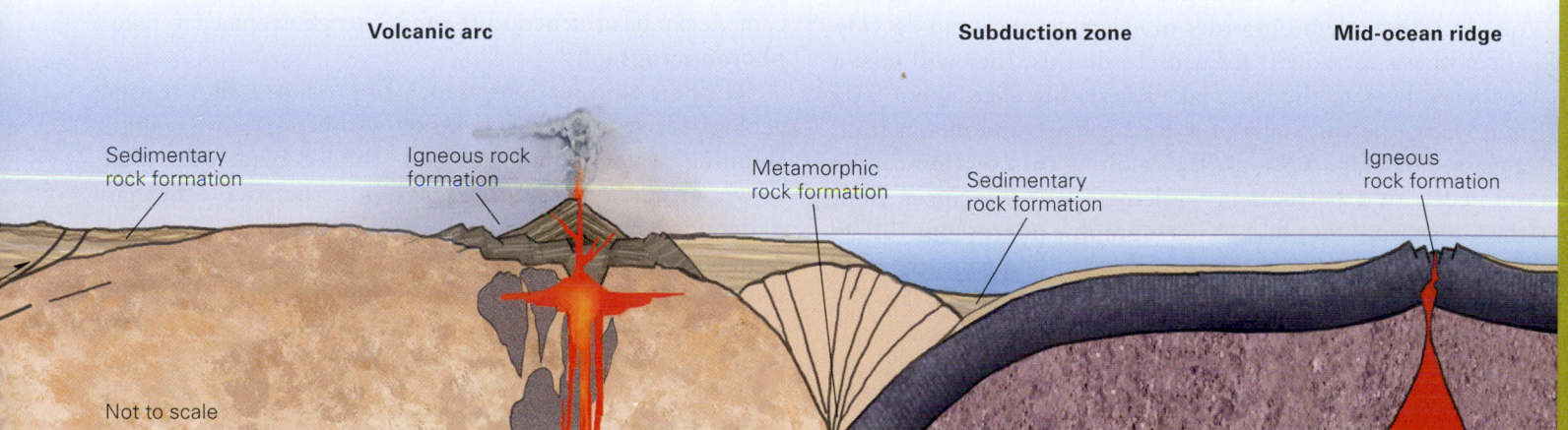

(a) Bedding in a sedimentary rock, here defined by alternating layers of coarser and finer grains, as exposed on a cliff along an Oregon beach. Older beds were tilted before younger ones were deposited.

(b) Foliation in this outcrop of metamorphic rock near Mecca, California, is defined by alternating light and dark layers. The color of the layers depends on the minerals comprising the layers.

rocks around it and will allow you to detect layering. Geologists carefully record observations about an outcrop, then use a hammer to break off a **hand specimen** (**Fig. A.8a–c**)—a fist-sized piece of a rock—that they can examine more closely with a *hand lens*, a type of high-quality magnifying glass (**Fig. A.8d**). (Breaking rocks with hammers can be dangerous and should only be done with appropriate eye protection.) Observation with a hand lens enables geologists to identify sand-sized or larger mineral grains and may enable characterization of the rock's texture.

Thin-Section Study

Geologists may have to examine rock composition and texture in minute detail in order to identify a rock and develop a hypothesis for how it formed. To do this, they will take a specimen back to the lab, make a very thin slice (about 0.03 mm thick, the thickness of a human hair), and mount it on a glass slide (**Fig. A.9a–c).** The resulting **thin section** can be studied with a petrographic microscope (*petro* comes from the Greek word for rock). A **petrographic microscope** differs from an ordinary microscope in that it illuminates the thin section with transmitted polarized light. This means that the illuminating light beam first passes through a special filter that makes all the light waves in the beam vibrate in the same plane, and then the light passes up through the thin section before entering a polarized eye piece. An observer, therefore,

looks through the thin section as if it were a stained glass window. When illuminated with transmitted polarized light, each type of mineral grain displays a unique suite of colors (**Fig. A.9d**). The specific color the observer sees depends on both the identity of the grain and its orientation with respect to the waves of polarized light, for a crystal interferes with polarized light and allows only certain wavelengths to pass through.

The brilliant colors and strange shapes visible in a thin section rival the beauty of an abstract painting. By examining a thin section with a petrographic microscope, geologists can identify most of the minerals constituting the rock and can describe the way in which the grains connect to each other. To convey information visible in thin section, a camera can be attached to the microscope eyepiece to take a **photomicrograph**.

How are thin sections made? To produce a thin section, geologists use a special rock saw with a very thin, rapidly spinning, water-cooled diamond-studded blade. The blade slowly grinds a very thin groove into the rock—the hard diamonds embedded in the saw blade scratch and pulverize minerals as the saw blade rubs against the rock. By making four cuts, geologists can produce a small rectangular block, or "chip," of rock (see Fig. A.7). Geologists then cement the chip to a glass microscope slide with an epoxy adhesive (glue) and cut off the excess rock with the rock saw. By pressing the chip facedown against a *lap,* a spinning plate

FIGURE A.8 Basic tools for studying rocks in the field.

(b) A rock hammer.

(a) A field geologist examining a hand specimen (on a cold day).

(c) A hand specimen.

(d) A hand lens.

coated with abrasive, geologists grind the chip down until only a thin slice, still cemented to the glass slide, remains, ready for examination with a petrographic microscope.

High-Tech Analytical Equipment

Beginning in the 1950s, high-tech electronic instruments became available that have enabled geologists to examine rocks on an even finer scale than is possible with a petrographic microscope. Modern research laboratories typically obtain instruments such as *scanning electron microscopes* (SEMs), which can image the surface of a rock chip at extremely high magnification and can map the distribution of elements in the chip; *electron microprobes*, which can focus a beam of electrons on a small part of a grain to create a signal that defines the chemical composition of the mineral (**Fig. A.10**); *mass spectrometers*, which analyze the proportions of different isotopes of elements contained in a rock; and *X-ray diffractometers*, which identify minerals by looking at the way X-ray beams pass through crystals in a rock. Such instruments, in conjunction with optical examination, can provide geologists with highly detailed characterizations of rocks, which in turn help them understand how the rocks formed and where the rocks came from. This information enables geologists to use the study of rocks as a basis for deciphering Earth history.

FIGURE A.9 Studying rocks in thin section.

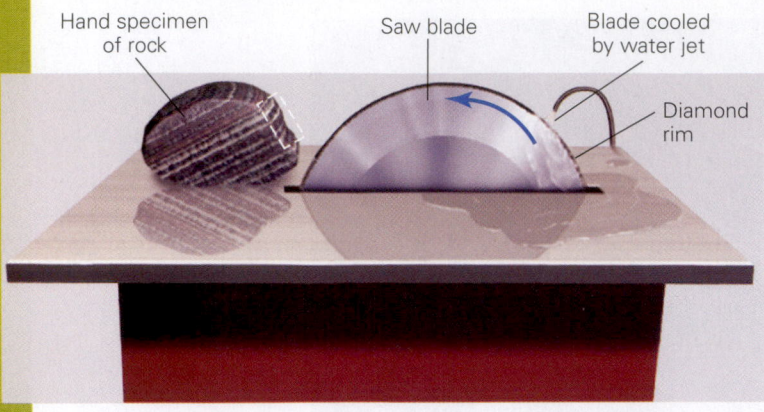

Hand specimen of rock

Saw blade

Blade cooled by water jet

Diamond rim

(a) Using a special saw, a geologist cuts a thin chip of a rock specimen.

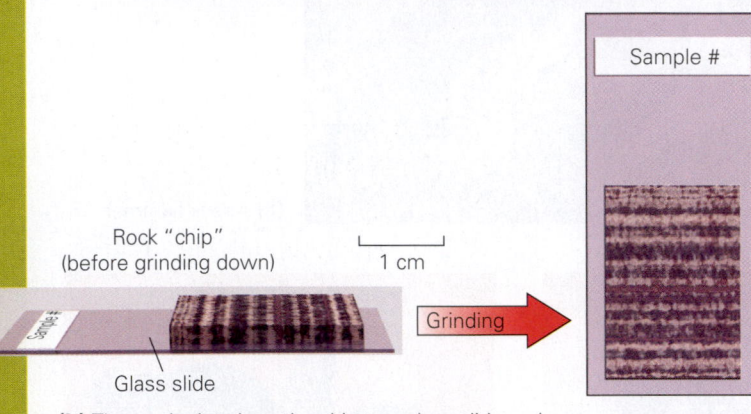

Rock "chip" (before grinding down)

1 cm

Glass slide

Grinding

Sample #

(b) The geologist glues the chip to a glass slide and grinds it down until it is so thin that light can pass through it.

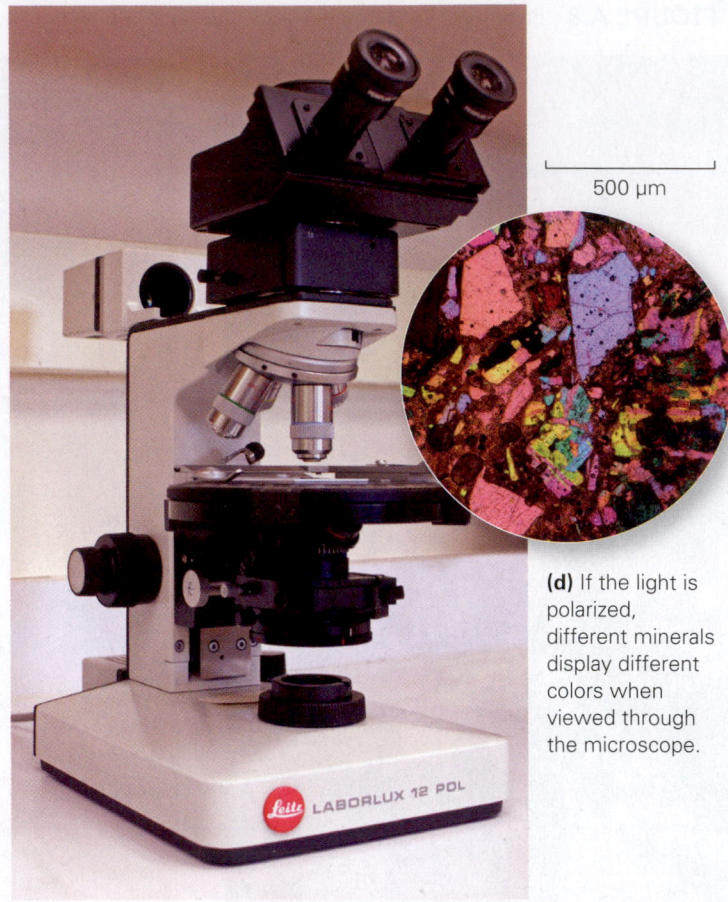

500 μm

(d) If the light is polarized, different minerals display different colors when viewed through the microscope.

(c) With a petrographic microscope, it's possible to view thin sections with light that shines through the sample from below.

FIGURE A.10 High-tech equipment for analyzing rocks.

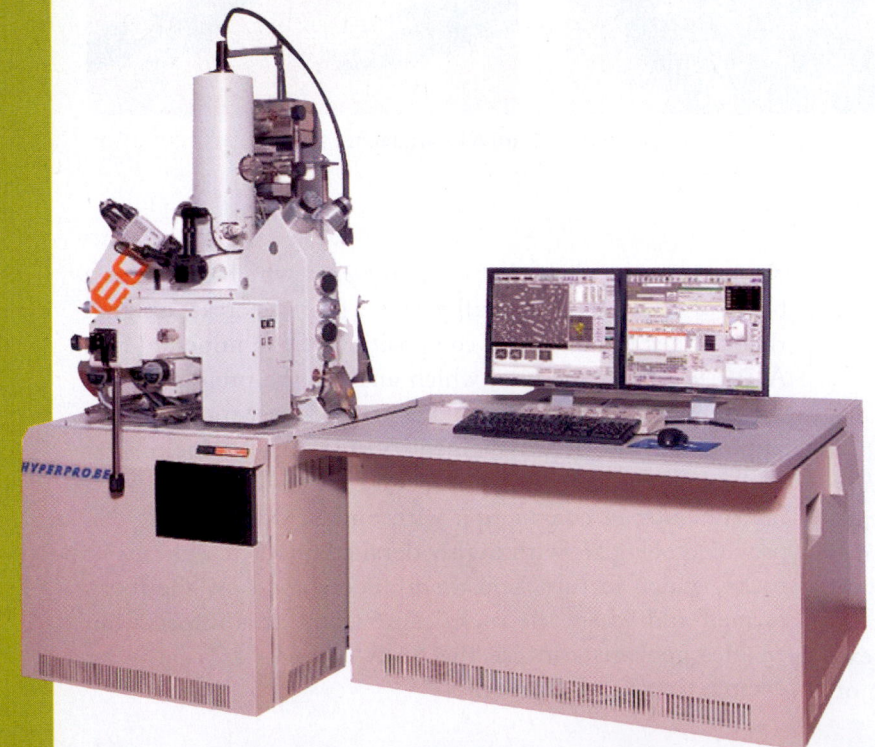

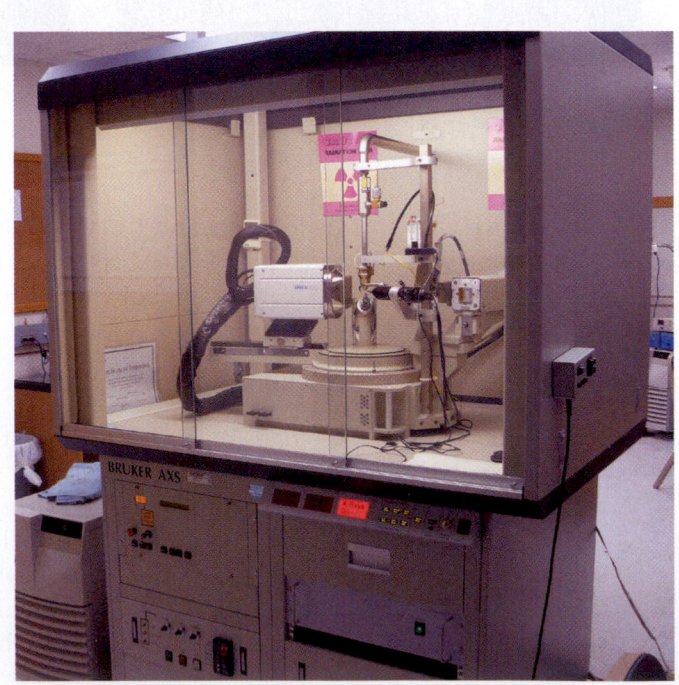

(a) An electron microprobe uses a beam of electrons to analyze the chemical composition of minerals.

(b) An X-ray diffractometer can define the crystal structure of minerals in rocks.

INTERLUDE SUMMARY

- Rock is a coherent, naturally occurring solid, consisting of an aggregate of minerals or of a mass of glass. Non-glassy rocks can be classified as crystalline or clastic.
- Bedrock consists of in-place rock that is still connected to the underlying crust. Outcrops, exposures of bedrock at the Earth's surface, can be natural or manmade.
- Geologists classify a rock as igneous, sedimentary, or metamorphic based on how the rock formed.

- A variety of characteristics prove helpful in describing rocks. Examples include: grain size, shape, composition, texture, and the nature of layering.
- Hand lenses, microscopes (after making thin sections), and sophisticated electronic equipment help geologists interpret the origin of rocks.

GUIDE TERMS

bedding (p. 147)
bedrock (p. 144)
cement (p. 142)
clastic rock (p. 142)
crystal (p. 142)
crystalline rock (p. 142)

equant (p. 146)
grain (p. 142)
hand specimen (p. 148)
igneous rock (p. 144)
inequant (p. 146)

metamorphic foliation (p. 147)
metamorphic rock (p. 144)
outcrop (p. 144)
petrographic microscope (p. 148)

photomicrograph (p. 148)
rock (p. 142)
rock composition (p. 146)
sedimentary rock (p. 144)
thin section (p. 148)

REVIEW QUESTIONS

1. What is the geologist's definition of the term, rock? Can a brick be considered to be a rock? Explain your answer.
2. Explain the difference between a clastic and crystalline rock?
3. Give examples of different kinds of rock outcrops. Can you find outcrops everywhere? Explain your answer.
4. What is the principle basis that geologists use to classify rocks into three classes? What are these classes?
5. Explain the difference between an equant and an inequant grain.
6. Where are two examples of layering that occur in rock?
7. What are thin sections, how are they examined, and what do they allow you to see?
8. What kinds of high-tech equipment can be used to study rocks?

These peaks in the Wind River Mountains of Wyoming consist of granite, formed by the cooling of molten rock over a long period of time, at depth in the crust. The granite, which formed over 2.5 billion years, did not reach the surface of the Earth until kilometers of overlying rock were eroded away.

CHAPTER 6

Up from the Inferno: Magma and Igneous Rocks

> Granite—it seems inevitable to begin with granite, even though so many
> people have ended with it, lying under those glossy pinkish slabs labeled
> in gold or black . . .
>
> —Jacquetta Hawkes (British archaeologist and writer; 1910–1996)

LEARNING OBJECTIVES

By the end of this chapter, you should understand . . . .

- the difference between magma and lava, and the characteristics of each.
- why melting only occurs at special places in the Earth.
- how melt moves to locations where it solidifies and how solidification takes place.
- why there are different kinds of igneous rocks and how to describe and classify them.
- where and why igneous activity happens, in the context of plate tectonics theory.

6.1 Introduction

Every now and then hot molten rock, or "melt," fountains from a fissure or hole on the big island of Hawaii. The incandescent liquid, which has a temperature of 1,100° to 1,200°C when first disgorged from the Earth's interior, may accumulate in a seething pool around the vent or may spill downhill as a syrupy red-yellow stream (**Fig. 6.1a**). Along the way, the melt stream burns its way through forest, submerges roads, and incinerates houses (**Fig. 6.1b**). In places close to the vent where the slope is steep and the melt follows a channel, it can move swiftly, cascading over escarpments at speeds of up to 60 km per hour, but farther from its source, where the melt has cooled and/or has spread onto gentler slopes, it slows to less than 10 km per hour. As it cools, the melt's surface darkens and crusts over, forming a carapace that insulates the still hot, sticky mass oozing below. Finally, the melt stops moving entirely and within days or weeks freezes into a hard black solid through and through (**Fig. 6.1c**). New **igneous rock**—rock made by the freezing of a melt—has formed. Considering the fiery heat of the melt from which igneous rock solidifies, the name *igneous*, from the Latin *ignis*, meaning fire, makes sense. Since we can see certain types of igneous rocks forming at the surface, it may seem like the process happens only at the surface. But that's not the case. In fact, the vast majority of igneous rock forms by cooling and solidification underground, hidden from view.

It may seem strange to speak of "freezing" in the context of forming rock, for most people think of freezing as the transformation of liquid water to solid ice when the temperature drops below 0°C (32°F). Nevertheless, the freezing of liquid melt to form solid igneous rock represents the same phenomenon—the solidification of a liquid by forming crystals and/or glass. But unlike water, igneous rock freezes at high temperatures, between 650°C and 1,100°C. To put such temperatures in perspective, keep in mind your home oven attains a maximum temperature of only 260°C (500°F).

A great variety of igneous rocks exist in the Earth's crust—they make up all of the ocean crust, beneath the thin veneer of sea-floor sediment, as well as vast volumes of the continental crust. To understand why and how these rocks form, and why there are so many different kinds, this chapter begins by discussing why melt forms, why it rises, how it flows, and how it freezes in intrusive and extrusive environments. We then examine the scheme that geologists use to classify igneous rocks. Finally, we relate the formation of igneous rocks to plate tectonics theory.

6.2 Why Do Melts Form?

Discussing Molten Rock and Its Occurrences

In discussions of molten rock in the Earth system, geologists use special names for molten rock and for the settings in which it occurs. Specifically, we refer to melt that's underground as **magma** and to melt that has emerged at the Earth's surface as **lava**. The vent from which lava emerges is a **volcano**, and an event during which lava emerges is a volcanic eruption. We also use the word *volcano* in reference to a hill or mountain built from the products of eruption. Eruptions can produce a *lava fountain*, in which lava forcefully rises meters to hundreds of meters into the air; a *lava lake*, in which lava pools over the vent; or a **lava flow**, in which lava flows down a slope. We also refer to the rock sheet or mound formed when lava cools to form rock as a lava flow. Some volcanic eruptions explosively send clouds of shattered pre-existing igneous rock as well as frozen droplets of lava skyward. This fragmental

FIGURE 6.1 Formation and evolution of lava flows.

Lava fountain

Active lava flow

As the lava cools, it darkens.

(a) Lava erupts as a fountain from a volcanic vent on Hawaii. A fast-moving river of lava then flows downslope.

Smoke comes from burning vegetation.

STOP

Time

(b) At a distance from the vent, the lava has completely crusted over with new rock, but the interior of the flow remains molten.

(c) Eventually, the flow cools completely and becomes a layer of new rock. This flow engulfed a road on Hawaii.

material and dust is called **pyroclastic debris**, from the Greek word *pyro*, meaning fire, and *klastos*, meaning broken (**Fig. 6.2**). There's a lot more to say about volcanoes and their eruptions. We explore the topic further in Chapter 9.

Geologists also distinguish between two main categories of igneous rock based on where it solidifies (**Fig. 6.3a**). Rock that forms by the freezing of lava above ground, after it spills out (*extrudes*) onto the surface of the Earth and comes into contact with air or water, or forms by the cementing or welding together of pyroclastic debris, is **extrusive igneous rock** (**Fig. 6.3b**). As we've noted, vastly more igneous rock forms from magma that had pushed its way, or *intruded*, into pre-existing rock, or **wall rock**, and solidified out of view underground. We refer to such rock as intrusive igneous rock and a body of such rock as an **igneous intrusion** (**Fig. 6.3c**). In the intrusive realm, magma may accumulate in an irregularly shaped zone called a **magma chamber**, or in a chimney-like column, or along planar cracks, or in thin sheets between pre-existing layers. Each of these processes yields a different shape of intrusion.

Why Is It Hot inside the Earth?

Clearly, if the Earth were not hot inside, igneous processes could not take place. Where does our planet's internal heat come from? Some of the heat was left over from the Earth's early days. Recall that, according to the nebula theory, our planet formed from the collision and merging of countless planetesimals. Every time a collision occurred, the kinetic energy (energy of motion) transformed into heat energy. (You can simulate this phenomenon by banging a hammer repeatedly on a nail—the head of the nail becomes quite warm.) As the Earth grew, gravity pulled matter inward until eventually the weight of overlying material squeezed the matter

inside tightly together. Such compression made the Earth's insides even hotter, just as compressing a gas by a piston makes it hotter. Eventually, the Earth became hot enough for iron inside to melt, and the dense iron sank to the center to form the core. Friction between sinking iron and its surroundings generated still more heat, just as rubbing your hands together generates heat. Soon after Earth's formation, but probably after its differentiation into the mantle and core, a Mars-sized object collided with the Earth (see Chapter 1). This collision generated vast amounts of heat. And even after the Earth had grown to become a planet, intense bombardment continued to add heat energy—this bombardment didn't cease until about 3.9 billion years ago. Taken together, collisions and differentiation made the early Earth so hot that it was at least partially molten throughout, and its surface may, at times, have been an ocean of lava.

FIGURE 6.2 A volcanic eruption can produce both lava and pyroclastic debris.

Ever since the heat-producing catastrophes of its early days, the Earth has radiated heat into space and, therefore, has slowly cooled. Eventually, the sea of lava on its surface solidified and formed igneous rock. Therefore, we can think of igneous rock as the first rock of this planet's crust to form. Significantly, if no heat had been added to the Earth after the era of intense bombardment, the Earth might have become too cold by now

FIGURE 6.3 The intrusive and extrusive realms.

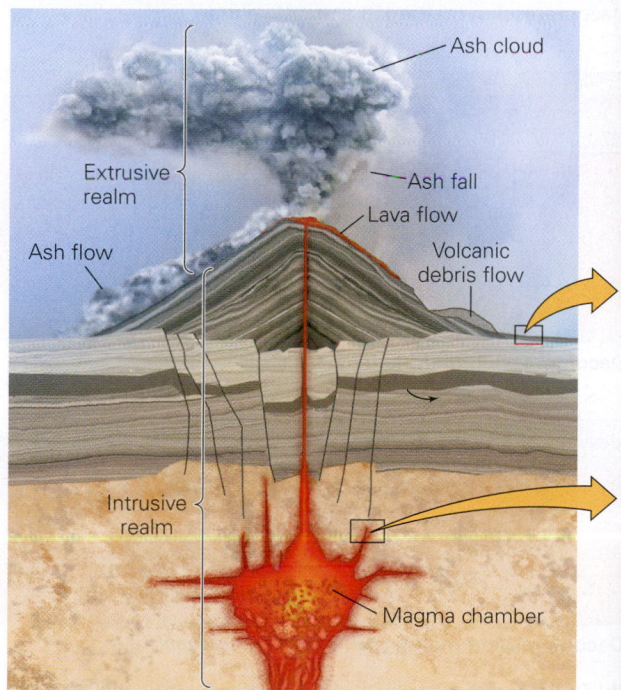

(a) The intrusive realm lies underground and the extrusive realm lies above ground. Lava flows, as well as various types of ash eruptions, all produce extrusive rocks.

(b) Extrusive rocks include lava flows and pyroclastic layers.

(c) An intrusion of basalt (dark rock) cuts across a mass of granite (light rock).

for igneous activity to take place today. Such cooling hasn't happened because of the presence of radioactive elements in the Earth (primarily in the crust). Decay of a single radioactive atom produces only a tiny amount of heat, but the cumulative heat of radioactive decay in the Earth has been sufficient to slow the cooling of this planet overall. Thus, Earth remains very hot today, with temperatures at the base of the lithosphere reaching almost 1,300°C, and temperatures at the planet's center exceeding 4,700°C. (By comparison, the surface of our Sun is about 5,700°C.)

Causes of Melting

As we've seen, the Earth is very hot inside. But despite the high temperatures, the common image that the solid crust of the Earth floats on a sea of molten rock is not correct. In fact, magma forms only in special places where pre-existing solid rock melts, for under the very high pres-

> **Did you ever wonder...**
>
> whether Earth's crust floats on a magma sea?

sures that exist inside the Earth, rock remains solid except in special places where local conditions allow some melt to form. Such conditions can develop both in the upper part of the asthenosphere and in the lower part of the crust. Let's now discuss each of the special physical conditions that lead to melting.

Melting Due to a Decrease in Pressure (Decompression) We can portray how temperature changes with increasing depth in the Earth with a **geotherm**, a curving line on a graph that plots temperature on the horizontal axis and pressure on the vertical axis. The graph emphasizes that temperature always increases with depth but not always at the same rate. By reading the graph, we see that beneath typical oceanic crust temperatures comparable to those of lava exist in the upper mantle (**Fig. 6.4a**). But, as we've

FIGURE 6.4 Decompression melting.

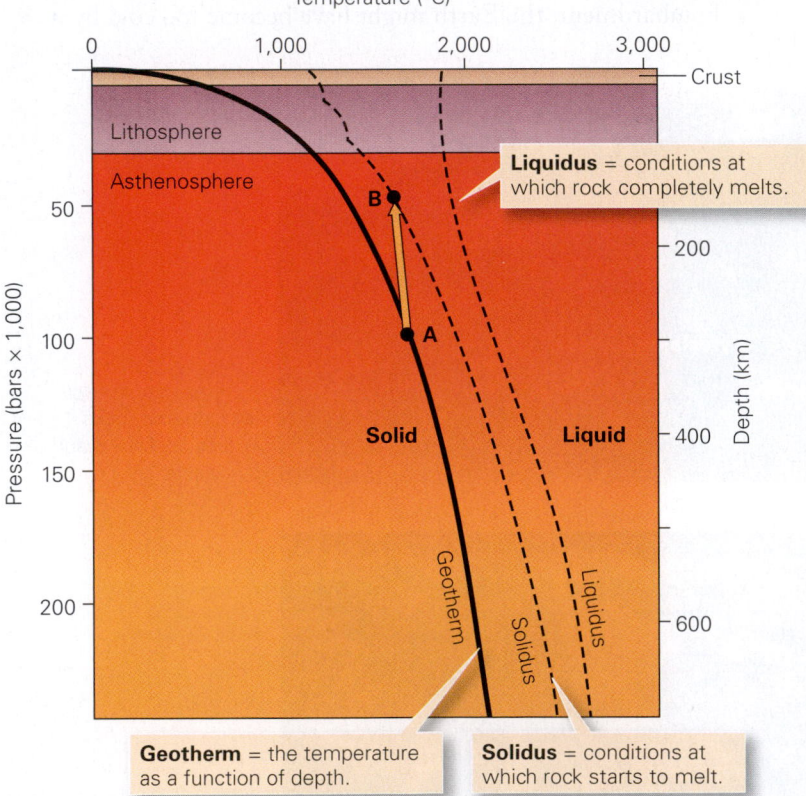

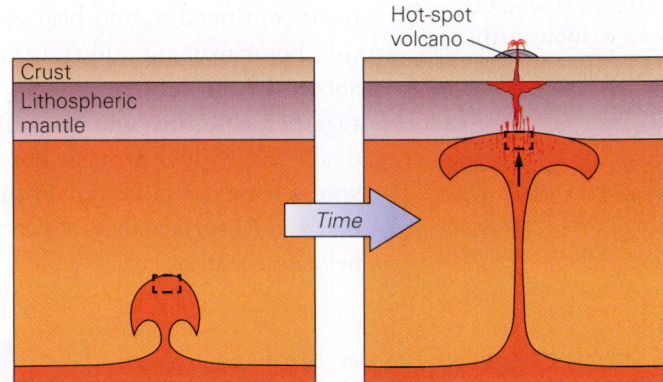

Decompression melting in a mantle plume

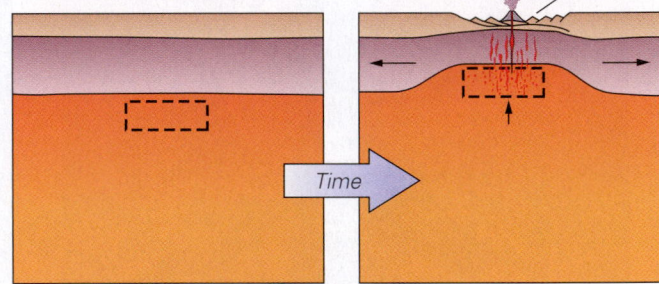

Decompression melting beneath a rift

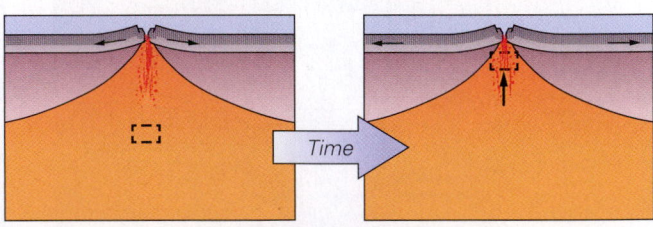

Decompression melting beneath a mid-ocean ridge

(a) Decompression takes place when the pressure acting on hot rock decreases. As this graph of pressure and temperature conditions in the Earth shows, when rock rises from point A to point B, the pressure decreases a lot, but the rock cools only a little, so the rock begins to melt.

(b) The conditions leading to decompression melting occur in several different geologic environments. In each case, a volume of hot asthenosphere (outlined by dashed lines) rises to a shallower depth, and magma (red dots) forms.

noted, even though the upper mantle is very hot, its rock stays solid because it is also under great pressure from the weight of overlying rock. Simplistically, pressure squeezes atoms together so they can't easily break free from solid mineral crystals.

Because pressure prevents melting, a decrease in pressure can permit melting. Specifically, if the pressure affecting hot mantle rock decreases while the temperature remains nearly unchanged, magma may form. This kind of melting, called **decompression melting**, occurs in places where hot mantle rock rises to shallower depths in the Earth. As we'll see, such movement occurs in mantle plumes, beneath rifts, and beneath mid-ocean ridges (**Fig. 6.4b**).

Melting as a Result of the Addition of Volatiles

Magma also forms at locations where chemicals called volatiles have the opportunity to mix with hot mantle rock. Recall that "volatiles" are substances such as water (H_2O) and carbon dioxide (CO_2) that evaporate easily and can exist in gaseous forms at the Earth's surface. When volatiles mix with hot rock, they help break chemical bonds. So if you add volatiles to a solid, hot, dry rock, the rock begins to melt. In effect, adding volatiles decreases a rock's melting temperature. Melting due to addition of volatiles is sometimes called **flux melting**. Of the common volatiles, water plays the most important role in triggering melting. We'll see that flux melting happens in the mantle above subducting oceanic crust (**Fig. 6.5a**).

Melting as a Result of Heat Transfer from Rising Magma

When very hot magma from the mantle rises up into the crust, it brings substantial amounts of heat with it. This heat can conduct into the wall rock surrounding an intrusion and thus can raise the temperature of the wall rock. In some cases, the added heat may be sufficient to cause the wall rock to begin melting. To picture the process, imagine injecting hot fudge into ice cream—the fudge conveys heat to the ice cream, raises its temperature, and causes it to melt. We call such melting **heat-transfer melting**, because it results from the movement of thermal energy from a hotter material to a cooler one. As we'll see, heat-transfer melting can happen in rifts, along convergent-plate boundaries, and at hot spots, all places where magma from the mantle can stall while rising into the crust, and transfer heat into rocks that have a lower melting temperature (**Fig. 6.5b**).

Take-Home Message

Molten rock underground is called magma, and molten rock that has come out of a vent at the Earth's surface is lava. Solidification of magma produces intrusive rocks; solidification of lava or pyroclastic debris produces extrusive igneous rock. Temperature increases with depth, at a rate defined by the geotherm. Nevertheless, the mantle and crust are mostly solid, so production of magma only occurs in special locations where there is decompression, addition of volatiles, or heat transfer.

QUICK QUESTION: Why hasn't the Earth cooled sufficiently to become entirely solid, over geologic time?

FIGURE 6.5 Flux melting and heat-transfer melting.

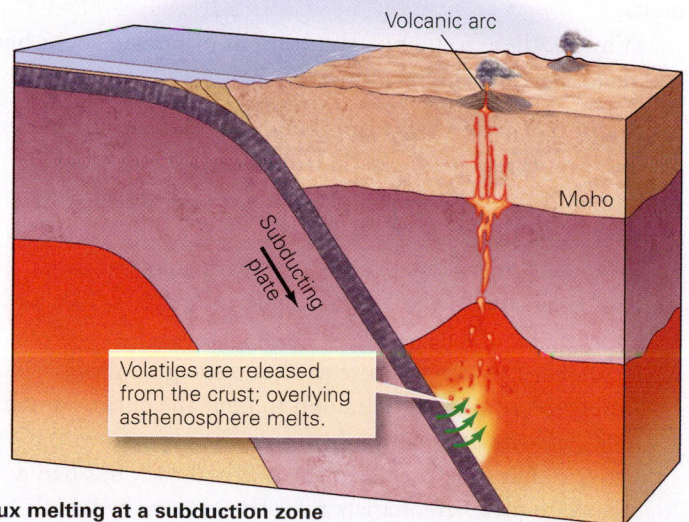

Flux melting at a subduction zone

(a) Flux melting occurs where volatiles enter hot mantle; this happens at subduction zones.

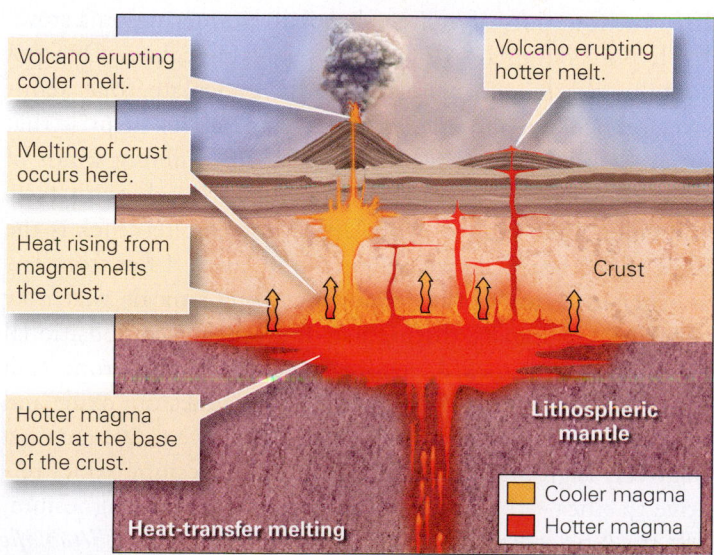

(b) Heat-transfer melting occurs when rising magma brings heat up with it and melts overlying or surrounding rock. (We'll see that the cooler magma is felsic, and the hotter magma is mafic.) (Not to scale.)

6.3 What Is Molten Rock Made of?

Molten Rock as a Chemical Soup

In a general sense, molten rock (magma or lava) is just a hot liquid consisting of many chemicals. For convenience, geologists tend to describe the chemical composition of molten rock in terms of the proportions of "oxides" that it contains—an oxide, in this context, is simply a molecule consisting of an element bonded to oxygen. The most common oxides in magma or lava are silica (SiO_2), aluminum oxide (Al_2O_3), iron oxide (FeO or Fe_2O_3), calcium oxide (CaO), magnesium oxide (MgO), sodium oxide (Na_2O), and potassium oxide (K_2O). Because molten rock is a liquid, oxide molecules do not lie in an orderly crystalline lattice but are bonded together instead in clusters or short chains that can move with respect to one another.

Geologists distinguish between "dry" melts, which contain no volatiles, and "wet" melts, which do. Wet melts may include up to 15% dissolved volatiles such as water (H_2O), carbon dioxide (CO_2), nitrogen (N_2), hydrogen (H_2), and sulfur dioxide (SO_2). These volatiles come out of the Earth at volcanoes in the form of gas (**Fig. 6.6**). Usually water constitutes at least half of the gas erupting at a volcano. Note that because it can release volatiles, magma and lava provide not only the molecules that make up rocks but also the molecules that become Earth's liquid water or air.

The Major Chemical Types of Molten Rock

Imagine four pots of molten chocolate simmering on a stove. Each pot contains a different type of chocolate. One pot contains white chocolate, one milk chocolate, one semi-sweet chocolate, and one dark chocolate. It's no surprise that different kinds of molten chocolate yield different kinds of solid chocolate; each type differs from the others in taste and color. Like molten chocolate, not all molten rock is the same. Specifically, melts (magmas and lavas) differ from one another in terms of the proportions of chemicals that they contain. Geologists distinguish four major compositional types of molten rock depending, overall, on the proportion of silica relative to the combination of magnesium oxide and iron oxides that it contains (**Table 6.1**). *Mafic melts* contain a relatively high proportion of iron and magnesium oxides relative to silica—the "ma-" in the word stands for magnesium, and the "-fic" comes from the Latin word for iron. *Ultramafic melts* have an even higher proportion of magnesium and iron oxides, relative to silica. *Felsic melts* have a relatively high

FIGURE 6.6 A cloud of gas, mixed with ash, rising above a volcano in the Aleutian Islands, Alaska.

> Volcanoes also erupt gas.

TABLE 6.1 The Four Categories of Magma

Felsic (or silicic) magma	66–76% silica*
Intermediate magma	52–66% silica
Mafic magma	45–52% silica
Ultramafic magma	38–45% silica

*The numbers provided are "weight percent," meaning the proportion of the magma's weight that consists of silica.

proportion of silica relative to magnesium and iron oxides. (Occasionally, geologists use the term *silicic* interchangeably with *felsic*.) *Intermediate melts* are so named because their composition lies partway between that of mafic and felsic melts.

Why are there so many compositions of molten rock? Several factors play a role, including:

- *Source-rock composition:* The composition of a melt reflects the composition of the solid from which it was derived. Not all melts form from the same source rock, so not all melts have the same composition.
- *Partial melting:* Under the temperature and pressure conditions that occur in the Earth, only 2% to 30% of an original rock can melt to produce magma at a given location—the temperature at sites of magma production simply never gets high enough to melt the entire source rock, and the magma tends to migrate away from the site of melting before all of the original rock has had a chance to melt. Geologists refer the process by which only part of an original rock melts to produce magma as **partial melting** (**Fig. 6.7a**). Magmas formed by partial

FIGURE 6.7 Phenomena that can affect the composition of magma.

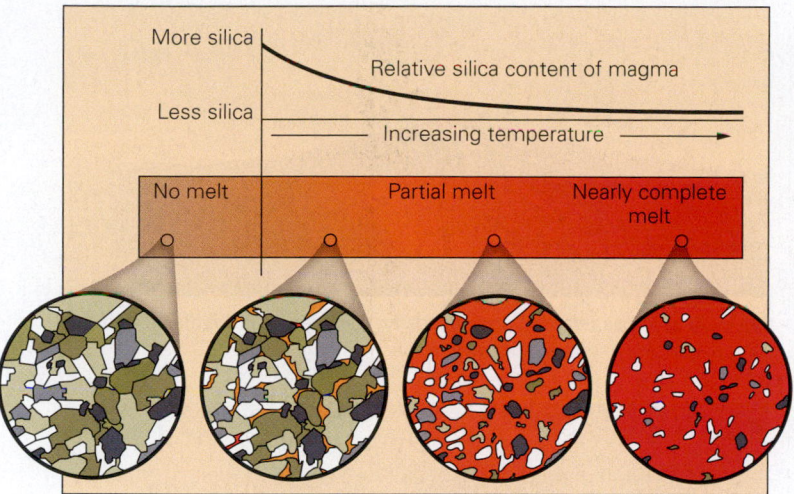

(a) Partial melting: The first-formed melt will be richer in silica than the original rock. As melting continues, magma becomes increasingly mafic.

Partial melting of wall rock produces new magma that mixes with magma from below.

Blocks of rock fall into magma and dissolve; this process is assimilation.

Deep magma rises

(b) Mixing and assimilation: Heat provided by deep magma partially melts wall rock; the new magma may then mix with deep magma. Also, blocks of wall rock can dissolve (assimilate) in the deep magma, and the wall rock may chemically react with the magma.

melting are more felsic than the source rock from which they were derived because relatively more silica enters the liquid, as melting begins, than remains behind in the still-solid source. Thus, for example, partial melting of an ultramafic rock produces a mafic magma.

- *Assimilation:* As magma sits in a magma chamber before completely solidifying, it may incorporate chemicals dissolved from the wall rocks of the chamber or from blocks that detached from the wall and sank into the magma (**Fig. 6.7b**). This process is called contamination or **assimilation**.

- *Magma mixing:* Different magmas formed in different locations from different sources may enter a magma chamber. In some cases, the originally distinct magmas mix, or mutually dissolve in each other, to produce a new, different magma. Thoroughly mixing a felsic magma with a mafic magma in equal proportions produces an intermediate magma.

Take-Home Message

Molten rock contains many chemicals, of which the dominant component is silica. Geologists distinguish among four types of molten rock, based on composition, as defined by the proportion of silica to magnesium and iron oxides. The composition of a given magma depends on many factors, both at the site where melting took place and in the magma chamber where it resides.

QUICK QUESTION: Why doesn't a magma formed by melting a given source rock have the same composition as the source rock?

6.4 Movement and Solidification of Molten Rock

If magma stayed put once it formed, new igneous rocks would not develop in or on the crust. But it doesn't stay put—magma tends to move upward, away from where it formed. It may eventually stop and freeze underground, but as we've seen, some molten rock reaches the Earth's surface to erupt as lava at volcanoes. Rise of molten rock serves an important role in the Earth System in that it transfers material from deeper parts of the Earth upward and provides the raw material from which new rock, as well as the atmosphere and ocean, forms.

Why Does Magma Rise?

Magma in the Earth rises for two reasons. First, since magma is less dense than surrounding rock, a buoyancy force acts on it to drive it upward. A similar process happens when you submerge a piece of Styrofoam in a bucket of water—because Styrofoam is less dense than water, it will rise upward through the water. Second, since rock has substantial weight, pressure develops at depth, and this pressure squeezes magma upward. A similar process happens when you step in a mud puddle while barefoot, and your weight squeezes mud up between your toes.

What Controls the Speed of Molten Rock Flows?

The resistance to flow, or **viscosity**, of a liquid affects the speed with which the liquid moves. We can say, for example, that molasses is more viscous than water because it flows more slowly than water. All molten rock (magma or lava) is more viscous than molasses, but not all molten rock has the same viscosity. The viscosity of molten rock depends primarily on temperature, volatile content, and silica content. Specifically, hotter melt is less viscous than cooler melt, because thermal energy breaks bonds and allows atoms or molecules to move more easily. A melt containing more volatiles is less viscous than a dry (volatile-free) melt because the volatiles also tend to break apart silicate molecules. Finally, mafic melt is less viscous than felsic melt because relatively more silicon-oxygen tetrahedra occur in felsic melt, and the tetrahedra tend to link together to create long chains that can't move past each other easily. With these relationships in mind, it's not surprising that a very hot mafic lava has relatively low viscosity and can flow in thin sheets over wide regions, but cool felsic lava is very viscous and clumps up at the volcanic vent to form a bulbous mound (**Fig. 6.8**).

Transforming Melt into Rock

If a melt stayed at its point of origin, and nothing in its surroundings changed, it would stay molten. But melts don't last forever. Rather, they eventually solidify or "freeze." Most commonly, freezing takes place simply because a melt cools below its freezing temperature. Cooling happens when magma rises because the temperature of the crust decreases upward. If magma becomes trapped underground as an intrusion, it slowly loses heat to surrounding wall rock, drops below its freezing temperature, and solidifies. If a magma reaches the Earth's surface and extrudes as lava at the ground surface, it cools because it comes in contact with much cooler air or water. In some cases, magmas freeze, in part due to loss of volatiles—we saw earlier that the addition of volatiles to hot rock can cause it to melt, so the subtraction of volatiles from magma can cause it to freeze. Volatiles bubble out of magma as the magma rises and pressure acting on it decreases, so "devolatilization" may accelerate the freezing that is already happening due to the cooling of a magma.

The time it takes for a magma to cool depends on how fast it can transfer heat into its surroundings. To see why, think about the process of cooling coffee. If you pour hot coffee into a thermos bottle and seal it, the coffee stays hot for hours because the insulation of the bottle allows the coffee to lose heat to the air outside only very slowly. Like a thermos bottle, wall rock surrounding an intrusion acts as insulation, so magma in an intrusive environment cools relatively slowly. In contrast, if you spill coffee on a

table, it cools quickly because it loses heat to the cold air. Thus, lava that erupts at the ground surface cools relatively quickly.

Not all magma trapped in the intrusive realm cools at the same rate. Three factors can control the cooling time of such magma.

- *The depth of intrusion:* Magma intruded deep in the crust, where it is surrounded by hot wall rock, cools more slowly than does magma intruded near the ground surface, where wall rock is cool.
- *The shape and size of a magma body:* Heat escapes from magma at an intrusion's surface, so the greater the surface area for a given volume of intrusion, the faster it cools. Thus, a body of magma roughly with the shape of a pancake cools faster than one with the shape of a melon. And since the ratio of surface area to volume increases as size decreases, a body of magma the size of a car cools faster than one the size of a ship (**Fig. 6.9a**).
- *The presence of circulating groundwater:* Water passing through wall rock carries away heat, much like the coolant that flows around an automobile engine. Thus, magma that interacts with circulating groundwater cools faster than does magma that intruded dry rock (**Fig. 6.9b**).

The same factors that control cooling of magma also control cooling of lava. Specifically, a thin lava flow cools faster than a

FIGURE 6.8 Viscosity affects lava behavior.

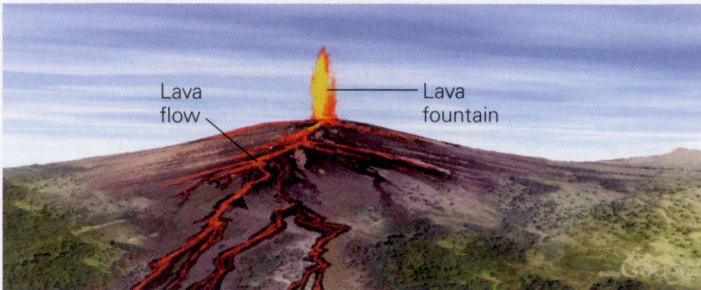

Lava flow — Lava fountain

(a) Mafic lava has relatively low viscosity. It can erupt in fountains, move long distances, and form thin lava flows.

Lava dome

(b) Felsic to intermediate lava is very viscous. When it erupts, it may form a mound-like lava dome around the volcano's vent.

FIGURE 6.9 Factors affecting the cooling rate of intrusions.

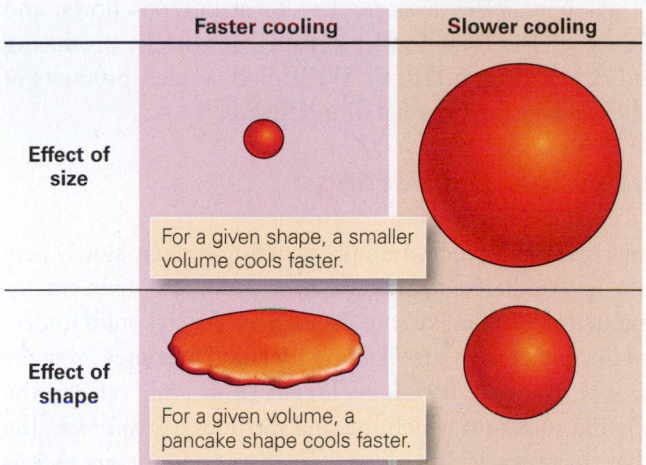

	Faster cooling	Slower cooling
Effect of size	For a given shape, a smaller volume cools faster.	
Effect of shape	For a given volume, a pancake shape cools faster.	

(a) The shape and size of an intrusion affects the surface area to volume ratio.

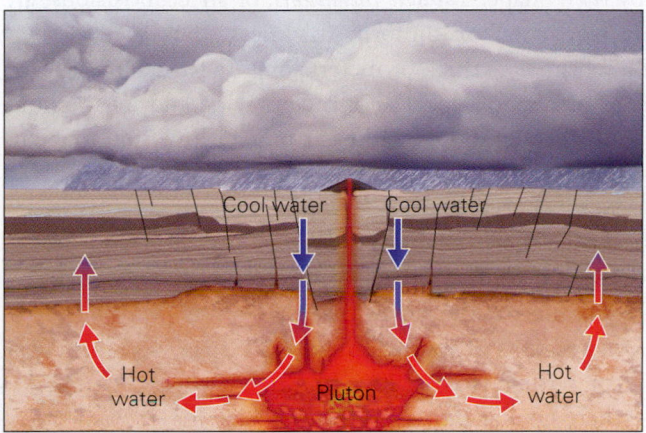

Cool water   Cool water

Hot water   Pluton   Hot water

(b) Circulating groundwater can cause a pluton to cool more quickly.

thick one because the surface area relative to volume is greater for a thin flow than it is for a thick one, and tiny droplets of lava sprayed into the air in an explosive eruption freeze much faster than does a coherent lava flow. Also of note, lava immersed in water cools faster than lava immersed in air, because heat moves faster through water than it does through air.

Changes in Molten Rock during Cooling: Fractional Crystallization

Most people are familiar with the process of forming ice out of liquid water—cool the water to a temperature of 0°C and crystals of ice start to form. Keep the temperature cold enough for long enough and all the water becomes solid, composed entirely of one type of mineral—water ice. The process of freezing magma or lava is much more complicated because molten rock, unlike water, contains many different compounds, so during freezing of molten rock, many different minerals form. Further, not all of the minerals that grow in a freezing magma or lava form at the same time.

To gain insight into the complexity of magma freezing, let's look at an example. When a mafic magma starts to freeze, mafic minerals (meaning, iron- and magnesium-rich minerals), such as olivine and pyroxene, crystallize first. These solid crystals are denser than the remaining liquid magma, so they start to sink (**Fig. 6.10**). Some of the crystals react chemically with the remaining magma as they sink, but some reach the floor of the magma chamber and become isolated from the magma. This process of sequential crystal formation and settling is called **fractional crystallization**. It extracts iron and magnesium from the magma so that the remaining magma becomes a bit more felsic. If a magma freezes completely before very much fractional crystallization has occurred, the magma becomes mafic igneous rock. If the process of fractional crystallization continues, more and

FIGURE 6.10 Factors that affect the freezing of molten rock.

Small drops cool very quickly.

A thin flow cools quickly.

Cooler wall rock

Increasing temperature

Warmer wall rock

A large blob cools slowly.

A shallow sheet cools faster than a deep sheet.

(a) The cooling rate of molten rock depends on the size and shape of the magma or lava body, and on its depth.

Fractional crystallization

After mafic minerals settle out, the remaining magma becomes more felsic.

Time 1   Time 2

The original melt is mafic.

Decreasing temperature

(b) The process of fractional crystallization results in a progressive change in magma composition during freezing.

more iron and magnesium will be extracted from the magma, so the magma evolves to become progressively more felsic. Freezing of the magma that remains after lots of fractional crystallization has taken place can, therefore, yield a felsic igneous rock. In the early 20th century, an American geologist, N. L. Bowen, conducted a series of laboratory experiments that allowed him to work out the specific sequence in which minerals form (**Box 6.1**).

Take-Home Message

Magma rises because it's buoyant and because pressure due to overlying rocks squeezes it upward. The rate of melt movement is affected by viscosity, which depends primarily on composition and temperature. When molten rock enters a cooler environment, it freezes. The rate of cooling depends on both the temperature of the surroundings and on the shape of the magma body. Magma composition may evolve as the magma cools.

QUICK QUESTION: Which cools faster—a large blob of magma intruded at depth or a thin flow extruded at the surface? Why?

6.5 Comparing Extrusive and Intrusive Environments

Extrusive Igneous Settings

Not all volcanic eruptions are the same, so not all extrusive rocks are the same. Some volcanoes erupt streams of low-viscosity lava that cover broad swaths of the countryside with relatively thin lava flows (**Fig. 6.11a**). Others erupt viscous masses of lava that pile into mounds of angular blocks. And still others erupt in cataclysmic explosions, which forcefully eject turbulent clouds of volcanic ash and debris that can rise several kilometers into the sky (**Fig. 6.11b**). The debris includes larger fragments, known as **lapilli**, that fall like hail on or near the volcano, as well as finer, dust-sized material, known as **ash**, which may be carried by the wind for great distances, before it drifts down from the sky like snow. An explosive eruption may also produce a fast-moving **pyroclastic flow**, a hot avalanche of ash and other debris that races down the surface of the volcano, destroying everything in its path.

Whether an eruption produces mostly sheets of lava, mounds of lava, or clouds of pyroclastic debris depends largely on a lava's viscosity and volatile content. Mafic lavas tend to have low viscosity and can spread in broad, thin flows (**Fig. 6.11c**); gas-poor felsic lavas tend to form bulbous flows; and volatile-rich felsic lavas tend to erupt explosively, producing ash and debris (**Fig. 6.11d, e**). We'll discuss such products of extrusive settings in more detail in Chapter 9.

Intrusive Igneous Settings

Magma rises and intrudes into pre-existing rock by slowly percolating upward between grains and/or by forcing open cracks. Magma that doesn't make it to the surface freezes solid underground in contact with pre-existing rock and becomes *intrusive igneous rock*. As we've noted, geologists commonly refer to the pre-existing rock into which magma intrudes as *wall rock*. The boundary between wall rock and an intrusive igneous rock is an **intrusive contact**.

Geologists distinguish among different types of intrusions by their shape. *Tabular intrusions*, or sheet intrusions, are roughly planar and of uniform thickness. Smaller ones may be only centimeters thick or meters long, but the largest reach widths of meters to hundreds of meters wide and tens to hundreds of kilometers long. A **dike** is a tabular intrusion that cuts across pre-existing layering (bedding or foliation), whereas a **sill** is a tabular intrusion that injects between layers (**Fig. 6.12**). Dikes may radiate outward from a volcano, or they may develop over broad regions of crust during the process of rifting (**Fig. 6.13a, b**). In places where tabular intrusions intrude rock that does not have layering, we generally refer to a vertical, wall-like example as a dike and a nearly horizontal, tabletop-shaped example as a sill. In some places, the magma injecting between layers gets blocked and can't spread laterally very far, so the magma accumulates to form a blister-shaped intrusion known as a **laccolith**, which pushes overlying strata upward into a dome.

Plutons are irregular or blob-shaped intrusions that range in size from tens of meters across to tens of kilometers across (**Fig. 6.14**; **Fig. 6.15**). Intrusion of numerous plutons in a region creates a vast composite body that may be several hundred kilometers long and over 100 km wide; such an immense mass of igneous rock is called a **batholith**. The rugged high peaks of the Sierra Nevada

FIGURE 6.11 Examples of eruptions and extrusive volcanic materials.

(a) This volcano is producing lava flows and fountains.

Some of the lava freezes in mid-air.

Lava flow

(b) This volcanic explosion produced two styles of ash eruption.

Ash cloud

Pyroclastic flow

(c) A stack of over 50 thin lava flows, capped by debris, visible from inside Mt. Vesuvius, Italy.

A group of people

(d) A close-up of a layer of ash containing a piece of lapilli.

A golf-ball-sized fragment of lapilli

(e) Thick layers of ash deposited by explosive eruptions in New Mexico, about 1.14 Ma. Note the highway, for scale.

Range in California expose intrusive rocks of a batholith formed from plutons that intruded between 145 and 80 Ma (million years ago).

Where does the space for intrusions come from? It's relatively easy to visualize how space for tabular intrusions can be produced. For example, igneous dikes form in regions where the crust is being stretched horizontally in response to rifting or sea-floor spreading, so as magma intrudes vertical cracks can simply open up horizontally (**Fig. 6.16a**). In effect, the magma fills space being generated by stretching. Intrusion of sills occurs near the surface of the Earth, so the pressure of the magma simply pushes overlying rock and the Earth's surface upward to make room (**Fig. 6.16b**).

Understanding how the space for plutons develops remains a subject for further research. Some plutons may have originated as "diapirs," meaning they rose upward through the crust

BOX 6.1 CONSIDER THIS . . .

Bowen's Reaction Series

FIGURE Bx6.1 Bowen's reaction series indicates the succession of crystallization in cooling magma.

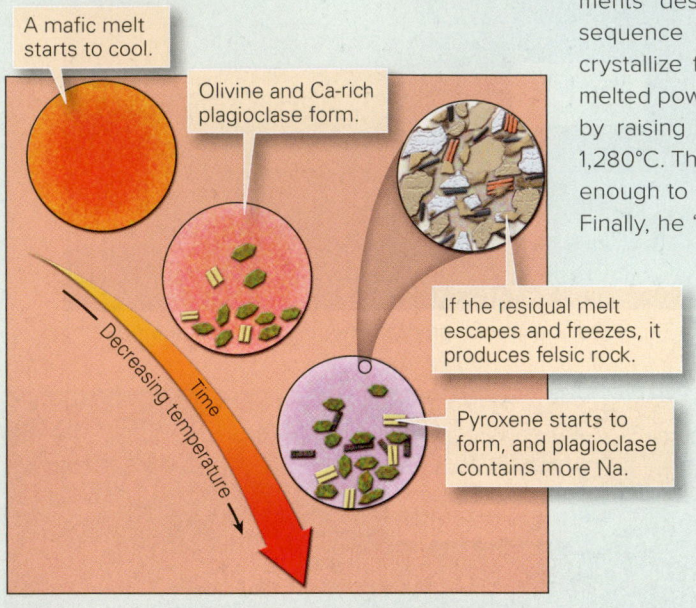

A mafic melt starts to cool.

Olivine and Ca-rich plagioclase form.

If the residual melt escapes and freezes, it produces felsic rock.

Pyroxene starts to form, and plagioclase contains more Na.

Decreasing temperature

Time

(a) With decreasing temperature, fractional crystallization begins and the composition of the remaining magma becomes more felsic.

In the 1920s, Norman L. Bowen began a series of laboratory experiments designed to determine the sequence in which silicate minerals crystallize from a melt. First, Bowen melted powdered mafic igneous rock by raising its temperature to about 1,280°C. Then he cooled the melt just enough to cause part of it to solidify. Finally, he "quenched" the remaining melt by submerging it quickly in cold mercury. Quenching, which means sudden cooling to form a solid, transformed any remaining liquid into glass. The glass trapped the earlier-formed crystals within it. Bowen identified mineral crystals

formed before quenching with a microscope, and he analyzed the chemical composition of the remaining glass to determine the composition of the magma.

After experiments at different temperatures, Bowen found that as new crystals form they extract certain chemicals preferentially from the melt (**Fig. Bx6.1a**). Thus, the chemical composition of the remaining melt progressively changes as the melt cools. Bowen described the specific sequence of mineral-producing reactions that take place in a cooling, initially mafic, magma. This sequence is now called **Bowen's reaction series** in his honor.

Let's examine the sequence more closely. In a cooling melt, olivine and calcium-rich plagioclase form first. The Ca-plagioclase reacts with the melt to form more plagioclase, which contains more sodium (Na). Meanwhile, some olivine crystals react

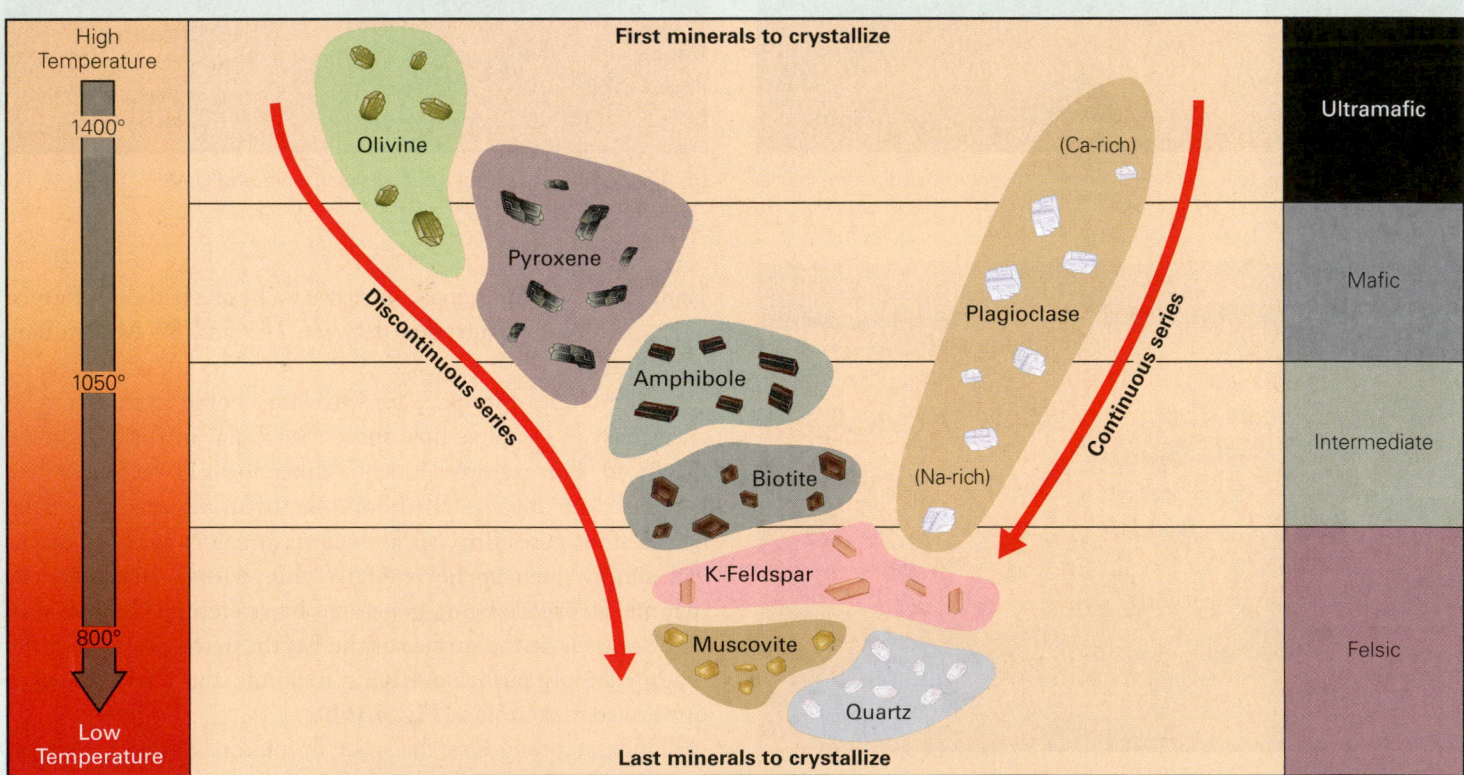

High Temperature

1400°

1050°

800°

Low Temperature

First minerals to crystallize

Olivine

Pyroxene

Amphibole

Biotite

Muscovite

Quartz

K-Feldspar

Plagioclase

(Ca-rich)

(Na-rich)

Discontinuous series

Continuous series

Last minerals to crystallize

Ultramafic

Mafic

Intermediate

Felsic

(b) This chart displays the discontinuous and continuous reaction series. Rocks formed from minerals at the top of the series are mafic, whereas rocks formed from the bottom of the series are felsic.

with the remaining melt to produce pyroxene, which may encase olivine crystals or even replace them. However, some of the olivine and Ca-plagioclase crystals settle out of the melt, taking iron, magnesium, and calcium atoms with them. By this process, the remaining melt becomes enriched in silica. As the melt continues to cool, plagioclase continues to form, with later-formed plagioclase having progressively more sodium (Na) than earlier-formed plagioclase. Pyroxene crystals react with melt to form amphibole, and then amphibole reacts with the remaining melt to form biotite. All the while, crystals continue to settle out, so the remaining melt continues to become more felsic. At temperatures between 650°C and 850°C, only about 10% melt remains, and this melt has a high silica content. At this stage, the final melt freezes, yielding quartz, K-feldspar, and muscovite.

On the basis of his observations, Bowen realized that there are two tracks to the reaction series. The "discontinuous reaction series" refers to the sequence olivine, pyroxene, amphibole, biotite, K-feldspar/muscovite/quartz: each step yields a different class of silicate mineral. The "continuous reaction series" refers to the progressive change from calcium-rich to sodium-rich plagioclase: the steps yield different versions of the same mineral (**Fig. Bx6.1b**). It's important to note that not all minerals listed in the series appear in all igneous rock. For example, a mafic magma may completely crystallize before felsic minerals such as quartz or K-feldspar have a chance to form.

Also note that the succession of minerals in the discontinuous series is not random—it begins with minerals having isolated tetrahedra (olivine) and progresses to those having single chains of tetrahedra (pyroxene), then double chains (amphibole), and finally sheets (mica) or 3-D networks (quartz). Put another way, the minerals that crystallize later in the discontinuous series have more Si-O-Si bonds and smaller O/Si than those that crystallize earlier.

FIGURE 6.12 Igneous sills and dikes, examples of tabular intrusions.

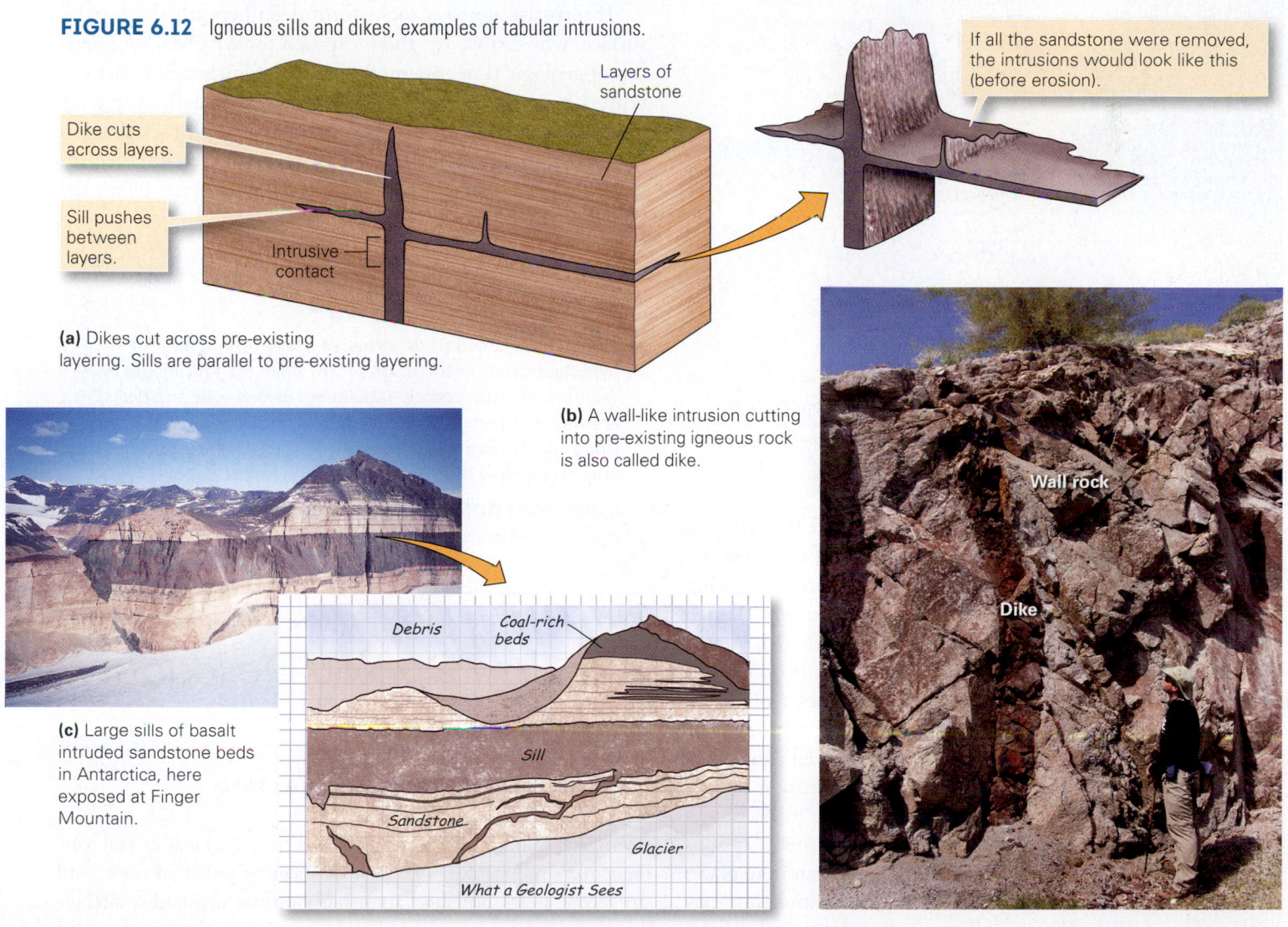

If all the sandstone were removed, the intrusions would look like this (before erosion).

Layers of sandstone

Dike cuts across layers.

Sill pushes between layers.

Intrusive contact

(a) Dikes cut across pre-existing layering. Sills are parallel to pre-existing layering.

(b) A wall-like intrusion cutting into pre-existing igneous rock is also called dike.

Wall rock

Dike

(c) Large sills of basalt intruded sandstone beds in Antarctica, here exposed at Finger Mountain.

Debris

Coal-rich beds

Sill

Sandstone

Glacier

What a Geologist Sees

FIGURE 6.13 Examples of igneous dikes.

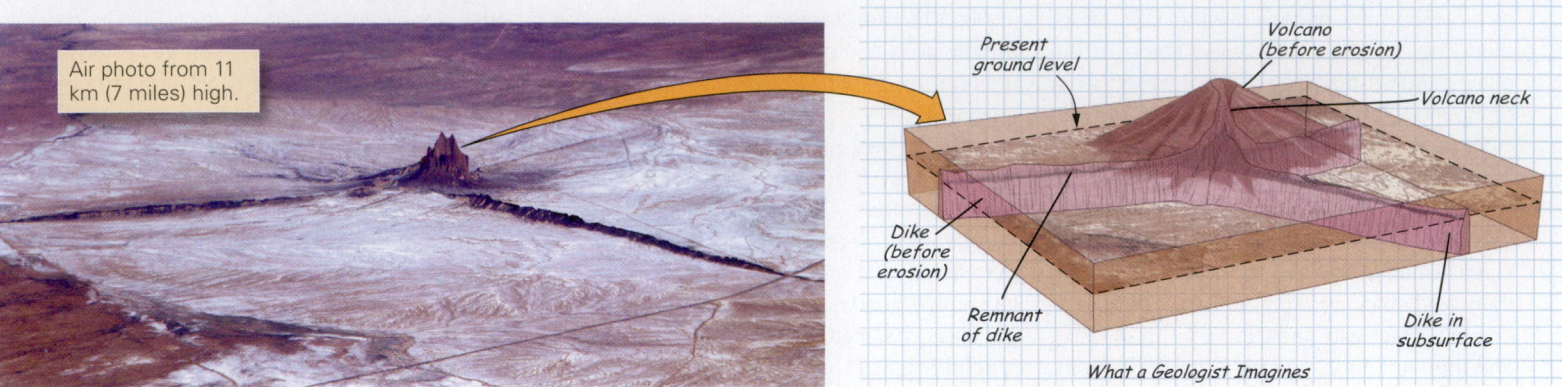

Air photo from 11 km (7 miles) high.

Present ground level

Volcano (before erosion)

Volcano neck

Dike (before erosion)

Remnant of dike

Dike in subsurface

What a Geologist Imagines

(a) Shiprock, in northwestern New Mexico, is the remnant of an eroded volcano. Large dikes radiate from the central hill. The central hill is the frozen magma chamber under what was once a volcano. The volcano has eroded away.

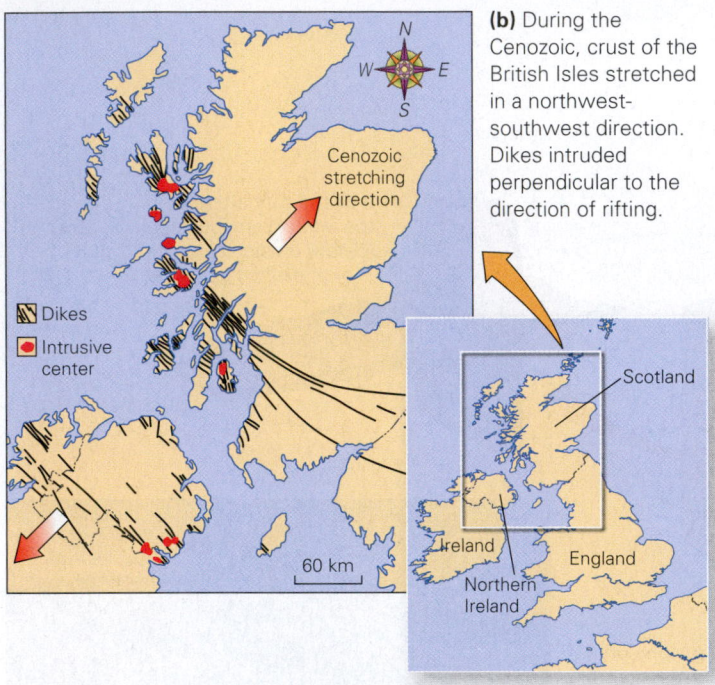

(b) During the Cenozoic, crust of the British Isles stretched in a northwest-southwest direction. Dikes intruded perpendicular to the direction of rifting.

Cenozoic stretching direction

Dikes

Intrusive center

60 km

Scotland

Ireland

England

Northern Ireland

as a buoyant, light-bulb-shaped blob of magma that pierced overlying rock and pushed it aside as it rose (**Fig. 6.17a, b**). Pluton intrusion may also involve **stoping**, a process during which magma assimilates wall rock, and/or blocks of wall rock break off and sink into the magma (**Fig. 6.17c**). If a stoped block does not melt entirely but rather becomes surrounded by new igneous rock, it is a **xenolith**, after the Greek word *xeno*, meaning foreign (**Fig. 6.17d**). Recent work has questioned the general applicability of diapiric and stoping models and leads to the alternative view that plutons form by intrusion of several superimposed dikes or sills, which coalesce to become a single, massive body. If the high temperature of the intrusions is maintained for a long time so that diffusion can take place, the rock may gradually recrystallize and its composition may evolve (**Fig. 6.17e**). The overall space for plutons may form

because of crustal stretching and/or uplift and erosion of overlying rock.

If intrusive igneous rocks form deep beneath the Earth's surface, why can we see them exposed today? Over long periods of geologic time, mountain building slowly uplifts belts of crust. Erosion by water, wind, and ice can gradually strip away the thick, overlying rock and expose the intrusive rock that has formed below. Some intrusive rocks exposed in mountain cliffs today first solidified kilometers to tens of kilometers below the surface.

Take-Home Message

Molten rock can extrude either as a lava flow or as pyroclastic debris. Intrusions form when magma injects into wall rock and then cools. Tabular intrusions (sills and dikes) are wall-like intrusions. Blob-shaped intrusions are plutons. Huge batholiths consist of many plutons. Debate continues concerning how the space for plutons forms.

QUICK QUESTION: Intrusions occupy space—where does the room for plutons come from?

6.6 How Do You Describe an Igneous Rock?

Characterizing Color and Texture

Wander around a city and you'll find that the facades and lobbies of many buildings consist of slices of polished rock, and in recent years polished rock has become a popular surfacing material for home kitchen countertops. Architects refer

FIGURE 6.14 Igneous plutons, "blob-shaped" intrusions.

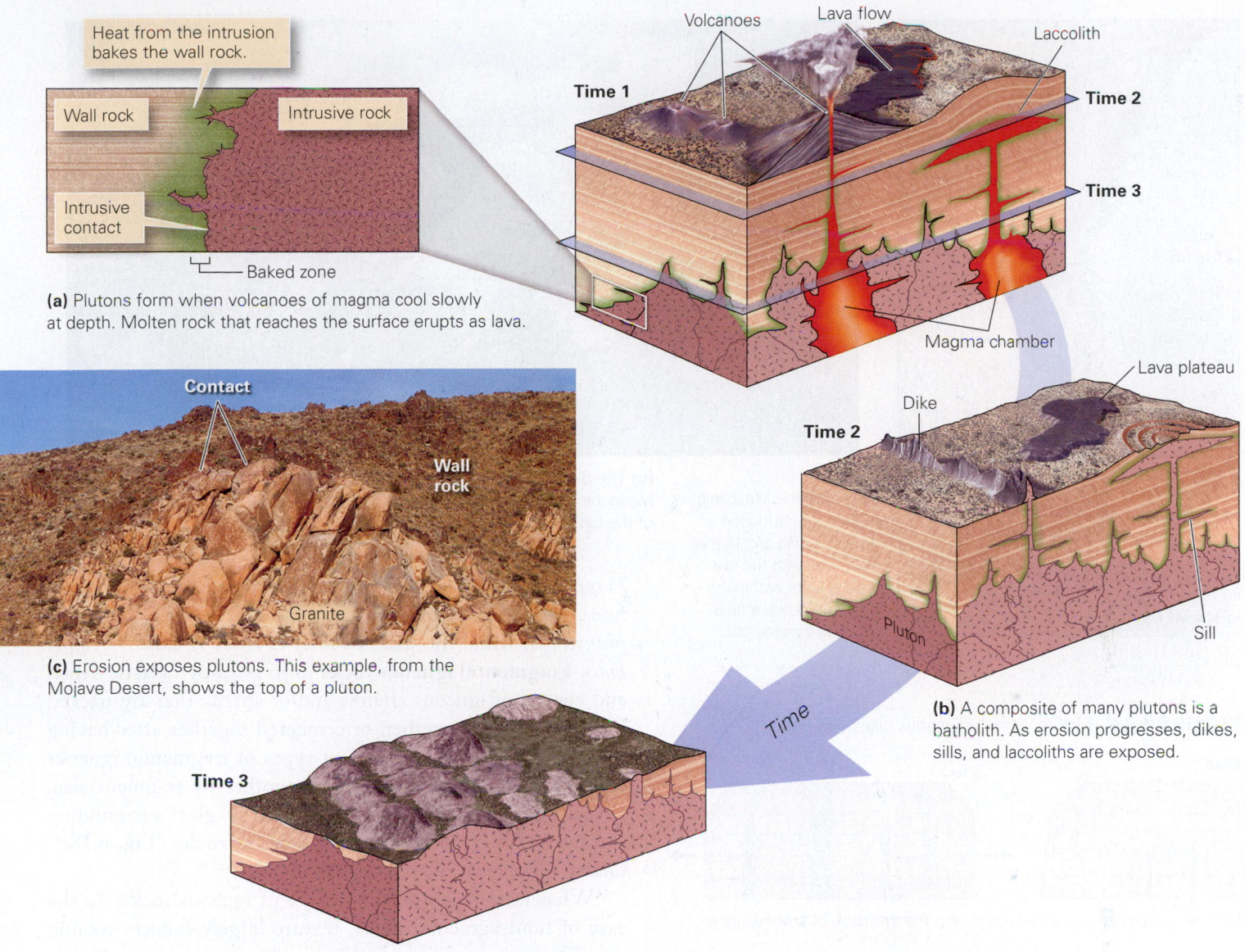

Heat from the intrusion bakes the wall rock.

Wall rock

Intrusive rock

Intrusive contact

Baked zone

(a) Plutons form when volcanoes of magma cool slowly at depth. Molten rock that reaches the surface erupts as lava.

Volcanoes

Lava flow

Laccolith

Time 1

Time 2

Time 3

Magma chamber

Lava plateau

Dike

Time 2

Pluton

Sill

Contact

Wall rock

Granite

(c) Erosion exposes plutons. This example, from the Mojave Desert, shows the top of a pluton.

Time

(b) A composite of many plutons is a batholith. As erosion progresses, dikes, sills, and laccoliths are exposed.

Time 3

to this rock in a generic way as "granite" and like to use it because it's both beautiful and durable. Its durability comes from the minerals it contains—most architectural "granite" contains feldspar and other minerals with high numbers on the Mohs hardness scale—and the way grains interlock. But, if you look at architectural granite closely, you'll discover that it doesn't all look the same. In fact, by the end of this section you'll discover that the word *granite*, as a geologist would use the term, has a relatively limited applicability, and most architectural granite is actually a different kind of rock.

Imagine that you wanted to describe an example of the stone in a facade to a friend—what adjectives would you use?

Did you ever wonder...

why "granite" used in floors and countertops is so hard?

You would probably start by noting the rock's color. Overall, is the rock dark or light? More specifically, is it gray, pink, white, or black? Describing color may not be easy because some igneous rocks contain many different minerals, each with a different color—but even so, you'll probably be able to characterize the overall hue of the rock. What physical factors control the color of igneous rock? Generally, the color reflects the rock's composition, but it isn't always so simple because color may also be influenced by grain size and by the presence of trace amounts of impurities. (For example, the presence of a small amount of iron oxide gives rock a reddish tint.)

Next, you would probably characterize the rock's texture. A description of igneous texture indicates whether the rock consists of crystals, fragments, or solid glass. **Crystalline igneous rocks** consist of mineral crystals that intergrow when the melt solidifies and thus fit together like pieces of a jigsaw

FIGURE 6.15 Mesozoic batholiths of western North America.

(a) During the Mesozoic, subduction produced a huge volcanic arc. In the crust, beneath the arc, large granite batholiths formed. They are now exposed by erosion.

(b) The Sierra Nevada Mountains of California provide exposures of one the Mesozoic batholiths. The huge granite cliffs, popular with climbers, are part of the batholith.

FIGURE 6.16 Making room for igneous dikes and sills.

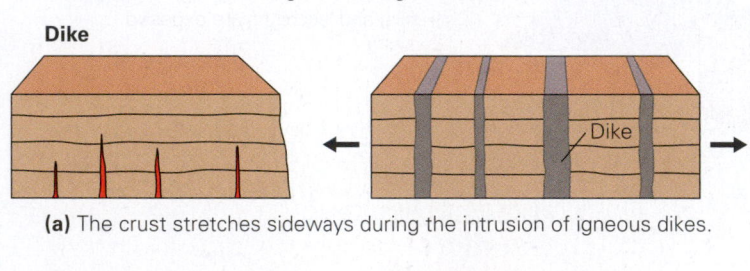

(a) The crust stretches sideways during the intrusion of igneous dikes.

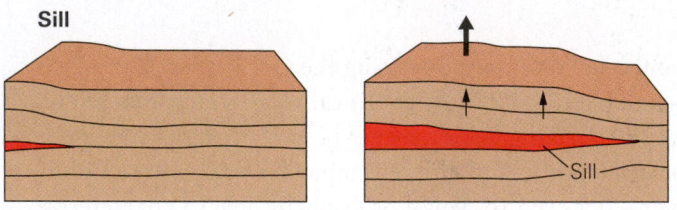

(b) The Earth's surface rises to make room for sills.

puzzle (**Fig. 6.18a**). The interlocking of crystals in these rocks develops because once earlier-formed grains exist, later-formed grains fill in the space in between and grow around the earlier-formed grains. Geologists distinguish subcategories of crystalline igneous rocks according to the size of the crystals. Coarse-grained (*phaneritic*) rocks have crystals large enough to be identified with the naked eye. Fine-grained (*aphanitic*) rocks have crystals too small to be identified with the naked eye.

Porphyritic rocks have larger crystals surrounded by a mass of fine crystals. In a porphyritic rock, the larger crystals are called *phenocrysts*, while the mass of finer crystals is called *groundmass*. **Fragmental igneous rocks** form from pyroclastic debris and consist of igneous chunks and/or shards that are packed together, welded together, or cemented together after having solidified (**Fig. 6.18b**). Different types of fragmental igneous rocks are distinguished from one another by fragment size. Rocks made of a solid mass of glass or of glass surrounding isolated small crystals are **glassy igneous rocks** (**Fig. 6.18c**). Glassy rocks typically fracture conchoidally.

What factors control the texture of igneous rocks? In the case of nonfragmental rocks, texture largely reflects cooling rate. The presence of glass indicates that cooling happened so quickly that the atoms within a lava didn't have time to arrange into crystal lattices. Crystalline rocks form when a melt cools more slowly. A melt that cools rapidly, but not rapidly enough to make glass, forms fine-grained rock because many crystal seeds form but none has time to grow large. A melt that cools very slowly forms a coarse-grained rock because relatively few seeds form and each crystal has time to grow large.

Because of the relationship between cooling rate and texture, lava flows, dikes, and sills tend to be composed of fine-grained igneous rock. In contrast, plutons tend to be composed of coarse-grained rock. Plutons that intrude into hot wall rock at great depth cool very slowly and thus tend to have larger crystals than plutons that intrude into cool wall rock at shallow depth. Porphyritic rocks form when a melt cools in two stages. First, the melt cools at depth slowly enough that phenocrysts form. Then the melt erupts and the remainder cools quickly, so groundmass forms around the phenocrysts.

FIGURE 6.17 Making room for igneous plutons.

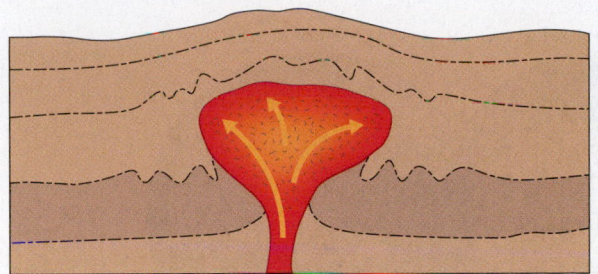

(a) During diapiric rise of a mass of magma, the magma is a blob that forces its way up through wall rock, which plastically deforms to move out of the way.

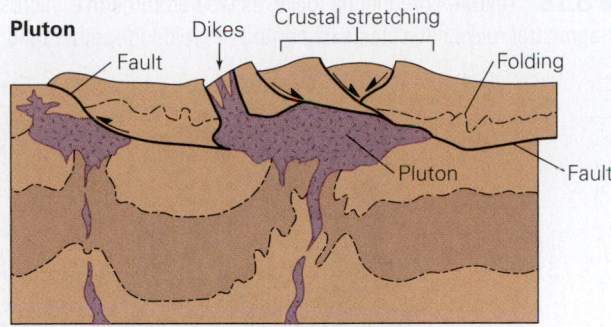

(b) Faulting associated with crustal stretching may accommodate the emplacement of a diapir.

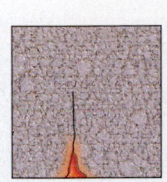  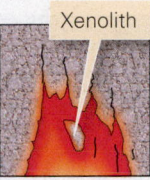 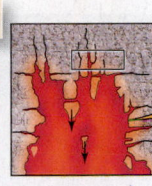

Xenolith

Time

(c) Magma pushes up into cracks, and blocks may be incorporated in the melt by stoping.

The white rock (granite) is intruding into the dark wall rock.

Xenolith

(d) A xenolith of darker rock surrounded by light-colored granite.

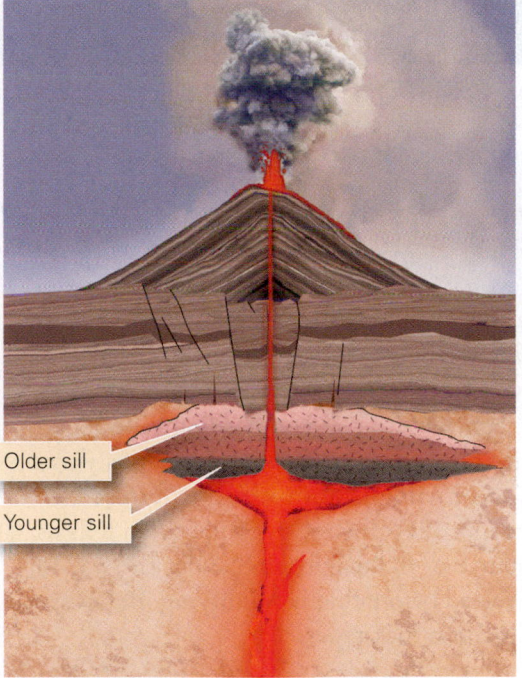

Older sill

Younger sill

(e) Plutons may intrude as a succession of sill-like sheets.

There is, however, an exception to the standard cooling-rate and grain-size relationship. A very coarse-grained igneous rock called **pegmatite** doesn't necessarily cool slowly. Pegmatite, which can contain crystals up to tens of centimeters across, typically occurs in dikes. Because pegmatite occurs in dikes, which generally cool relatively quickly, the coarseness of the rock may seem surprising. Researchers have shown that pegmatites are coarse because they form from water-rich melts in which atoms can move around so rapidly that large crystals can grow very quickly.

FIGURE 6.18 Textures of igneous rocks, as viewed through a microscope. The field of view is about 3 mm. Crystalline rocks have interlocking crystals, fragmental rocks have clasts cemented or welded together, and glassy rocks contain glass (black material) as well as isolated crystals.

Crystalline   Fragmental   Glassy

(a) Granite is crystalline.   **(b)** Tuff is fragmental.   **(c)** Obsidian is glassy.

Classifying Igneous Rocks

Because melts can have a variety of compositions and can freeze to form igneous rocks in many different environments above and below the surface of the Earth, a wide spectrum of distinct igneous rock types occurs on Earth. We can classify these types according to their texture and composition. The rock's texture provides clues to the rate at which it cooled, as we've seen, and therefore the environment in which it formed (see **Geology at a Glance**, pp. 171). The rock's composition tells us about the original source of the magma and about the way in which the magma evolved before finally solidifying. Below we introduce some key igneous rock types.

Crystalline Igneous Rocks The scheme for classifying the principal types of crystalline igneous rocks is quite simple. The different compositional classes are distinguished on the basis of silica content—*ultramafic*, *mafic*, *intermediate*, or *felsic*—whereas the different textural classes are distinguished according to whether the grains are coarse or fine. The chart in **Figure 6.19** gives the texture and composition of the most commonly used crystalline rock names. As a rough guide, the color of an igneous rock reflects its composition: mafic rocks

tend to be black or dark gray, intermediate rocks tend to be lighter gray or greenish gray, and felsic rocks tend to be tan to pink or maroon. **Figure 6.20** provides images of some of these rocks. Note that the basalt, andesite, and rhyolite could have come from the same magmas as gabbro, diorite, and granite, respectively. But the three fine-grained rocks likely cooled quickly in lava flows or near-surface dikes and sills, while the three coarse-grained rocks likely cooled more slowly in plutons.

Glassy Igneous Rocks Glassy texture develops more commonly in felsic igneous rocks because the high concentration of silica inhibits the easy growth of crystals. But basaltic and intermediate lavas can form glass if they cool rapidly enough. In some cases, a rapidly cooling lava freezes while it still contains gas bubbles—these bubbles remain as open holes known as **vesicles**. Geologists distinguish among several different kinds of glassy rocks.

- **Obsidian** is a mass of vesicle-free felsic glass that tends to be black or brown (see Fig. 6.18c). Because it breaks conchoidally, sharp-edged

Did you ever wonder . . .

how black glass once used for arrowheads formed?

Formation of Igneous Rocks

Igneous rock forms by the cooling of magma underground or of lava at the surface. Igneous rocks that solidify underground are intrusive, whereas those that solidify at the surface are extrusive. The type of igneous rock that forms depends on the composition of the melt and the environment of cooling.

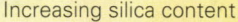

Stratified volcanic tuff

EXTRUSIVE ENVIRONMENT

Pyroclastic flow

Lava flow

Dike swarm

Sills

Lava dome

Laccolith

Ring dikes

Volcanic neck

Irregular stock

Increasing silica content

Fast cooling

Slow cooling

MAFIC	FELSIC
Scoria (glassy)	Obsidian (glassy)
Basalt (fine grained)	Rhyolite (fine grained)
Gabbro (coarse grained)	Granite (coarse grained)

In the extrusive environment, melt may cool quickly and have a glassy texture. Melt that explodes into the air forms ash and other debris with fragmental texture.

In the intrusive environment, crystals grow together to form an interlocking texture. Slower cooling makes coarser grains.

Minerals in an igneous rock crystallize in succession as the melt cools.

Cooler

Hotter

Pluton

INTRUSIVE ENVIRONMENT

Magma chamber

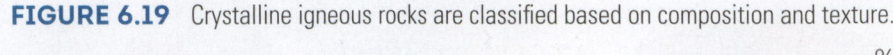

FIGURE 6.19 Crystalline igneous rocks are classified based on composition and texture.

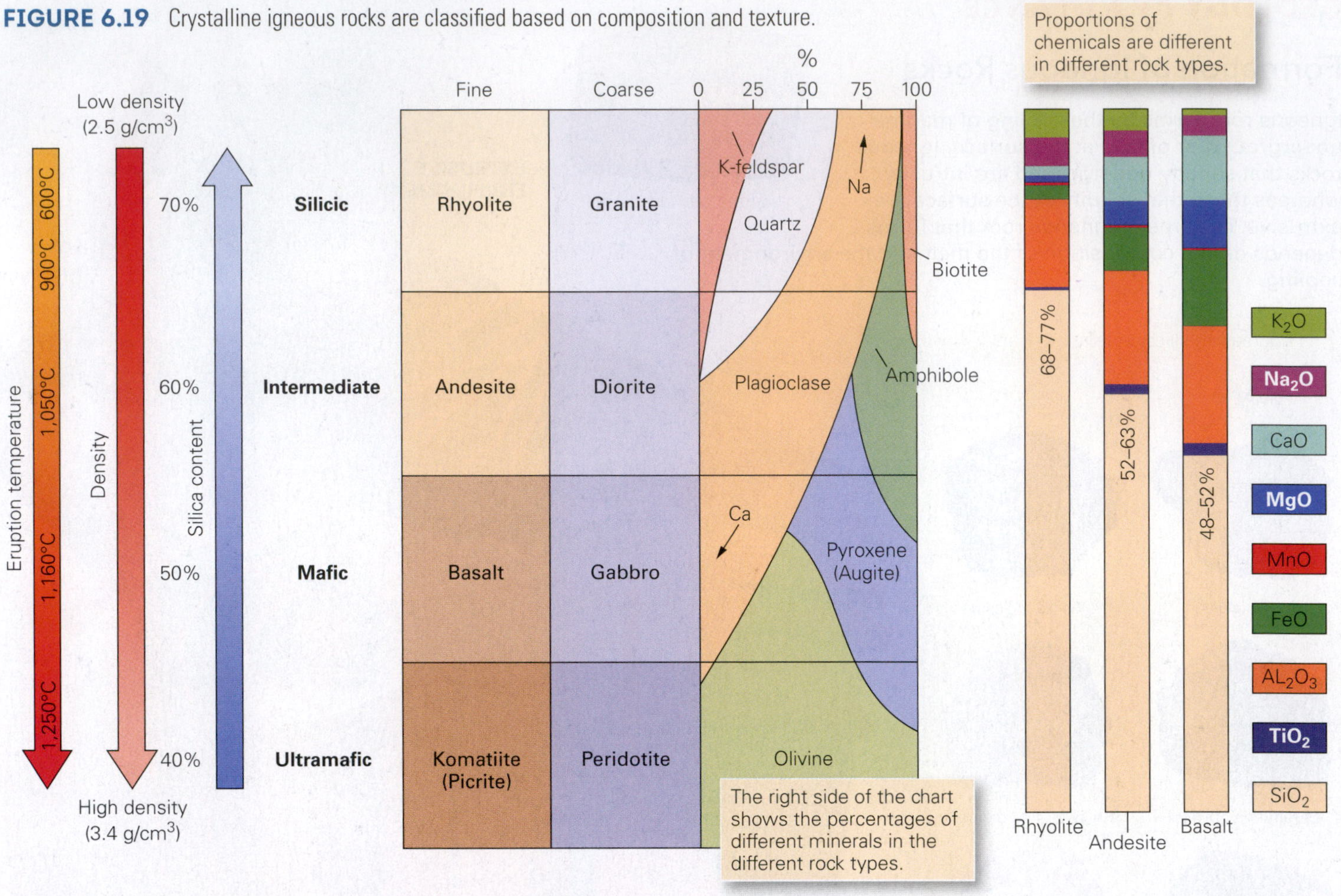

Eruption temperature

900°C / 600°C
1,050°C
1,160°C
1,250°C

Density

Low density (2.5 g/cm³)
High density (3.4 g/cm³)

Silica content

70% — **Silicic**
60% — **Intermediate**
50% — **Mafic**
40% — **Ultramafic**

	Fine	Coarse
Silicic	Rhyolite	Granite
Intermediate	Andesite	Diorite
Mafic	Basalt	Gabbro
Ultramafic	Komatiite (Picrite)	Peridotite

%
0 25 50 75 100

K-feldspar
Na
Quartz
Biotite
Plagioclase
Amphibole
Ca
Pyroxene (Augite)
Olivine

The right side of the chart shows the percentages of different minerals in the different rock types.

Proportions of chemicals are different in different rock types.

Rhyolite 68–77%
Andesite 52–63%
Basalt 48–52%

K₂O
Na₂O
CaO
MgO
MnO
FeO
AL₂O₃
TiO₂
SiO₂

---

pieces split off its surface when you hit a sample with a hammer. Pre-industrial people worldwide used such pieces for scrapers, arrowheads, and knife blades.

- **Tachylite** is a vesicle-free mass consisting of more than 80% mafic glass. This rock is relatively rare in comparison with obsidian.
- **Pumice** is a felsic volcanic rock that contains abundant vesicles (75% to 90%), giving it the appearance of a sponge (**Fig. 6.21a**). Pumice forms by the quick cooling of frothy lava, which occurs when lots of gas bubbles come out of a solution but can't escape because the felsic magma is so sticky. The felsic "froth" from which pumice forms resembles the head of foam on a glass of beer. In some cases, a chunk of pumice contains so many air-filled pores that it can actually float on water, like Styrofoam. Ground-up pumice makes the grainy abrasive that manufacturers use to "stonewash" blue jeans. Pumice tends to be light gray to tan in color.
- **Scoria** is a mafic volcanic rock that contains abundant vesicles (more than about 30%). Generally, the vesicles in scoria are bigger than those in pumice, there's more glassy

rock between individual vesicles, and the rock has an overall darker appearance (**Fig. 6.21b**).

Pyroclastic Igneous Rocks As we have noted, when some volcanoes erupt, they spew out droplets and clots of still-molten lava that freezes in midair as it flies. Explosive eruptions may spray out a fine mist of lava, pulverize recently solidified pumice that had frozen in the vent of the volcano, or blast apart pre-existing volcanic rocks that formed the volcano and send them flying, too. Volcanic fragments come in a great range of sizes—dust-sized specks or flakes of glass or pulverized rock comprise *ash*; pea- to marble-sized pellets are called *lapilli*; and still larger chunks form *bombs* (if streamlined) or *blocks* (if angular; see Chapter 9).

Accumulations of fragmental volcanic debris are called pyroclastic deposits. When the material in these deposits consolidates into a solid mass, due either to welding together of still-hot clasts or to later cementation by minerals precipitating from groundwater, it becomes a **pyroclastic rock**. Geologists distinguish among several types of pyroclastic rocks based on grain size. We introduce just a few below.

- **Tuff** is a fine-grained pyroclastic rock composed of volcanic ash or of ash mixed with lapilli-sized pumice fragments (**Fig. 6.21c**). Tuff can form from debris that settled out of the air like snow, accumulating in beds that blanket the landscape. Or it may form from debris that rushed down the volcano's side in a pyroclastic flow. Of note, ash in the interior of a pyroclastic flow may be so hot that it welds together, when the flow stops moving, to produce "welded tuff."
- **Volcanic agglomerate** consists of accumulations of lapilli or bombs.
- **Volcanic breccia** consists of angular fragments of volcanic debris that have been cemented together (**Fig. 6.21d**). A special type of breccia, called **hyaloclastite**, forms when lava erupts under water or ice and violently shatters into glassy fragments that react chemically with the surrounding water. Of note, not all volcanic breccias form from debris falling from the sky or tumbling down a volcano. Some develop when the crust of a lava flow breaks up as the still-liquid interior continues to move.

When we investigate volcanic eruptions further in Chapter 9, we will see that pyroclastic debris may mix with water, after deposition, and move down the slope of a volcano before eventually getting buried and turned to rock. Geologists use the term **volcaniclastic rock** to refer to any rock that contains a large proportion of volcanic fragments; this category includes both pyroclastic rocks and rocks formed from water-transported volcanic debris.

FIGURE 6.21 Glassy and fragmental igneous rocks.

(a) Pumice is a felsic glassy rock with tiny vesicles.

(b) Scoria is a mafic glassy rock with many vesicles.

(c) Tuff is a fragmental rock composed of ash and, locally, pumice fragments.

(d) Volcanic breccia consists of angular fragments.

Take-Home Message

Geologists classify and assign names to igneous rocks based on texture and composition. The grain size of crystalline rocks commonly reflects cooling rate. Mafic rocks tend to be darker than felsic rocks. Glassy rocks do not have grains but may contain vesicles, gas bubbles trapped when the lava froze. Pyroclastic rocks consists of fragmental debris.

QUICK QUESTION: What kind of rock forms from felsic magma cooled in a large pluton at depth? Could you make a sharp axe out of this rock?

6.7 Plate Tectonic Context of Igneous Activity

Look at a map showing the distribution of **igneous activity**—the formation, movement, and in some cases eruption of molten rock—around the world (**Fig. 6.22**), and compare the map to a map showing the location of plate boundaries. You'll realize that most igneous activity occurs in the volcanic arc of convergent-plate boundaries, along the mid-ocean ridge axis of divergent-plate boundaries, or within continental rifts (where continents are stretching and pulling apart and a new plate boundary may form). Clearly, the processes that lead to melt production in the Earth are governed largely by plate tecton-ics. But the map also shows that igneous activity happens at isolated hot spots, some of which occur on plate boundaries and some of which occurs in plate interiors. Let's look at the settings where igneous activity occurs more closely, so we can see why the particular setting leads to melt production and why different types of melts form at different settings.

Products of Subduction

A chain of volcanoes, called a **volcanic arc** (or, commonly, just an arc), develops along all convergent-plate boundaries. Recall that at convergent boundaries a downgoing plate of oceanic lithosphere subducts beneath an overriding plate that can consist either of oceanic or continental lithosphere. You will find the volcanic arc on the edge of the overriding plate, generally at a distance of 150 to 300 km from the axis of the trench (**Fig. 6.23**). Geologists use the word *arc* to emphasize that many of these chains define a curve on a map. *Continental arcs*, such as the Andean arc of South America and the Cascade arc of northwestern United States, lie on the edge of a continent, where the oceanic lithosphere subducts beneath a continent. *Island arcs*, such as the Aleutian arc of Alaska and the Mariana arc of the western Pacific, protrude from the ocean at localities where one oceanic plate subducts beneath another. Igneous activity at volcanic arcs produces both extru-sive rocks and intrusive rocks—in the crust beneath each vol-cano, dikes, sills, and plutons have developed or are develop-ing. It takes millions of years of exhumation—uplift of rocks at depth as erosion removes rocks at the surface—to expose these intrusive rocks at the ground surface.

How does subduction trigger melting? When the theory of plate tectonics was first being developed, geologists speculated that all the magma came from melting of the subducted plate.

FIGURE 6.22 The tectonic setting of igneous rocks.

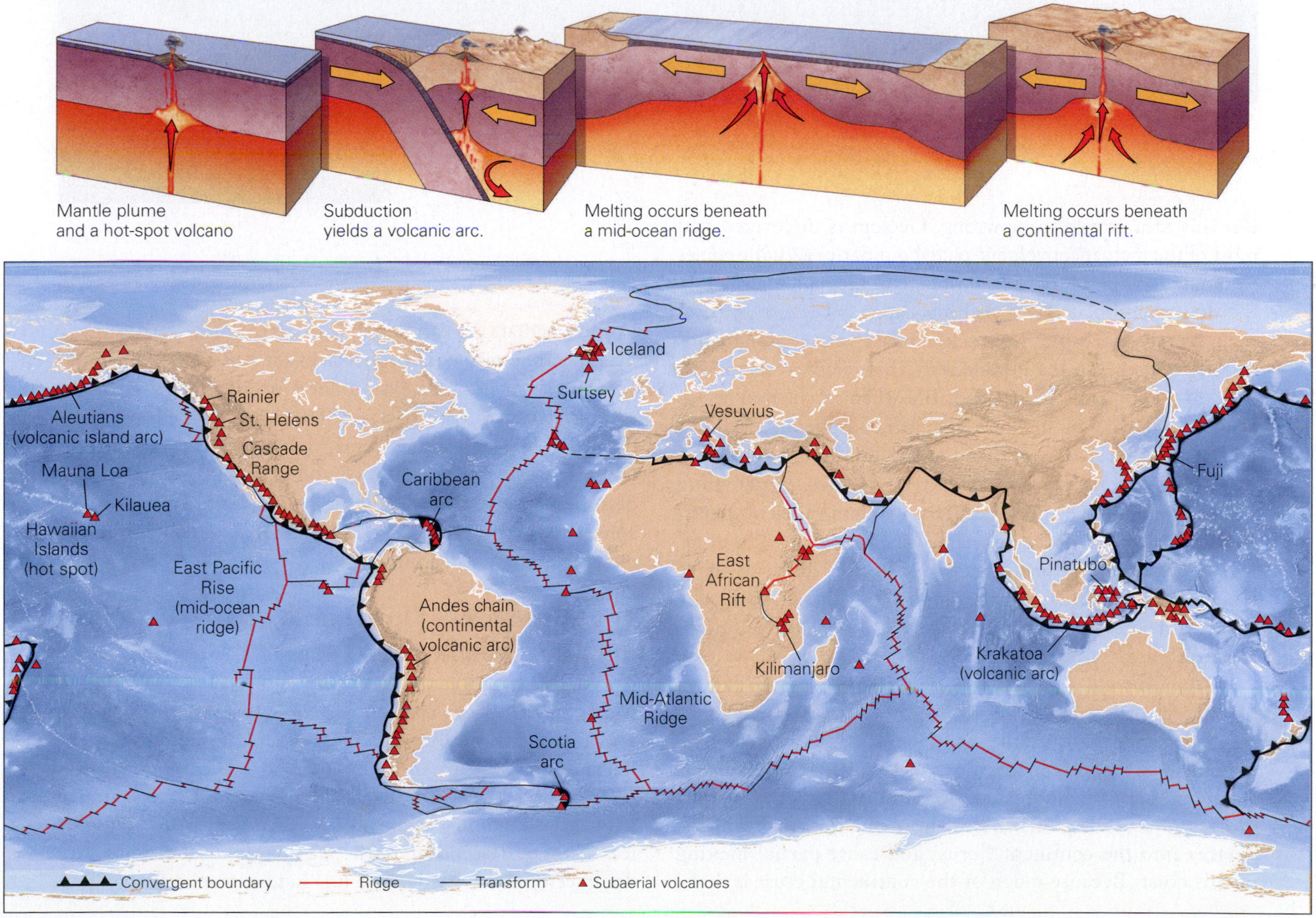

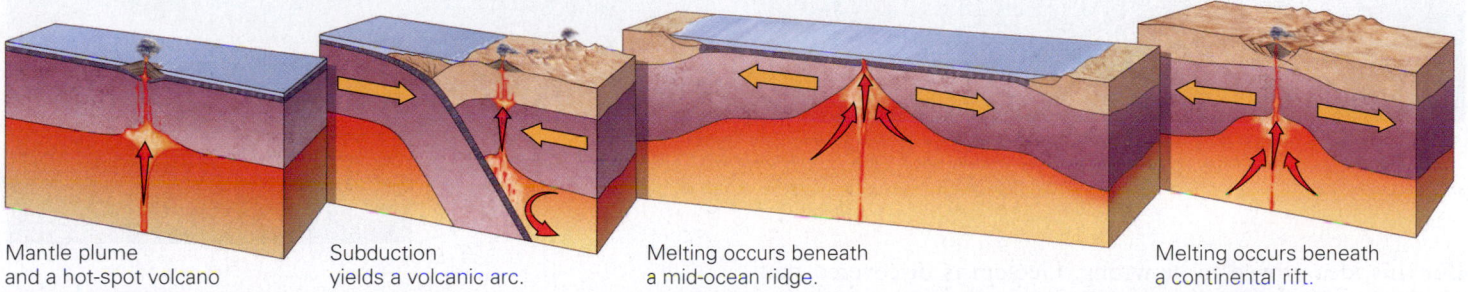

Mantle plume and a hot-spot volcano

Subduction yields a volcanic arc.

Melting occurs beneath a mid-ocean ridge.

Melting occurs beneath a continental rift.

Iceland
Surtsey
Vesuvius
Aleutians (volcanic island arc)
Rainier
St. Helens
Cascade Range
Mauna Loa
Kilauea
Hawaiian Islands (hot spot)
Caribbean arc
Fuji
East Pacific Rise (mid-ocean ridge)
Andes chain (continental volcanic arc)
East African Rift
Pinatubo
Kilimanjaro
Krakatoa (volcanic arc)
Mid-Atlantic Ridge
Scotia arc

▲▲▲ Convergent boundary　── Ridge　── Transform　▲ Subaerial volcanoes

FIGURE 6.23 The Aleutian volcanic arc of Alaska. The trench and volcanic islands are clearly visible. The enlargement shows Umnak Island hosting two volcanoes.

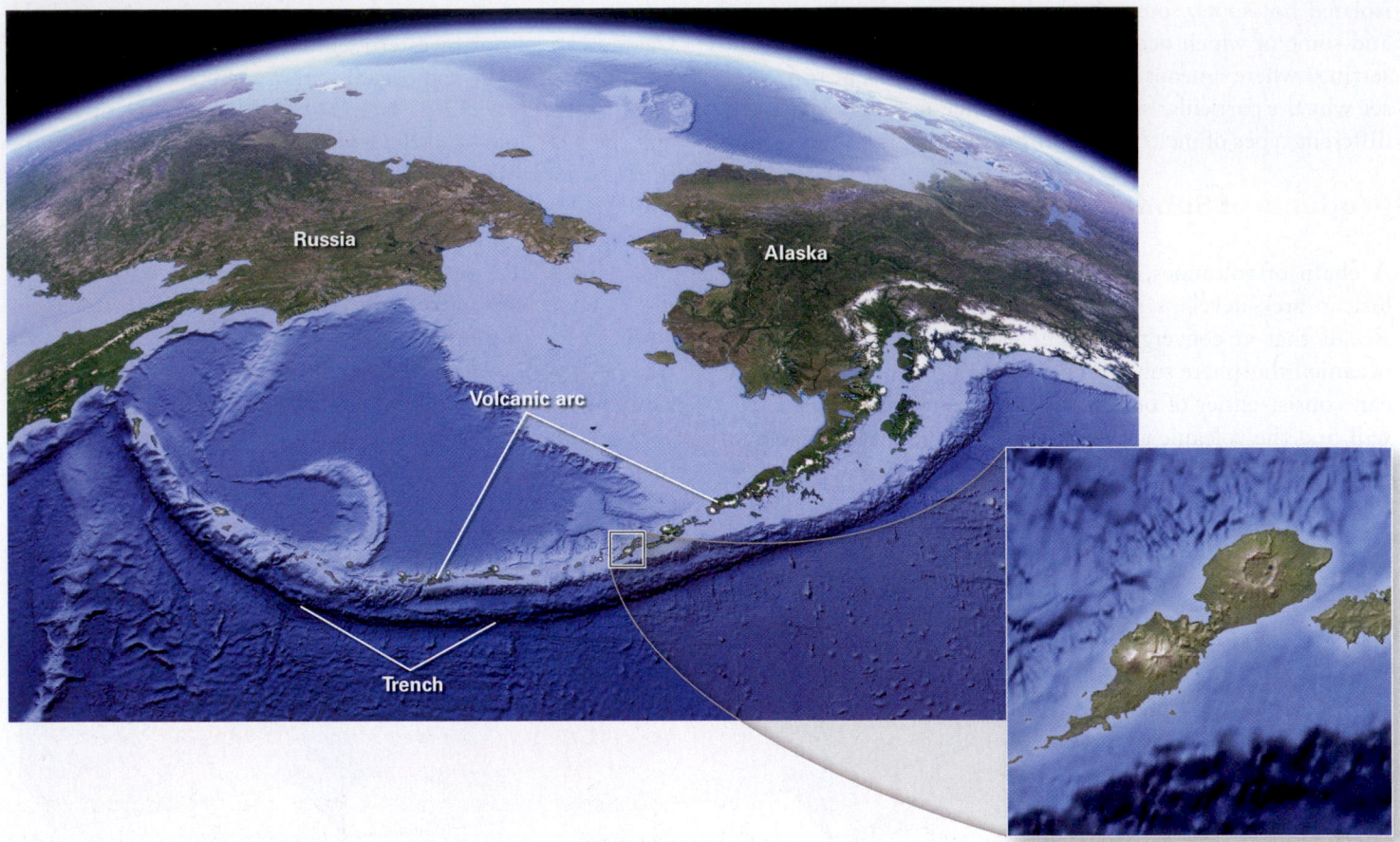

But this idea proved to be wrong. Geologists discovered that most of the melt at convergent-plate boundaries actually comes from flux melting of the asthenosphere in the region just above the downgoing plate. This melting happens because some of the minerals in oceanic crust rocks contain volatile compounds (mostly water). At shallow depths, the volatiles chemically bond to the minerals in the crust. But when subduction brings oceanic crust down into the hot asthenosphere, the "wet" crustal rocks start to warm, and at a depth of about 150 km, the crustal rock has become so hot that volatiles separate from minerals and diffuse upward into the overlying asthenosphere. Addition of volatiles triggers partial melting of the hot ultramafic rock in the asthenosphere, a process that yields mafic magma. Some of this magma rises to form basaltic sills and dikes in the crust, and some makes it all the way to the surface to extrude as basaltic lava.

In continental volcanic arcs, not all the mantle-derived basaltic magma rises directly to the surface—some gets trapped at the base of the continental crust and some in magma chambers deep in the crust. When this happens, fractional crystallization may occur, so part of the magma evolves into a felsic melt. Also, heat can transfer from the magma into the continental crust and cause partial melting of this crust. Because much of the continental crust is mafic to intermediate in composition to start with, the resulting magmas are intermediate to felsic in composition—remember that partial melting always produces magma that is richer in silica than was the source rock. This magma rises, leaving the basalt behind, and either cools higher in the crust to form plutons or rises to the surface and erupts. For this reason, granitic plutons, as well as felsic and andesite lavas, form at continental arcs (**Fig. 6.24**). The word *andesite* come from the name of the Andes Mountains.

The Formation of Igneous Rocks at Mid-Ocean Ridges

Most *subaerial volcanoes* on Earth—those that rise above sea level and erupt into the air producing visible displays of igneous activity—are in island arcs. So it may seem surprising that, in fact, most igneous activity at the Earth's surface happens at the mid-ocean ridges of divergent-plate boundaries. Without descending to the sea floor in a submersible, we don't see this activity because it is hidden from our view by a

FIGURE 6.24 The rounded boulders shown here are a consequence of in-place weathering of a homogeneous granite pluton in Joshua Tree National Monument, California.

2-km-thick layer of ocean water. Think about it—the entire oceanic crust, a 7- to 10-km-thick layer of basalt and gabbro that covers 70% of the Earth's surface, forms at mid-ocean ridges. And this entire volume gets subducted and replaced by new crust about every 200 million years.

Igneous magmas form at mid-ocean ridges because of decompression melting; as sea-floor spreading occurs and oceanic lithosphere plates drift away from the ridge, hot asthenosphere rises beneath the ridge axis. As this astheno-

sphere rises, the pressure caused by overlying rock progressively decreases, which leads to partial melting and the generation of basaltic magma. As noted in Chapter 4, some of this magma rises into the crust and collects in a shallow magma chamber, forming a mush of liquid and crystals. Slow cooling along the margins of the magma chamber produces gabbro. Not all the melt stays in the magma chamber—about 30% eventually rises still further. Of this volume, two-thirds intrudes and freezes within vertical cracks that propagate as newly formed crust splits apart and forms basalt dikes (see Fig. 4.7a). The remaining third reaches the sea floor and extrudes as lava on the surface of the sea floor, commonly in the form of pillow basalt (**Fig. 6.25**).

Igneous Rocks at Rifts

Rifts are places where continental lithosphere is being stretched horizontally and, therefore, thinned vertically. As this process takes place, the weight of rock overlying the asthenosphere decreases, so pressure in the asthenosphere decreases and decompression melting takes place, producing basaltic magma. Some of this magma intrudes as sills or dikes in the crust, and some makes it to the Earth's surface and erupts as basaltic lava. But, unlike mid-ocean ridges, part of the magma produced in the asthenosphere beneath rifts gets trapped at the base of the continental crust or even in the continental crust itself, and this magma transfers enough heat

FIGURE 6.25 Pillow basalt forms when lava erupts under water and cools very quickly.

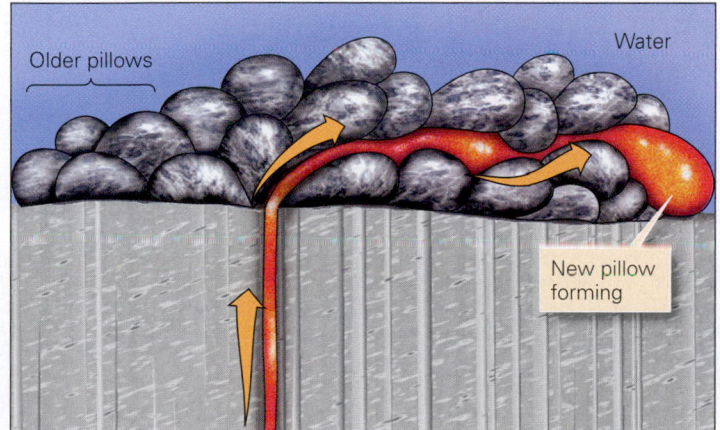

(a) The formation of pillow basalt. Lava squeezes through a conduit and emerges at the surface like toothpaste.

(b) This pillow basalt forms part of an ophiolite, a slice of sea floor that was pushed up onto the surface of a continent during mountain building.

to the crust to cause partial melting of the continental crust. This partial melting yields felsic magmas that erupt mostly as rhyolitic ash. Thus, a sequence of extrusive rocks in a rift generally includes both basaltic flows and sheets of rhyolitic tuff.

Products of Hot Spots

Hawaii and numerous other South Pacific island volcanoes are *oceanic hot-spot volcanoes*, isolated volcanoes that are not a consequence of plate-boundary interactions. Most oceanic hot-spot volcanoes erupt in the interior of oceanic plates, away from convergent or divergent boundaries. But some, such as Iceland, sit astride a divergent boundary. (Geoscientists associate Iceland with a hot spot because its volcanoes generate far more lava than normal mid-ocean ridge volcanoes do.) Not all hot-spot volcanoes are oceanic, however—some erupt in the interior of continents. Eruptions from such *continental hot-spot volcanoes* produced colorful sulfur- and iron-stained layers of volcanic ash, the "yellow stone" of Yellowstone National Park in northwestern Wyoming and adjacent states.

As we learned in Chapter 4, most researchers associate hot-spot igneous activity with mantle plumes, columns of hot mantle rock rising from deep in the mantle, though this idea is not universally accepted. According to the plume hypothesis, the plume itself does *not* consist of magma—it's composed of solid rock but rock that is hot and soft enough to flow plastically at rates of a few centimeters a year. When the hot rock of a plume reaches the

base of the lithosphere, decompression causes partial melting, a process that generates mafic magma. (Geologists who don't agree with the plume hypothesis have suggested alternatives.) At oceanic hot spots, much of the mafic magma erupts at the surface as basalt. At continental hot spots, part of the mafic magma erupts to form basalt, but some transfers heat to the continental crust, which then partially melts itself, as happens beneath rifts, producing felsic magmas that erupt to form rhyolite.

Large Igneous Provinces

In many places on Earth, particularly voluminous quantities of mafic magma have erupted and/or intruded (**Fig. 6.26**). Some of these regions occur along the margins of continents, some in the interior of oceanic plates, and some in the interior of continents. The largest of these, the Ontong Java Oceanic Plateau of the western Pacific, covers an area of about 5,000,000 km² of the sea floor and has a volume of about 50,000,000 km³. Such provinces also occur on land. It's no surprise that geologists refer to an occurrence of such a huge volume of igneous rock as a **large igneous province (LIP)**. More recently, this term has also been applied to areas covered by immense eruptions of felsic rocks, so Yellowstone Park can also be called a LIP.

Mafic LIPs may form when the bulbous head of a mantle plume first reaches the base of the lithosphere. More partial melting can occur in a plume head than in the asthenosphere rising from relatively shallow depths beneath normal mid-ocean ridges or rifts, because the rock in the plume head came from great depth in the mantle, so its temperature is higher. Thus, an unusually large quantity of unusually hot basaltic magma forms in the plume head. If the lithosphere

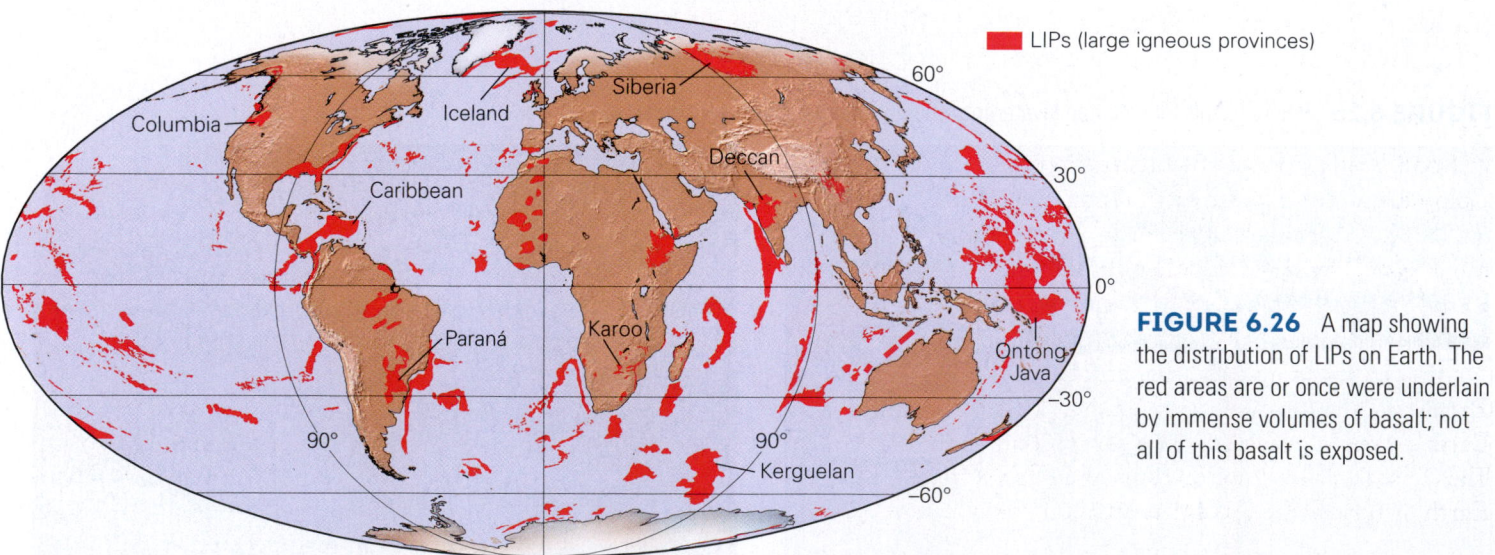

FIGURE 6.26 A map showing the distribution of LIPs on Earth. The red areas are or once were underlain by immense volumes of basalt; not all of this basalt is exposed.

FIGURE 6.27 Flood basalts form when vast quantities of low-viscosity mafic lava "floods" over the landscape and freezes into a thin sheet. Accumulation of successive flows builds a flat-topped plateau.

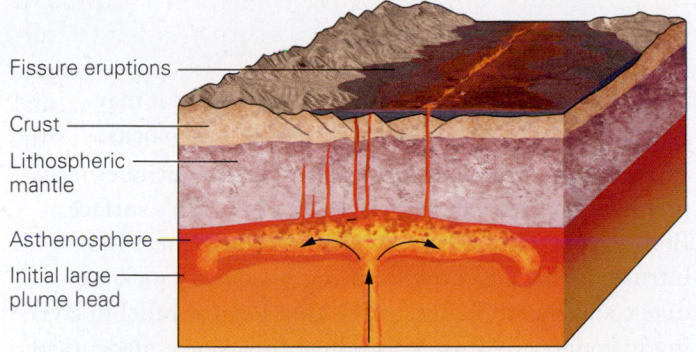

(a) The plume model for forming flood basalts.

Labels: Fissure eruptions; Crust; Lithospheric mantle; Asthenosphere; Initial large plume head

(b) Flood basalts form the layers exposed in Palouse Canyon, Washington.

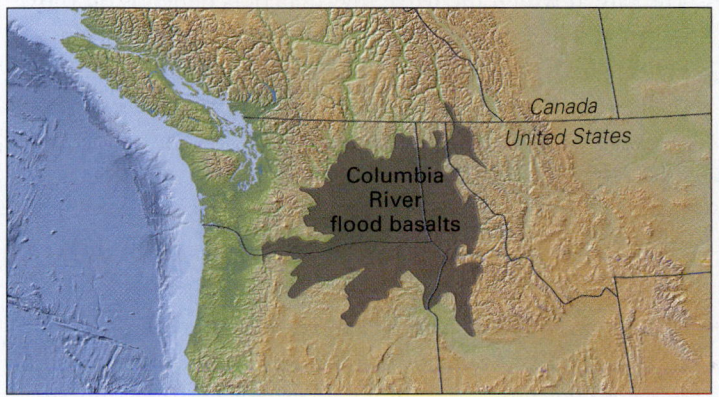

(c) Flood basalts underlie the Columbia River Plateau in Washington and Oregon, the dark area of this map.

(d) Iguaçu Falls, on the Brazil-Argentina border. The falls flow over the huge flood basalt sheet (the black rock) of the Paraná Plateau. Flood basalt underlies all of the region in view.

over the plume head starts to stretch and rift, huge quantities of basaltic lava can rise and spew out of the ground. The particularly hot basaltic lava that erupts at such localities has such low viscosity that it can flow tens to hundreds of kilometers across the landscape. Geoscientists refer to such flows as **flood basalts**. Flood basalts make up the bedrock of the Columbia River Plateau in Oregon and Washington (**Fig. 6.27a, b**), the Palouse Canyon in Washington (**Fig. 6.27c**), the Paraná Plateau in southeastern Brazil (**Fig. 6.27d**), the Karoo region of southern Africa, and the Deccan region of southwestern India.

The volume of material that erupted at the biggest LIPs on Earth may have exceeded the amount that erupted along the Earth's entire mid-ocean ridge system during the same time. Thus, it seems that eruptions of such LIPs are special events in Earth history. Some geologists attribute them to **superplumes** in the mantle—plumes that bring up vastly more hot asthenosphere than do normal plumes.

Take-Home Message

Igneous activity only occurs in special locations, where melting takes place. Flux melting takes place just above subducting plates, and decompression melting takes place beneath rifts, mid-ocean ridges, and hot spots. When igneous activity takes place beneath continental crust, heat-transfer melting may trigger partial melting of the continental crust.

QUICK QUESTION: Explain why both mafic and felsic igneous rocks form at continental rifts and at continental hot spots.

CHAPTER SUMMARY

- Magma is liquid rock (melt) under the Earth's surface. Lava is melt that has erupted from a volcano at the Earth's surface.

- Magma forms when hot rock in the Earth partially melts. This process occurs only under certain circumstances—when the pressure decreases (decompression), when volatiles (such as water or carbon dioxide) are added to hot rock, and when heat is transferred to adjacent rock by magma rising from below.

- Magma occurs in a range of compositions: felsic (silicic), intermediate, mafic, and ultramafic. The composition of magma is determined in part by the original composition of the rock from which the magma formed and in part by the way the magma evolves by such processes as assimilation and fractional crystallization.

- During partial melting, only part of the source rock melts to form magma. Magma tends to be more felsic than the rock from which it was extracted.

- Magma rises from depth because of its buoyancy and because pressure caused by the weight of overlying rock squeezes magma upward.

- Magma viscosity (its resistance to flow) depends on its composition. Felsic magma is more viscous than mafic magma.

- Geologists distinguish between two types of igneous rocks. Extrusive igneous rocks form from lava that erupts out of a volcano and freezes in contact with air or water. Intrusive igneous rocks develop from magma that freezes inside the Earth.

- Lava may solidify to form flows or domes, or it may explode into the air to form ash, lapilli, and blocks.

- Intrusive igneous rocks form when magma intrudes into pre-existing rock (country rock) below Earth's surface. Blob-shaped intrusions are called plutons. Sheet-like intrusions that cut across layering in country rock are dikes, and sheet-like intrusions that form parallel to layering in country rock are sills. Huge intrusions, made up of many plutons, are known as batholiths.

- The rate at which intrusive magma cools depends on the depth at which it intrudes, on the size and shape of the magma body, and on whether circulating groundwater is present. The cooling time controls the texture of an igneous rock.

- Crystalline (nonglassy) igneous rocks are classified according to texture and composition. Glassy igneous rocks are classified according to texture (a solid mass is obsidian; ash that has cemented or welded together is tuff).

- The origin of igneous rocks can readily be understood in the context of plate tectonics. Magma forms at continental or island volcanic arcs along convergent margins, mostly because of the addition of volatiles to the asthenosphere above the subducting slab. Igneous rocks form at hot spots, owing to the decompression melting of a rising mantle plume. Igneous rocks form at rifts as a result of decompression melting of the asthenosphere below the thinning lithosphere or of heat transfer from mantle melts into crustal rocks. Igneous rocks form along mid-ocean ridges because of decompression melting of the rising asthenosphere.

GUIDE TERMS

ash (p. 162)
assimilation (p. 159)
batholith (p. 162)
Bowen's reaction series (p. 164)
crystalline igneous rock (p. 167)
decompression melting (p. 157)
dike (p. 162)

extrusive igneous rock (p. 154)
flood basalt (p. 179)
flux melting (p. 157)
fractional crystallization (p. 161)
fragmental igneous rock (p. 168)
geotherm (p. 156)

glassy igneous rock (p. 168)
heat-transfer melting (p. 157)
hyaloclastite (p. 173)
igneous activity (p. 174)
igneous rock (p. 153)
intrusive contact (p. 162)
igneous intrusion (p. 154)
laccolith (p. 162)
lapilli (p. 162)

large igneous province (LIP) (p. 178)
lava (p. 153)
lava flow (p. 153)
magma (p. 153)
magma chamber (p. 154)
obsidian (p. 170)
partial melting (p. 158)
pegmatite (p. 169)

pluton (p. 162)
pumice (p. 172)
pyroclastic debris (p. 154)
pyroclastic flow (p. 162)
pyroclastic rock (p. 172)

scoria (p. 172)
sill (p. 162)
stoping (p. 166)
superplume (p. 179)
tachylite (p. 172)

tuff (p. 173)
vesicle (p. 170)
viscosity (p. 160)
volcanic agglomerate (p. 173)
volcanic arc (p. 175)

volcanic breccia (p. 173)
volcaniclastic rock (p. 173)
volcano (p. 153)
wall rock (p. 154)
xenolith (p. 166)

REVIEW QUESTIONS

1. How is the process of freezing magma similar to that of freezing water? How is it different?
2. What is the source of heat in the Earth? How did the first igneous rocks on the planet form?
3. Describe the three processes that are responsible for the formation of magmas.
4. Why are there so many different types of magmas?
5. Why do magmas rise from depth to the surface of the Earth?
6. What factors control the viscosity of a melt?
7. What factors control the cooling time of a magma within the crust?

8. How does grain size reflect the cooling time of a magma?
9. What does the mixture of grain sizes in a porphyritic igneous rock indicate about its cooling history?
10. Describe the way magmas are produced in subduction zones.
11. What process in the mantle may be responsible for causing hot-spot volcanoes to form?
12. Describe how magmas are produced at continental rifts.
13. What is a large igneous province (LIP)? How might the formation of LIPs have affected the Earth System?
14. Why does melting take place beneath the axis of a mid-ocean ridge?

ON FURTHER THOUGHT

15. If you look at the Moon, even without a telescope, you see broad areas where its surface appears relatively darker and smoother. These areas are individually called *mare* (plural: *maria*), from the Latin word for sea. The term is misleading, for they are not bodies of water but rather plains of igneous rock formed after huge meteors struck the Moon and formed very deep craters. These impacts occurred early in the history of the Moon. Propose a cause for the igneous activity, and suggest the type of igneous rock that fills the mare. (Hint: Think about how the presence of a deep crater affects pressure in the region below the crater, and think about the viscosity of a magma that could spread over such a broad area.)
16. The Cascade volcanic chain of the northwestern United States is only about 800 km long (from the southernmost volcano in California to the northernmost one in Washington State). The volcanic chain of the Andes is several thousand kilometers long. Look at a map showing the Earth's plate boundaries, and explain why the Andes volcanic chain is so much longer than the Cascade volcanic chain.
17. The photograph of **Figure Ft6.1** shows a nearly horizontal layer of basalt that now crops out as a 250-mile-high cliff on the western shore of the Hudson River, across from New York City. It is parallel to layers of sandstone above and below, and is younger than all of these layers. The basalt formed around 190 Ma, as Pangaea was beginning to break apart. What is this body of basalt, and in what geologic setting did it form?

FIGURE Ft6.1 Explaining mineral distribution in the Palisades.

(a) The Palisades sill in New Jersey rises above the Hudson River.

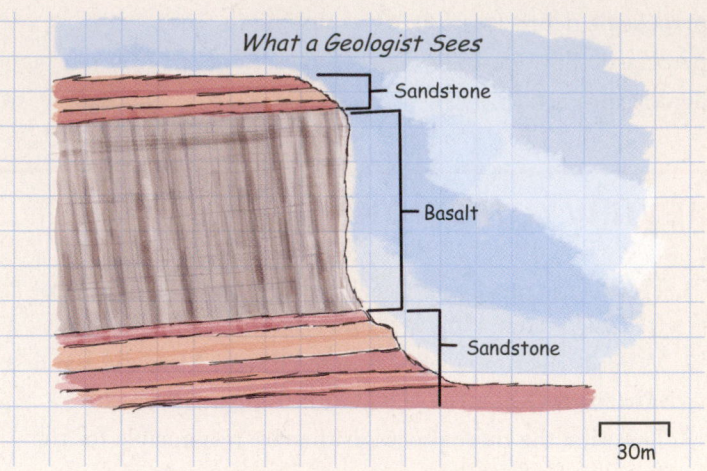

What a Geologist Sees

Sandstone

Basalt

Sandstone

30m

(b) What a geologist sees, as portrayed in cross section.

smartwork   smartwork.wwnorton.com

This chapter's Smartwork features:

- Interactive activities on decompression melting.
- What A Geologist Sees exercises on igneous rock composition.
- In-depth explorations of the characteristics of magma.

GEOTOURS

This chapter's GeoTour exercise (D) features:

- Batholiths and Laccoliths
- Dikes

Another View   The lighter-colored elliptical areas in this satellite image of a region called the Pilbara block, in northwestern Australia, are complex batholiths consisting of Archean granite. The darker rock surrounding the batholiths consists of basalt, komatiite, and sedimentary rock. The Pilbara block is about 300 km measured side to side.

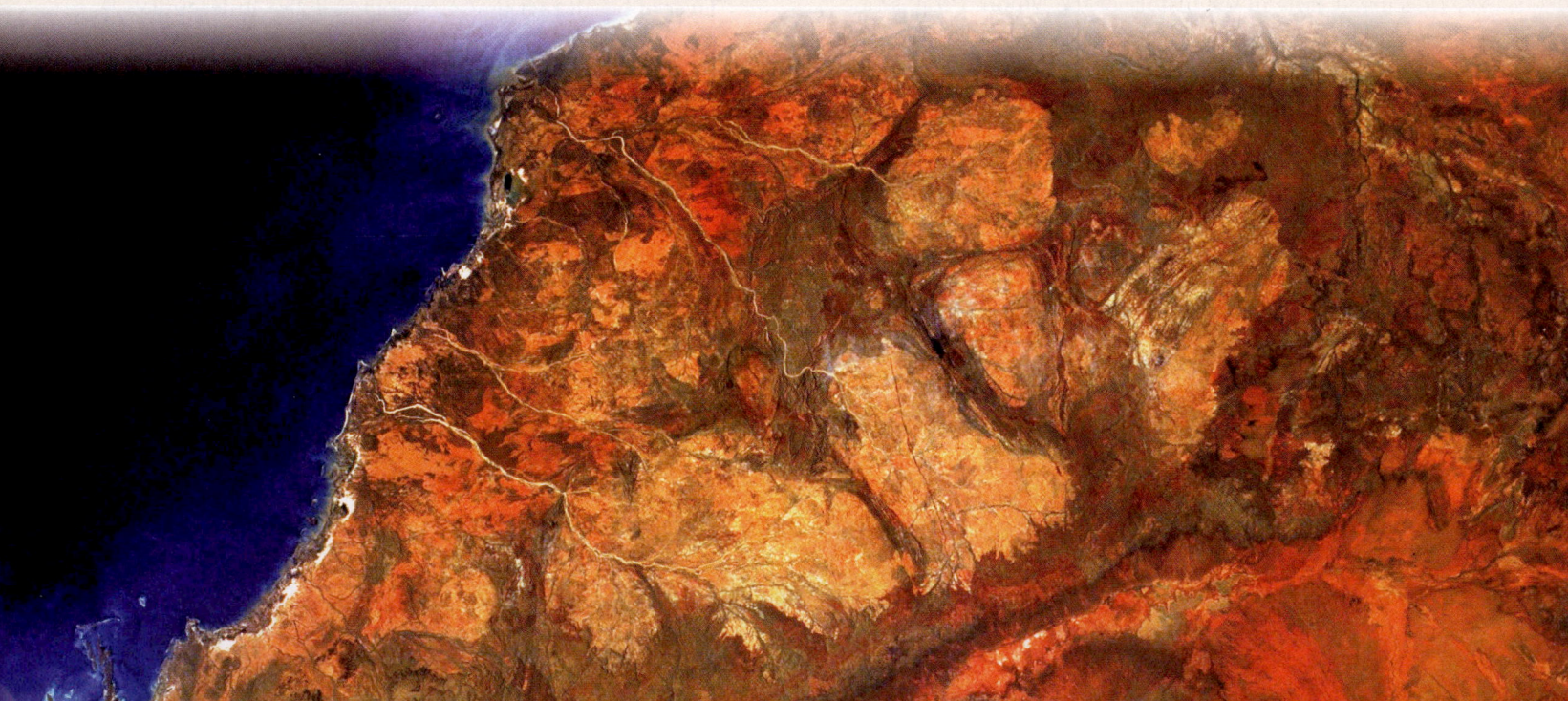

Here at the Earth's surface, in a small canyon in Indiana, a layer of sediment and soil covers bedrock. In stream cuts, bedrock peeks out from under this layer.

INTERLUDE B

INTERLUDE B

A Surface Veneer: Sediments and Soils

LEARNING OBJECTIVES

By the end of this interlude, you should understand . . .

- how rocks undergo change at and near the Earth's surface due to weathering.
- how weathering produces sediment.
- the difference between sediment and soil.
- factors that affect the character and thickness of soil.

B.1 Introduction

In the 1950s, the government of Egypt decided to build the Aswan High Dam to trap water of the Nile River in a huge reservoir before the water could reach the Mediterranean Sea. In the process of identifying a good site for the dam's foundation, geologists discovered that the present-day Nile flows on top of a 1,500-m-thick wedge of gravel, sand, and mud that fills what was once a canyon as large as the Grand Canyon (**Fig. B.1a**). Evidently, at some time in the geologic past, the Nile had carved a canyon 1,500 m deep, and subsequently this canyon filled with debris. Geologists of the 1950s couldn't understand how such a sequence of events could have happened. As a general rule, the surface of a river where it empties into the sea can't be lower than sea level—otherwise, the river would have

FIGURE B.1 The Nile River flows on the top of a sediment-filled canyon. This canyon formed when water evaporated from the Mediterranean, and the river cut down to that level.

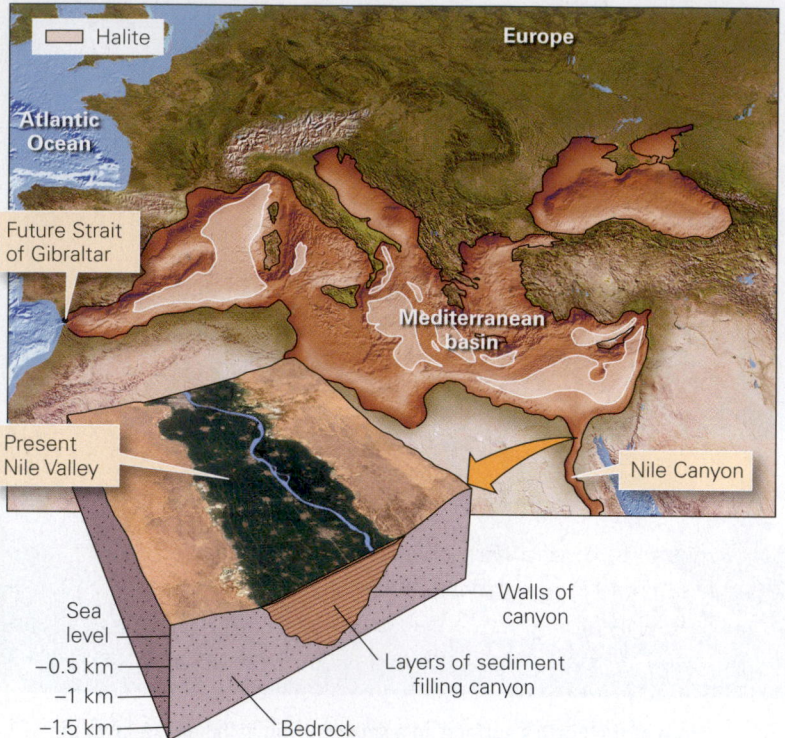

(a) The floor of the Mediterranean basin lies about 1.5 km below present sea level.

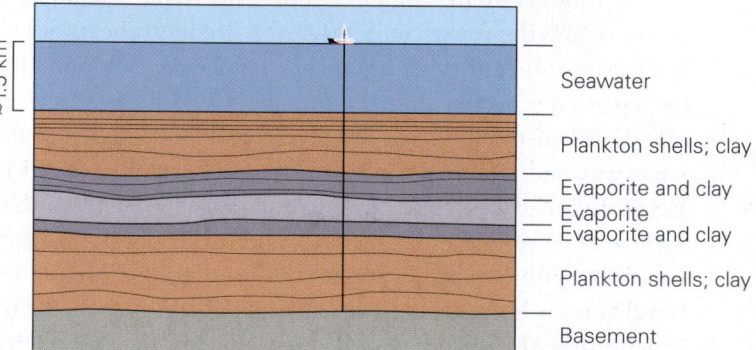

(b) Evidence for the past evaporation of the Mediterranean Sea was the result of drilling, which revealed a thick evaporite layer.

to flow upstream. Since the floor of the Nile's channel lies only about 15 m below the river's surface, it seemed impossible for the river to cut a canyon 100 times deeper.

The origin of the sub-Nile canyon remained a mystery until the summer of 1970, when geologists began to study the Mediterranean Sea's floor by drilling into it from a ship, as part of a 15-year-long program of global sub-seafloor exploration called the Deep Sea Drilling Project. The researchers expected to find layers consisting of the shells of plankton (tiny floating organisms) that had settled out of the water and of clay that rivers had carried to the sea (**Fig. B.1b**). The drill did penetrate such materials, but to their surprise, the drill also penetrated a 2-km-thick layer of halite and gypsum. Such minerals form only when seawater dries up and thus are known as evaporites. Since only about 3.5% of seawater consists of dissolved minerals, drying up the 1,500-m-deep Mediterranean would form a layer of evaporate less than 100 m thick. Thus, formation of a 2-km-thick evaporite layer meant that the entire Mediterranean must have dried up slowly over a long time period, and that it probably dried up completely several times, with the sea refilling after each drying event.

This discovery solved the mystery of the pre-Nile canyon. When the Mediterranean Sea dried up, the Nile River would have been flowing down into a basin whose floor was 1,500 m below present-day sea level. Over time, the river would have cut a canyon down to the low elevation of the dry sea floor. Later, when the sea refilled with water, the river could no longer cut down, and the flooded canyon filled with gravel, sand, and mud brought in from upstream.

Why did the Mediterranean Sea dry up? Only 10% of its water comes from rivers, so for the Mediterranean Sea to remain full, water must flow in from the Atlantic Ocean through the Strait of Gibraltar. Put another way, if the passage to the Atlantic were to be blocked, water would evaporate from the Mediterranean at a rate 10 times faster than it would flow into the Mediterranean from rivers. About 6 million years ago, the northward-drifting African Plate collided with the European Plate, forming a natural dam separating the Mediterranean from the Atlantic. At times when global sea level dropped, water stopped flowing over this dam from the Atlantic, and the Mediterranean evaporated. The salt that had been dissolved in its water accumulated as a solid deposit of halite and gypsum on the floor of the resulting basin, and the pre-Nile canyon formed. When sea level rose, water flooded in from the Atlantic into the Mediterranean, filling the basin again. This process was repeated many times. About 5.5 million years ago, the Mediterranean rose to its present level, and gravel, sand, and mud carried by the Nile River filled the canyon to its present level.

Geologists refer to the kinds of deposits just described—sand, mud, gravel, halite and gypsum crystals, and shell

FIGURE B.2 Various types of sediments form and accumulate at the surface of the Earth.

(a) Gravel, sand, and mud have washed or tumbled into a valley floor in California.

(b) Shells accumulate on a beach in Florida and later break into smaller pieces.

(c) Salt precipitates on fence posts protruding from the Dead Sea, which is a very salty lake.

fragments—as sediment. **Sediment**, broadly defined, consists of loose fragments of rocks or minerals broken off of bedrock, mineral crystals that precipitate directly out of water, and shells or shell fragments (**Fig. B.2**).

Sediments are produced by *weathering*, the physical and chemical breakdown of pre-existing rock at or near the Earth's surface. They form a surface veneer, or cover, on bedrock (**Fig. B.3**). This cover ranges from 0 km in thickness, at places where bedrock crops out at the Earth's surface, to 20 km in thickness, where deep, sediment-filled depressions called *sedimentary basins* have developed. Interaction with water and organisms may modify sediment over time and transform it into *soil*, an essential material for life. In this Interlude, we'll now look at how weather-ing produces sediment and how soils form and evolve. This Inter-lude provides important background for Chapter 7, which looks at how sediment can be transformed into sedimentary rock and why there are so many different kinds of sedimentary rock.

B.2 Weathering: Forming Sediment

If you excavate a road cut deep into bedrock, you'll find that the rock contains the minerals and textures that it has had since it first formed. Geologists informally refer to such rock as "fresh" rock (**Fig. B.4**). In the surface and near-surface realm of our planet, fresh rock doesn't last forever, both because the Earth System has an atmosphere and hydrosphere that contain water and other chemicals that react with minerals in rocks and be-cause the Earth hosts living organisms that can interact with Earth materials. As a result, rocks in the surface or near-sur-face realm undergo weathering. In a general sense, weathering is the combination of processes that break up and corrode solid rock and may eventually transform it into loose debris. (Of note, geologists may refer to an accumulation of debris as *detritus*, to emphasize that it is what's left over from the breakup of once-solid rock, and to a layer of debris as **regolith**. Regolith can include both loose sediment and soil.) Weathering represents one type of change that can take place in rock. It happens, in part, because the original assemblage of minerals comprising rock may not be stable in the presence of air and water at the low pressures and temperatures at the Earth's surface; because of weathering, rock brought to the surface sooner or later will crumble away. Note that these processes don't happen on the Moon because the Moon has no atmosphere, hydrosphere, or

FIGURE B.3 A layer of unconsolidated sediment (sand, clay, and cobbles), topped by dark soil, overlies bedrock in this outcrop along the coast of western Ireland.

Soil

Unconsolidated sediment

Bedrock

FIGURE B.4 The contrast between fresh and weathered granite in an Arizona road cut. Weathered granite can break apart. Grains fall off and collect as regolith at the base of the outcrop. The inset photos show how weathering visibly changes the rock.

Weathered granite

Fresh granite

Regolith derived from granite

Weathered

Fresh

biosphere. (Later, in Chapter 8, we'll learn about another kind of change, called metamorphism, which can happen to rock when it is subjected to conditions found deep in the crust.)

Just as a plumber can unclog a drain by using physical force (with a plumber's snake) or by causing a chemical reaction (with a dose of liquid drain opener), nature can attack rocks in two ways. So geologists distinguish between two types of weathering: physical and chemical. We'll discuss each of these separately.

Physical Weathering

Physical weathering, sometimes also referred to as *mechanical weathering*, breaks intact rock into unconnected **clasts** (grains or chunks). Each size range of clasts has a name (**Table B.1**). Many different phenomena contribute to physical weathering, as described below.

Jointing Rocks buried deep in the Earth's crust endure enormous pressure due to the weight of overlying rock, or overburden. Rocks at depth are also warmer than rocks nearer the surface because of the Earth's geothermal gradient. Over long periods of time, moving water, air, and ice at the Earth's surface grind away and remove overburden, so rock formerly at depth rises closer to the Earth's surface. Geologists refer to this overall process as **exhumation**. As exhumation progresses, the pressure squeezing rock rising from depth decreases, and the rock becomes cooler. This change in pressure and temperature can cause a rock's shape to change slightly—specifically, as the weight acting on the rock decreases, it stretches by a few percent vertically and shortens slightly horizontally, and as it cools, it contracts. Such changes can generate forces sufficient to break rock apart. Natural cracks that form in rocks due to removal of overburden or due to cooling (and for other reasons as well; see Chapter 11) are known as **joints**.

TABLE B.1 Clasts Are Classified by Grain Diameter

Boulders	More than 256 mm
Cobbles	Between 64 mm and 256 mm
Pebbles	Between 2 mm and 64 mm
Sand	Between 1/16 mm and 2 mm
Silt	Between 1/256 mm and 1/16 mm
Mud	Less than 1/256 mm

Cobbles

Sand

Almost all rock outcrops contain joints. Some joints are fairly planar, some are curving, and some are irregular. Large granite plutons may split into onion-like sheets along joints that lie parallel to the mountain face, a process known as *exfoliation*, while sedimentary rock beds tend to break into rectangular blocks (**Fig. B.5a**). Regardless of their orientation, the formation of joints turns formerly intact bedrock into separate blocks. Gravity pulls on these blocks, and eventually they fall from the outcrop at which they formed. After a while, they may collect in an apron of **talus**, the rock rubble at the base of a slope (**Fig. B.5b**).

Frost Wedging Freezing water can burst pipes and shatter bottles because water expands when it freezes and pushes the walls of the container apart. The same phenomenon happens in rock. When water trapped in a joint freezes, it forces the joint open and may cause the joint to grow. Such **frost wedging** helps break blocks free from intact bedrock (**Fig. B.6a**).

FIGURE B.5 Joints (natural cracks) break bedrock into blocks and sheets, which can tumble down a slope.

(a) When buried deeply, rocks are subjected to a large downward pressure. Later, after erosion removes overburden, rocks expand and crack.

Downward pressure

Sedimentary rock layers

Granite pluton

Time

Exfoliation joints are parallel to the ground surface, so rock peels off like layers of an onion.

Joints in granite, California.

Vertical joints intersect bedding to form rectangular blocks.

Vertical joints   Bedding

100-m-high sandstone cliff, Ireland.

Intact rock

Talus

(b) Joint-bounded blocks tumble from a cliff and accumulate in talus aprons near Mt. Snowdon in Wales.

Root Wedging Have you ever noticed how the roots of an old tree can break up a sidewalk? Even though the wood of roots doesn't seem very strong compared to rock or concrete, as roots grow and expand they apply forces to their surroundings. Tree roots that grow into joints can push joints open in a process known as **root wedging** (**Fig. B.6b**).

Salt Wedging In arid climates, dissolved salt in groundwater precipitates and grows as crystals in open pore spaces in rocks. This process, **salt wedging**, pushes apart the surrounding grains and weakens the rock, so when exposed to wind and rain the rock disintegrates into separate grains.

The same phenomenon happens on outcrop faces in coastal areas, where salt spray percolates into rock and then dries (**Fig. B.6c**).

Thermal Expansion When the heat of an intense forest fire bakes a rock, the outer layer of the rock expands. On cooling, the layer contracts. This change creates forces in the rock sufficient to make the outer part of the rock break off in sheet-like pieces. Recent research suggests that the intense heat of the Sun's rays sweeping across dark rock in a desert may, over time, cause the rock to fracture into thin slices.

FIGURE B.6 Wedging is one type of physical (mechanical) weathering.

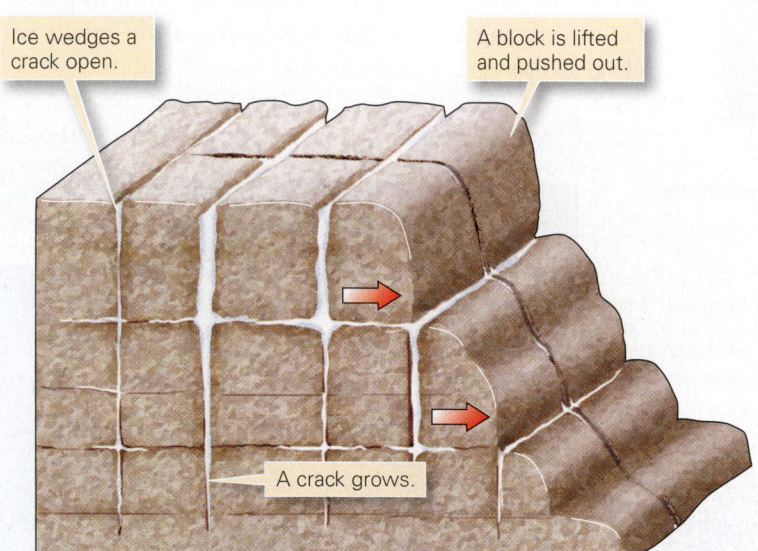

(a) When the water that fills cracks freezes, it expands and wedges the cracks open.

(c) Salt wedging led to disintegration of gravestones in Whitby, England.

(b) Root wedging pushes open a joint, slowly separating a block from the cliff.

FIGURE B.7 Examples of chemical weathering.

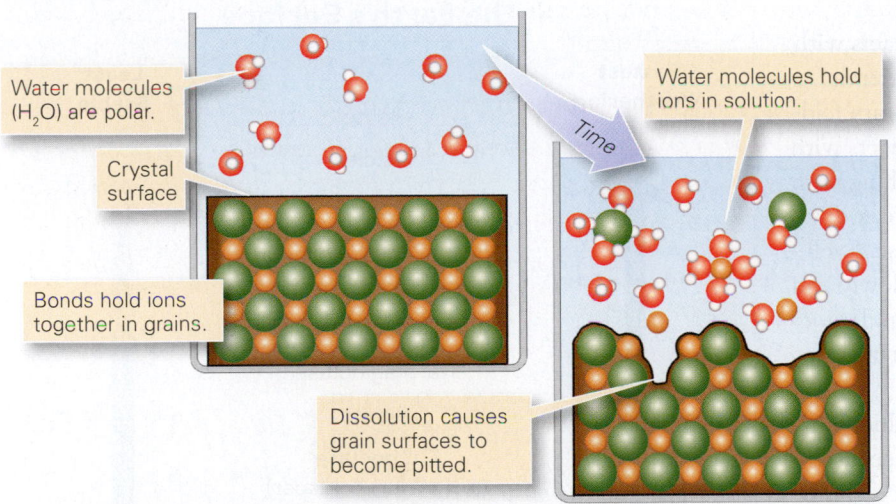

Water molecules (H₂O) are polar.

Crystal surface

Bonds hold ions together in grains.

Time

Water molecules hold ions in solution.

Dissolution causes grain surfaces to become pitted.

Water seeping into joints in limestone produced troughs.

(a) Dissolution occurs when water molecules pluck ions off of grain surfaces.

Weathered pyrite crystals

(b) Pyrite (FeS₂) is a sulfide mineral that reacts with air to form iron oxide.

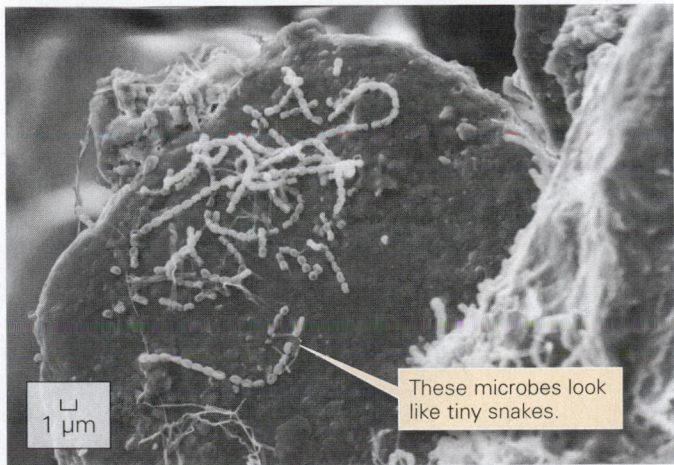

1 μm

These microbes look like tiny snakes.

(c) Certain types of microbes obtain their life energy from the chemical bonds in minerals.

Animal Attack Animal life also contributes to physical weathering. For example, burrowing creatures, from earthworms to gophers, move rock fragments. Recently, humans have become perhaps the most energetic agent of physical weathering on the planet. When we excavate quarries, foundations, mines, or roadbeds by digging and blasting, we shatter and displace rock that might otherwise have remained intact for millions of years more.

Chemical Weathering

Up to now we've taken the plumber's-snake approach to breaking up rock. Now let's look at the liquid-drain-opener approach. **Chemical weathering** refers to the chemical reactions that alter or destroy minerals when rock comes in contact with water solutions or air. Common reactions involved in chemical weathering include the following.

- *Dissolution:* Chemical weathering during which minerals dissolve into water is called **dissolution**. Dissolution primarily affects salts and carbonate minerals (**Fig. B.7a**), which dissolve relatively easily. But even quartz can dissolve slightly. Some minerals, such as halite, dissolve rapidly in pure rainwater. But others, such as calcite, dissolve rapidly only when the water is *acidic*, meaning that it contains an excess of hydrogen ions (H^+). Acidic water reacts with calcite to form a solution of Ca^+ and CO_3^{2-}, and releases CO_2 gas (see Chapter 5).

  How does the water in rock near the surface of the Earth become acidic? As rainwater falls, it dissolves carbon dioxide gas in the atmosphere, and as the water sinks down through soil containing organic debris, it reacts with the debris. Both processes yield carbonic acid. Because of the solubility of calcite, limestone and marble—two types of rock composed of calcite—dissolve sufficiently to yield underground caverns.

- *Hydrolysis:* During hydrolysis, water chemically reacts with minerals and breaks them down to form other minerals (*lysis* means loosen in Greek). For example, potassium feldspar (orthoclase), a common mineral in granite, reacts with acidic water to produce kaolinite (a type of clay) along with a variety of dissolved ions. Hydrolysis reactions break down not only feldspars but many other silicate minerals as well— amphibole, pyroxene, mica, and olivine all react slowly and transform into various types of clay. Quartz also undergoes hydrolysis but does so at such a slow rate that quartz grains survive such weathering in most climates.

- *Oxidation:* Chemists refer to a reaction during which an element loses electrons as an *oxidation reaction*, because such a loss commonly takes place when elements combine with oxygen. The oxidation, or rusting, of iron serves as an example. Oxidation reactions in rocks transform iron-bearing minerals (such as biotite and pyrite) into a rusty-brown mixture of various iron-oxide and iron-hydroxide minerals (**Fig. B.7b**).

- *Hydration:* Hydration, the absorption of water into the crystal structure of minerals, causes some minerals, such as certain types of clay, to expand. Such expansion weakens rock.

Not all minerals undergo chemical weathering at the same rates (**Table B.2**). Some weather in a matter of months or years, whereas others remain unweathered for millions of years. In temperate climates, for example, calcite weathers faster than most silicate minerals. Of the silicate minerals, those that crystallize at the highest temperatures are generally less stable under the cool temperatures of the Earth's surface than those that

TABLE B.2 Relative Stability of Minerals at the Earth's Surface

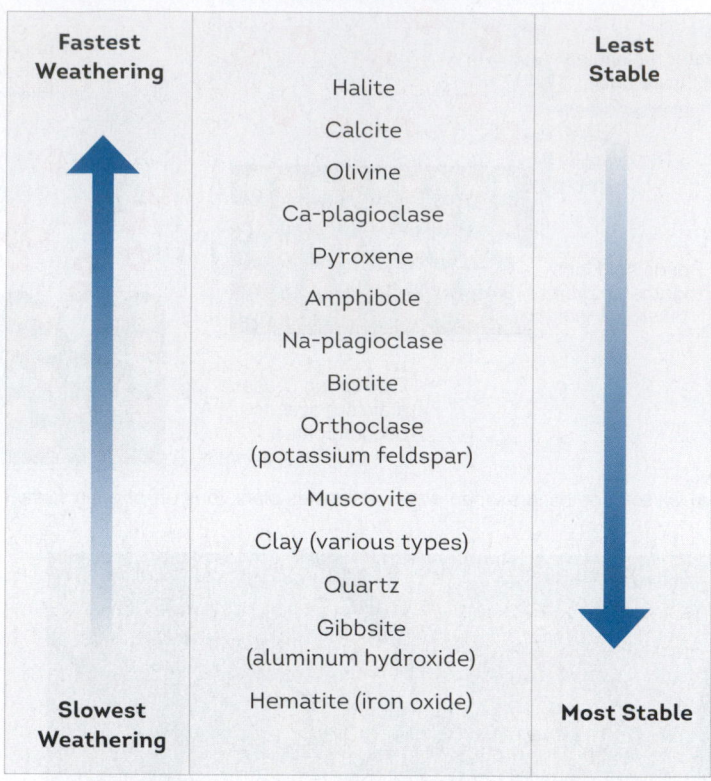

Fastest Weathering		Least Stable
	Halite	
	Calcite	
	Olivine	
	Ca-plagioclase	
	Pyroxene	
	Amphibole	
	Na-plagioclase	
	Biotite	
	Orthoclase (potassium feldspar)	
	Muscovite	
	Clay (various types)	
	Quartz	
	Gibbsite (aluminum hydroxide)	
Slowest Weathering	Hematite (iron oxide)	Most Stable

Note that minerals that form early in Bowen's reaction series (see Box 6.1) are among the least stable minerals at the Earth's surface. Minerals that are the products of weathering reactions (e.g., hematite) are among the most stable minerals at the Earth's surface. Mafic minerals weather by oxidation, felsic minerals by hydrolysis, and carbonates and salts by dissolution; oxide minerals don't weather at all.

These slopes have smoothed surfaces because the rock they are made of has weathered.

crystallize at lower temperatures. The difference depends partly on crystal structure and partly on chemical composition. Specifically, minerals with fewer linkages between silicon-oxygen tetrahedra tend to have weaker structures and thus weather faster than do minerals with more linkages. And minerals containing iron, magnesium, sodium, potassium, and aluminum tend to weather faster than minerals without these elements. Quartz (pure SiO_2), for example, is a 3-D network silicate mineral with strong bonds in all directions, and tends to be very stable. When a granite (a rock that contains quartz, mica, and feldspar) undergoes chemical weathering, everything but quartz transforms to clay. That's why beaches typically consist of quartz sand; quartz is the most common mineral left after the other minerals have turned to clay and washed away.

Until fairly recently, geologists tended to think of chemical weathering as a strictly inorganic chemical reaction, occurring entirely independently of life forms. But we now realize that organisms play a major role in the chemical-weathering process. For example, the roots of plants, fungi, and lichens secrete organic acids that help dissolve minerals in rocks, in order to extract nutrients from the minerals. *Microbes*, microscopic single-celled organisms (including bacteria and archaea), are amazing in that they literally eat minerals for lunch (**Fig. B.7c**). Microbes can pluck off molecules of a mineral's surface and use the energy from broken chemical bonds to supply their own life force.

Physical and Chemical Weathering Working Together

So far we've looked at the processes of chemical and physical weathering separately, but in the real world they happen together, aiding each other in disintegrating rock to form sediment (see **Geology at a Glance**, pp. 192–193).

Physical weathering speeds up chemical weathering. To understand why, keep in mind that chemical-weathering reactions take place at the surface of a material. Thus, the overall rate at which chemical weathering occurs depends on the ratio of surface area to volume—the greater the surface area, the faster the volume as a whole can chemically weather. When jointing (physical weathering) breaks a large block of rock into smaller pieces, the surface area increases, so chemical weathering happens faster (**Fig. B.8a**).

Similarly, chemical weathering speeds up physical weathering by dissolving away grains or cements that hold a rock together, by transforming hard minerals (like feldspar) into soft minerals (like clay), and by causing minerals to absorb water and expand. These phenomena make rock weaker, so it can disintegrate more easily (**Fig. B.8b**). If you drop a block of fresh granite on the ground, it will most likely stay intact, but if you drop a block of chemically weathered granite on the ground, it will crumble into a pile of debris.

FIGURE B.8 Physical and chemical processes work together during the weathering process.

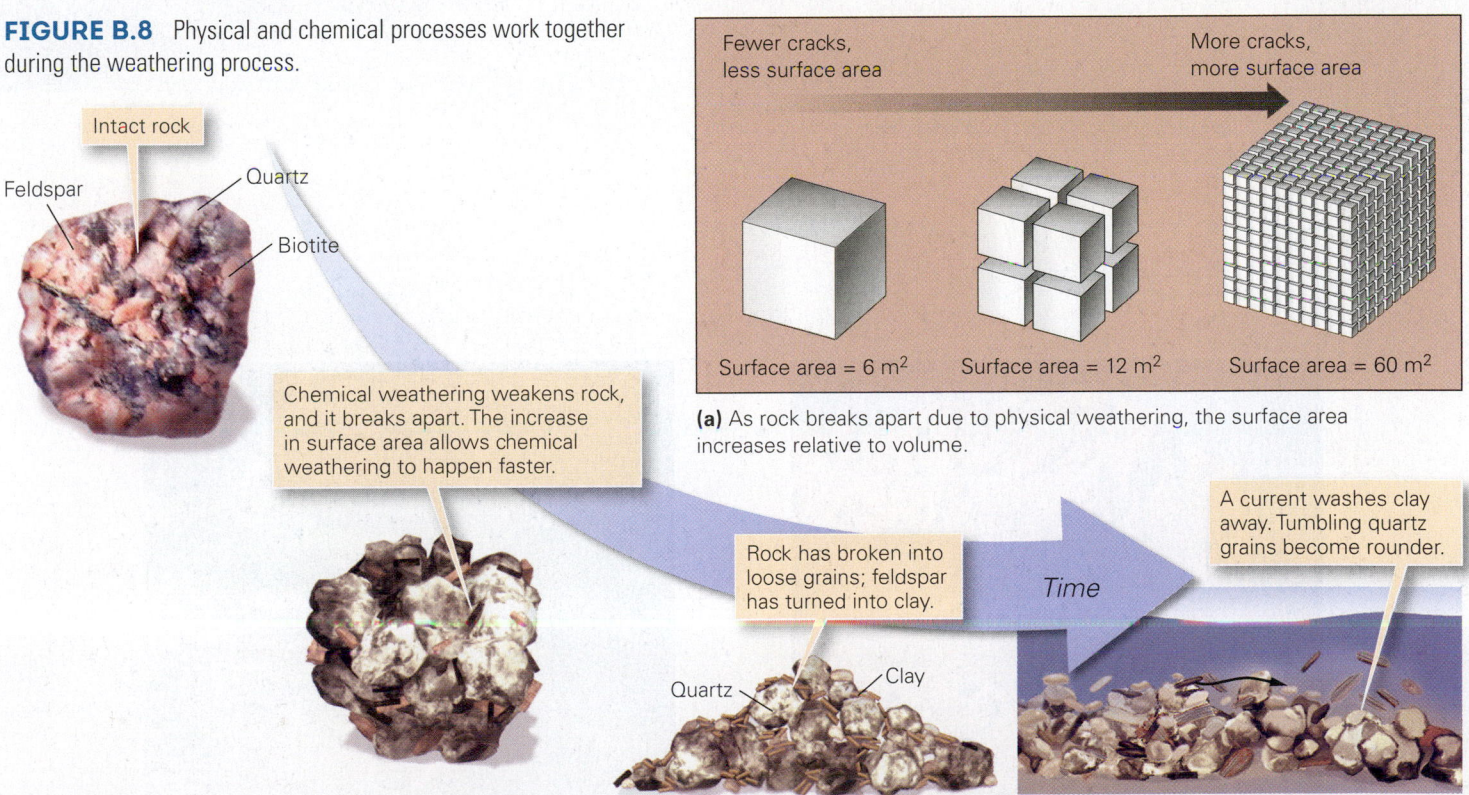

(a) As rock breaks apart due to physical weathering, the surface area increases relative to volume.

(b) Chemical weathering weakens rock, so it breaks apart. As this happens, the surface area increases, so chemical weathering happens still faster. Eventually, the rock completely disaggregates to form sediment. Weathering of granite produces quartz sand and clay.

GEOLOGY AT A GLANCE

Weathering, Sediment, and Soil Production

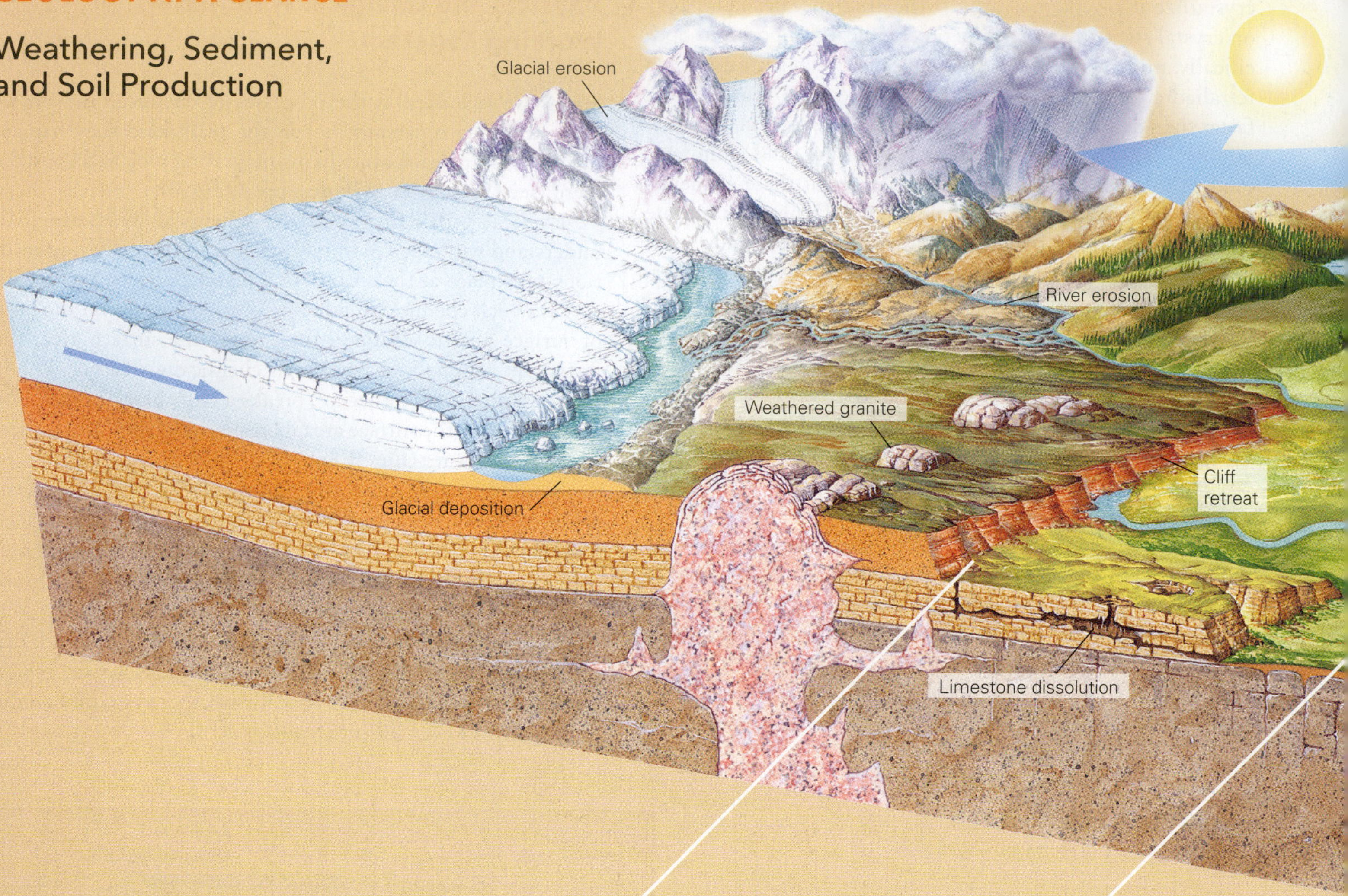
Glacial erosion
River erosion
Weathered granite
Cliff retreat
Glacial deposition
Limestone dissolution

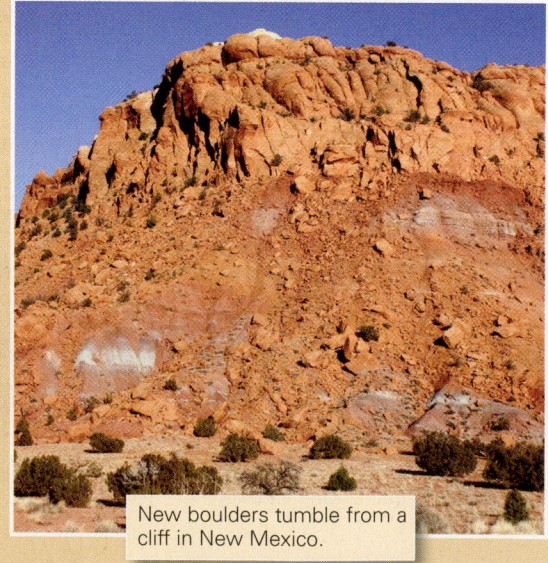

New boulders tumble from a cliff in New Mexico.

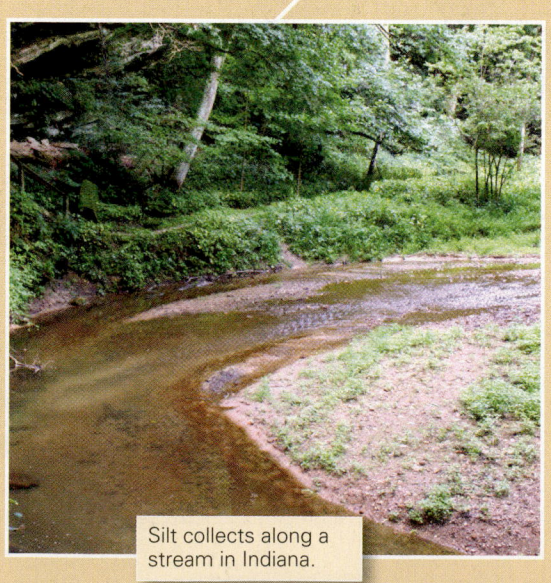

Silt collects along a stream in Indiana.

Tectonic processes uplift the land surface above sea level. Once exposed, rock interacts with air and water and undergoes chemical and physical weathering, ultimately breaking down to produce sediment. Convection in the atmosphere generates wind, rain, and snow. Flowing water, ice, and air erode and transport sediment to sites of deposition. Leaching by downward-percolating rainwater, along with the addition of organic material, produces soil.

Wind

Coastal erosion

River deposition

Soil formation

Coastal deposition

Soil forms on bedrock of chalk in southern England.

Waves move sand on a beach in Brazil.

Erosion carves the coastal cliffs of Ireland.

Weathering happens faster at edges, and even faster at the corners of broken blocks, because weathering attacks a flat face from only one direction, an edge from two directions, and a corner from three directions. Thus, with time, edges of blocks become blunt and corners become rounded (**Fig. B.9a**). In rocks such as granite, which do not contain layering that can affect weathering rates, rectangular blocks transform into a spheroidal shape (**Fig. B.9b**).

When different rocks in an outcrop undergo weathering at different rates, we say that the outcrop has undergone *differential weathering*. As a result of differential weathering, cliffs composed of a variety of rock layers take on a stair-step or sawtooth shape (**Fig. B.9c**). Weak layers may weather away beneath a more resistant layer, so the resistant layer becomes an overhanging ledge. Similarly, the rate at which the land surface weathers depends on the rock type, so valleys tend to develop over weak rocks, while strong rocks hold up hills.

You can easily see the consequences of differential weathering if you walk through a graveyard. The inscriptions on some headstones are sharp and clear, whereas those on other stones have become blunted or have even disappeared (**Fig. B.9d**). That's because the minerals in these different stones have different resistances to weathering. Granite, an igneous rock with a high quartz content, retains inscriptions the longest. In contrast, marble, a metamorphic rock composed of calcite, dissolves away relatively rapidly in acidic rain.

FIGURE B.9 Differential weathering.

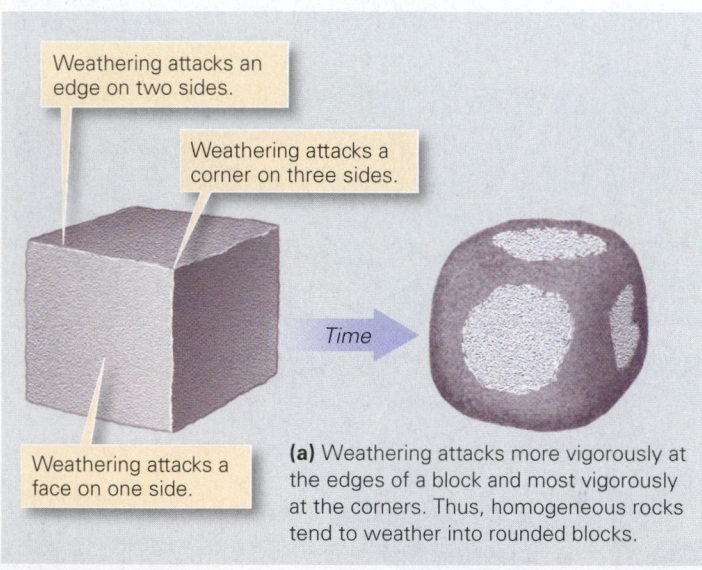

Weathering attacks an edge on two sides.

Weathering attacks a corner on three sides.

Weathering attacks a face on one side.

Time

(a) Weathering attacks more vigorously at the edges of a block and most vigorously at the corners. Thus, homogeneous rocks tend to weather into rounded blocks.

(b) Weathering along cracks in granite of the Mojave Desert led to the formation of rounded blocks. This style is called spheroidal weathering.

Weak shale

Strong sandstone

(c) Sawtooth weathering profiles develop in sequences of alternating strong and weak layers on this exposure in New Mexico. Weak layers are indented, whereas strong layers protrude.

(d) Inscriptions on a granite headstone (left) last for centuries, but those on a marble headstone (right) may weather away in decades. These gravestones are in the same cemetery and are about the same age.

B.3 Soil

If you've ever had the chance to dig in a garden, you've seen firsthand that the material in which flowers grow looks and feels different from beach sand or potter's clay. We call the material in a garden dirt or, more technically, soil. In a general sense, **soil** consists of rock or sediment that has been modified by physical and chemical interaction with organic material and rainwater, over time, to produce a substrate that can support the growth of plants. Soil is one of our planet's most valuable resources, for without it there could be no agriculture, forestry, ranching, or home gardening.

How Does Soil Form?

Three processes taking place at or just below the surface of the Earth contribute to soil formation.

- *Debris production:* Chemical and physical weathering produces loose debris, new mineral grains (such as clay), and ions in solution.
- *Interaction with water:* When rain falls, some of the water percolates through the debris and carries dissolved ions and clay flakes downward. We refer to the region in which this downward transport occurs as the **zone of leaching**,

because *leaching* means extracting, absorbing, and removal. Deeper down, new mineral crystals precipitate out of the downward-percolating water or form by reaction of the water with debris. Also, the water leaves behind its load of fine clay. The region in which new minerals grow and clay collects is the **zone of accumulation** (**Fig. B.10a**).

- *Interaction with organisms:* Soils serve as home for a remarkable number of organisms—they are an important realm in which biologic and physical components of the Earth System interact profoundly. For example, a single cubic centimeter of moist soil in a warm region hosts over 1 billion microbe cells, and over 1.5 million earthworms wriggle through each acre of such soil. The microbes, fungi, plants, and animals in soil produce acids that weather grains, absorb nutrient atoms, leave behind organic waste, and also physically churn and break up the soil. When organisms die, they rot and transform into organic carbon that mixes in with the mineral debris of soil. An accumulation of rotted organic debris is called **humus**.

As a consequence of the above processes, regolith and rock evolve into soil, and the soil's "character"—its texture and composition—becomes very different from that of the starting material. Because different soil-forming processes operate at different depths, soils typically develop distinct zones, known as **soil horizons**, arranged in a vertical sequence called a **soil profile** (**Fig. B.10b, c**). Let's look at an idealized soil profile,

FIGURE B.10 Formation of soil horizons.

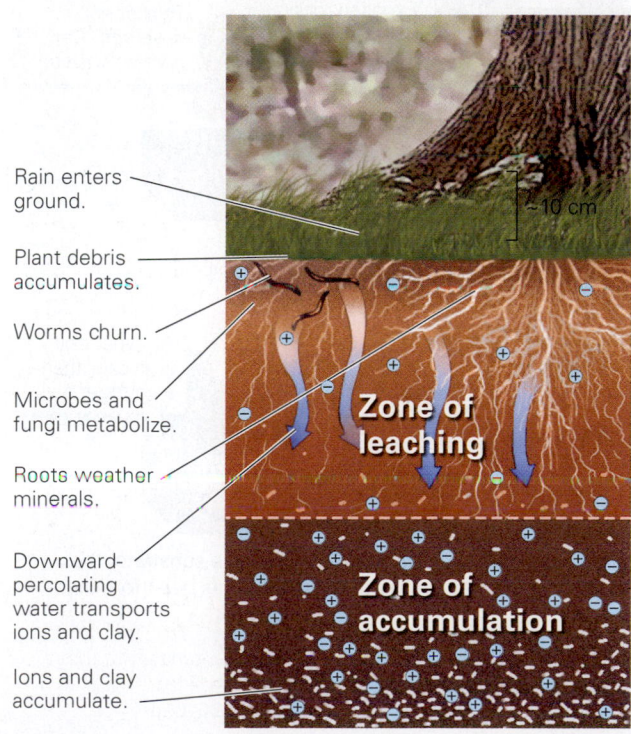

Rain enters ground.

Plant debris accumulates.

Worms churn.

Microbes and fungi metabolize.

Roots weather minerals.

Downward-percolating water transports ions and clay.

Ions and clay accumulate.

Zone of leaching

Zone of accumulation

(a) Soil horizons develop as water percolates downward, carrying ions and clay with it, and as organisms interact with the soil.

Horizon designation

O

A

E

B

C

Topsoil

Transition

Subsoil

Weathered bedrock

Solid bedrock

(b) Distinct soil horizons develop, each with a characteristic composition and texture.

Grass

Topsoil

Transition

Subsoil

10 cm

Weathered bedrock

(c) Soil horizons exposed on the wall of a gully in eastern Brazil.

from top to bottom, using a soil formed in a temperate forest as our example. The highest horizon is the *O-horizon* (the prefix stands for organic), so called because it consists almost entirely of organic matter and contains barely any mineral matter. Below the O-horizon we find the *A-horizon*, in which humus has decayed further and has mixed with mineral grains (clay, silt, and sand). Water percolating through the A-horizon causes chemical weathering reactions to occur and produces ions in solution and new clay minerals. The downward-moving water eventually carries soluble chemicals and fine clay deeper into the subsurface. The O- and A-horizons constitute dark-gray to blackish-brown **topsoil**, the fertile portion of soil that farmers till for planting crops. In some places, the A-horizon grades downward into the *E-horizon*, a soil level that has undergone substantial leaching but has not yet mixed with organic material. Ions and clay accumulate in the *B-horizon*, or **subsoil**. (Note from our description that the O-, A-, and E-horizons lie within the zone of leaching, whereas the B-horizon lies in the zone of accumulation.) Finally, at the base of a soil profile we find the *C-horizon*, which consists of material derived from the substrate that's been chemically weathered and broken apart, but has not yet undergone leaching or accumulation. The C-horizon grades downward into unweathered bedrock or into unweathered sediment.

Farmers, foresters, and ranchers well know that soil in one locality differs greatly from soil in another in terms of composition, thickness, and texture. And crops that grow well in one type of soil may wither and die in another. Such diversity exists because the makeup of a soil depends on several *soil-forming factors* (**Fig. B.11**).

- *Climate:* The total rainfall, the distribution of rainfall during the year, and the range and average of temperature during the year determine the rate and amount of chemical weathering and leaching that take place at a given location. Large amounts of rainfall and warm temperatures accelerate chemical weathering and cause most of the soluble elements to be leached. In regions with small amounts of rainfall and cooler temperatures, soils take a long time to develop and can retain unweathered minerals and soluble components. Climate seems to be the single most important factor in determining the nature of soils that develop.

- *Substrate composition:* Some soils form on basalt, some on granite, some on volcanic ash, and some on recently deposited quartz sand. These different substrates consist of different materials, so the soils formed on them end up with different chemical compositions. For example, a soil formed on basalt tends to be richer in iron than a soil formed on granite, because basalt contains more iron than does granite. Also, soils tend to develop faster on "unconsolidated material" (loose ash or sediment) than on hard, solid bedrock.

FIGURE B.11 Factors that control the character of soil.

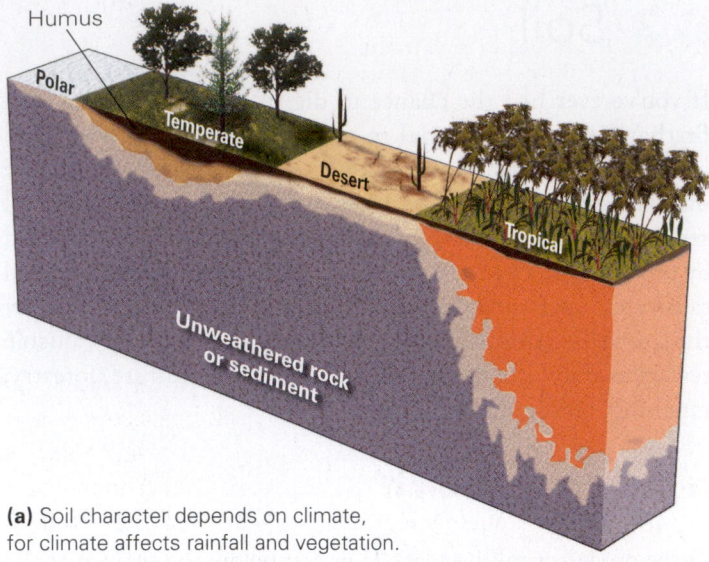

(a) Soil character depends on climate, for climate affects rainfall and vegetation.

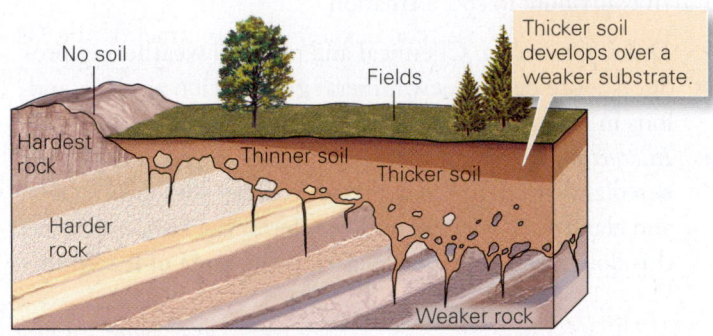

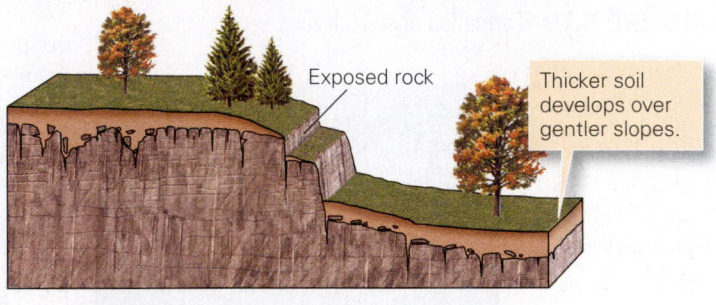

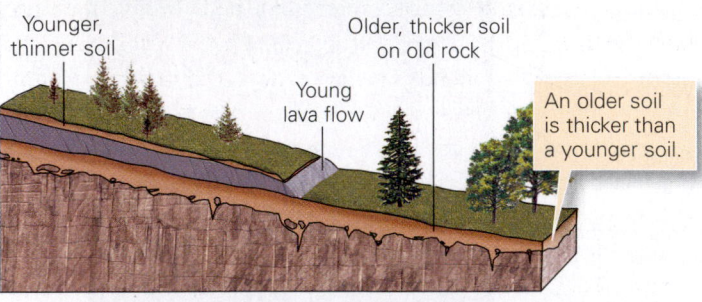

(b) Soil character also depends on the strength of the substrate, the steepness of the slope, and the length of time soil has been forming.

- *Slope steepness:* A thick soil can accumulate under land that lies flat. But on a steep slope, regolith may wash away before it can evolve into a soil. Thus, all other factors being equal, soil thickness increases as the slope angle decreases.
- *Wetness:* Depending on the details of local topography and on the depth below the surface at which groundwater occurs, some soil is wetter than other soil in the same region. Wet soils tend to contain more organic material than do dry soils.
- *Time:* Because soil formation is an evolutionary process, a younger soil tends to be thinner and less developed than an older soil in the same location. The rate of soil formation varies greatly with environment. In a protected, moist, warm region, soils may develop over the course of a few years to a few decades. But in an exposed, cold, dry region, soils may take thousands of years or more to develop. In temperate regions, soil forms at a rate of 0.02 to 0.20 mm per year, thereby producing 1 meter of soil in about 10,000 years.
- *Vegetation type:* Different kinds of plants extract or add different nutrients and quantities of organic matter to a soil. Also, some plants have deeper root systems than others and help prevent soil from washing away, so that it builds into a thicker layer.

Soil Classification

Depending on their evolution and composition, soils come in a variety of textures, structures, and colors. *Soil texture* reflects the relative proportions of sand, silt, and clay-sized grains in the soil. For many crops, farmers prefer to sow in **loam**, a type of soil consisting of 10% to 30% clay and the rest silt and sand. In loam, pores (open spaces) can remain between grains so that water and air can pass through and roots can easily penetrate. In soils with too much clay, the clay packs together and prevents water movement. *Soil structure* refers to the degree to which soil grains clump together to form lumps or clods, which soil scientists refer to as "peds" (from the Latin *pedo*, meaning soil). The structure changes as a soil develops, because structure depends on clay content and organic content, both of which change with time. These materials give soil its stickiness. Soil color reflects its composition: organic-rich soil tends to be gray or black, organic-poor and calcite-rich soil is whitish, and iron-rich soil is red or yellow.

Soil scientists worldwide have struggled mightily to develop a rational scheme for *soil classification*. Not all schemes utilize the same criteria, and even today there is not worldwide agreement on which works best. In the United States, a country that includes many climates at mid-latitudes, many soil scientists use the U.S. Comprehensive Soil Classification System, which distinguishes among 12 orders of soil based on the physical characteristics and environment of soil formation (**Table B.3** and **Fig. B.12**). Canadians use a different scheme focusing only on soils that develop in cooler, high-latitude climates.

Rainfall and vegetation play a key role in determining the type of soil that forms. For example, in deserts, where there is very little rainfall and sparse vegetation, an *aridisol* forms (**Fig. B.13a**). (In older classifications, these were known as pedocal soils.) Aridisols have no O-horizon (because there is so little organic material), and the A-horizon is thin. Soluble minerals, specifically calcite, that would be washed away entirely if there were more rainfall instead accumulate in the B-horizon. In fact, capillary action may bring calcite up from deeper down as water evaporates at the ground surface. Calcite cements clasts together in the B-horizon to form a rock-like mass called *caliche* or *calcrete*. In temperate environments, an *alfisol* forms—this soil has an O-horizon, and because of moderate amounts of rainfall, materials

TABLE B.3 Soil Orders: U.S. Comprehensive Soil Classification System
(see also Fig. B.12)

Alfisol	Gray/brown, has subsurface clay accumulation and abundant plant nutrients. Forms in humid forests.
Andisol	Forms in volcanic ash.
Aridisol	Low in organic matter, has carbonate horizons. Forms in arid environments.
Entisol	Has no horizons. Formed very recently.
Gelisol	Underlain with permanently frozen ground.
Histosol	Very rich in organic debris. Forms in swamps and marshes.
Inceptisol	Moist, has poorly developed horizons. Formed recently.
Mollisol	Soft, black, and rich in nutrients. Forms in subhumid to subarid grasslands.
Oxisol	Very weathered, rich in aluminum oxide and iron oxide, low in plant nutrients. Forms in tropical regions.
Spodosol	Acidic, low in plant nutrients, ashy, has accumulations of iron and aluminum. Forms in humid forests.
Ultisol	Very mature, strongly weathered soils, low in plant nutrients.
Vertisol	Clay-rich soils capable of swelling when wet, and shrinking and cracking when dry.

FIGURE B.12 U.S. Department of Agriculture map of soil types around the world.

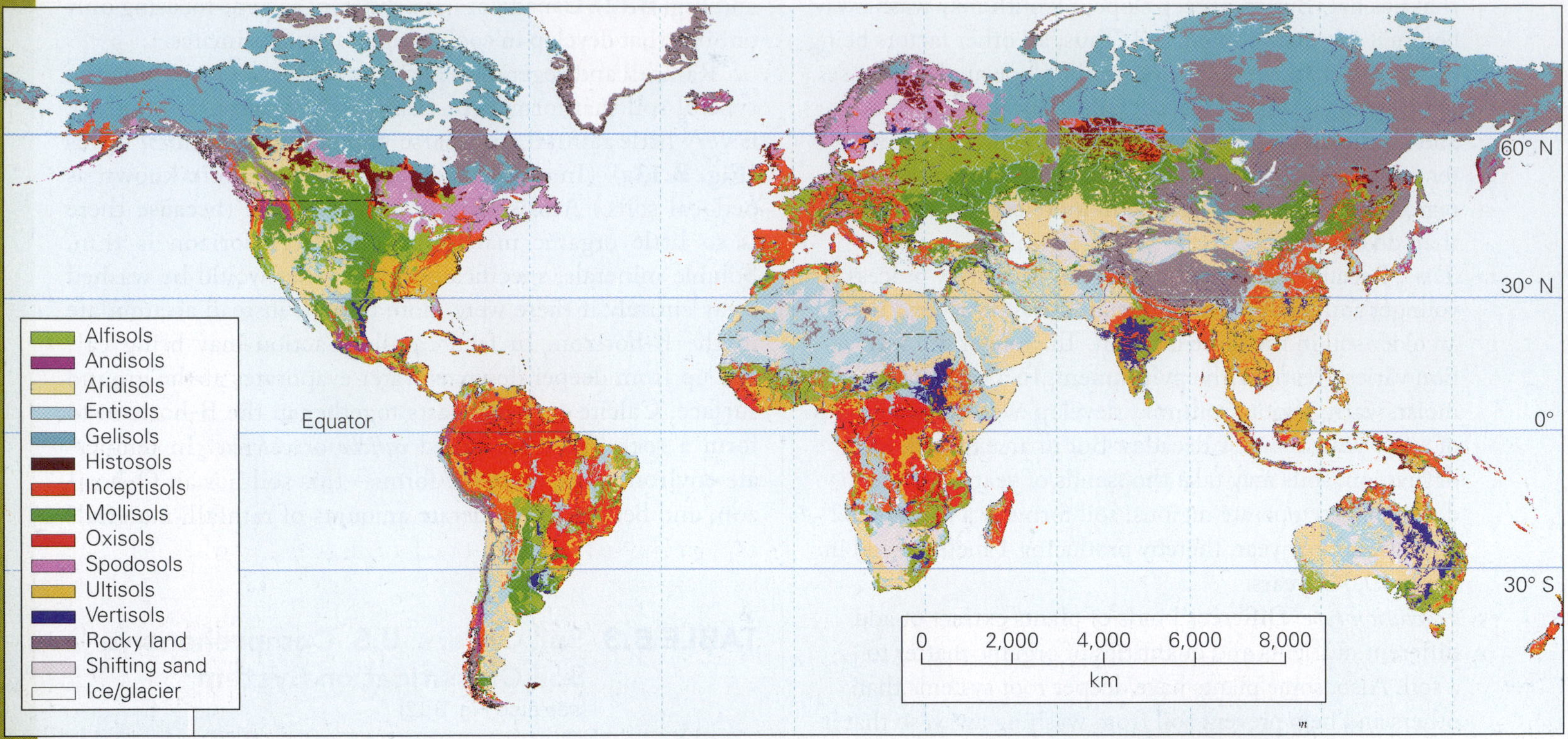

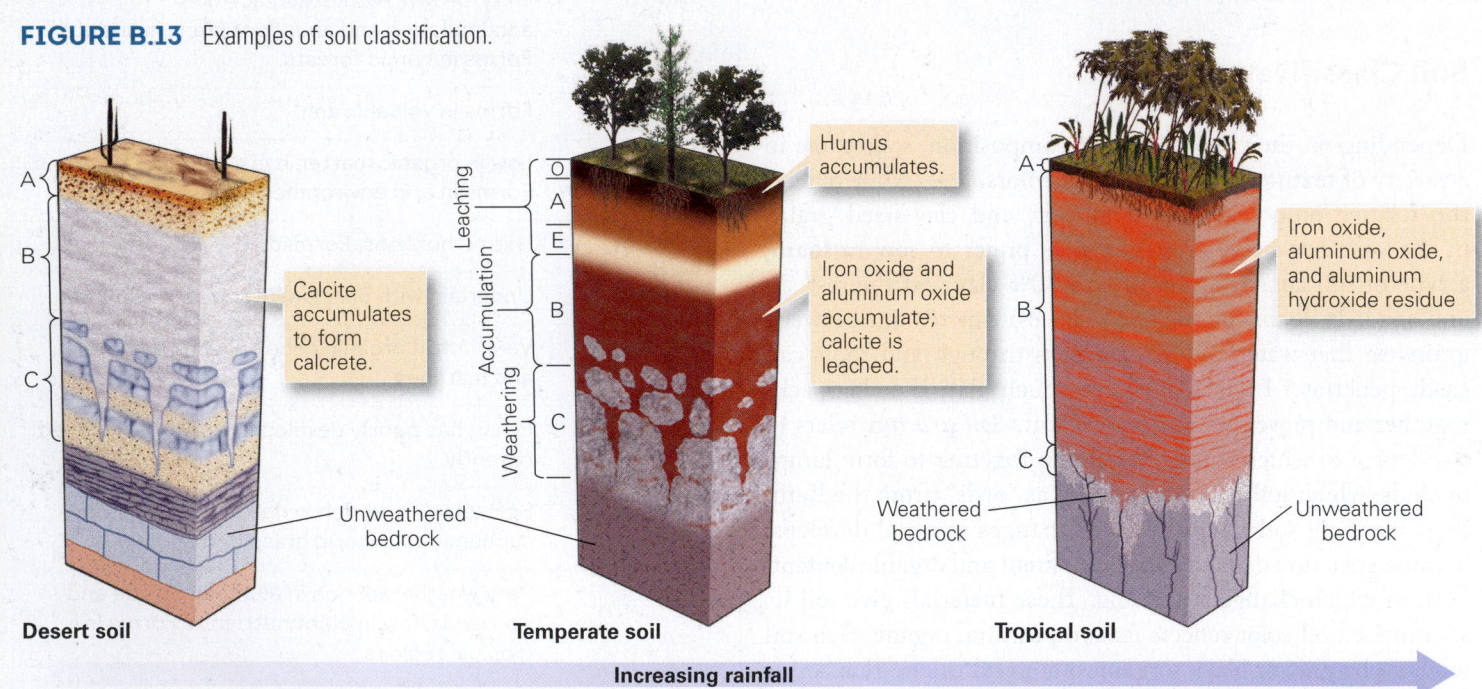

- Alfisols
- Andisols
- Aridisols
- Entisols
- Gelisols
- Histosols
- Inceptisols
- Mollisols
- Oxisols
- Spodosols
- Ultisols
- Vertisols
- Rocky land
- Shifting sand
- Ice/glacier

Equator

60° N
30° N
0°
30° S

0   2,000   4,000   6,000   8,000
km

FIGURE B.13 Examples of soil classification.

Humus accumulates.

Calcite accumulates to form calcrete.

Leaching
Accumulation
Weathering

Iron oxide and aluminum oxide accumulate; calcite is leached.

Iron oxide, aluminum oxide, and aluminum hydroxide residue

A
B
C

Unweathered bedrock

Desert soil

Unweathered bedrock

Temperate soil

Weathered bedrock

Unweathered bedrock

Tropical soil

Increasing rainfall

(a) Aridisol forms in deserts. Rainfall is so low that no O-horizon forms, and soluble minerals accumulate in the B-horizon.

(b) Alfisol forms in temperate climates. An O-horizon forms, and less-soluble materials accumulate in the B-horizon.

(c) Oxisol forms in tropical climates where percolating rainwater leaches all soluble minerals, leaving only iron- and aluminum-rich residues.

leached from the A-horizon accumulate in the B-horizon (**Fig. B.13b**). (In older classifications, these were known as pedalfer soils.) In a tropical climate, *oxisols* develop. Here so much rainfall percolates down into the ground that all reactive minerals in the soil undergo chemical weathering, producing ions and clay that flush downward. This process leaves an A-horizon that contains substantial amounts of stable iron-oxide, aluminum-oxide, and aluminum-hydroxide residues (**Fig. B.13c**). The resulting soil tends to be brick red. So much water flushes down through some oxisols that a B-horizon cannot develop.

Laterite is a type of oxisol that tends to be very rich in iron oxides and can be hard enough to be broken into blocks that can be used for construction—in fact, the name comes from the Latin word *later*, which means brick (**Fig. B.14**). Typically, laterite forms in regions that have a distinct dry season, during which capillary action brings additional oxide minerals back up into the A-horizon, where they accumulate. Laterite that develops on iron-rich rock can contain so much iron that it can be an economic source for iron, and oxisols that develops from an aluminum-rich rock (e.g., granite) may accumulate so much aluminum hydroxide that they can be an economic source of aluminum; aluminum-rich laterite is also known as bauxite (see Chapter 15).

Soil Destruction

As we have seen, soils take time to form, so soils capable of supporting crops or forests should be considered a natural resource worthy of protection. However, human activities such as agriculture, overgrazing, and clear-cutting have led to the destruction of soil. This destruction involves two phenomena: nutrient removal and soil erosion.

Nutrient Removal   In a natural ecosystem, plants remove *nutrients* (specific chemicals, such as nitrogen and phosphorus, necessary for life) from soil, but when the plants die and transform into humus, they return the nutrients to the soil. Agriculture, by its nature, involves planting, growth, and then harvesting (removal) of organic material. The nutrients incorporated in the crop are removed during harvest. As a result, a few seasons of farming may deplete the concentration of nutrients in the soil sufficiently to inhibit the growth of successive crops. Farmers can overcome this problem by fertilizing the soil—fertilizer containing the missing nutrients can be mixed into the soil to add the nutrients back. Fertilizer is expensive, so it adds significantly to the cost of food production.

The consequences of rainforest destruction on soil nutrients are particularly profound. In an established rainforest, lush growth provides sufficient organic debris so that trees can grow. But if the forest is logged or cleared for agriculture, the humus rapidly disappears, leaving laterite, which contains few nutrients. Crop plants consume the remaining nutrients so rapidly that the soil becomes infertile after only a year or two, useless for agriculture and unsuitable for regrowth of rainforest trees.

Soil Erosion   When the natural plant cover disappears, the surface of the soil becomes exposed to wind and water. Actions such as the impact of falling raindrops or the rasping of a plow break up the soil at the surface, with the result that it can wash away in water or blow away as dust. When this happens, **soil erosion**, the removal of soil by wind or by running water, takes place (**Fig. B.15**). Soil erosion can remove up to six tons of soil from an acre of land in a year. Human activities increase rates of soil erosion by 10 to 100 times, which far exceeds the rate of soil formation. When soil erosion becomes severe, once-clear

FIGURE B.14   Laterite in Brazil.

(a) Farm fields, after harvest, are like deserts in that they have no plant cover. Plowing a dry field sends clouds of soil into the air as dust.

(b) When vegetation doesn't protect soil, water erosion carves a "badlands landscape" of closely spaced gullies and carries the soil away.

streams flowing through the affected area turn brown because of all the sediment that they carry.

Droughts exacerbate the situation. For example, during the 1930s a succession of droughts killed off so much vegetation in the American plains that wind stripped the land of soil and caused devastating dust storms. Large numbers of people were forced to migrate away from the Dust Bowl of Oklahoma and adjacent areas. Deforestation in temperate and tropical climates can also trigger significant soil erosion.

Nutrient removal and soil erosion are two of several problems that face society. The overuse of fertilizers, pesticides, and herbicides, as well as spills of a great variety of toxic chemicals, causes *soil contamination*. And too much irrigation in arid climates can make soils too saline for plant growth, for irrigation water contains trace amounts of salts that get left behind when irrigation water evaporates. Fortunately, people have begun to realize the fragility of soil and have been working on ways to promote *soil conservation*.

INTERLUDE SUMMARY

- Sediment consists of loose fragments, derived from pre-existing rock, precipitated from water, or formed by the breakup of shells.
- Rock at or near the Earth's surface weathers over time. Physical weathering breaks larger rock bodies into smaller pieces. Chemical weathering involves a variety of chemical reactions that dissolve and/or alter minerals.

- Soil is regolith that underwent change over time when water percolated down through it. Soil can be modified by interacting with organisms. The character of soil depends on the composition of its source material, as well as on climate and time.
- Soil is an essential resource to society, for it provides the basis for agriculture. Various phenomena, some caused by humans, can lead to the loss of soil.

GUIDE TERMS

chemical weathering (p. 189)

clast (p. 186)

dissolution (p. 189)

exhumation (p. 186)

frost wedging (p. 187)

humus (p. 195)

joints (p. 186)

laterite (p. 199)

loam (p. 197)

physical weathering (p. 186)

regolith (p. 185)

root wedging (p. 188)

salt wedging (p. 188)

sediment (p. 185)

soil (p. 195)

soil erosion (p. 199)

soil horizon (p. 195)

soil profile (p. 195)

subsoil (p. 196)

talus (p. 187)

topsoil (p. 196)

zone of accumulation (p. 195)

zone of leaching (p. 195)

REVIEW QUESTIONS

1. Explain the difference between physical and chemical weathering.
2. What processes can cause originally solid rock to break into pieces?
3. What are the various reactions that can contribute to chemical weathering?
4. Why doesn't weathering take place on the Moon?
5. Explain the process of soil formation.
6. Why do soils develop distinct horizons?
7. What factors determine the character (e.g., thickness, texture, types of horizons, etc.) of a soil?
8. How does a soil that forms in a tropical climate differ from one that forms in an arid climate?
9. Explain why soil erosion has been exacerbated by human activity.

Another View In this image, we can see that the lack of natural plant coverage has led to severe soil erosion by wind. Similar conditions produced the Dust Bowl of the 1930s.

Look closely at this cliff in southern Nevada. You'll see layer upon layer of sedimentary strata. The ledges exposed resistant sandstone. Slopes are underlain by weaker shale. These rocks formed when sediments were buried and lithified long ago.

CHAPTER 7

Pages of Earth's Past: Sedimentary Rocks

> In every grain of sand there is a story of Earth.
>
> —*Rachel Carson (American conservationist, 1907–1964)*

LEARNING OBJECTIVES

By the end of this chapter, you should understand . . .

- what distinguishes various classes of sedimentary rocks from one another.
- how clastic sedimentary rocks form and how to recognize and name major types.
- the role of life in the production of rocks such as limestone and coal.
- how the layering (bedding) in sedimentary rock forms.
- how shapes and textures preserved in sedimentary rocks reflect depositional environments.
- why thick accumulations of sedimentary rock can be found only in certain locations.

7.1 Introduction

In this day when *Google Earth™* can take you to every nook and cranny of our planet's surface at the touch of a computer mouse, it's hard to imagine a world in which vast regions were blanks on a map. But only a little over a century ago, that was the state of affairs that members of the 1910–13 British Antarctic Expedition were seeking to change. Led by Robert Falcon Scott, a team of explorers from the expedition set out to be the first to reach the South Pole. The first part of the journey took them over the Ross Ice Shelf, a broad plain of ice not far above sea level. But to reach the pole, they had to haul their heavy sledges up the Beardmore Glacier, a river of ice that had cut its way down through the rugged Transantarctic Mountains (**Fig. 7.1a**), for the South Pole lies on the Polar Plateau at an elevation of about 3 km. The cliffs overlooking the glacier, as is the case along much of the length of the Transantarctic Mountains, expose layer upon layer of a light-colored grainy rock (**Fig. 7.1b**). One of the expedition's members, Edward Wilson, a physician and naturalist, served as the team's geologist, and during the journey he collected numerous specimens of these rocks. Scott, Wilson, and the others succeeded in reaching the South Pole, on January 17, 1912. But when they arrived, they found to their profound disappointment that the Norwegian explorer Roald Amundsen had beaten them there by 34 days.

On their return journey, all of the British explorers perished in the blizzards and cold of the southernmost continent. When rescuers eventually came upon Scott's last campsite, they found Wilson's specimens, some of which contained fossils of *Glossopteris*, the fossil whose distribution Alfred Wegener would use as evidence of continental drift.

What are the grainy, layered rocks that Wilson collected? They are a type of sedimentary rock called sandstone. Formally defined, **sedimentary rock** is rock that forms at or near the surface of the Earth in one of several ways: by the cementing together of loose **clasts** (fragments or grains) that had been produced by physical or chemical weathering of pre-existing rock, by the growth of shell masses or by the cementing together of shells and shell fragments, by the accumulation and subsequent alteration of organic matter derived from living organisms, or by the precipitation of minerals directly from surface-water solutions. Layers, or "beds," of sedimentary rock are like the pages of a book, recording tales of ancient events and ancient environments on the ever-changing face of the Earth. They occur only in the upper part of the crust, and form a "cover" that buries the underlying "basement" of igneous and/or metamorphic rock (**Fig. 7.2**).

In Interlude B, we introduced the concept of weathering and showed how it attacks solid rock and breaks it down into ions and loose sediment grains. In this chapter, we see how these materials can be buried and transformed into sedimentary rock. We also introduce various specific types of sedimentary rock and show how geologists use the study of sedimentary rocks to characterize Earth System history. Finally, we discuss the special settings, called sedimentary basins, in which particularly thick successions of sedimentary rock accumulate.

7.2 Classes of Sedimentary Rocks

Geologists divide sedimentary rocks into four major classes, based on their mode of origin: (1) **clastic sedimentary rock** consists of cemented-together *clasts*, solid fragments and grains broken off of pre-existing rocks (the word *clastic* comes from the Greek *klastos*, meaning broken); (2) **biochemical sedimentary rock** consists of shells grown by organisms; (3) **organic sedimentary rock** consists of carbon-rich relicts of cellular material from plants or other organisms; and (4) **chemical sedimentary rock** comes from minerals that precipitated

FIGURE 7.1 Robert Falcon Scott and his companions, including naturalist Edward Wilson, traversed the Transantarctic Mountains in 1912 to reach the Polar Plateau. The mountains expose sedimentary rocks.

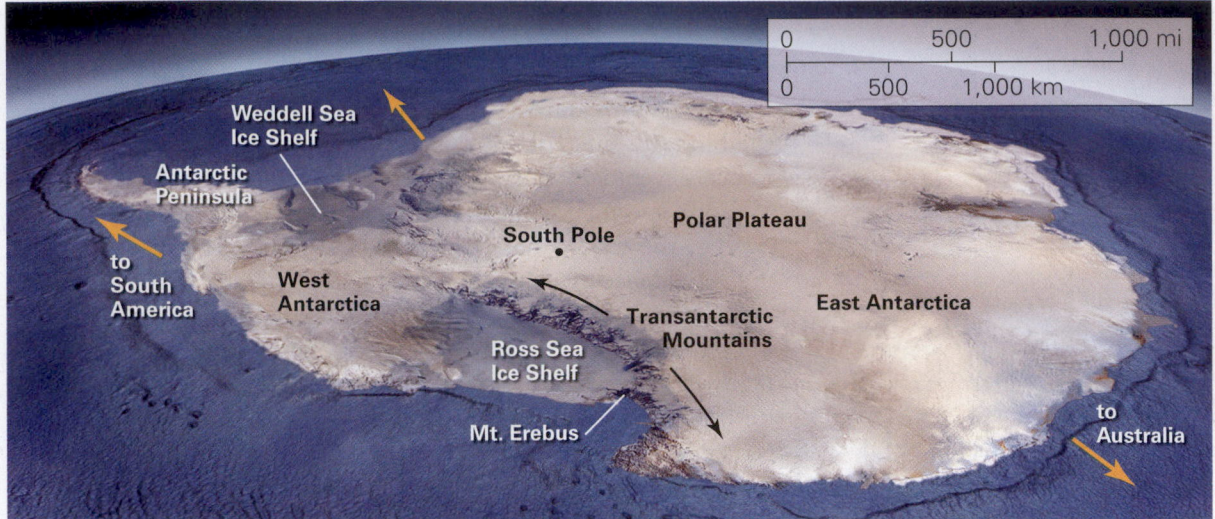

(a) Scott started near Mt. Erebus and crossed the Ross Ice Shelf and the Transantarctic Mountains to reach the South Pole. Roald Amundson beat him there.

(b) The white rock of the high cliffs in the Antarctic Dry Valleys is sedimentary. The black rock is a sill of basalt.

directly from surface-water solutions. Geologists sometimes also use various adjectives to characterize sedimentary-rock composition—*siliceous rocks* contain mostly quartz, *argillaceous rocks* contain mostly clay minerals, and *carbonate rocks* contain mostly calcite and/or dolomite. By some estimates, 70% to 85% of all the sedimentary rocks on Earth are siliceous or argillaceous clastic rocks, and 15% to 25% are carbonate biochemical or chemical rocks. Other kinds of sedimentary rocks occur only in minor quantities.

Let's now look at the four major classes in more detail. For each class, we explain how the rocks form and introduce the names of common examples.

Clastic Sedimentary Rocks

Formation Nine hundred years ago, a thriving community of Native Americans inhabited the high plateau of Mesa Verde, Colorado. In hollows beneath huge overhanging ledges, they built multistory stone-block buildings that have survived to this day (**Fig. 7.3**). Clearly, the blocks are solid and durable—they are, after all, rock. But if you were to rub your thumb along one, it would feel gritty, and small grains of quartz would break free and roll under your thumb, for the blocks consist of quartz sand grains cemented together. Geologists call such rock a **sandstone**.

Sandstone is an example of a clastic sedimentary rock (also referred to as a detrital sedimentary rock), because it consists of loose clasts (detritus) that have been stuck together to form a solid mass. The clasts, or *grains*, can consist of individual minerals, such as fragments of quartz or flakes of clay minerals, or of chunks of rock, such as pebbles of granite. Production of a clastic sedimentary rock involves five steps (**Fig. 7.4a**).

- *Weathering:* The grains from which clastic rocks form come from the disintegration of pre-existing bedrock into separate grains due to physical and chemical weathering. The dissolved ions that eventually precipitate as new minerals hold the grains together in sedimentary rock are also a product of weathering.

- *Erosion:* Once formed, grains produced by weathering do not stay in place forever. Gravity may cause them to fall off the outcrop, or they may be removed from the outcrop by flowing water or ice, or by wind. **Erosion** refers to the combination of processes that separate clasts from their

FIGURE 7.2 Sedimentary strata form a blanket that covers basement.

(a) During the formation of the Grand Canyon, erosion cut through this cover into the basement.

(b) The contact between the sedimentary cover and underlying basement lies at the top of the inner gorge.

original substrate. Dissolution of outcrop faces in water, producing dissolved ions, is also a kind of erosion.

- *Transportation:* Once produced by weathering, and removed by erosion, clasts and dissolved ions can be carried away in a "transporting medium" (wind, water, or ice). The ability of a medium to carry sediment depends on its viscosity and velocity. Solid ice, for example, can transport clasts of any size, regardless of how slowly the ice moves. Very fast-moving, turbulent water can transport very coarse clasts (cobbles and boulders), moderately fast-moving water can carry only sand and gravel, and slowly moving water carries only silt and mud. Strong winds can move sand and dust, but gentle breezes carry only dust.
- *Deposition:* A transporting medium does not carry sediment forever. Eventually, the sediment undergoes **deposition**, the process by which sediment falls out of the medium (**Fig. 7.4b**).

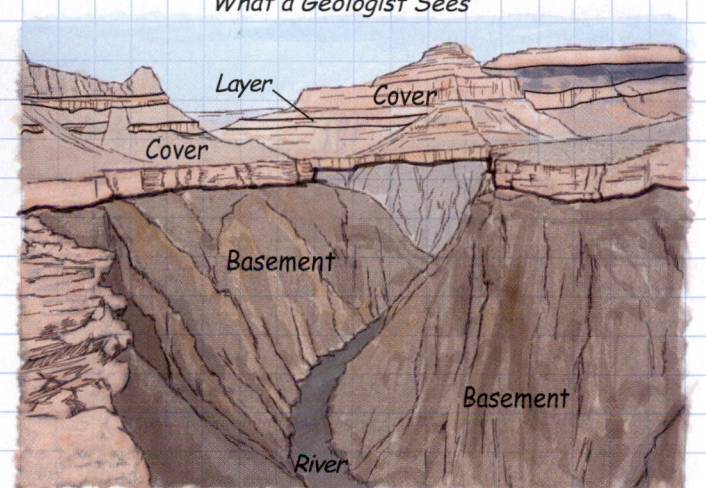

(c) A geologist's sketch emphasizes the contact. Here the basement consist of metamorphic and igneous rock.

FIGURE 7.3 The cliff dwellings nestled beneath a ledge at Mesa Verde, Colorado, are made of sandstone blocks. The inset shows the grainy character of the rock.

FIGURE 7.4 The five steps in clastic sedimentary rock formation.

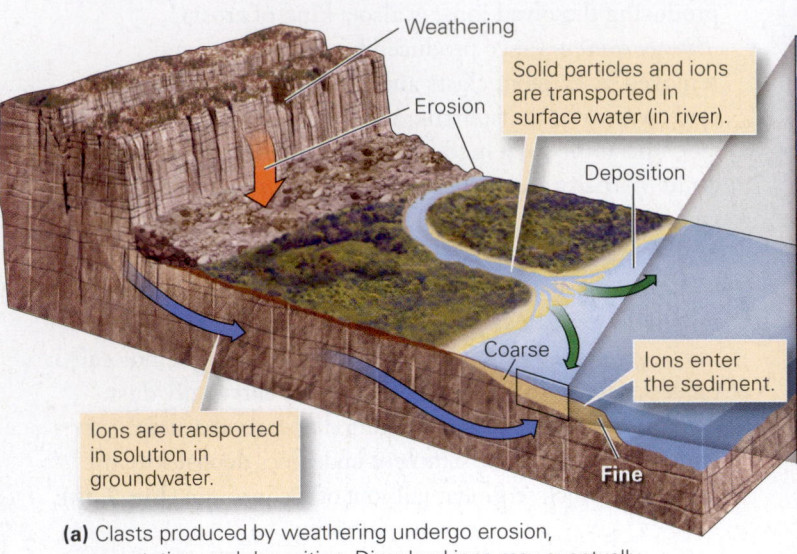

(a) Clasts produced by weathering undergo erosion, transportation, and deposition. Dissolved ions may eventually become cement.

(b) The process of lithification takes place during progressive burial.

(c) Over time, cement fills the spaces between grains.

Sediment settles out of wind or moving water when these fluids slow, because as the velocity decreases, the fluid no longer has the ability to carry sediment. Sediment carried by ice accumulates when the ice melts.

- *Lithification:* Geologists refer to the last stage of producing a clastic sedimentary rock, namely the transformation of loose sediment into solid rock, as **lithification**. Lithification of clastic sediment involves two steps. First, when the sediment has been buried, pressure generated by the weight of overlying material squeezes out the water and air that had been trapped between clasts, and clasts press together tightly, a process called **compaction**. Second, during and/or after compaction, sediment may be bound in place to make coherent sedimentary rock by the process of **cementation** (**Fig. 7.4c**). During cementation, minerals (commonly quartz or calcite) precipitate from groundwater and fill spaces between clasts. The resulting **cement** acts like glue and holds grains together.

Classifying Clastic Sedimentary Rocks Say that you pick up a clastic sedimentary rock and want to describe it sufficiently so that, from your words alone, another person can picture the rock. What characteristics should you mention? Geologists find the following characteristics most useful:

- *Clast size:* Geologists refer to the diameter of the grains making up a clastic sedimentary rock as the *clast size* or *grain size*. Names used for clast size, listed in order from coarsest to finest, are boulder, cobble, pebble, sand, silt, and clay (see Table B.1 in Interlude B). Geologists informally use the term *gravel* for an accumulation of pebbles

and cobbles and the term *mud* for an accumulation of wet clay and/or very fine silt. In this context, *clay* refers to all grains less than 0.004 mm in diameter—these grains are mostly clay minerals (several different types of sheet-silicate minerals) but also include tiny specks of quartz and other minerals.

- *Clast composition:* Not all clasts consist of the same mineral or rock fragments. Composition refers to the makeup of clasts in sedimentary rock. Larger clasts (pebbles or larger) typically consist of rock fragments, meaning the clasts themselves are an aggregate of many mineral grains, whereas smaller clasts (sand, silt, or clay) typically consist of individual minerals. In some cases, chips of fine-grained rock may be mixed in with sand grains. Such chips are called *lithic clasts*. Some sedimentary rocks contain only clasts of one composition, but others contain a variety of different kinds of clasts.

- *Angularity and sphericity:* Angularity indicates the degree to which clasts have smooth or angular corners and edges (**Fig. 7.5a, b**). Sphericity, in contrast, refers to the degree to which a clast is equidimensional, or resembles the shape of a sphere.

FIGURE 7.5 Grain characteristics, and their evolution with increasing transport and weathering.

Grain size

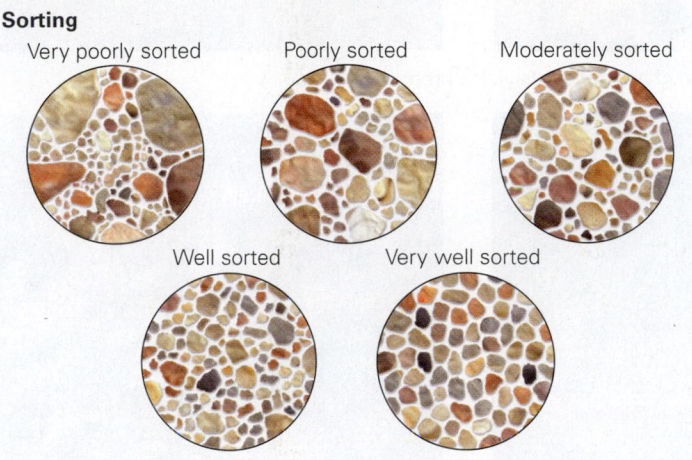

Closer to source ⟶ Farther from source

(a) As the amount of transport of a sediment (by a stream or by waves) increases, the sediment tends to become finer grained.

Sorting

Very poorly sorted Poorly sorted Moderately sorted

Well sorted Very well sorted

(c) If transport sifts grains, carrying smaller ones farther and leaving coarser ones behind, grains in a sediment tend to be the same size.

Angularity

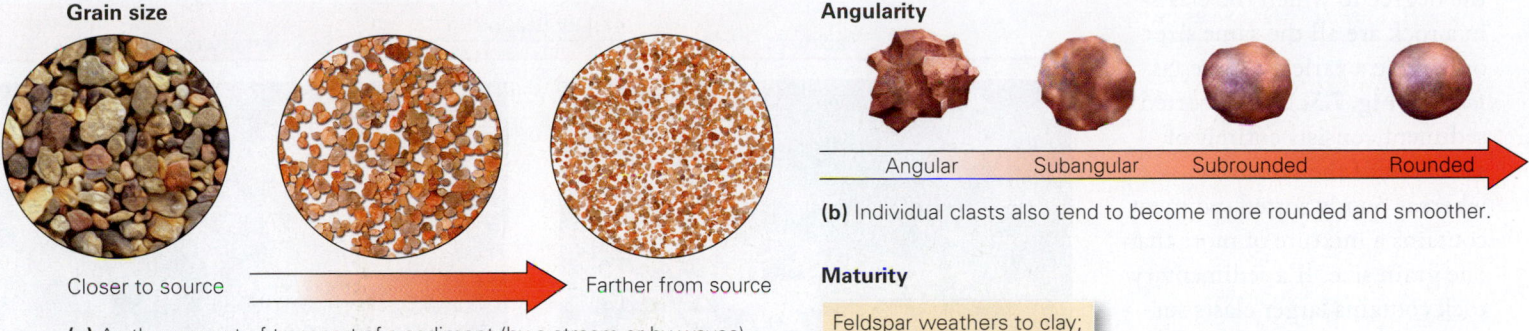

Angular Subangular Subrounded Rounded

(b) Individual clasts also tend to become more rounded and smoother.

Maturity

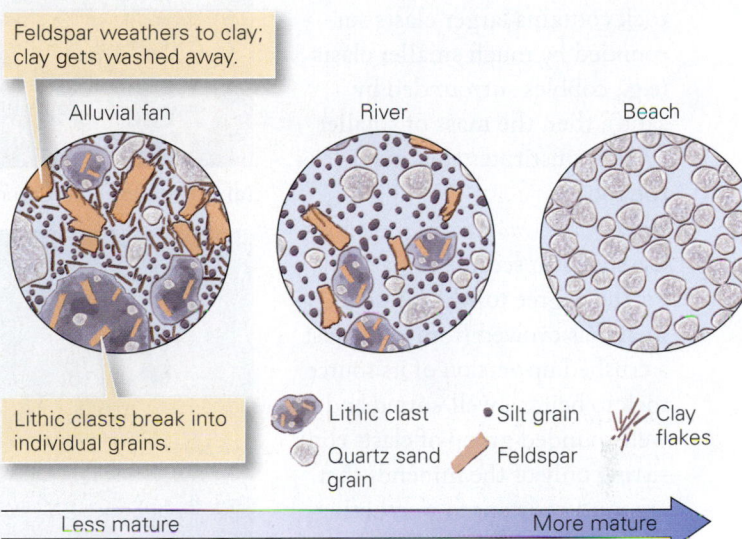

Feldspar weathers to clay; clay gets washed away.

Alluvial fan River Beach

Lithic clasts break into individual grains.

Lithic clast · Silt grain ✦ Clay flakes
Quartz sand grain ▬ Feldspar

Less mature ⟶ More mature

(d) A less mature sediment consists of fragments of the original rock and contains both resistant and nonresistant minerals. A mature sediment contains only well-sorted resistant minerals.

TABLE 7.1 Classification of Clastic Sedimentary Rocks

Clast Size	Clast Character	Rock Name (Alternate Name)
Coarse to very coarse (> 2 mm)	Rounded pebbles and cobbles	**Conglomerate**
	Angular clasts	**Breccia**
	Large clasts in muddy matrix	**Diamictite**
Medium to coarse (0.06–2 mm)	Sand-sized grains	**Sandstone**
	• Quartz grains only	• Quartz sandstone (quartz arenite)
	• Quartz and feldspar sand	• Arkose
	• Sand-sized rock fragments	• Lithic sandstone
	• Sand and rock fragments in a clay-rich matrix	• Wacke (informally called graywacke)
Fine (0.004–0.06 mm)	Silt-sized clasts	**Siltstone**
Very fine (< 0.004 mm)	Clay and/or very fine silt	**Shale**, if it breaks into platy sheets
		Mudstone, if it doesn't break into platy sheets

- *Sorting:* Geologists refer to the degree to which the clasts in a rock are all the same size or include a variety of sizes as *sorting* (**Fig. 7.5c**). Well-sorted sediment consists entirely of sediment of the same size, whereas poorly sorted sediment contains a mixture of more than one grain size. If a sedimentary rock contains larger clasts surrounded by much smaller clasts (e.g., cobbles surrounded by sand), then the mass of smaller grains constitutes the "matrix" of the rock.

- *Sedimentary maturity:* Geologists use the term sediment maturity for the degree to which a sediment has evolved from being just a crushed up version of its source rock to being a well-sorted and well-rounded group of clasts consisting only of the minerals that are most resistant to weathering (**Fig. 7.5d**).

- *Character of cement:* Not all clastic sedimentary rocks have the same kind of cement. In some, the cement consists predominantly of quartz, whereas in others it consists predominantly of calcite. Other kinds of mineral cements do occur, but they are rarer.

With these characteristics in mind, we can distinguish among several common types of clastic sedimentary rocks (**Table 7.1**). This table provides common rock names—specialists sometimes use other, more precise names based on more complex classification schemes. Though no single characteristic serves as a complete basis for classifying clastic sedimentary rocks, grain size is the most important one. Geologists further distinguish among different kinds of sandstone (quartz sandstone, arkose, wacke) on the basis of clast composition and/or sorting, and they distinguish between shale and mudstone on the basis of the way in which the rock breaks—shale splits into thin sheets, whereas mudstone does not.

FIGURE 7.6 Different kinds of clasts lithify into different kinds of sedimentary rocks.

Sediment ———— Lithification ————▶ Sedimentary rock

(a) Lithification of an accumulation of angular clasts yields breccia.

(b) Layers of river gravel lithify into conglomerate.

Alluvial fan

(c) Sediment deposited in an alluvial fan, close to its source, can be feldspar rich. Lithification of this sediment yields arkose.

Origin of Various Types of Clastic Rocks Characteristics of a sedimentary rock provide clues to the source of the sediment, and to the environment of deposition. To see how, let's follow the fate of rock fragments as they gradually move from a cliff face in the mountains via a river to the seashore. Different kinds of sediment develop along the route. Each kind, if buried and lithified, would yield a different type of sedimentary rock.

Sediment → Lithification → **Sedimentary rock**

(d) Layers of beach or dune sand lithify into sandstone.

Sandstone

Shale

(e) Layers of mud, exposed beneath marsh grass, lithify to form shale. Here the thin-bedded shale is interbedded with sandstone.

(f) A rock formed from material containing large clasts surrounded by a fine-grained matrix is a diamictite.

To start, imagine that large blocks of rock tumble off a cliff and slam into other blocks already at the bottom. The impact shatters the blocks, producing a pile of angular fragments with sharp edges. If these fragments were to be cemented together, the resulting rock would be **breccia** (**Fig. 7.6a**). Later, a storm causes the fragments (clasts) to slide downslope into a turbulent river. In the moving water, clasts bang into each other and into the riverbed, a process that shatters them into still smaller pieces and breaks off their sharp edges. Angular clasts gradually become rounded clasts. When the river water slows, pebbles and cobbles stop moving and form a mound, or bar, of gravel. Burial and lithification of these rounded clasts produces **conglomerate** (**Fig. 7.6b**).

If the gravel stays put for a long time, it undergoes chemical weathering. As a consequence, cobbles and

> **Did you ever wonder...**
>
> where beach sand comes from?

pebbles of the rock break apart into individual mineral grains. For example, disintegration of granite would yield a mixture of quartz, feldspar, and clay. Clay is so fine that even slowly moving water can carry it away, leaving sand containing a mixture of quartz and some feldspar grains—this sediment, if buried and lithified, becomes *arkose* (**Fig. 7.6c**). Over time, feldspar grains in sand continue to weather into clay so that gradually, during successive events that wash the sediment downstream, the sand loses its feldspar and ends up being composed almost entirely of durable quartz sand grains. Some of the sand may make it to the sea, where waves carry it to beaches, and some may end up in desert dunes. Such sediment, when buried and lithified, becomes quartz sandstone (**Fig. 7.6d**). Meanwhile, silt and clay may accumulate in the flat areas bordering streams, regions called *floodplains* (see Chapter 17), which become submerged only during floods. And some silt and mud settles in a wedge, called a *delta*, at the mouth of the river, or in quiet lagoons (protected bodies of quiet water) or mudflats (broad, quiet water areas exposed at low tide) along the shore. The silt, when lithified, becomes **siltstone**, and the mud, when lithified, becomes **shale** or **mudstone** (**Fig. 7.6e**).

Two of the rock names in Table 7.1 did not appear in the above narrative. *Diamicton* is a very poorly sorted sediment that contains clasts of all sizes. For example, diamicton may contain cobbles or boulders surrounded by a matrix of sand, silt, and clay. A rock made from a diamicton is a *diamictite* (**Fig. 7.6f**). Research suggests that diamictites can form from the lithification of *debris flows* (viscous slurries consisting of mud mixed with larger clasts) both on land and underwater, or of *glacial till*, the debris left behind as glacial ice melts. Another sedimentary rock type, known as *wacke*, is a poorly sorted sedimentary rock that consists of sand grains and lithic fragments suspended in a matrix of mud. Wackes can form from the deposits of submarine avalanches; most wacke has a grayish color, and thus geologists informally refer to it as graywacke. You may have sensed from our narrative that as sediment moves downstream, it becomes more mature.

Biochemical Sedimentary Rocks

Numerous organisms have evolved the ability to extract dissolved ions from seawater to make solid shells. Some of these organisms are anchored to the sea floor, while others crawl or burrow on the sea floor or float in the water above. When the organisms die, the solid material in their shells survives, and this material, when lithified, constitutes *biochemical sedimentary rock*. The formation of such rock represents a significant type of interaction between the physical and living components of the Earth System. Geologists recognize several different types of biochemical sedimentary rocks, two of which we now describe.

Biochemical Limestone
A snorkeler gliding above a reef sees an incredibly diverse community of coral and algae, around which creatures such as clams, oysters, snails (gastropods), and lampshells (brachiopods) live, and above which plankton, including coccolithophores and forams, float (**Fig. 7.7a**). All of these organisms make solid shells out of calcium carbonate ($CaCO_3$), either as calcite or as its polymorph, aragonite. When the organisms die, the shells remain and may accumulate. Rocks formed dominantly from this material are the **biochemical limestone**. Since the principal compound making up limestone is $CaCO_3$, geologists refer to limestone as a type of *carbonate rock*.

Limestone comes in a variety of textures because the material that forms it accumulates in a variety of ways. For example, limestone can originate from reef builders (such as coral) that grew in place (**Fig. 7.7b**), from shell debris that was broken up and transported, or from carbonate mud consisting of plankton shells that settled like snow out of water. Because of this vari-

ety, we can distinguish among *fossiliferous limestone*, containing visible fossil shells or shell fragments (**Fig. 7.7c**); *micrite*, consisting only of very fine carbonate mud (**Fig. 7.7d**); and *chalk*, consisting of plankton shells. (Experts recognize many other types as well and use a more precise, but complex, terminology for limestone classification.)

Typically, limestone is a massive light-gray to dark-bluish-gray rock that breaks into chunky blocks—it doesn't look much like a pile of shell fragments (**Fig. 7.7e**). That's because several processes take place that change the texture of the rock over time. Prior to lithification, organisms may burrow into recently formed or deposited shells and break them up. Later, water passing through the rock not only precipitates cement but also dissolves some carbonate grains and causes new ones to grow. Thus, original crystals may be replaced by new ones. Typically, all the aragonite that was originally in shells transforms into calcite, a more stable mineral, and smaller crystals of calcite are replaced by larger ones.

Biochemical Chert
If you walk beneath the northern end of the Golden Gate Bridge in California, you will find outcrops of reddish, almost porcelain-like rock occurring in 3- to 15-cm-thick layers (**Fig. 7.8a**). Hit it with a hammer, and the rock cracks, almost like glass, creating smooth, spoon-shaped (conchoidal) fractures. It's made from cryptocrystalline quartz (*crypto* is the Greek word for hidden), quartz grains that are too small to be seen without the extreme magnification of a scanning electron microscope. Such chert formed from plankton, such as radiolaria and diatoms, that produce shells composed of SiO_2 (silica), which had accumulated along with clay to form a siliceous "ooze" on the sea floor. Gradually, the shells dissolved, forming silica solutions, which fill pores in the surrounding clay. Eventually, as tiny crystals grow from these solutions, the sediment lithifies to become chert. Geologists call this rock *biochemical chert*, to emphasize that it formed from the shells of organisms, or *bedded chert*, to emphasize that it occurred as a succession of layers. Chert can come in a variety of colors, depending on the impurities that it contains. For example, chert containing traces of iron oxide tends to be red—such chert is also known as *jasper*.

Organic Sedimentary Rocks

We've seen how the mineral shells of organisms can accumulate and lithify to become *biochemical* sedimentary rocks. What happens to the "guts" of the organisms—the cellulose, fat, carbohydrate, protein, and other organic compounds that make up living tissue? Commonly, organic debris gets eaten by other organisms or decays at the Earth's surface. But in some environments, such as oxygen-poor quiet water in swamps, lagoons, or lakes, the debris settles along with other sediment and eventually gets buried and preserved. At the elevated

FIGURE 7.7 The formation of carbonate rocks (limestone).

(a) In this modern coral reef, corals produce shells. If buried and preserved, these become limestone.

Relict of a small reef

(b) A Vermont quarry shows the gray color of 400-Ma limestone. The white mounds are relics of small reefs.

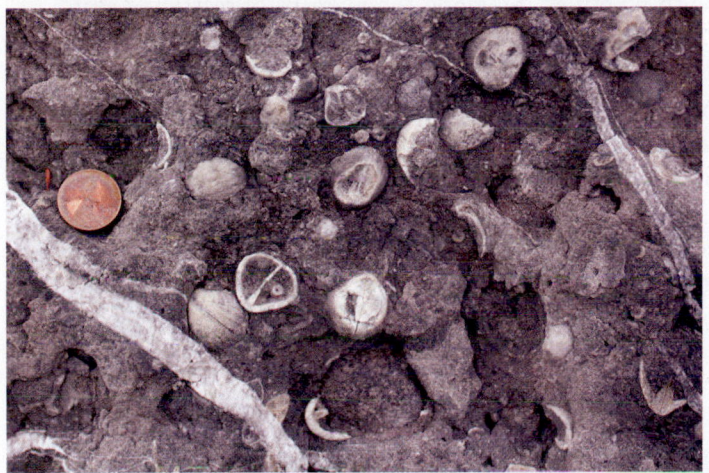

(c) Fossil shells of ~415-Ma branchipods protrude from an outcrop of limestone in New York.

(d) The grains of this micrite are almost too small to see. This micrite contains very thin bedding.

(e) This roadcut near Kingston, New York, exposes beds of limestone. The vertical stripes are drillholes.

temperatures and pressures that exist at depth below the Earth's surface, organic matter undergoes chemical reactions that slowly transform it into *organic sedimentary rock*, distinct from other sedimentary rock in that it contains a high proportion of organic chemicals. Since the dawn of the industrial revolution in the early 19th century, organic sedimentary rock has provided the fuel of modern industry and transportation (see Chapter 14), for organic chemicals can burn to produce energy. We'll briefly consider two types of organic sedimentary rock—coal and oil shale.

Coal is a black, combustible rock containing between 40% and 90% carbon; the remainder consists of clay and quartz. The carbon of coal occurs in large, complicated molecules called *macerals*—a typical maceral consists of about 85% carbon and 15% oxygen, nitrogen, and hydrogen. As discussed further in Chapter 14, coal forms from plant remains that have been buried deeply. Under the pressure and temperature conditions found at depth, the organic material of the plants becomes tightly compacted, and volatile molecules such as hydrogen, water, carbon dioxide, and ammonia break free and escape. As this happens, the carbon atoms reorganize into macerals (**Fig. 7.8b**).

Oil shale is a shale that contains not only clay but also between 15% and 30% organic material in a form called *kerogen*. The kerogen in oil shale comes from the fats and proteins that made up the living part of plankton or algae. If the tiny organisms settle in an environment where they do not immediately rot away or get eaten, they mix with the clay minerals in mud. When the mud gets buried and lithified, to form shale, the organic material transforms into kerogen. The presence of organic material colors oil shale black.

Chemical Sedimentary Rocks

The colorful terraces, or mounds, that grow around the vents of hot-water springs; the immense layers of salt that underlie the floor of the Mediterranean Sea; the smooth, sharp point of an ancient arrowhead—these materials all have something in common. They all consist of rock formed primarily by the precipitation of minerals directly out of water solutions. We call such rocks *chemical sedimentary rocks*. They typically have a crystalline texture, partly formed during their original precipitation and partly when, at a later time, new crystals grow at the expense of old ones through the process of recrystallization. In some chemical sedimentary rocks, crystals are coarse, whereas in others they are too small to see. Geologists distinguish among many types of chemical sedimentary rocks, primarily on the basis of composition.

Evaporites—Products of Saltwater Evaporation In 1965, two daredevil drivers in jet-powered cars battled to be the first to surpass a speed of 600 mph on land. On November 7, the *Green Monster* peaked at 576.127 mph. Eight days later, the *Spirit of America* reached 600.601 mph. Traveling at such speeds, a driver must maintain an absolutely straight line, for the slightest turn will catapult the vehicle out of control. Thus, high-speed trials must take place on extremely long and flat racecourses. Not many places can provide such conditions—the Bonneville Salt Flats of Utah do. The salt flats formed from the evaporation of an ancient salt lake. Under the heat of the Sun, the water turned to vapor and drifted up into the atmosphere, but the salt that had been dissolved in the water stayed behind. Such salt precipitation occurs wherever saturated salt water develops—in desert lakes with no outlet and along the margins of restricted seas (**Fig. 7.9**). For thick deposits of salt to form, large volumes of water must evaporate. This may happen when plate tectonic movements temporarily cut off arms of the sea (as we saw in the case of the Mediterranean Sea; see Interlude B), or during continental rifting, when seawater first begins to spill into the rift valley.

Because salt deposits form as a consequence of evaporation, geologists refer to them as **evaporites**. The specific type of salt

FIGURE 7.8 Examples of biochemical and organic sedimentary rocks.

(a) This bedded chert developed on the deep-sea floor by the deposition of plankton that secrete silica shells.

Sandstone and shale

Coal layer

(b) Coal is deposited in layers (beds), just like other kinds of sedimentary rocks.

minerals comprising an evaporite depends on the amount of evaporation. For example, when 80% of the seawater trapped in a basin evaporates, gypsum forms, and when 90% of the water evaporates, halite precipitates. If seawater were to evaporate entirely, the resulting evaporite would consist of 80% halite, 13% gypsum, and the remainder of other salts and carbonates.

Travertine (Chemical Limestone) Most limestone is biochemical, in that it forms from the shells of organisms. But one type, called *travertine*, consists of crystalline calcium carbonate ($CaCO_3$) that precipitates directly from groundwater that has seeped out at the ground surface either in hot- or cold-water springs or on the walls of caves. What causes this precipitation? It happens, in part, when the groundwater "degasses," meaning that some of the carbon dioxide that had been dissolved in the groundwater bubbles out of solution, for removal of carbon dioxide decreases the ability of the water to hold dissolved carbonate. Precipitation also occurs when water evaporates, thereby increasing the concentration of carbonate. Recent research suggests that various kinds of microbes can accelerate the process of precipitation.

Travertine produced at springs forms terraces and mounds that are meters or even hundreds of meters thick (**Fig. 7.10a**). Spectacular terraces of travertine occur at Mammoth Hot Springs in Yellowstone National Park. Amazing column-like mounds of travertine grew up from the floor of Mono Lake, California (**Fig. 7.10b**), where hot springs seeped into the cold water of the lake. Travertine also grows on the walls of caves where groundwater seeps out (**Fig. 7.10c**). In cave settings, travertine builds up beautiful and complex growth forms called *speleothems* (see Chapter 19).

Travertine has been quarried for millennia to make building stones and decorative stones. The rock's beauty comes in part because in thin slices it is translucent and in part because it typically displays colored growth bands. Bands develop in response to changes in the composition of groundwater or in the environment into which the water drains. Some travertines (a type called tufa) contain abundant large pores (open spaces).

Dolostone Not all carbonate rock consists of pure calcite. A variety of carbonate rock called *dolostone* differs from limestone in that it contains the mineral dolomite ($CaMg[CO_3]_2$).

FIGURE 7.9 The formation of evaporite deposits.

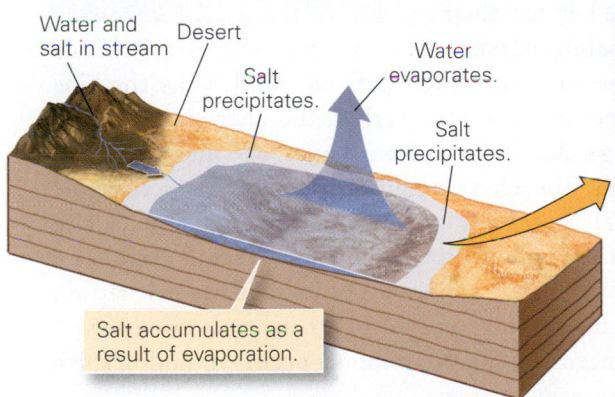

Salt on the floor of Death Valley, California

Water and salt in stream — Desert — Water evaporates. — Salt precipitates. — Salt precipitates. — Salt accumulates as a result of evaporation.

(a) In lakes with no outlet, tiny amounts of salt brought in by streams stay behind as the water evaporates. When the water evaporates entirely, a white crust of salt remains.

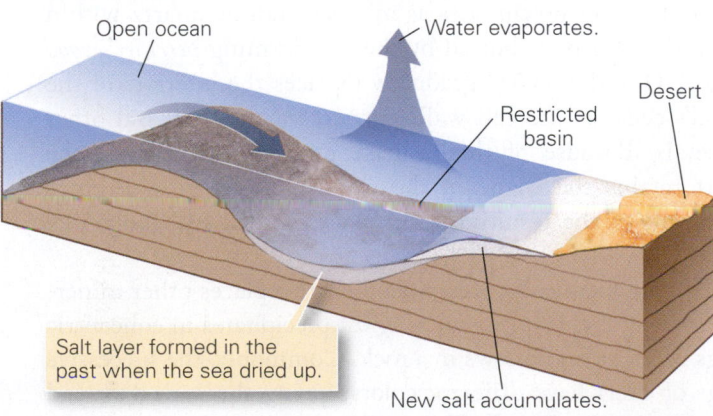

Open ocean — Water evaporates. — Desert — Restricted basin — Salt layer formed in the past when the sea dried up. — New salt accumulates.

(b) Salt precipitation can also occur along the margins of a restricted marine basin, if salt water evaporates faster than it can be resupplied.

(c) Thick layers of salt may be buried deeply. Here salt is being mined deep underground.

FIGURE 7.10 Examples of travertine (chemical limestone) deposits.

Terrace of new travertine

2 m

(a) Travertine accumulates in terraces at Mammoth Hot Springs in Yellowstone Park, Wyoming.

(b) Columns of tufa precipitated from springs that were once under the saline water of Mono Lake, California, when the lake level was higher.

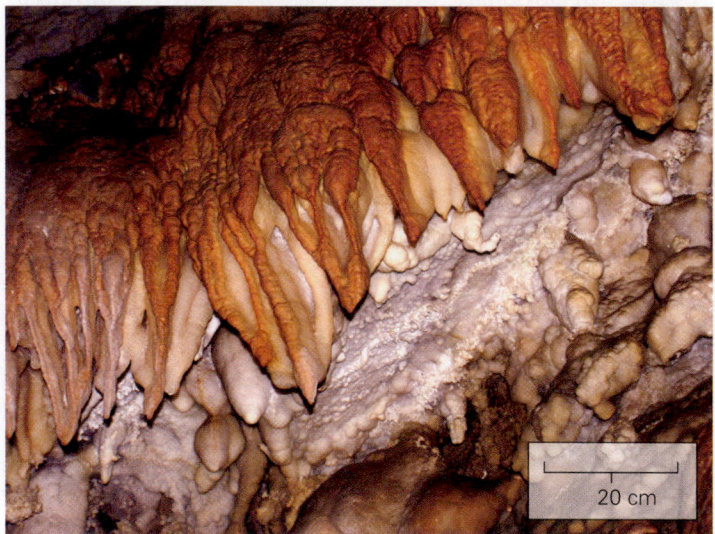

20 cm

(c) Travertine speleothems form as calcite-rich water drips from the ceiling of Timpanogos Cave in Utah.

Where does the magnesium in dolomite come from? Dolostone forms by a chemical reaction between solid calcite and magnesium-bearing groundwater. This change may take place beneath lagoons along a shore soon after the limestone formed or, a long time later, after the limestone has been buried deeply.

Chemically Precipitated Chert A tribe of Native Americans, the Onondaga, once lived off the land in eastern New York State. Here outcrops of limestone contain *nodules* (small, rounded lumps or lenses) of a black chert (**Fig. 7.11a, b**). Because of the way it breaks, the tribe's artisans could fashion sharp-edged tools (arrowheads and scrapers) from this chert, so the Onondaga collected it for their own tool-making indus-try and for use in trade with other people. Unlike the bio-chemical chert we described earlier, the chert collected by the Onondaga didn't form from layers of silica plankton shells. Rather, the chert nodules grew when microscopic quartz crystals gradually precipitated and replaced calcite crystals within a bed of limestone, long after the limestone was originally deposited. Because of its mode of formation, geologists refer to this type of chert as *replacement chert*. Because replacement chert commonly occurs in nodules, it's also known as *nodular chert*.

Did you ever wonder...

how flint used for arrowheads first formed?

Replacement chert doesn't only form in limestone. Some grows in buried silica-rich volcanic ash beds, when the silica in groundwater precipitates as microcrystalline quartz within wood that has been buried by the ash, forming *petrified wood* (**Fig. 7.11c**) The quartz gradually replaces the interiors of the wood's cells, as the cell walls convert into carbon and other minerals. Because of this delicate replacement process, the chert retains the shape of the wood and the growth rings within it, so the resulting solid block of rock still looks like wood.

Not all chemically precipitated chert replaces other miner-als. A type of chert, known as *agate*, precipitates in concentric rings inside open hollows in a rock. Commonly, the successive rings of chert have different colors, giving the rock a striped appearance (**Fig. 7.11d**). These colors are caused by variations in the type and concentration of impurities.

FIGURE 7.11 Examples of chert that precipitated in place.

Layer of black chert

Tilted limestone beds

(a) Replacement chert forms as layers of nodules between tilted limestone beds in New York.

(b) Close-up of a black chert nodule surrounded by white chalk, exposed on a cliff in southern England. The coin is for scale.

(c) This 14 cm-diameter log of petrified wood from Wyoming formed about 50 Ma.

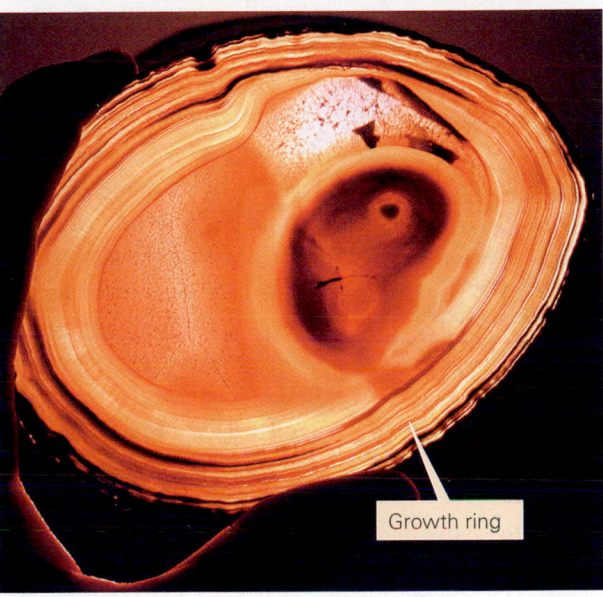

Growth ring

(d) A thin slice of Brazilian agate, lit from the back, shows growth rings.

Take-Home Message

Geologists distinguish among many types of sedimentary rocks based on the mode of formation. Clastic sedimentary rocks consist of grains weathered and eroded from pre-existing rocks and transported by wind, water, or ice to a site of deposition where they were buried and lithified. Clastic rocks can be classified by grain size. Biochemical sedimentary rocks, such as limestone, consist of the shells of organisms, and organic sedimentary rocks, such as coal, form from the organic remains of organisms. Chemical sedimentary rocks precipitate from water solutions.

QUICK QUESTION: Do all sedimentary rocks have the same composition? Why or why not?

7.3 Sedimentary Structures

One of the first things that you may notice when looking at a large outcrop of sedimentary rock is that rock in the outcrop is not generally homogeneous. The outcrop likely contains distinct layers, perhaps of different sedimentary rock types. On the surface of layers you may see small ridges and/or indentations, and within the layers the grains may be sorted or oriented so as to define notable textures. These features, which range from subtle to obvious, formed during the process of deposition and, as we will see, provide key clues to the environment in which the sediments were deposited. Geologists use the term **sedimentary structure** for such features. Here we

examine some of the more important types of sedimentary structures.

Bedding and Stratification

Let's start by introducing the jargon geologists use for discussing sedimentary layers. A single layer of sediment or sedimentary rock with a recognizable top and bottom is called a **bed**; the boundary between two beds is a bedding plane; several beds together constitute **strata** (singular stratum, from the Latin *stratum*, meaning pavement); and the overall arrangement of sediment into a sequence of beds is **bedding**, or stratification. From the word *strata*, we derive other words, such as *stratigrapher* (a geologist who specializes in studying strata) and *stratigraphy* (the study of the record of Earth history preserved in strata).

When you examine strata in a region with good exposure, the bedding generally stands out clearly—it looks like bands or stripes across a cliff face (**Fig. 7.12a**). Typically, contrasts in rock type distinguish one bed from adjacent beds. Each bed has a definable thickness (from a couple of centimeters to tens of meters) and may display a contrast in composition, color, and/or grain size that distinguishes it from its neighbors. For example, if a sequence of strata contains a bed of sandstone overlain by a bed of shale and overlain again by a bed of siltstone, the surfaces separating one rock type from another define bedding. In many cases, however, adjacent beds all have the same overall composition, and bedding may be defined by subtle changes in grain size, by surfaces that represent interruptions in deposition and an interval of weathering, or by cracks that formed parallel to bed surfaces.

Why does bedding form? To find the answer, we need to think about how sediment accumulates. Changes in the climate, water depth, current velocity, or the sediment source control the type of sediment deposited at a location at a given time. For example, on a normal day a slow-moving river may carry only silt, which collects on the riverbed (**Fig. 7.12b, c**). During a flood, the river flows faster and carries sand and pebbles, so a layer of sandy gravel forms over the silt layer. Then, when the flooding stops, more silt buries the gravel. If this succession of sediments become lithified and exposed for you to see, they

appear as alternating beds of siltstone and sandy conglomerate. Bedding is not always well preserved. In some environments, burrowing organisms disrupt the layering. Worms, clams, and other creatures churn sediment and may leave behind burrows; this process is called *bioturbation*.

During geologic time, long-term changes in a depositional environment can take place. Thus, a given sequence of strata may differ markedly from sequences of strata above or below. A sequence of strata that is distinctive enough to be traced as a package across a fairly large region is called a **stratigraphic formation** or, simply, a *formation* (**Fig. 7.13a**). For example, a region may contain a succession of alternating sandstone and shale beds deposited by rivers, overlain by beds of marine limestone deposited later when the region was submerged by the sea. A stratigrapher might identify the sequence of sandstone and shale beds as one formation and the sequence of limestone beds as another. Formations are often named after the locality where they were first found and studied. A map that portrays the distribution of stratigraphic formations is called a **geologic map** (**Fig. 7.13b**).

Ripple Marks, Dunes, and Cross Bedding: Consequences of Deposition in a Current

Many clastic sediments accumulate in moving fluids (wind, rivers, or waves). Fascinating sedimentary structures develop at the interface between the sediment and the fluid. These structures are called *bedforms*. Bedforms that develop at a given location reflect factors such as the velocity of the flow and the size of the clasts. Though there are many types of bedforms, we'll focus on only two—ripple marks and dunes. The growth of both produces cross bedding, a special type of lamination within beds.

Ripple marks are relatively small (generally no more than a few centimeters high) elongated ridges that form on a bed surface at right angles to the direction of current flow. If the current always flows in the same direction, the ripple marks tend to be asymmetric, with a steeper slope on the downstream (lee) side (**Fig. 7.14a**). Along the shore, where water flows back and forth due to wave action, ripples tend to be symmetric. The crest (the high ridge) of a symmetric ripple is a sharp ridge, whereas the trough between adjacent ridges is a smooth, concave-up curve (**Fig. 7.14b**). You can find ripples on modern beaches and preserved on bedding planes of ancient rocks (**Fig. 7.14c, d**). A **dune** looks like a ripple, only it's larger. For example, dunes on the bed of a stream may be tens of centimeters to several meters high, and wind-formed dunes formed in deserts may be tens of meters to over 100 meters high.

If you examine a vertical slice cut into a ripple or dune, you will find distinct internal laminations that are inclined at an angle to the boundary of the main sedimentary layer.

FIGURE 7.12 Bedding in sedimentary rocks.

(a) Differential weathering makes stratification stand out. It's clear—even from a distance—that these hills near Las Vegas, Nevada, consist of sedimentary rock.

(b) An example of the formation of bedding during deposition of sediment by a stream.

A layer of silt, deposited during normal river flow

Silt

Basement

A layer of gravel, deposited during flood

Gravel

Later, another layer of silt accumulates.

Silt

Gravel

Time

After burial, the sediment turns to beds of rock.

(c) Beds of sedimentary rock exposed along a road in Utah.

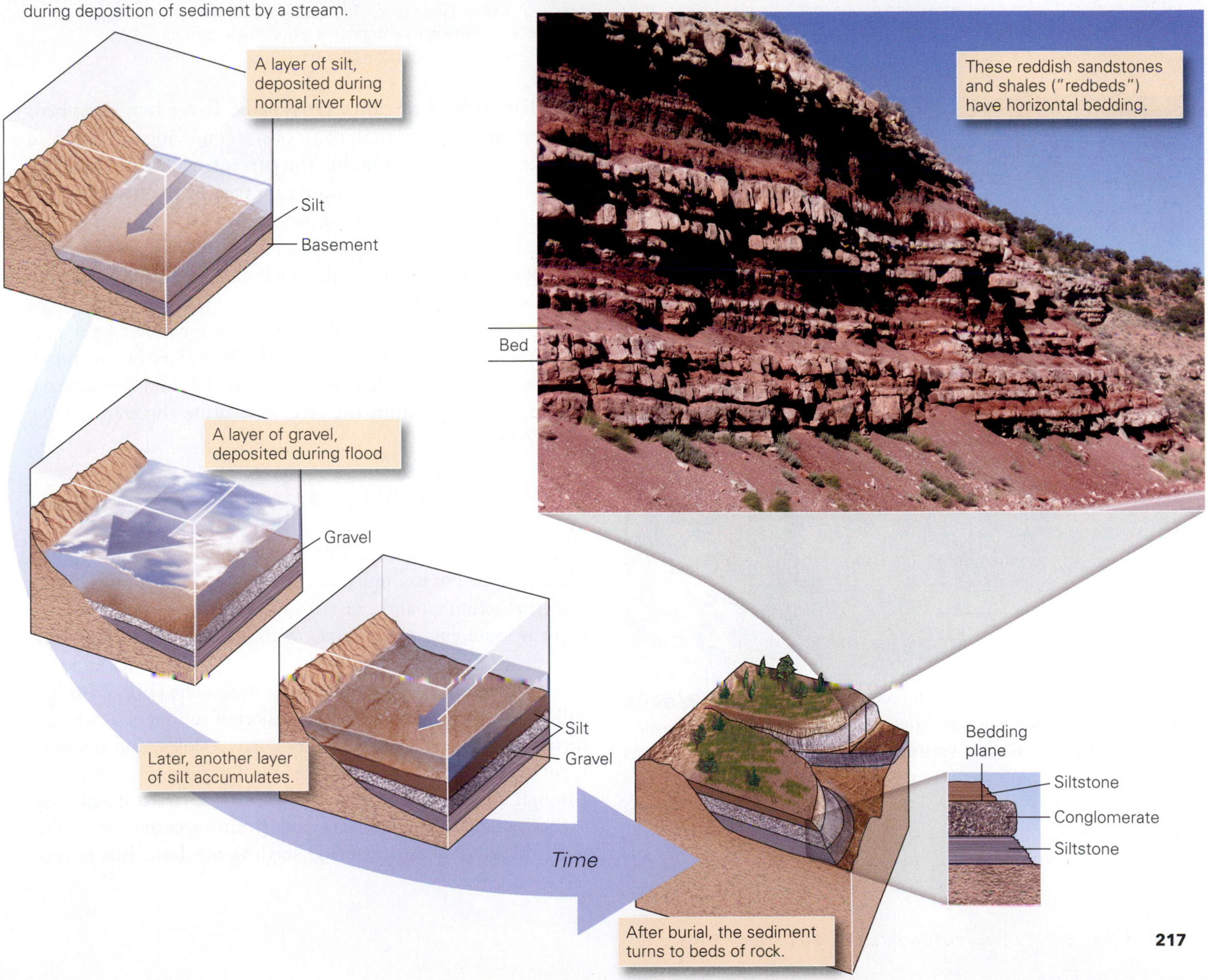

These reddish sandstones and shales ("redbeds") have horizontal bedding.

Bed

Bedding plane

Siltstone

Conglomerate

Siltstone

FIGURE 7.13 The concept of a stratigraphic formation.

The surface between two units is called a contact.

Kaibab Limestone
Toroweap Formation
Coconino Sandstone
Hermit Shale
Supai Group

Toroweap — Kaibab
Coconino
Hermit
Supai
Redwall
Muave
Bright Angel
Tapeats — Plateau on Tapeats
Vishnu

(a) The names of formations consisting of one rock type may indicate the rock type (e.g., Kaibab Limestone). The name of a formation including more than one rock type includes the word *formation* (Toroweap Formation). Several related formations comprise a group (Supai Group).

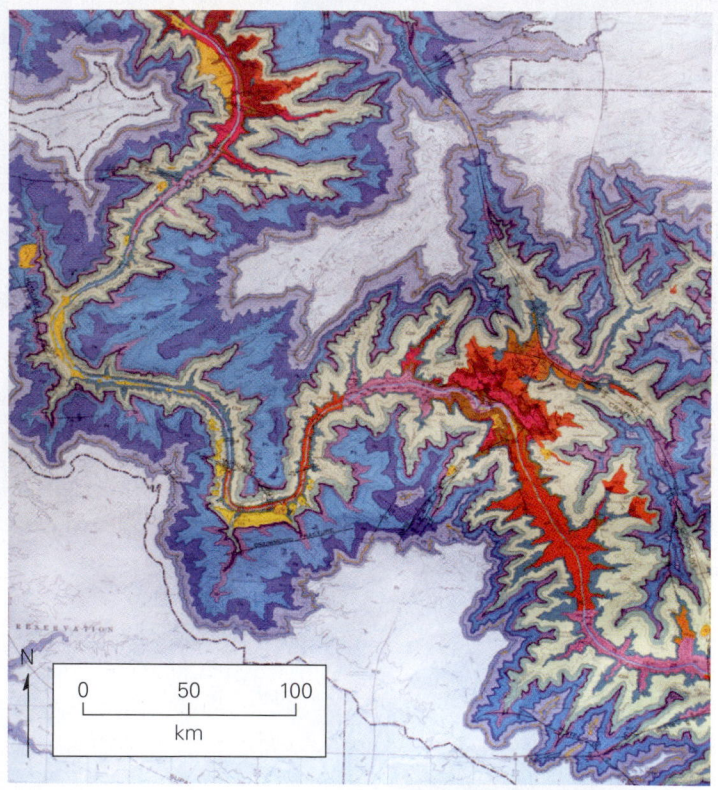

N
0   50   100
km

(b) A geologic map portrays the distribution of formations in a portion of the Grand Canyon. Each color band is a specific formation.

Such laminations are called **cross beds**. To see how cross beds develop, imagine a current of air or water moving uniformly in one direction (**Fig. 7.15a, b**). The current erodes and picks up clasts from the upwind or upstream part of the bedform and deposits them on the leeward part. Sediment builds up on the leeward side until gravity causes it to slip down. With time, the leeward side of the bedform builds in the downstream direction. The surface of the slip face establishes the shape of the cross beds. Eventually, a new cross-bedded layer may build out over a pre-existing one. The boundary between two successive layers is called the *main bedding*, and the internal curving surfaces within the layer constitute the *cross bedding* (**Fig. 7.15c, d**).

Turbidity Currents and Graded Beds

Sediment deposited on a submarine slope might not stay in place forever. For example, an earthquake or storm might disturb this sediment and cause it to slip downslope. If the sediment is loose enough, it mixes with water to create a murky, turbulent cloud. This cloud is denser than clear water and thus flows downslope like an underwater avalanche (**Fig. 7.16**). We call this moving submarine suspension of sediment a *turbidity current*. Downslope, the turbidity current slows, and the sediment that it carried starts to settle out. Larger grains sink faster through a fluid than do finer grains, so the coarsest sediment settles out first. Progressively finer grains accumulate on top, with the finest sediment (clay) settling out last. This process

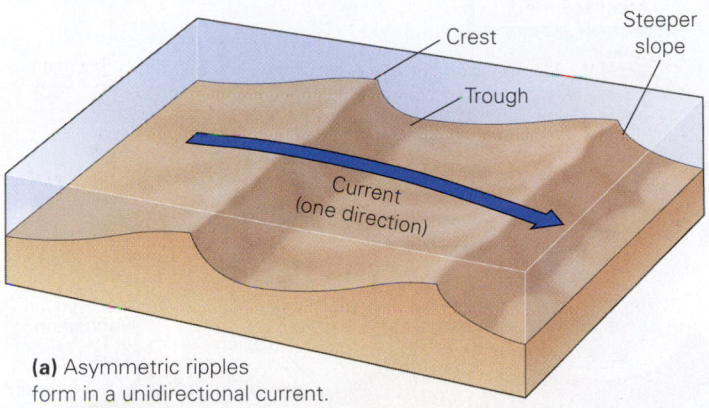

(a) Asymmetric ripples form in a unidirectional current.

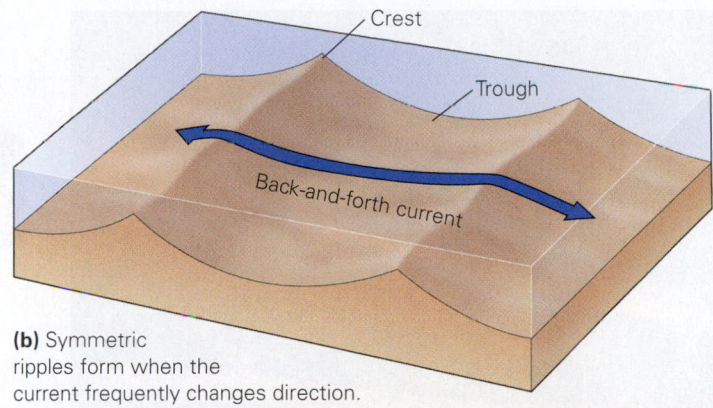

(b) Symmetric ripples form when the current frequently changes direction.

(c) Modern ripples exposed at low tide along a sandy beach on the shore of Cape Cod, Massachusetts.

(d) These 145-Ma ripples are preserved on a tilted bed of solid sandstone at Dinosaur Ridge, Colorado.

forms a *graded bed*—that is, a layer of sediment in which grain size varies from coarse at the bottom to fine at the top. Geologists refer to a deposit from a turbidity current as a *turbidite*.

Bed-Surface Markings

A number of features appear on the surface of a bed as a consequence of events that happen during deposition or soon after. These bed-surface markings include the following.

- *Mud cracks:* If a mud layer dries up after deposition, it cracks into roughly hexagonal plates that typically curl up at their edges. We refer to the openings between the plates as **mud cracks**. If buried, mud cracks can be preserved (**Fig. 7.17**).
- *Scour marks:* As currents flow over a sediment surface, they may erode small troughs, called *scour marks*, parallel to the current flow. These indentations can be buried and preserved.

- *Fossils:* Fossils are relics of past life. Some fossils are shell imprints, footprints, or feeding traces on a bedding surface (see Interlude E).

Why Study Sedimentary Structures?

Sedimentary structures are not just a curiosity; they also provide important clues that help geologists understand the environment in which clastic sedimentary beds were deposited (**Geology at a Glance**, pp. 222–223). For example, the presence of ripple marks and cross bedding indicates that layers were deposited in a current; the presence of mud cracks indicates that the sediment layer was exposed to the air and dried out on occasion; and graded beds indicate deposition by turbidity currents. Also, fossil types can tell us whether sediment was deposited along a river or in the deep sea, for different species of organisms live in different environments. In the next section of this chapter, we examine these environments in greater detail.

FIGURE 7.15 Cross breeding, a type of sedimentary structure within a bed.

(a) Large sand dunes formed in a strong wind.

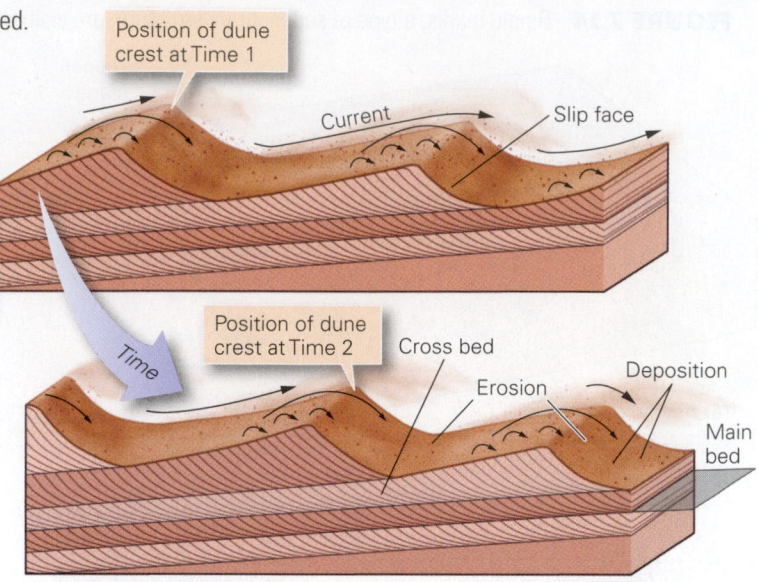

Position of dune crest at Time 1

Current

Slip face

Time

Position of dune crest at Time 2

Cross bed

Erosion

Deposition

Main bed

(b) Cross beds form as sand blows up the windward side of a dune or ripple and then accumulates on the slip face. With time, the dune crest moves.

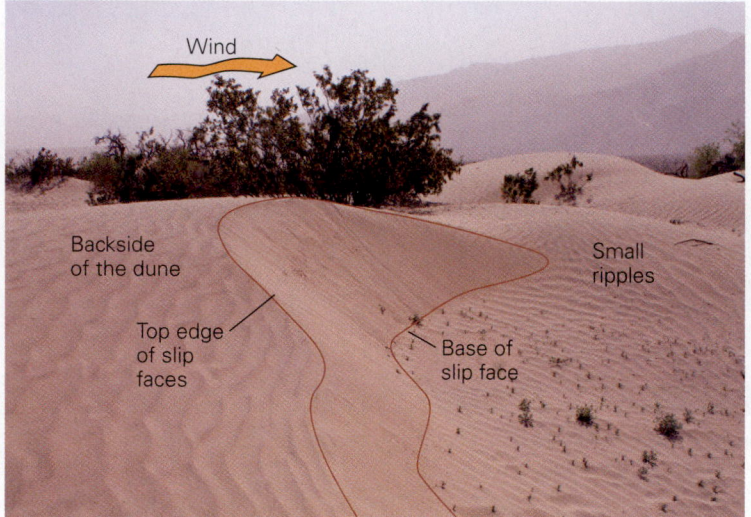

Wind

Backside of the dune

Top edge of slip faces

Base of slip face

Small ripples

(c) Slip face of a small sand dune, Death Valley. The edges of the slip face are highlighted.

Cross-bed orientation indicates wind direction at the time of deposition.

Wind

Cross bedding

Main bedding

(d) A cliff face in Zion National Park, Utah, displays large cross beds formed between 200 and 180 Ma, when the region was a desert with large sand dunes.

Take-Home Message

Exposures of sedimentary rocks typically contain a variety of sedimentary structures, features or textures formed during deposition. Examples include bedding, ripple marks on bed surfaces, and cross-bedding within beds. Features like these provide clues to the environment of deposition. For example, cross beds and dunes form only in a current.

QUICK QUESTION: What is the difference between a bed and a stratigraphic formation?

7.4 How Do We Recognize Depositional Environments?

Imagine that you could travel anywhere on Earth and took the opportunity to visit a variety of natural settings where sediment accumulates. In each setting, if you looked closely, you'd see that the character of sediment being deposited was

FIGURE 7.16 The development of graded bedding from turbidity currents.

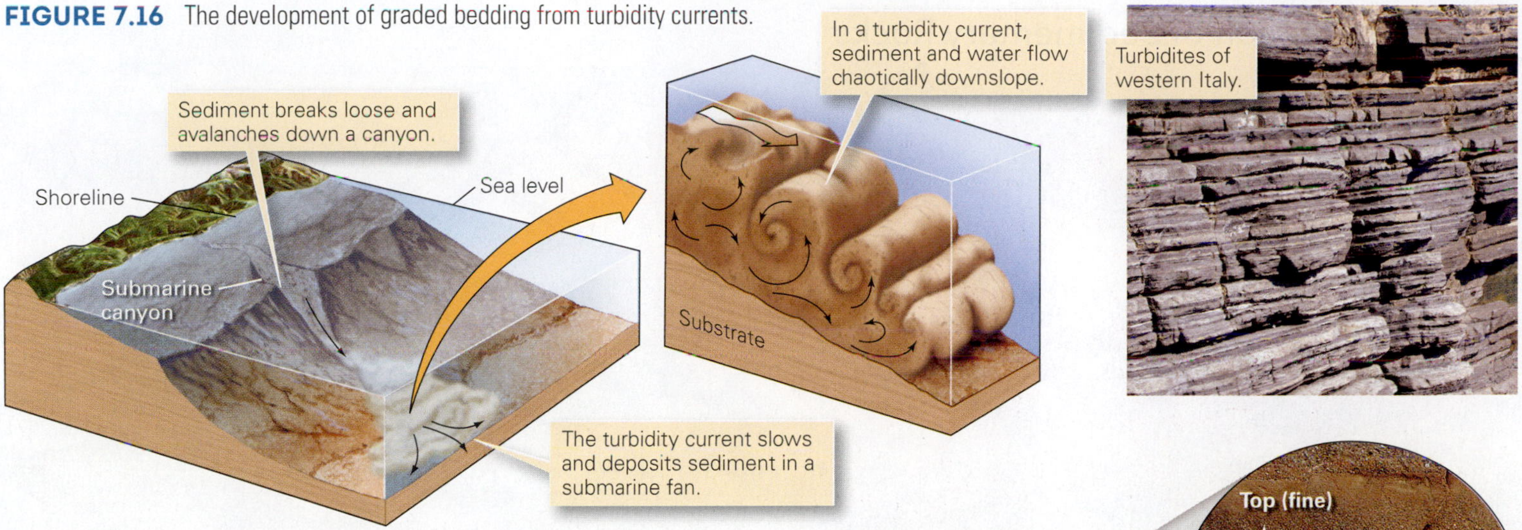

Sediment breaks loose and avalanches down a canyon.

In a turbidity current, sediment and water flow chaotically downslope.

Turbidites of western Italy.

Shoreline

Sea level

Submarine canyon

Substrate

The turbidity current slows and deposits sediment in a submarine fan.

(a) An earthquake or storm triggers an underwater avalanche (turbidity current).

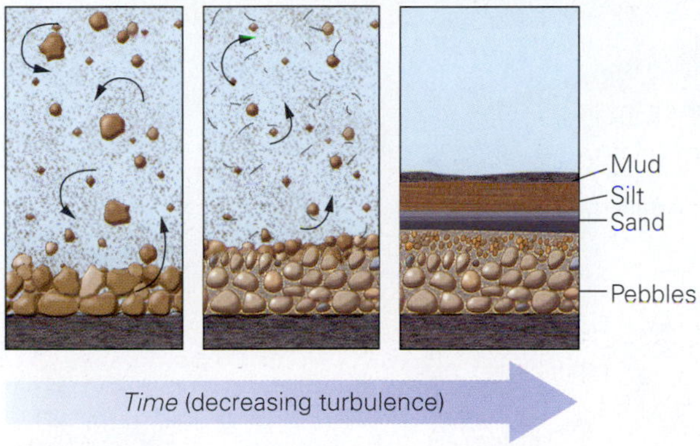

Mud
Silt
Sand

Pebbles

Time (decreasing turbulence)

(b) As the turbidity current slows, larger grains settle first, followed by progressively finer grains.

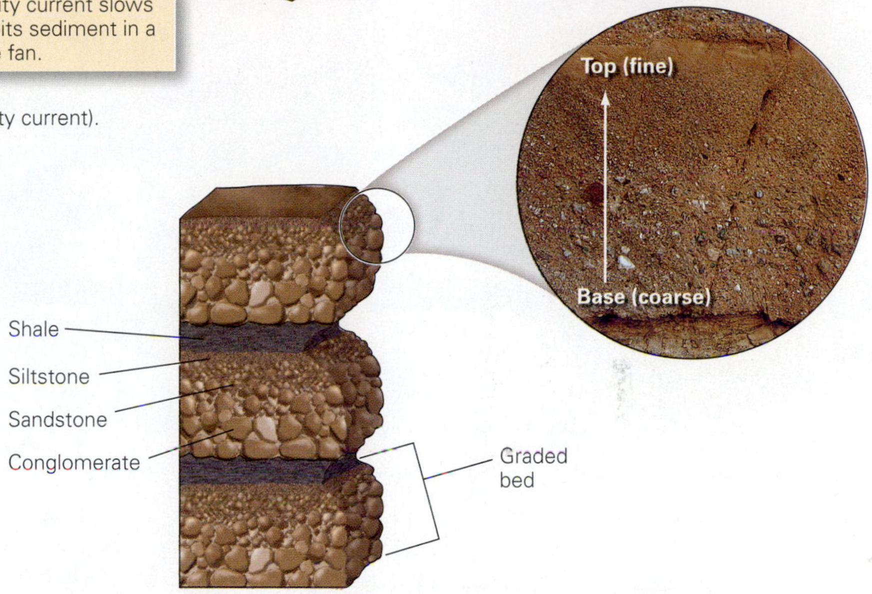

Top (fine)

Base (coarse)

Shale
Siltstone
Sandstone
Conglomerate

Graded bed

(c) As the process repeats, a succession of graded beds accumulates.

FIGURE 7.17 Mud cracks, a sedimentary structure formed by the drying out of mud.

(a) Mud cracks in red mud at Bryce Canyon, Utah. Note how the edges of the mud plates curl up.

(b) Mud cracks visible on the surface of a 410-Ma bed exposed on a cliff in New York.

The Formation of Sedimentary Rocks

Glacial environment

Estuary

Beach

Bar

Continental shelf

Coastal erosion

Turbidity current

Submarine fan

Deep-sea current

Bedding

Redbeds

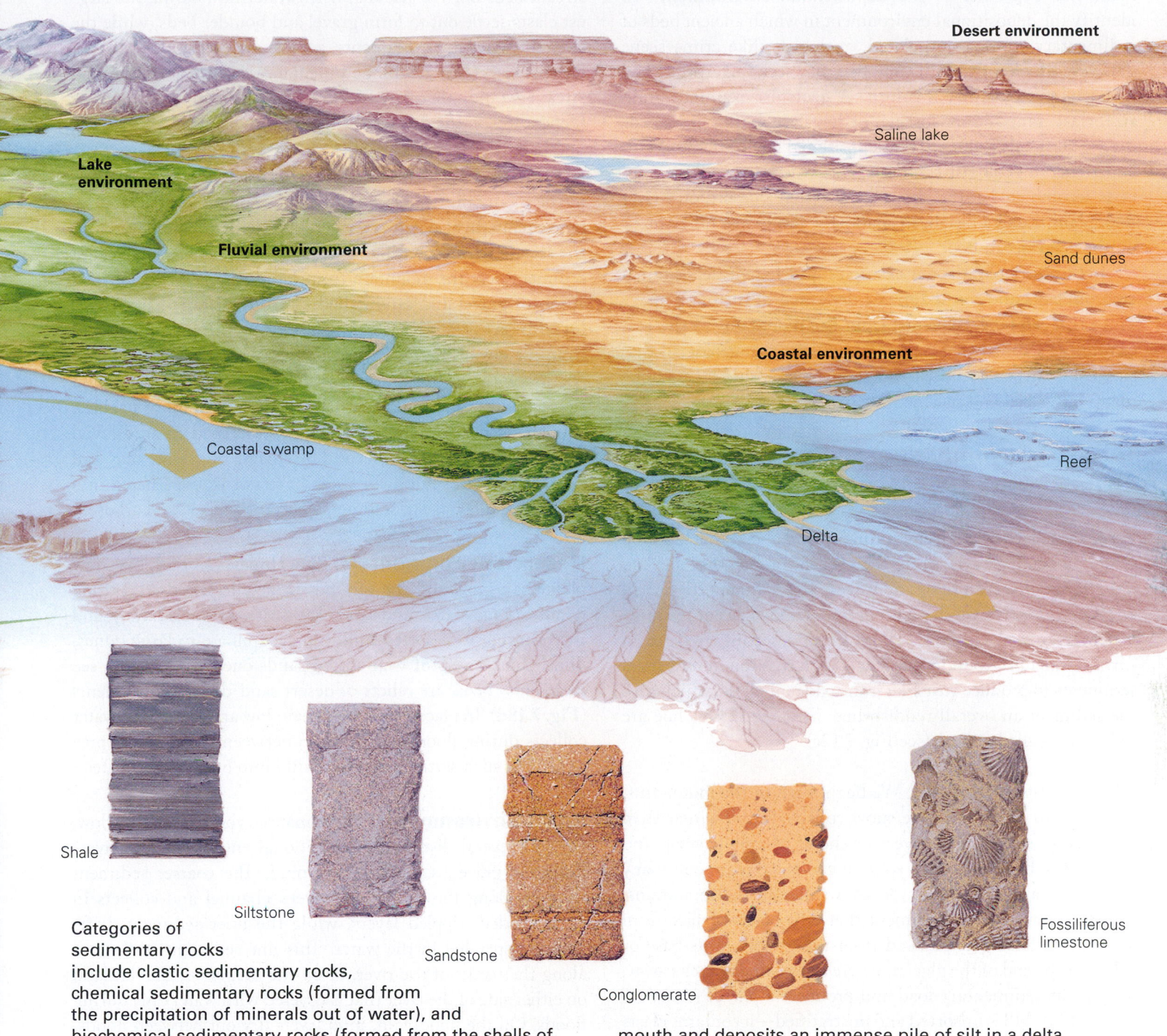

Desert environment

Saline lake

Sand dunes

Lake environment

Fluvial environment

Coastal environment

Coastal swamp

Reef

Delta

Shale

Siltstone

Sandstone

Conglomerate

Fossiliferous limestone

Categories of sedimentary rocks include clastic sedimentary rocks, chemical sedimentary rocks (formed from the precipitation of minerals out of water), and biochemical sedimentary rocks (formed from the shells of organisms). Clastic sedimentary rocks develop when grains (clasts) break off pre-existing rock by weathering and erosion and are transported to a new location by wind, water, or ice; the grains are deposited to create sediment layers, which are then cemented together. We distinguish among types of clastic sedimentary rocks on the basis of grain size.

The character of a sedimentary rock depends on the composition of the sediment and on the environment in which it accumulated. For example, glaciers carry sediment of all sizes, so they leave deposits of poorly sorted (different-sized) till; streams deposit coarser grains in their channels and finer ones on floodplains; a river slows down at its

mouth and deposits an immense pile of silt in a delta. Fossiliferous limestone develops on coral reefs. In desert environments, sand accumulates into dunes and evaporates precipitate in saline lakes. Offshore, submarine canyons channel avalanches of sediment, or turbidity currents, out to the deep-sea floor.

Sedimentary rocks tell the history of the Earth. For example, the layering, or bedding, of sedimentary rocks is initially horizontal. So where we see layers bent or folded, we can conclude that the layers were deformed during mountain building. Where horizontal layers overlie folded layers, we have an unconformity: for a time, sediment was not deposited, and/or older rocks were eroded away.

distinct. For example, the sediment carried by desert winds doesn't look like the sediment settling on the floor of a tropical lagoon. Geologists refer to the conditions in which sediment was deposited as the **depositional environment**. To identify the depositional environment in which ancient beds of sedimentary strata accumulated, geologists, like crime-scene detectives, look for clues. Detectives may seek fingerprints and bloodstains to identify a culprit. A geologist examines grain size, clast composition, sorting, bed-surface marks, cross bedding, and fossils to identify a depositional environment. With experience, a geologist can determine whether a succession of beds accumulated at the end of a glacier, in a desert, on a river floodplain, along a beach, in shallow seawater just offshore, or on the deep-ocean floor.

Let's now look at some examples of different depositional environments and the sediments deposited in them, by imagining that we take a journey from the mountains to the sea, examining sediments as we go. We will see that geologists distinguish between two basic categories of depositional environment: terrestrial and marine.

Terrestrial (Nonmarine) Sedimentary Environments

Terrestrial depositional environments are those that develop inland, far enough away from the ocean shoreline that they are not affected by ocean tides and waves. The sediments of terrestrial deposits settle on dry land or under and adjacent to freshwater streams, glaciers, and lakes. In some settings, oxygen in surface water or groundwater reacts with the iron in terrestrial sediments to produce rust-like iron-oxide minerals, which give the sediment an overall reddish hue. Strata with this hue are informally called *redbeds* (see Fig. 7.12c).

Glacial Environments We begin high in the mountains, where it's so cold that more snow collects in the winter than melts away, so glaciers—rivers or sheets of ice—develop and slowly flow. Because ice is a solid, it can move sediment of any size. So as a glacier moves down a valley in the mountains, it carries along *all* the sediment that falls on its surface from adjacent cliffs or gets plucked from the ground at its base or sides. At the end of the glacier, where the ice finally melts away, it drops its sedimentary load and produces a pile of *glacial till* (**Fig. 7.18a**). Till is unsorted and unstratified—it contains clasts ranging from clay size to boulder size all mixed together—and thus becomes a type of diamicton. We provide further details on glacial sediments in Chapter 22.

Mountain Stream Environments As we walk down beyond the end of the glacier, we enter a realm where tur-

bulent streams rush downslope in steep-sided valleys. This fast-moving water has the power to carry large clasts—in fact, during floods, boulders and cobbles can tumble down the stream bed. Between floods, when water flow slows, the largest clasts settle out to form gravel and boulder beds, while the stream carries finer sediments, such as sand and mud, farther downstream (**Fig. 7.18b**). Sedimentary deposits of a mountain stream would, therefore, include breccia and conglomerate (depending on the degree of rounding).

Alluvial-Fan Environments Our journey now takes us to the mountain front, where the fast-moving stream empties onto a plain. In arid regions, where there is not enough water for the stream to flow continuously, the stream deposits its load of sediment near the mountain front, producing a wedge-shaped apron of gravel and sand called an **alluvial fan** (**Fig. 7.18c**). Deposition takes place here because when the stream emerges from a canyon mouth, it spreads out over a broader area, so friction with the ground causes the water to slow down, and slow-moving water does not have the power to move pebbles, cobbles, or coarse sediment. Notably, the sand in an alluvial fan whose sediment was derived by erosion of granite may still contain feldspar grains, for these have not yet weathered into clay. Thus, alluvial-fan sediments become breccia and arkose.

Desert Environments If the climate is very dry, few plants can grow and the ground surface lies exposed. Strong winds can move dust and sand. The dust gets carried away, and the resulting well-sorted sand can accumulate in large dunes. Thus, thick layers of well-sorted sandstone, in which we see large cross beds, are relics of desert sand-dune environments (**Fig. 7.18d**). In places, deserts contain low areas in which water collects during floods but dries up between floods. Salts precipitate in such settings and can build into beds of evaporites.

River Environments In climates where streams flow, we find several distinctive depositional environments. Rivers transport gravel, sand, silt, and mud. The coarser sediment tumbles along the bed in the river's channel and collects in cross-bedded, rippled layers, while the finer sediment drifts along, suspended in the water. This fine sediment settles out along the banks of the river, or on the *floodplain*, the flat land on either side of the river that is covered with water only during floods. On the floodplain, mud layers dry out between floods, leading to the formation of mud cracks. River sediments lithify to form sandstone, siltstone, and shale. Typically, channels of coarser sediment are surrounded by layers of fine-grained floodplain deposits; in cross section, the channel has a lens-like shape (**Fig. 7.18e**). Geologists commonly refer to river deposits as fluvial sediments, from the Latin word *fluvius*, for river.

FIGURE 7.18 Examples of nonmarine depositional environments.

(a) Glacial till at the end of a glacier in France.

(b) Boulders and cobbles deposited by a mountain stream in Colorado.

(c) An alluvial fan in Death Valley, California.

(d) Sand dunes in Brazil.

(e) Deposits of an ancient river channel in Indiana. Note how the floor of the channel cuts across older strata. The geologist's sketch emphasizes the relationship.

Edge of photo | What a Geologist Sees

Younger floodplain deposits

Channel fill

Older floodplain deposits

(Talus)

(f) Laminated mud from a lake bed.

Lake Environments In temperate climates, where water remains at the surface throughout the year, lakes form. In lakes, the relatively quiet water can't move coarse sediment; any coarse sediment brought into the lake by a stream settles out at the stream's outlet. Only fine clay makes it out into the center of the lake, where it settles to form mud on the lake bed. Thus, lake sediments typically consist of finely laminated shale (**Fig. 7.18f**).

At the mouths of streams that empty into lakes, small deltas may form. A **delta** is a wedge of sediment that accumulates where moving water enters standing water. Deltas were so named because the map shape of some deltas resembles the Greek letter *delta* (Δ), as we discuss further in Chapter 17. In 1885, an American geologist named G. K. Gilbert showed that small deltas contain three components (**Fig. 7.19**): topset beds of gravel, foreset beds of gravel and sand, and bottomset beds of silt.

Coastal and Marine Sedimentary Environments

Along the seashore, a variety of distinct coastal environments occur; the character of each reflects the nature of the sediment supply and the climate. Marine environments start at the high-tide line and extend offshore, to include the deep-ocean floor. The type of sediment deposited at a location depends on the water temperature, water clarity, water depth, and the supply of clasts. That's because temperature and clarity determine the species of organisms that can live in the water, and the availability of clasts and the degree to which wave motion moves water affects the size of the clasts that accumulate.

Marine Delta Deposits After following the river downstream for a long distance, we reach its mouth, where it empties into the sea. Here the river water stops flowing, so water velocity slows and sediment settles out to build a delta out into the sea. Large marine deltas are much more complex than the small lake examples that Gilbert studied, for marine deltas host many different sedimentary environments, including swamps, channels, floodplains, and submarine slopes. Sea-level changes may cause the position of the different environments to move inshore or offshore with time. Thus, deposits of an ocean-margin delta produce a great variety of sedimentary rock types (**Fig. 7.20a**).

Coastal Beach Sands Now we leave the delta and wander along the coast (**Fig. 7.20b**). Oceanic currents transport sand along the coastline. The sand washes back and forth in the surf, so it becomes well sorted (waves winnow out mud and silt) and well rounded, and because of the back-and-forth

movement of ocean water, the sand surface may become rippled. Thus, if you find well-sorted, medium-grained sandstone, perhaps with ripple marks, you may be looking at the remnants of a beach environment.

Shallow-Marine Clastic Deposits From the beach, we proceed offshore. In deeper water, where wave energy does not stir the sea floor, finer sediment can accumulate. Because the water here may be only meters to a few tens of meters deep, geologists refer to this depositional setting as a *shallow-marine environment*. Clastic sediments that accumulate in this environment tend to be fine-grained, well-sorted, and well-rounded silt. A great variety of organisms such as mollusks, gastropods, and worms live in or on this sediment. Thus, if you see beds of siltstone and mudstone containing marine fossils, you may be looking at shallow-marine clastic deposits.

Shallow-Water Carbonate Environments In shallow-marine settings where relatively little sand and mud enters the water, warm, clear, nutrient-rich water can host an abundance of organisms with carbonate shells, which becomes carbonate sediment (**Fig. 7.21**). Beaches collect sand composed of shell fragments; lagoons are sites where carbonate mud accumulates; and reefs consist of coral and coral debris. Farther offshore of a reef, we can find a sloping apron of reef fragments. Shallow-water carbonate environments transform into various kinds of limestone.

Deep-Marine Deposits We conclude our journey by sailing far offshore. Along the transition between coastal regions and the deep ocean, turbidity currents deposit graded beds. In the deep-ocean realm, only fine clay and plankton provide a source for sediment. The clay eventually settles out onto the deep-sea floor, forming deposits of finely laminated mudstones, and plankton shells settle to form chalk (from calcite shells; **Fig. 7.22**) or chert (from siliceous shells). Thus, deposits of mudstone, chalk, or bedded chert indicate a deep-marine origin.

FIGURE 7.19 A simple "Gilbert-type" delta formed where a stream enters a lake.

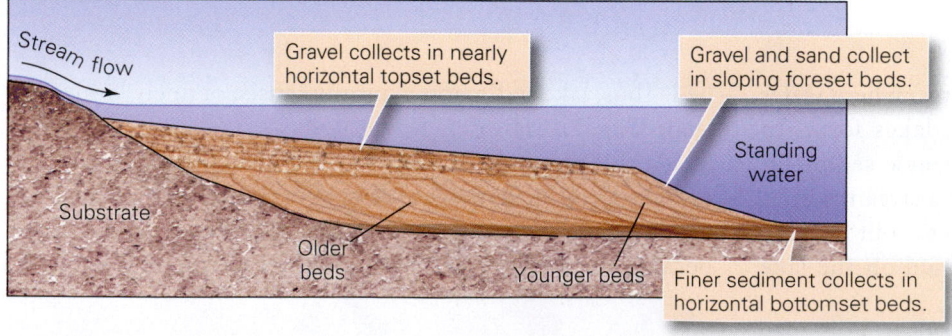

Stream flow

Gravel collects in nearly horizontal topset beds.

Gravel and sand collect in sloping foreset beds.

Standing water

Substrate

Older beds

Younger beds

Finer sediment collects in horizontal bottomset beds.

FIGURE 7.20 Examples of coastal depositional environments.

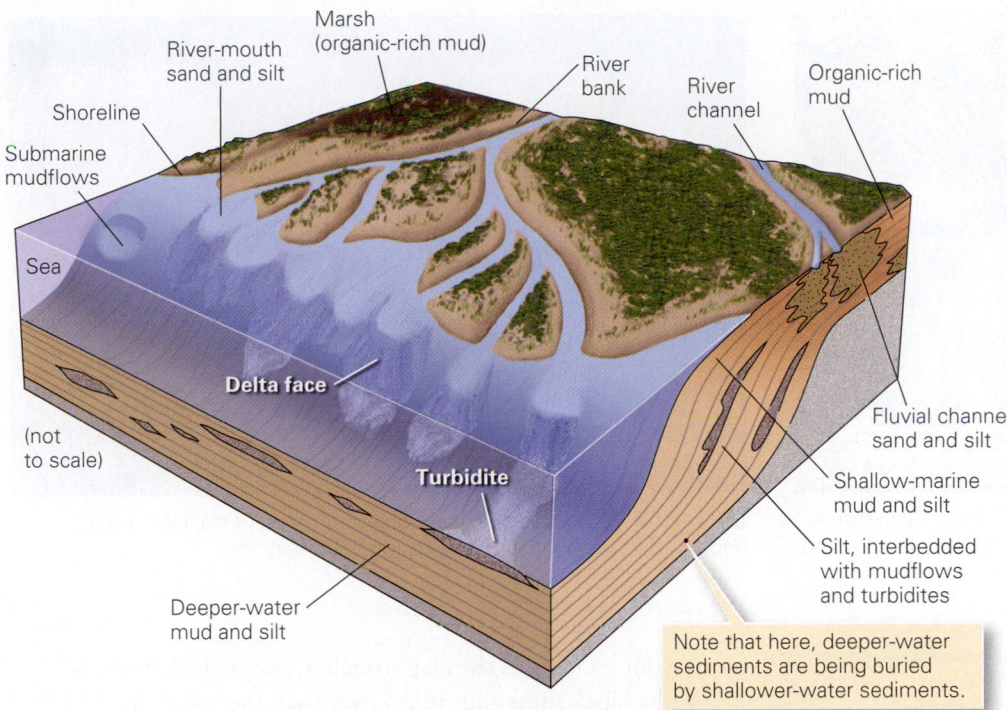

Shoreline
Submarine mudflows
Sea
River-mouth sand and silt
Marsh (organic-rich mud)
River bank
River channel
Organic-rich mud
Delta face
(not to scale)
Turbidite
Deeper-water mud and silt
Fluvial channel sand and silt
Shallow-marine mud and silt
Silt, interbedded with mudflows and turbidites

Note that here, deeper-water sediments are being buried by shallower-water sediments.

(a) A major river delta along an ocean coast is a complex depositional environment. Sea-level changes affect locations of depositional settings.

(b) Waves on this California beach wash and sort the sand.

FIGURE 7.21 Reef environments for the deposition of carbonate rocks.

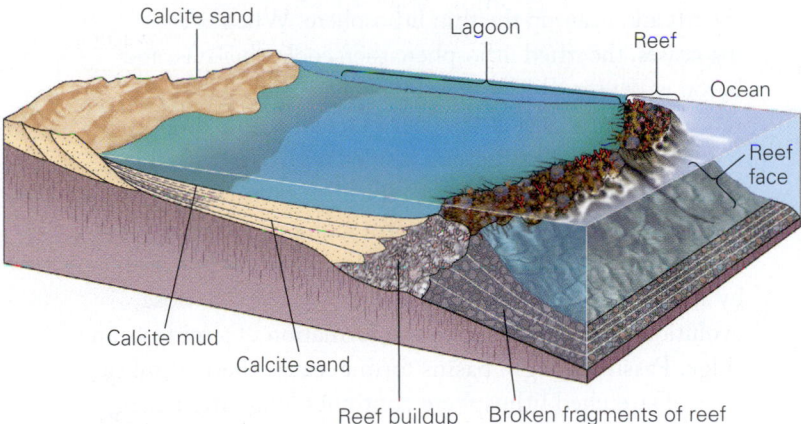

Calcite sand
Lagoon
Reef
Ocean
Reef face
Calcite mud
Calcite sand
Reef buildup
Broken fragments of reef

(a) Carbonate reefs form along shorelines in warm-water environments. In detail, reefs include many distinct depositional environments.

(b) A dramatic reef surrounds an island in the tropical Pacific. Deeper water is darker. Note the surf along the edge of the reef.

Take-Home Message

Different types of sedimentary rocks accumulate in different depositional environments. Thus, strata deposited along a river differ from strata deposited by ocean waves, by glaciers, or in the deep sea. By studying sedimentary rocks at a location, geologists can distinguish between various marine and terrestrial depositional settings, and they can deduce environments that existed at the locality in the past.

QUICK QUESTION: How can you distinguish sediment deposited in an alluvial fan from sediment deposited in a shallow-marine environment?

FIGURE 7.22 Examples of deep-marine sediment.

(a) These plankton shells, which make up some kinds of deep-marine sediment, are so small that they could pass through the eye of a needle.

(b) The chalk cliffs of southeastern England consist of plankton shells deposited on the sea floor tens of millions of years ago.

7.5 Sedimentary Basins

The sedimentary veneer on the Earth's surface varies greatly in thickness. If you stand in central Siberia or Canada, you will find yourself on igneous and metamorphic basement rocks that are over a billion years old—sedimentary rocks are nowhere in sight. Yet if you stand along the southern coast of Texas, you would have to drill through over 15 km of sedimentary beds before reaching igneous or metamorphic basement. Thick accumulations of sediment form only in special regions where the surface of the Earth's lithosphere sinks, providing space in which sediment can collect. Geologists use the term **subsidence** to refer to the process by which this sinking takes place and the term **sedimentary basin** for the resulting sediment-filled depression. In what geologic settings do sedimentary basins form? Plate tectonics theory provides a key.

Categories of Basins in the Context of Plate Tectonics Theory

Geologists distinguish among different kinds of sedimentary basins on the basis of the region of a lithospheric plate in which the basin formed, as defined in the context of plate tectonics theory (**Fig. 7.23**). Let's consider a few examples.

- *Rift basins:* These form in continental rifts, regions where the lithosphere has been stretched. During the early stages of rifting, the surface of the Earth subsides simply because crust becomes thinner as it stretches. (To picture this process, imagine pulling on either end of a block of clay with your hands—as the clay stretches, the central region of the block thins and sinks lower than the ends.) As the rift grows, slip on faults drops blocks of crust down, producing low areas bordered by narrow mountain ridges. Alluvial-fan deposits form along the base of the mountains, and salt flats or lakes develop in the low areas between the mountains.

  Thinning is not the only reason that rifted lithosphere subsides. During rifting, warm asthenosphere rises beneath the rift and heats up the thin lithosphere. When rifting ceases, the rifted lithosphere then cools, thickens, and becomes denser. This heavier lithosphere sinks down, causing more subsidence, just as the deck of a tanker ship drops to a lower elevation when the ship is filled with ballast. Sinking due to cooling of the lithosphere is called thermal subsidence.

- *Passive-margin basins:* These form along the edges of continents that are not plate boundaries. They are underlain by stretched lithosphere, the remnants of a rift whose evolution successfully led to the formation of a mid-ocean ridge. Passive-margin basins form because thermal subsidence of stretched lithosphere continues long after rifting ceases and sea-floor spreading begins. They fill with sediment carried to the sea by rivers and with carbonate rocks formed in coastal reefs. Sediment in a passive-margin basin can reach an astounding thickness of 15 to 20 km.

- *Intracontinental basins:* These develop in the interiors of continents, initially because of subsidence over a rift. They may continue to subside in pulses even hundreds of millions of years after they first formed, for reasons that are not yet well understood. Illinois and Michigan are each underlain by an intracontinental basin—the Illinois basin and the Michigan basin, respectively—in which up to 7 km of sediment has accumulated. Most of this sediment

FIGURE 7.23 The geologic setting of sedimentary basins.

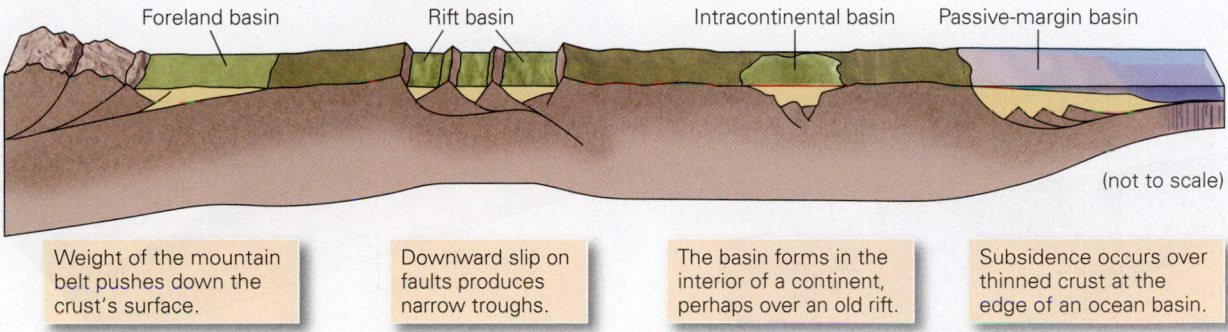

Foreland basin   Rift basin   Intracontinental basin   Passive-margin basin

(not to scale)

Weight of the mountain belt pushes down the crust's surface.

Downward slip on faults produces narrow troughs.

The basin forms in the interior of a continent, perhaps over an old rift.

Subsidence occurs over thinned crust at the edge of an ocean basin.

is fluvial, deltaic, or shallow marine. At times, extensive swamps formed along the shoreline in these basins. The plant matter of these swamps was buried to form coal.

- *Foreland basins:* These form on the continent side of a mountain belt because the forces produced during convergence or collision push large slices of rock up faults and onto the surface of the continent. The weight of these slices pushes down on the surface of the lithosphere, producing a wedge-shaped depression adjacent to the mountain range that fills with sediment eroded from the range. Fluvial and deltaic strata accumulate in foreland basins.

Transgression and Regression

Sea-level changes, relative to the land surface, can control the succession of sediments that we see in a sedimentary basin (**Fig. 7.24**). At times during Earth history, sea level has risen by as much as a couple of hundred meters, creating shallow seas that submerge the interiors of continents, and at other times sea level has fallen by a couple of hundred meters, exposing the continental shelves to air. Global sea-level changes may be due to a number of factors, including climate change, that control the amount of ice stored in polar ice caps and cause changes in the volume of ocean basins. Sea level at a location may also be due to the local uplift or sinking of the land surface.

When relative sea level rises, the shoreline migrates inland. During this process, which is called **transgression**, terrestrial sediments are progressively buried by coastal sediments, and then coastal sediments are buried by deeper-water sediment. Note that the extensive layer of beach sand laid down during a transgression may look like a blanket of sand that was deposited all at once, but in fact the sand deposited at one location differs in age from the sand deposited at another location. When sea level falls, the coast migrates seaward. We call this process **regression**. Typically, the record of a regression will not be well preserved, because as sea level drops areas that had been sites of deposition become exposed to erosion. A succession of strata deposited during a cycle of transgression and regression is called a *depositional sequence*.

Diagenesis

Earlier in this chapter we discussed the process of lithification, by which sediment hardens into rock. Lithification is an aspect of a broader phenomenon called diagenesis. Geologists use the term **diagenesis** for all the physical, chemical, and biological processes that transform sediment into sedimentary rock and that alter characteristics of sedimentary rock after the rock has formed.

In sedimentary basins, sedimentary rocks may become buried very deeply. As a result, the rocks endure higher pressures and temperatures and come in contact with warm groundwater. Diagenesis, under such conditions, can cause chemical reactions in the rock that produce new minerals and can also cause cement to dissolve or precipitate. As temperature and pressure increase still deeper in the subsurface, the changes that take place in rocks become more profound. At sufficiently high temperature and pressure, metamorphism begins: a new assemblage of minerals forms, and/or mineral grains become aligned parallel to each other. The transition between diagenesis and metamorphism in sedimentary rocks is gradational and occurs between temperatures of 150°C and 300°C. In the next chapter, we enter the realm of true metamorphism.

Take-Home Message

In certain geologic settings, Earth's surface sinks (subsides) to form a depression that fills with sediment. The depression with its thick fill of sediment is a sedimentary basin. As sea level rises and falls, the coast, and therefore depositional environments, can migrate inland or offshore, respectively. Once sediment has been deposited, it undergoes various changes, known as diagenesis, in response to pressure and to interaction with groundwater.

QUICK QUESTION: What is the thickest that the fill of a sedimentary basin can become?

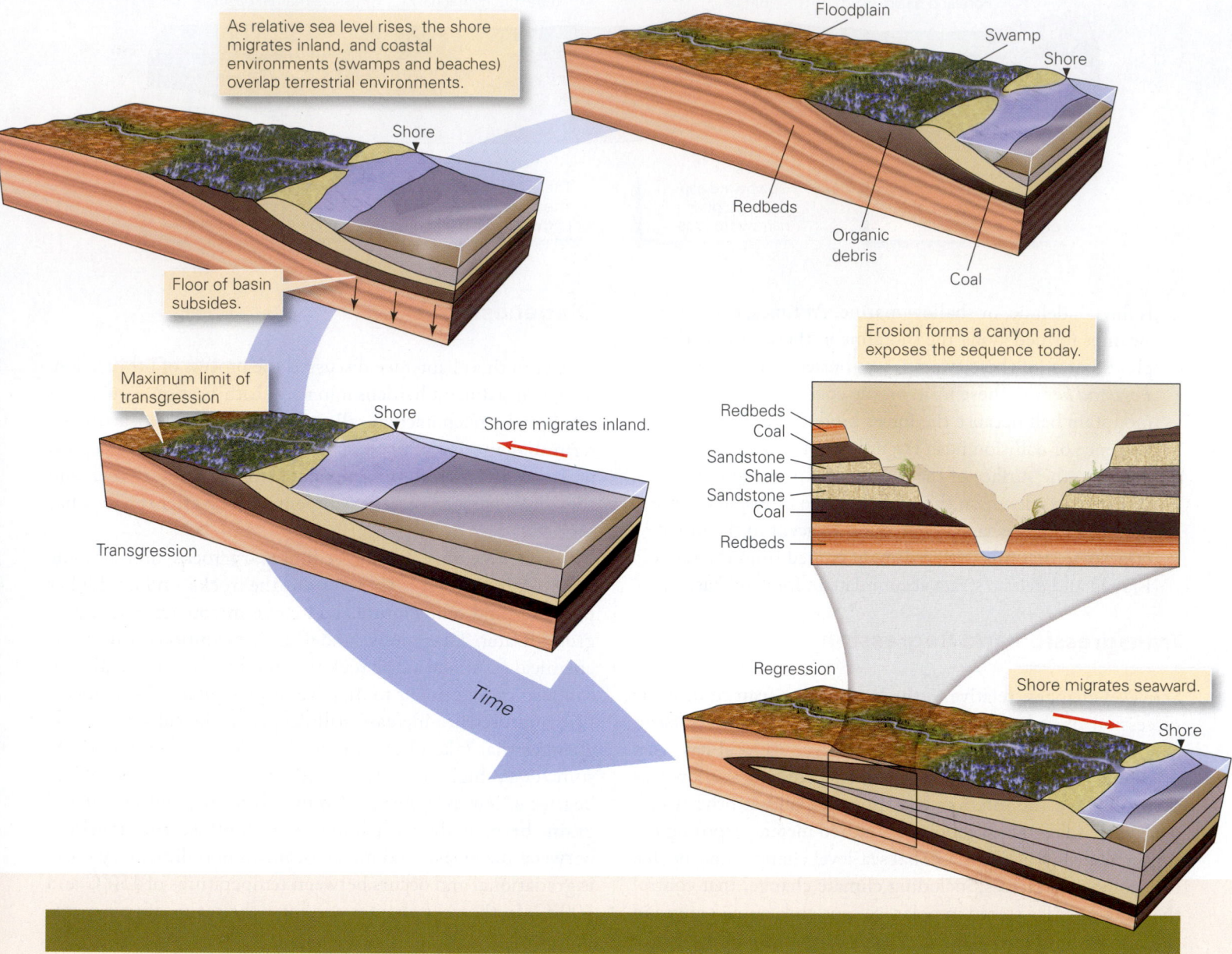

FIGURE 7.24 The concept of transgression and regression during deposition of sedimentary sequence.

As relative sea level rises, the shore migrates inland, and coastal environments (swamps and beaches) overlap terrestrial environments.

Shore

Floodplain

Swamp

Shore

Redbeds

Organic debris

Coal

Floor of basin subsides.

Maximum limit of transgression

Shore

Shore migrates inland.

Erosion forms a canyon and exposes the sequence today.

Redbeds
Coal
Sandstone
Shale
Sandstone
Coal
Redbeds

Transgression

Time

Regression

Shore migrates seaward.

Shore

CHAPTER SUMMARY

- Geologists recognize four major classes of sedimentary rocks. Clastic rocks form from cemented-together grains that were first produced by weathering and were then transported, deposited, and lithified. Biochemical rocks develop from the shells of organisms. Organic rocks consist of plant debris or of altered plankton remains. Chemical rocks precipitate directly from water.

- Formation of clastic rocks begins when grains erode from pre-existing rock. Moving water, air, or ice transport these grains to a site of deposition, where they accumulate. Lithification, involving compaction and cementation, converts loose sediment into rock.

- Clastic rocks are classified based primarily on grain size. Characteristics (roundness, sorting, composition) help to distinguish sediment source and depositional setting.

- Sedimentary structures include bedding, cross bedding, graded bedding, ripple marks, dunes, and mud cracks. They serve as clues to depositional settings.

- Biochemical and organic rocks form from materials produced by living organisms.

- Limestone consists predominantly of calcite; chert forms from silica, coal from carbon, shale from clay, and sandstone from quartz grains.

- Evaporites consist of minerals precipitated from saline water.

- Glaciers, streams, alluvial fans, deserts, rivers, lakes, deltas, beaches, shallow seas, and deep seas each accumulate a different, distinctive assemblage of sedimentary strata.
- Thick piles of sedimentary rocks accumulate in sedimentary basins, regions where the lithosphere sinks.

- Transgressions occur when sea level rises and the coastline migrates inland. Regressions occur when sea level falls and the coastline migrates seaward.
- Diagenesis involves processes leading to lithification and processes that alter sedimentary rock once it has formed.

GUIDE TERMS

alluvial fan (p. 224)
bed (p. 216)
bedding (p. 216)
biochemical sedimentary rock (p. 203)
biochemical limestone (p. 210)
breccia (p. 209)
cement (p. 206)
cementation (p. 206)
chemical sedimentary rock (p. 203)

clastic sedimentary rock (p. 203)
clasts (p. 203)
coal (p. 211)
compaction (p. 206)
conglomerate (p. 209)
cross bed (p. 216)
delta (p. 225)
deposition (p. 205)
depositional environment (p. 224)
diagenesis (p. 229)

dune (p. 216)
erosion (p. 204)
evaporite (p. 212)
geologic map (p. 216)
lithification (p. 206)
mud crack (p. 219)
mudstone (p. 210)
oil shale (p. 212)
organic sedimentary rock (p. 203)
regression (p. 229)
ripple mark (p. 216)

sandstone (p. 204)
sedimentary basin (p. 228)
sedimentary rock (p. 203)
sedimentary structure (p. 215)
shale (p. 210)
siltstone (p. 210)
strata (p. 216)
stratigraphic formation (p. 216)
subsidence (p. 228)
transgression (p. 229)

REVIEW QUESTIONS

1. Describe how a clastic sedimentary rock forms from its unweathered parent rock.
2. Explain how biochemical sedimentary rocks form.
3. How do grain size and shape, sorting, sphericity, and angularity change as sediments move downstream?
4. Describe the two different kinds of chert. How are they similar? How are they different?
5. What kinds of conditions produce evaporites?
6. How does dolostone differ from limestone, and how does dolostone form?

7. What are cross beds, and how do they form? How can you read the current direction from cross beds?
8. Describe how a turbidity current forms and moves. How does it produce graded bedding?
9. Compare deposits of an alluvial fan with those of a deep-marine deposit.
10. What kinds of sediments accumulate in river and delta systems?
11. Why don't sediments accumulate everywhere? What types of tectonic conditions are required to create basins?
12. What kinds of changes may take place during diagenesis?

ON FURTHER THOUGHT

13. Recent exploration of Mars by robotic vehicles suggests that layers of sedimentary rock cover portions of the planet's surface. On the basis of examining images of these layers, some researchers claim that the layers contain cross bedding and relicts of gypsum crystals. At face value, what do these features suggest about depositional environments on Mars in the past?

14. The Gulf Coast of the United States is a passive-margin basin that contains a very thick accumulation of sediment. Drilling reveals that the base of the sedimentary succession in this basin consists of redbeds. These are overlain by a thick layer of evaporite. The evaporite, in turn, is overlain by deposits composed predominantly of sandstone and shale. In some intervals, the sandstone occurs in channels

and contains ripple marks, and the shale contains mud cracks. In other intervals, the sandstone and shale contain fossils of marine organisms. The sequence contains hardly any conglomerate or arkose. Be a sedimentary detective, and explain the succession of sediment in the basin.

15. Examine the Bahamas (at Lat 23°58′40.98″N Long 77°30′20.37″W) with *Google Earth™*. Note that broad expanses of very shallow water surround the islands, that white-sand beaches occur along the coast of the islands, and that small reefs occur offshore. What does the sand consist of, and what rock will it become if it eventually becomes buried and lithified? Compare the area of shallow water in the Bahamas area with the area of Florida. The bedrock of Florida consists mostly of shallow-marine limestone. What does this observation suggest about the nature of the Florida peninsula in the past? Keep in mind that sea level on Earth changes over time and that most of the land surface of Florida lies at less than 50 m above sea level.

smartwork.wwnorton.com

This chapter's Smartwork features:

- An art-based ranking activity on sediment grain sorting.
- Interactive exercises on the formation of sedimentary rock.
- Visual simulations of metamorphic processes.

GEOTOURS

This chapter's GeoTour exercise (F) features:

- Sedimentary rocks exposed in the Grand Canyon
- Tilted and folded sedimentary rocks
- Arid depositional environments
- Fluvial depositional environments

Another View These cliffs, near Bryce Canyon (Utah), expose beds of sedimentary rock deposited in lakes, and by streams, over 40 Ma. Present-day erosion has produced aprons of debris at the base of the cliffs.

In the stark mountains of northwestern Scotland, outcrops of Lewisian Gneiss poke up from the scruffy grass. The rock formed when pre-existing rocks were subjected to high pressures and temperatures, as well as compression and shear, during Precambrian mountain building.

CHAPTER 8

Metamorphism:
A Process of Change

> Nothing in the world lasts, save eternal change.
>
> —Honorat de Bueil (French poet, 1589–1670)

8.1 Introduction

Cool winds sweep across Scotland for much of the year. In this blustery climate, vegetation has a hard time taking hold, so the landscape provides countless outcrops of barren rock. During the mid-18th century, James Hutton, the man who would come to be known as the "father of geology," examined these outcrops in great earnest, hoping to learn how rock formed. Hutton discovered that many features in the outcrops resembled the products of present-day sediment deposition or volcanic activity, and from this observation he came to an understanding of how sedimentary and igneous rock form. But Hutton also recognized that some rocks contained minerals and textures quite different from those visible in sedimentary and igneous samples. He described this puzzling rock as "a mass of matter which had evidently formed originally in the ordinary manner . . . but which is now extremely distorted in its structure . . . and variously changed in its composition."

The rock that so puzzled Hutton is now known as metamorphic rock, from the Greek words *meta*, meaning change, and *morphe*, meaning form. In modern terms, a **metamorphic rock** is one that forms when a pre-existing rock, or **protolith**, undergoes a solid-state change in response to the modification of its environment at depth in the Earth. This process of change is called **metamorphism**.

Let's look at the components of our definition more closely. By *solid-state*, we mean that a metamorphic rock does not form by solidification of magma—remember that geologists consider rocks that solidified from magma to be igneous. By change, we mean that metamorphism produces new minerals that did not occur in the protolith, and/or produces a new *texture* (arrangement of mineral grains) that is distinct from that of the protolith. By modification of environment, we mean that metamorphism takes place when a protolith endures a rise or fall in temperature and/or pressure, undergoes compression and shear, or reacts with *hydrothermal fluids* (very hot-water solutions). And by using the term at depth, geologists are implying that metamorphism takes place at higher pressures and temperatures than those responsible for diagenesis (see Chapter 6). Pressures, temperatures, and fluids together define the *metamorphic conditions*.

Hutton did more than just note the existence of metamorphic rock—he also tried to understand why metamorphism takes place. Because he found metamorphic rocks adjacent to igneous intrusions, he concluded that metamorphism can take place when heat from an intrusion "cooks" the rock into which it intrudes. And because he found that metamorphic rocks can occur over broad regions in the absence of intrusions, he speculated that metamorphism can also take place when rock becomes deeply buried, as occurs during mountain building.

From Hutton's day to the present, geologists have undertaken field studies, laboratory experiments, and theoretical calculations to better characterize metamorphism. In this chapter, we introduce the results of their work. We begin by explaining the causes of metamorphism and the basis for classifying metamorphic rocks. We conclude by discussing the geologic settings in which these rocks form. As you will see, Hutton's speculations on the origin of metamorphic rock were basically correct, but they represented only part of the story—the rest of the story could not take shape until the theory of plate tectonics was proposed.

Did you ever wonder . . .

if, once formed, rocks ever change?

8.2 Consequences and Causes of Metamorphism

What Is a Metamorphic Rock?

If someone were to put a rock on a table in front of you, how would you know that it is metamorphic? First, metamorphic rocks can possess **metamorphic minerals**, new minerals that grow in place within the solid rock only under metamorphic conditions. In fact, metamorphism can produce a group of minerals that together make up a *metamorphic mineral assemblage*. Second, metamorphic rocks can have **metamorphic texture** defined by the arrangement of mineral grains. The texture can be manifested by **metamorphic foliation**, characterized by the parallel alignment of platy minerals (such as mica) and/or the presence of alternating light-colored and dark-colored layers (**Fig. 8.1**). Because of its metamorphic minerals and textures, a metamorphic rock can be as different from its protolith as a butterfly is from a caterpillar. For example, metamorphism of red shale can yield a metamorphic rock consisting of aligned mica flakes and brilliant garnet crystals (**Fig. 8.2a**); metamorphism of limestone composed of cemented-together fossil fragments can yield a metamorphic rock consisting of large interlocking crystals of calcite (**Fig. 8.2b**); and metamorphism of granite, a rock with randomly oriented crystals, can produce a rock with crystals that align parallel to one another

(**Fig. 8.2c**). As seen through a microscope, the change in the grains of a rock due to metamorphism is every bit as dramatic as the change in the cells of a caterpillar becoming a butterfly through metamorphosis.

The formation of metamorphic minerals and textures takes place very slowly—it may take thousands to millions of years—and it involves several processes, which sometimes occur alone and sometimes together. The most common processes include the following.

- *Recrystallization*, which changes the shape and size of grains without changing the identity of the mineral making up the grains. For example, during recrystallization of sandstone, a tightly fitting mosaic of irregularly shaped, large quartz grains may replace a cluster of cemented-together, small, round quartz grains (**Fig. 8.3a**).
- *Phase change*, which transforms one mineral into another mineral with the same composition but a different crystal structure. For example, the transformation of quartz into a denser mineral called coesite represents a phase change, for these minerals have the same formula (SiO_2) but different crystal structures. On an atomic scale, phase change involves the rearrangement of atoms.
- *Metamorphic reaction*, or *neocrystallization* (from the Greek *neos*, for new), which results in the growth of new mineral crystals that differ from those of the protolith (**Fig. 8.3b**). During neocrystallization, chemical reactions digest minerals of the protolith and yield new minerals of the metamorphic rock. For this to take place, atoms *diffuse* (migrate) through solid crystals, a very slow process, and/or dissolve and reprecipitate at grain boundaries.
- *Pressure solution*, which happens when a wet rock is squeezed more strongly in one direction than in the other. Mineral grains dissolve where their surfaces are pressed against other grains, producing ions that migrate through the water to precipitate elsewhere. Precipitation may take place on faces where the grains are squeezed together less strongly. Thus, pressure solution can cause grains to become shorter in one direction and new growth to occur in another (**Fig. 8.3c**). Pressure solution takes place only under conditions in which liquid water can exist—thus, it happens during diagenesis and during relatively low-temperature metamorphism.
- *Plastic deformation*, which happens when a rock is squeezed or sheared at elevated temperatures and pressures, conditions during which minerals behave like soft

FIGURE 8.1 An outcrop of 2.7-billion-year-old metamorphic rock in Ontario, Canada, shows a distinct foliation, in this case defined by alternating bands of light and dark minerals.

plastic and can change shape without breaking (**Fig. 8.3d**). The atomic-scale processes causing plastic deformation are complex, so we must leave the explanation of them to more advanced books.

Caterpillars undergo metamorphosis because of hormonal changes in their bodies. Rocks undergo metamorphism when they are subjected to heat, pressure, compression and shear, and/or very hot water. Metamorphism can change the grains inside of a rock just as profoundly as metamorphosis changes the cells inside a caterpillar (**Fig. 8.2d**). Let's now consider the details of how these *agents of metamorphism* operate.

Metamorphism Due to Heating

When you heat cake batter sufficiently, the batter transforms into a new material—cake. Similarly, when you heat a rock sufficiently, its ingredients transform into a new material—metamorphic rock. Why? Think about what happens to atoms in a mineral grain as the grain warms. Heat causes the atoms to vibrate rapidly, so the chemical bonds that lock atoms to their neighbors stretch and bend. If bonds stretch and bend too far, they break, causing atoms to detach from their original neighbors, move slightly, and form new bonds with other atoms. Repetition of this process—trillions of times—leads to enough rearrangement of atoms within grains, or enough migration of atoms into and out of grains, that recrystallization and/or neocrystallization takes place, and a new metamorphic mineral assemblage grows in the rock.

Metamorphism takes place at temperatures between those at which diagenesis occurs (which modifies the rock without producing metamorphic minerals and textures) and those that cause melting. Roughly speaking, this means that most metamorphic rocks you find in outcrops on continents formed at temperatures of between 250° and 850°C.

FIGURE 8.2 Metamorphism causes changes in mineral makeup and texture.

(a) A specimen of red shale (left) contains clay, quartz, and iron oxide. When this rock undergoes intense metamorphism, it may change into a gneiss containing different minerals. This gneiss sample (right) contains biotite, quartz, feldspar, and purple garnet.

(b) In a limestone, individual fossils are visible. During metamorphism, the texture changes profoundly.

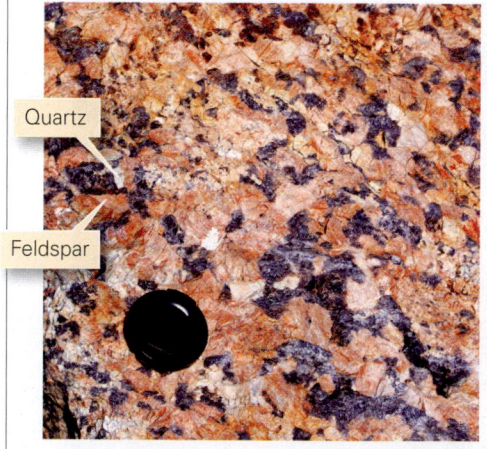

(c) Metamorphism can transform a rock in which crystals are randomly oriented (left) into one in which they are aligned (right).

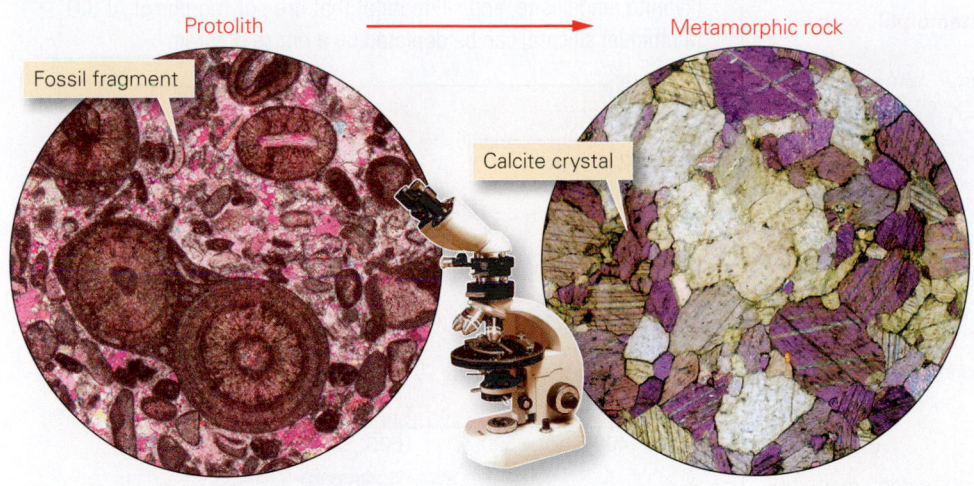

FIGURE 8.2 cont.

Protolith → Metamorphic rock

Fossil fragment

Calcite crystal

(d) The contrast in texture between a protolith of fossiliferous limestone and a marble formed by metemorphism is evident when viewed through a microscope.

12 kbar (= 3,000 to 12,000 bars, where 1 bar is the air pressure at Earth's surface). Pressure increases at about 300 bars/km due to the weight of overlying rock, so these pressures occur at depths of between 10 and 40 km. However, in a few locations geologists have found *ultrahigh-pressure* (*UHP*) *metamorphic rocks*, which appear to have formed at pressures of up to almost 30 kbar, meaning depths of 80 to 100 km. Rocks subjected to ultrahigh pressure contain grains of coesite, a phase of SiO_2 that is much denser than familiar quartz. In fact, some of these rocks contain tiny grains of diamond, a phase of carbon that forms only under very high pressure.

However, melting temperature depends on composition and water content (see Chapter 6)—wet granite, for example, can melt at temperatures of less than 650°C, and very dry peridotite can melt at temperatures as high as 1,200°C. So the upper limit of the metamorphic realm actually ranges between 650° and 1,200°C, depending on rock composition and water content.

The depth in the Earth at which metamorphic temperatures exist depends on the geothermal gradient, which, in turn, reflects the geologic setting. For example, near a hot, igneous intrusion, metamorphic temperatures can develop near the Earth's surface. But in the upper part of average continental crust, away from intrusions, the geothermal gradient is about 25°C/km and a temperature of 500°C occurs at a depth of about 20 km.

Metamorphism Due to Pressure.

As you swim underwater in a swimming pool, water squeezes against you equally from all sides—in other words, your body feels **pressure**. Pressure can cause a material to collapse inward. For example, if you pull an air-filled balloon down to a depth of 10 m in a lake, the balloon becomes significantly smaller. Pressure can have the same effect on minerals. Near the Earth's surface, minerals with relatively open crystal structures (meaning, more space between atoms) can be stable. However, if you subject these minerals to extreme pressure, the atoms pack more closely together (resulting in less space between atoms) and denser minerals tend to form. Such transformations involve phase changes and/or neocrystallization.

Most metamorphic rocks that occur in outcrops on continents were metamorphosed at pressures of between 3 and

Changing Both Pressure and Temperature

So far, we've considered changes in pressure and temperature as separate phenomena. However, in reality, pressure and temperature in the Earth change together with increasing depth. For example, at a depth of 8 km, temperature in the crust reaches about 200°C and pressure reaches about 2.3 kbar. If a rock slowly becomes buried to a depth of 20 km, as can happen during mountain building, temperature in the rock increases to more than 500°C, and pressure increases to 5.5 kbar. Experiments and calculations show that the "stability" of certain minerals (the ability of a mineral to form and survive) and of mineral assemblages depends on both pressure and temperature. Thus, a metamorphic rock formed at 8 km does not contain the same minerals as one formed at 20 km.

We can illustrate the relationship of mineral stability to pressure and temperature by studying the behavior of Al_2SiO_5 (aluminum silicate) as portrayed on a *phase diagram*, a graph with temperature indicated by one axis and pressure indicated by the other (**Fig. 8.4**). Al_2SiO_5 can exist as three different minerals (also called polymorphs, or phases): kyanite, andalusite, and sillimanite. Each of these minerals exists only under a specific range of temperatures and pressures, indicated by an area called a *stability field*, on the phase diagram. If a protolith containing the elements necessary to produce Al_2SiO_5 is taken to a depth in the Earth where the pressure is 2 kbar and the temperature is 450°C (Point X), then andalusite grows. If the temperature stays at 450°C but pressure on the rock increases to 5 kbar (Point Y), then andalusite becomes unstable and kyanite grows. And if the pressure stays at 5 kbar but the temperature increases to 650°C (Point Z), then sillimanite grows. So the presence of one of these polymorphs is a clue to the pressure and temperature at which metamorphism occurred.

FIGURE 8.3 Metamorphic processes, as seen through a microscope.

Protolith ⟶ Metamorphic rock

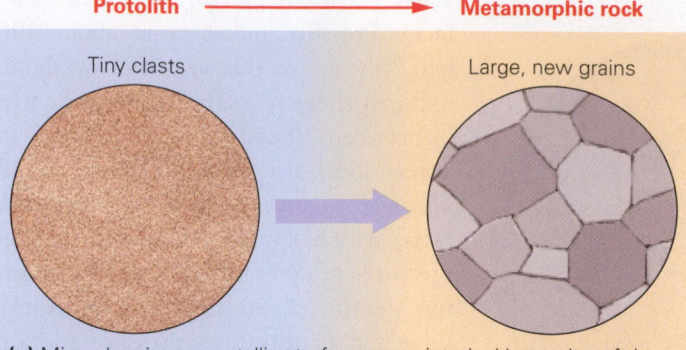

(a) Mineral grains recrystallize to form new, interlocking grains of the same mineral. Typically, grains get larger.

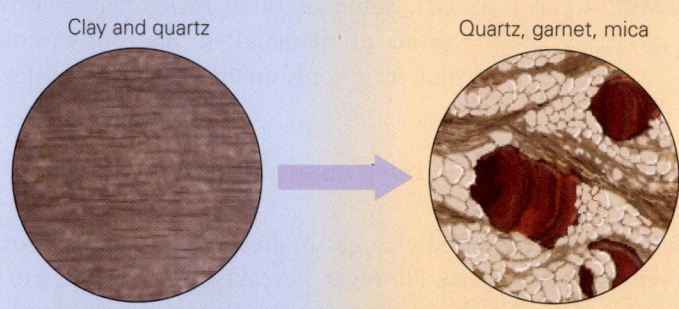

(b) Chemical reactions change the original assemblage of minerals into a new, metamorphic assemblage of minerals.

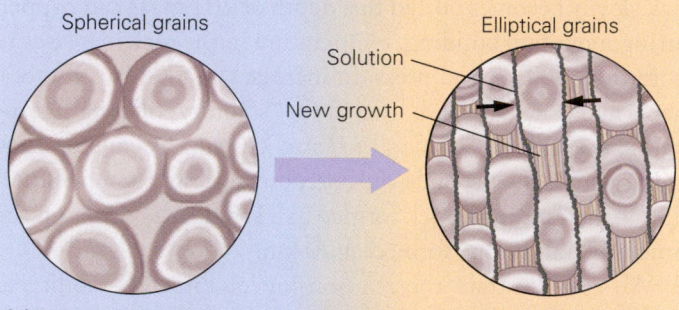

(c) Pressure solution dissolves grains on the sides undergoing more pressure and precipitates new mineral material where the pressure is lower. Arrows indicate the squeezing direction.

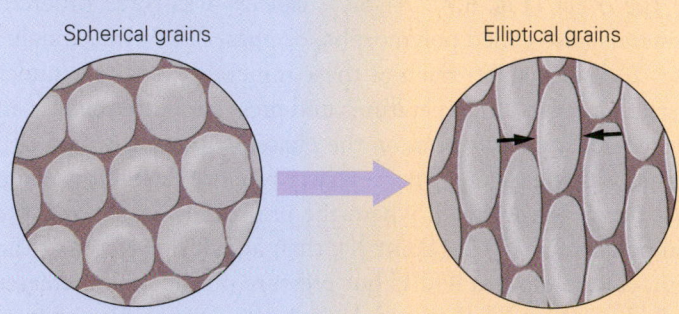

(d) Plastic deformation changes the shape of grains—without breaking them—as a result of squeezing or shear at high temperatures.

FIGURE 8.4 The stability fields for three metamorphic minerals (kyanite, andalusite, and sillimanite) that are polymorphs of Al_2SiO_5 (aluminum silicate) can be depicted on a phase diagram.

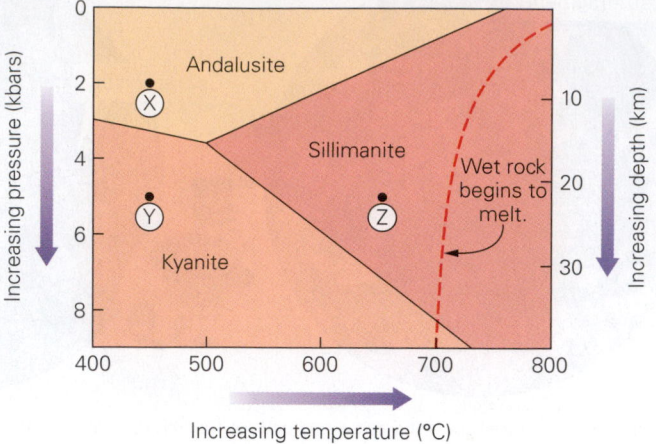

Compression, Shear, and Development of Preferred Orientation

Imagine that you have just built a house of cards, and being in a destructive mood, you step on it. The structure collapses because the downward push you apply with your foot exceeds the push provided by air in other directions. If a material is squeezed (or stretched) unequally from different sides, we say that it is subjected to **differential stress**. (Note that differential stress differs from pressure—in the latter, the push is the same in all directions.) In other words, under conditions of differential stress, the push or pull in one direction differs in magnitude from the push or pull in another direction (**Fig. 8.5a**). We can distinguish two kinds of differential stress.

- *Normal stress:* Normal stress pushes or pulls perpendicular to a surface. We call a push *compression* and a pull *tension*. Compression flattens a material (**Fig. 8.5b**), whereas tension stretches a material.
- *Shear stress:* Shear stress, or shear, moves one part of a material sideways, relative to another. If you place a deck of cards on a table, then set your hand on top of the deck and move your hand parallel to the table, you've sheared the deck (**Fig. 8.5c**).

Compression and shear at metamorphic temperatures and pressures may cause a body of rock to change shape without breaking. During the process, a **preferred orientation** develops, in that pancake-shaped (platy) grains become roughly parallel with one another, and cigar-shaped (elongate) grains become aligned with one another. Platy and elongate grains are examples of *inequant grains* that have different dimensions in different directions (**Fig. 8.5d, e**); in

FIGURE 8.5 Compression and shear change rock grains during metamorphism.

Before   After

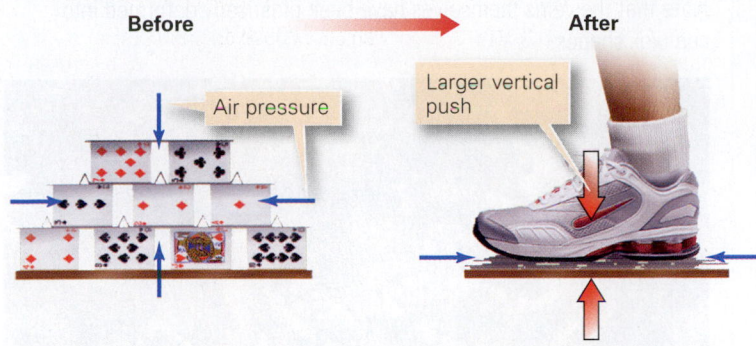

(a) A standing house of cards is subject only to air pressure, which is equal from all sides. When you step on the cards, they are pushed downward because of the vertical compression.

Horizontal compression

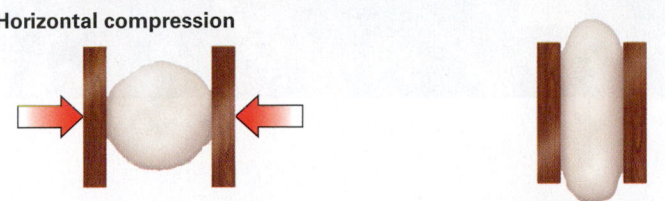

(b) Here horizontal compression flattens a dough ball between wood blocks.

Shear

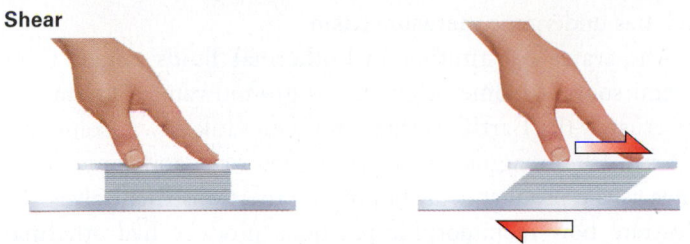

(c) A shear stress acts parallel to a surface. Here shear smears out a deck of cards parallel to a table top.

Before   After

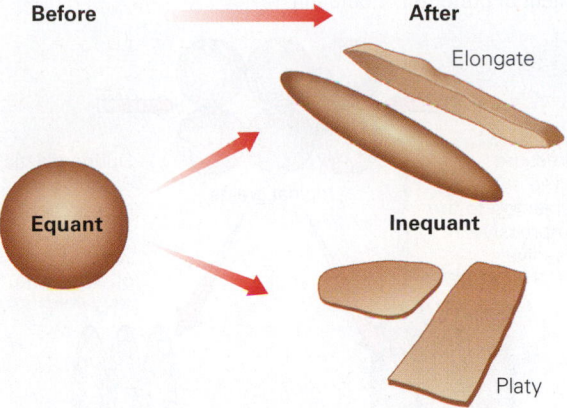

(d) Compression and shear can transform equant grains into inequant grains. Inequant grains can be elongate (cigar-shaped) or platy (pancake-shaped).

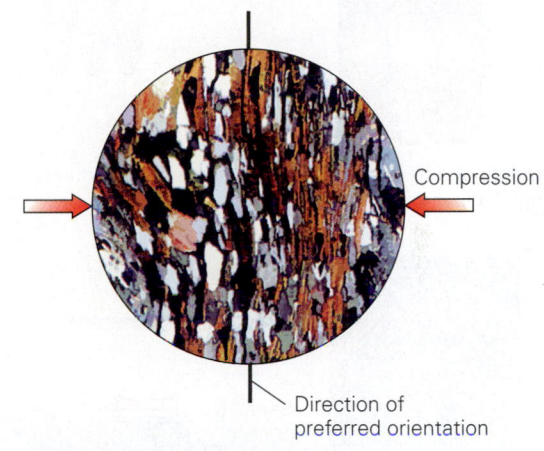

(e) In metamorphic rock, inequant grains may be aligned to form a preferred orientation. As seen through a microscope, the flat planes of grains are perpendicular to the compression direction.

contrast, *equant grains* have roughly the same dimensions in all directions.

Let's look more closely at how preferred orientation forms. In wet rocks at relatively low temperatures, pressure solution dissolves grains on faces perpendicular to the direction of compression. So, as a result of pressure solution, grains become shorter in the direction of compression. (Sometimes, precipitation of new mineral crystals takes place on faces where compression is less, so the grains may effectively get longer in the direction perpendicular to compression.) At relatively high temperatures, weaker grains flatten due to differential stress by means of plastic deformation (**Fig. 8.6a**); the grains become narrower in one direction and become longer in the other. Also, as a rock changes shape, stronger inequant grains distributed throughout a softer matrix rotate into parallelism, much the way logs scattered in a flowing river align with the current (**Fig. 8.6b**). Weaker grains may change shape, and be

smeared into parallelism. Growth of new metamorphic minerals during shear contributes to the development of preferred orientation—the new minerals grow faster in the direction in which a rock is stretching than in other directions.

The Role of Hydrothermal Fluids

Metamorphic reactions commonly take place in the presence of hydrothermal fluids. We initially defined hydrothermal fluids simply as very hot-water solutions. In fact, they actually can include hot water, steam, and so-called supercritical fluid. (A supercritical fluid is a substance that forms under high temperatures and pressures and has characteristics of both liquid and gas; it can permeate rock, seeping into every conceivable opening.) Hydrothermal fluids chemically react with rock by dissolving, transporting, and providing ions and also by providing water molecules that can become incorporated in minerals.

FIGURE 8.6 Changes in grain shape and orientation, and the development of preferred orientation.

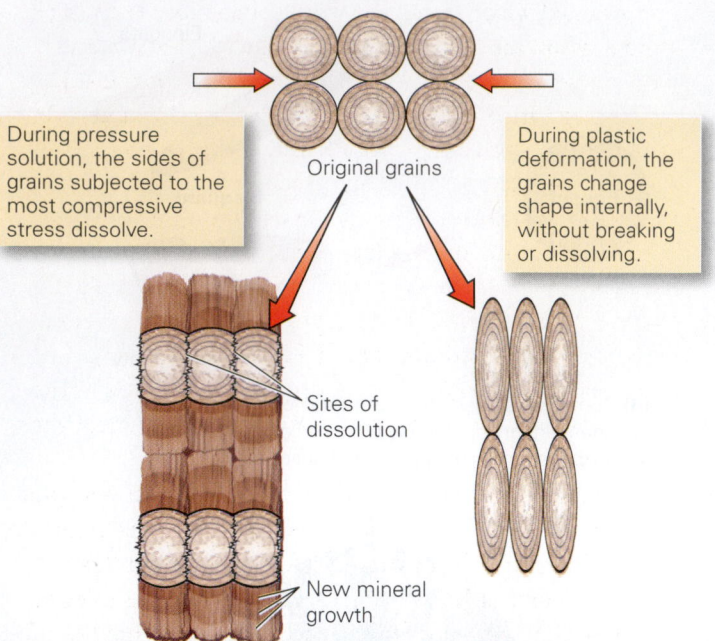

During pressure solution, the sides of grains subjected to the most compressive stress dissolve.

Original grains

During plastic deformation, the grains change shape internally, without breaking or dissolving.

Sites of dissolution

New mineral growth

(a) Application of differential stress (directed pressure) during metamorphism can change the shape of grains in two ways.

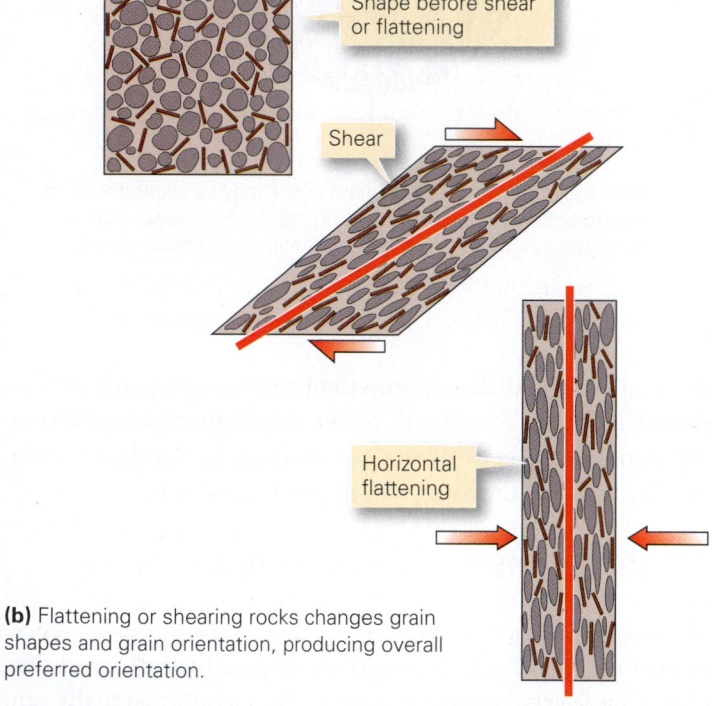

Shape before shear or flattening

Shear

Horizontal flattening

(b) Flattening or shearing rocks changes grain shapes and grain orientation, producing overall preferred orientation.

Locally, minerals dissolved in the fluids precipitate in cracks so that the cracks fill with new minerals. Such mineral-filled cracks, or **veins**, commonly consist of milky-white quartz (**Fig. 8.7**). In some cases, hydrothermal fluids passing through a rock during metamorphism pick up some ions of one element and

FIGURE 8.7 Veins in metamorphic rock on the coast of Wales. Note that the veins themselves have been plastically deformed into complex shapes.

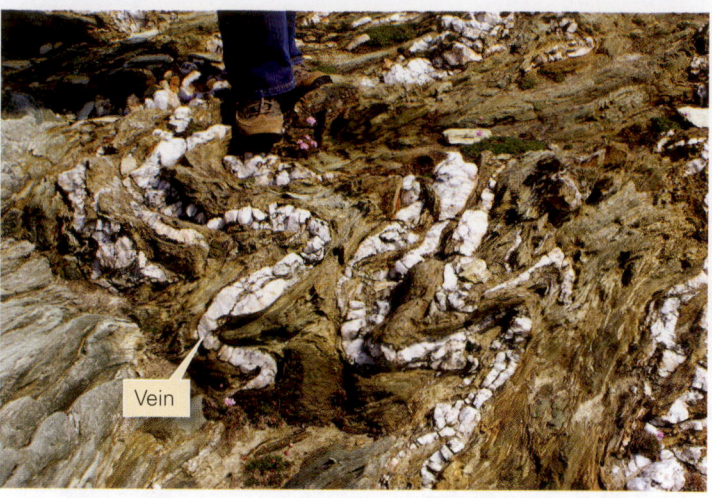

Vein

drop off ions of another, thereby changing the overall chemical composition of the rock. When this happens, we say that the rock has undergone **metasomatism**.

The water constituting hydrothermal fluids comes from several sources. Some originates as groundwater that entered the crust at the Earth's surface and then sank down, some was released from magma when the magma rose, and some originates as the product of metamorphic reactions themselves. To illustrate how metamorphic reactions produce hydrothermal fluids, consider the metamorphism of muscovite and quartz at high temperature:

$$KAl_3Si_3O_{10}(OH)_2 + SiO_2 \rightarrow KAlSi_3O_8 + Al_2SiO_5 + H_2O$$

muscovite   quartz   K-feldspar   sillimanite   water

This formula indicates that muscovite and quartz of the protolith decompose while new crystals of K-feldspar and sillimanite grow and new water molecules are released.

Take-Home Message

Metamorphic rocks form in response to changes in temperature, pressure, application of compression and shear, and/or interaction with hydrothermal fluids. During metamorphism, new mineral grains grow as pre-existing ones disappear, and rock may develop foliation due to the alignment of inequant grains. A variety of processes, such as recrystallization, chemical reaction, and plastic deformation may be involved.

QUICK QUESTION: Can the overall composition of rock change during metamorphism?

8.3 Types of Metamorphic Rocks

Coming up with a way to classify and name the great variety of metamorphic rocks on Earth hasn't been easy. After decades of debate, geologists have found it most convenient to divide metamorphic rocks into two fundamental classes: foliated rocks and nonfoliated rocks. Each class contains several rock types, as we now see.

Foliated Metamorphic Rocks

To understand this class of rocks, we first need to discuss the nature of foliation in more detail. The word comes from the Latin *folium*, for leaf. Geologists use *foliation* to refer to an assemblage of parallel planar surfaces and/or layers in a metamorphic rock. Foliation can give metamorphic rocks a striped or streaked appearance in an outcrop, and/or give them the ability to split into thin sheets. A foliated metamorphic rock has foliation because it contains inequant mineral crystals that are aligned parallel to one another, defining preferred mineral orientation, and/or because the rock has alternating dark-colored and light-colored layers.

Foliated metamorphic rocks can be distinguished from one another according to their composition, their grain size, and the nature of their foliation. The most common types are:

- *Slate:* The finest-grained foliated metamorphic rock, **slate**, forms by metamorphism of shale or mudstone—rocks composed of clay—under relatively low pressures and temperatures. Slate contains a type of foliation called *slaty cleavage*, which allows it to split into thin sheets that make excellent roofing shingles (**Fig. 8.8a**). Slaty cleavage develops when compression causes clay flakes to reorient and regrow into an orientation perpendicular to the direction of compression. For example, horizontal compression of a sequence of horizontal shale beds produces vertical slaty cleavage (**Fig. 8.8b**). Commonly, such compression also causes the layers to bend into curves called folds. The development of aligned clay in slate is primarily a consequence of pressure solution and recrystallization—grains lying at an angle to the cleavage plane dissolve, whereas grains parallel to the cleavage plane grow. In addition, during this process, less-soluble inequant grains may passively rotate into the plane of cleavage.

> **Did you ever wonder . . .**
>
> why slate makes such nice roofing shingles?

- *Phyllite:* **Phyllite** is a fine-grained metamorphic rock with a foliation caused by the preferred orientation of very fine-grained white mica. The word comes from the Greek word *phyllon*, meaning leaf, as does the word *phyllo*, the flaky dough in Greek pastry. The parallelism of translucent fine-grained mica gives phyllite a silky sheen known as phyllitic luster (**Fig. 8.9a**). Phyllite forms when slate is subjected to a temperature high enough to produce a new assemblage of metamorphic minerals (fine-grained white mica and chlorite) out of clay.

- *Metaconglomerate:* Under the metamorphic conditions that produce slate or phyllite, a protolith of conglomerate becomes **metaconglomerate**. Typically, pressure solution and plastic deformation flatten pebbles and cobbles of the rock into pancake-like shapes. The alignment of inequant clasts defines a foliation (**Fig. 8.9b**).

- *Schist:* **Schist** is a medium- to coarse-grained metamorphic rock that possesses a type of foliation, called schistosity, defined by the preferred orientation of large mica flakes (muscovite and/or biotite; **Fig. 8.9c**). Schist forms at even higher temperatures than those needed to form phyllite, and it differs from phyllite in that the mica grains are larger. Typically, schists also contain other minerals such as quartz, feldspar, garnet, and amphibole—the specific minerals that grow depending on the chemical composition of the protolith. Schist can form from a shale but also from a great variety of other protoliths as long as the protolith contains the appropriate elements to make mica. In some cases, certain mineral grains in schists grow to be much larger than surrounding minerals. For example, garnet crystals in schist may become many times larger than those of other minerals (see Fig. 8.3b). Especially large crystals that grow in a metamorphic rock are called *porphyroblasts*. Smaller grains surrounding porphyroblasts constitute a rock's matrix.

- *Gneiss:* **Gneiss** is a compositionally layered metamorphic rock, typically comprised of alternating dark-colored and light-colored layers that range in thickness from millimeters to meters (**Fig. 8.10**). This compositional layering, or *gneissic banding*, gives gneiss a striped appearance. The contrasting colors represent contrasting compositions. Light-colored layers contain predominantly felsic minerals such as quartz and feldspar, whereas the dark-colored layers contain predominantly mafic minerals such as amphibole, pyroxene, and biotite. If gneiss contains mica, the mica-rich layers may have schistosity. Of note, gneiss that formed at very high temperatures does not contain mica, because at these temperatures mica reacts to form other minerals.

  How does the banding in gneiss form? Banding in some examples of gneiss evolved directly from the original bedding in a rock. For example, metamorphism of a

FIGURE 8.8 Slate is a foliated metamorphic rock that forms at relatively low temperatures and pressures.

(a) A block of slate splits easily along cleavage planes, which may be at a high angle to the bedding planes. Workers split slate to produce roof shingles that, when overlapped, make a watertight surface (see inset).

(b) Slaty cleavage forms in response to compressive stress. In this example, layers also bend to form folds, as slaty cleavage develops. The cleavage tends to be oriented parallel to the axial plane, an imaginary surface that, simplistically, divides the fold in half.

protolith consisting of alternating beds of sandstone and shale produces a gneiss consisting of alternating beds of quartzite and mica. Gneissic banding can also form when the protolith undergoes an extreme amount of shearing under conditions in which the rock can flow like soft plastic (**Fig. 8.11a**). Such flow stretches, folds, and smears out pre-existing compositional contrasts in the rock and transforms them into aligned sheets. To picture this process, imagine slowly stirring vanilla batter in which there are blobs of chocolate batter—eventually you will see thin, alternating layers of dark and light batter. Similarly, imagine what happens if you take a rolling pin and flatten a ball consisting of two different colors of dough, then fold it in half and flatten it again—you end up with thin parallel layers of contrasting colors. Finally, banding in

some gneisses can develop by an incompletely understood process called *metamorphic differentiation*. Differentiation may involve dissolution of minerals in some layers and migration of the chemical components of those minerals to other layers, where new minerals then grow. In effect, chemical reactions segregate different minerals into different layers (**Fig. 8.11b**).

• *Migmatite:* Under certain conditions, gneiss may begin to melt, producing felsic magma and residual, still-solid mafic rock. If the melt freezes again before flowing out of the source area, a mixture of igneous rock and relict metamorphic rock forms. This mixture is called **migmatite** (**Fig. 8.12**). Plastic flow during the process can contort the light-colored felsic rock and the dark-colored mafic rock into complex shapes.

FIGURE 8.9 Examples of foliated metamorphic rocks formed at higher temperatures and pressures.

(a) During formation of phyllite, clay recrystallizes to form tiny mica flakes that reflect light, giving the rock a sheen.

(b) In metaconglomerate, pebbles and cobbles flatten into a pancake shape without cracking.

(c) A schist contains coarse mica flakes, along with other metamorphic minerals.

FIGURE 8.10 The foliation (gneissic banding) in gneiss comes at a huge range of scales.

(a) Huge (> 100 m high) cliffs in Greenland consist of gneiss. The bands of light and dark are tens of meters across.

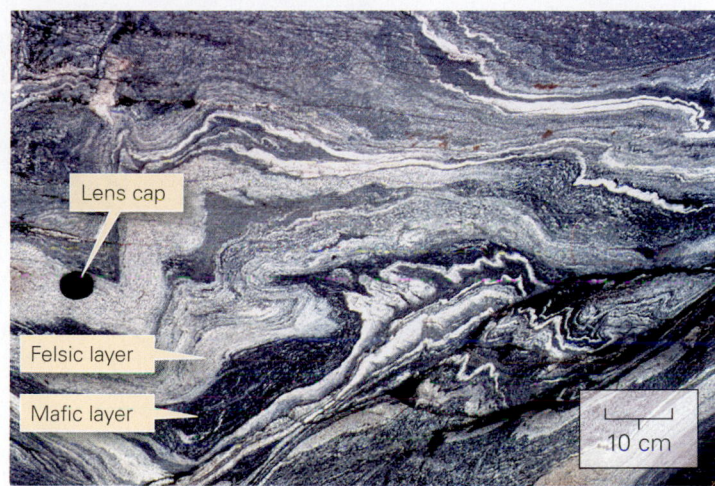

Lens cap

Felsic layer

Mafic layer

10 cm

(b) In this outcrop of gneiss in Brazil, some of the layers are only centimeters across.

Nonfoliated Metamorphic Rocks

Nonfoliated metamorphic rocks contain minerals that recrystallized or grew during metamorphism but have no foliation. The lack of foliation means either that metamorphism occurred in the absence of compression and shear or that most of the new crystals are equant. We list below some of the rock types that can occur without foliation.

- *Hornfels:* **Hornfels** is a fine-grained nonfoliated rock that contains a variety of metamorphic minerals (some equant and some inequant). The specific mineral assemblage in a hornfels depends on the composition of the protolith and on the temperature and pressure of metamorphism. Inequant minerals that grow in hornfels are randomly oriented—in other words, they do not display a preferred orientation.

- *Quartzite:* Most **quartzite** forms by the metamorphism of pure quartz sandstone. During metamorphism, pre-existing quartz grains recrystallize, creating new, larger grains. In the process, the distinction between cement and grains disappears, open pore space disappears, and the grains become interlocking. When quartzite cracks, the fracture cuts across grain boundaries—in contrast, fractures in sandstone curve *around* grains. Quartzite looks glassier than sandstone and does not have the grainy, sandpaper-like surface characteristic of sandstone (**Fig. 8.13a**). Depending on the impurities it contains, quartzite can vary in color from white to gray, purple, or green.

Most quartzite is nonfoliated because it does not contain aligned mica or compositional layering. In some cases, however, quartz grains deform plastically and become pancake shaped. The alignment of pancakes yields a foliation. (To avoid confusion, quartzite with this texture should be called "foliated quartzite.")

- *Marble:* The metamorphism of limestone yields **marble** (**Fig. 8.13b**). During the formation of marble, calcite composing the protolith recrystallizes, so fossil shells, pore space, and the distinction between grains and cement disappear. Thus, marble typically consists of a fairly uniform mass of interlocking calcite crystals. It may also contain other, less-familiar minerals formed from the reaction of calcite with quartz, clay, and iron oxide, if these minerals had existed in the protolith.

Sculptors love to work with marble because the rock is relatively soft and has a uniform texture that gives it the cohesiveness and homogeneity needed to fashion large, smooth, highly detailed sculptures. Marble comes in a variety of colors—white, pink, green, and black—depending on the impurities it contains. (Michelangelo, one of the great Italian Renaissance artists, sought large, unbroken blocks of creamy white marble from quarries in western Italy for his masterpieces). If the original protolith contained layers with different impurities, the resulting marble has color banding that makes it a prized decorative stone (**Fig. 8.13c**). Because marble is a relatively weak rock, it flows like soft plastic under metamorphic conditions, and this flow can smear out different-colored portions of marble into beautiful contorted, curving bands.

- *Granofels:* A granofels is a coarse-grained nonfoliated rock consisting primarily of quartz and feldspar.
- *Amphibolite:* Metamorphism of mafic rocks (basalt or gabbro) can't produce quartz and muscovite when metamorphosed, for these rocks don't contain the right mix of

FIGURE 8.11 The formation of gneiss, which takes place at very high temperatures and pressures.

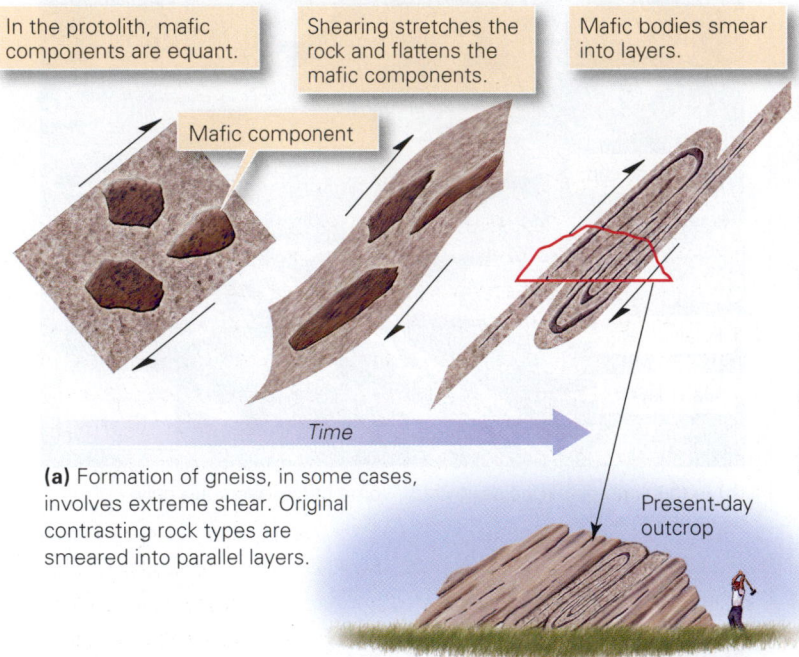

In the protolith, mafic components are equant.

Shearing stretches the rock and flattens the mafic components.

Mafic bodies smear into layers.

Mafic component

Time

Present-day outcrop

(a) Formation of gneiss, in some cases, involves extreme shear. Original contrasting rock types are smeared into parallel layers.

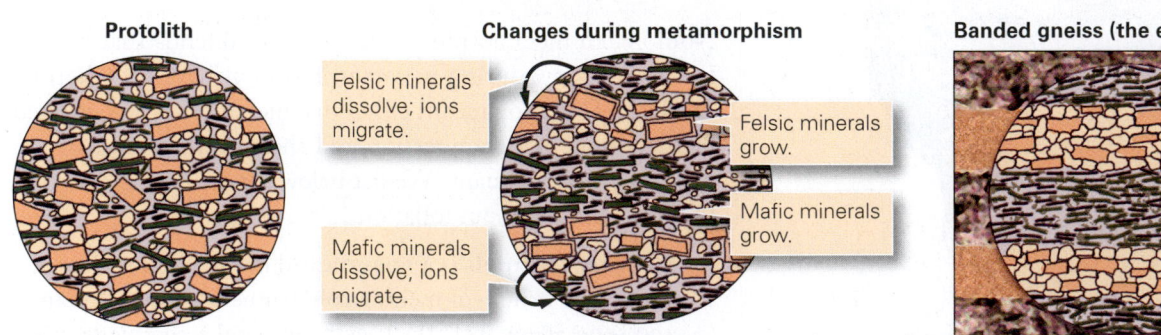

Protolith

Changes during metamorphism

Banded gneiss (the end product)

Felsic minerals dissolve; ions migrate.

Felsic minerals grow.

Mafic minerals grow.

Mafic minerals dissolve; ions migrate.

Mafic band

Felsic band

Mafic band

Felsic band

Time

(b) Gneiss may also form by metamorphic differentiation, during which chemical reactions cause felsic and mafic minerals to grow in distinct, separate layers.

FIGURE 8.12 An outcrop of migmatite in northern Michigan contains both light-colored (felsic) igneous rock and dark-colored (mafic) metamorphic rock. The mixture of the two rock types makes migmatite resemble marble cake.

Light (felsic) bands formed from melt.

Dark (mafic) bands remained solid.

chemicals to yield such minerals. Rather, they transform into amphibolite, a dark-colored metamorphic rock containing predominantly hornblende (a type of amphibole) and plagioclase (a type of feldspar), and in some cases, a tiny bit of biotite. Where subjected to differential stress, amphibolites can develop a foliation, but the foliation tends to be poorly defined because the rock contains very little mica.

Chemical Composition of Metamorphic Rocks

Up to this point, we've emphasized the importance of temperature and pressure in determining the mineral assemblage in a metamorphic rock. Let's not forget that composition plays a key role in determining which minerals form as well. For example, it's impossible to form a biotite-rich schist from a pure quartz sandstone because biotite contains elements, such as iron, that do not occur in quartz.

To distinguish among different compositions of metamorphic rock, geologists use the following terms. (1) *Pelitic* metamorphic rocks form from sedimentary protoliths such as shale, which contain a relatively high proportion of aluminum. Metamorphism of these rocks produces aluminum-rich metamorphic minerals such as muscovite. (2) *Mafic* metamorphic rocks contain relatively little silica and an abundance of iron and magnesium. During metamorphism, minerals such as biotite and hornblende grow. (3) *Calcareous* metamorphic rocks form from calcium-rich sedimentary rocks (limestone) and contain calcite ($CaCO_3$). (4) *Quartzo-feldspathic* metamorphic rocks form from protoliths (such as granite) that contain mostly quartz and feldspar.

Take-Home Message

Geologists divide metamorphic rocks into two main classes based on whether the rock contains foliation. Foliated rocks include slate, schist, and gneiss. Nonfoliated rocks include marble and quartzite. The type of metamorphic rock that forms depends on the conditions of metamorphism.

QUICK QUESTION: Why don't builders use gneiss to make roofing shingles?

8.4 Defining Metamorphic Intensity

Not all metamorphism takes place under the same physical conditions. For example, rocks carried to a great depth beneath a mountain range undergo more intense metamorphism than do rocks closer to the surface. Geologists use the term **metamorphic grade** in a somewhat informal way to indicate the intensity of metamorphism, meaning the amount or degree of metamorphic change (**Fig. 8.14a**). Classification of metamorphic grade depends primarily on temperature because temperature plays the dominant role in determining the extent of recrystallization and neocrystallization during metamorphism. Metamorphic rocks that form at relatively low temperatures (between about 250° and 400°C) are *low-grade* rocks, and metamorphic rocks that form at relatively high temperatures (over 600°C) are *high-grade* rocks. *Intermediate-grade* rocks form at temperatures between these two extremes. To provide a more complete indication of the intensity of metamorphism, geologists use the concept of *metamorphic facies* (**Box 8.1**).

Different grades of metamorphism yield different metamorphic mineral assemblages (**Fig. 8.14b**). As grade increases, recrystallization and neocrystallization tend to produce coarser grains and new mineral assemblages that are stable at higher temperatures and pressures. Metamorphism that occurs while temperature and pressure progressively increase is called *prograde metamorphism*. As grade increases, metamorphic reactions release water, so high-grade rocks tend to be "drier" than low-grade rocks. This means that minerals in high-grade rocks do not contain minerals with $-OH$ in their chemical formula, whereas minerals in lower-grade rocks can.

FIGURE 8.13 Examples of quartzite and marble—typically, but not always, these are nonfoliated metamorphic rocks.

Quartz sandstone—the protolith of quartzite.

(a) This unfoliated maroon quartzite from Wisconsin breaks on smooth fractures that cut across grains.

Italian marble quarry sliced into a mountain

(b) Michelangelo used nonfoliated white marble for his spectacular sculptures.

To understand prograde metamorphism, consider the changes that a shale undergoes when it starts near the Earth's surface and ends up at great depth beneath a mountain range (**Fig. 8.14c**). The clay flakes in shale lie more or less parallel to the bedding. Under low-grade metamorphic conditions and differential stress, shale transforms into slate. In slate, the clay flakes are a bit larger, are better formed, and align parallel to cleavage. As metamorphic grade increases a little more, the clay flakes decompose and new crystals of fine-grained white mica as well as new crystals of chlorite and quartz grow, transforming the rock into phyllite. Under intermediate-grade conditions, the minerals in phyllite react and decompose, yielding atoms that combine to produce large crystals of mica (such as muscovite and biotite) as well as other minerals such as garnet. The reactions also release water, which escapes. In our example, this metamorphism is taking place under differential stress, so the micas grow with a preferred orientation and the rock becomes a schist. During metamorphic reactions under high-grade conditions, yet another assemblage of minerals forms. High-grade rocks do not contain much mica, if any, because mica contains –OH and tends to decompose and release water at high temperatures. In fact, high-grade rock typically includes water-free minerals, such as feldspar, quartz, pyroxene, and garnet. As mica disappears, the rock loses its schistosity but can develop gneissic layering.

Metamorphism that takes place when temperatures and pressures progressively decrease is known as *retrograde metamorphism*. For retrograde metamorphism to occur, hydrothermal fluids must enter the rock to add back water. In fact, under cold and dry conditions, retrograde metamorphic reactions either cannot proceed or take place too slowly to cause much change, because atoms cannot diffuse easily at low temperature. It is for this reason that high-grade rocks formed early during Earth history have survived and can be exposed at the surface of the Earth today.

In this unmetamorphosed limestone, fossils and bedding are visible.

(c) The marble floor in the inset photo has color banding inherited from original bedding like that seen in the limestone to the left. The layers became contorted during metamorphism.

FIGURE 8.14 Intensity of metamorphism is indicated by metamorphic grade.

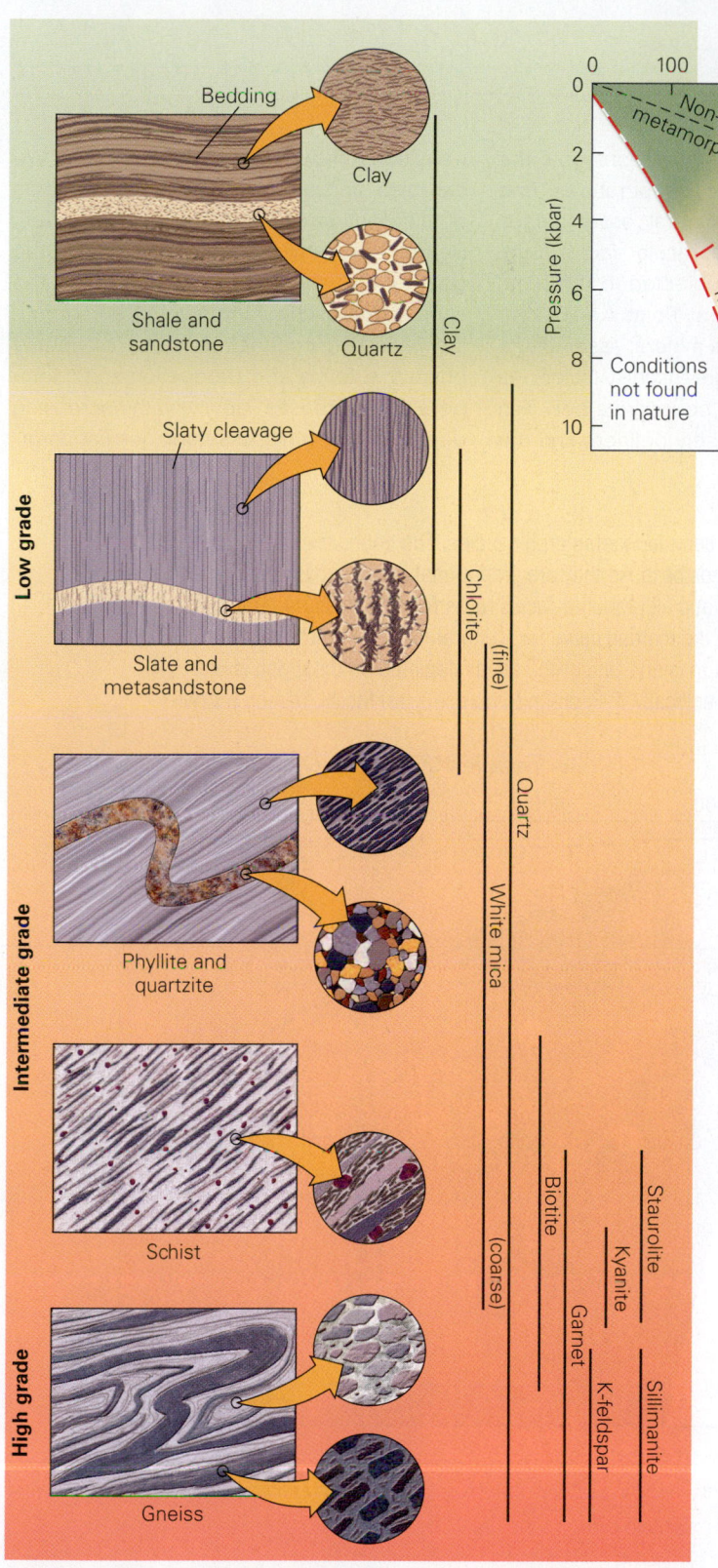

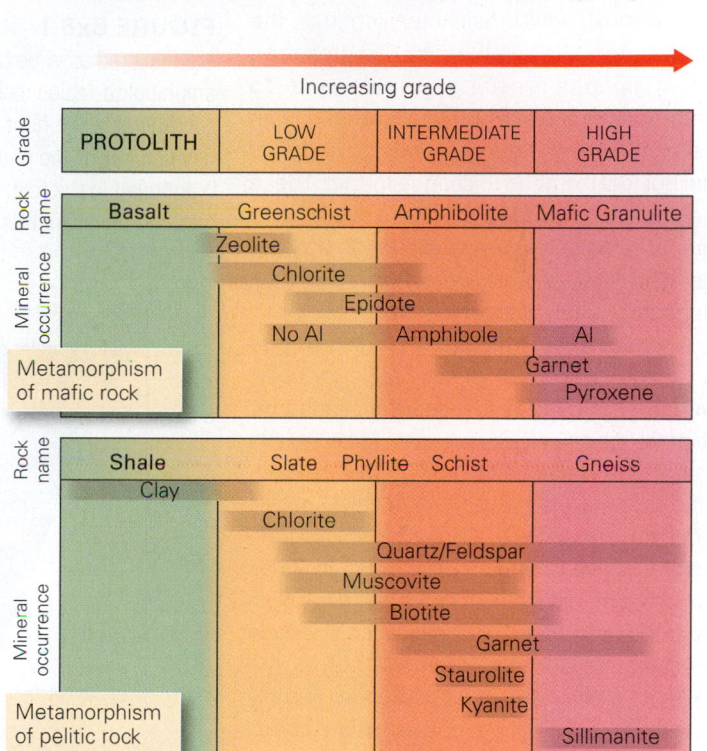

(a) This graph depicts the approximate temperatures and pressures of metamorphic grades. Different conditions occur in different geologic settings.

(b) The metamorphic minerals that form in a given rock depend on grade and composition. This chart contrasts important metamorphic minerals that form, at different grades, from a basalt protolith with those formed from a shale protolith.

(c) Here we see the consequences of the progressive metamorphism of shale and sandstone from low grade to high grade during mountain building. Lines on the right indicate the range of grade in which key metamorphic minerals form.

BOX 8.1 CONSIDER THIS . . .

Metamorphic Facies

In the early years of the 20th century, geologists working in Scandinavia, where erosion by glaciers has left beautiful, nearly unweathered outcrops, came to realize that metamorphic rocks, in general, do not consist of a hodgepodge of minerals formed at different times and in different places but rather consist of a distinct assemblage of minerals that grew in association with each other at a certain pressure and temperature. It seemed that the mineral assemblage present in a rock more or less represented a condition of chemical equilibrium, meaning that the chemicals making up the rock had organized into a group of mineral grains that were—to anthropomorphize a bit—comfortable with each other and their surroundings and thus did not feel the need to change further. These geologists concluded that the specific mineral assemblage in a rock depends both on pressure and temperature conditions and on the composition of the protolith.

This discovery led the geologists to propose the concept of metamorphic facies. A **metamorphic facies** is a set of metamorphic mineral assemblages indicative of a certain range of pressure and temperature. Each specific assemblage in a facies reflects a specific protolith composition. According to this definition, a given metamorphic facies includes several different kinds of rocks that differ from each other in terms of chemical composition and, therefore, mineral content—but all the rocks of a given facies formed under roughly the same temperature and pressure conditions. Geologists recognize several facies, of which the major ones are zeolite, hornfels, greenschist, amphibolite, blueschist, eclogite, and granulite. The names of the different facies are based on a distinctive feature or mineral found in some of the rocks of the facies.

We can represent the approximate conditions under which metamorphic facies formed by using a pressure-temperature graph (**Fig. Bx8.1**). Each area on the graph,

labeled with a facies name, represents the approximate range of temperatures and pressures in which mineral assemblages characteristic of that particular facies form. For example, a rock subjected to the pressure and temperature at Point A (4.5 kbar and 400°C) develops a mineral assemblage characteristic of the greenschist facies. As the graph implies, the boundaries between facies cannot be precisely defined, and the

transitions between facies are gradual. We can also portray the geothermal gradients of different crustal regions on the graph. Beneath mountain ranges, for example, the geothermal gradient passes through the zeolite, greenschist, amphibolite, and granulite facies. In contrast, in the accretionary prism that forms at a subduction zone, temperature increases slowly with increasing depth, so blueschist assemblages can form.

FIGURE Bx8.1 The common metamorphic facies. The boundaries between the facies are depicted as wide bands because they are gradational and approximate. Note that some amphibolite-facies rocks and all granulite-facies rocks form at pressure-temperature conditions to the right of the melting curve for wet granite. Thus, such metamorphic rocks develop only if the protolith is dry. One of the facies depicted on the graph is not mentioned in the text: specifically, P-P (prehnite-pumpellyite) facies, named for two metamorphic minerals.

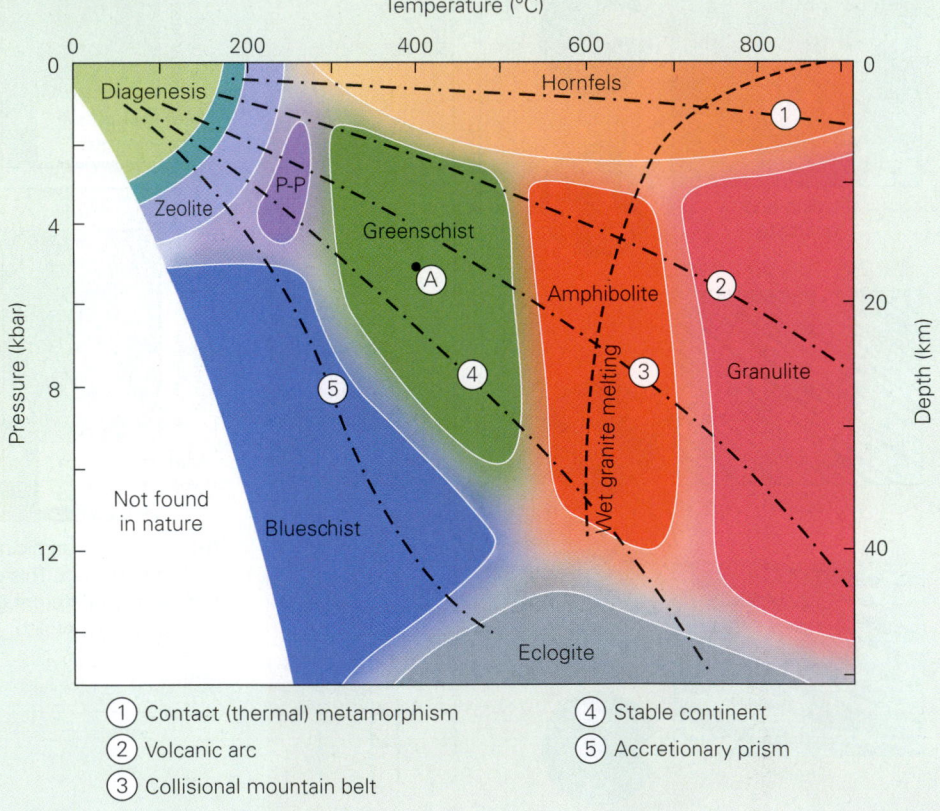

1. Contact (thermal) metamorphism
2. Volcanic arc
3. Collisional mountain belt
4. Stable continent
5. Accretionary prism

We can represent the concept of prograde and retrograde metamorphism as a path plotted on a pressure-temperature graph (see Fig. 8.14a). When a rock gets buried progressively

deeper, temperature and pressure increase; the rock follows a prograde path on the graph. (In this particular example, the rock was buried quickly, so pressure increased faster than temperature; rocks are good insulators and thus heat up very slowly.) Later, when the rock moves back toward the Earth's surface, because uplift raises the rock and erosion strips away overlying rock, it follows the retrograde path. Geologists have developed techniques to determine the times at which a rock reached particular locations along the pressure-temperature path. This information defines a *P-T-t path* (pressure-temperature-time path) for the rock (**Fig. 8.15**). Considering these factors helps geologists interpret the geologic history of the rock.

The presence of certain minerals known as **index minerals** in a rock indicates the approximate metamorphic grade of the rock. The line on a map along which an index mineral first appears is called an *isograd* (from the Greek *iso*, meaning equal). All points along an isograd have approximately the same metamorphic grade. **Metamorphic zones** are regions between two isograds; zones are named after an index mineral that was not present in the previous, lower-grade zone. To compare rocks of different grades, you could take a hike from central New York State eastward into central Massachusetts in the eastern United States. Your path starts in a region where rocks were not metamorphosed, and it takes you into the internal part of the Appalachian Mountain belt, where rocks were intensely metamorphosed. As a consequence, you cross several metamorphic zones (**Fig. 8.16**).

Take-Home Message

The mineral assemblage in a metamorphic rock depends on temperature and pressure and on protolith composition. Metamorphic grade informally specifies the intensity of metamorphism; high-grade rocks form at higher temperatures, and low-grade rocks form at lower temperatures. The mineral assemblage that a metamorphic rock contains reflects its grade. A region in which a specific grade of rock occurs is a metamorphic zone, and the boundary between zones is an isograd. A metamorphic facies is a set of mineral assemblages indicative of certain pressure and temperature conditions.

QUICK QUESTION: The geothermal gradient on Mars (~ 8°C/km) is less than that on Earth, and in places the crust of Mars is only 30 km thick. Do high-grade metamorphic rocks exist in this crust?

8.5 Where Does Metamorphism Occur?

So far, we've discussed the nature of changes that occur during metamorphism, the agents of metamorphism (heat, pressure, differential stress, and hydrothermal fluids), the rock types that

FIGURE 8.15 A P-T-t path for a hypothetical rock. The ages in parentheses indicate the time at which the specified pressure and temperature conditions were achieved. P_m is the peak pressure and T_m is the peak temperature.

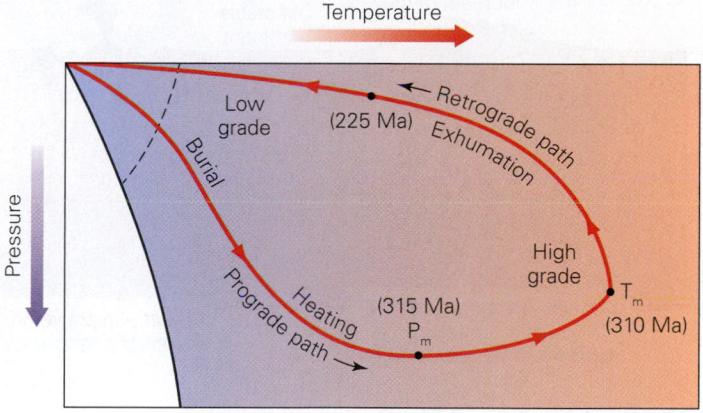

FIGURE 8.16 Metamorphic zones and isograds.

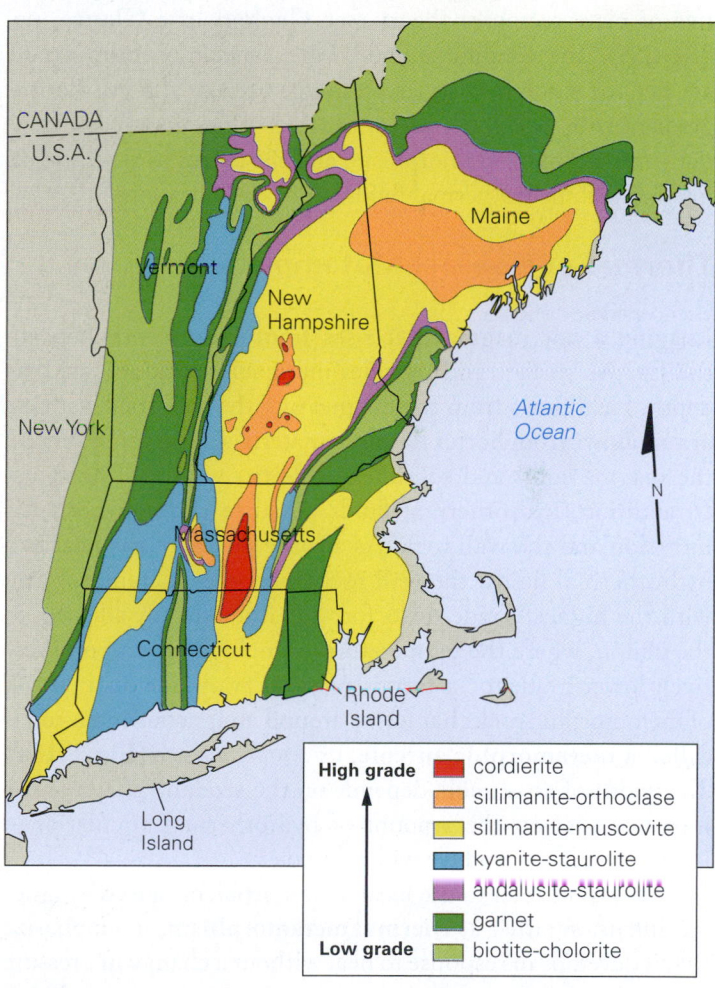

form as a result of metamorphism, and the concepts of metamorphic grade and metamorphic facies. With this background, let's now examine the wide range of geologic settings on Earth

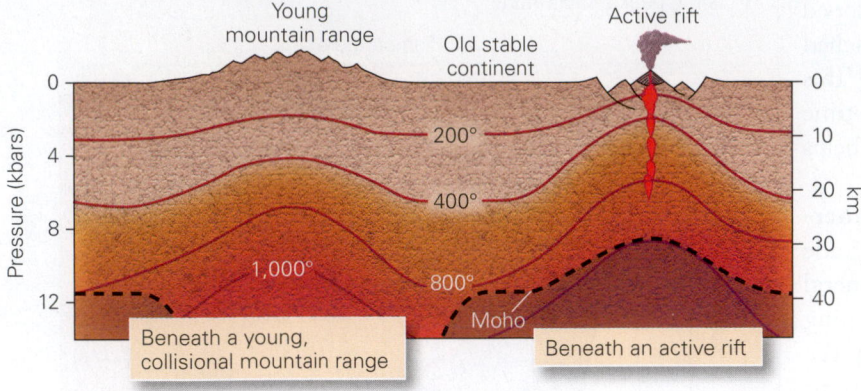

FIGURE 8.17 The change in temperature with depth at different locations in a continent. The solid lines are isotherms (in °C)—they connect locations where the temperature has the same value.

where metamorphism takes place, as viewed from the perspective of plate tectonics theory (see **Geology at a Glance**, pp. 254–255). The conditions under which metamorphism occurs are not the same in all these settings, because the geothermal gradient (**Fig. 8.17**), the extent to which rocks endure differential stress during metamorphism, and the extent to which rocks interact with hydrothermal fluids varies with location.

Thermal or Contact Metamorphism

Imagine a hot magma that rises from great depth beneath the Earth's surface and intrudes into cooler rock at a shallow depth. Heat flows from the magma into the wall rock, for heat always flows from hotter to colder materials. As a consequence, the magma cools and solidifies while the wall rock heats up. In addition, hydrothermal fluids circulate through both the intrusion and the wall rock. As a consequence of the heat and hydrothermal fluids, the wall rock undergoes metamorphism, with the highest-grade rocks forming immediately adjacent to the pluton, where the temperatures were highest, and progressively lower-grade rocks forming farther away. The distinct belt of metamorphic rock that forms around an igneous intrusion is called a **metamorphic aureole**, or *contact aureole* (**Fig. 8.18a**). The width of an aureole depends on the size and shape of the intrusion and on the amount of hydrothermal circulation—larger intrusions produce wider aureoles.

Geologists refer to the local metamorphism caused by igneous intrusion either as **thermal metamorphism**, to emphasize that it develops in response to heat without a change in pressure and without differential stress, or as **contact metamorphism**, to emphasize that it develops adjacent to the contact of an intrusion with its wall rock (**Box 8.2**). Because this metamorphism takes place without application of compression or shear,

aureoles contain hornfels, a nonfoliated metamorphic rock. Contact metamorphism occurs anywhere that the intrusion of plutons occurs. In the context of plate tectonics theory, plutons intrude into the crust at convergent-plate boundaries, in rifts, and during the mountain building that takes place when continents collide.

You can see a classic example of contact metamorphism by traveling to the state of Maine in the northeastern United States. Here you will find a 14-km-long by 4-km-wide granitic pluton, named the Onawa Pluton, which formed about 400 million years ago when an 850°C magma intruded into wall rock comprising 300°C slate, several kilometers below the surface of the Earth. Heat from the magma transformed the slate into hornfels in an aureole that reaches a maximum width of 2 km. Subsequently, erosion stripped off overlying rock, so outcrops of the granite and hornfels can be seen today (**Fig. 8.18b–f**).

Burial Metamorphism

As sediment gets buried in a subsiding sedimentary basin, the pressure increases due to the weight of overburden, and the temperature increases due to the geothermal gradient. In the upper few kilometers, temperatures and pressures are low enough that the changes taking place represent diagenesis. But at depths greater than 8 to 15 km, depending on the geothermal gradient, temperatures may be high enough for metamorphic reactions to begin, and low-grade metamorphic rocks form. Metamorphism due only to the consequences of very deep burial is called **burial metamorphism**. Of note, burial metamorphism destroys the organic molecules of oil; for this reason, oil drillers stop drilling when the bottom of the hole reaches depths at which burial metamorphism has begun.

Dynamic Metamorphism

Faults are surfaces on which one piece of crust slides, or shears, past another. Near the Earth's surface (in the upper 10 to 15 km) this movement can fracture rock, breaking it into angular fragments or even crushing it to a powder. But at greater depths, rock is so warm that it behaves like soft plastic as shear along the fault takes place. During this process, the minerals in the rock recrystallize. We call this process **dynamic metamorphism** because it occurs as a consequence of shearing alone under metamorphic conditions, without requiring a change in temperature or pressure. The resulting rock, called a **mylonite**, is extremely fine grained and has a strong foliation that roughly parallels the fault (**Fig. 8.19**). (More advanced geology books

FIGURE 8.18 Contact ("thermal") metamorphism occurs in an aureole adjacent to an igneous intrusion. This type of metamorphism produces hornfels. The grade of hornfels decreases away from the contact.

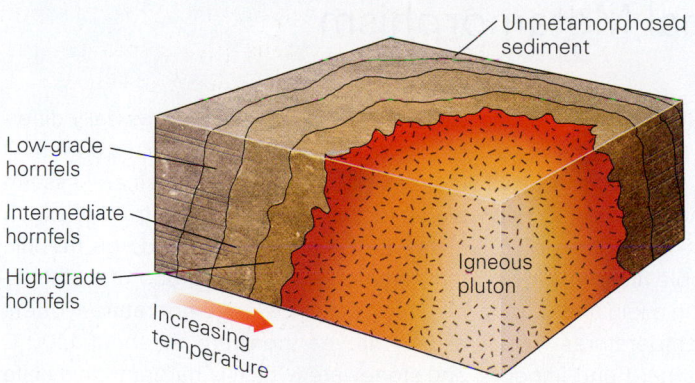

(a) Heat radiated from a large pluton can produce a metamorphic aureole, in which hornfels develops. Grade decreases progressively away from the pluton contact.

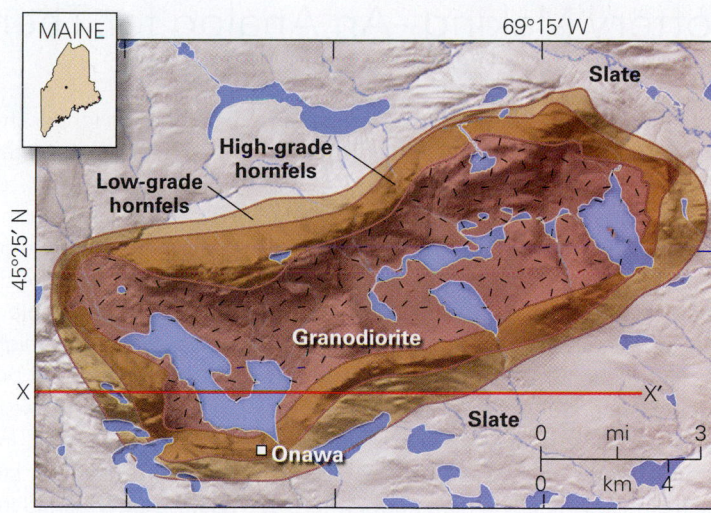

(b) A metamorphic aureole, on a map, appears as a ring around a pluton. This example is the Onawa Pluton in Maine.

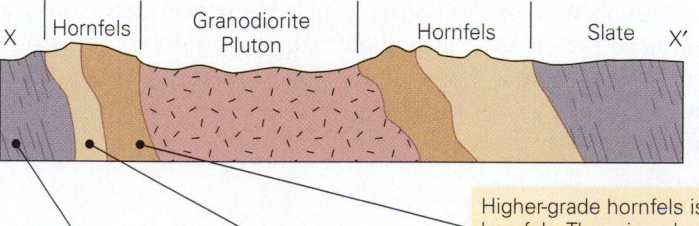

(c) A cross section shows the aureoles in the subsurface. Grade decreases away from the contact of the pluton as indicated by the arrows.

Higher-grade hornfels is coarser than lower-grade hornfels. The mineral assemblage depends on grade.

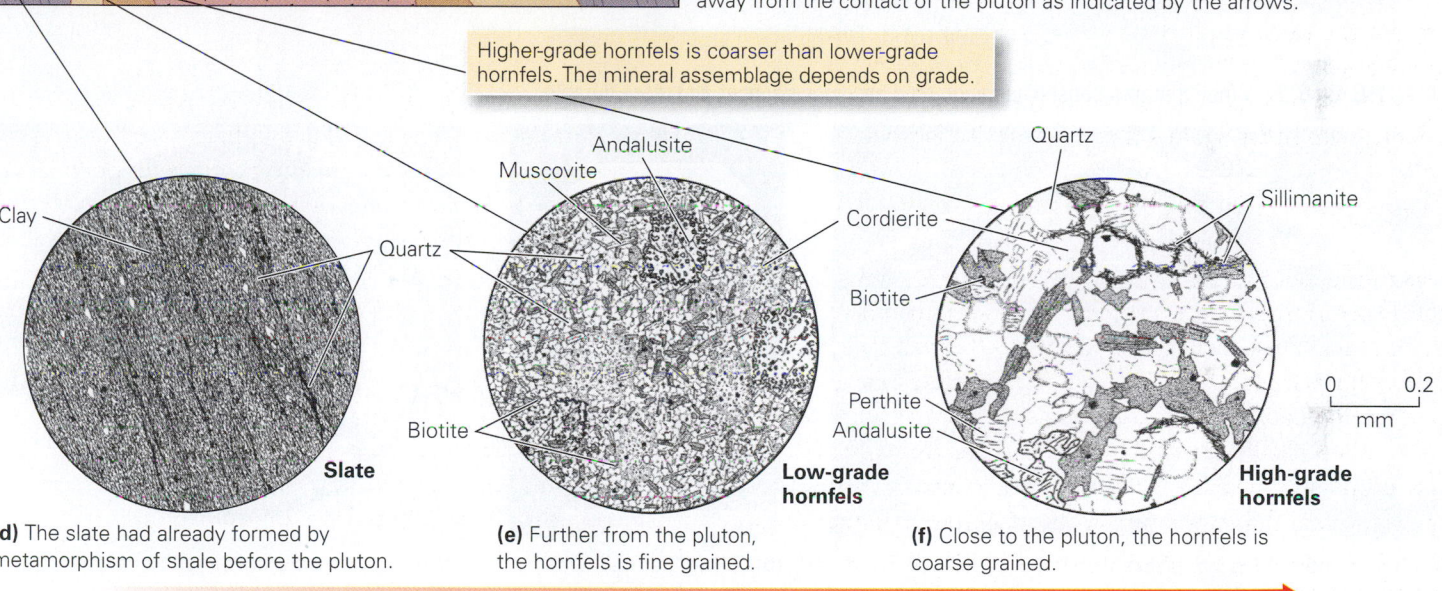

(d) The slate had already formed by metamorphism of shale before the pluton.

(e) Further from the pluton, the hornfels is fine grained.

(f) Close to the pluton, the hornfels is coarse grained.

Increasing grade

explain how the fine grains form from coarser ones.) Dynamic metamorphism takes place anywhere that faulting occurs at depth in the crust. Thus, mylonites can be found at all plate boundaries, in rifts, and in collision zones.

Dynamothermal or Regional Metamorphism

During the development of large mountain ranges, in response to either convergent-margin tectonics or continental collision, broad regions of crust undergo compression and large slices of continental crust slip up and over other portions of the crust along faults. As a consequence, rock that was once near the Earth's surface along the margin of a continent may end up at great depth beneath the mountain range (**Fig. 8.20**). In this environment, three changes happen: (1) the protolith heats up because of the geothermal gradient and because of igneous activity; (2) the protolith endures greater pressure because of the weight of overburden; and (3) the protolith undergoes compression and shearing. As a result of these changes, the protolith transforms into foliated metamorphic

BOX 8.2   CONSIDER THIS . . .

Pottery Making—An Analog for Thermal Metamorphism

A brick for the wall of an adobe house, an earthenware pot, a stoneware bowl, or a translucent porcelain teacup may all be formed from the same lump of soft clay, scooped from the surface of the Earth and shaped by human hands (**Fig. Bx8.2**). This pliable and slimy muck is a mixture of very fine clay minerals and quartz grains formed during the chemical weathering of rock and water. Fine potter's clay for making white china contains a particular clay mineral called *kaolinite*, named after the locality in China (Kauling, meaning high ridge) where it was originally discovered.

People in arid climates make adobe bricks by forming damp clay into blocks, which they then dry in the sun: Such bricks can be used for construction only in arid climates, because if it rains heavily the bricks will rehydrate and turn back into sticky muck—drying clay in the sun does not change the structure of the clay minerals.

To make a more durable material, brick makers place clay blocks in a kiln and "fire," or bake, them at high temperatures. This process makes the bricks hard and impervious to water. Potters use the same process to make jugs. In fact, fired clay jugs that were used for storing wine and olive oil have been found intact in sunken Greek and Phoenician ships that have rested on the floor of the Mediterranean Sea for thousands of years! Clearly, the firing of a clay pot fundamentally and permanently changes clay in a way that makes it physically different. In other words, firing causes a thermal metamorphic change in the mineral assemblage that composes pottery. The extent of the transformation depends on the kiln temperature, just as the grade of metamorphic rock depends on temperature. Potters usually fire earthenware at about 1,100°C and stoneware (which is harder than a knife or fork) at about 1,250°C. To produce porcelain—fine china—the clay must partially melt at even higher temperatures. Just as it begins to melt, the potter cools it quickly. Such quenching of the melt creates glass, which gives porcelain its translucent, vitreous (glassy) appearance.

FIGURE Bx8.2 When metamorphosed, common mud becomes stronger and more durable.

(a) Mud can be shaped into blocks that, when dried, were used to build this house in Peru.

(b) Baking the mud turns it into much harder brick.

(c) At high temperatures, mud turns into porcelain, as in this plate from China.

rock. The type of foliated rock that forms depends on the grade of metamorphism—slate forms at shallower depths, whereas schist and gneiss form at greater depths. Since the metamorphism we've just described involves not only heat but also compression and shearing, we can call it **dynamothermal metamorphism**. Since such metamorphism tends to affect a broad region (tens to even hundreds of kilometers across and hundreds to thousands of kilometers long), geologists also call it **regional metamorphism**. Erosion eventually removes the mountains, exposing a belt of metamorphic rock that once lay at depth.

Hydrothermal Metamorphism at Mid-Ocean Ridges

Hot magma rises beneath the axis of mid-ocean ridges, so when cold seawater sinks through cracks down into the oceanic crust along ridges, it heats up and transforms into hydrothermal fluid.

FIGURE 8.19 The formation of mylonite during dynamic metamorphism in a shear zone.

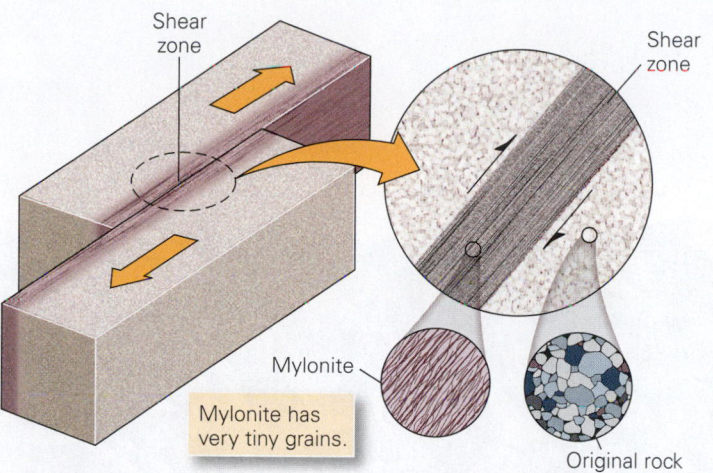

(a) Shearing of a rock under plastic conditions causes original crystals to divide into tiny crystals without breaking to form a mylonite.

(b) Typically, a mylonite has a strong, very fine grain and a strong foliation, as in this example from Ontario.

FIGURE 8.20 The formation of dynamothermal metamorphic rocks during the development of mountain belts. The process is also called regional metamorphism.

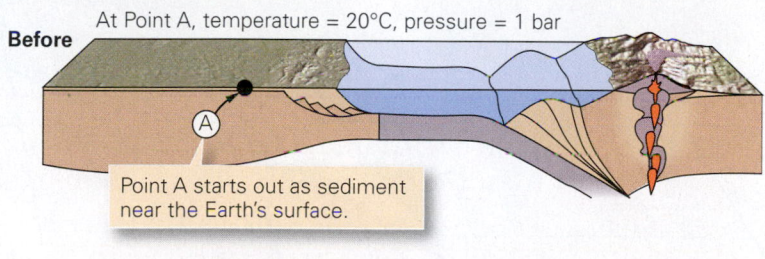

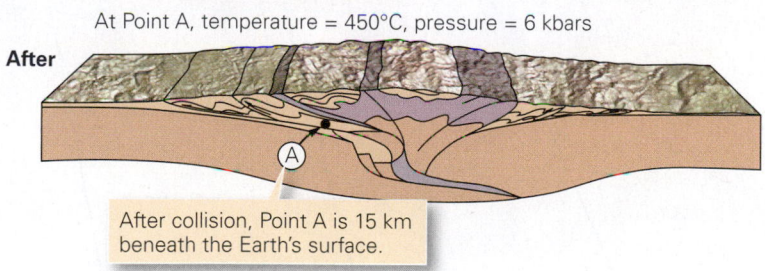

This fluid then rises through the crust, near the ridge, causing **hydrothermal metamorphism** of ocean-floor basalt (**Fig. 8.21a, b**). Eventually, the fluid escapes through vents back into the sea; these vents are called black smokers (see Chapter 2).

Metamorphism in Subduction Zones

Blueschist is a relatively rare rock that contains an unusual blue-colored amphibole. Laboratory experiments indicate that formation of this mineral requires very high pressure but relatively low temperature. Such conditions do not develop in continental crust—usually at the high pressure needed to produce blue amphibole, temperature in continental crust is also high (see Box 8.1). So, to figure out where blueschist forms, we must search for a geologic setting where high pressures can develop at relatively low temperatures.

Plate tectonics theory provides the answer to this puzzle. Researchers found that blueschist occurs only in the accretionary prisms that form at subduction zones (see Geology at a Glance, pp. 254–255). They realized that because prisms grow to be over 20 km thick, rock at the base of the prism feels high pressure (due to the weight of overburden). But because the subducted oceanic lithosphere beneath the prism is cool, temperatures at the base of the prism remain relatively low. Under these conditions, blue amphibole can form. Because of shear between the subducting plate and the overriding plate, blueschist develops a foliation.

Shock Metamorphism

When large meteorites slam into the Earth, a vast amount of kinetic energy instantly transforms into heat, and a pulse of extreme compression (a shock wave) propagates into the Earth. The heat may be sufficient to melt or even vaporize rock at the impact site, and the extreme compression of the shock wave causes the crystal structure of quartz grains in rocks below the impact site to suddenly undergo a phase change to a denser mineral called coesite. The changes in rock due to the passage of a shock wave are called **shock metamorphism**. When astronauts sampled the Moon, they discovered that the regolith covering the lunar surface contains the products of shock metamorphism produced by countless impacts.

Environments of Metamorphism

Metamorphic rocks form when a pre-existing rock (a protolith) undergoes changes in texture and/or mineral content in the solid state in response to changes in temperature, pressure, and/or differential stress. Metamorphism may also reflect interaction with hydrothermal fluids. Some metamorphic rocks are nonfoliated (they do not have metamorphic layering), whereas others are foliated (they do

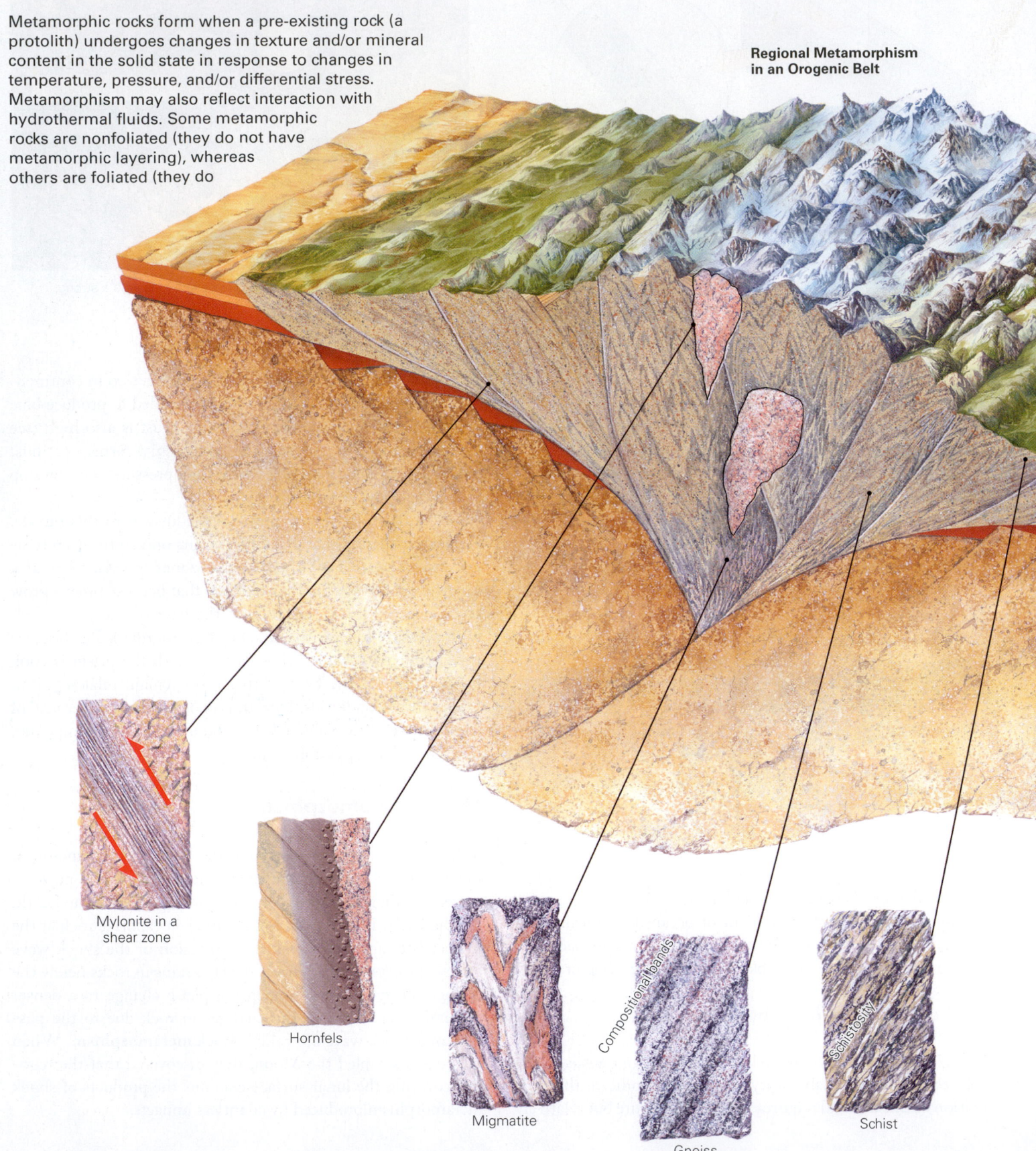

Regional Metamorphism in an Orogenic Belt

Mylonite in a shear zone

Hornfels

Migmatite

Compositional bands

Gneiss

Schistosity

Schist

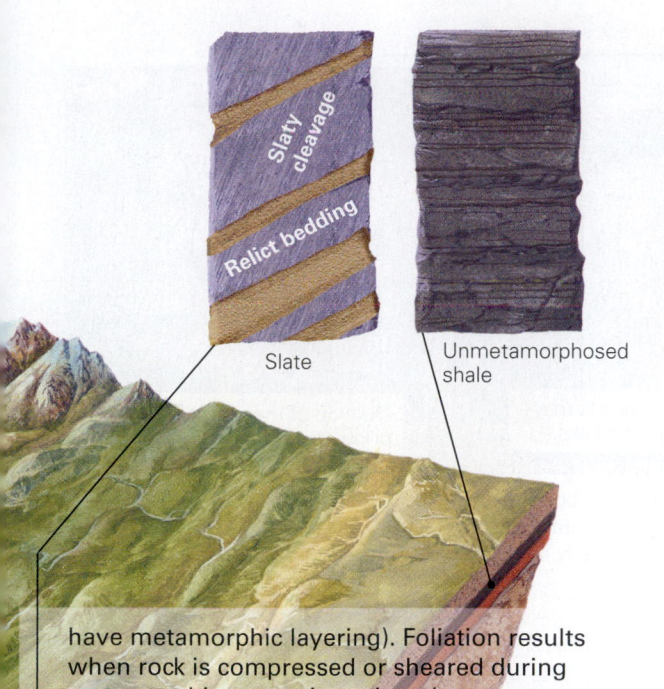

Slate

Unmetamorphosed shale

Slaty cleavage

Relict bedding

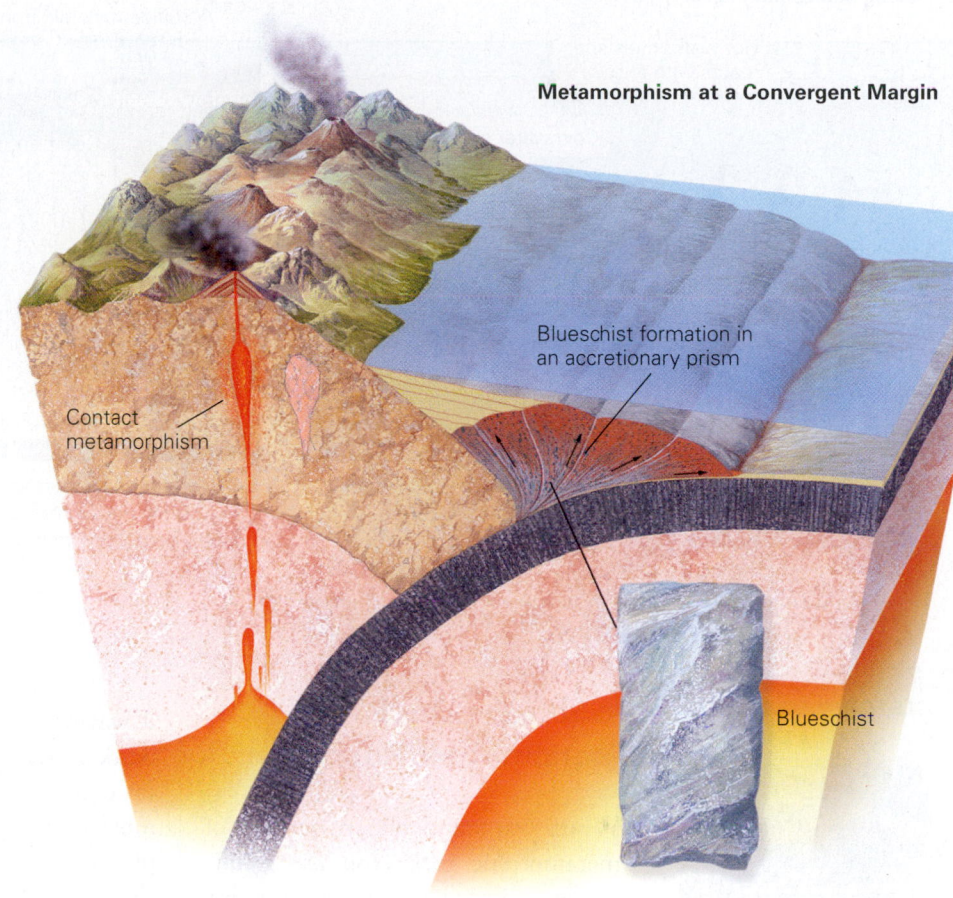

Metamorphism at a Convergent Margin

Contact metamorphism

Blueschist formation in an accretionary prism

Blueschist

have metamorphic layering). Foliation results when rock is compressed or sheared during metamorphism, causing minerals to grow or rotate into parallelism with each other. Dynamothermal (regional) metamorphism occurs during mountain building. Contact metamorphism takes place around an igneous intrusion, or pluton, caused by the heat released by the pluton.

Geologists distinguish among metamorphic rocks according to the type of foliation and the mineral assemblage a rock contains. Hornfels is unfoliated and forms as a result of contact metamorphism. Mylonite develops when shearing creates a foliation but not necessarily a change in types of minerals. Slate, which forms from shale, contains slaty cleavage; clay flakes are typically aligned at an angle to bedding. Schist contains coarse grains of mica (muscovite and/or biotite) aligned parallel to each other. Gneiss has compositional banding. (Migmatite forms when part of the rock melts, and thus it is a mixture of metamorphic and igneous rock.) Quartzite is composed predominantly of quartz (it is metamorphosed sandstone), whereas marble is composed predominantly of calcite or dolomite (it is metamorphosed limestone or dolostone). Quartzite and marble are usually unfoliated.

The types of minerals and foliation in a metamorphic rock indicate the rock's grade. High-grade rocks, such as gneiss, form at higher temperatures and pressures, whereas low-grade rocks, such as schist, form at lower pressures and temperatures. Blueschist is an unusual metamorphic rock that develops under relatively high pressures but relatively low temperatures—the environment of an accretionary prism.

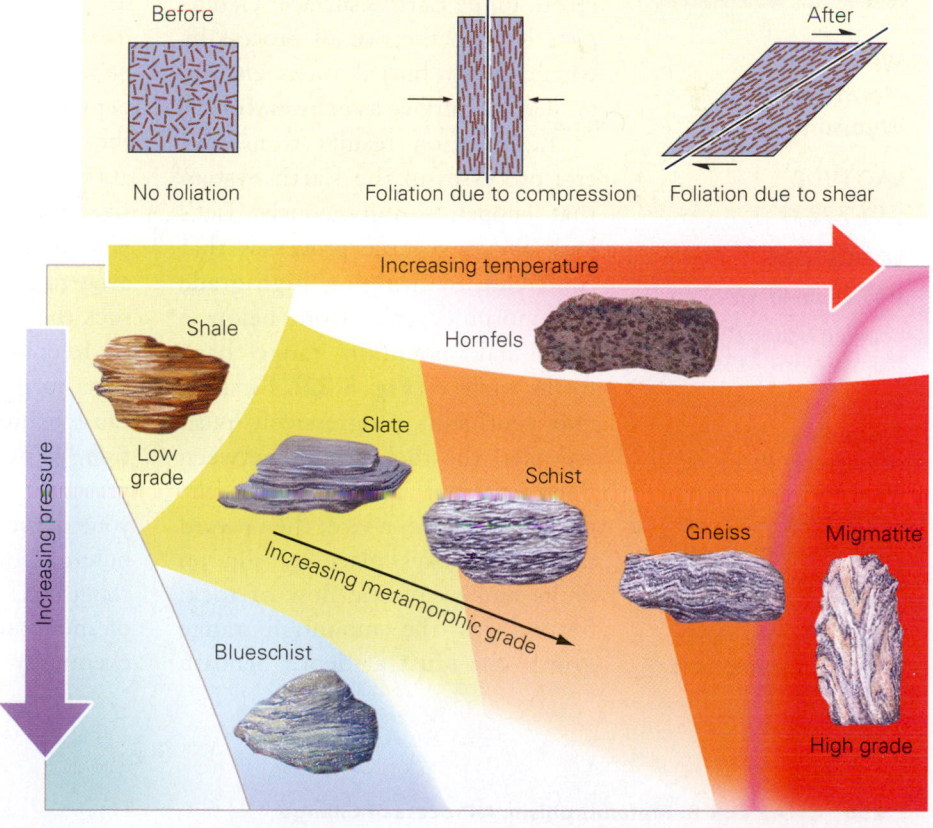

Foliation resulting from deformation

Before

After

No foliation

Foliation due to compression

Foliation due to shear

Increasing temperature

Increasing pressure

Increasing metamorphic grade

Shale

Hornfels

Slate

Schist

Gneiss

Migmatite

Low grade

Blueschist

High grade

FIGURE 8.21 Metamorphism due to hydrothermal circulation along mid-ocean ridges.

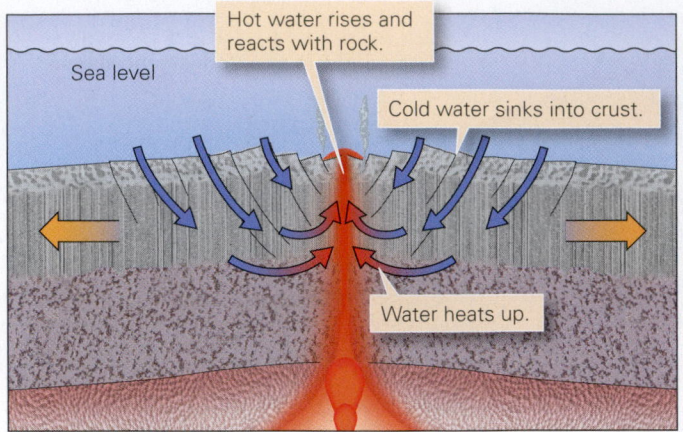

(a) The rising magma at a ridge axis heats water, which then convects. The hot water reacts with the crust and forms metamorphic minerals.

(b) Hydrothermal metamorphism in oceanic crust produces greenish minerals (chlorite and epidote). Sulfide minerals typically precipitate from water mostly at the surface.

Wind River Mountains, Wyoming

LATITUDE
43°6'7.22"N

LONGITUDE
109°21'36.45"W

Looking obliquely from 20 km (~ 12.5 mi).

Faulting uplifted the Precambrian basement of western Wyoming 40 and 80 mya. You can see that the overlying cover of sedimentary strata was warped into a huge fold. The main part of the range contains metamorphic rock.

Where Do You Find Metamorphic Rocks?

When you stand on an outcrop of metamorphic rock, you are standing on material that once lay many kilometers beneath the surface of the Earth. How does metamorphic rock return to the Earth's surface? Geologists refer to the overall process by which deeply buried rocks end up back at the surface as **exhumation**.

Exhumation results from several processes in the Earth System that happen simultaneously. Let's look at the specific processes that contribute to bringing high-grade metamorphic rocks from below a collisional mountain range back to the surface (**Fig. 8.22**). First, as two continents progressively push together, the rock caught between them squeezes upward, much like dough pressed in a vise; the upward movement takes place by slip on faults and by plastic flow of rock. Second, as the mountain range grows, the crust at depth beneath it warms up and becomes softer and weaker. Eventually, the range starts to collapse under its own weight, much like a block of soft cheese placed in the hot sun. As a result of this collapse, the upper crust spreads out laterally. Stretching of the upper part of the crust in the horizontal direction causes it to become thinner in the vertical direction, and as the upper part of the crust becomes thinner the deeper crust ends up closer to the surface. Third, erosion takes place at the surface; weathering, landslides, river flow, and glacial flow together play the role of a giant rasp, stripping away rock at the surface and exposing rock that was once below the surface.

Keeping in mind the processes that form metamorphic rock and cause exhumation, let's ask the question "Where are metamorphic rocks presently exposed?" You can start your quest to find metamorphic rock outcrops by hiking into a mountain range. As we've seen, the process of mountain building produces and eventually exhumes metamorphic

Canadian Shield

LATITUDE
61°09'50.15"N

LONGITUDE
76°42'43.88"W

Zoom to an elevation of 225 km (~ 140 mi) and look straight down.

The Canadian Shield east of Hudson Bay is covered by sparse vegetation. Large areas of metamorphic rock are exposed. Differential erosion of a gneiss defines a band of foliation.

FIGURE 8.22 Processes that bring metamorphic rock back to Earth's surface. Three phenomena contribute to exhumation of rocks at depth. Here the red dot (representing metamorphic rocks formed at the base of a mountain range) gets progressively closer to the surface over time.

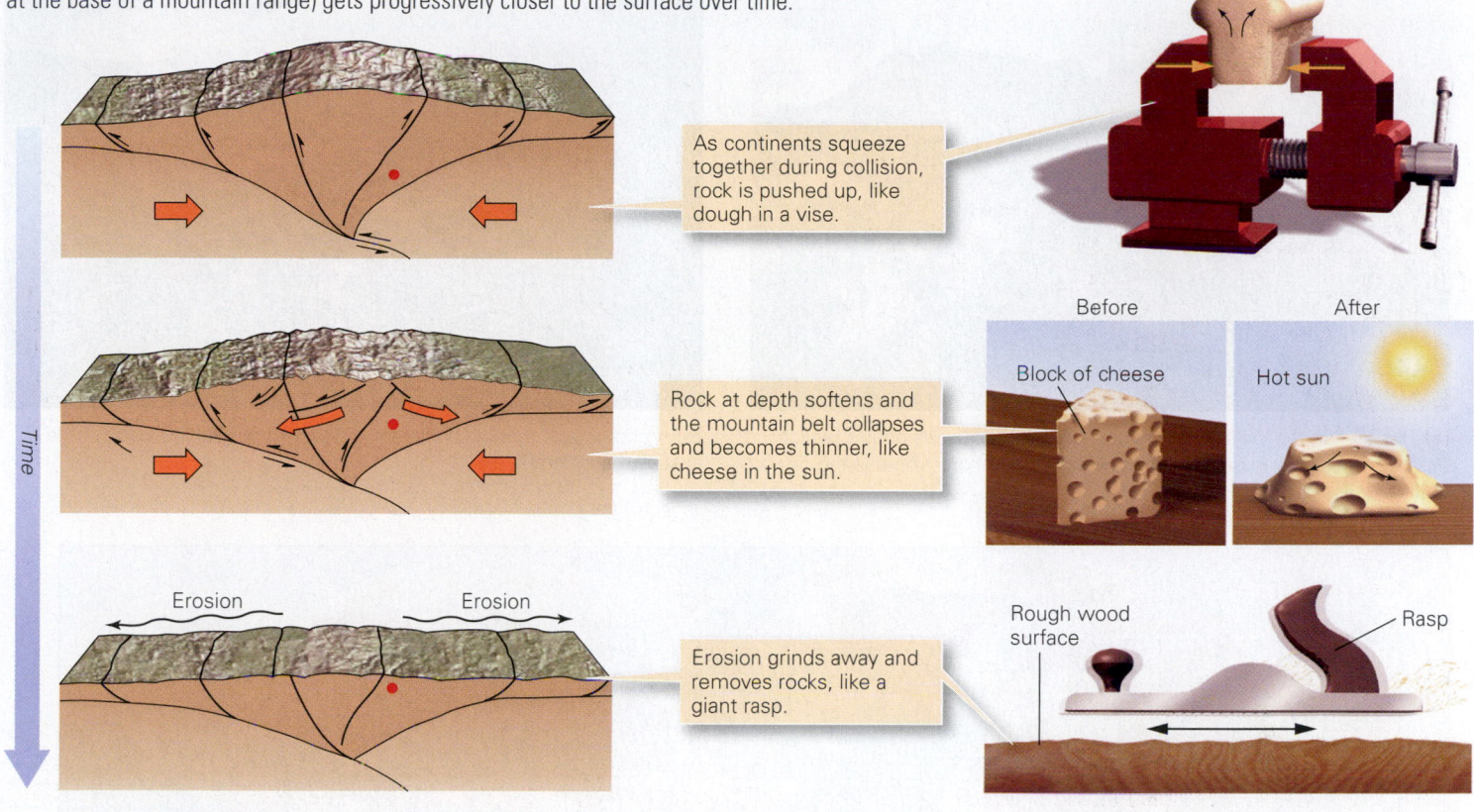

As continents squeeze together during collision, rock is pushed up, like dough in a vise.

Rock at depth softens and the mountain belt collapses and becomes thinner, like cheese in the sun.

Before — Block of cheese
After — Hot sun

Erosion grinds away and removes rocks, like a giant rasp.

Rough wood surface — Rasp

Time

Erosion   Erosion

rocks. The towering cliffs in the interior of a mountain range typically reveal schist, gneiss, and quartzite (**Fig. 8.23a**). Even after the peaks have eroded away, the record of mountain building remains in the form of a belt of metamorphic rock at the ground surface. Vast expanses of metamorphic rock crop out in continental shields. A **shield** is a broad region of long-lived, stable continental crust where Phanerozoic sedimentary cover either was not deposited or has been eroded away so that Precambrian rocks are exposed (**Fig. 8.23b, c**). These rocks were metamorphosed during a succession of Precambrian mountain-building events that led to the original growth of continents.

Take-Home Message

Thermal (contact) metamorphism develops around igneous intrusions due to heat from a pluton, and dynamothermal (regional) metamorphism develops beneath mountain ranges where rock undergoes compression and shear at high temperatures and pressures. Metamorphism can also happen in response to deep burial, reaction with hydrothermal fluids, and meteorite impact. Erosion and uplift eventually expose metamorphic rock in mountain ranges or continental shields.

QUICK QUESTION: Could you find a layer of metamorphic rock between the layers of sedimentary rock in a sedimentary basin? Why or why not?

FIGURE 8.23 Examples of rock exposures consisting of Precambrian metamorphic rocks.

(a) This cliff, in the Wasatch Mountains of Utah, exposes gneiss, which has been intruded by granite. The gneiss has foliation, whereas the granite does not.

(b) A photograph from an airplane window of the flat landscape of the eastern Canadian Shield.

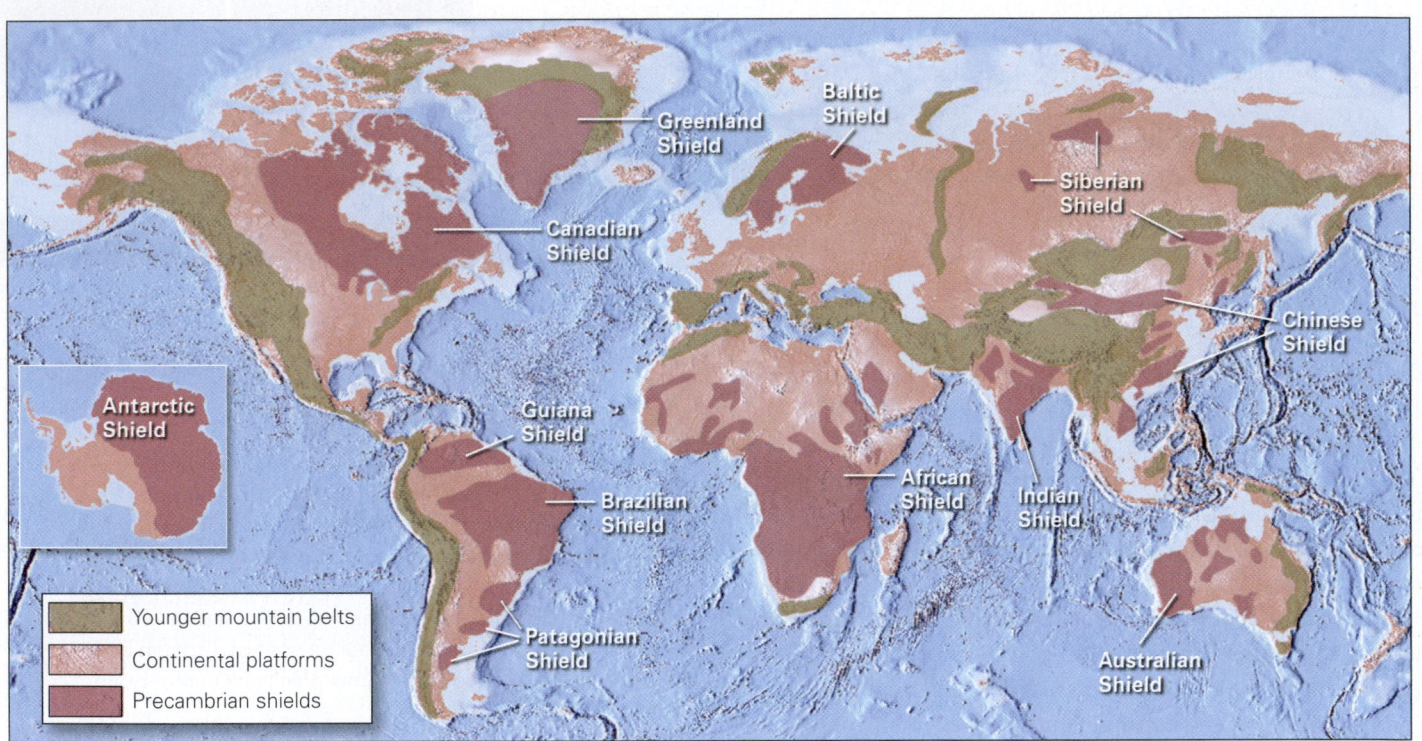

(c) A map showing the distribution of shields, areas where broad expanses of Precambrian crust, including Precambrian metamorphic rocks, crop out.

CHAPTER SUMMARY

- Metamorphism refers to changes in a rock that result in the formation of a metamorphic mineral assemblage, and/or metamorphic texture, in response to change in temperature and/or pressure, to the application of differential stress, and to interaction with hydrothermal fluids (hot-water solutions).

- Metamorphism involves recrystallization, phase changes, metamorphic reactions (neocrystallization), pressure solution, and/or plastic deformation. If hydrothermal fluids bring in or remove elements, we say that metasomatism has occurred.

- Metamorphic foliation can be defined either by preferred mineral orientation (aligned inequant crystals) or by compositional banding. Preferred mineral orientation develops where differential stress causes the compression and shearing of a rock so that its inequant grains align parallel with each other.

- Geologists separate metamorphic rocks into two classes, foliated rocks and nonfoliated rocks, depending on whether the rocks contain foliation.

- The class of foliated rocks includes slate, phyllite, metaconglomerate, schist, and gneiss. The class of nonfoliated rocks includes hornfels, quartzite, and marble. Migmatite, a mixture of igneous and metamorphic rock, forms under conditions where melting begins.

- Rocks formed under relatively low temperatures are known as low-grade rocks, whereas those formed under high temperatures are known as high-grade rocks. Intermediate-grade rocks develop between these two extremes. Different metamorphic mineral assemblages form at different grades.

- Geologists track the distribution of different grades of rock by looking for index minerals. Isograds indicate the location at which index minerals first appear. A metamorphic zone is the region between two isograds.

- A metamorphic facies is a group of metamorphic mineral assemblages that develop under a specified range of temperature and pressure conditions.

- Thermal metamorphism (also called contact metamorphism) occurs in an aureole surrounding an igneous intrusion. Burial metamorphism occurs at depth in a sedimentary basin. Dynamically metamorphosed rocks form along faults, where rocks undergo plastic shearing. Dynamothermal metamorphism (also called regional metamorphism) results when rocks undergo heating and shearing during mountain building. Hydrothermal metamorphism takes place due to the circulation of hot water in oceanic crust at mid-ocean ridges. Shock metamorphism happens during the impact of a meteorite.

- We find belts of metamorphic rocks in mountain ranges. Blueschist forms in accretionary prisms. Shields expose broad areas of Precambrian metamorphic rocks.

GUIDE TERMS

burial metamorphism (p. 250)
contact metamorphism (p. 250)
differential stress (p. 238)
dynamic metamorphism (p. 250)
dynamothermal metamorphism (p. 252)
exhumation (p. 256)
gneiss (p. 241)
hornfels (p. 243)

hydrothermal metamorphism (p. 253)
index mineral (p. 249)
marble (p. 244)
metaconglomerate (p. 241)
metamorphic aureole (p. 250)
metamorphic facies (p. 248)
metamorphic foliation (p. 235)
metamorphic grade (p. 245)
metamorphic mineral (p. 235)

metamorphic rock (p. 234)
metamorphic texture (p. 235)
metamorphic zone (p. 249)
metamorphism (p. 234)
metasomatism (p. 240)
migmatite (p. 242)
mylonite (p. 250)
phyllite (p. 241)
preferred orientation (p. 238)
pressure (p. 237)
protolith (p.234)

quartzite (p. 244)
regional metamorphism (p. 252)
schist (p. 241)
shield (p. 257)
shock metamorphism (p. 253)
slate (p. 241)
thermal metamorphism (p. 250)
vein (p. 240)

REVIEW QUESTIONS

1. How are metamorphic rocks different from igneous and sedimentary rocks?

2. What two features characterize most metamorphic rocks?

3. What phenomena can cause metamorphism?

4. What is metamorphic foliation, and how does it form?

5. How does slate differ from a phyllite? How does phyllite differ from a schist? How does schist differ from a gneiss?

6. Why is hornfels nonfoliated?

7. What is a metamorphic grade, and how can it be determined? How does grade differ from facies?

8. Describe the geologic settings where thermal, dynamic, and dynamothermal metamorphism take place.

9. Why does metamorphism happen at the site of meteor impacts and along mid-ocean ridges?

10. How does plate tectonics explain the combination of low-temperature but high-pressure minerals found in a blueschist?

11. Where would you go if you wanted to find exposed metamorphic rocks, and how did such rocks return to the surface of the Earth after being at depth in the crust?

ON FURTHER THOUGHT

12. Do you think that you would be likely to find a broad region (hundreds of kilometers across and hundreds of kilometers long) in which the outcrop consists of high-grade hornfels? Why or why not? (Hint: Think about the causes of metamorphism and the conditions under which a hornfels forms.)

13. Would we likely find broad regions of gneiss and schist on the Moon? Why or why not?

smartwork smartwork.wwnorton.com

This chapter's Smartwork features:

- Ranking activities to identify the geologic settings of metamorphism.
- Interactive simulations covering earthquakes.
- Visual identification tasks on plate boundaries.

GEOTOURS

This chapter's GeoTour exercise (G) features:

- Zones of metamorphic grade
- Types of metamorphism
- Metamorphism: past and present

Another View This outcrop of marble (metamorphosed limestone) forms the wall of Mosaic Canyon, near Death Valley, California. During metamorphism, the rock flowed somewhat like plastic, producing the horizontal foliation. During modern floods, sediment-laden water has sculpted and polished the face of the outcrop.

~15 cm

Sedimentary strata, Utah

Metamorphic rock, Utah

Igneous rock forming, Hawaii

The Rock Cycle in the Earth System

On our dynamic planet, the atoms in a rock of one class may, over time, be incorporated in rock of another class, and then another. Such transformations comprise the rock cycle.

LEARNING OBJECTIVES

By the end of this interlude, you should understand . . .

- that rocks don't always last forever because the Earth System is dynamic, so phenomena such as uplift, weathering, burial, heating, and/or melting can occur.
- how components of a given rock may, over time, become incorporated in other rocks or even other rock classes, a progression of change called the rock cycle.
- how pathways through the rock cycle reflect geologic settings, in the context of plate tectonics.
- that steps of the rock cycle can happen at vastly different rates.
- how energy from inside the Earth, from the Sun, and from gravity drive the rock cycle.

C.1 Introduction

"Stable as a rock." This familiar expression implies that a rock, once formed, is a permanent entity. But it isn't necessarily. In the time frame of Earth history, a span of over 4.54 billion years, components making up a rock of one class may later be rearranged or moved elsewhere to form another rock of the same class or a different rock class altogether. In some places, this process of change has happened many times. Geologists refer to the progressive transformation of Earth materials from one rock to another over time as the **rock cycle** (Fig. C.1). Discussion of the rock cycle illustrates important relationships among the rock classes described in the previous three chapters. In this interlude, we'll illustrate how to apply the concept of the rock cycle and will conclude by showing how it represents one of many important cycles in the Earth System.

C.2 Pathways through the Rock Cycle

The Variety of Paths Reflect Geologic History

By following the arrows in Figure C.1, you can see that there are many paths around or through the rock cycle. For example, igneous rock formed by solidification of a melt that rose from the mantle may undergo weathering and erosion to produce sediment. Later, burial and lithification transforms the sediment into a new sedimentary rock. This sedimentary rock may, in turn, become buried so deeply that it transforms into metamorphic rock. Extreme heating of the metamorphic rock might cause it to partially melt and produce new magma. This new magma might later solidify to form a new igneous rock. We can express this path as follows:

igneous → sedimentary → metamorphic → igneous

FIGURE C.1 The stages of the rock cycle showing various alternative pathways.

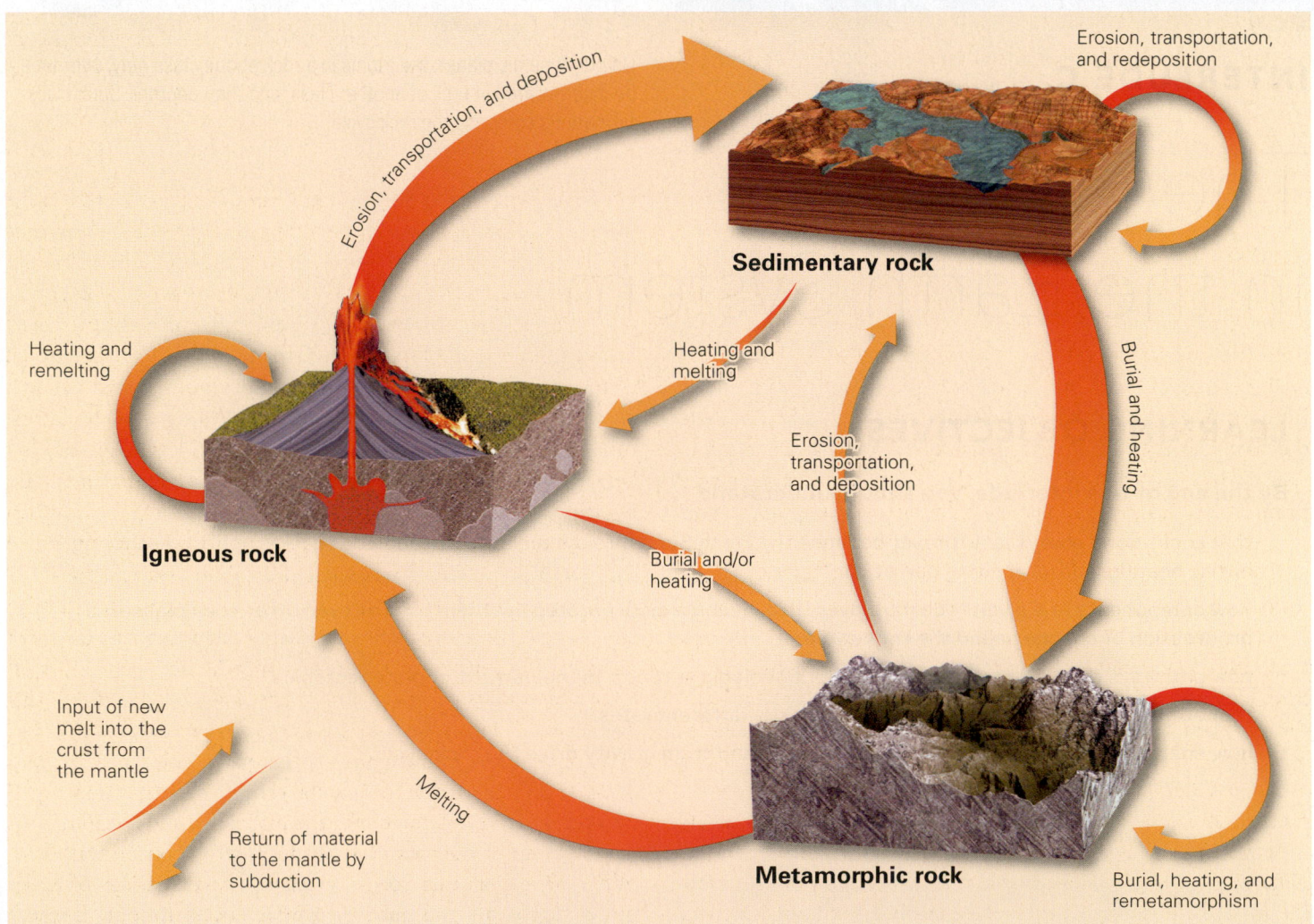

Erosion, transportation, and deposition

Erosion, transportation, and redeposition

Sedimentary rock

Heating and remelting

Heating and melting

Burial and heating

Erosion, transportation, and deposition

Burial and/or heating

Igneous rock

Input of new melt into the crust from the mantle

Return of material to the mantle by subduction

Melting

Metamorphic rock

Burial, heating, and remetamorphism

Given local geologic conditions, a rock at one stage in the cycle could follow another path. For example, the metamorphic rock, once formed, could itself be uplifted and eroded to form new sediment and, later, new sedimentary rock, without melting. This path takes a shortcut through the cycle that we can illustrate as follows:

igneous → sedimentary → metamorphic → sedimentary

Likewise, the original igneous rock might have been deeply buried before it was eroded and could undergo metamorphism directly, without first turning to sediment. The resulting metamorphic rock might then be uplifted and eroded to produce sediment that becomes sedimentary rock, defining another shortcut path:

igneous → metamorphic → sedimentary

Not all steps have to yield a new rock type. For example, if a sedimentary rock, once formed, were uplifted and eroded to form a new sediment and then buried and lithified, the pathway would be:

sedimentary → sedimentary

To get a clearer sense of how the rock cycle works, let's look at a case study of the rock cycle, in the context of plate tectonics.

C.3 A Case Study of the Rock Cycle

Material may originally enter the rock cycle when magma rises from the mantle. Suppose the magma erupts and forms basalt (an igneous rock) at a continental hot-spot volcano (**Fig. C.2a**). Interaction with wind, rain, and vegetation gradually weathers the basalt, physically breaking it into smaller fragments and chemically weathering it to yield clay. Water washes the newly formed clay away and transports it downstream—if you've ever seen a brown-colored river, you've seen clay traveling to a site of deposition. Eventually the river reaches the sea, where the water slows down and the clay settles out.

Let's imagine, in our example, that the clay accumulates as a deposit of mud along the margin of Continent X. Gradually, through time, the mud becomes buried to a depth of 6 km and the clay flakes pack tightly together to yield a new sedimentary rock, shale. The shale resides 6 km below the continental shelf for millions of years, until Continent X collides with Continent Y. As the two continents squeeze together, slip on faults transports the edge of Continent Y up and over the shale of Continent X as mountains grow. The shale that had once been only 6 km below the surface ends up at a depth of 30 km below the surface, and

under the pressure and temperature conditions existing at this depth, it metamorphoses into schist (**Fig. C.2b**).

The story's not over. Once mountain building stops, erosion grinds away the mountain range, and exhumation brings some of the schist back up to the ground surface. Some of this schist erodes to form sediment, which is carried off and deposited elsewhere to form new sedimentary rock (perhaps sandstone, siltstone, and shale). The rest remains preserved below the surface (**Fig. C.2c**). Eventually, continental rifting takes place at the site of the former mountain range, and the crust containing the schist begins to split apart. Injection of new, hot magma may cause some of the schist to partially melt and a new felsic magma forms. This felsic magma rises to the surface of the crust and freezes into rhyolite, a new igneous rock (**Fig. C.2d**). In terms of the rock cycle, we're back at the beginning, having once again made igneous rock.

Note that atoms, as they pass through the rock cycle, do not always stay within the same mineral. In our example, a silicon atom in a pyroxene crystal of the basalt may become part of a clay crystal in the shale, part of a mica crystal in the schist, or part of a quartz crystal in the rhyolite. Similarly, atoms that were adjacent in the starting rock don't necessary end up near each other at a later stage in the rock cycle. Moving water or wind, for example, may carry one atom of what was originally a single mineral crystal to a location hundreds or even thousands of kilometers away from what was originally an adjacent atom.

Rates of Transfer

We have seen that not all atoms pass through the rock cycle in the same way. Similarly, not all atoms pass through the rock cycle at the same rate, and for that reason we find rocks of many different ages at the surface of the Earth—some rocks have remained in one form for less than a few million years, while others have stayed unchanged for most of Earth's history. For example, rocks exposed in the Appalachian Mountains today may have passed through stages of the rock cycle many times during the past several hundred million years, for the eastern margin of North America has been subjected to multiple events of basin formation, mountain building, and rifting during this time. In contrast, Precambrian chert beds now in the interior of a continent may have remained unchanged since they first lithified over 3 billion years ago.

Most atoms that comprise continental rocks never return to the mantle because continental crust is buoyant and does not subduct. However, a small amount of sediment that erodes off a continent ends up in deep-ocean trenches, and some of this does eventually get carried back into the mantle by subduction. Also, recent research suggests that metamorphic and igneous rocks at the base of the continental crust locally may be scraped off and transported down into the mantle by subduction.

Our tour of the rock cycle has focused on continental rocks. What about oceanic rocks? Oceanic crust consists of igneous rock (basalt and gabbro) overlain by sediment. Because

FIGURE C.2 An example of the rock cycle in the context of plate tectonics.

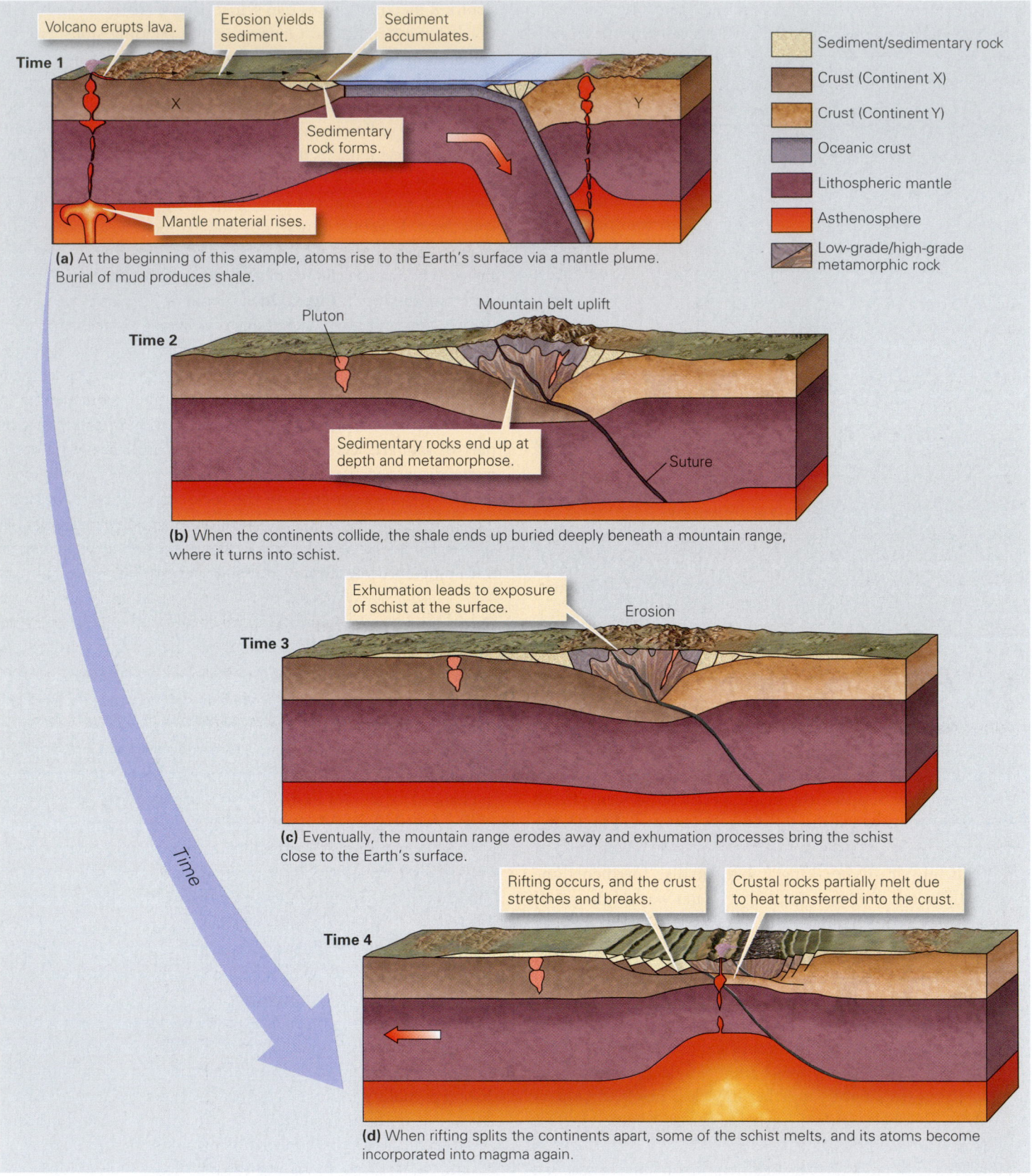

Volcano erupts lava.

Erosion yields sediment.

Sediment accumulates.

Time 1

Sedimentary rock forms.

Mantle material rises.

Sediment/sedimentary rock

Crust (Continent X)

Crust (Continent Y)

Oceanic crust

Lithospheric mantle

Asthenosphere

Low-grade/high-grade metamorphic rock

(a) At the beginning of this example, atoms rise to the Earth's surface via a mantle plume. Burial of mud produces shale.

Mountain belt uplift

Pluton

Time 2

Sedimentary rocks end up at depth and metamorphose.

Suture

(b) When the continents collide, the shale ends up buried deeply beneath a mountain range, where it turns into schist.

Exhumation leads to exposure of schist at the surface.

Erosion

Time 3

(c) Eventually, the mountain range erodes away and exhumation processes bring the schist close to the Earth's surface.

Time

Rifting occurs, and the crust stretches and breaks.

Crustal rocks partially melt due to heat transferred into the crust.

Time 4

(d) When rifting splits the continents apart, some of the schist melts, and its atoms become incorporated into magma again.

a layer of water blankets the crust, erosion does not affect it, so oceanic crustal rock generally does not follow the path into the sedimentary loop of the rock cycle. But sooner or later, oceanic crust subducts. When this happens, the rock of the crust undergoes metamorphism, for as it sinks it endures progressively higher temperatures and pressures. Most oceanic igneous rock eventually gets carried down into the mantle. A small part of this subducted rock may melt and be incorporated in magma that rises and becomes new igneous rock, but most eventually mixes back into the mantle.

What Drives the Rock Cycle?

The rock cycle occurs because the Earth is a dynamic planet and there are many different rock-forming environments (see **Geology at a Glance**, pp. 266–267). Our planet's internal heat and gravitational field together ultimately drive plate movements and plume-associated hot spots. Plate interactions, in turn, cause the uplift of mountain ranges, a process that exposes rock to weathering and erosion, which leads to sediment production. Plate interactions also generate the geologic settings in which pre-existing rock melts and produces magma, metamorphism occurs, and sedimentary basins develop.

At the surface of the Earth, heat from the Sun, together with gravity, drive wind, rain, ice, and currents. These agents of weathering and erosion grind away at the surface of the Earth and send material into the sedimentary loop of the cycle. In the Earth System, life also plays a key role in the rock cycle by adding corrosive oxygen to the atmosphere and by directly contributing to weathering. In sum, external energy (solar heat), internal energy (Earth's internal heat), gravity, and life all play a role in driving the rock cycle by keeping the mantle, crust, atmosphere, and oceans in motion and by making portions of the system chemically reactive.

C.4 Cycles of the Earth System

In a general sense, a **cycle**, from the Latin word *cyclus*, meaning circle, is a series of interrelated events or steps that occur in succession and can be repeated. During a *temporal cycle*—such as the phases of the Moon, the seasons of the year, or the tides—events happen according to a timetable, but the materials involved do not necessarily change. During a **mass-transfer cycle**, materials physically transfer among different regions of the Earth System. We refer to each region that holds the material as a **reservoir**—you can think of a reservoir as a holding tank that incorporates the material for a period of time. Examples of reservoirs include the ocean, rock, the atmosphere, the sea, and living organisms. Geologists refer to the time during which a component stays within a given reservoir as the **residence time** in the reservoir. Residence times can vary from hours, as is the case for cycles involving transfer of volatile elements between land and air, to millions or billions of years, as is the case for the rock cycle.

Some mass-transfer cycles are considered to be *geochemical cycles* in that they involve primarily physical components in the Earth System, whereas others are *biogeochemical cycles* in that they involve both physical and living components (**Fig. C.3**). Further, some cycles involve many different reservoirs, above, below, and

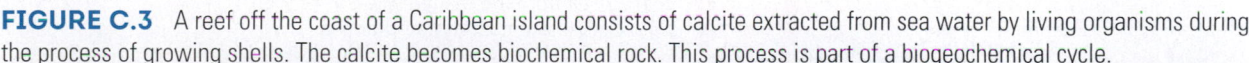

FIGURE C.3 A reef off the coast of a Caribbean island consists of calcite extracted from sea water by living organisms during the process of growing shells. The calcite becomes biochemical rock. This process is part of a biogeochemical cycle.

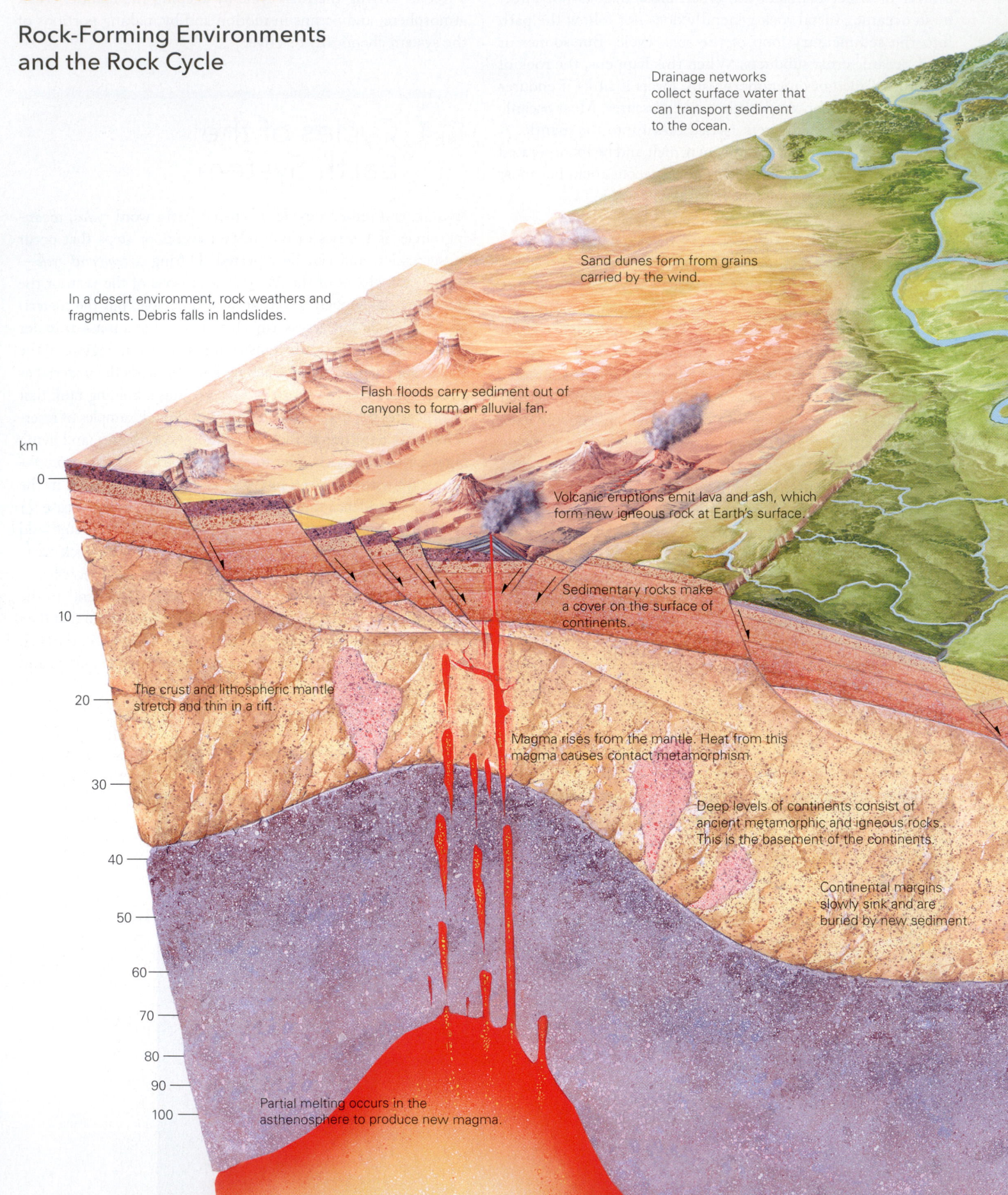

GEOLOGY AT A GLANCE

Rock-Forming Environments and the Rock Cycle

Drainage networks collect surface water that can transport sediment to the ocean.

Sand dunes form from grains carried by the wind.

In a desert environment, rock weathers and fragments. Debris falls in landslides.

Flash floods carry sediment out of canyons to form an alluvial fan.

Volcanic eruptions emit lava and ash, which form new igneous rock at Earth's surface.

Sedimentary rocks make a cover on the surface of continents.

The crust and lithospheric mantle stretch and thin in a rift.

Magma rises from the mantle. Heat from this magma causes contact metamorphism.

Deep levels of continents consist of ancient metamorphic and igneous rocks. This is the basement of the continents.

Continental margins slowly sink and are buried by new sediment.

Partial melting occurs in the asthenosphere to produce new magma.

km

0

10

20

30

40

50

60

70

80

90

100

Glaciers erode rock and can transport sediment of all sizes.

In a region of continental collision, rocks that were near the surface are deeply buried and metamorphosed.

In humid climates, thick soils develop.

Magma that cools and solidifies underground forms igneous intrusions.

Along coastal plains, rivers meander. Sediment collects in the channel and floodplain.

Where a river enters the sea, sediment settles out to form a delta.

Reefs grow from calcite-secreting organisms. These will eventually turn into limestone.

Many different kinds of sediment accumulate along coastlines, building out a continental shelf.

Underwater avalanches carry a cloud of sediment that settles to form a submarine fan.

Fine clay and plankton shells settle on the oceanic crust.

The oceanic crust consists of igneous rocks formed at a mid-ocean ridge.

Rocks form in many different environments. Igneous rocks develop where melt rises from depth and cools; sedimentary rocks form at or near the surface in many different environments; and metamorphic rocks form deep underground due to either heating by plutons or deep burial beneath a mountain belt. Since the Earth is dynamic—because of plate motion and the circulation of water, ice, and air at the surface—environments at a given location change over time, so atoms don't necessarily stay within rock type for all of geologic time. This progressive shift of material over time, from one rock to another, is the rock cycle. There are many pathways through the rock cycle; which pathway (or succession of pathways) takes place depends on the geologic setting.

on the surface of the Earth, whereas others involve just a few. We'll be discussing several cycles later in this book. Examples include the *hydrological cycle* (movement of water from sea to air to rain to stream and/or to and from ice and living tissue) and the *carbon cycle* (movement of carbon between living material, water solutions, air, calcite or fossil fuel).

We can consider the rock cycle to be either a geochemical or biogeochemical cycle in the Earth System, depending on circumstances. For example, transfer of chemicals from the sedimentary rock reservoir to the metamorphic rock reservoir may be entirely physical, without the involvement of life—the

process could happen simply due to heating and/or burial. In fact, such a transfer doesn't necessarily involve long-distance movement of atoms but could just entail a rearrangement of atoms in place. In some cases, however, the transfer of material from one reservoir to another involves life and may occasion movement of materials from one location to another far away. For example, the weathering and erosion of an igneous rock produces clasts and ions in solution. The ions may be carried to the sea, where they become incorporated in the shells of living organisms that later settle to become part of a new sedimentary rock.

INTERLUDE SUMMARY

- A given rock doesn't necessarily last forever. The atoms in a rock may, over time, be incorporated in different rock types. This transfer of atoms progressively from one rock type to another, over time, is the rock cycle.
- Not all atoms follow the same path through the rock cycle. For example, an igneous rock could later become eroded and turned into sediment, which becomes a sedimentary

rock that may eventually be metamorphosed. Or the igneous rock could be metamorphosed directly.
- The rock cycle happens because the Earth is dynamic, and there are internal and external sources of energy driving melting, uplift, faulting, weathering, erosion, and burial.
- The rock cycle is one of many cycles in the Earth System.

GUIDE TERMS

cycle (p. 265)   rock cycle (p. 262)   residence time (p. 265)
mass-transfer cycle (p. 265)   reservoir (p. 265)

REVIEW QUESTIONS

1. Once formed, does a rock necessarily last for all of Earth's history?
2. Define the rock cycle, and give three examples of pathways through it.

3. Have all rocks on Earth passed through the rock cycle the same number of times? Explain your answer.
4. Is there a rock cycle on the Moon? Why or why not?

Another View   In this view up a canyon in the Rocky Mountains of Colorado, we see the rock cycle at work. The cliff consists of metamorphosed sandstone and shale. It breaks into blocks that tumble downslope. Some make it into the stream below, where they break up further, eventually becoming sand and clay that will be deposited downstream as sediment. The sediment might later be buried and turn into new sedimentary rock.

PART III

TECTONIC ACTIVITY OF A DYNAMIC PLANET

Cultures around the world, recognizing the dynamic nature of the Earth, have conjured myriad imaginative explanations for different earth-shaking events. The ancient Greeks and Romans attributed volcanic eruptions to sparks and smoke flying from the fire god's forges. The Maori believed that movements of the god Rūaumoko within the Earth Mother's womb caused earthquakes. In Norse folklore, Odin and his brothers slew the frost giant Ymir and fashioned the mountains from his bones. We now know that these phenomena are all a result of plate movements and interactions. In Part III, each chapter focuses on one of the dramatic consequences of such tectonic activity in the Earth System: volcanoes (Chapter 9), earthquakes (Chapter 10), and mountains (Chapter 11). Interlude D explores how seismic (earthquake) waves reveal the composition of Earth's interior and how that composition generates the magnetic and gravity fields surrounding the planet. Why does molten rock rise like a fountain out of the ground? Why can the ground shake enough to topple a city? What processes cause the land surface to rise to form mountain belts? How do rocks bend, squash, stretch, and break? Read on, and you will be able to answer these questions and understand some of the deadly natural hazards that threaten society.

9 **The Wrath of Vulcan: Volcanic Eruptions**

10 **A Violent Pulse: Earthquakes**

D **The Earth's Interior, Revisited: Seismic Layering, Gravity, and the Magnetic Field**

11 **Crags, Cracks, and Crumples: Crustal Deformation and Mountain Building**

These mountains, in Utah, emphasize the continuing interplay between plate-tectonic forces displacing the Earth's surface, and erosive forces grinding it away.

An eruption of the Batu Tara Volcano in Indonesia spews glowing bombs of lava skyward. Volcanoes are an important component of the Earth System, serving to transfer rock and gas from our planet's interior to its surface. But, while beautiful to watch from a distance, the force of eruptions can threaten life and property.

CHAPTER 9

The Wrath of Vulcan: Volcanic Eruptions

> Glowing waves rise and flow, burning all life on their way, and freeze into
> black, crusty rock which adds to the height of the mountain and builds the land,
> thereby adding another day to the geologic past. . . . I became a geologist forever,
> by seeing with my own eyes: the Earth is alive!
>
> —*Hans Cloos (geologist, 1886–1951), on seeing Mt. Vesuvius erupt*

9.1 Introduction

Every few hundred years, one of the hills on Vulcano, an island in the Mediterranean Sea off the western coast of Italy, rumbles and spews out molten rock, glassy cinders, and dense "smoke" (actually a mixture of various gases, fine ash, and very tiny liquid droplets). Ancient Romans thought that such eruptions happened when Vulcan, the god of fire, fueled his forges beneath the island to manufacture weapons for the other gods. Geologic study suggests instead that eruptions take place when hot magma, formed by melting inside the Earth, rises through the crust and emerges at the surface. No one believes the Roman myth anymore, but the island's name evolved into the English word **volcano**, which geologists use to designate either an erupting vent through which molten rock reaches the Earth's surface or an edifice (hill or mountain) built from the products of eruption.

On the main peninsula of Italy, not far from Vulcano, another volcano, Mt. Vesuvius, towers over the Bay of Naples. Nearly 2,000 years ago, a prosperous Roman resort and trading town named Pompeii sprawled at the foot of Vesuvius. One morning in 79 C.E., earthquakes signaled the mountain's awakening. At 1:00 P.M. on August 24, a ferociously turbulent, mottled cloud boiled up above Mt. Vesuvius's summit to a height of 27 km. The cloud soon drifted over Pompeii, turning day into night. Blocks and pellets of rock fell like hail, while fine ash and choking fumes enveloped the town (**Fig. 9.1**). People frantically rushed to escape, but for most it was too late. As the growing weight of falling volcanic debris began to crush buildings, a scalding, turbulent avalanche of ash mixed with pumice fragments surged down the flank of the volcano and

FIGURE 9.1 The eruption of Vesuvius buried Pompeii and nearby Herculaneum in 79 C.E.

(a) In this 1817 painting, the British artist J.M.W. Turner depicted the cataclysmic explosion.

(b) An artist's interpretation of the early phase of the eruption when roof-crushing debris rained on the town.

swept over Pompeii. When the next day dawned, the town and its neighbor, Herculaneum, had vanished beneath a 6-m-thick gray-black blanket of debris (**Fig. 9.2a–d**). This covering pro-tected the ruins of Pompeii and Herculaneum so well that when archaeologists started to excavate the towns 1,800 years later, they found artifacts and structures that gave an amazingly

FIGURE 9.2 The burial of Pompeii.

Pre-79 C.E. volcano
20th-century volcano
1.5X vertical exaggeration

(a) As seen looking southeast from the air today, it's evident that Vesuvius was once much bigger. The red dot shows the location of Pompeii.

Pre-79 C.E. profile

(b) Excavations exposed ruins of Pompeii with Vesuvius in the distance. The dashed line shows the volcano's profile prior to its eruption.

A modern suburb of Naples has been built on top of the ash.

The ash layer here is about 25 m thick.

Excavation exposes Roman columns.

(c) Volcanic debris still buries much of Herculaneum.

You can see ruts carved into streets by chariots.

(d) Streets and buildings of Pompeii are well preserved.

(e) Casts of people and animals convey the terror of that awful day.

complete picture of Roman daily life. In addition, they discovered odd-shaped open spaces in the debris covering Pompeii. Out of curiosity, they filled the spaces with plaster and then dug away the surrounding ash. The spaces turned out to be fossil casts of Pompeii's unfortunate inhabitants, contorted in agony or huddled in despair (**Fig. 9.2e**).

Clearly, volcanoes are unpredictable and dangerous. Volcanic activity can build a towering mountain, or it can blast one apart. The materials produced by this activity can provide the fertile soil and mineral deposits that enable a civilization to thrive, or they can provide a rain of destruction that can snuff one out. Because of the diversity and consequences of volcanic activity, this chapter sets out ambitious goals. We first build on Chapter 6 by looking more closely at the products of volcanic eruptions and at the basic characteristics of volcanoes. Then we consider the different kinds of volcanic eruptions on Earth and why they occur where they do. Finally, we review the hazards posed by volcanoes, the efforts made by geoscientists to predict eruptions and help minimize the damage they cause, and the possible influence of eruptions on climate and civilization.

9.2 The Products of Volcanic Eruptions

The drama of a volcanic eruption serves an important role in the Earth System because it transfers materials from inside the Earth to our planet's surface. Products of an eruption come in three forms—lava flows, pyroclastic debris, and gas. (Note that we use the term *lava flow* for both a molten, moving layer of lava and for the solid layer of rock that forms when the lava freezes.) We'll examine each of these products in turn.

Lava Flows

Sometimes lava races down the side of a volcano like a fast-moving, incandescent stream, sometimes it builds into a rubble-covered mound at a volcano's summit, and sometimes it oozes into blobs like a sticky but scalding paste. Clearly, not all lava behaves in the same way when it rises out of a volcano, so not all lava flows look the same. Why?

The character of a lava primarily reflects its **viscosity** (resistance to flow), and not all lavas have the same viscosity. Differences in viscosity depend on a variety of factors, including chemical composition, temperature, gas content, and crystal content. Silica content (the proportion of SiO_2, one of many chemicals making up lava) plays a particularly key role in controlling viscosity. Specifically, silica-poor (basaltic) lava is less viscous and thus flows farther than silica-rich (rhyolitic) lava (**Fig. 9.3**). That's because if silica is abundant in lava, silicon-oxygen

FIGURE 9.3 The characteristics of a lava flow depends on its viscosity.

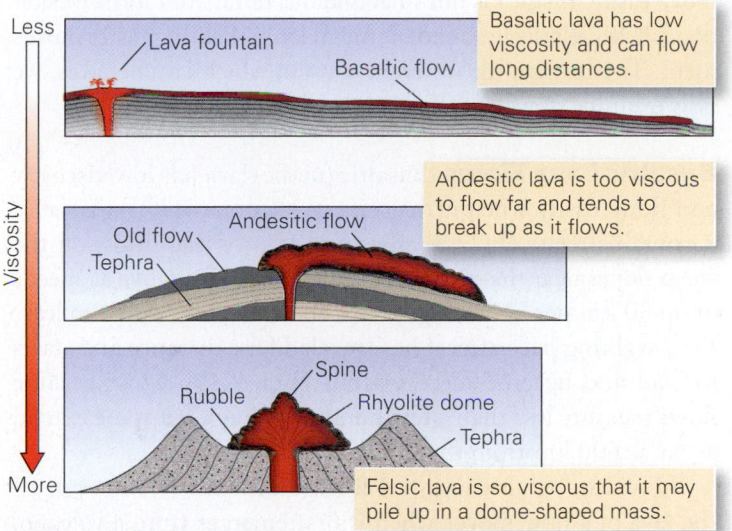

Basaltic lava has low viscosity and can flow long distances.

Andesitic lava is too viscous to flow far and tends to break up as it flows.

Felsic lava is so viscous that it may pile up in a dome-shaped mass.

(a) Three different kinds of flows.

(b) A satellite view of Hawaii shows that numerous lava flows have come from fissures and vents on Kilauea.

(c) This rhyolite dome formed about 650 years ago, in Panum Crater, California. Tephra (cinders) accumulated around the vent.

tetrahedra link in strongly bonded chains or networks that can't move easily. If silica is not so abundant, tetrahedra form weaker bonds with relatively abundant metal ions, allowing faster movement. To illustrate the different ways in which lava behaves, we now examine flows of different compositions.

Basaltic Lava Flows Basaltic (mafic) lava has low viscosity and flows easily when it first emerges from a volcano because it contains relatively little silica and is very hot. Thus, on the steep slopes near the summit of a volcano, it can move at speeds of up 30 km per hour (**Fig. 9.4a**). The lava slows down to less-than-walking pace after it has traveled for a distance and starts to cool and become more viscous (**Fig. 9.4b**). Most basaltic flows measure less than a few kilometers long, but some extend as far as 600 km from the source.

How can basaltic lava travel large distances? Although all the lava of a flow moves when it first emerges from a volcano, rapid cooling causes the surface of the flow to harden after the flow has moved a short distance from the source. The solid crust serves as insulation, allowing the hot interior of the flow to remain liquid and continue to move. New molten lava injects between the original ground surface and the new solid crust—the addition of this lava effectively inflates the flow, jacking up the hardened crust and making the overall flow thicker. As time progresses, part of the flow's interior solidifies, so eventually molten lava moves only through a tunnel-like passageway, or **lava tube**, within the flow. The largest lava tubes may be tens of meters in diameter (**Fig 9.4c**). In some cases, they eventually drain and become empty tunnels (**Fig. 9.4d**).

The character of a basaltic lava flow's surface reflects the timing of freezing. Flows that have warm, pasty surfaces wrinkle into smooth, glassy, rope-like ridges (**Fig. 9.4e**)—geologists refer to such flows by its Hawaiian name, **pahoehoe** (pronounced pa-hoy-hoy). If the surface layer of the lava freezes, but then breaks up due to the continued movement of lava underneath, it becomes a jumble of sharp, angular fragments, creating a rubbly flow also known by a Hawaiian name, **a'a'** (pronounced ah-ah) (**Fig. 9.4f**). Footpaths made by people living in basaltic volcanic regions follow the smooth surface of pahoehoe flows rather than the rough, foot-slashing surface of a'a' flows.

During the final stages of cooling, lava flows contract because rock shrinks as it loses heat and may fracture into polygonal columns. This type of fracturing is called **columnar jointing** (**Fig. 9.5a**). Columnar jointing typically terminates in the rubble that occurs at the top and bottom of a flow.

Basaltic flows that erupt underwater look different from those that erupt on land because the lava cools so much more quickly in water. Because of rapid cooling, submarine basaltic lava can travel only a short distance before its surface freezes, producing a glass-encrusted blob, or "pillow" (**Fig. 9.5b**). The rind of a pillow momentarily stops the flow's advance, but soon the pressure of the lava squeezing into the pillow breaks the rind, and a new blob of lava squirts out, freezes, and produces another pillow. In some cases, successive pillows add to the end of previous ones, forming worm-like chains. Geologists refer to a lava flow consisting of a pile of such blobs as **pillow lava**.

Andesitic and Rhyolitic Lava Flows Because of its greater viscosity, andesitic lava cannot flow as easily as basaltic lava. When erupted, andesitic lava forms a mound above the vent. This mound advances slowly down the volcano's flank at only 1 to 5 m a day, becoming a lumpy flow with a bulbous snout. Typically, andesitic flows are less than a few kilometers long, though unusually hot flows may travel farther. Because the lava moves so slowly, the outside of the flow has time to solidify, so as it moves, the surface breaks up into angular blocks, and the whole flow looks like a jumble of rubble called *blocky lava*. On steep slopes, the blocks may tumble downhill, so the flow may evolve into a landslide of blocks.

Rhyolitic lava is the most viscous of all lavas because it has the highest silica concentration and the coolest temperature. Therefore, it tends to accumulate either above the vent in a bulbous mass called a *lava dome* (see Fig. 9.3c). Sometimes rhyolitic lava freezes while still in the vent and then pushes upward as a column-like *lava spire* or lava spine rising up to 100 m above the vent. Rhyolitic flows, where they do form, are rarely more than 1 to 2 km long and have broken and blocky surfaces.

Volcaniclastic Deposits

On a mild day in February 1943, as Dionisio Pulido prepared to sow the fertile soil of his field 330 km (200 miles) west of Mexico City, an earthquake jolted the ground, as it had dozens of times in the previous days. But this time, to Dionisio's amazement, the surface of his field visibly bulged upward by a few meters and then cracked. Ash and sulfurous fumes filled the air, and Dionisio fled. When he returned the following morning, his rich land lay buried beneath a 40-m-high mound of gray cinders—Dionisio had witnessed the birth of Paricutín, a new volcano. During the next several months, Paricutín erupted continuously, at times blasting clots of lava into the sky like fireworks. By the following year, it had become a steep-sided cone over 300 m high. Nine years later, when the volcano ceased all activity, its lava and debris covered 25 square km, and Dionisio's farm and those of his neighbors were gone.

This description of Paricutín's eruption, and that of Vesuvius at the beginning of this chapter, emphasizes that volcanoes produce large quantities of fragmental material. Geologists

> **Did you ever wonder . . .**
>
> if anyone has ever seen a brand-new volcano appear?

FIGURE 9.4 Features of basaltic lava flows. They have low viscosity and thus can flow long distances. Their surface and interior can be complex.

(a) A fast-moving flow coming from Mt. Etna, Sicily.

Lava flow

Highway

(b) A basaltic lava flow covers a highway in Hawaii.

A "skylight" into an active lava tube

(c) In a lava tube, still-molten lava flows beneath a crust of solid basalt.

An ancient lava tube in a road cut, Hawaii

(d) A drained lava tube exposed in a road cut on Hawaii.

"Ropes" of pahoehoe

0.5 m

(e) Pahoehoe from a recent lava flow in Hawaii. Note the coin for scale.

(f) The rubbly surface of an a'a' flow, Sunset Crater, Arizona.

FIGURE 9.5 Examples of structures within lava flows.

(a) Columnar jointing develops when the interior of a flow cools and cracks. Such jointing develops in dikes and sills. This example is Devils Postpile in California.

(b) Pillow basalt develops when lava erupts underwater. Later uplift may expose pillows above sea level, as in this Oregon outcrop.

use the term *volcaniclastic deposit*, in a general sense, for any accumulation of this material. Volcaniclastic deposits include **pyroclastic debris** (from the Greek *pyro*, meaning fire), which is specifically the debris forcefully ejected from a volcano during an eruption. Pyroclastic debris includes: fragments formed from lava ejected into the air while still molten, so they freeze in midair or soon after they land; fragments of all sizes ejected when an eruption blasts apart either recently solidified lava or pumice of the volcano's throat; or pre-existing volcanic rock surrounding the volcano's vent. But volcaniclastic deposits also include debris that tumbled down the flank of a volcano in landslide, or has mixed with water to form a muddy slurry, or has been carried and sorted by streams. Let's look at these components in more detail—you'll see that different types form in association with different kinds of eruptions.

Pyroclastic Debris from Basaltic Eruptions

Basaltic lava rising in a volcano may contain dissolved volatiles, such as water. As such lava approaches the surface, the volatiles form bubbles, and in basaltic magma the bubbles can rise faster than the magma itself. When the bubbles reach the surface of the lava, they burst and eject clots and drops of molten lava upward to form dramatic fountains (**Fig. 9.6a**). To picture this process, think of the droplets that spray from a just-opened bottle of soda.

Geologists recognize several different types of fragments formed from frozen clots or drops of lava. Pea-sized fragments of glassy lava and scoria comprise a type of **lapilli**, from the Latin word for little stones; they are informally known as *cinders*. Rarely, flying droplets may trail thin strands of lava,

which freeze into filaments of glass known as *Pelé's hair*, after the Hawaiian goddess of volcanoes, and the droplets themselves freeze into tiny streamlined glassy beads known as *Pelé's tears*. Apple- to refrigerator-sized fragments, called **blocks** (**Fig. 9.6b**), may consist of already-solid volcanic rock, broken up during the eruption. Blocks tend to be angular and chunky. In some cases, however, blocks form from soft lava squirting out of the vent—such blocks, also known as **bombs**, have streamlined, polished surfaces (**Fig. 9.6c**).

Pyroclastic Debris from Andesitic or Rhyolitic Eruptions

Andesitic or rhyolitic lava is more viscous than basalt and tends to be more gas rich. Eruptions of these lavas also tend to be explosive. Volcanic explosions can produce immense quantities of pyroclastic debris, much more than can come from a basaltic volcano (**Fig. 9.7a**). Debris ejected from explosive eruptions includes: **ash**, which consists of glassy particles less than 2 mm in diameter formed when frothy lava or recently formed pumice explosively breaks up during an eruption, or when pre-existing volcanic rock gets pulverized by the force of an explosion; *pumice lapilli*, which consists of angular pumice fragments; and *accretionary lapilli*, which consists of snowball-like lumps of ash formed when ash mixes with water in the air and then sticks together to form small balls (**Fig. 9.7b–d**).

Unconsolidated deposits of pyroclastic grains, regardless of size or composition, constitute **tephra** (**Fig. 9.8a**). Ash, or ash mixed with lapilli, becomes **tuff** when buried and transformed into coherent rock (**Fig. 9.8b**). In some cases the coherence forms during deposition, because grains are so hot that they

FIGURE 9.6 Examples of pyroclastic debris from a basaltic eruption on Hawaii.

(a) A fountain of basaltic lapilli spouts from a vent on Hawaii.

(b) Blocks and lapilli on the flank of a Hawaiian volcano.

(c) A bomb has a smooth, streaked surface.

FIGURE 9.7 Pyroclastic debris from andesitic or rhyolitic eruptions.

(a) Pyroclastic debris billowing from the 2008 eruption of Chaitén in Chile.

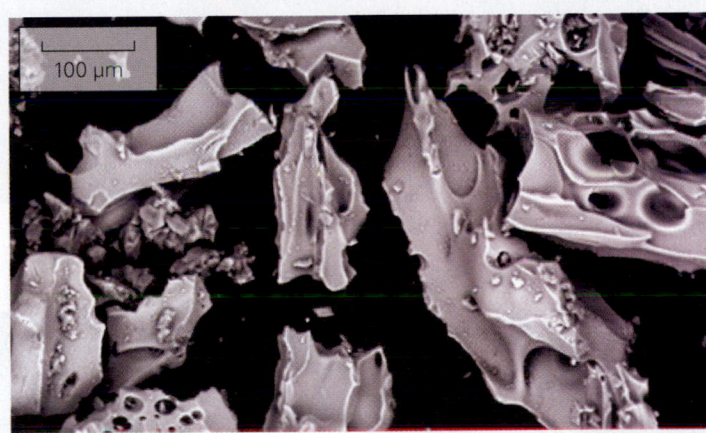

100 μm

(b) Electron photomicrograph of ash.

5 cm

(c) Pumice lapilli.

1 cm

(d) Accretionary lapilli.

FIGURE 9.8 Examples of volcaniclastic deposits (debris flows and lahars).

(a) Recent tephra on the flank of a volcano in Hawaii.

(b) A cliff of ~1 Ma tuff in New Mexico.

The landslide stripped away the forest.

(c) Water-soaked volcanic debris slid down the side of this volcano in Nicaragua.

Debris flow deposits

Stratified ash

Paleo-channel wall

(d) Deposits of a debris flow that accumulated about 35 Ma in Utah. Note that the debris flow filled a channel cut into finer, stratified ash.

(e) A lahar fills a river bed in New Zealand after an eruption in 2007.

(f) Deposits of a lahar from Mt. St. Helens, 20 years after its 1980 eruption, include logs ripped off hill slopes.

weld together. But more commonly, the coherence comes when the debris gets cemented together either by minerals precipitated from groundwater or by minerals that grow in the ash as it reacts with groundwater.

Other Volcaniclastic Deposits The nature and origin of debris produced during and after eruptions continue to be the subject of active research, for some of the deposits prove to be difficult to interpret. As we've noted, geologists use the term *volcaniclastic deposit* for any material that consists of volcanic igneous fragments, and they recognize three categories:

1. *Pyroclastic* deposits, as we have just seen, consist of fragments that ejected during an eruption, and they accumulate directly from the clouds of debris ejected into the sky or sent in avalanches down the flank of the volcano. The fragments in such deposits have not moved, subsequent to their original deposition.

2. *Volcani-sedimentary* deposits consist of volcanic material (lava and pyroclastic debris) that later moved downslope and was redeposited elsewhere, subsequent to accumulating after an eruption. Some of this material tumbles in landslides, breaking up to varying degrees as it moves. If volcanoes are covered with snow and ice or are drenched with rain, water mixes with debris to form a *volcanic debris flow* that moves downslope like wet concrete (**Fig. 9.8c–f**). Very wet, ash-rich debris flows down in a relatively fast-moving slurry called a **lahar**, which can reach speeds of 50 km per hour. Lahars tend to follow river channels and may travel for tens of kilometers away from the volcano. When debris flows and lahars stop moving, they yield a layer consisting of volcanic blocks suspended in ashy mud. Rivers may eventually sort and transport some volcanic sediment. Where this material accumulates, perhaps far downstream, it forms deposits of *volcanic sandstone* and/or volcanic conglomerate.

3. *Fragmental lava* deposits consist of debris produced when lava breaks up into angular clasts while flowing, without ever being ejected into the air.

As we've seen, fragmentation (or brecciation) happens when the inside of a lava flow continues to move after its surface has frozen—the crust breaks up due to the movement. But fragmentation may also happen when lava freezes very quickly and shatters upon erupting into water or ice. The resulting material, hyalocastite, consists of glassy fragments embedded in ash that has reacted with hot water.

Volcanic Gas

Most magma contains dissolved gases, including water (H_2O), carbon dioxide (CO_2), sulfur dioxide (SO_2), and hydrogen sulfide (H_2S). Generally, felsic lavas can contain more dissolved gas than mafic lavas—in fact, up to 9% by weight of a felsic magma consists of volatiles. As we've seen, these dissolved gases come out of solution when the magma approaches the Earth's surface. This process happens for two reasons, the first of which is that the ability of a liquid to hold dissolved gas decreases as the pressure acting on the liquid decreases. You see this phenomenon when you pop off the top of a carbonated beverage—the beverage was injected with CO_2 under pressure when it was bottled, and because popping the top decreases the pressure, bubbles form and give the beverage its sparkle. Second, gas comes out of solution as a side effect of crystallization. Gases can't fit easily into growing crystals, so they remain in the liquid magma, causing the concentration of gas in the liquid to increase until it exceeds the capacity of the liquid to keep it in solution. When this happens, gas bubbles form.

The sulfurous gases emitted by some volcanoes smell like rotten eggs. Incorporation of these gases dissolved in tiny water droplets yields corrosive sulfuric acid. This acid occurs in the form of an **aerosol**, meaning a haze of droplets, or solid particles, that are so small that they can remain suspended in air for a long time.

The fate of bubbles in magma depends on the viscosity of magma. For example, in low-viscosity mafic magma, gas bubbles can rise faster than the magma moves, and thus most reach the surface of the magma and enter the atmosphere before the lava does. Thus, some volcanoes may, for a while, produce large quantities of steam, without much lava (**Fig. 9.9a**). The last bubbles to form, however, remain as holes when the lava freezes around them. As we discussed in Chapter 6, these holes are called **vesicles** (**Fig. 9.9b**), and mafic rock in which more than 50% of the rock's volume consists of bubbles is called *scoria*. In high-viscosity felsic magmas, the gas has trouble escaping because bubbles can't push through the very sticky lava. When such magma reaches shallower depths, it effectively becomes a foam. As this foam approaches the Earth's surface, and the weight of overlying lava decreases, the gas expands so that in some cases bubbles may account for as much as 50% to 75% of the volume of the magma.

FIGURE 9.9 The gas component of volcanic eruptions.

(a) A volcano in Alaska erupting large quantities of steam.

(b) Gas bubbles frozen in lava produce vesicles, as in this block from Sunset Crater, Arizona.

Pumice forms when this material freezes—the thin films of melt between bubbles turn into glass. In some cases, pumice has such low density that it can float.

Take-Home Message

Volcanoes erupt lava, pyroclastic debris, and gas. The character of a lava flow—whether it has low viscosity and spreads over a large area or has high viscosity and builds a bulbous mound over the vent—depends largely on its silica content. Pyroclastic debris includes ash, lapilli, blocks, and bombs. Magma contains dissolved gas, which comes out of solution when the magma approaches the Earth's surface.

QUICK QUESTION: How can lava travel tens of kilometers or more from a volcanic vent without freezing?

9.3 Structure and Eruptive Style

Volcanic Architecture

As we saw in Chapter 6, melting in the upper mantle and lower crust produces magma, which rises into the upper crust. Typically, this magma accumulates underground in a **magma chamber**, a zone of open spaces and/or fractured rock that can contain a large quantity of magma and/or a mush of magma mixed with crystals. Some of the magma may solidify in the magma chamber and transform into intrusive igneous rock, but the rest rises along a pathway, or *conduit*, to the Earth's surface and erupts from an opening, or **vent**. The conduit may have the

shape of a chimney or may be a long crack called a **fissure** (**Fig. 9.10**). At times, vents at the top of chimney-shaped conduits erupt tall fountains of lava, whereas those along fissures erupt long curtains of lava.

Over time, new igneous rock (lava and/or pyroclastic debris) builds up around the vent to form a *volcanic edifice*. Eruptions that take place at the top of the edifice are called *summit eruptions*, whereas those that break through along the

FIGURE 9.10 Crater eruptions and fissure eruptions come from conduits of different shapes.

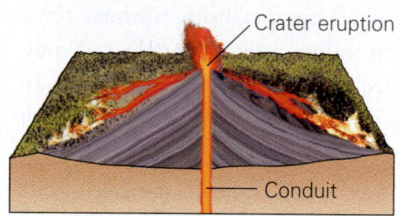

Crater eruption

Conduit

(a) At a crater eruption, lava spouts from a chimney-shaped conduit.

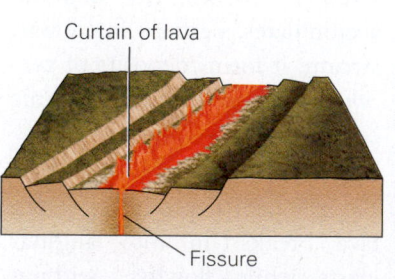

Curtain of lava

Fissure

(b) At a fissure eruption, lava comes out in a curtain, along the length of a crack.

FIGURE 9.11 The formation of volcanic calderas.

(a) The crater of Santa Ana Volcano in El Salvador.

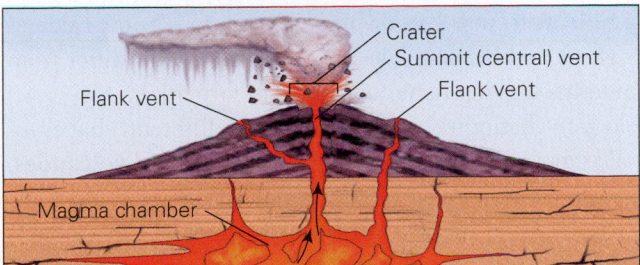

(b) As an eruption begins, the magma chamber inflates with magma. There can be a central vent and one or more flank vents.

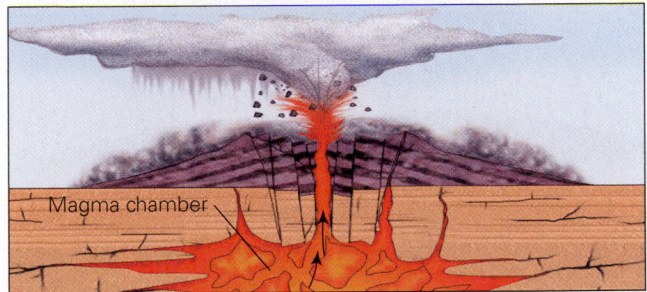

(c) During an eruption, the magma chamber drains, and the central portion of the volcano collapses downward.

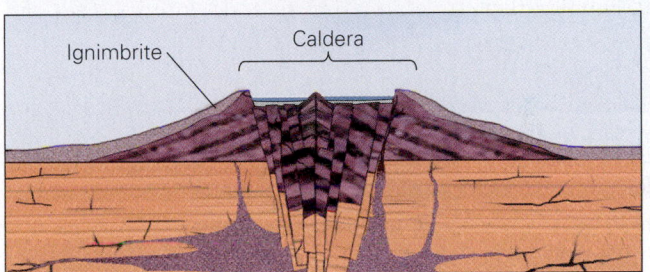

(d) The collapsed area becomes a caldera. Later, a new volcano may begin to grow within the caldera.

Time

sides, or flanks, of the volcano are *flank eruptions*. In some cases, the vent lies at the floor of a circular depression called a **crater**, shaped like a bowl, up to 500 m across and 200 m deep (**Fig. 9.11a**). Craters form either because material accumulates around the vent during eruption or because the top part of the edifice collapses into the drained conduit when the eruption ceases.

During some major eruptions, a large portion of the volcanic edifice collapses into the drained magma chamber below, producing a **caldera**, a large circular depression up to thousands of meters across and up to several hundred meters deep (**Fig. 9.11b–e**). Typically, a caldera has steep walls and a fairly flat floor and may be

(e) This caldera in Oregon formed about 7,700 years ago. Afterward, it filled with water to become Crater Lake. Wizard Island, protruding from the lake, is a small volcano that grew on top of the caldera floor.

partially filled with new lava or pyroclastic debris. Some calderas fill with water and become lakes. Note that calderas differ from craters in terms of size, shape, and mode of formation.

Geologists distinguish among several different shapes of "subaerial" (above sea level) volcanic edifices. **Shield volcanoes**, broad, gentle domes, are so named because they resemble a soldier's shield lying on the ground. They form when the products of eruption have low viscosity and thus cannot pile up around the vent but rather spread out over large areas. The volcanoes of Hawaii, which produce layer upon layer of low-viscosity basaltic lava, are shield volcanoes (**Fig. 9.12a**), as are some volcanoes that erupt successive ignimbrites. **Cinder cones**, also known as *scoria cones*, consist of cone-shaped piles of basaltic lapilli and blocks, sometimes from a single eruption (**Fig. 9.12b**). **Stratovolcanoes**,

FIGURE 9.12 Different shapes of volcanoes.

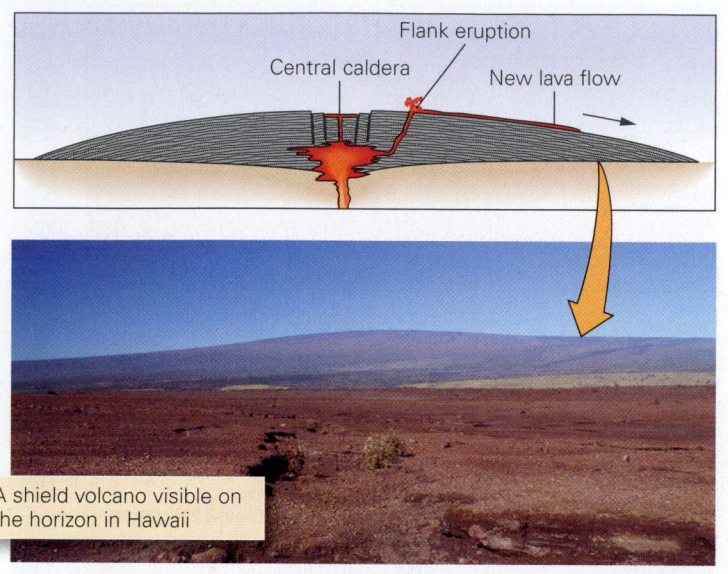

A shield volcano visible on the horizon in Hawaii

(a) A shield volcano, made from successive flows of low-viscosity basalt, has very gentle slopes.

(b) A cinder cone on the flank of a larger volcano in Arizona. The pile of cinders has assumed the angle of repose. A lava flow covers the land surface in the distance.

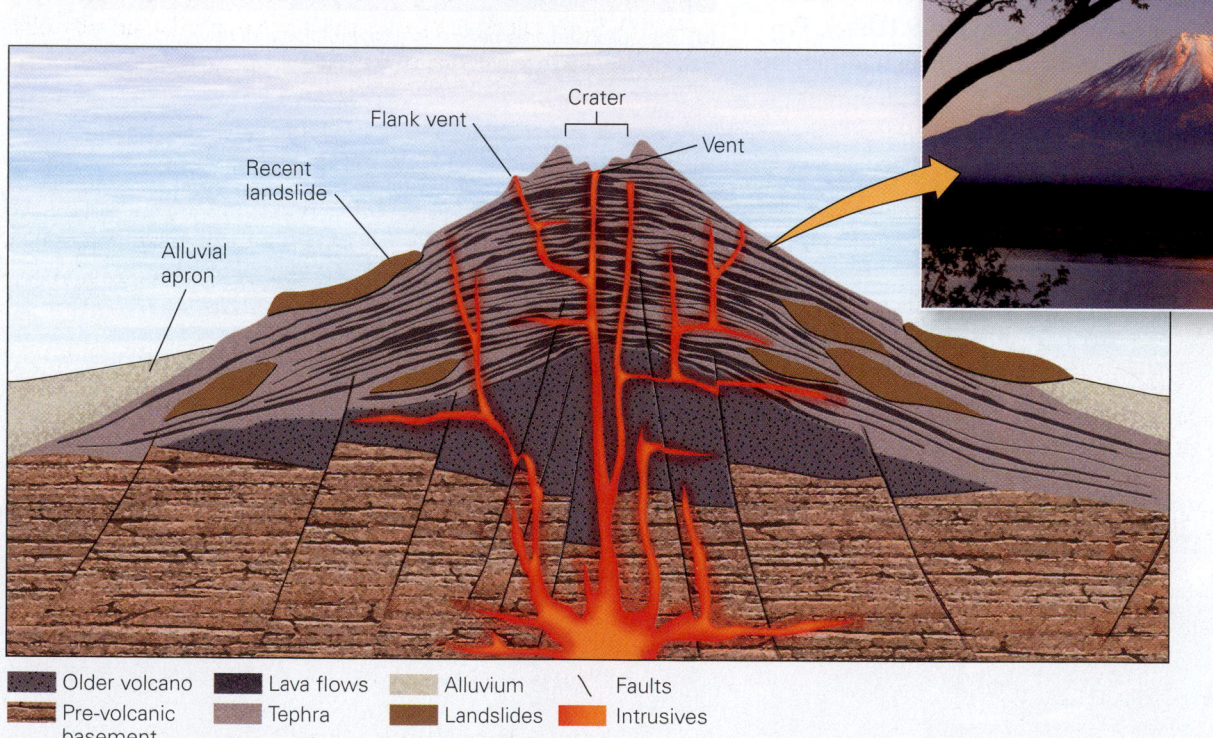

Mt. Fuji, a composite volcano in Japan, last erupted in 1707.

▨ Older volcano	▨ Lava flows	▨ Alluvium	\ Faults
▨ Pre-volcanic basement	▨ Tephra	▨ Landslides	▨ Intrusives

(c) A composite volcano consists of layers of tephra and lava. Volcanic debris flows and ash avalanches modify slopes and contribute to the development of a classic cone-like shape.

also known as *composite volcanoes*, are large (up to 3 km high) and cone-shaped, generally with steeper slopes near the summit, and consist of interleaved contrasting layers (hence the prefix *strato–*) of lava, tephra, and volcaniclastic debris (**Fig. 9.12c**). Their shape, exemplified by Japan's Mt. Fuji, serves as the classic image that most people have of a volcano.

Let's look at the nature of stratovolcanoes a little more closely. Their edifices build from the products of many eruptions over an extended period of time. Not all eruptions produce the same kind of material. Specifically, eruptions producing pyroclastic debris produce layers of tephra, eruptions of andesite produce blocky flows that breakup into bouldery landslides on the volcano's slopes, and eruptions of low-viscosity lava produce flows that cascade down the flanks of the volcano. Lava flows resist erosion and thus armor the underlying tephra, preventing it from being washed away. Over time, the volcano builds into a symmetrical cone. But, for many reasons, this shape doesn't last indefinitely. For example, large landslides carry masses of rock down the slopes and build broad aprons around the base of the volcano. Heavy rains or spring snow melt can trigger debris flows, which transport volcaniclastic sediment out to alluvial fans surrounding the volcano. Finally, explosive eruptions can blast away a large portion of the volcano's edifice.

The hills or mountains resulting from volcanism come in a great range of sizes (**Fig. 9.13**). Shield volcanoes tend to be the largest, followed by stratovolcanoes. Cinder cones tend to be relatively small and are often found on the flanks of larger volcanoes.

The Concept of Eruptive Style: Will It Flow or Will It Blow?

Kilauea, a volcano on Hawaii, produces rivers of lava that cascade down the volcano's flanks. Mt. St. Helens, a volcano near the Washington-Oregon border, exploded catastrophically in 1980 and blanketed the surrounding countryside with tephra. Clearly, different volcanoes erupt differently and, as we've noted, successive eruptions from the same stratovolcano may differ markedly in character from one another. Geologists refer to the character of an eruption as **eruptive style**. Below we describe several distinct eruptive styles and explore why the differences occur (see **Geology at a Glance**, pp. 286–287).

Effusive Eruptions

The term *effusive* comes from the Latin word *effundere*, to pour out, and indeed that's what happens during an **effusive eruption**—lava pours out from a vent or fissure. This lava may fill a *lava lake* around the crater and/or spill down the side of the mountain in a sheet or in a channelized flow (**Fig. 9.14a, b**). Geologists refer to small- to moderate-sized effusive eruptions as *Hawaiian eruptions*, because they are common on Hawaii. Successive eruptions build layer upon layer of basalt (**Fig. 9.14c**).

Effusive eruptions occur where the magma feeding the volcano is hot and mafic and therefore has low viscosity. Pressure, applied to the magma chamber by the weight of overlying rock, squeezes magma upward and out of the vent; in some cases, the pressure is great enough to drive the magma up into a *lava fountain* spewing out of the vent (see Fig. 9.10a). As we have seen, when the magma rises, gas comes out of solution and forms bubbles. The presence of bubbles decreases the overall density and viscosity of the magma, allowing it to rise faster. If the conduit through which the magma rises narrows in the throat of the volcano, the velocity of the rising magma increases even more, much like the velocity of water coming out of a hose increases if you pinch the end of the hose. Such pinching of conduits may play a role in producing very high magma fountains spurting up to 500 m into the air above the vent.

Explosive Eruptions

Volcanoes that forcefully emit significant quantities of pyroclastic debris are called **explosive eruptions**. Geologists recognize a variety of different types based on the size of the explosion and the products of the explosion, as we now see.

FIGURE 9.13 These profiles emphasize that volcanoes come in different sizes. Large shield volcanoes, like those on Hawaii, are many times larger than cinder cones.

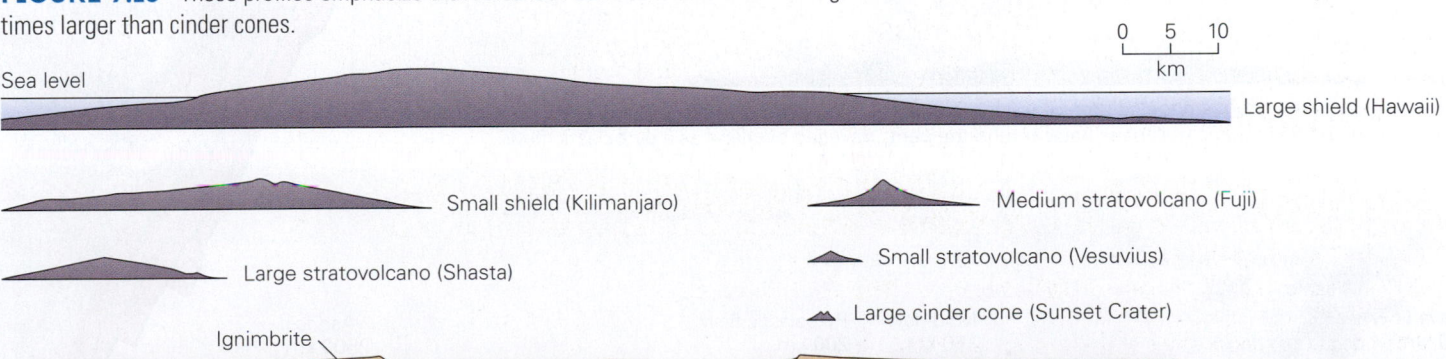

0   5   10
km

Sea level

Large shield (Hawaii)

Small shield (Kilimanjaro)

Medium stratovolcano (Fuji)

Large stratovolcano (Shasta)

Small stratovolcano (Vesuvius)

Large cinder cone (Sunset Crater)

Ignimbrite

Large caldera (Yellowstone)

GEOLOGY AT A GLANCE

Volcanoes

Beneath a volcano, magma rises to fill a pervasively cracked region of crust and forms a magma chamber. Some of the magma erupts at a surface vent. Once molten rock has erupted at the surface, it is called lava. Some lava spills down the side of the volcano in lava flows. Some fountains out of a vent to form scoria fragments that pile up in a cone around the vent. Eruptions may eject larger chunks as blocks or bombs. The nature of eruptions depends on the viscosity of the lava, which in turn depends on lava composition. Volcanic explosions blast up a cloud of ash and lapilli, and may pulverize pre-existing

Cinder cone

Caldera

Vulcanian eruptions occur when a buildup of gas and magma explodes.

Strombolian crater explosions frequently burst through thinly crusted lava.

Hawaiian fountain explosions are caused by escaping gas.

Side vent

Eroded cone

Lava cone

Lava flow

Sills

Dikes

Cinder cones

Lava pavement (cracked/broken)

Plinean explosions shoot a huge column of pumice fragments up to 50 km into the atmosphere. The ash fall rains down and the column collapses back around the vent, traveling overland as a pyroclastic flow.

Lava flow
50 km

Mud flow
150 km

Pyroclastic flow
200 km

Ash fall
2500 km

The distance volcanic hazards can travel from an eruption.

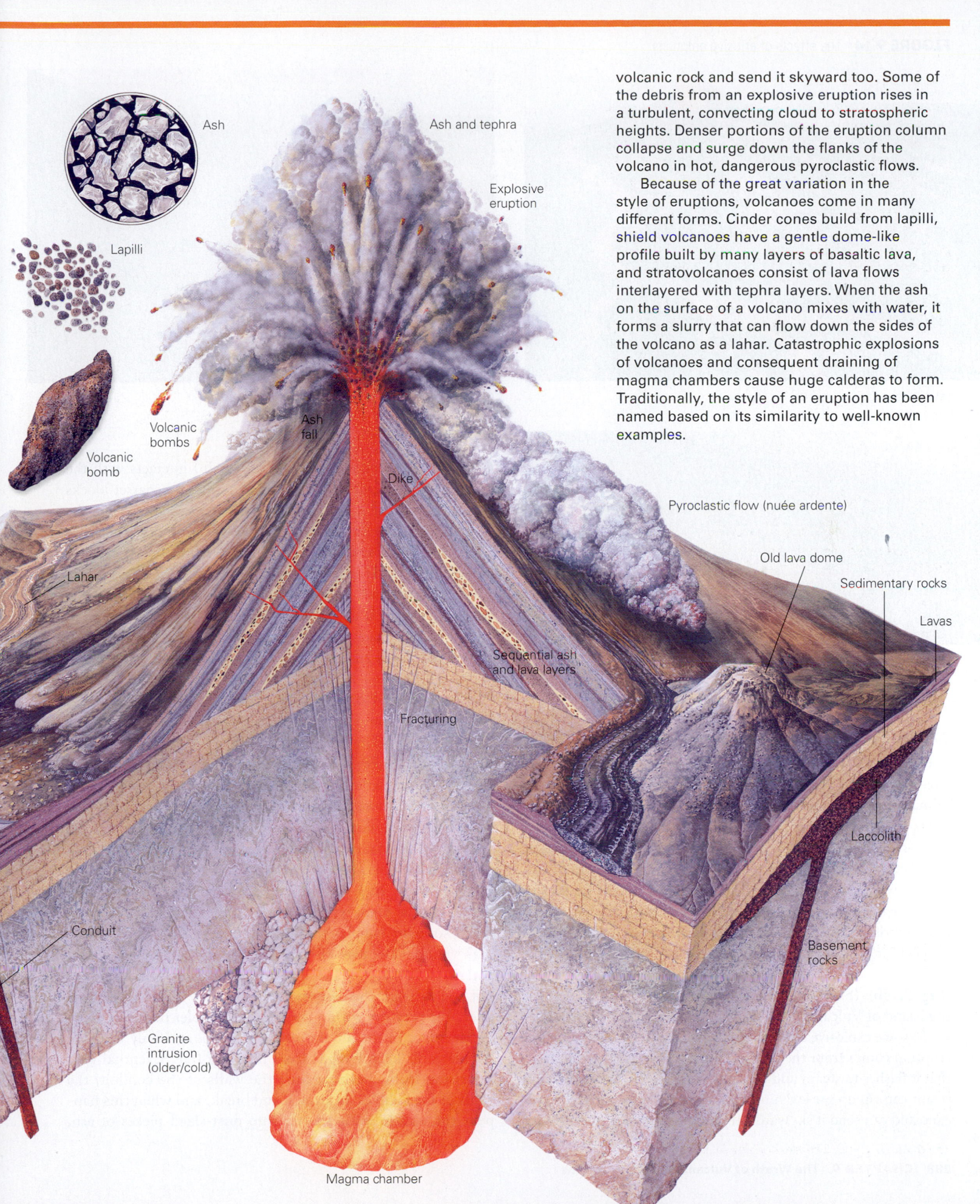

volcanic rock and send it skyward too. Some of the debris from an explosive eruption rises in a turbulent, convecting cloud to stratospheric heights. Denser portions of the eruption column collapse and surge down the flanks of the volcano in hot, dangerous pyroclastic flows.

Because of the great variation in the style of eruptions, volcanoes come in many different forms. Cinder cones build from lapilli, shield volcanoes have a gentle dome-like profile built by many layers of basaltic lava, and stratovolcanoes consist of lava flows interlayered with tephra layers. When the ash on the surface of a volcano mixes with water, it forms a slurry that can flow down the sides of the volcano as a lahar. Catastrophic explosions of volcanoes and consequent draining of magma chambers cause huge calderas to form. Traditionally, the style of an eruption has been named based on its similarity to well-known examples.

Ash

Lapilli

Volcanic bombs

Volcanic bomb

Ash and tephra

Explosive eruption

Ash fall

Dike

Pyroclastic flow (nuée ardente)

Old lava dome

Sedimentary rocks

Lavas

Lahar

Sequential ash and lava layers

Fracturing

Conduit

Granite intrusion (older/cold)

Laccolith

Basement rocks

Magma chamber

FIGURE 9.14 The effects of effusive eruptions.

(a) A 1986 effusive eruption on Hawaii.

(b) A lava lake in the caldera of Kilauea, Hawaii.

(c) Layer upon layer of basaltic lava flows, exposed on the wall of a caldera in Hawaii.

The Diversity of Explosive Types Ancient Romans referred to the island of Stromboli as "the lighthouse of the Mediterranean" because it has erupted about every 10 to 20 minutes through recorded history, and the red-hot clots that it ejects trace out glowing arcs of light in the night sky. Occasionally, Stromboli also produces basaltic lava flows, but most of the material it erupts comes out in the form of scoria lapilli and blocks, which build into a cone around the vent (**Fig. 9.15a**). (Not surprisingly, geologists refer to eruptions that produce a fountain of basaltic lapilli as a *Strombolian eruption*, regardless of where it occurs.) Somewhat larger explosive eruptions emit both a fountain of lava and a dense plume of pyroclastic debris (**Fig. 9.15b**). (Such events are *Vulcanian eruptions*, named for the island of Vulcano.)

In some explosive eruptions, much of the pressure driving the eruption comes from the sudden heating of water by magma so that it flashes to steam and expands very rapidly. The expanding steam can rip up pre-existing rock from the conduit of the volcano and can send it skyward. In *phreatic eruptions*, groundwater (possibly from melting snow or heavy rain) interacts with the magma and the eruption blasts steam, ash, and coarser blocks skyward, but little if any lava appears at the surface. When the vent lies in relatively shallow seawater—heating the water produces prodigious amounts of steam that billow out of the sea, along with fountains of wet ash (**Fig. 9.15c**). (Such eruptions are called *Surtseyan eruptions*, for the island of Surtsey off the coast of Iceland.) If seawater suddenly gains access via cracks to a large magma chamber, the resulting flash to steam may blast the entire volcano apart in a huge explosion.

Plinean Eruptions Really huge explosive eruptions of stratovolcanoes are known as *Plinean eruptions*, named for the Roman scholar, Pliny the Younger, who observed and described the eruption of Vesuvius. Plinean eruptions (**Fig. 9.15d**) can eject many cubic kilometers of material into the atmosphere and may destroy a substantial part of the stratovolcano's edifice itself. In fact, the explosion can completely change the profile of the volcano such that it no longer has a cone-like shape (**Box 9.1**).

Plinean eruptions take place when andesitic and rhyolitic magmas rising in a volcano contain very large quantities of gas. As we've seen, when this gas comes out of solution, it forms so many bubbles that they comprise most of the magma's volume. Because of their high silica content, andesitic and rhyolitic magmas are so viscous that the gas bubbles cannot rise through the magma and escape as the gassy magma rises. The pressure within the trapped bubbles becomes much greater than in the air above the volcano, and only the thin bubble walls keep the gas from bursting free. Eventually, as the rising froth shears against the walls of the conduit, the walls of the bubbles do stretch and break, and when this happens the bubble walls shatter into dust-sized pieces of ash,

FIGURE 9.15 Examples of explosive eruptive styles. No two eruptions are exactly alike.

(a) A large Strombolian eruption on Mt. Etna, Sicily.

(b) A lava fountain in the plume during an eruption of Mt. Etna.

Sea surface

(c) A Surtseyan eruption of a subsea volcano near Tonga.

Convective cloud

(d) The Plinean eruption of Mt. Pinatubo in the Philippines.

and these surround the pumice lapilli (larger chunks that still contain both bubbles and bubble walls). Suddenly the very hot gas that had been held in by bubble walls is released and it expands violently, producing immense pressure that pushes surrounding debris upward. If the pressure cracks the lava dome capping the vent, the mixture of fragments and gas burst out of the volcano's vent at very high velocity (over 300 km per hour), like a giant shotgun blast. The sudden release of this material decreases the pressure on magma deeper in the conduit, allowing bubbles in the deeper magma to expand and shatter, so more and more pyroclastic debris erupts until the magma chamber drains. The ejected debris forms a huge **eruption column**.

Geologists recognize several distinct regions within a Plinean eruptive column (**Fig. 9.16a**). The blast from an explosive

eruption can propel debris upward for hundreds of meters to a couple of kilometers and forms the lower part, or *gas-thrust region*, of the eruption column. But the eruption column of huge eruptions is much taller than the gas-thrust region. That's because the mixture of hot ash, hot volcanic gas, and air is lighter than cooler air around it and thus, like a hot-air balloon, is buoyant. The debris continues to rise as a turbulent, billowing cloud or *convective plume*. At stratospheric heights, 10 to 50 km high, the convective plume spreads out into a broad *ash umbrella*—the mushroom head of an overall mushroom cloud (**Fig. 9.16b**). The formation of an umbrella occurs where the plume has cooled enough so that it is no longer buoyant.

Debris from the convective cloud and ash umbrella of a huge volcanic explosion falls from the air, like hail and snow, and blankets the countryside near the volcano. Strong winds of the

BOX 9.1   CONSIDER THIS . . .

Volcanic Explosions to Remember

A volcanic explosion generates an enduring image of destruction, for an explosion can rip a volcano apart and devastate the surrounding region. To compare volcanic explosions, geologists sometimes use the **volcanic explosivity index** (VEI), a logarithmic scale on which the largest known eruption has been assigned a VEI of 8; larger ones could conceivably take place. This index takes into account the volume of debris ejected, the height of the eruptive column, and an estimate of the energy released during the explosion. The historic record shows that there have been about 100 eruptions with VEIs in the range of 4 to 6 since 1800 C.E., and the largest-observed eruption in recorded history (Tambora, in 1815) ranks as a 7. The geologic record shows that "mega-colossal" explosions with a VEI of 8 have taken place during the past few million years. One of these formed the caldera of Yellowstone National Park, Wyoming, about 600,000 years ago (**Fig. Bx9.1a**). To get a sense of the

consequences of a volcanic explosion, let's look at three notable examples.

Mt. St. Helens, a snow-crested strato-volcano in the Cascade Mountains of north-western United States, had not erupted since 1857. However, geologic evidence suggested that the mountain had a violent past, punctuated by many explosive eruptions. On March 20, 1980, an earthquake announced that the volcano was awakening. A week later, a crater 80 m in diameter burst open at the summit and began emitting gas and pyroclastic debris. Geologists who set up monitoring stations to observe the volcano noted that its north side was beginning to bulge markedly, suggesting that the volcano was filling with magma and was expanding like a balloon. Their concern that an eruption was imminent led local authorities to evacuate people in the area.

The climactic eruption came suddenly. At 8:32 A.M. on May 18, a geologist, David Johnston, monitoring the volcano from a distance

of 10 km, shouted over his two-way radio, "Vancouver, Vancouver, this is it!" An earthquake had triggered a huge landslide that caused 3 cubic km of the volcano's weakened north side to slide away. The sudden landslide released pressure on the magma inside the volcano, causing a sudden and violent expansion of gases that blasted through the side of the volcano (**Fig. Bx9.1b**). Rock, steam, and ash screamed north at the speed of sound and flattened a forest and everything in it over an area of 600 square km (**Fig. Bx9.1c**). Tragically, Johnston, along with 60 others, vanished forever. Seconds after the sideways blast, a vertical column carried about 540 million tons of ash (about 1 cubic km) 25 km into the sky, where the jet stream carried it away; the ash was able to circle the globe. In towns near the volcano, a blizzard of ash choked roads and buried fields. Water-saturated ash formed viscous slurries, or lahars, that flooded river valleys, carrying away everything in their path. When the eruption was finally over, the once

FIGURE Bx9.1   Examples of explosive eruptions.

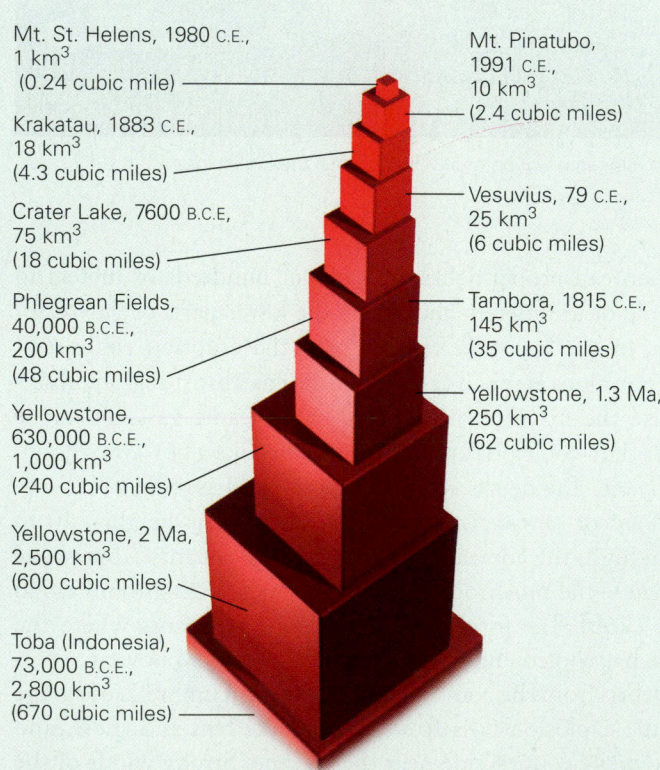

Mt. St. Helens, 1980 C.E.,
1 km³
(0.24 cubic mile)

Krakatau, 1883 C.E.,
18 km³
(4.3 cubic miles)

Crater Lake, 7600 B.C.E,
75 km³
(18 cubic miles)

Phlegrean Fields,
40,000 B.C.E.,
200 km³
(48 cubic miles)

Yellowstone,
630,000 B.C.E.,
1,000 km³
(240 cubic miles)

Yellowstone, 2 Ma,
2,500 km³
(600 cubic miles)

Toba (Indonesia),
73,000 B.C.E.,
2,800 km³
(670 cubic miles)

Mt. Pinatubo,
1991 C.E.,
10 km³
(2.4 cubic miles)

Vesuvius, 79 C.E.,
25 km³
(6 cubic miles)

Tambora, 1815 C.E.,
145 km³
(35 cubic miles)

Yellowstone, 1.3 Ma,
250 km³
(62 cubic miles)

(a) The relative amounts of pyroclastic debris (in cubic km) ejected during major explosive eruptions.

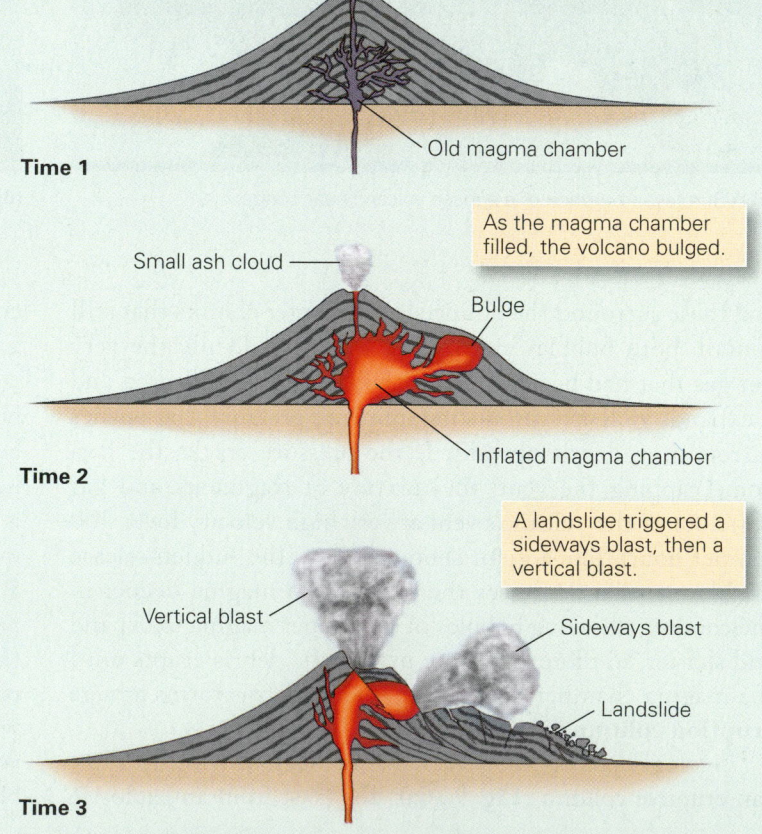

Time 1

Old magma chamber

As the magma chamber filled, the volcano bulged.

Small ash cloud

Bulge

Inflated magma chamber

Time 2

A landslide triggered a sideways blast, then a vertical blast.

Vertical blast

Sideways blast

Landslide

Time 3

(b) Stages during the eruption of Mt. St. Helens, 1980.

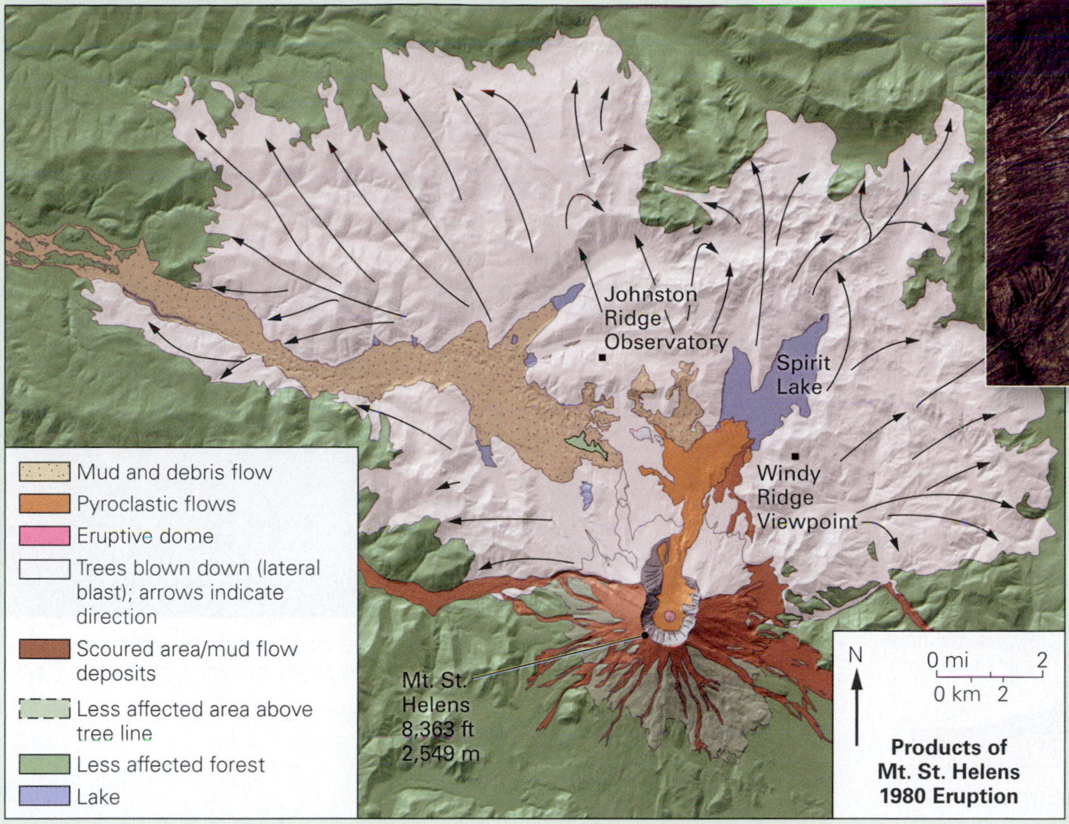

The blast knocked trees down as if they were toothpicks.

Thirty years later, the downed trees remain.

(c) A map shows the dimensions of the region destroyed by the eruption of Mt. St. Helens. The arrows indicate the blast direction. The neighboring forest was flattened by a blast of rock, steam, and ash.

Legend:
- Mud and debris flow
- Pyroclastic flows
- Eruptive dome
- Trees blown down (lateral blast); arrows indicate direction
- Scoured area/mud flow deposits
- Less affected area above tree line
- Less affected forest
- Lake

Johnston Ridge Observatory

Spirit Lake

Windy Ridge Viewpoint

Mt. St. Helens 8,363 ft 2,549 m

N

0 mi 2
0 km 2

Products of Mt. St. Helens 1980 Eruption

cone-shaped peak of Mt. St. Helens had disappeared—the summit now lay 440 m lower, and the once snow-covered mountain was a gray mound with a large gouge in one side. The volcano came alive again in 2004, but it did not explode.

The death toll due to the 1902 explosion of Mt. Pelée, on the Caribbean island of Martinique, was much higher. On the morning of May 8, the 100-m-high rhyolite spire stuck in the conduit of the volcano suddenly disintegrated. The immense gas pressure that had been building beneath the obstruction was suddenly released, and in the same way champagne bursts out of a bottle when the cork is pulled, a cloud of ash and pumice lapilli spewed out of Mt. Pelée. Collapse of the ash column formed a pyroclastic flow—at a temperature of 200°C to 450°C—that swept down Pelée's flank. This flow rode on a cushion of air and reached speeds of up to 300 km per hour before slamming into the busy port town of St. Pierre. Within moments, the town's buildings were flattened and its 28,000 inhabitants were dead of incineration or asphyxiation. Only two people survived—one was a prisoner who was protected by the stout walls of his underground cell.

An even greater explosion happened in 1883. Krakatau, a volcano between Indonesia and Sumatra, had grown to become a 9-km-long island rising 800 m (2,600 feet) above the sea. On May 20, the island began to erupt with a series of large explosions, yielding ash that settled as far as 500 km away. Smaller explosions continued through June and July, and steam and ash rose from the island, forming a huge black cloud that rained ash into the surrounding straits. Ships sailing by couldn't see where they were going, and their crews had to shovel ash off the decks. Krakatau's demise came at 10 A.M. on August 27, perhaps when the volcano cracked and the magma chamber suddenly flooded with seawater. The resulting blast, 5,000 times greater than the Hiroshima atomic bomb

explosion, could be heard as far as 4,800 km away, and subaudible sound traveled around the globe seven times. Giant sea waves pushed out by the explosion slammed into nearby coastal towns, killing over 36,000 people. Near the volcano, a layer of ash up to 40 m thick accumulated. When the air finally cleared, Krakatau was gone, replaced by a submarine caldera some 300 m deep (**Fig. Bx9.1d**). All told, the eruption shot 20 cubic km of rock into the sky. Some ash reached elevations of 27 km. Because of this ash, people around the world could view spectacular sunsets during the next several years.

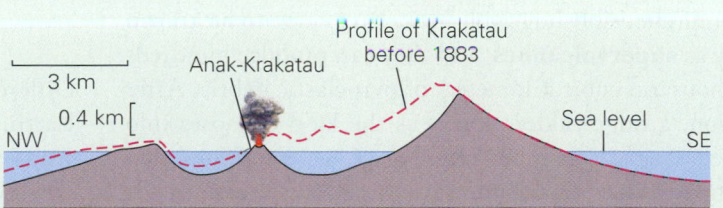

Profile of Krakatau before 1883

Anak-Krakatau

3 km

0.4 km

NW

Sea level

SE

(d) Profile of Krakatau before and after the eruption. Note that a new volcano (Anak-Krakatau) has formed.

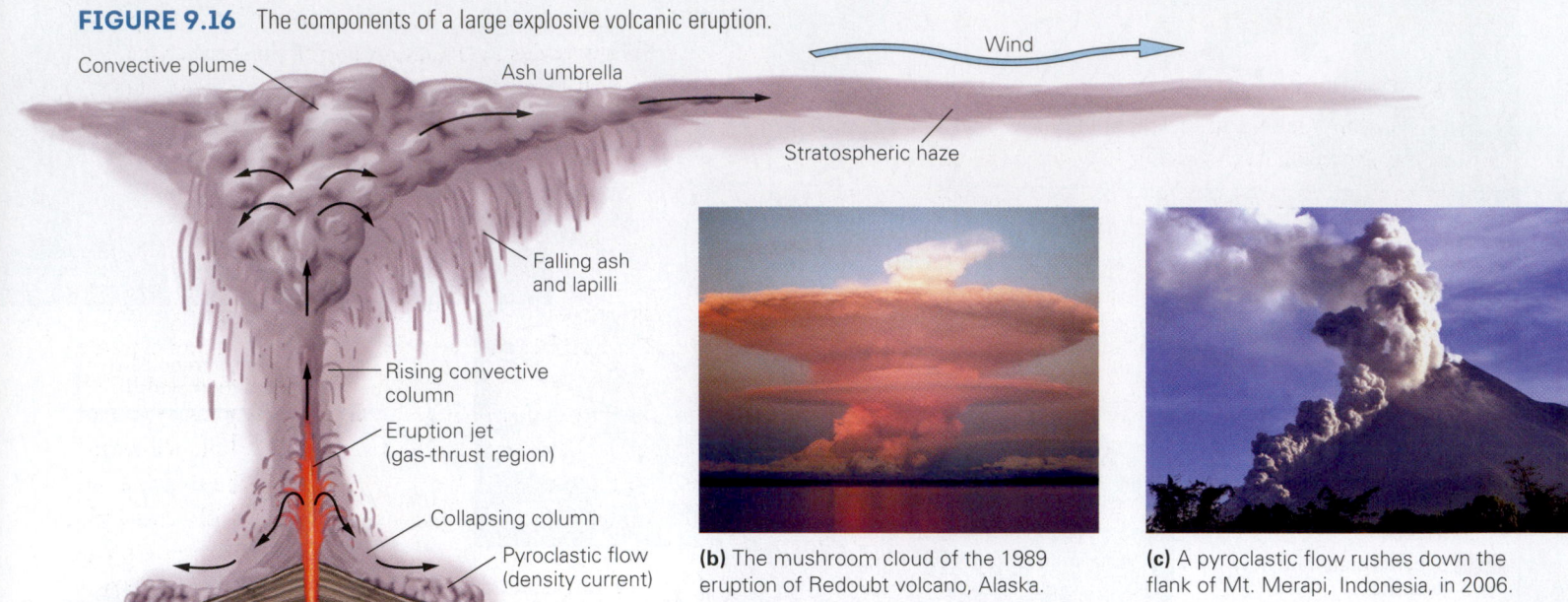

FIGURE 9.16 The components of a large explosive volcanic eruption.

Convective plume

Ash umbrella

Wind

Stratospheric haze

Falling ash and lapilli

Rising convective column

Eruption jet (gas-thrust region)

Collapsing column

Pyroclastic flow (density current)

(a) A large explosive eruptive cloud (Plinean-type) contains several components.

(b) The mushroom cloud of the 1989 eruption of Redoubt volcano, Alaska.

(c) A pyroclastic flow rushes down the flank of Mt. Merapi, Indonesia, in 2006.

jet stream may waft substantial amounts of ash in the umbrella great distances—hundreds or even thousands of kilometers—away from the volcano. In some cases, gravity can cause heavier debris at the top of the gas-thrust region to collapse downward, forming a mixture of air, hot ash, and pumice lapilli that rushes down the side of a volcano in a scalding avalanche. This surge of air and debris stays relatively close to the ground because it is denser than the clear air above (**Fig. 9.16c**). Such avalanches of hot ash are known as **pyroclastic flows** (or as pyroclastic density currents, by analogy to underwater turbidity currents). In older literature, a pyroclastic flow was called a *nuée ardente*, French for glowing cloud. Tuff formed from ash and/or pumice lapilli that fell like snow from the sky is called *air-fall tuff*, whereas a sheet of tuff that formed from a pyroclastic flow is an **ignimbrite**. Ash and pumice lapilli in an ignimbrite is sometimes so hot that it welds together to form a hard mass.

Supervolcanoes In recent years, geologists have realized that some volcanic explosions of the geologic past dwarf any that have been observed during human history (see Box 9.1). Such incomprehensibly huge volcanoes have come to be known informally as **supervolcanoes**, and they can produce hundreds to a few thousand cubic kilometers of pyroclastic debris. After the eruption, a huge caldera forms as the land collapses into the drained magma chamber (see Fig. 9.11). Eruptions that occurred hundreds of thousands of years ago in the area that is now Yellowstone Park serve as an example—in the aftermath of one of these explosions, a caldera 72 km across formed!

Take-Home Message

At a volcano, lava rises from a magma chamber and erupts from chimney-like conduits or from crack-like fissures. During effusive eruptions, low-viscosity basalt lava flows build low, dome-like shield volcanoes. Fountaining basalt spatters to build tephra cones. Volcanoes erupting underwater produce huge bursts of steam and clots of muddy ash. Successive, alternating eruptions of felsic or intermediate pyroclastic debris and lava build stratovolcanoes, and drainage of lava beneath the surface or explosions produce calderas. Immense explosions of supervolcanoes produce giant calderas.

QUICK QUESTION: How can you distinguish between a volcanic caldera and a meteorite crater?

9.4 Geologic Settings of Volcanism

Different styles of volcanism occur at different locations on Earth. Most eruptions occur along plate boundaries, but major eruptions also occur at hot spots and in rifts (**Fig. 9.17a**). We'll now look at the settings in which eruptions occur, in the context of plate tectonics theory, and see why different kinds of volcanoes form in different settings.

Mid-Ocean Ridge Submarine Eruptions

Products of mid-ocean ridge volcanism cover 70% of our planet's surface. We don't generally see this volcanic activity, however, because the ocean hides most of it beneath a blanket of water. Mid-ocean ridge volcanoes, which develop along fissures parallel to the ridge axis, are not all continuously active. Each one turns on and off in a time scale measured in tens to hundreds of years. They erupt basalt, formed when hot mantle rock rises from great depth to shallow depths beneath the ridge and undergoes decompression melting (see Chapter 6). This basalt, because it cools so quickly underwater, forms pillow-lava mounds or aprons. The pillow basalts commonly occur in association with hyaloclastites. Water that heats up as it circulates through the crust near the magma chamber bursts out of hydrothermal (hot-water) vents along these mounds, producing black smokers (see Chapter 4).

Volcanic Arcs at Convergent Boundaries

Most of the subaerial volcanoes on Earth lie along convergent-plate boundaries (subduction zones). The volcanoes form when volatiles rise from the rock of the subducting plate into the overlying hot asthenosphere, causing flux melting in the asthenosphere. The resulting magma then rises and eventually erupts along the edge of the overriding plate. Some of these volcanoes grow on oceanic crust and become volcanic island arcs, such as the Marianas of the western Pacific and the Aleutians of the northern Pacific (**Fig. 9.17b**). Others grow on continental crust, building continental volcanic arcs such as the Cascade volcanic chain of Washington and Oregon or the Andes chain of South America. Typically, individual volcanoes in volcanic arcs lie about 50 to 100 km apart. Subduction zones border over 60% of the Pacific Ocean, creating a 20,000-km-long chain of volcanoes known as the *Ring of Fire*.

In island arcs, where magma rises from the mantle through oceanic crust, volcanoes initially primarily produce basalt, formed by partial melting of the mantle. This lava builds an edifice of pillows and hyaloclastites that eventually rises above the sea level. During its initial appearance above sea level, the eruption produces blasts of steam and ash. When the vent rises entirely above sea level, a shield of basalt starts to grow. Processes such as fractional crystallization and assimilation (see Chapter 6) may eventually yield andesitic lava, so the volcano can evolve into a stratovolcano.

In continental arcs, some basalt rises to the surface, but andesitic and rhyolitic eruptions are more common because more of the magma undergoes fractional crystallization and assimilation within the crust, and in addition, heat transfer partially melts some of the continental crust and produces felsic magma (see Chapter 6). Because many different kinds of magma form at volcanic arcs, these volcanoes sometimes have effusive eruptions and sometimes pyroclastic eruptions, and build stratovolcanoes, which occasionally explode. Examples include the elegant symmetrical cone of Mt. Fuji (see Fig. 9.12c) and the blasted-apart hulk of Mt. St. Helens (see Box 9.1).

Volcanism of Continental Rifts

The igneous activity of rifts happens because thinning of the continental lithosphere allows the underlying asthenosphere to rise to shallower depths, where it undergoes decompression and partially melts to produce basaltic magma. Some of this magma rises straight to the surface and erupts as basalt, but some gets trapped at the base or within the continental crust and undergoes fractional crystallization and/or assimilates surrounding crust to produce intermediate or felsic magmas. Trapped magma can also partially melt continental crust through heat transfer, producing rhyolitic magma.

Because of the diversity of magmas that can form beneath rifts, rifts can host both basaltic fissure eruptions, in which curtains of lava fountain up or linear chains of cinder cones develop, and explosive rhyolitic volcanoes. Thus, you may find both large basalt flows and immense sheets of rhyolites in rifts. In some locations, eruptions build stratovolcanoes such as Mt. Kilimanjaro in Africa.

Oceanic Hot-Spot Volcanism

Oceanic hot-spot volcanoes form where asthenosphere undergoes decompression melting and produces voluminous amounts of basaltic magma. Most oceanic hot-spot volcanoes, such as the ones that produced Hawaii, occur in the interior of plates, away from plate boundaries. A few, such as the ones that produced Iceland, sit astride a mid-ocean ridge. Most geologists favor the hypothesis that hot-spot volcanoes lie above mantle plumes, localized upwellings from the deep mantle, but some argue for alternative interpretations.

When a hot-spot volcano first forms on oceanic lithosphere, basaltic magma erupts at the surface of the seafloor. At first, such submarine eruptions yield a mound of pillow lava and hyaloclastite. With time, the volcano grows up above the sea surface and becomes an island. After the volcano emerges from the sea, the basalt lava that erupts no longer freezes so quickly and thus flows as a thin sheet over a great distance. Thousands of thin basalt flows pile up, layer upon layer, to build a broad, dome-shaped shield volcano with gentle slopes (**Fig. 9.18**). As the volcano grows, portions of it can't resist the pull of gravity and slip seaward, creating large submarine slumps.

The big island of Hawaii, the tallest oceanic hot-spot volcano on Earth today, currently consists of five shield volcanoes, each built around a different vent. The island now

FIGURE 9.17 Volcanoes of the world.

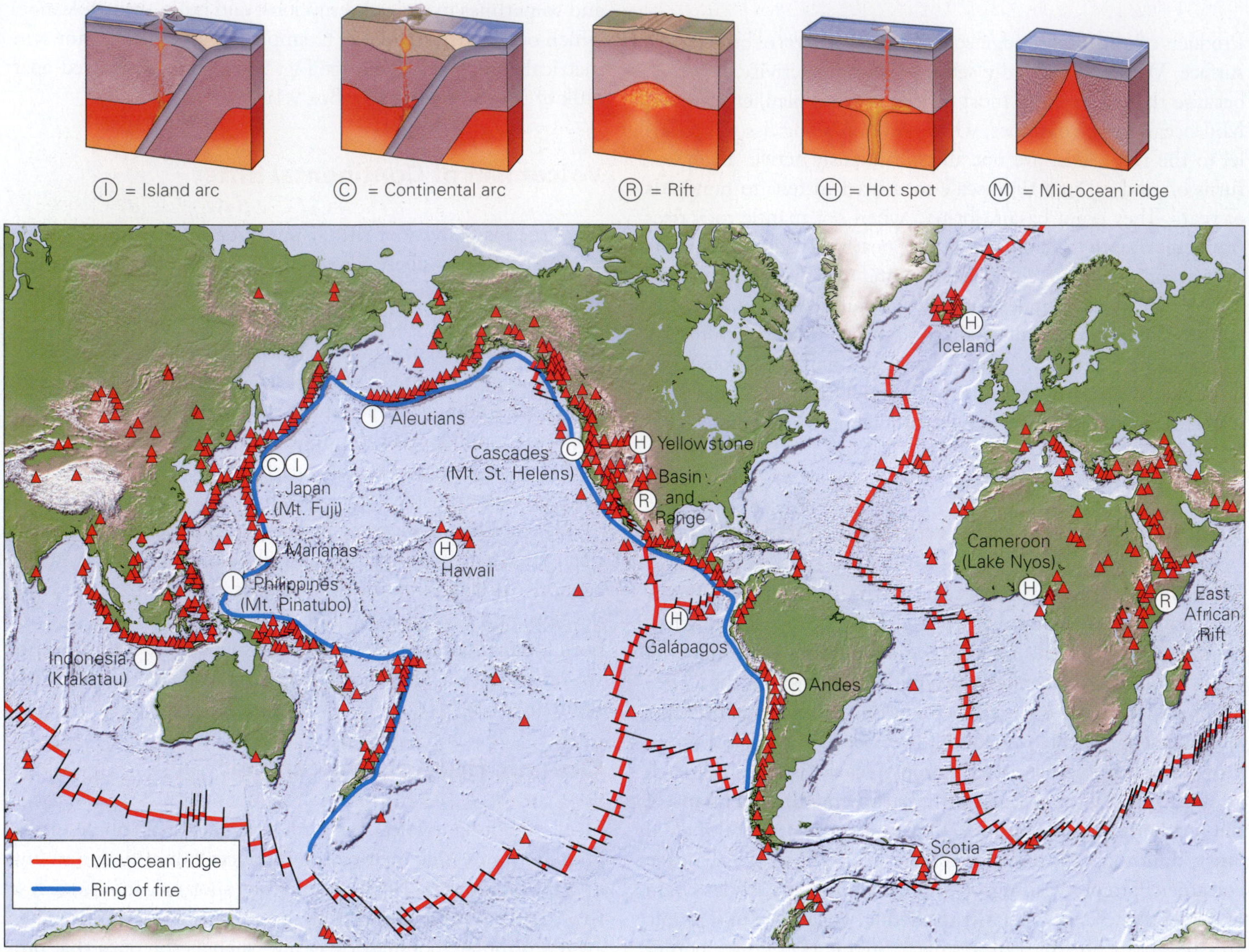

Ⓘ = Island arc Ⓒ = Continental arc Ⓡ = Rift Ⓗ = Hot spot Ⓜ = Mid-ocean ridge

(a) A map showing the distribution of volcanoes around the world and the basic geologic settings in which volcanoes form, in the context of plate tectonics theory.

(b) The Aleutian Arc forms the northern edge of the Pacific Plate. It displays a distinct curvature.

(c) An oblique view of the Aleutian arc, as seen looking northwest.

FIGURE 9.18 The interior of an oceanic hot-spot volcano is complicated. Initially, eruption produces pillow basalts. When the volcano emerges above sea level, it becomes a shield volcano. The margins of the island frequently undergo slumping, and the weight of the volcano pushes down the surface of the lithosphere. The Hawaiian Islands exemplify this architecture.

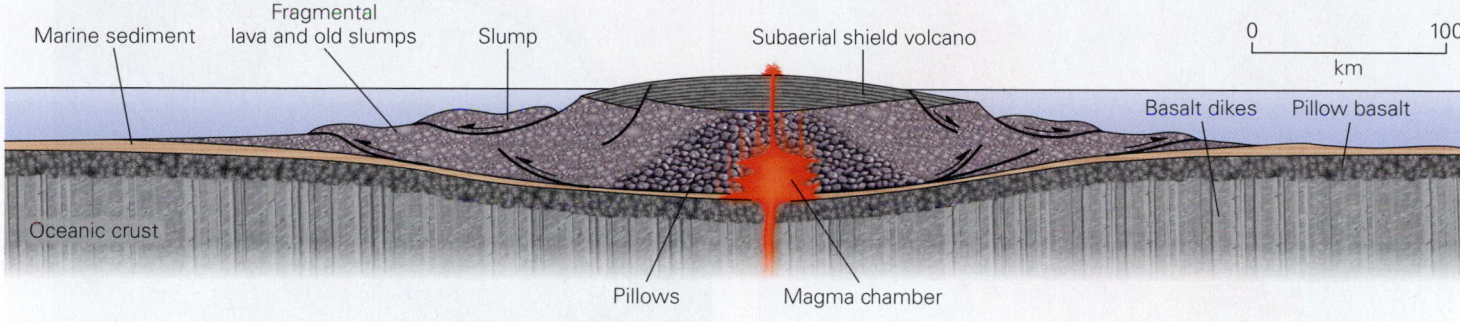

towers over 9 km above the adjacent ocean floor (about 4.2 km above sea level), the greatest relief from base to top of any mountain on Earth—by comparison, Mt. Everest rises 8.85 km above the plains of India. Calderas up to 3 km wide have formed at the summit. Vast sheets of basaltic lava extrude from chimney-shaped conduits and fissures, both at the summit and on the flanks of the volcanoes (see Fig. 9.3b).

Iceland also formed over a hot spot, one that lies beneath the Mid-Atlantic Ridge—the presence of this hot spot means that far more lava erupted here than beneath other places along the ridge. As a result, Iceland sits on a broad oceanic plateau. Because Iceland straddles a divergent-plate boundary, it is being stretched apart, with faults forming as a consequence. Indeed, the central part of the island is a narrow rift, in which the youngest volcanic rocks of the island have erupted (**Fig. 9.19**). This rift is the trace of the Mid-Atlantic Ridge. Faulting cracks the crust and so provides a conduit to a magma chamber. Thus, eruptions on Iceland tend to begin as fissure eruptions, spewing curtains of lava. Eventually, the curtains die out and are replaced by more localized eruptions that form linear chains of cinder cones.

Not all volcanic activity on Iceland occurs subaerially. Some eruptions take place under glaciers. During 1996, for example, an eruption at the base of a 600-m-thick glacier melted the ice and produced a column of steam that rose several kilo-

meters into the air. Meltwater accumulated under the ice for six days, until it burst through the edge of the glacier and became a flood (called a *jökulhlaup*, in Icelandic) that lasted two days and destroyed roads, bridges, and telephone lines. The 2010 eruption of the Eyjafjallajokull Volcano caused similar problems, and was also disruptive in other ways, as we will see.

Some of Iceland's volcanic activity occurs off the coast. Such activity produced the island of Surtsey. The birth of Surtsey was heralded by huge quantities of steam bubbling up from the ocean. Eventually, steam pressure explosively ejected ash as high as 5 km into the atmosphere. Surtsey finally emerged from the sea on November 14, 1963, building up a cone of ash and lapilli that rose almost 200 m above sea level in just three months. Waves could easily have eroded the cinder cone away, but the island has survived because lava erupted from the vent and flowed over the cinders, effectively encasing them in an armor-like blanket of solid rock.

Continental Hot-Spot Volcanism

Yellowstone National Park lies at the northeast end of a string of calderas, known as the Yellowstone hot-spot track, whose remnants crop out in the Snake River Plain of Idaho (**Fig. 9.20a**). The oldest of these calderas, at the southwest end of the track, erupted 16 million years ago (Ma). Recent studies have found evidence of a mantle plume beneath Yellowstone, adding support to the hypothesis that the Yellowstone hot-spot track formed as the North American plate moved over a plume.

Ongoing activity beneath Yellowstone has yielded fascinating landforms, volcanic rock deposits, and geysers. Eruptions at the Yellowstone hot spot differ from those in Hawaii in an important way: unlike Hawaii, the Yellowstone hot spot erupts both basaltic lava and rhyolitic pyroclastic debris. This happens because heat transfer from the rising basaltic magma partially melts the continental crust to produce felsic magma.

FIGURE 9.19 Iceland, a hot spot on the Mid-Atlantic Ridge.

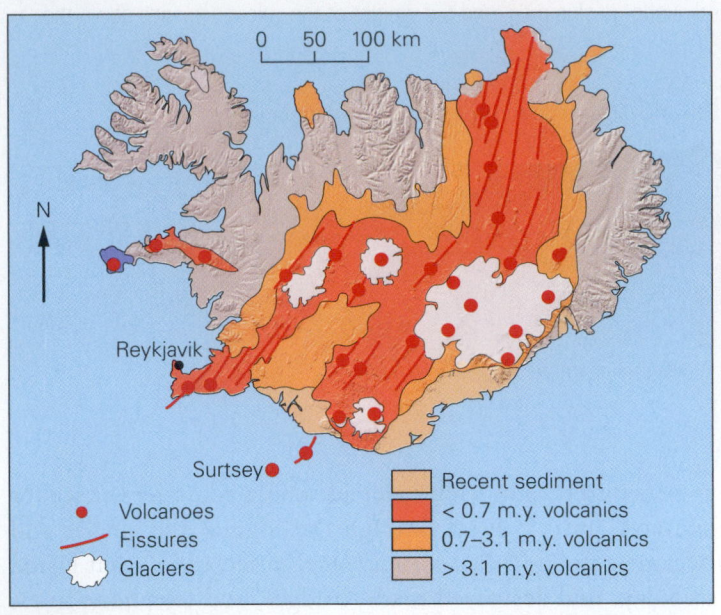

(a) A geologic map of Iceland shows how the youngest volcanoes occur in the central rift, effectively the on-land portion of the Mid-Atlantic Ridge.

(b) A bathymetric map shows that Iceland sits atop a huge plateau straddling the Mid-Atlantic Ridge. Light blue is shallower water; dark blue is deeper.

(c) A rift cutting through basalt, on Iceland.

About 630,000 years ago (0.63 Ma), immense pyroclastic flows and convective clouds of ash and pumice lapilli blasted out of the Yellowstone region. Close to the eruption, numerous ignimbrites built up, and ash and lapilli from the giant cloud sifted down over the United States as far east as the Mississippi River (**Fig. 9.20b**). The 0.63-Ma eruption produced an immense caldera, up to 72 km across, that overlaps earlier calderas (**Fig. 9.20c**). When the debris settled, it blanketed an area of 2,500 square km with tuffs that, in the park, reached a thickness of 400 m—Yellowstone was a supervolcano! The park's name reflects the brilliant color of volcaniclastic debris exposures in the park's canyons (**Fig. 9.20d**). Eruptive activity, producing basalt and rhyolite lavas

and more pyroclastic debris, continued until about 70,000 years ago. Magma remains in the crust beneath the park today. It's the energy radiating from this magma that heats the water filling hot springs and spurting out of geysers.

Flood-Basalt Eruptions

In several locations around the world, huge sheets of low-viscosity lava erupted and spread out in vast sheets. Geologists refer to the lava of these sheets as **flood basalt** (**Fig. 9.21a**). Over time, many successive eruptions of flood basalt can build up a broad *basalt plateau*. What causes flood-basalt eruptions? A popular hypothesis suggests that flood basalts form when a mantle plume starts to rise beneath a region that is undergoing rifting (Fig. 6.27a). As the plume reaches the base of the lithosphere, it has a bulbous head containing a large amount of partially molten rock. Stretching and thinning of the overlying lithosphere results in further decompression of the plume head and causes even more melt to form. The melt intrudes along fissures that form in the rift and erupts spectacularly at the surface. Once the plume head no longer exists, the volume of eruption decreases, and "normal" hot-spot volcanism (with less magma production) takes place.

The aggregate volume of rock in a basalt plateau may be so great (over 175,000 km^3) that geologists also refer to the region as a **large igneous province** (**LIP**; see Fig. 6.26). (The term has also been used for regions of immense rhyo-

FIGURE 9.20 Hot-spot volcanic activity in Yellowstone National Park.

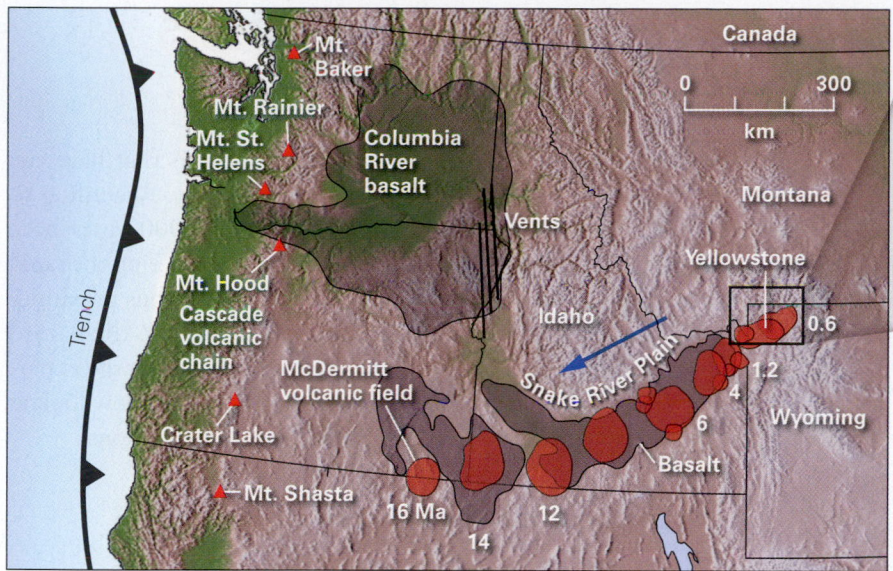

(a) Yellowstone lies at the end of a continental hot-spot track. Progressively older calderas follow the Snake River Plain. The blue arrow indicates plate motion.

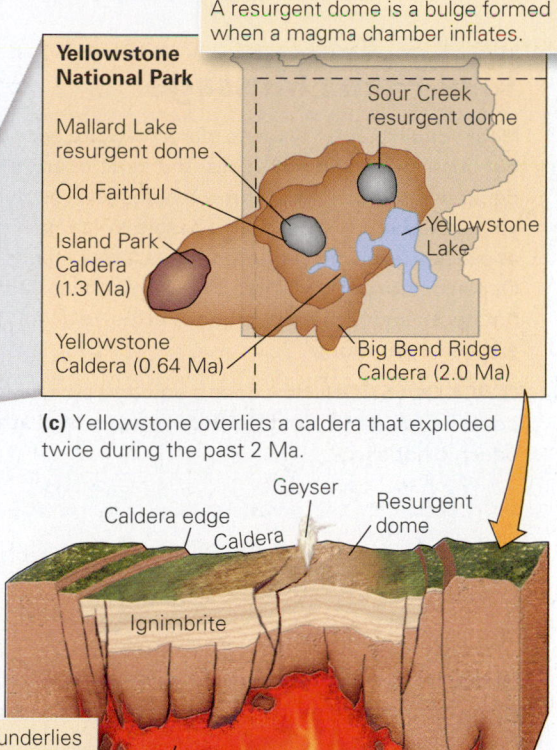

(c) Yellowstone overlies a caldera that exploded twice during the past 2 Ma.

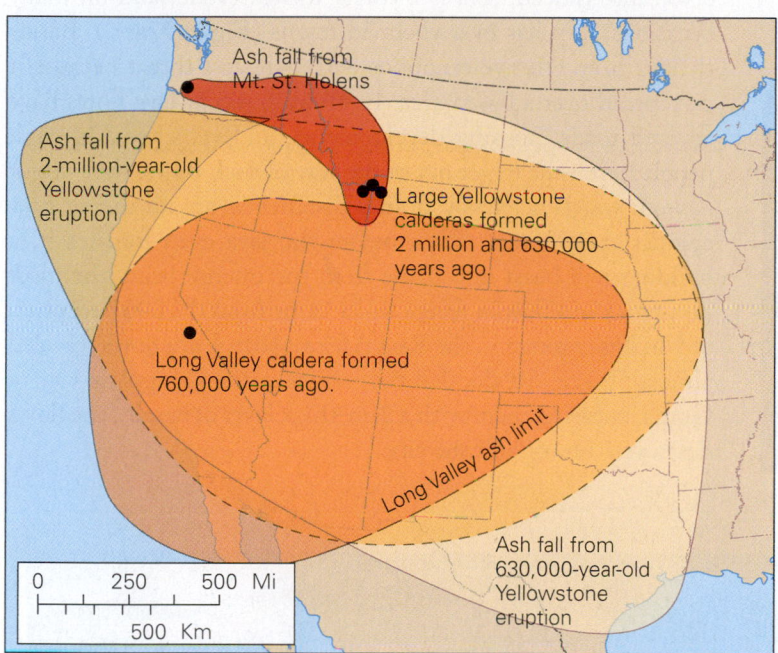

(b) The ash produced by explosions of the Yellowstone calderas covered vast areas—much more than Mt. St. Helens.

(d) Felsic tuffs form the colorful walls of Yellowstone Canyon.

lite eruption, such as the Yellowstone region.) An example of a LIP, the Columbia River Plateau, occurs in Washington and Oregon (see Fig. 6.27b). The basalt here, which erupted around 15 million years ago, reaches a thickness of 3.5 km. Geologists have identified about 300 individual flows in the Columbia River Plateau. Lava in some of these flows traveled great distances—up to 600 km—from its source. Eventually, basalt covered an area of 220,000 km². Even larger

flood-basalt provinces occur in eastern Siberia (an occurrence known as the Siberian traps; **Fig. 9.21b**), the Deccan Plateau of India, the Paraná region of Brazil, and the Karoo Plateau of South Africa.

FIGURE 9.21 According to one hypothesis, flood basalts erupt when the head of a plume reaches the base of rifting lithosphere.

(a) Flood-basalt layers exposed on the wall of a canyon in Idaho.

9.5 Beware: Volcanoes Are Hazards!

Like earthquakes, volcanoes are natural hazards that have the potential to cause great destruction to humanity. According to one estimate, volcanic eruptions in the last 2,000 years have caused about a quarter of a million deaths. Considering the rapid expansion of cities, far more people live in dangerous proximity to volcanoes today than ever before, so if anything, the hazard posed by volcanoes has gotten worse—imagine if a large explosion were to occur next to a major city today. Let's now look at the different kinds of threats posed by volcanic eruptions.

Hazards due to Eruptive Materials

Threat of Flows When you think of an eruption, perhaps the first threat that comes to mind is the lava that flows from a volcano. Indeed, lava is a threat to real estate, and on many occasions, lava has overwhelmed towns (**Fig. 9.22a–d**). Basaltic lava from effusive eruptions is the greatest threat because it can spread over a broad area. In Hawaii, recent lava flows have covered roads, housing developments, and vehicles. Although people have time to get out of the way of such flows, they might have to watch helplessly from a distance as an advancing flow engulfs their homes. Even before the lava even touches it, a building will burst into flame from the intense heat. The most disastrous lava flow in recent times came from the 2002 eruption of Mt. Nyiragongo in the East African Rift. Lava flows traveled almost 50 km and flooded the streets in the Congolese city of Goma, encasing them with a 2-m-thick layer of basalt. The flows destroyed almost half the city.

(b) An area the size of Europe was covered by flood basalts (the "Siberian traps") and associated tuffs in Siberia. The time of the eruption, about 250 Ma, coincides with the end of the Paleozoic, when there was mass extinction. The map shows the maximum extent of the volcanic rock before erosion.

Pyroclastic flows can move extremely fast (100 to 300 km per hour) and are so hot (500° to 1,000°C) that they represent a profound hazard to humans and the environment (**Fig. 9.22e, f**). Even relatively small examples, such as the flow that struck St. Pierre on Martinique, can flatten towns and devastate fields even though they leave only a few centimeters of ash and lapilli behind. People caught in the direct path of such flows may be incinerated, and even those protected from the ash itself may die from inhaling toxic, superhot gases.

Threat of Falling Ash and Lapilli During a pyroclastic eruption, large quantities of ash and lapilli erupt into the air, later to fall back to the ground (**Fig. 9.22g**). Close to the volcano, pumice and lapilli tumble out of the sky and can accumulate to form a blanket up to several meters thick. The mass, especially when saturated with rainwater, causes roofs and power lines to collapse. Winds can carry fine ash over a broad region. In the Philippines, for example, a typhoon spread heavy air-fall ash from the 1991 eruption of Mt. Pinatubo so that it covered a 4,000-square-km area. An ash fall buries crops, coats the leaves of trees, and may spread toxic chemicals that poison the soil. Ash also insidiously infiltrates machinery, causing moving parts to wear out.

Threat to Aircraft Fine ash from an eruption can also present a hazard to airplanes. Like a sandblaster, the sharp, angular shards of ash abrade turbine blades, greatly reducing engine efficiency. The ash, along with sulfuric acid formed from the volcanic gas, scores windows and damages the fuselage. Also, when heated inside a jet engine, the ash melts, creating a liquid that coats interior parts of the engine and freezes to glass, coating temperature sensors, which falsely indicate that the engines are overheating so they automatically shut down.

Encounters between airliners and volcanic plumes have led to terrifying incidents. In 1982, a British Airways 747 flew through the ash cloud above a volcano in Java. The windshield turned opaque and all four engines failed. For 13 minutes, the plane silently glided earthward, dropping from 11.5 km (37,000 ft). The pilot frantically tried to restart the engines to no avail and prepared to ditch at sea. Finally, at 3.7 km (12,000 ft), the engines had cooled sufficiently and suddenly roared back to life and the plane headed to Jakarta for an emergency landing. There, without functioning instruments, the pilot squinted out an open side window to see the runway and brought the plane safely to a halt with only his toes touching the pedals. A similar event happened between a KLM 747 and the eruptive cloud of Redoubt Volcano in Alaska in 1989.

Because of the lessons learned from such incidents, the 2010 eruption of the Eyjafjallajokull Volcano in Iceland had a profound impact on air traffic. All told, the eruption sent about 0.25 cubic km of pyroclastic material up into the air. The jet stream, a high-altitude current of rapidly moving air, was passing over Iceland at the time of the eruption and thus dispersed the ash throughout European air space. Because of concern that this ash could damage planes, officials shut down almost all air traffic across Europe for six days. The closure directly cost airlines $200 million/day, disrupted travel plans for countless passengers, and halted shipment of everything from electronic goods to flowers, thus impacting economies worldwide.

Other Hazards Related to Eruptions

Threat of the Blast Most exploding volcanoes direct their fury upward. But some, like Mt. St. Helens, explode sideways. The forcefully ejected gas and ash, like the blast of a bomb, flattens everything in its path. In the case of Mt. St. Helens, the region around the volcano had been a beautiful pine forest. But after the eruption, the once-towering trees, stripped of bark and needles, lay flattened onto the hill slopes, all pointing in the direction away from the blast (see Box 9.1).

Threat of Landslides Eruptions commonly trigger large landslides along a volcano's flanks. The debris, composed of ash and solidified lava that erupted earlier, can move quite fast (250 km per hour) and far. During the eruption of Mt. St. Helens, 8 billion tons of debris took off down the mountainside, careened over a 360-m-high ridge, and tumbled down a river valley, until the last of it finally came to rest over 20 km from the volcano.

Threat of Lahars When volcanic ash and other debris mix with water, the result is a lahar, an ashy slurry that resembles wet concrete. A lahar can move downslope at speeds of over 50 km per hour. Because lahars are denser and more viscous than clear water, they pack more force than clear water and literally carry away everything in their path. The lahars of Mt. St. Helens traveled along existing drainages for more than 40 km from the volcano. When they had passed, they left a gray and barren wake of mud, boulders, broken bridges, and crumpled houses, as if a giant knife had scraped across the landscape. The lahars generated during the 1991 eruption of Pinatubo in the Philippines were even more devastating, for the eruption was bigger and coincided with the drenching rains of a typhoon.

Lahars may develop in regions where snow and ice cover an erupting volcano, for the eruption melts the snow and ice, thereby creating a supply of water. Perhaps the most destructive lahar of recent times accompanied the eruption of the snow-crested Nevado del Ruiz in Colombia on the night of November 13, 1985. The lahar surged down a valley like a 40-m-high wave, hitting the sleeping town of Armero, 60 km from the volcano. Ninety percent of the buildings in the town vanished, replaced by a 5-m-thick layer of mud, which now entombs the bodies of 25,000 people (**Fig. 9.22h**).

FIGURE 9.22 Hazards due to lava and ash from volcanic eruptions.

Lava Flows

(a) A lava flow reaches a house in Hawaii and sets it on fire.

(b) Lava from Mt. Etna threatens a town and olive grove in Sicily.

(c) Residents rescue household goods after a lava flow filled the streets of Goma, along the East African Rift.

(d) This empty school bus was engulfed by lava in Hawaii.

Pyroclastic Debris

(e) A pyroclastic flow from the 1991 eruption of Mt. Pinatubo chases a fleeing vehicle.

(f) A blizzard of ash fell from the cloud erupted by Mt. Pinatubo in the Philippines.

(g) Lapilli falls from an eruption in Iceland.

Lahar

(h) A lahar submerges farmland in Colombia.

Threat of Earthquakes Earthquakes accompany almost all major volcanic eruptions, for the movement of magma breaks rocks underground. Such earthquakes may trigger landslides on the volcano's flanks and can cause nearby buildings to collapse and dams to rupture, even before the eruption itself begins.

Threat of Tsunamis Where explosive eruptions occur in an island arc, the blast and the underwater collapse of a caldera can generate huge sea waves, or tsunamis, tens of meters high. Most of the 36,000 deaths attributed to the 1883 eruption of Krakatau were not due to ash or lava but rather to tsunamis that slammed into nearby coastal towns. Tsunamis may also be generated by huge submarine landslides that occur when part of a volcanic island suddenly slumps into the sea.

Threat of Gas We have already seen that volcanoes erupt not only solid material but also large quantities of gases such as water vapor, carbon dioxide, sulfur dioxide, and hydrogen sulfide. Usually the gas eruption accompanies the lava and ash eruption, with the gas contributing only a minor part of the calamity. For example, sulfur gases, mixing with moisture, produce sulfuric acid aerosols that can cause respiratory problems in people who live downwind. But occasionally the gas alone snuffs out life in its path without causing any other damage. Such an event occurred in 1986 near Lake Nyos in western Africa.

Lake Nyos is a small but deep lake filling the crater of an active volcano in Cameroon. Though only 1 km across, the lake reaches a depth of over 200 m. Because of its depth, the cool bottom water of the lake does not mix with warm surface water. Carbon-dioxide gas slowly bubbles out of cracks in the floor of the crater and dissolves in the cool bottom water, eventually saturating the water with CO_2. On August 21, 1986, perhaps because a landslide or wind disturbed the water, the water within the lake overturned and the saturated bottom water rose to the surface (**Fig. 9.23a, b**). As it rose, the pressure

FIGURE 9.23 The CO_2 gas disaster, Lake Nyos, Cameroon.

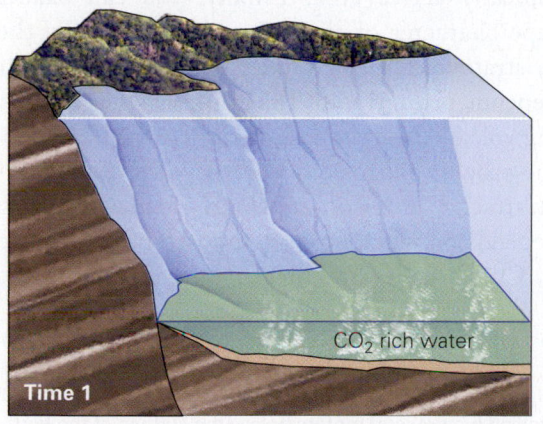
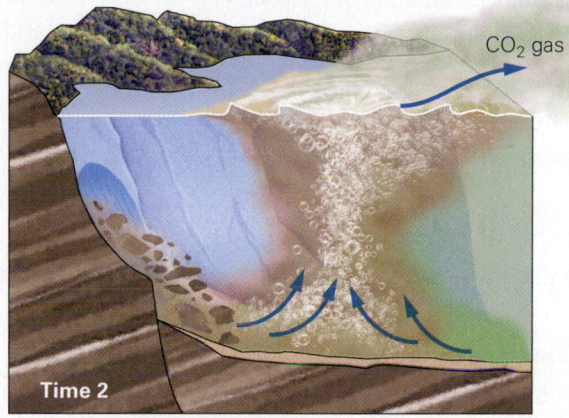

(a) CO_2 dissolved in the colder bottom water (purple). When a landslide or wind disturbed the water, it rose and the CO_2 came out of solution and, in gas form, flowed out of the crater.

(b) Lake Nyos, after the disaster. The crater lake has been discolored by turbulence.

(c) The CO_2 suffocated cattle on the slopes below the volcano.

acting on the water decreased, the CO_2 came out of solution, and the lake suddenly expelled a forceful froth of CO_2 bubbles. Because it is denser than air, this invisible gas flowed down the flank of the volcano and spread out over the countryside for a distance of about 23 km before dispersing. Although not toxic, carbon dioxide cannot provide oxygen for metabolism or oxidation. When the gas cloud engulfed the village of Nyos, it quietly put out cooking fires and suffocated the sleeping inhabitants. The next morning, the landscape looked exactly as it had the day before, except for the lifeless bodies of 1,742 people and about 6,000 head of cattle (**Fig. 9.23c**).

Take-Home Message

Volcanoes can be dangerous! The lava flows, pyroclastic debris, explosions, mudflows (lahars), landslides, earthquakes, and tsunamis that can be produced during eruptions can destroy cities and farmland. Ash that enters the air can be a hazard for air travel.

QUICK QUESTION: Why can a lahar do so much more damage than an equivalent-sized flood of clear water?

9.6 Protection from Vulcan's Wrath

Volcanic eruptions are a natural hazard of extreme danger. Can anything be done to protect lives and property from this danger? The answer is yes. Below we first examine the evidence that geologists use to determine whether a volcano has the potential to erupt, and then we consider the suite of observations that may allow geologists to predict the timing of an impending eruption.

Active, Dormant, and Extinct Volcanoes

Geologists refer to volcanoes that are erupting, have erupted recently, or are likely to erupt soon as **active volcanoes** and distinguish them from **dormant volcanoes**, which have not erupted for hundreds to thousands of years but may erupt again in the future. Volcanoes that were active in the geologic past but have shut off entirely and will never erupt again because the geologic cause of volcanism no longer exists are called **extinct**

volcanoes. As examples, geologists consider Hawaii's Kilauea to be active, for it currently is erupting and has erupted frequently during recorded history. In contrast, Mt. Rainier in the Cascades last erupted centuries to millennia ago, but since subduction continues along the western edge of Oregon and Washington, the volcano could erupt in the future, and so it is considered dormant. Devils Tower, in eastern Wyoming, is the remnant of a shallow igneous intrusion, formed beneath a volcano and subsequently exposed by erosion (**Fig. 9.24**). This was a volcanic region at the time the magma of Devils Tower intruded, tens of millions of years ago. Volcanism will not happen again where the geologic causes for volcanism no longer exist, so we can say that volcanism in the Devils Tower area is extinct. Similarly, when the volcanoes that built the island of Kauai lay over the Hawaiian hot spot, they were active, but now that the island has moved off, its volcanoes have become extinct.

How do you determine whether a volcano is active, dormant, or extinct? One way is to examine the historical record. Another is to determine the age of erupted rocks and to search for evidence that the volcano still lies within a tectonically active area. Finally, you can examine the landscape character of the volcano. Specifically, the shape (shield, stratovolcano, or cinder cone) of an erupting volcano depends primarily on the eruptive style, because at an erupting volcano, the process of construction happens faster than the process of erosion. Once a volcano stops erupting, erosion attacks. The rate at which erosion destroys a volcano depends on whether it's composed of pyroclastic debris or lava. Cinder cones and ash piles can wash away quickly.

FIGURE 9.24 Devils Tower, Wyoming formed as an intrusion into sedimentary rocks beneath a volcano. It solidified hundreds of meters below the surface of the Earth and has been exposed by erosion. Cooling produced spectacular columnar joints.

The vertical lines are columnar joints.

In contrast, composite or shield volcanoes, which have been armor-plated by lava flows, can withstand the attack of water and ice for quite some time. In the end, however, erosion wins out, and you can distinguish a dormant volcano from an active volcano by the extent to which river or glacial valleys have been carved into its flanks (**Fig. 9.25**). In some cases, the softer exterior of a volcano completely erodes away, leaving behind the plug of harder frozen magma that once lay within or beneath the volcano, as well as the network of dikes that radiate from this plug. You can see good examples of such landforms at Shiprock, New Mexico (see Fig. 6.13).

Predicting Eruptions

Predicting the time of an eruption decades or even years in advance is impossible. The only way to constrain a long-term prediction is to determine the recurrence interval (the average time between eruptions). Geologists determine recurrence intervals by determining the age of erupted layers comprising the volcano's edifice. For example, Mt. Fuji has erupted about 65 times in the last 10,000 years. So its eruptive recurrence interval is about 150 years. Note that a recurrence interval does not indicate periodicity—some of Fuji's eruptions were decades apart while others were centuries apart. The recurrence interval just gives a sense of the probability, and surprises happen.

Did you ever wonder . . .

if a volcano could erupt beneath London, England?

Unlike earthquakes, volcanic eruptions can (in general) be predicted. The 2014 eruption of Mt. Ontake in Japan, however, serves as a rare exception. The microseismicity, hinting that magma was rising in the throat of the volcano, did not start until an hour before a significant eruption of gas and ash. Sadly, the ash covered and killed about 30 hikers who had climbed the volcano to enjoy the Fall foliage. Some volcanoes send out distinct warning signals announcing that an eruption may take place very soon, for as magma squeezes into the magma chamber, it causes a number of changes that geologists can measure.

- *Earthquake activity:* Movement of magma generates vibrations in the Earth. When magma flows into a volcano, rocks surrounding the magma chamber crack, and blocks slip with respect to each other. Such cracking and shifting cause earthquakes. Thus, in the days or weeks preceding an eruption, the region between 1 and 7 km beneath a volcano becomes seismically active.
- *Changes in heat flow:* The presence of hot magma increases the local heat flow, the amount of heat passing up through rock. In some cases, the increase in the heat flow melts snow or ice on the volcano, triggering floods and lahars even before an eruption occurs.

FIGURE 9.25 The shape of a volcano changes as it is eroded.

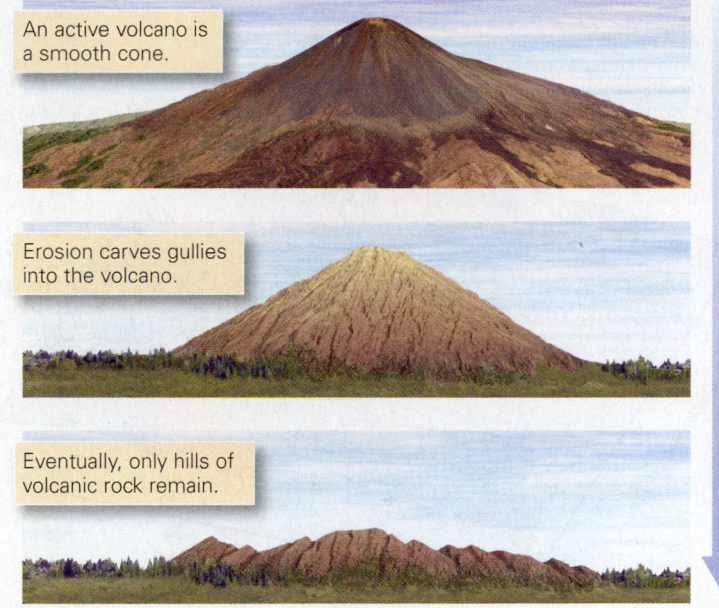

An active volcano is a smooth cone.

Erosion carves gullies into the volcano.

Eventually, only hills of volcanic rock remain.

Time

- *Changes in shape:* As magma fills the magma chamber inside a volcano, it pushes outward and can cause the surface of the volcano to bulge; the same effect happens when you blow into a balloon. Geologists now use laser sighting, accurate tiltmeters, surveys using global positioning systems, and a technique called satellite interferometry (InSAR, which uses radar beams emitted by satellites to measure elevation changes) to detect modification of a volcano's shape due to the rise of magma (**Fig. 9.26a**).
- *Increases in gas and steam emission:* Even though magma remains below the surface, gases bubbling out of the magma, or steam formed by the heating of groundwater by the volcano, percolate upward through cracks in the Earth and rise from the volcanic vent. So an increase in the volume of gas emission, or the birth of new hot springs, may indicate that magma has entered the ground below (**Fig. 9.26b**).

Mitigating Volcanic Hazards

Danger Assessment Maps Let's say that a given active volcano has the potential to erupt in the near future. What can we do to prevent the loss of life and property? Since we can't prevent the eruption, the first and most effective precaution is to define the regions that can be directly affected by the eruption—to compile a **volcanic-hazard assessment map** (**Fig. 9.27**). These maps delineate areas that lie in the path of potential lava flows, lahars, or pyroclastic flows. River valleys originating on the flanks of a volcano are particularly dangerous places to be because lahars may flow down them.

Before the 1991 eruption of Mt. Pinatubo in the Philippines, geologists had defined areas potentially in the path of

FIGURE 9.26 Measuring the activity of volcanoes.

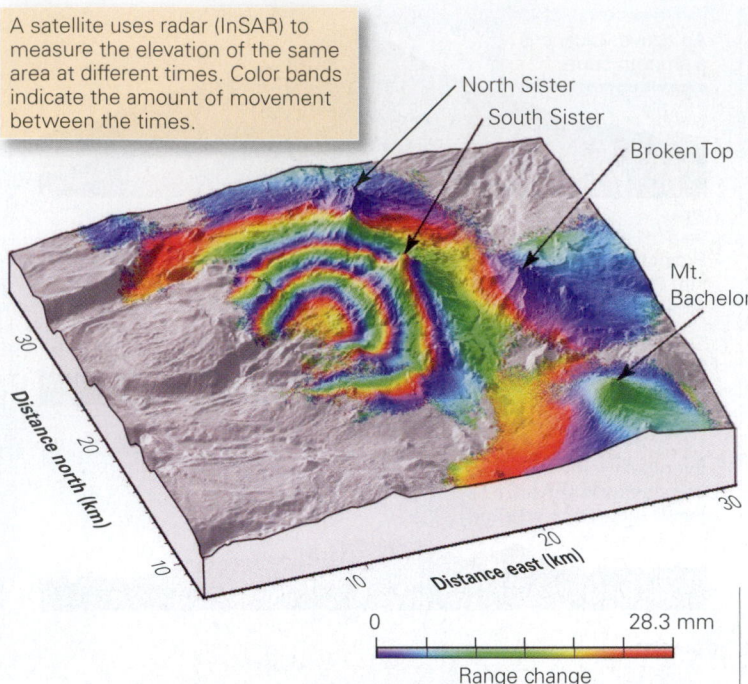

A satellite uses radar (InSAR) to measure the elevation of the same area at different times. Color bands indicate the amount of movement between the times.

North Sister
South Sister
Broken Top
Mt. Bachelor

Distance north (km)
Distance east (km)

0 ——————— 28.3 mm
Range change

(a) An InSAR map of the Three Sisters Wilderness in Oregon. Satellites can map and detect movement of the ground surface over time.

(b) Geologists can detect changes in gas composition.

pyroclastic flows and had predicted which river valleys were likely hosts for lahars. Although the predicted pyroclastic-flow paths proved to be accurate, the region actually affected by lahars was much greater. Nevertheless, many lives were saved by evacuating people in areas thought to be under threat.

Volcano Monitoring and Evacuation Because geologists can determine when magma has moved into the magma chamber of a volcano, government agencies now send monitoring teams to a volcano at the first sign of activity. These teams set up instruments to record earthquakes, measure the heat flow, determine changes in the volcano's shape, and analyze emissions. If an eruption seems imminent, they may issue a warning and call for area residents to evacuate.

Unfortunately, because of the uncertainty of prediction, the decision about whether or not to evacuate is a hard one. In the case of Mt. St. Helens in 1980, hundreds of lives were saved by timely evacuation, but in the case of Mt. Pelée in 1902, thousands of lives were lost because warning signs were ignored. In the Philippines, evacuation of people from the danger area around Mt. Pinatubo, over many years, succeeded in saving thousands of lives, so the prediction was a success. In 1976, however, there was a fierce debate over the need for an evacuation around a volcano on Guadeloupe, in the French West Indies. Eventually, the population of a threatened town was evacuated, but as months passed, the volcano did not erupt. Instead, tempers did, and anger at the cost of the evacuation translated into lawsuits; the prediction was not a success.

Diverting Flows Sometimes people have used direct force to change the direction of a flow or even to stop it. For example, during a 1669 eruption of Mt. Etna, an active volcano on the Italian island of Sicily, basaltic lava formed a glowing orange river that began to spill down the side of the mountain. When the flow approached the town of Catania, 16 km from

FIGURE 9.27 A volcanic-hazard assessment map for Mt. Rainier, Washington, showing the regions that might be affected by flows and lahars.

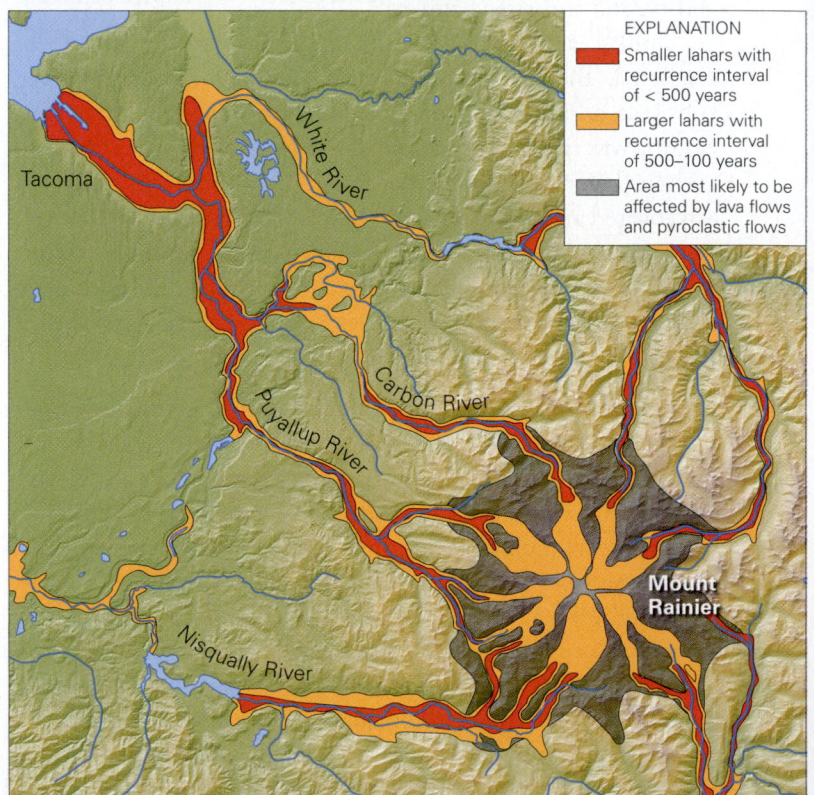

EXPLANATION

🟥 Smaller lahars with recurrence interval of < 500 years

🟧 Larger lahars with recurrence interval of 500–100 years

⬛ Area most likely to be affected by lava flows and pyroclastic flows

Tacoma
White River
Carbon River
Puyallup River
Nisqually River
Mount Rainier

FIGURE 9.28 Efforts to divert lava flows away from inhabited locations.

(a) Workers spray the lava flow to solidify it and use a bulldozer to build an embankment to divert it on the flanks of Mt. Etna.

(b) Firefighters pumping 6 million cubic meters of water on a lava flow, in an effort to freeze it and stop it.

the summit, 50 townspeople protected by wet cowhides boldly hacked through the chilled side of the flow to create an opening through which the lava could exit. They hoped thereby to cut off the supply of lava feeding the end of the flow, near their homes. Their strategy worked, and the flow began to ooze through the new hole in its side. Unfortunately, the diverted flow began to move toward the neighboring town of Paterno. Five hundred men of Paterno then chased away the Catanians so that the hole would not be kept open, and eventually the flow once again headed toward Catania, burying part of the town.

More recently, people use high explosives to blast breaches in the flanks of flows and use bulldozers to build dams and channels to divert flows. Major efforts to divert flows from a 1983 eruption of Mt. Etna, and again in 1992, were successful (**Fig. 9.28a**). Inhabitants of Iceland used a particularly creative approach in 1973 to stop a flow before it overran a town—they sprayed cold seawater onto the flow to freeze it in its tracks (**Fig. 9.28b**). The flow did stop short of the town, but whether this was a consequence of the cold shower it received remains unknown.

> **Did you ever wonder . . .**
>
> whether people could redirect a lava flow?

Take-Home Message

Volcanoes don't erupt continuously and don't last forever, so we can distinguish among active, dormant, and extinct volcanoes. Once a volcano ceases to erupt, erosion destroys its eruptive shape. Geologists can provide near-term predictions of eruptions so that people can take precautions.

QUICK QUESTION: Using the Web, discover how many times Vesuvius has erupted since 79 C.E. and calculate its recurrence interval. Is it wise to continue building housing on its flanks?

9.7 Effect of Volcanoes on Climate and Civilization

Can Eruptions Affect Climate?

In 1783, Benjamin Franklin was living in Europe, serving as the American ambassador to France. The summer of that year seemed to be unusually cool and hazy. Franklin, who was an accomplished scientist as well as a statesman, couldn't resist seeking an explanation for this phenomenon and learned that in June of 1783 a huge volcanic eruption had taken place in Iceland. He wondered if the "smoke" from the eruption had prevented sunlight from reaching the Earth, thus causing the cooler temperatures. Franklin reported this idea at a meeting, and by doing so, he may well have been the first scientist ever to suggest a link between eruptions and climate.

Franklin's idea seemed to be confirmed in 1815, when Mt. Tambora in Indonesia exploded. Tambora's explosion ejected over 100 cubic km of ash and pumice into the air (compared with 1 cubic km from Mt. St. Helens). Ten thousand people were killed by the eruption and the associated tsunami. Another 82,000 died of starvation. The sky became so hazy that stars dimmed by a full magnitude. Temperatures dipped so low in the northern hemisphere that 1816 became known as "the year without a summer." The unusual weather of that year inspired artists and writers. For example, memories of fabulous sunsets and the hazy glow of the sky inspired the luminous and atmospheric quality that made the landscape paintings of the English artist J. M. W. Turner so famous (see Fig. 9.1a), and Byron's 1816 poem "Darkness" contains the gloomy lines "The bright Sun was extinguish'd, and the stars / Did wander darkling in the eternal space . . . / Morn came and went—and came, and brought no day." Two years later, Mary Shelley, trapped indoors by bad weather, wrote Frankenstein, with its numerous scenes of gloom and doom.

Geoscientists have witnessed other examples of eruption-triggered coolness more recently. In the months following the 1883 eruption of Krakatau and the 1991 eruption of Pinatubo, global temperatures noticeably dipped. Classical literature provides more evidence of the volcanic impact on climate. For example, Plutarch wrote around 100 C.E., "there was . . . after Caesar's murder . . . the obscuration of the Sun's rays. For during all the year its orb rose pale and without radiance . . . and the fruits, imperfect and half ripe, withered away." Similar conditions appear to have occurred in China the same year, as described in records from the Han dynasty, and may have been a consequence of volcanic eruption.

To study the effect of volcanic activity on climate even further in the past, geologists have studied ice from the glaciers of Greenland and Antarctica. Glacial ice has layers, each of which represents the snow that fell in a single year. Some layers contain concentrations of sulfuric acid, formed when sulfur dioxide from volcanic gas dissolves in the water from which snow forms. These layers indicate years in which major eruptions occurred. Years in which ice contains acid correspond to years during which the thinness of tree rings elsewhere in the world indicates a cool growing season.

How can a volcanic eruption create these cooling effects? As a result of a large explosive eruption, fine ash and sulfuric-acid aerosols enter the stratosphere. It takes only about two weeks for the ash and aerosols to circle the planet (**Fig. 9.29**),

FIGURE 9.29 A map showing the global concentration of aerosols two months after the Mt. Pinatubo eruption in 1991.

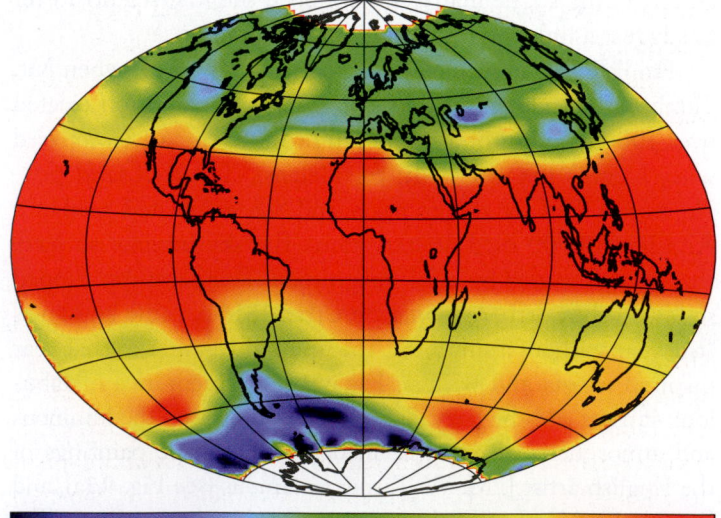

Low concentration · High concentration

and they stay suspended in the stratosphere for many months to years because they float above the weather and do not get washed away by rainfall. The resulting haze causes cooler average temperatures because it reflects and/or scatters incoming visible solar radiation during the day but not the infrared radiation that rises from the Earth's surface at night. Thus, it keeps energy from reaching the Earth, but unlike greenhouse gases (such as CO_2), it does not prevent heat from escaping. A Krakatau-scale eruption can lead to a drop in global average temperature of 0.3° to 1°C. According to some calculations, a series of large eruptions over a short period of time could cause a global average temperature drop of 6°C. The temperature dip seems to last up to a few years.

Some researchers are currently exploring the possibility that the eruption of LIPs may have longer-lasting effects, perhaps changing climate so much that species go extinct. The apparent coincidence of the eruption of the Siberian traps with the extinctions that define the end of the Paleozoic may be an example.

Volcanoes and Civilization

Not all volcanic activity is bad. Over time, volcanic activity has played a major role in making the Earth a habitable planet. Eruptions and underlying igneous intrusions produced the rock making up the Earth's crust, and gases emitted by volcanoes provided the raw materials from which the atmosphere and oceans formed. The black smokers surrounding vents along mid-ocean ridges may have served as a birthplace for life, and volcanic islands in the oceans have hosted isolated populations whose evolution adds to the diversity of life on the planet. Volcanic activity continues to bring nutrients (potassium, sulfur, calcium, and phosphorus) from Earth's interior to the surface and to provide fertile soils that nurture plant growth. And in more recent times, people have exploited the mineral and energy resources generated by volcanic eruptions. Thus, volcanoes and people have lived in close association since the first human-like ancestors walked the Earth 3 million years ago. In fact, one of the earliest relics of human ancestors consists of footprints fossilized in a volcanic ash layer in East Africa.

But as we have seen, volcanic eruptions also pose a hazard. Eruptions may even lead to the demise of civilizations. The history of the Minoan people, who inhabited several islands in the eastern Mediterranean during the Bronze Age, illustrates this possibility. Beginning around 3000 B.C.E., the Minoans built elaborate cities and prospered. Then their civilization waned and disappeared (**Fig. 9.30a**). Geologists have discovered that the disappearance of the Minoans came within 150 years of a series of explosive eruptions of the Santorini Volcano in the first half of the 17th century B.C.E. Remnants of

(a) The ruins of palaces left by the Minoans.

(b) From space, we can see the 11-km-wide caldera, all that is left of Santorini.

(c) The inner wall of the caldera reveals the typical geology of a stratovolcano; we see both lava and tuff layers.

the volcano now constitute Thera, one of the islands of Greece (**Fig. 9.30b**). After a huge eruption, the center of the volcano collapsed into the sea, leaving only a steep-walled caldera (**Fig. 9.30c**). Archaeologists speculate that pyroclastic debris from the eruptions periodically darkened the sky, burying Minoan settlements and destroying crops. In addition, related earthquakes crumbled homes, and tsunamis generated by the eruptions damaged Minoan seaports. Perhaps the Minoans took these calamities as a sign of the gods' displeasure, became demoralized, and left the region. Or perhaps trade was disrupted, and bad times led to political unrest. Eventually the Mycenaeans moved in, bringing the culture that evolved into that of classical Greece. The Minoans, though, were not completely forgotten. Plato, in his dialogues, refers to a lost city, home of an advanced civilization that bore many similarities to that of the Minoans. According to Plato, this city, which he named Atlantis, disappeared beneath the waves of the sea. Per-

haps this legend evolved from the true history of the Minoans, as modified by Egyptian scholars who passed it on to Plato.

Numerous cultures living along the Pacific Ring of Fire have evolved religious practices that are based on volcanic activity—no surprise, considering the awesome might of a volcanic eruption in comparison with the power of humans. In some cultures, this reverence took the form of sacrifice in hopes of preventing an eruption that could destroy villages and bury food supplies. In traditional Hawaiian culture, Pelé, goddess of the volcano, created all the major landforms of the Hawaiian Islands. She gouged out the craters that top the volcanoes, her fits and moods bring about the eruptions, and her tears are the smooth, glassy lapilli ejected from the lava fountains.

The largest eruption to have happened in the last million years of Earth's history took place at the Toba Volcano in Indonesia about 73,000 years ago. The explosion put out huge quantities of ash, some of which was nearly white and covered much of southern Asia like a snowfall. In addition, the eruption injected a huge dose of aerosols into the atmosphere. The aerosols and ash diminished the solar radiation reaching the Earth, and the white ash on the ground reflected some of the radiation that did reach the planet back into space. The resulting cooling event may have cooled the atmosphere for a decade, possibly as long as 1,000 years. The duration of the cooling may have been sufficient to cause species of organisms to go extinct. Anthropologists have speculated that the event killed off all but 3,000 to 10,000 humans on Earth. Genetic studies suggest that a bottleneck in evolution occurred about the same time as the Toba eruption, in that all modern humans are descended from a single ancestor who lived about that time. If so, then the Toba eruption could have significantly impacted human evolution.

The observed effect of volcanic eruptions on the climate provides a model with which to predict the consequences of a nuclear war. Researchers have speculated that so much dust and gas would be blown into the sky in the mushroom clouds of nuclear explosions that a "nuclear winter" would ensue.

Take-Home Message

The ash, gases, and aerosols produced by explosive eruptions can be blown around the globe. This material can cause significant global cooling. Climatic effects, as well as other consequences of eruptions, may have impacted human evolution and civilization.

QUICK QUESTION: Not all eruptions of equivalent size (defined by the volume of material erupted) trigger equivalent amounts of global cooling. Why? (*Hint:* Think about eruptive style.)

FIGURE 9.31 Volcanism on other planets and moons in the Solar System.

(a) Maria of the Moon were seas of basaltic lava.

(b) A volcano rises above the plains of Venus.

(c) Olympus Mons rises 27 km above the surface of Mars. It's the largest volcano in our Solar System.

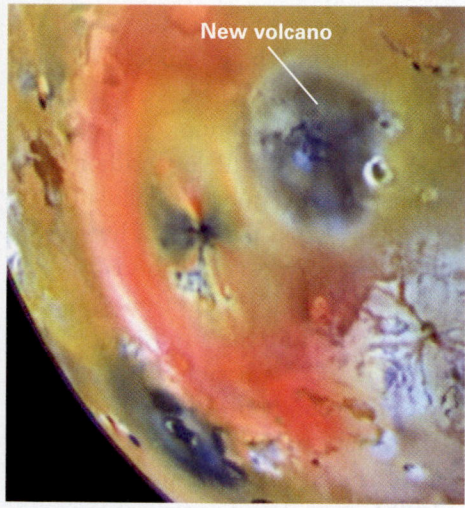

(d) An active volcano erupts sulfur on Io, a moon of Jupiter.

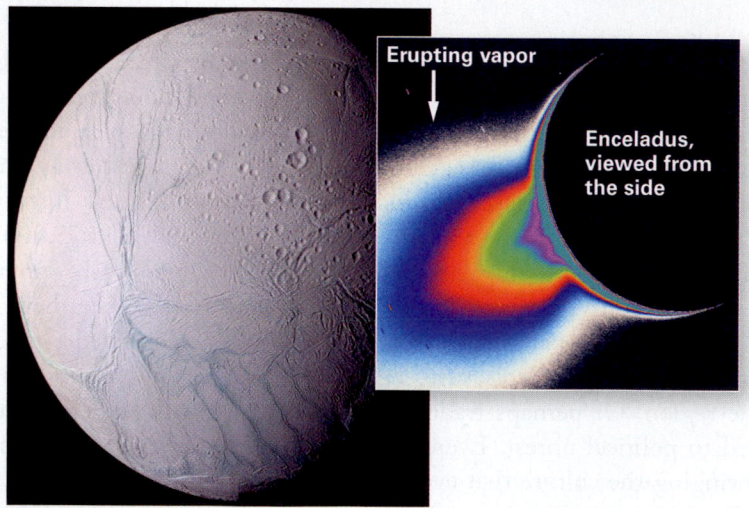

(e) Large cracks at the southern end of Enceladus, a moon of Saturn, erupt water vapor.

9.8 Volcanoes on Other Planets

We conclude this chapter by looking beyond the Earth, for our planet is not the only one in the Solar System to have hosted volcanic eruptions. We can see the effects of volcanic activity on our nearest neighbor, the Moon, just by looking up on a clear night. The broad darker areas of the Moon, the **maria** (singular *mare*, after the Latin word for sea), consist of flood basalts that erupted more than 3 billion years ago (**Fig. 9.31a**). Geologists propose that the flood basalts formed when huge meteors collided with the Moon, blasting out giant craters. Crater formation decreased the pressure in the Moon's mantle so that it underwent partial melting, producing basaltic magma that rose to the surface and filled the craters.

On Venus, about 22,000 volcanic edifices have been identified. Some of these even have calderas at their crests (**Fig. 9.31b**). Though no volcanoes currently erupt on Mars, the planet's surface displays a record of a spectacular volcanic past. The largest known mountain in the Solar System, Olympus Mons (**Fig. 9.31c**), is an extinct shield volcano on Mars. The base of Olympus Mons is 600 km across, and its peak rises 27 km above the surrounding plains.

Active volcanism currently occurs on Io, one of the many moons of Jupiter. Cameras in the *Galileo* spacecraft have recorded these volcanoes in the act of spraying plumes of sulfur gas into space (**Fig. 9.31d**) and have tracked immense, moving basaltic lava flows. Different colors of erupted material make the surface of this moon resemble a pizza. Researchers have proposed that the volcanic activity is due to tides: the gravitational pull exerted by Jupiter and by other moons alternately stretches and then squeezes Io, generating sufficient friction to keep Io's mantle hot. Geologists have also detected eruptions from moons of Saturn (**Fig. 9.31e**).

Take-Home Message

Space exploration reveals that volcanism occurs not only on Earth but has also left its mark on other terrestrial planets and on the moons of giant planets. Satellites have detected active eruptions on moons of Jupiter and Saturn.

QUICK QUESTION: Since most volcanic activity on Earth is a result of plate tectonics, what does the lack of volcanic activity on the Moon tell us about whether or not plate tectonics happens on the Moon?

<aside>
Did you ever wonder...

if the Earth is the only planet with volcanoes?
</aside>

CHAPTER SUMMARY

- Volcanoes are vents at which molten rock (lava), pyroclastic debris, gas, and aerosols erupt at the Earth's surface. A hill, mountain, crater, or caldera depression created by volcanism is also called a volcano.
- The characteristics of a lava flow depend on its viscosity, which in turn depends on its temperature and composition.
- Basaltic lava can flow great distances. Pahoehoe flows have smooth, ropy surfaces, whereas a'a' flows have rough, rubbly surfaces. Andesitic and rhyolitic lava flows tend to pile into mounds at the vent.
- Pyroclastic debris includes powder-sized ash, marble-sized lapilli, and apple- to refrigerator-sized blocks and bombs. Some falls from the air, whereas other debris forms incandescent pyroclastic flows (density currents) that rush away from the volcano.

- Eruptions may occur at a volcano's summit or from fissures on its flanks. The summit of an erupting volcano may collapse to form a bowl-shaped depression called a caldera.
- A volcano's shape depends on the type of eruption. Shield volcanoes are broad, gentle rises. Cinder cones are steep-sided, symmetrical hills composed of tephra. Composite volcanoes (stratovolcanoes) can become quite large and consist of alternating layers of pyroclastic debris and lava.
- The type of eruption depends on several factors, including the lava's viscosity and gas content. Effusive eruptions produce only flows of lava, whereas explosive eruptions produce clouds and flows of pyroclastic debris.
- Different kinds of volcanoes form in different geologic settings as defined by plate tectonics theory.
- Volcanic eruptions pose many hazards: lava flows overrun roads and towns, ash falls blanket the landscape, pyroclastic

flows incinerate towns and fields, landslides and lahars bury the land surface, earthquakes topple structures and rupture dams, tsunamis wash away coastal towns, and invisible gases suffocate people and animals.

- Eruptions can be predicted by earthquake activity, changes in heat flow, changes in shape of the volcano, and the emission of gas and steam.

- We can minimize the consequences of an eruption by avoiding construction in danger zones and by drawing up evacuation plans.
- Immense flood basalts cover portions of the Moon. The largest known volcano, Olympus Mons, towers over the surface of Mars. Satellites have documented evidence for eruptions on moons of Jupiter and Saturn.

GUIDE TERMS

a'a' (p. 276)
aerosol (p. 281)
active volcano (p. 302)
ash (p. 278)
block (p. 278)
bomb (p. 278)
caldera (p. 283)
cinder cone (p. 284)
columnar jointing (p. 276)
crater (p. 283)
dormant volcano (p. 302)

effusive eruption (p. 285)
eruption column (p. 289)
eruptive style (p. 285)
explosive eruption (p. 285)
extinct volcano (p. 302)
fissure (p. 282)
flood basalt (p. 296)
ignimbrite (p. 292)
lahar (p. 281)
lapilli (p. 278)

large igneous province (LIP) (p. 296)
lava tube (p. 276)
magma chamber (p. 282)
mare (p. 309)
pahoehoe (p. 276)
pillow lava (p. 276)
pyroclastic debris (p. 278)
pyroclastic flow (p. 292)
shield volcano (p. 284)
stratovolcano (p. 284)

supervolcano (p. 292)
tephra (p. 278)
tuff (p. 278)
vent (p. 282)
vesicle (p. 281)
viscosity (p. 275)
volcanic explosivity index (VEI) (p. 290)
volcanic-hazard assessment map (p. 303)
volcano (p. 273)

REVIEW QUESTIONS

1. Describe the three different kinds of material that can erupt from a volcano.
2. Describe the differences between a pyroclastic flow and a lahar.
3. Describe the differences among shield volcanoes, stratovolcanoes, and cinder cones. How are these differences explained by the composition of their lavas and other factors?
4. Why do some volcanic eruptions consist mostly of lava flows, whereas others are explosive and do not produce flows?

5. Describe the activity in the mantle that leads to hot-spot eruptions.
6. How do continental-rift eruptions form flood basalts?
7. Contrast an island volcanic arc with a continental volcanic arc. What is a hot-spot volcano?
8. Identify some of the major volcanic hazards, and explain how they develop.
9. How do geologists predict volcanic eruptions?
10. Explain how steps can be taken to protect people from the effects of eruptions.

ON FURTHER THOUGHT

11. The Long Valley Caldera, near the Sierra Nevada range, exploded about 700,000 years ago and produced a huge ignimbrite (a lapilli tuff deposited from vast pyroclastic density currents) called the Bishop Tuff. About 30 km to the northwest lies Mono Lake (see photo), with an island in the middle and a string of craters extending south from its south shore. Hot springs and tufa deposits can be found along the lake. You can see the lake on *Google Earth*™ at Lat 37° 59′ 56.58″ N Long 119° 2′ 18.20″ W. Explain the origin of Mono Lake. Do you think that it represents a volcanic hazard?

12. Mt. Fuji is a 3.6-km-high stratovolcano in Japan formed as a consequence of subduction (see photo below). With *Google Earth*™ you can reach the volcano at Lat 35° 21′ 46.72″ N Long 138° 43′ 49.38″ E. It contains volcanic rocks with a range of compositions, including some andesitic rocks. Why do andesites erupt at Mt. Fuji? Very little andesite occurs on the Marianas Islands, which are also subduction-related volcanoes. Why?

13. The city of Albuquerque lies along the Rio Grande in New Mexico. Within the valley, numerous volcanic features crop out. Using *Google Earth*™, you can fly to Albuquerque and then along the river to find many examples. Many of the volcanoes are basaltic, but in places you will see huge caldera remnants. In fact, the city of Los Alamos lies atop thick ignimbrites. What causes the volcanism in the Rio Grande Valley, and why are there different kinds of volcanism? Look at the photo of the volcanic cluster (see photo below). These occur north of Santa Fe, at Lat 36° 45′ 27.51″ N Long 105° 47′ 24.85″ W—use *Google Earth*™ to get a closer view. Judging from the character of these volcanoes, would you say they are active? Why? How would you evaluate the volcanic hazard of this region? (*Hint:* Use the Web to find a map of seismicity for the Rio Grande Valley region, and think about its implications.)

smartwork   smartwork.wwnorton.com

This chapter's Smartwork features:

- Art exercise on lava flows and viscosity.
- Simulation exercises on basin and range formation and sedimentation.
- Composite volcano labeling activity.

GEOTOURS

This chapter's GeoTour exercise (E) features:

- Shield volcanoes
- Composite cone volcanoes
- Cinder cone volcanoes

Another View Left: the ash cloud from the 2011 eruption of a volcano in the Puyehue-Cordón Caulle chain of Chile. The ash circled the globe within two weeks, disrupting air traffic throughout the southern hemisphere. Right: A large ash plume with lava erupting from the Eyjafjallajokull Volcano in Iceland, April 2010. The ash cloud from the eruption grounded air traffic around the world.

A 20-story apartment building tipped over and broke in two during a major earthquake (magnitude 8.8) that struck Chile in 2010. Violent earthquakes, which are inevitable on our dynamic planet, can have catastrophic consequences.

CHAPTER 10

A Violent Pulse: Earthquakes

We learn geology the morning after the earthquake.

—*Ralph Waldo Emerson (American poet, 1803–1882)*

LEARNING OBJECTIVES

By the end of this chapter, you should understand . . .

- the causes of earthquakes and why earthquakes occur where they do.
- how to determine the size and location of an earthquake.
- the many ways in which earthquakes cause damage.
- the limitations on people's ability to predict earthquakes.
- the steps that people can take to prevent earthquake damage.

10.1 Introduction

It was mid-afternoon of March 11, 2011, and in the many seaside towns along the eastern coast of Honshu, the northern island of Japan, fishing fleets unloaded their catch, factories churned out goods, shoppers browsed the stores, and office workers tapped at their computers, unaware that their surroundings would suddenly be changing forever. This coast is near the convergent-plate boundary at which the Pacific Plate slips beneath the edge of Japan and sinks back into the mantle. Averaged over time, the relative movement across this boundary takes place at a rate of about 8 cm per year. But the motion doesn't take place smoothly. Rather, for a while, rocks adjacent to the boundary quietly bend and distort to accommodate the motion, until suddenly, like a stick that snaps after you bend it too far, the rocks break and a large amount of slip on a fracture takes place in a matter of seconds to minutes (**Fig. 10.1**). On March 11, at 2:46 P.M., such a "snap" started about 130 km (80 miles) east of Japan's coast, at a depth of about 24 km (15 miles) below the seafloor; the Pacific Plate lurched westward by several meters, relative to Japan, and the seafloor moved up by about 30 cm. The stage had been set for a disaster.

Geologists refer to a fracture on which such sliding takes place as a **fault**. In the case of the March 11 event, the fault delineates the boundary between Honshu and the Pacific Plate. The sudden sliding and breaking of rock along a fault produces energy that propagates through the crust in the form of vibrations. In essence, these waves resemble the shock waves that travel through a breaking stick to your hands. In the Earth, vibrations can race through the crust at an average speed of 11,000 km (7,000 miles) per hour, 10 times the speed of sound in air. When the vibrations produced on March 11 reached Japan, the land surface lurched back and forth and bounced up and down for over a minute. People became disoriented, panicked, and even seasick—some lost their balance and crouched or fell, and some heard a dull rumbling or thumping. Bottles and plates flew off shelves and crashed to the floor, buildings twisted and swayed, ceilings and facades fell in a shower of debris, dust rose from the ground to create a fog-like mist, power lines stretched and sparked, and weaker buildings collapsed (**Fig. 10.2a**). In addition, several natural gas tanks and pipes broke, sending flammable vapors into the air—locally, the gas ignited in billows of flame that set fire to damaged buildings. A major

FIGURE 10.1 What happens during an earthquake?

(a) Before deformation, rock layers in this example are not bent.

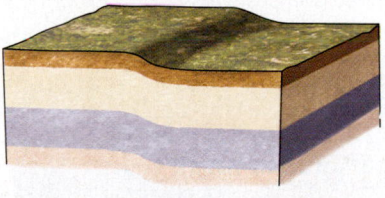 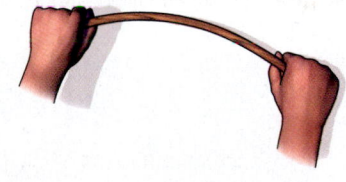

(b) Before an earthquake, rock bends elastically, like a stick that you arch between your hands. The drawing exaggerates the amount of bending.

(c) Eventually, the rock breaks, and sliding suddenly occurs on a fault. This break generates vibrations. You feel such vibrations when you break a stick.

FIGURE 10.2 The great Tōhoku earthquake and tsunami in Japan, 2011.

(a) Some buildings collapsed due to ground shaking.

(b) The tsunami filled harbors and spilled over sea walls.

(c) The tsunami washed over low coastal areas. In this photograph, the wave is moving from left to right and is just about to wash over a canal.

earthquake—an episode of ground shaking—had occurred. Geologists now refer to this event as the Tōhoku earthquake, named for the eastern province of Honshu.

In many cases, the societal calamity due to an earthquake is a direct consequence of ground shaking, because it causes buildings to collapse and crush their inhabitants and sends debris careening down slopes. Building collapse and landslides triggered by the May 12, 2008, earthquake of Sichuan, in central China, leveled cities—70,000 people died, and millions were left homeless. And in February 2011, sudden rupture on a fault near Christchurch, New Zealand, toppled buildings and the walls of a stalwart stone cathedral. But in Japan, a country struck by earthquakes fairly frequently, officials have established stringent building codes that successfully prevented widespread building collapse. Unfortunately, even strong buildings could not resist what happened minutes after the earthquake shock. The sudden displacement of the seafloor off Japan's coast displaced a massive amount of ocean water and produced immense water waves known as *tsunamis*. When these reached land, they became a roaring jumble of water and debris that submerged low-lying land as far as 10 km in from the shoreline (**Fig. 10.2b, c**). To make matters worse, the tsunamis destroyed power sources used to run the pumps that cooled reactors in the Fukushima nuclear power plant, ultimately leading to the release of radioactivity into the environment. In the end, the immensity of the total devastation due to the Tōhoku earthquake was almost beyond comprehension.

Earthquakes are a fact of life on planet Earth—almost 1 million detectable earthquakes happen every year. Most are a consequence of plate movement—they punctuate each step in the growth of mountains, the drift of continents, and the opening and closing of ocean basins. Fortunately, most cause no damage or casualties, either because they are too small or they occur in unpopulated areas. But a few hundred earthquakes per year rattle the ground sufficiently to crack or topple buildings and injure their occupants, and every 5 to 20 years, on average, a great earthquake triggers a horrific calamity. In fact, during the past two millennia, ground shaking, tsunamis, landslides, fires, and other phenomena caused by earthquakes have killed over 3.5 million people (**Table 10.1**).

What geologic phenomena trigger earthquakes? Why do earthquakes take place where they do? How do they cause damage? Can we predict when earthquakes will happen or even prevent them from happening? Many of these questions have been addressed by the hard work of **seismologists** (from the Greek word *seismos*, for shock or earthquake), geoscientists who study earthquakes, during the past century. In this chapter, we present some of the answers that they have obtained.

TABLE 10.1 Some Notable Earthquakes

Year	Location	Deaths
2011	Tōhoku, Japan (tsunami)	20,000
2011	Christchurch, New Zealand	180
2010	Haiti	230,000
2010	Concepcion, Chile	1,000
2008	Sichuan, China	70,000
2005	Pakistan	80,000
2004	Sumatra (tsunami)	230,000
2003	Bam, Iran	41,000
2001	Bhuj, India	20,000
1999	Calaraca/Armenia, Colombia	2,000
1999	Izmit, Turkey	17,000
1995	Kobe, Japan	5,500
1994	Northridge, California	51
1990	Western Iran	50,000
1989	Loma Prieta, California	65
1988	Spitak, Armenia	24,000
1985	Mexico City	9,500
1983	Turkey	1,300
1978	Iran	15,000
1976	T'ang-shan, China	255,000
1976	Caldiran, Turkey	8,000
1976	Guatemala	23,000
1972	Nicaragua	12,000
1971	San Fernando, California	65
1970	Peru	66,000
1968	Iran	12,000
1964	Anchorage, Alaska	131
1963	Skopje, Yugoslavia	1,000
1962	Iran	12,000
1960	Agadir, Morocco	12,000
1960	Southern Chile	6,000
1948	Turkmenistan, USSR	110,000
1939	Erzincan, Turkey	40,000
1939	Chillan, Chile	30,000
1935	Quetta, Pakistan	60,000
1932	Gansu, China	70,000
1927	Tsinghai, China	200,000
1923	Tokyo, Japan	143,000
1920	Gansu, China	180,000
1915	Avezzano, Italy	30,000
1908	Messina, Italy	160,000
1906	San Francisco	500
1896	Japan	22,000
1886	Charleston, South Carolina	60
1866	Peru and Ecuador	25,000
1811–12	New Madrid, Missouri (3 events)	Few
1783	Calabria, Italy	50,000
1755	Lisbon, Portugal	70,000
1556	Shen-shu, China	830,000

10.2 What Causes Earthquakes?

What causes earthquake activity, or **seismicity**? Traditional cultures commonly attributed it to the action of supernatural beasts. In Japanese folklore, for example, seismicity happened when a giant catfish, Namazu, which lived in the mud below the surface of the ground, started to thrash about. Similarly, in Indian folklore, seismicity happened when one of eight elephants holding up the Earth shook its head. Scientific study instead associates seismicity with any of the following:

- the sudden formation of a new fault
- sudden slip on an existing fault
- a phase change, when atoms in minerals suddenly rearrange
- movement of magma in, or explosion of, a volcano
- a giant landslide
- a meteorite impact
- an underground nuclear-bomb test

In this chapter, we will focus our attention on fault-generated earthquakes, as they are by far the most common.

When describing the location of an earthquake, seismologists use two terms. First, we define the point within the Earth at which rock starts to rupture and slip on a fault as the *hypocenter*, or **focus**, of an earthquake (**Fig. 10.3a**). Simplistically, we can picture the focus as the point on the fault from which earthquake energy, in the form of vibrations, begins to propagate. (In reality, however, the area of a fault that slips during an earthquake extends beyond the focus, so not all of the energy produced during an earthquake actually originates right at the focus.) Second, we define the point on the surface of the Earth that lies directly above the focus as the earthquake's **epicenter**. We can portray the position of an epicenter as a point on a map (**Fig. 10.3b**).

Faults in the Crust

At first glance, a fault may look simply like a break that cuts across rock or sediment. But on closer examination, you may be able to see evidence of the sliding that occurred on a fault. For example, the rock adjacent to the fault may be broken up into angular fragments or may be pulverized into tiny grains, due to the crushing and grinding that can accompany slip, and the surface of a fault may be polished and grooved as if scratched by a rasp. In some localities, a fault cuts through a distinct *marker*, such as a distinct sedimentary bed, an igneous dike, or a fence. Where this happens, the marker on one side of the fault no longer lies adjacent to the marker on the other side

FIGURE 10.3 Earthquake hypocenters and epicenters.

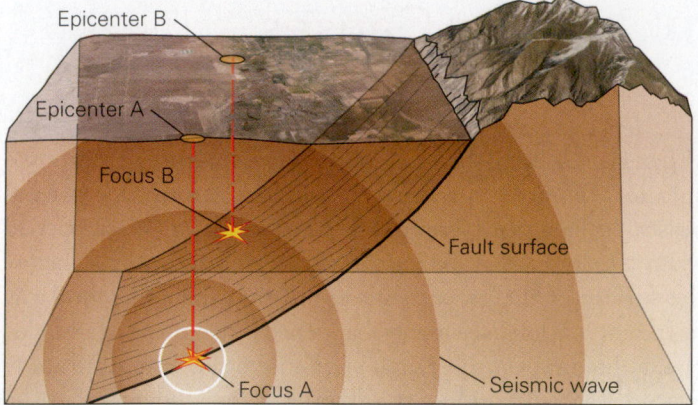

(a) The focus is the point on the fault where slip begins. Seismic energy starts radiating from it. The epicenter is the point on the Earth's surface directly above the focus. Earthquake A just happened; earthquake B happened a while ago.

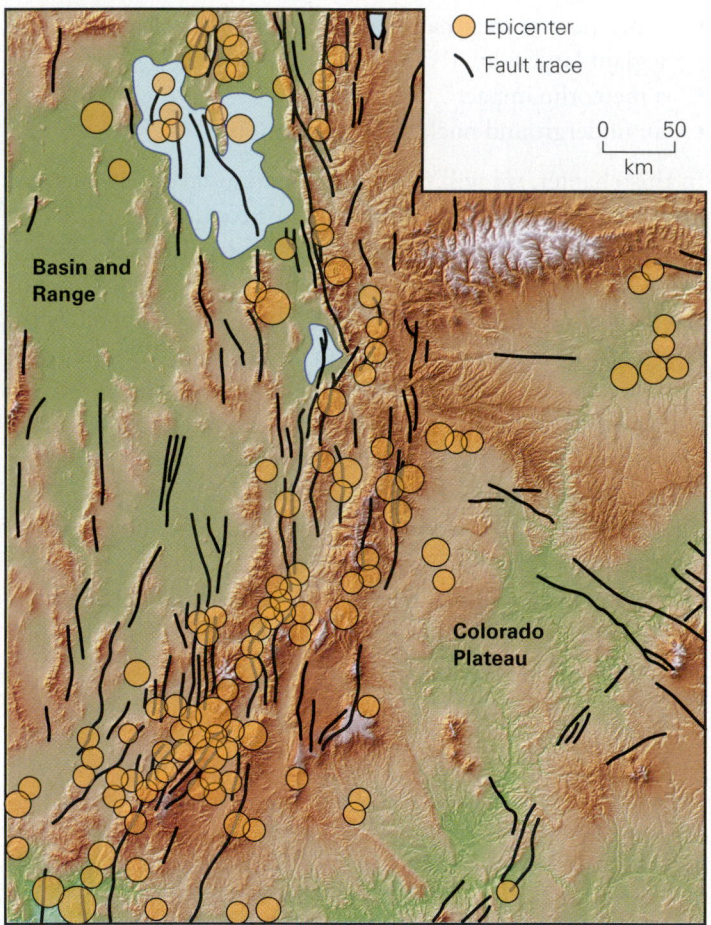

Legend:
- Epicenter
- Fault trace

(b) A map of Utah showing the distribution of earthquake epicenters, recorded over several years. The map indicates that seismic activity happens mostly in a distinct belt following the boundary between the Basin and Range rift and the Colorado Plateau.

after fault slip. The distance between two ends of the marker, as measured along the fault surface in the direction of slip, is the fault's **displacement** (**Fig. 10.4**). Many faults do not cut

the surface while active. Such *blind faults* are completely underground while active, but they may become visible if exposed by erosion of overlying rock. But some faults intersect and offset the ground surface, producing a step called a **fault scarp** (**Fig. 10.5**). Geologists refer to the ground surface exposure of a fault, either because it cut the surface or was later exposed by erosion, as the fault line or the **fault trace**.

In the 19th century, miners who encountered faults in mine tunnels referred to the rock mass above a sloping fault plane as the *hanging wall*, because it hung over their heads, and the rock mass below the fault plane as the *footwall*, because it lay beneath their feet. These miners described the direction in which rock masses slipped on a sloping fault by specifying the direction that the hanging wall moved relative to the footwall, and we still use their terms today (see **Geology at a Glance**, pp. 320–321). Specifically, a fault whose hanging wall slipped down the slope of the fault is a *normal fault*, and a fault whose hanging wall slipped up the slope is a *reverse fault* (if steep) or a *thrust fault* (if shallowly sloping). A *strike-slip fault* is near-vertical fracture on which slip occurs parallel to an imaginary horizontal line, called a strike line, on the fault plane. No up-or-down motion takes place on a strike-slip fault. Note that though we depict faults as single surfaces in Figure 10.5, in reality, major faults commonly consist of a zone including several smaller faults.

The direction of slip on a fault reflects the nature of crustal movements causing the fault. For example, normal faults form in response to stretching or extension of the crust, whereas reverse faults and thrust faults develop in response to squeezing (compression) and shortening of the crust. Strike-slip faults form where one block of crust slides past another laterally. Slip on strike-slip faults tends to form narrow bands of low ridges and narrow depressions because faults are not perfect planes, and at bends along the fault the ground pulls apart or squeezes together (see Fig. 10.4b).

Faults can be found in many locations—but don't panic! Not all faults are likely to be the source of earthquakes. Faults that have moved recently or are likely to move in the near future are called *active faults* (and if they generate earthquakes, news media sometimes refer to them as "earthquake faults"). Most active faults occur along plate boundaries or in currently developing collision zones and rifts. Faults that last moved in the distant past and probably won't move again in the near future are called *inactive faults*. Some faults have been inactive for billions of years and may never slip again.

Generating Earthquake Energy: Stick Slip

Imagine that you grip each side of a brick-shaped block of rock with a clamp. Apply an upward push on one of the clamps and a downward push on the other. By doing so, you have applied a stress to the rock—simplistically, we can think of **stress** as a push, pull, or shear. At first, the rock bends slightly but doesn't

FIGURE 10.4 Examples of fault displacement on the San Andreas fault in California.

Displacement

Fault trace

What a Geologist Sees

These ruptures formed where the fault broke the ground surface.

Dirt road

(a) A wooden fence built across the fault was offset during the 1906 San Francisco earthquake. The displacement indicates that the motion was strike slip.

(b) An aerial photograph shows offset of a dirt road by slip on the fault in 1999.

FIGURE 10.5 The basic types of faults. Fault types are distinguished from one another by the direction of slip relative to the fault surface.

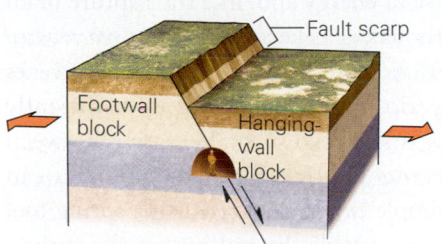

Fault scarp

Footwall block

Hanging-wall block

(a) Normal faults form during extension of the crust. The hanging wall moves down.

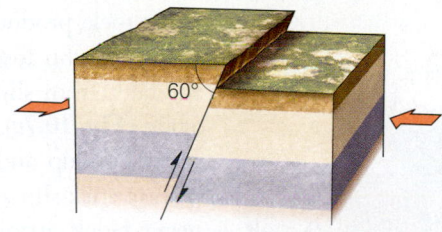

60°

(b) Reverse faults form during shortening of the crust. The hanging wall moves up and the fault is steep.

30°

(c) Thrust faults also form during shortening. The fault's slope is gentle (less than 30°).

Strike-slip faults tend to be vertical.

(d) On a strike-slip fault, one block slides laterally past another, so no vertical displacement takes place.

start to form in the rock, typically in a diagonally zone. Eventually the cracks connect to one another to form a fracture that cuts across the entire block of rock (**Fig. 10.6b**). The instant that such a throughgoing fracture forms, the block breaks in two, and the rock on one side of the fracture suddenly slides past the rock on the other side. When this happens, any elastic strain that had built up in the rock gets released, so the rock straightens out (**Fig. 10.6c**). When sliding occurs, the fracture has become a fault. Such fault formation in a previously intact rock releases energy and generates earthquake vibrations.

A new fault can't slip forever, for friction eventually slows and stops the movement, just as friction slows and stops a book sliding across a table. **Friction**, defined as the force that resists sliding on a surface, exists because fault surfaces are not perfectly smooth. The small protrusions, or asperities, on a fault surface act like tiny anchors and dig into the opposing surface (**Fig. 10.7a, b**).

We've just seen how earthquakes can happen when intact rock ruptures. Earthquakes can also happen by the sudden reactivation of sliding on pre-existing faults. In this case, stress builds up in rock until it's sufficient to overcome friction, either

break (**Fig. 10.6a**). In fact, if you were to stop applying stress before the rock breaks, the rock would return to its original shape. Geologists refer to such a phenomenon as **elastic behavior**—the same phenomenon happens when you stretch a spring and then let go. The change in shape due to elastic bending, stretching, or shortening is called *elastic strain*. Now repeat the experiment, but bend the rock even more. If you bend the rock far enough, a number of small cracks or breaks

FIGURE 10.6 A model representing the development of a new fault. Rupturing can generate earthquake-like vibrations.

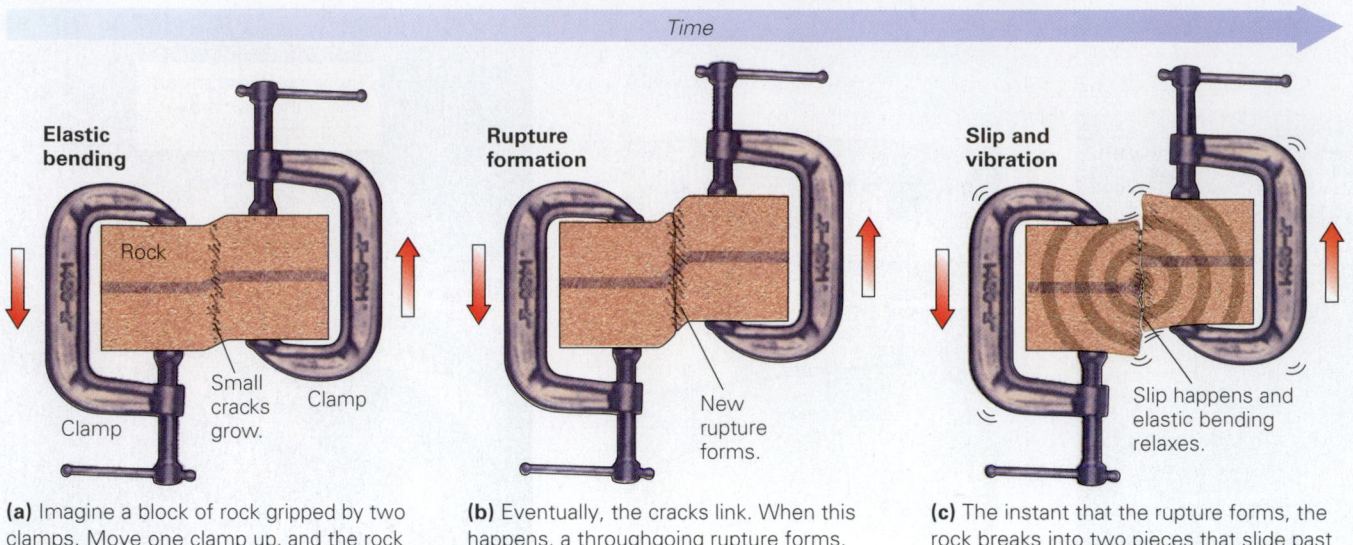

Time

Elastic bending

Rock

Small cracks grow.

Clamp

Clamp

(a) Imagine a block of rock gripped by two clamps. Move one clamp up, and the rock starts to bend. Small cracks develop along the bend.

Rupture formation

New rupture forms.

(b) Eventually, the cracks link. When this happens, a throughgoing rupture forms.

Slip and vibration

Slip happens and elastic bending relaxes.

(c) The instant that the rupture forms, the rock breaks into two pieces that slide past each other. The energy that is released generates vibrations (earthquake energy).

FIGURE 10.7 The concept of friction and stick-slip behavior. On a microscopic scale, fault surfaces are rough and friction temporarily inhibits sliding.

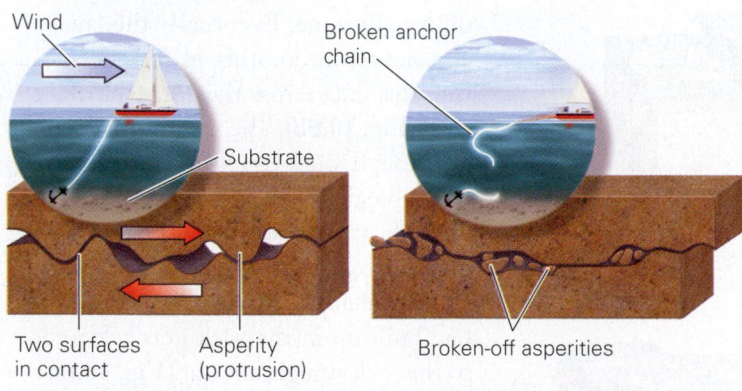

Wind

Broken anchor chain

Substrate

Two surfaces in contact

Asperity (protrusion)

Broken-off asperities

(a) Before movement, protrusions lock together, causing friction that prevents sliding. Similarly, an anchor keeps a boat in place.

(b) When a fault slips, the protrusions break off, and like a ship moving when its anchor chain breaks, the block slides.

by breaking off asperities or by having asperities plow a groove into the opposing fault surface. The breaking or plowing of asperities causes the release of energy and, like the rupture of an intact rock, produces earthquakes. Like a new fault, a *reactivated fault* can't slip forever. Friction stops the motion and prevents the fault from slipping again until stress builds up sufficiently again (**Fig. 10.7c**). Geologists refer to such alternation between stress buildup and slip events as **stick-slip behavior**. You can picture stick slip with a simple experiment. Attach a spring to a heavy block sitting on a rough table. By pulling on the spring, you elastically stretch it and apply a stress to the block. The more you stretch the spring, the greater the stress. Suddenly the block slips, and when this happens, the spring relaxes, so the stress decreases. You have to pull on the spring once again to increase stress enough to cause another sliding event.

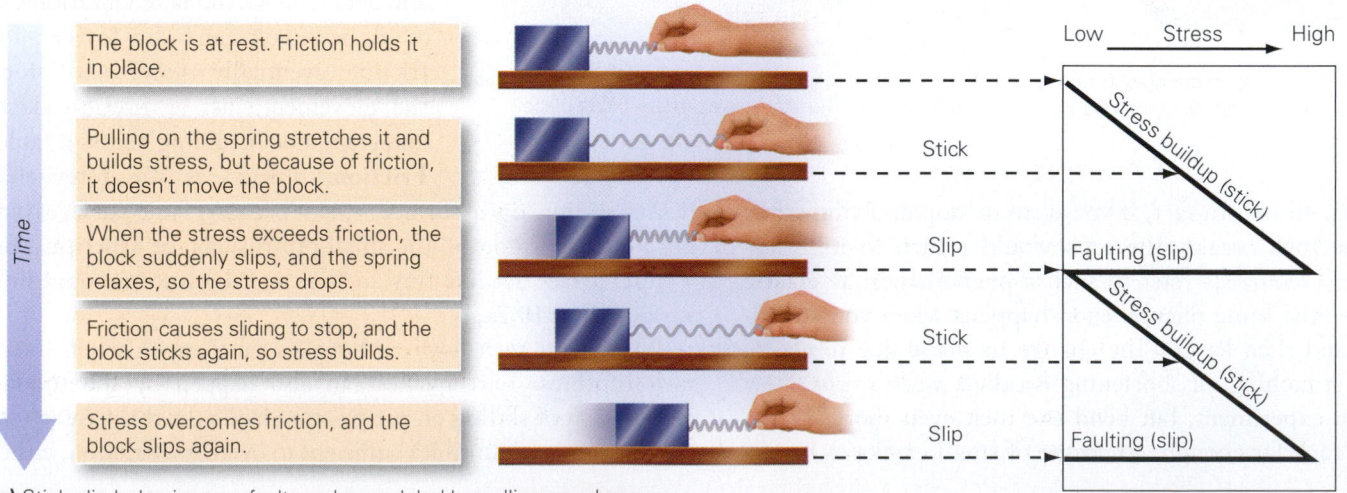

The block is at rest. Friction holds it in place.

Pulling on the spring stretches it and builds stress, but because of friction, it doesn't move the block.

When the stress exceeds friction, the block suddenly slips, and the spring relaxes, so the stress drops.

Friction causes sliding to stop, and the block sticks again, so stress builds.

Stress overcomes friction, and the block slips again.

Time

Low   Stress   High

Stick

Slip

Stick

Slip

Stress buildup (stick)

Faulting (slip)

Stress buildup (stick)

Faulting (slip)

(c) Stick-slip behavior on a fault can be modeled by pulling on a heavy block with a spring. As the spring stretches, it applies stress to the block.

In sum, we see that earthquakes happen because stresses build up, causing rock adjacent to the fault to develop elastic strain until either intact rock breaks or a preexisting fault reactivates. When the movement takes place, the once-bent rocks adjacent to the fault "twang" back to their original, unbent shape, thereby relieving the elastic strain. Geologists refer to this overall concept as the **elastic-rebound theory** of earthquake generation. Notably, the stress necessary to reactivate a fault tends to be less than the stress necessary to break intact rock, so most earthquakes probably represent slip on pre-existing faults.

Foreshocks and Aftershocks

Of note, a major earthquake, or *mainshock*, along a fault may be preceded by smaller ones, called **foreshocks**. These possibly result from the development of the smaller cracks in the vicinity of what will become the major rupture. Smaller earthquakes, called **aftershocks**, follow a major earthquake—the largest of these are ten times smaller than the mainshock, and most are much smaller. Most aftershocks, especially larger ones, take place soon after the mainshock. But they can occur sporadically for weeks or even years. Their distribution defines the overall area of the fault that slipped during the mainshock. Aftershocks of the Tōhoku earthquake indicate that the faults slipped over an area that measured 600 km long, as measured parallel to the coast, by 200 km, as measured perpendicular to the coast (**Fig. 10.8**). Aftershocks happen because slip during the mainshock does not leave the fault in a perfectly stable configuration. For example, after the mainshock, irregularities on one side of the fault surface, in their new position, may push into the opposing side and generate new stresses. Such stresses may become large enough to cause a small portion of the fault around the irregularity to slip again or may trigger slip in a nearby fault.

Sizes and Amount of Slip on Faults

Faults come in a huge range of sizes. Small ones may only be a few centimeters or a few meters from top to bottom, as measured in cross section, and it's possible for you to see the "fault tip," the place where the displacement on the fault dies out. Beyond the fault

FIGURE 10.8 The distribution of aftershocks outlines the area of the main slip area of a major earthquake.

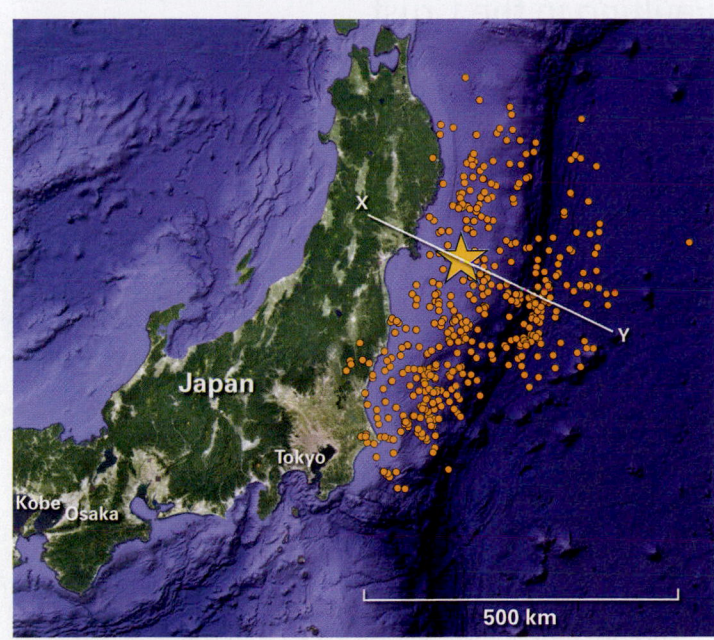

(a) The dots represent epicenters of aftershocks after the 2011 Tōhoku earthquake. The star is the epicenter of the main earthquake.
Source: Adapted from Lou Estey (UNAVCO's Jules Verne Voyager data).

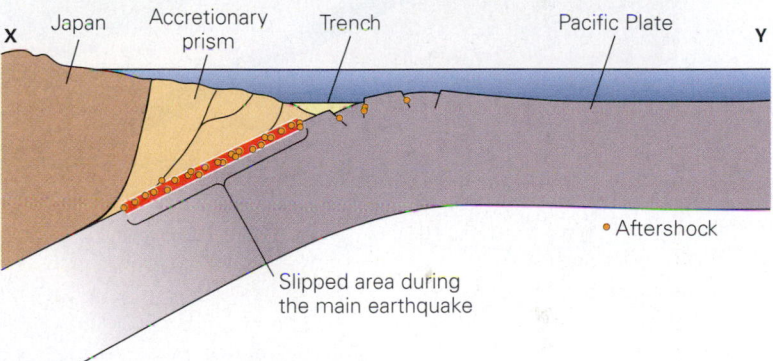

(b) A cross section, drawn along the section line in Fig. 10.8a, shows that the aftershocks are primarily along the main slip area, which lies at the base of the accretionary prism.

tip, rock is intact and unbroken, though it may be bent and distorted. Such small faults are likely only centimeters to meters long, measured in map view. But some faults are immense, and in cross section they can extend entirely through the crust and even into the lithospheric mantle. Such faults may be hundreds to thousands of kilometers long as measured on a map. For example, the San Andreas fault, a transform-plate boundary, extends from the Earth's surface to the base of the lithosphere and has a trace that runs from the Salton Sea to Cape Mendocino, a distance of about 1,400 km. Convergent-plate boundaries are, in effect, thrust faults that extend for the entire length of the plate boundary.

Faulting in the Crust

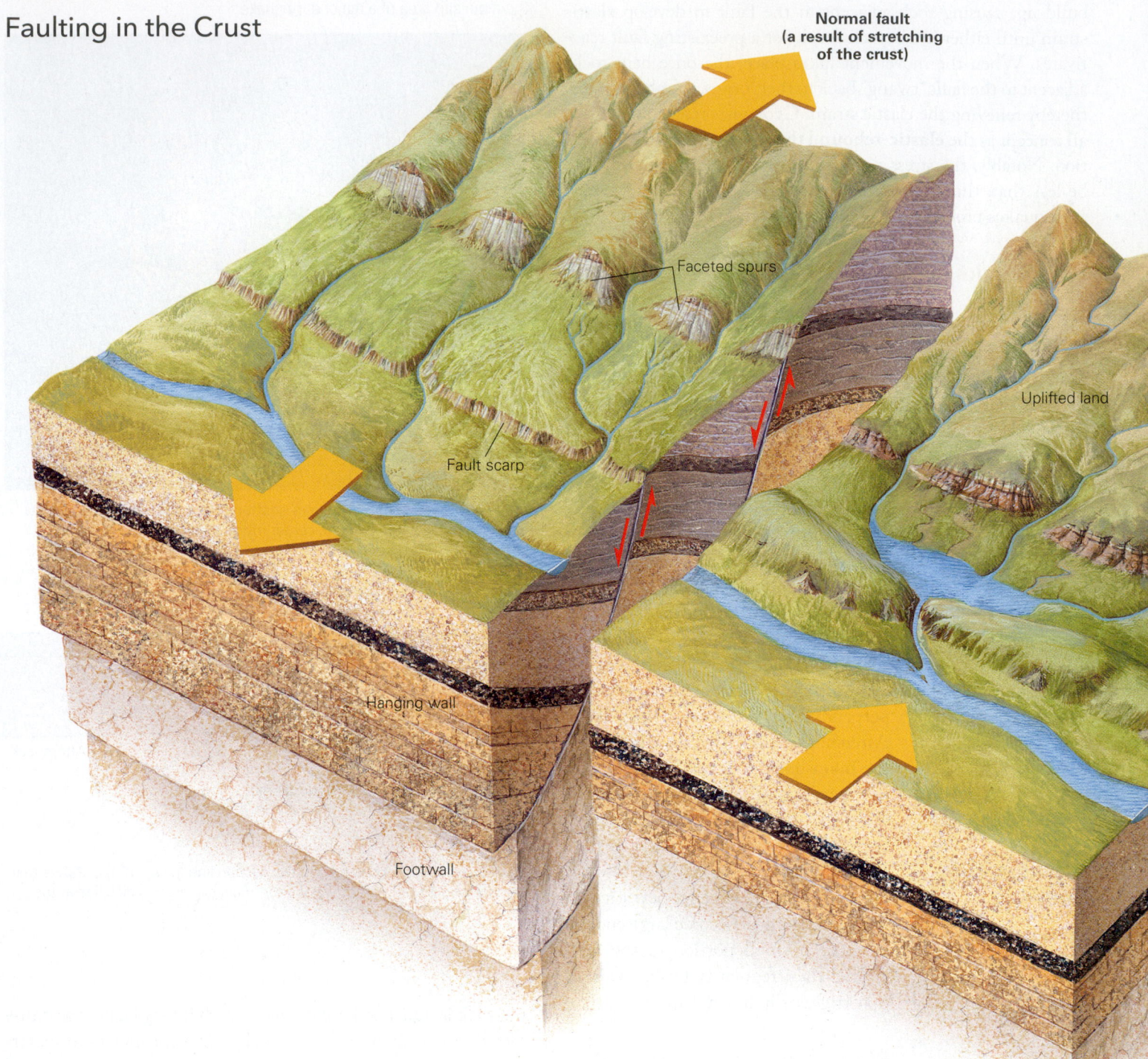

Normal fault
(a result of stretching
of the crust)

Faceted spurs

Uplifted land

Fault scarp

Hanging wall

Footwall

Faults are fractures along which one block of crust slides past another block. Sometimes movement takes place slowly and smoothly, without earthquakes, but other times the movement is sudden, and rocks break as a consequence. The sudden breaking of rock sends seismic waves through the crust, creating vibrations at the Earth's surface—an earthquake.

Geologists recognize three types of faults. If the hanging-wall block (the rock above a fault plane) slides down the fault's slope relative to the footwall block (the rock below the fault plane), the fault is a normal fault. (Normal faults form where the crust is being stretched apart, as in a continental rift.) If the

hanging-wall block is being pushed up the slope of the fault relative to the footwall block, then the fault is a reverse fault. (Reverse faults develop where the crust is being compressed or squashed, as in a collisional mountain belt.) If one block of rock slides past another and there is no up or down motion, the fault is a strike-slip fault. Strike-slip fault planes tend to be nearly vertical. If a fault displaces the ground surface, it creates a ledge called a fault scarp. Where fault scarps cut a system of rivers and valleys, the ridges are truncated to make triangular facets. Strike-slip faults may offset ridges and streams sideways.

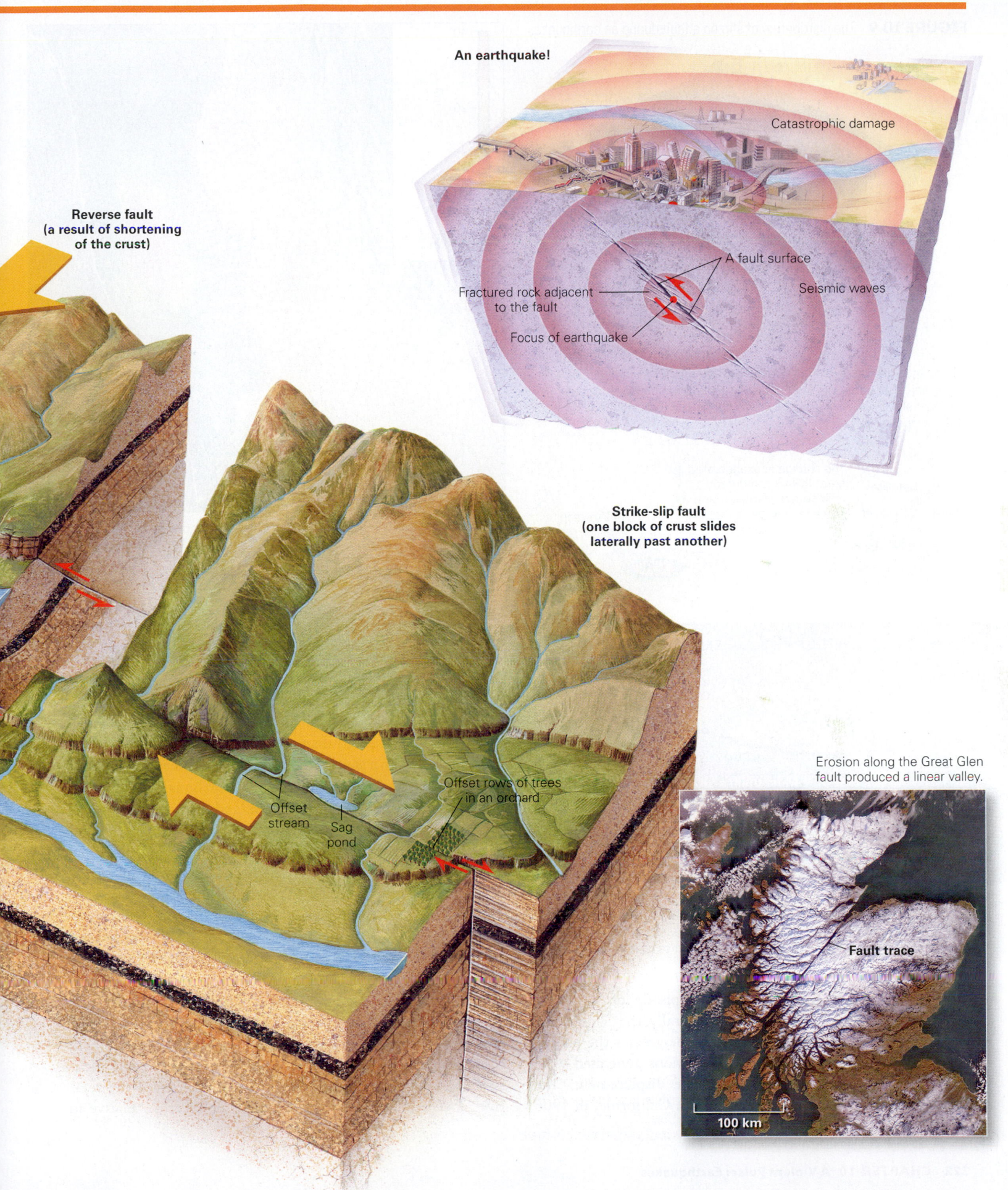

An earthquake!

Catastrophic damage

A fault surface

Fractured rock adjacent to the fault

Seismic waves

Focus of earthquake

Reverse fault (a result of shortening of the crust)

Strike-slip fault (one block of crust slides laterally past another)

Offset stream

Sag pond

Offset rows of trees in an orchard

Erosion along the Great Glen fault produced a linear valley.

Fault trace

100 km

FIGURE 10.9 The distribution of slip on a fault during an earthquake.

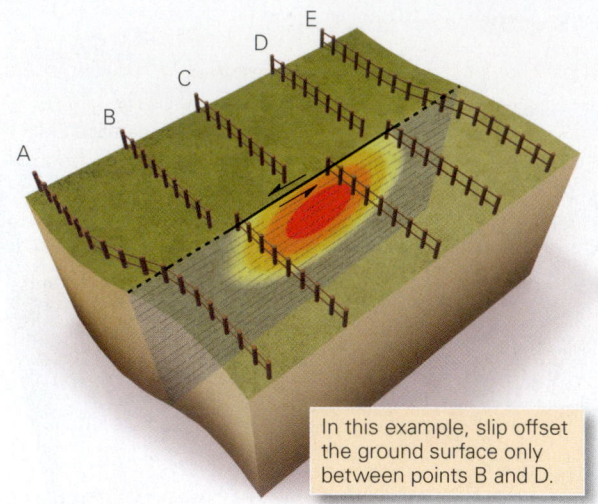

In this example, slip offset the ground surface only between points B and D.

(a) The area of a pre-existing fault can be larger than the area that slips during an earthquake. Slip starts at a point and dies out. Red indicates the most slip. Rock of the crust is bent, not broken, beyond the ends of the slipped area.

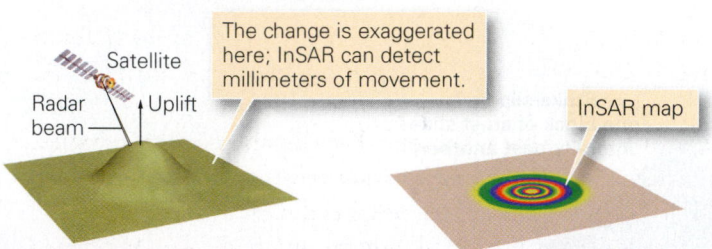

Satellite

Radar beam — Uplift

The change is exaggerated here; InSAR can detect millimeters of movement.

InSAR map

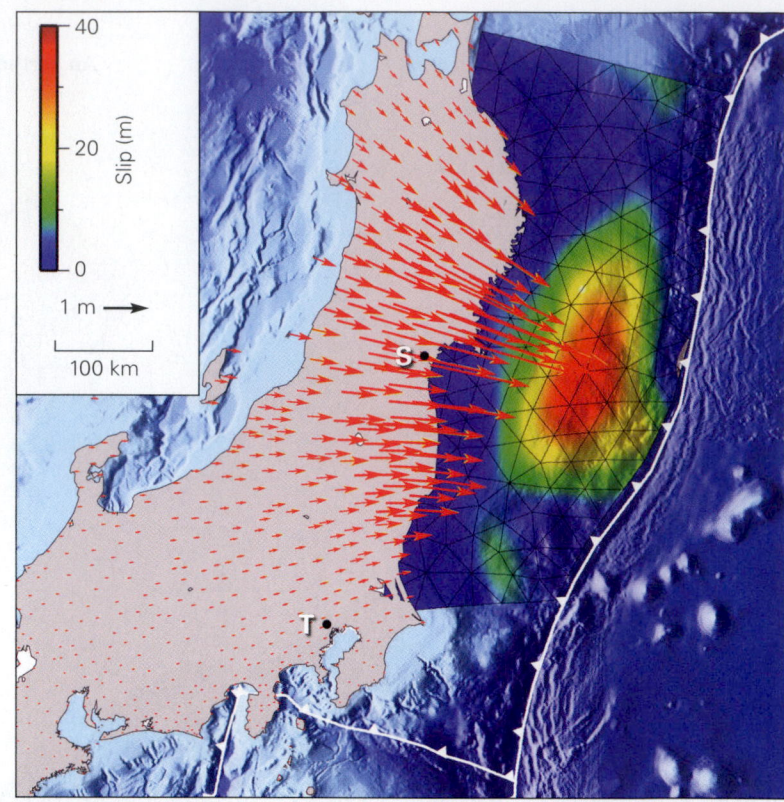

(b) A map of the slipped area during the Tōhoku earthquake. Each red arrow indicates the motion of Japan relative to the Pacific Plate, at the point beneath the arrow.

Source: Created by Lou Estey with UNAVCO's Jules Verne Voyager.

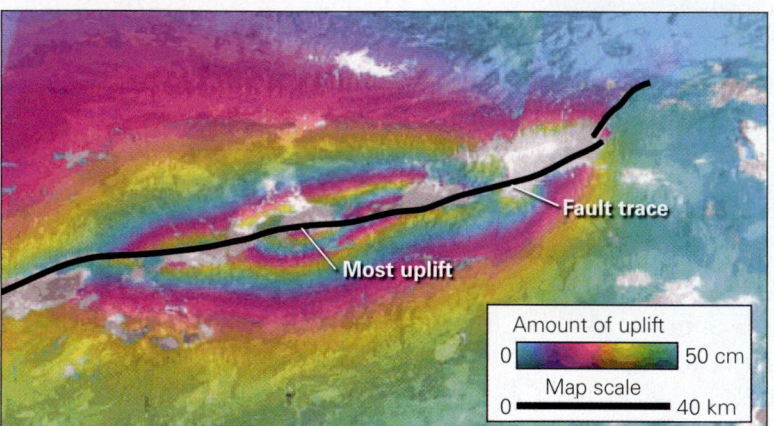

Fault trace

Most uplift

Amount of uplift

0 ——————— 50 cm

Map scale

0 ———————————— 40 km

(c) Slip on a fault locally warps the Earth's surface adjacent to a fault. Measurements made using InSAR, a type of satellite-based radar, portray the warping of Earth's surface by color bands on a map. The span of each rainbow indicates a specified amount of uplift.

During an earthquake-generating episode of slip, only part of a fault may slip (**Fig. 10.9a**). Generally, the "larger" the earthquake—meaning, as we will see, the greater the energy released and the more intense the vibrations generated—the larger the slip area and the greater the displacement. The earthquake that destroyed San Francisco, California, in 1906 ruptured a 430-km-long (measured parallel to the Earth's surface) by 15-km-deep (measured perpendicular to the Earth's surface) segment of the San Andreas fault.

Large earthquakes can result in meters of slip. For example, the maximum observed displacement during the 1906 earthquake was 7 m, in a strike-slip sense. Displacement on a thrust fault that caused the 1964 Good Friday earthquake in southern Alaska reached a maximum of 12 m; and displacement on the Tōhoku earthquake was as much as 30 m (**Fig. 10.9b**). Smaller earthquakes, such as the one that hit Northridge, California, in 1994, resulted in only about 0.5-m slip—even so, this earthquake toppled homes, ruptured pipelines, and killed 51 people. The smallest-felt earthquakes result from displacements measured in millimeters to centimeters. Slip on a fault varies with location along a fault. Slip tends to decrease with increasing distance from the focus, until it eventually decreases to zero, but not necessarily in a symmetrical pattern.

The amount of displacement during a slip event can be studied using a variety of techniques, ranging from field observation of offset markers to sophisticated satellite measurements involving GPS, which directly measure the relative movements of one point on the Earth's crust relative to another (see Fig. 10.9b), and InSAR measurements, which use radar beams to detect subtle vertical displacement of the crust (**Fig. 10.9c**).

Although the cumulative movement on a fault during a human life span may not amount to much, over geologic time the cumulative movement becomes significant. For example, if earthquakes occurring on a strike-slip fault cause 1 cm of displacement per year, on average, the fault's movement will yield 10 km of displacement after 1 million years. Thus, earthquakes mark the incremental movements that create mountains.

Can Faults Slip without Earthquakes?

When a material breaks along fractures or cracks and separates into pieces, we say that it has undergone *brittle deformation*. For example, a glass plate shattering on the floor is a type of brittle deformation. The faulting that generates earthquakes also represents brittle deformation. Under certain conditions, a rock flows, very slowly, without breaking. We call such behavior *plastic deformation*. If the glass plate were heated to a high temperature, it could be molded into another shape plastically, without breaking. Some materials, like glass and rock, are brittle at lower temperatures and plastic at higher temperatures, whereas others, like chewing gum, can be plastic even at low temperatures. Plastic deformation changes the shape of rocks and accommodates crustal movements; therefore, it does not cause earthquakes.

Because the temperature of the Earth increases with depth, most brittle deformation and, therefore, earthquake-generating faulting in continental crust generally occur only in the upper 15 to 20 km of the crust. At greater depths, shear and movement can take place by plastic deformation without generating earthquakes. For that reason, earthquake generation on the San Andreas fault happens only in the upper 15 km even though the fault is a plate boundary that cuts clear through the lithosphere. In oceanic plates that have been subducted, earthquakes can happen much deeper, as we'll see later in this chapter.

Even though most upper-crustal continental rock is brittle, geologists have found that some movement on faults in the upper crust appears to take place slowly and steadily, without generating earthquakes. When movement on a fault happens without generating earthquakes, we call the movement **fault creep**. Seismologists do not completely understand fault creep but speculate that it occurs in particularly weak rock or when the fault surface has a coating of soft clay. Where fault creep happens relatively near the Earth's surface, geologists can detect it either by using satellite data, including GPS measurements, or by InSAR.

Take-Home Message

Most earthquakes happen when stress builds to cause either sudden formation of a new fault or slip on a pre-existing fault. Faults display stick-slip behavior in that stress builds until slip takes place and then builds again. Elastic rebound of rock when fracturing or slip takes place generates earthquake energy. The focus is the point in the Earth where the slip begins, and the epicenter is the point on the surface of the Earth directly above the focus. The size of faults and the amount of displacement during a fault vary greatly.

QUICK QUESTION: Why might foreshocks and aftershocks take place?

10.3 Seismic Waves and Their Measurement

How does the energy produced at the focus of an earthquake travel to the surface or even pass through the entire Earth? As we've noted, earthquake energy travels through rock and sediment in the form of vibrations. We call these **seismic waves**, or earthquake waves. You feel such waves when you touch one end of a brick and strike the other end with a hammer.

Seismologists distinguish among different types of seismic waves on the basis of where and how the waves move. **Body waves** pass through the interior of the Earth, whereas **surface waves** travel along the Earth's surface. Waves that cause particles of material to move back and forth parallel to the direction in which the wave itself moves are called **compressional waves**. As a compressional wave passes, the material first compresses (or squeezes together), then dilates (or expands). To see this kind of motion in action, push on the end of a spring and watch as the little pulse of compression moves along the length of the spring. Waves that cause particles of material to move back and forth *perpendicular* to the direction in which the wave itself moves are called **shear waves**. To see shear-wave motion, jerk the end of a rope up and down and watch how the up-and-down motion travels along the rope. With these concepts in mind, we can define four basic types of seismic waves (**Fig. 10.10**):

- P-waves (*P* stands for primary) are compressional body waves.
- S-waves (*S* stands for secondary) are shear body waves.
- L-waves (*L* stands for Love, the name of a seismologist) are surface waves that cause the ground to ripple back and forth, producing a snake-like movement.
- R-waves (*R* stands for Rayleigh, the name of a physicist) are surface waves that cause the ground to ripple up and down.

FIGURE 10.10 Different types of earthquake waves.

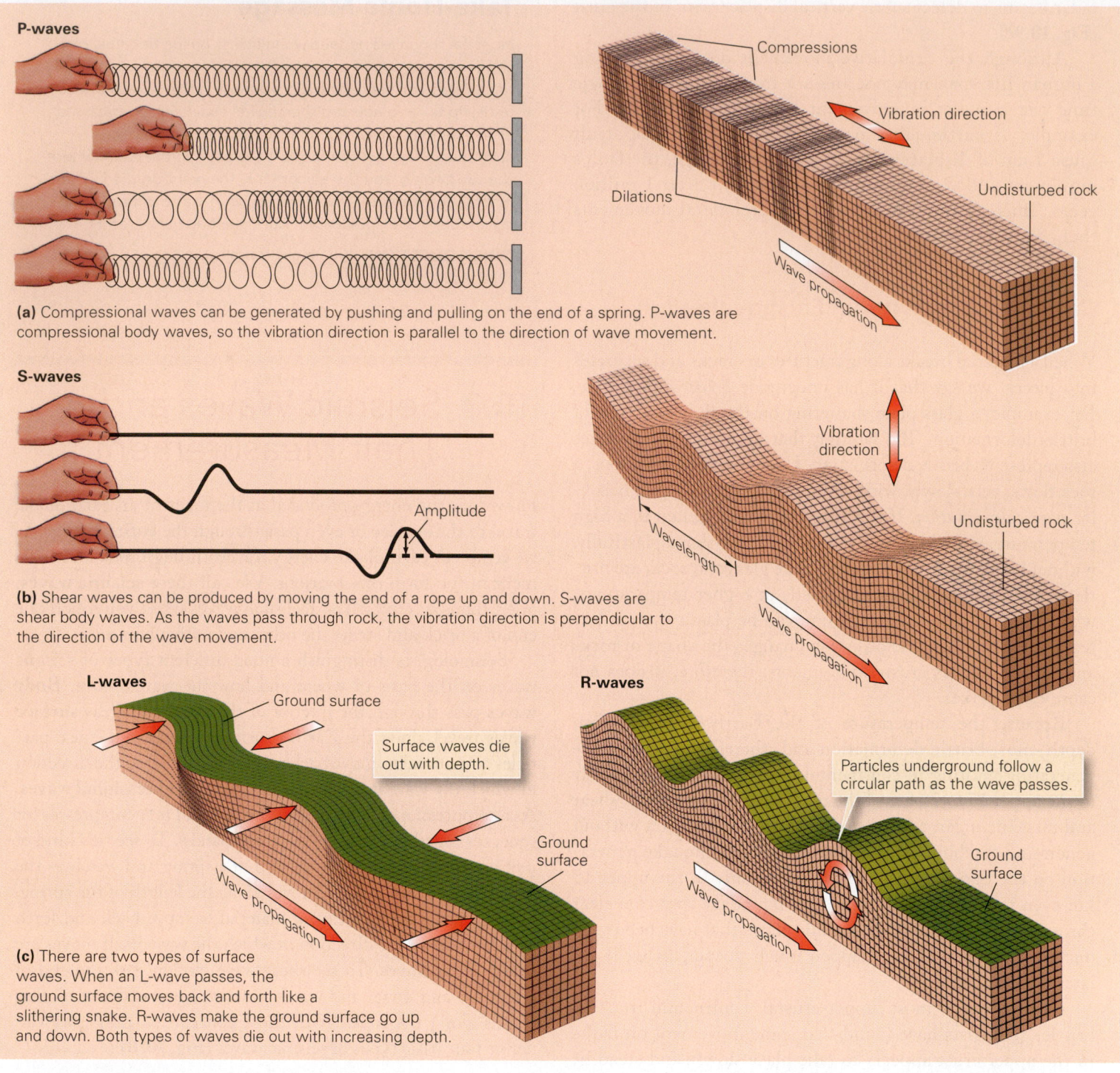

P-waves

(a) Compressional waves can be generated by pushing and pulling on the end of a spring. P-waves are compressional body waves, so the vibration direction is parallel to the direction of wave movement.

Compressions

Vibration direction

Dilations

Undisturbed rock

Wave propagation

S-waves

Amplitude

(b) Shear waves can be produced by moving the end of a rope up and down. S-waves are shear body waves. As the waves pass through rock, the vibration direction is perpendicular to the direction of the wave movement.

Vibration direction

Wavelength

Undisturbed rock

Wave propagation

L-waves

Ground surface

Surface waves die out with depth.

Ground surface

Wave propagation

(c) There are two types of surface waves. When an L-wave passes, the ground surface moves back and forth like a slithering snake. R-waves make the ground surface go up and down. Both types of waves die out with increasing depth.

R-waves

Particles underground follow a circular path as the wave passes.

Ground surface

Wave propagation

The different types of seismic waves travel at different velocities. P-waves travel the fastest and thus arrive first. S-waves travel more slowly, at about 60% of the speed of P-waves, so they arrive later. Surface waves are the slowest of all.

Friction absorbs energy as waves pass through a material, and waves bounce off layers and obstacles in the Earth, so the amount of energy carried by seismic waves decreases the farther they travel. Thus, people near the epicenter may be thrown off their feet by a large earthquake, but those 200 km away barely feel it. Similarly, an earthquake caused by slip on a fault deep in the crust causes less damage than one caused by slip on a fault near the surface.

Seismometers and the Record of an Earthquake

How do seismologists detect and measure seismic waves? The principle behind the key tool now used to accomplish this task was discovered in the 19th century and culminated in the construction of an instrument formerly known as a seismograph, now called a **seismometer**. Seismologists use two basic configurations of seismometers, one for measuring vertical (up-and-down) ground motion and the other for measuring horizontal (back-and-forth) ground motion.

A traditional, mechanical vertical-motion seismometer consists of a heavy weight (like a pendulum) suspended from a spring (**Fig. 10.11a**). The spring connects to a sturdy frame that has been bolted to the ground. A pen extends sideways from the weight and touches a vertical revolving cylinder of paper that has been connected to the seismometer frame. If the ground is steady, the pen traces out a straight reference line. But when an earthquake wave arrives and causes the ground surface to move up and down, it makes the seismometer frame and the attached paper cylinder move up and down as well. The weight, however, because of its *inertia* (the tendency of an object at rest to remain at rest), remains fixed in space. As the revolving paper roll moves up and down with respect to the fixed pen, the line traced by the pen on the paper deflects away from the reference line. The deflection of the pen, therefore, represents the up-and-down movement. Note that if the paper cylinder did not revolve, the pen would simply move back and forth in the same place on the paper, but since the paper revolves under the pen, the pen traces out curves that resemble seismic waves. A mechanical horizontal-motion seismometer works on the same principle, except that the paper cylinder is horizontal and the weight hangs from a wire (**Fig. 10.11b**). Sideways back-and-forth movement of this seismometer causes the pen to trace out waves.

In sum, the key to a seismometer is the presence of a weight that stays fixed in space while everything else moves around it. Typically, seismologists place these instruments in vaults below the ground surface, away from traffic, trains, and swaying trees that together can produce "noise," or non-earthquake vibrations (**Fig. 10.11c**). The entire configuration comprises a *seismometer station*. Modern electronic seismometers work on the same principle as traditional mechanical ones, except the weight is a magnet that moves relative to a wire coil, producing an electric signal that can be recorded digitally (**Fig. 10.12**). Such seismometers are so sensitive that they can record ground movements of a millionth of a millimeter (only 10 times the diameter of an atom)—movements that people can't feel. This allows the

FIGURE 10.11 The basic operation of a seismometer.

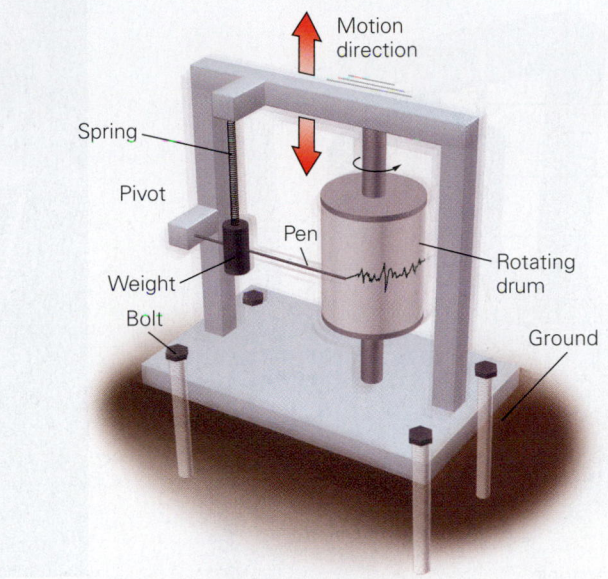

(a) A vertical-motion seismometer records up-and-down ground motion.

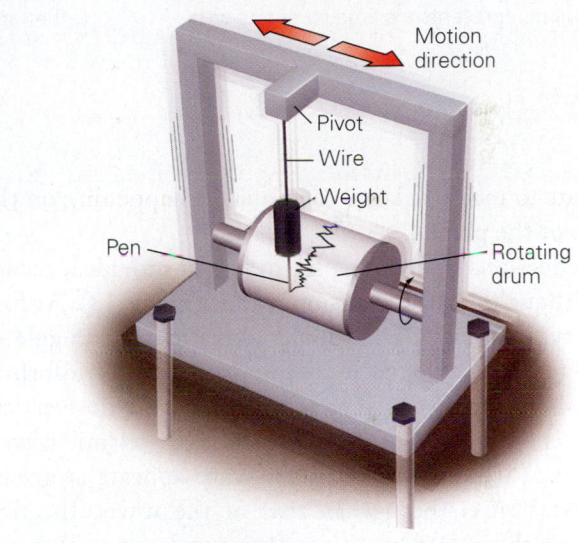

(b) A horizontal-motion seismometer records back-and-forth ground motion.

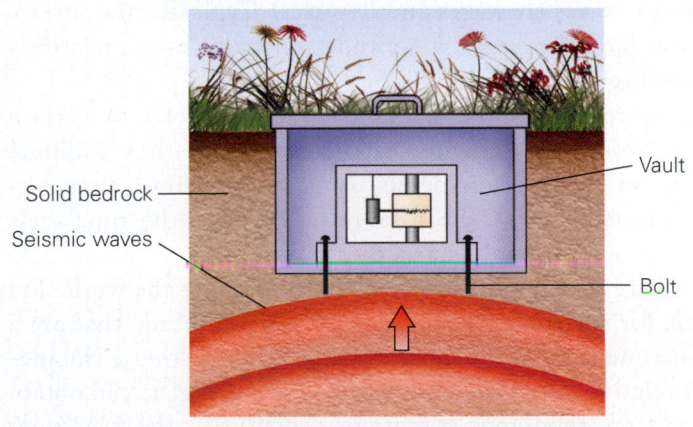

(c) Seismometers are bolted to bedrock in a protected shelter or vault.

FIGURE 10.12 Modern electronic seismometers.

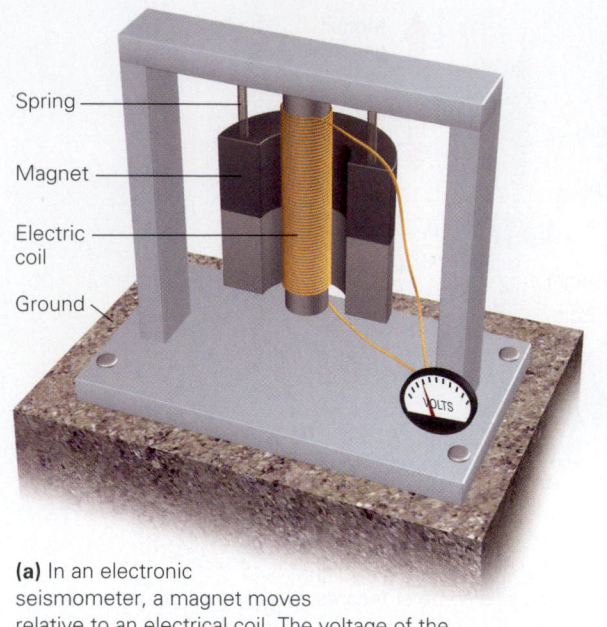

Spring

Magnet

Electric coil

Ground

(a) In an electronic seismometer, a magnet moves relative to an electrical coil. The voltage of the electrical current represents the amount of motion.

(b) Modern seismometers are very compact. Inside, they are very complex.

can also detect nuclear-bomb tests, so seismic records can allow governments to verify compliance with nuclear test-ban treaties.

Finding the Epicenter Using Seismograms

How do we find the location of an earthquake's epicenter? The key to this problem comes from measuring the difference between the P-wave arrival time and the S-wave arrival time at seismometer stations. P-waves and S-waves pass through the interior of the Earth at different velocities, so the delay between P-wave and S-wave arrival times increases as the distance from the epicenter increases (**Fig. 10.14a**). To picture why, imagine a car race. If one car travels faster than another, the distance between the two cars increases as the race proceeds.

instruments to measure large earthquakes happening on the other side of the planet.

The deflections traced by a seismometer provide a record of the earthquake called a **seismogram** (**Fig. 10.13**). At first glance, a typical seismogram looks like a messy squiggle of lines, but to a seismologist it contains a wealth of information. The horizontal axis represents time, and the vertical axis represents the *amplitude* (the size) of the seismic waves. The instant at which an earthquake wave appears at a seismometer station is the *arrival time* of the wave. The first squiggles on the record represent P-waves, because P-waves travel the fastest. Next come the S-waves and finally the surface waves (R-waves and L-waves). Typically, the surface waves have the largest amplitude and arrive over a relatively long interval of time.

Seismologists the world over have agreed to certain standards for measuring earthquakes, and they calibrate time on their seismometers using very accurate time signals broadcast by GPS satellites. This way, the time scale on all seismograms is exactly the same, so it's possible to compare seismograms from different parts of the world and look for variations in arrival time and amplitude that are a consequence of differences in the location of the seismometer relative to the epicenter. Today, seismologists can obtain data from thousands of stations constituting the *worldwide seismic network*. Governments have supported this network because in addition to recording natural earthquakes, it

We can represent the time delay between P- and S-waves on a graph where the horizontal axis indicates distance from the epicenter and the vertical axis indicates time. A curve on the graph is called a *travel-time curve*—it shows how the duration of time that it takes for the wave to move from its origin to a seismometer station increases as the distance between the epicenter and the seismometer station increases (**Fig. 10.14b**). Since P-waves travel faster than S-waves, the travel-time curve for an S-wave differs from that of a P-wave. The gap between the two curves (as measured along a vertical line) represents the time delay between the P-wave arrival and the S-wave arrival.

To use a graph of travel-time curves for determining the distance to an epicenter, start by measuring the time difference between the P- and S-waves on your seismogram; this is called the *S – P time* (pronounced S minus P time). Then draw a line segment on a piece of tracing paper to represent this amount of time, at the scale used for the vertical axis of the graph. Orient the line segment parallel to the vertical time axis and move it back and forth until one end lies on the P-wave curve and the other end lies on the S-wave curve (this gives the S – P time). You have now identified the gap between the two travel-time curves for the seismic station. Extend the line down to the horizontal distance axis, and simply read off the distance to the epicenter.

The analysis of one seismogram tells you only the distance between the epicenter and the seismometer station—it does not tell you in which direction from the station the epicenter lies. To

FIGURE 10.13 The nature of seismograms.

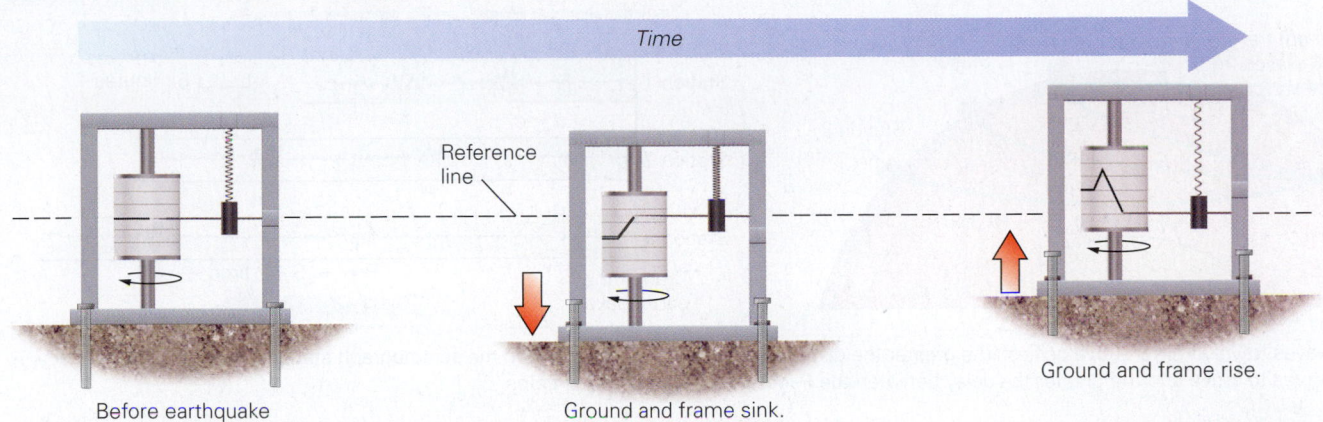

(a) Before an earthquake, the pen traces a straight line. During an earthquake, the paper roll moves up and down while the pen stays in place.

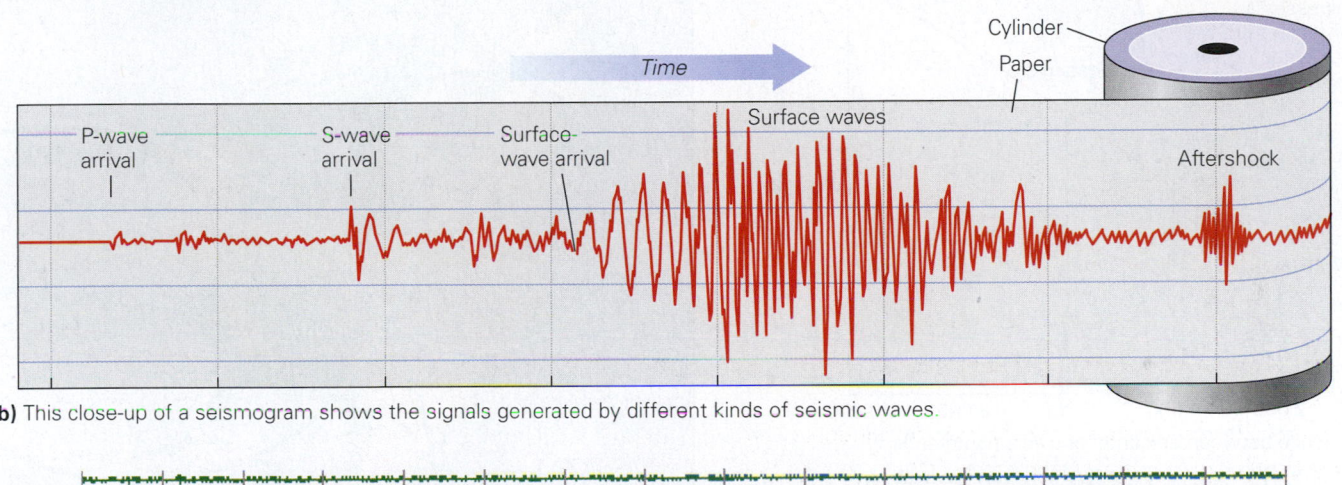

(b) This close-up of a seismogram shows the signals generated by different kinds of seismic waves.

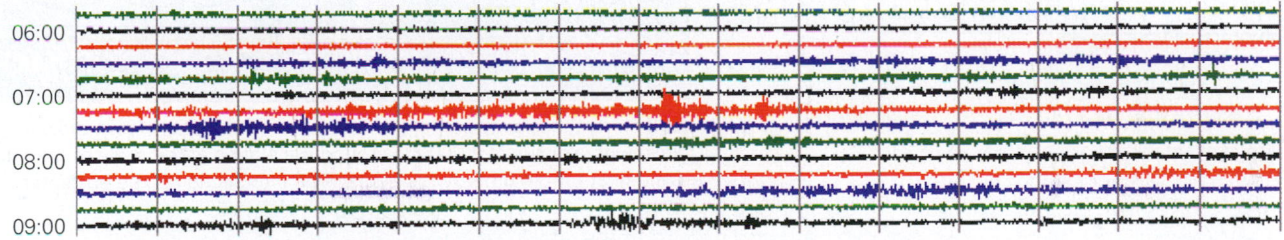

(c) A digital seismic record from a seismometer station in Arkansas. The space between vertical lines represents 1 minute. Colors have no meaning but make the figure more readable. Each color band represents the record of 15 minutes.

determine the map position of the epicenter, we use a method called triangulation, by plotting the distance from the epicenter to three stations. For example, imagine that you use the method in the previous paragraph to determine that the epicenter lies 2,000 km from Station 1, 4,000 km from Station 2, and 6,000 km from Station 3. On a map, you can draw a circle around each station, such that the radius of the circle represents the distance between the station and the epicenter, at the scale of the map. The epicenter lies at the intersection of the three circles, for this is the only point at which the epicenter has the appropriate measured distance from all three stations (**Fig. 10.14c**).

Take-Home Message

Earthquake energy travels as seismic waves. Body waves travel through the interior of the Earth, whereas surface waves travel along the surface. Ground shaking, due to arrival of waves, generally decreases with distance from the focus. We can detect an earthquake using a seismometer. A seismogram is the record of an earthquake.

QUICK QUESTION: How can you determine the location of an earthquake's epicenter?

FIGURE 10.14 The method for locating an earthquake epicenter.

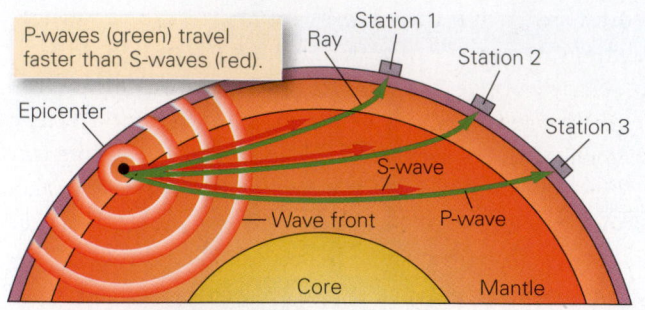

P-waves (green) travel faster than S-waves (red).

Station 1

Ray

Station 2

Station 3

Epicenter

S-wave

Wave front   P-wave

Core   Mantle

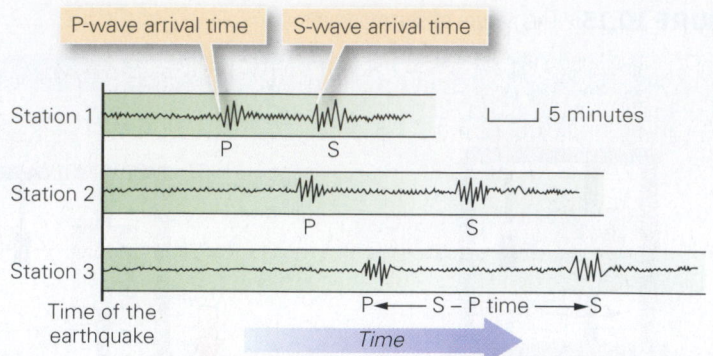

P-wave arrival time   S-wave arrival time

Station 1   P   S   5 minutes

Station 2   P   S

Station 3   P ◄— S – P time —► S

Time of the earthquake

Time

(a) Seismic waves travel at different velocities. The greater the distance between the epicenter and the seismograph station, the longer it takes for earthquake waves to arrive and the greater the delay between the P-wave and S-wave arrival times.

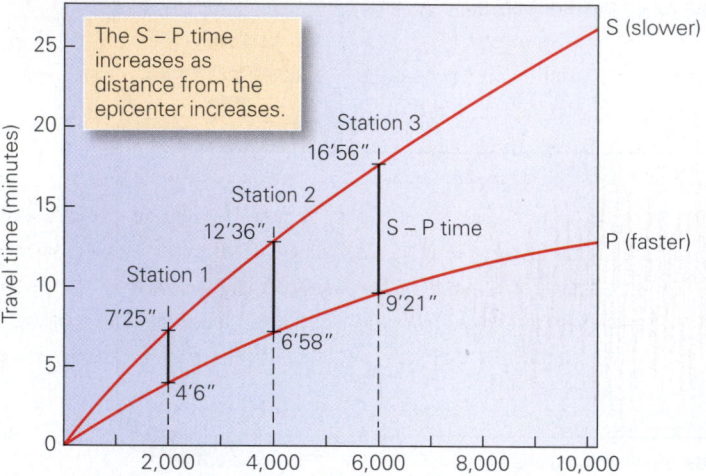

The S – P time increases as distance from the epicenter increases.

S (slower)

Travel time (minutes)

Station 3
16'56"

Station 2
12'36"   S – P time

P (faster)

Station 1
7'25"   9'21"

6'58"

4'6"

Distance between epicenter and seismometer (km)

(b) We can represent the different arrival times of P-waves and S-waves on a graph of travel-time curves.

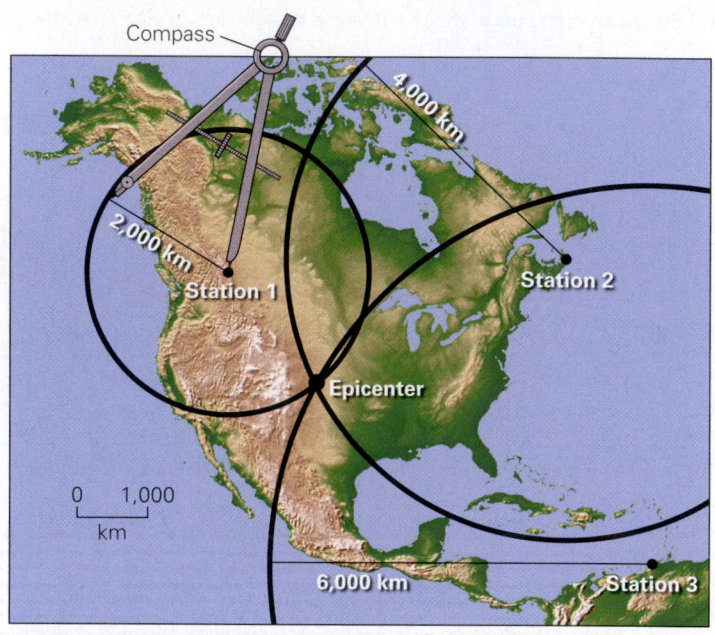

Compass

4,000 km

2,000 km

Station 1

Station 2

Epicenter

0   1,000
km

6,000 km   Station 3

(c) If an earthquake epicenter lies 2,000 km from Station 1, we draw a circle with a radius of 2,000 km around the station at the scale of the map. We repeat for the other two stations. The intersection is the epicenter.

10.4 Defining the "Size" of Earthquakes

Some earthquakes shake the ground violently, whereas others can barely be felt—in other words, some are bigger than others. Seismologists have developed two scales to define "size" in a meaningful way so that they can systematically describe and compare earthquakes. The first scale, called the Mercalli Intensity Scale, is subjective in that it depends on human perception of ground shaking and the damage resulting from it at a given locality. The second scale, called the magnitude scale, focuses on the amount of ground motion, as measured by a seismometer, at a defined distance from the epicenter.

Mercalli Intensity Scale

The **intensity** of an earthquake refers to the effect or consequence of a given earthquake's ground shaking at a locality on the Earth's surface. In 1902, an Italian scientist named Giuseppe Mercalli devised a scale for defining intensity by systematically assessing human perception of shaking and of the damage that the earthquake caused. A version of this scale, called the **Modified Mercalli Scale (MMI)**, continues to be used today (**Table 10.2**). The Modified Mercalli Scale uses Roman numerals ranging from I (low intensity) to XII (extremely high intensity). To use the scale, seismologists visit the affected region to interview residents and observe the damage. For example, if an earthquake at a location causes everyone to notice shaking but does not damage buildings, then the earthquake had an intensity of III at that location. In contrast,

TABLE 10.2 Modified Mercalli Intensity Scale

MMI	Destructiveness (Perceptions of the Extent of Shaking and Damage)
I	Detected only by seismic instruments; causes no damage.
II	Felt by a few stationary people, especially in upper floors of buildings; suspended objects, such as lamps, may swing.
III	Felt indoors; standing automobiles sway on their suspensions; it seems as though a heavy truck is passing.
IV	Shaking awakens some sleepers; dishes and windows rattle.
V	Most people awaken; some dishes and windows break, unstable objects tip over; trees and poles sway.
VI	Shaking frightens some people; plaster walls crack, heavy furniture moves slightly, and a few chimneys crack, but overall little damage occurs.
VII	Most people are frightened and run outside; a lot of plaster cracks, windows break, some chimneys topple, and unstable furniture overturns; poorly built buildings sustain considerable damage.
VIII	Many chimneys and factory smokestacks topple; heavy furniture overturns; substantial buildings sustain some damage, and poorly built buildings suffer severe damage.
IX	Frame buildings separate from their foundations; most buildings sustain damage, and some buildings collapse; the ground cracks, underground pipes break, and rails bend; some landslides occur.
X	Most masonry structures and some well-built wooden structures are destroyed; the ground severely cracks in places; many landslides occur along steep slopes; some bridges collapse; some sediment liquefies; concrete dams may crack; facades on many buildings collapse; railways and roads suffer severe damage.
XI	Few masonry buildings remain standing; many bridges collapse; broad fissures form in the ground; most pipelines break; severe liquefaction of sediment occurs; some dams collapse; facades on most buildings collapse or are severely damaged.
XII	Earthquake waves cause visible undulations of the ground surface; objects are thrown up off the ground; there is complete destruction of buildings and bridges of all types.

if ground shaking feels significant and can destroy poorly constructed buildings, but causes only minor damage to substantial buildings, then the earthquake had an intensity of VIII.

Note that the specification of earthquake intensity depends on a subjective assessment of perception and damage, not on a direct measurement with an instrument. Also, the Mercalli intensity value varies with location for a given seismic event—we cannot assign a single Mercalli number to a given earthquake. Typically, the intensity is greater near the epicenter and decreases progressively away from the epicenter. To illustrate how intensity varies over a region for a given earthquake, seismologists draw lines, called contours, which delimit zones where the earthquake had a given intensity (**Fig. 10.15**). Generally, seismologists refer to the area within the intensity II contour line as the "felt area." Of note, the maximum intensity of a "large" earthquake is greater than that of a "small" earthquake, and the distance from the epicenter to a given contour tends to be greater for a large earthquake than for a small one. In detail, the spacing between intensity contours for a given earthquake depends on the strength of the crust in the region where the earthquake occurred. Where the crust is strong, it can transmit earthquake energy more efficiently, so the contours are farther apart. In contrast, in regions where the crust is weak (because it contains lots of fractures and/or consists of soft rock), the contours are close together and the area affected by the earthquake is smaller.

FIGURE 10.15 This map shows Modified Mercalli Intensity contours for the 1886 Charleston, South Carolina, earthquake. Note that near the epicenter ground shaking reached MMI of X, and in New York City ground shaking reached MMI of II to III.

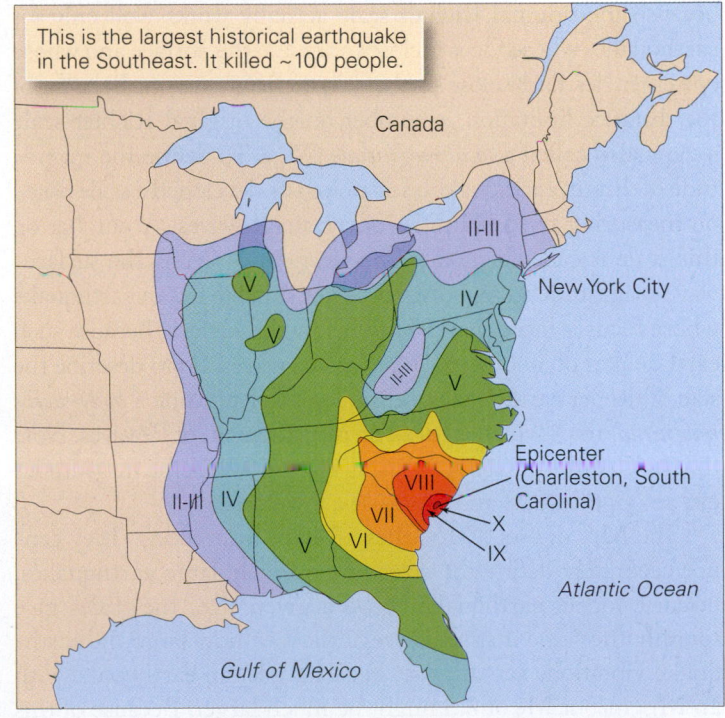

This is the largest historical earthquake in the Southeast. It killed ~100 people.

Earthquake Magnitude Scales

When you read a report of an earthquake disaster in the news, you will likely come across a phrase that states something like "An earthquake with a magnitude of 7.2 struck the city yesterday at noon." What does this mean? The **magnitude** of an earthquake is a number that indicates the maximum amplitude of ground motion recorded by a seismometer were the instrument at a specific, standard distance from the focus. By "amplitude of ground motion," we mean the amount of up-and-down or back-and-forth motion of the ground. The larger the ground motion, the greater the deflection of a seismometer pen as it traces out a seismogram. Since seismologists take into account the distance between the epicenter and the seismometer when calculating a magnitude, a magnitude value does not depend on this distance. Use of the magnitude scale, therefore, allows seismologists to define the size of an earthquake objectively.

The American seismologist Charles Richter developed an early method for defining and measuring earthquake magnitude in 1935. The scale he proposed came to be known as the **Richter scale** and is based on the maximum amplitude of motion as it would be recorded at a seismic station 100 km from the epicenter. Because the amount of deflection depends on the distance between the seismometer and the epicenter, and since most seismic stations do not happen to lie exactly 100 km from the epicenter, seismologists use a chart to adjust for distance of the station from the epicenter (**Fig. 10.16**).

Richter's scale became so widely used that news reports often include such wording as "The earthquake registered a 7.2 on the Richter scale." These days, however, seismologists actually use several different magnitude scales, not just the Richter scale, because the original Richter scale actually works well only for earthquakes whose focus is close to the Earth's surface and whose epicenter lies fairly close to the seismometer station. Because of the distance limitation, a number on the original Richter scale is now also called a *local magnitude* (M_L). To define the magnitude of distant earthquakes, seismologists developed a scale based on measuring the amplitudes of certain R-waves. A number on this scale is called a *surface-wave magnitude* (M_S). The surface-wave magnitude scale, however, is not suitable for an earthquake whose focus is more than 50 km below the surface, because such earthquakes do not create large surface waves. So to describe the size of deeper earthquakes, seismologists determine a *body-wave magnitude* (m_b), which is based on measurement of P-waves. Note that by an unfortunately confusing convention, some magnitudes use an uppercase M and some use a lowercase m.

The M_L, m_b, and M_S scales all have limitations. They cannot accurately define the sizes of extremely large earthquakes, because for an earthquake above a given size, the scales give roughly the same magnitude regardless of how large the earthquake vibrations actually are. For example, an earthquake with an M_L, m_b, or M_S of 8.3 might be much larger. Because of this

FIGURE 10.16 Using the Richter magnitude scale.

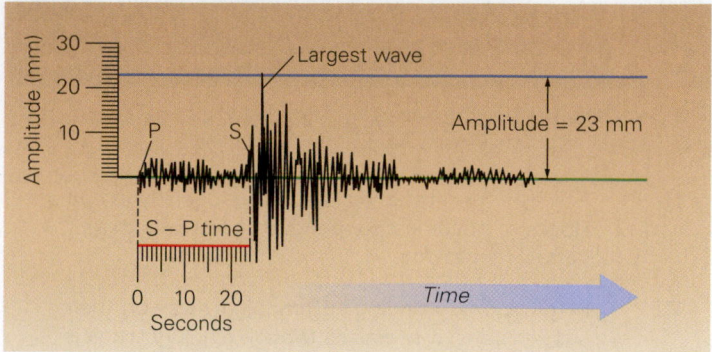

(a) To calculate the Richter magnitude from a seismograph, first measure the S – P (S minus P) time, to determine the distance to the epicenter. Then measure the amplitude of the largest wave.

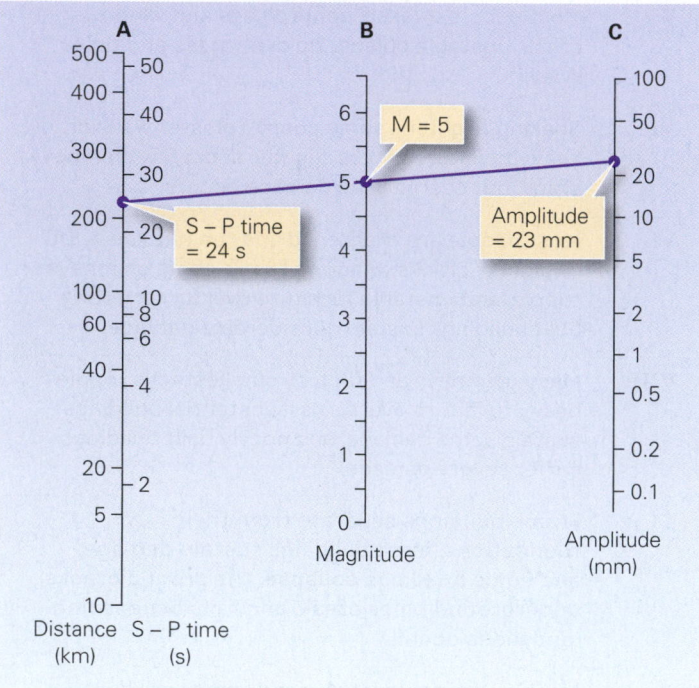

(b) Draw a line from the point on Column A representing the S – P time or the distance to the epicenter, to the point on Column C representing the wave amplitude. Read the Richter magnitude off Column B.

problem, seismologists developed the **moment-magnitude scale** (M_W). The moment magnitude scale provides the most accurate representation of an earthquake's size. To calculate the moment magnitude, seismologists measure the amplitude of several different seismic waves, determine the dimensions of the slipped area on the fault, and determine the displacement that occurred. The largest recorded earthquake in history, the great 1960 Chilean quake, registered as an 8.5 on the M_L scale and as a 9.5 on the M_W scale. The larger number makes more sense because this earthquake was indeed much larger than other known events for which M_L = 8.5. Of note, the catastrophic 2011 Tōhoku earthquake had a magnitude of M_W = 9.0.

What magnitude is given in modern reports of earthquakes in the newspaper? For early reports, seismologists report a preliminary magnitude, such as an M_L, which can be calculated quickly. Later on, after they have had the chance to collect the necessary data, seismologists report an M_W, which becomes the number generally used for archival records.

All magnitude scales are logarithmic, meaning that an increase of one integer of magnitude represents a 10-fold increase in the maximum ground motion. Thus, a magnitude 8 earthquake results in ground motion that is 10 times greater than that of a magnitude 7 earthquake, and 1,000 times greater than that of a magnitude 5 earthquake. To make discussion easier, seismologists use familiar adjectives to describe the size of an earthquake, as listed in **Table 10.3**.

Energy Release by Earthquakes

As we've pointed out, an earthquake releases energy. Seismologists have compared the energy release from an earthquake

TABLE 10.3 Adjectives for Describing Earthquakes

Adjective	Magnitude	Approximate maximum intensity at epicenter	Effects
Great	> 8.0	X to XII	Major to total destruction
Major	7.0 to 7.9	IX to X	Great damage
Strong	6.0 to 6.9	VII to VIII	Moderate to serious damage
Moderate	5.0 to 5.9	VI to VII	Slight to moderate damage
Light	4.0 to 4.9	IV to V	Felt by most; slight damage
Minor	> 3.9	III or smaller	Felt by some; hardly any damage

FIGURE 10.17 Measuring the energy released by earthquakes.

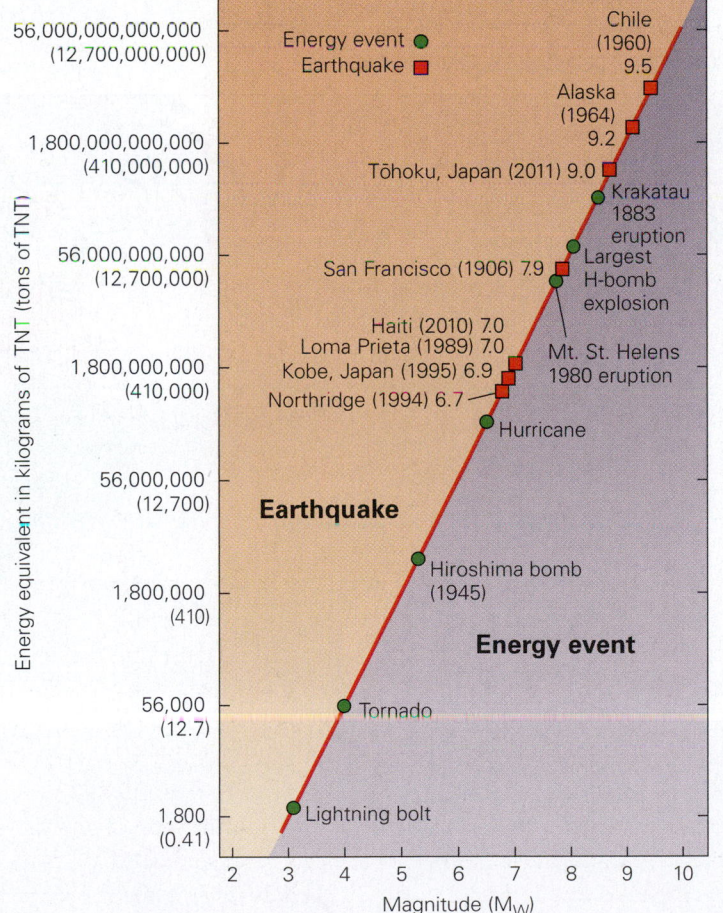

(a) Energy released by earthquakes increases dramatically with magnitude. Great earthquakes release vastly more energy than the largest bombs.

to that released by other events. According to some estimates, a magnitude 5.3 earthquake releases about as much energy as the Hiroshima atomic bomb, and a magnitude 9 earthquake releases significantly more energy than the largest hydrogen bomb ever detonated and even more than the explosion of Krakatau. Notably, an increase in magnitude by one integer represents approximately a 32-fold increase in energy. Thus, a magnitude 8 earthquake releases about 1 million times more energy than a magnitude 4 earthquake (**Fig. 10.17a**). In fact,

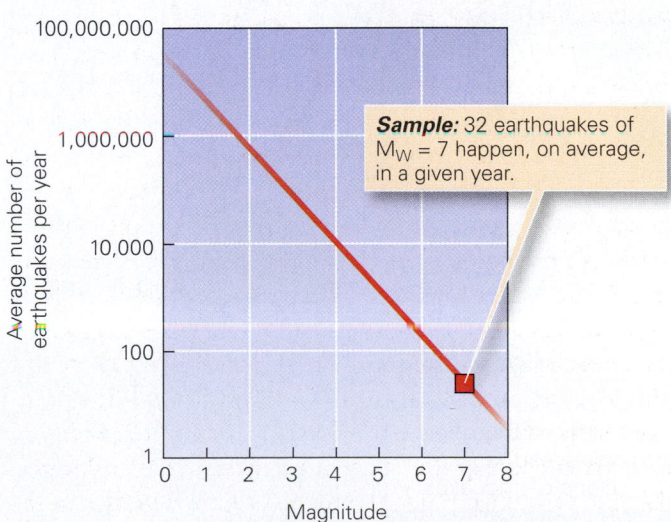

(b) The number of earthquakes per year of a given magnitude decreases with increasing magnitude.

a single magnitude 8.9 earthquake releases as much energy as the entire average global annual release of seismic energy coming from all other earthquakes combined! Fortunately, such large earthquakes occur much less frequently than small earthquakes. There are about 100,000 magnitude 3 earthquakes every year, but a magnitude 8 earthquake happens only about once or twice a year (**Fig. 10.17b**).

Take-Home Message

We can specify earthquake size by intensity (based on a perception of shaking and damage caused) or by magnitude (based on a measurement of ground motion on a seismogram). An increase of one magnitude number represents a 10-times increase in shaking and a 32-times increase in energy release. A large earthquake releases more energy than a giant hydrogen bomb.

QUICK QUESTION: Can you specify the "size" of an earthquake by giving just one intensity number? How about a single magnitude number?

10.5 Where and Why Do Earthquakes Occur?

Earthquakes do not take place everywhere on the globe. By plotting the distribution of earthquake epicenters on a map, seismologists

Did you ever wonder . . .

if an earthquake might happen near where you live?

find that most, but not all, earthquakes occur in fairly distinct **seismic belts**, or *seismic zones*. Most seismic belts coincide with plate boundaries, and earthquakes within these belts are called *plate-boundary earthquakes*. Earthquakes that occur away from plate boundaries are called *intraplate earthquakes*; the prefix *intra-* means within (**Fig. 10.18**). Eighty percent of the earthquake energy released on Earth comes from the plate-boundary earthquakes in the belts surrounding the Pacific Ocean.

Earthquakes do not occur at random depths in the Earth. Seismologists distinguish three classes of earthquakes based on focus depth: *shallow earthquakes* occur in the top 60 km of

FIGURE 10.18 A map of epicenters emphasizes that most earthquakes occur in distinct belts along plate boundaries.

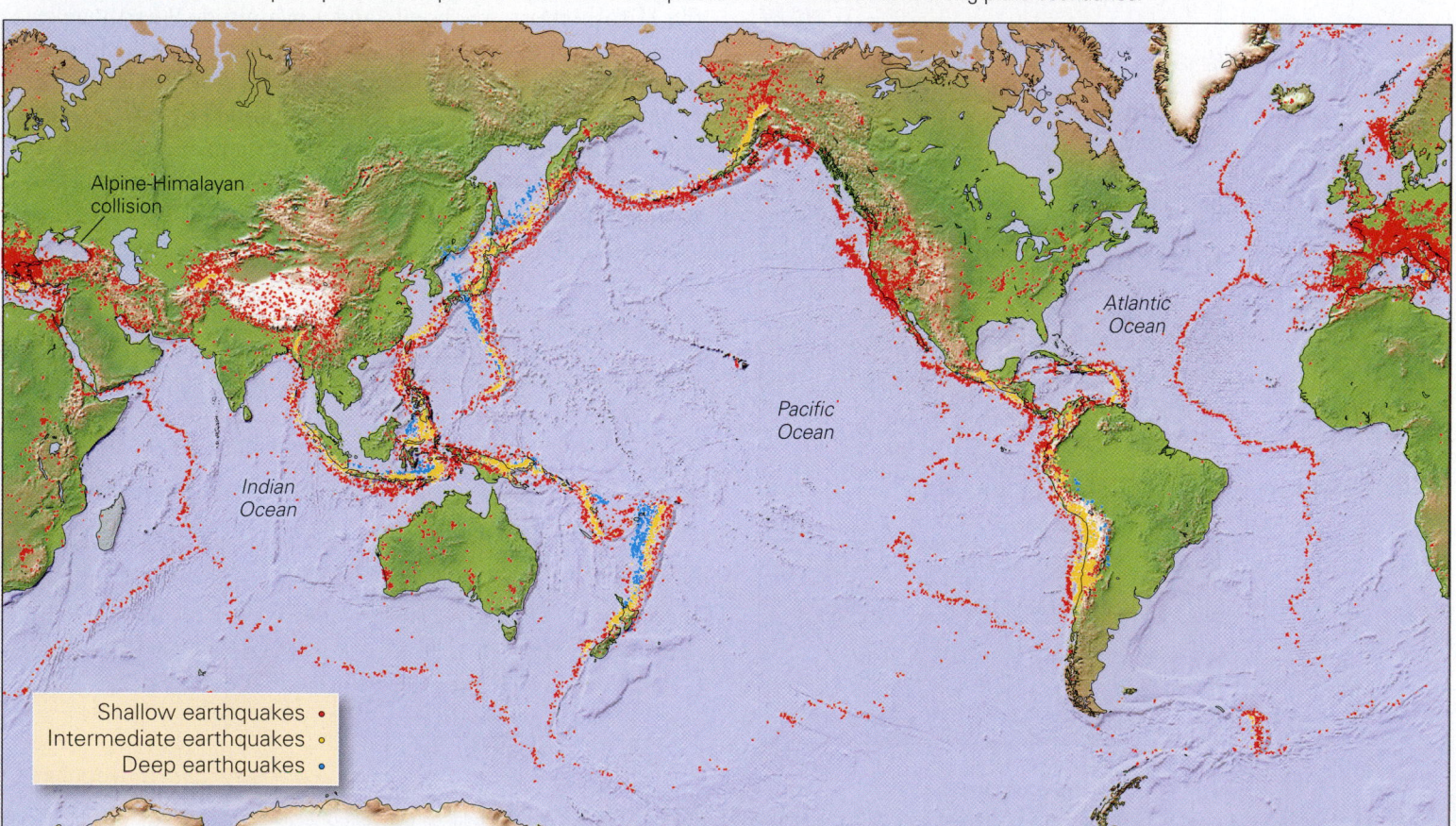

Alpine-Himalayan collision

Atlantic Ocean

Pacific Ocean

Indian Ocean

Shallow earthquakes •
Intermediate earthquakes •
Deep earthquakes •

the Earth, *intermediate earthquakes* take place between 60 and 300 km, and *deep earthquakes* occur down to a depth of about 660 km. Earthquakes do not happen below a depth of about 660 km. Let's look at the characteristics of earthquakes in various geologic settings and learn why earthquakes take place both where they do and at the depths that they do.

Earthquakes at Plate Boundaries

The majority of earthquakes happen at faults along plate boundaries, for the relative motion between plates causes slip on faults. We find different kinds of faulting at different types of plate boundaries.

Divergent-Plate-Boundary Seismicity

At divergent-plate boundaries (mid-ocean ridges), two oceanic plates form and move apart. Divergent boundaries are broken into spreading segments linked by transform faults. Therefore, two kinds of faults develop at divergent boundaries. Along spreading segments, stretching generates normal faults, whereas along transform faults strike-slip displacement occurs (**Fig. 10.19**). All earthquakes along mid-ocean ridges are shallow—they generally represent slip at depths of less than 10 km. Since most ridges lie out in the ocean, far away from settled areas, mid-ocean ridge earthquakes don't cause damage. Only a few populated localities (such as Iceland) lie astride divergent-plate boundaries.

FIGURE 10.19 The distribution of earthquakes at a mid-ocean ridge. Note that normal faults occur along the ridge axis, and strike-slip faults occur along active transform faults. Earthquakes don't take place along inactive fracture zones.

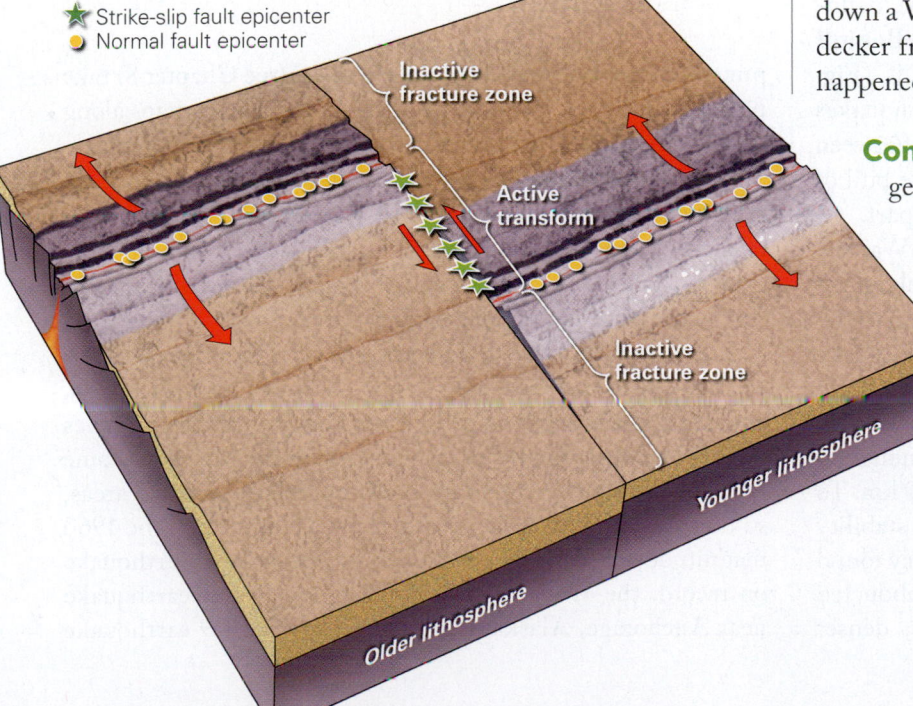

★ Strike-slip fault epicenter
● Normal fault epicenter

Transform-Plate-Boundary Seismicity

At transform-plate boundaries, where one plate slides past another without the production or consumption of oceanic lithosphere, most faulting results in strike-slip motion. The majority of transform faults in the world link segments of oceanic ridges, as we've just discussed. But a few, such as the San Andreas fault of California, the Alpine fault of New Zealand, and the Anatolian fault in Turkey, cut across continental lithosphere. All transform-fault earthquakes have a shallow focus, so the larger ones on land can cause disaster.

The San Francisco earthquake of 1906 serves as an example of a continental transform-fault earthquake (**Fig. 10.20a**). In the wake of the gold rush, San Francisco was a booming city with broad streets and numerous large buildings. But it was built on the transform boundary along which the Pacific Plate moves north at an average rate of 6 cm per year, relative to North America. Because of the stick-slip behavior of the fault, this movement happens in sudden jerks, each of which causes an earthquake. At 5:12 A.M. on April 18, the fault near San Francisco slipped by as much as 7 m, and minutes later seismic waves struck the city. Witnesses watched in horror as the streets undulated, buildings swayed and banged together, laundry lines stretched and snapped, and weaker buildings collapsed. Fire followed soon after, consuming huge areas of the city (**Fig. 10.20b**). Judging from the damage, seismologists estimate that the earthquake would have registered as a magnitude 7.9.

The San Francisco earthquake has not been the only one to strike along the San Andreas and nearby related faults during historic time. Over a dozen major earthquakes have happened on these faults during the past two centuries, including the 1857 magnitude 7.7 earthquake just east of Los Angeles and the 1989 magnitude 7.1 Loma Prieta earthquake, which occurred 100 km south of San Francisco but nevertheless shut down a World Series game and caused the collapse of a double-decker freeway (**Fig. 10.20c**). Even deadlier earthquakes have happened on other transform faults.

Convergent-Plate-Boundary Seismicity

Convergent plate boundaries are complicated regions at which several different kinds of earthquakes take place. Specifically, as the downgoing plate begins to subduct, it bends and scrapes along the base of the overriding plate. Large thrust faults define the contact between the downgoing and overriding plates, and shear on these faults can produce disastrous, shallow earthquakes. Thrust faults also develop in the accretionary prism. Bending causes normal faults to develop in the downgoing plate, seaward of the trench. In some cases, interaction between the downgoing plate and the overriding plate also triggers

FIGURE 10.20 The San Andreas fault system, a continental transform.

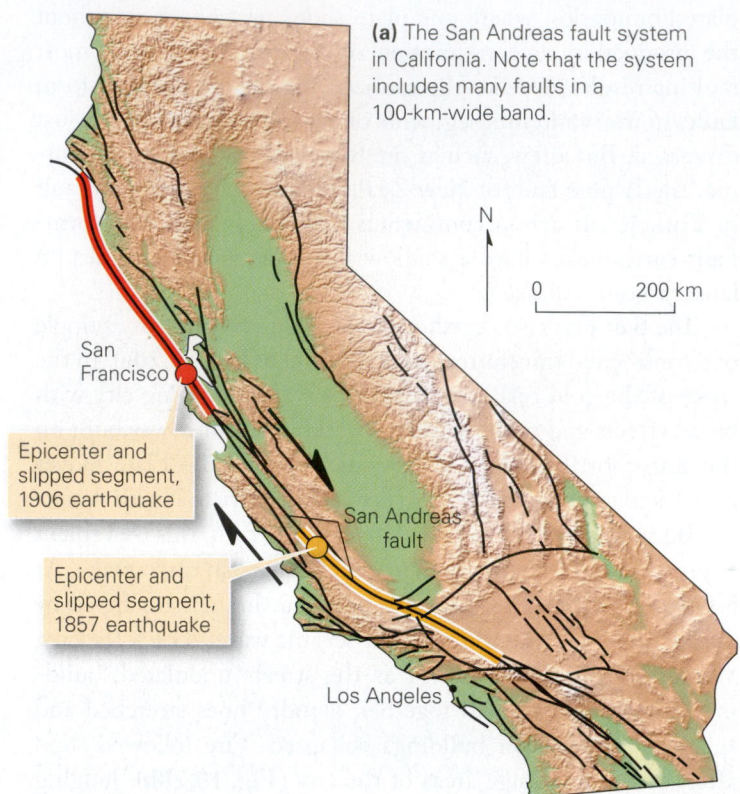

(a) The San Andreas fault system in California. Note that the system includes many faults in a 100-km-wide band.

N

0   200 km

San Francisco

Epicenter and slipped segment, 1906 earthquake

San Andreas fault

Epicenter and slipped segment, 1857 earthquake

Los Angeles

(b) A street in San Francisco after the 1906 earthquake. Huge fires swept through the city.

(c) The 1989 Loma Prieta earthquake caused a double-decker bridge to collapse, as support columns gave way.

shallow faulting in the overriding plate within and on both sides of the volcanic arc.

In contrast to other types of plate boundaries, where only shallow-focus earthquakes occur, convergent-plate boundaries also host intermediate- and deep-focus earthquakes. These occur in the downgoing slab as it sinks into the mantle, defining the sloping band of seismicity called a **Wadati-Benioff zone**, after the seismologists who first recognized it (**Fig. 10.21a**; see also Fig. 4.11c). Intermediate and deep earthquakes happen partly in response to stresses caused by shear between the downgoing plate and the mantle and partly by the pull of the sinking, deeper part of the plate on the shallower part.

Why can intermediate and deep earthquakes of a Wadati-Benioff zone take place? Shouldn't the rock of a subducted plate at these depths be too warm and soft to break brittlely? To answer these questions, seismologists studied the rate at which a subducting slab warms up as it sinks down through hot asthenosphere. They determined that rock is such a good insulator that the interior of a plate actually remains cool enough to fracture seismically, even down to a depth of about 300 km. To explain deeper earthquakes, seismologists studied the stability of minerals comprising the rock in the lithosphere. They found that at the extreme pressures developed in deeply subducted lithosphere, certain minerals collapse to form new, denser

minerals. As such sudden "phase changes" (see Chapter 8) take place, the minerals abruptly decrease in volume, perhaps along a fault-like band. This process could generate an earthquake. The fate of subducted lithosphere below a depth of 660 km remains uncertain. Some of the subducted plate may accumulate at 660 km, whereas some sinks still deeper. But at depths greater than 660 km, processes that generate earthquakes in subducted plate can no longer take place.

Earthquakes in southern Alaska, eastern Japan, the western coast of South America, the coast of Oregon and Washington, and along island arcs in the western Pacific serve as examples of convergent-boundary earthquakes (see Fig. 10.21a). Some of these earthquakes are large and occur near populated areas, so they can be devastating. Notable examples include the 1960 magnitude 9.5 earthquake of Chile, the largest earthquake on record; the 1964 magnitude 9.2 Good Friday earthquake near Anchorage, Alaska; the 1995 magnitude 6.9 earthquake

FIGURE 10.21 An example of a convergent-plate-boundary earthquake.

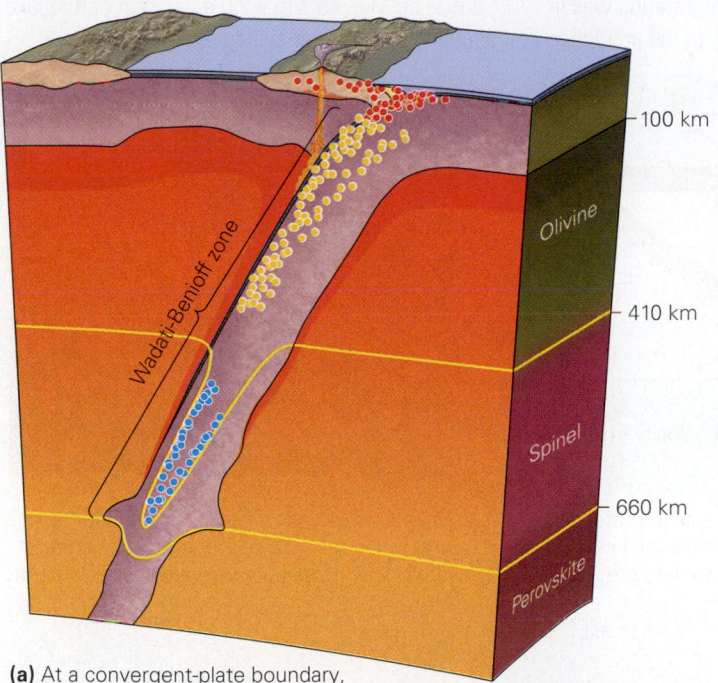

(a) At a convergent-plate boundary, earthquakes occur along the contact between the two plates, as well as in the downgoing plate and overriding plate. In the asthenosphere, the phase changes happen at 410 km and at 660 km. Its change to spinel may cause deep earthquakes.

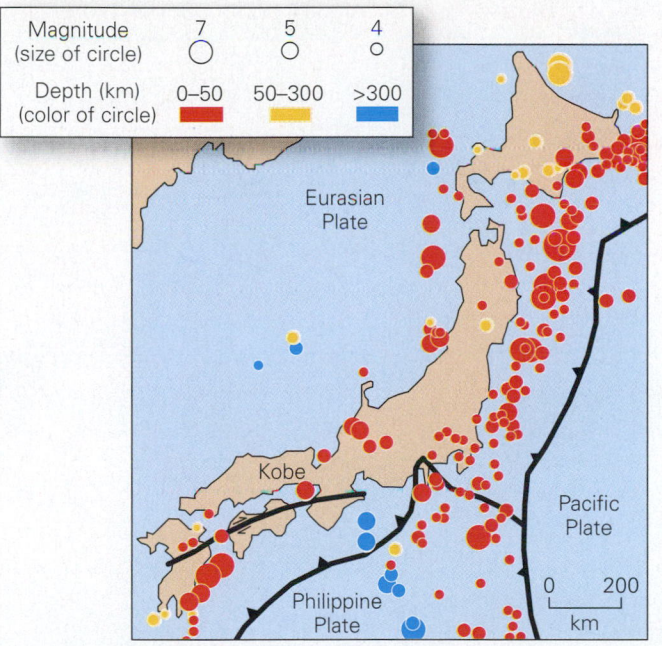

(b) Map of earthquake epicenters and subduction zones in and near Japan.

(c) A collapsed building after the Kobe earthquake in Japan.

of Kobe, Japan, which devastated the city (**Fig. 10.21b, c**); the 2004 magnitude 9.3 Sumatra earthquake, which triggered the giant Indian Ocean tsunami that killed 230,000 people; the 2010 magnitude 8.8 Chilean earthquake; and the 2011 magnitude 9.0 Tōhoku earthquake.

Recently, geologists have recognized that in 1700 a huge earthquake, with a magnitude of 8.7–9.2, accompanied slip on the Cascadia subduction zone off the now densely populated coast of Oregon and Washington. An earthquake of similar size today would be disastrous to the region. GPS measurements indicate that the crust of the region is moving and thus that stresses are likely building (**Fig. 10.22**).

Earthquakes due to Continental Rifting and Collision

Continental Rifts The stretching of continental crust at continental rifts generates normal faults. Active rifts today include the East African Rift (**Fig. 10.23a**), the Basin and Range Province (mostly in Nevada, Utah, and Arizona), and the Rio Grande Rift (in New Mexico). In all these places, shallow earthquakes occur, similar in nature to the earthquakes at mid-ocean ridges. But in contrast to mid-ocean ridges, these seismic zones occur on land and thus can be located under or near populated areas.

Collision Zones Two continents collide when the oceanic lithosphere that once separated them has been completely subducted. Such collisions produce great mountain ranges such as the Alpine-Himalayan chain (**Fig. 10.23b**). Though a variety of earthquakes happen in collision zones, the most common earthquakes result from movement on thrust faults.

An example of a collision-zone earthquake happened in October 2005, when compression resulting from the northward push of the Indian subcontinent into Asia caused a magnitude 7.6 earthquake in the Kashmir region along the border of India and Pakistan. In this region of poorly constructed homes,

FIGURE 10.22 GPS measurements made as part of the EarthScope program show that the crust is actively shortening in western Oregon and Washington due to subduction. The last earthquake to occur because of this deformation was in 1700; it was an $M_W = 8.7$ to 9.2 (i.e., a great earthquake) and caused a major tsunami. The observation that movement is occurring leads to the prediction that another earthquake may happen in the future.

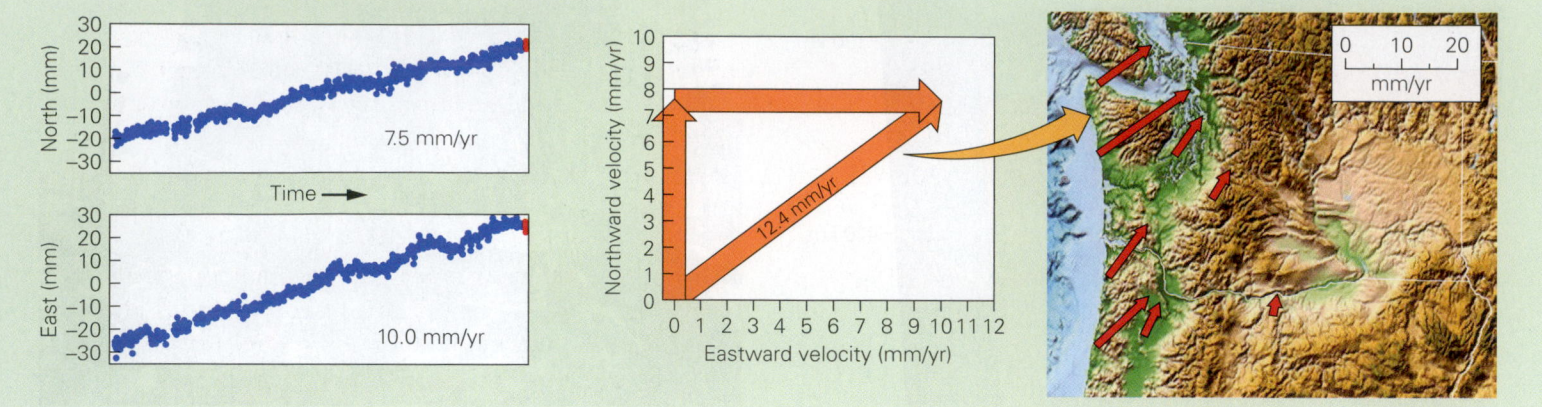

(a) The position of the survey point at Neah Bay, Washington, changes over time as indicated by GPS measurements. The top graph gives the north component, and the bottom gives the east component. Adding these components indicates the overall movement is 12.4 mm per year to the northeast.

(b) Each arrow indicates the velocity of the point at the rear end of the arrow. Note that the velocity decreases progressively eastward.

FIGURE 10.23 Earthquakes of rifts and collision zones.

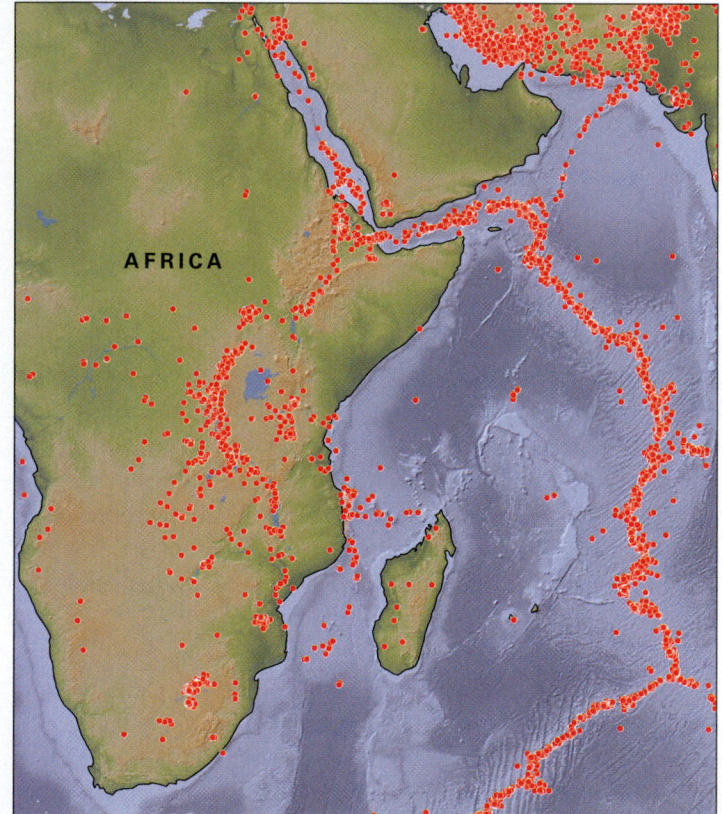

(a) Earthquakes in Africa occur mostly along the East African Rift.

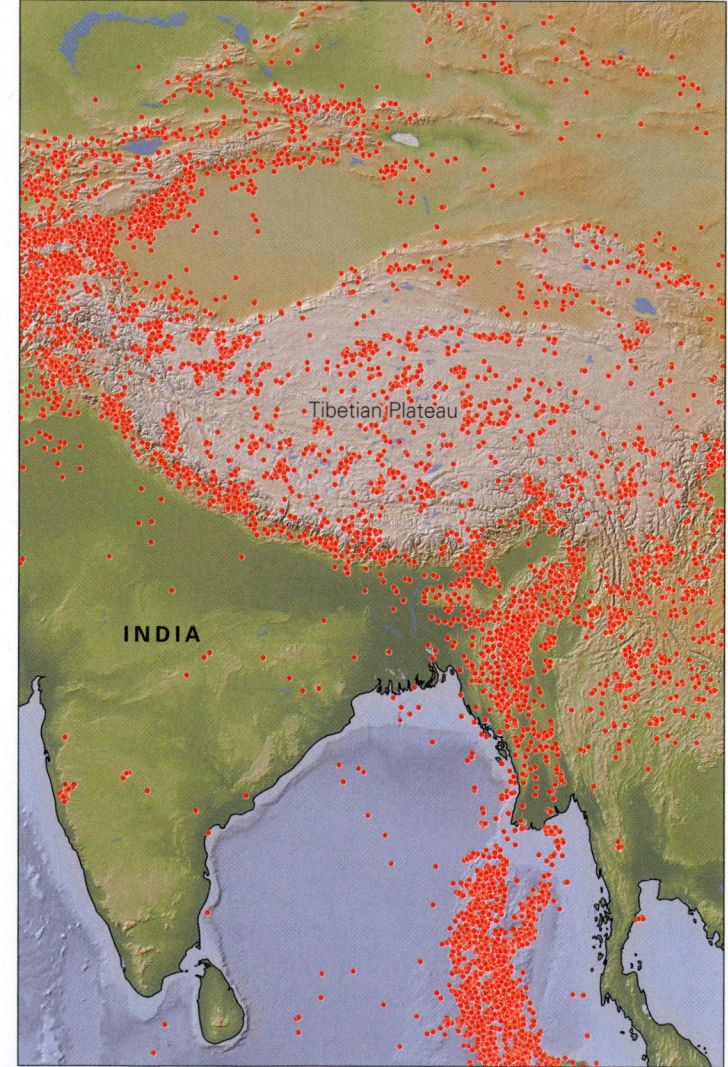

(b) Earthquakes in southern Asia occur primarily in crust deforming due to the collision of India.

ground shaking resulted in the collapse of whole towns. Associated landslides ripped out roads, denying access to potential rescuers. At least 86,000 people perished, and another 4 million were left homeless.

Intraplate Earthquakes

Some earthquakes occur in the interiors of plates and are not associated with plate boundaries, active rifts, or collision zones (**Fig. 10.24**). These **intraplate earthquakes**, almost all of which have a focus that lies at a depth of less than 20 km, account for only about 5% of the earthquake energy released in a year. Most intraplate earthquakes happen in continents, and because some can occur under populated areas, large ones have the potential to cause significant damage.

What causes intraplate earthquakes? Most seismologists favor the idea that stress applied to continental lithosphere causes pre-existing fault zones in the crust to suddenly slip. Many of these faults may have first formed during Precambrian rifting events and thus represent long-lived weak "scars" in the crust. The source of the stress driving fault reactivation remains controversial. Some seismologists attribute the stress to the push acting on plate boundaries, whereas others suggest that it comes from the shear between the lithosphere and the underlying asthenosphere or to changes in the shape of continents due to the unloading that happens when glaciers on the surface melt.

Intraplate earthquakes happen on all continents but are not uniformly distributed. In North America, for example, intraplate earthquakes occur in the vicinities of New Madrid, Missouri; Charleston, South Carolina; eastern Tennessee; Montreal, Quebec; and the Adirondack Mountains in New York. A magnitude 7.3 earthquake occurred near Charleston in 1886, ringing church bells up and down the coast and vibrating buildings as far away as Chicago. In Charleston itself, over 90% of the buildings were damaged, and 60 people died. In 2011, a magnitude 5.9 earthquake rattled central Virginia, abruptly reminding residents of the eastern United States that the region is not immune to seismicity. The tremor was felt from the Carolinas to New England, and people evacuated buildings in Washington, D.C., where some damage occurred to the Washington Monument and other structures, and in New York City.

The largest intraplate earthquakes to affect the United States took place in the early 19th century, near New Madrid, which lies near the Mississippi River in southernmost Missouri. At the time, the region was inhabited by a small population of Native Americans and an even smaller population of European descent. During the winter of 1811–12, three magnitude 7 to 7.4 earthquakes struck the region. The ground motion temporarily reversed the flow of the Mississippi River and toppled cabins (**Fig. 10.25a**). The earthquakes resulted from slip on thrust and strike-slip faults that underlie the Mississippi Valley (**Fig. 10.25b**). St. Louis, Missouri, and Memphis, Tennessee, lie close to the epicenter, so if large earthquakes were to happen in the New Madrid region again, they could cause significant damage.

Induced Seismicity

Most earthquakes reflect geologic phenomena independent of human activity. But the timing of some earthquakes relative to human-caused events suggests that in certain cases people indeed can influence seismicity. **Induced seismicity**, meaning seismic events caused by actions of people, generally occurs in response to changes in groundwater pressure. That's because the pressure of groundwater can

Did you ever wonder . . .

if people could trigger earthquakes?

FIGURE 10.24 The tectonic settings in which earthquakes occur in continental lithosphere. Subduction-related earthquakes in continental crust are not shown.

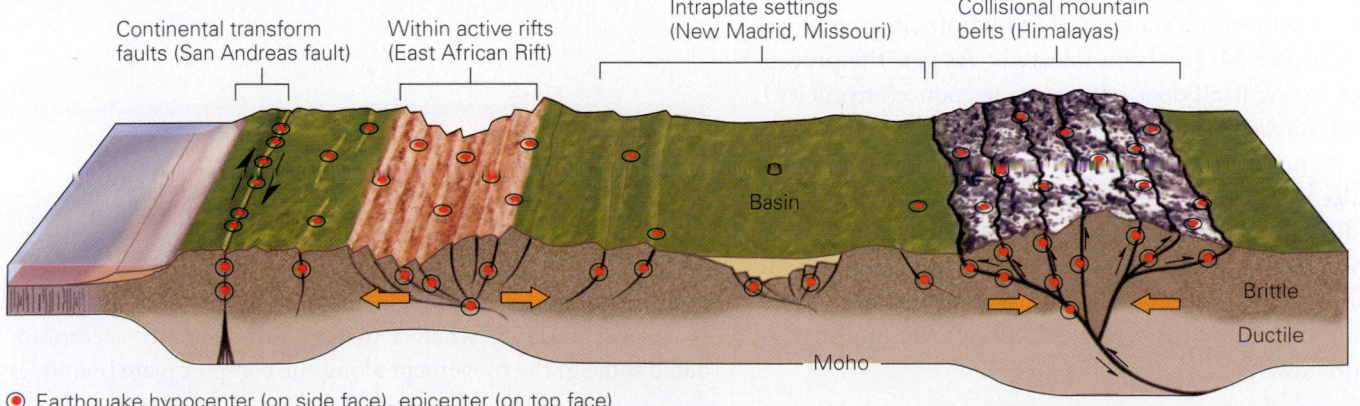

Continental transform faults (San Andreas fault)
Within active rifts (East African Rift)
Intraplate settings (New Madrid, Missouri)
Collisional mountain belts (Himalayas)
Basin
Brittle
Ductile
Moho

⊙ Earthquake hypocenter (on side face), epicenter (on top face)

FIGURE 10.25 New Madrid, Missouri, is an example of intraplate seismic activity.

(a) The earthquakes of 1811–12 destroyed cabins and disrupted the Mississippi River.

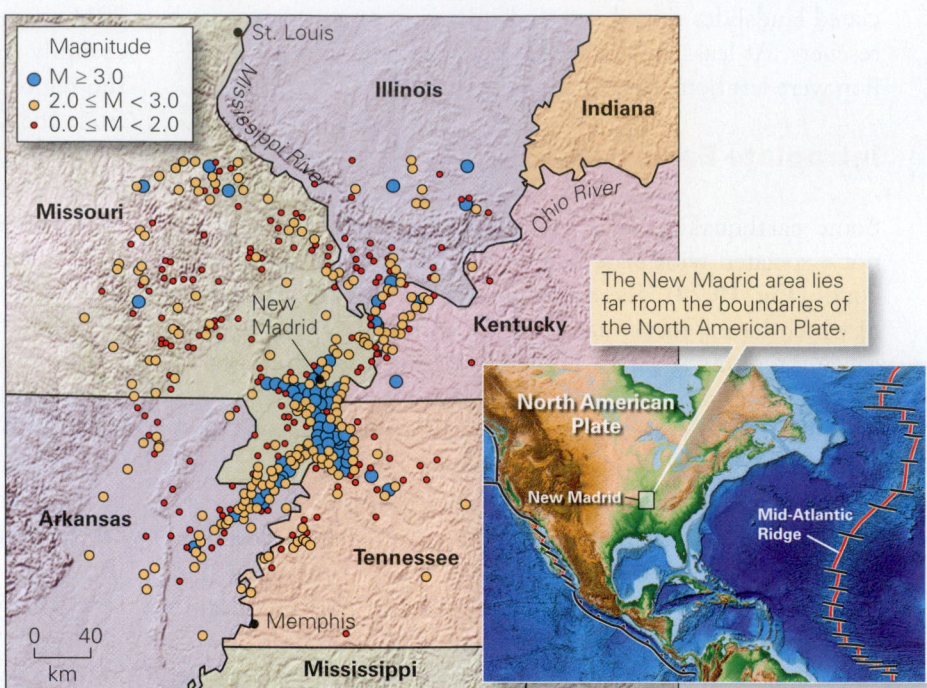

The New Madrid area lies far from the boundaries of the North American Plate.

(b) The epicenters of recent small earthquakes in the New Madrid area, as recorded by modern seismic instruments. The region remains active.

slightly push apart the opposing surfaces of faults and thus effectively decrease the friction that resists slip on them. So when people increase groundwater pressure by pumping lots of water underground in a region containing an active fault, the fault may slip under regional stress conditions that might not have otherwise led to slip. Seismologists observed such a relationship near Denver, Colorado, when engineers pumped wastewater from a military installation down a deep well—as soon as the pumping began, small earthquakes started in the region. A similar phenomenon seems to be happening in Oklahoma, where seismicity has increased markedly since 2009. The observed "earthquake swarm" lies near high-volume disposal wells into which water used in hydrofracturing of gas wells (see Chapter 14) has been injected. (As yet, the process of hydrofracturing itself does not appear to induce seismicity.)

Induced seismicity is a particular danger when people build dams and create large reservoirs in valleys overlying active faults. Faulting generally breaks up rock, making it more erodible by rivers, so it is no surprise that deep river valleys commonly form over large faults. When a reservoir fills over a fault, water seeps down into the fault and, under the pressure caused by the water column above, might trigger earthquakes.

Take-Home Message

Most, but not all, earthquakes happen along plate boundaries; thrust faults dominate at convergent boundaries, strike-slip faults at transforms, and normal faults at mid-ocean ridges. Normal faults also occur commonly in rifts and thrust faults collision zones. Earthquakes occasionally happen in plate interiors, probably along long-lived weak faults.

QUICK QUESTION: On a global basis, why are earthquakes in continental crust more dangerous to society than those along mid-ocean ridges?

10.6 How Do Earthquakes Cause Damage?

The great Lisbon earthquake happened on All Saints' Day, November 1, 1755, when a thrust fault suddenly accommodated some of the movement along the complex plate boundary

FIGURE 10.26 Types of ground motion during earthquakes. The ground can shake in many ways at once, causing surface structures to move.

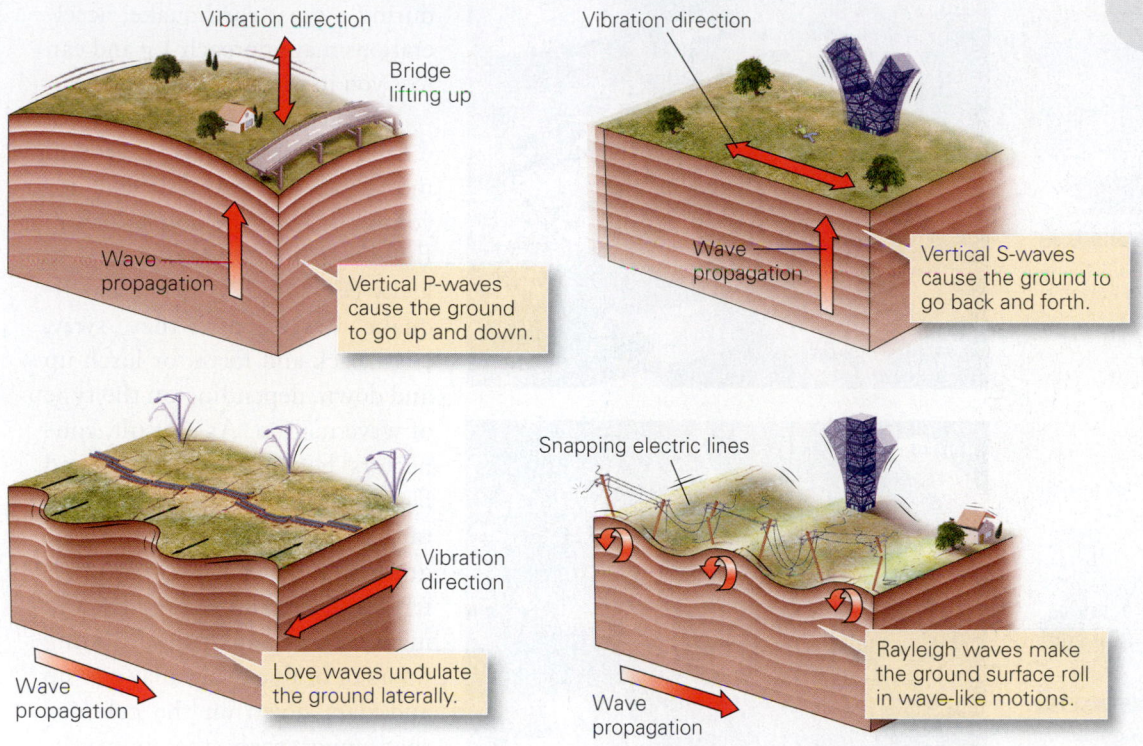

Vibration direction

Bridge lifting up

Wave propagation

Vertical P-waves cause the ground to go up and down.

Vibration direction

Wave propagation

Vertical S-waves cause the ground to go back and forth.

Vibration direction

Wave propagation

Love waves undulate the ground laterally.

Snapping electric lines

Wave propagation

Rayleigh waves make the ground surface roll in wave-like motions.

the aftershocks. The duration depends both on how long it took for slip on the earthquake-generating fault to take place and on the distance between the focus and the surface location; since different seismic waves travel at different velocities, the difference between their arrival times increases as distance from the focus increases. The intensity of an earthquake at a given location depend on four factors: (1) the magnitude of the earthquake, because larger-magnitude events release more energy; (2) the distance from the focus, because the amplitude of vibrations decreases as waves pass through the Earth; (3) the nature of the substrate at the location (i.e., the character and thickness of different materials beneath the ground surface), because earthquake waves tend to be amplified in weaker substrate; and (4) the "frequency" of the earthquake waves (where frequency equals the number of oscillations that pass a point in a specified interval of time)—high-frequency vibrations are not as dangerous as low-frequency vibrations because the former don't cause buildings to sway as much.

Different kinds of earthquake waves cause different kinds of ground motion (**Fig. 10.26**). Generally, P-waves are almost perpendicular to the ground surface when they arrive and cause the ground to buck up and down. Next come the S-waves, which also reach the surface at a steep angle. These waves are more complicated but tend to cause back-and-forth motion parallel to the ground surface. Almost immediately afterward, L-waves, the first surface waves, arrive and cause snake-like side-to-side motion. Finally, the R-waves arrive and cause a rolling motion as particles near the surface of the ground follow elliptical movements. Interference among the different kinds of waves causes motion to be anything but regular. In great earthquakes, the ground's movement can have an amplitude of as much as 1 m, but in moderate earthquakes, motions fall in the range of a few centimeters or less. Ground accelerations caused by moderate earthquakes lie in the range of 10%

between the Eurasian and African plates. It didn't take long for seismic waves from the earthquake, which is estimated to have had a magnitude of between 8.5 and 9.0, to reach Lisbon, 440 km to the northeast. Lisbon, the capital of Portugal, was one of the great port cities and cultural centers of the day. The resulting ground shaking toppled 85% of the city's buildings, and fires set by overturned stoves then consumed much of the wreckage. Forty minutes later, a tsunami inundated the coast and washed away Lisbon's harbor. When it was all over, more than 50,000 people had lost their lives, and an irreplaceable library, which housed all the records of Portuguese exploration and countless Renaissance artworks, was gone forever. The event led philosophers such as Voltaire (1694–1778) and Kant (1724–1804) to question long-held beliefs and set the stage for the Age of Enlightenment.

Visitors to Lisbon in the months after the 1755 earthquake attested that an area ravaged by a major earthquake is a heartbreaking sight. Let's consider the many factors that, sadly, can contribute to earthquake devastation. Understanding them, as we will see, may help prevent such devastation in the future.

Ground Shaking and Displacement

A large earthquake at a location on the Earth's surface may last from a few seconds to several minutes, not including

FIGURE 10.27 Examples of earthquake damage due to vibration.

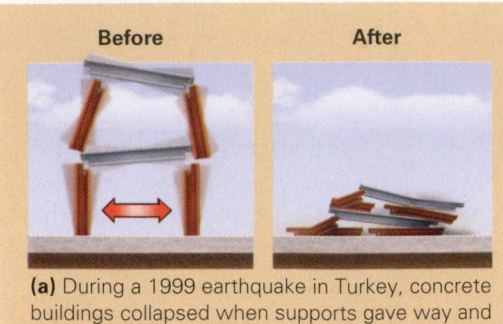

(a) During a 1999 earthquake in Turkey, concrete buildings collapsed when supports gave way and floors piled on each other like pancakes.

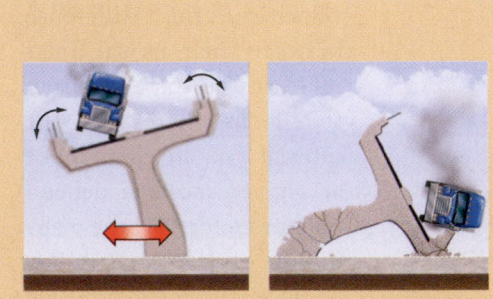

(b) An elevated bridge tipped over during the 1995 Kobe, Japan, earthquake.

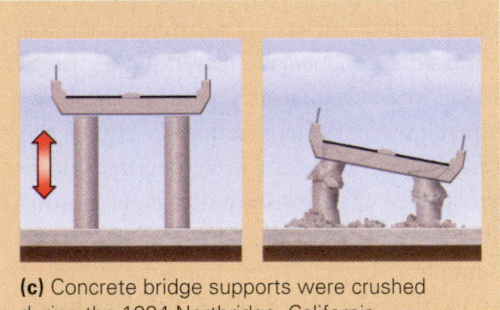

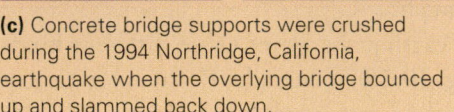

(c) Concrete bridge supports were crushed during the 1994 Northridge, California, earthquake when the overlying bridge bounced up and slammed back down.

(d) A neighborhood of masonry buildings in Armenia collapsed during a 1999 earthquake because the walls broke apart.

to 20% of *g* (where *g* is the acceleration resulting from gravity), and during a great earthquake, accelerations may approach 1 *g* and can toss you in the air.

If you're out in an open field during an earthquake, ground motion alone won't kill you, for your body is too flexible to break. Buildings and bridges aren't so lucky (**Fig. 10.27**). When earthquake waves pass, they sway, twist back and forth, or lurch up and down, depending on the type of wave motion. As a result, connectors between the frame and facade of a building may separate, so the facade crashes to the ground. The flexing of walls shatters windows and wall board and makes roofs collapse. Building floors or bridge decks may rise up and slam down on the columns that support them, thereby crushing the columns. Some buildings collapse with their floors piling on top of one another like pancakes in a stack, whereas others simply tip over.

The majority of earthquake-related deaths and injuries happen when people are hit by debris or are crushed beneath falling walls or roofs. Aftershocks worsen the problem, because they may topple already weakened buildings, trapping rescuers. During earthquakes, roads, rail lines, and pipelines may also buckle and rupture. If a building, fence, road, pipeline, or rail line straddles a fault, slip on the fault can crack the structure and separate it into two pieces.

Occasionally, ground motion causes water in lakes, bays, reservoirs, and pools to slosh back and forth, in some cases thousands of kilometers from the epicenter. The water's rhythmic movement, known as a *seiche*, can build up waves almost 10 m high and can last for hours. Seiches capsize small boats and flood shoreline homes. And if they occur in reservoirs, seiches may wash over and weaken retaining dams.

Landslides

The shaking of an earthquake can cause ground on steep slopes or ground underlain by weak sediment to give way. This movement results in a *landslide*, the tumbling and flow of soil and rock downslope (see Chapter 16). Seismically-triggered landslides occur along the coast of California, for movement on faults has rapidly uplifted this coastline in the past few million years, resulting in the development of steep cliffs. When earthquakes take place, the cliffs collapse, often carrying expensive homes down to the beach below (**Fig. 10.28**). Such events lead to the misperception that "California will someday fall into the sea." Although small portions of the coastline do collapse, the state as a whole remains firmly attached to the continent, despite what Hollywood scriptwriters say.

Did you ever wonder . . .

if California will fall into the sea?

Earthquake-triggered landslides that transport massive amounts of debris into reservoirs, lakes, or bays may cause huge waves. One of the biggest known examples happened in 1958, when a magnitude 8.3 earthquake in southeastern Alaska triggered a landslide at the head of Lituya Bay. The splash from the displaced water washed the forest off the opposite wall of the bay up to an elevation of 516 m (nearly 1,700 feet)—this wall of water was 25% higher than the Empire State Building!

Sediment Liquefaction

In 1964, a magnitude 7.5 earthquake struck Niigata, Japan. A portion of the city had been built on land underlain by wet sand. During the ground shaking, foundations of over 15,000 buildings sank into their substrate, causing walls and roofs to crack. Several four-story-high buildings in a newly built apartment complex tipped over (**Fig. 10.29a**). In 2011, an earthquake in Christchurch, New Zealand, caused sand to erupt and produce small, cone-shaped mounds called *sand volcanoes* or *sand blows* on the ground surface (**Fig. 10.29b**). The transfer of sand from underground to the surface led to the formation of depressions large enough to swallow cars (**Fig. 10.29c**).

The above examples are manifestations of a phenomenon called **sediment liquefaction**. Liquefaction in beds of wet sand or silt happens because ground shaking causes the sediment grains to try and settle together. But because the spaces (pores) between grains are filled with water, water pressure in the pores increases and pushes the grains apart, and the wet silt or sand becomes a slurry—quicksand. As the material

FIGURE 10.28 Examples of landslide damage triggered by earthquakes.

(a) Shaking triggers landslides, which caused a steep slope along the coast of California to collapse, carrying part of a home with it.

(b) During the 1964 Alaska earthquake, slumping caused the land to give way beneath parts of Anchorage.

above the liquefied sediment settles downward, pressure squeezes the sand upward and out onto the ground surface. The result being sand volcanoes, as we've seen. The settling of sedimentary layers down into a liquefied layer can also disrupt bedding and can lead to formation of open fissures of the land surface (**Fig. 10.29d, e**).

The 1964 Good Friday earthquake in Alaska caused liquefaction beneath the Turnagain Heights neighborhood of Anchorage. The neighborhood was built on a small terrace of uplifted sediment. The edge of the terrace was a 20-m-high escarpment that dropped down to Cook Inlet, a bay of the Pacific Ocean. Here, as the ground shaking began, a layer of wet clay beneath the neighborhood liquefied. In this case, liquefaction allowed the overlying terrace, along with the houses built on top of it, to slide seaward. In the process, the terrace

FIGURE 10.29 Examples of liquefaction triggered by earthquakes.

(a) Liquefaction under their foundations caused these apartment buildings in Niigata, Japan, to tip over during a 1964 earthquake.

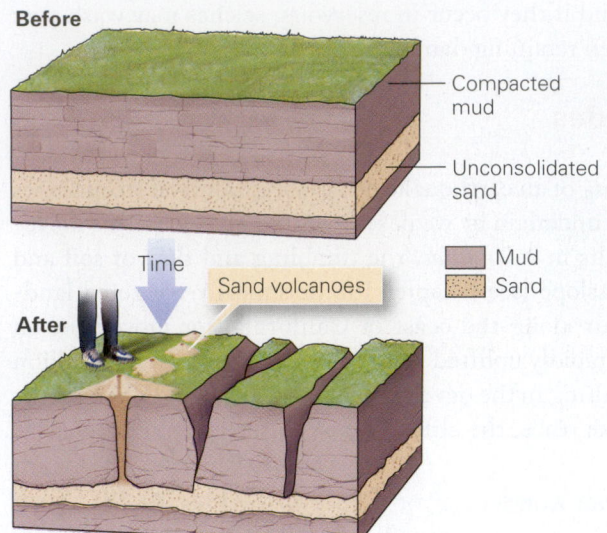

Before

Compacted mud

Unconsolidated sand

Time

Sand volcanoes

After

☐ Mud
☐ Sand

(d) Liquefaction of a sand layer causes the ground to crack and sand volcanoes to erupt.

(b) A sand volcano (sand blow) formed during the 2011 Christchurch earthquake in New Zealand.

(e) Liquefaction of a sand layer beneath the dry soil of this field caused the soil to crack and fissures to develop.

(c) During the 2011 Christchurch earthquake, liquefied sand spurted out and spread over the pavement. The process produced open space underground, so the pavement collapsed to form a sinkhole.

broke into separate blocks that tilted, turning the landscape into a chaotic jumble (**Fig. 10.30**). Liquefaction beneath Turnagain Heights happened because in wet clay the clay flakes stick together only because of the "surface tension" of the water between the flakes. When still, wet clay can behave like a solid gel, but when shaken, the weak bonds between water molecules break and the clay transforms into a viscous liquid. Clay that displays this behavior is called thixotropic clay, or *quick clay*.

Fires Associated with Earthquakes

The shaking during an earthquake can tip over lamps, stoves, or candles with open flames, and it may break wires or topple power lines, generating sparks. As a consequence, areas already turned to rubble, and even areas not so badly damaged, may be consumed by fire. Ruptured gas pipelines and oil tanks feed the flames, sending columns of fire erupting

FIGURE 10.30 The 1964 Turnagain Heights disaster.

The dark area in this aerial photograph is the slump that was Turnagain Heights.

(a) A landslide carried a neighborhood out to sea.

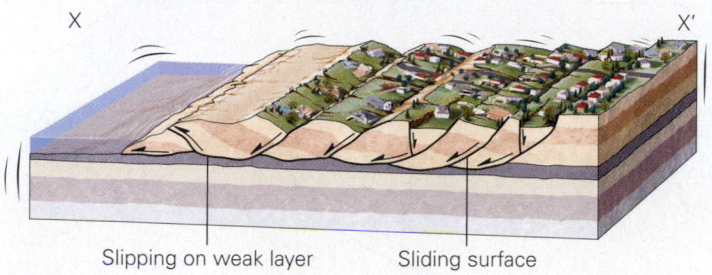

(b) Slip occurred on a weak layer.

FIGURE 10.31 Fire sometimes follows an earthquake.

(a) Broken gas tanks erupt in fountains of flame after the 2011 Tōhoku, Japan earthquake.

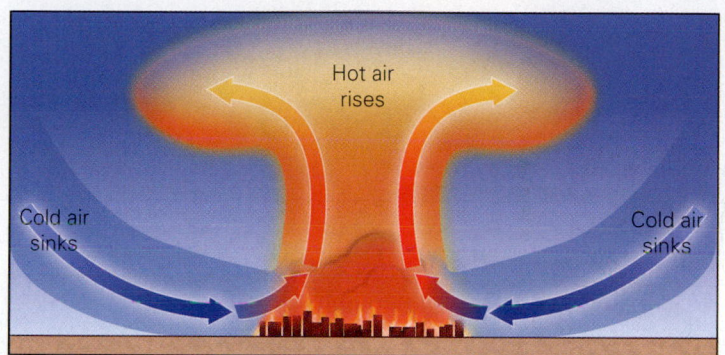

(b) A firestorm develops when cool air rushes in to replace rising hot air above a huge fire. The cool air stokes the blaze, making it larger and hotter.

skyward (**Fig. 10.31a**). Firefighters might not even be able to reach the fires because the doors to the firehouse won't open or rubble blocks the streets. Moreover, firefighters may find themselves without water, for ground shaking and landslides damage water lines.

Once a fire starts to spread, it can become an unstoppable inferno. Most of the destruction of the 1906 San Francisco earthquake resulted from fire. For three days, the blaze spread through the city until firefighters contained it by blasting a firebreak. By then, 500 blocks of structures had turned to ash, causing 20 times as much financial loss as the shaking itself. When a large earthquake hit Tokyo in September 1923, fires set by cooking stoves spread quickly through the wood-and-paper buildings, creating an inferno that heated the air above the city. The hot air rose like a balloon, and when cool air rushed in, creating wind gusts of over 100 mph, the wind stoked the blaze, which engulfed 120,000 people (**Fig. 10.31b**).

Tsunamis

The azure waters and palm-fringed islands of the Indian Ocean's east coast hide one of the most seismically active plate boundaries on Earth—the Sunda Trench. Along this convergent boundary, the Indian Ocean floor subducts at about 4 cm per year, leading to the accumulation of a large elastic strain, during the "stick"

phase of a stick-slip cycle. Just before 8:00 A.M. on December 26, 2004, the crust above a 1,300-km-long by 100-km-wide portion of one of these faults slipped and lurched westward by as much as 15 m. The rupture started at the focus and then propagated north at 2.8 km per second; thus, the rupturing process, over-all, took about 9 minutes. This slip triggered a magnitude 9.3 earthquake—a great earthquake—and pushed the seafloor up by tens of centimeters. The rise of the seafloor, in turn, shoved up overlying ocean water (**Fig. 10.32**). Because the area that rose was so broad, the volume of displaced water was immense. As a consequence, tragedy of an unimaginable extent was about to unfold. Gravity caused the water that had been pushed up to begin moving outward at speeds of about 800 km per hour (500 mph)—almost the speed of a jet plane. Geologists refer to such a wave, caused by the sudden displacement of a large vol-ume of water, as a *tsunami*. This Japanese word translates literally as harbor wave, an apt name because tsunamis can be particu-larly damaging to harbor towns. Though we most often hear of tsunamis generated by displacement due to an earthquake, they

FIGURE 10.32 Formation of a tsunami; an example from a convergent-plate boundary.

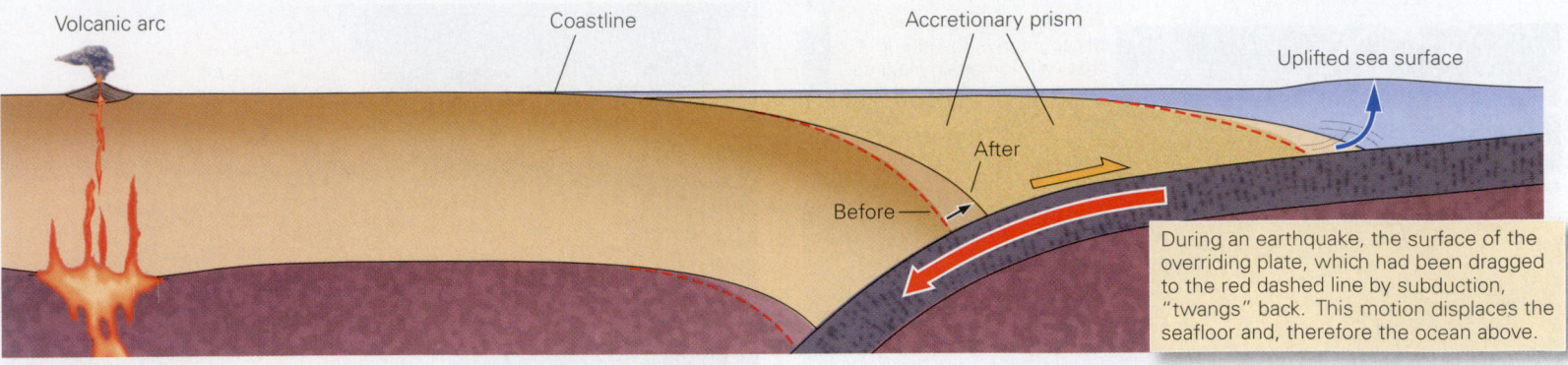

Volcanic arc | Coastline | Accretionary prism | Uplifted sea surface

After

Before

During an earthquake, the surface of the overriding plate, which had been dragged to the red dashed line by subduction, "twangs" back. This motion displaces the seafloor and, therefore the ocean above.

(a) A schematic showing how subduction leads to the buildup of elastic strain. Release of this strain by fault slip displaces the seafloor.

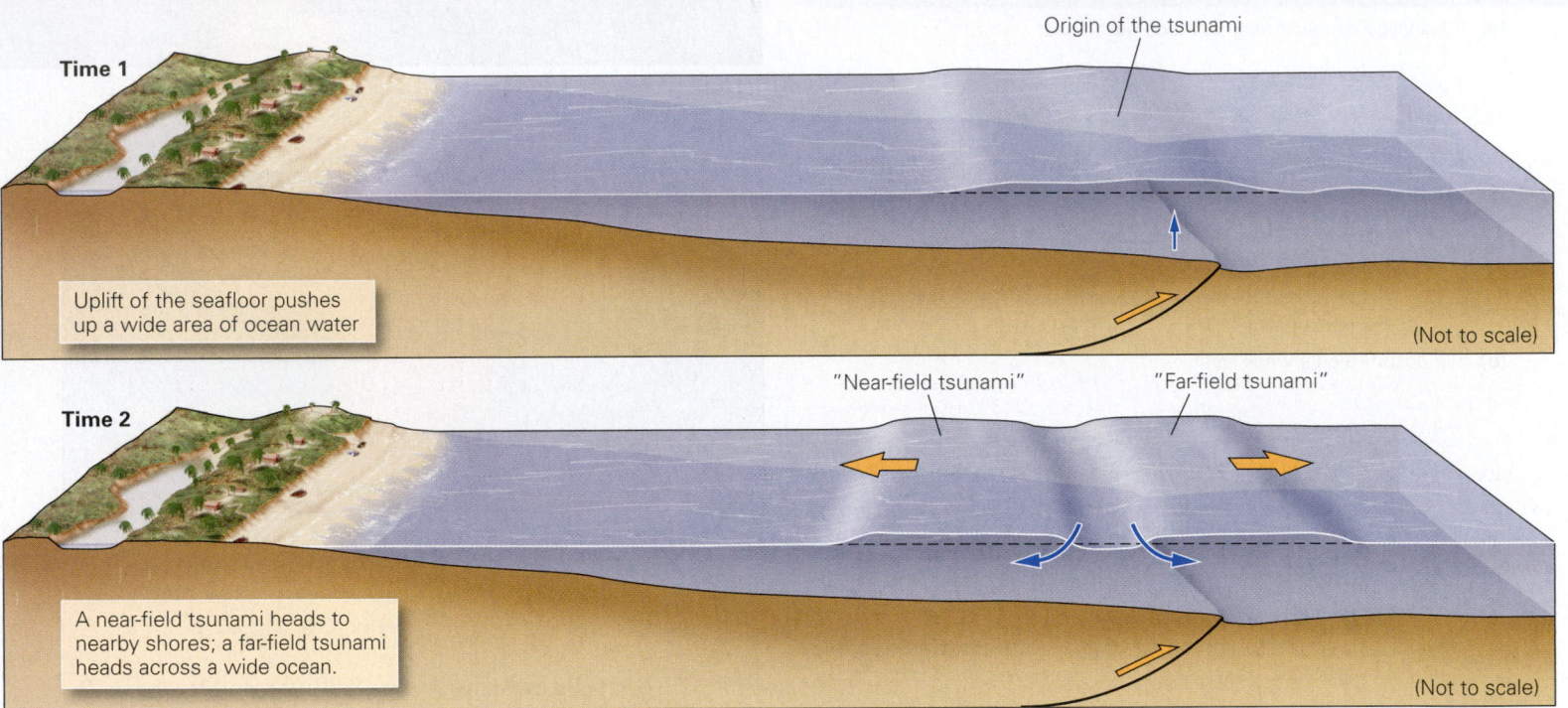

Origin of the tsunami

Time 1

Uplift of the seafloor pushes up a wide area of ocean water

(Not to scale)

"Near-field tsunami"   "Far-field tsunami"

Time 2

A near-field tsunami heads to nearby shores; a far-field tsunami heads across a wide ocean.

(Not to scale)

(b) After initial uplift of a mound of water, gravity causes the wave to spread out. The waves travel almost as fast as a jet plane.

can also be triggered by submarine landslides (see Chapter 16) or by explosion of volcanic islands (see Chapter 9).

In older literature, tsunamis were called tidal waves because when one arrives, as we will see, water rises as if a huge tide were coming in. But the waves have nothing to do with the Earth's daily tidal cycles, so the name is misleading. Though in popular media the word tsunami tends to be associated with giant or very high waves, in fact tsunamis come in all sizes—some may be just centimeters high and are barely noticeable, whereas others are colossal, reaching heights of 30 m when they reach the shore.

Regardless of cause, tsunamis are very different from familiar, wind-driven storm waves (**Fig. 10.33**). Large wind-driven waves can reach heights of 10 to 30 meters in the open ocean

or as they crash on beaches. But even such monsters have wavelengths (the distance between adjacent wave crests) of only tens of meters and thus only contain a relatively small volume of water. In contrast, although a tsunami in deep water may cause a rise in the sea surface of, at most, only a few tens of centimeters—a ship crossing one wouldn't even notice it—tsunamis have wavelengths of tens to hundreds of kilometers and an individual crest can be several kilometers wide, as measured perpendicular to the wave front. Thus, the wave contains an immense volume of water. In simpler terms, we can think of the width of a tsunami, in map view, as being more than 100 times the width of a wind-driven wave. Because of this difference, a storm wave and a tsunami have very different effects when they strike the shore.

FIGURE 10.33 A comparison of tsunami to storm waves (not to scale).

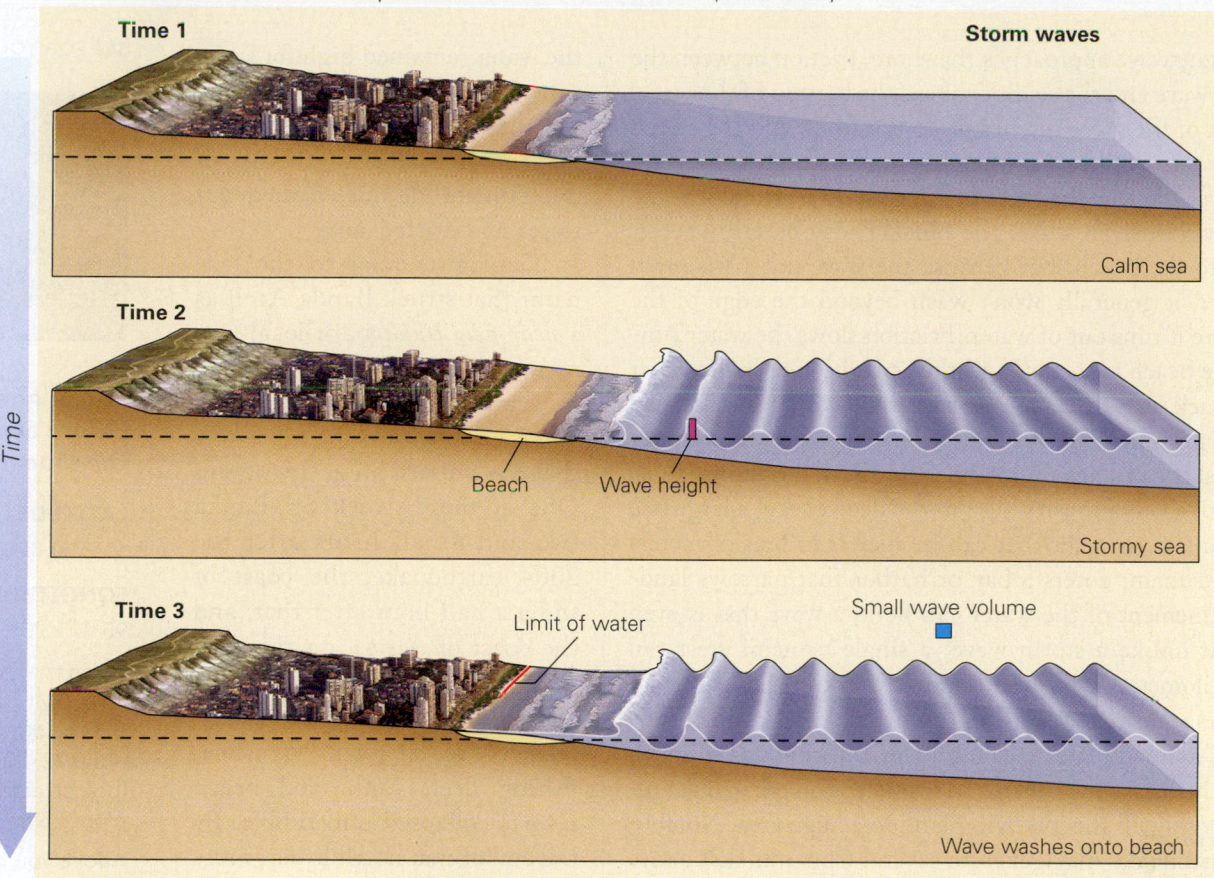

Storm waves

Time 1

Calm sea

Time 2

Beach
Wave height

Stormy sea

Time 3

Limit of water

Small wave volume

Wave washes onto beach

Time

(a) Storm waves can be high, but because they have short wavelengths, they contain relatively little water. Most waves runs out of water by the upslope edge of the beach.

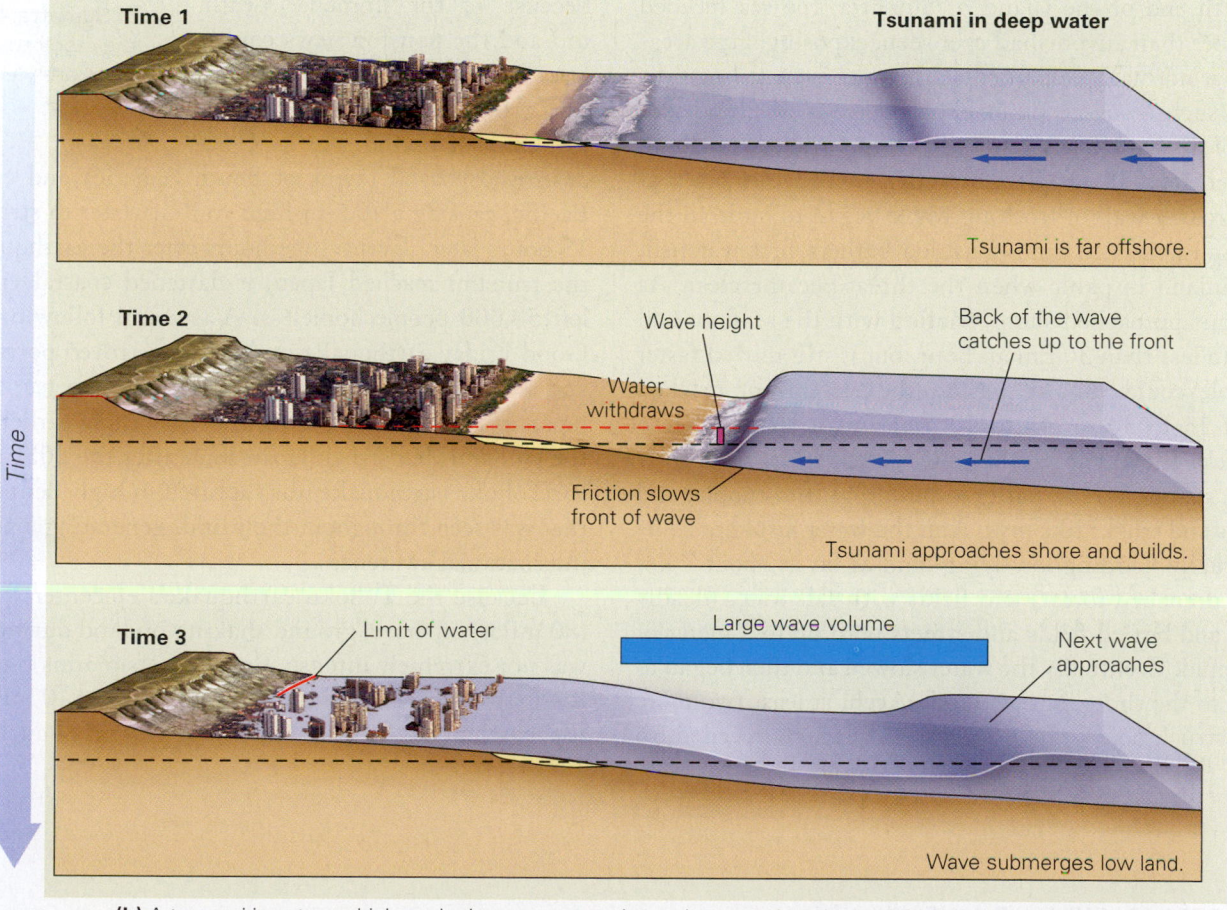

Tsunami in deep water

Time 1

Tsunami is far offshore.

Time 2

Wave height

Water withdraws

Friction slows front of wave

Back of the wave catches up to the front

Tsunami approaches shore and builds.

Time 3

Limit of water

Large wave volume

Next wave approaches

Wave submerges low land.

Time

(b) A tsunami is not very high out in the open ocean, but as it approaches the land, friction slows the front of the wave, so the rear catches up and the wave grows. It is so wide that it can cover a wide area of low-lying land.

When any wave approaches the shore, friction between the base of the wave and the seafloor slows the bottom of the wave, so the back of the wave catches up to the front, and the added volume of water builds the wave higher. The top of the wave may fall over the front of the wave and cause a breaker. In the case of a wind-driven wave, the breaker may be tall when it washes onto the beach, but because the wave doesn't contain much water, it generally won't wash beyond the edge of the beach before it runs out of water. Friction slows the water running up the beach to a stop, and then gravity causes the water to recede back seaward. As is the case for a storm wave, when a tsunami approaches the shore, friction slows it down so water farther offshore catches up to the water near the shore. Thus, even though a tsunami isn't high in the deep ocean, a large one can build into a monster that can be meters to tens of meters high. If a tsunami enters a bay or harbor that narrows landward, confinement of the water can build a wave that is even higher. But unlike a storm wave, a single tsunami crest can be many kilometers wide and hundreds of kilometers long. If you picture a storm wave as a narrow ridge, you can picture a tsunami as a broad plateau. Thus, a tsunami contains so much water that it crosses the beach and just keeps on going, eventually submerging all the low-lying land in a huge area. Notably, a single earthquake may generate several tsunamis that arrive on distant shores as much as an hour apart.

Tsunami damage can be beyond catastrophic (**Fig. 10.34**). When the December 2004 wave struck Banda Aceh, a city at the north end of the island of Sumatra, the sea receded much farther than anyone had ever seen, exposing large areas of reefs that normally remained submerged even at low tide. Although such a severe pullback of water is an important warning of an impending tsunami, people walked out onto the exposed reefs in wonder. But then, with a rumble that grew to a roar, a wall of frothing water began to build in the distance and approach land. Puzzled bathers first watched, then ran inland in panic when the threat became clear. As the tsunami approached shore, friction with the seafloor had slowed it to less than 30 km an hour, but it still moved faster than people could run. In places, the wave front reached heights of 15 to 30 m (45 to 100 feet) as it slammed into Banda Aceh. The impact of the water ripped boats from their moorings, snapped trees, battered buildings into rubble, and tossed cars and trucks like toys. And the water just kept coming, eventually flooding low-lying land as far as about 7 km inland. It drenched forests and fields with salt water (deadly to plants) and buried fields and streets with up to a meter of sand and mud. Eventually the water slowed and then began to rush back to the shore, but at Banda Aceh, at least two more tsunamis struck before the first one had entirely receded, so

the water remained high for some time. When the water level finally returned to normal, a jumble of flotsam, as well as the bodies of unfortunate victims, floated out to sea and drifted away.

Geologists refer to the tsunami that struck Banda Aceh as a *near-field tsunami*, or local tsunami, because of its proximity to the earthquake. A *far-field tsunami*, or distant tsunami, is one that has crossed an entire ocean. One of these struck Sri Lanka two and a half hours after the 2004 earthquake, the coast of India a half hour after that, and the coast of Africa, on the west side of the Indian Ocean, five and a half hours after the earthquake. Coastal towns vanished, fishing fleets sank, and beach resorts collapsed into rubble. By the end of that horrible day, more than 230,000 people had died.

The 2004 Indian Ocean event remains etched in people's minds because of the immense death toll and the nonstop news coverage. But it is not unique. Tsunamis generated by the magnitude 9.5 Chilean earthquake in 1960 destroyed coastal towns of South America and crossed the Pacific, causing a 10.7-m-high wall of water to strike Hawaii 15 hours later. Twenty-one hours after the earthquake, when the tsunami reached Japan, it flattened coastal villages and left 50,000 people homeless. A tsunami following the 1964 Good Friday earthquake in Alaska destroyed ports at Valdez and Kodiak (**Fig. 10.35**). And a devastating tsunami struck Japan in 2013, as we have seen. Unlike these earlier examples, the tsunami that struck Japan soon after the 2011 magnitude 9.0 Tōhoku earthquake was captured in high-definition video that was seen throughout the world, generating a new level of international awareness.

Because the Tōhoku earthquake's epicenter was 130 km (80 miles) offshore, ground shaking on land during the event was not extremely intense. But, since tsunamis travel so fast, the first waves reached the shore only about 10 minutes after the earthquake struck, and there was not much time for

FIGURE 10.34 The great Indian Ocean tsunami of 2004.

(a) A devastating tsunami was triggered by an earthquake off Sumatra. Three hours later, the leading wave struck the coasts of Sri Lanka and India. A computer model shows the wave.

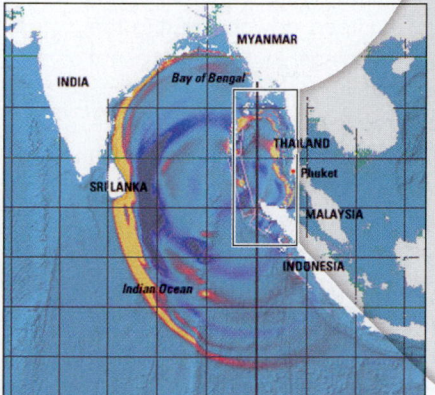

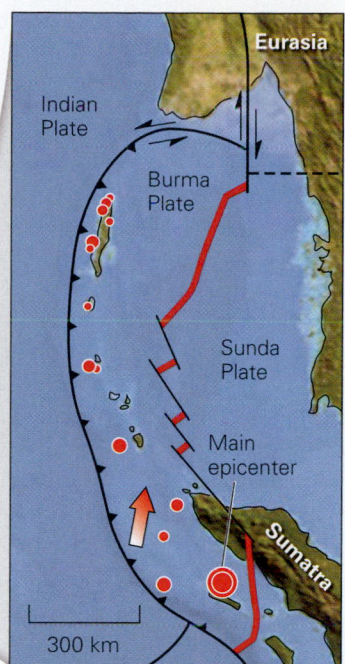

Colors represent wave height—yellow is highest. There were several waves.

The earthquake was caused by subduction at a trench. Red dots are epicenters.

(b) This snapshot shows the wave rushing toward the coast of Sumatra. Recession of water in advance of the wave exposed a reef.

(c) The wave blasts through a grove of palm trees as it strikes the coast of Thailand.

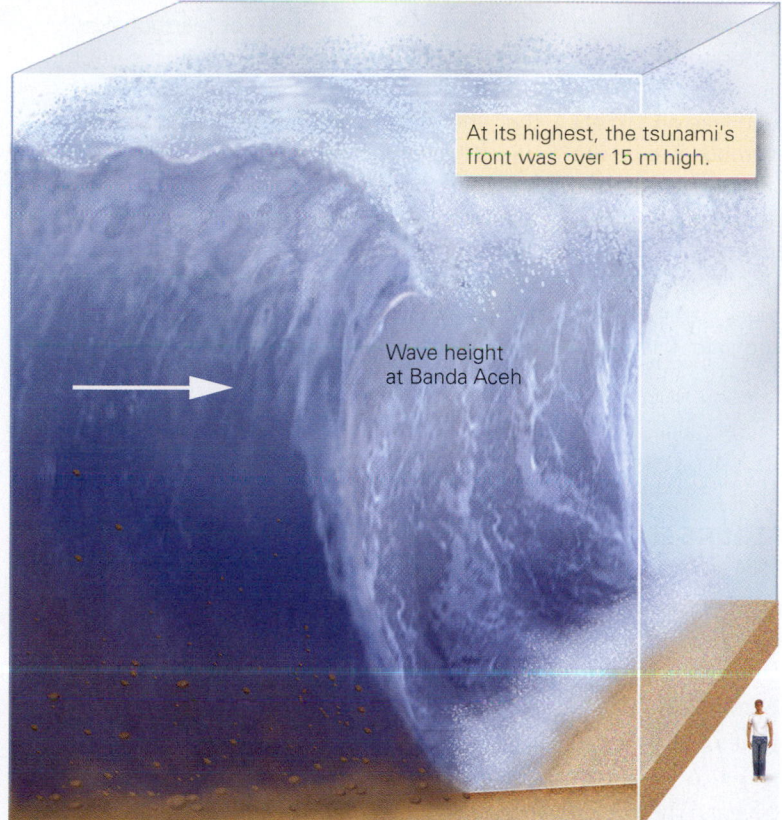

At its highest, the tsunami's front was over 15 m high.

Wave height at Banda Aceh

(d) Satellite photos of the Indonesian province of Aceh before and after the tsunami struck. Note that the city was washed away and the beach vanished.

BOX 10.1 CONSIDER THIS . . .

The 2010 Haiti Catastrophe

The history of Haiti changed forever on the sunny afternoon of January 12, 2010. At 4:53 P.M., a 70-km-long segment of a strike-slip fault suddenly slipped by an average amount of 1.8 m and, locally, by as much as 4 m. The motion began at a focus only 25 km west-southwest of Port-au-Prince, and 13 km beneath the ground surface, so the shock waves of the resulting magnitude 7 earthquake reached the capital city in a matter of seconds, causing the ground to lurch violently over a duration of 35 seconds. The event released as much energy as a 32-megaton nuclear weapon, 1,000 times the energy released by the Nagasaki atomic bomb.

Ground shaking caused insufficiently reinforced structures to crack and crumble (**Fig. Bx10.1a, b**). Unable to hold up the shifting weight of the building above, columns gave way, bringing floors down into pancake-like stacks. Similarly, brick or block walls broke apart, roads buckled, and hill slopes slumped. And beneath the harbor, sediment liquefied, causing the wharfs to sink into the sea and the giant crane used to unload ships to topple sideways. When the shaking, which reached an intensity of IX on the Mercalli scale (**Fig. Bx10.1c, d**), finally stopped, most of Port-au-Prince had collapsed. As a dense cloud of white dust slowly rose over the rubble, survivors began the frantic scramble to dig out victims, a task made more hazardous by aftershocks, of which there were over 50 with magnitudes between 4.5 and 6.1—the renewed shaking caused still-standing but weakened walls to collapse on rescuers. Sadly, only about 150 people were eventually pulled from the rubble. No one will ever know exactly how many people died during the earthquake or of injuries in the weeks afterward, but some estimates place the death toll at 230,000, about 2.5% of the country's population. More than a million people, in a country of 9 million, lost their homes.

Why did the earthquake occur? Haiti sits astride the transform-plate boundary along which the North American Plate moves westward at about 2 cm per year, relative to the Caribbean Plate. So earthquakes in Haiti are inevitable. Slip-accommodating plate motion in Haiti is divided among a few large faults; the southernmost of these accommodates about half the slip and was responsible for the disaster. The last major earthquakes on this plate boundary happened about 240 years ago, so stress has been building on the fault for quite some time.

The impact of an earthquake on society depends not only on its size but also on the nature of the substrate, on the steepness of the slopes, on construction practices, and on the quality of emergency services in the affected area. Much of Port-au-Prince was built on a basin of weak sediment—as earthquake waves passed into this basin from regions of harder bedrock, they were amplified and therefore caused particularly large ground movements. In addition, many of the city's neighborhoods perch on steep slopes, which slumped downhill during the quake. Sadly, buildings in Haiti were not designed to withstand ground vibration, and local emergency services were overwhelmed. The country did not have enough rescue workers, doctors, or supplies to cope with the catastrophe, and debris made streets impassable. To make things worse, the country has only a small airport, slowing down delivery of supplies by air, and the destruction of the port prevented ships from bringing relief. Continued poor sanitation contributed to the spread of cholera over the next three years.

What does the future hold? One possibility is that the earthquake may have released enough stress buildup that another large event will not happen for another century or two. (The last major earthquakes near Port-au-Prince took place in the 18th century.) But some seismologists worry that the 2010 earthquake might be the beginning of a chain of earthquakes along the plate boundary. Because nothing can stop the movement of plates, the safety of the region's inhabitants will depend on the strength of the new buildings to rise from the rubble and on efforts to reinforce buildings in other regions along the fault.

---

residents of coastal towns to escape when the warning sirens went off. The 10-m-high seawalls that fringe the coast were not high enough to stop the advance of the wave that, in places, built to a height of 30 m when it reached shore. The rising sea picked up boats and ships in the harbor and flung them over the seawalls and in some cases onto the roofs of buildings. It crossed the beach traveling at a speed of 30 km per hour, and once on land, it smashed through houses and tumbled cars as if they were pebbles. As the churning wave picked up dirt and debris, it became a viscous slurry, resembling a volcanic lahar, moving with such force that nothing could withstand its impact (see Fig. 10.2c). When the

FIGURE 10.35 A tsunami inundated the shore of Valdez, Alaska, in 1964 and washed away the snow cover. The town's port was destroyed.

FIGURE Bx10.1 The disastrous January 2010 earthquake in Haiti, and its geologic setting.

(a) Survivors salvage what they can in Port-au-Prince, the capital of Haiti, after the devastating earthquake of January 2010.

(b) Ground shaking during the 2010 Haiti earthquake caused most of the houses in this residential neighborhood to collapse.

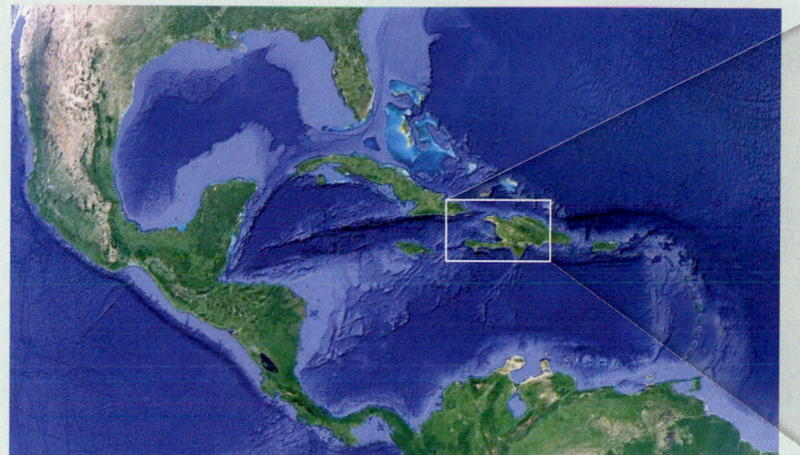

(c) The Caribbean Plate has complex boundaries, delineated by bathymetric features. The white rectangle shows the location of Haiti.

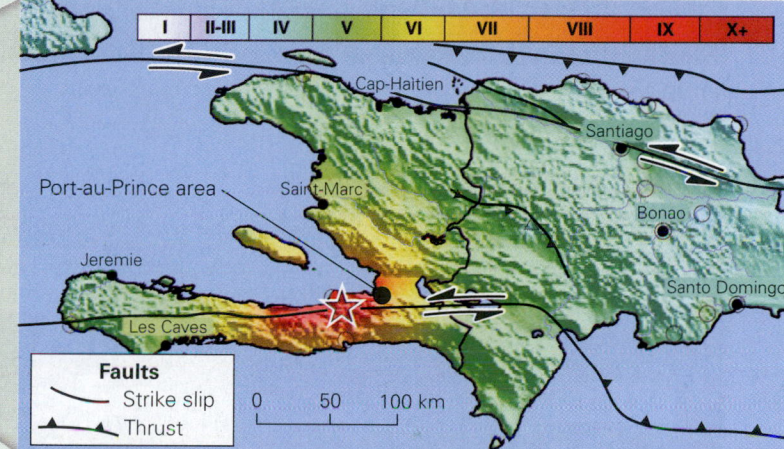

(d) A map showing the Mercalli intensity of shaking in Haiti. The star marks the epicenter of the quake.

wave finally ran out of water, it receded to the sea, carrying debris and victims with it and leaving behind a wasteland (**Fig. 10.36a**).

But the catastrophe was not over. The wave had also hit the Fukushima nuclear power plant. Though the plant had withstood ground shaking of the earthquake and had automatically shut down, its radioactive core still needed to be cooled by water in order to remain safe. The tsunami not only destroyed power lines, cutting the plant off from the electrical grid, but it also drowned the backup diesel generators, so the cooling pumps stopped functioning. Eventually, water surrounding the heat-producing radioactive core of the

reactors, as well as the water cooling spent fuel, boiled away. Some of the superheated water separated into hydrogen and oxygen gas, which then exploded, thereby breaching the integrity of the reactor structure and releasing radioactivity into the environment (**Fig. 10.36b**).

Because tsunamis are so dangerous, predicting their arrival can save thousands of lives. A tsunami warning center in Hawaii keeps track of earthquakes around the Pacific and uses data relayed from tide gauges and seafloor pressure gauges to determine whether a particular earthquake has generated a tsunami. If observers detect a tsunami, they flash warnings to authorities in communities around the Pacific.

FIGURE 10.36 Damage due to the 2011 Tōhoku tsunami.

(a) The complete destruction of a Japanese coastal town by a tsunami, which followed the Tōhoku, Japan, earthquake of March 2011.

Before

The power plant was built next to the shore.

Reactor building 4

After

Reactor building 4

(b) Each cubic building houses a reactor of the Fukushima nuclear power plant. The tsunami washed over the seawalls and inundated the plant to a depth of 14 m (46 ft), destroying power to the cooling pumps. Hydrogen explosions destroyed the reactor buildings.

Disease

Once the ground shaking and fires have stopped, disease may still threaten lives in an earthquake-damaged region. Earthquakes destroy housing, leaving victims exposed to the weather; sever water and sewer lines, contaminating clean-water supplies and exposing the public to bacteria; and cut transportation lines, preventing food and medicine from reaching the area. The severity of such problems depends on the ability of emergency services to cope (**Box 10.1**).

Take-Home Message

Earthquakes cause devastation in many ways. Ground shaking, landslides, sediment liquefaction, and tsunamis can topple buildings and disrupt the land. Fire and disease may follow.

QUICK QUESTION: Is ground shaking the major cause of loss of life in all earthquakes?

10.7 Can We Predict the "Big One"?

We have seen that large earthquakes occurring near population centers cause catastrophe. Needless to say, many lives could be saved if we could determine which regions are earthquake prone so that building codes could require stronger structures. Many more could be saved if we knew exactly when and where a specific earthquake will happen, so people could evacuate dangerous buildings and potentially unstable slopes. Even if we had seconds to minutes of advance warning, it might be enough to shut off gas lines, electrical circuits, and water valves. Can seismologists predict the next great earthquake, the proverbial "big one?" The answer depends on the time frame of the prediction. With our present understanding of the distribution of seismic zones and the frequency at which earthquakes occur, we can make *long-term predictions* (on the time scale of decades to centuries). For example, with some certainty, we can say that a major earthquake will probably rattle California during the next 100 years and that a major earthquake probably won't strike central Canada during the next 10 years. But despite extensive research, seismologists cannot make *short-term predictions* (on the time scale of hours to weeks or even years). Thus, we cannot say, for example, that an earthquake will happen in Montreal next month. But new technologies may permit warnings to be sent in advance of the arrival of seismic waves if an earthquake does happen.

In this section, we look at the scientific basis of long-term and short-term predictions and consider the consequences of a prediction. We also will introduce earthquake early warning systems. Seismologists refer to studies leading to predictions as seismic-risk assessment, or *seismic–hazard assessment*.

Long-Term Predictions

When making a prediction, we use the word *probability* because a prediction only gives the likelihood of an event. For example, a seismologist may say, "The probability of a major earthquake occurring in the next 20 years in this state is 20%." This sentence implies that there's a one-in-five chance that the earthquake will happen during the 20-year period. Urban planners and civil engineers can use long-term predictions to help create building codes for a region—codes requiring stronger, more expensive buildings make sense for regions with greater seismic risk. Planners may also use predictions to determine whether to build vulnerable structures such as nuclear power plants, hospitals, or dams in potentially seismic areas. Seismologists base long-term earthquake predictions on two pieces of information: the identification of **seismic zones** (places where earthquakes have happened fairly frequently in the past) and the **recurrence interval** (the average time between successive events) of earthquakes along a given fault.

To identify a seismic zone, seismologists produce a map showing the epicenters of earthquakes that have happened during a set period of time (say, 30 years). Clusters or belts of epicenters define the seismic zone. The basic premise of long-term earthquake prediction can be stated as follows: a region in which there have been many earthquakes in the past will likely experience more earthquakes in the future. Seismic zones, therefore, are regions of greater seismic risk. This doesn't mean that a disastrous earthquake can't happen far from a seismic zone—they can and do—but the risk that an event will happen in a given time window is less.

Epicenter maps can be produced with data from only the past 60 years or so, because before that time seismologists did not have enough seismometers to locate epicenters accurately.

Fortunately, geologists can also gain insight into seismic risk by examining landforms for evidence of recent faulting. For example, the presence of a distinct fault scarp in a landscape indicates that faulting has happened so recently that erosion has not yet had time to grind away the evidence (**Fig. 10.37**).

To determine the recurrence interval for large earthquakes at a location, seismologists must determine when large earthquakes happened at the location in the past—this type of research is called *paleoseismology*. Since the historical record does not provide information far enough back in time, they study geologic evidence for great earthquakes. For example, a

FIGURE 10.37 Identifying recent fault movement.

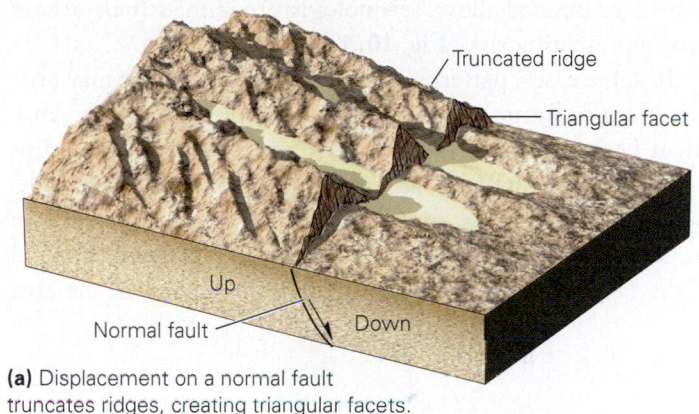

(a) Displacement on a normal fault truncates ridges, creating triangular facets.

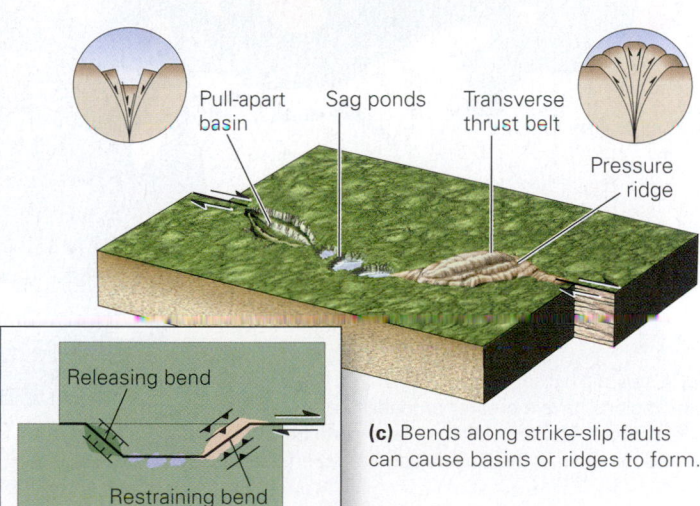

(b) A strike-slip fault may offset ridges and streams.

(c) Bends along strike-slip faults can cause basins or ridges to form.

trench cut into sedimentary strata near a fault may reveal layers of sand volcanoes and disrupted bedding in the stratigraphic record. Each layer, whose age can be determined by using radiocarbon dating of plant fragments, records the time of an earthquake (**Fig. 10.38**). By calculating the number of years between successive events and taking the average, seismologists obtain the recurrence interval. As an example, imagine that disrupted layers formed 260, 820, 1,200, 2,100, and 2,300 years ago. We can say that the recurrence interval between events is about 510 years. Note again that a recurrence interval does not specify the exact number of years between events, only the average number. Since stress builds up over time on a fault, the probability that an earthquake will happen in any given year probably increases as time passes. Information on a recurrence interval allows seismologists to refine *seismic-hazard maps* representing risk (**Fig. 10.39**).

In some cases, patterns of seismicity along a fault may provide clues to future seismicity. For example, the North Anatolian fault, a large strike-slip fault along which Turkey slips westward (**Fig. 10.40a**), has been the site of numerous earthquakes in historic time. Since 1939, 11 major earthquakes have occurred along the fault—each rupturing a different portion of the fault (**Fig. 10.40b**). By mapping the extent of the area

FIGURE 10.38 Evidence of paleoseismicity, used to determine recurrence interval. The block shows the walls of two trenches cut into the ground at a fault zone.

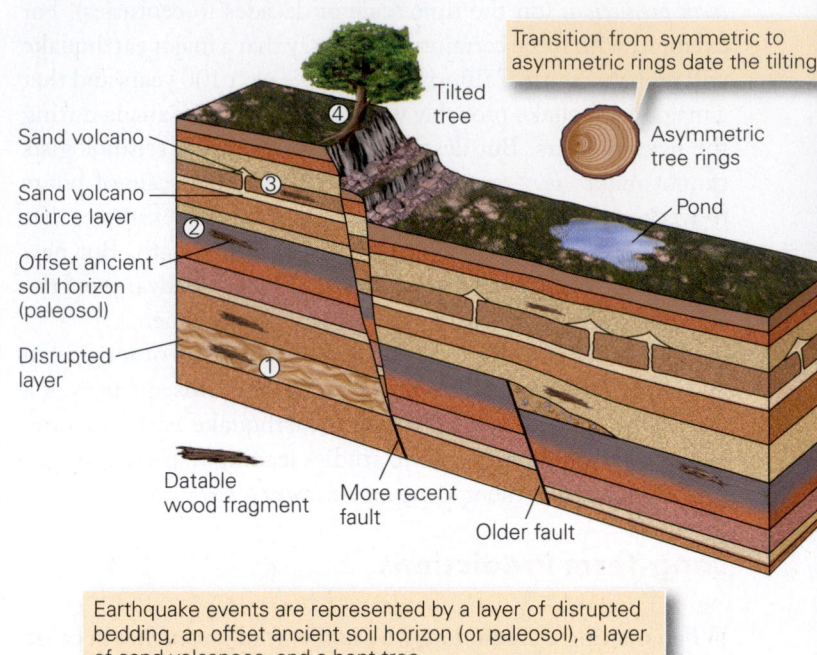

Transition from symmetric to asymmetric rings date the tilting.

Tilted tree

Asymmetric tree rings

Sand volcano

Pond

Sand volcano source layer

Offset ancient soil horizon (paleosol)

Disrupted layer

Datable wood fragment

More recent fault

Older fault

Earthquake events are represented by a layer of disrupted bedding, an offset ancient soil horizon (or paleosol), a layer of sand volcanoes, and a bent tree.

FIGURE 10.39 Examples of seismic-hazard maps.

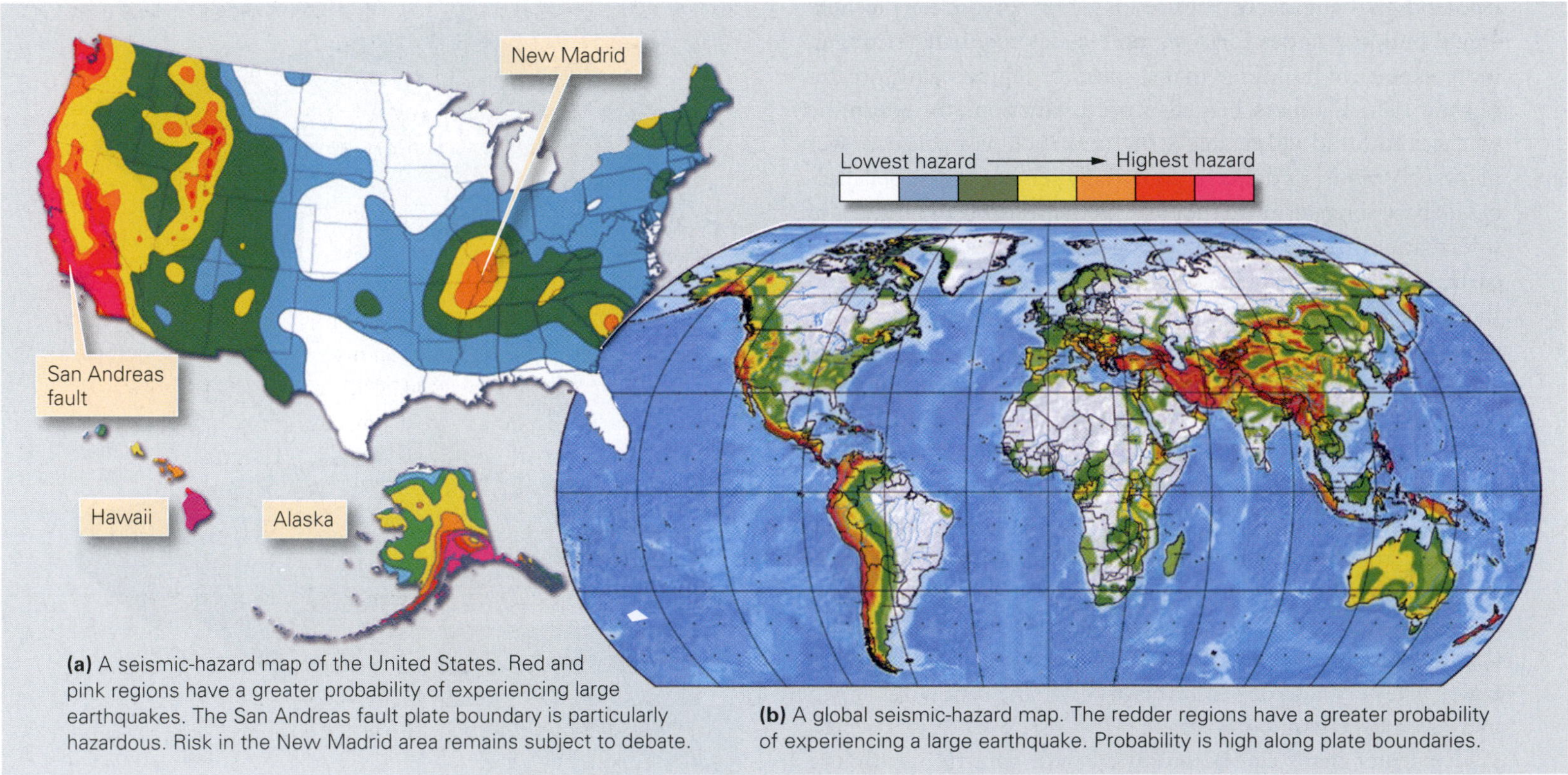

New Madrid

San Andreas fault

Hawaii

Alaska

Lowest hazard ——→ Highest hazard

(a) A seismic-hazard map of the United States. Red and pink regions have a greater probability of experiencing large earthquakes. The San Andreas fault plate boundary is particularly hazardous. Risk in the New Madrid area remains subject to debate.

(b) A global seismic-hazard map. The redder regions have a greater probability of experiencing a large earthquake. Probability is high along plate boundaries.

FIGURE 10.40 Major earthquakes of the last century along the Anatolian fault of Turkey have occurred roughly in sequence from east to west.

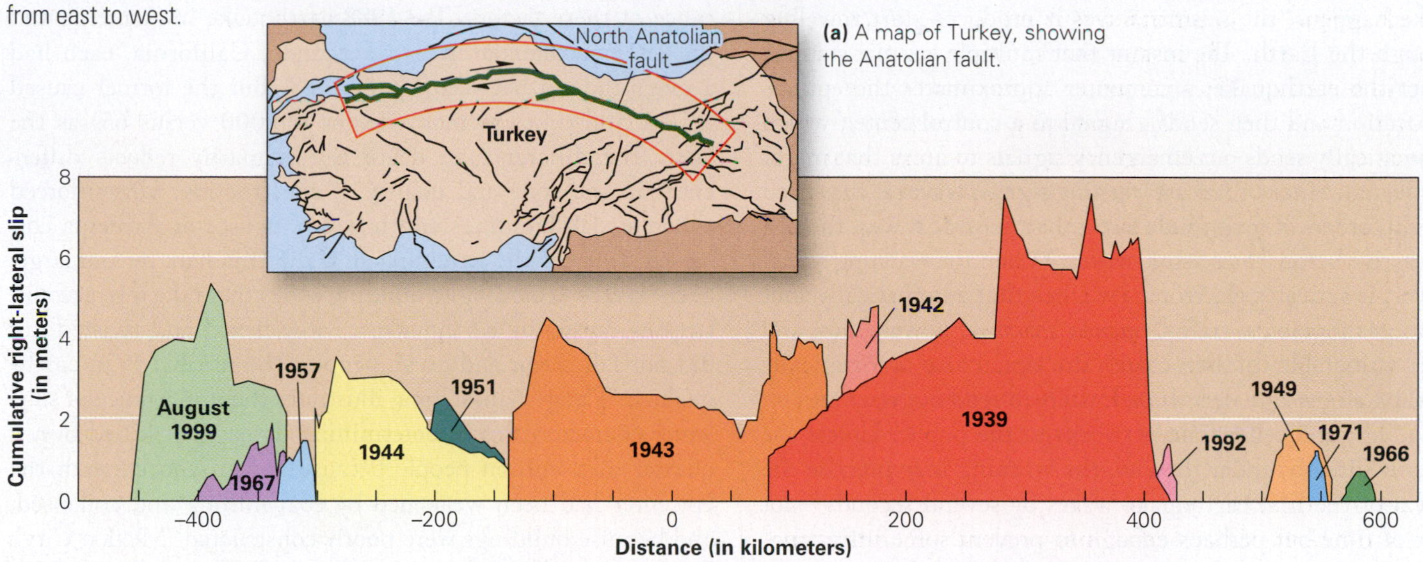

(a) A map of Turkey, showing the Anatolian fault.

(b) A graph representing regions of the fault. The vertical axis represents the amount of slip, and the horizontal axis represents location along the fault.

that slipped during each earthquake, seismologists recognize a general westward progression in the faulting and thus predict that the next major earthquake on the North Anatolian fault has a higher probability of happening at the west end of the fault than elsewhere.

Some seismologists suspect that places called *seismic gaps*, where a known active fault has not slipped for a long time, may be particularly dangerous. In a seismic gap, either the fault moves nonseismically or stress is building up to be released by a major earthquake at some time in the future.

Short-Term Predictions

Short-term predictions, which could lead to such precautions as evacuating dangerous buildings, shutting off gas and electricity, and readying emergency services, are not and may never be reliable. Seismologists have explored a number of possible clues to imminent earthquakes, but none have yet led to an accurate prediction, and often, possible clues can be recognized only in hindsight. For example, since rocks start to crack before a throughgoing rupture forms and slips, recognition of a "swarm" (a cluster of events during a short period of time) of foreshocks may be a clue. But foreshocks do not always occur, and even if they do, they may be indistinguishable from other small earthquakes.

Similarly, since the crust elastically deforms prior to seismic slip (according to the elastic-rebound theory) and the local volume of rock may increase as open cracks start to develop, precise surveying of the ground (using lasers or InSAR) may detect upwarping or downwarping of the land surface, which

may hint at an upcoming earthquake. But again, such warping may be recognized only in hindsight. Recently, geologists have begun to use computer models of stress to predict where stress buildups may lead to earthquakes, but these models have not yet been proven to be a reliable predictor of events.

Other changes that have been explored but have not been confirmed as precursors of earthquakes include changes in the water level in wells; appearance of gases, such as radon or helium, in wells; changes in the electrical conductivity of rock underground; and unusual animal behavior. Believers in these proposed clues suggest that they all reflect the occurrence of cracking in the crust prior to an earthquake, but most investigators remain skeptical.

As long as short-term predictions remain questionable, emergency service planners must ask, What should be done if someone makes a prediction? Should schools and offices be shut? Should millions of dollars be spent to evacuate people and leave cities open to looters? Should the public be notified, or should only officials be notified, creating a potential for rumor? If the prediction proves wrong, can seismologists be sued? No one knows the answers to such questions.

Earthquake Early Warning Systems

Even though seismologists cannot provide weeks to years of advance notice that an earthquake will happen, they have successfully developed an *earthquake warning system* in locations where there are many seismic stations so that some can detect the earthquake before the seismic waves have had time to reach populated areas.

An early warning system works as follows. When an earthquake happens, the seismic waves it produces start traveling through the Earth. The instant that multiple seismic stations detect the earthquake, a computer approximates the epicenter location and then sends a signal to a control center, which automatically sends out emergency signals to areas that might be affected. Since broadcast warning signals travel at the speed of light, orders of magnitude faster than seismic waves, the signals arrive before the seismic waves. When the warning signal arrives, it activates electronic switches that automatically shut down gas pipelines, trains, nuclear reactors, power lines, and other vulnerable infrastructure. The signal also automatically activates sirens and alerts broadcasters to send out warnings on radio, TV, and cell-phone networks to the public. Unless the focus is directly under the city, the warning may precede the arrival of the first earthquake waves by several seconds—not a lot of time but perhaps enough to prevent some infrastructure damage and to allow people to seek a safer location. Some earthquake-prone regions such as Japan and California have already installed commercial earthquake warning systems.

Take-Home Message

Researchers can determine regions where earthquakes are more likely and can estimate the recurrence interval (average time) between successive major earthquakes. Seismic hazard is greater where seismicity has happened more frequently in the past. It's not possible to predict exactly when and where an event will occur. Early warning systems can, however, provide seconds of warning by sending out signals that travel faster than seismic waves.

QUICK QUESTION: What is the relation between recurrence interval and the likelihood that an earthquake will happen in a given location in a given year?

10.8 Earthquake Engineering and Zoning

The destruction from an earthquake of a given size varies widely and depends on a number of factors. The most important factors include the size of the earthquake, the proximity of the epicenter to a population center, the depth of the focus, the style of construction in the epicentral region, whether the earthquake occurred in a region of steep slopes or along the coast, whether building foundations are on solid bedrock or on weak substrate, whether the earthquake happened when people were outside or inside, and whether the government was able to provide emergency services promptly.

A comparison of two earthquakes illustrates the significance of these factors. The 1988 earthquake in Armenia and the 1971 earthquake in San Fernando, California, each had a magnitude of between 6.6 and 6.8. But the former caused almost 500 times as many deaths (24,000 versus 65) as the later. The difference in death toll primarily reflects differences in the style and quality of construction. Unreinforced concrete-slab buildings and masonry houses of Armenia collapsed, whereas the structures in California had, by and large, been erected according to building codes that take into account stresses caused by earthquakes. Most flexed and twisted but did not fall down and crush people. The terrible 1976 earthquake in T'ang-shan, China, illustrates the importance of substrate characteristics in determining damage. It killed over a quarter of a million people because the ground beneath the epicenter had been weakened by coal mining and collapsed, and because buildings were poorly constructed. Mexico City's 1985 earthquake proved disastrous because the city lies over a sedimentary basin whose composition and bowl-like shape focused seismic energy, and the vibrations had a frequency that caused certain buildings to resonate (**Box 10.2**). During the 1989 Loma Prieta quake in California, portions of Route 880 in Oakland that were built on a weak substrate collapsed, whereas portions built on bedrock remained standing.

Communities can mitigate, or diminish, the consequences of earthquakes by taking sensible precautions. Clearly, *earthquake engineering* (the designing of buildings that can withstand shaking) and earthquake zoning (the determination of where land is stable and where it might collapse) can help save lives and property. In regions prone to large earthquakes, buildings and bridges should be constructed so they are able to withstand vibrations without collapsing (**Fig. 10.41a, b**). They should be somewhat flexible so that ground motions can't crack them, but they should have sufficient bracing so movements don't become too severe. Also, supports should be strong enough to maintain loads far in excess of the loads caused by their static (nonmoving) weight. Wrapping steel cables around bridge support columns makes them many times stronger. Bolting the bridge spans to the top of a column prevents the spans from bouncing off. Bolting buildings to foundations keeps them in place, and adding diagonal braces to frames keeps them from twisting and shearing too much. Earthquake construction techniques have been improved greatly through the use of *shaking tables*, platforms that shake in response to computer-activated pistons; shaking tables have facilitated the testing of different building designs.

Certain kinds of construction should be avoided in seismic zones. For example, concrete-block, unreinforced-concrete, and brick buildings crack and tumble under conditions in which wood-frame, steel-girder, or reinforced-concrete buildings remain standing. Traditional heavy, brittle tile roofs shatter and

BOX 10.2 CONSIDER THIS . . .

When Earthquake Waves Resonate—Beware!

When you shine a flashlight, you produce white light. If you aim that beam at a prism, the light spreads into a spectrum of different colors, with each color representing light waves of a different frequency. Thus, the original beam of white light contained all these frequencies. Similarly, when an earthquake occurs, it produces seismic waves with a variety of frequencies. As the waves travel away from the hypocenter, however, the Earth acts like a filter in that high-frequency waves (waves with short wavelengths) lose energy more rapidly than low-frequency waves (waves with long wavelengths). You've experienced this phenomenon if you've ever heard a car stereo playing loud rap music—when you stand near the car, you hear all the sound frequencies and can make out soprano voices and high-frequency guitar notes, but if you are far away, all you can hear are the low-frequency thump-thump-thumps of the bass guitar and bass drum. Because of this effect, the frequency content (the variety of wave frequencies) of the earthquake changes with distance from the hypocenter.

Why is frequency content important? Different frequencies of waves cause different amounts of ground acceleration. The frequency of waves is also important because of a phenomenon called resonance. **Resonance** happens when each new wave arrives at just the right time to add more energy to a system. To understand resonance, picture a boy on a swing. If the boy pumps his legs at just the right time, he swings higher; but if not, he slows down. The same phenomenon occurs if you slide a block of Jell-O that is resting on a plate back and forth on a table. If you move the plate too fast, the Jell-O merely trembles, but if your motion is at just the right frequency, resonance begins and the block sways wildly.

Resonance played a major role in accentuating the damage during an earthquake in Mexico City. On September 19, 1985, a magnitude 8.1 earthquake occurred 350 km away, on the convergent-plate boundary along which the Cocos Plate grinds beneath North America. Though the epicenter was far away, ground movements in Mexico City were very intense. That's because Mexico City sits on a thick sequence of unconsolidated lake-bed sediments, exposed when the Spanish conquistadors drained Lake Texcoco. The sedimentary basin, somewhat like a lens, focused earthquake energy on the city. Because of the distance from the epicenter, high-frequency waves had weakened, but because of the nature of the sedimentary basin, low-frequency waves were amplified. These waves had just the right frequency to make buildings between 8 and 18 stories high begin to resonate. In some cases, neighboring buildings slammed together like clapping hands. Engineers had not designed the buildings to accommodate such motion, so many buildings collapsed. Between 8,000 and 30,000 people died, and another 250,000 were left homeless.

---

bury the inhabitants inside, whereas sheet-metal or asphalt-shingle roofs do not. Loose decorative stone and huge open-span roofs also do not fare well when vibrated. Of note, inadequate structures can be made safer by *seismic retrofitting*, the process of strengthening existing buildings in potentially seismically hazardous areas. Examples of retrofitting include adding shock absorbers to foundations, adding braces, jacketing support columns, and coating masonry with resins. In some cases, substrates can be strengthened by draining water.

Similarly, developers should avoid construction on land underlain by weak sediment that could liquefy. They should not build on top of, on, or at the base of steep escarpments because the escarpments could fail and produce landslides, and they should avoid locating large population centers downstream of dams (which could crack and collapse, causing a flood). And they should also avoid constructing vulnerable buildings directly over active faults, because fault movement could crack and destroy the buildings. Cities in seismic zones need to draw up emergency plans to deal with disaster. Communication centers should be situated in safe localities, and strategies need to be implemented for providing supplies under circumstances where roads may be impassable. In coastal areas, tsunami warning systems need to be implemented (**Fig. 10.41c**).

Finally, individuals should learn to protect themselves during an earthquake. In your home, keep emergency supplies accessible, bolt bookshelves to walls, strap the water heater in place, install locking latches on cabinets, know how to shut off the gas and electricity, know how to find the exit, have a fire extinguisher handy, and know where to go to find family members. Schools and offices should have earthquake-preparedness drills. When an earthquake strikes, stay away from buildings. If you are trapped inside, a heavy table or solid door frame may provide protection (**Fig. 10.41d**). As long as lithosphere plates continue to move, earthquakes will continue to shake. But we can learn to live with them.

Take-Home Message

Earthquakes are a fact of life on this dynamic planet. People in regions facing high seismic risk should build on stable ground, avoid unstable slopes, and design construction that can survive shaking. Individuals should learn what to do in the event of an earthquake.

QUICK QUESTION: What factors influence the degree of devastation during an earthquake?

FIGURE 10.41 Preventing damage and injury during an earthquake.

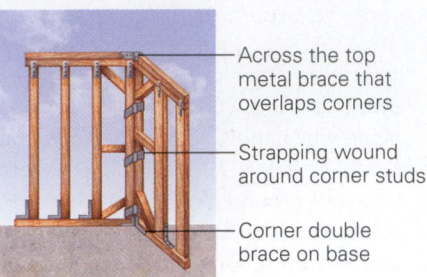

- Across the top metal brace that overlaps corners
- Strapping wound around corner studs
- Corner double brace on base

Adding corner struts, braces, and connectors can substantially strengthen a wood-frame house.

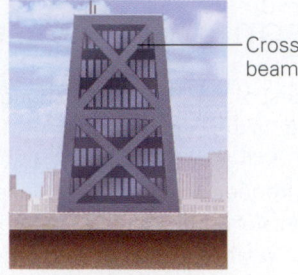

- Cross-beam

Buildings are less likely to collapse if they are wider at the base and if crossbeams are added for strength.

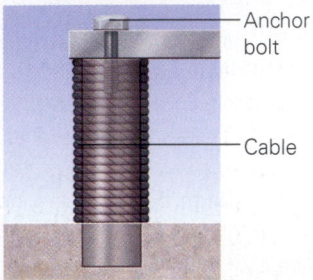

- Anchor bolt
- Cable

Wrapping a bridge's support columns in cable and bolting the span to the columns will prevent the bridge from collapsing so easily.

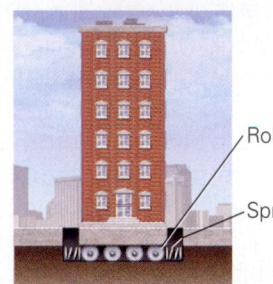

- Rollers
- Spring

Placing buildings on rollers or shock absorbers lessens the severity of the vibrations.

(a) Damage can be prevented if buildings are designed to withstand vibration.

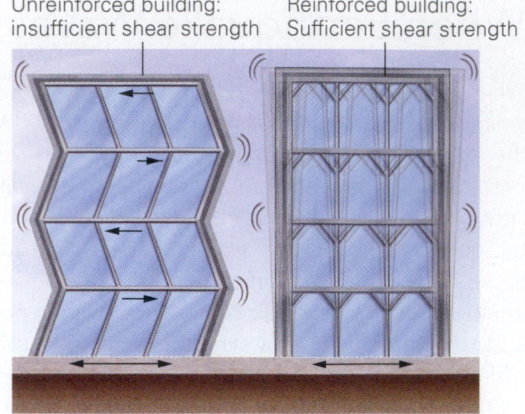

Unreinforced building: insufficient shear strength

Reinforced building: Sufficient shear strength

(b) An unreinforced building will shear side to side in a way that causes floors to shift out of alignment.

(c) Buoys can detect a tsunami in the open ocean, so people on land can be warned.

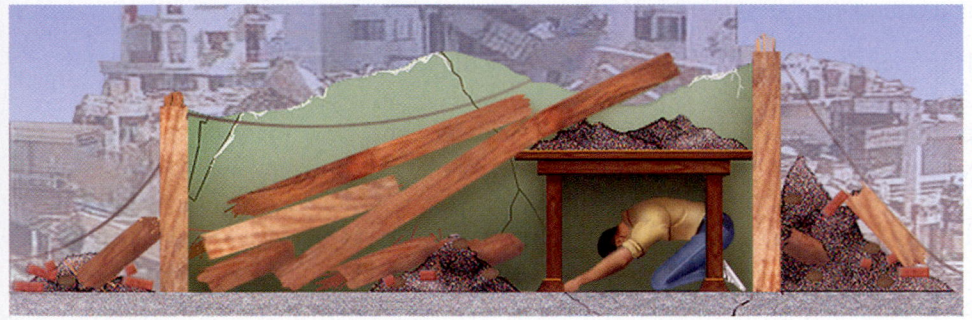

(d) If an earthquake strikes, take cover under a sturdy table near a wall.

CHAPTER SUMMARY

- Earthquakes are episodes of ground shaking. Earthquake activity is called seismicity.
- Most earthquakes happen when rock slips during faulting. The place where rock begins to break and release seismic energy is the focus (hypocenter) of an earthquake, and the point on the ground directly above the hypocenter is the epicenter.

- Active faults are faults on which movement is likely. Inactive faults ceased being active long ago but can still be recognized because of the displacement across them. Displacement on active faults that intersect the ground surface may yield a fault scarp.
- During fault formation, rock elastically bends, then cracks. Eventually, cracks link to form a throughgoing

rupture on which sliding occurs. When this happens, the rock breaks and vibrates, and this generates an earthquake.

- Once formed, faults exhibit stick-slip behavior in that they move in sudden increments. Most earthquakes happen when stress overcomes friction on a preexisting fault, and the fault slips again.
- Earthquake energy travels in the form of seismic waves. Body waves, which pass through the interior of the Earth, include P-waves and S-waves. Surface waves, which pass along the surface of the Earth, include R-waves and L-waves.
- We can detect earthquake waves by using a seismometer. In principle, a seismometer consists of a weight whose inertia keeps it in place, while the Earth around it moves.
- Seismograms demonstrate that different earthquake waves arrive at different times because they travel at different velocities. Using the difference between P-wave and S-wave arrival times, seismologists can pinpoint the epicenter location.
- The Mercalli Intensity Scale characterizes earthquake size by documenting human perception of ground shaking and of damage caused by an earthquake.
- Magnitude scales, such as the Richter scale, are based on measuring the amount of ground motion, as indicated by

traces of waves on a seismogram. The moment-magnitude scale takes into account the amount of slip, the length and depth of the rupture, and the strength of the ruptured rock.

- A magnitude 8 earthquake yields about 10 times as much ground motion as a magnitude 7 earthquake and releases about 32 times as much energy.
- Most earthquakes occur in seismic belts, the majority of which lie along plate boundaries. Intraplate earthquakes happen in the interior of plates. Different kinds of earthquakes happen at different kinds of plate boundaries.
- Earthquake damage results from ground shaking (which can topple buildings), landslides (set loose by vibration), sediment liquefaction (the transformation of sediment into a muddy slurry), fire, and tsunamis (water waves).
- Seismologists predict that earthquakes are more likely in seismic zones than elsewhere and can determine the recurrence interval (the average time between successive events) for great earthquakes. But it may never be possible to pinpoint the exact time and place at which an earthquake will happen.
- Earthquake hazards can be reduced with better construction practices and zoning, and by educating people about what to do during an earthquake.

GUIDE TERMS

aftershock (p. 319)
body wave (p. 323)
compressional wave (p. 323)
displacement (p. 316)
earthquake (p. 314)
elastic behavior (p. 317)
elastic-rebound theory (p. 319)
epicenter (p. 315)
fault (p. 313)
fault creep (p. 323)

fault scarp (p. 316)
fault trace (p. 316)
focus (p. 315)
foreshock (p. 319)
friction (p. 317)
induced seismicity (p. 337)
intensity (p. 328)
intraplate earthquake (p. 337)
magnitude (p. 330)

Modified Mercalli Scale (p. 328)
moment magnitude scale (p. 330)
recurrence interval (p. 351)
resonance (p. 355)
Richter scale (p. 330)
sediment liquefaction (p. 341)
seismic belt (p. 332)
seismic zone (p. 351)

seismicity (p. 315)
seismic wave (p. 323)
seismogram (p. 326)
seismologist (p. 314)
seismometer (p. 325)
shear wave (p. 323)
stick-slip behavior (p. 318)
stress (p. 316)
surface waves (p. 323)
Wadati-Benioff zone (p. 334)

REVIEW QUESTIONS

1. Compare normal, reverse, and strike-slip faults.
2. Describe elastic-rebound theory and the concept of stick-slip behavior.
3. What are the four types of seismic waves? Which are body waves, and which are surface waves?
4. Explain how the vertical and horizontal components of an earthquake motion are detected on a seismometer.

5. Explain the differences among the scales used to describe the size of an earthquake.
6. How does seismicity on mid-ocean ridges compare with seismicity at convergent or transform boundaries? Do all earthquakes occur at plate boundaries?
7. What is a Wadati-Benioff zone, and why was it important in understanding plate tectonics?

8. Describe the types of damage caused by earthquakes.

9. What is a tsunami, and why does it form?

10. Explain how liquefaction occurs in an earthquake and how it can cause damage.

11. How are long-term and short-term earthquake predictions made? What is the basis for determining a recurrence interval, and what does a recurrence interval mean?

12. What types of structure are most prone to collapse in an earthquake? What types are most resistant to collapse?

ON FURTHER THOUGHT

13. Is seismic risk greater in a town on the west coast of South America or in one on the east coast? Explain your answer.

14. The northeast-trending Ramapo fault crops out north of New York City near the east coast of the United States. Precambrian gneiss forms the hills to the northwest of the fault, and Mesozoic sedimentary rock underlies the lowlands to the southeast. (You can see the fault on *Google Earth*™ by going to Lat 41° 10′ 21.12″ N Long 74° 5′12.36″ W. Once you're there, tilt the image and fly northeast along the fault.) Where the fault crosses the Hudson River, there is an abrupt bend in the river. A nuclear power plant was built near this bend. Geologic studies suggest that the Ramapo fault first formed during the Precambrian, was reactivated during the Paleozoic, and was the site of major displacement during the Mesozoic rifting that separated North America from Africa. Imagine that you are a geologist with the task of determining the seismic risk of the fault. What evidence of present-day or past seismic activity could you look for?

15. On the seismogram of an earthquake recorded at a seismic station in Paris, France, the S-wave arrives six minutes after the P-wave. On the seismogram obtained by a station in Mumbai, India, for the same earthquake, the difference between the P-wave and S-wave arrival times is 4 minutes. Which station is closer to the epicenter? From the information provided, can you pinpoint the location of the epicenter? Explain.

 **smartwork**   smartwork.wwnorton.com

This chapter's Smartwork features:

- Interactive exercises on identifying features of an earthquake.
- Video questions on stratigraphic cross sections.
- Multipart problems on body waves and surface waves.

GEOTOURS

This chapter's GeoTour exercise (H) features:

- Earthquake activity along a plate boundary
- Intraplate earthquake activity
- Stress transfer and earthquake prediction
- Tsunami devastation

Another View   A model showing the predicted height of the tsunami in the Pacific Ocean generated by the 2011 Tōhoku earthquake.

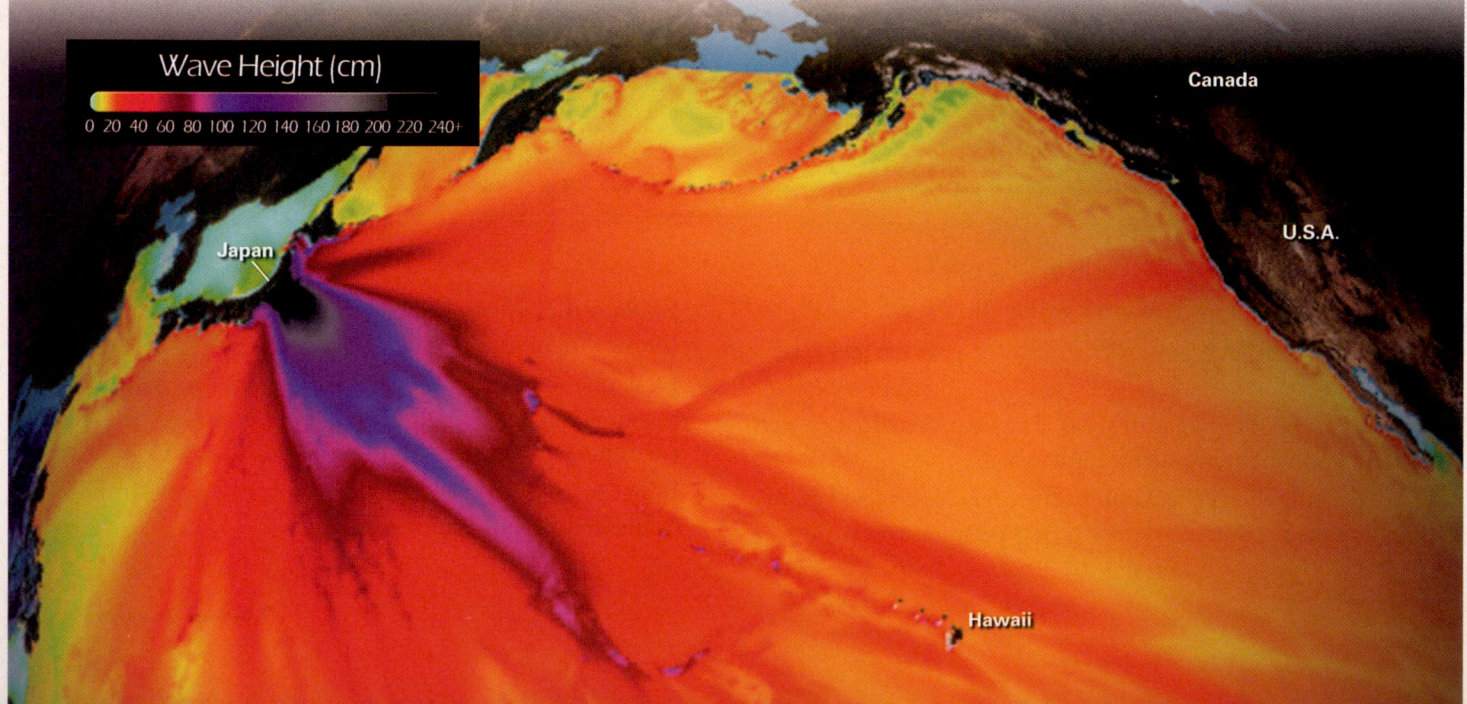

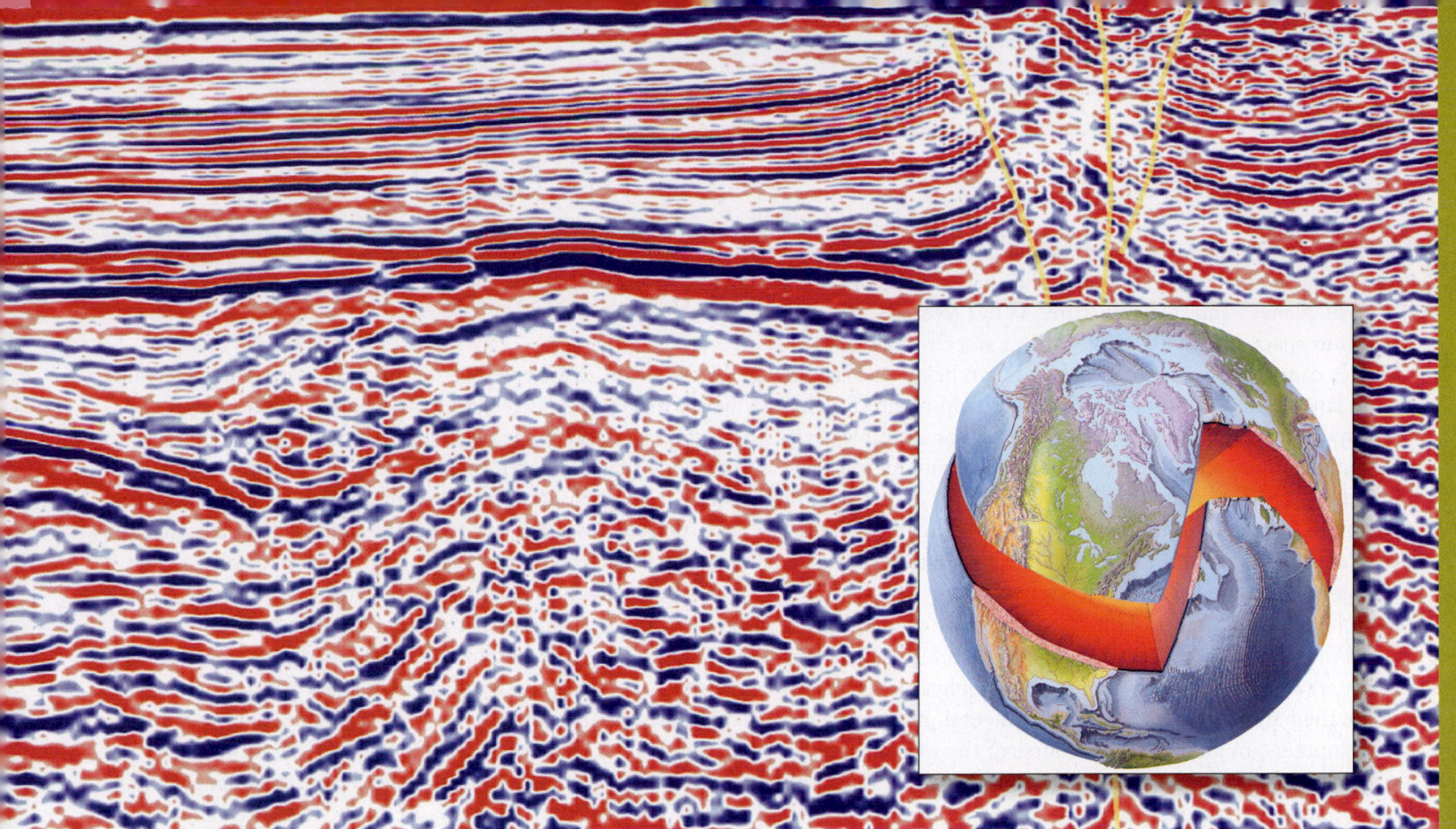

Using geophysical techniques, researchers obtain images of the Earth's interior. Here we see a seismic-reflection profile of sedimentary strata in the upper crust.

The Earth's Interior, Revisited: Seismic Layering, Gravity, and the Magnetic Field

LEARNING OBJECTIVES

By the end of this interlude, you should understand . . .

- how seismic waves behave as they pass through the Earth's interior.
- what the study of seismic waves can tell us about layering inside the Earth.
- that the lithosphere, overall, floats at an appropriate level to achieve isostasy.
- that Earth's gravitational attraction varies with location and what these variations mean.
- why Earth's magnetic field exists and why its strength varies with location.

D.1 Introduction

Strange as it may seem, researchers had an understanding of our Solar System's structure long before they had a clear concept of our own planet's internal structure. Why? We can see light years into space just by looking up. But since rock is opaque, our eyes cannot see even a centimeter down below the surface of the land. Tunnels and drillholes don't help much, for they literally only scratch the planet's surface. For example, even the deepest mine, a gold mine in South Africa which reaches a depth of 3.9 km below the surface, provides access to only the top 0.06% of the Earth, and the deepest drillhole, which penetrates 12.3 km into the crust beneath northwest Russia, samples only the top 0.19%. Thus, to study the Earth's interior, researchers have sought insight from the study of **geophysics**, the subdiscipline of geoscience that focuses on understanding seismic waves, gravity, and magnetism. Geophysicists typically use mathematical calculations, instrumental measurements, and computer simulations in the course of their work.

This interlude builds on concepts introduced earlier in this book (Chapters 2 through 6 and Chapter 10) to give a sense of what geophysics can tell us about the Earth's interior. We begin by examining how seismic waves interact with layer boundaries, in a general sense, and then discuss how an understanding of this interaction led first to the discovery of specific layer boundaries within the Earth and then later to an image of three-dimensional variations within each layer. We also discuss this planet's gravity field and examine why gravitational pull varies with location. This interlude concludes by reconsidering the Earth's magnetic field, this time with a focus on understanding why the magnetic field exists.

D.2 The Basis for Seismic Study of the Interior

Setting the Stage

As discussed in Chapter 2, the first clues used to determine what's inside our planet came from measurements of the Earth's overall mass and shape. These measurements led geologists to conclude that the Earth consists of three concentric layers that differ from each other in terms of density (**Fig. D.1a**). From the surface down, the layers are the crust, of low density; the mantle, of intermediate density; and the core, of high density.

To gain further insight into the nature of the crust, geologists examined samples of rocks exposed on land and samples dredged and drilled from the seafloor. To characterize the mantle, geologists analyzed rocks formed from magmas that originated in the mantle, as well as chunks of the mantle brought up in the magmas. Further insight came from the study of meteorites thought to have come from planetesimals, which had differentiated into a mantle and core before fragmenting. In fact, study of meteorites that came from the cores of planetesimals is the only way for us to "see" a core first hand.

Overall, this work led geologists to conclude that beneath a veneer of sediment, oceanic crust consists predominantly of mafic rock (basalt and gabbro), whereas continental crust consists of a variety of igneous and metamorphic rocks ranging from mafic to felsic in composition (**Fig. D.1b**). (In detail, the average chemical composition of the continental crust resembles that of granodiorite, a rock with a silica content partway between that of granite and diorite.) The mantle consists of ultramafic rock (peridotite), so its composition differs markedly from that of both kinds of crust. The core does not consist of rock at all but rather of a metallic iron alloy.

To go beyond the basic understanding we've just reviewed, researchers searched for a tool that could provide an actual image of the interior. Study of seismic waves provides that tool. By measuring how fast seismic waves travel through the Earth and how the waves bend and/or reflect as they travel,

FIGURE D.1 Simplified images of the Earth's interior.

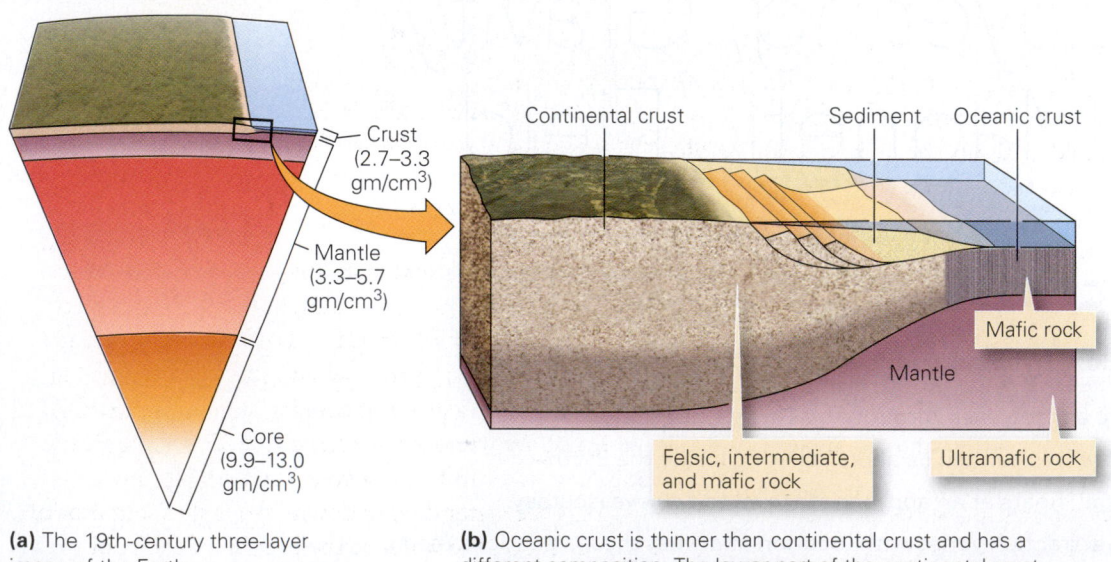

Crust (2.7–3.3 gm/cm³)

Mantle (3.3–5.7 gm/cm³)

Core (9.9–13.0 gm/cm³)

(a) The 19th-century three-layer image of the Earth.

Continental crust Sediment Oceanic crust

Mafic rock

Mantle

Felsic, intermediate, and mafic rock

Ultramafic rock

(b) Oceanic crust is thinner than continental crust and has a different composition. The lower part of the continental crust tends to be more mafic than the upper.

researchers could determine the exact depth to the crust-mantle boundary and to the mantle-core boundary, and they identified sublayers within the crust, mantle, and core, as we'll now see.

Seismic Wave Fronts and Travel Times

Energy travels from one location to another in the form of waves. For example, the impact of a pebble on the surface of a pond produces water waves that eventually cause a stick meters away to bob up and down, and the takeoff of a jet plane produces sound waves that travel through air and rattle windows of nearby houses. Similarly, the energy produced by a sudden rupture of intact rock or by the sudden frictional slip of rock on a fault produces seismic waves that travel through rock or metal in the Earth. These waves transmit energy outward from the earthquake's focus (hypocenter) in all directions at once, eventually rattling distant seismometers.

The boundary between the rock through which a wave has passed and the rock through which it has not yet passed is called a **wave front**. In three dimensions, a wave front expands outward from the earthquake focus like a growing bubble (**Fig. D.2a**). We can represent a succession of waves in a drawing by a series of concentric wave fronts. The changing position of an imaginary point on a wave front as the front moves through rock is called a **seismic ray**. Seismic rays are lines drawn perpendicular to wave fronts; each point on a curving wave front follows a slightly different ray. The time it takes for a seismic wave to travel from the focus to a seismometer along a given ray is the **travel time** along that ray. For example, though P-waves in rock travel about 10 to 25 times faster than sound waves in air, they take about 20 minutes to travel along the Earth's diameter from one side of the planet to the other. Thus, we can say that the travel time for a P-wave along this ray is 20 minutes. (See Chapter 10 for a description of the various types of seismic waves.)

The ability of a seismic wave to travel through a material, and the velocity at which it travels, depend on the character of the material. Factors such as *density* (mass per unit volume), *rigidity* (how stiff or resistant to bending a material is), and *compressibility* (how easily a material's volume changes in response to squashing) all affect seismic-wave velocity. As a result, seismic waves exhibit the following traits:

- Seismic waves travel at different velocities in different rock types (**Fig. D.2b**). For example, P-waves travel at 3.5 km per second in sandstone (a porous sedimentary rock) but at 8 km per second in peridotite (an ultramafic igneous rock).
- The velocity of seismic waves can change when the waves pass from one rock type into another.
- In general, seismic waves travel faster in a solid than in a liquid. For example, seismic waves travel more slowly in molten iron alloy than in solid iron alloy (**Fig. D.2c**).

FIGURE D.2 The propagation of earthquake waves.

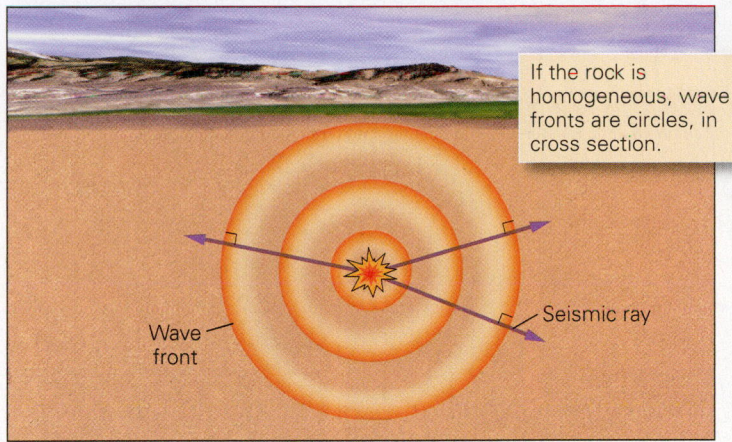

If the rock is homogeneous, wave fronts are circles, in cross section.

Wave front — Seismic ray

(a) An earthquake sends out waves in all directions. Seismic rays are perpendicular to wave fronts.

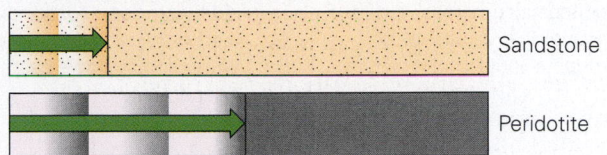

Sandstone

Peridotite

(b) Seismic waves travel at different velocities in different rock types. After a given time, the wave will have traveled farther in peridotite than in sandstone.

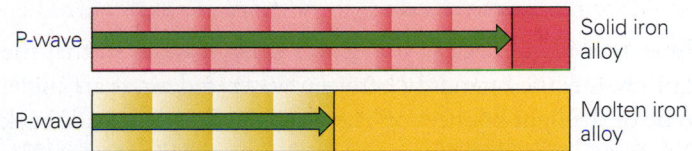

P-wave — Solid iron alloy

P-wave — Molten iron alloy

(c) P-waves travel faster in solid iron alloy than in liquid, such as molten iron alloy.

S-waves stop at this boundary.

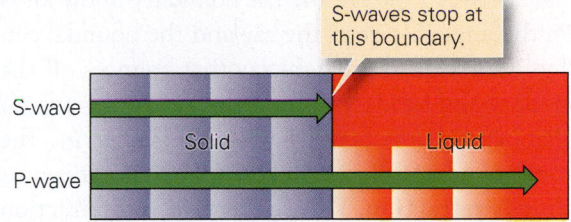

S-wave — Solid — Liquid

P-wave

(d) Both P-waves and S-waves can travel through a solid, but only P-waves can travel through a liquid.

- Both P-waves (compressional waves) and S-waves (shear waves) can travel through a solid, but only P-waves can travel through a liquid (**Fig. D.2d**). To see why, picture what happens if you push down on the water surface in a pool—you send a pulse of compression to the bottom of the pool. Now move your hand sideways through the water. The water in front of your hand simply slides or flows past the water deeper down—your shearing motion has no effect on the water at the bottom of the pool (**Fig. D.3**).

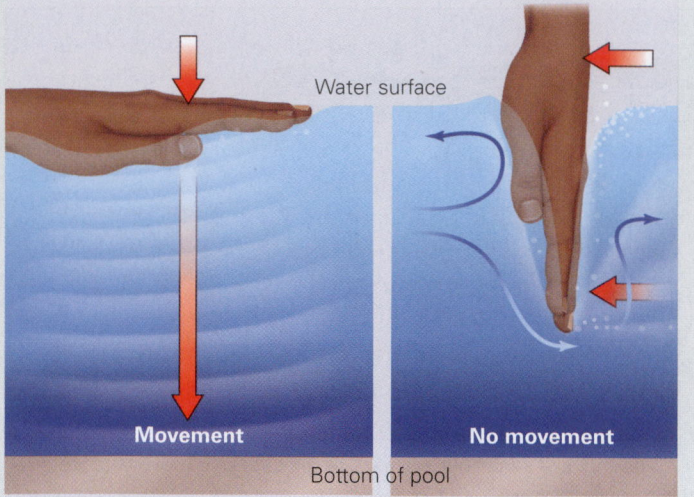

(a) Pushing down on a liquid produces a compressive pulse (P-wave) that can travel far through a liquid.

(b) Moving your hand sideways does *not* generate a shear wave; the moving water simply flows past deeper water.

Reflection and Refraction of Seismic Wave Energy

Shine a flashlight or laser into a container of water so that the light ray hits the boundary between water and air at an angle. Some of the light bounces off the water surface and heads back up into the air, while some enters the water (**Fig. D.4a**). The light ray that enters the water bends at the air-water boundary, so that the angle between the ray and the boundary in the air is different from the angle between the ray and the boundary in the water. Physicists refer to the light ray that bounces off the air-water boundary and heads back into the air as the *reflected ray*; the ray that bends at the boundary is the *refracted ray*. The phenomenon of bouncing off is **reflection**, and the phenomenon of bending is **refraction**. Wave reflection and refraction take place at the interface between two materials if the wave travels at different velocities in the two materials.

The angle at which a reflected wave bounces off a boundary is always that same as the angle at which the incoming, or incident, wave strikes the surface. The angle by which a refracted wave bends at a boundary, however, depends on the contrast in wave velocity between the two materials in contact at the boundary and on the angle at which a wave hits the interface. As a rule, if a wave enters a material through which it will travel more slowly, the ray representing the wave bends down and away from the interface (**Fig. D.4b**). For example, the light ray in Figure D.4b bends down when hitting the air-water boundary because light travels more slowly in water. (To see why, picture a car driving from a paved surface diagonally

FIGURE D.4 Refraction and reflection of waves.

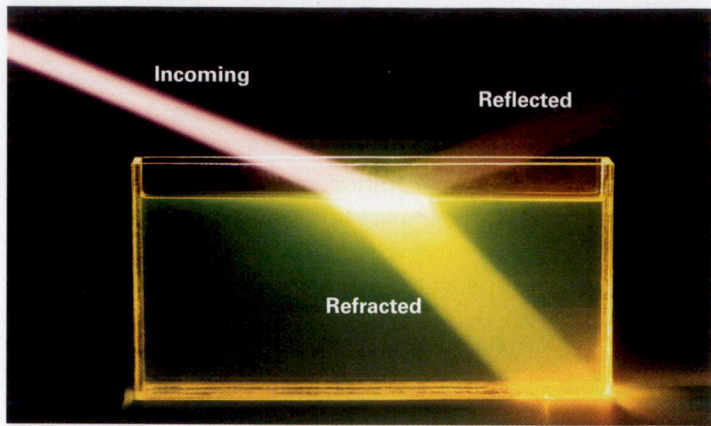

(a) A lab experiment showing how a ray of light reflects and partly refracts when it crosses the boundary between two different materials.

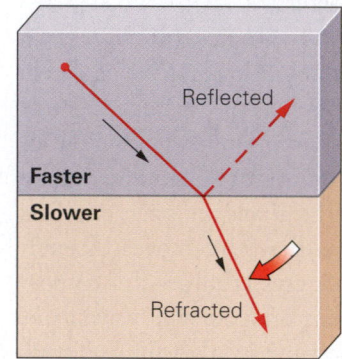
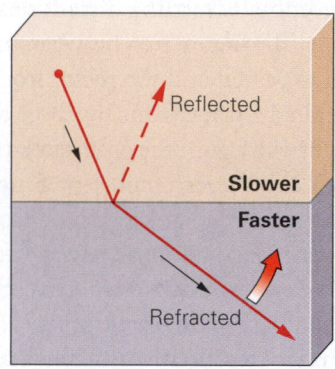

(b) A ray that enters a material through which it travels more slowly bends away from the boundary. A ray that enters a faster medium bends toward the boundary.

onto a sandy beach—the wheel that rolls onto the sand first slows down relative to the wheel still on the pavement, causing the car to turn toward the sand.) Alternatively, if the ray were to pass from a layer in which it travels slowly into one in which it travels more rapidly, the rays representing the waves would bend up and toward the interface (Fig. D.4b).

D.3 Results from Seismic Study of Earth's Interior

Let's now utilize your knowledge of seismic velocity, refraction, and reflection to see how each of the major layer boundaries inside the Earth was discovered.

Discovering the Crust-Mantle Boundary

The concept that seismic waves refract at boundaries between different layers led to the identification of the core-mantle boundary's depth. In 1909, Andrija Mohorovičić, a Croatian seismologist, noted that P-waves arriving at seismometers less than 200 km from the epicenter traveled at an average speed of 6 km per second, whereas P-waves arriving at seismometers more than 200 km from the epicenter traveled at an average speed of 8 km per second. To explain this observation, he suggested that P-waves reaching nearby seismometers followed a shallow path through the crust, in which they traveled relatively slowly, whereas P-waves reaching distant seismometers followed a deeper path through the mantle, in which they traveled relatively rapidly.

To understand Mohorovičić's proposal, examine **Figure D.5a**, which shows P-waves, depicted as rays, generated by an earthquake in the crust. Ray C, the shallower wave, travels through the crust directly to a seismometer. Ray M, the deeper wave, heads downward, refracts at the crust-mantle boundary, curves through the mantle, refracts again at the crust-mantle boundary, and then proceeds through the crust up to the seismometer. At seismometers less than 200 km from the epicenter, Ray C arrives first, because it has a shorter distance to travel. But at seismometers more than 200 km from the epicenter (**Fig. D.5b**), Ray M arrives first, even though it has farther to go, because it travels faster for much of its length. Calculations based on this observation require the crust-mantle boundary beneath most continental regions to be at a depth of about 35 to 50 km. Later studies showed that the crust-mantle boundary beneath the ocean floor lies at a depth of 7 to 10 km, and beneath some mountain belts it reaches a depth of 70 km. As we learned in Chapter 2, the crust-mantle boundary is now called the **Moho**, in honor of Mohorovičić.

Subsequent studies indicate that seismic velocities in the lower continental crust are faster than those in the upper continental crust. To interpret these observations, geologists have measured velocities in rock samples of various compositions under laboratory conditions. Such experiments suggest that the upper crust has, on average, felsic to intermediate composition, whereas the lower continental crust has, on average, mafic composition, as indicated in Figure D.1b.

Defining the Structure of the Mantle

If the density, rigidity, and compressibility of mantle peridotite were exactly the same at all depths, seismic velocities would be the same everywhere in the mantle, and seismic rays passing through the mantle would be straight lines. But, by studying travel times, seismologists have determined that seismic waves travel at different velocities at different depths. Let's now look at these variations and how they affect seismic-ray paths.

Beneath oceanic crust, seismic velocity in the mantle increases down to a depth of about 100 km. Beneath 100 km, and continuing down to a depth of 200 km, seismic velocities become slower (**Fig. D.6a**). Seismologists propose that this **low-velocity zone (LVZ)** occurs because at the temperature and pressure conditions in this depth range mantle peridotite melts by 1% to 6%. The melt, a liquid, coats solid grains and fills voids between grains. Because seismic waves travel more slowly through liquids than through solids, the coatings of melt slow seismic waves down. In the context of plate tectonics theory, the LVZ represents the upper part of the asthenosphere and is the weak layer on which oceanic lithosphere plates move. Seismologists do not find a distinct LVZ beneath continents, suggesting that there is no partial melt in the upper asthenosphere beneath continents.

Below the LVZ under ocean lithosphere, and throughout the mantle under continents, seismic-wave velocities in the mantle increase progressively with depth. Seismologists interpret this increase to mean that mantle peridotite becomes progressively less compressible, more rigid, and denser with depth. This proposal makes sense intuitively, considering that the weight of overlying rock increases with depth, and as pressure increases, the atoms making up rock squeeze together more tightly and are not so free to move. Because of refraction, the increase in seismic velocity with depth causes seismic rays to curve in the mantle when traveling a long distance (**Fig. D.6b**). So, at the scale of the mantle, the expanding "bubble" of a wave front is not spherical but rather is somewhat elliptical.

To understand the shape of a curved ray, imagine representing a portion of the mantle by a series of imaginary layers,

FIGURE D.5 Discovery of the Moho.

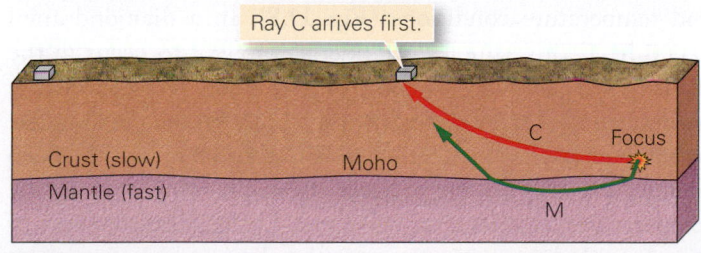

(a) Seismic waves traveling only in the crust reach a nearby seismometer first.

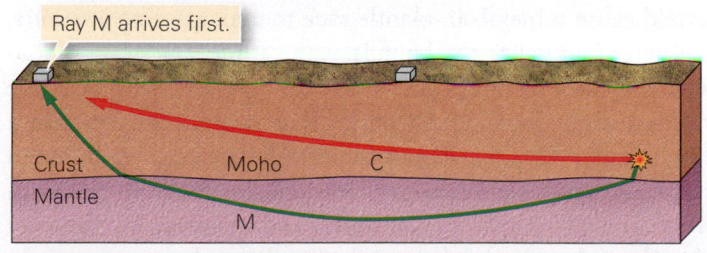

(b) Seismic waves traveling for most of their path in the mantle reach a distant seismometer first.

FIGURE D.6 The velocity of P-waves in the mantle changes because the physical properties of the mantle change with depth.

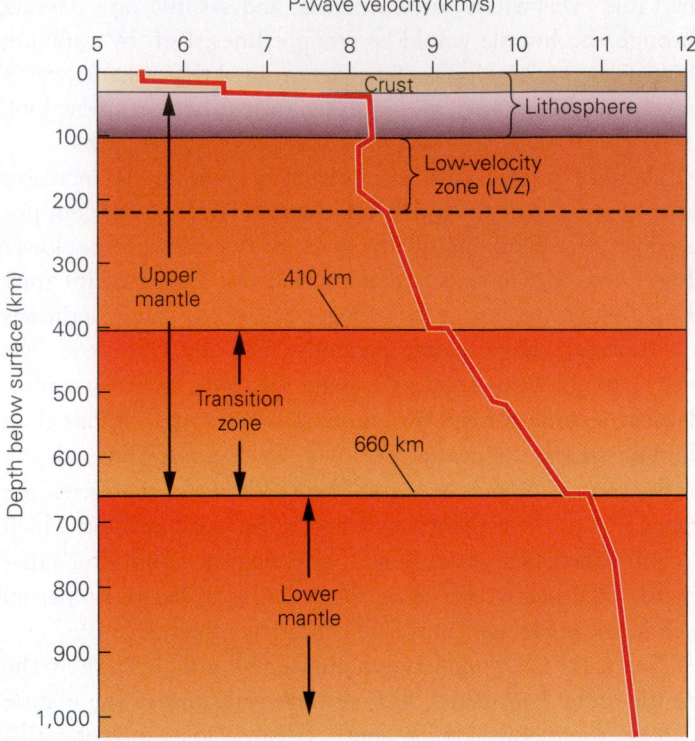

(a) The velocity of P-waves changes with depth in the mantle.

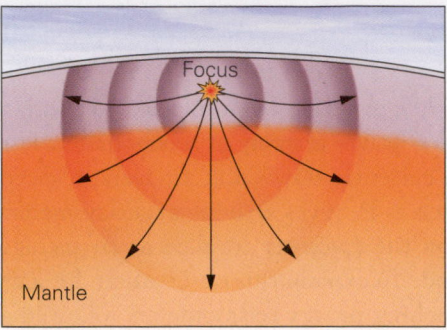

(b) The velocity of seismic waves increases with depth in the mantle, so rays curve and wave fronts are oblong.

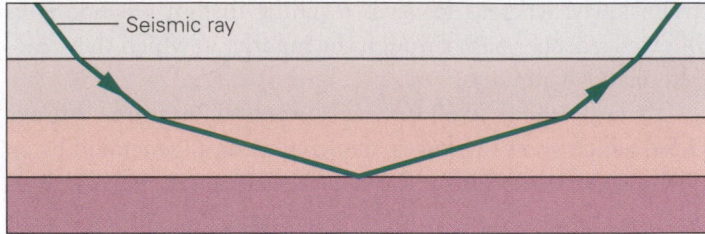

(c) In a stack of discrete layers, rays bend at each boundary. If the velocity is progressively faster in each lower layer, the ray eventually bends back and returns to the surface.

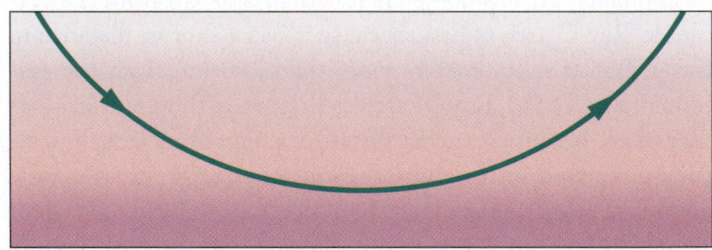

(d) In a material in which the velocity increases gradually with depth, rays curve smoothly and eventually return to the surface.

each of which has a slightly greater seismic-wave velocity than the layer above (**Fig. D.6c**). Every time a seismic ray crosses the boundary between adjacent layers, it refracts a little toward the boundary. After the ray has crossed several layers, it has bent so much that it begins to head back up toward the top of the stack. If we replace the stack of distinct layers with a single layer in which velocity increases with depth at a constant rate, the wave follows a smoothly curving path (**Fig. D.6d**).

As shown in Figure D.6a, at depths between 410 km and 660 km, seismic velocity increases in a series of abrupt steps, so in this interval, the stack of layers in Figure D.6c is actually a somewhat realistic image. Experiments suggest that these **seismic-velocity discontinuities** occur at depths where pressure would cause minerals in mantle rock to undergo a *phase change*, meaning that below the boundary, the atoms are crowded into a denser, more tightly packed configuration than they are above the boundary. This contrast means that, overall, the rock above the boundary has a different compressibility and rigidity than the rock below the boundary. Specifically, above the 410-km discontinuity, silicon, oxygen, iron, and magnesium, the most abundant elements of the mantle, bond together to form crystals of the mineral olivine, but below the discontinuity, the

atoms bond together to form crystals of a mineral called magnesium spinel. Magnesium spinel can be stable down to a depth of about 660 km. Below this depth, that atoms are bonded to form an even denser crystal structure known as perovskite. Because of these seismic-velocity discontinuities, seismologists subdivide the mantle into the **upper mantle**, above 660 km, and the **lower mantle**, below 660 km (**Fig. D.7**). The region between 410 km and 660 km, where several velocity discontinuities occur, is the mantle's **transition zone**.

At this point, you may be wondering whether there is any way to test ideas about the identity of minerals that occur deep in the mantle. The answer is yes. Researchers use a laboratory device called a *diamond anvil* to simulate deep-mantle pressure and temperature conditions (**Fig. D.8**). In a diamond-anvil experiment, tiny samples of minerals thought to occur in the mantle are squeezed between two diamonds. Diamond is the hardest mineral known, and the pressure between the two diamonds can reach values comparable to those found in the deep mantle. Electric heaters or lasers heat the sample to mantle temperatures as it is being squeezed. Using measurements of how laser light passing through the diamond changes as it interacts with sample, researchers can calculate seismic veloci-

FIGURE D.7 The layering of the mantle, drawn to scale, beneath the ocean floor. Note that the upper mantle is divided into several parts and that, overall, it is much thinner than the lower mantle.

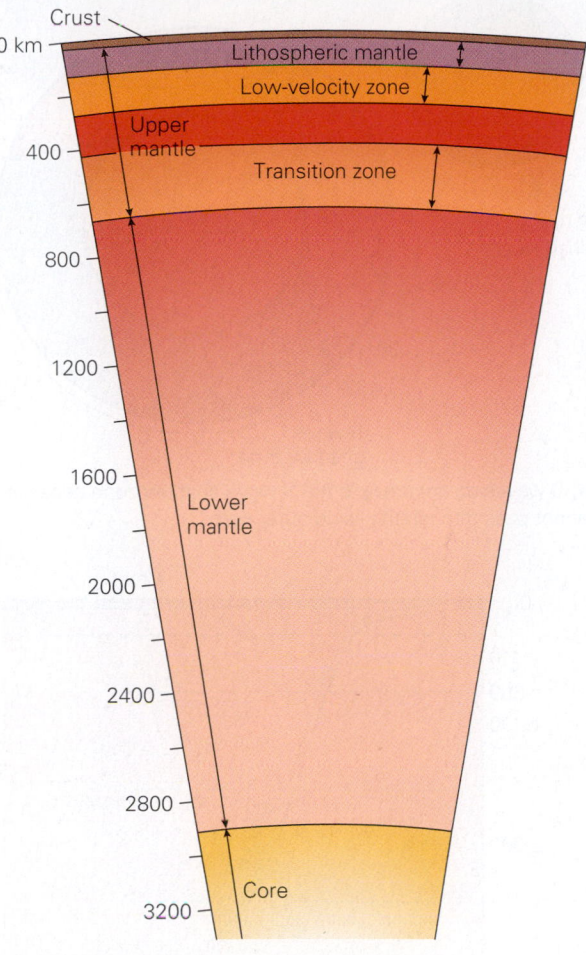

FIGURE D.8 A laboratory apparatus for studying the characteristics of minerals under very high pressures and temperatures. The green laser beam is probing a microscopic sample being squeezed between two diamonds hidden at the center of the metal cylinder.

the surface. Thus, the presence of a shadow zone means that deep in the Earth a major interface exists where seismic waves abruptly refract down, implying that the velocity of seismic waves suddenly decreases. This interface, now called the **core-mantle boundary**, lies at a depth of about 2,900 km.

To see why the P-wave shadow zone exists, follow the two seismic rays labeled A and B in Figure D.9a. Ray A curves smoothly in the mantle (we are ignoring seismic-velocity discontinuities in the mantle, for simplicity) and passes just above the core-mantle boundary before returning to the surface. It reaches the surface at 103° from the epicenter. In contrast, Ray B penetrates the boundary and refracts down into the core. Ray B then curves through the core and refracts again when it crosses back into the mantle. As a consequence, Ray B intersects the surface at more than 143° from the epicenter.

ties in the sample. By comparing the velocities measured in the laboratory with those observed for the real mantle, researchers gain insight into whether or not the laboratory sample could be a mantle mineral.

Discovering the Core-Mantle Boundary

During the first decade of the 20th century, seismologists installed seismometers at many stations around the world, expecting to be able to record waves produced by a large earthquake anywhere on Earth. In 1914, one of these seismologists, Beno Gutenberg, noticed that P waves from an earthquake do not arrive at seismometers lying in a band that lies at a distance of between 103° and 143° from the earthquakes epicenter, as measured along the surface of the Earth. This band is now called the **P-wave shadow zone** (**Fig. D.9a**). If the density of the Earth increased gradually with depth all the way to the center, the shadow zone would not exist because rays passing into the interior would curve up and reach every point on

Discovering the Nature of the Core

The iron alloy of the core contains about 85% iron, 5% nickel, and 10% of a lighter element (probably oxygen, silicon, and/ or sulfur). Notably, this iron alloy is much denser than mantle peridotite—in fact, the density contrast across the core-mantle boundary is greater than the density contrast between the Earth's crust and water! Seismological study provides insight into the physical state of iron alloy in the core. Specifically, the downward bending of P-waves when they pass from the mantle down into the core indicates that P-wave velocity, at least in the outer part of the core, is slower than in the mantle. Thus, even though the core is deeper and denser than the mantle, at least the outer part of the core must be less rigid

FIGURE D.9 Shadow zones and the discovery of the Earth's core. (Note that the circumference of a circle is 360°, so it is 180° from a given point to a locality on the other side of the planet.)

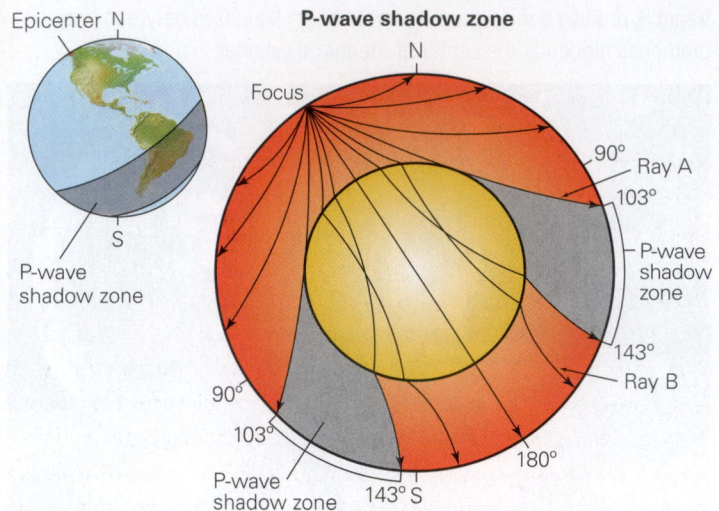

(a) P-waves do not arrive in the P-wave shadow zone because they refract at the core-mantle boundary.

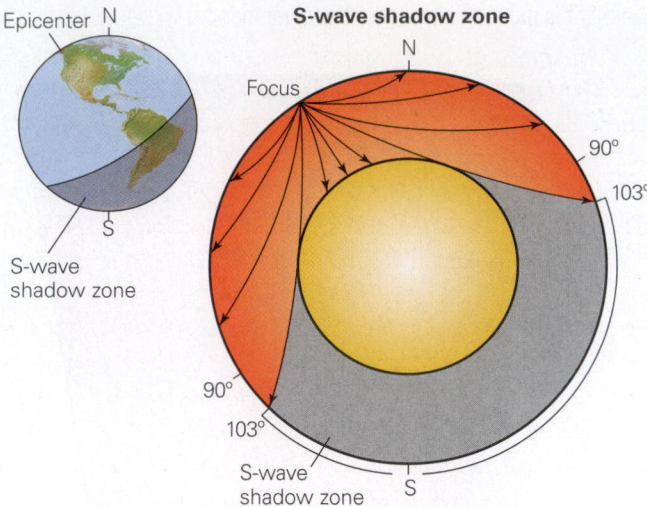

(b) S-waves do not arrive in the S-wave shadow zone because they cannot pass through the liquid outer core.

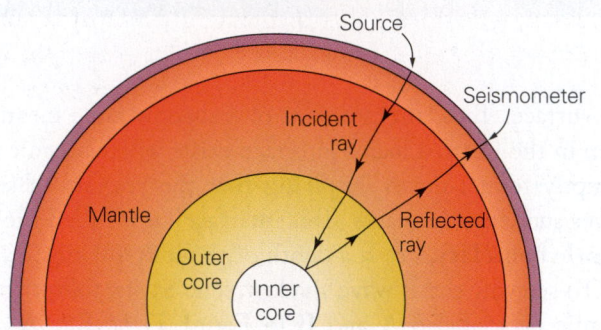

(c) Seismic waves reflect off the inner core–outer core boundary.

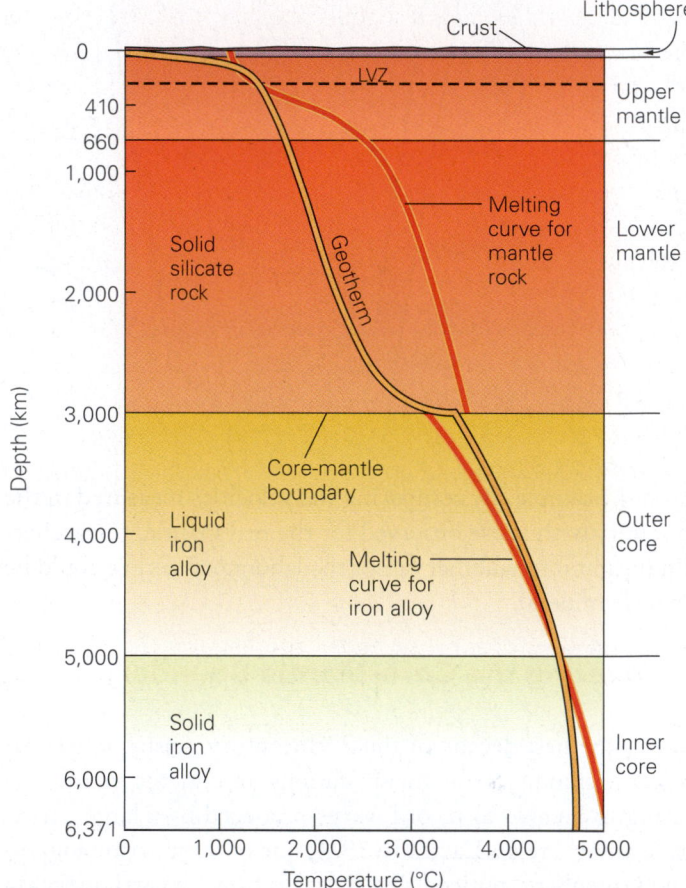

(d) A graph of the geotherm and melting curve for the Earth. Note that the melting temperature is less than the Earth's temperature in the outer core, so the outer core is molten.

than the mantle. How can this be? The answer to this question came from examining how S-waves pass through the interior. Seismologists found that S-waves do not arrive at any stations located between 103° and 180° from the epicenter, a band called the **S-wave shadow zone**. The presence of this shadow zone indicates that S-waves cannot pass through the core at all. Keeping in mind that S-waves are shear waves, which by their nature can travel only through solids, the fact that S-waves do not pass through the core means that the core, or at least the outer part of it, consists of liquid (**Fig. D.9b**).

At first, seismologists thought that the entire core might be liquid iron alloy. But in 1936, a Danish seismologist, Inge Lehmann, discovered that P-waves passing into the core reflect off a boundary within the core. Based on this observation, she proposed that the core consists of two parts: an **outer core**, made of liquid iron alloy, and an **inner core** made of solid iron alloy. Lehmann's work defined the existence of the inner core but could not locate the depth at which the inner core–outer core interface occurs. This depth was eventually located by measuring the exact time it took for seismic waves generated

by nuclear explosions to penetrate the Earth, bounce off the inner core–outer core boundary, and return to the surface (**Fig. D.9c**). The measurements showed that this boundary lies at a depth of about 5,155 km.

Why does the core have two layers—an outer liquid one and an inner solid one? An examination of **Figure D.9d** provides some insight. This graph shows two curves: the *geotherm*, which indicates how temperature changes with increasing depth in the Earth, and the *melting curve*, which indicates how the temperature at which materials start to melt changes with increasing depth in the Earth. As the graph shows, the geotherm lies to the left of the melting curve through most of the mantle and in the inner core. This means that temperatures in most of the mantle and in the inner core are not high enough to cause melting, under the intense pressures found in these regions, so these regions are solid. But the geotherm lies to the right of the melting curve in the low-velocity zone of the mantle, as we discussed earlier, and in the outer core, so these regions contain molten material.

Before we leave our discussion of the core, let's look again at a particularly important characteristic of the liquid outer core—its flow. As noted earlier in the book, the liquid outer core undergoes convection, and this convection generates Earth's magnetic field. Geologists suggest that convection in the outer core cannot be explained by thermal contrasts between the top and bottom of the core, for the outer core consists of metal, which is a very good heat conductor so temperature differences would be hard to maintain. Recent research suggests that the density differences leading to convection are largely due to differences in chemical composition between the top and bottom of the outer core. These compositional differences arise because the solid inner core is slowly growing as the Earth, overall, cools. The new crystals of solid iron that form along the surface of the inner core do not have room in their crystal structure for low-density elements such as silicon, sulfur, or oxygen. So these elements are expelled into the base of the molten outer core. Thus, at any given time, the base of the outer core has a lower density than the upper parts, and it starts to rise convectively.

A Modern Image of Earth's Layers

By the middle of the 20th century, seismologists had identified the crust-mantle boundary (the Moho), the layering within the mantle, the core-mantle boundary, and the layering within the core. In other words, the image of the Earth's interior as being a layered onion-like sequence of concentric zones had been established. Beginning in the 1950s, seismologists began working to refine this image, ironically a task made easier by the Cold War. Because of their desire to detect nuclear explosions, governments funded the installation of an array of seismometer stations scattered around the world. They used the data from this array to develop a graph now known as a **velocity-versus-depth curve**. We've already presented the upper part of the curve (see Fig. D.6a)—**Figure D.10** presents the rest of it. This curve shows the depths at which seismic velocity suddenly changes. These changes define the principal layers and sublay-

ers in the Earth. Note that the graph does not show a velocity for S-waves in the outer core because S-waves cannot travel through a liquid.

Seismic Tomography

In recent years, seismologists have developed a technique, called **seismic tomography**, to produce three-dimensional images of variation in seismic velocities in the Earth's interior. This technique resembles the method used to produce three-dimensional CAT (computerized axial tomography) scans of the human body. In seismic tomography studies, researchers compare the observed travel time of seismic waves following a specific ray path with the predicted travel time that waves following the same path would have if the average velocity-versus-depth model of Figures D.6 and D.10 were completely correct. They found that waves following some paths take more time than predicted, whereas waves following other paths take less time than predicted. By repeating the measurements for many different wave paths in many different directions, researchers can outline three-dimensional regions of the mantle in which waves travel unexpectedly fast or unexpectedly slow.

FIGURE D.10 The velocity-versus-depth profile of the whole Earth.

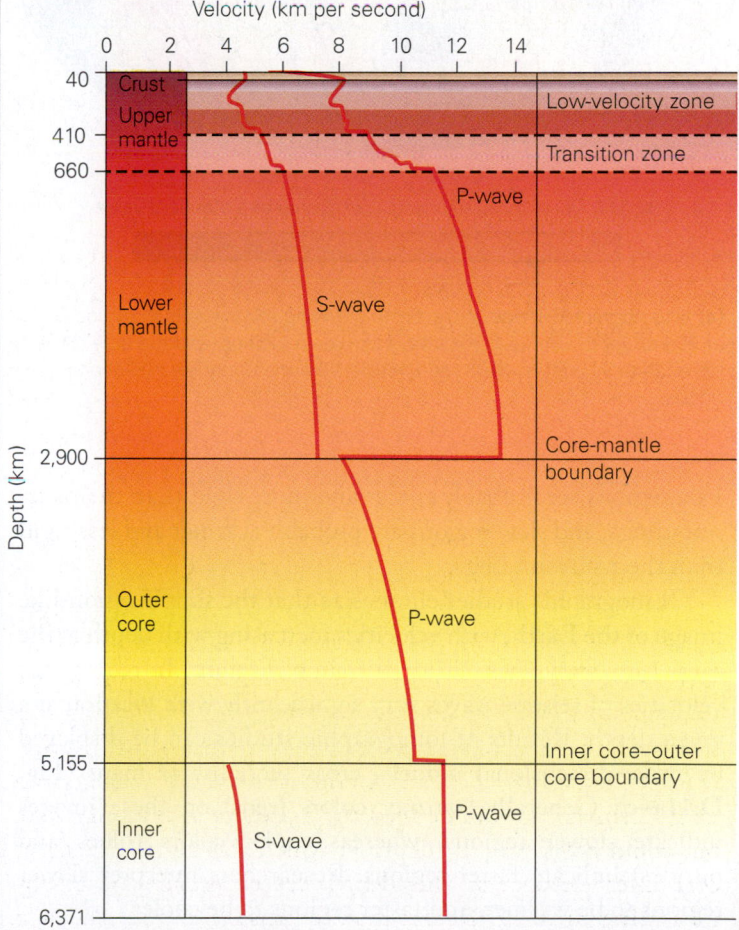

FIGURE D.11 Tomographic images of the Earth's interior and their interpretation.

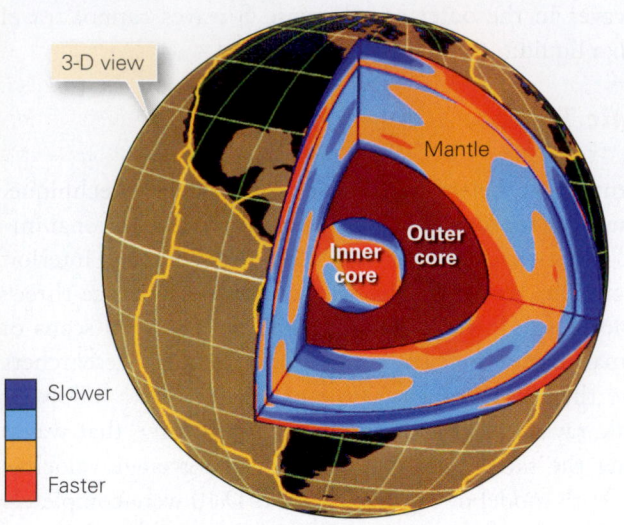

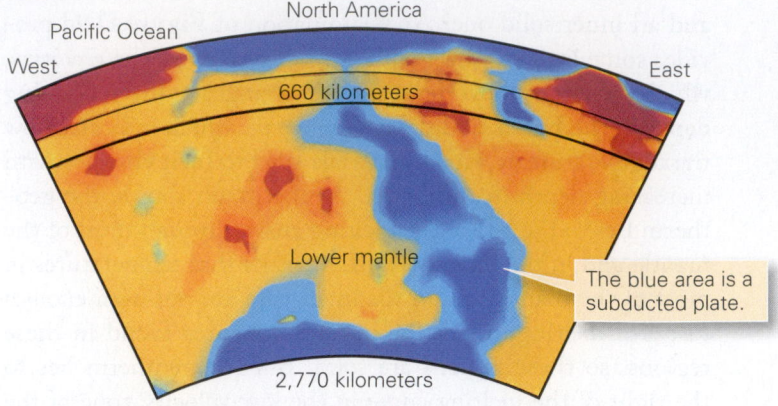

(b) Close-up tomographic image of the mantle beneath North America and the Pacific Ocean. This is a cross-section (vertical slice).

(a) Tomographic image of the Earth. Slower areas may be relatively warmer than their surroundings, and faster areas may be relatively cooler.

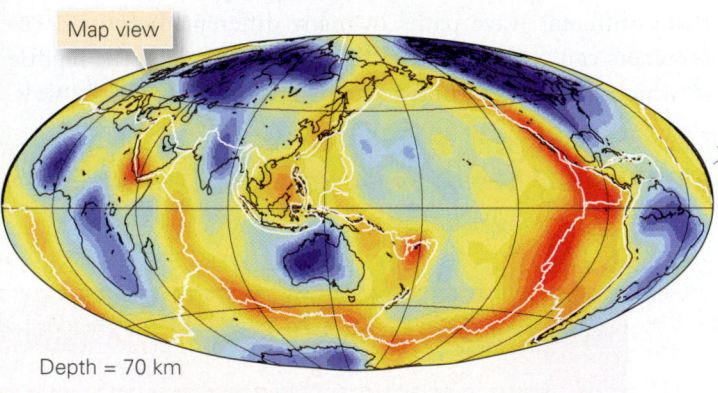

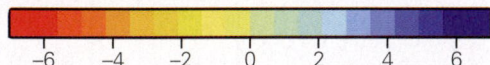

% difference in seismic velocity, relative to a standard value

(c) A tomographic image of the Earth showing relative seismic velocities at 70-km depth. The red areas have slower velocities, and the blue areas have faster velocities. Note red (warmer) mantle underlies mid-ocean ridges.

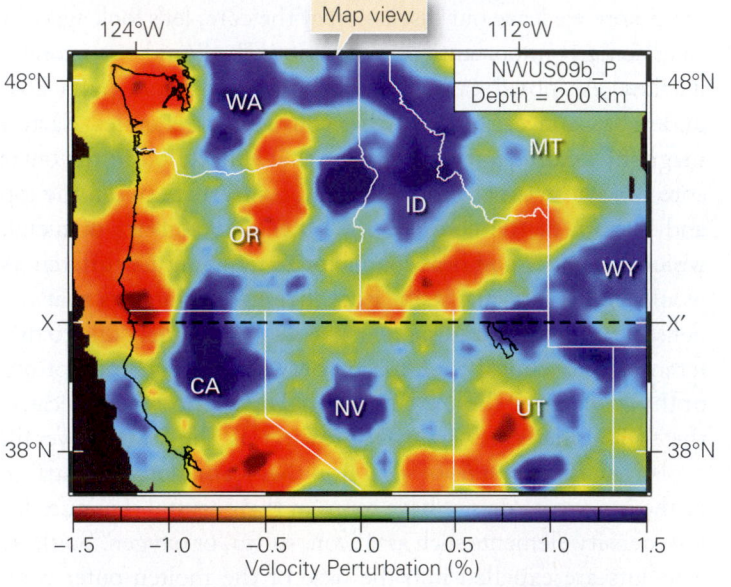

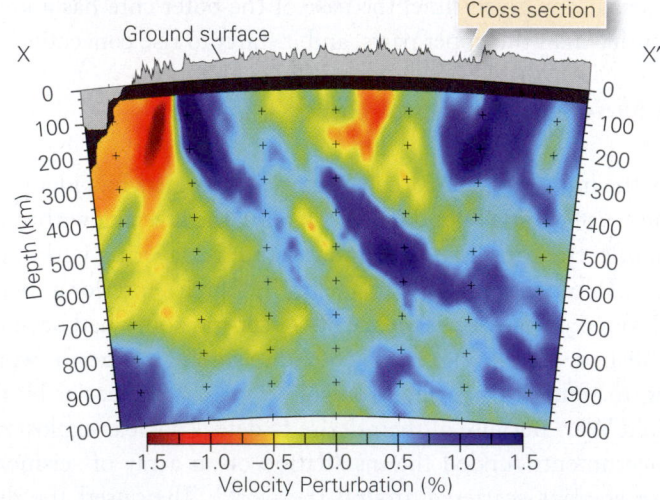

Fast regions are probably cooler and more rigid than their surroundings, and slow regions are probably warmer and less rigid than their surroundings.

Tomographic studies emphasize that the simple onion-like image of the Earth, with velocities increasing with depth at the same rate everywhere, is an oversimplification. In reality, the velocities of seismic waves vary significantly with location at a given depth. Results of tomographic studies can be displayed by three-dimensional models, cross sections, or maps (**Fig. D.11a–c**). Generally, warmer colors (reds) on these images indicate slower regions, whereas cooler colors (blues and purples) indicate faster regions. Researchers interpret slower regions to be warmer, and faster regions to be cooler.

(d) A map (of a surface at a depth of 200 km) and cross section of the Earth beneath the western United States, constructed using tomographic data from the EarthScope array. Red areas have faster-than-expected velocities and probably are warmer, whereas blue areas are cooler.

Seismic tomography studies provide new insight into the Earth's interior. For example, the studies show that a region of fast (cool) mantle lies in the asthenosphere beneath North America (see Fig D.11b). This region may represent the remnants of oceanic lithosphere that was subducted during the Mesozoic. The observation that the fast region extends into the lower mantle suggests that some subducted plates may sink down into the lower mantle. In fact, some researchers suggest that there is a "plate graveyard" at the base of the mantle, where denser portions of subducted plates sink and accumulate. But others maintain that most subducted plate material remains in the upper mantle. Tomographic images of the uppermost mantle (70-km depth) show a region of slow (warm) mantle that occurs in the region beneath mid-ocean ridges (Fig. D.11d). This result supports the concept that warm asthenosphere rises beneath ridges.

Tomography studies focused on the very deep earth have defined a 200-km-thick layer of warmer mantle just above the core-mantle boundary. This layer, called the D″ layer (pronounced dee double prime), may represent the region in which the mantle has absorbed heat radiating from the core. Some researchers speculate that it may be the source of mantle plumes, but this proposal remains a subject for debate. And

studies of the inner core have shown that the inner core contains a distinct pattern of fast and slow regions (see D.11a). Significantly, the orientation of this pattern changes over time, relative to the orientation of this pattern in the mantle. Researchers interpret this to mean that the inner core rotates slightly faster than the rest of the Earth—it makes an extra rotation about once every 20 to 25 years. This process has come to be known as **core superrotation**.

On a global scale, seismologists have been able to outline broad areas of **mantle upwelling**, where warmer and less dense mantle is rising, and broad areas of **mantle downwelling**, where cooler and denser mantle is sinking. The existence of these zones characterizes the large-scale complex pattern of mantle convection. Recently, aided by the installation of arrays of more closely spaced seismometers (**Box D.1**), seismologists may even be able to see localized mantle plumes associated with hot-spot volcanoes. Tomographic studies of the heterogeneities in the mantle have inspired researchers to develop supercomputer models of what this convection looks like (**Fig. D.12a, b**). Even though many important questions remain, tomography has led geologists to picture the Earth's insides as a dynamic place, not just a region of static, concentric shells (**Fig. D.12c**).

FIGURE D.12 Images of convection in the Earth's interior.

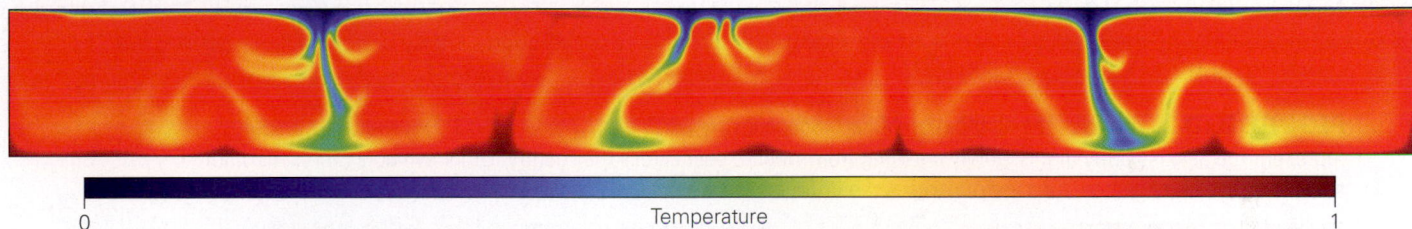

0   Temperature   1

(a) A computer model that simulates convective cells formed in part of the mantle. Blue areas are cooler, downwelling regions, whereas dark red areas are warmer, upwelling regions. Computer models provide insight into how the real Earth might work.

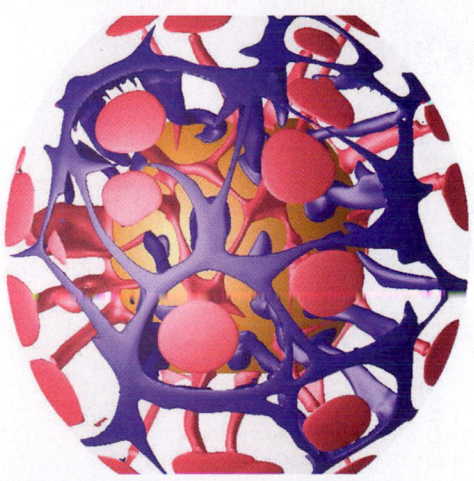

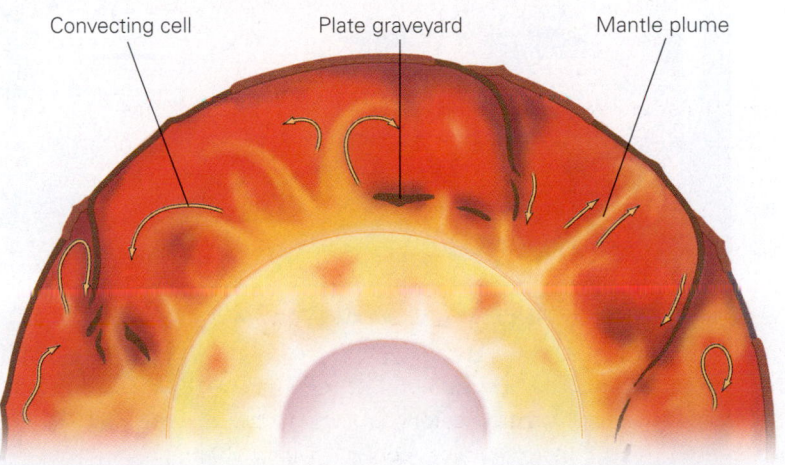

Convecting cell   Plate graveyard   Mantle plume

(b) A 3-D computer model showing plumes (in red) and subduction zones (in blue).

(c) An artist's rendition of a modern view of the complex and dynamic Earth interior. Note the convecting cells, the mantle plumes, and the graveyards of subducted plates.

Resolving the Details of Earth's Interior with EarthScope

Unlike film cameras, digital cameras record images by using a sensor that consists of an array of points, called pixels, each of which detects the light that strikes it. The *resolution* of an image refers to the detail or sharpness of an image—a low-resolution photograph of a billboard might show a blurry image of dark and light areas, while a high resolution of the same image might show distinct letters. Resolution depends, in part, on the number of pixels used on the sensor. In the late 1990s, digital cameras first began to be widely used by the public. The early versions could take 1-megapixel images. This means that the sensor on the camera had about 1 million points on the surface of its light-sensitive sensor. While this sounds like a lot, it actually yields a relatively low-resolution image. Today, even compact digital cameras can take 18-megapixel images, sharp enough to be blown up to the size of poster. Resolution counts, if you want to see detail in an image.

Just like the resolution of a digital photograph depends on the spacing of pixels on a sensor in the camera, the resolution of tomographic images of the Earth's interior depends on the spacing of seismometers on the Earth's surface. As a rough rule of thumb, the size of features visible on a tomographic image is roughly the same as spacing between seismometers. So, if seismometers are hundreds of kilometers apart, they can only detect features (blue areas or red areas) that are hundreds of kilometers across. Data from such widely spaced instruments cannot resolve distinct features in the lithosphere. To overcome this problem, the U.S. National Science Foundation of the United States funded a project called **USArray**, which consists of about 400 seismometers arranged in a grid-like array within a belt that extends from the northern to southern borders of the United States (**Fig. BxD.1**). At any given time, the array is about 700 to 800 km across, in an east-to west direction. Each seismometer of the array remains in place for about 2 years, but a team of technicians constantly removes instruments from the west side of the array and installs them on the east, so the array as a whole moves eastwards across the country like a giant bulldozer tread. Instruments in the array are spaced about 70 km apart, and thus can see features inside the Earth that are about 70 km across. Individual research groups have installed additional instruments, with closer spacing (10 km) in smaller areas to obtain even higher resolution images of specific structures. The initial results of studies using records from USArray have revealed features in the mantle that researchers previously didn't know existed. Knowledge of these features has provided new insight into the convection of the mantle, the nature of hot spots, and the causes of uplift and sinking (subsidence) of the Earth's surface.

USArray is one component of an even larger project called **EarthScope**. Other projects operating under the banner of EarthScope include the Plate-Boundary Observatory. The array of GPS stations and other instruments document the exact distribution of motion between the North American and Pacific plates in order to gain insight into the process by which the movement occurs and to potentially better constrain seismic risk.

FIGURE BxD.1 A map of the distribution of EarthScope instruments, as of January 2013. The small red triangles represent the seismic array that is slowly crawling from west to east across the country. Other symbols represent other seismic arrays or magnetotelluric instruments—the latter measure electrical conductivity in the crust and can detect magma.

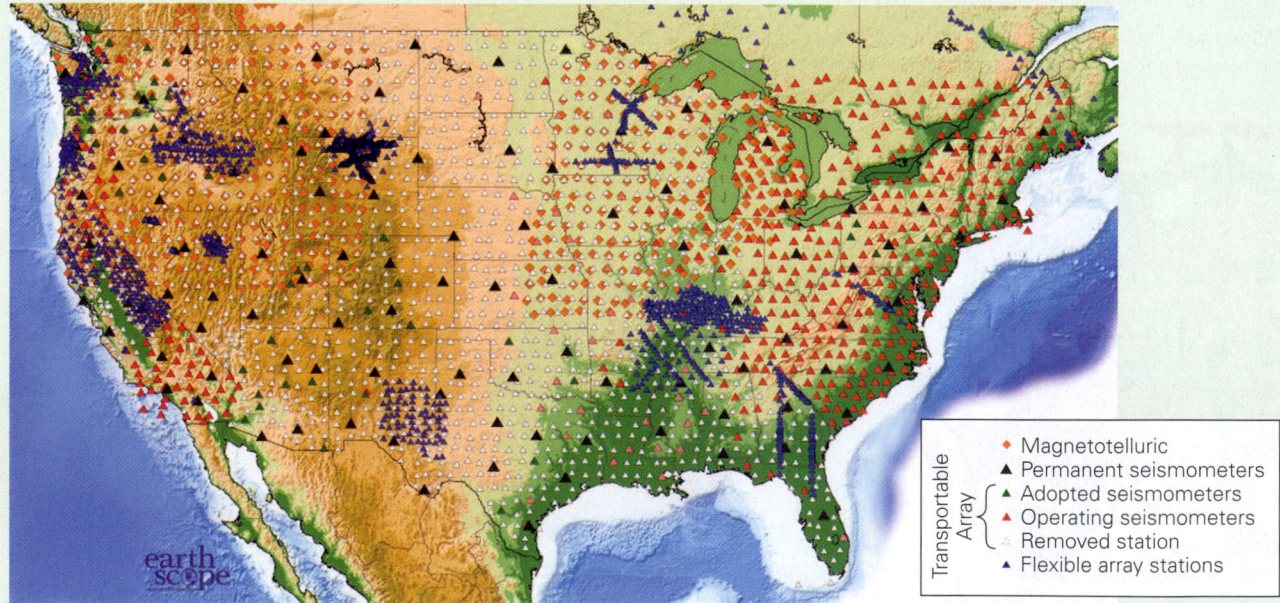

Seismic-Reflection Profiling

So far, we've focused our discussion in this interlude on how seismological research provides a picture of the deep interior of the Earth. Seismic techniques also help geologists fine-tune our image of the upper crust. During the past half century, geologists have found that by exploding dynamite, by banging large weights against the Earth's surface, or by releasing bursts of compressed air into the water, it's possible to create artificial seismic waves that propagate down into the Earth and reflect off the boundaries between different layers of rock in the crust.

By recording the travel time of the reflected waves, geologists can determine the depth to these boundaries. With this information, they can produce a cross-sectional view of the crust called a **seismic-reflection profile** (Fig. D.13a–c). This image can define subsurface bedding and stratigraphic formation contacts and can also reveal the presence of subsurface folds (bends in layers) and faults. Oil and gas companies purchase many seismic-reflection profiles, despite the high cost of the profiles, to identify likely locations of energy resources underground. Research geologists at universities have used the profile to obtain images of the lower crust, the Moho, and even the upper mantle.

FIGURE D.13 Seismic-reflection profiling.

(a) Trucks thumping on the ground to generate the signal needed for making a seismic-reflection profile.

(b) Geologists at Shell Oil Company process seismic data.

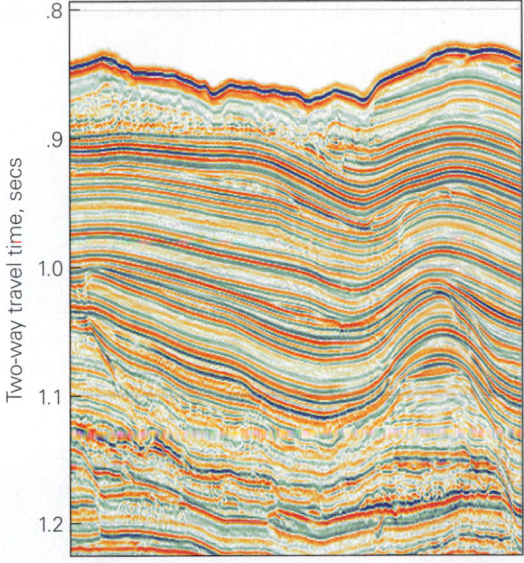

(c) After computer analysis, the data yield a seismic-reflection profile. Color bands represent horizons in the stratigraphic sequence.

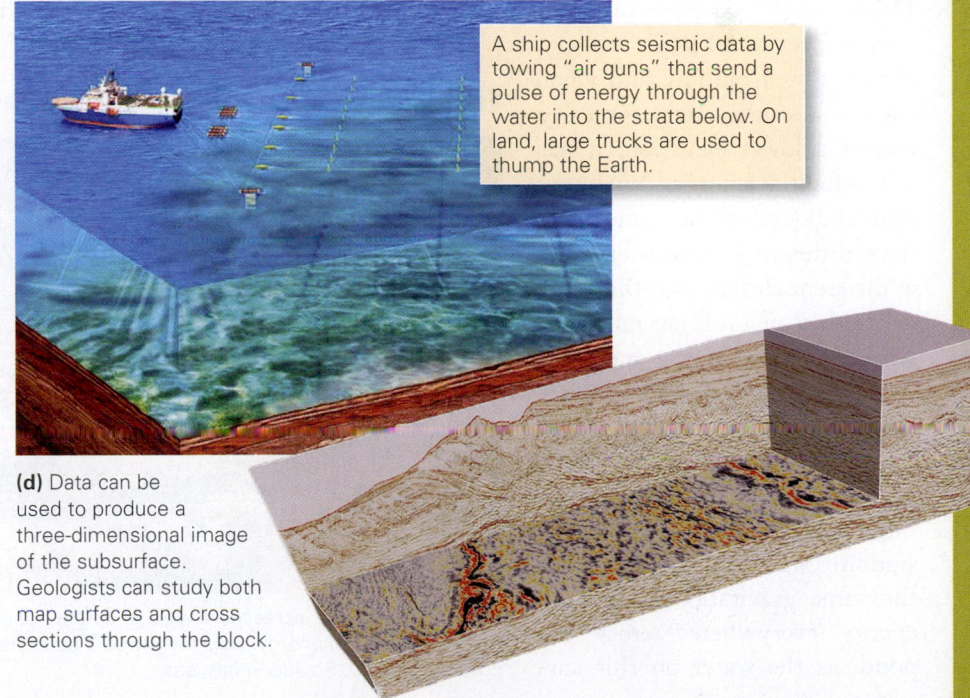

(d) Data can be used to produce a three-dimensional image of the subsurface. Geologists can study both map surfaces and cross sections through the block.

A ship collects seismic data by towing "air guns" that send a pulse of energy through the water into the strata below. On land, large trucks are used to thump the Earth.

In the past decade, data-processing techniques have become so sophisticated, and computers have become so fast, that geologists can now produce three-dimensional seismic-reflection images of the crust (**Fig. D.13d**). These provide so much detail that they can even show individual ribbons of sandstone, representing the channel of an ancient stream, buried kilometers below the Earth's surface.

D.4 Earth's Gravity

We can develop additional understanding of the complexity of the Earth's interior from the study of gravity and magnetism, the two field forces that emanate from our planet. Recall that a **field force** can apply a push or pull across a distance. In other words, a field force exerted by one object can cause another object to *accelerate* (change speed and/or direction) without ever touching it. In this section we discuss gravity, and in the next, we'll turn our attention to magnetism.

The Geoid

Gravity is an attractive field force that one mass exerts on another. As Isaac Newton first recognized, the magnitude of gravitational pull depends on the size of the masses and on the distance between them—larger masses exert stronger pulls than do smaller ones, and closer-together masses produce stronger pulls than do farther-apart ones. When gravity acts on an object but can't make the object move, then the object has **gravitational potential energy**. For example, a boulder sitting at rest on a hill slope has gravitational potential energy (**Fig. D.14a**). If the boulder starts to tumble, the potential energy transforms into kinetic energy, the energy of motion, and when the boulder ultimately comes to rest at a lower elevation, it has less potential energy. Note that two boulders of the same mass have different potential energies at different elevations—the boulder higher on a hill has more potential energy than the one lower on a hill—and the same gravitational potential energy when at the same elevation.

With the above concepts in mind, we see that the surface of standing water in a still pond has the same gravitational potential energy everywhere across the pond, as the water on this sur-face all lies at the same elevation. If you throw a pebble in the water, the energy of impact temporarily forms waves (**Fig. D.14b**), but the waves eventually disappear because water is so weak that gravity eventually evens out highs and lows of the surface (**Fig. D.14c**). A surface on which all points have the same potential energy, such as the surface of standing water in a pond, is called an **equipotential surface**.

What does the gravitational equipotential surface of the Earth look like? Let's start to address this question by imagining what sea level would look like (not what it really is) if an ocean completely surrounded a stationary (nonrotating), perfectly smooth, spherical globe on which there are no winds or currents. This equipotential surface would itself be a sphere. In reality, however, the Earth's real equipotential surface is not a perfect sphere. Part of this deviation is due to our planet's spin on its axis. Rotation produces centrifugal force, which flattens the Earth into a *spheroid*. The imaginary spheroid that most closely has the shape of the Earth has a slight equatorial bulge so that its radius at the equator is 6,378 km (3,963 miles) while its radius at the pole is 6,357 km (3,950 miles). Geologists refer to this imaginary equipotential surface as the **reference geoid** (**Fig. D.15a**).

While the reference geoid is a starting point for describing the Earth's gravity, it doesn't provide a completely accurate picture for many reasons, including (1) the density of material within the Earth's layers is not uniform throughout the layer; (2) the surface of the Earth is not smooth, for the land surface lies higher than the seafloor, and both have mountains and valleys; and (3) rock making up the outermost layer of the Earth, the lithosphere, is strong enough to hold up heavy loads such as volcanoes or glaciers, or to allow depressions such as trenches to persist for a long time. Because of these factors, and others, there can be more mass or less mass at a location

FIGURE D.14 The concept of gravitational potential energy.

(a) Potential energy increases as you go up the hill. Boulder 1 has the most potential energy. Boulders 2 and 3 have the same potential energy. Boulder 4 has less.

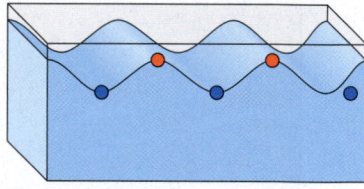

(b) The crests of waves on a pond have higher potential energy than the troughs.

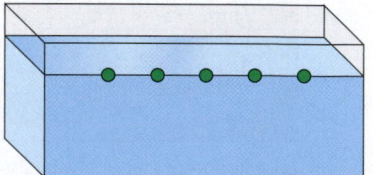

(c) When there are no waves, the surface of the water is an equipotential surface.

FIGURE D.15 Representations of the geoid, the shape of an equipotential surface representing Earth's gravity. Colors represent *elevations* of the geoid relative to the reference geoid.

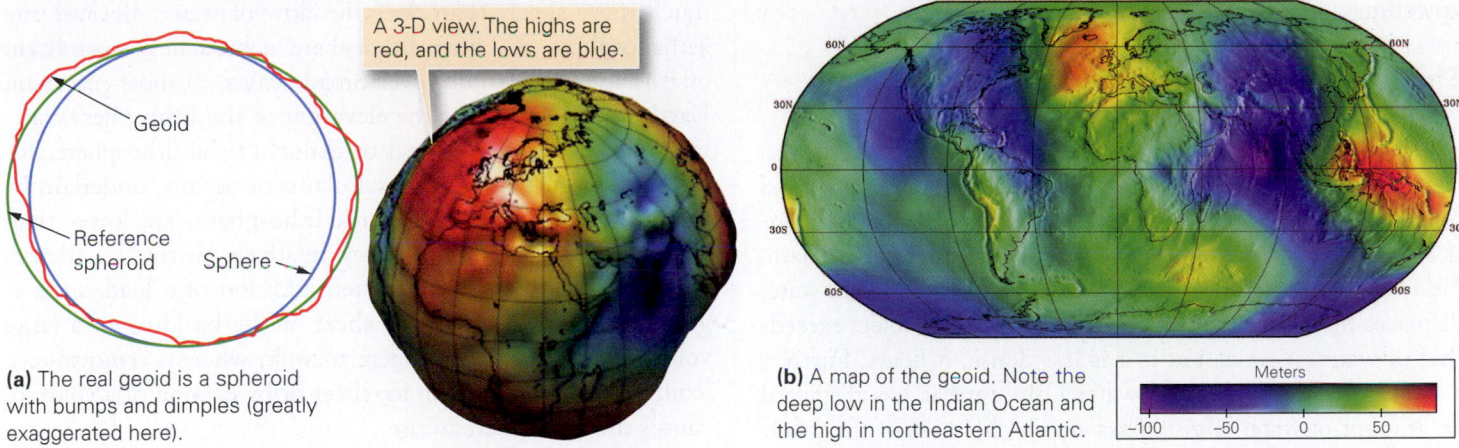

A 3-D view. The highs are red, and the lows are blue.

Geoid

Reference spheroid   Sphere

(a) The real geoid is a spheroid with bumps and dimples (greatly exaggerated here).

(b) A map of the geoid. Note the deep low in the Indian Ocean and the high in northeastern Atlantic.

Meters

−100   −50   0   50

than the reference geoid predicts, so the real equipotential surface for the Earth has bumps and dimples. Geologists refer to this irregular surface, which provides the best representation of the Earth's gravity, simply as the **geoid** (**Fig. D.15b**). The surface of the geoid differs from that of the reference geoid only slightly—the highest point on the geoid lies 85 m above the reference geoid, and the lowest point lies 107 m below the reference geoid. But even these small differences can noticeably influence the orbits of satellites or the accuracy of land surveys.

Gravity Anomalies

Geologists can measure the pull of gravity at a point on the Earth by using an instrument called a **gravimeter**, which consists of a weight hanging from a delicate spring—stronger gravity pulls the weight down more; weaker gravity pulls it down less. Traditionally, measurements of gravity required placing a gravimeter on the ground and letting it sit long enough to stabilize before making a measurement. In recent years, researchers have developed methods that allow satellites (such as the GRACE satellite, launched by NASA) to detect tiny variations in the Earth's gravitational pull at the surface over broad areas very rapidly. On-land and satellite-based measurements reveal that the actual surface of the Earth is not an equipotential surface, meaning that the pull of gravity at the surface varies with location.

How can we describe variations in the pull of gravity? Recall that gravity is a force, so it causes an unrestrained object to accelerate, or increase its velocity. Researchers describe the acceleration caused by the Earth's gravity using a unit called the *Gal* (named for the Italian astronomer Galileo, 1564–1642), where 1 Gal = 1 cm/s^2. On average,

the acceleration due to Earth's gravity is 981 Gals ($\approx$ 9.8 m/s^2), but it ranges from 976 Gals to 983 Gals. Differences in the gravitational pull between one location and an adjacent one may be so tiny that geologists must specify them in milliGals (mGal), where 1 mGal = 1/1,000 Gal. Geologists refer to a deviation of the observed (real) geoid from the reference geoid as a **gravity anomaly**. Over a *positive gravity anomaly*, the pull is stronger, whereas over a *negative gravity anomaly*, the pull is weaker. A positive anomaly indicates that there is extra mass below the site, perhaps due to a body of particularly dense rock underground, whereas a negative anomaly means that there is a deficit of mass, perhaps due to the presence of less-dense rock (**Fig. D.16**).

As a practical point, study of gravity anomalies helps geologists to find reserves of valuable metal ores (rocks that can be mined and processed to yield metal) underground, because

FIGURE D.16 A gravity anomaly map of the United States. On this map, colors represent variations in the magnitude of gravitational pull. Red is stronger, blue is weaker.

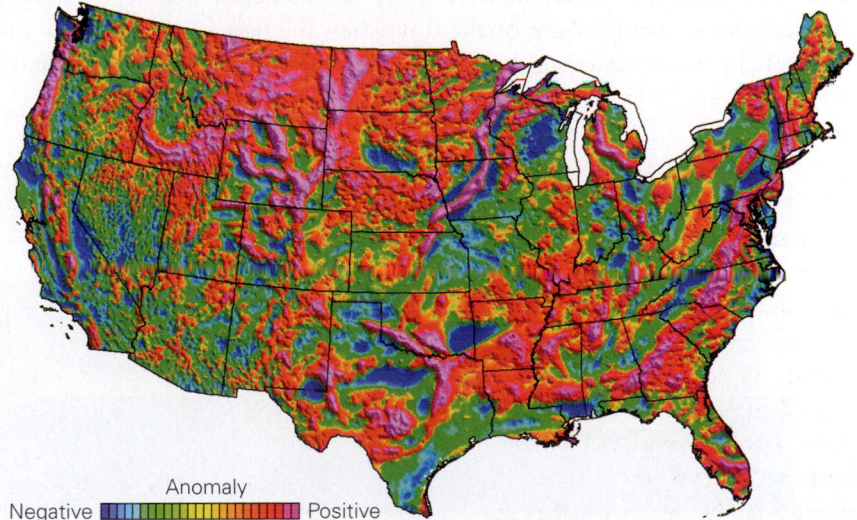

Anomaly

Negative ▐▐▐▐▐▐▐▐▐▐▐▐▐▐▐▐ Positive

metal ore tends to be denser than other rock. And from an academic perspective, gravity anomalies help geologists to detect upwelling and downwelling zones in the mantle.

The Concept of Isostasy

Archimedes (ca. 287–212 B.C.E.), a Greek mathematician, had been puzzling over the question of why some objects float and others sink, until one day when, according to legend, he noticed the water level rise in a tub as he stepped in to take a bath. He realized that this meant that an object sinking into water displaces the water and that if the density of the object exceeds that of water, it sinks, but if it is less dense, it floats. Further, a floating object sinks into water only until it has displaced an amount of water whose mass equals the mass of the whole object. This concept is now known as **Archimedes' principle**.

According to Archimedes' principle, a cargo ship anchored in a harbor floats at just the right level so that the mass of the water displaced by the ship equals the mass of the whole ship. A ship, in effect, is just a steel-sheathed bubble—it floats because its overall average density, taking into account the fact that most of a ship's volume consists of air, is less than that of water, even though its hull consists of heavy steel. When the ship remains empty, the distance that its deck lies above the water, its so-called *freeboard*, is large (**Fig. D.17**). Addition of heavy cargo causes the ship's hull to settle down deeper into the water, so the freeboard decreases. Such movement can happen only if the water beneath the ship can flow out of the way. If the ship's keel rests on the harbor's solid floor, addition of cargo will not change freeboard, because the floor would hold the ship up. When the freeboard of the ship is just right for a given cargo, so that ship's freeboard does not "want" to rise or sink, we say that the ship is in **isostatic equilibrium**, or that a condition of **isostasy** exists.

As we learned in Chapter 4, the outer layer of the Earth, the lithosphere, can be thought of as a relatively rigid shell that "floats" on the underlying, relatively soft asthenosphere. Asthenosphere can flow out of the way when the base of the lithosphere moves downward, or it can flow back in to keep the space beneath rising lithosphere filled, just like water does beneath a ship. But the flow of asthenosphere happens much, much more slowly than does the flow of water. Because the asthenosphere can flow, lithosphere approaches a condition of isostatic equilibrium, over broad scales, in most places on Earth. This means that the elevation of the lithosphere's surface reflects the density and thickness of the lithosphere. It's for this reason that the abyssal plains of oceans, underlain by relatively dense, cold, and thick lithosphere, are lower than mid-ocean ridges, underlain by relatively warm, less-dense, and thinner lithosphere. Further, addition of a load, such as growth of a large glacial ice sheet or the building of a large volcano, causes the lithosphere to sink, whereas removal of a load, say by melting of an ice sheet or by erosion of a volcano, causes the lithosphere to rise.

At the local scale, however, lithosphere is not always in isostatic equilibrium. First, deviations from isostasy develop because asthenosphere flows so slowly, so there are places where the lithosphere hasn't had time to sink enough or rise enough to accommodate for a change in loading. As an example, regions that were covered by a huge continental ice sheets during the most recent ice age, which ended only about 10,000 years ago, are still in the process of rising and are not yet in full isostatic equilibrium. Second, since the lithosphere itself is relatively strong, it can hold up relatively small loads without sinking. To picture this, imagine that the lithosphere is like a surface of a trampoline. If a heavy person steps in it, the trampoline's surface bends down. But if a small block of wood is placed on it, the surface supports the load and doesn't bend.

In locations where isostatic equilibrium has not been achieved, we observe gravity anomalies. Examples include deep-sea trenches, where the lithosphere has been bent and held down by the process of subduction—since the density of water is much less than that of rock, there is a deficit of mass over a trench and, therefore, a negative gravity anomaly. In places where a mountain consists of particularly dense rock but is small enough to be held up by the strength of the lithosphere, there may be an excess of mass and therefore a positive gravity anomaly. Similarly, rapid eruption of a volcano may produce a

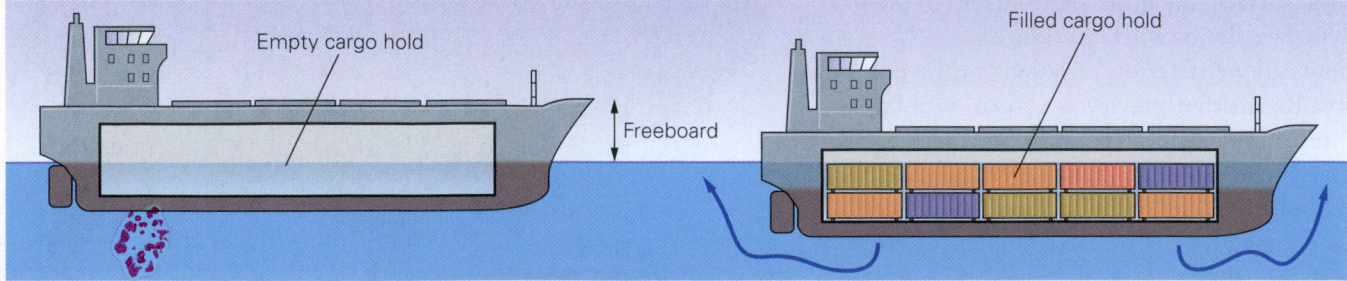

FIGURE D.17 A ship analog for the concept of isostasy. At isostatic equilibrium, the freeboard reflects Archimedes' principle, so the freeboard decreases as the mass of the ship increases. Water flows out of the way as the ship moves down (right image).

load on the lithosphere so quickly that the lithosphere does not have time to sink—here again, isostasy has not been achieved and you'll detect a positive gravity anomaly. We'll discuss isostasy further in Chapter 11 to see how it is involved in explaining the height of mountain ranges.

D.5 Earth's Magnetic Field, Revisited

As we learned in Chapters 2 and 3, the Earth behaves like a weak magnet and thus produces a magnetic field. Here we look more closely at the magnetic field, focusing on why the field exists and on why there are variations in the field strength over regions. To set the stage, we review the fundamentals of magnetism.

A Brief Review of Earth's Magnetism

If you hold a magnet over steel paper clips, it lifts the clips by exerting an attraction or pull on them that, at close range, exceeds the force of gravity. A magnet can also produce a repulsion, capable of pushing an object away. Recall that the push or pull exerted by a magnet is its *magnetic force*, and the region around a magnet in which we can detect a magnetic force is a *magnetic field*. Magnetic fields can be represented by curving lines, called *magnetic field lines* (**Fig. D.18**); a compass needle placed in the field aligns with these lines. All magnets have two *magnetic poles*, north at one end and south at the other. Opposite poles attract, and like poles repel—thus, the north end of one magnet attracts the south end of another, but it repels the north end of another. The direction from the north

FIGURE D.18 The geometry of Earth's dipole field.

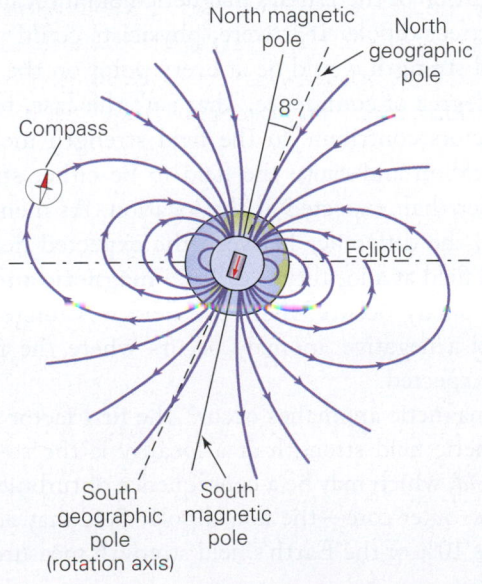

pole to the south pole is represented by an arrow called a **dipole**. Magnetic field lines form a continuous loop through the magnet's dipole.

In Chapter 2, we learned that our planet has a magnetic field, which deflects solar wind and traps deadly cosmic rays. (Of note, not all planets have fields; in fact, neither Mars nor Venus, the two planets most like the Earth, have measurable fields.) The dipole representing most of the Earth's field currently lies about 10° from the Earth's spin axis—thus, in the northern hemisphere, the *magnetic pole* (the point where the magnetic dipole intersects the planet's surface) currently lies about 855 km from the *geographic pole* (the point where the Earth's spin axis intersects the Earth's surface). Magnetic field lines are parallel to the planet's surface at the magnetic equator and are vertical at the magnetic poles. The angle between the magnetic dipole and the spin axis changes over time, so the location of the magnetic pole on the surface of the Earth changes. Currently, the magnetic pole is moving at about 50 km per year. Measurements suggest that on a time scale of thousands of years Earth's magnetic poles follow looping paths but don't stray more than about 15° of latitude from the geographic pole (see Fig. 2.3).

Geologists, by convention, refer to the tail end of the dipole, which lies close to the north geographic pole, as the **north magnetic pole**; the arrowhead end of the dipole, which lies near the south geographic pole, is the **south magnetic pole**. (This convention can be a bit confusing because, to a physicist, the north end of a magnet corresponds to the arrowhead end, and the arrowhead representing the Earth's dipole currently lies in the south. We use the convention because it means that the north-seeking end of a compass needle points toward the north.) As we've seen, the polarity of the Earth field suddenly changes every now and then—during times of reversed polarity, the dipole arrow points toward our planet's north geographic pole. The last reversal happened about 730,000 years ago.

Origin of the Earth's Magnetic Field

How can magnetic fields be produced? In an **electromagnet**, the field develops when a current runs through a wire. In a **permanent magnet**, the magnetism comes from the arrangement of atoms. Each atom can be pictured as a tiny dipole whose magnetism arises, simplistically, from the orbiting of its electrons. In nonmagnetic materials, these dipoles are randomly oriented and thus the fields they produce cancel one another out (**Fig. D.19**). But in a magnetic material, like solid iron, the dipoles are locked into parallelism with one another, so that the field produced by each adds to the fields produced by the others to create an overall measurable magnetic field.

Why does the Earth have a magnetic field? Though books commonly portray the field as emanating from a giant permanent magnet in the interior, that's not the real situation, for

FIGURE D.19 The origin of permanent magnetization in a material.

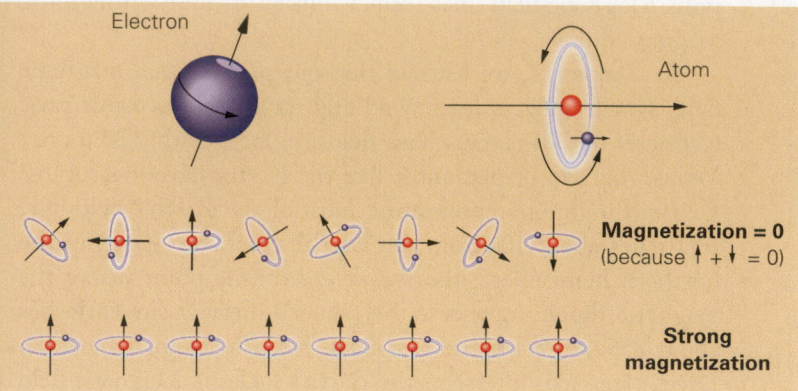

FIGURE D.20 An artist's representation of the spiral convection cells in the Earth's outer core that may cause the magnetic field.

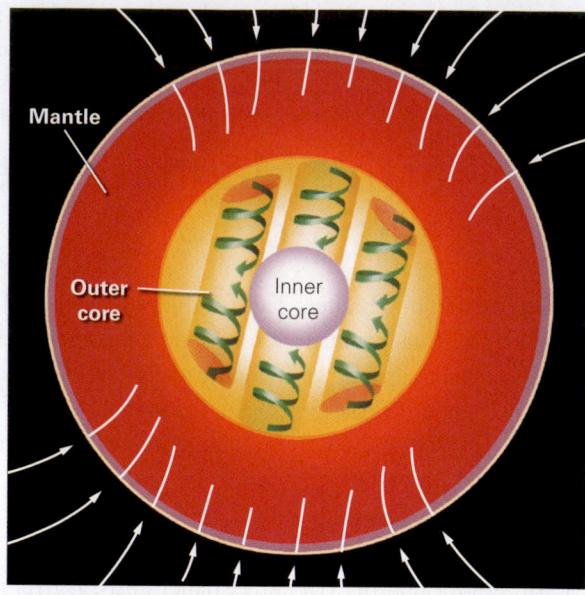

at the temperatures that occur deep in the Earth, even iron cannot behave as a permanent magnet—the atoms in a hot material tumble chaotically, so their dipoles cannot align and lock into the same orientation. The path toward discovering the origin of the Earth's field started in 1926, when researchers learned that our planet's outer core consists of liquid iron alloy. Since flow of metal can produce an electric current, it seemed that the flow of the outer core must somehow be responsible for the magnetic field. But how?

Clues to decoding the production of the Earth's magnetic field come from understanding how an electric power plant works. In a power plant, water or wind power spins a wire coil (an electrical conductor) around an iron bar (a permanent magnet). Such an apparatus is called a **dynamo**. Because an electric current in a wire produces a magnetic field, movement of a wire relative to a magnetic field produces an electric current. The Earth's interior is effectively a dynamo. In the case of the Earth, flow in the outer core serves the role of the spinning wire coil. But the inner core can't serve the role of the permanent magnet because it is so hot. Researchers suggest instead that the Earth is a **self-exciting dynamo**. Evidently, during the Earth's early history, flow in the outer core took place in the presence of a magnetic field. This flow generated an electric current, and once the current existed it generated a magnetic field. Continued flow in the presence of the generated magnetic field produced more electric current that, in turn, maintains the magnetic field. Once started, the system perpetuates itself, as long as there's an input of energy to keep the outer core in motion.

The key to understanding the orientation of the Earth's magnetic dipole comes from understanding the geometry of the convection cells in the outer core. Researchers suggest that the Coriolis force (see Chapter 18) causes convective cells in the outer core to become spirals that align roughly with the Earth's spin axis (**Fig. D.20**). The spirals wobble, so at any given time the overall dipole axis they generate is not exactly parallel to the spin axis. Notably, the spirals apply a force to the inner core,

causing it to rotate slightly faster than the rest of the Earth, the phenomenon we introduced earlier as "superrotation."

Why does the polarity of the Earth's field reverse every now and then? Computer models suggest that reversals happen because circulation spirals in the outer core are unstable. Over time, they slow and fade away, as they interact, causing the magnetic field to weaken or disappear temporarily. As new spirals become established, the field reappears, possibly with a different polarity. The details of this process remain a subject of active research.

Magnetic Anomalies, Revisited

Close examination of the Earth's magnetic field indicates that it is not a perfect dipole. If it were, physicists could predict what the field strength would be at every point on the planet with a high degree of confidence. That isn't the case, because two other factors contribute to the field strength measured at a given location and cause the field to be either stronger than or weaker than expected at the location. As mentioned in Chapter 3, the difference between the expected field and the measured field at a locality is called a **magnetic anomaly**. A positive anomaly occurs where the field is stronger than expected, and a negative anomaly occurs where the field is weaker than expected.

Why do magnetic anomalies occur? The first factor affecting the magnetic field strength at a locality is the so-called *non-dipolar field*, which may be a consequence of turbulence in the flow of the outer core—the non-dipolar field may account for as much as 10% of the Earth's field strength measured at a given location. The second factor is the magnetization of rock in the crust—some rocks contain minerals, such as magnetite and hematite, that behave as permanent magnets. These make the rock, overall, behave like a weak magnet. The presence of magnetic rocks in the crust will produce local magnetic anomalies that are independent of the Earth's internal magnetic field.

We noted in Chapter 3 that magnetic anomalies due to the magnetization of rocks take the form of parallel stripes on the seafloor, because the basalt comprising the upper layer of oceanic crust was produced at the mid-ocean ridge axis by seafloor spreading during successive polarity reversals. A positive anomaly stripe occurs over basalt with normal polarity because the magnetization of the basalt adds to the Earth's field, whereas a negative anomaly stripe occurs over basalt with reversed polarity because the magnetization of this basalt subtracts from that of the Earth's field. On continents, in contrast, the pattern of magnetic anomalies is much more complex, for the distribution of rock types is more complex (**Fig. D.21**). We see anomalies due to igneous intrusions (which contain magnetic minerals), lava flows, concentrations of iron-rich sediments, and variations in the depth to basement. The shapes of the anomalies may reflect the shape of intrusions or extrusions, or the shape of iron-rich sedimentary layers.

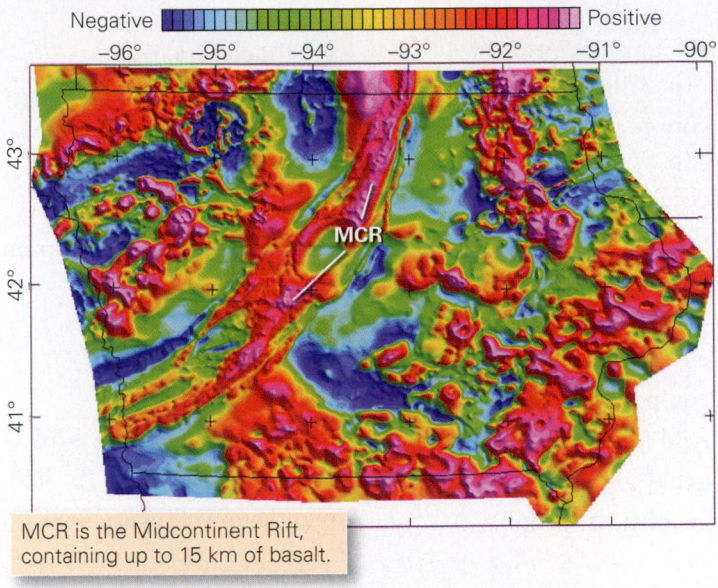

FIGURE D.21 Magnetic anomaly map of Iowa. Red areas have stronger magnetization and blue areas have weaker magnetization.

MCR is the Midcontinent Rift, containing up to 15 km of basalt.

INTERLUDE SUMMARY

- Details about the depth to Earth's internal layers, and about divisions within the layers, come from the study of seismic waves passing through the Earth. This is because waves refract and reflect at boundaries.

- The Moho was discovered because seismic waves travel faster through the mantle than through the crust. Seismic-velocity discontinuities define boundaries in the mantle. Several discontinuities occur in the transition zone.

- Seismic shadow zones define the depth to the core. S-waves cannot pass through the core, so part of it must be liquid. The reflection of waves off the surface of the solid, inner core defines its depth.

- Seismic tomography reveals that the onion-like model of the Earth's interior is an oversimplification, as the layers are inhomogeneous. Modern EarthScope studies can outline subducted plates.

- Seismic-reflection, using artificially produced seismic waves, allows geologists to identify layering and structures in the sedimentary strata of the upper crust.

- The geoid, the representation of the surface on which the pull of gravity is the same, is not a perfect sphere. Local anomalies may indicate the presence of particularly dense rocks below the surface.

- The elevation of the lithosphere's surface regionally reflects isostatic equilibrium. Local bending or loading, however, yields places that are not in equilibrium.

- The Earth's magnetic dipole exists because of convective circulation in the outer core. Local anomalies may exist where rocks contain magnetic minerals.

GUIDE TERMS

Archimides' principle (p. 374)

core-mantle boundary (p. 365)

core superrotation (p. 369)

dipole (p. 375)

dynamo (p. 376)

EarthScope (p. 370)

electromagnet (p. 375)

equipotential surface (p. 372)

field force (p. 372)

geoid (p. 373)

geophysics (p. 360)

gravimeter (p. 373)

gravitational potential energy (p. 372)

gravity anomaly (p. 373)

inner core (p. 366)

isostasy (isostatic equilibrium) (p. 374)

lower mantle (p. 364)

low-velocity zone (LVZ) (p. 363)

magnetic anomaly (p. 376)

mantle downwelling (p. 369)

mantle upwelling (p. 369)

Moho (p. 363)

north magnetic pole (p. 375)

outer core (p. 366)

permanent magnet (p. 375)

P-wave shadow zone (p. 365)

reference geoid (p. 372)

reflection (p. 362)

refraction (p. 362)

seismic ray (p. 361)

seismic-reflection profile (p. 371)

seismic tomography (p. 367)

seismic-velocity discontinuity (p. 364)

self-exciting dynamo (p. 376)

south magnetic pole (p. 375)

S-wave shadow zone (p. 366)

transition zone (p. 364)

travel time (p. 361)

upper mantle (p. 364)

USArray (p. 370)

velocity-versus-depth curve (p. 367)

wave front (p. 361)

REVIEW QUESTIONS

1. What basic layers of the Earth were recognized before the use of seismology?

2. Do seismic waves travel at the same velocities in all rock types?

3. What is the difference between refraction and reflection of waves?

4. What observation led to the discovery of the Moho?

5. How and why do seismic waves bend as they pass through the mantle?

6. What are seismic-velocity discontinuities, and what do they tell us?

7. What are P-wave and S-wave shadow zones, and what do they tell us?

8. Is seismic velocity constant at a given depth? Explain how seismologists learned the answer to this question and what the answer tells us about the mantle.

9. What can we learn from seismic-reflection surveys?

10. What is the principle of isostasy, and why is the lithosphere able to be in isostatic compensation, in general?

11. Explain what the geoid is and why it is so bumpy. What is a gravity anomaly?

12. Why might the Earth's magnetic field exist? What causes magnetic anomalies?

Look closely at this small cliff face in Arizona. You can see layers that have been contorted into toothpaste-like curves and have been broken along jagged cracks. We're seeing geologic structures, evidence that these rocks were subjected to deformation during mountain building.

CHAPTER 11

Crags, Cracks, and Crumples: Crustal Deformation and Mountain Building

Innumerable peaks, black and sharp, rose grandly into the dark blue sky. . . . [Mountains] are nature's poems carved on tables of stone. . . . How quickly these old monuments excite and hold the imagination!

—*John Muir (1838–1914), American naturalist*

LEARNING OBJECTIVES

By the end of this chapter, you should understand . . .

- how rocks deform (crack; bend; flow) during mountain building in response to stresses.
- the characteristics of basic geologic structures (joints, faults, folds, foliations) and how to describe them.
- where and why mountain belts form and what processes cause uplift.

11.1 Introduction

Geographers call the peak of Mt. Everest "the top of the world," for this mountain, which lies in the Himalayas of southern Asia, rises higher than any other on Earth. The cluster of flags on Mt. Everest's summit flap at 8.85 km (29,029 feet) above sea level—almost the cruising height of modern jets. No one can survive very long at the top, for the air there is too thin to breathe. In 1953, Sir Edmund Hillary, from New Zealand, and Tenzing Norgay, a Nepalese guide, became the first to reach the summit. Over 6,000 other people have also succeeded, and now hundreds make the ascent every year—but more than 200 climbers and guides have died trying. Success depends not just on the skill of the climber but also on the path of the jet stream, a 200-km-per-hour current

FIGURE 11.1 Digital map of world topography, showing the locations of major mountain ranges.

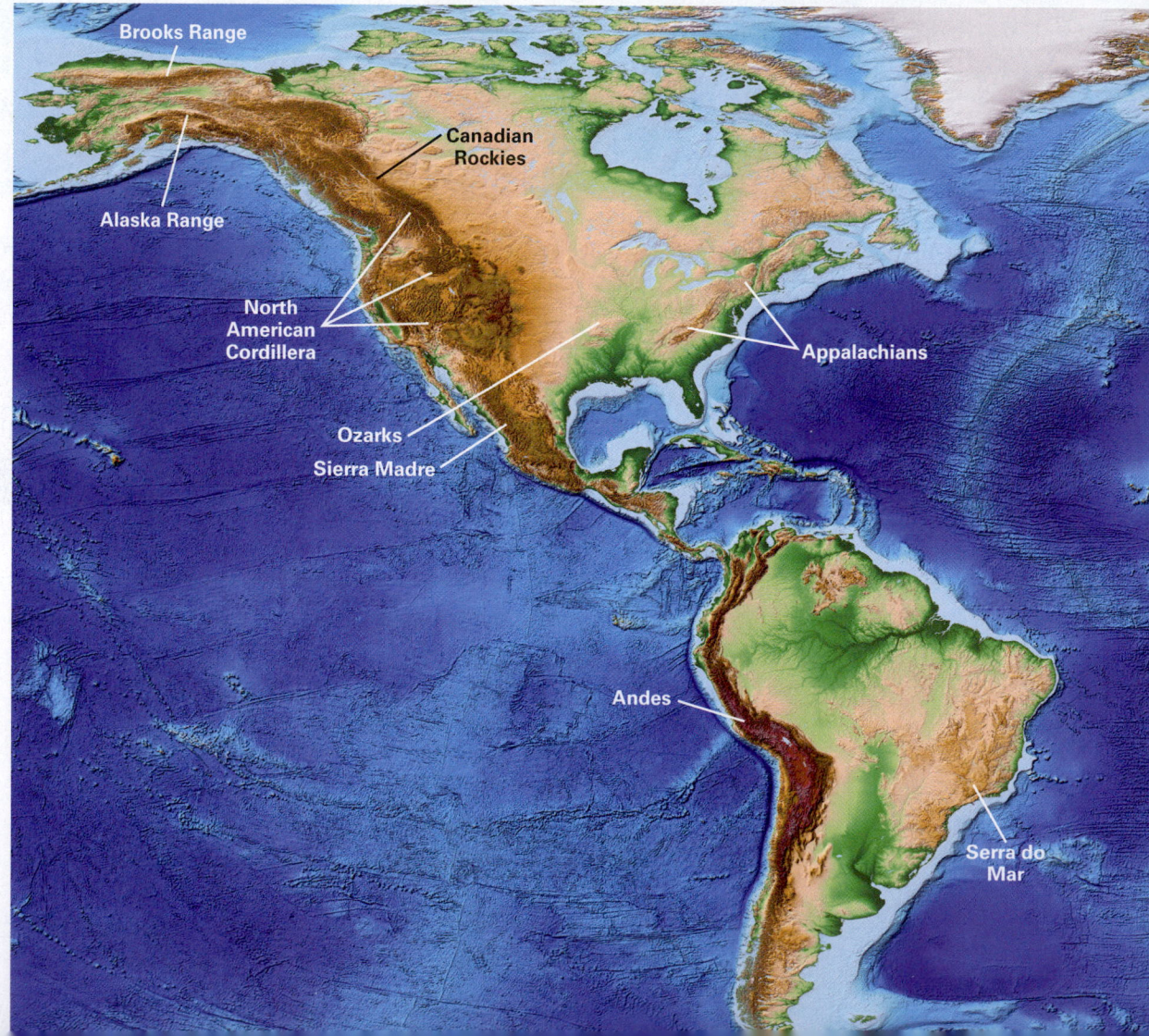

of air that flows at high elevations. If the jet stream crosses the summit, it engulfs everyone there in heat-robbing winds that can freeze a person's face, hands, and feet, even if they're swaddled in high-tech clothing.

Mountains draw nonclimbers as well, for everyone loves a vista of snow-crested peaks. The stark cliffs, clear air, meadows, forests, streams, and glaciers of mountains provide a refuge from the mundane. Geologists have a particular fascination with mountains, for high peaks provide one of the most obvious indications of dynamic activity on Earth. Think about it . . . to make a mountain, cubic kilometers of rock must be pushed skyward, against the pull of gravity. With the exception of the large volcanoes formed over hot spots, mountains do not occur in isolation but rather as part of a linear chain or range called a **mountain belt**, or **orogen** (from the Greek words *oros*, meaning mountain, and *genesis*, meaning formation). Geographers define about a dozen major mountain belts and numerous smaller ones worldwide (**Fig. 11.1**).

Mountain building, or **orogeny**, the process of forming a mountain belt, not only raises the surface of the crust, a process called *uplift*, but it also causes rocks to undergo *deformation*, a process by which rocks bend, break, or flow in response to "stress" (compression, tension, or shearing). Deformation yields *joints* (cracks), *faults* (fractures on which one body of rock slides past another), *folds* (bends, curves, or wrinkles of rock layers), and *foliations* (a new fabric or layering in rock)—all of these features are examples of *geologic structures*. Orogeny also can cause metamorphism and igneous activity.

Mountain building happens on the Earth for a variety of reasons. Some mountains develop along convergent-plate boundaries, some are due to the collision of continents, and some are a consequence of continental rifting. A particular event may last for tens of millions of years. Significantly, as the land surface rises and relief—the elevation difference between a higher area and a lower area—develops, erosion starts to grind away the land surface. This process produces immense volumes of sediment and sculpts awesome, jagged topography (**Fig. 11.2**). When uplift ceases, erosion can eventually bevel a mountain range down to near sea level over a period of tens of millions of years. But even after the high peaks are gone, a distinct belt of fractured, contorted, and metamorphosed rock

Did you ever wonder . . .

if mountain belts last forever?

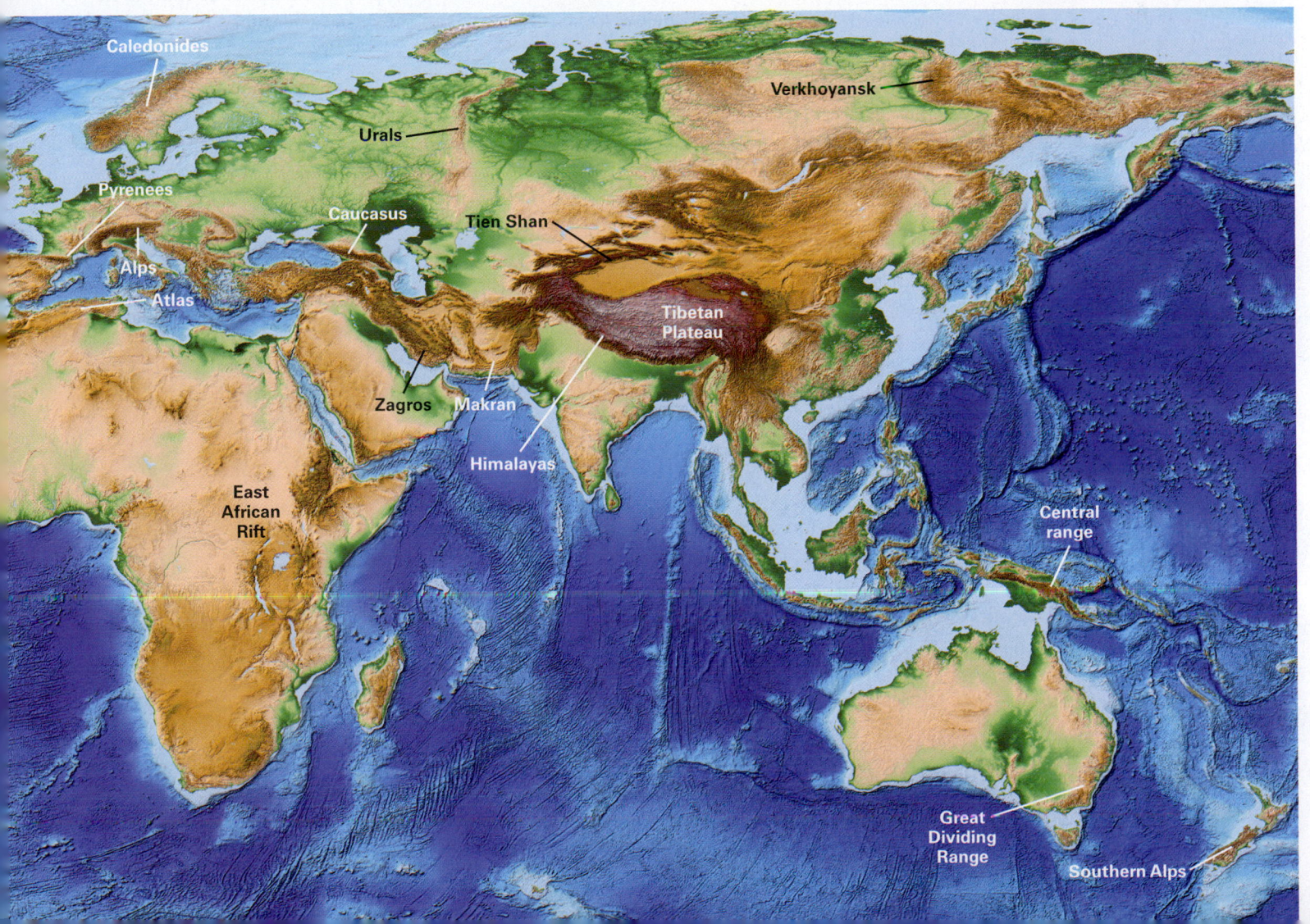

remains. Such crustal scars serve as a permanent monument to what had once been a region of high peaks.

In this chapter, we'll learn about deformation, uplift, and other phenomena that happen during mountain building and will discuss ways in which we can describe and interpret geologic structures. You'll find that, for better or worse, geologists have invented a lot of jargon for geologic structures, in order to streamline the description of these structures. The chapter concludes by addressing the broad question of why mountains form, in the context of plate tectonics theory.

11.2 Rock Deformation in the Earth's Crust

Deformation and Strain

As we've just noted, deformation, meaning the bending, breaking, shortening, stretching, or shearing of rock, produces a variety of geologic structures. To get a visual sense of deformation, let's compare a road cut along a highway in a region that has not undergone orogeny with a cliff face in a region that has.

This road cut, which lies at an elevation of only about 200 m (about 650 ft) above sea level, exposes nearly horizontal beds of sandstone, shale, and limestone—these beds have the same orientation that they had when first deposited (**Fig. 11.3a**). Sand grains in sandstone beds of this outcrop have a nearly spherical shape (the same shape they had when deposited), and clay flakes in the shale lie roughly parallel to the bedding because of compaction. Rock of such outcrops are essentially undeformed, meaning that they contain no geologic structures other than a few joints (**Fig. 11.3b**).

Rocks of the mountain cliff, exposed at an elevation of 3 km (about 10,000 ft), look very different. Here we find layers of quartzite, slate, and marble, the metamorphic equivalents of sandstone, shale, and limestone, respectively. The layers are cut by joints, but they have also been contorted into folds (**Fig. 11.3c, d**). On close examination, we also note that the grains in the quartzite aren't spherical, like those of a sandstone, but have been flattened so that they are somewhat elliptical. And in the slate, clay flakes are aligned to form a type of foliation called slaty cleavage (see Chapter 8). Finally, if we try tracing the quartzite and slate layers along the outcrop face, we find that they abruptly terminate at a sloping surface along which rock has been fragmented. This surface is a fault. In our example, thick layers of marble lie below the fault, so we can infer that the quartzite and slate

FIGURE 11.2 The Matterhorn of the Alps along the Swiss-Italian border. The strata of the peak lay the seafloor during the late Mesozoic. Mountain building thrust the strata upwards so they now reach an elevation of 4.5 km. Erosion continuously sculpts uplifted land, to yield jagged peaks.

must have moved some distance along the fault from where they first formed to get to their present location, juxtaposed against the marble.

Clearly, the beds in the mountain cliff have been deformed, and as a result the cliff exposes a variety of geologic structures. Note that because of deformation, beds no longer have the same shape and position that they had when first formed, and even the shape and orientation of grains has changed. This example illustrates that deformation includes one or more of the following (**Fig. 11.4**): change in location (displacement); change in orientation (rotation); and change in shape (distortion).

Geologists refer to a measure of the change in shape that deformation causes as **strain**. We distinguish among different kinds of strain according to the nature of the shape change. If a layer of rock becomes longer, it has undergone *stretching*, but if the layer becomes shorter, it has undergone *shortening* (**Fig. 11.5a–c**). If a change in shape involves the movement of one part of a rock body past another so that angles between features in the rock change, the result is called *shear strain* (**Fig. 11.5d, e**).

Brittle versus Plastic Deformation

Imagine that a plate tumbles off a table and lands on a hard floor—the plate breaks and smashes into pieces. Similarly, if you strike a glass window with a ball, the window cracks and may even shatter. Such phenomena serve as familiar examples of **brittle deformation** (**Fig. 11.6a, b**). Now imagine that you squeeze a ball of soft dough between a book and a tabletop—

FIGURE 11.3 Deformation changes the character and configuration of rocks.

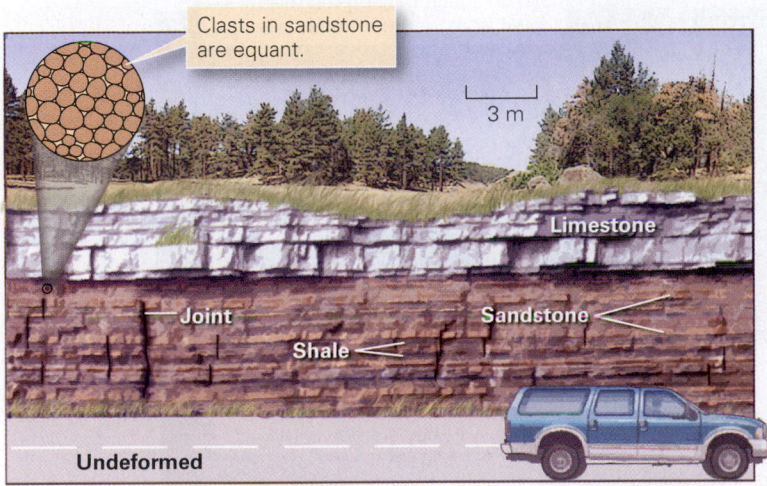

Clasts in sandstone are equant.

Limestone

Joint — Sandstone

Shale

Undeformed

(a) Flat-lying beds of strata along a highway in the Great Plains of North America are essentially undeformed. A few joints, formed when overlying rock eroded away, are visible.

(b) Another example of undeformed, horizontal beds. These form a 100-m-high sea cliff in western Ireland. The vertical lines are joints.

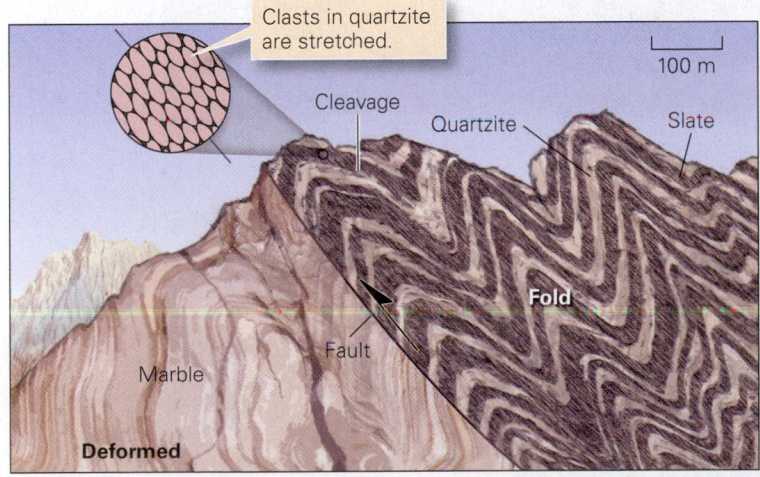

Clasts in quartzite are stretched.

Cleavage

Quartzite — Slate

Fold

Fault

Marble

Deformed

(c) In a mountain belt, deformation may cause layers to undergo folding and faulting. In addition, foliation (such as slaty cleavage) and stretched clasts may develop.

(d) Another example of deformed beds. These have arched into a fold, along the coast of Wales. Note how the beds look "wrinkled."

the dough flattens into a pancake. Similarly, if you bend a stick of chewing gum, it changes from a plane into a curve. During such **plastic deformation** objects change shape without visibly breaking (**Fig. 11.6c, d**). Informally, geologists also refer to plastic deformation as *ductile deformation*, though the terms have slightly different meanings to specialists.

What actually happens within mineral grains during these two different kinds of deformation? Recall that the atoms that make up mineral grains are connected by chemical bonds. During brittle deformation, large numbers of bonds break and stay broken, leading to the formation of a permanent crack across which material no longer connects. During plastic deformation, simplistically, some bonds break, but new ones quickly

form. In this way, the atoms within grains rearrange, and the grains change shape without permanent cracks forming.

Why do rocks inside the Earth sometimes deform brittlely and sometimes plastically? The behavior of a rock depends on a number of variables.

- *Temperature:* Warmer rocks tend to deform plastically, whereas colder rocks tend to deform brittlely. To see this contrast, try an experiment with a candle. Chill a candle in a freezer, then press its middle against the edge of a table—the candle will brittlely snap in two. But if you first warm the candle in an oven, it will plastically bend without breaking when pressed against the table. Heat makes materials softer.

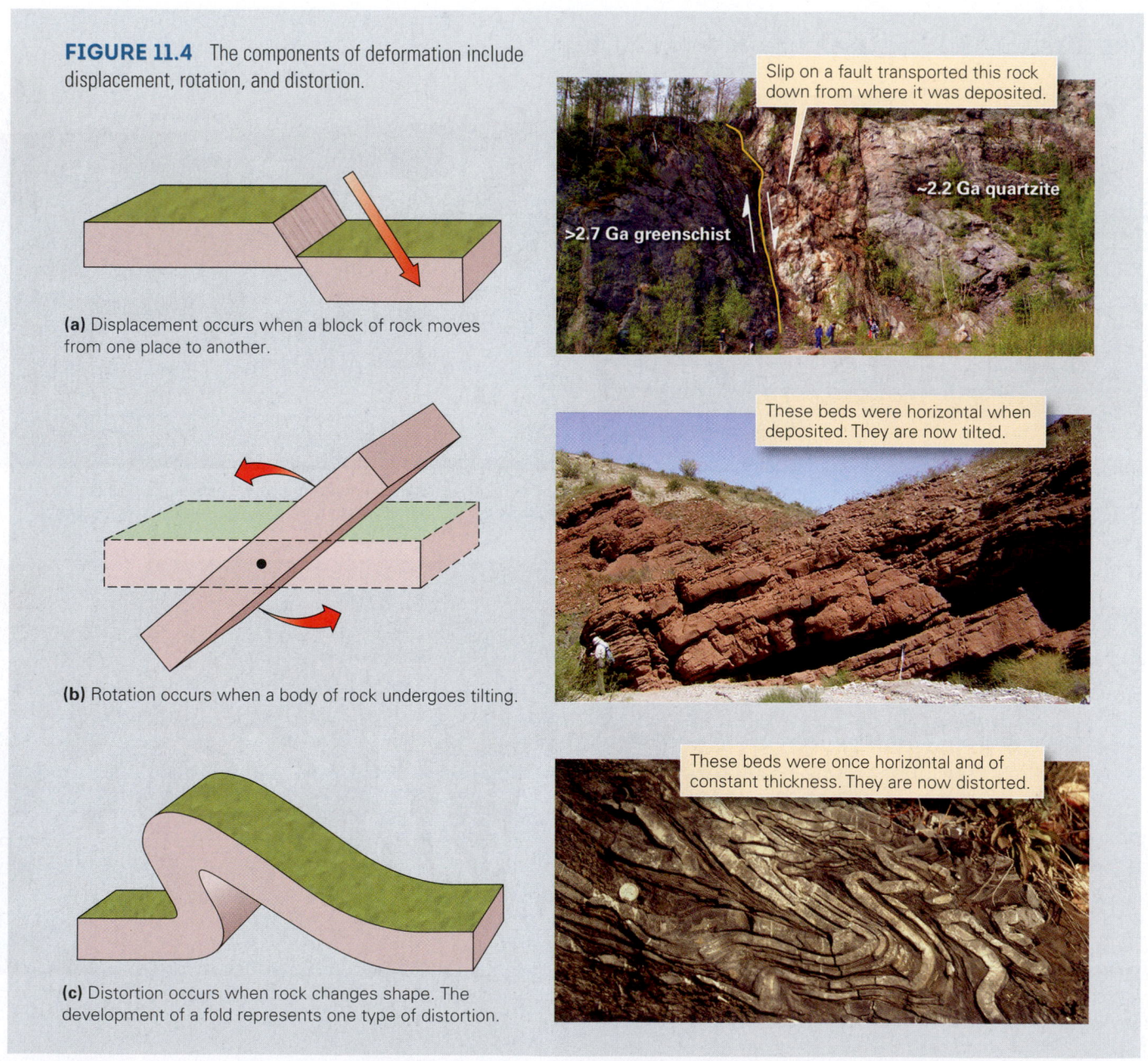

FIGURE 11.4 The components of deformation include displacement, rotation, and distortion.

(a) Displacement occurs when a block of rock moves from one place to another.

Slip on a fault transported this rock down from where it was deposited.

~2.2 Ga quartzite

>2.7 Ga greenschist

(b) Rotation occurs when a body of rock undergoes tilting.

These beds were horizontal when deposited. They are now tilted.

(c) Distortion occurs when rock changes shape. The development of a fold represents one type of distortion.

These beds were once horizontal and of constant thickness. They are now distorted.

FIGURE 11.5 Different kinds of strain in rock. Strain is a measure of the distortion, or change in shape, that takes place in rock during deformation.

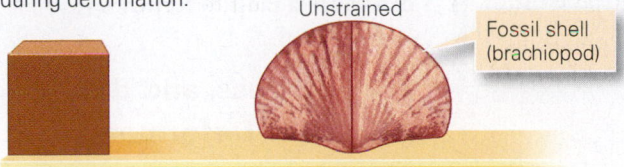

Unstrained

Fossil shell (brachiopod)

(a) An unstrained cube and an unstrained fossil shell (brachiopod).

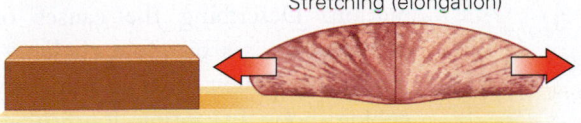

Stretching (elongation)

(b) Horizontal stretching changes the cube into a horizontal brick and elongates the shell.

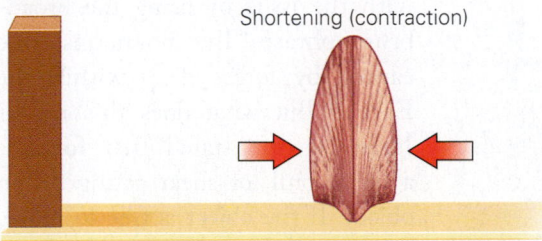

Shortening (contraction)

(c) Horizontal shortening changes the cube into a vertical brick and makes the shell narrower.

Shear

(d) Shear strain tilts the cube and transforms it into a parallelogram and changes angular relationships in the shell.

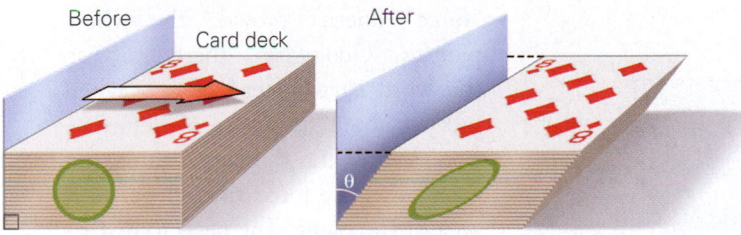

Before   After

Card deck

(e) You can produce shear strain by moving a deck of cards so that each card slides a little with respect to the one below.

- *Pressure:* Under great pressures deep in the Earth, rock behaves more plastically than it does under low pressures near the surface. Pressure effectively prevents rock from separating into fragments.
- *Deformation rate:* A sudden change in shape causes brittle deformation, whereas a slow change in shape causes plastic deformation. For example, if you hit a marble bench with

a hammer, it shatters, but if you leave the bench alone for a century, gravity causes it to gradually sag, without breaking.

- *Composition:* Some rock types are softer than others; for example, halite (rock salt) deforms plastically under conditions in which granite deforms brittlely.

Considering that pressure and temperature both increase with depth in the Earth, geologists find that in typical continental crust, rocks generally behave brittly above 10 to 15 km and plastically below this depth. The depth at which this change in behavior takes place is called the *brittle-plastic transition.* Earthquakes in continental crust happen only above this depth because these earthquakes involve brittle breaking.

FIGURE 11.6 Brittle versus plastic deformation.

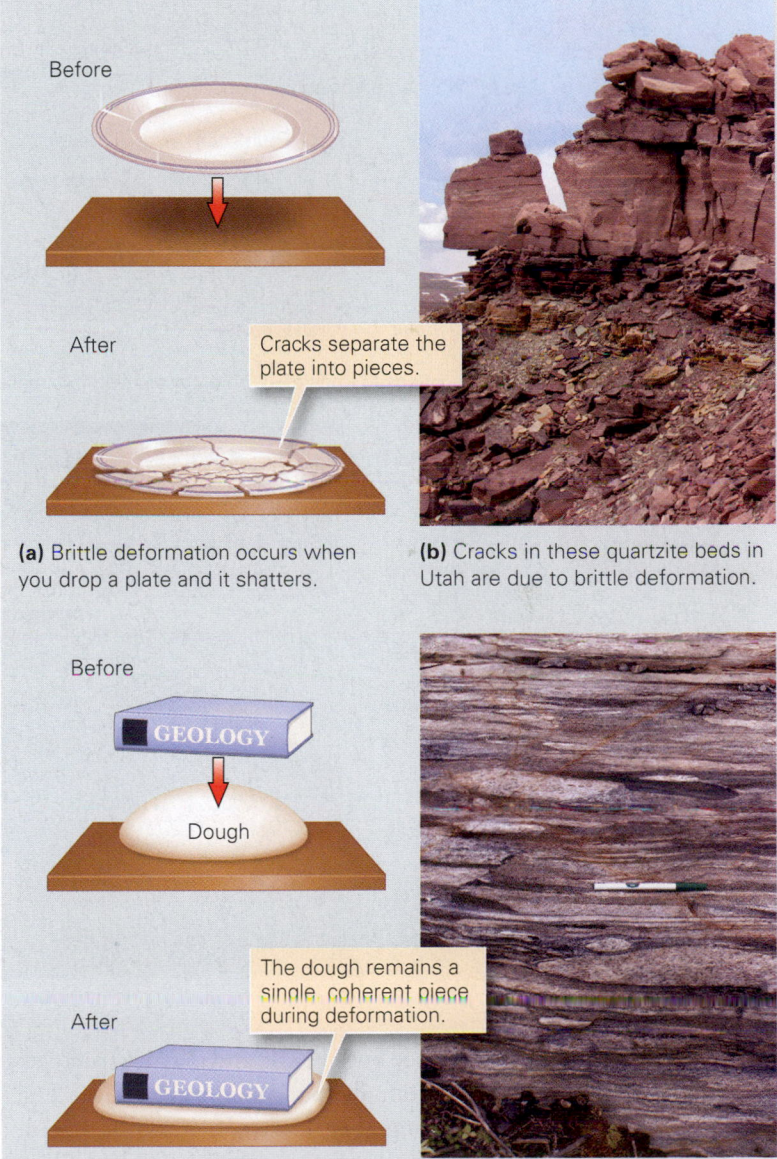

Before

After

Cracks separate the plate into pieces.

(a) Brittle deformation occurs when you drop a plate and it shatters.

(b) Cracks in these quartzite beds in Utah are due to brittle deformation.

Before

GEOLOGY

Dough

The dough remains a single, coherent piece during deformation.

After

GEOLOGY

(c) Plastic deformation occurs when you squash a ball of dough.

(d) The quartzite cobbles in the conglomerate were flattened plastically.

FIGURE 11.7 There are several kinds of stress.

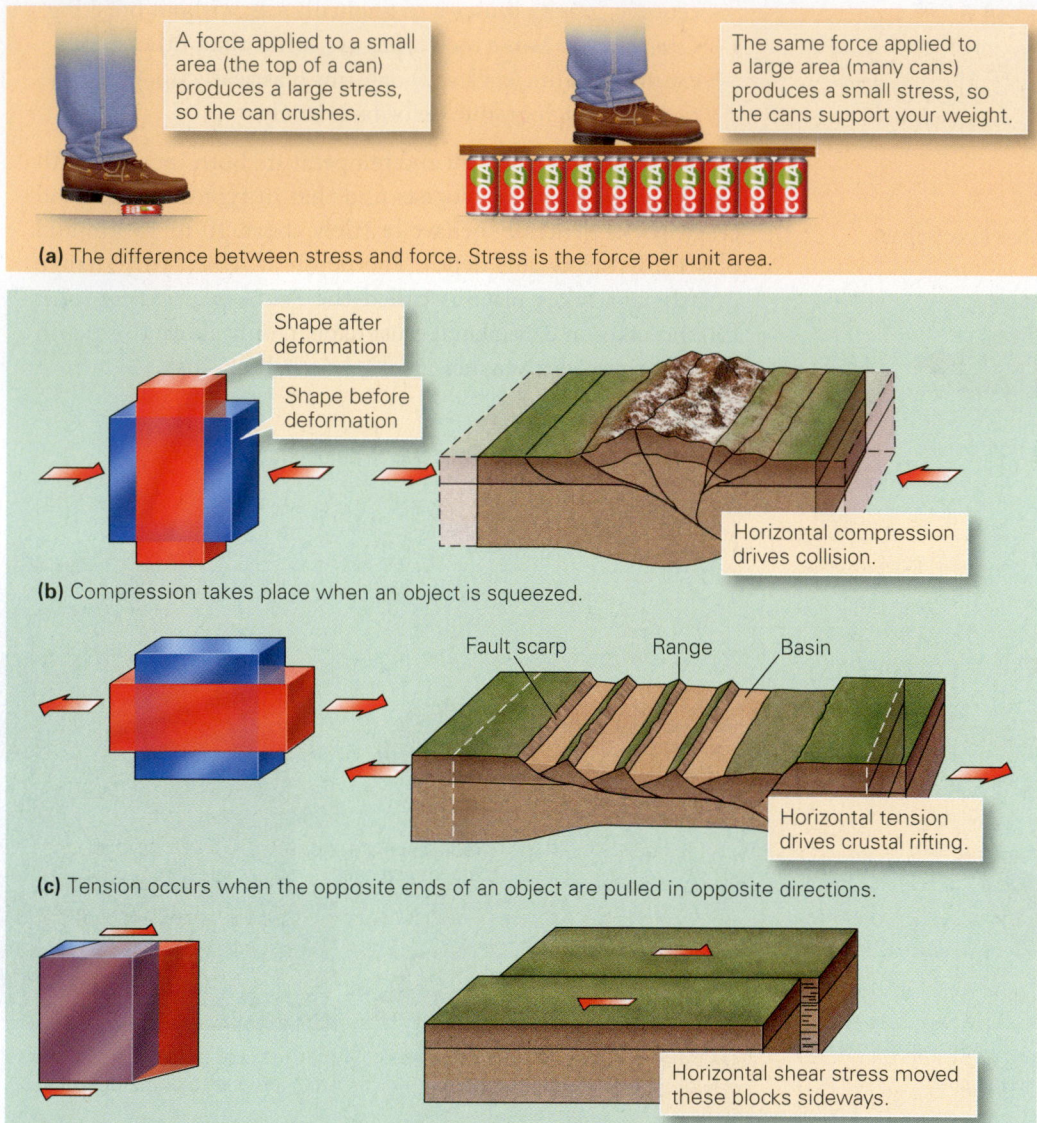

A force applied to a small area (the top of a can) produces a large stress, so the can crushes.

The same force applied to a large area (many cans) produces a small stress, so the cans support your weight.

(a) The difference between stress and force. Stress is the force per unit area.

Shape after deformation

Shape before deformation

Horizontal compression drives collision.

(b) Compression takes place when an object is squeezed.

Fault scarp   Range   Basin

Horizontal tension drives crustal rifting.

(c) Tension occurs when the opposite ends of an object are pulled in opposite directions.

Horizontal shear stress moved these blocks sideways.

(d) Shear stress develops when one surface of an object slides relative to the other surface.

A diver underwater feels pressure.

(e) Pressure occurs when an object feels the same stress on all sides.

In some cases, you can see both brittle and plastic structures in the same outcrop. For example, our mountain cliff (see Fig. 11.3c) displays both faulting (brittle deformation) and folding (plastic deformation). Such an occurrence may seem like a paradox at first. But a juxtaposition of different styles can happen because of changes in the deformation rate during orogeny.

Slow deformation yielded the folds, whereas a pulse of rapid deformation can cause a fault to form.

Force, Stress, and the Causes of Deformation

Up to this point, we've focused on picturing the consequences of deformation. Describing the causes of deformation is a bit more challenging in the context of an introductory geology book. In captions for displays about mountain building, museums and national parks typically dispense with the issue by using the broad-brush phrase "The mountains were caused by forces deep within the Earth." But what does this mean? Isaac Newton stated that **force** is a push, pull, or shear acting on an object. If the object is free to move, a force will cause the object to speed up, slow down, rotate, or change direction. In the context of geology, plate interactions and continent-continent collisions apply forces to rock and thus cause rock to change location, orientation, or shape. In other words, the application of forces in the Earth indeed causes deformation.

But describing the cause of deformation in terms of the action of a "force" doesn't provide the whole picture. Geologists, therefore, use the word *stress* instead of *force* when talking about the cause of deformation. We define the **stress** acting on a plane as the force applied per unit area of the plane. The need to distinguish between stress and force arises because the actual consequences of applying a force depend not just on the amount of force but also on the area over which the force acts. A pair of simple experiments shows why (**Fig. 11.7a**). Experiment 1: Stand on a single, empty aluminum can. All of your weight—a force—focuses entirely on the can, and the can crushes. Experiment 2: Place a board atop 100 cans and stand on the board. In this case, your weight spreads across 100 cans, and the cans don't crush. In both experiments, the force caused by the weight of

your body was the same, but in Experiment 1 the force was applied over a small area so a large stress developed, whereas in Experiment 2 the same force was applied over a large area so only a small stress developed—a large stress could crush a can, but a small one could not. How does this concept apply to geology? During mountain building, the force of one plate interacting with another applies across the broad area of contact between the two plates, so the deformation resulting at any specific location actually reflects the stress developed at that location, not on the total force produced by the plate interaction.

Different kinds of stress occur in rock bodies (**Fig. 11.7b–e**). **Compression** takes place when a rock is squeezed, *tension* occurs when a rock is pulled apart, and *shear stress* develops when one part of a rock body moves sideways past another. **Pressure** refers to a special stress condition that happens when the same push acts on all sides of an object.

Note that *stress* and *strain* have very different meanings to geologists, even though people tend to use these words interchangeably in everyday English: "stress" specifically refers to the amount of force applied per unit area of a rock, whereas strain specifically refers to a change in shape of a rock. Thus, stress causes strain—compression causes shortening, tension leads to stretching, and shear stress produces shear strain. Pressure can cause an object to become smaller but will not cause it to move, rotate, or change shape. With our knowledge of stress and strain, we can now look at the nature and origin of various classes of geologic structures and discuss their significance.

11.3 Brittle Structures

Joints and Veins

If you look at the photographs of rock outcrops in this book, you'll notice thin dark lines that cross the rock faces. These lines represent traces of natural cracks along which the rock broke and separated into two pieces during brittle deformation. Geologists refer to such natural cracks as **joints** (**Fig. 11.8a, b**). Rock bodies do not slide past each other on joints. Since joints are roughly planar structures, we define their orientation by their *strike* and *dip* (**Box 11.1**).

FIGURE 11.8 Examples of joints and veins.

(a) Prominent vertical joints cut red sandstone beds in Arches National Park, Utah, as seen from the air.

(b) Vertical joints on a cliff face in shale near Ithaca, New York.

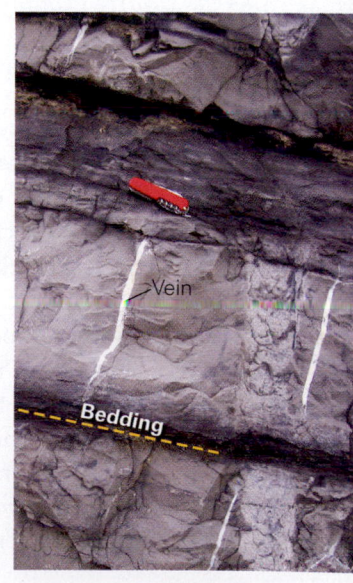

(c) Milky white quartz veins cut across gray limestone beds.

BOX 11.1   CONSIDER THIS . . .

Describing the Orientation of Geologic Structures

When discussing geologic structures, it's important to be able to communicate information about their orientation. For example, does a fault exposed in an outcrop at the edge of town continue beneath the nuclear power plant 3 km to the north, or does it go beneath the hospital 2 km to the east? If we knew the fault's orientation, we might be able to answer this question. To describe the ori- entation of a geologic structure, geologists picture the structure as a simple geometric shape, then specify the angles that the shape makes with respect to a horizontal plane (a flat surface parallel to sea level), a vertical plane (a flat surface perpendicular to sea level), and the north direction (a line of longitude).

Let's start by considering planar struc- tures such as faults, beds, and joints. We call these structures planar because they resem- ble a geometric plane. A planar structure's orientation can be specified by its **strike** and **dip**. The strike is the angle between an imaginary horizontal line (the strike line) on the structure and the direction to true north (**Fig. Bx11.1a, b**). We measure the strike with a special type of compass that has a level bubble so that we can be sure that the compass surface is exactly horizontal (**Fig. Bx11.1c**). The dip is the angle of the struc- ture's slope—more precisely, it is the angle between a horizontal plane and the dip line (an imaginary line parallel to the steep- est slope on the structure), as measured in a vertical plane perpendicular to the strike

FIGURE Bx11.1 Specifying the orientation of planar and linear structures.

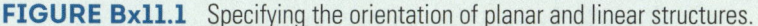

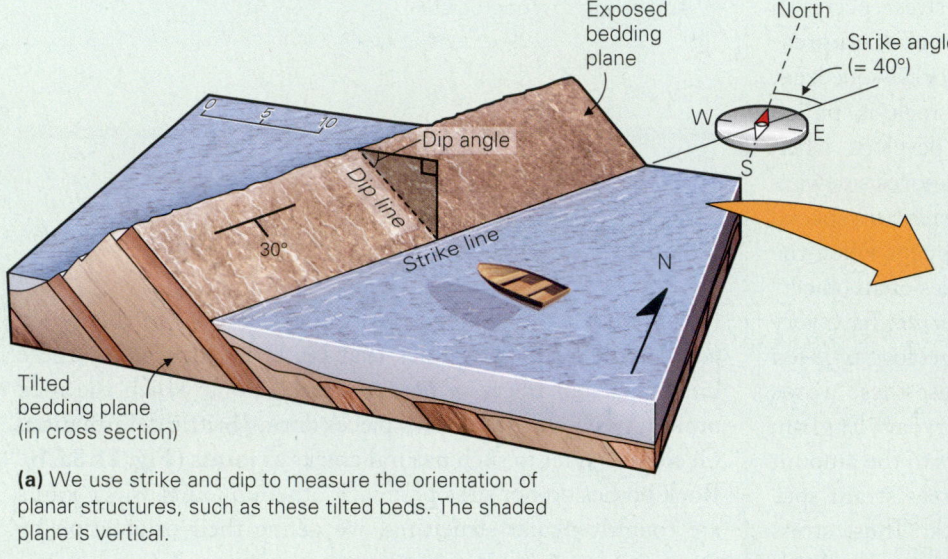

(a) We use strike and dip to measure the orientation of planar structures, such as these tilted beds. The shaded plane is vertical.

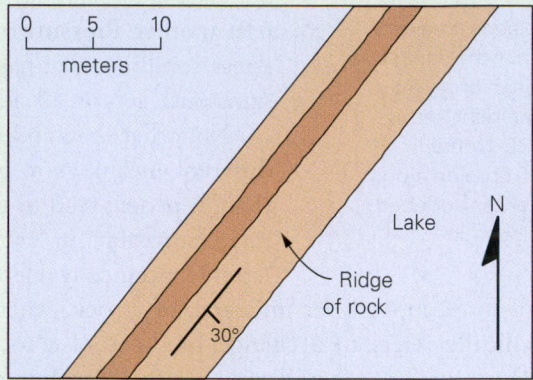

(b) On a map, the line segment represents the strike direction and the tick on the segment represents the dip direction. The number indicates the dip angle as measured in degrees.

Joints develop in response to tensile stress in brittle rock: a rock splits open because it has been pulled slightly apart. Joints may form for a variety of geologic reasons. For example, some joints form when a rock cools and contracts, because contraction makes one part of a rock pull away from the adjacent part. Oth- ers develop when rock formerly at depth undergoes a decrease in pressure as overlying rock erodes away and thus changes shape slightly by expanding vertically and contracting horizontally. Still others form when relatively brittle rock layers bend.

Rock bodies may contain two categories of joints. *Systematic joints* are long planar cracks that occur fairly regularly through a rock body, whereas *nonsystematic joints* are short cracks that occur in a range of orientations, are randomly spaced, and may be irregular or curved. A group of systematic joints constitutes a *joint set*, a spectacular example of which can be seen in sand- stone beds of Arches National Park in Utah (see Fig. 11.8a). Erosion along the joints produced narrow gullies. In sedimen- tary rocks, systematic joints typically are vertical planes (see Fig. 11.8b). If groundwater seeps through joints underground for a long period of time, minerals such as quartz or calcite may precipitate and fill the joint. Such mineral-filled joints are a type of **vein** and look like white stripes cutting across a body of rock (**Fig. 11.8c**).

Geotechnical engineers, people who study the geologic set- ting of construction sites, pay close attention to jointing when recommending where to put roads, dams, and buildings. Water

(**Fig. Bx11.1d**). We measure the dip angle with a clinometer, a type of protractor. (A geologist's protractor typically contains a built-in compass.) A horizontal plane has a dip of 0°, and a vertical plane has a dip of 90°. We represent strike and dip on a geologic map using the symbol shown in Figure Bx11.1b.

A linear structure resembles a geometric line rather than a plane; examples of linear structures include scratches or grooves on a fault surface. Geologists specify the orientation of linear structures by giving their **plunge** and **bearing** (**Fig. Bx11.1e**). The plunge is the angle between a line and horizontal in the vertical plane that contains the line. A horizontal line has a plunge of 0°, and a vertical line has a plunge of 90°. The bearing is the compass heading of the line—more precisely, the angle between the projection of the line on the horizontal plane and the direction to true north.

The edge of the compass is parallel to the strike.

The intersection of the water and the rock is horizontal.

(c) Geologists use a Brunton compass to measure strike and dip.

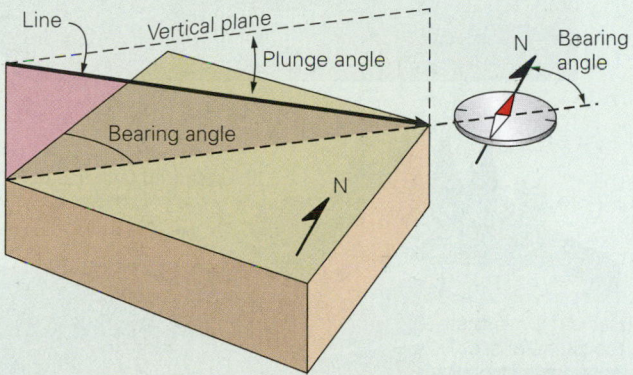

(e) To specify the orientation of a line, we use plunge and bearing.

Dipping beds intersect a calm sea.

Bedding

(d) The intersection of dipping beds with the horizontal water surface is a strike line. The slope of the bed surface is the dip.

flows much more easily through joints than it does through solid rock, so it would be a bad investment to situate a water reservoir over rock containing lots of joints—the water would leak down into the joints. Also, building a road on a steep cliff composed of jointed rock could be risky, for joint-bounded blocks separate easily from bedrock, and the cliff might collapse.

Faults: Surfaces of Slip

After the San Francisco earthquake of 1906, geologists found a rupture that seemingly had torn the land surface near the city. Where this rupture crossed orchards, it offset rows of trees, and where it crossed a fence, it broke the fence in two—

the western side of the fence moved northward by about 2 m (see Fig. 10.4a). The rupture represents the trace of the San Andreas fault. As we have seen, a **fault** is a fracture on which sliding occurs, and slip events, or *faulting*, can generate earthquakes. Faults, like joints, are more or less planar structures, so we represent their orientation by strike and dip (see Box 11.1).

Faults have formed throughout Earth history. Some are currently *active* in that sliding has been occurring on them in recent geologic time, but most are *inactive*, meaning that sliding on them ceased long ago. Some faults, such as the San Andreas, intersect the ground surface and thus displace the ground when they move. Others, known as *blind faults*, accommodate the sliding of rocks in the crust at depth and remain

FIGURE 11.9 The different categories of faults.

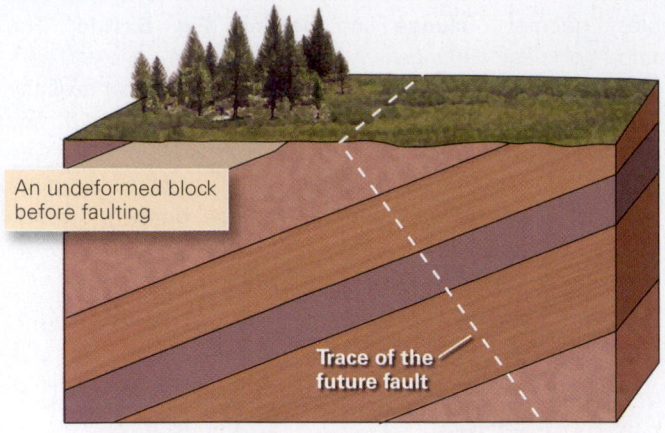

An undeformed block before faulting

Trace of the future fault

(a) The block of crust above a non-vertical fault is the hanging wall, whereas the block below is the footwall.

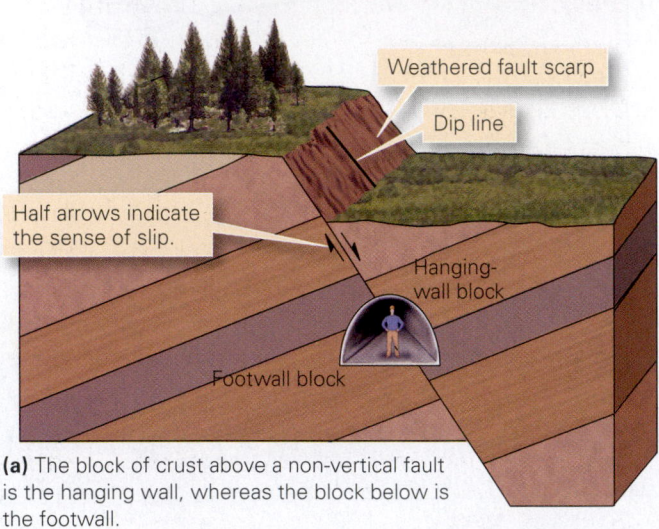

Weathered fault scarp

Dip line

Half arrows indicate the sense of slip.

Hanging-wall block

Footwall block

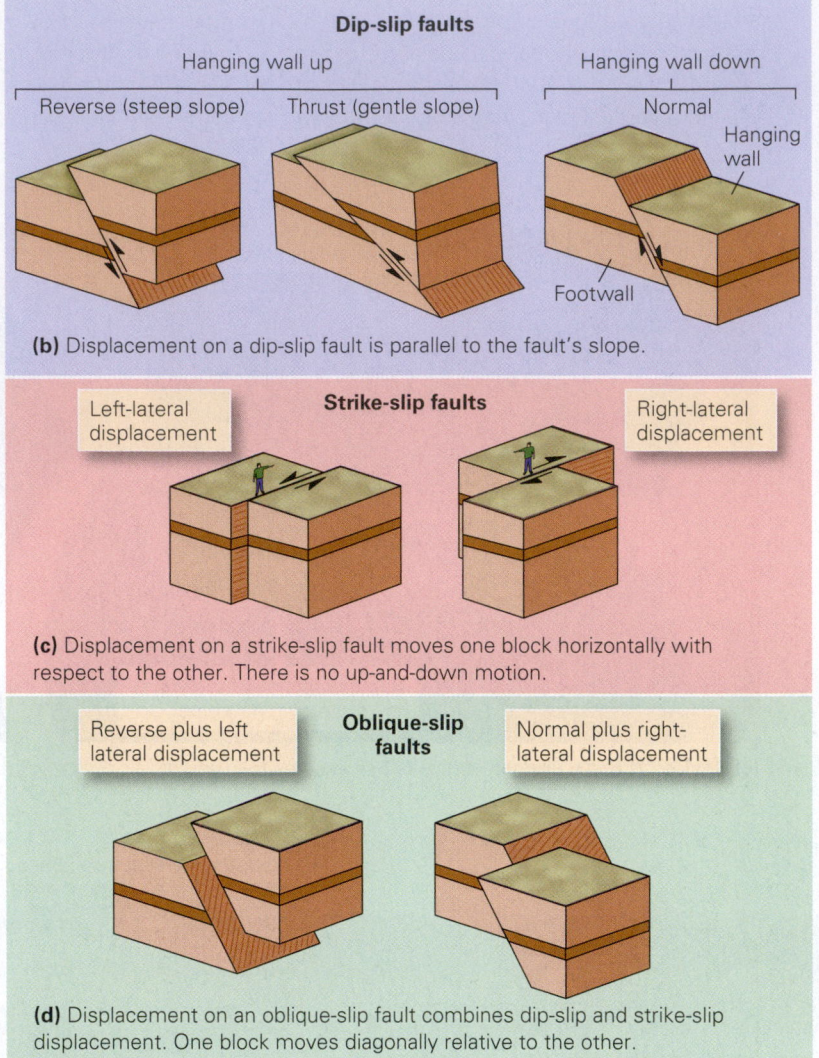

Dip-slip faults

Hanging wall up

Reverse (steep slope)   Thrust (gentle slope)

Hanging wall down

Normal

Hanging wall

Footwall

(b) Displacement on a dip-slip fault is parallel to the fault's slope.

Strike-slip faults

Left-lateral displacement

Right-lateral displacement

(c) Displacement on a strike-slip fault moves one block horizontally with respect to the other. There is no up-and-down motion.

Oblique-slip faults

Reverse plus left lateral displacement

Normal plus right-lateral displacement

(d) Displacement on an oblique-slip fault combines dip-slip and strike-slip displacement. One block moves diagonally relative to the other.

invisible at the surface unless they are later exposed by erosion. Geologists refer to the intersection of a fault with the land surface, regardless of whether the fault is active and has offset the surface, or if it was blind and later exposed by erosion, as the **fault trace**, or *fault line*.

Geologists study faults not only because the movement on some faults causes earthquakes but also because faults juxtapose bodies of rock that did not originally lie adjacent to each other and thus complicate the arrangement of rocks in the crust. For example, in our mountain cliff (see Fig. 11.3c), movement on a fault placed quartzite and slate beds against marble beds. Geologists must understand these rearrangements in order to interpret orogenic events and to predict where resources lie underground.

Fault Classification

Not all faults result in the same kind of crustal deformation—some accommodate horizontal shear, some accommodate

shortening, and some accommodate stretching. It's important for geologists to distinguish among different kinds of faults in order to interpret their tectonic significance. Fault classification focuses on two characteristics of faults: (1) the dip, or slope, of the fault surface (see Box 11.1)—the dip can be vertical, horizontal, or any angle in between; and (2) the *shear sense* across the fault—by shear sense, we mean the direction that material on one side of the fault moved relative to the material on the other side. With this concept in mind, let's consider the principal kinds of faults. (This description builds on that of Chapter 10 and adds an additional category.)

- *Dip-slip faults:* On a dip-slip fault, movement is parallel to the dip line, a line on the fault surface that is oriented down the steepest slope of the fault surface. (*Dip* refers to the angle that this line makes relative to horizontal; see Box 11.1.) We distinguish among different types of dip-slip faults based on whether the *hanging-wall* block, meaning the material above the fault surface, slides up or down relative to the *footwall* block, the material below the fault surface (**Fig. 11.9a, b**). If the hanging-wall block slides up, the fault is a **reverse fault**. Geologists refer to a reverse fault with a gentle dip (<30°) as a **thrust fault**. Reverse faults and thrust faults accommodate shortening of the crust, as happens during continental collision. If the hanging-wall block slides down, the fault is a **normal fault**. Normal faults accommodate stretching of the Earth's crust, as happens during rifting. Note that because of the strain that develops during movement on dip-slip faults, the overall length of a crustal block will be shorter after reverse faulting and longer after normal faulting.

- *Strike-slip faults:* A **strike-slip fault** is a fault on which the slip direction is parallel to a strike line, a line drawn parallel to the strike of the fault. (*Strike* is the compass trend of a horizontal line on a surface; see Box 11.1). This means that the block on one side of the fault slips sideways relative to the block on the other side, and there is no up-or-down motion (**Fig. 11.9c**). The faults that occur at transform-plate boundaries are strike-slip faults. Most strike-slip faults have a steep to vertical dip. Geologists describe the motion on strike-slip faults based on the shear sense as viewed when you are facing the fault and looking across it. If the block across the fault slipped to your left, the fault is a *left-lateral* strike-slip fault, and if the block slipped to the right, the fault is a *right-lateral* strike-slip fault.

- *Oblique-slip faults:* On an oblique-slip fault, sliding occurs diagonally. In effect, an oblique-slip fault is a combination of a strike-slip and a dip-slip fault (**Fig. 11.9d**).

Recognizing Faults

How do you recognize a fault when you see one? The most obvious criterion is the occurrence of **displacement**, or *offset*, meaning the amount of movement across a fault plane. After displacement has occurred, layers on one side of a fault are not continuous with layers on the other side. If the layers are distinctive, you easily can see the displacement in an outcrop (**Fig. 11.10a, b**). In some cases, faults juxtapose two different rock units (see Fig. 11.3c). Typically, thrust or reverse faults cutting sedimentary beds place older beds on younger ones, whereas normal faults place younger beds on older. In some cases, layers of rock cut by a fault undergo folding during or just before slip. Geologists informally refer to the resulting folds adjacent to a fault as *drag folds* (see Fig. 11.10a)

Faults may also leave their mark on the landscape. Those that intersect the ground surface while they are active can displace natural landscape features, such as stream valleys or glacial moraines (**Fig. 11.10c**), and human-made features, such as highways, fences, or rows of trees in orchards. Displacement of the land surface on a dip-slip or oblique-slip fault yields a small step called a **fault scarp** along the fault trace (**Fig. 11.11a**). Faults that are no longer active may still be readily identifiable in the landscape; because faults tend to break up rock, their trace can be more easily eroded. If this happens, the fault trace will be marked by a linear valley (see the inset photo of the Great Glen fault on p. 321). And, if a fault juxtaposes two different rock units with different resistances to erosion, the land surface may, over time, erode to be lower on the side of the weaker rock, so the fault trace will be marked by a noticeable escarpment.

Faulting under brittle conditions may crush or break adjacent rock. If this shattered rock consists of visible angular fragments, it is called *fault breccia* (**Fig. 11.11b**), but if it consists of a fine powder, it is called *fault gouge*. Because fault gouge is so fine grained, it may quickly undergo chemical weathering and turn into clay. Some fault surfaces are polished and grooved by the movement of the hanging wall past the footwall. Polished fault surfaces are **slickensides**, and linear grooves on fault surfaces are **slip lineations** (**Fig. 11.11c**).

The shear on some faults takes place under plastic conditions at depth in the crust. Where this happens, rock does not break up into breccia or gouge along the fault zone, because the rock is too soft; rather, it shears plastically to form a band of fine-grained foliated rock called mylonite. The fine grain size of mylonite results not from brittle fracturing during shear but instead from a type of metamorphic recrystallization that subdivides large grains into small ones. Geologists commonly refer to faults whose movement occurred plastically as a **shear zone** (**Fig. 11.12**).

Fault Systems

An array of several related faults comprises a **fault system**. In systems of dip-slip faults, individual faults in the system typically merge at depth with a nearly horizontal **detachment fault**. Displacements in a *thrust-fault system* shorten the crust because slip makes slices of the crust overlap like shingles (**Fig. 11.13a, b**). During the development of a thrust system, strata within the *thrust sheets*, the bodies of rock transported by a thrust fault, tend to become folded. Geologists commonly refer to areas in which a thrust system and associated folds develop as a **fold-thrust belt**. The Canadian Rocky Mountains serve as an example of a fold-thrust belt—the rugged cliffs and peaks of this mountain belt formed when compressive stress shoved eastward rocks from the west (**Fig. 11.13c**). The process resembles the building of a pile of snow or sand in front of a plow as the plow moves forward and pushes it.

Displacements in a *normal-fault system*, on the other hand, stretch the crust as slivers of crust drop down (**Fig. 11.13d**). Many normal-fault systems consist of parallel faults that curve to shallower dips at depth, where they merge with the

FIGURE 11.10 Recognizing fault displacement in the field.

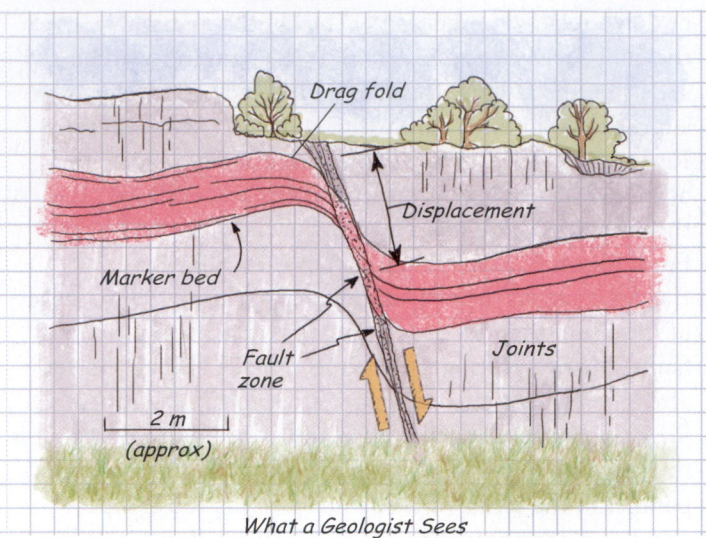

What a Geologist Sees

(a) A steep normal fault has displaced a distinctive red bed (a marker layer). Note that the displacement formed a 0.5-m-wide fault zone of broken rock. Drag folds developed adjacent to the fault.

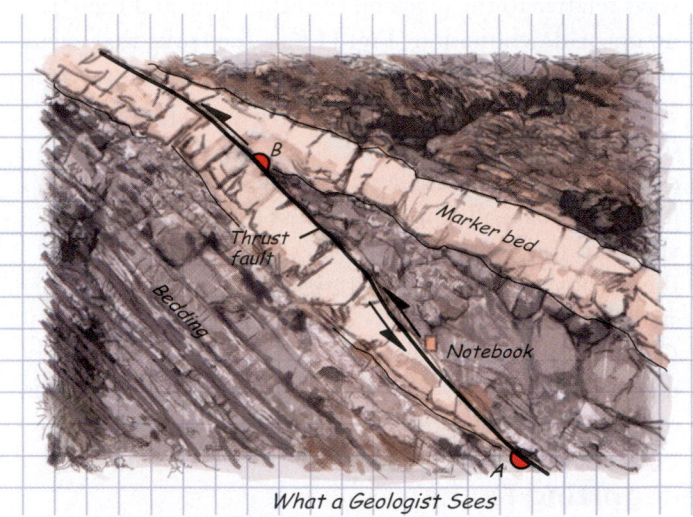

What a Geologist Sees

(b) Slip on a thrust fault caused one part of the light-colored marker bed to be shoved over another part, as emphasized in the drawing. Note that the beds are tilted to the right. The distance between the red dots is the displacement.

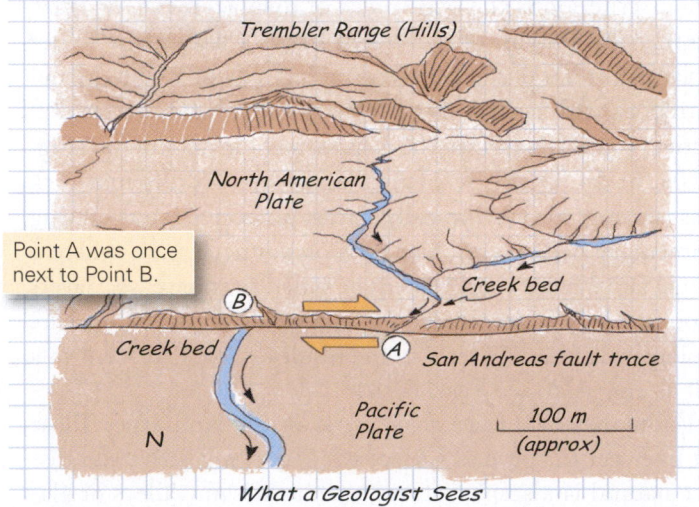

What a Geologist Sees

(c) An aerial photograph of a portion of the San Andreas Fault. As emphasized by the sketch, the fault offsets a stream channel in a right-lateral sense.

FIGURE 11.11 Features of exposed fault surfaces.

A scarp due to slip on a normal fault in Nevada.

(a) A fault scarp develops where faulting displaces the land surface.

(b) This fault breccia consists of irregular fragments of light-colored rock.

Striation orientation

(c) Slip lineations or striations on the surface of a strike-slip fault may look like grooves or scratches.

FIGURE 11.12 An outcrop of a small shear zone in granite. Note that the boundaries of the shear zone are gradational. Note that a foliation has developed in the shear zone. A portion of the edge is highlighted.

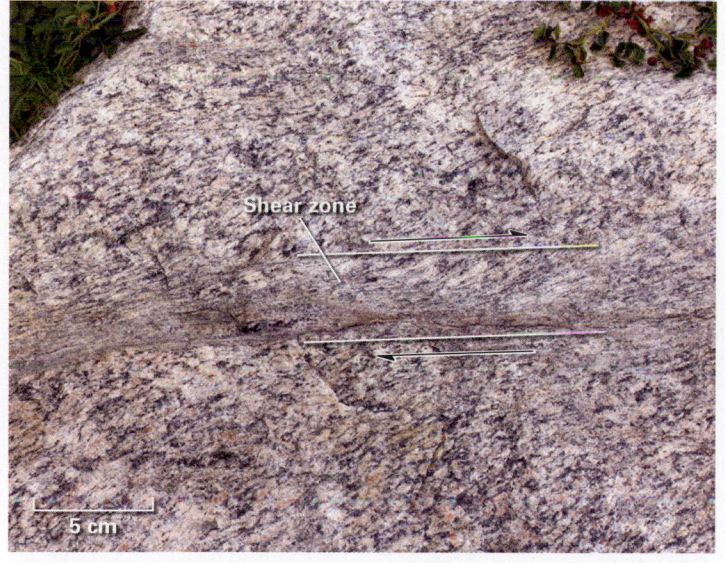

Shear zone

5 cm

detachment. As slip progresses on these curved faults, the hanging-wall block rotates, and a wedge-shaped space between the fault surface and the tilted top of the hanging-wall block develops. This space, a *half graben*, fills with sediments to form a basin. Locally, two normal faults dip toward each other—the keystone-shaped block that drops down between them is a *graben*, whereas the high block between adjacent grabens is a *horst* (**Fig. 11.13e**). Normal fault systems develop in rifts.

Take-Home Message

Brittle structures include joints (cracks), veins (mineral-filled cracks), and faults (fractures on which sliding occurs). Geologists distinguish among different kinds of faults based on the relative displacement across the fault. Slip shatters or pulverizes rock and can polish and scratch a fault's surface. You can identify faults by searching for offset layers or by steps in the landscape. Slip on systems of faults can accommodate significant crustal stretching or shortening.

QUICK QUESTION: How can you distinguish between a joint and a fault in the field?

11.4 Folds and Foliations

The Geometry of Folds

Imagine a carpet lying flat on the floor. Push on one end of the carpet, and it will wrinkle or contort into a series of wave-like curves. Stresses developed during mountain building can similarly warp or bend bedding, foliation, veins, or other planar features in rock. The result—a curving surface—is called a **fold**.

Not all folds look the same—some look like arches, some like troughs, and some have other shapes. To describe these

FIGURE 11.13 Examples of fault systems, arrays of related faults formed in sequence during a deformation event.

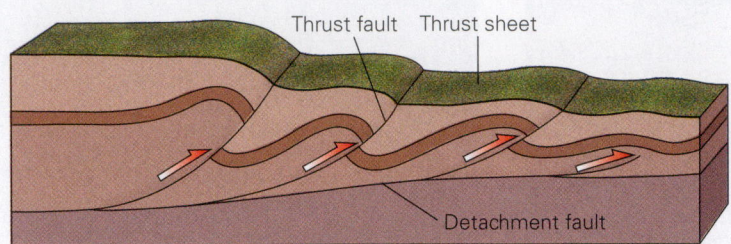

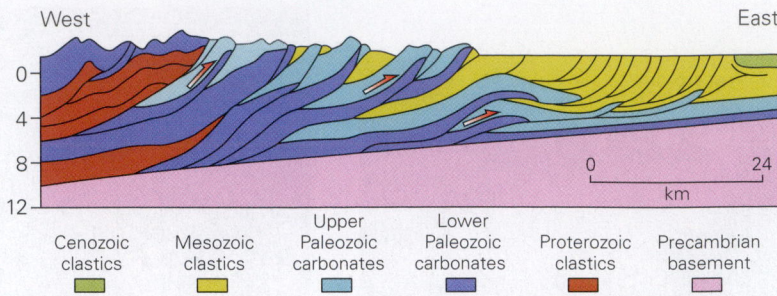

West   East

		Upper	Lower		
Cenozoic	Mesozoic	Paleozoic	Paleozoic	Proterozoic	Precambrian
clastics	clastics	carbonates	carbonates	clastics	basement

(a) In this thrust-fault system, several related thrust faults merge at depth with a detachment fault. Note that displacement on the thrusts shortens the layers of rock above the detachment.

(b) A cross-section of the Canadian Rockies, showing that the range is made of many thrust sheets, that were pushed toward the east.

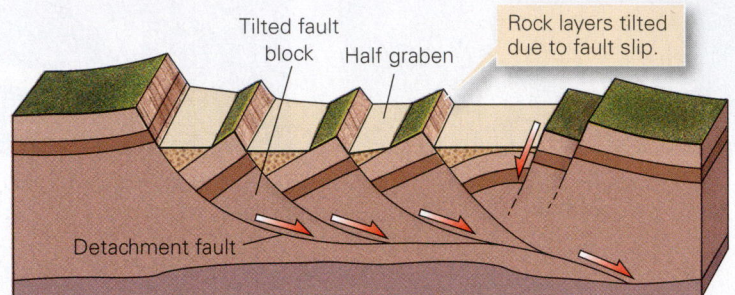

(c) A close-up photo of the folding within one of the thrust slices of the Canadian Rockies. Snow highlights curving layers.

(d) A simplified diagram showing a normal-fault system. Here, several faults merge with a detachment at depth.

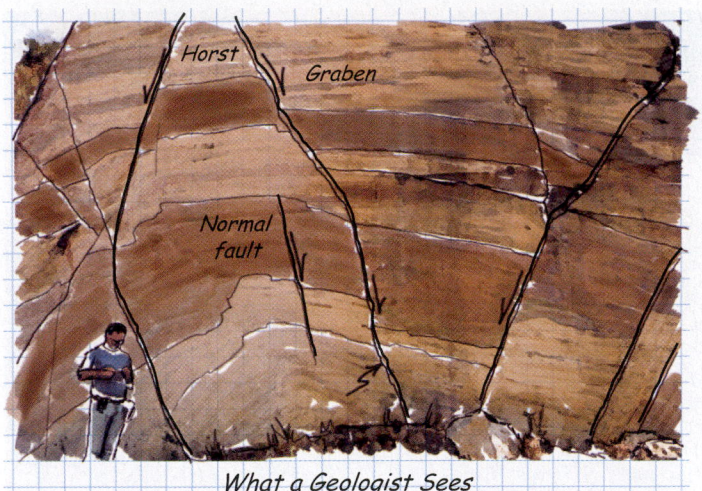

(e) An example of horsts and grabens exposed in the wall of a marble quarry in Brazil.

What a Geologist Sees

---

shapes efficiently, we must first label the parts of a fold. The **hinge** refers to a line along which the curvature is greatest, and the **limbs** are the sides of the fold that display less curvature. For example, if the fold looks like an arch, the hinge lies at the top. Geologists refer to the imaginary plane that contains the hinges of successive layers as the **axial plane** of the fold. With these terms in hand, we can distinguish among the following characteristics.

- *Anticlines, synclines, and monoclines:* Folds that have an arch-like shape in which the limbs dip away from the hinge are called **anticlines** (**Fig. 11.14a**), whereas folds with a trough-like shape in which the limbs dip toward the hinge are called **synclines** (**Fig. 11.14b**). Where a fold intersects the ground surface, the oldest beds crop out near the hinge and the youngest away from the hinge in an anticline; the opposite is true in a syncline. A

FIGURE 11.14 Geometric characteristics of folds.

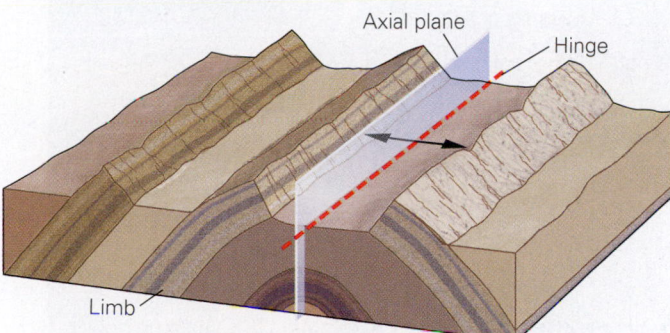

(a) An anticline looks like an arch. The beds dip away from the hinge.

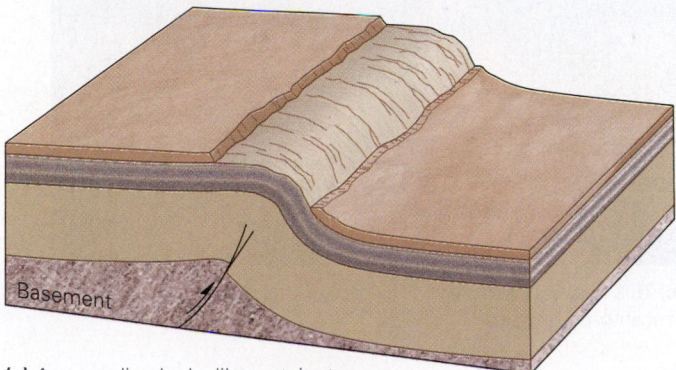

(c) A monocline looks like a stair step and is commonly draped over a fault block.

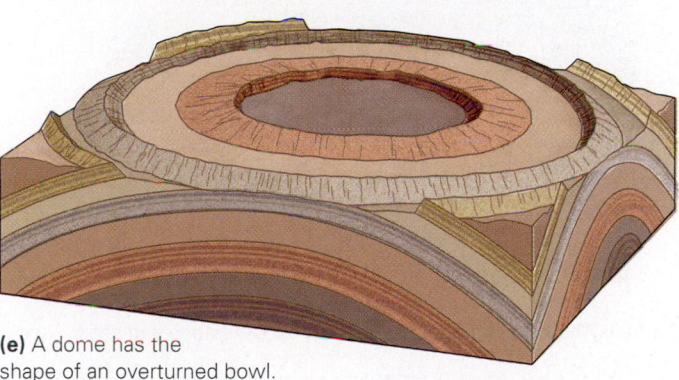

(e) A dome has the shape of an overturned bowl.

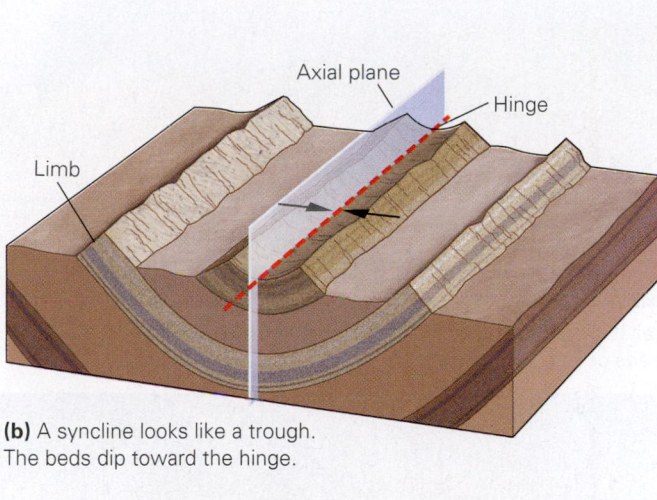

(b) A syncline looks like a trough. The beds dip toward the hinge.

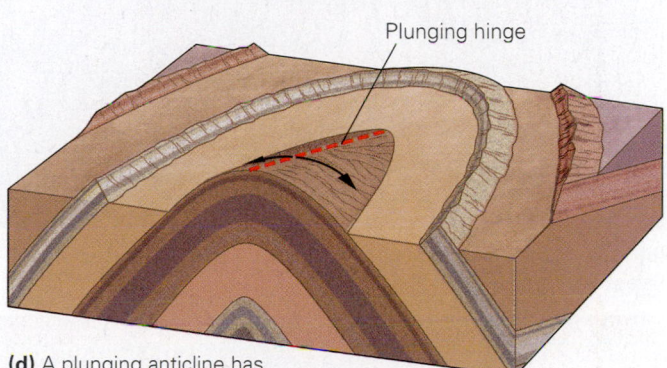

(d) A plunging anticline has a tilted hinge.

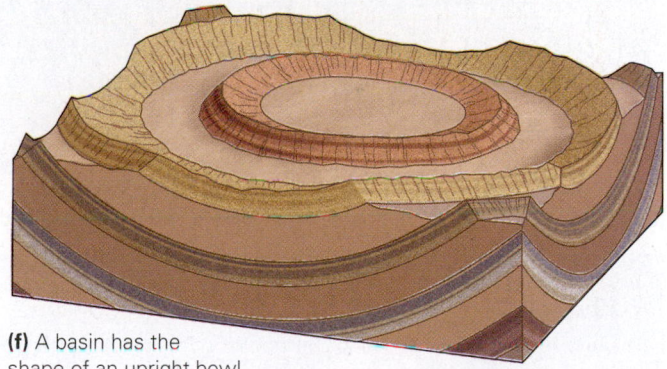

(f) A basin has the shape of an upright bowl.

monocline has the shape of a carpet draped over a stair step (**Fig. 11.14c**).

- *Nonplunging and plunging folds:* If the hinge is horizontal, the fold is a nonplunging fold, but if the hinge tilts, the fold is called a plunging fold (**Fig. 11.14d**).
- *Domes and basins:* A fold with the shape of an overturned bowl is a **dome**, whereas a fold shaped like an upright bowl is a **basin** (**Fig. 11.14e, f**). Domes and basins that intersect

the ground surface both display circular outcrop patterns that look like bull's-eyes—the oldest layer crops out in the center of a dome, whereas the youngest layer crops out in the center of a basin.

Using the geological terms for describing folds we've just presented, see if you can identify the various folds shown in **Figure 11.15**.

FIGURE 11.15 Examples of folds on outcrops and in the landscape.

(a) This anticline, exposed in a road cut near Kingston, New York, involves beds of Paleozoic limestone.

(b) This syncline, exposed in a road cut in Maryland, involves beds of Paleozoic sandstone and shale.

(c) This fold, exposed along the coast of Brazil, occurs in Precambrian gneiss.

(d) A train of folds exposed in sea cliffs in eastern Ireland includes anticlines and synclines. The folds affect beds of Paleozoic sandstone and shale.

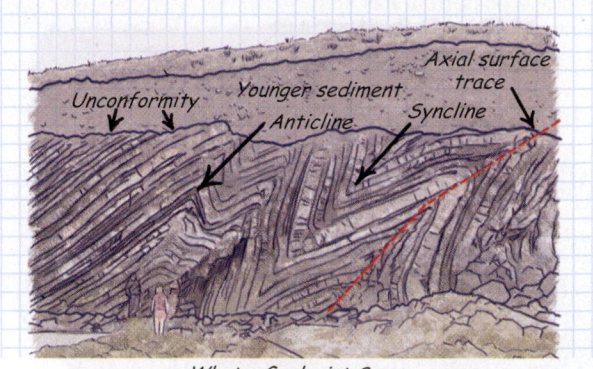

What a Geologist Sees

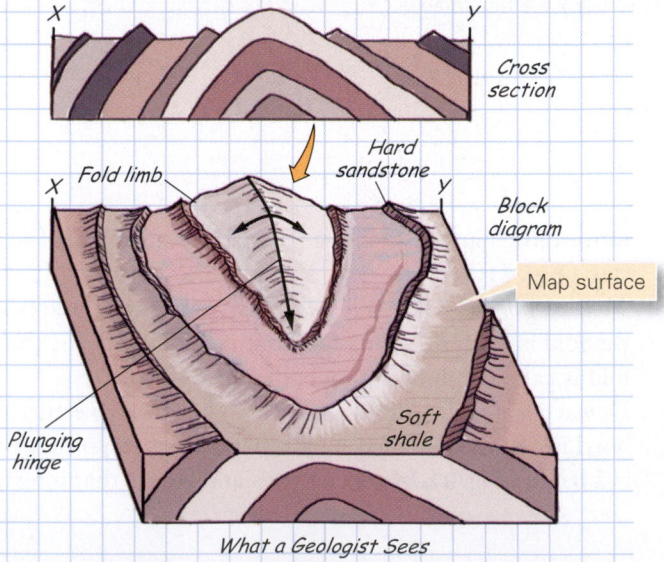

What a Geologist Sees

(e) The plunging anticline of Sheep Mountain, Wyoming, is easy to see because of the lack of vegetation. Resistant rock layers (sandstone) stand out as ridges, whereas weak rock layers (shale) erode away. A block diagram shows how surface exposures relate to underground structure.

FIGURE 11.16 Fold development in flexural-slip and passive-flow folding.

(a) During the formation of flexural-slip folds, layers maintain constant thickness, so for the fold to form, layers must bend. To accommodate the bending, each bed slips relative to its neighbor.

(b) During the formation of passive-flow folds, the rock slowly flows. Different points along a marker line flow at different rates, causing the layer to become folded. Note that the layer's thickness doesn't stay constant during folding.

(c) A fold resulting from shearing in South Australia.

You can recognize folds not only in a *cross section* (a vertical slice through the crust) but also by the pattern of rock layers in map view (a horizontal plane representing the ground surface). For example, a nonplunging anticline involving sedimentary layers appears as a series of parallel stripes, with the oldest layer in the center and progressively younger layers away from the center; the stripes are symmetrically positioned around the hinge. Layers in a plunging fold have a U-shape on the ground surface (see Fig. 11.15e). We can represent the hinge of the fold by a heavy line bordered by outward-pointing arrows for an anticline and inward-pointing arrows for a syncline. Note that because some layers erode more easily than others, the shapes of folds may be indicated by ridges and valleys in the landscape.

Formation of Folds

Folds develop in two principal ways. During formation of *flexural-slip folds*, a stack of layers bends, and slip occurs between the layers (**Fig. 11.16a, c**). The same phenomenon happens when you bend a deck of cards—to accommodate the change

FIGURE 11.17 Folding is caused by several different processes.

Before ⟶ After

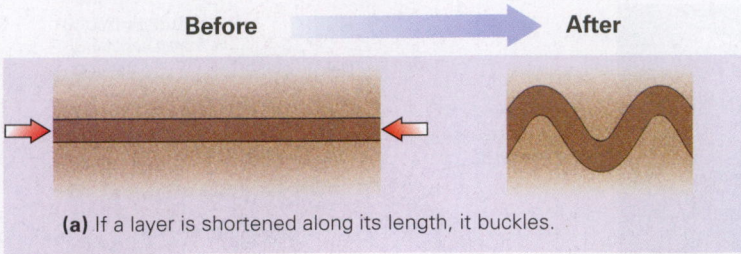

(a) If a layer is shortened along its length, it buckles.

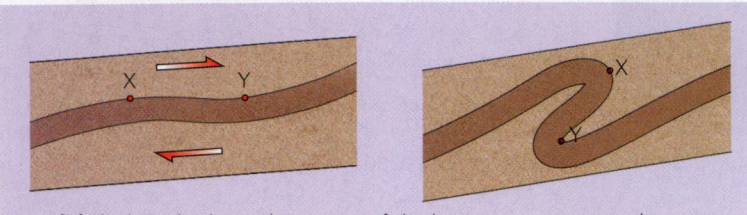

(b) If a layer is sheared, one part of the layer moves over another part to produce a fold.

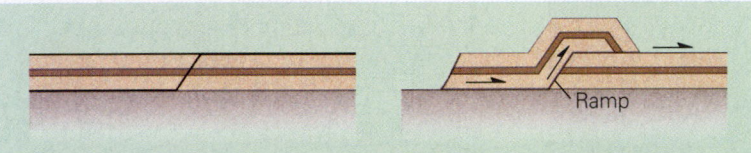

Ramp

(c) When layers move up and over step-shaped faults, they must bend into folds. The step, where the fault cuts layers, is a "ramp."

Axial plane trace

Bedding

What a Geologist Sees

(d) A ramp anticline along a highway in New York State.

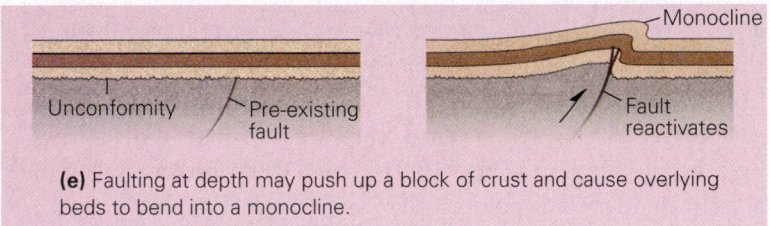

Monocline

Unconformity   Pre-existing fault   Fault reactivates

(e) Faulting at depth may push up a block of crust and cause overlying beds to bend into a monocline.

in shape, the cards slide with respect to each other. *Passive-flow folds* form when the rock, overall, is so soft that it behaves like weak plastic and slowly flows; these folds develop simply because different parts of the rock body move at different rates (**Fig. 11.16b, d**).

Why do folds form? Some layers wrinkle up, or buckle, in response to end-on compression (**Fig. 11.17a**). Others form where shear stress gradually shifts one part of a layer up and over another part (**Fig. 11.17b**). Still others develop where rock layers move up and over step-like bends, called ramps, in an underlying fault, so the layers must curve to conform to the fault's shape (**Fig. 11.17c, d**). Finally, some folds form when new slip on a fault causes a block of basement to move up so that the overlying sedimentary layers must warp (**Fig. 11.17e**).

Tectonic Foliation in Rocks

In an undeformed sandstone, the grains of quartz are roughly spherical, and in an undeformed shale, clay flakes press together into the plane of bedding so that shales tend to split parallel to the bedding. During deformation, however, internal changes take place in a rock that gradually modify the original shape and arrangement of grains. For example, plastic deformation of quartz grains may transform them into cigar shapes, elongate ribbons, or tiny pancakes, and clay flakes may recrystallize or reorient so that they lie at an angle to the bedding. Overall, deformation can produce inequant grains, meaning grains that have different orientations in different directions, and can cause them to align parallel to each other. We refer to layering developed by the alignment of grains in response to deformation as **tectonic foliation**.

We introduced foliation, such as slaty cleavage, schistosity, and gneissic layering, in Chapter 8 while discussing the effects of metamorphism. Here we add to the story by noting that such foliation forms in response to flattening and shearing in plastically deforming rocks—in other words, foliation indicates that the rock has developed a strain (**Fig. 11.18**). For example, in rocks with slaty cleavage, the cleavage planes lie perpendicular to the direction of the shortening strain, so the cleavage may be parallel to the axial plane of folds. In schists and gneisses, the foliation commonly lies parallel to or at a slight angle to planes on which shear took place because shear smears grains out (**Fig. 11.19**).

FIGURE 11.18 The development of tectonic foliation in rock.

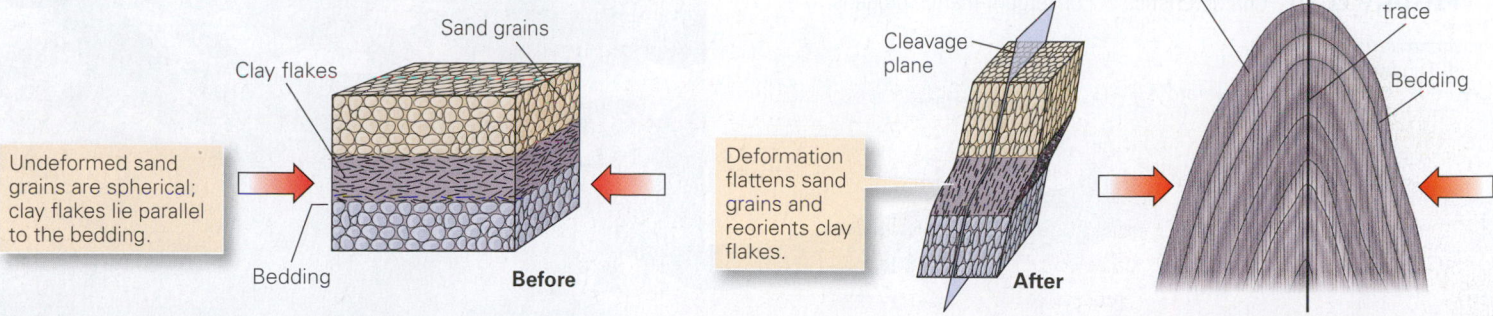

Undeformed sand grains are spherical; clay flakes lie parallel to the bedding.

Clay flakes

Sand grains

Bedding

Before

Deformation flattens sand grains and reorients clay flakes.

Cleavage plane

After

Slaty cleavage

Axial plane trace

Bedding

(a) Compression shortens beds, flattens sand grains, and reorients clay flakes. Clay flakes were originally parallel to bedding, but they become parallel to slaty cleavage during deformation. Folding may accompany cleavage formation.

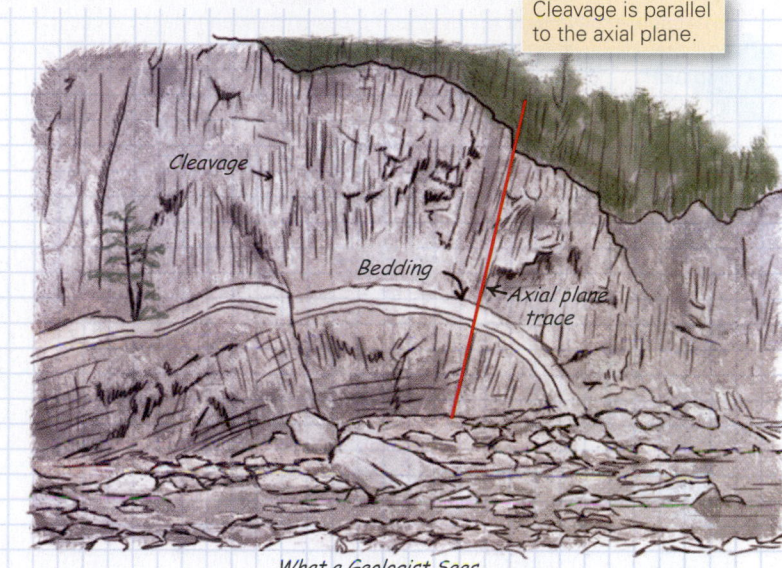

Cleavage is parallel to the axial plane.

Cleavage

Bedding

Axial plane trace

What a Geologist Sees

(b) An example of slaty cleavage developed in Paleozoic strata exposed in a stream cut in New York. Relict bedding is still visible, but note that the rock breaks more easily on the cleavage. This cleavage formed in association with folding.

FIGURE 11.19 Development of a foliation due to shearing. In this rock, quartz and mica-rich layers have become very fine grained and strongly foliated. The foliation is parallel to the shearing direction.

Arrows give shear direction.

Large feldspar grains have been flattened and have "tails."

Take-Home Message

Folds are bends or curves defined by the shape of rock layers. Arch-like folds are anticlines, and trough-like folds are synclines. Folds form for a variety of reasons; for example, buckling of layers due to end-on compression and crustal shortening can yield folds. Deformation can change the shape and orientation of grains, aligning them to produce a planar fabric called foliation. Foliation commonly develops when rocks simultaneously undergo metamorphism and deformation.

QUICK QUESTION: What mechanisms allow layers to undergo folding?

FIGURE 11.20 Characteristics of convergent-margin orogens.

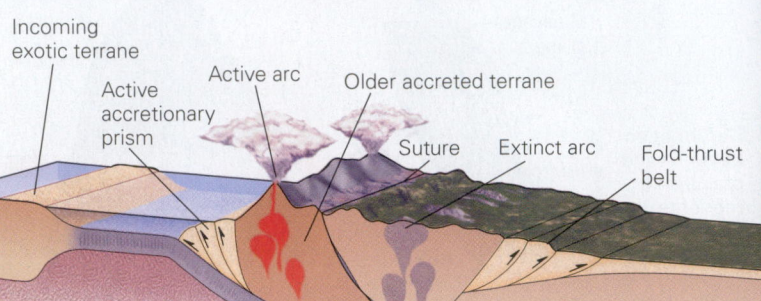

(a) At some convergent margins, compression between the downgoing and overriding plates uplifts a mountain range in which volcanism occurs. Subduction may bring in exotic crustal blocks that collide and become incorporated in the orogen.

(b) The Andes in Chile formed along a convergent margin.

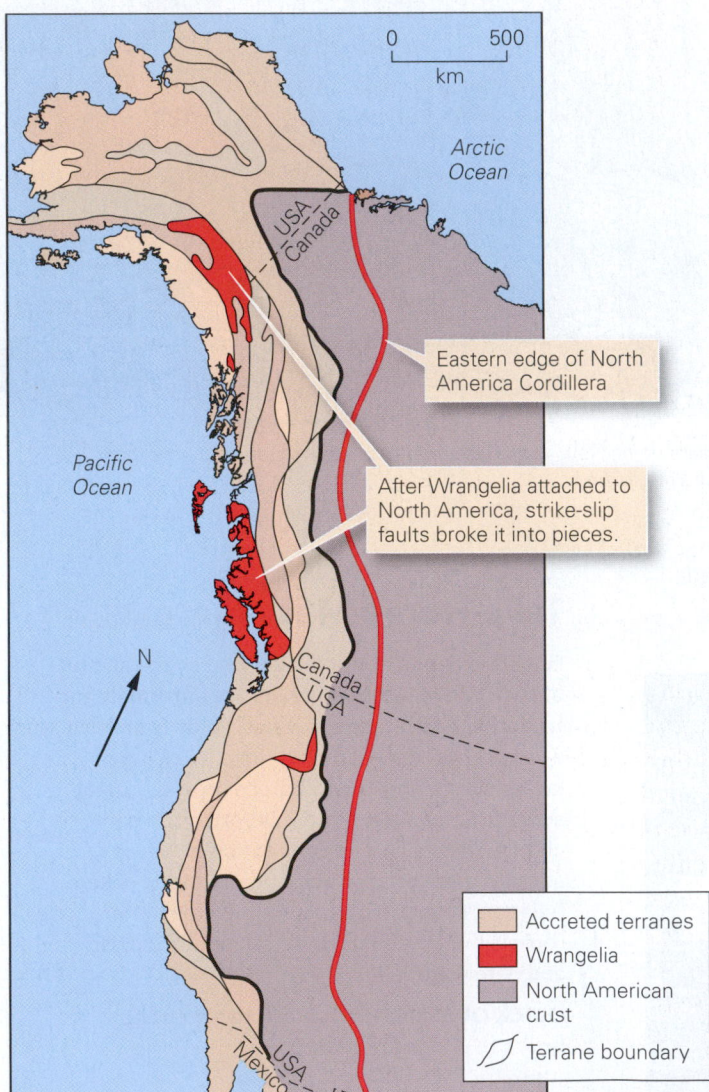

Eastern edge of North America Cordillera

After Wrangelia attached to North America, strike-slip faults broke it into pieces.

Accreted terranes
Wrangelia
North American crust
Terrane boundary

(c) The western portion of the North American Cordillera consists of accreted terranes that attached to the continent during the Mesozoic. During docking, a distinct terrane called Wrangelia (highlighted in red) was sliced into pieces that were displaced by strike-slip faults.

11.5 Causes of Mountain Building

Before plate tectonics theory became established, geologists were just plain confused about how mountains formed. In the context of plate tectonics, however, the many processes driving mountain building became clear: mountains form in response to convergent-boundary deformation, continental collisions, and rifting. Since collision zones, rifts, and plate boundaries are elongate, mountain belts are elongate. Below we look at these different settings and the types of mountains and geologic structures that develop in each one.

Did you ever wonder...

why mountains occur in distinct belts?

Mountains Related to Subduction

As we saw in Chapter 9, subduction at a convergent-plate boundary produces a volcanic arc. But that's not the only feature to develop in response to plate interactions at such boundar-

FIGURE 11.21 Characteristics of collisional orogens.

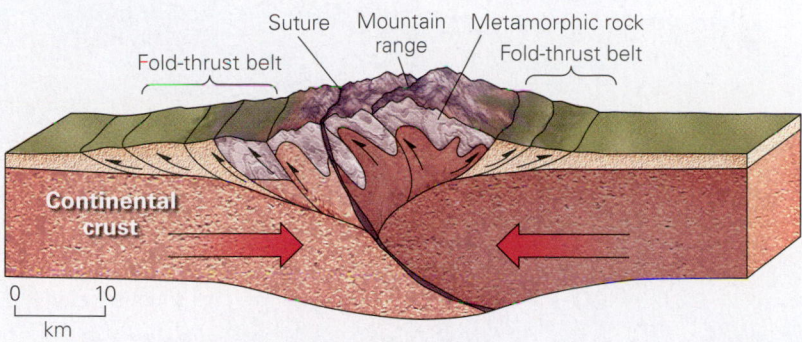

(a) During collision, continents squeeze together and deform. Thrusting brings metamorphic rock up to shallower levels.

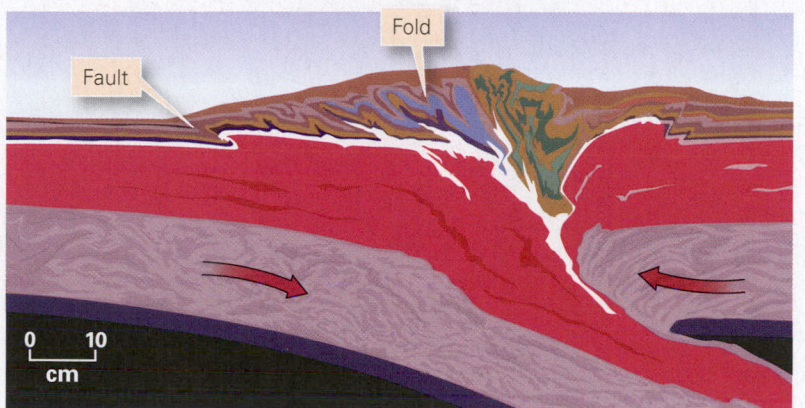

(b) Geologists simulate collision in the laboratory using layers of colored sand. Dragging the left side of the model under the right produces structures and uplift, as shown in this sketch of a model.

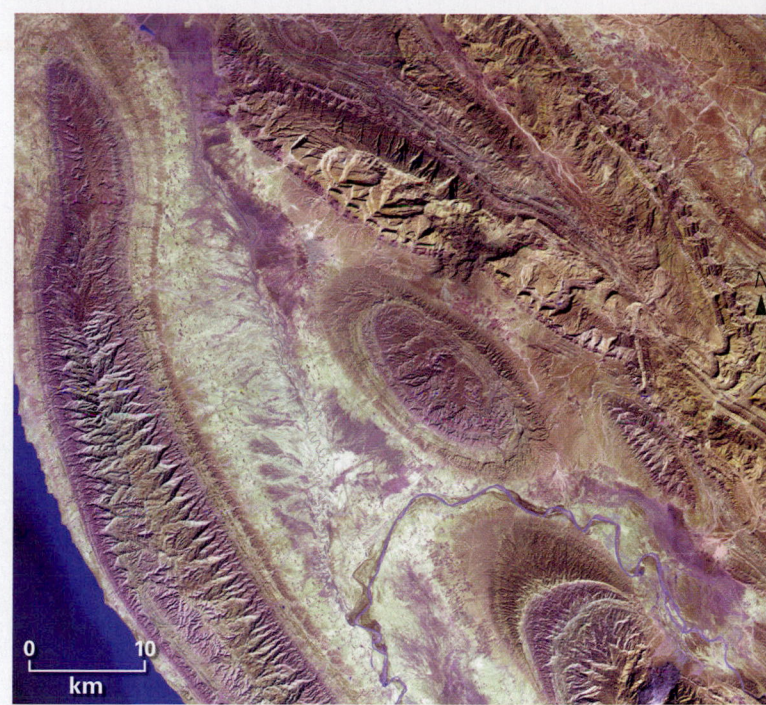

(c) In this satellite photo the ridges and valleys of the Zagros Mountains along the coast of Iran represent eroded folds of a collisional orogen. Ridges consist of durable sandstone beds.

ies. At some convergent-plate boundaries, compressional stress develops and drives crustal shortening in the overriding plate. Such shortening produces a fold-thrust belt (**Fig. 11.20a**). The Andes orogen of western South America displays the rugged topography that can develop in a compressional convergent-margin orogen (**Fig. 11.20b**).

If subduction continues over a long time, offshore volcanic arcs, oceanic plateaus, and small fragments of continents (called *microcontinents*) may drift into the convergent margin (see Fig. 11.20a). Such crustal blocks are too buoyant to subduct and sink back into the mantle, so instead they collide with the overriding plate and attach to the edge of the overriding plate. Geologists refer to the fault at which two once separate pieces of crust have attached as a **suture** and to the process of building out a continent by attaching new crustal fragments as **accretion**. The buoyant crustal block is called an *accreted terrane* once it has attached to the overriding plate. Once accretion occurs, the convergent-plate boundary may jump to the seaward side of the accreted terrane so that subduction can continue and perhaps bring in additional crustal slivers. The process of accretion

can add substantial new crust to a convergent-margin orogen. For example, the western part of North American Cordillera, the huge (7,000 km long and 600 to 1,500 km wide) orogen that extends from Alaska through Mexico, consists of accreted terranes that attached to the continent during between 250 and 150 million years ago (Ma) (**Fig. 11.20c**). At its widest, the belt of accreted crust is 500 km wide, as measured in an east-west direction.

Mountains Related to Continental Collision

Once the oceanic lithosphere between two continents completely subducts, the continents themselves collide with each other. Continental collision results in the formation of large mountain ranges, such as the present-day Himalayas of Asia, the Alps of Europe, and the Paleozoic Appalachian Mountains of eastern North America (see **Geology at a Glance**, pp. 402–403). The final stage in the growth of the Appalachians happened when Africa and North America collided.

During collision, intense compression generates fold-thrust belts on the margins of the orogen (**Fig. 11.21**). In the interior of the orogen, where one continent overrides the edge of the other, high-grade metamorphism occurs, accompanied by formation of passive-flow folds and tectonic foliation. During this process, *crustal thickening* takes place as the crust shortens horizontally and thickens vertically, and thrust faults place slices of crust on top of one another.

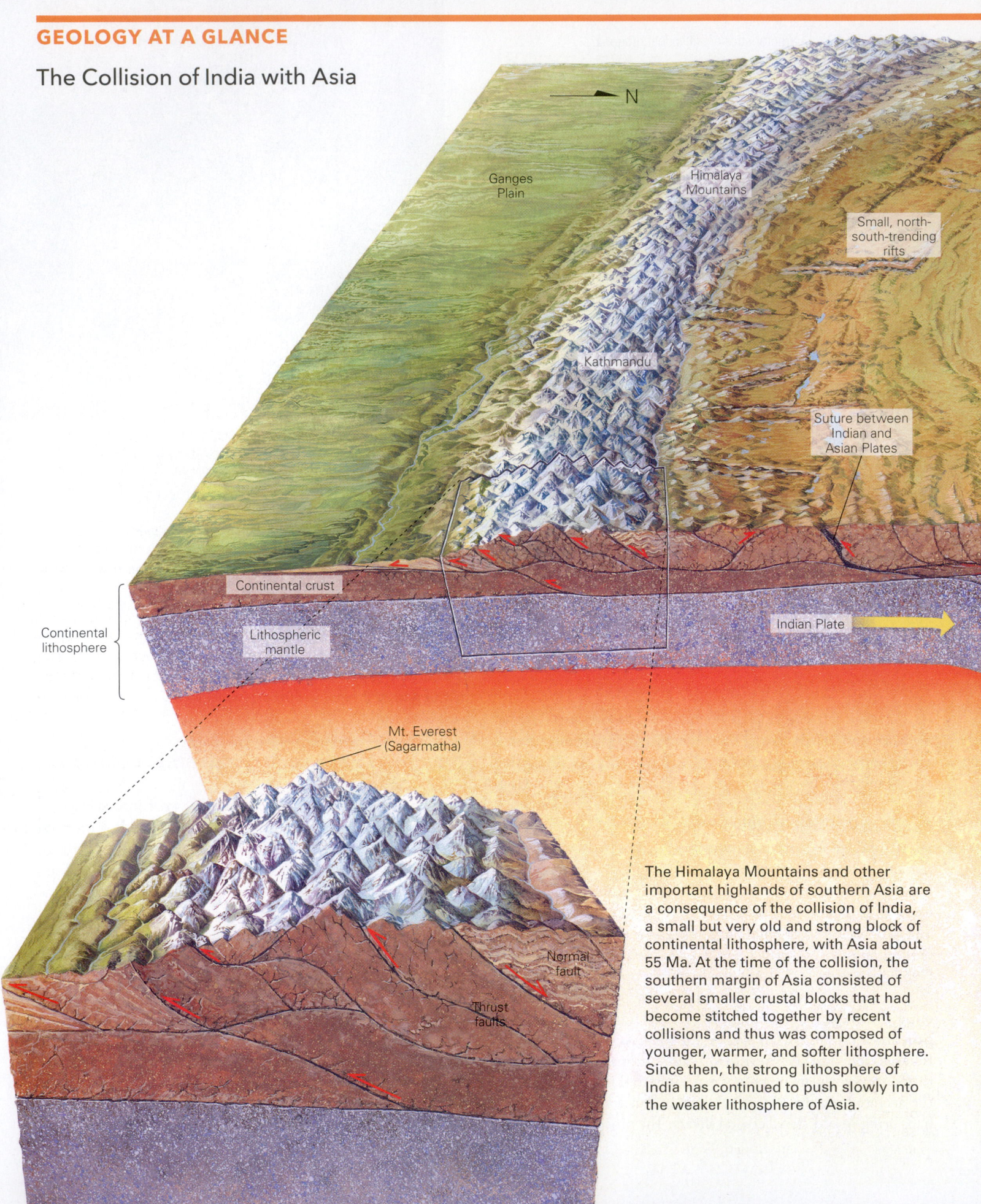

The Collision of India with Asia

N

Ganges
Plain

Himalaya
Mountains

Small, north-
south-trending
rifts

Kathmandu

Suture between
Indian and
Asian Plates

Continental crust

Continental
lithosphere

Lithospheric
mantle

Indian Plate

Mt. Everest
(Sagarmatha)

Normal
fault

Thrust
faults

The Himalaya Mountains and other
important highlands of southern Asia are
a consequence of the collision of India,
a small but very old and strong block of
continental lithosphere, with Asia about
55 Ma. At the time of the collision, the
southern margin of Asia consisted of
several smaller crustal blocks that had
become stitched together by recent
collisions and thus was composed of
younger, warmer, and softer lithosphere.
Since then, the strong lithosphere of
India has continued to push slowly into
the weaker lithosphere of Asia.

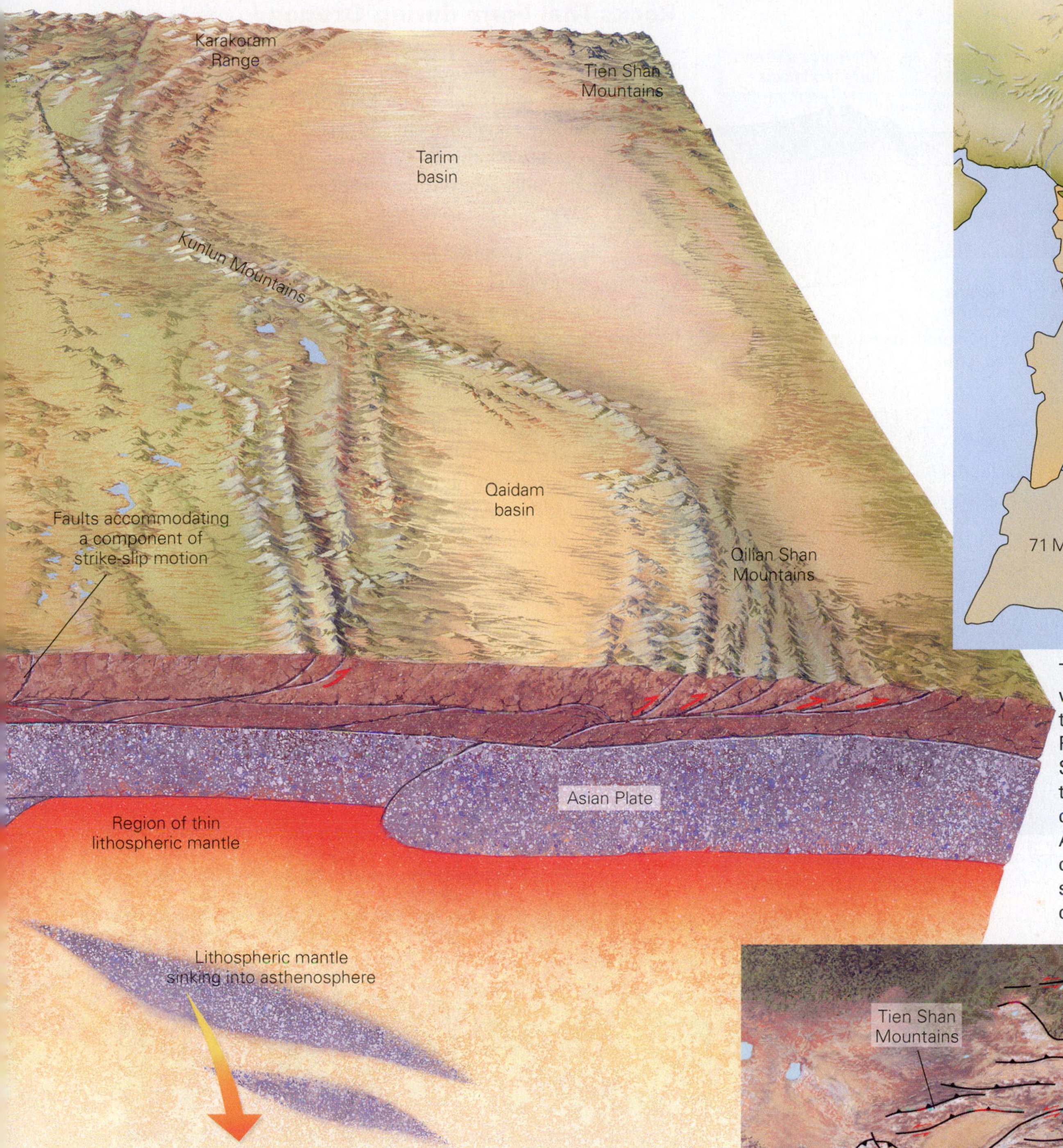

Karakoram
Range

Tien Shan
Mountains

Tarim
basin

Kunlun Mountains

Qaidam
basin

Qillan Shan
Mountains

Faults accommodating
a component of
strike-slip motion

Asian Plate

Region of thin
lithospheric mantle

Lithospheric mantle
sinking into asthenosphere

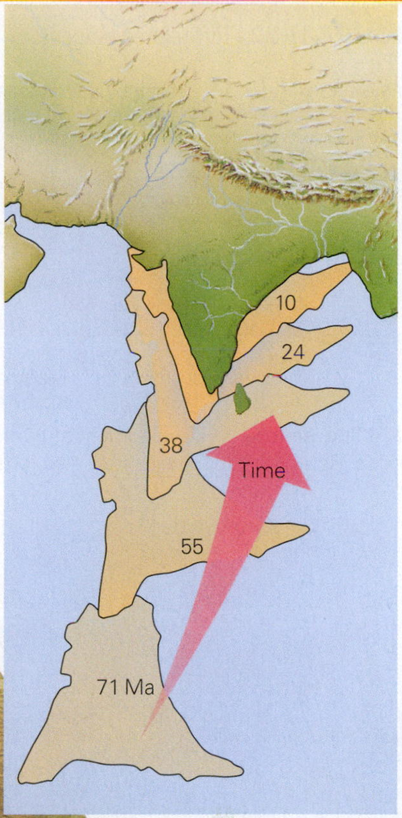

The collision of India
with Asia has uplifted
the Himalayas and Tibet.
Portions of China and
Southeast Asia have slipped
to the east to "escape" the
collision. Faults in central
Asia have become active,
causing the uplift of ranges
such as the Tien Shan, as
compressive forces build up.

The development of large thrust faults has uplifted the curving
Himalayan chain where Asia begins to thrust over India. Why the broad
plateau of Tibet has risen remains something of a mystery. In part,
the uplift may be a consequence of the thickening of the crust as it is
squashed horizontally, continental crust is relatively weak and so may
spread laterally (like soft cheese in the sun), leading to the formation of
normal faults and small rifts in the upper crust and a plastic-like flow in
the deep crust. The uplift may also be due to the heating of the region
when slabs of the underlying lithospheric mantle drop off and sink and
are replaced by hot asthenosphere.

As India has pushed into Asia, it may have squeezed blocks of
China and Southeast Asia sideways, toward the east; this motion is
accommodated by slip on strike-slip faults. The collision may also have
caused reverse faults in the interior of Asia to become active, uplifting
a succession of small mountain ranges, such as the Tien Shan.

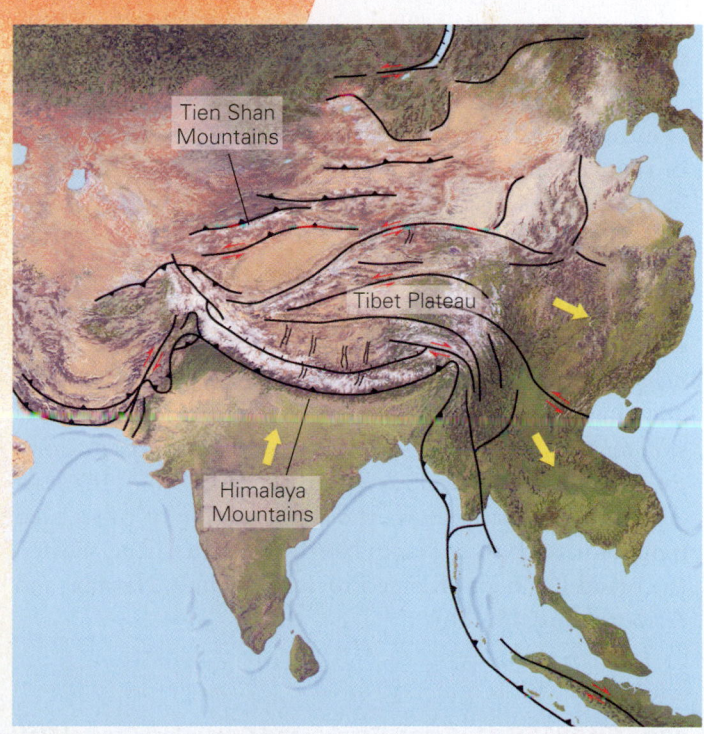

Tien Shan
Mountains

Tibet Plateau

Himalaya
Mountains

FIGURE 11.22 Rift-related orogens.

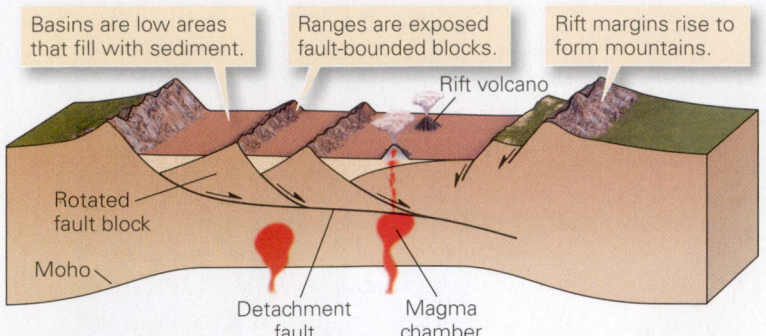

Basins are low areas that fill with sediment.

Ranges are exposed fault-bounded blocks.

Rift margins rise to form mountains.

Rift volcano

Rotated fault block

Moho

Detachment fault

Magma chamber

(a) Rifting leads to the development of numerous narrow mountain ranges. Rift-margin mountains may also form.

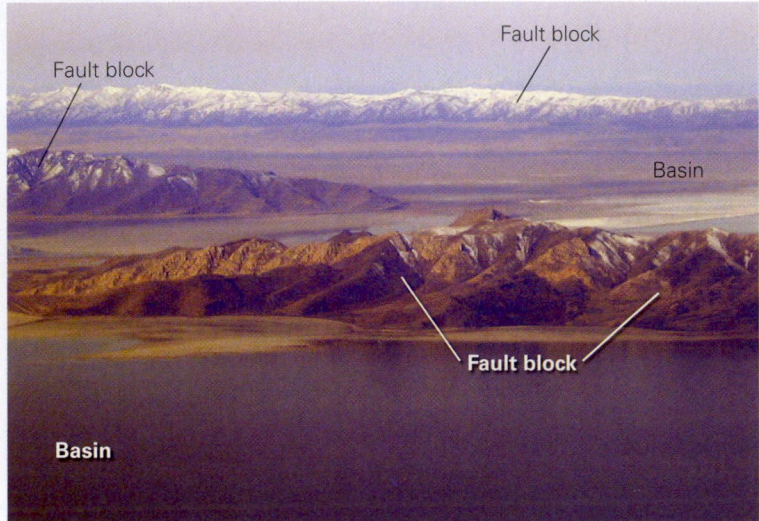

Fault block

Fault block

Fault block

Basin

Fault block

Basin

(b) The Basin and Range Province of the western United States formed during Cenozoic rifting.

Mountains Related to Continental Rifting

Continental rifts are places where continents are splitting in two. During rifting, stretching causes normal faulting in the brittle crust (**Fig. 11.22a**). Movement on the normal faults drops down blocks of crust, producing deep, sediment-filled basins (grabens and half grabens) separated by narrow, elongate mountain ranges that contain tilted crustal blocks. These ranges are sometimes called *fault-block mountains*. Stretching during rifting thins the lithosphere, allowing hot asthenosphere to rise and undergo decompression melting (see Chapter 6). This process produces magmas that rise to form volcanoes within the rift. Today, the East African Rift clearly shows the configuration of rift-related mountains and volcanoes. In North America, rifting yielded the broad Basin and Range Province of Utah, Nevada, and Arizona (**Fig. 11.22b**).

Rocks That Form during Orogeny

We've discussed how orogeny produces a variety of geologic structures and uplifts elongate belts of crust. Here we point out that the process of orogeny also establishes geologic conditions appropriate for the formation of a great variety of rocks. We'll briefly consider examples from all three rock categories (**Fig. 11.23a**):

- *Igneous activity during orogeny:* At convergent-plate boundaries, melting takes place in the mantle above the subducting plate and produces large volumes of magma that rise into the crust. Some of this magma erupts from volcanic arcs, but much of it solidifies underground to form numerous plutons that together form batholiths (see Chapter 6). Thus, orogens formed along convergent margins may include dramatic cliffs of granitic rocks. Stretching and thinning of lithosphere along rifts causes decompression melting of the underlying mantle, producing magmas the erupt as to add layers of basaltic lava and sheets of rhyolitic ash to the surface of the crust. And during continental collision, melting may take place where deep portions of the crust undergo heating, producing granitic magmas the rise to form plutons. Clearly, igneous activity contributes significant volumes of new rock to orogens.
- *Sedimentation during orogeny:* Weathering and erosion in mountain belts generate vast quantities of sediment. This

FIGURE 11.23 An example of the various rocks formed during orogeny.

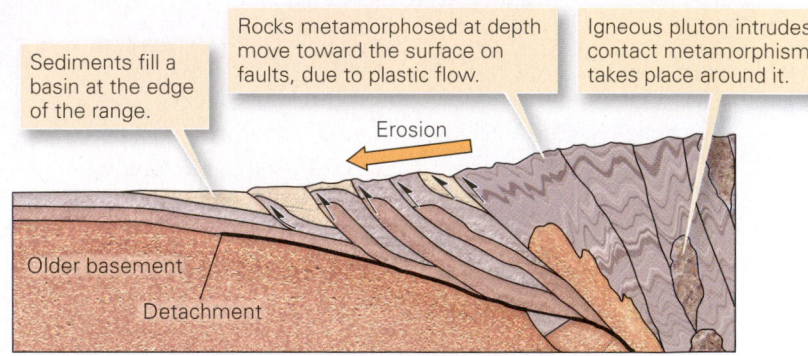

Sediments fill a basin at the edge of the range.

Rocks metamorphosed at depth move toward the surface on faults, due to plastic flow.

Igneous pluton intrudes; contact metamorphism takes place around it.

Erosion

Older basement

Detachment

(a) In the internal zone of the range, metamorphic rocks and igneous rocks form. At the edge of the range, sedimentary rocks form.

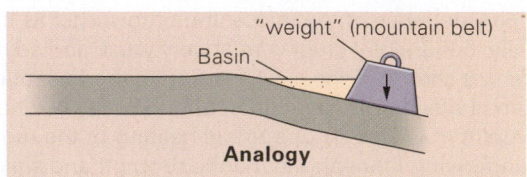

"weight" (mountain belt)

Basin

Analogy

(b) The sedimentary basin develops because the mountain range acts as a weight that pushes down the surface of the lithosphere.

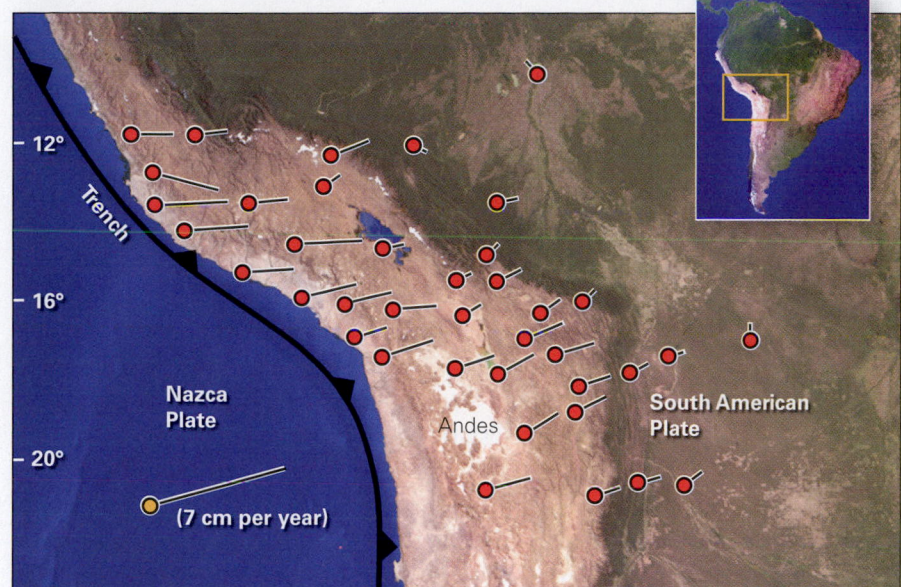

geologists can determine where coastal areas have been rising relative to the sea level by locating ancient beaches that now lie high above the water. And they can tell where the land surface has risen relative to a river by identifying places where a river has recently carved a new valley. In addition, geologists now use the **global positioning system (GPS)** to measure rates of uplift and horizontal shortening in orogens. Though standard handheld GPS devices provide locations with accuracies of only about 2 m, research-quality GPS systems can specify locations to within 2 mm. By comparing the position of a location within an orogen to a location outside an orogen over a time period of a few years, it is possible to detect crustal motion. Thus, we can "see" the Andes shorten horizontally at a rate of a couple of centimeters per year (**Fig. 11.24**), and we can "watch" as mountains along this convergent boundary rise by a couple of millimeters per year.

sediment tumbles down slopes and gets carried away by glaciers or streams that transport it to low areas where it accumulates in large fans or deltas. Large collisional or convergent mountain belts act as heavy loads on the top of the lithosphere and thus push down the surface of the lithosphere, thereby producing deep sedimentary basin along the border of the range (**Fig. 11.23b**). In rifts, the basins that form as the crust stretches fill with thick wedges of sediment.

- *Metamorphism during orogeny:* We just noted that igneous activity commonly happens in orogens. So contact metamorphic aureoles form adjacent to igneous intrusions in orogens in new metamorphic rocks form (see Chapter 8). Regional metamorphism (dynamothermal metamorphism) occurs where mountain building substantially thickens the crust, for when this happens, rocks that may have first formed near the Earth's surface end up at great depth where they are subjected to high temperature and pressure. Because deformation accompanies this process, the resulting metamorphic rocks contain tectonic foliation.

Measuring Mountain Building in Progress

Not all mountains are just "old monuments," as John Muir mused. The rumblings of earthquakes and the eruptions of volcanoes attest to present-day, continuing movements in some ranges. Geologists can measure the rates of these movements through field studies and satellite technology. For example,

---

Take-Home Message

Mountain belts form in association with convergence, collision, and rifting. During continental collisions, and at some convergent margins, the crust thickens, large thrust faults and folds form, and regional metamorphism develops. Tilting of crustal blocks during rifting yields fault-block mountains.

QUICK QUESTION: Could fold-thrust belts develop in rifts? Why or why not?

11.6 Mountain Topography

Leonardo da Vinci, the Renaissance artist and scientist, enjoyed walking in the mountains, sketching ledges and examining the rocks he found there. In the process, he discovered marine shells (fossils) in limestone beds cropping out a kilometer above sea level, and he suggested that the rock containing the fossils had risen from below sea level up to its present elevation. Modern geologists agree with Leonardo and now refer to the process by which the surface of the Earth moves vertically from a lower to a higher elevation as **uplift**. Mountain building requires substantial uplift of the Earth's surface (**Fig. 11.25**). Think about it—to form a mountain range, on the order of 1 million km³ of rock rises up.

FIGURE 11.25 Mountain cliffs tower above a lake in the Canadian Rockies.

because asthenosphere flows only very slowly, and because lithosphere is strong enough to hold up loads, isostasy does not exist everywhere; see Interlude D.)

To picture isostasy, imagine placing a block of wood into a bathtub full of water. If the block is less dense than water, it floats, with part of the block remaining above the water surface and most of the block submerged below. Now place a denser block of the same thickness next to the first block. The top of the denser block sits lower than that of the less-dense block of the same thickness. Similarly, the top surface of a thicker block sits higher than the top surface of a thinner block of the same density. If you were to add another block of wood on top of one that is already floating, the lower block would sink adjusting for the addition so as to maintain isostatic equilibrium.

From our bathtub experiment, we can deduce that any phenomenon that changes the thickness and/or density of a floating block will affect the elevation of the block's surface above the water surface. Since the lithosphere floats on the asthenosphere, the elevation of the lithosphere's surface at a location depends on the thickness and density of the lithosphere beneath. So to answer the question of why mountain belts can rise, we must identify geologic processes that can change the thickness and/or density of layers in the lithosphere. Let's consider some examples of how these charges take place.

Crustal Shortening and Thickening During collisional orogeny or certain types of convergent-margin orogeny, horizontal compression causes the crust to shorten horizontally and thicken vertically. In fact, the folding, faulting, and plastic flow that take place during such events can almost double the crust's thickness. For example, the crust beneath the tallest range, the Himalayas, is 70 km thick, whereas crust beneath the plains of the central United States averages 40 km thick (**Fig. 11.26a**; see also Geology at a Glance, pp. 402–403). To isostatically compensate for the thickening of the crust (the geologic equivalent to adding another block of low-density wood to the top of a floating block), the base of the crust and underlying lithospheric mantle sink or subside (**Fig. 11.26b**). Indeed, since the Himalayas are approximately 8 km high, most of the thickened crust extends downward beneath the range. This downward protrusion of crust is called a **crustal root**. Simply stated, the higher the mountains in a collisional or convergent orogen, the deeper the root. We can illustrate this relationship in a bathtub model by lining up a row of floating blocks of different thickness—if all the blocks have the same density, the thicker blocks rise farther above water and protrude deeper below the surface (**Fig. 11.26c**).

Where did the idea that mountains have roots come from? The discovery began with an observation by Sir George Everest,

What kinds of distances are we talking about when referring to uplift in mountain ranges? As noted earlier, Mt. Everest rises 8.85 km above sea level. Although this distance may seem monumental—that's equal to 5,000 people standing one on top of another—it represents only about 0.06% of the Earth's diameter. In fact, if the Earth were shrunk to the size of a billiard ball, its surface (mountains and all) would feel smoother than that of an actual billiard ball. In general, the individual peaks that you see in a mountain range represent only a fraction of the range's total height, for the plains at the base of the mountains may be significantly higher than sea level. Nevertheless, mountain heights are spectacular, and in this section we will look at why uplift occurs, how erosion carves rugged landscapes out of uplifted crust, and why Earth's mountains can't get much higher than Mt. Everest.

Why Are Mountains High?

What processes can cause the surface of the Earth to rise to mountainous heights? There are many because, as we have seen, mountain building happens in numerous different geologic settings. To understand how the variety of uplift processes work, we must begin by remembering the concept of isostasy, introduced in Interlude D.

The lithosphere, which consists of relatively rigid crust and lithospheric mantle, "floats" on the softer asthenospheric mantle below. As a consequence, the elevation of the top surface of the lithosphere, over a broad region, represents a balance between buoyancy force pushing lithosphere up and gravitational force pulling the lithosphere down. Geologists refer to the condition that exists when this balance has been achieved as **isostasy**, or isostatic equilibrium. Put another way, isostasy exists where the elevation of the Earth's surface reflects the level at which the lithosphere naturally floats. (Note that

FIGURE 11.26 The concept of isostasy as applied to collisional mountain ranges.

(a) The Himalayas are the world's highest mountain range. The crust beneath them is almost twice the normal thickness.

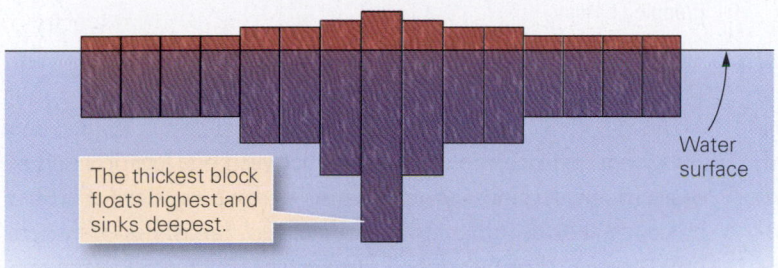

The thickest block floats highest and sinks deepest.

Water surface

(c) Because of isostasy, blocks of wood floating in water sink to a depth such that the mass of the water displaced is the same as the mass of the block.

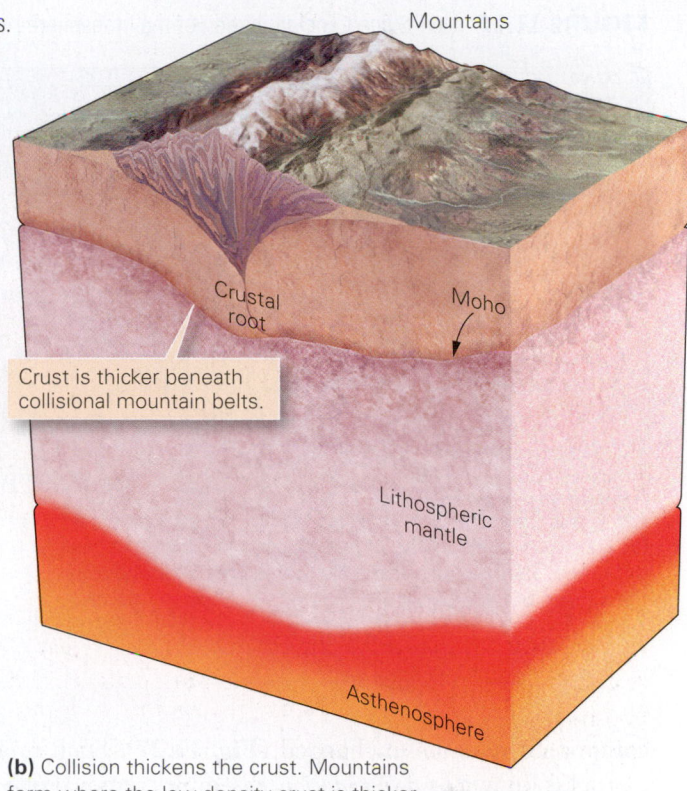

Mountains

Crustal root

Moho

Crust is thicker beneath collisional mountain belts.

Lithospheric mantle

Asthenosphere

(b) Collision thickens the crust. Mountains form where the low-density crust is thicker.

after whom the world's highest mountain was named. When surveying in India during the mid-1800s, Everest discovered that the gravitational attraction due to the mass of the Himalayas was large enough to deflect a plumb bob (a lead weight at the end of a string) away from vertical. Subsequent calculations demonstrated that the amount of deflection was actually less than expected, given the size of the range. A British scientist, George Airy, came up with an explanation. Realizing that crustal rocks are less dense than mantle rocks, Airy speculated that if the Himalayas have a low-density crustal root, the range would have less mass than expected and thus would not exert as much gravitational pull on a plumb bob. Note that Airy's proposal resembles the image shown in Figure 11.26c, so this image is known as the Airy model of isostasy.

Adding Igneous Rock to the Crust When a volcano erupts, a thick pile of pyroclastic debris and/or lava may build up on the surface of the crust. Successive eruptions over millions of years, and granitic intrusions within and below the volcanic sequence, can produce a range whose peaks rise kilometers above sea level. Of note, the elevation of the range is not simply the thickness of the new igneous rock accumulation—adding

the load to the surface of the Earth bends the lithosphere down to maintain isostasy, so the heights end up being less than the thickness of the accumulation.

Adding volcanic material to the surface of the crust and granitic rock to the upper crust is not the only way that igneous activity causes uplift. Geologists suspect that in some locations, basaltic magma formed by partial melting of the mantle gets trapped in sills at or near the base of the crust and accumulates. Addition of this hidden basalt thickens the crust relative to the lithospheric mantle. Even though basalt (a mafic rock) is denser than granite (a felsic rock), it is less dense than the peridotite (an ultramafic rock) that makes up the lithospheric mantle. Thus, adding basalt to the base of the crust causes the land surface to rise and the base of the crust to sink to maintain isostasy.

Removal of Lithospheric Mantle Lithospheric mantle consists of very dense rock (peridotite). The weight of this rock pulls the lithosphere down, just like heavy ballast makes a ship settle deeper into the water. Removal of some or all of the lithospheric mantle from the base of a plate, therefore, causes the surface of the remaining lithosphere to rise to maintain isostasy, even if the thickness of the crustal

FIGURE 11.27 Uplift, due to delamination of the lithosphere root, may happen after collision.

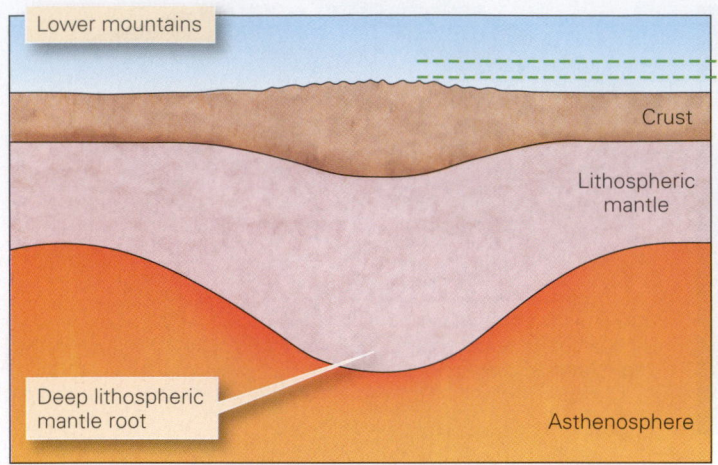
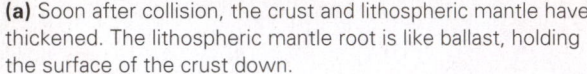

(a) Soon after collision, the crust and lithospheric mantle have thickened. The lithospheric mantle root is like ballast, holding the surface of the crust down.

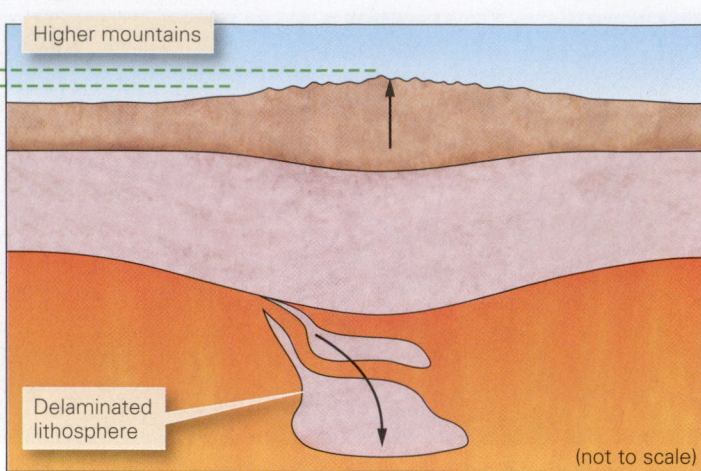

(b) If the lithospheric mantle root detaches and sinks, a process called delamination, the surface of the lithosphere may rise, like a balloon dropping ballast.

component remains unchanged (**Fig. 11.27**). Such removal, a process known as *delamination*, resembles removal of ballast from the hold of a ship—as the weight of the ballast disappears, the deck of the ship rises. Several data sources suggest that the region of the Tibet Plateau in Asia underwent an episode of uplift when some of the underlying lithospheric mantle dripped or peeled off the base of the plate and sank down into the asthenosphere.

Thinning and Heating the Lithosphere In rifts, the lithosphere undergoes stretching and thinning. As a result, relatively less-dense asthenosphere rises beneath the rift, and the remaining, thinned lithosphere heats up. Heating causes rocks to expand and therefore causes the density of the thinned lithosphere to decrease. Replacing denser lithospheric mantle with less-dense asthenosphere, together with heating the remaining thinned lithosphere, causes the overall region of the rift, as well as the margin of the rift, to rise in order to maintain isostasy. It's for this reason that the region of the East African Rift hosts dramatic cliffs. Such a process may also be responsible for the Wasatch Range, the chain of mountains that lies along the east edge of the Basin and Range rift. Uplift of rifts is not necessarily permanent. When the process of stretching stops, rifted lithosphere ages, thereby thickening and becoming denser, so the surface of the lithosphere may subside.

What Goes Up Must Come Down

When the land surface rises significantly, for whatever reason, it doesn't remain a smooth welt or bulge on the Earth's surface.

As soon as a difference in elevation between one location and an adjacent one develops, gravity begins to drive a variety of erosive processes. For example, as slopes steepen, landslides of various types cause rock and debris to tumble from higher to lower elevations (see Chapter 16); when rain falls, streams form and sculpt valleys and canyons (see Chapter 17); and if temperatures remain cold enough, glaciers grow and flow, carving pointed peaks at their origin and widening and deepening valleys downslope (see Chapter 22). The net result of all these processes is to grind away elevated areas and produce the jagged landscapes that we associate with mountain terrains (**Fig. 11.28**). Glaciers appear to be particularly effective in keeping mountains lower than they might otherwise grow—so efficient, in fact, that geologists informally refer to glacial erosion in mountains as the "glacial buzz saw."

It's important to keep in mind that uplift and erosion happen simultaneously in active mountain belts. So, if the elevation of a range increases over time, the rate of uplift must exceed the rate of erosion. However, if the erosion rate exceeds the uplift rate, then the range's elevation decreases over time, even if the tectonic processes driving uplift continue to operate. And, if the rate of erosion equals the rate of uplift, the elevation of the range stays the same, maintaining a "steady-state" elevation. Notably, the balance between erosion rate and uplift rate evolves during the history of an orogen, either due to a change in rates of tectonic processes (e.g., a change in relative plate motion) or due to a change in climate. For example, if rainfall or snowfall increases, so rivers and glaciers grow larger and flow faster, the erosion rate increases and a mountain range may start to erode away even if the uplift rate hasn't changed.

When tectonic processes driving uplift eventually stop because convergence, collision, or rifting ceases, erosion con-

(a) Glaciers carved these rugged peaks in Switzerland.

(b) Streams cut valleys into weathered bedrock in Brazil.

tinues and will eventually bevel a mountain range back to near sea level. But the process takes a long time, at least 30 to 50 million years. It takes so long, in part, because of isostasy. Specifically, removal of rock from the top of a mountain range is like taking cargo off the deck of a ship—removing cargo causes the ship to float higher, and erosion of rock from the range causes isostatic rebound (uplift). Roughly speaking, for every 3 cm eroded, the range rises by 1 cm.

The highest point on Earth, the peak of Mt. Everest, lies 8.85 km above sea level. Can our planet's mountain ranges get significantly higher? Probably not. Mountains as high as Olympus Mons on Mars, which rises 27 km above the plain at its base, couldn't form on Earth because of the relatively high geothermal gradient (the rate of increase in temperature with depth) in Earth's crust. Due to the gradient, quartz-rich crustal rocks at mid-crustal depths (15–30 km) become so warm and weak that they can flow plastically. When this flow begins, overlying mountains above begin to collapse under their own weight, and spread laterally like soft cheese that has been left out in the summer sun (**Fig. 11.29**). Geologists call this process orogenic collapse. During orogenic collapse, the upper crust breaks and a system of normal faults develop to accommodate the horizontal stretching.

Because erosion at the surface takes place as uplift continues, rock intruded or metamorphosed at depth eventually becomes exposed at the surface. This process of revealing deeper rocks by removal of the overlying crust is called unroofing, or **exhumation**. In large mountain ranges, rocks that were metamorphosed at depths of over 20 km become exhumed, and can eventually be exposed in outcrops at the surface. Exhumation

rates, averaged over time, range from less than 1mm per year up to about 10 mm per year. This may seem slow, but even at a rate of just 1 mm per year, rock rises from depth by about 1 km every 1 million years, and therefore, rock from 10 km below the surface reaches the surface in 10 million years.

Take-Home Message

Mountains exist where major uplift occurs. Uplift occurs for a variety of reasons but ultimately reflects the tendency of lithosphere to achieve isostasy (meaning to "float" at the appropriate level). Crustal thickening causes uplift in collisional and convergent mountain ranges. Uplift of rifts happens because less dense asthenosphere rises and replaces denser lithospheric mantle. Once uplifted, erosion sculpts rugged topography. Mountain belts may eventually collapse under their own weight.

QUICK QUESTION: How do rocks metamorphosed at 20 km below the surface eventually become exposed in a mountain range?

11.7 Basins and Domes in Cratons

A **craton** consists of crust that has not been affected by orogeny for at least the last 1 billion years and, as a result, has become quite cool and therefore relatively strong and stable.

FIGURE 11.29 The concept of orogenic collapse.

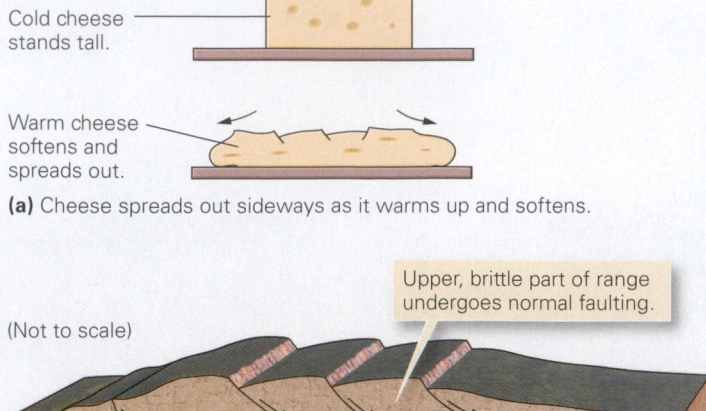

Cold cheese stands tall.

Warm cheese softens and spreads out.

(a) Cheese spreads out sideways as it warms up and softens.

(Not to scale)

Upper, brittle part of range undergoes normal faulting.

Moho

Deep, ductile part of range flows sideways.

(b) Similarly, mountain belts spread out sideways once they reach a certain thickness. The ductile crust at depth flows, whereas the upper (brittle) crust is broken by normal faults.

Each continent has one or more cratons. North America, for example, contains one large craton, bounded on the west by the North American Cordillera, an orogen that formed in the Mesozoic and Cenozoic, and bounded on the east and south by the Appalachian-Ouachita orogen, which developed during the Paleozoic (**Fig. 11.30**). South America contains several smaller cratons, which were sutured together in Late Precambrian and Early Paleozoic time, and are now bounded to the west by the Andean orogen (Fig. 11.24). Geologists divide cratons into two provinces: **shields**, where Precambrian metamorphic and igneous rocks crop out at the ground surface, and *cratonic platforms*, where a relatively thin layer of Paleozoic and locally younger sedimentary strata cover the Precambrian rocks (Fig. 11.30).

Shields tend to be broad, low-lying regions in which we can find widespread exposures of intensively deformed metamorphic rocks with abundant examples of flow folds and tectonic foliation. That's because the crust making the cratons was deformed during a succession of orogenies in the Precambrian. These orogens are so old that erosion has worn away the original topography, in the process exhuming deep crustal rocks.

FIGURE 11.30 North America's craton consists of a shield, where Precambrian rock is exposed, and a platform, where Paleozoic sedimentary rock covers the Precambrian.

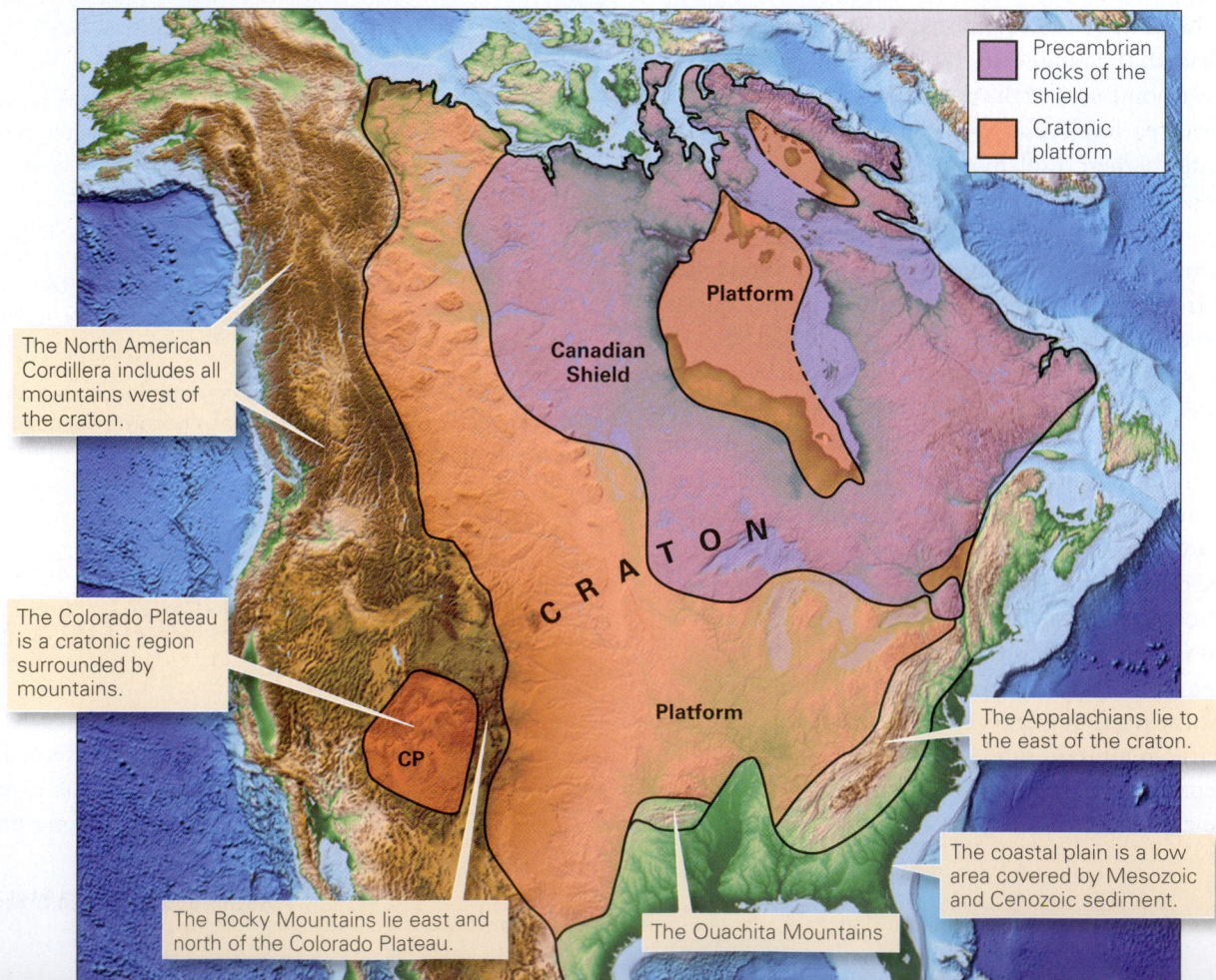

Precambrian rocks of the shield

Cratonic platform

Platform

Canadian Shield

CRATON

The North American Cordillera includes all mountains west of the craton.

The Colorado Plateau is a cratonic region surrounded by mountains.

CP

Platform

The Appalachians lie to the east of the craton.

The coastal plain is a low area covered by Mesozoic and Cenozoic sediment.

The Rocky Mountains lie east and north of the Colorado Plateau.

The Ouachita Mountains

Cratonic platforms also tend to be expansive plains. Outcrops in these regions expose nearly flat-lying and nearly undeformed strata. But at a regional scale, the pattern of contacts between stratigraphic formations on a geologic map defines regional-scale domes, basins, and arches. Cratonic basins are regions that over geologic time gradually subsided so that stratigraphic units are warped down into the shape of a broad bowl, whereas domes and arches have stayed high or have risen, and thus stratigraphic units have the shape of an overturned bowl or elongate bowl (**Fig. 11.31**).

As examples of basins and domes, let's look at the central Midwest of the United States. In the Illinois basin, strata warp downward into a huge bowl that is also about 300 km across and up to 7 km deep. Significantly, strata get progressively thicker toward the basin center, indicating that the floor of the basin was subsiding as sediment was accumulating—there was more room for sediment to accumulate where the basin subsided the most, so the sediment layer became thinner. On a geologic map, the basin also has a bull's-eye shape, with the youngest strata exposed in the center. Note that because of thickening toward the interior of the dome and because the dome is so much wider than it is deep, bed dips are so gentle as to be hardly noticeable at an individual outcrop. The Ozark dome of Missouri also has a diameter of about 300 km. Individual sedimentary layers thin toward the crest of the dome—since the crest was high during the time of deposition, less sediment could accumulate there than in adjacent basins. The dome also looks like a bull's-eye on a map, but in the case of a dome, the oldest rocks (Precambrian granite, in the case of the Ozark dome), are exposed near the center. Geologists refer to the broad vertical movements that generate broad mid-continent domes and basins as **epeirogeny**. The causes of epeirogeny are not well understood.

We noted earlier that strata in cratonic platforms are nearly undeformed, but they are not completely undeformed. Locally, faults cut the strata, and movement on faults, like the rise of a trapdoor beneath a carpet, causes overlying strata to bend into monoclines. Geologists refer to the faults as *intracratonic faults*. (The prefix *intra* means within, so *intracratonic* means within the craton.) Most of the faults appear to have originated during the Precambrian, perhaps in response to episodes of rifting. They reactivated in pulses during the Paleozoic, generally at times when orogenies were active along the margins of the continent. Evidently, stress generated during the orogenies was sufficient to cause slip on pre-existing faults in the craton but was not sufficient to generate the pervasive folds and tectonic foliation found within orogens. Most intracratonic faults are blind faults in that they die out up-dip, before they reach the ground surface.

FIGURE 11.31 Domes and basins of the North American cratonic platform.

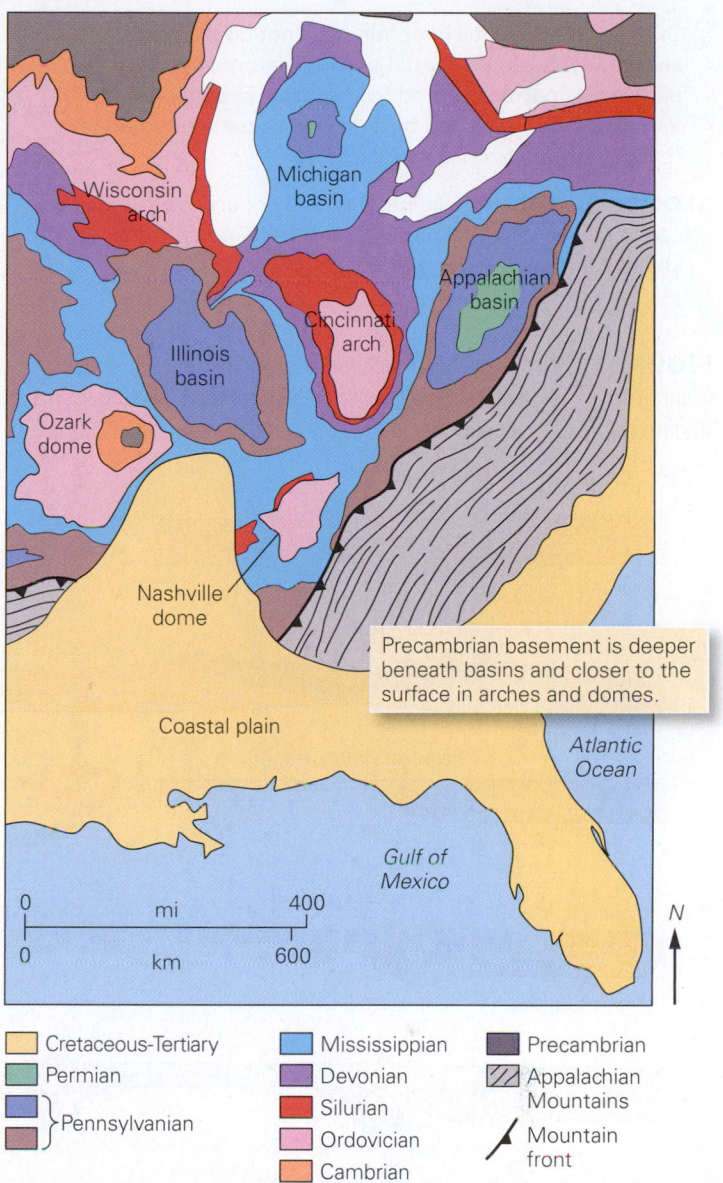

Precambrian basement is deeper beneath basins and closer to the surface in arches and domes.

Cretaceous-Tertiary	Mississippian	Precambrian
Permian	Devonian	Appalachian Mountains
Pennsylvanian	Silurian	Mountain front
	Ordovician	
	Cambrian	

(a) A geologic map of the U.S. mid-continent platform region showing the locations of basins and domes.

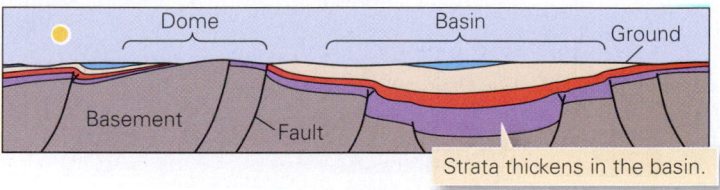

Strata thickens in the basin.

(b) A cross section showing how strata thin toward the crest of a dome and thicken toward the center of a basin. The cross section is vertically exaggerated. This cross section symbolically represents the Ozark dome and Illinois basin.

FIGURE 11.32 These idealized stages show the tectonic evolution of the Appalachian Mountains. Note that although mountains do form during rifting events, geologists traditionally assign names only to the collisional or convergent events.

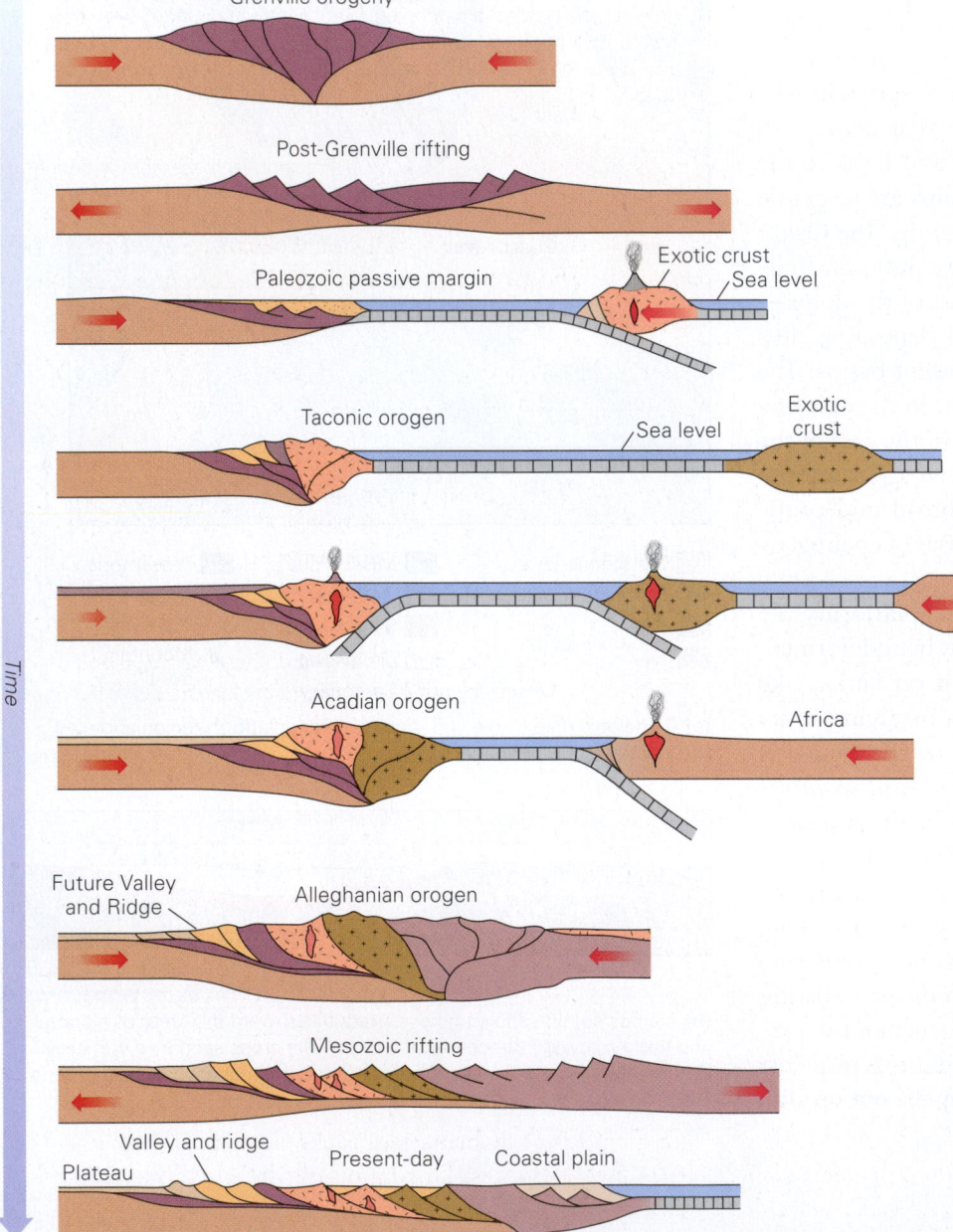

11.8 Life Story of a Mountain Range: A Case Study

Perhaps the easiest way to bring together all the information in this chapter is to look at the life story of one particular mountain range. We'll take the Appalachian Mountains of eastern North America as our example. (We've simplified the story a bit, for ease of reading.) Geologists have constructed the range's life story by studying structures, by determining the ages of igneous and metamorphic rocks, and by searching for strata formed from sediment eroded from the range.

The story begins at about 1.1 billion years ago, when the eastern margin of North America was involved in a massive collision with another continent (**Fig. 11.32**). This event, called the *Grenville orogeny*, yielded a belt of deformed and metamorphosed rocks that underlie the eastern fifth of the continent. For a while after the Grenville event, eastern North America lay in the middle of a supercontinent. But this supercontinent rifted apart around 600 Ma. Eventually, new ocean formed to the east, and the former rifted margin of eastern North America cooled, subsided, and evolved into a passive margin. (Recall that a passive margin is a continental edge that is not a plate boundary.) From 600 to about 420 Ma, the passive margin slowly sank to become a basin that gradually filled with a thick sequence of sediment washed off the adjacent continent.

Between 420 and 370 Ma, two collisions took place along the eastern margin of North America. During the first convergent event, called the *Taconic orogeny*, a volcanic arc collided with eastern North America, and during the second convergent event, the *Acadian orogeny*, continental crustal slivers accreted to the continent. At times between these events, the margin of North America was a convergent-plate boundary. The accretion of these terranes deformed the sediment that had accumulated in the passive-margin basin and made the continent grow eastward. At 270 Ma, the Taconic and Acadian orogens

FIGURE 11.33 The relationship between deformation and topography.

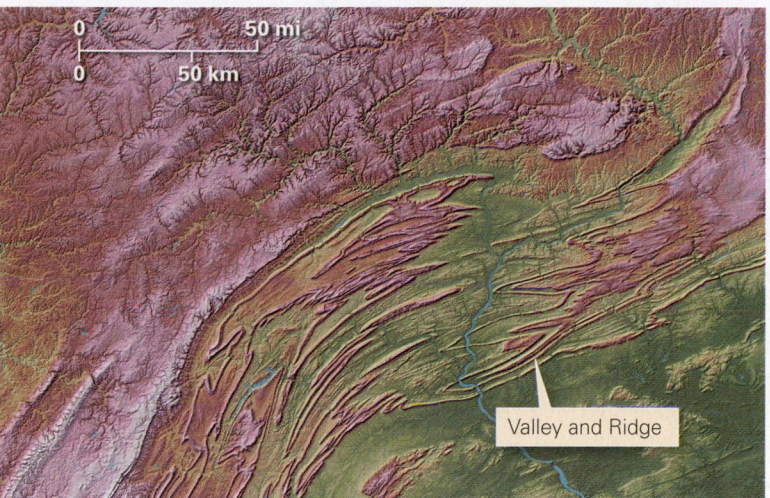

(a) The topography of eastern Pennsylvania, here shown in shaded relief, shows how erosional patterns reveal the folds of the Appalachians, producing a region called the Valley and Ridge.

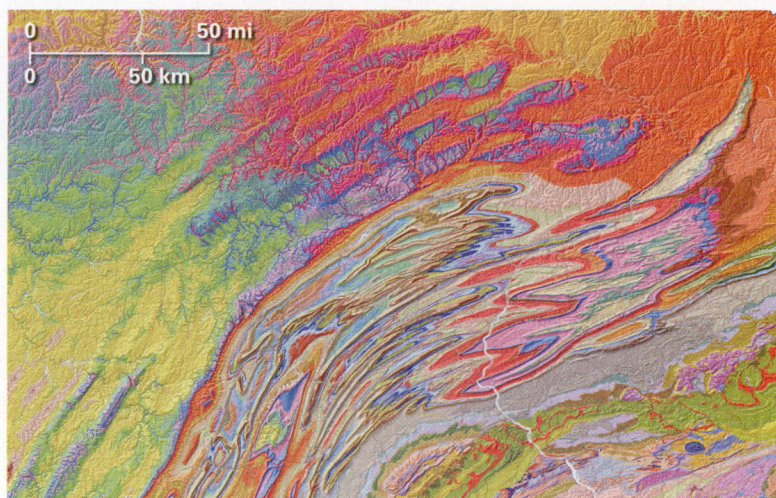

(b) A geologic map of the same area emphasizes the relation between the distribution of folded stratigraphic formations (indicated by different colors) and the topographic features. A resistant sandstone formation forms the ridges.

were caught in a gigantic vice as the large continent of Africa collided with North America. This event, the *Alleghanian orogeny*, yielded a huge mountain range resembling the present-day Himalayas and generated a wide fold-thrust belt along the mountains' western margin. Eroded folds of this belt make up the topography of the present Valley and Ridge Province in Pennsylvania (**Fig. 11.33**).

When the Alleghanian orogeny ceased, the Appalachian region once again lay in the interior of a supercontinent (Pangaea), where it remained until about 180 Ma (Mesozoic time). At that time, rifting split the region open again, creating the Atlantic Ocean and a new passive margin that forms the east coast of North America today. As you can see from this example, major ranges such as the Appalachians incorporate the products of multiple orogenies and reflect the opening and closing of ocean basins. The succession of opening an ocean

basin to form a passive margin, to be followed later by closing the ocean basin to form a collisional orogen, has come to be known as the *Wilson cycle* after J. Tuzo Wilson, the Canadian geophysicist who first recognized the steps.

Take-Home Message

The Appalachian Mountains expose remnants of structures and rocks formed during three distinct orogenic events in the Paleozoic, each associated with a collision. The present eastern margin of the continent formed subsequently. Orogenic belts clearly can have complex histories.

QUICK QUESTION: How does the Appalachian orogen differ from the cratonic platform of the United States? See if you can find the boundary between the two on *Google Earth™*.

CHAPTER SUMMARY

- Mountains occur in linear ranges called mountain belts, orogenic belts, or orogens. An orogen forms during an orogeny, or mountain-building event.
- Mountain building causes rocks to undergo deformation during which they can change their location, orientation,

and shape. Geologic structures are the product of deformation.
- During brittle deformation, rocks crack and break into two or more pieces. During plastic deformation, rocks change shape without breaking.

- Stress (compression, tension, and shear) causes deformation to take place. Strain is a measure of the change of shape, or distortion, that rocks undergo when subjected to a stress. Simply put, stress causes strain.

- Joints are natural cracks in rock, formed in response to tension under brittle conditions. Veins can form when minerals precipitate out of water passing through joints.

- Faults are fractures on which there has been shearing. Geologists distinguish among normal, reverse, strike-slip, and oblique faults, based on the sense of slip. Thrust faults are gently dipping reverse faults.

- Folds are curved layers of rock. Anticlines are arch-like, synclines are trough-like, monoclines resemble the shape of a carpet draped over a stair step, basins are shaped like a bowl, and domes are shaped like an overturned bowl.

- Tectonic foliation forms when grains flatten, rotate, or grow so that they align parallel with one another.

- Uplift in mountains, over broad regions, is controlled largely by isostasy, meaning that the elevation of the Earth's surface reflects the level at which lithosphere naturally floats.

- Large collisional mountain ranges are underlain by buoyant roots. Within these ranges, folds, faults, and foliations develop and crustal thickening takes place.

- Fold-thrust belts form on the continental edge of collisional and convergent-margin orogens.

- Mountain areas may also rise if part of the underlying lithospheric mantle detaches and sinks into the asthenosphere.

- With modern GPS technology, it is now possible to measure the slow shortening and uplift of mountains.

- Once uplifted, mountains are sculpted by erosion. When crust thickens during mountain building, the deep crust may eventually become warm and weak, leading to orogenic collapse.

- Mountain belts formed by convergent-margin tectonism may incorporate accreted terranes; as a result, new crustal material adds to the edge of a continent.

- Rifting produces tilted blocks of crust that become narrow, elongate mountain ranges. Heating of the lithosphere causes uplift of rifts.

- Cratons are the old, relatively stable parts of continental crust. They include shields and platforms. Broad regional domes and basins form in platform areas.

- Large orogens, such as the Appalachians, may record multiple events of collision and rifting.

GUIDE TERMS

accretion (p. 401)
anticline (p. 394)
axial plane (p. 394)
basin (p. 395)
bearing (p. 389)
brittle deformation (p. 383)
compression (p. 387)
craton (p. 409)
crustal root (p. 406)
detachment fault (p. 391)
dip (p. 388)
displacement (p. 391)
dome (p. 395)

epeirogeny (p. 411)
exhumation (p. 409)
fault (p. 389)
fault scarp (p. 391)
fault system (p. 391)
fault trace (p. 390)
fold (p. 393)
fold-thrust belt (p. 391)
force (p. 386)
global positioning system (GPS) (p. 405)
hinge (p. 394)
isostasy (p. 406)

joint (p. 387)
limb (of fold) (p. 394)
monocline (p. 395)
mountain belt (p. 381)
mountain building (p. 381)
normal fault (p. 391)
orogen (p. 381)
orogeny (p. 381)
plastic deformation (p. 384)
plunge (p. 389)
pressure (p. 387)
reverse fault (p. 391)
shear zone (p. 391)

shield (p. 410)
slickenside (p. 391)
slip lineation (p. 391)
strain (p. 383)
stress (p. 386)
strike (p. 388)
strike-slip fault (p. 391)
suture (p. 401)
syncline (p. 394)
tectonic foliation (p. 398)
thrust fault (p. 391)
uplift (p. 405)
vein (p. 388)

REVIEW QUESTIONS

1. What changes do rocks undergo during formation of an orogenic belt?

2. What is the difference between brittle and plastic deformation?

3. What factors determine whether a rock will behave brittlely or plastically?

4. How are stress and strain different?

5. How is a fault different from a joint?

6. Compare the motion of normal, reverse, and strike-slip faults.

7. How do you recognize faults in the field?

8. Describe the differences among an anticline, a syncline, and a monocline.

9. Discuss the relationship between foliation and deformation.

10. Discuss the processes by which mountain belts form in convergent margins, in continental collisions, and in continental rifts.

11. Describe the principle of isostasy and how it affects the elevation of a mountain range.

12. What processes are involved in sculpting the rugged topography of mountain belts? How can rocks from depth beneath a mountain range be exhumed?

13. How does a craton differ from an orogenic belt?

14. Can we attribute all the structures of the Appalachian orogen to a single mountain-building event? Explain your answer.

ON FURTHER THOUGHT

15. Imagine that a geologist sees two outcrops of resistant sandstone, as depicted in the cross-section sketch below. The region between the outcrops is covered by soil. A distinctive bed of cross-bedded sandstone occurs in both outcrops, so the geologist correlated the western outcrop (on the left) with the eastern outcrop. The curving lines in the bed indicate the shape of the cross beds.

(a) Keeping in mind how cross beds form (see Chapter 7), sketch how the cross-bedded bed connected from one outcrop to the other before erosion. What geologic structure have you drawn?

(b) Is the bedrock directly beneath the geologist older than or younger than the sandstone bed of the outcrop?

 smartwork.wwnorton.com

This chapter's Smartwork features:

- What A Geologist Sees exercises on fault characteristics.
- Labeling exercises on characteristics of orogens.
- What A Geologist Sees activities on fossil identification.

GEOTOURS

This chapter's GeoTour exercise (I) features:

- Calculating strike and dip from flatirons
- Brittle structures: joints
- Brittle structures: faults
- Ductile structures: folds

PART IV

HISTORY BEFORE HISTORY

Perhaps the most important contribution that the science

of geology has made to humanity's understanding of the Earth is the demonstration that our planet existed long, long before humans took their first steps. In Part IV, we peer back into this history. First, in Interlude E, we take an in-depth look at fossils, remnants of past life that allow geologists to correlate life's evolution with that of Earth. Then, in Chapter 12, we learn how geologists gaze into "deep time"—geologic time, or the time since Earth formed—first by determining the relative ages of geologic features (whether one feature is older or younger than another) and then by learning how to calculate numerical ages (ages in years)

E Memories of Past Life:
   Fossils and Evolution
12 Deep Time: How Old Is Old?
13 A Biography of Earth

based on ratios of radioactive elements to their daughter products in minerals. With the background provided in Chapter 12, we're ready for Chapter 13's brief synopsis of Earth history, from the birth of the planet to the present. We see how plate tectonics has redistributed continents and built mountains, how sea level has risen and fallen, and how Earth's climate has changed over time.

Motorists speeding down this highway in Utah are passing a panorama of time. The layers of brightly colored rock exposed on the cliffs near the road illustrate the narrative of our planet's past, when the landscape at this location would have looked very different.

This fossil of a fish was carefully exposed by removing the overlying layer of rock. Fine details of the fish's bones and fins have been preserved for millions of years, allowing us to see a species that no longer exists.

Memories of Past Life: Fossils and Evolution

LEARNING OBJECTIVES

By the end of this interlude, you should understand . . .

- what a fossil is and how fossils form.
- why relatively few organisms become fossils.
- how to recognize some of the more common fossils.
- how the study of fossils contributes to an understanding of life evolution.

E.1 The Discovery of Fossils

If you look closely at outcrops of sedimentary strata or of air-fall tuff, you might occasionally find features that resemble shells, bones, leaves, or footprints, either on the surfaces of layers or within the rock itself (**Fig. E.1a–c**). Especially impressive examples were displayed in ancient Greek temples as trophies, "dragon" (dinosaur) bones were used in traditional Chinese medicine, and mythical creatures such as griffins and the Native American thunderbird may have been inspired by skeletons embedded in rock. The origin of these features mystified early thinkers. Some thought that they had somehow grown underground, in already solid rock. Today such **fossils**—from the Latin word *fossilis*, which means dug up—are considered to be the remnants or traces of ancient living organisms that were buried with the material from which the rock formed and were preserved after lithification. Surprisingly, this interpretation, though proposed by the Greek historian Herodotus in 450 B.C.E. and revived by Leonardo da Vinci in 1500 C.E., did not become widely accepted until the 1669 publication of a book in which a Danish physician named Nicolas Steno (1638–86) argued that components of organisms could be incorporated in rock without losing their distinctive shape. A British contemporary of Steno, Robert Hooke (1635–1703), emphasized that fossils provide insight into the nature of life that existed earlier in Earth history, because most fossils represent *extinct species*, meaning species that lived in the past but can no longer be found alive today. The concept of extinction remained controversial until Georges Cuvier (1769–1832), a French zoologist, carefully demonstrated that the skeletons and teeth of fossil organisms differ from those of any modern ones. Cuvier was also the first to classify fossils using the same systematic approach that biologists had developed for classifying modern organisms. Subsequently, **paleontologists**, researchers who study fossils, collected vast numbers of fossil specimens, which they named, classified, and cataloged. Museums accumulated extensive collections of fossils, which, in turn, could be used as a basis for comparison in the analysis of new discoveries (**Fig. E.1d**). Thus, the 19th century saw **paleontology**, the study of fossils, ripen into a science.

Paleontology went beyond being merely an exercise in description when William Smith, a British engineer who supervised canal construction in England during the 1830s, noted that fossil species in the lower layers of strata differ from those in the higher layers within a sequence of sedimentary rocks. Smith eventually realized that sequences of strata, in fact, contain a predictable succession of fossils, from base to top and that a given fossil species can only be found in a specific interval of the strata. His discovery made it possible for geologists to use fossils as a basis for determining the age of one interval of sedimentary strata relative to another. Fossils, therefore, have become an indispensable tool for studying geologic history and for documenting the evolution of life.

How do fossils form? Where can we find them? What can they tell us about how life has changed over Earth's history? To address these questions, this interlude provides a brief overview

FIGURE E.1 Examples of fossils and fossil collections.

(a) Fossil shells in 400-Ma sandstone. (Ma means "million-year old.")

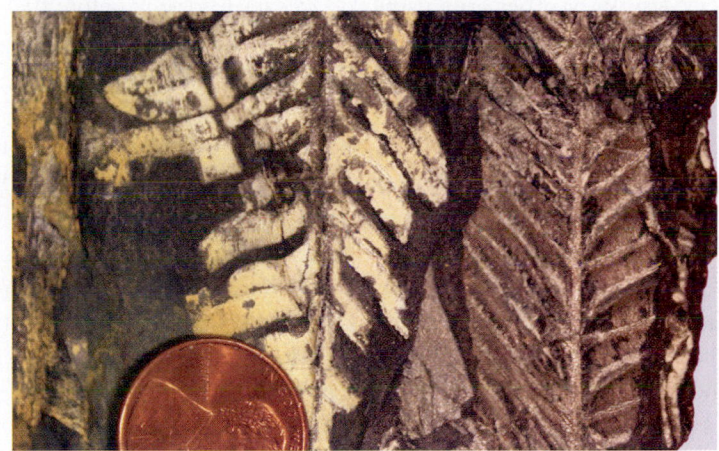

(b) Fossil leaves in 300-Ma shale.

(c) Fossil skeleton in 200-Ma sandstone.

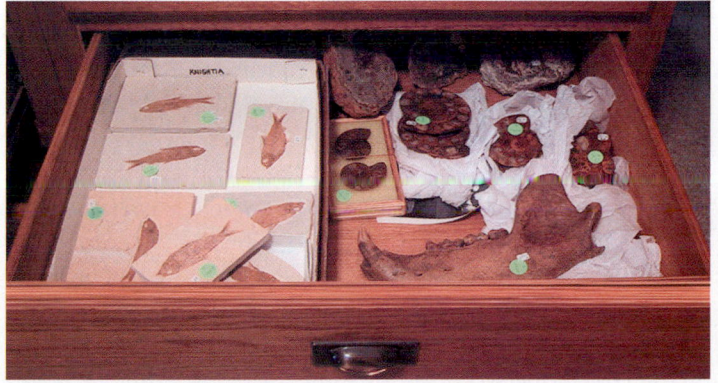

(d) A drawer of fossil specimens in a museum.

of fossils. We'll use the background provided here in the discussion of geologic time and Earth history covered by the next two chapters.

E.2 Fossilization

What Kinds of Rocks Contain Fossils?

Fossils form when organisms die and become buried by sediment or air-fall ash, or when organisms travel over or through these materials and leave imprints or debris. The vast majority of fossils occur in sedimentary rocks, though some important examples have been found in volcanic tuffs (**Fig. E.2**). Fossils cannot survive the recrystallization, new mineral growth, and shearing that accompany all but the lowest grades of metamorphism and thus do not occur in metamorphic rocks. Similarly, fossils cannot be found in intrusive igneous rocks because organisms do not live in intrusive environments and could not survive in igneous heat. And, with one exception, they are not found in lava flows or hot pyroclastic flows because they would be incinerated before the material solidified into rock. (The exception is where very low-viscosity basaltic lava flows around tree trunks or other organisms and freezes into rock before the organisms have completely burned up—the space left in the lava, where the organism once stood is, effectively, a fossil.)

Forming Fossils

Paleontologists refer to the process of forming a fossil as **fossilization**. To see how a fossil develops, let's consider an example (**Fig. E.3a**). Imagine an elderly dinosaur searching for food along the muddy shore of a lake on a scalding summer day in the geologic past. The hungry dinosaur succumbs to the heat and collapses dead into the mud. Soon after, scavengers strip the skeleton of meat and scatter the bones. But before the bones have had time to weather away, it rains heavily and streams draining nearby hills dump silty water into the lake, so the lake level rises and submerges the carcass with that water. After the storm, silt settles out of the quiet water and buries the bones along with the dinosaur's footprints. More sediment from succeeding floods buries the bones and prints still deeper, permanently protecting them from being destroyed by currents or by burrowing organisms. Much later, sea level rises and a thick sequence of marine sediment accumulates over the beds of mud and silt until, eventually, the sediment containing the bones and footprints becomes so deeply buried that it turns into sedimentary rock—the silt turns into siltstone and mud turns into shale. The dinosaur's footprints remain outlined by the contact between the siltstone and shale while its bones reside within the siltstone. Over time, minerals precipitating from groundwater replace some of the chemicals constituting the bones, until the bones themselves become rock-like. The buried bones and footprints are now fossils.

How do fossils end up back at the Earth's surface, where they can be found? As time passes, the region containing the dinosaur-fossil-bearing beds stops subsiding and starts to undergo uplift. Erosion gradually strips away overlying strata until rocks containing the bones and footprints become exposed in an outcrop. If a team of paleontologists observes the fossils protruding from the outcrop, they may undertake an excavation, taking care to avoid breaking the specimens (**Fig. E.3b**). If they're lucky, they can uncover enough bones to reconstruct a skeleton so that the dinosaur can rise again, though this time in a museum (**Fig. E.3c**).

FIGURE E.2 The famous fossil footprints at a site called Laetoli, in Olduvai Gorge, Tanzania. They were left when an adult and child of a human ancestor, *Australopithecus*, walked on two feet over ash that had recently been erupted by a nearby volcano. The ash had been dampened by rain when a second ash eruption buried it and thereby preserved the footprints.

FIGURE E.3 From a living organism to a museum display.

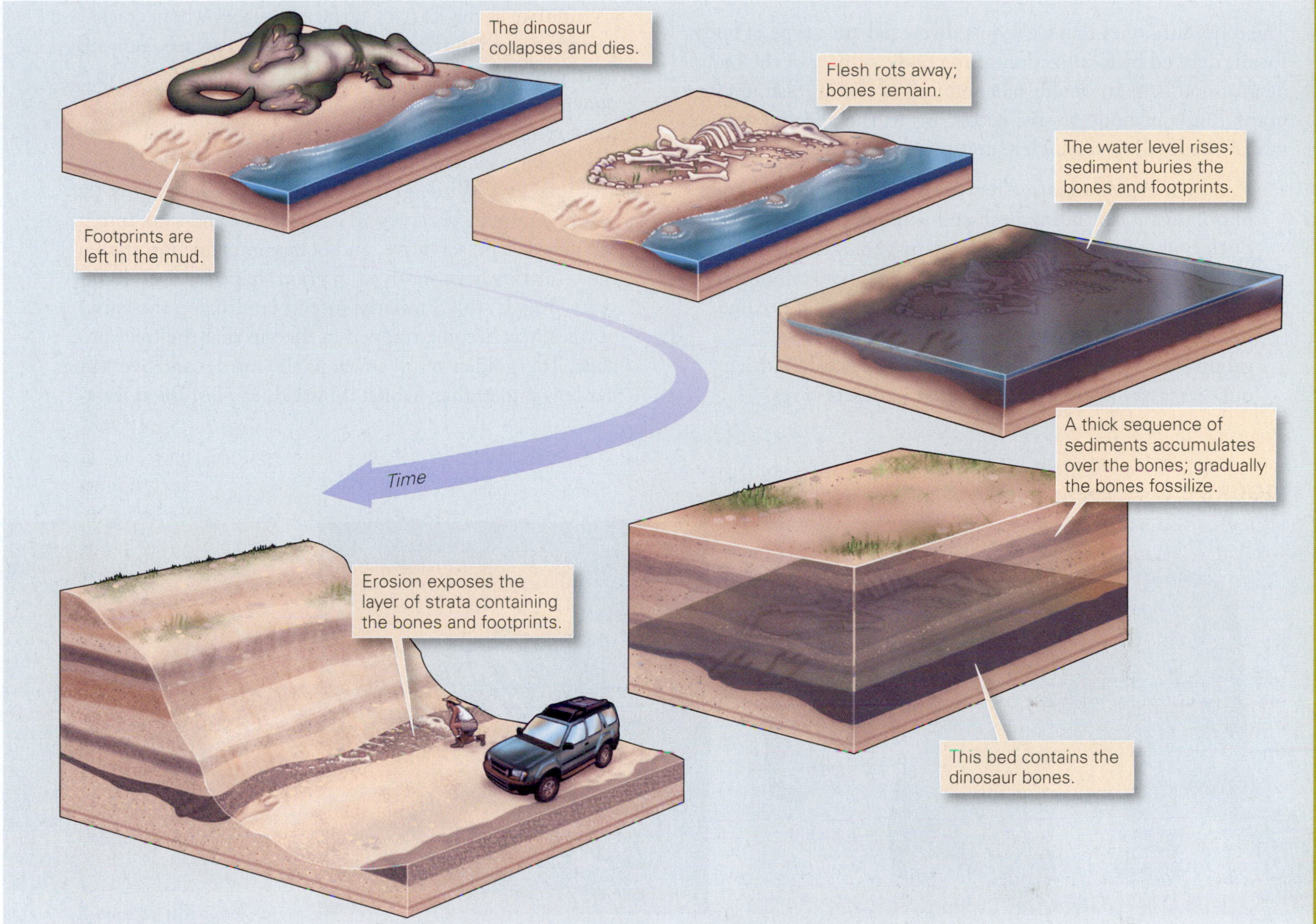

The dinosaur collapses and dies.

Footprints are left in the mud.

Flesh rots away; bones remain.

The water level rises; sediment buries the bones and footprints.

Time

A thick sequence of sediments accumulates over the bones; gradually the bones fossilize.

Erosion exposes the layer of strata containing the bones and footprints.

This bed contains the dinosaur bones.

(a) The stages in fossilization of a dinosaur.

(b) A paleontologist collecting specimens.

(c) A display of dinosaur fossils (or plaster casts of fossils) in a museum.

Paleontologists may also quarry out slabs of rock containing footprints. In recent years, bidding wars have made some fossil finds extremely valuable. For example, a skeleton of a *Tyrannosaurus rex*, a 67-million-year-old dinosaur, sold at auction in 1997 for $7.6 million. The specimen, named Sue, after its discoverer, now stands in the Field Museum of Chicago.

The Many Different Kinds of Fossils

The dinosaur bones that we've just discussed are a type of **body fossil**, derived by fossilization of the body, or part of the body, of an organism. In detail, paleontologists distinguish among many kinds of body fossils, according to the specific way in which the fossil formed. Here are some examples:

- *Frozen or dried fossils:* In a few environments, whole bodies of organisms, including flesh and skin, may be preserved. Such body fossils are fairly young, by geologic standards—their ages can be measured in thousands, not millions, of years. Examples include woolly mammoths that became incorporated in the permafrost (permanently frozen ground) of Siberia (**Fig. E.4a**) and the desiccated (dried out) carcasses of creatures that died in desert caves.

- *Fossils preserved in tar or amber:* Small pools of oil locally accumulate at the Earth's surface in places where cracks provide a conduit through which the oil can seep upward from underground. Once at the surface, volatile components of the oil evaporate, and bacteria degrade what remains, leaving behind sticky tar. At the La Brea Tar Pits in Los Angeles, California, tar accumulated in a swampy area long ago. While grazing, drinking, or hunting at the swamp, animals became mired in the tar and sank into it. Tar acts as a preservative, and bones embedded in the La Brea Tar Pits have survived in great shape for over 40,000 years (**Fig. E.4b**). Similarly, insects landing on the bark of trees may become trapped in the sap that the trees produce. This golden syrup envelops the insects and over time hardens into **amber**, which if buried, can last for at least

FIGURE E.4 Examples of different kinds of fossils.

(a) This 1-m long baby mammoth, found in Siberia, died 37,000 years ago.

(b) A fossil skeleton of a 2-m-high giant ground sloth from the La Brea Tar Pits in California.

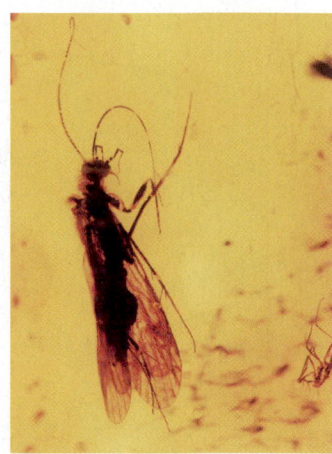

(c) This insect became embedded in amber about 200 million years ago.

(d) Fossil shells, the "hard parts" of invertebrates.

(e) The carbonized impressions of fern fronds in a shale.

(f) Petrified wood from Arizona. It is so hard that it remains after the rock that surrounded it has eroded away.

200 million years (**Fig. E.4c**). Amber that contains insect fossils has become popular as jewelry.

- *Preserved hard parts:* Paleontologists refer to the bones (internal skeletons) of vertebrate animals and the shells (external skeletons) of invertebrate animals informally as hard parts. Bones and shells are hard because they consist of durable minerals, such as calcite or silica. In some cases, the original minerals of a bone or shell may survive in a rock long after the rock lithifies. More commonly, the minerals of the hard parts recrystallize during diagenesis of sedimentary rock (see Chapter 7). But even when this happens, the shape of the bone or shell can be preserved as a fossil (**Fig. E.4d**).

- *Molds and casts:* As sediment compacts around the hard parts of an organism, it conforms to their shape. If the hard parts later dissolve away, the shape may still remain as an indentation called a **mold** (see Fig. E.1a)—sculptors use the same word to refer to the receptacle into which they pour bronze or plaster. The sediment that fills the mold forms a **cast** (see Fig. E.4d), which also depicts the shape of the hard part. Molds and casts can appear as indentations or protrusions from bedding surfaces, respectively, and since they are a relict of an organism's body, they are considered to be a type of body fossil.

- *Carbonized impressions of bodies:* Impressions are simply flattened molds or casts created when soft or semi-soft organisms (such as leaves, insects, shell-less invertebrates, sponges, feathers, or jellyfish) get pressed between layers of sediment. Chemical reactions eventually remove most of the organic material, leaving only a thin film of carbon on the surface of the impression (**Fig. E.4e**).

- *Permineralized fossils:* **Permineralization** refers to the process by which minerals precipitate in porous material, such as wood or bone, underground. The ions from which the minerals grow come from groundwater solutions that slowly seeped into the pores. **Petrified wood**, an example of a permineralized fossil, forms when volcanic ash buries a forest. Groundwater passing through the ash dissolves silica and carries it into the wood, where the silica replaces cell interiors. Eventually, the wood completely transforms into hard chert. (The word *petrified* literally means turned to stone.) During the process, the cell walls of the wood transform into organic films that survive permineralization, so the fine detail of the wood's cell structure and bark can be seen in a petrified log (**Fig. E.4f**). The color of petrified logs come from impurities such as iron or carbon in the chert.

Not all visible fossils are body fossils. Paleontologists refer to a fossilized feature formed by the action of an organism as a **trace fossil**. In detail, we can distinguish among several types of trace fossils, including the following:

- *Footprints:* We've already noted that organisms may leave distinctive footprints in the mud. These can be buried and preserved in rock (**Fig. E.5a, b**). Paleontologists can study assemblages of fossil footprints to determine how organisms moved and even how they interacted with one another.

- *Burrows:* These are sediment-filled holes or tunnels left behind when an organism digs its way through sediment at or below the surface. Sediment within the burrow may have a slightly different texture or permeability than surrounding unburrowed sediment and thus remain visible in rock.

- *Feeding marks:* These form when an organism disrupts the surface of a sedimentary bed in order to eat organisms that live in the sediment (**Fig. E.5c**).

FIGURE E.5 Examples of trace fossils.

(a) Molds of dinosaur footprints. These push down into the top of a bed.

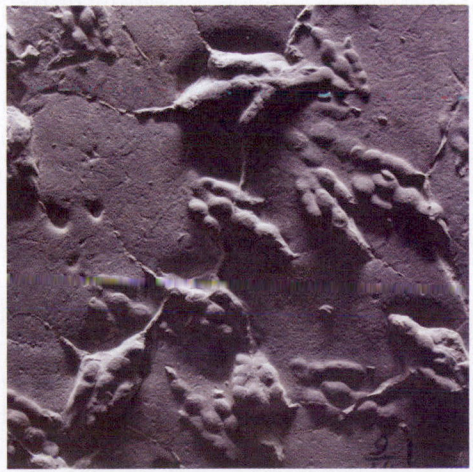

(b) Casts of dinosaur footprints. These protrude from the base of a bed.

(c) Feeding traces on the surface of a bed of sandstone in Ireland.

- *Coprolites:* Organisms leave behind excrement that has passed through their digestive track. This material, when buried and fossilized by the same processes that lead to fossilization of the organism itself, becomes fossils called *coprolites*.

Some fossils are not visible, but instead are chemical. These **chemical fossils**, which can be studied only with laboratory instruments, include:

- *Molecular fossils (biomarkers):* The life activity of an organism produces organic chemicals. When the organism dies and becomes buried in sediment, these organic chemicals tend to break up. In some cases, specific segments of the chemicals may be durable enough to undergo lithification of sediment. Such distinctive, durable molecular segments are called molecular fossils, or **biomarkers**.
- *Distinctive isotope ratios:* An *isotope ratio* refers to the relative proportion of a lighter isotope to a heavier isotope of the same element. (Recall that lighter and heavier isotopes have the same atomic number, but the lighter one has a smaller atomic weight than does the heavier one.) The life activity of organisms tends to utilize lighter isotopes slightly more than heavier ones, so the isotopic ratio found in strata deposited at the time the organisms lived differs from that found in strata when they didn't. For example, photosynthesis preferentially uses ^{12}C instead of ^{13}C, so enrichment of ^{12}C in strata serves as fossil evidence for the appearance of photosynthetic organisms. Isotope ratios are a type of trace fossil, because they are a consequence of the organisms' life activity.

We've seen how paleontologists describe a body fossil based on how the fossil formed. It's also useful to distinguish among different fossils on the basis of the fossil's size. Paleontologists, therefore, distinguish between **macrofossils**, which are ones that are large enough to be seen with the naked eye, and **microfossils**, which can be seen only with a microscope. Microfossils include the remnants of plankton, algae, bacteria, and pollen (**Fig. E.6**).

Fossil Preservation

Not all living organisms become fossils when they die. In fact, only a small percentage do, for it takes special circumstances to produce a fossil and allow it to survive. Examples of conditions that have an effect on the degree to which a recognizable fossil can be preserved include the following.

- *How fast burial takes place:* If an organism dies in a depositional environment where sediment can accumulate rapidly, it may be buried before it has time to rot, oxidize, or be eaten and thus has a better chance of being preserved. Organisms that lie exposed for a long time before burial will be less likely to be preserved.

FIGURE E.6 Examples of fossil plankton shells. Because of their size, geologists refer to these as microfossils or nanofossils.

- *The energy of the depositional setting:* Specimen preservation will be better for organisms deposited along with sediment in quiet water, a so-called low-energy environment, for the fossils remain intact as they are buried. In high-energy environments, where there are strong currents or waves, organisms tumble about, break up, and are dispersed as fragments before being buried.
- *The presence of hard parts:* Organisms without durable hard parts (shells or bones) usually don't get fossilized, for soft flesh decays long before hard parts do under most depositional conditions. For this reason, there are more examples of bivalves (a class of organisms, including clams and oysters, with strong shells) in the fossil record than there are of jellyfish (which have no shells) or spiders (which have very fragile shells).
- *Oxygen content of the depositional environment:* A dead squirrel by the side of the road won't become a fossil. As time passes, birds, dogs, or other scavengers may come along and eat the carcass. And if that doesn't happen, maggots, bacteria, and fungi infest the carcass and gradually digest it. Flesh that has not been eaten or does not rot can oxidize (react with oxygen), a reaction that breaks down organic chemicals. The remaining skeleton weathers in air and turns to dust. Thus, before roadkill can become incorporated in sediment, it has vanished. If, however, a carcass settles into an oxygen-poor environment (*anoxic conditions*), such as occurs in a stagnant lagoon, oxidation reactions happen slowly, scavenging organisms aren't abundant, and bacterial metabolism takes place very slowly. In such environments, the organism won't rot away before it has a chance to be buried and preserved, so the likelihood that the organism becomes fossilized increases.

By carefully studying modern organisms and by taking into account the concepts we've just described, paleontologists can provide rough estimates of the **preservation potential** of organisms, meaning the likelihood that an organism will be buried and eventually transformed into a fossil. For example, in a typical modern-day shallow-marine environment, such as the mud-and-sand sea floor close to a beach, about 30% of the organisms have sturdy shells and thus a high preservation potential, 40% have fragile shells and a low preservation potential, and the remaining 30% have no hard parts at all and are not likely to be fossilized except in special circumstances. Of the 30% with sturdy shells, though, few happen to die in a depositional setting where they are buried fast enough and actually *do* become fossilized. Thus, fossilization is the exception rather than the rule.

Extraordinary Fossils: A Special Window to the Past

Although, as we've just seen, only hard parts survive in most fossilization environments, paleontologists have discovered a few special locations where relicts of soft parts have as well; such fossils are known as **extraordinary fossils**. Extraordinary fossils include insects preserved in amber and frozen or desiccated organisms. The category also includes fossils that settled on the anoxic floor of quiet-water lakes or lagoons or the deep ocean. In these settings, the soft parts have disappeared but not before leaving distinct carbonized impressions.

One of the most famous sources of extraordinary fossils occurs in a small quarry near Messel, Germany. By carefully prying apart thin beds in the quarry, paleontologists have been able to extract extraordinary fossils of 49-million-year-old mammals, birds, fish, and amphibians that died in a shallow-water lake (**Fig. E.7a**). Bird fossils from the quarry include the delicate imprints of feathers, bat fossils come complete with impressions of ears and wings, and other mammal fossils have an aura of carbonized fur. In southern Germany, exposures of the Solnhofen Limestone, an approximately 150-million-year-old rock made of calcite mud deposited in a stagnant lagoon, contain extraordinary fossils of about 600 species, including *Archaeopteryx*, one of the earliest birds (**Fig. E.7b**). And exposures of the Burgess Shale in the Canadian Rockies of British Columbia have yielded a plentitude of fossils showing what shell-less invertebrates that inhabited the sea floor about 510 million years ago looked like (**Fig. E.7c, d**). Significantly, the *Burgess Shale fauna* is so strange—for example, it includes organisms with circular jaws—that it has been hard to determine how these organisms are related to present-day ones.

In a few cases, extraordinary fossils include actual tissue, a discovery that has led to a research race to find the oldest preserved DNA. (**DNA**, deoxyribonucleic acid, is the complex molecule, shaped like a double helix, that contains the code that guides the growth and development of an organism. Individual components of this code are called *genes*.) Paleontologists have isolated small segments of DNA from amber-encased insects that are over 40 million years old. The amounts are not enough, however, to clone extinct species, as suggested by the popular 1993 film *Jurassic Park*.

E.3 Taxonomy and Identification

Since the days of Georges Cuvier, the classification of fossils has followed the same principles that Carolus Linnaeus, a Swedish biologist, developed in the 18th century for the classification of living organisms. The study of how to classify organisms is now referred to as **taxonomy**.

Linnaeus's scheme for taxonomy has a hierarchy of divisions. First, all life is divided into three *domains*, named Archaea, Bacteria, and Eukarya. The domains differ from one another based on fundamental characteristics of the genes in their DNA. **Archaea** include a vast array of tiny single-celled microorganisms that grow not only in the mild environments of oceans, soils, and wetlands but also in the harsh environments of hot springs, black smokers, salt lakes, very acidic sediments, and saline groundwater (**Fig. E.8a**). Organisms that can survive in harsh environments are known as *extremophiles*; many extremophiles do not need light or air but rather live off the energy stored in the chemical bonds of minerals. **Bacteria** are also tiny single-celled organisms, species of which inhabit almost all livable environments on Earth (**Fig. E.8b**). Visually, it may be difficult to distinguish bacteria from archaea, but on a genetic level they are profoundly different. Nevertheless, both archaea and bacteria are **prokarya**, meaning that their cells do not contain a nucleus (a distinct membrane-surrounded region of a cell containing all the cell's DNA). In this regard, archaea and bacteria differ from **eukarya**, organisms whose cells do contain a nucleus. Taxonomists divide the Eukarya domain into several kingdoms.

- *Protista:* various unicellular and simple multicellular organisms
- *Fungi:* mushrooms and yeast
- *Plantae:* trees, grasses, and ferns
- *Animalia:* sponges, corals, snails, dinosaurs, ants, and people

Each kingdom consists of one or more phyla. A phylum, in turn, includes several classes, a class includes several orders, an order

FIGURE E.7 Extraordinary fossils. These fossils are particularly well preserved.

(a) A 50-million-year-old mammal fossil was chiseled from oil shale near Messel, Germany. It still contains the remains of skin.

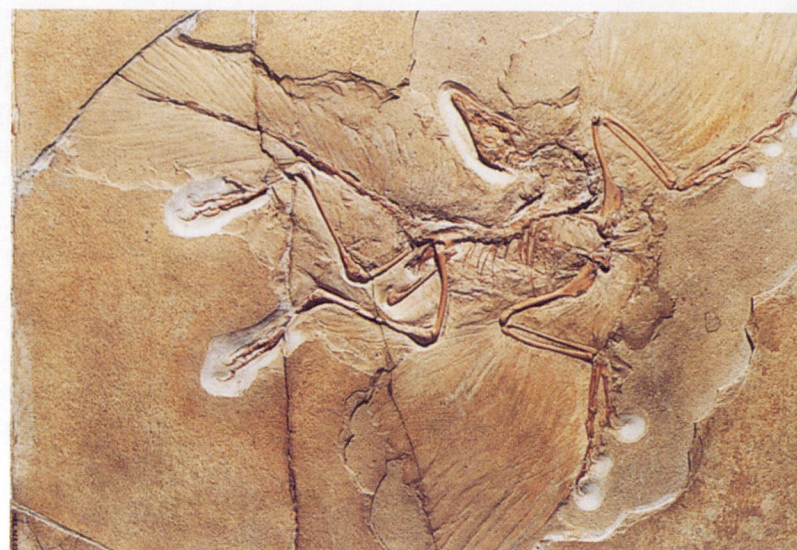

(b) *Archaeopteryx* from the 150-million-year-old Solnhofen Limestone of Germany. The imprints of feathers are clearly visible.

(c) The Burgess Shale of the Canadian Rockies contains unique Cambrian arthropods, such as Marella, shown here.

(d) An artist's reconstruction of a Cambrian ecosystem. The paintings are based on Burgess Shale fossils.

includes several families, a family includes several genera, and a genus includes one or more species (**Fig. E.9**). Kingdoms, therefore, represent the broadest category of eukarya, and species represent the narrowest.

Keeping in mind the concepts of taxonomy, you'll find that there's nothing magical about identifying fossils. Typically, you can recognize common fossils in the field by examining their *morphology* (form or shape). If the fossil is well preserved and has distinctive features, the process can be straightforward so even beginners can distinguish the major groups of fossils from one another on sight. But if fossils are broken into fragments

and parts are missing, or if the fossil shares many characteristics with many different species, identification can be challenging. In many cases classification of a given organism has remained controversial.

Many fossil organisms resemble modern organisms, and this similarity provides a starting point for classification. For example, to identify a fossil, paleontologists look at the nature of the skeleton (is it internal or external?); the symmetry of the organism (is it bilaterally symmetric, like a mammal, so that one side is a mirror image of the other, or does it have fivefold symmetry like a starfish?); the design of the shell (in the case of

FIGURE E.8 The simplest forms of life.

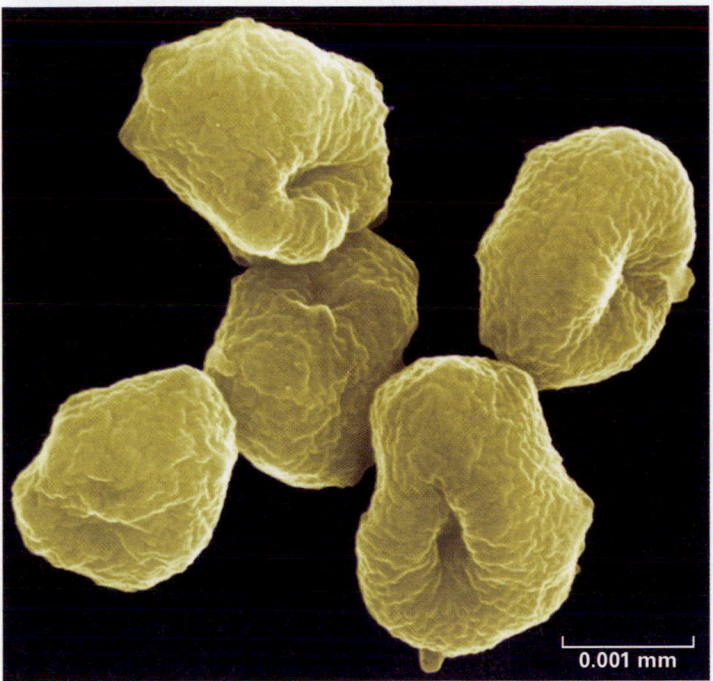

(a) Archaea cells.

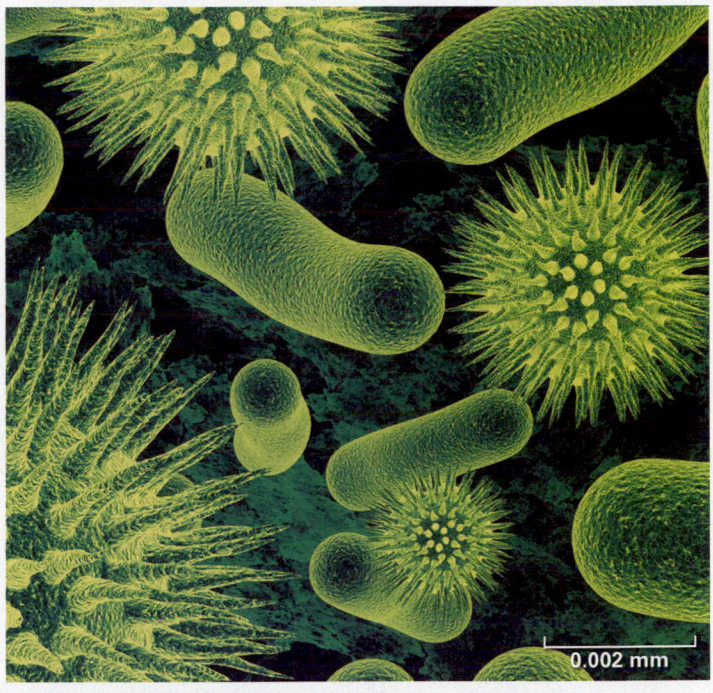

(b) Bacteria cells.

invertebrates); and the design of the jaws, teeth, or feet (in the case of vertebrates). A fossil mammal looks like a mammal and not a fish, and a fossil clam (class Bivalvia) looks like a clam and not a snail (class Gastropoda). Similarly, a fossil organism with a spiral shell that does not contain internal chambers is a member of the class Gastropoda (the snails), whereas an organism with a chambered, spiral shell is a member of the class Cephalopoda (**Fig. E.10a**). At taxonomic levels below class, identification may involve focusing on details, such as the number of ridges on the surface of the shell.

Paleontologists have found countless species of fossils that resemble living organisms only at the level of orders or even higher. For example, a group of extinct organisms called trilobites have no close living relatives (**Fig. E.10b**). But they were clearly segmented invertebrate animals, and as such, they resemble arthropods such as insects and crustaceans. Thus, they are considered to be a distinct class of the phylum Arthropoda. In the case of fossils that contain preserved DNA, paleontologists may someday be able to determine relationships among organisms more accurately by specifying the percentages of shared genes within the DNA.

Figure E.11 provides simplified sketches of some of the common types of invertebrate fossils. With this figure, you should be able to identify many of the fossils you'll find in a typical bed of limestone. Some of the notable characteristics of these fossils include the following.

- *Trilobites:* These have a segmented shell that is divided lengthwise into three parts. They are a type of arthropod.
- *Gastropods (snails):* Most fossil specimens of gastropods have a spiral shell that does not contain internal chambers.
- *Bivalves (clams and oysters):* These have a shell that can be divided into two similar halves. The plane of symmetry is parallel to the plane of the shell.
- *Brachiopods (lamp shells):* The top and bottom parts of these shells have different shapes, and the plane of symmetry is perpendicular to the plane of the shell. Examples typically have ridges radiating out from the hinge.
- *Bryozoans:* These are colonial animals. Their fossils resemble a screen-like grid of cells. Each cell is the shell of a single animal.
- *Crinoids (sea lilies):* These organisms look like a flower but actually are animals. They have a stalk consisting of numerous circular plates stacked one on top of the other.
- *Graptolites:* These look like tiny saw blades. They are remnants of colonial animals that floated in the sea.
- *Cephalopods:* These include ammonites, with a spiral shell, and nautiloids, with a straight shell. Their shells contain internal chambers and have ridged surfaces. These organisms were squid-like.
- *Corals:* These include colonial organisms that form distinctive mounds or columns in tropical reefs. Paleozoic examples include solitary examples with a cone-like shell.

FIGURE E.9 The taxonomic subdivisions. The center column shows how the names apply to human beings.

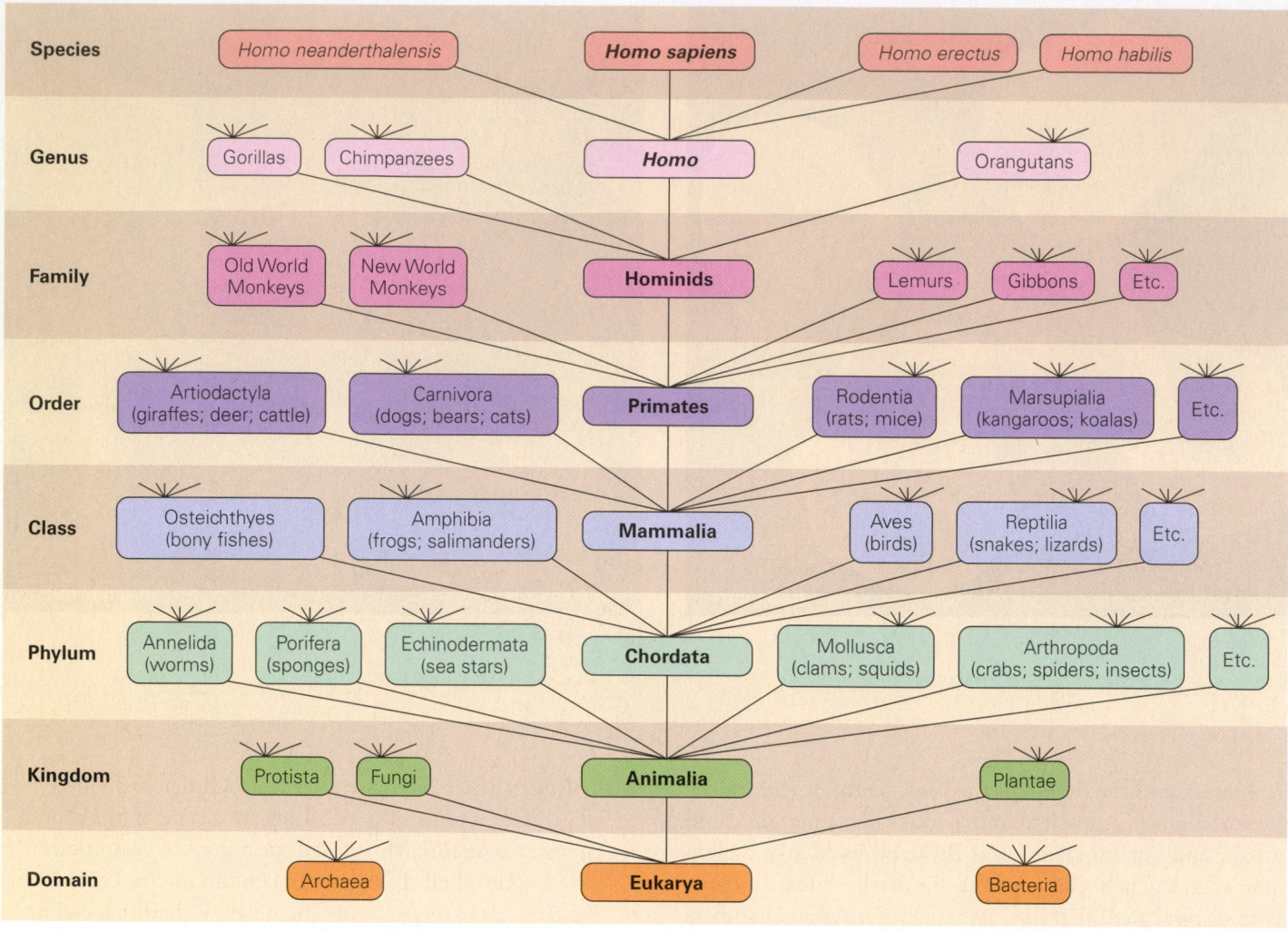

E.4 The Fossil Record

A Brief History of Life

Based on laboratory experiments conducted in the 1950s, researchers speculated that reactions in concentrated "soups" of chemicals that formed when seawater evaporated in shallow, coastal pools led to the formation of the earliest protein-like organic chemicals (proto-life). More recent studies suggest, instead, that such reactions took place in warm groundwater beneath the Earth's surface or at hydrothermal vents (black smokers) on the sea floor. Presently, researchers are paying particular attention to the hydrothermal vent environment as the cradle of life. Vent chimneys are hollow and have walls

through which fluids can pass, so the walls resemble membranes that separate different chemical environments and across which weak electrical currents develop. In this environment, chemical reactions do yield protein-like chemicals. While these chemicals are non-biotic, researchers speculate that they could at some point have developed into the building blocks of life.

While the nature of proto-life remains a mystery, the study of fossils in the oldest-known sedimentary rocks has started to provide an image of early life. Specifically, archaea and bacteria fossils appear in rocks as old as about 3.7 billion years. For the first billion years or so of life history, cells of these organisms were the only types of life on Earth. Then, at about 2.5 Ga (billion years ago), organisms of the Protista kingdom first appeared. Early multicellular organisms, shell-less invertebrates of the animal kingdom, and fungi came into existence at perhaps 1.0 to 1.5 Ga. Complex multicellular organisms began

FIGURE E.10 Examples of fossil classification.

(a) Examples of ammonites, a type of cephalopod. The inset shows segmentation inside a shell.

(b) Examples of trilobites. Note that each specimen has three parts (hence the prefix *tri-*).

FIGURE E.11 Common types of invertebrate fossils.

Trilobite Gastropod Bivalve

Brachiopod Bryozoan

Crinoid

Graptolite Ammonite Coral

to populate marine environments by about 635 Ma (million years ago), and the first shelly fauna can be found in rocks as old as 542 Ma. The great variety of shelly invertebrate classes whose descendants are alive today appeared during the next 20 million years or so, a relatively short period of time compared to the age of the Earth. Paleontologists refer to this event as the **Cambrian explosion**, after the geologic time interval in which it occurred. Within each class, life "radiated" (diversified) into different orders. Eventually, during the next few hundred million years, fish, land plants, amphibians, reptiles, and finally birds and mammals appeared successively.

Researchers have been working hard to understand *phylogeny* (evolutionary relationships) among organisms, using both the morphology of organisms and, more recently, the study of genetic material. We can portray ideas about which groups radiated from which ancestors in a chart called the tree of life, or, more formally, the **phylogenetic tree** (**Fig. E.12**). Study of DNA is now enabling researchers to understand relationships between molecular processes and evolutionary change.

Is the Fossil Record Complete?

By some estimates, more than 250,000 species of fossils have been collected and identified to date, by thousands of investigators and collectors working on all continents over the past two centuries. These fossils define the framework of life's evolution. But the record is far from complete—known fossils cannot account for every intermediate step in the evolution of every organism. As many as 8.7 million eukaryotic species may

FIGURE E.12 Phylogenetic trees.

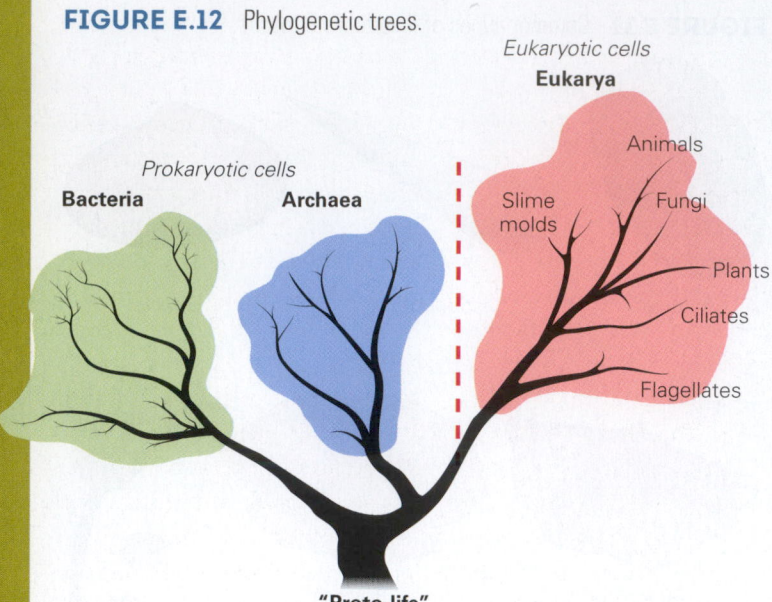

(a) The "tree of life" symbolically illustrates relations among different domains of organisms on Earth. Archaea and bacteria are both prokaryotes and don't have nuclei. All other life forms are eukaryotes because they have cells with a nucleus.

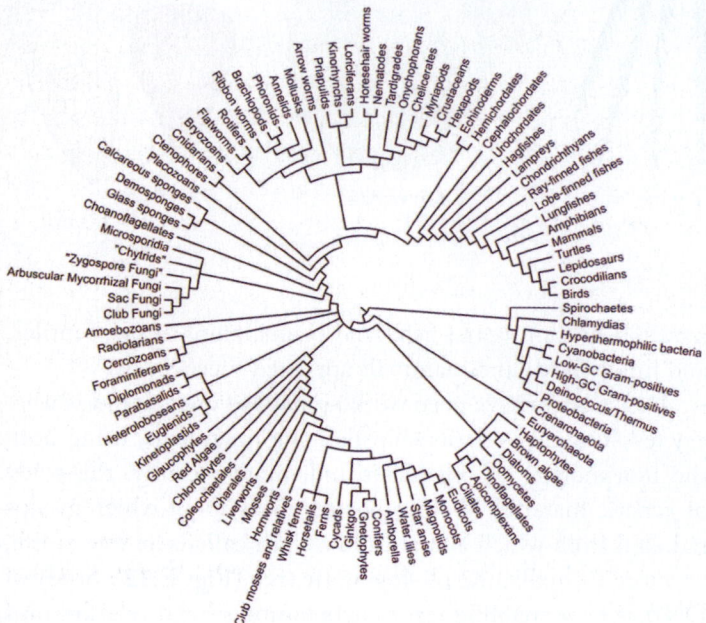

(b) Example of a phylogenetic tree in biology. The position of each branch is based on the study of DNA.

be living on Earth today, and over the billions of years that life has existed there may have been 5 billion to 50 billion species of life in the three domains. Clearly, known fossils represent at most a very tiny percentage of these species. Why is the record so incomplete?

First, despite all the fossil-collecting efforts of the past two centuries, paleontologists have not even come close to sampling every cubic centimeter of exposed sedimentary rock on Earth, much less the vast volumes of sedimentary rock that remain below the surface. Just as biologists have not yet identified every living species of insect, paleontologists have not yet identified every species of fossil. New species and even genera of fossils continue to be discovered every year.

Second, not all organisms are represented in the rock record because not all organisms have a high preservation potential. As noted earlier, fossilization occurs only under special conditions, and thus only a minuscule fraction of the organisms that have lived on Earth have left a fossil record. There may be few, if any, fossils of a vast number of extinct species, so we have no way of knowing what they looked like or even that they ever existed.

Finally, as we will learn in Chapter 12, the sequence of sedimentary strata that exists on Earth does not account for every minute of time since the formation of our planet. Sediments accumulate only in environments where conditions are appropriate for deposition and not for erosion—sediments do not accumulate, for example, on the dry great plains or on mountain peaks, but they do accumulate in the sea and in the floodplains and deltas of rivers. Because Earth's climate changes through time and because the sea level rises and falls, certain locations on continents are sometimes sites of deposition and sometimes aren't, and on occasion they may become sites of erosion. Therefore, strata only accumulate episodically.

In sum, a rock sequence provides an incomplete record of Earth history, organisms have a low probability of being preserved, and paleontologists have found only a small percentage of the fossils preserved in rock. So the incompleteness of the fossil record comes as no surprise.

E.5 Evolution and Extinction

Darwin's Grand Idea

As a young man in England in the early 19th century, Charles Darwin had been unable to settle on a career but had developed a strong interest in natural history. Therefore, he jumped at the opportunity to serve as a naturalist aboard HMS *Beagle* on an around-the-world surveying cruise. During the five years of the cruise, from 1831 to 1836, Darwin made detailed observations of plants, animals, and geology in the field and amassed an immense specimen collection from South America, Australia, and Africa. Just before Darwin departed on the voyage, a friend gave him a copy of the first geology textbook, Charles Lyell's 1830 publication *Principles of Geology*, which argued in favor of James Hutton's proposal that the Earth had a long history and that geologic time extended much farther into the past than did human civilization.

A visit to the Galápagos Islands, off the coast of Peru, led to a turning point in Darwin's thinking during the voyage. The naturalist was most impressed with the variability of Galápagos finches. He marveled not only at the fact that different varieties of the bird occurred on different islands but at how each variety had adapted to utilize a particular food supply. With Lyell's writings in mind, Darwin developed a hypothesis that the finches had begun as a single species that later branched into several different species when populations of the birds became isolated on different islands. This can happen because offspring can differ from their parents and new traits can be transferred to succeeding generations. If enough change accumulates over many generations, the living population ends up being so different from its distant ancestors that the population can be classified as a new species. During the course of evolution, old species vanish and new species appear. The accumulation of changes may eventually yield a population that is so different from its ancestors that taxonomists consider it to be a new genus. Greater accumulations of differences resulted in new classes, orders, or even phyla earlier in Earth's history. Change in a population over a succession of generations, due to the transfer of inheritable characteristics, is the process of **evolution**.

Darwin and his contemporary, Alfred Russel Wallace, not only proposed that evolution took place but also came up with an explanation for why it occurs. The crux of their explanation is simply this: A population of organisms cannot increase in number forever because it is limited by competition for scarce resources in the environment. In nature, only organisms capable of survival can pass on their characteristics to the next generation. In each new generation, some individuals have characteristics that make them more fit, whereas some have characteristics that make them less fit. The fitter organisms are more likely to survive long enough to produce offspring. Thus, the beneficial characteristics that they possess get passed on to the next generation. Darwin called this process **natural selection**, because it occurs on its own in nature. According to Darwin, when natural selection takes place over long periods of time, it eventually produces new organisms that differ so significantly from their distant ancestors that the new organisms can be considered to constitute a new species. If environmental conditions change, or if competitors enter the environment, species that do not evolve and become better adapted to survive eventually die off and become extinct.

Darwin's view of evolution has been successfully supported by many observations and so far has not been definitively disproven by any observation or experiment. Also, it can be used to make testable predictions. Thus, scientists now refer to Darwin's idea as the **theory of evolution by natural selection** (often abbreviated "the theory of evolution"; see Box P.1 in the Prelude for the definition of a theory).

In the century and a half since Darwin published his work, *genetics* (the study of genes) has developed and provided insight into how evolution works. Progress began in the late 19th century when an Austrian monk, Gregor Mendel, studied peas in the garden of his monastery and showed that genetic mutations led to new traits that could be passed on to offspring. Traits that make an organism less likely to survive are not passed on, either because the organism dies before it has offspring or because the offspring themselves cannot survive, but traits that make an organism better suited to survival are passed on to succeeding generations. With the discovery of DNA in 1953, biologists began to understand the molecular nature of genes and mutations, and thus of evolution. And with the genome projects of the 21st century, which define the detailed architecture of DNA molecules for a given species, it is now possible to pinpoint the exact arrangement of genes responsible for specific traits.

The theory of evolution by natural selection provides a conceptual framework in which to understand paleontology. By studying fossils in sequences of strata, paleontologists have observed progressive changes in species through time, and they can document that some species have died out and others have appeared during Earth history. But because of the incompleteness of the fossil record, many questions remain as to the rates at which evolution takes place during the course of geologic time. As a result, different researchers have suggested different concepts of rates. For example, Cuvier, back in the early 19th century, thought that extinction happened primarily during catastrophes, such as giant floods, and that a whole new assemblage of life took over the earth after each catastrophe—Cuvier's view came to be known as **catastrophism**. In contrast, James Hutton and his followers, such as Lyell and Darwin, assumed that evolution happened at a constant, slow rate—their view came to be called **gradualism**. (Gradualism reflects a strict interpretation of a broader idea, called uniformitarianism, which we'll discuss in Chapter 12.)

More recently, researchers have suggested that evolution takes place in fits and starts: evolution occurs very slowly for quite a while, and then during a relatively short period, it takes place very rapidly. This concept is called **punctuated equilibrium**. Factors that could cause sudden accelerations in the rate of evolution include (1) a sudden *mass-extinction event* during which many organisms disappear, leaving ecological niches open for new species to colonize; (2) a relatively rapid change in the Earth's climate that puts stress on organisms—new species that can survive the new climate survive, whereas those that can't become extinct; (3) formation of new environments, as may happen when rifting splits apart a continent and generates a new ocean with new coastlines; and (4) the isolation of a breeding population. The punctuated evolution concept takes into account that sudden events leading to widespread extinction do take place during Earth's history but that evolution can happen steadily during time intervals between these events.

Regardless of the process of evolution, the survivability of different kinds of organisms is not all the same. Some populations are very durable in that they survive as an identifiable genus or even species for long intervals of geologic time (10s to 100s of million years). But others appear and then disappear within a relatively short interval of geologic time (less than a few million years).

Extinction: When Species Vanish

As we've noted, **extinction** occurs when the last members of a species die, so there are no parents to pass on their genetic traits to offspring. Some species become extinct as a population evolves into new species, whereas other species just vanish, leaving no hereditary offspring. These days, we take for granted that species become extinct because a great number have, unfortunately, vanished from the Earth during human history. Before the 1770s, however, few geologists thought that extinction occurred; they thought fossils that didn't resemble known species must have living relatives somewhere on the planet. Considering that large parts of the Earth remained unexplored, this idea wasn't so far-fetched. However, by the end of the 18th century, in light of Cuvier's work and the completion of the map of the Earth by explorers, it became clear that fossil organisms did not have modern-day counterparts. The bones of mastodons and woolly mammoths, for example, were too different from those of elephants to be of the same species, but the animals were too big to hide. Twentieth-century studies concluded that many different phenomena can contribute to extinction. Some of the geologic factors that may cause extinction include the following.

- *Global climate change:* At times, the Earth's mean temperature has been significantly colder than today's, whereas at other times it has been much warmer. Because of a change in climate, an individual species may lose its habitat, and if it cannot adapt to the new habitat or migrate to stay with its old one, the species will disappear.
- *Tectonic activity:* Tectonic activity causes both vertical movement of the crust over broad regions and changes in sea-floor spreading rates. These phenomena can modify the distribution and area of habitats. Species that cannot adapt die off.
- *Asteroid or comet impact:* Many geologists have concluded that impacts of large meteorites with the Earth have been catastrophic for life. A large impact would send dust and debris into the atmosphere that could blot out the Sun and plunge the Earth into darkness and cold (see Chapter 23). Such a change, though relatively short lived, could interrupt the food chain.
- *Voluminous volcanic eruption:* Several times during Earth history, incredible quantities of lava have spilled out on the surface and/or incredible volumes of ash and gas have spewed into the air. These eruptions, perhaps due to the rise of superplumes in the mantle, were accompanied by the release of enough greenhouse gas into the atmosphere to alter the climate.
- *The appearance of a new predator or competitor:* Some extinctions may happen simply because a new predator appears on the scene and kills individuals of a given species at a faster rate than new individuals can be born. (For example, when humans first arrived in North America, they killed off most species of giant mammals.) Similarly, if a more efficient competitor appears, the competitor steals an ecological niche from the weaker species, whose members can't obtain enough food and thus die off.

Some extinctions happen over long time intervals, when the replacement rate of a population simply becomes lower than the mortality rate, but others happen suddenly, when a cataclysmic event leads to the rapid extermination of many organisms. For example, in 1870 the population of passenger pigeons in North America exceeded 3 billion. Due to widespread hunting by people, the population dropped rapidly during the next two decades and the last representative of the species died in 1914.

Of note, paleontologists have found that the number of different genera of fossils, a representation of **biodiversity** (the overall variation of life), changes over time and has abruptly decreased at specific times during Earth history. A worldwide abrupt decrease in the number of fossil genera is called a **mass-extinction event**. At least five major mass-extinction events have happened during the past half-billion years (**Fig. E.13**).

FIGURE E.13 This graph shows how the diversity of life has changed with time. Sudden drops indicate periods when mass extinctions occurred.

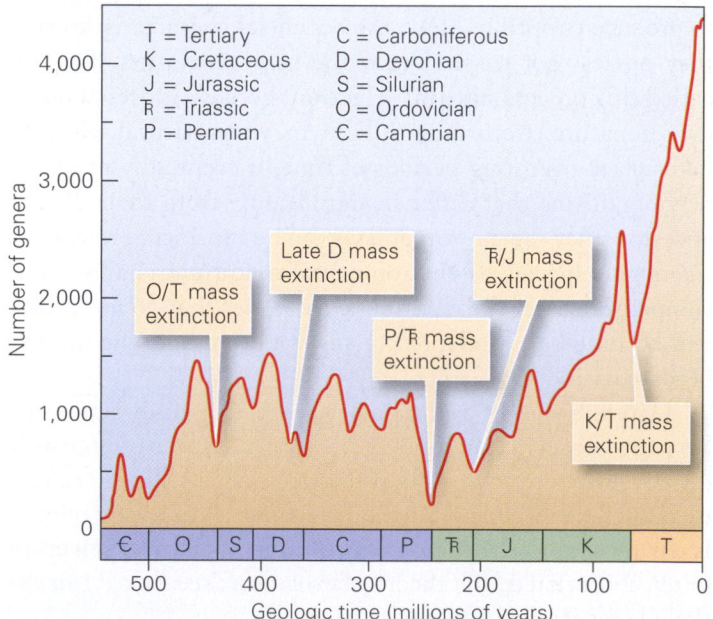

These events define the boundaries between some of the major intervals into which geologists divide time. For example, a major extinction event marks the end of the Cretaceous Period, 65 Ma. During this event, all dinosaur species (with the exception of their modified descendants, the birds) vanished, along with most marine invertebrate species. A huge extinction event also brought the Permian to a close. Significantly, the rate at which species have been disappearing during the past few centuries has been so rapid that some researchers and writers refer to our present time as the *sixth extinction*.

INTERLUDE SUMMARY

- Fossils are records of past life on Earth. Shapes of materials derived from the body of an organism are body fossil, whereas imprints due to the activity of an organism are trace fossils.
- Fossilization involves several steps, and not all organisms have the same likelihood of being preserved. Those with hard parts have a higher preservation potential, while organisms with soft parts can only be preserved under special circumstances.
- Some fossils are molds or casts of hard parts; some form when minerals replace organic materials. Chemical fossils are remnants of molecules formed by living organisms.

- Taxonomy classifies organisms into subdivisions, from domains down to species. Generally, fossils are classified based on morphology of the fossil or organism. It's fairly easy to recognize classes of common fossil organisms.
- The fossil record is incomplete, yet paleontologists can track the evolution of many organisms and can define the "tree of life." Whether evolution happens gradually, or in fits and starts, remains unclear.
- Extinction can happen for many reasons. During mass extinction events, huge numbers of species go extinct all at once and, therefore, biodiversity is decreased.

GUIDE TERMS

amber (p. 422)	chemical fossil (p. 424)	macrofossil (p. 424)	phylogenetic tree (p. 429)
archaea (p. 425)	DNA (p. 425)	mass-extinction event (p. 432)	preservation potential (p. 425)
bacteria (p. 425)	eukarya (p. 425)	microfossil (p. 424)	prokarya (p. 425)
biodiversity (p. 432)	evolution (p. 431)	mold (p. 423)	punctuated equilibrium (p. 431)
biomarker (p. 424)	extinction (p. 432)	natural selection (p. 431)	
body fossil (p. 422)	extraordinary fossil (p. 425)	paleontologist (p. 419)	taxonomy (p. 425)
Cambrian explosion (p. 429)	fossil (p. 419)	paleontology (p. 419)	theory of evolution by natural selection (p. 431)
cast (p. 423)	fossilization (p. 420)	permineralization (p. 423)	
catastrophism (p. 431)	gradualism (p. 431)	petrified wood (p. 423)	trace fossil (p. 423)

REVIEW QUESTIONS

1. What is a fossil, and how can fossils form?
2. Do fossil bones and shells always have the same chemical composition as do modern ones? Explain your answer.
3. What is meant by the "preservation potential" of an organism?
4. Explain the basis for classifying fossils.

5. What is the theory of evolution by natural selection, and how does the study of fossils provide evidence that can test its validity?
6. What are the alternative ideas that paleontologists have had concerning the rate of evolution over geologic time.
7. What is a mass-extinction event, and what phenomena might cause such an event?

The view from an airplane window of the Colorado Plateau region in Arizona and Utah reveals layer upon layer of strata deposited long ago and now exposed by erosion. These layers preserve a record of the Earth's very long history.

CHAPTER 12

Deep Time: How Old Is Old?

> If the Eiffel Tower were now representing the world's age, the skin of paint on the pinnacle-knob at its summit would represent man's share of that age; and anybody would perceive that that skin was what the tower was built for. I reckon they would, I dunno.
>
> —*Mark Twain (1835–1910)*

12.1 Introduction

In May of 1869, a one-armed Civil War veteran named John Wesley Powell set out with a team of nine geologists and scouts to explore the previously unmapped expanse of the Grand Canyon, the greatest gorge on Earth. Though Powell and his companions battled fearsome rapids and the pangs of starvation, most managed to emerge from the mouth of the canyon three months later (**Fig. 12.1**). During their voyage, seemingly insurmountable walls of rock both imprisoned and amazed the explorers and led them to pose important questions about the Earth and its history, questions that even casual tourists to the canyon ponder today: Did the Colorado River sculpt this marvel, and if so, how long did it take? When did the rocks making up the walls of the canyon form? Was there a time before the colorful layers accumulated? Such questions pertain to **geologic time**, the span of time since the Earth's formation.

In this chapter, we first learn the geologic principles that allowed geologists to develop the concept of geologic time and thus to create a frame of reference for describing the *relative age* of rocks, fossils, structures, and landscapes. This information

sets the stage for discussing the *geologic column*, the chart that geologists use to divide time into intervals. Then we describe how geologists can obtain a rock's *numerical age* (its age in years), by a procedure called *isotopic dating* (also known as *radiometric dating*). Isotopic dating permitted the establishment of the *geologic time scale*, which provides numerical ages for intervals on the geologic column. With the concept of geologic time in mind, a hike down a trail into the Grand Canyon becomes a trip into what some authors have called *deep time*. Humans are obsessed with time (**Box 12.1**), so the geologic discovery that our planet's history extends billions of years into the past changed humanity's perception of its place in the Universe as profoundly as did the astronomical discovery that the limit of space extends billions of light years beyond the edge of our Solar System.

FIGURE 12.1 Woodcut illustration of the "noonday rest in Marble Canyon," from J. W. Powell's *The Exploration of the Colorado River and Its Canyons* (1895). "We pass many side canyons today that are dark, gloomy passages back into the heart of the rocks."

BOX 12.1 CONSIDER THIS . . .

Time: A Human Obsession

When you plan your daily schedule, you have to know not only where you need to be but when you need to be there. Because time assumes such significance in human consciousness today, we have developed elaborate tools to measure it and formal scales to record it. We use a *second* as the basic unit of time measurement. What exactly is a second? From 1900 to 1968, we defined the second as 1/31,556,925.9747 of the year 1900, but now we define it as the duration of time that it takes for a cesium atom to change back and forth between two energy states 9,192,631,770 times. This change is measured with a device called an atomic clock, which is accurate to about 1 second per 30 million years. We sum 60 seconds into 1 minute, 60 minutes into 1 hour, and 24 hours into 1 day, about the time it takes for Earth to spin once on its axis.

In the pre-industrial era, each locality kept its own time, setting noon as the moment when the Sun reached the highest point in the sky. But with the advent of train travel and telegraphs, people needed to calibrate schedules from place to place. So in 1883, countries around the globe agreed to divide the world into 15°-wide bands of longitude called time zones—in each time zone, all clocks keep the same standard time. The times in each zone are set in relation to Greenwich mean time (GMT), the time at the astronomical observatory in Greenwich, England. Today the world standard for time is determined by a group of about 200 atomic clocks that together define Coordinated Universal Time (abbreviated as UTC, based on the French *Temps universel coordonné*). UTC is the basis for the global positioning system (GPS), used for precise navigation. The time that appears on your cell phone is UTC.

12.2 The Concept of Geologic Time

Setting the Stage for Studying the Past

Until relatively recently, people in most cultures believed that geologic time began about the same time that human history began and that our planet has been virtually unchanged since its birth. With this concept in mind, an Irish archbishop named James Ussher (1581–1656) tallied successions of lineages and reigns described in the Old and New Testaments to determine the age of the Earth and, in 1654, stated his conclusion: the birth of the Earth took place on October 23, 4004 B.C.E.

Not long after Ussher had implied that, in effect, the Earth has existed for only about 250 human generations (6,000 years), Nicolas Steno (1638–86) proposed an idea that established the foundation for a very different approach to thinking about geologic time. Steno was serving as a physician in a nobleman's court in Florence, Italy, when he realized that unusual triangular-shaped rocks, known as *tongue stones* by the local people who had chiseled them out of nearby outcrops, resembled the teeth of sharks (**Fig. 12.2**). He speculated that the tongue stones were not the tongues of dragons, as the locals thought, but rather were shark teeth that had been buried with sediment and seashells on the seafloor and that the sediment and teeth together had later transformed into rocks that were uplifted above sea level when the mountains formed. Steno eventually concluded that the presence of shark teeth, along with other **fossils** (remnants of ancient life preserved in rock; see Interlude E) of marine organisms in rocks now exposed in mountains, implied that the Earth could change over time. This realization set the stage for the founding of the science of geology, by James Hutton, a century later.

Hutton (1726–97), a Scottish gentleman farmer and doctor, lived during the Age of Enlightenment when, sparked by the discovery of physical laws by Sir Isaac Newton, scientists

FIGURE 12.2 A fossilized shark's tooth. Before Steno explained the origin of such fossils, they were thought to be dragons' tongues.

began to seek natural rather than supernatural explanations for features of the world around them. While wandering in the highlands of Scotland, a region where rocks are well exposed, Hutton noted that many features (such as ripple marks and cross beds) found in sedimentary rock types resembled features he could see forming today in modern depositional environments. These observations led Hutton to propose that the formation of rocks and landscapes, in general, were a consequence of processes that he could see happening today.

Hutton's idea, discussed in his 1785 book called *The Theory of the Earth*, came to be known as the principle of **uniformitarianism**. According to this principle, physical processes we observe today also operated in the past at roughly the same rates, and these processes were responsible for the formation of geologic features that we now see in outcrops. More concisely, the principle can be stated as "the present is the key to the past." Because the rates of most geologic processes taking place today are so slow, Hutton deduced that the development of individual geologic features takes a very long time. Further, he deduced that not all features formed at the same time, so the Earth has a history that includes a succession of slow events, and thus that the Earth existed for a long time before human history began. In fact, he speculated that we could see "no vestige of a beginning, nor prospect of an end."

Hutton was not a particularly clear writer, and it took the efforts of subsequent geologists to clarify the implications of the principle of uniformitarianism and to publicize them. Once this had been accomplished, geologists around the world began to apply their growing understanding of geologic processes to define and interpret geologic events of the Earth's past.

Relative versus Numerical Age

Like historians, geologists strive to establish both the sequence of events that created an array of geologic features—such as rocks, structures, and landscapes—and, when possible, the date on which each event happened (**Fig. 12.3**). We specify the age of one feature with respect to another in a sequence as its **relative age** and

the age of a feature given in years as its **numerical age** (or, in older literature, its *absolute age*). Recall that we can abbreviate numeral ages by using the units Ka, for thousands of years, Ma for millions of years, and Ga for billions of years; *K* stands for kilo-, *M* for mega-, and *G* for giga-. Geologists learned how to determine relative age long before they could determine numerical age, so we will look next at the principles leading to relative-age determination next.

Take-Home Message

The principle of uniformitarianism ("the present is the key to the past") implies that the Earth must be very old, for geologic processes happen slowly, and that we can interpret events in Earth's history. Geologists distinguish between relative age (is one event older or younger than another?) and numerical age (how many years ago did an event happen?).

QUICK QUESTION: What observations led Hutton to propose uniformitarianism?

FIGURE 12.3 The difference between relative and numerical age.

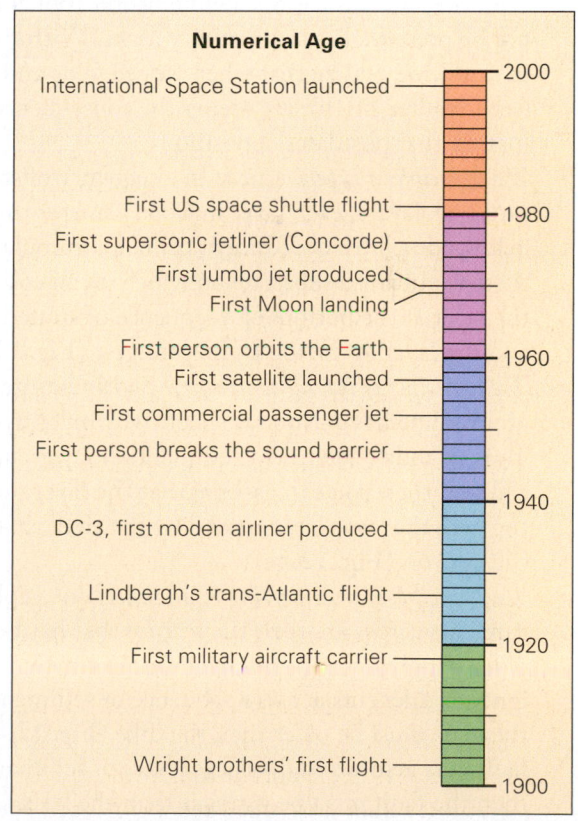

Relative Age
International Space Station launched
First US space shuttle flight
First supersonic jetliner (Concorde)
First jumbo jet produced
First Moon landing
First person orbits the Earth
First satellite launched
First commercial passenger jet
First person breaks the sound barrier
DC-3, first moden airliner produced
Lindbergh's trans-Atlantic flight
First military aircraft carrier
Wright brothers' first flight

(a) The relative ages of important moments in aviation and space flight.

(b) The numerical ages of these same events. Clearly this chart provides more information, for it displays the dates of the events and indicates the amount of time between the events.

12.3 Geologic Principles Used for Defining Relative Age

Building from the work of Steno, Hutton, and others, the British geologist Charles Lyell (1797–1875) laid out a set of formal, usable geologic principles in the first modern textbook of geology (*Principles of Geology*, published between 1830 and 1833). These principles, defined below, continue to provide the basic framework within which geologists read the record of Earth history and determine relative ages.

- *The principle of uniformitarianism*: As noted earlier, physical processes we observe operating today also operated in the past, at roughly comparable rates (**Fig. 12.4a, b**); in other words, the present is the key to the past.
- *The principle of original horizontality:* Sediments on Earth settle out of fluids in a gravitational field, and the surfaces on which sediments accumulate (such as a floodplain or the bed of a lake or sea) are fairly flat. Therefore, layers of sediment when originally deposited are fairly horizontal (**Fig. 12.4c**). If sediments collect on a steep slope, they typically slide downslope before lithification and so will not be preserved as sedimentary rocks. With this principle in mind, we realize that when we see folds and tilted beds (see Chapter 11), we are seeing the consequences of deformation that postdates deposition.
- *The principle of superposition:* In a sequence of sedimentary rock layers, each layer must be younger than the one below, for a layer of sediment cannot accumulate unless there is already a substrate on which it can collect. Thus, the layer at the bottom of a sequence of strata is the oldest, and the layer at the top is the youngest (**Fig. 12.4d**).
- *The principle of lateral continuity:* Sediments generally accumulate in continuous sheets within a given region. Thus, if today you find a sedimentary layer cut by a canyon, then you can assume that the layer once spanned the area that was later eroded by the river that formed the canyon (**Fig. 12.4e**).
- *The principle of cross-cutting relations:* If one geologic feature cuts across another, the feature that has been cut is older. Applying this principle, we can conclude that if an igneous dike cuts across a sequence of sedimentary beds, the beds must be older than the dike (**Fig. 12.4f**), and if a fault cuts across and displaces layers of sedimentary rock, then the fault must be younger than the layers. In contrast, if a layer of sediment buries a fault, the sediment must be younger than the fault.

- *The principle of baked contacts:* During the formation of an igneous intrusion, hot magma injects into cooler rock. As a consequence, heat from the intrusion "bakes" (metamorphoses) surrounding rocks. Thus, rock that has been baked by an intrusion must be older than the intrusion (see Fig. 12.4f). Note that since an intrusion loses heat to its surroundings at its margins, the margin of an intrusion cools more rapidly. A fine-grained chilled margin occurs within the younger intrusion (**Fig. 12.4g**).
- *The principle of inclusions:* This principle states that an "inclusion" (a fragment of one rock incorporated in another) is always older than the rock that contains it. Thus, a layer of younger sediment deposited on older rock may contain inclusions (clasts) of the older rock, whereas a younger intrusion into an older wall rock may contain inclusions (xenoliths) of the wall rock. Note that we can use this concept to distinguish sills from lava flows (**Fig. 12.4h**).

Geologists apply the above geologic principles to determine the relative ages of rocks, structures, and other geologic features at a given location. We can then go further by interpreting the development of each feature to be the consequence of a specific "geologic event." Examples of geologic events include deposition of sedimentary beds, erosion of the land surface, intrusion or extrusion of igneous rocks, deformation (folding and/or faulting), and episodes of metamorphism. The succession of events, in order of relative age, that have produced the rock, structure, and landscape of a region is called the *geologic history* of the region.

To see how to work out the geologic history of an area, let's work through an example (**Fig. 12.5a**). The principle of superposition requires that the oldest sedimentary layer of the figure is Bed 1 while the youngest is Bed 7, and the principle of original horizontality means that folding postdates deposition of beds. The principles of cross-cutting relations, inclusions, and baked contacts allow us to determine the relative ages of the intrusions and faults relative to the beds and to each other. Thus, in this example, we can propose the following geologic history for this region (**Fig. 12.5b**): deposition of the sedimentary sequence in order from Beds 1 to 7, intrusion of the sill, folding of the sedimentary beds and the sill, intrusion of the granite pluton, faulting, intrusion of the dike, erosion to form the present-day land surface.

Adding Fossils to the Story: Fossil Succession

As Britain entered the industrial revolution in the late 18th and early 19th centuries, new factories demanded coal to fire their steam engines. Various companies decided to build a network of canals to transport coal and iron, and hired an

FIGURE 12.4 Major geological principles used for determining relative ages.

Present-day mudcracks form in clay-rich sediment.

Ancient mudcracks in solid rock

(a) Uniformitarianism: The processes that formed cracks in the dried-up mud puddle on the left also formed the mudcracks preserved in the ancient, solid rock on the right. We can see these ancient mudcracks because erosion removed the adjacent bed.

Present-day volcanism produces molten lava.

Layers of basalt formed during volcanic activity.

(b) Uniformitarianism (cont.): We can observe lava flows forming today, so we can infer that solid lava flows represent the products of volcanic eruptions in the past.

Modern sediment, exposed at low tide, on the coast of France.

Horizontal sandstone beds in Wisconsin

Youngest bed

Bedding plane

Cross beds

Oldest bed

What a Geologist Sees

(c) Original horizontality: Gravity causes sediments to accumulate in horizontal sheets.

FIGURE 12.4 *(continued)*

Time 1   Time 2   Time 3

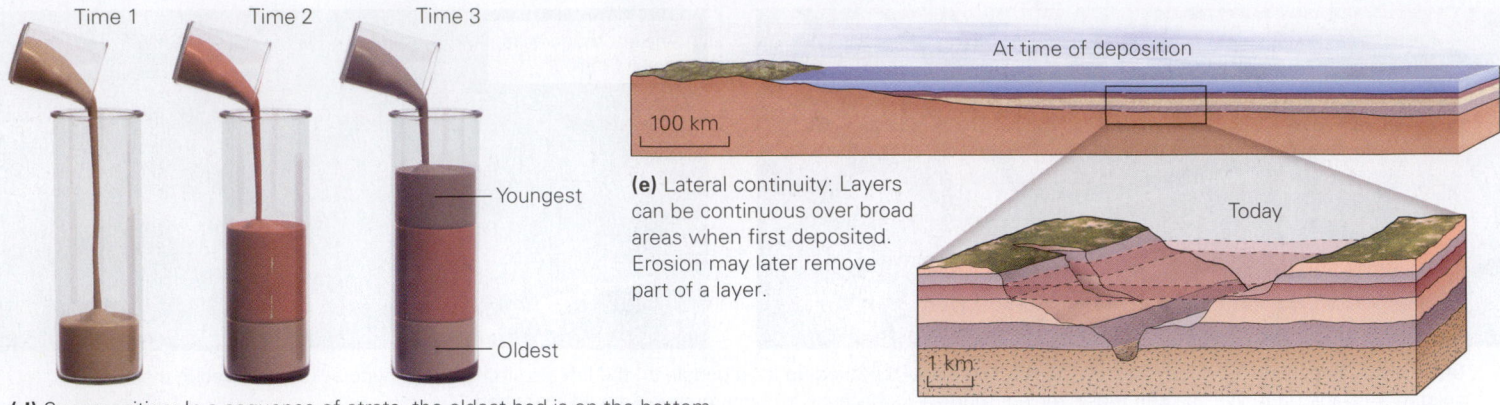

— Youngest

— Oldest

(d) Superposition: In a sequence of strata, the oldest bed is on the bottom, and the youngest on top. Pouring sand into a glass illustrates this point.

(e) Lateral continuity: Layers can be continuous over broad areas when first deposited. Erosion may later remove part of a layer.

At time of deposition

100 km

Today

1 km

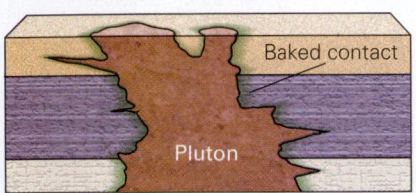

Baked contact

Pluton

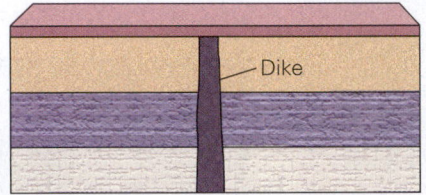

Dike

(f) By the principle of cross-cutting relations, the pluton is younger than the beds it cuts across (a baked contact—metamorphic aureole—forms in the strata next to the pluton), and the dike is younger than the beds it cuts across. The sediment layer that buries the dike is younger than the dike.

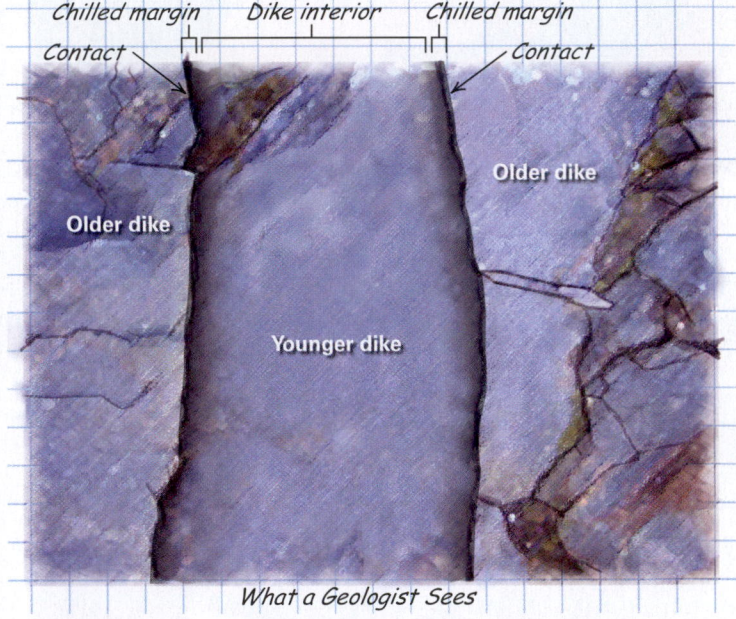

Chilled margin   Dike interior   Chilled margin

Contact   Contact

Older dike   Older dike

Younger dike

What a Geologist Sees

(g) Chilled margins can be used to determine age relationships. In this outcrop in California, a younger dike intrudes an older one. The younger dike is finer-grained and darker where it cooled faster at the contact with the older one.

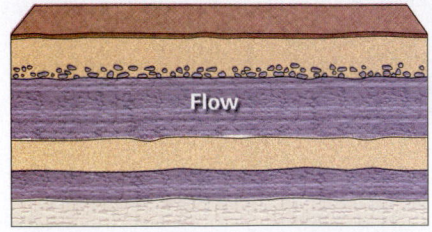

Flow

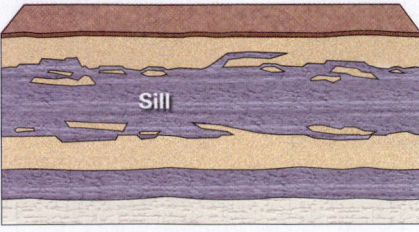

Sill

(h) By the principle of inclusions, the pebbles of basalt in a conglomerate must be older than the conglomerate and xenoliths of sandstone must be older than the basalt containing them.

FIGURE 12.5 Interpreting the geologic history of a region, using geologic principles as a guide.

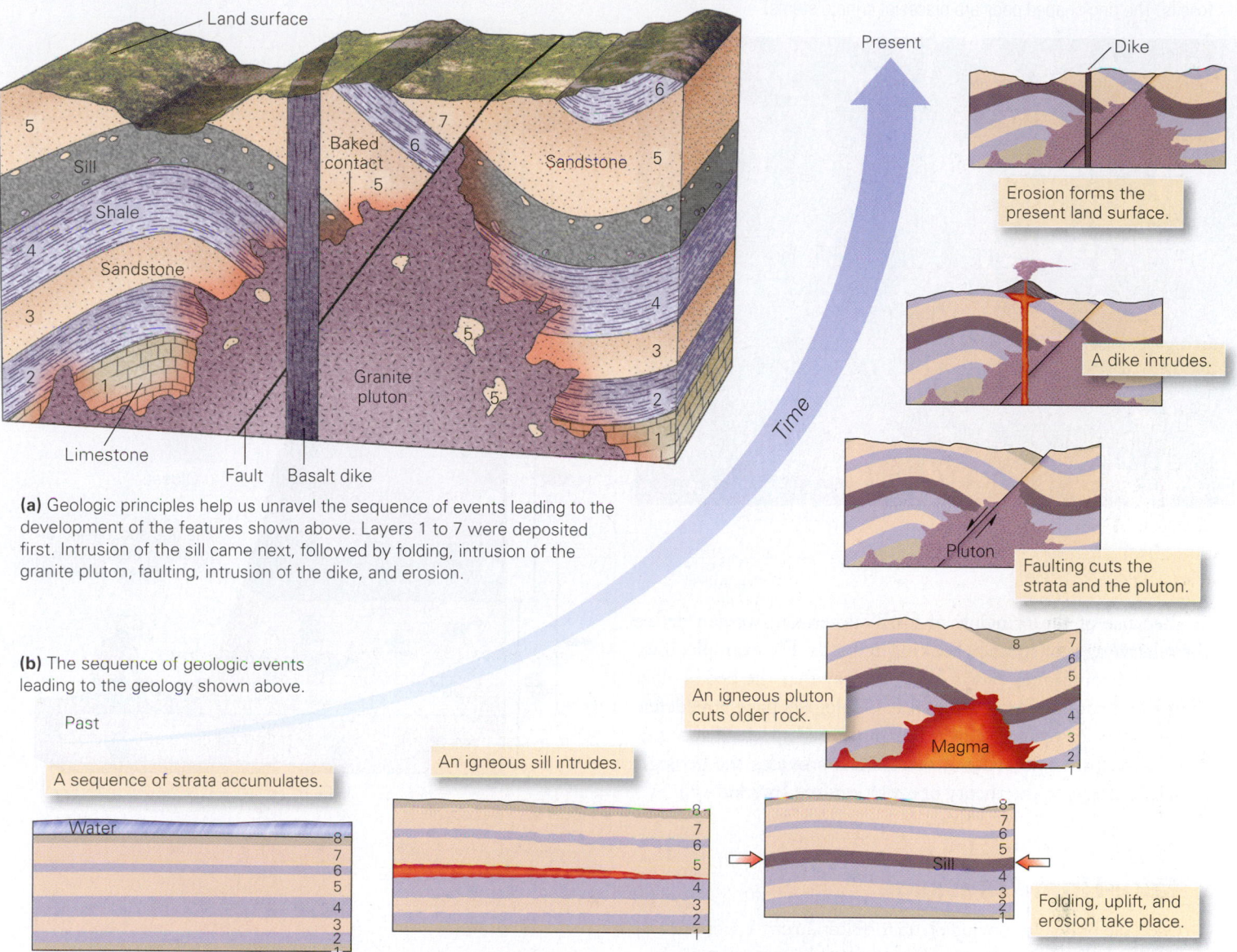

(a) Geologic principles help us unravel the sequence of events leading to the development of the features shown above. Layers 1 to 7 were deposited first. Intrusion of the sill came next, followed by folding, intrusion of the granite pluton, faulting, intrusion of the dike, and erosion.

(b) The sequence of geologic events leading to the geology shown above.

Erosion forms the present land surface.

A dike intrudes.

Faulting cuts the strata and the pluton.

An igneous pluton cuts older rock.

An igneous sill intrudes.

A sequence of strata accumulates.

Folding, uplift, and erosion take place.

engineer named William Smith (1769–1839) to survey the excavations. Canal digging provided fresh exposures of bedrock, which previously had been covered by vegetation. Smith learned to recognize distinctive layers of sedimentary rock and to identify the **fossil assemblage**—the group of fossil species—that they contained (**Fig. 12.6**). As we noted in Interlude E, he also realized that a particular fossil assemblage can be found only in a limited interval of strata and not in beds above or below this interval. Thus, once a fossil species disappears at a horizon in a sequence of strata, it never reappears higher in the sequence. In other words, "extinction is forever." Smith's observation has been repeated at thousands of locations around the world and has been codified as the *principle of fossil succession*.

To see how this principle works, examine **Figure 12.7**, which depicts a sequence of strata. Bed 1 at the base contains fossil species A, Bed 2 contains A and B, Bed 3 contains B and C, Bed 4 contains C, and so on. From these data, we can define the *range* of specific fossils in the sequence, meaning the interval in the sequence in which the fossils occur. Note that the sequence contains a definable succession of fossils (A, B, C, D, E, F), that the range in which a particular species occurs may overlap with the range of other species, and that once a species vanishes, it does not reappear higher in the sequence. Some species can be found over a broad region, but existed only for a short interval of geologic time, and thus can be diagnostic of a precise time interval in rocks at many different locations. The fossils of such species are called **index fossils.**

FIGURE 12.6 A close-up photo of a bedding surface showing many fossils. The ring-shaped ones are pieces of crinoid stems.

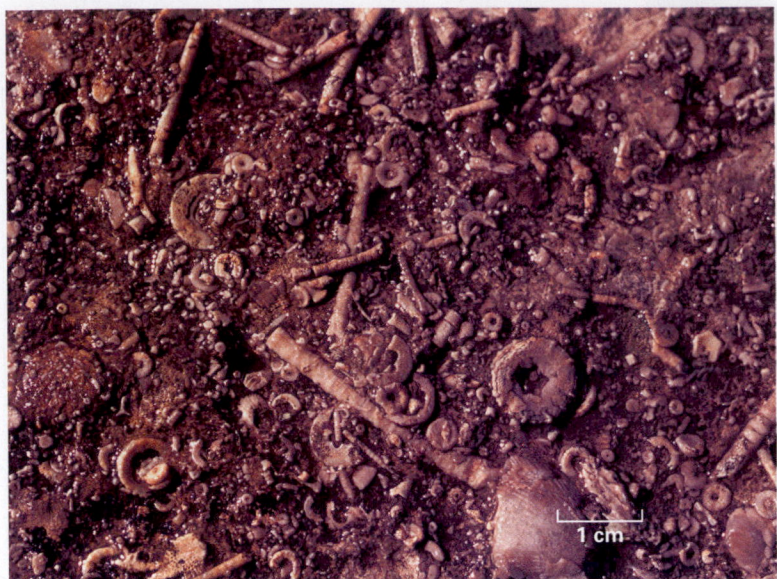

1 cm

FIGURE 12.7 The principle of fossil succession.

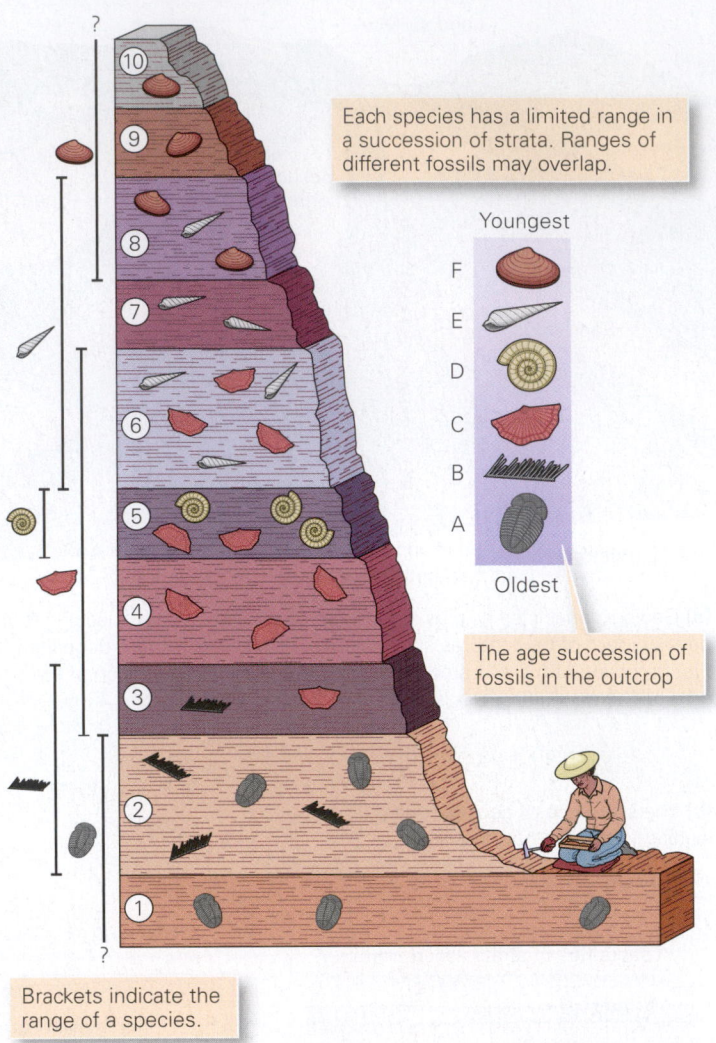

Each species has a limited range in a succession of strata. Ranges of different fossils may overlap.

Youngest

F
E
D
C
B
A

Oldest

The age succession of fossils in the outcrop

Brackets indicate the range of a species.

Because of the principle of fossil succession, we can define the relative ages of strata by looking at fossils. For example, if we find a bed containing Fossil A, we can say that the bed is older than a bed containing, say, Fossil F. Geologists have now determined the relative ages of over 200,000 fossil species. Recognition of the principle of fossil successions provides the geologic underpinnings for the theory of evolution (see Interlude E).

Take-Home Message

Geologic principles (including uniformitarianism, superposition, cross-cutting relations, and fossil succession) provide the basis for determining the relative ages of rocks and other geologic features. By working out relative ages, we can reconstruct the geologic history of a region.

QUICK QUESTION: If a fault forms a fault scarp, what is the relative age of the fault and the landscape surface?

12.4 Unconformities: Gaps in the Record

James Hutton used a boat to explore the seaside of Scotland, because shore cliffs provided great exposures of rock, stripped of soil and shrubbery. He was particularly puzzled by an outcrop he found at Siccar Point on the east coast. Here he saw a

sequence of strata consisting of red sandstone and conglomerate that rested on a distinctly different sequence, one that consisted of gray sandstone and shale (**Fig. 12.8a**). Further, the beds of gray sandstone and shale have a nearly vertical dip (slope), whereas the beds of red sandstone and conglomerate have a dip of less than 15°—the gently dipping layers seemed to lie across the truncated ends of the vertical layers, like a handkerchief resting on a row of books. (See Box 11.1 for the definition of *dip*.) Perhaps as Hutton sat and stared at this odd geometric relationship, the tide came in and deposited a new layer of sand on top of the rocky shore. With the principle of uniformitarianism in mind, Hutton suddenly realized the significance of the outcrop—the gray sandstone–shale sequence had been deposited, turned into rock, tilted, and truncated by erosion before the red sandstone–conglomerate beds had been deposited. Thus, the surface between the gray and red rock sequences represented a time interval during which new strata

FIGURE 12.8 Examples of unconformities, as visible in outcrops.

(a) James Hutton found this unconformity along the east coast of Scotland. He deduced that the layers above were deposited long after the beds below had been tilted. Geologists have since determined that strata above are about 80 million years younger than strata below.

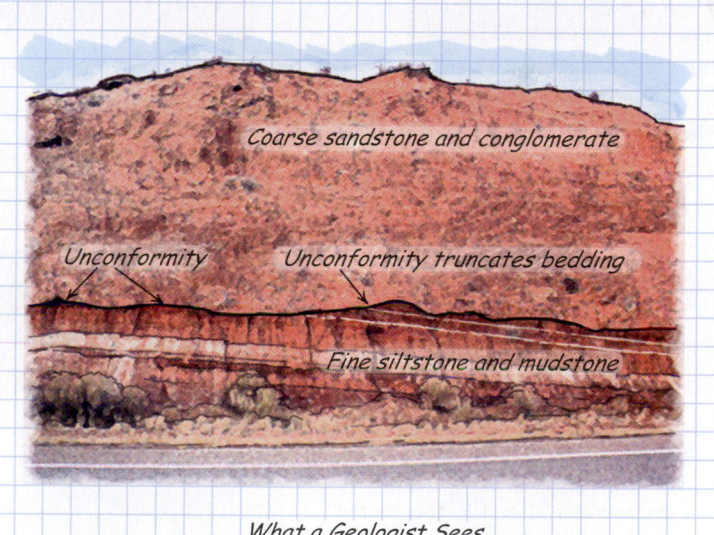

(b) A roadcut in Utah reveals an unconformity between a coarse sandstone and conglomerate above and a fine siltstone and mudstone below. Note that the irregularities along this unconformity locally truncate bedding of the siltstone and mudstone.

had not been deposited at Siccar Point and the older strata had been eroded away. Hutton realized that Siccar Point exposed the record of a long and complex saga of geologic history.

We now refer to a boundary surface between two units, which represents a period of nondeposition and possibly erosion, as an **unconformity** (**Fig. 12.8b**). The gap in the geologic record that an unconformity represents is called a *hiatus*. Geologists recognize three main types of unconformities.

• *Angular unconformity:* An **angular unconformity** represents an erosional surface that cuts across previously tilted or folded underlying layers, such that the orientation of layers below an unconformity differs from that of the layers above (**Fig. 12.9a**)—Siccar Point, shown in Figure 12.8a, exposes an angular unconformity. Angular unconformities form where rocks were deformed before being exposed at the Earth's surface.

FIGURE 12.9 The three kinds of unconformities and their formation.

Time →

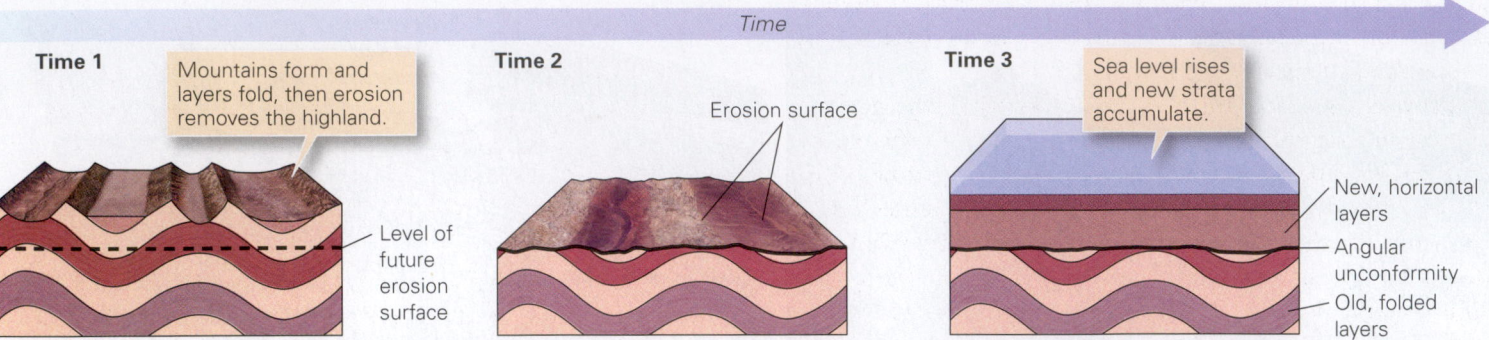

Time 1 — Mountains form and layers fold, then erosion removes the highland. — Level of future erosion surface

Time 2 — Erosion surface

Time 3 — Sea level rises and new strata accumulate. — New, horizontal layers — Angular unconformity — Old, folded layers

(a) An angular unconformity: (1) layers undergo folding; (2) erosion produces a flat surface; (3) sea level rises and new layers of sediment accumulate.

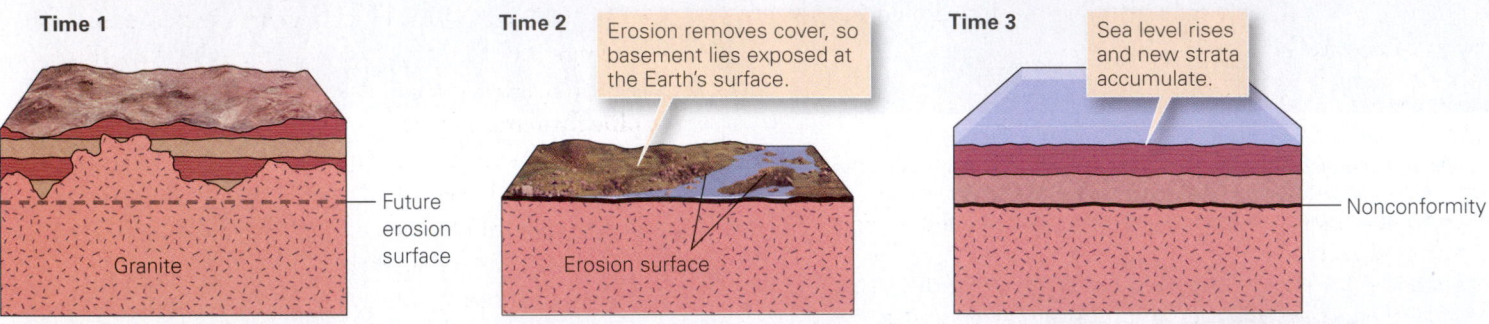

Time 1 — Granite — Future erosion surface

Time 2 — Erosion removes cover, so basement lies exposed at the Earth's surface. — Erosion surface

Time 3 — Sea level rises and new strata accumulate. — Nonconformity

(b) A nonconformity: (1) a pluton intrudes; (2) erosion cuts down into the crystalline rock; (3) new sedimentary layers accumulate above the erosion surface.

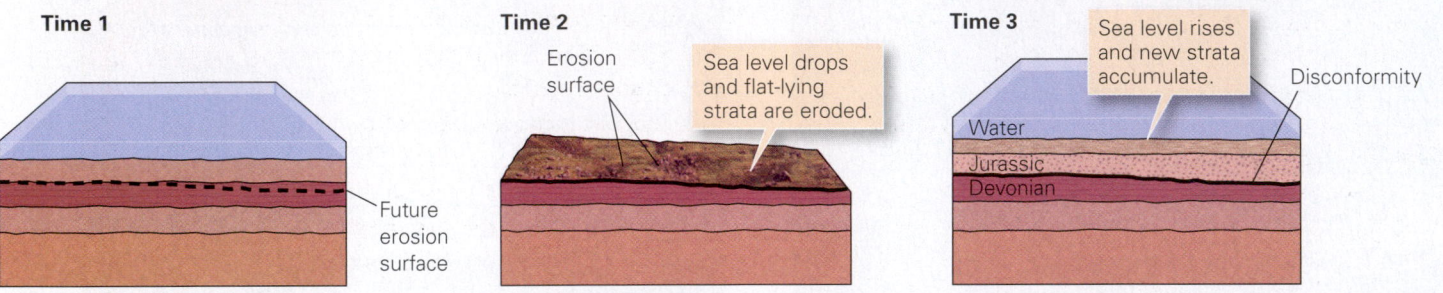

Time 1 — Future erosion surface

Time 2 — Erosion surface — Sea level drops and flat-lying strata are eroded.

Time 3 — Sea level rises and new strata accumulate. — Disconformity — Water — Jurassic — Devonian

(c) A disconformity: (1) layers of sediment accumulate; (2) sea level drops and an erosion surface forms; (3) sea level rises and new sedimentary layers accumulate.

• *Nonconformity:* A **nonconformity** is a type of unconformity at which sedimentary rocks overlie older intrusive igneous rocks and/or metamorphic rocks (**Fig. 12.9b**). The igneous or metamorphic rocks underwent cooling, uplift, and erosion prior to becoming the substrate or "basement" on which the "cover" of new sediments accumulated.

• *Disconformity:* Imagine that a sequence of sedimentary beds has been deposited beneath a shallow sea. Then sea level drops, exposing the beds for some time. During this time, no new sediment accumulates, and some of the pre-existing sediment gets eroded away. Later, sea level rises, and a new sequence of sediment accumulates over the old. The boundary between the two sequences is a **disconformity** (**Fig. 12.9c**). Even though the beds above

and below the disconformity are parallel, the boundary between them represents an interruption in deposition. The unconformity in the outcrop shown in Figure 12.8b is a disconformity.

The succession of strata at a particular location provides a record of Earth history there. But because of unconformities, the record preserved in the rock layers is incomplete. It's as if geologic history were being chronicled by a video recorder that turns on only intermittently—when it's on (= times of deposition), the

Did you ever wonder . . .

if the strata of the Grand Canyon represent all of Earth's history?

rock record accumulates, but when it's off (= times of non-deposition and possibly erosion), an unconformity develops. Because of unconformities, no single location on Earth contains a complete record of Earth history.

How can you recognize an unconformity in the field? At angular unconformities, the strata above have a different dip than the strata below, and at nonconformities, the juxtaposition of cover over basement serves as a clue. Disconformities may be hard to recognize since beds above and below are parallel. A disconformity may be indicated by a gap in the fossil succession and/or by the presence of a surface of erosion and weathering. If the surface was exposed at the Earth's surface for a while, a pebbly layer of debris might occur just above it, and/or a *paleosol* (a remnant of a soil horizon that has been lithified) may be visible just below it (**Fig. 12.10**).

Take-Home Message

Unconformities represent time intervals of nondeposition and possibly erosion. Strata above and below an unconformity may be parallel, so a gap in the fossil record or evidence of erosion represents the unconformity. In cases where rock below the unconformity was folded or tilted before deposition of strata above, an angular discordance marks the unconformity. Some unconformities juxtapose strata above with basement below.

QUICK QUESTION: Is there any one place on the surface of the Earth where the exposed stratigraphic succession represents all of geologic time? Explain.

FIGURE 12.10 This road cut in Utah shows a sand-filled channel cut down into floodplain mud. The mud was exposed between floods, and a soil formed on it. When later buried, all the sediment turned into rock; the channel floor is now an unconformity, and the ancient soil is now a paleosol. Note that the channel cut across the paleosol. The paleosol also represents an unconformity, a time during which deposition did not occur.

Channel (unconformity surface)

Paleosol horizon

12.5 Stratigraphic Formations and Their Correlation

The Concept of a Formation

When William Smith first began to explore the strata exposed along the newly dug canals of England, he realized that distinctive sets of beds, with distinctive assemblages of fossils, could be found at many locations. Smith, and the generations of geologists that have followed, now routinely divide thick successions of strata into recognizable units, called stratigraphic formations, which others can recognize and identify. Formally defined, a **stratigraphic formation** (formation, for short) is an interval of strata composed of a specific rock type or group of rock types that together can be traced over a fairly broad region. A formation represents the products of deposition during a definable interval of time. The boundary surface between two formations is a type of **geologic contact**, or contact, for short. (Fault surfaces, unconformities, and the boundary between an igneous intrusion and its wall rock are also types of contacts.)

Geologists summarize information about the sequence of sedimentary strata at a location by drawing a **stratigraphic column**. Typically, stratigraphic columns are constructed to scale so that the relative thicknesses of beds or formations portrayed on the column are in proportion to the thicknesses of these units in outcrop. Geologists may represent the relative resistance to erosion of beds (or formations) by making the right side of the column irregular to symbolize the way the units might erode on a cliff face—units that stick out further are more resistant.

Let's see how the concept of a stratigraphic formation and of a stratigraphic column applies to the Grand Canyon. The walls of the canyon look striped because they expose a variety of rock types that differ in color and in resistance to erosion. Geologists identify major

contrasts distinguishing one interval of strata from another and use them as a basis for dividing the strata into formations, each of which may consist of many beds (**Fig. 12.11**). Note that some formations consist of beds of a single rock type, whereas others include interlayered beds of two or more rock types. Also, not all formations have the same thickness. Commonly, geologists name a formation after a locality where it was first identified or first studied. For example, the Schoharie Formation was first defined based on exposures in Schoharie Creek of eastern New York. If relevant, geologists can depict unconformities in the section by a wavy line. If a formation consists of only one rock type, we may incorporate that rock type in the name (such as the Kaibab Limestone), but if a formation contains more than one rock type, we use the word formation in the name (such as the Toroweap Formation). Note that in the formal name of a formation, all words are capitalized. Several adjacent formations in a succession may be lumped together as a **stratigraphic group** (or simply a group).

Correlating Strata

How does the stratigraphy of a sedimentary succession exposed at one locality relate to that exposed at another? Stated another way, can we determine the relative age of strata at one location to that of strata at another? The answer is yes. Geologists determine such relations by a process called **stratigraphic correlation** (correlation, for short). Geologists use two approaches for correlating intervals of strata.

When correlating formations among nearby regions, we can simply look for similarities in successions of rock type. We call this method *lithologic correlation*. For example, the sequence of strata on the southern rim of the Grand Canyon clearly correlates with the sequence on the northern rim, because they contain the same rock types in the same order. In some cases, a sequence contains a key bed, or **marker bed**, which is a particularly unique layer that provides a definitive basis for correlation.

Lithologic correlation doesn't necessarily work over broad areas because the depositional setting and the source of sediments can change from location to location. Therefore, beds deposited at one location during a given time interval may look quite different from the beds deposited at another location during the same time interval. To correlate rock units over *broad* areas, we must rely on fossils to define the relative ages of sedimentary units. We call this method *fossil correlation*. If fossils of the same relative age occur at both locations, we can say that the strata at the two locations correlate. Note that the fossils in correlative units are not necessarily the same species—they won't be if the depositional environments are different—but they lived during the same time interval. Fossil correlation can allow geologists to determine whether or not beds of the same rock type at different locations represent the same formation.

As an example, imagine that the Santuit Sandstone and Oswaldo Sandstone look the same but are of different ages (**Fig. 12.12**). In Location A, the units are separated by the Milo Limestone, but at Location C the units are in direct contact. At first glance, the combination of the two units at Location C may look like a single unit. But a sharp-eyed geologist, looking at the strata closely, will find that the fossils are of significantly different age and will depict the contact between the two units by an unconformity.

Let's now apply correlation principles to the challenge of determining the relative ages of formations exposed in the Grand Canyon to those exposed in the mountains near Las Vegas, Nevada, 150 km to the west (**Fig. 12.13a**). Near Las Vegas, we find a sequence of sedimentary rocks that includes a limestone formation called the Monte Cristo Limestone. The Monte Cristo Limestone contains fossils of the same relative age as those of the Redwall Limestone of the Grand Canyon, but the Monte Cristo Limestone is much thicker than the Redwall Limestone. Because the formations contain fossils of the same relative age, we conclude that they were deposited during the same time interval, and thus we say that they correlate with one another. Note also that not only are the units thicker in the Las Vegas area than in the Grand Canyon area, but there are more of them. Thus, the contact beneath the Grand Canyon's Redwall Limestone is an unconformity.

Why does the stratigraphy of Las Vegas differ from that of the Grand Canyon? During part of the time when thick sediments were accumulating near Las Vegas, no sediments accumulated near the Grand Canyon because the region of Las Vegas lay below sea level, whereas the region of the Grand Canyon was dry land. Geologists studying this contrast concluded that in the geologic past the location of Las Vegas was part of a passive-margin basin that sank (subsided) rapidly and remained submerged below the sea almost continuously; the Grand Canyon column, on the other hand, accumulated on the crust of a craton that episodically emerged above sea level (**Fig. 12.13b**).

Geologic Maps

Once William Smith succeeded in correlating stratigraphic formations throughout central England, he faced the challenge of communicating his ideas to others. One way would be to create a table that compared stratigraphic columns from different locations. But since Smith was a surveyor and worked with maps, it occurred to him that he could outline and color in areas on a map to represent areas in which strata of a given relative age occurred. He did this using the data he had collected, and in 1815 he produced the first modern geologic map. In general, a **geologic map** portrays the spatial distribution of rock units at the Earth's surface.

FIGURE 12.11 The stratigraphic column and stratigraphic formation: examples from the Grand Canyon in Arizona.

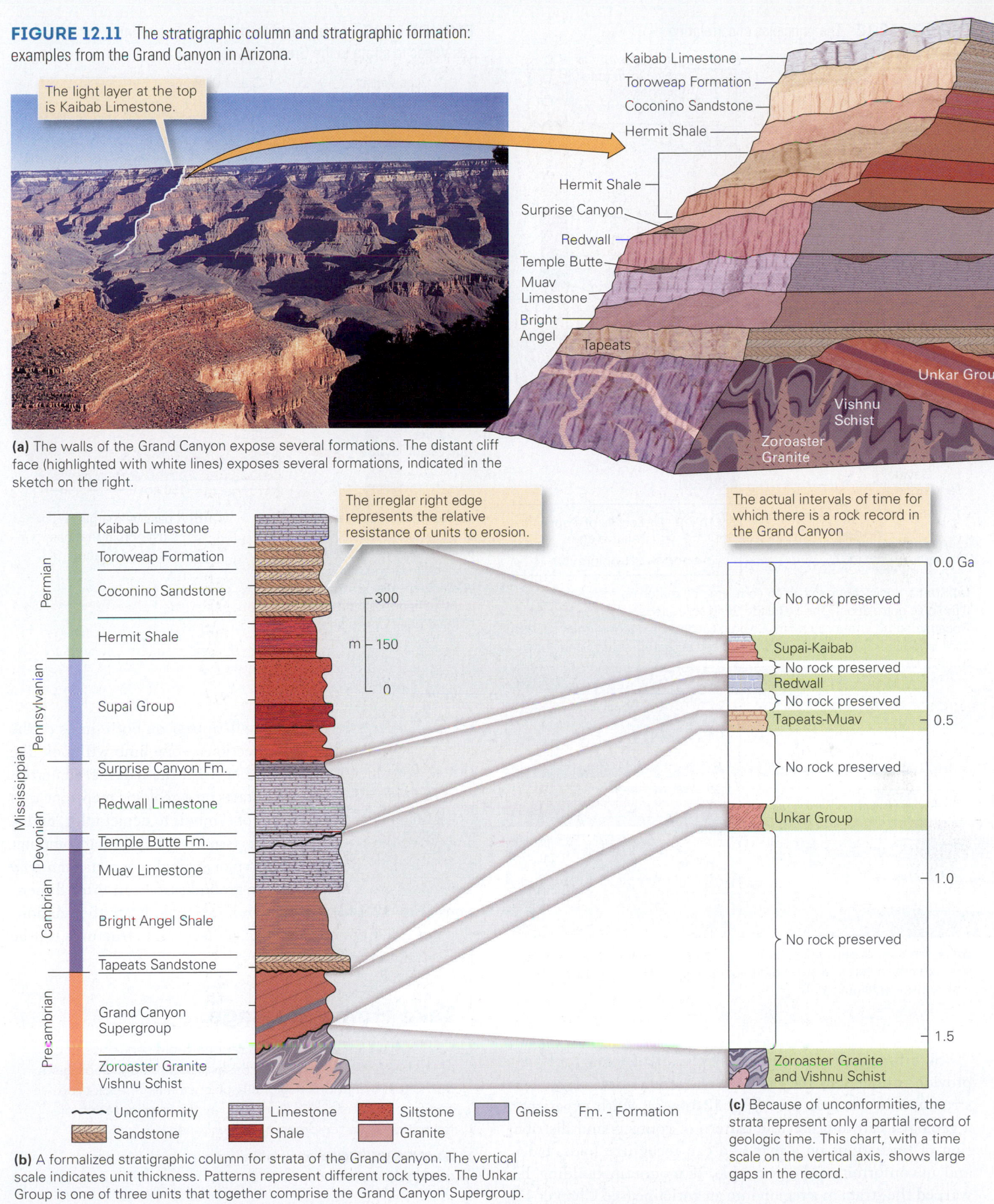

The light layer at the top is Kaibab Limestone.

Kaibab Limestone
Toroweap Formation
Coconino Sandstone
Hermit Shale

Hermit Shale
Surprise Canyon
Redwall
Temple Butte
Muav Limestone
Bright Angel
Tapeats

Unkar Group
Vishnu Schist
Zoroaster Granite

(a) The walls of the Grand Canyon expose several formations. The distant cliff face (highlighted with white lines) exposes several formations, indicated in the sketch on the right.

The irreglar right edge represents the relative resistance of units to erosion.

The actual intervals of time for which there is a rock record in the Grand Canyon

Permian
Kaibab Limestone
Toroweap Formation
Coconino Sandstone
Hermit Shale

Pennsylvanian
Supai Group

Mississippian
Surprise Canyon Fm.
Redwall Limestone

Devonian
Temple Butte Fm.
Muav Limestone

Cambrian
Bright Angel Shale
Tapeats Sandstone

Precambrian
Grand Canyon Supergroup
Zoroaster Granite
Vishnu Schist

m — 300
m — 150
0

0.0 Ga

No rock preserved

Supai-Kaibab
No rock preserved
Redwall
No rock preserved
Tapeats-Muav   0.5

No rock preserved

Unkar Group

1.0

No rock preserved

1.5

Zoroaster Granite and Vishnu Schist

Unconformity   Limestone   Siltstone   Gneiss   Fm. - Formation
Sandstone   Shale   Granite

(b) A formalized stratigraphic column for strata of the Grand Canyon. The vertical scale indicates unit thickness. Patterns represent different rock types. The Unkar Group is one of three units that together comprise the Grand Canyon Supergroup.

(c) Because of unconformities, the strata represent only a partial record of geologic time. This chart, with a time scale on the vertical axis, shows large gaps in the record.

FIGURE 12.12 The principles of correlation.

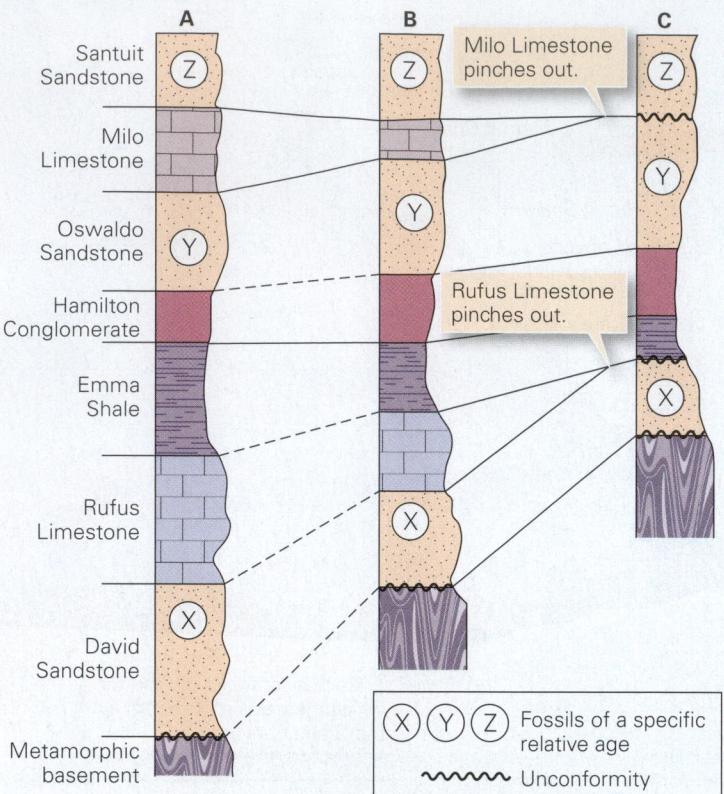

(a) Stratigraphic columns can be correlated by matching rock types (lithologic correlation). The Hamilton Conglomerate is a marker horizon. Because some strata pinch out, Column C contains unconformities. Fossil correlation indicates that the youngest beds in C are Santuit Sandstone.

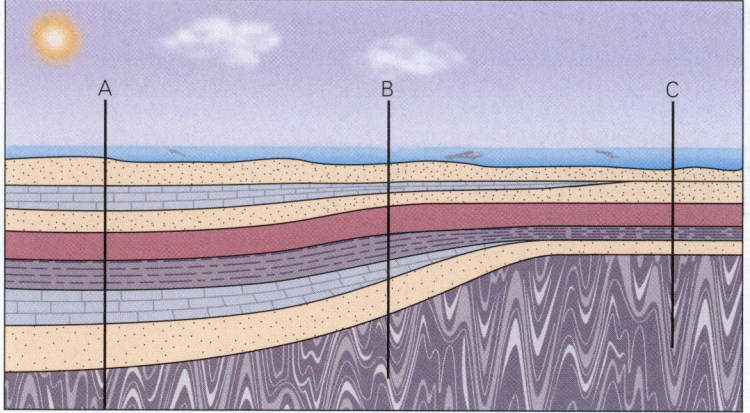

(b) At the time of deposition, locations A, B, and C (which correlate with the columns in part a) were in different parts of a basin. The basin floor was subsiding fastest at A.

FIGURE 12.13 An application of correlation to relate strata near Las Vegas to strata in the Grand Canyon.

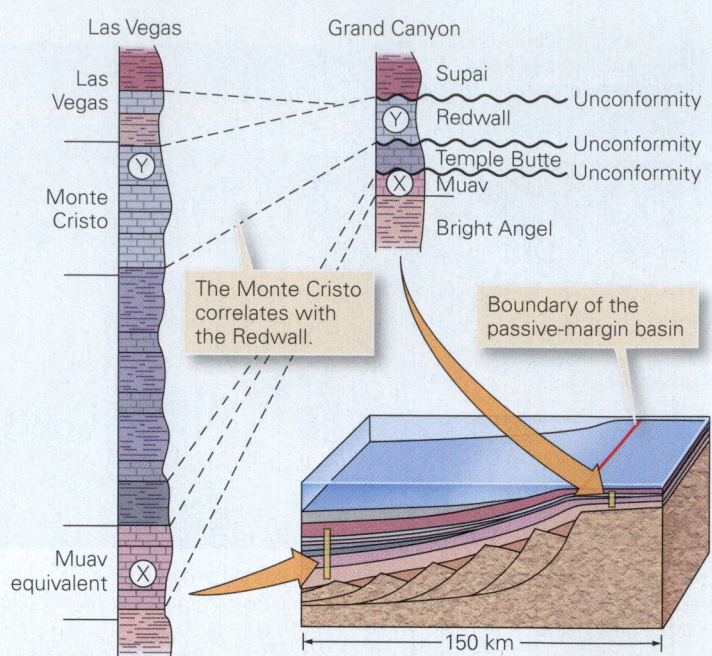

(a) The section between fossils of age X and fossils of age Y is much thicker near Las Vegas than in the Grand Canyon. The edge of a passive-margin basin lay between the two localities at the time of deposition.

(b) A passive-margin basin forms over crust that has stretched and thinned.

Significantly, the pattern displayed on a geologic map can provide insight into the presence and orientation of geologic structures in the map area (**Fig. 12.14a–c**). With experience, a geologist can interpret the pattern of contacts and distribution of formations on a map and can recognize folds, faults, and unconformities. For example, if mountain building has warped the strata in a region into an anticline (see Chapter 11), the same succession of strata will appear on both limbs of the fold but with opposite dip directions—one limb will look like the mirror reflection of the other, with the oldest strata cropping out along the hinge. Contacts on a geologic map appear as lines. Geologists use a variety of symbols to depict folds, faults, and the strike-and-dip of layers (see Box 11.1). Using computer technology, it's now possible to plot geologic contacts on *digital elevation maps* that portray the ground surface in three dimensions (**Fig. 12.14d**). *Geologic cross sections* indicate the relationships of underground geologic contacts and structures on the plane of a vertical slice.

Take-Home Message

A stratigraphic formation is a recognizable sequence of beds that can be mapped across a broad region. Geologists correlate formations regionally on the basis of rock type and fossil content, and they portray the configuration of formations in a region on a geologic map.

QUICK QUESTION: What will a syncline look like on a geologic map?

FIGURE 12.14 A geologic map depicts the distribution of rock units and structures.

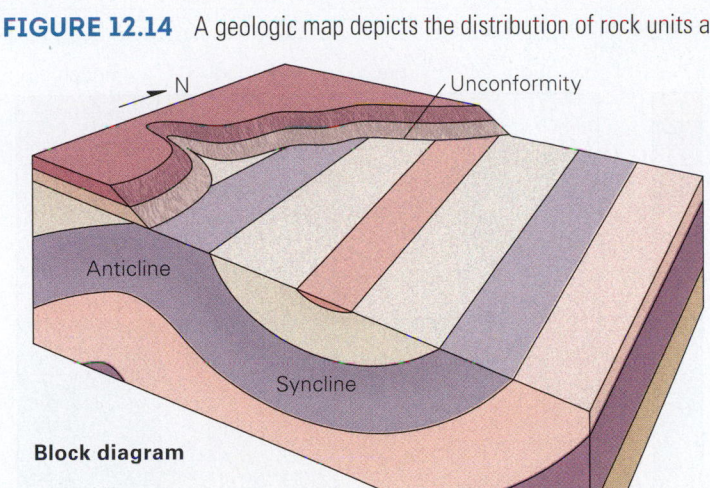

Block diagram

(a) A block diagram provides a three-dimensional representation. Here we see an angular unconformity over folded strata.

A geologic map shows the view looking straight down.

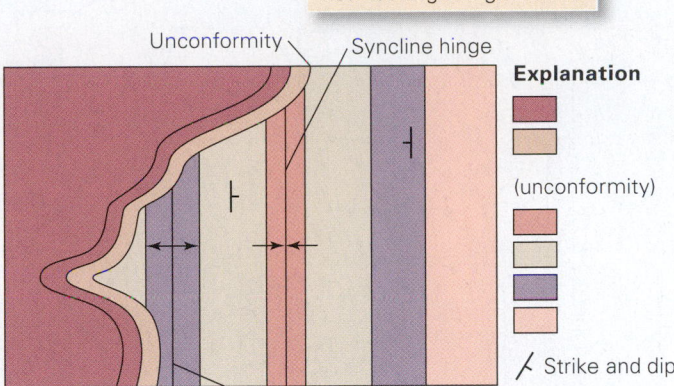

Explanation

(unconformity)

⊬ Strike and dip

5 km

Geologic map

(b) A geologic map shows the distribution of units. Contacts occur between units. Note that the map also shows geologic structures.

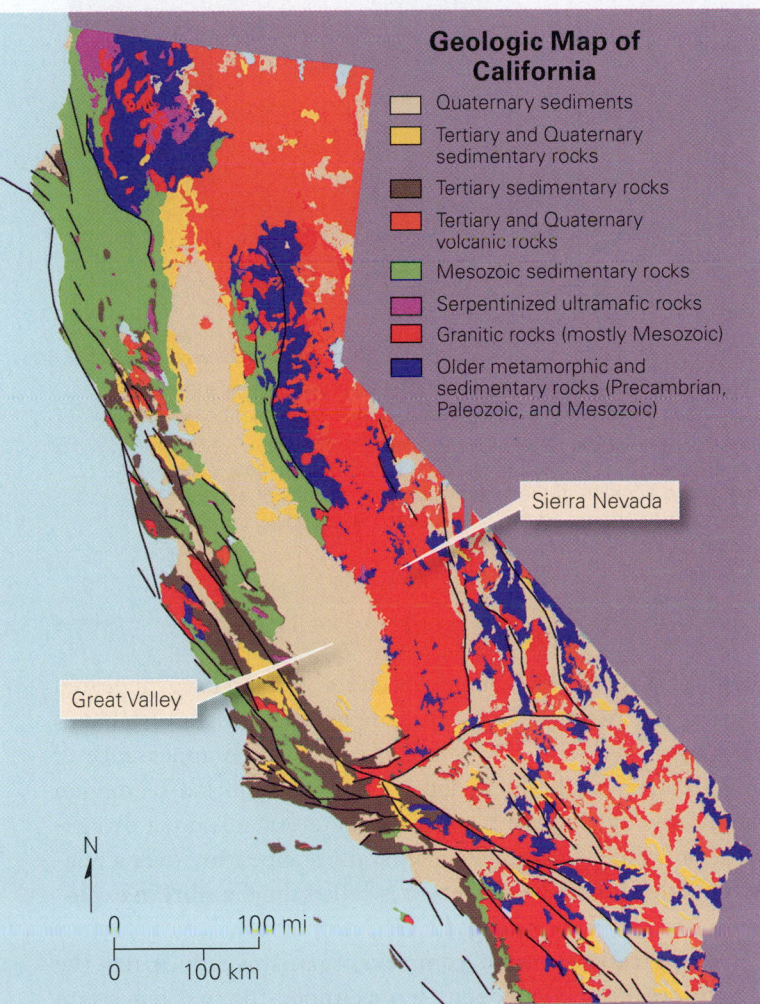

Geologic Map of California

- Quaternary sediments
- Tertiary and Quaternary sedimentary rocks
- Tertiary sedimentary rocks
- Tertiary and Quaternary volcanic rocks
- Mesozoic sedimentary rocks
- Serpentinized ultramafic rocks
- Granitic rocks (mostly Mesozoic)
- Older metamorphic and sedimentary rocks (Precambrian, Paleozoic, and Mesozoic)

Sierra Nevada

Great Valley

N

0   100 mi
0   100 km

(c) A geologic map of California indicates that the state is underlain by many different rock units. Granite underlies the Sierras, and Quaternary sediments underlie the Great Valley. The black lines are fault traces.

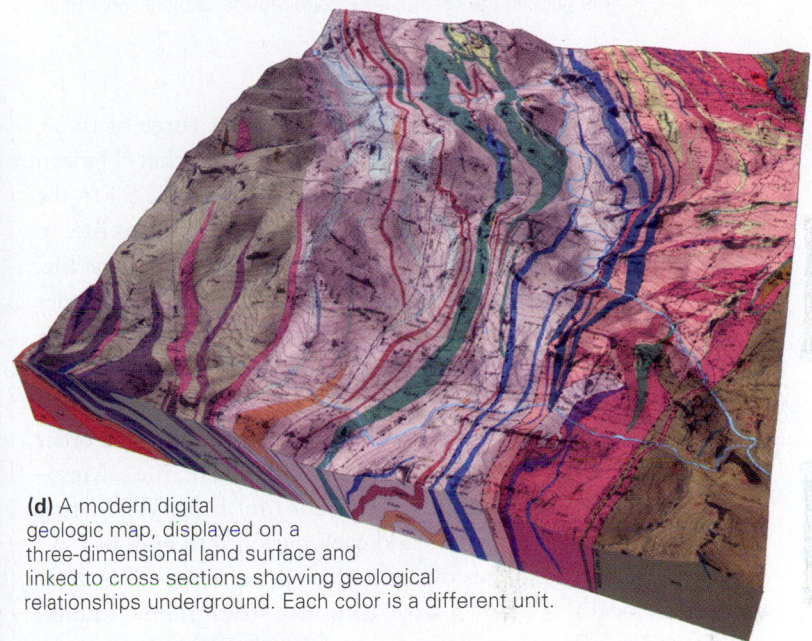

(d) A modern digital geologic map, displayed on a three-dimensional land surface and linked to cross sections showing geological relationships underground. Each color is a different unit.

12.6 The Geologic Column

Dividing Time

As stated earlier, no one locality on Earth provides a complete record of our planet's history, because stratigraphic columns can contain unconformities. But by correlating strata from locality to locality at millions of places around the world, geologists have pieced together a composite stratigraphic column, called the **geologic column**, that represents the entirety of Earth history (**Fig. 12.15**). The column is divided into segments, each of which represents a specific interval of time. The largest subdivisions break Earth history into the Hadean, Archean,

FIGURE 12.15 Global correlation of strata led to the development of the geologic column. (Not to scale.)

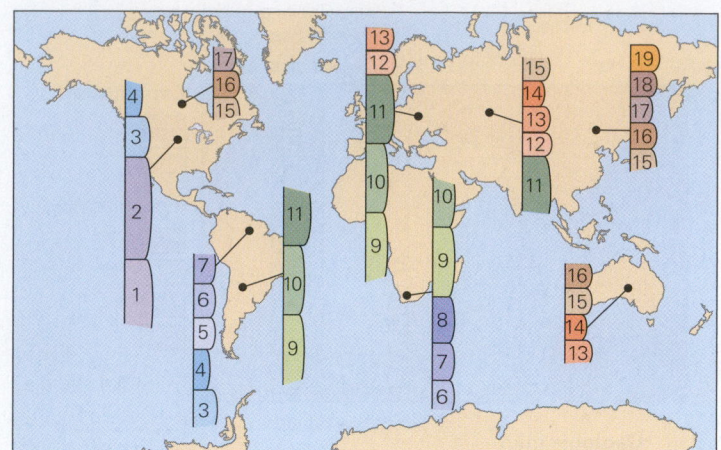

(a) Each of these small columns represents the stratigraphy at a given location. By correlating these columns, geologists determined their relative ages, filled in the gaps in the record, and produced the geologic column.

Eon	Era	Period	Epoch
Phanerozoic	Cenozoic	Quaternary	Holocene
			Pleistocene
		Neogene	Pliocene
		Tertiary	Miocene
		Paleogene	Oligocene
			Eocene
			Paleocene
	Mesozoic	Cretaceous	
		Jurassic	
		Triassic	
	Paleozoic	Permian	Pennsylvanian
		Carboniferous	Mississippian
		Devonian	
		Silurian	
		Ordovician	
		Cambrian	
Precambrian	Proterozoic		
	Archean		

(b) By correlation, the strata from localities around the world were stacked in a chart representing geologic time to create the geologic column. Geologists assigned names to time intervals, but since the column was built without knowledge of numerical ages, it does not depict the duration of these intervals. Subdivisions of eons in the Precambrian are not shown. The Hadean is not shown because rocks do not preserve a record of it.

Proterozoic, and Phanerozoic **Eons**—the first three of these, together, constitute the **Precambrian**. (Note that Hadean rocks don't exist, so they are not shown in Figure 12.5.) In the names of the two youngest eons, the suffix *–zoic* means life, so Phanerozoic means visible life, and Proterozoic means first life. These names can be a bit confusing, though, because decades after the eons had been named, geologists discovered that the earliest life, cells of bacteria and archaea, actually appeared during the Archean Eon. Eons, in turn, can be subdivided into **eras**. The Phanerozoic Eon, for example includes, in order from oldest to youngest, the Paleozoic (ancient life), Mesozoic (middle life), and Cenozoic (recent life) Eras. We further divide each era into **periods** and each period into **epochs**.

Where do the names of the periods come from? They refer either to localities where a fairly complete stratigraphic column representing that time interval was first identified (e.g., rocks representing the Devonian Period crop out near Devon, England) or to a characteristic of the time (rocks from the Carboniferous Period contain a lot of coal). The terminology was not set up in a planned fashion that would make it easy to learn. Instead, time divisions and their naming were established haphazardly in the years between 1760 and 1845, when geologists first began to refine their understanding of geologic history and fossil succession. Also, because the divisions were defined before numerical ages could be determined, they are all of different durations.

Life Evolution in the Context of the Geologic Column

The succession of fossils preserved in strata of the geologic column defines the course of life's evolution throughout Earth history (**Fig. 12.16**). Simple bacteria and archaea appeared during the Archean Eon, but complex shell-less invertebrates did not evolve until the late Proterozoic. The appearance of invertebrates with shells defines the Precambrian-Cambrian boundary. During the Cambrian, there was a sudden diversification in life, with many new families appearing over a relatively short interval—this event is called the **Cambrian explosion** (see Interlude E; **Fig. 12.17**).

Progressively more complex organisms populated the Earth during the Paleozoic. For example, the first fish swam in Ordovician seas, land plants started to spread over the continents during the Silurian, and amphibians appeared during the Devonian. (Note that prior to the Silurian, the land surface was completely unvegetated and would have been a stark, dusty desert, even at the equator.) Though

FIGURE 12.16 Life evolution in the context of the geologic column. The Earth formed at the beginning of the Hadean Eon.

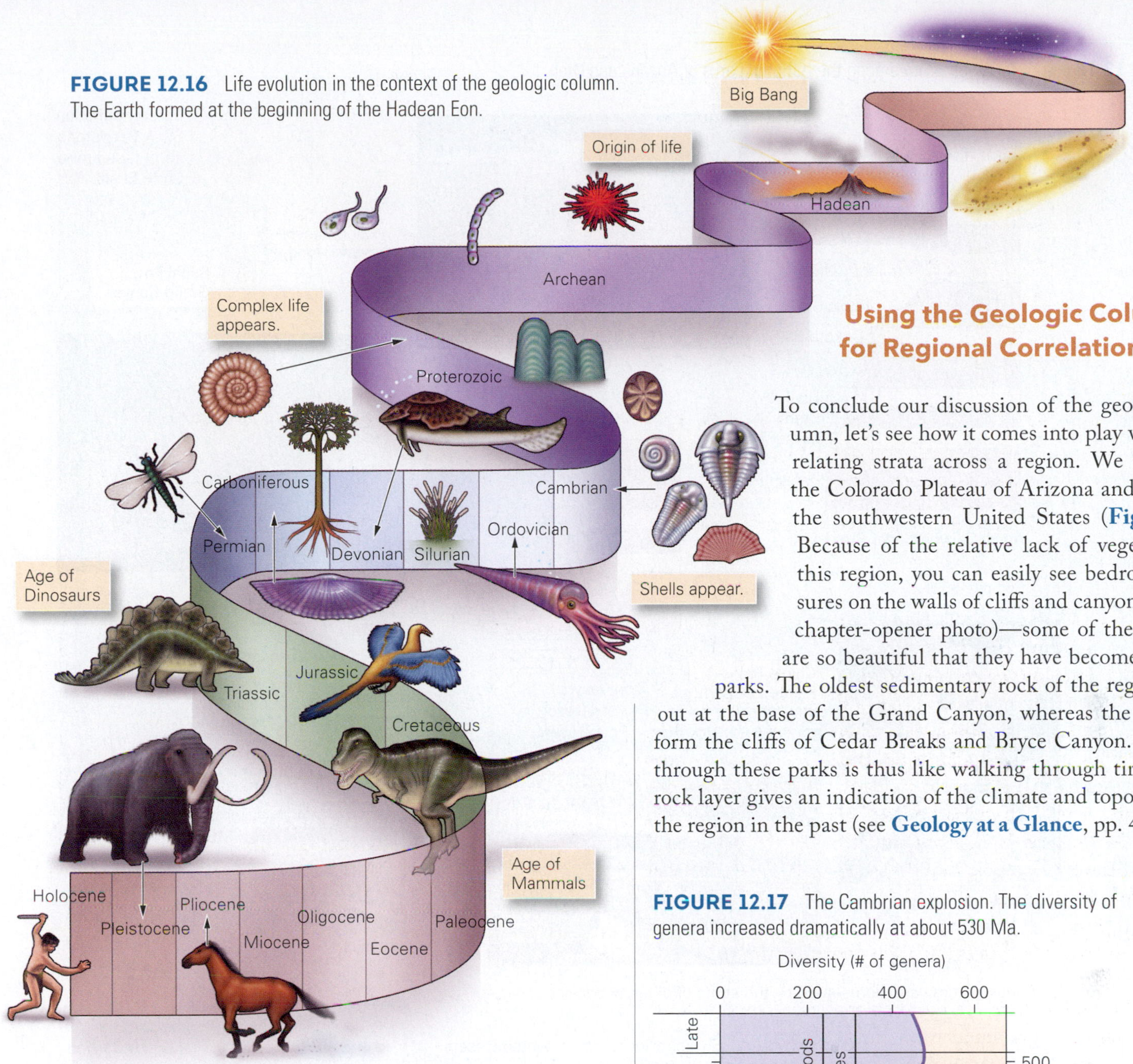

Big Bang

Origin of life

Hadean

Archean

Complex life appears.

Proterozoic

Carboniferous

Cambrian

Ordovician

Permian

Devonian Silurian

Shells appear.

Age of Dinosaurs

Jurassic

Triassic

Cretaceous

Age of Mammals

Holocene

Pliocene

Oligocene

Paleocene

Pleistocene

Miocene

Eocene

Using the Geologic Column for Regional Correlation

To conclude our discussion of the geologic column, let's see how it comes into play when correlating strata across a region. We return to the Colorado Plateau of Arizona and Utah, in the southwestern United States (**Fig. 12.18**). Because of the relative lack of vegetation in this region, you can easily see bedrock exposures on the walls of cliffs and canyons (see the chapter-opener photo)—some of these locales are so beautiful that they have become national parks. The oldest sedimentary rock of the region crops out at the base of the Grand Canyon, whereas the youngest form the cliffs of Cedar Breaks and Bryce Canyon. Walking through these parks is thus like walking through time—each rock layer gives an indication of the climate and topography of the region in the past (see **Geology at a Glance**, pp. 454–455).

FIGURE 12.17 The Cambrian explosion. The diversity of genera increased dramatically at about 530 Ma.

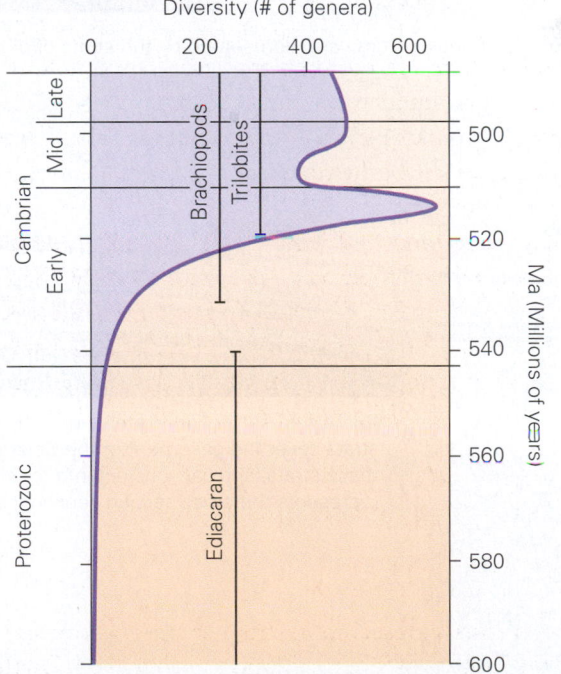

Diversity (# of genera)

Brachiopods

Trilobites

Ediacaran

Cambrian — Late / Mid / Early

Proterozoic

Ma (Millions of years)

reptiles appeared during the Pennsylvanian Period, the first dinosaurs did not stomp across the land until the Triassic. Dinosaurs continued to inhabit the Earth until their sudden extinction at the end of the Cretaceous Period. For this reason, geologists refer to the Mesozoic Era as the *Age of Dinosaurs*. Small mammals appeared during the Triassic Period, but the diversification (development of many different species) of mammals to fill a wide range of ecological niches did not happen until the beginning of the Cenozoic Era, so geologists call the Cenozoic the *Age of Mammals*. Birds also appeared during the Mesozoic—specifically, at the beginning of the Cretaceous Period—but underwent great diversification in the Cenozoic Era.

FIGURE 12.18 Correlation of strata among the national parks of Arizona and Utah.

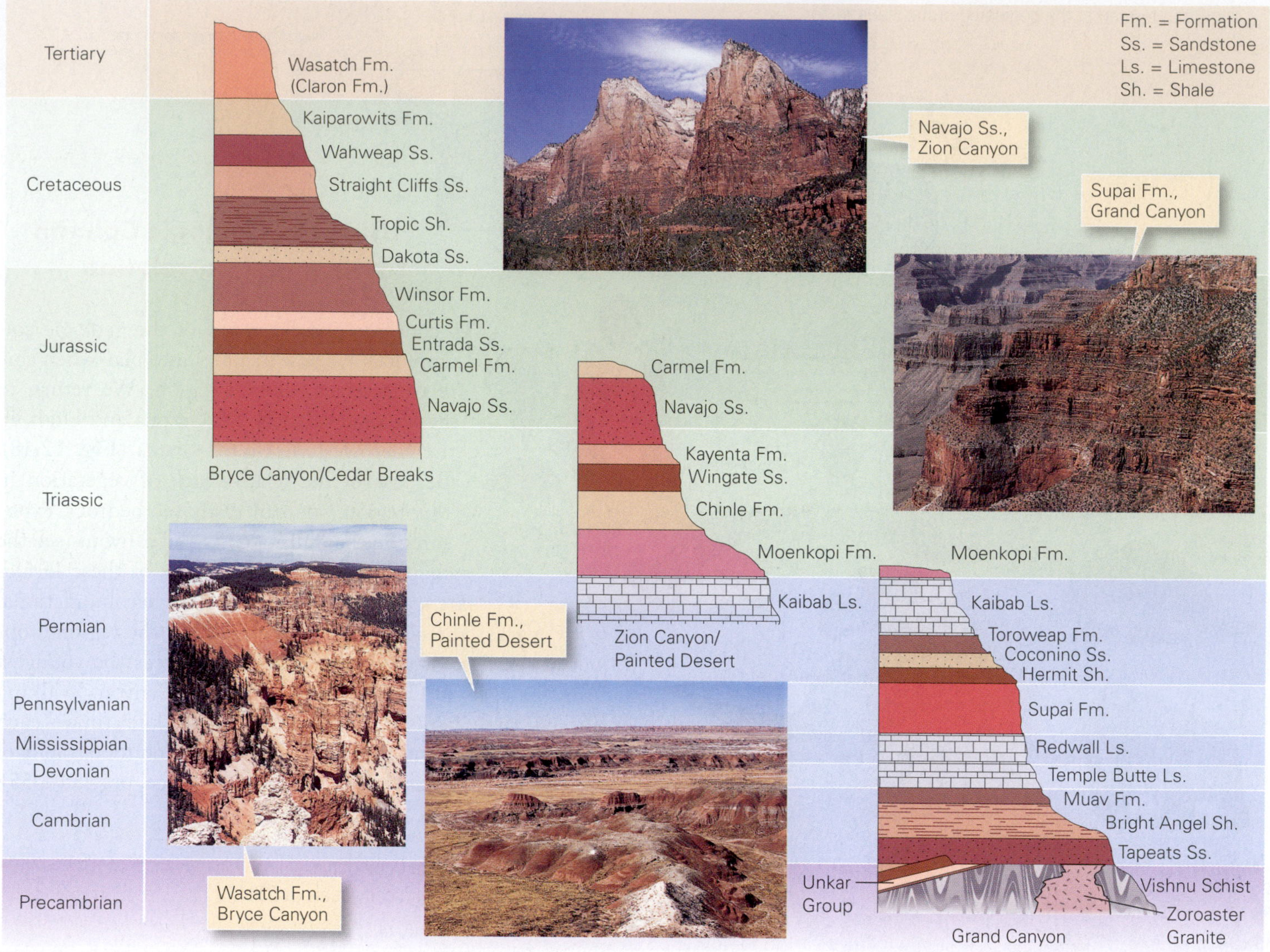

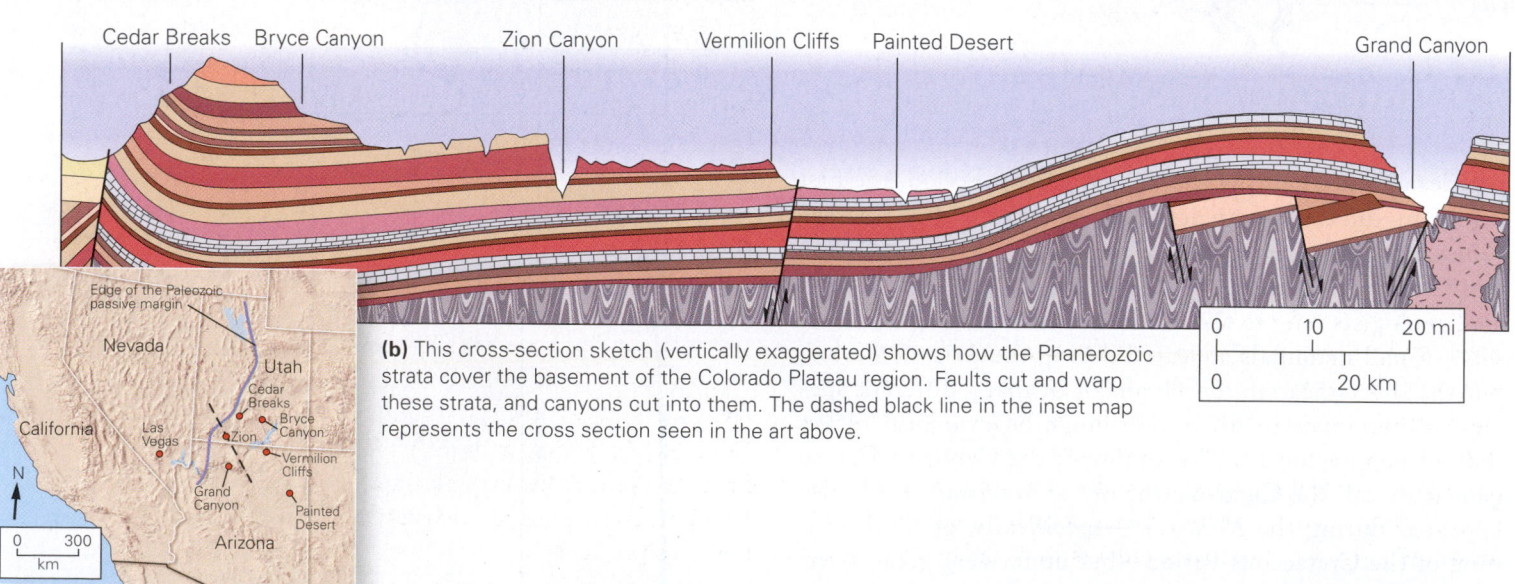

Fm. = Formation
Ss. = Sandstone
Ls. = Limestone
Sh. = Shale

Tertiary

Wasatch Fm. (Claron Fm.)

Navajo Ss., Zion Canyon

Cretaceous

Kaiparowits Fm.
Wahweap Ss.
Straight Cliffs Ss.
Tropic Sh.
Dakota Ss.

Supai Fm., Grand Canyon

Jurassic

Winsor Fm.
Curtis Fm.
Entrada Ss.
Carmel Fm.
Navajo Ss.

Carmel Fm.
Navajo Ss.

Kayenta Fm.
Wingate Ss.
Chinle Fm.

Bryce Canyon/Cedar Breaks

Triassic

Moenkopi Fm.

Moenkopi Fm.

Kaibab Ls.

Kaibab Ls.

Permian

Chinle Fm., Painted Desert

Zion Canyon/ Painted Desert

Toroweap Fm.
Coconino Ss.
Hermit Sh.

Pennsylvanian

Supai Fm.

Mississippian

Redwall Ls.

Devonian

Temple Butte Ls.

Cambrian

Muav Fm.
Bright Angel Sh.
Tapeats Ss.

Precambrian

Wasatch Fm., Bryce Canyon

Unkar Group

Vishnu Schist
Zoroaster Granite

Grand Canyon

(a) Different intervals of geologic time are represented by the strata of different parks.

Cedar Breaks   Bryce Canyon   Zion Canyon   Vermilion Cliffs   Painted Desert   Grand Canyon

| 0 | 10 | 20 mi |
| 0 | 20 km | |

Edge of the Paleozoic passive margin

Nevada

Utah

Cedar Breaks
Bryce Canyon
Las Vegas
Zion
Vermilion Cliffs
Grand Canyon
Painted Desert

California

N

Arizona

| 0 | 300 |
| km | |

(b) This cross-section sketch (vertically exaggerated) shows how the Phanerozoic strata cover the basement of the Colorado Plateau region. Faults cut and warp these strata, and canyons cut into them. The dashed black line in the inset map represents the cross section seen in the art above.

For example, when the Precambrian metamorphic and igneous rocks exposed in the inner gorge of the Grand Canyon first formed, the region was a high mountain range, perhaps as dramatic as the Himalayas today. When the fossiliferous beds of the Kaibab Limestone at the rim of the canyon first developed, the region was a Bahama-like carbonate reef and platform, bathed in a warm, shallow sea. And when the rocks making up the towering red cliffs of sandstone in Zion Canyon were deposited, the region was a Sahara-like desert, blanketed with huge sand dunes.

Take-Home Message

Correlation of stratigraphic sequences from around the world led to the production of a chart, the geologic column, that represents the entirety of Earth history. The column, developed using only relative-age relations, is subdivided into eons, eras, periods, and epochs.

QUICK QUESTION: What feature of living organisms appeared at the Precambrian/Cambrian boundary?

12.7 How Do We Determine Numerical Ages?

Geologists since the days of Hutton could determine the relative ages of geologic events, but they had no way to specify *numerical ages* (called absolute ages in older literature), meaning the age in years of an event. Thus, they could not define a timeline for Earth history or determine the duration of events. This situation changed with the discovery of radioactivity. Simply put, a **radioactive element** is one that decays so that its atoms transform into atoms of another element at a constant rate. The rate can be measured in the lab and can be specified in years. In the 1950s, geologists first developed techniques for using measurements of radioactive elements to calculate the ages of rocks. Geologists originally referred to this process of determining the numerical age of rocks as *radiometric dating*; more recently, it has come to be known as **isotopic dating**. We refer to the overall determination and interpretation of numerical ages as **geochronology**. The technique of isotopic dating has been vastly improved over the years. We begin our discussion of this technique by first learning more about radioactive decay.

Radioactive Decay

As we've noted earlier, different versions of an element, called **isotopes** of the element, have the same atomic number but a different atomic weight (see Box 1.3). To see how this terminology applies, let's consider two isotopes of uranium. All uranium atoms have 92 protons, but the uranium-238 isotope (abbreviated ^{238}U) has an atomic weight of 238 and thus has 146 neutrons, whereas the ^{235}U isotope has an atomic weight of 235 and thus has 143 neutrons. Some elements have only one isotope, which is stable in that all atoms of the element could last until the end of the Universe—unless they become fuel for the nuclear reactions in stars. But many elements have two or more isotopes. Of these, some are *stable*, but some are not. The *unstable* isotopes, by definition, are called *radioactive isotopes*. Unstable in this context means that the isotopes undergo a change called *radioactive decay*, which converts them into a different element. Radioactive decay can take place by a variety of reactions, but regardless of the details, all these reactions change the atomic number of the nucleus and, therefore, the identity of the element. We refer to the isotope that undergoes decay as the *parent isotope* and the decay product as the *daughter isotope*.

Physicists cannot specify how long an individual radioactive isotope will survive before it decays, but they can measure how long it takes for half of a group of parent isotopes to decay. This time is called the **half-life** of the isotope. **Figure 12.19** can help you visualize the concept of a half-life. Imagine a crystal containing 16 radioactive parent isotopes. (Note that in a real crystal, the number of atoms would be immensely larger.) After one half-life, 8 isotopes have decayed, so the crystal now contains 8 parent and 8 daughter isotopes. After a second half-life, 4 of the remaining parent isotopes have decayed, so the crystal contains 4 parent and 12 daughter isotopes. And after a third half-life, 2 more parent isotopes have decayed, so the crystal contains 2 parent and 14 daughter isotopes. For a given decay reaction, the half-life is a constant, measured in years.

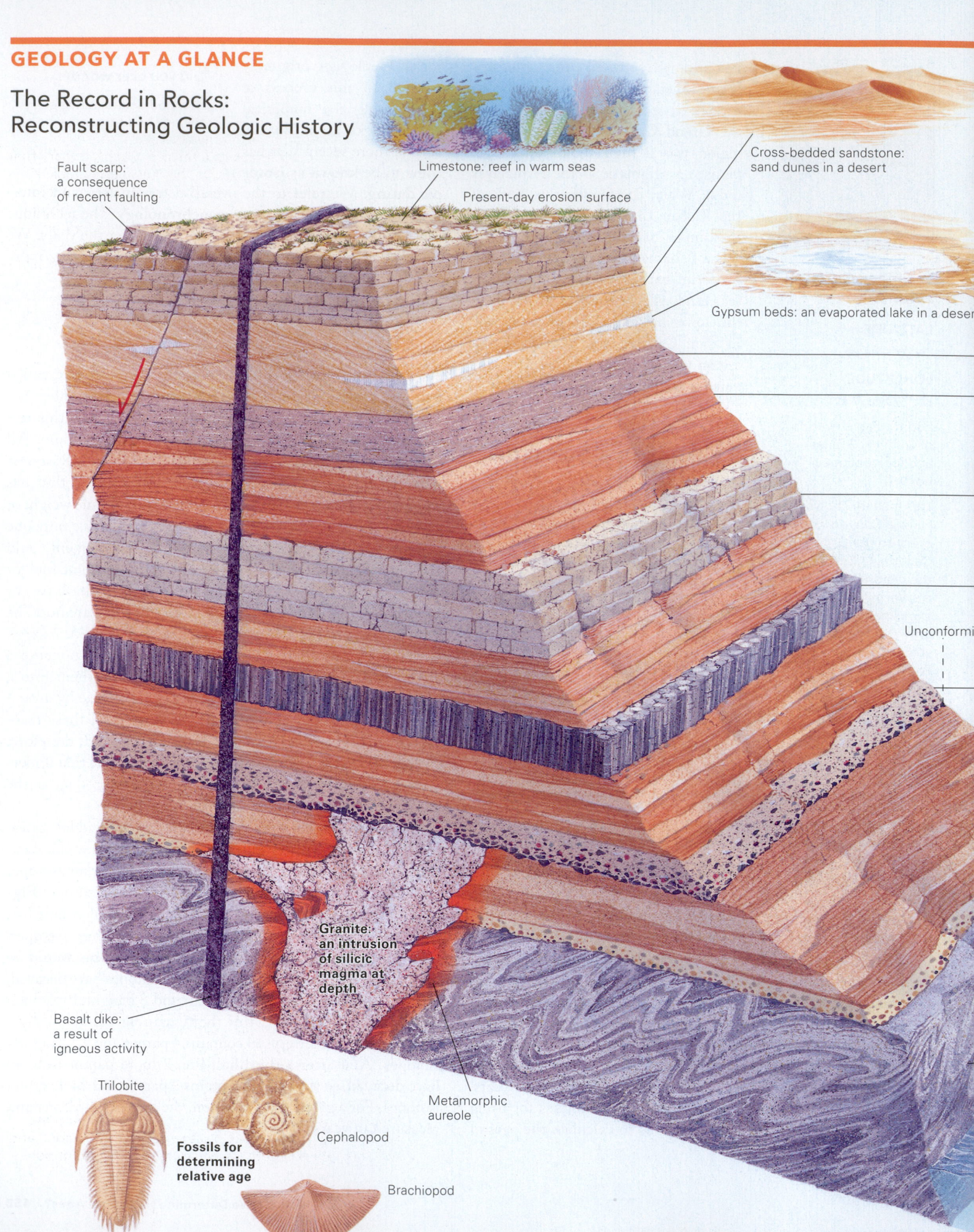

The Record in Rocks: Reconstructing Geologic History

Fault scarp: a consequence of recent faulting

Limestone: reef in warm seas

Present-day erosion surface

Cross-bedded sandstone: sand dunes in a desert

Gypsum beds: an evaporated lake in a desert

Unconformity

Granite: an intrusion of silicic magma at depth

Basalt dike: a result of igneous activity

Metamorphic aureole

Trilobite

Cephalopod

Fossils for determining relative age

Brachiopod

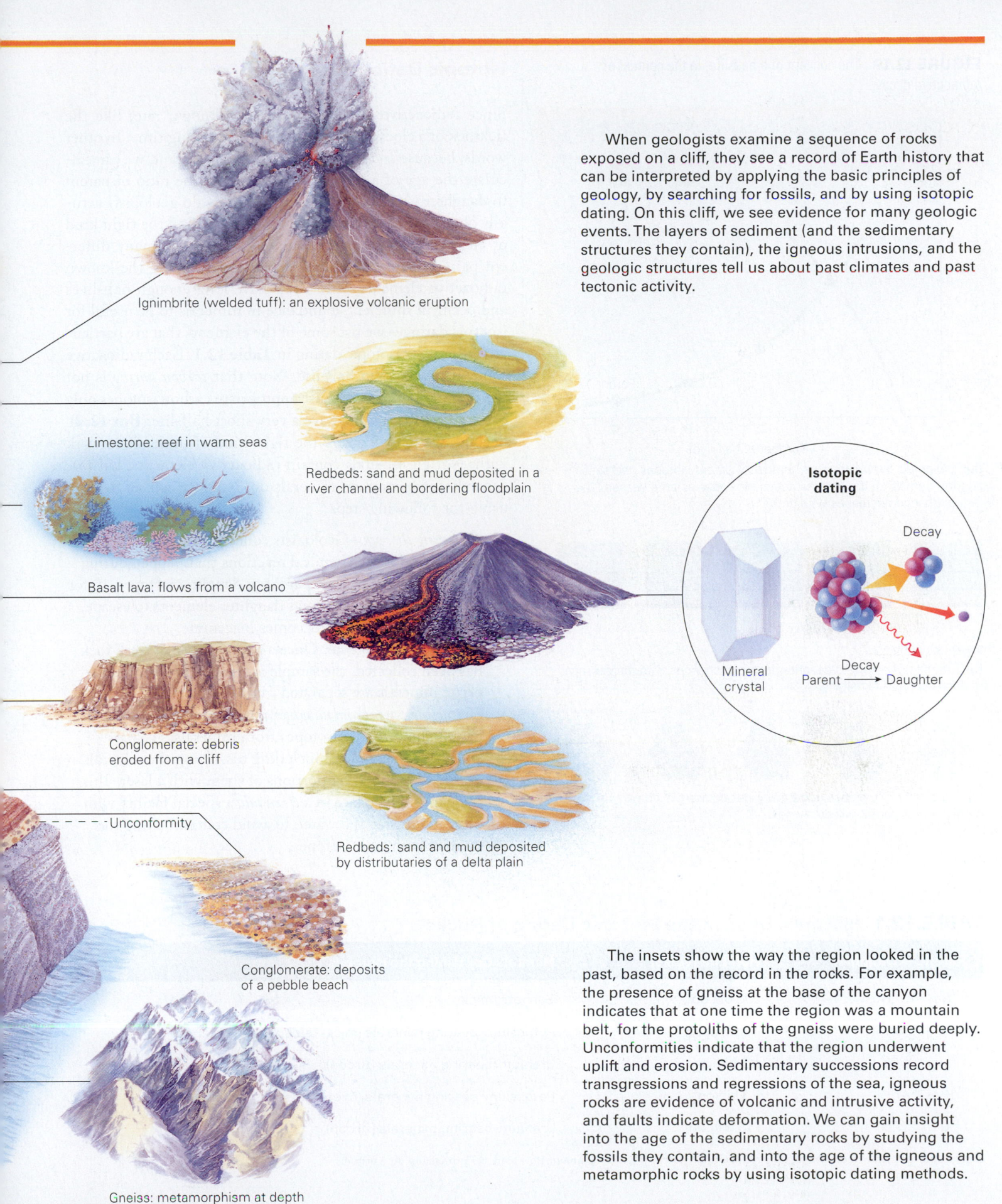

When geologists examine a sequence of rocks exposed on a cliff, they see a record of Earth history that can be interpreted by applying the basic principles of geology, by searching for fossils, and by using isotopic dating. On this cliff, we see evidence for many geologic events. The layers of sediment (and the sedimentary structures they contain), the igneous intrusions, and the geologic structures tell us about past climates and past tectonic activity.

Ignimbrite (welded tuff): an explosive volcanic eruption

Limestone: reef in warm seas

Redbeds: sand and mud deposited in a river channel and bordering floodplain

Basalt lava: flows from a volcano

Conglomerate: debris eroded from a cliff

Unconformity

Redbeds: sand and mud deposited by distributaries of a delta plain

Conglomerate: deposits of a pebble beach

Gneiss: metamorphism at depth beneath a mountain belt

Isotopic dating

Mineral crystal

Decay

Decay

Parent ⟶ Daughter

The insets show the way the region looked in the past, based on the record in the rocks. For example, the presence of gneiss at the base of the canyon indicates that at one time the region was a mountain belt, for the protoliths of the gneiss were buried deeply. Unconformities indicate that the region underwent uplift and erosion. Sedimentary successions record transgressions and regressions of the sea, igneous rocks are evidence of volcanic and intrusive activity, and faults indicate deformation. We can gain insight into the age of the sedimentary rocks by studying the fossils they contain, and into the age of the igneous and metamorphic rocks by using isotopic dating methods.

FIGURE 12.19 The concept of a half-life, in the context of radioactive decay.

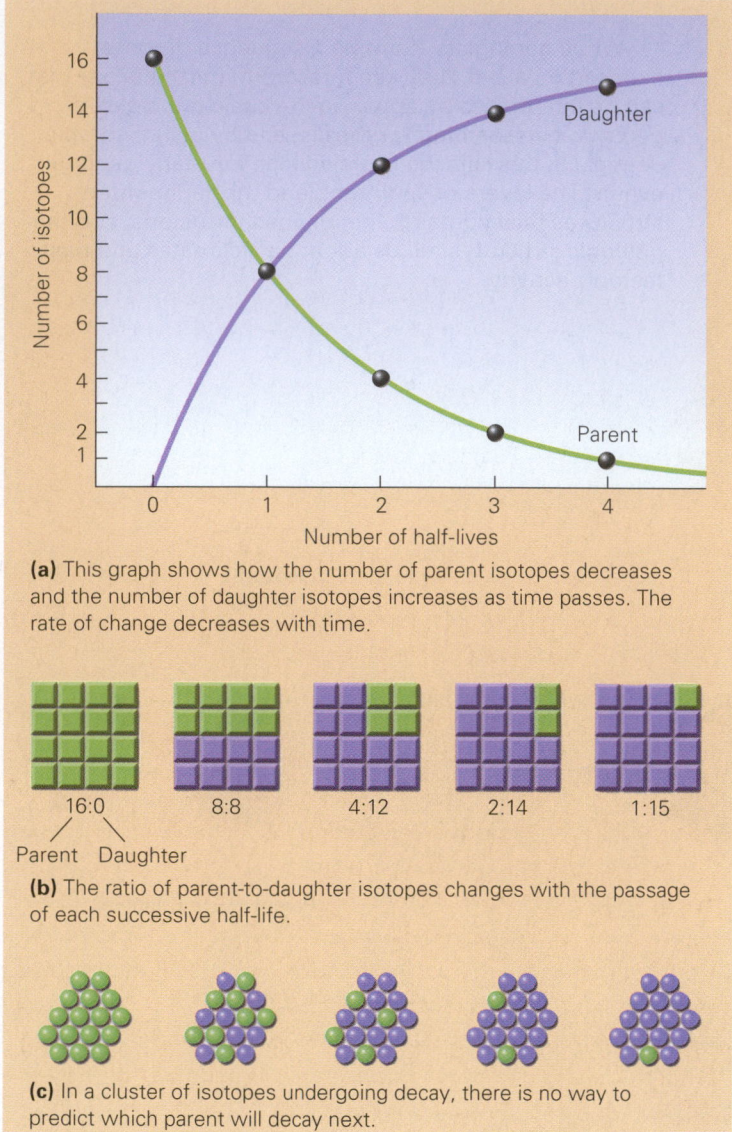

(a) This graph shows how the number of parent isotopes decreases and the number of daughter isotopes increases as time passes. The rate of change decreases with time.

16:0   8:8   4:12   2:14   1:15

Parent   Daughter

(b) The ratio of parent-to-daughter isotopes changes with the passage of each successive half-life.

(c) In a cluster of isotopes undergoing decay, there is no way to predict which parent will decay next.

Isotopic Dating Technique

Since radioactive decay proceeds at a known rate, like the ticktock of a clock, it provides a basis for telling time. In other words, because an element's half-life is a constant, we can calculate the age of a mineral by measuring the ratio of parent to daughter isotopes in the mineral. How do geologists actually obtain an isotopic date? First, we must find the right kind of elements to work with. Although there are many different pairs of parent and daughter isotopes among the known radioactive elements, only a few have long enough half-lives and occur in sufficient abundance in minerals to be useful for isotopic dating—we list some of the elements that are particularly useful for isotopic dating in **Table 12.1**. Each radioactive element has its own half-life. Note that *carbon dating* is not used for dating rocks because appropriate carbon isotopes only occur in organisms and have a very short half-life (**Box 12.2**). Second, we must identify the right kind of minerals to work with. Not all minerals contain radioactive elements, but fortunately some common minerals do. Now we can set to work using the following steps.

- *Collecting the rocks:* Geologists collect unweathered rocks for dating, for the chemical reactions that happen during weathering may remove parent or daughter isotopes. If, for example, weathering allows daughter elements to escape, then the isotopic clock becomes inaccurate.
- *Separating the minerals:* Once a good sample of fresh rock has been collected, the sample is crushed and the appropriate minerals are separated from the debris.
- *Extracting parent and daughter isotopes:* To separate out the parent and daughter isotopes from minerals, geologists use several techniques, including dissolving the minerals in acid or evaporating portions of them with a laser. This stage must take place in a *clean lab*, a special facility with ultra-filtered air and water, to avoid contaminating the samples with stray isotopes.

TABLE 12.1 Isotopes Used in the Isotopic Dating of Rocks

Parent → Daughter	Half-Life (years)	Minerals Containing the Isotopes
$^{147}Sm \rightarrow ^{143}Nd$	106 billion	Garnets, micas
$^{87}Rb \rightarrow ^{87}Sr$	48.8 billion	Potassium-bearing minerals (mica, feldspar, hornblende)
$^{238}U \rightarrow ^{206}Pb$	4.5 billion	Uranium-bearing minerals (zircon)
$^{40}K \rightarrow ^{40}Ar$	1.3 billion	Potassium-bearing minerals (mica, feldspar, hornblende)
$^{235}U \rightarrow ^{207}Pb$	713 million	Uranium-bearing minerals (zircon)

Sm = samarium, Nd = neodymium, Rb = rubidium, Sr = strontium, U = uranium, Pb = lead, K = potassium, Ar = argon.

BOX 12.2 CONSIDER THIS . . .

Carbon-14 Dating

Many people who have heard of carbon-14 (^{14}C) dating assume that it can be used to define the numerical age of rocks. But this is not the case. Rather, ^{14}C dating tells us the ages of organic materials—such as wood, cotton fibers, charcoal, flesh, bones, and shells—that contain carbon originally extracted from the atmosphere by photosynthesis. ^{14}C, a radioactive isotope of carbon, forms naturally in the atmosphere when cosmic rays (charged particles from space) bombard atmospheric nitrogen-14 (^{14}N) atoms. When plants consume carbon dioxide during photosynthesis, or when animals consume plants, they ingest a tiny amount of ^{14}C along with ^{12}C, the more common isotope of carbon. After an organism dies and can no longer exchange carbon with the atmosphere, the ^{14}C in its body begins to decay back to ^{14}N. Thus, the ratio of ^{14}C to ^{12}C changes at a rate determined by the half-life of ^{14}C.

We can use ^{14}C dating to determine the age of prehistoric fire pits or of organic debris in sediment. ^{14}C has a short half-life—only 5,730 years. Thus, the method cannot be used to date anything older than about 70,000 years, for after that time essentially no ^{14}C remains in the material. But this range makes it a useful tool for geologists studying sediments of the last ice age and for archaeologists studying ancient cultures or prehistoric peoples. Again, since rocks do not contain organic carbon, and may be significantly older than 70,000 years, we cannot determine the age of rocks by using the ^{14}C dating method.

- *Analyzing the parent-daughter ratio*: Geologists pass the isotopes through a mass spectrometer, a sophisticated instrument that uses a strong magnet to separate isotopes from one another according to their respective weights (**Fig. 12.20**). The instrument can count the number of atoms of specific isotopes separately.

At the end of the laboratory process, geologists can define the ratio of parent-to-daughter isotopes in a mineral and from this ratio determine the age of the mineral. Needless to say, the

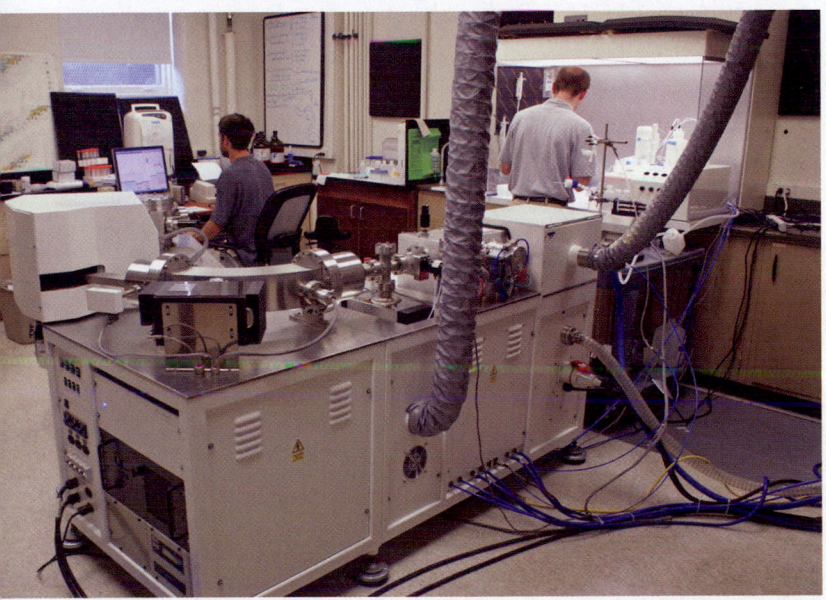

FIGURE 12.20 In an isotopic dating laboratory, samples are analyzed using a mass spectrometer. This instrument measures the ratio of parent-to-daughter isotopes.

description of the procedure here has been simplified—in reality, obtaining an isotopic date requires complex calculations and can be time-consuming and expensive.

What Does an Isotopic Date Mean?

At high temperatures, atoms in a crystal lattice vibrate so rapidly that chemical bonds break and reattach relatively easily. As a consequence, isotopes escape from or move into crystals, so parent-daughter ratios are meaningless. Because isotopic dating is based on the parent-daughter ratio, the "isotopic clock" starts only when crystals become cool enough for isotopes to be locked into the lattice. The temperature below which isotopes are no longer free to move is called the **closure temperature** of a mineral. When we specify an isotopic date for a rock, we are defining the time at which a specific mineral in the rock cooled below its closure temperature.

With the concept of closure temperature in mind, we can interpret the meaning of isotopic dates. In the case of igneous rocks, isotopic dating of minerals with high closure temperature (e.g., >650°C) tells you when a magma or lava cooled to form a solid, cool igneous rock. In the case of metamorphic rocks, an isotopic date from minerals with high closure temperature tells you when a rock cooled from a metamorphic temperature (e.g., >450°C) above the closure temperature to a temperature below. Not all minerals have the same closure temperature, so different minerals in a rock that cools very slowly will yield different dates.

In recent years, geologists have started to carry out dating using isotopic systems that have very low closure temperatures (as low as 75°C). Dating these low-closure temperature minerals constrains the time of exhumation, meaning the time at which uplift and erosion led to the return of a rock that was once, say, 20 km below the surface back to within a few kilometers of the surface.

Can we isotopically date a clastic sedimentary rock directly? No. If we date minerals in a sedimentary rock, we're determining only when these minerals first crystallized as part of an igneous or metamorphic rock, not the time when the minerals were deposited as sediment nor the time when the sediment lithified to form a sedimentary rock. For example, if we date the feldspar grains contained within a granite pebble in a conglomerate, we're dating the time the granite cooled below feldspar's closure temperature, not the time the pebble was deposited by a stream.

The age of mineral grains in sediment, however, can be useful. In recent years, geologists have used *detrital geochronology*, determination of the ages of detrital (clastic) grains, to learn the age of the rocks in the sediment's source region and thus determine where the sediment came from. Such ages also put an upper limit on the age of the sedimentary rock, for according to the principle of inclusions, the sedimentary rock must be younger than the grains it contains.

Other Methods of Determining Numerical Age

Isotopic dating is not the only means of constraining the age of a geological material or of events that have happened on Earth. Let's consider a few additional techniques.

Counting Changes of Season
The changes in seasons affect a wide variety of phenomena, including the following.

- *The growth rate of trees:* Trees grow seasonally, with different growth rates in different seasons. These variations produce alternating dense (dark) and less-dense (light) bands, for growth rate affects cell size which, in turn, affects wood density.
- *The organic productivity of lakes and seas:* During the winter, less light reaches the Earth, water temperatures decrease, and the supply of nutrients decreases. Therefore, more organic material will be deposited in summer than in winter. So alternations between organic-rich and organic-poor sediment indicate the seasons.
- *The growth rate of chemically precipitated sedimentary rocks:* Factors that control precipitation rates, such as the rate of groundwater flow and temperature, change seasonally in some locations. Such changes may form distinct bands in chemical sedimentary rocks such as travertine.
- *The growth rate of shell-secreting organisms:* Organisms grow and produce new shell material on a seasonal basis. Such variations can produce ridges in shells of mollusks and alternating dark and light bands in the heads of coral colonies.
- *The layering in glaciers:* During the snowy season, more snow falls relative to dust than during the dry season, so

annual snowfall includes a dusty layer and a clean layer. These contrasts are preserved as visible banding in ice.

As a consequence of these seasonal changes, **growth bands** develop in trees, travertine deposits, and shelly organisms, and *rhythmic layering* develops in sedimentary accumulations and glacier ice (**Fig. 12.21a–c**). By counting bands or layers, geologists can determine how long an organism survived and how long a sedimentary accumulation took to form. If rings or layers have developed right up to the present, we can count backward and determine how long ago they began to form.

Dendrochronologists, scientists who study and date **tree rings**, the growth bands in trees, have found that not only do such rings count time, but they also preserve a record of the changing climate, for during warm, rainy years, trees grow faster than they do during drought years. In fact, climate variations over time yield distinctive patterns of tree rings, much like the bar codes on products in supermarkets. By correlating the older rings of a still-living tree with the youngest rings measured in a log from a dead tree, by using these patterns, it's possible to extend the record of tree rings back before living trees started to grow. For example, the oldest living trees, bristlecone pines, are almost 4,000 years old—by correlating the older rings of these trees with rings in preserved logs, dendrochronologists have extended the tree-ring record back for many thousands of years more (**Fig. 12.21d**).

Glacier ice also preserves a valuable record of past climate, for the ratio of different oxygen isotopes in the water molecules making up the ice reflects the global temperature at the time the snow fell to create the ice, as we'll discuss further in Chapter 22. Ice cores drilled through the thick Greenland ice cap contain a continuous record of climate in polar latitudes back through 750,000 years. Geologists are racing to sample the record of past climate recorded in glaciers on mountain peaks in temperate and equatorial regions before the glaciers melt away entirely.

Magnetostratigraphy
As we discussed in Chapter 3, the polarity of Earth's magnetic field flips every now and then through geological time. Geologists have determined when the reversals took place and have constructed a reference column showing the succession of reversals through time (**Fig. 12.22**). By comparing the pattern of the reversals in a sequence of strata with the pattern of reversals in a reference column, a study known as *magnetostratigraphy*, geologists can determine the age of the sequence.

Fission-Track Dating
In certain minerals, the ejection of an atomic particle during the decay of a radioactive isotope damages the nearby crystal, creating a line called a **fission track**. This track resembles the line of crushed grass left behind when you ride a bike across a lawn. As time passes, more atoms undergo fission, so the number of fission tracks in

FIGURE 12.21 Using layering and rings as a basis for dating.

(a) Dust settling during the summer highlights boundaries between snow layers, now turned to ice in this Oregon glacier.

(b) The ridges in this giant clamshell from the Red Sea form during successive seasons of growth.

(c) Each ring in this slice of wood represents the growth of one year. The width of each ring depends on the temperature and rainfall during growth. Patterns of rings are like bar codes—the pattern in a living tree can be correlated with the pattern in a dead tree, which can be correlated with the pattern in an ancient, buried tree. Such correlation extends the tree-ring record back in time.

the crystal increases (**Fig. 12.23**). Therefore, the number of fission tracks in a given volume of a crystal represents the age of the crystal. Geologists have been able to measure the rate at which fission tracks are produced and thus can determine the age of a mineral grain by counting the fission tracks within it. The closure temperature for fission tracks is fairly low, so fission-track dating is used primarily to help constrain rates of exhumation.

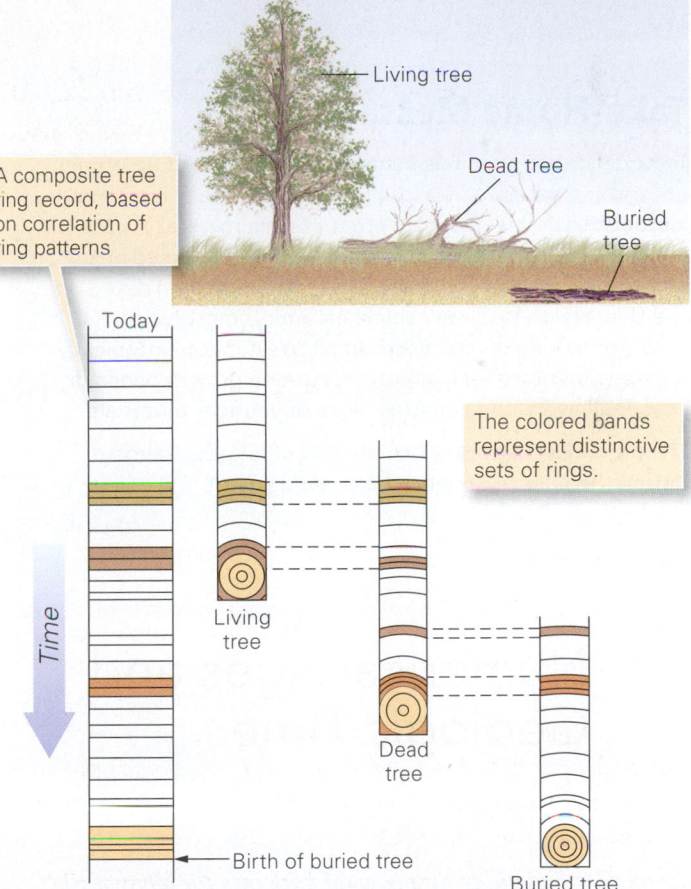

A composite tree ring record, based on correlation of ring patterns

Living tree

Dead tree

Buried tree

The colored bands represent distinctive sets of rings.

Today

Time

Living tree

Dead tree

Birth of buried tree

Buried tree

(d) Dendrochronology is based on the correlation of tree rings. Each of the columns in the diagram represents a core drilled out of a tree. Distinctive clusters of closely spaced rings indicate dry seasons. By correlation, researchers extend the climate record back in time before the oldest living tree started to grow.

FIGURE 12.22 Magnetostratigraphy involves comparing the sequence of polarity reversals in strata with the sequence of polarity reversals in a global reference column to determine the age of the strata.

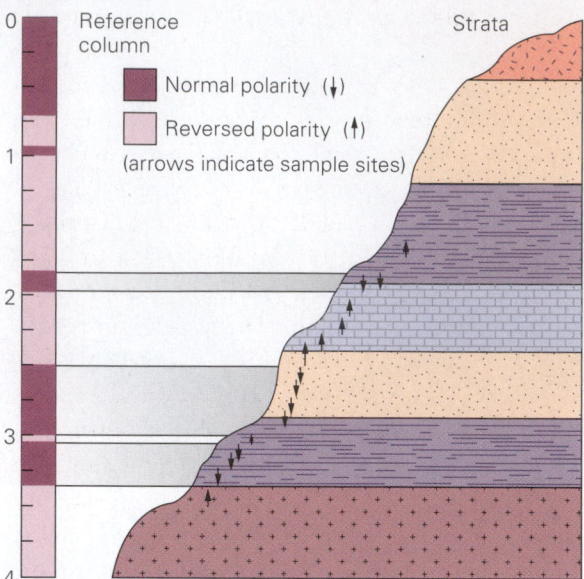

FIGURE 12.23 The method of fission-track dating.

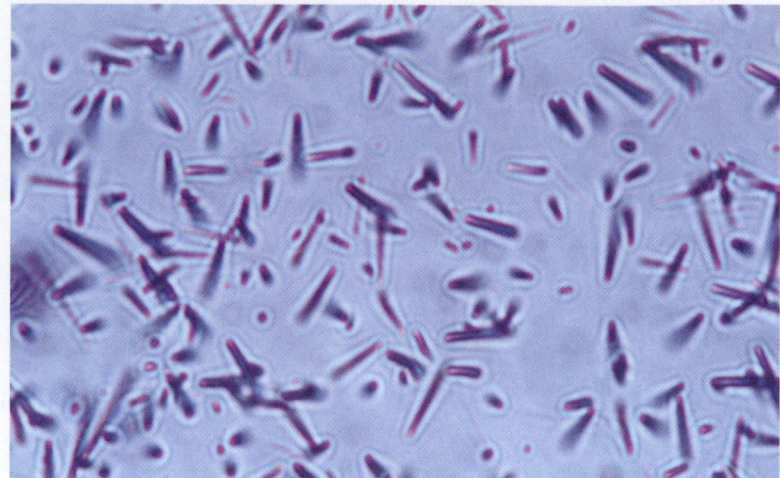

(a) This high-magnification photomicrograph shows fission tracks in a crystal. When formed, each track is only a few atoms wide. Treating the sample with acid enlarges the tracks so they are visible.

Time

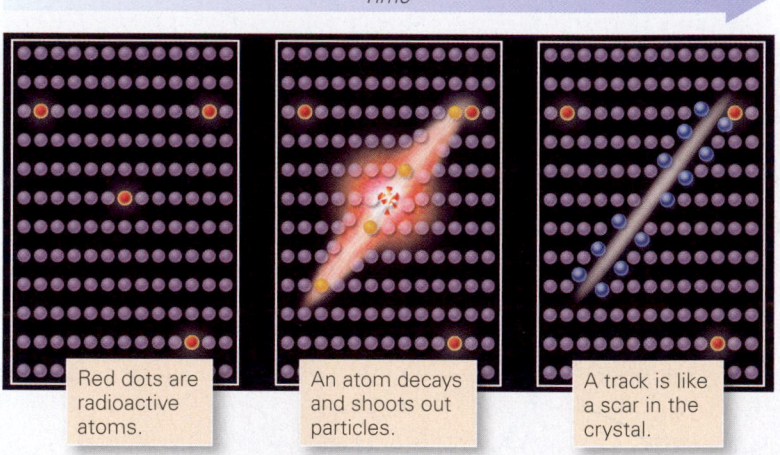

Red dots are radioactive atoms.

An atom decays and shoots out particles.

A track is like a scar in the crystal.

(b) A fission track forms when a radioactive atom decays and blasts out particles that disrupt the crystal structure in a narrow band.

Take-Home Message

Isotopic dating specifies numerical ages in years. To obtain an isotopic date, we measure the ratio of parent radioactive isotopes to stable daughter products in a mineral. A date for a mineral gives the time at which the mineral cooled below a closure temperature. Using this method, we can date the time an igneous rock solidifies, a metamorphic rock cools, or a body of rock is exhumed. We cannot isotopically date sedimentary rock directly. Counting growth bands or seasonal layers can constrain ages of younger materials.

QUICK QUESTION: How can we date changes in climate during the past few hundred thousand years?

12.8 Numerical Ages and Geologic Time

Dating Sedimentary Rocks

The mind grows giddy gazing so far back into the abyss of time.
—John Playfair (British geologist; 1747–1819)

We have seen that isotopic dating can be used to date the time when igneous rocks formed and when rocks metamorphosed

but not when sedimentary rocks were deposited. So how do we determine the numerical age of a sedimentary rock? We must answer this question if we want to add numerical ages to the geologic column—remember, the column was originally constructed by studying only the relative ages of fossil-bearing sedimentary rocks and did not specify dates.

Geologists obtain dates for sedimentary rocks by studying cross-cutting relationships between sedimentary rocks and datable igneous or metamorphic rocks. For example, if we find a sequence of sedimentary strata deposited unconformably on a datable granite, the strata must be younger than the granite. If a datable basalt dike cuts the strata, the strata must be older than the dike. And if a datable volcanic ash buried the strata, then the strata must be older than the ash.

Let's see how cross-cutting relations can help us constrain the numerical age of an imaginary bed of sandstone that contains fossilized dinosaur bones (**Fig. 12.24**). Geologists assign the beds to the Cretaceous Period by correlating the bones with fossils in known Cretaceous strata from elsewhere. Field study shows that the sandstone was deposited unconformably over an eroded pluton consisting of granite that has a numerical age, determined by uranium-lead (U-Pb) dating, of 125 Ma. Further study shows that a basalt dike whose numerical age, based on potassium-argon (K-Ar) dating, is 80 Ma, cuts across the bed. These measurements mean that this Cretaceous sandstone bed was deposited sometime between 125 Ma and 80 Ma. Note that the data provide an age range, not an exact age. Thus, from this observation we can only conclude that the Cretaceous Period includes, but is not limited to, the time interval of 125 to 80 Ma.

The Geologic Time Scale

Geologists have searched the world for localities where they can recognize cross-cutting relations between datable igneous rocks and sedimentary rocks or for layers of datable volcanic rocks interbedded with sedimentary rocks. By isotopically dating the igneous rocks, they have provided numerical ages for the boundaries between all geologic periods. For example, a compilation of many studies from around the world has led to

FIGURE 12.24 The Cretaceous sandstone bed was deposited on the granite, so it must be younger than 125 Ma. The dike cuts the bed, so the bed must be older than 80 Ma. Thus, the Cretaceous bed was deposited between 125 and 80 Ma. The Paleocene sandstone was unconformably deposited over the dike and lies beneath a 50-million-year-old layer of ash. Therefore, it must have been deposited between 80 and 50 Ma.

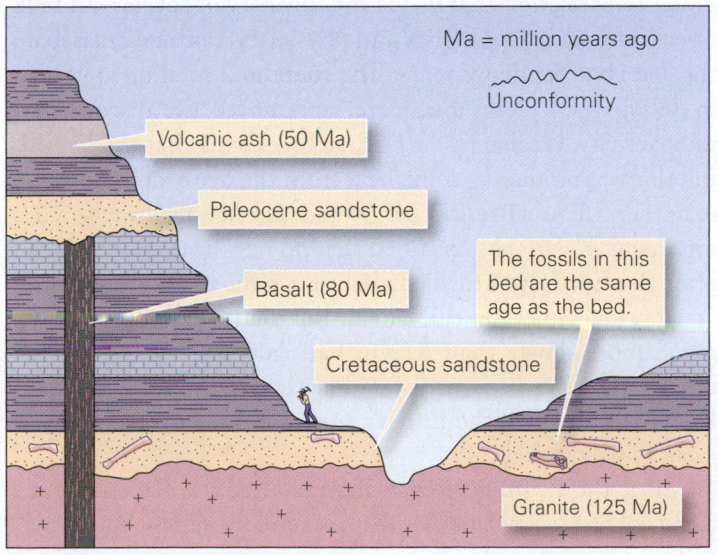

Ma = million years ago

Unconformity

Volcanic ash (50 Ma)

Paleocene sandstone

Basalt (80 Ma)

The fossils in this bed are the same age as the bed.

Cretaceous sandstone

Granite (125 Ma)

a refinement of the numerical ages assigned to the Cretaceous Period—currently, the beginning of the Cretaceous has been placed at 145 Ma and the end at 66 Ma. (With this information in hand, we can say that the bed in Figure 12.24 was deposited during the middle part of the Cretaceous, not at the beginning or end.)

Because numerical ages depend on the documentation of datable cross-cutting relations, discovery of new data can require changes in the numerical ages assigned to period boundaries. (That's why we prefer the term *numerical age* to *absolute age*.) For example, around 1995, new dates on rhyolite ash layers above and below the Cambrian/Precambrian boundary showed that this boundary occurred closer to an age of 542 Ma—previous, less-definitive studies had placed the boundary at 570 Ma. More recently, the boundary has been moved to 541 Ma. Also, because repeated measurements of the same sample might yield slightly different results, a given numerical age is not perfectly "precise." To be technically accurate, a reported age has an uncertainty (plus or minus) value attached, which indicates the precision of the measurement. For example, researchers would report the start of the Cambrian as 541 ± 1 Ma. (We are leaving off precision specifications in this book for the sake of simplicity.) Note that the terms *precision* and *accuracy* have different meanings in scientific discussion. *Precision* refers to the uncertainty of a result that reflects the repeatability of a measurement. *Accuracy* refers to the closeness of the result to the real or true value.

The chart in **Figure 12.25** gives the numerical ages of periods and eras in the geologic column, as reported by the International Commission on Stratigraphy in 2013. We refer to such a chart, on which intervals of the geologic column have been assigned numerical ages, as the **geologic time scale**. Because of the numerical constraints provided by the geologic time scale, when we say that the first dinosaurs appeared during the Triassic Period, we are implying that dinosaurs appeared after 252 Ma. (The oldest dinosaur fossil known actually comes from strata that are 240 to 245 Ma.)

What Is the Age of the Earth?

During the 18th and 19th centuries, before the discovery of isotopic dating, scientists came up with a great variety of clever solutions to the question, "How old is the Earth?" All of these have since been proven wrong. Lord William Kelvin, a 19th-century physicist renowned for his discoveries in thermodynamics, made the most influential scientific estimate of the Earth's age of his time. Kelvin calculated how long it would take for the Earth to cool down from a temperature as hot as the Sun's and concluded (in 1862) that our planet is about 20 million years old.

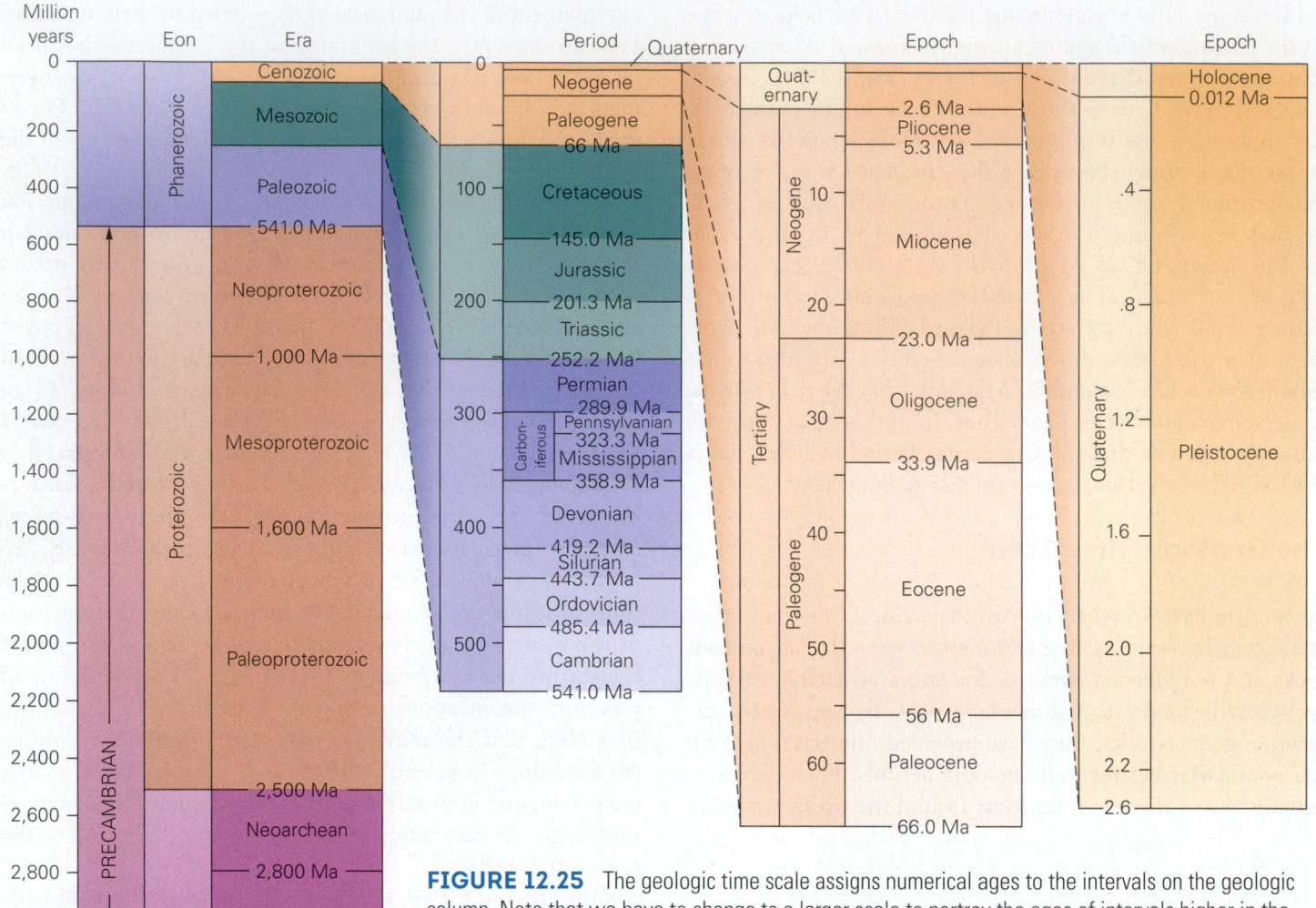

FIGURE 12.25 The geologic time scale assigns numerical ages to the intervals on the geologic column. Note that we have to change to a larger scale to portray the ages of intervals higher in the column, because these are shorter subdivisions. This time scale utilizes numbers favored by the International Commission on Stratigraphy.

that shape the Earth and form its rocks, as well as the process of natural selection that yields the diversity of species, all take a very long time. Geologists and physicists continued to debate the age issue for many years. The route to a solution appeared in 1896, when Henri Becquerel announced the discovery of radioactivity. Geologists immediately realized that the Earth's interior was producing some heat from the decay of radioactive material. This realization uncovered the key flaw in Kelvin's argument: Kelvin had assumed that no new heat was produced after the Earth first formed. Because radioactivity constantly generates new heat in the Earth, the planet has cooled down much more slowly than Kelvin had calculated and could be much older. The discovery of radioactivity not only invalidated Kelvin's estimate of the Earth's age, it also led to the development of isotopic dating.

Since the 1950s, geologists have scoured the planet to identify its oldest rocks. Samples from several localities (Wyoming,

Kelvin's estimate contrasted with those being promoted by followers of Hutton, Lyell, and Darwin, who argued that if the concepts of uniformitarianism and evolution were correct, the Earth must be much older. They held that physical processes

Canada, Greenland, and China) have yielded dates as old as 4.03 Ga (**Fig. 12.26**). Individual clastic grains of the mineral zircon have yielded dates of up to 4.4 Ga, indicating that by 4.4 Ga solid crust existed, at least for a while. Isotopic dating of Moon rocks yields dates of up to 4.50 Ga, and dates on the meteorites thought to reflect the most primitive solids of the Solar System have yielded ages as old as 4.57 Ga. Geologists consider 4.57 Ga meteorites to be fragments of very early, undifferentiated planetesimals. Meteorites from the oldest differentiated planetesimals (ones that had separated into a core and mantle) are slightly younger, and these ages are taken to be the same as the age of the Earth itself. Based on these ages, therefore, researchers estimate that the Earth itself formed at 4.54 Ga.

> **Did you ever wonder . . .**
>
> how old the Earth's oldest rock is?

Why don't we find whole rocks with ages between 4.03 and 4.54 Ga in the Earth's crust? Geologists have come up with several ideas to explain the lack of extremely old rocks. One idea comes from calculations defining how the temperature of this planet's interior has changed over time. These calculations indicate that during the first half-billion years of its existence, the Earth might have been so hot that rocks in the crust remained above the closure temperature for minerals, and isotopic clocks could not start "ticking." More recent studies, looking at isotope ratios in the oldest (4.4 Ga)

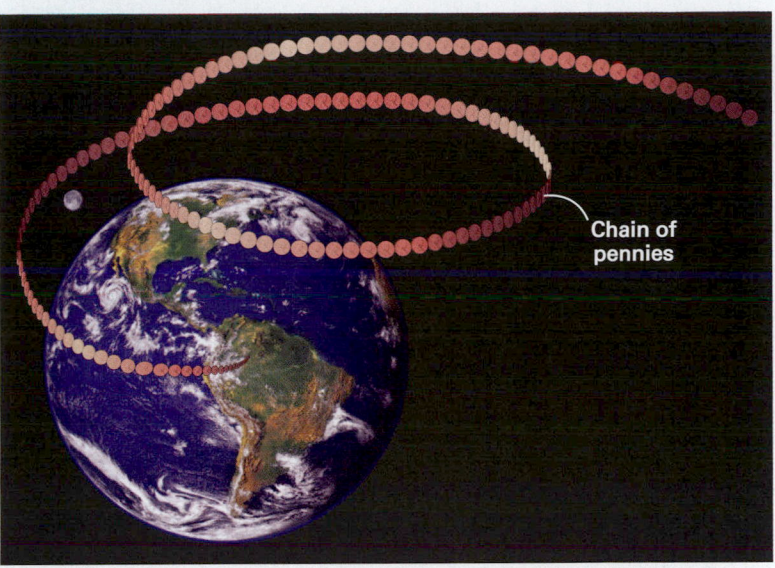

FIGURE 12.27 We can use the analogy of distance to represent the duration of geologic time.

Chain of pennies

zircons, suggest that the Earth had cooled sufficiently to host oceans of water within only a couple of hundred million years of its formation. So an alternative view is that intense bombardment of the Earth by meteorites just prior to 4.03 Ga destroyed or remelted any crust that existed and vaporized the earliest oceans. As noted earlier, geologists have named the time interval between the birth of the Earth and the origin of the oldest isotopically dated rock as the Hadean Eon, to emphasize that conditions at the surface, at times, resembled literary images of Hades.

Picturing Geologic Time

The number 4.57 billion is so staggeringly large that we can't really begin to comprehend it. If you lined up this many pennies in a row, they would make an 87,400-km-long line that would wrap around the Earth's equator more than twice (**Fig. 12.27**). Notably, at the scale of our penny chain, human history is only about 100 city blocks long.

Another way to grasp the immensity of geologic time is to equate the entire 4.57 billion years to a single calendar year. On this scale, the oldest rocks preserved on Earth date from early February, the first archaea appear in the sea on February 21, the first shelly invertebrates burrowed through the mud on October 25, and the first amphibians crawled onto the land about November 20. On December 7, the continents coalesce into the supercontinent of Pangaea. The first mammals and birds appear about December 15, along with the dinosaurs, and the Age of Dinosaurs ends on December 25. The last week of December

FIGURE 12.26 An outcrop of the Acasta Gneiss, in Canada, showing rocks of different ages. The oldest have yielded isotopic ages of 4.03 Ga.

represents the last 66 million years of Earth history, including the entire Age of Mammals. The first human-like ancestor appears on December 31 at 3 P.M., and our species, *Homo sapiens*, shows up an hour before midnight. The last ice age ends a minute before midnight, and all of recorded human history takes place in the last 30 seconds. To put it another way, human history occupies the last 0.0001% of Earth history. The Earth is so old that there has been more than enough time for the rocks, mountains, and life forms of Earth to have formed and evolved.

Take-Home Message

Numerical dates for sedimentary rocks come from isotopic dating of cross-cutting datable rocks. Such work led to the geologic time scale, which assigns dates to periods. The oldest rock of Earth's crust is about 4.0 Ga. Dating of meteorites indicates the Earth is 4.54 Ga.

QUICK QUESTION: Why don't all periods on the geologic time column have the same duration in years?

CHAPTER SUMMARY

- Geologic time refers to the time span since the Earth's formation.

- Relative age specifies whether one geologic feature is older or younger than another; numerical age provides the age of a geologic feature in years.

- Using such principles as uniformitarianism, original horizontality, superposition, and cross-cutting relations, we can construct the geologic history of a region.

- The principle of fossil succession states that the assemblage of fossils in strata changes from base to top of a sequence. Once a species becomes extinct, it never reappears.

- Strata are not necessarily deposited continuously at a location. An interval of nondeposition and/or erosion is called an unconformity. Geologists recognize three kinds: angular unconformity, nonconformity, and disconformity.

- A stratigraphic column shows the succession of strata in a region. The process of determining the relationship between strata at one location and strata at another is called correlation.

- A given succession of strata that can be traced over a fairly broad region is called a stratigraphic formation. A geologic map shows the distribution of formations.

- A composite chart that represents the entirety of geologic time is the geologic column. The column's largest subdivisions, each of which represents a specific interval of time, are eons. Eons are subdivided into eras, eras into periods, and periods into epochs.

- The numerical age of rocks can be determined by isotopic (radiometric) dating. This is because radioactive elements decay at a rate characterized by a known half-life.

- The isotopic date of a mineral specifies the time at which the mineral cooled below a closure temperature. We can use isotopic dating to determine when an igneous rock solidified and when a metamorphic rock cooled from high temperatures. To date sedimentary strata, we must examine cross-cutting relations among dated igneous or metamorphic rock and the strata.

- Other methods for dating materials include counting growth rings in trees and seasonal layers in glaciers.

- From the isotopic dating of meteors and Moon rocks, geologists conclude that the Earth formed about 4.54 billion years ago. Our species, *Homo sapiens*, has been around for only a tiny fraction of geologic time.

GUIDE TERMS

angular unconformity
 (p. 443)
Cambrian explosion (p. 450)
closure temperature (p. 457)
disconformity (p. 444)
eon (p. 450)
epoch (p. 450)
era (p. 450)
fission track (p. 458)
fossil (p. 436)

fossil assemblage (p. 441)
geochronology (p. 453)
geologic column (p. 449)
geologic contact (p. 445)
geologic map (p. 446)
geologic time (p. 435)
geologic time scale (p. 461)
growth bands (p. 458)
half-life (p. 453)
index fossil (p. 441)

isotope (p. 453)
isotopic dating (p. 453)
marker bed (p. 446)
nonconformity (p. 443)
numerical age (p. 437)
period (p. 450)
Precambrian (p. 450)
radioactive element (p. 453)
relative age (p. 437)
stratigraphic column (p. 445)

stratigraphic correlation
 (p. 446)
stratigraphic formation
 (p. 445)
stratigraphic group (p. 446)
tree rings (p. 458)
unconformity (p. 443)
uniformitarianism (p. 437)

REVIEW QUESTIONS

1. Contrast numerical age with relative age.
2. Describe the principles that allow us to determine the relative ages of geologic events.
3. How does the principle of fossil succession allow determination of relative ages?
4. How does an unconformity develop? Describe the three kinds of unconformities.
5. Describe two different methods of correlating rock units. How was correlation used to develop the geologic column? What is a stratigraphic formation?
6. What does the process of radioactive decay entail?
7. How do geologists obtain an isotopic date? What does the age of an igneous rock mean? What does the age of a metamorphic rock mean?
8. Why can't we date sedimentary rocks directly? How do we assign numerical ages to intervals on the geologic column, to produce a geologic time scale?
9. How are growth rings and ice layering useful in determining the ages of geologic events?
10. What is the age of the oldest rocks on Earth? What is the current estimate of the numerical age of the Earth? Why is there a difference?

ON FURTHER THOUGHT

11. Imagine an outcrop exposing a succession of alternating sandstone and conglomerate beds. A geologist studying the outcrop notes the following.

   • The sandstone beds contain fragments of land plants, but the fragments are too small to permit identification of species.
   • A layer of volcanic ash overlies the sandstone bed. Isotopic dating indicates that this ash is 300 Ma.
   • A paleosol occurs at the base of the ash layer.

   • A basalt dike, dated at 100 Ma, cuts both the ash and the sandstone-conglomerate sequence.
   • Pebbles of granite in the conglomerate yield radiometric dates of 400 Ma.

   On the basis of these observations, how old is the sandstone and conglomerate? (Specify both the numerical age range and the period or periods of the geologic column during which it formed.) If the igneous rocks were not present, could you still specify the maximum or minimum ages of the sedimentary beds? Explain.

12. Examine the photograph and the "What a Geologist Sees" interpretation of an outcrop in eastern New York state below. Write a brief geologic history that explains the relationships displayed in this outcrop. The strata directly above the unconformity are Late Silurian (Rondout Formation), whereas the strata below the unconformity are Middle Ordovician (Austin Glen Formation).

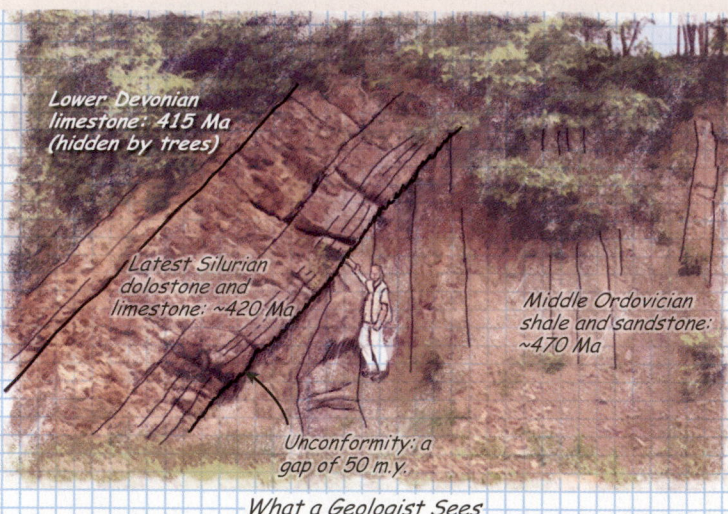

Lower Devonian limestone: 415 Ma (hidden by trees)

Latest Silurian dolostone and limestone: ~420 Ma

Middle Ordovician shale and sandstone: ~470 Ma

Unconformity: a gap of 50 m.y.

What a Geologist Sees

13. Examine the photograph and the "What a Geologist Sees" interpretation of an outcrop in Missouri below. Note that the strata thin close to the unconformity. Write a brief geologic history that explains the relationships displayed in this outcrop. The strata above the unconformity are of the Cambrian-age Lamotte Sandstone.

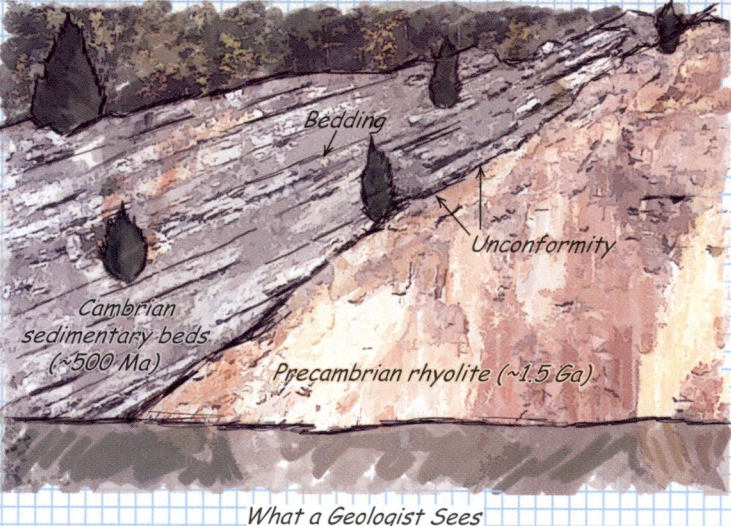

Bedding

Unconformity

Cambrian sedimentary beds (~500 Ma)

Precambrian rhyolite (~1.5 Ga)

What a Geologist Sees

smartwork

smartwork.wwnorton.com

This chapter's Smartwork features:

- Art-based problems on unconformities.
- What A Geologist Sees exercises on strata of the Grand Canyon.
- Geologic time labeling activity.

GEOTOURS

This chapter's GeoTour exercise (J) features:

- Relative age dating and unconformities
- Stratigraphic formations in southern Utah
- Rock layers and monoclines, Circle Cliffs, Utah

The presence of these fossiliferous limestone beds in Illinois tell us that the middle of North America was once covered by a shallow sea. The dip of the strata tells us that long after the sediment turned to rock, tectonics deformed this part of the continent. Much later, uplift and erosion exposed the rock. Clearly, the Earth has a history!

CHAPTER 13

A Biography of Earth

> [T]he man who should know the true history of the bit of chalk which every carpenter carries about in his breeches pocket, though ignorant of all other history, is likely, if he will think his knowledge out to its ultimate results, to have a truer and therefore a better conception of this wonderful universe and of man's relation to it than the most learned student who [has] deep-read the records of humanity [but is] ignorant of those of nature.
>
> —Thomas Henry Huxley, from *On a Piece of Chalk* (1868)

LEARNING OBJECTIVES

By the end of this chapter, you should understand . . .

- how many geologic clues indicate that Earth changes over time, so the Earth has a history.

- that the earliest crust and oceans may date to 4.4 Ga but were destroyed by bombardment.

- that oceans and life have existed continuously since at least 3.8 Ga, but that the atmosphere began to accumulate oxygen at only about 2.5 Ga.

- when supercontinents formed and then rifted apart during Earth history.

- that the fossil record indicates that early life consisted of archaea and bacteria; after oxygen levels increased, complex multicellular life became possible, and after shells evolved, organisms diversified.

- that due to sea-level rise and fall, continental interiors sometimes hosted shallow seas.

- when and why mountain belts formed in the past.

13.1 Introduction

In 1868, a well-known British scientist, Thomas Henry Huxley, presented a public lecture on geology to an audience in Norwich, England. Seeking a way to convey his fascination with Earth history to people with no geologic background, he focused his audience's attention on the piece of chalk he had been writing with (see the epigraph above). And what a tale the chalk has to tell! Chalk, a type of limestone, consists of microscopic marine algae shells and shrimp feces. The specific chalk that Huxley held came from beds deposited in Cretaceous time (the name Cretaceous, in fact, derives from the Latin word for chalk). These beds now form the white cliffs bordering the shore of England (**Fig. 13.1**). Geologists in Huxley's day knew of similar chalk beds in outcrops throughout much of Europe and had discovered that the chalk contains not only plankton shells but also fossils of bizarre swimming reptiles, fish, and invertebrates—species absent in the seas of today. Clearly, when the chalk was deposited, warm seas holding unfamiliar creatures covered some of what is dry land today.

Clues in his humble piece of chalk allowed Huxley to demonstrate to his audience that the landscape features and living inhabitants of the Earth in the past differ markedly from those today and thus that the Earth has a history. Geologic research of the past few centuries has led to the conclusion that physical and biological components of the Earth System have interacted pervasively during this long history, in ways that have transformed a formerly barren, crater-pocked surface into countless environments and landscapes supporting a diversity of life.

In this chapter, we offer a concise geologic biography of our planet, from its birth 4.54 billion years ago to the present. We illustrate how continents came into existence and have waltzed across the globe ever since. We also describe mountain-building events, changes in Earth's climate and sea level through time, and the evolution of life. To simplify the discussion, remember that we use the following abbreviations: Ga for billion years ago, Ma for million years ago, and Ka for thousand years ago.

13.2 Methods for Studying the Past

When historians outline human history, they describe daily life, wars, economics, governments, leaders, inventions, and explorations. When geologists outline Earth history, they describe the changes of depositional environments, mountain-building events (orogenies), past climates, the rise and fall in sea level, life evolution, the past configuration of plate boundaries and continents, and changes in the composition of the atmosphere and oceans. Historians collect data by reading written accounts, examining relics and monuments, and for more recent events, listening to recordings or watching videos. Geologists collect data by examining rocks, geologic structures, and fossils and, for more recent events, by studying sediments, ice cores, and tree rings.

Figuring out Earth's past hasn't been an easy task for geologists. The record that we can see isn't complete, because the materials that hold the record of the past don't form continuously through time and many important clues have been eroded away and/or covered by younger rocks. Also, it can be very challenging to obtain and interpret accurate isotopic ages of old rocks. Nevertheless, enough of the record exists to outline major geologic events of the past. Following are a few examples of how geologists use observational data to study Earth history.

- *Identifying ancient orogens:* We identify present-day orogens (mountain belts) by finding regions of high, rugged peaks. However, since it takes as little as 50 million years to erode a mountain range entirely away, we cannot identify orogens of the past simply by studying topography. Rather, we look for the rock record that orogeny leaves behind. Orogeny causes igneous activity, deformation, and metamorphism. Thus, a belt of crust containing these features represents an ancient orogen (**Fig. 13.2**). We can determine the age of an orogeny by isotopically dating the metamorphic and igneous rocks that crop out in the orogen. Orogeny also leads to the development of unconformities, for uplift exposes rocks to erosion. It also leads to the formation of sedimentary basins in which the

detritus, produced by erosion of mountains, accumulates. These basins, which border the orogen, develop because the weight of the orogen pushes the lithosphere's surface down, forming a depression that traps sediment.

- *Recognizing the growth of continents:* Not all continental crust formed at the same time. To determine how a continent grew, geologists find the ages of different regions of the crust by using isotopic dating techniques. They can figure out not only when rocks originally formed from magmas rising out of the mantle but also when the rocks were metamorphosed during a subsequent orogeny. The identities of the rock types making up the crust indicate the tectonic environment in which the crust formed.
- *Recognizing past depositional environments:* The environment at a particular location changes through time. To learn about these changes, we study successions of sedimentary rocks, for depositional environment controls both the type of sediment accumulating at a location and the type of organisms that live there.
- *Recognizing past changes in relative sea level:* We can determine when sea level has gone up or down by looking for changes in the depositional environment. For example, a marine limestone above an alluvial-fan conglomerate indicates a rise in sea level.
- *Recognizing positions of continents in the past:* To help us find out where a continent was located in the past, we have three sources of information. First, apparent polar-wander paths give us a sense of continental movement (see Chapter 3; **Fig. 13.3a**). Second, marine magnetic anomalies

FIGURE 13.1 Horizontal chalk beds exposed along the coast of southern England, formed from layers of deep-sea sediment. The thin, dark beds, between white chalk layers, consist of chert. The chalk erodes easily, so the pebbles on the beach all consist of chert.

Bedding

Person

FIGURE 13.2 Evidence of mountain building in the past. Even after the topography of a mountain belt has eroded away, a record of mountain building remains; deformed and metamorphosed rock defines a distinct belt.

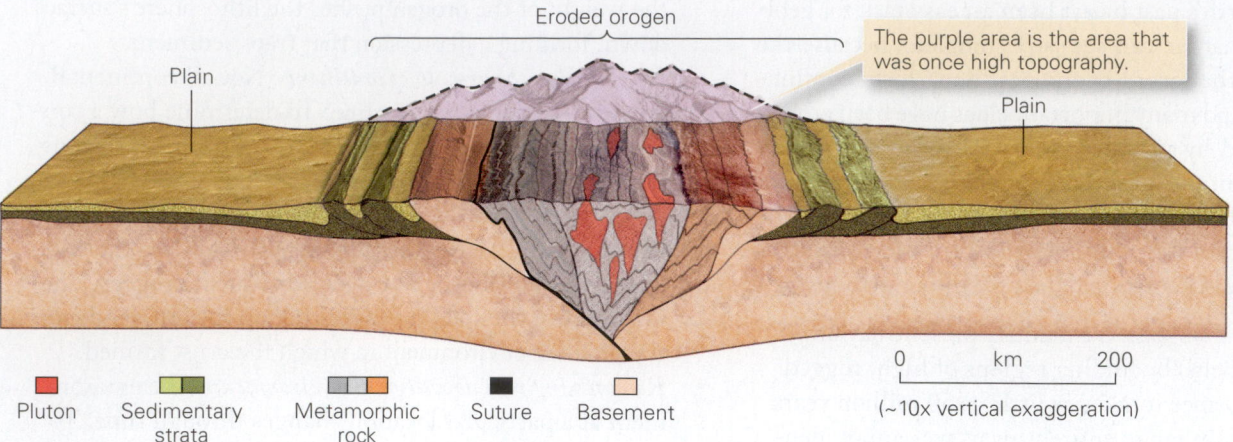

Eroded orogen

Plain

The purple area is the area that was once high topography.

Plain

■ Pluton   ■ Sedimentary strata   ■ Metamorphic rock   ■ Suture   ■ Basement

0   km   200

(~10x vertical exaggeration)

define changes in the ocean basin width between continents over time (**Fig. 13.3b**). Third, comparison of rocks and/or fossils from different continents permits correlations that indicate whether continents were adjacent.

- *Recognizing past climates:* We can gain insight into past climates by looking at fossils and rock types that formed at given latitudes. For example, if organisms requiring semitropical conditions lived near the poles during a given time period, then the atmosphere overall must have been warmer. Geologists have also learned how to use the ratios of certain isotopes in fossil shells as an indication of past temperatures.

- *Recognizing life evolution:* Progressive changes in the assemblage of fossils in a sequence of strata represent changes in the assemblage of organisms inhabiting Earth through time and thus characterize life's evolution.

As you read the following sections describing specific events in Earth's history, try to think about how geologists used the tools that we've just described to identify and characterize the event.

Take-Home Message

We can reconstruct geologic history from a variety of clues. Outcrops of deformed and metamorphosed rocks indicate where mountain belts once existed, the character and distribution of sedimentary strata provide clues to the timing of sea-level rise and fall and to the nature of past environments, study of paleomagnetism and correlation of rock units allows researchers to constrain the past positions of continents relative to one another, and the fossil record provides a window into the evolution of life.

QUICK QUESTION: How can you determine the numerical age of an ancient orogen?

13.3 The Hadean and Before

Formation of the Earth, Revisited

The current scientific interpretation of the Solar System's formation places the beginning of the process at about 4.570 Ga, when a supernova explosion sent shock waves, as well as atoms of heavier elements, into a nebula within our region of the Milky Way Galaxy. (Note: We're adding an extra decimal to the ages in our discussion here, so as to distinguish among events that happened fairly soon after one another. The ages given have a precision of about ± 0.003 Ga, but their accuracy remains a subject of research.) Perhaps stimulated by the shock waves, this nebula collapsed rapidly to form the Sun, which ignited around 4.567 Ga.

According to the nebular theory of Solar System formation that we discussed in Chapter 1, during the next million years a protoplanetary disc formed, and dust and ice condensed in the disc. Based on the isotopic dating of meteorites, thought to be remnants of the earliest planetesimals, the planetesimals soon began to grow and between 4.560 and 4.540, several had grown into sizable protoplanets by sweeping in material from their orbit. Once they got large enough and hot enough, protoplanets underwent **differentiation**, during which iron within the body began to melt and gravity pulled the iron down to the center of the planet where it accumulated to form the core. Core formation left behind a mantle composed of ultramafic rock. (Recall that the heat leading to differentiation came from the compression of matter into a dense ball, the transfer of kinetic energy of colliding

FIGURE 13.3 Paleomagnetic tools that geologists used to determine the past positions of the continents.

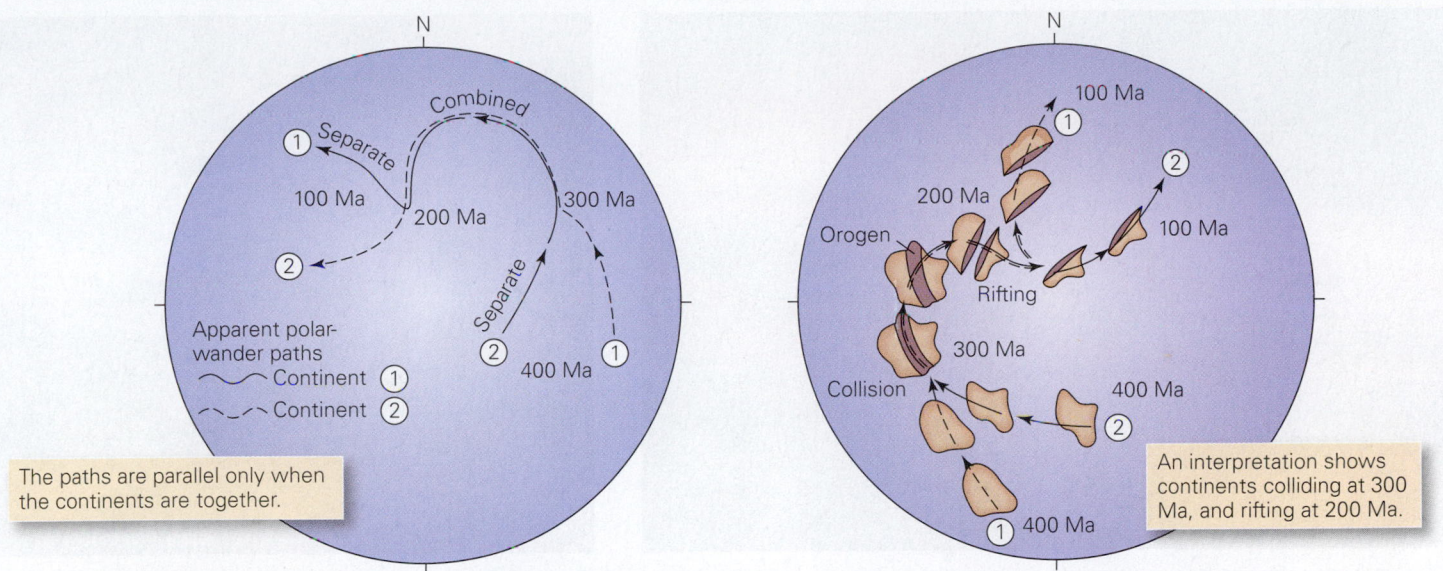

(a) Comparison of the apparent polar-wander paths for two continents indicates when they moved together and when they moved separately.

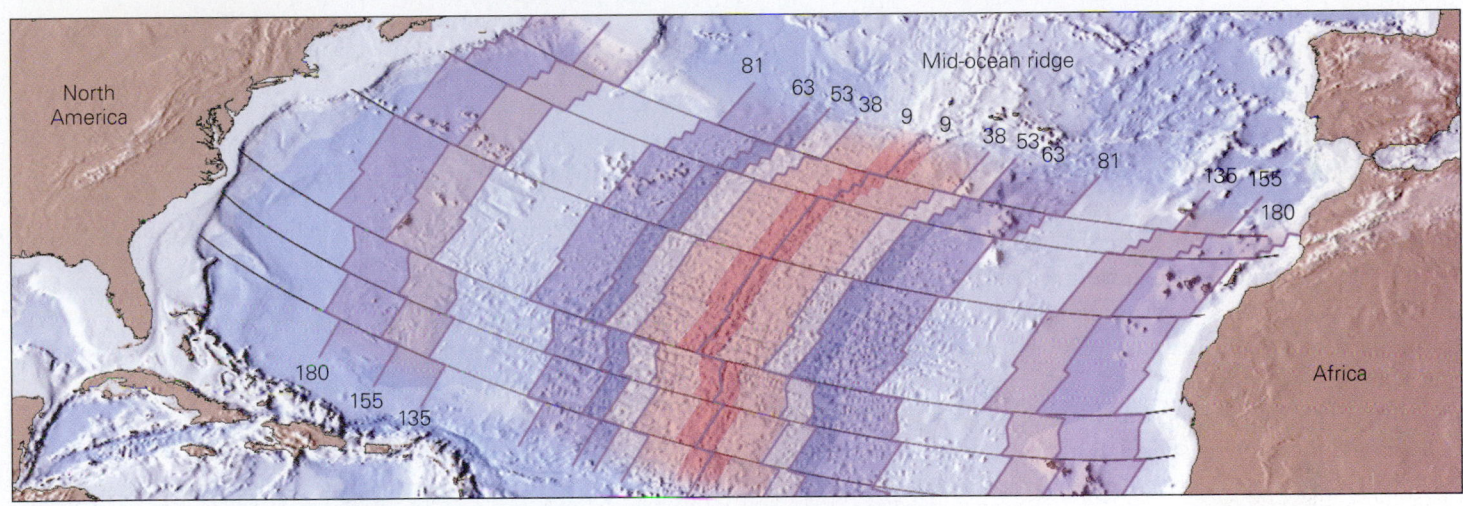

(b) Marine magnetic anomalies indicate how the distance between continents separated by a mid-ocean ridge changes over time. On this map, the age of anomaly boundaries is indicated in Ma.

objects into thermal energy, and the production of heat by decay of relatively abundant radioactive elements.) By about 4.540 Ga, the largest protoplanets had differentiated, as indicated by isotopic ages of meteorites thought to be remnants of differentiated planetesimals, and had swept their orbits clear of debris, thus attaining the status of "planet" (see Chapter 1). In the context of this scenario, geologists use 4.540 Ga as the birth date of the Earth.

Events of the Hadean

The newborn Earth started to cool and stabilize, but it didn't remain unscathed for long. At about 4.533, it cataclysmically collided with a protoplanet named *Theia*. This event blasted much of the Earth's mantle into orbit and may have added new material from the colliding object into the Earth's remaining mantle and core. The orbiting debris quickly coalesced into the Moon, which, at the time of its formation, was only 20,000 km from the Earth—by comparison, the Moon is 384,000 km from Earth today, and it continues to move farther away at about 3.8 cm per year.

In the wake of Moon formation, the Earth was so hot that much of its surface was probably an ocean of seething magma (**Fig. 13.4a**). But temperature rapidly diminished as heat radiated into space and the supply of new heat from radioactive decay diminished (because elements with short half-lives had

FIGURE 13.4 Visualizing the early Earth is a challenge, but by using geologic interpretations, artists have provided useful images.

(a) At a very early stage, during or after differentiation, the surface might have been largely molten. The loss of heat to space would allow patches of solid ultramafic solid crust to form. Meteorite impacts could have then destroyed the crust.

(b) When Earth's surface fell below the boiling point of water, water from the atmosphere rained onto the surface, submerging it with early oceans.

decayed). Rafts of solid rock formed on the surface of the magma ocean, but most of these eventually sank and remelted to be recycled into new rocks. In addition, rapid outgassing took place, meaning that volatile (gassy) elements or compounds originally incorporated in mantle minerals were released and erupted at the Earth's surface, along with lava. These gases accumulated to constitute a toxic atmosphere consisting mostly of water (H_2O), methane (CH_4), ammonia (NH_3), hydrogen (H_2), nitrogen (N_2), carbon dioxide (CO_2), and sulfur dioxide (SO_2). Some researchers speculate that gases from comets colliding with Earth may have contributed additional gases to the early atmosphere.

By about 4.4 Ga, the Earth might have become cool enough for a solid crust and even liquid water to exist at its surface (**Fig. 13.4b**). The evidence for this statement comes from grains of a durable mineral called zircon, which has been extracted from much younger sandstone beds exposed in Western Australia. The zircons, which yield isotopic ages of 4.4 Ga, formed originally in an igneous rock, indicating the Earth was cool enough for rock to solidify. Further, isotopic ratios of oxygen in the zircon indicate that it possibly was altered by interaction with surface water, hinting at the presence of early oceans. What did the Earth's surface look like at this time? An observer from space probably would have found small, barren landmasses spotted with volcanoes poking up above an acidic sea. But both land and sea would have been obscured by murky, dense (H_2O-, CO_2-, and SO_2-rich) air.

Take-Home Message

The Earth formed not long after the birth of the Sun. It differentiated and swept its orbit clear of debris by about 4.54 Ga, a date taken as the birth of the Earth. Very little record remains of the first half-billion years of Earth history. During the Hadean (4.57–3.85 Ga), the Earth differentiated and the Moon formed. A rock record of this eon doesn't exist because our planet's surface may have been partially molten and rapidly recycled by remelting.

QUICK QUESTION: When might the earliest liquid water on the Earth have accumulated? What's the evidence?

13.4 The Archean Eon: Birth of Continents and Life

The Transition from Planet Formation to Planet Evolution

Though mineral grains as old as 4.4 Ga exist, most rocks are younger than 3.85 Ga. What destroyed most, if not all, of the pre-3.85-Ga rock (and any oceans, if they existed) of the Earth? The answer may come from studies of cratering on the Moon. These studies suggest that the Moon—and, therefore, all inner planets of the Solar System—underwent a period of intense meteorite bombardment, called the *late heavy bombardment*, between about 4.0 and 3.85 Ga. Researchers speculate that this event would have pulverized and/or melted almost all crust that had existed on Earth at the time and would have destroyed the existing atmosphere and ocean. Only after the bombardment ceased could longer-lasting crust, atmosphere, and oceans begin to form. In addition to bombardment, convective movements involving mantle and crust may have continuously recycled crust—in effect, soon after an area of crust formed, it might have been subducted and remelted.

The discovery of 3.85-Ga marine sedimentary rocks in Greenland suggests that the appearance or reappearance of land, sea, and an atmosphere happened at about 3.85 Ga. The end of the Hadean was once placed at 3.85 Ga, the time at which the late heavy bombardment ceased, and the rock record becomes fairly complete. Now geologists consider the end of the Hadean to be at 4.0 Ga, essentially the age of the oldest-known rock and thus the time at which planet formation processes had ceased and the record of the rocks begins. The next eon of Earth history is the **Archean**, from the Greek words *arkhaios* meaning ancient and *arche* meaning beginning.

Land Appears

With the advent of the Archean, the Earth's crust was cool and stable enough for isotopic clocks to start ticking; the rate of crustal recycling decreased, and marine strata started to be preserved. Thus, the Earth of the Archean clearly had land and sea, a situation that has persisted ever since. Geologists still argue about whether plate tectonics in the form that occurs today operated in the early part of the Archean Eon. Most researchers picture an early Archean Earth with rapidly moving small plates, numerous volcanic island arcs, and abundant hot-spot volcanoes—the rates of plate-tectonic processes may have been faster in the Archean than they are today because the Earth's interior was hotter (due to the availability of more radioactive atoms and to leftover heat from planet formation). Others propose that early Archean lithosphere was too warm and buoyant to subduct and that plate tectonics could not have operated until the later part of the Archean; these authors argue that plume-related volcanism was the main source of new crust until the late Archean. Regardless of which model ultimately proves more correct, it is clear that the Archean was a time during which significant volumes of new continental crust came into existence.

What processes produced continental crust? According to one model, early crust formed from mafic igneous rocks that originally extruded or intruded at convergent plate boundaries and/or at hot-spot volcanoes. When the arcs and oceanic plateaus collided with one another, they sutured together to form larger, relatively buoyant blocks that remained at the Earth's surface.

The development of convergent-plate boundaries along the margins of these blocks, and of rifts and hot-spot volcanoes within the blocks, led to production of flood basalts. Partial melting of basaltic crust, perhaps at or near its base, yielded felsic and intermediate magmas, which rose and solidified in the upper crust or were extruded at the surface. Thus, with time, continents differentiated into a more mafic lower crust and a more felsic upper crust. As collisions continued, the blocks coalesced into still larger *protocontinents* (**Fig. 13.5**), which slowly cooled and became stronger. As a result of these processes, the first long-lived blocks of durable continental crust came into existence between 3.2 and 2.7 Ga, and by the end of the Archean Eon about 80% of the Earth's continental area had formed (**Fig. 13.6**). Since that time, relatively little "juvenile" (newly extracted from the mantle) rock has formed. Most rock that is younger than about 2.7 Ga is "recycled" in the sense that it either has gone through various stages of the rock cycle in the crust (see Interlude C) or it has been carried back into the mantle where it subsequently became incorporated in new magma.

A clear stratigraphic record of marine sediment deposition has been preserved in remnants of Archean crust, indicating that oceans filled in the Archean and have existed ever since. Permanent oceans could survive only after the Earth's surface had cooled below the boiling point of water. Prior to that time, gaseous H_2O saturated the atmosphere—in fact, prior to ocean formation, H_2O and CO_2 were the dominant gases of the atmosphere. Once the oceans formed, however, the atmosphere lost most of its H_2O. And once liquid water existed, substantial amounts of atmospheric CO_2 dissolved into it. Thus, the Archean saw the atmosphere transform from a foggy mixture of H_2O and CO_2 into a transparent gas dominantly composed of N_2 gas. Since N_2 is inert, meaning it doesn't chemically react with or dissolve in other materials, it was left behind and became the major component of the atmosphere.

Archean cratons contain five principal rock types: *gneiss*, relics of Archean metamorphism in collisional zones; *greenstone*, metamorphosed relics of ocean crust trapped between colliding blocks of continental crust, as well of basalts that had filled early continental rifts, basalts produced at hot spots, and basaltic volcanoes of island arcs; *granite*, formed from magmas generated by the partial melting of the crust in continental volcanic arcs or above hot spots; *graywacke*, a mixture of sand and clay eroded from the volcanic areas and dumped into the ocean; and *chert*, formed by the precipitation of silica in the deep sea. Archean shallow-water sediments are rare, either because continents were so small that depositional environments in which such sediments could accumulate didn't exist or because any that were present have eroded away.

Once land areas had formed, rivers flowed over their stark, unvegetated surfaces. Geologists reached this conclusion because sedimentary beds from this time contain clastic grains

FIGURE 13.5 A model for crust formation during the Archean Eon.

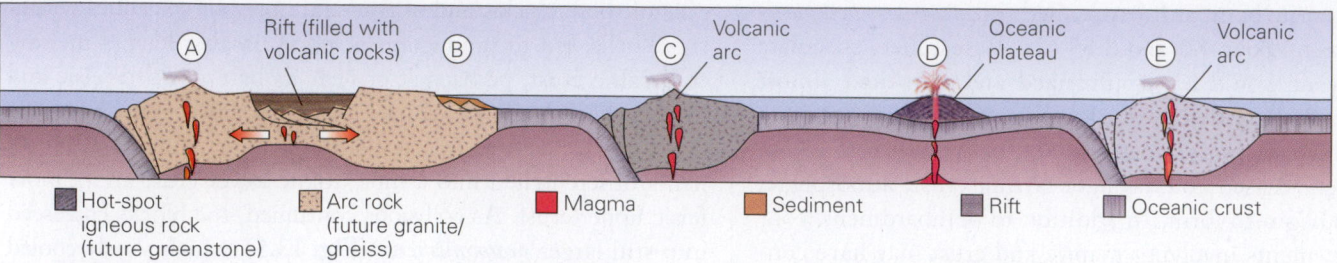

Hot-spot igneous rock (future greenstone) | Arc rock (future granite/ gneiss) | Magma | Sediment | Rift | Oceanic crust

(a) In the Archean, island arcs and hot-spot volcanoes built small blocks of buoyant crust. Rifting of these blocks may have produced flood basalts, and erosion of the blocks produced sediment. (Not to scale.)

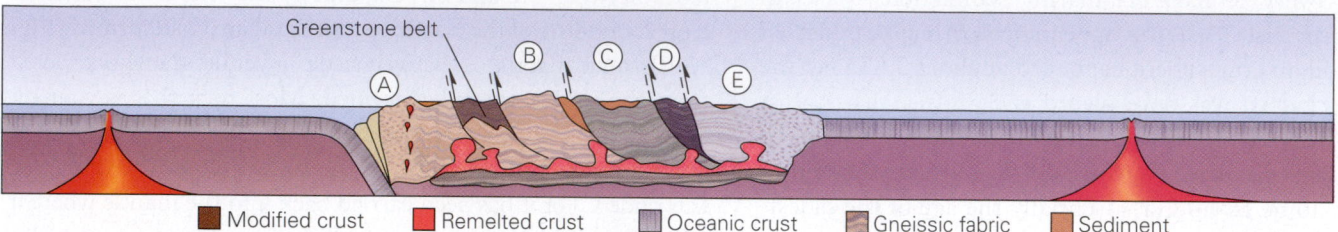

Modified crust | Remelted crust | Oceanic crust | Gneissic fabric | Sediment

(b) Buoyant blocks collided and sutured together, forming protocontinents. Melting at depth produced granite. Eventually, regions of crust cooled, stabilized, and became cratons. (Not to scale.)

(c) An exposure of Archean rock in the Upper Peninsula of Michigan. This rock, migmatite, was once buried so deeply that it started to melt.

that were clearly rounded by transport in liquid water. Salts that weathered out of rock and were transported to the sea by rivers made the oceans salty.

The First Life

Clearly, the Archean Eon saw many firsts in Earth history. Not only did the first continents appear during the Archean but probably also the first life. Geologists use three sources of evidence to identify early life.

- *Chemical (molecular) fossils, or biomarkers:* These are durable chemicals that represent pieces of larger molecules produced by the metabolism of living organisms.

- *Isotopic signatures:* By analyzing the ratio of ^{12}C to ^{13}C in carbon-rich sediment, geologists can determine if the sediment once contained the bodies of organisms because organisms preferentially incorporate ^{12}C.

- *Fossil forms:* Given appropriate depositional conditions, fossils of bacteria or archaea cells can be preserved in rock. However, identification of such fossil forms remains controversial—similar shapes can result from inorganic crystal growth.

The search for the earliest evidence of life continues to make headlines in the popular media (**Box 13.1**). Most geologists currently conclude that life has existed on Earth since at least 3.5 Ga, and perhaps since 3.8 Ga, for rocks of this age contain isotopic signatures of organisms. The oldest undisputed body fossils of bacteria and archaea occur in 3.2-Ga

FIGURE 13.6 As time progressed, the area of the Earth covered by continental crust increased. Most crust had formed by the beginning of the Proterozoic.

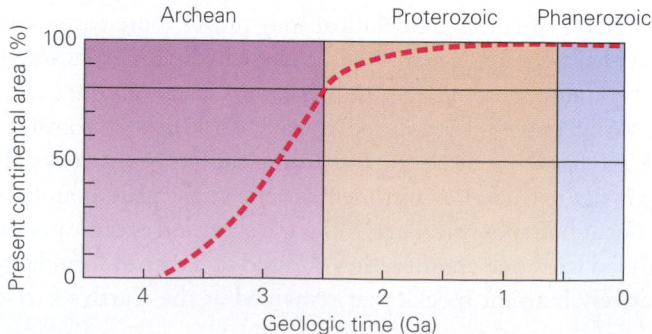

FIGURE 13.7 Archean life forms.

rocks (**Fig. 13.7a**)—shapes resembling such organisms occur in rocks as old as 3.4 to 3.5 Ga, but their identity remains less certain. Some rocks of this age contain **stromatolites**, distinctive mounds of sediment, some of which were produced by mats of cyanobacteria. Such stromatolites form because cyanobacteria secrete a mucus-like substance to which sediment settling from water sticks. As the mat gets buried, new cyanobacteria colonize the top of the sediment, building a mound upward (**Fig. 13.7b**); modern examples locally occur in shallow, tropical waters (**Fig. 13.7c**). Biomarkers in Archean sediments indicate that photosynthetic organisms appeared by 2.7 Ga.

By the end of the Archean, and perhaps as early as 3.5 Ga, organisms similar to cyanobacteria evolved the ability to carry out photosynthesis and moved into shallower, well-lit water. Though these organisms produced oxygen, very little of it accumulated in the atmosphere, for it was either dissolved in the sea or absorbed by weathering reactions with rocks. Thus, though the composition of the Earth's atmosphere at the end of the Archean differed significantly from that of the atmosphere that existed at the start of the Hadean, it was still unbreathable. It probably consisted mostly of about 75% N_2 gas and 25% CO_2 gas, with only traces of oxygen.

As the Archean Eon came to a close, the first continents had formed, and life colonized not only the depths of the sea but also the shallow-marine realm. Plate tectonics had commenced, continental drift was taking place, collisional mountain belts were forming, and erosion was occurring. Oxygen was beginning to enter the air but had not yet accumulated in any significant quantity—the air would not be breathable. The stage was set for another major change in the Earth System.

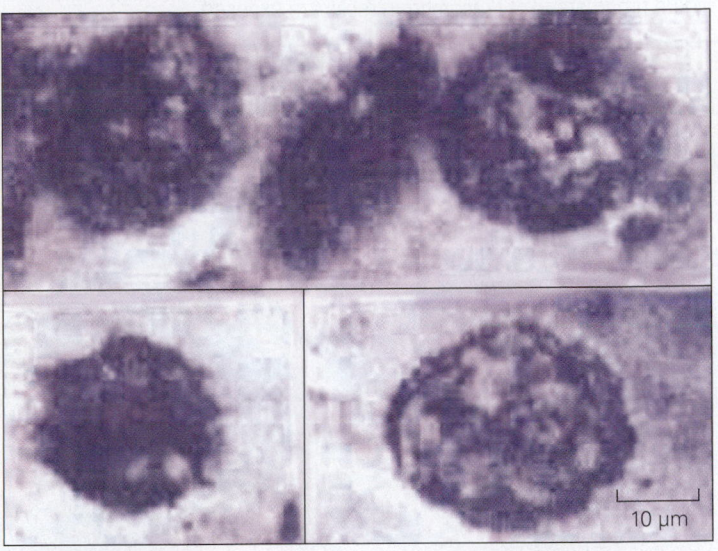

(a) These shapes in 3.2-Ga chert from South America are thought to be fossil bacteria or archaea.

(b) This weathered outcrop of 1.85-Ga dolostone near Marquette, Michigan, reveals the layer-like structure of stromatolites. The delicate ridges represent the fossilized remnants of bacterial mats. Similar stromatolites also occur in exposures of Archean rocks.

Take-Home Message

The Archean (4.0–2.5 Ga) began with the late heavy bombardment, during which almost all crust was destroyed. A solid crust, reformed by cooling of the mantle's skin, appeared soon after, and by 3.8 Ga, oceans formed and have existed ever since. Plate-tectonic-like activity probably began in the Archean; there were also huge mantle plumes. During this eon, the first continental crust formed from colliding volcanic arcs and hot-spot volcanoes, the atmosphere changed, and life appeared. By the end of the eon, the first continents existed.

QUICK QUESTION: What kinds of rock form the blocks of Archean crust that remain today?

(c) Modern stromatolites in Sharks Bay, Western Australia.

FIGURE 13.8 Major crustal provinces of Earth. The black lines indicate the borders of regions underlain by Precambrian crust. Shields are regions where broad areas of Precambrian rocks are exposed.

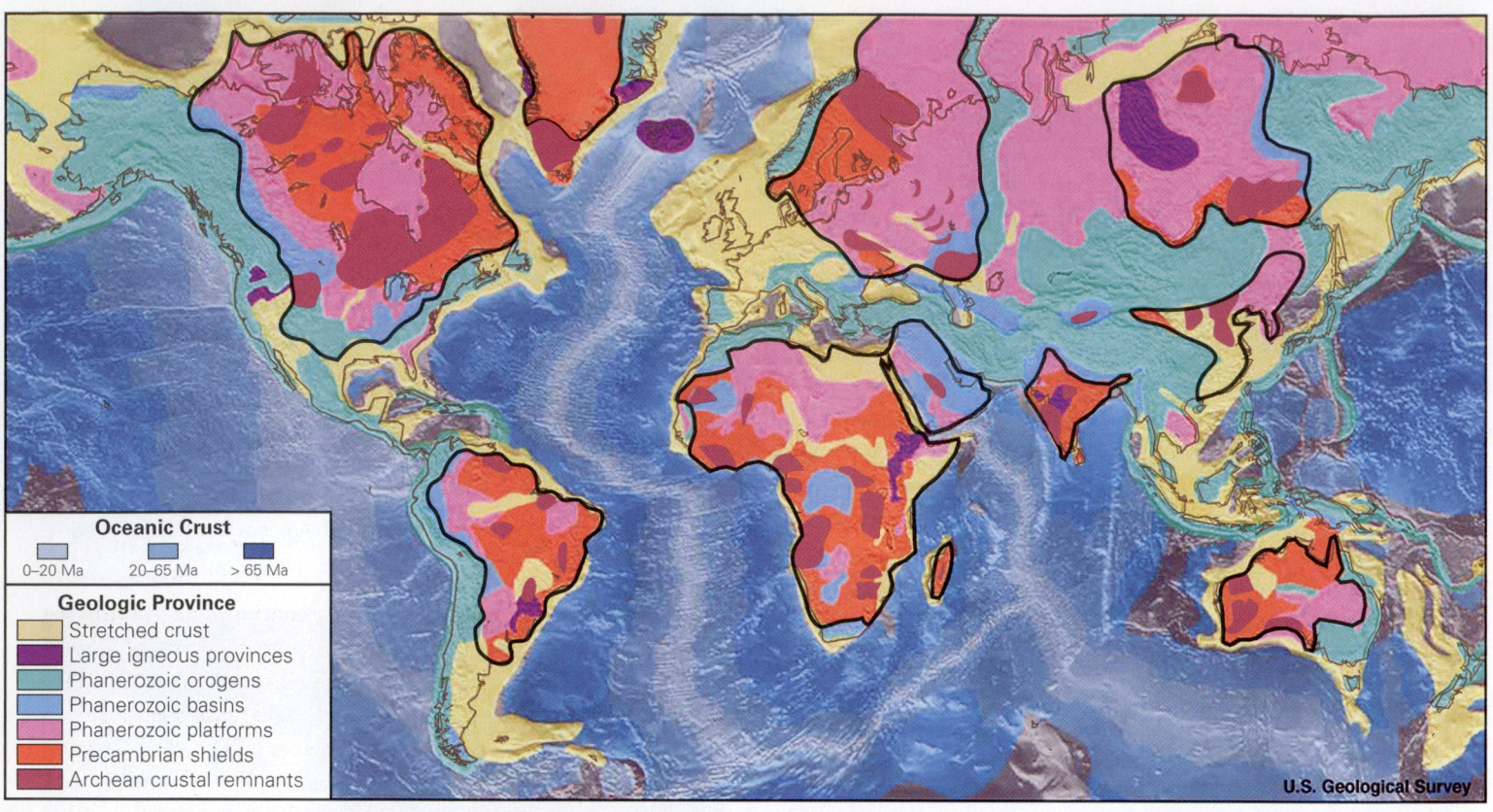

Oceanic Crust
0–20 Ma   20–65 Ma   > 65 Ma

Geologic Province
- Stretched crust
- Large igneous provinces
- Phanerozoic orogens
- Phanerozoic basins
- Phanerozoic platforms
- Precambrian shields
- Archean crustal remnants

U.S. Geological Survey

13.5 The Proterzoic Eon: The Earth in Transition

Continued Growth of Continents

The **Proterozoic Eon** (from the Greek meaning earlier life) spans roughly 2 billion years, from about 2.5 Ga to the beginning of the Cambrian Period at 541 Ma—thus, it encompasses almost half of Earth's history. During Proterozoic time, Earth's surface environment changed from being an unfamiliar world of small, fast-moving plates, small continents, and an oxygen-free atmosphere, to the more familiar world of mostly large, slow-moving plates, large continents, and an oxygenated atmosphere.

First, let's look at changes to the continents. New continental crust continued to form during the Proterozoic Eon but at progressively slower rates, and by the middle of the eon over 90% of the Earth's continental crust had formed. In addition, collisions between Archean continental blocks, and between these blocks and volcanic island arcs or hot-spot volcanoes, gradually assembled larger continents. Size matters when it comes to the geologic behavior of continents, for the interior of

a larger continent can be isolated from heating by subduction-related igneous activity that happened along its margins. Such interior regions, therefore, slowly cool and strengthen until they become rigid and durable. The resulting region of cold, relatively stable continental crust, as we have seen, is called a **craton**. All cratons that exist today had formed by about 1 Ga (**Fig. 13.8**), meaning that the crust of cratons ranges from 3.85 Ga to about 1 Ga. Thus, cratons are the old, long-lived parts of continents.

To understand the character of a craton, let's examine North America's craton a bit more closely. We see that it consists of two regions (**Fig. 13.9**). Throughout the **shield**, outcrops expose Precambrian "basement," which consists of igneous and metamorphic rocks older than about 1 Ga. The landscape of the shield tends to have fairly low relief—there are small hills and valleys but no dramatic mountain ranges. Most of North America's shield lies in Canada, so geologists refer to it as the Canadian Shield. Throughout the **cratonic platform**, which surrounds the shield and also underlies Hudson Bay, a blanket or "cover" of Paleozoic or Mesozoic strata overlies the Precambrian basement. In the eastern platform of North America, Paleozoic strata are the youngest bedrock, whereas in the western platform most of the Paleozoic strata are covered

BOX 13.1   CONSIDER THIS . . .

Where Was the Cradle of Life?

What specific environment on the Archean Earth served as the cradle of life? Laboratory experiments conducted in the 1950s led many researchers to think that life began in warm pools of surface water, beneath a methane- and ammonia-rich atmosphere streaked by bolts of lightning (see Interlude E). The only problem with this hypothesis is that more recent evidence suggests that the early atmosphere consisted mostly of CO_2 and N_2, with relatively little methane and ammonia. Thus, some researchers suggest instead that submarine hot-water vents, so-called black smokers, served as the hosts of the first organisms. These vents emit clouds of ion-charged solutions from which sulfide minerals precipitate and build chimneys. The chimney's are hollow, and their surfaces act like membranes in keeping two chemically distinct environments separate. This difference creates a weak electric charge, perhaps providing energy for "proto-life" molecules, molecules that have structures that resemble proteins of living organisms, to start developing. The earliest life in the Archean Eon may well have been thermophilic (heat-loving) bacteria or archaea that originated at deep-sea hydrothermal vents and dined on pyrite at dark depths in the ocean alongside these vents.

---

by Mesozoic and Cenozoic strata derived from sediments eroded from mountains to the west.

By using isotopic dating on samples from both outcrops and drill holes, geologists have been able to subdivide the Precambrian basement of North America's craton into distinct provinces, each of which has been given a name (**Fig. 13.10**). It

FIGURE 13.9   On this map of North America, we see four different geologic provinces: shield areas, where Precambrian rocks of the craton crop out; platform areas, where Phanerozoic sedimentary rocks have buried Precambrian rocks. Phanerozoic orogenic belts composed of rocks deformed during the past half-billion years, and the coastal plain, which is underlain by Cretaceous and Cenozoic strata.

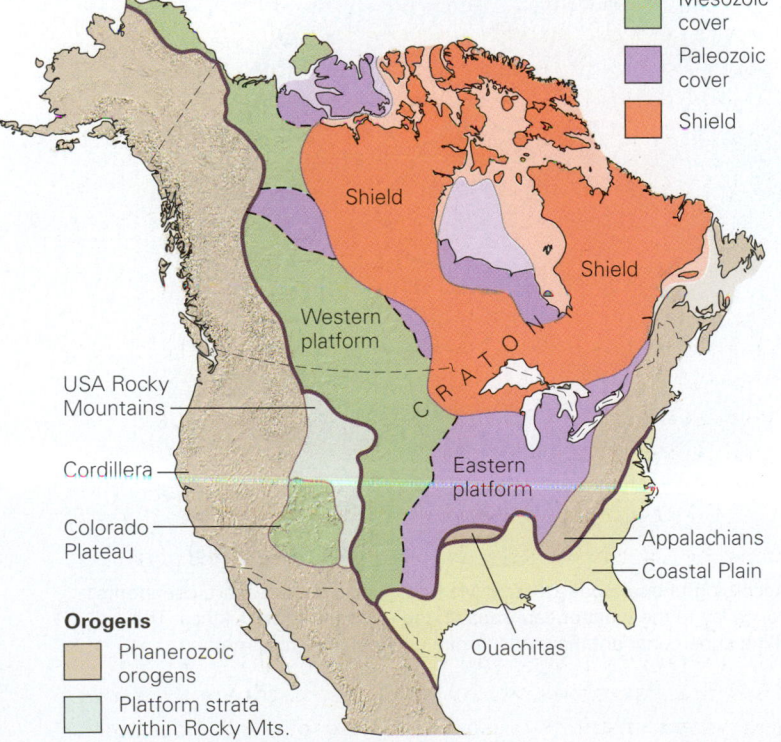

appears that the Canadian Shield consists of several Archean crustal blocks sutured together by Proterozoic orogens. The basement of the cratonic platform in the United States, in contrast, grew when a series of volcanic island arcs and continental slivers accreted, or attached, to the margin of the Canadian Shield between 1.8 and 1.6 Ga—these accreted belts are known as the Yavapai and Mazatzal Provinces, respectively. In the Midwest, granite plutons intruded much of this accreted region, and rhyolite ash flows covered it, due to widespread felsic igneous activity between 1.5 and 1.3 Ga.

Successive collisions ultimately brought together most continental crust on Earth into a single supercontinent, named **Rodinia**, by around 1 Ga. The last major collision during the formation of Rodinia was the **Grenville orogeny**. The resulting Grenville orogen was likely as huge as the present-day Himalayas. If you look at a popular (though not universally accepted) reconstruction of Rodinia, you can identify the crustal provinces that would eventually become the familiar continents of today (**Fig. 13.11a**). Several studies suggest that sometime between 800 and 600 Ma, Rodinia effectively turned inside out in that Antarctica, India, and Australia broke away from western North America and later collided with the future South America, possibly forming a short-lived supercontinent that some geologists refer to as *Pannotia* (**Fig. 13.11b**).

Life Becomes More Complex

The map of the Earth clearly changed radically during the Proterozoic. But that's not all that changed—fossil evidence suggests that this eon also saw important steps in the evolution of life. When the Proterozoic began, most life was *prokaryotic*, meaning that it consisted of single-celled organisms (archaea and bacteria) without a nucleus. Studies of chemical fossils (biomarkers; see Interlude E) hint that *eukaryotic* life, consisting of cells that have nuclei, originated as early as 2.7

FIGURE 13.10 The North American craton consists of a collage of different belts and blocks stitched together during collisional and accretionary orogenies of Precambrian time.

Explanation

G = Grenville; M = Mazatal; Y = Yavapai;

P = Penokean; THO = Trans-Hudson orogen;

WY = Wyoming; WT = Wopmay; T = Thelon;

S = Superior; M = Mojave; RH = Rae and Hearn;

SL = Slave

— Edge of the craton

— Pre-1.8 and post-1.8 Ga crust boundary

▨ Proterozeroic rifts of various ages

▨ Grenville orogen (1.3 – 1.0 Ga)

▨ Granite-rhyolite province (1.5 – 1.3 Ga)

▨ Mazatzal accreted crust (1.7 – 1.6 Ga)

▨ Yavapai accreted crust (1.8 – 1.7 Ga)

▨ Proterozoic collisional orogens (1.9 – 1.8 Ga)

▨ Proterozoic accreted crust (2.0 – 1.8 Ga)

▨ Archean provinces (> 2.5 Ga)

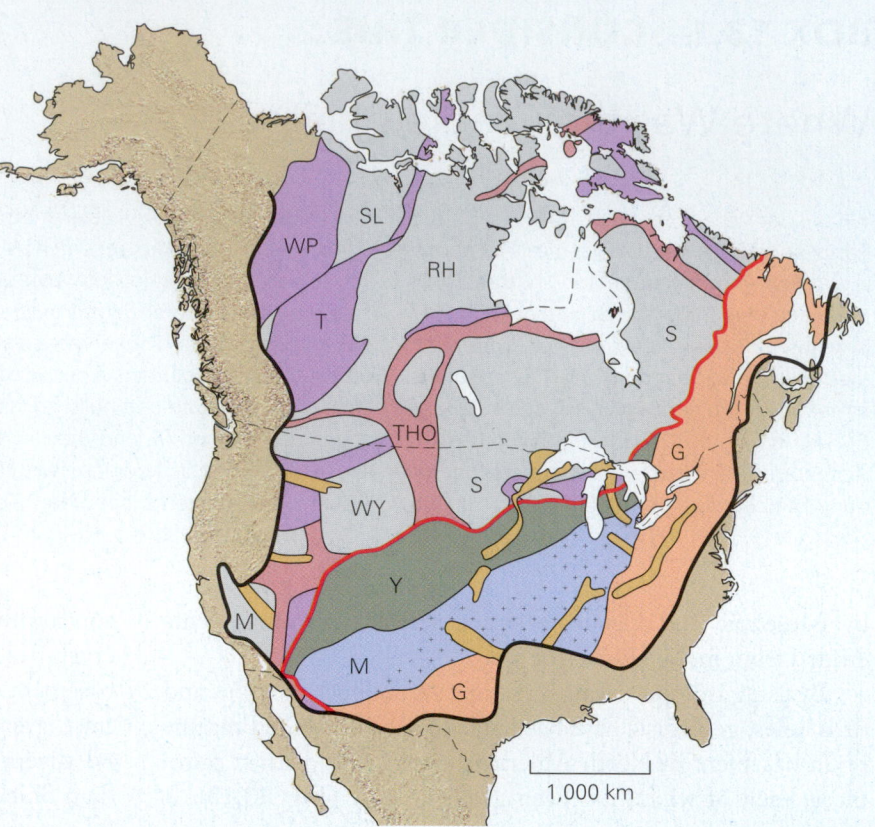

FIGURE 13.11 Supercontinents in the late Precambrian.

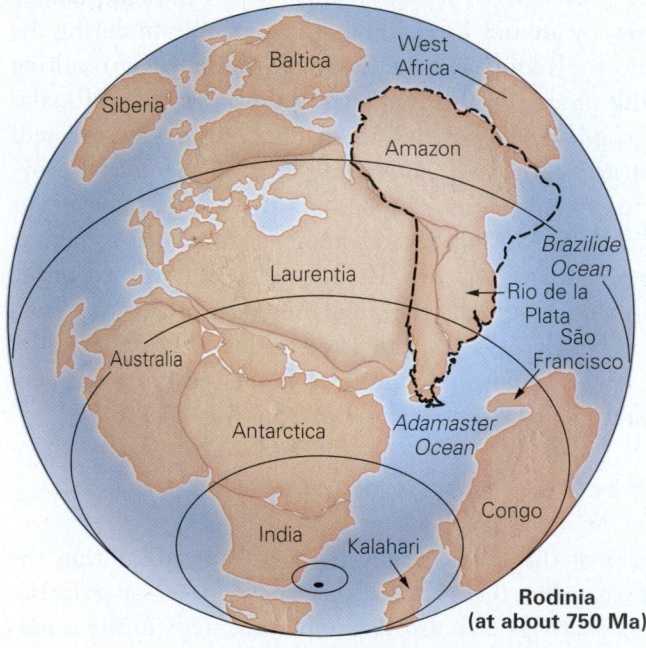

(a) Rodinia formed around 1 Ga and lasted until about 700 Ma. North America and Greenland together comprise Laurentia.

(b) According to one model, by 570 Ma Rodinia had broken apart; continents that once lay to the west of Laurentia ended up to the east of Africa. The resulting supercontinent, Pannotia, broke up soon after it formed.

Ga, but the first body fossils of eukaryotic organism appear in 2.1 Ga rocks, and abundant body fossils of eukaryotic organisms can be found only in rocks younger than about 1.5 Ga. Thus, the proliferation of eukaryotic life, the foundation from which complex organisms eventually evolved, took place during the Proterozoic.

The last half-billion years of the Proterozoic Eon saw the remarkable transition from simple organisms into complex ones. Ciliate protozoans (single-celled organisms coated with fibers that give them mobility) appear at about 750 Ma. A great leap forward in complexity of organisms occurred during the next 150 million years of the eon, for sediments deposited perhaps as early as 620 Ma and certainly by 565 Ma contain several types of multicellular animals that together constitute the **Ediacaran fauna**, named for a region in southern Australia where fossils of these organisms were first found. Ediacaran species survived into the beginning of the Cambrian before becoming extinct. Their fossil forms suggest that some of these invertebrate organisms resembled jellyfish, while others resembled worms (**Fig. 13.12a**).

FIGURE 13.12 Major changes in the Earth System during the Proterozoic Eon.

(a) *Dicksonia*, a fossil of the Ediacaran fauna. These complex, soft-bodied marine organisms appeared in the late Proterozoic.

(b) An outcrop of BIF in the Iron Ranges of Michigan's Upper Peninsula. The red stripes are jasper (red chert) and the gray stripes are hematite. The rock was folded during a mountain-building event long after deposition. The hammer indicates scale.

Strata contain large clasts surrounded by mudstone, for glaciers can carry clasts of all sizes.

Bedding

(c) Layers of Proterozoic glacial till crop out in Africa, indicating that low-latitude landmasses were glaciated during the Proterozoic.

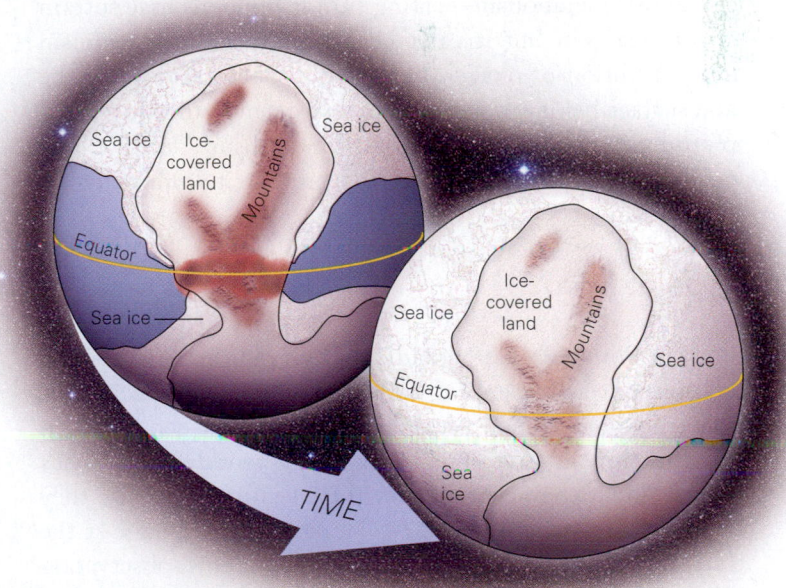

Sea ice
Ice-covered land
Sea ice
Mountains
Equator
Sea ice

Sea ice
Ice-covered land
Mountains
Sea ice
Equator
Sea ice

TIME

(d) This planet may have frozen over completely to form "snowball Earth." Glaciers first grew on land, and eventually the sea surface froze over.

The evolution of life played a key role in the evolution of Earth's atmosphere. Before life appeared, there was hardly any free oxygen (O_2) in the atmosphere. With the appearance of photosynthetic organisms, oxygen began to enter the atmosphere. But it was not until about 2.4 Ga that the concentration of oxygen in the atmosphere increased dramatically. This event, called the **great oxygenation event**, happened when the land and sea were no longer able to react with, absorb, or dissolve all the oxygen produced by organisms, so the oxygen began to accumulate as a gas in air.

One of the important consequences of this change is that the oceans became oxidizing environments (meaning that they contained atoms, such as oxygen, that could transfer electrons to other atoms). Iron atoms, in particular, can pick up electrons and thus become *oxidized*. Oxidized iron is not soluble in water, so when the oceans became oxidizing, they could no longer contain large quantities of dissolved iron. Between 2.4 Ga and 1.8 Ga, as the atmosphere became oxidizing, huge amounts of iron settled out of the ocean to form colorful sedimentary beds known as banded iron formation (BIF). *BIF* consists of alternating layers, or bands, of iron oxide minerals (hematite or magnetite) and jasper (red chert) (**Fig. 13.12b**). Though BIF did form in the Archean, most of the world's BIF, humanity's primary source of iron for making steel, accumulated between 2.4 and 1.8 Ga, implying that the transition to an oxygenated atmosphere was complete by about 1.8 Ga. Other geological evidence reinforces this idea, as described in **Box 13.2**.

The great oxygenation event profoundly influenced the evolution of life in the Earth System. Life could become more complex because oxygen-dependent (aerobic) metabolism can produce energy much more efficiently than can oxygen-free (anaerobic) metabolism—eating sulfide minerals may sustain an archaea cell, but it can't keep a multicellular organism alive. Addition of oxygen to the atmosphere also made the land surface habitable, because some oxygen in the air reacted to produce ozone (O_3) molecules, which absorb deadly ultraviolet radiation from the Sun and prevent it from reaching the surface.

Snowball Earth

Study of Proterozoic strata indicates that radical climate shifts took place near the end of the Proterozoic Eon. Specifically, researchers have found accumulations of glacial sediments in late Proterozoic stratigraphic sequences worldwide. What's strange about the occurrence of these sediments is that they occur even in regions that were located at the equator at the time they were deposited (**Fig. 13.12c**). This observation implies that the entire planet was cold enough at the end of the Proterozoic for glaciers to grow at all latitudes. Geologists still are debating the character and history of these global ice ages (see Chapter 22), but in one model, glaciers covered all land, and the entire ocean surface froze. The ice-covered globe that our planet may have become at the end of the Proterozoic has come to be known as **snowball Earth** (**Fig. 13.12d**).

Did you ever wonder . . .

if the oceans have ever frozen over entirely?

The shell of ice covering snowball Earth would have cut off the oceans from the atmosphere, and many life forms died off. Earth might have remained a snowball forever were it not for volcanic CO_2. The icy sheath covering the oceans prevented atmospheric CO_2 from dissolving in seawater, but it did not prevent volcanic activity from continuing to add CO_2 to the atmosphere. CO_2 is a greenhouse gas, meaning that it traps heat in the atmosphere much as glass panes trap heat in a greenhouse (see Chapter 23), so as the CO_2 concentration increased, Earth warmed up and eventually the glaciers melted. Life may have survived snowball Earth conditions only near submarine black smokers and near hot springs. When the ice vanished, life rapidly expanded into new environments, where new species, such as those of the Ediacaran fauna, evolved.

Transition to the Phanerozoic Eon

As the Proterozoic came to a close, Earth's climate warmed and rifting broke apart the late Proterozoic supercontinent. As continents drifted apart, life evolved and diversified to occupy the many new environments that formed. Over a relatively short period of time, shells appeared and the fossil record became much more complete. This event defines the end of the Proterozoic Eon, and therefore of the Precambrian, and the start of the Phanerozoic Eon. Of note, geologists recognized the significance of this event long before they could assign it a numerical age (currently 541 Ma).

The **Phanerozoic Eon** (Greek for visible life) encompasses the last 541 million years of Earth history. Its name reflects the appearance of diverse organisms with hard shells or skeletons that became the well-preserved fossils you can find easily in sedimentary rock outcrops. The Phanerozoic Eon consists of three eras—the Paleozoic (Greek for ancient life), the Mesozoic (middle life), and the Cenozoic (recent life). Geologists have divided the Mesozoic and Cenozoic each into three periods and the Paleozoic into six periods. In the sections that follow, we consider changes in the map of our planet's surface (its *paleogeography*), as manifested by the distribution of continents, seas, and mountain belts, as well as life evolution that happened during the three eras.

BOX 13.2 CONSIDER THIS . . .

The Evolution of Atmospheric Oxygen

Without oxygen, the great variety of life that exists on Earth could not survive. Presently, the atmosphere contains 21% oxygen, but this has not always been the case. Throughout most of the Archean Eon and into the beginning of the Proterozoic Eon, the atmosphere contained less than 1% oxygen. Several lines of evidence lead geologists to conclude that a transformation from an oxygen-poor to an oxygen-rich atmosphere occurred during the great oxygenation event, a period lasting from about 2.4 to about 1.8 Ga, the early part of the Proterozoic.

As noted earlier, one line of evidence comes from studying the deposition of BIFs, a process that could only have happened in an oxygenated environment. Another line of evidence comes from examining clastic grains in sandstones. In sediments deposited before 1.8 Ga, well-formed pyrite (iron sulfide) occurs as clasts in sediment. This could only be pos-

sible if the atmosphere before 1.8 Ga contained very little oxygen, for in an oxygen-rich atmosphere, pyrite rapidly undergoes chemical weathering (oxidation) and doesn't survive long enough at the Earth's surface to become a sedimentary clast. A third line of evidence comes from studying the age of *redbeds*, clastic sedimentary rocks colored by the presence of bright red hematite (iron oxide). Redbeds form when oxygen-rich groundwater flows through sediment during lithification, and such rocks appear in the geologic record only after 1.8 Ga.

Oxygen concentration in the atmosphere has changed considerably over Earth's history. As we have noted, it constituted less than 1% of the atmosphere during the Archean. By 1.8 Ga, after the great oxygenation event, it had risen to about 3% and stayed at roughly that level until about 0.6 Ga (**Fig. Bx13.2**). Then, because all the materials on land and in the

sea that could absorb oxygen had become *saturated* and could hold no more O_2, atmospheric concentrations gradually grew, reaching about 12% at the end of the Proterozoic. The proportion grew substantially when photosynthetic organisms began to prosper on land, and it has oscillated between 35% and 15% during the past half-billion years. Oxygen concentration has remained at 21% since about 25 Ma. Keeping in mind that atmospheric oxygen comes from photosynthetic organisms, geologists suggest that increases reflect proliferation of photosynthetic life (such as land-based plants) and that decreases reflect mass-extinction events. The concentration of O_2 can't get higher than about 35%, because if it did, land plants would become explosively combustible, since oxygen feeds fire, and so much vegetation would burn that the amount of photosynthesis would decrease until oxygen levels decreased.

FIGURE Bx13.2 The change in the proportion of oxygen content in the atmosphere over time. Oxygen level remained low until the end of the Protoerozoic.

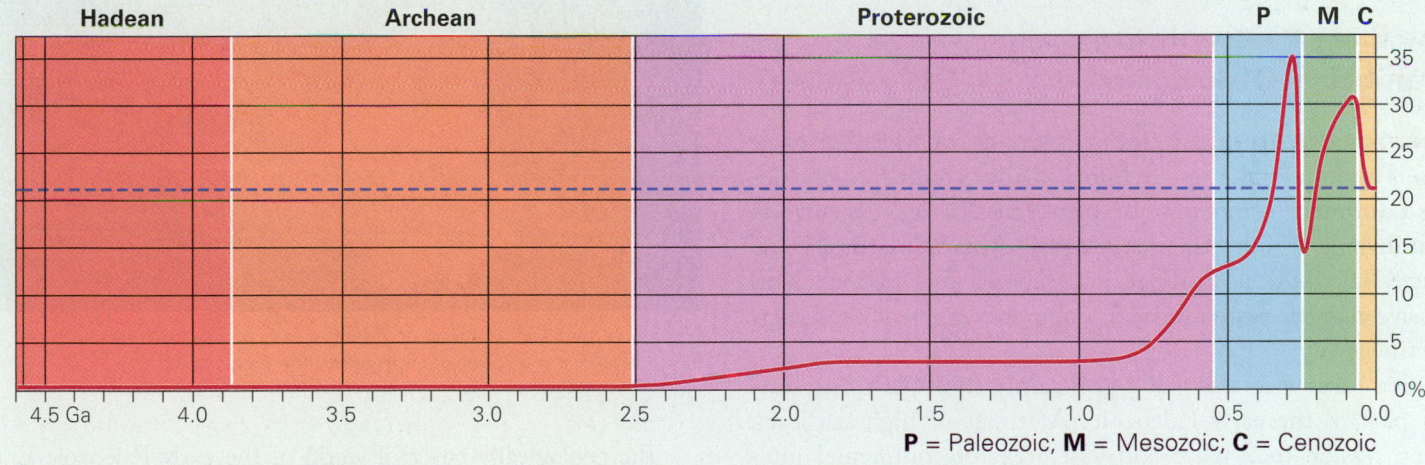

P = Paleozoic; M = Mesozoic; C = Cenozoic

Take-Home Message

During the Proterozoic (2.5–0.54 Ga), cratons formed and then sutured together to form continents and, eventually, supercontinents. The atmosphere began to accumulate significant oxygen, and the chemical behavior of the seas changed. Near the end of the eon, the Earth may have been completely ice covered. When the eon ended, the climate warmed, supercontinents rifted apart, and multicellular organisms appeared.

QUICK QUESTION: What evidence suggests that oxygen began to accumulate in the air during the Proterozoic?

FIGURE 13.13 Land and sea in the early Paleozoic Era.

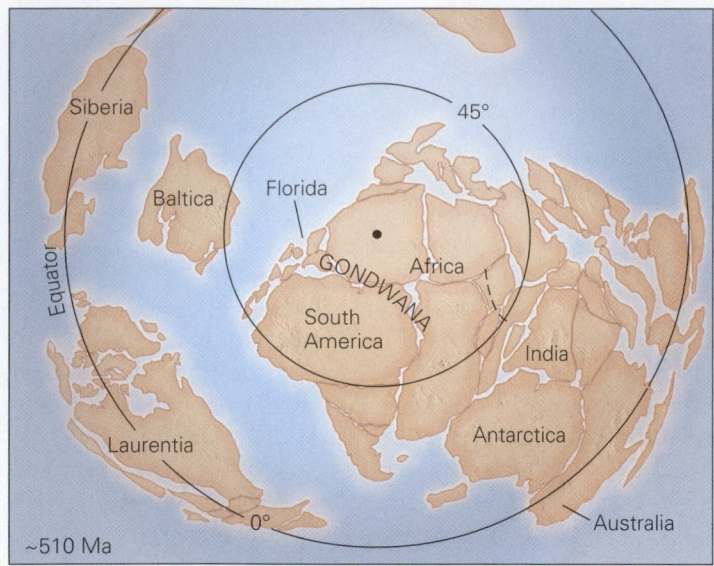

(a) The distribution of continents in the Cambrian Period (510 Ma), as viewed looking down on the South Pole.

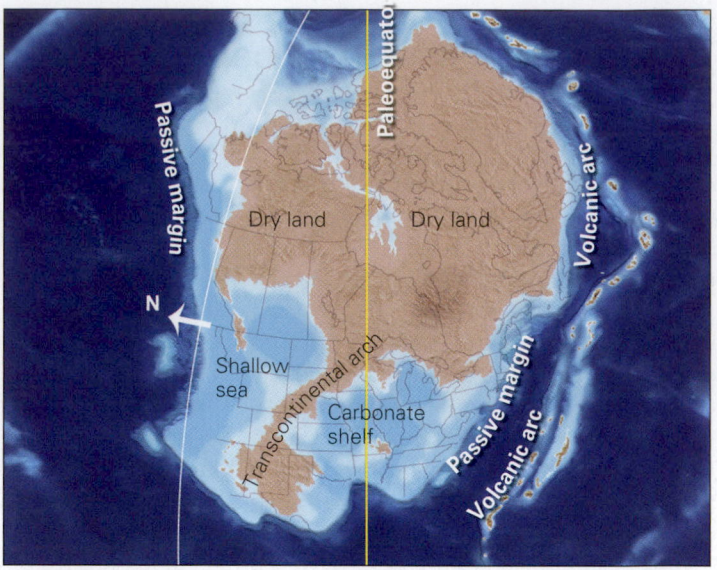

(b) A paleogeographic map of North America shows the regions of dry land and shallow sea in the Late Cambrian Period.

13.6 The Paleozoic Era: Continents Reassemble, and Life Gets Complex

The Early Paleozoic Era (Cambrian-Ordovician Periods, 541–444 Ma)

Paleogeography At the beginning of the Paleozoic Era, Pannotia broke up, yielding smaller continents, including **Laurentia** (composed of North America and Greenland), **Gondwana** (South America, Africa, Antarctica, India, and Australia), Baltica (Europe), and Siberia (**Fig.13.13a**). New passive-margin basins formed along the edges of these new continents.

Sea level rose and fell significantly multiple times during part of the early Paleozoic. At times of high sea level, transgression took place and vast areas of continental interiors became shallow seas, known as *epicontinental seas* (**Fig. 13.13b**). These regions are now cratonic platforms. In many places, water depths in epicontinental seas reached only a few meters, creating a well-lit marine environment in which life abounded, so deposition in these seas yielded layers of fossiliferous sediment. When sea level dropped, regression took place and unconformities formed. Thus, the craton was covered by unconformity-bounded sequences of sediment—the layer cake of strata in the Grand Canyon formed from such sediment (**Box 13.3**).

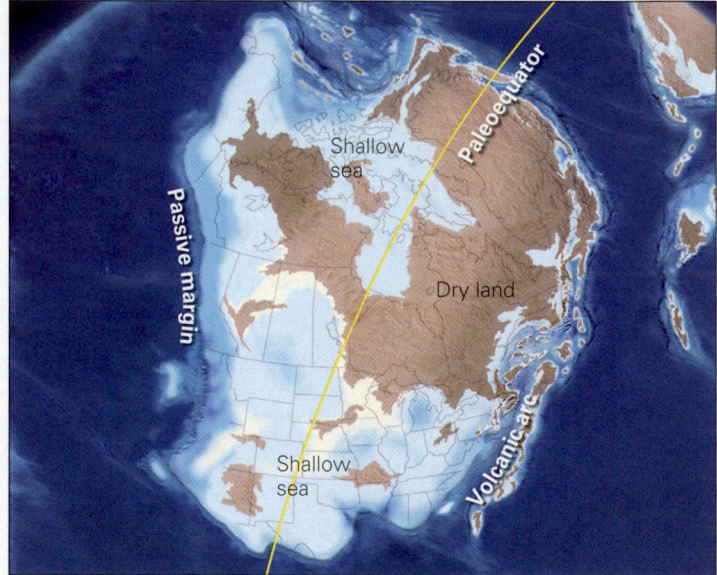

(c) During the Middle Ordovician Period, shallow seas covered much of North America. A volcanic arc formed off the east coast.

The geologically peaceful world of the early Paleozoic Era in Laurentia abruptly came to a close in the Middle Ordovician Period, for at this time its eastern margin rammed into a volcanic island arc and other crustal fragments. The resulting collision, called the *Taconic orogeny*, deformed and metamorphosed strata of the continent's margin and produced a mountain range in what is now the eastern part of the Appalachians (**Fig. 13.13c**).

Life Evolution The fossil record indicates that soon after the Cambrian began, life underwent remarkable diversification. This event, which paleontologists refer to as the **Cambrian**

explosion, took several million years (see Interlude E). What caused this event? No one can say for sure, but considering that it occurred roughly at the time a supercontinent broke up, it may have had something to do with the production of new ecological niches and the isolation of populations that resulted when small continents formed and drifted apart.

The first animals to appear in the Cambrian Period had simple tube- or cone-shaped shells, but soon thereafter the shells became more complex. Shells may have evolved as a means of protection against predation by organisms such as conodonts, small, eel-like organisms with hard parts that resemble teeth. By the end of the Cambrian, trilobites were grazing the seafloor. Trilobites shared the environment with mollusks, brachiopods, nautiloids, gastropods, graptolites, and echinoderms (**Fig. 13.14**; see Interlude E). Thus, a complex food chain arose, which included plankton, bottom feeders, and at the top, predators. Many of the organisms crawled over or swam around reefs composed of mounds of sponges with mineral skeletons. The Ordovician Period saw the first crinoids and the first vertebrate animals, jawless fish.

Although the sea teemed with organisms during the early Paleozoic Era, there were no land organisms for most of this time, so the land surface was a stark landscape of rock and sediment, subjected to rapid erosion rates. Our earliest record of primitive land plants and green algae comes from the Late Ordovician Period, but these plants were very small and occurred only along bodies of water. As we mentioned earlier, the invasion of the land could begin only when there was enough ozone in the atmosphere to protect the land surface from UV radiation. At the end of the Ordovician, mass

extinction took place, perhaps because of a brief ice age and associated sea-level lowering of the time.

The Middle Paleozoic Era (Silurian-Devonian Periods, 444-359 Ma)

Paleogeography As the world entered the Silurian Period, the start of the middle Paleozoic, global climate warmed (leading to so-called greenhouse conditions), sea level rose, and the continents flooded once again. In some places, where water in the epicontinental seas was clear and could exchange with water from the oceans, huge reef complexes grew, forming a layer of fossiliferous limestone on the continents. Also, several orogenies took place, yielding new mountain belts during the middle Paleozoic Era. For example, collisions on the eastern side of Laurentia during Silurian and Devonian time produced the *Caledonian orogen* (affecting eastern Greenland, western Scandinavia, and Scotland) and the *Acadian orogen* in the region that is now the Appalachians (**Fig. 13.15a**).

Throughout much of the middle Paleozoic, the western margin of North America continued to be a passive-margin basin. But finally, in the Late Devonian, the quiet environment of the basin ceased, possibly because of a collision with an island arc. This event, known as the *Antler orogeny*, was the first of many orogenies to affect the western margin of the continent. The Caledonian, Acadian, and Antler orogenies all shed deltas of sediment onto the continents—these deposits formed thick successions of redbeds, such as those visible today in the Catskill Mountains (**Fig. 13.15b, c**).

Life Evolution During the middle Paleozoic Era, new species of trilobites, gastropods, crinoids, and bivalves appeared in the sea, replacing species that had disappeared during the mass extinction at the end of the Ordovician Period. On land, vascular plants with woody tissues, seeds, and veins (for transporting water and food) rooted for the first time. With the evolution of veins and wood, plants could grow much larger, and by the Late Devonian Period the land surface hosted swampy forests with tree-sized relatives of club mosses and ferns. Also at this time, spiders, scorpions, insects, and crustaceans began to exploit both dry-land and freshwater habitats, and jawed fish, including sharks and bony fish, began to cruise the oceans. Finally, at the very end of the Devonian Period, the first amphibians crawled out onto land and inhaled air with lungs (**Fig. 13.15d**).

The Late Paleozoic Era (Carboniferous-Permian Periods, 359-251 Ma)

Paleogeography The climate cooled significantly in the late Paleozoic (leading to so-called icehouse conditions). Seas

FIGURE 13.14 A museum diorama illustrates what early Paleozoic marine organisms may have looked like.

Coral

Trilobite

Nautiloid

Brachiopod

FIGURE 13.15 Paleogeography and fossils of Silurian and Devonian time.

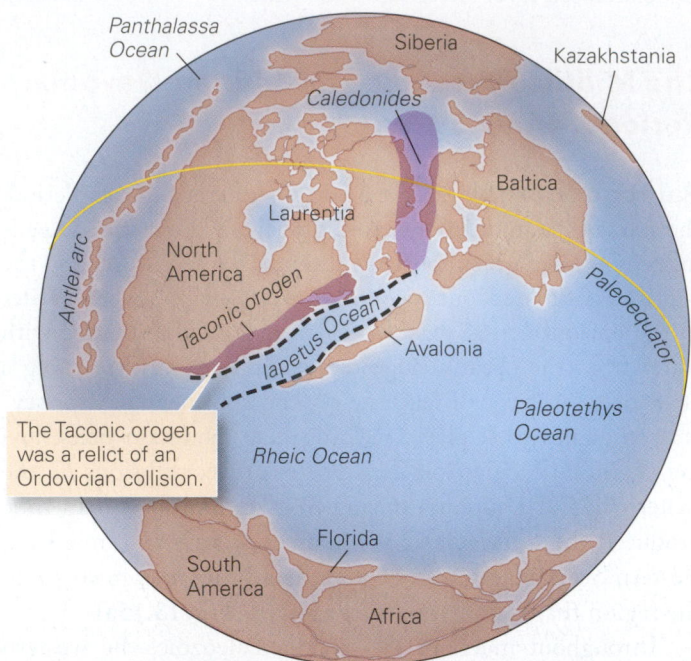

(a) During the Silurian and Devonian Periods, Laurentia collided with Baltica, Avalonia, and South America in succession, as oceans in between were consumed. The Antler arc formed off the west coast.

The Taconic orogen was a relict of an Ordovician collision.

Terranes that eventually attach to Laurentia.

(b) During the Devonian Period, the Acadian orogeny shed sediments into a shallow sea to form the Catskill Delta on the east coast of Laurentia. The Antler orogeny shed sediments in the west.

(c) A road cut exposing Devonian redbeds (sandstone and shale) in New York.

~ 20 cm

(d) A Late Devonian fossil skeleton of *Tiktaalik*; this lobe-finned fish was one of the first animals to walk on land.

gradually retreated from the continents, so that during the Carboniferous Period, regions that had hosted the limestone-forming reefs of epicontinental seas now became coastal areas and river deltas in which sand, shale, and organic debris accumulated. In fact, during the Carboniferous Period, Laurentia lay near the equator, so it enjoyed tropical and semitropical conditions that favored lush growth in swamps. This growth left thick piles of plant debris that transformed into coal after burial. Much of Gondwana and Siberia, in contrast, lay at high latitudes and, by the Permian Period, had become covered by ice sheets.

The late Paleozoic Era also saw a succession of continental collisions, culminating in the formation of a single supercontinent, Pangaea (**Fig. 13.16a**). The largest collision occurred during Carboniferous and Permian time, when Gondwana rammed into Laurentia and Baltica, causing the *Alleghanian orogeny* of North America (**Fig. 13.16b**). During this event, the final stage in the development of the Appalachians, eastern North America rammed against northwestern Africa, and what is now the Gulf Coast region of North America squashed against the northern margin of South America. A vast mountain belt grew, in which deformation generated huge faults and folds. We now see the eroded remnants of rocks deformed during this event in the Appalachian and Ouachita Mountains of North America. By the end of the collisions, nearly all land on Earth had

FIGURE 13.16 Paleogeography at the end of the Paleozoic Era.

Approximate area of part (b)

(a) At the end of the Paleozoic, almost all land had combined into a single supercontinent called Pangaea.

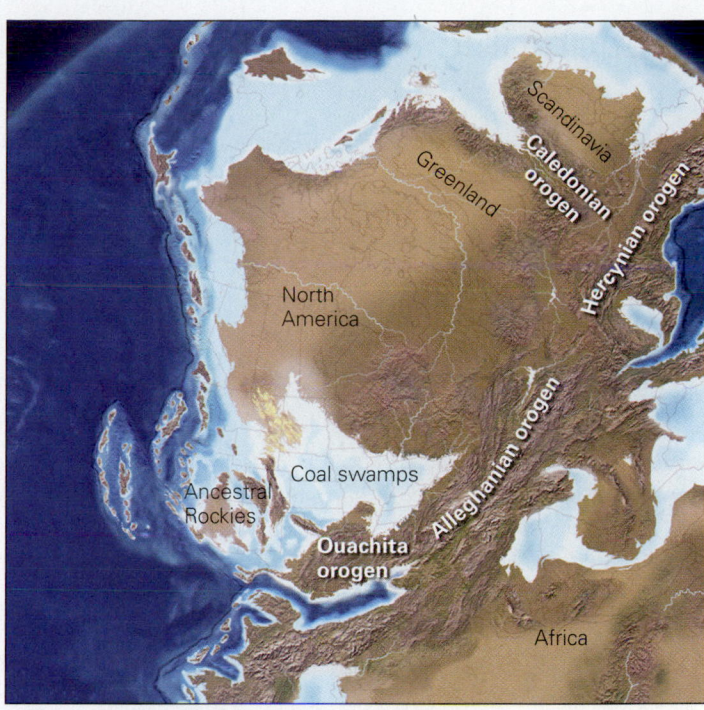

(b) During the Alleghanian and Hercynian orogenies, a huge mountain belt formed. The Caledonian orogeny had formed earlier. Coal swamps bordered interior seas, and the Ancestral Rockies rose.

combined into the giant supercontinent, Pangaea, so named by Alfred Wegener who, more than a century ago, first proposed its existence. Imagine how different the world would be if you could walk from the middle of North America to the middle of Africa or Asia.

Along the continental side of the Alleghanian orogen, a wide band of deformation called the **Appalachian fold-thrust belt** formed. In this province, a distinctive style of deformation, called *thin-skinned deformation*, took place. In a thin-skinned fold-thrust belt, an array or system of thrust faults cuts across what was originally a layer cake of sedimentary strata (**Fig. 13.17**). Movement on the faults displaces the strata and results in the formation of large folds. At depth, the thrust faults merge with a near-horizontal sliding surface, called a detachment, that lies just above the Precambrian basement. The adjective *thin-skinned* highlights the fact that the thrust faults do not cut into the basement but occur only in the overlying cover or "skin" of sedimentary strata.

Stresses generated during the Alleghanian orogeny were so strong that pre-existing faults in the continental crust clear across North America became active again. The movement produced *basement uplifts* (local high areas, cored by Precambrian metamorphic and igneous rocks) and sediment-filled basins in the Midwest and in the region of the present-day Rocky Mountains (Colorado, New Mexico, and Wyoming).

Geologists refer to the late Paleozoic basement uplifts of the Rocky Mountain region as the **Ancestral Rockies**.

The assembly of Pangaea involved a number of other collisions around the world as well. Notably, Africa collided with southern Europe during the *Hercynian orogeny*. Also, a rift or small ocean in Russia closed, leading to the uplift of the Ural Mountains, and parts of China along with other fragments of Asia attached to southern Siberia.

Life Evolution The fossil record indicates that during the late Paleozoic Era, plants and animals continued to evolve toward more familiar forms. In coal swamps, fixed-wing insects such as huge dragonflies flew through a tangle of ferns, club mosses, and scouring rushes, and by the end of the Carboniferous Period insects such as the cockroach, with foldable wings, appeared (**Fig. 13.18**). Forests containing gymnosperms ("naked seed" plants, such as conifers) and cycads (trees with a palm-like stalk peaked by a fan of fern-like fronds) became widespread in the Permian Period. Amphibians and, later, reptiles populated the land. The appearance of reptiles marked the evolution of a radically new component in animal reproduction: eggs with a protective shell. By producing such eggs, reptiles could reproduce without returning to the water and thus could populate previously uninhabitable environments on land.

BOX 13.3 CONSIDER THIS . . .

Stratigraphic Sequences and Sea-Level Change

As we have seen in this chapter, there have been intervals in geologic history when the interior of North America was submerged beneath a shallow sea and times when it was high and dry. This observation implies that sea level, relative to the surface of the continent, rises and falls through geologic time. When the continental surface was dry and exposed to the atmosphere, weathering and erosion ground away at previously deposited rock and, as a result, created a continent-wide unconformity. In the early 1960s, an American stratigrapher named Larry Sloss introduced the term **stratigraphic sequence** to refer to the strata deposited on the continent during periods when continents were submerged. Such sequences are bounded above and below by regional unconformities.

We can picture the deposition of an *idealized* sequence as follows. As the starting condition, much of the surface of a continent lies above sea level, and an unconformity develops. Then, as a transgression takes place, the shoreline migrates inland and the continent's interior progressively floods. Thus, the pre-existing unconformity gets progressively buried, and the bottom layers of the newly deposited sequence simplistically tend to be older near the margin of the continent than toward the interior.

During transgression, the depositional environment at a location changes as the sea deepens. Thus, the base of a sequence consists of terrestrial strata (river alluvium). This sediment is buried by nearshore sediment, then by deeper-water sediment (**Fig. Bx13.3a**). Eventually, nearly the entire width

of the continent, with the exception of highlands and mountain belts, lies below sea level. As regression proceeds, the interior of the continent is exposed first and then the margins. (In detail, the location and behavior of individual sedimentary basins influences

the local positions of shorelines.) So a new unconformity forms first in the interior and then later along the margins. Many shorter-duration transgressions and regressions may happen during a single long-duration rise or fall.

FIGURE Bx13.3 The rise and fall of sea level and its manifestation in the stratigraphic record.

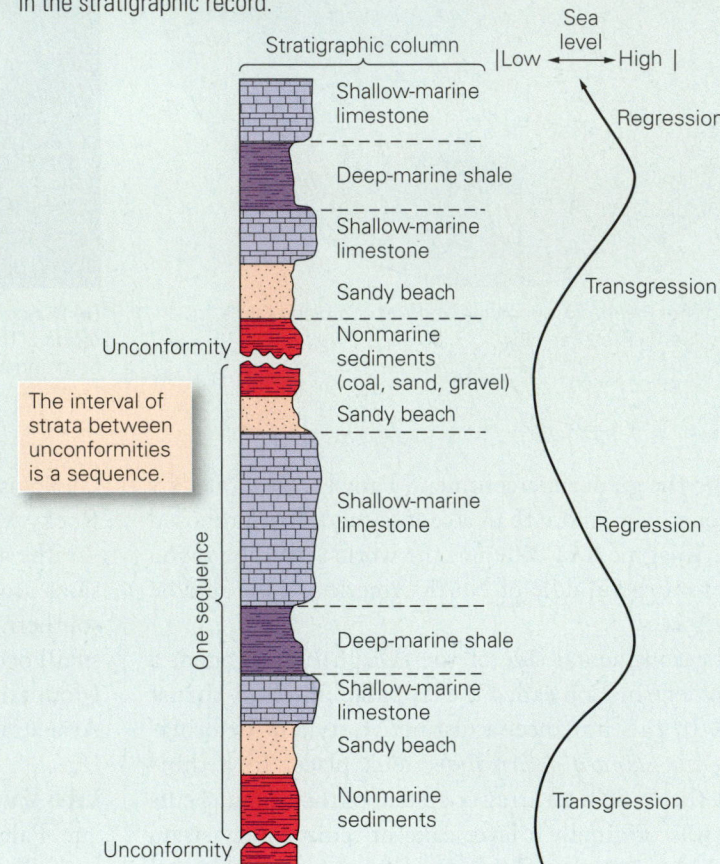

(a) Different types of strata are deposited as sea level rises and falls. Unconformities develop when sea level is low.

The late Paleozoic Era came to a close with two major mass-extinction events, during which over 95% of marine species disappeared. Why these particular events occurred remains a subject of debate. According to one hypothesis, the terminal **Permian mass extinction** occurred as a result of an episode of extraordinary volcanic activity in the region that is now Siberia—basalt sheets extruded during the event are known as the Siberian traps. Eruptions could have clouded the atmosphere, acidified the oceans, and disrupted the food chain. Another hypothesis relates the mass extinction to a huge meteorite impact.

Take-Home Message

Life diversified radically at the beginning of the Paleozoic. During the Paleozoic, life moved onto land, and by the end of it, reptiles roamed through forests of trees. Shallow seas transgressed and regressed over the continents, building a layer cake of Paleozoic strata, and a series of collisions produced orogens. At the end of the eon, collisions led to the assembly of Pangaea.

QUICK QUESTION: What event, indicated by the fossil record, marks the end of the Paleozoic?

Sloss recognized six major stratigraphic sequences in North America, and he named each after a nation of Native Americans (**Fig. Bx13.3b**). Each sequence represents deposition during an interval of time lasting tens of millions of years. The transgressions and regressions could have been caused by global (eustatic) sea-level change, or they could reflect mountain-building processes, or they could reflect the subsidence or uplift of continents.

In more recent decades, studies of strata along continental passive margins have provided an even more detailed record of sequences, because there is less erosion in these regions.

Taken together, this information may provide insight into the history of the global rise and fall of sea level, though this interpretation remains controversial (**Fig. Bx13.3c**). Some sea-level changes may reflect changes in seafloor-spreading rates and thus in the volume of mid-ocean ridges; some may reflect hot-spot activity; some may reflect variations in Earth's climate that caused the formation or melting of ice sheets; others may represent changes in areas of continents accompanying continental collisions; and still others may reflect the warping of continental surfaces as a result of mountain building.

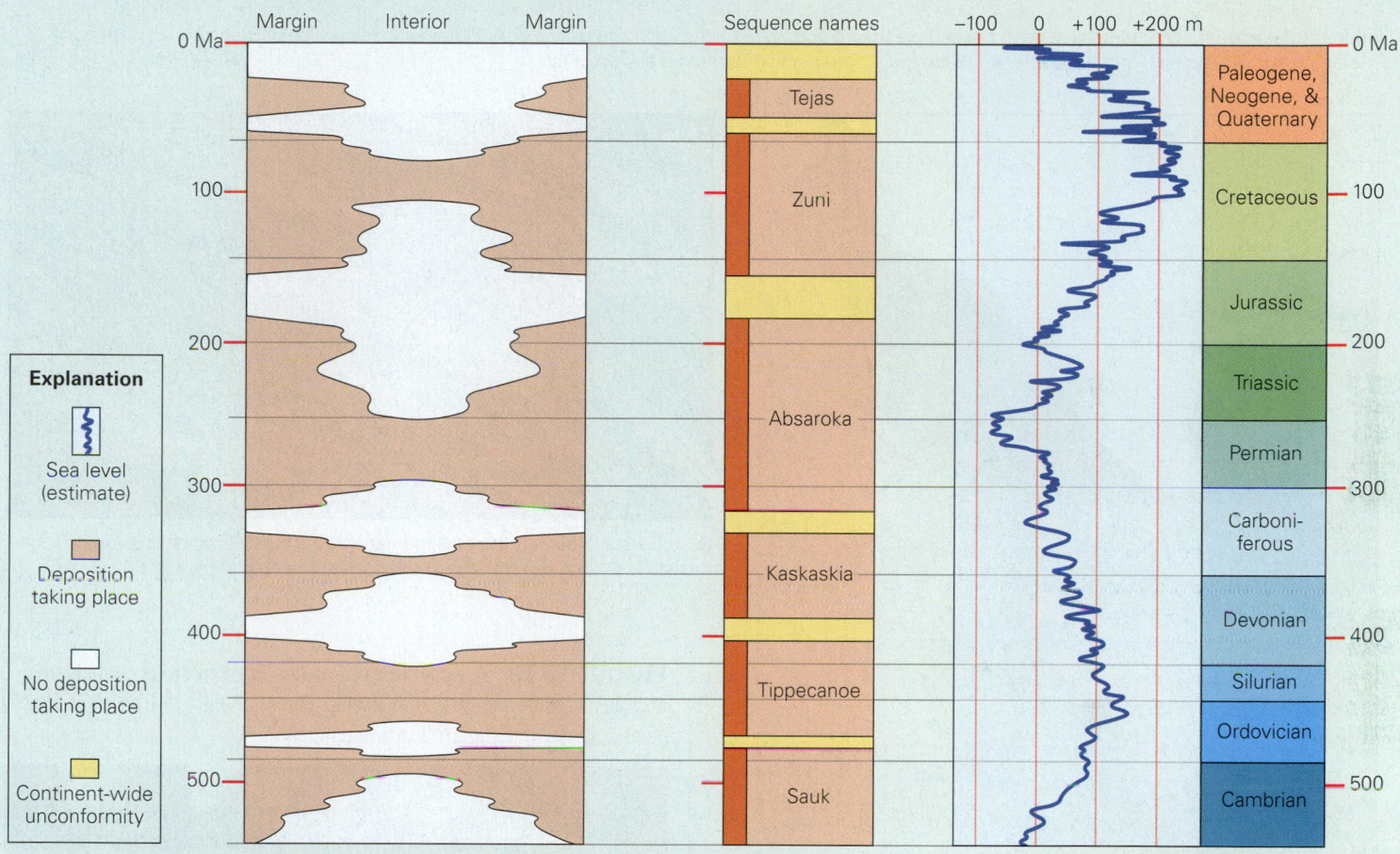

(b) Stratigraphic sequences of North America, as named by Sloss (1962). Symbolically, the chart on the left shows that, overall, continental margins were sites of deposition while the interior was not. At times, the whole continent was a not a site of deposition—the resulting continent-wide unconformities separate the sequences, as shown by the chart in the middle.

(c) An interpretation of global sea-level change, based on analysis of stratigraphy in passive-margin basins. The interpretation remains controversial. (0 = present day)

13.7 The Mesozoic Era: When Dinosaurs Ruled

The Early and Middle Mesozoic Era (Triassic–Jurassic Periods, 251–145 Ma)

Paleogeography Pangaea existed for about 100 million years, until the Late Triassic. Then rifts developed and the supercontinent began to break apart, and by the end of the Jurassic Period, the Mid-Atlantic Ridge formed, the North Atlantic Ocean and Gulf of Mexico started to grow, and North America split away from Europe and Africa (**Fig. 13.19a**). During the early stages of its formation, the North Atlantic was narrow and shallow, and evaporation made its water so salty that thick evaporite deposits accumulated along its margins. These evaporites now lie buried beneath the younger strata in the passive-margin basins that rim the North Atlantic—they are particularly thick beneath along the Gulf Coast region of the United States.

The Earth, overall, had a warm climate during the Triassic and Early Jurassic. But during the Late Jurassic and Early

FIGURE 13.17 Features of the Appalachian Mountains in the eastern United States.

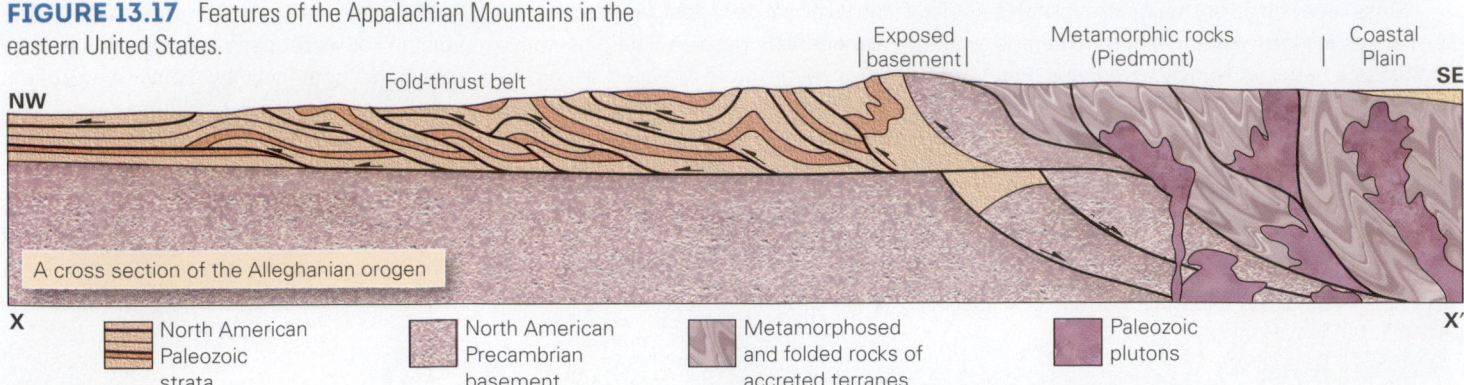

A cross section of the Alleghanian orogen

| | North American Paleozoic strata | | North American Precambrian basement | | Metamorphosed and folded rocks of accreted terranes | | Paleozoic plutons |

(a) In the fold-thrust belt, strata have been pushed westward and were folded and faulted. The Blue Ridge exposes a slice of Precambrian basement. Metamorphic and plutonic rocks underlie the Piedmont. Cross section location is shown in (b).

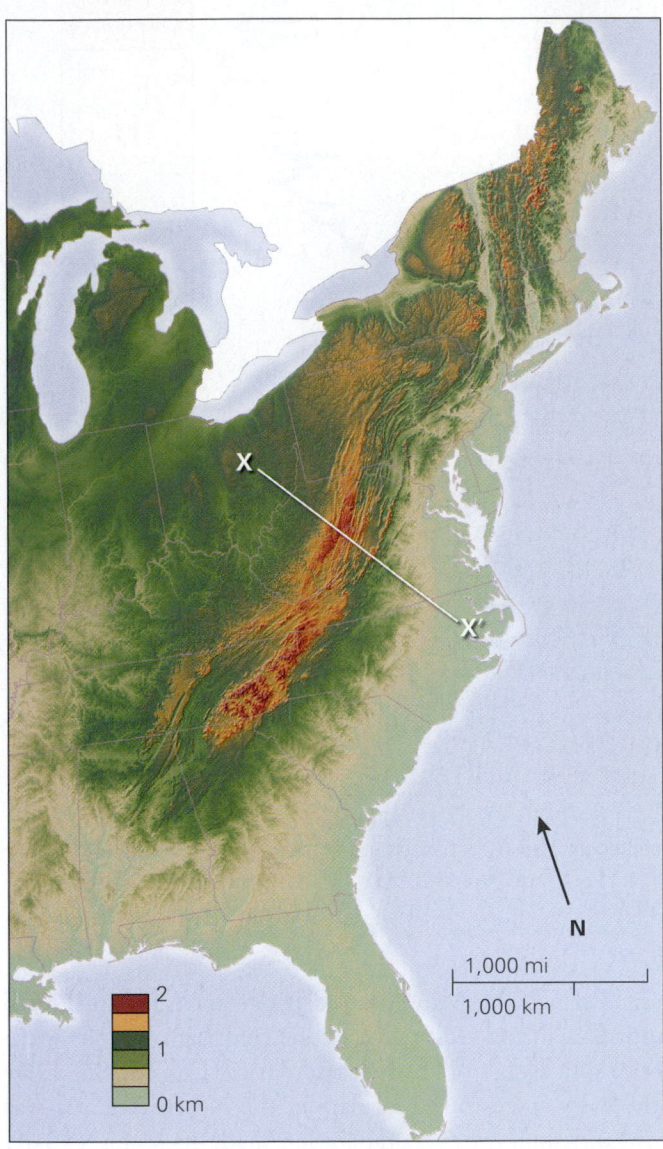

(b) The eroded remnants of the Appalachian orogeny stand out in the eastern United States. The white line shows the approximate cross-section position.

(c) Folds stand out in this satellite image of the Pennsylvania Valley and Ridge. Resistant sandstone layers form the ridges. The field of view is 80 km wide.

FIGURE 13.18 A museum diorama of a Carboniferous coal swamp includes a giant dragonfly with a wingspan of about 1 m. The inset photo gives a sense of its size relative to a human.

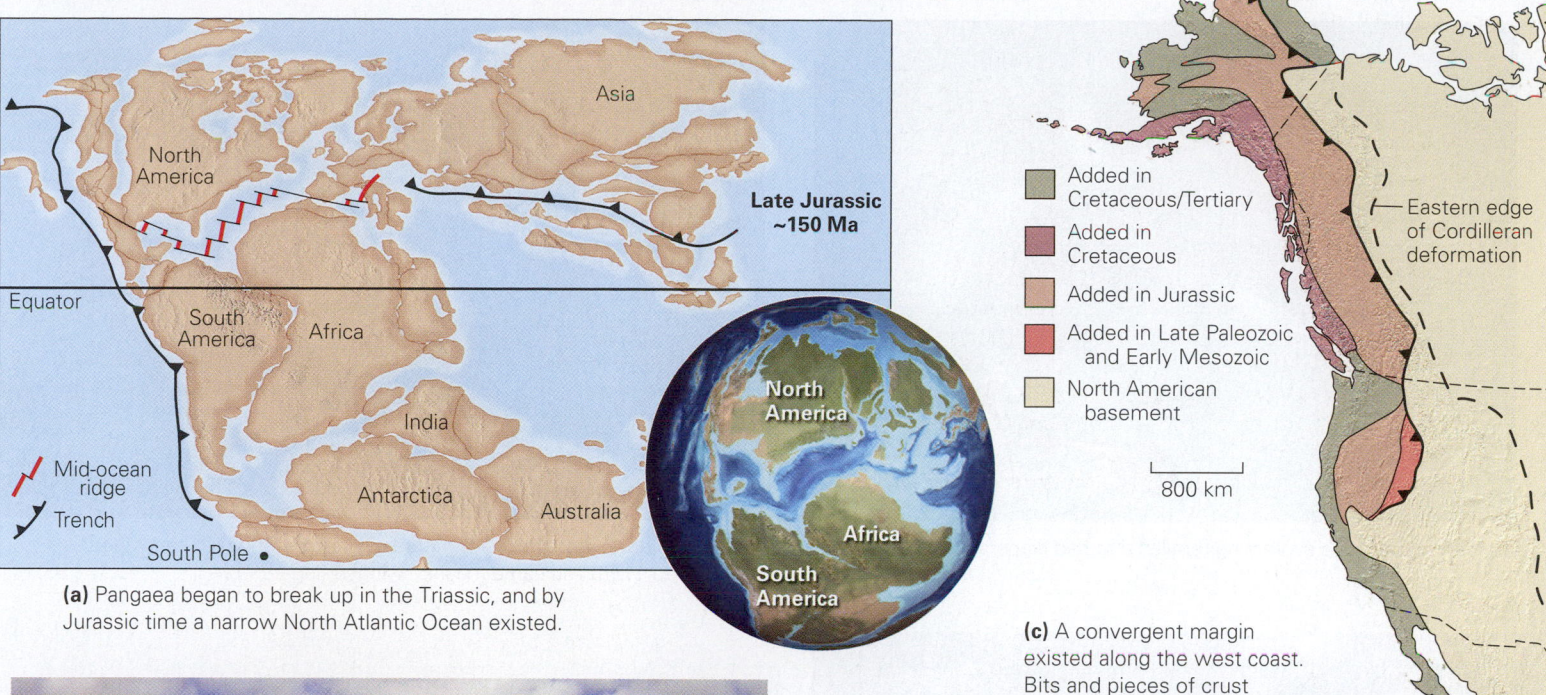

(a) Pangaea began to break up in the Triassic, and by Jurassic time a narrow North Atlantic Ocean existed.

Late Jurassic ~150 Ma

Equator

Mid-ocean ridge

Trench

South Pole

North America, Asia, South America, Africa, India, Antarctica, Australia

Legend (c):
- Added in Cretaceous/Tertiary
- Added in Cretaceous
- Added in Jurassic
- Added in Late Paleozoic and Early Mesozoic
- North American basement

Eastern edge of Cordilleran deformation

800 km

(c) A convergent margin existed along the west coast. Bits and pieces of crust attached as accreted terranes.

(b) During the Jurassic, immense sand dunes blanketed the southwestern United States. Sandstone beds in Zion Park are the relicts of these dunes.

Cretaceous, the climate cooled. Large areas of North America's interior were nonmarine environments in which thick deposits of red sandstones and shales, now exposed in the spectacular cliffs of Zion National Park, were deposited (**Fig. 13.19b**). Then, as the Middle Jurassic Period began, sea level began to rise until eventually a shallow sea submerged much of what is now the Rocky Mountain region of the western United States.

On the western margin of North America, convergent-margin tectonics became the order of the day. Beginning with Late Permian and continuing through Mesozoic time, subduction generated volcanic island arcs and caused them, along with microcontinents and oceanic plateaus (the product of hot-spot volcanism), to collide with North America. Thus,

North America grew in land area by the "accretion" (addition) of crustal fragments onto its western margin (**Fig. 13.19c**). Because these fragments consist of crust that formed elsewhere, not originally on or adjacent to the continent, geologists refer to them as **exotic terranes**. At the end of the Jurassic, subduction of Pacific Ocean floor beneath North America began, an event that produced a major continental volcanic arc, the Sierran arc. This arc continued to erupt through the Cretaceous—we'll learn more about it later in this chapter.

Life Evolution During the early Mesozoic Era, a variety of new plant and animal species appeared, filling the ecological niches left vacant by the Late Permian mass extinction. Reptiles, such as plesiosaurs, begin to swim in the oceans (**Fig. 13.20a**), and new kinds of corals became the predominant reef builders. On land, gymnosperms and reptiles diversified, and the Earth saw its first turtles and flying reptiles (such as pterodactyls; **Fig. 13.20b**). And at the end of the Triassic Period, the first true dinosaurs evolved. Dinosaurs differed from other reptiles in that their legs were positioned under their bodies rather than off to the sides, and they were possibly warm-blooded. By the end of the Jurassic Period, gigantic sauropod dinosaurs (weighing up to 100 tons), along with other familiar beasts such as stegosaurus, thundered across the landscape (**Fig. 13.20c**), and the

Did you ever wonder...

when the dinosaurs lived?

FIGURE 13.20 Reptiles take to the land and sea.

(a) Plesiosaurs were swimming reptiles that had flippers instead of legs.

(b) There were many species of pterodactyl, some of which had wing spans up to 11 m. The name means "winged finger."

(c) During the Jurassic, giant dinosaurs roamed the land. This painting shows several species.

first feathered birds, such as archaeopteryx, took to the skies. The earliest ancestors of mammals appeared at the end of the Triassic Period, in the form of small, rat-like creatures.

The Late Mesozoic Era (Cretaceous Period, 145–65 Ma)

Paleogeography During the Cretaceous Period, several more *exotic* arcs accreted along the west coast (**Fig. 13.21a**). Also, the Earth's climate continued to shift to warmer greenhouse conditions, and sea level rose significantly, reaching levels that had not been attained for the previous 200 million years. Great shallow seas flooded most of the continents. In fact, during the latter part of the Cretaceous Period, a shark could have swum from the Gulf of Mexico to the Arctic Ocean in the Western Interior Seaway or, similarly, across much of western Europe (**Fig. 13.21b**).

In western North America, the *Sierran arc*, a large continental volcanic arc that was initiated at the end of the Jurassic Period, continued to be active. This arc resembled the present-day Andean arc of western South America. Though the volcanoes of the Sierran arc have long since eroded away, we can see their roots in the form of the plutons that now constitute the granitic batholith of the Sierra Nevada range. A thick accretionary prism, formed from sediments and debris scraped off the subducting oceanic plate, piled up to the west of the Sierran arc and now crops out in the Coast Ranges of California. Compressional stresses along the western North American convergent boundary activated large thrust faults east of the arc, an event geologists refer to as the **Sevier orogeny**. This orogeny produced a thin-skinned fold-thrust belt whose remnants you can see today in the Canadian Rockies and in western Wyoming (**Fig. 13.21c**).

FIGURE 13.21 Cretaceous paleogeography.

Convergent boundary · Thrust fault · Volcanic arc

(a) In Early Cretaceous, several exotic island arcs collided with western North America.

Early Cretaceous

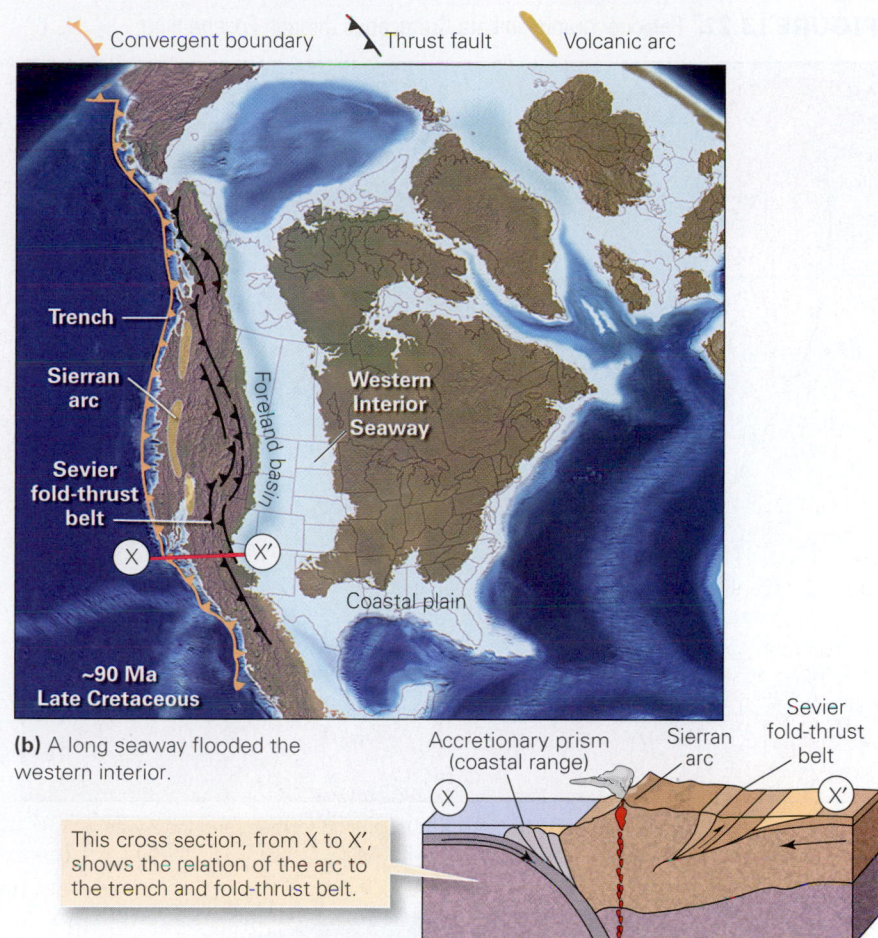

Trench
Sierran arc
Sevier fold-thrust belt
Foreland basin
Western Interior Seaway
Coastal plain

~90 Ma
Late Cretaceous

(b) A long seaway flooded the western interior.

This cross section, from X to X', shows the relation of the arc to the trench and fold-thrust belt.

Accretionary prism (coastal range) · Sierran arc · Sevier fold-thrust belt

(c) In Late Cretaceous, a continental volcanic arc formed. A fold-thrust belt formed to the east, as did a transcontinental seaway.

The weight of the fold-thrust belt helped push the surface of the continent down, forming a wide foreland basin that filled with sediment (see Chapter 7). The depth of the basin may have been enhanced by the downward pull of the subducting plate beneath it. This foreland basin constituted the western part of the Western Interior Seaway.

The breakup of Pangaea continued through the Cretaceous Period, with the opening of the South Atlantic Ocean and the separation of South America and Africa from Antarctica and Australia. India broke away from Gondwana and headed rapidly northward toward Asia (Fig. 13.22a, b). Along the continental margins of the newly formed Mesozoic oceans, large passive-margin basins developed, which filled with great thicknesses of sediments. The passive-margin basin sediment along the Gulf Coast of the United States, eventually accumulated a wedge of sediment that's over 15 km thick.

At the end of the Cretaceous Period, continued compression along the convergent boundary of western North America caused deformation to sweep eastwards (Fig. 13.22c). Slip occurred on the large reverse faults in the region of Wyoming,

Colorado, eastern Utah, and northern Arizona. In contrast to the faults of fold-thrust belts, these faults penetrated deep into the Precambrian basement rocks of the continent and thus movement on them generated **basement uplifts** (Fig. 13.22d) by bringing basement rocks in the hanging-wall block up and over Paleozoic strata in the footwall. In the process, layers of Paleozoic strata overlying the uplifting basement warped over the fault to form large *monoclines*, folds whose shape resembles the drape of a carpet over a step. This event, which happened during the **Laramide orogeny**, formed the structure of the present Rocky Mountains in the United States (Fig. 13.23a). Some geologists have suggested that the contrast in the location of faulting between the Sevier and Laramide orogenies may reflect contrasts in the dip of the subducting plate. During the Laramide orogeny, the subducting plate entered the mantle at a shallower angle and therefore scraped along and applied stress to the base of the continent farther inland. The change in subduction angle may have been due to the presence of an oceanic plateau, a region of thicker oceanic crust, in the downgoing slab. Significantly, this change in dip angle did not occur

FIGURE 13.22 Paleogeography in Late Cretaceous through Eocene time.

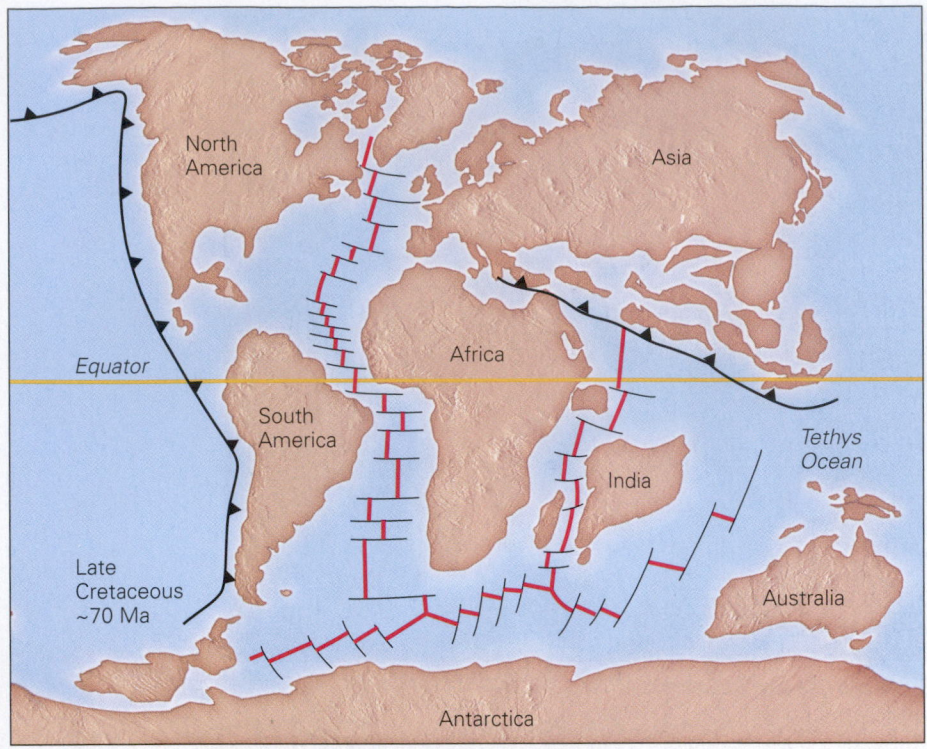

(a) By the Late Cretaceous Period, the Atlantic Ocean had formed, and India was moving rapidly northward to eventually collide with Asia.

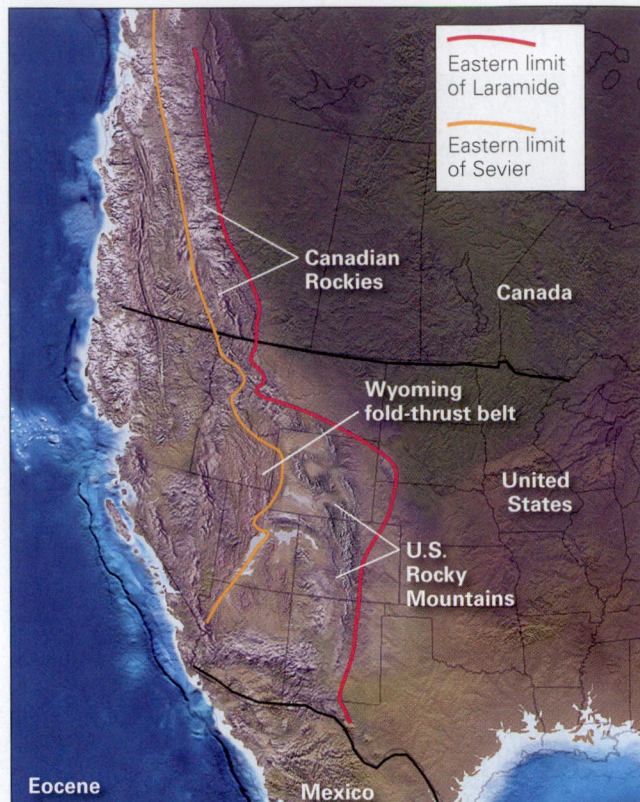

(c) During the Laramide orogeny, deformation shifted eastward in the United States, moving from the Sevier belt to the Rocky Mountains, and the style of deformation changed.

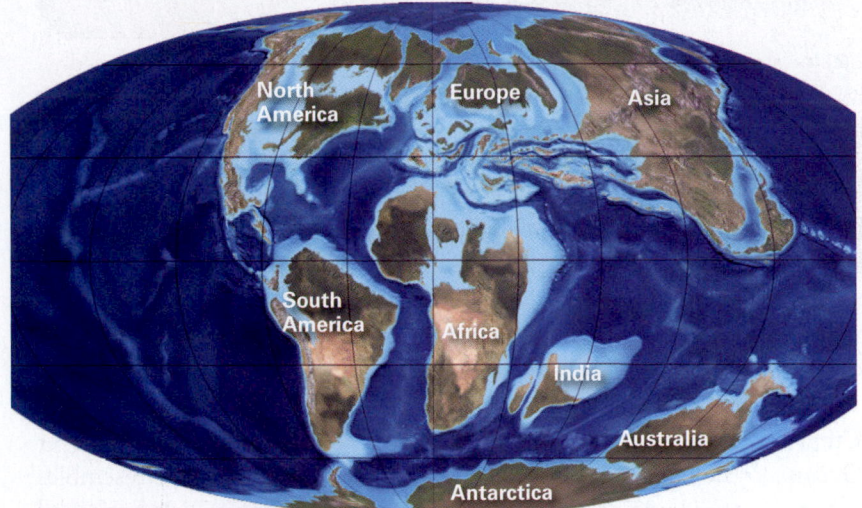

(b) In this Late Cretaceous paleogeographic reconstruction, southern Europe and Asia are beginning to form from a collage of many crustal blocks.

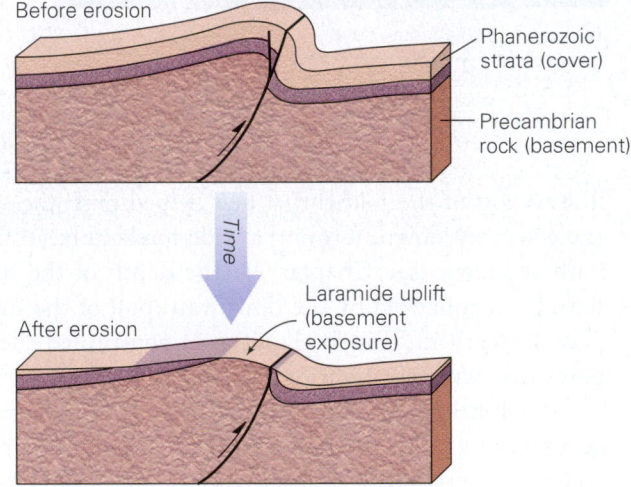

(d) The Laramide orogeny produced "basement-cored uplifts." In these, faults lifted up blocks of basement, causing the overlying strata to bend into a stair-step-like fold.

in Canada, so during the Laramide orogeny, the Canadian Rockies simply continued to grow eastward as a thin-skinned fold-thrust belt (**Fig. 13.23b**).

Geologists have determined that seafloor-spreading rates may have been as much as three times faster during the Cretaceous than they are today. As a result, more of the oceanic crust was younger and warmer than it is today, and since young seafloor lies at a shallower depth than does older seafloor (due to isostasy; see Chapter 11), Cretaceous mid-ocean ridges occupied more volume than they do today. Also during the Cretaceous, huge submarine plateaus formed from basalts erupted at hot-spot volcanoes. The existence of these plateaus implies that particularly active mantle plumes, or **superplumes**, reached the base of the lithosphere. Melting at the top of such plumes

FIGURE 13.23 Examples of mountains formed during the Laramide orogeny.

Unconformity between basement and cover had erosion not taken place

Exposed Precambrian basement   Dipping strata

(a) An air photo of the Wind River Mountains in Wyoming, showing the basement that has been uplifted. The fault is on the left (southwest) side of the range.

(b) The Laramide portion of the Canadian Rocky Mountains consists of thrust sheets containing dipping strata.

produced immense quantities of magma, which erupted and built up the plateaus. Growth of submarine plateaus also displaced seawater. The combination of having broader mid-ocean ridges and large oceanic plateaus, by displacing seawater, may have caused the sea-level rise that happened during the Cretaceous. During this time, the sea submerged what is now the coastal plain of eastern and southern United States, and submerged much of England and Europe, creating the setting in which the chalk deposits we mentioned at the beginning of the chapter could accumulate.

Notably, volcanism associated with extra-rapid seafloor spreading, as well as with submarine plateau growth, likely released large quantities of CO_2, a greenhouse gas, into the atmosphere. Geologists hypothesize that this increased atmospheric CO_2 concentration led to a global rise in atmospheric temperature. Rising temperatures would cause seawater to expand and polar ice sheets to melt, both phenomena that would make sea level go up even more. Considering all the phenomena that caused sea level to rise during the Cretaceous, it's no surprise that the continents flooded and that large epicontinental seas formed during this era.

Life Evolution In the seas of the late Mesozoic world, modern fish appeared and became dominant. In contrast with earlier fish, new fish had short jaws, rounded scales, symmetrical tails, and specialized fins. Huge swimming reptiles and gigantic turtles (with shells up to 4 m across) preyed on the fish. On land, cycads largely vanished, and angiosperms (flowering plants), including hardwood trees, began to compete successfully with conifers for dominance of the forest. Dinosaurs reached their peak of success

(a) An artist's image of the 13-km-wide object as it hit.

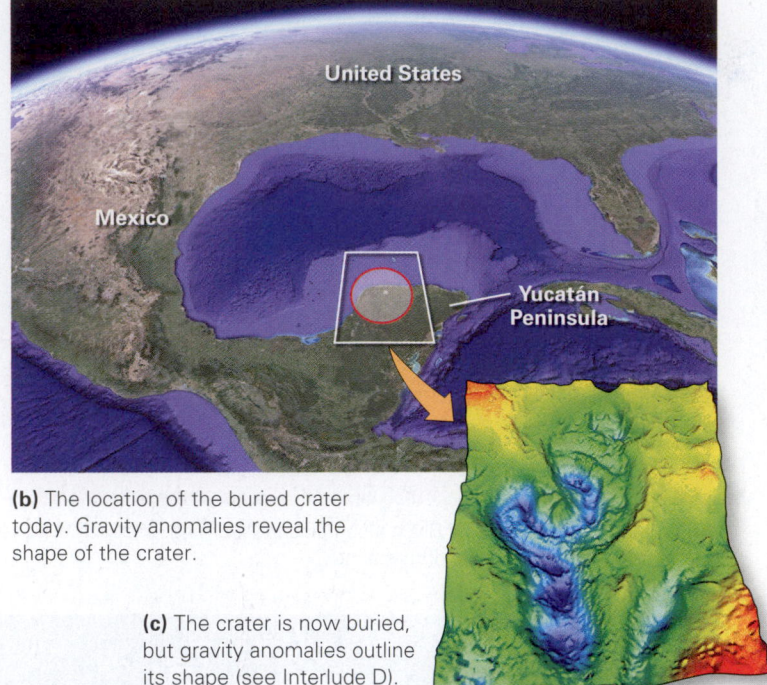

(b) The location of the buried crater today. Gravity anomalies reveal the shape of the crater.

(c) The crater is now buried, but gravity anomalies outline its shape (see Interlude D).

at this time, inhabiting almost all environments on Earth. Social herds of grazing dinosaurs roamed the plains, preyed on by the fearsome *Tyrannosaurus rex* (a Cretaceous, not a Jurassic, dinosaur, despite what Hollywood says!). Pterosaurs, with wingspans of up to 11 m, soared overhead, and birds began to diversify. Mammals also diversified and developed larger brains and more specialized teeth, but for the most part they remained small and rat-like.

The "K-T Boundary Event"

Geologists first recognized the K-T boundary (K stands for Cretaceous and T for Tertiary, an older term for Cenozoic time before the Quaternary) from 18th-century studies that identified an abrupt global change in fossil assemblages. Until the 1980s, most geologists assumed the faunal turnover took millions of years. But modern dating techniques indicate that this change happened almost instantaneously and that it signaled a sudden mass extinction of most species on Earth. The dinosaurs, rulers of the planet for over 150 million years, simply vanished, along with 90% of some plankton species in the ocean and up to 75% of plant species. What catastrophe could cause such a sudden and extensive mass extinction? From data collected in the 1970s and 1980s, most geologists have concluded that the Cretaceous Period came to a close, at least in part, as a result of the impact of a 13-km-wide meteorite at the site of the present-day Yucatán Peninsula in Mexico (**Fig. 13.24a, b**).

The discoveries leading up to this conclusion provide fascinating insight into how science works. The story began when Walter Alvarez, a geologist studying strata in Italy, noted that a thin layer of clay interrupted the deposition of deep-sea limestone precisely at the K-T boundary. Cretaceous plankton shells constituted the limestone below the clay layer, whereas Cenozoic plankton shells made up the limestone just above the clay. Apparently, for a short interval of time at the

K-T boundary, all the plankton died, so only clay settled out of the sea. When Alvarez, his father, Luis (a physicist), and other colleagues analyzed the clay, they learned that it contained *iridium*, a very heavy element found only in extraterrestrial objects. Soon geologists were finding similar iridium-bearing clay layers at the K-T boundary all over the world. Further study showed that the clay layer contained other unusual materials, such as tiny glass spheres (formed from the flash freezing of molten rock), wood ash, and shocked quartz (grains of quartz that had been subjected to intense pressure). Only an immense impact could explain all these features. The glass spherules formed when melt sprayed in the air from the impact site, the iridium came from fragments of the colliding object, and the shocked quartz grains were produced and scattered by the force of the impact. The wood ash resulted when forests were set ablaze at the time—this conceivably happened because the impact ejected super-hot debris at such high velocity that the debris almost went into orbit and could reach forests worldwide. The impact also generated 2-km-high tsunamis that inundated the shores of continents and generated a blast of super-hot air.

Researchers suggest that the impact caused unfathomable destruction because the dust, ash, and aerosols that it produced lofted into the atmosphere and transformed the air into a murky haze that reflected incoming sunlight. As a result, photosynthesis became difficult, so all but the hardiest plants died, and winter-like cold might have lasted all year, perhaps for years. Finally, sulfur-bearing aerosols could have combined with water to produce acid rain, acidifying the ocean and land. These conditions could have broken the food chain and, therefore, could have triggered extinctions.

Geologists suggest that the meteorite responsible for the K-T boundary event landed on the northwestern coast of the Yucatán Peninsula, because beneath the reefs and sediments of this tropical realm lies a 100-km-wide by 16-km-deep scar called the Chicxulub crater (**Fig. 13.24c**). A layer of glass spherules up to 1 m thick occurs at the K-T boundary in strata near the site. And radiometric dating indicates that igneous melts in the crater formed at the time of the K-T boundary event. The discovery of this event has led geologists to speculate that other such collisions may have punctuated the path of life evolution throughout Earth history and has led to modern-day efforts to track asteroids that pass close to the Earth.

Take-Home Message

The Mesozoic began with the breakup of Pangaea and the formation of the Atlantic. A convergent-plate boundary formed along the west coast of North America, yielding the Sierran arc and eventually leading to the uplift of the Rocky Mountains. Dinosaurs ruled the planet and modern forests appeared. A huge meteorite impact marks the K-T boundary event, the end of the era—it may have caused the extinction of the dinosaurs.

QUICK QUESTION: Did all of today's continents break off of Pangaea at the same time?

13.8 The Cenozoic Era: The Modern World Comes to Be

Paleogeography Plate tectonics doesn't stop, so during the last 66 million years the map of the Earth has continued to change, gradually producing the configuration of continents we see today. The final stages of the Pangaea breakup separated Australia from Antarctica and Greenland from North America and formed the North Sea between Britain and continental Europe. The Atlantic Ocean continued to grow because of seafloor spreading on the Mid-Atlantic Ridge, and thus the Americas moved relatively westward, away from Europe and Africa. Meanwhile, the continents that once constituted Gondwana drifted northward as the intervening ocean was consumed by subduction. Several volcanic island arcs and microcontinents started colliding with Asia beginning around 50 to 60 million years ago, and then about 40 Ma India collided. The overall result of this protracted period of collisional

tectonics uplifted the Himalayas and the Tibetan Plateau. Meanwhile, Africa along with some volcanic island arcs and microcontinents collided with Europe to produce the Alps of southern Europe and the Zagros Mountains of Iran. Finally, collision between Australia and New Guinea led to orogeny in Papua New Guinea. Thus, collisions of the former Gondwana continents with the southern margins of Europe and Asia resulted in the formation of the largest orogenic belt on Earth today, the **Alpine-Himalayan chain** (**Fig. 13.25**).

As the Americas moved westward, convergent-plate boundaries evolved along their western margins. In South America, convergent-boundary activity built the Andes, which remains an active orogen to the present day. In North America, convergent-boundary activity continued without interruption until about 40 Ma (the Eocene Epoch) yielding, as we have seen, the Laramide orogeny. Then the Farallon-Pacific ridge reached the continental margin, and the configuration of plates along the western shore of North America began to evolve. By 25 Ma, a transform boundary had replaced part of the convergent boundary in the western part of the continent (**Fig. 13.26**). Where this happened, volcanism and compression ceased in western North America, and strike-slip faulting took over. This led to the formation of the San Andreas fault system in California, and the Queen Charlotte fault system off the west coast of Canada. Along the San Andreas and Queen Charlotte faults today, the Pacific Plate moves northward with respect to North America at a rate of about 6 cm per year. In the western United States, convergent-boundary tectonics continues only in Washington, Oregon, and northern California, where subduction of the Juan de Fuca Plate generates the volcanism of the Cascade volcanic chain. At depth beneath North America, the subducted plate peeled off the base of North America and began to sink deeper into the mantle.

As compression associated with convergent tectonics ceased in the western United States south of the Cascades, the region began to undergo extension in a roughly east-west direction. The result was the formation of the **Basin and Range Province**, a broad continental rift. Continued extension in this rift, over the past 20

FIGURE 13.25 The two main active continental orogenic systems on the Earth today. The Alpine-Himalayan system formed when Africa, India, and Australia collided with Asia (inset). The Cordilleran and Andean systems reflect the consequences of convergent-boundary tectonism along the eastern Pacific Ocean.

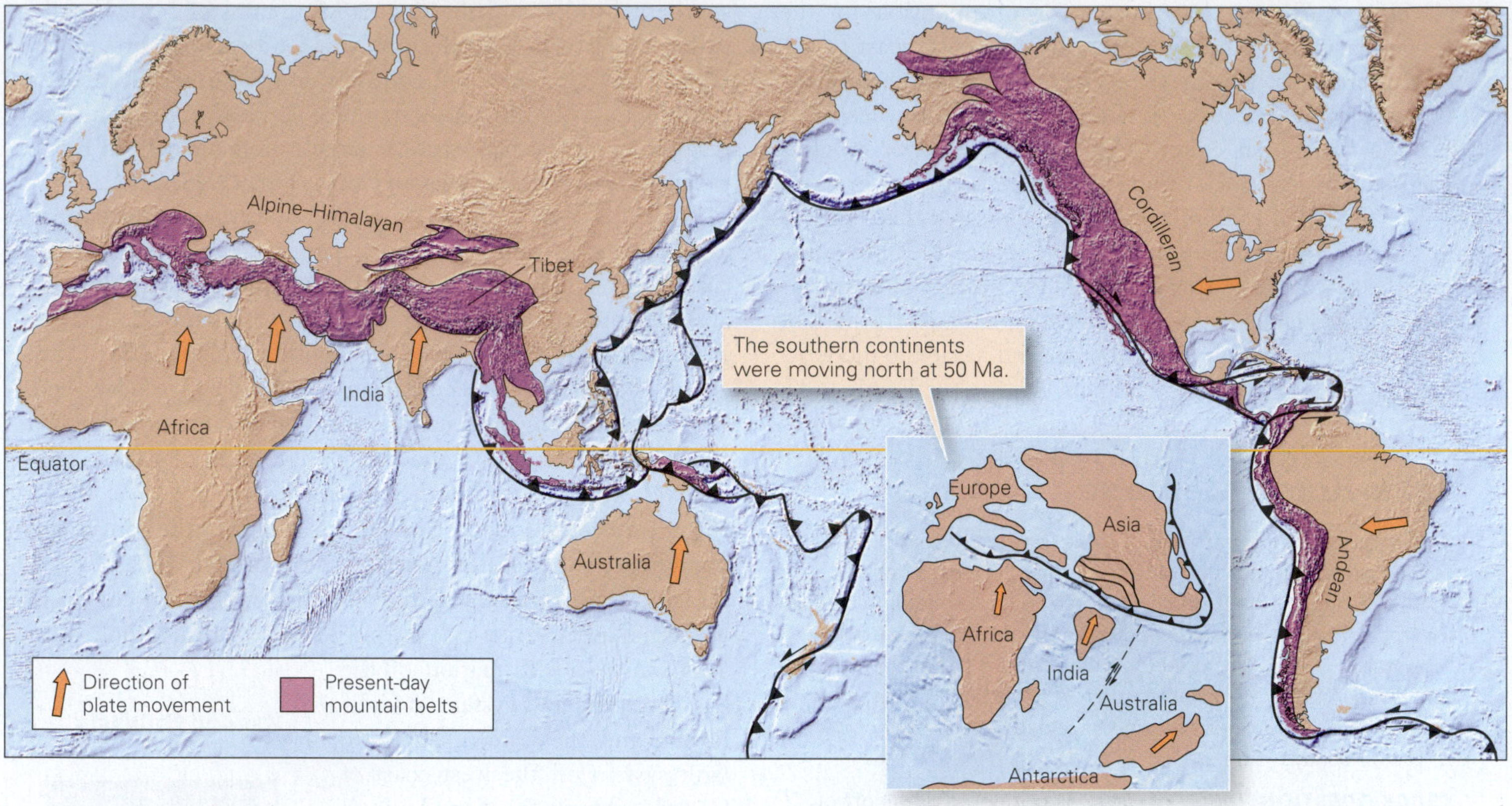

FIGURE 13.26 The western margin changed from a convergent-plate boundary into a transform-plate boundary after subduction of the Farallon Ridge. Then the San Andreas fault developed. To the east, the Basin and Range rift was developed.

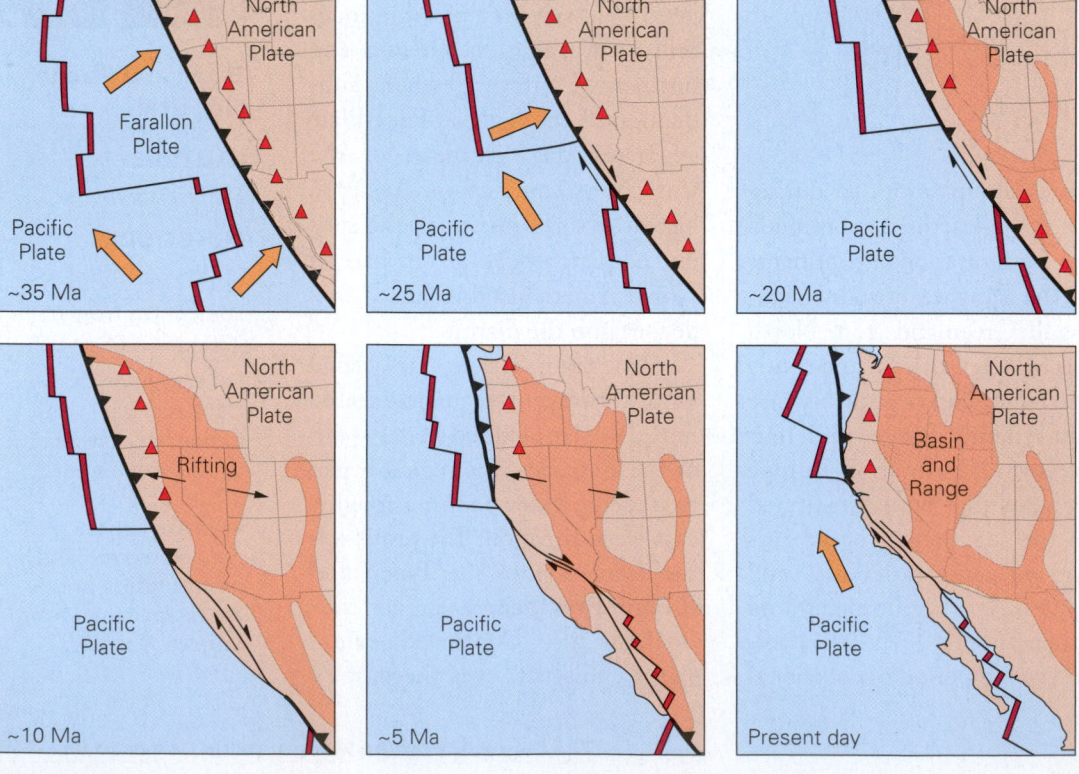

Ma, has caused the region to stretch to twice its original width (**Fig. 13.27**). The Basin and Range Province's name reflects its topography—the province contains long, narrow mountain ranges separated from each other by flat, sediment-filled basins. This topography formed when the crust of the region was broken up by normal faults (see Chapter 11). Blocks of crust above these faults slipped down and tilted, producing narrow, wedge-shaped depressions. The upward protruding crests of the tilted blocks form the ranges. The depressions between the ranges rapidly filled with sediment eroded from the ranges and became the basins. The Basin and Range Province terminates just north of the Snake River Plain, the track of the hot spot

FIGURE 13.27 The Basin and Range Province is a rift. Its opening caused rotation of the Sierra Nevada. The opening of the Rio Grande Rift caused rotation of the Colorado Plateau, which is an unrifted block of cratonic crust. The inset shows a cross section along the red line.

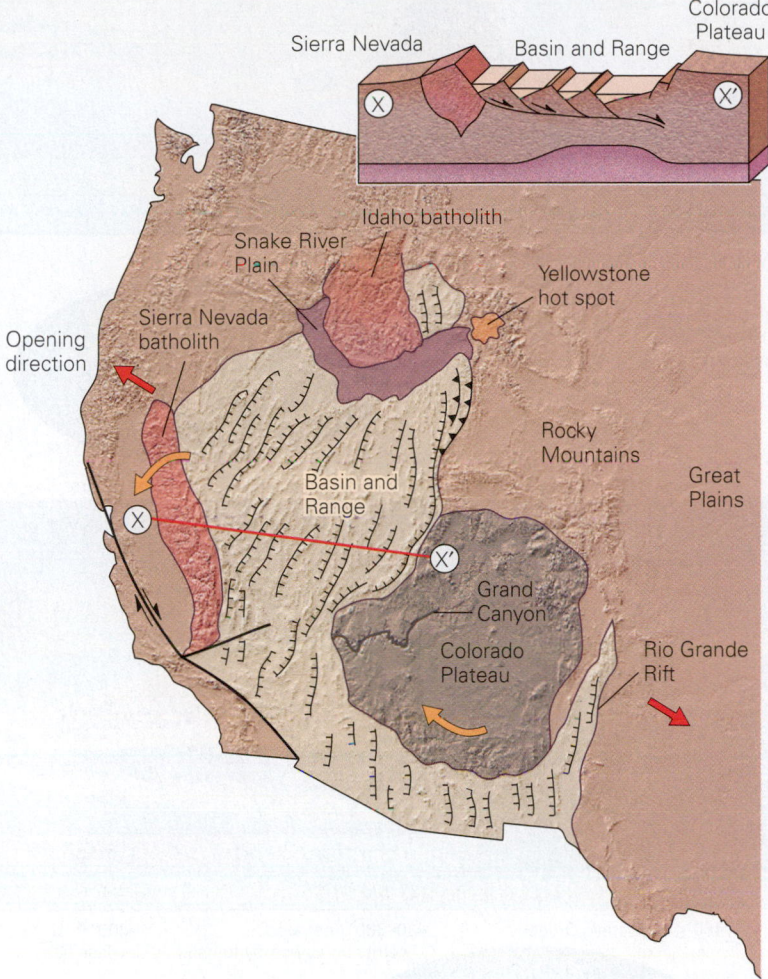

this name is no longer "official" but remains widely used.) The Quaternary begins at 2.6 Ma and continues to the present. Between 2.6 Ma and 12 Ka, an interval of time called the Pleistocene Epoch, continental glaciers expanded and retreated across northern continents at least 20 times, an event known as the **Pleistocene Ice Age** (**Fig. 13.28**). During each *glaciation*, the time intervals when the glaciers grew, sea level fell so much that the continental shelf became exposed to air, and at times a **land bridge** formed across the Bering Strait, west of Alaska. Animals and people migrated across this bridge from Asia into North America. A partial land bridge also formed from Southeast Asia to Australia, making human migration to Australia easier. Erosion and deposition by the glaciers created much of the landscape we see today in northern temperate regions. During *interglacials*, intervals of time when glaciers retreated, the land bridges and broad areas of continental shelves were submerged again. About 12,000 years ago, the climate warmed, and we entered the interglacial time interval we are still experiencing today (see Chapter 22). This most recent interval of time is the **Holocene Epoch**.

Life Evolution Within a few million years of the K-T boundary catastrophe, plant life recovered, and forests of both angiosperms and gymnosperms reappeared. The grasses, which first appeared in the Cretaceous, spread across the plains in temperate and subtropical climates by the middle of the Cenozoic Era, transforming them into vast grasslands. The dinosaurs, except

FIGURE 13.28 The maximum advance of the Pleistocene ice sheet in North America.

that now lies beneath Yellowstone National Park. As North America drifts westward, volcanic calderas have formed along the Snake River Plain; Yellowstone National Park straddles the most recent caldera (see Chapter 9).

Recall that during the Cretaceous Period, the world experienced greenhouse conditions and sea level rose so that extensive areas of continents were submerged. During the Cenozoic Era, however, the global climate rapidly shifted to icehouse (cooler) conditions, and by the early Oligocene Epoch Antarctic glaciers reappeared for the first time since the Triassic. The climate continued to grow colder through the Late Miocene Epoch, leading to the formation of grasslands in temperate climates. In the Pliocene, when the Isthmus of Panama formed, land separated the Atlantic completely from the Pacific, and the configuration of oceanic currents changed. The change in currents, in turn, may have decreased the transport of oceanic heat to polar regions and may have triggered the formation of the sea ice that covered the Arctic Ocean.

The Cenozoic consists of three periods—the Paleogene, the Neogene, and the Quaternary. (Until recently, the Paleogene and Neogene together comprised the Tertiary Period;

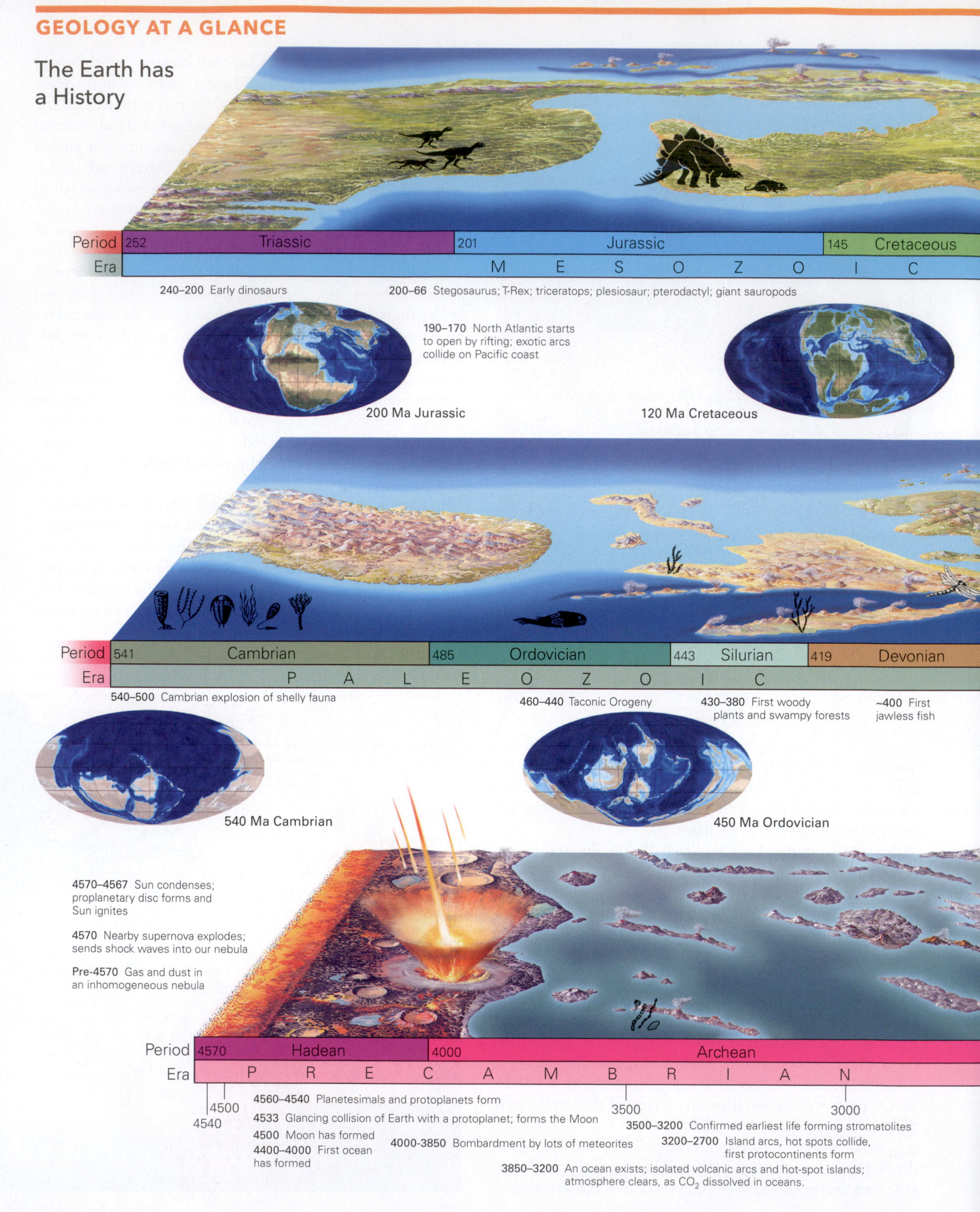

GEOLOGY AT A GLANCE

The Earth has a History

Period	252	Triassic	201	Jurassic	145	Cretaceous
Era			M E S O Z O I C			

240–200 Early dinosaurs

200–66 Stegosaurus; T-Rex; triceratops; plesiosaur; pterodactyl; giant sauropods

190–170 North Atlantic starts to open by rifting; exotic arcs collide on Pacific coast

200 Ma Jurassic

120 Ma Cretaceous

Period	541	Cambrian	485	Ordovician	443	Silurian	419	Devonian
Era				P A L E O Z O I C				

540–500 Cambrian explosion of shelly fauna

460–440 Taconic Orogeny

430–380 First woody plants and swampy forests

~400 First jawless fish

540 Ma Cambrian

450 Ma Ordovician

4570–4567 Sun condenses; proplanetary disc forms and Sun ignites

4570 Nearby supernova explodes; sends shock waves into our nebula

Pre-4570 Gas and dust in an inhomogeneous nebula

Period	4570	Hadean	4000	Archean
Era		P R E C A M B R I A N		

4500

4540

4560–4540 Planetesimals and protoplanets form

4533 Glancing collision of Earth with a protoplanet; forms the Moon

4500 Moon has formed

4400–4000 First ocean has formed

4000-3850 Bombardment by lots of meteorites

3500

3000

3500–3200 Confirmed earliest life forming stromatolites

3200–2700 Island arcs, hot spots collide, first protocontinents form

3850–3200 An ocean exists; isolated volcanic arcs and hot-spot islands; atmosphere clears, as CO_2 dissolved in oceans.

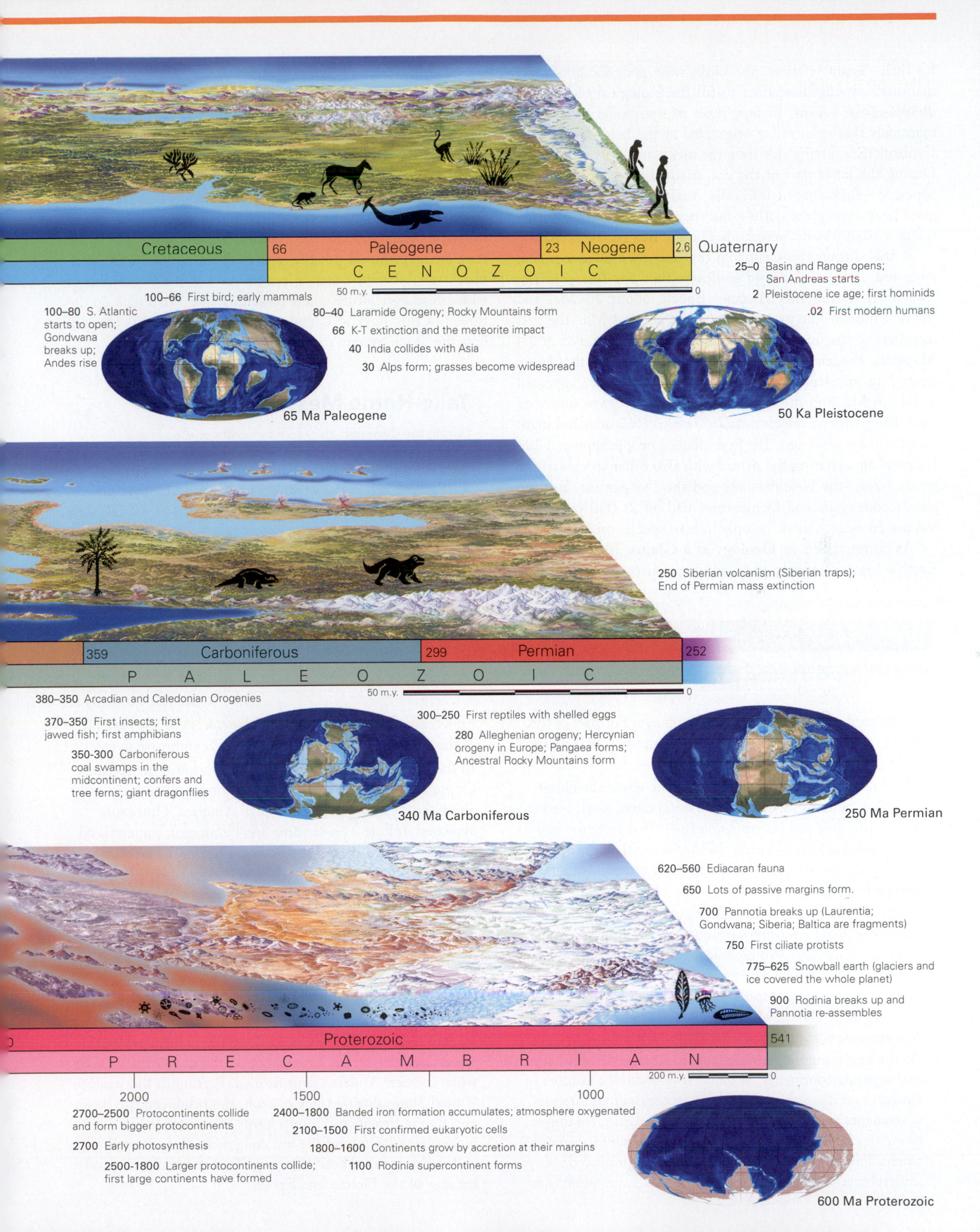

Cretaceous | 66 | Paleogene | 23 | Neogene | 2.6 | Quaternary

C E N O Z O I C

50 m.y. _____ 0

100–66 First bird; early mammals

100–80 S. Atlantic starts to open; Gondwana breaks up; Andes rise

80–40 Laramide Orogeny; Rocky Mountains form

66 K-T extinction and the meteorite impact

40 India collides with Asia

30 Alps form; grasses become widespread

25–0 Basin and Range opens; San Andreas starts

2 Pleistocene ice age; first hominids

.02 First modern humans

65 Ma Paleogene

50 Ka Pleistocene

359 | Carboniferous | 299 | Permian | 252

P A L E O Z O I C

50 m.y. _____ 0

380–350 Arcadian and Caledonian Orogenies

370–350 First insects; first jawed fish; first amphibians

350-300 Carboniferous coal swamps in the midcontinent; confers and tree ferns; giant dragonflies

300–250 First reptiles with shelled eggs

280 Alleghenian orogeny; Hercynian orogeny in Europe; Pangaea forms; Ancestral Rocky Mountains form

250 Siberian volcanism (Siberian traps); End of Permian mass extinction

340 Ma Carboniferous

250 Ma Permian

Proterozoic | 541

P R E C A M B R I A N

200 m.y. _____ 0

2000   1500   1000

620–560 Ediacaran fauna

650 Lots of passive margins form.

700 Pannotia breaks up (Laurentia; Gondwana; Siberia; Baltica are fragments)

750 First ciliate protists

775–625 Snowball earth (glaciers and ice covered the whole planet)

900 Rodinia breaks up and Pannotia re-assembles

2700–2500 Protocontinents collide and form bigger protocontinents

2700 Early photosynthesis

2500-1800 Larger protocontinents collide; first large continents have formed

2400–1800 Banded iron formation accumulates; atmosphere oxygenated

2100–1500 First confirmed eukaryotic cells

1800–1600 Continents grow by accretion at their margins

1100 Rodinia supercontinent forms

600 Ma Proterozoic

for their distant relatives, the birds, were gone for good, and mammals rapidly diversified to fill the ecological niches that dinosaurs left vacant. In fact, most of the modern groups of mammals that exist today originated at the beginning of the Cenozoic Era, giving this time the nickname *Age of Mammals*. During the latter part of the era, particularly huge mammals appeared—including mammoths, mastodons, giant beavers, giant bears, and giant sloths—but these became extinct during the past 10,000 years, perhaps because of hunting by humans.

It was during the Cenozoic that our own ancestors first appeared. The fossil record indicates that ape-like primates diversified during the Miocene Epoch, about 20 Ma, and the first human-like primates appeared about 4 Ma. The first members of the human genus, *Homo*, have been found in 2.4 Ma strata. Evidence from studies in Africa indicates that *Homo erectus*, an ancestor capable of making stone axes, appeared about 1.6 Ma, and the line leading to *Homo sapiens* (our species) diverged from *Homo neanderthalensis* (Neanderthal man) about 500 Ka years ago. The first modern people appeared 200 Ka years ago, sharing the planet with two other species of the genus *Homo*, the Neanderthals and the Denisovans. But the last Neanderthals and Denisovans died off 25,000 years ago, leaving *Homo sapiens* as the only human species on Earth.

As summarized in **Geology at a Glance** (pp. 498–499), Earth's history reflects the complex consequences of plate interactions, sea-level changes, atmospheric changes, life evolution, and even meteorite impact. In the past few millennia, humans have had a huge effect on the planet, causing changes significant enough to be obvious in the geologic record of the future. In fact, geologists now informally refer to the more recent portion of the Holocene, specifically, the time during which human activities have had a major impact on the Earth System, as the **Anthropocene**. Different researchers assign different start dates to the Anthropocene—some place the start at the beginning of the industrial revolution, a few centuries ago, whereas others place its start at the beginning of widespread agriculture, a few millennia ago. We'll pick up the thread of this story in Chapter 23, where we discuss ideas of how the Earth System may change in the future.

Take-Home Message

During the Cenozoic, the mountain belts of today rose, and modern plate boundaries became established. After the K-T mass extinction, mammals diversified. During the Pleistocene, glaciers covered large areas of continents, and humans appeared.

QUICK QUESTION: What interval of time does the Anthropocene refer to?

CHAPTER SUMMARY

- Earth formed about 4.54 billion years ago. For part of the first 600 million years, the Hadean Eon, the planet was so hot that its surface was a magma ocean.

- The Archean Eon began about 4.00 Ga, when the oldest rock that remains formed. Continental crust, assembled out of volcanic arcs and hot-spot volcanoes that were too buoyant to subduct, grew during the Archean. The atmosphere contained little oxygen, but the first life forms—bacteria and archaea—appeared.

- In the Proterozoic Eon, which began at 2.5 Ga, Archean cratons collided and were sutured together along orogenic belts, forming large Proterozoic cratons. Photosynthesis by organisms added oxygen to the atmosphere. By the end of the Proterozoic, complex shell-less marine invertebrates populated the planet. Most continental crust accumulated to form a supercontinent called Rodinia at 1 Ga.

- At the beginning of the Paleozoic Era, rifting yielded several separate continents. Sea level rose and fell a number of times, creating sequences of strata in continental interiors. Continents began to collide and coalesce again, leading to orogenies and, by the end of the era, to another supercontinent, Pangaea. Early Paleozoic evolution produced many invertebrates with shells, and jawless fish. Land plants and insects appeared in the middle Paleozoic. And, by the end of the eon, there were land reptiles and gymnosperm trees.

- In the Mesozoic Era, Pangaea broke apart and the Atlantic Ocean formed. Convergent-boundary tectonics dominated along the western margin of North America. Dinosaurs appeared in Late Triassic time and became prominent land animals through the Mesozoic Era. During the Cretaceous Period, sea level was very high, and the continents flooded. Angiosperms appeared at this time, along with modern fish. A huge mass-extinction event, which wiped out the dinosaurs, occurred at the end of the Cretaceous Period, probably because of the impact of a large meteorite.

- In the Cenozoic Era, continental fragments of Pangaea collided again. The collision of Africa and India with Asia and Europe formed the Alpine-Himalayan orogen. Convergent tectonics has persisted along the margin of South America, creating the Andes, but ceased in North America when the San Andreas fault formed. Rifting in the western United States during the Cenozoic Era produced the Basin and Range Province. Various kinds of mammals filled niches left vacant, and the human genus, *Homo*, appeared and evolved throughout the radically shifting climate and ice ages of the Pleistocene Epoch.

KEY TERMS

Alpine-Himalayan chain (p. 495)

Ancestral Rockies (p. 485)

Anthropocene (p. 498)

Appalachian fold-thrust belt (p. 485)

Archean Eon (p. 473)

basement uplifts (p. 491)

Basin and Range Province (p. 495)

Cambrian explosion (pp. 482–83)

craton (p. 476)

cratonic platform (p. 476)

differentiation (p. 470)

Ediacaran fauna (p. 479)

exotic terrane (p. 489)

Gondwana (p. 482)

great oxygenation event (p. 480)

Grenville orogeny (p. 477)

Holocene Epoch (p. 497)

land bridge (p. 497)

Laramide orogeny (p. 491)

Laurentia (p. 482)

Permian mass extinction (p. 486)

Phanerozoic Eon (p. 480)

Pleistocene Ice Age (p. 497)

Proterozoic Eon (p. 476)

Rodinia (p. 477)

Sevier orogeny (p. 490)

shield (p. 476)

snowball Earth (p. 480)

stratigraphic sequence (p. 486)

stromatolite (p. 475)

superplume (p. 492)

REVIEW QUESTIONS

1. Why are there no whole rocks on Earth that yield isotopic dates older than 4 billion years?

2. Describe the condition of the crust, atmosphere, and oceans during the Hadean Eon.

3. How did the atmosphere and tectonic conditions change during the Proterozoic Eon?

4. What evidence do we have that the Earth nearly froze over twice during the Proterozoic Eon?

5. How did the Cambrian explosion of life change the nature of the living world?

6. How did the Alleghanian and Ancestral Rockies orogenies affect North America?

7. What are the major types of organisms that appeared during the Paleozoic?

8. Describe the plate-tectonic conditions that led to the formation of the Sierran arc and the Sevier thrust belt. What happened during the Laramide orogeny?

9. What life forms appeared during the Mesozoic?

10. What may have caused the flooding of the continents during the Cretaceous Period?

11. What could have caused the K-T extinctions?

12. What continents formed as a result of the breakup of Pangaea?

13. What caused the Himalayas and the Alps to form?

14. What major tectonic provinces formed in the western United States during the Cenozoic?

15. What major climatic and biologic events happened during the Pleistocene?

ON FURTHER THOUGHT

16. During intervals of the Paleozoic, large areas of continents were submerged by shallow seas. Using *Google Earth*™, tour North America from space. Do any present-day regions within North America consist of continental crust that was submerged by seawater? What about regions offshore? (Hint: Look at the region just east of Florida.)

17. Geologists have concluded that 80% to 90% of Earth's continental crust had formed by 2.5 Ga. But if you look at a geological map of the world, you find that only about 10% of the Earth's continental crustal surface is labeled "Precambrian." Why?

PART V

EARTH RESOURCES

The earliest humans were hunter-gatherers and needed only food and water to survive. But then people discovered that fires made food easier to eat and campsites more comfortable, weapons made hunting more successful, shelters made daily life more pleasant, and farming made food supplies more reliable—and humanity's needs expanded. Specifically, people began to require energy resources (sources of heat and/or power) and mineral resources (materials from which metals and other chemicals can be derived). In a general sense, we use the term resource for any item that can be employed for a useful purpose. Modern society requires resources obtained directly from the Earth System in order to survive and prosper. Think about it . . . metals, oil, plastics, wallboard, brick, coal, pottery, cement, uranium, and more all come from the Earth's crust. To appreciate the value and cost of such Earth resources, it's important to know how they form, where they can be found, how they're extracted, whether or not they're sustainable, and how their use can impact the environment. In Chapter 14, we focus on the energy resources that come from the Earth. These include fossil fuels (oil and coal) as well as nuclear fuel and moving water. Chapter 15 focuses on nonenergy resources, particularly the mineral deposits from which we obtain metals.

14 Squeezing Power from a Stone: Energy Resources

15 Riches in Rock: Mineral Resources

An exposed fracture surface reveals a weathered coating of malachite, a beautiful green mineral sometimes used for jewelry. Malachite is also a copper ore mineral, meaning that its crystals contain a significant proportion of copper atoms. The copper we use for wires and coins comes from this and other ore minerals. The Earth provides many of the resources essential to society.

These lumps of coal, piled near a coal mine in Indiana, formed from vegetation that accumulated in swamps of central North America about 310 Ma and was then buried deeply where it transformed into rock. Burning one of the larger chunks would light a lightbulb for a day.

CHAPTER 14

Squeezing Power from a Stone: Energy Resources

> To keep a lamp burning, we have to keep putting oil in it.
>
> —*Mother Teresa (Nobel Peace Prize winner, 1910–1997)*

14.1 Introduction

The extreme chill of an arctic midwinter doesn't stop a wolf from stalking its prey. The wolf's legs move through the snow, its heart pumps, its lungs inhale and exhale, and its body radiates heat—all these activities require energy. **Energy**, as defined by a physicist, is the capacity to do work, to cause a change in a physical or biological system. This means that energy can raise the temperature of a material, can drive chemical reactions, can cause an object to move or change shape, can generate light and/or magnetism, or can change the state of a material (from solid to liquid or gas). A wolf's energy comes from the metabolism of sugar, protein, and carbohydrates in its body. These chemicals, in turn, come from the food the wolf catches and eats—thus, we can think of mice, rabbits, and deer as an **energy resource**, a source of materials that yield energy, for a wolf.

Early humans, like wolves, could supply their energy needs entirely from their food, so they could survive by hunting and gathering. But when people discovered how to use fire, tools, and weapons, their need for energy resources began to exceed that of other animals. Before the advent of civilization, wood and dried dung could supply humanity's energy resource needs—these materials could serve as **fuel**, an energy resource in a usable and transportable form (**Fig. 14.1a**). As people began to congregate in towns, however, they began to use energy for agriculture and transportation. At first, animal power, wind, and flowing water met society's additional energy demand, but as populations grew, and new industries such as iron smelting emerged, energy resource needs began to outpace the supplies available at the Earth's surface. In fact, to feed the smelting industry, 17th-century woodcutters devastated European forests (**Fig. 14.1b**). The industrial revolution could

FIGURE 14.1 Nongeologic sources of energy dominated in the past.

(a) Women collecting dung to burn for cooking in India (ca. 1977).

(b) A painting of an ironworks in Norway from 1800 by John Edy (1760–1820).

not begin unless new fuel supplies could be found, and society turned to coal, the fossilized remains of woody plants, to meet the demand. Coal could be transported easily and contains about 1.7 times more energy per kilo than wood, so coal kept 18th-century cooking and heating stoves hot, and it powered the steam engines of factories and trains around the industrializing world.

Since the industrial revolution, society's hunger for energy has increased unabated (**Fig. 14.2**). In the United States today, for example, an average city dweller uses more than 110 times the amount of energy that a prehistoric hunter did. Most energy for human consumption in the industrial world now comes from oil, natural gas, and coal, but we still use wind and flowing water. In the last half century we've added nuclear energy, geothermal energy, and solar energy to the list of energy resources, and there has been growing interest in expanding the use of biofuels from plant crops. In 2007, the world's energy picture changed significantly when efforts to extract natural gas from shale began to burgeon, and natural gas has now started to substitute for other fuels. But use of energy brings with it myriad challenges to society—the

distribution of supplies can spark discord among nations, production and consumption both can have undesirable consequences for the environment and climate, and issues concerning long-term sustainability of resources remain incompletely answered.

Why does a geology book include a chapter devoted to energy resources? Because most of these resources originate in geologic materials or are the result of geologic processes. Thus, to understand the source and limitations of energy resources and to find new resources, we must understand their geological context and geological consequences. (That's why the energy industry employs tens of thousands of geologists.) To help you to understand the geology of energy, this chapter begins by surveying the various types of energy resources on Earth. Then we focus on *fossil fuels* (oil, gas, and coal), which are combustible materials derived from organisms that lived in the past. The chapter continues with a survey of other energy resources and concludes by outlining the dilemmas that society faces as conventional energy resources begin to run out and products of energy consumption enter our environment.

FIGURE 14.2 The proportion of different sources of energy that people use has change over time, and the amount of energy used almost continuously increases.

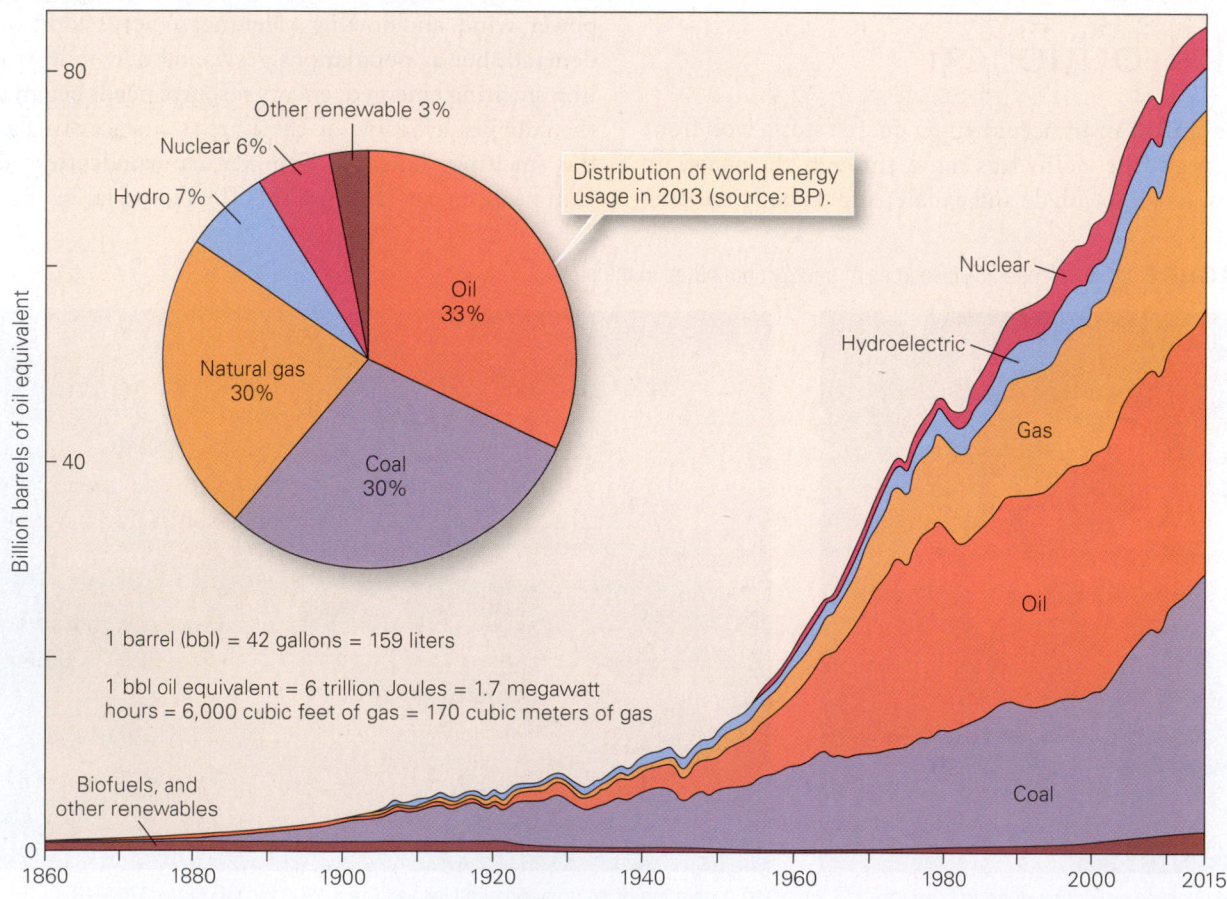

14.2 Sources of Energy in the Earth System

What comes to mind when someone asks you to name an energy resource? Perhaps you think about the gasoline that fills the tanks of cars, or the mounds of coal piled outside of a power plant, or the flowing stream that turns a water-wheel. Alternatively, you may think of windmills or arrays of solar panels, because they are appearing on the landscape with increasing frequency. Let's step back and consider where the energy in these energy resources comes from in the first place (**Fig. 14.3**):

- *Energy directly from the Sun:* Solar energy, resulting from nuclear fusion reactions in the Sun, bathes the Earth's surface. It may be converted directly into electricity, using solar-energy panels, or it may be used to heat water in tanks.
- *Energy directly from gravity:* The gravitational attraction of the Moon, and to a lesser extent the Sun, causes ocean *tides*, the daily up-and-down movement of the sea surface. The flow of water in and out of channels during tidal changes can drive turbines.
- *Energy involving both solar energy and gravity:* Solar radiation heats the air, which becomes buoyant and rises. As this happens, gravity causes cooler air to sink. The resulting air movement, *wind*, powers sails and windmills. Solar energy also evaporates water, which enters the atmosphere. When the water condenses, it rains and falls on the land, where it accumulates in streams that flow downhill in response to gravity. This moving water powers water-wheels and turbines.
- *Energy via photosynthesis:* Algae and green plants absorb some of the solar energy that reaches the Earth's surface. Their green color comes from a pigment called *chlorophyll*. With the aid of chlorophyll, plants produce sugar through a chemical reaction called *photosynthesis*. In chemist's shorthand, we can write the reaction in this way:

$$6CO_2 + 12H_2O + Light \rightarrow 6O_2 + C_6H_{12}O_6 + 6H_2O$$

carbon dioxide   water   oxygen   sugar   water

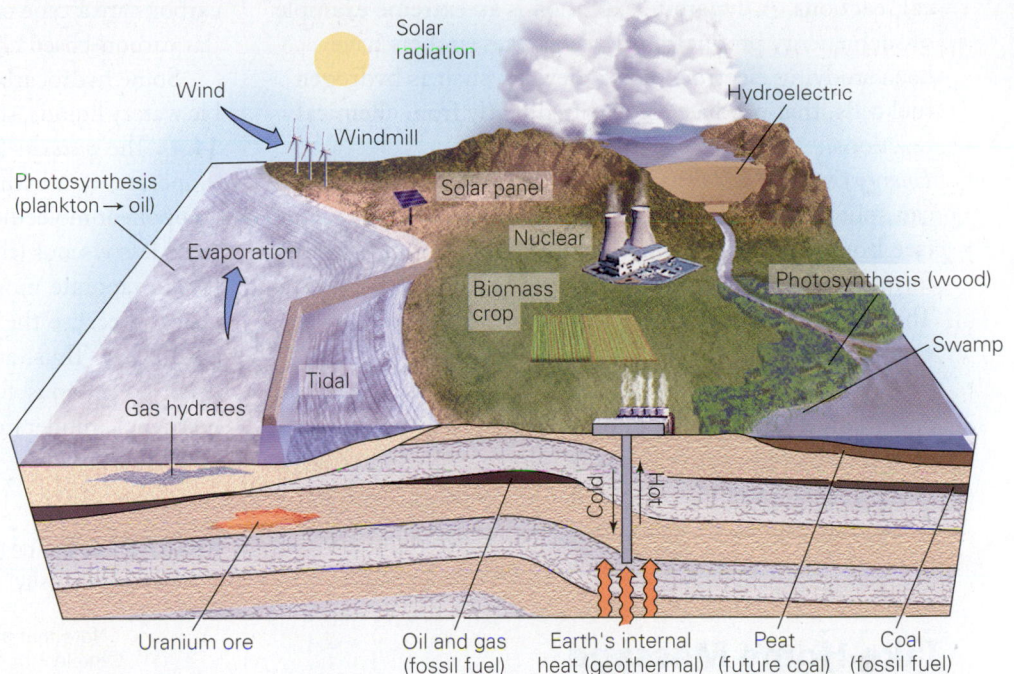

FIGURE 14.3 The diverse sources of energy on Earth. Surface sources are, ultimately, driven by heat from the Sun. Subsurface sources include fossil fuels, heat rising from hot rocks at depth, and radioactive minerals.

Plants use the sugar produced by photosynthesis to manufacture more complex chemicals, or they metabolize it directly to provide themselves with energy. Burning plant matter in a fire releases potential energy stored in the chemical bonds of organic chemicals. During burning, the organic molecules react with oxygen and break apart to produce carbon dioxide, water, and carbon (soot):

$$\text{plant molecules} + O_2 \xrightarrow{\text{burning}} CO_2 + H_2O + C + \text{other gases} + \text{heat energy}$$

The flames you see in fire consist of glowing gases released and heated by this reaction.

People have burned wood to produce energy for centuries. More recently, plant material (*biomass*) from corn, sugar cane, switchgrass, and algae has been used to produce ethanol, a flammable alcohol.

- *Energy from fossil fuels:* Oil, natural gas, and coal come from organisms that lived long ago and thus store solar energy that reached the Earth long ago. We refer to these substances as **fossil fuels**, to emphasize that they were derived from ancient organisms, the remains of which have been preserved in rocks over geologic time.

> **Did you ever wonder . . .**
>
> what the "fossils" are in fossil fuel?

- *Energy from chemical reactions:* A number of inorganic chemicals can burn to produce light and energy. The energy results from "exothermic" (heat-producing) chemical reactions. A dynamite explosion is an extreme example of such energy production. Recently, researchers have been studying electrochemical devices, such as hydrogen fuel cells, that produce electricity directly from chemical reactions.
- *Energy from nuclear fission:* Atoms of radioactive elements can split into smaller pieces, a process called *nuclear fission* (see Box 1.3). During fission, a tiny amount of mass transforms into a large amount of energy, called *nuclear energy*. This type of energy runs nuclear power plants and nuclear submarines.
- *Energy from Earth's internal heat:* Some of Earth's internal energy dates from the birth of the planet, while some is produced by radioactive decay in minerals. This internal energy heats water underground. The resulting hot water, when transformed to steam, provides *geothermal energy* that can drive turbines or heat buildings.

Take-Home Message

Energy comes from several sources—solar radiation, gravity, chemical reactions, radioactive decay, and the Earth's internal heat. Solar radiation and gravity together drive wind and water movement. Products of living organisms can be preserved in rocks as fossil fuels.

QUICK QUESTION: What kinds of energy sources are available on the Moon?

14.3 Introducing Hydrocarbon Resources

What Are Oil and Gas?

For reasons of economics and convenience, industrialized societies today rely primarily on familiar products derived from oil (such as gasoline, jet fuel, kerosene, and diesel) and various kinds of natural gas (such as methane and propane) for their energy needs. Why are these fuels so popular? They have a relatively high **energy density**, meaning that they contain a relatively large amount of energy per unit of weight. For example, 1 g of oil provides twice as much energy as 1 g of coal and about 500 times more energy than a battery of comparable weight provides. So an airplane can cross an ocean on a tank of jet fuel but wouldn't even be able to get off the ground if it had to run on batteries. Chemically, both oil and natural gas are

hydrocarbons, because they consist of chain-like or ring-like molecules made of carbon and hydrogen atoms. For example, bottled gas (propane) has the chemical formula C_3H_9. Hydrocarbons are a type of organic chemical, so-named because similar carbon-based chemicals make up living organisms.

Some hydrocarbons are gaseous and invisible, some resemble watery liquids, some appear syrupy, and some are solid (**Fig. 14.4**). The *viscosity* (ability to flow) and the *volatility* (ability to evaporate) of a hydrocarbon product depend on the size of its component molecules. Products composed of short chains tend to be less viscous (they can flow more easily) and more volatile (they evaporate more easily) than products composed of long chains, because the long chains tend to tangle and bond with each other. Thus, at room temperature short-chain molecules occur in gaseous form (such as cooking gas, which is 95% methane), moderate-length-chain molecules occur in liquid

FIGURE 14.4 The diversity of hydrocarbon products in order of increasing viscosity.

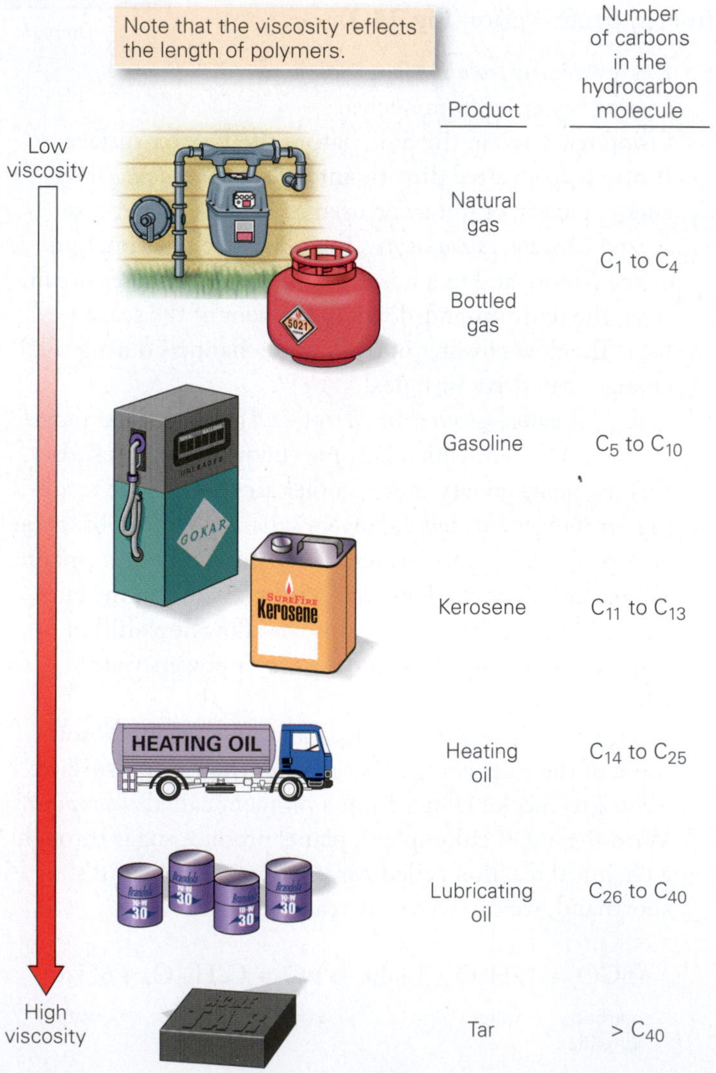

Note that the viscosity reflects the length of polymers.

	Product	Number of carbons in the hydrocarbon molecule
Low viscosity	Natural gas	C_1 to C_4
	Bottled gas	
	Gasoline	C_5 to C_{10}
	Kerosene	C_{11} to C_{13}
	Heating oil	C_{14} to C_{25}
	Lubricating oil	C_{26} to C_{40}
High viscosity	Tar	$> C_{40}$

form (such as gasoline and oil), and long-chain molecules occur in solid form (tar).

Why can we use hydrocarbons as fuel? Simply because hydrocarbons, like wood, burn, meaning they react with oxygen to form carbon dioxide, water, and heat. As an example, we can describe the burning of gasoline by the following reaction:

$$2C_8H_{18} + 25O_2 \rightarrow 16CO_2 + 18H_2O + \text{heat and light}$$

During such reactions, potential energy stored in the chemical bonds of the hydrocarbon molecules converts into heat and light (**Fig. 14.5**). This energy can be used to run engines or to transform water into the steam that drives generators in a power plant.

Hydrocarbon Generation in Source Rocks

Popular media often incorrectly imply that oil and gas are derived from buried trees or the carcasses of dinosaurs. In fact, the hydrocarbon molecules form from organic chemicals, such as fatty molecules called lipids, that were once in plankton. *Plankton* is made up of very tiny floating organisms including single-celled and very small multicellular plants (algae) as well as protists and microscopic animals. Typically, most planktonic organisms range in size from 0.02 to 2.0 mm in diameter. Note that it is the organic cells of plankton that can transform into hydrocarbons, not the mineral shells.

When the organisms die, they sink to the floor of the lake or sea that they lived in, and if the water is relatively "quiet" (nonflowing), they accumulate (**Fig. 14.6**). In most locations, relatively little organic matter settles out of the water column, because most plankton gets consumed by organisms higher in the food chain. But if surface waters are nutrient-rich and receive a lot of sunlight, plankton blooms and a significant amount can settle out. Commonly, the seafloor or lake-floor hosts an oxygen-rich environment populated by scavengers

FIGURE 14.6 The formation of oil. The process begins when organic debris settles with sediment. As burial depth increases, heat and pressure transform the sediment into black shale in which organic matter becomes kerogen. At appropriate temperatures, kerogen becomes oil, which then seeps upward.

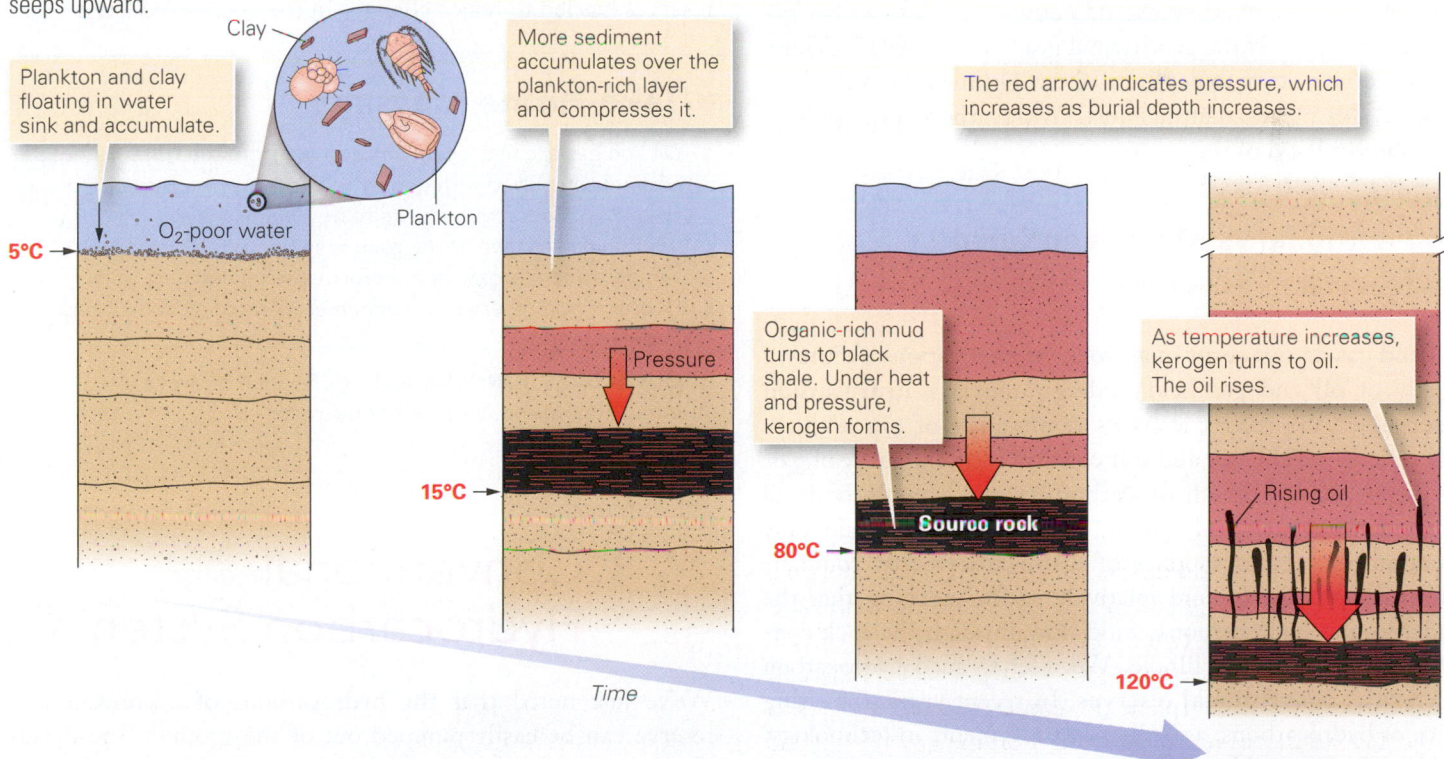

and microbes. As a result, dead plankton cells are consumed, decay, or oxidize before being buried. So, the organic matter in their cells transforms into CH_4 or CO_2 that bubbles away. But in oxygen-poor waters, the organic material can survive long enough to mix with clay and form an organic-rich, muddy ooze, which can then become buried by still more sediment so that it becomes preserved. Eventually, pressure from the weight of overlying sediment squeezes out the water, and the ooze becomes compacted and lithified to become black *organic shale*. (Shale that does not contain organic matter, in contrast, tends to be gray, tan, or red.) Organic shale contains the raw materials from which hydrocarbons form, so we refer to it as a **source rock**.

If organic shale is buried deeply enough (2 to 4 km), it becomes warmer, since temperature increases with depth in the Earth. Chemical reactions that take place in warm source rocks slowly transform the organic material in the shale into a variety of large waxy molecules that together comprise **kerogen**. Shale containing more than 25% to 75% kerogen is called **oil shale**. If oil shale warms to temperatures of greater than about 90°, kerogen molecules break into smaller oil and natural gas molecules, a process known as **hydrocarbon generation**. At temperatures over about 160°, any remaining oil breaks down to form natural gas. And at temperatures over 225°, organic matter loses all its hydrogen and transforms into graphite (pure carbon). Thus, oil itself forms only in a relatively narrow range of temperatures, called the **oil window** (**Fig. 14.7**). For regions with a geothermal gradient of 25°C/km, the oil window lies at depths of 3.5 to 6.5 km. Since gas survives to a higher temperature, the gas window extends down to 9 km, so the gas window is larger. If the geothermal gradient is low (15°C/km), oil can survive down to depths of about 11 km and gas down to 15 km. This means that, at most, hydrocarbons exist only in the topmost third of the crust.

Conventional vs. Unconventional Hydrocarbon Reserves

Oil and gas do not occur in all rocks at all locations. A known supply of oil and gas held underground is a **hydrocarbon reserve**—if the reserve consists dominantly of oil, it's usually called an *oil reserve* and if it consists predominantly of gas, it's a *gas reserve*. Some hydrocarbon reserves contain both oil and gas. Until relatively recently, oil companies could only economically obtain supplies of hydrocarbons that could be pumped from the ground relatively easily, meaning that the underground hydrocarbons could flow through the rock containing them to the drillhole. We refer to such hydrocarbon reserves as **conventional reserves**. In recent years, the rising price of hydrocarbons, as well as improvements in technology, have made it possible to obtain supplies that previously were

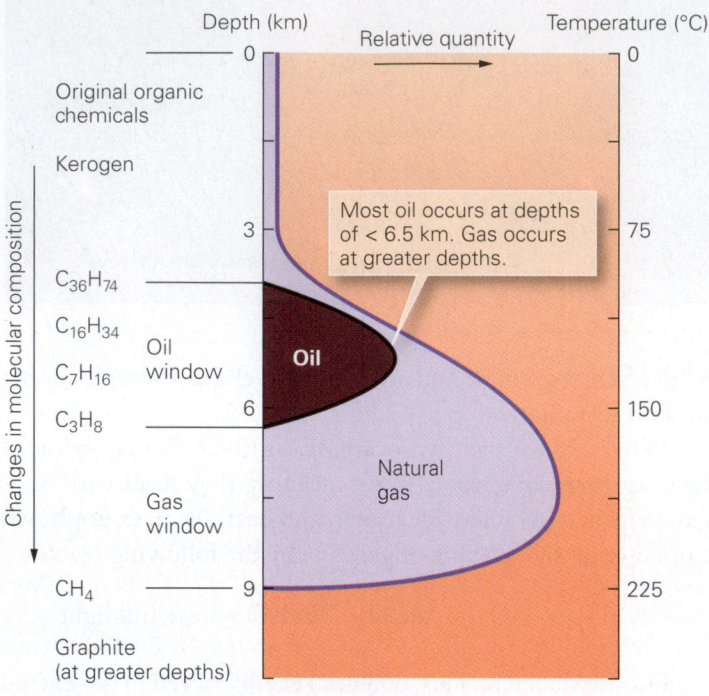

FIGURE 14.7 The "oil window" indicates subsurface conditions in which oil can form and survive. Deeper down, oil breaks down to form gas. At greater depth, metamorphism transforms organic material to graphite.

not easy to pump out simply by drilling a hole into the hydrocarbon-bearing rock. These hard-to-get supplies are known as **unconventional reserves**. The ability to access unconventional reserves has led to major changes in the energy industry.

Take-Home Message

Oil and gas are hydrocarbons derived from the remains of plankton buried with clay in an organic ooze. Burial transforms ooze into source rock, a black shale containing kerogen. If the rock is subjected to appropriate temperatures, organics transform into oil and gas. Hydrocarbon reserves are accumulations of oil and/or gas underground.

QUICK QUESTION: What is the difference between a conventional and unconventional reserve?

14.4 Conventional Hydrocarbon Systems

We've just noted that the hydrocarbons of a conventional reserve can be easily pumped out of the ground. The development of such a reserve requires a specific association of materials, conditions, and time. Geologists refer to this association as a **conventional hydrocarbon system** (**Fig. 14.8**). We already discussed the first two components of the system, namely the formation of a source rock of organic shale and the existence of thermal conditions that transform the kerogen in the shale into smaller hydrocarbon molecules. Let's now look at the remaining components, the migration of oil into a reservoir rock that lies within a configuration of rocks called an oil trap.

Reservoir Rocks and Hydrocarbon Migration

The clay flakes of an oil shale fit together tightly and thus prevent kerogen and any hydrocarbons forming within the rock or from moving easily through the rock. Therefore, you can't simply drill a hole into a source rock and pump out oil—the oil won't flow into the well fast enough to make the process cost efficient. To extract oil or gas from a conventional reserve, energy companies instead drill into **reservoir rocks**, rocks that contain, or could contain, accessible oil or gas. By "accessible," we mean oil or gas that can flow through rock and be sucked into a well fairly easily by pumping.

> **Did you ever wonder...**
>
> whether there are actually lakes or pools of oil underground?

To be a reservoir rock, a body of rock must have space in which the oil or gas can reside and must have channels through which the oil or gas can move. The space can be in the form of openings, or **pores**, between clastic grains (which exist because the grains didn't fit together tightly and because cement didn't fill all the spaces during cementation) or in the form of cracks and fractures that developed after the rock formed. In some cases, groundwater passing through rock dissolves minerals and creates new pores. **Porosity** refers to the amount of open space in a rock. Not all rocks have the same porosity (**Fig. 14.9**). For example, shale typically has a low porosity (less than 10%), whereas poorly cemented sandstone has a high porosity (up to 35%). That means that about a third of a block of porous sandstone actually consists of open space. The oil or gas in a reservoir rock occurs in the pores (just as water fills the holes in a sponge) so it is distributed through the rock—it does not occur in open pools underground. **Permeability** refers to the degree to which pore spaces connect to one another. In a permeable rock, the pores and cracks are linked, so a fluid is able to flow slowly through the rock, following a tortuous pathway. Note that even if a rock has high porosity, it is not

FIGURE 14.8 Stages of a conventional hydrocarbon system.

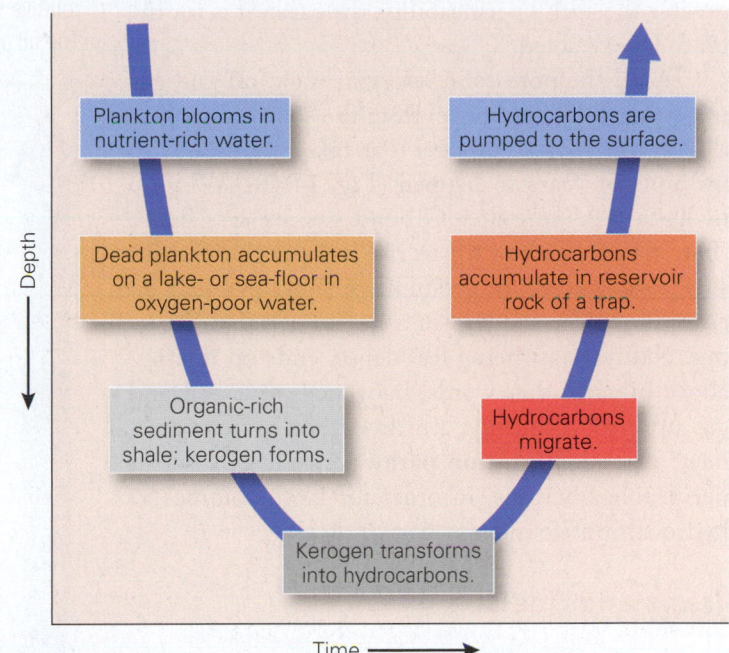

necessarily permeable (see Fig. 14.9), for if the pores aren't connected fluid can't move from one to another.

Keeping the concepts of porosity and permeability in mind, we can see that a poorly cemented sandstone makes a good reservoir rock because it is both porous and permeable. A highly fractured rock can be porous and permeable, even if there is no pore space between individual grains. A limestone can be permeable if groundwater has dissolved the surfaces of cracks in the rock. The greater the porosity, the greater

FIGURE 14.9 Porosity and permeability in sedimentary rocks. Rocks with high porosity and high permeability make the best reservoir rocks.

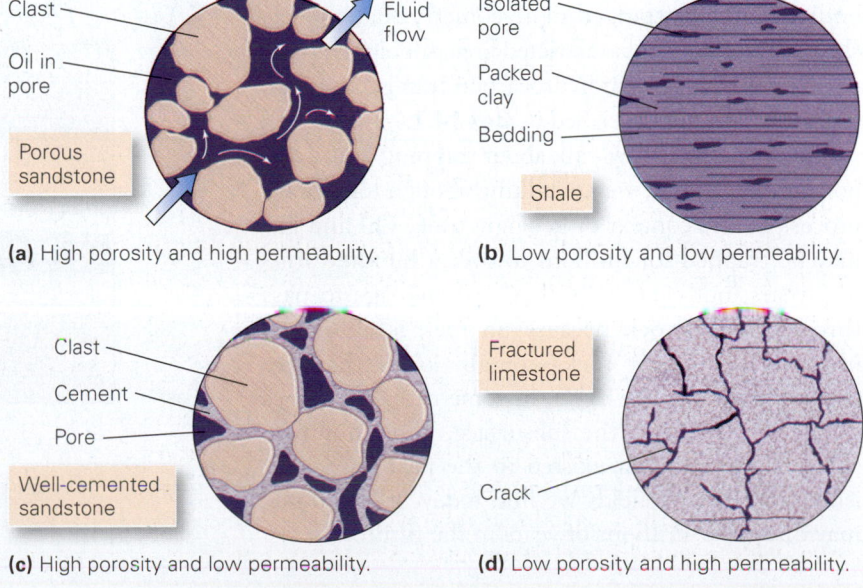

(a) High porosity and high permeability.

(b) Low porosity and low permeability.

(c) High porosity and low permeability.

(d) Low porosity and high permeability.

the capacity of a reservoir rock to hold oil, and the greater the rock's permeability, the easier it is for the oil to be extracted.

To fill the pores of a reservoir rock, oil and gas must first **migrate** (move) from the source rock into a reservoir rock, a process that takes thousands to millions of years to happen (**Fig. 14.10**). Why do hydrocarbons migrate? Oil and gas are less dense than water, so they try to rise toward the Earth's surface to get above groundwater, just as salad oil rises above the vinegar in a bottle of salad dressing. Natural gas, being less dense, ends up floating above oil. In other words, buoyancy drives oil and gas upward. Typically, a hydrocarbon system must have a good **migration pathway**, such as a set of permeable fractures, in order for large volumes of hydrocarbons to move.

Traps and Seals

The existence of reservoir rock alone does not create a conventional reserve, because if hydrocarbons can flow into a reservoir rock, they can also flow out. If oil or gas escape from the reservoir rock and ultimately reach the Earth's surface, where they can leak away at a **oil seep** or *gas seep* (**Fig. 14.11**), there will be none left underground to extract. Thus, for an oil reserve to exist, oil and gas must be held underground in the reservoir rock by means of a geologic configuration called a **trap**.

An oil or gas trap has two components. First, a **seal rock**, a relatively impermeable rock such as shale, salt, or unfractured limestone, must lie above the reservoir rock and stop the hydrocarbons from rising further. Second, the seal and reservoir rock bodies must be arranged in a geometry that collects the hydrocarbons in a restricted area. Geologists recognize several types of hydrocarbon trap geometries, four of which are described in **Box 14.1**.

Note that when we talk about trapping hydrocarbons underground, we are talking about a temporary process in the context of geologic time. Oil and gas may be trapped for millions to over a hundred million years, but eventually they may manage to pass through a seal rock because no rock is absolutely impermeable—most rocks contain joints that can provide permeability. Also, in some cases, microbes eat hydrocarbons in the subsurface. Thus innumerable oil reserves that existed in the past have vanished, and the oil fields we find today, if left alone, may disappear millions of years in the future.

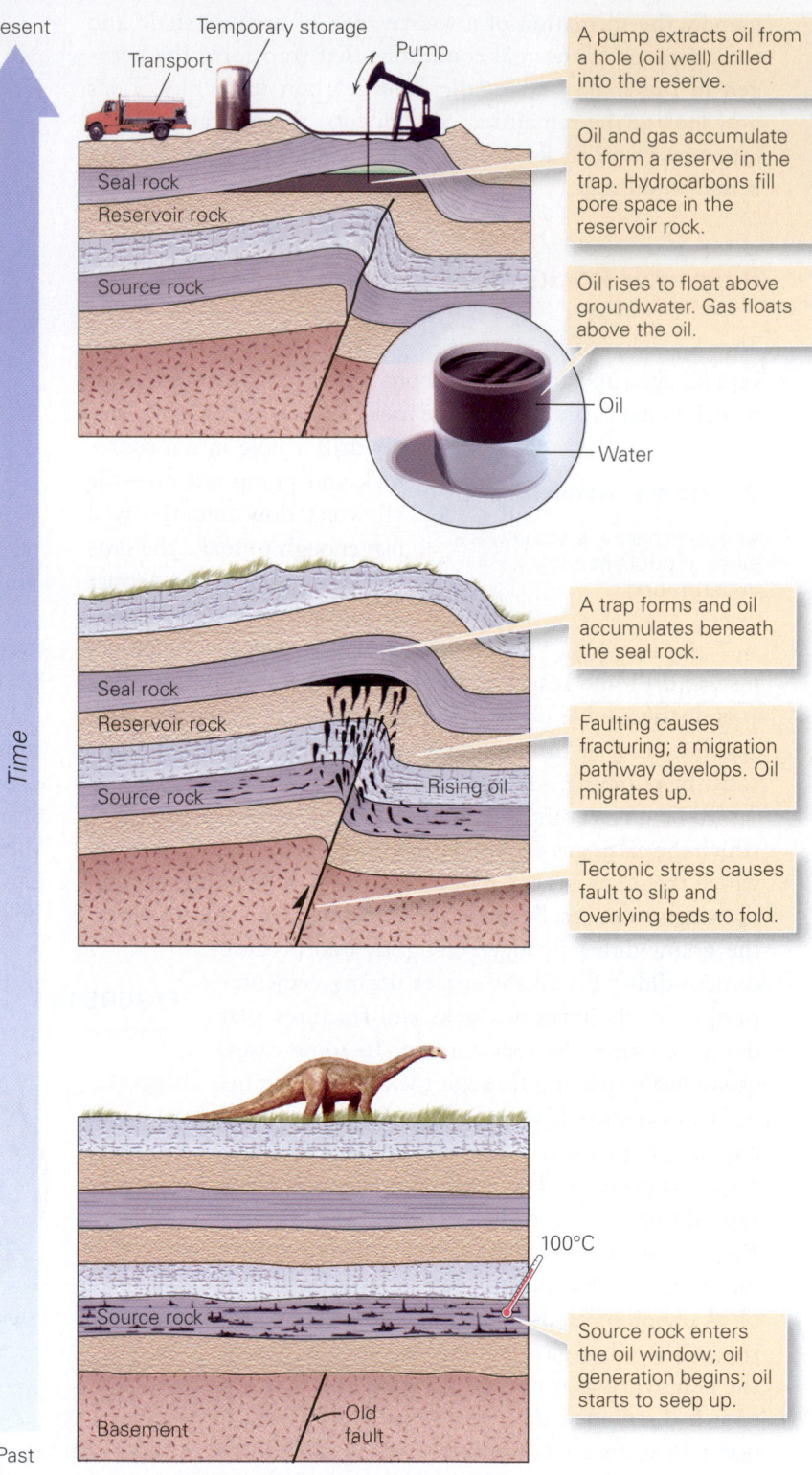

FIGURE 14.10 Initially, oil resides in the source rock. Because it is buoyant relative to groundwater, the oil migrates into the overlying reservoir rock. The oil accumulates beneath a seal rock in a trap.

A pump extracts oil from a hole (oil well) drilled into the reserve.

Oil and gas accumulate to form a reserve in the trap. Hydrocarbons fill pore space in the reservoir rock.

Oil rises to float above groundwater. Gas floats above the oil.

A trap forms and oil accumulates beneath the seal rock.

Faulting causes fracturing; a migration pathway develops. Oil migrates up.

Tectonic stress causes fault to slip and overlying beds to fold.

Source rock enters the oil window; oil generation begins; oil starts to seep up.

FIGURE 14.11 Examples of hydrocarbon seeps at the Earth's surface.

(a) The La Brea tar pit is an oil seep that now forms the centerpiece of a small park in Los Angeles, California.

(b) The Darvasa gas crater (known as the "door to hell") in Turkmenistan formed in 1971 when a drilling rig tapped into a cavern filled with natural gas. Engineers lit a fire hoping to burn off the gas, but it's above a seep, so the fire has burned ever since.

Birth of the Conventional Oil Industry

People have used oil since the dawn of civilization—as a cement, as a waterproof sealant, and even as a preservative to embalm mummies. In the United States, during the first half of the 19th century, "rock oil" (later called petroleum, from the Latin words *petra*, meaning rock, and *oleum*, meaning oil), collected at seeps, was used to grease wagon axles and to make patent medicines. But such oil was rare and expensive. In 1854, George Bissell, a New York lawyer, realized that oil might have broader uses, particularly as fuel for lamps, to replace increasingly scarce whale oil. Bissell and a group of investors hired Edwin Drake, a colorful character who had drifted among many professions, to find a way to drill for oil in rocks beneath a hill near Titusville, Pennsylvania, where oily films floated on the water of springs. Using the phony title "Colonel" to add respectability to his name, Drake hired drillers and obtained a steam-powered drill. Work was slow and the investors became discouraged, but the very day that Drake received a letter ordering him to stop drilling, his drillers discovered that the hole, which had reached a depth of 21.2 m, had filled with oil. They set up a pump, and on August 27, 1859, for the first time in history, pumped oil out of the ground. No one had given much thought to the question of how to store the oil, so workers dumped it into empty whisky barrels. (These days, a **barrel of oil** [**bbl**] has become a standard unit of measurement: 1 bbl = 42 gallons = 159 liters. Oil may also be sold by weight: 1 metric ton (1 tonne) of oil = 6.5 barrels.)

Within a few years, thousands of oil wells had been drilled in many states, and by the turn of the 20th century, civilization had begun its addiction to oil. Initially, most oil went into the production of kerosene for lamps. Later, when electricity took over from kerosene as the primary source for illumination, gasoline derived from oil became the fuel of choice for the newly invented automobile. Oil was also used to fuel electric power plants. In its early years, the oil industry was in perpetual chaos. When "wildcatters" discovered a new oil field, there would be a short-lived boom during which the price of oil could drop to pennies a barrel. In the midst of this chaos, John D. Rockefeller established the Standard Oil Company, which monopolized the production, transport, and marketing of oil. In 1911, the Supreme Court broke Standard Oil down into several companies (including Exxon, Chevron, Mobil, Sohio, Amoco, Arco, Conoco, and Marathon), some of which have recombined in recent decades. Oil became a global industry governed by the complex interplay of politics, profits, supply, and demand.

The Modern Search for Oil

Wildcatters discovered the earliest *oil fields*, the land area above an oil-filled trap, either by blind luck or by searching for surface seeps. But in the 20th century, when most known seeps had been drilled and blind luck became too risky, oil companies realized that finding new oil fields would require systematic exploration. The modern-day search for oil is a complex, sometimes dangerous, and often exciting endeavor with many steps.

Source rocks are always sedimentary, as are most reservoir and seal rocks, so geologists begin exploration by looking for a region containing appropriate sedimentary rocks. Then they compile a geologic map of the area, showing the distribution of rock units. From this information, it may be possible to construct a preliminary cross section depicting the geometry of the sedimentary layers underground as they would appear on an imaginary vertical slice through the Earth.

BOX 14.1 CONSIDER THIS . . .

Types of Oil and Gas Traps

Geologists who work for oil companies spend much of their time trying to identify underground traps. No two traps are exactly alike, but we can classify most into the following four categories.

- *Anticline trap:* In some places, sedimentary beds are not horizontal, as they are when originally deposited, but have been bent by the forces involved in mountain building. These bends, as we have seen, are called folds. An anticline is a type of fold with an arch-like shape (**Fig. Bx14.1a**). If the layers in the anticline include a source rock overlain by a reservoir rock that is overlain by a seal rock, then we have the recipe for an oil reserve. The oil and gas rise from the source rock, enter the reservoir rock, and are trapped in the crest of the anticline.

- *Fault trap:* If the slip on a fault crushes and grinds the adjacent rock to make an impermeable layer along the fault, then oil and gas may migrate upward along bedding in the reservoir rock until they stop at the fault surface (**Fig. Bx14.1b**). A fault trap may also develop if the slip juxtaposes a seal rock against a reservoir rock.

- *Salt-dome trap:* In some sedimentary basins, the sequence of strata contains a thick layer of salt, deposited when the basin was first formed and seawater covering the basin was shallow and very salty. Sandstone, shale, and limestone overlie the salt. The salt layer is not as dense as sandstone, limestone, or shale, so it is buoyant and tends to rise up slowly through the overlying strata. Once the salt starts to rise, the weight of surrounding strata squeezes the salt out of the salt layer and up into a growing, bulbous *salt dome*. As the dome rises, it bends up the adjacent layers of sedimentary rock. Oil and gas in reservoir rock layers migrate upward until they are trapped against the boundary of the salt dome, for salt is impermeable (**Fig. Bx14.1c**).

- *Stratigraphic trap:* In a stratigraphic trap, a tilted reservoir rock bed "pinches out" (thins and disappears up-dip) between two impermeable layers. Oil and gas migrating upward along the bed accumulate at the pinch-out (**Fig. Bx14.1d**).

FIGURE Bx14.1 Examples of oil traps. A trap is a configuration of a seal rock over a reservoir rock in a geometry that keeps the oil underground.

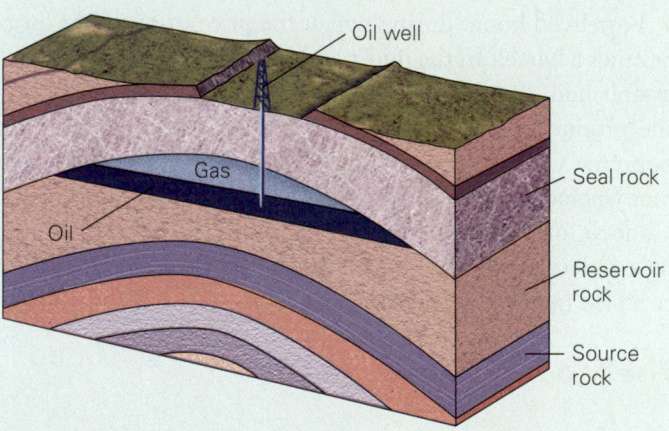

(a) Anticline trap. Oil and gas rise to the crest of the fold.

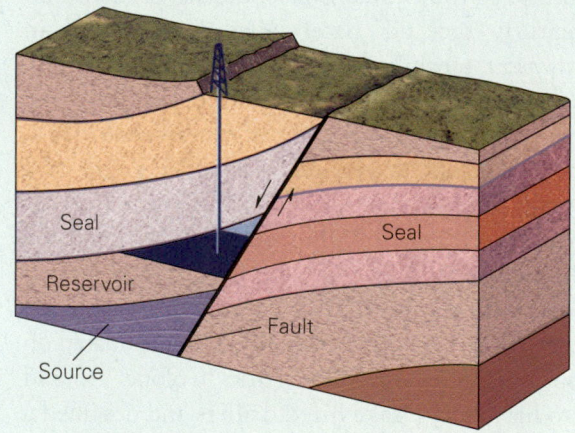

(b) Fault trap. Oil and gas collect in tilted strata adjacent to the fault.

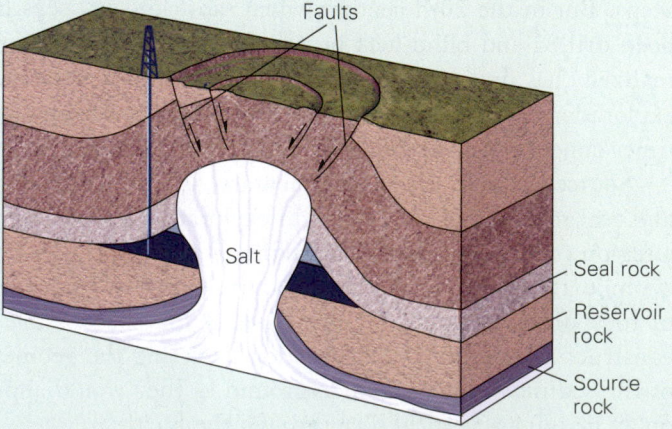

(c) Salt-dome trap. Oil and gas collect in strata on the flanks of the dome, beneath salt.

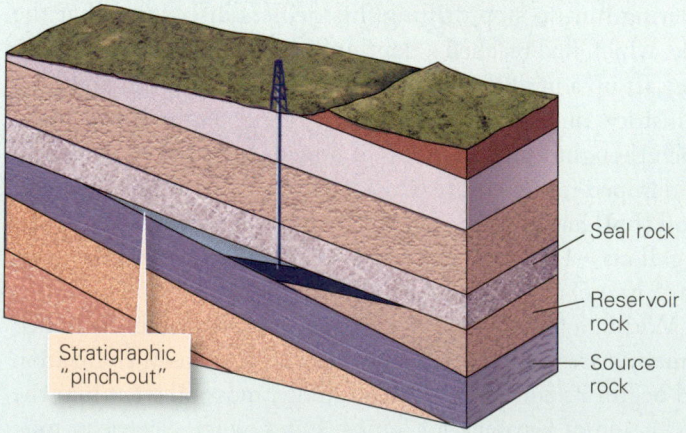

(d) Stratigraphic trap. Oil and gas collect where the reservoir layer pinches out.

To add detail to the cross section, explorationists obtain a **seismic-reflection profile** of the region. On land, this is done by using a special vibrating truck or by setting off dynamite explosions that send seismic waves into the ground (**Fig. 14.12a**; see Interlude D). The seismic waves reflect off contacts between rock layers, just as sonar waves sent out by a submarine reflect off the bottom of the sea. Reflected seismic waves then return to the ground surface, where sensitive seismometers (geophones) stuck into the ground record their arrival. A computer measures the time between the generation of a seismic wave and its return and, based on this information, defines the depth to the contacts at which the wave reflected (**Fig. 14.12b**). To explore for oil in sedimentary layers below the sea floor, the exploration company uses an "air gun" dragged behind a ship to send pulses of compressed air into the water (**Fig. 14.12c**). The pulses carry enough energy to produce seismic waves in the strata beneath the water. Reflected signals are received by geophones dragged behind the ship. Using the data collected by geophones, computer programs construct an image of the configuration of underground rock layers and, in some cases, can even detect reserves of oil or gas. Technological advances now enable geologists to create 3-D seismic-reflection data blocks of the subsurface both under land and underwater (**Fig. 14.12d**). Such blocks are expensive—just one may cost millions of dollars to create.

Drilling and Refining

If geologic studies identify a trap, as well as good source rocks and reservoir rocks, and a likelihood that strata have been heated into the oil or gas window, then geologists make a recommendation to drill. They do not make such recommendations lightly, as drilling a deep well may cost tens of millions of dollars. If management accepts the geologists' recommendation, drillers go to work.

These days, drillers use rotary drills to grind a hole down through rock. A rotary drill consists of a pipe tipped by a rotating bit, a bulb of metal studded with hard metal prongs (**Fig. 14.13a**). As the bit rotates, it scratches and gouges and

FIGURE 14.12 Using seismic-reflection profiling to describe the character of underground beds and help locate reservoirs.

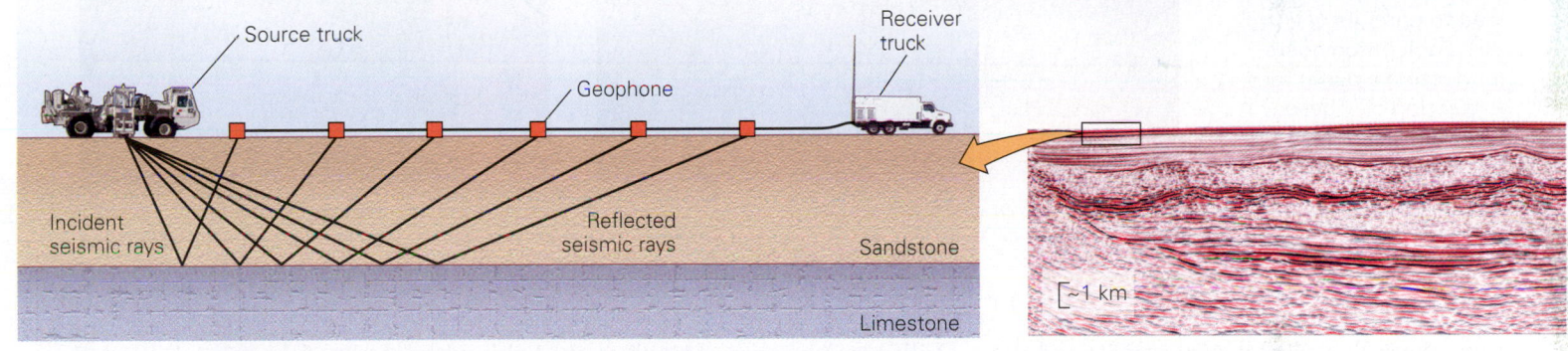

(a) Source truck sends a vibration into the Earth. The vibration reflects off layer boundaries up to geophones on the surface. The time it takes for the vibration to travel indicates the depth to the reflector.

(b) A two-dimensional seismic profile can reveal the presence of structure underground.

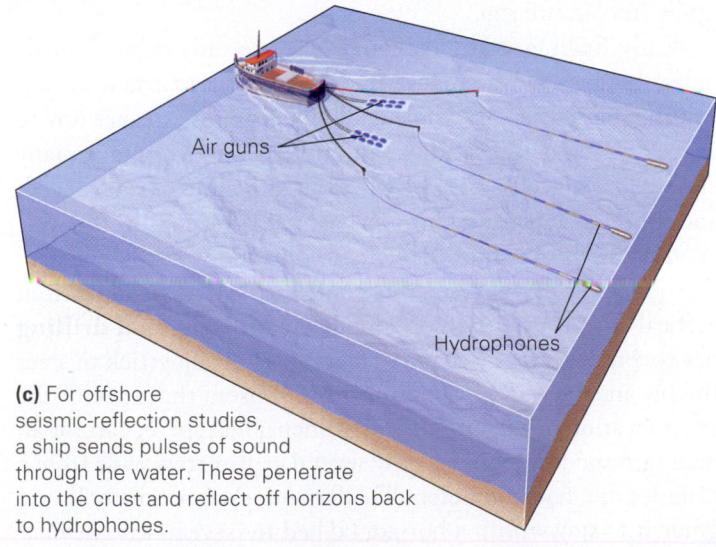

(c) For offshore seismic-reflection studies, a ship sends pulses of sound through the water. These penetrate into the crust and reflect off horizons back to hydrophones.

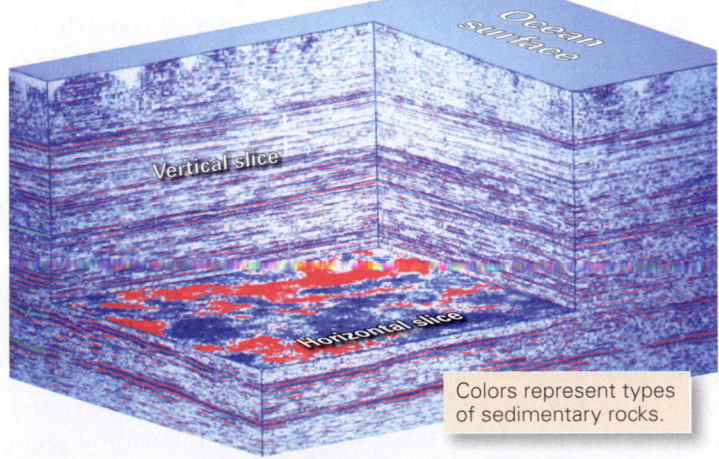

Colors represent types of sedimentary rocks.

(d) Modern techniques produce three-dimensional "blocks" of data, so geologists can study both vertical and horizontal slices through the subsurface.

Oil Field near Lamesa, Texas

LATITUDE
32°33'18.42"N

LONGITUDE
101°46'55.39"W

Looking down from 8 km up (~5 mi).

This view shows a grid of farm fields, 1.6 km (1 mi) wide, within which small roads lead to patches of dirt. Each patch hosts (or hosted) a pump for extracting oil. These wells are tapping an oil reserve in Permian strata in Texas.

FIGURE 14.13 Drilling rigs and oil pumps.

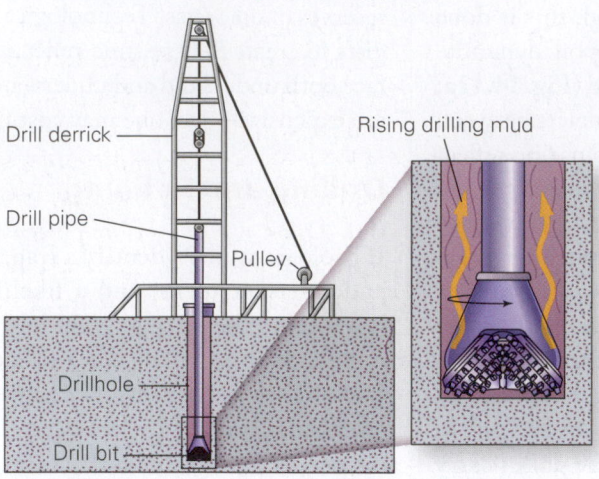

Drill derrick
Drill pipe
Pulley
Drillhole
Drill bit
Rising drilling mud

(a) An on-shore drilling rig. The inset shows a close-up of the drill bit. Drilling mud carries the cuttings up and out of the hole.

(b) The Lakeview gusher, in California, spilled 9 million bbl in 1910.

(c) An aerial view of an oil field with many, closely-spaced wells.

(d) An off-shore drilling rig. The derrick is constructed on top of a platform. The flare burns off gas.

turns the rock it contacts into powder and chips. Drillers pump **drilling mud**, a slurry of water mixed with clay and other materials, down the center of the pipe. The mud flows down, past a propeller that rotates the drill bit, and then squirts out of holes at the end of the bit. The extruded mud cools the bit head, which otherwise would heat up due to friction as it grinds against rock, then flows up the hole on the outside of the drill pipe. As it rises, the mud carries *rock cuttings* (fragments of rock that were broken up by the drill bit) up and out of the hole. Mud also serves another very important purpose—its weight counters the natural pressure of the oil and gas in underground reservoir rocks. By doing so, it prevents hydrocarbons from entering the hole until drilling stops and the hole has been "completed." Completing a hole involves removing the drill pipe, inserting a casing of steel pipe, filling the space between the casing and the walls of the drillhole with concrete, and capping the hole. Were it not for the mud, the natural pressure in the reservoir rock would drive oil and/or gas into the hole. And if the pressure were great enough, the hydrocarbons

would rush up the hole and spurt out of the ground as a *gusher* or *blowout* (**Fig. 14.13b**). Gushers and blowouts can be disastrous, because they spill oil onto the land and, in some cases, ignite into an inferno.

Early drilling methods could produce only vertical drillholes, because it wasn't just the drill bit that turned during drilling, but rather the whole drill pipe from the surface down, as well as the drill bit. Thus, an oil field might contain many closely spaced wells (**Fig. 14.13c**). But as technology advanced, and the drill bit became the only part to rotate, engineers developed methods that allow drillers to control the path of the drill bit so the hole can curve and become inclined at an angle from vertical or can even be horizontal. Such **directional drilling** has become so precise that a driller, by using a joystick to steer the bit and by watching output from sensors that specify the exact location of the bit in three-dimensional space, can hit an underground target that is only several centimeters wide from a distance of a few kilometers. Drillers can even bend a hole and guide it to stay within a horizontal bed for several kilometers.

Drillers must use derricks (towers) to hoist the heavy drill pipe. To drill in an offshore hydrocarbon reserve, one that occurs in strata beneath the continental shelf, the derrick must be constructed on an *offshore-drilling facility* (**Fig. 14.13d**). These may be built on huge towers rising from the seafloor or on giant submerged pontoons. Using directional drilling, it's possible to reach multiple targets from the same platform. On completion of a hole, workers remove the drilling rig and set up a pump. Some pumps resemble a bird pecking for grain; their heads move up and down to pull up oil that has seeped out of pores in the reservoir rock into the drillhole (**Fig. 14.14a**).

You may be surprised to learn that simple pumping gets only about 30% of the oil in a reservoir rock out of the ground. Thus oil companies use *secondary recovery techniques* to coax out more oil (as much as 20% more). For example, a company may drive oil toward a drillhole by forcing steam into holes in the ground nearby—the steam heats the oil in the ground, making it less viscous, and pushes it along. In some cases, drillers create artificial fractures in rock around the hole by pumping a high-pressure water and chemical mixture into a portion of the hole. This process, called **hydrofracturing** (*hydraulic fracturing* or simply *fracking*), creates new fractures and opens up pre-existing ones—the fractures provide easy routes for the oil to follow from the rock to the well. We'll discuss hydrofracturing in more detail later in this chapter.

Once extracted directly from the ground, "crude oil" flows first into storage tanks and then into a pipeline or tanker, which transports it to a refinery (**Fig. 14.14b–d**). At a refinery, workers distill crude oil into several separate components by first heating it to a temperature of about 400°C and then by pumping it into a vertical pipe called a *distillation column* (**Fig. 14.14e**). Lighter molecules rise to the top of the column, while heavier molecules stay at the bottom. Outlets at different levels in the column allow removal of molecules of different sizes—tar comes out the bottom, oil above that, gasoline above that, and propane from the top. The heat may also "crack" larger molecules to make smaller ones. Chemical factories buy the largest molecules left at the bottom and transform them into plastics.

Where Do Conventional Reserves of Oil Occur?

Conventional reserves are not randomly distributed around the Earth. Currently, countries bordering the Persian Gulf contain the world's largest reserves in 25 *supergiant fields* (**Fig. 14.15a**). In fact, this region has almost 60% of the world's conventional reserves, whereas the United States, the largest consumer of oil, has only a few percent of the total.

What determines where conventional reserves occur? To start with, a region must have been in an environment where plankton could grow well, so that sediments being deposited in the region are rich in organic content. Much of the region that is now the Middle East was situated in tropical areas between latitude 20° S and 20° N between the Jurassic (135 million years ago [Ma]) and the Late Cretaceous (66 Ma), where biological productivity was high (**Fig. 14.15b**). So sediments deposited at that time were rich in organic matter. Thick successions of porous sandstone buried the source rocks of the Middle East, and crustal compression due to the collision between Africa and Asia folded the strata to produce excellent traps.

The Middle East is not the only source of conventional oil (see Fig. 14.15a). Reserves also occur in sedimentary basins formed along passive continental margins, such as the Gulf Coast of the United States and the Atlantic coasts of Africa and Brazil, as well as in the intracratonic and foreland basins on continents (see Chapter 7).

Take-Home Message

Conventional hydrocarbon reserves are ones in which oil has migrated from a source rock into a porous and permeable reservoir rock, situated within an oil trap, so that the oil or gas can be pumped fairly easily. The first oil drilling took place in 1859. Modern methods for finding, drilling, producing, and refining oil are very complex and expensive.

QUICK QUESTION: Where do most of the conventional oil reserves in the world occur today?

14.5 Unconventional Hydrocarbon Reserves

In an **unconventional hydrocarbon reserve**, the material (rock or sediment) contains significant quantities of hydrocarbons, but either the material does not have adequate permeability or the hydrocarbons themselves are too viscous to flow. Thus, the hydrocarbons cannot be extracted simply by drilling into the material and pumping. Extraction of unconventional hydrocarbons was not possible until engineers developed new technologies and the price per barrel of hydrocarbons became high enough for the extraction to be profitable. These conditions were met 10 to 15 years ago, and since then efforts to extract hydrocarbons from, specifically, shale gas and tar sand have grown at a very rapid rate, and now such reserves provide a significant portion of the global energy. Let's examine the geologic context of these reserves.

FIGURE 14.14 Pumping, transporting, and refining oil.

The head of the pump goes up and down.

Storage tanks

Hole

(a) When drilling is complete, the derrick is replaced by a pump, which sucks oil out of the ground. This design is called a pumpjack.

(d) Distilling columns of an oil refinery transform crude oil into gasoline and other hydrocarbon products.

(b) This oil supertanker is 330 meters long and can carry 300,000 metric tons. When the ship is loaded, the red painted area is almost completely underwater.

(c) The Trans-Alaska Pipeline transports oil from fields on the Arctic coast to a tanker port on the southern coast of Alaska.

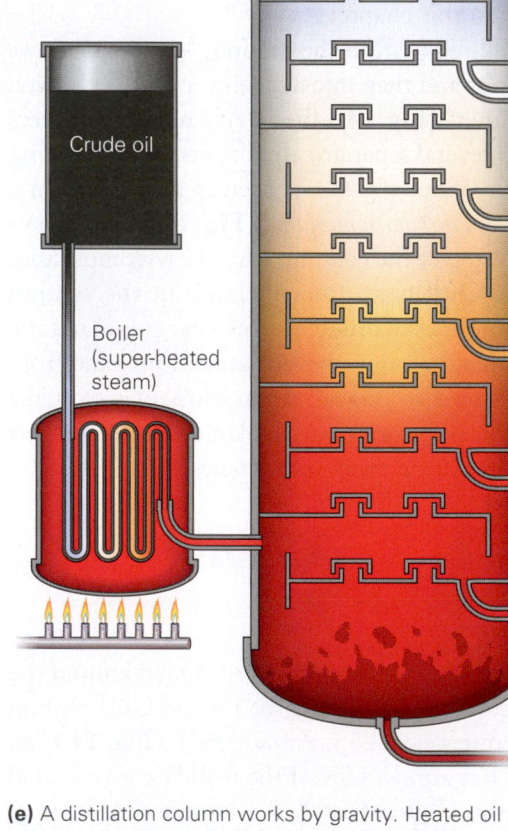

Fractionating column

<30°C
Bottled gas (C_1–C_4)

30°–100°C
Naptha (C_5–C_{12})

100°–120°C
Gasoline (C_5–C_{12})

120°–160°C
Kerosene (C_{11}–C_{13})

160°–250°C
Diesel/Heating (C_{14}–C_{25})

250°–320°C
Lubricating (C_{25}–C_{40})

340°–600°C
Bitumen/Residual (>C_{70})

320°–340°C
Fuel oil (C_{40}>C_{70})

Crude oil

Boiler (super-heated steam)

(e) A distillation column works by gravity. Heated oil separates into bubbles of lighter hydrocarbons and droplets of heavier ones. Heavier ones sink and light ones rise.

FIGURE 14.15 The distribution of oil reserves around the world.

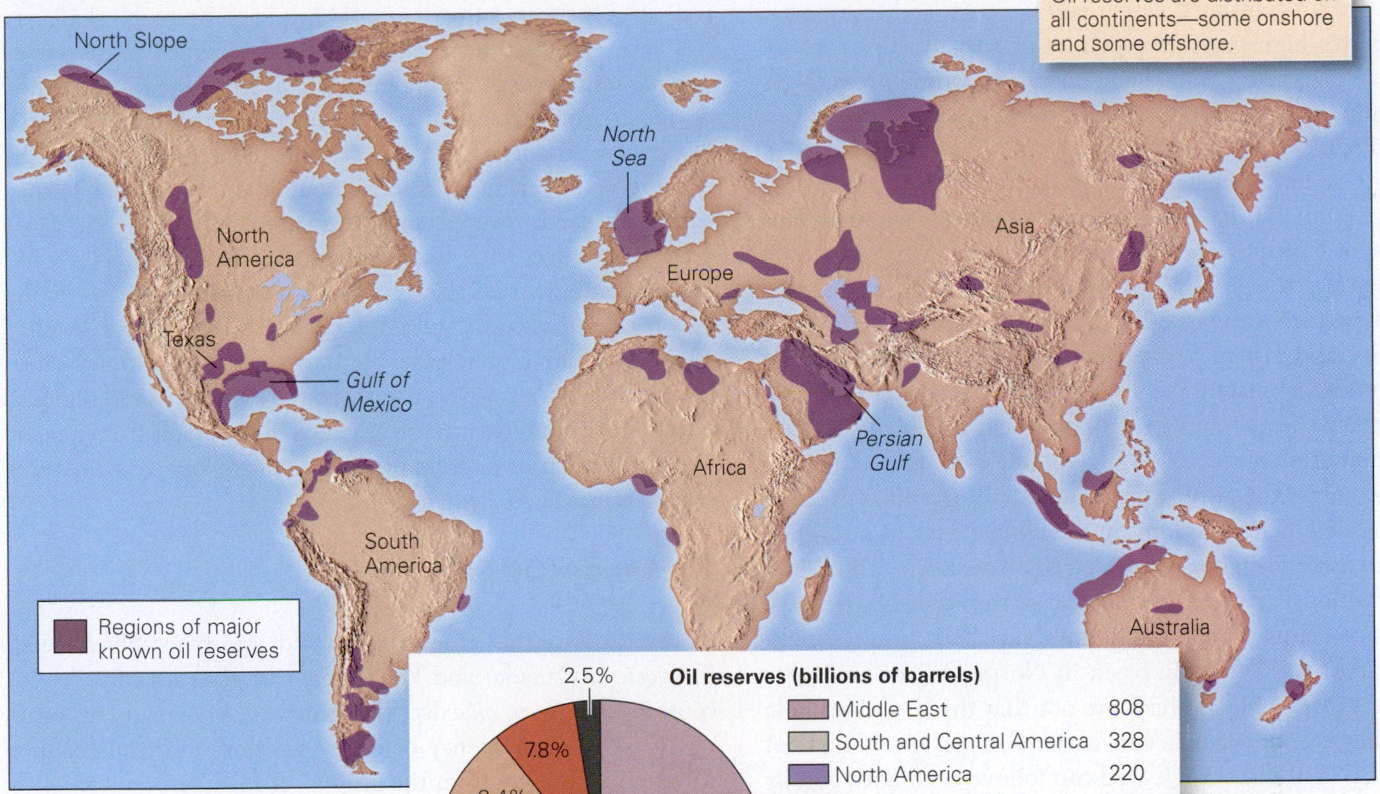

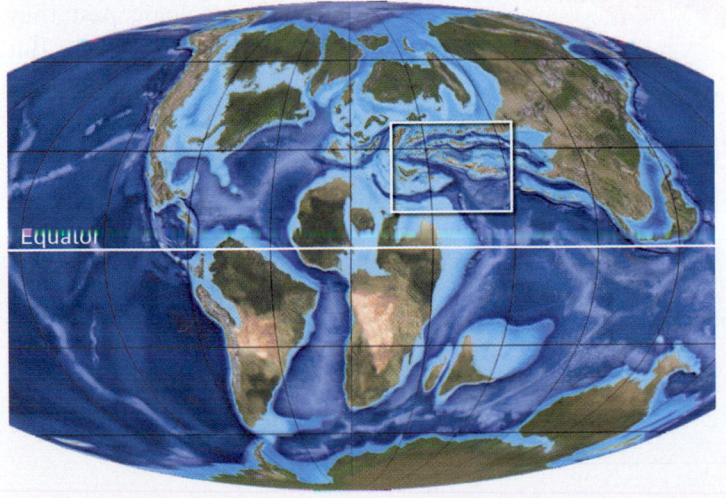

Oil reserves are distributed on all continents—some onshore and some offshore.
North Slope
North America
Texas
Gulf of Mexico
North Sea
Europe
Asia
Africa
Persian Gulf
South America
Australia
Regions of major known oil reserves

(a) Oil reserves are distributed on all continents—some onshore and some offshore.

Oil reserves (billions of barrels)
2.5%
7.8%
8.4%
13.2%
19.7%
48.4%
Middle East 808
South and Central America 328
North America 220
Europe and Eurasia 141
Africa 130
Asia Pacific 42
World total 1,669*
* includes ~414 of tar sands
Oil reserves are typically specified in barrels: 1 barrel (bbl) = 42 gallons (159 liters). Data source: BP Statistical Review of World Energy (2013).

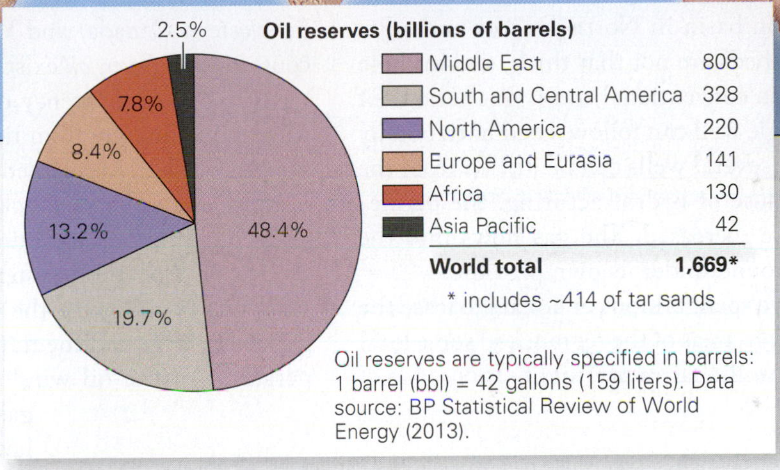

(b) A map of the Late Cretaceous world during the time when the deposition of sediment from which source rocks in the Mideast are forming. The box shows the region that will become the Middle East oil fields of today. Note that the region lay in the warm subtropics in which marine plankton could thrive.

Equator

Shale Gas

As we've seen, natural gas consists of volatile short-chain hydrocarbon molecules (methane, ethane, propane, and butane). Gas burns more cleanly than oil, in that combustion of gas produces only carbon dioxide and water, while the burning of oil not only produces carbon dioxide and water but also complex organic pollutants. Thus, natural gas has long been a preferred fuel for home cooking and heating. And when the cost of natural gas decreased, many electrical generating stations were converted to burn natural gas. It can also be used to fuel cars and trucks if the vehicles have been appropriately modified.

If gas is the main hydrocarbon of a conventional reserve, it may be economical to extract and transport the gas. Gas often occurs in association with oil, or in place of oil, in conventional reserves. In many of these reserves, the volume of gas is too small to be economically produced because gas must be compressed for transport, which takes energy, and transportation requires high-pressure pipelines or special ships that are expensive to operate. So oil-field operators commonly vent gas from a pipe and burn it as a flare where it enters the air (see Fig. 14.13d).

Huge quantities of natural gas remain in source rocks (organic shale) that have been heated to the gas window. This resource is called **shale gas**. Until recently, this gas could not be extracted economically by pumping because of shale's very low permeability. In 2008, Terry Engelder of Penn State University and others pointed out that previous reports had greatly underestimated the amount of shale gas available in sedimentary basins that lie near populated areas (**Fig. 14.16**). The combination of increased gas-volume estimates and of the ability to access the gas by directional drilling and hydrofracturing has led to a "gas boom" in the United States and worldwide. For example, thousands of wells have been drilled into the Marcellus Shale, a Devonian formation that lies beneath portions of Pennsylvania, Ohio, and New York, and the Bakken Shale of the Williston basin in North Dakota and adjacent states. The shale gas beds are not that thick, and in these basins they do not occur in conventional traps. But because of directional drilling, a single well can follow a bed horizontally for many kilometers, and several wells can be drilled from the same platform. And because of hydrofracturing, the permeability of the rock can be increased. The gas flows into the drillhole and up to the ground under its own pressure.

Energy companies have spent billions of dollars to lease the right to obtain gas from large areas of the region, and some local landowners have become millionaires overnight. Thousands of people have obtained jobs working for energy companies as well as in the towns that house the workers. So much shale gas has been extracted in recent years that it has led some power companies to switch their generators to natural gas, and it has decreased the amount of overseas oil that the United States has had to import. The boom remains controversial, though, because of lingering questions about the amount of gas that can really be recovered and because of environmental concerns associated with hydrofracturing (**Box 14.2**). Not only are residents worried that chemicals in hydrofracturing fluids may contaminate water supplies, but there are also worries that fracturing may also release gas into sources of groundwater. (Natural gas does occur in the groundwater of some regions, but it's not always clear if the gas entered the water from new or enlarged fractures or if its presence reflects long-term natural seepage from shale beds.)

Tar Sands (Oil Sands)

In several locations around the world, most notably Alberta (in western Canada) and Venezuela, vast reserves of very viscous, tar-like *heavy oil* exist. This heavy oil, known also as bitumen, has the consistency of gooey molasses and thus cannot be pumped directly from the ground. It fills the pore spaces of sand or of poorly cemented sandstone, constituting up to 12% of the sediment or rock volume. Sand or sandstone containing high concentrations of bitumen is known as **tar sand** or *oil sand*.

The hydrocarbon system that leads to the generation of tar sands begins with the production and burial of a source rock in a large sedimentary basin. When subjected to temperatures of the oil window, the source rock yields oil and gas, which migrate into sandstone layers and then up the dip of tilted layers to the edge of the basin, where they become caught in stratigraphic traps (see Box 14.1). Initially, these hydrocarbons have relatively low viscosity, so in the geologic past they could have been pumped easily. But over time, microbes attacked the oil reserve underground, digested lighter, smaller hydrocarbon molecules, and left behind only the larger molecules, whose presence makes the remaining oil so viscous. Geologists refer to such a transformation process as *biodegradation*. Generation of tar sand by biodegradation is yet another example of the interaction between physical and biological components of the Earth System.

FIGURE 14.16 Map of basins with assessed shale oil and shale gas formations, as of May 2013.

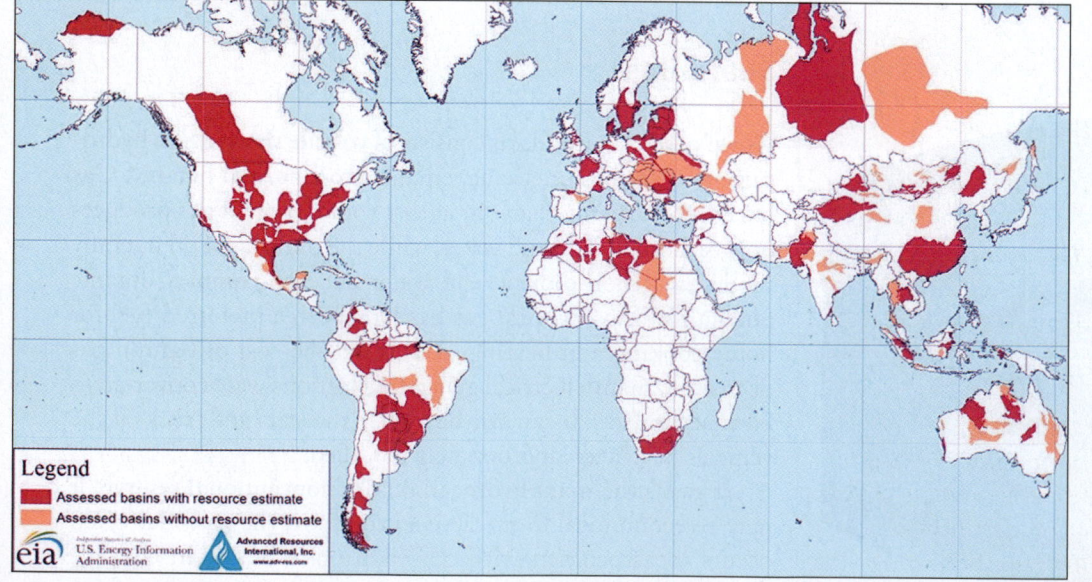

Legend
■ Assessed basins with resource estimate
■ Assessed basins without resource estimate

Production of usable oil from tar sand is difficult and expensive but not impossible. It takes about two tons of tar sand, and a lot of energy, to produce one barrel of oil. Oil companies mine near-surface deposits in vast open-pit mines and then heat the tar sand in a furnace to extract the oil (**Fig. 14.17a**). Producers then crack the heavy oil molecules to produce smaller, more usable molecules. Trucks dump the drained sand back into the mine pit. To extract oil from deeper deposits of tar sand, oil companies drill wells and pump steam or solvents down into the sand to liquefy the oil enough so that it can be pumped out.

Oil Shale

Vast reserves of organic shale have not been subjected to temperatures of the oil window, or if they were, they did not stay within the oil window long enough to complete the transformation to oil. Such rock still contains a high proportion of kerogen. Shale that contains at least 15 to 30% kerogen is called **oil shale**. Oil shale is not the same as coal because the organic matter within it exists in the form of waxy hydrocarbon molecules, not as elemental carbon.

Lumps of oil shale can be burned directly and thus have been used as a fuel since ancient times (**Fig. 14.17b**). In the 1850s, researchers developed techniques to produce liquid oil from oil shale. The process involves heating the oil shale to a temperature of 500°C; at this temperature, the shale decomposes and the kerogen transforms into liquid hydrocarbon and gas. Large supplies of oil shale occur in Estonia, Scotland, China, and Russia, and in the Green River basin of Wyoming in the United States. As is the case with tar sand, production of oil from oil shale is possible but very expensive. In addition to the expense of mining and environmental reclamation, producers must pay for the energy needed to heat the shale. It takes about 40% of the energy yielded by a volume of oil shale to produce the oil itself.

Gas Hydrate

Gas hydrate is a chemical compound consisting of methane (CH_4) molecules surrounded by a cage-like arrangement of water molecules. An accumulation of gas hydrate occurs as a whitish solid that resembles ordinary water ice (**Fig. 14.17c**). Gas hydrate forms when anaerobic bacteria (bacteria that live in the absence of oxygen) eat organic matter, such as dead plankton that have been incorporated into sediments of the seafloor. When the bacteria digest organic matter, they produce methane as a by-product, and the methane bubbles into the cold seawater that fills pore spaces in sediments. Under pressures found at depth in the ocean, the methane dissolves in the pore water and produces gas-hydrate molecules—the

FIGURE 14.17 Examples of alternative hydrocarbon sources.

(a) An open-pit mine in Canada for digging up tar sand. Trucks haul the sand to a plant where it is heated so hydrocarbons can be extracted.

(b) Oil shale can be set ablaze.

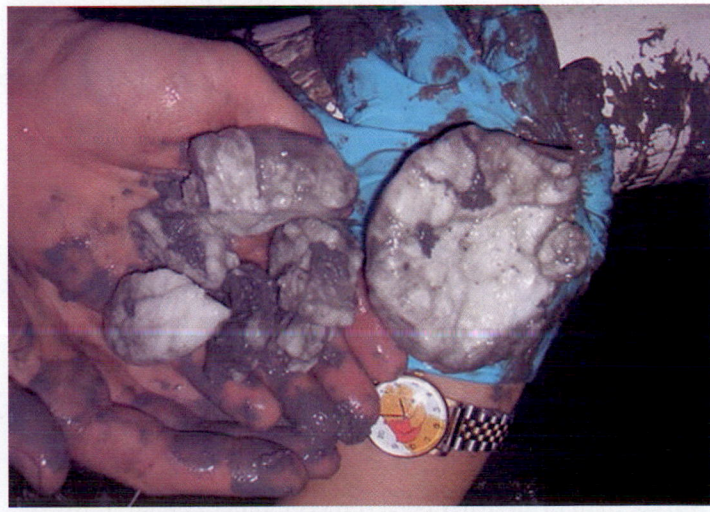

(c) Gas hydrate samples (white material) dug up from the muddy sea floor.

BOX 14.2   CONSIDER THIS . . .

Hydrofracturing (*Fracking*)

Hydrofracturing (*hydraulic fracturing* or *fracking*), a technology first utilized in the 1950s, has been in the news quite a bit in recent years, where it's commonly referred to as *fracking*. Drillers use the technique to open and propagate existing joints in rocks at depth as well as to generate new cracks in the rock. These fractures provide a permeability pathway through which hydrocarbons can flow to reach a drill hole and be extracted. Originally, hydrofracturing was used in conventional vertical wells as a means to enhance secondary recovery. Today it is also being widely used in horizontal wells following gas shale beds (**Fig Bx14.2a, b**). What is hydrofracturing, how does it work, and what risks are potentially involved in its use?

To hydraulically fracture a hole, drillers start by sealing off a length of a well with *packers*, which are simply inflatable balloons that when filled with high-pressure fluid press tightly against the wall of the hole (**Fig Bx14.2c**). Then the drillers insert a pipe through one of the packers into the sealed-off section of the well and pump *fracking fluid* into this section under high pressure. When the pressure generated by the fluid within the sealed section becomes great enough, it forces open existing joints that intersect the hole and may cause these joints to lengthen at their tip (**Fig Bx14.2d**). The process may also generate new cracks in the rock adjacent to the hole. The area affected by hydrofracturing can extend tens of meters out from the drillhole. Once fracturing has been completed, drillers pump out the fluid and

complete the hole. Hydrocarbons then flow along the fractures into the drillhole and up to the surface, where drillers capture it and compress it for transport (**Fig Bx14.2e**). The volume of fluid used to hydrofracture a section of hole is roughly equivalent to the volume in an Olympic swimming pool—usually several sections of a hole may be subjected to hydrofracturing, so the process is repeated several times in the hole.

What's in fracking fluid? A typical example consists of about 95% water, 9.5% quartz sand, and 0.5% other chemicals. The sand is necessary because it props the holes open once the fluid has been removed—without the sand, the cracks opened by hydrofracturing would close up tightly under the pressure applied by surrounding rock, and thus could not serve as permeability pathways. The 0.5% portion of fracking fluid that is not water or sand is a mixture of many chemicals, including oils that make the fluid more slippery so it can inject farther into the rock; acid, which dissolves cement between grains and increases porosity; detergent, which lowers the surface tension of water so it doesn't stick to grains; guar gum to make the fluid more viscous so that it can carry more sand; antifreeze to prevent scale buildup; and biocides, which prevent the growth of bacteria that could clog pores. Hydrofracturing requires a lot of equipment and materials, so at a drilling site where it's taking place, there will be lots of trucks carrying water, sand, and other chemicals as well as trucks carrying portable pumps and

giant mixing vats. The drill site may also have holding tanks or retaining ponds for storing fluid that has been removed from the ground once hydrofracturing has been finished.

Concerns about hydrofracturing have become the subject of intense public debate. The most common concern is that fracking fluid can contaminate drinking water underground. To understand the nature of this risk, it's necessary to understand how groundwater changes with depth. As we'll discuss further in Chapter 19, *groundwater* is water that fills or saturates pores and cracks in rock or sediment underground, beneath a surface called the *water table*—above the water table, pores and cracks contain some air. Typically, groundwater in the upper several hundred to a few thousand meters can be fresh and drinkable, but below that depth groundwater tends to be saline (**Fig Bx14.2f**). If the section of the hole subjected to hydrofracturing lies deeper than the saline boundary, the fluids from the section probably won't mix with drinkable groundwater, for they are denser than groundwater and thus are not buoyant. Leakage from the portion of the vertical hole above the saline boundary, however, can be problematic, so it is important that this portion of the hole be cased and sealed thoroughly before fracking fluid is pumped in. Leakage at the surface, from tanks or holding ponds, or from transporting trucks is also of concern—to avoid contamination, handling the fluids at the surface must be very carefully monitored.

---

reaction can occur in water as shallow as 90 m in colder, polar water but not until depths of about 300 m in warmer, equatorial water. Exploration tests suggest that gas hydrate occurs as layers interbedded with sediment and/or as a cement holding together the sediment at depths of between 90 and 900 m beneath the seafloor. Geologists estimate that an immense amount of methane lies trapped in gas-hydrate layers. In fact, worldwide there may be more organic carbon stored in gas hydrate than in all other reservoirs combined! So far, however, techniques for safely recovering gas hydrate from the seafloor have not been devised.

Take-Home Message

Gas shale, tar sand, and oil shale are unconventional reserves, because it is difficult and expensive to extract hydrocarbons from them. Development of new technologies, however, has made extraction feasible. In particular, directional drilling and hydrofracturing has led to a huge increase in production of shale gas.

QUICK QUESTION: How can usable fuels be obtained from oil shale and tar sand?

FIGURE Bx14.2 Directional drilling and hydrofracturing. New technologies have led to economic production of shale gas.

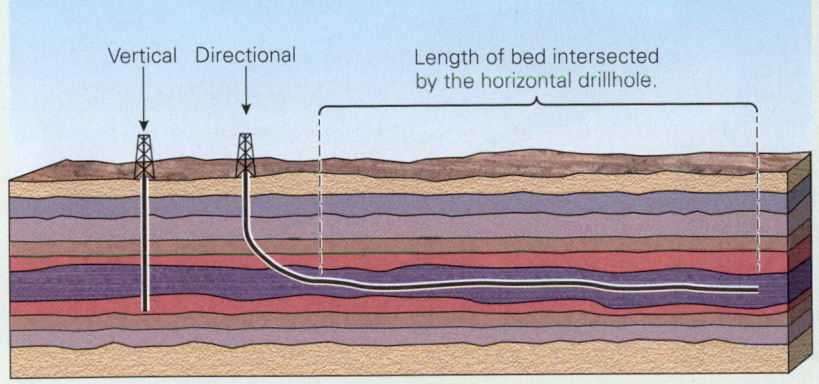

(a) Directional drilling permits the drillhole to follow a relatively thin bed for many kilometers, so the amount of shale gas accessed can be large. A vertical hole intersects the gas shale for only a short distance.

(b) A drilling site. The trucks and the holding pond are used during hydrofracturing. Many holes can be drilled from this site, like spokes of a wheel.

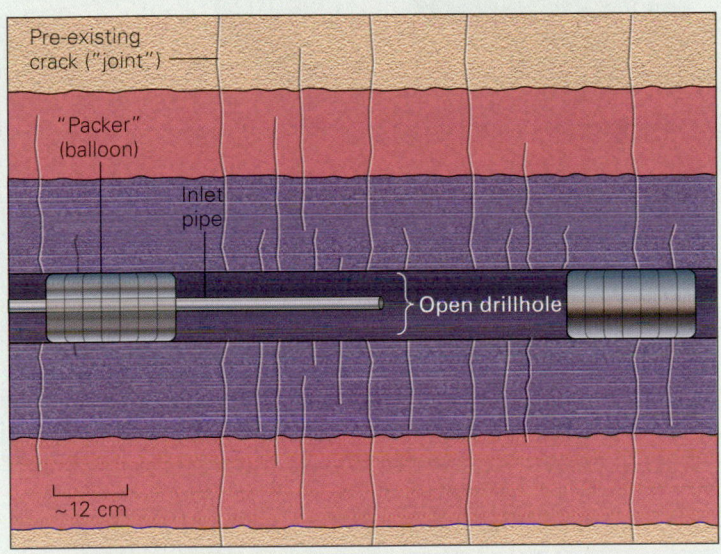

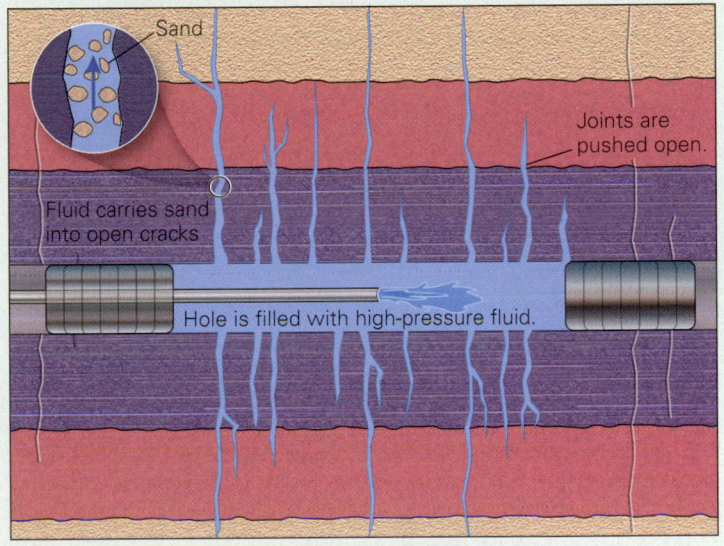

(c) The first step in hydrofracturing. After the hole has been drilled, packers seal off a portion, and a pipe is inserted through one of the packers.

(d) High-pressure fluid is pumped into the segment of hole. The pressure pushes open cracks and forms new ones. Sand injected with the fluid keeps the cracks from closing.

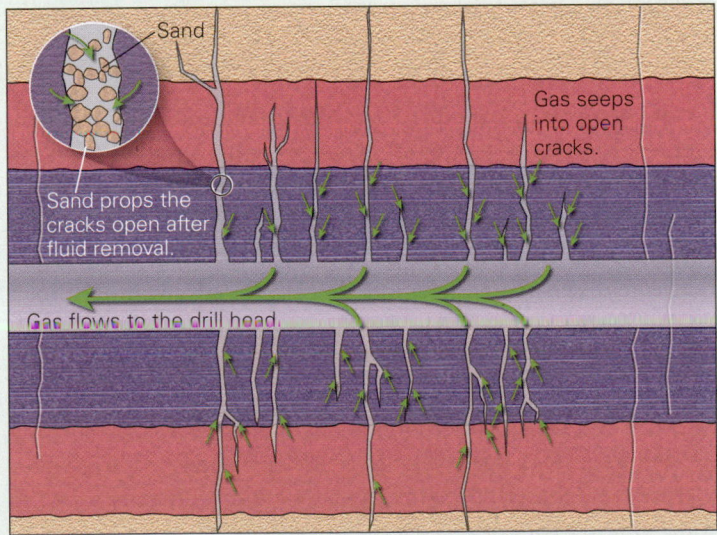

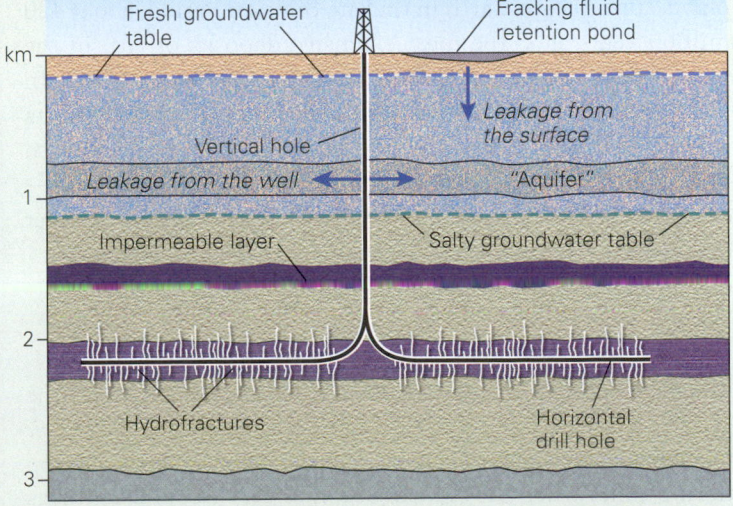

(e) After the packers and the fluid are removed, gas can seep into the pipe and flow to the drill head.

(f) Potential for the contamination of groundwater by hydrofracturing.

FIGURE 14.18 The formation of coal. Coal forms when plant debris becomes deeply buried.

(a) A museum diorama depicting a Carboniferous coal swamp.

(b) If there is a transgression, peat formed in the coal swamp can be buried and preserved.

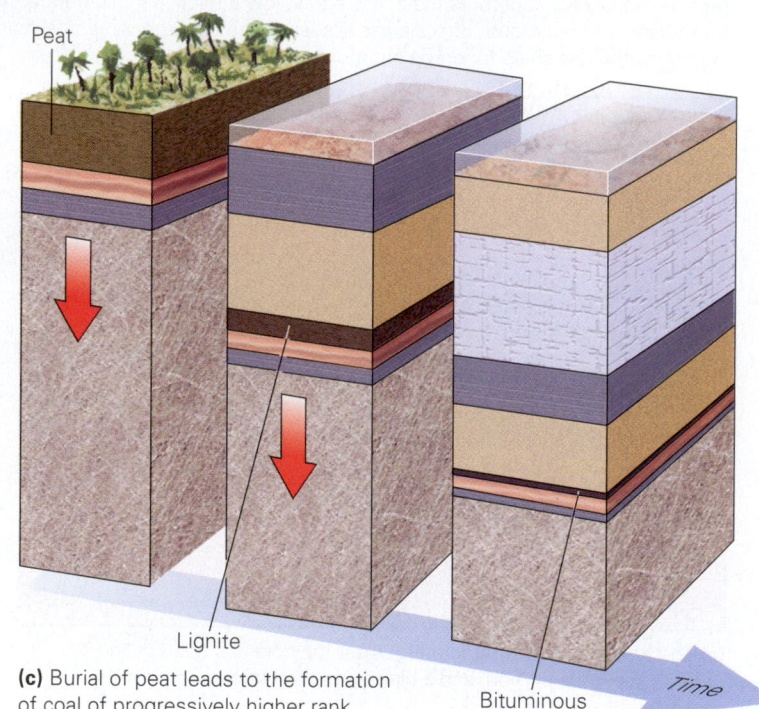

(c) Burial of peat leads to the formation of coal of progressively higher rank.

14.6 Coal: Energy from the Swamps of the Past

Coal, a black, brittle, sedimentary rock that burns, consists of elemental carbon mixed with minor amounts of organic chemicals, quartz, and clay. Typically, the carbon atoms in coal have bonded together, forming large, complicated molecules called coal *macerals*. Like oil and gas, coal is a fossil fuel because it stores solar energy that reached Earth long ago. But coal does not have the same composition or origin as oil or gas—coal contains carbon, not hydrocarbons—and in contrast to oil, coal forms from plant material (wood, stems, leaves), not plankton. Coal commonly occurs in beds, called *coal seams*, that may be centimeters to meters thick and may be traceable over very large regions.

Significant coal deposits could not form until vascular land plants appeared on Earth in the late Silurian Period, about 420 million years ago. The most extensive deposits of coal in the world occur in Carboniferous-age strata (359 to 299 Ma). In fact, geologists coined the name *Carboniferous* because strata representing this interval of the geologic column contain so much coal. Not all coal reserves, however, are Carboniferous—during the Cretaceous (145 to 66 Ma), large areas of freshwater coal swamps developed in Wyoming and adjacent states.

The Formation of Coal

The development of broad, continuous coal seams requires the accumulation of a substantial amount of vegetation in an *anoxic* (oxygen-poor) *environment*. Such accumulations can develop during times when the climate becomes very wet and warm, as happened in the Carboniferous, when large areas of continents

lay in equatorial latitudes, and in the Cretaceous, when superplumes dumped large quantities of greenhouse gases into the atmosphere. Wet climates cause the water table to rise until broad plains become saturated with water, and lower areas become submerged by shallow water. If it's warm, vegetation flourishes, so such wet regions become **coal swamps**, broad lowlands that resemble the wetlands and rainforests of modern tropical to semitropical regions (**Fig. 14.18a**). In the stagnant water of the swamp, oxygen from the air doesn't mix into the water, and microorganisms completely consume any oxygen that had been dissolved in the water, so the water becomes anoxic. Thus, vegetation that dies and falls into the water doesn't rot entirely away but rather gets buried by more dead vegetation until a compacted and partially decayed layer of organic matter, called **peat**, accumulates. The peat may be buried by fluvial (river) deposits, or if sea level then rises, and a transgression takes place (see Chapter 7), layers of marine sediments—such as sand, mud, or carbonate debris—bury and preserve the peat. If buried deeply enough, the succession of sediment turns into sedimentary strata, with the coal occurring as a sedimentary bed (**Fig. 14.18b**).

How do the remains of plants transform into coal? As we've seen, the first step involves the formation of peat, which contains about 50% carbon. (Notably, peat itself serves as a fuel in many parts of the world, where thick deposits formed from moss and grasses in bogs during the last several thousand years. This peat can easily be cut out of the ground and, once dried, burns.) To transform peat into coal, the peat must be buried to a depth of 4 to 10 km by overlying sediment. Such deep burial can happen where the surface of the continent gradually sinks, creating a sedimentary basin (see Chapter 7). At depth in the pile, the weight of overlying sediment compacts the peat and squeezes out any remaining water. Then, because temperature increases with depth in the Earth, deeply buried peat gradually heats up. Heat accelerates chemical reactions that gradually destroy plant fiber and release molecules such as H_2O, CO_2, CH_4, and N_2. These gases seep out of the reacting peat layer, leaving behind a residue concentrated with carbon. Once the proportion of carbon in the residue exceeds about 60%, the deposit formally becomes coal. With further burial and higher temperatures, chemical reactions yield progressively higher concentrations of carbon (**Fig. 14.18c**).

The Classification of Coal

Geologists classify coal according to the concentration of carbon, which in turn reflects the temperature to which the coal has been subjected underground. At temperatures of less than 100°C, coal is a soft, dark brown material called *lignite*. At higher temperatures (100° to 200°C), lignite transforms into dull, black *bituminous coal*. At still higher temperatures (200° to 300°C), bituminous coal transforms into shiny, black *anthracite* (also called hard coal). The progressive transformation of peat to anthracite, which occurs as the coal layer is buried more deeply and becomes warmer, reflects the completeness of chemical reactions that remove hydrogen, oxygen, and nitrogen atoms from the organic chemicals of the peat and leave behind carbon (**Fig. 14.18c**). Thus, lignite contains only about 60% carbon, bituminous about 70%, and anthracite about 90%. As the carbon content of coal increases, we say the **coal rank** increases—lignite is low-rank coal, bituminous is intermediate-rank coal, and anthracite is high-rank coal.

The burning of coal is a chemical reaction: $C + O_2 \rightarrow CO_2$. Therefore, the different ranks of coal produce different amounts of energy when burned. Anthracite contains more carbon per kilogram than lignite contains, so it produces more energy when burned. Specifically, burning a kilogram of anthracite yields about five times as much energy as burning a kilogram of peat.

Notably, the temperatures necessary to form anthracite develop only on the borders of mountain belts. Here mountain-building processes can push thick sheets of rock up along thrust faults and over the coal-bearing sediment, so the sediment ends up at depths of 8 to 10 km, where temperatures reach 300°C. In addition, mountain uplift drives very hot groundwater from great depth up through the coal and can cause its rank to increase. Metamorphism in the interiors of mountain belts leads to the expulsion of all elements except carbon from organic layers, and the remaining carbon atoms rearrange to form graphite, the gray mineral used to make pencils. Thus, coal cannot be found in metamorphic rocks.

Finding, Mining, and Using Coal

Because the vegetation that eventually becomes coal was initially deposited in a sequence of sediment, coal occurs as sedimentary beds (seams, in mining parlance) interlayered with other sedimentary rocks (**Fig. 14.19**). To find coal, geologists search for sequences of strata that were deposited in tropical to semitropical shallow-marine to terrestrial environments—the environments in which a swamp could exist. The sedimentary strata of continents contain huge quantities of discovered coal, or **coal reserves**. For example, economic seams (beds of coal 1 to 3 m thick, thick enough to be worth mining) of Cretaceous age occur in the U.S. and Canadian Rocky Mountain region, while economic seams of Carboniferous age are found throughout the midwestern United States. Coal is found widely in Europe, Asia, and Australia (**Fig. 14.20a, b**).

FIGURE 14.19 An example of coal beds interlayered with beds of sandstone and shale.

FIGURE 14.20 The distribution of coal, and its consumption.

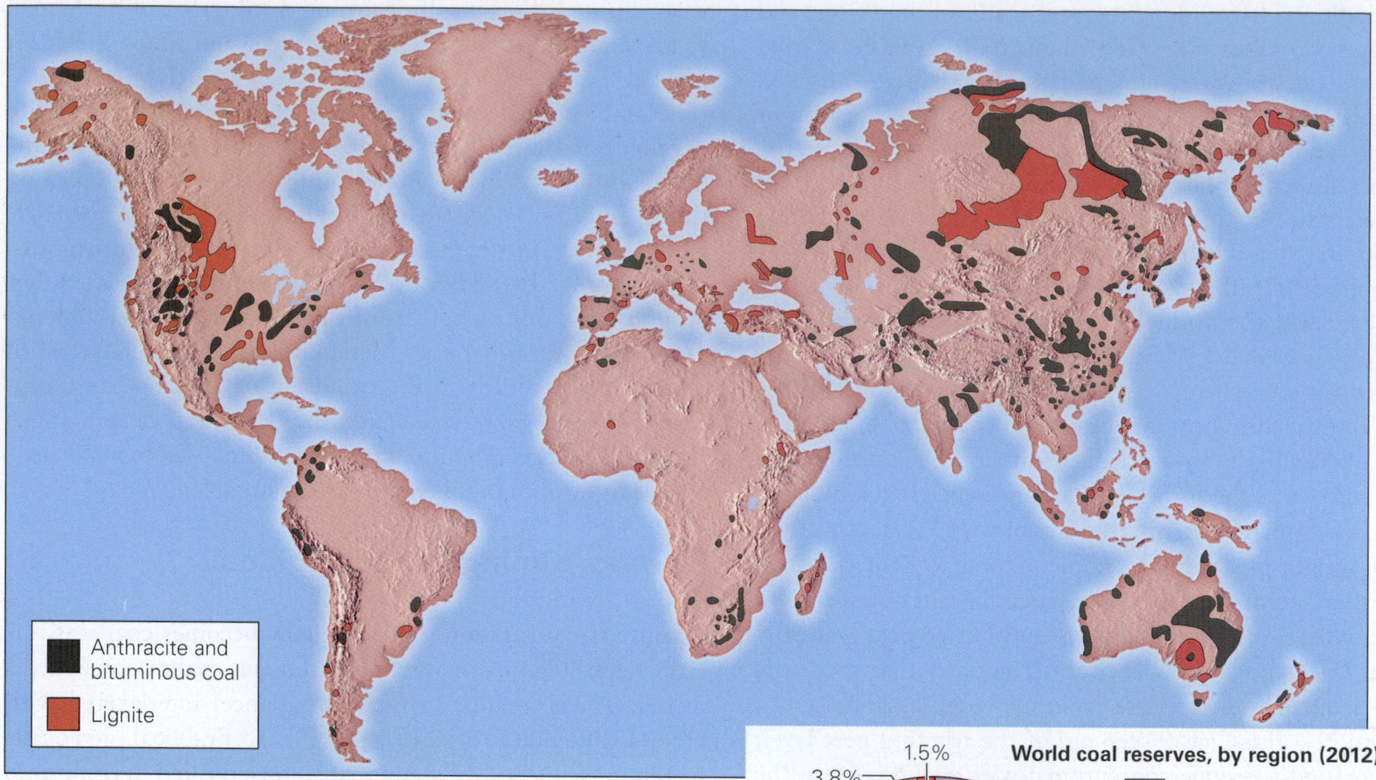

![World map showing global distribution of coal reserves with legend: Anthracite and bituminous coal (black), Lignite (red)]

Anthracite and bituminous coal

Lignite

(a) A map showing global distribution of coal reserves. Most coal accumulated in continental interior basins.

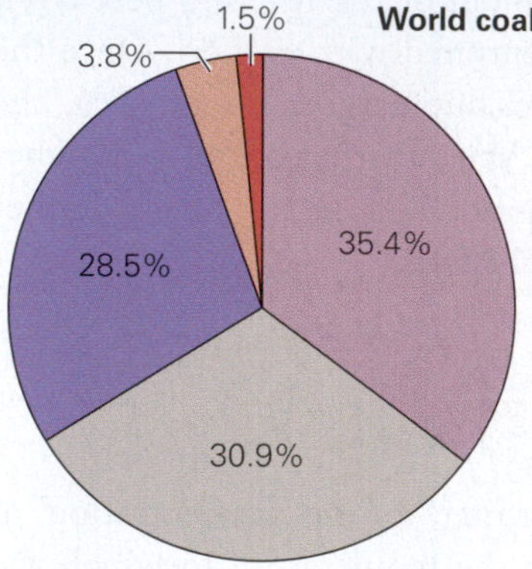

World coal consumption, by region (2013)

Asia Pacific	2,609
Europe and Eurasia	517
North America	469
Africa	98
South and Central America	28
Middle East	10
World total	**3,730***

Pie chart: 70%, 14%, 13%, 7.8%, 1%, 1%

* 1 tonne = 1,000 kg = 2,240 pounds = 1.12 tons

In the United States, a tonne is called a "metric ton."

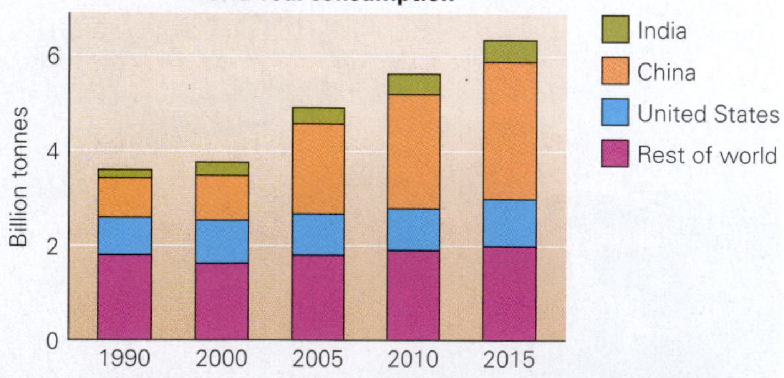

World coal reserves, by region (2012)

Europe and Eurasia	
Asia Pacific	
North America	
Middle East and Africa	
South and Central America	

Pie chart: 35.4%, 30.9%, 28.5%, 3.8%, 1.5%

World total = 860,938 million tonnes

(b) A graph illustrating the distribution of coal reserve quantities, by region. Data source: BP Statistical Review of World Energy (2013).

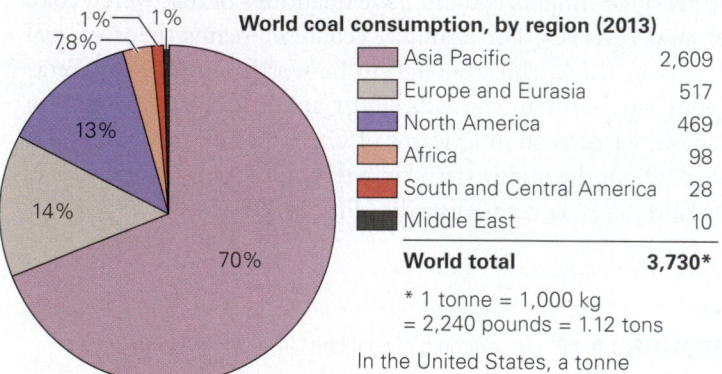

World coal consumption

Legend: India, China, United States, Rest of world

Y-axis: Billion tonnes (0, 2, 4, 6)
X-axis: 1990, 2000, 2005, 2010, 2015

c) Coal consumption varies greatly and is changing rapidly. Growth of consumption in China has more than tripled in the last 25 years. Data source: U.S. Energy Information Administration (EIA).

The United States was the largest consumer of coal until about 1986, when China's rate of consumption surpassed that of the United States. China's consumption has tripled since 2005, reflecting the country's rapid industrialization—in 2012, China burned 55% of the coal mined, the United States about 12%, India about 8%, and the rest of the world about 25% (**Fig. 14.20c**). All told, society is now consuming about 8 billion tons of coal a year, about 70% of which fuels electricity-generating stations. In fact, coal currently provides about 21% of the world's energy supply.

The way in which companies mine coal depends on the depth of the coal seam. If the coal seam lies within about 100 m of the ground surface, *strip mining* proves to be most economical. In strip mines, miners use a giant shovel, called a dragline, to scrape off soil and layers of sedimentary rock above the coal seam (**Fig. 14.21a**). Draglines are so big that the shovel could swallow a two-car garage without a trace. Once the dragline has exposed the seam, smaller machinery scrapes out the coal and dumps it into trucks or onto a conveyor belt where it is carried to storage piles (**Fig. 14.21b, c**). Before modern environmental awareness took hold, strip mining left huge scars on the landscape. Without topsoil, the rubble and exposed rock of the mining operation remained barren of vegetation. Coal beds that formed from coastal swamps tend to contain pyrite (FeS), which weathers when exposed to air and water to form sulfuric acid; sulfuric acid contributes to making the soil unsuitable for growth. In many contemporary mines, however, the dragline operator separates out and preserves soil. Then, when the coal has been scraped out, the operator fills the hole with the rock that had been stripped to expose the coal and covers the rock

back up with the saved soil, on which grass or trees may eventually grow. Within years to decades, the former mine site can become a pasture or a forest. In hilly areas, however, miners may use a practice called *mountaintop removal*, during which they blast off the top of the mountain and dump the debris into adjacent valleys. This practice disrupts the landscape permanently.

Deep coal can be obtained only by *underground mining*. To develop an underground mine, miners dig a shaft down to the depth of the coal seam and then create a maze of tunnels, using huge grinding machines that chew their way into the coal (**Fig. 14.21d**). Depending on circumstances, miners either leave columns of coal behind to hold up the mine roof, or they let the mined area collapse, so the ground level above sinks once mining is finished.

Underground coal mining can be very dangerous not only because the sedimentary rocks forming the roof of the mine are

FIGURE 14.21 Coal can be mined in strip mines or in underground mines.

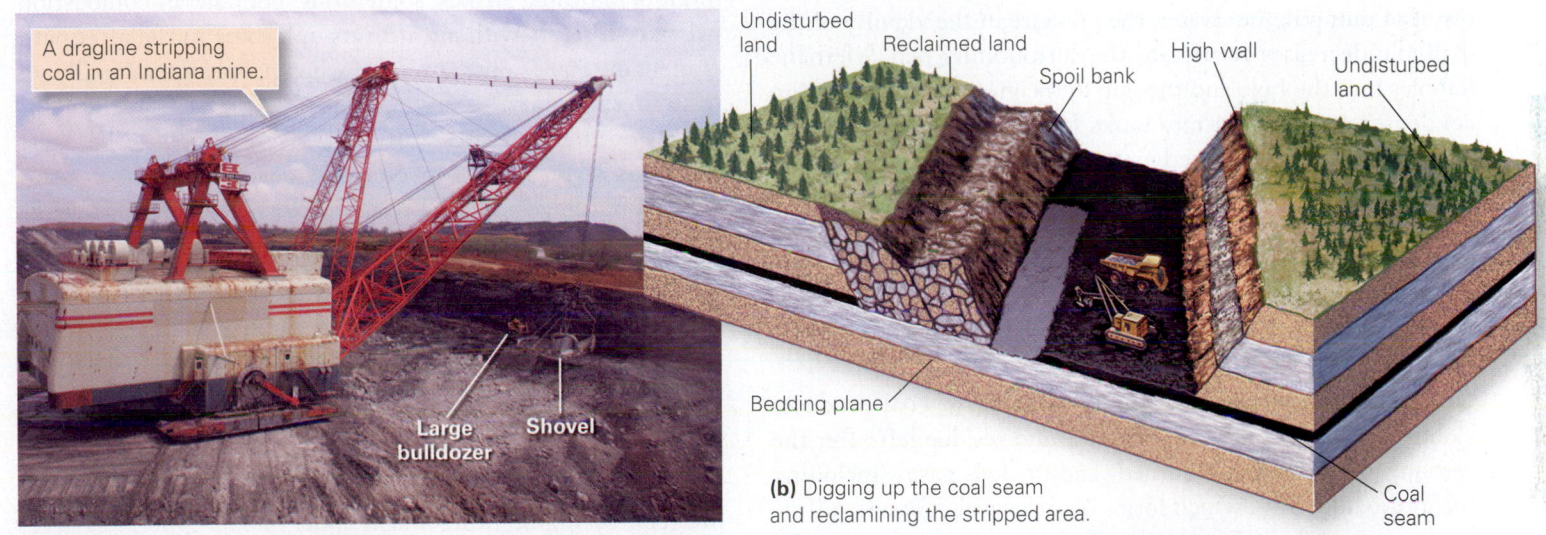

(a) A dragline stripping overburden.

(b) Digging up the coal seam and reclaiming the stripped area.

(c) Piles of recently mined coal.

(d) Underground coal mining.

weak and can collapse but also because methane gas released by continuing chemical reactions in coal can accumulate in the mine, leading to the danger of a small spark triggering a deadly mine explosion. Unless they breathe through filters, underground miners also risk contracting black lung disease from the inhalation of coal dust. The dust particles wedge into tiny cavities of the lungs and gradually cut off the oxygen supply or cause pneumonia.

Gas from Coal

Coal-Bed Methane As we've noted, the natural process by which coal forms underground yields methane, a type of natural gas. Over time, some of the gas escapes to the atmosphere, but vast amounts remain within the coal in pores or bonded to coal molecules. Such **coal-bed methane**, trapped in strata too deep to be reached by mining, is an energy resource that has become a target for extraction in many regions of the world.

Obtaining coal-bed methane from deep layers of strata involves drilling rather than mining. Drillers penetrate a coal bed with a hole and then start pumping out groundwater. As a result of pumping out water, the pressure in the vicinity of the drillhole decreases relative to the surrounding bed. Methane bubbles into the hole and then up to the ground surface, where condensers compress it into tanks for storage. Disposal of the water produced by coal-bed methane extraction can be a major problem. If the water is pure, it can be used for irrigation, but in deep coal beds the water may be saline and thus cannot be used for crops. Producers either pump this water back underground or evaporate it in large ponds, so that they can extract and collect the salt.

Coal Gasification. Traditional burning of coal produces clouds of smoke containing *fly ash* (solid residue left after the carbon in coal has been burned) and noxious gases (including sulfur dioxide, SO_2, which forms because coal contains sulfur-bearing minerals such as pyrite). Today smoke can be partially cleaned by expensive scrubbers, but pollution remains a major problem—the abundance of coal-burning plants in China has played a major role in producing the country's air pollution. Alternatively, solid coal can be transformed into various gases, as well as solid by-products, before burning. The gases burn relatively cleanly. The process of producing relatively clean-burning gases from solid coal is called **coal gasification**; the process was invented in the late 18th and early 19th centuries and was used extensively to produce fuels during World War II.

Coal gasification involves the following steps. First, pulverized coal is placed in a large container. Then a mixture of steam and oxygen passes through the coal at high pressure. As a result, the coal heats up to a high temperature but does not ignite. Under these conditions, chemical reactions break down and oxidize the molecules in coal to produce hydrogen (H_2) and other gases such as carbon monoxide (CO), H_2O,

and CO_2. Solid ash, as well as sulfur and mercury, concentrate at the bottom of the container and can be removed before the gases are burned so that the contaminants do not go up the chimney and into the atmosphere. The CO gas and H_2 gas can then be combined, through a chemical reaction, to produce hydrocarbons. Alternatively, the CO gas can react with water to produce more H_2 (plus CO_2). Of note, the hydrogen can be concentrated to produce fuel cells (which we will discuss later).

Underground Coal-Bed Fires

Coal will burn not only in furnaces but also in surface and subsurface mines as long as the fire has access to oxygen. Coal mining of the past two centuries has exposed much more coal to the air and has provided many more opportunities for fires to begin. Once started, such a *coal-bed fire* progresses underground by sucking in oxygen from joints and pore spaces in surrounding rock. The fires may be difficult or impossible to extinguish because they are inaccessible. Some fires begin as a result of lightning strikes, some from spontaneous combustion (when coal reacts with air, it heats up), some in the aftermath of methane explosions, and some when people intentionally set trash fires in mines.

Major coal-bed fires can be truly disastrous. The most notorious of these fires in North America began as a result of trash burning in a mine near the town of Centralia, Pennsylvania (**Fig. 14.22a**). For the past 50 years, the fire has progressed underground, eventually burning coal seams beneath the town itself. The fire produces toxic fumes that rise through the ground and make the overlying landscape uninhabitable, and it also causes the land surface to collapse and sink. Eventually, inhabitants had to abandon many neighborhoods of the town. Much longer-lived fires occur elsewhere—one in Australia may have been burning for a few thousand years. Satellite imagery, by highlighting warm spots on the ground surface, indicates that thousands of coal-bed fires are currently burning. Many of the fires occur in northern China—recent estimates suggest that 200 million tons of coal burn in China every year, an amount equal to approximately 20% of the annual national production of coal in China (**Fig. 14.22b**).

Take-Home Message

Coal forms from the accumulations of plant material over time in anoxic coal swamps. When buried deeply, this organic material undergoes reactions that concentrate carbon. Higher-rank coal contains more carbon. Coal occurs as beds, called seams, in sedimentary successions and must be mined underground or in open pits.

QUICK QUESTION: Why do underground coal fires start, and why are they so hard to stop?

FIGURE 14.22 Coal mine fires—long-lived disasters.

(a) A coal-bed fire beneath Centralia, Pennsylvania, produces noxious gas and has forced the evacuation of many homes.

(b) A burning coal bed in China, exposed in a mine wall, glows red.

14.7 Nuclear Power

How Does a Nuclear Power Plant Work?

So far we have looked at fuels, such as oil, gas, and coal, that release energy when they undergo *chemical burning*. During such burning, a chemical reaction between the fuel and oxygen releases the potential energy stored in the chemical bonds of the fuel. The energy that drives a nuclear power plant comes from a totally different process—*nuclear fission*, the breaking of the nuclear bonds that hold protons and neutrons together in the nucleus. Fission splits an atom into smaller pieces, and during this process, a small amount of mass transforms into thermal and electromagnetic energy, as Einstein characterized in his famous equation, $E = mc^2$ (where E is energy, m is mass, and c is the speed of light).

A **nuclear reactor**, the heart of the plant, commonly lies within a containment building made of reinforced concrete (**Fig. 14.23**). The reactor contains nuclear fuel, consisting of pellets of concentrated uranium oxide or a comparable radioactive material that is packed into metal tubes called *fuel rods*. Fission occurs when a speeding neutron strikes a radioactive isotope, causing it to split. For example, radioactive uranium-235 (^{235}U) splits into barium-141 (^{141}Ba) and krypton-92 (^{92}Kr) plus three neutrons. The neutrons released during the fission of one atom strike other atoms, thereby triggering more fission in a self-perpetuating process called a **chain reaction**, which overall produces lots of energy, including heat. Pipes carry water close to the heat-generating fuel rods, and the heat transforms the water into high-pressure steam. The pipes then carry this steam to a turbine, where it rotates fan blades—the rotation drives a dynamo that generates electricity. Eventually the steam goes into cooling towers, where it condenses back into water that can be reused in the plant or returned to the environment.

Compared with fossil fuels, the energy density of enriched uranium is vast—1 g of reactor fuel contains almost 80,000 times as much energy as 1 g of gasoline, and yields no air pollution or greenhouse gases. Thus, nuclear power plants could produce a large proportion of global energy supply. But they don't, at present, because of societies' concerns about handling nuclear fuel and nuclear waste.

The Geology of Uranium

Where does uranium come from? The Earth's radioactive elements, including uranium, probably developed during the explosion of a supernova that happened before the existence of our Solar System. Uranium atoms from this explosion became part of the nebula out of which the Earth formed and thus were incorporated into the planet. Large atoms like uranium don't fit well into the crystal structure of minerals, so when melts form, they preferentially enter the melt and rise with the melt. Eventually, uranium atoms were carried into the upper crust by rising granitic magma.

Even though granite contains uranium, it does not contain very much. But nature has a way of concentrating uranium. Hot water circulating through a pluton after intrusion dissolves the uranium and precipitates it as the mineral pitchblende (UO_2) in veins. Uranium may be further concentrated once plutons, and the associated uranium-rich veins, weather and erode at the ground surface. Sand derived from a weathered pluton washes down a stream, and as it does so, uranium-rich grains stay behind because they are so heavy relative to quartz and feldspar grains. The world's richest uranium deposits, in fact, occur in ancient streambed gravels. Uranium deposits may also form when groundwater percolates through uranium-rich

FIGURE 14.23 Producing electricity at a nuclear power plant.

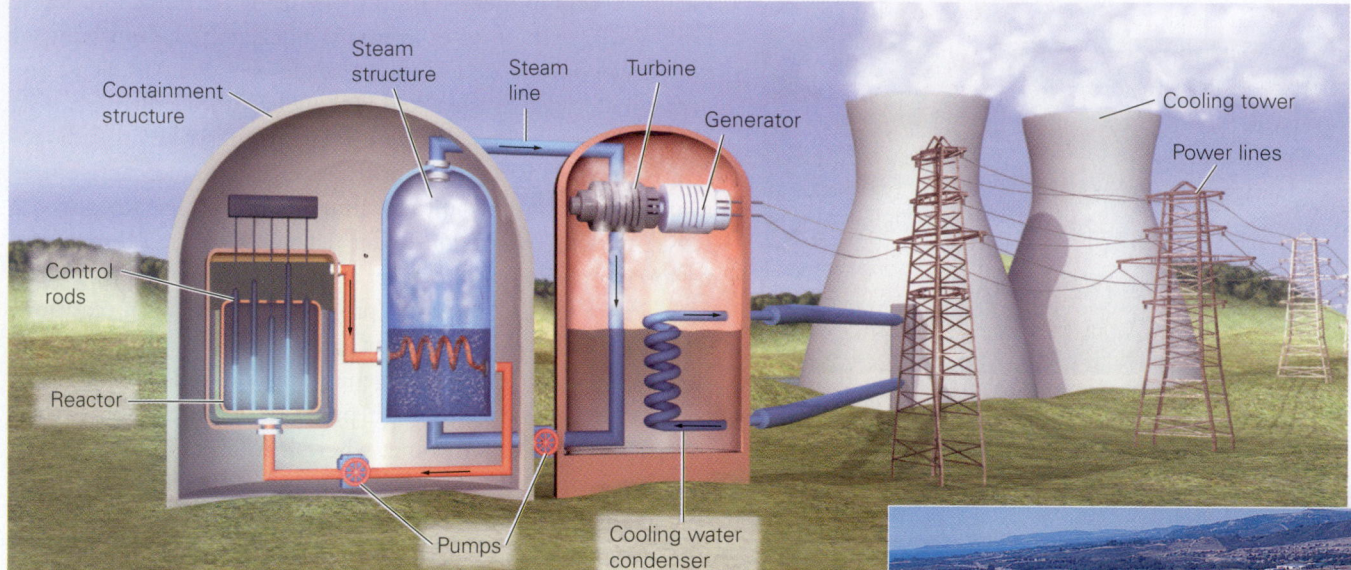

(a) In a nuclear power plant, a reactor heats water, which produces high-pressure steam. The steam drives a turbine that in turn drives a generator to produce electricity. A condenser transforms the steam back into water.

(b) This nuclear power plant in California has two reactors, each in its own containment building.

sedimentary rocks, for the uranium dissolves in the water and moves with the water to another location where the chemical environment is different, causing new uranium-bearing minerals to precipitate out of solution and fill the pores of the host sedimentary rock.

You can't just mine uranium and put it into a reactor. That's because ^{235}U, the isotope of uranium that serves as the most common fuel for conventional nuclear power plants, accounts for only about 0.7% of naturally occurring uranium—most uranium consists of ^{238}U. Thus, to make a fuel suitable for use in a power plant, the ^{235}U concentration in a mass of natural uranium must be increased by a factor of 2 or 3, an expensive process called *uranium enrichment*.

Challenges of Using Nuclear Power

Maintaining safety at nuclear power plants requires hard work. In conventional reactors, operators must ensure that circulating water constantly cools the nuclear fuel, and the rate of nuclear fission must be regulated by the insertion of *control rods*, made of materials that absorb neutrons and thus decrease the number of collisions between neutrons and radioactive atoms. Without control rods, the number of neutrons dashing around in the nuclear fuel would progressively increase, causing the rate of fission and accompanying heat production to increase. Eventually, the fuel would become so hot that it would melt. Such

a **meltdown** might cause a steam explosion, or it could generate such high temperatures that water molecules break to form hydrogen gas and oxygen gas, a mixture that can be very explosive. If the explosion is large enough, it could breach the containment building and scatter radioactive debris into the air. Note that a meltdown is not the same as an atomic bomb explosion. An atomic bomb explosion can only occur if there is a sufficient "critical mass" (quantity) of highly enriched uranium (90% ^{235}U), in which fission reactions happen so quickly that the fuel itself explodes. (By comparison, "reactor-grade" uranium is only 3% to 4% ^{235}U.)

The first nuclear power plant designed to generate electricity for the public became operational in 1954. Today, about 435

> **Did you ever wonder . . .**
>
> whether a nuclear power plant could explode like an atomic bomb?

nuclear power plants are in operation, and about 70 are under construction. Over the past 60 years, there have been three significant accidents. The first occurred in 1979 at the Three Mile Island plant in Pennsylvania, when a stuck valve allowed coolant water to escape, causing the reactor to overheat. Eventually, some of that radioactive coolant leaked out into the environment, but contamination due to the leak was limited. The next, and much more serious nuclear accident occurred at the power plant in Chernobyl, Ukraine, in April 1986, while engineers were conducting a test. The fuel pile became too hot, triggering a hydrogen explosion that ruptured the roof of the containment building and spread fragments of the reactor and its fuel around the plant grounds, and within six weeks, 20 people had died from radiation sickness. In addition, some of the radioactive material entered the atmosphere and dispersed over eastern Europe and Scandinavia, but no one yet knows whether this fallout affected the health of exposed populations. The reactor has been entombed in concrete, and the surrounding region has remained closed. In 2011, a disaster second only to Chernobyl in terms of the amount of radiation released, occurred at the multireactor Fukushima power plant along the east coast of northern Japan. This disaster occurred when the catastrophic tsunami that followed the magnitude 9.0 Tōhoku earthquake knocked out both the power lines providing electricity to pumps that circulated cooling water and the backup diesel generators (see Chapter 10). As a result, some of the reactors overheated, and they suffered partial meltdowns and hydrogen gas explosions. Long-term health effects are not known.

One of the biggest challenges to the nuclear industry pertains to the storage of **nuclear waste**, the radioactive material produced in a nuclear plant. It includes spent fuel, which contains radioactive elements, as well as water and equipment that have come in contact with radioactive materials. Radioactive elements emit gamma rays and X-rays, which can damage living organisms and cause cancer. Some radioactive material decays quickly (in decades to centuries), but some remains dangerous for thousands of years or more. Nuclear waste cannot just be stashed in a warehouse or buried in a town landfill. Some of the waste is hot enough that it needs to be cooled with water. Even cooler waste has the potential to leak radioactive elements into municipal water supplies or nearby lakes or streams. Ideally, waste should be sealed in containers that will last for thousands of years (the time needed for the short-lived radioactive atoms to undergo decay) and stored in a place where it will not come in contact with the environment. Finding an appropriate place is not easy, and so far, experts disagree about which is the best way to dispose of nuclear waste. For some years the U.S. government favored storing waste at Yucca Mountain in the Nevada desert, but this is no longer likely and most nuclear waste remains on the site of the power plant that produced it.

Take-Home Message

Controlled fission in reactors produces nuclear power. The fuel consists of uranium or other elements obtained by mining. Reactors run the risk of meltdown but cannot explode like an atomic bomb. They also yield radioactive waste, which is difficult to store.

QUICK QUESTION: What is the difference between a meltdown in a reactor and the explosion of an atomic bomb?

14.8 Other Energy Sources

Geothermal Energy

As the name suggests, **geothermal energy** comes from heat in the Earth's crust. We can distinguish between high-temperature geothermal energy, which is used for producing heat and electricity at a commercial scale, and low-temperature geothermal energy (also known as *ambient geothermal energy*), which is used for warming and cooling the water of an individual household.

High-temperature geothermal energy exists because the crust becomes progressively hotter with increasing depth, at a rate defined by the geothermal gradient. In active volcanic areas, the increase happens so fast that a temperature of 100°C or more can be attained within only several hundred meters of the Earth's surface. Thus, at relatively shallow depths, bedrock, and any groundwater contained in bedrock, is at or near the boiling point of water. Power companies can use this geothermal energy in many ways. For example, in some places, they simply pump hot groundwater through pipes to heat houses or spas directly. If the groundwater is hot enough, it turns to steam when it rises to the Earth's surface and decompresses. This steam can drive

FIGURE 14.24 Geothermal power plants utilize hot groundwater. In some cases, the water is so hot that it becomes steam when it rises and undergoes decompression. Most geothermal plants are in volcanic areas.

Steam turns turbines to generate electricity.

Geothermal well

Rain

Hot groundwater

Cold groundwater

Magma

Natural heat warms groundwater.

Hot water rises in a well and turns to steam.

Steam-filled fracture

A geothermal power plant in New Zealand

turbines and generate electricity (**Fig. 14.24**). In Iceland and New Zealand, which sit astride volcanic areas, geothermal energy provides a substantial portion of electricity needs. But on a global basis, the impact is much smaller.

Ambient geothermal energy takes advantage of the fact that below a depth of a few meters, the ground temperatures remain nearly constant all year (at about the average annual temperature of the air above). By installing pipes in the ground, typically down to a depth of 5 m, home owners can cool the water used in their house during the summer, and can heat the water used in their house in the winter, before running it through a powered cooling system or heating system, respectively. This decreases the amount of energy needed to heat or cool the water and lower costs.

Biofuels

In recent years, farmers have begun to produce rapidly growing crops specifically for the purpose of producing biomass for fuel production. The resulting liquids are called **biofuels**. The most commonly used biofuel is ethanol, a type of alcohol, which can substitute for gasoline in car engines. The process of producing ethanol from corn includes the following steps: First, producers grind corn into a fine powder, mix it with water, and cook it to produce a mash of starch. Then they add an enzyme to the mash, which converts the mash into sugar. The sugar, when mixed with yeast, ferments. Fermentation produces ethanol and CO_2. Finally, the fermented mash is distilled to concentrate the ethanol. While corn is the main source of ethanol in North America, ethanol in Brazil is produced directly, without fermentation, from sugar extracted from sugarcane.

Researchers have also begun to develop processes that yield ethanol from cellulose, permitting perennial grasses and the stalks of other plants to become a source of liquid fuel, or from algae, which naturally produces fatty chemicals (lipids) from which hydrocarbons can be produced. Such processes could help make ethanol a renewable energy source. And recent technologies have also led to the commercial production of *biodiesel*, a fuel produced by chemical modification of fats and vegetable oils. Biodiesel can be mixed with petroleum-derived diesel to run trucks and buses.

Hydroelectric and Wind Power

For millennia, people have used flowing water and air to produce energy. In fact, many towns were established next to rivers, where streams could rotate the waterwheels that powered mills and factories. And in agricultural areas, farmers used windmills to pump water for irrigation. In the past century, engineers have begun to employ the same basic technology to drive generators that produce electricity. Energy derived from flowing fluids is clean, or "green" in the sense that its production does not release chemical or radioactive pollutants, and is renewable in that its production does not consume limited resources. But its use does impact the environment.

In a modern hydroelectric power plant, the potential energy of water is converted into kinetic energy as the water flows from a higher elevation to a lower elevation. The flowing water turns turbine blades placed in a pipe, and the turbine drives an electrical generator. Most hydroelectric plants rely on water from a reservoir held back by a dam. The largest of these is the Three Gorges Dam on the Yangtze River in China

(**Fig. 14.25a**). Dam construction increases the available potential energy of the water because the water level in a filled reservoir is higher than the level of the valley floor that the dam spans. Hydroelectric energy is clean, and reservoirs may have the added benefit of providing flood control, irrigation water, and recreational opportunities. But the construction of dams and reservoirs may bring unwanted changes to a region, so the benefits of their construction must be weighed against potential harm. Damming a river may flood a spectacular canyon, eliminate exciting rapids, or destroy an ecosystem. Further, reservoirs trap sediment and nutrients, thus disrupting the supply of these materials to downstream floodplains or deltas, a process that may adversely affect agriculture.

Not all hydroelectric power generation utilizes flowing river water—engineers have been developing new means to tap **tidal power** (the daily rise and fall of the sea; see Chapter 18). One approach involves building a dam, called a *tidal barrage,* across the entrance to a bay or estuary (the flooded mouth of a river) in which there are large tides. When the tide rises, water spills into the enclosed area through openings in the dam. When the tide outside drops, water flows back to the sea via a pipe that carries it through a power-generating turbine (**Fig. 14.25b**). More recently, engineers have developed technologies to place huge fan blades underwater in nearshore regions where tidal currents naturally flow; the blades slowly turn in the current and run generators.

When you think of wind energy, you may picture a classic Dutch windmill driving a water pump. Modern efforts to harness the wind are on a much larger scale. To produce wind-generated electricity, engineers identify regions, either onshore or just offshore, with steady breezes. In these regions, they build wind farms that consist of numerous towers, each of which holds a wind turbine, a giant fan blade that turns even in a gentle breeze (**Fig. 14.25c**). Some towers are on the

FIGURE 14.25 Hydroelectric and tidal power.

(a) Gravity causes water held back by the Three Gorges Dam to flow through turbines and generate electricity.

(c) A wind farm in southwestern England. The towers are about 50 m high.

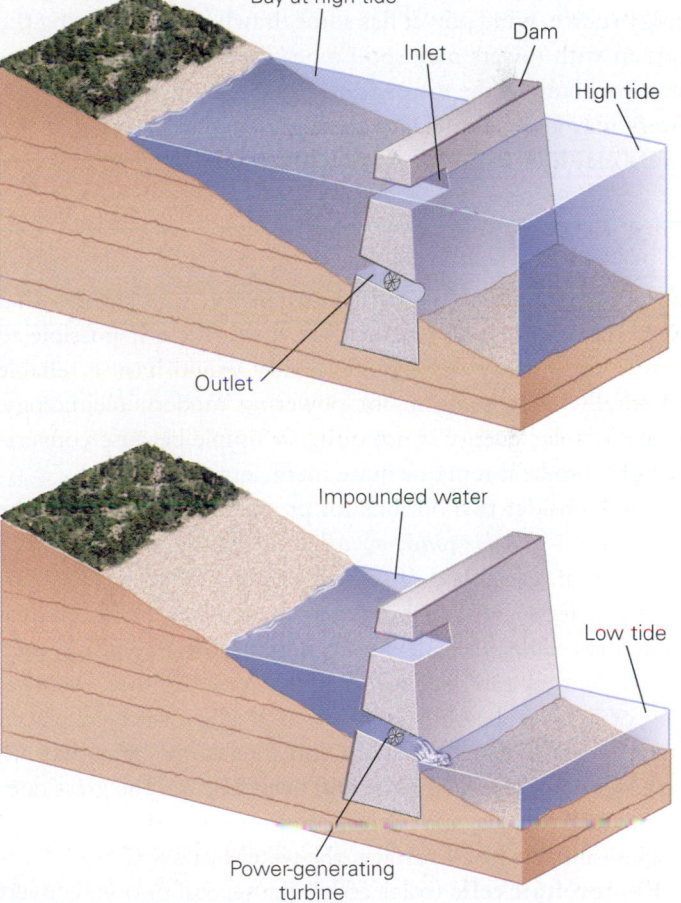

(b) To produce tidal power, a dam traps seawater at high tide; the water can flow through a turbine at low tide.

FIGURE 14.26 Solar energy and fuel cells.

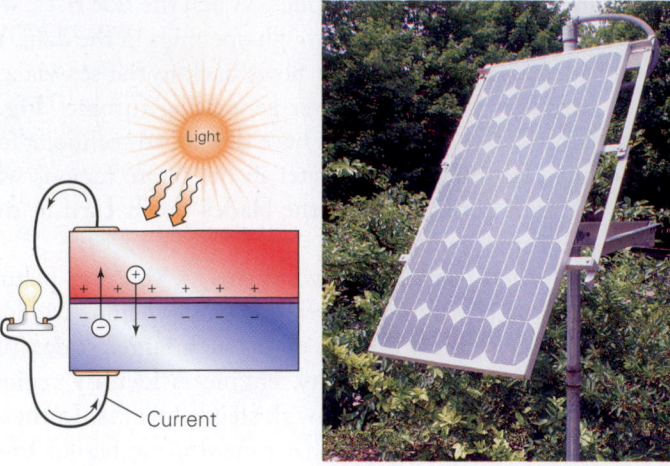

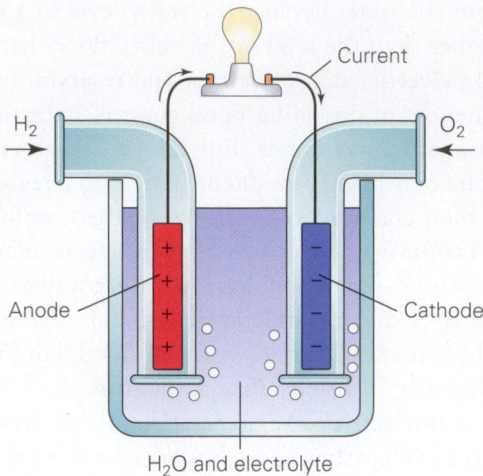

(a) A photovoltaic cell produces electricity directly from solar radiation.

(b) A hydrogen fuel cell.

order of 100 m (300 ft) tall, with fan blades that are over 40 m (120 ft) long. Large wind farms can host hundreds of towers. Wind energy produces no pollutants, but as with any type of energy source, wind power has some drawbacks. Cluttering the horizon with towers may spoil a beautiful view, and the constant loud hum of the towers can disturb nearby residents. The towers may also be a hazard to migrating birds, and offshore towers may interfere with marine life.

Solar Energy

The Sun drenches the Earth with energy in quantities that dwarf the amounts stored in fossil fuels. Were it possible to harness this energy directly, humanity would have a reliable and totally clean solution for powering modern technology. But using solar energy is not quite so simple because converting light into heat remains quite inefficient.

Let's consider two options for producing solar energy. The first option is a *solar collector*, a device that collects energy to produce heat. One class of solar collectors includes mirrors or lenses that focus light striking a broad area into a smaller area. On a small scale, such devices can be used for cooking; on a large scale, they can produce steam to drive turbines. Another class of solar collectors consists of a black surface placed beneath a glass plate. The black surface absorbs light that has passed through the glass plate and heats up, and the glass does not let the heat escape. When a consumer runs water between the glass and the black surface, the water heats up.

Photovoltaic cells (solar cells), the second option, convert light energy directly into electricity (**Fig. 14.26a**). Most photovoltaic cells consist of two wafers of silicon pressed together. One wafer also contains atoms of arsenic, and the other wafer includes atoms of boron. When light strikes the cell, arsenic atoms release electrons that flow over to the boron atoms. If a

wire loop connects the back side of one wafer to the back side of the other, this phenomenon produces an electrical current. The production costs of solar cells have decreased substantially in recent years, and thus their use has increased substantially.

Fuel Cells

In a fuel cell, chemical reactions produce electricity directly. Let's consider a *hydrogen fuel cell*. Hydrogen gas flows through a tube across an anode (a strip of platinum) that has been placed in a water solution containing an electrolyte (a substance that enables the solution to conduct electricity). At the same time, a stream of oxygen gas flows onto a separate platinum cathode that has also been placed into the solution. A wire connects the anode and the cathode to provide an electrical circuit (**Fig. 14.26b**). In this configuration, hydrogen reacts with oxygen to produce water and electricity. Fuel cells are efficient and clean. Their limitation lies in the need for a design that will protect the cells from damage by impact and that will enable them to store hydrogen, an explosive gas, in a safe way. Also, it takes a significant amount of energy from other sources to produce the hydrogen used in fuel cells.

Take-Home Message

A variety of alternative energy resources are now under development. Biomass can be transformed into burnable alcohol and biodiesel. Geothermal energy utilizes groundwater that has been warmed by heat from Earth's interior. Flowing water (in rivers or tides), flowing air, solar panels, and fuel cells can also produce energy.

QUICK QUESTION: Is there any way that home owners can use heat stored in the Earth on their property to lower their energy costs?

14.9 Energy Choices, Energy Problems

The Age of Oil and the Oil Crunch

Energy usage in industrialized countries grew with dizzying speed through the mid-20th century, and during this time people came to rely increasingly on oil (see Fig. 14.2). Oil remains the single largest source of energy globally, accounting for about 33% of global energy consumption. The United States, European Union, and Japan combined used more oil than the rest of the world until about 2006. Then consumption in these industrialized nations leveled off and started to decrease, while usage in the rest of the world has increased as countries in Asia and South America industrialize. According to statistics published in 2013, the Asia-Pacific region uses about 33% of global energy, North America as a whole now uses about 26%, Europe and Eurasia together use about 21%, and Africa uses only 4%. The United States still uses more oil per capita (about 22 bbl/year) than any country except Saudi Arabia—by comparison, per capita use of oil in China is about 3 bbl/year.

During the latter half of the 20th century, conventional oil supplies within the borders of industrialized countries could no longer match the demand, and these countries began to import more oil than they produced themselves. Through the 1960s, oil prices remained low (about $1.80 a barrel), so this was not a problem (**Fig. 14.27a**). In 1973, however, a complex tangle of politics and war led the Organization of Petroleum Exporting Countries (OPEC) to limit its oil exports. In the United States, fear of an oil shortage turned to panic, and motorists began lining up at gas stations, in many cases waiting for hours to fill their tanks. The price of oil rose to $18 a barrel, and newspaper headlines proclaimed, "Energy Crisis!" Governments in industrialized countries instituted new rules to encourage oil conservation. During the last two decades of the 20th century, the oil market stabilized, though political events occasionally led to price jumps and short-term shortages. Since 2004, oil prices rose overall, passing the $147/bbl mark in 2008. During the Great Recession of 2008, the price then collapsed, but in recent years, it has hovered around $100/bbl.

Will a day come when shortages of conventional oil arise not because of an embargo or limitations on refining capacity, but because there is no more oil to produce? Already, evidence suggests that the rate of new discoveries can no longer keep up with the rate of consumption, so society is tapping into reserves faster than reserves are increasing (**Fig. 14.27b**). To address such a question, we must keep in mind the distinction between renewable and nonrenewable energy resources (**see Geology at a Glance**, pp. 536–537). We can call a particular resource renewable if nature can replace it within a short time relative to a human life span (in months or, at most, decades), whereas a resource is nonrenewable if nature takes a very long time (hundreds to perhaps millions of years) to replenish it. Oil is a nonrenewable resource in that the rate at which humans consume it far exceeds the rate at which nature replenishes it, so we will inevitably run out of oil. The question is, when?

Historians in the future may refer to our time as the Oil Age because so much of our economy depends on oil. How

FIGURE 14.27 Price, discovery, and consumption of oil.

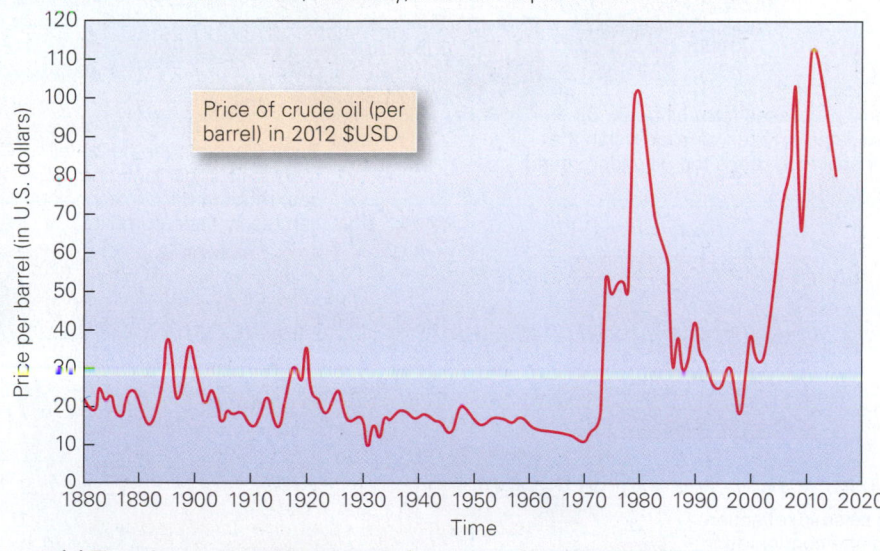

(a) The price of oil held fairly steady for almost 100 years. Starting in 1970, it has risen and fallen dramatically. Data source: BP Statistical Review of World Energy (2013).

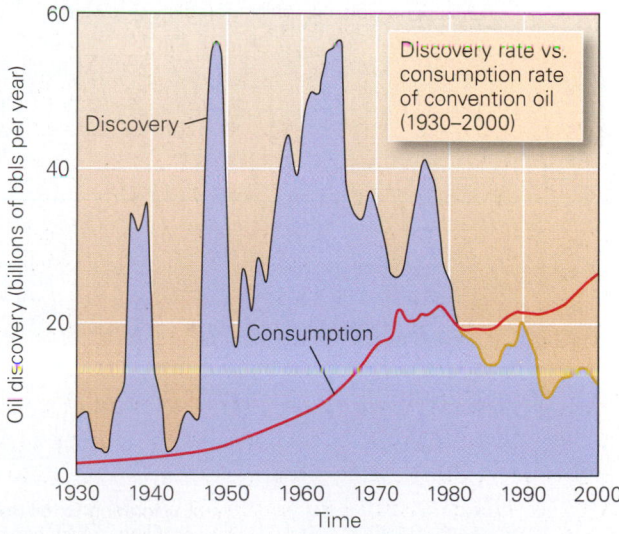

(b) Despite short-term dips, consumption of oil has continued to grow, but the rate of new discovery has not. Data source: ExxonMobil; consumption based on data from EIA.

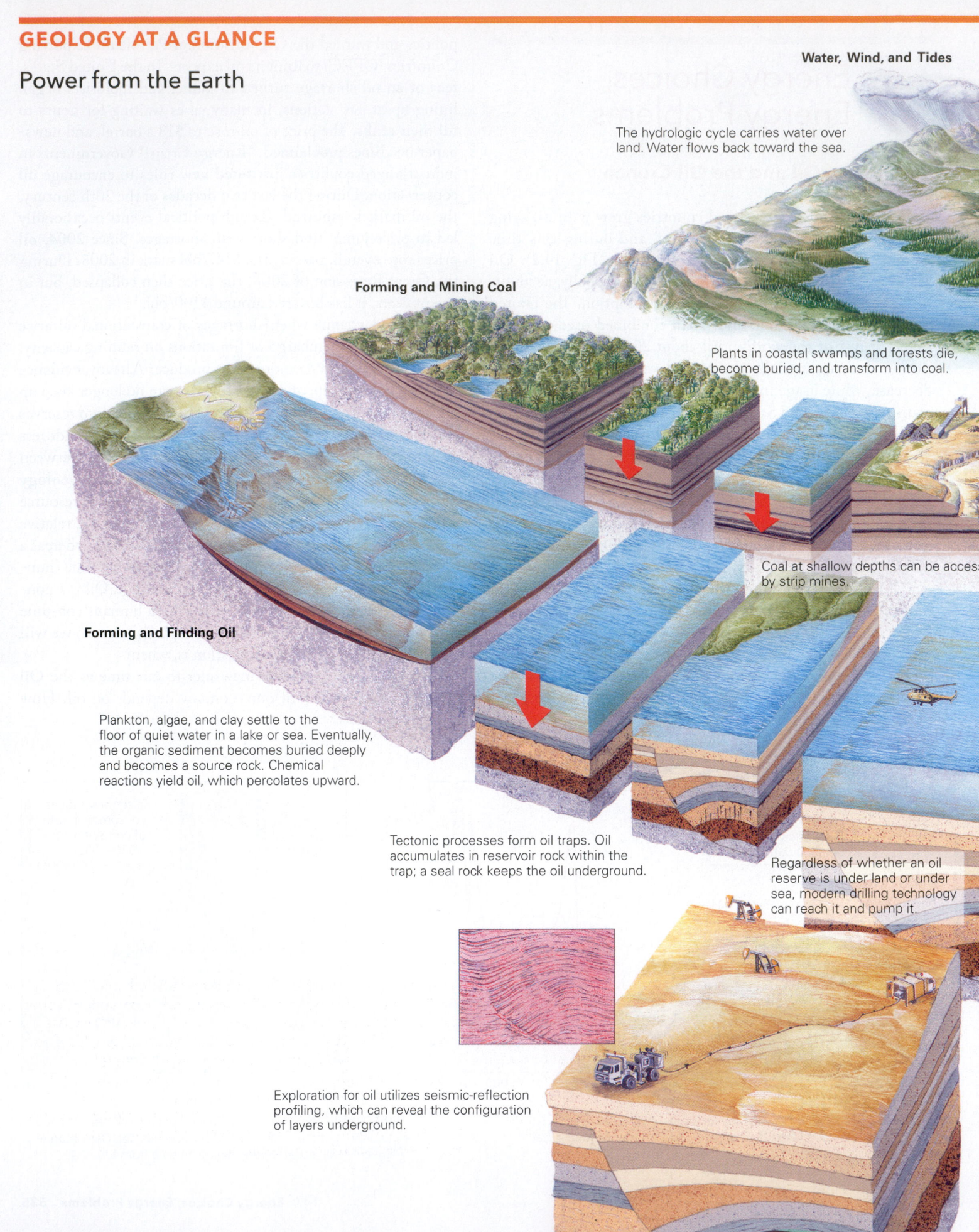

GEOLOGY AT A GLANCE

Power from the Earth

Water, Wind, and Tides

The hydrologic cycle carries water over land. Water flows back toward the sea.

Forming and Mining Coal

Plants in coastal swamps and forests die, become buried, and transform into coal.

Coal at shallow depths can be accessed by strip mines.

Forming and Finding Oil

Plankton, algae, and clay settle to the floor of quiet water in a lake or sea. Eventually, the organic sediment becomes buried deeply and becomes a source rock. Chemical reactions yield oil, which percolates upward.

Tectonic processes form oil traps. Oil accumulates in reservoir rock within the trap; a seal rock keeps the oil underground.

Regardless of whether an oil reserve is under land or under sea, modern drilling technology can reach it and pump it.

Exploration for oil utilizes seismic-reflection profiling, which can reveal the configuration of layers underground.

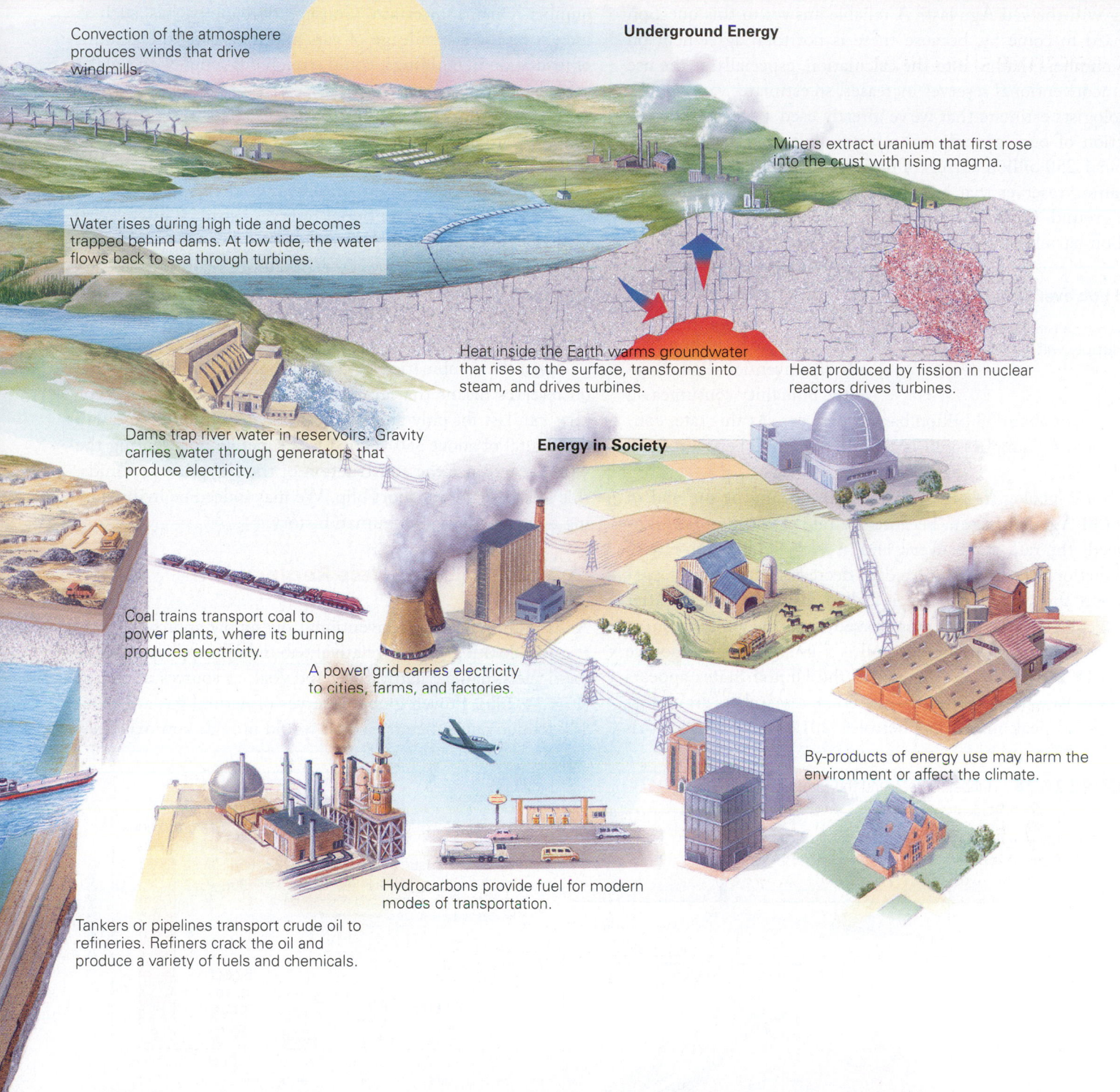

Convection of the atmosphere produces winds that drive windmills.

Underground Energy

Miners extract uranium that first rose into the crust with rising magma.

Water rises during high tide and becomes trapped behind dams. At low tide, the water flows back to sea through turbines.

Heat inside the Earth warms groundwater that rises to the surface, transforms into steam, and drives turbines.

Heat produced by fission in nuclear reactors drives turbines.

Dams trap river water in reservoirs. Gravity carries water through generators that produce electricity.

Energy in Society

Coal trains transport coal to power plants, where its burning produces electricity.

A power grid carries electricity to cities, farms, and factories.

By-products of energy use may harm the environment or affect the climate.

Hydrocarbons provide fuel for modern modes of transportation.

Tankers or pipelines transport crude oil to refineries. Refiners crack the oil and produce a variety of fuels and chemicals.

Modern society, for better or worse, uses vast amounts of energy to produce heat, to drive modes of transportation, and to produce electricity. This energy comes either from geologic materials stored in the Earth or from geologic processes happening at our planet's surface. For example, oil and gas fill the pores of reservoir rocks at depth below the surface, coal occurs in sedimentary beds, and uranium concentrates in ore deposits. A hydroelectric power plant taps into the hydrologic cycle, windmills operate because of atmospheric convection, and geothermal energy comes from hot groundwater.

Ultimately, the energy in the sources just listed comes from the Sun, from gravity, from Earth's internal heat, and/or from nuclear reactions. Oil, gas, and coal are fossil fuels because the energy they store first came to Earth as sunlight, long ago.

As energy usage grows, easily obtainable energy resources dwindle, the environment can be degraded, and the composition of the atmosphere changes. The pattern of energy use that forms the backbone of society today may have to change radically in the not-so-distant future if we wish to avoid a decline in living standards.

long will the Oil Age last? A reliable answer to this question is hard to come by, because there is not total agreement on the numbers that go into the calculation, especially as the use of unconventional reserves increases, so estimates vary widely. Geologists estimate that we've already used a substantial proportion of our conventional reserves, but that there are still about 1,250 billion barrels of proven conventional oil reserves, meaning reserves that have been documented and are still in the ground. Optimistically, there may be an additional 2,000 billion barrels of unproven conventional reserves, meaning oil that has not yet been found but might exist. Thus, the world possibly holds between 1,250 and 3,350 billion barrels of conventional oil. Presently, humanity consumes oil at a rate of about 33 billion barrels per year. At this rate, conventional oil supplies will last until sometime between 2050 and 2150.

Some geologists argue that the beginning of the end of the Oil Age has begun, because the rate of consumption now exceeds the rate of discovery, and in many regions the rate of production has already started to decrease. The peak of production for a given reserve is called **Hubbert's Peak**, after the geologist who first emphasized that the production of reserves must decline because oil is a nonrenewable resource (**Fig. 14.28a**). Hubbert's Peak for the United States appears to have been passed in the 1970s. Some researchers argue that the global peak may occur between 2012 and 2015, but this number remains uncertain. Conservation approaches, such as increasing the gas mileage of cars and increasing the amount of insulation in buildings, could stretch out supplies and make them last decades longer.

Of course, the picture of oil reserves changes significantly if unconventional reserves are included in estimates. Currently, there are an estimated 414 billion barrels of proven unconventional oil reserves, and there may be more than a trillion additional barrels of unconventional reserves yet to be discovered. But wide disagreement remains concerning whether it's fair to include all of these reserves in estimates of hydrocarbon supplies because a significant proportion would be so difficult and expensive to access that they may never really be an economical energy source.

Even the combination of conventional and unconventional oil reserves means that at current rates of consumption supplies can last for only another 200 years, so the Oil Age will last a total of about 350 years. On a timeline representing the 4,000 years since the construction of the Egyptian pyramids, this looks like a very short blip. We may indeed be living during a unique interval of human history.

Can Other Fossil Fuels Replace Oil?

As true limits to the conventional oil supply approach, societies are looking first at relatively abundant supplies of other fossil fuels, namely natural gas and coal, as sources of energy (**Fig. 14.28b**). Proven global reserves of natural gas are about 187 trillion cubic meters, which would provide approximately

FIGURE 14.28 Hubbert's Peak and the future of energy.

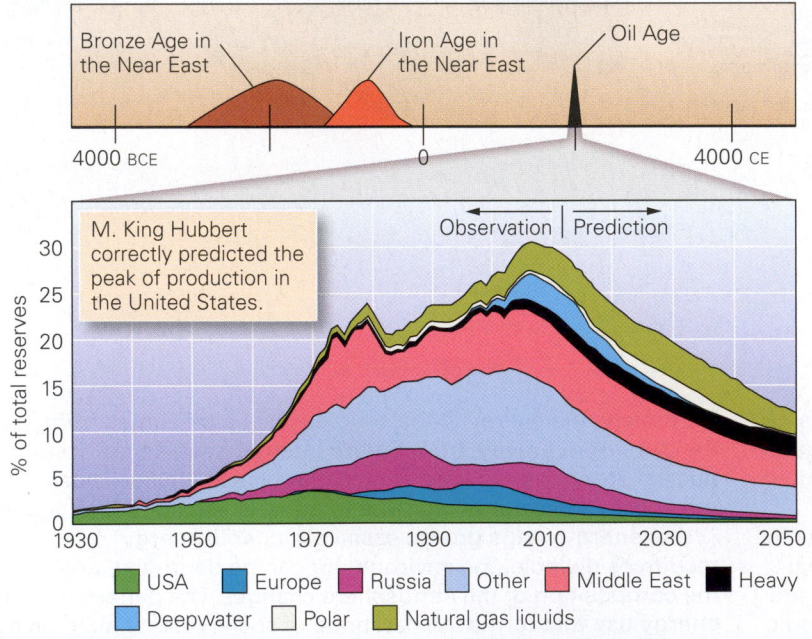

(a) A plot of production vs. time suggests that some time in the early 21st century, the production of oil globally will start to decrease.

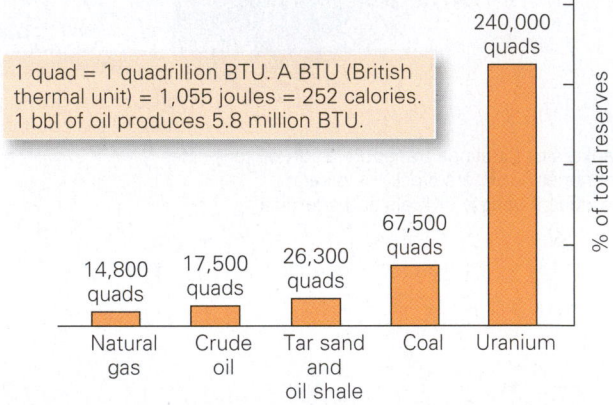

(b) Estimated nonrenewable energy resources on Earth.

the same amount of energy as about 1.2 trillion bbl. Tapping into this gas supply requires expensive technologies for extraction and transport, but given the current price of oil, it has become economical not only to produce conventional gas but also to produce shale gas. In the United States, production of shale gas has increased from less than 33 billion cubic meters per year in 2006 to 306 billion cubic meters per year in 2013. This change has significantly altered the dependence of the United States on imported oil. Similarly, worldwide coal reserves are estimated to be about 850 trillion tons, which contains approximately the same amount of energy as 11 trillion bbl. But the stated number for coal reserves does not distinguish clearly between accessible coal, which can be mined, and inaccessible coal, which is too deep to be mined.

Environmental Issues of Fossil Fuel Use

Environmental concerns about energy resources begin right at the source. Oil drilling requires substantial equipment, the use of which can damage the land. And as demonstrated by the 2010 Gulf of Mexico offshore well blowout, oil drilling can lead to tragic loss of life and disastrous marine oil spills (**Box 14.3**). Oil spills from pipelines or trucks sink into the subsurface and contaminate groundwater, and oil spills from ships and tankers create slicks that spread over the sea surface and foul the shoreline (**Fig. 14.29**). Coal and uranium mining also scar the land and can lead to the production of **acid mine runoff**, a dilute solution of sulfuric acid that forms when sulfur-bearing minerals such as pyrite (FeS_2) in mines react with rainwater. The runoff enters streams and kills fish and plants. Collapse of underground coal mines may cause the ground surface to sink.

Numerous air-pollution issues also arise from the burning of fossil fuels, which sends soot, carbon monoxide, sulfur dioxide, nitrous oxide, and unburned hydrocarbons into the air. Coal, for example, commonly contains sulfur, primarily in the form of pyrite, which enters the air as sulfur dioxide (SO_2) when coal is burned. This gas combines with rainwater to form dilute sulfuric acid (H_2SO_4), or **acid rain**. For this reason, many countries now regulate the amount of sulfur that coal can contain when it is burned.

But even if pollutants can be decreased, the burning of fossil fuels still releases carbon dioxide (CO_2) into the atmosphere. CO_2 is important because it is a **greenhouse gas**, meaning that it traps heat in the Earth's atmosphere much like glass traps heat in a greenhouse (see Chapter 23). A global increase of CO_2 will lead to a global increase in atmospheric temperature (global warming), which in turn may alter the distribution of climatic belts and lead to a rise in sea level, among many other effects. Because of concern about CO_2 production, research efforts are under way to replace fossil fuels with biofuels derived from perennial plants, for these biofuels are "carbon neutral" in that, since plants absorb CO_2 as they grow, the CO_2 added to the air by burning fuels derived from plants will be absorbed by the growth of new plants. There are also efforts to develop techniques to capture CO_2 at power plants, liquefy it, and pump it into reservoir rocks deep underground. This process is called *carbon sequestration*. We'll learn more about this issue in Chapter 23.

Alternatives to Hydrocarbons

Can nuclear power or hydroelectric power replace oil? Vast supplies of uranium, the fuel of traditional nuclear plants, remain

FIGURE 14.29 Marine oil spills. These can come from drilling rigs, or from tankers.

(a) An oil tanker leaking oil on the sea surface.

(b) Oil spills can contaminate the shore and can be very difficult to clean.

BOX 14.3 CONSIDER THIS . . .

Offshore Drilling and the Deepwater Horizon Disaster

A substantial proportion of the world's oil reserves reside in the sedimentary basins that underlie the continental shelves of passive continental margins. To access such reserves, oil companies build offshore drilling platforms. Such drilling requires complex technology. In water less than 600 m (2,000 ft) deep, companies position fixed platforms on towers resting on the seafloor. In deeper water, semi-submersible platforms float on huge submerged pontoons (**Fig. Bx14.3a**). With these, oil companies can now access fields lying beneath 3 km (10,000 ft) of water.

North America's largest offshore fields occur in the passive-margin basin that fringes the coast of the Gulf of Mexico (**Fig. Bx14.3b**). This basin started forming in Jurassic time, subsequent to the breakup of Pangaea, and its floor has been slowly sinking ever since. Up to 15 to 20 km (9 to 12 miles) of sediment (mud, salt, sand, marine shells) fill the basin. Some of the mud was rich in organic matter; when buried deeply, this converted to hydrocarbons that now fill pores of reservoir rocks in salt-dome traps and stratigraphic traps. More than 3,500 platforms operate in the Gulf at present, together yielding up to 1.7 million bbl/day.

During both onshore and offshore exploration, drillers worry about the possibility of a **blowout.** A blowout happens when the pressure within a hydrocarbon reserve penetrated by a well exceeds the pressure that drillers had planned for, causing the hydrocarbons (oil and/or gas) to rush up the well in an uncontrolled manner and burst out of the well at the surface in an oil gusher or gas plume, which can explode and burn if ignited. Blowouts are rare because although fluids below the ground are under great pressure due to the weight of overlying material, engineers always fill the hole with drilling mud (a mixture of water, clay, and other chemicals) with a density greater than that of clear water. The weight of drilling mud can counter the pressure of underground hydrocarbons and hold the fluids underground. But if drillers encounter a bed in which pressures are unexpectedly high, or if they remove the mud before the walls of the well have been sealed with a casing (a pipe, cemented in place by concrete, that lines the hole), a blowout may happen.

A catastrophic blowout occurred on April 20, 2010, when drillers on the Deepwater Horizon, a huge semi-submersible platform leased by BP, were completing a 5.5-km-long (18,000 ft) hole in 1.5-km-deep (5,000 ft) water south of Louisiana. Due to a series of errors, the casing was not sufficiently strong when workers began to replace the drilling mud with clear water. Thus, the high-pressure, gassy oil in the reservoir that the well had punctured rushed up the drill hole. A backup safety device called a blowout preventer failed, so the gassy oil reached the platform and sprayed 100 m (328 ft) into the sky. Sparks from electronic gear triggered an explosion, and the platform became a fountain of flame and smoke that killed 11 workers. An armada of fireboats could not douse the conflagration (**Fig. Bx14.3c**), and after 36 hours the still-burning platform tipped over and sank.

Robot submersibles sent to the seafloor to investigate found that the twisted mess of bent and ruptured pipes at the well head was billowing oil and gas (**Fig. Bx14.3d**). Estimates as to the amount of hydrocarbons released vary wildly, but most data suggest that on the order of 50,000 to 62,000 bbl/day of hydrocarbons entered the Gulf's water from the well. (By comparison, natural oil seeps at many localities around the Gulf together yield about 1,000 to 4,000 bbl/day.) Stopping this underwater gusher proved to be an immense challenge, and initial efforts to block the well or to put a containment dome over the well head failed. It was not until July 15, almost three months after the blowout, that the flow was finally stopped, and it was not until September 19 that a new relief well intersected the blown well and provided a conduit to pump concrete down to block the original well permanently.

All told, about 4.2 million bbl of hydrocarbons contaminated the Gulf from the Deepwater Horizon blowout. The more volatile components of the spill evaporated, but enough liquid hydrocarbons remained to form a slick that, at times, covered an area of over 100,000 sq km (**Fig. Bx14.3e**). Thousands of people labored to contain the damage—floating booms were set up to contain the oil, skimmer ships traversed the slick to suck up oil, sandbag barriers were placed along the coast from Louisiana to Florida to keep the oil out of wetlands and beaches. Also, planes spread chemical dispersants on the floating oil to break it into tiny particles. In the near term, the spill was devastating to wetlands, wildlife, and the fishing and tourism industries. Over the longer term, natural processes—microbes that eat the oil droplets—will conquer the spill.

---

untapped. Further, nuclear engineers have designed alternative plants, powered by breeder reactors, that essentially produce new fuel. But many people view nuclear plants with concern because of issues pertaining to radiation, accidents, terrorism, and waste storage, and these concerns have slowed the industry. A substantial increase in hydroelectric power production is not likely, as most major rivers have already been dammed, and industrialized countries have little appetite for taming any more. Similarly, the growth of geothermal-energy output

seems limited. Coal supplies are fairly abundant, but burning large quantities of coal contributes to air pollution and adds greenhouse gases to the atmosphere.

Because of the potential problems that might result from relying more on coal, hydroelectric, and nuclear energy, researchers have been increasingly exploring clean energy options. One possibility is solar power, and the cost of solar power is steadily decreasing. Similarly, we can turn to wind power for relatively small-scale energy production. Because

FIGURE Bx14.3 Offshore drilling and the Deepwater Horizon disaster.

(a) The Deepwater Horizon oil-drilling platform as it appeared in the Gulf of Mexico in July of 2009.

(c) Fireboats dousing the burning rig before the rig sank.

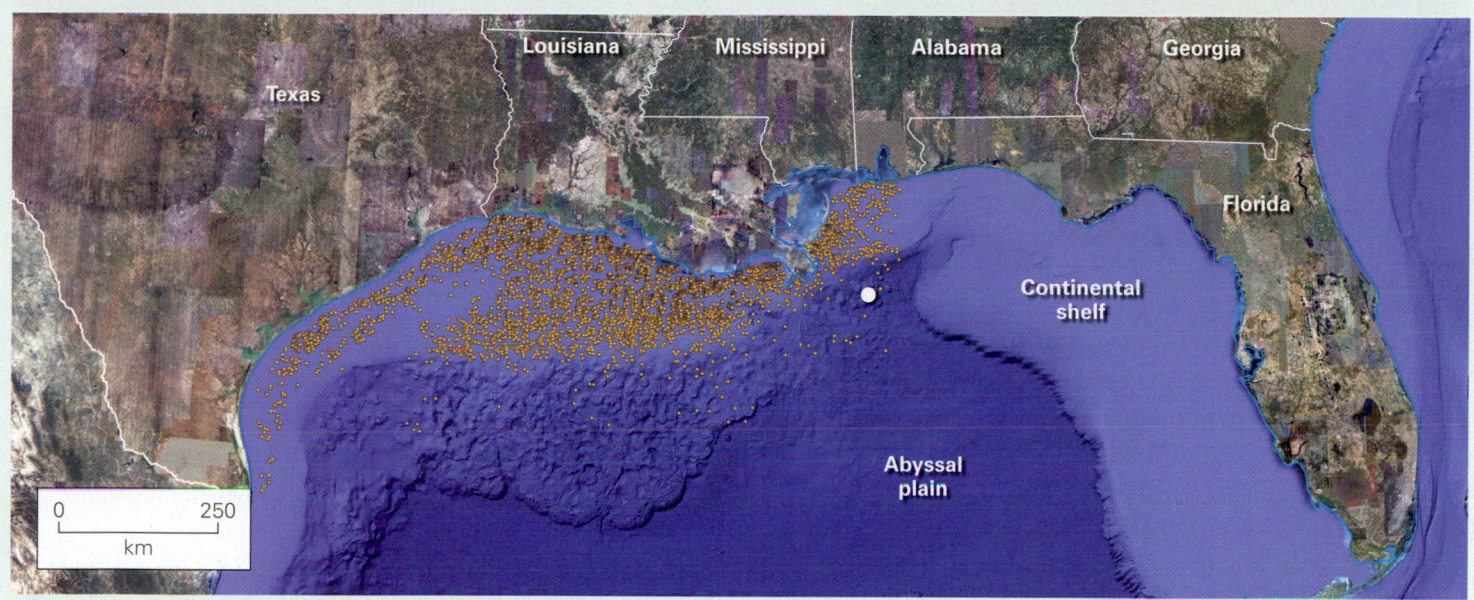

(b) Drilling on the Gulf Coast margin. Most drill sites are on the continental shelf, and more recently deeper water sites have been explored. The roughness of the slope to the abyssal plain is due to salt domes. Each yellow dot is a drilling platform. The location of the Deepwater Horizon platform is indicated by the larger, white dot.

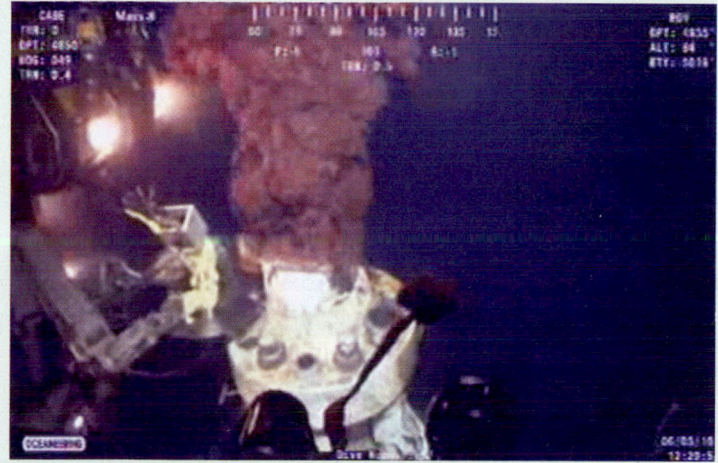

(d) A plume of oil billows into the water from the well head, as viewed by underwater cameras.

(e) A satellite image showing the oil slick in the Gulf of Mexico, southeast of the Mississippi Delta.

sunlight reaching the Earth changes during the day and with the amount of overcast, and because wind speed varies, solar power and wind power can't provide the steady supply of energy that the present-day *energy grid* (the interconnected network of power lines and transformers that distribute energy around the country) requires. This problem might someday be overcome with "smart grids," which can accommodate for rapid fluctuations in energy input. Fusion power may be possible some day, but physicists and engineers have not yet figured out a way to harness it. Clearly, society will be facing difficult choices in the not-so-distant future about where to obtain energy, and we will need to invest in the research required to discover new alternatives.

Take-Home Message

Oil is a nonrenewable resource and conventional supplies may run out in less than a century—we live in the Oil Age, which may, when it ends, have lasted less than 350 years. Unconventional hydrocarbon supplies may last longer, but use of fossil fuel has environmental consequences.

QUICK QUESTION: What is "Hubbert's Peak," and when was it reached in the United States?

CHAPTER SUMMARY

- Energy resources come in a variety of forms: energy directly from the Sun; energy from tides, flowing water, or wind; energy stored by photosynthesis (either in contemporary plants or in fossil fuels); energy from inorganic chemical reactions; energy from nuclear fission; and geothermal energy.

- Oil and gas are hydrocarbons, a type of organic chemical. The viscosity and volatility of a hydrocarbon depend on the length of its molecules.

- Oil and gas originally develop from the dead bodies of plankton and algae, which settle out in a quiet-water, oxygen-poor depositional environment and form black organic shale. Later, chemical reactions at elevated temperatures convert the dead plankton into kerogen and then oil.

- For a conventional oil reserve to be usable, oil must migrate from a source rock into a porous and permeable rock called a reservoir rock. Unless the reservoir rock is overlain by an impermeable seal rock, the oil will escape to the ground surface. The subsurface configuration of strata that ensures the entrapment of oil in a good reservoir rock is called an oil trap.

- Substantial volumes of hydrocarbons also exist in unconventional reserves, such as tar sand, oil shale, and gas hydrate.

- For coal to form, abundant plant debris must be deposited in an oxygen-poor environment so that it does not completely decompose. Compaction near the ground surface creates peat, which, when buried deeply and heated, transforms into coal. Coal has a high concentration of carbon.

- Geologists rank coal based on the amount of carbon it contains: lignite (low rank), bituminous (medium rank), and anthracite (high rank).

- Coal occurs in beds, interlayered with other sedimentary rocks. Coal beds can be mined by either strip mining or underground mining.

- Coal-bed methane and coal gasification provide additional sources of energy.

- Nuclear power plants generate energy by using the heat released from the nuclear fission of radioactive elements. The heat turns water into steam, and the steam drives turbines.

- Some economic uranium deposits occur as veins in igneous rock bodies; some are found in sedimentary beds.

- Nuclear reactors must be carefully controlled to avoid overheating or meltdown. The disposal of radioactive nuclear waste can create environmental problems.

- Geothermal energy uses Earth's internal heat to transform groundwater into steam that drives turbines; hydroelectric power uses the potential energy of water; and solar energy uses solar cells to convert sunlight to electricity.

- We now live in the Oil Age, but oil supplies may last only for another century or two. Natural gas may become an increasingly important energy supply in the near future.

- Most energy resources have environmental consequences. Oil spills pollute the landscape and sea, and the sulfur associated with some fuel deposits causes acid mine runoff. The burning of coal can produce acid rain, and the burning of coal and hydrocarbons produces smog and can contribute to global warming.

GUIDE TERMS

acid mine runoff (p. 539)
acid rain (p. 539)
barrel of oil (bbl) (p. 513)
biofuel (p. 532)
blowout (p. 540)
chain reaction (p. 529)
coal (p. 524)
coal-bed methane (p. 528)
coal gasification (p. 526)
coal rank (p. 525)
coal reserve (p. 525)
coal swamp (p. 524)
conventional hydrocarbon
 system (p. 511)
conventional reserve (p. 510)

directional drilling (p. 516)
drilling mud (p. 516)
energy (p. 505)
energy density (p. 508)
energy resource (p. 505)
fossil fuel (p. 507)
fuel (p. 503)
gas hydrate (p. 521)
geothermal energy (p. 531)
greenhouse gas (p. 539)
Hubbert's Peak (p. 538)
hydrocarbon (p. 508)
hydrocarbon generation
 (p. 510)
hydrocarbon reserve (p. 510)

hydrofracturing (pp. 517, 522)
kerogen (p. 510)
meltdown (p. 530)
migrate (p. 512)
migration pathway (p. 512)
nuclear reactor (p. 529)
nuclear waste (p. 531)
oil seep (p. 512)
oil shale (pp. 510, 521)
oil window (p. 510)
peat (p. 525)
permeability (p. 511)
photovoltaic cell (p. 534)
pore (p. 511)
porosity (p. 511)

reservoir rock (p. 511)
seal rock (p. 512)
seismic-reflection profile
 (p. 515)
shale gas (p. 520)
source rock (p. 510)
tar sand (p. 520)
tidal power (p. 533)
trap (p. 512)
unconventional hydrocarbon
 reserve (p. 517)

REVIEW QUESTIONS

1. What are the fundamental sources of energy?
2. How does the length of a hydrocarbon chain affect its viscosity and volatility?
3. What is the source of the organic material in oil?
4. What is the oil window, and what happens to oil at temperatures higher than the oil window?
5. How is organic matter trapped and transformed to yield an oil reserve?
6. What are the different kinds of oil traps?
7. What are tar sand and oil shale, and how can oil be extracted from them?
8. What are gas hydrates, and where do they occur?
9. How do porosity and permeability affect the oil-bearing potential of a rock?
10. Where is most of the world's oil found? At present rates of consumption, how long will oil supplies last?
11. How is coal formed, and what class of rock is coal considered to be?
12. What conditions cause coal to transform in rank from peat to anthracite?
13. What are some of the environmental drawbacks of mining and burning coal?
14. What is coal-bed methane, and how is it extracted?
15. Describe the nuclear reactions in a nuclear reactor and the means that engineers use to control reaction rates.
16. Where does uranium form in the Earth's crust? Where does it usually accumulate in minable quantities?
17. What are some of the drawbacks of nuclear energy?
18. Discuss the pros and cons of alternative energy sources.
19. What is geothermal energy? What limits its use?
20. What is the difference between renewable and nonrenewable resources?
21. What are the major environmental consequences of producing and burning fossil fuels?
22. What is the likely future of hydrocarbon production and use in the 21st century?

ON FURTHER THOUGHT

23. Much of the oil production in the United States takes place at offshore platforms along the coast of the Gulf of Mexico. Consider the geologic setting of the Gulf Coast, in the context of the theory of plate tectonics, and explain why an immensely thick sequence of sediment accumulated in this region and why so many salt-dome traps formed.

24. Ethanol can potentially be used as an alternative to petroleum as a liquid fuel. Ethanol can be produced by processing corn, sugarcane, or certain perennial grasses (e.g., switchgrass or Miscanthus). What factors should be considered in determining which of these crops would be the most appropriate for use as a source of ethanol in North America?

 smartwork.wwnorton.com

This chapter's Smartwork features:

- Labeling exercise on types of oil traps.
- Animation problem on oil formation and trapping.
- Activities on formation and types of coal.

GEOTOURS

This chapter's GeoTour exercise (L) features:

- Hydrocarbon resources
- Coal resources
- Other energy resources

Another View Solar-panel arrays are beginning to carpet the landscape in sunny regions. This example (the Beneixama photovoltaic power plant in Spain) generates 20 megawatts of electricity at peak production and covers 500,000 m^2 (= 0.5 km^2, or 123 acres). By comparison, a typical nuclear power plant produces 1,000 megawatts of electricity (enough for a city of 1.2 million). A solar array large enough to produce as much electricity as a nuclear power plant would have to cover an area of about 25 square km.

A bronze artifact made in ancient China contains metals extracted from ore deposits that formed in rocks in the crust. Through the ages, society has tapped into Earth's resources for metals and many other materials.

CHAPTER 15

Riches in Rock: Mineral Resources

*Truth, like gold, is to be obtained not by its growth,
but by washing away from it all that is not gold.*

—Leo Tolstoy (Russian author, 1828–1910)

15.1 Introduction

In June 1845, James Marshall arrived by horse at Sutter's Fort in central California to make a new life. Having just finished a stint as a rancher and a few months as a soldier, Marshall decided to go into the lumber trade, and he convinced Captain John Sutter to finance the construction of a sawmill in the foothills of the Sierra Nevada mountain range. Marshall's scruffy crew finished the mill by the beginning of 1848. As Marshall stood admiring the new building, he noticed a glimmer of metal in the gravel that littered the bed of the adjacent stream, picked it up, banged it between two rocks to test its hardness, and shouted, "Boys, by God, I believe I have found a gold mine!" For a short while, Marshall and Sutter managed to keep the discovery secret. But word of the gold soon spread, and within weeks the workers at Marshall's mill had disappeared into the mountains to seek their own fortunes.

Gold fever eventually reached the East Coast, and as a result, 1849 saw 40,000 prospectors head to California. These "forty-niners," as they were called, had abandoned their friends and relatives on the gamble that they could strike it rich (**Fig. 15.1**). Some traveled by land, some by sea, to reach San Francisco, a town of mud streets and plank buildings. In many cases, crews abandoned their ships in the harbor to join the scramble to the gold fields. Although $20 million worth of gold was mined in 1849, few of the forty-niners actually became rich. Most lost whatever wealth they found in the saloons, stores, and gambling halls that sprouted around the gold fields.

Gold is but one of many **mineral resources**, meaning minerals extracted from the Earth's crust for use by people. Without these resources, industrialized societies could not function. Geologists divide mineral resources into two categories: *metallic mineral resources* (materials containing a concentration of gold, copper, aluminum, iron, or other metals) and *nonmetallic mineral resources* (building stone, gravel, sand, gypsum, phosphate, and salt, which are materials used in construction or for chemical production). In this chapter, we look at the nature of such mineral resources, the geologic phenomena responsible for their formation, and the ways people mine them. We conclude by considering the sustainability of mineral reserves in coming years.

FIGURE 15.1 The Gold Rush of 1849.

(a) An 1849 poster promoting travel to the gold regions of California.

(b) A prospector, or "forty-niner," mining in the Sierra Nevada range.

FIGURE 15.2 What makes a metal, a metal?

(a) A copper crystal consists of wafer-like sheets of copper atoms.

(b) Bending of the crystal can take place because the rows can easily slide past each other. Metals can be rolled into thin sheets without breaking.

Pennies

Sheets of rolled copper are stamped to make coins.

15.2 Metals and Their Discovery

What Is Metal?

A **metal** is an opaque, shiny, smooth solid that can conduct electricity and can be bent, drawn into wire, or hammered into thin sheets. Metals look and behave quite differently from wood, plastic, meat, or rock because the atoms that make up metals are held together by *metallic bonds*, so electrons can flow from atom to atom fairly easily, whereas the atoms in the other materials are connected by covalent or ionic bonds and thus cannot provide roving electrons (see Chapter 5). Despite the mobility of their electrons, metals are solids, so their atoms lie fixed in a regular lattice defining a crystal structure. Atoms in pure copper, for example, sit in wafer-like layers (**Fig. 15.2a**).

Not all metals behave the same way. Some have great strength, whereas others are particularly *malleable* (meaning they can be bent or molded). The behavior of a metal depends on the strength of bonds between atoms and on the architecture of its crystalline structure. For example, the wafer-like layers of atoms in copper can slip past each other quite readily, so copper can be bent or stretched easily (**Fig. 15.2b**). The shape and dimensions of crystals in a piece of metal reflect the rate at which the metal cools. Metal can be changed by *tempering* (alternate heating and cooling) or by *cold working* (manipulating the shape of the metal after it is cooled). Before the development of the modern science of metallurgy, metalworkers learned how to obtain the qualities they wanted in a metal object simply by trial and error. Skilled metalworkers passed trade secrets from master to apprentice for generations. But now engineers with a detailed knowledge of chemistry are able to control how a metal product behaves.

The Discovery of Metals

Certain substances—namely, copper, silver, gold, and mercury—can occur in rock as native metals (**Fig. 15.3**). A **native metal** consists only of metal atoms and thus looks and behaves like metal. Prehistoric hunters collected nuggets of native metal from stream beds and pounded them with stone hammers to make tools, and because native metals are rare and durable, people began to use them as money. Gold attained particular popularity as a currency because of its unique warm yellow glow and its resistance to tarnish or rust. Gold flakes and nuggets are simply pieces of native metal that have been eroded free of bedrock. Though iron does not occur as a native metal in bedrock on Earth, chunks of it fall to Earth from space in the form of meteorites; such meteorite iron was highly valued for toolmaking.

If we had to rely on native metals as our only source of metal, we would have access to but a tiny fraction of our current metal supply. Most of the metal atoms in materials we use today originally occurred as ions bonded to nonmetallic elements in a great variety of minerals that themselves look nothing like metal. Only because of the chance discovery by some prehistoric genius that metals can be extracted from certain minerals do we now have the ability to produce sufficient metal for the needs of industrialized society. The process of extracting metal from mineral is called **smelting**—the decomposition of minerals during smelting yields metal plus a nonmetallic residue called *slag*.

Of the principal metals in use today (copper, iron, and aluminum), copper began to be used first, because copper smelting from sulfide minerals is relatively easy. Copper implements appeared as early as 4000 B.C.E. Pure copper has limited value for toolmaking or weapon making because it is too soft to retain a sharp edge. By chance, around 2800 B.C.E., Sumerian craftsmen discovered that copper could be mixed with tin to produce bronze, an **alloy** (a compound containing two or more

FIGURE 15.3 Examples of native metals.

Gold bracelets on display in a Kuwaiti marketplace

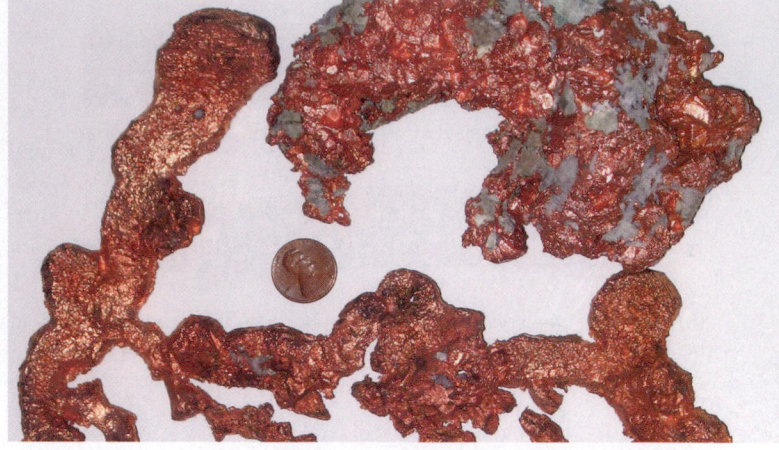

(a) Gold occurs as native metal within quartz veins. The quartz breaks up to form sand, leaving nuggets of gold.

(b) Native copper occurs in complex shapes, which remain when the surrounding rock weathers away. Note the penny for scale.

metals) whose strength exceeds that of either metal alone, and warriors came to rely on bronze for their swords.

Iron proves superior to copper or bronze for many purposes because of its strength, hardness, and abundance. Still, people didn't start using iron widely until 1,500 years after they had begun to use bronze. The delay was due, in part, to the fact that iron has a very high melting temperature that can be difficult to work with. But, in addition, the metal generally occurs in iron-oxide minerals (such as hematite, Fe_2O_3, or magnetite, Fe_3O_4), and the liberation of iron metal from oxide minerals requires a chemical reaction. Thus, the widespread use of iron became possible only after someone discovered that metallic iron can be produced by heating iron-oxide minerals in

> **Did you ever wonder ...**
>
> where the iron in the steel bodies of cars comes from?

the presence of carbon monoxide (CO) gas, which comes from burning charcoal (**Fig. 15.4**). Chemists describe this reaction by the formula $Fe_2O_3 + 3CO \rightarrow 2Fe + 3CO_2$. More recently, people learned to make steel, an alloy of iron and carbon, and stainless steel, an alloy of iron and chromium, which resists corrosion.

You can find aluminum in many minerals of the crust. Aluminum, for many uses, is preferable to iron because it weighs less, but it does not occur in native form. Extracting aluminum from

FIGURE 15.4 The smelting of iron ore in a blast furnace requires temperatures of 1,250°C to 1,400°C. Workers use very hot air to heat iron ore, coke (carbon), and limestone. A chemical reaction ($Fe_2O_3 + 3CO \rightarrow 2Fe + 3CO_2$) produces iron metal.

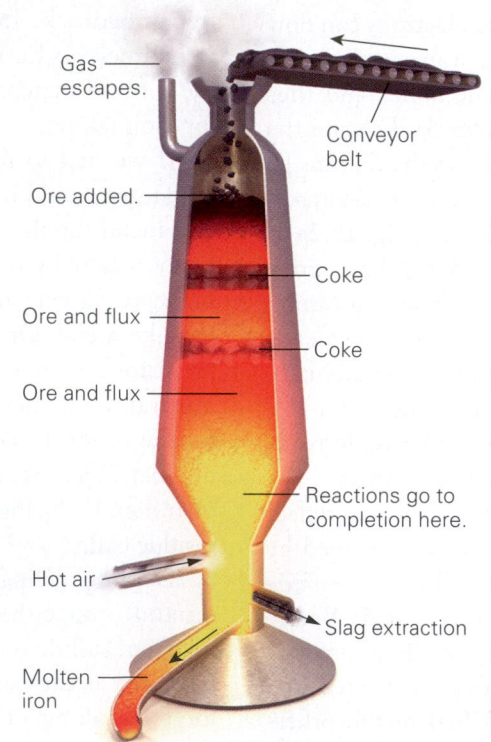

Gas escapes.

Conveyor belt

Ore added.

Coke

Ore and flux

Coke

Ore and flux

Reactions go to completion here.

Hot air

Slag extraction

Molten iron

(a) Molten iron sinks to the bottom and "slag" (silicates and oxides) floats. "Flux" decreases the melting point of ore.

(b) In a factory, huge tubs pour the end product, molten iron.

minerals requires complex methods that use lots of electricity, and have become economically practical only since the 1880s.

These days, in addition to iron, copper, tin, and aluminum, we use a vast array of different metals. Some are known as *precious metals* (gold, silver, and platinum) and others as *base metals* (copper, lead, zinc, and tin) because of the difference in their price. Surprisingly, of the 63 or so metals in use today, people knew of only nine (gold, copper, silver, mercury, lead, tin, antimony, iron, and arsenic) before the year 1700.

> ## Take-Home Message
>
> Metals are malleable materials in which metallic bonds hold atoms together. Though some occur in pure form as native metals, most occur in other minerals and have to be extracted by smelting. Copper was the first metal to be used by humans. Iron and aluminum are more difficult to extract.
>
> **QUICK QUESTION:** Why did the Bronze Age come before the Iron Age in human history?

15.3 Ores, Ore Minerals, and Ore Deposits

What Is an Ore?

As we have seen, metals occur in two forms—as native elements (copper, silver, or gold) or as ions bonded to nonmetallic elements in a great variety of minerals. Many of these minerals can be found in common rocks. For example, if you pick up a chunk of common granite and analyze its mineral content, you will find that it consists mostly of quartz (SiO_2), plagioclase ($[Na,Ca]AlSi_3O_8$), and potassium feldspar ($KAlSi_3O_8$). Looking at these chemical formulas, you can see feldspars contain about 8% aluminum, so common granite includes aluminum ions. But we don't mine granite to produce aluminum. Why? Simply because it's not economical—overall, granite contains relatively little aluminum, and it's difficult to separate aluminum atoms from feldspar, so processing costs would exceed selling costs. To obtain the metals needed for industrialized society, we mine **ore**, rocks containing a significant concentration of native metals or ore minerals.

Ore minerals (or *economic minerals*) are minerals that contain metal both in sufficiently high concentrations and in a form that can be easily extracted. For example, galena (PbS) contains about 50% lead, and lead can be separated from sulfur fairly easily, so we consider galena to be an ore mineral of lead (**Fig. 15.5a**). Similarly, we obtain most of our iron from oxide minerals such as hematite and magnetite, so these minerals are iron ore minerals. Geologists have identified many different kinds of ore minerals (**Table 15.1**). As you can see from the chemical formulas, many ore minerals are sulfides, in which the metal occurs in combination with sulfur (S), or oxides, in which the metal occurs in combination with oxygen (O). Numerous ore minerals are colorful and come in interesting shapes (**Fig. 15.5b**), and some have a metallic luster.

To be an **ore**, a rock must contain a sufficient amount of ore minerals to make the rock worth mining. For example, iron constitutes about 6.2% of the continental crust's weight, whereas it makes up 30% to 60% of iron ore (**Fig. 15.6**). We refer to the concentration of a useful metal in an ore as the **grade** of the ore—the higher the concentration, the higher the grade. Whether or not ore of a certain grade is worth mining depends on the price of metal in the market. For example,

FIGURE 15.5 Examples of ore minerals.

(a) This lead ore, from Missouri, consists of galena (PbS) crystals that grew in dolostone.

(b) Most copper comes from ore minerals that look nothing like metallic copper. This ore consists of azurite (blue) and malachite (green).

TABLE 15.1 Some Common Ore Minerals

Metal	Mineral Name	Chemical Formula
Copper	Chalcocite	Cu_2S
	Chalcopyrite	$CuFeS_2$
	Bornite	Cu_5FeS_4
	Azurite	$Cu_3(CO_3)2(OH)_2$
	Malachite	$Cu_2(CO_3)(OH)_2$
Iron	Hematite	Fe_2O_3
	Magnetite	Fe_3O_4
Tin	Cassiterite	SnO_2
Lead	Galena	PbS
Mercury	Cinnabar	HgS
Zinc	Sphalerite	ZnS
Aluminum	Kaolinite	$Al_2Si_2O_5(OH)_4$
	Corundum	Al_2O_3
Chrome	Chromite	$(Fe,Mg)(Cr,Al,Fe)2O_4$
Nickel	Pentlandite	$(Ni,Fe)_9S_8$
Titanium	Rutile	TiO_2
	Ilmenite	$FeTiO_3$
Tungsten	Sheelite	$CaWO_4$
Molybdenum	Molybdenite	MoS_2
Magnesium	Magnesite	$MgCO_3$
	Dolomite	$CaMg(CO_3)_2$
Manganese	Pyrolusite	MnO_2
	Rhodochrosite	$MnCO_3$

FIGURE 15.6 The difference between ore and other rock. Iron in a block of granite is not worth extracting.

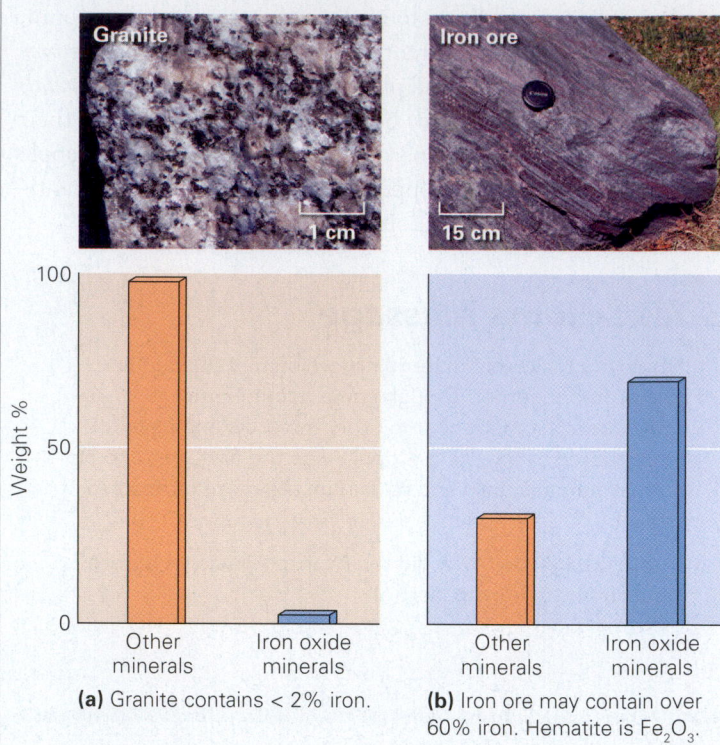

(a) Granite contains < 2% iron.

(b) Iron ore may contain over 60% iron. Hematite is Fe_2O_3.

minerals in ore deposits. Simply put, an **ore deposit** is an economically significant occurrence of ore. Geologists distinguish among many different kinds of ore deposits, which differ from each other in terms of which ore minerals they contain and which kind of rock body they occur in. Below we introduce a few examples.

Magmatic Deposits As a magma cools, sulfide ore minerals crystallize early and then may accumulate to form a **magmatic deposit**. When the magma freezes solid, the resulting igneous body contains concentrations of sulfide minerals. Because of their composition, these concentrations make up a type of **massive sulfide deposit** (**Fig. 15.7a**).

Hydrothermal Deposits Hydrothermal activity involves the circulation of hot-water solutions through a magma or through the rocks surrounding an igneous intrusion. These fluids dissolve metal ions. When a solution enters a region of lower pressure, lower temperature, different acidity, and/or different availability of oxygen, the metals come out of solution and form ore minerals that precipitate in fractures and pores, creating a **hydrothermal deposit** (**Fig. 15.7b**). Such deposits may form within an igneous intrusion or in surrounding country rock. If the resulting ore minerals disperse through the intrusion, we call the deposit a *disseminated deposit*, but if they precipitate to fill cracks in pre-existing

in 1880, copper-bearing rocks needed to contain at least 3% copper to be considered "economic ore" (ore worth mining), but due to improvements in technology and a rise in price, rock containing only about 0.3% copper can now be considered economic.

How Do Ore Deposits Form?

Ore minerals do not occur uniformly through rocks of the crust. If they did, we would not be able to extract them economically. Fortunately for humanity, geologic processes concentrate these

FIGURE 15.7 Various processes that form ore deposits.

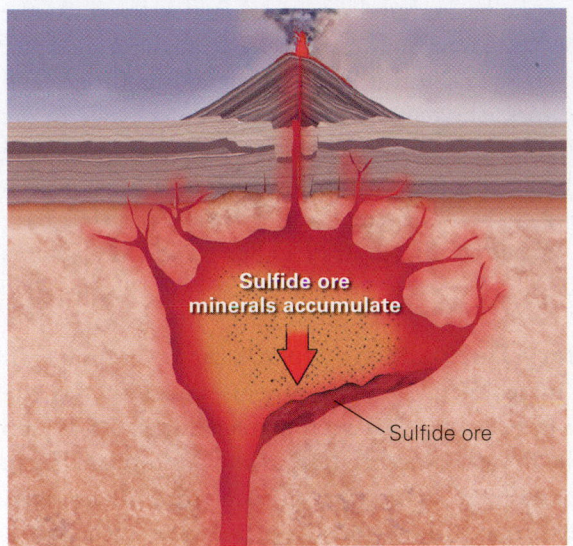

(a) Massive sulfide deposits can form when sulfide ore minerals form concentrations in a magma chamber.

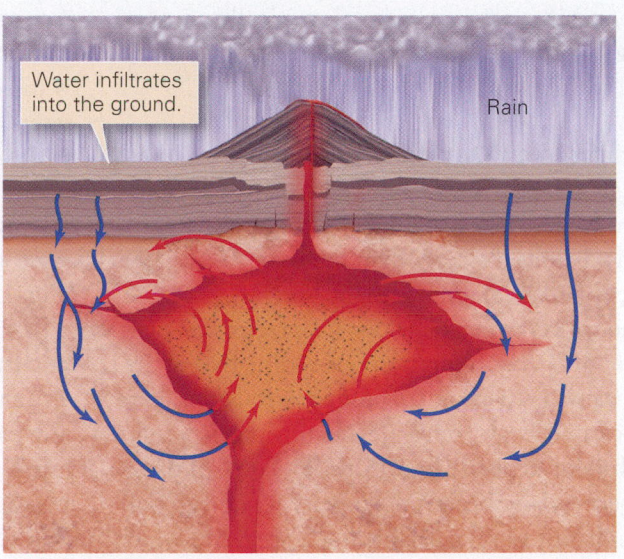

(b) Hydrothermal deposits form when water circulating around and through magma dissolves and redistributes metals (arrows indicate flowing water).

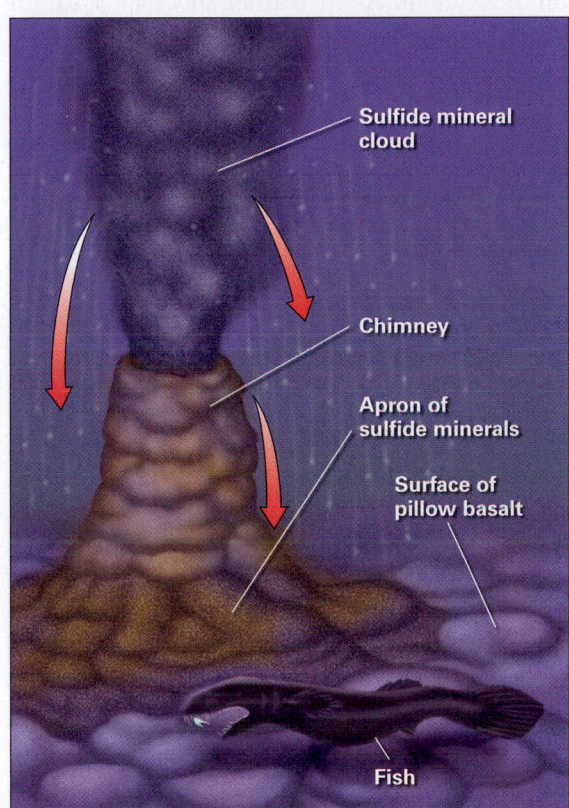

(c) Massive sulfide deposits also form when ore minerals precipitate around hydrothermal vents (black smokers) along a mid-ocean ridge.

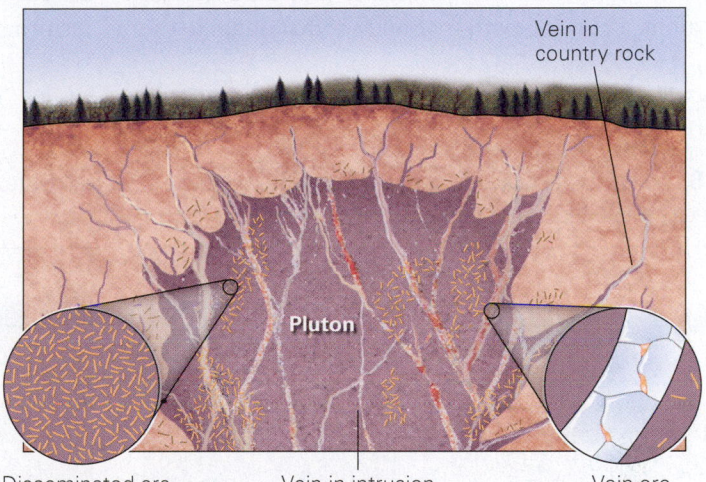

(d) Disseminated ore consists of a rock in which ore minerals are dispersed. Veins consist of minerals precipitated in cracks.

rock, we call the deposit a *vein deposit* (veins are mineral-filled cracks; **Fig. 15.7c**). Hydrothermal copper deposits commonly occur in porphyritic igneous intrusions—these examples are known as *porphyry copper deposits*. Typically, vein deposits include quartz in addition to the ore minerals. For example, native gold commonly appears as flakes in milky-white quartz veins.

Submarine-Vent Deposits Hydrothermal activity at the submarine volcanoes along mid-ocean ridges leads to the eruption of hot water out of vents. This water contains high concentrations of dissolved metal and sulfur, and when the hot vent water comes in contact with cold seawater, the dissolved components precipitate as tiny crystals of metal-sulfide minerals (**Fig. 15.7d**). The erupting water, therefore, looks like a black cloud, so the vents are called black smokers (see Chapter 4). The minerals in the cloud eventually sink and form a chimney of ore minerals around the vent and a layer of minerals on the surrounding seafloor. Since the ore minerals typically are sulfides, the resulting hydrothermal deposits constitute another type of massive sulfide deposit.

Secondary-Enrichment Deposits Sometimes groundwater passes through ore-bearing rock long after the rock first formed. This groundwater dissolves some of the ore minerals and carries away the dissolved ions. When the water eventually flows into a different chemical environment (for instance, one with a different amount of oxygen or acidity), it precipitates new ore minerals, commonly in concentrations exceeding that of the original deposit. A new ore deposit containing metals that had been dissolved and carried away from a pre-existing ore deposit is a **secondary-enrichment deposit** (**Fig. 15.8a, b**). Some of these deposits contain spectacularly beautiful copper-bearing carbonate minerals, such as azurite and malachite (see Fig. 15.5b).

MVT Ores Rain falling along one margin of a large sedimentary basin may sink into the subsurface and then flow as groundwater along a curving path that takes it first down to the bottom of the basin, and then eventually back up to the opposite margin of the basin, hundreds of kilometers away. At the bottom of the basin, temperatures increase enough for the water to dissolve metals. When the water returns to the surface and enters cooler rock, the metals that it carries precipitate in ore minerals. Ore deposits formed in this way commonly contain lead- and zinc-bearing minerals. Many such deposits appear in dolomite beds of the Mississippi Valley region and thus have come to be known as **Mississippi Valley–type (MVT) ores** (**Fig. 15.8c**).

Sedimentary Deposits of Metals Some ore minerals accumulate in sedimentary environments under special circumstances. For example, between about 2.5 and 1.8 billion years ago, the oxygen concentration in the atmosphere began to increase, and more oxygen began to dissolve in seawater. This change affected the chemistry of seawater such that it could no longer hold large quantities of dissolved iron in solution. This iron precipitated as iron-oxide minerals that settled as sediment on the seafloor. As we learned in Chapter 13, the resulting iron-rich sedimentary deposits are known as **banded**

FIGURE 15.8 Secondary enrichment occurs when oxidizing groundwater leaches (extracts) metals from a rock and moves them elsewhere into reducing conditions where they precipitate.

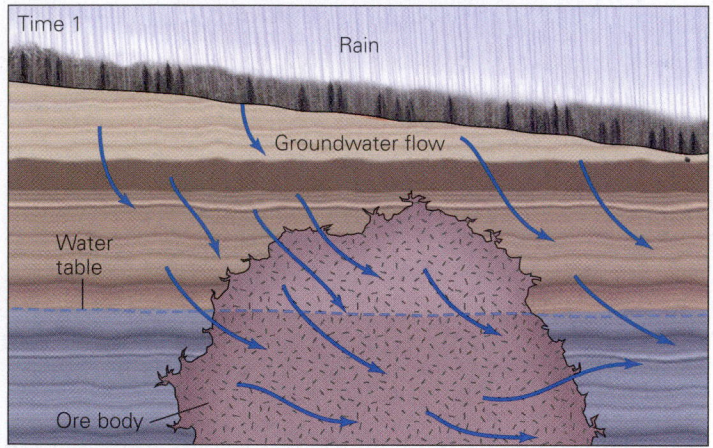

(a) Downward-flowing water dissolves ore minerals. The metal-rich solution moves down below the water table.

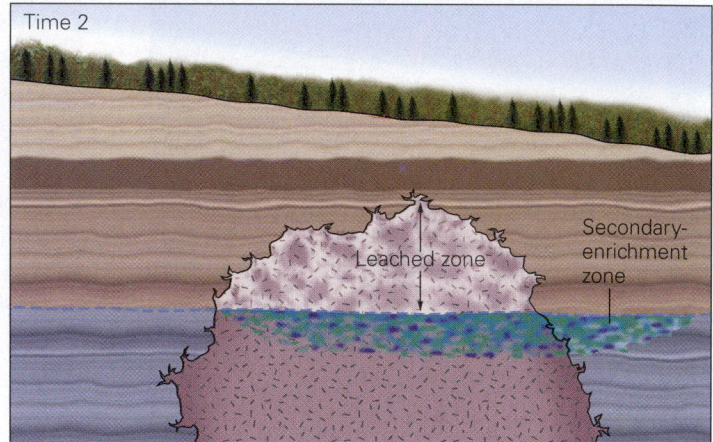

(b) Below the water table, the chemical environment changes, and new ore minerals precipitate.

(c) Mississippi Valley-type (MVT) deposits form when groundwater sinks deeply beneath a mountain range, dissolves metals, and then transports hot water across a sedimentary basin to a cooler environment where the metals precipitate.

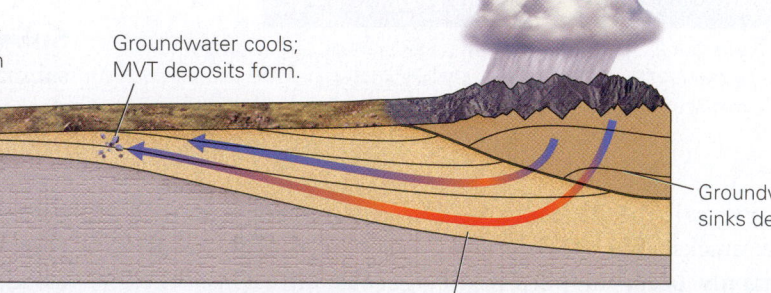

Groundwater cools; MVT deposits form.

Groundwater sinks deeply.

Deep groundwater picks up heat and dissolves metals.

Mountain belt
Sedimentary basin
Basement

iron formations (BIFs) (Fig. 15.9a) because after lithification they consist of alternating beds of gray iron oxide (magnetite or hematite) and red beds of jasper (iron-rich chert). Microbes may have participated in the precipitation process.

The chemistry of seawater in some parts of the ocean today leads to the deposition of manganese-oxide minerals on the sea floor. These minerals grow into lumpy accumulations known as **manganese nodules** (Fig. 15.9b). Mining companies have begun to explore technologies for vacuuming up these nodules because according to some estimates the worldwide supply of nodules contains 720 years' worth of copper and 60,000 years' worth of manganese, at current rates of consumption.

Residual Mineral Deposits Recall from Interlude B that as rainwater sinks into the Earth, it leaches (dissolves) certain elements and leaves behind others as part of the process of forming soil. In rainy, tropical environments, the residuum left behind in soils after leaching includes concentrations of iron or aluminum. Locally, these metals become so concentrated that the soil itself becomes an ore deposit (Fig. 15.10a, b). We refer to such deposits as **residual mineral deposits**. Most of the aluminum ore mined today comes from *bauxite*, a residual mineral deposit created by the extreme leaching of rocks (such as granite) containing aluminum-bearing minerals (Fig. 15.10c).

Placer Deposits Ore deposits may develop when rocks containing native metals erode, producing a mixture of sand grains and metal flakes or nuggets (pebble-sized fragments). By this process, for example, gold can accumulate in sand or gravel bars along the course of rivers, for the moving water carries away lighter mineral grains (quartz and feldspar) but can't move the heavy metal grains (gold) so easily. Concentrations of metal grains in stream sediments are a type of **placer deposit** (Fig. 15.11a). (The term is also used for concentrations of diamonds.) Panning further concentrates gold flakes or nuggets—swirling water in a pan causes the lighter sand grains to wash away, leaving the gold behind (Fig. 15.11b). Placer deposits may eventually be buried and lithify to become part of a new sedimentary rock, which could be mined. By tracking placer deposits upstream, prospectors may find the "mother lode," the bedrock source of the gold.

Where Are Ore Deposits Found?

The Inca Empire of 15th-century Peru built elaborate cities and temples, decorated with fantastic masks, jewelry, and sculptures made of gold. Then, around 1532, Spanish ships arrived, led by conquistadors who quipped, "We Spaniards suffer from a disease that only gold can cure." The Incas, already weakened by civil war, were no match for the armor-clad Spaniards with their guns and horses. Within six years, the Inca Empire had vanished, and Spanish ships were transporting Inca treasure back to Spain. Why did the Incas possess so much gold? Or to ask the broader question, what geologic factors control the distribution of ore? Once again, we can find the answer by considering the consequences of plate tectonics.

Several of the ore-deposit types we've mentioned occur in association with igneous rocks. As we learned in Chapter 6, igneous activity does not happen randomly around the Earth but rather concentrates along convergent-plate boundaries (specifically, in the overriding plate of a subduction zone),

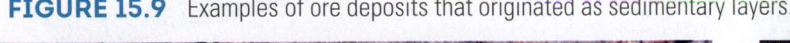

FIGURE 15.9 Examples of ore deposits that originated as sedimentary layers.

Magnetite layer

Jasper layer

(a) Precambrian BIF from northern Michigan consists of hematite interbedded with jasper.

(b) A view from a submersible of seafloor on which polymetallic nodules have grown in the sediment.

FIGURE 15.10 The formation of residual mineral deposits by intense weathering and leaching.

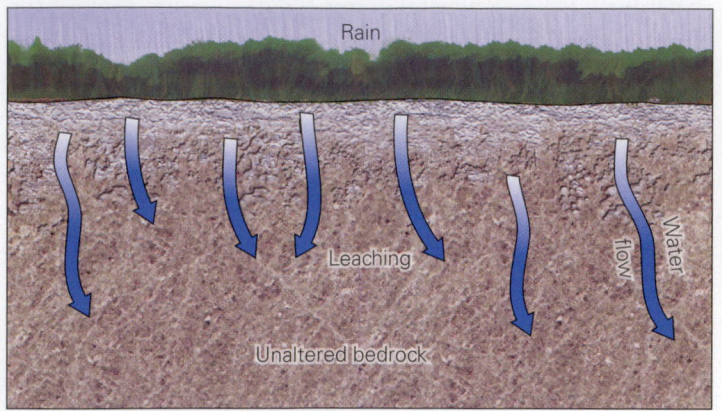

(a) When water sinks through the ground, bedrock weathers and slowly transforms into soil.

(b) If the amount of water is great, it leaches (dissolves and removes) many elements, leaving behind a residuum rich in iron or aluminum.

(c) Bauxite ore looks like reddish soil. It's typically obtained from near-surface open-pit mines.

divergent-plate boundaries (along mid-ocean ridges), continental rifts, or hot spots. Thus, magmatic and hydrothermal deposits (and secondary-enrichment deposits derived from these) form along plate boundaries, along rifts, or at hot spots. Placer deposits accumulate in sediments eroded from magmatic or hydrothermal deposits and thus occur downstream of the igneous bodies. Inca gold, for example, came to the surface from the Earth's interior with magmas that fed the Andean volcanic arc along a convergent-plate boundary where the Pacific Ocean floor subducts beneath the South American Plate. As the mountains rose, erosion stripped away surface rocks to expose the hydrothermal deposits associated with plutons that had intruded into the continental crust beneath. Inca miners quarried gold-bearing veins in the plutons and surrounding rock or panned for gold downstream. Plutons that contain similar ore deposits developed in the western United States during the Mesozoic and Cenozoic Eras.

As noted earlier, some massive sulfide deposits accumulate from black smokers along a mid-ocean ridge system and are interlayered with sea-floor basalt. Miners can gain access

FIGURE 15.11 Placer deposits.

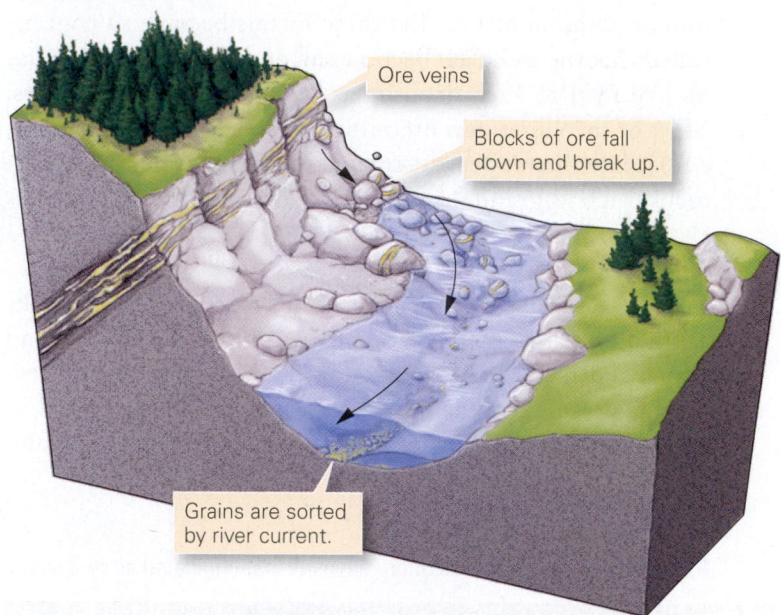

Ore veins

Blocks of ore fall down and break up.

Grains are sorted by river current.

(a) Placer deposits form where erosion produces clasts of native metals. Sorting by the stream concentrates the metals.

(b) A young woman panning for gold in the Mekong River, Laos.

to such seafloor deposits only in places where the collision of continents slides a sliver of seafloor up along a fault and onto continental crust. A slice of seafloor thrust up onto continents is known as an *ophiolite*—some ophiolites contain massive sulfide deposits.

Not all ore deposits result directly from plate tectonics activity, and thus not all occur in association with present-day or ancient plate boundaries. For example, BIF formed along passive continental margins during the Precambrian. Its occurrence today reflects the present distribution of preserved Precambrian sedimentary rocks, and thus most major exposures occur in shield areas of continents. Bauxite forms where aluminum-rich bedrock occurs, and extreme leaching takes place during soil formation. Thus, some bauxite deposits form on Precambrian granite bedrock in stable continental areas that now lie in tropical regions.

Take-Home Message

Ores (rocks, or sediments that can be processed economically to produce useful metals) form in many ways, including crystallizing in a melt, precipitating from hot water during igneous activity, accumulating in sediment, interacting with groundwater, and extreme weathering. The distribution of ores can be explained by plate tectonics.

QUICK QUESTION: Why don't we use fresh granite as a source for aluminum?

15.4 Ore-Mineral Exploration and Production

Imagine an old prospector of days past clanking through the desert with a worn-out donkey. To find bedrock ore, such prospectors would eye hillsides for a "show," visible evidence on the ground surface that ore lies below. What does a show look like? It may be an outcropping of milky-white quartz veins, for veins could indicate the presence of hydrothermal ore deposition. It may be the sparkle of minerals with metallic luster disseminated through the outcrop. Or, it may be the presence of orangish, yellowish, or bluish stains in outcrops, for these stains could result either from the presence of brightly colored ore minerals, or from the rusting (oxidation) of oxide or sulfide minerals (**Fig. 15.12a**). To find placer deposits, prospectors would "pan" a pile of sand or gravel from a streambed.

On finding a possible ore, a prospector would take samples back to town for an *assay*, a test to determine how much extractable metal the samples contain. If the assay indicated a significant concentration of metal, the prospector might "stake

a claim" by literally marking off an area of ground with stakes. In some locations, a claim gave the prospector rights to all the mineral deposits on or under the land and the opportunity to develop mines. When one prospector would find ore, word would quickly spread and others would rush to stake neighboring claims.

These days, large mining companies employ geologists to survey ore-bearing regions systematically. The geologists focus their studies on rocks that developed in settings appropriate for ore formation. Once such a region has been identified, they may measure the local strength of Earth's magnetic field and the local pull of gravity. These measurements may lead them to ore bodies, because ore minerals tend to be denser and more magnetic than average rocks (**Fig. 15.12b**). Geologists also sample rocks and soils to test for metal content and may even analyze plants in the area to detect traces of metals, for plants absorb traces of metal through their roots. Once geologists have identified a possible ore deposit, they drill holes to sample subsurface rock and to determine the ore deposit's shape and extent (**Fig. 15.12c**). Ore-mineral exploration sometimes takes geologists into jungles, deserts, and tundras worldwide. The largest gold deposit now being mined, at the Grasberg Mine, on a 4-km-high mountain in Papua (Indonesia), was found in 1988 by a geologist who helicoptered from mountaintop to mountaintop looking for shows.

If calculations indicate that mining an ore deposit will yield a profit, and if environmental concerns can be properly addressed, a company develops a mine. Mines can be below or above ground, depending on how close the ore deposit lies to the surface. To develop an **open-pit mine** (**Fig. 15.12d**), workers first drill a series of holes into the solid bedrock and then fill the holes with high explosives. They must space the holes carefully and set off the charges in a precise sequence, so that the bedrock shatters into appropriate-sized blocks for handling. When the dust settles, large front-end loaders dump the ore into giant trucks, which can

FIGURE 15.12 Finding and mining ore deposits.

The blue indicates the presence of copper ore.

20 cm

(a) Prospectors look for traces of ore minerals on rock outcrops.

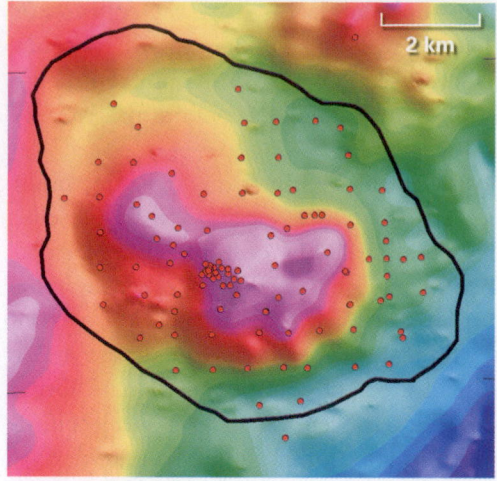

Stronger field

Weaker field

(b) A map of anomalies in the magnetic field may hint at ore bodies because metal is magnetic. The black line outlines the ore body and the red dots are drill-hole locations for sampling.

2 km

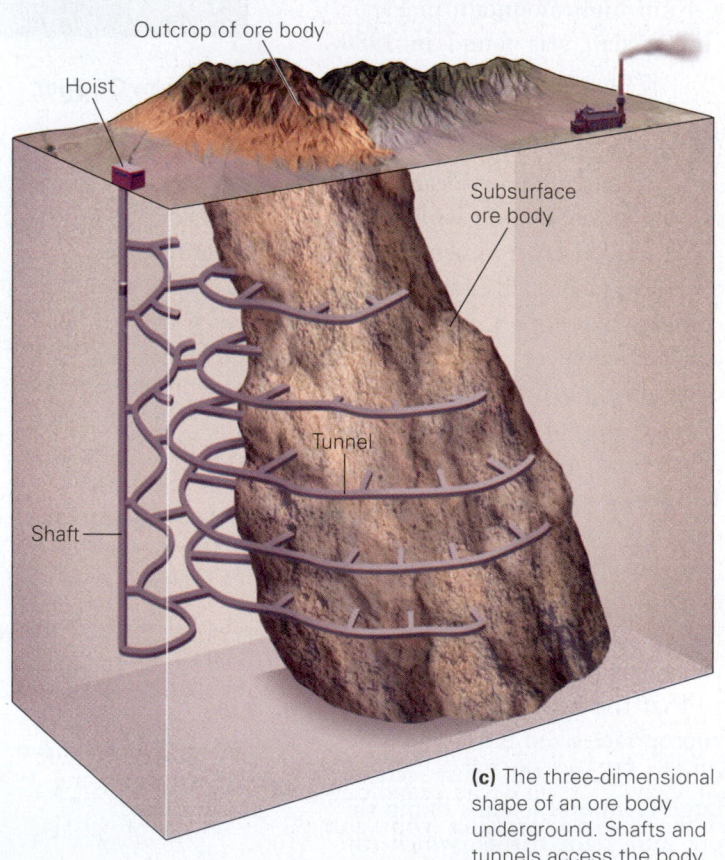

Outcrop of ore body

Hoist

Subsurface ore body

Tunnel

Shaft

(c) The three-dimensional shape of an ore body underground. Shafts and tunnels access the body.

(d) Open-pit mining utilizes ore that lies fairly close to the ground surface. Terracing the mine walls helps to stabilize them.

carry as much as 200 tons of rock in a single load. (In comparison, a loaded cement mixer weighs about 70 tons.) Tires on these mine trucks are so huge that a tall person comes up only to the base of the hub. The trucks transport waste rock or *tailings* (rock that doesn't contain ore) to a tailings pile and the ore to a crusher, a giant set of moving steel jaws that smash the ore into small fragments. Workers then separate ore minerals from other minerals and send the ore-mineral concentrate to a processing plant, where it undergoes smelting to separate metal atoms from other atoms. At the end of the process, workers melt the metal and then pour it into molds to make ingots (brick-shaped blocks) for transport to a manufacturing facility.

To develop an **underground mine**, miners either dig a tunnel into the side of a mountain (the entrance to the tunnel is an *adit*), or they sink a vertical shaft in which they install an elevator. In some cases, they cut a spiral tunnel downward, to provide a gentle ramp for trucks to carry ore up to the surface. At the level in the crust where the ore body appears, they build a maze of tunnels into the ore by drilling holes into the rock and then blasting. The rock removed must be conveyed back to the surface. Rock columns between the tunnels hold up the ceiling of the mine. The deepest mine on the planet, located in South Africa, currently reaches a depth of 3.5 km, where temperatures exceed 55°C, making mining there a very uncomfortable occupation. Miners face danger from mine collapse and rock falls (**Box 15.1**). Some miners have been killed or injured by "rock bursts," sudden explosions of rock off the ceiling or walls of a tunnel. These explosions happen because the rock surrounding the adit is under such great pressure that it sometimes spontaneously fractures.

Take-Home Message

Geologists study outcrops and drillholes for ore shows, and they make maps of gravity and rock magnetism in order to discover ore deposits. Mining takes place either in open-pit mines or in shafts and tunnels underground. The process can be expensive and dangerous.

QUICK QUESTION: How can geologists confirm that an ore deposit lies deep beneath a given location?

15.5 Nonmetallic Mineral Resources

So far, this chapter has focused on resources that contain metal. But society uses many other geological materials, commonly known as **industrial minerals**, as well. For example, the stone used to make roadbeds and buildings, the chemicals composing fertilizers, the gypsum in drywall, the salt filling saltshakers, and the sand used to make glass, all come from the Earth. This section looks at a few of these materials and explains their origin (**Geology at a Glance**, pp. 562–563).

Dimension Stone

The Parthenon, a colossal stone temple rimmed by 46 carved columns, has stood atop a hill overlooking the city of Athens for almost 2,500 years. No wonder—*stone*, an architect's word for rock, outlasts nearly all other construction materials. We use stone to make facades, roofs, curbs, steps, countertops, and floors. We value stone for its visual appeal as well as its durability. The names that architects give to various types of stone may differ from the formal rock names that geologists use. For example, architects refer to any polished rock containing carbonate minerals as "marble," whether or not it has been metamorphosed. Likewise, they refer to any rocks containing silicate minerals as "granite," regardless of whether the rock has an igneous or a metamorphic texture, or a felsic or mafic composition.

To extract intact slabs and blocks of rock—known as **dimension stone** for architectural purposes—workers must carefully cut rock out of the walls of quarries (**Fig. 15.13a, b**). (Note that a *quarry* provides stone, whereas a *mine* supplies ore.) Quarry operators split rock blocks from bedrock either by hammering a series of wedges into the rock until a new crack propagates or by cutting directly through the bedrock with a wireline saw, a thermal lance, or a water jet. A *wireline saw* consists of a loop of braided wire moving between two pulleys. In some cases, as the wire moves along the rock surface, the quarry operator spills abrasive (sand or garnet grains) and water onto the wire. The movement of the wire drags the abrasive along the rock and grinds a slice into it. Alternatively, the quarry operator may use a diamond-coated wire, cooled with pure water. A *thermal lance* looks like a long blowtorch—it produces an ultra-intense flame of burning diesel fuel stoked by high-pressure air that can cut a slot in rock. More recently, quarry operators have begun using abrasive water jets, which squirts out water and abrasives at very high pressure, to cut rock.

Crushed Stone and Concrete

Crushed stone forms the substrate of highways and railroads and serves as the raw material for manufacturing cement, concrete, and asphalt. In crushed-stone quarries (**Fig. 15.13c, d**), operators use high explosives to break up bedrock into rubble that they then transport by truck to a jaw crusher. This reduces the rubble into usable-size chunks.

Most of the buildings and highways constructed in the past two centuries consist of walls, floors, columns, and roads made of concrete that has been spread into a layer or poured into a form, or they are made of bricks attached to each other by mortar. Both mortar and concrete start out as a slurry, but when allowed to set they harden into a hard, rock-like substance (**Box 15.2**). The slurry from which mortar and concrete form consists of *aggregate* (sand and/or gravel) mixed with water and cement. **Cement** consists of minerals that grow in the slurry, after the slurry has been left to set. These minerals bind the already solid grains in the slurry together. (In this regard, the cement in concrete or mortar plays the same role as the cement holding together the clasts of a sedimentary rock.) Before mixing, the material from which cement forms is a powder that consists of lime (CaO), quartz (SiO_2), aluminum oxide (Al_2O_3), and iron oxide (Fe_2O_3)—typically, lime accounts for 66% of cement, silica for 25%, and the remaining chemicals for about 9%. When this powder mixes with water, the chemicals comprising

BOX 15.1 CONSIDER THIS...

The Amazing Chilean Mine Rescue of 2010

The San José mine, near Copiapó in northern Chile, penetrates an ore-bearing intrusive body of diorite (intermediate igneous rock) that formed due to Cenozoic Andean convergent-margin tectonism. Like many other mines in the Andes, the mine produces copper and gold. The ore body extends downward to great depth, and over the years miners have been extracting ore from progressively deeper levels. The mining operation uses a long, gently sloped ramp that spirals downward to provide access in and out of the mine for miners, supplies, and ore. The ramp intersects the ore body at a depth of between 150 and 800 m (500 to 2,600 feet) below the ground surface. Off of this ramp, miners have cut numerous horizontal tunnels from which they remove ore. They also have enlarged some areas to produce "rooms" for repair shops and for shelters.

The rocks in the mine are under great stress due to the weight of overlying rock, so when miners cut tunnels they must set up supports that are strong enough to resist the overlying load and prevent rock from collapsing. In addition, workers may need to bolt, cement, or fence loose rocks to keep them from falling, even in places where the tunnel, overall, has adequate support.

On August 5, 2010, something went wrong. At a depth of around 500 m (1,600 feet) below the surface, a catastrophic rock fall suddenly filled a portion of the ramp with thousands of tons of debris. The fall isolated a group of 33 miners who were working near the bottom of the mine. Choking dust from the rock fall blocked visibility and made breathing difficult, but the miners managed to retreat to a refuge room at a depth of over 688 m (2,300 feet) below the surface. Immediately, in an amazing example of self-reliance and strength, the miners organized into a functioning group under the leadership of their foreman, carefully rationed emergency food supplies, dug small wells to provide water, and established a daily routine to wait until rescue.

For 17 days the miners sweltered in the humid, 35°C (95°F) air of their refuge, while unbeknownst to them, rescuers were frantically drilling exploratory holes in hopes of reaching the miners to provide ventilation and communication. Finally, a drill broke through the ceiling of the refuge. The miners tapped on the pipe to indicate they were there and taped a paper message to the end of the drill bit announcing to the world, *Estamos bien en el refugio los 33*. ("We are all right in the shelter—all 33.") With contact established, rescuers drilled a wider hole allowing them to send down supplies, letters, and even a video link. But the wait had to continue, for engineers estimated that bringing the men out might be four months away.

Immediately, mining engineers at the surface set to work drilling rescue holes wide enough for miners to fit in. The effort was not easy, for the igneous rocks of the mine are very hard, and drilling would have to avoid earlier workings. Three different plans (A, B, and C), using different drilling technologies, began the race through solid rock to reach the miners. In the end, Plan B reached them first, weeks before the predicted rescue date. Plan B utilized a percussion drilling machine manufactured by a Pennsylvania supplier—four hammers at the end of a device sent down one of the ventilation holes bashed into the surrounding rock and widened the hole. The drilling progressed at a rate of 40 m (130 ft) per day and sent a cascade of rock debris down the ventilation hole into the workroom. To

keep the workroom open, the trapped miners had to haul away the debris—all 700 tons of it.

Once the access hole was complete, rescuers sent down a special rescue cylinder, or capsule, named the Phoenix 2, in which one miner at a time could fit. Before strapping into the cylinder, which came equipped with oxygen and other safety devices, each miner donned sunglasses, so the brightness of light at the Earth's surface would not harm his eyes. One by one, they were pulled to safety by a large winch at a rate of about 1 m/s (2 mph), on a journey that took up to 18 minutes. As a huge crowd, including the Chilean president, waited at the surface, and an audience of 1 billion people watched on TV worldwide, the miners emerged on October 13, after having been underground for 69 days (**Fig. Bx15.1**).

FIGURE Bx15.1 The rescue capsule emerges at the surface.

FIGURE 15.13 Stone production in quarries.

(a) An active quarrying operation in Missouri that produces large blocks of cut dimension stone.

(b) Sheets of cut dimension stone being measured for cutting to become a kitchen countertop.

(c) A large crushed-stone quarry in Silurian limestone of Illinois. Drillers are working on the shelf in the distance.

(d) A large truck hauls debris from a quarry blast to the crusher.

Did you ever wonder . . .

how concrete differs from rock?

it dissolve. Mortar and concrete set when these chemicals react and precipitate to form an assemblage of new mineral crystals.

The ancient Romans probably were the first people to use cement—they made it from a mixture of volcanic glass and limestone. In the 18th and early 19th centuries, workers produced the powder for making cement by heating specific types of limestone in a kiln up to a temperature of about 1,450°C. The heating releases CO_2 gas and produces "clinker," chunks consisting of lime and other oxide compounds. Manufacturers crush the clinker into cement powder and pack it in bags for transport. The special limestone from which such *natural limestone* could be produced is fairly rare, for it has to contain calcite, clay, and quartz in just the right proportions to yield the chemicals needed to make the minerals in cement. Therefore, most cement used today is **Portland cement**, made by physically mixing crushed limestone, sandstone, and shale from different geologic formations in the correct ratios to provide the chemicals needed to produce cement. Isaac Johnson, an English engineer, came up with the recipe for Portland cement in 1844, and named it after the town of Portland, England, because he thought that, when set, it resembled rock exposed there.

BOX 15.2 CONSIDER THIS . . .

The Sidewalks of New York

Untold tons of concrete have gone into the construction of New York City. In fact, with the exception of a few city parks, most of the walking space in the city consists of concrete (**Fig. Bx15.2**). And concrete skyscrapers tower above the concrete plain. Where does all this concrete come from?

Much of the sand used in New York concrete was deposited during the last ice age. As vast glaciers moved southward over 14,000 years ago, they ground away the igneous and metamorphic rocks that constituted central and eastern Canada. These ancient rocks contained abundant quartz, and since quartz lasts a long time (it does not undergo chemical weathering easily), the sediment transported by the glaciers retained a large amount of quartz. Glaciers deposited this sediment in huge piles called moraines (see Chapter 22). As the glaciers melted, fast-moving rivers of meltwater washed the sediment, sorting sand from mud and pebbles. The sand was deposited in bars in the meltwater rivers, and these relict bars now provide thick lenses of sand that can be economically excavated.

What about the cement? Cement contains a mixture of lime, derived from limestone, and other elements (such as silica) derived from shale and sandstone. The bedrock of New York, though, consists largely of schist and gneiss, not sedimentary rocks. Fortunately, a source of rocks appropriate for making cement lies up the Hudson River. A Silurian-age rock unit, exposed in low hills just west of the river and called the Rosendale Formation, naturally contains exactly the right mixture of lime and silica needed to make durable cement. Beginning in the late 1820s, workers began quarrying the Rosendale Formation for cement. Quarry operators followed the Rosendale beds closely, making horizontal mine tunnels where the beds were horizontal, tilted mine tunnels where the beds tilted, and vertical mine tunnels where the beds were vertical. They then dumped the excavated rock into nearby kilns and roasted it to produce lime mixed with other oxides. The resulting powder was packed into barrels, loaded onto barges, and shipped downriver to New York. As demand for cement increased, operators eventually dug open-pit quarries from which they excavated Devonian limestone and shale units, mixing them together in the correct proportion to make Portland cement.

The rocks making up the strata that provide the source for cement consist of shell fragments and small, reef-like colonies of organisms. In other words, the lime in the concrete of New York sidewalks was originally extracted from seawater by living organisms—brachiopods, crinoids, and bryozoans—over 400 million years ago (Ma).

FIGURE Bx15.2 The production of concrete effectively transforms natural rock into human-made stone.

(a) A quarry of limestone. The trucks are carrying rock to a crusher.

(b) The heat of a kiln transforms limestone into lime (CaO).

(c) A concrete mixer pouring wet concrete.

(d) A sidewalk in New York.

Nonmetallic Minerals for Homes and Farms

We use an astounding variety of nonmetallic geologic resources (**Table 15.2**) without ever realizing where they come from. Consider the materials in a typical house or apartment. We've already talked about how ore deposits provide metals, such as the copper in a house's electrical wiring or pipes, and the iron in household tools and appliances, and we've also seen how concrete in the foundation, floors, and walls comes from limestone mixed with sand or gravel. Let's now look at some other components.

TABLE 15.2 Common Nonmetallic Resources

Limestone	Sedimentary rock made of calcite; used for gravel or cement
Crushed stone	Any variety of coherent rock (limestone, quartzite, granite, gneiss)
Siltstone	Beds of sedimentary rock; used to make flagstone
Granite	Coarse igneous rock; used for dimension stone
Marble	Metamorphosed limestone; used for dimension stone
Slate	Metamorphosed shale; used for roofing shingles
Gypsum	A sulfate salt precipitated from salt water; used for wallboard
Phosphate	From the mineral apatite; used for fertilizer
Pumice	Frothy volcanic rock; used to decorate gardens and paths
Clay	Very fine mica-like mineral in sediment; used to make bricks or pottery
Sand	From sandstone, beaches, or riverbeds; quartz sand is used for construction and for making glass
Salt	From the mineral halite, formed by evaporating saltwater; used for food seasoning, and for melting ice on roads
Sulfur	Occurs either as native sulfur, typically above salt domes, or in sulfide minerals; used for fertilizer and chemicals

The bricks in the exterior walls originated as clay, formed from the chemical weathering of silicate rocks and perhaps dug from the floodplain of a stream. To make bricks, workers mold wet clay into blocks and then bake it. Baking drives out water and causes metamorphic reactions that recrystallize the clay (**Fig. 15.14a**). Clay also serves as the raw material of pottery, porcelain, and other ceramic materials (see Chapter 8). The glass used to glaze windows consists largely of silica, formed by first melting and then freezing pure quartz sand from a beach deposit or a sandstone formation. Quartz may also be used in the construction of photovoltaic cells for solar panels. Gypsum board (drywall), used to construct interior walls, comes from a slurry of water and the mineral gypsum sandwiched between sheets of paper. Gypsum ($CaSO_4 \cdot 2H_2O$) occurs in evaporite strata precipitated from seawater or saline lake water. Evaporites provide other useful materials as well, such as halite (for

FIGURE 15.14 Clay and salt can be used for many products.

(a) These bricks consist of clay baked at a high temperature.

(b) Lithium-bearing salts being mined in Bolivia.

Forming and Processing Earth's Mineral Resources

Mining and processing ore has environmental consequences, including acid runoff, acid rain, and groundwater contamination.

Ore deposits can be obtained either in strip mines or in underground mines.

Circulating groundwater may extract and concentrate metals to form ore deposits.

Clay, when formed into blocks and baked, becomes brick.

Gravel itself may be quarried for construction purposes.

Mud, a mixture of clay minerals and water, accumulates in beds.

Ore minerals may collect on the bottom of a magma chamber.

Miners pan for gold in placer deposits where metal flakes and nuggets occur in sand and gravel.

From Mud to Brick

Hydrothermal vents (black smokers) produce accumulations of massive sulfides.

From Magma to Metal

Erosion tears down mountains and produces gravel and sand.

From Stream Channel to Roadbed

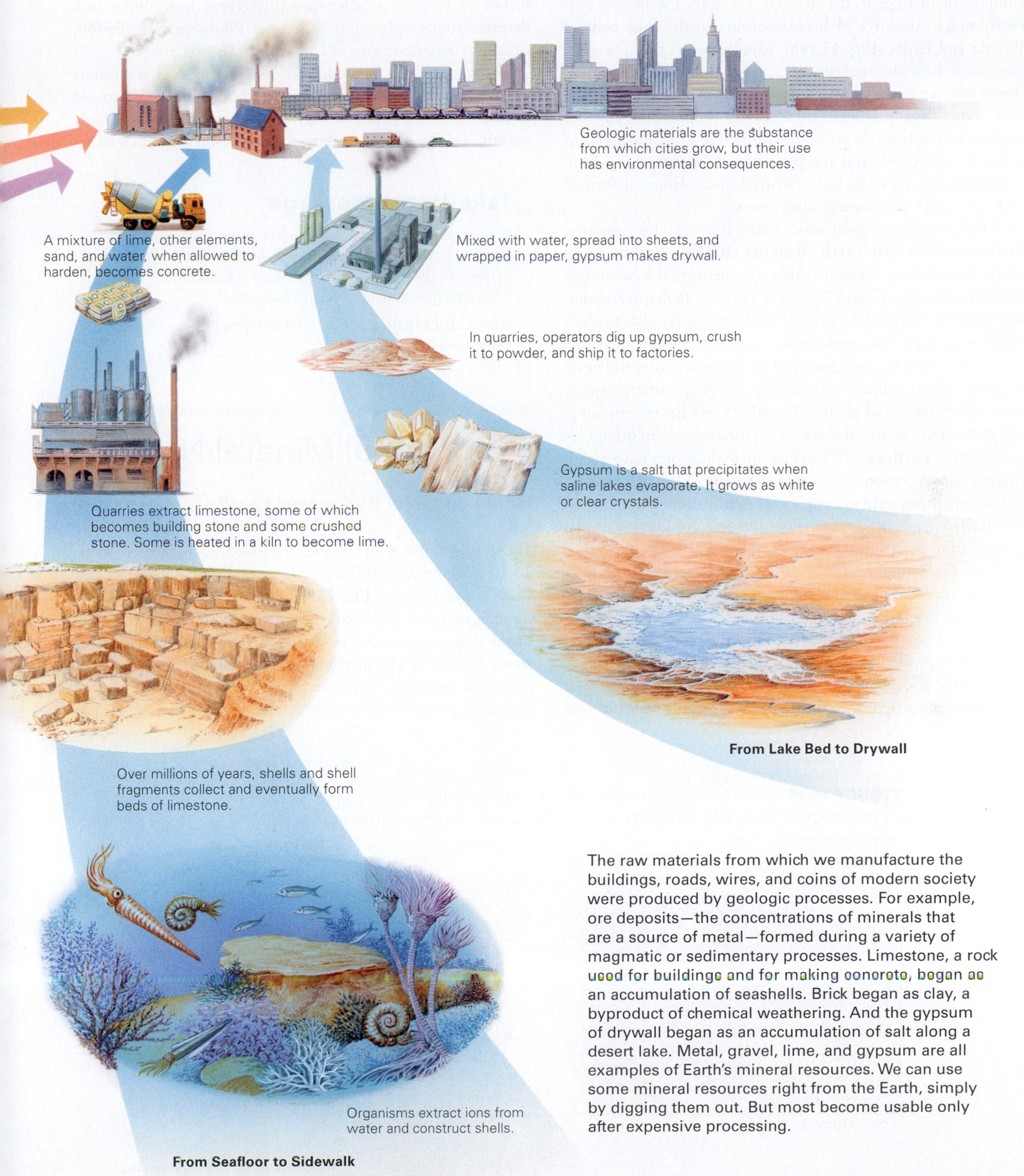

Geologic materials are the substance from which cities grow, but their use has environmental consequences.

A mixture of lime, other elements, sand, and water, when allowed to harden, becomes concrete.

Mixed with water, spread into sheets, and wrapped in paper, gypsum makes drywall.

In quarries, operators dig up gypsum, crush it to powder, and ship it to factories.

Quarries extract limestone, some of which becomes building stone and some crushed stone. Some is heated in a kiln to become lime.

Gypsum is a salt that precipitates when saline lakes evaporate. It grows as white or clear crystals.

Over millions of years, shells and shell fragments collect and eventually form beds of limestone.

From Lake Bed to Drywall

Organisms extract ions from water and construct shells.

From Seafloor to Sidewalk

The raw materials from which we manufacture the buildings, roads, wires, and coins of modern society were produced by geologic processes. For example, ore deposits—the concentrations of minerals that are a source of metal—formed during a variety of magmatic or sedimentary processes. Limestone, a rock used for buildings and for making concrete, began as an accumulation of seashells. Brick began as clay, a byproduct of chemical weathering. And the gypsum of drywall began as an accumulation of salt along a desert lake. Metal, gravel, lime, and gypsum are all examples of Earth's mineral resources. We can use some mineral resources right from the Earth, simply by digging them out. But most become usable only after expensive processing.

seasoning food and thawing ice on roads) and lithium (a key component of high-tech batteries). Of note, the largest currently mined resource of lithium occurs in dry lake beds of Bolivia and Chile (**Fig. 15.14b**). Finally, recall that asbestos, once used to make roof shingles, floor tiles, and break pads comes from serpentine, a rock created by the reaction of olivine with water. (The olivine, in turn, occurs in ophiolites, slices of oceanic lithosphere thrust onto continental crust during continental collisions.) And the plastic used in everything from countertops to pipes to light fixtures comes from oil formed and extracted from underground reserves.

Modern technological innovations have greatly increased the demand for **rare earth elements** (**REEs**), a group of 17 elements including the lanthanides (elements that have atomic numbers between 57 and 71 on the periodic table), scandium, and yttrium. While the names of REEs are unfamiliar to most people—and are also quite hard to pronounce—the elements themselves have become essential in the production of lasers, magnets, X-ray tubes, night-vision goggles, camera lenses, high-tech lamps, and chemical catalysts. REEs are not actually that rare in terms of their abundance relative to other elements in the Earth's crust. But localities where ores have a high enough concentration to be mined are rare. Most REEs used today come from strip mining either granitic plutons that contain REE-rich veins, or sediments and soils (residual deposits) derived from such plutons.

Chemicals employed for agricultural purposes also come from the ground. For example, potash (K_2CO_3) comes from the minerals in evaporite deposits. Phosphate (PO_4^{-3}) comes from the mineral apatite, which crystallizes in organic-rich muds that were deposited in shallow, oxygen-free (anoxic) seawater. Much of the phosphate now used in the United States has been extracted from strip mines accessing 10-Ma shale beds (formed from organic-rich muds) that lie about 15 m below the land surface of Florida. Smaller quantities come from mines in a Permian stratigraphic unit called the Phosphoria Formation, which underlies portions of the western United States.

As you can see, geologic processes acting over thousands to billions of years provide many of the material goods used in homes, farms, and industry. Truly, without the geologic resources of the Earth, modern society would grind to a halt.

> ## Take-Home Message
>
> Society uses a great variety of nonmetallic geologic materials. These include dimension stone, crushed stone, cement (made from roasted limestone), evaporites (including gypsum), and clay (to make bricks).
>
> **QUICK QUESTION:** What is the difference between natural cement and Portland cement?

15.6 Global Mineral Needs

How Long Will Resources Last?

The average citizen of an industrialized country uses about 20 kg of aluminum, 10 kg of copper, and 500 kg of iron and steel in a year's time (**Fig. 15.15; Table 15.3**). If you combine these figures with the quantities of energy resources and nonmetallic geologic resources a person uses, you get a total of about 15,000 kg (15 metric tons) of non-fuel resources used per capita each year. Thus, the population of the United States consumes about 4 billion metric tons of geologic material per

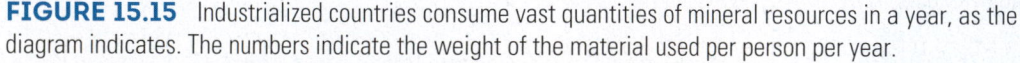

FIGURE 15.15 Industrialized countries consume vast quantities of mineral resources in a year, as the diagram indicates. The numbers indicate the weight of the material used per person per year.

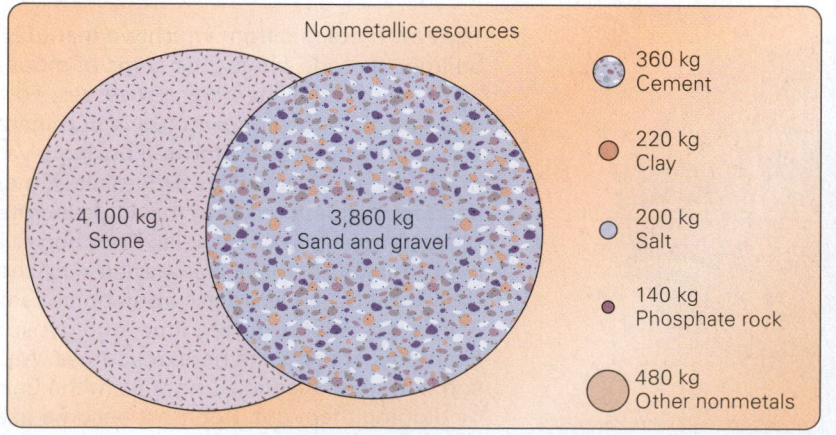
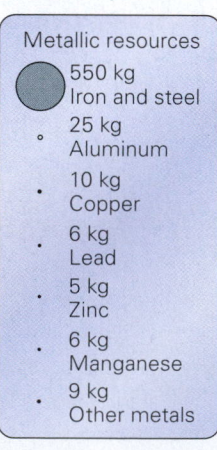

Nonmetallic resources

4,100 kg Stone
3,860 kg Sand and gravel

360 kg Cement
220 kg Clay
200 kg Salt
140 kg Phosphate rock
480 kg Other nonmetals

Metallic resources

550 kg Iron and steel
25 kg Aluminum
10 kg Copper
6 kg Lead
5 kg Zinc
6 kg Manganese
9 kg Other metals

TABLE 15.3 Per Capita Usage of Earth Materials in the USA

Material	Weight Used in a Year
Stone	4,100 kg
Sand and gravel	3,860 kg
Petroleum	3,050 kg
Coal	2,650 kg
Natural gas	1,900 kg
Iron and steel	550 kg
Cement	360 kg
Clay	220 kg
Salt	200 kg
Phosphate	140 kg
Aluminum	25 kg
Copper	10 kg
Lead	6 kg
Zinc	5 kg

1 kg = 2.205 pounds.

year. To create this supply, workers must mine, quarry, or pump 18 billion metric tons of Earth materials. By comparison, the Mississippi River transports 190 million metric tons of sediment per year. If you were to sum up consumption over a lifetime and add to it the weight of fossil fuels used, the average citizen of the United States will use 1.9 to 1.5 million kg (2.5 to 3.0 million pounds) of Earth materials in a lifetime.

Mineral resources, like oil and coal, are nonrenewable. Once mined, an ore deposit or a limestone hill disappears forever. Natural geologic processes do not happen fast enough to replace the deposits as quickly as we use them. Geologists have calculated **reserves** (known quantities of a commodity still in the ground) for various mineral deposits just as they have for oil. Based on current definitions of reserves, which depend on today's prices, and rates of consumption, supplies of some metals may run out in only decades to centuries (**Table 15.4**). But these estimates may change as supplies become depleted and prices rise, making previously uneconomical deposits worth mining. And supplies could increase if geologists discover new reserves or if new ways of mining become available. Further, increased efforts at conservation and recycling can cause a

dramatic decrease in rates of consumption and thereby stretch the lifetime of existing reserves.

Ore deposits do not occur everywhere because their formation requires special geologic conditions. As a result, some countries possess vast supplies, whereas others have none. In fact, no single country owns all the mineral resources it needs, so nations must trade with each other to maintain supplies, and global politics inevitably affects prices. Many wars have their roots in competition for mineral reserves, and it is no surprise that the outcomes of some wars have hinged on who controls these reserves.

The United States worries in particular about supplies of so-called **strategic minerals**, which include manganese, platinum, chromium, and cobalt—metals alloyed with iron to make the special-purpose steels needed in the aerospace industry. At present, the country must import 100% of its supply of many strategic minerals. Principal reserves of these metals lie in the crust of countries that have not always practiced open trade with the United States. As a defense precaution, the United States stockpiles these metals in case supplies are cut off.

We've already noted that many sources of lithium occur in nonindustrialized countries, and that industrialized countries are dependent on this material. Resources of REEs have become a subject of international tension in recent years. Currently, about 90% of the mined supply occurs in China. But China has started to limit its exports, which has led other countries to reactivate mines. Mining REE is challenging,

TABLE 15.4 Expected Lifetimes of Currently Known Ore Resources (in Years)

Metal	World Resources	U.S. Resources
Iron	120	40
Aluminum	330	2
Copper	65	40
Lead	20	40
Zinc	30	25
Gold	30	20
Platinum	45	1
Nickel	75	less than 1
Cobalt	50	less than 1
Manganese	70	0
Chromium	75	0

because it often occurs in association with radioactive elements, which should require miners to use special precautions to prevent environmental contamination.

Mining and the Environment

Mining leaves a big footprint in the Earth System. Some gaping holes that open-pit mining creates in the landscape have become so big that astronauts can see them from space. Both open-pit and underground mining yield immense quantities of waste rock, which miners dump in tailings piles (**Fig. 15.16a**). Some tailings piles grow into artificial hills 200 meters high and many kilometers long. Lacking soil, tailings piles tend to remain unvegetated for a long time. Mining also exposes ore-bearing rock to the atmosphere, and since many ore minerals are sulfides, they react with rainwater to produce **acid mine runoff**, which can severely damage vegetation downstream (**Fig. 15.16b**). In many cases, mining companies douse tailings with acidic solutions to leach out more metals; these acids sometimes escape into the environment.

FIGURE 15.16 Environmental consequences of producing metallic mineral resources.

(a) The tailings piles of the Bingham open-pit copper mine in Utah. The pit is 1.2 km deep and 4 km wide. During its century of operation, the mine has yielded about 17 million tons of copper, 700 tons of gold, and 6,000 tons of silver. The ore formed due to hydrothermal fluid circulation that accompanied magmatism about 36 Ma.

Much of the color comes from bacteria and archaea living in the water.

(b) The orange color in this mine runoff is due to iron and sulfide in the water.

Nickel smelter

The "superstack" is 380 m (1,270 ft) high.

Tailings pile

(c) Acidic smelter smoke killed off vegetation near Sudbury, Ontario, in the 1970s. A large tailings pile can be seen in the distance.

Ore processing tends to release noxious chemicals, which can mix with rain, forming acidic aerosols that spread over the countryside, and can harm vegetation. Before the installation of modern environmental controls, smoke from ore-processing plants caused severe air pollution. Plumes of smoke from the old smelters in Sudbury, Ontario, for example, created a wasteland for many kilometers downwind (**Fig. 15.16c**).

Recent years have seen efforts to reclaim mining spoils, and new technologies have been developed to extract metals in ways that are less deleterious to the environment and that treat waste more efficiently. Clearly, mining and ore processing can potentially become a scar on the landscape, the size of which depends on the willingness of producers and consumers to minimize damage and the degree to which regulations are successfully designed and enforced.

Take-Home Message

People use a vast quantity of mineral resources during the course of a lifetime. Mineral resources are nonrenewable, so minerals have limited reserves, and because reserves are not distributed uniformly around the planet, all supplies are not accessible to all consumers. Thus, the trade in economic minerals is politically charged. Also, mineral extraction and utilization has significant environmental consequences that are challenging to address.

QUICK QUESTION: If one country restricts the export of a strategic mineral, would that change the minimum concentration of the mineral necessary to make an ore deposit elsewhere worth mining? Why?

CHAPTER SUMMARY

- Industrial societies use many types of minerals, all of which must be extracted from the upper crust. We distinguish two general categories: metallic resources and non-metallic resources.
- Metals are materials in which atoms are held together by metallic bonds. They are malleable and make good conductors.
- Metals come from ore. An ore is a rock containing native metals or ore minerals (sulfide, oxide, or carbonate minerals that contains a high proportion of metal) in sufficient quantities to be worth mining. An ore deposit is an accumulation of ore.
- Magmatic deposits form when sulfide ore minerals accumulate during solidification of magma. In hydrothermal deposits, ore minerals precipitate from hot-water solutions. Secondary-enrichment deposits form when groundwater carries metals away from a pre-existing deposit. MVT deposits precipitate from groundwater that has passed long distances through sedimentary basins. Sedimentary deposits precipitate out of the ocean. Residual mineral deposits are the result of severe leaching in tropical soils. Placer deposits develop when heavy metal grains accumulate in sediment along a stream.

- Many ore deposits are associated with igneous activity in subduction zones, along mid-ocean ridges, along continental rifts, or at hot spots.
- Nonmetallic resources include dimension stone for decorative purposes, crushed stone for cement and asphalt production, clay for brick making, sand for glass production, and many other materials. A large proportion of materials in your home have a geological ancestry.
- Mineral resources are nonrenewable. Many are now or may soon be in short supply.
- The production and processing of mineral resources can harm the environment if not done carefully.

GUIDE TERMS

acid mine runoff (p. 566)

alloy (p. 547)

banded iron formation (BIF) (p. 553)

cement (p. 557)

dimension stone (p. 557)

grade (p. 549)

hydrothermal deposit (p. 550)

industrial mineral (p. 557)

magmatic deposit (p. 550)

manganese nodule (p. 553)

massive-sulfide deposit (p. 550)

metal (p. 547)

mineral resource (p. 546)

Mississippi Valley–type (MVT) ore (p. 552)

native metal (p. 547)

open-pit mine (p. 555)

ore (p. 549)

ore deposit (p. 550)

ore mineral (economic mineral) (p. 549)

placer deposit (p. 553)

Portland cement (p. 559)

rare earth element (REE) (p. 564)

reserve (p. 565)

residual mineral deposit (p. 553)

secondary-enrichment deposit (p. 552)

smelting (p. 547)

strategic mineral (p. 565)

underground mine (p. 556)

REVIEW QUESTIONS

1. Why did people use stone weapons before using bronze weapons?

2. What's the difference between an ore mineral and other minerals and between an ore and other kinds of rock?

3. How is the formation of certain types of ore minerals associated with igneous activity?

4. Explain how ores can occur or develop in sedimentary rocks.

5. In what geologic settings do massive sulfide deposits form?

6. What procedures are used to locate and mine mineral resources today?

7. How can dimension stone be obtained from a quarry?

8. What are the ingredients of cement? How is Portland cement made?

9. Name materials in your home that come from Earth materials.

10. How many kilograms of Earth materials does the average person in an industrialized country use in a year?

11. What are strategic minerals, and why have they become a political issue?

12. What are some environmental hazards of large-scale mining?

ON FURTHER THOUGHT

13. The costs of mining can be immense. To get a rough sense of this expense, imagine that an ore deposit of a certain metal contains 0.6% grade ore. This means that 0.6% by weight of a block of ore consists of the metal. On the open market, the pure metal sells for $8,000/ton. It costs $15/ton to mine the ore, $15/ton to transport the ore to the processing plant, and $15/ton to process the ore and produce pure metal. Start-up costs (building the mine and building the processing factory) are about $100 million. How much profit does the company make when it sells a ton of metal? How much ore (in tons) does the operation have to mine to pay back the start-up costs? Considering that a giant dump truck in a mine can carry 200 tons of ore at a time, how many dump-truck loads will have been transported at the break-even point? If the mine has eight trucks that can each make six loads a day, about how many years will it take to break even?

14. An ore deposit at a location in Arizona has the following characteristics: One portion of the ore deposit is an intrusive igneous rock in which tiny grains of copper sulfide minerals are dispersed among the other minerals of the rock. Another nearby portion of the ore deposit consists of limestone in which malachite fills cavities and pores in the rock. What types of ores are these? Describe the geologic history that led to the formation of these deposits.

smartwork smartwork.wwnorton.com

This chapter's Smartwork features:

- What A Geologist Sees exercises covering Hurricane Sandy.
- In-depth reading comprehension activities on mineral resources.
- Labeling activity on hydrothermal vents.

GEOTOURS

This chapter's GeoTour exercise (L) features:

- Mineral resources
- Bingham Copper Mine, Utah

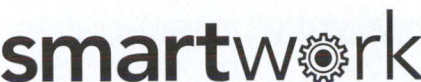

Another View Aerial view of a copper mine in British Columbia, Canada.

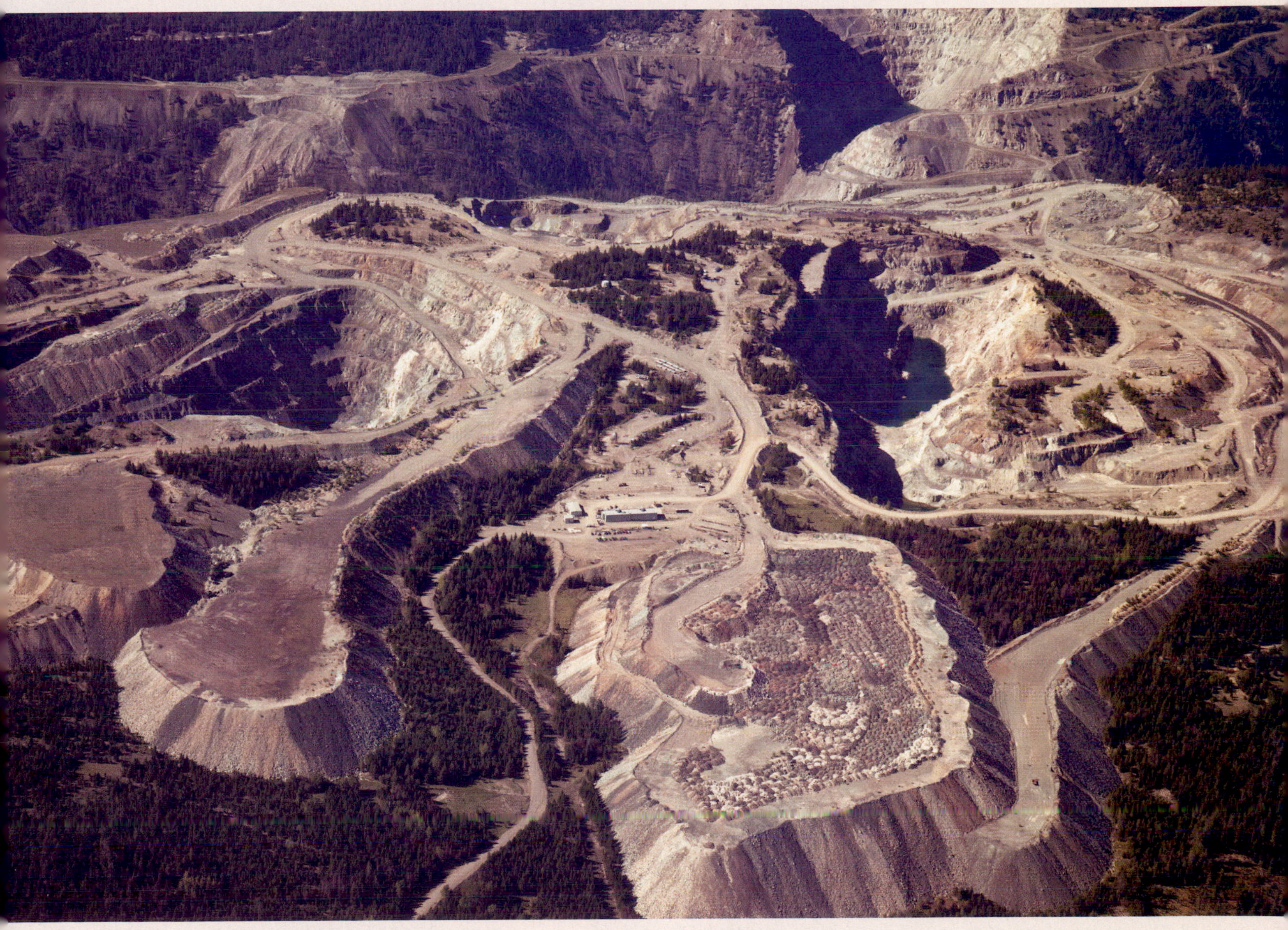

PART VI

PROCESSES AND PROBLEMS AT THE EARTH'S SURFACE

In the last part of this book, we focus on Earth's surface and near-surface realms. This portion of the Earth System—which encompasses the interface among the lithosphere, hydrosphere, atmosphere, and biosphere—displays great variability, for the dynamic interplay between *internal processes* (driven by Earth's internal heat) and *external processes* (driven by the warmth of the Sun) under the influence of Earth's gravitational field has resulted in a diverse array of landscapes. In Chapters 16 through 22, we examine five of these landscapes, plus groundwater and the atmosphere. Finally, in Chapter 23, we see how forces at work in the Earth System—including human activities—cause the planet to change over time. These changes affect the *critical zone*, the realm that supports life.

F Ever-Changing Landscapes and the Hydrologic Cycle

16 Unsafe Ground: Landslides and Other Mass Movements

17 Streams and Floods: The Geology of Running Water

18 Restless Realm: Oceans and Coasts

19 A Hidden Reserve: Groundwater

20 An Envelope of Gas: Earth's Atmosphere and Climate

21 Dry Regions: The Geology of Deserts

22 Amazing Ice: Glaciers and Ice Ages

23 Global Change in the Earth System

Mountains tower above sand dunes in this landscape of the eastern Mojave Desert, California. We're seeing the products of processes that modify landscapes at the Earth's surface. Erosion and landslides change the slopes, floods carry sediment into alluvial fans at the base of the mountains, and winds carry sand into drifting mounds on which it's hard for life to anchor.

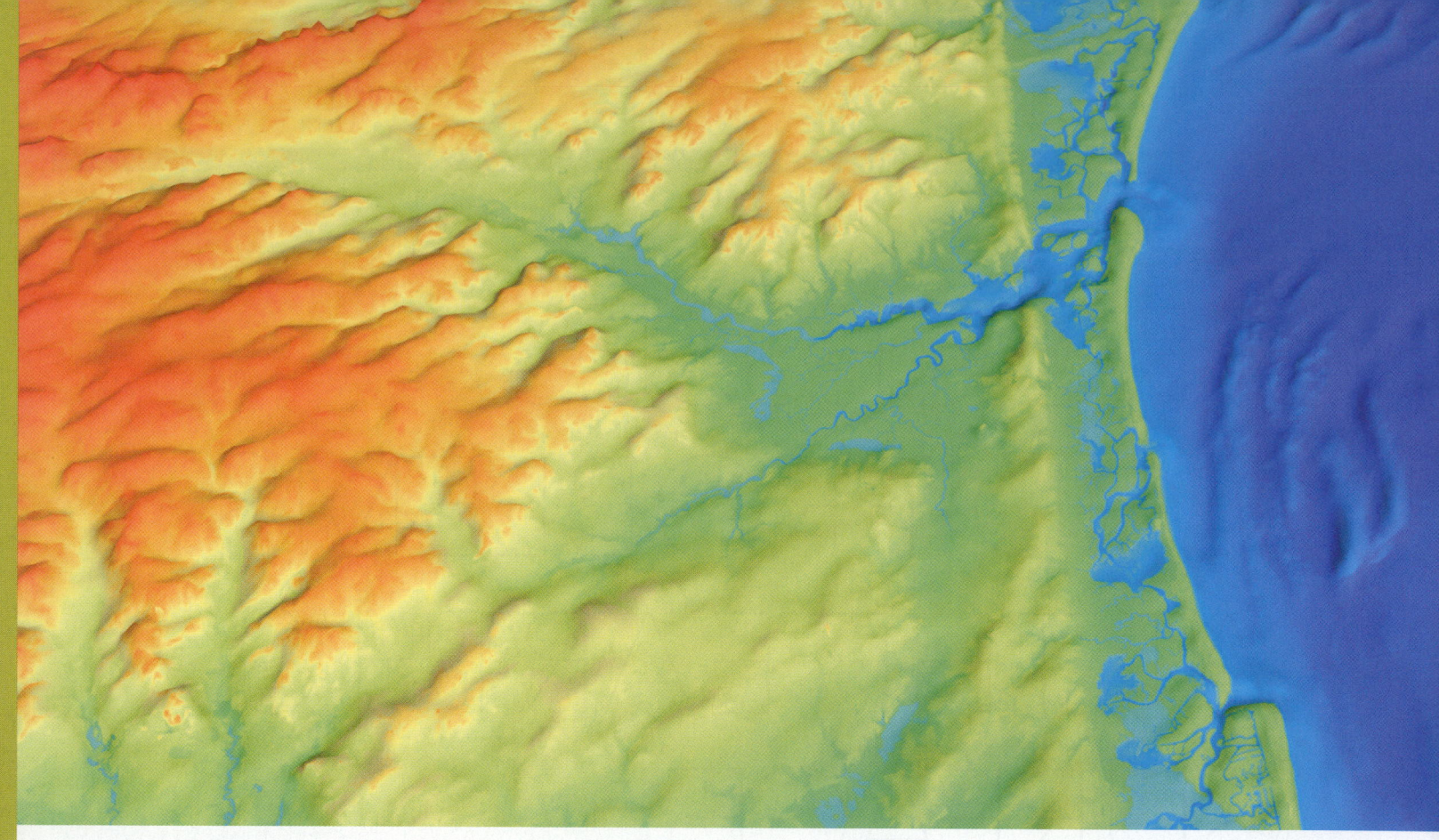

This digital elevation model of the Atlantic City, New Jersey, region reveals the details of a landscape modified by the ocean surf on the east and stream erosion inland. Water cycles among the ocean, air, land surface, and subsurface in this region.

Ever-Changing Landscapes and the Hydrologic Cycle

LEARNING OBJECTIVES

By the end of this interlude, you should understand . . .

- what a landscape is and what questions geologists ask about landscapes.
- the difference between uplift and subsidence and the forces that drive them.
- the contrasts between internal and external energy in the Earth System.
- how erosional and depositional landscapes differ.
- the reservoirs and exchange processes of the hydrologic cycle.

Nothing that is can pause or stay—
The moon will wax, the moon will wane,
The mist and cloud will turn to rain,
The rain to mist and cloud again,
Tomorrow be today.
— *Henry Wadsworth Longfellow (1807–1882)*

F.1 Introduction

The Earth's surface is at once a place of endless variety and intricate detail. Observe the height of its mountains, the expanse of its seas, the desolation of its deserts, and you may be inspired, frightened, or calmed. It's no wonder that artists and writers from across the ages have sought inspiration from the **landscape**—the character and shape of the land surface in a region—for landscapes encompass the diversity of human emotion (**Fig. F.1**). Geologists, like artists and writers, savor the impression of a dramatic landscape. But on seeing one, they can't help but ask, "How did it come to be, and how will it change in the future?"

This interlude introduces the general driving forces behind landscape development and sets the stage for interpreting distinct **landforms**, the individual shapes (such as valleys, cliffs, fans, mesas, and beaches) that make up landscapes, which we will be discussing in Part VI. We also introduce the *hydrologic cycle*—the

FIGURE F.1 Examples of the great variety of landscapes on Earth.

(a) Rounded "sugarloaf" mountains surround Rio de Janeiro, Brazil.

(b) Glaciated peaks of the Alps, France.

(c) Buttes of sandstone, Monument Valley, Arizona.

(d) Cliffs rise from the forest in the Blue Mountains, Australia.

(e) The Amazon jungle, Peru.

(f) Sandy beaches of Cape Cod, Massachusetts.

pathway water molecules follow as they move from ocean to air to land and back to ocean—for in Part VI you will see that water, both in its liquid and solid forms, serves many roles in surface and near-surface processes on the Earth. We conclude by taking a glimpse at landscapes found on other planets.

F.2 Shaping the Earth's Surface

Uplift and Subsidence; Erosion and Deposition

If the Earth's surface were totally flat, the great diversity of landscapes that embellish our vistas would not exist. But the surface isn't flat, because a variety of geologic processes can cause one portion of the surface to move up or down relative to an adjacent region. We refer to the relative upward movement of a region as **uplift** and the relative downward movement of a region as **subsidence**. Both uplift and subsidence occur for a variety of reasons, as outlined in **Table F.1**.

Cartographers and geologists refer to variations in land-surface elevation as **topography**. We can represent variations in elevation on a *topographic map* (**Box F.1**), or by a *shaded-relief map*, which conveys the impression of three dimensions by shading appropriate slopes to appear as if they are in shadows cast when the Sun is low in the sky (**Fig. F.2a**).

TABLE F.1 Causes of Uplift and Subsidence

Causes of Uplift

- **Thickening of the crust.** At convergent and collisional boundaries, compression causes the crust to shorten horizontally (by development of folds, faults, and foliations) and thicken in the vertical direction. Because of isostasy (see Chapter 11), lithosphere with thickened crust floats relatively higher on the asthenosphere, with the result that the surface of the crust in mountain belts rises. Intrusion or extrusion of igneous rocks thickens the crust or builds volcanoes on top of the surface, and also can cause uplift.

- **Heating of the lithosphere.** Heating decreases the thickness and density of lithosphere, so to maintain isostatic equilibrium, lithosphere floats higher.

- **Rebound due to unloading.** Removal of a heavy load (such as a glacier or mountain) from the surface causes the Earth's surface to rise in a manner similar to the way a trampoline's surface rises when you step off it.

- **Delamination.** If dense lithospheric mantle separates from the base of the plate and sinks into the mantle, the surface of the lithosphere rises. The effect resembles the consequence of unloading ballast from a ship.

Causes of Subsidence

- **Thinning of the crust due to stretching.** In rifts, where the crust undergoes horizontal stretching, the axis of the rift drops down by slip on normal faults.

- **Cooling of the lithosphere.** Cooling thickens the lithospheric mantle and makes it denser, so to maintain isostatic equilibrium, the lithosphere sinks down and its surface lies at a lower elevation.

- **Sinking due to loading.** Where a heavy load (such as a glacier or volcano) forms on the Earth's surface, the lithosphere warps downward, somewhat like the surface of a trampoline warps down when you stand on it.

FIGURE F.2 Portraying the shape of the Earth's surface.

(a) A hand-painted shaded-relief map of Europe.

(b) An oblique digital elevation model of Oahu, Hawaii.

BOX F.1 CONSIDER THIS . . .

Topographic Maps and Profiles

We can distinguish one landform from another by its shape, as manifested by variations in elevation within a region. For example, as you will see in succeeding chapters, a river-carved valley simply does not look like a glacially carved valley. As we've noted, geologists use the term *topography* to refer to variations in elevation, effectively the shape of the land surface.

How can we convey information about topography—a three-dimensional feature—on a two-dimensional sheet of paper? Cartographers and geologists do this by means of a **topographic map**, which uses contour lines to represent variations in elevation (**Fig. BxF.1a**). A **contour line** is an imaginary line along which all points have the same elevation. For example, if you walk along the 200-m contour line on a hill slope, you stay at exactly the same elevation (200 m). As another example, the shoreline on a flat, calm body of water is a contour line. In other words, you can picture a contour line as the intersection between the land surface and an imaginary horizontal plane (**Fig. BxF.1b**). The elevation difference between two adjacent contour lines on a topographical map is the **contour interval**. For a given topographic map, the contour interval is constant, so the spacing between contour lines represents the steepness of a slope. Specifically, closely spaced contour lines represent a steep slope, whereas widely spaced contour lines represent a gentle slope.

We can also represent variations in elevation by means of a **topographic profile**, the trace of the ground surface as it would appear on a vertical plane that sliced into the ground. Put another way, a profile represents the shape of the ground surface as viewed from the side (**Fig. BxF.1c**). If we add a representation of geologic features under the ground surface, then we have a **geologic cross section**. In some cases, geologists gain insight into subsurface geology simply by looking at the shape of a landform (**Fig. BxF.1d**). For example, a steep cliff in a region of dipping sedimentary strata may indicate the presence of a resistant (difficult to erode) layer; low areas may be underlain with nonresistant (easy to erode) layers.

FIGURE BxF.1 Topographic maps and profiles.

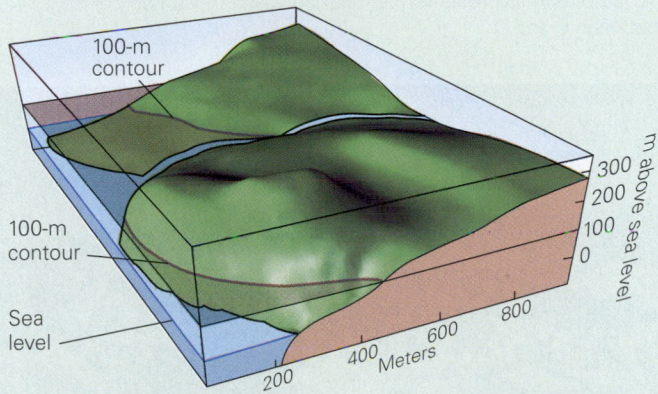

(a) A contour line is the intersection of a horizontal plane with the land surface. This block diagram shows the map area.

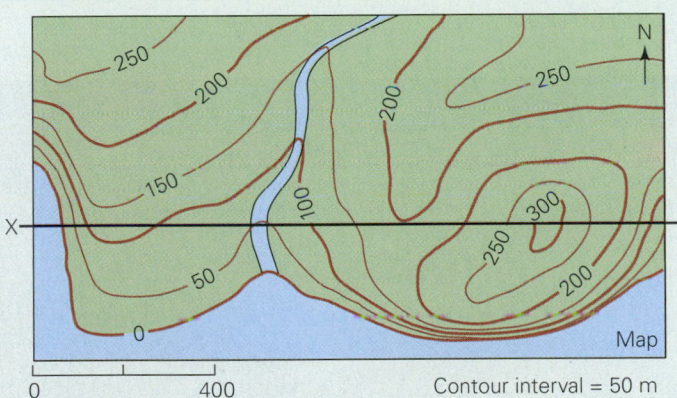

(b) A topographic map depicts the shape of the land surface through the use of contour lines. The difference in elevation between two adjacent lines is the contour interval.

Contour interval = 50 m

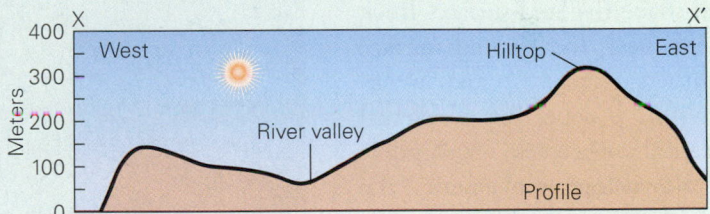

(c) A topographic profile (along section line X-X′) shows the shape of the land surface as seen in a vertical slice.

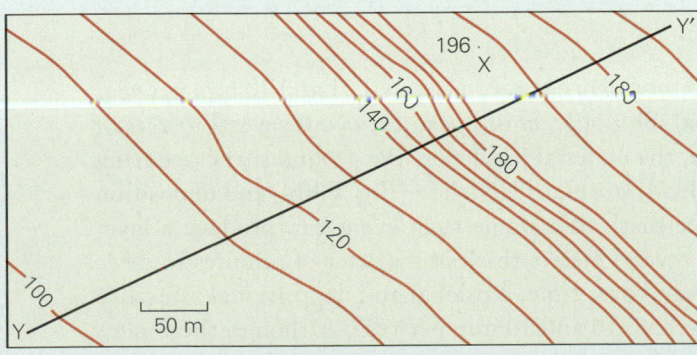

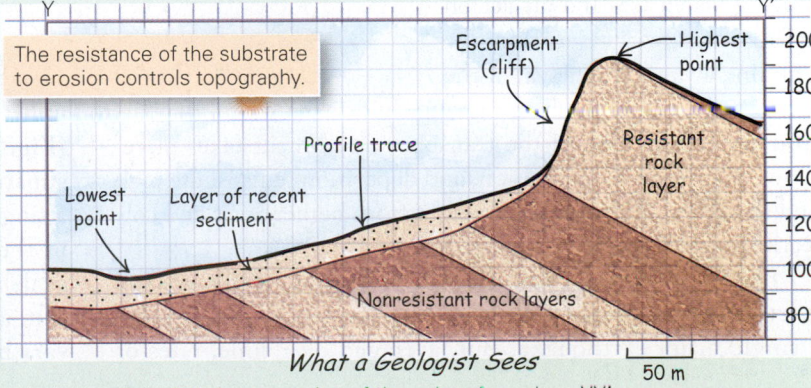

The resistance of the substrate to erosion controls topography.

What a Geologist Sees

(d) This topographic map shows a distinct cliff. A cross-section (right) depicts a geologist's interpretation of the subsurface along YY′. The cliff is the edge of a resistant rock layer.

In recent years, geologists have used radar beamed from satellites to produce highly detailed digital data sets of elevation variations on the Earth's surface. The resulting images, called **digital elevation models** (DEMs), can be analyzed with a computer to produce a digital elevation map, shaded and colored to give the impression of three dimensions. The digital image can also be tilted to portray an oblique view (**Fig. F.2b**).

The occurrence of uplift and/or subsidence generates **relief**, the elevation difference between two points separated by a specified horizontal distance on a map (**Fig. F.3a**). In discussion, we can say that a region has "high relief" if there are large elevation differences and that a region has "low relief" if there is a small elevation difference—for example, a region of tall mountains cut by deep valleys has high relief, whereas a plain with a nearly flat land surface has low relief (**Fig. F.3b, c**).

Whenever relief develops, various components of the Earth System kick into action to modify and shape the land surface. Rock at or near the ground surface weathers, fractures, and weakens. On a slope, this weakened material becomes susceptible to **downslope movement**, the gravity-driven tumbling or sliding of debris from higher elevations to lower ones. In addition, moving water, ice, and air act on uplifted land to cause **erosion**, the grinding away and removal of the Earth's surface. And where moving fluids slow down, **deposition** or accumulation of the transported sediment takes place. Downslope movement, erosion, and deposition redistribute rock and sediment, ultimately stripping it from higher areas and collecting it in low areas.

How rapidly do uplift and subsidence take place? The Earth's surface can rise or sink by as much as 3 m during a single major earthquake. But, averaged over time, rates of uplift and subsidence range between 0.01 and 10 mm per year (**Fig. F.4a**). Similarly, erosion can carve out several meters of **substrate**, the material just below the ground surface, during a single flood, storm, or landslide (**Fig. F.4b**), and deposition in the aftermath of a single such event can produce a layer of debris several meters thick in a matter of minutes to days. But, averaged over time, erosional and depositional rates also vary between 0.10 and 10 mm per year. Although these rates

FIGURE F.3 The concept of relief.

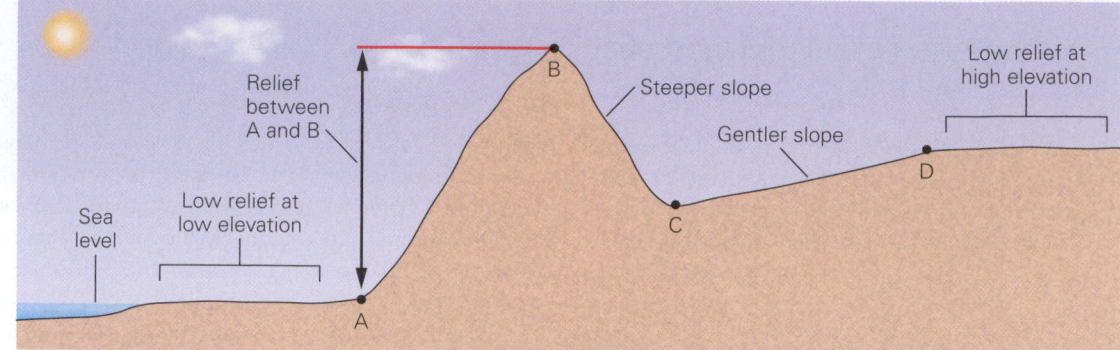

(a) The elevation difference between Points A and B, on this profile, is the relief between those two points. The relief is steeper between B and C, than between C and D.

(b) This mountainous area in Alaska has high relief.

(c) This plain in Ontario has low relief.

FIGURE F.4 The processes of uplift, subsidence, erosion, and deposition can be slow or rapid.

(a) Uplifted beach terraces form where the coast is rising relative to sea level. Present-day wave erosion is forming a new terrace and cutting a cliff on the edge of the old one.

(b) So much erosion can take place during a single hurricane that houses built along the beach become undermined.

seem small, a change in surface elevation of just 0.5 mm (the thickness of your fingernail) per year can yield a net change of 5 km in 10 million years. Uplift can build a mountain range, and erosion can whittle one down to near sea level—it just takes time!

What Drives Landscape Evolution?

The energy that drives landscape evolution comes from three sources: **internal energy**, the heat within the Earth, which ultimately keeps the asthenosphere hot and plastic enough so that plates can move and interact, and mantle plumes can rise; **external energy**, energy coming to the Earth from the Sun, which warms the atmosphere and ocean; and **gravitational energy**, which exerts a downward pull on material at higher elevation and, along with external energy, causes the convection of water and air that yields winds and waves. In fact, we can think of landscape evolution as a "battle" between tectonic processes such as collision, convergence, rifting, and volcanism, caused by plate interactions and hot spots, which build relief by driving uplift or subsidence, and processes such as downslope movement, erosion, and deposition, which destroy relief by removing material from high areas and depositing it in low ones.

If, in a particular region, the rate of uplift exceeds the rate of erosion, the land surface rises, whereas if the rate of subsidence exceeds the rate of deposition, the land surface sinks. Without uplift and subsidence, erosion and deposition would have long ago transformed Earth's surface into a flat plain. And without erosion and deposition, high and low areas would have lasted for the entirety of Earth history.

F.3 Factors Controlling Landscape Development

Imagine traveling across a continent. On your journey, you pass plains, swamps, hills, valleys, mesas, and mountains. Some of these features are **erosional landforms**, which result from the breakdown and removal of rock or sediment and develop where **agents of erosion**, such as water, ice, or air, carve into the substrate. Other features are **depositional landforms**, which result from the deposition of sediment where the medium carrying the sediment evaporates, slows down, or melts. The Part VI opening photo shows both erosional and depositional landforms in the same area. The specific landforms that develop at a given locality and that together make up the landscape reflect several factors.

- *Eroding or transporting agents:* Water, ice, and wind all cause erosion and transport sediment. But the shapes of landforms produced by each are different because of differences in the abilities of these agents to carve into the substrate and to carry debris. Of these three agents, water has the greatest impact, on a global basis.
- *Relief:* The elevation difference, or relief, between adjacent places in a landscape determines the height and steepness of slopes (see Fig. F.3a). Steepness, in turn, controls the velocity of ice or water flow and determines whether rock or soil stays in place or tumbles downslope.
- *Climate:* The average mean temperature, the volume of precipitation, and the distribution of precipitation through

the year (in other words, the *climate*), determines whether running water, flowing ice, or wind serves as the main agent of erosion or deposition in a region. Climate also affects the way in which substrate weathers.

- *Substrate composition:* The material comprising the substrate determines how the substrate responds to erosion. For example, strong rocks can stand up to form steep cliffs, while soft sediment collapses to generate gentle slopes.
- *Life activity:* Some life activity weakens the substrate (by burrowing, wedging, or digesting), while some holds it together (by binding it with roots).
- *Time:* Landscapes evolve through time, in response to continued erosion and/or deposition. For instance, a gully that has just started to form in response to the flow of

a stream does not look the same as a deep canyon that develops after the same stream has existed for a long time.

Although water, wind, and ice are responsible for the development of most landscapes, human activities have had an increasingly important impact on the Earth's surface. We have dug pits (mines) where once there were mountains, have built hills (tailings piles and landfills) where once there were valleys, and have made steep slopes gentle and gentle slopes steep (**Fig. F.5**). By constructing concrete walls, we modify the shapes of coastlines, change the courses of rivers, and fill new lakes (reservoirs). In cities, buildings and pavements completely seal the ground and cause water that might once have seeped down into the ground to spill into streams instead, increasing their flow. The area of land covered by pavement or buildings in the United States now exceeds the area of Ohio! And in the

FIGURE F.5 Human influence on a geologic scale.

(a) The pyramids of Egypt are human-made hills that rise above the desert sands. They have lasted for thousands of years.

(b) In the process of making highway cuts, deep valleys are cut through high ridges. This example borders a highway near Denver.

(c) This stone dam holds back a reservoir in Colorado. Think about how long it would take a glacier to pile up so much sediment.

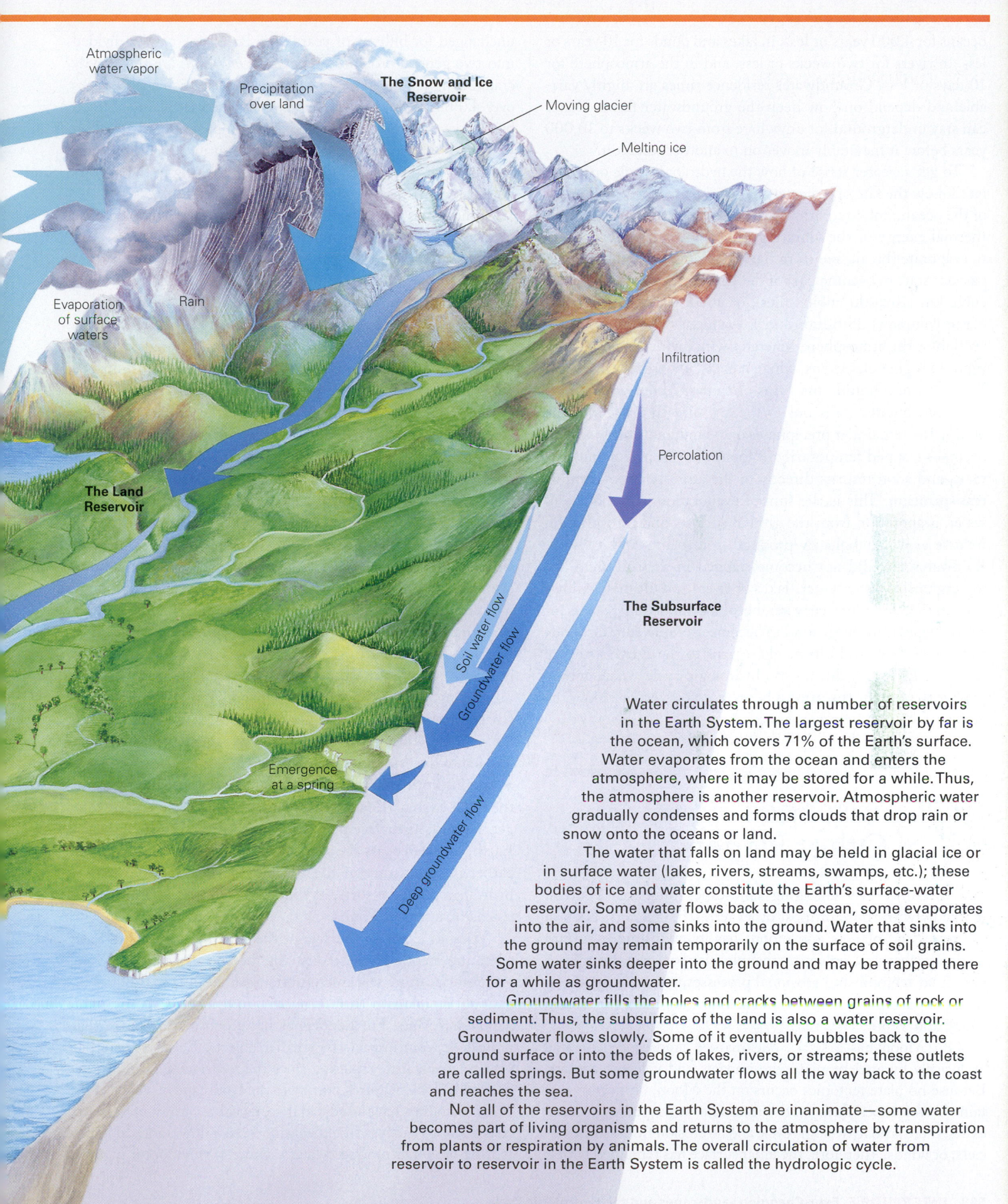

Atmospheric water vapor

Precipitation over land

The Snow and Ice Reservoir

Moving glacier

Melting ice

Evaporation of surface waters

Rain

Infiltration

Percolation

The Land Reservoir

Soil water flow

Groundwater flow

The Subsurface Reservoir

Emergence at a spring

Deep groundwater flow

Water circulates through a number of reservoirs in the Earth System. The largest reservoir by far is the ocean, which covers 71% of the Earth's surface. Water evaporates from the ocean and enters the atmosphere, where it may be stored for a while. Thus, the atmosphere is another reservoir. Atmospheric water gradually condenses and forms clouds that drop rain or snow onto the oceans or land.

The water that falls on land may be held in glacial ice or in surface water (lakes, rivers, streams, swamps, etc.); these bodies of ice and water constitute the Earth's surface-water reservoir. Some water flows back to the ocean, some evaporates into the air, and some sinks into the ground. Water that sinks into the ground may remain temporarily on the surface of soil grains. Some water sinks deeper into the ground and may be trapped there for a while as groundwater.

Groundwater fills the holes and cracks between grains of rock or sediment. Thus, the subsurface of the land is also a water reservoir. Groundwater flows slowly. Some of it eventually bubbles back to the ground surface or into the beds of lakes, rivers, or streams; these outlets are called springs. But some groundwater flows all the way back to the coast and reaches the sea.

Not all of the reservoirs in the Earth System are inanimate—some water becomes part of living organisms and returns to the atmosphere by transpiration from plants or respiration by animals. The overall circulation of water from reservoir to reservoir in the Earth System is called the hydrologic cycle.

oceans for 4,000 years or less, in lakes and ponds for 10 years or less, in rivers for two weeks or less, and in the atmosphere for 10 days or less. Groundwater residence times are highly variable and depend on how deep the groundwater flows. Water can stay underground for anywhere from two weeks to 10,000 years before it inevitably moves on to another reservoir.

To get a clearer sense of how the hydrologic cycle operates, let's follow the fate of seawater that has just reached the surface of the ocean. Solar radiation heats the water, and the increased thermal energy of the vibrating water molecules allows them to evaporate (break free from the liquid) and drift upward in a gaseous state to become part of the atmosphere. About 417,000 cubic km (102,000 cubic miles), or about 0.03% of the total ocean volume (1.35 billion km^3), evaporates every year. Convection of the atmosphere generates wind, which carries water vapor to higher elevations, where it cools, undergoes condensation to form a liquid, and rains or snows. About 76% of this water precipitates (falls out of the air) directly back into the ocean. The remainder precipitates onto land; most of this water becomes trapped temporarily in the soil, or in plants and animals, and soon returns directly to the atmosphere by **evapotranspiration**. This is the sum of evaporation from bodies of water, evaporation from the ground surface, and transpiration (release as a metabolic by-product) from plants and animals. Rainwater that did not become trapped in the soil or in living organisms either enters lakes or rivers and ultimately flows back to the sea as surface water, becomes trapped in glaciers, or sinks deeper into the ground to become groundwater. Groundwater also flows and ultimately returns to the Earth's surface reservoirs. In sum, during the hydrologic cycle, water moves among the ocean, the atmosphere, reservoirs on or below the land surface, and living organisms.

F.5 Landscapes of Other Planets

The dynamic, ever-changing landscapes of Earth contrast markedly with those of most other terrestrial planets. Each of the terrestrial planets and moons has its own unique surface landscape features, reflecting the interplay between the object's particular tectonic and erosional processes, either currently, or in the far-distant past. Let's look at a few examples: the Moon, Mars, and Venus.

Our Moon has a static, pockmarked landscape generated exclusively by ancient meteorite impacts and volcanic activity. Because no plate tectonics occurs on the Moon, no new mountains or volcanoes form; and because no atmosphere or ocean exists, there is no hydrologic cycle and no erosion from rivers, glaciers, or winds. Therefore, the lunar surface has remained largely unchanged for billions of years. The landscape can be divided into two general provinces. The *lunar highlands* are the heavily cratered, light-colored regions of the Moon, which expose rocks over 4.0 billion years old. The *mare* are the vast plains of flood basalt, possibly formed in response to impacts over 3.8 billion years ago—these impacts could have excavated such huge craters that that they caused decompression melting in the Moon's mantle and extrusion of flood basalts (**Fig. F.7a, b**).

Landscapes on Mars differ from those of the Moon because Mars does have an atmosphere, though much less dense than that of Earth. Martian winds generate huge dust storms that can obscure nearly the entire surface of the planet for months at a time. The landscapes of Mars also differ from the Moon's because Mars once had surface water (**Box F.2**). Thus, the Martian surface appears to expose four kinds of materials: volcanic flows and deposits (primarily of basalt), debris from impacts, windblown sediment, and water-laid sediment. There is even evidence that soil-forming processes affected surface materials. Of note, Martian winds not only deposit sediment, but they also slowly erode impact craters and polish surface rocks.

Landscapes on Mars also differ from those on Earth, because like the Moon, Mars does not have plate tectonics. So, unlike Earth, Mars has no mountain belts or volcanic arcs. In fact, most landscape features on Mars, with the exception of wind-related ones, are over 3 billion years old. There is, however, significant relief on Mars (**Fig. F.7c**). Long ago, a huge mantle plume formed, causing the uplift of a 9-km-high bulge (the Tharsis Ridge) that covers an area comparable to that of North America. Rifting of the Tharsis Ridge produced an immense canyon, the Valles Marineris, that is 4,000 km (2,500 miles) long, 200 km (120 km) wide, and 7 km (4.4 miles) deep. (By comparison, the Grand Canyon on earth is about 230 km long, 30 km wide, and 1.8 km deep.) Martian thermal activity also led to the eruption of gargantuan hot-spot volcanoes, such as the 22-km-high Olympus Mons, the highest mountain in the Solar System. Because Mars has no vegetation and no longer has rain, its surface does not weather and erode like that of Earth, so it still bears the scars of impact by swarms of meteors earlier in the history of the Solar System. Mars does have a hydrologic cycle, of sorts, in that it has ice caps that grow and recede on a seasonal basis.

Venus is closer to the size of the Earth and may still have operating mantle plumes. Virtually the entire surface of Venus was resurfaced by volcanic eruptions about 300 to 1,600 Ma, making the planet's surface much younger than those of the Moon and Mars. Further, Venus has a dense atmosphere that protects it from impacts by smaller objects. Because there has been relatively little cratering since the resurfacing event, volcanic and tectonic features dominate the landscape of Venus (**Fig. F.7d**). Satellites have used radar to reveal a variety of volcanic constructions (such as shield volcanoes, lava flows, and calderas). Rifting on Venus produced faults, some of which occur in asso-

FIGURE F.7 Landscapes of other planets.

(a) The heavily cratered surface of Earth's moon.

(b) A close-up view of the lunar landscape, with the lunar rover and an astronaut for scale.

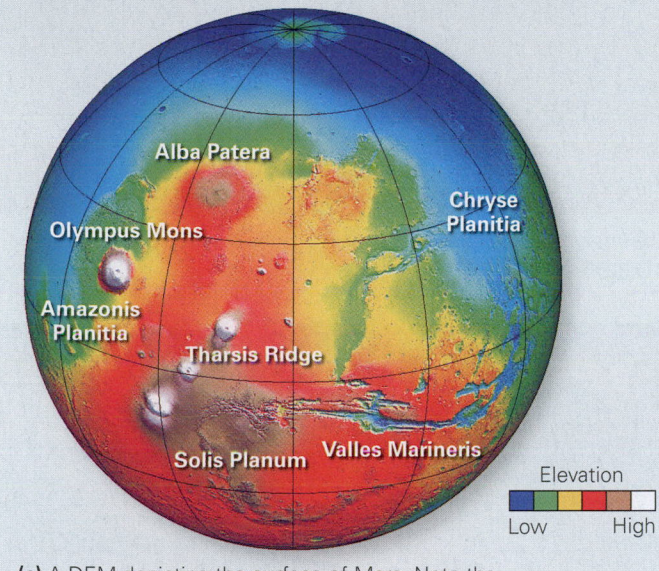

Alba Patera

Olympus Mons

Chryse Planitia

Amazonis Planitia

Tharsis Ridge

Solis Planum   Valles Marineris

Elevation

Low   High

(c) A DEM depicting the surface of Mars. Note the huge bulge of Tharsis Ridge, the giant Olympus Mons volcano, and the deep Valles Marineris canyon.

(d) A radar image of Venus. A thick blanket of clouds obscures the planet's surface, so it can't be seen through a telescope. Red areas are higher, blue areas lower.

ciation with volcanic features. Liquid water cannot survive the scalding temperatures of Venus's surface, so no hydrologic cycle operates there and no life exists. Because of the density of the atmosphere, winds are too slow to cause much erosion or deposition, thus leaving volcanic landforms virtually unchanged.

During the past two decades, spacecraft visiting the moons of the outer planets have sent home amazing images of surface features that differ markedly from any found on Earth. As an example, consider Enceladus, a 500-km-diameter moon of Saturn (**Fig. F.8a**). Much of Enceladus's ice-covered surface is cracked and wrinkled and largely crater-free, suggesting that tectonic movements rifted and folded this moon's crust subsequent to the intense meteorite bombardment episodes of early Solar System history. The still-cratered terrains may be older and stabler regions of the crust. Io, a 3,600 km-diameter moon of Jupiter, has significant ongoing volcanic activity, producing mafic and ultramafic flows as well as coatings of multicolored sulfur-rich ash (**Fig. F.8b**).

BOX F.2   CONSIDER THIS . . .

Water on Mars?

In 1877, an Italian astronomer named Giovanni Schiaparelli studied the surface of Mars with a telescope and announced that long, straight *canali* crisscrossed the planet's surface. *Canali* should have been translated into the English word *channel*, but perhaps because of the recent construction of the Suez Canal, newspapers of the day translated the word into the English *canal*, with the implication that the features had been constructed by intelligent beings. An eminent American astronomer began to study the "canals" and suggested that they had been built to carry water from polar ice caps to Martian deserts.

Late 20th-century satellite mapping of Mars showed that the so-called canals do not exist—they were simply optical illusions. There are no lakes, oceans, or flowing rivers on the surface of Mars today. The atmosphere of Mars has such low density and thus exerts so little pressure on the planet's surface that any liquid water released at the surface would quickly evaporate. Thus, Mars has no hydrologic cycle the way the Earth does. But three crucial questions remain: Does liquid water ever form, even for short periods, on the Martian surface today? Did Mars ever have a significant amount of running water or standing water in the past? And, if the planet once had significant water, where is the water now? The question of the presence of water lies at the heart of an even more basic question: given that the simplest life as we know it requires liquid water, is there, or was there, life on Mars?

The case for liquid water on Mars is very strong. Much of the evidence comes from comparing landforms on the planet's surface with landforms of known origin on Earth. High-resolution images of Mars reveal a number of landforms that look as if they formed in response to flowing water. Examples include networks of channels resembling river networks on Earth (**Fig. BxF.2**), scour features, deep gullies, and streamlined deposits of sediment. Studies by the *Odyssey* satellite in 2003, and by the Mars rovers (*Spirit* and *Opportunity*) that landed on the planet in 2004, added intriguing new data to the debate. *Odyssey* detected hints that hydrogen, an element in water, exists beneath the surface of the planet over broad regions, and the Mars rovers have documented the existence of hematite and gypsum, minerals that form in the presence of water. The rovers also have found sedimentary deposits that appear to have been deposited in water. The *Phoenix* lander, in 2008, confirmed the existence of water ice by digging into the surface to expose some, and starting in 2012, the *Curiosity* rover has found further supporting evidence. Researchers speculate that Mars was much wetter in its past, perhaps billions of years ago. But since the atmosphere became less dense, the water evaporated and now lies hidden underground or trapped in polar ice caps.

FIGURE BxF.2   Evidence for water on Mars, earlier in the planet's history.

(a) A DEM of the boundary between the Acidalia Plantia lowlands (blue, green, yellow) and the Tempe Terra Plateau (red and brown), based on data from the Mars Express satellite. Several stream channels are visible.

(b) An oblique view showing stream channels on Mars resembling channels cut by rivers on Earth.

(c) Layers of strata, interpreted to be water deposited, in Chasma Canyon, Mars. Note the elongate islands in the channel, resembling islands that have been shaped by rivers on Earth.

FIGURE F.8 The variety of moons surrounding gas-giant planets. The moons come in a variety of compositions and with different amounts of cratering. Left: Enceladus is an icy moon of Saturn with very little cratering, indicating more recent geologic activity. Right: Io, a moon of Jupiter, is volcanically active.

INTERLUDE SUMMARY

- The character and shape of the land surface in a region is a landscape. Individual shapes are landforms. Topographic maps and DEMs portray the shape of landscapes.
- Land can undergo uplift or subsidence, to yield relief. Debris formed by the weathering of uplifted land undergoes downslope movement, and collects in lower areas.
- The energy driving landscape evolution comes from three sources: Earth's internal energy, gravitational energy, and energy radiating from the Sun.

- The nature of a landscape depends on climate, time, relief, slope angles, elevation, the activity of organisms, substrate composition, and the rate of tectonic movement.
- Water moves among various reservoirs during the hydrologic cycle. Landscape evolution involves erosion by flowing water or ice.
- Landscapes on other planets differ markedly from those on Earth.

GUIDE TERMS

agents of erosion (p. 577)
contour interval (p. 575)
contour line (p. 575)
deposition (p. 576)
depositional landform (p. 577)
digital elevation model (DEM) (p. 576)

downslope movement (p. 576)
erosion (p. 576)
erosional landform (p. 577)
evapotranspiration (p. 582)
external energy (p. 577)
geologic cross section (p. 575)

gravitational energy (p. 577)
hydrologic cycle (p. 579)
hydrosphere (p. 579)
internal energy (p. 577)
landform (p. 573)
landscape (p. 572)
relief (p. 576)

residence time (p. 579)
subsidence (p. 574)
substrate (p. 576)
topographic map (p. 575)
topographic profile (p. 575)
topography (p. 574)
uplift (p. 574)

REVIEW QUESTIONS

1. What is the difference between uplift and subsidence?
2. Why do landscapes on the Earth change over geologic time, while they remain static on the Moon?
3. What is topography, and how can we portray it on a sheet of paper?
4. What are the principle agents of erosion on Earth?

5. What factors affect the character of erosional or depositional landforms that develop in a region?
6. Explain the steps in the hydrologic cycle.
7. How do landscapes of other planets differ from those of Earth?

An 2010 earthquake triggered a landslide on a rain-soaked hillslope in Taiwan. The debris buried a highway. Where there are slopes, unstable ground can develop and become a natural hazard.

CHAPTER 16

Unsafe Ground: Landslides and Other Mass Movements

> Gravity is a habit that is hard to shake off.
>
> —*Terry Pratchett (British author)*

LEARNING OBJECTIVES

By the end of this chapter, you should understand . . .

- the characteristics and consequences of different types of mass movements.

- factors that determine whether a slope is stable or unstable.

- the events that can trigger a mass-movement event.

- why some regions are more susceptible to mass movements than are others.

- how landslide hazards can be evaluated and, in some cases, prevented.

16.1 Introduction

It was Sunday, May 31, 1970, a market day, and thousands of people had crammed into the Andean town of Yungay, Peru, to shop. Suddenly they felt the jolt of an earthquake that was strong enough to topple some masonry houses. But worse was yet to come. Shocks from the earthquake also caused an 800-m-wide ice slab to break off the end of a glacier at the top of Nevado Huascarán, a nearby 6.6-km-high mountain peak. Gravity instantly pulled the ice slab down the mountain's steep slopes. As it tumbled down over 3.7 km, the ice disintegrated into a chaotic avalanche of chunks traveling at speeds of over 300 km per hour. Near the base of the mountain, most of the avalanche channeled into a valley and thickened into a moving layer as high as a ten-story building, ripping up rocks and soil along the way. Friction transformed the ice into water, which when mixed with rock and dust created 50 million cubic meters of a muddy slurry viscous enough to buoy along boulders larger than houses. This mass, sometimes riding on a compressed air cushion that allowed it to pass by without disturbing the grass below, traveled over 14.5 km in less than 4 minutes.

On rounding a curve near the mouth of the valley, part of the mass shot up the sides and flew over the ridge between the valley and Yungay. As the town's inhabitants and visitors stumbled out of earthquake-damaged buildings, they heard a deafening roar and looked up to see a churning mud cloud descending on them. Moments later, the town was completely buried under several meters of mud and rock—only the top of the church and a few palm trees remained visible to show where Yungay once lay (**Fig. 16.1**). Over 18,000 people are forever entombed beneath the resulting debris layer. Today the site is a grassy meadow, spotted with memorials left by mourning relatives.

Could the Yungay tragedy have been prevented? Perhaps. A few years earlier, climbers had recognized the instability of glacial ice on Nevado Huascarán, and Peruvian newspapers had published a warning, but alas, no one took notice. In the aftermath of the event, geologists discovered that

FIGURE 16.1 The May 1970 Yungay landslide disaster in Peru.

(a) Before the landslide, the town of Yungay perched on a hill near the ice-covered mountain Nevado Huascarán.

(b) The landslide completely buried the town beneath debris. A landslide scar is visible on the mountain in the distance.

Yungay had been built on ancient layers of debris from past calamities. Peru subsequently prevented new construction in the danger zone.

People often assume that the ground beneath them is *terra firma*, a solid foundation on which they can build their lives. But the catastrophe at Yungay says otherwise. The **substrate** (material below the surface) underlying sloping regions of the Earth's surface—regardless of whether it consists of rock or *regolith* (loose, sediment, debris, and soil)—is inherently "unstable" in the sense that under the relentless pull of gravity it will eventually move downslope. Geologists refer to the downslope transport of rock, regolith, snow, and ice as **mass movement**, or *mass wasting*. Like earthquakes, volcanic eruptions, storms, and floods, mass movements are a type of **natural hazard**, meaning a dangerous aspect of the environment that can cause damage to life and property. Unfortunately, mass movement becomes more of a threat every year because as the world's population grows, cities expand into areas of unsafe ground. In addition to representing a hazard that we must address, mass movement also plays a critical role in the rock cycle, for it serves as the first step in the transportation of sediment. And it plays a critical role in the evolution of landscapes in that it modifies the shapes of slopes.

In this chapter, we look at the types, causes, and consequences of mass movement and the precautions society can take to protect people and property from its dangers. You might want to consider this information when selecting a site for your home or when voting on land-use propositions for your community.

16.2 Types of Mass Movement

In general discussion, most people refer to any mass movement of rock and/or regolith down a slope as a **landslide**. Geologists and engineers, however, find it useful to distinguish among different kinds of mass movements based on four features: (1) the type of material involved (rock or regolith); (2) the velocity of movement (slow, intermediate, or fast); (3) the character of the moving mass (coherent or chaotic; wet or dry); and (4) the environment in which the movement takes place (subaerial or submarine). Similarly, most people think of an **avalanche** as a mass movement of snow—geologists and engineers, however, apply the term more broadly to include any mass movement that moves like a turbulent cloud.

Why bother classifying mass movements? We make these distinctions because different types of mass movements have different consequences and therefore represent different kinds of hazard—by characterizing mass movements more completely,

we can better prepare for them. Below we examine mass movements that occur on land, roughly in order from slow to very fast. We then briefly introduce submarine mass movements.

Creep and Solifluction

Creep (also known as *soil creep*) refers to the slow, gradual downslope movement of regolith on a slope. Creep happens when regolith alternately expands and contracts in response to freezing and thawing or wetting and drying. During freezing or wetting, the regolith expands and its particles move outward, perpendicular to the slope. During the thawing or drying, the regolith contracts and gravity makes the particles sink vertically and thus migrate downslope slightly (**Fig. 16.2a, b**). You can't see creep by staring at a hillslope because it occurs too slowly, but over a period of years, creep causes trees, fences, gravestones, walls, and foundations built on a hillside to tilt downslope. Notably, trees that continue to grow after they have been tilted display a pronounced curvature at their base (**Fig. 16.2c**).

In Arctic or high-elevation regions, regolith freezes solid to great depth during the winter. In the brief summer thaw, only the uppermost 1 to 3 m of the ground thaws. Since meltwater cannot sink into the underlying *permafrost* (permanently frozen ground), the melted layer becomes soggy and weak and slowly flows downslope in overlapping sheets. Geologists refer to this kind of creep, characteristic of cold, treeless tundra regions, as **solifluction** (**Fig. 16.2d**).

Rock Glaciers

Slow mass movement also takes place in a **rock glacier**, a body made of rock fragments embedded in a matrix of ice (**Fig. 16.2e**). Rock glaciers differ from more familiar ice-dominated glaciers in terms of the proportion of rock fragments to ice—in a rock glacier, most of the volume consists of rock, whereas in an ice-dominated glacier, most of the volume consists of ice. In effect, rock glaciers are breccias cemented by ice. Since ice is weak, the combined mass of rock and ice in a rock glacier can slowly flow downslope, as does the relatively pure ice of an ice-dominated glacier.

Rock glaciers form in two ways. First, they develop where snow or rain percolates down into a pile of rock debris that has accumulated above permafrost at the base of a cliff. This water can't infiltrate the frozen ground below the debris pile, so it freezes to form ice in the pores between clasts within the debris pile. Second, rock glaciers develop where an ice-dominated glacier already containing abundant rock debris begins to melt. As melting progresses, the proportion of rock to ice in the glacier increases, and if enough melting takes place, the glacier eventually contains more rock than ice.

FIGURE 16.2 The process and consequences of slow mass movements (creep and solifluction).

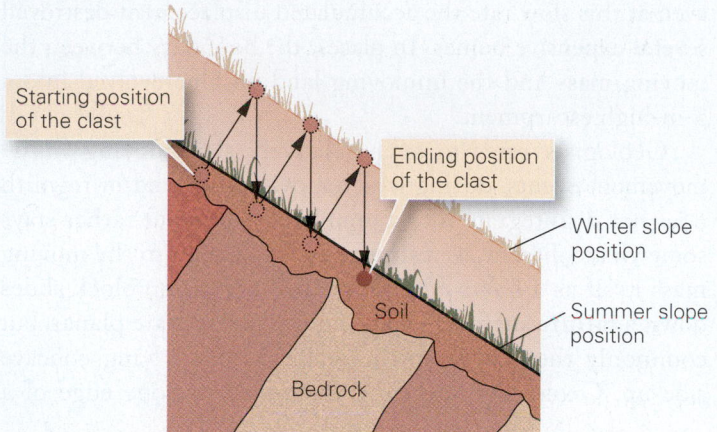

(a) Creep due to freezing and thawing: The clast rises perpendicular to the ground during freezing and sinks vertically during thawing. After 3 years, it migrates to the position shown.

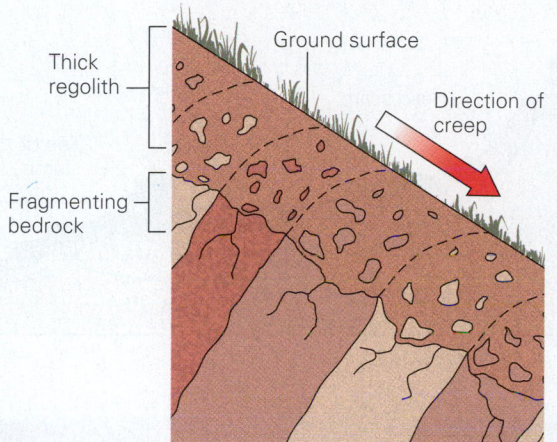

(b) As rock layers weather and break up, the resulting debris creeps downslope.

(c) Soil creep causes walls to bend and crack, building foundations to sink, trees to bend, and power poles and gravestones to tilt.

(d) Solifluction on a hillslope in the tundra.

(e) A rock glacier in Alaska. Note how the flow of the ice below wrinkles the rock layer on the glacier's surface.

Slumps

Near the Pacific Palisades, along the coast of southern California, Highway 1 runs between the beach and a 120-m-high cliff. On March 31, 1958, a gash developed about 200 m inland of the cliff's edge, and a semicoherent mass of sediment and rock began to move downslope. Four days later, when the movement finally stopped, a 1-km-long stretch of the coastal highway had been buried—it took weeks for bulldozers to make this stretch passable again. A similar event happened in northern New York State during the summer of 2011 when, after weeks of drenching rains, a 1.5-km-wide portion of a slope began to move down and out into the floor of Keene Valley. The mass

moved at only centimeters to tens of centimeters per day, but even at this slow rate the accumulated displacement destroyed several expensive homes. In places, the boundary between the moving mass and the unmoving land upslope evolved into a 5-m-high escarpment.

Geologists refer to such relatively slow-moving mass-movement events, during which moving rock and/or regolith does not disintegrate into a jumble of debris but rather stays somewhat coherent, as a **slump**, and they refer to the moving mass itself as a *slump block* (**Fig. 16.3**). A slump block slides down a **failure surface**—some failure surfaces are planar, but commonly they curve and resemble a spoon lying concave side up. Geologists refer to the exposed upslope edge of a

FIGURE 16.3 The process of slumping on a hillslope. Note the scarps that form at the head of the slump.

(a) A head scarp on the hillslope.

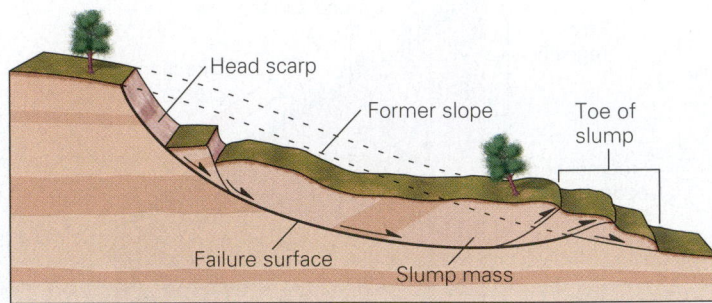

(b) Cross section of a slump.

(c) Slumping dumped sediment into this river in Costa Rica.

(d) A slump beginning to form along a highway in Utah.

failure surface as the **head scarp** and the downslope end of a slump block as the block's "toe." The upslope and downslope ends of a slump block may break into a series of discrete slices, each separated from its neighbor by a small sliding surface. On a hillslope, slices of the toe can move up and over the pre-existing land surface to form curving ridges. If the toe ends up on a beach or river bank, moving water eventually erodes it away. Slumps come in all sizes, from only a few meters across to tens of kilometers across. They move at speeds from millimeters per day to tens of meters per minute. Structures (such as houses, patios, and swimming pools) built on slump blocks or across the head scarp crack and fall apart, whereas those build at the toe may be knocked over or buried.

Mudflows, Debris Flows, and Lahars

Rio de Janeiro, Brazil, originally occupied only the flatlands bordering beautiful crescent beaches that had formed between steep hills. But in recent decades, the city's population has grown so much that in many places densely populated communities of makeshift shacks have been built on steep slopes. These communities, many of which have no storm drains, were built on the thick regolith that resulted from long-term weathering of bedrock in Brazil's tropical climate. Particularly heavy rains can saturate the regolith, transforming it into a viscous slurry of mud, resembling wet concrete, that flows downslope. The communities built on the regolith can disappear overnight, replaced by a muddle of mud and debris. And at the base of the cliffs, the flowing mud may knock over and bury buildings of all sizes. In unpopulated areas of Brazil, similar mass movements rip away forests (**Fig. 16.4a, b**). Geologists refer to a moving slurry of mud as a **mudflow**, or *mudslide*, and a slurry consisting of a mixture of mud mixed with larger, pebble- to boulder-sized fragments as a **debris flow**, or *debris slide* (**Fig. 16.4c; Box 16.1**).

Mudflows are not just phenomena of tropical regions—any slope underlain by poorly consolidated material can give way in a mudflow or debris flow during or following a heavy rain, and if people live nearby, the

FIGURE 16.4 Examples of mudflows and lahars.

(a) A 2011 mudslide destroyed a high-rise building at the base of the hill.

(b) Mudslides of 2011 stripped away forests on hillslopes in Brazil.

(c) A recent debris flow in Utah. Note the chaotic mixture of rock chunks and mud.

What Goes Up Must Come Down

Along the shore of California, as waves slowly cut into the land to produce low, flat areas called *wave-cut benches* (see Chapter 18), tectonic motions of this active plate boundary are constantly at work. As a result, the land surface slowly rises such that former wave-cut benches become small plateaus, or terraces. One such terrace lies at an elevation of 180 m above sea level, about 500 m east of the present-day beach at La Conchita. While the surface of the terrace is flat, the west face of this terrace has become a cliff-like bluff (**Fig. Bx16.1a**). Relatively little vegetation covers the bluff or the terrace above, so rain infiltrates into the ground, sinks down, and saturates clay and debris on the terrace and its bluff, making the material very weak. Every now and then, the weight of surface material causes the bluff to give way, and a mass of mud and debris flows downslope at rates of up to 10 m per second.

If the region of La Conchita were uninhabited, such mass wasting would just be part of the natural process of landscape evolution—gravity brings down land that had been raised by tectonic activity. But when downslope movements take place in La Conchita, it makes headlines because on the modern wave-cut bench between the shore and the base of the bluff developers built a community housing 350 people. In 1995, a mud and debris flow overwhelmed 9 houses at the base of the bluff. An even more devastating flow happened in 2005, burying 13 houses, damaging 23, and killing 10 people (**Fig. Bx16.1b, c**).

FIGURE Bx16.1 The 2005 La Conchita mudslide along the coast of California.

(a) A housing development was built in a narrow strip between the beach and steep cliffs.

(c) Rescuers at the toe of the mudslide.

(b) During heavy rains, the slope gave way and heavy mud flowed down, burying houses and taking several lives.

FIGURE 16.5 The Oso, Washington, mudslide of 2014.

(a) The slide surged over the river, damming it temporarily, and tragically buried a small community on the opposite bank.

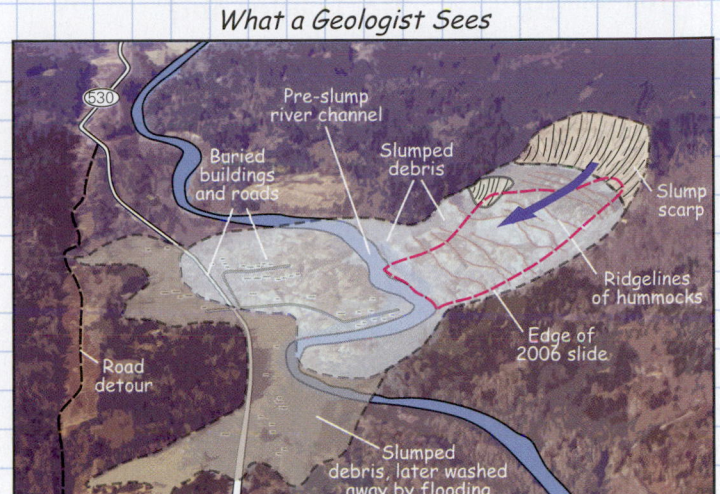

(b) The mudslide happened in an area where previous slides had taken place. The mountain slope was being undercut by the river.

movement has tragic consequences. In March 2014, near Oso, Washington, in the temperate northwest, a forested hillslope bordering a river gave way and within a few minutes buried a 2.5-sq-km (1-sq-mile) area of the river valley below with mud and wet sand (**Fig. 16.5**).

The speed at which material moves in a debris flow or mud flow depends on the slope angle and on the water content. Flows move faster if they contain more water so they are less viscous, and if they move on steeper slopes. On a gentle slope, drier mudflows move like molasses, but on a steep slope, very wet mud may move at over 100 km per hour. Because mud and debris flows have greater viscosity than clear water, they can carry large rock chunks as well as houses and cars. They typically follow channels downslope and at the base of the slope will spread out into a broad lobe.

Particularly devastating mudflows spill down the river valleys bordering volcanoes. These mudflows, known as **lahars**, form when volcanic ash from an ongoing or previous eruption mixes with water from the snow and ice that melts in a volcano's heat or from heavy rains (**Fig. 16.6**; see Chapter 9). A lahar occurred on November 13, 1985, in the Andes Mountains in Colombia. That night, a major eruption melted a volcano's thick snowcap, creating hot water that mixed with ash. A scalding lahar rushed down river valleys and swept over the nearby town of Armero while most inhabitants were asleep. Of the 25,000 residents, 20,000 perished.

FIGURE 16.6 Lahars develop when volcanic ash mixes with water from rain or melting snow and ice.

(a) A lahar that rushed down the side of Mt. St. Helens, Washington.

(b) Lahars can overrun populated areas far from the volcano.

Rockslides and Debris Slides

In the early 1960s, engineers built a huge new dam across a river on the northern side of Monte Toc in the Italian Alps to create a reservoir for generating electricity. This dam, the Vaiont Dam, was an engineering marvel, a concrete wall rising 260 m (as high as an 85-story skyscraper) above the valley floor (**Fig. 16.7a**). Unfortunately, the dam's builders did not appreciate the hazard posed by nearby Monte Toc. The side of Monte Toc facing the reservoir was underlain by a succession of limestone beds overlying a horizon of weak shale. These beds dipped parallel to the surface of the mountain and curved under the reservoir. As the reservoir filled, the flank of the mountain cracked, shook, and rumbled. Local residents began to call Monte Toc *la montagna che cammina* (the mountain that walks).

After several days of rain, Monte Toc began to rumble so much that on October 9, 1963, engineers lowered the water level in the reservoir. They thought the wet ground might slump a little into the reservoir, with minor consequences, so no one ordered the evacuation of the town of Longarone, a few kilometers down the valley below the dam. Unfortunately, the engineers underestimated the problem. At 10:30 that evening, a huge chunk of Monte Toc—600 million tons of rock—detached from the mountain and slid downslope along the weak shale horizon into the reservoir. Some debris rocketed up the opposite wall of the valley to a height of 260 m above the original reservoir level (**Fig. 16.7b**). The displaced water of the reservoir spilled over the top of the dam and rushed down into the valley below. When the flood had passed, nothing of Longarone and its 1,500 inhabitants remained. Though the dam itself still stands, it holds back only debris and has never provided any electricity.

Geologists refer to such a sudden movement of rock and debris down a nonvertical slope as a **rockslide** if the mass consists only of rock or as a **debris slide** if it consists mostly of regolith. Once a slide has taken place, it leaves a scar on the slope and forms a debris pile at the base of the slope. Slides happen when bedrock and/or regolith detaches from a slope, slips rapidly downhill on a failure surface, and breaks up into a chaotic jumble. Slides are more likely to occur where a weak layer of rock or sediment at depth below the ground parallels the land surface. (At the Vaiont Dam, the plane of weakness that would become the failure surface was a weak shale bed.) Slides may move at speeds of up to 300 km per hour—they are particularly fast when a cushion of air gets trapped beneath, in which case there is virtually no friction between the slide and its substrate, and the mass moves like a hovercraft. Rock and debris slides sometimes have enough momentum to climb the opposite side of the valley into which they fall. Slides, like slumps, come at a variety of scales. Most are small, involving blocks up to a few meters across. Some, such as the Vaiont slide, are large enough to cause a catastrophe.

Avalanches

In the winter of 1999, an unusual weather system passed over the Austrian Alps. First it snowed. Then the temperature warmed and the snow began to melt. But then the weather turned cold again, and the melted snow froze into a hard, icy crust. This cold snap ushered in a blizzard that blanketed the ice crust with tens of centimeters (1 to 2 ft) of new snow. With the frozen snow layer underneath acting as a failure surface, 200,000 tons of new snow began to slide down the mountain. As it accelerated, the mass transformed into a **snow avalanche**, a chaotic jumble of snow surging downslope. At the bottom of the slope, the avalanche overran a ski resort, crushing and carrying away buildings, cars, and trees and killing more than 30 people. It took searchers and their specially trained dogs many days to find buried survivors and victims under the 5- to 20-m-thick pile of snow that the avalanche deposited (**Fig. 16.8a**).

Snow avalanches display a variety of behaviors, depending on both the temperature of the snow and on the steepness of the slope down which the snow moves. Specifically, *wet-snow avalanches*, which involve snow that has started to melt and thus contains some liquid water, behave like a viscous slurry in that they hug the slope as they move and entrain relatively little air. Wet-snow avalanches generally travel at speeds of less than 30 km per hour and can pick up rock debris and vegetation along their path. *Dry-snow avalanches* contain cold, powdery snow that on tumbling down a steep slope disintegrates into a turbulent, air-rich cloud that rushes along at hurricane speeds of up to 250 km per hour (**Fig. 16.8b**). Both wet-snow and dry-snow avalanches can become powerful enough to flatten forests in their paths (**Fig. 16.8c**).

What triggers snow avalanches? Some happen when a *cornice*, a large drift of snow that builds up on the lee side of a windy mountain summit, suddenly gives way and falls onto slopes below where it knocks free additional snow. Others happen when a broad slab of snow on a moderate slope detaches

FIGURE 16.7 The Vaiont Dam disaster—a catastrophic landslide that displaced the water in a reservoir with rock debris.

(a) Before the landslide, the north flank of Mt. Toc was forested. When the reservoir filled, the slope became unstable. A shale bed a few hundred meters below the ground surface became a failure surface.

(b) Thirty-three million cubic meters of debris slid and displaced water in the reservoir. The water surged over the dam and swept away a village in the valley below.

from its substrate along an icy failure surface. Avalanches tend to affect the same localities year after year, because of the characteristics of snow buildup and because of the occurrence of *avalanche chutes*, shallow valleys running down the slope that channel the tumbling snow. To protect populated areas downslope of known avalanche chutes, experts may use special explosives to trigger small, controlled avalanches before the snow piles deep enough to become a dangerous hazard.

As we've noted, geologists commonly use the word *avalanche* in a broader sense for any mass movement during which the solid fragments are suspended in so much fluid (air or water) that the flowing mixture behaves like a turbulent cloud. Thus, a *debris avalanche* contains rock fragments and regolith mixed with air, and a *submarine avalanche* consists of sediment suspended in water. Avalanches of all types flow downslope because the mixture of solid and fluid in an avalanche is denser than the surrounding pure fluid—for this reason, geologists refer to various kinds of avalanches as *density currents*.

Rockfalls and Debris Falls

Rockfalls and debris falls, as their names suggest, occur when a mass free-falls from a cliff (**Fig. 16.9a, b**). Commonly, rockfalls happen when a body of rock separates from a cliff face along

a *joint*, a natural crack in rock across which a block of rock no longer connects to bedrock (see Chapter 11). Some joints are vertical, so after the rock fall, a new vertical cliff forms. Others, called exfoliation joints, are parallel to the hillslope. Most rockfalls involve only a few blocks detaching from a cliff face and dropping downslope. But some falls dislodge immense quantities of rock. In September 1881, a 600-m-high crag of slate, undermined by quarrying, suddenly collapsed onto the town of Elm in a valley of the Swiss Alps. Over 10 million cubic meters of rock fell to the valley floor, burying Elm and its 115 inhabitants to a depth of 10 to 20 m.

Friction and collision with other rocks may bring some blocks that have fallen to a halt before they reach the bottom of the slope; these blocks pile up to form a **talus**, a sloping apron or fan of rocks along the base of the cliff (**Fig. 16.9c**). Debris that has fallen a long way can reach speeds of 300 km per hour and may have so much momentum that it keeps moving as an avalanche-like cloud of fragments mixed with air when it reaches the base of a cliff. Large, fast rockfalls push the air in front of them, creating a blast of hurricane-like wind. For example, the wind in front of a 1996 rockfall in Yosemite National Park flattened over 2,000 trees.

Rockfalls happen fairly frequently along steep highway road cuts, leading to the posting of "falling-rock zone" signs

FIGURE 16.8 Examples of avalanches.

(a) Aftermath of a 1999 avalanche in the Austrian Alps. Masses of snow buried several homes.

(b) Avalanche chutes down the side of a mountain in the Canadian Rockies. Recent avalanches have flattened trees.

Origin of the avalanche

Toe of the avalanche

(c) A dry-snow avalanche in Alaska is a turbulent cloud.

Did you ever wonder...

why earthquakes don't occur everywhere?

(Fig. 16.9d). Such rockfalls take place because highway cuts are, effectively, new cliffs. In some cases, construction of the road cut weakens the material underlying the road cut, allowing frost wedging and root wedging to more easily break it free.

Submarine Mass Movements

So far, we've focused on mass movements that occur subaerially, for these are the ones we can see and are affected by most.

But mass wasting also happens underwater. The sedimentary record contains abundant evidence of submarine mass movements, because after they take place they tend to be buried by younger sediments and are preserved.

Geologists distinguish three types of submarine mass movements, according to whether the mass remains coherent or disintegrates as it moves (**Fig. 16.10**). In **submarine slumps**, semicoherent blocks slip downslope on weak detachments—the layers constituting the blocks become contorted as they move, like a tablecloth that slides off a table. In **submarine debris flows**, the moving mass breaks apart to form a slurry containing larger clasts (pebbles to boulders) suspended in a mud matrix. And in **turbidity currents**, sediment disperses in water to create a turbulent cloud of suspended sediment that rushes downslope as an avalanche. Turbidity currents commonly flow down submarine canyons—their movement, in fact, erodes the seafloor so it contributes to the formation of the canyon. When the turbidity current starts to slow, its entrained sediment settles out in a sequence, with coarser grains at the base and finer grains at the top. The resulting deposits, therefore, consist of

FIGURE 16.9 Examples of rockfalls.

(a) Successive rockfalls have littered the base of this sandstone cliff with boulders. Note the talus at the base of the cliff.

(c) A boulder apron in the Canadian Rockies; boulders fall from the high cliffs and accumulate in the apron.

(b) A rockslide buried the forest bordering a lake in the Uinta Mountains, Utah. Fresh rock exposed by the slide has a lighter color.

(d) A rockfall along a highway in Vermont. Note that the blocks separated from the wall along joints.

graded beds (see Chapter 7)—typically, the deposits accumulate in a fan at the mouth of a submarine canyon.

In recent years, geologists have used satellites as well as ship-board *side-scan sonar* (sonar that analyzes a 60-km-wide swath of the seafloor, instead of just a line, as it moves) to map out the extent of submarine landslides (**Fig. 16.11a**). The shapes of slumps and landslide deposits stand out on the resulting new generation of high-resolution bathymetric maps. Geologists have found that submarine slopes bordering both hot-spot volcanoes and active plate boundaries are scalloped by immense slumps, because tectonic activity jars these areas with earthquakes that set masses of material in motion. For example, slumps up to 200 km long and 100 km wide have substantially modified the flanks of the Hawaiian Islands (**Fig. 16.11b**). Some

of the slumping events even carried away large chunks of the subaerial parts of the islands—in fact, the steep portions of the islands' coasts are the head scarps of huge slumps. Studies suggest that huge slumping events off Hawaii happen, on average, about once every 100,000 years. Significantly, passive-margin coasts are not immune to slumping, and immense slumps have been mapped along the coasts of the Atlantic Ocean.

Since a submarine slump can develop fairly quickly, and since its movement can displace a large area of the seafloor, it can trigger a tsunami. A submarine slump set in motion by a 1998 earthquake in Papua New Guinea generated a tsunami that devastated a 40-km-long stretch of coast and killed 2,100 people. A prehistoric tsunami triggered by a slump off the coast of Norway left its trace all around the North Sea (**Box 16.2**).

FIGURE 16.10 Submarine mass movements.

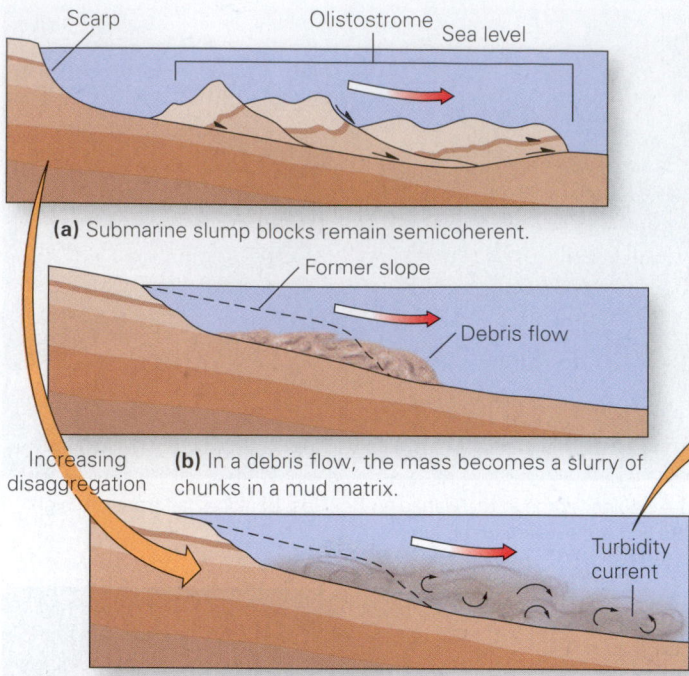

(a) Submarine slump blocks remain semicoherent.

(b) In a debris flow, the mass becomes a slurry of chunks in a mud matrix.

Increasing disaggregation

A laboratory model of a turbidity current. The cloud consists of fine clay suspended in water.

(c) A turbidity current is a cloud of sediment suspended in water; it flows near the seafloor because it is denser than clear water.

FIGURE 16.11 Examples of huge submarine slumps and debris flows.

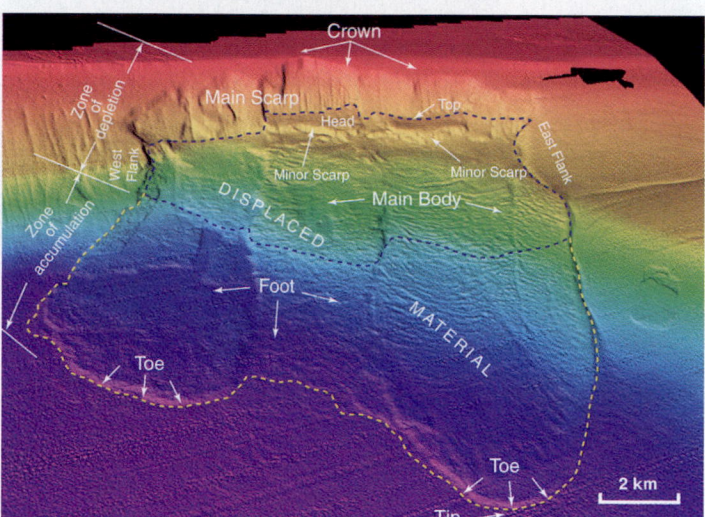

(a) A digital bathymetric map of a slip along the coast of California. The parts of the slump are labeled.

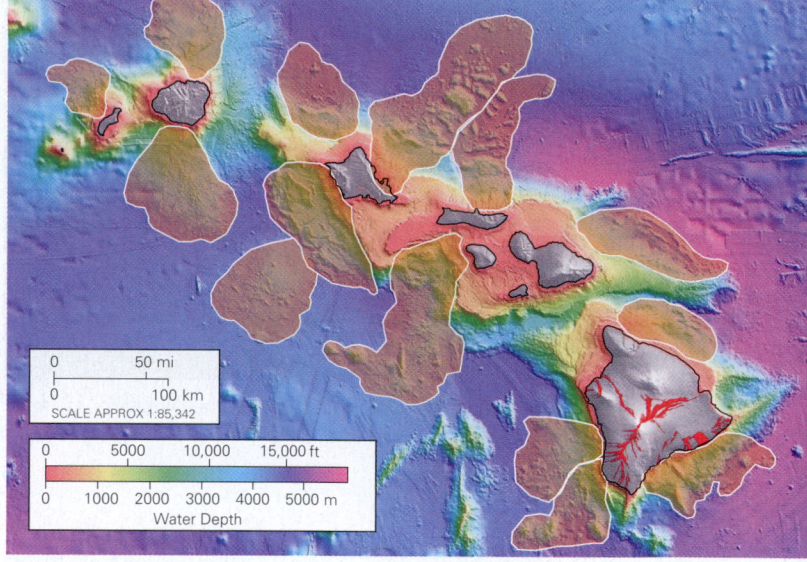

(b) A bathymetric map of the area around Hawaii shows several huge slumps, shaded in tan.

Take-Home Message

Mass movements differ from one another based on speed and character. Creep, slumping, and solifluction are slow. Mudflows and debris flows move faster, and avalanches and rockfalls move the fastest. Movements occur on land and underwater.

QUICK QUESTION: In what way is a snow avalanche like a turbidity current?

16.3 Why Do Mass Movements Occur?

We've seen that mass movements travel at a range of different velocities, from slow (creep) to faster (slumps, mud and debris flows, and rock and debris slides) to fastest (snow avalanches and rock and debris falls; see **Geology at a Glance,** pp. 602–603).

BOX 16.2   CONSIDER THIS . . .

The Storegga Slide and North Sea Tsunamis

The Firth of Forth, a long inlet of the North Sea, forms the waterfront of Edinburgh, Scotland. At its western end, it merges with a broad plain in which mud and peat have been accumulating since the last ice age. Around 1865, geologists investigating this sediment discovered an unusual layer of sand containing bashed seashells, marine plankton, and torn-up fragments of substrate. This sand layer lies sandwiched between mud layers at an elevation of up to 4 m above the high-tide limit and 80 km inland from the shore. How could the sand layer have been deposited? The mystery baffled geologists for many decades. During this time, layers of shelly sand, similar to the one found in Scotland, were discovered at many other localities along both sides of the North Sea (**Fig. Bx16.2**). In some cases, the layer occurred 20 m above the high-tide limit. Geologists studying the coast also found locations where coastal cliffs appeared to have been eroded by wave action at elevations well out of reach of normal storm waves.

While land-based geologists puzzled about these unusual sand beds and coastal erosional features, marine geologists investigating the continental shelf off the western coast of Norway discovered a region of very irregular sea floor underlain with a jumble of chaotic blocks, some of which are 10 km by 30 km across and 200 m thick. When mapped out, they indicate that a 290-km-long sector of the continental shelf had collapsed in a series of submarine slides that together involve about 5,580 cubic km of debris—the overall feature is called the Storegga Slide.

Further studies show that the Storegga Slide formed during three movement events: one occurred 30,000 years ago, the second about 7,950 years ago, and the third about 6,000 years ago. Now the pieces of the puzzle were in place. Immense submarine slides displace enough ocean water to create tsunamis. The wave produced by the second Storegga Slide may coincide with the disappearance of Stone Age tribes along the North Sea coast, suggesting that the tribes were effectively washed away.

If such a calamity happened in the past, could it happen again, with submarine-slide-generated tsunamis inundating coastal cities with large populations? Geologists now realize that submarine landslides can trigger tsunamis every bit as devastating as earthquake- and volcano-generated tsunamis. A slide in 1929 along the coast of Newfoundland, for example, not only created a turbidity current that broke the trans-Atlantic telephone cable but also generated tsunamis that washed away houses and boats along the coast of Newfoundland at elevations of up to 27 m above sea level. With a bit of looking, geologists have found huge boulders flung by tsunamis onto the land, layers of sand and gravel deposited well above the high-tide limit, and erosional features high up on shoreline cliffs along many coastal areas, even in areas (such as the Bahamas and southeastern Australia) far from seismic or volcanic regions. And submarine mapping shows that many large slumps occur all along continental shelves.

FIGURE Bx16.2   Map of the North Sea region showing the location of sites (red dots) where marine sand layers occur significantly above the high-tide limit. These were caused by tsunamis generated by movement of the Storegga Slide. The map shows the estimated position of the tsunamis at 2 hours, 4 hours, and 6 hours.

FIGURE 16.12 Jointing broke up this thick sandstone bed along a cliff in Utah. Blocks of sandstone break free along joints and tumble downslope.

The answer comes from looking at the strength of the attachments holding materials together. A mass of intact bedrock is relatively strong because the chemical bonds within its interlocking grains, or within the cements between grains, can't be broken easily. Weathered rock tends to be weaker, because strong bonds between the original grains of the rock have been replaced by weaker bonds between weathering products, such as clays and iron oxides. Regolith is relatively weak because the grains are held together only by friction (caused by roughness along the contact between two grains), electrostatic attraction (the weak force that develops between clay flakes because the surfaces of the flakes are charged), and/or the surface tension of water (the weak force caused by the attraction of water molecules to one another). All of these forces combined are weaker than chemical bonds holding together the atoms in the minerals of intact rock. To picture this contrast, think how much easier it is to bust up a sand castle (whose strength comes primarily from the surface tension of water films on the sand grains) than it is to bust up a granite sculpture of a castle.

The velocity depends on the steepness of the slope and the water or air content of the mass. Why do mass movements take place? The stage must be set by the following phenomena: (1) fracturing and weathering of the substrate, which weakens the substrate so it cannot hold up against the pull of gravity; (2) the development of relief, which provides slopes down which masses move; and (3) an event that sets mass in motion. Let's look at these phenomena more closely.

Weakening the Substrate: Fragmentation and Weathering

If the Earth's surface were covered by completely unfractured rock, mass movements would be of little concern, for intact rock has great strength and could form stalwart mountain faces that would not tumble. But, in reality, the rock of the Earth's upper crust has been fractured by jointing and faulting and has undergone chemical weathering. Also, in many locations long-term weathering and/or deposition has covered bedrock with regolith (sediment or soil). Regolith and fractured rock are much weaker than intact rock and can indeed collapse in response to gravitational pull (**Fig. 16.12**). Thus jointing, faulting, and weathering ultimately make mass movements possible.

Why are regolith and fractured rocks weaker than intact bedrock?

Slope Stability

Mass movements do not take place on all slopes, and even on slopes where such movements are possible, they occur only occasionally. Geologists distinguish between *stable slopes*, on which sliding is unlikely, and *unstable slopes*, on which sliding will likely happen. When material starts moving on an unstable slope, we say that **slope failure** has occurred. Whether a slope fails or not depends on the balance between two forces—the *downslope force*, caused by gravity, and the *resistance force*, which inhibits sliding. If the downslope force exceeds the resistance force, the slope fails and mass movement results.

Let's examine the battle between downslope forces and resistance forces more closely by imagining a block sitting on a slope. We can represent the gravitational attraction between the

FIGURE 16.13 Forces that trigger downslope movement.

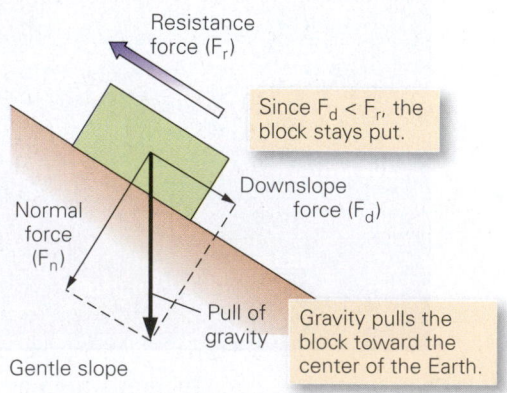

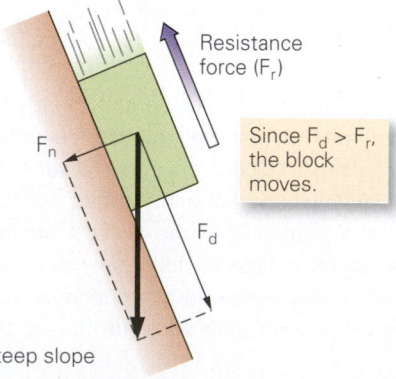

(a) Gravity can be divided into a normal force and a downslope force. If the resistance force, caused by friction, is greater than the downslope force, the block does not move.

(b) If the slope angle increases, the downslope force due to gravity increases. If the downslope force becomes greater than the resistance force, the block starts to move.

block and the Earth by an arrow (a vector) that points straight down, toward the Earth's center of gravity. This arrow can be separated into two components—the downslope force parallel to the slope and the normal force perpendicular to the slope. Note that for a given mass the magnitude of the downslope force increases as the slope angle increases, so downslope forces are greater on steeper slopes. We can symbolize the resistance force by an arrow pointing uphill. (As we have seen, resistance force comes from chemical bonds, electrostatic attraction, friction, and surface tension.) If the downslope force is larger than the resistance force, then the block moves—otherwise, it stays in place (**Fig. 16.13**).

Because of resistance force, granular debris tends to pile up to produce the steepest slope it can without collapsing. The angle of this slope is called the **angle of repose**, and for most dry, unconsolidated materials (such as dry sand) it generally has a value of between 30° and 37°. The angle depends partly on the shape and size of grains, which determine the amount of friction across grain boundaries. For example, steeper angles of repose (up to 45°) characterize talus slopes composed of large, irregularly shaped grains (**Fig. 16.14**).

In many locations, the resistance force is less than might be expected because a weak surface exists at some depth below ground level. If downslope movement begins on the weak surface, the weak surface becomes a failure surface. Geologists recognize several different kinds of weak surfaces that are likely to become failure surfaces (**Fig. 16.15**). These include wet clay layers; wet, unconsolidated sand layers; joints; weak bedding planes, such as shale beds or evaporite beds; and metamorphic foliation planes.

Weak surfaces that dip parallel to the land surface slope are particularly likely to become failure surfaces. An example of such failure occurred in Madison Canyon, southwestern Montana, on August 17, 1959. That day, shock waves from a strong earthquake jarred the region. Metamorphic rock with a strong foliation formed the bedrock of the canyon's southern wall. When the ground vibrated, rock detached along a foliation plane and tumbled downslope. Unfortunately, 28 campers lay sleeping on the valley floor. They were probably awakened by the hurricane-like winds blasting in front of the moving mass but seconds later were buried under 45 m of rubble.

FIGURE 16.14 The angle of repose is the steepest slope that a pile of unconsolidated sediment can have and remain stable. The angle depends on the shape and size of grains.

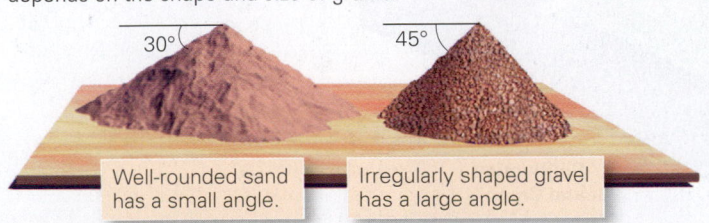

30°

45°

Well-rounded sand has a small angle.

Irregularly shaped gravel has a large angle.

FIGURE 16.15 Different kinds of weak surfaces can become failure surfaces.

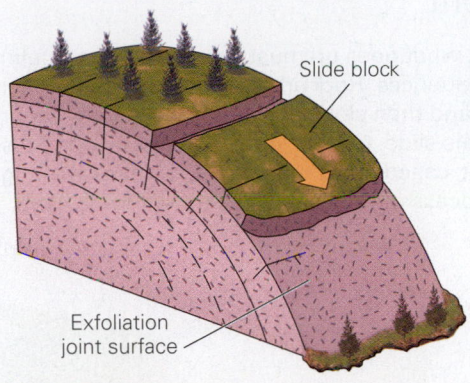

(a) Exfoliation joints form parallel to slope surfaces in granite and become failure surfaces.

Slide block

Exfoliation joint surface

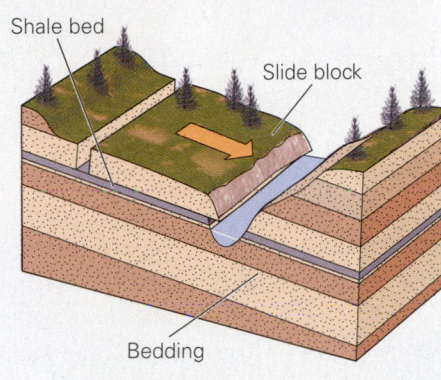

(b) In sedimentary rocks, bedding planes (particularly in weak shale) become failure surfaces.

Shale bed

Slide block

Bedding

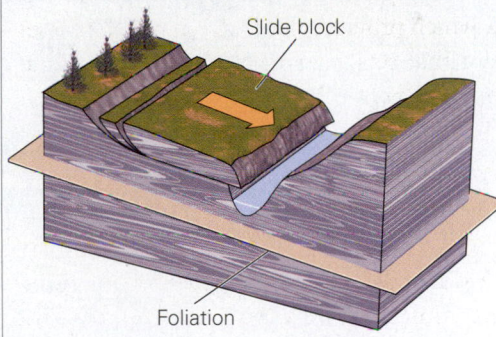

(c) In metamorphic rock, foliation planes (particularly in mica-rich schist) become failure surfaces.

Slide block

Foliation

Fingers on the Trigger: What Causes Slope Failure?

What triggers an individual mass-wasting event? In other words, what causes the balance of forces to change so that the downslope force exceeds the resistance force and a slope suddenly fails? Here we look at various phenomena—natural and human-made—that trigger slope failure.

Shocks, Vibrations, and Liquefaction Earthquake tremors, storms, the passing of large trucks, or blasting in construction sites may cause a mass that was on the verge of moving actually to start moving. For example, an earthquake-triggered slide dumped debris into southeastern Alaska's Lituya Bay in 1958. The debris displaced the water in the bay,

Mass Movement

In Earth's gravity field, what goes up must come down—sometimes with disastrous consequences. Rock and regolith are not infinitely strong, so every now and then slopes or cliffs give way in response to gravity, and materials slide, tumble, or career downslope. This downslope movement, called mass movement, or mass wasting, is the first step in the process of erosion and sediment formation.

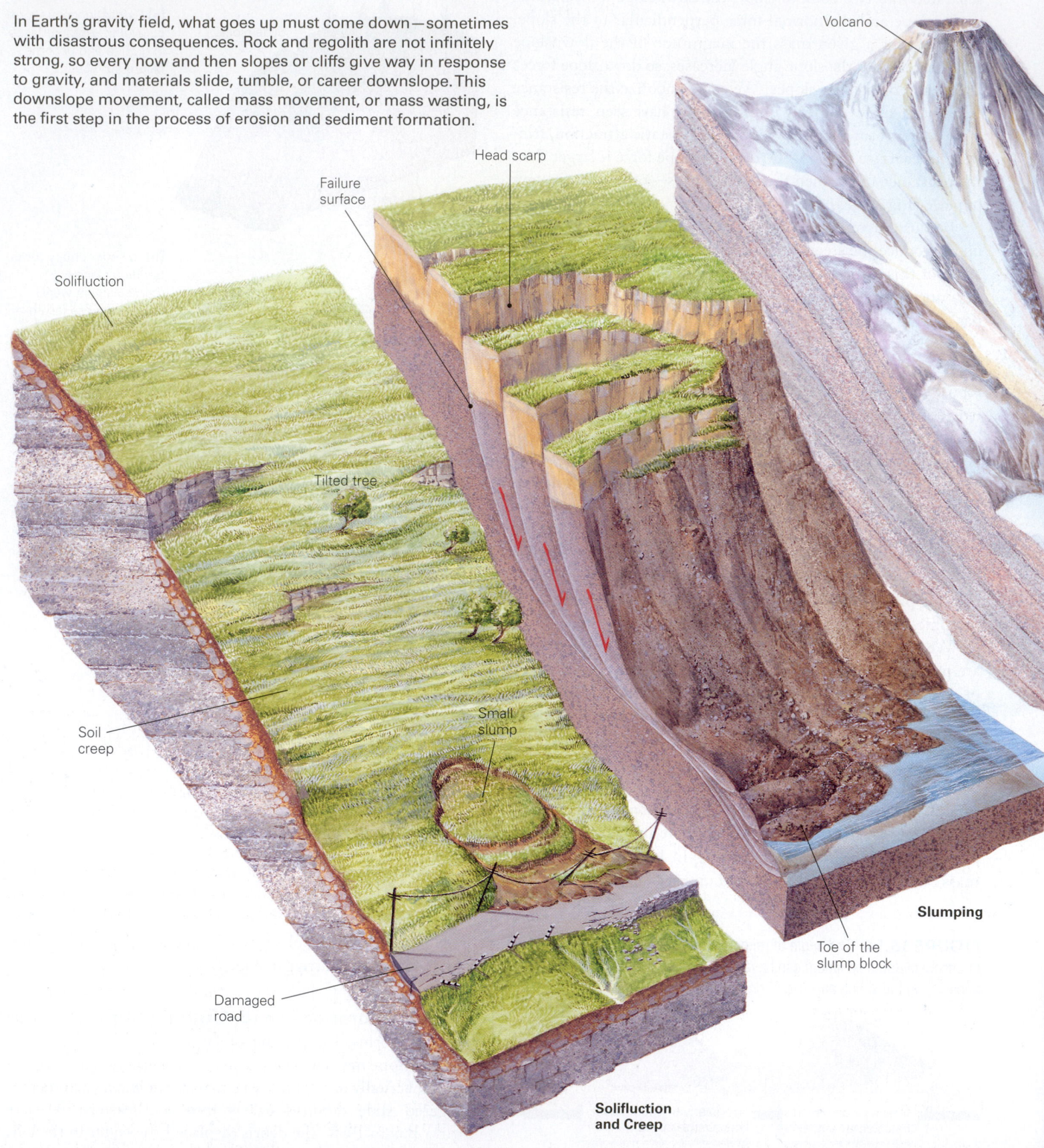

Volcano

Head scarp

Failure surface

Solifluction

Tilted tree

Soil creep

Small slump

Damaged road

Slumping

Toe of the slump block

Solifluction and Creep

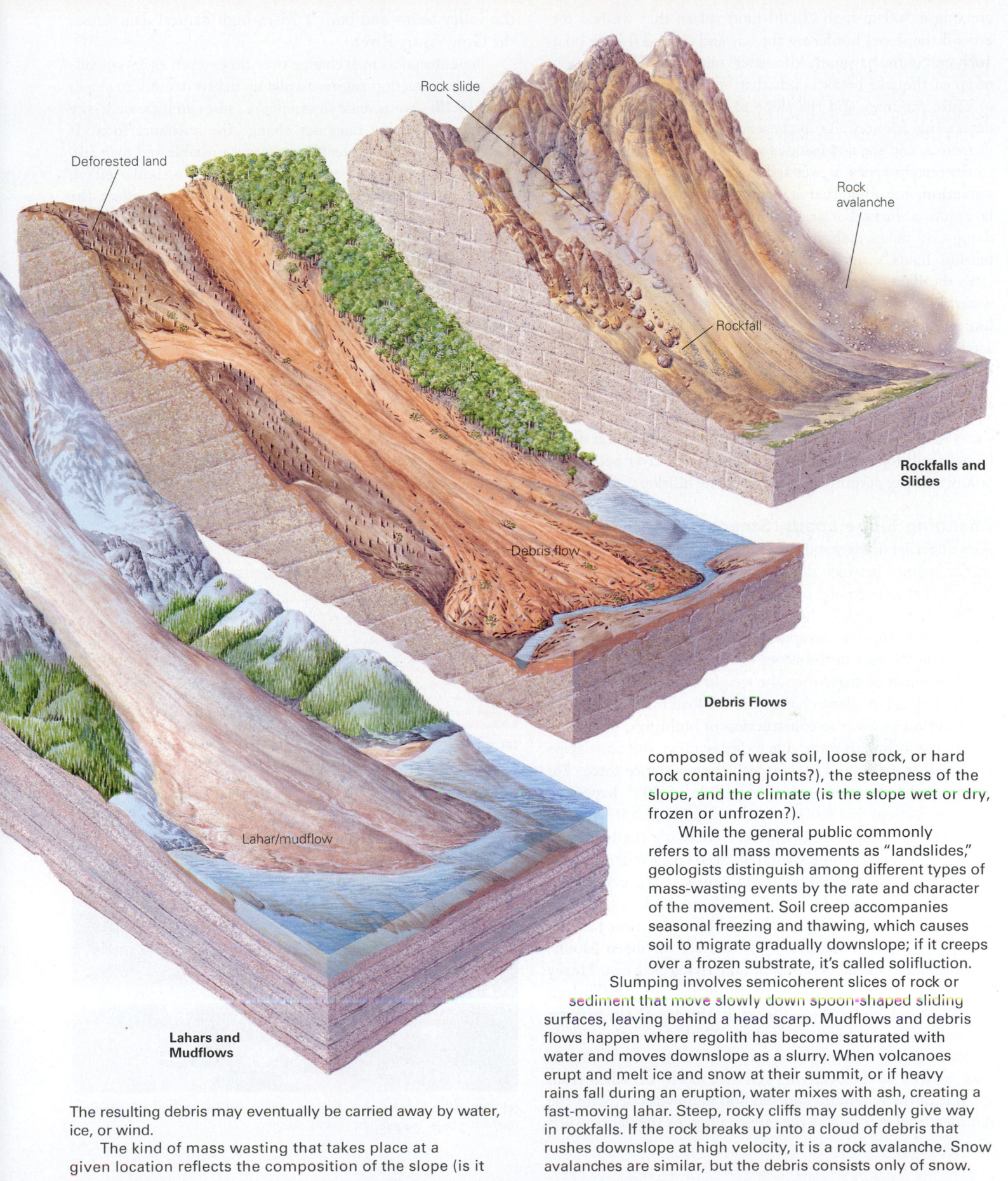

Deforested land

Rock slide

Rock avalanche

Rockfalls and Slides

Rockfall

Debris flow

Debris Flows

Lahar/mudflow

Lahars and Mudflows

composed of weak soil, loose rock, or hard rock containing joints?), the steepness of the slope, and the climate (is the slope wet or dry, frozen or unfrozen?).

While the general public commonly refers to all mass movements as "landslides," geologists distinguish among different types of mass-wasting events by the rate and character of the movement. Soil creep accompanies seasonal freezing and thawing, which causes soil to migrate gradually downslope; if it creeps over a frozen substrate, it's called solifluction. Slumping involves semicoherent slices of rock or sediment that move slowly down spoon-shaped sliding surfaces, leaving behind a head scarp. Mudflows and debris flows happen where regolith has become saturated with water and moves downslope as a slurry. When volcanoes erupt and melt ice and snow at their summit, or if heavy rains fall during an eruption, water mixes with ash, creating a fast-moving lahar. Steep, rocky cliffs may suddenly give way in rockfalls. If the rock breaks up into a cloud of debris that rushes downslope at high velocity, it is a rock avalanche. Snow avalanches are similar, but the debris consists only of snow.

The resulting debris may eventually be carried away by water, ice, or wind.

The kind of mass wasting that takes place at a given location reflects the composition of the slope (is it creating a 300-m-high (1,200-foot) splash that washed forests off the slopes bordering the bay and carried fishing boats anchored in the bay many kilometers out to sea. The vibrations of an earthquake break bonds that hold a mass in place and/ or cause the mass and the slope to separate slightly, thereby decreasing friction. As a consequence, the resistance force decreases, and the downslope force sets the mass in motion.

In certain types of wet sediment, shaking can cause **liquefaction**, meaning that it turns what seems to be a solid layer into a slurry. For example, **quick clay**, which consists of damp clay flakes, behaves like a solid when still, for surface tension holds water-coated flakes together—shaking separates the flakes from one another and suspends them in the water, thereby transforming the clay into wet mud that flows like a fluid (**Fig. 16.16**). And in wet sand, shaking causes the sand grains to shift slightly relative to one another, which then causes the water pressure in pores between the grains to increase, thus pushing the grains apart such that they become suspended in water. As a result, the wet sand becomes a sandy slurry with virtually no strength. If a layer of sediment at depth below a hillslope undergoes liquefaction, the layer becomes a failure surface, permitting collapse of the hillslope.

Changing Slope Loads, Steepness, and Support

As we have seen, the stability of a slope at a given time depends on the balance between downslope force and resistance force. Factors that change one or the other of these forces can lead to failure. Examples include changes in slope loads, failure-surface strength, slope steepness, and the support provided by material at the base of the slope.

Slope loads change when the weight of the material above a potential failure plane changes. If the load increases, due to the addition of fill, the construction of buildings, or saturation of regolith with water due to heavy rains, the downslope force will increase and may exceed the resistance force. For example, the devastating Oso mudslide of 2014 happened after a 6-week period during which rainfall was 200% greater than average, so the sandy ground underlying the hillslopes of the region had become saturated. Seepage of water into the ground may also trigger failure in bedrock. An example of such failure triggered the huge Gros Ventre Slide, which took place in 1925 on the flank of Sheep Mountain, near Jackson Hole, Wyoming (**Fig. 16.17**). The surface of Sheep Mountain is parallel to the bedding in underlying bedrock. Heavy rains saturated the permeable sandstone beds of the Tensleep Formation, which lay above the weak Amsden Shale, making the bedrock much heavier. On June 27, the downslope force exceeded the restraining force, and 40 million cubic meters of rock, as well as the overlying soil and forest, detached from the side of the mountain and slid 600 m downslope, with the Amsden Shale serving as the failure surface. The debris filled the valley below and built a 75-m-high natural dam across the Gros Ventre River.

Slope steepness may change over time, when rivers cut valleys, or construction engineers pile up debris or cut into slopes (**Fig. 16.18**). An increase in steepness causes an increase in the downslope force but does not change the resistance force. If the slope becomes too steep, it becomes unstable and may fail. Removing support at the base of a slope has a similar effect, and plays a major role in triggering slope failures. In effect, the material at the base of a slope acts like a retaining wall or "dam" holding back the material farther up the slope. When natural erosion, or bulldozer excavations, takes away this "dam," the upslope material can start to move. Such a process contributed to setting the stage for both Oso Mudslide (see Fig. 16.5), and the Gros Ventre Slide—in both cases, the river flowing at the base of what would become the slide had been cutting into the base of the hill. Cutting terraces into hillslopes for road building can have the same effect, as can wave erosion at the base of a slope along a coast. In some cases, erosion by a river or by waves eats into the base of a cliff and produces an overhang. When such **undercutting** has occurred, rock making up the overhang eventually breaks away from the slope and falls (**Fig. 16.19**).

FIGURE 16.16 Quick clay mudslides.

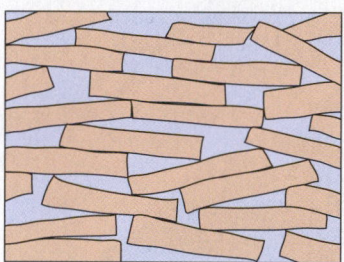

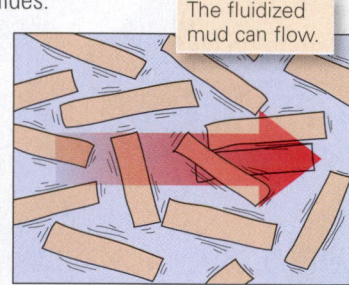

The fluidized mud can flow.

(a) Before shaking, clay flakes stick together.

(b) During shaking, clays separate and become suspended in water.

(c) A mudslide near Namsos, Norway, destroyed ten houses. Movement may have been triggered by nearby blasting.

FIGURE 16.17   Stages leading to the 1925 Gros Ventre Slide in Wyoming.

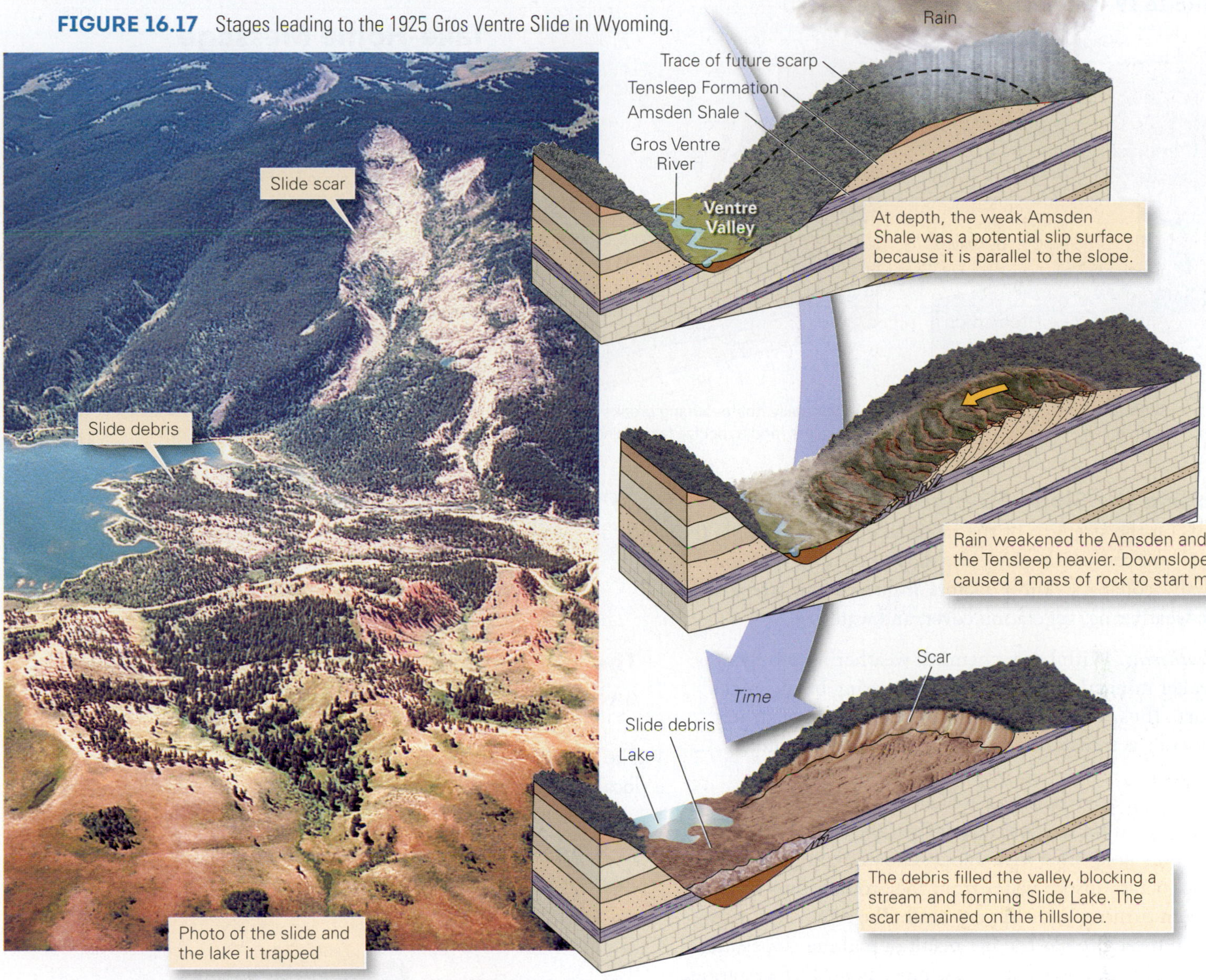

Slide scar

Slide debris

Photo of the slide and the lake it trapped

Rain

Trace of future scarp

Tensleep Formation

Amsden Shale

Gros Ventre River

Ventre Valley

At depth, the weak Amsden Shale was a potential slip surface because it is parallel to the slope.

Rain weakened the Amsden and made the Tensleep heavier. Downslope force caused a mass of rock to start moving.

Time

Scar

Slide debris

Lake

The debris filled the valley, blocking a stream and forming Slide Lake. The scar remained on the hillslope.

FIGURE 16.18   Processes that can steepen slopes and make them unstable.

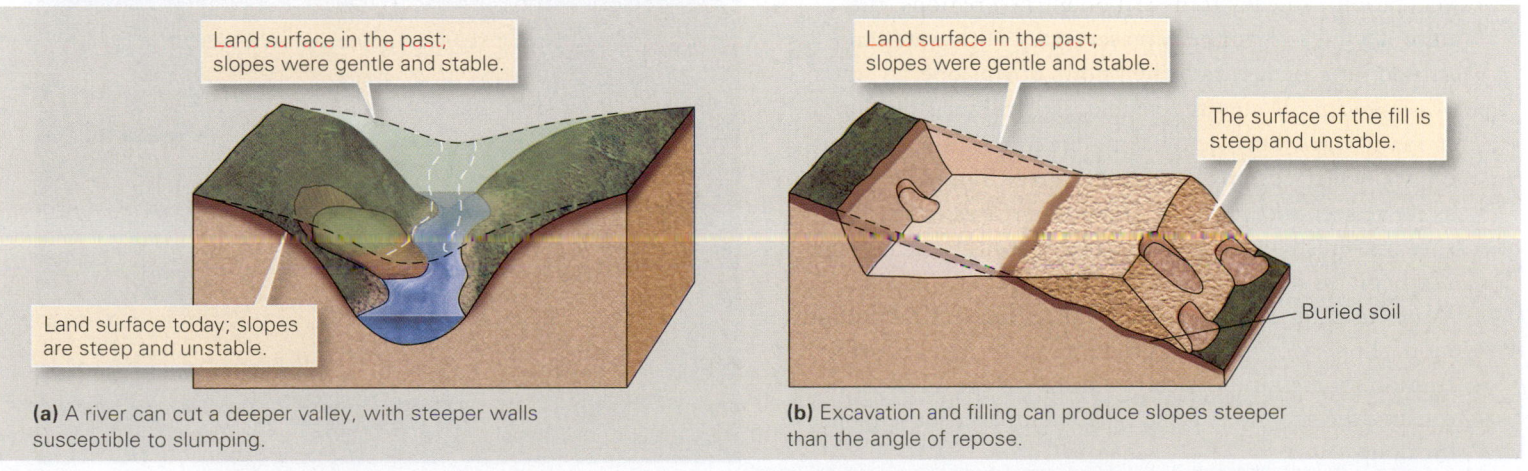

Land surface in the past; slopes were gentle and stable.

Land surface today; slopes are steep and unstable.

Land surface in the past; slopes were gentle and stable.

The surface of the fill is steep and unstable.

Buried soil

(a) A river can cut a deeper valley, with steeper walls susceptible to slumping.

(b) Excavation and filling can produce slopes steeper than the angle of repose.

FIGURE 16.19 Undercutting and collapse of a sea cliff.

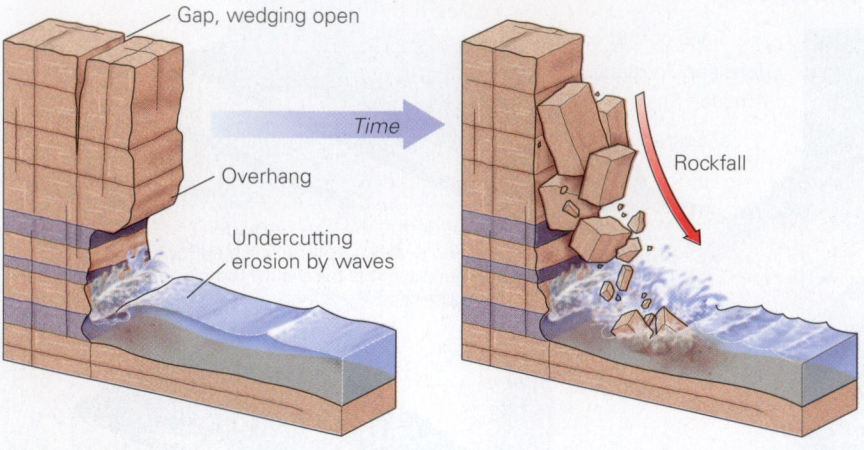

Gap, wedging open

Overhang

Undercutting erosion by waves

Time

Rockfall

(a) Undercutting by waves removes the support beneath an overhang.

(b) Eventually, the overhang breaks off along joints, and a rockfall takes place.

Take-Home Message

Weathering and fragmentation weaken slope materials and make them more susceptible to mass movement. The stability of a slope reflects the relative sizes of downslope force and resistance force. Failure occurs when downslope force exceeds resistance force, perhaps due to shocks, changing slope angles and strength, changing water content, and changing slope support.

QUICK QUESTION: If the slope of a sand pile is less than the angle of repose, will the slope of the pile fail?

Changing the Slope Strength The stability of a slope depends on the strength of the material constituting it. If the material weakens with time, the slope becomes weaker and eventually collapses. Three factors influence the strength of slopes: weathering, vegetation cover, and water:

- *Weathering:* With time, chemical weathering produces weaker minerals, and physical weathering breaks rocks apart. Thus, a formerly intact rock composed of strong minerals is transformed into a weaker rock or into regolith.
- *Vegetation cover:* In the case of slopes underlain by regolith, vegetation tends to strengthen the slope because the roots hold otherwise unconsolidated grains together. Also, plants absorb water from the ground, thus keeping it from turning into slippery mud. The removal of vegetation therefore has the net result of making slopes more susceptible to downslope mass movement. Deforestation, for example, can lead to catastrophic mass wasting of the forest's substrate (**Fig. 16.20**). And terrifying wildfires, stoked by strong winds, can destroy ground-covering vegetation. If heavy rains fall on burned regions, the unprotected soil can became saturated with water and turned into mud, which then flows downslope.
- *Water content:* Water affects materials comprising slopes in many ways. Surface tension, due to the film of water on grain surfaces, may help hold regolith together. But, as we've seen, if the water content increases, water pressure may push grains apart so that regolith liquefies and can begin to flow or may make the substrate heavier. Water infiltration may make weak surfaces underground more slippery or may push surfaces apart and decrease friction. Some kinds of clays absorb water and expand, causing the ground surface to rise and, as a consequence, break up.

16.4 Where Do Mass Movements Occur?

The Importance of Relief, Climate, and Substrate

The single most important factor in determining whether a locality is susceptible to mass movements is the relief (elevation

FIGURE 16.20 Deforestation decreases slope stability. A large slump has formed on this deforested hill in Brazil.

Slump scar

difference) of a region, for mass wasting does not occur without slopes. Further, regions with steeper slopes tend to be more susceptible to mass movements than are regions with gentle slopes. Thus, locations where rivers or glaciers have cut valleys, mountains have risen, or ocean waves have eaten into the shores to form sea cliffs are places where mass movements are more likely.

Slope is not the only factor affecting the character and frequency of mass movements, though. Climate also affects mass movements. For example, regions with heavy seasonal rainfalls are subject to mudflows and landslides, especially if fire or deforestation has removed vegetation. In these regions, the climate has led to deep weathering and weakening of the substrate, and rainfalls increase slope loads and weaken failure surfaces. Desert and alpine regions, where slopes tend to be underlain by unweathered bedrock, tend to host rare but dramatic rock falls.

The Importance of the Tectonic Setting

Most unstable ground on Earth ultimately owes its existence to the activity of plate tectonics. As we've seen, plate tectonics causes uplift, generates relief, and causes faulting, which fragments the crust. And, of course, earthquakes on plate boundaries trigger devastating landslides. Spend a day along the steep slopes of the Southern Alps, a mountain range now uplifting along plate boundary that transects New Zealand, and you can hear mass movement in progress—during heavy rains, rockfalls and landslides clatter with astounding frequency, as if the mountains were falling down around you.

A Case Study: Southern California

To see the interplay of plate tectonics and other factors, let's consider a case study in southern California. Contractors have built expensive homes on cliffs overlooking the Pacific so that homeowners can enjoy spectacular sunset views from their backyard patios. But the landscape is not ideal from the standpoint of stability. Slumps and mudflows on coastal cliffs have consumed some of these homes over the years, with a cost to their owners (or insurance companies) of millions of dollars. What is special about southern California that makes it so susceptible to mass wasting?

First, California lies along an active plate boundary, the San Andreas fault. Repeated slip events along the fault over millions of years have shattered the rock of California's crust. Fractures not only act as planes of weakness but can provide paths for water to seep into bedrock and cause chemical weathering, producing slippery clay that weakens the rock. Also, the rocks in many areas are weak to begin with

Did you ever wonder . . .

why parts of the California coast slip into the sea?

because they formed as part of an accretionary prism, a chaotic mass of sediment that was scraped off subducting oceanic lithosphere during the Mesozoic Era.

Though most of the movement on the boundary between the North American and Pacific plates involves strike-slip displacement, there is a component of compression across the boundary. This compression leads to uplift and slope formation in California (see Box 16.1). Since the uplifted region borders the coast, wave erosion steepens and, in some places, undercuts cliffs. And because it is a plate boundary, numerous earthquakes rock the region, thus shaking regolith loose.

California is also susceptible to mass movements because of its climate. In general, the region is hot and dry and thus supports only semidesert flora. Brush fires remove much of this cover, leaving large areas with no dense vegetation. But since the region lies on the Pacific coast, it endures occasional heavy winter rains. The water sinks quickly into the sparsely vegetated ground, adds weight to the mass on the slope, and weakens failure surfaces.

Development of cities and suburbs also contributes to triggering mass movements in southern California. Development has oversteepened and overloaded slopes and has caused the water content of regolith to change. The consequences can be seen in the Portuguese Bend Slide of the Palos Verdes area near Los Angeles (**Fig. 16.21**). The Portuguese Bend region is underlain with a thick, seaward-dipping layer of weak volcanic ash (now altered to weak clay) resting on shale. The weak ash has served as a failure plane a few times during the past few thousand years. In 1956, developers deposited a 23-m-thick layer of fill over the ground surface and built homes on top. Residents began to water their lawns and to use septic tanks that were susceptible to leaking. The water seeped into the ground and decreased the

SEE FOR YOURSELF . . .

Slumps near San Francisco

LATITUDE
37°40'47.33"N

LONGITUDE
122°29'46.14"W

Zoom to 2.5 km (~8200 ft.) and rotate the view so north is on the left (top photo). Then tilt, to look obliquely east (bottom photo) from 550 m.

You are seeing the coast near San Francisco, where waves erode sea cliffs at the base of a terrace. Occasionally, slumps along the cliffs carry debris to the beach. The bowls cut into the terrace are places where slumps have occurred in the past.

strength of the ash layer. Because of the decrease in strength, the added weight, and the erosion of the toe of the hill by the sea, the upper 30 m of land began to move once again. Between 1956 and 1985, the Portuguese Bend Slide moved at rates of up to 2.5 cm per day. Eventually, portions of a 260-acre region slid by over 200 m, and in the process more than 150 homes were destroyed.

Take-Home Message

Mass movement can occur where slopes have developed, particularly if the substrate has undergone weathering and fragmentation, for these processes weaken slope materials. Failure occurs when the downslope force exceeds resistance force due to shocks (such as earthquakes), changing slope angles and strength, changing water content, and changing slope support. Climate can impact the susceptibility of a region to mass wasting because it affects the character of vegetation and the amount and distribution of rainfall.

QUICK QUESTION: Are mass movements likely to be frequent on the surface of the Moon? Why or why not?

16.5 How Can We Protect against Mass-Movement Disasters?

Identifying Regions at Risk

Clearly, landslides, mudflows, and slumps are natural hazards that we cannot ignore. Too many of us live in regions where mass wasting has the potential to kill people and destroy property. In many cases, the best solution is avoidance: don't build, live, or work on slopes or below slopes where mass movement is likely to take place. But avoidance is possible only if we know where the hazards are greater.

To pinpoint regions that are particularly susceptible to mass movements, and thus are regions of elevated risk, geologists look for landforms known to result from mass movements, for where mass movements have happened in the past, they might happen again in the future. Features such as slump head scarps, swaths of forest in which trees have been tilted, piles of loose debris at the base of hills, and hummocky land surfaces all indicate recent mass wasting.

Geologists may also be able to detect regions that are beginning to move (**Fig. 16.22**). For example, roads, buildings, and pipes begin to crack over unstable ground. Power lines may be too tight or too loose because the poles to which they are attached move together or apart. Visible cracks form on the ground at the potential head of a slump, while the ground may bulge up at the toe of the slump. In some cases, subsurface cracks may drain the water from an area and kill off vegetation, whereas in other areas land may sink and form a swamp. Even if mass wasting takes place too slowly to be perceptible to people, it can be documented with sensitive instruments that can detect a subtle tilt of the ground or changes in distance between nearby points. Specifically, measurements with satellite data, with laser surveys, and with tiltmeters permit geologists to identify movements of just a few millimeters that, while small, may indicate the reactivation of a slump.

If various clues indicate that a landmass is beginning to move, and if conditions make accelerating movement likely (e.g., persistent rain, rising floodwaters, or continuing earthquake aftershocks), then officials may order an evacuation. Evacuations have saved lives, and ignored warnings have cost lives.

FIGURE 16.21 The Portuguese Bend Slide viewed from the air.

Head scarp

Approximate edge of slump

Hummocky slump surface

FIGURE 16.22 Surface features warn that a large slump is beginning to develop. Cracks that appear at the head scarp may drain water and kill trees. Power-line poles tilt and the lines become tight. Fences, roads, and houses on the slump begin to crack.

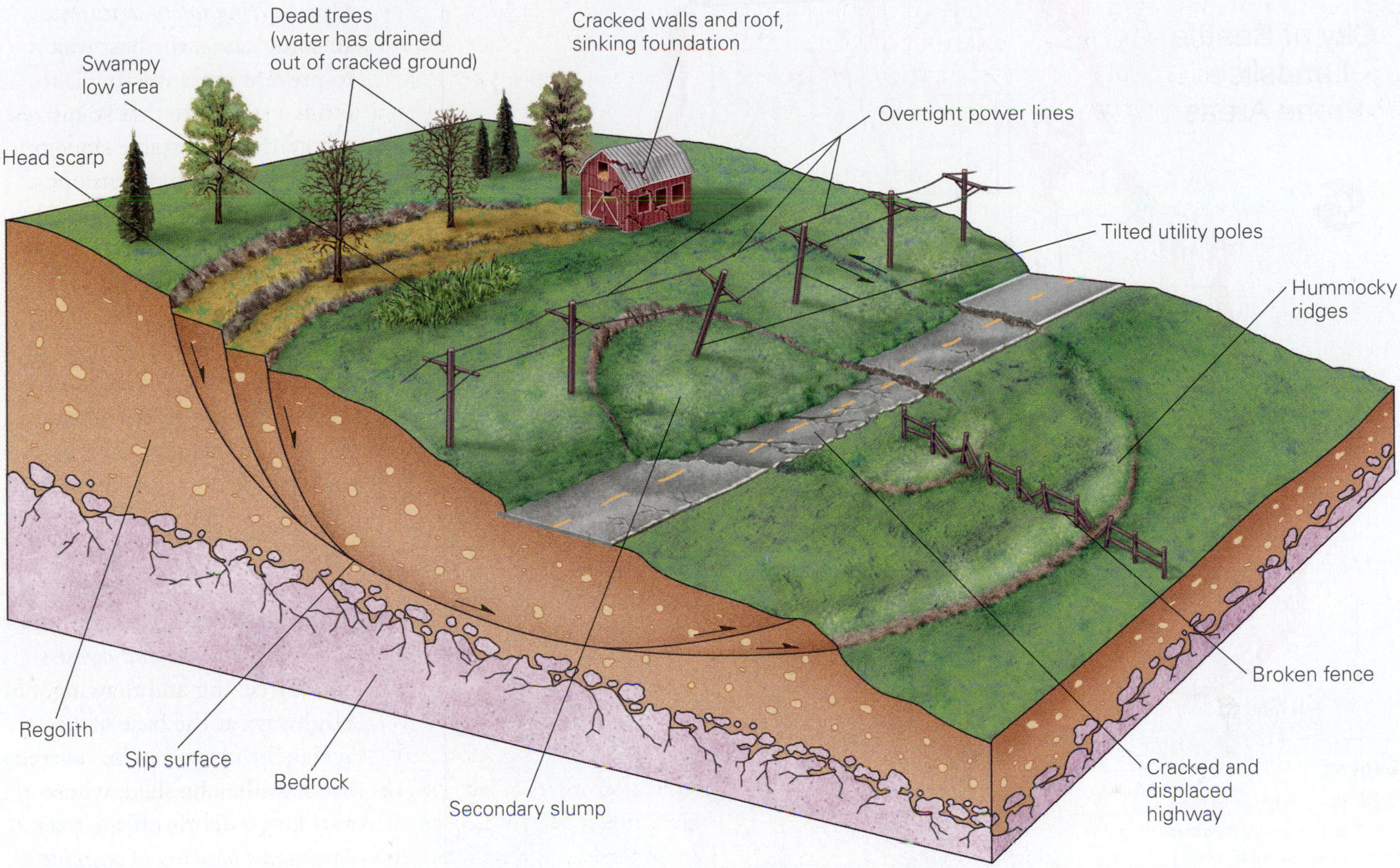

Swampy low area

Dead trees (water has drained out of cracked ground)

Cracked walls and roof, sinking foundation

Overtight power lines

Head scarp

Tilted utility poles

Hummocky ridges

Broken fence

Regolith

Slip surface

Bedrock

Secondary slump

Cracked and displaced highway

But unfortunately, some mass movements happen without any warning, and some evacuations prove costly but unnecessary.

In places that are being directly monitored, warnings can be quite precise. For example, an immense landslide that happened within the Bingham Canyon Mine of Utah was predicted with enough advance warning to evacuate the miners before the landslide happened. Thus, the debris flow, which carried material almost 2 km at speeds of up to 100 km per hour (**Fig. 16.23**), destroyed some buildings and equipment but did not cause any injury.

Even if there is no evidence of recent movement, a danger may still exist: just because a steep slope hasn't collapsed in the recent past doesn't mean it won't in the future. In recent years, geologists have begun to identify such potential hazards by using computer programs that evaluate factors that trigger mass wasting. With these data, they create maps that portray the degree of risk for a certain location. These factors include the following: slope steepness; strength of substrate; degree of water saturation; orientation of bedding, joints, or foliation relative to the slope; nature of vegetation cover; potential for heavy rains; potential for undercutting to occur; and likelihood of earthquakes. From such hazard-assessment studies, geologists compile **landslide-potential maps**, which rank regions according to the likelihood that a mass movement will occur (**Fig. 16.24**). Officials may restrict construction in landslide-prone areas.

FIGURE 16.23 A huge landslide took away one of the walls of the Bingham open-pit copper mine in Utah. No lives were lost because engineers detected the instability of the slope and evacuated the mine in time.

FIGURE 16.24 A landslide hazard map of the Seattle area.

City of Seattle
Landslide
Prone Areas

Legend

ECA - Forty% steep slope

ECA - Potential Slide Areas

Forty% steep slope areas and potential slide areas overlap and are coincidental in many areas.

No warranties of any sort, including accuracy, fitness, or merchantability accompany this product. Copyright 2008, All Rights Reserved, City of Seattle

Preventing Mass Movements

In areas where a hazard exists, people can take certain steps to remedy the problem and stabilize the slope (**Fig. 16.25**).

- *Revegetation:* Since bare ground is more vulnerable to downslope movement than is vegetated ground, stability in deforested areas will be greatly enhanced if land owners replant the region with vegetation that sends down deep roots.
- *Regrading:* An oversteepened slope can be regraded or terraced so that it does not exceed the angle of repose.
- *Reducing subsurface water:* Because water weakens material beneath a slope and adds weight to the slope, an unstable situation may be remedied either by improving drainage so that water does not enter the subsurface in the first place or by removing water from the ground.
- *Preventing undercutting:* In places where a river undercuts a cliff face, engineers can divert the river. Similarly, along coastal regions they may build an offshore breakwater or pile **riprap** (loose boulders or concrete) along the beach to

absorb wave energy before it strikes the cliff face.
- *Constructing safety structures:* In some cases, the best way to prevent mass wasting is to build a structure that stabilizes a potentially unstable slope or protects a region downslope from debris if a mass movement does occur. For example, civil engineers can build retaining walls or bolt loose slabs of rock to more coherent masses in the substrate in order to stabilize highway embankments. The danger from rock falls can be decreased by covering a road cut with chain-link fencing or by spraying road cuts with "shotcrete," a cement that coats the wall and prevents water infiltration and consequent freezing and thawing. Highways at the base of an avalanche chute can be covered by an avalanche shed, whose roof keeps debris off the road.
- *Controlled blasting of unstable slopes:* When it is clear that unstable ground or snow threatens a particular region, the best solution may be to blast the unstable ground or snow loose at a time when its movement can do no harm.

Clearly, the cost of preventing mass-wasting calamities is high, and people might not always be willing to pay the price. In such cases, they have a choice of avoiding the risky area, taking the chance that a calamity will not happen while they are around, buying appropriate insurance, or counting on relief agencies to help if disaster does strike. Once again, geology and society cross paths.

Take-Home Message

Various features of the landscape may help geologists to identify unstable slopes and estimate risk. Systematic study allows production of landslide-potential maps. Engineers can use a variety of techniques to stabilize slopes.

QUICK QUESTION: Why is mass movement of major concern during production of large road cuts?

FIGURE 16.25 A variety of remedial steps can stabilize unstable ground.

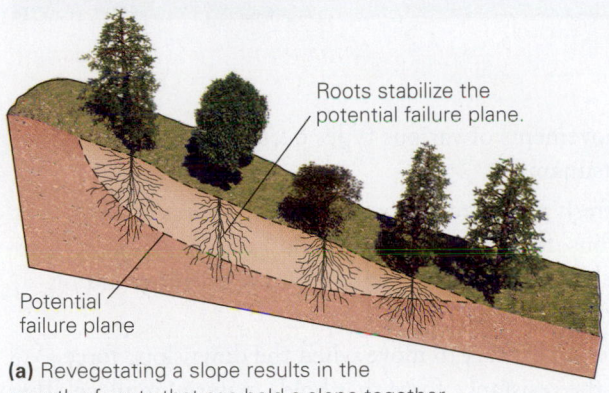

Roots stabilize the potential failure plane.

Potential failure plane

(a) Revegetating a slope results in the growth of roots that can hold a slope together.

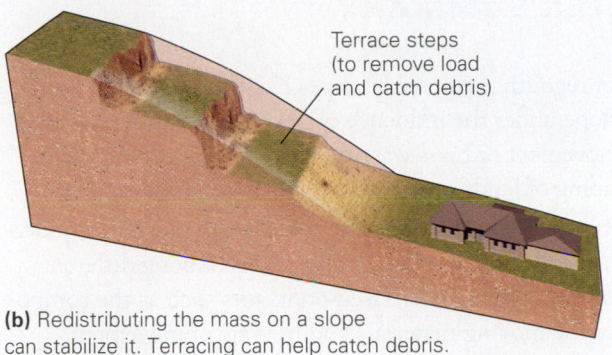

Terrace steps (to remove load and catch debris)

(b) Redistributing the mass on a slope can stabilize it. Terracing can help catch debris.

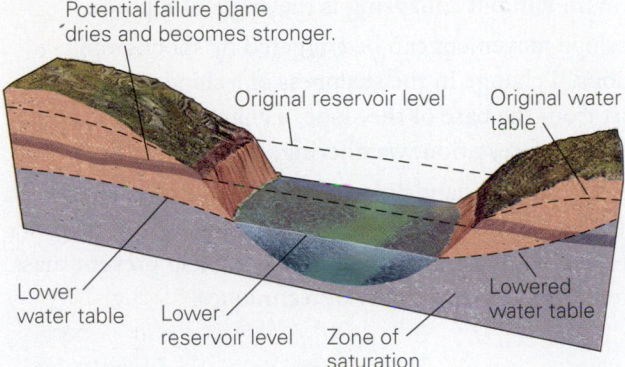

Potential failure plane dries and becomes stronger.

Original reservoir level

Original water table

Lower water table

Lower reservoir level

Zone of saturation

Lowered water table

(c) Lowering the level of the water table can strengthen a potential failure surface.

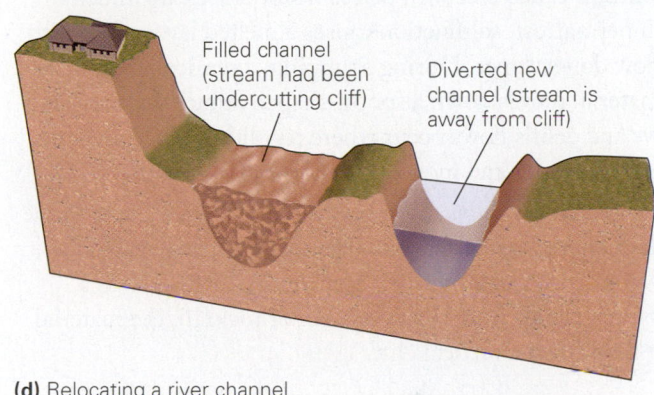

Filled channel (stream had been undercutting cliff)

Diverted new channel (stream is away from cliff)

(d) Relocating a river channel can prevent undercutting.

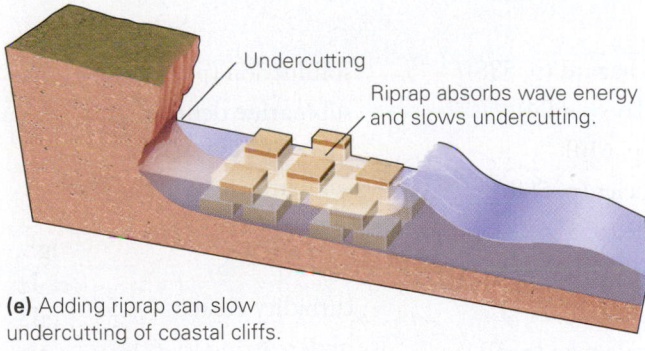

Undercutting

Riprap absorbs wave energy and slows undercutting.

(e) Adding riprap can slow undercutting of coastal cliffs.

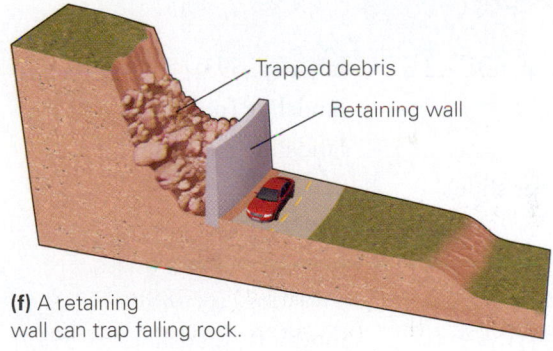

Trapped debris

Retaining wall

(f) A retaining wall can trap falling rock.

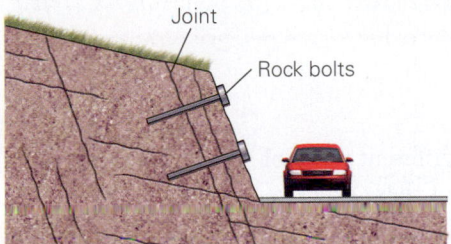

Joint

Rock bolts

(g) Bolting or screening a cliff face can hold loose rocks in place.

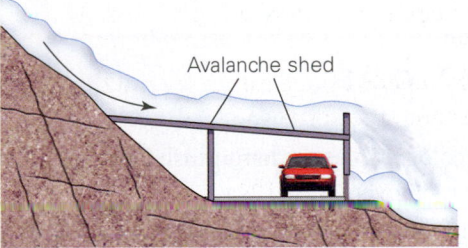

Avalanche shed

(h) An avalanche shed diverts debris or snow over a roadway.

CHAPTER SUMMARY

- Rock or regolith on unstable slopes has the potential to move downslope under the influence of gravity. This process, called mass movement or mass wasting, plays an important role in the shaping of landforms and transport of sediment.

- Though in everyday language people refer to most mass movements as landslides, geologists distinguish among different types of mass movements based on factors such as the composition of the moving materials and the rate of movement.

- Slow mass movement, caused by the freezing and thawing of regolith, is called creep. In places where slopes are underlain with permafrost, solifluction causes a melted layer of regolith to flow down slopes. During slumping, a semicoherent mass of material moves down a spoon-shaped failure surface. Mudflows and debris flows occur where regolith has become saturated with water and moves downslope as a slurry.

- Rock and debris slides move very rapidly down a slope; the rock or debris breaks apart and tumbles. During avalanches, snow or debris mixes with air and moves downslope as a turbulent cloud. And in a debris fall or rockfall, the material free-falls down a vertical cliff.

- Mass movements of various types occur undersea and can trigger tsunamis.

- Intact, fresh rock is too strong to undergo mass movement. Thus, for mass movement to be possible, rock must be weakened by fracturing (joint formation) and/or weathering.

- Unstable slopes start to move when the downslope force exceeds the resistance force that holds material in place. The steepest angle at which a slope of unconsolidated material can remain without collapsing is the angle of repose.

- Downslope movement can be triggered by shocks and vibrations, a change in the steepness of a slope, removal of support from the base of the slope, a change in the strength of a slope, deforestation, weathering, or heavy rain.

- Geologists produce landslide-potential maps to identify areas susceptible to mass movement. Engineers can help detect incipient mass movements and can help prevent mass movements by using a variety of techniques.

GUIDE TERMS

angle of repose (p. 601)
avalanche (p. 588)
creep (p. 588)
debris flow (debris slide) (p. 591)
debris slide (p. 594)
failure surface (p. 590)
head scarp (p. 591)

lahar (p. 593)
landslide (p. 588)
landslide-potential map (p. 609)
liquefaction (p. 604)
mass movement (mass wasting) (p. 588)
mudflow (mudslide) (p. 591)

natural hazard (p. 588)
quick clay (p. 604)
riprap (p. 610)
rock glacier (p. 588)
rockslide (p. 594)
slope failure (p. 600)
slump (p. 590)
snow avalanche (p. 594)

solifluction (p. 588)
submarine debris flow (p. 596)
submarine slump (p. 596)
substrate (p. 588)
talus (p. 595)
turbidity current (p. 596)
undercutting (p. 604)

REVIEW QUESTIONS

1. What factors do geologists use to distinguish among various types of mass movements?

2. Identify the key differences among a slump, a debris flow, a lahar, an avalanche, a rockslide, and a rockfall.

3. Explain the process of creep, and discuss how it differs from solifluction.

4. Distinguish among different types of submarine mass movements. Which of these types can trigger a major tsunami, and why?

5. Why is intact bedrock stronger than fractured bedrock? Why is it stronger than regolith?

6. Explain the difference between a stable and unstable slope. What factors determine the angle of repose of a material? What features are likely to serve as failure surfaces?

7. Discuss the variety of phenomena that can cause a stable slope to become so unstable that it fails.

8. How can ground shaking cause fairly solid layers or sand or mud to become weak slurries capable of flowing?

9. Discuss the role of vegetation in slope stability. Why can fires and deforestation lead to slope failure?

10. Identify the various factors that make the coast of California susceptible to mass movements.

11. What factors do geologists take into account when producing a landslide-potential map, and how can geologists detect the beginning of mass movement in an area? What steps can people take to avoid landslide disasters?

ON FURTHER THOUGHT

12. Imagine that you have been asked by the World Bank to determine whether it makes sense to build a dam in a steep-sided, east-west-trending valley in a small central Asian nation. The local government has lobbied for the dam because the climate of the country has gradually been getting drier and the farms of the area are running out of water, and construction of the dam would employ thousands of now-jobless people. Initial investigation shows that the rock of the valley floor consists of schist containing a strong foliation that dips south. Outcrop studies reveal that abundant fractures occur in the schist along the valley floor; the surfaces of most fractures are coated with slickensides. Moderate earthquakes have rattled the region. What would you advise the bank? Explain the hazards and what might happen if the reservoir were filled.

smartwork.wwnorton.com

This chapter's Smartwork features:

- Labeling exercise on types of mass movements.
- What A Geologist Sees activity on slump features.
- Video questions on microgravity measurement.

GEOTOURS

This chapter's GeoTour exercise (M) features:

- Portuguese Bend Landslide, CA
- La Conchita Mudslide, CA
- Gros Ventre Slide, WY
- Vaiont Dam Slide, Italy

Another View   The huge Ganges Chasma landslide, which formed along the Valles Marineris, a canyon wall on Mars, has taken part of a meteorite crater with it. The field of view is about 60 km across.

This winter view from an airplane illustrates how running water, flowing down slopes in response to the pull of gravity, cuts channels into the landscape to produce drainage networks.

CHAPTER 17

Streams and Floods: The Geology of Running Water

> Nothing in the world is more flexible and yielding than water.
> Yet when it attacks the firm and the strong, none can withstand it.
>
> —*Laozi (legendary ancient Chinese philosopher)*

17.1 Introduction

Wind swooping northward across warm oceans can pick up vast quantities of water vapor. It can then transport that water as moist air into interior of a continent. On a September day in 2013, such a mass of moist air collided with a mass of cold air stalled along the eastern edge of the Rocky Mountains in Colorado. The moist air rose; the moisture in it condensed; and drenching rains began to fall. In Boulder County, 430 mm (17 in) of rain fell over a period of a few days—normally, the county receives 525 mm (21 in) over the course of a whole year! Some of the water soaked into the ground, but most ended up in the numerous stream channels that flow from the mountains toward the plains. At times, the Big Thompson River carried 30 times more water than normal. The water rushing down the river became faster, more turbulent, and muddier, and eventually it spilled over the river's banks. The roaring torrent washed away houses and trees, scoured away roads (**Fig. 17.1a**), destroyed rail lines, inundated farmland, broke pipelines, overwhelmed sewage treatment plants, and undermined bridges. By the time the storms subsided and water levels returned to normal, the tragedy had claimed eight lives and destroyed $2 billion of property. During the same year, flooding in Alberta had the same effect in the eastern Canadian Rockies

(**Fig. 17.1b**). Where this water spilled out onto the plains, it submerged portions of Calgary (**Fig. 17.1c**).

The people of Colorado and Alberta had, unfortunately, experienced the immense power of *running water*, water that flows down the surface of sloping land in response to the pull of gravity. Geologists use the term **stream** for any body of running water that flows in a **channel**, an elongate depression or trough—the edges of the channel are the stream's *banks*, the floor of the channel is the *streambed*, and a length of the channel is called a *reach*. (In everyday English, we refer to large streams as rivers and medium-sized ones as creeks.) Water in a stream, overall, flows from the *upstream region*, closer to the source or **headwaters** of the stream, to the *downstream region*, closer to the end or **mouth** of the stream. Streams drain water from the landscape and carry it into lakes or to the sea, much as culverts drain water from a parking lot. In the process, streams relentlessly erode the landscape and transport sediment and debris to sites of deposition. Generally, a stream stays within the confines of its channel, but when the supply of water entering a stream exceeds the channel's capacity, water spills out and covers the surrounding land, thereby causing a **flood**, such as the ones that devastated parts of Colorado and Alberta.

The Earth is the only planet in the Solar System that currently hosts flowing streams. Streams are of great importance to human society not only because of how they modify the landscape, especially during floods, but also because they provide avenues for travel and commerce, nutrients and sediment for agriculture, water for irrigation and consumption, and energy to produce electricity and drive factories. In this chapter, we examine how streams operate in the Earth System. First, we learn about the origin of running water and about how the water draining a region organizes into networks of streams. Then we look at the process of stream erosion and deposition and at the landscapes that form as a consequence of to these processes. Finally, we focus on the nature and consequences of flooding and address the question of how people can mitigate flooding risk.

17.2 Draining the Land

Forming Streams and Drainage Networks

Where does the water in a stream come from? Recall that water enters the hydrologic cycle by evaporating from the Earth's surface and rising into the atmosphere (see Interlude F). After a

FIGURE 17.1 Examples of flooding damage illustrate the power of running water.

(a) A stream roaring down a canyon in Colorado undercut and carried away part of a highway.

(b) Heavy rains in the Canadian Rockies caused extensive flooding in Alberta.

relatively short residence time, atmospheric water condenses and falls back to the Earth's surface as rain or snow, which accumulates in various reservoirs. Some rain or snow remains in relatively nonflowing accumulations on the land (puddles, swamps, lakes, snowfields, and glaciers), some flows down the land surface as a thin film called **sheetwash**, and some sinks into the ground where it either becomes trapped in soil (as soil moisture) or descends below the water table to become groundwater. (As we will discuss in Chapter 19, the *water table* is the level below which groundwater fills all the pores and cracks in subsurface rock or sediment.) The flowing water in a stream can come from any or all of these reservoirs (**Fig. 17.2**). Specifically, it can spill from the outlet of a lake or swamp, it can melt from the end of a glacier or snowfield, it can bubble up from springs that tap into groundwater, it can seep out of the soil, and it can spill into a stream as sheetwash.

(c) The water inundated parts of Calgary, including the stadium for the city's Stampede (rodeo).

FIGURE 17.2 Excess surface water (runoff) comes from rain, melting ice or snow, and groundwater springs. On flat ground, water accumulates in puddles or swamps, but on slopes it flows downslope in streams.

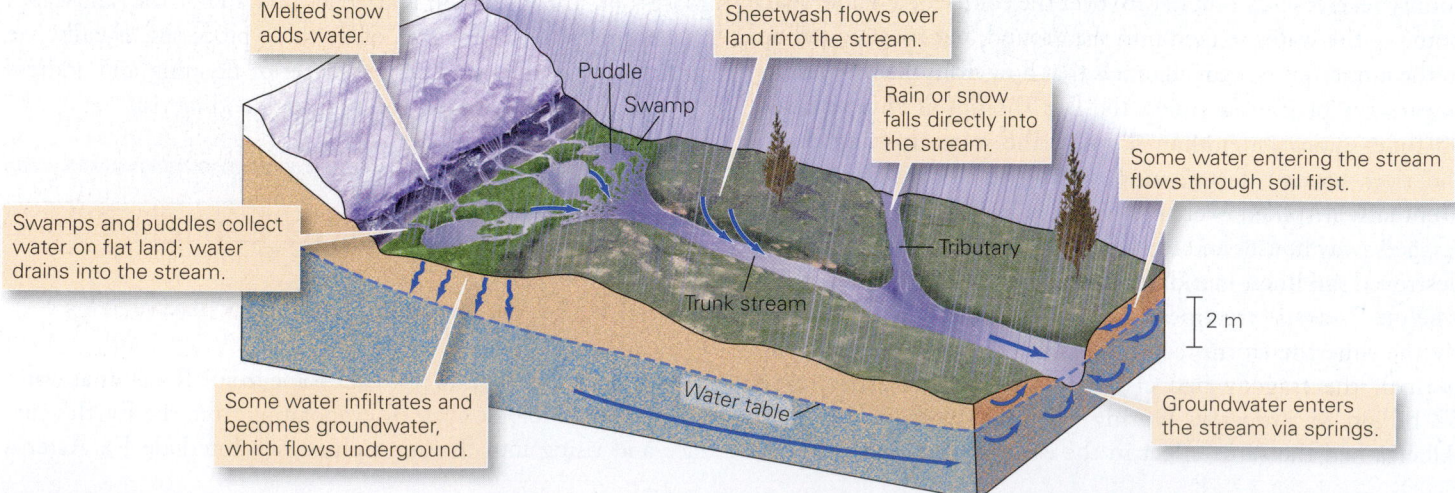

Runoff, water flowing on the surface of the Earth, can collect in stream channels because a channel is lower than the surrounding area and gravity always moves material from higher to lower elevation. How does a stream channel form in the first place? In some cases, the channel initiates at the mouth of a spring, where groundwater seeps out and starts to flow on the surface. The process of channel formation can also begin when sheetwash starts flowing downslope (**Fig. 17.3a**), for like any flowing fluid, sheetwash erodes its substrate, the material it flows over. The efficiency of such erosion depends on the velocity of the flow, (faster flows have more power and thus erode more rapidly) and on the resistance of the substrate to erosion. Where the flow happens to be a bit faster, the substrate is a little weaker, or protective groundcover a bit more sparse, erosion scours (digs) a channel. Since the channel is lower than the surrounding ground, sheetwash in adjacent areas starts to head toward it. With time, the extra flow deepens the channel, a process called **downcutting**. More sheetwash diverts into the channel in which, therefore, flow velocity increases and erosion takes place even more rapidly. If this process continues for enough time, a distinct stream forms. Note that downcutting is, in effect, a "positive feedback" process—once downcutting begins to produce a channel, water flow focuses into the channel, so the channel deepening accelerates.

As time passes, a stream channel lengthens by digging into the land at the head of the channel, a process called **headward erosion** (**Fig. 17.3b**). Headward erosion occurs for two reasons. First, it happens when the surface flow converging at the head of a channel has sufficient erosive power to downcut. Second, it happens at locations where groundwater seeps out of the ground and enters the head of the stream channel. Such seepage, called *groundwater sapping*, gradually weakens and undermines the soil or rock just upstream of the channel's endpoint until the material collapses into the channel—the collapsed debris eventually washes away during a flood. Each increment of collapse makes the channel longer.

As downcutting deepens the main channel, the surrounding land surfaces start to slope toward the channel. Thus, new

FIGURE 17.3 The formation of stream channels and drainage networks.

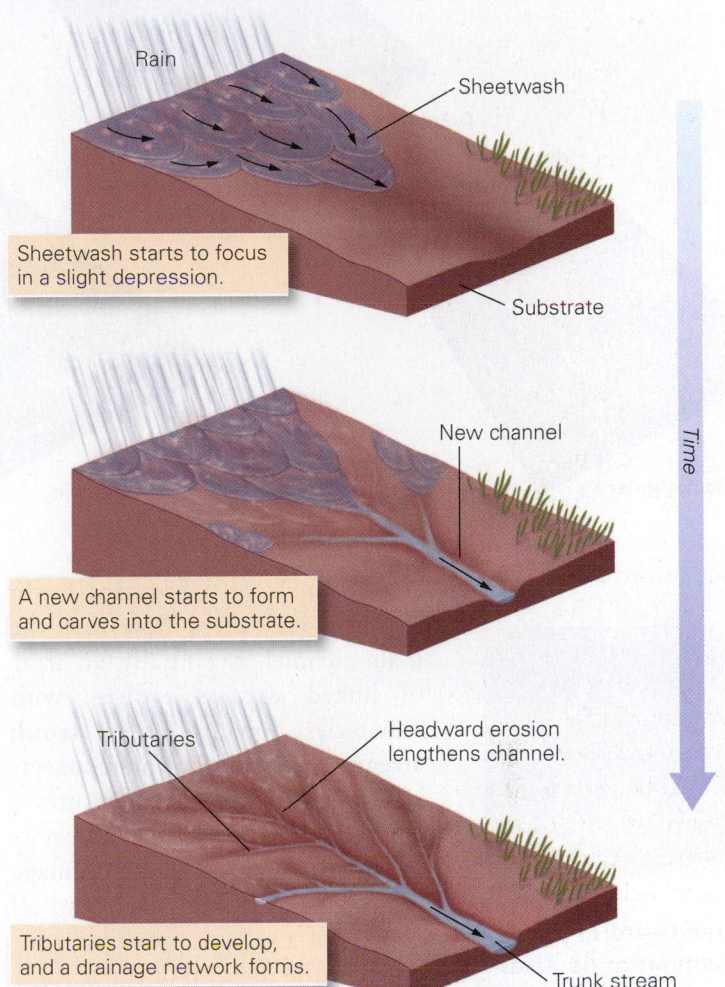

(a) Progressive stages during the development of a network of stream channels.

(b) As seen in this air photo, headward erosion is cutting into a plateau.

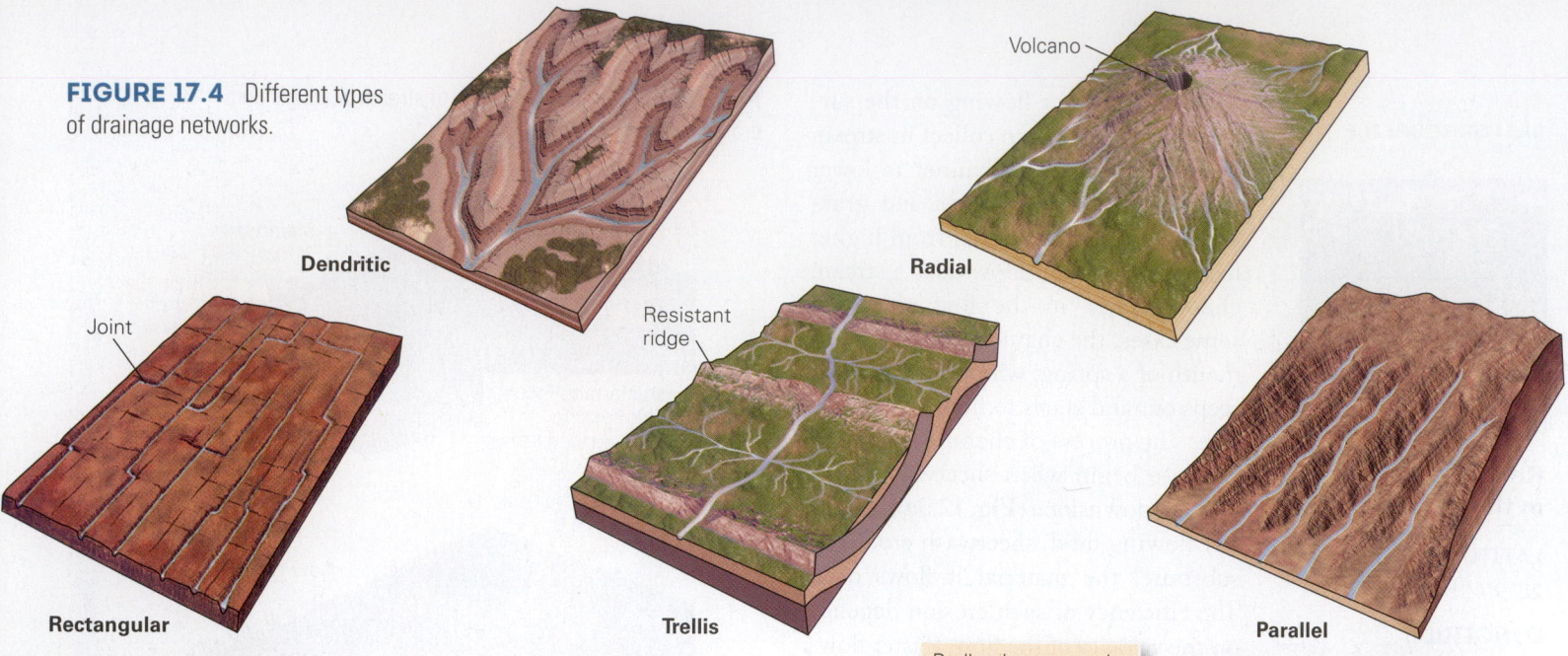

FIGURE 17.4 Different types of drainage networks.

Dendritic

Radial

Volcano

Joint

Resistant ridge

Rectangular

Trellis

Parallel

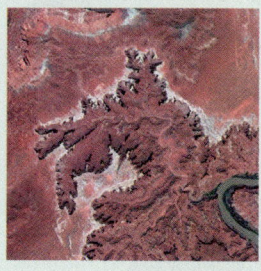

Headward Erosion, Canyonlands, Utah

LATITUDE
38°17'58.16"N

LONGITUDE
109°50'18.76"W

Looking down from 8 km (~5 mi).

Tributaries flow down canyons into the Colorado River in Canyonlands National Park. The tributaries are downcutting into horizontal strata. Note the steep escarpments at the head of each canyon. Here, groundwater sapping causes collapse of debris into the canyon, resulting in the headward erosion.

side channels, or **tributaries**, begin to form, and these flow into the main channel. Eventually, an array of linked streams evolves, with tributaries flowing into a **trunk stream**. The array of interconnecting streams together constitutes a **drainage network**. Like transportation networks of roads, drainage networks of streams reach into all corners of a region, providing conduits for the removal of runoff.

The configuration of tributaries and trunk streams defines the map pattern of a drainage network. The pattern that develops in a given region depends on the shape of the landscape and the character of the substrate. Geologists recognize several distinct geometries of drainage networks (**Fig. 17.4**).

- *Dendritic:* When rivers flow over a fairly uniform substrate with a fairly gentle slope, a dendritic network develops. It looks like the pattern of branches connecting to the trunk of a deciduous tree. In fact, the word *dendritic* comes from the Greek *dendros*, meaning tree.
- *Radial:* Drainage networks forming on the surface of a cone-shaped mountain, such

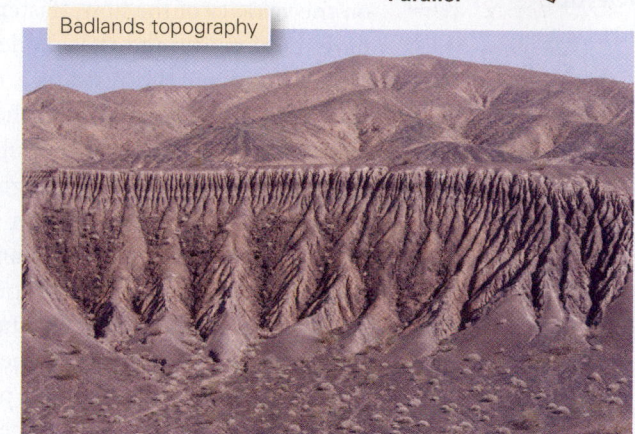

Badlands topography

as a volcano, flow outward from the mountain peak, like spokes on a wheel. Such a pattern defines a radial network.

- *Rectangular:* In places where a rectangular grid of fractures (vertical joints) breaks up the ground, channels form along the pre-existing fractures, and streams join one another at right angles, creating a rectangular network.
- *Trellis:* In places where a drainage network develops across a landscape of parallel valleys and ridges, major tributaries flow down a valley and join a trunk stream that cuts across the ridges. The place where a trunk stream cuts across a resistant ridge is a *water gap*. The resulting map pattern resembles a garden trellis, so the arrangement of streams constitutes a trellis network.
- *Parallel:* On a uniform, fairly steep slope, several streams with parallel courses develop simultaneously. The group constitutes a parallel network. Downslope, the streams merge into fewer, but still parallel, channels. Parallel networks typically form on the sides of steep escarpments of weak substrate—if the surface of the substrate is bare of sediment, the resulting landscape is called *badlands topography*.

FIGURE 17.5 Drainage divides and basins.

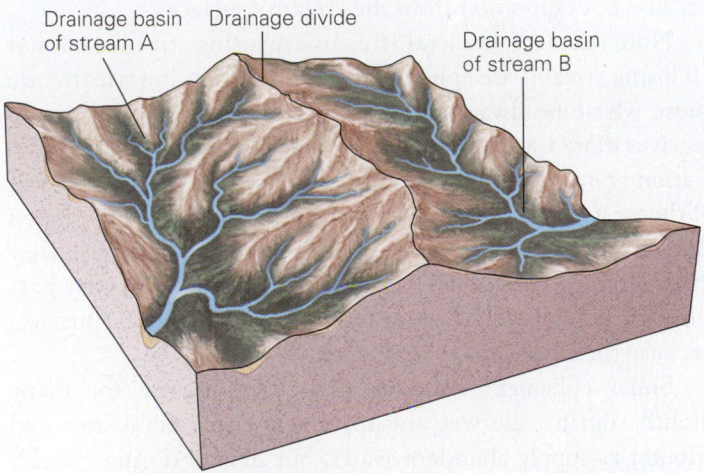

Drainage basin of stream A

Drainage divide

Drainage basin of stream B

(a) A drainage divide is a relatively high ridge that separates two drainage basins.

(c) The major drainage basins of North America.

Main drainage divide

A trunk stream

Secondary drainage divide

A small drainage basin

What a Geologist Sees

(b) An air view shows numerous small drainage basins formed during erosion of this ridge in Oregon.

Drainage Basins and Divides

A drainage network collects water from a region, variously called a *drainage basin*, *catchment*, or **watershed**. The highland, or ridge, that separates one watershed from another is called a **drainage divide** (**Fig. 17.5a**). You can recognize a local divide on a topographic map or from the view out an airplane window (**Fig. 17.5b**). Regional divides separate two large drainage basins—for example, an important divide runs follows the Appalachians and separates Atlantic Ocean drainage from Gulf of Mexico drainage. A **continental divide** separates drainage that flows into one ocean from drainage that flows into another. In North America, one continental divide runs the length of the North American Cordillera and separates watersheds that ultimately drain into the Atlantic or Gulf of Mexico from those that drain to the Pacific (**Fig. 17.5c**). Another continental divide separates watersheds that drain into Hudson Bay and the Arctic Ocean from water that drains into other oceans. The largest drainage basin in North America, the Mississippi, covers much of the interior of the United States, and the largest in the world, the Amazon, extends across almost the entire width of South America (**Fig. 17.6**).

Streams That Last and Streams That Don't

Since streams can serve as water supplies for agriculture, industry, or municipalities, it's important to distinguish between types of streams based on whether the volume of water in

FIGURE 17.6 The Amazon watershed of South America stretches from the crest of the Andes to the Atlantic Ocean. Note that the drainage network is dendritic and that it flows to the east.

Atlantic

Pacific

N

500 km

the stream increases or decreases along its length. In a *gaining stream*, the volume of water increases in the downstream direction, whereas in a *losing stream*, the volume of water decreases in the downstream direction. Whether a stream is gaining or losing depends on two factors: the amount of water entering the stream from tributaries, because water coming from tributaries can replace water lost to evaporation or to infiltration into the streambed, and the depth of the water table relative to the streambed, because if the streambed lies below the water table, then groundwater flows up into the stream through springs.

Planners also find it useful to distinguish between types of streams based on the duration of flow—**permanent streams** flow all year long, whereas **ephemeral streams** flow for only part of the year. What determines whether a stream is permanent or ephemeral? Streams in temperate or tropical climates tend to be permanent because the water table generally lies above the streambed (**Fig. 17.7a, b**). Such streams also tend to be gaining streams. Most streams in arid (dry) climates, however, tend to be ephemeral because they are losing streams that lose so much water along their length that they dry out. This happens because the water table lies at depth below the streambed, so

water sinks down through the streambed (**Fig. 17.7c**), and also because of evaporation from the stream's surface.

Note that all ephemeral streams are losing streams, but not all losing streams are ephemeral. Permanent losing streams are those whose headwaters lie in a wetter region, so the stream receives more water from its upper reaches than it loses to infiltration or evaporation in its lower reaches. The Colorado River of the western United States and the Nile River of northeastern Africa serve as examples of permanent losing streams—they flow all year, even though for much of their lengths they pass through deserts where water tables lie far below the surface, because their headwaters drain very wet regions.

Some ephemeral streams flow continuously for many months during the wet season, when water tables rise and tributaries supply abundant water, but dry up during the dry season, when the water table sinks below the streambed and tributaries don't supply water. In contrast, some ephemeral streams flow only for a brief time after a heavy rain and are dry most of the year. The dry channel of these ephemeral streams is called a *dry wash*, wadi, or arroyo (**Fig 17.7d**).

Flow Taking Place in the Ground around the Channel

When we look at a stream, we see only the flow taking place within the channel. But if a permanent stream flows over a permeable substrate, it's important to keep in mind that water is also moving through the substrate in the direction of the stream flow, down to a depth of tens of centimeters to tens of meters below the streambed. The region affected by this flow is called the *hyporheic zone* (**Fig. 17.8**). In this zone, flow is much slower than in the open channel, because the water has to find its way around clasts and along cracks, but it is significantly faster than in groundwater outside of the zone. Notably, many organisms spawn within the hyporheic zone, where eggs can hatch without being washed away or eaten.

Take-Home Message

Streams drain water from the land. Their channels form by downcutting and lengthen by headward erosion, and they carry water from sheetwash, lakes, springs, and melting snow or ice. In a region, a drainage network of tributaries feeding a trunk stream carries water from a watershed to the sea. Divides separate watersheds. Permanent streams flow all year, while ephemeral streams flow only intermittently. The permanence of a stream depends on climate and on the supply of water from upstream.

QUICK QUESTION: Why do streams develop distinct channels?

FIGURE 17.7 The contrast between permanent and ephemeral streams.

(a) The floor of a permanent stream in a temperate climate lies below the water table. Springs add water from below, so the stream contains water even between rains.

An empty stream channel is called a dry wash.

(c) The channel of an ephemeral stream lies above the water table, so the stream flows only when water enters the stream faster than it can infiltrate into the ground.

(b) An example of a permanent stream in the Wind River Mountains, Wyoming.

(d) An example of a dry wash in the Buckskin Mountains, Arizona.

FIGURE 17.8 Since streambeds are permeable, water from a permanent stream mixes with groundwater in a region beneath the streambed called the hyporheic zone. Water in this zone flows in the same direction as the stream, but not as fast.

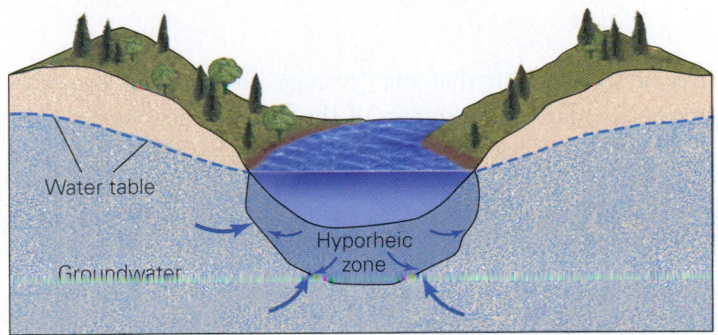

17.3 Describing Flow in Streams: Discharge and Turbulence

Imagine two streams—a larger one in which water flows slowly and a smaller one in which water flows rapidly. Which stream carries more water? The answer is not obvious. To answer this question completely, geologists and engineers must calculate the streams' discharge. Technically speaking, we define **discharge** as the volume of water passing a reference point in a given time. We can calculate discharge (D) by using a simple equation: $D = A_C \times v_A$. In this equation, A_C is the cross-sectional area of the stream as measured in a vertical plane oriented perpendicular to the stream bank, and v_A is the average velocity at which water moves in the downstream direction. If, for example, a stream's cross-sectional area is 10 m^2 and its

FIGURE 17.9 Flow velocity and its measurement in streams.

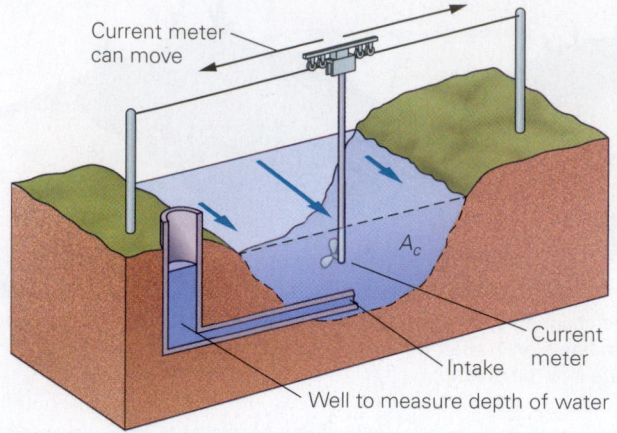

(a) At a stream-gauging station, geologists measure the cross-sectional area (A_c), the depth, and the average velocity of the stream. Velocity is slower near the banks.

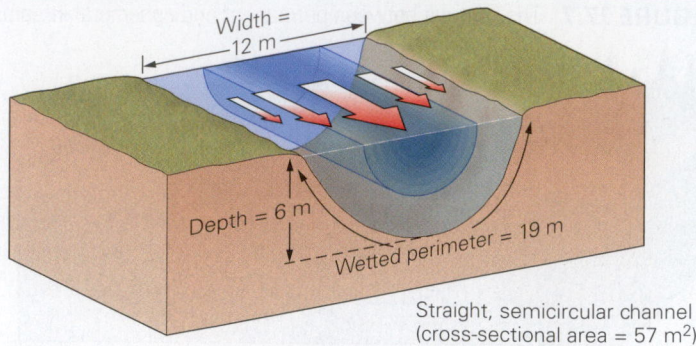

Straight, semicircular channel (cross-sectional area = 57 m²)

(b) In a straight channel, the fastest velocity occurs near the surface of the stream, equidistant from the banks.

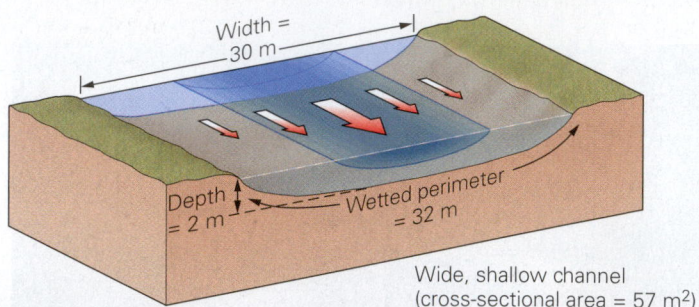

Wide, shallow channel (cross-sectional area = 57 m²)

(c) Velocity in a wide, shallow channel is less than that of a semicircular channel with the same cross-sectional area because the wetted perimeter is greater.

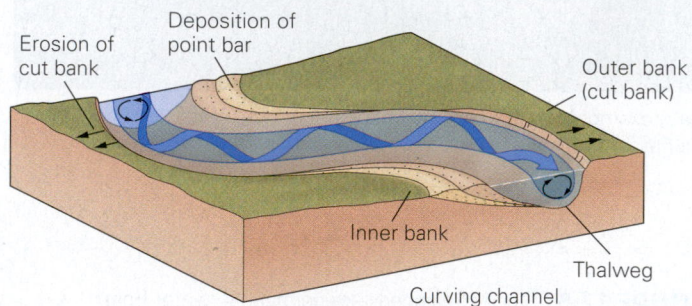

(d) In a curved channel, the fastest flow shifts to the outer edge of the stream, and water follows a spiral-like path.

water's average velocity is 5 m/s, then the discharge is 50 m³/s. Note that we specify discharge by volume per unit time, using units such as cubic meters per second (m³/s), where 1 m³/s = 35 cubic feet per second (ft³/s). We can measure discharge at a *stream-gauging station*, using instruments that analyze velocity and depth at various points across the stream (**Fig. 17.9a**).

Discharge depends on the size of the watershed, on the amount of rain or snow falling in the watershed, and on whether the stream is gaining or losing. As we've seen, the discharge of a gaining stream in a tropical or temperate region increases in the downstream direction, whereas the discharge of a losing stream decreases downstream, as more and more water seeps into the ground or evaporates. Discharge can also be affected by human activity. For example, if people divert the river's water for irrigation, the river's discharge decreases downstream. We can define an average discharge of a stream for a given location along the stream by taking the average of the discharge as measured daily over several years. The average discharge at a given location varies with time: a stream's discharge during the wet season may be double or triple the amount during the dry season, and during a flood, discharge may increase to more than 100 times normal.

Measurements of the average discharge at the mouth of a trunk stream allow us to compare different watersheds. The rivers that drain the two largest rainforests on Earth have the largest discharges. Specifically, the Amazon River has an average discharge of about 200,000 m³/s, roughly 15% of global river discharge, whereas the Congo River has an average discharge of 40,000 m³/s. In contrast, the "mighty" Mississippi, temperate North America's largest river, has a discharge of only 17,000 m³/s.

Calculations of discharge, as we have seen, are based on measurements of the average velocity at a point along the

stream. It turns out that the "average velocity" can be difficult to calculate, because not all the water in a stream moves at the same velocity past a given point on the bank. Velocity variations happen, in part, because friction along the sides and bed of the stream slows the flow. In general, water near the stream banks or the streambed moves more slowly than water in the middle of the flow, and the fastest-moving part of the stream flow lies near the surface in the center of the channel (**Fig. 17.9b**). The amount by which friction slows the flow depends both on the roughness of the walls and bed and on the channel shape. Rougher walls slow the floor more, and a wide, shallow stream channel has a larger *wetted perimeter* (the area in which water touches the channel walls) than does a semicircular channel, so water flows more slowly in the former than in the latter (**Fig. 17.9c**).

In a curving stream channel, the fastest flow shifts toward the outside curve, somewhat as a car swerves to the outer edge of a curve on a highway. Therefore, the deepest part of a channel, its **thalweg**, lies near the outside curve. In fact, as the water flows toward the outside wall of a curving channel, it starts to follow a spiral path because as surface water moves toward the outer bank, water deeper down must flow toward the inner bank to replace the surface water (**Fig. 17.9d**).

Velocity variations in a stream also happen because of **turbulence**, the twisting, swirling motion that on a large scale can create *eddies* (whirlpools) in which water curves and actually flows upstream or circles in place (**Fig. 17.10**). Turbulence develops because the shearing motion of one water volume against its neighbor causes the neighbor to spin and because obstacles such as boulders on the streambed deflect volumes, forcing them to move in a different direction.

Take-Home Message

Stream discharge, the amount of water passing through a cross section of the stream in a given time, depends on such factors as watershed area and climate. Discharge changes along the stream's length, and seasonally. Water velocity varies across a stream due to turbulence and also friction with the streambed.

QUICK QUESTION: Relative to a given point along the bank of a stream, where is the velocity of flow in a stream fastest? Why?

17.4 The Work of Running Water

How Do Streams Erode?

The energy that makes running water move comes from gravity. As water flows downslope from a higher to a lower elevation, the gravitational potential energy stored in water transforms into kinetic energy. About 3% of this energy goes into the work of eroding the walls and beds of stream channels. Running water causes erosion in four ways.

- *Scouring:* Running water can remove loose fragments of sediment, a process called **scouring**.
- *Breaking and lifting:* In some cases, the push of flowing water can break chunks of solid rock off the channel floor or walls. In addition, the flow of a current over a clast can cause the clast to rise or lift off the substrate.

FIGURE 17.10 Turbulence in streams.

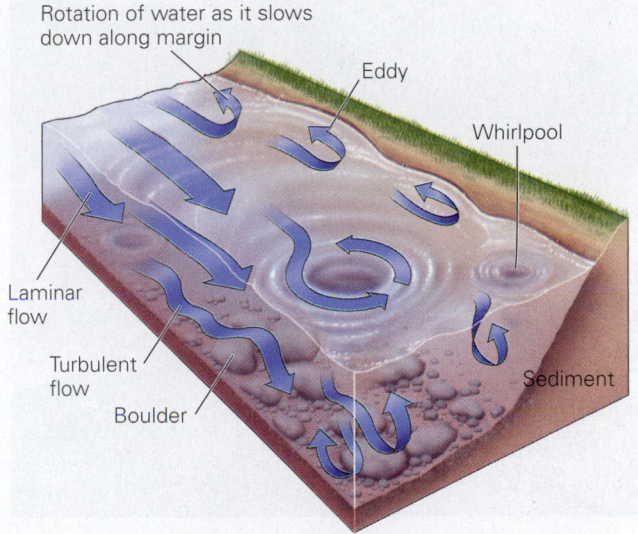

(a) Water in a stream doesn't usually follow a straight path. It swirls and twists, producing turbulence.

(b) The turbulence of this river in Rochester, New York, is indicated by the roughness of the surface, and the chaos of the bubble trails.

- *Abrasion:* Clean water has little erosive effect, but sand-laden water acts like sandpaper and grinds or rasps away at the channel floor and walls, a process called **abrasion**. In places where turbulence produces long-lived whirlpools, abrasion by sand or gravel carves a bowl-shaped depression, called a **pothole**, into the floor of the stream (**Fig. 17.11a, b**).
- *Dissolution:* Running water dissolves soluble minerals as it passes and carries the minerals away in solution.

FIGURE 17.11 Erosion and transportation in streams.

(a) A pothole in the bed of a stream near Ithaca, New York.

(b) This slot canyon in Arizona formed when many potholes linked together.

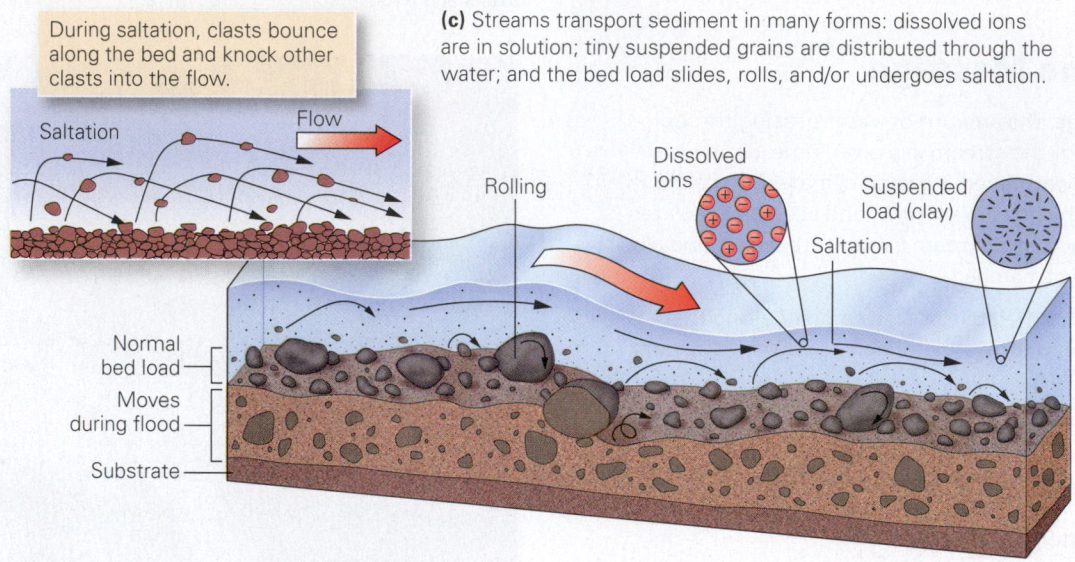

(c) Streams transport sediment in many forms: dissolved ions are in solution; tiny suspended grains are distributed through the water; and the bed load slides, rolls, and/or undergoes saltation.

The efficiency of erosion depends on the velocity and volume of water and on its sediment content. A large volume of fast-moving, turbulent, sandy water causes more erosion than does a trickle of quiet, clear water. Thus, most erosion takes place during floods, which supply streams with large volumes of fast-moving, sediment-laden water.

How Do Streams Transport Sediment?

The Mississippi River received the nickname "Big Muddy" for a reason—its water can become chocolate brown because of all the clay and silt it carries. All streams carry sediment, though not the same amount. Geologists refer to the total volume of sediment carried by a stream as its *sediment load*. The sediment load consists of three components (**Fig. 17.11c**):

- *Dissolved load*: Running water dissolves soluble minerals in the sediment or rock of its substrate, and groundwater seeping into a stream through the channel walls brings dissolved minerals with it. The ions of these dissolved minerals constitute a stream's **dissolved load**.
- *Suspended load*: The **suspended load** of a stream consists of tiny solid grains (silt or clay sized) that swirl along with the water without settling to the floor of the channel; this sediment makes the water brown (**Fig. 17.12a–c**).
- *Bed load*: The **bed load** of a stream consists of large particles, such as sand, pebbles, or cobbles, that bounce or roll along the stream floor (**Fig. 17.12d, e**). Bed-load movement commonly involves **saltation**. During saltation, a multitude of grains bounce along in the direction of flow within a zone that extends up from the surface of the streambed

FIGURE 17.12 Sediment, carried and deposited by streams. The clast size depends on stream velocity.

(a) On a sunny day, this stream in Switzerland is fairly clear; you can see cobbles on the bed.

(b) On a rainy day, the stream's discharge increases and the faster, more turbulent water becomes brown due to its load of sediment.

(c) An air photo shows a muddy river emptying into the Black Sea. Currents carry the sediment away.

(d) Gravel deposited within the channel of a stream in the Wasatch Mountains, Utah.

(e) Gravel in a streambed of a mountain stream in Denali National Park, Alaska. The large clasts were carried during floods.

(f) Point bars of mud deposited along a gentle, slowly moving stream in Brazil.

for a distance of several centimeters to several tens of centimeters. Each saltating grain in this zone follows a curved trajectory up through the water and then back down to the bed. When it strikes the bed, it knocks other grains upward and thus supplies grains to the saltation zone.

When describing a stream's ability to carry sediment, geologists specify its competence and capacity. The **competence** of a stream refers to the maximum particle size it carries—a stream with high competence can carry large particles, whereas one with low competence can carry only small particles. A fast-moving, turbulent stream has greater competence than does a slow-moving stream, and a stream in flood has greater competence than does a stream with normal flow. In fact, the huge boulders that litter the bed of a mountain creek move only during floods. The **capacity** of a stream refers to the total quantity of sediment it can carry. A stream's capacity depends on both its competence and discharge.

Depositional Processes

A raging torrent of water can carry coarse and fine sediment—the finer clasts rush along with the water as suspended load, whereas the coarser clasts may bounce and tumble as bed load. If the flow velocity decreases, either because the slope of the streambed decreases or because the channel broadens out and friction between the streambed and the water increases, then the competence of the stream decreases and sediment settles out. The size of the clasts that settle at a particular locality depends on how slow the flow has become. Thus, coarser sediment tends to settle out farther upstream, where water flows faster, whereas finer grains settle out farther downstream, where the water flows more slowly, and the finest sediment settles out when the stream flows into a standing body of water. Because of this process of *sediment sorting*, stream deposits tend to be segregated by size—gravel, sand, silt, and mud can collect in different locations.

Geologists refer to sediments transported by a stream as *fluvial deposits* (from the Latin *fluvius*, meaning river), or **alluvium**. Alluvium may accumulate along the streambed in elongate mounds called **bars** (see Fig. 17.12d). Some stream channels follow broad curves, which, as we'll see, are called meanders. Water slows along the inner edge of a meander, so crescent-shaped **point bars** bordering the shoreline develop along the inner edge (**Fig. 17.12f**). During floods, a stream may overtop the banks of its channel and spread out over its floodplain, the broad, flat area bordering the stream. Friction slows the water on the floodplain, so a sheet of silt and mud settles out to comprise floodplain deposits. Where a stream empties at its mouth into a standing body of water, the water slows and a wedge of sediment, called a **delta**, accumulates. We'll discuss meanders, floodplains, and deltas in more detail later in this chapter.

Take-Home Message

Streams erode by scouring, breaking and lifting, abrasion, and dissolution. They carry sediment as dissolved, suspended, or bed loads. Competence, the ability to carry sediment, depends on flow velocity. Where the velocity decreases, sediment settles out.

QUICK QUESTION: Why do point bars form on the inner arc of a curve in a stream?

17.5 How Do Streams Change along Their Length?

Longitudinal Profiles

In 1803, under President Thomas Jefferson's leadership, the United States bought the Louisiana Territory, a vast tract of land encompassing the western half of the Mississippi drainage basin. At the time, the geography of the territory was a mystery. To fill the blank on the map, Jefferson asked Meriwether Lewis and William Clark to lead a voyage of exploration across the Louisiana Territory to the Pacific.

Lewis and Clark, along with a crew of 40, began their expedition at the mouth of the Missouri River, where it joins the Mississippi. At this juncture, the Missouri is a wide, languid stream of muddy water. The group found the Missouri's downstream reach, where the river's channel is deep and the water smooth, to be easy going. But the farther upstream they went, the more difficult their voyage became, for the **stream gradient**, meaning the slope of the stream, became progressively steeper and the stream's discharge became less. When Lewis and Clark reached the site of what is now Bismarck, North Dakota, they had to abandon their original boats and haul smaller vessels up *rapids*, where turbulent water plunges over a steep, bouldery bed, and occasionally they had to carry their boats around *waterfalls*, where water drops over an escarpment. When they reached what is now southwestern Montana, they abandoned these boats as well and trudged along the stream valley on foot or on horseback, struggling up steep gradients until they reached the continental divide.

If Lewis and Clark had been able to plot a graph showing their elevation above sea level relative to their distance along the Missouri, they would have found that the **longitudinal profile** of the Missouri, a cross-sectional image showing the variation in the river's elevation along its length, is roughly a concave-up curve (**Fig. 17.13**). This curve illustrates that stream gradient is steeper near its headwaters than near its mouth. Real longitudinal profiles are not perfectly smooth curves but rather display little plateaus and steps, representing interruptions by lakes or waterfalls. Near its headwaters, an idealized stream flows down deep valleys or canyons, whereas near its mouth, it flows over nearly horizontal plains.

The Concept of a Base Level

Streams progressively deepen their channels by downcutting, but there is a depth below which a stream cannot downcut any further. The lowest elevation that a stream can attain at a

FIGURE 17.13 Drainage basins and the change in character of a stream along its longitudinal profile.

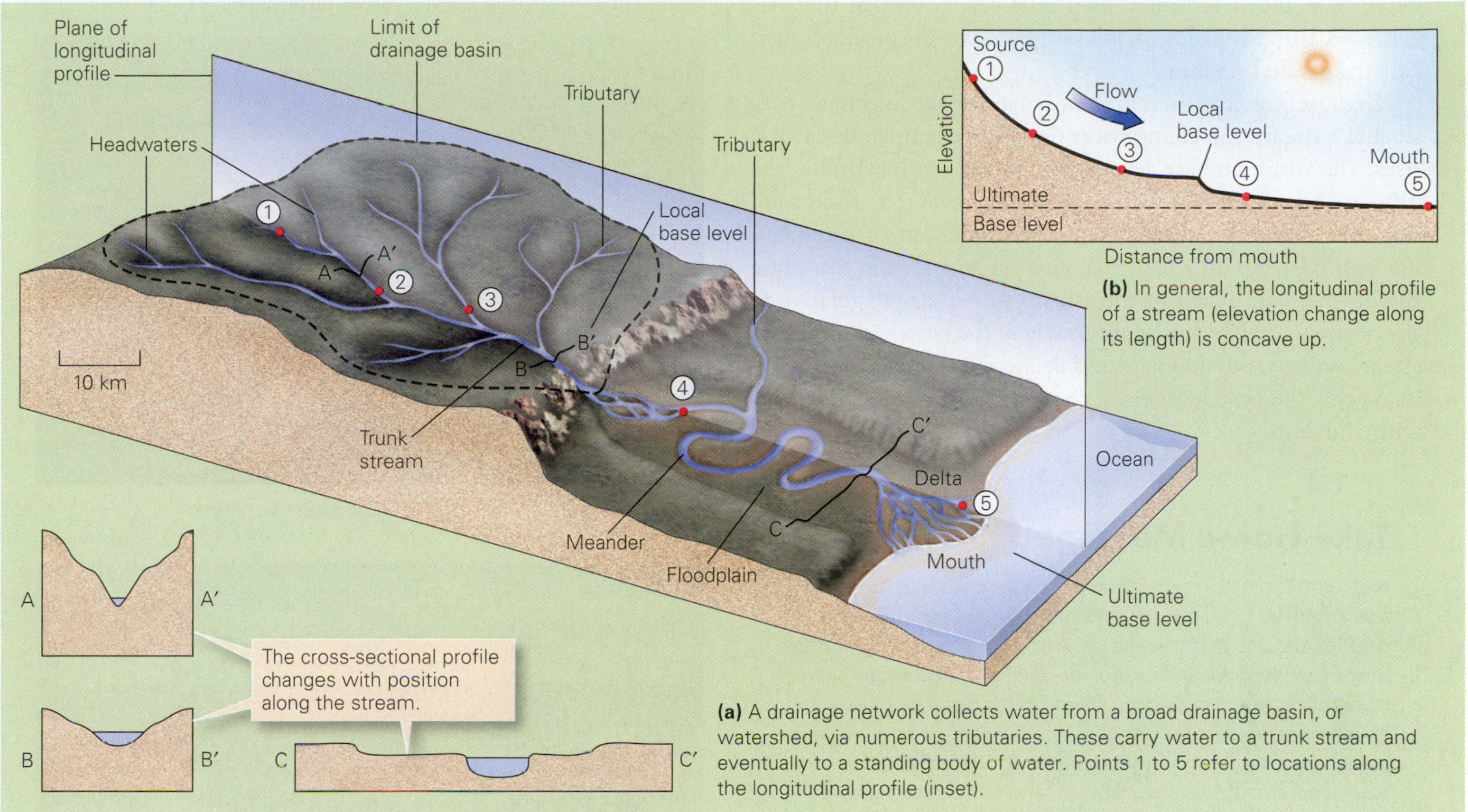

Plane of longitudinal profile

Limit of drainage basin

Tributary

Tributary

Headwaters

Local base level

Trunk stream

Meander

Floodplain

Delta

Mouth

Ocean

Ultimate base level

10 km

(b) In general, the longitudinal profile of a stream (elevation change along its length) is concave up.

Source

Flow

Local base level

Mouth

Ultimate Base level

Elevation

Distance from mouth

The cross-sectional profile changes with position along the stream.

(a) A drainage network collects water from a broad drainage basin, or watershed, via numerous tributaries. These carry water to a trunk stream and eventually to a standing body of water. Points 1 to 5 refer to locations along the longitudinal profile (inset).

locality is the **base level** of the stream. A *local* base level is one that occurs upstream of a stream's mouth, and the *ultimate* base level (i.e., the lowest possible elevation along the longitudinal profile) is sea level. The surface of the trunk stream where it enters the sea cannot be lower than sea level, for if it were, the stream would have to flow upslope.

Lakes or reservoirs act as local base levels along a stream, for where the stream enters such standing bodies of water, it slows almost to a halt and cannot downcut further (**Fig. 17.14a**). A ledge of resistant rock can also act as a local base level, for the stream level cannot drop below the ledge until the ledge erodes away (**Fig. 17.14b**). Finally, where a tributary joins a larger stream, the channel of the larger stream acts as the base level for the tributary. (Thus, the mouth of a tributary tends to lie at the same elevation as the stream that it joins, at the point of intersection.) Local base levels do not last forever, because running water eventually removes the obstructions that create them.

The steps or ledges defining local base levels along the stream represent the steps and deviations from the concave-up shape of an ideal longitudinal profile. Over time, streams erode away the steps, or deposit sediments to fill in low areas, until any point along the stream approaches a condition such that there is no net erosion or deposition—the stream can

FIGURE 17.14 The concept of local base levels in stream profiles.

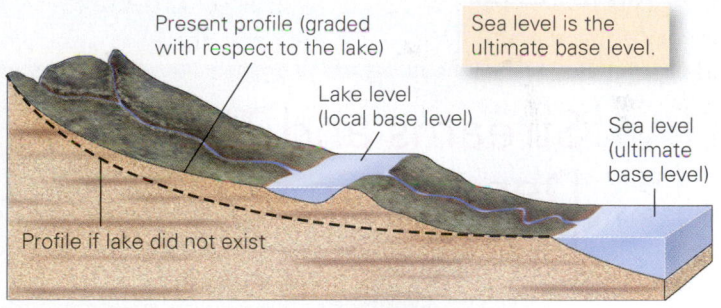

Present profile (graded with respect to the lake)

Sea level is the ultimate base level.

Lake level (local base level)

Sea level (ultimate base level)

Profile if lake did not exist

(a) A lake acts as a local base level. The stream profile uphill of the lake lies above the profile that would form if the lake didn't exist.

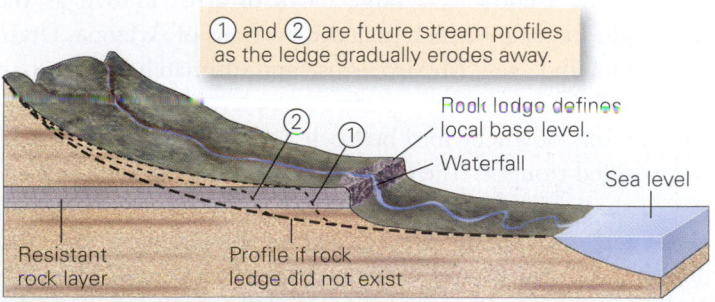

① and ② are future stream profiles as the ledge gradually erodes away.

Rock ledge defines local base level.

Waterfall

Sea level

Resistant rock layer

Profile if rock ledge did not exist

(b) A resistant rock ledge can form a local base level. Headward erosion gradually cuts back into the ledge.

carry all the sediment that has been supplied to it, and it deposits as much sediment as it removes. A stream that has achieved this condition displays a concave-up profile and is called a **graded stream**.

Because the Earth remains dynamic, the ultimate base level of a stream can change over time. For example, if sea level rises, the ultimate base level rises, and if sea level falls, the ultimate base level may fall. A local base level may rise or fall due to tectonic movements or to accumulation of sediment. For example, slip on a fault that cuts across a stream can cause the downstream reach of the stream to rise or fall relative to the upstream reach. In situations where the base level falls, the stream will start to downcut and deepen the valley that it flows in. When the base level rises, the stream will fill in its valley with alluvium.

Take-Home Message

A stream typically has a steeper gradient toward its source and a gentler gradient near its mouths, so longitudinal profiles tend to be concave up. A stream's mouth cannot be lower than its base level, and sea level is the ultimate base level for drainage networks. Over time, streams achieve an equilibrium so the amount of sediment carried into a reach is the same as the amount carried out. Changes in the base level disrupt this equilibrium and can cause renewed downcutting or deposition within a reach.

QUICK QUESTION: Why can't a stream downcut more deeply than its base level?

17.6 Streams and Their Deposits in the Landscape

Valleys and Canyons

During the Cenozoic, a large block of crust known as the Colorado Plateau—which includes portions of Arizona, Utah, Colorado, and New Mexico—rose and ultimately attained an average elevation of 2 km, relative to sea level. As the land went up, streams downcut into bedrock and produced spectacular, steep-sided troughs. The Colorado River flows down the largest of these, the Grand Canyon. In places, 1.6 km of vertical relief separates the river surface from the Canyon's rim. The formation of the Grand Canyon illustrates a general phenomenon. In regions where the land surface lies well above the base level, a stream can carve a deep trough, much deeper than the

FIGURE 17.15 Canyons and valleys form when uplift of the land, relative to a stream's base level, causes downcutting.

(a) Part of the Grand Canyon, Arizona, a stair-step canyon.

(b) Vertical walls of Canyon de Chelly, Arizona.

(c) V-shaped valleys in the Andes of Peru.

FIGURE 17.16 The shape of a canyon or valley depends on the resistance of its walls to erosion slumping.

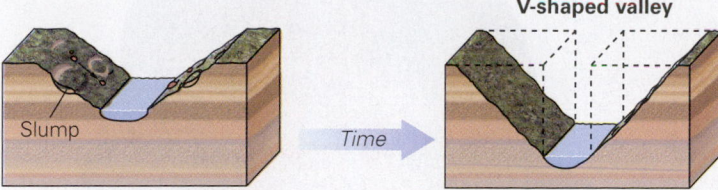

(a) If mass wasting takes place as fast as downcutting occurs, a V-shaped valley develops.

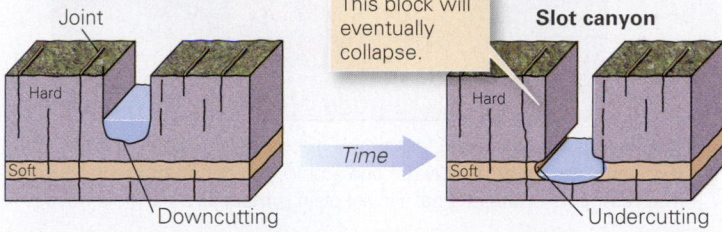

(b) If downcutting by the stream happens faster than mass wasting on the walls, a slot canyon forms. The canyon widens as the stream undercuts the walls.

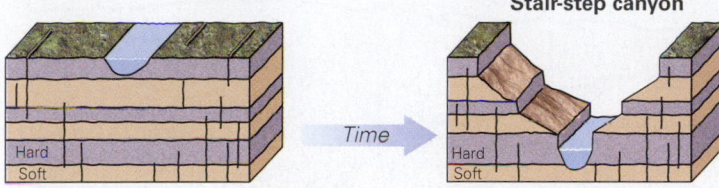

(c) Downcutting through alternating hard and soft layers produces a stair-step canyon.

channel itself. If the walls of the trough slope relatively gently, the landform is a **valley**. If they slope relatively steeply, the landform is a **canyon** (**Fig. 17.15**).

Whether stream erosion produces a valley or a canyon depends on the rate at which downcutting takes place relative to the rate at which mass wasting causes the walls on either side of the stream to collapse. In places where the walls collapse as fast as the stream downcuts, landslides and slumps gradually cause the slope of the walls to approach the angle of repose. When this happens, the stream channel lies at the floor of a valley whose cross-sectional shape resembles the letter V (**Fig. 17.16a**; see 17.15c); this landform is called a *V-shaped valley*. In places where a stream downcuts through its substrate faster than the walls of the stream collapse, erosion creates a *slot canyon* (a relatively narrow, steep-walled canyon). Such canyons typically form in hard rock, which can hold up steep cliffs for a long time (**Fig. 17.16b**; see 17.15b). Where the walls of the stream consist of alternating layers of hard and soft rock, the walls develop a stair-step shape such as that of the Grand Canyon (**Fig. 17.16c**; see 17.15a).

In places where active downcutting occurs, the valley floor remains relatively clear of sediment, for the stream—especially when it floods—carries away sediment that has fallen or slumped into the channel from the stream walls. But if the stream's base level rises, its discharge decreases, or its sediment load increases, the valley floor can fill with sediment, creating an *alluvium-filled valley* (**Fig. 17.17a, b**). The surface of the alluvium becomes a broad floodplain. If the stream's base level later drops again and/or the discharge increases, the stream will start to cut down into the alluvium, a process that generates **stream terraces** bordering the present floodplain (**Fig. 17.17c**).

Rapids and Waterfalls

When Lewis and Clark forged a path up the Missouri River, they came to reaches that could not be navigated by boat because of **rapids**, particularly turbulent water with a rough surface (**Fig. 17.18a**). Rapids form where water flows over steps or large clasts in the channel floor, where the channel abruptly narrows, or where its gradient abruptly changes. The turbulence in rapids produces eddies, waves, and whirlpools that roil and churn the water surface, in the process creating *whitewater*, a mixture of bubbles and water. Modern-day whitewater rafters or kayakers thrill to the unpredictable movement of rapids (**Fig. 17.18b**).

A **waterfall** forms where the gradient of a stream becomes so steep that the water literally free-falls down the streambed (**Fig. 17.18c, d**). The energy of falling water may scour a depression, called a *plunge pool*, at the base of the waterfall. Some waterfalls develop where a stream crosses a resistant ledge of rock, and some develop as a result of faulting because displacement produces an escarpment. Waterfalls also occur where glacial erosion has deepened a trunk valley relative to tributary valleys to form a *hanging valley* (**Fig. 17.18d**) whose mouth is much higher than the floor of the trunk valley (see Chapter 22).

Although a waterfall may appear to be a permanent feature of the landscape, all waterfalls eventually disappear because headward erosion slowly eats back the resistant ledge until the stream reaches grade. We can see this process taking place at Niagara Falls, where water flowing from Lake Erie to Lake Ontario drops over a 55-m-high ledge of resistant Silurian dolostone that overlies a weak shale. Extreme turbulence of the water in the plunge pool erodes the shale and causes undercutting of the dolostone. Gradually, the overhang of dolostone becomes unstable and collapses in a rock fall, with the result that the position of the waterfall migrates upstream. Before the industrial age, the edge of Niagara Falls cut upstream at an average rate of 1 m per year; but since then, the diversion of water from the Niagara River into a hydroelectric power station has decreased the discharge over the falls, cutting the rate of headward erosion in half. Nevertheless, geologists estimate that Niagara Falls will cut all the way back to Lake Erie in about 60,000 years (**Fig. 17.19**).

> **Did you ever wonder . . .**
>
> why a waterfall forms and whether it will always be there?

FIGURE 17.17 The evolution of alluvium-filled stream valleys and the development of terraces.

Time 1

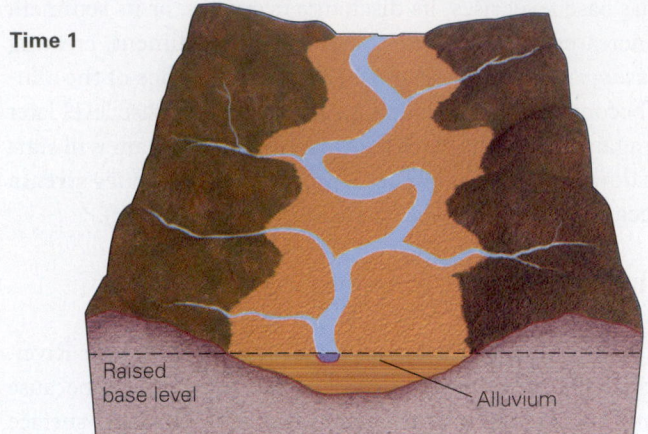

(a) A rise in the base level or a decrease in discharge causes the valley to fill with alluvium.

Time 2

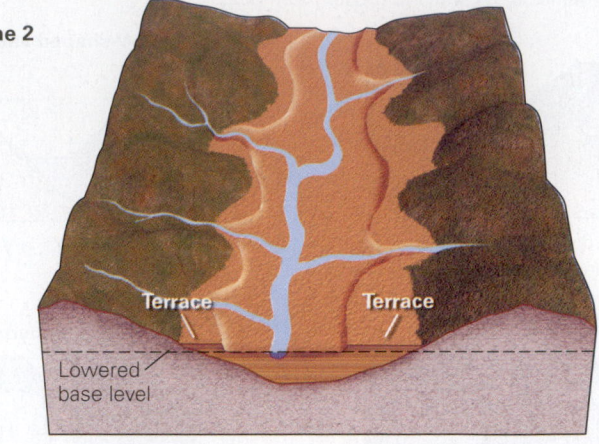

(b) Later, if the base level falls or the discharge increases, the stream downcuts through the alluvium and a new, lower floodplain develops. The remnants of the original alluvial plain remain as a pair of terraces.

(c) In this view of a valley in Utah, we can see a floodplain and two terrace levels.

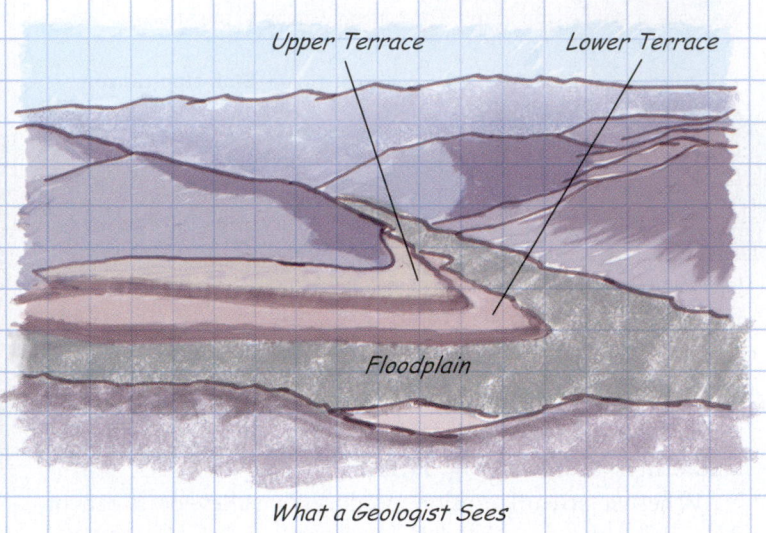

What a Geologist Sees

Alluvial Fans and Braided Streams

Where a fast-moving ephemeral stream abruptly emerges from a mountain canyon onto an open plain, water that had been confined to a narrow channel spreads over a broader surface, slows, and deposits a lens of sediment. The sediment decreases the gradient at the mouth of the outlet that it came from, so the next time the stream flows, it follows a different, steeper path, to one side of the previous lens. Each new flow event deposits its load in a remaining lower area. Over time, therefore, deposition episodes build a broad, gently sloping, wedge-shaped apron of sediment called an **alluvial fan** (**Fig. 17.20a**). Particularly large flows may spread debris over the whole fan and smooth the fan's surface. Individual smaller flows may cut local channels into the fan and carry sediment further downslope.

In some localities, streams carry abundant coarse sediment during floods but cannot carry this sediment during normal flow. Thus, during normal flow, the sediment settles out and chokes the channel. Because the gravelly sediment can't stick together, the stream cannot cut a single deep channel with steep banks—the channel walls simply collapse. As a consequence, the stream divides into numerous strands weaving back and forth between elongate bars of gravel and sand. The result is a **braided stream**—the name emphasizes that the strands entwine like hair in a braid (**Fig. 17.20b**).

Meandering Streams and Their Floodplains

A riverboat cruising along the lower reaches of the Mississippi River cannot sail in a straight line, for the river channel winds back and forth in a series of snake-like curves each of which is called a **meander** (**Fig. 17.21**). In fact, the boat has to go 500 km along the river channel to travel 100 km as the crow flies. A **meandering stream** is one with many meanders. Effectively,

FIGURE 17.18 Examples of rapids and waterfalls.

(a) These rapids in the Grand Canyon formed when a flood from a side canyon dumped debris into the channel of the Colorado River.

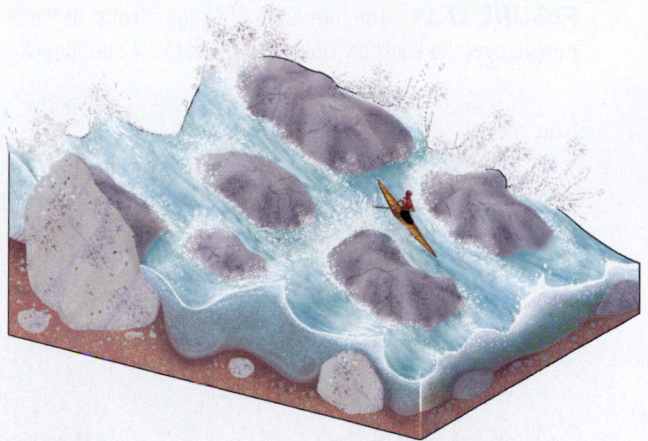

(b) Some rapids are extremely difficult to navigate.

(c) Iguaçu Falls, at the Brazil-Argentina border, spills across layers of basalt. The basalt acts as a resistant ledge.

(d) Waterfall spilling into Milford Sound, New Zealand, from a hanging valley.

the development of meanders increases the volume of water the stream can carry by increasing the stream's length.

How do meanders evolve? Even if a stream starts out with a straight channel, natural variations in the water depth and associated friction cause the fastest-moving current to swing back and forth. The water erodes the side of the stream more effectively where it flows faster, so it begins to cut away faster on the outer arc of the curve. Thus, each curve begins to migrate sideways and grow more pronounced until it becomes a meander. On the outside edge of a meander, erosion continues to eat away at the channel wall and a steep *cut bank* develops. In contrast, on the

inside edge of the meander, water slows down so that its competence decreases and sediment accumulates, forming a wedge-shaped deposit called a *point bar*, as noted earlier. (Mark Twain, who worked as a riverboat pilot on the Mississippi River before writing such classic books as *Huckleberry Finn*, took his pen name from the signals that the mate of a paddle-wheel riverboat called out to the skipper to indicate water depth; *mark twain* means that the water is 2 fathoms, or about 4 m, deep.) With continued erosion, a meander may curve through more than 180°, so that the cut banks of adjacent meanders approach each other, leaving a *meander neck*, a narrow isthmus of land separating two meanders.

FIGURE 17.19 The formation of Niagara Falls, at the border between Ontario, Canada, and New York State. The falls tumble over the Lockport Dolomite, a relatively strong rock layer.

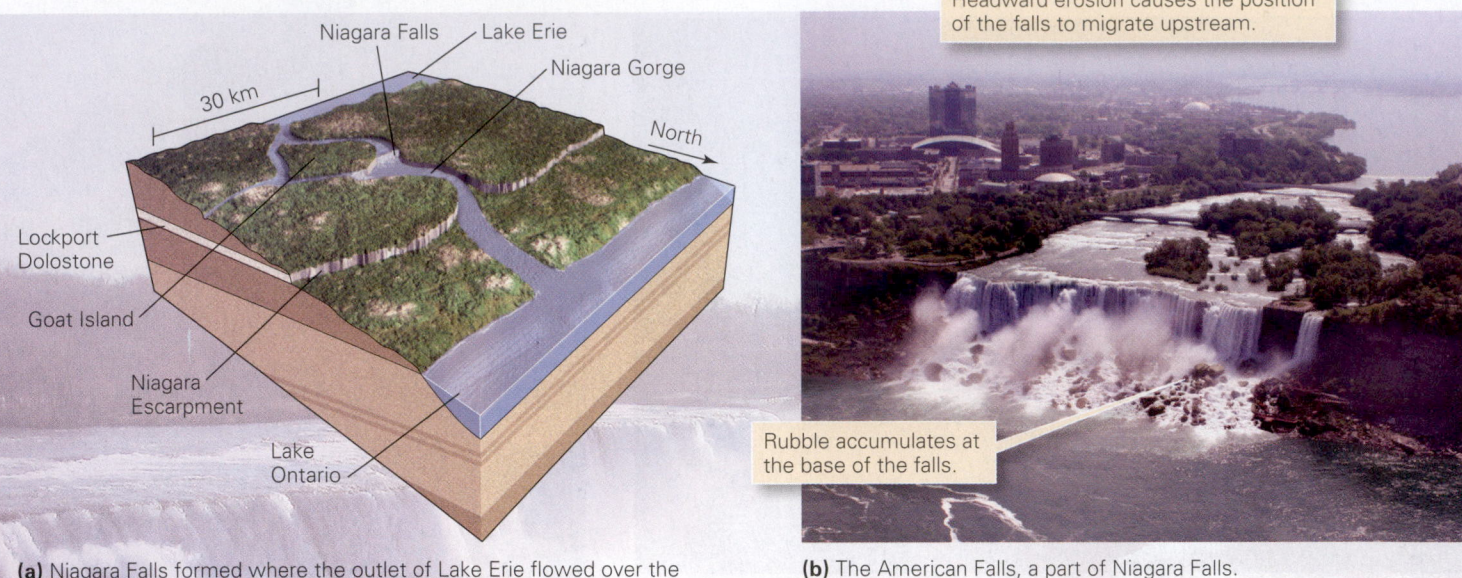

Headward erosion causes the position of the falls to migrate upstream.

Rubble accumulates at the base of the falls.

(a) Niagara Falls formed where the outlet of Lake Erie flowed over the Niagara escarpment.

(b) The American Falls, a part of Niagara Falls.

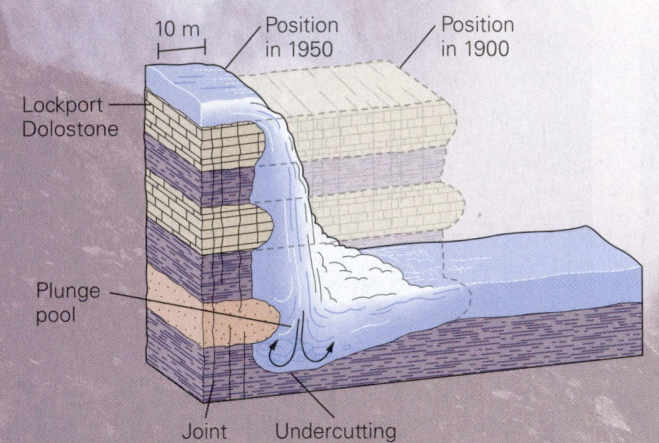

(c) The face of the falls retreats over time. The lower shale erodes, and stronger layers above break off at joints.

(d) In the 1960s, the water was diverted from the falls, making the ledge of dolomite visible.

FIGURE 17.20 Examples of depositional landforms produced from stream sediment.

(a) An alluvial fan in Death Valley, California, consists of sand, gravel, and debris flows. The curving black line is a road.

(b) A braided stream, carrying meltwater from a glacier near Denali, Alaska, deposits elongate bars of gravel.

FIGURE 17.21 The character and evolution of meandering streams and floodplains.

(a) A meandering stream wanders across a floodplain.

(b) The flat land of this floodplain hosts farm fields. Trees delineate the channel.

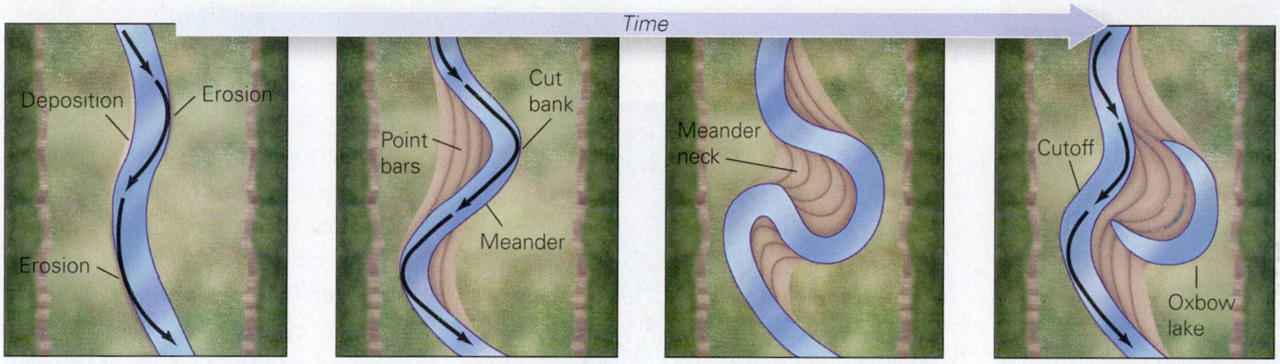

Time

(c) Meanders evolve because erosion occurs faster on the outer bank of a curve, and deposition takes place on the inner curve. Eventually, a cutoff isolates an oxbow lake.

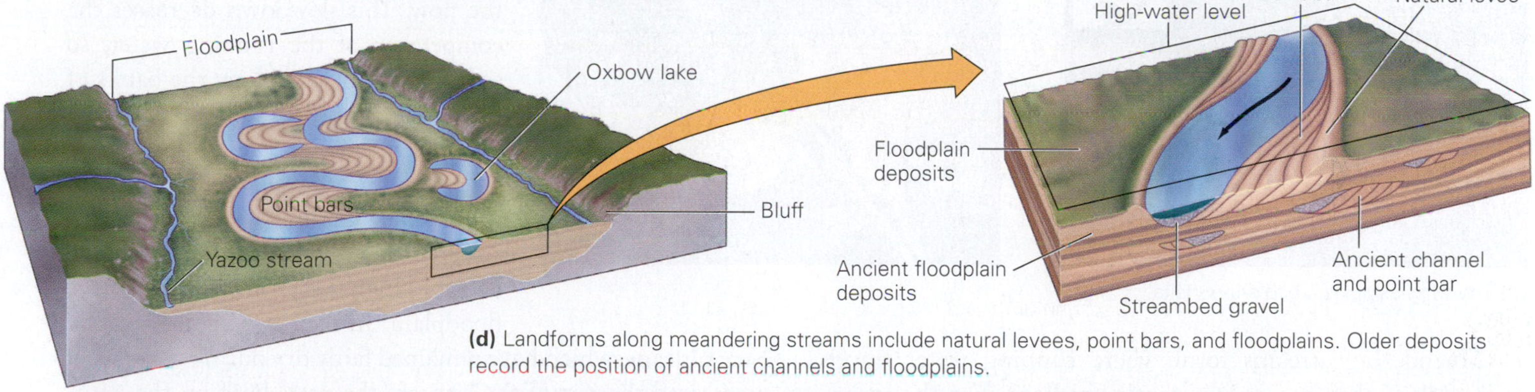

(d) Landforms along meandering streams include natural levees, point bars, and floodplains. Older deposits record the position of ancient channels and floodplains.

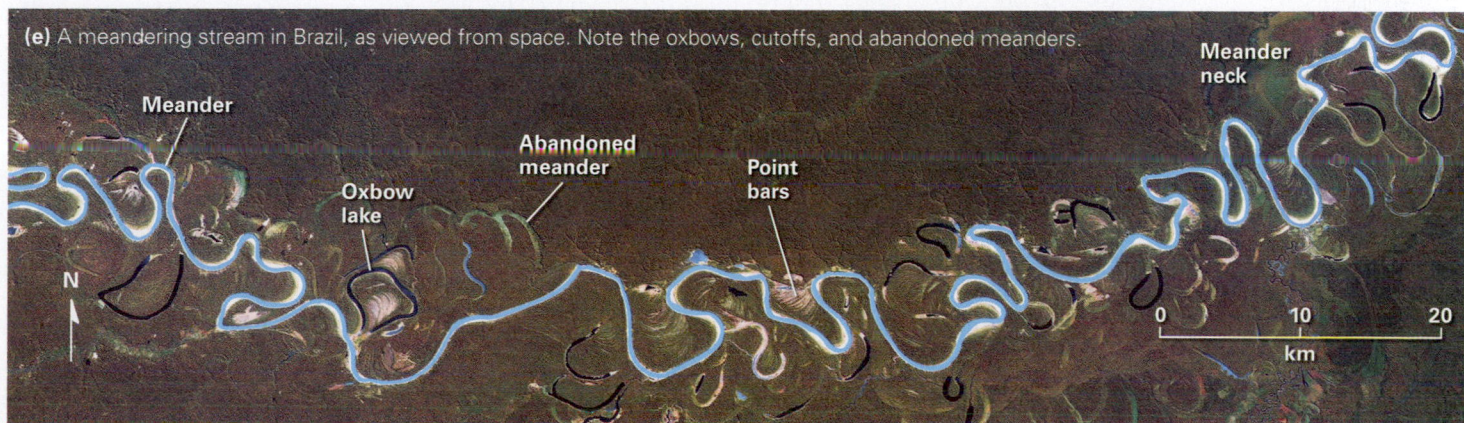

(e) A meandering stream in Brazil, as viewed from space. Note the oxbows, cutoffs, and abandoned meanders.

FIGURE 17.22 Formation of deltas and their variety of shapes.

(a) The Nile is a Δ-shaped delta.

Mediterranean Sea
Africa
Sand
Delta plain

(c) The Niger is an arc-like delta.

Atlantic Ocean
Africa

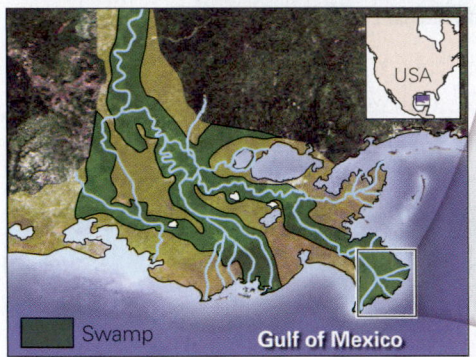

(d) The Mississippi is a bird's-foot delta.

USA
Swamp
Gulf of Mexico

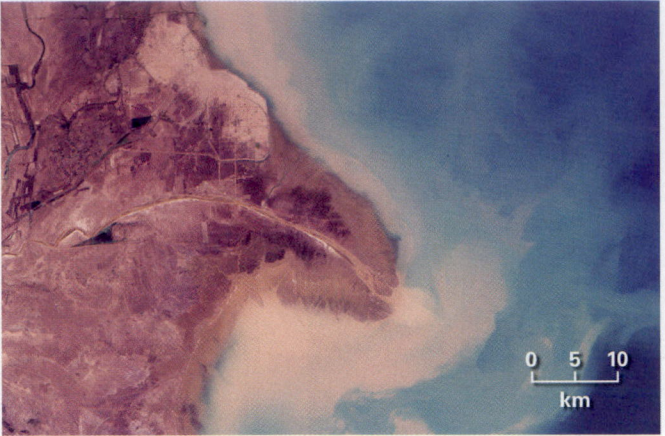

(b) The sediment of the Yellow River, China, settles out where it enters the sea, as seen from space.

0 5 10
km

Mouth of the Mississippi seen from space.

Natural levee

and the establishment of a new one, a process that geologists call an *avulsion*, can take place during a single flood. The meander that has been cut off is called an **oxbow lake** if it remains filled with water or an *abandoned meander* if it dries out.

Most meandering stream channels, during normal flow, cover only a relatively small portion of their broad, gently sloping **floodplain**. Floodplains, as we've noted earlier, are so named because during a flood they become submerged because water overtops the channel bank. In many cases, a floodplain terminates at its sides along a bluff, or escarpment; large floods may cover the entire region from bluff to bluff. As the water leaves the channel, friction between the ground and the thin sheet of water moving over the floodplain slows down the flow. This slowdown decreases the competence of the running water, so sediment settles out along the banks of the channel. Over time, the accumulation of this sediment creates a low ridge called a **natural levee**, on either side of the stream. Natural levees may grow so large that the floor of the channel becomes higher than the surface of the floodplain. In fact, the higher areas of New Orleans, which have remained fairly dry during floods that submerged the rest of the city, are the parts built on the natural levees. In places where large natural levees exist, the region between the bluffs and the levees may become a low, marshy swamp. Also because of the levees, small tributaries may be blocked from joining the trunk stream—these tributaries, called *yazoo streams*, flow in the floodplain and trend parallel to the main river.

Deltas: Deposition at the Mouth of a Stream

Along most of its length, only a narrow floodplain—covered by green, irrigated farm fields—borders the Nile River in Egypt. But at its mouth, the trunk stream of the Nile divides into

Meandering streams form where running water travels over a plain that has only a gentle gradient. For the meander curves to evolve, the substrate of a stream must be strong enough for cut banks to hold up. In unconsolidated substrates, this strength comes from plant roots. Recent research suggests that meandering streams did not develop until abundant land plants appeared during the Silurian—before that all streams on plains may have been braided.

People building communities along a riverbank may assume that the shape of a meander remains fixed for a long time. It doesn't. In a natural meandering river system, the river channel migrates back and forth across the floodplain. When erosion eats through a meander neck, a new straight reach called a *cut-off* develops. The abandonment of a portion of a river channel a fan of smaller streams, called **distributaries**, and the area of green agricultural lands broadens into a triangular patch. The Greek historian Herodotus noted that this triangular patch resembles the shape of the Greek letter delta (Δ), and so the region became known as the Nile Delta (**Fig. 17.22a**). In general, a delta develops wherever sediment-laden water of a stream enters standing water, because as the current slows, the stream loses competence, so sediment settles out (**Fig. 17.22b**). Small deltas form from small streams entering lakes. Huge ones develop where large river systems enter the sea.

Relatively few deltas display the simple triangular shape of the Nile Delta. Some curve out into the sea, whereas others, called *bird's-foot deltas*, consist of many elongate lobes (**Fig. 17.22c, d**). The shape of a delta depends on many factors. Deltas that form where the strength of the river current exceeds that of ocean currents have a bird's-foot shape, since the sediment can be carried far offshore. In contrast, deltas that form where the ocean currents are strong have a Δ shape, for the ocean currents redistribute sediment in bars running parallel to the shore. And in places where waves and currents are strong enough to remove sediment as fast as it arrives, a river has no delta at all.

Why do rivers divide into distributaries at their mouths? When a river reaches standing water, its velocity slows. The sediment settles out at the mouth to form a midstream bar (**Fig. 17.23a**). The presence of the bar causes the stream to split into two channels. Similar bars created at the mouths of these two subsidiary channels cause each of them to separate in turn, until eventually numerous distributary channels have formed (**Fig. 17.23b**).

With time, the sediment of a large marine delta compacts, and the lithosphere beneath the delta subsides (see Chapter 7). As a consequence, the surface of a delta slowly sinks. In a natural delta, distributaries provide sediment that fills the resulting space so that the delta's entire surface remains at or just above sea level, forming a broad flat area called a *delta plain*.

But if people build up artificial levees to constrain the river to its channel, sediment bypasses the interior part of a delta and gets carried directly to the seaward edge of the delta. Thus, the delta's interior "starves" (does not receive sediment), and when this happens, the land surface drops below sea level. Because of this process, much of New Orleans lies below sea level, so high waters from Hurricane Katrina in 2005 caused the city to flood extensively (see Chapter 20).

The position of a delta at the mouth of a major meandering river may change over time, because the main course of the river can shift over time. Sometimes these shifts are a result of an avulsion associated with a meander cutoff just upstream of the delta. And sometimes shifts happen when a toe of a bird's-foot delta builds so far out into the sea that the gradient of the stream becomes too gentle to allow the river to flow. At this point, the river overflows a natural levee upstream, an avulsion takes place, and the stream begins to flow along a new channel. The distinct lobes of the Mississippi Delta, a bird's-foot delta, reflect avulsions that have happened several times during the past 9,000 years (**Fig. 17.24**). New Orleans, built along one of the Mississippi's distributaries, may eventually lose its riverfront, for a break in a levee upstream of the city could divert the Mississippi into the Atchafalaya River channel.

Take-Home Message

Erosion carves valleys and canyons, with shapes that depend on the balance between slope-collapse and downcutting rates. Streams choked with sediment become braided; those following snake-like paths across floodplains are meandering. Meandering streams do not stay in the same place over time, due to the formation of cutoffs. Deltas build where a stream empties into standing water.

QUICK QUESTION: What factors cause the formation of rapids and waterfalls?

FIGURE 17.23 The formation of distributaries on a delta.

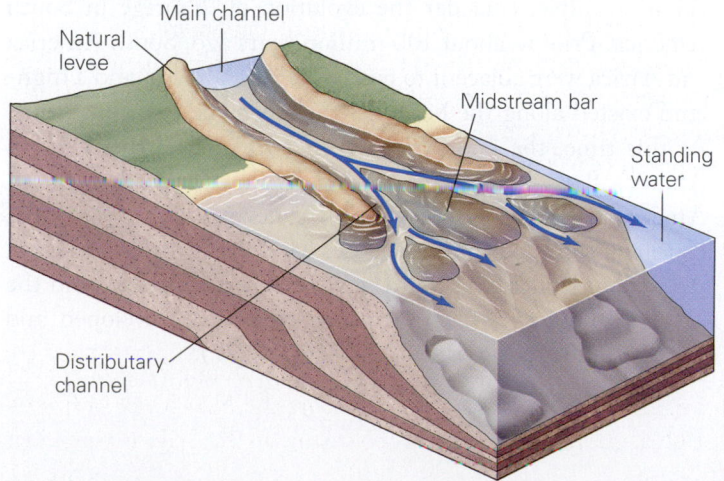

(a) Distributaries form because the river deposits sediment in its mouth when it reaches standing water.

(b) A small delta forming in Lake Tekapo, New Zealand, shows several distributaries.

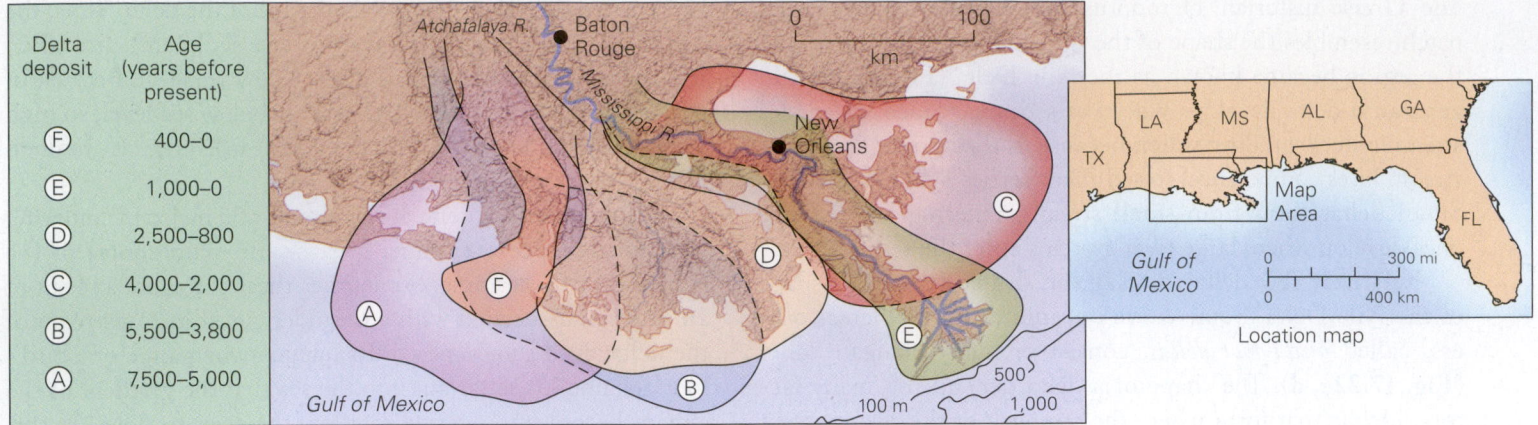

FIGURE 17.24 A map showing ancient lobes of the Mississippi Delta. A major flood could divert water from the Mississippi into the channel of the Atchafalaya.

Delta deposit	Age (years before present)
F	400–0
E	1,000–0
D	2,500–800
C	4,000–2,000
B	5,500–3,800
A	7,500–5,000

17.7 The Evolution of Drainage

Beveling Topography

Over time, *fluvial landscapes*—those modified by the erosion and/or deposition by streams—gradually evolve (**Fig. 17.25**). To see the progression, imagine that a region undergoes significant uplift. As soon as relief develops, rivers will get to work and a drainage network forms. In these "young" fluvial landscapes, streams of the network have steep gradients and their water churns down rapids and tumbles over waterfalls as they carve deep valleys or canyons. But with time, mass wasting along valley walls, combined with the removal of debris by running water, transforms rugged mountains into low, rounded hills, and once-deep, narrow valleys broaden into wide floodplains with gentler gradients. Eventually, streams that had followed straight courses down steep gradients begin to meander on the surfaces of nearly horizontal floodplains. As more time passes, even low hills erode away, so the landscape becomes a "mature" one with low relief, and its elevation lies close to the elevation of the stream's base level. (Some geologists refer to such a low-relief landscape as a *peneplain*, from the Latin *paene*, which means almost.) Through these stages, extensive *denudation*—the removal of rock and regolith from the Earth's surface—can take place. Of course, this idealized model, like many models of components in the Earth System, is an oversimplification. Typically, tectonic processes cause the region to undergo uplift or subsidence again, and/or

global sea level rises or falls over time, so the base level of the drainage network likely changes before a peneplain ever has a chance to form.

Stream Piracy and Drainage Reversal

Stream piracy sounds like pretty violent stuff. In reality, it's just a natural process that happens when headward erosion by one stream causes the stream to intersect the course of another stream. When this happens, the pirate stream "captures" the water of the stream that it has intersected, so that the water of the captured stream starts to flow down the channel of the pirate stream. Because of piracy, the channel of the captured stream, downstream of the point of capture, dries up (**Fig. 17.26**).

We tend to think of a given drainage network as always flowing in the same direction, because gravity moves water from a higher to lower elevation. But tectonic processes, acting over geologic time, can change the slope of the land on a continental scale. If the overall tilt of the land reverses, a **drainage reversal** will take place; that is, a drainage network reorganizes so that water flows, overall, in the opposite direction. As an example, consider the evolution of drainage in South America. Prior to about 100 million years ago, South America and Africa were adjacent to each other in Pangaea, and a highland existed along the boundary between the two continents. At this time, the main drainage network of northern South America flowed westward to the Pacific. Later, when South America rifted away from Africa, the northeastern coast of the continent subsided, and the Andes Mountains rose to the west. As a consequence, westward flow became impossible, and the eastward-flowing Amazon drainage network developed and has continued to the present day (**Fig. 17.27**).

FIGURE 17.25 Evolution of a fluvial landscape when base level drops.

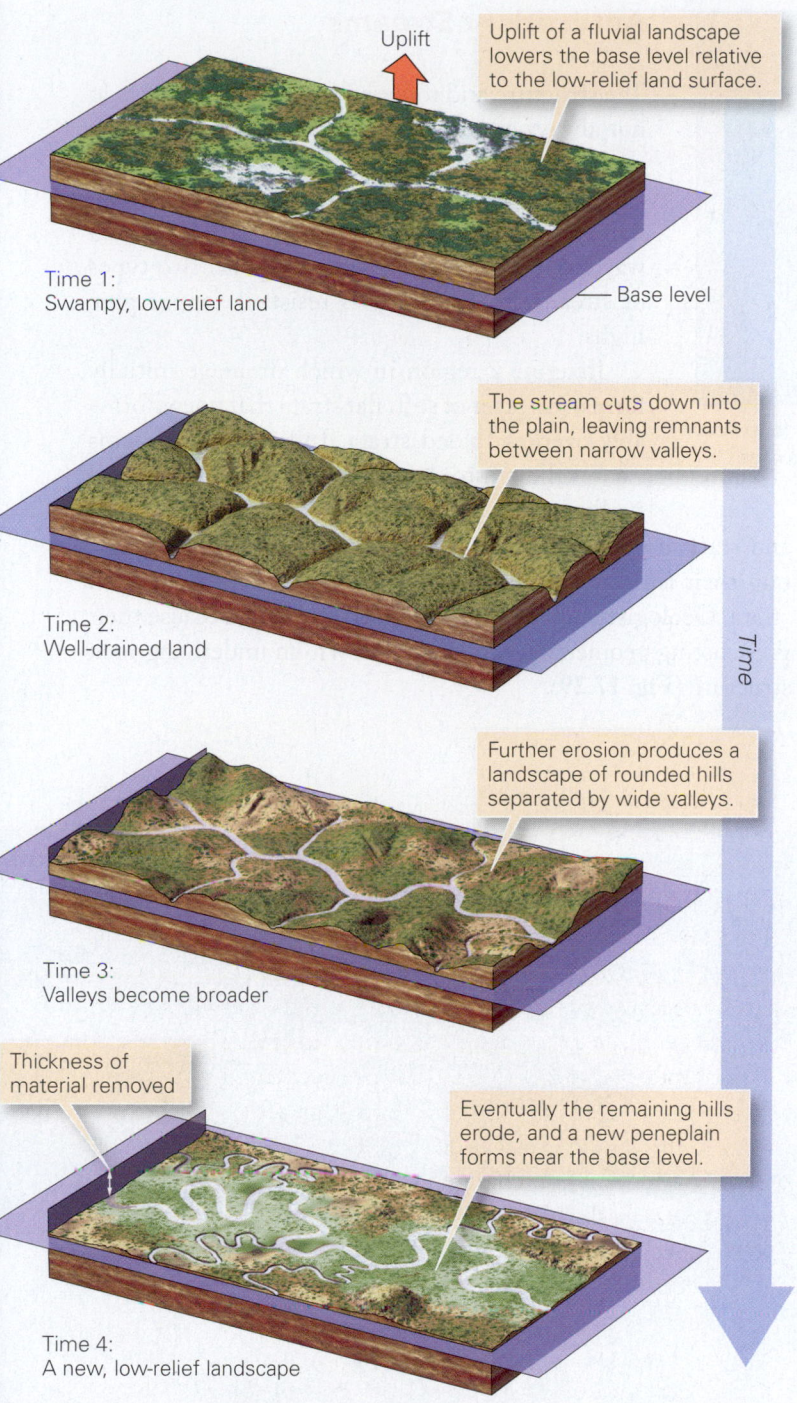

Uplift

Uplift of a fluvial landscape lowers the base level relative to the low-relief land surface.

Base level

Time 1:
Swampy, low-relief land

The stream cuts down into the plain, leaving remnants between narrow valleys.

Time 2:
Well-drained land

Further erosion produces a landscape of rounded hills separated by wide valleys.

Time 3:
Valleys become broader

Thickness of material removed

Eventually the remaining hills erode, and a new peneplain forms near the base level.

Time 4:
A new, low-relief landscape

Time

Stream Rejuvenation

Where streams cut down into a landscape of low relief that was originally near the stream's base level, **stream rejuvenation** has occurred. Rejuvenation happens when the base level of a stream drops, when land rises beneath a stream, or when the discharge of a stream increases. What's the evidence that rejuvenation took place at a given locality? In the case of a stream flowing in an alluvium-filled valley, renewed downcutting allows the stream to create a new floodplain at a lower elevation than the original one (see Fig. 17.17b). As we have seen, the younger floodplain tends to be narrower than the older, and the surface of the older floodplain becomes a terrace on either side of the new floodplain. In the case of a stream flowing on bedrock, a drop of the base level causes the stream to "incise" (cut down into) the underlying bedrock

FIGURE 17.26 The concept of stream capture or piracy.

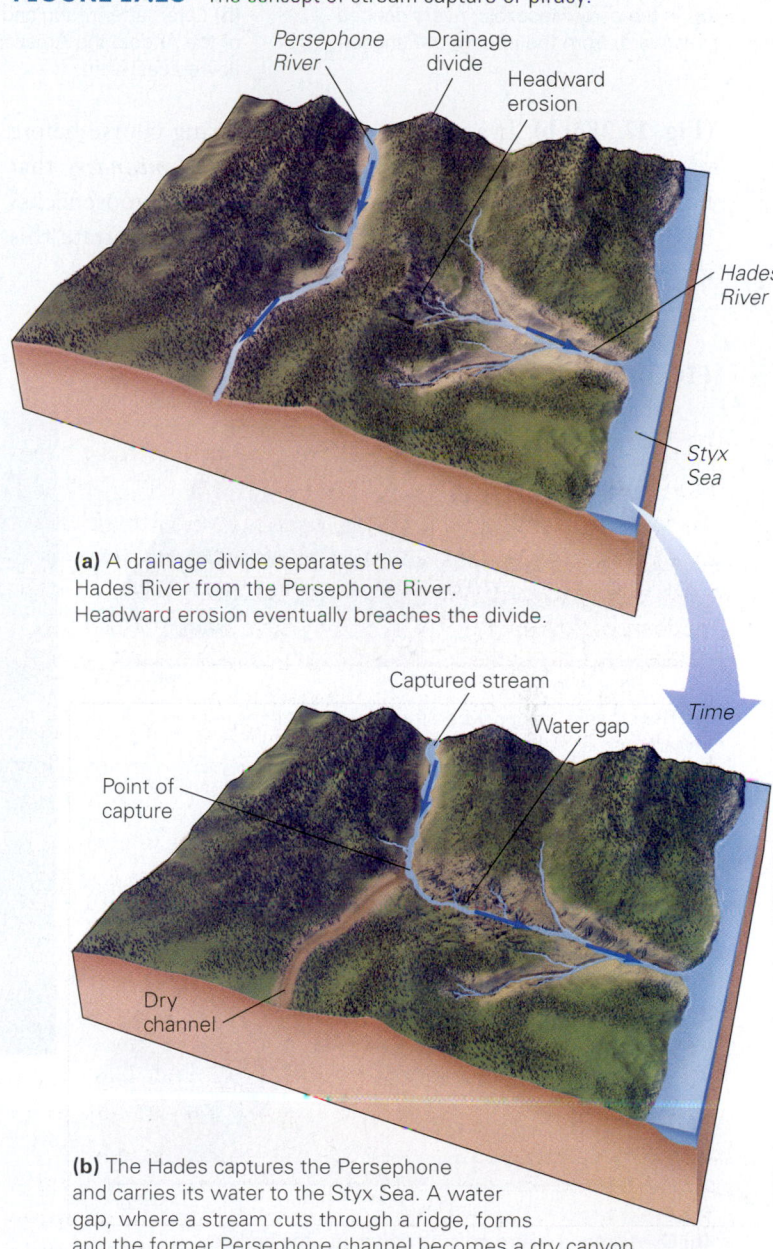

Persephone River   Drainage divide   Headward erosion

Hades River

Styx Sea

(a) A drainage divide separates the Hades River from the Persephone River. Headward erosion eventually breaches the divide.

Captured stream

Point of capture

Water gap

Time

Dry channel

(b) The Hades captures the Persephone and carries its water to the Styx Sea. A water gap, where a stream cuts through a ridge, forms and the former Persephone channel becomes a dry canyon.

FIGURE 17.27 An example of continental-scale drainage reversal in South America.

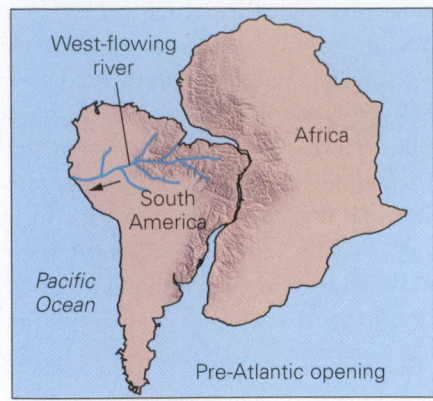

(a) In the early Mesozoic, rivers drained westward, from the interior of Pangaea.

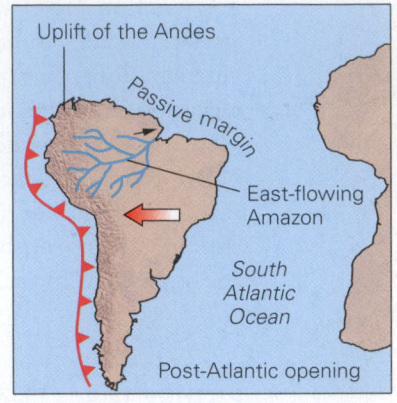

(b) Later, after rifting and the formation of the Andes, the Amazon drainage flowed eastward.

Superposed and Antecedent Streams

The structure and topography of the landscape do not always appear to control the path, or course, of a stream. For example, imagine a stream that carves a deep canyon straight across a strong mountain ridge—why didn't the stream find a way around the ridge? We distinguish two types of streams that cut across resistant topographic highs.

Imagine a region in which drainage initially forms on a layer of soft, flat strata that unconformably overlies folded strata. Initially, the streams carve valleys into the flat strata. When they eventually erode down through the unconformity and start to downcut into the folded strata, they may maintain their earlier course, ignoring the structure of the folded strata. Geologists call these **superposed streams**, because their pre-existing geometry has been laid down on underlying rock structure (**Fig. 17.29**).

(**Fig. 17.28a, b**). If a stream had a meandering course before rejuvenation, it will downcut to form *incised meanders* that lie at the bottom of a steep-walled canyon. The "goosenecks" of the San Juan, a landform in southern Utah, illustrate this geometry (**Fig. 17.28c**).

FIGURE 17.28 The formation of incised meanders.

(a) The process begins when the base level drops relative to a stream that is meandering on a plain.

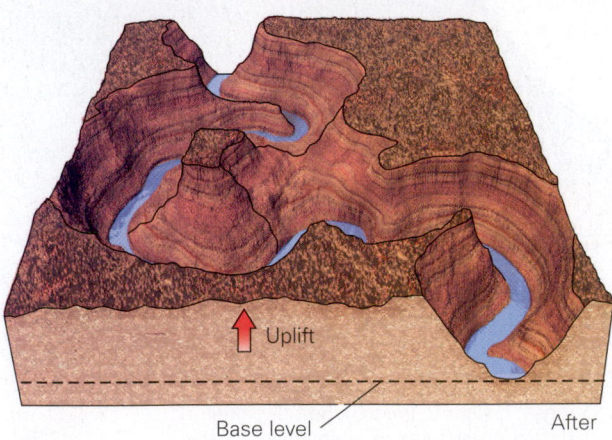

(b) Over time, the stream cuts down into the bedrock. Meanders continue to evolve while this happens.

(c) The "goosenecks" of the San Juan River, Utah, are incised meanders.

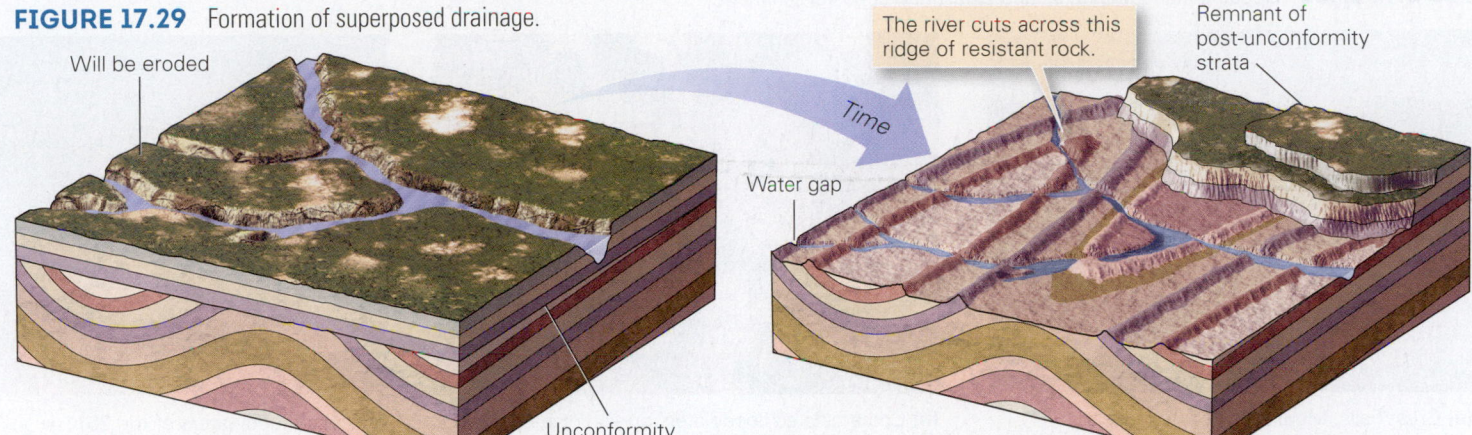

FIGURE 17.29 Formation of superposed drainage.

Will be eroded

The river cuts across this ridge of resistant rock.

Remnant of post-unconformity strata

Time

Water gap

Unconformity

(a) A superposed stream establishes its geometry while flowing over a uniform substrate above an unconformity.

(b) When erosion exposes underlying rock with a different structure, the river is superposed on the structure. As a result, it cuts across resistant ridges instead of flowing around them.

In some cases, tectonic activity, such as subduction or collision, causes localized uplift, so a mountain range rises beneath the course of an already established stream. If the stream downcuts as fast as the range rises, it can maintain its course and will cut right across the range. Geologists call such streams **antecedent streams**, from the Greek *ante*, meaning before, to emphasize that they existed before the range uplifted. Note that if the range rises faster than the stream can downcut, the new highlands divert the stream's course so that it starts to flow parallel to the range face (**Fig. 17.30**).

FIGURE 17.30 Development of antecedent and diverted streams.

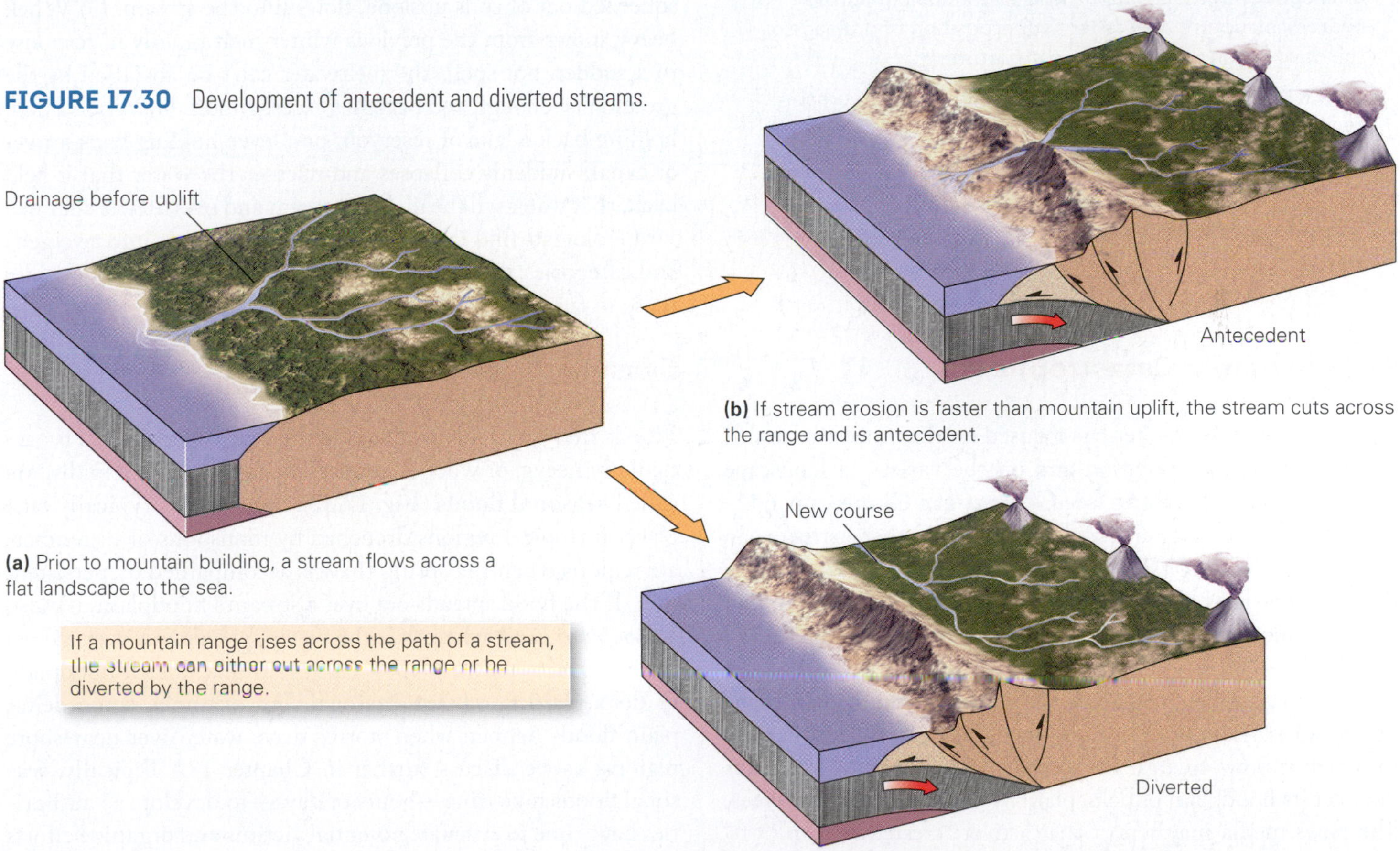

Drainage before uplift

Antecedent

(b) If stream erosion is faster than mountain uplift, the stream cuts across the range and is antecedent.

New course

(a) Prior to mountain building, a stream flows across a flat landscape to the sea.

If a mountain range rises across the path of a stream, the stream can either cut across the range or be diverted by the range.

Diverted

(c) If uplift happens faster than erosion, the stream is diverted and flows along the edge of the range.

FIGURE 17.31 Flooding inundates fields and cities, and deposits sediment.

(a) Great Falls, Montana, was submerged by floodwaters in 1975.

(b) Contaminated floodwaters spread disease during a 2007 flood in Indonesia.

(c) When floodwaters of the 2011 Missouri River flood in Nebraska receded, they left a layer of sediment on farm fields as well as desert-like conditions.

Take-Home Message

Stream-carved landscapes evolve over time as gradients diminish and ridges between valleys erode away. Eventually, the landscape may attain low relief. Subsequent changes in base level can trigger rejuvenation. Over time, some streams capture the flow of other streams. And if the regional tilt of the land surface changes, the direction of drainage reverses. Locations where streams cut across bedrock structure may reflect superposition of drainage or the downcutting through rising structure.

QUICK QUESTION: What can cause a drainage-reversal event to take place?

17.8 Raging Waters

The Inevitable Catastrophe

Up until now, this chapter has focused on the process of drainage formation and evolution and on the variety of landscape features formed by streams (see **Geology at a Glance**, pp. 642–643). Now we turn our attention to the havoc that a stream can cause when flooding takes place. Floods can be catastrophic—they can strip land of forests and buildings, they can bury land in mud and silt, and they can submerge communities (**Fig. 17.31**). As noted earlier, a flood takes place when the volume of water flowing down a stream exceeds the volume of the stream channel, so water rises out of its normal channel and fills the canyon in which it flows to a greater depth than normal, or it spreads out over its floodplain or delta plain by breaking through levees. The news media may report that a river "crested at 3 m (9 ft) above flood stage at 10 p.m." This means that the water surface in the stream was 3 m higher than the top of the normal

channel at 10 p.m., and after that time it became lower—the *flood crest* is the highest level that the stream reaches.

Floods can happen for several reasons: (1) During abrupt, heavy rains, water falls on the ground faster than it can infiltrate deep into the subsurface. The excess becomes runoff that can fill the stream channel beyond capacity. (2) After a long period of continuous rain, the ground becomes saturated with water and can hold no more. Any additional rain, along with water squeezed out of soils upslope, flows into the stream. (3) When heavy snows from the previous winter melt rapidly in response to a sudden hot spell, the meltwater can't be absorbed by the ground fast enough and becomes excess runoff. (4) When a dam holding back a lake or reservoir, or a levee holding back a river or canal, suddenly collapses and releases the water that it held back, that water will head downstream and overfill the channel.

Geologists find it convenient to divide floods into two general categories: seasonal floods and flash floods. Let's consider these in turn.

Seasonal Floods

Floods that occur during times of the year when rainfall is particularly heavy, or when winter snow starts to melt rapidly, are called **seasonal floods** (**Fig. 17.32**). Such floods typically take place in tropical regions drenched by monsoons or in temperate regions when the spring thaw is accompanied by persistent rain. If the flood spreads out over a stream's floodplain, it's also a *floodplain flood*, whereas if the water spreads out over a delta plain, it's also a *delta-plain flood*. (Note that not all floodplain or delta-plain floods are seasonal. For example, some delta-plain floods happen when storms drive water over near-shore regions, as we discuss further in Chapter 18.) Typically, seasonal floods take time—hours or days—to develop, so authorities have time to evacuate potential victims and organize efforts to protect property. But because preparation and/or evacuation doesn't always take place, and because so many people live on

FIGURE 17.32 Examples of seasonal floods, in a floodplain.

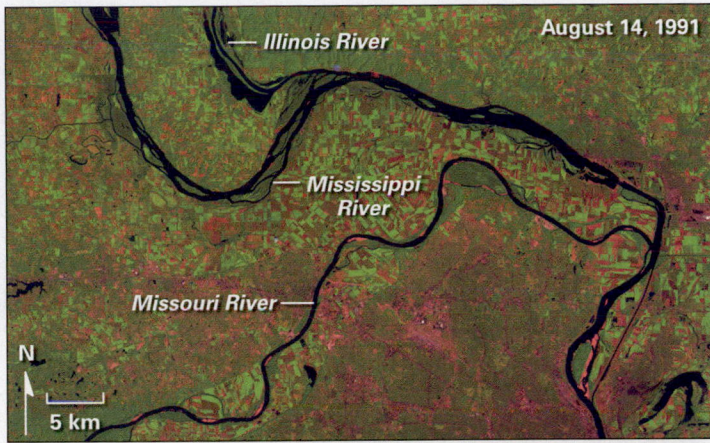

Illinois River — August 14, 1991
— Mississippi River
Missouri River —
N
5 km

August 19, 1993
N
5 km

(a) Satellite images show how the rivers in the Midwestern United States covered their floodplans.

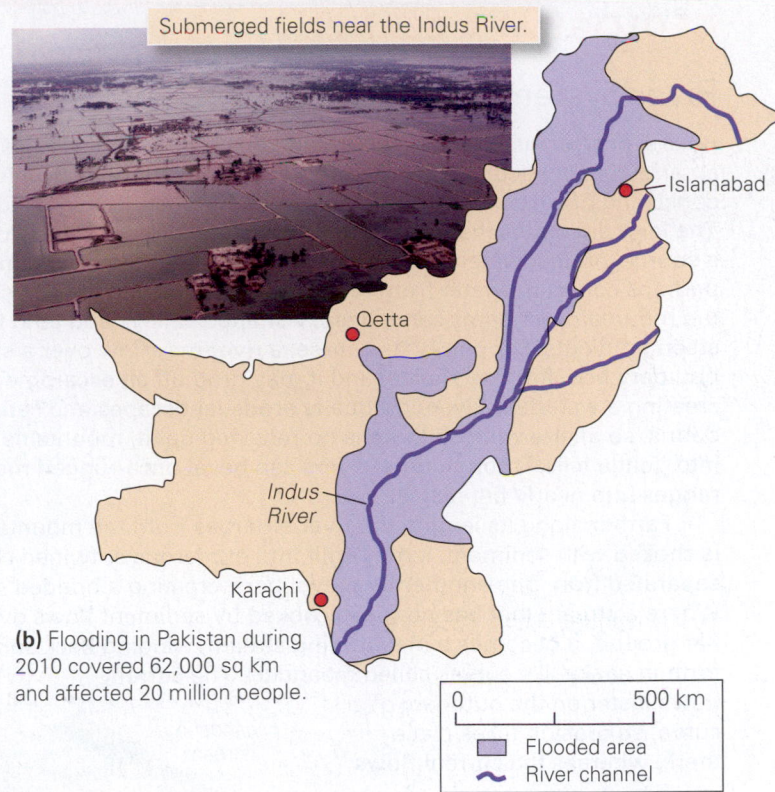

Submerged fields near the Indus River.

Islamabad
Qetta
Indus River
Karachi

(b) Flooding in Pakistan during 2010 covered 62,000 sq km and affected 20 million people.

0   500 km

Flooded area
River channel

floodplains and delta plains, these floods can cause a staggering loss of life and property.

Seasonal floods happen every year, somewhere. Indeed, along many rivers, seasonal flooding serves an important role in the sustainability of ecosystems and societies. For example, seasonal floods along the Nile River and its delta played a major role in replenishing nutrients to agricultural areas of Egypt. Unfortunately, some seasonal floods cause so much damage and loss of life that they make a mark in history, either locally or globally. For example, an 1887 flood of China's Hwang (Yellow) River—so named because of the yellow silt it carries—killed as many as 2.5 million people, and a 1931 flood of the Yangtze River in China led to a famine that killed 3.7 million people. During the 1990 monsoon season in Bangladesh, rain fell almost continuously for weeks; the delta plain became inundated, and the resulting flood killed 100,000 people. Seasonal floods struck Indonesia in 2007, killing dozens of people and displacing almost half a million, nearly half of whom became sick from contact with filthy water and mud that submerged 60% of the capital and hundreds of square kilometers of farmland. One of the most devastating floods of recent times began in July 2010, when a seasonal flood fed by

intense monsoonal rains submerged floodplains of the Indus River drainage system in Pakistan (see Fig. 17.32b). On some days, it had rained up to 40 cm (1.3 ft) in 24 hours! The floods put almost 70,000 km² of the country underwater and severely impacted the lives of over 20 million people (12% of the population)—many people lost all they owned. Crops growing in the fertile floodplain floated away or rotted, and over 200,000 cattle drowned. Due to the destruction of clean-water supplies and of road and rail networks, survivors were stranded for weeks or more, and sadly, disease spread despite relief efforts by organizations from around the world.

Seasonal flooding happens frequently in the Mississippi Valley of the central United States, and in some years, the floods are disasterous. One such year was 1993, which can serve as a case study of such an event. High-altitude (10- to 15-km-high) winds drifted southward and for weeks formed an invisible wall that trapped warm, moist air from the Gulf of Mexico over the central United States. This air rose to higher elevations and cooled, and the water it held condensed and fell as rain, rain, and more rain. In fact, almost a whole year's supply of rain fell during the spring of 1993, and some regions received 400% more than usual. The ground became saturated and could no longer absorb additional water, so the excess entered the region's streams, which carried it into the Missouri and Mississippi Rivers. Eventually, the water in these rivers rose above the height of levees and spread out over the floodplain. By July, parts of nine states were under water (Fig. 17.32a).

River Systems

Rivers, or streams, drain the landscape of surface runoff. Typically, an array of connected streams called a drainage network develops, consisting of a trunk stream into which numerous tributaries flow. The land drained is the network's watershed. A stream starts from a source, or headwaters. Some headwaters are in the mountains, perhaps collecting water from rainfall or from melting ice and snow. In the mountains, streams carve deep, V-shaped valleys and tend to have steep gradients. For part of its course, a river may flow over a steep, bouldery bed, forming rapids, and it may drop off an escarpment, creating a waterfall. Rivers gradually erode landscapes and carry away debris, so after a while, if there is no renewed uplift, mountains evolve into gentle hills. Through time, rivers can bevel once-rugged mountain ranges into nearly flat plains.

Farther along its length, the river emerges from the mountains. If it is choked with sediment, it may split into numerous entwined channels separated from one another by gravel bars, creating a braided stream. Where a stream that has not been choked by sediment flows over flat ground, it becomes a meandering stream, winding back and forth in snake-like curves called meanders. The current flows faster on the outer arc of a curve, so erosion takes place there, whereas the current flows

Developing drainage networks.

Transportation along the channel

Rapids

Braided channel

Meandering stream

Cut bank

Deposition

Terraced floodplain

(present floodplain)

(oldest floodplain)

Bank erosion

Deposition of point bar

Back swamps

Yazoo stream

Wide meanders

Neck

Oxbow lake

Cutoff

Natural levees

Wide floodplain

Point bars forming on inner curves.

Meanders, abandoned meanders, and cutoffs.

A small delta in a mountain lake.

Headward erosion

Valleys with
high relief

Glaciers

Melting
ice

Lake

Dendritic
drainage

Rapids

Waterfall

**Collection of water
in watershed**

Streams contribute to carving mountains.

Waterfall in Hawaii spilling over a basalt ledge.

more slowly on the inner arc, where it drops sediment. Because of erosion and deposition, a meandering stream changes shape over time. Occasionally a meander may be cut off, leaving a curving lake called an oxbow lake. A broad floodplain, covered with water only during floods, may develop on either side of the stream. Natural levees build up between the channel and the floodplain from sediment dropped as a flooding river starts to spill out of its channel. Eventually, a river reaches a standing body of water and slows down, and the sediment it carries gets deposited to form a delta. On a delta, the trunk stream divides into many smaller channels called distributaries.

**Deposition
at mouth**

Delta

Distributaries

Natural levees

Swamps and
marsh

Tidal flats

Bar

Banks

The roiling, muddy flood of 1993 uprooted trees, swept cars away, and even unearthed coffins (which floated out of the inundated graveyards). All barge traffic along the Mississippi came to a halt, bridges and roads were undermined and washed away, and towns along the river were submerged in muddy water. For example, in Davenport, Iowa, the riverfront district and baseball stadium were covered with 4 m (14 feet) of water. In Des Moines, Iowa, 250,000 residents lost their supply of drinking water when floodwaters contaminated the municipal water supply with raw sewage and chemical fertilizers. Rowboats replaced cars as the favored mode of transportation in towns where only the rooftops remained visible. In St. Louis, Missouri, the river crested 14 m (47 ft) above flood stage. When the water finally subsided, it left behind a thick layer of silt and mud, filling living rooms and kitchens in floodplain towns and burying crops in floodplain fields. For 79 days, the flooding continued. In the end, more than 40,000 km² of the floodplain had been submerged, 50 people died, at least 55,000 homes were destroyed, and countless acres of crops were buried. Officials estimated that the flood caused over $12 billion in damage. Comparable flooding happened again in the spring of 2011.

Flash Floods

Events during which the floodwaters rise so fast that it may be difficult to escape from the path of the water are called **flash floods** (Fig. 17.33). A flash flood may be the harbinger of a seasonal flood in that the high waters may arrive rapidly but remain for a long time, or it may be a short-lived event in the wake of an intense rainstorm, a dam collapse (Box 17.1), or a levee failure, in which the high waters subside in a matter of minutes to hours. The water may rise so fast that there's no time for it to sink into the ground even if the ground is unsaturated. Flash floods can be particularly unexpected in arid or semiarid regions, where water from an isolated rainstorm may suddenly fill the channel of an otherwise dry wash. Such a flood may even affect areas downstream that had not received a drop of rain. If a flash flood affects a drainage basin in which streams have cut steep-sided valleys or canyons without a floodplain, the flood may arrive as a wall of water, slamming downstream with great force, and water may quickly fill the canyon or stream to meters to tens of meters above normal in a matter of minutes.

The historic Big Thompson River flood of July 31, 1976, illustrates the power of a flash flood. This river, which drains the Front Range of the Rocky Mountains, north of Denver, Colorado, normally seems quite harmless. It carries clear water, dripping from rains or from melting ice and snow. The stream has a steep gradient, so this water froths over and around boulders and cobbles of the streambed. The channel of the river does not occupy the whole valley floor, so a road follows the course of the river, and in places, vacation cabins and campgrounds dot the land between the river and the road. On July 31, warm, moist air blew toward the Rocky Mountain front. As this air rose over the mountains, towering thunderheads built up, and at 7:00 P.M., rain began to fall. It poured, in quantities that even old-timers couldn't recall. In a little over an hour, 19 cm (7.5 in.) of rain drenched the watershed of the Big Thompson River. The river's discharge grew to more than four times the maximum recorded during the previous century, and the water level rose by several meters.

Turbulent currents swirled down the canyon at up to 8 m per second and churned up so much sand and mud that the once clear river became a viscous slurry. Slides of rock and soil tumbled down the steep slopes bordering the river and

FIGURE 17.33 Flash floods can occur after torrential rains.

(a) A flash flood in a desert region of Israel has washed over a highway, forcing the evacuation of truckers.

(b) During the 1976 Big Thompson River flash flood, this house was carried off its foundation and dropped on a bridge.

BOX 17.1 CONSIDER THIS . . .

The Johnstown Flood of 1889

By the 1880s, Johnstown, built along the Conemaugh River in scenic western Pennsylvania, had become a significant industrial town. Recognizing the attraction of the surrounding hills as a summer retreat, speculators built a mud-and-gravel dam across the river, upstream of Johnstown, to trap a pleasant reservoir of cool water. A group of industrialists and bankers bought the reservoir and established the exclusive South Fork Hunting and Fishing Club, a cluster of lavish 15-room "cottages" on the shore. Unfortunately, the dam had been poorly designed, and debris blocked its spillway (a passageway for surplus water), setting the stage for a monumental tragedy. On May 31, 1889, torrential rain drenched Pennsylvania, and the reservoir filled until water began to flow over the dam. Despite frantic attempts to strengthen the dam, the soggy structure abruptly collapsed, and the reservoir emptied into the Conemaugh River Valley. A 20-m-high wall of water roared downstream and slammed into Johnstown, transforming bridges and buildings into twisted wreckage (**Fig. Bx17.1**). When the water subsided, 2,300 people were dead, and Johnstown became the focus of national sympathy. The recently founded Red Cross set to work building dormitories, and citizens nationwide donated everything from clothes to beds. Nevertheless, it took years for the town to recover, and many residents simply picked up and left.

FIGURE Bx17.1   During the 1889 Johnstown flood, raging waters could tumble large houses.

fed the torrent with even more sediment. The raging water undercut foundations and washed the houses and bridges away. Parts of the road were eroded away, and other parts were buried by debris. Boulders that had stood like landmarks for generations bounced along in the torrent like beach balls, striking and shattering other rocks along the way; the largest rock known to be moved by the flood weighed 275 tons. Cars drifted downstream until they finally wrapped like foil around obstacles. When the flood subsided, the canyon had changed forever, and 144 people had lost their lives. As we described at the beginning of this chapter, flooding devastated the area again in 2013.

Ice-Age Torrents

Perhaps the greatest floods chronicled in the geologic record happen when natural ice dams burst. The Great Missoula Flood of about 11,000 years ago illustrate this phenomenon. This flood occurred at the end of the last ice age, when a glacier acted like a dam, holding back a large lake called Glacial Lake

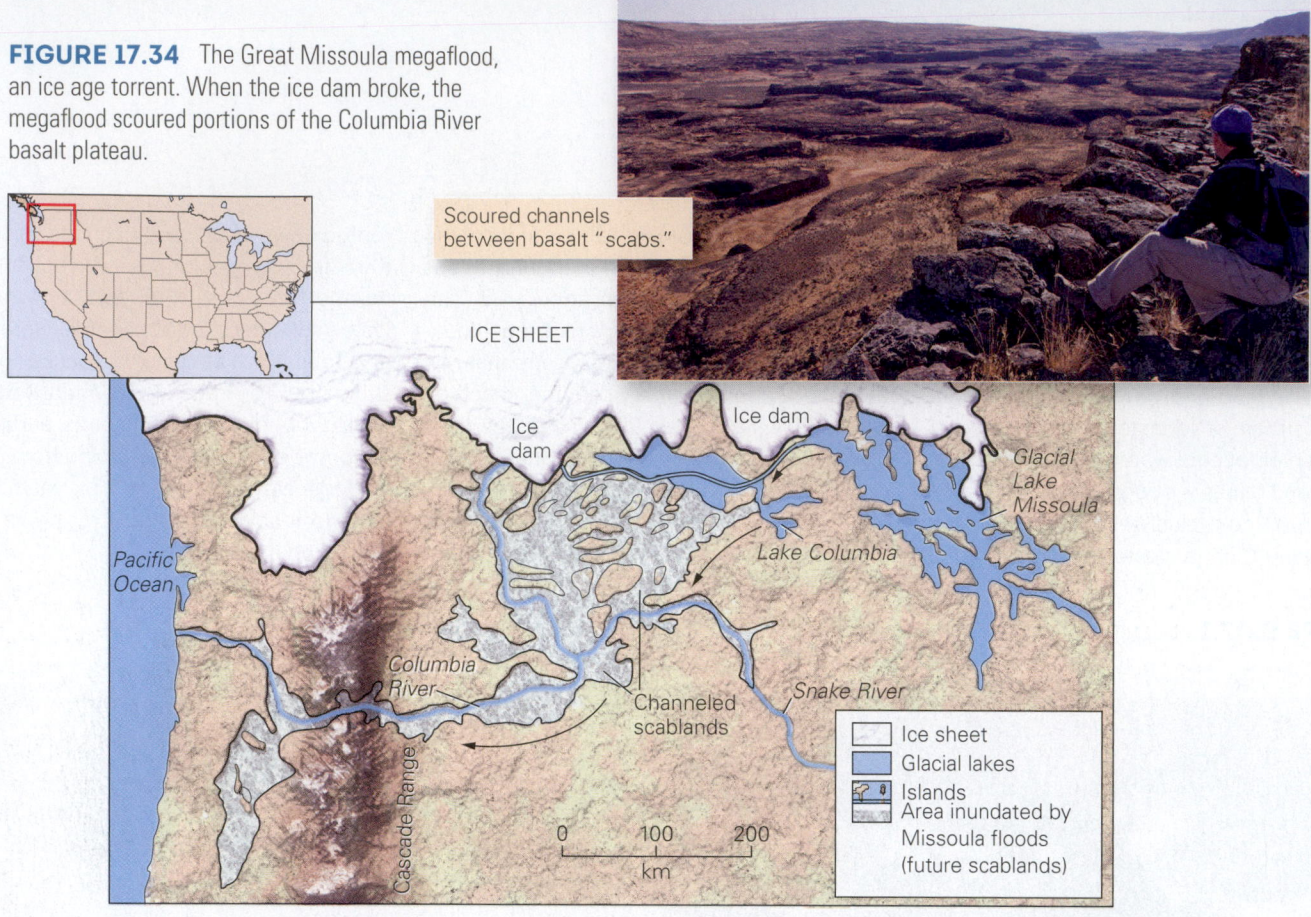

FIGURE 17.34 The Great Missoula megaflood, an ice age torrent. When the ice dam broke, the megaflood scoured portions of the Columbia River basalt plateau.

Scoured channels between basalt "scabs."

ICE SHEET

Ice dam

Ice dam

Glacial Lake Missoula

Pacific Ocean

Lake Columbia

Columbia River

Channeled scablands

Snake River

Cascade Range

0   100   200
km

Ice sheet
Glacial lakes
Islands
Area inundated by Missoula floods (future scablands)

Missoula. When the glacier melted and the dam suddenly broke, the lake abruptly drained, and water roared over what is now eastern Washington, eventually entering the Columbia River Valley and flowing on out to the Pacific Ocean. The glacier then grew again, and the dam re-formed, trapping a new lake, which drained during a subsequent failure. These floods stripped off the soil and regolith covering the dark basaltic bedrock of eastern Washington, leaving barren, craggy landscape now known as the *channeled scablands* (**Fig. 17.34**).

The hypothesis that the channeled scablands formed as a consequence of catastrophic flooding was first proposed by J. Harlan Bretz, who studied the landscape of the region in the 1920s. Initially, other geologists ridiculed Bretz because his idea seemed to violate the well-accepted principle of uniformitarianism (see Chapter 12). But Bretz steadily fought back, demonstrating that the scablands are littered with boulders too large to have been carried by normal rivers, that hills in the region were giant ripples a thousand times larger than the ripples typically found in a streambed, that the now-dry Grand Coulee was once a giant waterfall hundreds of times larger than Niagara Falls, and that deep pits were scoured by whirlpools. Ultimately, the geologic community accepted the reality of the Great Missoula Flood.

Now the consequences of numerous other *glacial torrents*, ice-age megafloods formed due to the sudden failure of a glacial dam, have been recognized. Glacial torrents can carve valleys that are much wider than could have been carved by present-day rivers, even during a large flood. For example, the northern reaches of the Illinois River Valley have very wide valleys bordered by cliffs that were scoured up to an elevation of a hundred meters above the present-day stream level.

Living with Floods

Flood Control Mark Twain once wrote of the Mississippi that we "cannot tame that lawless stream, cannot curb it or confine it, cannot say to it, 'go here or go there,' and make it obey." Was Twain right? Since ancient times, people have attempted to confine rivers to set courses so as to prevent undesired flooding. In the 20th century, flood-control efforts intensified as the population living along rivers increased. For example, since the passage of the 1927 Mississippi River Flood Control Act, drafted after a cataclysmic flood took place that year, the U.S. Army Corps of Engineers has labored to control the Mississippi. First, engineers built about 300 dams along the river's tributaries so that excess runoff could be stored in reservoirs and later be released slowly. Second, they built *artificial levees* of sand and mud, and flood walls of concrete, along the river (**Fig. 17.35**). These structures effectively increase the channel's

FIGURE 17.35 Flood control structures.

High-water marks of past floods.

(a) Artificial levees built to protect the downtown of Galena, Illinois, which is built along a tributary of the Mississippi.

(b) Floodwalls, which can be closed to protect Cape Girardeau, Missouri, from Mississippi River floods.

volume and isolate discrete areas of the floodplain to prevent them from being inundated.

Although the Corps' strategy worked for floods up to a certain size, it was insufficient to handle the 1993 and 2011 floods when the reservoirs filled to capacity and additional runoff headed downstream. The river rose until it spilled over the tops of some levees and undermined others (**Fig. 17.36**). *Levee undermining* occurs when rising floodwaters increase the water pressure on the river side of the levee, forcing water down through sand under the levee. In susceptible areas, water begins to spurt out of the ground on the dry side of the levee, thereby washing away the levee's support. The levee finally becomes so weak that it collapses, and water rushes through the breech.

Using lessons learned from 1993, the Army Corps of Engineers undertook a number of proactive but controversial steps that reduced the impact of the 2011 Mississippi and Missouri River floods. The 2011 floodwaters simply didn't fit in the space bounded by existing levees and floodwalls, so at a few places, the Corps had to choose between flooding fields or flooding towns on the floodplain. Generally, it chose the former. For example, to protect Cairo, Illinois, the Corps blasted a 3-km-long gap in a levee upstream of the town. This action diverted enough water into a 530 sq km area of farmed floodplain so that the river level remained lower than the levees at Cairo and water stayed out of the town's streets. Farther downstream, the Corps opened floodgates to divert water gradually into the Atchafalaya basin. Without this action, the river might have broken through natural levees, and if its new course did not reach the port of New Orleans, that could have had huge economic implications.

Because of expense, it's just not feasible to build levees high enough to handle all conceivable floods. And since building high levees confines the river to the channel, levees cause the water

height in the channel to become higher than it would have been if no levees were present, and this may cause greater flooding further downstream. Those who live on floodplains must face the reality of flooding risk and be willing to accept that flooding damage will occasionally happen, to build temporary levees of sandbags to protect local property, or to consider alternative ways to use floodplains that won't make property susceptible to damage (**Fig. 17.37**). The cost of flood damage has quadrupled in recent years despite the billions of dollars that have been spent on flood control, because more people have settled in floodplains. This has led to the challenge of figuring out ways to insure floodplain properties in ways that are fair to both floodplain inhabitants and to others who share in the cost of insurance.

Some communities are looking at new ways to mitigate flooding risk. For example, instead of building levees to isolate portions of the floodplain, the portions can be transformed into natural wetlands, for wetlands absorb water like a sponge and thus their presence can diminish flooding hazards. Property may also be kept safe by moving levees farther away from the river, so as to define **floodways**, regions likely to be flooded and thus off limits for building homes or businesses (**Fig. 17.38**). The existence of a floodway effectively increases the river, so the rising waters during a flood would not overtop the levees.

Evaluating Flooding Hazard When making decisions about investing in flood-control measures, mortgages, or insurance, planners need a basis for defining the hazard or risk posed by flooding. If floodwaters submerge a locality every year, a bank officer would be ill advised to approve a loan that would encourage building there. But if floodwaters submerge the locality only very rarely, then the loan may be worth the risk. Geologists characterize the risk of flooding in two ways. The

FIGURE 17.36 The failure of levees.

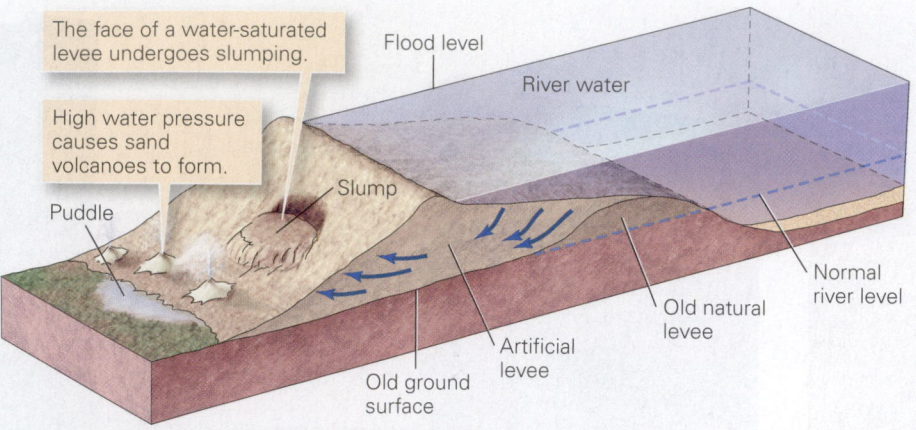

The face of a water-saturated levee undergoes slumping.

High water pressure causes sand volcanoes to form.

Flood level

River water

Puddle

Slump

Old ground surface

Artificial levee

Old natural levee

Normal river level

(a) Levees can be undermined if the river floods.

(b) Breaching of a levee lets water spill into the floodplain.

Flooded flood plain

Breached inner levee

Outer levee

Still-dry floodplain

(c) Astronauts in the Space Station could see the flood. Inner levees have been breached, but outer levees held.

annual probability of flooding—more formally known as the "annual exceedance probability"—indicates the likelihood that a flood of a given size or larger will happen at a specified locality during any given year. For example, if we say that a flood of a given size has an annual probability of 1%, then we mean there is a 1 in 100 chance that a flood of at least this size will happen in any given year. The **recurrence interval** of a flood of a given size is defined as the average number of years between successive floods of at least this size. For example, if a flood of a given size happens once in 100 years, on average, then it is assigned a recurrence interval of 100 years and is called a 100-year flood. Floods with a shorter recurrence interval are

Did you ever wonder . . .

what newscasters mean by a "100-year flood"?

more likely to occur than floods with a longer recurrence interval (**Fig. 17.39a**). Note that annual probability and recurrence interval are related:

$$\text{annual probability} = \frac{1}{\text{recurrence interval}}$$

For example, the annual probability of a 50-year flood is 1/50, which can also be written as 0.02 or 2%. To learn how to calculate annual probabilities and recurrence intervals of floods in more detail, see **Box 17.2**.

Unfortunately, some people may be misled by the meaning of recurrence interval and think that they do not face future flooding hazard if they buy a home within an area just after a 100-year flood has occurred. Their confidence comes from making the incorrect assumption that because such flooding

FIGURE 17.37 Emergency measures can protect local property.

(a) People increased the height of levees by adding sandbags.

(b) Some individual homeowners succeeded in protecting their houses.

FIGURE 17.38 By building a floodway, the stream can hold more water before flooding property.

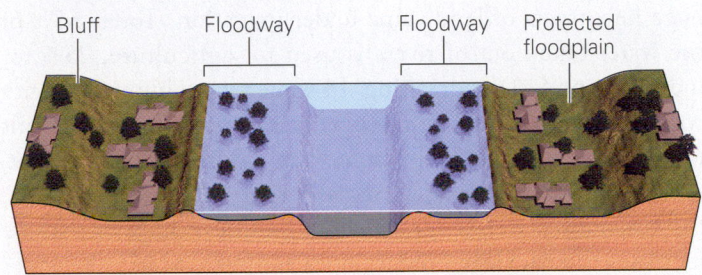

FIGURE 17.39 The conceptual relationship between flood size and probability.

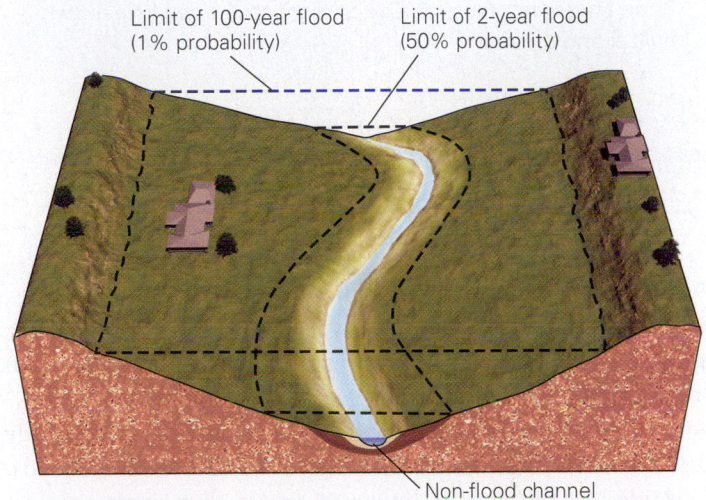

(a) A 100-year flood covers a larger area than a 2-year flood and occurs less frequently.

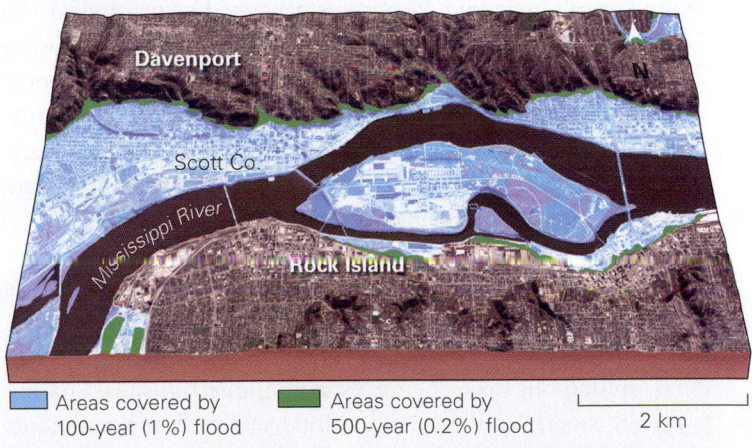

Areas covered by 100-year (1%) flood	Areas covered by 500-year (0.2%) flood	2 km

(b) A flood-hazard map shows areas likely to be flooded. Here, near Rock Island, Illinois, even large floods are confined to the floodplain.

just happened, it can't happen again until "long after I'm gone." They may regret their decision, because two 100-year floods can occur in consecutive years or even in the same year (alternatively, the interval between such floods could be, say, 210 years). Because the term *recurrence interval* can lead to confusion, it may be better to report risk in terms of annual probability.

As an example of how to think about flood hazards, let's consider the case of Nashville, Tennessee. The "home of country music" endured a 500-year flood on May 1 and 2 of 2010. Put another way, the likelihood of such a flood, based on previous discharge records, is only 0.2% in any given year. But comparable floods had happened in 1927 and 1937. The 2010 disaster began when a storm system stalled and warm, moist air from the Gulf of Mexico channeled over the region. Ferocious storms spawned at least 13 tornadoes, and in a 2-day period, 0.34 m (13.5 in) of rain fell over Nashville. (Not far from Nashville, almost half a meter, or 19 inches, of rain fell.) The Cumberland River overtopped its banks and rose 3 m (12 ft) above flood stage, putting it 15 m (over 52 ft) above its normal height. Much of the downtown, as well as the Grand Ole Opry, a famous performance venue, lay beneath water.

Knowing the discharge during a flood of a specified annual probability, and knowing the shape of the river channel and the elevation of the land bordering the river, geologists can predict the extent of land that will be submerged by such a flood (Fig. 17.39a). Such data, in turn, permit production of **flood-hazard maps**. In the United States, the Federal Emergency Management Agency (FEMA) produces maps that show the 1% annual probability (100-year) flood area and the 0.2% annual probability (500-year) flood risk zones (**Fig. 17.39b**).

> ## Take-Home Message
>
> Seasonal floods submerge floodplains and delta plains at certain times of the year. Flash floods arrive quickly and can be short-lived. People try to protect land in floodplains by building dams, levees, and floodwalls, but these don't always work. We can specify the probability that a flood of a certain size will happen in a given year and, from this information, can produce flooding risk maps.
>
> **QUICK QUESTION:** Why might the flood level (height above flood stage) increase for a given discharge when levees are built along the river?

17.9 Vanishing Rivers

As *Homo sapiens* evolved from hunter-gatherers into farmers, areas along rivers became attractive places to settle. Rivers serve as avenues for transportation and are sources of food, irrigation water, drinking water, power, recreation, and (unfortunately) waste disposal. Further, their floodplains provide particularly fertile soil for fields, replenished annually by seasonal floods. Considering the multitudinous resources that rivers provide, it's no coincidence that ancient cultures developed in river valleys and on floodplains. The civilization of Mesopotamia arose around the Tigris and Euphrates Rivers, Egypt around the Nile, India in the Indus Valley, and China along the Hwang River. Over the millennia, rivers have killed millions of people in floods, but they have been the lifeblood for hundreds of millions more. Nevertheless, over time, humans have increasingly tended to abuse or overuse the Earth's rivers. Here we note four pressing environmental issues.

Pollution The capacity of some rivers to carry pollutants has long been exceeded, transforming them into deadly cesspools. Pollutants include raw sewage and storm drainage from urban areas, spilled oil, toxic chemicals from industrial sites, floating garbage, excess fertilizer, and animal waste. Some pollutants directly poison aquatic life, some feed algae blooms that strip water of its oxygen, and some settle out to be buried along with sediments. River pollution has become overwhelming in countries with inadequate waste-treatment facilities.

Dam Construction In 1950, there were about 5,000 large (over 15-m-high) dams worldwide, but today there are over 48,000. Damming rivers has both positive and negative results. Reservoirs provide irrigation water and hydroelectric power, and they trap some floodwaters and create popular recreation areas. But in some locations their construction destroys "wild rivers" (whitewater streams of hilly and mountainous areas) and alters the ecosystem of a drainage network by forming barriers to migrating fish, by decreasing the nutrient supply to organisms downstream, and by removing the source of sediment and nutrients for the floodplain and delta.

Overuse of Water Because of growing populations, our thirst for river water continues to increase, but the supply of water does not. The use of water has grown especially in response to the "green revolution" of the 1960s, during which huge new tracts of land came under irrigation. Today 65% of the water taken out of rivers is used for agriculture, 25% for industry, and 9% for drinking. In some places human activity consumes the entire volume of a river's water, and as a result the channel contains little more than a saline trickle, if that, at its mouth. For example, except during unusually wet years, the Colorado River's channel contains almost no water where it crosses the Mexican border; pipes and canals carry the water instead to growing cities such as Phoenix and Los Angeles (**Fig. 17.40**).

Effects of Urbanization and Agriculture on Streams When it rains in a naturally vegetated region, much of the water that falls from the sky either soaks into the ground or gets absorbed by plants. Some of the soil moisture or groundwater eventually seeps into a nearby stream, but the remainder flows elsewhere underground. As a result, the amount of water that reaches nearby streams after a storm is less than the total amount of precipitation, and a significant *lag time* occurs between when the rain falls and when the stream's discharge increases. Urbanization changes this picture. When developers transform fields and forests into parking lots, roads, and buildings, a layer of impermeable concrete and asphalt prevents rainfall from infiltrating, the amount of living biomass available to absorb water is less, and storm sewers divert water directly to streams. So not only does the overall volume of water entering the streams increase, but the lag time decreases. This change can be illustrated by diagrams, called *hydrographs*, that show how discharge varies with time (**Fig. 17.41**). Unfortunately, the change in volume and lag time can lead to flash flooding.

BOX 17.2 CONSIDER THIS . . .

Calculating the Threat Posed by Flooding

How do we calculate the probability that a flood of a given size at a locality along a stream will happen in a given year? (Note that "size" in this context is indicated by the stream's discharge, as measured in cubic feet per second or cubic meters per second). First, researchers collect data on the stream's discharge at the locality for at least 10 to 30 years to get a sense of how the discharge varies during a year and from year to year. Then, they pick the largest, or peak, discharge for each year and make a table listing the peak discharges. The largest peak discharge is given a rank of 1, the second-largest discharge is given a rank of 2, and so on. Researchers can then calculate the recurrence interval for each different discharge by using a simple equation:

$$R = (n + 1) \div m$$

where R is the recurrence interval in years, n is the total number of years for which there is a record, and m is the rank.

Once the recurrence interval for each peak discharge has been calculated, the researchers plot a graph: the vertical axis represents peak discharge, and the horizontal axis represents recurrence interval. In order for all the data to fit on a reasonable-size graph, the horizontal axis must be logarithmic. Typically, the data for a stream plot roughly along a straight line (**Fig. Bx17.2a**). In the example shown in this figure, a flood with a peak discharge of about 460 cubic feet per second has a recurrence interval of 10 years (meaning an annual probability of 10%). We can extend the line beyond the data points (the dashed line on Fig. Bx17.2a)

to make predictions about the recurrence interval and, therefore, the annual probability (= 1/R), of floods with discharges larger than the ones that have been measured. As more data become available, the graph may need to be modified. In this example, note that the graph predicts that a 1% probability flood (a 100-year flood) will have a discharge of about 650 cubic feet per second.

The peak annual discharge of the Mississippi River at St. Louis has been measured almost continuously since 1850. A bar graph of these data shows that floods characterized as 100-year floods (meaning 1% probability floods) or larger happened in 1844, 1903, and 1993 (**Fig. Bx17.2b**). Note that the time between "100-year floods" is not exactly 100 years.

FIGURE Bx17.2 Flood frequency and peak discharge graphs.

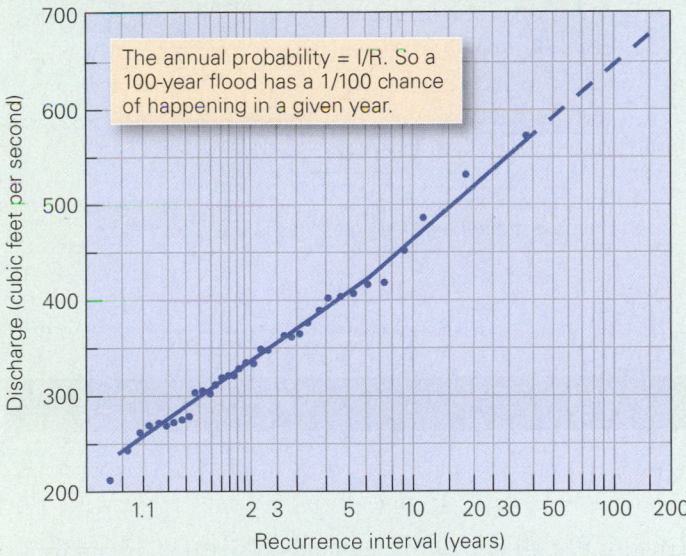

The annual probability = I/R. So a 100-year flood has a 1/100 chance of happening in a given year.

(a) A flood-frequency graph shows the relationship between the recurrence interval and the discharge for an idealized river.

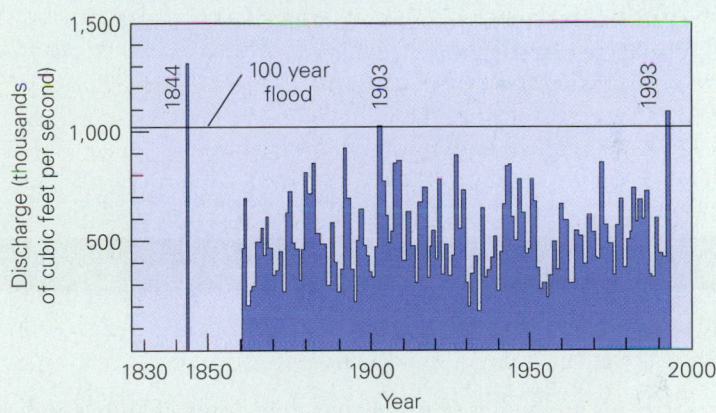

(b) A peak discharge graph for the Mississippi at St. Louis. Each bar represents the largest discharge for a given year.

Agriculture can have a similarly profound effect on streams. Though covered with green during the growing season, fields of many crops actually host relatively little vegetation because farmers keep the space between rows of the crop plants free of weeds. Further, if the crop consists of annual plants, such as corn or soybeans, the fields are completely bare of vegetation during the winter. As a consequence, the runoff from farm fields can be greater than that from forests or from naturally vegetated fields, and runoff can carry significant amounts of soil into streams, thus increasing the sediment load that the stream carries.

FIGURE 17.40 The Central Arizona Project canal shunts water from the Colorado River to Phoenix.

FIGURE 17.41 Hydrographs, showing discharge as a function of time, are affected by urbanization.

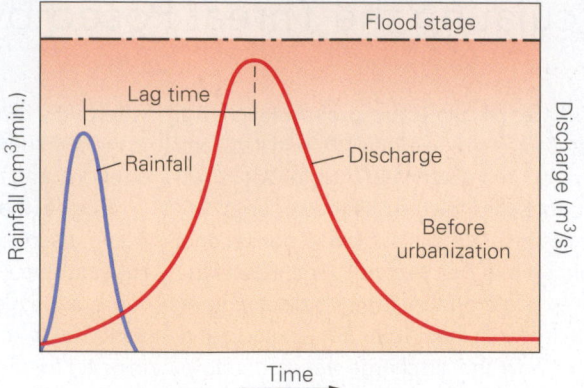

(a) Before urbanization, rain infiltrated the ground so discharge was smaller and peak runoff occurred after a long lag time.

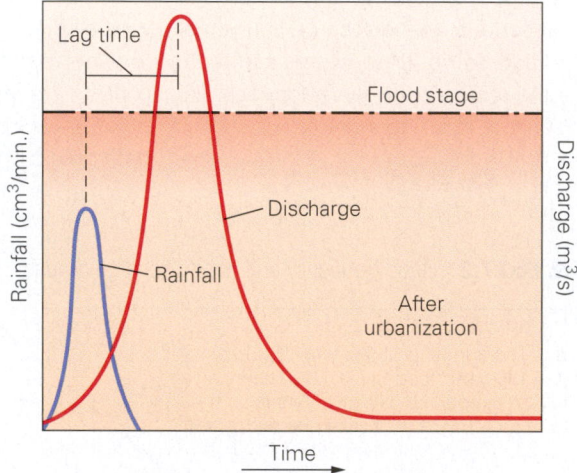

(b) After urbanization, water flows directly into streams, discharge is greater, and lag time is less.

Take-Home Message

Society depends on streams for water supplies, irrigation, energy, and transport, but the growth of populations can negatively affect streams. Diversion of flow may decrease stream discharge to a trickle, dam construction can change flow, pollution can foul the water, and urbanization or runoff can change discharge and sediment load.

QUICK QUESTION: Why can urbanization decrease the discharge of a stream?

CHAPTER SUMMARY

- Streams are bodies of water that flow down channels and drain the land surface. Channels develop when running water cuts into the substrate and sculpts a trough; they grow by headward erosion. Streams carry water out of a drainage basin. A drainage divide separates two adjacent catchments.

- Drainage networks consist of many tributaries that flow into a trunk stream. Several different patterns of these networks exist.

- The flow in gaining streams increases downstream, whereas that of losing streams decreases. Permanent

streams last all year, while ephemeral streams dry up for part or most of the year. Whether a stream is gaining and permanent, or not, depends in large part on whether the streambed lies above or below the water table.

- The discharge of a stream is the total volume of water passing a point along the bank in a second. Flow velocity slows where water comes in contact with the stream banks or streambed. Most streams are turbulent, so their water swirls in complex patterns.

- Streams erode the landscape by scouring, lifting, abrading, and dissolving. The resulting sediment provides dissolved loads, suspended loads, and bed loads. The total quantity of sediment carried by a stream is its capacity. Capacity differs from competence, the maximum particle size a stream can carry. When stream water slows, it deposits alluvium.

- The longitudinal profile of a stream is concave up. Typically, a stream has a steeper gradient at its headwaters than near its mouth. Streams cannot cut below the base level.

- Whether streams cut valleys or canyons depends on the rate of downcutting relative to the rate at which the slopes on either side of the stream undergo mass wasting. Where a stream flows down steep gradients and has a bed littered with large rocks, rapids develop, and where a stream plunges off a vertical face, a waterfall forms.

- A meandering stream curves back and forth across a floodplain. It erodes its outer bank and builds out sediment into a point bar on the inner bank. Eventually, a meander may be cut off and turn into an oxbow lake. Natural levees form on either side of the river channel. Braided streams consist of many entwined channels.

- Where streams or rivers flow into standing water, they deposit deltas. The shape of a delta depends on the balance between the amount of sediment supplied by the river and the amount of sediment redistributed or carried away by wave activity along the coast.

- With time, fluvial erosion can bevel landscapes to a nearly flat plain. If the base level drops or the land surface rises, stream rejuvenation causes the stream to start downcutting into the peneplain. The headward erosion of one stream may capture the flow of another.

- If an increase in rainfall or spring melting causes more water to enter a stream than the channel can hold, a flood results. Some floods are seasonal in that they accompany monsoonal rains. Some floods submerge broad floodplains or delta plains. Flash floods happen very rapidly. Officials try to prevent floods by building reservoirs and levees.

- Rivers are becoming a vanishing resource because of pollution, damming, and overuse.

GUIDE TERMS

abrasion (p. 623)
alluvial fan (p. 630)
alluvium (p. 626)
annual probability (p. 648)
antecedent stream (p. 639)
bar (p. 626)
base level (p. 627)
bed load (p. 624)
braided stream (p. 630)
canyon (p. 629)
capacity (p. 626)
channel (p. 615)
competence (p. 625)
continental divide (p. 619)
delta (p. 624)
discharge (p. 621)

dissolved load (p. 624)
distributary (p. 635)
downcutting (p. 617)
drainage divide (p. 619)
drainage network (p. 618)
drainage reversal (p. 636)
ephemeral stream (p. 620)
flash flood (p. 644)
flood (p. 615)
flood-hazard map (p. 650)
floodplain (p. 634)
floodway (p. 647)
graded stream (p. 628)
headward erosion (p. 617)
headwaters (p. 615)
longitudinal profile (p. 626)

meander (p. 630)
meandering stream (p. 630)
mouth (p. 615)
natural levee (p. 634)
oxbow lake (p. 634)
permanent stream (p. 620)
point bar (p. 626)
pothole (p. 623)
rapid (p. 629)
recurrence interval (p. 648)
runoff (p. 617)
saltation (p. 624)
scouring (p. 623)
seasonal flood (p. 640)
sheetwash (p. 616)
stream (p. 615)

stream gradient (p. 626)
stream piracy (p. 636)
stream rejuvenation (p. 637)
stream terrace (p. 629)
superposed stream (p. 638)
suspended load (p. 624)
thalweg (p. 623)
tributary (p. 618)
trunk stream (p. 618)
turbulence (p. 623)
valley (p. 629)
waterfall (p. 629)
watershed (p. 619)

REVIEW QUESTIONS

1. What role do streams serve during the hydrologic cycle? Indicate various sources of water in streams.

2. Describe the five different types of drainage networks. What factors are responsible for the formation of each?

3. What factors determine whether a stream is permanent or ephemeral, gaining or losing?

4. How does discharge vary according to the stream's length, climate, and position along the stream course?

5. Why is average downstream velocity always less than maximum downstream velocity?

6. Why does stream flow tend to become turbulent?

7. Describe how streams and running water erode the Earth.

8. What are three components of sediment load in a stream?

9. Distinguish between a stream's competence and its capacity.

10. Describe how a drainage network changes, along its length, from headwaters to mouth.

11. What factors determine the position of a local base level? What is the ultimate base level, and why?

12. What do lakes, rapids, waterfalls, and terraces indicate about the stream gradient and base level? Why do canyons form in some places and valleys in others?

13. How does a braided stream differ from a meandering stream?

14. Describe how meanders form, develop, are cut off, and then are abandoned.

15. Describe how deltas grow and develop. How do they differ from alluvial fans?

16. How does a stream-eroded landscape evolve as time passes?

17. What is stream piracy? What causes a drainage reversal?

18. How are superposed and antecedent drainages similar? How are they different?

19. Explain the difference between a seasonal flood and a flash flood, and describe the causes of each.

20. What human activities tend to increase flood risk and damage?

21. What is the recurrence interval of a flood, and how is it related to the annual probability? Why can't someone say, "The hundred-year flood happened last year, so I'm safe for another hundred years"?

22. How have humans abused and overused the resource of running water?

ON FURTHER THOUGHT

23. The northeastern two-thirds of Illinois, in the midwestern United States, was last covered by glaciers only 14,000 years ago. The rest of the state was last covered by glaciers over 100,000 years ago. Until the advent of modern agriculture, the recently glaciated area was a broad, grassy swamp, cut by very few stream channels. In contrast, the area that was glaciated over 100,000 years ago is not swampy and has been cut by numerous stream valleys. Why?

24. Records indicate that flood crests for a given amount of discharge along the Mississippi River have been getting higher since 1927, when a system of levees began to block off portions of the floodplain. Why?

25. The Ganges River carries an immense amount of sediment load, which has been building a huge delta in the Bay of Bengal. Look at the region using an atlas or *Google Earth*™, think about the nature of the watershed supplying water to the drainage network that feeds the Ganges, and explain why this river carries so much sediment.

26. Look closely at the graph in Box 17.2. What is the recurrence interval of a flood with a discharge of 650 cubic feet per second? In a given year, how much more likely is a flood with a discharge of 200 cubic feet per second than a flood of 400 cubic feet per second? For the sake of discussion, imagine that the floodplain of the river is completely covered when a flood with an annual probability of 1/300 occurs. Would you build a new home in the floodplain? Would it make a difference to you if the last flood with this probability happened 1 year ago? One hundred years ago?

smartwork.wwnorton.com

This chapter's Smartwork features:

- What A Geologist Sees exercise on river landscapes.
- Labeling activity on drainage networks.
- In-depth reading comprehension problems on stream characteristics.

GEOTOURS

This chapter's GeoTour exercise (N) features:

- Headward erosion
- Stream patterns
- Meandering stream features

The rocky coast of Brittany in northwestern France gets pounded by waves, which gradually carve headlands and bays. The ocean is constantly in motion.

CHAPTER 18

Restless Realm: Oceans and Coasts

> The three great elemental sounds in nature are the sound of rain, the sound of wind in a primeval wood, and the sound of the outer ocean on a beach.
>
> —*Henry Beston (American naturalist, 1888–1968)*

LEARNING OBJECTIVES

By the end of this chapter, you should understand . . .

- how tectonic processes produce bathymetric features of the seafloor.
- the nature and causes of surface currents and deep currents.
- the behavior of tides and the forces that cause tides to happen.
- why waves form and how waves behave as they approach the shore.
- how a great variety of different coastal landforms develop and evolve.
- that changes in sea level, wave erosion, and human activities affect the coast.

18.1 Introduction

A thousand kilometers from the nearest shore, two scientists and a pilot wriggle through the entry hatch of the research submersible *Alvin*, ready for a cruise to the floor of the ocean and, hopefully, back. *Alvin* consists of a superstrong metal sphere embedded in a cigar-shaped tube (**Fig. 18.1a**). The sphere protects its crew from the immense water pressures of the deep ocean, and the tube holds motors and oxygen tanks. When the hatch seals, *Alvin* sinks at a rate of 1.8 km per hour. Most of this journey takes place through utter darkness, for light penetrates only the top few hundred meters of ocean water. On reaching the bottom, at a depth of 4.5 km, the cramped explorers turn on outside lights to reveal a stark vista of loose sediment, black rock, and the occasional sea creature. For the next five hours they take photographs and use a robotic arm to collect samples. When finished, they release ballast and rise like a bubble, reaching the surface two hours later.

FIGURE 18.1 Modern oceanographic research vessels explore the surface and subsurface realms.

(a) *Alvin*, a 3-person submersible, explores mid-ocean ridges.

(e) *ABE* is a robotic submersible.

(b) The HMS *Challenger* (ca. 1876) was the first oceanographic research vessel.

(c) RV *Atlantis* can take 24 scientists to sea for 2 months.

(d) Submersibles use spotlights to see at depth.

Alvin dives began in the 1970s, but humans have explored the ocean for tens of centuries. In fact, Phoenician traders had circumnavigated Africa by 590 B.C.E., Polynesian sailors in outrigger canoes traveled among South Pacific islands beginning around 700 C.E., and Chinese naval ships may have circled the globe in the 15th century. European mapmakers have known that the ocean spanned the entire globe since the tattered remnants of Ferdinand Magellan's crew completed a round-the-world voyage in 1519–22. But no one could systematically map the ocean until the late 18th century, when navigators first obtained the equipment (an accurate chronometer) necessary to calculate longitude accurately. Subsequently, naval officers of many nations started to gather data on water depths in the ocean, and by 1839 data compilations demonstrated that the greatest ocean depths could swallow the highest mountains, without a trace.

A converted British navy ship, the HMS *Challenger*, made the first true ocean research cruise (**Fig. 18.1b**). Beginning in 1872, onboard scientists spent four years dredging rocks from the seafloor, analyzing water composition, collecting specimens of marine organisms, and measuring water depths and currents. But still our knowledge of the ocean remained spotty. In fact, we knew less about the ocean floor than we did about the surface of the Moon, for at least we could see the Moon with a telescope.

Then, in the latter half of the 20th century, the fields of *oceanography* (the study of ocean water and its movements), *marine geology* (the study of the ocean floor and shorelines), and *marine biology* (the study of life forms in the sea) expanded rapidly, as new technology became available and a fleet of oceanographic research ships crisscrossed the seas (**Fig. 18.1c–e**). It's now commonplace for ships to tow instruments, such as *side-scan sonar*, to generate detailed bathymetric maps along swaths

of the ocean floor (**Fig. 18.2a**). Some ships send high-energy bursts that penetrate the seafloor and reflect off layers in the subsurface to provide an image, a seismic-reflection profile, revealing the layering in the oceanic crust (**Fig. 18.2b**; see Interlude D). Other ships, such as the *JOIDES Resolution*, drill holes as deep as 4 km into the seafloor and bring up samples of the oceanic crust. In addition, satellites use remote sensing techniques to map the characteristics of the oceans globally. And while ships and satellites research the open ocean, land-based geologists continue to study the evolution of its margins.

When seen from space, Earth glows blue, for the oceans cover 70.8% of its surface. The ocean provides the basis for life, tempers Earth's climate, and spawns its largest storms. It is a vast reservoir for water and chemicals that cycle into the atmosphere and crust and for sediment washed off the continents. In this chapter, we first learn about the fundamental characteristics of ocean basins and seawater and the role they play in the Earth System. Then we focus on the landforms that develop along the **coast**, the region where the land meets the sea and where over 60% of the global population lives today. Finally, we consider the hazards of coastal areas and how people may confront them.

18.2 Landscapes beneath the Sea

If the surface of the lithosphere were completely smooth, an ocean would surround the Earth as a uniform 2.5-km-deep layer. But in reality about 30% of this planet's surface is dry land, and most seawater resides in distinct *ocean basins*, which have an average depth of around 4.5 km (**Fig. 18.3a**). These

FIGURE 18.2 Modern methods for surveying the seafloor.

(a) Side-scan sonar (also called multi-beam sonar) can map the detailed bathymetry of a swath of the seafloor rapidly.

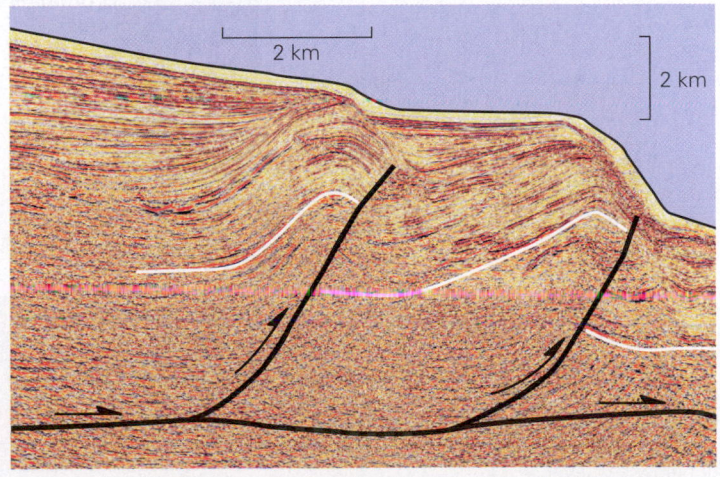

(b) Seismic-reflection profiles reveal layers and structures beneath the seafloor. This example shows the accretionary prism of southern Alaska.

basins exist for a reason! Due to isostasy, the surface of the denser but thinner oceanic lithosphere sinks deeper than the surface of the relatively buoyant, thicker continental lithosphere. It's the low areas over oceanic lithosphere that fill with water to become oceans—higher areas, underlain by continental lithosphere, remain as dry land (**Fig. 18.3b**). Cartographers delineate several distinct oceans and seas, which differ significantly in terms of their volumes (**Fig. 18.4**). Some of the boundaries between oceans are obvious—for example, the Americas separate the Pacific from the Atlantic—but some are rather arbitrary, defined by a line of latitude or longitude or by a chain of islands. Note that all oceans are, in effect, interconnected for water can ultimately flow from one ocean to another.

Have you ever wondered what the ocean floor would look like if all the water evaporated? Marine geologists can now provide a clear image of the ocean's **bathymetry**, or variation in depth. Depth measurements were first obtained by using a *plumb line*, a lead weight attached to a cable. Then, beginning in the mid-20th century, sonar scans became available, and today, satellites survey the ocean, measuring variations in the pull of gravity that can be translated into variations in water depth. Such studies indicate that the ocean contains broad *bathymetric provinces*, distinguished from each other by their water depth. Let's examine each of these provinces.

Continental Shelves, Slopes, and Rises

Imagine you're in a submersible that is cruising just above the floor of the western half of the North Atlantic (**Fig. 18.5a**). If you start at the shoreline of North America and head east,

you will cross the 200- to 500-km-wide **continental shelf**, a relatively shallow portion of the ocean that fringes the continent. Water depth over the continental shelf does not exceed 500 m—most of the major marine fisheries occur in the relatively shallow waters of the shelf. Across the width of the shelf, the ocean floor slopes seaward at only about 0.3°, an almost imperceptible angle. At its eastern edge, the continental shelf merges with the *continental slope*, which descends to depths of nearly 4 km at an angle of about 2°. From about 4 km down to about 4.5 km, a province called the *continental rise*, the angle decreases gradually until at a depth of 4.5 km you'll find yourself above a vast, nearly horizontal plain—the **abyssal plain**.

Broad continental shelves, like the one bordering eastern North America, form along **passive continental margins**. Recall that such margins are not plate boundaries and thus lack seismicity (see Fig. 18.5a and Chapter 4). Passive margins originate after rifting succeeded in breaking a continent in two, for when rifting ceases and seafloor spreading begins (along a newly formed mid-ocean ridge), the stretched lithosphere at the boundary between the ocean and continent gradually cools and sinks. Sand and mud that washed off the continent, along with the shells of marine creatures that grow on the seafloor or settle from the water above, bury the sinking crust, slowly producing a pile of sediment up to 15 or even 20 km thick. Geologists refer to the region along passive margins that accumulates thick wedges of sediment as a *passive-margin basin*. The flat surface of a passive-margin basin constitutes the continental shelf. Of note, in some locations, a thick evaporate layer underlies that strata of a passive-margin basin. The weak salt comprising this evaporite can serve as a failure horizon or detachment,

FIGURE 18.3 Contrasts between continental lithosphere and oceanic lithosphere.

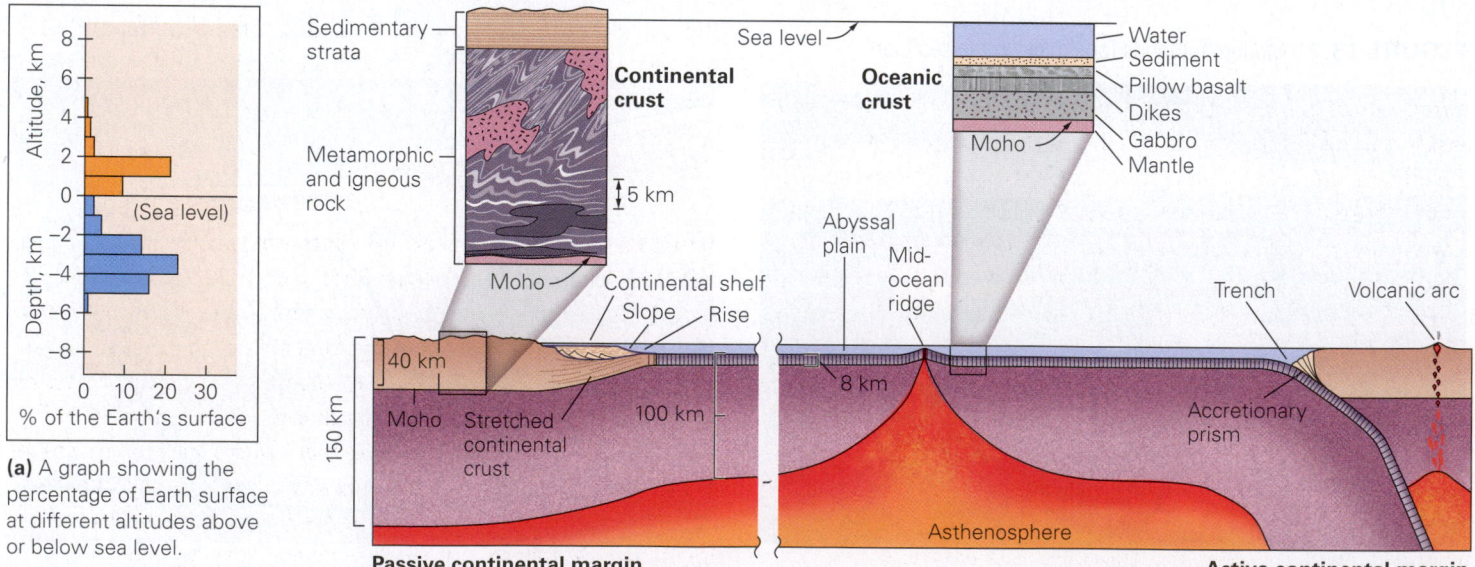

(a) A graph showing the percentage of Earth surface at different altitudes above or below sea level.

(b) The crustal portion of the continental lithosphere differs markedly from that of oceanic lithosphere.

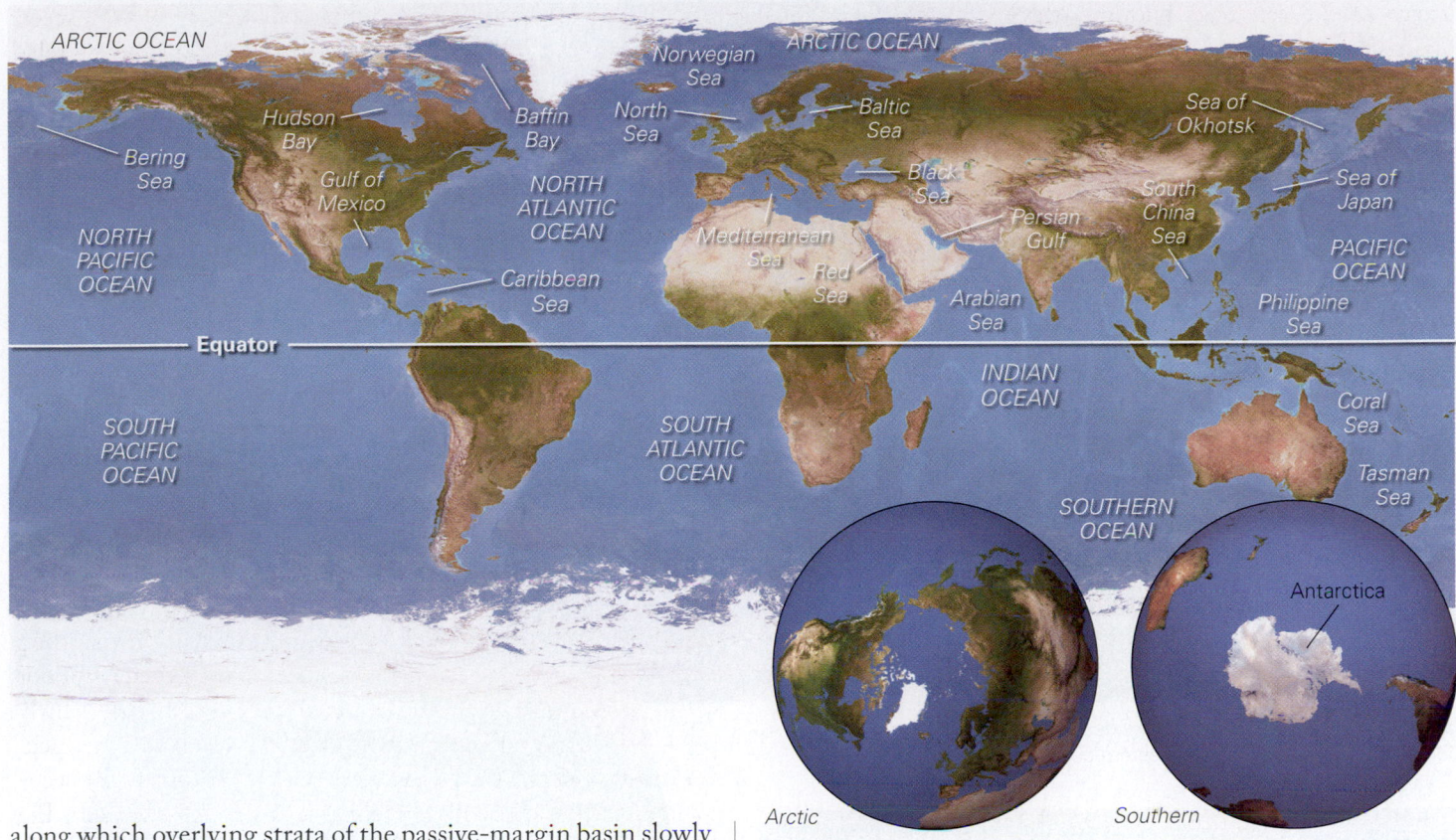

along which overlying strata of the passive-margin basin slowly slump seaward, movement that generates faults, folds, and salt domes in the strata.

If you were to take your submersible to the western coast of South America and cruise out into the Pacific, you would find a very different type of continental margin. This continental margin is an **active continental margin**, a margin that coincides with a plate boundary that hosts many earthquakes (**Fig. 18.5b**). The western South American coast is, in particular, a convergent-plate boundary where subduction takes place. Because of subduction, a 100- to 200-km-wide *accretionary prism*, consisting of contorted and broken-up sediment and basalt that has been scraped off the subducting plate, lies along the edge of the coast. The upper part of the accretionary prism may be buried by sediment eroded from the land to form a narrow continental shelf. The shelf ends at the continental slope, which corresponds to the face of the accretionary prism—its surface slopes at an angle of 3.5° down to the axis of the trench, which lies at a depth of 5 to 7.5 km.

At many locations, relatively narrow and deep valleys called **submarine canyons** downcut into continental shelves and slopes, both along active and passive margins (**Fig. 18.5c**). Some submarine canyons start offshore of major rivers, and for good reason—rivers cut into the continental shelf at times when sea level was low and the shelf was exposed. But river erosion cannot explain the total depth of these canyons—some

slice almost 1,000 m down into the continental margin, far deeper than the maximum sea-level change. Submarine exploration demonstrates that much of the erosion that cuts submarine canyons results from turbidity currents, avalanches of sediment mixed with water (see Chapters 7 and 16). When turbidity currents finally reach the base of the continental slope, the velocity of flow decreases, so sediments settle out and turbidites, composed of graded beds, accumulate and build into a *submarine fan*.

The Bathymetry of Oceanic Plate Boundaries

You can see all three types of plate boundaries by studying the bathymetry of the ocean floor (see Fig. 18.5a). Seafloor spreading at a divergent boundary yields a *mid-ocean ridge*, a 2-km-high submarine mountain belt. Because crust stretches and breaks as seafloor spreading continues, the axis of a ridge may be bordered by fault scarps due to slip on normal faults (see Chapter 11). *Oceanic transform faults*, strike-slip faults along which one plate shears sideways past another, typically link segments of mid-ocean ridges (see Chapter 4). Transforms are delineated by *fracture zones*, narrow belts of steep escarpments and broken-up rock. These fracture zones can be

FIGURE 18.5 Bathymetric features of the seafloor. The maps are produced by computer using measurements from satellites or submersibles.

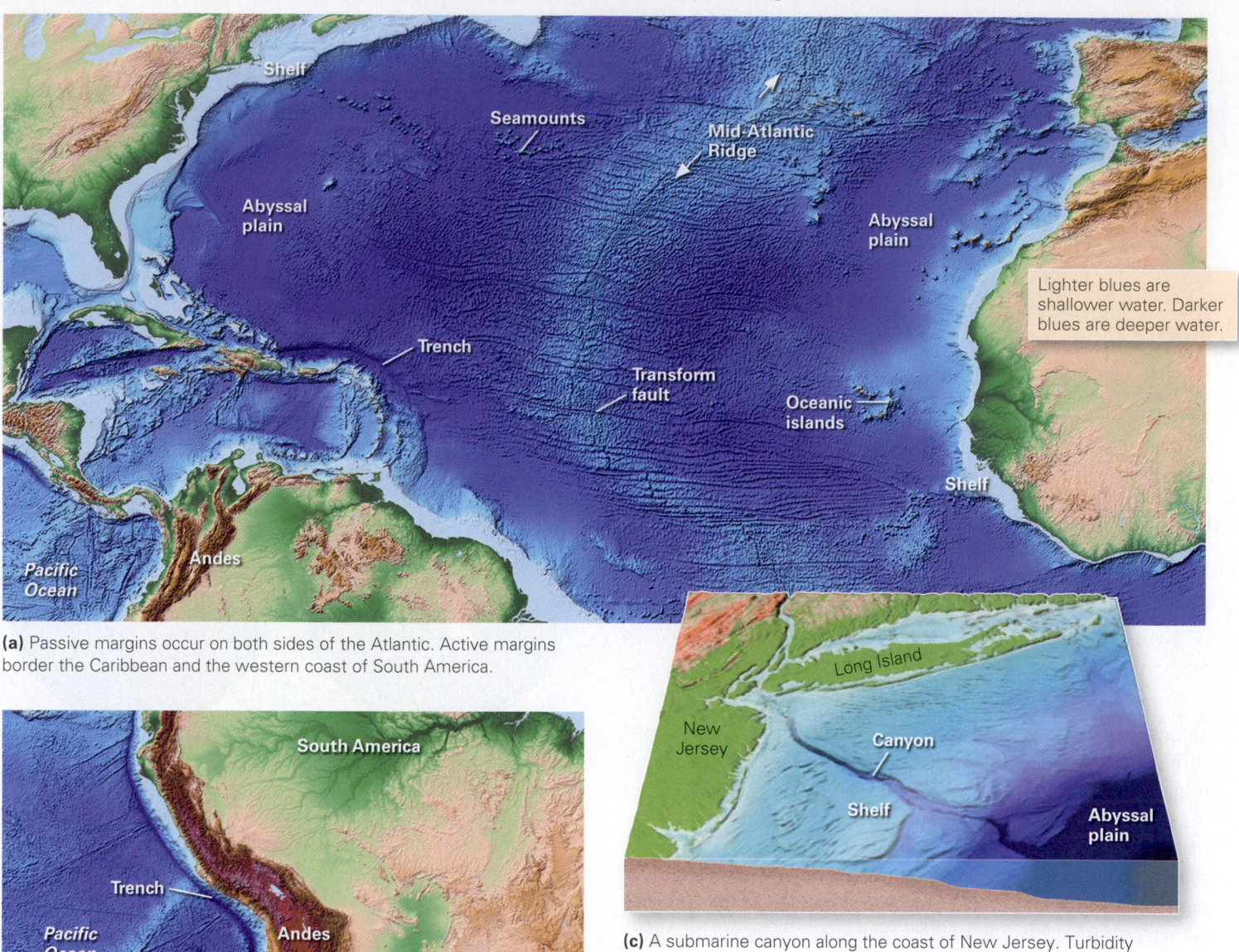

(a) Passive margins occur on both sides of the Atlantic. Active margins border the Caribbean and the western coast of South America.

(b) The western coast of South America is an active continental margin.

(c) A submarine canyon along the coast of New Jersey. Turbidity currents in submarine canyons carry sediment to the abyssal plain.

traced into the oceanic plate away from the ridge axis where they are not seismically active and, therefore, are not plate boundaries—the inactive portions of fracture zones juxtapose portions of plates that of different age. Subduction at convergent boundaries yields a *trench*, a deep, elongate trough bordering a volcanic arc. Some trenches reach depths of over 8 km. In fact, the deepest point in the global ocean, at −11,035 m, lies in the Mariana Trench of the western Pacific. Some trenches border continents as we described earlier; others border island arcs, which are curving chains of active volcanic islands.

Abyssal Plains, Seamounts, and Oceanic Plateaus

As oceanic crust ages and moves away from the axis of the mid-ocean ridge, two changes take place. First, the lithosphere cools and thickens, and as it does so, its surface sinks to maintain isostasy (see Interlude D). Second, a blanket of **pelagic sediment** gradually accumulates and covers the basalt of the oceanic crust (**Fig. 18.6a**). This sediment consists mostly of microscopic plankton shells and fine flakes of clay, which slowly fall like snow from the ocean water and settle on the

FIGURE 18.6 The sedimentary layer of the deep seafloor.

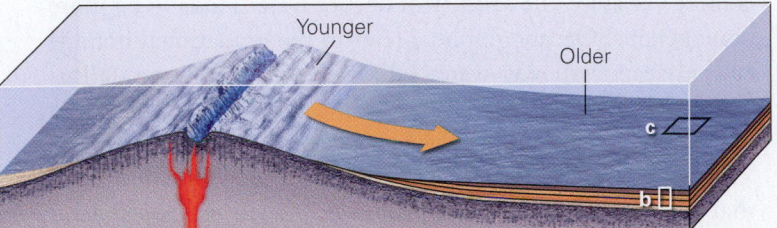

(a) The sedimentary layer of the seafloor thickens with increasing distance from the mid-ocean ridge axis.

(c) Cores extracted from drillholes shows bedding in the sediment.

(b) A photo of the abyssal plain surface showing mud and few organisms.

over time, the island gradually sinks. In addition, oceanic islands gradually undergo erosion, and every once in a while chunks of the island slump and tumble to greater depths. Overall, these processes end up causing the island to submerge beneath the waves and become a **seamount** (**Fig. 18.7**). Oceanic islands and seamounts that developed above the same hot spot line up in a chain (a hot-spot track) with the oldest seamount at one end and the youngest seamount or island at the other (see Chapter 4).

In several localities, there are broad areas of seafloor whose depths are less than what estimates based on the well-understood relation between ocean floor depth and lithosphere age predicts. Such a location is an **oceanic plateau** (see Fig. 18.7). Seismic-reflection profiles across oceanic plateaus demonstrate that they are underlain by unusually thick oceanic crust—whereas most oceanic crust is about 7 km thick, the crust beneath oceanic plateaus is 10 to 12 km thick. Geologists speculate that oceanic plateaus are regions where particularly voluminous hot-spot igneous activity once took place—in other words, they are large igneous provinces (LIPs; see Chapter 6), perhaps formed over superplumes in which much more melting takes place than is typical of a plume.

Take-Home Message

Ocean basins exist because oceanic and continental lithosphere differ in thickness and composition. Seafloor bathymetric features (ridges, trenches, and fracture zones) reflect plate-tectonic processes. Abyssal plains overlie old lithosphere, and continental shelves overlie passive-margin basins. Oceanic islands and plateaus are the products of hot-spot volcanism.

QUICK QUESTION: How do submarine canyons form?

seafloor. Because the ocean crust gets progressively older away from the ridge axis, the thickness of the sediment blanket increases away from the ridge axis, for sediment has had more time to accumulate on older seafloor (**Fig. 18.6b**). Eventually, the sediment buries the escarpments that had formed at the mid-ocean ridge, resulting in a flat, featureless surface of the abyssal plain (**Fig. 18.6c**).

Hot-spot igneous activity has produced unusually thick accumulations of basalt that build up on the seafloor. In places, the lava builds into a mound that protrudes above sea level to form an oceanic island. Oceanic islands that lie over hot spots host active volcanoes, whereas those that have moved off the hot spot are extinct. Notably, as the seafloor beneath an oceanic island gets older, it thickens and sinks due to isostasy. Thus,

FIGURE 18.7 Bathymetry of the southwestern Pacific, showing seamounts and oceanic plateau.

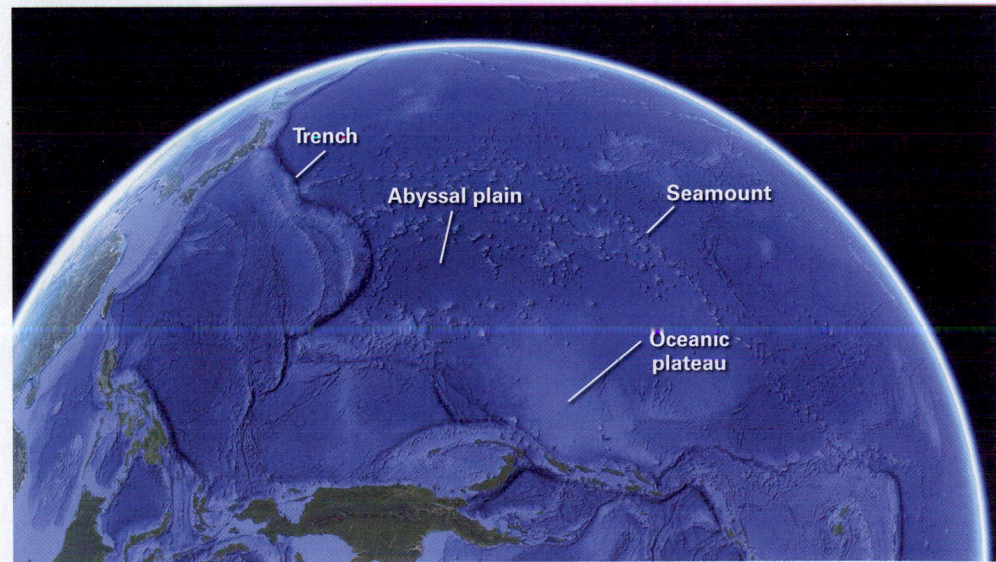

18.3 Ocean Water and Currents

Composition

If you've ever had a chance to swim in the ocean, you may have noticed that you float much more easily in ocean water than you do in freshwater. That's because ocean water contains an average of 3.5% dissolved salt (**Fig. 18.8**). Dissolved ions fit between water molecules without changing the volume of the water, so adding salt to water increases the water's density, and you float higher in a denser liquid. There's so much salt in the ocean that if all the water magically evaporated, a 60-m-thick layer of salt would coat the ocean floor. This layer would consist of about 75% halite (NaCl) with lesser amounts of gypsum ($CaSO_4 \cdot H_2O$), anhydrite ($CaSO_4$), and other salts. Oceanographers refer to the concentration of salt in water as the water's **salinity**.

Although ocean surface salinity averages 3.5%, measurements from around the world demonstrate that salinity varies with location, ranging from about 1.0% to about 4.1%. Ocean surface salinity reflects the balance between the addition of freshwater by rivers or rain and the removal of freshwater by evaporation. (When seawater evaporates, salt stays behind and only H_2O goes into the vapor phase.) Salinity may also depend on water temperature, for warmer water can hold more salt in solution than can cold water. Thus, salt concentration varies with latitude, proximity to the coast, and depth (**Fig. 18.9a, b**). As shown in Figure 18.9a, variations in salinity generally occur only in the upper 1 km of seawater, where evaporation

and addition of freshwater take place, and where temperature is more variable. The salinity of deeper water tends to be more homogenous. Oceanographers refer to the gradational boundary between shallower-water salinities and deep-water salinities as the *halocline*.

Where does the salt in seawater come from? Leonardo da Vinci, the famous Renaissance artist and scientist, speculated that sea salt came buried in salt layers, such as those found in salt mines. But modern studies demonstrate that most cations, or positive ions, in sea salt—such as Na^+, K^+, Ca^{2+}, and Mg^{2+}—come from chemical weathering of rocks and that the anions of negative ions—such as Cl^- and SO_4^{2-}—come from volcanic gases in the air. Dissolved ions get carried to the sea by groundwater and river water because, in fact, "freshwater" is not ion-free but rather is a dilute solution typically containing 0.01% to 0.05% salt. Rivers actually deliver over 2.5 billion tons of salt to the sea every year. Salt remains in the sea because, as we've noted, evaporation removes only H_2O molecules. Eventually, precipitation removes salt from seawater, and this salt becomes incorporated in new evaporite deposits.

Temperature

When the *HMS Titanic* sank after striking an iceberg in the North Atlantic, most of the unlucky passengers and crew who jumped or fell into the sea died within minutes because the seawater temperature at the site of the tragedy was about 1°C (34°F), and cold water removes heat from a body very rapidly. Yet swimmers can play for hours in the Caribbean, where sea-surface temperatures reach 29°C (84°F). Though the average global sea-surface temperature hovers around 17°C (63°F), it

FIGURE 18.8 The composition of average seawater. The expanded part of the graph shows the proportions of ions in the salt of seawater.

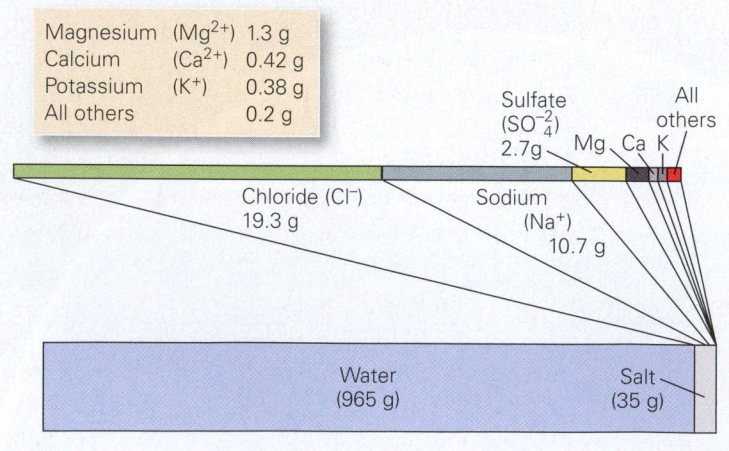

Magnesium	(Mg^{2+})	1.3 g
Calcium	(Ca^{2+})	0.42 g
Potassium	(K^+)	0.38 g
All others		0.2 g

Sulfate (SO_4^{-2}) 2.7 g
Mg Ca K
All others

Chloride (Cl⁻) 19.3 g
Sodium (Na⁺) 10.7 g

Water (965 g)
Salt (35 g)

Workers on Cape Verde Island collecting salt produced by evaporation of seawater.

FIGURE 18.9 Physical characteristics of the ocean.

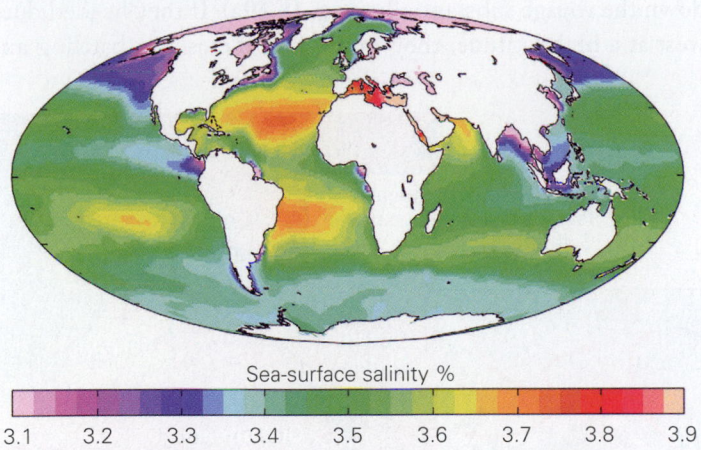

Sea-surface salinity %

3.1   3.2   3.3   3.4   3.5   3.6   3.7   3.8   3.9

(a) Regional variations in salinity, specified in percentages. Contour lines separate areas of different salinity.

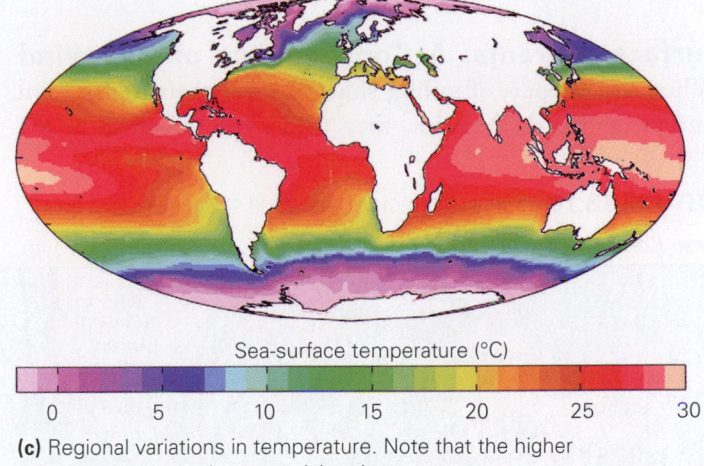

Sea-surface temperature (°C)

0   5   10   15   20   25   30

(c) Regional variations in temperature. Note that the higher temperatures occur in equatorial regions.

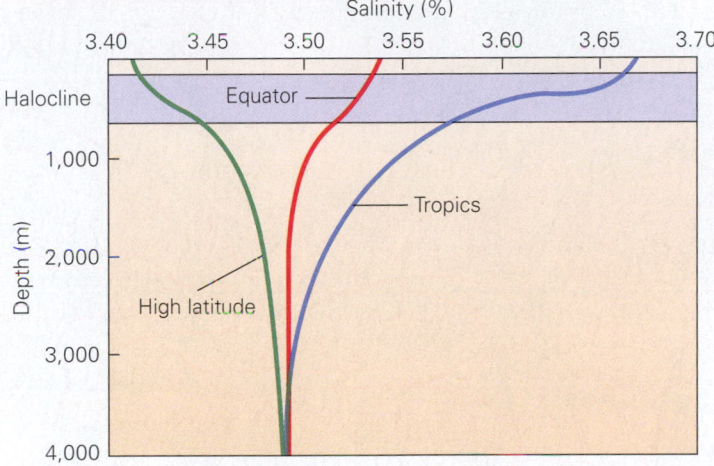

(b) Changes in salinity with depth. Note the halocline (blue band), an interval where salinity changes rapidly.

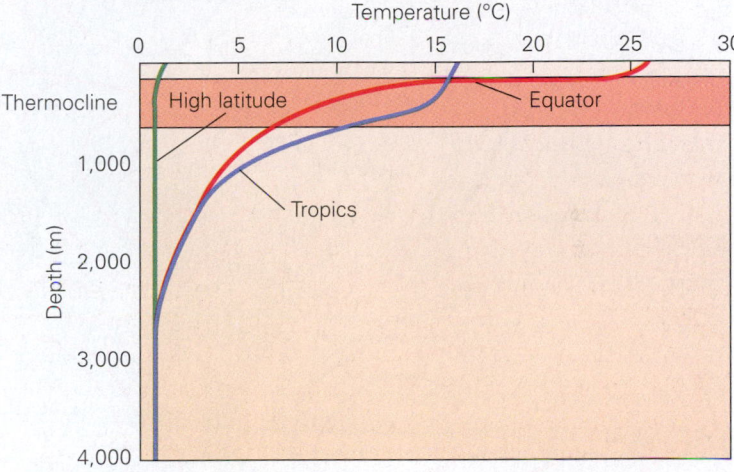

(d) Changes in temperature with depth. Note the thermocline (orange band), an interval where temperatures decrease rapidly.

ranges between −2°C (28°F) near the poles to almost 36°C (97°F) in restricted tropical seas (**Fig. 18.9c**). (Note that, because ocean water is salty, it freezes at a temperature below that of freshwater.) The general correlation of average temperature with latitude exists because the intensity of solar radiation varies with latitude. As we'll see, ocean temperature at a given latitude actually varies significantly because of currents.

Surface seawater temperature at a given locality varies with the season because the intensity of solar radiation varies with the season. But the difference is only around 2° in the tropics, 8° in the temperate latitudes, and 4° near the poles. Seasonal seawater temperature changes remain in a narrow range because water has a large *heat capacity*, meaning it can absorb or release large amounts of heat without changing temperature very much. Because of this characteristic, the ocean regulates the temperatures of coastal regions. Thus, the air temperature range in Vancouver, on the Pacific coast of Canada, is much less than that of Winnipeg, in the plains of central

Canada—Vancouver rarely drops below freezing, while a cold day in Winnipeg can reach −30°C (−22°F).

Water temperature in the ocean varies markedly with depth (**Fig. 18.9d**), for the solar energy does not penetrate deeply, and since warm water is less dense than cold water, it remains at the surface. Typically, a *thermocline* between warm water above and significantly colder water below appears at a depth of about 300 m in the tropics. A pronounced thermocline does not develop in polar seas since surface waters there are already so cold.

Currents: Rivers in the Sea

Since first setting sail on the open ocean, people have known that the water of the ocean does not stand still but rather flows or circulates at velocities of up to several kilometers per hour in fairly well-defined streams called **currents**. Oceanographic studies made since the *Challenger* expedition demonstrate that circulation in the sea occurs at two levels: *surface currents* affect the upper hundred meters of water, while *deep currents* can reach down to the seafloor.

Surface Currents: A Consequence of the Wind

When the skippers of sailing ships planned their routes from Europe to North America, they paid close attention to the directions of surface currents, for sailing against a current slowed down the voyage substantially (**Fig. 18.10a**). If they headed due west at a high latitude, they would find themselves battling an

FIGURE 18.10 Surface currents of the world's oceans.

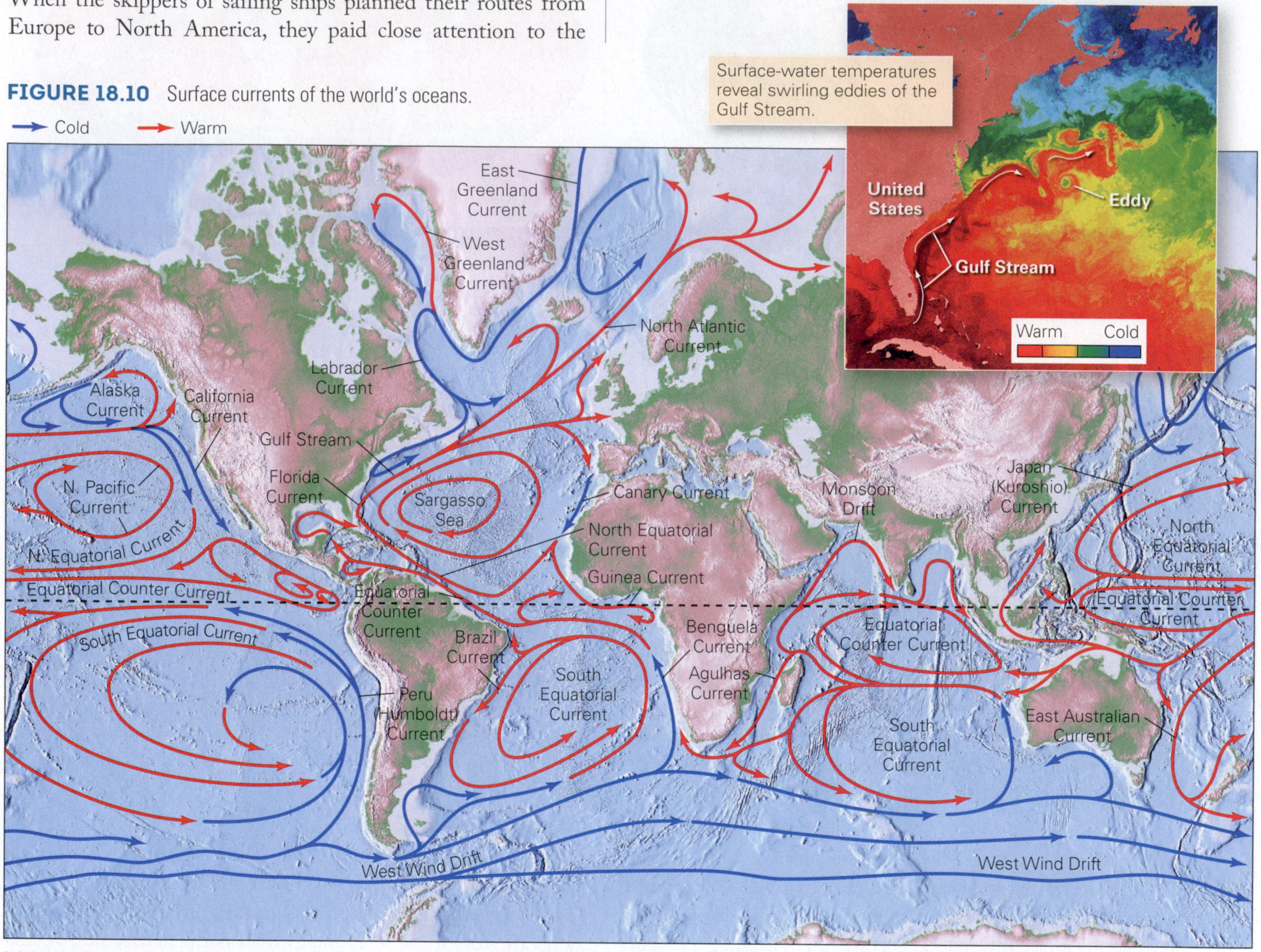

→ Cold   → Warm

Surface-water temperatures reveal swirling eddies of the Gulf Stream.

(a) A simplified map showing the major currents. The color key distinguishes cold from warm currents, and the inset shows a detail of the Gulf Stream, which carries warm water to the northeast.

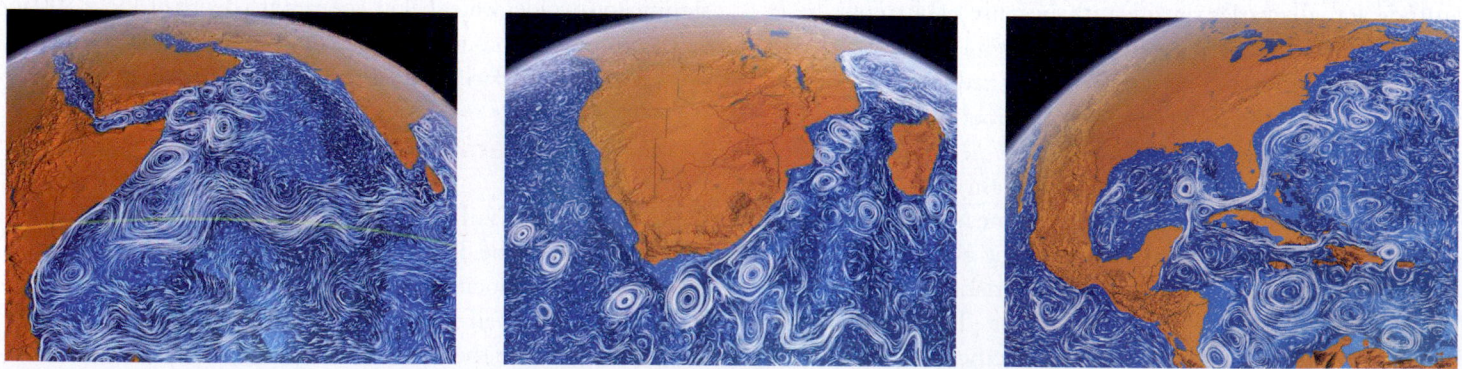

(b) Animations of ocean currents, prepared by NASA based on data collected over a two-year period, show the complexities that develop where currents shear against the margins of the continents or are squeezed between land masses. The circular flows are eddies.

eastward-flowing surface current, the Gulf Stream. Further, they found that the water moving in a surface current does not flow smoothly but displays turbulence. Isolated swirls or ring-shaped currents of water, called eddies, form along the margins of currents. Eddies also develop where regional currents squeeze between land masses or shear along a coastline (**Fig. 18.10b**).

Surface currents occur in all the world's oceans. They result from interaction between the sea surface and the wind, for as moving air molecules shear across the surface of the water, the friction between air and water drags the water along. But the movement of water resulting from wind shear does not exactly parallel the movement of the wind. Specifically, because of the **Coriolis effect**, a consequence of the Earth's rotation, surface currents in the northern hemisphere veer toward the right and surface currents in the southern hemisphere veer toward the left of the average wind direction (**Box 18.1**). Across the width of an ocean, the Coriolis effect causes surface currents to make a complete loop known as a **gyre**. Surface water may become trapped for a long time in the center of the gyre, where currents hardly exist, so these regions tend to accumulate floating plastics, sludge, and seaweed. The "Sargasso Sea," named for a kind of floating seaweed, lies at the center of the North Atlantic gyre, and the "Great Pacific Garbage Patch," an accumulation of floating plastic and trash, lies at the center of the North Pacific gyre.

In the past, when continents were in different positions, the geometry of ocean currents was quite different from what it is today. For example, the modern circum-Antarctic current did not appear until the Drake Passage, between South America and Antarctica, opened 25 million years ago. Currents that move from the poles to the equator bring cool water toward the equator, whereas currents that move from the equator toward the poles carry warm water poleward, which means that warm water can occur locally at fairly high latitudes. This transport of heat moderates the global climate, so changes in the pattern of currents through geologic time affect the climate of the Earth System.

Upwelling, Downwelling, and Deep Currents

Surface currents are not the only means by which water flows in the ocean; it also circulates in the vertical direction. Oceanographers have now identified *downwelling zones*, places where near-surface water sinks, and *upwelling zones*, places where subsurface water rises. Let's consider some of the causes of upwelling and downwelling.

Downwelling and upwelling locally takes place where winds push surface water toward or away from the coast. If near-surface water moves toward the coast, then an oversupply of water develops along the shore and excess water must sink, causing downwelling (**Fig. 18.11a**). Alternatively, if near-surface water moves away from the coast, then a deficit of water

FIGURE 18.11 Upwelling and downwelling can happen along coasts.

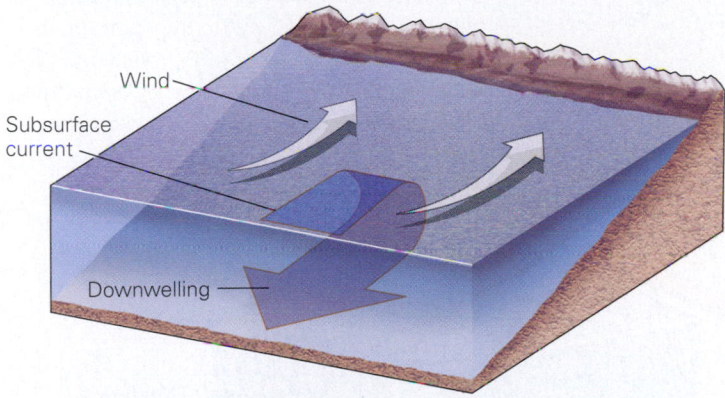

(a) A wind blowing toward shore can cause local downwelling.

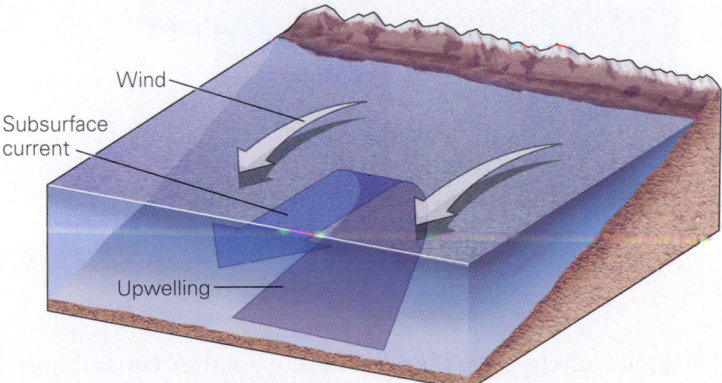

(b) A wind blowing offshore can cause local upwelling.

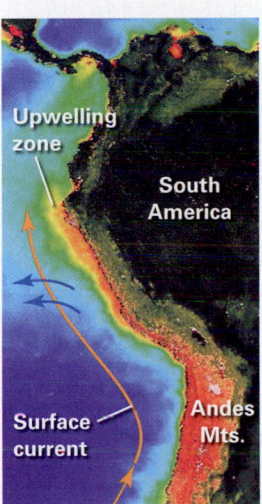

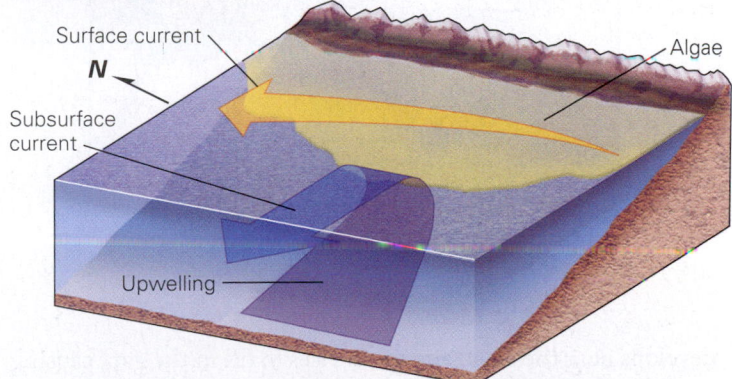

(c) Due to Ekman transport (see Box 18.1), the north-flowing Humbolt Current causes upwelling. This upwelling brings up nutrients that feed near-surface algae.

BOX 18.1   CONSIDER THIS . . .

The Coriolis Effect

Imagine that you have a huge cannon—aim it due south and fire a projectile from the North Pole to a target on the equator (**Fig. Bx18.1a**). If the Earth were standing still, the shot would follow a line of longitude. But the Earth isn't standing still. It rotates counterclockwise around its axis, an imaginary line that passes through the planet's center and its geographic poles, as viewed looking down on the North Pole. To an observer in space, an object at the pole doesn't move at all as the Earth spins because it is sitting on the axis, but an object at the equator moves at about 1,665 km per hour (1,035 mph). Because of this difference,

the target on the equator will have moved by the time the projectile reaches it. In fact, to an observer standing on the Earth and moving with it, the projectile follows a curving trajectory. The same phenomenon happens if you place the cannon on the equator and fire the projectile due north (**Fig. Bx18.1b**)—the projectile's path curves because, as it moves north, the projectile moves eastward progressively faster than the land beneath. This behavior is called the *Coriolis effect*, after the French engineer who, in 1835, described its consequences. You'll see the Coriolis effect in the southern hemisphere, too, but in the

opposite direction. Because of the Coriolis effect, north-flowing surface currents in the northern hemisphere deflect by about 45° to the east of the wind direction (**Fig. Bx18.1c**), and south-flowing surface currents deflect about 45° to the west of the wind direction.

In 1902, a Swedish researcher named Vagn Ekman thought further about the effect of the Coriolis effect on surface-water movement and suggested that the upper 100 m of the ocean could be pictured as a series of layers, each shearing the layer below. Because of the Coriolis effect, the deflection of each successive deeper layer moves at an angle of 45° relative to the layer above, so arrows reflecting the motion of successively deeper layers define a spiral known as an *Ekman spiral* (Fig. Bx18.1c). The length of successively deeper arrows is shorter because friction between layers slows the flow. Ekman showed that the net result of this movement pattern causes net water transport in the upper 100 m to be about 90°, relative to the wind direction. This somewhat surprising movement is now known as *Ekman transport*.

FIGURE Bx18.1 The Coriolis effect occurs because the velocity of a point at the equator, in the direction of the Earth's spin, is greater than that of a point near the pole.

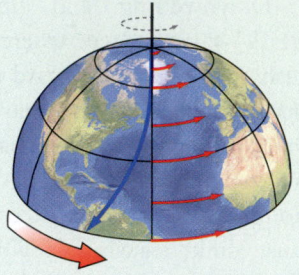

(a) A projectile shot from the pole to the equator deflects to the west.

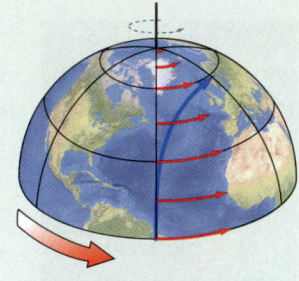

(b) A projectile shot from the equator to the pole deflects to the east.

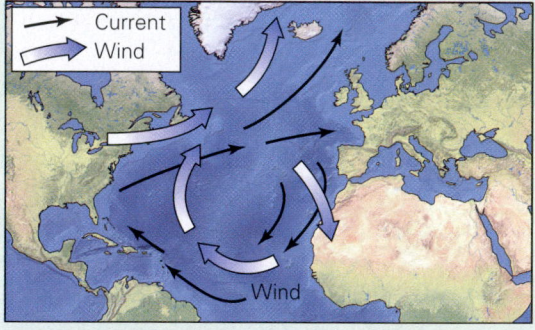

(c) Due to the Coriolis effect, currents deflect clockwise, relative to the wind, in the North Atlantic.

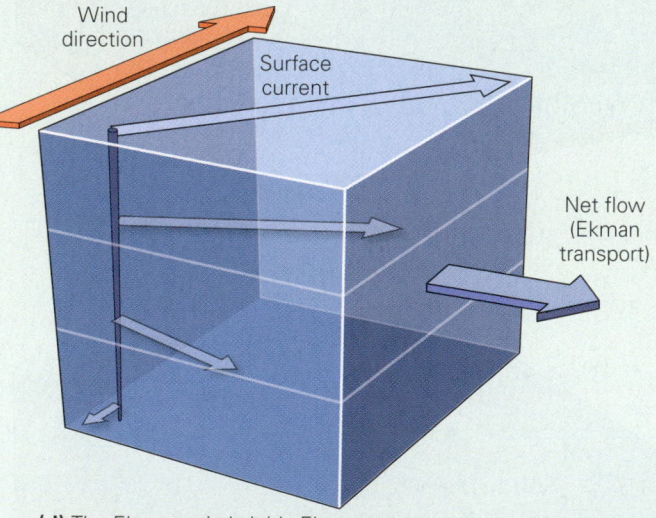

(d) The Ekman spiral yields Ekman transport.

develops near the coast and water rises to fill in the gap, causing upwelling (**Fig. 18.11b**). Where a strong surface current flows along a coast, it can cause upwelling or downwelling depending on its direction (see Box 18.1). For example, the north-flowing

Humbolt Current along the west coast of South America causes upwelling, which is important for the fishing industry because rising water brings up nutrients that encourage growth of the plankton on which fish feed (**Fig. 18.11c**). Similarly, upwelling

FIGURE 18.12 Global-scale upwelling and downwelling of ocean water.

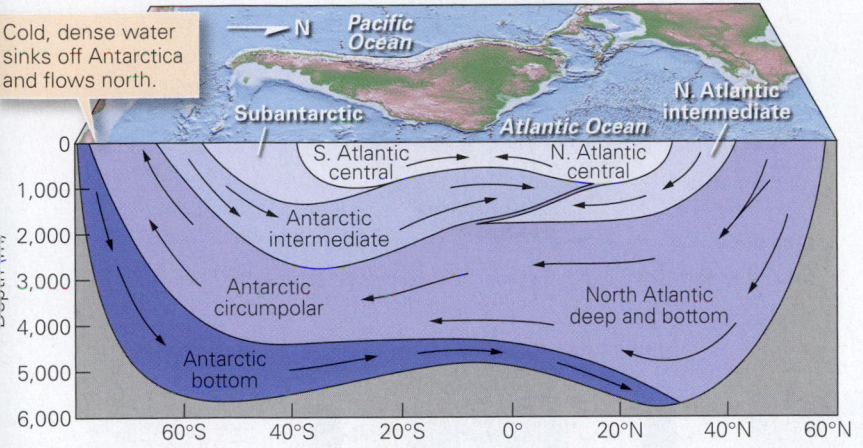

Cold, dense water sinks off Antarctica and flows north.

(a) Due to variations in density, the oceans are stratified into distinct water masses.

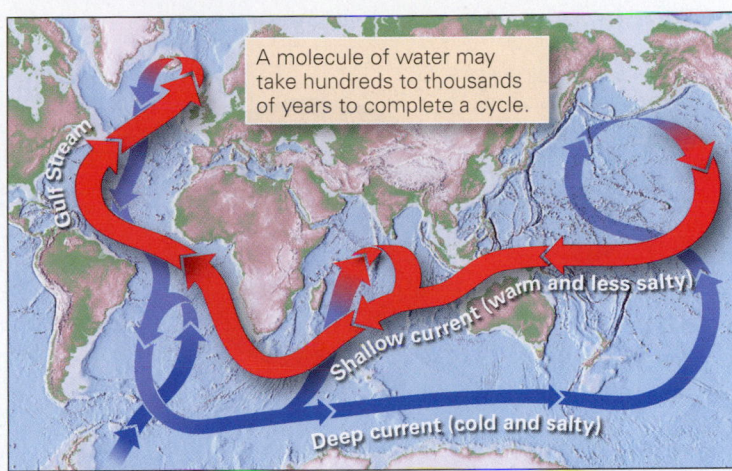

A molecule of water may take hundreds to thousands of years to complete a cycle.

(b) Thermohaline circulation results in a global-scale "conveyor belt" that circulates water throughout the entire ocean system. The ocean mixes entirely in a 1,500-year period.

occurs along the equator because the winds blow steadily from east to west—the Coriolis effect causes the surface water to deflect to the right in the northern hemisphere and to the left in the southern hemisphere, resulting in a net deficit of water along the equator. Upwelling replaces this deficit and causes the surface water at the equator to be cooler and rich in the nutrients that foster an abundance of life in equatorial water.

Contrasts in water density, caused by differences in temperature and salinity, can also drive upwelling and downwelling. We refer to the rising and sinking of water driven by such density contrasts as **thermohaline circulation**. During such circulation, denser water (cold and/or saltier) sinks, whereas water that is less dense (warm and/or less salty) rises. As a result, the cold water in polar regions sinks and flows back along the bottom of the ocean toward the equator. This process divides the ocean vertically into a number of distinct *water masses*, which mix only very slowly with one another. In the Atlantic Ocean, for example, the "Antarctic Bottom Water" sinks along the coast of Antarctica, and the "North Atlantic Deep Water" sinks in the north polar region (**Fig. 18.12a**). The combination of surface currents and thermohaline circulation, like a conveyor belt, moves water and heat among the various ocean basins (**Fig. 18.12b**).

Take-Home Message

Ocean salinity and temperature vary with depth and location. The wind drives surface currents, forming large gyres. Upwelling and downwelling develop due to wind-driven water surplus or deficit or due to density variations related to temperature and salinity.

QUICK QUESTION: Why does downwelling of ocean water occur at polar latitudes?

18.4 Tides

A ship captain seeking to float a ship over reefs, a fisherman hoping to set sail from a shallow port, marines planning to attack a beach from the sea, a tourist eager to harvest shellfish from nearshore mud—all must pay attention to the rise and fall of sea level, a vertical movement called a **tide**, if they are to be successful (**Fig. 18.13a–d**). If the tide is too low, ships run aground or stay trapped in harbors, and if the tide is too high, a beach may become too narrow to permit access. During a rising tide, or flood tide, the *shoreline* (the boundary between water and land) moves inland, whereas during the falling tide, or ebb tide, the shoreline moves seaward. The vertical difference between sea level at high tide and sea level at low tide is called the **tidal range**. Tidal range varies with location. For example, out in the open ocean, the range averages about 0.6 m, but along coasts it gets higher—the largest tidal reach on Earth, 16.8 m (54.6 feet), occurs in the Bay of Fundy, on the east coast of Canada. The region submerged at high tide and exposed at low tide, is called the *intertidal zone*, and it is a fascinating ecological niche populated by organisms that must be able to survive out of water for hours at a time.

The horizontal distance over which the shoreline migrates between high and low tides depends on both the tidal range and the slope of the shore surface. Where the tidal range is large and the slope is gentle, the position of the shore can move a long way during a tidal cycle, leaving a broad *tidal flat* exposed to the air in the intertidal zone at low tide (**Fig. 18.13e, f**). Some tidal flats extend out from the high-tide line for over a kilometer (**Fig. 18.14a**). Wide tidal flats can be dangerous, for when the tide turns, the arrival of a flood tide can create a *tidal bore*, a

FIGURE 18.13 Ocean tides and their manifestation.

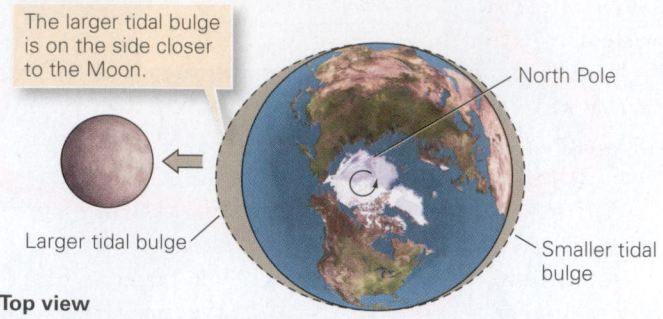

The larger tidal bulge is on the side closer to the Moon.

North Pole

Larger tidal bulge

Smaller tidal bulge

Top view

(a) Tides develop as the Earth spins relative to the two tidal bulges.

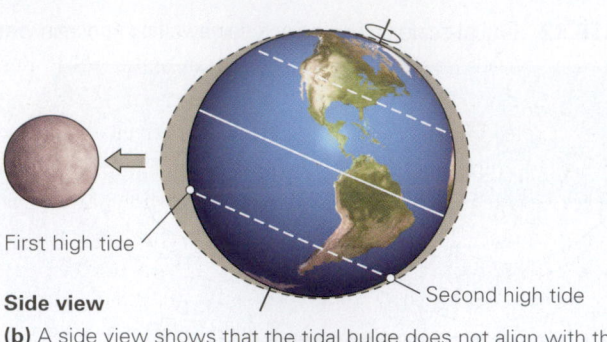

First high tide

Second high tide

Side view

(b) A side view shows that the tidal bulge does not align with the equator.

Spring tide

Solar tide

Lunar tide

Full moon

New moon

Sun

Extra-high tides are spring tides.

Neap tide

Lunar tide

Solar tide

Sun

Extra-low tides are neap tides.

(d) Gravitational pull of the Sun can add to that of the Moon to cause extra-high (spring) tides. If the Sun's pull is at right angles to that of the Moon, the tides are particularly low (neap tides).

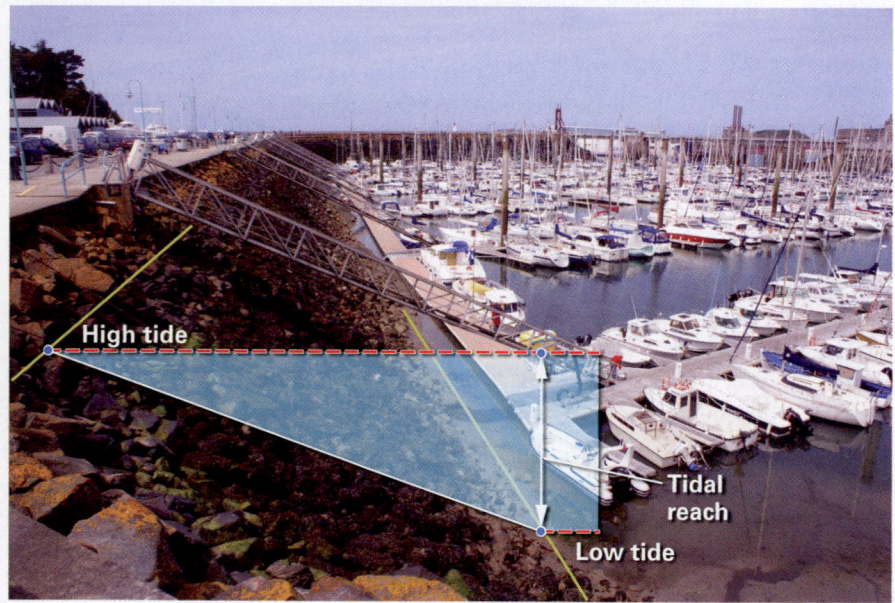

High tide

Tidal reach

Low tide

(c) The tidal reach is the vertical distance between low tide and high tide. This photo of a French harbor was taken at low tide.

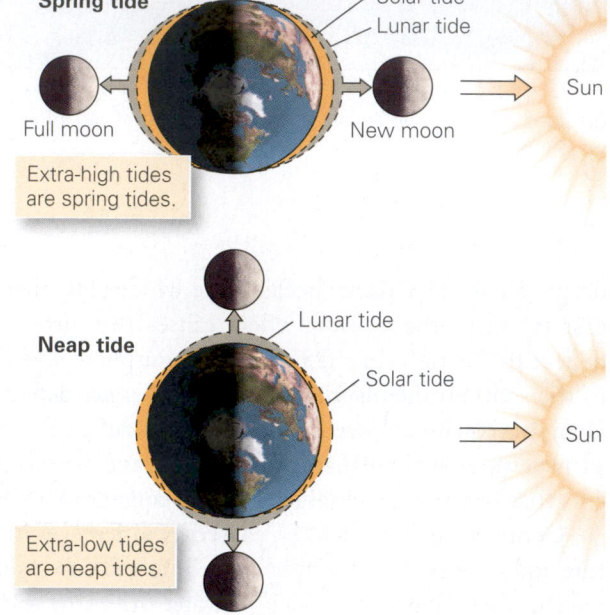

(e) High tide at Perranporth in Cornwall, England (August 28, 2007; high water: 6:00 pm).

(f) Low tide at Perranporth (August 29, 2007; low water: 11:20 pm). Note that the rocks in the foreground are the same ones that the man was fishing from in part (e).

visible wall of water ranging from a few centimeters to a couple of meters high, which moves inland at speeds of up to 35 km per hour—faster than a person can run (**Fig. 18.14b**). In February 2004, for example, 15 shellfish hunters lost their lives along the coast of northwestern England because they were far offshore searching for cockles in the mud when the flood tide came in.

Tides are caused by a **tide-generating force**, which is due partly to the gravitational attraction of the Sun and Moon and partly to centrifugal force caused by the revolution of the Earth-Moon system around its center of mass. (To understand the meaning of this complex statement, see **Box 18.2**.) Overall, tide-generating forces create two bulges in the global ocean (see Fig. 18.13a, b). One bulge, the sublunar bulge, lies on the side of the Earth closer to the Moon—it forms because the Moon's gravitational attraction is greatest at this point. The other bulge, the secondary bulge, lies on the opposite (far) side of the Earth (12,000 km—the diameter of the Earth—farther from the Moon); it forms because the Moon's gravitational attraction is weakest at this point. Here centrifugal force can push water outward (see Box 18.2). A depression in the global ocean surface separates the two bulges. Simplistically, when a shore location lies under a tidal bulge, it experiences a high tide, and when it passes under a depression it experiences low tide.

If the Earth's solid surface were smooth and completely submerged beneath the ocean so that there were no continents or islands, the timing of tides would be fairly simple to understand. Because the Earth spins on its axis once a day, we would predict two high tides and two low tides at a given point per day. But the story isn't quite that simple—many other factors affect the timing and magnitude of tides. These include the following.

- *Tilt of the Earth's axis:* Because the spin axis of the Earth is not perpendicular to the plane of the Earth-Moon system, a given point passes between a high part of one bulge during one part of the day and through a lower part of the other bulge during another part of the day, so the two high tides at the given point are not the same size (see Fig 18.3b).

- *The Moon's orbit:* The Moon progresses in its 28-day orbit around the Earth in the same direction as the Earth rotates. High tides arrive 50 minutes later each day because of the difference between the time it takes for Earth to spin on its axis and the time it takes for the Moon to orbit the Earth.

- *The Sun's gravity:* When the angle between the direction to the Moon and the direction to the Sun is 90°, we experience extra-low tides (neap tides) because the Sun's gravitational attraction counteracts the Moon's. When the Sun is on the same side as the Moon, we experience extra-high tides (spring tides) because the Sun's attraction adds to the Moon's (see Fig. 18.13d).

- *Focusing effect of bays:* In a bay that narrows toward the shore, the flood tide brings a large volume of water into a small area, so the bay head experiences an especially large high tide.

- *Basin shape:* The shape of the basin containing a portion of the sea influences the sloshing of water back and forth within the basin as tides rise and fall. Depending on the timing and magnitude of this sloshing, this effect can locally add to the global tidal bulge or subtract from it, and thus it can affect the rhythm and magnitude of tides. In some locations, the net effect is to cancel one of the

FIGURE 18.14 Tidal flats and tidal bores.

(a) A broad tidal flat is exposed at low tide around Mont-Saint-Michel, on the coast of France.

(b) This large tidal bore, entering the mouth of a river along the coast of China, is a tourist attraction.

BOX 18.2 CONSIDER THIS . . .

The Forces Causing Tides

Fundamentally, tides result from interaction between two forces: gravitational attraction exerted by the Moon and Sun on the Earth and centrifugal force caused by the revolution of the Earth around the center of mass of the Earth-Moon system. To explain this statement, we must review some key terms from physics.

- *Gravitational pull* is the attractive force that one mass exerts on another. The magnitude of gravitational pull depends on the amount of mass in each object and on the distance between the two masses.
- *Centrifugal force* is the apparent outward-directed ("center-fleeing") force that material on or in an object feels when the object spins or moves in orbit around a point. Note that centrifugal force differs from centripetal force, the "center-seeking" force; this distinction can be confusing. To experience centripetal force, tie a ball to a string and swing it around your head. The string exerts an inward-directed centripetal force on the ball—if the string breaks, the centripetal force ceases to exist and the ball heads off in a straight-line path. To picture centrifugal force, imagine that the ball is hollow and that you've placed a marble inside. As you twirl the ball around your head, the marble moves to the outer edge of the ball. The apparent force pushing the marble outward is the centrifugal force. But as such, centrifugal force is not a real force—it is simply a manifestation of inertia, and it exists only from the perspective, or reference frame, of the orbiting object. (A physics book explains this contrast in greater detail.)
- *Earth-Moon system* refers to this pair of objects viewed as a unit as they move together through space.
- *Center of mass* is the point within an object, or a group of objects, about which

mass is evenly distributed; put another way, it is the location of the average position, or the balance point, of the total mass in a single object or a group of objects. Because the Earth is 81 times more massive than the Moon, the center of mass of the Earth-Moon system actually lies 1,700 km below the surface of the Earth.

With these terms in mind, let's first consider the origin of centrifugal force in the Earth-Moon system. To do this, we must examine the way in which the Earth-Moon system moves. The center of the Earth itself does not follow a simple orbit around the Sun. Rather, it is the center of mass of the Earth-Moon system that follows this trajectory; the Earth actually spirals around this trajectory as it speeds around the Sun. To picture this motion, imagine that the Earth-Moon system is a pair of dancers, one of them much heavier than the other. The dancers face each other, hold hands, and whirl in a circle as they drift across the dance floor (**Fig. Bx18.2a**). Each dancer's head orbits the center of mass.

Revolution of the Earth around the Earth-Moon system's center of mass generates centrifugal forces on both the Earth and the Moon that would cause the Earth and the Moon to fly away from each other were it not for the gravitational attraction holding them together. We can see this by looking again at our dancer analogy (**Fig. Bx18.2b**)—the centrifugal force acting on each dancer points outward, away from his or her partner, and is the same for all points on each dancer. We can represent the direction and magnitude of centrifugal force by an arrow called a *vector*. (A vector is a quantity that has magnitude and direction.) In this case, the length of the arrow represents the magnitude of the force, and the orientation of the arrow indicates the direction of the force. If we think of the dancers as the Earth and the Moon,

then centrifugal force vectors at all points on the surface of the Earth point away from the Moon (**Fig. Bx18.2c**). On the Earth, therefore, centrifugal force causes the surface of the ocean to bulge outward, away from the center of mass of the Earth-Moon system, on the far side of the Earth.

Now let's consider how the force of gravity comes into play in causing tides. To simplify this discussion, we examine only the effect of the Moon's gravity on Earth. (This is reasonable because gravitational pull by the Moon contributes most of the tide-generating force. The Sun, even though it is larger, is so far away that its contribution is only 46% that of the Moon.) Vectors representing the magnitude and direction of the Moon's gravitational pull at any point on the surface of the Earth all point toward the center of the Moon. Because the magnitude of gravity depends on distance, the Moon exerts more attraction on the near side of the Earth than at the Earth's center, and it exerts less attraction on the far side of the Earth than at the Earth's center. Gravity, therefore, causes the surface of the ocean on the near side of the Earth to bulge toward the Moon.

In the Earth-Moon system, both centrifugal force and gravitational pull operate at the same time. How do they interact? If we draw vectors representing both centrifugal force and gravitational force at various points on or in the Earth, we see that the vectors representing centrifugal force do not have the same length as those representing gravitational attraction except at the Earth's center. Moreover, the vectors representing centrifugal force do not point in the same direction as the vectors representing gravitational attraction. The force that the ocean water feels is the sum of the two forces acting on the water. You can determine the sum of two vectors by drawing the vectors so they

daily tides entirely; as a result, the locality experiences only one high tide and one low tide in a day.
- *Air pressure:* The weight of the overlying atmosphere produces pressure that pushes down on the sea surface.

Air pressure can be affected by temperature and circulation in the atmosphere and thus can vary over time (see Chapter 20). When air pressure decreases, sea level rises slightly. And when air pressure increases, touch head to tail—the sum is the vector that completes the triangle. This sum is the *tide-generating force*.

The magnitude and direction of the tide-generating forces vary with location on the Earth. For example, on the side of the Earth closer to the Moon, gravitational vectors are larger than centrifugal force vectors, so adding the two gives a net tide-generating force that pulls the sea surface to bulge toward the Moon. On the side of the Earth farther from the Moon, the centrifugal force vectors are larger, so centrifugal force caused by the orbiting of the Earth-Moon system around the center of mass causes the surface of the sea to bulge outward, away from the Moon (**Fig. Bx18.2d**). Thus, the ocean has two tidal bulges—one on the side close to the Moon and one on the opposite side of the Earth. The bulge closer to the Moon is larger.

FIGURE Bx18.2 The concept of the center of mass of the Earth-Moon system and the tide-generating force.

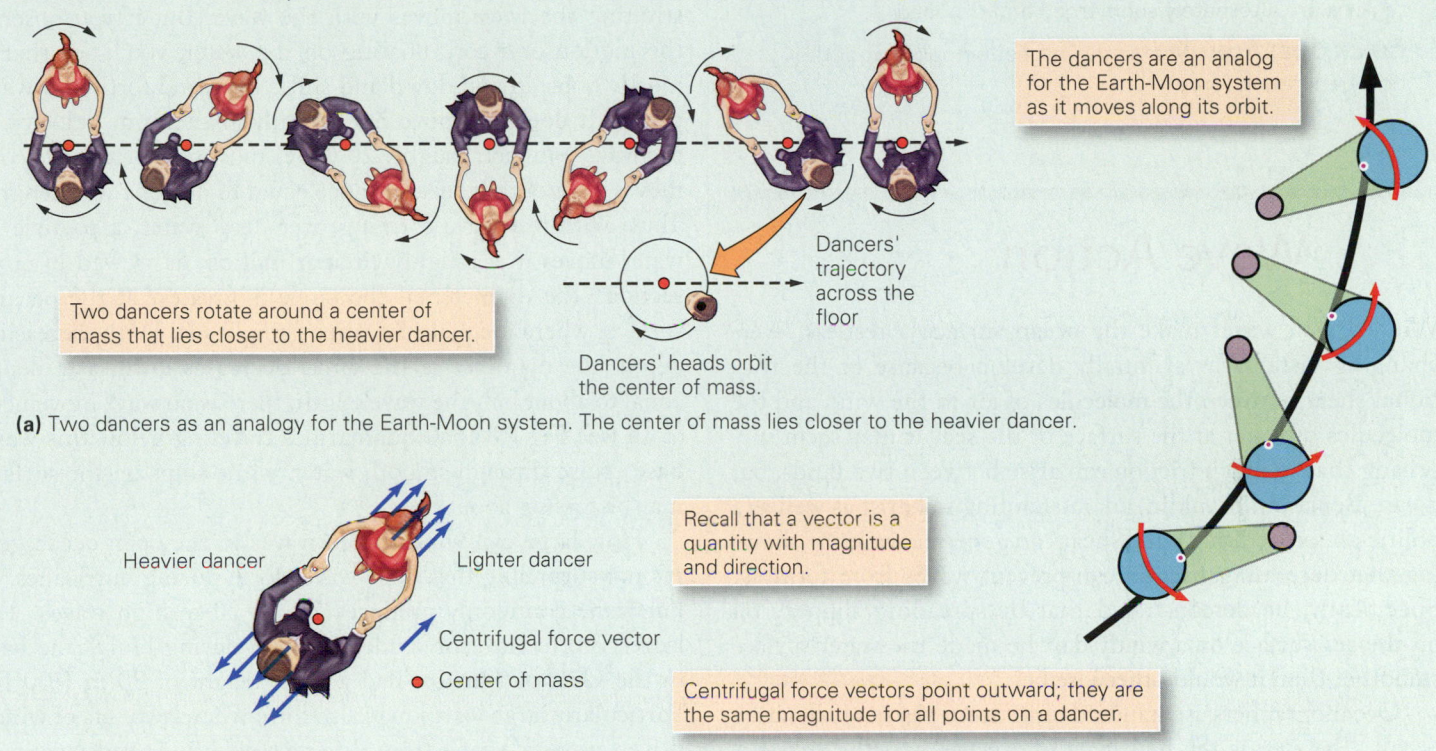

The dancers are an analog for the Earth-Moon system as it moves along its orbit.

Two dancers rotate around a center of mass that lies closer to the heavier dancer.

Dancers' trajectory across the floor

Dancers' heads orbit the center of mass.

(a) Two dancers as an analogy for the Earth-Moon system. The center of mass lies closer to the heavier dancer.

Heavier dancer   Lighter dancer

Recall that a vector is a quantity with magnitude and direction.

↑ Centrifugal force vector

• Center of mass

Centrifugal force vectors point outward; they are the same magnitude for all points on a dancer.

(b) Each point on each dancer feels an outward-directed centrifugal force.

Earth

Moon

← Centrifugal force
↘ Gravitational attraction

(c) Each point on the Earth's surface feels the same centrifugal force due to the spin of the Earth-Moon system around its center of mass but feels a different gravitational pull from the Moon because gravitational force depends on

Sea surface (greatly exaggerated)

Center of mass

↗ Tide-generating force

(d) The tide-generating force is the sum of the centrifugal force vector and the gravitational force vectors. On the side closer to the Moon, the gravitational force vector dominates, producing a bulge toward the Moon. On the other side, centrifugal force dominates,

sea level drops slightly. Normally, this effect is relatively small, but during a hurricane air pressure can drop so much that sea level rise can contribute to coastal flooding.

Because of the complexity of factors contributing to tides, the timing and magnitude of tides vary significantly along a coast. Nevertheless, at a given location the tides are periodic and can be predicted. Tides gave early civilizations a rudimentary way to tell time. In fact, in some languages the word for tide is the same as the word for time.

Take-Home Message

Gravitational attraction by the Moon and Sun, as well as centrifugal force in the Earth-Moon System, cause two tidal bulges that move around the Earth. Tidal range, the difference between high and low tides, varies significantly with location. Because of the rise and fall of tides, intertidal regions are alternately submerged and exposed.

QUICK QUESTION: What is the largest component of the tide-generating force?

18.5 Wave Action

Wind-driven waves make the ocean surface a restless, ever-changing vista. Waves initially develop because of the frictional shear between the molecules of air in the wind and the molecules of water at the surface of the sea. It may seem surprising that so much friction can arise between two fluids, but it can. Benjamin Franklin, an outstanding scientist as well as a politician, proved that wind shear can generate waves by showing that decreasing friction can prevent waves from forming. Specifically, he demonstrated that by spreading slippery oil on the sea surface on a windy day, he made the water surface smoother than it would otherwise be.

Oceanographers use standard terms from physics to describe waves. The *crest* is the high part of a wave, and the *trough* is the low part. The total vertical distance from trough to crest is the *wave height*, the *amplitude* is half the wave height, and the *wavelength* is the horizontal distance between two adjacent wave troughs or adjacent wave crests (**Fig. 18.15a**).

Wave height in the open ocean depends on the wind velocity and on the *fetch*, the horizontal distance of the ocean over which the wind blows without changing direction. When the wind first begins to blow, it creates *ripples* in the water surface, pointed wavelets with small amplitude and wavelength. With continued blowing of a long fetch, larger waves will begin to build.

Swells initiate where the wind has piled up water beneath a storm. **Swells** are sets of large waves (typically with wave heights of between 2 and 10 m and wavelengths of between 40 and 700 m) that can travel hundreds to thousands of kilometers across the ocean, far beyond the region where they formed. Physicists call them *gravity waves*—oscillations that develop when water undergoes vertical uplift in the presence of gravity. To illustrate, a storm's strong winds and low atmospheric pressure can cause water to pile up. Gravity will then pull the uplifted surface of the water back down and inertia takes the sinking water below the equilibrium level (the level of a flat calm sea). Since the water seeks equilibrium, it rises back up, its inertia sending it again above the equilibrium line, only to have it pulled down again by gravity, continuing the cycle. This process resembles the rise and fall of a weight suspended from a spring. Each successive oscillation sends water out sideways as a swell, allowing the waves to travel great distances.

When you watch a wave travel across the open ocean, you may get the impression that the whole mass of water constituting the wave moves with the wave. But if you observe the motion of a cork floating on the water, you'll see that it mostly bobs up and down and shifts back and forth as a wave passes. It does not move horizontally at the same velocity as the wave—in fact, roughly 20 waves must pass before the cork moves horizontally by a distance equal to just one wavelength. Thus, within a wave forming over deep water, a particle of water moves in a roughly circular motion, as viewed in cross section. The diameter of the circle is greatest at the ocean's surface, where it equals the wave's amplitude. With increasing depth, the diameter of the circle decreases until, at a depth equal to about half the wavelength, there is no wave movement at all (see Fig. 18.15a). Submarines traveling below this **wave base** cruise through smooth water, while ships on the surface may be tossing about.

How large can wind-driven waves in the open ocean get? It's not surprising that huge waves form during hurricanes. A hurricane commonly produces 15- to 20-m-high waves. The largest storm waves, recorded in 2005 during Hurricane Ivan in the Gulf of Mexico, had a wave height of 30 m (100 ft). Particularly large waves may also form where two sets of wind-driven waves coming from different directions constructively interfere in such a way that wave crests add to each other. This happened in 1979, when waves generated by an east-blowing gale collided with waves generated by a west-blowing gale in the waters off Ireland during the Fastnet Yacht Race. Waves with amplitudes of over 15 m developed, and 23 of the 300 sailboats in the race capsized.

Wave interference, the interaction of wind-driven waves with strong currents, and focusing due to the shape of the coastline or seafloor can lead to the formation of **rogue waves**, defined as waves that are two to five times higher than other large waves passing a location during a period of time. Long thought to exist only in the imagination of sailors, rogue waves now have been documented numerous times. In fact, wave-measuring instruments on oil platforms in the North Sea recorded almost 500 encounters with rogue waves during a 10-year period, and radar studies from satellites demonstrate that at any given time there are about 10 rogue waves in the world's oceans. The decks of large ships—including famous cruise ships—have been swamped by rogue waves in the open ocean (**Fig. 18.15b**). In 1995, for example, a 29-m-high wave

FIGURE 18.15 Ocean waves build in response to the shear of wind blowing over the water surface.

Beneath the wave base, water molecules are not affected by the wave.

(a) Within a deep-ocean wave, water molecules follow a circular path. The radius of the circle decreases with depth. Note that the wave height is twice the amplitude.

(b) This oil tanker was sailing through 7.5 m (25 ft) high waves when, suddenly, a rogue wave that was 20 m (65 ft) high struck it broadside, completely submerging the deck, which normally lies 18 m above sea level.

Breakers along a California beach.

(c) As the wave approaches shore, friction slows its base. Water motion in the wave becomes more elliptical, and the wave becomes a breaker.

struck the *Queen Elizabeth II*, and in 1933 a military ship encountered one at least 34 m (112 ft) high. If a rogue wave reaches the shore, it can wash unsuspecting bystanders off shoreside piers or beaches.

Waves have no effect on the ocean floor as long as the floor lies below the wave base. However, near the shore, where the wave base just touches the floor, it causes a slight back-and-forth motion of sediment. Closer to shore, as the water gets shallower, friction between the wave and the seafloor slows the deeper part of the wave, and the motion of water in the wave becomes more elliptical. Eventually, water at the top of the wave curves over the base, and the wave becomes a *breaker*, ready for surfers to ride (**Fig. 18.15c**). Breakers crash onto the shore in the surf zone, sending a surge of water up the beach. This upward surge, or **swash**, continues until friction brings motion to a halt. Then gravity draws the water back down the beach as **backwash**.

Waves may make a large angle with the shoreline as they're coming in, but they bend as they approach the shore, a phenomenon called **wave refraction**. When they reach the shore their crests are rarely at an angle of more 5° with respect to the shoreline (**Fig. 18.16**). To understand why wave refraction takes place, imagine a wave approaching the shore with a crest that makes an angle of 45° with the shoreline. The end of the wave closer to the shore touches bottom first and slows down because of friction, whereas the end farther offshore continues to move at its original velocity, swinging the whole wave around so that it is more parallel with the shoreline.

Although wave refraction decreases the angle at which waves move close to shore, it generally does not eliminate the angle. Where waves do arrive at the shore obliquely, water in

Did you ever wonder...

why the breakers that surfers love form near the shore?

FIGURE 18.16 Wave refraction and its consequences along the shore.

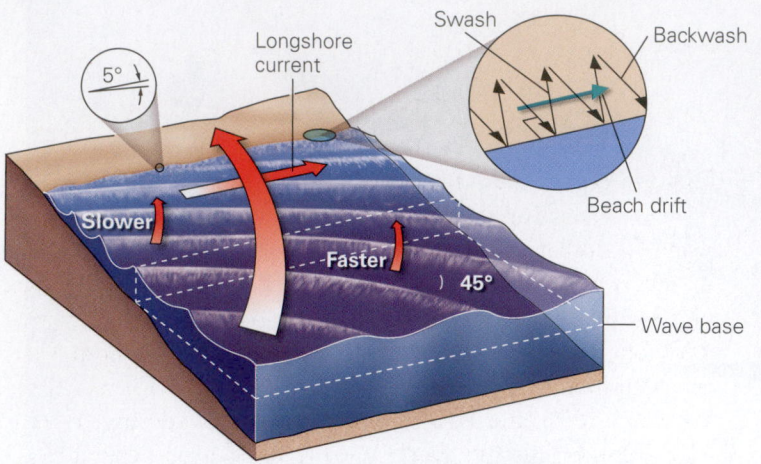

Wave refraction occurs where coastlines are not straight.

(a) Wave refraction occurs when waves approach the shore at an angle. If the wave reaches the beach at an angle, it causes a longshore current and beach drift of sand.

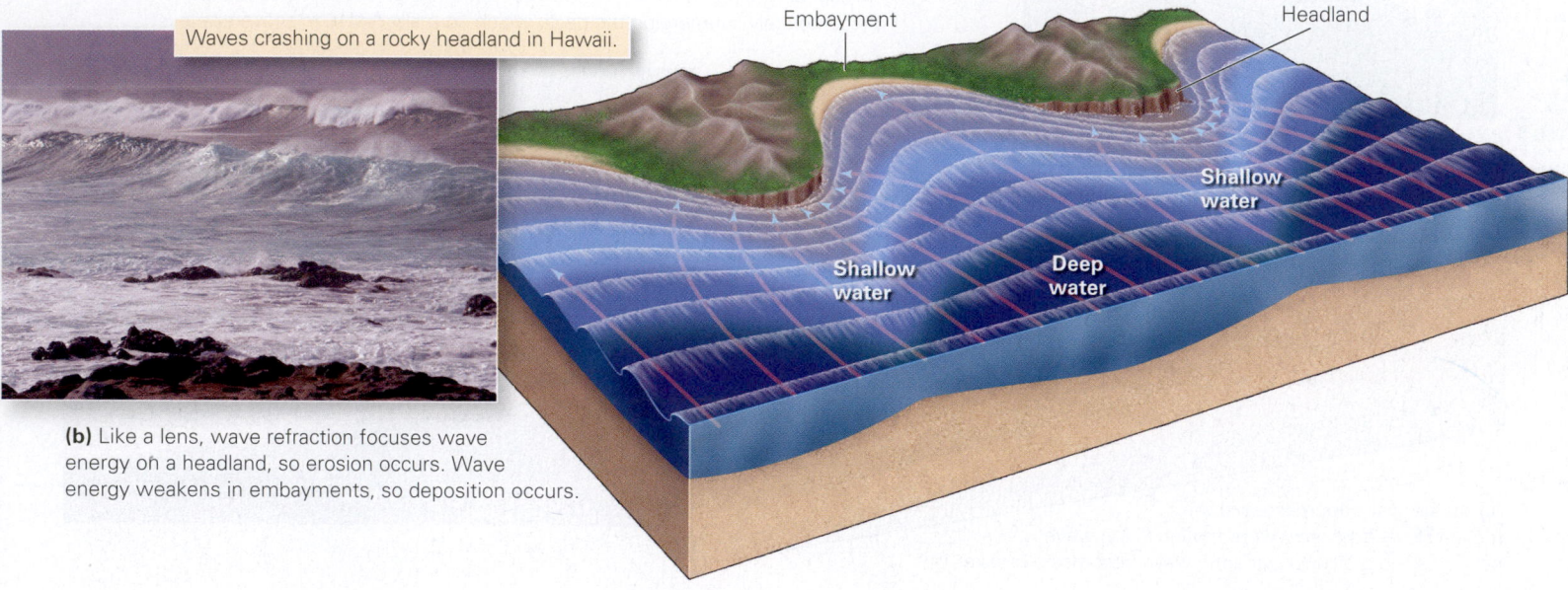

Waves crashing on a rocky headland in Hawaii.

(b) Like a lens, wave refraction focuses wave energy on a headland, so erosion occurs. Wave energy weakens in embayments, so deposition occurs.

the nearshore region has a component of motion that trends parallel to the shore. This **longshore current** causes swimmers floating in the water just offshore to drift gradually in a direction parallel to the beach. And where waves roll onto the shore at an angle, sediment in the surf follows a sawtooth pattern of movement that results in a gradual net transport of beach sediment parallel to the beach—such movement is called **longshore drift** (or beach drift). This sawtooth pattern happens because the swash of a wave moves perpendicular to the wave crest, so an oblique wave carries sediment diagonally up the beach, but the backwash must flow straight down the slope of the beach due to gravity.

Waves pile water up on the shore incessantly. As the excess water moves back to the sea, it may localize into a strong seaward flow perpendicular to the beach called a rip current (**Fig. 18.17**). Rip currents cause many drownings every year along beaches, because they carry unsuspecting swimmers away from the beach and cause swimmers to panic. If caught in a rip current, your best strategy is to swim calmly parallel to the shore.

Take-Home Message

The friction of the wind against the sea surface causes waves to start forming. Within a wave, water moves roughly in a circle; the amount of motion decreases with depth to the wave base. Large storms generate large waves. Near the shore, friction between the wave and the seafloor causes water to pile up into breakers. Waves refract when they approach the shore, but if waves wash ashore obliquely, longshore currents and longshore drift develop.

QUICK QUESTION: What's the difference between swash and backwash?

FIGURE 18.17 Waves bring water up on shore. The water may return to sea in a narrow rip current perpendicular to the shore.

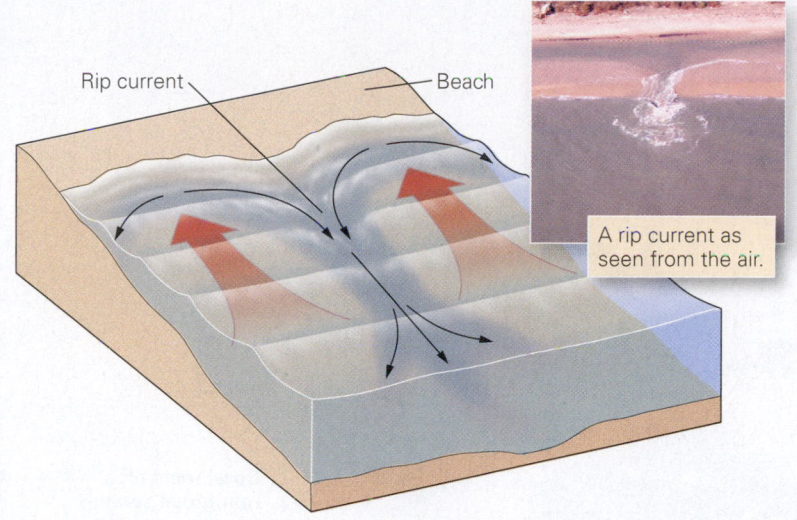

Rip current

Beach

A rip current as seen from the air.

18.6 Where Land Meets Sea: Coastal Landforms

Tourists along the Amalfi Coast of Italy thrill to the sound of waves crashing on rugged rocky shores. But in the Virgin Islands sunbathers can find seemingly endless white sand beaches. Large dome-like mountains rise directly from the sea in Rio de Janeiro, Brazil, but a 100-m-high vertical cliff marks the boundary between the Nullarbor Plain of southern Australia and the Great Southern Ocean (**Fig. 18.18**). As these examples illustrate, *coasts*, the belts of land bordering the sea, vary dramatically in terms of topography and associated landforms (**Fig. 18.19**).

FIGURE 18.18 Spectacular coastal scenery from localities around the world.

(a) Rocky cliffs form the Amalfi Coast of western Italy.

(c) Rounded "sugar loaf" mountains rise from the bays of Rio de Janeiro.

(b) A sandy beach along the coast of St. John, U.S. Virgin Islands.

(d) The abrupt edge of the Nullarbor Plain forms the south coast of Australia.

FIGURE 18.19 Examples of different kinds of coasts.

FIGURE 18.19 Examples of different kinds of coasts.

Uplifted terraces

Glacial fjords

Coastal plains
and offshore
sandbars

Coral reefs off a
mangrove swamp

Drowned river
valleys (estuaries)

A swampy
delta

Coastal
sand dunes
and a wide beach

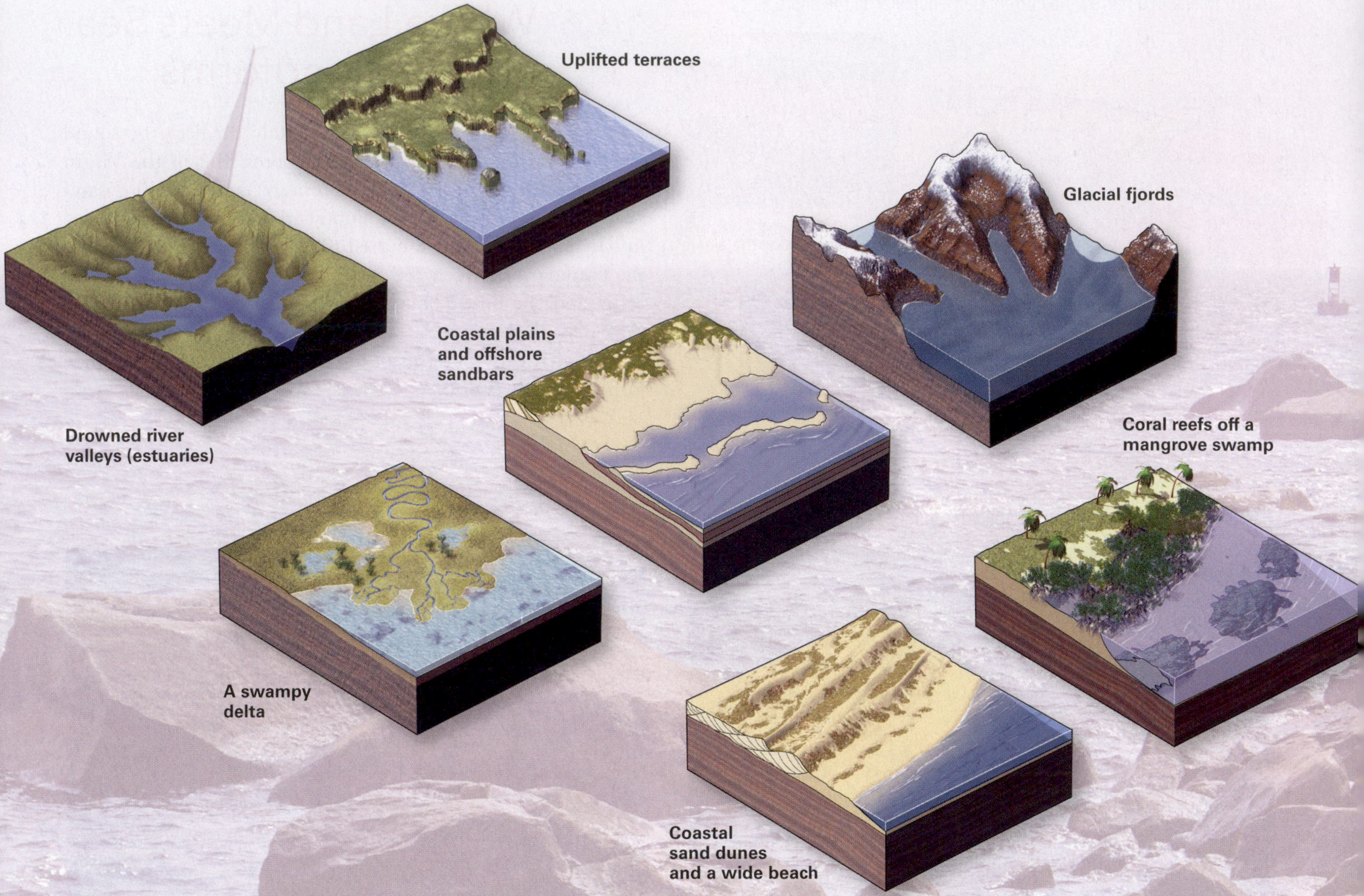

Beaches and Tidal Flats

For many vacationers, the ideal holiday includes a trip to a **beach**, a gently sloping fringe of sediment along the shore. Some beaches consist of rounded pebbles or boulders, but most are made up of a blanket of sand grains (**Fig. 18.20a, b**). The character of sediment on a beach is no accident, because beaches are subject to the constant swash and backwash of waves. If the beach lies close to cliffs that shed coarse debris, wave action breaks up larger blocks, grinds away edges and corners, rounds the clasts, and produces a beach of cobbles. If the beach forms from poorly sorted sediment

Did you ever wonder...

why beautiful sandy beaches don't form along all coasts?

deposited by rivers or glaciers, wave action winnows out finer sediment, such as silt and mud, and carries it offshore to quieter water, leaving sand behind. If the beach forms from shells or coral, wave action breaks up the shells until they the size of sand grains. Wave action can't make sand grains smaller, because even when tossed in a current, the grains can't collide with enough energy to break up.

The composition of sand itself varies from beach to beach, because different sands come from different sources. Sands derived from the weathering and erosion of silicic-to-intermediate rocks consist mainly of quartz; other minerals in these rocks chemically weather to form clay, which washes away in waves. Beaches made from the erosion of coral reefs and shells consist of carbonate sand. And beaches derived from the erosion of basalt boast black sand, made of tiny basalt grains.

FIGURE 18.20 Characteristics of beaches, barrier islands, and tidal flats.

(a) A gravel beach along the Olympic Peninsula, Washington. The clasts were derived from erosion of adjacent cliffs.

(b) A sand beach on the western coast of Puerto Rico. Wave action has carried away finer sediment.

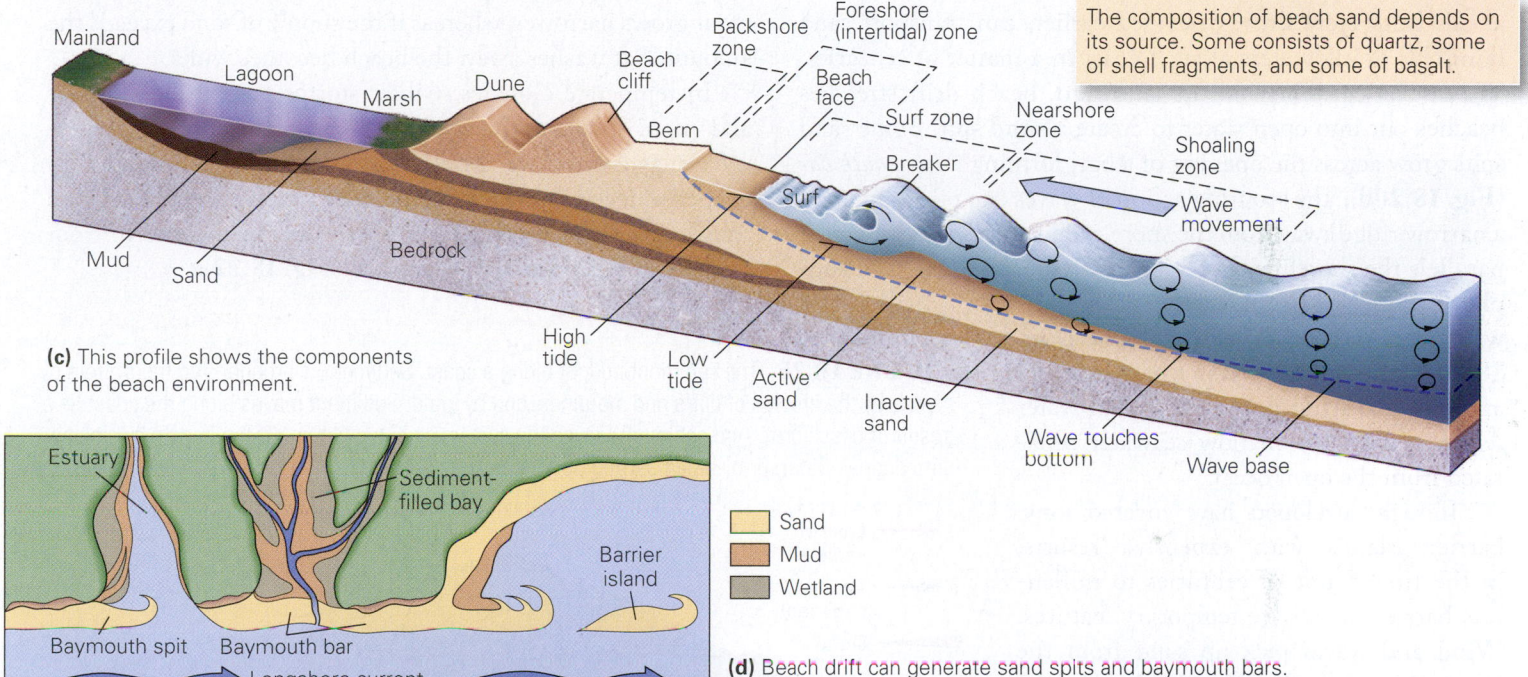

(c) This profile shows the components of the beach environment.

The composition of beach sand depends on its source. Some consists of quartz, some of shell fragments, and some of basalt.

(d) Beach drift can generate sand spits and baymouth bars. Sedimentation eventually fills in the region behind a baymouth bar.

Sand
Mud
Wetland

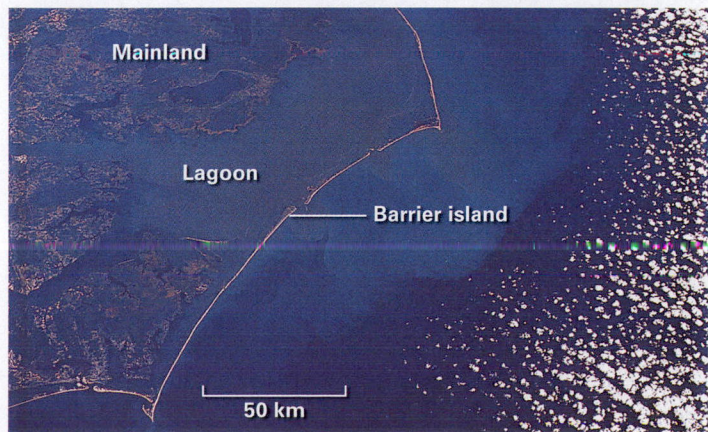

(e) A satellite image showing barrier islands off the coast of North Carolina.

(f) At low tide, boats rest in the mud of a tidal flat on the coast of Brittany.

A **beach profile**, a cross section drawn perpendicular to the shore, illustrates the shape of a beach (**Fig. 18.20c**). Starting from the sea and moving landward, a beach consists of a *foreshore zone*, the intertidal zone across which the tide rises and falls. The *beach face*, a steeper, concave-up part of the foreshore zone, forms where the swash of the waves actively scours the sand. The *backshore zone* extends from a small step, cut by high-tide swash to the front of the dunes or cliffs that lie farther inshore. The backshore zone includes one or more *berms*, horizontal to landward-sloping terraces that received sediment during a storm.

Geologists commonly refer to beaches as "rivers of sand," to emphasize that beach sand moves along the coast over time—it is not a permanent substrate. Wave action at the shore moves an active sand layer on the seafloor on a daily basis. Inactive sand, buried below this layer, moves only during severe storms or not at all. Longshore drift, discussed earlier, can transport sand hundreds of kilometers along a coast in a matter of centuries. Where the coastline indents landward, beach drift stretches beaches out into open water to create a **sand spit**. Some sand spits grow across the opening of a bay, forming a *baymouth bar* (**Fig. 18.20d**). The scouring action of waves can pile sand up in a narrow ridge away from the shore called an *offshore bar*, which parallels the shoreline. In regions with an abundant sand supply, offshore bars rise above the mean high-water level and become *barrier islands* (**Fig. 18.20e**). The water between a barrier island and the mainland becomes a quiet-water *lagoon*, a body of shallow seawater separated from the open ocean.

Though developers have covered some barrier islands with expensive resorts, in the time frame of centuries to millennia, barrier islands are temporary features. Wind and waves pick up sand from the ocean side of the barrier island and drop it on the lagoon side, causing the island to migrate landward. Storms may breach barrier islands and create an inlet (a narrow passage of water). Finally, beach drift gradually transports the sand of barrier islands and modifies their shape.

Tidal flats, regions of mud and silt exposed or nearly exposed at low tide but totally submerged at high tide, develop in regions protected from strong wave action, as we noted earlier (see Fig. 18.14a; **Fig. 18.20f**). They are typically found along the margins of lagoons or on shores protected by barrier islands. Here mud and silt accumulate to form thick, sticky layers. In tidal flats that provide a home for burrowing organisms such as

clams and worms, **bioturbation** (stirring by living organisms) mixes sediments together.

Because of the movement of sediment, the "sediment budget" (the difference between sand supplied and sand removed) plays an important role in determining the long-term evolution of a beach. Let's look at how the budget works for a small segment of beach (**Fig. 18.21**). Sand may be supplied to the segment from local rivers or by wind from nearby dune fields. It may also be brought from just offshore by waves or from far away by beach drift. In fact, the large quantity of sand along beaches of the southeastern United States may have been brought to the region by glaciers during the ice age. Some of the sand from a stretch of beach may be removed by beach drift, whereas some gets carried offshore by waves, where it either settles locally or tumbles down a submarine canyon into the deep sea. If the lost sand cannot be replaced, the beach segment grows narrower, whereas if the supply of sand exceeds the amount that washes away, the beach becomes wider.

In temperate climates, winter storms tend to be stronger and more frequent than summer ones. The larger, shorter-wavelength waves of winter storms wash beach sand into deeper water and thus make the beach narrower, whereas the smaller, longer-wavelength summer waves bring sand in from offshore and deposit it on the beach (**Fig. 18.22**).

FIGURE 18.21 The sediment budget along a coast. Sediment is brought into the system by rivers, by the erosion of cliffs and moraines, and by wind. Sediment moves along the coast as a result of beach drift. And sediment leaves the system by being blown off the beach, by sinking into deeper water, or by being carried out by the longshore current.

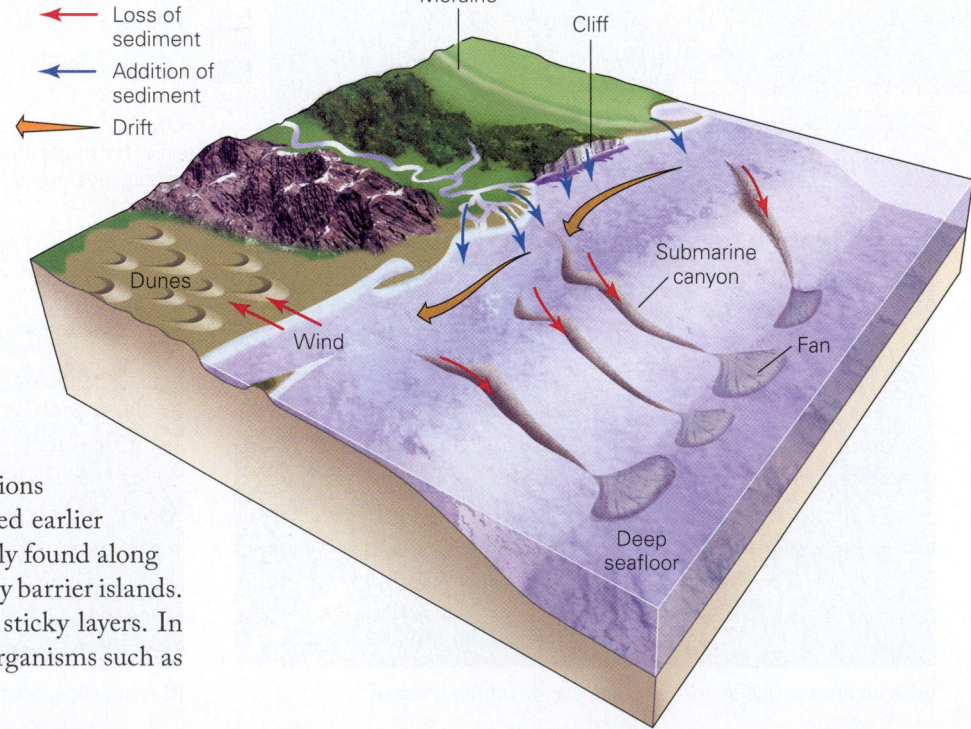

FIGURE 18.22 Contrasts between winter and summer beaches.

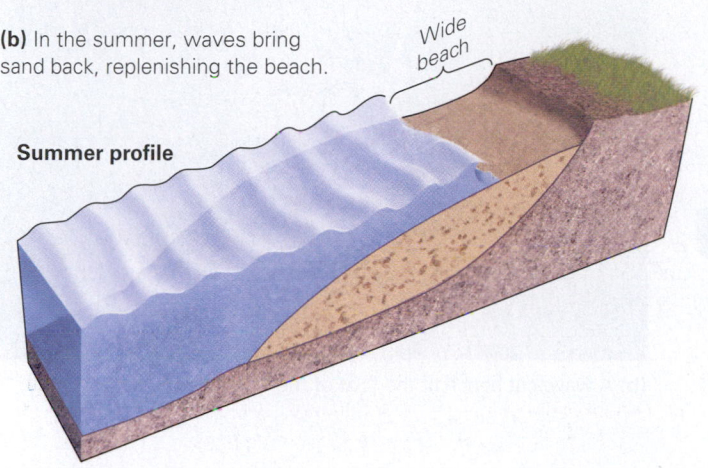

(a) In the winter, waters are stormier so sediment moves farther offshore.

Winter profile

Narrow beach

Berm

Gravel

A berm on a beach.

(b) In the summer, waves bring sand back, replenishing the beach.

Summer profile

Wide beach

Rocky Coasts

More than one ship has met its end, smashed and splintered in the spray and thunderous surf of a rocky coast, where bedrock cliffs rise directly from the sea. Lacking the protection of a beach, rocky coasts feel the full impact of ocean breakers. The water pressure generated during the impact of a breaker can pick up boulders and smash them together until they shatter, and it can squeeze air into cracks, creating enough force to split off fragments. Further, because of its turbulence, the water hitting a cliff face carries suspended sand and thus can abrade the cliff. The combined effects of shattering, wedging, and abrading, together called *wave erosion*, gradually undercut a cliff face and make a *wave-cut notch* (**Fig. 18.23a**). Undercutting continues until the overhang becomes unstable and breaks away at a joint, creating a pile of rubble at the base of the cliff that waves immediately attack and break up. In this process, wave erosion cuts away at a rocky coast such that the cliff gradually migrates inland. Such cliff retreat leaves behind a *wave-cut bench*, or platform, which becomes visible at low tide (**Fig. 18.23b**).

Other processes besides wave erosion also break up the rocks along coasts. For example, salt spray coats the cliff face above the waves and infiltrates into pores. When the water evaporates, salt crystals grow and push apart the grains, thereby weakening the rock. Plants and animals in the intertidal zone also contribute to erosion; by boring into the rocks to create rootholds or burrows, they weaken the rock.

Many rocky coasts start out with an irregular coastline, with *headlands* protruding into the sea and *embayments* set back from the sea. Such irregular coastlines tend to be temporary features in the context of geologic time, for wave refraction focuses wave energy on headlands and disperses it in embayments. As a result, waves preferentially remove debris at headlands and deposit sediment in embayments (**Fig. 18.23c**). Of note, a headland erodes in stages (**18.23d**). Because of refraction, waves curve and attack the sides of a headland, slowly eating through it to create a *sea arch* connected to the mainland by a narrow bridge. Eventually the arch collapses, leaving isolated *sea stacks* just offshore (**Fig. 18.23e**). Once formed, a sea stack protects the adjacent shore from waves. Therefore, sand may collect in the lee of the stack, slowly building a *tombolo*, a narrow ridge of sand that links the sea stack to the mainland.

Estuaries

Along some coastlines, a relative rise in sea level causes the sea to flood river valleys that merge with the coast, resulting in **estuaries**, where seawater and river water mix. You can recognize an estuary on a map by the dendritic pattern of its river-carved coastline (**Fig. 18.24**). Oceanic and fluvial waters interact in two ways within an estuary. In quiet estuaries, protected from wave action or river turbulence, the water becomes stratified, with denser oceanic saltwater flowing upstream as a wedge beneath less-dense fluvial freshwater. Such a *saltwater wedge*

FIGURE 18.23 Erosion landforms of rocky shorelines.

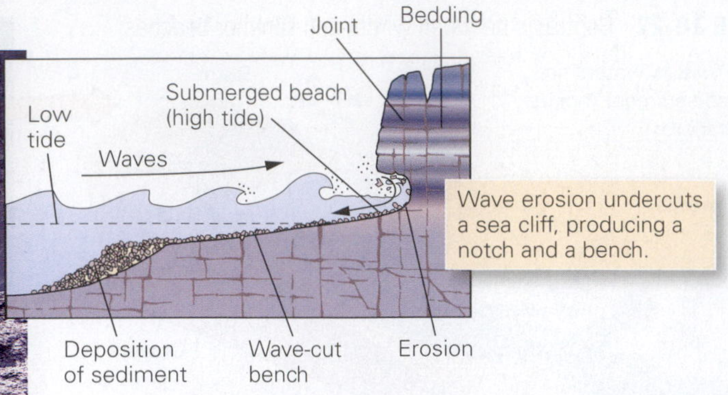

(a) A wave-cut notch.

Joint
Bedding
Submerged beach (high tide)
Low tide
Waves
Wave erosion undercuts a sea cliff, producing a notch and a bench.
Deposition of sediment
Wave-cut bench
Erosion

Wave-cut bench exposed at low tide.

(b) A wave-cut bench at the foot of the cliffs at Étretat, France.

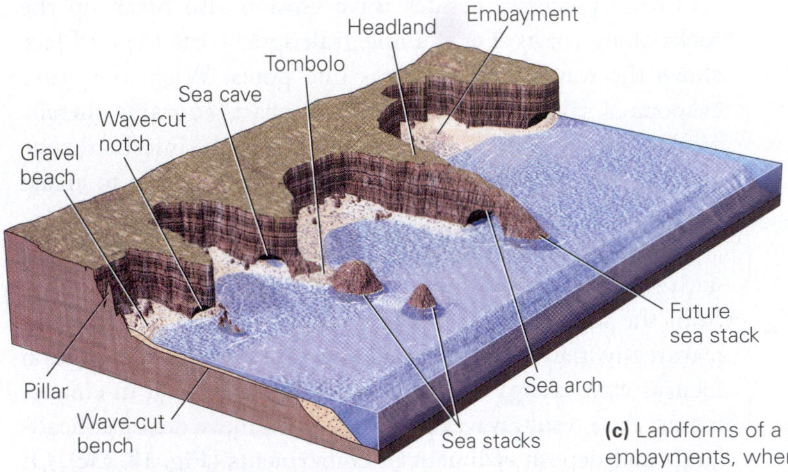

Headland
Embayment
Tombolo
Sea cave
Wave-cut notch
Gravel beach
Pillar
Wave-cut bench
Sea stacks
Future sea stack
Sea arch

(c) Landforms of a rocky shore. Beaches collect in embayments, whereas erosion concentrates at headlands.

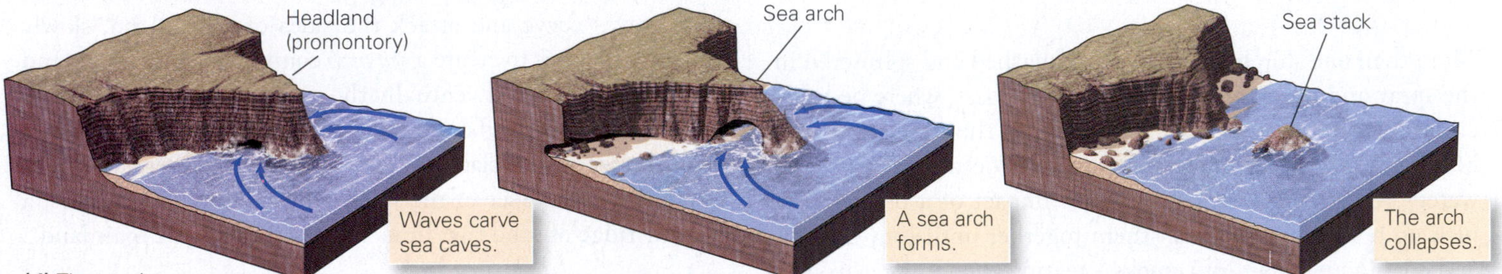

Headland (promontory)
Sea arch
Sea stack
Waves carve sea caves.
A sea arch forms.
The arch collapses.

(d) The erosion of a headland.

Time

(e) Coastal erosion along Australia's southern coast produced a sea arch (left). Eventually, the bridge will collapse, and only sea stacks will remain (right). These sea stacks are among several that together are known locally as the Twelve Apostles.

FIGURE 18.24 The Chesapeake Bay estuary formed when the sea flooded river valleys. The region is sinking relative to other coast areas because it overlies a buried meteor crater.

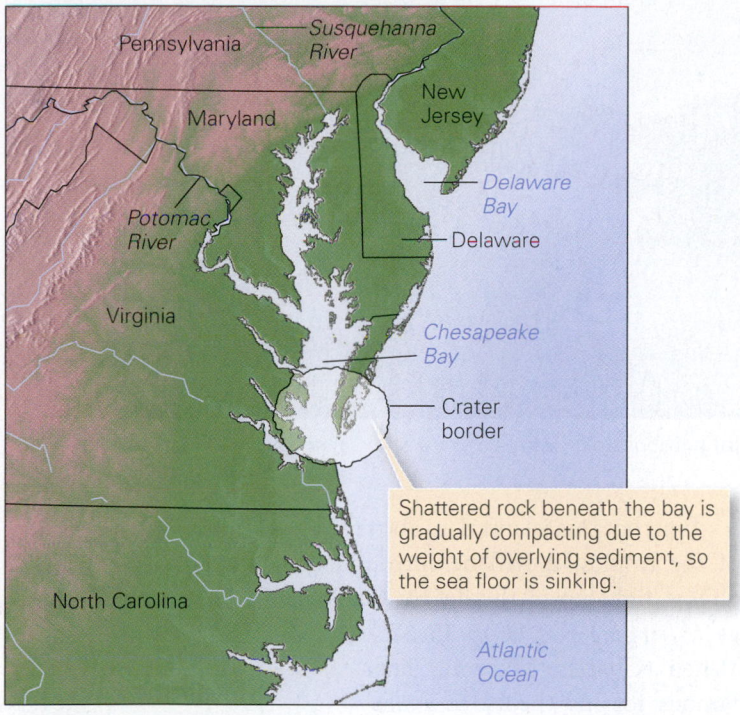

Shattered rock beneath the bay is gradually compacting due to the weight of overlying sediment, so the sea floor is sinking.

migrates about 100 km up the Hudson River in New York, and about 40 km up the Columbia River in Oregon. In turbulent estuaries, such as the Chesapeake Bay, oceanic and fluvial water combine to create nutrient-rich *brackish water*, with a salinity between that of oceans and rivers. Estuaries are complex ecosystems inhabited by unique species of shrimp, clams, oysters, worms, and fish that can tolerate large changes in salinity.

Fjords

During the last ice age, glaciers carved deep valleys in coastal mountain ranges. When the ice age came to a close, the glaciers melted away, leaving deep, U-shaped valleys (see Chapter 22). The water stored in the glaciers, along with the water within the vast ice sheets that covered continents during the ice age, flowed back into the sea and caused sea level to rise. The rising sea filled the deep valleys, creating **fjords**, or flooded glacial valleys. Coastal fjords are fingers of the sea surrounded by mountains. Because of their deep blue water and steep walls of polished rock, they are distinctively beautiful (**Fig. 18.25**). Some of the world's most spectacular fjords decorate the western coasts of Norway, British Columbia, and New Zealand. Smaller examples appear along the coast of Maine and southeastern Canada.

Organic Coasts

Coasts in which living organisms control landforms along the shore are called **organic coasts**. The nature of an organic coast depends on the type of organisms that live there, which, in turn, depends on climate.

A **coastal wetland** is a vegetated, flat-lying stretch of coast that floods at high tide but does not feel the impact of strong waves. In temperate climates, coastal wetlands include *swamps* (wetlands dominated by trees), *marshes* (wetlands dominated by grasses; **Fig. 18.26a**), and *bogs* (wetlands dominated by moss and shrubs). So many marine species spawn in wetlands that despite their relatively small area when compared with the oceans as a whole wetlands account for 10% to 30% of marine organic productivity.

FIGURE 18.25 Fjord landscapes form where relative sea-level rise drowns glacially carved valleys.

Glacial valleys have a U shape, so fjords have steep sides.

A fjord in Norway

FIGURE 18.26 Examples of coastal wetlands.

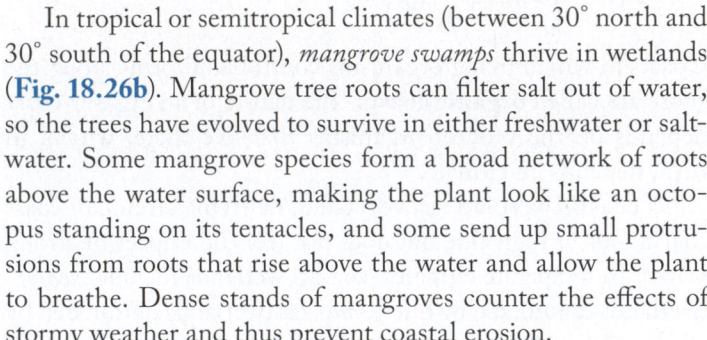

(a) A salt marsh along the coast of Cape Cod.

(b) A mangrove growing along the shore of southern Florida.

In tropical or semitropical climates (between 30° north and 30° south of the equator), *mangrove swamps* thrive in wetlands (**Fig. 18.26b**). Mangrove tree roots can filter salt out of water, so the trees have evolved to survive in either freshwater or saltwater. Some mangrove species form a broad network of roots above the water surface, making the plant look like an octopus standing on its tentacles, and some send up small protrusions from roots that rise above the water and allow the plant to breathe. Dense stands of mangroves counter the effects of stormy weather and thus prevent coastal erosion.

Along the azure coasts of Hawaii, visitors swim through colorful growths of living coral. Some corals look like brains, others like elk antlers, still others like delicate fans (**Fig. 18.27a**). Sea anemones, sponges, and clams grow on and around the coral. Though at first glance coral looks like a plant, it is actually a colony of tiny invertebrates. An individual coral animal, or polyp, has a tube-like body with a head of tentacles. Corals obtain part of their livelihood by filtering nutrients out of seawater; the remainder comes from algae that live on the corals' tissue. Corals have a symbiotic (mutually beneficial) relationship with the algae in that the algae photosynthesize and provide nutrients and oxygen to the corals while the corals provide carbon dioxide and nutrients for the algae.

Coral polyps secrete calcite shells, which gradually build into a mound of solid limestone whose top surface lies from just below the low-tide level down to a depth of about 60 m. At any given time, only the surface of the mound lives—the mound's interior consists of shells from previous generations of coral. The realm of shallow water underlain by coral mounds, associated organisms, and debris comprises a **coral reef**. Reefs absorb wave energy and thus serve as a living buffer zone that protects coasts from erosion. Corals need clear, well-lit, warm (18° to 30°C) water with normal oceanic salinity, so coral reefs grow only along clean coasts at latitudes of less than about 30° (**Fig. 18.27b**).

If an oceanic island rises above sea level in the warm waters of the tropics, it will be surrounded by a reef that extends almost right up to the beach. Geologists refer to such a reef as a *fringing reef* (**Fig. 18.27c**). When Charles Darwin, later famous for his theory of evolution, was a young naturalist on the *HMS Beagle*, he noticed that some islands of the tropics had fringing reefs, whereas others were surrounded by a *barrier reef* that lay offshore, separated from the island by a quiet lagoon. And in some locations, such as the Marshall Islands of the western Pacific, one can see a roughly circular coral reef, called an *atoll*, surrounding a lagoon, but no island remains in the middle of the lagoon. Darwin correctly surmised that fringing reefs evolved into barrier reefs, which, in turn, evolved into an atoll due to the slow subsidence of an island. As we've seen, this happens because of the sinking of the surface of the lithosphere beneath the island as the lithosphere ages and because the island erodes and slumps away. Had Darwin been able to use sonar, invented more than a century after his historic voyage, he would have identified one more step in the evolution of

FIGURE 18.27 The character and evolution of coral reefs.

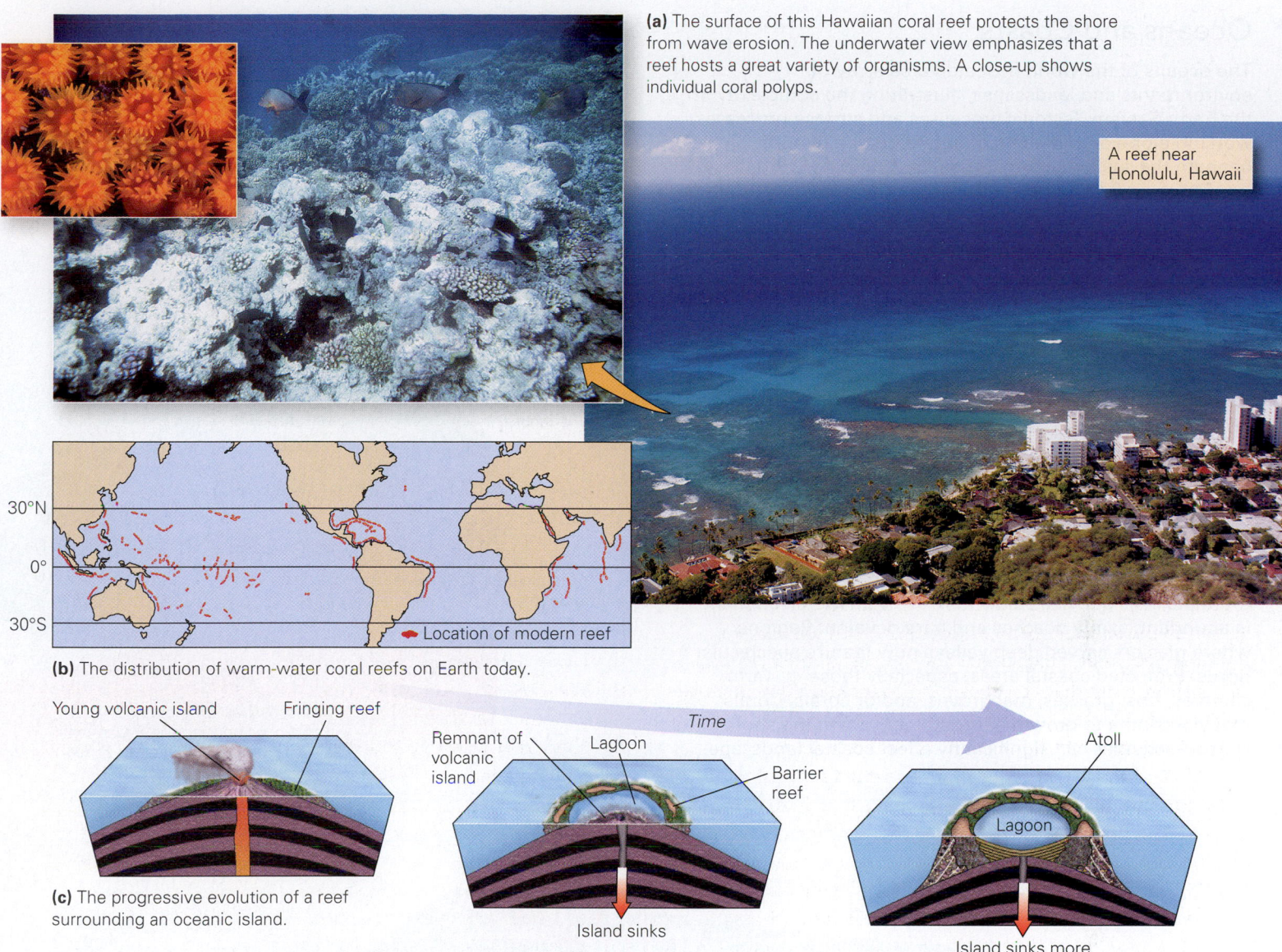

(a) The surface of this Hawaiian coral reef protects the shore from wave erosion. The underwater view emphasizes that a reef hosts a great variety of organisms. A close-up shows individual coral polyps.

A reef near Honolulu, Hawaii

30°N

0°

30°S

Location of modern reef

(b) The distribution of warm-water coral reefs on Earth today.

Young volcanic island   Fringing reef

Time

Remnant of volcanic island   Lagoon   Barrier reef   Atoll

Lagoon

(c) The progressive evolution of a reef surrounding an oceanic island.

Island sinks

Island sinks more

a reef—eventually, even the reef itself sinks below sea level, for the rate of coral growth can't keep up with the rate of subsidence. When this happens, the resulting seamount has a flat top—geologists refer to a flat-topped seamount as a **guyot.**

Take-Home Message

Beaches form from wave-washed and rounded sediment. Wave action may build bars, and beach drift may produce spits. Rocky coasts evolve due to wave erosion. Submergence of coastal valleys produces fjords and estuaries. Some coasts host wetlands and reefs.

QUICK QUESTION: Does the sand of a beach stay in the same location for a long time? Why?

18.7 Causes of Coastal Variability

Plate Tectonic Setting

Why does the character of the coast vary so much from place to place around the world? We can start to address this question from the perspective of plate tectonics theory. Specifically, the tectonic setting of a coast plays a role in determining whether the coast displays steep-sided mountain slopes or a broad lowland (see **Geology at a Glance**, pp. 684–685). For example, along some active margins, compressive stresses cause crustal

Oceans and Coasts

The oceans of the world host a diverse array of environments and landscapes, illustrating the complexity of the Earth System. Tectonic processes and surface processes working alone or in tandem generate unique features beneath the sea and along its coasts.

The major structures of the ocean floor are the result of plate-tectonic activity. For example, mid-ocean ridges define divergent boundaries, fracture zones form along transform faults, and trenches mark subduction zones. Oceanic islands, seamounts, and plateaus build above hot spots. Along passive continental margins, broad continental shelves develop, locally incised by submarine canyons.

Within the ocean, currents circulate water due to regional variations in both water salinity and temperature and to the traction between wind and the water surface. Tides cause sea level to rise and fall, and wind builds waves that churn the sea surface, erode shorelines, and transport sediment.

Coastal landscapes reflect variations in sediment supply, relative sea-level rise or fall, and climate. For example, where the supply of sediment is low and the landscape is rising relative to sea level, rocky shores with dramatic cliffs and sea stacks may evolve. Where sediment is abundant, sandy beaches and bars develop. Regions where glaciers carved deep valleys now feature spectacular fjords. Protected coastal areas, especially those in warm climates, host grasses, mangroves, and/or corals. Corals may contribute to growth of broad reefs along the shore. Human activities can significantly affect coastal landscape.

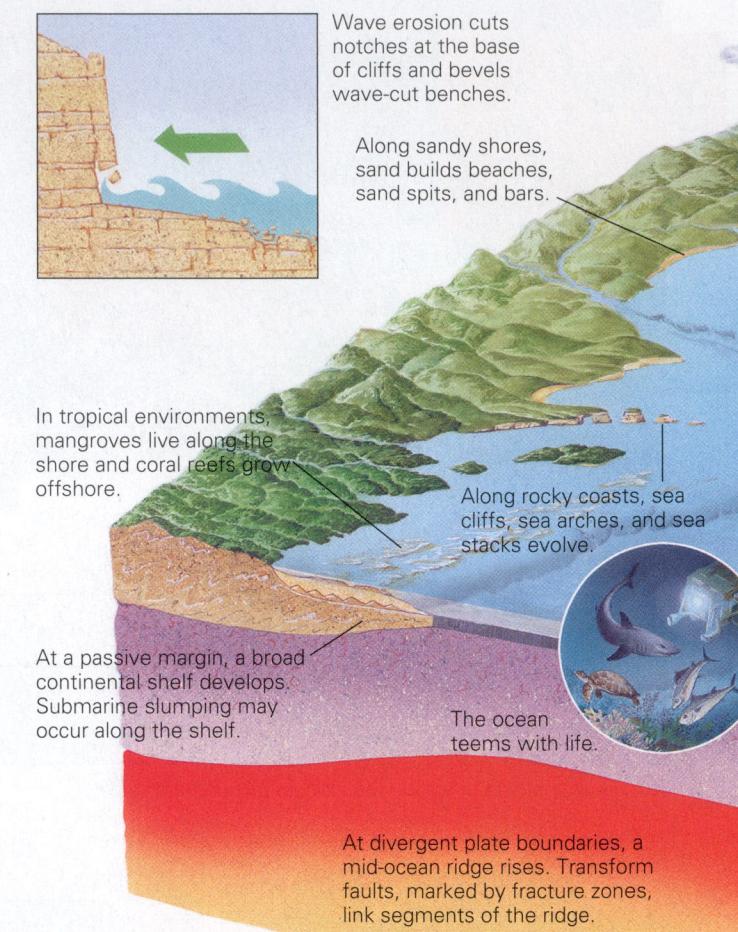

Wave erosion cuts notches at the base of cliffs and bevels wave-cut benches.

Along sandy shores, sand builds beaches, sand spits, and bars.

In tropical environments, mangroves live along the shore and coral reefs grow offshore.

Along rocky coasts, sea cliffs, sea arches, and sea stacks evolve.

At a passive margin, a broad continental shelf develops. Submarine slumping may occur along the shelf.

The ocean teems with life.

At divergent plate boundaries, a mid-ocean ridge rises. Transform faults, marked by fracture zones, link segments of the ridge.

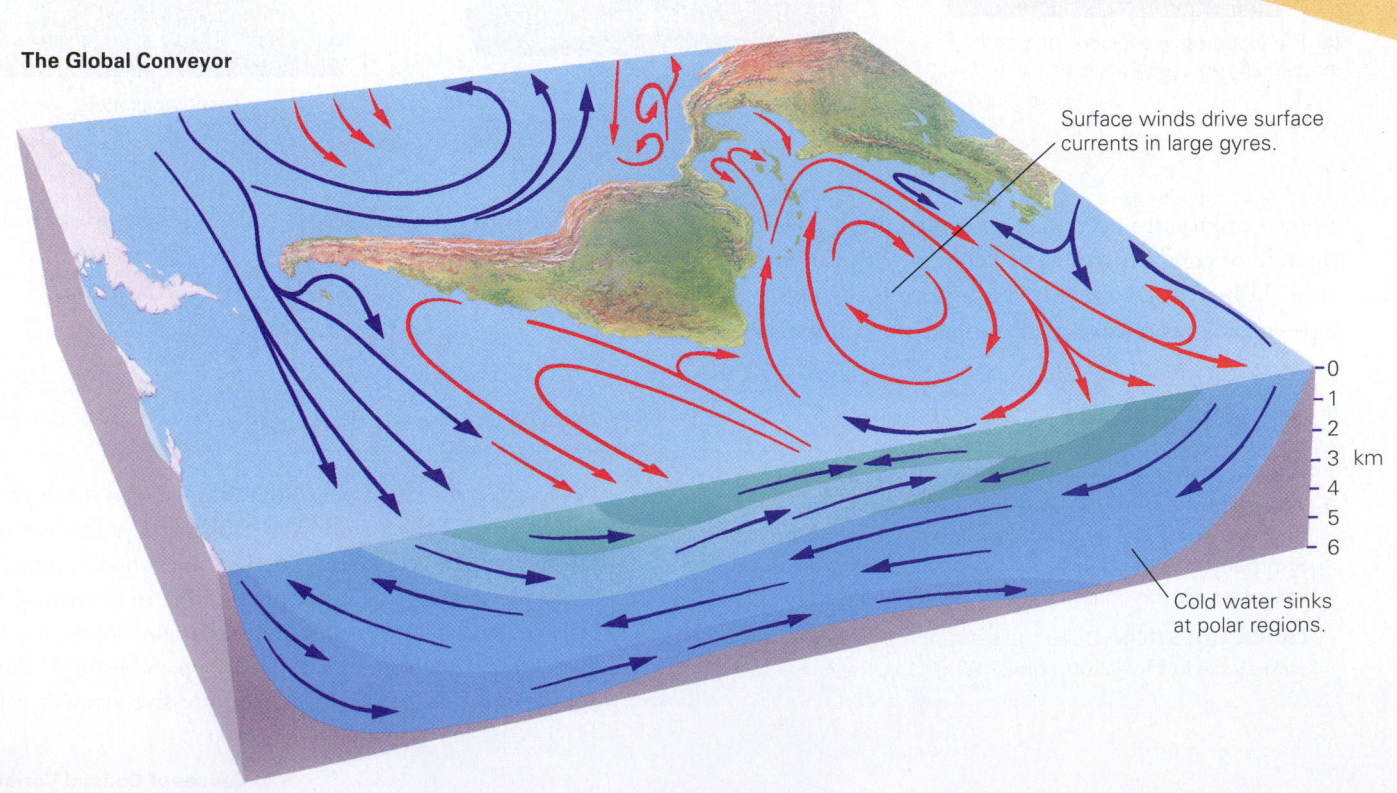

The Global Conveyor

Surface winds drive surface currents in large gyres.

0
1
2
3 km
4
5
6

Cold water sinks at polar regions.

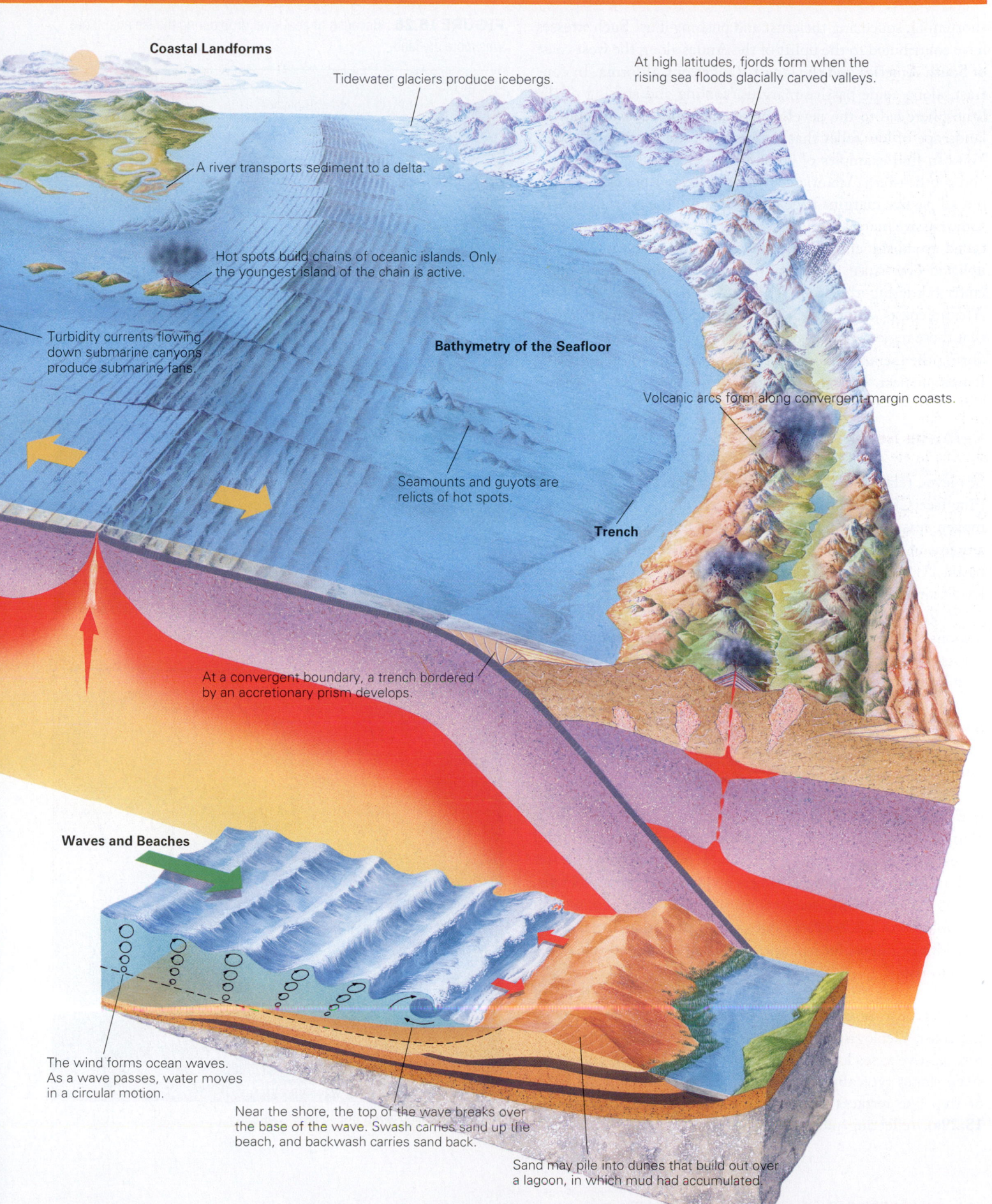

Coastal Landforms

Tidewater glaciers produce icebergs.

At high latitudes, fjords form when the rising sea floods glacially carved valleys.

A river transports sediment to a delta.

Hot spots build chains of oceanic islands. Only the youngest island of the chain is active.

Bathymetry of the Seafloor

Turbidity currents flowing down submarine canyons produce submarine fans.

Volcanic arcs form along convergent-margin coasts.

Seamounts and guyots are relics of hot spots.

Trench

At a convergent boundary, a trench bordered by an accretionary prism develops.

Waves and Beaches

The wind forms ocean waves. As a wave passes, water moves in a circular motion.

Near the shore, the top of the wave breaks over the base of the wave. Swash carries sand up the beach, and backwash carries sand back.

Sand may pile into dunes that build out over a lagoon, in which mud had accumulated.

shortening, squeezing the crust and pushing it up. Such stresses have contributed to the uplift of the Andes along the west coast of South America and the Coast Ranges of California. In contrast, along some passive margins, cooling and sinking of the lithosphere led to the development of a broad **coastal plain**, a landscape of low relief that merges with the continental shelf. You can find examples of coastal plains along the Gulf Coast and southeastern Atlantic coast of the United States. Of note, not all passive margins have coastal plains. The coastal areas of some passive margins uplifted during the rifting event that preceded establishment of the passive margin and have remained uplifted ever since. Such uplifts produced present-day highlands bordering the Red Sea, southeastern Brazil, southern African coasts, and eastern Australia. Recent research suggests that these regions may have undergone pulses of renewed uplift during the Cenozoic, long after rifting ceased, for reasons that remain unclear.

Relative Sea-Level Changes

Sea level, relative to the land surface, changes during geologic time (see Chapter 23). Some of these changes reflect vertical movements of the land, whereas others reflect changes in the amount of water held in the ocean or in the volume of ocean basins. Let's consider a few causes of changes in relative sea level more closely.

Regional changes in land surface elevation may develop in response to plate-tectonic processes. For example, compression across the coast of California has caused the land to rise. They may also be due to flow in the asthenosphere, which can push the lithosphere up or pull it down from below. Addition or removal of large glaciers from the surface of continents will cause the surface to sink or rise, respectively, in order to maintain isostasy (see Interlude D).

Geologists refer to sea-level changes resulting from a global rise or fall of the ocean surface as *eustatic sea-level changes*. Such changes may happen in association with the advances and retreats of large glaciers during ice ages, for glaciers store water on land (**Fig. 18.28**), so as glaciers on land grow, sea level falls, and as glaciers shrink, sea level rises. Others reflect changes in the volume of mid-ocean ridges—an increase in the number or width of ridges, for example, displaces water and causes sea level to rise.

Relative sea-level changes, either due to regional rise or fall of the land, or to eustatic sea-level change, affect landforms along the coast. Geologists refer to coasts where the land is rising or rose relative to sea level as **emergent coasts**. At emergent coasts, steep slopes typically border the shore. In some cases, a series of step-like terraces form along some emergent coasts (**Fig. 18.29a**), reflecting pulses of uplift between times of erosion.

FIGURE 18.28 Because of sea-level drop during the ice age, there was more dry land.

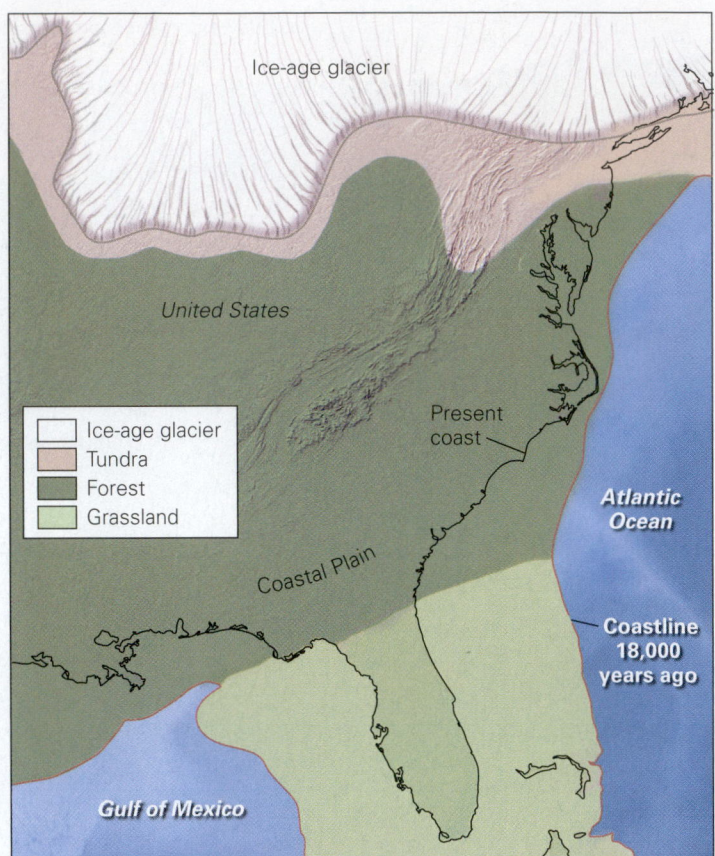

Ice-age glacier
United States
Present coast
Atlantic Ocean
Coastal Plain
Coastline 18,000 years ago
Gulf of Mexico

Legend:
☐ Ice-age glacier
☐ Tundra
☐ Forest
☐ Grassland

(a) Exposed land along eastern North America.

Iceland
Ice-age glacier
Sea ice
Atlantic Ocean
Tundra
Cold-weather conifer forest and steppe
Grass
Mediterranean Sea

(b) Exposed land along the coast of Europe.

Coasts at which the land sinks relative to sea level become **submergent coasts** (Fig. 18.29b). At submergent coasts, landforms include estuaries and fjords that developed when the sea flooded coastal valleys. Many of the coastal landforms of eastern North America are the consequences of submergence.

Sediment Supply and Climate

The amount of sediment supplied to a shore also affects its character. At *erosional coasts*, waves wash sediment away faster than it can be supplied. These coastlines recede landward and may become rocky, whereas coastlines that receive more sediment than that which erodes away, so-called *accretionary coasts*, grow seaward and develop broad beaches.

Regardless of whether a coast is emergent or submergent, erosional or accretionary, what it looks like ultimately reflects climate. Shores that enjoy generally calm weather erode less rapidly than those constantly subjected to ravaging storms, so a sediment supply large enough to generate an accretionary coast in a calm environment may be insufficient to prevent the development of an erosional coast in a stormy environment. The climate also affects biological activity along coasts. For example, in the warm water of tropical climates, mangrove swamps flourish along the shore, and coral reefs form offshore. The reefs may build into a broad carbonate platform such as that which appears in the Bahamas today. In cooler climates, salt marshes develop, whereas in arctic regions, the coast may be a stark environment of lichen-covered rock and barren sediment.

FIGURE 18.29 Features of emergent coastlines (relative sea level is falling) and submergent coastlines (relative sea level is rising).

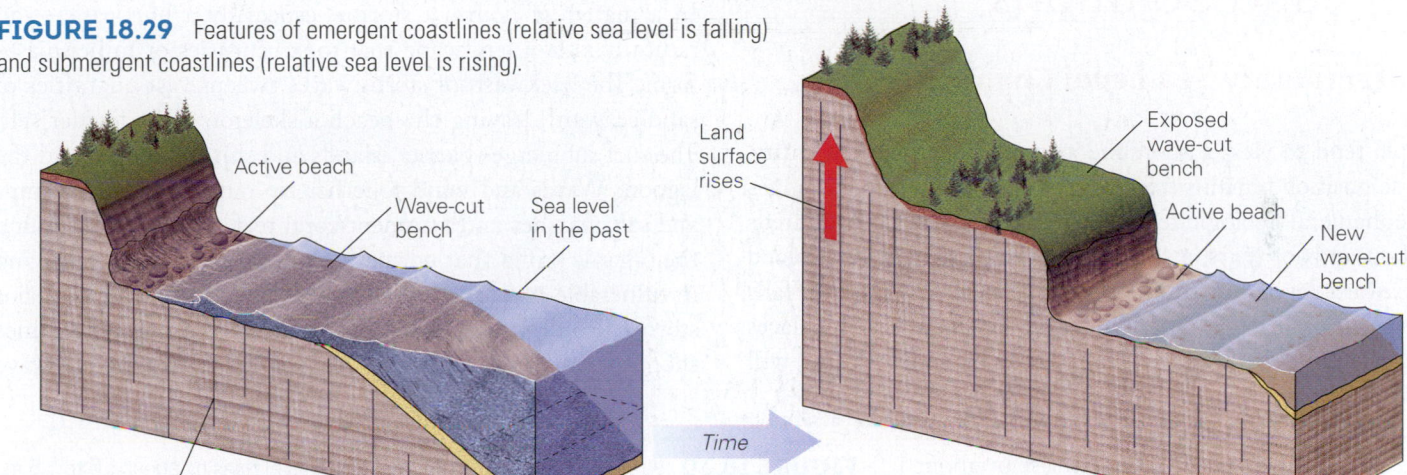

(a) *Emergent coasts*: Wave erosion produces a wave-cut bench along an emergent coast. As the land rises, the bench becomes a terrace, and a new wave-cut bench forms.

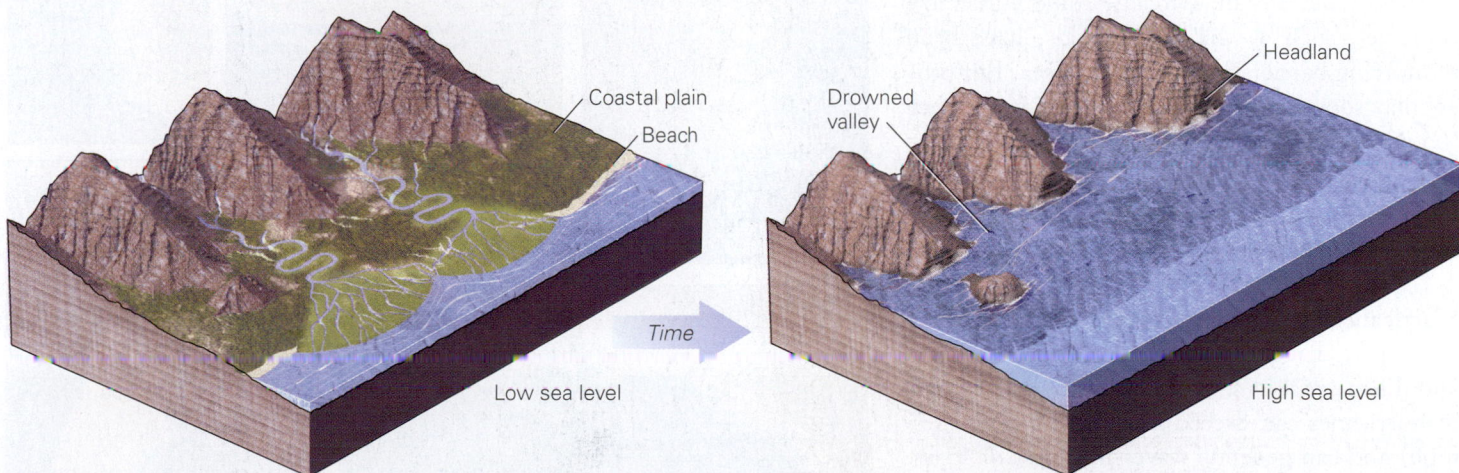

(b) *Submergent coasts*: A coast before sea level rises. Rivers drain valleys and deposit sediment on a coastal plain. As a submergent coast forms, sea level rises and floods the valleys, and waves erode the headlands.

the overlying atmosphere) decreases substantially in the region beneath a hurricane, and without the downward push of the air, the sea surface (level) rises still further. When this combination of locally high sea level and powerful waves reaches the shore, it causes a **storm surge** that inundates low areas along the shore, and the waves batter both the shore and offshore reefs. In regions where the coast is a low-lying delta plain, the land can be submerged for days or more. Such catastrophic flooding has taken a dreadful toll on the Ganges Delta in Bangladesh and on the Gulf Coast of the United States. Flooding during Hurricane Sandy in 2012 caused tragic loss of life and tens of billions of dollars in damage in the northeastern United States. We discuss hurricanes and their consequences more fully in Chapter 20.

18.8 Coastal Problems and Solutions

Contemporary Sea-Level Changes

People tend to view a shoreline as a rather permanent entity, whose position remains fixed over time. But in fact, shorelines are ephemeral geologic features. On a time scale of hundreds to thousands of years, a shoreline can move kilometers inland or seaward depending on whether relative sea level rises or falls or whether sediment supply increases or decreases. In places where relative sea level now rises today, shoreline towns will eventually be submerged. For example, sea level is currently rising at a rate of about 3.3 mm/year. If this rate continues, sea level will rise by about 1 m in the next few centuries. A 1-m rise will have a significant effect on coastal cities (**Fig. 18.30**) and could submerge about 15% of the islands in the Pacific Ocean. Already, some vulnerable countries, such as the Netherlands, have been reinforcing barriers to hold out the sea. But others may soon have to begin such projects or face the potential for loss of life and immense property damage in low-lying coastal areas.

Hurricanes and Coastal Floods

Hurricanes are immense storms that grow over the waters of equatorial oceans. Some are born and die at sea, but some move onto land. Winds in hurricanes can exceed 250 km per hour (155 mph) and can generate waves in excess of 15 m. They can also push a bulge of water landward. Further, because air rises beneath a hurricane, atmospheric pressure (caused by the weight of

Beach Destruction—Beach Protection?

In a matter of hours, a storm—especially a hurricane—can radically alter a landscape that took centuries or millennia to form. The backwash of storm waves sweeps vast quantities of sand seaward, leaving the beach a skeleton of its former self. The surf submerges barrier islands and shifts them toward the lagoon. Waves and wind together rip out mangrove swamps and salt marshes and fragment coral reefs, thereby destroying the organic buffer that normally protects the coast and leaving it vulnerable to erosion for years to come. Of course, major storms also destroy human constructions: erosion undermines shoreside buildings, causing them to collapse into the sea; wave

FIGURE 18.30 Areas that may be submerged if sea level rises by about 1.0 to 1.5 m.

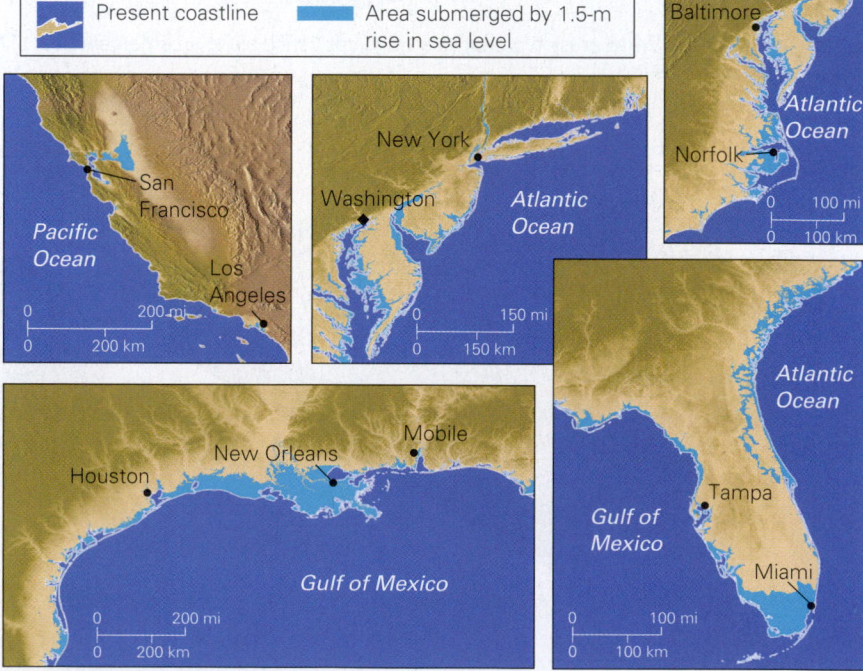

impacts smash buildings to bits; and the storm surge floats buildings off their foundations (**Fig. 18.31a**).

But even less-dramatic events, such as the loss of river sediment, a gradual rise in sea level, a change in the shape of a shoreline, or the destruction of coastal vegetation, can alter the balance between sediment accumulation and sediment removal on a beach, leading to **beach erosion** (**Fig. 18.31b**). In some places, beaches retreat landward at rates of 1 to 2 m per year, forcing home owners to pick up and move their houses. Even large lighthouses have been moved to keep them from washing away or tumbling down eroded headlands.

In many parts of the world, beachfront property has great value; but if a hotel loses its beach sand, it probably won't stay in business. Thus property owners often construct artificial barriers to protect their stretch of coastline or to shelter the mouth of a harbor from waves. These barriers alter the natural movement of sand in the beach system and thus change the shape of the beach, sometimes with undesirable results. For example, people may build *groins*, concrete or stone walls protruding perpendicular to the shore, to prevent beach drift from removing sand (**Fig. 18.32a, d**). Sand accumulates on the updrift side of the groin, forming a long triangular wedge, but sand erodes away on the downdrift side. Needless to say, the property owner on the downdrift side doesn't appreciate this process. Harbor engineers may build a pair of walls called *jetties* to protect the entrance to a harbor (**Fig. 18.32b**). But jetties erected at the mouth of a river channel effectively extend the river into deeper water and thus may lead to the deposition of an offshore sandbar. Engineers may also build an offshore wall called a *breakwater*, parallel or at an angle to the beach, to prevent the full force of waves from reaching a harbor. With time, however, sand builds up in the lee of the breakwater and the beach grows seaward, clogging the harbor (**Fig. 18.32c**). To protect expensive shoreside homes, people build *seawalls* out of riprap (large stone or concrete blocks) or reinforced concrete (**Fig. 18.32e**). But seawalls reflect wave energy, which crosses the beach, back to sea, so this process increases the rate of erosion at the foot

SEE FOR YOURSELF . . .

Groins of Chicago's shore

LATITUDE
41°55'5.52"N

LONGITUDE
87°37'37.24"W

Looking down from 3 km (~9,800 ft).

Below can be seen the sandy beaches that fringe the coastline of Lake Michigan. Waves strike the shore obliquely. As a result, beach drift carries sand southwards. The city constructed a series of groins in an attempt to prevent beach erosion.

FIGURE 18.31 Examples of beach erosion.

September 9, 2008

September 15, 2008

(a) When Hurricane Ike hit Galveston, Texas, in September 2008, many beachfront homes were washed away.

(b) Wave erosion has completely removed the beach and has started to erode a beach cliff along the coast of Cape Cod.

FIGURE 18.32 Techniques used to preserve beaches.

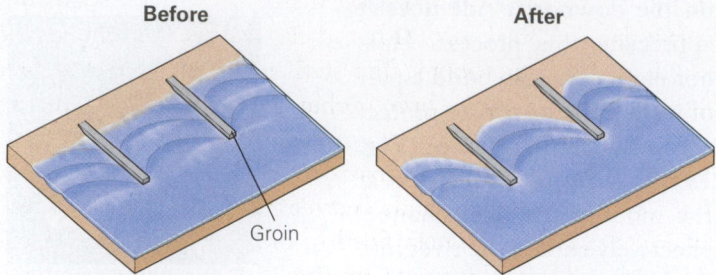

Before   **After**

Groin

(a) The construction of groins may produce a sawtooth beach.

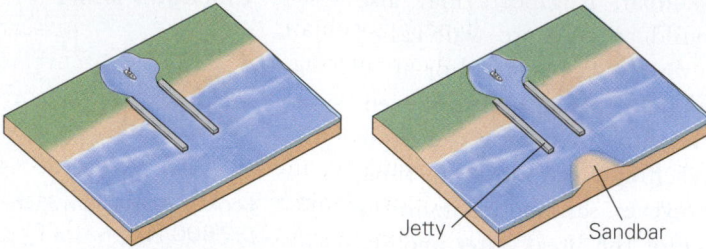

Jetty   Sandbar

(b) Jetties extend a river farther into the sea but may cause a sandbar to form at the end of the channel.

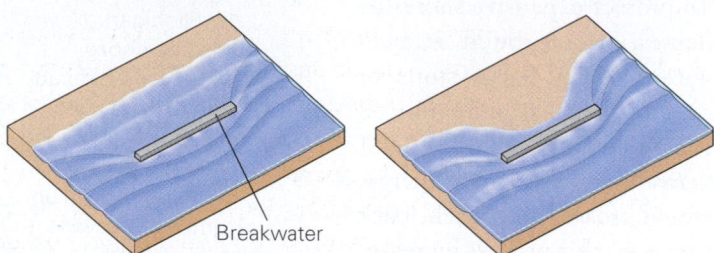

Breakwater

(c) A beach may grow seaward behind a breakwater.

(d) Groins spaced along a beach in southern England are intended to prevent loss of sediment.

(e) Riprap will slow erosion of a parking lot along a California beach.

of the seawall. During a large storm, the seawall may be undermined so that it collapses (**Fig. 18.33**).

In some places, people have given up trying to decrease the rate of beach erosion and instead have worked to increase the rate of sediment supply. To do this, they pump sand from farther offshore or truck in sand from elsewhere to replenish a beach. This procedure, called *beach nourishment*, can be hugely expensive and at best provides only a temporary fix, for the backwash and beach drift that removed the sand in the first place continue unabated as long as the wind blows and the waves break. Clearly, beach management remains a controversial issue, for beachfront properties are expensive, but protecting them can be even more expensive.

Coastal Pollution

Bad cases of beach pollution create headlines. Because of beach drift, garbage dumped in the sea in an urban area may drift along the shore and be deposited on a tourist beach far from its point of introduction. For example, hospital waste from New York City has washed up on beaches tens of kilometers to the south. Oil spills, either from ships that flush their bilges or from tankers that have run aground or foundered in stormy seas or blowouts on offshore oil rigs, have contaminated shorelines at several places around the world.

The influx of nutrients, from sewage and agricultural runoff, into coastal waters can create dead zones along coasts. A *dead zone* is a region in which water contains so little oxygen that fish and other organisms within it die. Dead zones form when the concentration of nutrients rises enough to stimulate an algae bloom, for overnight respiration by algae depletes dissolved oxygen in the water, and the eventual death and decay of plankton depletes oxygen even more. Dead zones are particularly common at the mouths of rivers that drain agricultural areas. One of the world's largest dead zones occurs in the Gulf of Mexico, offshore of the Mississippi River's mouth.

FIGURE 18.33 A seawall protects the sea cliff under most conditions, but during a severe storm the wave energy reflected by the seawall helps scour the beach. As a result, the wall may be undermined and collapse.

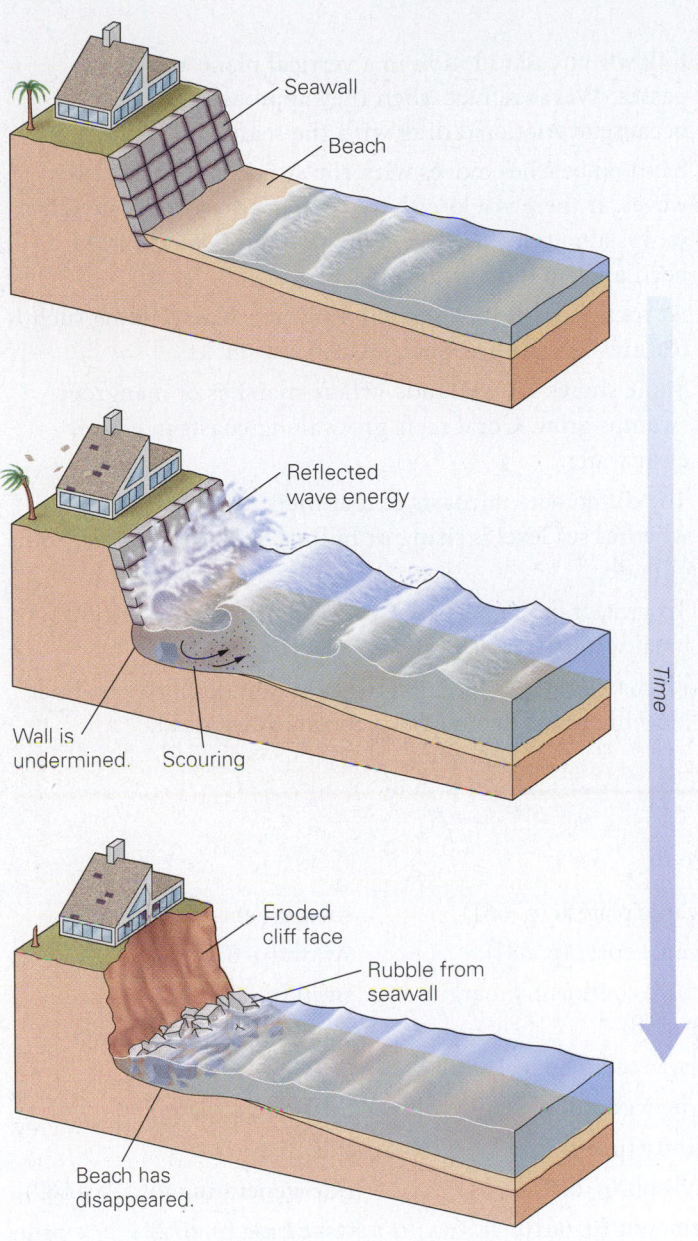

Seawall

Beach

Reflected wave energy

Wall is undermined.   Scouring

Time

Eroded cliff face

Rubble from seawall

Beach has disappeared.

Wetland and Reef Destruction

Coastal wetlands and coral reefs are particularly susceptible to changes in the environment, and many of them have been destroyed in recent decades. Their loss increases a coast's vulnerability to erosion and, because these areas provide spawning grounds for marine organisms, their loss disrupts the global food chain of the ocean. The statistics of wetland and reef destruction worldwide are frightening—ecologists estimate that between 20% and 70% of wetlands have already been destroyed, and along some coasts 90% of reefs have died.

Destruction of wetlands and reefs happens for many reasons. Wetlands have been filled or drained to serve as farmland, housing developments, resorts, or garbage dumps. Reefs have been destroyed by boat anchors, dredging, the activities of divers, dynamite explosions intended to kill fish, and quarrying operations intended to obtain construction materials. Chemicals and particulates entering coastal water from urban, industrial, and agricultural areas can cause havoc in wetlands and reefs, for these materials cloud water and/or trigger algal blooms, killing filter-feeding organisms. Toxic chemicals in such runoff can also poison plankton and burrowing organisms and, therefore, other organisms progressively up the food chain.

Global climate change, which will be more fully described in Chapter 23, also impacts the health of organic coasts. For example, transformation of once-vegetated regions into deserts means that the amount of dust carried by winds from the land to the sea has increased. This dust can interfere with coral respiration and can bring dangerous viruses. A global increase in seawater temperature may be contributing to *reef bleaching*, the loss of coral color due to the death of the algae that live in coral polyps. Reef bleaching and other reef diseases have transformed once-living landscapes hosting diverse species of colorful coral into mounds and piles of whitish rubble.

Take-Home Message

In a geologic context the position and character of a coastal landscape is temporary. Sea-level rise and storms can flood coastal areas, and beaches and bars erode—especially during hurricanes—despite intense efforts to build protective structures or replenish sand. Pollution and a rise in ocean water temperature can destroy wetlands and coral reefs.

QUICK QUESTION: What problems can be caused by construction of groins or seawalls?

CHAPTER SUMMARY

- The landscape of the seafloor depends on the character of the underlying crust. Particularly wide continental shelves form over passive-margin basins. Continental shelves may be cut by submarine canyons. Abyssal plains develop on old, cool oceanic lithosphere. Seamounts and guyots form above hot spots.

- The salinity, temperature, and density of seawater vary with location and depth.

- Water in the oceans circulates in currents. Surface currents are driven by the wind and are deflected in their path by the Coriolis effect. The vertical upwelling and downwelling of water create deep currents. Some of this movement is thermohaline circulation, a consequence of variations in temperature and salinity.

- Tides—the daily rise and fall of sea level—are caused by a tide-generating force. The largest contribution to this force comes from the gravitational pull of the Moon.

- Waves are caused by friction where the wind shears across the surface of the ocean. Water particles approximately

follow a circular motion in a vertical plane as a wave passes. Waves refract when they approach the shore because of frictional drag with the seafloor.

- Sand on beaches moves with the swash and backwash of waves. If there is a longshore current, the sand gradually moves along the beach and may extend outward from headlands to form sand spits.

- At rocky coasts, waves grind away at rocks, yielding such features as wave-cut benches and sea stacks.

- Some shores are wetlands, where marshes or mangrove swamps grow. Coral reefs grow along coasts in warm, clear water.

- The differences in coasts reflect their tectonic setting, whether sea level is rising or falling, sediment supply, and climate.

- To protect beach property, people build groins, jetties, breakwaters, and seawalls.

- Human activities have led to the pollution of coasts. Reef bleaching has become dangerously widespread.

GUIDE TERMS

abyssal plain (p. 658)

active continental margin (p. 659)

backwash (p. 673)

bathymetry (p. 658)

beach (p. 676)

beach erosion (p. 689)

beach profile (p. 678)

bioturbation (p. 678)

coast (p. 657)

coastal plain (p. 686)

coastal wetland (p. 681)

continental shelf (p. 658)

coral reef (p. 682)

Coriolis effect (p. 665)

current (p. 663)

emergent coast (p. 686)

estuary (p. 679)

fjord (p. 681)

guyot (p. 683)

longshore current (p. 674)

longshore drift (p. 674)

oceanic plateau (p. 661)

organic coast (p. 681)

passive continental margin (p. 658)

pelagic sediment (p. 660)

rogue wave (p. 672)

salinity (p. 662)

sand spit (p. 678)

seamount (p. 661)

storm surge (p. 688)

submarine canyon (p. 659)

submergent coast (p. 687)

swash (p. 673)

swell (p. 672)

thermohaline circulation (p. 667)

tidal range (p. 667)

tide (p. 667)

tide-generating force (p. 669)

wave base (p. 672)

wave refraction (p. 673)

REVIEW QUESTIONS

1. How much of the Earth's surface is covered by oceans? What proportion of the world's population lives near a coast?

2. Describe the typical topography of a passive continental margin, from the shoreline to the abyssal plain. How does

the lithosphere beneath a passive margin differ from that beneath an abyssal plain?

3. How do the shelf and slope of an active continental margin differ from those of a passive margin?

4. Where does the salt in the ocean come from? How does the salinity in the ocean vary? How does the temperature of the oceans vary?

5. What factors control the direction of surface currents in the ocean? What is the Coriolis effect, and how does it affect oceanic circulation? Explain thermohaline circulation.

6. What causes the tides? Why do the range and reach of tides vary with location?

7. Describe the motion of water molecules in a wave. How does wave refraction cause longshore currents?

8. Describe the components of a beach profile.

9. How does beach sand migrate as a result of longshore drift? Explain the sediment budget of a coast.

10. Describe how waves affect a rocky coast and how such coasts evolve.

11. What is an estuary? What is the difference between an estuary and a fjord?

12. Discuss the different types of coastal wetlands. Describe the different kinds of reefs and how a reef surrounding an oceanic island changes with time.

13. How do plate tectonics, sea-level changes, sediment supply, and climate change affect the shape of a coastline? Explain the difference between emergent and submergent coasts. What is the difference between an erosional and depositional coast?

14. In what ways do people try to modify or "stabilize" coasts? How do the actions of people threaten the natural systems of coastal areas?

ON FURTHER THOUGHT

15. In 1789, the crew of the HMS *Bounty* mutinied. Near Tonga, in the Friendly Islands (approximately 20° S and 175° W), the crew, led by Fletcher Christian, forced the ship's commanding officer, Lieutenant Bligh, along with those crewmen who remained loyal to Bligh, into a rowboat and set them adrift in the Pacific Ocean. The castaways survived, and 47 days later, they landed at Timor, 6,700 km to the west. Why did they end up where they did? Considering how long the journey took them, how fast were they moving?

16. A hotel chain would like to build a new beachfront hotel along a north-south-trending stretch of beach where a strong longshore current flows from south to north. The neighbor to the south has constructed an east-west-trending groin on the property line. Will this groin pose a problem? If so, what solutions could the hotel try?

17. Observations made during the last decade suggest that sea level is rising in response to global warming. Much of southern Florida lies at elevations of less than 6 m above sea level. What changes will take place to the region as sea level rises? To answer, keep in mind the concepts of emergent and submergent coasts, and assume that southern Florida will continue to lie in the subtropical realm.

smartwork   smartwork.wwnorton.com

This chapter's Smartwork features:

- Labeling exercise on continental and oceanic lithosphere.
- What A Geologist Sees exercise on glacial features.
- In-depth reading comprehension problems on groundwater characteristics.

GEOTOURS

This chapter's GeoTour exercise (O) features:

- Seafloor bathymetry
- Coral reefs
- Barrier islands and spits
- Coastlines and Sea-level rise

Steaming pools and travertine ledges have developed where groundwater bubbles to the surface from the hot springs of Rotorua, New Zealand.

CHAPTER 19

A Hidden Reserve: Groundwater

> When the rain falls and enters the earth, when a pearl drops into the depth of the sea, you can dive in the sea and find the pearl, you can dig in the earth and find the water.

> —*Mei Yao-ch'en (Chinese poet, 1002–1060)*

LEARNING OBJECTIVES

By the end of this chapter you should understand . . .

- what groundwater is, where it resides in the Earth, and how its composition varies.
- the difference between porosity and permeability and how these features form.
- the difference between aquifers and aquitards and the nature of the water table.
- the various types of wells and springs that provide access to groundwater.
- how hot springs and geysers originate.
- how groundwater supplies can be damaged or depleted and how to address these problems.
- how caves and karst landscapes originate and evolve.

19.1 Introduction

Imagine Mae Rose Owens's surprise when, on May 8, 1981, she looked out her window and discovered that a large sycamore tree in the backyard of her Winter Park, Florida, home had suddenly disappeared. It wasn't a particularly windy day, so the tree hadn't blown over—it had just vanished! When Owens went outside to investigate, she found that more than the tree had disappeared. Her whole backyard had become a deep, gaping pit. The pit continued to grow for a few days until finally it swallowed Owens's house and six other buildings, as well as the municipal swimming pool, part of a road, and several expensive Porsches in a car dealer's lot (**Fig. 19.1a**).

What had happened in Winter Park? The bedrock beneath the town consists of limestone. **Groundwater**, the liquid water that resides in sediment or rock under the surface of the Earth, had gradually dissolved the limestone over time, carving open rooms, or *caverns*, underground. On May 8, the roof of a cavern underneath Owens's backyard began to collapse, forming a circular depression called a **sinkhole** (**Fig. 19.1b**). The sycamore tree and the rest of the neighborhood simply dropped down into the sinkhole. It would have taken too much effort to fill in the sinkhole with soil, so the community allowed it to fill with water, and now it's a circular lake—Lake Rose—the centerpiece of a municipal park. Similar lakes spot the landscape throughout central Florida (**Fig. 19.1c**), and they continue to form, sometimes with tragic loss of life and/or property (**Fig. 19.1d**).

Sinkholes serve as one of the more dramatic reminders that significant quantities of water reside underground. Though we can easily see Earth's *surface water* (in lakes, rivers, streams, marshes, glaciers, and oceans) and *atmospheric water* (in clouds and rain), *groundwater* lies hidden beneath the surface. This hidden reserve contains about 23,400,000 km³ (5,600,000 mi³) of water—about 123 times as much water as all the visible lakes, rivers, and swamps combined. Thus, it's not surprising that, globally, groundwater accounts for about two-thirds of the Earth's freshwater resources used by agriculture, industry, and homes.

In this chapter, we examine the nature of groundwater, discuss how it enters the subsurface, where it resides in the subsurface, and how it slowly moves underground until it returns to the surface through springs or wells. We also examine how human activities impact subsurface water. The chapter concludes with a brief survey of landscape features, such as hot springs and caves, that form as a consequence of the interaction between groundwater and its surroundings.

SEE FOR YOURSELF . . .

Sinkholes in Central Florida

LATITUDE
28°37'50.59"N

LONGITUDE
81°23'13.60"W

Looking down from 20 km (~12.5 mi).

The image shows several sinkholes that range in size from 100 to 800 m across. These sinkholes lie within suburban developments. Since the water table lies very close to the surface, the sinkholes have filled with water and are now lakes.

FIGURE 19.1 Development of sinkholes in central Florida.

(a) The Winter Park sinkhole, as seen from a helicopter.

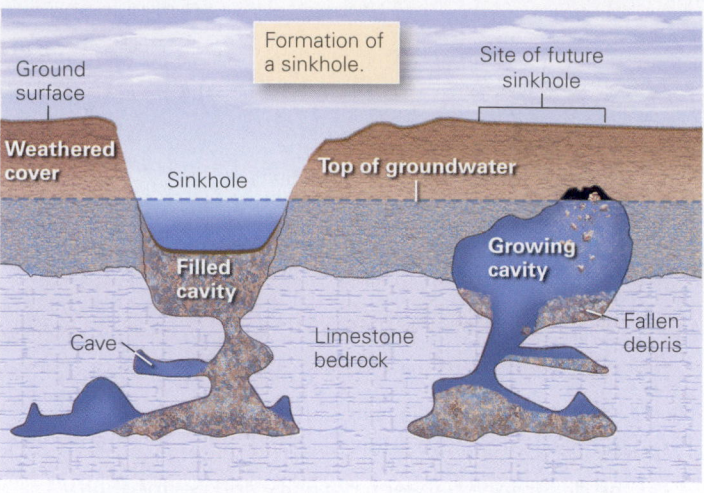

(b) As overburden slowly washes into underlying caves, a cavity forms. When the roof of this cavity collapses, a sinkhole forms (not to scale).

(c) An airplane view of Florida sinkholes that have become lakes.

(d) A new apartment building collapsed when a sinkhole formed under it.

19.2 Where Does Groundwater Reside?

As we saw in Interlude F, water moves among various reservoirs (the ocean, the atmosphere, rivers and lakes, groundwater, living organisms, soil, and glaciers) during the *hydrologic cycle*. Of the water that falls on land, some evaporates directly back into the atmosphere, some gets trapped in glacial ice, and some becomes runoff that enters a network of streams and lakes that drains to the sea. The remainder sinks or percolates downward, by a process called *infiltration*, into the ground. In effect, the upper part of the crust behaves like a giant sponge that can soak up water. Infiltration can take place because

most Earth materials (sediment and rocks) are not perfectly solid but rather contain some interconnected open space. Let's examine these openings and their connections in more detail.

Porosity: The Open Space in Rock and Regolith

The existence of liquid water underground requires the existence of open space underground. What is the nature of this space? When asked this question, many people picture networks of large caves containing spooky lakes and inky streams. Indeed, as we see later in this chapter, caves do provide room for subsurface water to drip and flow, but only locally. Contrary to popular belief, only a small proportion of underground water actually occurs in caves. Most groundwater resides in

FIGURE 19.2 Porosity—where water resides underground. Groundwater can fill microscopic pores (holes) and cracks between grains, or it can fill gaps formed along joints.

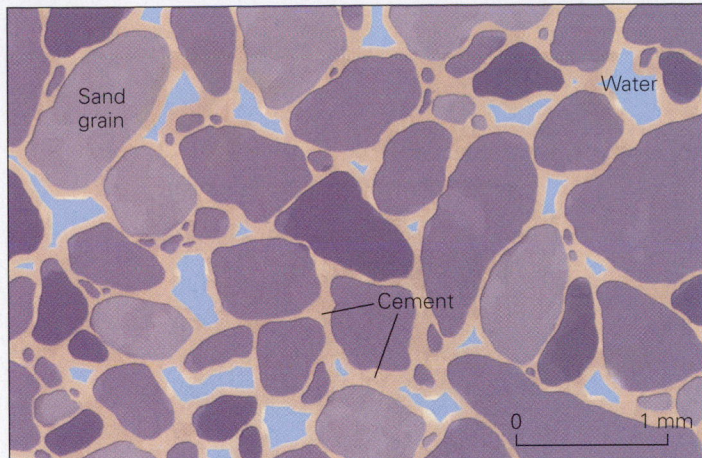

(a) Isolated pores in a sandstone occur in the spaces between grains. Water or air can fill pores.

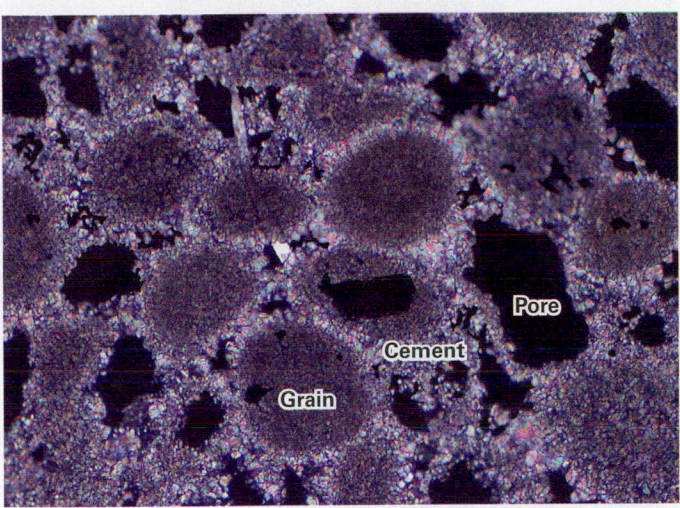

(b) This photo of a sedimentary rock, as seen through a microscope, shows grains, cement, and pores. The field of view is about 3 mm.

These fractures have been enlarged by dissolution.

(d) Limestone outcrop on the coast of Ireland contains abundant fractures that provide secondary porosity.

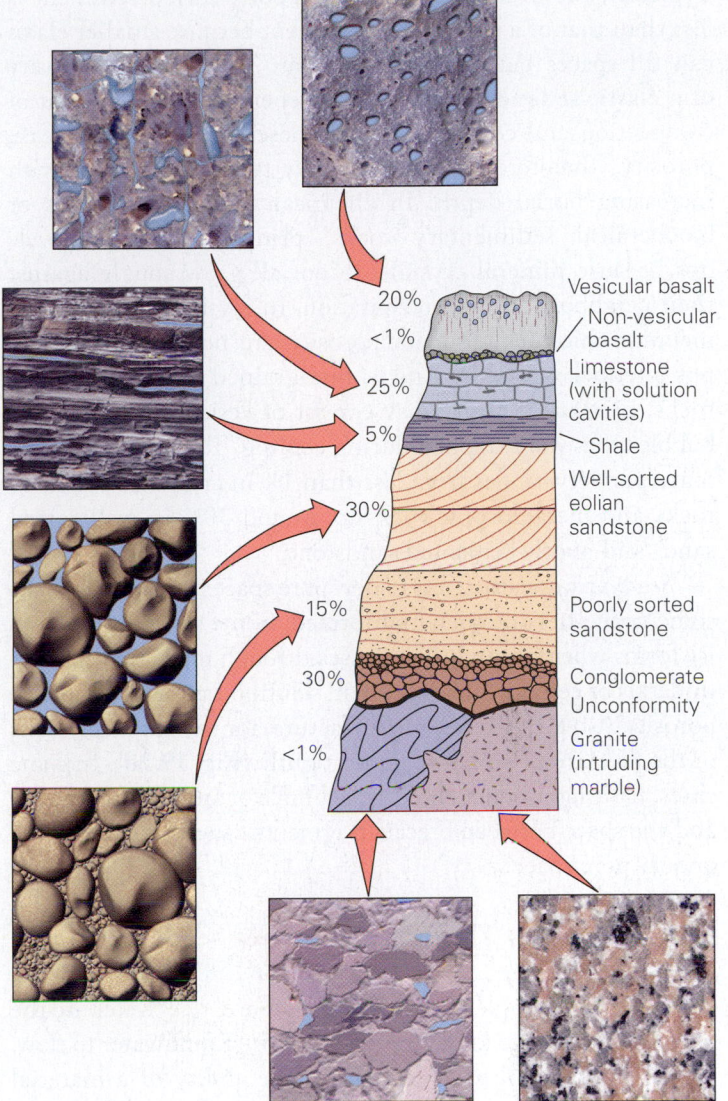

20% — Vesicular basalt
<1% — Non-vesicular basalt
25% — Limestone (with solution cavities)
5% — Shale
30% — Well-sorted eolian sandstone
15% — Poorly sorted sandstone
30% — Conglomerate
Unconformity
<1% — Granite (intruding marble)

(c) Different kinds and amounts of porosity occur in different kinds of rocks. Well-sorted and poorly cemented sandstone contains high porosity, whereas granite contains low porosity.

tiny open spaces or voids within sediment and within seemingly solid rock. As we first noted in Chapter 14, geologists use the term **pore** for any open space within a volume of sediment, or within a body of rock, and the term **porosity** for the total amount of open space within a material, specified as a percentage. For example, if we say that a block of rock has 30% porosity, then 30% of the block consists of pores. What makes up pores? We distinguish between two basic kinds of porosity—primary and secondary.

Primary porosity develops during sediment deposition and during rock formation (**Fig. 19.2a–c**). In clastic sedimentary rocks, it includes the gaps between grains. These gaps exist because the grains don't fit together tightly during deposition. Typically, the overall porosity of a poorly sorted sediment is less than that of a well-sorted sediment because smaller clasts can fill spaces between larger grains. The primary porosity of a clastic sedimentary rock also depends on the amount of compaction and cementation, for these phenomena decrease porosity. Therefore, primary porosity tends to decrease with increasing burial depth. In chemical sedimentary rocks or biochemical sedimentary rocks, primary porosity develops because mineral crystals do not all grow snugly against their neighbors during precipitation. In crystalline igneous or metamorphic rock, primary porosity can persist if grains do not interlock perfectly. And in fine-grained or glassy igneous rocks, primary porosity may consist of vesicles, relicts of air bubbles that were trapped during cooling. The amount of primary porosity ranges from less than 1% in crystalline igneous rocks and metamorphic rocks to around 30% in well-sorted sands and poorly cemented sandstone.

Secondary porosity refers to new pore space produced in rocks some time after the rock first formed. Some secondary porosity forms when groundwater passes through rock and dissolves minerals or cement, creating small solution cavities. Secondary porosity also forms when rocks fracture, for the opposing walls of the fracture do not fit together tightly (**Fig. 19.2d**). In some cases, faulting produces breccia, a jumble of angular fragments, and the space between breccia fragments can also serve as secondary pore space.

Permeability: The Ease of Flow

If solid rock completely surrounds a pore, the water in the pore cannot flow to another location. For groundwater to flow, pores must be linked by conduits. The ability of a material to allow fluids to pass through an interconnected network of pores is a characteristic known as **permeability**. To develop an intuitive image of permeability, take a sturdy glass jar and fill it with gravel composed of rounded pebbles. If you look closely at this gravel through the side of the jar, you will see air-filled pores between the pebbles, because the pebbles don't fit together perfectly. If you pour water into the jar, the water can trickle between grains down to the bottom of the jar, where it displaces air and fills the pores (**Fig. 19.3**). Water flows easily through a *permeable material*, whereas it flows slowly or not at all through an *impermeable material*. Informally, we can describe materials as having "high permeability," "low permeability," or "no permeability" to convey the relative ease of fluid movement. The permeability of a material depends on several factors.

FIGURE 19.3 Permeability—the ability of water to flow through rock from pore to pore or along joints.

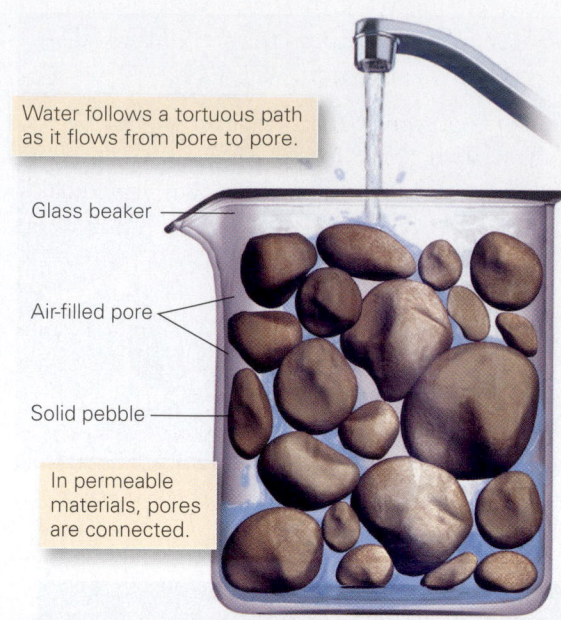

Water follows a tortuous path as it flows from pore to pore.

Glass beaker

Air-filled pore

Solid pebble

In permeable materials, pores are connected.

(a) Gravel contains pore space because clasts don't fit together tightly. The connection of pores produces permeability, so water can flow through gravel.

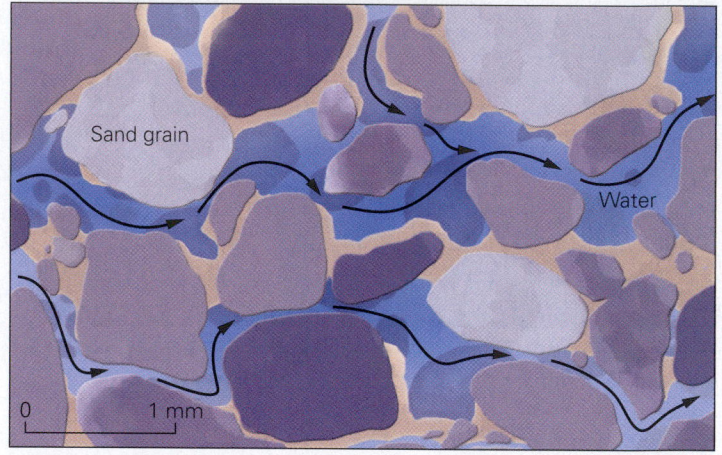

Sand grain

Water

0   1 mm

(b) Even at a microscopic scale, if passages connect pores, then water can flow through and the rock has permeability.

- *Number of available conduits:* As the number of conduits increases, permeability increases, for there are more routes through which water can move.
- *Size of the conduits:* Greater volumes of fluids can travel through wider conduits than through narrower ones. They can travel faster, because there is less friction with the conduit walls.

- *Straightness of the conduits:* Water flows more rapidly through straight conduits than it does through crooked ones. In crooked channels, the distance a water molecule actually travels may be many times the straight-line distance between the two end points.

Note that the factors that control permeability in rock or sediment resemble those that control the ease with which traffic moves through a city. Specifically, traffic can flow quickly through cities with many straight, multilane boulevards, whereas it flows slowly through cities with only a few narrow, crooked streets.

Porosity and permeability are not the same feature. A material whose pores are isolated from each other can have high porosity but low permeability. Cork exhibits this behavior—it has high porosity (so it floats) but low permeability (so it can plug up a bottle). In cork, a type of tree bark, woody cell walls isolate adjacent pores (empty cells) and prevent communication between them. Vesicular basalt is an example of a rock that can have high porosity but low permeability.

Aquifers and Aquitards

With the concept of permeability in mind, *hydrogeologists* (researchers who study groundwater) distinguish between **aquifers**, sediment or rocks that hold a lot of groundwater and transmit it easily because they have both high porosity and permeability, and **aquitards**, sediment or rocks that have low permeability and do not transmit groundwater easily. Generally, aquitards do not hold much groundwater because they have low porosity, too. (The less-used term, *aquiclude*, refers to a geologic material that does not transmit water at all.) Coarse gravels, poorly cemented sandstones, and highly fractured and partially dissolved limestones typically make good aquifers. Shales, evaporites, and very well-cemented sandstones serve as aquitards.

As we learned in Chapter 7, the type of sediment deposited at a location can vary over time as the depositional environment changes. Therefore, successions of sedimentary strata typically contain interlayered aquifers and aquitards. From the Earth's surface, water can infiltrate directly down into an *unconfined aquifer*. Water cannot migrate directly down into a *confined aquifer*, for an aquitard isolates the water in a confined aquifer from the Earth's surface (**Fig. 19.4**).

To get a sense of the diversity of characteristics that aquifers can have, and the diversity of ways in which aquifers form, let's look at examples of specific aquifers that provide important groundwater supplies in the United States. Each aquifer has a name, based on the region in which it occurs or on the stratigraphic unit that hosts the groundwater.

FIGURE 19.4 An aquifer is a high-porosity, high-permeability rock. Some aquifers are unconfined, and some are confined.

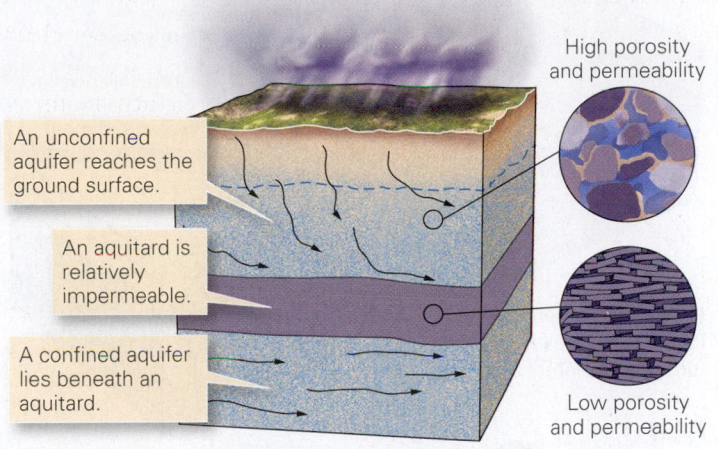

An unconfined aquifer reaches the ground surface.

An aquitard is relatively impermeable.

A confined aquifer lies beneath an aquitard.

High porosity and permeability

Low porosity and permeability

- *Mahomet aquifer, Illinois:* When glaciers melted during the Pleistocene Ice Age, meltwater streams carried coarse gravel southward. This gravel filled stream valleys that had previously been cut into the bedrock of central Illinois. Younger glacial sediment later buried the filled valleys. The porous and permeable gravel of the largest buried valley comprises the Mahomet aquifer, named for a nearby town (**Fig. 19.5a**).

- *Phoenix basin aquifer, Arizona:* The city of Phoenix resides in the Basin and Range Province, a continental rift. During rifting, downward slip of crustal blocks on normal faults produced deep wedge-shaped basins that filled with gravels and sands eroded from adjacent ranges (**Fig. 19.5b**). These sediments serve as aquifers that provide groundwater for irrigation and drinking in Phoenix.

- *The Dakota Sandstone aquifer:* During the Cretaceous, rivers flowing from mountains in the western United States dumped sediment into an inland sea, forming a vast sheet of sand that later was buried by mud. The sand layer lithified to become the Dakota Sandstone, an aquifer, and the mud layer became a shale aquitard. Continued mountain building warped the strata into a syncline. Water infiltrates into the western limb of the fold and slowly flows eastward (**Fig. 19.5c**). The Dakota Sandstone serves as a groundwater source where it returns to the ground surface on the eastern limb of the syncline. Note that it is a confined aquifer.

- *The High Plains (Ogallala) aquifer:* Erosion of the Rocky Mountains 3 million years ago led to the deposition of huge alluvial fans over what is now the High Plains region. The resulting deposits comprise the Ogallala

FIGURE 19.5 Examples of different types of aquifers. The aquifers are stippled.

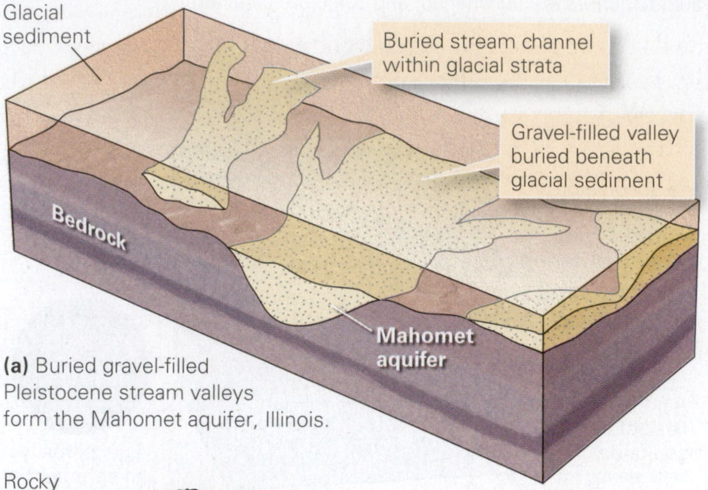

(a) Buried gravel-filled Pleistocene stream valleys form the Mahomet aquifer, Illinois.

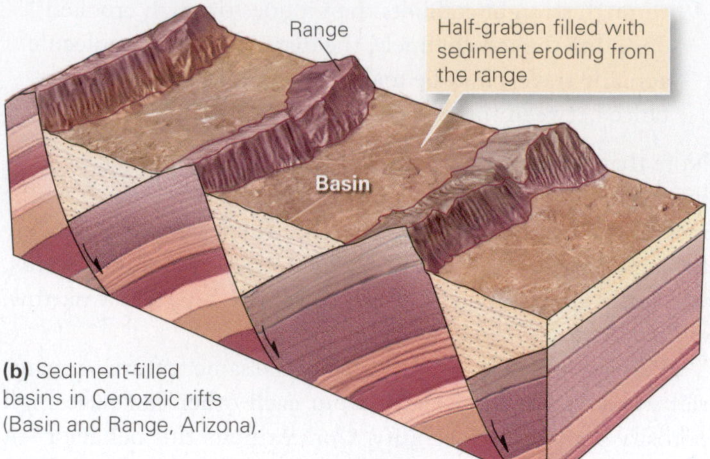

Glacial sediment

Buried stream channel within glacial strata

Gravel-filled valley buried beneath glacial sediment

Bedrock

Mahomet aquifer

(b) Sediment-filled basins in Cenozoic rifts (Basin and Range, Arizona).

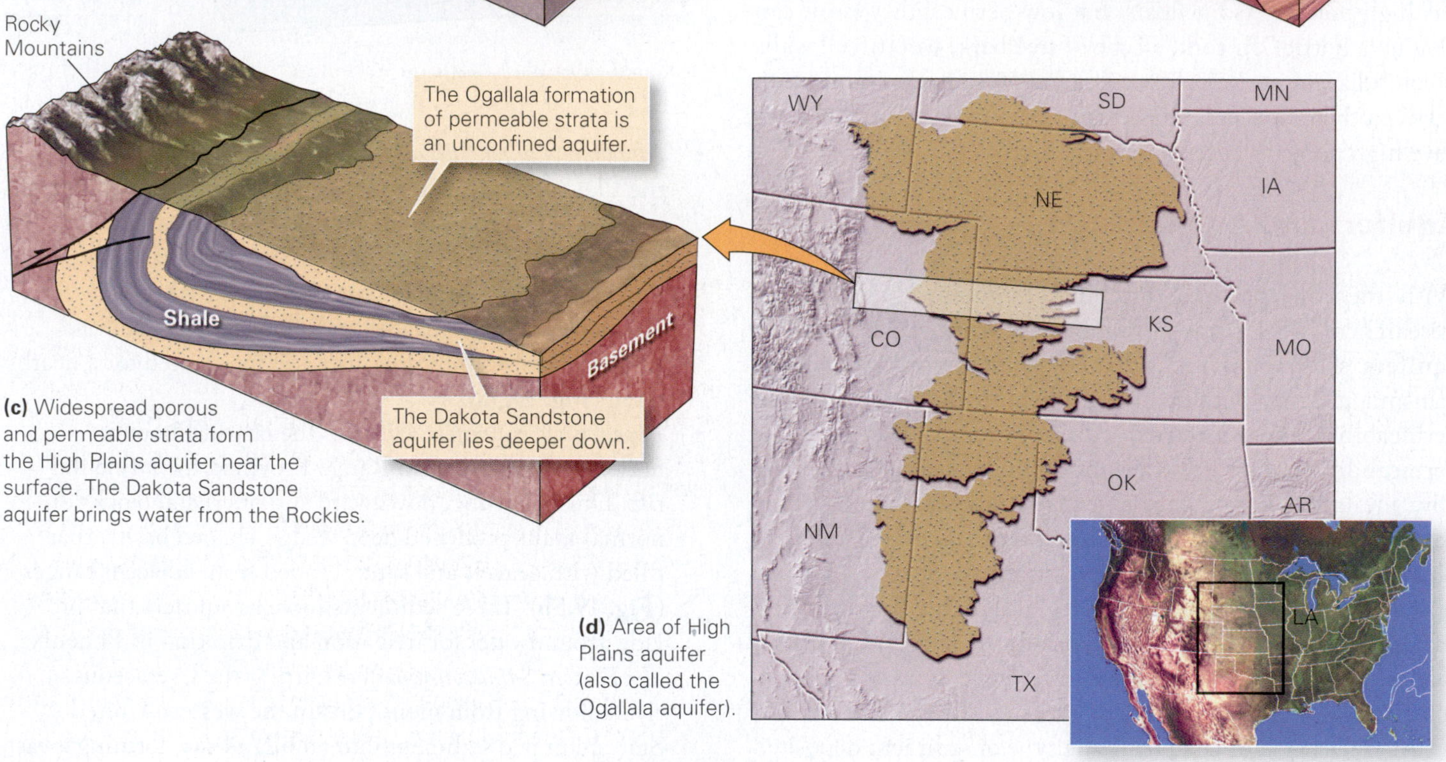

Range

Half-graben filled with sediment eroding from the range

Basin

(c) Widespread porous and permeable strata form the High Plains aquifer near the surface. The Dakota Sandstone aquifer brings water from the Rockies.

Rocky Mountains

The Ogallala formation of permeable strata is an unconfined aquifer.

Shale

The Dakota Sandstone aquifer lies deeper down.

Basement

(d) Area of High Plains aquifer (also called the Ogallala aquifer).

Formation, a sheet of porous and permeable strata (fluvial sandstones and conglomerates, and paleo-desert sands) that remains just below the ground surface (**Fig. 19.5d**). Water can infiltrate this unconfined aquifer over a broad area. Note that the Ogallala aquifer lies above the Dakota Sandstone aquifer over much of its extent.

- *Floridan limestone aquifers:* Sheets of Cenozoic limestone, deposited when the region was submerged by a shallow sea, underlie the entire peninsula of Florida as well as adjacent areas of Georgia and Alabama. Fractures in the limestone have been widened by dissolution and now provide spaces that hold large quantities of groundwater.

Take-Home Message

Most groundwater fills pores and cracks in rock or sediment. Porosity refers to the total amount of open space within a material, whereas permeability indicates the degree to which pores connect. Aquifers have high porosity and permeability whereas aquitards don't. An unconfined aquifer is one that connects to the ground surface, while a confined aquifer is one that lies beneath an aquitard. There are many different kinds of aquifers, each with distinct geologic characteristics.

QUICK QUESTION: Why do poorly cemented sandstones make good aquifers?

19.3 Characteristics of the Water Table

So far, we've used the term *groundwater* in a fairly general sense for all water underground. But now that we understand the concept of porosity and permeability, and have been introduced to the various kinds of aquifers, we can look more closely at the distribution of subsurface water and see that we can define a boundary, called the water table, that marks the upper limit of what hydrogeologists formally define as groundwater.

Nearer the ground surface, water may only partially fill the pores, leaving some space that remains filled with air (**Fig. 19.6**). The region of the subsurface in which water only partially fills pores is called the *unsaturated zone* (also known as the *vadose zone*). If the top of the unsaturated zone includes a soil, we can also refer to the water wetting the surfaces of grains and organic material making up the soil as **soil moisture**—this water tends to evaporate back into the atmosphere, seep into streams, or get sucked up by the plant roots and transpire back into the atmosphere. Below the unsaturated zone, water completely fills, or saturates, the pores, yielding the *saturated zone* (also called the *phreatic zone*). In a strict sense, hydrogeologists restrict use of the term *groundwater* to subsurface water in the saturated zone. The **water table** is the horizon that separates the unsaturated zone above from the saturated zone below (see Fig. 19.6). We can picture the water table as the top boundary of groundwater in an unconfined aquifer. Surface tension, the electrostatic attraction of water molecules to one another and to mineral surfaces, causes water to seep upward from the saturated zone to form a layer called the **capillary fringe** just above the water table. Capillary fringes typically have a thickness of between 5 and 30 cm.

The depth of the water table in the subsurface varies greatly with location. In some places, the water table defines the surface of a permanent stream, lake, or marsh, and thus effectively lies at or above the ground level (**Fig. 19.7a**). (In such places, there is no unsaturated zone, and the soil itself can be saturated, so we can't really distinguish soil moisture from groundwater.) Elsewhere, the water table lies hidden below the ground surface. In humid regions, it typically lies at depths of only meters to tens of meters, but in arid regions it may lie hundreds of meters below the surface. Thus, permanent lakes or streams cannot exist in arid regions, unless fed by a water supply from elsewhere, because water can infiltrate down through the lake or stream bed to the water table far below (**Fig. 19.7b**; see Chapter 17). Rainfall rates affect the water table depth in a given locality (**Fig. 19.7c**). Specifically, the water table drops during the dry season and rises during the wet season. Streams or ponds that hold water during the wet season may dry up during the dry season.

We've defined the water table as the top of groundwater in the crust. Can we define the base of groundwater? Put another way, how deep down in the crust can we find groundwater? Recent research shows that liquid groundwater does circulate in basement igneous and metamorphic rock, perhaps to depths of over 15 km. Most of this deep circulation takes place along fractures. The ultimate base of groundwater in crust may be taken as the depth where water endures such high temperatures and pressures that it becomes a "supercritical" hydrothermal fluid involved in metamorphic reactions, and rock becomes weak enough to flow plastically and close up pores. The depth at which this transition takes place depends on the geothermal gradient, but roughly speaking it lies at a depth of between 12 and 20 km. Thus, groundwater occurs only in the upper crust.

Topography of the Water Table

In hilly regions, if the subsurface has low to moderate permeability, the water table is not a planar surface. Rather, its shape mimics, in a subdued way, the shape of the overlying topography (**Fig. 19.8a**). This means that the water table lies at a higher elevation beneath hills than it does beneath valleys, but the relief (the vertical distance between the highest and lowest elevations) of the water table is not as great as that of the overlying

FIGURE 19.6 The water table is the top of the groundwater reservoir in the subsurface. It separates the unsaturated (vadose) zone above from the unsaturated zone below. A capillary fringe forms at the boundary.

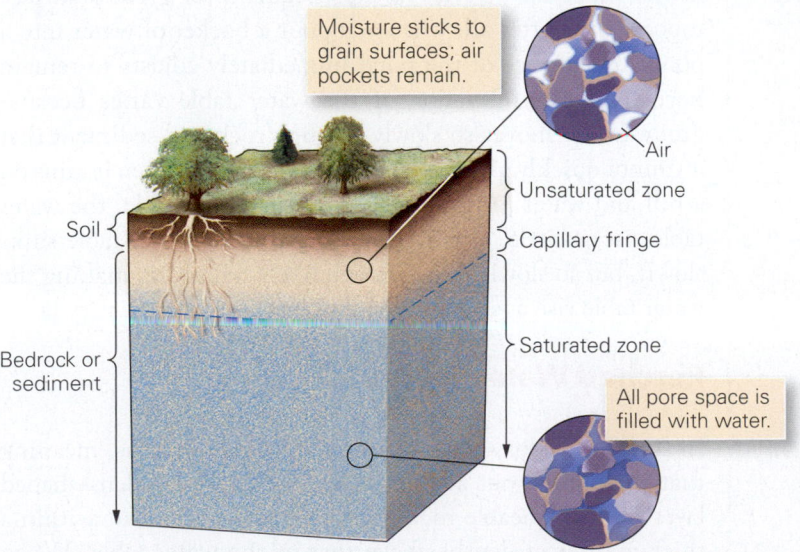

Moisture sticks to grain surfaces; air pockets remain.

Air

Unsaturated zone

Soil

Capillary fringe

Bedrock or sediment

Saturated zone

All pore space is filled with water.

FIGURE 19.7 The depth of the water table depends on the water supply from above.

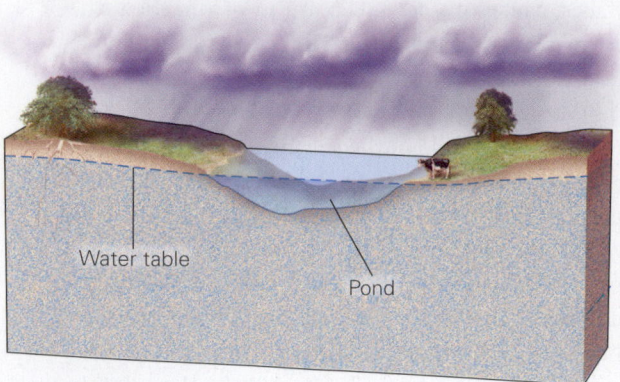

The water table lies above the ground in this swamp.

(a) Where the water table lies close to the ground surface, ponds remain filled—the water table is the surface of the pond.

Water seeps down until it reaches the water table.

In a drought, the water table sank and this pond dried up.

(b) In dry regions, the water table sinks deep below the surface. Water that collects temporarily in low areas sinks into the subsurface.

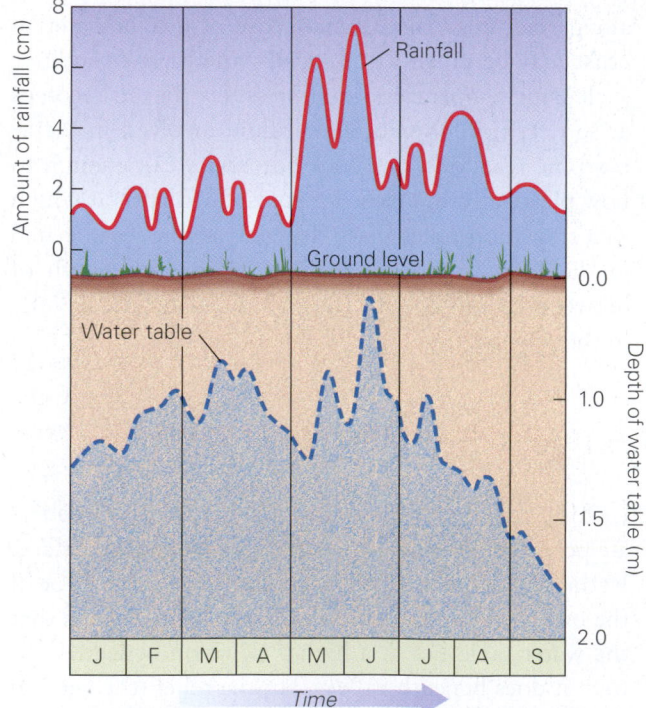

(c) This graph illustrates how the water table rises and falls depending on the amount of rainfall.

land. Thus, the surface of the water table tends to be smoother than that of the landscape.

At first thought, it may seem surprising that the elevation of the water table varies as a consequence of ground-surface topography. After all, when you pour a bucket of water into a pond, the surface of the pond immediately adjusts to remain horizontal. The elevation of the water table varies because groundwater moves so slowly through rock and sediment that it cannot quickly assume a horizontal surface. When it rains on a hill and water infiltrates down to the water table, the water table rises a little. When it doesn't rain, the water table sinks slowly, but so slowly that rain will likely fall again, making the water table rise again, before it has had time to sink very far.

Perched Water Tables

In some locations, layers of strata are discontinuous, meaning that they pinch out at their sides. As a result, a lens-shaped layer of impermeable rock (such as shale) may occur within a thick aquifer at depths above that of the water table. When

FIGURE 19.8 Factors that influence the position of the water table.

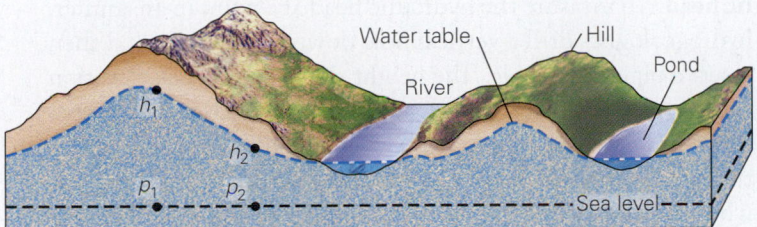

(a) The shape of a water table beneath hilly topography. Point h_1 on the water table is higher than Point h_2, relative to a reference elevation (sea level). The pressure at p_1, is, therefore, more than the pressure at p_2.

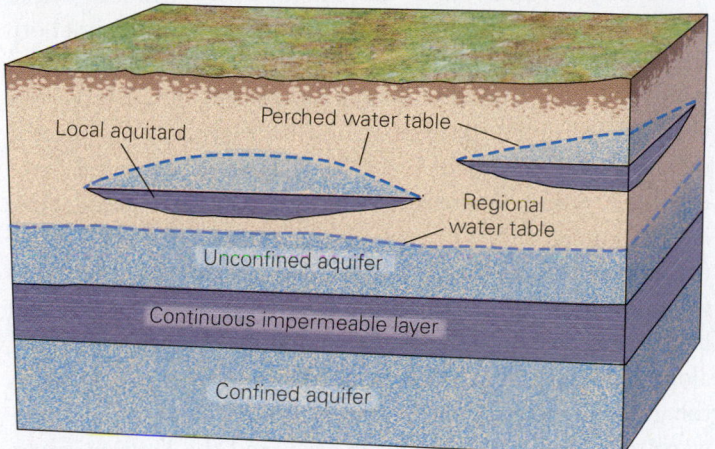

(b) A perched water table occurs where a mound of groundwater becomes trapped above a localized aquitard that lies above the regional water table.

this happens, a mound of groundwater accumulates above each aquitard lens because the lens prevents the groundwater from sinking down to the regional water table (**Fig. 19.8b**). The top of each mound is called a **perched water table** because it lies above the regional water table.

Take-Home Message

Groundwater is water that resides underground in the pores of rock or sediment. Below the water table, in the saturated zone, water fills available pore space; above the water table, pores are partially or entirely filled with air. The water table is higher in regions with wetter climates than in regions with drier climates; it can rise or fall depending on the amount of rainfall. In hilly areas, the water table itself has topography. Where the water table intersects valleys or depressions, streams or lakes form.

QUICK QUESTION: Why is the water table higher beneath hills than beneath valleys?

19.4 Groundwater Flow

What happens to groundwater over time? Does it just sit, unmoving, like the water in a stagnant puddle, or does it flow and eventually find its way back to the surface? Countless measurements confirm that groundwater enjoys the latter fate—groundwater indeed flows, and in some cases it moves great distances underground, taking anywhere from a few weeks to tens of thousands of years before returning to the surface to pass once again into other reservoirs of the hydrologic cycle. First, we examine factors that drive groundwater flow and determine the path that this flow takes. Then we examine the rate (velocity) at which groundwater moves.

Groundwater Flow Paths

In the unsaturated zone—the region between the ground surface and the water table—water percolates down, like the water passing through a drip coffeemaker, for this water moves only in response to the downward pull of gravity. But in the zone of saturation—the region below the water table—water flow is more complex, for in addition to the downward pull of gravity, water responds to differences in pressure. Pressure can cause groundwater to flow sideways or even upward. (If you've ever watched water spray from a fountain, you've seen pressure pushing water upward.) Thus, to understand the nature of a *groundwater flow path*, the overall trajectory that groundwater follows over time, we must first understand the origin of pressure in groundwater. For simplicity, we'll consider only the case of groundwater in an unconfined aquifer.

Pressure in groundwater at a specific point underground is caused by the weight of all the overlying water from that point up to the water table. (The weight of overlying rock does not contribute to the pressure exerted on groundwater, for the contact points between mineral grains bear the rock's weight.) Thus, a point at a greater depth below the water table feels more pressure than does a point at lesser depth. If the water table is horizontal, the pressure acting on an imaginary horizontal reference plane at a specified depth below the water table is the same everywhere. But if the water table is not horizontal (see Fig. 19.8a), the pressure at points on a horizontal reference plane at depth changes with location. For example, the pressure acting at point p_1, which lies below the hill in Figure 19.8a, is greater than the pressure acting at point p_2, which lies below the valley, even though both p_1 and p_2 are at the same elevation (sea level, in this case).

Both the elevation of a volume of groundwater and the pressure within the water provide energy that, if given the chance, will cause the water to flow. Physicists refer to such stored energy as *potential energy*. To understand why elevation provides potential energy, imagine a bucket of water high on a hill; the

FIGURE 19.9 The flow of groundwater.

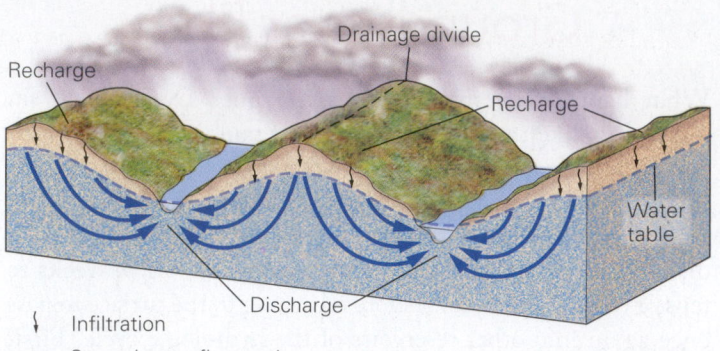

(a) Groundwater flows from recharge areas to discharge areas. Typically, the flow follows curving paths.

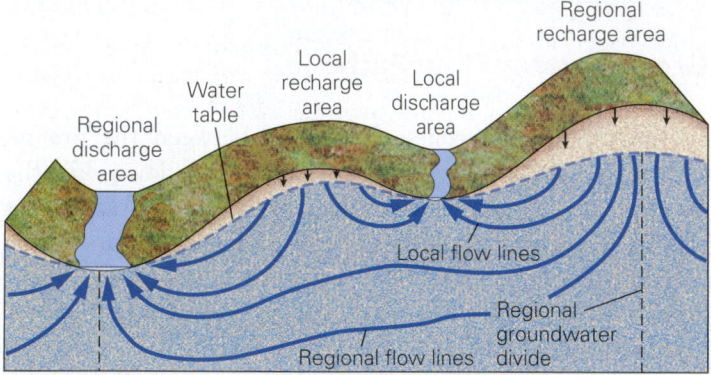

(b) In a region of complicated topography, local flow may be different from regional flow.

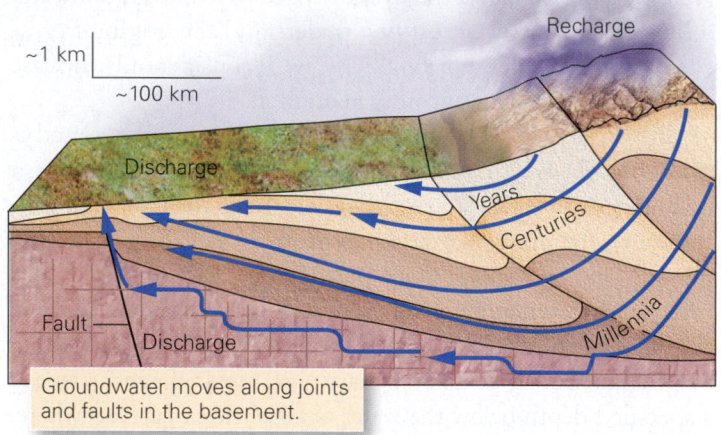

Groundwater moves along joints and faults in the basement.

(c) The large hydraulic head resulting from uplift of a mountain belt may drive groundwater hundreds of kilometers across regional sedimentary basins. Deeper flow paths take longer.

water has potential energy due to Earth's gravity, which will cause it to flow downslope if the bucket were suddenly to rupture. To understand why pressure provides potential energy, imagine a water-filled plastic bag sitting on a table; if you puncture the bag and then squeeze the bag to exert pressure, water spurts out. The potential energy available to drive the flow of a

given volume of groundwater at a location is called the **hydraulic head**. To measure the hydraulic head at a point in an aquifer, hydrogeologists drill a vertical hole down to the point and then insert a pipe in the hole. The height above a reference elevation (such as sea level) to which water rises in the pipe represents the hydraulic head. Thus, water rises higher in the pipe where the head is higher. As a rule, groundwater flows from regions where it has higher hydraulic head to regions where it has lower hydraulic head. This statement generally implies that groundwater, regionally, flows from locations where the water table is higher to locations where the water table is lower.

Hydrogeologists have calculated how hydraulic head changes with location underground by taking into account both the effect of gravity and the effect of pressure. These calculations reveal that groundwater flows along concave-up curved paths (**Fig. 19.9a**). (Specialized books on hydrogeology provide the details of why flow paths have this shape.) These curved paths eventually take groundwater from regions where the water table is high (under a hill) to regions where the water table is low (below a valley), but because of flow-path shape, some groundwater may flow deep down into the crust along the first part of its path and then may flow back up, toward the ground surface, along the final part of its path. The location where water enters the ground, meaning the region where the flow has a downward trajectory, is the **recharge area**, and the location where groundwater flows back up to the surface is the **discharge area** (see Fig. 19.9a). Because landscapes are complex, not all water entering the ground a given recharge zone ends up in the same discharge zone. Some flow follows longer paths to regional discharge areas, and some follows shorter paths to local discharge areas (**Fig. 19.9b**). We can define a *groundwater divide* as the vertical plane separating flow that goes to different discharge zones, and these too can be local or regional.

Groundwater following short paths close to the Earth's surface travels tens of meters to a few kilometers before returning to the surface. Such *local flow* has a residence time (stays underground) for only hours to weeks. Groundwater following paths of several kilometers to tens of kilometers constitutes *intermediate flow* and has a residence time of weeks to years. Groundwater following paths that carry it hundreds of kilometers across a large sedimentary basin constitutes *regional flow* and stays underground for centuries to millennia (**Fig. 19.9c**).

Rates of Groundwater Flow

Flowing water in an ocean current moves at up to 3 km per hour (over 26,000 km per year), and water in a steep river channel can reach speeds of up to 30 km per hour (over 260,000 m per year). In contrast, groundwater moves at less than a snail's pace—hydrogeologists have found that typical rates range between 0.01 and 1.4 m per day (only about 4 to 500 m per year). They can measure the rate (velocity) of groundwater flow in a region by determining how long it takes a "tracer," such as a chemical dye or some radioactive isotopes, injected in one well to arrive at another well a known distance away. Groundwater moves much more slowly than surface water for two reasons. First, groundwater must percolate through a complex, crooked network of tiny conduits, so it must travel a much greater distance than it would if it could follow a straight path. Second, the surface tension of water makes it stick to the solid material around it and also slows its escape from pores. Even in larger conduits, friction between groundwater and conduit walls slows the water flow.

Simplistically, the velocity of groundwater flow depends on the slope of the water table and on the permeability of the material through which the groundwater is flowing. Groundwater flows faster through high-permeability rocks than it does through low-permeability rocks, and it flows faster in regions where the water table has a steep slope than it does in regions where the water table has a gentle slope. Thus, groundwater flows relatively quickly (about 30 m per year) through high-permeability gravel under a steep hillslope, but it flows relatively slowly (about 2 m per year) through a low-permeability, well-cemented sandstone under the Great Plains. In detail, hydrogeologists use Darcy's law to determine flow rates at a location (**Box 19.1**).

Take-Home Message

Gravity and pressure cause groundwater to flow slowly from recharge to discharge areas. Simplistically, the rate of flow depends on the water table's slope and on permeability. In detail, flow is driven by differences in hydraulic head. Groundwater can follow curving flow paths that take it deep into the crust. It typically flows at rates of 4 m to 0.5 km per year, so it can stay underground for thousands of years. Darcy's law allows calculation of flow rates.

QUICK QUESTION: Does steepening the water table increase or decrease the rate of groundwater flow?

19.5 Tapping Groundwater Supplies

All living organisms rely on access to water. Of course, while marine life, coastal plants, and certain types of microbes survive in saline water, most terrestrial organisms, such as our own species, require freshwater. Groundwater contains the second-largest reservoir of freshwater on the planet, after the glaciers of Antarctica and Greenland. Although lakes and streams provide access to freshwater, they don't occur everywhere, so even our distant prehistoric ancestors, as well as many other plants and animals, need to utilize **springs**, natural outlets from which groundwater

flows, in order to survive. Springs don't occur everywhere, either, so almost 10,000 years ago, people learned to dig *wells*, artificial holes that provide access to groundwater. The oldest-known wells, found on the island of Cyprus, date from about 7500 B.C.E. In this section, we first examine the variability of natural groundwater, and then focus on the nature of springs and wells.

Natural Groundwater Quality

Shallow, new groundwater has a composition similar to the rainwater or snow melt that served as its source—pure H_2O. But as groundwater percolates downward, it reacts with the soil, sediment, or rock through which it passes, and these reactions transform it into a chemical solution. Deep in sedimentary basins, groundwater derived from rain or snow may mix with water left in pore spaces from the time of the original deposition of sediment millions of years ago. If the sediment was deposited in a marine environment, this ancient deep groundwater can be saline.

The concentration of dissolved ions in groundwater—meaning the quantity of ions dissolved per unit volume of water—depends on the temperature, pressure, acidity, availability of oxygen, and residence time of the groundwater. Warmer groundwater, for example, can carry more ions in solution than can cooler groundwater. And the longer that groundwater has to react with its surroundings, the greater the concentration can become. When groundwater attains "saturation," the water contains as many dissolved ions as possible under the local environmental conditions.

Because groundwater flows, it eventually enters different environments. If groundwater enters a new environment where it has the capacity to contain more ions, it may dissolve surrounding rock or sediment and create secondary porosity. Alternatively, if groundwater enters an environment in which it cannot hold as many dissolved ions, some of the ions bond together and become solid mineral grains that precipitate to form cement or vein fill.

Shallow groundwater can be crystal clear and pure enough to drink right out of the ground, for rocks and sediment are natural filters capable of removing suspended solids—these solids get trapped in tiny pores or stick to the surfaces of clay flakes. In fact, the commercial distribution of bottled groundwater, so-called spring water, has become a major business worldwide. Small amounts of dissolved minerals may add flavor to water, so it can be sold as "mineral water." Dissolved CO_2 can give some groundwater a natural fizziness.

Did you ever wonder...

where bottled "spring water" comes from?

In too high a quantity, however, dissolved chemicals make natural groundwater undesirable. For example, groundwater that has passed through salt-containing strata or has mixed with old saline pore water may become salty and unsuitable

BOX 19.1 CONSIDER THIS . . .

Darcy's Law for Groundwater Flow

The rate at which groundwater flows at a given location depends on the permeability of the material containing the groundwater—groundwater flows faster in a more permeable material than it does in a less permeable material. The rate also depends on the **hydraulic gradient**, meaning the change in hydraulic head per unit of distance between two locations, as measured along the flow path. To calculate the hydraulic gradient, we divide the difference in hydraulic head between two points by the distance between the two points as measured along the flow path. This can be written as a formula:

$$\text{hydraulic gradient} = \frac{h_1 - h_2}{j}$$

where $h_1 - h_2$ is the difference in head (given in meters or feet, because head can be represented as an elevation) between two points along the water table, and j is the distance between the two points as measured along the flow path (**Fig. Bx19.1**). A hydraulic gradient exists anywhere that the water table is not horizontal. Typically, the slope of the water table is so small that the path length is almost the same as the horizontal distance between two points. So, in general, the hydraulic gradient is roughly equivalent to the slope of the water table.

In 1856, a French engineer named Henry Darcy carried out a series of experiments designed to characterize factors that control the velocity of groundwater flow

between two locations (1 and 2), each of which has a different hydraulic head (h_1 and h_2). Darcy represented the velocity of flow by a quantity called the discharge (Q), meaning the volume of water passing through an imaginary vertical plane, perpendicular to the groundwater's flow path, in a given time. He found that the discharge depends on the hydraulic head ($h_1 - h_2$); the area (A) of the imaginary plane through which the groundwater is passing; and a number called the hydraulic conductivity (K). The hydraulic conductivity represents the ease with which a fluid can flow through a material and depends on many factors, such as the

viscosity and density of the fluid, but mostly it reflects the permeability of the material. The relationship that Darcy discovered, now known as **Darcy's law**, can be written in the form of an equation as:

$$Q = \frac{KA(h_1 - h_2)}{j}$$

The equation states that if the hydraulic gradient increases, discharge increases, and that as conductivity increases, discharge increases. Put in simpler terms, the flow rate of groundwater increases as the permeability increases and as the slope of the water table gets steeper.

FIGURE Bx19.1 The level to which water rises in a drillhole is the hydraulic head (h). The hydraulic gradient (HG) is the difference in head divided by the length of the flow path.

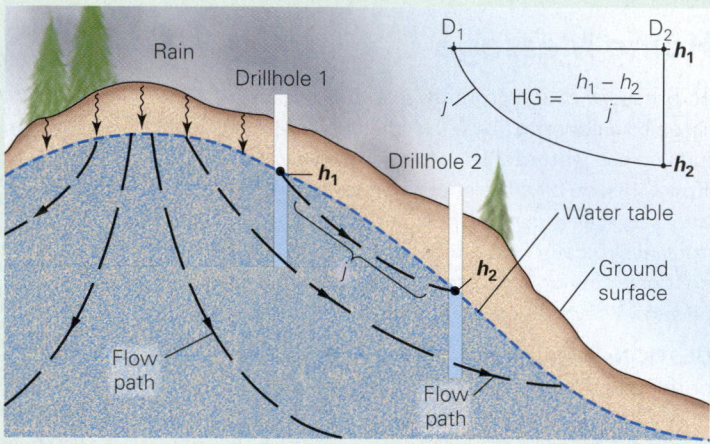

for irrigation or drinking. Groundwater that passed through limestone or dolomite may become saturated in calcium (Ca^{2+}) and magnesium (Mg^{2+}) ions—such water, called **hard water,** can be a problem because carbonate minerals precipitate from it to form "scale," which clogs pipes, and prevent soap from developing a lather. And groundwater that has passed through iron-bearing rocks may contain dissolved iron oxide that precipitates to form rusty stains.

Unfortunately, some groundwater contains dangerous chemicals. Examples include hydrogen sulfide, which comes out of solution when the groundwater rises to the surface to form a poisonous gas with a rotten-egg smell, and arsenic, a highly toxic chemical, which enters groundwater when arsenic-bearing minerals dissolve. In the United States, government

standards require drinking water to contain less than 10 micrograms of arsenic per liter, but surveys suggest that about 5% of groundwater supplies contain 20 or more micrograms per liter. Natural gas may become a problem where bedrock contains organic matter. The gas dissolves in the water under pressure deeper below the surface but bubbles out of solution when it comes out of the tap.

The quality of groundwater can vary with depth in a given region because typically the age of groundwater increases with increasing depth. Thus, groundwater in a deep sedimentary basin may become more saline (brackish) and harder with increasing depth. In fact, in deep basins, a boundary at a depth a few hundred meters up to a few thousand meters divides drinkable groundwater above from undrinkable groundwater below.

FIGURE 19.10 Geologic settings in which springs form.

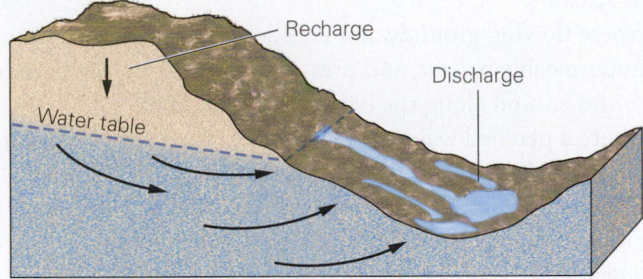

(a) Groundwater reaches the ground surface in a discharge zone.

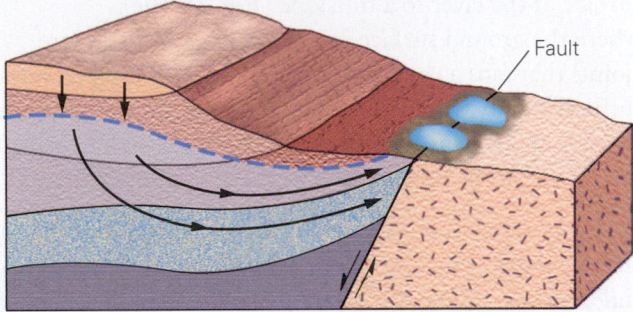

(b) Where groundwater reaches an impermeable barrier, it rises.

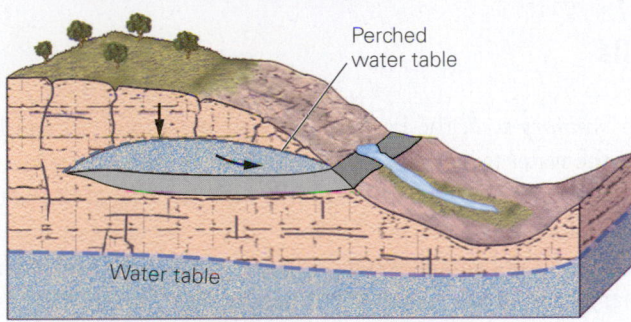

(c) Groundwater seeps where a perched water table intersects a slope.

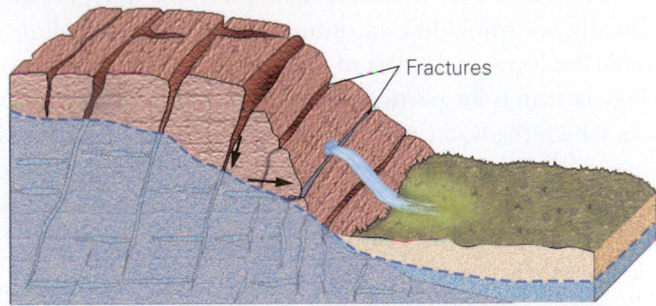

(d) A network of interconnected fractures channel water to the surface of the hill.

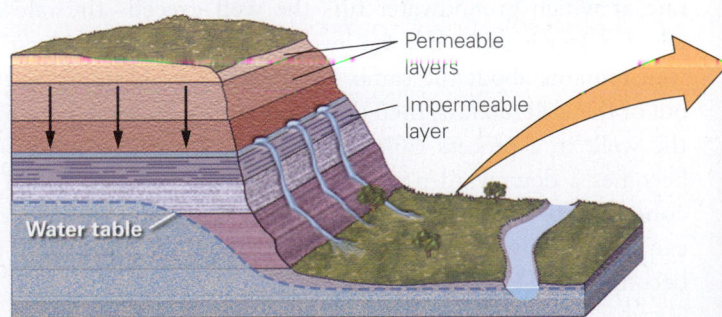

(e) Groundwater seeps out of a cliff face at the top of a relatively impermeable bed.

Springs

More than one town has grown up around a spring, a place where groundwater naturally flows or seeps onto the Earth's surface, for springs can provide fresh, clear groundwater for drinking or irrigation, without the expense of drilling or digging. Some springs spill water onto dry land. Others bubble up through the bed of a stream or lake. Springs form under a variety of conditions:

- where the ground surface intersects the water table in a discharge area (**Fig. 19.10a**); such springs typically occur

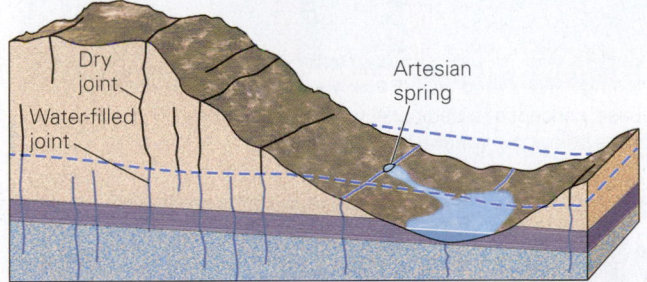

(f) In cases where water under pressure lies below an aquitard, a crack may provide a pathway for an artesian spring to form.

A spring on the wall of the Grand Canyon

FIGURE 19.11 The nature of ordinary wells.

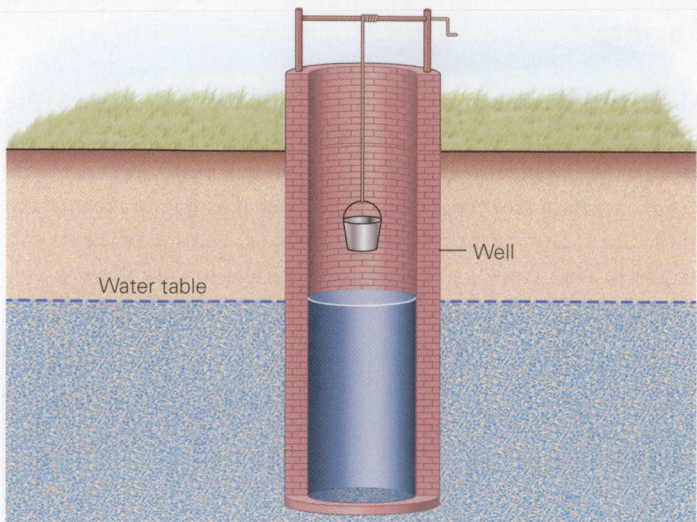

(a) The basic concept of a traditional well. It's a vertical cylindrical hole that extends below the water table.

(b) At a traditional ordinary well, villagers obtain water by lifting it with a bucket.

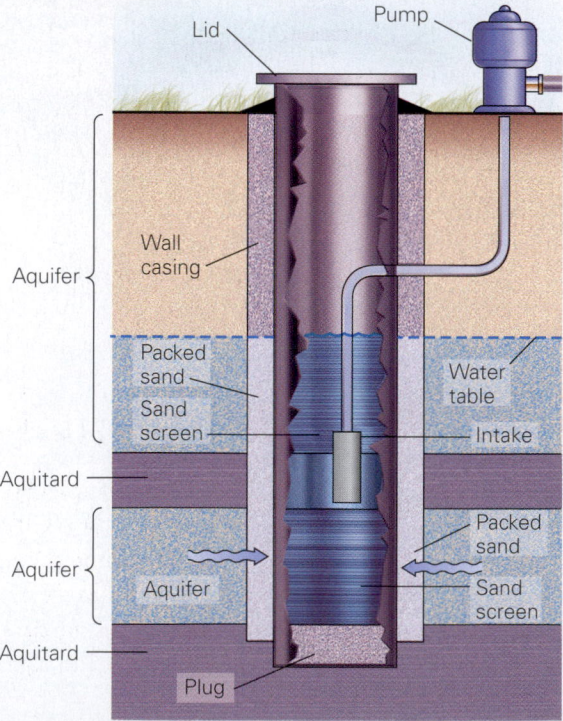

(c) A modern ordinary well sucks up water with an electric pump. The packed sand filters the water.

in or near valley floors, where they may add water to lakes or streams.

- where flowing groundwater collides with a steeply dipping impermeable barrier, and pressure pushes it up the barrier to the ground along the barrier (**Fig. 19.10b**).
- where a perched water table intersects the surface of a hill (**Fig. 19.10c**).
- where a network of interconnected fractures channels groundwater to the surface of a hill (**Fig. 19.10d**).
- where downward-percolating water runs into a relatively impermeable layer (aquitard) and migrates along the top surface of the layer to a hillslope (**Fig. 19.10e**).
- where the ground surface intersects a natural fracture (joint) that taps a confined aquifer in which the pressure is sufficient to drive the water to the surface; such an occurrence is an **artesian spring** (**Fig. 19.10f**)—we'll explain the source of the pressure in our discussion of wells.

Springs can provide water in regions that would otherwise be uninhabitable. For example, oases in deserts may develop around a spring. An **oasis** is a wet area, where plants can grow, in an otherwise bone-dry region (**Box 19.2**).

Wells

In an *ordinary well*, the base of the well penetrates an aquifer below the water table (**Fig. 19.11**). Water from the pore space in the aquifer seeps into the well and fills it to the level of the water table. Drilling into rock that lies above the water table, or into an aquitard, will not supply water and thus yields a *dry well*. Some ordinary wells are seasonal and function only during the rainy season when the water table rises. During the dry season, the water table lies below the base of the well, so the well goes dry.

Ideally, we would like an ordinary well to be as shallow as possible (to decrease the cost of digging or drilling). So hydrogeologists search for particularly porous and permeable aquifers in which the water table lies near the surface. Contrary to legend, a "dowser" cannot find water simply by using a forked stick—when dowsers do strike water, either they have enjoyed dumb luck or they have prior knowledge of the water table in the area of their search.

To obtain water from an ordinary well, you can either pull water up in a bucket or pump the water out. As long as the rate at which groundwater fills the well exceeds the rate at which water is removed, the level of the water table near the well remains about the same. However, if users pump water out of the well too fast, then the water table sinks down around the well, in a process called *drawdown*, and the water table becomes a downward-pointing, cone-shaped surface called a **cone of depression** (**Fig. 19.12a**). Drawdown in a deep well can lower the water table, overall, causing shallower wells to become dry (**Fig. 19.12b**).

BOX 19.2 CONSIDER THIS . . .

Oases

The Sahara of northern Africa is now one of the most barren and desolate places on Earth, for it lies in a climatic belt where rain seldom falls. But it wasn't always that way. During the last ice age, when glaciers covered parts of northern Europe on the other side of the Mediterranean, the Sahara enjoyed a more temperate climate, and the water table was high enough that permanent streams dissected the landscape. These streams are long gone, a victim of the drier, warmer climate of recent millennia, but vast reserves of groundwater fill a vast underground aquifer composed of porous sandstone. In general, the water of the aquifer can be obtained only by drilling deep wells, but locally, water spills out at the surface—either because tectonic folding brings the aquifer particularly close to the ground so that valley floors intersect the water table or because artesian pressure pushes groundwater up along joints or faults (**Fig. Bx19.2a**). In either case, the aquifer feeds springs that quench the thirst of plants and create an *oasis*, an island of green in the sand sea (**Fig. Bx19.2b**). Oases became important stopping points along caravan routes, allowing both people and camels to replenish water supplies.

In some oases, people settled and used the groundwater to irrigate date palms and other crops. For example, the Bahariya Oasis, about 400 km southwest of Cairo, Egypt, hosted a town of perhaps 30,000 between 300 B.C.E. and 300 C.E. During that time, the water table lay only 5 m below the ground and could easily be reached by shallow wells. Today, as a result of changing climates and centuries of use, the water table lies 1,500 m below the ground, almost out of reach. Bahariya's glorious past came to light in 1996, quite by accident. A man was riding his donkey in the desert near the oasis when the ground beneath the donkey suddenly caved in. The man had inadvertently opened the roof into a tomb filled with over 150 mummies, along with thousands of well-preserved artifacts. The site has since come to be known as the Valley of the Mummies.

FIGURE Bx19.2 The formation of oases in deserts.

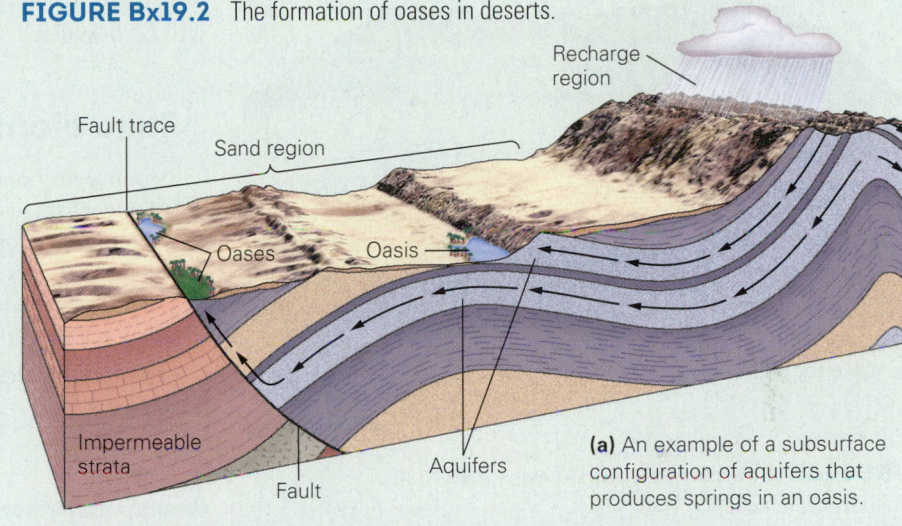

(a) An example of a subsurface configuration of aquifers that produces springs in an oasis.

(b) A small oasis in the Sahara Desert.

An **artesian well,** named for the province of Artois in France, penetrates confined aquifers in which water is under enough pressure to rise on its own to a level above the surface of the aquifer. If this level lies below the ground surface, the well is a *nonflowing artesian well*. But if the level lies above the ground surface, the well is a *flowing artesian well*, and water actively fountains out of the ground (**Fig. 19.13a**). (We've seen that the same phenomenon happens at an artesian spring, where a natural fracture cuts across an aquitard to provide access to the pressurized water.)

We can understand why artesian wells exist if we look first at the configuration of a city water supply (**Fig. 19.13b**). City water companies pump water into a high tank that has a significant hydraulic head relative to the surrounding areas. If the water were connected by a water main to a series of vertical pipes, pressure caused by the elevation of the water in the high

FIGURE 19.12 Pumping groundwater from a normal well can affect the water table.

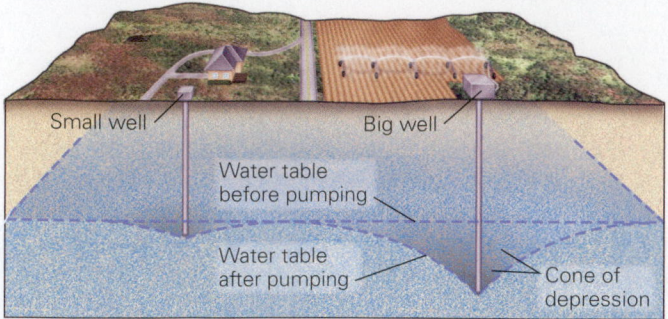

(a) Pumping forms a cone of depression in the water table.

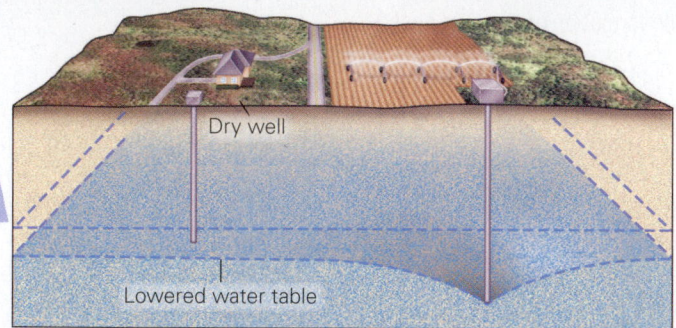

(b) Pumping by the big well may be enough to make the small well run dry.

FIGURE 19.13 Artesian wells, where water rises from the aquifer without pumping.

(a) A flowing artesian well in Wisconsin; the water rises from underground in the corrugated pipe without the need of pumping, and it continuously flows out of the pipe.

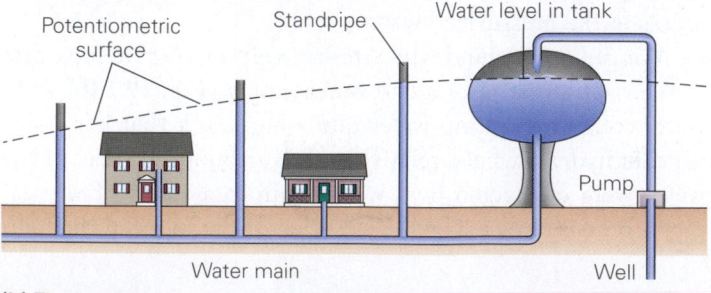

(b) The configuration of a city water supply. Water rises in vertical pipes up to the level of the potentiometric surface.

tank would make the water rise in the pipes until it reached an imaginary surface, called a *potentiometric surface*, that lies above the ground. This pressure drives water through water mains to household water systems without requiring pumps. In an artesian system, water enters a tilted, confined aquifer that intersects the ground in the hills of a high-elevation recharge area (**Fig. 19.13c**). The confined groundwater flows down to the adjacent plains, which lie at a lower elevation. The potentiometric surface to which the water would rise were it not confined lies above this aquifer. Pressure in the confined aquifer pushes water up an artesian well or spring. Where the potentiometric surface lies underground, the well will be non-flowing, but where the surface lies above the ground, the well will be flowing.

Take-Home Message

Groundwater can be obtained at springs (natural outlets) or at wells (built by people). Springs are found in many geologic settings. In ordinary wells, water must be lifted to the surface, but in artesian wells and springs, it rises due to its natural pressure. Pumping groundwater lowers the water table and can create a cone of depression.

QUICK QUESTION: Why do some springs emerge along the side of a hill?

19.6 Hot Springs and Geysers

Hot springs, springs that emit water ranging in temperature from 30° to 104°C, develop in two geologic settings. First, they occur where groundwater, as it slowly flows from recharge area to discharge area, follows a flow path that takes it many kilometers down in the crust, where bedrock is naturally warm due to the geothermal gradient. This groundwater absorbs heat from the bedrock and carries it back up to the surface. Such

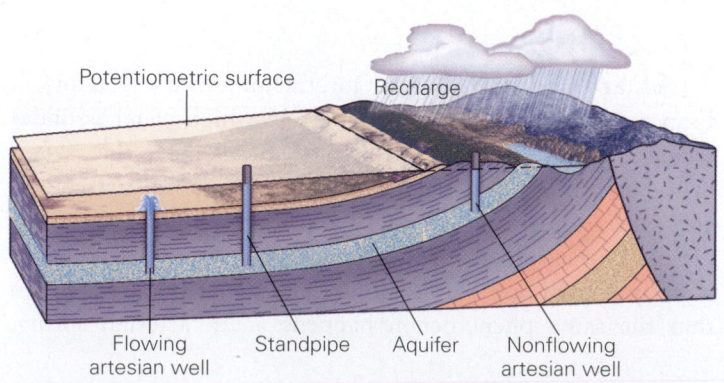

(c) The configuration of a regional artesian system.

water loses some heat as it enters shallower crustal levels but in some localities may still be warm when it seeps from a spring. Second, hot springs develop in **geothermal regions**, places where volcanism is currently taking place or has occurred recently (**Fig. 19.14**). In geothermal areas, that magma and/ or very hot rock resides close to the Earth's surface, so even shallow groundwater can become very hot. Since it only has to rise a short distance before returning to the surface at a discharge area, it comes out of the ground while still at a high temperature.

Notably, hot groundwater dissolves minerals from rock that it passes through, because water becomes a more effective solvent when hot. So water emitted at hot springs can contain a high concentration of dissolved minerals. People use the water emitted at hot springs to fill relaxing mineral baths (**Fig. 19.15a**). The minerals in hot water feed microbes, so natural pools of geothermal water may become brightly colored—the

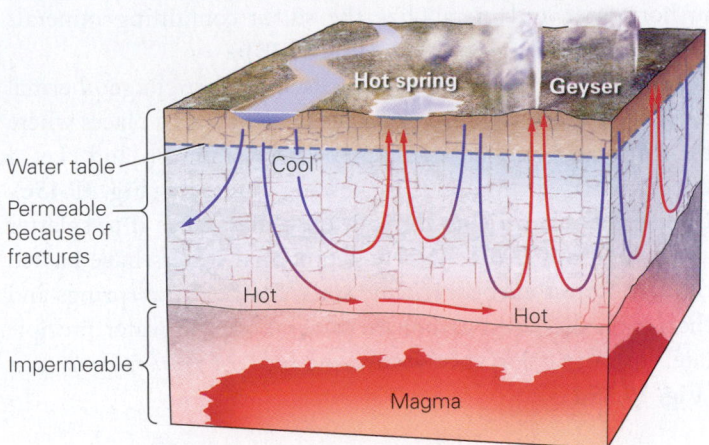

FIGURE 19.14 Geothermal springs form where very hot rock, or magma, at depth heats groundwater which then rises.

FIGURE 19.15 Various features of hot springs.

(a) Hot springs in Iceland, warmed by magma below, attract tourists from around the world.

(b) Colorful bacteria- and archaea-laden pools, Yellowstone National Park, Wyoming.

(c) Mudpots form where boiling water mixes with ash. The ash changes into clay and forms a muddy soup.

(d) Terraces of minerals precipitated at Mammoth Hot Springs, Yellowstone.

gaudy greens, blues, and oranges of these pools come from "thermophyllic" (heat-loving) bacteria and archaea that thrive in hot water and metabolize the sulfur-containing minerals dissolved in the groundwater (**Fig. 19.15b**).

Numerous distinctive geologic features form in geothermal regions as a result of the eruption of hot water. In places where the hot water rises into soils rich in volcanic ash and clay, a viscous slurry forms and fills bubbling mud pots (**Fig. 19.15c**). Bubbles of steam rising through the slurry cause it to splatter about in goopy drops. Where geothermal waters have passed through limestone bedrock and spill out of natural springs and then cool, dissolved carbonate minerals in the water precipitate, forming colorful travertine mounds, or *travertine terraces* (**Fig. 19.15d**).

FIGURE 19.16 Geysers form where steam erupts from the ground.

(a) A vent in Iceland begins to bubble.

Time

(b) Water spurts out, releasing pressure on super-hot water below.

(c) Water at depth transforms into vapor, which rises and pushes overlying water out of the vent.

Under special circumstances, in geothermal regions a fountain of steam and boiling hot water erupts every now and then from a vent in the ground (**Fig. 19.16**). An episodic fountain of hot, steamy water is a **geyser**—the name comes from the Icelandic *geysir*, a word that means gusher. To understand why a geyser erupts, we first need to picture the underground plumbing beneath one. Beneath a geyser lies a network of irregular fractures in very hot rock, one or more of which connects to the ground surface. Groundwater sinks and fills these fractures and absorbs heat from the rock. Since the boiling point of water, the temperature at which it vaporizes, increases with growing pressure, hot groundwater at depth can remain in liquid form even if its temperature has become greater than the boiling point of water at the Earth's surface (100°C or 212°F). When such "superheated" groundwater begins to rise through a conduit toward the surface, pressure in it decreases until eventually some of the water transforms into steam and expands. The resulting expansion causes water higher up in the fracture to spill out of the conduit at the ground surface. When this spill happens, the pressure deeper in the conduit, due the weight of overlying water, suddenly decreases. This sudden drop in pressure causes the superhot water at depth to turn into steam instantly, and this steam quickly rises, ejecting all the water and steam above it out of the conduit in a geyser eruption. Once the conduit empties, the eruption ceases, and the conduit fills once again with water that gradually heats up, starting the eruptive cycle all over again.

Hot springs are found in many localities around the world, such as at Hot Springs, Arkansas, where deep groundwater rises to the surface; in Yellowstone National Park, above a continental hot spot; around the Salton Sea in southern

(d) The Old Faithful geyser in Yellowstone erupts predictably.

California, and in the Geysers Geothermal Field of California, both places where rifting has triggered local volcanism; in Iceland, which has grown on top of an oceanic hot spot along the Mid-Atlantic Ridge; and in Rotorua, New Zealand, which lies in an active volcanic field above a subduction zone.

People live in some geothermal regions, though the areas have inherent natural hazards. In Rotorua, signs along the road warn "STEAM!," which can obscure visibility, and steam indeed spills out of holes in backyards and parking lots. But all this hot water does offer a benefit—in Rotorua, waters circulate through pipes to provide home heating. In geothermal areas worldwide, as we've seen, steam provides a relatively inexpensive means of generating electricity.

Take-Home Message

At hot springs, warm or even boiling water emerges at the ground surface. Some hot springs form where groundwater rises from warm rock several kilometers down; others form where igneous activity heats water near the surface. At a geyser, a fountain of steam and scalding water spurts from the ground episodically.

QUICK QUESTION: Why do geysers generally erupt episodically instead of continuously?

19.7 Groundwater Problems

Since prehistoric times, groundwater has been an important resource that people have relied on for drinking, irrigation, and industry. Groundwater feeds the lushness of desert oases in the Sahara, the amber grain in the North American high plains, and the growing cities of sunny arid regions. The proportion of groundwater that contributes to the water supply of a region depends on the local climate, the availability of other local supplies (such as lakes or rivers), local demand, and politics. While we tend to think of groundwater as being primarily consumed for drinking, on a global basis, agricultural and industrial usage actually accounts for most groundwater usage—roughly 80% to 95% of the groundwater used by society goes into crop irrigation or into various industrial applications. So, as once-empty land comes under cultivation and countries become increasingly industrialized, demands on the groundwater supply soar. Currently, groundwater provides only 20% to 30% of the water we use worldwide, but this percentage has been increasing as surface-water resources decrease. Problems inherent in groundwater management are important to society. In this section, we examine some of these problems.

Depletion of Groundwater Supplies

Is groundwater a renewable resource? In the context of geologic time, the answer is yes, for stages in the hydrologic cycle will eventually resupply depleted reserves, and new reserves will form as old ones disappear. But in a time frame of decades to millennia—the span of a human lifetime or a civilization—groundwater in many regions of the Earth must be viewed as a nonrenewable resource. Whether a particular groundwater reservoir is renewable or not depends on the balance between the rate of recharge and the rate of depletion. In the era before modern agriculture and industry and rapid population growth, the water table rose and sank with the season and would change depending on whether it was a particularly wet or dry year, but averaged over decades or centuries it would not change by much. Two changes have taken place since the dawn of the industrial era that have had a major impact on this balance.

First, the huge increase in human utilization of groundwater has made groundwater extraction a phenomenon of geologic significance. The *green revolution*, which involves using irrigation and fertilization to improve food production, has greatly increased the amount of groundwater used for agriculture, and farmers now plow areas once too dry for cultivation (**Fig. 19.17a, b**). Consider that it takes 10 liters (2.6 gallons)

FIGURE 19.17 Irrigation can turn the dry areas green.

(a) Irrigating a hay field in Utah consumes vast quantities of water.

(b) A view from an airplane shows irrigated crop circles in an otherwise dry landscape.

FIGURE 19.18 This map shows the combination of surface water and recharge, as measured in cubic meters per capita per year. It emphasizes where freshwater supplies are becoming a problem.

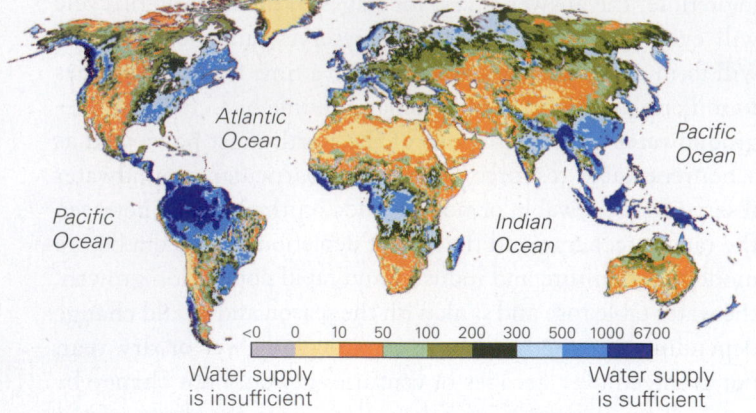

Atlantic
Ocean

Pacific
Ocean

Pacific
Ocean

Indian
Ocean

<0   0   10   50   100   200   300   500   1000 6700

Water supply
is insufficient

Water supply
is sufficient

of water to manufacture one sheet of paper. Industrialization, which has been increasing rapidly in developing countries, has also increased groundwater use. Finally, higher standards of living lead to an increase in the volume of water used per capita for household use, and since human population has grown almost exponentially since the beginning of the industrial revolution, the total consumption of groundwater for households has also dramatically increased. Because of increased demand, withdrawal of groundwater significantly exceeds recharge in many regions, and the total amount of freshwater entering a region, the sum of surface flow and recharge, is not sufficient (**Fig. 19.18**). According to some estimates, we are depleting over 20% of the world's aquifers, and in some places we use groundwater at a rate that is 50 times the rate of recharge.

Second, global climate change (see Chapter 23) is causing a decrease in the amount of recharge in key agricultural areas. As climate belts shift, the amount of water from rain and/or melting snow can't keep up with water removed by society. Since groundwater is underground and we can't see it, groundwater loss is, in effect, a "hidden disaster" in the making, one that policy makers and the public must learn to mitigate by changing practices of land use and/or by water recycling.

When we extract groundwater from wells at a rate faster than it can be resupplied by nature, a cone of depression forms locally around the well. Then the water table gradually becomes lower in a broad region. As a consequence, existing wells, springs, and rivers dry up (**Fig. 19.19a, b**), and to continue tapping into the water supply, we must drill progressively deeper. During the past decade, researchers have used data from GRACE satellites to study groundwater depletion. These satellites provide precise measurements of how the magnitude

of Earth's gravitational pull in a given region varies over time. In some regions, a very small but detectable decrease in gravitational pull reflects groundwater depletion because when air replaces water in subsurface pores, the mass of the near-surface realm becomes smaller. By calibrating the magnitude of gravity decrease to the amount of groundwater depletion, researchers can produce maps depicting how much the water table has dropped (**Fig. 19.19c, d**).

Notably, the water table can also drop when people divert surface water from the recharge area. Such a problem has developed in the Everglades of southern Florida, a huge swamp where, before the expansion of Miami and the development of agriculture, the water table lay at the ground surface (**Fig. 19.19e, f**). Diversion of water from the Everglades' recharge area into canals has significantly lowered the water table, causing parts of the Everglades to dry up.

Other Consequences of Groundwater Use

Saline Intrusion In many regions, fresh groundwater lies in a layer above saline (salty) water, because freshwater is less dense than saline water so it floats above the saline water. The boundary between fresh and saline water may be hundreds or even a few thousand meters below the surface of a large sedimentary basin, but it may lie at relatively shallow depth along the coast, where water enters an aquifer from the adjacent ocean (**Fig. 19.20a, b**). If people pump water out of a well too quickly, the boundary between the saline water and the fresh groundwater rises. And if this boundary rises above the base of the well, then the well will start to yield useless saline water. Geologists refer to this phenomenon as *saline intrusion*.

Pore Collapse and Land Subsidence When groundwater fills the pore space of a rock, it holds the grains of the rock or regolith apart, for water cannot be compressed. The extraction of water from a pore eliminates the support holding the grains apart, because the air that replaces the water can be compressed. As a result, the grains pack more closely together. Such *pore collapse* permanently decreases the porosity and permeability of a rock and thus lessens its value as an aquifer (**Fig. 19.20c, d**).

Pore collapse also decreases the volume of the aquifer, with the result that the ground above the aquifer sinks. Such *land subsidence* may cause fissures at the surface to develop and the ground to tilt (**Fig. 19.21a**). Buildings constructed over regions undergoing land subsidence may themselves tilt, or their foundations may crack. In the San Joaquin Valley of California, the land surface subsided by 9 m between 1925 and 1975, because water was removed to irrigate farm

FIGURE 19.19 Effects of human modification of the water table.

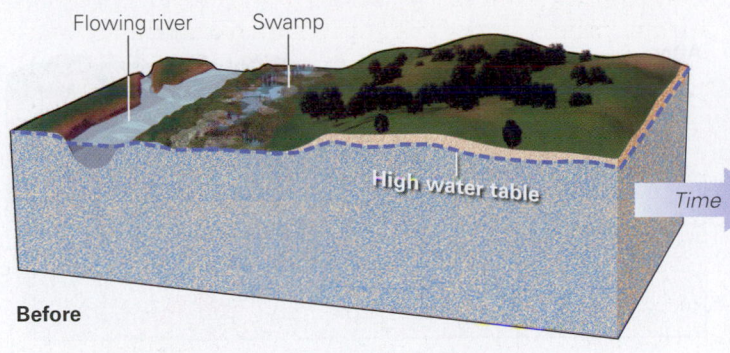

(a) Before humans start pumping groundwater, the water table is high. A swamp and permanent stream exist.

(b) Pumping for consumers in a nearby city causes the water table to sink in, so the swamp dries up.

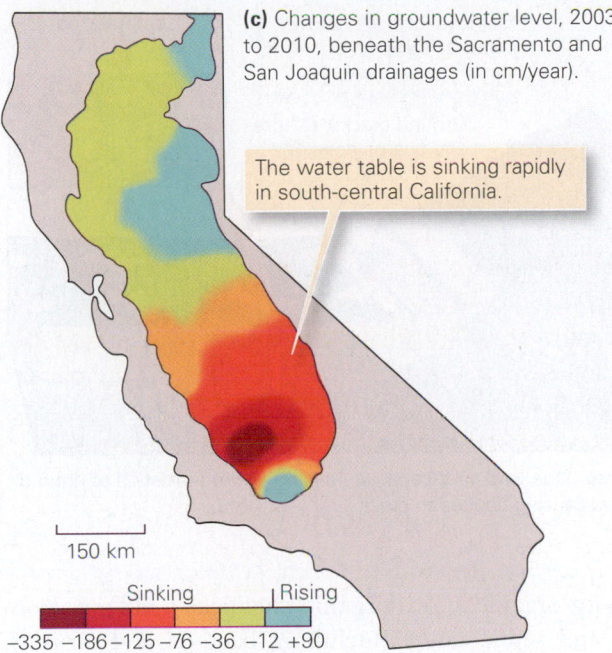

(c) Changes in groundwater level, 2003 to 2010, beneath the Sacramento and San Joaquin drainages (in cm/year).

The water table is sinking rapidly in south-central California.

150 km

Sinking | Rising
−335 −186 −125 −76 −36 −12 +90

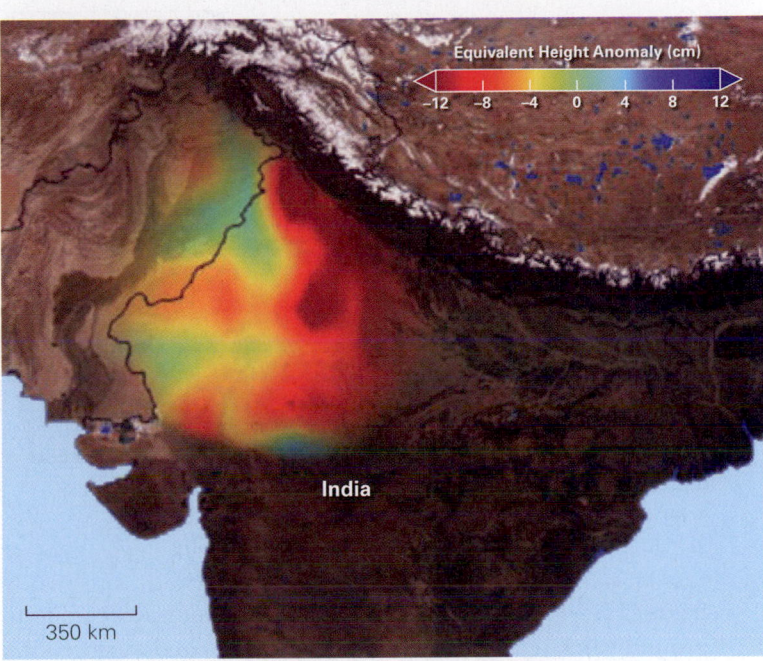

Equivalent Height Anomaly (cm)
−12 −8 −4 0 4 8 12

India

350 km

(d) Changes in groundwater level in northwest India, based on GRACE satellite data.

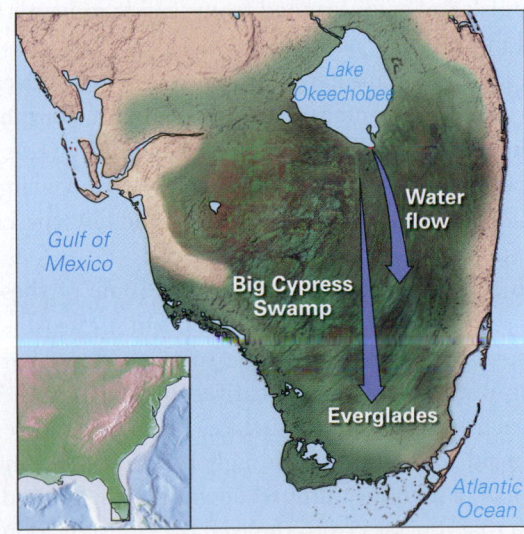

Lake Okeechobee

Water flow

Gulf of Mexico

Big Cypress Swamp

Everglades

Atlantic Ocean

(e) The Florida Everglades before the advent of urban growth and intensive agriculture.

~1700 C.E.   Today

Time

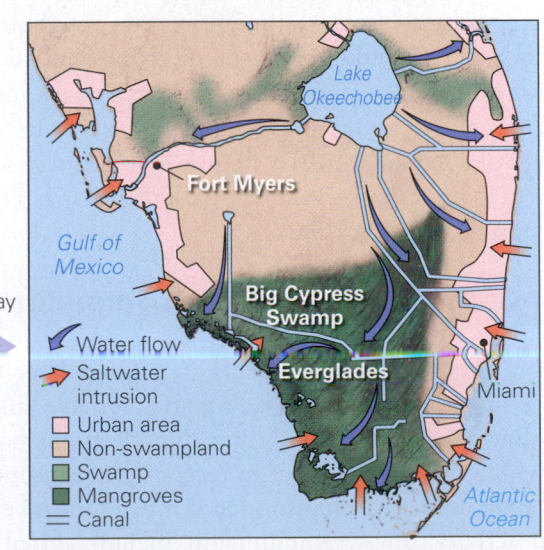

Lake Okeechobee

Fort Myers

Gulf of Mexico

Big Cypress Swamp

Everglades

Miami

Water flow
Saltwater intrusion
Urban area
Non-swampland
Swamp
Mangroves
Canal

Atlantic Ocean

(f) Channelization and urbanization have removed water from recharge areas, disrupting flow paths.

FIGURE 19.20 Some causes of groundwater problems.

Before

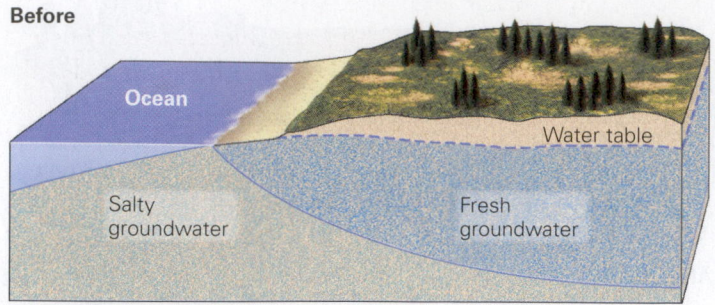

(a) Before pumping, fresh groundwater forms a lens below the ground.

After

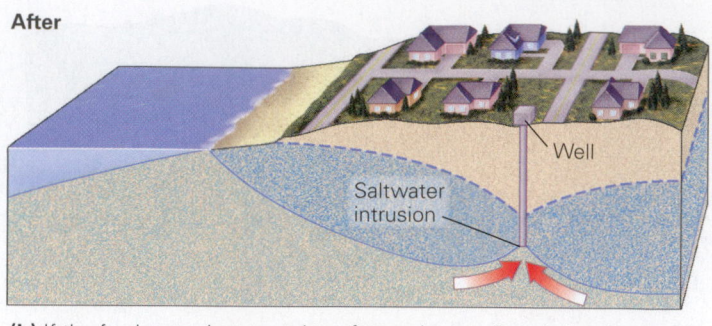

(b) If the freshwater is pumped too fast, saltwater from below is sucked up into the well. This is saltwater intrusion.

Before

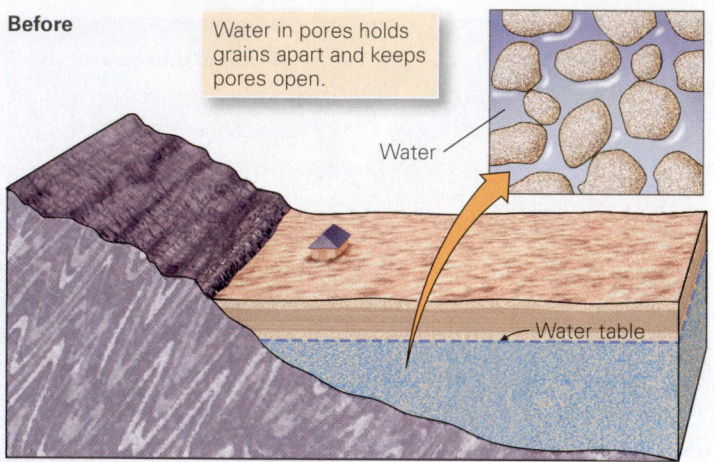

(c) When intensive irrigation removes groundwater, pore space in an aquifer collapses.

After

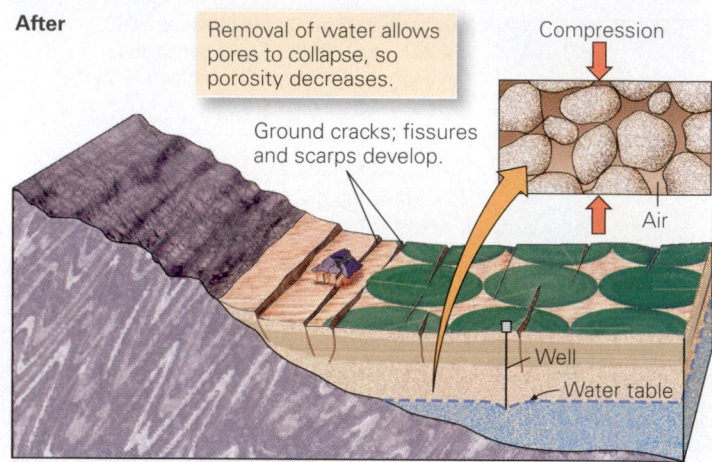

(d) As a result, the land surface sinks, leading to the formation of ground fissures and causing houses to crack.

fields (**Fig. 19.21b**). In coastal areas, land subsidence due to groundwater removal may even make the land surface sink below sea level. The flooding of Venice, Italy, for example, is due to land subsidence accompanying the withdrawal of groundwater beneath Venice (**Fig. 19.21c**).

To avoid such problems, communities have sought to prevent groundwater depletion either by directing surface water into recharge areas or by pumping surface water back into the ground. For example, some communities excavate to lower the land surface of a park and then configure storm sewers so that they drain onto its grassy surface. The park then acts as a catchment for storm water (**Fig. 19.22**).

Human-Caused Groundwater Contamination

As we've noted, some contaminants in groundwater occur naturally—sulfur, iron, calcium carbonate, methane, and salt can all be introduced to groundwater directly from the rock through which it is flowing. But in recent decades, contaminants have increasingly been introduced into aquifers because of human activity. These contaminants include agricultural waste (pesticides, fertilizers, and animal sewage); industrial waste (both organic and inorganic chemicals); effluent from landfills and septic tanks (including bacteria and viruses); petroleum products and other chemicals that do not dissolve in water (together referred to as nonaqueous-phase liquids); radioactive waste (from weapons manufacture, power plants, and hospitals); and acids leached from sulfide minerals in coal and metal mines.

The cloud of contaminated groundwater that moves away from the source of contamination is called a **contaminant plume** (**Fig. 19.23a, b**). These plumes flow in the same direction as groundwater. Note that the development of a large cone of depression around a big well can cause the flow direction of a nearby plume to change (**Fig. 19.23c, d**).

Where do contaminants come from? Some of these contaminants seep into the ground from leaks in surface or subsurface tanks; some infiltrate from the surface when downward-percolating water dissolves and carries chemicals with it; and some comes from spills on the surface. Finally, some are intentionally forced into aquifers through *injection wells*, meaning wells in which a liquid is pumped down into the ground under pressure so that it passes from the well back into the pore space of the rock or regolith. Some injection wells go deep enough

FIGURE 19.21 Consequences of groundwater removal and subsurface collapse.

(a) Ground fissures in Arizona.

The dates indicate ground elevation in the past.

(b) Sinking ground surface, California.

(c) Sinking and submergence continue to cause damage in Venice, Italy.

in 2014 when an organic chemical used to wash coal spilled from a storage tank onto the ground and then migrated underground to the nearby Elk River, which provides the water supply for the city of Charleston and surrounding regions. Three hundred thousand residents temporarily lost their water supply.

The best way to avoid **groundwater contamination** is to prevent contaminants from entering groundwater in the first place. This can be done by locating potential sources of contamination on impermeable bedrock so that they are isolated from aquifers. If such a site is not available, the storage area should be lined with a thick layer of clay, for the clay not only acts as an aquitard but it can hold on to contaminants or over an impermeable plastic sheet. Government agencies have studied various options for safely storing the containers containing contaminants. One option involves stockpiling the containers in tunnels cut into salt domes, because salt is impermeable. Another option involves placing waste containers in tunnels above the water table. For example, officials considered a proposal to store American nuclear waste in a network of tunnels 300 m beneath Yucca Mountain, Nevada, a hill that consists of dry, fairly impermeable tuff high above the water table.

In some cases, natural processes can clean up groundwater contamination. For example, chemicals may be absorbed by clay, oxygen in the water may oxidize the chemicals, and bacteria in the water may metabolize the chemicals, thereby turning them into harmless substances. If natural processes don't solve the problem, what can we do? First, environmental engineers drill test wells to determine which way and how fast the contaminant plume is flowing. Once they know the flow path, they can close wells in the path to prevent consumption

FIGURE 19.22 Sketch of an enhanced recharge catchment in a city.

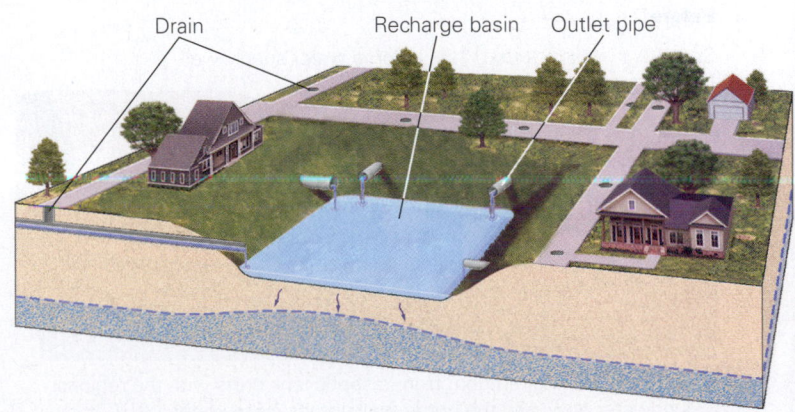

Drain   Recharge basin   Outlet pipe

that they push wastes into undrinkable saline water. The long-term consequences of this injection remain a subject of study.

Sadly, staggering quantities of contaminating liquids (trillions of gallons in the United States alone) enter the groundwater system every year. Contaminants can be carried with local or regional groundwater flow up to tens of kilometers from the source. In some cases, local groundwater flow carries contaminants into a nearby stream that may serve as a municipal water supply. Such a situation developed in West Virginia

of contaminated water. Engineers may also attempt to clean the groundwater by drilling a series of extraction wells to pump it out of the ground. If the contaminated water does not rise fast enough, engineers drill injection wells to force clean water or steam into the ground beneath the contaminant plume (**Fig. 19.24**). The injected fluids then push the contaminated water up into the extraction wells.

More recently, environmental engineers have begun exploring two techniques that may help remediate groundwater contamination. The first technique, *bioremediation*, involves microbes. To carry out bioremediation, environmental geologists inject oxygen and nutrients into a contaminated aquifer to foster growth of bacteria. The bacteria consume and break down contaminant molecules. The second involves the insertion of *permeable reactive barriers* underground. These are, in effect, subsurface walls of materials such as iron filings that react with contaminants chemically to transform the contaminants into safer, less-soluble materials. To produce a barrier, engineers dig a trench in the path of the flow, fill it with the reactive material, and bury it. When the plume flows through the barrier, the reactions take place. Needless to say, cleaning techniques are expensive and may be only partially effective.

Unwanted Effects of Rising Water Tables

We've seen the negative consequences of sinking water tables, but what happens when the water table rises? Is that necessarily good? Sometimes but not always. If the water table rises above

FIGURE 19.23 Contamination plumes in groundwater.

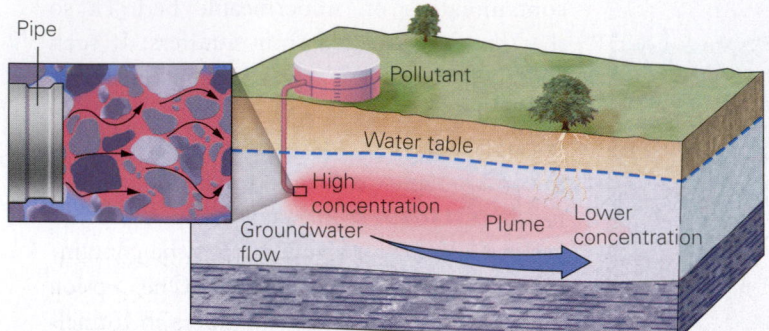

(a) A contaminant plume as seen in cross section. The darker the color, the greater the concentration of contaminant.

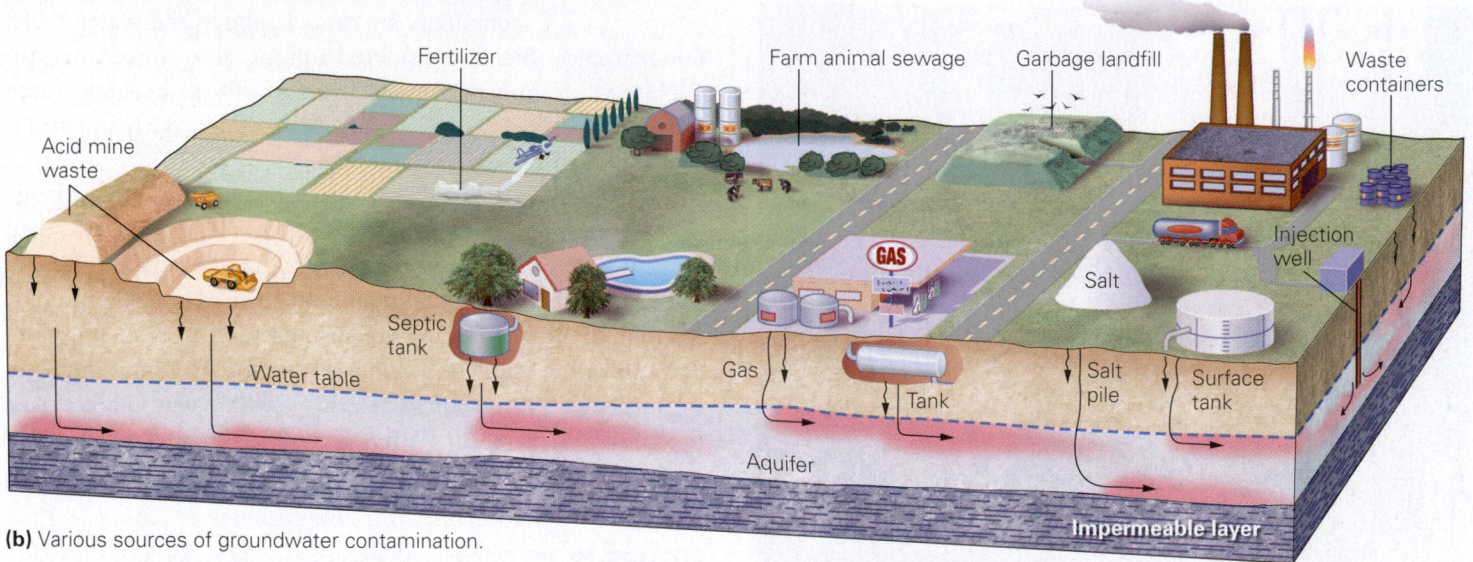

(b) Various sources of groundwater contamination.

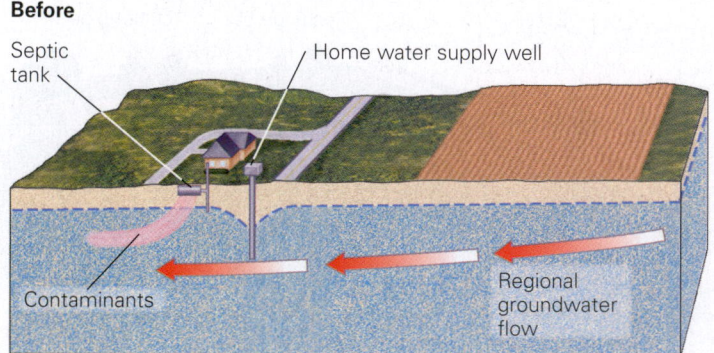

(c) Before pumping, effluent from a septic tank drifts with the regional groundwater flow, and the home well pumps clean water.

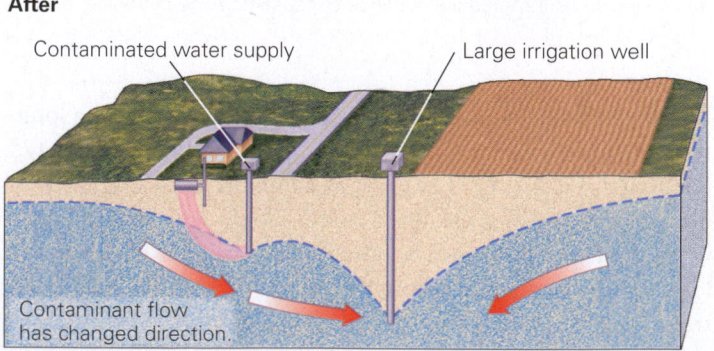

(d) After pumping by a nearby irrigation well, effluent flows into the home well in response to the new local slope of the water table.

FIGURE 19.24 Steam injected beneath the contamination drives the contaminated water upward in the aquifer, where pumping wells remove it.

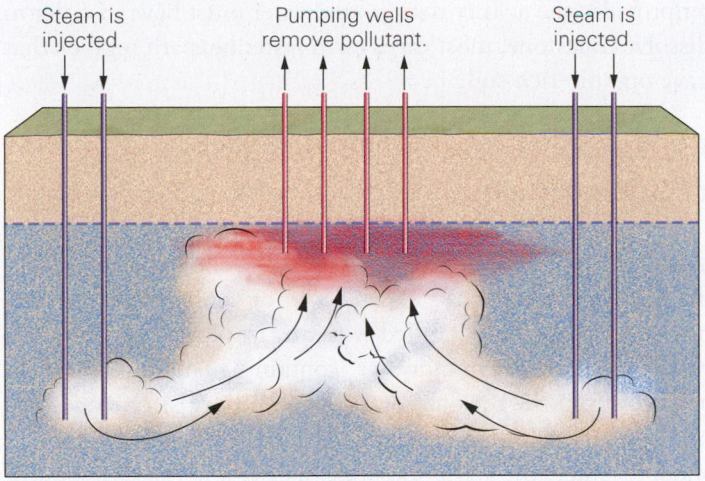

Steam is injected.

Pumping wells remove pollutant.

Steam is injected.

Take-Home Message

Groundwater usage can cause problems, and growing evidence shows that in many locations, globally, people are extracting groundwater at rates far in excess of natural recharge. Too much pumping lowers the water table, causing land subsidence and/or saltwater intrusion. Contamination can ruin a groundwater supply. Remediation of a contaminant plume can be extremely expensive.

QUICK QUESTION: What is bioremediation?

the level of a house's basement, water seeps through the foundation and floods the basement floor. As we noted in Chapter 16, catastrophic damage may occur when a rising water table weakens the base of a hillslope or an underground failure surface triggers landslides and slumps (**Fig. 19.25**).

FIGURE 19.25 When the water table rises, material above a weak sliding surface begins to slump, and a landslide may result.

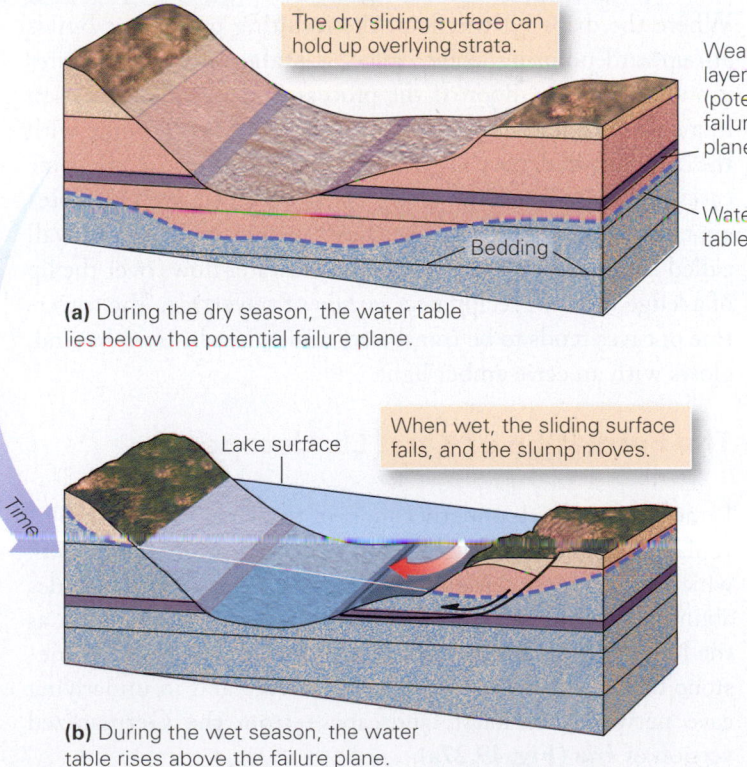

The dry sliding surface can hold up overlying strata.

Weak layer (potential failure plane)

Water table

Bedding

(a) During the dry season, the water table lies below the potential failure plane.

Lake surface

When wet, the sliding surface fails, and the slump moves.

Time

(b) During the wet season, the water table rises above the failure plane.

19.8 Caves and Karst

The Development of Caves

In 1799, as legend has it, a hunter by the name of Houchins was tracking a bear through the wooded hills of Kentucky when the bear suddenly disappeared. Baffled, Houchins plunged through the brambles trying to sight his prey. Suddenly he felt a draft of surprisingly cool air flowing downslope from uphill. Now curious, Houchins climbed up the hill and found a dark portal into the hillslope beneath a ledge of rocks. Bear tracks were all around—was the creature inside? He returned later with a lantern and cautiously stepped into the passageway. After walking a short distance, he found himself in a large, underground room. Houchins was the first settler of European descent to enter Mammoth Cave. This is an immense maze of natural tunnels and large subterranean openings called a *cave network*. A walk through the entire network would extend for 630 km!

Most large cave networks, such as Mammoth Cave, develop in limestone bedrock because limestone dissolves relatively easily in acidic groundwater. Generally, such groundwater is dilute carbonic acid (H_2CO_3), which forms when water absorbs carbon dioxide (CO_2). The CO_2 in groundwater comes from two sources—a small amount dissolves in rain as it falls through the sky, but most dissolves when water percolates down through organic-rich soil on its way down to the water table. Carbonic acid is corrosive, in that when it comes in contact with the calcite ($CaCO_3$) in limestone, it reacts to produce HCO_3^{1-} and Ca^{2+} ions, which then dissolve. The reaction releases CO_2 back into the air—you've seen this reaction if you used the acid test to identify calcite when studying minerals in Chapter 5.

Did you ever wonder...

why huge underground caverns form?

In recent years, geologists have discovered that about 5% of the limestone caves around the world formed due to reactions with sulfuric-acid-bearing water—Carlsbad Caverns in New Mexico serves as an example. Such caves form where limestone overlies strata containing oil, because microbes convert the sulfur in the oil to hydrogen sulfide (H_2S) gas. This gas rises and reacts with oxygen and water to produce sulfuric acid, which in turn eats into limestone and reacts to produce gypsum and CO_2 gas.

Geologists debate about the depth at which limestone cave networks form. Some limestone dissolves above the water table, particularly along joints, which act as conduits for water to flow into the subsurface, and some limestone may dissolve deep below the water table. But it appears that most cave growth, or *speleogenesis*, takes place in limestone that lies just below the water table. Here groundwater acidity remains high, the mixture of groundwater and newly added rainwater is undersaturated (meaning it has the capacity to dissolve more ions), and groundwater flow is fastest. If the water table goes up or down, the depth at which speleogenesis takes place goes up or down, so a region may hold several levels of caves.

The Character of Cave Networks

Cave networks include rooms, or *chambers*, which are large, open spaces sometimes with cathedral-like ceilings, and *passages*, which are tunnel- or slot-shaped corridors (see **Geology at a Glance**, pp. 724–725). Some chambers may host underground lakes, and some passages may serve as conduits for underground streams. The shape of the cave network reflects variations in permeability and in the composition of the rock from which the caves formed. Rooms develop where the limestone was most soluble and where groundwater flow was fastest. Thus, in a sequence of strata, caves develop preferentially in the more soluble limestone beds. Passages in cave networks typically follow pre-existing joints, which provide secondary porosity along which groundwater can flow faster (**Fig. 19.26a**). Because joints commonly occur in orthogonal systems (consisting of two sets of joints oriented at right angles to each other; see Chapter 11), cave passages may form a grid.

Why do extensive cave networks, with large rooms and abundant passages, develop in some locations but not in others? There are several reasons. First, most caves form in limestone, so without a thick layer of limestone in the subsurface, extensive networks can't form. Second, the dissolution that forms caves occurs primarily in freshwater at the water table, so unless the water table lies above sea level and below the land surface, extensive networks can't form. Third, for caves to develop, sufficient liquid water must be present. Thus extensive networks form preferentially in temperate or tropical regions, drenched by rain—they do not develop in polar regions, where ice covers the ground and subsurface water remains permanently frozen, nor do they develop in desert regions, where water is scarce. Finally, since percolation through organic matter provides the acidity that groundwater must have in order to dissolve limestone, most caves form faster beneath regions that have organic-rich soil.

Precipitation and the Formation of Speleothems

When the water table drops below the level of a cave that has developed, speleogenesis slows down or ceases, and the cave becomes an open space filled with air. In places where downward-percolating groundwater containing dissolved calcite emerges from the rock above the cave and drips from the ceiling, the surface of the cave gradually changes. As soon as this water re-enters the air, it evaporates a little and releases some of its dissolved carbon dioxide. As a result, calcite precipitates out of the water and produces a type of travertine called *dripstone*. The various intricately shaped formations that grow in caves by the accumulation of dripstone are called **speleothems**.

Cave explorers (spelunkers) and geologists have developed a detailed nomenclature for different kinds of speleothems (**Fig. 19.26b**). Where water drips from the ceiling of the cave, initially calcite precipitates around the outside of the drip, forming a delicate, hollow stalactite called a *soda straw*. But eventually, the soda straw fills up and water migrates down the margin of the cone to form a more massive, icicle-like cone called a **stalactite**; stalactites grow down from the ceiling. Where the drips hit the floor, the resulting precipitate builds an upward-pointing cone called a **stalagmite**; stalagmites grow up from the floor. If the process of dripstone formation in a cave continues long enough, a stalagmite will merge with the overlying stalactite to create a *column* of travertine. In some cases, groundwater flows along the surface of a wall and precipitates to produce cloth-like sheets of travertine on the wall called *flowstone* (**Fig. 19.26c**). If groundwater flows over the lip of a ledge, it may precipitate a *curtain* of travertine. The travertine of caves tends to be translucent and, when lit from behind, glows with an eerie amber light.

The Formation of Karst Landscapes

Limestone bedrock underlies most of the Kras Plateau in Slovenia, along the east coast of the Adriatic Sea. The name *kras*, which means rocky ground, is apt because this region includes abundant rock exposures. Geologists refer to regions such as the Kras Plateau, where surface landforms develop when limestone bedrock dissolves both at the surface and in underlying cave networks, as karst landscapes—from the Germanized version of *kras* (**Fig. 19.27a**).

FIGURE 19.26 Development of karst and dripstone.

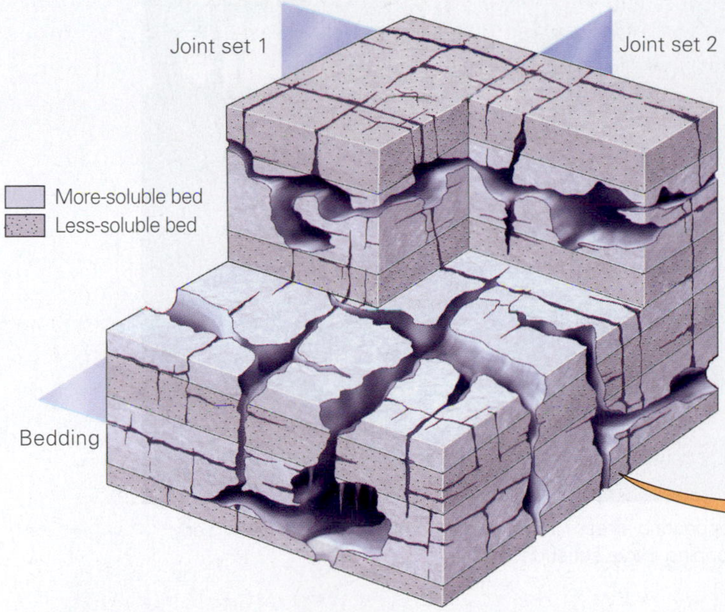

Joint set 1   Joint set 2

More-soluble bed
Less-soluble bed

Bedding

(a) Joints act as conduits for water in cave networks. Caves and passageways follow joints and preferentially form in more-soluble beds.

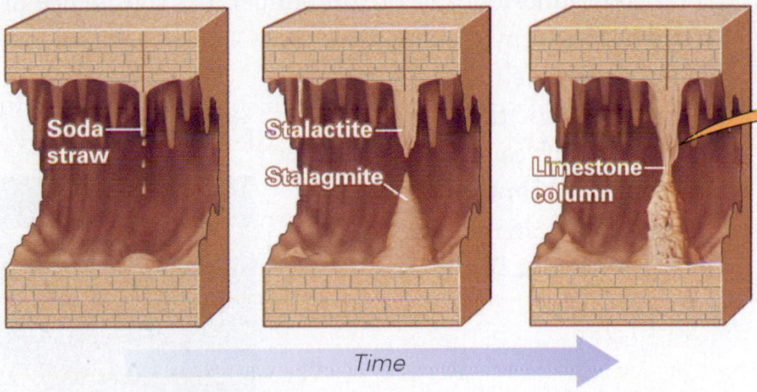

Soda straw

Stalactite
Stalagmite

Limestone column

Time

(b) The evolution of a soda straw stalactite into a limestone column.

Squeezing through a joint-controlled passage, Utah

Stalactites and stalagmites in a cave in Italy

1m

Karst landscapes typically display a number of distinct landforms. Perhaps the most widespread are *sinkholes*, which as we've seen are circular depressions that form either when the ground collapses into an underground cave below or when surface bedrock dissolves in acidic water on the floor of a bog or pond. Not all of the caves or passageways beneath a karst landscape have collapsed, and this situation leads to unusual drainage patterns. Specifically, where surface streams intersect cracks (joints) or holes that link to caverns or passageways below, the water cascades downward into the subsurface and disappears (**Fig. 19.27b**). Such **disappearing streams** may flow through passageways underground and re-emerge from a cave entrance downstream. In cases where the ground collapses over a long, joint-controlled passage, sinkholes may be

(c) Flowstone on the wall of a cave in Vietnam.

FIGURE 19.27 Features of karst landscapes.

Sinkholes of the Kras Plateau.

(a) Karst terrains typically have a rough, rocky surface.

(b) A small disappearing stream in the Hudson Valley region of New York; the water is dropping into a subsurface cave.

elongate and canyon-like. Remnants of cave roofs remain as *natural bridges*. Ridges or walls between adjacent sinkholes tend to be steep-sided, for they were originally joint controlled. Over time, the walls erode, leaving only jagged, isolated limestone *spires*—a karst landscape dominated by such spires is called *tower karst*. The surreal collection of pinnacles constituting the tower karst landscape in the Guilin region of China have inspired generations of artists who portray them in scroll paintings (**Fig. 19.28**).

Karst landscapes typically take a long time to form. Their development involves a series of stages (**Fig. 19.29**):

- *The establishment of a water table in limestone:* The story of a karst landscape begins after the formation of a thick interval of limestone, as a region undergoes subsidence in a depositional environment where limestone forms. If, later in geologic time, erosion and uplift cause exhumation of the limestone, a water table can develop in the limestone below the ground surface.

FIGURE 19.28 Features of karst landscapes.

The landscape is treeless today, a consequence of industrialization policies in the 1950s.

Chinese artists painted scrolls depicting forested towers of karst.

FIGURE 19.29 The progressive formation of caves and a karst landscape.

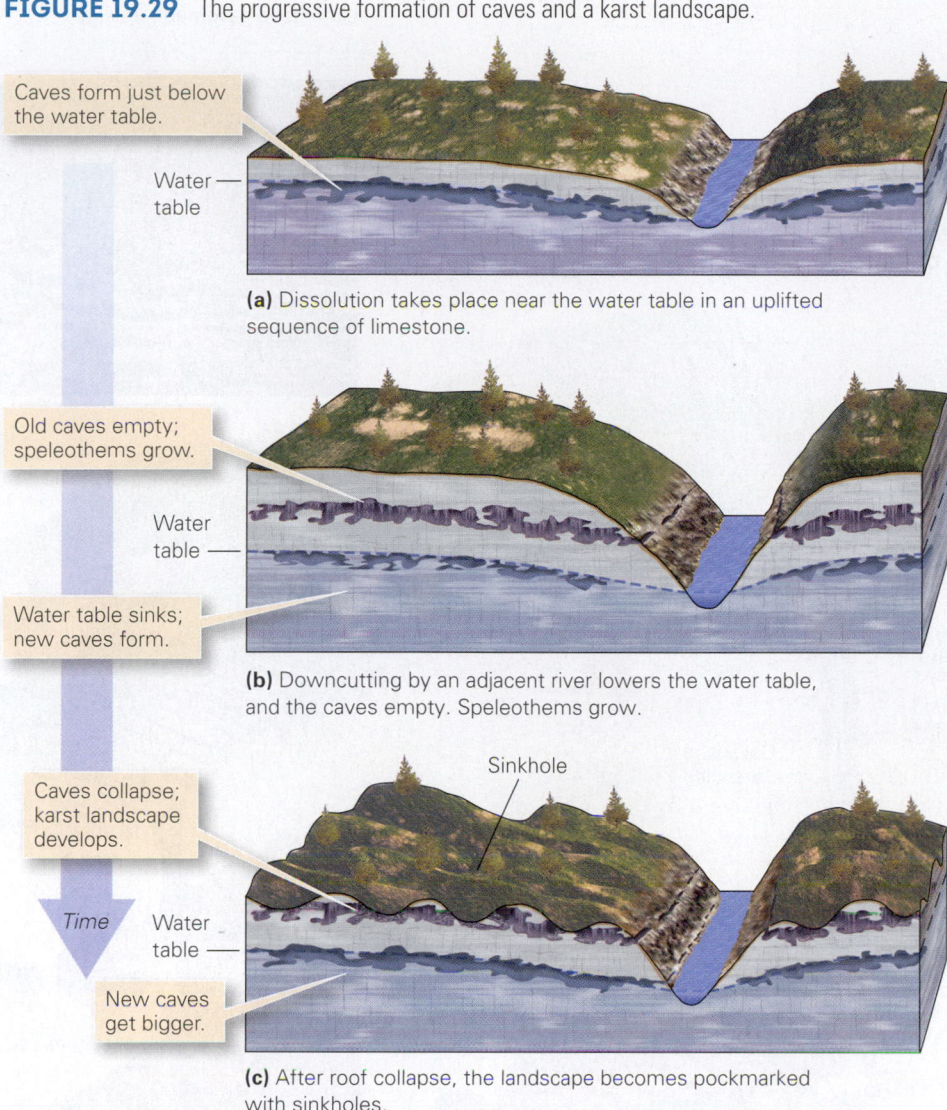

Caves form just below the water table.

Water table

(a) Dissolution takes place near the water table in an uplifted sequence of limestone.

Old caves empty; speleothems grow.

Water table

Water table sinks; new caves form.

(b) Downcutting by an adjacent river lowers the water table, and the caves empty. Speleothems grow.

Sinkhole

Caves collapse; karst landscape develops.

Time

Water table

New caves get bigger.

(c) After roof collapse, the landscape becomes pockmarked with sinkholes.

- *The formation of a cave network:* Once the water table has been established, dissolution begins and a cave network develops.
- *A drop in the water table:* If the water table later becomes lower, either because of a decrease in rainfall or because nearby rivers cut down through the landscape and drain the region, newly formed caves dry out. Downward-percolating groundwater emerges from the roofs of the caves; dripstone and flowstone precipitate.
- *Roof collapse:* If rocks fall off the roof of a cave for a long time, the roof eventually collapses. Such collapse creates sinkholes and troughs, leaving behind hills, ridges, and natural bridges of limestone.

Life in Caves

Despite their lack of light, caves are not sterile, lifeless environments. Caves that are open to the air provide a refuge for bats as well as for various insects and spiders. Similarly, fish and crustaceans enter caves where streams flow in or out. Species living in caves have developed some unusual characteristics. For example, some of the fish species that evolved in caves lost their pigment and, in some cases, their eyes (**Fig. 19.30a**). Explorers

FIGURE 19.30 Unusual organisms have evolved in the darkness of caves.

(a) A blind fish—there's no need for eyes in a cave.

(b) Gobs of bacteria drip from the ceiling of a cave to form snottites.

Caves and Karst Landscapes

Limestone is soluble in acidic water. Much of the water that falls to the ground as rain or seeps through the ground as groundwater tends to be acidic, so in regions of the Earth where bedrock consists of limestone, we find signs of dissolution. Underground openings that develop by dissolution are called caves or caverns. Some of these

Limestone pavement, Ireland

Disappearing stream

Sinkhole

Collapsed breccia

Dissolved joint

Stalagmite

Stalactite

Soda straw

Flowstone

Cavern

Stalactite

Limestone column

Underground stream

Underground pool

Corridor

Sinkholes

Underground pool, Mexico

Emerging spring

Natural Bridge, Virginia

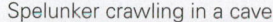

Spelunker crawling in a cave

may be large, open rooms, whereas others are long, narrow passages. Underground lakes and streams may cover the floor. A cave's location depends on the orientation of bedding and joints, for these features localize the flow of groundwater.

In many locations, groundwater drips from the ceiling of a cave or flows along its walls. As the water evaporates and loses its acidity, new calcite precipitates. Over time, this calcite builds into cave formations, or speleothems, such as stalactites, stalagmites, columns, and flowstone. Distinctive landscapes, called karst landscapes, develop at the Earth's surface over limestone bedrock that has undergone dissolution.

have discovered caves in Mexico in which warm, mineral-rich groundwater currently flows. Colonies of bacteria metabolize sulfur-containing minerals in this water and create thick mats of living ooze in the complete darkness of the cave. Long gobs of these bacteria slowly drip from the ceiling. Because of the mucus-like texture of these drips, they have come to be known as "snottites" (**Fig. 19.30b**).

Take-Home Message

Reaction with natural acids dissolves limestone underground to form caverns. Most dissolution takes place near the water table. If the water table sinks, relative to the cave, dripping water in caves can produce speleothems. Collapse of a cavern network produces karst terrain.

QUICK QUESTION: Can organisms live in the pitch black of caves? If so, how?

CHAPTER SUMMARY

- During the hydrologic cycle, water infiltrates the ground and fills the pores and cracks in rock and sediment. This subsurface water is called groundwater. The amount of open space in rock or sediment determines its porosity; the degree to which pores are interconnected so that water can flow through defines its permeability.

- Geologists classify rock and sediment according to their permeability. Aquifers are relatively porous and permeable, and aquitards are relatively impermeable.

- The water table is the surface in the ground above which pores contain some air and below which pores are filled with water. The shape of a water table is a subdued imitation of the shape of the overlying land surface.

- Groundwater flows wherever the water table has a hydraulic gradient, meaning it moves from where it's under more pressure to where it's under less pressure. It moves slowly from recharge areas to discharge areas. Darcy's law shows that flow rate depends on permeability and on the hydraulic gradient.

- Groundwater contains dissolved ions. These ions may come out of solution to form the cement or veins.

- Groundwater can be obtained at a spring, where groundwater exits the ground on its own. There are many geologic configurations that lead to spring formation.

- An ordinary well penetrates below the water table, but in an artesian well, water rises on its own. Pumping water out of a well too fast yields a cone of depression.

- Hot springs and geysers release hot water to the Earth's surface. This water may have been heated by passing deep in the crust or by the proximity of a magma chamber or of recently formed and still hot igneous rock.

- Groundwater is a precious resource, used for municipal water supplies, industry, and agriculture. In recent years, some regions have lost their groundwater supply because of overuse or contamination.

- When limestone dissolves just below the water table, underground caves are the result. Soluble beds and joints determine the location and orientation of caves. If the water table drops, caves empty out. Limestone precipitates out of water dripping from cave roofs and creates speleothems, such as stalagmites and stalactites.

- Regions where abundant caves have collapsed to form sinkholes are karst landscapes. These terrains contain sinkholes, natural bridges, and disappearing streams.

GUIDE TERMS

aquifer (p. 699)
aquitard (p. 699)
artesian spring (p. 708)
artesian well (p. 709)
capillary fringe (p. 701)
cone of depression (p. 708)

contaminant plume (p. 716)
Darcy's law (p. 706)
disappearing stream (p. 721)
discharge area (p. 704)
geothermal region (p. 711)
geyser (p. 712)

groundwater (p. 695)
groundwater contamination (p. 717)
hard water (p. 706)
hot spring (p. 710)
hydraulic gradient (p. 706)

hydraulic head (p. 704)
oasis (p. 708)
perched water table (p. 703)
permeability (p. 698)
pore (p. 697)
porosity (p. 697)

recharge area (p. 704)   soil moisture (p. 701)   spring (p. 705)   stalagmite (p. 720)

sinkhole (p. 695)   speleothem (p. 720)   stalactite (p. 720)   water table (p. 701)

REVIEW QUESTIONS

1. What is groundwater, and where does it reside on Earth?

2. How do porosity and permeability differ? Give examples of substances with high porosity but low permeability.

3. What is a water table, and what factors affect the level of the water table? What factors affect the flow direction of the water below the water table?

4. How does the rate of groundwater flow compare with that of moving ocean water or river currents?

5. What does Darcy's law tell us about rates of discharge?

6. How does the chemical composition of groundwater change with time? Why is "hard water" hard, and saline water saline?

7. How does excessive pumping affect the local water table?

8. Why do natural springs form?

9. How is an artesian well different from an ordinary well?

10. Explain why hot springs form and what makes a geyser erupt.

11. Is groundwater a renewable or nonrenewable resource? Explain your answer.

12. Describe some of the ways in which human activities adversely affect the water table.

13. What are some sources of groundwater contamination? How can it be prevented?

14. Describe the process leading to the formation of caves and the speleothems within caves.

15. Describe the various features of a karst landscape, and explain how they evolve.

ON FURTHER THOUGHT

16. The population of Desert Paradise (DP; a fictitious town in the southwestern United States) has been doubling every seven years. Most of the new inhabitants are "snowbirds," people escaping the cold winters of more northerly latitudes. There are no permanent streams or lakes anywhere near DP. In fact, the only standing water in the town occurs in the ponds of the many golf courses that have been built recently. The water in these ponds needs to be replenished almost constantly, for without supplementing it, the water seeps into the ground quickly and the ponds dry up. The golf courses and yards of the suburban-style developments of DP all have lawns of green grass. DP has been growing on a flat, sediment-filled basin between two small mountain ranges. Much of the water supply of DP comes from wells. What do you predict will happen to the water table of the area in coming years, and how might the land surface change as a consequence? Is there a policy that you might suggest to the residents of DP that could slow the process of change?

17. You are part of a cave-exploration team that is trying to map a cave network in a temperate region of flat-lying limestone beds that underlie a plateau. A set of northwest-southeast-trending systematic joints cuts the limestone beds. A river cuts through the region, and the entrance to the cave is along the valley wall. The surface of the river lies about 400 m below the surface of the plateau. What do you predict will be the trend of tunnels in the cave network, and how far below the surface of the plateau do you think that you can explore the cave network without scuba tanks?

smartwork.wwnorton.com

GEOTOURS

This chapter's Smartwork features:

- Labeling activity on the water table.
- Reading comprehension questions on biogeochemical cycles.
- Art exercise on understanding aquifers.

This chapter's GeoTour exercise (P) features:

- Irrigation in the Saudi Desert
- Surface and groundwater flow in the Everglades
- Hot springs in Yellowstone National Park
- Karst features around the world

Towering clouds above hint at stormy weather below . . . and a rather bumpy airplane ride. The gaseous outer layer of the Earth System, our atmosphere, is a dynamic place that makes this planet habitable.

CHAPTER 20

An Envelope of Gas: Earth's Atmosphere and Climate

Who has seen the wind?
Neither you nor I:
But when the trees bow down their heads,
The wind is passing by.

—*Christina Rossetti (British poet, 1830–1894)*

LEARNING OBJECTIVES

By the end of this chapter, you should understand . . .

- how the Earth's atmosphere has evolved over time and where its gases come from.
- that the atmosphere contains layers and that temperature and pressure vary with elevation.
- why the atmosphere circulates regionally and what causes prevailing winds.
- what causes fronts and clouds to form and how they relate to weather.
- why thunderstorms, tornadoes, and hurricanes originate and how they cause damage.
- the difference between weather and climate, and why climate varies with location.

20.1 Introduction

On March 1, 1999, Bertrand Piccard, a Swiss psychiatrist, and Brian Jones, a British balloon instructor, silently rose from a launch site in Switzerland aboard the *Breitling Orbiter 3* (**Fig. 20.1**), an airtight gondola suspended from a giant silvery helium-filled balloon. Their gondola provided life support—it contained a heater, bottled oxygen, food and water, and solar-powered instruments for navigation and communication, and it could even float if they had to ditch at sea—and the balloon provided lift. Nineteen days later, more than two centuries after the first documented manned balloon flight, they became the first people to circle the globe nonstop by balloon.

The voyage of the *Breitling Orbiter 3* could take place only because an **atmosphere**, a layer consisting of a unique mixture of gases called **air**, surrounds the Earth. Any volume of material placed in the atmosphere feels a *buoyancy force*, and if the volume is less dense than the surrounding air, then the volume will rise because air is weak enough to flow out of its way. Note that if the Earth, like the Moon, were surrounded by a vacuum, balloon travel would be impossible, because no buoyancy force exists in a vacuum. The *Breitling Orbiter 3* was carried aloft by helium, whose density is about 15 times less than that of air at sea level, but most recreational balloons use hot air, which is less dense than cool air because gases expand when warmed.

The complexity of Piccard and Jones's equipment provides insight into the character and behavior of the atmosphere. Balloonists control their vertical movements by changing either the buoyancy of the balloon or the weight of the payload, but they cannot directly control their horizontal motions. They can travel horizontally only because the atmosphere circulates, resulting in the flow of air that we know as the *wind*—balloons float with the wind, like sticks float with the current in a stream. To reach their destination before running out of supplies, long-distance balloonists must change elevation to find wind flowing rapidly in the correct direction—this can be challenging because atmospheric circulation is complex. To stay safe, Piccard and Jones had to pay close attention to the **weather**, the overall physical conditions (temperature, pressure, moisture content, wind velocity, and wind direction) of the atmosphere at any given time and location. They had to avoid **storms**, episodes of extreme and/or turbulent wind, accompanied by rain, ice, snow, or lightning. To stay alive, Piccard and Jones had to monitor conditions inside their gondola, for conditions of pressure and temperature vary with elevation.

FIGURE 20.1 The balloon and gondola used by Piccard and Jones during their successful attempt to circle the globe in March 1999.

Despite all the challenges they faced, Piccard and Jones's journey went quite smoothly. On leaving their launching point in the Swiss Alps, they drifted southwest until, after crossing the Mediterranean, they entered strong, steady winds flowing from west to east. These winds carried the balloon entirely around the planet to touch down in Egypt on March 21.

In this chapter, we explore the envelope of air—the atmosphere—through which Piccard and Jones traveled. We begin by learning where the gases came from and how the atmosphere evolved in the context of the Earth System. Then we look at the structure of the atmosphere and the global-scale and local-scale circulation of its lowermost layer. This circulation controls the weather and leads to the growth of storms. We conclude by introducing the climate, the average of weather conditions for a region over many years.

20.2 The Formation of the Atmosphere

The First and Second Atmospheres

When the Earth formed 4.54 billion years ago (Ga), it was initially surrounded by gas molecules gravitationally attracted to its surface from the protoplanetary disk out of which our planet had grown (see Chapter 1). This *first atmosphere*, which consisted mostly of hydrogen and helium, with traces of other gases, survived only a short time. Heat from the Sun caused lightweight atoms to move so rapidly that they eventually achieved escape velocity and zoomed into space, and atoms that didn't escape on their own were carried away by intense solar winds that reached the Earth's surface.

As the Earth's primary atmosphere was disappearing, volcanic activity yielded new gases. Once the Earth had differentiated and had developed a magnetic field capable of deflecting the solar wind, these gases began to accumulate to form the *second atmosphere*. Where do volcanic gases come from? Large quantities of volatile elements reside inside the Earth, bonded to solid minerals. During melting, the volatile elements separate from minerals and dissolve in magma. Near the Earth's surface, where pressure decreases, the gases come out of solution and form bubbles that burst out of lava and into the air. Volcanic gas consists of 70% to 90% water (H_2O), with smaller amounts of carbon dioxide (CO_2) and sulfur dioxide (SO_2), along with traces of other gases including nitrogen (N_2) and ammonia (NH_3). So the second atmosphere consisted of these gases plus other gases, such as methane (CH_4) and carbon monoxide (CO), some of which were brought to Earth by comets.

The Earth's second atmosphere evolved over geologic time. When the Earth cooled sufficiently for water to condense, rain filled oceans, lakes, and streams, or sank into the shallow crust to become groundwater. Transfer of water from a gaseous state in the atmosphere to a liquid state on or near the surface caused the proportion of water in the atmosphere to decrease radically. As we noted in Chapter 13, the timing of the first liquid oceans remains a subject of debate—growing evidence suggested that ephemeral oceans had formed as early as 4.4 Ga, but permanent oceans appear to have existed only since 3.85 Ga.

The accumulation of liquid water on or near the surface also led to a radical drop in the concentration of CO_2 in the atmosphere for two reasons. First, CO_2 dissolves in water to form carbonate ions, which in turn can combine with calcium ions to form solid carbonate minerals. Second, CO_2 reacts with certain silicate rocks during chemical weathering, in the presence of water, to produce carbonate ions that also can become incorporated in solid carbonate minerals. Formation of carbonate minerals effectively traps CO_2 in rock and keeps it out of the atmosphere.

With removal of large quantities of H_2O and CO_2 from the air, what was left? The second atmosphere also initially contained NH_3. Ultraviolet radiation from the Sun split molecules of NH_3 into nitrogen and hydrogen atoms. The lightweight hydrogen atoms eventually escaped into space, but the nitrogen atoms combined to form N_2 molecules. Molecular nitrogen is an inert (stable) gas that can remain in the air for a long time, and eventually the proportion of nitrogen in the atmosphere increased.

Notably, the small amount of CO_2 that remained in the air after the oceans formed serves an essential role in making and keeping the Earth habitable. CO_2 is a **greenhouse gas**, meaning that it lets solar radiation (light) from space pass through but prevents infrared radiation (heat) rising from the Earth's surface to escape back into space. Like other greenhouse gases (such as CH_4), CO_2 effectively traps heat in the atmosphere. The tiny amount of CO_2 and other greenhouse gases that remain in the air regulate atmospheric temperature and have kept the surface from becoming cold enough to freeze over, except perhaps for a relatively brief time in the Proterozoic.

> **Did you ever wonder . . .**
>
> if our atmosphere has always been breathable?

What would have happened if the Earth's surface had been too hot for liquid water to accumulate on it? Our atmosphere would have had a totally different history—CO_2 would not have been removed from the atmosphere, so instead of having about 0.04% CO_2, the Earth's atmosphere today would resemble the atmosphere of Venus, which currently contains 96.5% CO_2. Because of its high concentration of CO_2, Venus's atmosphere is so hot that lead can melt at the planet's surface.

FIGURE 20.2 Stages in the evolution of Earth's atmosphere over time (not to scale).

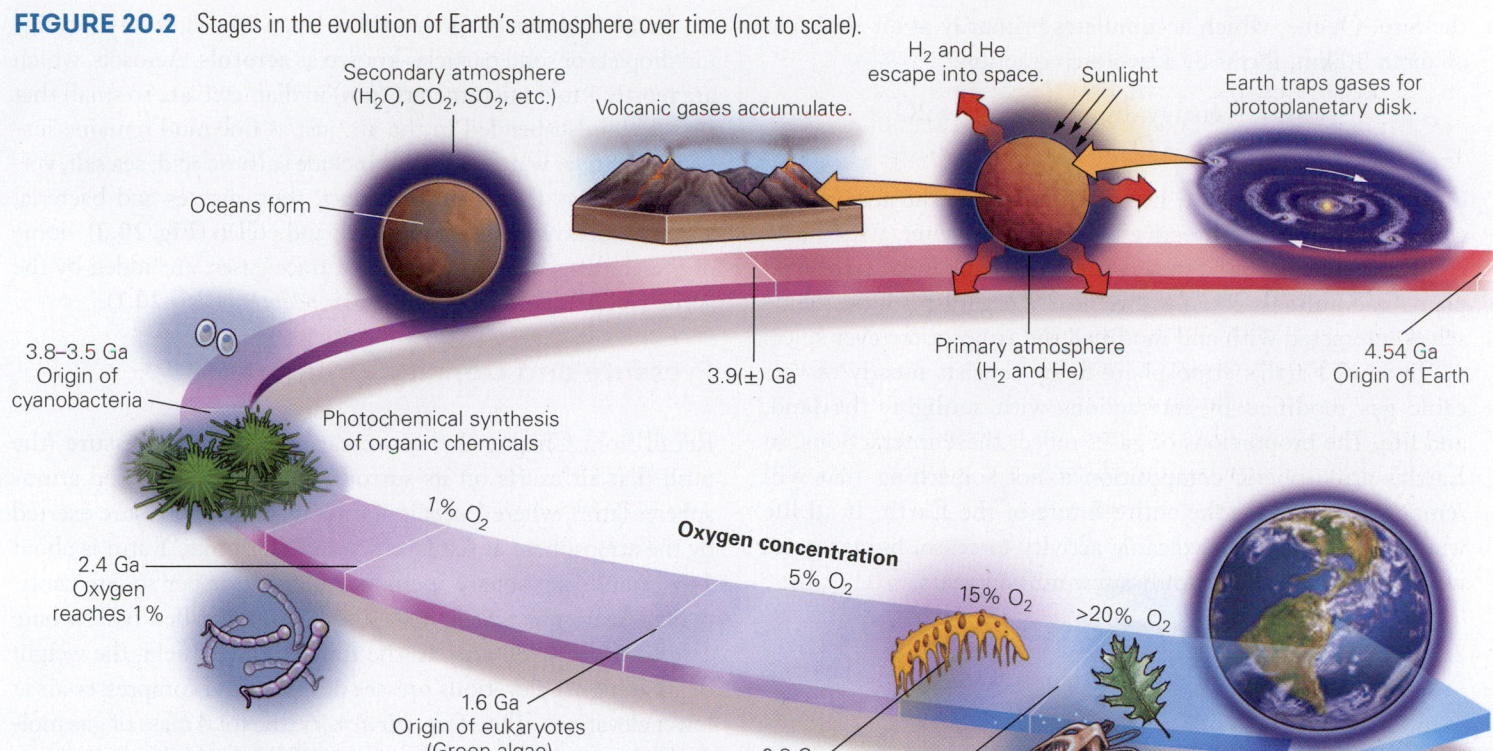

(a) The composition of the atmosphere has changed profoundly during geologic time.

The Third Atmosphere

If you were suddenly to travel back through time and appear on Earth 3.8 Ga, you would instantly suffocate, for the atmosphere back then contained virtually no molecular oxygen (O_2). It wasn't until the first photosynthetic organisms, cyanobacteria, appeared on Earth between 3.8 and 3.5 Ga that significant O_2 began to be produced. At first, virtually all of this O_2 was absorbed, by dissolution in water or by reactions with rocks. But by about 2.5 Ga, land and ocean oxygen "reservoirs" became saturated, and O_2 began to accumulate in the air, so that by about 2.4 Ga Earth's atmosphere contained 1% of its present oxygen level. This transition—from an oxygen-free atmosphere to an oxygen-containing atmosphere—is known as the *Great Oxygenation Event*. Oxygen concentrations increased slowly during the Proterozoic. At about 1.2 Ga, there may have been a boost in the production of O_2 with the appearance of photosynthetic algae, but only by about 600 million years ago (Ma) did oxygen levels in the air become substantial, ushering Earth's *third atmosphere*—the modern atmosphere dominated by N_2 and O_2—into existence.

Oxygen levels during the Phanerozoic remains the subject of research. The occurrence of charcoal in Silurian strata indicates that there must have been at least 13% O_2 in the air by

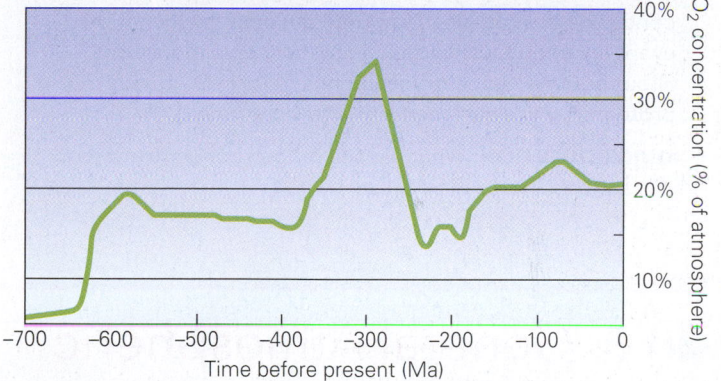

(b) During the Phanerozoic, oxygen composition has varied. It was particularly high during Caboniferous and Permian time.

about 425 Ma when vascular plants appeared, for vegetation cannot burn in air with less than 13% O_2. Oxygen concentration appears to have reached a peak of about 35% during the late Paleozoic coal age and has fluctuated since then, declining since the end of the Mesozoic to the present value of 21% (**Fig. 20.2**). If oxygen concentration exceeded 35%, plants would become so combustible that oxygen-producing forests would burn up.

Oxygen is important not only because it allows complex multicellular organisms to breathe but also because it supplies the raw components for the production of *ozone* (O_3), a gas that absorbs harmful ultraviolet (short-wavelength) radiation from the Sun. Ozone, which accumulates primarily at an elevation of about 30 km, forms by a two-step reaction:

(1) O_2 + energy (from the Sun) $\rightarrow$ 2O
(2) O_2 + O $\rightarrow$ O_3

Only when enough ozone had accumulated in the atmosphere could life leave the protective blanket of seawater, which also absorbs ultraviolet radiation. When this happened, terrestrial plants and animals could evolve. These organisms have themselves interacted with and modified the atmosphere ever since.

In sum, Earth's atmosphere today consists mostly of volcanic gas modified by interactions with sunlight, the land, and life. The proportions of gases reflect these interactions, so Earth's atmospheric composition is not something that will remain the same for the entire future of the Earth. If all life were to vanish and all volcanic activity to cease, lighter gases would leak into space in only a few million years.

Take-Home Message

The composition of the atmosphere has changed radically over Earth's history. Initially, it consisted of gases left over from planet formation, then of volcanic gases. When the ocean formed, water precipitated out to became liquid, and CO_2 dissolved in the oceans. The remaining atmosphere became rich in N_2. O_2 comes from photosynthesis. Its concentration remained low until the great oxygenation event and didn't rise to significant concentrations until about 600 Ma. Some oxygen bonds to form O_3, which protects Earth's surface from ultraviolet rays.

QUICK QUESTION: What caused oxygen concentrations in the atmosphere to increase?

20.3 General Atmospheric Characteristics

Atmospheric Composition

As we'll see later in this chapter, the moisture content (concentration of H_2O) of air can vary widely. So when atmospheric scientists analyze atmospheric composition, they first remove all the water. Completely dry, clean air near the Earth's surface consists almost entirely of two gases, 78% N_2 and 21% O_2. The remaining 1%, the trace gases of air, includes argon (Ar), CO_2, neon (Ne), CH_4, He, H_2, and O_3. Some of these trace gases serve an important function in the Earth System, as we've seen—CO_2 and CH_4 are greenhouse gases that regulate Earth's atmospheric temperature, and O_3 protects the surface from ultraviolet radiation.

Because air moves and has mass, it can keep aloft very tiny liquid droplets or solid particles known as **aerosols**. Aerosols, which are mostly 1 to 4 micrometers (µm) in diameter, are so small that they remain suspended in the air, just as fine mud remains suspended in river water. Aerosols include sulfuric acid, sea salt, volcanic ash, clay flakes, mineral dust, soot, viruses and bacteria, specks of decayed organic material, and pollen (**Fig. 20.3**). Some of these, along with some types of trace gases, are added by the activity of humans and comprise *air pollution* (**Box 20.1**).

Pressure and Density Variations

Recall from Chapter 2 that we can measure **air pressure** (the push that air exerts on its surroundings) in units called atmospheres (atm), where 1 *atm* is approximately the pressure exerted by the atmosphere at sea level. In familiar units, 1 atm is about 14.7 pounds per square inch, or 1,035 grams per square centimeter. Atmospheric scientists also use units called bars, where 1 *bar* is about 0.986 atm. In the Earth's gravity field, the weight of air at higher elevations presses down on and compresses air at lower elevations. Therefore, *air density* (the total mass of gas molecules in a unit of volume) increases from high elevation down toward the surface of the Earth (**Fig. 20.4**). In fact, the air pressure in the atmosphere at sea level is about three times that of the atmosphere on top of Mt. Everest, so a gulp of air at sea level contains about three times as many gas molecules as does a gulp of air on the top of Mt. Everest. Humans cannot breathe easily at elevations above about 5 km, so the cabin of an airliner must be pressurized to provide adequate oxygen for normal breathing.

Because of the decrease in air density with elevation, air is not uniformly distributed in the atmosphere. Specifically, 50% of the atmosphere's molecules lie below an elevation of 5.6 km,

FIGURE 20.3 A recent forest fire produces smoke that adds aerosols (e.g., soot), as well as CO_2 gas, to the atmosphere.

BOX 20.1 CONSIDER THIS . . .

Air Pollution

FIGURE Bx20.1 Air pollution in Beijing

Human activities (energy production, resource processing, manufacturing, and transportation) have changed concentrations of trace gases and have added new gases and aerosols to the air. We refer to these materials, which would not be in air (in the concentrations observed) due to natural phenomena alone, as *air pollution*, or **pollutants** (**Fig. Bx20.1**). Pollutants include SO_2, sulfate ($-SO_4$), nitrous oxide (N_2O), nitrogen dioxide (NO_2), CO, O_3, molecular carbon (soot), organic chemicals, chlorofluorocarbons (CFCs), metals (lead, uranium, mercury), CH_4, and CO_2, among many other chemicals.

The addition of some pollutants can have major effects on **air quality**, the clarity and health impact of air. For example, in the 19th century, the burning of coal for home heating in large cities created clouds of soot (carbon aerosols), which mixed with fog to produce *smog*. Today automobile exhaust creates chemicals that react in sunlight to produce a dangerous mix of gases and aerosols called *photochemical smog* over cities. Sulfur emitted into the air can have impacts that extend far beyond the source. Sulfur, in its various forms, dissolves in atmospheric moisture to produce sulfuric acid aerosols. These can become incorporated in raindrops to produce **acid rain**. Where acid rain falls, lakes, streams, and the ground become more acidic and thus toxic to fish and vegetation (particularly coniferous trees). Similarly, the emission of CFCs into the air can have global consequences. In 1985, researchers discovered that CFCs react with ultraviolet light from the Sun to release chlorine atoms, which, in turn, react with ozone and break it down. These reactions appear to happen mainly in high clouds above polar regions during certain times of the year, thus preferentially removing ozone from these regions to create an *ozone hole*. Fortunately, global regulation of CFCs seems to be diminishing the problem. Finally, as we discuss in Chapter 23, the burning of fossil fuels has significantly increased the amount of CO_2 in the atmosphere.

To give a sense of the severity of air pollution at a locality, governments now routinely issue measurements of air quality for regions. The U.S. Environmental Protection Agency defines air quality on a scale, the *air quality index* (AQI), from 0 to 500. If the AQI value exceeds 100, the air is potentially unhealthful, and if it is over 300, it is hazardous. Air quality in rapidly industrializing cities has become a major problem.

FIGURE 20.4 This graph shows air pressure versus elevation on the Earth. Climbers on top of Mt. Everest breathe an atmosphere that contains only about 33% of the atmospheric gases at sea level.

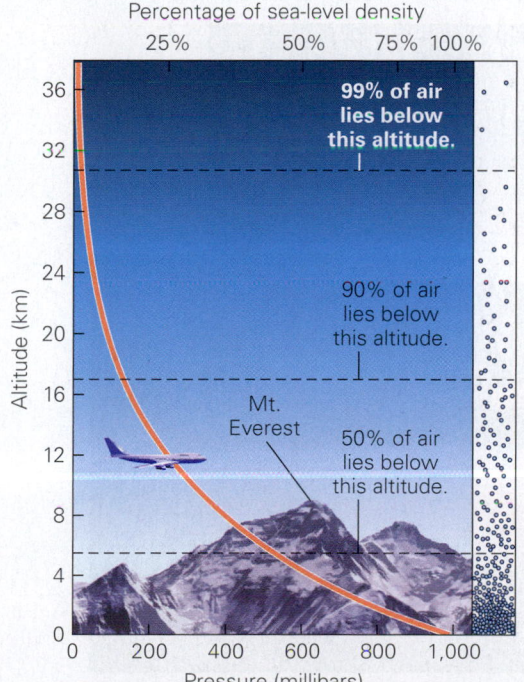

90% lie below 16 km, and 99.99997% lie below 100 km. Thus, even though the outer edge of the atmosphere, a vague boundary where the gas density becomes the same as that of interplanetary space, technically lies as far as 10,000 km from the Earth's surface, most of the atmosphere's molecules lie within a shell only 0.5% as wide as the solid Earth. Though thin, the atmospheric shell contains sufficient gas to turn the sky blue when viewed during the day (**Box 20.2**).

Heat and Temperature in Air

The molecules that constitute the atmosphere, or any gas, do not stand still but are constantly moving. We refer to the total kinetic energy (energy of motion) resulting from the movement of molecules in a gas as its *thermal energy*, or *heat*. Note that heat and temperature are not the same. A gas's *temperature* is a measure of the average kinetic energy of its molecules (see Chapter 2). Thus, a volume of gas with a small number of very rapidly moving molecules has a higher temperature but may contain less heat than a volume with a larger number of slowly moving molecules. And if we add heat to a gas in a given volume, its molecules move faster and its temperature rises.

Where does the heat in the atmosphere come from? Contrary to popular belief, only a small amount (12%) of this energy

BOX 20.2 CONSIDER THIS . . .

Why Is the Sky Blue?

When astronauts standing on the Moon looked up, even during the day, they saw a black sky filled with stars and a nearly white Sun. On Earth, when we look up during the day, we see a blue sky when there are no clouds and a white sky when there are. The color we see in the sky is the result of the dispersal of energy that occurs when light interacts with air molecules and other particles in the atmosphere, a process called *scattering*. When light is scattered, a single ray divides into countless beams, each heading off in a different direction. It's similar to what happens when you shine a spotlight on a mirror ball over a dance floor.

Sunlight consists of a broad spectrum of electromagnetic radiation. Different components of the atmosphere scatter different wavelengths of this light because the ability of a particle to scatter light depends on the particle's size relative to the light's wavelength. Aerosols (soot, water droplets, dust, etc.) are larger than all wavelengths of light, so all wavelengths reflect off these particles. As a result, light scattered from these particles appears white—that's why clouds are white.

The interaction of light with tiny gas molecules in the air is different. The molecules absorb some wavelengths and shortly afterward, re-emit them. Blue light (with shorter wavelengths) is absorbed and re-emitted more than red light (with longer wavelengths). Put another way, blue light is scattered more effectively by air, while red light passes through more easily. Since scattered light heads in all directions, some of it returns back to space; this light is called backscattered light. Because of backscattering, the intensity of light received on Earth's surface is less than it would be if Earth had no atmosphere. It's also because of scattering that shadows are not completely dark. Scattering sends light into regions (beneath trees, for example) that are blocked from the Sun.

When you look at an object, the color you see is the color of light either emitted from the object or reflected off it. A plum appears violet because it reflects violet light and absorbs other wavelengths, and a traffic light appears green because it emits green light. On a clear day, with the Sun high in the sky, the gas in the atmosphere scatters primarily blue light. When we look up, we are seeing the blue light scattered off the gas molecules (**Fig. Bx20.2a**). Because blue light is scattered, not all of it reaches the Earth (some has been backscattered to space), so the Sun appears yellow rather than white—if you subtract a little blue light from white light, you get yellow light. On a cloudy or hazy day, there are more aerosols (such as water droplets) in the atmosphere, and these scatter all wavelengths of light. Therefore, the sky appears white unless the clouds are so dense that they do not let light through, in which case they appear gray.

At the beginning or end of the day, when the Sun lies close to the horizon, light passes through a thicker amount of the atmosphere, and so much of the blue light scatters back to space that the sunlight reaching Earth contains mostly red wavelengths—if you subtract a lot of blue light from white light, you are left with red light. Thus, the Sun appears red, as does the light that reflects off clouds (**Fig. Bx20.2b**).

FIGURE Bx20.2 Atmospheric color depends on the thickness of the atmosphere that light passes through.

(a) Scattering of light by gas molecules leads to the brilliant blue of a clear sky at noon on top of a volcano in Hawaii.

(b) So much scattering happens when sunlight enters the atmosphere at a low angle during sunrise that all that's left is red light.

comes directly from incoming sunlight, because solar radiation consists predominantly of short-wavelength energy that air molecules cannot absorb. Most of the heat in air actually comes from long-wavelength (infrared) radiation energy that was absorbed by solids or liquids on the Earth's surface and then re-radiated and absorbed by greenhouse gases in the air. In fact, about 70% of the heat in the atmosphere comes from rising infrared energy, meaning, in effect, that the atmosphere is baked from below rather than broiled from above. Of the remaining heat, 4% comes directly from conduction of heat from contact with the Earth's surface, and 14% comes from the condensation of water in the air (when a gas changes state to become a liquid, it releases heat).

Relations between Pressure and Temperature

Adding heat to a given volume of air will cause the air to expand, for as the molecules in air start to move faster, they try to move outward. Similarly, removing heat from a given volume of air causes the air to contract, since its molecules slow down. Because of this behavior, a sealed balloon filled with air expands when you place it outside on a hot day because the molecules within it exert more outward force on the balloon's skin. Similarly, the balloon contracts if you place it in the refrigerator, where the molecules do not exert as much outward force.

You can see the same relationship between pressure and temperature when you change air pressure without adding or subtracting thermal energy. For example, when a given volume of air rises from a region of greater pressure to a region of less pressure, without adding or subtracting thermal energy, it expands, and when this happens, its temperature decreases (the total kinetic energy per unit volume decreases). Such a process is called **adiabatic cooling**, from the Greek *adiabatos*, meaning impassable. Air cools adiabatically at a rate of 6° to 10°C per kilometer that it rises, depending on its moisture content; dry air cools faster. (That's why it's cooler on a mountain peak than at its base.) Similarly, if air moves from a region of less pressure to a region of greater pressure, without adding or subtracting heat, **adiabatic heating** takes place as the air contracts and its temperature increases.

We'll see that adiabatic cooling and heating are important processes in the atmosphere because air pressure changes with elevation, and volumes of air move up in *updrafts*, or down in *downdrafts*. Air in an updraft cools as it rises, while air in a downdraft warms as it sinks.

Water in the Air

Air contains varying amounts of water vapor—from 0.3% above a hot desert to 4% in a rainforest before a heavy downpour. The moisture content in air changes with time at a given location, and with location at a given time. It affects your comfort level as well as the likelihood of rain or snow. How do we quantitatively distinguish wet from dry air?

Meteorologists, scientists who study the weather, specify the water content of air by a number called the **relative humidity**, the ratio between the measured water vapor content and the maximum possible amount of water vapor that the air could hold, expressed as a percentage. (Note that relative humidity differs from *absolute humidity*. The latter, which refers to the mass of water in a volume of air, is given in grams per cubic meter.) The maximum vapor content varies with temperature because warmer air can hold more water vapor than can colder air. Air that contains as much water vapor as possible is *saturated*, whereas air that doesn't is *unsaturated*. If we say that air at a given temperature has a relative humidity of 20%, we mean the air contains only 20% of the water that it could hold at that temperature when saturated. Such air feels dry. Air with a relative humidity of 100% is saturated and feels very humid, or damp. The higher the relative humidity, the more difficult it is for your body to cool by perspiring, so a hot, humid day is less comfortable than a hot, dry day.

Because cold air can't hold as much water vapor as warm air, unsaturated air may become saturated when cooled, without the addition of any new water. The temperature at which the air becomes saturated is called the *dewpoint temperature*; dew forms when unsaturated air cools at night and becomes saturated, causing liquid water to condense on surfaces. (When the temperature is below freezing, the moisture turns into frost.) When saturated air rises and adiabatically cools, its moisture condenses to form a **cloud**, a mist of tiny water droplets or tiny ice crystals (**Fig. 20.5**). Clouds contain about 50% vapor and 50% liquid or solid water in droplets.

As we noted earlier, when water in the air changes in state from liquid to gas, or vice versa, the temperature of the air changes. That's because when water evaporates, it absorbs heat, allowing molecules to break free from the liquid, and when it condenses, it releases heat as molecules slowly get locked into position and slow down. The heat released during condensation was "hidden" in the sense that it comes only from the change of state and does not require an external energy source. Thus, it is called the **latent heat of condensation**.

Take-Home Message

Air consists of 78% N_2, 21% O_2, and 1% trace gases. Air pressure decreases upward, so 90% of air lies below an elevation of 16 km. Heat in the air comes mostly from infrared radiation rising from the Earth's surface. As air rises, it expands and cools, and as it sinks it compresses and warms. Air contains variable amounts of water, as represented by its relative humidity. The amount of water that air can contain depends on temperature. When this water condenses, it releases latent heat.

QUICK QUESTION: What is air pollution?

FIGURE 20.5 As air rises and enters regions of lower pressure, it expands and adiabatically cools. Here we see moist air rising and becoming less dense; its moisture condenses at elevations above 3 km and produces a cloud.

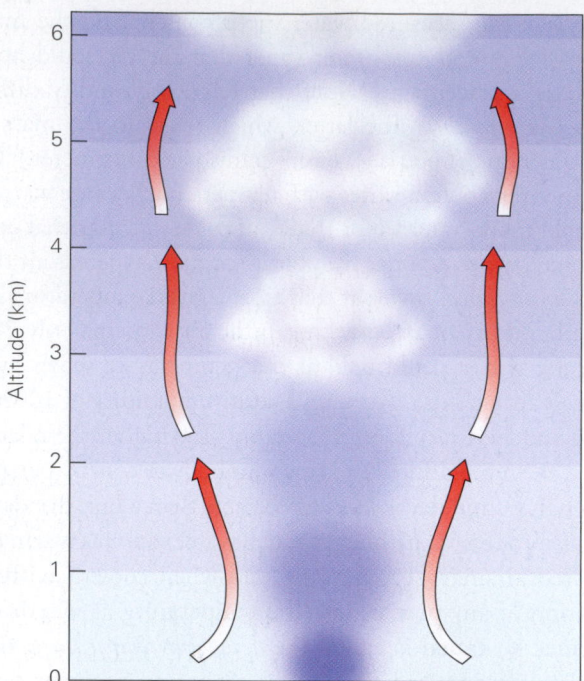

20.4 Atmospheric Layers

Layers Based on Temperature Variation

While air pressure in the atmosphere decreases uniformly from sea level upward, air temperature does not. In fact, starting from the surface and going up, temperature decreases, then increases, then decreases, then increases. Elevations where temperatures stop decreasing and start increasing, or vice versa, divide the Earth's atmosphere into four layers. Let's consider these, starting from the base (**Fig. 20.6**):

- *Troposphere:* The layer of air that lies between the surface of the Earth and an elevation of 5 km at the poles and 18 km at the equator is called the **troposphere**. Within this layer, the temperature decreases gradually from an average of 18°C at the surface to about −55°C at the top, a boundary called the *tropopause*. The name troposphere comes from the Greek *tropos*, which means turning. It is an appropriate name, because air in the troposphere constantly undergoes convection in that warm air rises in updrafts and cold air sinks in downdrafts. Why does

FIGURE 20.6 The principal layers of the atmosphere.

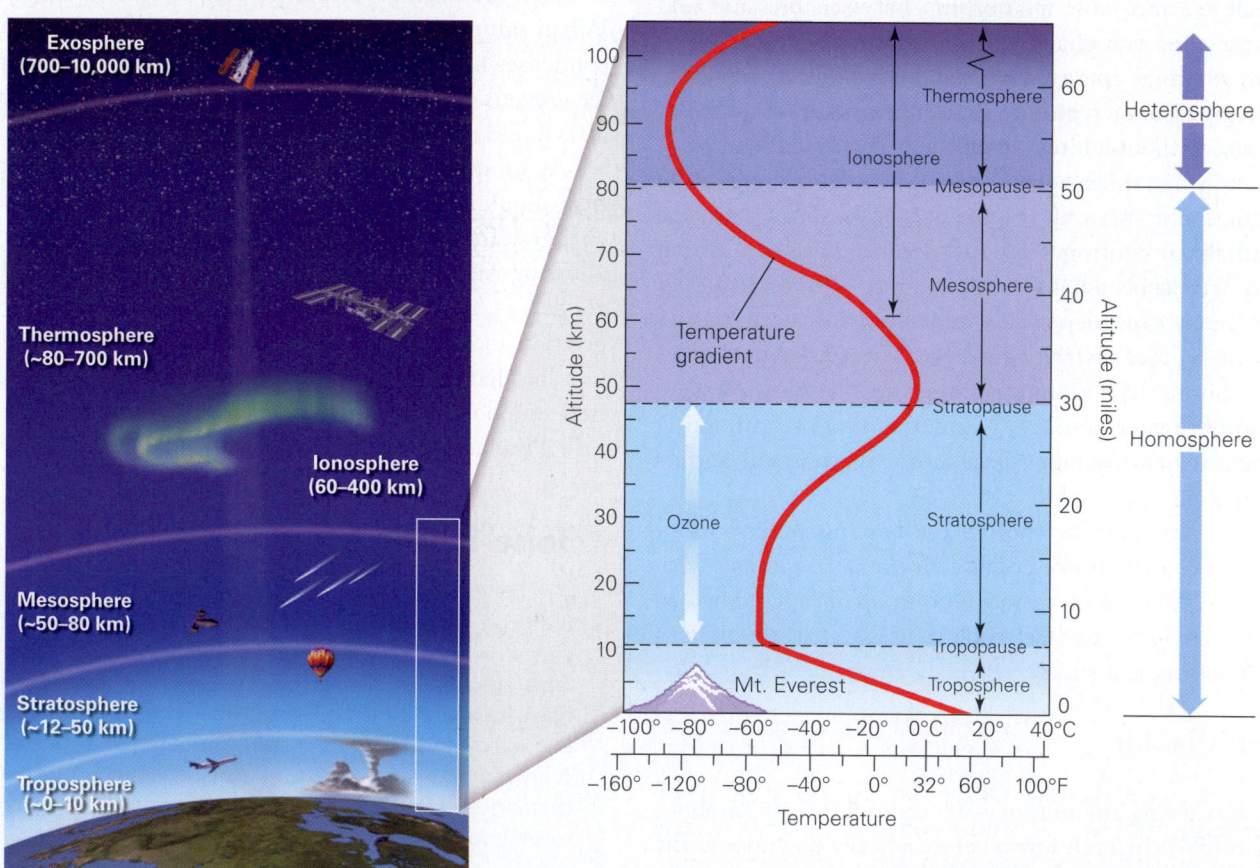

(Not to scale)

this convection take place? As we noted, most of the heat in the air comes from infrared radiation rising from the surface, meaning air gets heated from below. As soon as it forms, near-surface warm air becomes buoyant, so it rises. Cold air must sink to take the place of the rising air. As we will see, air movement in the troposphere drives most weather phenomena, so the troposphere can also be thought of as the "weather layer."

- *Stratosphere*: Beginning at the tropopause and continuing up for about 10 km, the temperature stays about the same. Then it slowly rises, reaching a maximum of about 0°C at an elevation of about 47 km, a boundary called the *stratopause*. The layer between the tropopause and the stratopause is the **stratosphere**, so named because it doesn't convect much and thus remains relatively stable and stratified. The stratosphere generally doesn't mix with the underlying troposphere because at the tropopause hotter (less dense) air already lies on top of cooler (denser) air, so there's no need for convection. Most of the ozone in Earth's atmosphere resides in the stratosphere; heating in the stratosphere happens because ozone absorbs ultraviolet radiation directly from the Sun.

- *Mesosphere:* Temperature decreases in the layer called the **mesosphere**, which lies between 47 and 82 km. At the *mesopause*, the top of the mesosphere, the temperature has dropped to −85°C. The mesosphere does not absorb much solar energy and thus cools with increasing distance from the hotter stratosphere below. Most meteors (so-called shooting stars) begin burning in the mesosphere and have vaporized by the time they reach an altitude of 25 km.

- *Thermosphere:* The outermost layer of the atmosphere, which lies between an elevation of 82 and 700 km, is called the **thermosphere**. It contains very little of the atmosphere's gas—less than 1%. Temperature increases with elevation in this layer because its gases absorb short-wavelength solar energy. In fact, the temperature at the top becomes very high. But because the thermosphere has so little gas, it contains very little heat, so an astronaut walking in space at an elevation of 200 km doesn't feel hot.

- *Exosphere:* The layer between 700 and 10,000 km is called the **exosphere**. It represents that gradual transition between the atmosphere and space. At 10,000 km, the gas concentration becomes the same as interplanetary space.

Layering Based on Composition

The density of gas in the lower three layers of the atmosphere is great enough that moving atoms and molecules frequently collide. Like billiard balls, they bounce off one another and shoot off in different directions. This constant, chaotic motion stirs the gas sufficiently to make a homogeneous mixture such that the air in the lower three layers has essentially the same proportion of different gases regardless of location. For this reason, atmospheric scientists refer to the troposphere, stratosphere, and mesosphere together as the *homosphere*. In contrast, atoms and molecules in the low-density thermosphere collide so infrequently that this layer does not homogenize. Rather, the air of the homosphere separates into distinct layers based on composition, with the heaviest gas (nitrogen) on the bottom, followed in succession by oxygen, helium, and at the top, hydrogen. To emphasize this composition, atmospheric scientists refer to the thermosphere as the *heterosphere*.

So far, we've distinguished atmospheric layers according to their thermal structure (troposphere, stratosphere, mesosphere, and thermosphere) and according to the degree their gases mix (homosphere and heterosphere). We need to add one more "sphere" to our discussion. The **ionosphere** is the interval between 60 and 400 km and thus includes most of the mesosphere and the lower part of the thermosphere. It was given its name because in this layer short-wavelength solar energy strips nitrogen molecules and oxygen molecules of their electrons and transforms them into positive ions. The ionosphere plays an important role in modern communication in that, like a mirror, it reflects radio transmissions from Earth back down so that they can be received over great distances.

The ionosphere also hosts a spectacular atmospheric phenomenon, the **aurorae** (*aurora borealis* in the northern hemisphere and *aurora australis* in the southern), which look like undulating, ghostly curtains of varicolored light in the night sky (**Fig. 20.7**). They appear when charged particles ejected from the Sun, especially when solar flares erupt, reach the Earth and interact with the ions in the ionosphere, making them release energy. Aurorae occur primarily at high latitudes because Earth's magnetic field traps solar particles and carries them to the poles.

FIGURE 20.7 The splendor of an aurora borealis lights up the night sky in Arctic Canada.

Take-Home Message

The atmosphere can be divided into layers based on temperature variation with elevation. From base to top, these are troposphere, stratosphere, mesosphere, and thermosphere. Weather happens in the troposphere. Researchers also distinguish layering based on composition. The lower well-mixed portion is the homosphere and the upper chemically layered interval is the heterosphere. The ionosphere is the layer containing abundant charged ions.

QUICK QUESTION: What causes the aurorae?

20.5 Wind and Global Circulation in the Atmosphere

A gusty breeze on a summer day, the steady currents of air that blew clipper ships across the oceans, and a fierce hurricane all are examples of **wind**, the horizontal component of air movement. We can feel wind because of the impacts of air molecules as they strike us. The existence of wind emphasizes that the lower part of the atmosphere constantly moves, swirling and overturning at rates of between a fraction of a kilometer and a few hundred kilometers per hour. This circulation happens on two scales, local and global. *Local circulation* refers to the movement of air over a distance of tens to hundreds of kilometers. *Global circulation* refers to the movement of volumes of air in paths that ultimately carry it around the entire planet. To understand both kinds of circulation, we must first see what drives air from one place to another and examine how energy inputs into the atmosphere vary with location.

Lateral Pressure Changes and the Cause of Wind

The air pressure of the atmosphere not only changes vertically, it also changes horizontally at a given elevation. The rate of pressure change over a given horizontal distance, called a *pressure gradient*, can be represented by the slope of a line on a graph plotting pressure on the vertical axis and map distance on the horizontal axis (**Fig. 20.8a**). We can use a map to illustrate how air pressure at a given elevation varies with location. A line on a map along which the air has a specified pressure is called an **isobar** (**Fig. 20.8b**); the pressure is the same at all points on an isobar. Isobars can never touch or cross because they represent different values of pressure.

Winds form wherever a pressure gradient exists. In a general sense, air accelerates from a high-pressure region to a low-pressure region; in other words, it starts flowing down a pressure gradient. To see why, step on one end of a balloon filled with air; you momentarily increase the pressure at that end, so the air flows toward the other end (**Fig. 20.8c**). A difference in pressure exists between one isobar and the next, so the air starts flowing perpendicular to these lines. The Coriolis effect modifies wind direction, so in general, flow is not perpendicular to isobars—in fact, in many locations air flow becomes subparallel to isobars, following the path that represents the balance between the force caused by the flow of air from higher to lower pressure realms and the force caused by the Coriolis effect.

Regional Circulation in the Atmosphere

Solar energy constantly bathes the Earth. Of this energy, 30% reflects back to space (off clouds, water, and land); air or clouds absorb 19%; and land and water absorb 51%. As we've seen, the energy absorbed by land and water re-radiates as infrared

FIGURE 20.8 The concept of a pressure gradient.

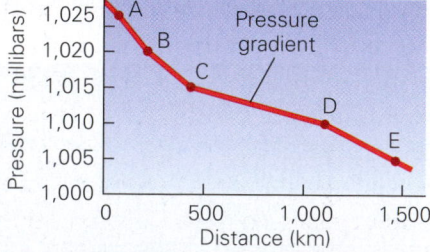

(a) A graph of pressure changes from Point A to E. The gradient is greater where the slope is steeper.

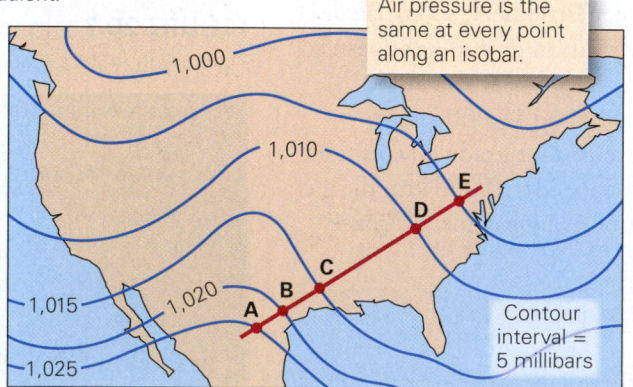

(b) The air pressure decreases between Points A and E. Note that the rate of change, the pressure gradient, is greater between Points A and C than between Points C and D. Spacing of isobars reflects the gradient.

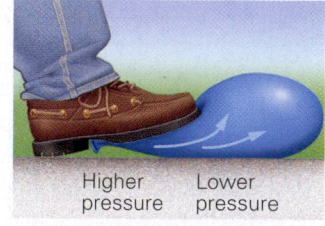

(c) Air flows from a region of high pressure to a region of low pressure.

radiation and thus bakes the atmosphere from below. Because the Earth is a sphere, not all latitudes receive the same amount of incoming solar energy, or **insolation**. Specifically, portions of the Earth's surface hit by direct rays of the Sun receive more energy per square meter than portions hit by oblique rays. We can illustrate this contrast with a flashlight. If you point a flashlight beam straight down, you get a small but bright spot of light on the ground, but if you aim the beam so that it hits the ground at an angle of 45°, the spot covers a broader area but does not appear as bright (**Fig. 20.9**). Higher latitudes thus receive less energy than lower latitudes. Because of the tilt of Earth's axis, the amount of solar radiation that any point on the surface receives changes during the year, which is why we have seasons (**Box 20.3**). The contrast in the amount of solar radiation received by different latitudes means that polar regions are cooler than equatorial regions.

In 1735, George Hadley, a British mathematician, realized that latitudinal contrasts in temperature could drive air circulation over broad areas. Specifically, he suggested

that very warm air at the equator would rise up, and then at high elevations flow toward the pole, to be replaced by cool polar air, which would flow to the equator at lower elevations (**Fig. 20.10a**). Hadley's proposal of a *hemispheric-scale circulation cell*, however, did not take into account the Earth's rotation and the resulting Coriolis effect. The Coriolis effect, as we learned in Chapter 18, causes an object moving from the equator to the pole above a rotating sphere to deflect relative to the ground (**Fig. 20.10b**). Because of the Coriolis effect, northward-moving high-altitude air in the northern hemisphere deflects to the east, so by the latitude of 30° N, it basically moves due east and cannot make it the rest of the way to the pole. Further, by the time it reaches this latitude, the air has also cooled significantly and thus starts to sink. When the sinking air reaches low elevations, it divides, some moving back toward the equator near the surface and some moving north near the surface. The southern flow heads back toward the equator, gradually deflecting to the southwest until it is flowing due west at the equator, where it warms and rises.

If Hadley's single hemispheric circulation cells can't exist, then what does global air circulation look like? Atmospheric scientists initially proposed a simplified model, in which three *circulation cells* develop within each hemisphere. The low-latitude cells, extending from the equator to a latitude of about 30°, were called *Hadley cells*, in honor of George Hadley; the mid-latitude cells were named *Ferrel cells*, in honor of the American meteorologist William Ferrel, who proposed them; and the high-latitude cells were called *polar cells* (**Fig. 20.10c**). In the context of this simplistic model, where the air of one circulation cell approaches another at the surface, a surface **convergence zone** develops—at this zone, the air has nowhere to go but up. Alternatively, where the air of one cell moves away from that of another, near the surface, a surface **divergence zone** develops. In the three-cell model, an *intertropical convergence zone* occurs at the equator, the southern edge of the Hadley cell; a *subtropical divergence zone* forms at subtropical latitudes (about 30°); and a *polar convergence zone* forms at the southern edge of polar cell (about 60°), a boundary also known as the **polar front**.

While the three-cell model described above gives an intuitive sense of how air moves globally, overall, in detail it is far from an accurate representation of the Earth's real atmospheric circulation. This complexity has many causes: the shear between cells; the fact that land and sea are not uniformly distributed and absorb and release heat and moisture at different rates; the variation of Earth's land surface topography, which forces horizontally moving air to rise or sink; seasonal variations of insolation; and transport of heat across latitudinal boundaries by ocean currents. All of these complexities impart turbulence to the global circulation patterns. Thus, while the Hadley cell tends to be recognizable, the other cells are substantially overprinted by spinning eddies, and the boundaries where visible have large wave-like curves in map view (**Fig. 20.10d**).

FIGURE 20.9 The amount of insolation depends on the angle at which sunbeams strike the Earth, so the poles are colder.

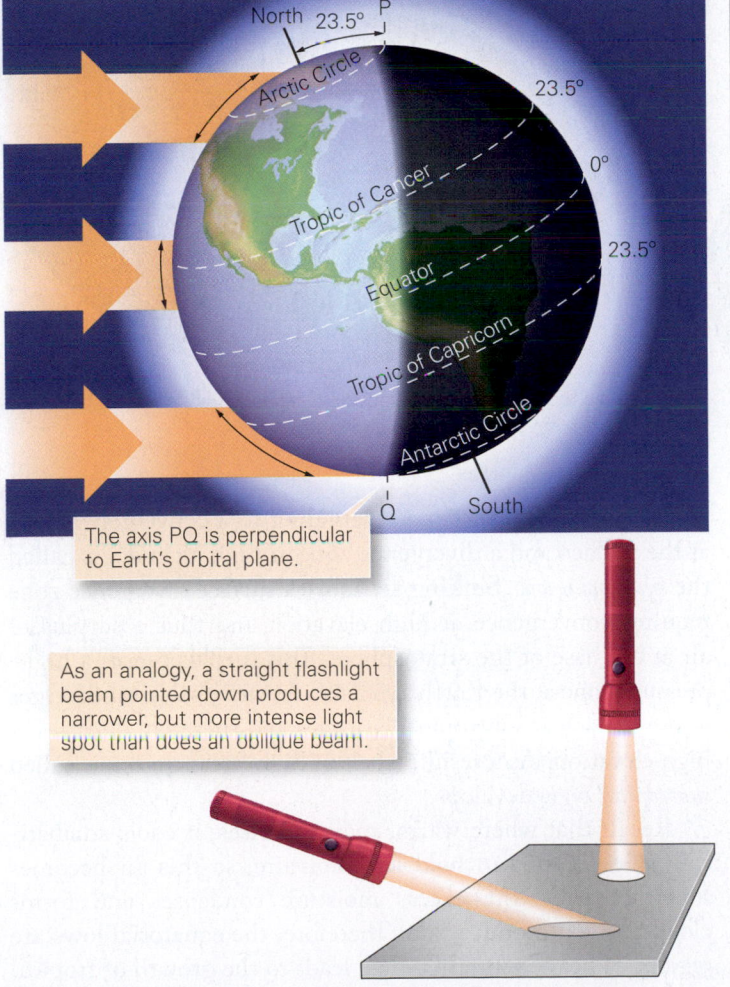

The axis PQ is perpendicular to Earth's orbital plane.

As an analogy, a straight flashlight beam pointed down produces a narrower, but more intense light spot than does an oblique beam.

FIGURE 20.10 Global circulation patterns in the atmosphere.

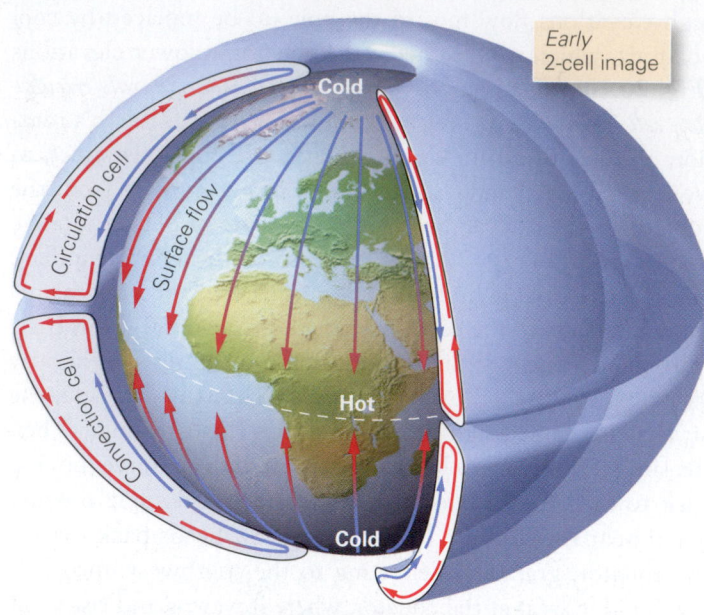

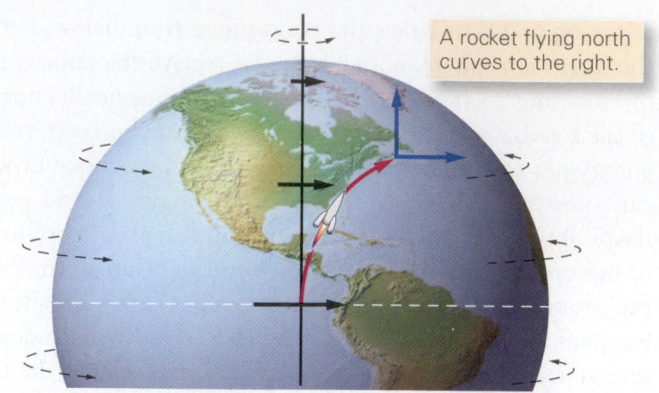

(a) If the Earth did not rotate, two simple circulation cells would exist, each stretching from the equator to the pole.

(b) But, because the Earth spins, the Coriolis effect influences the motion of moving objects and materials.

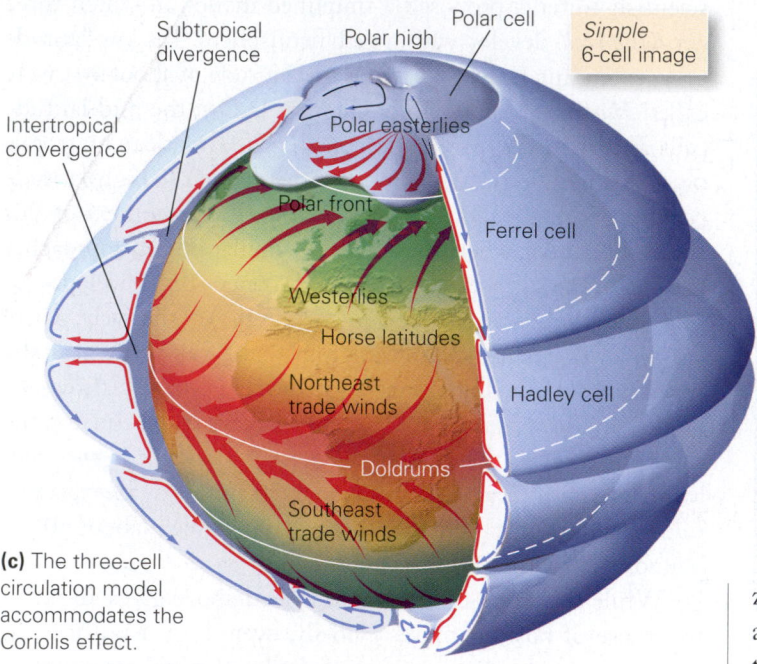

(c) The three-cell circulation model accommodates the Coriolis effect.

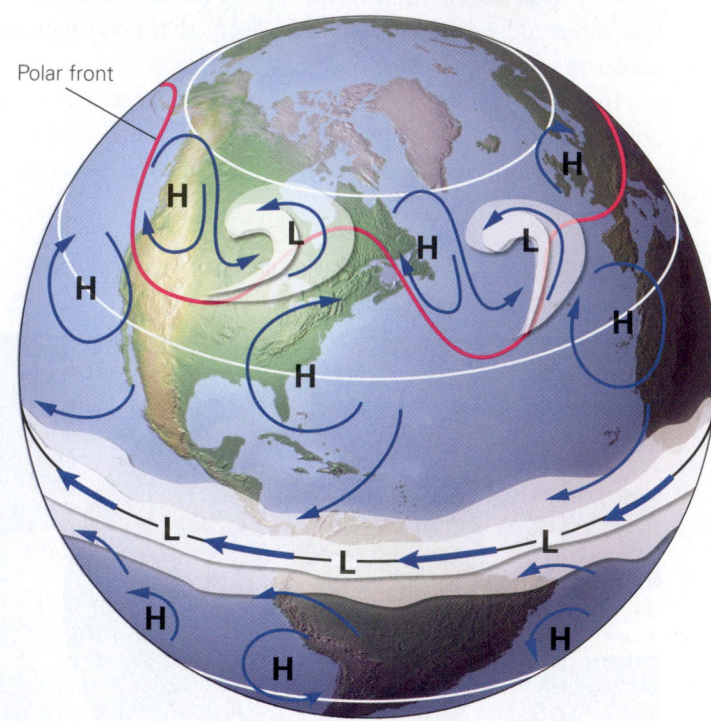

(d) The modern image of global circulation is very complicated; there are several large, moving eddies.

High-Pressure and Low-Pressure Zones

Significantly, the vertical movement of air above convergence and divergence zones affects surface air pressure below. For example, where air rises from the surface (at a surface convergence zone), it must diverge or spread out at high elevation (the base of the stratosphere). This high-altitude divergence produces a deficit in the total mass of air in a vertical column extending up from the surface and, as a result, an area of low atmospheric pressure at the Earth's surface. The low-pressure zone along the equator, corresponding to a convergence zone at the surface and a divergence zone at high altitude, is called the *equatorial low*. Sinking air above a surface divergence zone requires convergence at high elevation and thus a surplus of air at the base of the stratosphere. This surplus causes a high-pressure zone at the Earth's surface. As we've seen, air diverges at the surface at subtropical latitudes and therefore diverges at high elevation. As a result, a belt of high-pressure zones called *subtropical highs* develops.

Recall that where warm, moist air rises, it cools adiabatically. Cooler air can hold less moisture, so this air becomes supersaturated. The excess moisture condenses and forms clouds, which produce rain. Therefore, the equatorial lows are regions of heavy rainfall, which leads to the growth of tropical rainforests. In contrast, in high-pressure belts where cool, dry air sinks, air contracts and heats adiabatically. The resulting hot air can absorb moisture, so that it rains only rarely. Thus, as we will see in Chapter 21, the subtropical regions at latitude 30° include some of the Earth's major deserts.

Prevailing Surface Winds

Despite its complexities, overall circulation of the atmosphere creates belts in which surface air often moves in a consistent direction. Such airflows are called **prevailing winds**. Note

that when describing these winds, meteorologists label them according to the direction the air comes from. Thus, a "westerly wind" blows from west to east.

Let's start our tour of prevailing winds at the base of the Hadley cell in the northern hemisphere. Near-surface winds start to flow from 30° N to the south and are deflected west. Thus, between the equator and 30° N, surface winds come out of the northeast and are called the northeast **trade winds**, so named because they once carried trading ships westward from Europe to the Americas. Trade winds in the southern hemisphere, which start flowing northward and then deflect to the west, end up flowing from southeast to northwest and are called the southeast trade winds. Where the southeast and northeast trade winds merge at the equator, they flow almost due west (**Fig. 20.11a**). But winds along the equator tend to be very slow, because the air mostly rises. Ships tended to be becalmed in this belt, which came to be known as the **doldrums**.

In mid-latitudes, surface air starts to move toward the north, but because of the Coriolis effect it curves to the east. Thus, throughout much of North America and Europe, the prevailing surface winds come out of the west or southwest and are known as the *surface westerlies*. Winds, of course, shift as waves in the polar front pass by and as eddies pass over regions. In subtropical highs, where air flows primarily downward, winds are weak and tend to shift in different directions. In the past, these conditions inhibited the progress of sailing ships. Perhaps because so many horses being transported by ship died of heat exhaustion while the vessels traversed the subtropical high, the region came to be known as the *horse latitudes*. Finally, in polar latitudes, surface air starts by flowing from the pole southward but deflects to the west. The resulting prevailing winds are known as the *polar easterlies*. These winds converge with the westerlies of mid-latitudes at the polar front.

FIGURE 20.11 General surface-wind patterns.

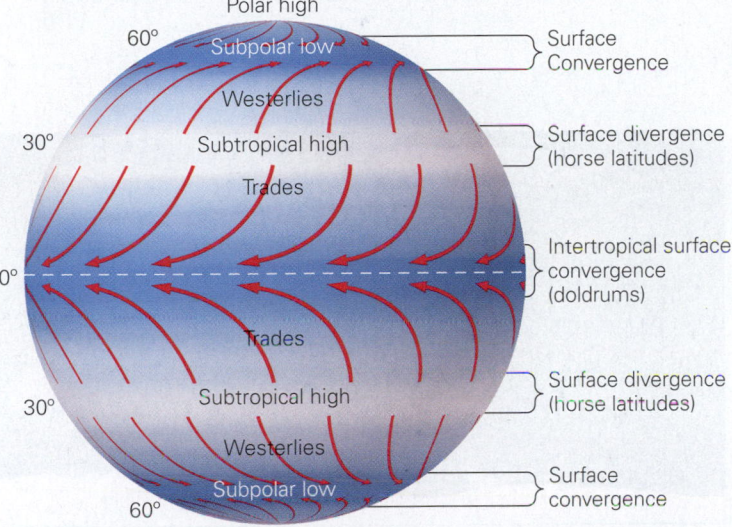

(a) If the Earth had a uniform surface, distinct high- and low-pressure zones would form on its surface. But surface winds flowing from high-pressure zones to low-pressure zones are deflected by the Coriolis effect, so for much of their course they flow almost parallel to pressure zones.

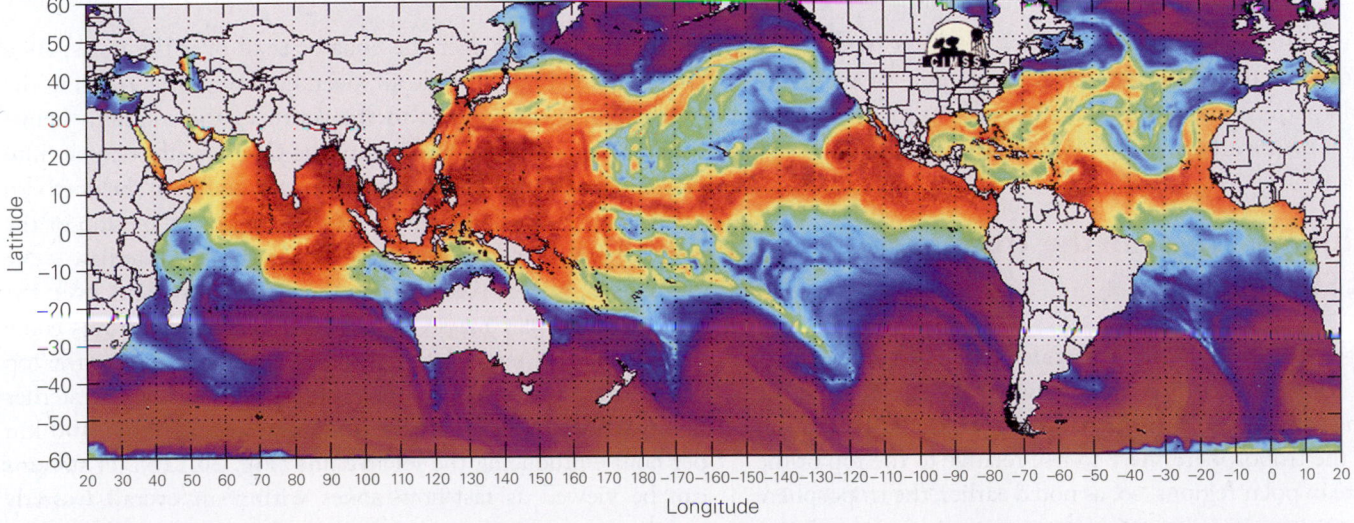

(b) A satellite map showing the amount of "precipitable moisture" (water that can fall as rain or snow) in the air. The pattern of colors gives a sense of the true complexity of air flow in the near-surface atmosphere.

BOX 20.3   CONSIDER THIS . . .

The Earth's Tilt: The Cause of Seasons

A satellite view emphasizes that snow cover distribution changes over the course of the year (**Fig. Bx20.3a**). This change occurs because the Earth has seasons. Seasons exist because the Earth's axis tilts (currently at 23.5°) relative to the plane of its orbit. As our planet moves around the Sun, the direction of tilt relative to the Sun varies, so the insolation at a given latitude and, therefore, the average monthly temperature, changes during the course of a year (**Fig. Bx20.3b**).

For example, when the northern hemisphere tilts toward the Sun, it receives more radiation, warms up, and enjoys summer, but when the northern hemisphere tilts away from the Sun it receives less radiation, cools down, and endures winter.

Any moment in time is day for half the Earth but night for the other half. The boundary between these two hemispheres is the *terminator*. Because of the Earth's tilt, the terminator does not always pass through the

North and South Poles (see Fig. Bx20.2b). On June 21, a special day called a *solstice*, the terminator lies 23.5° away from the poles, as measured along the Earth's surface. We refer to the line of latitude at this position as the Arctic Circle in the northern hemisphere and the Antarctic Circle in the southern hemisphere. On June 21, a person standing on the Arctic Circle can see the midnight sun, meaning it's light all day. Anyone south of the Antarctic Circle sees night for a full 24

FIGURE Bx20.3   Because of the tilt of Earth's axis, we have seasons.

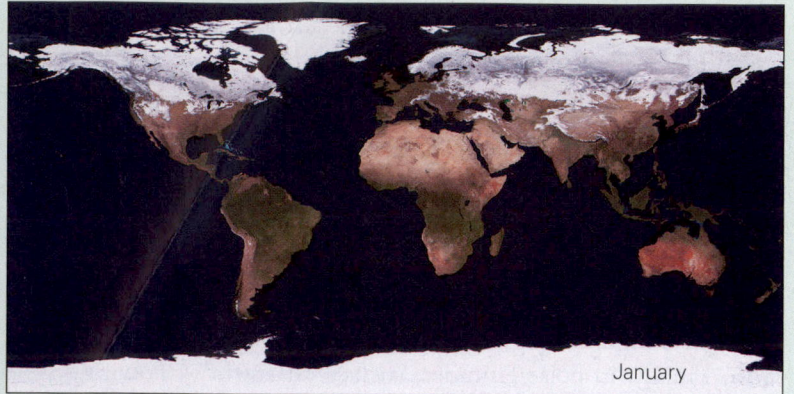
January

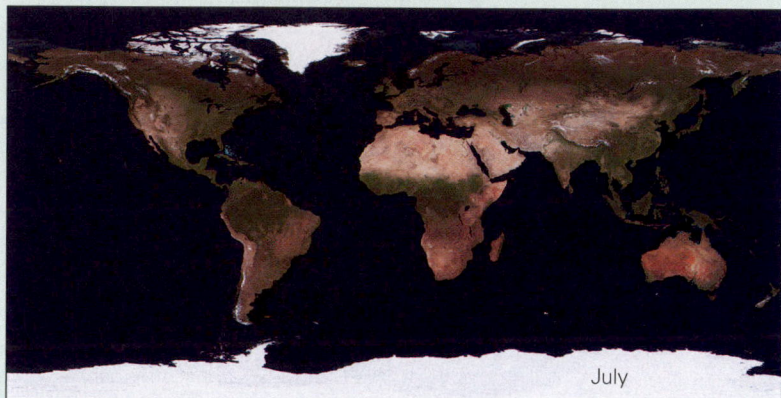
July

Satellite photos emphasize the contrast in snow cover of the northern hemisphere as seasons change.

Keep in mind that the concept of prevailing winds is a simplification. While it provides a general basis for picturing global wind patterns, in reality, air flow follows more complex patterns and swirls around many eddies (**Fig. 20.11b**).

High-Altitude Winds in the Troposphere: The Jet Streams

In the upper atmosphere, a global-scale pressure gradient exists because of temperature differences between the equator and the pole. At the equator, air is warmer and thus expands. This causes the top of the troposphere there to rise relative to the top of the troposphere in polar regions, so, as noted earlier, the troposphere

tends to be thicker over the equator than over the poles. As a consequence, high-altitude air, overall, flows to the north in the northern hemisphere and to the south in the southern hemisphere (**Fig. 20.12a**). Once again, the Coriolis effect comes into play and makes this air deflect to the east, and so in the northern hemisphere we have generally westerly winds at the top of the troposphere. These are called the high-altitude westerlies.

In two special places, over the polar front and over the horse latitudes, air masses of very different temperatures come in contact. This temperature difference causes a step at the top of the troposphere. Along these steps, high-altitude westerlies flow particularly fast—at speeds of between 200 and 400 km per hour—producing the **jet streams** (**Fig. 20.12b**). Jet streams can be viewed as fast-flow zones within an overall westerly hours. During the northern summer, regions north of the Arctic Circle see the midnight sun for more than a day, and the North Pole itself experiences the midnight sun for a full 6 months. Similarly, regions south of the Antarctic Circle experience 24-hour nights for more than a day, and the South Pole itself sees night for a full 6 months. On December 21, the other solstice, a person at the Antarctic Circle sees the midnight sun, whereas all regions north of the Arctic Circle have perpetual night. During the southern summer, the South Pole sees daylight for 6 months.

On the June 21 solstice, the Sun's rays are exactly perpendicular to the Earth (the Sun is directly overhead) at latitude 23.5° N, the Tropic of Cancer, and on December 21 the Sun's rays are exactly perpendicular to the Earth at 23.5° S, the Tropic of Capricorn. On September 22 and March 20, each known as an *equinox*, the Sun is directly overhead at the equator. The two solstices and two equinoxes divide the year into four astronomical seasons.

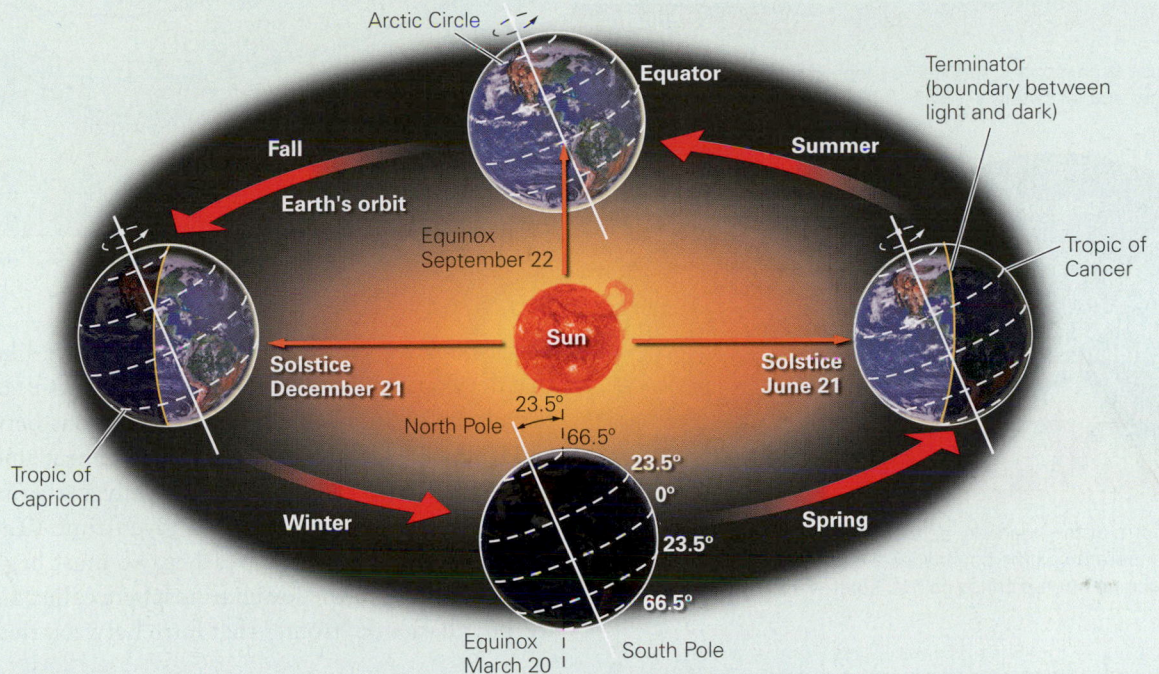

(b) In the northern hemisphere's summer solstice, the noonday Sun appears directly overhead at the Tropic of Cancer, and the region above the Arctic Circle sees the midnight sun, whereas at the winter solstice, the noonday Sun appears directly overhead at the Tropic of Capricorn and the region south of the Antarctic Circle sees the midnight sun. At the equinoxes, the noonday sun appears directly overhead at the equator.

high-altitude flow. The polar jet stream, which tends to be the stronger of the two, typically flows at 7 to 11 km above the surface, whereas the subtropical jet stream flows at 9 to 14 km above the surface. Thus, airliners commonly encounter the jet streams. Planes flying east have shorter flying times than those flying west, because the former are helped along by strong tailwinds, whereas the latter battle headwinds that slow them down.

As we've noted, the polar front tends to develop large, wave-like undulations. Thus, as viewed on a map, the positions of the polar front and therefore of the jet stream follow large, curving trajectories that sometimes bring the jet stream down to southern latitudes of North America and sometimes to northern latitudes of Canada. Further, the average position varies with the seasons (**Fig. 20.12c**).

Take-Home Message

Where air pressure changes horizontally, a pressure gradient exists. Air starts to move from high-pressure to low-pressure regions, producing a wind. The Coriolis force modifies the flow direction. Due to latitudinal variation in insolation, regional-scale air circulation cells develop in each hemisphere. But in detail, air circulation is much more complicated, and only the southern cell, the Hadley cell, is well defined. Regional variations in pressure drive prevailing surface winds. Jet streams form at high elevation, in association with elevation steps in the tropopause.

QUICK QUESTION: What are doldrums, and why do they form?

FIGURE 20.12 Jet streams form just above the tropopause and follow wavy paths that go around the Earth.

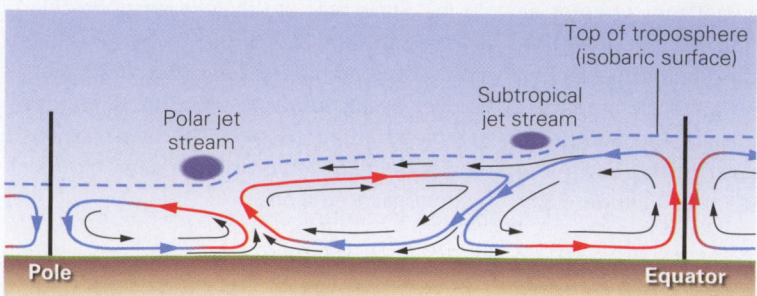

(a) The tropopause is an isobaric surface. At steps in this surface, high-altitude winds are stronger and jet streams form.

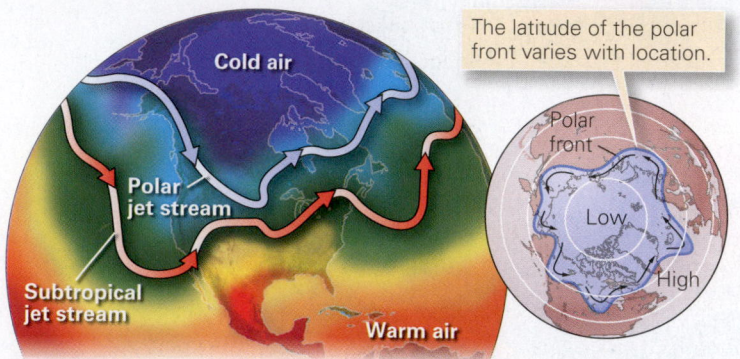

(b) Jet streams form at the boundary between regions of the troposphere with different temperatures. The boundaries are wavy, so the jet streams are wavy. The polar jet stream brings cold air southward.

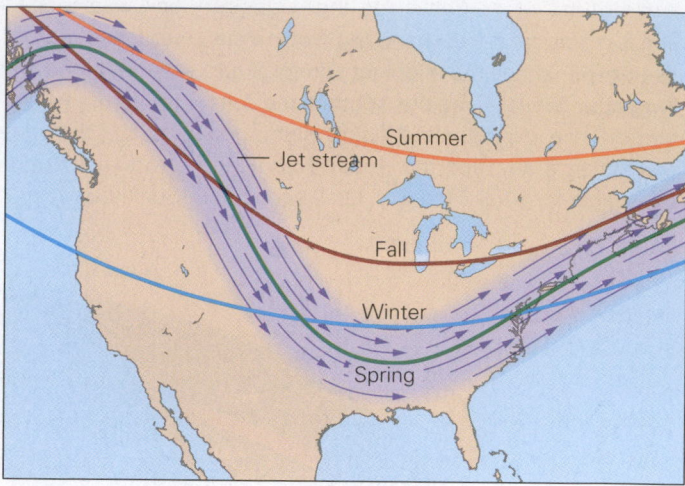

(c) The position of the jet stream changes over time.

fairly simple as well. But because of the wave-like undulations that develop along convergence and divergence zones; the local turbulence of atmospheric flow; interactions between air flow and landscape features; and exchanges of heat among land, sea, and air, weather can vary dramatically with time at a given location, and with location at a given time (**Table 20.1**). To understand what controls weather, we must begin by characterizing regions of the lower atmosphere called air masses, and the boundaries or "fronts" that form between them.

Air Masses and Fronts

Air that remains or passes over a certain region for a length of time takes on characteristics that reflect both its interaction with the Earth's surface and the amount of insolation it receives. For example, air that has hovered over a warm sea tends to become warm and moist, whereas air stuck over cold land becomes cold and dry. A body of air, on the order of 1,500 km across, with recognizable physical characteristics (temperature and moisture content), is called an **air mass**. Air masses move within

20.6 Weather and Its Causes

Weather refers to the local-scale conditions as defined by temperature, air pressure, relative humidity, rainfall or snow, cloud cover, and wind speed. If the surface of the Earth were a perfectly uniform sphere, airflow might follow the simple global circulation patterns and weather patterns, therefore, might be

TABLE 20.1 Weather Extremes

Lowest recorded air pressure at sea level	870 millibars (25.7 in.); during Typhoon Tip (1979), Pacific Ocean
Highest air pressure at sea level	1,084 millibars (32.02 in.); Agata, Siberia
Lowest air temperature (world)	−89°C (−129°F); Vostok, Antarctica
Lowest air temperature (North America)	−63°C (−81°F); Yukon, Canada
Highest air temperature (world)	58°C (136°F); Libya
Highest air temperature (North America)	57°C (134°F); Death Valley, California
Highest recorded wind speed (world), not in a tornado	372 km per hour (231 mph); Mt. Washington, New Hampshire

global-scale circulation of the atmosphere, and their paths are controlled by prevailing winds. Simplistically, distinct air masses, which have been named by meteorologists (**Fig. 20.13**), persistently develop in given regions; they are controlled by the locations of land and sea and by latitude. The weather of a location can change drastically when one air mass replaces another. For example, a summer day in a midwestern state will be cool and dry under a continental polar air mass but hot and humid under a maritime tropical air mass.

The boundary between two air masses is called a **front**. Meteorologists recognize several kinds of fronts, of which we introduce three. At a *cold front*, a cold air mass pushes underneath a warm air mass (**Fig. 20.14a**). As a consequence, the warm, moist air flows up to higher elevations, where it expands and cools adiabatically. The moisture it contains then condenses to form large clouds from which heavy rains may fall. As a *warm front* moves into a region, the warm air slowly rises over the cool air (**Fig. 20.14b**). Again, the rising air expands and cools, and its moisture condenses, so clouds develop over the boundaries between the two air masses. Note that a front is a sloping surface and that cold fronts typically are steeper than warm fronts. We can portray the trace of the intersection between a front and the land surface by a line on a map and can represent the movement of a front by the movement of this line.

Not all air masses move at the same velocity. Typically, cold fronts move faster than warm fronts and overtake them. Where this happens, the cold front lifts up the base of the warm front, so the warm front no longer intersects the ground surface. Meteorologists refer to the geometry that results when a cold front pushes underneath a warm front as an *occluded front* (**Fig. 20.14c**).

Cyclonic and Anticyclonic Flow

Imagine that the air rises persistently over a given region. When this air reaches the base of the stratosphere, it spreads out or, more formally, undergoes high-elevation divergence. Such movement results in a net loss of air molecules above the region and, therefore, in the development of a *low-pressure system* at the ground surface below. By definition, a pressure gradient exists between the low-pressure system and its surroundings, as defined by concentric isobars. As we noted earlier, air will flow from high pressure to low pressure, so winds carry air from the surroundings toward the center of the low-pressure system. If the Earth were not spinning, the wind direction would trend roughly perpendicular to the isobars. But because of the Coriolis effect, the winds deflect to follow paths around the center of

FIGURE 20.14 The formation of fronts between air masses.

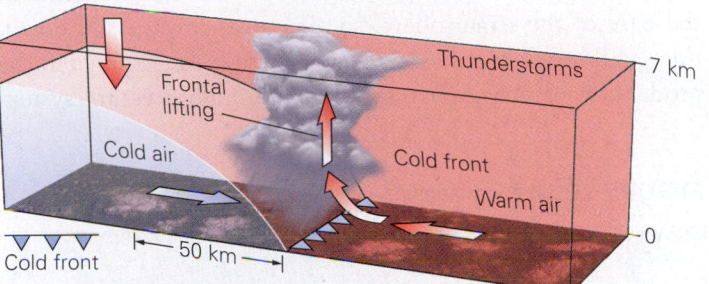

(a) A cold front develops where a cold air mass moves under a warm air mass.

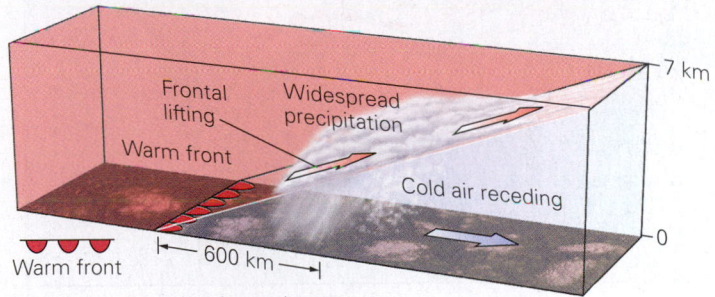

(b) A warm front develops where a warm air mass moves over a cold air mass.

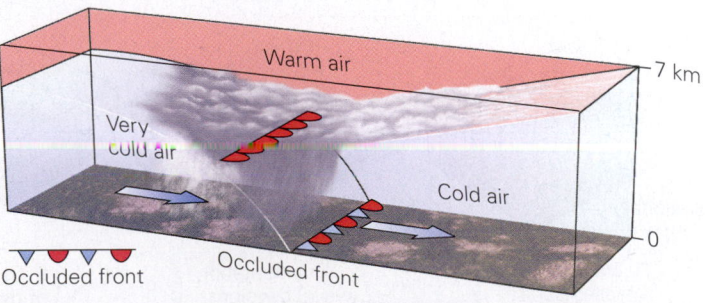

(c) An occluded front develops where a fast-moving cold front overtakes a warm front and lifts the base of the warm front off the ground. Note the symbols used to represent fronts.

FIGURE 20.13 Air masses that form in different places have been assigned different names. The arrows indicate the average directions in which the air masses move.

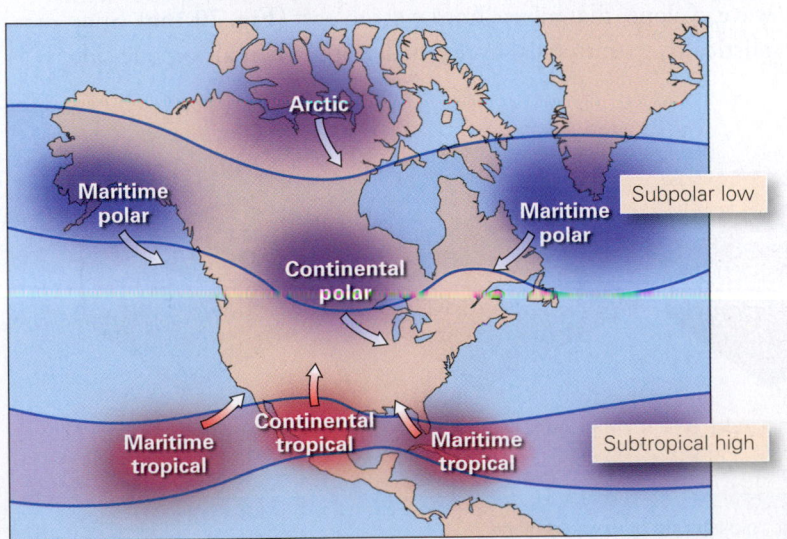

the low pressure. In fact, if there were no friction between the air and the Earth's surface, a balance would eventually develop between the pressure-gradient force and the Coriolis force, and the path of air flowing around the low-pressure center would be nearly circular. Friction, however, slows the wind and decreases the Coriolis force, so the pressure-gradient force dominates, and winds follow a spiral path toward the interior of the low-pressure system.

The rotational flow around a low-pressure system is called **cyclonic flow** (**Fig. 20.15**). In the northern hemisphere, this flow is counterclockwise. Note that because of the inward spiral of winds, convergence takes place in the center of a cyclonic low-pressure system at low elevations. The excess air upwells until it reaches the top of the troposphere where the divergence that we've already mentioned takes place. The upward spiraling air adiabatically cools, so the water vapor it contains condenses to form clouds that may produce rain. As a result, a low-pressure system tends to be associated with cloudy, rainy weather.

Not all spiral flows are cyclonic. If there is convergence at the base of the stratosphere, so air moves inward and down, there will be a net gain of molecules within the region and the production of a *high-pressure system*. A high-pressure system

FIGURE 20.15 Air spirals downward and clockwise (creating an anticyclone) at a high-pressure mass and upward and counterclockwise (creating a cyclone) at a low-pressure mass.

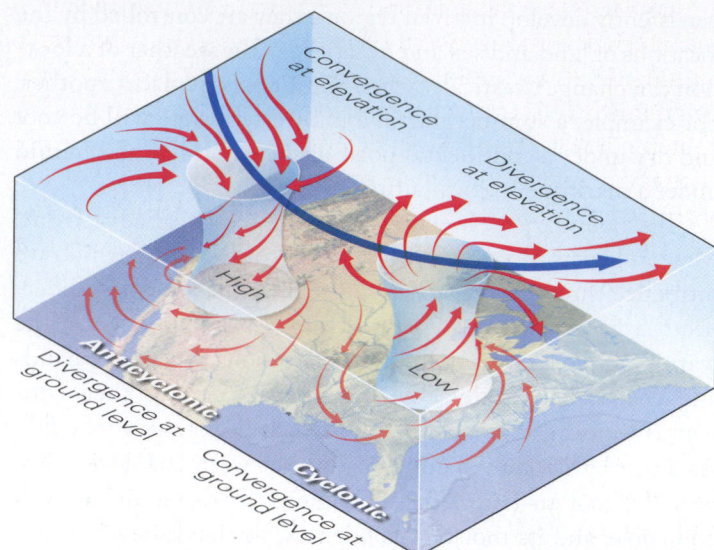

must also be surrounded by a pressure gradient and, therefore, winds. But the winds of a high-pressure system move outward at the base of the troposphere (near the Earth's surface) and, because of the Coriolis effect, follow a clockwise spiral path around the high-pressure system. Such movement is called an **anticyclonic flow** (see Fig. 20.15). As a consequence of anticyclonic flow, divergence takes place near the ground and air from above sinks down to fill the deficit. The sinking air compresses, warms, and undergoes a reduction in relative humidity. Thus, a high-pressure system tends to be associated with "fair" (clear and dry) weather.

In the mid-latitudes that encompass the United States and much of Canada as well as much of Europe, the weather commonly reflects the development of a large low-pressure system, known as an **extratropical cyclone**, a mid-latitude cyclone, or a wave cyclone, that moves from west to east (**Fig. 20.16a**). Simplistically, extratropical cyclones develop when air on one side

FIGURE 20.16 Formation of a mid-latitude (wave) cyclone.

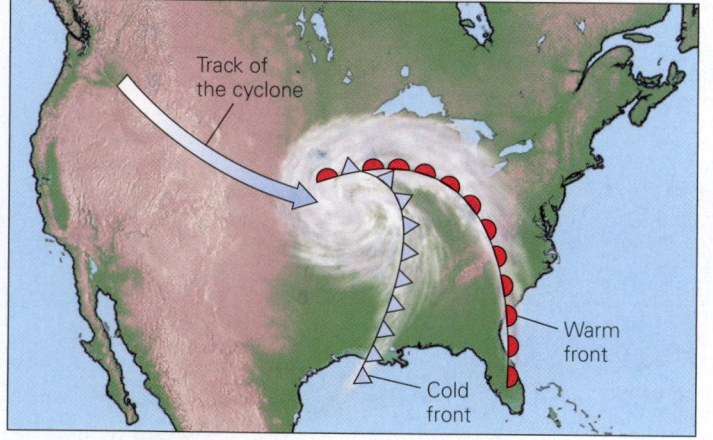

(a) An extratropical cyclone treks from west to east across North America.

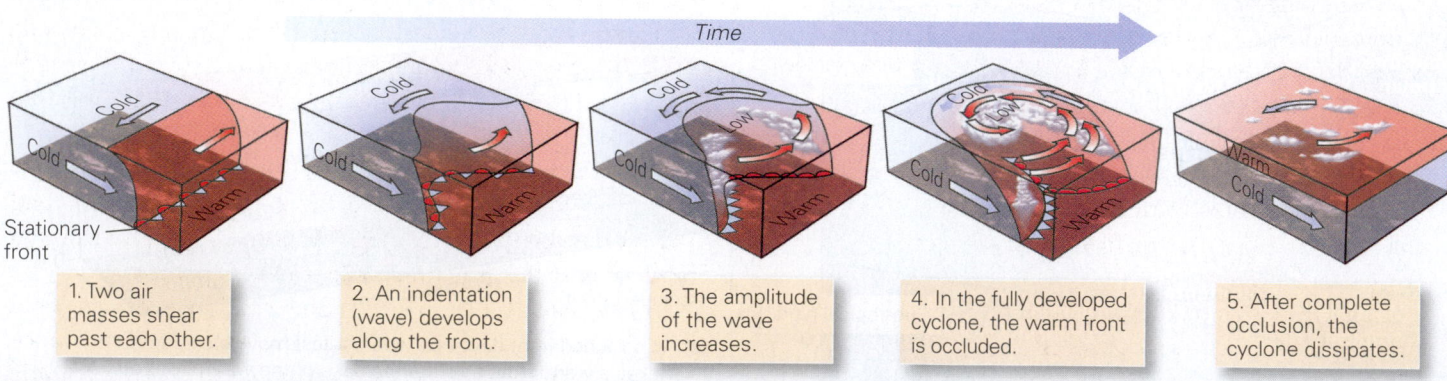

1. Two air masses shear past each other.

2. An indentation (wave) develops along the front.

3. The amplitude of the wave increases.

4. In the fully developed cyclone, the warm front is occluded.

5. After complete occlusion, the cyclone dissipates.

(b) Researchers can define many stages during the evolution of an extratropical (mid-latitude or wave) cyclone.

of a cold front shears sideways past air on the other side (**Fig. 20.16b**). The shear of air along the cold front warps the face of the front into the shape of a wave. When this happens, warm air starts to move north, up and over the cold air mass, whereas cold air circles around and starts to move south and downward, pushing the cold front forward. The two fronts meet at a V, the point of which lies near the center of the low-pressure mass. In a satellite image, a mid-latitude cyclone is a huge spiral mass of clouds, rotating counterclockwise; the clouds develop along both fronts and are centered on the low-pressure mass. The cold front of a wave cyclone tends to move faster than the warm front, so eventually the warm front becomes occluded and the cyclone dies out.

Forming Clouds and Precipitation

We've mentioned clouds several times in this chapter already. Let's now look at clouds a little more closely, ultimately to understand how to describe them and why they yield rain and snow (**Fig. 20.17**). As noted earlier, a cloud is a region of the atmosphere where about half of the atmospheric moisture exists in the form of tiny water droplets (about 20 μm across, less than a third of the diameter of a human hair; 1 μm = 0.001 mm) or in the form of tiny ice crystals. (Most clouds float above the Earth's surface; clouds that form at ground level make up *fog*.) Because clouds reflect and scatter incoming sunlight, they keep the ground cooler during the day, but at night they prevent infrared radiation from escaping and thus keep the ground warmer.

Droplets or ice grains in clouds form by condensation or deposition, respectively, around pre-existing aerosol particles called **condensation nuclei**. This takes place when the air becomes saturated with water vapor, either because evaporation from the Earth's surface provides additional water or because the air containing the water cools so that its capacity to hold water decreases. Cooling may take place at night, simply because of the loss of sunlight, or when air rises and expands. Meteorologists refer to conditions that cause air to rise as **lifting mechanisms**, and they recognize several different types.

- *Convective lifting:* This process occurs where air warms, becomes buoyant, and rises. This process tends to happen where the temperature of the Earth's surface varies with location. For example, land that heats during the day may cause convective lifting in the late afternoon when it starts radiating heat back into the base of the atmosphere. Similarly, convective lifting may develop over an island that releases more heat than does the surrounding sea (**Fig. 20.18a**).
- *Frontal lifting:* Frontal lifting takes place along the fronts between air masses, as we've seen (see Fig. 20.14). At cold fronts, warm air pushes up and over a steep wall of cold

air, rising rapidly to form large clouds. At warm fronts, the advancing warm air rides up the gentle slope of the front and condenses.
- *Convergence lifting:* Where air converges or pushes together near the ground surface, as happens when air spirals up into a low-pressure zone or when two winds that have been deflected around an obstacle meet again, the air has nowhere to go but up and thus rises and cools, forming clouds (**Fig. 20.18b**).
- *Orographic lifting:* This type of lifting happens when moisture-laden wind blows toward a mountain range and on meeting the mountain range can go no farther and must rise. As a result, clouds form above the mountain range (**Fig. 20.18c**).

Rain, snow, hail, and sleet fall from clouds in two ways, depending on the temperature of the cloud. In warm clouds, rain develops by **collision and coalescence**, during which the tiny droplets that compose the cloud collide and stick together to create a larger drop (**Fig. 20.19a**). Eventually, water drops that are too big to be held in suspension by circulating air start to fall, incorporating more droplets as they descend. Depending on how far the drops have fallen and on how much moisture is available, drops reach different sizes before they hit the ground. Typical raindrops have a diameter of 2 mm and fall at a velocity of about 20 km per hour. Any drops larger than 5 mm tend to break into smaller ones upon collision. If rain falls through colder air near the ground, it freezes to become *sleet*.

In cold clouds, the mist contains a mixture of very cold water droplets and tiny ice crystals. The water droplets evaporate faster than the ice (because water molecules are less tightly

FIGURE 20.18 Causes of cloud formation.

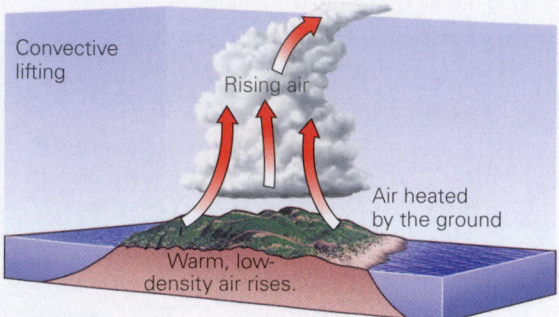

Convective lifting

Rising air

Air heated by the ground

Warm, low-density air rises.

(a) Convective lifting occurs where warm air starts to rise.

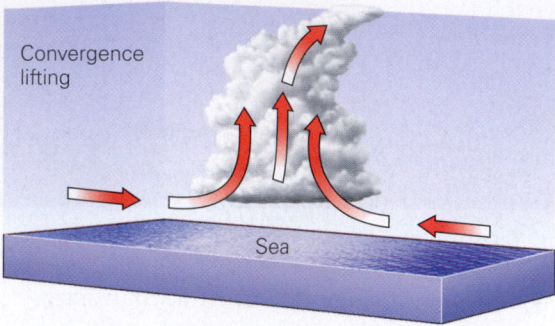

Convergence lifting

Sea

(b) Convergence lifting takes place where winds merge—the air has nowhere to go but up.

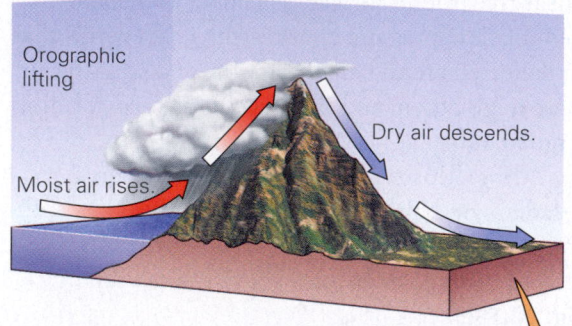

Orographic lifting

Dry air descends.

Moist air rises.

(c) Orographic lifting happens where moist winds run into a mountain range and are forced up.

Lifting forms clouds over a ridge in Utah.

FIGURE 20.19 Mechanisms of raindrop formation.

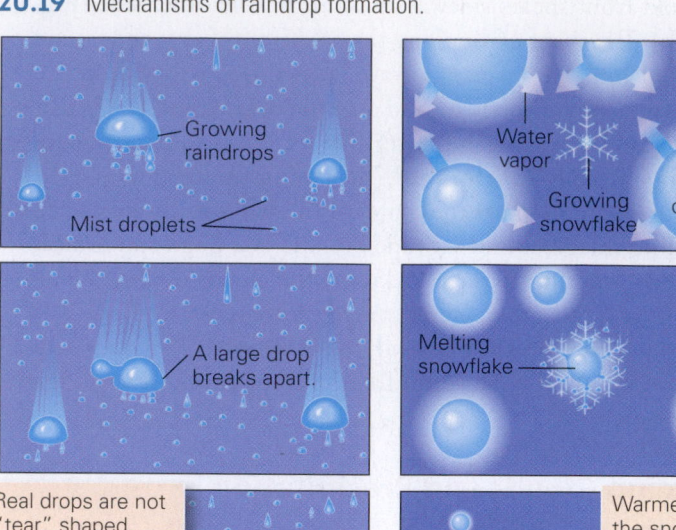

Time

Growing raindrops

Mist droplets

A large drop breaks apart.

Real drops are not "tear" shaped.

New drops grow.

Water vapor

Growing snowflake

Cloud droplet

Melting snowflake

Growing raindrop

Warmer air melts the snowflake.

(a) Rain can form when tiny droplets collide and coalesce. Once a drop becomes large enough to fall, it incorporates more drops on the way down. Really large drops can break apart.

(b) During the Bergeron process, water drops evaporate, releasing vapor that attaches to growing snowflakes, which then fall. At lower elevations, where air is warmer, the flakes melt.

bound to liquid than to solid) and provide moisture that condenses onto pre-existing ice crystals, leading to the growth of hexagonal snowflakes. If the air below the cloud is very cold, the snow falls as powder-like flakes, but if the air is close to the melting temperature, large wet clumps of snowflakes fall. And, if the lower air is warmer than 0°C, the snow transforms to rain before it hits the ground. This kind of formation, involving the growth of ice crystals in a cloud at the expense of water droplets, is called the *Bergeron process*, after Tor Bergeron, the Swedish meteorologist who discovered it (**Fig. 20.19b**).

Many kinds of clouds form in the troposphere. It wasn't until 1803, however, that Luke Howard, a British pharmacist, proposed a simple terminology for describing clouds (**Fig. 20.20**). First, we divide clouds into types based on their shape: puffy, cotton-ball- or cauliflower-shaped clouds are *cumulus* (from the Latin word for stacking). Clouds that occur in relatively thin, stable layers and thus have a sheet-like or layered shape are called *stratus*. High clouds that have a wispy shape and taper into delicate, feather-like curls are called *cirrus*. We can then add a prefix to the name of a cloud to indicate its elevation: high-altitude clouds (above about 7 km) take the prefix *cirro-*, mid-altitude clouds take the prefix *alto-*, and low-altitude clouds (below 2 km) do not have a prefix. Finally, we

FIGURE 20.20 The various types of clouds on Earth.

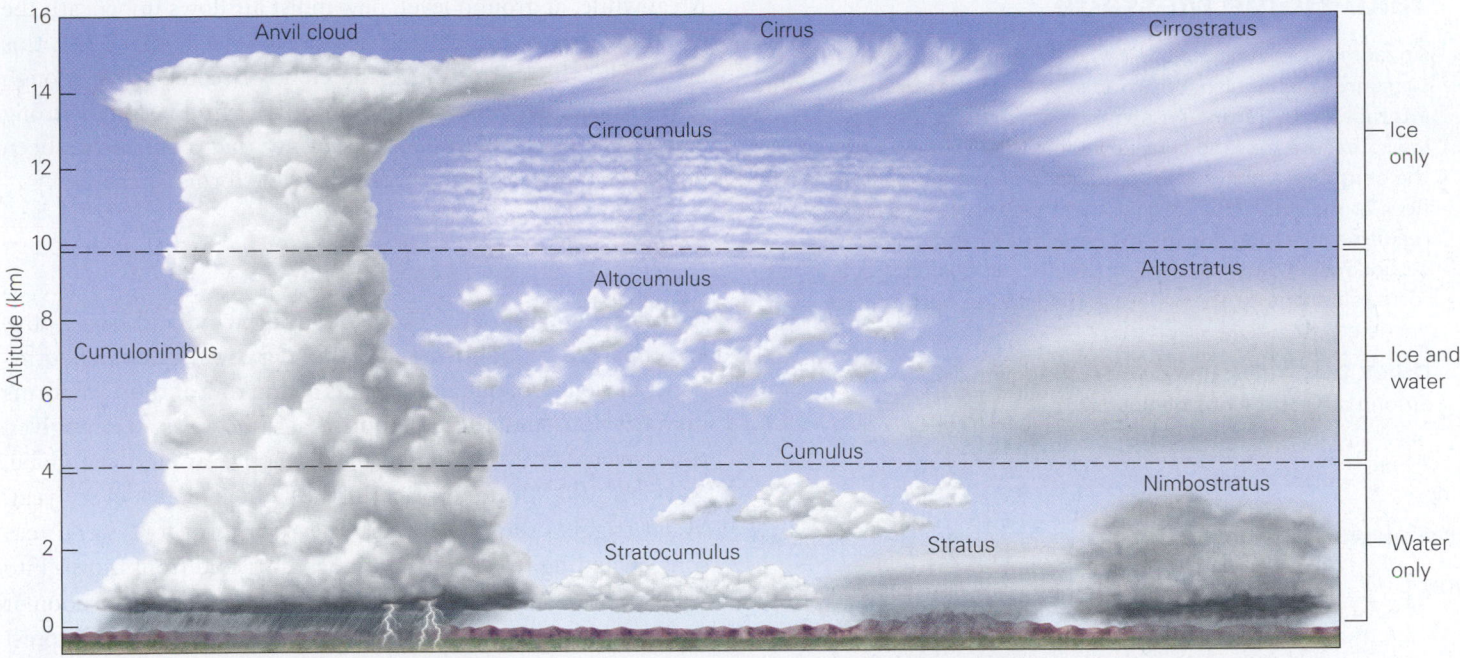

(a) The type of cloud that forms depends on the stability of the air, the temperature at which moisture condenses, and the wind speed.

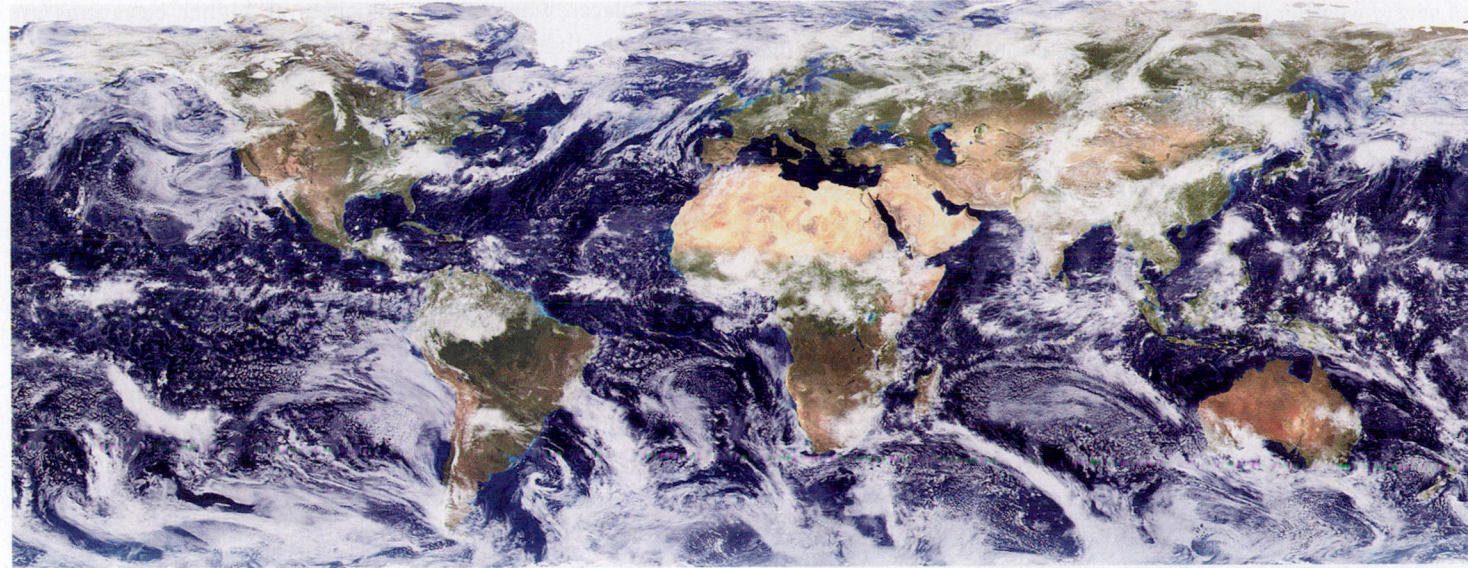

(b) A satellite photo of the Earth displays the distribution of clouds. At times, more than half of the surface has cloud cover.

add the suffix -*nimbus* or the prefix *nimbo-* if the cloud produces rain.

Applying this cloud terminology, we see that a nimbostratus is a layered, sheet-like rain cloud, and a cumulonimbus is a rain-producing puffy cloud. Cumulonimbus clouds can be truly immense, with their bases lying at less than 1 km high and their tops butting up against the tropopause at an elevation of over 14 km. Large cumulonimbus clouds spread laterally at the tropopause to form broad, flat-topped clouds called *anvil clouds* (see Fig. 20.17).

The differences in cloud types depend on whether the clouds develop in stable or unstable air. *Stable air* does not have a tendency to rise because it is colder than its surroundings—stratus clouds may form in such stable air. *Unstable air* has a tendency to rise in updrafts because it is warmer than its surroundings. Cumulus clouds, which in time-lapse photography look as if they're boiling, form in unstable air and contain intense updrafts, bordered by downdrafts. Plane flights through the unstable air of cumulus clouds can be intensely bumpy.

20.7 Storms: Nature's Fury

An overcast sky may be inconvenient for a picnic, but it won't threaten life or property—but a storm can! A storm is an episode of severe weather during which winds, rainfall, snowfall, and in some cases, lightning become strong enough to be bothersome and even dangerous (**Fig. 20.21**). Storms are commonly associated with large pressure gradients, which may exist across a steep front, for pressure gradients produce strong winds. Storms also form where local conditions cause relatively large quantities of warm, moist air to rise rapidly. Rising moist air can trigger a storm because when the air reaches higher elevations, it condenses and, as we have seen, releases latent heat. This heat warms the air, makes it more buoyant, and thus causes it to rise

FIGURE 20.21 A thunderstorm can drench an area with rain and strike it with lightning.

still further, until it becomes cool enough to produce clouds. Meanwhile, at ground level, new moist air flows in beneath the clouds to replace the air that has already risen; effectively, this air "feeds" the storm. Once the clouds become thick enough to start producing heavy rain, and/or the wind becomes strong enough to be troublesome, we can say that a storm has been born. We'll now look at various types of storms.

Thunderstorms

A **thunderstorm** is an episode of strong wind and heavy rain, accompanied by lightning and thunder. Some thunderstorms occur in isolation, but along fronts a long row of thunderstorms may develop, and these comprise a *squall line*. At any given time, over 2,000 thunderstorms are active around the globe, and over 100,000 take place in the United States every year. What triggers thunderstorms? They can form for several reasons, including frontal lifting, in which a cold front moves into a region of warm, moist air in temperate latitudes; afternoon or evening convective lifting in the tropics, where the rainforest can supply immense amounts of moisture; and orographic lifting, in which moist air is pushed up over mountains.

In places where wind velocities at higher elevations are faster than those at lower elevations, shear between the air at different elevations causes air in the storm to start rotating. Initially, this rotation has a horizontal axis, but updrafts eventually tilt the rotation axis down, so it becomes steep, and the whole storm starts to rotate around the updraft. Such a thunderstorm, which can grow to be huge, is called a **supercell**.

A typical thunderstorm has a relatively short lifespan, lasting from under an hour to a few hours (**Fig. 20.22**). The storm begins when a cumulonimbus cloud, fed by a steady supply of warm, moist air, grows large. The rising hot air, kept warm by the addition of energy from the latent heat of condensation, creates updrafts that cause the cloud to stack, or billow upward, toward the tropopause. When this air adiabatically cools, precipitation begins.

Once precipitation begins, a thunderstorm has reached its mature stage. Falling rain pulls air down with it, and evaporative cooling takes place, creating strong downdrafts. By this stage, the top of the cloud has reached the top of the troposphere and begins to spread laterally to form an anvil cloud. Because of the simultaneous occurrence of updrafts and downdrafts, a mature thunderstorm produces gusty winds and the greatest propensity for lightning. Eventually, downdrafts become the overwhelming wind; their cool air cuts off the supply of warm, moist air, so the thunderstorm dissipates.

Ice crystallizes in the higher levels of some thunderstorms, where temperatures are below freezing. Updrafts keep the ice particles aloft, allowing them to build into ice balls known as **hail**, or hailstones. As a hailstone tosses about high in the

FIGURE 20.22 Evolution of a thunderstorm takes place in three stages.

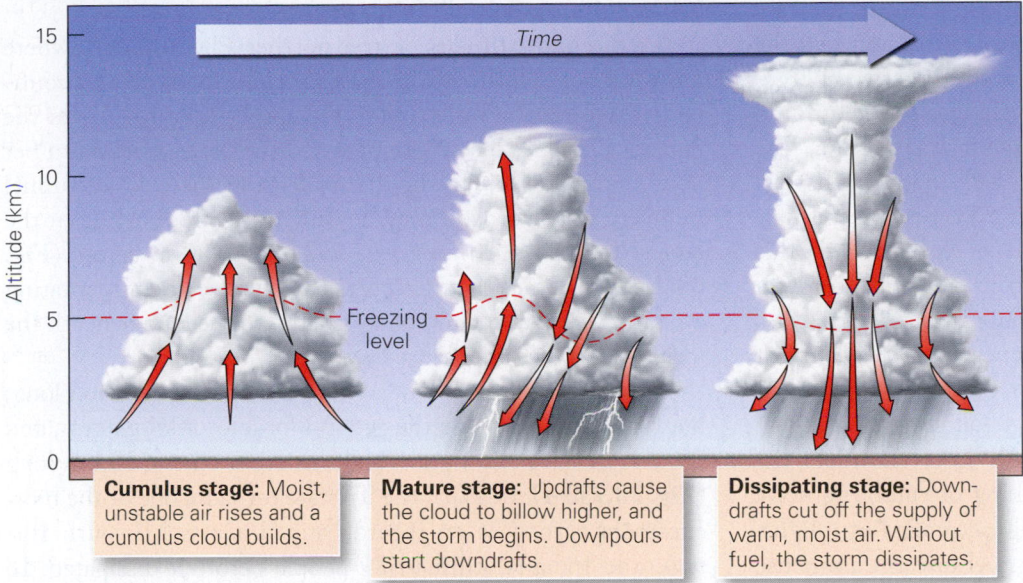

Cumulus stage: Moist, unstable air rises and a cumulus cloud builds.

Mature stage: Updrafts cause the cloud to billow higher, and the storm begins. Downpours start downdrafts.

Dissipating stage: Downdrafts cut off the supply of warm, moist air. Without fuel, the storm dissipates.

cloud, more ice deposits (attaches) to its surface, until the hailstone gets heavy enough to fall. A discrete shower of hail may tumble from a cloud over a few minutes to form a *hail streak* on the ground, typically 2 km by 10 km and elongated in the direction that the storm moves. Though most hailstones are pea-sized, the largest recorded hailstone reached a diameter of 14 cm and weighed 0.7 kilograms.

Thunderstorms are so named because of the crashes and rumbles of thunder that typically accompany them. The thunder is a consequence of lightning. A **lightning bolt** (or lightning stroke) is a giant spark that forms between two parts of a cloud, between two clouds, or between a cloud and the ground. It resembles the spark that shocks you when you shuffle along a rug and then touch a door handle. But the spark from the door handle is only about a millimeter long and a fraction of a millimeter in diameter, whereas a single lightning bolt may be up to 10 km long, up to 3 cm in diameter, and can contain enough energy to power a house for many months.

Lightning develops, in part, because of movement of various kinds of ice particles within storm clouds. Simplistically, the particles transfer electrons to each other when they come in contact, and different types of particles with different charges move to different parts of the storm cloud. Overall, positive charges accumulate toward the top of the cloud, and negative charges accumulate toward the bottom (**Fig. 20.23a**). Air is a good insulator, so the charge separation can become very large until a giant spark or pulse of current, the lightning stroke, jumps across the gap. Essentially, lightning is like a giant short circuit across which a huge (30-million-volt) pulse of electricity flows.

FIGURE 20.23 Formation of a lightning bolt happens because a charge separation forms in a cloud as rain falls.

Positive charge forms at the top of a cloud, negative at bottom.

(a) The negative charge at the cloud's base repels negative charges on the ground, so the ground develops a positive charge.

(b) As a leader descends from the cloud, positive charges flow upward from an object on the ground.

(c) When connection is complete, a return stroke from the ground to the cloud produces the main part of the flash.

As we noted, lightning strokes can jump from one part of a cloud to another or from a cloud to the ground. In the case of cloud-to-ground lightning, charge separation between the cloud and the ground develops because the negative charges at the base of the cloud repel negative charges on the ground below, creating a zone of positive charge on the ground. The stroke begins when electrons leak from the negatively charged base of the cloud incrementally downward across the insulating air gap, creating a conductive path, or *leader*. While this happens, positive charges flow upward to the cloud through conducting materials, such as trees or buildings (**Fig. 20.23b**). The instant that the charge flows connect, a strong current develops, allowing electrons to flow groundward. This current, the "return stroke," constitutes the main bolt (**Fig. 20.23c**).

We hear **thunder**, the cracking or rumbling noise that accompanies lightning, because the immense energy of a flash heats the surrounding air to a temperature of 8,000° to 33,000°C, causing it to expand almost instantly. This expansion, like an explosion, creates sound waves that travel through the air to our ears. Sound travels much more slowly than light, so we hear thunder after we see lightning. A 5-second time delay between the two means that the lightning flashed about 1.6 km (1 mile) away.

Over 80 people a year die from lightning strikes in the United States alone, and many more are seriously burned or shocked. Lightning that strikes trees heats the sap so quickly that the trees can literally explode. Lightning can spark devastating forest fires and set buildings on fire. You can reduce the hazard to buildings by installing lightning rods, upward-pointing iron spikes that conduct electricity directly to the ground so that it doesn't pass through the building.

Tornadoes

Thunderstorms that grow to be intense supercells may spawn one or more tornadoes. A **tornado** is a near-vertical, funnel-shaped cloud in which air rotates extremely rapidly around the axis (center line) of the funnel (**Fig. 20.24**). In other words, a tornado is an intense vortex beneath a severe thunderstorm. The word probably comes either from the Spanish *tonar*, meaning to turn, or *tronar*, meaning thunder (perhaps referring to the loud noise generated by a tornado). They form when a sudden, intense downdraft rushes down the back side of a supercell. When this downdraft reaches the ground, it forms an intense rotating cylinder of air whose axis is parallel to the ground. The updraft of the supercell then pulls up this cylinder so that the spinning air starts to spiral upward around a vertical axis. As the cylinder stretches up into the storm, it becomes narrower, and like a skater pulling her arms inward while spinning, the

cylinder spins faster, until eventually its wind speed can range between 100 and over 300 km per hour.

In the mid-latitudes of the northern hemisphere, where most tornadoes form, air in the funnel normally rotates counterclockwise around the center and spirals upward. Air in the fiercest tornadoes may move at speeds in excess of 322 km per hour (about 200 mph). The diameter of the base of the funnel in a very small tornado may be only 5 m across, while in the largest tornadoes it may be as wide as 1,500 m across (**Fig. 20.25a, b**). Because of the centrifugal force caused by rotating air, air diverges at the top of a tornado, so the land beneath the tornado becomes an intense low-pressure zone.

Small tornadoes may cut a swath less than a kilometer long, but large tornadoes raze the ground for tens of kilometers, and the largest have left a path of destruction up to 500 km long (**Fig. 20.25c–f**). In 1925, the Tristate tornado, one of the most enormous tornadoes on record, ripped across Missouri, Illinois, and Indiana, killing 689 people before it dissipated. In some cases, two or three tornadoes may erupt from a single thunderstorm. Massive thunderstorm fronts may produce a tornado swarm, or outbreak, dozens of tornadoes out of the same storm. In April 1974, a single thunderstorm system generated a swarm of at least 148 individual tornadoes, which killed 307 people over 11 states. On November 11, 2002, a chain of thunderstorms covering a belt from Ohio to Alabama spawned 66 tornadoes that left 66 people dead. The death toll would have been higher were it not for warnings broadcast by the U.S. National Weather Service, which sent people scrambling for safety. Between April 25 and April 28, 2011, a "super

FIGURE 20.24 The spiraling winds of a tornado's funnel.

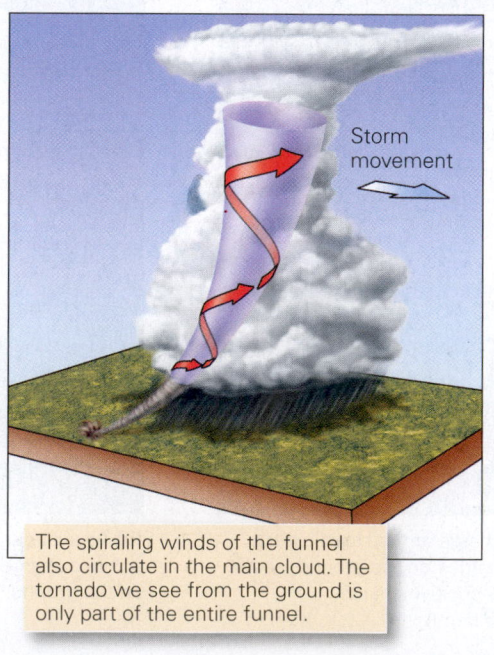

Storm movement

The spiraling winds of the funnel also circulate in the main cloud. The tornado we see from the ground is only part of the entire funnel.

outbreak" of tornadoes occurred in midwestern and southeastern states. At least 336 tornadoes were recorded. Particularly huge tornadoes struck Joplin, Missouri, in 2013 and Moore, Oklahoma, in 2014.

In North America, tornadoes drift with a thunderstorm from southwest to northeast because of the prevailing wind direction, traveling across the land at speeds of up to 100 km per hour. They tend to hopscotch across the landscape, touching down for

FIGURE 20.25 Tornadoes and the damage they cause.

(a) A moderate-sized tornado touches down near Mulvane, Kansas.

(b) A huge tornado near Eureka, Illinois.

(c) The swath of destruction left by a 1999 tornado in Oklahoma.

(d) Even downtowns are not immune—this tornado struck Miami in 1997.

(e) Close-up of the damage due to a 2007 tornado in Kansas.

(f) A helicopter view of the devastation track left by one of the April 2011 tornadoes that hit Tuscaloosa, Alabama.

a stretch, then rising up into the air for a while before touching down again. This characteristic leads to a bizarre incidence of damage—one house may be blasted off its foundation while its next-door neighbor remains virtually unscathed. Winds on one side of a tornado move in the same direction as the overall funnel drift, whereas those on the other side move in the opposite direction. Thus, for a northeast-moving counterclockwise funnel, damage is greater on the southeast side of the funnel.

Tornadoes cause damage because of the force of their rapidly moving wind. The wind lifts trucks and tumbles them for hundreds of meters, uproots trees, and flattens buildings. Particularly large tornadoes can even rip asphalt off a highway. In some cases, tornadoes cause strange kinds of damage: they have been known to drive straw through wood, lift cows and carry them unharmed for hundreds of meters, and raise railroad cars right off the ground. Because of the range of damage a tornado can cause, T. T. Fujita of the University of Chicago proposed a scale that distinguishes among tornadoes on the basis of damage (**Table 20.2**). The wind speeds in the **Enhanced Fujita Scale** are estimates, based on the damage assessment.

The special weather conditions that spawn tornadoes in the midwestern United States and Florida develop when cold polar air from Canada collides with warm tropical air from the Gulf of Mexico. These conditions happen most frequently during the months of March through September but sometimes occur at other times of the year. (Tornadoes, for example, raked across Illinois in November of 2014.) So many tornadoes occur during the summer in a belt from Texas to Indiana that this region has the unwelcome nickname "tornado alley" (**Fig. 20.26a**). During a 30-year span, the number of reported tornadoes per year ranged between 420 and 1,100 in the United States, with an average of 770 per year (fewer than 20 strike Canada annually). On average, about 80 people a year die in tornadoes, but in some years the toll may be in the hundreds. The 2011 super outbreak alone killed about 350 people because some of its tornadoes rampaged through cities (**Fig. 20.26b**).

Because of the threat tornadoes pose to life and property, meteorologists have worked hard to be able to forecast them. First they search for appropriate weather conditions. If these conditions exist, meteorologists issue a *tornado watch*. If observers spot an actual tornado or see one forming, they issue a **tornado warning** for the region in its general path (the exact path can't be predicted). If you hear a warning, it's best to take cover immediately in a basement or at least in an interior room away from windows. With the invention of Doppler radar, which uses the Doppler effect (see Chapter 1) to identify rain moving in strong winds, meteorologists may detect tornadoes without even going outside. Probable tornadoes appear in Doppler images as a distinct hook-like shape at the edge of a thunderstorm (**Fig. 20.26c**). The ball of debris carried by a tornado causes intense reflectivity on a Doppler image.

Extratropical Cyclones and Nor'easters

Extratropical (mid-latitude or wave) cyclones can produce incredible storms. In the Midwest, they can spawn intense thunderstorms and associated flash floods. In colder weather they can produce blizzards (snowstorms of immense proportions). A blizzard in 1888 dumped up to 1.5 m (60 in.) of snow, and a blizzard in 1996 buried New York under 1.2 m (48 in.) of snow. Large extratropical cyclones of North America that affect the Atlantic coast are called **nor'easters** because the

TABLE 20.2 Enhanced Fujita Scale for Tornadoes

Scale	Category	Wind Speed km per Hour (mph)	Average Path Length; Average Path Width	Typical Damage
EF0	Weak	104–137 (65–85)	0–1.6 km; 0–17 m	Branches and windows broken
EF1	Moderate	138–177 (86–110)	1.6–5.0 km; 18–55 m	Trees broken; shingles peeled off; mobile homes moved off their foundations
EF2	Strong	178–217 (111–135)	5–16 km; 56–175 m	Large trees broken; mobile homes destroyed; roofs torn off
EF3	Severe	218–266 (136–165)	16–50 km; 176–556 m	Trees uprooted; cars overturned; well-constructed roofs and walls removed
EF4	Devastating	267–322 (166–200)	50–160 km; 0.56–1.5 km	Strong houses destroyed; buildings torn off foundations; cars thrown; trees carried away
EF5	Incredible	over 322 (over 200)	160–500 km; 1.5–5.0 km	Cars and trucks carried more than 90 m; strong houses disintegrated; bark stripped off trees; asphalt peeled off roads

FIGURE 20.26 North American tornadoes are most common in "tornado alley," a band extending from Texas to Indiana, where the polar air mass collides with the Gulf Coast maritime tropical air mass. The storm systems move eastward.

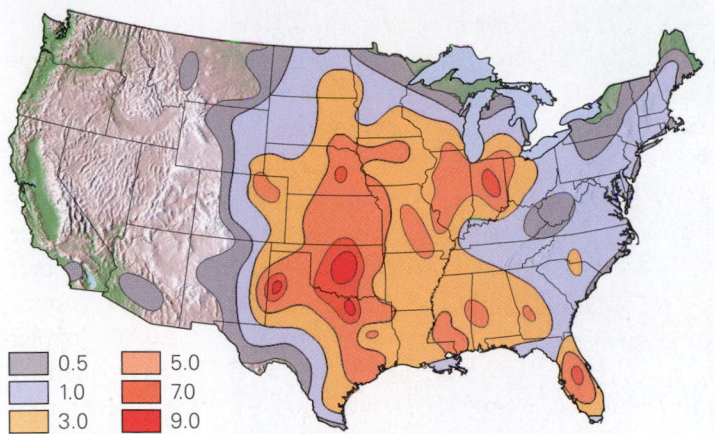

0.5	5.0
1.0	7.0
3.0	9.0

(a) Number of tornadoes per year (per 26,000 square km, for a 27-year period).

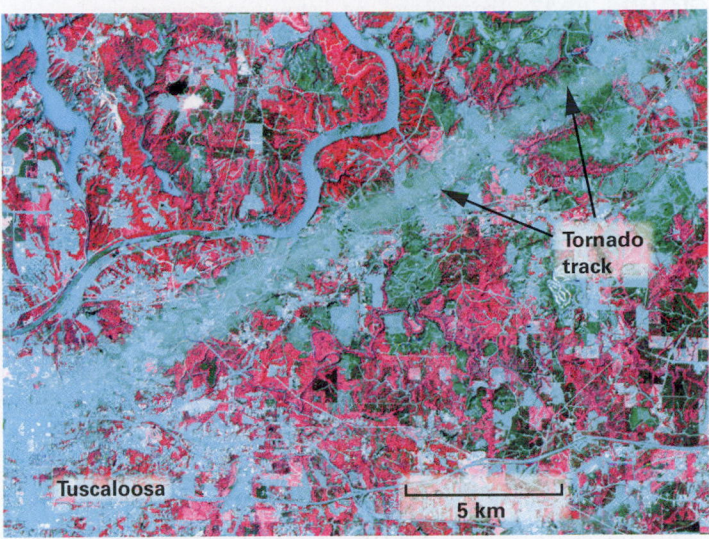

Tornado track

Tuscaloosa

5 km

(b) This false-color satellite image shows the track of a tornado near Tuscaloosa, Alabama. Red areas are forested—the winds ripped out the trees.

100 km

Tuscaloosa

Debris ball

The maroon is due to the intense reflectivity of debris being carried by the tornado.

Less ▮▮▮▮▮▮ More
Doppler reflectivity

(c) A radar image shows the characteristic "hook" of a tornado, as viewed by satellite, as it passes near Vilonia, Arkansas, in 2011.

cold, counterclockwise winds of these cyclones come out of the northeast. Some nor'easters are truly phenomenal storms. During some storms, winds are not as strong as those of a hurricane, but they cover such a large area that waves in the open ocean build to a height of over 11 m, a disaster for ships. When the waves reach shore, they erode huge stretches of beach.

Hurricanes—A Coastal Calamity

What Is a Hurricane? Global-scale convection of the atmosphere, influenced by the Coriolis effect, causes currents of warm air to flow steadily from east to west in tropical latitudes

(**Fig. 20.27a**). As the air flows over the ocean, it absorbs moisture. As we've seen, because air becomes less dense as it gets warmer, tropical air eventually begins to rise like a balloon. The rising air cools, and the water vapor it contains condenses to form clouds. Condensation during cloud formation releases latent heat, warming the clouds still further and causing them to billow still higher. If the air contains sufficient moisture, the clouds grow into a cluster of large thunderstorms, which consolidate to form a single, very large storm. Because of the Coriolis effect, this storm cluster starts rotating around a vertical axis and evolves into a broad swirl of clouds called a *tropical disturbance*. At the center of the swirl, air rises, following an upward spiral path, and at the top of the storm, this rising air spreads out, or diverges. The resulting loss of air molecules, at high elevation, over the storm creates low pressure at the base of the storm.

If a tropical disturbance remains over warm ocean water, as can happen in late summer and early fall, warm, moist air continues to feed the storm. Eventually the storm organizes into a spiral of rapidly circulating clouds, and the tropical disturbance becomes a *tropical depression*. Additional nourishment causes the tropical depression to spin even faster and grow broader, until it becomes a *tropical storm* and receives a name. If a tropical storm becomes powerful enough, it becomes a *tropical cyclone*. Formally defined, a **tropical cyclone** is a huge rotating storm, which forms in tropical latitudes, and in which winds exceed 119 km per hour (74 mph). In the northern hemisphere, it resembles a giant counterclockwise spiral of clouds—300 to 1,500 km (930 miles) wide—when viewed from space (**Fig. 20.27b**). Such a storm is a *hurricane* in the Atlantic, Caribbean, Gulf of Mexico, and eastern Pacific, a

FIGURE 20.27 The structure and distribution of hurricanes.

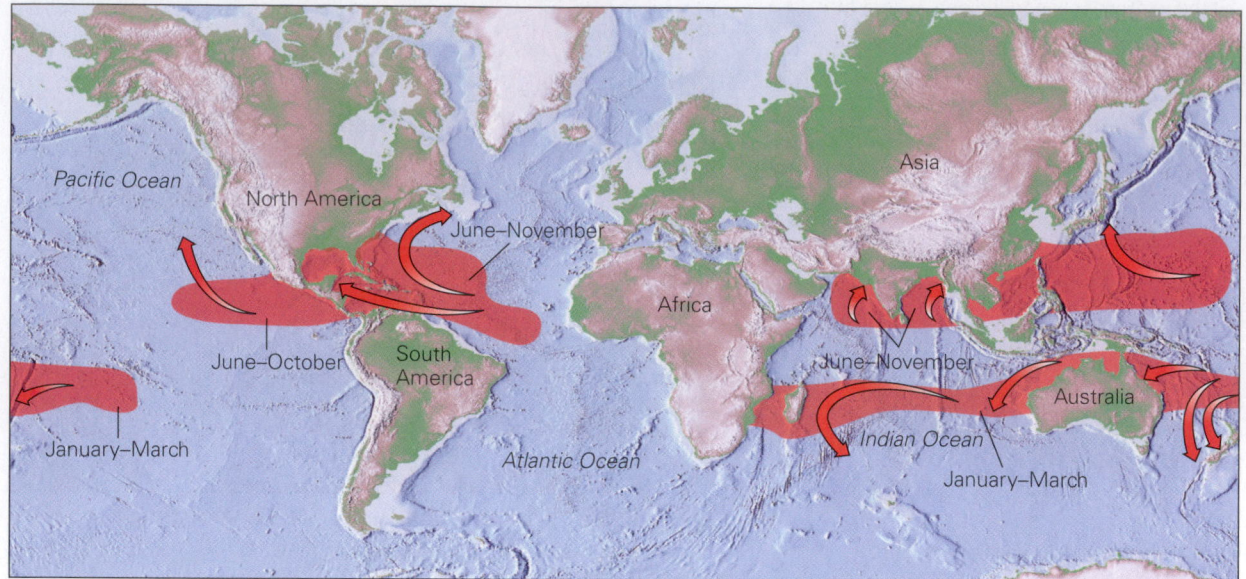

(a) Hurricanes form only in certain regions of the ocean, where water temperatures are high. They generally follow regional tracks and occur during certain times of the year.

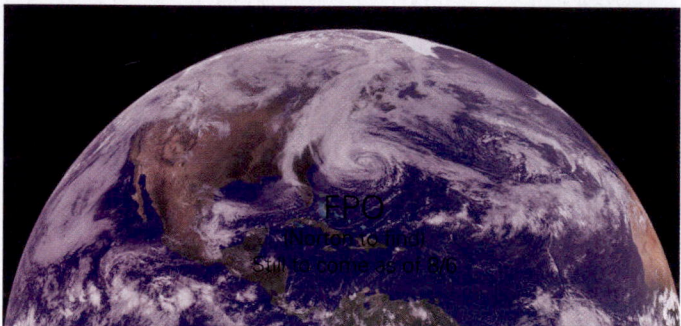

(b) Hurricane Sandy approaches the east coast of the United States in 2012, as seen from space.

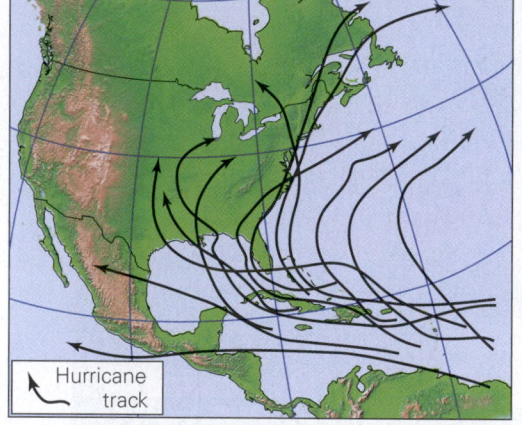

(c) Tracks of several Atlantic hurricanes show how most drift westward, then northward.

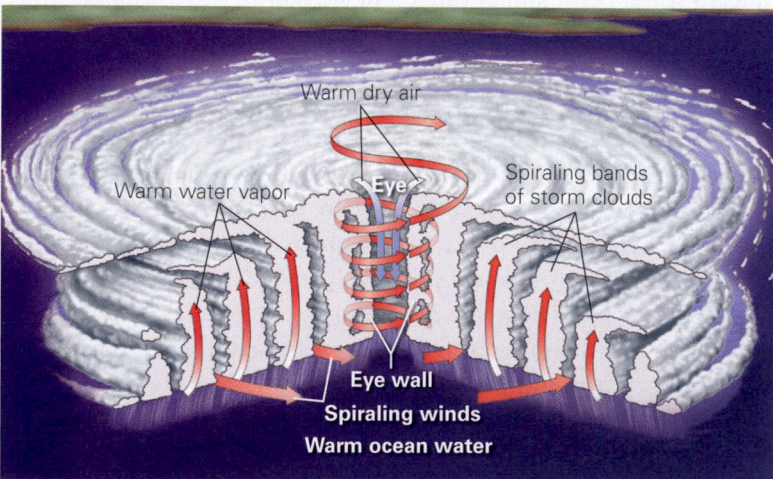

(d) In this cutaway drawing, we can see the spirals of clouds, the eye, and the eye wall of a hurricane. Dry air descends in the eye.

typhoon in the northwestern Pacific, and simply a *cyclone* in the southwestern Pacific and the Indian Ocean.

Atlantic hurricanes generally form in the ocean to the east of the Caribbean Sea, though some form in the Caribbean itself. They first drift westward at speeds of up to 60 km per hour (37 mph), then may eventually turn north and head into the North Atlantic or into the interior of North America, where they die when they run out of a supply of warm water (**Fig. 20.27c**). Weather researchers classify the strength of hurricanes using the **Saffir-Simpson Scale**, which runs from 1 to 5 (**Table 20.3**); somewhat different scales are used for typhoons and cyclones. On the Saffir-Simpson scale, a Category 5 hurricane has sustained winds of 250 km per hour or

TABLE 20.3 Saffir-Simpson Scale for Hurricanes

Scale	Category	Wind Speed km/h (mph)	Air Pressure in Eye (millibars)	Damage
1	Minimal	119–153 (74–95)	980 or more	Branches broken; unanchored mobile homes damaged; some flooding of coastal areas; no damage to buildings; storm surge of 1.2 to 1.5 m
2	Moderate	154–177 (96–110)	965–979	Some roofs, doors, and windows damaged; mobile homes seriously damaged; some trees blown down; small-boat moorings broken; storm surge of 1.6 to 2.4 m
3	Extensive	178–209 (111–130)	945–964	Some structural damage to small buildings; large trees blown down; mobile homes destroyed; structures along coastal areas destroyed by flooding and battering; storm surge of 2.5 to 3.6 m
4	Extreme	210–250 (131–155)	920–944	Some roofs completely destroyed; extensive window and door damage; major damage and flooding along coast; storm surge of 3.7–5.4 m. Widespread evacuation of regions within up to 10 km of the coast may be necessary.
5	Catastrophic	Over 250 (over 155)	Less than 920	Many roofs and buildings completely destroyed; extensive flooding; storm surge greater than 5.4 m. Widespread evacuation of regions within up to 16 km of the coast may be necessary.

greater (≥155 mph). The highest wind speed ever recorded during a hurricane was in excess of 300 km per hour.

A typical hurricane (or typhoon or cyclone) consists of several spiral arms, called *rain bands*, extending inward to a central zone of relative calm known as the *hurricane eye* (**Fig. 20.27d**). A rotating vertical cylinder of clouds, the eye wall, surrounds the eye. Winds spiral toward the eye and, similar to a tornado, they accelerate near the interior of the storm. In a hurricane, the fastest winds occur along the eye wall. Thus, hurricane-force winds of a given storm generally affect a belt that is only 15% to 35% as wide as the whole storm. On the side of the eye where winds blow in the same direction as the whole storm is moving, the ground speed of winds is greatest, because the storm's overall speed adds to the rotational motion.

The Damage Due to Hurricanes Hurricanes pose extreme danger in the open ocean, because their winds cause huge waves to build and thus have led to the foundering of countless ships. They also cause havoc in coastal regions, and even inland, though they die out rapidly after moving onshore. This damage happens for several reasons.

- *Wind:* Winds of weaker hurricanes tear off branches and smash windows (**Fig. 20.28a**). Stronger hurricanes uproot trees, rip off roofs, and collapse walls. Extreme hurricanes can carry away or flatten whole towns, causing wind damage as intense as that of a tornado but over a broader area.

- *Waves:* Winds shearing across the sea surface during a hurricane generate huge waves. In the open ocean, these waves can capsize ships. Near the shore, waves batter and erode beaches, rip boats from moorings, and destroy coastal property (**Fig. 20.28b**).

- *Storm surge:* Extremely low air pressure develops beneath a hurricane—in fact, the lowest sea-level air pressure ever recorded (0.87 bars) occurred during a typhoon in the western Pacific in 1970. This decrease in pressure causes the surface of the sea to bulge upward over an area with a diameter of 60 to 80 km. Sustained winds blowing in an onshore direction build this bulge even higher. When the hurricane reaches the coast, the bulge of water, or **storm surge**, swamps the land (**Fig. 20.28c**). If the bulge hits the land at high tide, the sea surface will be especially high and will affect a broader area. Storm waves develop on top of storm surge, so the height of the waves is the sum of the surge height plus the wave height.

- *Rain, stream flooding, and landslides:* Rain drenches the Earth's surface beneath a hurricane. In places, a half meter or more of rain falls in a single day. Rain causes streams to flood, even far inland, and can trigger landslides (**Fig. 20.28d**).

FIGURE 20.28 Examples of hurricane damage.

(a) Wind damage due to Hurricane Andrew, 1992.

(b) Waves pummel the shore during Hurricane Sandy, 2012.

(c) Damage due to storm surge from Typhoon Haiyan in the Philippines, 2013.

(d) Flooding caused by rains of Hurricane Irene, 2011, affected Vermont, far inland from the coast.

• *Disruption of social structure:* When the storm passes, the hazard is not over. By disrupting transportation and communication networks, breaking water mains, and washing away sewage-treatment plants, hurricane damage creates severe obstacles to search and rescue, and can lead to the spread of disease, fire, and looting.

Nearly all hurricanes that reach the coast cause death and destruction, but some are truly catastrophic. Storm surge from a 1970 cyclone making landfall on the low-lying delta lands of Bangladesh led to an estimated 500,000 deaths. In 1992, Hurricane Andrew leveled extensive areas of southern Florida, causing over $30 billion in damage and leaving 250,000 people homeless. Hurricane Katrina, in 2005, and 2013's Typhoon Haiyan serve as examples of particularly devastating storms, as we now see.

Case Studies of Hurricanes

Katrina, 2005 Tropical Storm Katrina came into existence over the Bahamas and headed west. Just before landfall in southeastern Florida, winds strengthened and the storm became Hurricane Katrina. This hurricane sliced across the southern tip of Florida, causing several deaths and millions of dollars in damage. It then entered the Gulf of Mexico and passed directly over the Loop Current, an eddy of summer-heated water from the Caribbean that had entered the Gulf of Mexico. Water in the Loop Current reaches temperatures of 32°C (90°F) and thus stoked the storm, injecting it with a burst of energy (from the latent heat of condensation) sufficient for the storm to morph into a Category 5 monster whose swath of hurricane-force winds reached a width of 325 km (200 miles) (**Fig. 20.29a–b**).

When it entered the central Gulf of Mexico, Katrina turned north and began to bear down on the Louisiana-Mississippi coast. The eye of the storm passed just east of New Orleans and then across the coast of Mississippi (**Fig. 20.29c**). Storm surges broke records, in places rising 7.5 m (25 feet) above sea level, and they washed coastal communities off the map along a broad swath of the Gulf Coast (**Fig. 20.30a, b**). In addition to the devastating wind and surge damage, Katrina led to the drowning of New Orleans.

To understand what happened to New Orleans, we must consider the city's geologic history. New Orleans grew on the Mississippi Delta between the banks of the Mississippi River on the south and Lake Pontchartrain (actually a bay of the Gulf of Mexico) on the north. The older parts of the town grew up on the relatively high land of the Mississippi's natural levee. Younger parts of the city, however, spread out over the topographically lower delta plain. As decades passed, people modified the surrounding delta landscape by draining wetlands, constructing artificial levees that confined the Mississippi River, and extracting groundwater. Sediment beneath the delta compacted, and the delta's surface has been starved of new sediment, so large areas of the delta sank below sea level. Today most of New Orleans lies in a bowl-shaped depression as much as 2 m (7 feet) below sea level—the hazard implicit in this situation had been recognized for years (**Fig. 20.30c**).

The winds of Hurricane Katrina ripped off roofs, toppled trees, smashed windows, and triggered the collapse of weaker buildings, but their direct consequences were not catastrophic. However, when the winds blew storm surge into Lake Pontchartrain, its water level rose beyond most expectations and pressed against the system of artificial levees and flood walls that had been built to protect New Orleans. Hours after the hurricane eye had passed, the high water found a weakness along the floodwall bordering a drainage canal and pushed out a section. Breaks eventually formed in a few

FIGURE 20.29 Hurricane Katrina's path through the warm waters of the Gulf of Mexico.

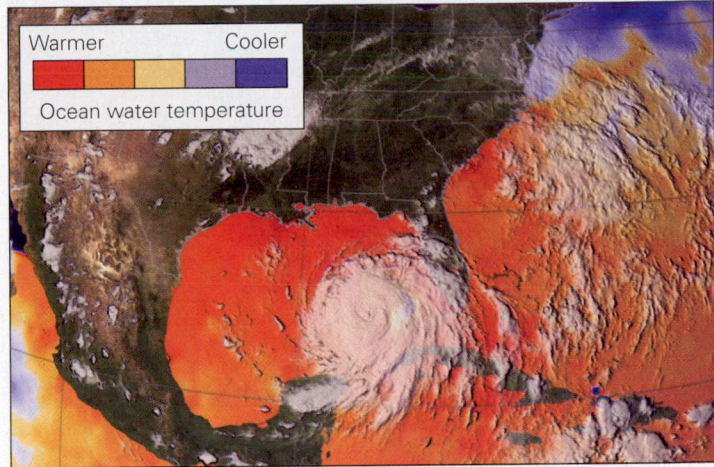

(a) A satellite photo of Hurricane Katrina over the hot water of the Gulf of Mexico.

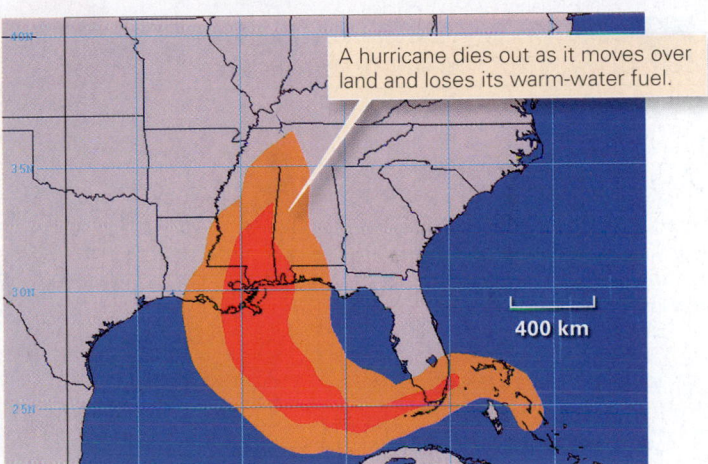

A hurricane dies out as it moves over land and loses its warm-water fuel.

(b) A wind-swath map of Hurricane Katrina. Red areas represent hurricane winds; orange areas represent tropical-storm winds.

(c) The track of the storm lay just east of New Orleans.

FIGURE 20.30 The devastation of coastal areas by Hurricane Katrina.

(a) Surge from the storm destroyed homes along the Alabama coast.

(b) Officials survey the storm damage.

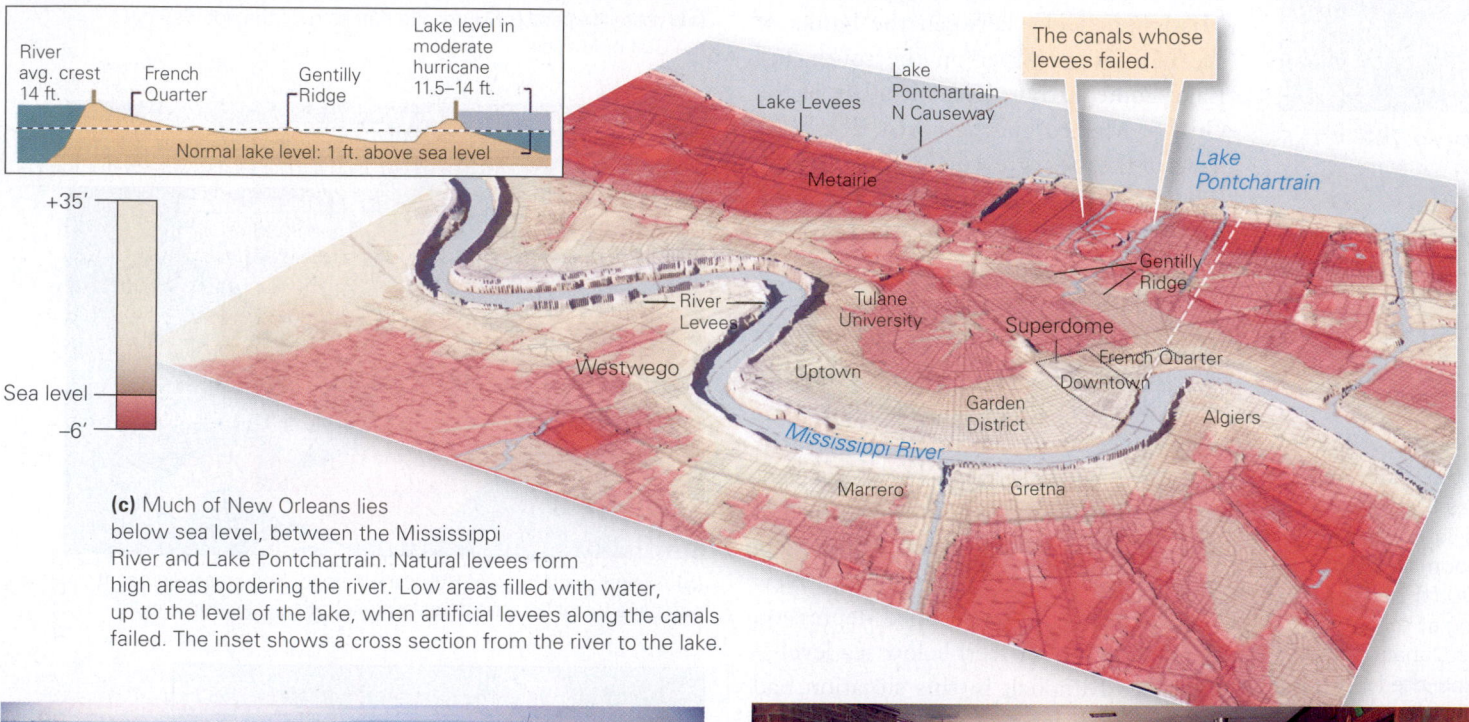

(c) Much of New Orleans lies below sea level, between the Mississippi River and Lake Pontchartrain. Natural levees form high areas bordering the river. Low areas filled with water, up to the level of the lake, when artificial levees along the canals failed. The inset shows a cross section from the river to the lake.

(d) Water flowing across the levees bordering the 17th Street Canal after the hurricane had passed.

(e) Damage due to flooding in a New Orleans home.

other locations as well. So, a day after the hurricane was over, New Orleans began to flood. As the water line climbed the walls of houses, residents fled first upstairs, then to their attics, and finally to their roofs. Water spread across the city until the bowl of New Orleans filled to the same level as Lake Pontchartrain, submerging 80% of the city (**Fig. 20.30d**).

Floodwaters washed some houses away and filled others with debris (**Fig. 20.30e**). The disaster took on national significance, as the trapped population sweltered without food, drinking water, or adequate shelter. With no communications, no hospitals, and few police, the city almost descended into anarchy. It took days for outside relief to reach the city, and by then, many had died and parts of New Orleans, a cultural landmark and major port, had become uninhabitable. It has taken years for the city to rebuild, and parts remain devastated.

Sandy, 2012 This storm began as a tropical depression on October 22, 2012, and became a hurricane just before hitting Jamaica. Then it curved north and struck New Jersey on October 28. When it hit, its maximum sustained wind was 185 km per hour (115 mph). But Sandy collided and merged with a large extratropical cyclone, to form the widest Atlantic hurricane ever documented, now known as "Superstorm Sandy." At its peak, it was 1,800 km (1,100 miles) wide. The storm surge washed away landmarks of the New Jersey shore and devastated expensive beachfront property. Its surge also inundated parts of New York City and even caused extensive flooding of the subway system. The $68 billion of damage that Sandy caused was a shock to the northeastern metropolitan areas and has led city governments to begin thinking about how to plan for the consequences of rising sea level.

Haiyan, 2013 Typhoon Haiyan began as a tropical depression in the western Pacific on November 2. As it drifted westward over very warm ocean, it grew to typhoon status within 2 days. By November 6, it had become a Category 5 behemoth and took aim at the Philippines. Measurements indicated that it had sustained winds of 315 km per hour (195 mph), the highest wind speeds ever recorded in a hurricane. In fact, at times, the wind gusted to 378 km per hour (235 mph). As a result, Haiyan has been called the strongest storm ever to reach land. In addition to intense wind and a record-breaking storm surge, the storm dumped as much as 282 mm (11 in.) of rain over the course of a few hours. By the time it had passed over the Philippines, more than 6,300 residents had lost their lives, whole towns had been flattened, and over a million homes sustained damage. Relief efforts struggled for weeks to reach people who were cut off by the destruction of roads, harbors, and airports.

Take-Home Message

Thunderstorms develop where convection causes warm, moist air to rise, resulting in billowing clouds and rain. Charge separation develops in clouds, resulting in lightning. The spiral of air in a tornado produces wind speeds of up to 500 km per hour. Extratropical cyclones can produce large storms. Hurricanes begin over warm seawater. Energy from latent heat of condensation enlarges the storm, producing winds ranging between 119 and 315 km per hour.

QUICK QUESTION: Why do hurricanes die after they cross onto land or head to higher latitudes?

20.8 Global Climate

Climate and weather are related but are not synonymous. Weather specifically refers to the atmospheric conditions at a certain location at a specified time. **Climate**, in contrast, refers to the average weather conditions (temperature, humidity, winds, and precipitation) of a region, the typical range of weather conditions for the region, the character of the region's storms, and the nature of the region's seasons, as observed over many decades. To parse this difference, let's consider an example. On a given summer day in Winnipeg in central Canada, the weather may be sunny, hot, and humid, and on a given winter day in Miami in the southeastern United States, it may be freezing and overcast. But averaged over time, the overall temperature of Winnipeg is less than that of Miami, the range of temperature change between winter and summer in Winnipeg is much greater than the range in Miami, and Miami has hurricanes while Winnipeg

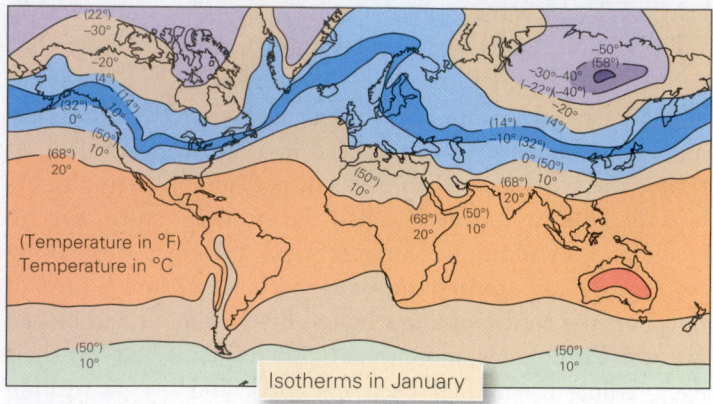

Isotherms in January

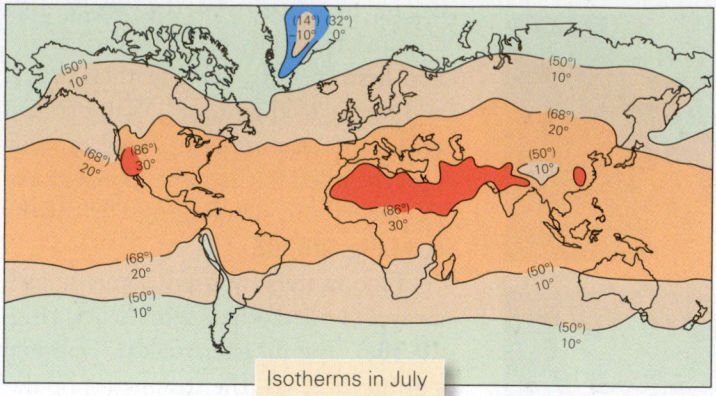

Isotherms in July

does not. Thus, Winnipeg and Miami have different climates, and for this reason, the native flora and fauna of Winnipeg differ significantly from those of Miami.

Climate Controls and Belts

Climatologists, scientists who study the Earth's climate, suggest that several distinct factors control the climate of a region.

- *Latitude:* Latitude determines the amount of solar energy a region receives as well as the contrasts among seasons and thus serves as the single most important factor in controlling climate. Polar regions, which receive much less solar radiation over the year, have colder climates than equatorial regions. And the contrast between winter and summer is greater in mid-latitudes than at the poles or the equator. We can easily see the influence of latitude by examining the global distribution of temperature, represented on a map by **isotherms**, lines along which the temperature is exactly the same at a point just above the ground surface (**Fig. 20.31**). Because land and sea do not heat up at the same rate, because the distribution of clouds is not uniform, and because ocean currents transfer heat across latitudes, isotherms are not perfect circles.
- *Altitude:* Because temperature decreases with elevation, cold climates exist at high elevations, even at the equator. Hiking from the base of a high mountain in the Andes to its summit takes you through the same range of climate belts you would pass through on a hike from the equator to the pole.
- *Proximity of water:* Land and water have very different *heat capacities* (ability to absorb and hold heat). Land absorbs or loses heat quickly, whereas water absorbs or loses heat slowly. Also, water can absorb and hold on to more heat than land can because water is semitransparent; sunlight

heats water to a depth of up to 100 m, whereas sunlight heats land to a depth of only a few centimeters. Thus, the proximity of the sea tempers the climate of a region. In fact, as a rule, locations in the interior of a continent experience a much greater range of weather conditions than do regions along the coast.

- *Proximity to ocean currents:* Where a warm current flows, it may heat the overlying air, and where a cold current flows, it may cool down the overlying air. For example, the Gulf Stream brings warm water north from the Gulf of Mexico and keeps Ireland, the United Kingdom, and Scandinavia much warmer than they would be otherwise.

FIGURE 20.32 Because of landmasses, atmospheric pressure belts, introduced in Figure 20.11, actually vary in width to create lens-shaped, semipermanent high- and low-pressure cells.

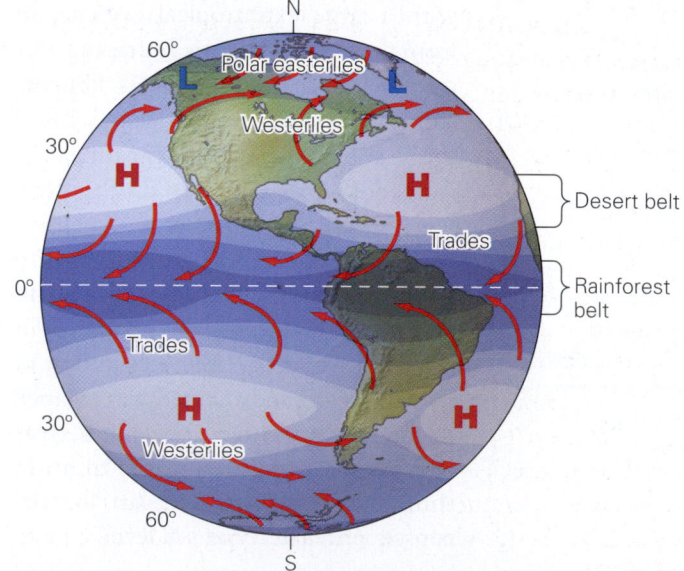

TABLE 20.4 Climate Types of the Earth

Climate Type	Regions and Characteristics
Tropical rainy	Tropical rainforests lie at equatorial latitudes and experience rain throughout the year. Rain commonly falls during afternoon thunderstorms. Tropical savanna (grasslands with brush and drought-resistant trees), which lie on either side of a rainforest, have a rainy season and a dry season. Rainforests and savannas may receive tropical monsoons.
Dry	Dry regions include deserts (regions with very little moisture or vegetation cover; vegetation that does exist has adapted to long periods without moisture) and steppes. Steppe regions (vast, grassy plains with no forest) border the desert and have somewhat more precipitation. Some steppe regions occur at high elevations, in latitudes where the climates would otherwise be more humid.
Humid mesothermal (temperate)	This category includes humid subtropical climates, with moist air and warm temperatures for much of the year, in which mixed deciduous-coniferous forest thrives; Mediterranean climates, coastal regions with most rainfall in the winter, very hot summers, and scrub forests; and marine west-coast climates, where the sea tempers the climate and may create a coastal temperate rainforest.
Humid microthermal (cold)	These higher-latitude temperate climates, which occur only in the northern hemisphere, include humid continental regions with long summers (as in the U.S. Midwest and mid-Atlantic states), in which deciduous forest thrives; humid continental regions with short summers, characterized by mixed deciduous-coniferous forest or coniferous-only forest; and subarctic climates with very short, cool summers and coniferous forest that becomes lower and scrubbier at higher latitudes.
Polar	These cold climates include tundra and ice caps. Tundra are regions with no summer and an extremely cold winter, in which only low, cold-resistant plants (moss, lichen, and grass) can survive. Much of the ground in tundra is permafrost (permanently frozen ground). In ice-cap regions, near the poles, the climate is subfreezing year-round, and any land not covered by ice has essentially no vegetation cover. Highlands are regions that lie at lower (nonpolar) latitudes but have such a high elevation that they have polar-like climates. When you enter a region above the tree line, you have entered a highland polar climate.

- *Proximity to orographic barriers:* An **orographic barrier** is a landform (such as a mountain range) that diverts airflow upward or laterally. This diversion affects the amount of precipitation and wind a region receives.
- *Proximity to high- or low-pressure cells:* Zones of high and low pressure, aligned roughly parallel to the equator, encircle the planet (see Fig. 20.11). Because land and sea have different heat capacities, they modify the zones, so high-pressure zones tend to be narrower over land. Meteorologists refer to the resulting somewhat elliptical regions of high or low pressure as semipermanent pressure cells (**Fig. 20.32**). These influence prevailing wind direction and relative humidity.

Climatologists, who have studied the distribution of climatic conditions around the globe, developed a classification scheme for climate based on such factors as the average monthly and annual temperatures and the total monthly and yearly amounts of precipitation. The vegetation of a region proves to be an excellent indicator of climate because plants are sensitive to temperature and to the amount and distribution of rainfall. **Table 20.4** lists the principal types of climate belts (**Fig. 20.33**).

Climate Variability: Monsoons and El Niño

The climate at a certain location may change during the course of one or more years. Here we look at two important examples of climate variability that affect human populations in notoriously significant ways.

Monsoons A **monsoon** is a major reversal in the wind direction that causes a shift from a very dry season to a very rainy season. In southern Asia, home to about half the world's population, people depend on monsoonal rains to bring moisture for their crops. The Asian monsoon develops primarily because Asia is so large that it includes vast tracts of land far from the sea. Further, a substantial part of this land, the Tibet Plateau, lies at a high elevation. During the winter, central Asia becomes very cold, much colder than coastal regions to the south. This coldness creates a stable high-pressure cell over central Asia. Dry air sinks and spreads outward from this cell and flows southward over southern Asia, pushing the intertropical convergence zone (ITCZ) out over the Indian Ocean, south of Asia (**Fig. 20.34a**). Thus, during the winter, southern Asia experiences a dry season. During the summer, central Asia warms up dramatically. As warm air rises and expands

FIGURE 20.33 Climate belts and their effect on vegetation.

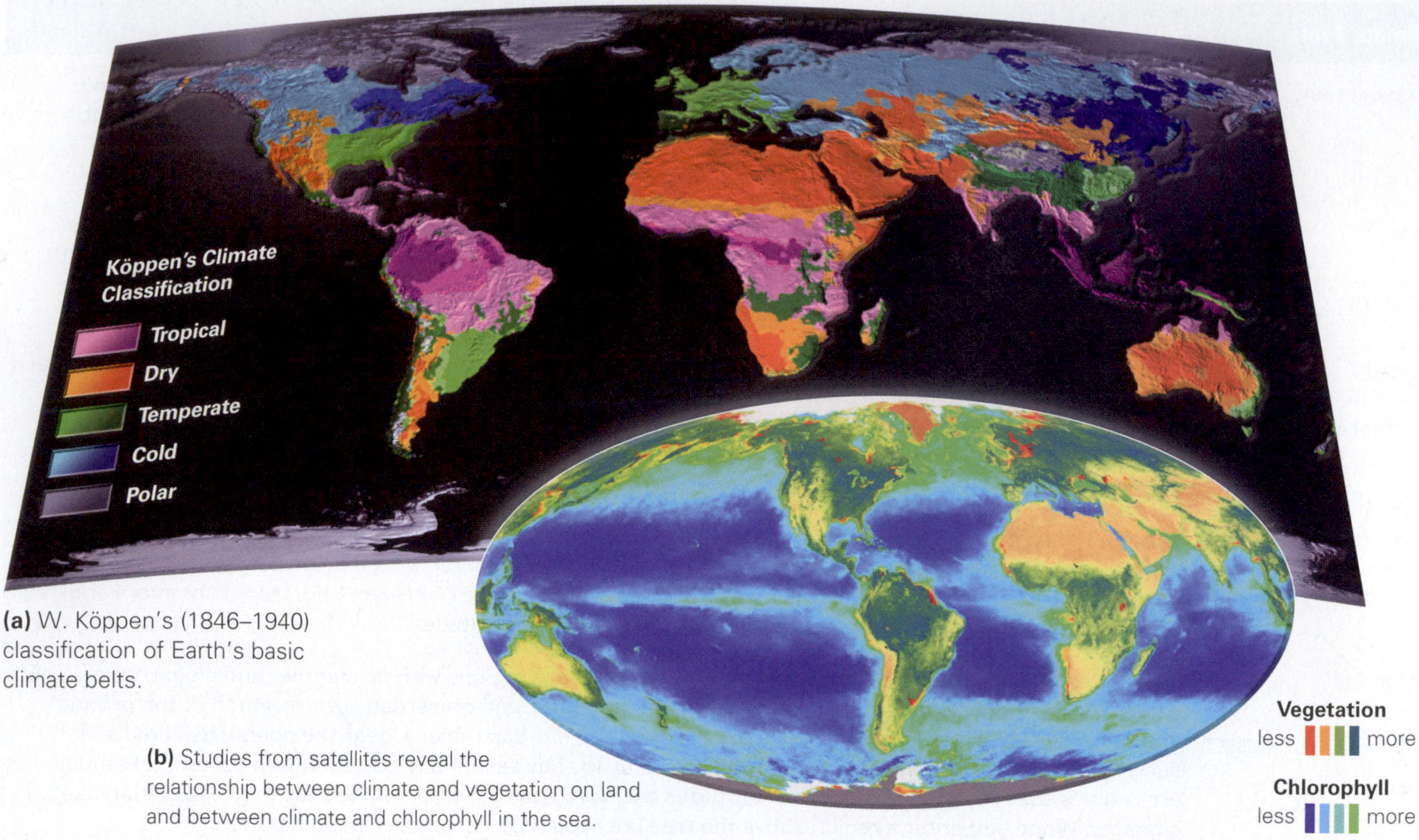

Köppen's Climate Classification

- Tropical
- Dry
- Temperate
- Cold
- Polar

(a) W. Köppen's (1846–1940) classification of Earth's basic climate belts.

(b) Studies from satellites reveal the relationship between climate and vegetation on land and between climate and chlorophyll in the sea.

Vegetation
less ▮▮▮▮ more

Chlorophyll
less ▮▮▮▮ more

FIGURE 20.34 The causes of monsoons. Because of monsoons, Asia has a wet season and a dry season every year.

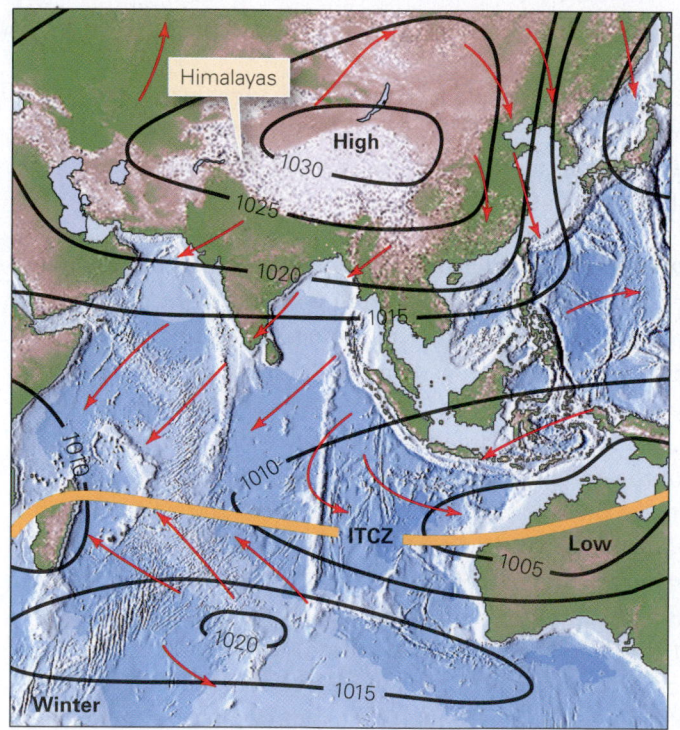

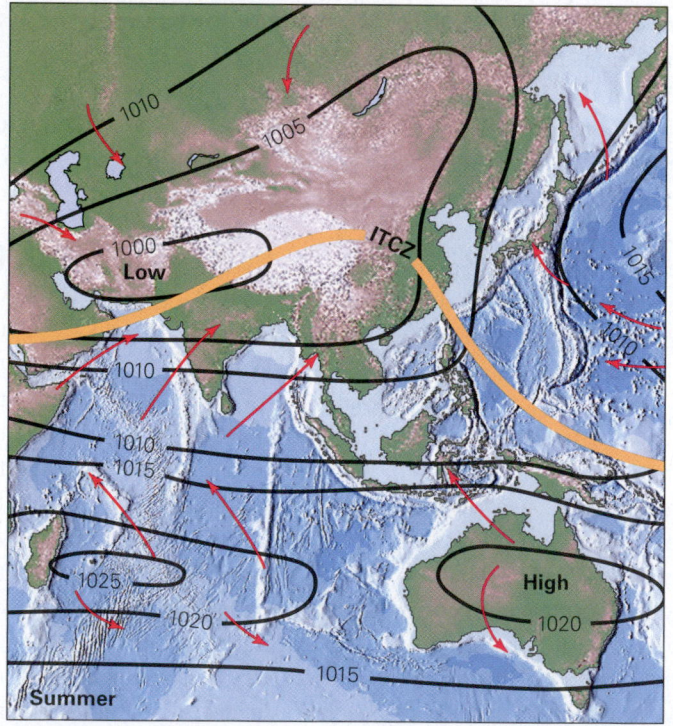

(a) In winter, the dry season, a high-pressure cell develops over central Asia, and the intertropical convergence zone lies south of Asia.

(b) In summer, the wet season, a low-pressure cell develops over Asia, so warm moisture-laden air flows landward. Orographic lifting over the Himalayas causes clouds to form, and intense rain begins.

over central Asia, a pronounced low-pressure cell develops, and the intertropical convergence zone moves north. When this happens, warm air flows northward from the Indian Ocean, bringing with it substantial moisture, and the summer rains begin. Rainfalls are especially heavy on the southern slope of the Himalayas, because orographic lifting leads to the production of huge cumulonimbus clouds (**Fig. 20.34b**).

El Niño Long before the modern science of meteorology became established, fishermen from Peru and Ecuador who ventured into the coastal waters west of South America knew that in late December the fish population that provided their livelihood diminished. Because of the timing of this event, it came to be known as **El Niño**, Spanish for little boy, or the Christ child. Why did the fish vanish? Fish are near the top of a food chain that begins with plankton, which live off nutrients

FIGURE 20.35 El Niño exists because of a change in winds and currents in the central Pacific.

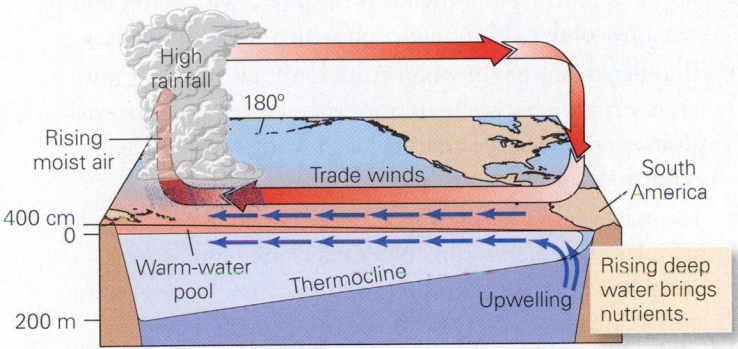

(a) During times between El Niño, low pressure lies over the western Pacific and surface winds blow west. Because these winds drive warm surface water away, cold water rises along the coast of South America.

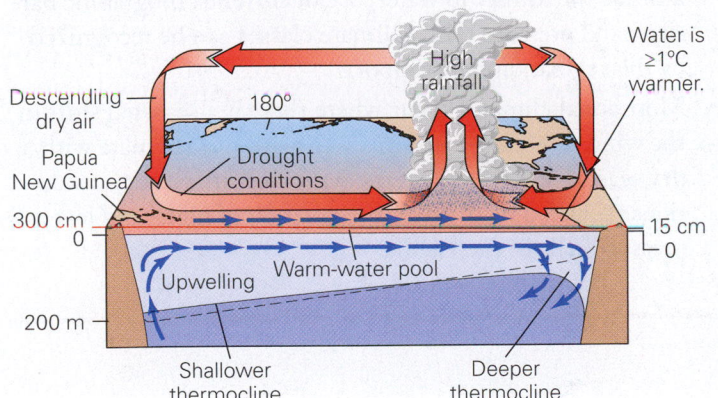

(b) During El Niño, the low-pressure cells move eastward and the westward flow stops, so upwelling ceases.

in the water. Along coastal South America, the nutrients to feed plankton come from upwelling deep water. During El Niño, warm-water currents flow eastward from the central Pacific, and the cold, nutrient-rich water that supports the marine food chain can't upwell. With fewer nutrients, there are fewer plankton, and without the plankton, the fish migrate elsewhere.

To understand why El Niño occurs, we need to look at atmospheric flow and related surface ocean currents in the equatorial Pacific (**Fig. 20.35**). When El Niño is not in progress, a major equatorial low-pressure cell exists in the western Pacific over Indonesia and Papua New Guinea, while a high-pressure cell forms over the eastern Pacific, along the coast of equatorial South America. This geometry (called La Niña, Spanish for little girl, when particularly intense) means that air rises in the western Pacific, flows east, sinks in the eastern Pacific, and then flows west at the surface. The easterly surface winds blow warm surface water westward, and cold water from the deep ocean rises along South America to replace the warm water that moved west. It is this rising cold water that brings nutrients to the surface, feeding the plankton, which in turn feed the fish. During El Niño, the low-pressure cell moves eastward over the central Pacific, and a high-pressure cell develops over Indonesia; so two circulation cells develop. As a result, surface winds start to blow from west to east in the western Pacific, driving warm surface water back to South America. This warm surface water prevents deep cold water from rising, and with fewer nutrients in the water, plankton quantities decrease and fish move elsewhere. In effect, pressure cells oscillate back and forth across the Pacific, an event now called the *southern oscillation*. Roughly speaking, El Niño events happen on roughly a 4-year cycle.

Take-Home Message

Climate refers to the long-term overall weather conditions in a region, as characterized by the range of weather conditions and the nature of seasons. In general, climate varies with latitude. In detail, factors such as proximity to the ocean or to orographic barriers affect local climate. Monsoons, seasonal reversals in the wind direction that trigger a shift from a dry to a rainy season, form because of changes in the position of the intertropical convergence zone. El Niño develops because of shifts in the position of air circulation at low latitudes.

QUICK QUESTION: How does El Niño affect fish supplies near the coast of South America?

CHAPTER SUMMARY

- The early atmosphere of the Earth contained high concentrations of water, carbon dioxide, and sulfur dioxide; these are gases erupted by volcanoes.

- When the oceans formed, much of the water and carbon dioxide was removed from the atmosphere. When photosynthetic organisms evolved, they produced oxygen, and the concentration of this gas gradually increased.

- Air now consists mostly of nitrogen (78%) and oxygen (21%). Several other gases occur in trace amounts. The atmosphere also contains aerosols. Air pressure decreases with elevation, so 90% of the air in the atmosphere occurs below an elevation of 16 km.

- When air rises, it expands and cools, a process called adiabatic cooling. If air sinks and undergoes compression, it heats up, a process called adiabatic heating.

- Air generally contains water. The ratio between the measured water content and the maximum possible amount of water that the air can hold is the air's relative humidity.

- The atmosphere is divided into layers. In the lowest layer, the troposphere, temperature decreases with elevation. The troposphere convects. The other layers are the stratosphere, the mesosphere, and at the top, the thermosphere.

- Air circulates on two scales, local and global. Winds blow because of pressure gradients: air begins to move from regions of higher pressure to regions of lower pressure, but its path is affected by the Coriolis effect.

- High latitudes receive less solar energy than low latitudes. This contrast initiates convection in the atmosphere. Because of the Coriolis effect, air moving north from the equator to the pole deflects to the east. Thus, a Hadley cell forms between the equator and a latitude of about 30°N. In detail, atmospheric circulation is very complicated.

- Prevailing surface winds—the northeast trade winds, the surface westerlies, and the polar easterlies—develop in latitudinal belts.

- Air pressure at a given latitude decreases from the equator to the pole, causing a poleward flow of air. In the northern hemisphere, the Coriolis effect deflects this flow to generate high-altitude westerlies. These winds, where particularly strong, are known as jet streams.

- Weather refers to the temperature, air pressure, wind speed, and relative humidity at a given time in a given location. Weather reflects the interaction of air masses. The boundary between two air masses is a front.

- Air sinks in high-pressure air masses and rises in low-pressure air masses. Because of the Coriolis effect, the air begins to rotate around the center of the mass, generating cyclones or anticyclones.

- Clouds, which consist of tiny droplets of water or tiny crystals of ice, form when the air is saturated with water and contains condensation nuclei on which water condenses.

- Thunderstorms begin when cumulonimbus clouds grow large. Friction between air and water molecules separates positive and negative charges. Lightning flashes when a giant spark jumps across charge separations.

- Tornadoes, rapidly rotating funnel-shaped clouds, develop in violent thunderstorms. Nor'easters are large storms associated with wave cyclones. Hurricanes, huge rotating storms, originate over warm oceans.

- Climate refers to the typical range of conditions, the nature of seasons, and the possible weather extremes of a region averaged over a long time. Climate is controlled by latitude, altitude, proximity to water, ocean currents, orographic barriers, and pressure cells. Climate classes can be recognized by the vegetation they support.

- Monsoonal climates occur where there is a seasonal shift in the wind direction, causing a wet season to alternate with a dry season. El Niño is a temporary shift in weather conditions triggered by shifts in the position of high- and low-pressure cells in the Pacific.

GUIDE TERMS

acid rain (p. 733)

adiabatic cooling, heating (p. 735)

aerosol (p. 732)

air (p. 729)

air mass (p. 744)

air pressure (p. 732)

air quality (p. 733)

anticyclonic flow (p. 746)

atmosphere (p. 729)

aurora (p. 737)

climate (p. 761)

cloud (p. 735)

collision and coalescence (p. 747)

condensation nuclei (p. 747)

convergence zone (p. 739)

cyclonic flow (p. 746)

divergence zone (p. 739)

doldrums (p. 741)

El Niño (p. 765)

Enhanced Fujita Scale (p. 754)

exosphere (p. 737)

extratropical cyclone (p. 746)

front (p. 745)

greenhouse gas (p. 730)

hail (p. 750)

insolation (p. 739)

ionosphere (p. 737)

isobar (p. 738)

isotherm (p. 762)

jet stream (p. 742)

latent heat of condensation (p. 735)

lifting mechanism (p. 747)

lightning bolt (p. 751)

mesosphere (p. 737)

monsoon (p. 763)

nor'easter (p. 754)

orographic barrier (p. 763)

polar front (p. 739)

pollutant (p. 733)

prevailing wind (p. 741)

relative humidity (p. 735)

Saffir-Simpson Scale (p. 756)

storm (p. 729)

storm surge (p. 757)

stratosphere (p. 737)

supercell (p. 750)

thermosphere (p. 737)

thunder (p. 752)

thunderstorm (p. 750)

tornado (p. 752)

tornado warning (p. 754)

trade wind (p. 741)

tropical cyclone (p. 755)

troposphere (p. 736)

weather (p. 729)

wind (p. 738)

REVIEW QUESTIONS

1. Describe the stages in the formation and evolution of Earth's atmosphere. Where does the ozone in the atmosphere come from, and why is it important?

2. Describe the composition of air (considering both its gases and its aerosols). Why are trace gases important?

3. How does air pressure change with elevation? Does the density of the atmosphere also change with elevation? Explain why or why not.

4. Describe the atmosphere's structure from base to top. What characteristics define the boundaries between layers?

5. What is the relative humidity of the atmosphere? What is the latent heat of condensation, and what is its relevance to the evolution of a thunderstorm or hurricane?

6. Explain the relation between the wind and variations in air pressure.

7. Why do changes in atmospheric temperature depend on latitude and the seasons?

8. Describe the pattern of global atmospheric circulation. Why don't convective cells extend from the equator to the pole?

9. Why do prevailing winds develop at the Earth's surface? Why do the jet streams form?

10. Explain the nature of cyclones and anticyclones, and note their relationship to high-pressure and low-pressure air masses. What is an extratropical (mid-latitude) cyclone?

11. How does a cold front differ from a warm front and from an occluded front?

12. Why do clouds form? (Include a discussion of lifting mechanisms.) What are the basic categories of clouds?

13. Under what conditions do thunderstorms develop? What provides the energy that drives clouds to the top of the troposphere? How do meteorologists explain lightning?

14. What conditions lead to the formation of a tornado? Where do most tornadoes appear?

15. Describe the stages in the development of a hurricane. Describe a hurricane's basic geometry.

16. What factors control the climate of a region? What special conditions cause monsoons and El Niño?

ON FURTHER THOUGHT

17. When Columbus set sail from Spain, his route first took him southward to the Canary Islands and then westward. His landfall in the new world, on an island southeast of Florida, was farther to the south than his point of departure. Why?

18. Explain why large rainforests occur in equatorial Africa (the Congo) and in equatorial South America (the Amazon).

19. A typhoon is moving due west and is approaching the east coast of Asia. Weather forecasters have predicted where the eye of the storm will make landfall in the middle of a narrow, north-south-trending island. People in the path of the storm have been told to evacuate but cannot head inland because they are on a narrow island. To escape the highest winds, should they move to the north or to the south of the eye?

smartwork   smartwork.wwnorton.com

GEOTOURS

This chapter's Smartwork features:

- Art question on atmospheric layers.
- Labeling exercises on thunderstorm and tornado formation.
- Comprehensive questions on the atmosphere.

There are no GeoTours for this chapter due to content.

Only particularly hardy shrubs can survive in this field of sand dunes, at the base of a barren rock ridge in the desert of Death Valley, California.

CHAPTER 21

Dry Regions: The Geology of Deserts

The bare hills are cut out with sharp gorges, and over their stone skeletons scanty earth clings. . . . A white light beat down, dispelling the last tract of shadow, and above hung the burnished shield of hard, pitiless sky.

—Clarence King (1842–1901; first director of the U.S. Geological Survey, describing a desert)

21.1 Introduction

For generations, nomadic traders have saddled camels to traverse the Sahara in northern Africa (**Fig. 21.1a**). The Sahara, the world's largest desert, receives so little rainfall that it has hardly any surface water or vegetation. So camels must be able to walk for up to 3 weeks without drinking or eating. They can survive these journeys because they sweat relatively little, thereby conserving their internal water supply. Also, they have the ability to metabolize their own body fat to produce new water, and they can withstand severe dehydration. Most mammals die after losing only 10 to 15% of their body fluid, but camels can survive 30% dehydration with no ill effects. Camels do get thirsty, though—after a marathon trek across the desert, a camel may guzzle up to 100 liters of water in less than 10 minutes.

The survival challenges faced by a camel emphasize that deserts are lands of extremes—extreme dryness, extreme heat, and extreme cold. But they can also be places of extreme beauty, for desert vistas include everything from sand seas

with sculpture-like dunes, to stony pavements spotted with flowers, to cactus-covered hills, to dramatic cliffs of colorful rock. Although less populated than other regions on Earth, deserts cover a significant percentage of the land surface—about 25%—and thus constitute an important component of the Earth System (**Fig. 21.1b**). In this chapter, we introduce the desert landscape. We learn why deserts occur where they do and how erosion and deposition shape their surface. We conclude by exploring life in the desert and by examining the problem of desertification, the gradual transformation of temperate lands into desert.

21.2 The Nature and Location of Deserts

What Is a Desert?

Formally defined, a **desert** is a region that is so arid (dry) that it contains no permanent streams, except for rivers that bring water in from temperate regions elsewhere, and supports vegetation on no more than 15% of its surface. In general, desert conditions exist where, on average, less than 25 cm of rain falls per year. But rainfall amounts alone do not determine the aridity of a region. Aridity also depends on rates of evaporation and on whether rainfall occurs only sporadically or more continuously during the year. If all the rain in a region drenches the land during isolated downpours only once every few years, the region becomes a desert because the intervals of drought last so long that plants and permanent streams cannot survive. Similarly, if high temperatures and dry air cause evaporation rates from the ground to exceed the rate at which rainfall wets the ground, then the region becomes a desert even if it receives more than 25 cm per year of rain.

Note that the definition of a desert depends on a region's aridity, not on its temperature. Geologists distinguish between *cold deserts*, where temperatures generally stay below about 20°C for the year, and *hot deserts*, where summer daytime temperatures exceed 35°C. Cold deserts exist at high latitudes, where the Sun's rays strike the Earth obliquely and thus don't provide

FIGURE 21.1 Deserts and their hardy inhabitants.

(a) Camels can survive the harsh conditions of the Arabian desert.

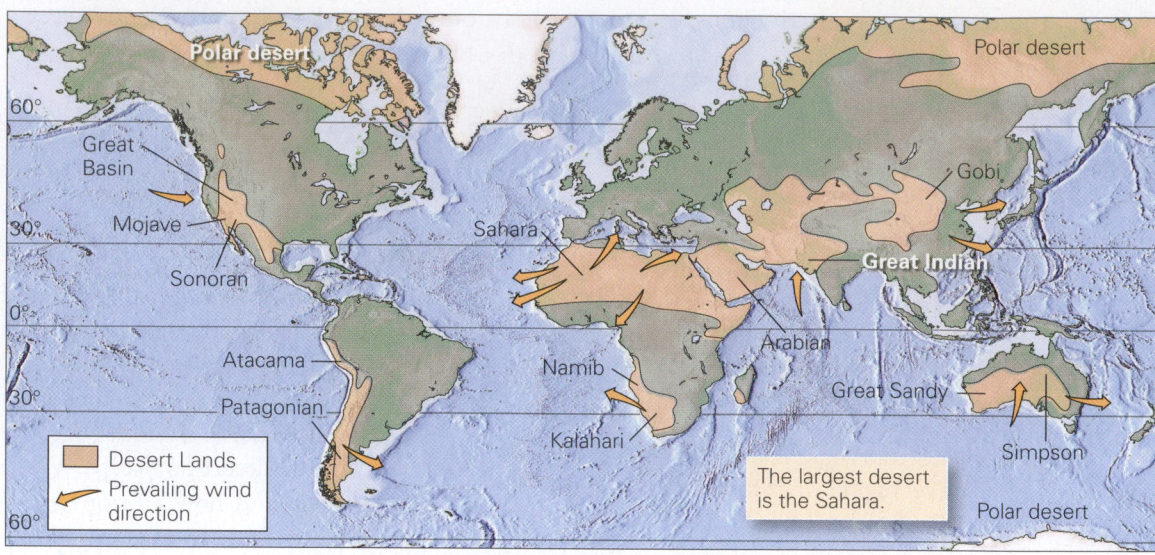

(b) The global distribution of deserts. Arid regions cover 25% of the land surface.

much energy; at high elevations, where the air is too thin to hold much heat; or in lands adjacent to cold oceans, where the cold water absorbs heat from the air above. Hot deserts develop at low latitudes, where the Sun's rays strike the desert at a high angle; at low elevations, where dense air can hold a lot of heat; and in regions distant from the cooling effect of cold ocean currents. The hottest recorded temperatures on Earth occur in low-latitude, low-elevation deserts—58°C (136°F) in Libya and 57°C (133°F) in Death Valley, California.

Did you ever wonder ...

how hot it can become in a desert?

Heat contributes to aridity by increasing the rate of evaporation. In fact, evaporation rates in hot deserts may be so great that even when it rains, the ground stays dry because falling raindrops evaporate in midair. In low latitudes, the bare ground of the desert absorbs so much energy from sunlight that a layer of very hot air (up to 77°C, or 170°F) forms just above the ground. This layer refracts sunlight, creating a *mirage*, a wavering pool of light, on the ground. Mirages make the dry sand of a desert wasteland look like a shimmering lake and distant mountains look like islands (**Fig. 21.2**). But even the hottest of hot deserts become cold at night because the dry air doesn't hold heat and there are no clouds or vegetation to trap heat. In fact, the air temperature at the ground surface in a desert may change by as much as 80°C in a single day.

Aridity causes weathering, erosion, and depositional processes in deserts to be different from those in temperate or tropical regions. Without plant cover, rain and wind batter and scour the ground, and during particularly heavy rains, water

accumulates into flash floods of immense power. In deserts, rocks and sediment do not undergo rapid chemical weathering, and humus (organic matter) does not collect on the ground surface. Thus, the desert land surface consists of any of the following: exposed bedrock, accumulations of clasts, relatively unweathered sediment, precipitated salt, or windblown sand. As we'll see, soils do develop in deserts, but they tend to be thinner and more mineralized than those of temperate or tropical regions.

FIGURE 21.2 The shimmering in this desert mirage may look like water, but it's not. Mirages result from the interaction of light with a thin layer of hot air just above the ground surface.

Types of Deserts

Each desert on Earth has unique characteristics of landscape and vegetation that distinguish it from others. Nevertheless, geologists can group deserts into five general classes, based on the setting in which the desert forms.

- *Subtropical deserts:* Subtropical deserts (e.g., the Sahara, Arabian, Kalahari, and Australian) form because of the pattern of air circulation in the atmosphere (see Chapter 20). At the equator, the air becomes warm and humid, for sunlight is intense and water rapidly evaporates from the ocean. The hot, moisture-laden air rises to great heights above the equator (**Fig. 21.3**). As this air rises, it expands and cools and can no longer hold so much moisture. Water condenses and falls in downpours that feed the lushness of the equatorial rainforest. The now-dry air high in the troposphere spreads laterally north or south. When this air reaches latitudes of 20° to 30°, a region called the subtropics, it has become cold and dense enough to sink. Because the air is dry, no clouds form, and intense solar radiation strikes the Earth's surface. The sinking, dry air condenses and heats up, soaking up any moisture present. Back at the surface, this hot, dry air sweeps back toward the equator. In the regions that it passes over, evaporation rates greatly exceed rainfall rates, so the land becomes parched.

    In subtropical deserts that border the sea, high tides may cause warm seawater to flood coastal tidal flats. Under the hot sun, the shallow seawater evaporates

and becomes saturated. Thus, salts precipitate onto the underlying organic-rich mud, producing a salty crust overlying the mud. Geologists refer to a salt-encrusted muddy tidal flat as a *sabkha.*

- *Deserts formed in rain shadows:* When moist air flowing landward reaches a coastal mountain range, the air must rise (**Fig. 21.4**). As the air rises, it expands and cools. The water it contains condenses and falls as rain on the seaward flank of the mountains, nourishing a coastal rainforest. Thus, when this air finally reaches the inland side of the mountains, it has lost all its moisture and can no longer provide rain. As a consequence, a **rain shadow** forms, and the land beneath becomes a *rain-shadow desert.* The landscapes flanking the Cascade Mountains of western Washington illustrate this pattern—the west side of the range is a temperate rainforest of dense vegetation, whereas the east side hosts a rain-shadow desert.

- *Coastal deserts formed along cold ocean currents:* In places where cold ocean currents flow along the margin of a continent, a desert can develop right along the coast. This happens because the cold water of the current cools the overlying air by absorbing heat, and this process decreases the capacity of the air to hold moisture. Such a *coastal desert* has formed along the western coast of South America, where the Humboldt Current, which carries icy water northward from Antarctica, cools the air that blows east over the coast. The air is so dry when it reaches the coast that rain rarely falls on the coastal areas of Chile and Peru. As a result, this coastal region hosts a desert landscape, including one of the driest landscapes in the world, the

FIGURE 21.3 Subtropical deserts form because the air that convectively flows downward in the subtropics warms and absorbs water as it sinks.

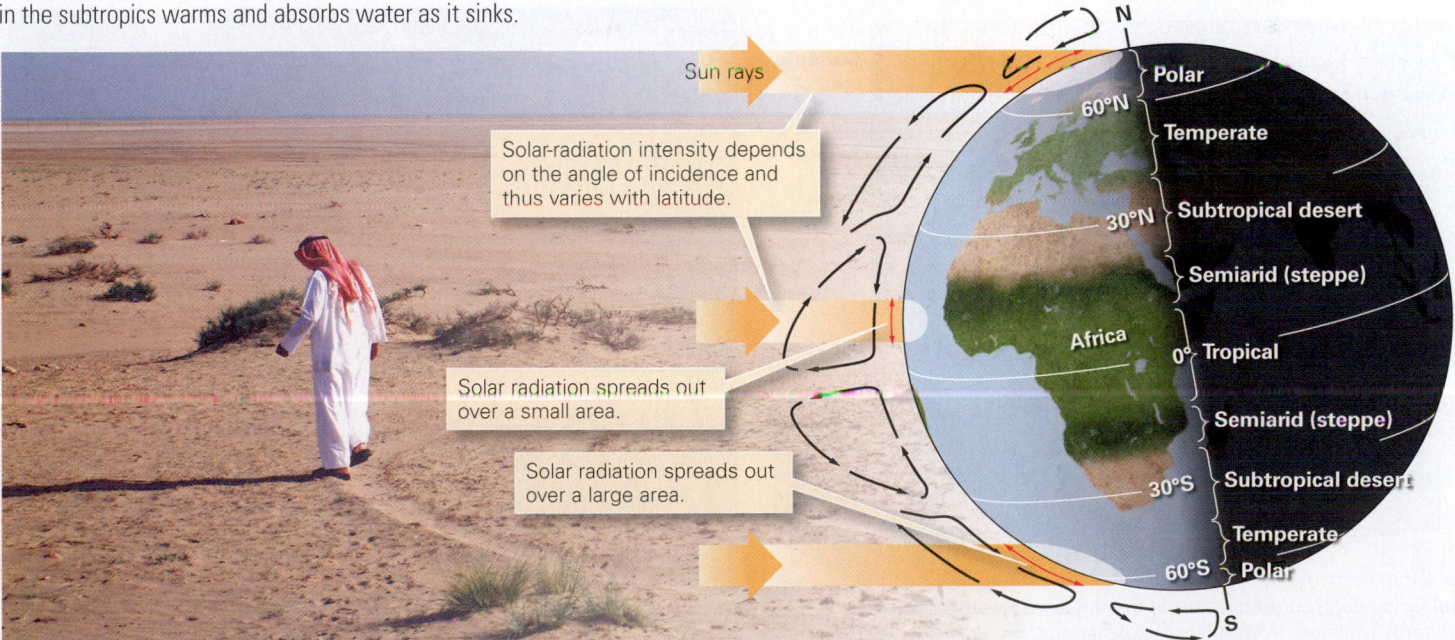

FIGURE 21.4 The formation of a rain-shadow desert.

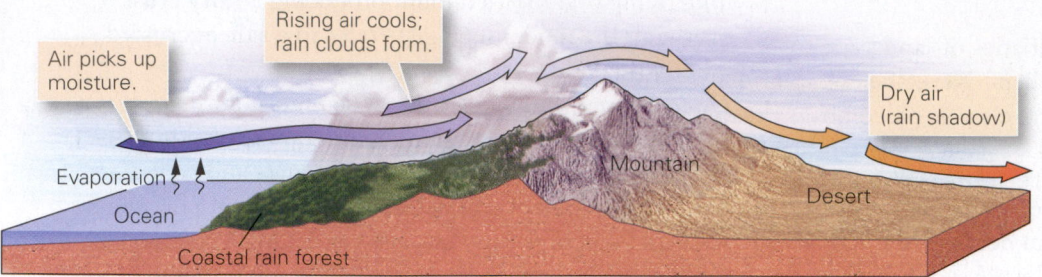

Air picks up moisture.

Rising air cools; rain clouds form.

Dry air (rain shadow)

Evaporation

Ocean

Mountain

Desert

Coastal rain forest

(a) Moist air rises and drops rain on the coastal side of the range. By the time the air has crossed the mountains, it is dry.

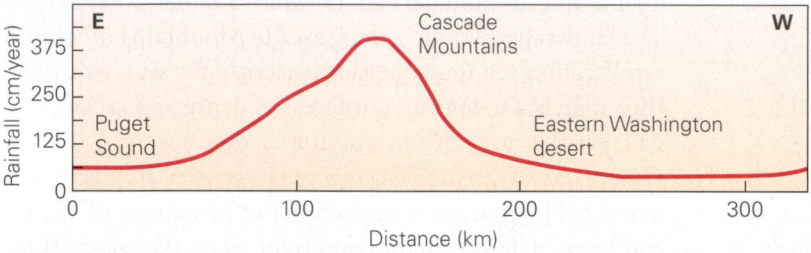

(b) A graph shows that measured rainfall is much greater on the western side of the Cascade Mountains in Washington than it is on the eastern side.

reaches the interior of a particularly large continent such as Asia, it has grown quite dry, so the land beneath becomes arid. The largest present-day example of such a *continental-interior desert*, the Gobi, lies in central Asia, over 2,000 km away from the nearest ocean.

• *Deserts of the polar regions:* So little precipitation falls in Earth's polar regions (north of the Arctic Circle, at 66°30′ N, and south of the Antarctic Circle, at 66°30′ S) that these areas are, in fact, arid. Polar regions are dry for two reasons. First, they are dry for the same reason that the subtropics are dry—the general pattern of global-scale air circulation causes the air flowing over these regions to be dry. Second, they are dry for the same reason that coastal areas along cold currents are dry—cold air holds relatively little moisture.

The distribution of deserts around the world through geologic time reflects the process of plate tectonics, for plate movements determine the latitude of landmasses, the position of landmasses relative to the coast, the proximity of landmasses to a mountain range, and the configuration of currents. Because of continental drift, the stratigraphic record shows that some regions that were deserts in the past are temperate or tropical regions now, and vice versa.

Atacama Desert (**Fig. 21.5**). Portions of this narrow (less than 200-km-wide) desert, which lies between the Pacific coast on the west and the Andes on the east, received no significant rain at all between 1570 and 1971.

• *Deserts formed in the interiors of continents:* As an air mass moves long distances across a continent, it gradually loses moisture by dropping rain. Thus, when the air mass

FIGURE 21.5 The formation of a coastal desert.

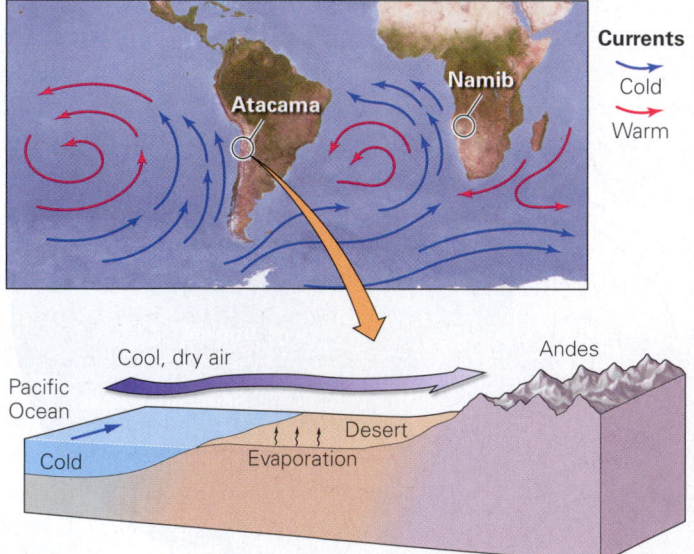

Currents
Cold
Warm

Namib
Atacama

Cool, dry air

Andes

Pacific Ocean

Cold

Desert

Evaporation

(a) Cold ocean currents cool and dry the air along the coast. This air absorbs moisture from adjacent coastal land.

(b) The Atacama Desert is the driest place on Earth.

21.3 Producing Desert Landscapes

Weathering and Soil Formation in Deserts

If you stand in the Mojave Desert of California and look around, you'll see barren cliffs exposing fractured rock, and slopes or plains littered with gravel (**Fig. 21.6**). Clearly, physical weathering happens in deserts—over time, blocks break off along joints and tumble downslope, perhaps shattering further on the way down because of collisions with other blocks. Once a block comes to rest on a desert landscape, it can sit unchanged for a long time, but not forever. Recent studies suggest that the daily alternation from midday heat to midnight cold can generate sufficient stress to break isolated blocks apart, in place.

Do chemical weathering reactions take place in deserts? Yes. Over time, the moisture from dew and occasional rain allows oxidation, hydrolosis, and dissolution reactions to destroy cements and to transform silicate minerals into clay, thereby causing rocks to disaggregate into pebble- or sand-sized debris. But without the presence of acidic organic matter, and without a steady supply of water to seep into rock or infiltrate down into the ground, these reactions take place relatively slowly.

Does soil form in deserts? Yes. But without roots to hold regolith in place, wind and water commonly move the regolith before a soil has time to evolve. Where sediment stays in place, soil can form, for infiltration of water after heavy rains leaches ions and transports fine clay downward (see Interlude D). In deserts, however, rains occur so infrequently that not enough water infiltrates to flush leached ions away entirely. Therefore, the ions precipitate to form new mineral cement not far below the ground surface. If the new cement consists of calcite, it can bind the regolith into a solid, rock-like material known as *caliche* or *calcrete* (**Fig. 21.7a**). Calcrete deposits can grow rapidly—some encase tools left behind by prospectors just a few decades ago.

Due to the lack of organic content, the black or brown colors of temperate soils don't develop in deserts, and variations in bedrock color tend to control the color of soil. Variations in the concentration of iron or in the degree of iron oxidation in adjacent beds result in spectacular color bands in rock layers and the thin soils derived from them. For example, iron-rich, well oxidized strata will be dark red or even maroon. Rocks with less iron will be lighter red or orange, and rocks in which iron concentration is low or has been chemically reduced can be tan, gray, or even greenish. The Painted Desert of northern Arizona earned its name from the brilliant and varied hues of oxidized iron in the region's shale bedrock (**Fig. 21.7b**).

Desert Varnish

Shiny **desert varnish**, a dark, rusty brown coating of iron oxide, manganese oxide, and clay, locally coats the surface of rocks in deserts (**Fig. 21.8a**). Desert varnish was once thought to form when water from rain or dew seeped into a rock, dissolved iron and magnesium ions, and carried the ions by capillary action back to the surface of the rock, where the ions precipitated. More recent studies, however, suggest that the minerals in desert varnish were not necessarily derived from the rocks they coat. Rather, desert varnish may form when windborne dust settles on the surface of the rock, for in the presence of moisture, microbes (bacteria and archaea) extract elements from the dust and transform it into iron or manganese oxide precipitates. Desert varnish doesn't form in humid climates because rain washes away dust before ions can be extracted from it.

Desert varnish takes a long time to form. In fact, measuring the thickness of a desert varnish layer can provide an estimate of how long a rock has been exposed at the ground surface. In past centuries,

FIGURE 21.6 Dark gravel, formed by physical weathering of rock exposed on the cliff, litters the gentle slope at the cliff's base, in the Mojave Desert.

FIGURE 21.7 Soils in deserts.

Stream erosion exposed this calcrete layer.

(a) Calcrete forms when calcium carbonate precipitates and binds together rock fragments.

(b) The red hues of the Painted Desert in Arizona are due to the oxidation of iron in the rock.

people have used desert-varnished rock as a medium for art—by chipping away the varnish to reveal the underlying lighter-colored rock, they were able to create figures or symbols on a dark background. The resulting drawings are called **petroglyphs** (**Fig. 21.8b**).

Water Erosion

Although rain rarely falls in deserts, when it does come it can radically alter a landscape in a matter of minutes. Since deserts lack plant cover, rainfall, sheetwash, and stream flow are all extremely effective agents of erosion. It may seem surprising, but water causes more erosion than does wind in most deserts (**Fig. 21.9a**). If the substrate eroded by water consists of soft,

homogeneous material, then a network of closely spaced parallel drainages, which merge downslope, develops—geologists refer to bare landscapes containing such drainage as *badlands* (**Fig. 21.9b**).

Water erosion begins with the impact of raindrops, which eject sediment from the ground into the air. On a hill, the ejected sediment lands downslope, so during a rain, sediment gradually migrates to lower elevations. The ground quickly becomes saturated with water during a heavy rain, so water starts flowing across the surface, carrying the loose sediment with it. Within minutes after a heavy downpour begins, dry stream channels fill with a turbulent mixture of water and sediment, which rushes downstream as a flash flood. When the rain stops, the water sinks into the streambed's gravel and

FIGURE 21.8 Desert varnish.

(a) A coating of desert varnish turns the surface of light tan sandstone beds to a dark, rusty brown.

(b) By chipping away desert varnish to reveal the lighter rock beneath, Native Americans created art and symbols.

FIGURE 21.9 Evidence of erosion by running water in deserts.

(a) These hills in the desert near Las Vegas, Nevada, are bone dry, but their shape indicates erosion by water. Note the numerous stream channels.

(b) Badlands topography develops where flowing water erodes a soft substrate in deserts.

(c) Flash floods carved this steep-walled channel bordered by polished rock in Mosaic Canyon, California.

(d) Gravel and sand are left behind on the floor of a dry wash after a flash flood in Death Valley. Erosion by the water is cutting a channel.

disappears. As we noted in Chapter 17, such drainages are called ephemeral streams; the channels of these streams are called **dry washes** or **arroyos** in North America and **wadis** in the Middle East and North Africa. Water in a flash flood can cause intense erosion, because the water moves so fast and can carry so much sediment. Thus, flash floods can polish bedrock that borders streams (**Fig. 21.9c**), cut steep-walled channels, and transport huge boulders downstream. As rocks roll and tumble along, they strike each other and shatter, creating smaller pieces that can be carried still farther—between floods, the floor of a dry wash consists of gravel littered with boulders (**Fig. 21.9d**).

Wind Erosion

In temperate and humid regions, plant cover protects the ground surface from the wind, but in deserts the wind has direct access to the ground. Wind, just like flowing water, can carry sediment both as suspended load and as surface load. **Suspended load** (fine-grained sediment such as dust and silt held in suspension) stays in the air for a long time and moves with it (**Fig. 21.10a**). The suspended sediment can be carried so high into the atmosphere (up to several kilometers above the Earth's surface) and so far downwind (tens to thousands of kilometers) that it may move completely out of its source region.

Of note, while a gentle breeze may be able to transport very fine grains as suspended sediment, it might not be able to break the grains free of the ground, for electrostatic charges on sediment grains keep the grains "stuck" to the ground. To break grains free of the ground and put it into suspension in the first place takes stronger breezes and/or turbulence. For example, on an otherwise calm day, tiny vortices locally develop due to the instability of the hot desert air, and these can churn up dust. In some cases, they become *dust devils* up to 100 m high that look like miniature tornadoes. Cars driving down dirt roads in deserts can also break dust free from the ground—that's why

(a) Dust storms can move huge volumes of sediment and cause intense erosion.

(b) A 2005 dust storm (haboob) approaches an army base in Iraq. The surging layer of dust-laden air is hundreds of meters high.

(c) A huge dust storm approaching Phoenix, Arizona.

a huge plume of suspended sediment develops in the wake of a car.

Particularly strong winds, such as the downdrafts in front of a thunderstorm, can generate a dramatic **dust storm** (known as a *haboob* in the Middle East) that can be 100 km long and up to 1.5 km (almost 1 mile) high (**Fig. 21.10b**). Dust storms can be dangerous, for they can cut visibility to almost nothing. They can shut down airports and bring highway traffic to a halt. They can also damage structures and infiltrate machinery with sand and grit, fouling the mechanisms. A particularly large dust storm rolled over Phoenix, Arizona, in July 2011—in photographs, it dwarfs the city below (**Fig. 21.10c**).

Surface loads, or bed loads, develop when winds become strong enough to cause **saltation**, a process during which sand grains roll and bounce along the ground (**Fig. 21.11a**). Saltation begins when turbulence caused by wind shearing along the ground surface lifts sand grains. The grains move downwind, following an asymmetric, arch-like trajectory. Eventually, gravity causes them to return to the ground, where they strike other sand grains, causing the new grains to bounce up and drift or roll downwind. The collisions between sand grains make the grains rounded and frosted. Saltating grains generally rise no more than 0.5 m, but where sand bounces on bedrock the grains may rise 2 m (6 ft).

The size of clasts that wind can transport depends on the wind velocity. Wind, therefore, does an effective job of sorting sediment, sending dust-sized particles skyward and sand-sized particles bouncing along the ground, while pebbles and larger grains remain behind. In some cases, wind carries away so much fine sediment that pebbles and cobbles become concentrated at the ground surface. An accumulation of coarser sediment left behind when fine-grained sediment blows away is called a **lag deposit** (**Fig. 21.11b**). Over time, in regions where the substrate consists of soft sediment, wind picks up and removes so much sediment that the land surface becomes lower. The process of lowering the land surface by wind erosion is called **deflation**. Shrubs can stabilize a small patch of sediment with their roots, so after deflation, forlorn shrubs with residual pedestals of soil stand isolated above a lowered ground surface (**Fig. 21.11c**). In some places, the shape of the land surface twists the wind into a local vortex that causes enough deflation to scour a deep, bowl-like depression called a *blowout*.

Just as sandblasting cleans the grime off the surface of a building, windblown sand and dust grind away at surfaces in the desert. Over long periods of time, such wind abrasion creates smooth faces, or facets, on pebbles, cobbles, and boulders. If a rock rolls or tips relative to the prevailing wind direction after it has been faceted on one side, or if the wind shifts direction, a new facet with a different orientation forms, and the

FIGURE 21.11 Moving the surface load during wind erosion.

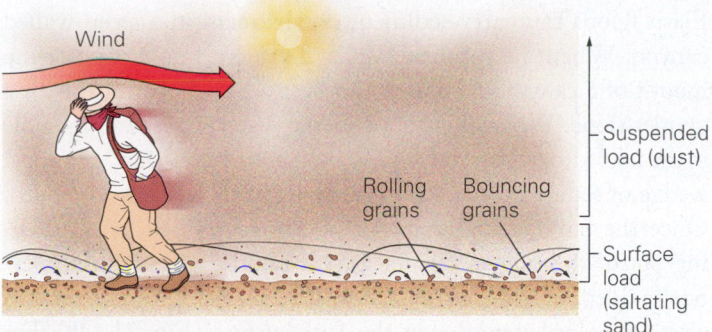

(a) Wind transports desert sediment as suspended load and surface load.

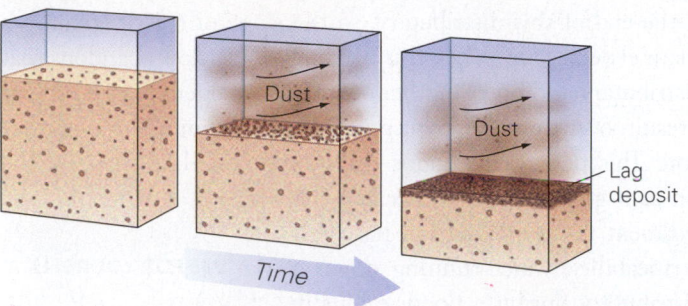

(b) A lag deposit develops when wind blows away finer sediment, leaving behind a layer of coarser grains.

(c) Deflation has removed the sediment between these shrubs in Death Valley, California.

two facets join at a sharp edge. Rocks whose surface has been faceted by the wind are called faceted rocks, or **ventifacts** (**Fig. 21.12a**). Wind abrasion also gradually polishes and bevels down irregularities on a desert pavement and polishes the surfaces of desert-varnished outcrops, giving them a reflective sheen. In places where a resistant layer of rock overlies a softer layer of rock, wind abrasion may create a formation consisting of a resistant block perched on an eroding mushroom-like column of softer rock. These unusual features are known as *yardangs*

(**Fig. 21.12b**). If a strong wind blows in only one direction, the yardangs become elongate and align with the wind direction.

Exploration by space probes over the last 20 years shows that wind affects the desert-like surface of Mars just as it does the deserts on Earth. During the Martian fall and spring, when differences in temperature between the ice-covered poles and the ice-free equator are greatest, strong winds blow from the poles to the equator, transporting so much dust that the planet's surface, as seen from Earth, visibly changes. At times, the entire planet becomes enveloped in a cloud of dust. Close-up images taken by spacecraft that have landed on Mars show that rocks have been abraded by saltating sand and that sand has accumulated on the lee (downwind) side of the rocks (**Fig. 21.13**).

FIGURE 21.12 The consequences of wind erosion.

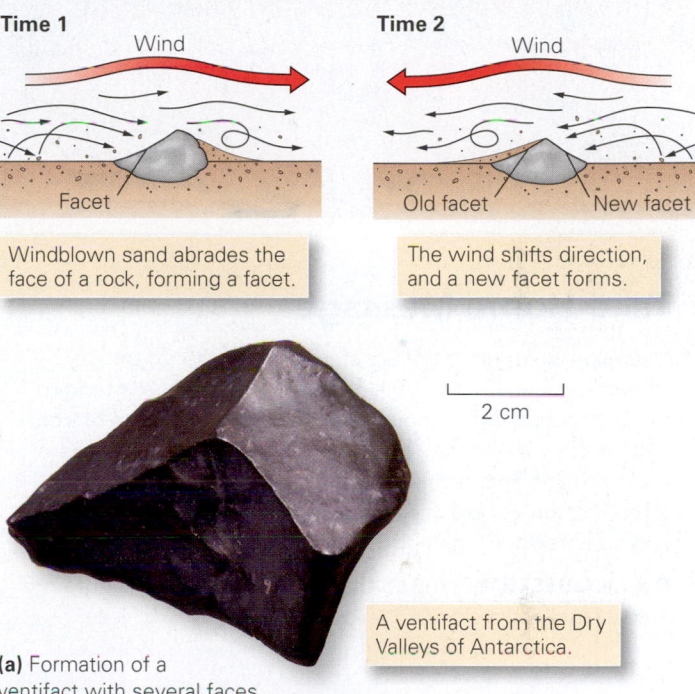

Time 1
Wind

Windblown sand abrades the face of a rock, forming a facet.

Facet

Time 2
Wind

The wind shifts direction, and a new facet forms.

Old facet   New facet

2 cm

A ventifact from the Dry Valleys of Antarctica.

(a) Formation of a ventifact with several faces.

(b) Yardangs develop when wind erodes a weaker layer beneath a stronger layer.

FIGURE 21.13 This NASA photo shows the wind abrasion of clasts and small sand dunes on the surface of Mars.

Take-Home Message

Chemical weathering occurs slowly in deserts. In the dry climate, unique soils form. Colors of rocks and soils tend to be controlled by the amount and degree of oxidation of iron, and desert varnish may coat rock surfaces. Water causes most erosion and sediment transport but rarely flows, so stream channels are dry washes. Wind transports sediment and can produce lag deposits and carve ventifacts.

QUICK QUESTION: Why do the unusual shapes of ventifacts and yardangs develop?

21.4 Deposition in Deserts

We've seen that erosion relentlessly eats away at bedrock and sediment in deserts. Where does the debris go? Below we examine the various desert settings in which sediment accumulates.

Talus

Over time, joint-bounded blocks of rock break off ledges and cliffs on the sides of hills. Under the influence of gravity, the resulting debris tumbles downslope and accumulates as **talus**, a pile of debris at the base of a hill (**Fig. 21.14a**). Talus can survive for a long time in desert climates, so we typically see talus aprons fringing the bases of cliffs in deserts. The angular clasts constituting talus gradually become coated with desert varnish.

Alluvial Fans

Flash floods can carry sediment downstream in a steep-walled canyon. When the turbulent water flows out into a plain at the mouth of a canyon, it spreads out over a broader surface with a gentler slope. When this happens, the water slows, and its sediment load drops out and builds into an **alluvial fan**, a conical wedge of sediment that builds outward from the canyon mouth. Once the fan's conical shape has been established, water spilling out of the canyon during a flash flood divides into numerous braided streams, called *distributaries*, that spread water and clasts radially outward over the fan's surface (**Fig. 21.14b**). For a period of time, one distributary of the fan may carry most of a flood's water flow. Eventually, however, sediment accumulates at the end of this distributary, so the slope of the distributary's channel decreases. When this happens, water flow abandons that distributary and finds another one with a steeper channel, and as a result, over time, different parts of the fan build out in succession. This process maintains the overall cone-like shape of the fan as it grows. Notably, because the sediment in an alluvial fan has high permeability, water running down a distributary during a flood gradually infiltrates into the fan. As a result, the proportion of water to sediment in the flood decreases downstream, and the flood evolves into a slurry-like debris flow that slows and finally comes to rest. Over time, alluvial fans emerging from adjacent valleys may merge and overlap along the front of a mountain range, producing an elongate wedge of sediment called a *bajada* (**Fig. 21.14c**).

Playas and Salt Lakes

During a particularly large storm or an unusually wet spring, a temporary lake may develop over the low part of a basin in a desert. In drier times, such desert lakes evaporate entirely, leaving behind a dry, flat, exposed lake bed known as a **playa** (**Fig. 21.15a, b**). Over time, a smooth crust of clay and various salts (halite, gypsum, borax, and other minerals) accumulate on the surface of playas. Some of these minerals have industrial uses and thus have been mined. In a few localities, isolated cobbles sit out on the surface of a playa, far from any slope down which they could have rolled. At the

SEE FOR YOURSELF . . .

Death Valley, California

LATITUDE
36°12′42.34″N

LONGITUDE
116°48′25.43″W

Looking down from 35 km (~21.8 mi).

You can see many desert landforms in Death Valley, a narrow basin whose floor lies below sea level. The white patch is a playa, a bajada forms the slope between the playa and the mountains to the west, and small alluvial fans spill into the basin on the east.

FIGURE 21.14 Production and transportation of debris and sediment in deserts.

(a) This talus apron along the base of a desert cliff formed from rocks that broke off and tumbled down the cliff.

(b) An alluvial fan has accumulated at the mouth of a small canyon along the edge of Death Valley.

(c) A bajada accumulating at the base of a mountain range in the Mojave Desert. Note that a bajada consists of overlapping fans.

best known example, Racetrack Playa of Death Valley, each cobble lies at the end of a groove formed when the cobble slid across the soft surface of the playa (**Fig. 21.15c**). Until recently, no one had ever actually seen the cobbles move, and geologists simply assumed that movement somehow occurred when the playa was wet and so slippery that strong winds could blow the cobbles along. In 2014, GPS-triggered cameras showed that,

in fact, the cobbles slide only under special circumstances. For sliding to occur, the playa must contain shallow water that freezes into a thin sheet during the night. When the air warms the next day, the ice breaks into broad plates, which move in response to strong winds and push the cobbles.

Where sufficient water flows from surrounding regions into a desert basin, a permanent lake will fill. If the basin occurs in an interior basin, meaning a basin that has no outlet to allow water to flow out, the lake becomes very salty. The salt accumulates because when water evaporates, only H_2O goes into the air—salt stays behind. The Great Salt Lake in Utah exemplifies this process (**Fig. 21.15d**). Even though the streams feeding the lake are fresh enough to drink, their water contains trace amounts of dissolved salt ions. Because the lake has no outlet, these ions have become concentrated in the lake over time, making it even saltier than the ocean.

Deposition from the Wind

As mentioned earlier, wind carries two kinds of sediment loads—a suspended load of dust-sized particles and a surface load of sand. Much of the dust is carried out of the desert and accumulates elsewhere. Locally, it builds into **loess**, layers of fine-grained sediment. Sand, however, cannot travel far, and it accumulates within the desert in piles called **dunes**, ranging in size from less than a meter to over 300 m high. In favorable locations, dunes accumulate to form vast sand seas hundreds of meters thick. We'll look at dunes in more detail later in this chapter.

> ### Take-Home Message
>
> Sediment carried by water and wind in deserts accumulates in a variety of landforms. Alluvial fans form at the outlets of canyons; playas form where water temporarily collects in basins, and dunes form where large amounts of sand are available. Lakes in interior basins become salty.
>
> **QUICK QUESTION:** How do bajadas develop?

21.5 Desert Landforms and Life

The popular media commonly portray deserts as endless seas of sand piled into dunes, which hide the occasional palm-studded oasis. In reality, immense sand seas are but one type of desert landscape. Some deserts are vast, rocky plains; others sport a stubble of cacti and other hardy desert plants; and still others contain intricate rock formations that look like

FIGURE 21.15 Playas form where a shallow, salty lake dries up.

An oblique air photo

Playa

Bajada

Range

(a) This playa in California formed at the base of a bajada.

A close-up of salt crystals.

(b) White salt crystals encrust the floor of a playa in Death Valley.

(c) Wind-driven ice pushed rocks, leaving tracks along the slippery clay surface of Racetrack Playa in Death Valley, California.

(d) The Great Salt Lake in Utah has no outlet. Sediments deposited along its shores are quite salty.

> **Did you ever wonder . . .**
>
> whether all deserts are completely covered by sand?

medieval castles. Explorers of the Sahara emphasized these differences by distinguishing among *hamada* (barren, rocky highlands), *reg* (vast, stony plains), and *erg* (sand seas in which large dunes form). As we've noted, desert landscapes, overall, tend to be harsher and more rugged than temperate or tropical ones (**Fig. 21.16a**). For example, if eastern North America, a temperate region, were instead a desert, the Appalachians would not be gentle, forested hills but rather would consist of stark, rocky ridges (**Fig. 21.16b**). In this section, we examine various desert landscapes.

Rocky Cliffs and Mesas

In hilly desert regions, the lack of soil exposes rocky ridges and cliffs. As cliffs erode when rocks split away along joints, the cliff face's position moves, but the face retains roughly the same shape. The process, commonly referred to as **cliff retreat**, or scarp retreat, occurs in fits and starts (**Fig. 21.17a**). A cliff may remain unchanged for decades or centuries, and then suddenly a wall of rock falls off it and crumbles into rubble at the foot of the cliff. Cliff height typically reflects bed thickness—in places where particularly thick, resistant beds crop out, tall cliffs develop, for joints tend to be widely spaced in thick beds,

FIGURE 21.16 Contrasts between desert hillslopes and temperate landscapes.

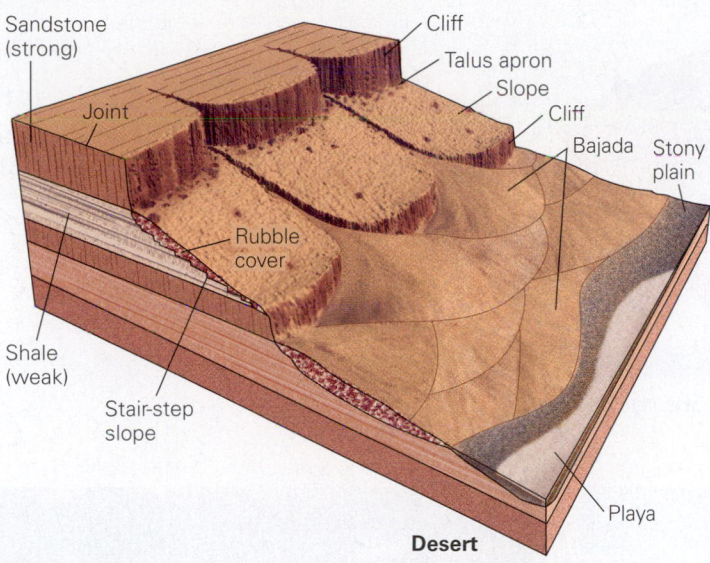

Desert

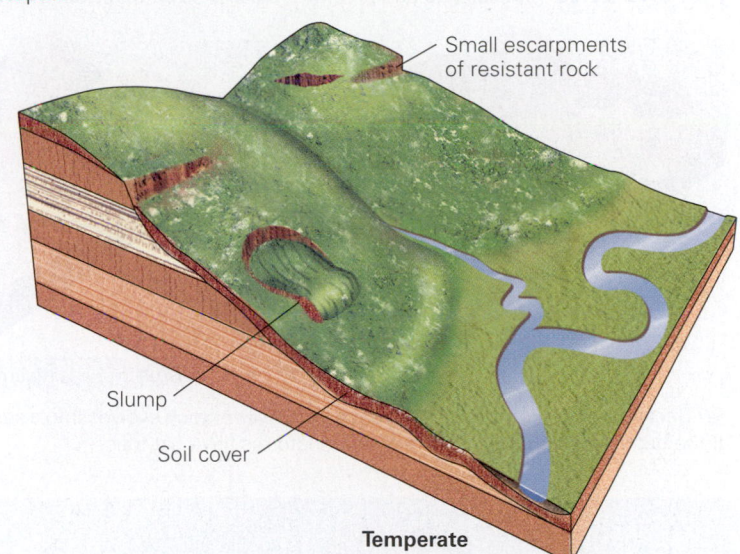

Temperate

(a) In a desert, steep cliffs form out of resistant rock, fans of debris form a bajada at the base of the cliffs, and playas form on the plain.

(b) In a temperate region, thick soils form, and slumping prevents steep slopes from developing. Vegetation covers the surface.

FIGURE 21.17 Cliff retreat in a desert environment.

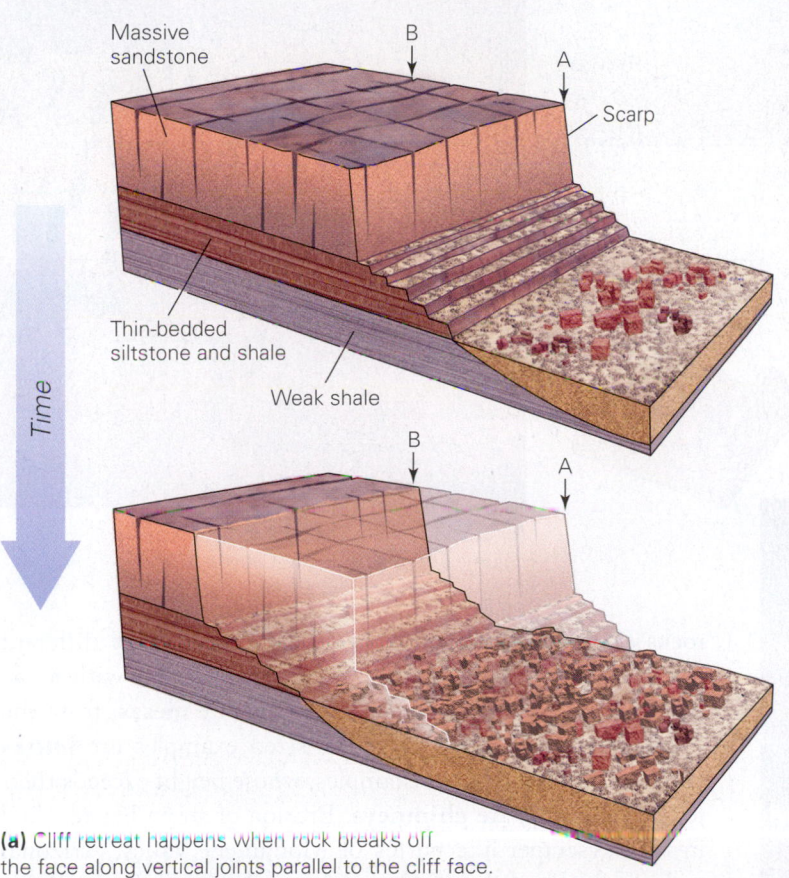

(a) Cliff retreat happens when rock breaks off the face along vertical joints parallel to the cliff face.

(b) On a hill in Utah, the strong sandstone holds up a cliff face, whereas weak shale forms a slope.

so the collapse of a portion of the wall generates huge new face. In thinly bedded shale, joints are small and closely spaced, so shale beds erode to make an overall gradual slope consisting of many tiny stair steps. Thus, cliffs formed from strata of contrasting strength develop a stair-step-like shape—strong layers (sandstone or limestone) become cliffs, whereas weak layers (shale) become slopes (**Fig. 21.17b**).

With continued erosion and cliff retreat, a plateau of rock slowly evolves into a cluster of isolated hills, ridges, or columns (**Fig. 21.18a**). Flat-lying strata or flat-lying layers of volcanic

FIGURE 21.18 Mesas and buttes form in deserts as cliffs retreat over time.

Butte
Chimney
Mesa

Time (increasing amount of erosion)

(a) Because of cliff retreat, a once-continuous layer of rock evolves into a series of isolated remnants. If the bedding is horizontal, the resulting landforms have flat tops.

(b) Buttes and mesas tower above the floor of Monument Valley, Arizona.

Joint
Preferential erosion along joints leaves walls ("fins").
Erosion through a fin produces an arch.
Eventually, the arches collapse.

(c) Erosion produced hoodoos, chimney-like columns of rock, in Bryce Canyon, Utah.

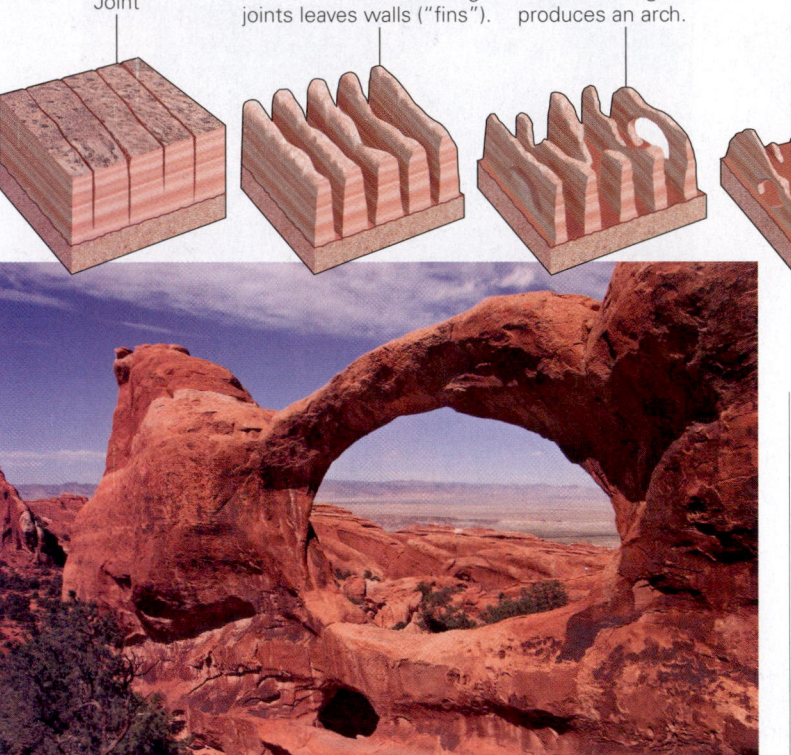

(d) Erosion of sedimentary beds containing large joints can produce natural arches.

rocks erode to make flat-topped hills. These go by different names, depending on their size. Large examples, with a top surface area of several square kilometers, are **mesas**, from the Spanish word for table. Medium-sized examples are **buttes** (**Fig. 21.18b**). And small examples, whose height exceeds their top surface area, are **chimneys**. Erosion of strata has resulted in the skyscraper-like buttes of Monument Valley, Arizona, and the stark cliffs of Canyonlands National Park. Bryce Canyon National Park in Utah contains countless chimneys of brightly colored shale and sandstone—locally, these chimneys are known as *hoodoos* (**Fig. 21.18c**). *Natural arches*, such as those of Arches National Monument, form when erosion removes the lower part of a wall of rock, while the upper part

FIGURE 21.19 Examples of erosional landscapes in deserts.

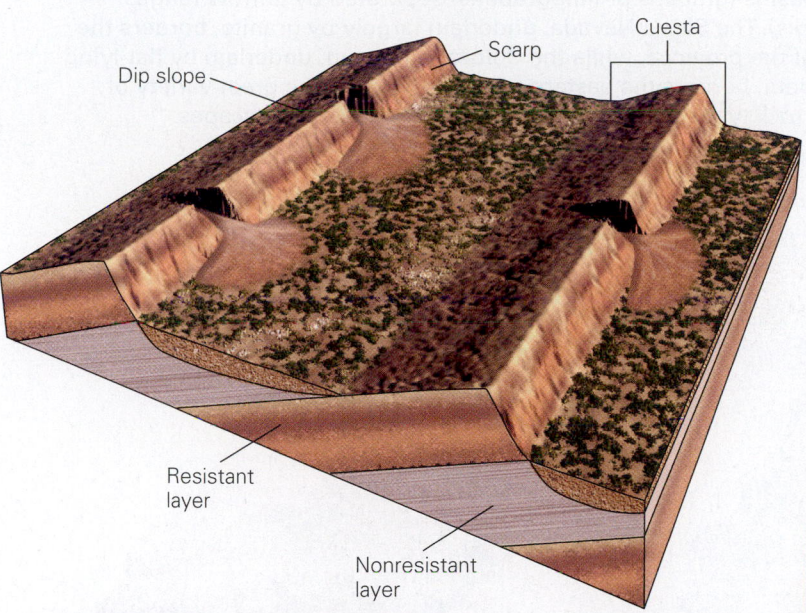

(a) Asymmetric ridges called cuestas develop where strata in a region are not horizontal.

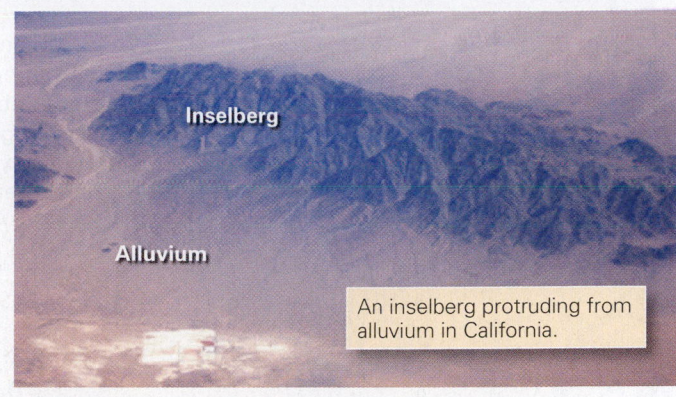

An inselberg protruding from alluvium in California.

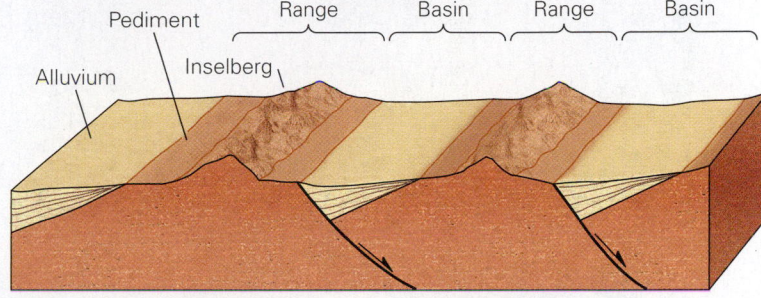

(c) In the Basin and Range Province of the southwestern U.S., tilted fault-block ranges evolve into inselbergs, bordered by sediment-filled basins.

(b) The homogeneous granite of a pluton breaks along joints into blocks, which erode into rounded boulders. This example crops out in southern Nevada.

remains. The walls of rock are bounded by joints and became walls because erosion preferentially removed rock along the joints (**Fig. 21.18d**; **Geology at a Glance**, pp. 784–785.)

In places where bedding dips at an angle, an asymmetric ridge called a *cuesta* develops. A joint-controlled cliff forms the steep front side of a cuesta, and the tilted top surface of a resistant bed forms the gradual slope on the backside (**Fig. 21.19a**). Because the angle of the gradual slope is the same as the dip angle of the bed (the angle the bed surface makes with respect to horizontal), it is known as a *dip slope*. If the bedding dip is steep to near vertical, a narrow symmetrical ridge, called a *hogback*, forms. If desert hills consist of homogeneous rock such as

granite rather than stratified rock, they typically erode to make a pile of rounded blocks (**Fig. 21.19b**).

With progressive erosion on all sides of a hill, finally all that remains of the hill is a relatively small island of rock, surrounded by alluvium-filled basins. Geologists refer to such islands of rock by the German word *inselberg* (island mountain; **Fig. 21.19c**). Depending on the rock type or the orientation of stratification in the rock, and on rates of erosion, inselbergs may be sharp-crested, plateau-like, or loaf-shaped with steep sides and a rounded crest. Inselbergs with a loaf geometry, as exemplified by Uluru (Ayers Rock) in central Australia (**Box 21.1**), are also known as *bornhardts*.

Desert Pavement

In many locations, the desert surface resembles a tile mosaic in that it consists of separate stones that fit together tightly, forming a fairly smooth surface layer above a soil composed of silt and clay. Such natural mosaics constitute **desert pavement** (**Fig. 21.20a, b**). Typically, desert varnish coats the top surfaces of the stones forming desert pavement. Geologists have proposed several explanations for the origin of desert pavements. Traditionally, pavements were thought to be lag deposits, formed when wind blows away the fine sediment between clasts, so that the clasts can settle down and fit together. Recently researchers have suggested instead that pavements form when windblown dust slowly sifts down onto the stones and then washes down between the stones. In this model, the pavement is "born at the surface," meaning that the stones forming the pavement were

GEOLOGY AT A GLANCE

The Desert Realm

The desert of the Basin and Range Province in Utah, Nevada, and Arizona consists of alternating basins (grabens or half-grabens) separated by narrow ranges (tilted fault blocks). The Sierra Nevada, underlain largely by granite, borders the western edge of the province, while the Colorado Plateau, underlain by flat-lying sedimentary strata, borders the eastern edge. Because of the great variety of elevations and rock types, the region hosts different desert landscapes.

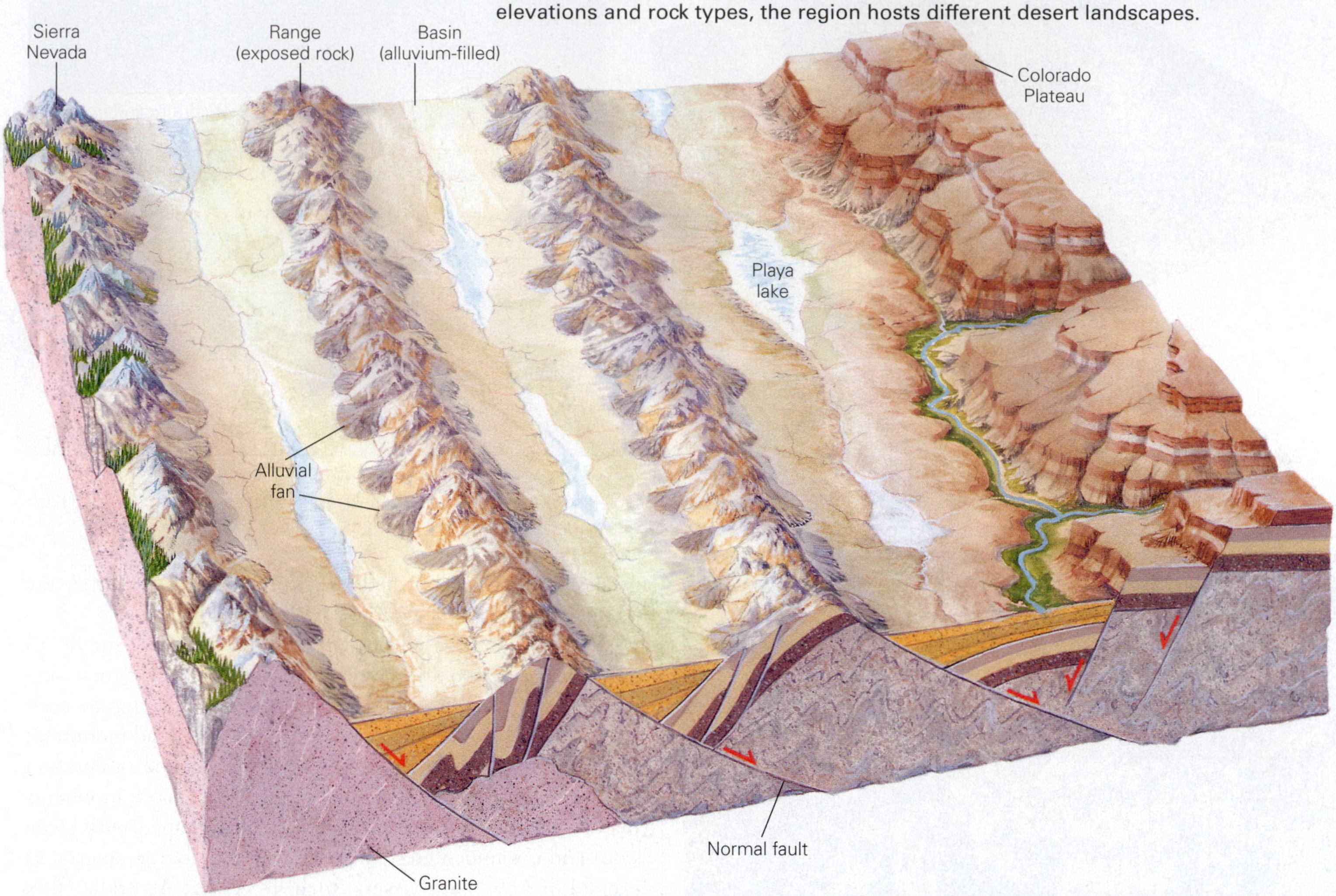

Sierra Nevada

Range (exposed rock)

Basin (alluvium-filled)

Colorado Plateau

Playa lake

Alluvial fan

Normal fault

Granite

Barchan dune

Cross beds

Most streams in deserts fill with water only during flash floods after heavy rains. The turbulent, muddy water of a flash flood can transport even large boulders. At other times, the stream channels are dry washes, arroyos, or wadis. Where there is a large supply of sand, a variety of sand dunes develop. The geometry of a particular sand dune depends on the sand supply and the wind. Inside sand dunes, we find cross beds.

Flash flood

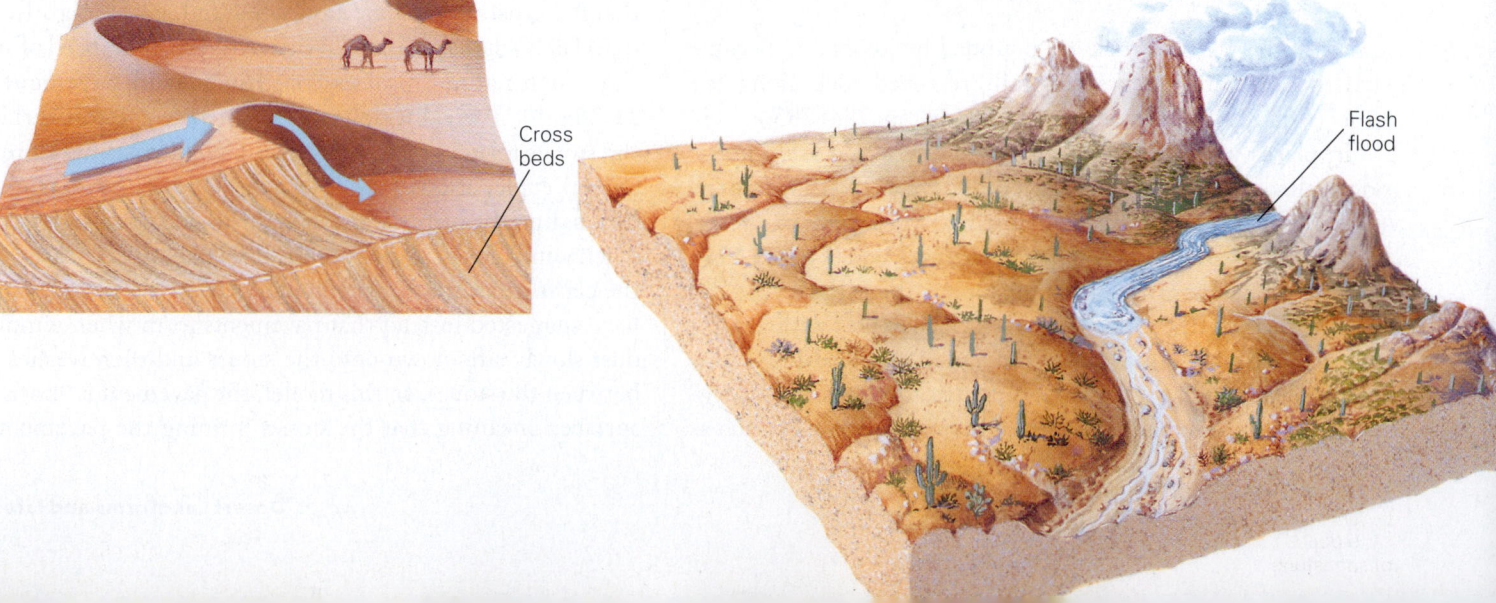

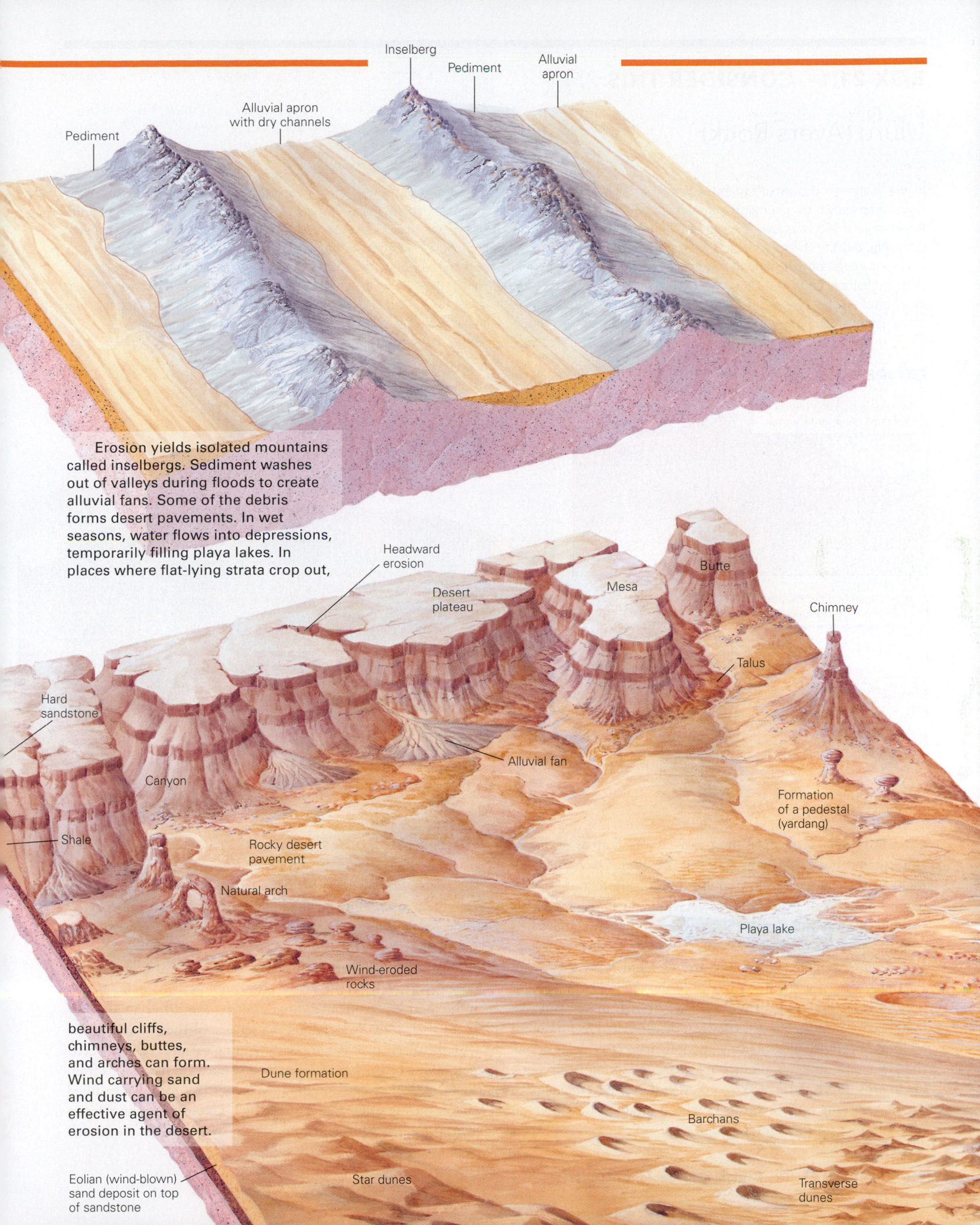

Pediment

Alluvial apron
with dry channels

Inselberg

Pediment

Alluvial
apron

Pediment

Erosion yields isolated mountains
called inselbergs. Sediment washes
out of valleys during floods to create
alluvial fans. Some of the debris
forms desert pavements. In wet
seasons, water flows into depressions,
temporarily filling playa lakes. In
places where flat-lying strata crop out,

Headward
erosion

Desert
plateau

Mesa

Butte

Chimney

Talus

Hard
sandstone

Alluvial fan

Formation
of a pedestal
(yardang)

Canyon

Shale

Rocky desert
pavement

Natural arch

Playa lake

Wind-eroded
rocks

beautiful cliffs,
chimneys, buttes,
and arches can form.
Wind carrying sand
and dust can be an
effective agent of
erosion in the desert.

Dune formation

Barchans

Eolian (wind-blown)
sand deposit on top
of sandstone

Star dunes

Transverse
dunes

BOX 21.1 CONSIDER THIS . . .

Uluru (Ayers Rock)

In the immense desert of central Australia, Uluru, also known by its English name, Ayers Rock, towers 360 m above a scrub-covered plain (**Fig. Bx21.1**). This rock mass, 3.6 km long and 2 km wide, consists of nearly vertical dipping sandstone beds. It makes up one limb of a huge regional syncline. The other limb is also a bornhardt, known locally as the Olgas.

Alluvium buries the entire area in between. Uluru formed because the sandstone that comprises it resisted erosion, whereas adjacent rock formations did not. Thus, over geologic time, alluvium buried the surrounding landscape, but Uluru remained high. Because its strata dip vertically, it has not developed the stair-step shape of mesas and buttes.

According to traditions of Australia's Aboriginal people, erosional features on the surface of the rock are scars from a fierce battle between ancient clans. In recent years, the rock has attracted tourists from around the world. Plaques at the base of the rock record the names of those who slipped and fell from its steep sides while climbing to the top.

FIGURE Bx21.1 The formation of Uluru (Ayers Rock) in central Australia.

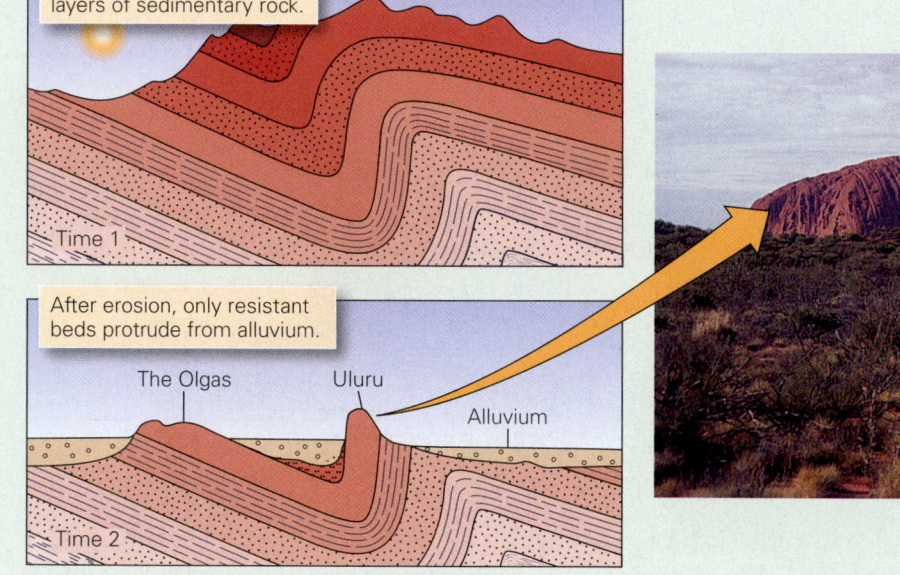

never buried but have been progressively lifted up as sediment collects and builds up beneath (**Fig. 21.20c**). Over time, the rocks at the surface crack, perhaps due to differential heating by the desert sun over time. Sheetwash, during downpours, may wash away fine sediment between fragments, and when soils dry and shrink between storms the clasts settle together, locking into a stable, jigsaw-like arrangement.

Desert pavements are remarkably durable and can last for hundreds or thousands of years if they are left alone. But like many features of the desert, they can be disrupted in a moment by human activity. For example, people driving vehicles across the pavement indent and crack its surface, making it susceptible to erosion. In parts of Arizona, vast desert pavements have become parking lots for campers who migrate to the desert in motor homes for the winter season.

Stony Plains and Pediments

The coarse sediment eroded from desert mountains and ridges ultimately washes into lowland plains to produce gravel-covered surfaces called stony plains. Portions of these stony plains evolve into desert pavements.

Not all stone-covered surfaces, however, are underlain by loose debris. When travelers began trudging through the desert of the southwestern United States during the 19th century, they found that in many locations the wheels of their wagons were actually rolling over flat or gently sloping surfaces of intact bedrock. These bedrock surfaces extended outward like ramps from the steep cliffs of a mountain range on one side to alluvium-filled valleys on the other (see Fig. 21.19c). Geologists now refer to such surfaces as **pediments**. Pediments develop when sheetwash

FIGURE 21.20 Desert pavement and a hypothesis for how it forms by building up a soil from below.

(a) A well-developed desert pavement in the Sonoran Desert, Arizona. The inset shows a close-up of the pavement.

(b) Students standing at the edge of a trench cut into desert pavement. Note the soil between the pavement and the underlying alluvium.

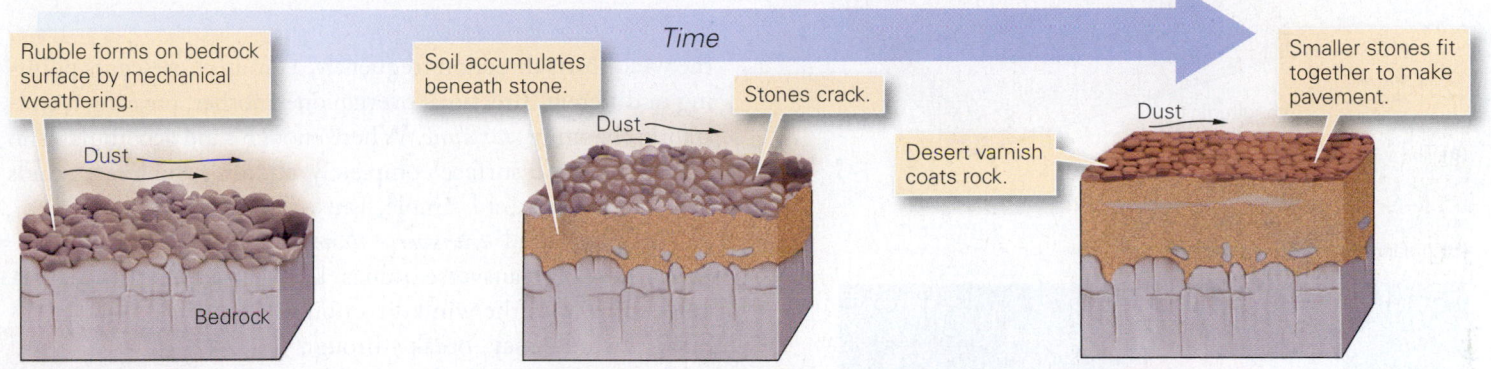

(c) Desert pavement forms in stages. First, loose pebbles and cobbles collect at the surface. Dust settles among the stones and builds up a soil layer below. The stones eventually crack into smaller pieces and settle to form a mosaic-like pavement.

during floods carries sediment away from the mountain front during cliff retreat. The moving sediment grinds away the bedrock that it tumbles over. Between erosional events, weathering weakens the surface of the pediment so that the next flood has more material to move. Alluvium that has been washed off pediments accumulates farther downslope and may eventually build up sufficiently to bury the pediment.

Seas of Sand: The Nature of Dunes

A **sand dune** is a pile of sand deposited by a moving current. Dunes can form due to currents of water on the beds of rivers as well as due to currents of air on the land surface of deserts.

Dunes in deserts start to form where sand becomes trapped on the windward side of an obstacle, such as a rock or a shrub. Gradually, the sand builds downwind into the lee of the obstacle. Once initiated, the dune itself affects the wind flow, and sand accumulates on the lee side of the dune. Here the sand eventually slides down the lee surface of the dune (**Fig. 21.21**).

In places where abundant sand accumulates, sand seas (ergs) bury the landscape. The wind builds the sand in these ergs into dunes that display a variety of shapes and sizes, depending on the character of the wind and the sand supply (**Fig. 21.22a**). Where the sand is relatively scarce and the wind blows steadily in one direction, beautiful crescents called *barchan dunes* develop, with the tips of the crescents pointing downwind. If

FIGURE 21.21 Progressive stages in the growth of a small sand dune.

Time 1

Time 2

Time 3

Time 4

FIGURE 21.22 The types of sand dunes and the cross beds within them.

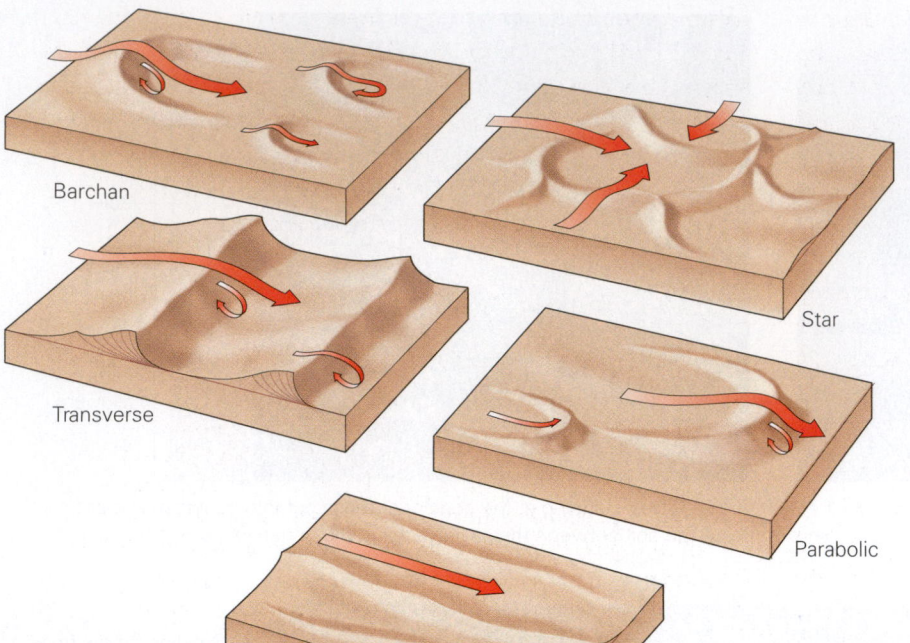

Barchan

Star

Transverse

Parabolic

Longitudinal

(a) The various kinds of sand dunes.

Main bed

Cross bed

(d) Cross beds preserved in Mesozoic sandstone beds of Zion National Park.

(b) A sand dune with surface ripples.

Wind

Windward side

Slip face

Dune

Surface ripples

Cross-bed surface

Slip face

Main bedding surface

(c) Cross bedding inside a dune.

the wind shifts direction frequently, a group of crescents pointing in different directions overlap one another, creating a constantly changing *star dune*. Where enough sand accumulates to bury the ground surface completely, and only moderate winds blow, sand piles into simple, wave-like shapes called *transverse dunes*. The crests of transverse dunes lie perpendicular to the wind direction. Strong winds may break through transverse dunes and change them into *parabolic dunes* whose ends point in the upwind direction. Finally, if there is abundant sand and a strong, steady wind, the sand streams into *longitudinal dunes* (also called *seif dunes* after the Arabic word for sword) whose axis lies parallel to the wind direction. In the southern third of the Arabian Peninsula, a region called the Empty Quarter because of its lack of population, a vast erg called the Rub al Khali contains seif dunes that stretch for almost 200 km and reach heights of over 300 m.

In a sand dune, sand saltates up the windward side of the dune, blows over the crest of the dune, and then settles on the steeper, lee face of the dune, where the air slows down. The slope of this face attains the angle of repose, the slope angle of a freestanding pile of sand. As

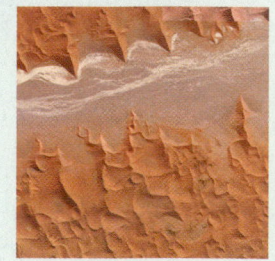

sand collects on this surface, it may become unstable and slide down the slope, so geologists refer to the lee side of a dune as the **slip face**. As more and more sand accumulates on the slip face, the crest of the dune migrates downwind, and former slip faces become preserved inside the dune. In cross section, these slip faces appear as cross beds (**Fig. 21.22b–d**). The surfaces of dunes are not, in general, smooth but rather with much smaller ripples.

With the exception of star and longitudinal dunes, sand dunes migrate downwind as the wind continuously picks up sand from the gently dipping windward slope and drops it onto the leeward side, or slip face. Rates of migration can exceed 25 m per year. Because of moving sand on an active dune, vegetation can't grow there. If a change in climate brings more rain or less wind, plant cover may grow and stabilize the dunes. At the end of the last ice age, for example, the Sand Hills region of western Nebraska was a vast active dune field, but in the past 11,000 years it has been covered by grasslands, and the dunes have become stabilized.

Life in the Desert

In the midst of a large erg, there seems to be nothing growing or moving at all. But most desert landscapes do include plants and animals. These organisms must possess special characteristics to enable them to survive in the desert: they must be able to withstand extremes of temperature—oppressive heat during the day and chilling cold at night—and to survive without abundant water.

Plants have evolved a number of different means to survive desert conditions. Some produce thick-skinned seeds that last until a heavy rainfall, then quickly germinate, grow, and generate new seeds only while water remains. The new generation of seeds then waits until the next rainfall to start the cycle over again. Other plants have evolved the ability to send long roots down to find deep groundwater. Still others have shallow root systems that spread over a broad area so they can efficiently soak up water when it does rain.

Many desert plants have thick, fleshy stems and leaves. These plants, known as *succulents*, can store water for long periods of time (**Fig. 21.23**). Because succulents may be the only source of water during a time of drought, they have developed threatening thorns or needles to keep away thirsty animals. Plant life tends to be more diverse in desert *oases*, the verdant islands that crop up where natural springs spill groundwater onto the surface (see Chapter 19). The nearly year-round supply of water in an oasis nourishes a variety of palms and other nonsucculent plants.

Animal life in the desert includes plant eaters, hunters, and scavengers. Animals face the same challenge as plants—they must be able to retain water and survive extreme temperatures. To accomplish these goals, desert animals have also evolved numerous strategies. Frogs, for example, burrow beneath the ground and remain dormant for months, waiting until the next rain. Reptiles escape the midday heat by crawling into dark cracks between rocks. Rodents forage for food only during the cool night. And kit foxes, jackrabbits, and mule deer have disproportionately large ears through which they efficiently lose body heat. Many desert mammals, such as camels, retain body water by not sweating, as we noted earlier.

Take-Home Message

A variety of erosional and depositional landscapes develop in deserts. Erosion of thick layers of horizontal sedimentary rock in deserts yields buttes and mesas, and erosion of tilted layers forms cuestas. Stony plains develop where finer sediment blows or washes away. Desert pavements form when fine sediment filters down between larger clasts, which settle together to form a mosaic. Regions with abundant sand contain many types of dunes. Particularly hardy organisms have evolved to survive in these harsh realms.

QUICK QUESTION: What factors control the shape, dimensions, and orientation of sand dunes?

21.6 Desert Problems

Humans can't live in the desert easily. The loss of body moisture in extreme heat can be so rapid that a person will die in less than 24 hours unless shaded from the Sun and supplied with at least 8 liters of water per day. Nevertheless, all but the most barren deserts are inhabited, though sparsely (**Fig. 21.24a**). Before technology provided water wells, pipelines, and mechanized transportation, desert peoples lived in small nomadic groups, spaced far enough apart that they could live off the land. In the past, nomadic desert dwellers either built temporary shelters out of local materials or traveled with tents. Locally, people carved underground dwellings in sediment or soft rock, for rock is such a good insulator that a few meters below the ground surface, it stays close to the region's mean temperature year round.

In recent times, more and more people have moved into desert regions. In fact, the population of the desert in the southwestern United States is growing faster than in any other part of the country, and as desert cities grow, environmental problems soon follow (**Fig. 21.24b**). As noted in Chapter 19, growing cities must either suck water out of the ground or bring in water via canals from rivers or reservoirs to meet their needs. As a result, water tables in deserts are dropping, rivers are drying up, the land surface is cracking, and vegetation is dying (**Fig. 21.25a**). People also have imported exotic plants and animals that invade the countryside and upset the ecological balance.

FIGURE 21.23 Plants of the Sonoran Desert, Arizona, are well adapted for dry conditions.

(a) Saguaro cactus can become huge. Don't try to hug one.

(b) "Teddy bear" cholla look soft, but the spines are extremely sharp and hard to remove.

(c) Cactus flowers stand out against brown sand and rock.

The modern era has seen a remarkable change in desert margins. Natural droughts (periods of unusually low rainfall), aggravated by overpopulation, overgrazing, careless agriculture, and diversion of water supplies, have transformed semiarid grasslands into true deserts, leading to tragic famines that have killed millions of people. **Desertification**, the process of transforming nondesert areas to desert, has accelerated.

The consequences of desertification have devastated portions of the Sahel, the belt of semiarid land that fringes the southern margin of the Sahara (**Fig. 21.25b**). In the past, the Sahel provided sufficient vegetation to support a small population of nomadic people and animals. But during the second half of the 20th century, large numbers of people migrated into the Sahel to escape overcrowding in central Africa. The immigrants began farming and maintained herds of cattle and goats. Plowing and overgrazing removed soil-preserving grass and

caused the soil to dry out. In addition, trampling by animals compacted the ground so it could no longer soak up water. In the 1960s and again in the 1980s, a series of natural droughts hit the region, bringing catastrophe (**Fig. 21.25c**). Wind erosion stripped off the remaining topsoil. Without vegetation, the air grew drier, and the semiarid grassland of the Sahel became desert, with mass starvation as the result. In some cases, desertification happens when winds blow in new dunes, and bury fields or forests.

Other arid regions on Earth are developing similar problems. The Aral Sea in Kazakhstan, for example, has almost entirely dried up. Rivers that once carried water into the sea have been diverted to provide water for irrigation, and now the rate at which the sea loses water to evaporation greatly exceeds the supply of new water, and the area of the sea has shrunk to a small fraction of what it once was. Boats that once plied the

FIGURE 21.24 Living in the desert is a challenge because of the need for water.

(a) A West African village struggles with blowing sand and the lack of vegetation.

(b) Suburbs of Las Vegas, Nevada, push into the desert. Residents use groundwater and water brought from the Colorado River to keep grass green.

FIGURE 21.25 Desertification is happening in parts of Africa.

(a) A woodland has dried out and died as huge dunes encroach.

(b) The Sahel is the semiarid land along the southern edge of the Sahara. Large parts have undergone desertification.

(c) Drought in the Sahel has brought deadly consequences. Here residents seek water from a dwindling pond.

waters of the Aral Sea now lie as rusting hulks in the salty dust (**Fig. 21.26**).

Desertification does not happen only in less-industrialized nations. People in the western Great Plains of the United States and Canada suffered from the problem beginning in 1933, the fourth year of the Great Depression. Banks had failed, workers had lost their jobs, the stock market had crashed, and hardship burdened all. No one needed yet another disaster—but that year, even nature turned hostile. All through the fall, so little rain fell in the plains of Texas and Oklahoma that the region's grasslands and croplands browned and withered, and the topsoil turned to powdery dust. Then, in November, strong storms blew eastward across the plains. Without vegetation to protect the ground, the wind lapped at it, stripped off the topsoil, and sent it skyward to form rolling black clouds that literally blotted out the sun (**Fig. 21.27**). People caught in the resulting dust storm found themselves choking and gasping for breath.

When the dust finally settled, it had buried houses and roads under huge drifts, and dirtied every nook and cranny. The dust blew east as far as New England, where it turned the snow brown. What had once been a rich farmland in the southwestern plains turned into a wasteland, the Dust Bowl.

For several more years the drought persisted, leading to starvation and bankruptcy. Many of the region's residents were forced to move to wherever they could find work. People from Texas and Oklahoma piled into jalopies and drove on old Route 66 out to California, looking for jobs in the state's still-green agricultural regions. Many people were subjected to exploitation once they arrived in California. John Steinbeck dramatized this staggering human tragedy, which came to symbolize the Depression, in his novel *The Grapes of Wrath*.

Why did the fertile soils of the southern Great Plains suddenly dry up? The causes were complex—some were natural and some were human-induced. The region, which has a semiarid climate, was settled in the 1880s and 1890s, which were unusually wet decades. Far more people moved into the region than it could sustain and farmed the land too intensively. Plowing destroyed the fragile grassland root systems that held the thin soil in place, so when the drought of the 1930s came, it brought catastrophe. Clearly, the Dust Bowl of the 1930s reminds us of how fragile the Earth's green blanket of vegetation really is. And while such desertification can be reversed, by irrigation and planting, water to nourish the plants has to come from somewhere, and when people obtain it by diverting rivers or by pumping groundwater, they can generate a new set of problems.

On a time scale of millennia, global climate change can shift climatic belts sufficiently to transform agricultural regions into deserts. For example, some 5,000 years ago, the swath of land between the Nile Valley in Egypt and the Tigris-Euphrates Valley of Mesopotamia was known as the fertile crescent. Here people first abandoned their nomadic ways and settled in agricultural communities. Now the original "land of milk and honey" hosts desert landscapes, and agriculture of the region requires intensive irrigation. The change in landscape reflects a change in climate—the beginning of Western civilization occurred during the warmest and wettest period of global climate since the time before the last ice age. So much water drenched the Middle East and North Africa that rivers flowed where Saharan sands now blow. In fact, geologists using ground-penetrating radar have mapped abandoned river channels buried beneath the sand. Unfortunately, as we discuss in Chapter 23, if current trends in climate change continue, our present agricultural belts could someday become new Saharas.

Desertification has an additional dangerous side effect—global transmission of chemicals and pathogens by blowing dust. As desert areas expand in response to desertification, and as desert pavement gets disrupted, windblown dust becomes more of a problem. Not only do winds have larger areas of dry, dusty land to churn, but the dust generated from lands that were once agricultural and are now desert may contain harmful chemicals (e.g., residues of herbicides and pesticides), fungi, and microbes that can themselves become windborne. In recent years, satellite images have revealed that windblown dust from deserts can travel across oceans and affect regions on the other side. For example, dust blown off the Sahara can traverse the Atlantic and settle over the Caribbean (**Fig. 21.28**). Geologists are concerned that this dust, along with the fungi, toxic chemicals, and microbes that it carries, may infect corals with disease or in some other way inhibit their life processes. Windblown dust from one part of the world may contribute to the destruction of coral reefs on the other side.

Take-Home Message

Human populations in arid regions have been burgeoning. Droughts and stresses caused by agriculture, grazing, and water diversion have led to desertification. Dust from desertified regions can carry pollutants and microbes and may potentially cause problems elsewhere in the world.

QUICK QUESTION: What factors transformed Oklahoma and Texas into a dust bowl during the 1930s?

FIGURE 21.26 Desertification of the Aral Sea in central Asia occurred when the Soviet Union diverted the rivers that flowed into the sea for irrigation projects. Once home to a fishing fleet, the sea has been reduced to a few ultra-salty ponds.

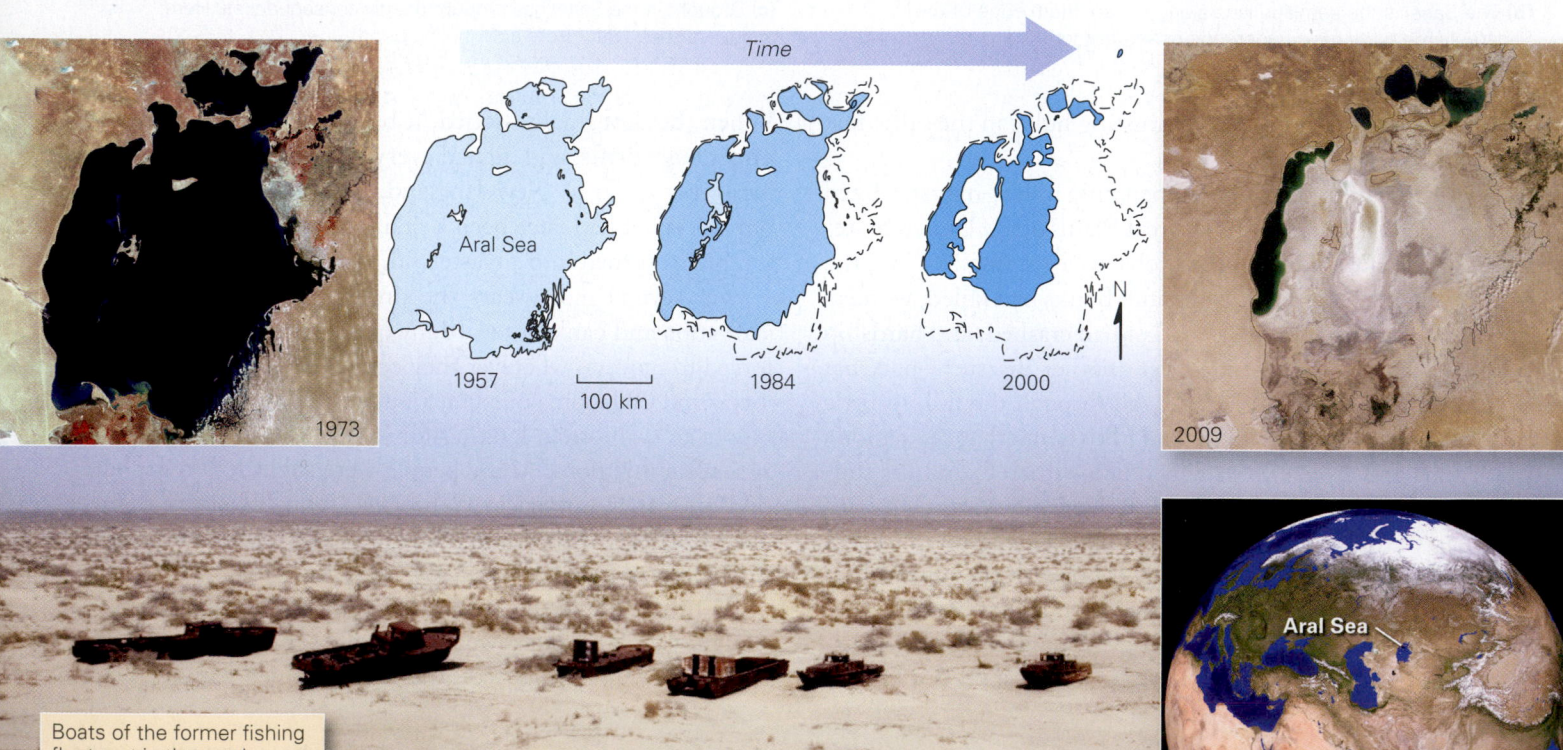

1973

Time

Aral Sea

100 km

1957   1984   2000

N

2009

Boats of the former fishing fleet rust in the sand.

Aral Sea

FIGURE 21.27 An iconic photograph of a farmer walking through a 1930s dust storm in Oklahoma. Much of the topsoil of the region was blown away.

FIGURE 21.28 In this satellite image, a huge dust cloud that originated in the Sahara blows across the Atlantic.

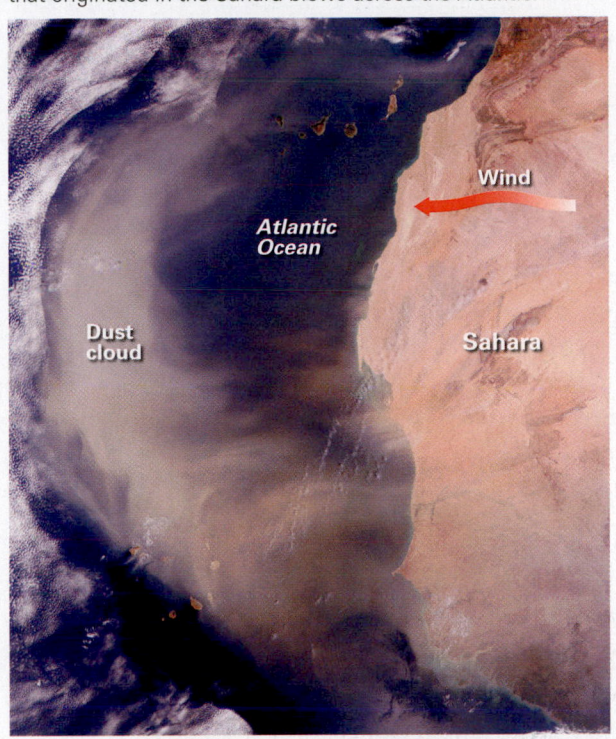

CHAPTER SUMMARY

- Deserts generally receive less than 25 cm of rain per year. Vegetation covers no more than 15% of their surface.

- Subtropical deserts form between latitudes of 20° and 30°, rain-shadow deserts occur on the inland side of mountain ranges, coastal deserts form on the land adjacent to cold ocean currents, continental-interior deserts exist in landlocked regions far from the ocean, and polar deserts develop at high latitudes.

- In deserts, chemical weathering happens slowly, so rock bodies tend to erode primarily by physical weathering. Desert varnish forms on rock surfaces, and soils tend to accumulate soluble minerals.

- Water causes significant erosion in deserts, mostly during heavy downpours. Flash floods carry large quantities of sediments down ephemeral streams. When the rain stops, these streams dry up, leaving dry washes.

- Wind can pick up dust and silt as suspended load and can cause sand to undergo saltation. Where wind blows away finer sediment, a lag deposit remains. Windblown sediment abrades the ground, creating ventifacts and yardangs.

- Desert pavements are mosaics of varnished stones armoring the surface of the ground.

- Talus piles form when rock fragments accumulate at the base of a slope. Alluvial fans form at a mountain front where water in ephemeral streams deposits sediment. When temporary desert lakes dry up, they leave playas.

- In some desert landscapes, erosion causes cliff retreat, eventually resulting in the formation of mesas, buttes, and inselbergs. Pediments of nearly flat or gently sloping bedrock surround some inselbergs.

- In places hosting abundant sand, the wind builds it into dunes. Common types include barchan, star, transverse, parabolic, and longitudinal (seif) dunes.

- Deserts contain a great variety of plant and animal species. All have adapted to survive extremes in temperature and without abundant water.

- Changing climates and land abuse may cause desertification, the transformation of semiarid land into deserts. Windblown dust, sometimes carrying microbes and toxins, may waft from deserts across oceans.

GUIDE TERMS

alluvial fan (p. 778)	desert pavement (p. 783)	loess (p. 779)	sand dune (p. 787)
arroyo (p. 775)	desert varnish (p. 773)	mesa (p. 782)	slip face (p. 789)
butte (p. 782)	desertification (p. 790)	pediment (p. 786)	surface load (p. 776)
chimney (p. 782)	dry wash (p. 775)	petroglyph (p. 774)	suspended load (p. 775)
cliff retreat (p. 780)	dune (p. 779)	playa (p. 778)	talus (p. 778)
deflation (p. 776)	dust storm (p. 776)	rain shadow (p. 771)	ventifact (p. 777)
desert (p. 769)	lag deposit (p. 776)	saltation (p. 776)	wadi (p. 775)

REVIEW QUESTIONS

1. What factors determine whether a region can be classified as a desert?
2. Explain the several settings that can cause deserts to form.
3. Have today's deserts always been deserts?
4. How do weathering processes in deserts differ from those in temperate or humid climates?
5. Describe how water modifies the landscape of a desert. Be sure to discuss both erosional and depositional landforms.
6. Explain the ways in which desert winds transport sediment.
7. Explain how the following features form: (a) desert varnish, (b) desert pavement, (c) ventifacts, and (d) yardangs.
8. Describe the process of formation of alluvial fans, bajadas, and playas.
9. Describe the process of cliff (scarp) retreat and the landforms that result from it.
10. What are the various types of sand dunes? What factors determine which type of dune develops at a particular location?
11. Discuss various adaptations that life forms have evolved in order to survive in desert climates.
12. What is the process of desertification, and what causes it? How can desertification in Africa affect the Caribbean?

ON FURTHER THOUGHT

13. Death Valley, California, lies to the east of a high mountain range, and its floor lies below sea level. During the summer, Death Valley is very hot and dry. Explain why it has such weather.
14. You are working for an international nongovernmental organization and have been charged with the task of providing recommendations to an African nation that wishes to slow or halt the process of desertification within its borders. What are your recommendations?
15. The Namib Desert lies to the north and west of the Kalahari Desert, in southern Africa. The reason that the former region is a desert is not the same as the reason that the latter is a desert. Explain this statement.

A small glacier, all that remains of what was once a large ice cap, clings to the side of a mountain in the Canadian Rockies. Flow of glaciers contributed to carving this rugged landscape.

CHAPTER 22

Amazing Ice:
Glaciers and Ice Ages

> I seemed to vow to myself that some day I would go to the region of ice and snow and go on and on till I came to one of the poles of the earth.
>
> —*Ernest Shackleton (British polar explorer, 1874–1922)*

22.1 Introduction

There's nothing like a good mystery, and one of the most puzzling in the annals of geology came to light in northern Europe early in the 19th century. When farmers of the region prepared their land for spring planting, they occasionally broke their plows by inadvertently running them into large boulders that were buried randomly through otherwise fine-grained sediment. Many of these boulders did not consist of local bedrock but rather came from outcrops hundreds of kilometers away. Because the boulders had apparently traveled so far, they came to be known as **erratics** (from the Latin *errare*, to wander).

The mystery of the wandering boulders became a subject of great interest to early-19th-century geologists, who realized that such deposits of extremely unsorted sediment, meaning sediment that contains a great variety of different clast sizes, could not be examples of typical stream alluvium, for running water sorts sediment by size. Most attributed the deposits to a vast flood, during which a slurry of boulders, sand, and mud spread across the continent. In 1837, however, a young Swiss geologist named Louis Agassiz proposed a radically different interpretation. Agassiz often hiked in the Alps near his home,

where he could study **glaciers**, rivers or sheets of recrystallized ice that last all year long and flow slowly under the influence of gravity. He observed that glacial ice could carry enormous boulders as well as sand and mud, because ice is a solid and has enough strength to support the weight of rock. Agassiz realized that because ice does not sort sediment as it flows, glaciers leave behind extremely unsorted sediment when they melt. On the basis of this observation, he proposed that the mysterious sediment and erratics of Europe were deposits left by **continental ice sheets**, vast glaciers that cover large areas of a continent (**Fig. 22.1**). In Agassiz's view, Europe had once been in the grip of an **ice age**, a time when the climate was significantly colder and glaciers grew to be immensely larger than they are today.

FIGURE 22.1 Agassiz's thoughts about the Ice Age.

(a) Agassiz found boulders protruding from the ground in places that are not currently glaciated. He proposed that the boulders are erratics left by now-vanished glaciers.

(b) Agassiz envisoned that vast areas of the northern hemisphere were once covered by vast ice sheets comparable to the one covering Antarctica today.

Agassiz's radical proposal faced intense criticism for the next two decades. But he didn't back down and instead challenged opponents to visit the Alps and examine the sedimentary deposits that alpine glaciers had left behind. By the late 1850s, most doubters had changed their minds, and the geological community concluded that the notion that Europe once had Arctic-like climates was correct. Later in life, Agassiz traveled to the United States and documented many glacier-related features in North America's landscape, proving that an ice age had affected vast areas of the planet.

Glaciers cover only about 10% of the land on Earth today, but during the most recent ice age, which ended less than 12,000 years ago, as much as 30% of continental land surface had a coating of ice. New York City, Montreal, and many of the great cities of Europe occupy land that once lay beneath hundreds of meters to a few kilometers of ice. The work of Louis Agassiz brought the subject of glaciers and ice ages into the realm of geologic study and led people to recognize that major climate changes have happened during Earth history. In this chapter, after considering the nature of ice, we see how glaciers form, why they move, and how they modify landscapes by erosion and deposition. A substantial portion of the chapter concerns the most recent ice age, known as the **Pleistocene Ice Age**, for its impact on the landscape can still be seen today, but we briefly introduce ice ages that happened earlier in Earth history, too. We conclude by considering hypotheses to explain why ice ages happen.

22.2 Ice and the Nature of Glaciers

What Is Ice?

Ice consists of solid water, formed when liquid water cools below its freezing point. We can consider a single ice crystal to be analogous to a mineral specimen: it is a naturally occurring, inorganic solid, with a definite chemical composition (H_2O) and a regular crystal structure. Ice crystals have a hexagonal form, so snowflakes grow into six-pointed stars (**Fig. 22.2a**). We can think of a layer of fresh snow as a layer of sediment, and a layer of snow that has been compacted so that its grains stick together as a bed of sedimentary rock (**Fig. 22.2b**). We can think of a coating of ice that appears on the surface of a pond in winter as an igneous rock, for it forms when molten ice—liquid water—solidifies. Glacier ice, in this context, is a metamorphic rock. It develops when pre-existing ice recrystallizes, meaning that the molecules in solid water rearrange to form new crystals (**Fig. 22.2c**).

Pure new ice has the transparency of glass, but if ice contains tiny air bubbles or cracks that disperse light, it becomes milky. Like glass, ice has a high **albedo**, meaning that it reflects light well—so well, in fact, that if you walk on ice without eye protection, you risk blindness from the glare. Ice differs from most other familiar materials in that its solid form is not as dense as its liquid form, for the architecture of an ice crystal holds water molecules apart. Ice, therefore, floats on water. This unusual characteristic prevents the oceans from freezing solid when it gets cold. If ice didn't float, ice in oceans would sink, leaving room for new ice to form above.

Ice also has the unusual property of being slippery—that's why skaters can skate! Surprisingly, researchers still don't completely understand this property. An older explanation—that skaters can glide on ice because a film of liquid water forms on the surface of the ice beneath their skates in response to frictional heating or to pressure-induced melting—can't explain how ice remains slippery even when it's so cold that water can't exist as liquid. Modern studies suggest that ice remains slippery at very low temperatures because the surface of ice consists of a layer of water molecules that are not completely fixed within a crystal lattice. The existence of unattached bonds permits the surface molecules to behave somewhat like a liquid, even while attached to the solid.

How Does a Glacier Form?

In order for a glacier to form, four conditions must be met. First, the local climate must be sufficiently cold that winter snow does not melt away entirely during the summer; second, there must be or must have been sufficient snowfall for a large amount of snow to accumulate; third, the surface on which the snow accumulates must have a gentle slope so that snow falling on it does not slide away in avalanches; and fourth, the area where snow falls must be protected enough so that snow doesn't blow away.

Glaciers develop in polar regions because even though relatively little snow falls today, temperatures remain so low that most ice and snow survive all year. Glaciers develop in mountains, even at low latitudes, because temperature decreases with elevation. Thus, at high elevations, the mean temperature stays low enough for ice and snow to survive all year. Since the temperature of a region depends on latitude, the specific elevation at which glaciers form in mountains depends on latitude. In Earth's present-day climate, glaciers form only at elevations above 5 km at latitudes of between 0° and 30°, but they can flow down to sea level at latitudes of between 60° to 90°. Thus, you can see high-latitude glaciers from a cruise ship, but at the equator you have to climb way up into the mountains to find glaciers. Mountain glaciers tend to develop on the side of mountains that receives less wind and on the side that receives less sunlight. Glaciers do not form on slopes greater than about 30°, because avalanches clear such slopes.

FIGURE 22.2 The nature of ice and the formation of glaciers. Snow falls like sediment and metamorphoses to ice when buried.

(a) The hexagonal shape of snowflakes. No two are alike.

A boundary between layers

The layers in the photo at left are part of this glacier in the Alps.

(b) Layers of snow accumulate. They recrystallize to become ice.

The wall of a tunnel bored into a glacier

(c) As revealed by a microscope, glacial ice has coarse grains and contains air bubbles.

Loose snow (90% air)

Granular snow (50% air)

Firn (25% air)

Fine-grained ice (< 20% air, in bubbles)

Coarse-grained ice (< 20% air, in bubbles)

10,000 years (250 m)

130,000 years (2,000 m)

(d) Snow compacts and melts to form firn, which recrystallizes into ice. Crystal size increases with depth.

The transformation of snow to glacier ice takes place slowly, as younger snow progressively buries older snow. Freshly fallen snow consists of delicate hexagonal crystals with sharp points. The crystals do not fit together tightly, so this snow contains about 90% air. With time, the points of the snowflakes become blunt because they either **sublimate** (evaporate directly into vapor) or melt, and the snow packs more tightly. As snow becomes buried, the weight of the overlying snow increases pressure, which causes any remaining points of contact between snowflakes to melt. This process of melting at points of contact, where the pressure is greatest, is another example of *pressure solution* (see Chapter 8). Gradually, the snow transforms into a packed granular material called **firn**, which contains only about 25% air (**Fig. 22.2d**). Melting of firn grains at contact points produces water that crystallizes in the spaces between grains until eventually the firn transforms into a solid mass of glacial ice composed of interlocking ice crystals. Such glacial ice, which may still contain up to 20% air trapped in bubbles, tends to absorb red light and thus has a bluish color (see Fig. 22.2c). The transformation of fresh snow to glacier ice can take as little as tens of years in regions with abundant snowfall or as long as thousands of years in regions with little snowfall.

Categories of Glaciers

Today glaciers highlight coastal and mountain scenery in Alaska, the Cordillera of western North America, the Alps of Europe, the Southern Alps of New Zealand, the Himalayas of Asia, and the Andes of South America, and they cover most of Greenland and Antarctica. Geologists distinguish between two main categories: mountain glaciers and continental glaciers.

Mountain glaciers (also called *alpine glaciers*) exist in or adjacent to mountainous regions (**Fig. 22.3a**). Overall, mountain glaciers flow from higher elevations to lower elevations. Mountain glaciers include *cirque glaciers*, which fill bowl-shaped depressions, or **cirques**, on the flank of a mountain; *valley glaciers*, rivers of ice that flow down valleys; mountain *ice caps*, mounds of ice that submerge peaks and ridges at the crest of a mountain range; and *piedmont glaciers*, fans or lobes of ice that form where a valley glacier emerges from a valley and spreads out into the adjacent plain (**Fig. 22.3b–f**). Mountain glaciers range in size from a few hundred meters to a few hundred kilometers long.

Continental glaciers are vast ice sheets that spread over thousands of square kilometers of continental crust. Today they exist only on Antarctica and Greenland (**Fig. 22.4**), but during ice ages they have covered other continental areas. Keep in mind that Antarctica is a continent, so the ice beneath the South Pole rests mostly on solid ground. We say *mostly* because new research reveals that at least three lakes underlie the glacier—the largest of these, Lake Vostok, has an area of 5,400 km^2.

The ice beneath the North Pole, in contrast, forms part of a thin sheet of sea ice floating on the Arctic Ocean. Continental glaciers flow outward from their thickest point (up to 3.5 km thick) and thin toward their margins, where they may be only a few hundred meters thick. The front edge of the glacier may divide into several tongue-shaped lobes, because not all of the glacier flows at the same speed. Of note, Earth is not alone in hosting polar ice sheets—Mars has them too (**Box 22.1**).

Geologists also find it valuable to distinguish between types of glaciers based on thermal conditions in the glacier. **Temperate glaciers** occur where atmospheric temperatures become warm enough for the glacial ice to be at or near its melting temperature during part or all of the year, so they contain some liquid water, in films between grains in the glacier, or in lenses and streams at the base of the glacier. Because of this water, a temperate glacier can also be called a *wet-based glacier*. **Polar glaciers** occur in regions where atmospheric temperatures stay so cold all year long that the glacial ice remains below melting temperature throughout the year—they are solid ice through and through. Geologists may also refer to a polar glacier as a *dry-based glacier*, because no liquid water collects at the base of the glacier.

The Movement of Glacial Ice

When Louis Agassiz became fascinated by glaciers, he decided to find out how fast the ice in them moved, so he hammered stakes into an Alpine glacier and watched the stakes change position during the year. More recently, researchers have observed glacial movement with the aid of time-lapse photography, which shows the evolution of a glacier over several years in a movie that lasts a few minutes. In such movies, the glacial ice seems to flow across the screen. How does this movement occur? Geologists have found that glacial flow involves two mechanisms—plastic deformation and basal sliding.

Did you ever wonder . . .

how a glacier moves?

At conditions found below depths of about 60 m in a glacier, ice deforms by **plastic deformation**, meaning the grains within it change shape very slowly, and/or new grains grow while old ones disappear (**Fig. 22.5a, b**). Simplistically, we can picture such changes to be a consequence of the rearrangement of water molecules within a crystal lattice as some chemical bonds break and new ones form. If ice becomes warm enough for thin water films to form along grain boundaries, plastic deformation may also involve the microscopic slip of ice grains past their neighbors along water films.

In some cases, significant quantities of liquid water collect at the base of a glacier. This water can occur as a lens of liquid under the ice, but commonly it mixes with subglacial sediment

FIGURE 22.3 A great variety of glaciers form in mountainous areas.

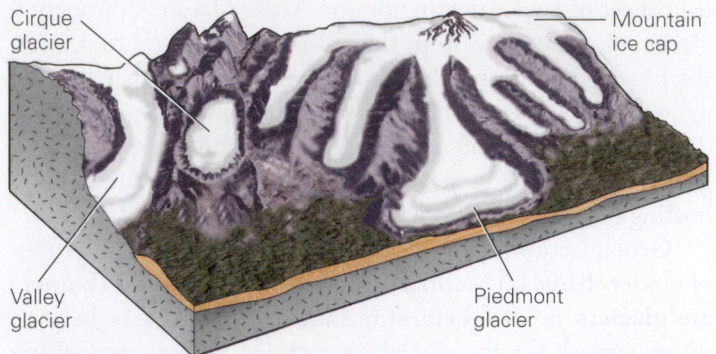

(a) Mountain glaciers are classified based on shape and position.

(d) A valley glacier and cirque glaciers in Switzerland.

(b) An ice cap in Alaska.

(e) A large trunk valley glacier and tributary glaciers in Pakistan.

(c) Valley glaciers draining a mountain ice cap in Alaska.

(f) A piedmont glacier near the coast of Greenland.

to form a slurry. The presence of liquid water or a wet slurry beneath a glacier allows the glacier to move by **basal sliding**. During this process, friction or bonding between the ice and its substrate has diminished the glacier so much that it effectively glides along on a wet cushion (**Fig. 22.5c**).

Where does the liquid water at the base of glaciers come from? Some forms when sunlight and atmospheric warming heats the glacier sufficiently to produce meltwater, either on the surface of the glacier or within the glacier. Recent studies show that surface meltwater ponds may drain in a matter of minutes to hours into cracks or tunnels that provide a conduit between the surface and the base of the glacier. Melting may also occur due to the trapping of heat rising from the ground beneath the glacier (for ice is an insulator) or due to the weight of overlying ice (for at elevated pressures, ice can melt even if its temperature remains below 0°C).

FIGURE 22.4 Two major continental glaciers exist today—one on Antarctica and one on Greenland.

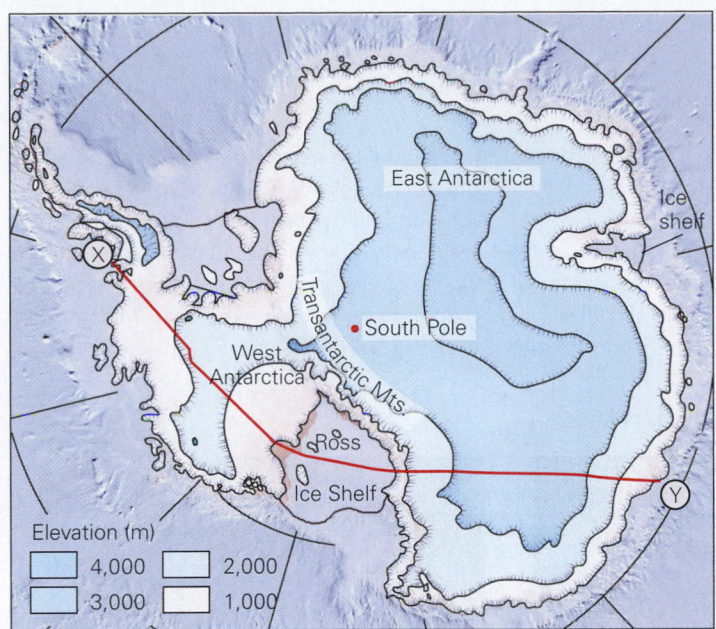

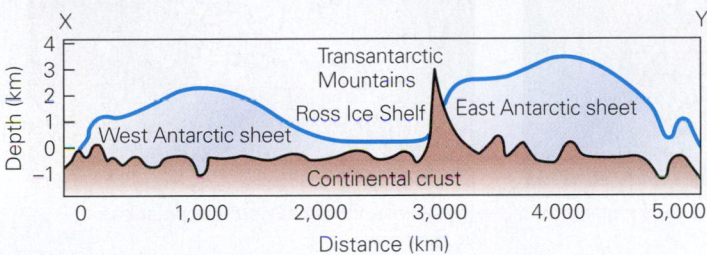

(a) A contour map of the Antarctic ice sheet. Valley glaciers carry ice from the ice sheet of East Antarctica down to the Ross Ice Shelf.

(b) A cross section X to Y of the Antarctic ice sheet. The Transantarctic Mountains separate East Antarctica from West Antarctica.

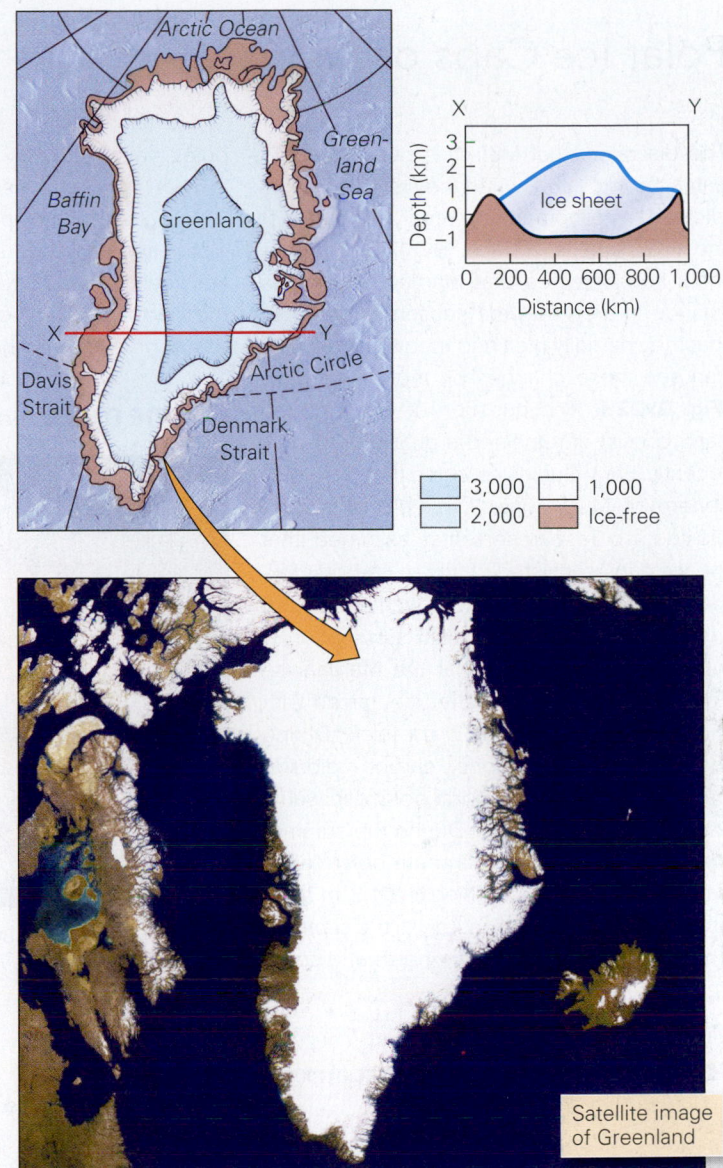

(c) Greenland is also covered by an ice sheet that at its thickest is 1 km thinner than Antarctica's.

In the case of polar glaciers, which are so cold that they have no internal water films and have a dry base, flow takes place generally by plastic deformation alone. In the case of temperate glaciers, which contain some intergranular water and/or have a wet base, flow can involve both plastic deformation and basal sliding. Note that not all parts of a given glacier necessarily flow in the same way. For example, imagine a continental glacier that originates in a very cold polar realm but eventually flows into temperate realms at lower latitudes. Near its cold origin, the glacier has a dry base and moves only by plastic deformation, but near its warmer margin, it becomes wet based and moves by both plastic deformation and basal sliding. Similarly, a long valley glacier may be dry based and flow only plastically at higher, colder elevations, but it may become wet based and flow by both plastic deformation and basal sliding at lower, warmer elevations.

As we noted earlier, plastic deformation takes place only at depths of greater than about 60 meters in a glacier—above this depth, known as the *brittle–plastic transition*, ice is too brittle to

flow. (Note that, by comparison, plastic deformation in silicate rocks of the Earth's crust occurs primarily under metamorphic temperatures greater than about 300°C; thus, the brittle–plastic transition in continental crust occurs at depths of about 10 to 15 km; see Chapter 11.) As a glacier overall undergoes movement, its upper 60 meters of ice deform predominantly by cracking. A crack that develops by brittle deformation of a glacier is called a **crevasse** (**Fig. 22.6**). In large glaciers, crevasses can be hundreds of meters long and tens of meters deep, and they may open into gashes that are many meters across. Tragically, explorers, hikers, and skiers have died by falling into crevasses, whose openings sometimes become covered

BOX 22.1 CONSIDER THIS . . .

Polar Ice Caps on Mars

The discovery that Mars has polar ice caps dates back to 1666, when the first telescopes allowed astronomers to resolve details of the red planet's surface. By 1719, astronomers had detected that Martian polar caps change in area with the season, suggesting that they partially melt and then refreeze. You can see these changes on modern images (**Fig. Bx22.1**). The question of what the ice caps consist of remained a puzzle until fairly recently. Early studies revealed that the atmosphere of Mars consists mostly of carbon dioxide, so researchers first assumed that the ice caps consisted of frozen carbon dioxide. But data from modern spacecraft led to the conclusion that this initial assumption is wrong. It now appears that the Martian ice caps consist mostly of water ice, mixed with dust, in layers from 1 to 3 km thick. During the winter, atmospheric carbon dioxide freezes and covers the north polar cap with a 1-m-thick layer of dry ice. During the summer this layer melts away. The south polar cap is different, for its dry ice blanket is 8 m thick and doesn't melt away entirely in the summer. The difference between the north and south poles may reflect elevation for the south pole is 6 km higher and therefore remains colder.

High-resolution photographs reveal that distinctive canyons, up to 10 km wide and 1 km deep, spiral outward from the center of the north polar ice cap. Why did this pattern form? Recent calculations suggest that if the ice sublimates (transforms into gas) on the sunny side of a crack and refreezes on the shady side, the crack will migrate sideways over time. If the cracks migrate more slowly closer to the pole, where it's colder, than they do farther away, they will naturally evolve into spirals.

FIGURE Bx22.1 The ice caps of Mars.

(a) During the winter, the ice caps expand to lower latitudes.

(b) A close-up of the northern polar cap in summer.

FIGURE 22.5 Mechanisms of glacial movement.

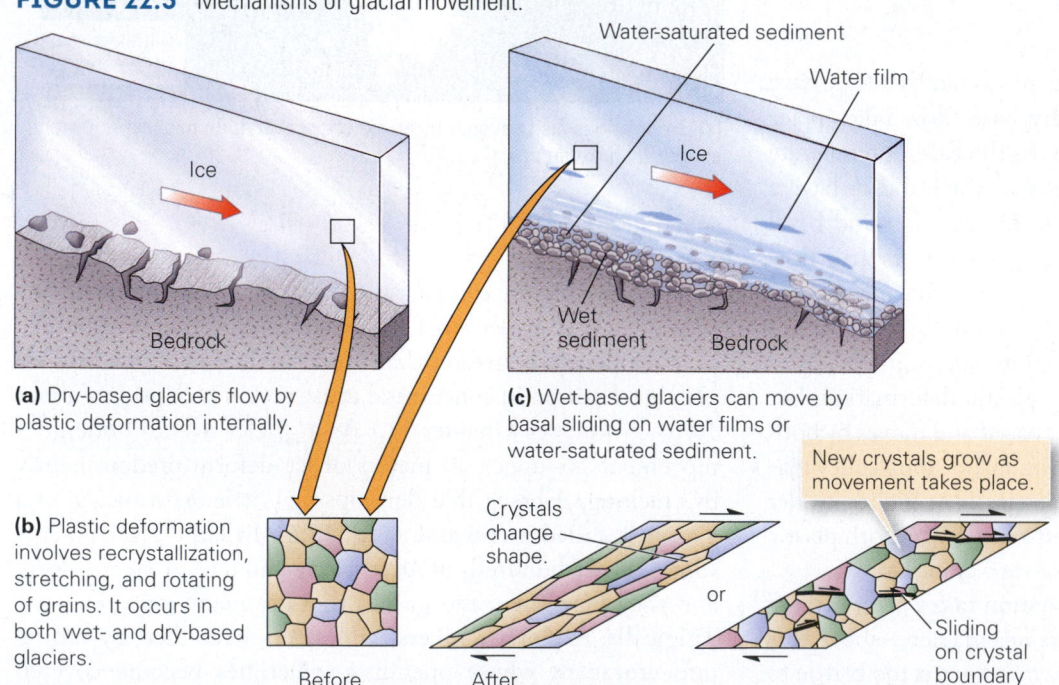

(a) Dry-based glaciers flow by plastic deformation internally.

(b) Plastic deformation involves recrystallization, stretching, and rotating of grains. It occurs in both wet- and dry-based glaciers.

(c) Wet-based glaciers can move by basal sliding on water films or water-saturated sediment.

Water-saturated sediment

Water film

New crystals grow as movement takes place.

Crystals change shape.

Before   After   or   Sliding on crystal boundary

by a bridge of weak, windblown snow. Crevasse formation typically localizes in regions where a glacier flows over steps or hills in the underlying bedrock surface, for the ice of the glacier must bend to accommodate the surface shape of the substrate.

Why do glaciers move? Ultimately, because the pull of gravity exceeds the strength of ice and can cause the ice to flow (**Fig. 22.7a**). A glacier flows in the direction in which its top surface slopes. Thus, valley glaciers flow down their valleys, and continental ice sheets spread outward from their thickest point. Note that it is the slope of the top surface that matters in

FIGURE 22.6 Crevasses form in the upper layer of a glacier, in which the ice is brittle. Commonly, cracking takes place where the glacier bends while flowing over steps or ridges in its substrate.

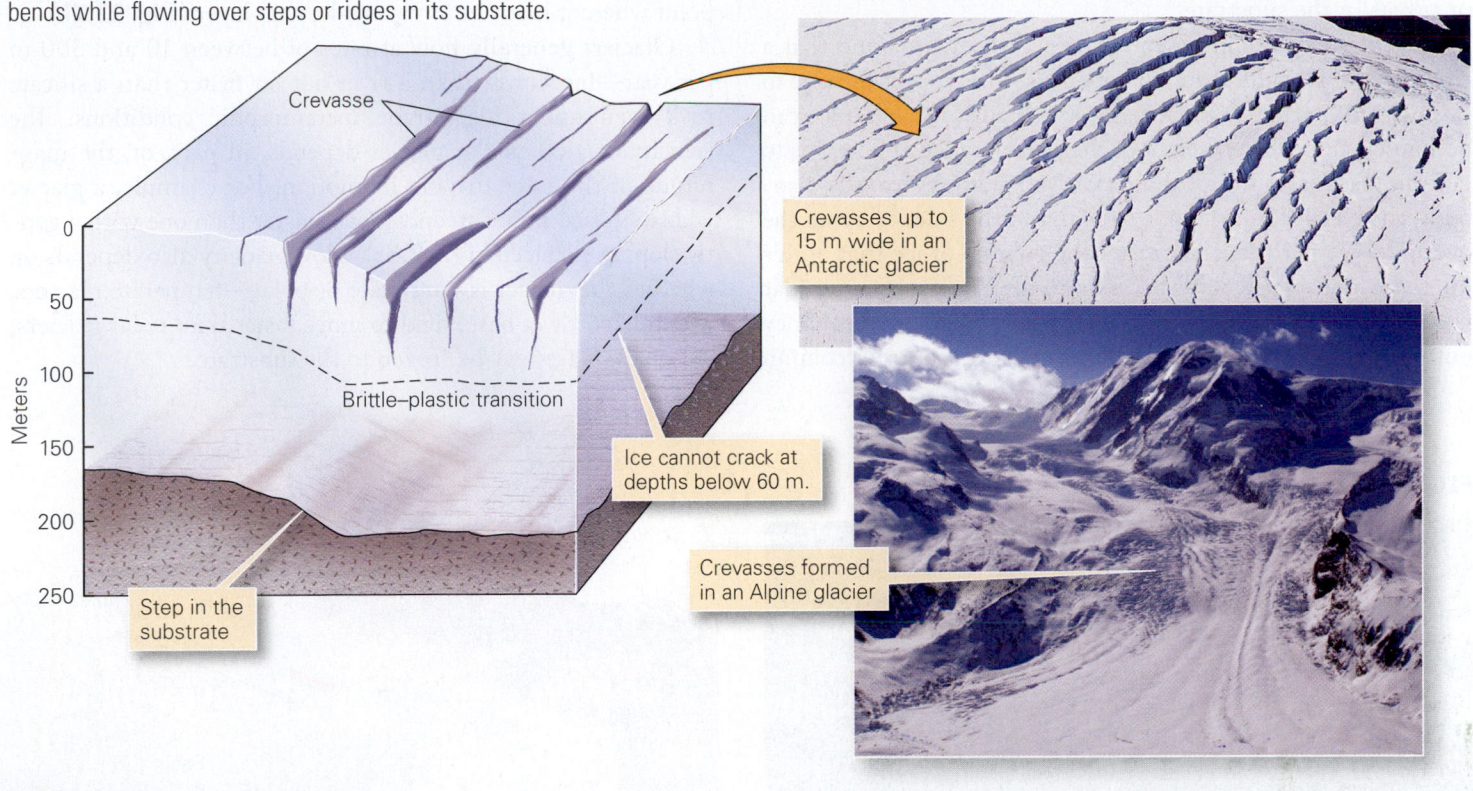

Crevasse

Meters

0

50

100

150

200

250

Brittle–plastic transition

Ice cannot crack at depths below 60 m.

Step in the substrate

Crevasses up to 15 m wide in an Antarctic glacier

Crevasses formed in an Alpine glacier

FIGURE 22.7 Forces that drive the movement of glaciers.

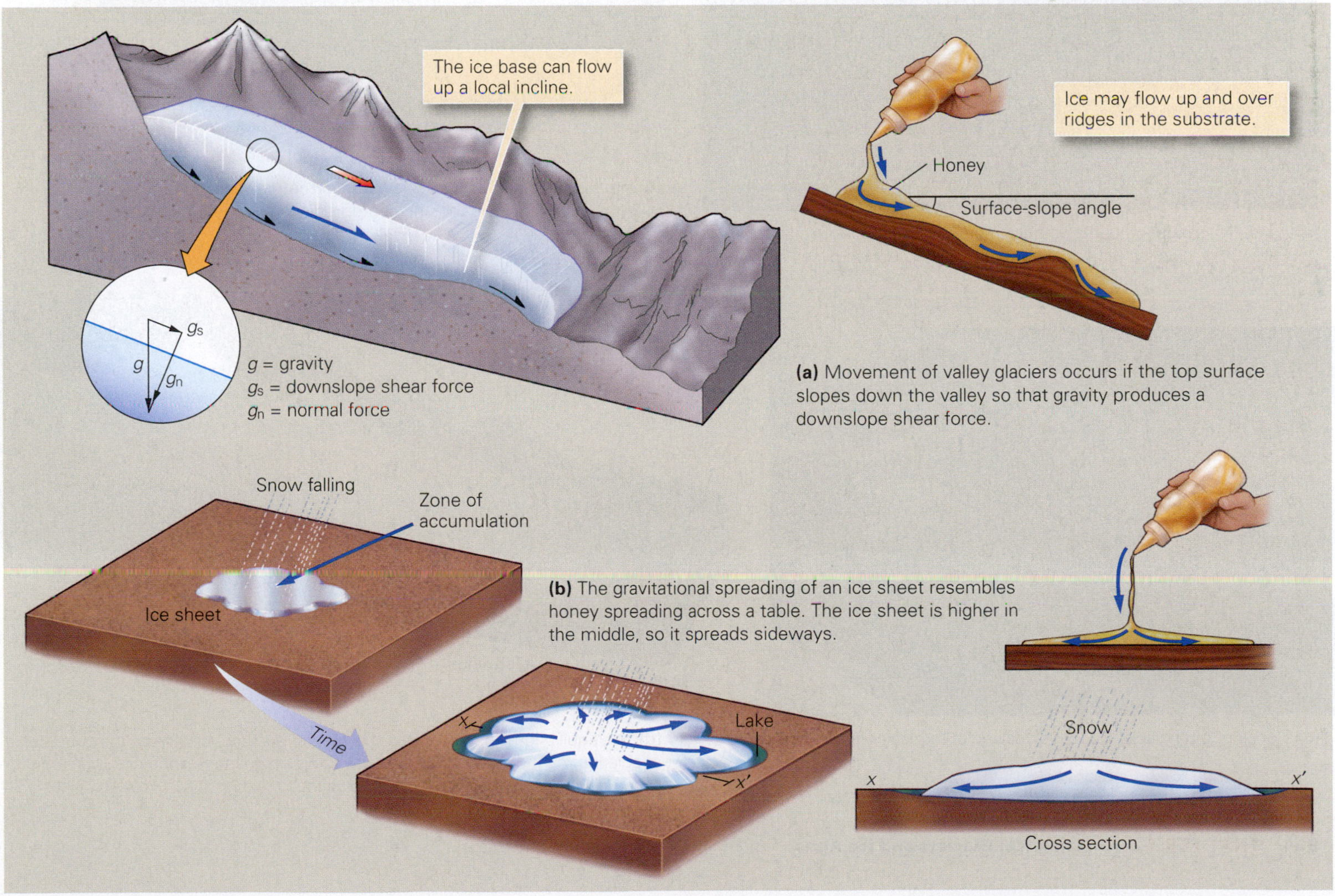

The ice base can flow up a local incline.

g = gravity
g_s = downslope shear force
g_n = normal force

g g_s g_n

Ice may flow up and over ridges in the substrate.

Honey

Surface-slope angle

(a) Movement of valley glaciers occurs if the top surface slopes down the valley so that gravity produces a downslope shear force.

Snow falling

Zone of accumulation

Ice sheet

Time

Lake

x

x′

(b) The gravitational spreading of an ice sheet resembles honey spreading across a table. The ice sheet is higher in the middle, so it spreads sideways.

Snow

x

x′

Cross section

driving ice forward—at its base, ice can flow up and over hills or ridges in the substrate.

To picture the movement of an ice sheet, imagine that a thick pile of ice builds up. Gravity causes the top of the pile to push down on the ice at the base. Eventually, the basal ice can no longer support the weight of the overlying ice and begins to deform plastically and/or slide on its substrate. When this happens, the basal ice starts squeezing out to the side, carrying the overlying ice with it. The greater the volume of ice that builds up, the wider the sheet of ice can become. You've seen a similar process of *gravitational spreading* if you've ever poured honey onto a plate. The honey can't build up into a narrow column because it's too weak; rather, it flows laterally away from the point where it lands to form a wide, thin layer (**Fig. 22.7b**).

Glaciers generally flow at rates of between 10 and 300 m per year—far slower than a river but far faster than a silicate rock even under high-grade metamorphic conditions. The velocity of a particular glacier depends, in part, on the magnitude of the force driving its motion. For example, a glacier whose surface slopes steeply moves faster than one with a gently sloping surface (**Fig. 22.8a**). Flow velocity also depends on whether the glacier is temperate or polar—temperate glaciers, which have a wet base, tend to move faster than polar glaciers, whose dry base may be frozen to the substrate.

FIGURE 22.8 Glacial flow, accumulation, and ablation.

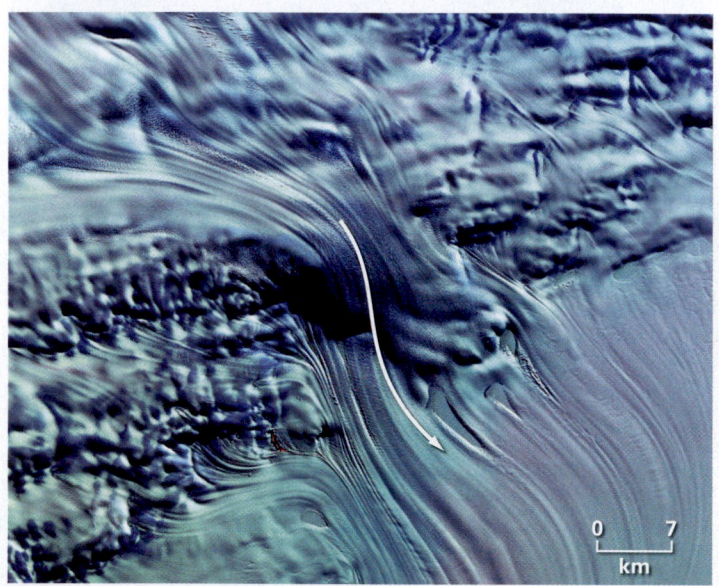

(a) A satellite image of ice flowing from the Polar Plateau of Antarctica, down a 400-m-high ice fall to the Lambert Glacier. Curving lines indicate the flow directions.

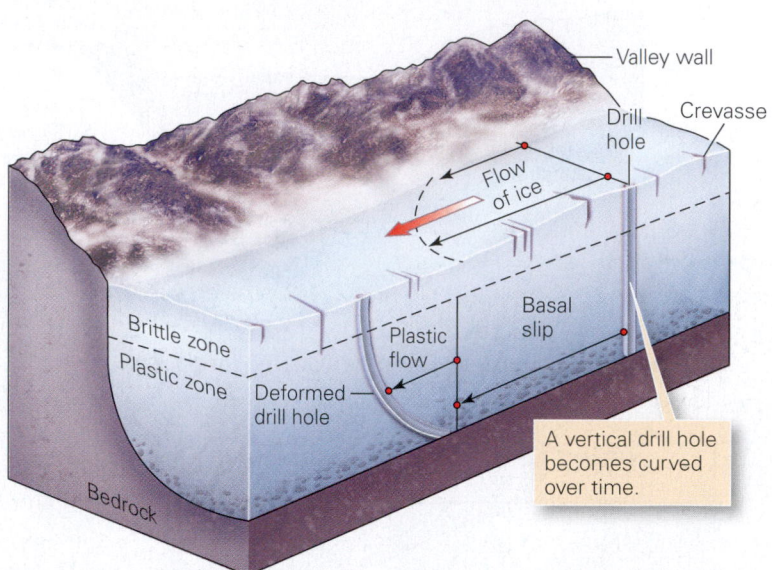

(b) Different parts of a glacier flow at different velocities due to friction with the substrate. The top and center regions flow fastest.

(c) Blocks of blue glacial ice, which calved off a glacier in the Alps.

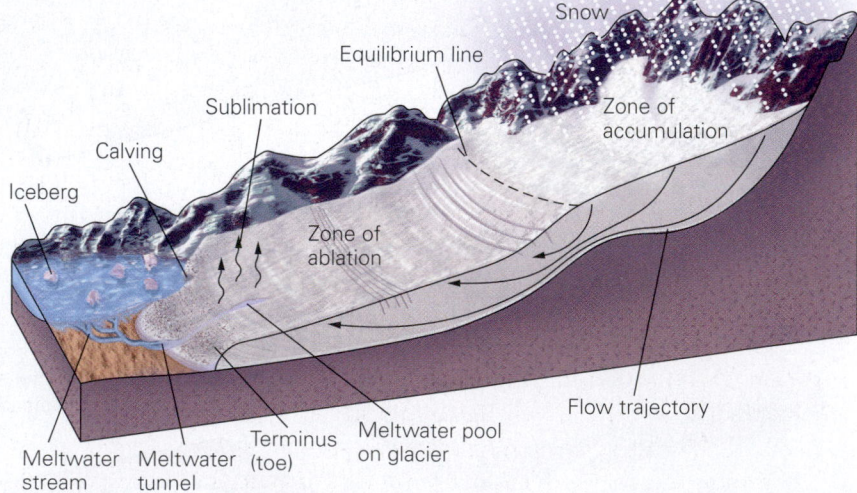

(d) The equilibrium line separates the zone of accumulation from the zone of ablation. As indicated by arrows, ice flows down in the zone of accumulation and up in the zone of ablation.

Not all parts of a glacier move at the same rate. For example, friction or bonding between rock and ice slows a glacier, so the center of a valley glacier moves faster than its margins and the top of a glacier moves faster than its base (**Fig. 22.8b**). Because water at the base of a glacier allows it to travel more rapidly, wet-based portions of a continental glacier can become *ice streams* that travel 10 to 100 times faster than adjacent dry-based portions of the glacier. The volume of water at the bottom of a wet-based glacier may change over time. If water suddenly builds up beneath a glacier to the point where a large area lifts the glacier off its substrate, basal sliding starts and the glacier undergoes a **surge** and flows much faster for a limited time (rarely more than a few months). During surges, glaciers have been clocked at speeds of 10 to 110 m per day! Sudden surges may generate **ice quakes**, because of the cracking that occurs in the brittle portion of the glacier. A surge stops when the water escapes, so basal sliding slows.

Glacial Advance and Retreat

Glaciers resemble bank accounts. Snowfall adds to the account, while **ablation**—the removal of ice—subtracts from the account. Ablation involves three processes: *sublimation* (the evaporation of ice into water vapor); *melting* (the transformation of ice into liquid water); and *calving* (the breaking off of chunks of ice at the end of the glacier) (**Fig. 22.8c**). Snowfall adds ice to a glacier in the **zone of accumulation**, whereas ablation subtracts ice from the glacier in the **zone of ablation**—the boundary between these two zones is the **equilibrium line** (**Fig. 22.8d**). The zone of accumulation occurs where the temperature remains cold enough year-round so that winter snow does not melt or sublimate away entirely during the summer. Therefore, elevation and latitude control the position of the equilibrium line.

The leading edge or margin of a glacier is called its **toe**, or *terminus* (**Fig. 22.9a**). If the rate at which ice builds up in the zone of accumulation exceeds the rate at which ablation occurs below the equilibrium line, then the toe moves forward into previously unglaciated regions, a change called a **glacial advance** (**Fig. 22.9b**). In mountain glaciers, the position of a toe moves downslope during an advance, and in continental glaciers, the toe moves outward, away from the glacier's origin. If the rate of ablation below the equilibrium line equals the rate of accumulation, then the position of the toe remains fixed. But if the rate of ablation exceeds the rate of accumulation, then the position of the toe moves back toward the origin of the glacier—such a change is called a **glacial retreat** (**Fig. 22.9c**). During a mountain glacier's retreat, the position of the toe moves upslope. But it's important to realize that when a glacier retreats, it's only the position of the toe that moves back toward the origin. Even during glacial retreat, ice continues to flow toward the toe as long as the surface of the glacier slopes

FIGURE 22.9 Glacial advance and retreat.

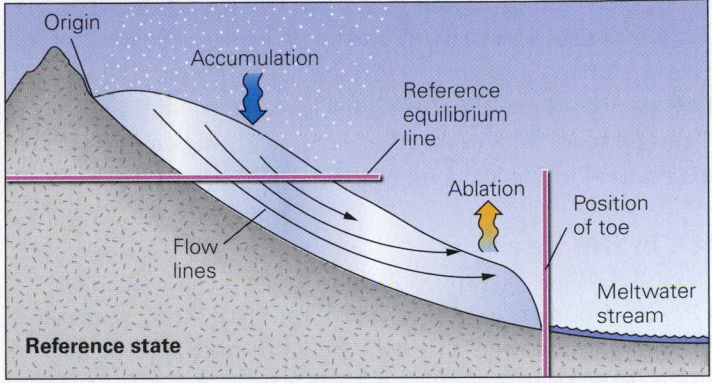

(a) The position of the toe represents a balance between addition by accumulation and loss by ablation.

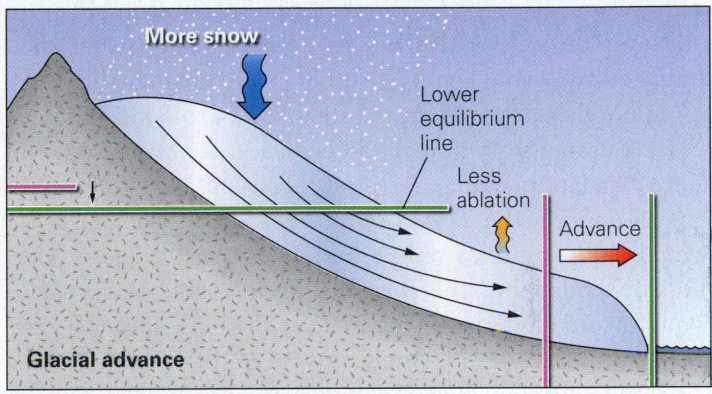

(b) If accumulation exceeds ablation, the glacier advances, the toe moves farther from the origin, and the ice thickens.

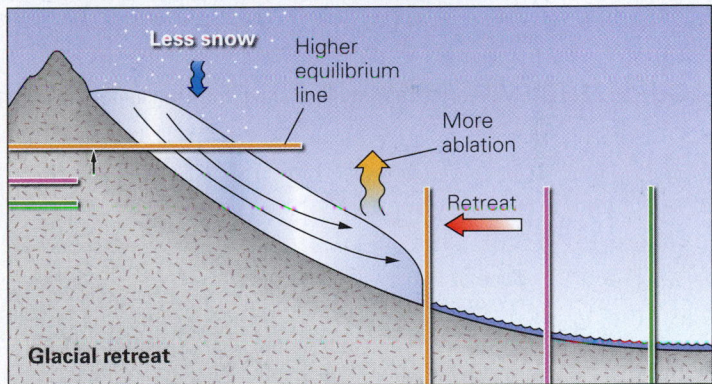

(c) If ablation exceeds accumulation, the glacier retreats and thins. The toe moves back, even though ice continues to flow toward the toe.

toward the toe—glacial ice cannot flow upslope, back toward the glacier's origin.

One final point before we leave the subject of glacial flow: Note that beneath the zone of accumulation a given volume of ice gradually moves down toward the base of the glacier as new ice accumulates above it. In contrast, beneath the zone of ablation, a given volume of ice gradually moves up toward the surface of the glacier as overlying ice ablates. Thus, as a glacier flows, ice volumes overall follow curved trajectories (see Fig. 22.9). For this reason, rocks picked up by ice at the base of the glacier slowly move to the surface near the toe. The upward flow of ice where the Antarctic ice sheet collides with the Transantarctic Mountains, for example, brings up meteorites long buried in the ice (**Fig. 22.10**).

Ice in the Sea

On the moonless night of April 14, 1912, the great ocean liner *Titanic* plowed through the calm but frigid waters of the North Atlantic on her maiden voyage from Southampton, England, to New York. Although radio broadcasts from other ships warned that **icebergs**, large blocks of ice floating in the water, had been sighted in the area and might pose a hazard, the ship sailed on, its crew convinced that they could see and avoid the biggest bergs and that smaller ones would not be a problem for the steel hull of this "unsinkable" vessel. But in a story now retold countless times, their confidence was fatally wrong. At 11:40 P.M., while first-class passengers danced, the *Titanic* struck an iceberg. Lookouts had seen the ghostly mass only minutes earlier and had alerted the ship's pilot, but the ship had been unable to turn fast enough to avoid disaster. The force of the blow split the steel hull spanning 5 of the ship's 16 watertight compartments. The ship could stay afloat if 4 compartments flooded, but the flooding of 5 meant it would sink. At about 2:15 A.M., the bow disappeared below the water, and the stern rose until the ship protruded nearly vertically from the water. Without water to support its weight, the hull buckled and split in two. The stern section fell back down onto the water and momentarily bobbed horizontally before following the bow, settling downward through over 3.5 km of water to the silent sea floor below. Because of an inadequate number of lifeboats, only 705 passengers survived; 1,500 expired in the frigid waters of the Atlantic. The *Titanic* remained lost until 1985, when a team of oceanographers located the sunken hull and photographed its eerie form.

Where do icebergs, such as the one responsible for the *Titanic*'s demise, originate? In high latitudes, mountain glaciers and continental ice sheets flow down to the shore. Glaciers that flow out into the sea along the coast become **tidewater glaciers**. Large valley glaciers may protrude several kilometers out into the ocean as elongate *ice tongues* (**Fig. 22.11a**). Continental glaciers entering the sea become broad, flat sheets called **ice shelves** (**Fig. 22.11b**). In shallow water, glacial ice remains grounded in that the base of the glacier rests on the sea floor (**Fig. 22.11c**). But where the water is deep enough, the ice floats with four-fifths of the ice below the water's surface. At the boundary between glacier and ocean, blocks of ice calve off and tumble into the water with an impressive splash, producing large waves. If a free-floating chunk rises 6 m above the water and is at least 15 m long, mariners refer to it as an *iceberg*. Smaller pieces, formed when ice blocks fragment before entering the water or after icebergs have had time to melt, include *bergy bits*, rising

FIGURE 22.10 Meteorites accumulate along the Transantarctic Mountains.

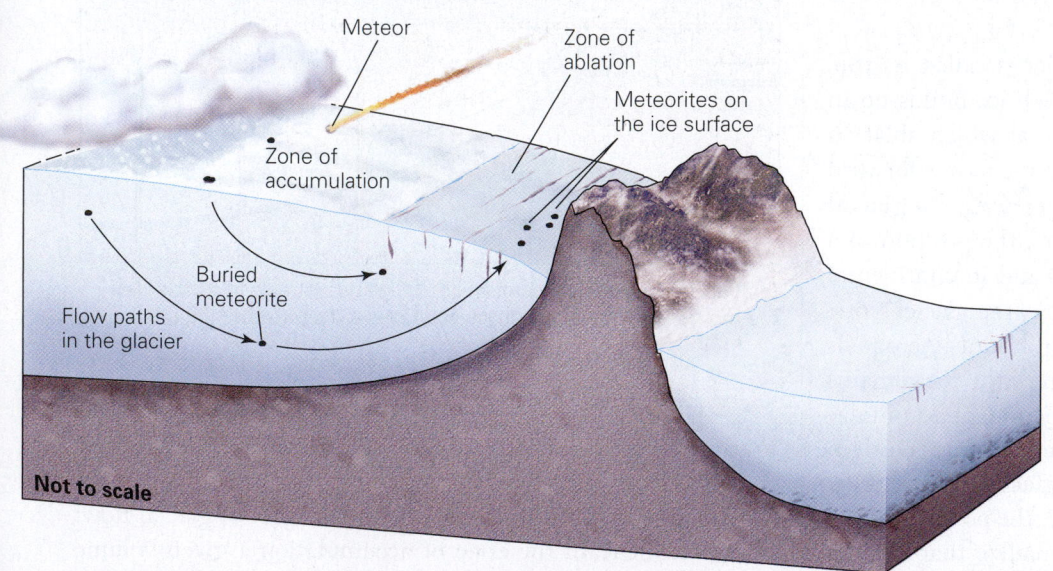

(a) Meteorites landing in the zone of accumulation are buried and incorporated in the flowing ice. They return to the ice surface in the zone of ablation.

(b) Researchers document a new meteorite discovery.

FIGURE 22.11 Ice along the edge of continents—shelves, tongues, and bergs.

(a) An ice tongue protruding into the Ross Sea along the coast of Antarctica.

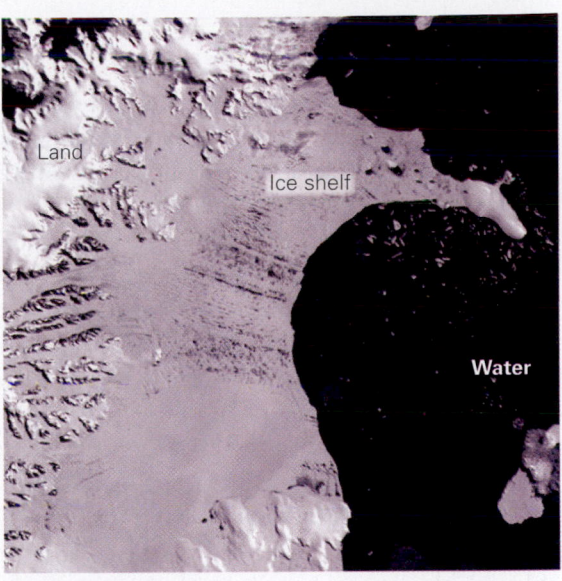

(b) The Larsen Ice Shelf along the coast of Antarctica, as viewed from a satellite in 2002.

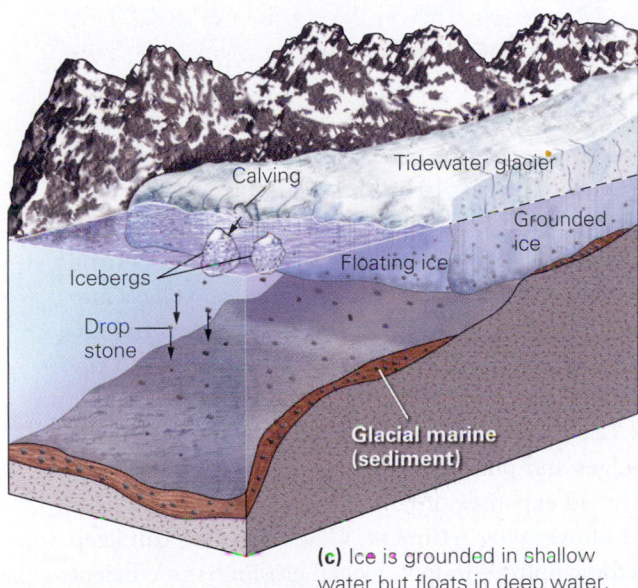

(c) Ice is grounded in shallow water but floats in deep water.

(e) This artist's rendition of an iceberg emphasizes that most of the ice is underwater.

(f) In summer, some of the sea ice of Antarctica breaks up to form tabular icebergs.

(d) A "growler" of floating ice off Alaska. Note the layering in the ice.

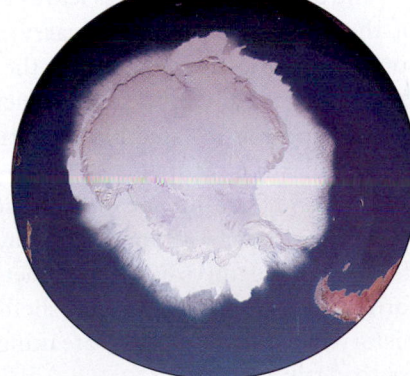

(g) Sea ice covers most of the Arctic Ocean (left) and surrounds Antarctica (right).

1 to 5 m above the water and covering an area of 100 to 300 m², and *growlers* (**Fig. 22.11d**), rising less than 1 m above the water and covering an area of about 20 m²—still big enough to damage a ship. Growlers get their name because of the sound they make as they bob in the sea and grind together.

Most large icebergs form along the western coast of Greenland or along the coast of Antarctica. Icebergs that calve off valley glaciers tend to be irregularly shaped with pointed peaks rising upward. Such glaciers are called castle bergs or *pinnacle bergs*—one of the largest on record protruded about 180 m above the sea. Since four-fifths of the ice lies below the surface of the sea, the base of a large iceberg may actually be a few hundred meters below the surface (**Fig. 22.11e**). Icebergs that originate in Greenland float into the "iceberg alley" region of the North Atlantic. These are the bergs that threaten ships, although the danger has diminished in modern times because of less ice and because of ice patrols that report the locations of floating ice. Blocks that calve off the vast ice shelves of Antarctica tend to have flat tops and nearly vertical sides—such glaciers are called *tabular bergs*. Some of the tabular bergs in the Antarctic are truly immense—air photos have revealed individual bergs over 160 km across.

Not all ice floating in the sea originates as glaciers on land. In polar climates, the surface of the sea itself freezes, forming **sea ice** (**Fig. 22.11f, g**). Some sea ice, such as that covering the interior of the Arctic Ocean, floats freely, but some protrudes outward from the shore (Fig. 22.11d). Icebreakers can crunch through sea ice that is up to 2.5 m thick; the icebreaker rides up on the ice, and its weight crushes the ice below. Vast areas of sea ice have been melting in recent years in association with global warming (see Chapter 23). For example, open regions develop in the Arctic Ocean during the summers, and the ice shelf in Antarctica has been decreasing rapidly in area. In some locations, large openings known as *polynyas* have developed in the sea ice of Antarctica. Some sea ice forms in winter and melts away in summer, but at high latitudes sea ice may last for several years. For example, in the Arctic Ocean, sea ice may last long enough to make the 7- to 10-year voyage around the Arctic Ocean, in response to currents, at least once.

The existence of icebergs leaves a record in the stratigraphy of the seafloor, for icebergs carry *ice-rafted sediment*. Larger rocks that drop from the ice to the sea floor are called **drop stones** (see Fig. 22.11c). In ancient glacial deposits, drop stones appear as isolated blocks surrounded by mud. Icebergs and smaller fragments also drop sand and gravel, derived by the erosion of continents, onto the seafloor. In cores extracted by drilling into seafloor sediment, horizons of such land-derived sediment, sandwiched between layers of sediment formed from marine plankton shells, indicate times in Earth history when glaciers were breaking up and icebergs became particularly abundant.

Take-Home Message

Glaciers form when buried snow lasts all year, turns to ice, and gradually recrystallizes. Glacial ice flows by plastic deformation or by basal sliding. Mountain glaciers form at high elevation and flow to lower elevations. Ice sheets form in high latitudes and spread over continents. The balance of accumulation to ablation controls glacial advance or retreat. Where glaciers reach the sea, they may spall off icebergs. In polar regions, sea ice covers large areas.

QUICK QUESTION: Does ice actually flow uphill during a glacial retreat?

22.3 Carving and Carrying by Ice

The Process of Glacial Erosion

The Sierra Nevada range of California consists largely of granite that formed during the Mesozoic Era in the crust beneath a volcanic arc. During the past 10 million years or more, the land surface slowly rose, and erosion stripped away overlying rock and yielded rounded, dome-like granite mountains. During the last ice age, valley glaciers cut deep, steep-sided valleys into the range. In the process, some of the domes were cut in half, leaving a rounded surface on one side and a steep cliff on the other. Half Dome in Yosemite National Park formed in this way (**Fig. 22.12a**). Such glacial erosion has also produced the knife-edge ridges and pointed spires of high mountains (**Fig. 22.12b**) and broad expanses of land where rock outcrops have been stripped of overlying sediment. Glacial erosion can keep pace with tectonic uplift—in fact, glacial erosion is so efficient at grinding away mountain peaks that geologists sometimes refer to the phenomenon as the "glacial buzz saw." Glaciers similarly strip material from the surface of continents—at least 30 m of rock was removed from the Canadian Shield during the last glaciation.

How does glacial erosion take place? In part, erosion in glaciated mountains takes place by landslides or rockfalls of debris onto glaciers, which then carry the debris away. Freezing and thawing in glacial environments may accelerate the mechanical weathering that sets the stage for such mass wasting. But erosion also takes place where the glacial ice flows along the land surface, for as ice moves, clasts embedded in the ice act like the teeth of a giant rasp and grind away the substrate. This process, **glacial abrasion**, produces very fine sediment called *rock flour*, just as sanding wood produces sawdust.

(a) Half Dome in Yosemite National Park, California.

(b) Examples of a cirque and an arête in the Swiss Alps.

(c) Glacially polished outcrop in Central Park, New York City.

(d) Small striations on an outcrop in Scotland.

(e) Glacial mega-striations in Victoria, British Columbia.

(f) Close-up of striations and chatter marks, Switzerland.

Rasping by embedded sand can smooth rock faces and produce **glacially polished surfaces** (**Fig. 22.12c**). Individual hard clasts protruding from moving ice yield grooves or scratches called **glacial striations** (1 to 10 cm across) in the bedrock below (**Fig. 22.12d**). These striations trend parallel to the flow direction of the glacier. In some cases, striations may be up to half a meter across and tens of meters long (**Fig. 22.12e**). Geologists don't fully understand the origin of such mega-striations, but they may be due to streamlined trains of sediment embedded in the ice, which can carve into bedrock along the same line for an extended period of time. Locally, when boulders entrained in the base of the ice strike bedrock below as the ice moves, asymmetric wedges of bedrock break off, leaving behind indentations called *chatter marks* (**Fig. 22.12f**). In regions of wet-based glaciers, sediment-laden water rushing through tunnels at the base of the glaciers can carve substantial subglacial channels.

Glaciers pick up fragments of their substrate in several ways. During *glacial incorporation*, ice surrounds loose debris so the debris starts to move with the ice (**Fig. 22.13a**). During *glacial plucking* (or glacial quarrying), a glacier breaks off fragments of

bedrock. Plucking occurs when ice freezes around rock that has just started to separate from its substrate; movement of the ice lifts off pieces of the rock. At the toe of an advancing glacier, ice may actually bulldoze sediment slightly before flowing over it (**Fig. 22.13b**).

Landforms Produced by Glacial Erosion

Let's now look more closely at the erosional features associated with mountain glaciers. If the glacier builds into an ice cap that completely covers the mountain, it smooths and rounds peaks (see the Chapter 6 opening photo). But, if the glacier's head (the top edge of the ice) lies below the peak of the mountain, then the ice carves rugged topography. Freezing and thawing during the fall and spring help fracture the rock bordering the head of the glacier. This rock falls on the ice or gets picked up at the base of the ice and moves downslope with the glacier. As a consequence, a bowl-shaped depression, or *cirque*, develops on the side of the mountain at the head of a glacier (see Fig. 22.12b). If the ice later melts, a lake called a tarn may remain at the base of the cirque, filling the base of the depression. An **arête** (French for "ridge"), a residual knife-edge ridge of rock, separates two adjacent cirques (see Fig. 22.12b), and a pointed mountain peak surrounded by at least three cirques is a **horn** (**Fig. 22.14a**). The Matterhorn, a famous peak in Switzerland, serves as a particularly beautiful example of a horn; each of its four faces originated as a cirque (**Fig. 22.14b**).

Glacial erosion severely modifies the shape of valleys. To see how, compare a river-eroded valley with a glacially eroded valley. If you look along the length of a river in unglaciated mountains, you'll see that it flows down a V-shaped valley, with the river channel forming the point of the V. The V develops because river erosion occurs only in the channel, and mass wasting causes the valley slopes to approach the angle of repose. But if you look down the length of a glacially eroded valley, you'll see that it resembles a U, with steep walls. A **U-shaped valley** (**Fig. 22.14c**) forms because the combined processes of glacial abrasion and plucking not only lower the floor of the valley but also bevel its sides. Remember that mountain faces above the ice level of a valley glacier erode as mechanical weathering breaks rock apart, and landslides carry debris onto the surface of the glacier below.

Glacial erosion in mountains also modifies the intersections between tributaries and the trunk valley. In a river system, the trunk stream serves as the local base level for tributaries (see Chapter 17), so the mouths of the tributary valleys lie at the same elevation as the trunk valley. The ridges (spurs) between valleys taper to a point when they join the trunk valley floor. During glaciation, tributary glaciers flow down side valleys into a trunk glacier. But the trunk glacier cuts the floor of its valley down to a depth that far exceeds the depth cut by the tributary glaciers. Thus, when the glaciers melt away, the mouths of the tributary valleys perch at a higher elevation than the floor of the trunk valley. Such side valleys are called **hanging valleys**. The water in post-glacial streams that flow down a hanging valley cascades over a spectacular waterfall to reach the post-glacial trunk stream (**Fig. 22.14d**). As they erode, trunk glaciers also remove the ends of spurs between valleys, producing *truncated spurs*.

FIGURE 22.13 The processes of incorporation and plowing.

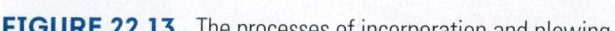

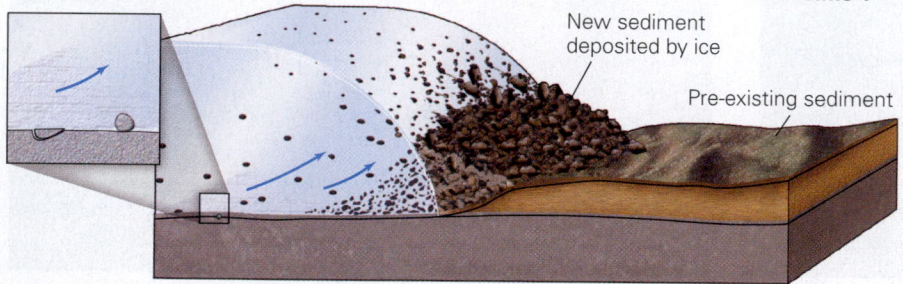

New sediment deposited by ice
Pre-existing sediment
Time 1

(a) Glacial ice can pluck, pick up, and incorporate chunks of rock that it flows over. The chunks then move with the ice, following the ice's flow trajectories, until deposition at the toe of the glacier.

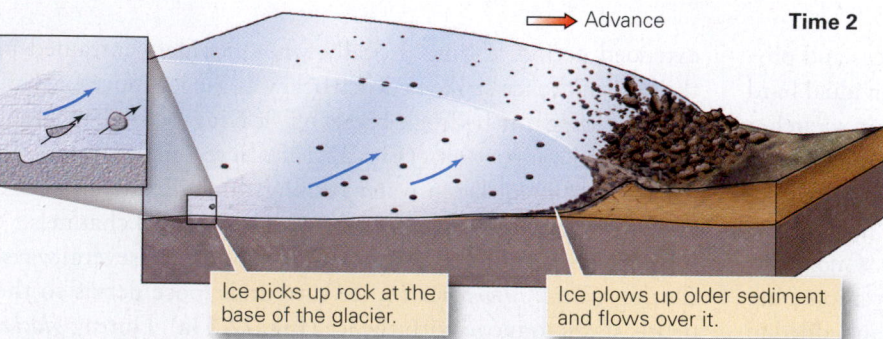

Advance   Time 2

Ice picks up rock at the base of the glacier.

Ice plows up older sediment and flows over it.

(b) At the toe of the glacier, ice can flow up and over pre-existing sediment. Locally, it bulldozes sediment, pushing it up into a ridge.

FIGURE 22.14 Landscape features formed by the glacial erosion of a mountainous landscape.

V-shaped valley

Tributary valley

Trunk valley

Trunk valley

Tributary valley

Before glaciation, valleys are V-shaped, and tributary mouths are the same elevation as the trunk stream.

Cirque

Tarn

Mt. Snowdon, Wales

During glaciation, the valleys fill with ice.

Time

After glaciation, the region contains U-shaped valleys, hanging valleys, truncated spurs, and horns.

Horn

Arête

Cirque

U-shaped valley

Hanging valley

Truncated spur

(a) Stages in the development of a glacially carved mountainous landscape.

(b) The Matterhorn in Switzerland. The first ascent was in 1865.

(c) A U-shaped glacial valley in the Tongass National Forest, Alaska.

(d) A waterfall spilling out of a U-shaped hanging valley in the Sierra Nevada.

FIGURE 22.15 The sculpting of hills by glacial erosion.

(a) Polished and striated bedrock in Ontario.

(b) Rounded hills in the highlands of Scotland formed when the entire region was covered by an ice sheet.

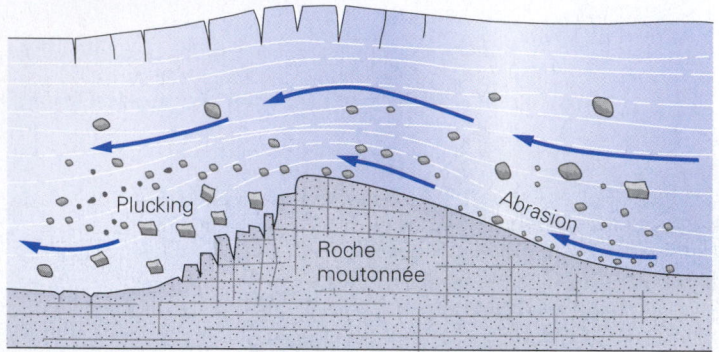

(c) Abrasion rasps the upstream side, and plucking carries away fracture-bounded blocks on the downstream side.

(d) An example of a roche moutonnée. The glacier flowed from right to left.

Now let's look at the erosional features produced by continental ice sheets. To a large extent, these depend on the nature of the pre-glacial landscape. Where an ice sheet spreads over a region of low relief, such as in Canada (**Fig. 22.15a**), glacial erosion creates a vast region of polished, flat, striated surfaces. Where an ice sheet spreads completely over a hilly area, it smooths hills (**Fig. 22.15b**). In Maine, for example, glaciers smoothed and streamlined the granite and metamorphic rock hills of Acadia National Park. Glacially eroded hills may become elongate in the direction of flow, and because glacial rasping smooths and bevels the upstream part of the hill, creating a gentle slope—whereas glacial plucking eats away at the downstream part, making a steep slope—the hills may become asymmetric. Ultimately, the hill's profile resembles that of a sheep lying in a meadow—such a hill is called a **roche moutonnée**, from the French for "sheep rock" (**Fig. 22.15c, d**). In some cases, a glacier erodes three sides of a hill, but deposits debris on the downstream or wake side of the hill. This process produces a *crag and tail*, with a steep cliff on the upstream side and a long ramp on the downflow side. The castle of Edinburgh, Scotland, was build on a crag. The steep scarps on three sides were easier to defend, while the tail provided a site for the growth of a town.

Fjords: Submerged Glacial Valleys

If the floor of a glacially carved valley lies below sea level along the coast, or beneath the water table inland, the floor of the valley becomes submerged with water. Geologists refer to any glacial valley that has filled partially or entirely with water as a **fjord**—marine fjords occur along the coast and have filled with seawater, whereas freshwater fjords lie inland (**Fig. 22.16**).

Spectacular examples of marine fjords can be found along the coasts of Norway, New Zealand, Chile, Alaska, and Greenland—in some cases, the walls of submerged U-shaped valleys rise straight from the sea as vertical cliffs up to 1,000 m high, and the

water depth just offshore may exceed a few hundred meters. How do such dramatic fjords develop? As noted earlier, where a valley glacier meets the sea, the glacier's base remains in contact with the ground until the water depth exceeds about four-fifths of the glacier's thickness. Further, during an ice age, water extracted from the sea becomes locked in the ice sheets on land, so sea level drops significantly. Therefore, the floors of valleys cut by coastal glaciers during the Pleistocene Ice Age could be cut much deeper than present sea level.

Take-Home Message

A glacier scrapes up and plucks rock from its substrate and carries debris that falls on its surface. Glacial erosion polishes and scratches rock and carves distinctive landforms, such as U-shaped valleys, cirques, and striations. An elongate bay or lake formed when water partially fills a glacial valley is a fjord.

QUICK QUESTION: Why do we find hanging valleys spilling waterfalls into trunk valleys, in regions that have been eroded by mountain glaciers?

22.4 Deposition Associated with Glaciation

The Glacial Conveyor and Glacial Moraines

Glaciers can carry sediment of any size and, like a conveyor belt, transport it in the direction of flow (**Fig. 22.17a**). Remember that ice flows toward the toe regardless of whether the glacier is advancing or retreating, so the transport of sediment always progresses in the direction of the toe. Where does the sediment come from? The sediment load either falls onto the surface of the glacier from bordering cliffs or gets plucked and lifted from the substrate and incorporated into the moving ice.

Sediment dropped on the glacier's surface from its margins becomes a stripe of debris, known as a **lateral moraine**, along the side edge of the glacier. When a glacier melts, its lateral moraines will be stranded along the side of the glacially carved valley, like bathtub rings. In places where two valley glaciers merge, the debris constituting two lateral moraines merges to become a **medial moraine**, running as a stripe down the interior of the composite glacier (**Fig. 22.17b, c**). Trunk glaciers created by the merging of many tributary glaciers contain several medial moraines. Sediment transported to a glacier's toe by the glacial conveyor accumulates in a pile

FIGURE 22.16 Examples of fjords.

(a) The Finger Lakes of central New York State are freshwater fjords.

(b) One of the many spectacular fjords of Norway. The water is an arm of the sea that fills a glacially carved valley. Tourists are standing on Pulpit Rock (Prekestolen).

at the toe and builds up to form an **end moraine**. If the glacier recedes, the end moraine will remain as a low ridge, outlining the former position of the toe. The name *moraine* originated from a term used by Alpine farmers and shepherds for any pile of rock and dirt. The word now applies exclusively to debris piles carried by or left by glaciers.

FIGURE 22.17 The glacial conveyor and the formation of lateral and medial moraines on glaciers.

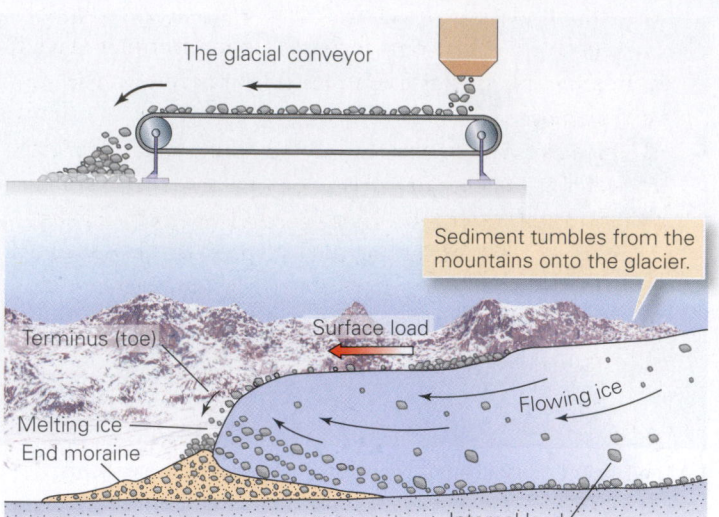

The glacial conveyor

(a) Sediment falls on a glacier from bordering mountains and gets plucked up from below. Glaciers are like conveyor belts, moving sediment toward the toe of the glacier.

(b) This glacier in the French Alps carries lots of sediment.

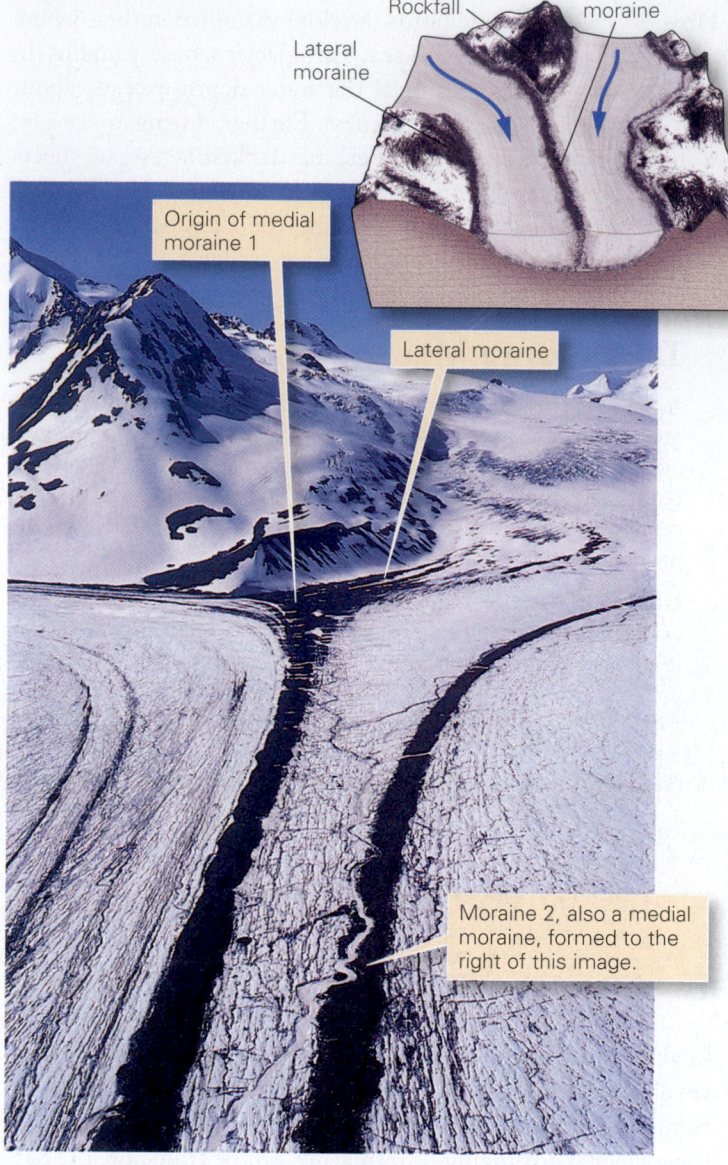

(c) A medial moraine forms where lateral moraines of two valley glaciers merge.

Types of Glacial Sedimentary Deposits

If you drill into the soil throughout much of the upper midwestern and northeastern United States and adjacent parts of Canada, the drill penetrates a layer of sediment deposited during the Pleistocene Ice Age. A similar story holds true for much of northern Europe. Thus, many of the world's richest agricultural regions rely on soil derived from sediment deposited by glaciers during the Ice Age. This sediment buries a pre–Ice Age landscape, as frosting fills the irregularities on a cake. Pre-glacial valleys may be completely filled with sediment.

Several different types of sediment can be deposited in glacial environments; all of these types together constitute **glacial drift**. (The term dates from pre-Agassiz studies of glacial deposits, when geologists thought that the sediment had "drifted" into place during an immense flood.) Glacial sediment, carried and deposited by ice, contains no layering, so geologists refer to it as *unstratified drift*. In contrast, glacial sediment that has been redistributed by flowing water, or settled through water, tends to contain layering and is called *stratified drift*. In detail, glacial drift includes the following features.

- *Till:* Sediment transported by ice and deposited beneath, at the side, or at the toe of a glacier is called **glacial till**. Glacial till is unsorted, so it's a type of diamicton (see Chapter 7), because the solid ice of glaciers can carry clasts of all sizes (**Fig. 22.18a**).
- *Erratics:* Boulders that have been dropped by a glacier are called *glacial erratics* (**Fig. 22.18b**). Some erratics protrude from till piles, and others rest on glacially polished surfaces.
- *Glacial marine:* Where a sediment-laden glacier flows into the sea, icebergs calve off the toe and raft clasts out to sea. As the icebergs melt, they drop the clasts, which settle into the muddy sediment on the seafloor. Pebble- and larger-size clasts deposited in this way, as we have seen, are called *dropstones*. Sediment consisting of ice-rafted clasts mixed with marine sediment makes up *glacial marine*. Glacial marine can also consist of sediment carried into the sea by water flowing at the base of a glacier.
- *Glacial outwash:* Till deposited by a glacier at its toe may be picked up and transported by meltwater streams that sort the sediment. The clasts are deposited by a braided stream network in a broad area of gravel and sandbars called an outwash plain. This sediment is known as **glacial outwash** (**Fig. 22.18c**).
- *Loess:* When the warmer air above ice-free land beyond the toe of a glacier rises, the cold, denser air from above the glacier rushes in to take its place; a strong wind, called *katabatic wind*, therefore blows at the margin of a glacier. This wind picks up fine silt and clay and transports it away from the glacier's toe. Where the winds die down, the sediment settles and forms a thick layer. This sediment, called **loess**, tends to stick together, so steep escarpments can develop by erosion of loess deposits (**Fig. 22.18d**).
- *Glacial lake-bed sediment:* Streams transport fine clasts, including rock flour, away from the glacial front. This sediment eventually settles in meltwater lakes, forming a thick layer of glacial lake-bed sediment. This sediment commonly contains varves. A **varve** is a pair of thin layers deposited during a single year. One layer consists of silt brought in during spring floods and the other of clay deposited in winter when the lake's surface freezes over and the water becomes still (**Fig. 22.18e**).
- *Kame deposits:* A *kame deposit* is an accumulation of sediment transported on the surface of a glacier by flowing meltwater. Because the sediment in a kame was in a current, the sediment in kame deposits tends to

be somewhat sorted. Some kame deposits form along the sides of glaciers, created by water sorting of lateral moraines, whereas others form in the interior of a glacier, as meltwater transports sediments into basins on the surface of the glacier.

- *Esker deposits:* In temperate glacial environments, the flowing water at the base of the glacier, moving through channels, transports much of the sediment load. Some of this water, along with its sedimentary load, exits the glacier through tunnels in the glacier's toe. But some never makes it out of the glacier and accumulates in sub-ice tunnels. This sediment is called an *esker deposit*.

Depositional Landforms of Glacial Environments

Picture a hunter, dressed in deerskin, standing at the toe of a continental glacier in what is now southern Canada, waiting for an unwary woolly mammoth to wander by. It's about 12,000 years ago, and the glacier has been receding for at least a millennium. Milky, sediment-laden streams gush from tunnels and channels at the base of the glacier and pour off the top as the ice melts. No mammoths venture by today, so the bored hunter climbs to the top of the glacier for a view. The climb isn't easy, partly because of the incessant katabatic wind and partly because deep crevasses interrupt his path. Reaching the top of the ice sheet, the hunter looks northward, and the glare almost blinds him. Squinting, he sees the white of snow, and where the snow has blown away, he sees the rippled, glassy surface of bluish ice (see Fig. 22.1b). Here and there, a rock protrudes from the ice. Now looking southward, he surveys a stark landscape of low, sinuous ridges separated by hummocky (bumpy) plains (see **Geology at a Glance**, pp. 820–821). Braided streams, which carry meltwater out across this landscape, flow through the hummocky plains and supply a number of lakes. Dust fills the air because of the wind.

All of the landscape features that the hunter observes as he looks southward were formed by deposition in glacial environments (**Fig. 22.19a, b**). The low, sinuous ridges that outline the former edge of the ice are end moraines, developed when the toe of a glacier stalls in one position for a while. The specific end moraine at the farthest limit of glaciation is called the **terminal moraine**. (The ridge of sediment that makes up Long Island, New York, and continues east-northeast into Cape Cod, Massachusetts, forms part of the terminal moraine of the ice sheet that covered New England and eastern Canada during the Pleistocene Ice Age; **Fig. 22.19c**.) End moraines that form when a glacier stalls temporarily as it recedes overall are **recessional moraines**.

FIGURE 22.18 Sedimentation processes and products associated with glaciation. Glacial sediment is distinctive.

(a) This glacial till in Ireland is unsorted, because ice can carry sediment of all sizes.

(b) Glacial erratics resting on a glacially polished surface in Wyoming.

(c) Braided streams choked with glacial outwash in Alaska. The streams carry away finer sediment and leave the gravel behind.

(d) Thick loess deposits underlie parts of the prairie in Illinois.

(e) In the quiet water of an Alaskan glacial lake, fine-grained sediments accumulate. Alternating layers in the sediment (varves), now exposed in an outcrop near Puget Sound, Washington, reflect seasonal changes.

FIGURE 22.19 The formation of depositional landforms associated with continental glaciation.

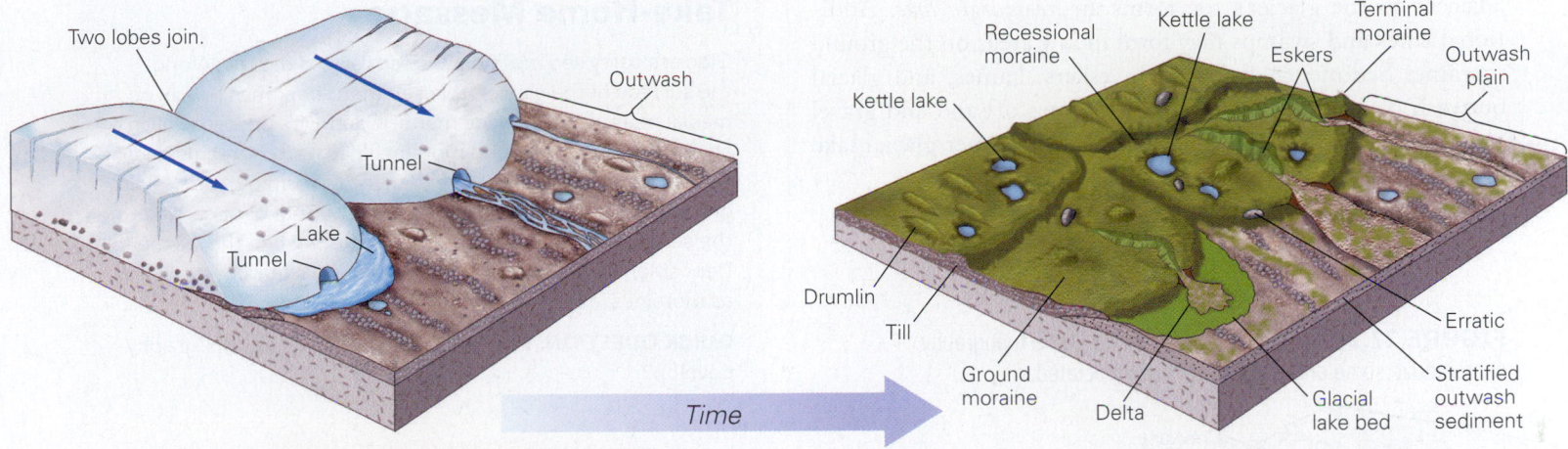

(a) The ice in continental glaciers flows toward the toe; sediment accumulates at the base and at the toe of the ice sheet.

(b) Several distinct depositional landforms form during glaciation; some developed under the ice and some at the toe.

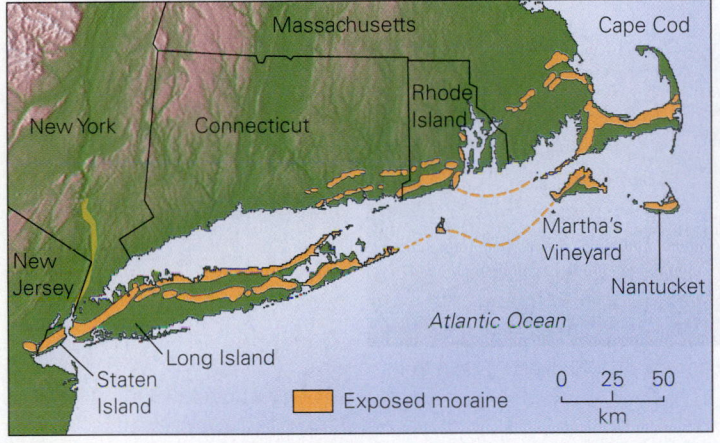

(c) Cape Cod, Long Island, and other landforms in the northeastern United States formed at the end of the continental ice sheet.

The till contains large boulders; finer sediment became soil.

(d) This glacial moraine, in Wyoming, formed when glaciers covered the region between the moraine and the mountains in the distance.

Till that has been released from the ice at the base of a flowing glacier and remains after the glacier has melted away makes up *lodgment till* (**Fig. 22.19d**). Clasts in lodgment till may be aligned and scratched during glacial flow. The till left behind during rapid recession forms a thin, hummocky layer on the land surface. This till, together with lodgment till, forms a landscape feature known as *ground moraine*. The flow of the glacier may mold till and other subglacial sediment into a streamlined, elongate hill called a **drumlin** (from the Gaelic word for hills). Drumlins tend to be asymmetric along their length, with a gentle downstream slope, tapered in the direction of flow, and a steeper upstream slope (**Fig. 22.20a, b**).

The hummocky surface of moraines reflects partly the variations in the amount of sediment supplied by the glacier and partly the formation of kettle holes, circular depressions made when blocks of ice calve off the toe of the glacier, become buried by till or kame deposits, and then melt to leave a depression (**Fig. 22.20c, d**). A land surface with many kettle holes separated by round hills of till displays *knob-and-kettle topography* (**Fig. 22.20e**).

As we've noted, ice does not directly deposit all of the sediment associated with glacial landscapes, for meltwater also carries and deposits sediment. Water-transported sediment, in contrast to till, tends to be sorted and stratified. When the underlying ice melts away, kame deposits become a small hill or ridge known as a *kame*; kames may comprise some of the knobs in knob-and-kettle topography. Esker deposits, filling meltwater tunnels beneath a glacier, may remain as a narrow sinuous ridge, known as an **esker**, when the glacier melts away; eskers tend to trend at a high angle to the toe of the glacier (**Fig. 22.21**). Braided meltwater streams that flow beyond the end of a glacier deposit layers of sand and gravel over a broad area, yielding a **glacial outwash plain**. Meltwater collecting adjacent to the glacier's toe forms an *ice-margin lake*. Additional lakes and swamps may form in low areas on the ground moraine. Sediments deposited in eskers, kames, and glacial outwash plains serve as important sources of sand and gravel for construction, and the fine sediment of former glacial lake beds evolves into fertile soil for agriculture.

Take-Home Message

Glaciers carry sediment of all sizes toward the toe. Along the surface of the glacier, this sediment becomes lateral or medial moraines. When ice melts, it deposits unsorted till in an end moraine or ground moraine. Meltwater streams and wind transport and sort the sediment to form outwash-plain gravels and loess deposits, respectively. Sediment carried to the sea by glaciers settles on the seafloor when the ice melts. Deposition by glaciers produces distinctive landforms, such as moraines, eskers, kames, and kettle holes.

QUICK QUESTION: How does knob-and-kettle topography develop?

FIGURE 22.20 Drumlins and knob-and-kettle topography characterize some areas that were once glaciated.

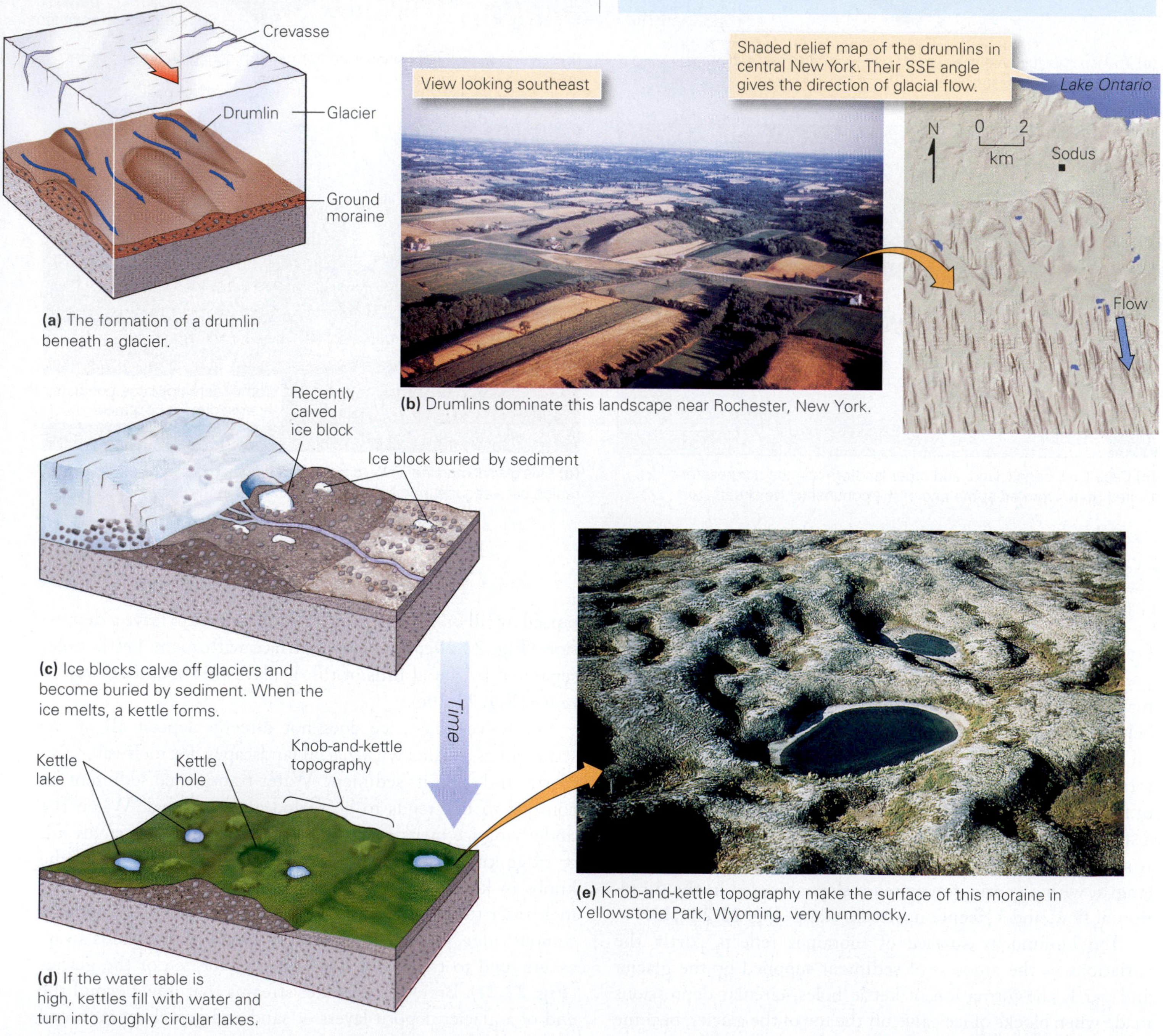

(a) The formation of a drumlin beneath a glacier.

Shaded relief map of the drumlins in central New York. Their SSE angle gives the direction of glacial flow.

View looking southeast

(b) Drumlins dominate this landscape near Rochester, New York.

(c) Ice blocks calve off glaciers and become buried by sediment. When the ice melts, a kettle forms.

(d) If the water table is high, kettles fill with water and turn into roughly circular lakes.

(e) Knob-and-kettle topography make the surface of this moraine in Yellowstone Park, Wyoming, very hummocky.

FIGURE 22.21 Eskers are snake-like ridges of sand and gravel that form when sediment fills meltwater tunnels at the base of a glacier.

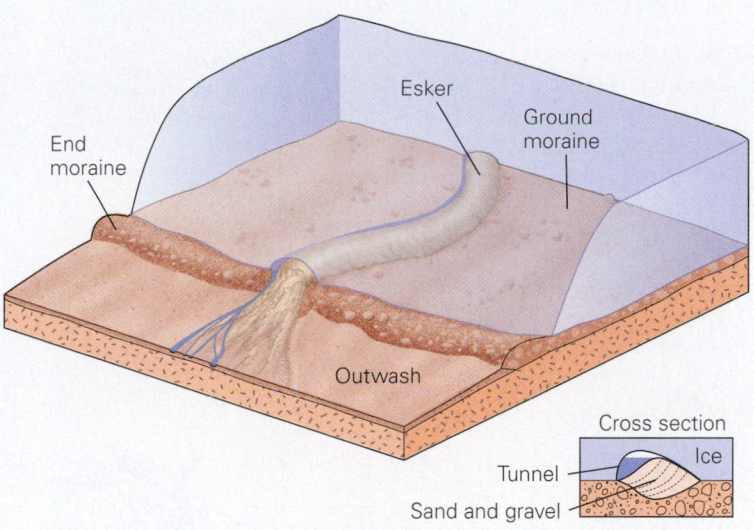

(a) At the time of formation, an esker develops beneath an ice sheet. In cross section (inset), wedges of sand accumulate in the tunnel.

(b) An example of an esker in an area once glaciated but now farmed.

22.5 Other Consequences of Continental Glaciation

Ice Loading and Glacial Rebound

When a large ice sheet (more than 50 km in diameter) grows on a continent, its weight causes the surface of the lithosphere to sink. In other words, ice loading causes **glacial subsidence**. Lithosphere, the relatively rigid outer shell of the Earth, can sink because the underlying asthenosphere is soft enough to flow slowly out of the way (**Fig. 22.22a**). As an analogy, imagine this simple experiment: Fill a bowl with honey and then place a thin rubber sheet over the honey. The rubber represents the lithosphere, and the honey represents the asthenosphere. If you place an ice cube on the rubber sheet, the sheet sinks because the weight of the ice pushes it down; the honey flows out of the way to make room. Because of ice loading, the rock surface underlying large areas of Antarctica's and Greenland's ice sheets now lie below sea level (see Fig. 22.4), so if the ice were instantly to melt away, these continents would be flooded by a shallow sea.

What happens when continental ice sheets do melt away? Gradually, the surface of the underlying continent rises back up to re-achieve isostasy (see Interlude D), by a process called **post-glacial rebound**. As this happens, the asthenosphere flows back underneath to fill the space (**Fig. 22.22b**). Where

rebound affects coastal areas, beaches along the shoreline rise several meters above sea level and become terraces (**Fig. 22.22c**). In the honey and rubber analogy, when you remove the ice cube, the rubber sheet slowly returns to its original shape. This process doesn't take place instantly because the honey can only flow slowly. Similarly, because the asthenosphere flows so slowly, it takes thousands of years for ice-depressed continents to rebound. Thus, glacial rebound is still taking place in some regions that were burdened by ice during the Pleistocene Ice Age. Recently, researchers in North America have documented this movement by using GPS measurements (**Fig. 22.22d**). Regions north of a line passing through the Great Lakes are now rising, relative to sea level.

Sea-Level Changes: The Glacial Reservoir

More of the Earth's surface and near-surface freshwater is stored in glacial ice than in any other reservoir. In fact, glacial ice accounts for 2.15% of Earth's total water supply, while lakes, rivers, soil, and the atmosphere together contain only 0.03%. The melting of glacial ice would transfer this water back into the ocean, causing sea level to rise. In fact, if today's ice sheets in Antarctica and Greenland were to melt, large areas of the coastal plain along the east coast and Gulf Coast of North America would become submerged, as would much of the Ganges Delta of Bangladesh.

During the last ice age, when glaciers covered almost three times as much land area as they do today, they held almost three times more water (70 million km^3, as opposed to 25 million km^3 today). In effect, during the ice age, water

Continental ice sheet

Crevasses

Ice shelf

Higher sea level

Lower sea level

Drop stones

Iceberg

Horn Valley glacier Lateral moraine Mountain ice cap

Cirque glacier

Arête

Medial
moraine

Meltwater lake

U-shaped valley

Erratic

Outwash
plain

Ground moraine

Kettle hole

Drumlin

Esker

Recessional
moraine

Braided stream

Note that the terminal moraine
here is not visible; it's offshore
and is submerged.

Striations Roche moutonnée

Glaciers are rivers or sheets of ice that last all year and slowly flow. Continental glaciers, vast sheets of ice up to a few kilometers thick, covered extensive areas of land during times when Earth had a colder climate. At the peak of the last ice age, ice sheets covered almost all of Canada, much of the United States, northern Europe, and parts of Russia.

The upper part of a sheet is brittle and may crack to form crevasses. Because ice sheets store so much of the Earth's water, sea level becomes lower during an ice age. When a glacier reaches the sea, it becomes an ice shelf. Rock that the glacier has plucked up along the way is carried out to sea with the ice; when the ice melts, the rocks fall to the sea floor as drop stones. At the edge of the shelf, icebergs calve off and float away.

Mountain or alpine glaciers grow in mountainous areas because snow can last all year at high elevations. During an

ice age, mountain glaciers grow and flow out onto the land surface beyond the mountain front. Glacial recession may happen when the climate warms, so ice melts away faster at the toe (terminus) of the glacier than it can be added at the source. Consequences of glacial erosion and deposition remain when a glacier melts away. Erosion features include striations on bedrock and roches moutonnées. Deposition features include glacial moraines, glacial outwash, and esker deposits. Even when the toe remains fixed in position for a while, the ice continues to flow and thus molds underlying sediment into drumlins. Ice blocks buried in till melt to form kettle holes. In the mountains, glaciers fill valleys or form ice caps. Sediment falling from the mountains creates lateral and medial moraines. Glaciers carve distinct landforms in the mountains, such as cirques, arêtes, horns, and U-shaped valleys.

FIGURE 22.22 The concept of subsidence and rebound due to continental glaciation and deglaciation. (Not to scale)

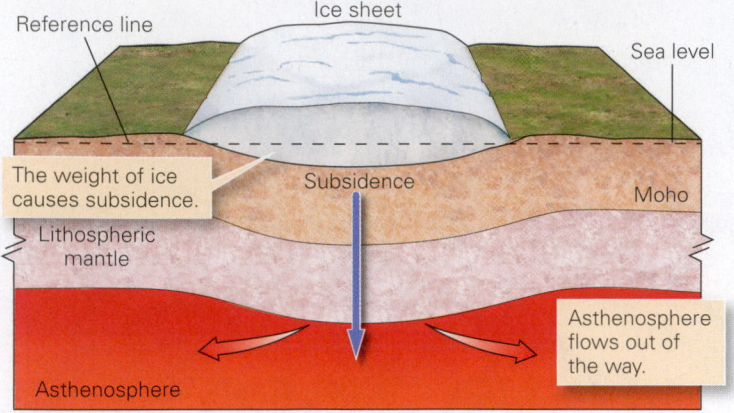

(a) The weight of the ice sheet causes the surface of the lithosphere to sink (subside).

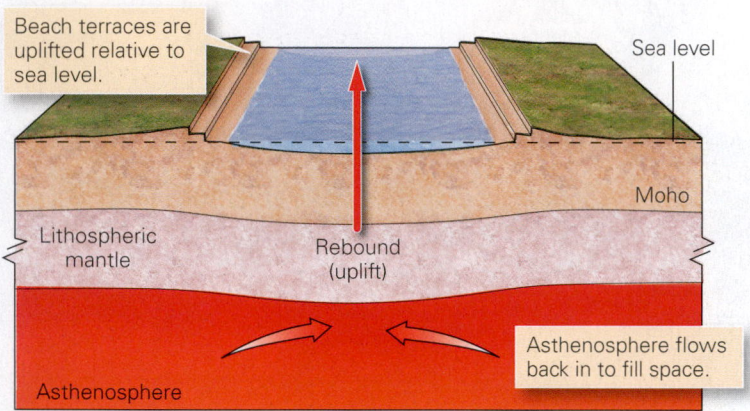

(b) After the glacier melts, the land surface rebounds and rises. This process uplifts beaches relative to sea level.

(c) Uplifted beaches along the coast of Arctic Canada form as the land undergoes post-glacial rebound.

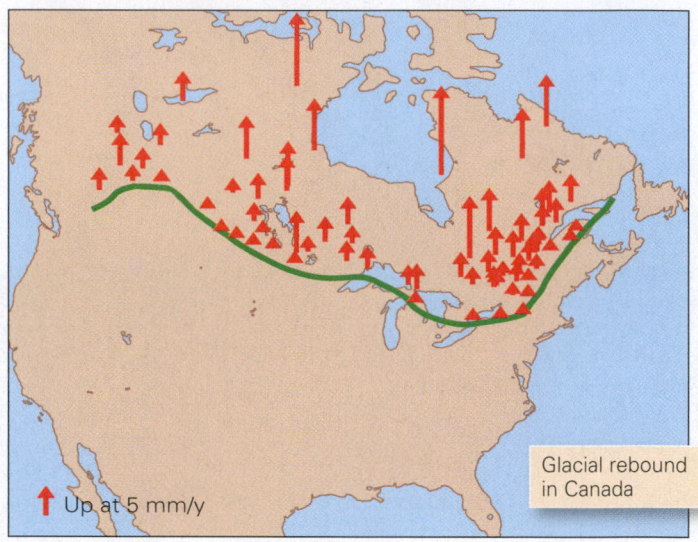

(d) GPS measurements show that the region north of the green line is rebounding. Different rates of uplift occur at different locations.

---

from the ocean reservoir was transferred from the ocean reservoir to the glacial reservoir, and remained trapped on land. As a consequence, sea level dropped by as much as 100 m, and extensive areas of continental shelves became exposed as the coastline migrated seaward, in places by more than 100 km (**Fig. 22.23**). People and animals migrated into the newly exposed ice-age coastal plains. In fact, fishermen dragging their nets along the Atlantic Ocean floor off New England today occasionally recover human artifacts. The drop in sea level also created land bridges across the Bering Strait between North America and northeastern Asia and between Australia and Indonesia, providing convenient migration routes for early people.

Effect on Drainage and Lakes

Continental glaciation can significantly modify the location and character of rivers and streams draining the land. For example, locally, the growth of a glacier, and/or the deposition of a moraine by glaciers, can block an individual stream. Diverted flow finds a new route and can carve a new valley. By the time the glacier melts away, these new streams have become so well established that pre-glacial channels remain abandoned.

At a regional scale, glaciation during the Pleistocene Ice Age profoundly modified North America's interior drainage. Before this ice age, several major rivers drained much of the interior of the continent to the north, into the Arctic Ocean (**Fig. 22.24a, b**). The ice sheet buried this drainage network and diverted the flow into the Mississippi–Missouri network, which became larger.

When ice-age glaciers receded, new lakes appeared on the land. For example, kettles in regions covered by knob-and-kettle topography turned into small lakes, as now occur in central and southern Minnesota by the thousands. In the Canadian Shield, scouring left innumerable depressions that have now become lakes.

FIGURE 22.23 The link between sea level and global glaciation: glaciers store water on land, so when glaciers grow, sea level falls, and when glaciers melt, sea level rises.

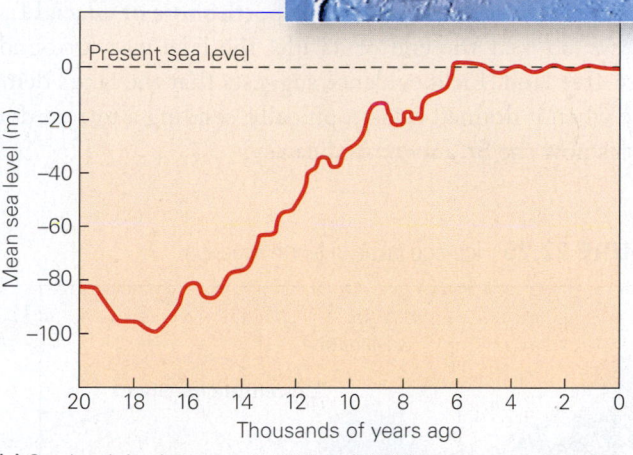

(b) Prehistoric people migrated across the Bering Strait land bridge.

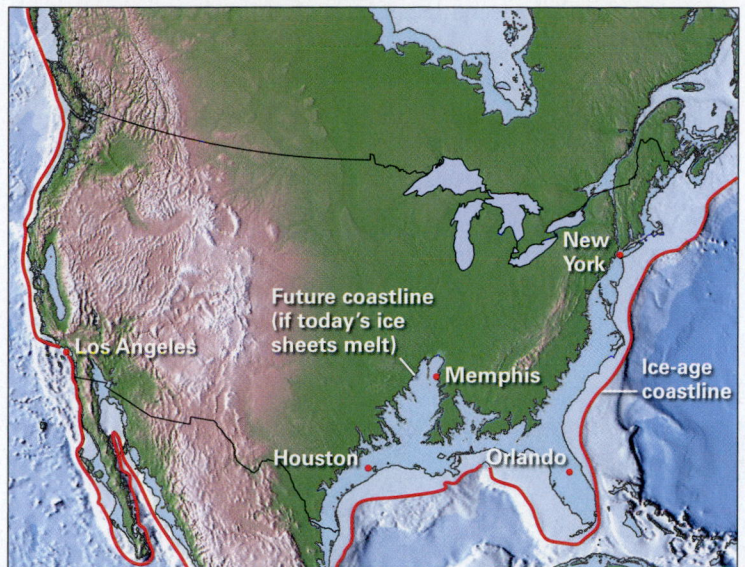

(a) The red line shows the coastline during the last ice age; much of the continental shelf was dry. If present-day ice sheets melt, coastal lands will flood.

(c) Sea-level rise between 17,000 and 7,000 B.C.E. was due to the melting of ice-age glaciers.

FIGURE 22.24 Ice-age glaciation changed the position of the divide between north-draining and south-draining river networks.

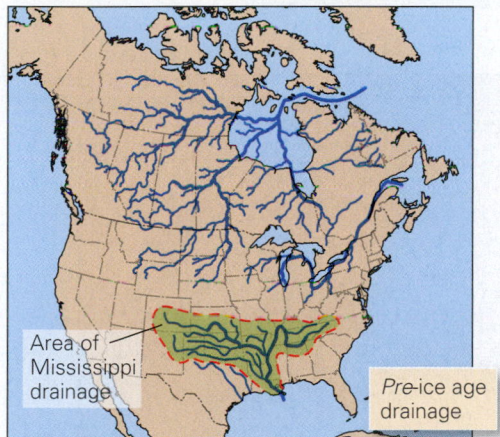

(a) Before the last ice age, more rivers flowed north; the Mississippi network was smaller.

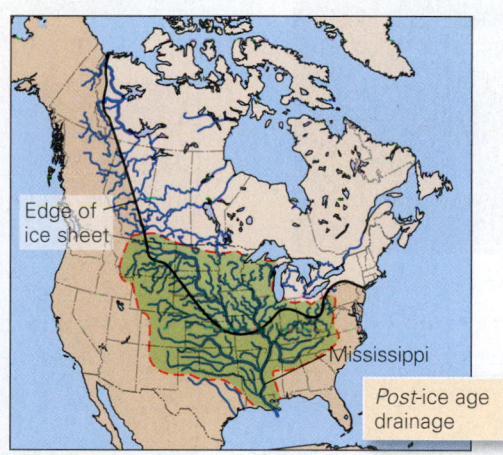

(b) Glaciation blocked northward drainage; the Mississippi network grew larger.

Erosion by glacial meltwater can carve valleys. Especially dramatic examples of this process result when the ice dams that held back large ice-margin lakes melted and broke. In a matter of hours to days, the contents of lakes could drain, yielding an immense flood called an outburst flood or *torrent*. Torrents can carve huge valleys and steep cliffs, strip the land of soil, and leave behind immense ripple marks (**Fig. 22.24c**). For example, when the ice dam holding back Glacial Lake Missoula in Montana broke, it released an immense torrent—known as the Great Missoula Flood—that scoured eastern Washington, creating a barren, soil-free landscape called the *channeled scablands* (see Chapter 17). Recent evidence suggests that this process repeated several times. Another torrent flowed down the channel of what is now the Illinois River and, in northern Illinois, carved a broad, steep-sided valley that is much too large to have been cut by the present-day Illinois River.

The largest known ice-margin lake covered portions of Manitoba and Ontario in south-central Canada and North Dakota and Minnesota in the United States (**Fig. 22.25a**).

(c) Giant ripple marks formed during the Great Missoula Flood.

This body of water, Glacial Lake Agassiz, existed between 11,700 and 9,000 years ago, a time during which the most recent phase of the last ice age came to a close and the continental glacier retreated north. At its largest, the lake covered over 250,000 square km (100,000 square miles), an area greater than that of all the present Great Lakes combined. Eventually, the ice sheet receded from the north shore of Glacial Lake Agassiz, so near the end of its life, the lake was surrounded by ice-free land. Field evidence suggests that the lake's demise came when it drained catastrophically, sending a torrent down what is now the St. Lawrence Seaway.

Pluvial Features

During the Pleistocene Ice Age, regions to the south of continental glaciers were wetter than they are today. Fed by enhanced rainfall, lakes accumulated in low-lying land even at a great distance from the ice front. The largest of these **pluvial lakes** (from the Latin *pluvia*, meaning rain) in North America flooded interior basins of the Basin and Range Province in Utah and Nevada (**Fig. 22.25b**). Examples include glacial Lake Bonneville, which covered almost a third of western Utah. When this lake suddenly drained after a natural dam holding it back broke,

FIGURE 22.25 Ice-age lakes in North America.

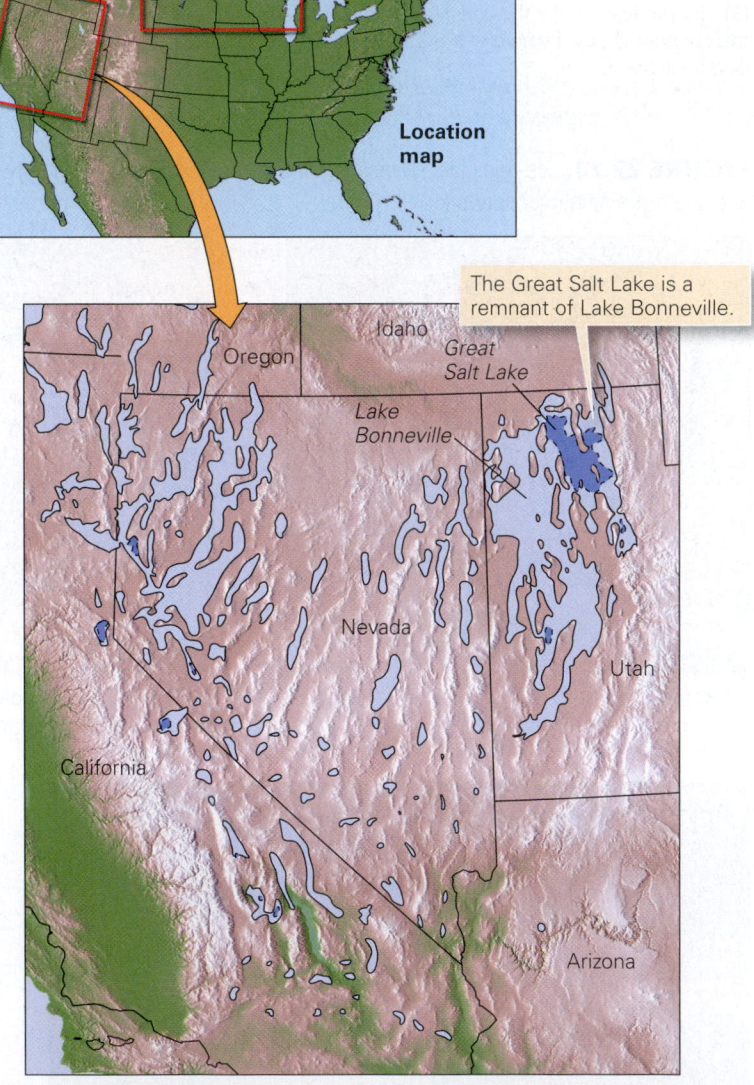

(a) Glacial Lake Agassiz was an ice-margin lake that formed near the end of the last ice age.

The Great Salt Lake is a remnant of Lake Bonneville.

(b) Pluvial lakes occurred throughout the Basin and Range Province during the last ice age due to the wetter climate. The largest of these was Lake Bonneville. Subtle horizontal terraces define the remnants of beaches, now over 100 m above the present level of the Great Salt Lake.

FIGURE 22.26 Periglacial regions are not ice covered but do include substantial areas of permafrost.

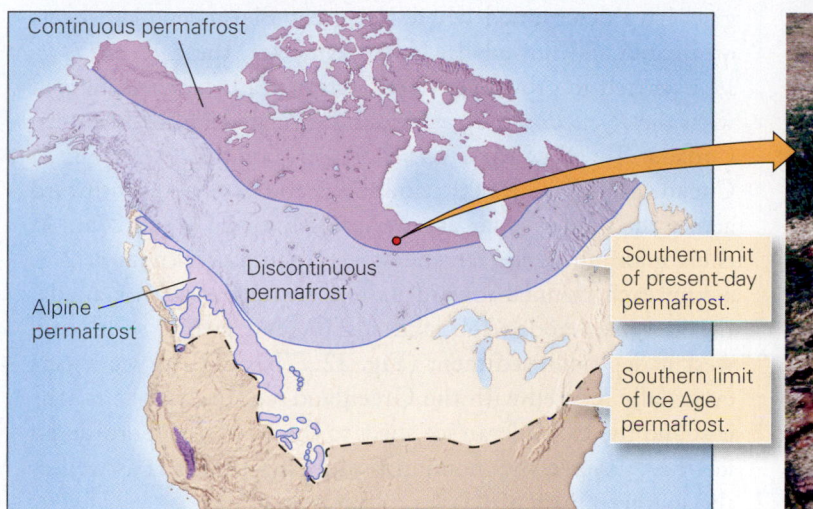

(a) The distribution of periglacial environments in North America.

(b) An example of patterned ground near a pond in Manitoba, Canada.

it left a bathtub ring of shoreline rimming the mountains near Salt Lake City (see Fig. 22.25b). Today's Great Salt Lake itself is but a small remnant of Lake Bonneville.

Periglacial Environments

In polar latitudes today, and in regions adjacent to the fronts of continental glaciers during the last ice age, the mean annual temperature stays low enough (below −5°C) that soil moisture and groundwater freeze and, except in the upper few meters, stay solid all year. Such permanently frozen ground, or **permafrost**, may extend to depths of 1,500 m below the ground surface. Regions with widespread permafrost that do not have a cover of snow or ice are called *periglacial environments* (the Greek *peri* means around, or encircling; periglacial environments appear around the edges of glacial environments; **Fig. 22.26a**).

The upper few meters of permafrost may melt during the summer months, only to refreeze again when winter comes. As a consequence of the freeze-thaw process, the ground of some permafrost areas splits into pentagonal or hexagonal shapes, creating a landscape called **patterned ground** (**Fig. 22.26b**). Water fills the gaps between the cracks and freezes to create wedge-shaped walls of ice. In some places, freeze-and-thaw cycles in permafrost gradually push cobbles and pebbles up from the subsurface. Because the expansion of the ground is not even, the stones gradually collect between adjacent bulges to form stone rings (**Fig. 22.26c**). Some stone rings may also form when mud at depth pushes up from beneath a permafrost layer and forces stones aside.

Permafrost presents a unique challenge to people who live in polar regions or who work to extract resources from these

(c) Stone circles of Spitzbergen form due to repeated freeze and thaw that separates gravel from silt.

regions. For example, heat from a building may warm and melt underlying permafrost, creating a mire into which the building settles. For this reason, buildings in permafrost regions must be placed on stilts so that cold air can circulate beneath them to keep the ground frozen. When geologists discovered oil on the northern coast of Alaska, oil companies faced the challenge of shipping the oil to markets outside of Alaska. After much debate over the environmental impact, the Trans Alaska Pipeline was built, and now it carries oil for 1,000 km to a seaport in southern Alaska (see Chapter 14). The oil must be warm during transport or it would be too viscous to flow; thus, to prevent the warm pipeline from melting underlying permafrost, it had to be built on a frame that holds it above the ground for most of its length.

22.6 The Pleistocene Ice Age

The Pleistocene Glaciers

Today most of the land surface in New York City lies hidden beneath concrete and steel, but in Central Park it's still possible to see land in a seminatural state. If you stroll through the park and study the rock outcrops, you'll find that their top surfaces are smooth and polished (see Fig. 22.12c) and in places have been grooved and scratched. You can also find numerous erratics. You are seeing evidence that an ice sheet once scraped along this now-urban ground. Geologists estimate that the ice sheet that overrode the New York City area may have been 250 m thick, enough to engulf the Empire State Building up to the 75th floor.

Glacial features such as those on display in Central Park first led Louis Agassiz to propose the idea that vast continental glaciers advanced over substantial portions of North America, Europe, and Asia during a great ice age. Since Agassiz's day, geologists, by mapping out the distribution of glacial deposits and landforms, have gradually defined the extent of ice-age glaciers and a history of their movement (**Box 22.2**).

The fact that glacially modified landscapes decorate the surface of the Earth today means that the most recent ice age occurred fairly recently during Earth history. This ice age, responsible for the glacial landforms of North America and Eurasia, happened mostly during the Pleistocene Epoch, which began about 2.6 million years ago (Ma) (see Chapter 13), so as we've noted earlier, it is commonly known as the *Pleistocene Ice Age*. Geologists use the term *Holocene* for about the last 12,000 years, the time since the last Pleistocene ice sheet melted away.

Based on mapping of glacial striations and deposits, geologists have determined where the great Pleistocene ice sheets originated and flowed. In North America, the *Laurentide ice sheet* started to grow over northeastern Canada, then merged with the *Keewatin ice sheet*, which originated in northwestern Canada. Together these ice sheets eventually covered all of Canada east of the Rocky Mountains and extended southward across the border as far as southern Illinois (**Fig. 22.27a**). At their maximum, the ice sheets attained a thickness of 2 to 3 km; each thinned toward its toe. In northeastern Canada, the ice sheet eroded the land surface, but farther south and west it deposited sediment (**Fig. 22.27b, c**). These ice sheets eventually merged with the Greenland ice sheet to the northeast and the Cordilleran ice sheet to the west. The Cordilleran ice sheet covered the mountains of western Canada as well as the southern third of Alaska.

During the Pleistocene Ice Age, mountain ice caps and valley glaciers also grew in the southern Rocky Mountains, the Sierra Nevada, and the Cascade Mountains, elevated regions to the south of the continental glacier. In Eurasia, a large ice sheet formed in northernmost Europe and adjacent Asia, and it gradually covered all of Scandinavia and northern Russia. This ice sheet flowed southward across France until it reached the Alps and merged with Alpine mountain glaciers. Ice also covered almost all of Ireland and the United Kingdom. Notably, even the highest mountains of Scotland were completely submerged beneath ice, so the peaks have been rounded by glacial erosion. A smaller ice sheet grew in eastern Siberia, and glaciers expanded in the mountains of central Asia. In the southern hemisphere, Antarctica remained ice covered, and mountain ice caps expanded in the Andes, but there were no continental glaciers in South America, Africa, or Australia.

In addition to continental ice sheets, sea ice in the northern hemisphere covered all of the Arctic Ocean and parts of the North Atlantic during the Pleistocene. Sea ice surrounded Iceland, approached Scotland, and also fringed most of western Canada and southeastern Alaska.

Life and Climate in the Pleistocene World

During the Pleistocene Ice Age, all climatic belts of the northern hemisphere shifted southward (**Fig. 22.28a, b**). Geologists can document this shift by examining fossil pollen, which can survive for thousands of years if preserved in the sediment of bogs. Presently, the southern boundary of North America's **tundra**, a treeless region supporting only low shrubs, moss, and lichen capable of living on permafrost, lies at a latitude of 68° N—during the Pleistocene Ice Age, it moved down to 48° N. Much of the interior of the United States, which now has temperate, deciduous forest, harbored cold-weather spruce and pine forest. Ice-age climates also changed the distribution

BOX 22.2   CONSIDER THIS . . .

So You Want to See Glaciation?

Though the Pleistocene continental ice sheet that once covered much of North America vanished about 6,000 years ago, you can find evidence of its power quite easily. The Great Lakes, along the U.S.–Canada border, the Finger Lakes and drumlins of New York, the low-lying moraines and outwash plains of Illinois, and the polished outcrops of the southern Canadian Shield all formed in response to the existence of this glacier. But if you want to see continental glaciers in action today, you must trek to Greenland or Antarctica.

Mountain glaciers are easier to reach. A trip to the mountains of western North America (including Alaska), the Alps of France or Switzerland, the Andes of South America, or the mountains of southern New Zealand will bring you in contact with active glaciers. You can even spot glaciers from the comfort of a cruise ship. Some of the most spectacular glacial landscapes in North America formed during the Pleistocene Epoch, when mountain glaciers were more widespread. These are now on display in national parks.

- **Glacier National Park (Montana):** This park, which borders Waterton Lakes National Park in Canada, displays giant cirques, U-shaped valleys, hanging valleys, and terminal moraines. In 1850, there were about 150 glaciers in the park, and some of these were quite large. Now only 25 active relicts of formerly larger glaciers remain, all in a mountainous terrain that reaches elevations of over 3 km. Unfortunately, these glaciers are melting away quickly and may vanish entirely by 2030.
- **Yosemite National Park (California):** A huge U-shaped valley carved into the Sierra Nevada granite batholith makes up the centerpiece of this park. Waterfalls spill out of hanging valleys bordering the valley.
- **Voyageurs National Park (Minnesota):** This park lacks the high peaks of mountainous parks but shows the dramatic consequences of glacial scouring and deposition on the Canadian Shield. The low-lying landscape, dotted with lakes, contains abundant polished surfaces, glacial striations, and erratics, along with moraines, glacial lake beds, and outwash plains.
- **Acadia National Park (Maine):** During the last ice age, the continental ice sheet overrode low bedrock hills and flowed into the sea along the coast of Maine. This park provides some of the best examples of the consequences. Its hills were scoured and shaped into large roches moutonnées by glacial flow. Some of the deeper valleys have now become small fjords.
- **Glacier Bay National Park (Alaska):** In Glacier Bay, huge tidewater glaciers fringe the sea, creating immense ice cliffs from which icebergs calve off. Cruise ships bring tourists up to the toes of these glaciers. More adventurous visitors can climb the coastal peaks and observe lateral and medial moraines, crevasses, and the erosional and depositional consequences of glaciers that have already retreated up the valley.

of rainfall on the planet. As we noted earlier, rainfall increased in North America, south of the glaciers, leading to the filling of pluvial lakes in Utah and Nevada. In contrast, rainfall decreased in equatorial regions, leading to shrinkage of the rainforest. Overall, the contrast between colder, glaciated regions and warmer, unglaciated regions created windier conditions worldwide. These winds sent glacial rock flour skyward, creating a dusty atmosphere (and, presumably, spectacular sunsets). The dust settled to create extensive deposits of loess. And because glaciers trapped so much water, as we have seen, sea level dropped.

Numerous species of now-extinct large mammals inhabited the Pleistocene world (**Fig. 22.28c**). Giant mammoths and mastodons, relatives of the elephant, along with woolly rhinos, musk oxen, reindeer, giant ground sloths, bison, lions, saber-toothed cats, and giant cave bears wandered forests and tundra in North America. Early human-like species were already foraging in the woods by the beginning of the Pleistocene Epoch, and by the end modern *Homo sapiens* lived on every continent except Antarctica and had discovered fire and invented tools.

Rapidly changing climates may have triggered a global migration of early humans, who gained access to the Americas, Indonesia, and Australia via land bridges that became exposed when sea level dropped.

Timing of the Pleistocene Ice Age

Louis Agassiz assumed that only one ice age had affected the planet. But close examination of the stratigraphy of glacial deposits on land revealed that *paleosols* (ancient soils preserved in the stratigraphic record), as well as beds containing fossils of warmer-weather animals and plants, separate distinct layers of glacial sediment. This observation suggested that between episodes of glacial deposition, glaciers receded and temperate climates prevailed. In the second half of the 20th century, when modern methods for dating geological materials became available, the difference in ages between the different layers of glacial sediment could be confirmed. Clearly, glaciers had advanced and then retreated more than once during the Pleistocene. Times during which the glaciers grew and

FIGURE 22.27 Pleistocene ice sheets and their consequences.

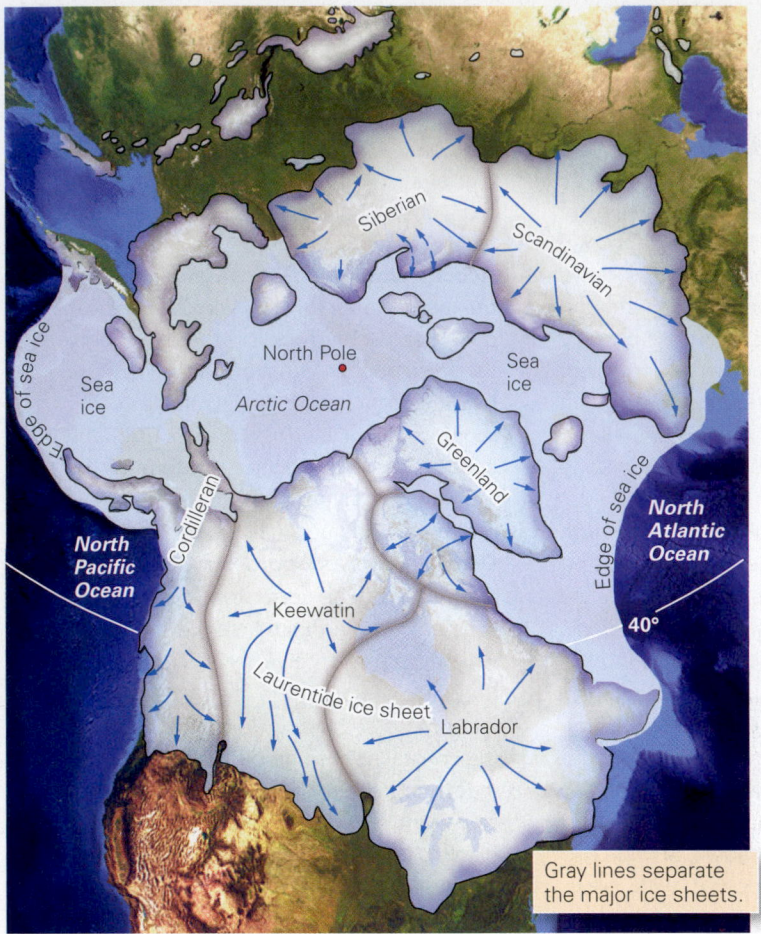

(a) During the Pleistocene, several distinct ice sheets formed. In several places, neighboring sheets came in contact.

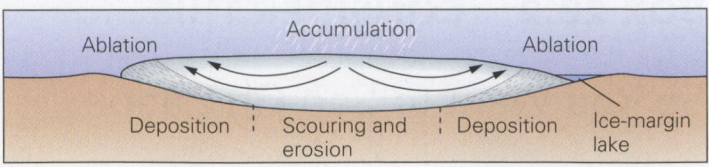

(b) Erosion dominates beneath the interior of the glacier, and deposition dominates along its margins.

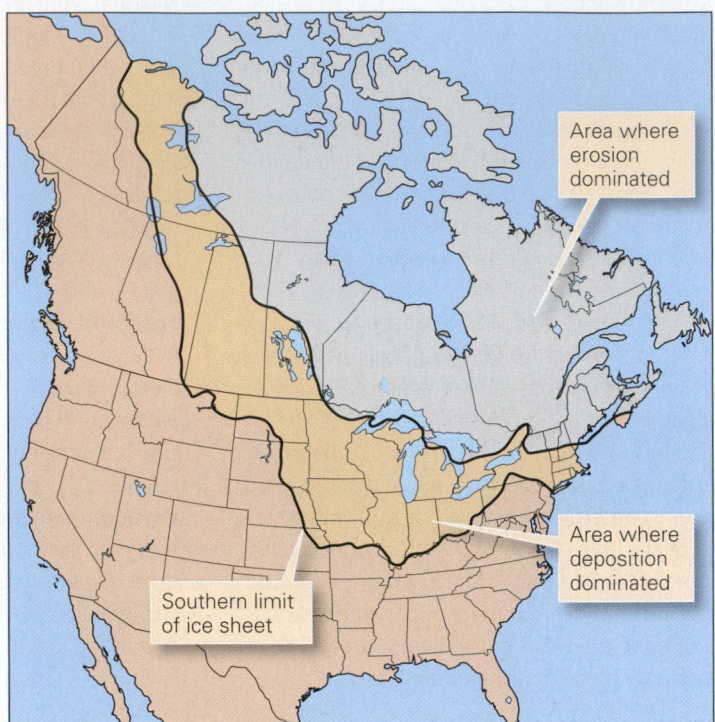

(c) Erosion dominated in northern and eastern Canada; deposition dominated in the Great Plains.

covered substantial areas of the continents are called glacial periods, or **glaciations**, and times between glacial periods are called interglacial periods, or **interglacials**.

Using the on-land sedimentary record, geologists recognized five Pleistocene glaciations in Europe (named, in order of increasing age, Würm, Riss, Mindel, Gunz, and Donau) and, traditionally, four in the midwestern United States (Wisconsinan, Illinoian, Kansan, and Nebraskan, named after the southernmost states in which their till was deposited; **Fig. 22.29**). Since the mid-1980s, geologists no longer distinguish the Nebraskan from the Kansan—they are lumped together as "pre-Illinoian." With the advent of radiometric dating in the mid-20th century, the ages of the younger glaciations were determined by dating wood trapped in glacial deposits. Geologists estimate the ages of the older glaciations by identifying fossils in the deposits.

The four- or five-stage chronology of glaciations was turned on its head in the 1960s, when geologists began to study submarine sediment containing the fossilized shells of microscopic marine plankton. Because the assemblage of plankton species living in warm water is not the same as the assemblage living in cold water, geologists can track changes in the temperature of the ocean by studying plankton fossils. Researchers found that in post-2.6-Ma sediment, assuming that cold water indicates a glacial period and warm water an interglacial period, there is a record of 20 to 30 different glacial advances during the Pleistocene Epoch. The four or five traditionally recognized glaciations possibly represent only the largest of these. Sediments deposited on land by other glaciations were eroded and redistributed during subsequent glaciations or were eroded away by streams and wind during interglacials.

Geologists refined their conclusions about the frequency of Pleistocene glaciations by examining the isotopic composition of fossil shells. Shells of many plankton species consist of calcite ($CaCO_3$). The oxygen in the shells includes two isotopes, a heavier one (^{18}O) and a lighter one (^{16}O). The ratio of

FIGURE 22.28 Climate belts during the Pleistocene.

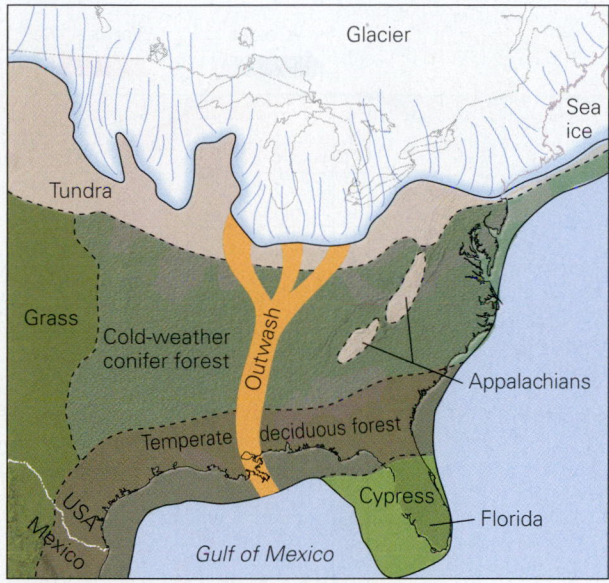

(a) Tundra covered parts of the United States, and southern states had forests like those of New England's today.

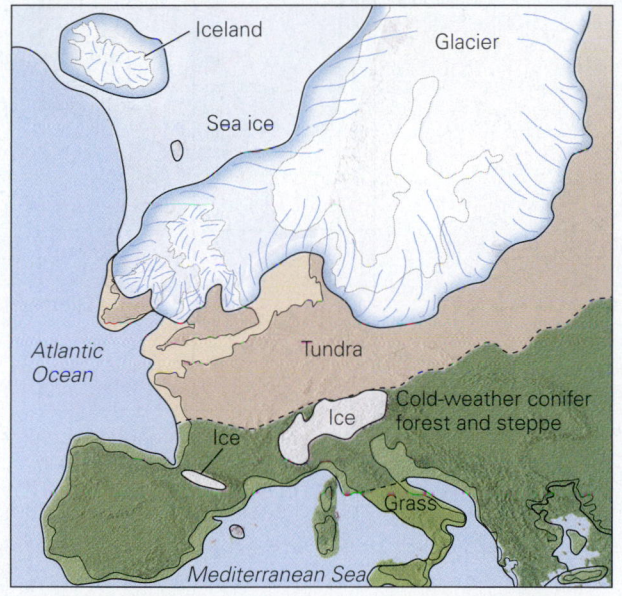

(b) Regions of Europe that support large populations today would have been barren tundra during the Pleistocene.

(c) Cold-adapted, now-extinct, large mammals roamed regions that are now temperate.

FIGURE 22.29 Pleistocene glacial deposits in the north-central United States. Curving moraines reflect the shape of glacial lobes.

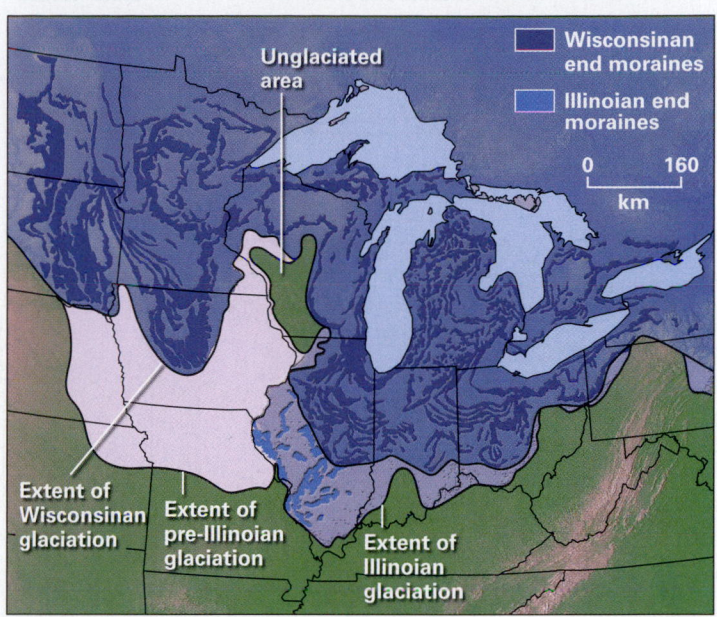

these isotopes tells us about the water temperature in which the plankton grew, for as water gets colder, plankton incorporate a higher proportion of ^{18}O into their shells (see Chapter 23). Thus, intervals in the stratigraphic record during which plankton shells have a large ratio of ^{18}O to ^{16}O define times when Earth had a colder, glacial climate. The isotope record confirms that 20 to 30 glaciations occurred during the last 2.6 Ma (**Fig. 22.30a**).

> **Did you ever wonder . . .**
>
> how many ice ages have happened during Earth history?

Older Ice Ages during Earth History

So far, we've focused on the Pleistocene Ice Age because of its importance in developing Earth's present landscape. Was this the only ice age during Earth history, or do ice ages happen frequently? To answer such questions, geologists study the stratigraphic record and search both for glacially striated and polished surfaces that have been buried by ancient strata, and for ancient deposits of till that have hardened into rock. Ancient, lithified till deposits are called **tillites** and consist of larger clasts distributed throughout a matrix of sandstone and mudstone.

By using the stratigraphic principles described in Chapter 12, geologists have determined that the most recent pre-Pleistocene widespread striated and polished surfaces and tillites formed in Permian time, about 280 Ma (**Fig. 22.30b**).

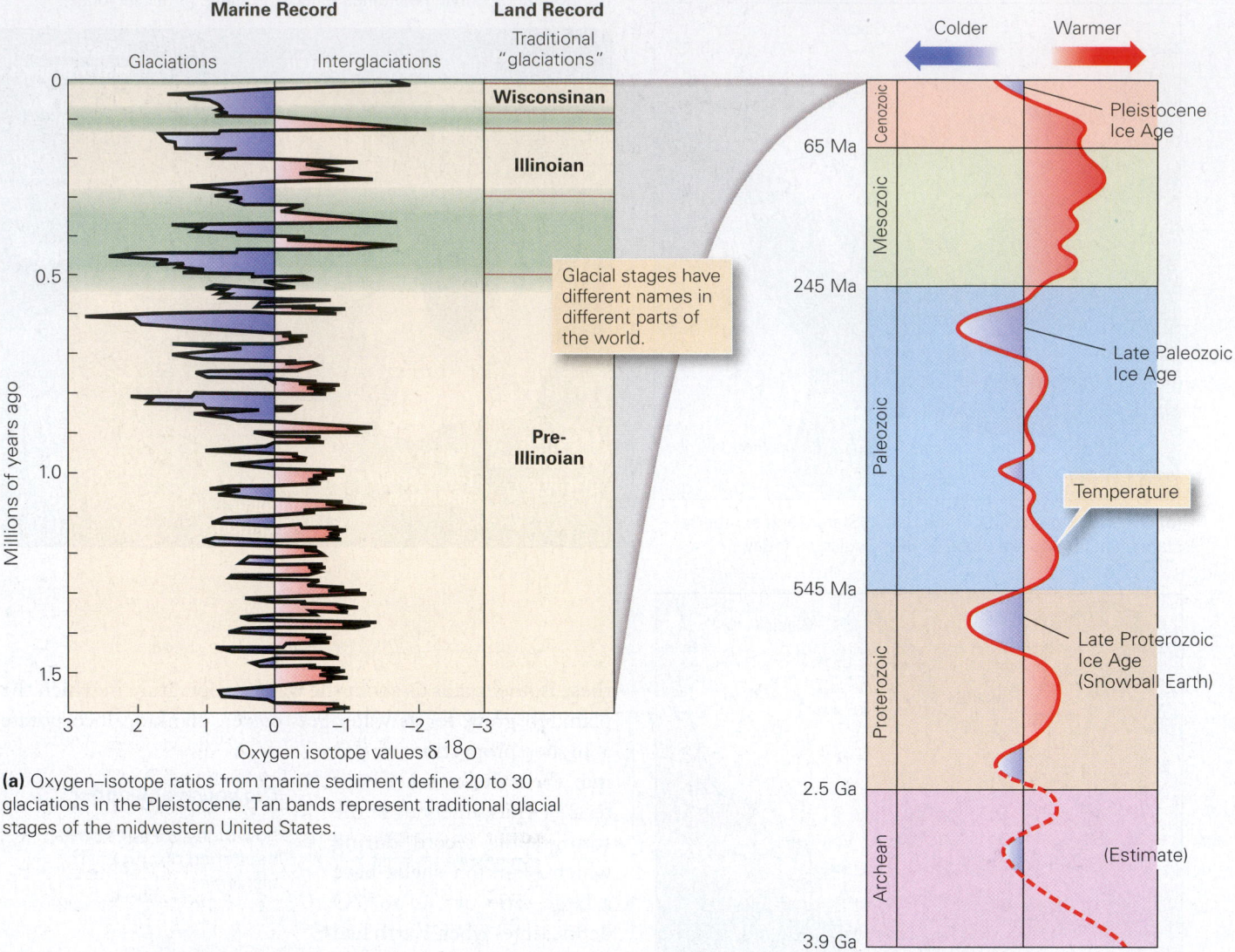

(a) Oxygen–isotope ratios from marine sediment define 20 to 30 glaciations in the Pleistocene. Tan bands represent traditional glacial stages of the midwestern United States.

(b) The Pleistocene is not the only ice-age time in Earth history. Glacial events happened in colder intervals of earlier eras, too.

These are the deposits Alfred Wegener studied when he developed the concept of continental drift, for on a reconstruction of Pangaea, the Permian glaciated areas are adjacent (see Chapter 3). Tillites were also deposited about 600 to 700 Ma (at the end of the Proterozoic Eon), about 2.2 billion years ago (Ga) (near the beginning of the Proterozoic), and perhaps about 2.7 Ga (in the Archean Eon). Strata deposited at other times in Earth history do not contain tillites. Thus, it appears that glacial advances and retreats have not occurred steadily throughout Earth history but rather are restricted to specific time intervals—*ice ages*—of which there were four or five: Pleistocene, Permian, late Proterozoic, early Proterozoic, and perhaps Archean.

Of particular note, some tillites of the late Proterozoic event were deposited at equatorial latitudes, suggesting that for at least a short time the continents worldwide were largely glaciated, and the sea may have been covered worldwide by ice. Geologists refer to the ice-encrusted planet as **snowball Earth** (see Chapter 13).

Take-Home Message

The most recent ice age, responsible for continental glacial landscapes of today, occurred during the past 2.6 Ma. The land record shows four to five discrete glaciations, but the marine record reveals that, in fact, ice sheets advanced and retreated about 20 to 30 times. Ice ages also happened earlier in Earth history. Proterozoic glaciers and sea ice may have covered all of "snowball Earth."

QUICK QUESTION: What's the evidence for multiple Pleistocene glaciations?

22.7 The Causes of Ice Ages

Ice ages occur only during restricted intervals of Earth history, hundreds of millions of years apart. But within an ice age, glaciers advance and retreat with a frequency measured in tens of thousands to hundreds of thousands of years. Thus, there must be both long-term and short-term controls on glaciation.

Long-Term Causes

Plate tectonics exercises long-term control over glaciation for several reasons. First, continental drift due to plate tectonics determines the distribution of continents relative to the equator. If all continents straddled the equator, none could become cold enough to host continental glaciations—ice ages can happen only when substantial continental area lies at high latitudes. Second, the distribution of continents relative to upwelling and downwelling zones of the mantle may influence overall land elevation and thus land-surface temperature. Third, the global volume of mid-ocean ridges, which reflects seafloor-spreading rates, influences global sea level—at times when continents are relatively low and sea level is relatively high, large areas of continents flood and cannot host glaciers. Finally, global climate can be affected by heat redistributed by oceanic currents—growth of island arcs and drift of continents can influence the configuration of currents and determine whether high-latitude regions can become cold enough to host ice sheet formation.

The concentration of carbon dioxide in the atmosphere may also play a key role in determining whether an ice age can or can't occur. As we've noted, carbon dioxide (CO_2) is a greenhouse gas—it traps infrared radiation rising from the Earth—so if the concentration of CO_2 increases, the atmosphere becomes warmer. Ice sheets cannot form during periods when the atmosphere has a relatively high concentration of CO_2, even if other factors favor glaciation. But what might cause long-term changes in CO_2 concentration? Possibilities include changes in the number of marine organisms that extract CO_2 to make shells; changes in the amount of chemical weathering on land, caused by growth of mountain ranges (weathering absorbs CO_2); changes in the amount of volcanic activity; and changes in the distribution and volume of photosynthetic organisms (these organisms remove CO_2 from the air). Of note, the widespread appearance of coal swamps may have triggered Permian glaciations of Pangaea.

Short-Term Causes

Now we've seen how the stage could be set for an ice age to occur, but why do glaciers advance and retreat periodically during an ice age? In 1920, Milutin Milanković, a Serbian astronomer and geophysicist, came up with an explanation. Milanković studied how the Earth's orbit changes shape and how its axis changes orientation through time, and he calculated the frequency of these changes. In particular, he evaluated three aspects of Earth's movement around the Sun.

- *Orbital eccentricity:* The Earth's orbit gradually changes from a more circular shape to a more elliptical shape. The degree to which an orbit deviates from a perfect circle is called the orbital eccentricity. Earth's eccentricity cycle takes around 100,000 years (**Fig. 22.31a**).
- *Tilt of Earth's axis:* We have seasons because the Earth's axis is not perpendicular to the plane of its orbit. Over time, the tilt angle varies between 22.5° and 24.5°, with a frequency of 41,000 years (**Fig. 22.31b**).
- *Precession of Earth's axis:* If you've ever set a top spinning, you've probably noticed that its axis gradually traces a conical path. This motion, or wobble, is called *precession* (**Fig. 22.31c**). The Earth's axis wobbles over the course of about 23,000 years. Right now, the Earth's axis points toward Polaris, making Polaris the North Star, but 12,000 years ago the axis pointed to Vega, a different star. Precession determines the relationship between the timing of the seasons and the position of Earth along its orbit around the Sun.

Milanković showed that precession, along with variations in orbital eccentricity and tilt, combine to affect the total annual amount of *insolation* (exposure to the Sun's rays) and the seasonal distribution of insolation that the Earth receives at the mid- to high-latitudes (such as 65° N) by as much as 25%. For example, such regions receive more insolation when the Earth's axis is almost perpendicular to its orbital plane than when its axis is greatly tilted. According to Milanković, glaciers tend to advance during times of cool summers at 65° N, which occur periodically (**Fig. 22.31d**). When geologists began to study the climate record, they found climate cycles with the frequency predicted by Milanković. These climate cycles, controlled by "orbital forcing," are now called **Milankovitch cycles**.

The discovery of Milankovitch cycles in the geologic record strongly supports the contention that changes in the Earth's orbit and tilt help trigger short-term advances and retreats during an ice age. But orbit and tilt changes cannot be the whole story because they could cause only about a 4°C temperature decrease (relative to today's temperature), and during glaciations the temperature decreased 5° to 7°C along coasts and 10° to 13°C inland. Geologists suggest that several other factors may come into play in order to trigger a glacial advance.

- *A changing albedo:* When snow remains on land throughout the year, or clouds form in the sky, the albedo (reflectivity) of the Earth increases, so Earth's surface reflects incoming sunlight and thus becomes even cooler.

FIGURE 22.31 Milankovitch cycles influence the amount of insolation received at high latitudes.

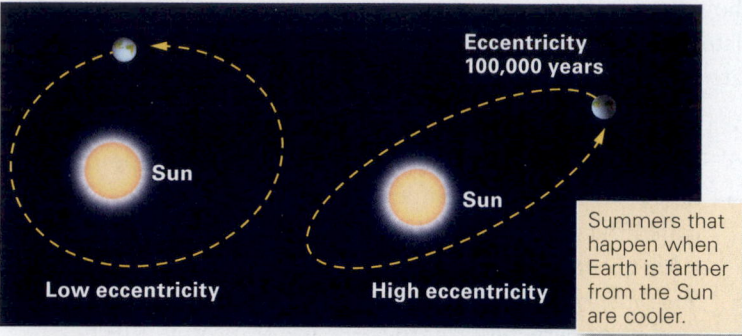

(a) Variations caused by changes in orbital shape.

Eccentricity 100,000 years

Sun
Low eccentricity

Sun
High eccentricity

Summers that happen when Earth is farther from the Sun are cooler.

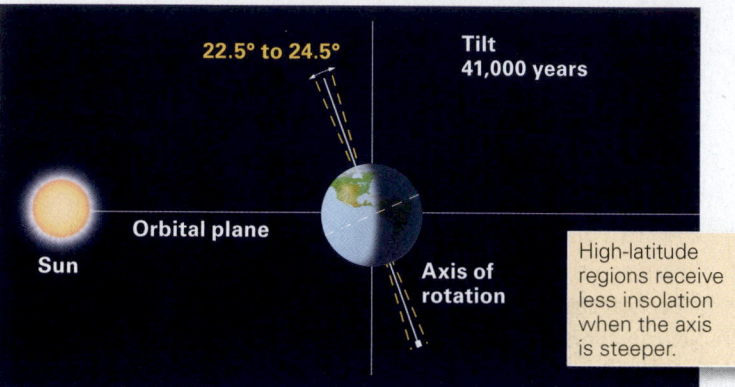

(b) Variations caused by changes in axis tilt.

22.5° to 24.5°
Tilt 41,000 years
Orbital plane
Sun
Axis of rotation

High-latitude regions receive less insolation when the axis is steeper.

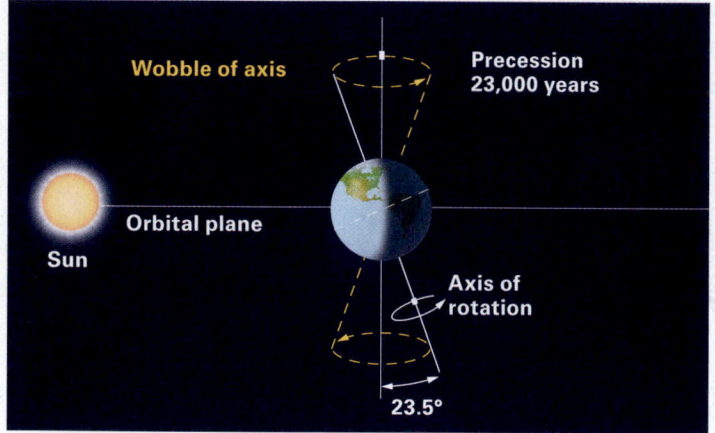

(c) Variations caused by the precession of Earth's axis.

Wobble of axis
Precession 23,000 years
Orbital plane
Sun
Axis of rotation
23.5°

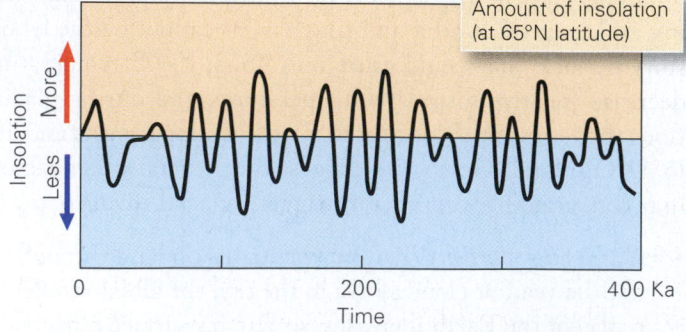

(d) Combining the effects of eccentricity, tilt, and precession produces distinct periods of more or less insolation.

Amount of insolation (at 65°N latitude)

Insolation — More / Less

0   200   400 Ka
Time

- *Interrupting the global heat conveyor:* As the climate cools, evaporation rates from the sea decrease, so seawater does not become as salty. Decreasing salinity might stop the system of thermohaline currents that brings warm water to high latitudes (see Chapter 18). Thus, the high latitudes become even colder than they would otherwise.

- *Biological processes that change CO_2 concentration:* Several kinds of biological processes may have amplified climate changes by altering the concentration of CO_2 in the atmosphere. For example, a greater amount of plankton growing in the oceans could absorb more CO_2 and thus remove it from the atmosphere. A bloom of plankton might reflect an addition of nutrients to the oceans, perhaps because of changing patterns of upwelling or changing amounts of runoff.

The above processes represent *positive feedback* in that they enhance the effects of the phenomenon that causes them. Because of positive feedback, the Earth could cool more than it would otherwise during the cooler stage of a Milankovitch cycle, and this could trigger a glacial advance. Variations in the radiation output of the Sun could also affect the amount of energy the Earth receives, but the long-term periodicity of such variability remains unclear, so the effect is hard to evaluate.

A Model for Pleistocene Ice Age History

Long-Term Cooling in the Cenozoic Era Taking all of the above causes into account, we can now propose a scenario for the events that led to the Pleistocene glacial advances. Our story begins in the Eocene Epoch, about 55 Ma (**Fig. 22.32**). At that time, climates were warm and balmy not only in the tropics but even above the Arctic Circle. At the end of the Middle Eocene (37 Ma), the climate began to cool, and by Early Oligocene time (33 Ma), Antarctica became glaciated. The Antarctic ice sheet came and went until the middle of the Miocene Epoch (15 Ma), when an ice sheet formed that has lasted ever since. Ice sheets did not appear in the Arctic, however, until 2 to 3 Ma, when the Pleistocene Ice Age began.

Cenozoic long-term climate changes may have been caused, in part, by changes in the pattern of oceanic currents that happened, in turn, because of plate tectonics. For example, in the Eocene, the collision of India with Asia cut off warm equatorial currents that had been flowing in the Tethys Sea. And in the Miocene and Oligocene, Australia and South America drifted away from Antarctica, allowing the cold circum-Antarctic current to develop. This new current prevented warm southward-flowing currents from reaching Antarctica, allowing ice to form and survive in the region. Without the warm currents, the climate of Antarctica overall underwent cooling,

FIGURE 22.32 Until recently, Earth's atmosphere has been gradually cooling, overall, since the Cretaceous.

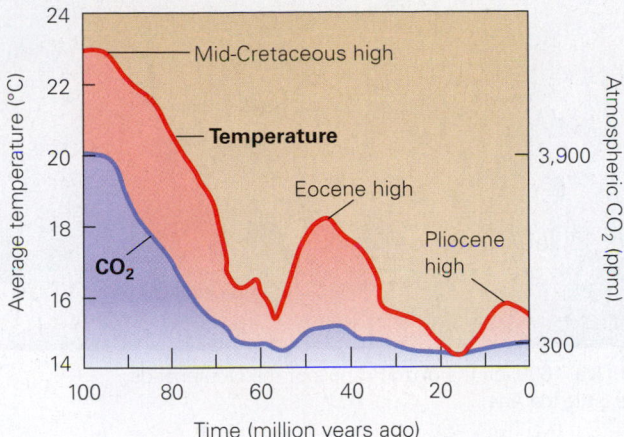

and this could have cooled the global ocean. Changes to atmospheric circulation and temperature may also have happened at this time. Models suggest that the uplift of the Himalayas and Tibet diverted winds in a way that cooled the climate. Further, this uplift exposed more rock to chemical weathering, perhaps leading to extraction of CO_2 from the atmosphere (as noted, chemical weathering reactions absorb CO_2). A decrease in the concentration of this greenhouse gas would contribute to atmospheric cooling.

So far, we've examined hypotheses that explain long-term cooling since about 40 Ma, but what caused the sudden appearance of the Laurentide ice sheet about 2.6 Ma? This event may coincide with other plate-tectonic events. For example, the gap between North and South America closed when the Isthmus of Panama grew and separated the waters of the Caribbean from those of the tropical Pacific for the first time. When this happened, warm currents that previously flowed out of the Caribbean into the Pacific were blocked and diverted northward to merge with the Gulf Stream. This current transfers warm water from the Caribbean up the Atlantic Coast of North America and ultimately to the British Isles. As the warm water moves up the Atlantic Coast, it generates warm, moisture-laden air that provides a source for the snow that falls over New England, eastern Canada, and Greenland. In other words, the Arctic has long been cold enough for ice caps, but until the Gulf Stream was diverted northward by the growth of Panama, there was no source of moisture to make abundant snow and ice needed for glacial growth.

Short-term Advances and Retreats in the Pleistocene Epoch

Once the Earth's climate had cooled overall, short-term processes such as the Milankovitch cycles led to periodic advances and retreats of the glaciers. To understand how, let's look at a possible case history of a single advance and

retreat of the Laurentide ice sheet. (Such models remain the subject of vigorous debate.)

- *Stage 1:* During the overall cooler climates of the late Cenozoic Era, the Earth reaches a point in the Milankovitch cycle when the average mean temperature in temperate latitudes drops. Because of glacial rebound, the ice-free surface of northern Canada has risen to an altitude of several hundred meters above sea level. With lower temperatures and higher elevations, not all of winter's snow melts away during the summer. Eventually, snow covers the entire region of northern Canada, even during the summer. Because of the snow's high albedo, it reflects sunlight, so the region grows still colder (a positive feedback) and even more snow accumulates. Precipitation rates are high, because evaporation off the Gulf Stream provides moisture. Finally, the snow at the base of the pile turns to ice, and the ice begins to spread outward under its own weight. A new continental glacier has been born.

- *Stage 2:* The ice sheet continues to grow as more snow piles up in the zone of accumulation. And as the ice sheet grows, the atmosphere continues to cool because of the albedo effect. But now the weight of the ice loads the continent and makes it sink, so the elevation of the glacier decreases and its surface approaches the equilibrium line. Also, the temperature becomes cold enough that in high latitudes the Atlantic Ocean begins to freeze. As the sea ice covers the ocean, the amount of evaporation decreases, so the source of snow is cut off and the amount of snowfall diminishes. The glacial advance basically chokes on its own success. The decrease in the glacier's elevation (leading to warmer summer temperatures) on the ice surface, as well as the decrease in snowfall, causes ablation to occur faster than accumulation, and the glacier begins to retreat.

- *Stage 3:* As the glacier retreats, temperatures gradually increase and the sea ice begins to melt. The supply of water to the atmosphere from evaporation increases once again, but with the warmer temperatures and lower elevations, this water precipitates as rain during the summer. The rain drastically accelerates the rate of ice melting, and the retreat progresses quite rapidly.

Will There Be Another Glacial Advance?

What does the future hold? Considering the periodicity of glacial advances and retreats during the Pleistocene Epoch, we may be living in an interglacial period. Pleistocene interglacials lasted about 10,000 years, and since the present interglacial began about 12,000 years ago, the time seems ripe for

FIGURE 22.33 The Little Ice Age and its demise. Glaciers that advanced between 1550 and 1850 have since retreated.

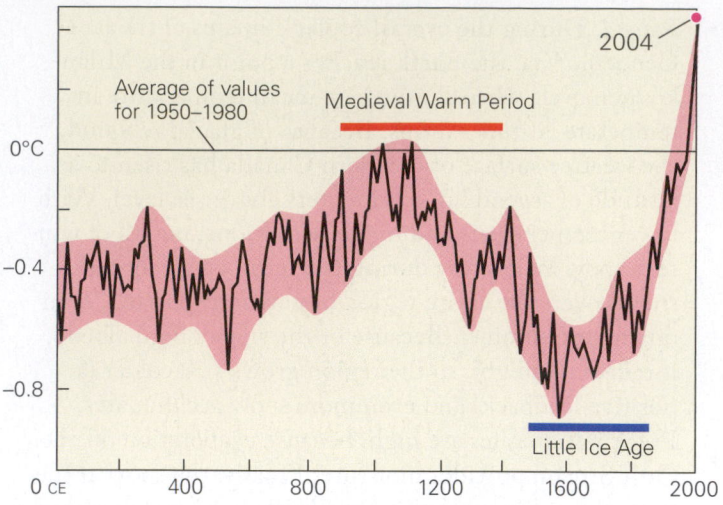

(a) A model of global temperature for the past 2,000 years. Overall trends display the Medieval Warm Period followed by the Little Ice Age. Since 1850, temperatures have warmed.

(b) Skaters (ca. 1600) on the frozen canals of the Netherlands during the Little Ice Age.

(c) During the Little Ice Age, a glacier filled this valley. In this 2003 photo, most of the glacier has vanished. Most of the retreat has happened in the last century.

a new glaciation. If a glacier on the scale of the Laurentide ice sheet were to develop, major cities and agricultural belts would be overrun by ice, and their populations would have to migrate southward. Long before the ice front arrived, though, the climate would become so hostile that northern cities would already be abandoned.

The Earth actually had a brush with ice-age conditions between the 1300s and the mid-1800s, when average annual temperatures in the northern hemisphere fell sufficiently for mountain glaciers to advance significantly. During this period, now known as the **Little Ice Age**, sea ice surrounded Iceland and canals froze in the Netherlands, leading to that country's tradition of skating (**Fig. 22.33a, b**). Some researchers speculate that the depopulation of the western hemisphere, in the wake of European conquest, which brought devastating epidemics, caused temporary reforestation, for without inhabitants, farmlands went untended and new forests, which absorbed CO_2, grew. This process caused atmospheric concentrations of CO_2 to decrease, leading to the cooler conditions that triggered the Little Ice Age. Others speculate that the change reflects increased cloud cover, not a change in CO_2 concentration. Researchers will likely propose additional ideas as work on this problem continues.

During the past 150 years, temperatures have warmed, and most mountain glaciers have retreated significantly (**Fig. 22.33c**). Large slabs have been calving off Antarctic ice shelves. In fact, the Larsen B Ice Shelf of Antarctica, an area larger than Rhode Island, disintegrated in 2002 over a period of only one month. Greenland's glaciers, in particular, are showing signs of accelerating retreat (**Fig. 22.34**). Large meltwater

ponds are forming on the surface of the ice sheet, and some of these drain abruptly through cracks to the base of the glacier a kilometer below. The vast majority of researchers suggest that this global-warming trend is due to increased CO_2 in the atmosphere from the burning of fossil fuels (see Chapter 23). Global warming could conceivably cause a "super-interglacial," meaning that the next glaciation could be substantially delayed or might never happen.

FIGURE 22.34 Greenland's melting glaciers. Melting has accelerated in the last few decades.

(a) Lakes of meltwater accumulate on the surface of the ice sheet during the summer.

(b) Lakes suddenly drain through cracks that carry the water to the base of the glacier, 1 km down. Addition of liquid water to the base allows the glacier to move faster, causing a "surge."

(c) Where glaciers meet the sea, huge masses calve off and crash into the water. This is happening so fast that the ice front is retreating.

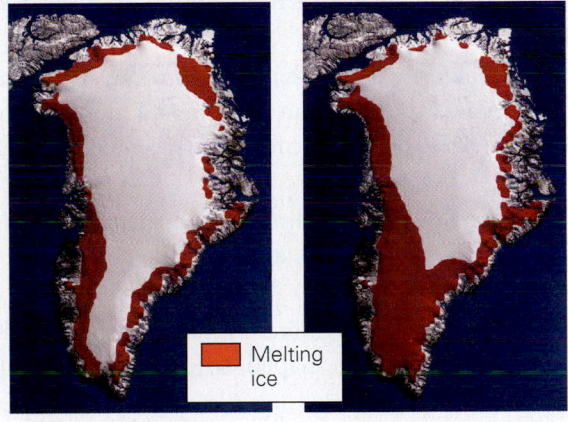

Melting ice

(d) The melting area has increased dramatically in recent years.

Current trends of global warming have led to concern that the ice sheet of West Antarctica might begin to float and then break up rapidly. If all of today's ice caps melted, global sea level would rise by 70 m (230 ft), extensive areas of coastal plains would be flooded, and major coastal cities such as New York, Miami, and London would be submerged (see Fig. 22.23). Instead of protruding from ice, the tip of the Empire State Building would protrude from the sea. Icehouse or greenhouse? We may not know which scenario will play out in the future until it happens. However, researchers have voiced concern that, at least in the near term, glacial melting will be the order of the day, as global temperatures seem to be rising. The next chapter addresses such change.

Take-Home Message

Ice ages occur when the distribution of continents, ocean currents, and the concentration of atmospheric CO_2 are appropriate. Advances and retreats during an ice age are controlled by Milankovitch cycles that take into account variations in Earth's orbit and rotation axis. Because of global warming, we may be living in a super-interglacial period.

QUICK QUESTION: How does positive feedback contribute to a glaciation during an ice age, and why does the glaciation eventually cease?

CHAPTER SUMMARY

- Glaciers are streams or sheets of recrystallized ice that survive for the entire year and flow in response to gravity. Mountain glaciers exist in high regions and fill cirques and valleys. Continental glaciers (ice sheets) spread over substantial areas of the continents.

- Glaciers form when snow accumulates over a long period of time. With progressive burial, the snow first turns to firn and then to ice.

- Ice in temperate glaciers melts during at least part of the year. Polar glaciers are frozen solid. Glaciers move by basal sliding over water or wet sediment, and/or by plastic deformation of ice grains. In general, glaciers move tens of meters per year.

- Glaciers move because of gravitational pull; they flow in the direction of their surface slope.

- Whether the toe of a glacier stays fixed in position, advances farther from the glacier's origin, or retreats back toward the origin depends on the balance between the rate at which snow builds up in the zone of accumulation and the rate at which glaciers melt or sublimate in the zone of ablation.

- Icebergs break off glaciers that flow into the sea. Continental glaciers that flow out into the sea along a coast make ice shelves. Sea ice forms where the ocean's surface freezes.

- As glacial ice flows over sediment, it incorporates clasts. The clasts embedded in glacial ice act like a rasp that abrades the substrate.

- Mountain glaciers carve numerous landforms, including cirques, arêtes, horns, U-shaped valleys, hanging valleys, and truncated spurs. Fjords are glacially carved valleys that filled with water.

- Glaciers can transport sediment of all sizes. Glacial drift includes till, glacial marine, glacial outwash, lake-bed mud, and loess. Lateral moraines accumulate along the sides of valley glaciers, and medial moraines form down the middle of two glaciers. End moraines accumulate at a glacier's toe.

- Glacial depositional landforms include moraines, knob-and-kettle topography, drumlins, kames, eskers, meltwater lakes, and outwash plains.

- Continental crust subsides as a result of ice loading. When the glacier melts away, the crust rebounds.

- When water is stored in continental glaciers, sea level drops. When glaciers melt, sea level rises.

- During past ice ages, the climate in regions south of the continental glaciers was wetter, and pluvial lakes formed. Permafrost (permanently frozen ground) exists in periglacial environments.

- During the Pleistocene Ice Age, large continental glaciers covered much of North America, Europe, and Asia.

- The stratigraphy of Pleistocene glacial deposits preserved on land records five European and four North American glaciations, times during which ice sheets advanced. The record preserved in marine sediments records 20 to 30 such events. The land record, therefore, is incomplete.

- Long-term causes of ice ages include plate tectonics and changes in the concentration of CO_2 in the atmosphere. Short-term causes include the Milankovitch cycles (caused by periodic changes in Earth's orbit and tilt).

GUIDE TERMS

ablation (p. 805)
albedo (p. 797)
arête (p. 810)
basal sliding (p. 800)
cirque (p. 799)
continental glacier (p. 799)
continental ice sheet (p. 796)
crevasse (p. 801)
drop stone (p. 808)

drumlin (p. 817)
equilibrium line (p. 805)
end moraine (p. 813)
erratic (p. 796)
esker (p. 817)
firn (p. 799)
fjord (p. 812)
glacial abrasion (p. 808)
glacial advance (p. 805)

glacial drift (p. 814)
glacially polished surface (p. 809)
glacial outwash (p. 815)
glacial outwash plain (p. 818)
glacial retreat (p. 805)
glacial striation (p. 809)
glacial subsidence (p. 819)
glacial till (p. 815)

glaciation (p. 828)
glacier (p. 796)
hanging valley (p. 810)
horn (p. 810)
ice age (p. 796)
iceberg (p. 806)
ice quake (p. 805)
ice shelf (p. 806)
interglacial (p. 828)

lateral moraine (p. 813)

Little Ice Age (p. 834)

loess (p. 815)

medial moraine (p. 813)

Milankovitch cycle (p. 831)

mountain (alpine) glacier (p. 799)

patterned ground (p. 825)

permafrost (p. 825)

plastic deformation (p. 799)

Pleistocene Ice Age (p. 797)

pluvial lake (p. 824)

polar glacier (p. 799)

post-glacial rebound (p. 819)

recessional moraine (p. 815)

roche moutonnée (p. 812)

sea ice (p. 808)

snowball Earth (p. 830)

sublimation (p. 799)

surge (p. 805)

temperate glacier (p. 799)

terminal moraine (p. 815)

tidewater glacier (p. 806)

tillite (p. 829)

toe (p. 805)

tundra (p. 826)

U-shaped valley (p. 810)

varve (p. 815)

zone of ablation (p. 805)

zone of accumulation (p. 805)

REVIEW QUESTIONS

1. What evidence did Louis Agassiz offer to support the idea of an ice age?

2. How do mountain glaciers and continental glaciers differ in terms of dimensions, thickness, and patterns of movement?

3. Describe the transformation from snow to glacial ice.

4. Explain how arêtes, cirques, and horns form.

5. Describe the mechanisms that enable glaciers to move, and explain why they move.

6. How fast do glaciers normally move? How fast can they move during a surge?

7. Explain how the balance between ablation and accumulation determines whether a glacier advances or retreats.

8. How can a glacier continue to flow toward its toe even though its toe is retreating?

9. How does a glacier transform a V-shaped river valley into a U-shaped valley? Discuss how hanging valleys develop.

10. Describe the various kinds of glacial deposits. Be sure to note the materials from which the deposits are made and the landforms that result from deposition.

11. How do the crust and mantle respond to the weight of glacial ice?

12. How was the world different during the glacial advances of the Pleistocene Ice Age? Be sure to mention the relation between glaciations and sea level.

13. How was the standard four-stage chronology of North American glaciations developed? Why is it so incomplete? How was it modified with the study of marine sediment?

14. Were there ice ages before the Pleistocene? If so, when?

15. What are some of the long-term causes that lead to ice ages? What are the short-term causes that trigger glaciations and interglacials?

ON FURTHER THOUGHT

16. If you fly over the barren cornfields of central Illinois during the early spring, you will see slight differences in soil color due to variations in moisture content—wetter soil is darker. These variations outline the shapes of polygons that are tens of meters across. What do these patterns represent, and how might they have formed? What do they tell us about the climate of central Illinois at the end of the last ice age?

17. An unusual late Precambrian rock unit crops out in the Flinders Range, a small mountain belt in South Australia near Adelaide. Structures in the belt formed at the beginning of the Paleozoic. This unit consists of clasts of granite and gneiss, in a wide range of sizes, suspended through a matrix of slate. What is this unusual rock?

smartwork   smartwork.wwnorton.com

This chapter's Smartwork features:

- Labeling exercise on identifying mountain glaciers.
- Interactive ranking activity on the transformation from ice to snow.
- Art-based, interactive exercises on glacial movements.

GEOTOURS

This chapter's GeoTour exercise (R) features:

- Continental glacier features in the northeastern U.S.
- Alpine-valley glacial features around the world
- Piedmont glacial features in Alaska

Rice paddies and villages cover the countryside near Shanghai, China. The landscape here would have looked vastly different before the arrival of humanity. Land-use change affects many aspects of the Earth System.

CHAPTER 23

Global Change in the Earth System

*All we in one long caravan
are journeying since the world began,
we know not whither, we know . . . all must go.*

—Bhartrihari (Indian poet, ca. 500 C.E.)

23.1 Introduction

Did the Earth's surface look the same in the Jurassic Period as it does today? Definitely not! Two hundred million years ago, the North Atlantic Ocean was just a narrow sea, and the South Atlantic Ocean didn't exist at all, so most dry land connected to form a single vast supercontinent (**Fig. 23.1**). Today the Atlantic is a wide ocean and the Earth has seven separate continents. Moreover, during the Jurassic, the call of the wild rumbled from the throats of dinosaurs, whereas today the largest land animals are mammals. In essence, what we see of the Earth today is just a snapshot, an instant in the life story of a constantly changing planet. This idea arguably stands as geology's greatest philosophical contribution to humanity's understanding of our Universe.

Why has the Earth changed so much over geologic time, and why does it continue to change? Ultimately, change happens because the Earth's internal heat keeps the asthenosphere weak enough to flow and because the Sun's heat keeps most of the Earth's surface at temperatures above the freezing point of water. Flow in the asthenosphere permits plate tectonics, which leads to continental drift, volcanism, and mountain building. Solar heat, together with gravity, keeps streams, glaciers, waves, and wind in motion, thereby causing erosion and deposition. Solar heat also makes the Earth's surface and near-surface regions hospitable to life. If the Earth did not have just the right mix of tectonic activity and solar heat, it would be a frozen dust bowl like Mars, a crater-pocked wasteland like the Moon, or a cloud-choked oven like Venus.

Many of the changes that take place on Earth reflect complex interactions among geologic and biological phenomena. For example, photosynthetic organisms affect the composition of the atmosphere by providing oxygen, and atmospheric composition, in turn, determines the nature of chemical weathering reactions that take place in rocks. We've referred to the global interconnecting web of physical and biological phenomena on Earth as the **Earth System** (see **Geology at a Glance**, pp. 840–841). We can now define **global change**, in a general sense, as transformations or modifications of physical and biological components in the Earth System through time.

FIGURE 23.1 The map of Earth's surface changes over time because of plate motions.

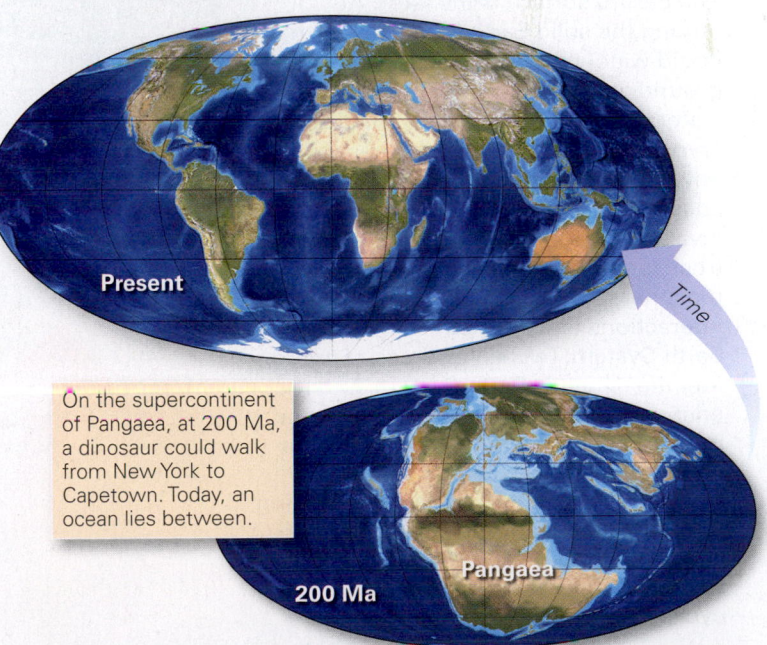

On the supercontinent of Pangaea, at 200 Ma, a dinosaur could walk from New York to Capetown. Today, an ocean lies between.

The Earth System

External energy

Sun

Thunderhead

Lightning

Rain and snow

Mountain uplift

Continental glacier

City

Ocean

Rocky coastline

Desert

Valley

Arid mountains

Mining

Lakes

Field pattern

Deciduous forest

Forested mountains

Beach

The Earth's surface is the interface among the solid Earth (lithosphere); the liquid water of oceans, lakes, streams, groundwater (the hydrosphere); the solid water of glaciers and permafrost (the cryosphere); and the planet's gaseous envelope (the atmosphere). Countless species of life, ranging from nearly invisible bacteria to giant whales and trees, populate complex ecosystems (the biosphere). These components, and the interactions among them, constitute the Earth System. Two key sources of energy fuel the dynamic Earth System. External energy comes from solar radiation, and internal energy comes from the Earth's interior and drives tectonic processes.

Various materials cycle among living and nonliving components of the Earth System during the hydrologic cycle, the rock cycle, and various biogeochemical cycles (such as the carbon cycle).

Tropical rain forest

Coral reef

Shark

Internal energy

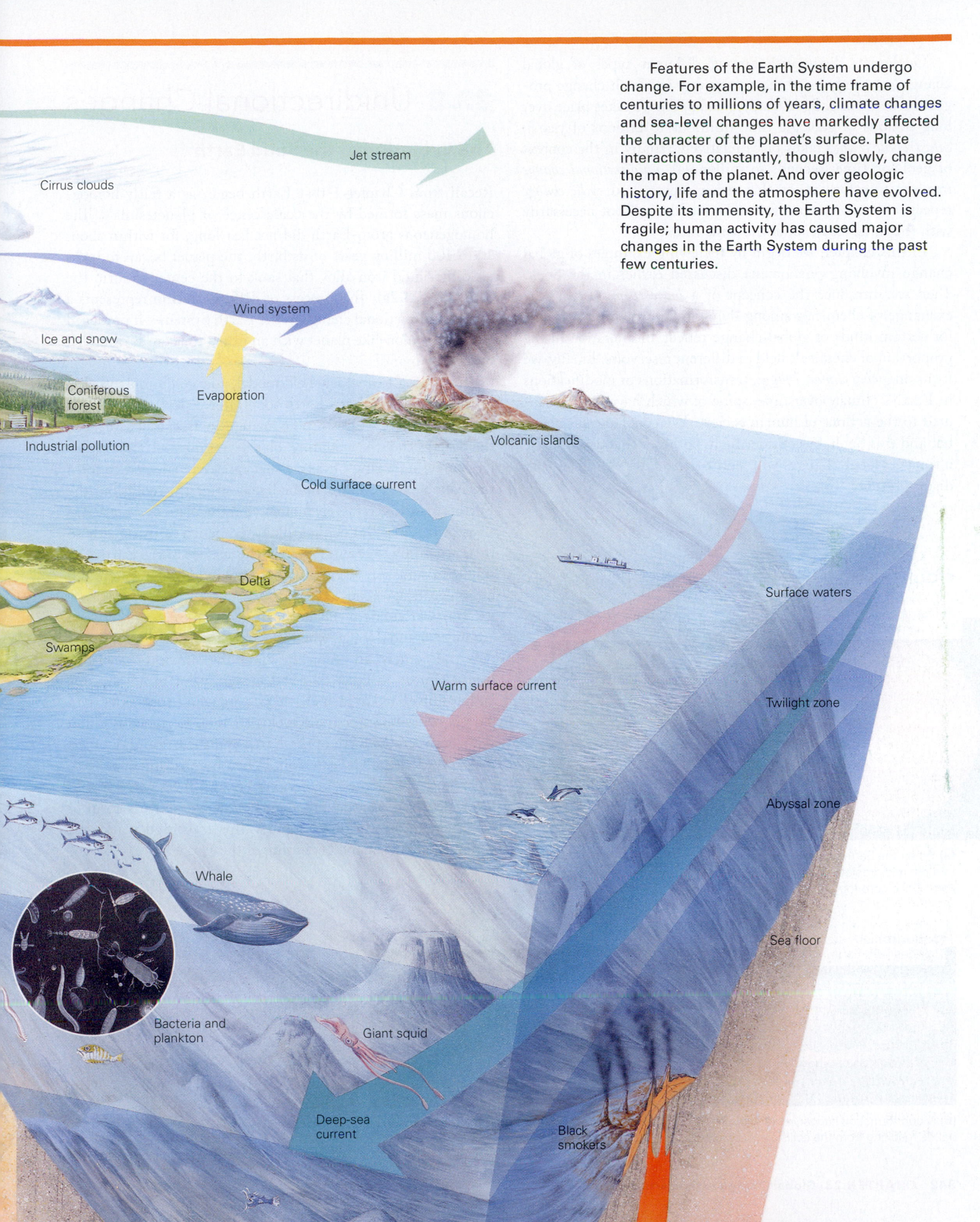

Features of the Earth System undergo change. For example, in the time frame of centuries to millions of years, climate changes and sea-level changes have markedly affected the character of the planet's surface. Plate interactions constantly, though slowly, change the map of the planet. And over geologic history, life and the atmosphere have evolved. Despite its immensity, the Earth System is fragile; human activity has caused major changes in the Earth System during the past few centuries.

Jet stream

Cirrus clouds

Wind system

Ice and snow

Coniferous forest

Evaporation

Industrial pollution

Volcanic islands

Cold surface current

Delta

Surface waters

Swamps

Warm surface current

Twilight zone

Abyssal zone

Whale

Sea floor

Bacteria and plankton

Giant squid

Deep-sea current

Black smokers

Geologists distinguish among different types of global change, on the basis of the rate or way in which change progresses with time. Specifically, *gradual change* takes place over long periods of geologic time (millions to billions of years); *catastrophic change* takes place relatively rapidly in the context of geologic time (seconds to millennia); *unidirectional change* involves transformations that never repeat; and *cyclic change* repeats the same steps over and over, though not necessarily with the same results or at the same rate.

In this chapter, we begin by reviewing examples of global change involving phenomena discussed earlier in the book. Then we introduce the concept of a *biogeochemical cycle*, the exchange of chemicals among living and nonliving reservoirs, for certain kinds of global change reflect an alteration in the proportion of chemicals held in different reservoirs. Finally, we focus on *global climate change*, transformations or modifications in Earth's climate over time, some of which have been attributed to the actions of human society. We conclude this chapter, and this book, by considering hypotheses that describe the ultimate global change—the end of the Earth—in the very distant future.

23.2 Unidirectional Changes

The Evolution of the Solid Earth

Recall from Chapter 1 that Earth began as a fairly homogenous mass formed by the coalescence of planetesimals. The homogeneous proto-Earth did not last long, for within about 10 to 100 million years of its birth, the planet began to melt, yielding liquid iron alloy that sank to the center to form the core (**Fig. 23.2a**). This process of **differentiation** represents a major unidirectional change in the Earth System—it produced a layered, onion-like planet with an iron alloy core surrounded by a rocky mantle.

According to a widely held model, a large protoplanet collided with the newborn Earth soon after differentiation. This collision caused a catastrophic change in that much of both the Earth and the colliding object fragmented and vaporized, creating a ring of debris that quickly coalesced to form the Moon (**Fig. 23.2b**). Immediately after this collision, the Earth's mantle was largely molten, and the planet's surface was a sea of magma. But cooling likely happened fairly quickly so that, according to recent research, the surface of the Earth had solidified and may even have hosted liquid water before 4.0 billion years ago (Ga).

The Earth underwent another catastrophic change between around 4.0 and 3.9 Ga, when it endured pummeling by asteroids and comets, an event known as the *late heavy bombardment*. Almost all crust that had formed prior to 3.9 Ga was largely pulverized or melted—in fact, no whole rock older than 4.0 Ga has yet been found. Geologists refer to the half-billion years between the birth of the Earth and the beginning of the rock record as the Hadean Eon. When bombardment slowed, our planet changed again, cooling enough for a new crust to form, new seas to accumulate, and probably an early form of plate tectonics to begin operating. This transition marks the end of the Hadean and beginning of the Archean.

FIGURE 23.2 Examples of major unidirectional change in Earth history.

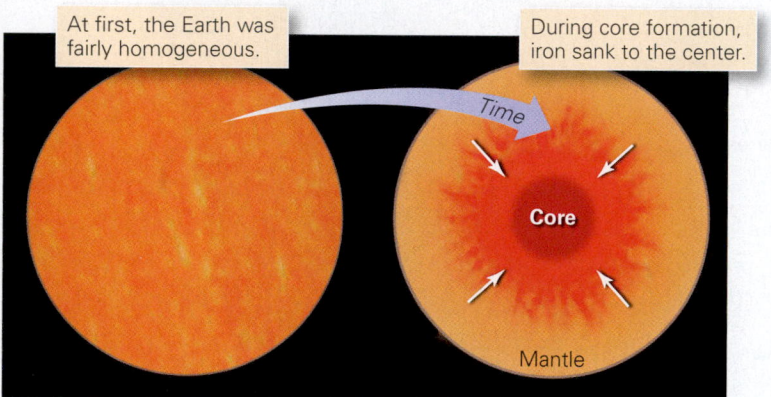

(a) When the Earth became hot enough inside, rock comprising it started to melt, and droplets of iron sank to the center, accumulating to form a core. Once core formation happened, it could not happen again.

(b) Moon formation happened early during Earth history but probably after core formation. According to a popular theory, the Moon coalesced from debris resulting from the collision of a Mars-sized protoplanet with the Earth.

During the Archean, another type of unidirectional change then began when partial melting, perhaps associated with subduction and/or the upwelling of mantle plumes, started to produce relatively low-density rocks, such as granite. Low-density rocks could not be subducted and thus remained at or near the surface to constitute relatively buoyant blocks of crust. Eventually, these blocks sutured together to form the first continents. Effectively, partial melting "distilled" the crust out of the mantle. The amount of continental crust progressively increased as the process continued until by the end of the Archean, continental crust covered about 25% of the Earth's surface. Subsequently, continental area has continued to grow but at a slower pace—today continental crust underlies about 30% of the surface.

Overall, therefore, the transition from the Hadean Eon to the Archean Eon saw a remarkable unidirectional change in the nature of the Earth. By early Archean time, our planet had distinct continents and ocean basins, and thus looked radically different from the way it did when it was first formed (see Chapter 13).

The Evolution of the Atmosphere and Oceans

The Earth's atmosphere has also undergone major unidirectional change over time. Changes in atmospheric composition are so profound that researchers distinguish among first, second, and third atmospheres. As discussed in Chapter 20, the "first atmosphere" consisted of gases from the protoplanetary disk that had been trapped by our planet's gravity. These gases eventually escaped and were replaced by gases belched from volcanoes, perhaps mixed with gases brought to the Earth by comets, and the Earth accumulated a "second atmosphere" composed dominantly of carbon dioxide (CO_2) and water (H_2O). Other gases, such as nitrogen (N_2), composed only a minor proportion of the second atmosphere. When the Earth's surface cooled, however, water condensed and fell as rain, collecting in low areas to form oceans. Gradually, CO_2 dissolved in the oceans and was absorbed by chemical-weathering reactions on land, so its concentration in the atmosphere decreased. N_2, which doesn't react with other chemicals, was left behind. Thus, the atmosphere's composition changed to become the "third atmosphere," dominated by N_2. Photosynthetic organisms appeared early in the Archean. But it probably wasn't until between 2.5 and 2.0 Ga, the early Proterozoic, that oxygen (O_2) became a significant proportion of the atmosphere, and it didn't reach breathable concentrations for another billion years.

> **Did you ever wonder . . .**
>
> whether the Earth's atmosphere has always been breathable?

The Evolution of Life

During its earliest stages, Earth's surface was probably lifeless, for carbon-based organisms could not survive the high temperatures of the time. The fossil record indicates that life had appeared at least by 3.8 Ga and has undergone evolution, a unidirectional change, in fits and starts ever since (see Interlude E). Though simple organisms such as archaea and bacteria still exist, life evolution during the late Proterozoic and early Phanerozoic yielded multicellular plants and animals (**Fig. 23.3**). Life now

FIGURE 23.3 New species of life have evolved over geologic time. Though some of the simplest still exist, more complex organisms have appeared more recently.

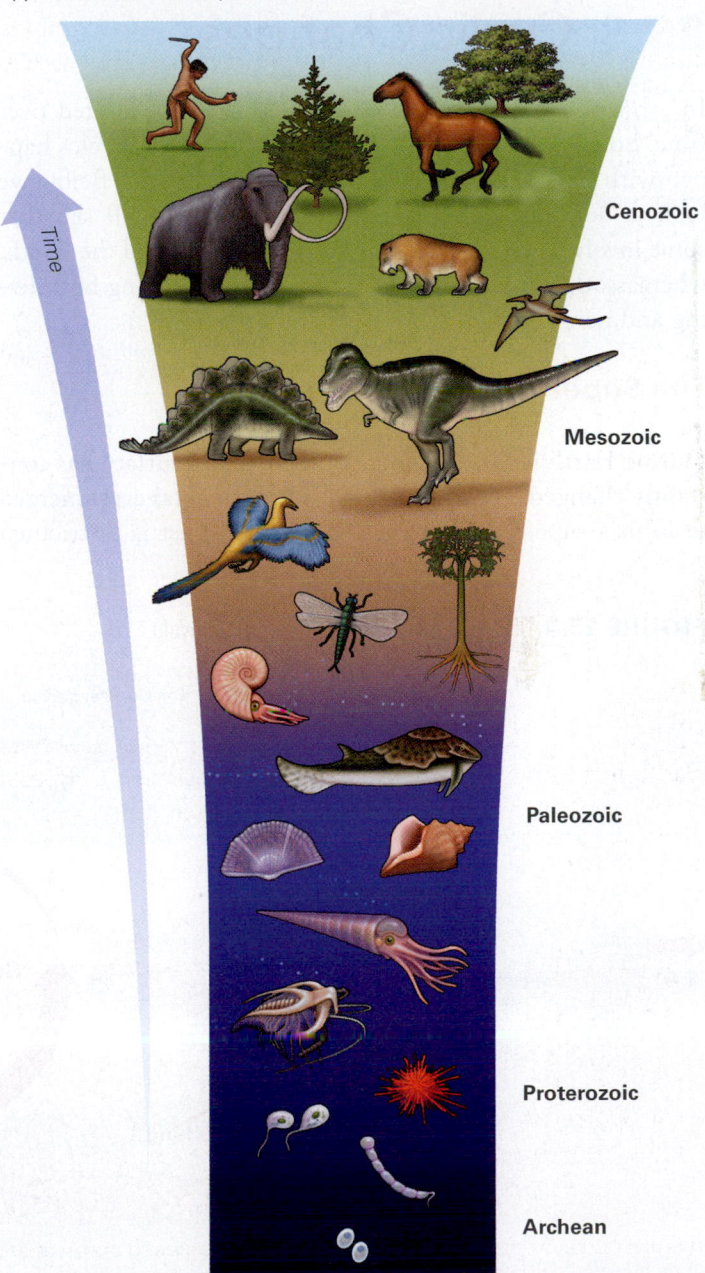

inhabits regions from a few kilometers below the surface to a few kilometers above, yielding a diverse and complex biosphere.

Take-Home Message

Since it first formed, the Earth has undergone major unidirectional changes, which will never repeat. Examples include internal differentiation (core formation), Moon formation, growth of continental crust, atmospheric evolution, ocean formation, and life evolution.

QUICK QUESTION: How are life evolution and atmosphere evolution linked?

23.3 Cyclic Changes

In *cyclic changes*, a sequence of stages may be repeated over time. Some cyclic changes are periodic in that the cycles happen with a definable frequency, but others are not. Below we look at several examples of cyclic change—you'll see that some involve movements of physical components of the Earth, whereas others involve transfer of chemicals among both living and nonliving reservoirs.

The Supercontinent Cycle

During Earth history, the map of the planet's surface has constantly changed. At times, almost all continental crust merged to form a supercontinent, but usually the crust is distributed among several smaller continents. The process of change during which a supercontinent forms and later breaks apart is called the **supercontinent cycle** (**Fig. 23.4**). Geologists have found evidence that supercontinents existed at least three or four times during the past 3 billion years of Earth history—no two included exactly the same arrangement of smaller continents. The most recent supercontinent, Pangaea, formed 300 million years ago (Ma) at the end of the Paleozoic Era and survived until it broke up to form today's continents, beginning about 200 Ma.

The Sea-Level Change Cycle

Global sea level rose and fell by as much as 300 m during the Phanerozoic and likely did the same in the Precambrian. When sea level rises, the shoreline migrates inland, and low-lying plains in the continental interiors become submerged. In fact, during periods of particularly high sea level, more than half of Earth's continental area was covered by shallow seas. At such times, shallow marine sediment buries continental regions (**Fig. 23.5a, b**). When sea level falls, the continents become dry again, and regional unconformities develop. We can see the record of this sea-level change cycle preserved in the sedimentary beds of the midwestern United States. Here a succession of strata record at least six continent-wide advances and retreats of the sea, each of which left behind a blanket of sediment called a **sedimentary sequence**. Unconformities define the boundaries between the sequences (**Fig. 23.5c**). Of note, the sequence deposited during the Pennsylvanian contains at least 30 shorter repeated intervals, called *cyclothems*,

FIGURE 23.4 The stages of the supercontinent cycle.

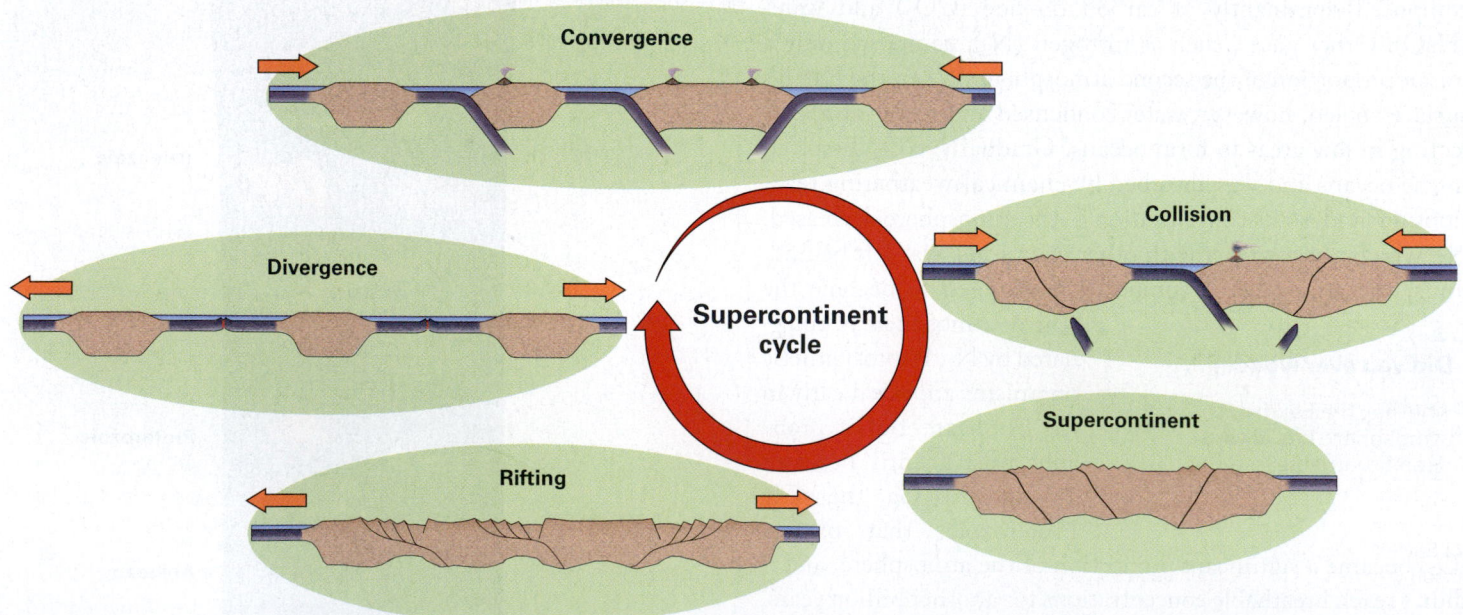

FIGURE 23.5 Sea-level change, and its manifestations, over geologic time.

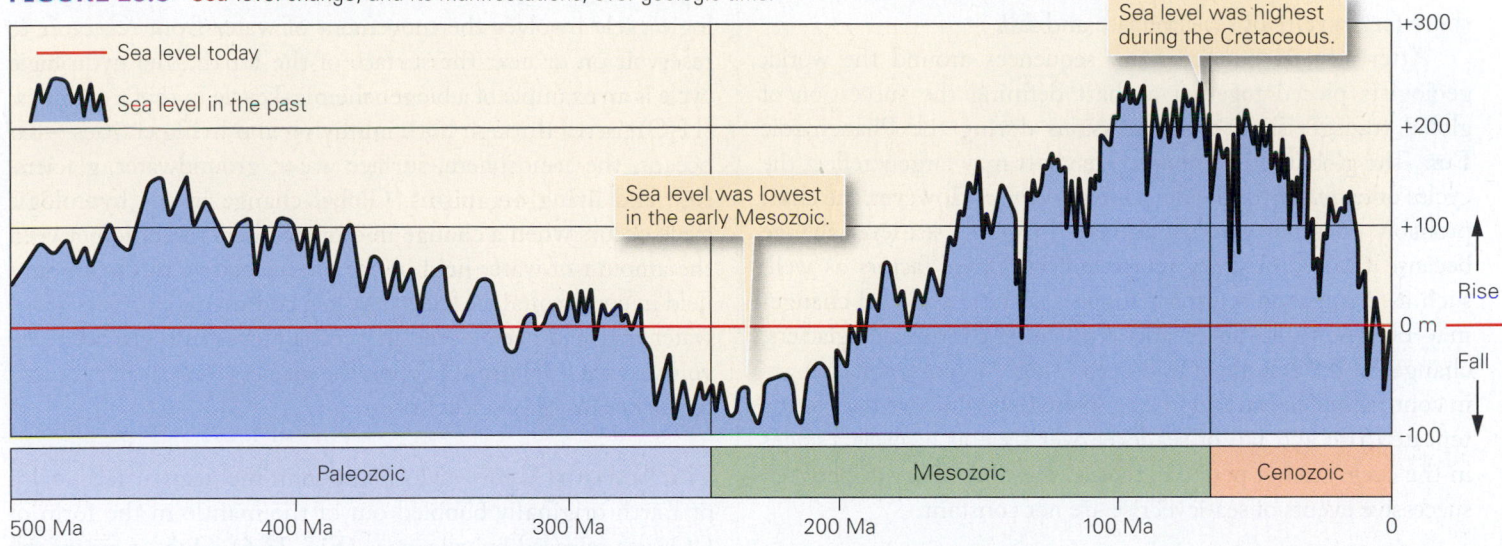

Sea level was highest during the Cretaceous.

Sea level was lowest in the early Mesozoic.

— Sea level today
Sea level in the past

Rise
Fall

+300
+200
+100
0 m
-100

Paleozoic | Mesozoic | Cenozoic

500 Ma | 400 Ma | 300 Ma | 200 Ma | 100 Ma | 0

(a) This chart provides one interpretation of sea-level change during the past half-billion years, based on the stratigraphic record. There is not full agreement about this interpretation; it remains the subject of research.

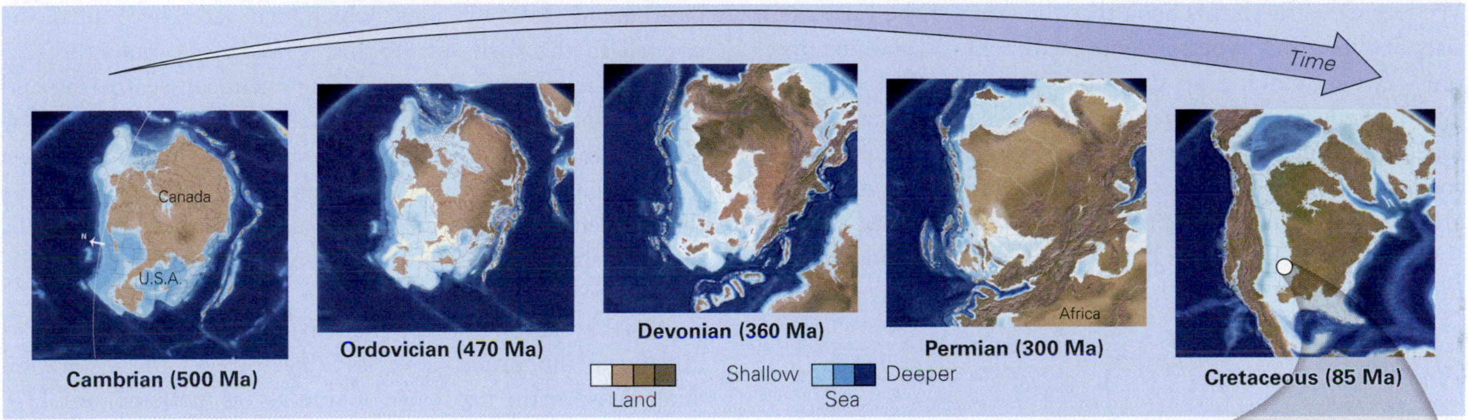

Time

Canada
U.S.A.

Cambrian (500 Ma)

Ordovician (470 Ma)

Devonian (360 Ma)

Permian (300 Ma)

Africa

Cretaceous (85 Ma)

Shallow | Deeper
Land | Sea

(b) When sea level is high, large parts of North America's interior became submerged by shallow seas.

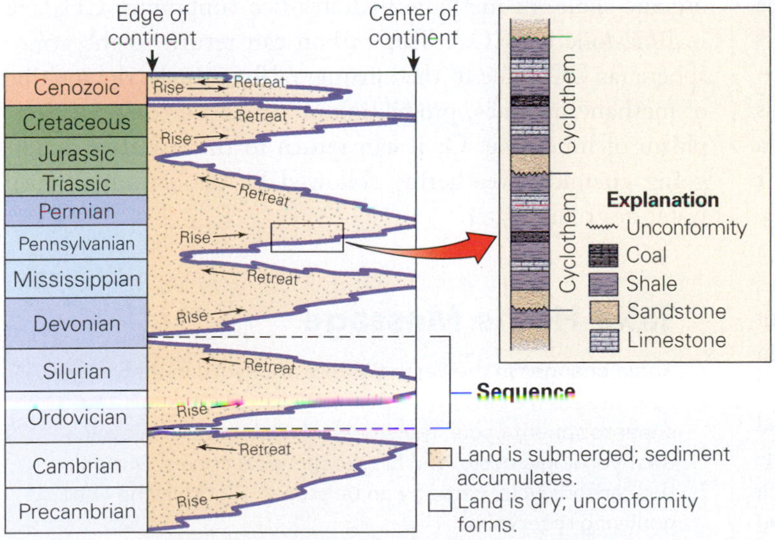

Edge of continent | Center of continent

Cenozoic — Rise — Retreat
Cretaceous — Retreat
Jurassic — Rise
Triassic
Permian — Retreat
Pennsylvanian — Rise
Mississippian — Retreat
Devonian — Rise
Silurian — Retreat
Ordovician — Rise
Cambrian — Retreat
Precambrian — Rise

Cyclothem | Cyclothem

Explanation
~~~~ Unconformity
▦ Coal
▨ Shale
▨ Sandstone
▦ Limestone

— **Sequence**

☐ Land is submerged; sediment accumulates.
☐ Land is dry; unconformity forms.

(c) Sea-level rise and fall left sedimentary sequences separated by unconformities. Pennsylvanian sequences contain cyclothems.

A plesiosaur searches for food in the Cretaceous seaway of North America.

each of which contains a layer of coal. Cyclothems represent short-term cycles of sea-level rise and fall.

After studying sedimentary sequences around the world, geologists pieced together a chart defining the succession of global transgressions and regressions during the Phanerozoic Eon. The global sedimentary cycle chart may largely reflect the cycles of *eustatic* (worldwide) *sea-level change*. However, the chart probably does not give us an exact image of sea-level change because the sedimentary record reflects other factors as well, such as changes in sediment supply. Eustatic sea-level changes may be due to advances and retreats of continental glaciers, changes in the volume of mid-ocean ridge systems, and changes in continental elevation and area. Note that while we refer to the repeated rise and fall of sea level over time as a "cycle," stages in the cycle are not periodic; that is, the time intervals between successive events of sea-level rise are not constant.

The Rock Cycle

We learned early in this book that the crust of the Earth consists of three rock types: igneous, sedimentary, and metamorphic. Atoms making up the minerals of one rock type may later become part of another rock type. As we learned in Interlude C, this process is the *rock cycle*. In effect, we can think of rocks as, simply, reservoirs of atoms—during the rock cycle, atoms move from reservoir to reservoir over time. Each stage in the rock cycle changes the Earth by redistributing and modifying material.

Biogeochemical Cycles

A **biogeochemical cycle** involves the passage of a chemical among nonliving and living reservoirs in the Earth System, mostly on or near the surface. Nonliving reservoirs include the atmosphere, the crust, and the ocean, whereas living reservoirs include plants, animals, and microbes. Some stages in a biogeochemical cycle may take only hours, some may take thousands of years, and others may take millions of years. The transfer of a chemical from reservoir to reservoir during these cycles doesn't really seem like a change in the Earth in the way that the movement of continents or the metamorphism of rock seems like a change, because for intervals of time biogeochemical cycles attain a **steady-state condition**, meaning the proportions of a chemical in different reservoirs remain fairly constant even though there is a constant flux (flow) of the chemical among reservoirs. When we speak of global change in a biogeochemical cycle, we mean a change in the relative proportions of a chemical held in different reservoirs at a given time—in other words, a change in the steady-state condition. Although a great variety of chemicals (water, carbon, oxygen, sulfur, ammonia, phosphorus, and nitrogen) participate in biogeochemical cycles, here we look at only two: water (H_2O) and carbon (C).

Hydrologic Cycle As we learned in Interlude F, the hydrologic cycle involves the movement of water from reservoir to reservoir on or near the surface of the Earth. The hydrologic cycle is an example of a biogeochemical cycle in that a chemical (H_2O) passes through both nonliving and living entities—the oceans, the atmosphere, surface water, groundwater, glaciers, soil, and living organisms. Global change in the hydrologic cycle occurs when a change in climate alters the ratio between the amount of water held in the ocean relative to the amount held in continental ice sheets. When continental glaciers grow, water that had been stored in oceans moves into glacial reservoirs, so sea level drops. When the glaciers melt, water returns to the oceans, so sea level rises.

The Carbon Cycle Most carbon in the near-surface realm of Earth originally bubbled out of the mantle in the form of CO_2 gas released by volcanoes (**Fig. 23.6**). Once it enters the atmosphere, it moves through various reservoirs of the Earth system in the **carbon cycle**. Some dissolves in seawater to form bicarbonate (HCO_3^-) ions, which may later become incorporated in the shells of organisms that settle onto the seafloor. Some reacts with rock during chemical weathering and becomes incorporated in minerals. And some gets absorbed by photosynthetic organisms (microbes and plants), which convert it into sugar and other organic chemicals—this carbon enters the food chain and ultimately makes up the flesh of animals. Of note, about 63 billion tons of carbon move from the atmosphere into life forms every year.

Of the carbon incorporated in organisms, some returns directly to the atmosphere (as CO_2) through the respiration of animals, by the flatulence of animals (as methane, or CH_4), or by the decay of dead organisms. But some can be stored for long periods of time in fossil fuels (oil, gas, and coal), in organic shale, in methane hydrates (ice containing CH_4), or in limestone ($CaCO_3$). This carbon can return to the atmosphere (as CO_2) due to the burning of fossil fuels, the melting of methane hydrates, production of cement, or the metamorphism of limestone. Or it can return to the sea after undergoing chemical weathering followed by dissolution in river water or groundwater.

Take-Home Message

Some changes in the Earth System are cyclic in that they have stages that may be repeated. Examples include the supercontinent cycle, the sea-level cycle, the rock cycle, the hydrologic cycle, and biogeochemical cycles. During the carbon cycle, carbon can be stored in both living and nonliving reservoirs.

QUICK QUESTION: What evidence indicates that sea level rises and falls over time?

FIGURE 23.6 In the carbon cycle, carbon transfers among various reservoirs at or near the Earth's surface. Red arrows indicate release to the air, and green arrows indicate absorption from air.

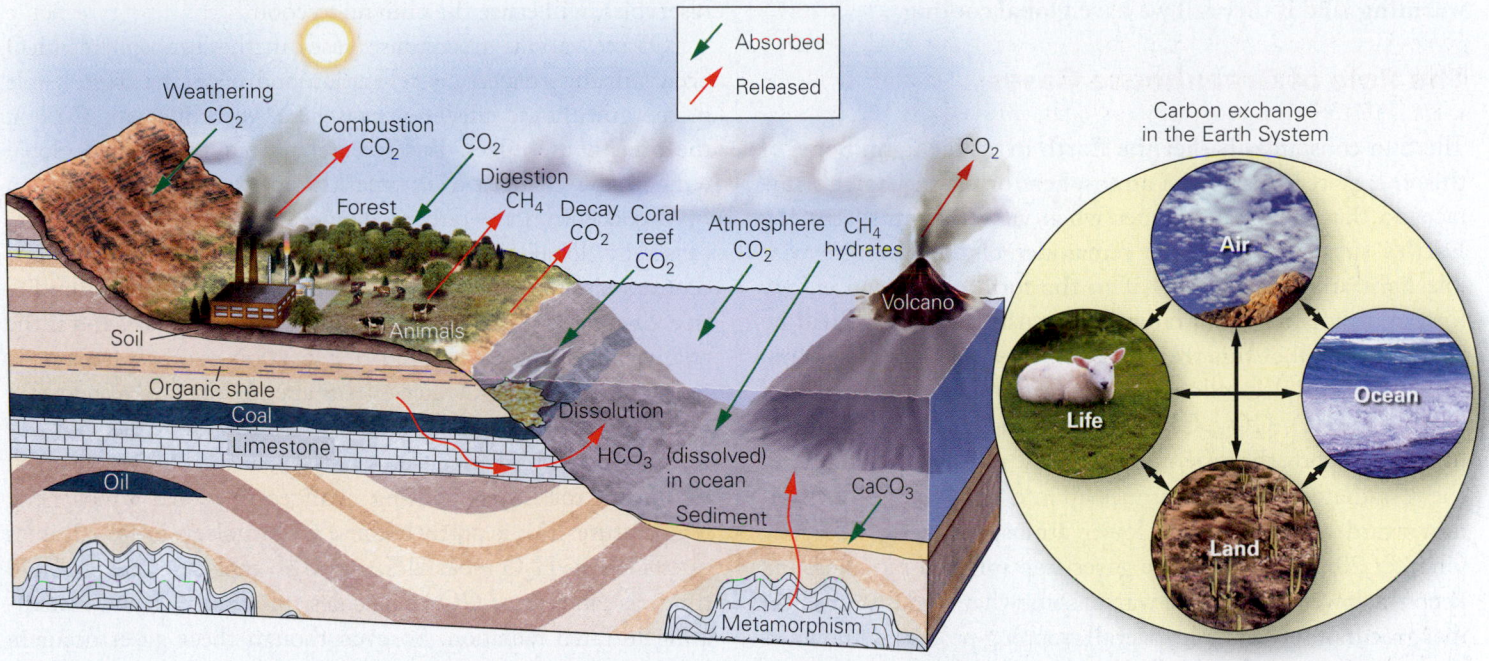

23.4 Global Climate Change

What Is Climate Change?

How often have you seen a newspaper proclaim "Record High Temperatures!" when thermometers register temperatures several degrees above "normal" for days on end. Do such headlines mean that "climate" is changing? To start addressing this question, we first need to distinguish between two common terms: weather and climate.

As discussed in Chapter 20, atmospheric conditions during a specific time interval in a given region define the region's **weather** for the time interval. The weather may be "windy, rainy, and cool" in the morning and "calm, sunny, and dry" in the afternoon. The term **climate**, in contrast, refers to overall range of weather conditions, as well as the typical daily to seasonal variability of weather conditions, as observed over a period of decades for a region. We can contrast, for example, a "tropical climate" with a "temperate climate"—the former tends to have hot, humid days all year, whereas the latter tends to have contrasting seasons and a wide range of temperature and humidity. So a newspaper's headline about a single hot spell or cold snap does not mean that the climate is changing. But if a new set of conditions—say, an overall increase in average temperature, a rising snow line, a longer growing season, or a change in storm frequency or intensity—becomes the new

norm for a region, then climate change has occurred. And if such changes happen worldwide, then **global climate change** has occurred.

The stratigraphic record clearly shows that global climate change has taken place repeatedly throughout Earth's long history for natural reasons. As we described in Chapter 13, for example, the Cretaceous was a time of relatively warm temperatures during which no polar ice caps existed, while the late Proterozoic may have seen "snowball Earth," when all oceans were frozen over and glaciers covered all continents. This well-established principle has been recognized since geologists first learned how to interpret the stratigraphic record, so it's not news. But global climate change has become a staple of the news media in recent years, and that's because research of the past few decades led to the conclusion that global climate change is now taking place, not at the slow pace typical of most geologic phenomena but at rates fast enough to have potential impacts in the next few decades. In other words, readers of this book may see the effects of contemporary global climate change in their lifetimes.

In this section, we begin our study of global climate change by first reviewing the fundamental role that greenhouse gases play in regulating atmospheric temperature, then by describing how researchers study past climates, and finally by defining different rates at which climate has changed over geologic time. For purpose of our discussion, we distinguish between *long-term climate change*, which takes place over millions to tens of millions of years, and *short-term climate change*, which takes place over tens to hundreds of thousands of years. If average atmospheric and sea-surface temperature rises, we have **global warming**, and if they fall we have **global cooling**.

The Role of Greenhouse Gases

The Sun constantly bathes the Earth in visible light. Some of this energy reflects off the atmosphere or off the Earth's surface, so that our planet shines when viewed from space. The Earth's surface absorbs the remainder of the incoming visible light and then releases it in the form of thermal energy (infrared radiation) that radiates upward. If the Earth had no atmosphere, all of this thermal energy would escape back into space. But our planet does have an atmosphere, and certain of its gases (H_2O, CO_2, CH_4, NO_2 [nitrogen dioxide], and O_3 [ozone]) absorb thermal radiation and re-radiate it. Some of the re-radiated energy continues up into space, but some heads downward and warms the lower atmosphere (**Fig. 23.7**; see Chapter 20). In effect, these gases trap infrared radiation and keep the lower atmosphere warm, somewhat as glass traps heat in a greenhouse. Thus, the overall trapping process is known as the **greenhouse effect**, and the gases that cause it are **greenhouse gases**.

Researchers estimate that were it not for the greenhouse effect, global average surface temperature of the Earth would be about 33°C (91.4°F) lower than it is today. Put another way, an Earth without greenhouse gases would have an average global temperature of about –19°C (–2.2°F), and our planet's surface would be a frozen wasteland—it is the presence of greenhouse gases that make the Earth habitable! Because greenhouse gases trap heat, any process that transfers these gases from underground, oceanic, or biomass reservoirs into the atmospheric reservoir will cause the climate to warm. Similarly, any process that removes greenhouse gases from the atmospheric reservoir and transfers them in biomass, or into oceanic or underground reservoirs, will cause the climate to cool.

Of the various greenhouse gases in the atmosphere, H_2O occurs in the greatest concentration and plays the biggest role in the greenhouse effect—it causes between 30% and 70% of the greenhouse effect. But researchers emphasize that global temperature changes over time are not driven by changes in H_2O concentration—rather, changes in H_2O concentration are caused by global temperature changes. That's because water concentration in the atmosphere, as represented by relative humidity (see Chapter 20), varies radically from place to place, and if the relative humidity gets too high at a location, excess water simply rains out. In fact, H_2O molecules that enter the atmosphere stay there for a very short time, generally less than nine days.

Of the other greenhouse gases, CO_2 and CH_4 play the most significant role in influencing changes in global atmospheric temperature. This is partly because CO_2 molecules are 20 times as efficient as H_2O molecules, and CH_4 molecules are about 70 times as efficient as CO_2 molecules, in absorbing and re-radiating infrared radiation. So even though these gases occur in very low concentrations, they can contribute significantly to the greenhouse effect (9% to 30%, and 4% to 9%, respectively). Also, CO_2 and CH_4 mix thoroughly with other gases in the lower atmosphere and remain in the atmosphere for a long time. So if new CO_2 or CH_4 enters the atmosphere from a surface or subsurface reservoir, it adds to the existing quantities of these gases and causes the overall concentration of these gases to increase—the excess doesn't simply rain back to Earth. The increase in the concentration of these gases causes an increase in the greenhouse effect and, therefore, global temperature increases.

Warming or cooling due to changes in the concentration of greenhouse gases can be amplified by positive or negative

FIGURE 23.7 The greenhouse effect shows how thermal energy can be trapped in the atmosphere.

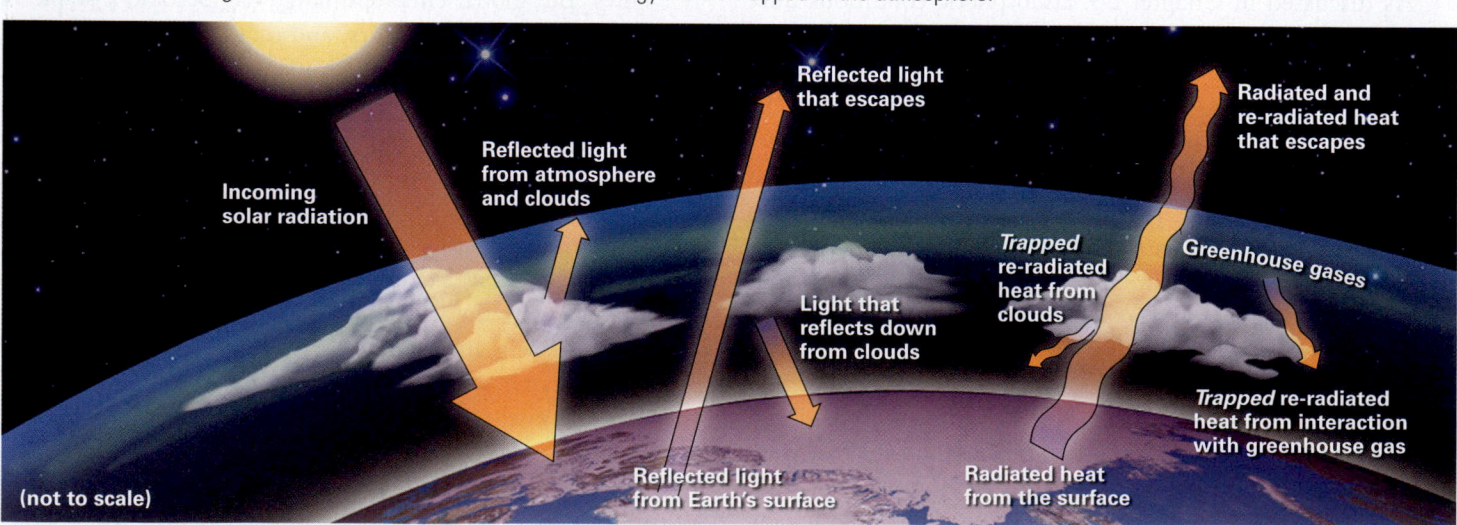

feedback mechanisms. *Negative* feedback slows a process down or even reverses it, whereas *positive* feedback enhances a process and amplifies its consequences. Let's consider an example of how positive feedback affects climate. Imagine that global average atmospheric temperature has increased due to an increase in CO_2. This increase will, in turn, cause the oceans to warm and evaporate more, so more water transfers into the atmospheric reservoir globally and increases the greenhouse effect due to water vapor. The increase also causes some of the CO_2 dissolved in the oceans to come out of solution and return to the atmosphere, and it causes CH_4 to enter the atmosphere from decay of organic matter in melting permafrost and perhaps from the melting of gas hydrates (ice containing dissolved CH_4). Thus, the warming due to the initial addition of CO_2 can cause the concentration of other greenhouse gases to increase overall, forcing the atmosphere to warm even more than it would have in the first place.

Methods of Studying Climate Change

Geologists and climatologists are working hard to define the nature of climate change, the rates at which change can take place, and the effects that change may have on our planet. There are three basic approaches to studying global climate change: (1) measure past climate change, as recorded by stratigraphy, to document the magnitude of changes that are possible and the rate at which such changes occurred; (2) conduct experiments and calculations to see how changes in the concentration of atmospheric components, such as CO_2 or dust, might affect climate; (3) develop computer programs (called *general circulation models*, or GCMs) to simulate how factors such as atmospheric composition, topography, ocean currents, and Earth's orbit affect the circulation of the atmosphere and, therefore, the distribution of climate belts. Researchers use GCMs to develop broader **climate-change models** that seek to provide insight into when and why changes took place in the past and whether they will happen in the future. Climate-change models try to predict changes in rainfall, sea level, ice cover, and other physical features that may be influenced by the warming or cooling of the atmosphere.

Did you ever wonder . . .

how researchers study past climates on Earth?

Let's look more closely at how geologists study the **paleoclimate** (past climate) so as to document climate changes throughout Earth history. Any feature whose character depends on the climate, and whose age can be determined, provides a clue to defining paleoclimate. Examples include:

- *Stratigraphic record:* The nature of sedimentary strata deposited at a certain location reflects the climate at that location. For example, an outcrop exposing cross-bedded sandstone, overlain successively by coal and glacial till, indicates that the site of the outcrop has endured different climates (desert, then tropical, then glacial) over time.

- *Paleontological evidence:* Different assemblages of species survive in different climatic belts. Thus, the succession of fossils in a sedimentary sequence provides clues to the changes in climate at that site. For example, a record of short-term climate change can be obtained by studying the succession of plankton fossils in seafloor sediments, because cold-water species of plankton are different from warm-water species. Fossil plant pollen preserved in the mud of bogs also provides information about paleoclimate since different plant species live in different climates (**Fig. 23.8a**). Pollen studies show, for example, that spruce forests, indicative of cool climates, have slowly migrated north since the last ice age (**Fig. 23.8b**).

- *Oxygen-isotope ratios:* Geologists have found that the ratio of ^{18}O to ^{16}O in glacial ice indicates the atmospheric temperature in which the snow that made up the ice formed: simplistically, the ratio is larger in snow that forms in warmer air, and smaller in snow that forms in colder air. Because of this relationship, the isotope ratio of the oxygen in H_2O measured in a succession of ice layers in a glacier indicates temperature change over time. Researchers have now obtained ice cores down to a depth of almost 3 km in Antarctica and in Greenland, a record that spans over 720,000 years (**Fig. 23.9a**). The $^{18}O/^{16}O$ ratio in the $CaCO_3$ making up plankton shells also gives geologists an indication of past temperatures. Thus, measurement of oxygen-isotope ratios in drill cores of marine sediment extends the record of temperature change back over millions of years (**Fig. 23.9b, c**).

- *Bubbles in ice:* Bubbles in ice trap the air present at the time the ice forms. By analyzing these bubbles, geologists can measure the concentration of CO_2 in the atmosphere back through time. This information can be used to correlate CO_2 concentration with past atmospheric temperature.

- *Growth rings:* If you've ever looked at a tree stump, you will have noticed the concentric rings visible in the wood. Each ring represents one year of growth, and the thickness of the ring indicates the rate of growth in a given year. Trees grow faster during warmer, wetter years and more slowly during cold, dry years (**Fig. 23.10a**). Thus, the succession of ring widths provides a calibrated record of climate during the lifetime of the tree. Bristlecone pines supply a record back through 4,000 years. To go further into the past, dendrochronologists (scientists who study tree rings) look at the record of rings in logs dated by the radiocarbon technique or in logs whose ages overlap with that of the oldest living tree. Growth rings in corals and shells can provide similar information.

FIGURE 23.8 Changes in the assemblage of pollen in sediment indicates a shift in climate belts.

Spruce pollen dominates when it's cooler.

Grass pollen dominates when it's warmer.

Tree pollen
Grass pollen

Receding Pleistocene ice sheet

Spruce forest 12,000 B.C.E.

Great Lakes

Spruce forest today

(a) The proportion of tree pollen relative to grass pollen can change in a sedimentary sequence through time. Researchers plot changes in the proportion of pollen types over time by examining samples from a column of sediment.

(b) Spruce forests (green) grew farther south 12,000 years ago than they do today.

• *Human history:* Researchers have been able to make careful, direct measurements of climate changes only for the past few decades. This record is not long enough to document long-term climate change. But history, both written and archaeological, contains important clues to climates at times centuries or millennia in the past. Periods of unusual cold or drought leave an impression on people, who record them in paintings, stories, and records of crop success or failure (**Fig. 23.10b; Box 23.1**).

Long-Term Climate Change

Using the variety of techniques described above, geologists have reconstructed an approximate record of global climate, represented by the Earth's mean atmospheric temperature, for geologic time. The record shows that, with the exception of snowball Earth intervals (see Chapter 13), temperature has stayed between the freezing point of water and the boiling point of water since the beginning of the Archean. But the temperature has not always been the same—at some times in the past, the Earth's atmosphere was significantly warmer than it is today, and at other times it was significantly cooler. The warmer periods have come to be known as **greenhouse** (or **hothouse**) **periods** and the colder as **icehouse periods**. (The more familiar term, *ice age*, refers to portions of an icehouse period when the Earth was cold enough for ice sheets to advance and cover substantial areas of the continents.) As the chart in **Figure 23.11a** shows, there have been at least five major icehouse periods during geologic time.

Let's look a little more closely at the climate record of the last 100 million years, for this time interval includes the transition between a greenhouse and an icehouse period. Paleontological and other data suggest that the climate of the Mesozoic Era, the Age of Dinosaurs, was much warmer than the climate of today. At the equator, average annual temperatures

FIGURE 23.9 The proportion of isotopes transferred between reservoirs during evaporation or precipitation depends on temperature. The $^{18}O/^{16}O$ ratio can be studied in glacial ice (H_2O) and fossil shells ($CaCO_3$).

4 cm

(a) A researcher examines an ice core in the field. Lab photos reveal annual layers.

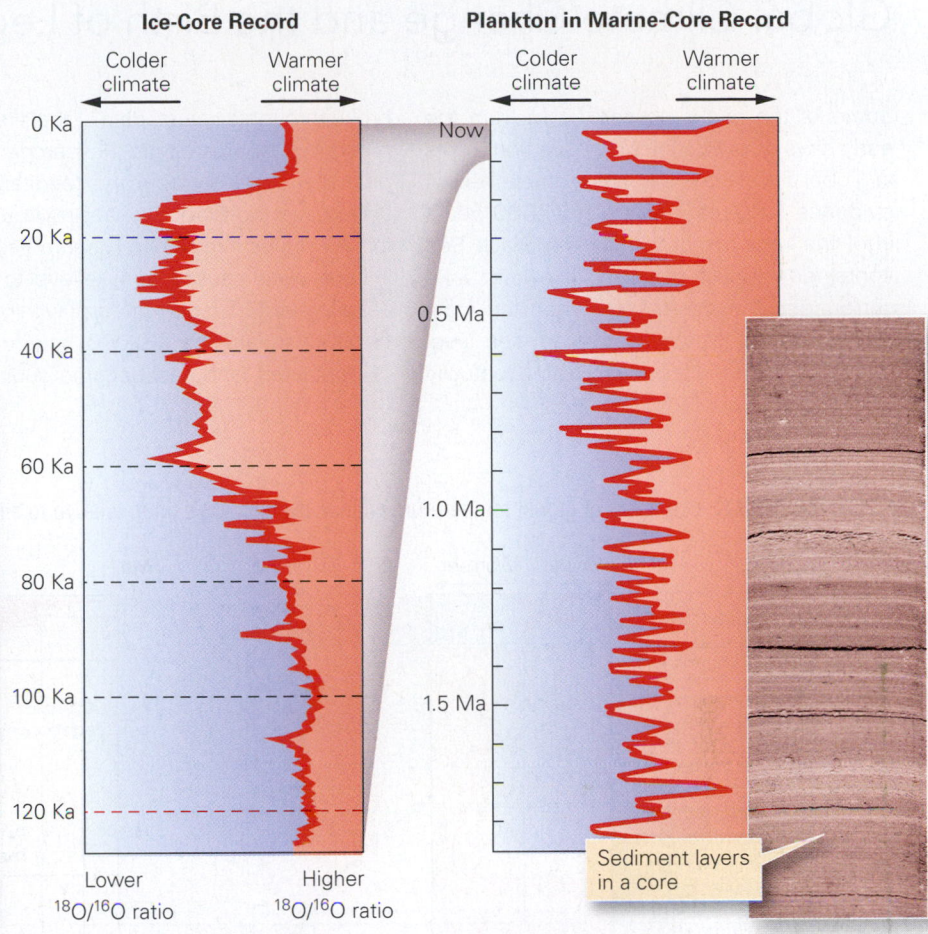

Ice-Core Record

Colder climate ← → Warmer climate

0 Ka
20 Ka
40 Ka
60 Ka
80 Ka
100 Ka
120 Ka

Lower $^{18}O/^{16}O$ ratio   Higher $^{18}O/^{16}O$ ratio

Plankton in Marine-Core Record

Colder climate ← → Warmer climate

Now
0.5 Ma
1.0 Ma
1.5 Ma

Sediment layers in a core

(b) The $^{18}O/^{16}O$ ratios in an ice core represent temperature changes.

(c) Studies of $^{18}O/^{16}O$ ratios in deep-sea cores also provide a climate record.

FIGURE 23.10 Records of recent climate change.

Researchers extract cores to see rings in living trees without damaging the tree.

(a) Tree rings provide a record of climate; more growth happens in wet years than in dry years.

(b) Historical archives provide records of floods and droughts.

BOX 23.1 CONSIDER THIS . . .

Global Climate Change and the Birth of Legends

Some of the myths passed down from the early days of civilization may have their roots in global climate change. For example, recent evidence suggests that before 7,600 years ago, the region that is now the Black Sea contained a much smaller freshwater lake surrounded by settlements. When the most recent ice-age glaciers retreated, sea level rose, and the Mediterranean Sea eventually broke through a natural dam at the site of the present Bosporus Strait. Researchers suggest that seawater from the Mediterranean spilled into the Black Sea basin via a waterfall 200 times larger than Niagara Falls. This influx of water caused the lake level to rise by as much as 10 cm per day, and within a year 155,000 square km (60,000 square miles) of populated land had become submerged beneath hundreds of meters of water. This traumatic flooding presumably forced many people to migrate, and its timing has led some researchers to speculate that it may have inspired the Epic of Gilgamesh (written in Babylonia ca. 2000 B.C.E.) and, later, the biblical epic of Noah's Ark.

FIGURE 23.11 Estimates of global temperature change over geologic time, relative to a reference value.

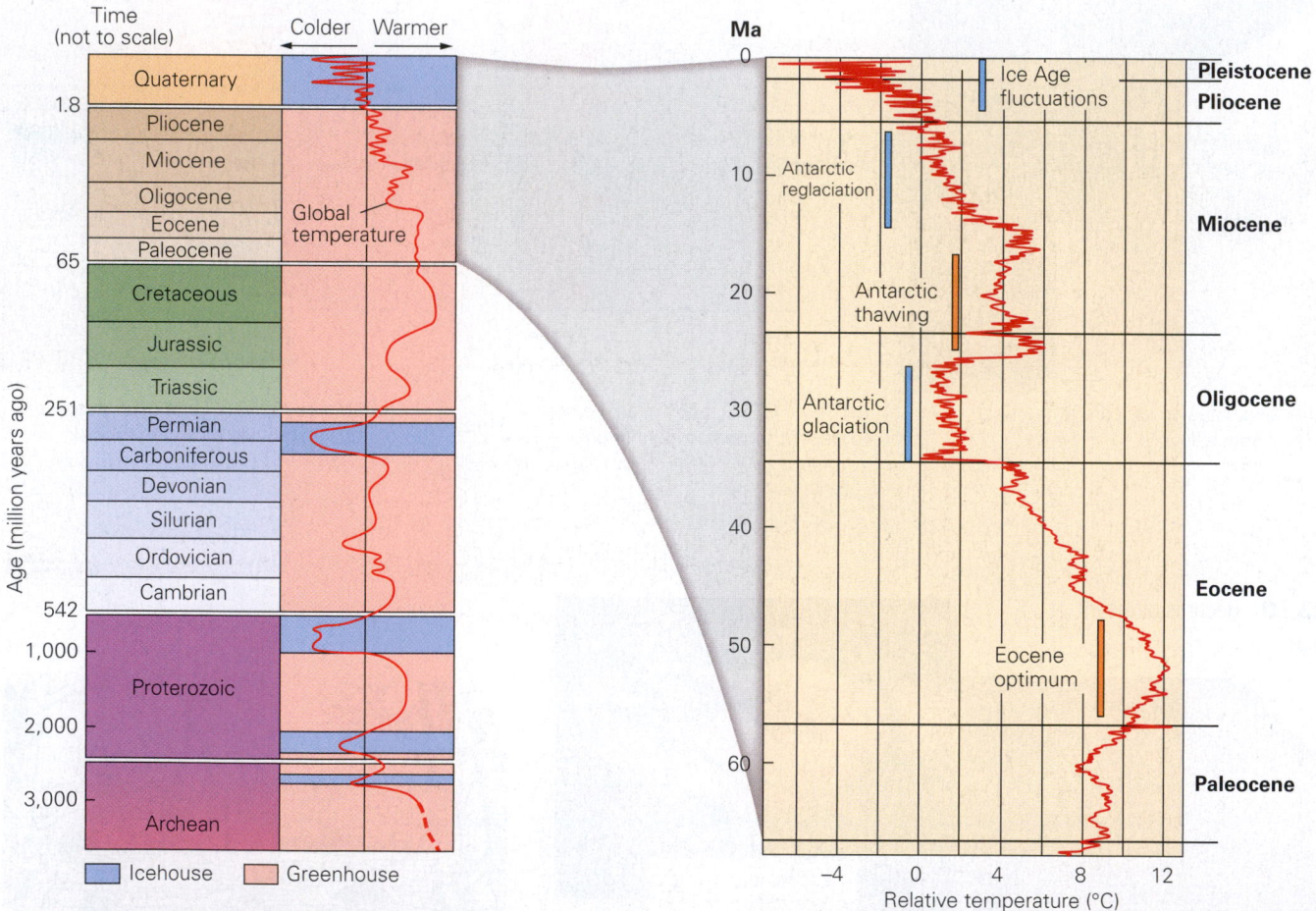

(a) Geologic data suggest the occurrence of icehouse and greenhouse phases at various times during Earth history.

(b) During the Cenozoic, there was an overall cooling trend; climate alternately warmed and cooled dramatically during the Pleistocene Ice Age.

may have been 2° to 6°C warmer, while at the poles temperatures may have been 20° to 60°C warmer. In fact, during the Cretaceous Period, dinosaurs could live at high latitudes, and there were no polar ice caps on Earth. But starting about 80 Ma, the Earth's atmosphere began to cool. The cooling trend continued, except for an interval of about 10 million years, a time now called the "Eocene climatic optimum" (**Fig. 23.11b**), and at about 33 Ma, our planet entered an icehouse period.

Temperatures in polar regions dropped below freezing, and the Antarctic ice sheet formed, and beginning at about 2.6 Ma, the Pleistocene Ice Age began.

What caused long-term global climate changes? The answer may lie in the complex relationships among the various solar, geologic, and biogeochemical cycles of the Earth System (**Box 23.2**). Factors that may have played a role include the following.

- *Positions of continents:* Continental drift influences the climate by controlling the pattern of oceanic currents, which redistribute heat around the planet's surface (**Fig. 23.12a**). Drift also determines whether the land is at high or low latitudes (and thus how much solar radiation strikes it), whether or not there are large continental interior regions where extremely cold winter temperatures can develop, and/or whether there is a lot of rainfall, which could cause weathering.

- *Volcanic activity:* A long-term global increase in volcanic activity may contribute to long-term global warming, if it increases the concentration of CO_2 in the atmosphere. For example, when Pangaea broke up during the Cretaceous Period, numerous rifts formed, and seafloor-spreading rates were particularly high, so volcanoes were more abundant than they are today. Such voluminous volcanic activity may have triggered Cretaceous greenhouse conditions.

- *Uplift of land surfaces:* Tectonic events that lead to the long-term uplift of the land affect atmospheric CO_2 concentration, because such events expose land to weathering, and chemical-weathering reactions absorb CO_2. Thus, uplift potentially decreases the greenhouse effect and causes global cooling. For example, uplift of the Himalayas and Tibet may have triggered Cenozoic icehouse conditions (**Fig. 23.12b**). Such uplift will also affect atmospheric circulation and rainfall rates (see Chapter 20).

FIGURE 23.12 Changes in the distribution and elevation of landmasses can affect climate.

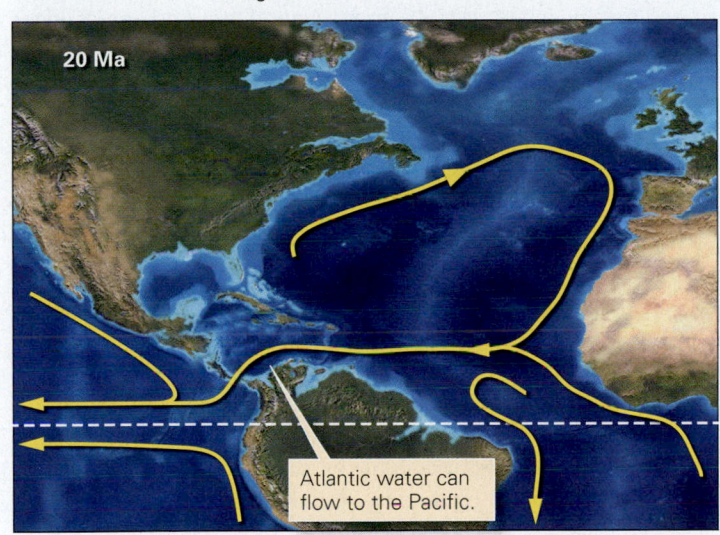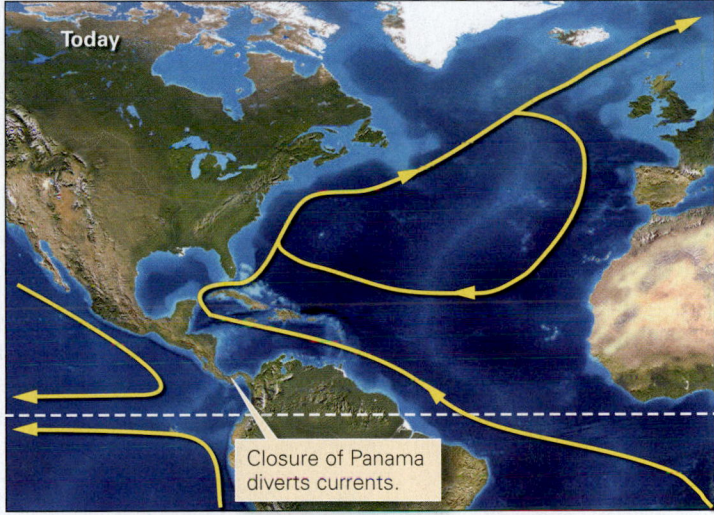

Atlantic water can flow to the Pacific.

Closure of Panama diverts currents.

(a) When the Isthmus of Panama, a volcanic arc, formed during the Miocene, the patterns of oceanic currents in the North Atlantic changed. The white dashed line is the equator.

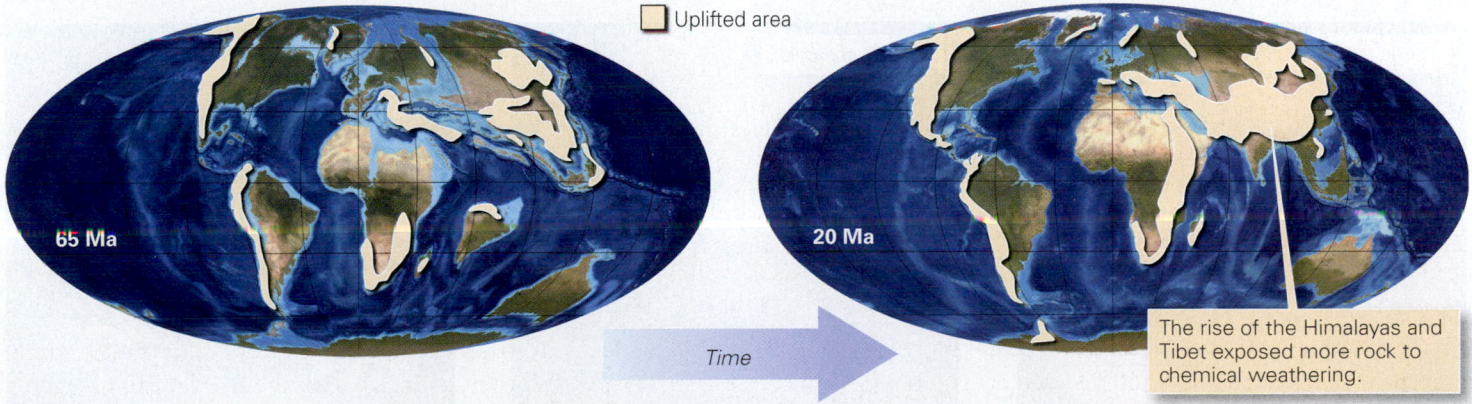

☐ Uplifted area

Time

65 Ma

20 Ma

The rise of the Himalayas and Tibet exposed more rock to chemical weathering.

(b) The percentage of elevated land increased between 65 Ma and 20 Ma because of orogeny. The exposure of more rock leads to more weathering, potentially affecting the concentration of greenhouse gases in the atmosphere.

BOX 23.2 CONSIDER THIS . . .

Goldilocks and the Faint Young Sun

Like Baby Bear's porridge in the tale of *Goldilocks and the Three Bears*, Earth is not too hot, and it's not too cold . . . it's just right for liquid water and, therefore, for life, to exist (**Fig. Bx23.2**). Researchers informally refer to the two related factors that make the Earth "just right" as the **Goldilocks effect**.

What factors keep Earth's surface habitable? The first is our planet's distance from the Sun, which determines the intensity of radiation that reaches the surface, and the second is the concentration of CO_2 and other greenhouse gases, which governs how much of that radiation remains trapped in the atmosphere. If the Earth orbited too close to the Sun, solar radiation would be so intense that water could not exist in liquid form, regardless of atmospheric composition. And without seas of liquid water, most CO_2 emitted by volcanoes could not have dissolved in the sea, so the atmosphere would contain so much water and CO_2 that the greenhouse effect would make surface temperatures on Earth way too hot for life. In contrast, if the Earth orbited too far from the Sun, temperatures would remain so cold that, regardless of atmospheric composition, any water present would freeze solid and life could not have evolved.

Astronomers define the distance from the Sun at which the Goldilocks effect is possible as the **habitable zone** of the Solar System.

The habitable zone currently extends roughly from 0.8 AU to 1.3 AU. (An AU, or astronomical unit, represents the mean distance between the Earth and the Sun.) Notably, the distance of the habitable zone from the Sun has increased over the history of the solar system. That's because the intensity of radiation emitted by the Sun has increased over time. This change has taken place because the Sun's energy comes from the fusion of four hydrogen atoms to form one helium atom, and one helium atom takes up less space than four hydrogen atoms. So, over time, production of helium has caused the Sun to contract. The resulting increase in internal pressure and temperature within the Sun, in turn, has caused the rate of fusion reactions to increase, and the Sun may be about 30% brighter today than it was when the Earth first formed.

If the Sun's intensity were the only factor controlling Earth's temperature, our planet should have been over 20°C cooler during the Archean than it is today, and all water should have been frozen. But this wasn't the case. Stratigraphic and fossil records indicate that water has existed in liquid form on our planet's surface at least since the early Archean (~ 3.8 Ga). Researchers refer to this apparent contradiction between the calculated temperature and the observed temperature of the early Earth as the **faint young Sun paradox**. Most researchers agree that

the paradox can be resolved by keeping in mind that earlier in Earth history, before the widespread appearance of photosynthetic life, the atmosphere contained more CO_2 than it does today. The greenhouse effect caused by the additional CO_2 increased the temperature of the Earth's atmosphere enough to counteract the lack of radiation from the faint young Sun, and this kept surface temperatures above freezing.

With the faint young Sun paradox in mind, astronomers speculate that when the Solar System was younger, Venus may also have orbited within the habitable zone and may have hosted liquid water. But as the Sun became brighter, Venus warmed up until its surface water evaporated. Addition of water to Venus's atmosphere caused the planet's surface to warm even more, leading to a drastic positive feedback that could not be stopped. Such a situation is called the *runaway greenhouse effect*. As a consequence, Venus's atmosphere eventually became so hot that water molecules broke apart, forming hydrogen atoms that escaped to space and oxygen atoms that reacted with rocks on the planet's surface to produce iron-oxide minerals. Volcanic CO_2 became the dominant gas of Venus's atmosphere, forming a dense blanket that now keeps the surface temperature of the planet at about 460°C, hot enough to melt lead.

FIGURE Bx23.2 The "Goldilocks effect," as applied to the Earth and its neighbors.

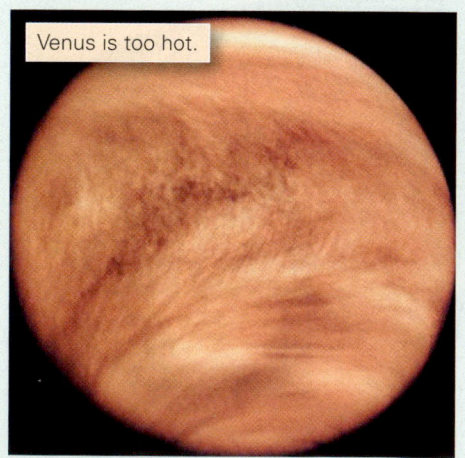
Venus is too hot.

Mars is too cold.

Earth is just right.

- *Formation of fossil fuels:* At various times during Earth history, environments suitable for the growth, accumulation, and burial of abundant organic material become particularly widespread. Once buried, the organic material transforms into coal or oil and can remain trapped underground. This overall process removes CO_2 from the atmosphere and thus may result in global cooling. The cooling that occurred in the late Paleozoic, a time when coal swamps were widespread, may be a manifestation of this process.

- *Life evolution:* The appearance or extinction of certain life forms may have affected climate significantly. For example, some researchers speculate that the appearance of lichens in the late Proterozoic may have decreased atmospheric CO_2 concentration and thus could have triggered icehouse conditions. Similarly, the appearance of grasses at around 30 to 35 Ma may have triggered Cenozoic icehouse conditions.

Natural Short-Term Climate Change

So far, we've focused on climate change that takes place on a time scale of millions to tens or even hundreds of millions of years. The geologic record of the past few million years provides enough detail to allow geologists to detect cycles of climate change that have durations of centuries to hundreds of thousands of years. Such *short-term climate change* must be a consequence of factors that can operate quickly in the context of geologic time. We can get a sense of short-term climate change by examining the Pleistocene stratigraphic record. Changes in fossil assemblages, sediment composition, and isotopic ratios indicate that continental glaciers advanced and retreated about 20 to 30 times in the northern hemisphere during the past 2.5 million years. Each advance (glaciation) represents an interval of global cooling, and each retreat (interglacial) represents an interval of global warming.

If we focus on the last 15,000 years, a period that includes all of the Holocene, we see trends of cooling or warming that last thousands of years, within which there are climate-change events whose duration lasts centuries or less (**Fig. 23.13a**). For example, the time between about 15,000 and 10,500 B.C.E. was a warming period during which the last ice-age glaciers retreated. At 10,500 B.C.E., the Earth entered the *Younger Dryas*, an interval of cooler temperatures named for an Arctic flower that became widespread at the time. Then climate warmed again, reaching a peak at 5,000 to 6,000 years ago, a period called the *Holocene maximum*, when average temperatures peaked at about 2°C above temperatures of today. This warming peak led to increased evaporation and therefore precipitation, making the Middle East region unusually wet and fertile—conditions that may partially account for the rise of civilization in Mesopotamia. The temperature dipped to a low about 3,000 years ago, before returning to a high during the Middle Ages, a time called the *Medieval Warm Period*. During this time, Vikings landed on the coast of Greenland and established settlements which could be self-supporting because of the relatively mild climate (**Fig. 23.13b**). The temperature dropped again from 1500 C.E. to about 1800 C.E., a period known as the *Little Ice Age*, when Alpine glaciers advanced and the canals of the Netherlands froze over in winter (**Fig. 23.13c**; see Fig. 22.33b). Overall, the climate has warmed since the end of the Little Ice Age, and today global temperature is comparable to that of the Medieval Warm Period.

What factors might control short-term climate change? There may be many, but geologists speculate that the following phenomena may have played the most important role.

- *Changes in Earth's orbit and tilt:* As Milanković first recognized in 1920, the tilt of Earth's axis changes over a period of 41,000 years, the Earth's axis undergoes precession over a period of 23,000 years, and the eccentricity of the Earth's orbit changes over a period of 100,000 years. Together these phenomena cause the amount of summer insolation and, therefore, the temperature in high latitudes to vary (see Chapter 22).

- *Changes in ocean currents:* Recent studies suggest that the configuration of currents can change quite quickly and that this configuration could affect the climate. The Younger Dryas, for example, may have resulted when a layer of freshwater from melting glaciers spread out over the North Atlantic and prevented thermohaline circulation in the ocean, thereby shutting down the Gulf Stream (see Chapter 18).

- *Large eruptions of volcanic aerosols:* Not all of the sunlight that reaches the Earth penetrates its atmosphere and warms the ground. Some gets reflected by the atmosphere. The degree of reflectivity, or **albedo**, of the atmosphere increases not only if cloud cover increases, as we have noted, but also if the concentration of volcanic aerosols in the atmosphere increases. Very large eruptions can emit enough aerosols (particularly SO_2) to affect global temperature for months to years. For example, the year following the 1815 eruption of Mt. Tambora in the western Pacific became known as the "year without a summer," for sulfur aerosols that erupted encircled the Earth and blocked the Sun. Snow fell in Europe throughout the spring, and the entire summer was cold. Recent studies suggest that the immense eruption of Toba (in Indonesia) 70,000 years ago disrupted climate for so many years that it caused worldwide extinctions and nearly eliminated *Homo sapiens*. (Note that the effect of volcanic aerosol emission is global cooling, the opposite of the effect of volcanic CO_2

FIGURE 23.13 Climate during the Holocene. Measurements suggest that temperature has varied significantly.

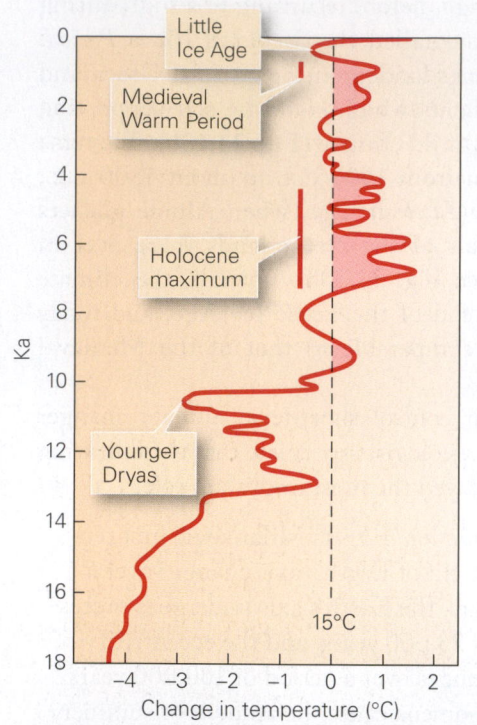

(a) There were several temperature highs and lows during the Holocene.

At the end of the Little Ice Age, this tributary glacier in France reached the main valley floor.

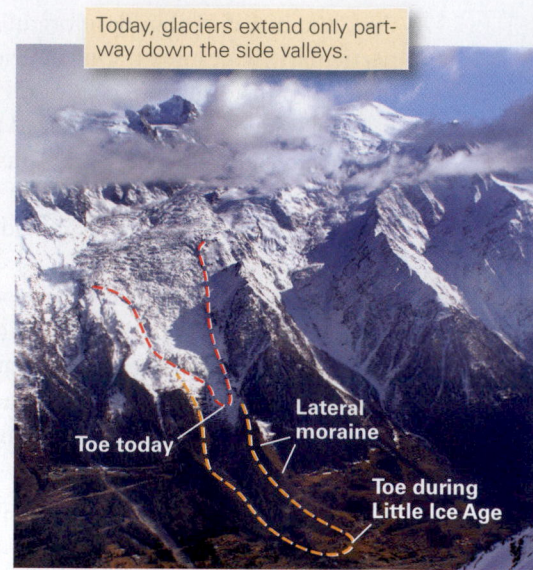

Today, glaciers extend only partway down the side valleys.

(c) Glaciers advanced in Europe during the Little Ice Age.

(b) During the Medieval Warm Period, Vikings settled in Greenland, where the climate was warm enough for agriculture.

emission, which causes global warming. Aerosols tend to impact climate for a relatively short time, but CO_2 can have effects for a relatively long time.)

- *Fluctuations in solar radiation:* The amount of energy produced by the Sun varies with the **sunspot cycle**. At peaks during the cycle, there are many sunspots (black spots thought to be magnetic storms on the Sun's surface), and at lulls in the cycle, there are few sunspots (**Fig. 23.14a, b**). Sunspots correlate with the amount of solar energy (irradiance) reaching the Earth, in that when there are many sunspots, more energy reaches the Earth (**Fig. 23.14c**).

A reliable record of sunspot activity can be taken back for about 400 years and reveals that cycles last 9 to 11.5 years, too fast to correlate with the rates of climate change observed in the geological record. There may be, however, longer-term cycles that have not yet been identified.

- *Fluctuations in cosmic rays:* Some researchers have speculated that changes in the rate of influx of cosmic rays may affect climate, perhaps by generating clouds. Specifically, recent research suggests that cosmic rays striking the atmosphere produce clusters of ions that become condensation nuclei around which water molecules congregate, thus forming the mist droplets making up clouds. But how cloud formation changes climate remains uncertain. High-elevation clouds could reflect incoming solar radiation and would cool the planet, whereas low-elevation clouds could absorb infrared rays rising from the Earth's surface and would warm the planet.

- *Changes in surface albedo:* Regional-scale changes in the nature of continental vegetation cover, and/or the proportion of snow and ice on the Earth's surface, and/or the sudden deposition of reflective volcanic ash could affect our planet's albedo. Increasing albedo causes cooling, whereas decreasing albedo causes warming. Of note, the Toba volcanic eruption covered vast areas of land with reflective white ash, which may have contributed to the global cooling that occurred after the event.

- *Abrupt changes in concentrations of greenhouse gases:* A relatively sudden change in greenhouse gas concentration in the atmosphere could affect climate. One such change might happen if sea temperature warmed or sea level

FIGURE 23.14 The abundance of sunspots varies with time and affects the radiation emitted by the Sun.

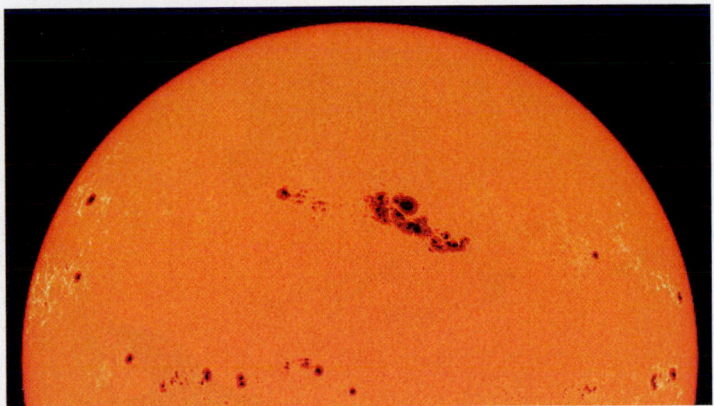

(a) Sunspots are magnetic storms that slow convection at the Sun's surface, producing a cooler area that appears as a dark patch.

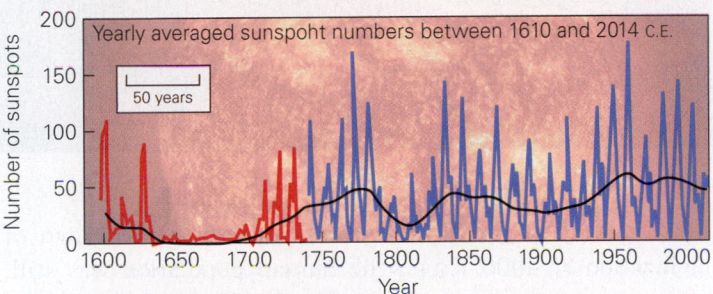

(b) The abundance of sunspots varies cyclically. When there are more sunspots, the Sun radiates less heat.

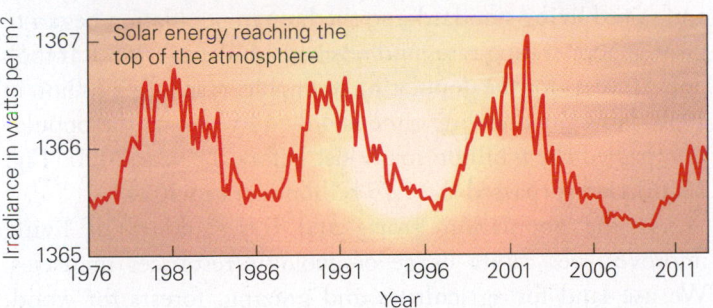

(c) Measurements since 1978 suggest that the solar energy reaching the atmosphere fluctuates periodically over time.

dropped, causing some of the methane hydrate that crystallized in sediment on the seafloor to melt suddenly. Such melting would release CH_4 to the atmosphere. Similarly, algal blooms and reforestation (or deforestation) conceivably could change CO_2 concentrations.

Catastrophic Climate Change and Mass-Extinction Events

Changes that happen on Earth almost instantaneously are called *catastrophic changes*. For example, a volcanic explosion,

an earthquake, a tsunami, or a landslide can change a local landscape in seconds or minutes. But such events affect only relatively small areas. Can such catastrophes happen on a global scale? In recent decades, geoscientists have come to the conclusion that the answer is yes. The stratigraphic record shows that Earth history includes several **mass-extinction events** (see Interlude E and Chapter 13), during which large numbers of species abruptly vanished (**Fig. 23.15a**). Some of these events define boundaries between geologic periods. A mass-extinction event decreases the *biodiversity*, the number of different species that exist at a given time, of life on Earth. It takes millions of years after a mass-extinction event for biodiversity to increase again, and the new species that appear differ from those that vanished, for evolution is unidirectional.

Geologists speculate that some mass-extinction events reflect a catastrophic change in the planet's climate, brought about by incredibly voluminous volcanic eruptions or by the impact of a comet or an asteroid with the Earth (**Fig. 23.15b**). Either of these events could eject enough debris into the atmosphere to block sunlight. Without the warmth of the Sun, winter-like or night-like conditions would last for weeks to years, long enough to disrupt the food chain. In addition, either event could eject aerosols that would turn into global acid rain, scatter hot debris that would ignite forest fires, or give off chemicals that when dissolved in the ocean would make the ocean either toxic, killing marine life, or so nutritious that oxygen-consuming algae could thrive.

Let's examine possible causes for two of the more profound mass-extinction events in Earth history. The first event marks the boundary between the Permian and Triassic periods and thus defines the boundary between the Paleozoic and Mesozoic Eras). During the Permian-Triassic extinction event, over two-thirds of the species on Earth became extinct. In fact, this boundary was first defined in the 19th century, precisely because the assemblage of fossils from rocks below the boundary differs so markedly from the assemblage in rocks above. Isotopic dating suggests that the extinction event roughly coincided with the eruption of vast quantities of basalt in Siberia. So much basalt erupted that geologists attribute its source to a "superplume," a mantle plume many times larger than the one currently beneath Hawaii. Because of the correlation between the time of the basalt eruptions and the time of the mass extinction, geologists suggest that the former caused the latter. Still controversial evidence suggests that, alternatively, a large asteroid collided with the Earth at the Permian-Triassic boundary and caused the mass extinction.

The second event, called the K-T boundary event, caused the mass extinction that marks the boundary between the Cretaceous and Tertiary Periods and, therefore, the boundary between the Mesozoic and Cenozoic Eras. The time of this event correlates well with the time at which an asteroid collided with the Earth at a site now called the Chicxulub crater

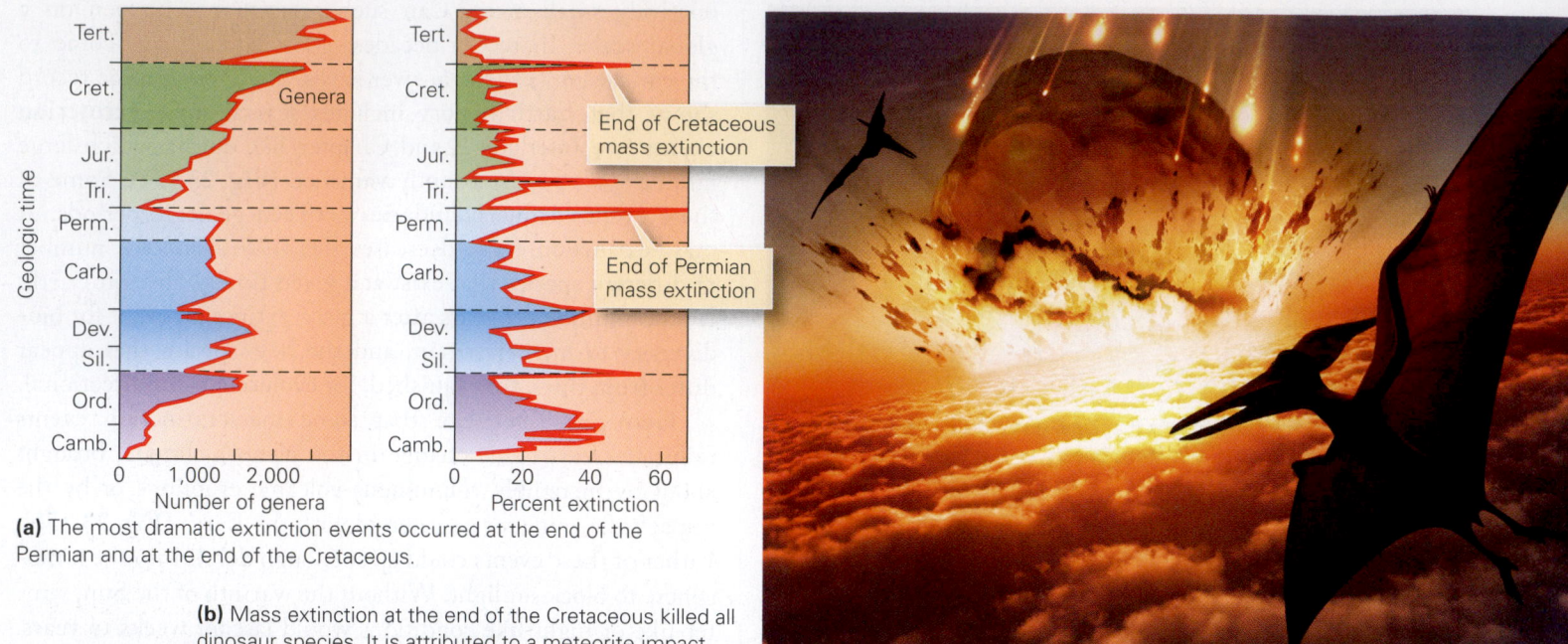

FIGURE 23.15 Life evolution has proceeded in fits and starts. During geologic time, there have been several catastrophic extinction events in which a larger percentage of the genera on Earth went extinct and biodiversity abruptly decreased.

(a) The most dramatic extinction events occurred at the end of the Permian and at the end of the Cretaceous.

(b) Mass extinction at the end of the Cretaceous killed all dinosaur species. It is attributed to a meteorite impact.

in Yucatán, Mexico. Thus, most geologists suggest that the mass extinction is the aftermath of this collision. A minority of researchers emphasize that the extinction is comparable in age to the eruption of extensive basalt flows in India and suggest that volcanic activity caused or at least contributed to causing the mass extinction.

Take-Home Message

Geologic study, using the stratigraphic record and fossils, shows that the climate has alternated between greenhouse and icehouse conditions over geologic time. Factors including life evolution, uplift of mountain belts, and continental drift may cause long-term change. Orbital cycles, eruption of volcanic aerosols, and perhaps solar radiation changes contribute to short-term change. Supervolcanoes or giant meteorite impacts can cause catastrophic change.

QUICK QUESTION: How does continental drift contribute to climate change?

23.5 Human Impact on Land and Life

According to some estimates, perhaps only 2,000 to 20,000 people lived on Earth after the catastrophic eruption of the Toba volcano about 70,000 years ago, and by the dawn of civilization at 4000 B.C.E., the human population was still, at most, a few tens of millions. But by the beginning of the 19th century, revolutions in industrial methods, agriculture, medicine, and hygiene had substantially lowered death rates and raised living standards, so the human population began to grow at accelerating rates and reached 1 billion in 1850. It took only 80 years for the population to double, reaching 2 billion in 1930. Now the doubling time is only 44 years, so the population passed the 6 billion mark just before the year 2000 (**Fig. 23.16**), and surpassed the 7.25 billion mark in 2014.

As the population grows and the standard of living improves, per capita usage of geologic resources increases. We use land for agriculture and grazing, forests for wood, rock and dirt for construction, oil and coal for energy or plastics, and ores for metals (see Chapter 12). Without a doubt, our usage of resources has affected the Earth System profoundly, and thus humanity has become a major agent of global change. Here we examine some of these anthropogenic (human-induced) impacts to the land surface, the environment, and finally to near-term climate.

The Modification of Landscapes

Every time we pick up a shovel and move a pile of rock or soil, we redistribute a portion of the Earth's crust, an activity that prior to humanity was accomplished only by rivers, the wind, rodents, and worms. In the last century, the pace of human-driven Earth movement has accelerated, for now we

FIGURE 23.16 Population now doubles about every 44 years. The Black Death pandemic caused an abrupt drop that lasted for a few decades.

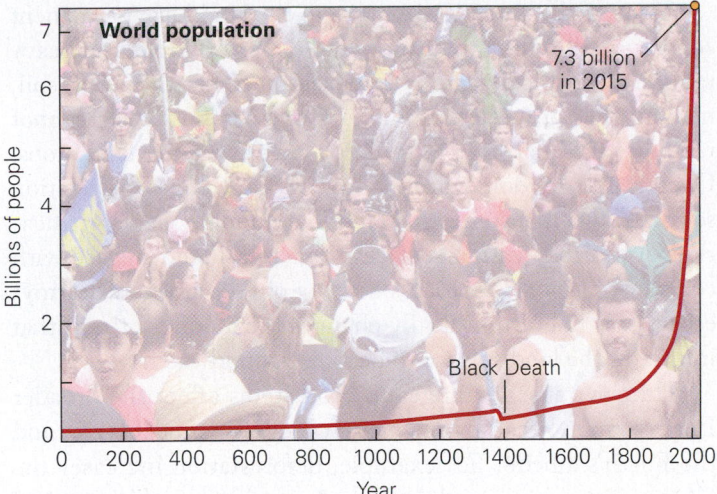

World population

7.3 billion in 2015

Black Death

have shovels in coal mines that can move 300 cubic meters of coal in a single scoop, trucks that can carry 200 tons of ore in a single load, and tankers that can transport 500 million liters (about 3 million barrels) of oil during a single journey. In North America, human activity now moves more sediment each year than rivers do. The extraction of rock during mining, the building of levees and dams along rivers or of sea walls along the coast, and the construction of highways and cities all involve the redistribution of Earth materials (**Fig. 23.17**). In addition, people clear and plow fields, drain and fill wetlands, and pave over the land surface. All these activities change the landscape, the water table, and the supply of sediment.

Landscape modification has side effects. For example, it may make the ground unstable and susceptible to landslides. It may expose the land to erosion, thereby changing the volume of sediment transported by natural agents, such as running water and wind. Locally, flood-control projects may diminish the sediment supply downstream, with unfortunate consequences. For example, the damming of the Nile by the Aswan High Dam cut off the sediment supply to the Nile Delta, so ocean waves along the Mediterranean coast of the delta now eat into the coastline by more than 1 m per year.

The Modification of Ecosystems

In undisturbed areas, the **ecosystem**—meaning an interconnected network of organisms, together with the physical environment in which they live—is the product of evolution for an extended period of time. The ecosystem's flora (plant life) include

FIGURE 23.17 Excavation, agriculture, and construction modify topography, drainage, infiltration, and ecology.

Humans move vast amounts of rock.

Agriculture eliminates diverse ecosystems.

Urbanization changes the water table.

species that have adapted to living together in that particular climate and on the substrate available, and its fauna (animal life) can survive local climate conditions and utilize local food supplies. Human-caused deforestation, overgrazing, agriculture, and urbanization disrupt ecosystems and lead to a decrease in biodiversity.

Archaeological studies have found that the earliest major example of human modification of an ecosystem took place in the Stone Age (pre-6,000 B.C.E.). Spear-bearing hunters, over a relatively short time, caused the mass extinction of mammoths and other large mammals. With the advent of agriculture and deforestation, and the growth of towns and cities, modification of the land accelerated, and today less than 5% of the landscapes in Europe and the United States retain their original ecosystems. In the developing world, landscape modification has claimed more than half of the area once occupied by tropical rainforests, and such forests are now disappearing at a rate of about 1.8% per year (**Fig. 23.18a, b**)—much of this loss comes from slash-and-burn agriculture, in which farmers and ranchers destroy forest to make open land for farming and grazing (**Fig. 23.18c, d**).

Some of the changes we make to the land have permanent consequences in a human time frame. For example, the heavy rainfall of tropical regions removes nutrients from the soil, making the soil useless in just a few years, so forests cannot regrow quickly even if farming or grazing of the land stops. Overgrazing by domesticated animals can remove vegetation so completely that some grasslands have undergone desertification. And urbanization replaces the natural land surface with concrete or asphalt, a process that not only completely destroys ecosystems but also radically changes the amount of rain that infiltrates the land surface to become groundwater.

Human-caused changes to ecosystems affect the broader Earth System because they modify biogeochemical cycles and the Earth's albedo. For example, deforestation increases the CO_2 concentration in the atmosphere, for the carbon stored

FIGURE 23.18 The area of forests has been shrinking. For example, tropical rainforests are being logged or burned.

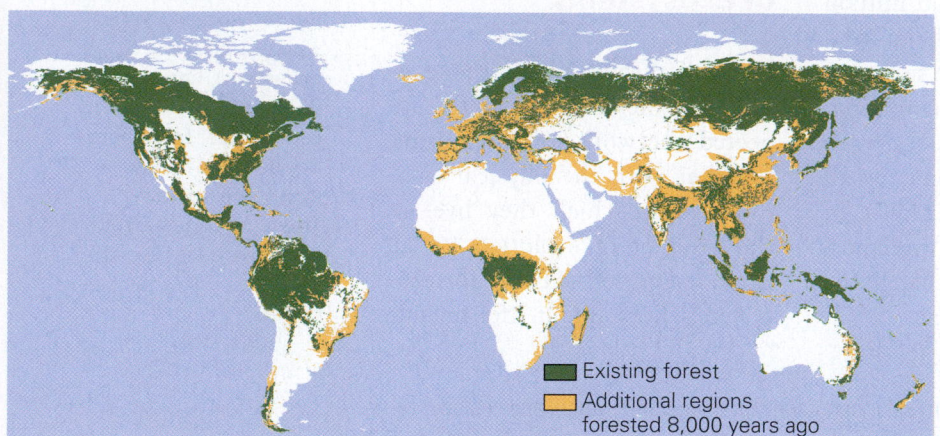

Existing forest

Additional regions forested 8,000 years ago

(a) Remaining forests today are much smaller than forests of 8,000 years ago.

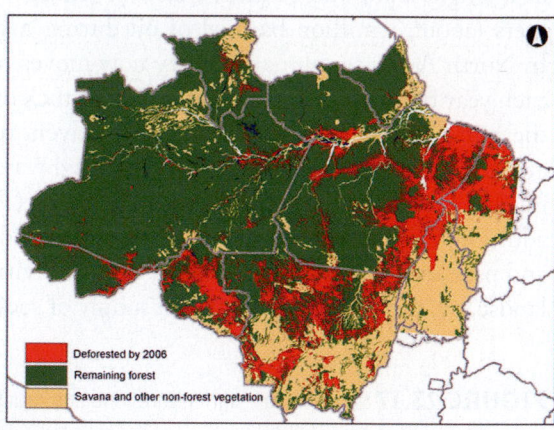

Deforested by 2006
Remaining forest
Savana and other non-forest vegetation

(b) A significant percentage of the Amazon rainforest has been lost in the past few decades.

(c) Part of the Amazon rainforest's destruction is due to slash-and-burn agriculture.

1975

2014

(d) Comparison of satellite imagery highlights areas that have undergone deforestation, such as this one near Ariquemes, Brazil.

in trees returns to the atmosphere when the trees burn. Further, the replacement of forest cover with concrete or fields increases the Earth's albedo.

Pollution

The environment has always contained various natural contaminants such as soot, dust, chemical run-off from sulfide minerals, and waste produced by organisms. In general, ecosystems could manage such contaminants naturally, by absorbing them, by breaking them down, or by locking them into accumulating strata so that the contaminants didn't damage the environment. The nearly exponential growth of human population, coupled with urbanization, industrialization, mining, expansion of farming and ranching, and a switch to engine-driven transportation has changed this situation. We've greatly increased both the quantity and the diversity of contaminants that enter the air, surface water, and groundwater. Some of these contaminants can be considered **pollution**, in that they yield poisonous or harmful effects. This pollution includes both natural and synthetic materials—in liquid, solid, or gaseous form—and has become a major problem because there is too much of it for the Earth System to accommodate. Pollution of the Earth System is a type of global change because it represents a redistribution and reformulation of materials. The following are key problems associated with this change.

- *Smog:* This term was originally coined to refer to the dank, dark air that resulted when smoke from the burning of coal mixed with fog in London and other industrial cities from the early 19th to mid-20th centuries. Beginning in the mid-20th century, another kind of smog, *photochemical smog*, has plagued cities. It forms when exhaust from cars and trucks reacts with air in the presence of sunlight to produce an ozone-rich brown haze. When this haze gets trapped in the lower kilometer or so of the air, it can make the air dangerous to breathe. In cities where photochemical smog has become a common problem, authorities publicize the *air-quality index*, a number representing the degree to which the air has become dangerous to breathe.
- *Water contamination:* Society dumps vast quantities of chemicals into surface water and groundwater. Examples include gasoline, other organic chemicals, radioactive waste, acids, fertilizers—the list could go on for pages. These chemicals can make groundwater supplies undrinkable and in some cases can kill off species in ecosystems where springs return the groundwater to the surface.
- *Acid runoff and acid rain:* When sulfide minerals remain buried in bedrock at depth, water and air can't reach them. But in mines and tailings piles, where rocks are exposed

and broken up, the minerals do come into contact with oxygen and water. The minerals may then react in the water to produce sulfuric acid. This *acid runoff*, which may flow out of mined areas, can be toxic to life. Similarly, when rain passes through air that contains sulfur-containing aerosols (emitted from power plants or factories), the water dissolves the sulfur and yields *acid rain*. Wind can carry aerosols far from their source, so acid rain can damage a broad region (**Fig. 23.19**).

- *Radioactive materials:* The by-products of producing nuclear weapons, nuclear energy, and medical-imaging equipment can all yield radioactive materials, including new radioactive isotopes, some of which have relatively short half-lives (see Chapter 14). Thus, society has changed the distribution and composition of radioactive material worldwide. Radioactive pollution arises when radioactive material escapes into the air or into rivers or groundwater.
- *Ozone depletion:* When emitted into the atmosphere, human-produced chemicals, most notably chlorofluorocarbons (CFCs), react with ozone in the stratosphere. This reaction, which happens most rapidly on the surfaces of tiny ice crystals in polar stratospheric clouds, destroys ozone molecules, thus creating an **ozone hole** over high-latitude regions, particularly during the spring (**Fig. 23.20**). Note that the "hole" is not really an area where no ozone is present but rather is a region where atmospheric ozone has been reduced substantially. The ozone hole is more prominent in the Antarctic than in the Arctic because a current of air circulates around the landmass of Antarctica and traps the air, with its CFCs, above the continent, preventing it from mixing with air from elsewhere. Ozone holes have dangerous consequences, for they affect the ability of the atmosphere to shield the Earth's surface from harmful ultraviolet radiation. In 1987, a summit conference in Montreal proposed a global reduction of ozone-destroying CFC emissions. As a consequence, reductions of such emissions have decreased, and the ozone hole has become smaller.

Take-Home Message

Human activities impact the landscape and ecosystems significantly by moving earth, paving the surface, and changing the character of land cover. Society also introduces contaminants into the environment at rates that cannot be accommodated by the Earth System, leading to pollution of land, air, surface water, and groundwater.

QUICK QUESTION: What is the ozone hole, and why did it form?

FIGURE 23.19 Acid rain forms when the sulfur dioxide in industrial smoke dissolves in water and produces sulfuric acid. Acid rain is a problem in all industrialized countries.

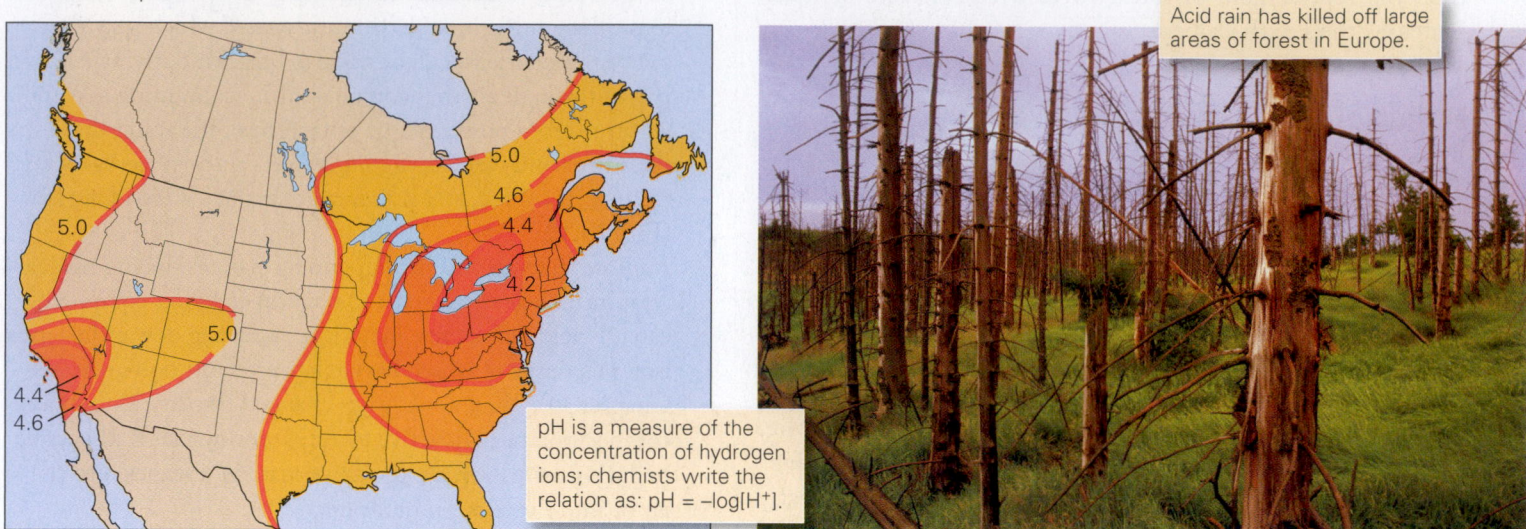

Acid rain has killed off large areas of forest in Europe.

pH is a measure of the concentration of hydrogen ions; chemists write the relation as: pH = –log[H⁺].

(a) Acid rain in North America occurs downwind of major industrial cities. We can specify the acidity of rainwater by stating its pH.

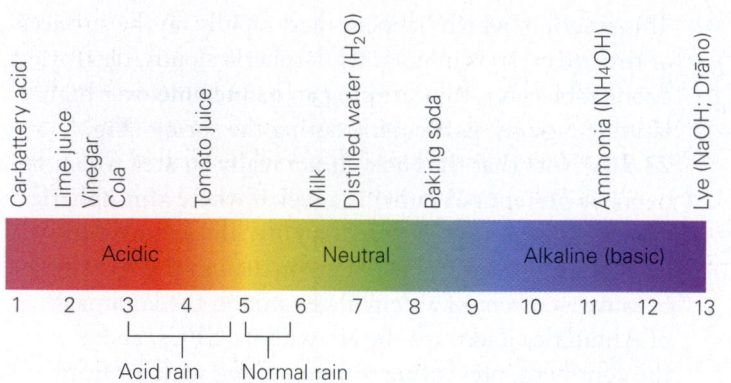

(b) Acid rain has a pH of between 3 and 5.

23.6 Recent Climate Change

Observed Changes in Atmospheric CO_2 and CH_4

We've seen that greenhouse gases, most notably CO_2 and CH_4, play a major role in the regulation of the Earth's surface temperatures—without these gases, the Earth could not be a home for life. Both CO_2 and CH_4 cycle through various biogeochemical reservoirs of the Earth System, and the rate of movement between reservoirs determines the amount in any given reservoir at any given time. For most of geologic time, the concentration of greenhouse gases in the atmosphere was governed by natural processes—volcanic eruptions, life-evolution events, forest fires, weathering of mountain belts, warming and cooling due to the Milankovitch cycles, changes in solar activity or cosmic-ray flux, and/or meteor impact. But beginning around 8,000 years ago, human society began to significantly modify the environment, first with the invention of agriculture and then, during the past two centuries, with the advent of industrialization.

Both industry and agriculture produce greenhouse gases and in effect transfer carbon that had been stored in underground reservoirs or biomass reservoirs into the atmospheric reservoir. For example: the burning of fossil fuels oxidizes vast quantities of carbon, which had previously been locked for millions of years in fossil fuel underground, to yield CO_2, which mixes into the atmosphere; the heating of calcite ($CaCO_3$) to produce the lime (CaO) of cement takes carbon that had been locked for millions of years in limestone underground and produces CO_2, which also mixes into the air; the clear-cutting of forests to make way for grazing land or fields replaces high-biomass vegetation (trees) with low-biomass vegetation (grasses or crops), thereby leaving CO_2 in the atmosphere; the decay of organic material in soggy rice paddies or melting permafrost, as well as the flatulence of cattle herds, produces significant quantities of CH_4 that would otherwise remain locked in biomass; and the melting of gas hydrates in the sediment of warming oceans releases CH_4 that would otherwise be locked in ice.

It may seem strange that human society's input of greenhouse gases can be so significant, but a comparison of human production of greenhouse gas relative to volcanic production shows that it indeed is. Specifically, researchers estimate that all volcanic eruptions together in a given year—including both submarine and subaerial eruptions—emit about 0.15 to 0.26 gigatons of CO_2. By comparison, activities of humanity emit about 35 gigatons of CO_2 every year (about 135 times as much).

FIGURE 23.20 The ozone hole over Antarctica.

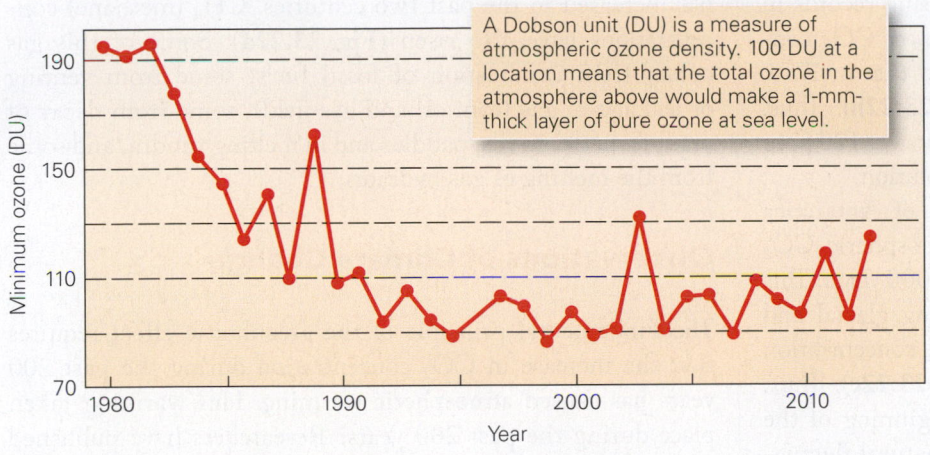

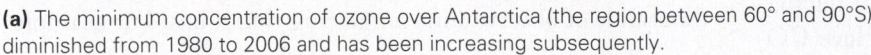

> A Dobson unit (DU) is a measure of atmospheric ozone density. 100 DU at a location means that the total ozone in the atmosphere above would make a 1-mm-thick layer of pure ozone at sea level.

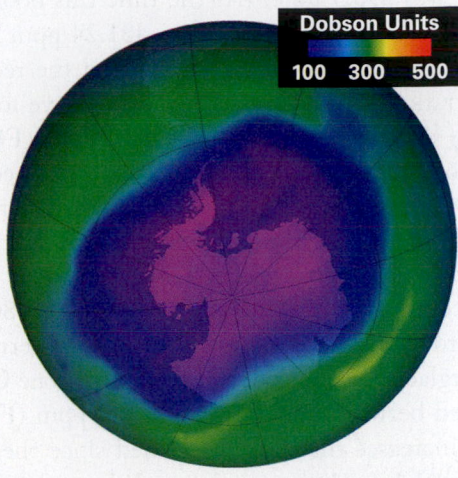

(a) The minimum concentration of ozone over Antarctica (the region between 60° and 90°S) diminished from 1980 to 2006 and has been increasing subsequently.

(b) A map showing the dimensions of the ozone hole in 2006—the largest hole ever recorded.

Put another way, human activities now produce more CO_2 in three days than do all of the volcanoes on Earth in a typical year. A large volcanic explosion, such as that of Mt. Pinatubo in 1991, emits only about as much CO_2 as society produces in one day, and a supervolcanic explosion (see Chapter 9), which happens only once every 100,000 to 300,000 years, produces about as much CO_2 as society produces in one year. About 85% of the CO_2 that we send into the atmosphere comes from burning fossil fuels and producing cement, while about 15% is a consequence of deforestation.

Significantly, not all of the greenhouse gases that society sends into the atmosphere stay there. In the case of CO_2, some dissolves in the ocean, some reacts with minerals in rocks during chemical weathering, and some gets incorporated into plants during photosynthesis. Before about 1900, natural "sinks" (the ocean, rock weathering, plants) could absorb most anthropogenic CO_2. But since then, the amount of CO_2 has exceeded the ability of natural sinks to absorb it (**Fig. 23.21**). In fact, calculations and isotopic studies suggest that only about 40% to 50% of the CO_2 that society produces—what researchers refer to as *anthropogenic CO_2*—gets reabsorbed by oceans, organisms or land during biogeochemical cycles. The remainder stays in the atmosphere.

Can we detect increases in atmospheric CO_2 concentration? In the early 1960s, a chemist named Charles Keeling decided to find out, and he set out to measure the concentration of CO_2 in the atmosphere using the most accurate methods available. To avoid areas with urban pollution, he collected air samples every month at the summit of Mauna Kea volcano in Hawaii. After completing many years of measurements, Keeling showed not only that distinct

seasonal variations occur (CO_2 concentration goes down in the warm summer when rates of photosynthesis increase, and it goes up in the winter when organic matter dies and decays) but also that the average annual concentration of CO_2 steadily rises (**Fig. 23.22a**). Specifically, he demonstrated that average yearly CO_2 concentration went from 320 parts per million (ppm) in 1965 to 360 ppm in 1995. Keeling died in 2005, but measurements at Mauna Kea have continued. On May 9, 2013, the daily CO_2 concentration surpassed 400 ppm for the first time in probably over 3 million years, and the average for the

FIGURE 23.21 Until about 1900, anthropogenic CO_2 could be absorbed by natural sinks. Since then, about 50% of anthropogenic CO_2 remains in the atmosphere.

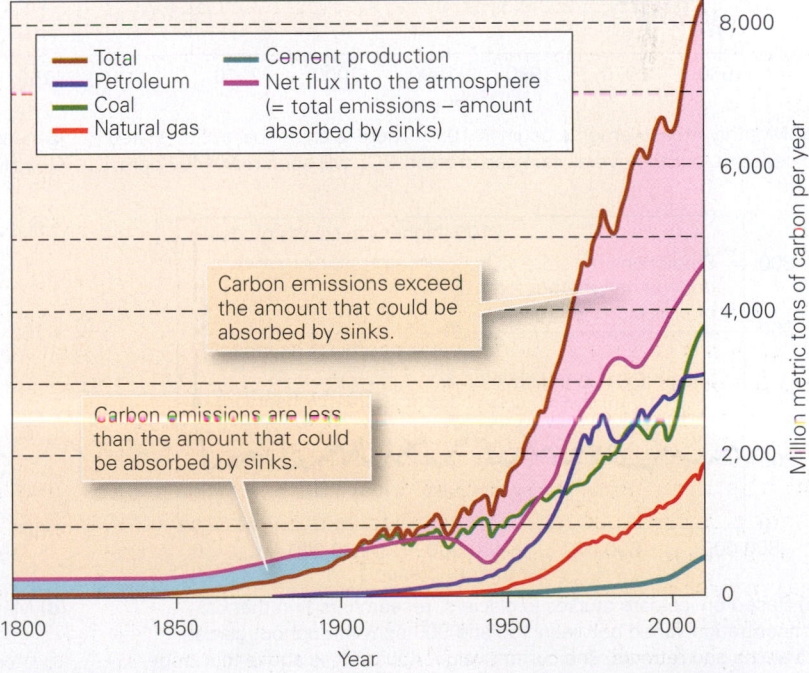

year was 396.5 ppm. (At the time this book went to press in 2014, CO_2 concentration was 401.24 ppm.) Using records in ice cores, researchers have extended the record of CO_2 concentration further back in time and have found that in 1750, CO_2 concentration was only 280 ppm (**Fig. 23.22b**). Thus, atmospheric CO_2 concentration has increased by over 120 ppm (40%) since the beginning of the industrial revolution.

Studies of gas bubbles trapped in glaciers of Antarctica allow researchers to extend the record of atmospheric CO_2 concentration back through almost the last 800,000 years. This record demonstrates that during the alternating glacial and interglacial periods of the Late Pleistocene CO_2 concentration varied between about 180 and 300 ppm (**Fig. 23.22c**). Thus, the increases that have happened since the beginning of the industrial revolution are beyond the range of natural fluctuations that occurred during the last 800,000 years. Have CO_2 concentrations ever been higher than they are today? Yes. For example, CO_2 concentration during the Eocene climatic optimum (49 to 56 Ma) was over 1,000 ppm, about 2.5 times what it is today.

CO_2 is not the only greenhouse gas whose concentration has increased in the past two centuries. CH_4 (methane) concentrations have also risen (**Fig. 23.22d**). Some of this gas comes from combustion of fossil fuels, some from venting or leakage of gas from oil and gas fields, some from decay of organic matter in rice paddies and in melting tundra, and some from the melting of gas hydrates.

Observations of Climate Change

The fundamental principle of the greenhouse effect requires that the increase in CO_2 concentration during the past 200 years has caused atmospheric warming. Has warming taken place during the past 200 years? Researchers have published thousands of observations suggesting that it has. For example:

*   Large ice shelves, such as the Larsen B Ice Shelf, along the Antarctic Peninsula, and the Ayles Ice Shelf, along Ellesmere Island in northernmost Canada, are breaking up rapidly (**Fig. 23.23a**).

FIGURE 23.22 Changes in carbon dioxide (CO_2) and methane (CH_4) concentrations over time.

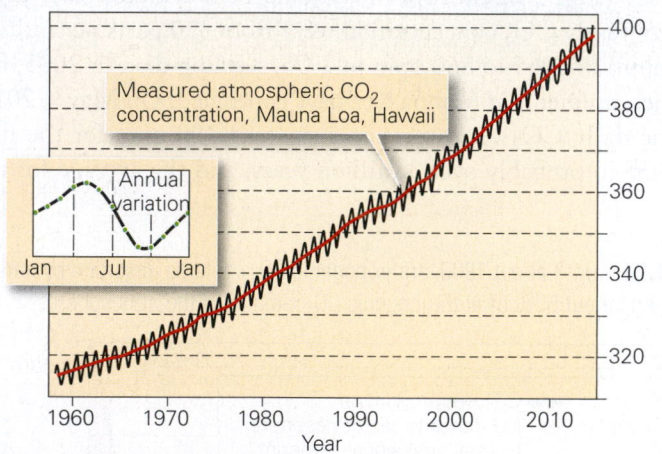

(a) Monthly measurements begin in 1960. There is an annual cycle, related to seasons, but the overall increase is clear; CO_2 will soon reach 402 ppm.

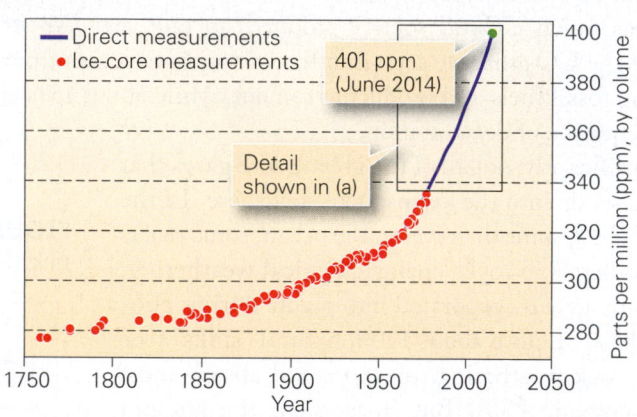

(b) Since the industrial revolution, CO_2 concentration has steadily increased.

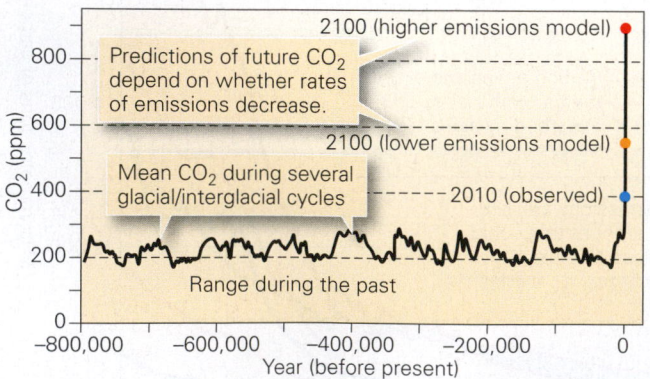

(c) Based on ice-core studies in glaciers, researchers find that CO_2 concentration varied between 180 and 300 ppm throughout glacial advances and retreats. The current value, 400 ppm, is above this range.

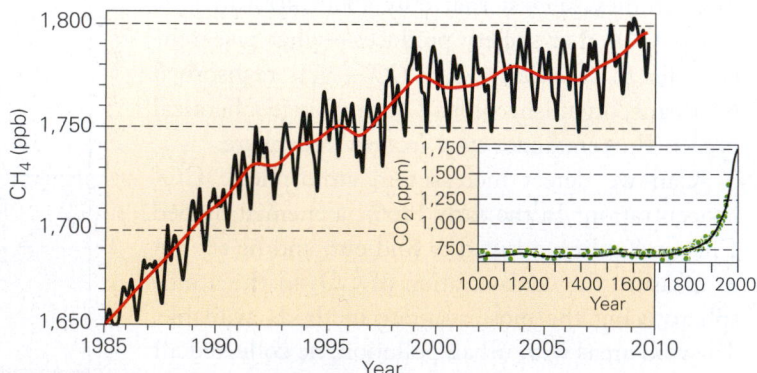

(d) Methane, another greenhouse gas, has been increasing, too.

- The area covered by sea ice in the Arctic Ocean has decreased substantially (**Fig. 23.23b**). Some estimates place the rate of ice-cover loss at about 3% per decade. But the trends are not simple—a graph of ice area from 1979 to 2013 shows many ups and downs lasting a few years (**Fig. 23.23c**), but the overall average appears to be in the direction of decreasing ice, and if this trend continues, it may be possible to sail across the Arctic Ocean within decades. Support for this prediction comes from studies of the age of the sea ice—the area of "old ice" (ice over 5 years old) has decreased by about 50% over the last 25 years (**Fig. 23.23d**).

- The Greenland ice sheet is melting at an accelerating pace. Studies suggest that the rate of ice loss has increased from 90 to 220 cubic km per year in the last 10 years and that in places the sheet is thinning by about 1 m per year. Warming has caused the number of days during the year when melting takes place to increase (**Fig. 23.24a**). In addition, the annual melt zone along the margins of the ice sheet has widened dramatically, because the elevation of the equilibrium line (see Chapter 22) has risen (**Fig. 23.24b**). Further indication of melting comes from observing the flow rates of valley glaciers draining the ice sheet—they flowed 50% faster in 2003 than they did in 1992.

- Valley glaciers worldwide have been retreating rapidly, so that areas that were once ice covered are now bare. The change is truly dramatic in many locations (**Fig. 23.25a**).

- Worldwide, glacial volumes have been diminishing. About 400 km³ of ice disappears every year (**Fig. 23.25b**).

- The area of permafrost in high latitudes has substantially decreased, and melt ponds have formed on the surface of once-frozen land (**Fig. 23.25c**). In fact, large regions that once stayed frozen all year are now melting in the summer.

- Average annual water vapor in the atmosphere has been increasing due to evaporation of warmer seas. This trend is hard to characterize, however, because humidity can vary significantly around the world at a given time.

- Biological phenomena that are sensitive to climate are being disrupted. For example: the time at which sap in the maple trees of

FIGURE 23.23 The melting of sea ice.

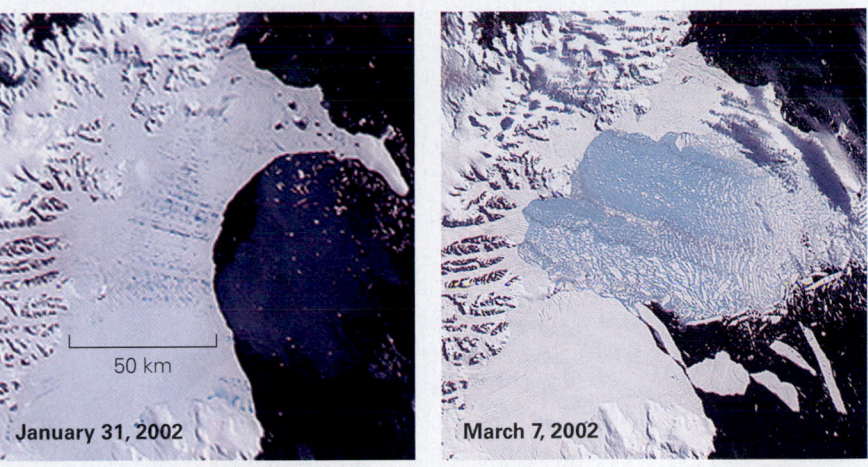

January 31, 2002   March 7, 2002

(a) In 2002, the Larson B Ice Shelf of Antarctica disintegrated over the course of a month.

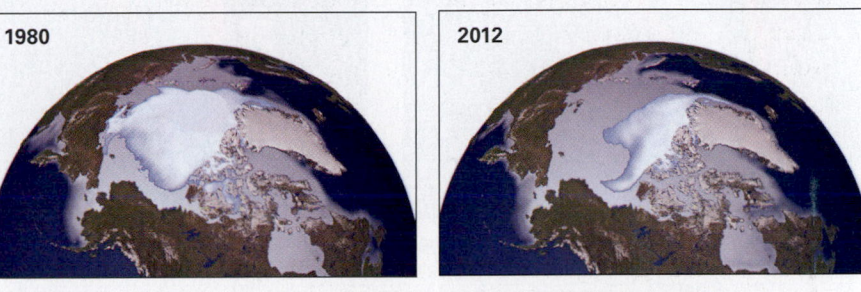

1980   2012

(b) Between 1980 and 2012, the coverage of sea ice in the Arctic decreased.

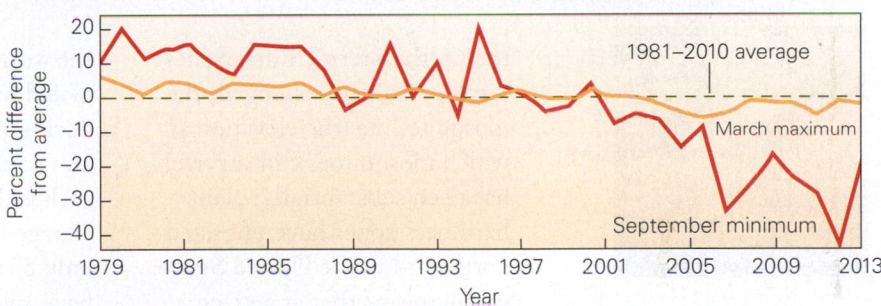

(c) The area of sea ice in the Arctic has gone up and down yearly, but on average it is decreasing.

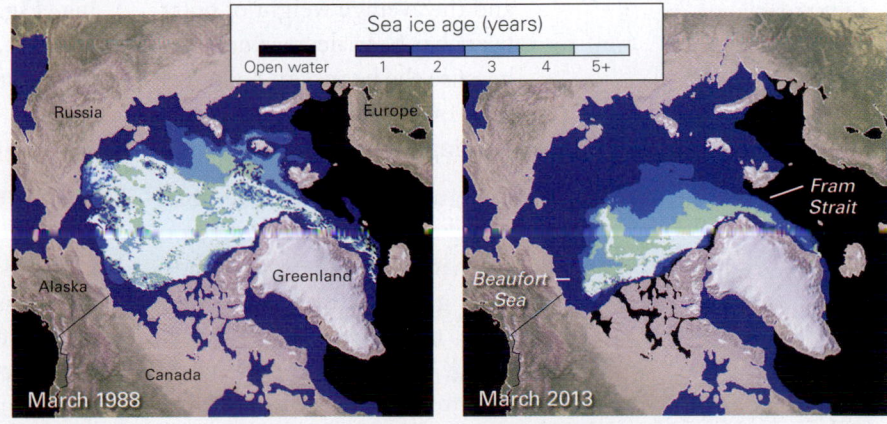

(d) The age of sea ice in the Arctic Ocean has decreased substantially during the past 25 years.

FIGURE 23.24 Melting of the Greenland Ice Sheet.

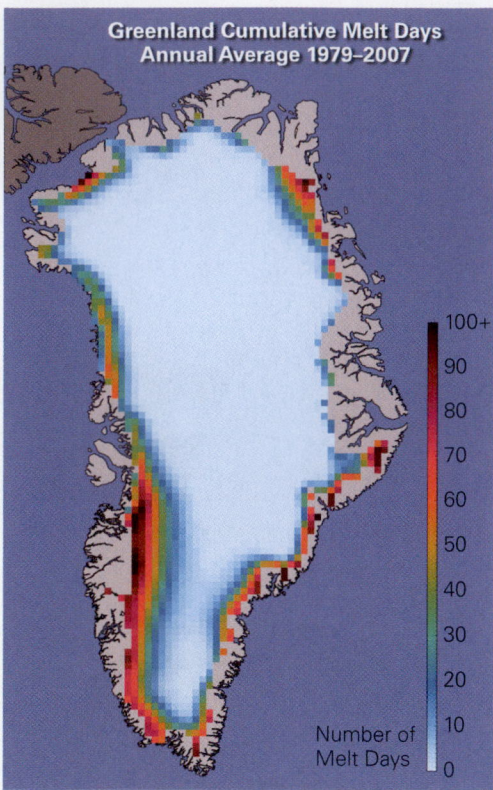

Greenland Cumulative Melt Days
Annual Average 1979–2007

100+
90
80
70
60
50
40
30
20
10
0

Number of
Melt Days

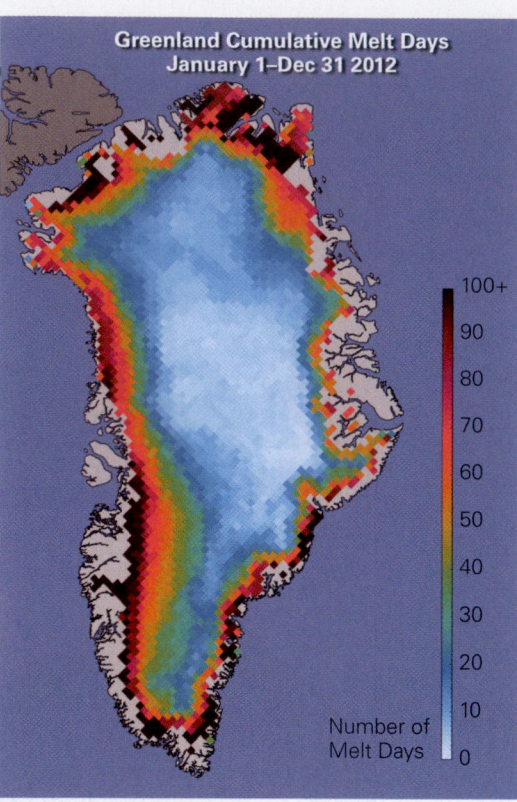

Greenland Cumulative Melt Days
January 1–Dec 31 2012

100+
90
80
70
60
50
40
30
20
10
0

Number of
Melt Days

The gray area, spotted with puddles, is the melt zone.

June 13, 2002

June 17, 2003

(a) The number of days during which melting of the Greenland Ice Sheet takes place has increased.

(b) The summer melt line has risen to higher elevations. Each photo shows the same area.

the northeastern United States starts to flow has changed; the mosquito line (the elevation at which mosquitoes can survive) has risen substantially; plant hardiness zones have migrated northward in the United States, which means that at a given location it's possible to plant earlier in the year (**Fig. 23.26**); and the average weight of polar bears has been decreasing, because the bears can no longer walk over pack ice to reach their hunting grounds in the sea.

Researchers consider the observations above to be substantive evidence that, averaged over decades, average global atmospheric temperature near the Earth's surface is rising, a phenomenon widely referred to as *global warming*. Direct measurements of temperatures, collected at recording stations around

the world since about 1880, support this proposal. Specifically, global mean atmospheric temperature has risen by almost 1°C during the last century and is higher now than it has been at any time during the past 2,000 years (**Fig. 23.27a, b**). Although such a change may seem small, the magnitude of temperature change between the last ice age and now, by comparison, was only 3° to 5°C. A small change in average temperature may have major consequences. Significantly, temperature change is not uniform around the world—some regions appear to be warming more than others, and some areas have been cooling (**Fig. 23.27c**). Because of the increase in atmospheric temperature, average measured values of near-surface ocean-water temperatures have also been rising (**Fig. 23.28**).

Interpretations and Potential Consequences of Climate Change

Climate data are not easy to obtain or to interpret, for individual measurements may have large uncertainties, and plots of measurements may show significant scatter. And, as is to be expected in any scientific endeavor, in some cases measurements and conclusions based on them do not stand the test of time and turn out to be incorrect, to be replaced by the results of newer studies. In the case of climate-change studies,

FIGURE 23.25 Global change in glaciers and permafrost.

1941

2004

(a) The Muir Glacier in Alaska retreated 12 km between 1941 and 2004.

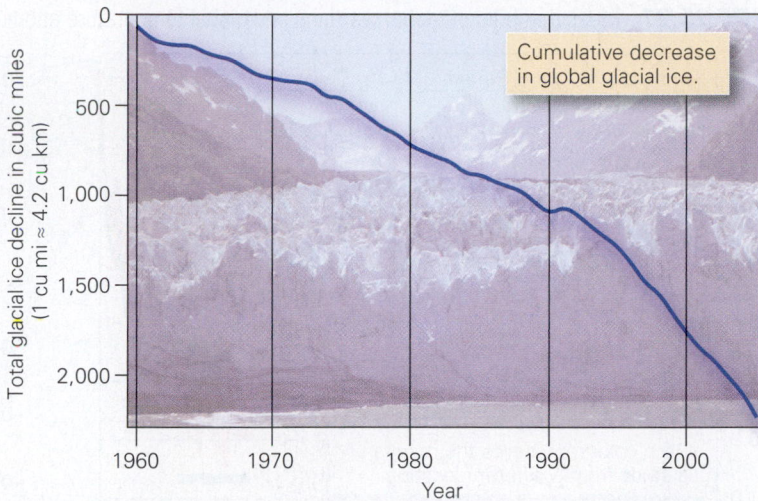

Cumulative decrease in global glacial ice.

Total glacial ice decline in cubic miles (1 cu mi ≈ 4.2 cu km)

(b) Global glacial ice volume has been decreasing by 400 km³ (about 100 mile³) per year.

(c) Lakes and puddles forming when permafrost melts along the coast of the Arctic Ocean.

the data show ups and downs from year to year and in some cases from decade to decade. This fact can lead to confusion in interpreting the data, because a trend that lasts for, say, 5 years, does not necessarily represent a trend that lasts for 100 years. Further, because so many variables control climate change, and because so many feedbacks are likely involved in climate change, it's not easy to associate specific individual effects with specific individual causes. In an attempt to make sense of overwhelming volumes of sometimes contradictory data and interpretations relevant to climate change, a group of leading scientists founded the *Intergovernmental Panel on Climate Change* (IPCC), whose purpose is to evaluate published climate studies from a broad perspective

FIGURE 23.26 Plant hardiness zones have been migrating northward.

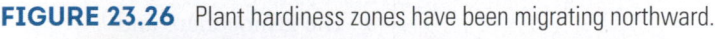

1990

2012

Zone

2 3 4 5 6 7 8 9 10

FIGURE 23.27 Measurements of global warming and global temperature anomalies.

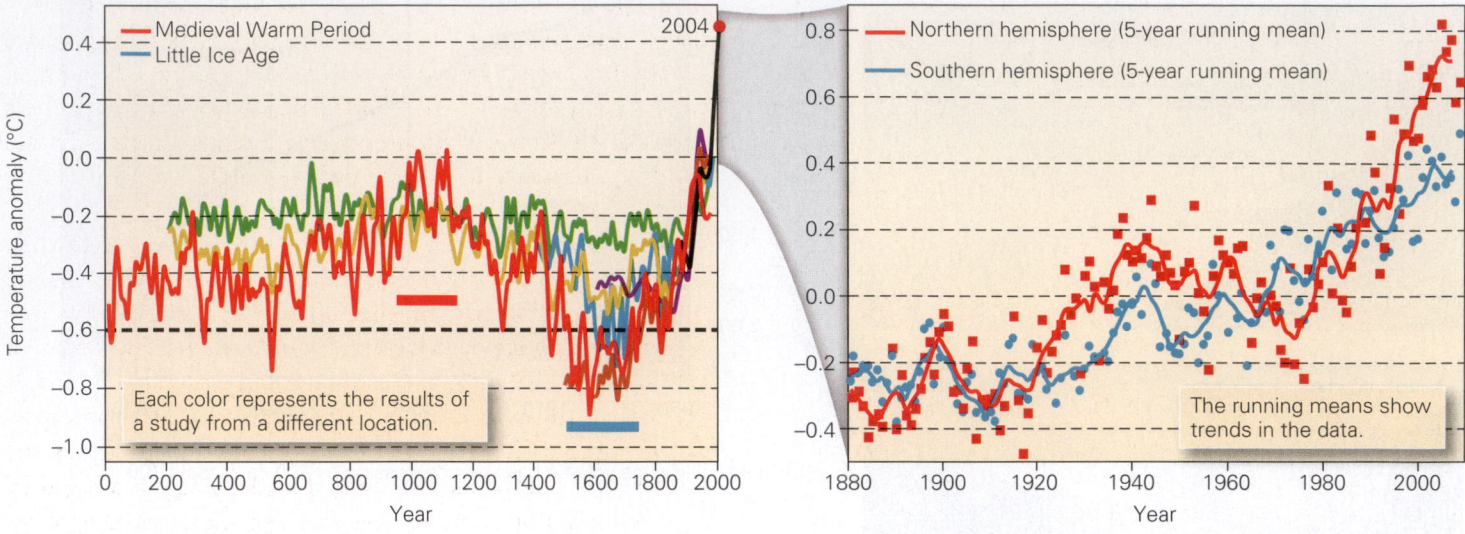

(a) Reconstructions of temperature during the past 2,000 years. Each color is a published estimate from a research team. The black line is the average of direct measurements.

(b) Graphs showing the change in global average temperature since 1880 relative to a reference value. Note that both the northern and southern hemispheres show a temperature increase.

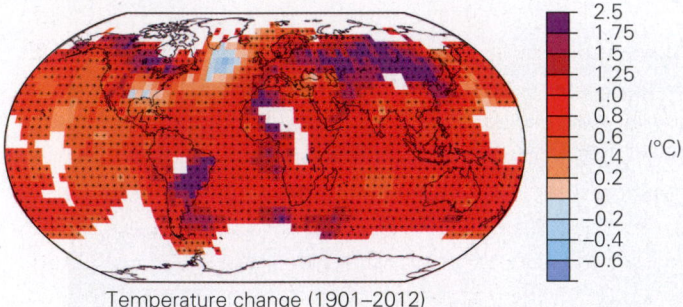

(c) Colors represent change in average atmospheric temperature at the Earth's surface, for the time period of 1901 to 2012. Redder areas are warm, and blue areas are cold.

and provide assessments of conclusions from the studies. This assessment has societal significance because climate change can have major implications for society and for global policy decisions. The IPCC, sponsored by the World Meteorological Organization and the United Nations, summarizes its conclusions in a report that is revised and published every 5 years. The language describing the likelihood that global warming is happening, and that humans have contributed significantly to causing it, has become progressively less equivocal in successive editions of the report. The *Fifth Assessment Report*, published in 2013, states:

Warming of the climate system is unequivocal, and since the 1950s, many of the observed changes are unprecedented over decades to millennia. The atmosphere and ocean have warmed, the amounts of snow and ice have diminished, sea level has risen, and the concentrations of greenhouse gases have increased . . . It is extremely likely that human influence has been the dominant cause of the observed warming since the mid-20th century.

FIGURE 23.28 The difference in global temperature of shallow ocean water relative to the average temperature between 1961 and 2006.

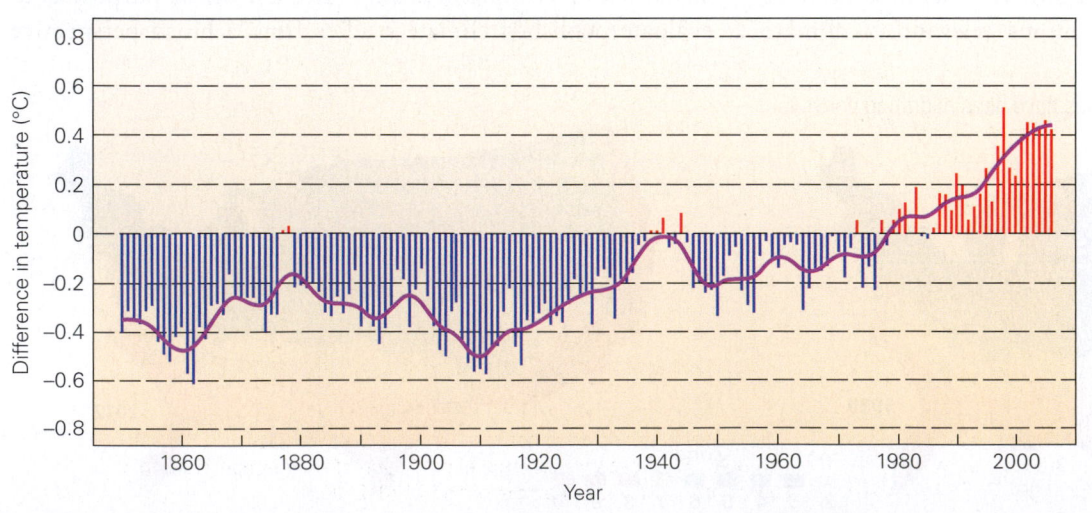

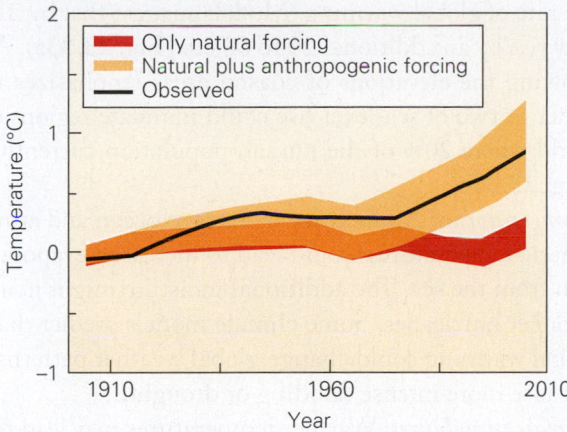

FIGURE 23.29 Model comparing the consequences of natural forcing both to natural plus anthropogenic forcing and to observed changes in temperature.

This wording means that the vast majority of climate researchers have concluded that global warming is real, and the activities of people (anthropogenic forcing)—the burning of fossil fuels, the production of cement, and changes in land use such as the cutting down of forests—have played a significant role in causing the change. Non-anthropogenic forcing, such as changes in solar radiation or cosmic-ray flux, do not appear to be of sufficient magnitude to have caused all observed warming. Calculations suggest that the rate of temperature change during the past 50 years is greater than the rate for the previous 50 years.

Some researchers suggest that human impact on climate became noticeable as far back as 8,000 years ago and that climate has been trending toward warmer conditions ever

since. Deviations from the warming trend have been attributed, speculatively, to times when human population abruptly decreased (due to pandemics). During these times, the production of greenhouse gas slowed and forests returned. But deviations have been relatively short-lived, so overall, researchers conclude that without anthropogenic forcing, Earth's climate would be significantly cooler, and we might even be heading toward another ice age (**Fig. 23.29**).

If the fifth assessment of the IPCC proves to be correct, then society faces the challenge of either slowing climate change or dealing with its consequences. The effects of global warming over the coming decades to centuries remain the subject of intense debate because predictions depend on computer models and not all researchers agree on how to construct or interpret these models. The role of clouds in climate change, for example, remains poorly understood and inadequately addressed by such models. But newer models, running on faster computers, provide increasingly reliable constraints on future trends. In a worst-case scenario, global warming will continue into the future at the present rate, so that by 2050—within the lifetime of many readers of this book—the average annual temperature will have increased in some parts of the world by 1.5° to 2.0°C. At these rates, by the end of the century, temperatures could be over 4°C warmer depending on the model used (**Fig. 23.30a**), and by 2150 global temperatures may be 5° to 11°C warmer than at present—the warmest since the Eocene Epoch, 40 million years ago. Models predict that warming will not be the same everywhere—the greatest impact will be in the Arctic (**Fig. 23.30b**). The effects of such a change are controversial, but according to many climate models, the following events might happen.

FIGURE 23.30 Model predictions of future global warming.

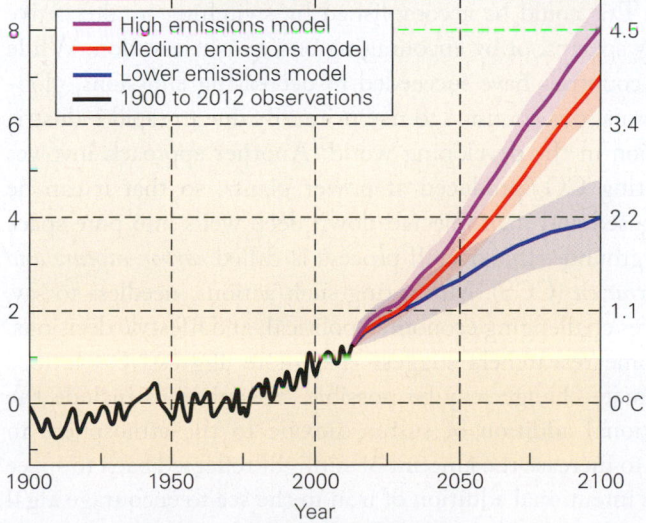

(a) Model calculations of global warming. Different models assume different rates of CO_2 emissions. All suggest significant global temperature increase by 2100.

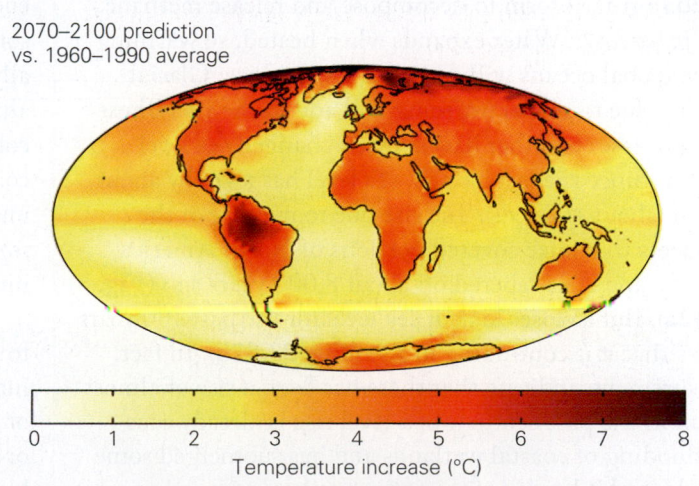

2070–2100 prediction vs. 1960–1990 average

(b) Global warming does not mean that all locations warm by the same amount. This map shows a model prediction of how temperature may vary with location for the time period 2070–2100.

- *Shift in climate belts:* As the climate overall warms, temperate and desert regions would occur at higher latitudes. Thus, regions that are now agricultural areas may dry out and become unfarmable (**Fig. 23.31a**). The change in climate would also affect the amount and distribution of precipitation (rain and snow) that falls. For example, in North America summers would become drier and winters would become wetter (**Fig. 23.31b**). One way to picture this change is to think of how the "climatic latitude" of a state in the United States might change over time. Models suggest that if current warming rates continue, the future climate of Illinois would be like the present climate of Texas, and the future climate of New Hampshire might be like the current climate of North Carolina (**Fig. 23.31c**). Models also suggest that global warming would dramatically increase the number of health-threatening *heat waves* (loosely defined as a prolonged period of excessively hot weather, i.e., above 32°C, or 90°F) that would affect particular regions (**Fig. 23.31d**).

- *Ice retreat and snow-line rise:* We've already seen evidence that the volume of glacial ice worldwide is decreasing. This is manifested by the shrinkage and retreat of most glaciers, exposing land that was once buried by ice. Of note, some glaciers are advancing. Current glacial advance could be a consequence either of surging due to addition of liquid water to the base of the glacier (see Chapter 22) or of local increases in snowfall due to the addition of moisture to the atmosphere as the oceans warm. As temperatures increase, the snow line in mountains rises to higher latitudes and higher elevations. This change affects ecosystems and may even impact tourism by shortening the ski season at resorts.

- *Melting permafrost:* Warming climates will cause areas of unglaciated land that previously had remained frozen most of the year to thaw during the summer. Such melting of permafrost will permit organic matter that had been stored in frozen form to decompose and release methane.

- *Rise in sea level:* Water expands when heated, so warming of the global oceans will cause sea level to rise. Glacial melting due to global warming will add to the rise. These two processes together have already changed sea level notably. Since the last ice age, sea level has risen by about 120 m (about 400 feet). The rise due to melting of the ice sheets that once covered portions of North America, Europe, and Asia tapered off about 8,000 years ago (**Fig. 23.32a**). But a closer look at sea level for the past 130 years shows that it is continuing to rise (**Fig. 23.32b**). In fact, measurements indicate that there has been a rise of almost 12 cm in the past century. Sea-level rise is already causing flooding of coastal wetlands and has submerged some islands, and it has led some nations to begin investing in new coastal flood-control measures. This rise correlates with warming of ocean water (**Fig. 23.32c**) and with the melting of glaciers (see Fig. 23.25b).

The amount of sea-level rise in the future depends on the rate of global warming. Models suggest that by 2100, it may rise by an additional 20 to 60 cm (**Fig. 23.33a**). A map showing the elevations of coastal areas emphasizes that a meter or two of sea-level rise could inundate regions of the world where 20% of the human population currently lives (**Fig. 23.33b**).

- *Stronger storms:* An increase in average ocean and atmospheric temperatures would lead to increased evaporation from the sea. The additional moisture might nourish stronger hurricanes. Some climate models predict that global warming could change global weather patterns and/or cause more intense flooding or drought.

- *Increase in wildfires:* Warmer temperatures may lead to an increase in the frequency of wildfires because the moisture content of plants is lower.

- *Interruption of the oceanic heat conveyor:* Oceanic currents play a major role in transferring heat across latitudes. According to some models, if global warming melts enough polar ice, the resulting freshwater would dilute surface ocean water at high latitudes. This water could not sink, and thus thermohaline circulation would be shut off (see Chapter 18), preventing the water from conveying heat.

The potential changes described above, along with other studies that estimate the large economic cost of global warming, imply that the issue needs to be addressed seriously and soon. But what can be done? The 160 nations that signed the 1997 Kyoto Accord, at a summit meeting held in Japan, propose that the first step would be to slow the input of greenhouse gases into the atmosphere by decreasing the burning of fossil fuels. This could be accomplished by switching to alternative energy sources, or by encouraging energy conservation. While some countries have succeeded in decreasing emissions, globally, emissions continue to rise primarily due to rapid industrialization in the developing world. Another approach involves collecting CO_2 produced at power plants, so that it can be condensed and then injected down deep wells into pore space underground—this overall process is called *carbon capture and sequestration* (CCS). Motivating such actions, needless to say, involves challenging economic, political, and lifestyle decisions.

Some researchers suggest that more aggressive solutions to climate change may be possible. Speculations include the intentional addition of sulfur dioxide to the atmosphere in order to increase the amount of sunlight reflected back to space or the intentional addition of iron to the sea to encourage algal blooms that would absorb CO_2. But the feasibility and potential risks of such approaches remains very uncertain.

FIGURE 23.31 Model predictions of changes that may happen if current global warming continues.

Tundra | Deciduous forest | Evergreen forest | Boreal forest | Shrub and grassland | Sparse vegetation

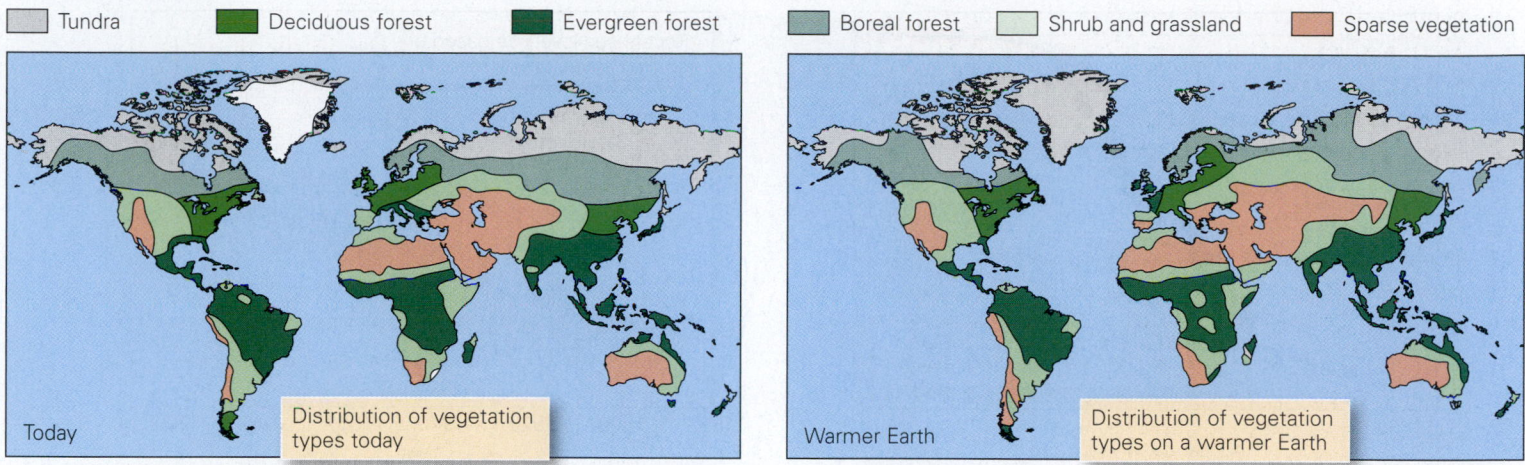

Distribution of vegetation types today

Today

Warmer Earth

Distribution of vegetation types on a warmer Earth

(a) Model predictions of changes that may happen if current global warming continues.

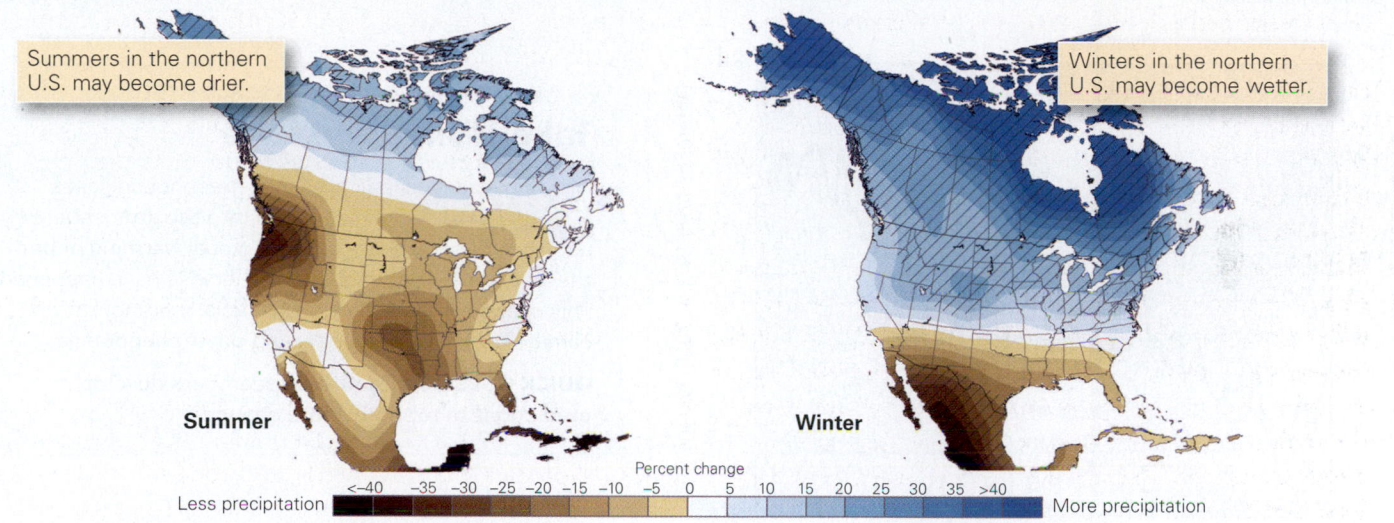

Summers in the northern U.S. may become drier.

Winters in the northern U.S. may become wetter.

Summer

Winter

Percent change

Less precipitation | <−40 −35 −30 −25 −20 −15 −10 −5 0 5 10 15 20 25 30 35 >40 | More precipitation

(b) Models suggest that the amount of precipitation (rain and snow) at locations in North America about 100 years from now will be different than it is today.

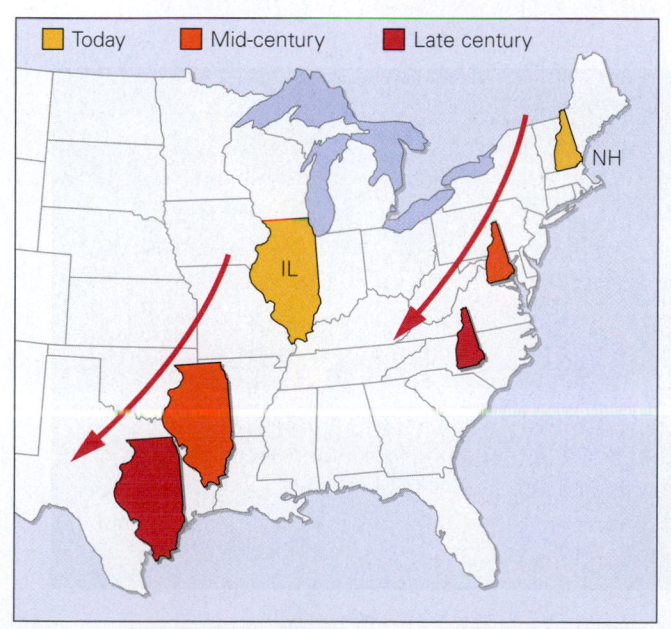

Today | Mid-century | Late century

NH

IL

(c) This map represents, symbolically, how the climate of a given state may change if the climate warms, according to models. At the end of the century, northern states (such as Illinois) may have climates that are like those of southern states today.

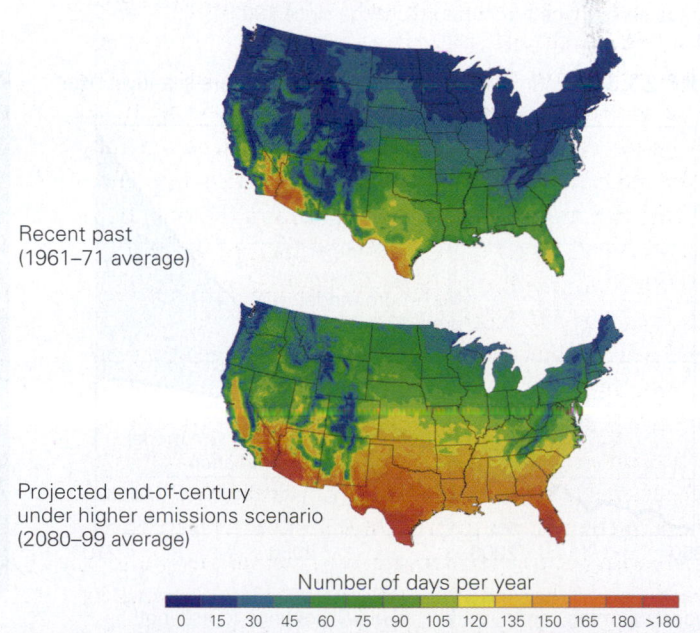

Recent past (1961–71 average)

Projected end-of-century under higher emissions scenario (2080–99 average)

Number of days per year

0 15 30 45 60 75 90 105 120 135 150 165 180 >180

(d) A prediction of the increase in the number of days above 32°C (90°F) at the end of the century, in comparison to today, in the United States.

FIGURE 23.32 Sea-level rise of the recent past and possible sea-level rise of the future.

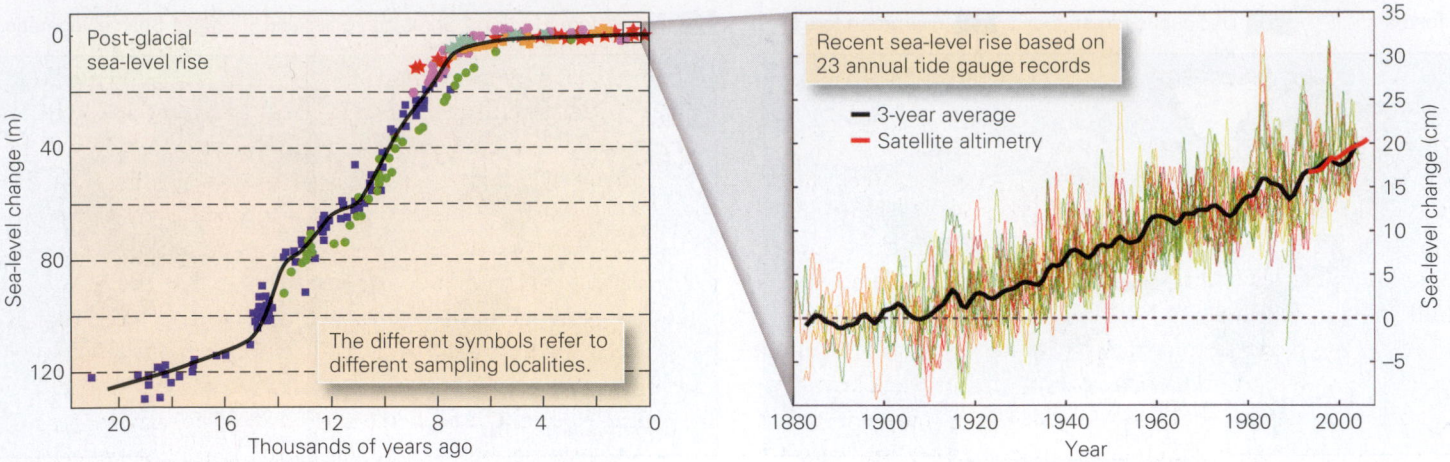

(a) Sea level rose rapidly after the last ice age due to melting of continental glaciers.

(b) Tide gauges document sea-level rise of the past 130 years.

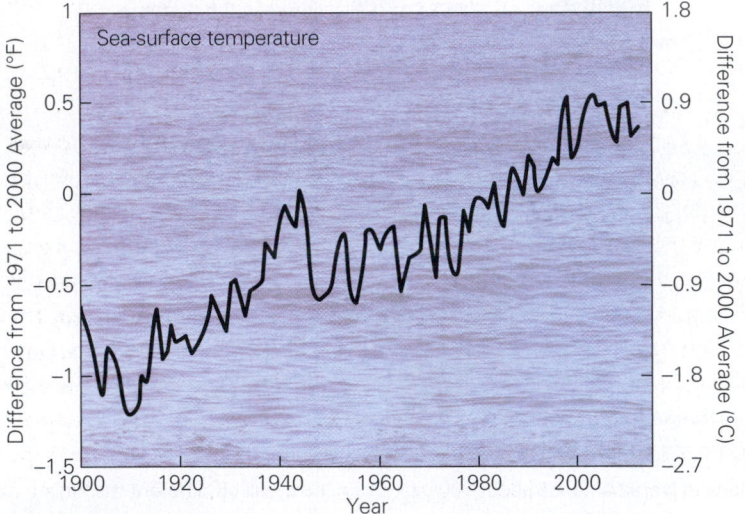

(c) Average sea-surface temperature change since 1900.

Take-Home Message

Growing evidence indicates that greenhouse gases (CO$_2$ and CH$_4$) being introduced into the atmosphere by activities of society are causing global warming of both the air and sea that might not otherwise have happened. This change is associated with glacial melting, shifting climate belts, sea-level rise, and other phenomena.

QUICK QUESTION: How do researchers develop predictions of future climate change?

FIGURE 23.33 Estimates and consequences of future sea-level rise.

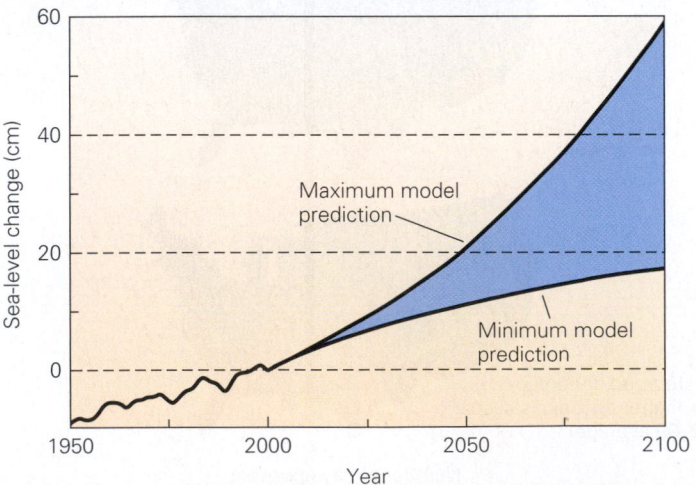

(a) If global warming continues, sea level will continue to rise, both because of the addition of glacial meltwater and the expansion of water that happens when water becomes warmer.

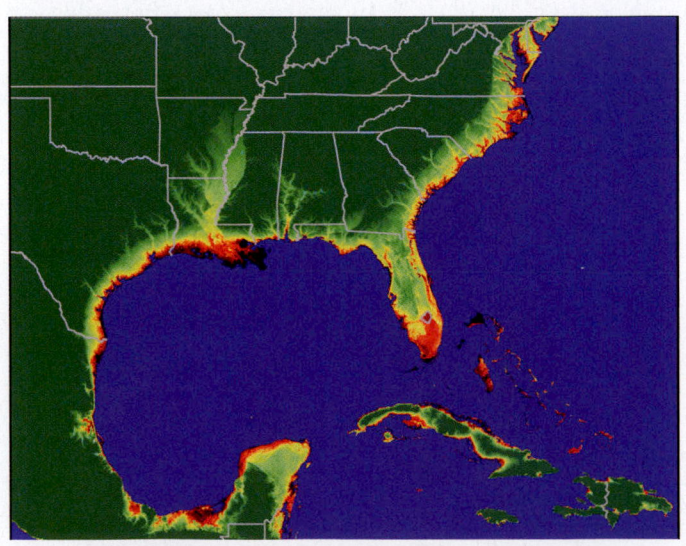

(b) The red and black areas of this map are so low that they could be flooded if sea level rose a few meters.

23.7 The Future of the Earth

Most of the discussion in this book has focused on the past, for the geologic record preserved in rocks tells us of earlier times. Let's now bring this book to a close by facing in the opposite direction and speculating what the world might look like in geologic time to come.

In the geologic near term, the future of the world likely depends largely on human activities. Whether the Earth System shifts to a new equilibrium, whether a mass-extinction event takes place, or whether society achieves **sustainable growth** (an ability to prosper within the constraints of the Earth System) will depend on our own foresight and ingenuity. Projecting thousands of years into the future, we might well wonder if the Earth will return to ice-age conditions, with glaciers growing over major cities and the continental shelf becoming dry land, or if the ice age is over for good because of global warming. No one really knows for sure. If we project millions of years into the future, it is clear that the map of the planet will change significantly because of the continuing activity of plate tectonics. For example, during the next 50 million years or so, the Atlantic Ocean will probably become bigger, the Pacific Ocean will shrink, and the western part of California will migrate northward. Eventually, Australia may crush against the southern margin of Asia, and the islands of Indonesia will be flattened in between.

Predicting the map of the Earth beyond that is hard, because we can't predict where new subduction zones will develop. Perhaps subduction of the Pacific Ocean will lead to the collision of the Americas with Asia, to produce a new supercontinent ("Amasia"). A subduction zone eventually will form on one side (or both sides) of the Atlantic Ocean, and the ocean will be consumed. As a consequence, the eastern margin of the Americas will collide with the western margin of Europe and/or Africa. The sites of major cities—New York, Miami, Rio de Janeiro, Buenos Aires, and London—will be incorporated in a collisional mountain belt and likely will be subjected to metamorphism and igneous intrusion before being uplifted and eroded. Shallow seas may once again cover the interiors of continents and then later retreat, and glaciers may once again cover the continents—it happened in the past, so it could happen again! And if the past is the key to the future, we *Homo sapiens* might not be around to watch our cities enter the rock cycle, for biological evolution may have introduced new species to the biosphere, and there is no way to predict what these species will be like. Perhaps 100 million years from now, the stratigraphic record of our time might be several centimeters of strata containing anomalous isotopes and unusual chemicals, trace fossils of concrete structures, and a record of widespread extinctions.

And what of the end of the Earth? Geologic catastrophes resulting from asteroid and comet collisions will undoubtedly occur in the future as they have in the past. We can't predict when the next strike will come, but unless the object can be diverted, Earth is in for another radical readjustment of surface conditions. It's not likely, however, that such collisions will destroy our planet. Rather, astronomers predict that the end of the Earth will occur some 5 billion years from now, when the Sun begins to run out of nuclear fuel. When this happens, outward-directed thermal pressure caused by fusion reactions will no longer be able to prevent the Sun from collapsing inward, because of the immense gravitational pull of its mass. Were the Sun a few times larger than it is, the collapse would trigger a supernova explosion that would blast matter out into space to form a new nebula, perhaps surrounding a black hole. But since the Sun is not that large, the thermal energy generated when

FIGURE 23.34 In about 5 billion years, the Sun will become a red giant.

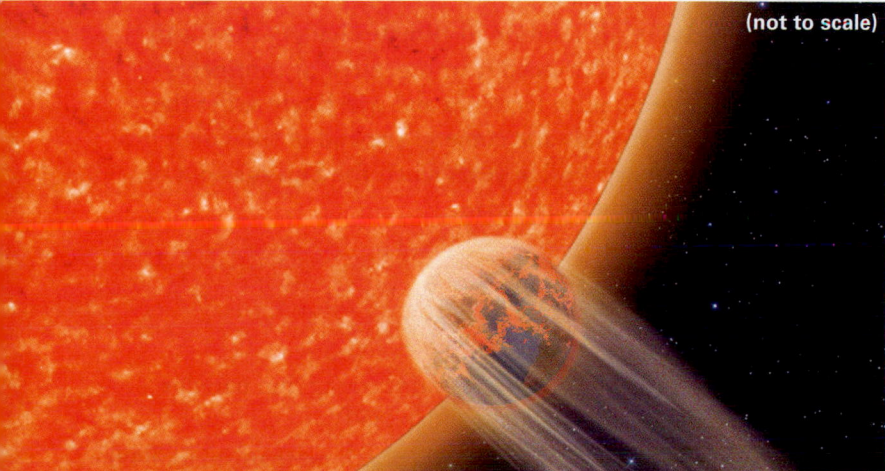

(a) As the surface of the red giant approaches the Earth, the Earth will evaporate like a giant comet.

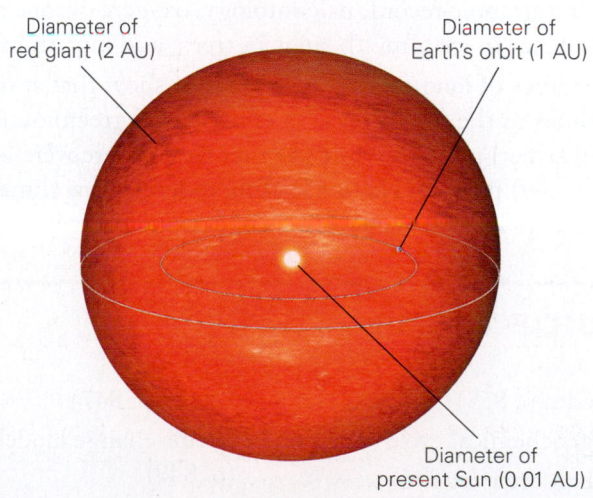

Diameter of red giant (2 AU)

Diameter of Earth's orbit (1 AU)

Diameter of present Sun (0.01 AU)

(b) At full size, the red giant will have a diameter twice that of the Earth's orbit, so atoms of the Earth will become part of the star.

its interior collapses inward will heat the gases of its outer layers sufficiently to cause them to expand. As a result, the Sun will become a red giant, a huge star whose radius would grow beyond the orbit of Earth (**Fig. 23.34**). Our planet will then vaporize, and its atoms will join an expanding ring of gas—the ultimate global change. If this happens, the atoms that once formed Earth and all its inhabitants through geologic time may eventually be incorporated in a future solar system, where the cycle of planetary formation and evolution will begin anew.

Take-Home Message

In the near term, Earth's surface will be affected by decisions of human society. Over longer time scales, the map of the Earth will change due to plate interactions and sea-level change.

QUICK QUESTION: What might happen to the atoms that make up the Earth long after the red-giant stage of Solar System evolution?

CHAPTER SUMMARY

- We refer to the global interconnecting web of physical and biological phenomena on Earth as the Earth System. Global change involves transformations or modifications of physical and biological components of the Earth System through time. Unidirectional change results in transformations that never repeat, whereas cyclic change involves repetition of the same steps over and over.

- Examples of unidirectional change include differentiation of the solid Earth, growth of continents, formation of the oceans, change in atmospheric composition, and life evolution.

- Examples of physical cycles that take place on Earth include the supercontinent cycle, the sea-level cycle, and the rock cycle.

- A biogeochemical cycle involves the passage of a chemical among nonliving and living reservoirs. Examples include the hydrologic cycle and the carbon cycle. Global change occurs when the relative proportions of the chemicals in different reservoirs change.

- Tools for documenting global climate change include the stratigraphic record, paleontology, oxygen-isotope ratios, bubbles in ice, growth rings in trees, and human history.

- Studies of long-term climate change show that at some times in the past the Earth experienced greenhouse (warmer) periods, while at other times there were icehouse (cooler) periods. Factors leading to long-term climate change include the positions of continents, volcanic activity, the uplift of land, and the formation of materials that remove CO_2.

- Short-term climate change can be seen in the record of the last million years. In fact, during only the past 15,000 years, we see that the climate has warmed and cooled several times. Causes of short-term climate change include changes in Earth's orbit and tilt, changes in surface albedo, changes in ocean currents, and perhaps fluctuations in solar radiation and cosmic rays.

- Mass extinction, a catastrophic change in biodiversity, may be caused by the impact of a comet or asteroid or by intense volcanic activity associated with a superplume.

- During the last two centuries, humans have changed landscapes and modified ecosystems and have added pollutants to the land, air, and water at rates faster than the Earth System can process.

- The addition of greenhouse gases (CO_2 and CH_4) to the atmosphere appears to be causing global warming, which could shift climate belts and lead to a rise in sea level. Sources for the added CO_2 include fossil-fuel burning, cement production, and deforestation.

- In the future, in addition to climate change, the Earth will witness a continued rearrangement of continents resulting from plate tectonics and will likely suffer the impact of asteroids and comets. The end of the Earth may come in about 5 billion years when the Sun runs out of fuel and becomes a red giant.

GUIDE TERMS

albedo (p. 855)
biogeochemical cycle (p. 846)
carbon cycle (p. 846)

climate (p. 847)
climate-change model (p. 849)

differentiation (p. 842)
Earth System (p. 839)
ecosystem (p. 859)

faint young Sun paradox (p. 854)
feedback mechanism (p. 849)

global change (p. 839)

global climate change
(p. 847)

global cooling (p. 848)

global warming (p. 848)

Goldilocks effect (p. 854)

greenhouse effect (p. 848)

greenhouse gas (p. 848)

greenhouse (hothouse) period
(p. 850)

habitable zone (p. 854)

icehouse period (p. 850)

mass-extinction event
(p. 857)

ozone hole (p. 861)

paleoclimate (p. 849)

pollution (p. 861)

sedimentary sequence
(p. 844)

steady-state condition
(p. 846)

sunspot cycle (p. 856)

supercontinent cycle (p. 844)

sustainable growth (p. 873)

weather (p. 847)

REVIEW QUESTIONS

1. What does the term *Earth System* refer to?

2. How have the Earth's interior, crust, and atmosphere changed since the planet first formed?

3. What processes control the rise and fall of sea level on Earth?

4. Describe the various reservoirs that play a role in the carbon cycle and how carbon transfers among these reservoirs.

5. How do paleoclimatologists study ancient climate change?

6. Contrast icehouse and greenhouse conditions. Have icehouse conditions happened prior to the most recent ice age?

7. What are the possible causes of long-term climatic change?

8. What factors explain short-term climatic change?

9. Give some examples of events that cause catastrophic change.

10. Give some examples of how humans have changed the solid Earth.

11. What are pollutants, and why are they a problem? What is the ozone hole, and why did it form?

12. What is the evidence that researchers use to argue that global warming has been taking place during the past few centuries in response to human activities? What effects might global warming have on the Earth System?

13. What approaches could be employed to decrease CO_2 emissions?

14. What are some likely scenarios for the long-term future of the Earth?

ON FURTHER THOUGHT

15. If global warming continues, how will the distribution of grain crops change? Might this affect national economies, and if so, why? How will the distribution of spruce forests change?

16. Currently, tropical rainforests are being cut down at a rate of 1.8% per year. At this rate, how many more years will the forests survive? In the eastern United States, the proportion of land with forest cover today has increased over the past century. In fact, most of the farmland that

existed in New York State in 1850 is forestland today. Why? How might this change affect erosion rates in the region?

17. Using the library or the Web, examine the change in the nature of world fisheries that has taken place in the last 50 years. Is the world's fish biomass sustainable if these patterns continue? What has happened to whale populations during the past 50 years?

smartwork

smartwork.wwnorton.com

This chapter's Smartwork features:

· Labeling exercise on the Carbon Cycle.
· Animation-based problem on the Milankovitch Cycles.
· Comprehensive questions dealing with climate change.

GEOTOURS

This chapter's GeoTour exercise (S) features:

· Effects of global warming
· Effects of deforestation
· Water use in arid regions
· Preservation of natural habitats

Additional Maps and Charts

This appendix contains several maps and charts for general reference. We list the purpose of each below.

The *Periodic Table of Elements* (**Fig. a.1**) has been provided as a reference since certain key topics of chemistry are an essential background to this text.

Mineral Identification Flowcharts: Geologists use these charts to identify unknown mineral specimens (**Fig. a.2a, b**). A mineral flowchart is simply an organized series of questions concerning the mineral's physical properties. The questions are arranged in a sequence such that appropriate answers ultimately lead you on a path to a specific mineral. To understand this concept, let's imagine that we are trying to identify a shiny, bronze- or gold-colored, metallic-looking mineral specimen. We start by observing the specimen's luster. It is metallic, so we follow the path on the chart for metallic-luster minerals (see Fig. a.2a). Next, we determine if the mineral is magnetic or nonmagnetic. If it is nonmagnetic, we follow the path for nonmagnetic minerals. Then, we look at the mineral's color. Since it is bronze- or gold-colored, our path ends at pyrite.

Notice that one of the flowchart questions in Fig. a.2b asks about the reaction of the specimen with hydrochloric acid (HCl). Only calcite and dolomite react, so the question allows definitive identification of these minerals. Another question pertains to striations, faint parallel lines on cleavage planes. Only plagioclase has striations.

World Magnetic Declination Map: This map shows the variation of magnetic declination with location on the surface of the Earth (**Fig. a.3a**). Declination exists because the position of the Earth's magnetic pole does not coincide exactly with that of the geographic pole. In fact, the magnetic pole location constantly moves, currently at a rate of about 20 km per year. It now lies off the north coast of Canada, and in the not too distant future, it may lie along the coast of Siberia. For a compass to give an accurate indication of direction, it must be adjusted to accommodate for the declination at the location of measurement.

US Magnetic Declination Map: This map shows the magnetic declination for the United States (**Fig. a.3b**).

The North America Tapestry of Time and Terrain: Another map provided by USGS superimposes a digital elevation map of North America over a geologic map of the same to get a full geologic picture of our continent (**Fig. a.4**).

Metric Conversion Chart: This chart shows the correlation between U.S. standard units and metric units for length, area, volume, mass, pressure, and temperature (including formulas for converting temperatures between Fahrenheit, Celsius, and Kelvin).

Legend:

Symbol	He	2	Atomic number
	Helium		Name
	4.002		Atomic weight

Alkali metals

H 1		
Hydrogen		
1.007		

Li 3	Be 4
Lithium	Beryllium
6.941	9.0121

Na 11	Mg 12
Sodium	Magnesium
22.989	24.305

K 19	Ca 20
Potassium	Calcium
39.098	40.078

Rb 37	Sr 38
Rubidium	Strontium
85.467	87.62

Cs 55	Ba 56
Cesium	Barium
132.905	137.327

Fr 87	Ra 88
Francium	Radium
223.019	226.025

Transition elements (metals)

Sc 21	Ti 22	V 23	Cr 24	Mn 25	Fe 26	Co 27	Ni 28	Cu 29	Zn 30
Scandium	Titanium	Vanadium	Chromium	Manganese	Iron	Cobalt	Nickel	Copper	Zinc
44.955	47.88	50.941	51.996	54.938	55.847	58.933	58.693	63.546	65.39

Y 39	Zr 40	Nb 41	Mo 42	Tc 43	Ru 44	Rh 45	Pd 46	Ag 47	Cd 48
Yttrium	Zirconium	Niobium	Molybdenum	Technetium	Ruthenium	Rhodium	Palladium	Silver	Cadmium
88.905	91.224	92.906	95.94	98.907	101.07	102.905	106.42	107.868	112.411

La 57	Hf 72	Ta 73	W 74	Re 75	Os 76	Ir 77	Pt 78	Au 79	Hg 80
Lanthanum	Hafnium	Tantalum	Tungsten	Rhenium	Osmium	Iridium	Platinum	Gold	Mercury
138.905	178.49	180.947	183.85	186.207	190.2	192.22	195.08	196.966	200.59

Ac   89
Actinium
227.027

Nonmetals

B 5	C 6	N 7	O 8	F 9	Ne 10
Boron	Carbon	Nitrogen	Oxygen	Fluorine	Neon
10.811	12.011	14.006	15.999	18.998	20.179

Al 13	Si 14	P 15	S 16	Cl 17	Ar 18
Aluminum	Silicon	Phosphorus	Sulfur	Chlorine	Argon
26.981	28.085	30.973	32.066	35.452	39.948

Ga 31	Ge 32	As 33	Se 34	Br 35	Kr 36
Gallium	Germanium	Arsenic	Selenium	Bromine	Krypton
69.723	72.61	74.921	78.96	79.904	83.80

In 49	Sn 50	Sb 51	Te 52	I 53	Xe 54
Indium	Tin	Antimony	Tellurium	Iodine	Xenon
114.82	118.710	121.757	12760	126.904	131.29

Tl 81	Pb 82	Bi 83	Po 84	At 85	Rn 86
Thallium	Lead	Bismuth	Polonium	Astatine	Radon
204.383	2072	208.980	208.982	209.987	222.017

Inert gases

He   2
Helium
4.002

Lanthanide series

Ce 58	Pr 59	Nd 60	Pm 61	Sm 62	Eu 63	Gd 64	Tb 65	Dy 66	Ho 67	Er 68	Tm 69	Yb 70	Lu 71
Cerium	Praseodymium	Neodymium	Promethium	Samarium	Europium	Gadolinium	Terbium	Dysprosium	Holmium	Erbium	Thulium	Ytterbium	Lutetium
140.115	140.907	144.24	144.912	150.36	151.965	157.25	158.925	162.50	164.930	16726	168.934	173.04	174.967

Actinide series

Th 90	Pa 91	U 92	Np 93	Pu 94	Am 95	Cm 96	Bk 97	Cf 98	Es 99	Fm 100	Md 101	No 102	Lr 103
Thorium	Protactinium	Uranium	Neptunium	Plutonium	Americium	Curium	Berkelium	Californium	Einsteinium	Fermium	Mendelevium	Nobelium	Lawrencium
232.038	231.035	238.028	237.048	244.064	243.061	247.070	247.070	251.079	252.083	257.095	258.10	259.100	262.11

FIGURE a.1   The modern periodic table of the elements. Each column groups elements with related properties. For example, inert gases are listed in the column on the right. Metals are found in the central and left parts of the chart.

FIGURE a.2 Simplified mineral identification flowcharts

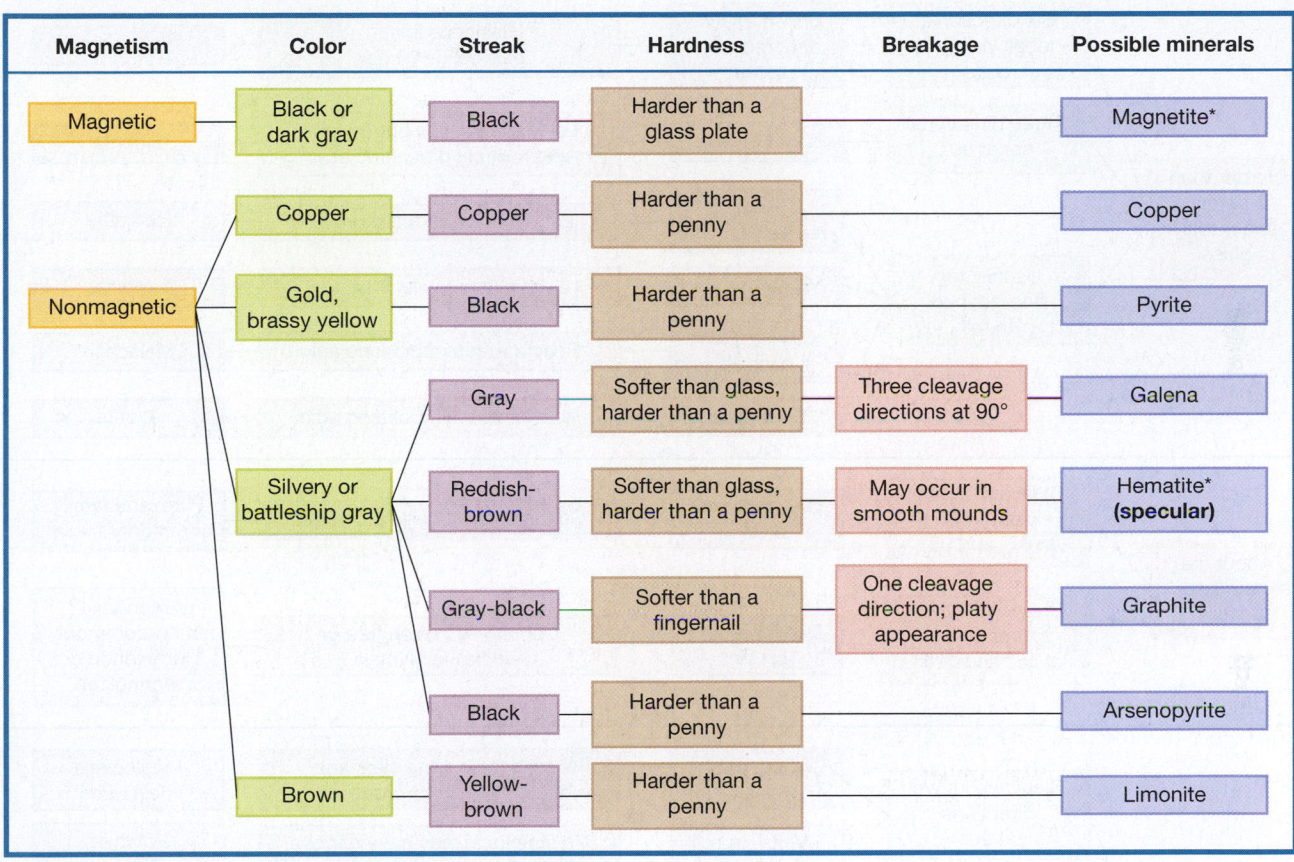

Magnetism	Color	Streak	Hardness	Breakage	Possible minerals
Magnetic	Black or dark gray	Black	Harder than a glass plate		Magnetite*
Nonmagnetic	Copper	Copper	Harder than a penny		Copper
	Gold, brassy yellow	Black	Harder than a penny		Pyrite
	Silvery or battleship gray	Gray	Softer than glass, harder than a penny	Three cleavage directions at 90°	Galena
		Reddish-brown	Softer than glass, harder than a penny	May occur in smooth mounds	Hematite* (specular)
		Gray-black	Softer than a fingernail	One cleavage direction; platy appearance	Graphite
		Black	Harder than a penny		Arsenopyrite
	Brown	Yellow-brown	Harder than a penny		Limonite

*Hematite is sometimes weakly magnetic.

(a) Minerals with metallic luster.

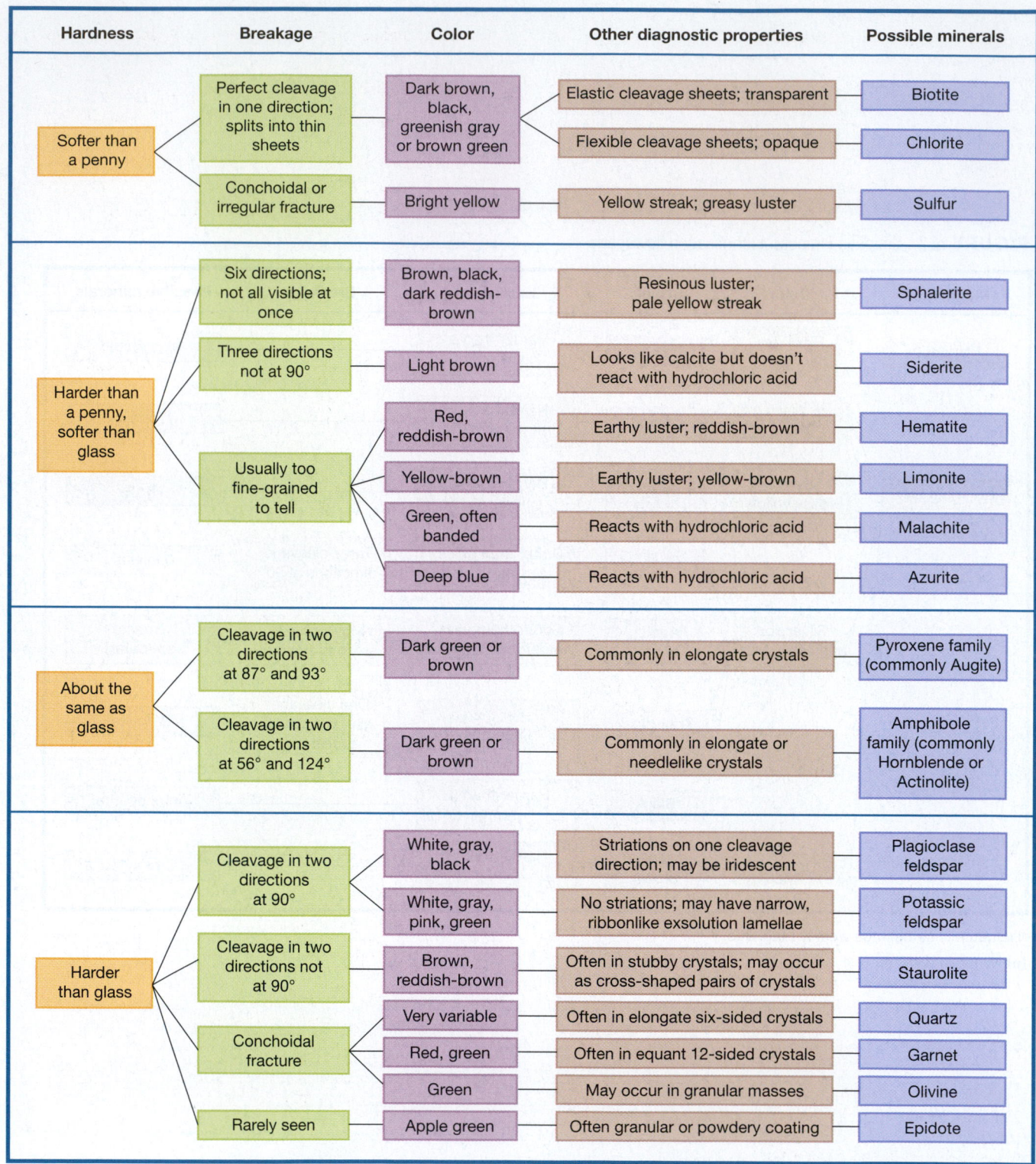

Hardness	Breakage	Color	Other diagnostic properties	Possible minerals
Softer than a penny	Perfect cleavage in one direction; splits into thin sheets	Dark brown, black, greenish gray or brown green	Elastic cleavage sheets; transparent	Biotite
			Flexible cleavage sheets; opaque	Chlorite
	Conchoidal or irregular fracture	Bright yellow	Yellow streak; greasy luster	Sulfur
Harder than a penny, softer than glass	Six directions; not all visible at once	Brown, black, dark reddish-brown	Resinous luster; pale yellow streak	Sphalerite
	Three directions not at 90°	Light brown	Looks like calcite but doesn't react with hydrochloric acid	Siderite
	Usually too fine-grained to tell	Red, reddish-brown	Earthy luster; reddish-brown	Hematite
		Yellow-brown	Earthy luster; yellow-brown	Limonite
		Green, often banded	Reacts with hydrochloric acid	Malachite
		Deep blue	Reacts with hydrochloric acid	Azurite
About the same as glass	Cleavage in two directions at 87° and 93°	Dark green or brown	Commonly in elongate crystals	Pyroxene family (commonly Augite)
	Cleavage in two directions at 56° and 124°	Dark green or brown	Commonly in elongate or needlelike crystals	Amphibole family (commonly Hornblende or Actinolite)
Harder than glass	Cleavage in two directions at 90°	White, gray, black	Striations on one cleavage direction; may be iridescent	Plagioclase feldspar
		White, gray, pink, green	No striations; may have narrow, ribbonlike exsolution lamellae	Potassic feldspar
	Cleavage in two directions not at 90°	Brown, reddish-brown	Often in stubby crystals; may occur as cross-shaped pairs of crystals	Staurolite
	Conchoidal fracture	Very variable	Often in elongate six-sided crystals	Quartz
		Red, green	Often in equant 12-sided crystals	Garnet
		Green	May occur in granular masses	Olivine
	Rarely seen	Apple green	Often granular or powdery coating	Epidote

(b) Minerals with nonmetallic luster, dark-colored.

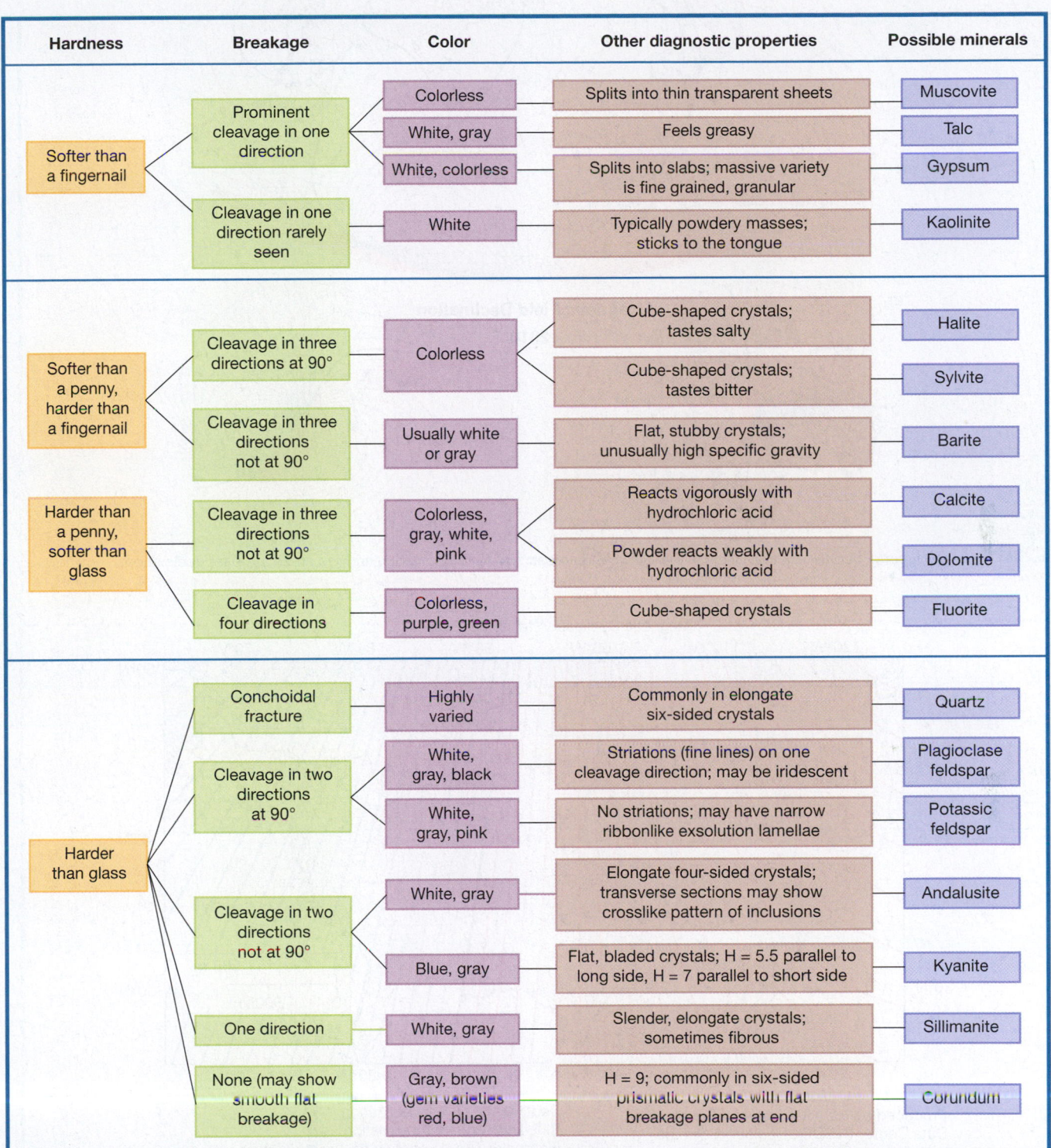

Hardness	Breakage	Color	Other diagnostic properties	Possible minerals
Softer than a fingernail	Prominent cleavage in one direction	Colorless	Splits into thin transparent sheets	Muscovite
		White, gray	Feels greasy	Talc
		White, colorless	Splits into slabs; massive variety is fine grained, granular	Gypsum
	Cleavage in one direction rarely seen	White	Typically powdery masses; sticks to the tongue	Kaolinite
Softer than a penny, harder than a fingernail	Cleavage in three directions at 90°	Colorless	Cube-shaped crystals; tastes salty	Halite
			Cube-shaped crystals; tastes bitter	Sylvite
	Cleavage in three directions not at 90°	Usually white or gray	Flat, stubby crystals; unusually high specific gravity	Barite
Harder than a penny, softer than glass	Cleavage in three directions not at 90°	Colorless, gray, white, pink	Reacts vigorously with hydrochloric acid	Calcite
			Powder reacts weakly with hydrochloric acid	Dolomite
	Cleavage in four directions	Colorless, purple, green	Cube-shaped crystals	Fluorite
Harder than glass	Conchoidal fracture	Highly varied	Commonly in elongate six-sided crystals	Quartz
	Cleavage in two directions at 90°	White, gray, black	Striations (fine lines) on one cleavage direction; may be iridescent	Plagioclase feldspar
		White, gray, pink	No striations; may have narrow ribbonlike exsolution lamellae	Potassic feldspar
	Cleavage in two directions not at 90°	White, gray	Elongate four-sided crystals; transverse sections may show crosslike pattern of inclusions	Andalusite
		Blue, gray	Flat, bladed crystals; H = 5.5 parallel to long side, H = 7 parallel to short side	Kyanite
	One direction	White, gray	Slender, elongate crystals; sometimes fibrous	Sillimanite
	None (may show smooth flat breakage)	Gray, brown (gem varieties red, blue)	H = 9; commonly in six-sided prismatic crystals with flat breakage planes at end	Corundum

(c) Minerals with nonmetallic luster, light-colored.

FIGURE a.3 Magnetic declination charts.

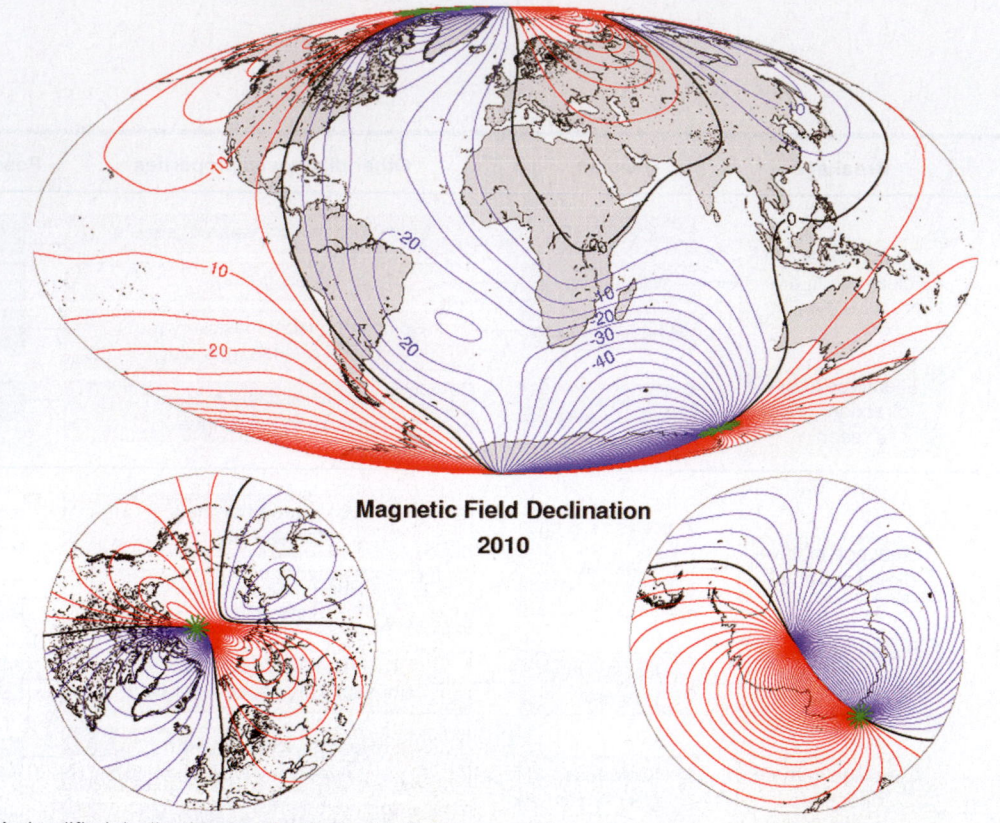

Magnetic Field Declination 2010

(a) A simplified declination map for the world. Blue areas are west declination and red areas are east declination.

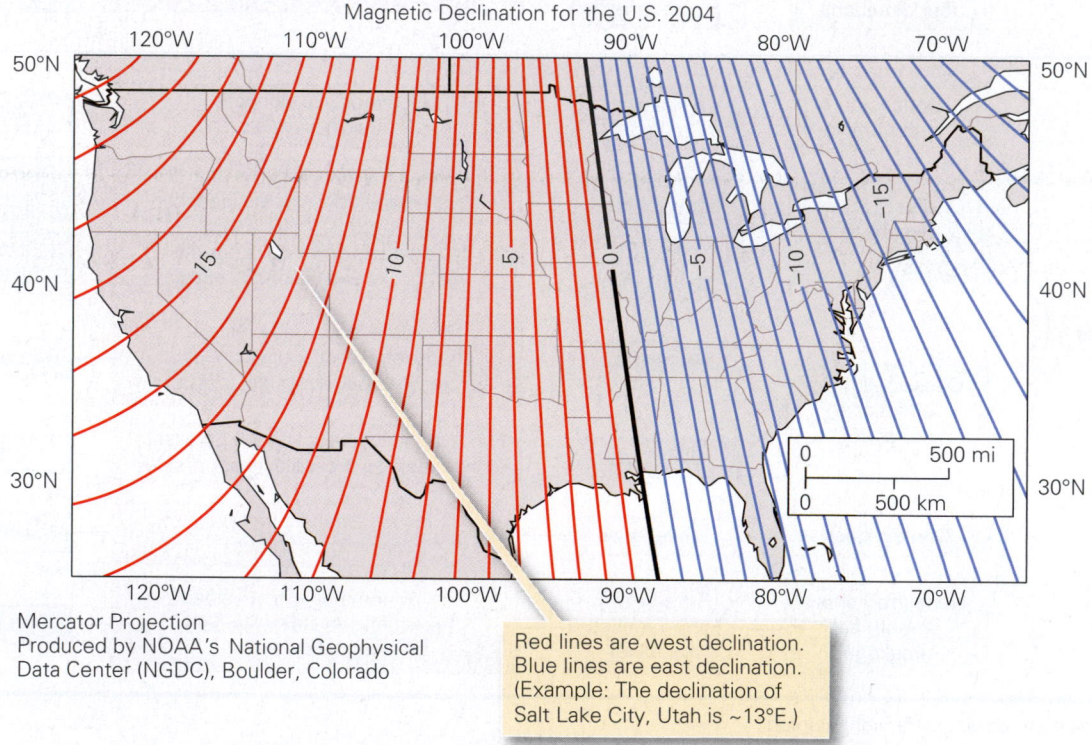

Magnetic Declination for the U.S. 2004

Mercator Projection
Produced by NOAA's National Geophysical
Data Center (NGDC), Boulder, Colorado

Red lines are west declination.
Blue lines are east declination.
(Example: The declination of
Salt Lake City, Utah is ~13°E.)

(b) A simplified declination map of the United States for 2004. Because of changes in the field, most of the lines are drifting westward at about 5′ (minutes) per year. (Note: 1° = 60′)

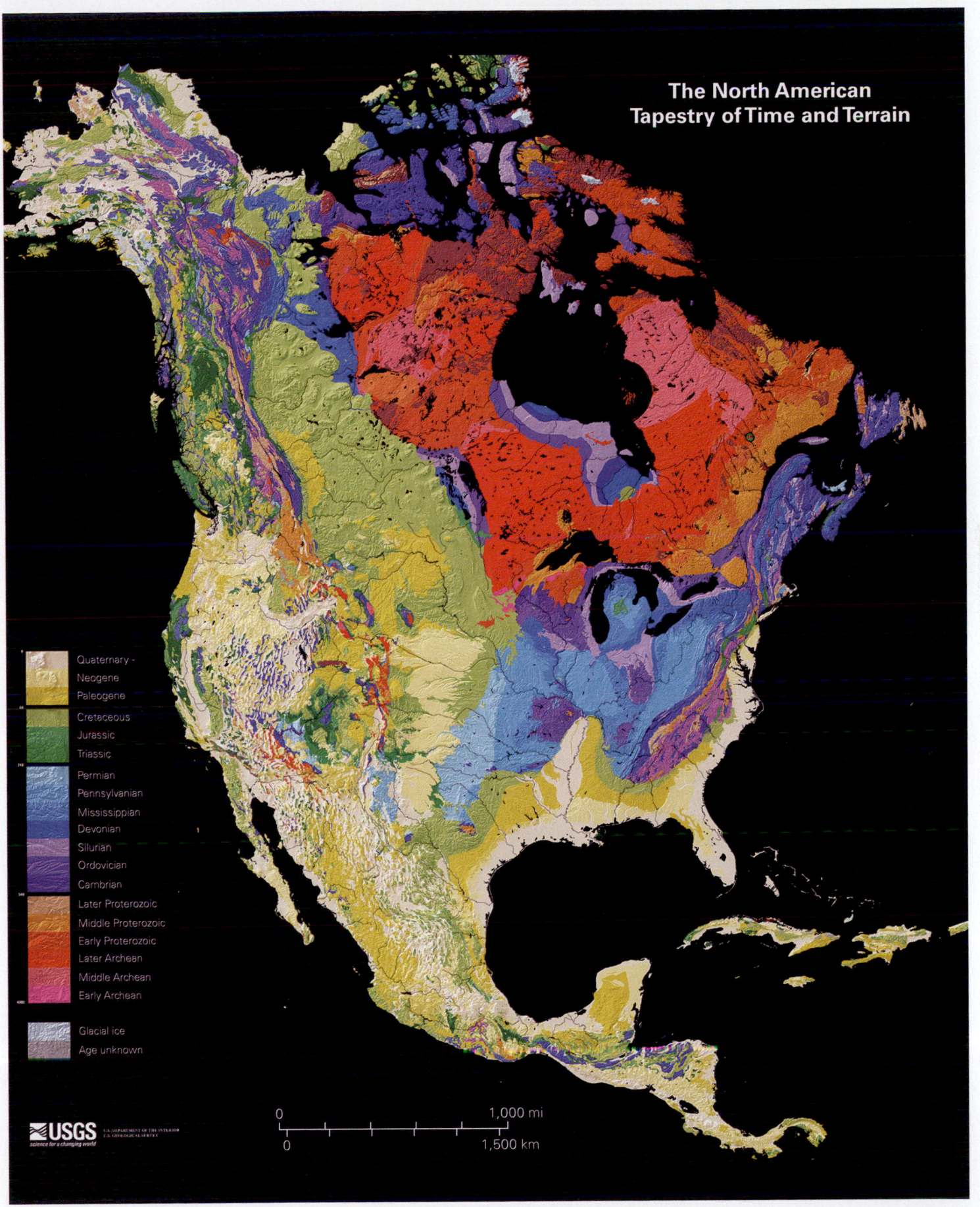

The North American
Tapestry of Time and Terrain

Quaternary -
Neogene
Paleogene

Cretaceous
Jurassic
Triassic

Permian
Pennsylvanian
Mississippian
Devonian
Silurian
Ordovician
Cambrian

Later Proterozoic
Middle Proterozoic
Early Proterozoic
Later Archean
Middle Archean
Early Archean

Glacial ice
Age unknown

0   1,000 mi

0   1,500 km

≈USGS
science for a changing world
U.S. DEPARTMENT OF THE INTERIOR
U.S. GEOLOGICAL SURVEY

FIGURE a.4 Geologic map of North America, superimposed on a digital elevation map.

Metric Conversion Chart

Length

1 kilometer (km) = 0.6214 mile (mi)
1 meter (m) = 1.094 yards = 3.281 feet
1 centimeter (cm) = 0.3937 inch
1 millimeter (mm) = 0.0394 inch
1 mile (mi) = 1.609 kilometers (km)
1 yard = 0.9144 meter (m)
1 foot = 0.3048 meter (m)
1 inch = 2.54 centimeters (cm)

Area

1 square kilometer (km^2) = 0.386 square mile (mi^2)
1 square meter (m^2) = 1.196 square yards (yd^2)
= 10.764 square feet (ft^2)
1 square centimeter (cm^2) = 0.155 square inch (in^2)
1 square mile (mi^2) = 2.59 square kilometers (km^2)
1 square yard (yd^2) = 0.836 square meter (m^2)
1 square foot (ft^2) = 0.0929 square meter (m^2)
1 square inch (in^2) = 6.4516 square centimeters (cm^2)

Volume

1 cubic kilometer (km^3) = 0.24 cubic mile (mi^3)
1 cubic meter (m^3) = 264.2 gallons
= 35.314 cubic feet (ft^3)
1 liter (1) = 1.057 quarts
= 33.815 fluid ounces
1 cubic centimeter (cm^3) = 0.0610 cubic inch (in^3)
1 cubic mile (mi^3) = 4.168 cubic kilometers (km^3)
1 cubic yard (yd^3) = 0.7646 cubic meter (m^3)
1 cubic foot (ft^3) = 0.0283 cubic meter (m^3)
1 cubic inch (in^3) = 16.39 cubic centimeters (cm^3)

Mass

1 metric ton = 2,205 pounds
1 kilogram (kg) = 2.205 pounds
1 gram (g) = 0.03527 ounce
1 pound (lb) = 0.4536 kilogram (kg)
1 ounce (oz) = 28.35 grams (g)

Pressure

1 kilogram per
square centimeter (kg/cm^2)* = 0.96784 atmosphere (atm)
= 0.98066 bar
= 9.8067×10^4 pascals (Pa)
1 bar = 0.1 megapascals (Mpa)
= 1.0×10^5 pascals (Pa)
= 29.53 inches of mercury (in a barometer)
= 0.98692 atmosphere (atm)
= 1.02 kilograms per square centimeter (kg/cm^2)
1 pascal (Pa) = 1 $kg/m/s^2$
1 pound per square inch = 0.06895 bars
= 6.895×10^3 pascals (Pa)
= 0.0703 kilogram per square
centimeter

Temperature

To change from Fahrenheit (F) to Celsius (C):
$$°C = \frac{(°F - 32°)}{1.8}$$
To change from Celsius (C) to Fahrenheit (F):
$$°F = (°C \times 1.8) + 32°$$
To change from Celsius (C) to Kelvin (K):
$$K = °C + 273.15$$
To change from Fahrenheit (F) to Kelvin (K):
$$K = \frac{(°F - 32°)}{1.8} + 273.15$$

---

*Note: Because kilograms are a measure of mass whereas pounds are a unit of weight, pressure units incorporating kilograms assume a given gravitational constant (g) for Earth. In reality, the gravitational for Earth varies slightly with location.

a'a A lava flow with a rubbly surface.

abandoned meander A meander that dries out after it was cut off.

ablation The removal of ice at the toe of a glacier by melting, sublimation (the evaporation of ice into water vapor), and/or calving.

abrasion The process in which one material (such as sand-laden water) grinds away at another (such as a stream channel's floor and walls).

absolute age Numerical age (the age specified in years).

absolute plate velocity The movement of a plate relative to a fixed point in the mantle.

absolute zero The lowest temperature possible (−273.15°C); at absolute zero, vibrations and movements of atoms in a material cease.

abyssal plain A broad, relatively flat region of the ocean that lies at least 4.5 km below sea level.

Acadian orogeny A convergent mountain-building event that occurred around 400 million years ago, during which continental slivers accreted to the eastern edge of the North American continent.

accreted terrane A block of crust that collided with a continent at a convergent margin and stayed attached to the continent.

accretionary coast A coastline that receives more sediment than erodes away.

accretionary lapilli Hailstone-like clumps of wet ash that fall from a volcanic eruptive cloud.

accretionary orogen An orogen formed by the attachment of numerous buoyant slivers of crust to an older, larger continental block.

accretionary prism A wedge-shaped mass of sediment and rock scraped off the top of a downgoing plate and accreted onto the overriding plate at a convergent plate margin.

acid mine runoff A dilute solution of sulfuric acid, produced when sulfur-bearing minerals in mines react with rainwater, that flows out of a mine.

acid rain Precipitation in which air pollutants react with water to make a weak acid that then falls from the sky.

active continental margin A continental margin that coincides with a plate boundary.

active fault A fault that has moved recently or is likely to move in the future.

active sand The top layer of beach sand, which moves daily because of wave action.

active volcano A volcano that has erupted within the past few centuries and will likely erupt again.

adiabatic cooling The cooling of a body of air or matter without the addition or subtraction of thermal energy (heat).

adiabatic heating The warming of a body of air or matter without the addition or subtraction of heat.

advection A process of heat transfer in which heat is carried into a solid by a liquid or gas moving through fractures or pores in the solid.

aerosols Tiny solid particles or liquid droplets that remain suspended in the atmosphere for a long time.

aftershocks The series of smaller earthquakes that follow a major earthquake.

air The mixture of gases that make up the Earth's atmosphere.

air-fall tuff Tuff formed when ash settles gently from the air.

air mass A body of air, about 1,500 km across, that has recognizable physical characteristics.

air pressure The push that air exerts on its surroundings.

albedo The reflectivity of a surface.

Alleghenian orogeny The orogenic event that occurred about 270 million years ago when Africa collided with North America.

alloy A metal containing more than one type of metal atom.

alluvial fan A gently sloping apron of sediment dropped by an ephemeral stream at the base of a mountain in arid or semiarid regions.

alluvium Sorted sediment deposited by a stream.

alluvium-filled valley A valley whose floor fills with sediment.

Alpine-Himalayan chain The largest orogenic belt on Earth today, formed by collisions of the former Gondwana continents with the southern margins of Europe and Asia.

amber Hardened (fossilized) ancient sap or resin.

amphibolite facies A set of metamorphic mineral assemblages formed under intermediate pressures and temperatures.

amplitude The height of a wave from crest to trough.

Ancestral Rockies The late Paleozoic uplifts of the Rocky Mountain region; they eroded away long before the present Rocky Mountains formed.

angiosperm A flowering plant.

angle of repose The angle of the steepest slope that a pile of uncemented material can attain without collapsing from the pull of gravity.

angularity The degree to which grains have sharp or rounded edges or corners.

angular unconformity An unconformity in which the strata below were tilted or folded before the unconformity developed; strata below the unconformity therefore have a different tilt than strata above.

anhedral grains Crystalline mineral grains without well-formed crystal faces.

anion A negatively charged ion.

annual probability The likelihood that a flood of a given size or larger will happen at a specified locality during any given year.

Antarctic bottom water mass The mass of cold, dense water that sinks along the coast of Antarctica.

antecedent stream A stream that cuts across an uplifted mountain range; the stream must have existed before the range uplifted and must then have been able to downcut as fast as the land was rising.

anthracite coal Shiny black coal formed at temperatures between 200°C and 300°C. A high-rank coal.

anticline A fold with an arch-like shape in which the limbs dip away from the hinge.

anticyclone The clockwise flow of air around a high-pressure mass.

anticyclonic flow A circulation of air around a high-pressure region in the atmosphere; it rotates clockwise in the northern hemisphere.

Antler orogeny The Late Devonian mountain-building event in which slices of deep-marine strata were pushed eastward, up and over the shallow-water strata on the western coast of North America.

anvil cloud A large cumulonimbus cloud that spreads laterally at the tropopause to form a broad, flat top.

aphanitic A textural term for fine-grained igneous rock.

apparent polar-wander path A path on the globe along which a magnetic pole appears to have wandered over time; in fact, the continents drift, while the magnetic pole stays fairly fixed.

aquiclude Sediment or rock that transmits no water.

aquifer Sediment or rock that transmits water easily.

aquitard Sediment or rock that does not transmit water easily and therefore retards the motion of the water.

archaea A kingdom of "old bacteria," now commonly found in extreme environments like hot springs. (Also called archaeobacteria.)

Archean Eon The middle Precambrian eon (4.0–2.5 Ga).

Archimedes' principle The mass of the water displaced by a block of material equals the mass of the whole block of material.

arête A residual knife-edge ridge of rock that separates two adjacent cirques.

argillaceous sedimentary rock Sedimentary rock that contains abundant clay.

arkose A clastic sedimentary rock containing both quartz and feldspar grains.

arroyo The channel of an ephemeral stream; dry wash; wadi.

artesian spring A location where the ground surface intersects a natural fracture (joint) that taps a confined aquifer in which the pressure can drive the water to the surface.

artesian well A well in which water rises on its own.

artificial levee A man-made retaining wall to hold back a river from flooding.

ash *See* Volcanic ash.

ash fall Ash that falls to the ground out of an ash cloud.

ash flow An avalanche of ash that tumbles down the side of an explosively erupting volcano.

assimilation The process of magma contamination in which blocks of wall rock fall into a magma chamber and dissolve.

asteroid One of the fragments of solid material, left over from planet formation or produced by collision of planetesimals, that resides between the orbits of Mars and Jupiter.

asthenosphere The layer of the mantle that lies between 100–150 km and 350 km deep; the asthenosphere is relatively soft and can flow when acted on by force.

atm A unit of air pressure that approximates the pressure exerted by the atmosphere at sea level.

atmosphere A layer of gases that surrounds a planet.

atoll A coral reef that develops around a circular reef surrounding a lagoon.

atom The smallest piece of an element that has the properties of the element; it consists of a nucleus surrounded by an electron cloud.

atomic mass The amount of matter in an atom; roughly, it is the sum of the number of protons plus the number of neutrons in the nucleus.

atomic number The number of protons in the nucleus of a given element.

atomic weight The number of protons plus the number of neutrons in the nucleus of a given element. (Also known as atomic mass.)

aurora australis The same phenomenon as the aurora borealis, but in the southern hemisphere.

aurora borealis A ghostly curtain of varicolored light that appears across the night sky in the northern hemisphere when charged particles from the Sun interact with the ions in the ionosphere.

avalanche A turbulent cloud of debris mixed with air that rushes down a steep hill slope at high velocity; the debris can be rock and/or snow.

avalanche chute A downslope hillside pathway along which avalanches repeatedly fall, consequently clearing the pathway of mature trees.

avulsion The process in which a river overflows a natural levee and begins to flow in a new direction.

axial plane The imaginary surface that encompasses the hinges of successive layers of a fold.

axial trough A narrow depression that runs along a mid-ocean ridge axis.

axis An imaginary line around which an object spins.

backscattered light Atmospheric scattered sunlight that returns to space.

backshore zone The zone of beach that extends from a small step cut by high-tide swash to the front of the dunes or cliffs that lie farther inshore.

backswamp The low marshy region between the bluffs and the natural levees of a floodplain.

backwash The gravity-driven flow of water back down the slope of a beach.

bacteria A type of tiny prokaryotic single-celled organism.

bajada An elongate wedge of sediment formed by the overlap of several alluvial fans emerging from adjacent valleys.

Baltica A Paleozoic continent that included crust that is now part of today's Europe.

banded-iron formation (BIF) Iron-rich sedimentary layers consisting of alternating gray beds of iron oxide and red beds of iron-rich chert.

bar (1) A sheet or elongate lens or mound of alluvium; (2) a unit of air-pressure measurement approximately equal to 1 atm.

barchan dune A crescent-shaped dune whose tips point downwind.

barrier island An offshore sand bar that rises above the mean high-water level, forming an island.

barrier reef A coral reef that develops offshore, separated from the coast by a lagoon.

basal sliding The phenomenon in which meltwater accumulates at the base of a glacier, so that the mass of the glacier slides on a layer of water or on a slurry of water and sediment.

basalt A fine-grained, mafic, igneous rock.

base level The lowest elevation a stream channel's floor can reach at a given locality.

basement Older igneous and metamorphic rocks making up the Earth's crust beneath sedimentary cover.

basement uplift Uplift of basement rock by faults that penetrate deep into the continental crust.

base metals Metals that are mined but not considered precious. Examples include copper, lead, zinc, and tin.

basin A fold or depression shaped like a right-side-up bowl.

Basin and Range Province A broad, Cenozoic continental rift that has affected a portion of the western United States in Nevada, Utah, and Arizona; in this province, tilted fault blocks form ranges, and alluvium-filled valleys are basins.

batholith A vast composite, intrusive, igneous rock body up to several hundred km long and 100 km wide, formed by the intrusion of numerous plutons in the same region.

bathymetric map A map illustrating the shape of the ocean floor.

bathymetric profile A cross section showing ocean depth plotted against location.

bathymetry Variation in depth.

bauxite A residual mineral deposit rich in aluminum.

baymouth bar A sandspit that grows across the opening of a bay.

beach A gently sloping fringe of sediment along the shore.

beach drift The gradual migration of sand along a beach.

beach erosion The removal of beach sand caused by wave action and longshore currents.

beach face A steeply concave part of the foreshore zone formed where the swash of the waves actively scours the sand.

bedding Layering or stratification in sedimentary rocks.

bed load Large particles, such as sand, pebbles, or cobbles, that bounce or roll along a streambed.

bedrock   Rock still attached to the Earth's crust.

Bergeron process   Precipitation involving the growth of ice crystals in a cloud at the expense of water droplets.

berm   A horizontal or landward-sloping terrace in the backshore zone of a beach that receives sediment during a storm.

Big Bang   A cataclysmic explosion that scientists suggest represents the formation of the Universe; before this event, all matter and all energy were packed into one volumeless point.

biochemical sedimentary rock   Sedimentary rock formed from material (such as shells) produced by living organisms.

biodiversity   The number of different species that exist at a given time.

biofuel   Gas or liquid fuel made from plant material (biomass). Examples of biofuel include alcohol (from fermented sugar), biodiesel from vegetable oil, and wood.

biogenic minerals   Substances that meet the definition of a mineral and are produced naturally by organisms (e.g., calcite in shells).

biogeochemical cycle   The exchange of chemicals between living and non-living reservoirs in the Earth System.

biomass   The amount of organic material in a specified volume.

bioremediation   The injection of oxygen and nutrients into a contaminated aquifer to foster the growth of bacteria that will ingest or break down contaminants.

biosphere   The region of the Earth and atmosphere inhabited by life; this region stretches from a few km below the Earth's surface to a few km above.

bioturbation   The mixing of sediment by burrowing animals such as clams and worms.

bituminous coal   Dull, black intermediate-rank coal formed at temperatures between 100°c and 200°C.

black-lung disease   Lung disease contracted by miners from the inhalation of too much coal dust.

black smoker   The cloud of suspended minerals formed where hot water spews out of a vent along a mid-ocean ridge; the dissolved sulfide components of the hot water instantly precipitate when the water mixes with seawater and cools.

blind fault   A fault that does not intersect the ground surface.

block   Large, angular pyroclastic fragments consisting of volcanic rock, broken up during the eruption.

blocking temperature   The temperature below which isotopes in a mineral are no longer free to move, so the radiometric clock starts.

blocky lava   Lava that is so viscous that it breaks into boulder-like blocks as it moves; typically, such lavas are andesitic or rhyolitic.

blowout   A deep, bowl-like depression scoured out of desert terrain by a turbulent vortex of wind.

blue shift   The phenomenon in which a source of light moving toward you appears to have a higher frequency.

body fossil   A relict of an organism's body, preserved in rock.

body waves   Seismic waves that pass through the interior of the Earth.

bog   A wetland dominated by moss and shrubs.

bolide   A solid extraterrestrial object such as a meteorite, comet, or asteroid that explodes in the atmosphere.

bornhardt   An inselberg with a loaf geometry, like that of Uluru (Ayers Rock) in central Australia.

Bowen's reaction series   The sequence in which different silicate minerals crystallize during the progressive cooling of a melt.

braided stream   A sediment-choked stream consisting of entwined subchannels.

breaker   A water wave in which water at the top of the wave curves over the base of the wave.

breakwater   An offshore wall, built parallel or at an angle to the beach, that prevents the full force of waves from reaching a harbor.

breccia   Coarse sedimentary rock consisting of angular fragments; or rock broken into angular fragments by faulting.

breeder reactor   A nuclear reactor that produces its own fuel.

brine   Water that is not fresh but is less salty than seawater; brine may be found in estuaries.

brittle deformation   The cracking and fracturing of a material subjected to stress.

brittle-ductile transition (brittle-plastic transition)   The depth above which materials behave brittlely and below which materials behave ductilely (plastically); this transition typically lies between a depth of 10 and 15 km in continental crustal rock, and 60 m deep in glacial ice.

buoyancy   The upward force acting on an object immersed or floating in fluid; the tendency of an object to float when placed in a fluid.

burial metamorphism   Metamorphism due only to the consequences of very deep burial.

butte   A medium-sized, flat-topped hill in an arid region.

caldera   A large circular depression with steep walls and a fairly flat floor, formed after an eruption as the center of the volcano collapses into the drained magma chamber below.

caliche   A solid mass created where calcite cements the soil together. (Also called calcrete.)

calorie   A unit of energy approximately equal to 4.2 joules; 1 calorie can raise the temperature of 1 gm of water by 1°C.

calving   The breaking off of chunks of ice at the edge of a glacier.

Cambrian explosion of life   The remarkable diversification of life, indicated by the fossil record, that occurred at the beginning of the Cambrian Period.

Canadian Shield   A broad, low-lying region of exposed Precambrian rock in the Canadian interior.

canyon   A trough or valley with steeply sloping walls, cut into the land by a stream.

capacity (of a stream)   The total quantity of sediment a stream can carry.

capillary fringe   The thin subsurface layer in which water molecules seep up from the water table by capillary action to fill pores.

carbonate rocks   Rocks containing calcite and/or dolomite.

carbon-14   A radioactive isotope of the element carbon; the ratio of C14 to C12 can provide an isotopic date of organic carbon.

carbon-14 dating   A radiometric dating process that can tell us the age of organic material containing carbon originally extracted from the atmosphere.

carbon sequestration   The process of extracting carbon dioxide from sources (e.g., power plants) and sending it back underground to keep it out of the atmosphere and diminish the greenhouse effect.

cast   Sediment that preserves the shape of a shell it once filled before the shell dissolved or mechanically weathered away.

catabatic winds   Strong winds that form at the margin of a glacier where the warmer air above ice-free land rises and the cold, denser air from above the glaciers rushes in to take its place.

catastrophic change   Change that takes place either instantaneously or rapidly in geologic time.

catchment   *See* Drainage network.

cation   A positively charged ion.

Celsius scale   A metric system measure of temperature in which the difference between the freezing point (0°C) and boiling point (100°C) of water is divided into 100 units; equivalent to centigrade scale.

cement   Mineral material that precipitates from water and fills the spaces between grains, holding the grains together.

cementation   The phase of lithification in which cement, consisting of minerals that precipitate from groundwater, partially or completely fills the spaces between clasts and attaches each grain to its neighbor.

Cenozoic Era   The most recent era of the Phanerozoic Eon, lasting from 66 Ma up until the present.

chain reaction   A self-perpetuating process in a nuclear reaction, whereby neutrons released during the fission trigger more fission.

chalk   Very fine-grained limestone consisting of weakly cemented plankton shells.

change of state   The process in which a material changes from one phase (liquid, gas, or solid) to another.

channel   A trough dug into the ground surface by flowing water.

channeled scablands   A barren, soil-free landscape in eastern Washington, scoured clean by a flood unleashed when a large glacial lake drained.

chatter marks   Wedge-shaped indentations left on rock surfaces by glacial plucking.

chemical   A material consisting of a distinct element or compound.

chemical bond   The invisible link that holds together atoms in a molecule and/or in a crystal.

chemical formula   The "recipe" that specifies the elements and their proportions in a compound.

chemical fossil   Distinctive molecules or molecular fragments, formed from the remains of living organisms, that can be preserved in rock.

chemical reaction   Interactions among atoms and/or molecules involving breaking or forming chemical bonds.

chemical sedimentary rocks   Sedimentary rocks made up of minerals that precipitate directly from water solution.

chemical weathering   The process in which chemical reactions alter or destroy minerals when rock comes in contact with water solutions and/or air.

chert   A sedimentary rock composed of very fine-grained silica (cryptocrystalline quartz).

Chicxulub crater   A circular excavation buried beneath younger sediment on the Yucatán Peninsula; geologists suggest that a meteorite landed there 65 Ma.

chimney   (1) A conduit in a magma chamber in the shape of a long vertical pipe through which magma rises and erupts at the surface; (2) an isolated column of strata in an arid region.

chron   The time interval between successive magnetic reversals.

cinder cone   A subaerial volcano consisting of a cone-shaped pile of tephra whose slope approaches the angle of repose for tephra.

cinders   Fragments of glassy rock ejected from a volcano.

cirque   A bowl-shaped depression carved by a glacier on the side of a mountain.

cirrus cloud   A wispy cloud that tapers into delicate, feather-like curls.

clast   A fragment or grain produced by the physical or chemical weathering of a pre-existing rock.

clastic (detrital) sedimentary rock   Sedimentary rock consisting of cemented-together detritus derived from the weathering of preexisting rock.

cleavage   (1) The tendency of a mineral to break along preferred planes; (2) a type of foliation in low-grade metamorphic rock.

cleavage planes   A series of surfaces on a crystal that form parallel to the weakest bonds holding the atoms of the crystal together.

cliff (scarp) retreat   The change in the position of a cliff face caused by erosion.

climate   The average weather conditions, along with the range of conditions, of a region over a year.

climate-change model   A computer-generated model designed to provide insight into how climate changed in the past and may change in the future, and what the consequences of climate change may be.

cloud   A mist of tiny water droplets in the sky.

coal   A black, organic rock consisting of greater than 50% carbon; it forms from the buried and altered remains of plant material.

coalbed methane   Natural gas created during the formation of coal, that gets trapped within the coal.

coal gasification   The process of producing relatively clean-burning gases from solid coal.

coal rank   A measurement of the carbon content of coal; higher-rank coal forms at higher temperatures.

coal reserve   The quantities of discovered, but not yet mined, coal in sedimentary rock of the continents.

coal swamp   A swamp whose oxygen-poor water allows thick piles of woody debris to accumulate; this debris transforms into coal upon deep burial.

coast   The belt of land bordering the sea.

coastal plain   Low-relief regions of land adjacent to the coast.

coastal wetland   A flat-lying coastal area that floods during high tide and drains during low tide, and hosts salt-resistant plants.

cold front   The boundary at which a cold air mass pushes underneath a warm air mass.

collision   The process of two buoyant pieces of lithosphere converging and squashing together.

color   The characteristic of a material due to the spectrum of light emitted or reflected by the material, as perceived by eyes or instruments.

columnar jointing   A type of fracturing that yields roughly hexagonal columns of basalt; columnar joints form when a dike, sill, or lava flow cools.

comet   A ball of ice and dust, probably remaining from the formation of the Solar System, that orbits the Sun.

compaction   The phase of lithification in which the pressure of the overburden on the buried rock squeezes out water and air that was trapped between clasts, and the clasts press tightly together.

competence (of a stream)   The maximum particle size a stream can carry.

composite volcano   *See* Stratovolcano.

compositional banding   A type of metamorphic foliation, found in gneiss, defined by alternating bands of light and dark minerals.

compound   A material composed of two or more elements that cannot be separated mechanically; the smallest piece is a molecule.

compressibility   The degree to which a material's volume changes in response to squashing.

compression   A push or squeezing felt by a body.

compressional waves   Waves in which particles of material move back and forth parallel to the direction in which the wave itself moves.

concentration   The proportion of one substance (the solute) dissolved within another (the solvent).

conchoidal fractures   Smoothly curving, clamshell-shaped surfaces along which materials with no cleavage planes tend to break.

condensation   The process of gas molecules linking together to form a liquid.

condensation nuclei   Preexisting solid or liquid particles, such as aerosols, onto which water condenses during cloud formation.

conduction   A process of heat transfer involving progressive migration of thermal energy from cooler to warmer regions in a material, without the physical flow of the material itself.

cone of depression   The downward-pointing, cone-shaped surface of the water table in a location where the water table is experiencing drawdown because of pumping at a well.

confined aquifer   An aquifer that is separated from the Earth's surface by an overlying aquitard.

conglomerate   Very coarse-grained sedimentary rock consisting of rounded clasts.

consuming boundary   *See* Convergent plate boundary.

contact   The boundary surface between two rock bodies (as between two stratigraphic formations, between an igneous intrusion and adjacent rock, between two igneous rock bodies, or between rocks juxtaposed by a fault).

contact metamorphism   *See* Thermal metamorphism.

contaminant plume   A cloud of contaminated groundwater that moves away from the source of the contamination.

continental crust   The crust beneath the continents.

continental divide   A highland separating drainage that flows into one ocean from drainage that flows into another.

continental-drift hypothesis   The idea that continents have moved and are still moving slowly across the Earth's surface.

continental glacier   A vast sheet of ice that spreads over thousands of square km of continental crust.

continental-interior desert   An inland desert that develops because by the time air masses reach the continental interior, they have lost all of their moisture.

continental lithosphere   Lithosphere topped by continental crust; this lithosphere reaches a thickness of 150 km.

continental margin   A continent's coastline.

continental rift   A linear belt along which continental lithosphere stretches and pulls apart.

continental rifting   The process by which a continent stretches and splits along a belt; if it is successful, rifting separates a larger continent into two smaller continents separated by a divergent boundary.

continental rise   The sloping sea floor that extends from the lower part of the continental slope to the abyssal plain.

continental shelf   A broad, shallowly submerged fringe of a continent; ocean-water depth over the continental shelf is generally less than 200 meters; the widest continental shelves occur over passive margins.

continental slope   The slope at the edge of a continental shelf, leading down to the deep sea floor.

continental volcanic arc   A long, curving chain of subaerial volcanoes on the margin of a continent adjacent to a convergent plate boundary.

contour lines   Lines on a map along which a parameter has a constant value; for example, all points along a contour line on a topographic map are at the same elevation.

control rod   Rods that absorb neutrons in a nuclear reactor and thus decrease the number of collisions between neutrons and radioactive atoms.

convection   Heat transfer that results when warmer, less dense material rises while cooler, denser material sinks.

convective cell   A distinct flow configuration for a volume of material that is moving during convective heat transport; simplistically, the material rises when warm and sinks when cool, and thus follows a loop-like path.

convergence zone   A place where two surface air flows meet so that air has to rise.

convergent margin   *See* Convergent plate boundary.

convergent plate boundary   A boundary at which two plates move toward each other so that one plate sinks (subducts) beneath the other; only oceanic lithosphere can subduct.

coral reef   A mound of coral and coral debris forming a region of shallow water.

core   The dense, iron-rich center of the Earth.

core-mantle boundary   An interface 2,900 km below the Earth's surface separating the mantle and core.

Coriolis effect   The deflection of objects, winds, and currents on the surface of the Earth owing to the planet's rotation.

cornice   A huge, overhanging drift of snow built up by strong winds at the crest of a mountain ridge.

correlation   The process of defining the age relations between the strata at one locality and the strata at another.

cosmic rays   Nuclei of hydrogen and other elements that bombard the Earth from deep space.

cosmology   The study of the overall structure of the Universe.

country rock (wall rock)   The preexisting rock into which magma intrudes.

covalent bonding   The attachment of one atom to another that develops when the atoms share electrons; one type of chemical bond.

crater   (1) A circular depression at the top of a volcanic mound; (2) a depression formed by the impact of a meteorite.

craton   A long-lived block of durable continental crust commonly found in the stable interior of a continent.

cratonic platform   A province in the interior of a continent in which Phanerozoic strata bury most of the underlying Precambrian rock.

creep   The gradual downslope movement of regolith.

crevasse   A large crack that develops by brittle deformation in the top 60 m of a glacier.

critical mass   A sufficiently dense and large mass of radioactive atoms in which a chain reaction happens so quickly that the mass explodes.

cross bed   Internal laminations in a bed, inclined at an angle to the main bedding; cross beds are a relict of the slip face of dunes or ripples.

cross section   A diagram depicting the geometry of materials underground as they would appear on an imaginary vertical slice through the Earth.

crude oil   Oil extracted directly from the ground.

crust   The rock that makes up the outermost layer of the Earth.

crustal root   Low-density crustal rock that protrudes downward beneath a mountain range.

crustal thickening   The process by which the continental crust increases in thickness, becoming up to 70 km thick (vs. normal thickness of about 35–40 km); it can occur during continental collision.

crystal   A single, continuous piece of a mineral bounded by flat surfaces that formed naturally as the mineral grew.

crystal face   The flat surfaces of a crystal, formed during the crystal's growth.

crystal form   The geometric shape of a crystal, defined by the arrangement of crystal faces.

crystal habit   The general shape of a crystal or cluster of crystals that grew unimpeded.

crystal lattice   The orderly framework within which the atoms or ions of a mineral are fixed.

crystal structure   The arrangement of atoms in a crystal.

crystalline   Containing a crystal lattice.

crystalline igneous   rock A rock that consists of minerals that grew when a melt solidified, and eventually interlock like pieces of a jigsaw puzzle.

cuesta   An asymmetric ridge formed by tilted layers of rock, with a steep cliff on one side cutting across the layers and a gentle slope on the other side; the gentle slope is parallel to the layering.

cumulonimbus cloud   A rain-producing, puffy cloud.

cumulus cloud   A puffy, cotton-ball-shaped cloud.

current   (1) A well-defined stream of ocean water; (2) the moving flow of water in a stream.

cut bank   The outside bank of the channel wall of a meander, which is continually undergoing erosion.

cutoff   A straight reach in a stream that develops when erosion eats through a meander neck.

cyanobacteria   Blue-green algae; a type of archaea.

cycle   A series of interrelated events or steps that occur in succession and can be repeated, perhaps indefinitely.

cyclone   (1) The counterclockwise flow of air around a low-pressure mass; (2) the equivalent of a hurricane in the Indian Ocean.

cyclonic flow   A circulation of air around a low-pressure region in the atmosphere; it rotates counterclockwise in the northern hemisphere.

cyclothem   A repeated interval within a sedimentary sequence that contains a specific succession of sedimentary beds.

Darcy's law A mathematical equation stating that a volume of water, passing through a specified area of material at a given time, depends on the material's permeability and hydraulic gradient.

daughter isotope The decay product of radioactive decay.

day The time it takes for the Earth to spin once on its axis.

debris avalanche An avalanche in which the falling debris consists of rock fragments and dust.

debris flow A downslope movement of mud mixed with larger rock fragments.

debris slide A sudden downslope movement of material consisting only of regolith.

decompression melting The kind of melting that occurs when hot mantle rock rises to shallower depths in the Earth so that pressure decreases while the temperature remains unchanged.

deep current An ocean current at a depth greater than 100 m.

deep-focus earthquake An earthquake that occurs at a depth between 300 and 670 km; below 670 km, earthquakes do not happen.

deflation The process of lowering the land surface by wind abrasion.

deformation A change in the shape, position, or orientation of a material, by bending, breaking, or flowing.

dehydration Loss of water.

delamination (plate tectonics) The process by which dense litho-spheric mantle separates from the base of a plate and sinks into the mantle.

delta A wedge of sediment formed at a river mouth when the running water of the stream enters standing water, the current slows, the stream loses competence, and sediment settles out.

delta plain The low, swampy land on the surface of a delta.

delta-plain flood A flood in which water submerges a delta plain.

dendritic network A drainage network whose interconnecting streams resemble the pattern of branches connecting to a deciduous tree.

dendrochronologist A scientist who analyzes tree rings to determine the geologic age of features.

density Mass per unit volume.

denudation The removal of rock and regolith from the Earth's surface.

deposition The process by which sediment settles out of a transporting medium.

depositional environment A setting in which sediments accumulate; its character (fluvial, deltaic, reef, glacial, etc.) reflects local conditions.

depositional landform A landform resulting from the deposition of sediment where the medium carrying the sediment evaporates, slows down, or melts.

desert A region so arid that it contains no permanent streams, except for those that bring water in from elsewhere, and has very sparse vegetation cover.

desertification The process of transforming nondesert areas into desert.

desert pavement A mosaic-like stone surface forming the ground in a desert.

desert varnish A dark, rusty-brown coating of iron oxide and magnesium oxide that accumulates on the surface of the rock.

detachment fault A nearly horizontal fault at the base of a fault system.

detritus The chunks and smaller grains of rock broken off outcrops by physical weathering.

dewpoint temperature The temperature at which air becomes saturated so that dew can form.

diagenesis All of the physical, chemical, and biological processes that transform sediment into sedimentary rock and that alter the rock after the rock has formed.

differential stress A condition causing a material to experience a push or pull in one direction of a greater magnitude than the push or pull in another direction; in some cases, differential stress can result in shearing.

differential weathering What happens when different rocks in an outcrop undergo weathering at different rates.

differentiation (of a planet) A process early in a planet's history during which dense iron alloy melted and sank downward to form the core, leaving less-dense mantle behind.

diffraction The splitting of light into many tiny beams that interfere with one another.

digital elevation map (DEM) A computer-produced portrayal of elevation differences commonly using shading to simulate shadows; the data used to produce the map assigns elevations to each point on the map.

dike A tabular (wall-shaped) intrusion of rock that cuts across the layering of country rock.

dimension stone An intact block of granite or marble to be used for architectural purposes.

dip The angle of a plane's slope as measured in a vertical plane perpendicular to the strike.

dipole A magnetic field with a north and south pole, like that of a bar magnet.

dipole field (for Earth) The part of the Earth's magnetic field, caused by the flow of liquid iron alloy in the outer core, that can be represented by an imaginary bar magnet with a north and south pole.

dip-slip fault A fault in which sliding occurs up or down the slope (dip) of the fault.

dip slope A hill slope underlain by bedding parallel to the slope.

directional drilling The process of controlling the trajectory of a drill bit to make sure that the drill hole goes exactly where desired.

disappearing stream A stream that intersects a crack or sinkhole leading to an underground cavern, so that the water disappears into the subsurface and becomes an underground stream.

discharge The volume of water in a conduit or channel passing a point in 1 second.

discharge area A location where groundwater flows back up to the surface and may emerge at springs.

disconformity An unconformity parallel to the two sedimentary sequences it separates.

displacement (offset) The amount of movement or slip across a fault plane.

disseminated deposit A hydrothermal ore deposit in which ore minerals are dispersed throughout a body of rock.

dissolution A process during which materials dissolve in water.

dissolved load Ions dissolved in a stream's water.

distillation column A vertical pipe in which crude oil is separated into several components.

distributaries The fan of small streams formed where a river spreads out over its delta.

divergence zone A place where sinking air separates into two flows that move in opposite directions.

divergent plate boundary A boundary at which two lithosphere plates move apart from each other; they are marked by mid-ocean ridges.

diversification The development of many different species.

DNA (deoxyribonucleic acid) The complex molecule, shaped like a double helix, containing the code that guides the growth and development of an organism.

doldrums A belt with very slow winds along the equator.

dome Folded or arched layers with the shape of an overturned bowl.

Doppler effect The phenomenon in which the frequency of wave energy appears to change when a moving source of wave energy passes an observer.

dormant volcano A volcano that has not erupted for hundreds to thousands of years but does have the potential to erupt again in the future.

downcutting The process in which water flowing through a channel cuts into the substrate and deepens the channel relative to its surroundings.

downdraft   Downward-moving air.

downgoing plate (slab)   A lithosphere plate that has been subducted at a convergent margin.

downslope force   The component of the force of gravity acting in the downslope direction.

downslope movement   The tumbling or sliding of rock and sediment from higher elevations to lower ones.

downwelling zone   A place where near-surface water sinks.

drag fold   A fold that develops in layers of rock adjacent to a fault during or just before slip.

drainage divide   A highland or ridge that separates one watershed from another.

drainage network (basin)   An array of interconnecting streams that together drain an area.

drainage reversal   When the overall direction of flow in a drainage network becomes the opposite of what it once had been.

drawdown   The phenomenon in which the water table around a well drops because the users are pumping water out of the well faster than it flows in from the surrounding aquifer.

drilling mud   A slurry of water mixed with clay that oil drillers use to cool a drill bit and flush rock cuttings up and out of the hole.

dripstone   Limestone (travertine in a cave) formed by the precipitation of calcium carbonate out of groundwater.

dropstone   A rock that drops to the sea floor once the iceberg that was carrying the rock melts.

drumlin   A streamlined, elongate hill formed when a glacier overrides glacial till.

dry-bottom (polar) glacier   A glacier so cold that its base remains frozen to the substrate.

dry wash   The channel of an ephemeral stream when empty of water.

dry well   (1) A well that does not supply water because the well has been drilled into an aquitard or into rock that lies above the water table; (2) a well that does not yield oil, even though it has been drilled into an anticipated reservoir.

ductile (plastic) deformation   The bending and flowing of a material (without cracking and breaking) subjected to stress.

dune   A pile of sand generally formed by deposition from the wind.

dust storm   An event in which strong winds hit unvegetated land, strip off the topsoil, and send it skyward to form rolling dark clouds that block out the Sun.

dynamic metamorphism   Metamorphism that occurs as a consequence of shearing alone, with no change in temperature or pressure.

dynamo   A power plant generator in which water or wind power spins an electrical conductor around a permanent magnet.

dynamothermal metamorphism   Metamorphism that involves heat, pressure, and shearing.

earthquake   A vibration caused by the sudden breaking or frictional sliding of rock in the Earth.

earthquake belt   A relatively narrow, distinct belt of earthquakes that defines the position of a plate boundary.

earthquake engineering   The design of buildings that can withstand shaking.

earthquake warning system   A communications network that provides an alert within microseconds after the first earthquake waves arrive at a seismograph near the epicenter, but before damaging vibrations reach population centers.

earthquake zoning   The determination of where land is relatively stable and where it might collapse because of seismicity.

Earth System   The global interconnecting web of physical and biological phenomena involving the solid Earth, the hydrosphere, and the atmosphere.

ebb tide   The falling tide.

eccentricity cycle   The cycle of the gradual change of the Earth's orbit from a more circular to a more elliptical shape; the cycle takes around 100,000 years.

ecliptic   The plane defined by a planet's orbit.

ecosystem   An environment and its inhabitants.

eddy   An isolated, ring-shaped current of water.

Ediacaran fauna   Multicellular invertebrate organisms that lived perhaps as early as 620 Ma and certainly by 565 Ma. They were named for a region in southern Australia.

effusive eruption   An eruption that yields mostly lava, not ash.

Ekman spiral   The change in flow direction of water with depth, caused by the Coriolis effect.

Ekman transport   The overall movement of a mass of water, resulting from the Eckman spiral, in a direction 90° to the wind direction.

elastic-rebound theory   The concept that earthquakes happen because stress builds up, causing rock adjacent to a fault to bend elastically until breaking and slip on a fault occurs; the slip relaxes the elastic bending and decreases stress.

elastic strain   A change in shape of a material; the change disappears instantly when stress is removed.

electromagnet   An electrical device that produces a magnetic field.

electron   A negatively charged subatomic particle that orbits the nucleus of an atom; electrons are about 0.0005 × the size of a proton.

electron microprobe   A laboratory instrument that can focus a beam of electrons on a small part of a mineral grain in order to create a signal that defines its chemical composition.

element   A material consisting entirely of one kind of atom; elements cannot be subdivided or changed by chemical reactions.

El Niño   The flow of warm water eastward from the Pacific Ocean that reverses the upwelling of cold water along the western coast of South America and causes significant global changes in weather patterns.

embayment   A low area of coastal land.

emergent coast   A coast where the land is rising relative to sea level or sea level is falling relative to the land.

end moraine (terminal moraine)   A low, sinuous ridge of till that develops when the terminus (toe) of a glacier stalls in one position for a while.

energy   The capacity to do work.

energy resource   Something that can be used to produce work; in a geologic context, a material (such as oil, coal, wind, flowing water) that can be used to produce energy.

eon   The largest subdivision of geologic time.

epeirogenic movement   The gradual uplift or subsidence of a broad region of the Earth's surface.

epeirogeny   An event of epeirogenic movement; the term is usually used in reference to the formation of broad mid-continent domes and basins.

ephemeral (intermittent) stream   A stream whose bed lies above the water table, so that the stream flows only when the rate at which water enters the stream from rainfall or meltwater exceeds the rate at which water infiltrates the ground below.

epicenter   The point on the surface of the Earth directly above the focus of an earthquake.

epicontinental sea   A shallow sea overlying a continent.

epoch   An interval of geologic time representing the largest subdivision of a period.

equant   A term for a grain that has the same dimensions in all directions.

equatorial low   The area of low pressure that develops over the equator because of the intertropical convergence zone.

equilibrium line (of a glacier)   The boundary between the zone of accumulation and the zone of ablation.

equinox One of two days out of the year (September 22 and March 21) in which the Sun is directly overhead at noon at the equator.

era An interval of geologic time representing the largest subdivision of the Phanerozoic Eon.

erg Sand seas formed by the accumulation of dunes in a desert.

erosion The grinding away and removal of Earth's surface materials by moving water, air, or ice.

erosional coast A coastline where sediment is not accumulating and wave action grinds away at the shore.

erosional landform A landform that results from the breakdown and removal of rock or sediment.

erratic A boulder or cobble that was picked up by a glacier and deposited hundreds of kilometers away from the outcrop from which it detached.

eruptive style The character of a particular volcanic eruption; geologists name styles based on typical examples (e.g., Hawaiian, Strombolian).

esker A ridge of sorted sand and gravel that snakes across a ground moraine; the sediment of an esker was deposited in subglacial meltwater tunnels.

estuary An inlet in which seawater and river water mix; created when a coastal valley is flooded because of either rising sea level or land subsidence.

Eubacteria The kingdom of "true bacteria."

euhedral crystal A crystal whose faces are well formed and whose shape reflects crystal form.

eukaryote An organism whose cells contain a nucleus; all plants and animals consist of eukaryotic cells.

eukaryotic cell A cell with a complex internal structure, capable of building multicellular organisms.

eustatic sea-level change A global rising or falling of the ocean surface.

evaporate To change from liquid to vapor.

evaporite Thick salt deposits that form as a consequence of precipitation from saline water.

evapotranspiration The sum of evaporation from bodies of water and the ground surface and transpiration from plants and animals.

exfoliation The process by which an outcrop of rock splits apart into onion-like sheets along joints that lie parallel to the ground surface.

exhumation The process (involving uplift and erosion) that returns deeply buried rocks to the surface.

exotic terrane A block of land that collided with a continent along a convergent margin and attached to the continent; the term exotic implies that the land was not originally part of the continent to which it is now attached.

expanding Universe theory The theory that the whole Universe must be expanding because galaxies in every direction seem to be moving away from us.

explosive eruptions Violent volcanic eruptions that produce clouds and avalanches of pyroclastic debris.

external process A geomorphologic process—such as downslope movement, erosion, or deposition—that is the consequence of gravity or of the interaction between the solid Earth and its fluid envelope (air and water). Energy for these processes comes from gravity and sunlight.

extinction The death of the last members of a species so that there are no parents to pass on their genetic traits to offspring.

extinct volcano A volcano that was active in the past but has now shut off entirely and will not erupt in the future.

extraordinary fossil A rare fossilized relict, or trace, of the soft part of an organism.

extratropical cyclone (wave cyclone, mid-latitude cyclone) A large, rotating storm system, in mid-latitudes, associated with a regional-scale low-pressure zone.

extrusive igneous rock Rock that forms by the freezing of lava above ground, after it flows or explodes out (extrudes) onto the surface and comes into contact with the atmosphere or ocean.

eye The relative calm in the center of a hurricane.

eye wall A rotating vertical cylinder of clouds surrounding the eye of a hurricane.

facet (of a gem) The ground and polished surface of a gem, produced by a gem cutter using a grinding lap.

facies (1) Sedimentary: a group of rocks and primary structures indicative of a given depositional environment; (2) metamorphic: a set of metamorphic mineral assemblages formed under a given range of pressures and temperatures.

Fahrenheit scale An English-system measure of temperature in which the difference between the freezing point (32°F) and the boiling point (212°F) is divided into 180 units.

failure surface A weak surface that forms the base of a landslide.

faint young Sun paradox The apparent contradiction implied by the fact that much of the Earth's surface temperature has remained above the melting point of water during the past 4 Ga, even though calculations indicate that the Sun produced much less energy when young.

fault A fracture on which one body of rock slides past another.

fault-block mountains An outdated term for a narrow, elongate range of mountains that develops in a continental-rift setting as normal faulting drops down blocks of crust or tilts blocks.

fault breccia Fragmented rock in which angular fragments were formed by brittle fault movement; fault breccia occurs along a fault.

fault creep Gradual movement along a fault that occurs in the absence of an earthquake.

fault gouge Pulverized rock consisting of fine powder that lies along fault surfaces; gouge forms by crushing and grinding.

faulting Slip events along a fault.

fault scarp A small step on the ground surface where one side of a fault has moved vertically with respect to the other.

fault system A grouping of numerous related faults.

fault trace (fault line) The intersection between a fault and the ground surface.

feedback mechanism A condition that arises when the consequence of a phenomenon influences the phenomenon itself.

felsic An adjective used in reference to igneous rocks that are rich in elements forming feldspar and quartz.

Ferrel cells The name given to the middle-latitude convection cells in the atmosphere.

fetch The distance across a body of water along which a wind blows to build waves.

field force A push or pull that applies across a distance (i.e., without contact between objects); examples are gravity and magnetism.

fine-grained A textural term for rock consisting of many fine grains or clasts.

firn Compacted granular ice (derived from snow) that forms where snow is deeply buried; if buried more deeply, firn turns into glacial ice.

fission A nuclear reaction during which the nucleus of a large atom splits to form two nuclei of smaller atoms; the process also releases neutrons and energy.

fission track A line of damage formed in the crystal lattice of a mineral by the impact of an atomic particle ejected during the decay of a radioactive isotope.

fissure A conduit in a magma chamber in the shape of a long crack through which magma rises and erupts at the surface.

fjord A deep, glacially carved, U-shaped valley flooded by rising sea level.

flank eruption An eruption that occurs when a secondary chimney, or fissure, breaks through the flank of a volcano.

flash flood A flood that occurs during unusually intense rainfall or as the result of a dam collapse, during which the floodwaters rise very fast.

flexing The process of folding in which a succession of rock layers bends and slip occurs between the layers.

flocculation The clumping together of clay suspended in river water into bunches that are large enough to settle out.

flood An event during which the volume of water in a stream becomes so great that it covers areas outside the stream's normal channel.

flood basalt Vast sheets of basalt that spread from a volcanic vent over an extensive surface of land; they may form where a rift develops above a continental hot spot, and where lava is particularly hot and has low viscosity.

flood-hazard map A representation of a portion of the Earth's surface that is designed to show how the danger of flooding varies with location.

floodplain The flat land on either side of a stream that becomes covered with water during a flood.

floodplain flood A flood during which a floodplain is submerged.

flood stage The stage when water reaches the top of a stream channel.

flood tide The rising tide.

floodway A mapped region likely to be flooded, in which people avoid constructing buildings.

flow fold A fold that forms when the rock is so soft that it behaves like weak plastic.

flowstone A sheet of limestone that forms along the wall of a cave when groundwater flows along the surface of the wall.

fluvial deposit Sediment deposited in a stream channel, along a stream bank, or on a floodplain.

flux Flow.

flux melting The transformation of hot solid to liquid that occurs when a volatile material injects into the solid.

focus The location where a fault slips during an earthquake (hypocenter).

fog A cloud that forms at ground level.

fold A bend or wrinkle of rock layers or foliation; folds form as a consequence of ductile deformation.

fold axis An imaginary line that, when moved parallel to itself, can trace out the shape of a folded surface.

fold-thrust belt An assemblage of folds and related thrust faults that develop above a detachment fault.

foliation Layering formed as a consequence of the alignment of mineral grains, or of compositional banding in a metamorphic rock.

foraminifera Microscopic plankton with calcitic shells, components of some limestones.

foreland sedimentary basin A basin located under the plains adjacent to a mountain front, which develops as the weight of the mountains pushes the crust down, creating a depression that traps sediment.

foreshocks The series of smaller earthquakes that precede a major earthquake.

foreshore zone The zone of beach regularly covered and uncovered by rising and falling tides.

formation *See* Stratigraphic formation.

fossil The remnant, or trace, of an ancient living organism that has been preserved in rock or sediment.

fossil assemblage A group of fossil species found in a specific sequence of sedimentary rock.

fossil correlation A determination of the stratigraphic relation between two sedimentary rock units, reached by studying fossils.

fossil fuel An energy resource such as oil or coal that comes from organisms that lived long ago and thus stores solar energy that reached the Earth then.

fossiliferous limestone Limestone consisting of abundant fossil shells and shell fragments.

fossilization The process of forming a fossil.

fractional crystallization The process by which a magma becomes progressively more silicic as it cools, because early-formed crystals settle out.

fracture zone A narrow band of vertical fractures in the ocean floor; fracture zones lie roughly at right angles to a mid-ocean ridge, and the actively slipping part of a fracture zone is a transform fault.

fragmental igneous rock A rock consisting of igneous chunks and/or shards that are packed together, welded together, or cemented together after having solidified.

frequency The number of waves that pass a point in a given time interval.

fresh rock Rock whose mineral grains have their original composition and shape.

friction Resistance to sliding on a surface.

fringing reef A coral reef that forms directly along the coast.

front The boundary between two air masses.

frost wedging The process in which water trapped in a joint freezes, forces the joint open, and may cause the joint to grow.

fuel rod A metal tube that holds the nuclear fuel in a nuclear reactor.

Fujita scale A scale that distinguishes among tornadoes on the basis of wind speed, path dimensions, and possible damage.

fusion A type of nuclear reaction during which nuclei collide and bond; fusion occurs in stars and hydrogen bombs.

Ga Billions of years ago (abbreviation).

gabbro A coarse-grained, intrusive, mafic igneous rock.

Gaia The term used for the Earth System, with the implication that it resembles a complex living entity.

galaxy An immense system of hundreds of billions of stars.

gem A finished (cut and polished) gemstone ready to be set in jewelry.

gemstone A mineral that has special value because it is rare and people consider it beautiful.

gene An individual component of the DNA code that guides the growth and development of an organism.

general circulation model A numerical calculation that simulates the flow of the atmosphere and resulting phenomena, due to changes in atmospheric temperature and other parameters.

genetics The study of genes and how they transmit information.

geocentric Universe concept An ancient Greek idea suggesting that the Earth sat motionless in the center of the Universe while stars and other planets and the Sun orbited around it.

geochronology The science of dating geologic events in years.

geode A cavity in which euhedral crystals precipitate out of water solutions passing through a rock.

geographical pole The locations (north and south) where the Earth's rotational axis intersects the planet's surface.

geologic column A composite stratigraphic chart that represents the entirety of the Earth's history.

geologic history The sequence of geologic events that has taken place in a region.

geologic map A map showing the distribution of rock units and structures across a region.

geologic time The span of time since the formation of the Earth.

geologic time scale A scale that describes the intervals of geologic time.

geology The study of the Earth, including our planet's composition, behavior, and history.

geotherm The change in temperature with depth in the Earth.

geothermal energy Heat and electricity produced by using the internal heat of the Earth.

geothermal gradient The rate of change in temperature with depth.

geothermal region A region of current or recent volcanism in which magma or very hot rock heats up groundwater, which may discharge at the surface in the form of hot springs and/or geysers.

geyser A fountain of steam and hot water that erupts periodically from a vent in the ground in a geothermal region.

giant planets The four outer, or Jovian, planets of our Solar System, which are significantly larger than the rest of the planets and consist largely of gas and/or ice.

glacial abrasion The process by which clasts embedded in the base of a glacier grind away at the substrate as the glacier flows.

glacial advance The forward movement of a glacier's toe when the supply of snow exceeds the rate of ablation.

glacial drift Sediment deposited in glacial environments.

glacial incorporation The process by which flowing ice surrounds and incorporates debris.

glacial marine Sediment consisting of ice-rafted clasts mixed with marine sediment.

glacial outwash Coarse sediment deposited on a glacial outwash plain by meltwater streams.

glacially polished surface A polished rock surface created by the glacial abrasion of the underlying substrate.

glacial plucking (glacial quarrying) The process by which a glacier breaks off and carries away fragments of bedrock.

glacial rebound The process by which the surface of a continent rises back up after an overlying continental ice sheet melts away and the weight of the ice is removed.

glacial retreat The movement of a glacier's toe back toward the glacier's origin; glacial retreat occurs if the rate of ablation exceeds the rate of supply.

glacial striation Grooves or scratches cut into bedrock when clasts embedded in the moving glacier act like the teeth of a giant rasp.

glacial subsidence The sinking of the surface of a continent caused by the weight of an overlying glacial ice sheet.

glacial till Sediment transported by flowing ice and deposited beneath a glacier or at its toe.

glaciation (glacial period) A portion of an ice age during which huge glaciers grew and covered substantial areas of the continents.

glacier A river or sheet of ice that slowly flows across the land surface and lasts all year long.

glass A solid in which atoms are not arranged in an orderly pattern.

glassy igneous rock Igneous rock consisting entirely of glass, or of tiny crystals surrounded by a glass matrix.

glide horizon The surface along which a slump slips.

global change The transformations or modifications of both physical and biological components of the Earth System through time.

global circulation The movement of volumes of air in paths that ultimately take it around the planet.

global climate change Transformations or modifications in Earth's climate over time.

global cooling A fall in the average atmospheric temperature.

global positioning system (GPS) A satellite system people can use to measure rates of movement of the Earth's crust relative to one another, or simply to locate their position on the Earth's surface.

global warming A rise in the average atmospheric temperature.

gneiss A compositionally banded metamorphic rock typically composed of alternating dark- and light-colored layers.

Gondwana A supercontinent that consisted of today's South America, Africa, Antarctica, India, and Australia. (Also called Gondwanaland.)

graben A down-dropped crustal block bounded on either side by a normal fault dipping toward the basin.

grade (of an ore) The concentration of a useful metal in an ore—the higher the concentration, the higher the grade.

graded bed A layer of sediment, deposited by a turbidity current, in which grain size varies from coarse at the bottom to fine at the top.

graded stream A stream that has attained an equilibrium longitudinal profile in which the sediment input into an area equals sediment removal.

gradualism The theory that evolution happens at a constant, slow rate.

grain A fragment of a mineral crystal or of a rock.

grain rotation The process by which rigid, inequant mineral grains distributed through a soft matrix may rotate into parallelism as the rock changes shape owing to differential stress.

granite A coarse-grained, intrusive, silicic igneous rock.

granulite facies A set of metamorphic mineral assemblages formed at very high pressures and temperatures.

gravitational spreading A process of lateral spreading that occurs in a material because of the weakness of the material; gravitational spreading causes continental glaciers to grow and mountain belts to undergo orogenic collapse.

gravity The attractive force that one mass exerts on another; the magnitude depends on the size of the objects and the distance between them.

graywacke An informal term used for sedimentary rock consisting of sand-sized up to small-pebble-sized grains of quartz and rock fragments all mixed together in a muddy matrix; typically, graywacke occurs at the base of a graded bed.

great oxygenation event The time in Earth's history, about 2.4 Ga, when the concentration of oxygen in the atmosphere increased dramatically.

greenhouse conditions (greenhouse period) Relatively warm global climate leading to the rising of sea level for an interval of geologic time.

greenhouse effect The trapping of heat in the Earth's atmosphere by carbon dioxide and other greenhouse gases, which absorb infrared radiation; somewhat analogous to the effect of glass in a greenhouse.

greenhouse gases Atmospheric gases, such as carbon dioxide and methane, that regulate the Earth's atmospheric temperature by absorbing infrared radiation.

greenschist facies A set of metamorphic mineral assemblages formed under relatively low pressures and temperatures.

greenstone A low-grade metamorphic rock formed from basalt; if foliated, the rock is called greenschist.

Greenwich mean time (GMT) The time at the astronomical observatory in Greenwich, England; time in all other time zones is set in relation to GMT.

Grenville orogeny The orogeny that occurred about 1 billion years ago and yielded the belt of deformed and metamorphosed rocks that underlie the eastern fifth of the North American continent.

groin A concrete or stone wall built perpendicular to a shoreline in order to prevent beach drift from removing sand.

ground moraine A thin, hummocky layer of till left behind on the land surface during a rapid glacial recession.

groundwater Water that resides under the surface of the Earth, mostly in pores or cracks of rock or sediment.

groundwater contamination Addition of chemicals or microbes (e.g., from agricultural and industrial activities, and landfills or septic tanks) to the groundwater supply.

group A succession of stratigraphic formations that have been lumped together, making a single, thicker stratigraphic entity.

growth ring A rhythmic layering that develops in trees, travertine deposits, and shelly organisms as a consequence of seasonal changes.

gusher A fountain of oil formed when underground pressure causes the oil to rise on its own out of a drilled hole.

guyot A seamount that had a coral reef growing on top of it, so that it is now flat-crested.

gymnosperm A plant whose seeds are "naked," not surrounded by a fruit.

gyre A large, circular flow pattern of ocean surface currents.

habitable zone (astronomy) The region in the Solar System where the intensity of radiation is sufficient to allow water to exist in liquid form on the surface of a planet.

Hadean Eon The oldest of the Precambrian eons; the time between Earth's origin and the formation of the first rocks that have been preserved.

Hadley cells The name given to the low-latitude convection cells in the atmosphere.

hail Falling ice balls from the sky, formed when ice crystallizes in turbulent storm clouds.

hail streak An approximately 2-by-10-km stretch of ground, elongate in the direction of a storm, onto which hail has fallen.

half-graben A wedge-shaped basin in cross section that develops as the hanging-wall block above a normal fault slides down and rotates; the basin develops between the fault surface and the top surface of the rotated block.

half-life The time it takes for half of a group of a radioactive element's isotopes to decay.

halocline The boundary in the ocean between surface-water and deep-water salinities.

hamada Barren, rocky highlands in a desert.

hanging valley A glacially carved tributary valley whose floor lies at a higher elevation than the floor of the trunk valley.

hanging wall The rock or sediment above an inclined fault plane.

hardness (of a mineral) A measure of the relative ability of a mineral to resist scratching; it represents the resistance of bonds in the crystal structure from being broken.

hard water Groundwater that contains dissolved calcium and magnesium, usually after passing through limestone or dolomite.

head (1) The elevation of the water table above a reference horizon; (2) the edge of ice at the origin of a glacier.

headland A place where a hill or cliff protrudes into the sea.

head scarp The distinct step along the upslope edge of a slump where the regolith detached.

headward erosion The process by which a stream channel lengthens up its slope as the flow of water increases.

headwaters The beginning point of a stream.

heat Thermal energy resulting from the movement of molecules.

heat capacity A measure of the amount of heat that must be added to a material to change its temperature.

heat flow The rate at which heat rises from the Earth's interior up to the surface.

heat-transfer melting Melting that results from the transfer of heat from a hotter magma to a cooler rock.

heliocentric Universe concept An idea proposed by Greek philosophers around 250 B.C.E. suggesting that all heavenly objects including the Earth orbited the Sun.

heliosphere A bubble-like region in space in which solar wind has blown away most interstellar atoms.

Hercynian orogen The late Paleozoic orogen that affected parts of Europe; a continuation of the Alleghenian orogen.

heterosphere A term for the upper portion of the atmosphere, in which gases separate into distinct layers on the basis of composition.

hiatus The interval of time between deposition of the youngest rock below an unconformity and deposition of the oldest rock above the unconformity.

high-altitude westerlies Westerly winds at the top of the troposphere.

high-grade metamorphic rocks Rocks that metamorphose under relatively high temperatures.

high-level waste Nuclear waste containing greater than 1 million times the safe level of radioactivity.

hinge The portion of a fold where curvature is greatest.

hogback A steep-sided ridge of steeply dipping strata.

Holocene The period of geologic time since the last glaciation.

Holocene climatic maximum The period from 5,000 to 6,000 years ago, when Holocene temperatures reached a peak.

homosphere The lower part of the atmosphere, in which the gases have stirred into a homogenous mixture.

hoodoo The local name for the brightly colored shale and sandstone chimneys found in Bryce Canyon National Park in Utah.

horn A pointed mountain peak surrounded by at least three cirques.

hornfels Rock that undergoes metamorphism simply because of a change in temperature, without being subjected to differential stress.

horse latitudes The region of the subtropical high in which winds are weak.

horst The high block between two grabens.

hot spot A location at the base of the lithosphere, at the top of a mantle plume, where temperatures can cause melting.

hot-spot track A chain of now-dead volcanoes transported off the hot spot by the movement of a lithosphere plate.

hot-spot volcano An isolated volcano not caused by movement at a plate boundary, but rather by the melting of a mantle plume.

hot spring A spring that emits water ranging in temperature from about 30°C to 104°C.

hummocky surface An irregular and lumpy ground surface.

hurricane A huge rotating storm, resembling a giant spiral in map view, in which sustained winds blow over 119 km per hour.

hurricane track The path a hurricane follows.

hyaloclastite A rubbly extrusive rock consisting of glassy debris formed in a submarine or sub-ice eruption.

hydration The absorption of water into the crystal structure of minerals; a type of chemical weathering.

hydraulic conductivity The coefficient K in Darcy's law; hydraulic conductivity takes into account the permeability of the sediment or rock as well as the fluid's viscosity.

hydraulic gradient The slope of the water table.

hydraulic head The potential energy available to drive the flow of a given volume of groundwater at a location; it can be measured as an elevation above a reference.

hydrocarbon A chain-like or ring-like molecule made of hydrogen and carbon atoms; petroleum and natural gas are hydrocarbons.

hydrocarbon generation A process in which oil shale warms to temperatures of greater than about 90°C so kerogen molecules transform into oil and natural gas molecules.

hydrocarbon reserve A known supply of oil and gas held underground.

hydrocarbon system The association of source rock, migration pathway, reservoir rock, seal, and trap geometry that leads to the occurrence of a hydrocarbon reserve.

hydrofracturing ("fracking") A process by which drillers generate new fractures or open preexisting ones underground, by pumping a high-pressure fluid into a portion of the drill hole, in order to increase the permeability of surrounding hydrocarbon-bearing rocks.

hydrogen bond The attraction of a hydrogen atom to a negatively charged atom or molecule (e.g., hydrogen bonds attach water molecules to each other).

hydrologic cycle The continual passage of water from reservoir to reservoir in the Earth System.

hydrolysis The process in which water chemically reacts with minerals and breaks them down.

hydrosphere The Earth's water, including surface water (lakes, rivers, and oceans), groundwater, and liquid water in the atmosphere.

hydrothermal deposit An accumulation of ore minerals precipitated from hot-water solutions circulating through a magma or through the rocks surrounding an igneous intrusion.

hydrothermal metamorphism When very hot water passes through the crust and causes metamorphism of rock.

hypocenter (focus) The place within the Earth where earthquake energy originates; commonly, the hypocenter is the place on a fault where slip took place.

hypsometric curve A graph that plots surface elevation on the vertical axis and the percentage of the Earth's surface on the horizontal axis.

ice age An interval of time in which the climate was colder than it is today, glaciers occasionally advanced to cover large areas of the continents, and mountain glaciers grew; an ice age can include many glacials and interglacials.

iceberg A large block of ice that calves off the front of a glacier and drops into the sea.

icehouse period A period of time when the Earth's temperature was cooler than it is today and ice ages could occur.

ice-margin lake A meltwater lake formed along the edge of a glacier.

ice-rafted sediment Sediment carried out to sea by icebergs.

ice sheet A vast glacier that covers the landscape.

ice shelf A broad, flat region of ice along the edge of a continent formed where a continental glacier flowed into the sea.

ice stream A portion of a glacier that travels much more quickly than adjacent portions of the glacier.

ice tongue The portion of a valley glacier that has flowed out into the sea.

igneous rock Rock that forms when hot molten rock (magma or lava) cools and freezes solid.

ignimbrite Rock formed when deposits of pyroclastic flows solidify.

inactive fault A fault that last moved in the distant past and probably won't move again in the near future, yet is still recognizable because of displacement across the fault plane.

inactive sand The sand along a coast that is buried beneath a layer of active sand and moves only during severe storms or not at all.

incised meander A meander that lies at the bottom of a steep-walled canyon.

index minerals Minerals that serve as good indicators of metamorphic grade.

induced seismicity Seismic events caused by the actions of people (e.g., filling a reservoir that lies over a fault with water).

industrial minerals Minerals that serve as the raw materials for manufacturing chemicals, concrete, and wallboard, among other products.

inequant A term for a mineral grain whose length and width are not the same.

inertia The tendency of an object at rest to remain at rest.

infiltrate Seep down into.

injection well A well in which a liquid is pumped down into the ground under pressure so that it passes from the well back into the pore space of the rock or regolith.

inner core The inner section of the core, extending from 5,155 km deep to the Earth's center at 6,371 km and consisting of solid iron alloy.

inselberg An isolated mountain or hill in a desert landscape created by progressive cliff retreat, so that the hill is surrounded by a pediment or an alluvial fan.

insolation Exposure to the Sun's rays.

intensity (seismology) A measure of the relative size of an earthquake (the severity of ground shaking) at a location, as determined by examining the amount of damage caused.

interglacial A period of time between two glaciations.

interior basin A basin with no outlet to the sea.

interlocking texture The texture of crystalline rocks in which mineral grains fit together like pieces of a jigsaw puzzle.

internal process A process in the Earth System, such as plate motion, mountain building, or volcanism, ultimately caused by Earth's internal heat.

intertidal zone The area of coastal land across which the tide rises and falls.

intertropical convergence zone The equatorial convergence zone in the atmosphere.

intraplate earthquakes Earthquakes that occur away from plate boundaries.

intrusive contact The boundary between country rock and an intrusive igneous rock.

intrusive igneous rock Rock formed by the freezing of magma underground.

ion A version of an atom that has lost or gained electrons, relative to an electrically neutral version, so that it has a net electrical charge.

ionic bond The attachment of one atom to another that happens when one atom transfers electrons to another; one type of chemical bond.

ionosphere The interval of Earth's atmosphere, at an elevation between 50 and 400 km, containing abundant positive ions.

iron catastrophe The proposed event very early in Earth history when the Earth partly melted and molten iron sank to the center to form the core.

isobar A line on a map along which the air has a specified pressure.

isograd (1) A line on a pressure-temperature graph along which all points are taken to be at the same metamorphic grade; (2) a line on a map making the first appearance of a metamorphic index mineral.

isostasy (isostatic equilibrium) The condition that exists when the buoyancy force pushing lithosphere up equals the gravitational force pulling lithosphere down.

isostatic compensation The process in which the surface of the crust slowly rises or falls to reestablish isostatic equilibrium after a geologic event changes the density or thickness of the lithosphere.

isotherm Lines on a map or cross section along which the temperature is constant.

isotopes Different versions of a given element that have the same atomic number but different atomic weights.

jet stream A fast-moving current of air that flows at high elevations.

jetty A man-made wall that protects the entrance to a harbor.

joints Naturally formed cracks in rocks.

joint set A group of systematic joints.

Jovian A term used to describe the outer gassy, Jupiter-like planets (gas-giant planets).

kame A stratified sequence of lateral-moraine sediment that's sorted by water flowing along the edge of a glacier.

karst landscape A region underlain by caves in limestone bedrock; the collapse of the caves creates a landscape of sinkholes separated by higher topography, or of limestone spires separated by low areas.

Kelvin (K) scale A measure of temperature in which 0 K is absolute zero and the freezing point of water is 273.15 K; divisions in the Kelvin scale have the same value as those in the Celsius scale.

kerogen The waxy molecules into which the organic material in shale transforms on reaching about 100°C. At higher temperatures, kerogen transforms into oil.

kettle hole A circular depression in the ground made when a block of ice calves off the toe of a glacier, becomes buried by till, and later melts.

knob-and-kettle topography A land surface with many kettle holes separated by round hills of glacial till.

K-T boundary event The mass extinction that happened at the end of the Cretaceous Period, 66 million years ago, possibly due to the collision of an asteroid with the Earth.

Kuiper Belt A diffuse ring of icy objects, remnants of Solar System formation, that orbit our Sun outside the orbit of Neptune.

laccolith A blister-shaped igneous intrusion that forms when magma injects between layers underground in a manner that pushes overlying layers upward to form a dome.

lag deposit The coarse sediment left behind in a desert after wind erosion removes the finer sediment.

lagoon A body of shallow seawater separated from the open ocean by a barrier island.

lahar A thick slurry formed when volcanic ash and debris mix with water, either in rivers or from rain or melting snow and ice on the flank of a volcano.

landslide A sudden movement of rock and debris down a nonvertical slope.

landslide-potential map A map on which regions are ranked according to the likelihood that a mass movement will occur.

land subsidence Sinking elevation of the ground surface; the process may occur over an aquifer that is slowly draining and decreasing in volume because of pore collapse.

La Niña Years in which the El Niño event is not strong.

lapilli Any pyroclastic particle that is 2 to 64 mm in diameter (i.e., marble-sized); the particles can consist of frozen lava clots, pumice fragments, or ash clumps.

Laramide orogeny The mountain-building event that lasted from about 80 Ma to 40 Ma, in western North America; in the United States, it formed the Rocky Mountains as a result of basement uplift and the warping of the younger overlying strata into large monoclines.

large igneous province (LIP) A region in which huge volumes of lava and/or ash erupted over a relatively short interval of geologic time.

latent heat of condensation The heat released during condensation, which comes only from a change in state.

lateral moraine A strip of debris along the side margins of a glacier.

laterite soil A hard, brick-red, soil formed from iron-rich rock in a tropical environment; it consists primarily of insoluble iron and aluminum oxide and hydroxide and forms due to extreme leaching.

Laurentia A continent in the early Paleozoic Era composed of today's North America and Greenland.

Laurentide ice sheet An ice sheet that spread over northeastern Canada during the Pleistocene ice age(s).

lava Molten rock that has flowed out onto the Earth's surface.

lava dome A dome-like mass of rhyolitic lava that accumulates above the eruption vent.

lava flows Sheets or mounds of lava that flow onto the ground surface or sea floor in molten form and then solidify.

lava lake A large pool of lava produced around a vent when lava fountains spew forth large amounts of lava in a short period of time.

lava tube The empty space left when a lava tunnel drains; this happens when the surface of a lava flow solidifies while the inner part of the flow continues to stream downslope.

leach To dissolve and carry away.

leader A conductive path stretching from a cloud toward the ground, along which electrons leak from the base of the cloud, and which provides the start for a lightning flash to the ground.

lightning bolt (lightning flash, lightning stroke) A giant spark or pulse of current that jumps across a gap of charge separation.

light year The distance that light travels in one Earth year (about 6 trillion miles or 9.5 trillion km).

lignite Low-rank coal that consists of 50% carbon.

limb The side of a fold, showing less curvature than at the hinge.

limestone Sedimentary rock composed of calcite.

liquefaction The process by which saturated, unconsolidated sediments are transformed into a substance that acts like a liquid as a result of ground shaking.

liquidus The lowest temperature at which all the components of a material have melted and transformed into liquid.

liquification The process by which wet sediment becomes a slurry; liquification may be triggered by earthquake vibrations.

lithification The transformation of loose sediment into solid rock through compaction and cementation.

lithologic correlation A correlation based on similarities in rock type.

lithosphere The relatively rigid, nonflowable, outer 100- to 150-km-thick layer of the Earth, constituting the crust and the top part of the mantle.

lithosphere plate One of many distinct pieces of the lithosphere (Earth's relatively rigid shell) that are separated from one another by breaks (plate boundaries).

little ice age A period of cooler temperatures, between 1500 and 1800 C.E., during which many glaciers advanced.

loam A type of soil consisting of roughly equal parts of sand, silt, and clay; it tends to be good for growth of crops.

local base level A base level upstream from a drainage network's mouth.

lodgment till A flat layer of till smeared out over the ground when a glacier overrides an end moraine as it advances.

loess Layers of fine-grained sediments deposited from the wind; large deposits of loess formed from fine-grained glacial sediment blown off outwash plains.

longitudinal (seif) dune A dune formed when there is abundant sand and a strong, steady wind, and whose axis lies parallel to the wind direction.

longitudinal profile A cross-sectional image showing the variation in elevation along the length of a river.

longshore current The flow of water parallel to the shore just off a coast, because of the diagonal movement of waves toward the shore.

longshore drift The movement of sediment laterally along a beach; it occurs when waves wash up a beach diagonally.

lower mantle The deepest section of the mantle, stretching from 670 km down to the core-mantle boundary.

low-grade metamorphic rocks Rocks that underwent metamorphism at relatively low temperatures.

low-velocity zone The asthenosphere underlying oceanic lithosphere in which seismic waves travel more slowly, probably because rock has partially melted.

luster The way a mineral surface scatters light.

L-waves (Love waves) Surface seismic waves that cause the ground to ripple back and forth, creating a snake-like movement.

Ma Millions of years ago (abbreviation).

macrofossil A fossil large enough to be seen with the naked eye.

mafic A term used in reference to magmas or igneous rocks that are relatively poor in silica and rich in iron and magnesium.

magma Molten rock beneath the Earth's surface.

magma chamber A space below ground filled with magma.

magma contamination The process in which flowing magma incorporates components of the country rock through which it passes.

magmatic deposit An ore deposit formed when sulfide ore minerals accumulate at the bottom of a magma chamber.

magnetic anomaly The difference between the expected strength of the Earth's magnetic field at a certain location and the actual measured strength of the field at that location.

magnetic declination The angle between the direction a compass needle points at a given location and the direction of true north.

magnetic dipole An imaginary vector that points from the north magnetic pole to the south magnetic pole of a magnetic field.

magnetic field The region affected by the force emanating from a magnet.

magnetic field lines The trajectories along which magnetic particles would align, or charged particles would flow, if placed in a magnetic field.

magnetic force The push or pull exerted by a magnet.

magnetic inclination The angle between a magnetic needle free to pivot on a horizontal axis and a horizontal plane parallel to the Earth's surface.

magnetic poles The ends of a magnetic dipole; all magnetic dipoles have a north pole and a south pole.

magnetic reversal The change of the Earth's magnetic polarity; when a reversal occurs, the field flips from normal to reversed polarity, or vice versa.

magnetic-reversal chronology The history of magnetic reversals through geologic time.

magnetism An attractive or repulsive field force generated by permanent magnets or by an electrical current.

magnetization The degree to which a material can exert a magnetic force.

magnetometer An instrument that measures the strength of the Earth's magnetic field.

magnetosphere The region protected from the electrically charged particles of the solar winds by Earth's magnetic field.

magnetostratigraphy The comparison of the pattern of magnetic reversals in a sequence of strata, with a reference column showing the succession of reversals through time.

magnitude (of an earthquake) The number that represents the maximum amplitude of ground motion that would be measured by a seismometer placed at a specified distance from the epicenter.

manganese nodules Lumpy accumulations of manganese-oxide minerals precipitated onto the sea floor.

mantle The thick layer of rock below the Earth's crust and above the core.

mantle plume A column of very hot rock rising up through the mantle.

marble A metamorphic rock composed of calcite and transformed from a protolith of limestone.

mare The broad, darker areas on the Moon's surface; they consist of flood basalts that erupted over 3 billion years ago and spread out across the Moon's lowlands.

marginal sea A small ocean basin created when sea-floor spreading occurs behind an island arc.

marine magnetic anomaly The difference between the *expected* strength of the Earth's main dipole field at a certain location on the sea floor and the *actual* measured strength of the magnetic field at that location.

maritime tropical air mass A mass of air that originates over tropical or subtropical oceanic regions.

marker bed A particularly unique layer that provides a definitive basis for correlation.

marsh A wetland dominated by grasses.

mass The amount of matter in an object; mass differs from weight in that its value does not depend on the strength of gravity.

mass-extinction event A time when vast numbers of species abruptly vanish.

mass movement (mass wasting) The gravitationally caused downslope transport of rock, regolith, snow, or ice.

matter The material substance of the universe; it consists of atoms and has mass.

matrix Finer-grained material surrounding larger grains in a rock.

meander A snake-like curve along a stream's course.

meandering stream A reach of stream containing many meanders (snake-like curves).

meander neck A narrow isthmus of land separating two adjacent meanders.

mean sea level The average level between the high and low tide over a year at a given point.

mechanical force A push, pull, or shear applied by one object on another; it can be applied only if the objects are in contact.

mechanical weathering *See* Physical weathering.

medial moraine A strip of sediment in the interior of a glacier, parallel to the flow direction of the glacier, formed by the lateral moraines of two merging glaciers.

Medieval Warm Period A period of high temperatures in the Middle Ages.

melt Molten (liquid) rock.

meltdown The melting of the fuel rods in a nuclear reactor that occurs if the rate of fission becomes too fast and the fuel rods become too hot.

melting curve The line defining the range of temperatures and pressures at which a rock melts.

melting temperature The temperature at which the thermal vibration of the atoms or ions in the lattice of a mineral is sufficient to break the chemical bonds holding them to the lattice, so a material transforms into a liquid.

meltwater lake A lake fed by glacial meltwater.

mesa A large, flat-topped hill (with a surface area of several square km) in an arid region.

mesopause The boundary that marks the top of the mesosphere of Earth's atmosphere.

mesosphere The cooler layer of atmosphere overlying the stratosphere.

Mesozoic Era The middle of the three Phanerozoic eras; it lasted from 252 Ma to 66 Ma.

metaconglomerate A metamorphic rock produced by metamorphism of a conglomerate; typically, it contains flattened pebbles and cobbles.

metal A solid composed almost entirely of atoms of metallic elements; it is generally opaque, shiny, smooth, malleable, and can conduct electricity.

metallic bond A chemical bond in which the outer atoms are attached to each other in such a way that electrons flow easily from atom to atom.

metamorphic aureole The region around a pluton, stretching tens to hundreds of meters out, in which heat transferred into the country rock and metamorphosed the country rock.

metamorphic facies A set of metamorphic mineral assemblages indicative of metamorphism under a specific range of pressures and temperatures.

metamorphic foliation A fabric defined by parallel surfaces or layers that develop in a rock as a result of metamorphism; schistocity and gneissic layering are examples.

metamorphic grade A representation of the intensity of metamorphism, meaning the amount or degree of metamorphic change.

metamorphic mineral New minerals that grow in place within a solid rock under metamorphic temperatures and pressures.

metamorphic mineral assemblage A group of minerals that form in a rock as a result of metamorphism.

metamorphic rock Rock that forms when preexisting rock changes into new rock as a result of an increase in pressure and temperature and/or shearing under elevated temperatures; metamorphism occurs without the rock first becoming a melt or a sediment.

metamorphic texture A distinctive arrangement of mineral grains produced by metamorphism.

metamorphic zone The region between two metamorphic isograds, typically named after an index mineral found within the region.

metamorphism The process by which one kind of rock transforms into a different kind of rock.

metasomatism The process by which a rock's overall chemical composition changes during metamorphism because of reactions with hot water that bring in or remove elements.

meteor A streak of bright, glowing gas created as a meteoroid vaporizes in the atmosphere due to friction.

meteoric water Water that falls to Earth from the atmosphere as either rain or snow.

meteorite A piece of rock or metal alloy that fell from space and landed on Earth.

micrite Limestone consisting of lime mud (i.e., very fine-grained limestone).

microfossil A fossil that can be seen only with a microscope or an electron microscope.

mid-ocean ridge A 2-km-high submarine mountain belt that forms along a divergent oceanic plate boundary.

migmatite A rock formed when gneiss is heated high enough so that it begins to partially melt, creating layers, or lenses, of new igneous rock that mix with layers of the relict gneiss.

Milanković cycles Climate cycles that occur over tens to hundreds of thousands of years because of changes in Earth's orbit and tilt.

mine A site at which ore is extracted from the ground.

mineral A homogenous, naturally occurring, solid inorganic substance with a definable chemical composition and an internal structure characterized by an orderly arrangement of atoms, ions, or molecules in a lattice. Most minerals are inorganic.

mineral classes Groups of minerals distinguished from each other on the basis of chemical composition.

mineralogist A geoscientist specializing in the study of minerals.

mineral resources The minerals extracted from the Earth's upper crust for practical purposes.

Mississippi Valley–type (MVT) ore An ore deposit, typically in dolostone, containing lead- and zinc-bearing minerals that precipitated from groundwater that had moved up from several km depth in the upper crust; such deposits occur in the upper Mississippi Valley.

mixture A material consisting of two or more substances that can be separated mechanically (i.e., without chemical reactions).

Modified Mercalli scale An earthquake characterization scale based on the amount of damage that the earthquake causes.

Moho The seismic-velocity discontinuity that defines the boundary between the Earth's crust and mantle. Named for Andrija Mohorovičić.

Mohs hardness scale A list of ten minerals in a sequence of relative hardness, with which other minerals can be compared.

mold A cavity in sedimentary rock left behind when a shell that once filled the space weathers out.

molecule The smallest piece of a compound that has the properties of the compound; it consists of two or more atoms attached by chemical bonds.

monocline A fold in the land surface whose shape resembles that of a carpet draped over a stair step.

monsoon A seasonal reversal in wind direction that causes a shift from a very dry season to a very rainy season in some regions of the world.

moon A sizable solid body locked in orbit around a planet.

moraine A sediment pile composed of till deposited by a glacier.

mountain front The boundary between a mountain range and adjacent plains.

mountain (alpine) glacier A glacier that exists in or adjacent to a mountainous region.

mountain ice cap A mound of ice that submerges peaks and ridges at the crest of a mountain range.

mouth The outlet of a stream where it discharges into another stream, a lake, or a sea.

mudflow A downslope movement of mud at slow to moderate speed.

mud pot A viscous slurry that forms in a geothermal region when hot water or steam rises into soils rich in volcanic ash and clay.

mudstone Very fine-grained sedimentary rock that will not easily split into sheets.

mylonite Rock formed during dynamic metamorphism and characterized by foliation that lies roughly parallel to the fault (shear zone) involved in the shearing process; mylonites have very fine grains formed by the nonbrittle subdivision of larger grains.

native metal A naturally occurring pure mass of a single metal in an ore deposit.

natural arch An arch that forms when erosion along joints leaves narrow walls of rock; when the lower part of the wall erodes while the upper part remains, an arch results.

natural hazard A natural feature of the environment that can cause injury to living organisms and/or damage to buildings and the landscape.

natural levees A pair of low ridges that appear on either side of a stream and develop as a result of the accumulation of sediment deposited naturally during flooding.

natural selection The process by which the fittest organisms survive to pass on their characteristics to the next generation.

neap tide An especially low tide that occurs when the angle between the direction of the Moon and the direction of the Sun is 90°.

nebula A cloud of gas or dust in space.

nebular theory of planet formation The concept that planets grow out of rings of gas, dust, and ice surrounding a newborn star.

negative anomaly An area where the magnetic field strength is less than expected.

negative feedback Feedback that slows a process down or reverses it.

neocrystallization The growth of new crystals, not in the protolith, during metamorphism.

neutron A subatomic particle, in the nucleus of an atom, that has a neutral charge.

Nevadan orogeny A convergent-margin mountain-building event that took place in western North America during the Late Jurassic Period.

nonconformity A type of unconformity at which sedimentary rocks overlie basement (older intrusive igneous rocks and/or metamorphic rocks).

nonflowing artesian well An artesian well in which water rises on its own up to a level that lies below the ground surface.

nonfoliated metamorphic rock Rock containing minerals that recrystallized during metamorphism but has no foliation.

nonmetallic mineral resources Mineral resources that do not contain metals; examples include building stone, gravel, sand, gypsum, phosphate, and salt.

nonplunging fold A fold with a horizontal hinge.

nonrenewable resource A resource that nature will take a long time (hundreds to millions of years) to replenish or may never replenish.

nonsystematic joints Short cracks in rocks that occur in a range of orientations and are randomly placed and oriented.

nor'easter A large, midlatitude North American cyclone; when it reaches the east coast, it produces strong winds that come out of the northeast.

normal fault A fault in which the hanging-wall block moves down the slope of the fault.

normal force The component of the gravitational force acting perpendicular to a slope.

normal polarity Polarity in which the paleomagnetic dipole has the same orientation as it does today.

normal stress The push or pull that is perpendicular to a surface.

North Atlantic deep-water mass The mass of cold, dense water that sinks in the north polar regions.

northeast tradewinds Surface winds that come out of the northeast and occur in the region between the equator and 30°N.

nuclear bond The force that attaches subatomic particles to each other within the nucleus of an atom.

nuclear fuel Pellets of concentrated uranium oxide or a comparable radioactive material that can provide energy in a nuclear reactor.

nuclear fusion The process by which the nuclei of atoms fuse together, thereby creating new, larger atoms.

nuclear reaction A process that results in changing the nucleus of an atom by breaking or forming nuclear bonds.

nuclear reactor The part of a nuclear power plant where the fission reactions occur.

nuclear waste The radioactive material produced as a byproduct in a nuclear plant that must be disposed of carefully due to its dangerous radioactivity.

nucleus The central ball of an atom that consists of protons and neutrons (except for hydrogen, whose nuclei contains only a proton).

nuée ardente *See* Pyroclastic flow.

numerical age (in older literature, "absolute age") The age of a geologic feature given in years.

oasis A verdant region surrounded by desert, occurring at a place where natural springs provide water at the surface.

oblique-slip fault A fault in which sliding occurs diagonally along the fault plane.

obsidian An igneous rock consisting of a solid mass of volcanic glass.

occluded front A front that no longer intersects the ground surface.

oceanic crust The crust beneath the oceans; composed of gabbro and basalt, overlain by sediment.

oceanic plateau A region of oceanic floor that is higher than surrounding areas; such regions have particularly thick oceanic crust and are relicts of submarine large igneous provinces.

oceanic lithosphere Lithosphere topped by oceanic crust; it reaches a thickness of 100 km.

offshore bar A narrow ridge of sand that forms off the shore of a beach; some offshore bars rise above sea level, and separate a lagoon on one side from the open ocean on the other.

oil (geology) A liquid hydrocarbon that can be burned as a fuel.

Oil Age The period of human history, including our own, so named because the economy depends on oil.

oil field A region containing a significant amount of accessible oil underground.

oil reserve The known supply of oil held underground.

oil shale Shale containing kerogen.

oil trap A geologic configuration that keeps oil underground in the reservoir rock and prevents it from rising to the surface.

oil window The narrow range of temperatures under which oil can form in a source rock.

olistotrome A large, submarine slump block, buried and preserved.

Oort Cloud A cloud of icy objects, left over from Solar System formation, that orbit the Sun in a region outside of the heliosphere.

ophiolite A slice of oceanic crust that has been thrust onto continental crust.

ordinary well A well whose base penetrates below the water table and can thus provide water.

ore Rock containing native metals or a concentrated accumulation of ore minerals.

ore deposit An economically significant accumulation of ore.

ore minerals Minerals that have metal in high concentrations and in a form that can be easily extracted.

organic carbon Carbon that has been incorporated in an organism.

organic chemical A carbon-containing compound that occurs in living organisms, or that resembles such compounds; it consists of carbon atoms bonded to hydrogen atoms along with varying amounts of oxygen, nitrogen, and other chemicals.

organic coast A coast along which living organisms control landforms along the shore.

organic sedimentary rock Sedimentary rock (such as coal) formed from carbon-rich relicts of organisms.

organic shale Lithified, muddy, organic-rich ooze that contains the raw materials from which hydrocarbons eventually form.

orogen (orogenic belt) A linear range of mountains.

orogenic collapse The process in which mountains begin to collapse under their own weight and spread out laterally.

orogeny A mountain-building event.

orographic barrier A landform that diverts air flow upward or laterally.

outcrop An exposure of bedrock.

outer core The section of the core, between 2,900 and 5,150 km deep, that consists of liquid iron alloy.

outwash plain A broad area of gravel and sandbars deposited by a braided stream network, fed by the melt water of a glacier.

overburden The weight of overlying rock on rock buried deeper in the Earth's crust.

overriding plate (slab) The plate at a subduction zone that overrides the downgoing plate.

oversaturated solution A solution that contains so much solute (dissolved ions) that precipitation begins.

oversized stream valley A large valley with a small stream running through it; the valley formed earlier when the flow was greater.

oxbow lake A meander that has been cut off yet remains filled with water.

oxidation reaction A reaction in which an element loses electrons; an example is the reaction of iron with air to form rust.

ozone O3, an atmospheric gas that absorbs harmful ultraviolet radiation from the Sun.

ozone hole An area of the atmosphere, over polar regions, from which ozone has been depleted.

pahoehoe A lava flow with a surface texture of smooth, glassy, ropelike ridges.

paleoclimate The past climate of the Earth.

paleomagnetism The record of ancient magnetism preserved in rock.

paleopole The supposed position of the Earth's magnetic pole in the past, with respect to a particular continent.

paleosol Ancient soil preserved in the stratigraphic record.

Paleozoic Era The oldest era of the Phanerozoic Eon (541–252 Ma).

Pangaea A supercontinent that assembled at the end of the Paleozoic Era.

Pannotia A supercontinent that may have existed sometime between 800 Ma and 600 Ma.

parabolic dunes Dunes formed when strong winds break through transverse dunes to make new dunes whose ends point upwind.

parallax The apparent movement of an object seen from two different points not on a straight line from the object (e.g., from your two different eyes).

parallax method A trigonometric method used to determine the distance from the Earth to a nearby star.

parent isotope A radioactive isotope that undergoes decay.

partial melting The melting in a rock of the minerals with the lowest melting temperatures, while other minerals remain solid.

passive margin A continental margin that is not a plate boundary.

passive-margin basin A thick accumulation of sediment along a tectonically inactive coast, formed over crust that stretched and thinned when the margin first began.

patterned ground A polar landscape in which the ground splits into pentagonal or hexagonal shapes.

pause An elevation in the atmosphere where temperature stops decreasing and starts increasing, or vice versa.

peat Compacted and partially decayed vegetation accumulating beneath a swamp.

pedalfer soil A temperate-climate soil characterized by well-defined soil horizons and an organic A-horizon.

pediment The broad, nearly horizontal bedrock surface at the base of a retreating desert cliff.

pedocal soil Thin soil, formed in arid climates. It contains very little organic matter, but significant precipitated calcite.

pegmatite A coarse-grained igneous rock containing crystals of up to tens of centimeters across and occurring in dike-shaped intrusions.

pelagic sediment Microscopic plankton shells and fine flakes of clay that settle out and accumulate on the deep-ocean floor.

Pelé's hair Droplets of basaltic lava that mold into long, glassy strands as they fall.

Pelé's tears Droplets of basaltic lava that mold into tear-shaped, glassy beads as they fall.

peneplain A nearly flat surface that lies at an elevation close to sea level; thought to be the product of long-term erosion.

perched water table A quantity of groundwater that lies above the regional water table because an underlying lens of impermeable rock or sediment prevents the water from sinking down to the regional water table.

percolation The process by which groundwater meanders through tiny, crooked channels in the surrounding material.

peridotite A coarse-grained ultramafic rock.

periglacial environment A region with widespread permafrost but without a blanket of snow or ice.

period An interval of geologic time representing a subdivision of a geologic era.

permafrost Permanently frozen ground.

permanent magnet A special material that behaves magnetically for a long time all by itself.

permanent stream A stream that flows year-round because its bed lies below the water table, or because more water is supplied from upstream than can infiltrate the ground.

permeability The degree to which a material allows fluids to pass through it via an interconnected network of pores and cracks.

permineralization The fossilization process in which plant material becomes transformed into rock by the precipitation of silica from groundwater.

petrified A term used by geologists to describe plant material that has transformed into rock by permineralization.

petroglyph Drawings formed by chipping into the desert varnish of rocks to reveal the lighter rock beneath.

petroleum *See* Oil.

phaneritic A textural term used to describe coarse-grained igneous rock.

Phanerozoic Eon The most recent eon, an interval of time from 542 Ma to the present.

phenocryst A large crystal surrounded by a finer-grained matrix in an igneous rock.

photochemical smog Brown haze that blankets a city when exhaust from cars and trucks reacts in the presence of sunlight.

photosynthesis The process during which chlorophyll-containing plants remove carbon dioxide from the atmosphere, form tissues, and expel oxygen back to the atmosphere.

phreatomagmatic eruption An explosive eruption that occurs when water enters the magma chamber and turns into steam.

phyllite A fine-grained metamorphic rock with a foliation caused by the preferred orientation of very fine-grained mica.

phyllitic luster A silk-like sheen characteristic of phyllite, a result of the rock's fine-grained mica.

phylogenetic tree A chart representing the ideas of paleontologists showing which groups of organisms radiated from which ancestors.

physical weathering The process in which intact rock breaks into smaller grains or chunks.

piedmont glacier A fan or lobe of ice that forms where a valley glacier emerges from a valley and spreads out into the adjacent plain.

pillow basalt Glass-encrusted basalt blobs that form when magma extrudes on the sea floor and cools very quickly.

placer deposit Concentrations of metal grains in stream sediment that develop when rocks containing native metals erode and create a mixture of sand grains and metal fragments; the moving water of the stream carries away lighter mineral grains.

planet An object that orbits a star, is roughly spherical, and has cleared its neighborhood of other objects.

planetesimal Tiny, solid pieces of rock and metal that collect in a planetary nebula and eventually accumulate to form a planet.

plankton Tiny plants and animals that float in sea or lake water.

plastic deformation The deformational process in which mineral grains behave like plastic and, when compressed or sheared, become flattened or elongate without cracking or breaking.

plate One of about 20 distinct pieces of the relatively rigid lithosphere.

plate boundary The border between two adjacent lithosphere plates.

plate-boundary earthquakes The earthquakes that occur along and define plate boundaries.

plate-boundary volcano A volcanic arc or mid-ocean ridge volcano, formed as a consequence of movement along a plate boundary.

plate interior A region away from the plate boundaries that consequently experiences few earthquakes.

plate tectonics *See* Theory of plate tectonics.

playa The flat, typically salty lake bed that remains when all the water evaporates in drier times; forms in desert regions.

Pleistocene Epoch The period of time from about 2 Ma to 14,000 years ago, during which the Earth experienced an ice age.

plunge pool A depression at the base of a waterfall scoured by the energy of the falling water.

plunging fold A fold with a tilted hinge.

pluton An irregular or blob-shaped intrusion; can range in size from tens of m across to tens of km across.

pluvial lake A lake formed to the south of a continental glacier as a result of enhanced rainfall during an ice age.

point bar A wedge-shaped deposit of sediment on the inside bank of a meander.

polar cell A high-latitude convection cell in the atmosphere.

polar easterlies Prevailing winds that come from the east and flow from the polar high to the subpolar low.

polar front The convergence zone in the atmosphere at latitude 60°.

polar glacier *See* Dry-bottom glacier.

polar high The zone of high pressure in polar regions created by the sinking of air in the polar cells.

polarity The orientation of a magnetic dipole.

polarity chron The time interval between polarity reversals of Earth's magnetic field.

polarity subchron The time interval between magnetic reversals if the interval is of short duration (less than 200,000 years long).

polarized light A beam of filtered light waves that all vibrate in the same plane.

polar wander The phenomenon of the progressive changing through time of the position of the Earth's magnetic poles relative to a location on a continent; significant polar wander probably doesn't occur—in fact, poles seem to remain fairly fixed, while continents move.

polar-wander path The curving line representing the apparent progressive change in the position of the Earth's magnetic pole, relative to a locality X, assuming that the position of X on Earth has been fixed through time (in fact, poles stay fixed while continents move).

pollen Tiny grains involved in plant reproduction.

pollution Natural and synthetic contaminant materials introduced to the Earth's environment by the activities of humans.

polymorphs Two minerals that have the same chemical composition but a different crystal lattice structure.

pore A small, open space within sediment or rock.

pore collapse The closer packing of grains that occurs when groundwater is extracted from pores, thus eliminating the support holding the grains apart.

porosity The total volume of empty space (pore space) in a material, usually expressed as a percentage.

porphyritic A textural term for igneous rock that has phenocrysts distributed throughout a finer matrix.

Portland cement Cement made by mechanically mixing limestone, sandstone, and shale in just the right proportions, before heating in a kiln, to provide the correct chemical makeup of cement.

positive anomaly An area where the magnetic field strength is stronger than expected.

positive-feedback mechanism A mechanism that enhances the process that causes the mechanism in the first place.

potentiometric surface The elevation to which water in an artesian system would rise if unimpeded; where there are flowing artesian wells, the potentiometric surface lies above ground.

pothole A bowl-shaped depression carved into the floor of a stream by a long-lived whirlpool carrying sand or gravel.

Precambrian Period The interval of geologic time between Earth's formation about 4.57 Ga and the beginning of the Phanerozoic Eon 542 Ma.

precession The gradual conical path traced out by Earth's spinning axis; simply put, it is the "wobble" of the axis.

precious metals Metals (like gold, silver, and platinum) that have high value.

precipitate (chemistry, n.) A solid substance formed when atoms attach and settle out of a solution, or attach to the walls of the container holding the solution; (chemistry, v.) the action of forming a solid substance from a solution; (meteorology, v.) the dropping of snow or rain from the sky.

precipitation (1) The process by which atoms dissolved in a solution come together and form a solid; (2) rainfall or snow.

preferred mineral orientation The metamorphic texture that exists where platy grains lie parallel to one another and/or elongate grains align in the same direction.

pressure Force per unit area, or the "push" acting on a material in cases where the push (compressional stretch) is the same in all directions.

pressure gradient The rate of pressure change over a given horizontal distance.

pressure solution The process of dissolution at points of contact, between grains, where compression is greatest, producing ions that then precipitate elsewhere, where compression is less.

prevailing winds Surface winds that generally flow in the same direction for long time periods.

primary porosity The space that remains between solid grains or crystals immediately after sediment accumulates or rock forms.

principal aquifer The geologic unit that serves as the primary source of groundwater in a region.

principle of baked contacts When an igneous intrusion "bakes" (metamorphoses) surrounding rock, the rock that has been baked must be older than the intrusion.

principle of cross-cutting relations If one geologic feature cuts across another, the feature that has been cut is older.

principle of fossil succession In a stratigraphic sequence, different species of fossil organisms appear in a definite order; once a fossil species disappears in a sequence of strata, it never reappears higher in the sequence.

principle of inclusions If a rock contains fragments of another rock, the fragments must be older than the rock containing them.

principle of original continuity Sedimentary layers, before erosion, formed fairly continuous sheets over a region.

principle of original horizontality Layers of sediment, when originally deposited, are fairly horizontal.

principle of superposition In a sequence of sedimentary rock layers, each layer must be younger than the one below, for a layer of sediment cannot accumulate unless there is already a substrate on which it can collect.

principle of uniformitariansim The physical processes we observe today also operated in the past in the same way, and at comparable rates.

product (chemistry, n.) materials produced or formed by a chemical reaction.

prograde metamorphism Metamorphism that occurs as temperatures and pressures are increasing.

prokaryote An organism whose cells do not contain a nucleus; archaea and bacteria consist of prokaryotic cells.

Proterozoic Eon The most recent of the Precambrian eons (2,500–541 Ma).

protocontinent A block of crust composed of volcanic arcs and hotspot volcanoes sutured together.

protolith The original rock from which a metamorphic rock formed.

proton A positively charged subatomic particle in the nucleus of an atom.

protoplanet A body that grows by the accumulation of planetesimals but has not yet become big enough to be called a planet.

protoplanetary nebula A ring of gas and dust that surrounded the newborn Sun, from which the planets were formed.

protostar A dense body of gas that is collapsing inward because of gravitational forces and that may eventually become a star.

pumice A glassy igneous rock that forms from felsic frothy lava and contains abundant (over 50%) pore space.

pumice lapilli Marble-sized chunks consisting of frothy, siliceous igneous rock that fall from a volcanic eruptive cloud.

punctuated equilibrium The hypothesis that evolution takes place in fits and starts; evolution occurs very slowly for quite a while and then, during a relatively short period, takes place very rapidly.

P-waves Compressional seismic waves that move through the body of the Earth.

P-wave shadow zone A band between 103° and 143° from an earthquake epicenter, as measured along the circumference of the Earth, inside which P-waves do not arrive at seismograph stations.

pycnocline The boundary between layers of water of different densities.

pyroclastic debris Fragmented material that sprayed out of a volcano and landed on the ground or sea floor in solid form.

pyroclastic flow A fast-moving avalanche that occurs when hot volcanic ash and debris mix with air and flow down the side of a volcano.

pyroclastic rock Rock made from fragments that were blown out of a volcano during an explosion and were then packed or welded together.

quarry A site at which stone is extracted from the ground.

quartzite A metamorphic rock composed of quartz and transformed from a protolith of quartz sandstone.

quenching A sudden cooling of molten material to form a solid.

quick clay Clay that behaves like a solid when still (because of surface tension holding the water-coated clay flakes together) but that flows like a liquid when shaken.

radial network A drainage network in which the streams flow outward from a cone-shaped mountain and define a pattern resembling spokes on a wheel.

radiation (physics) Electromagnetic energy traveling away from a source through a medium or space.

radioactive decay The process by which a radioactive atom undergoes fission or releases particles, thereby being transformed into a new element.

radioactive isotope An unstable isotope of a given element.

radiometric dating The science of dating geologic events in years by measuring the ratio of parent radioactive atoms to daughter product atoms.

rain band A spiraling arm of a hurricane radiating outward from the eye.

rain shadow The inland side of a mountain range, which is arid because the mountains block rain clouds from reaching the area.

range (for fossils) The interval of a sequence of strata in which a specific fossil species appears.

rapids A reach of a stream in which water becomes particularly turbulent; as a consequence, waves develop on the surface of the stream.

rare earth element One of a group of 17 elements including the lanthanides, scandium, and yttrium; they are essential in the production of high-tech devices.

reach A specified segment of a stream's path.

reactant (chemistry) The starting materials of a chemical reaction.

recessional moraine The end moraine that forms when a glacier stalls for a while as it recedes.

recharge area A location where water enters the ground and infiltrates down to the water table.

recrystallization The process in which ions or atoms in minerals rearrange to form new minerals.

rectangular network A drainage network in which the streams join each other at right angles because of a rectangular grid of fractures that breaks up the ground and localizes channels.

recurrence interval The average time between successive geologic events.

red giant A huge red star that forms when Sun-sized stars start to die and expand.

red shift The phenomenon in which a source of light moving away from you very rapidly shifts to a lower frequency; that is, toward the red end of the spectrum.

reef bleaching The death and loss of color of a coral reef.

reflected ray A ray that bounces off a boundary between two different materials.

refracted ray A ray that bends as it passes through a boundary between two different materials.

refraction The bending of a ray as it passes through a boundary between two different materials.

refractory materials Substances that have a relatively high melting point and tend to exist in solid form.

reg A vast stony plain in a desert.

regional metamorphism *See also* Dynamothermal metamorphism; metamorphism of a broad region, usually the result of deep burial during an orogeny.

regolith Any kind of unconsolidated debris that covers bedrock.

regression The seaward migration of a shoreline caused by a lowering of sea level.

relative age The age of one geologic feature with respect to another.

relative humidity The ratio between the measured water content of air and the maximum possible amount of water the air can hold at a given condition.

relative plate velocity The movement of one lithosphere plate with respect to another.

relief The difference in elevation between adjacent high and low regions on the land surface.

renewable resource A resource that can be replaced by nature within a short time span relative to a human life span.

reservoir rock Rock with high porosity and permeability, so it can contain an abundant amount of easily accessible oil.

residence time The average length of time that a substance stays in a particular reservoir.

residual mineral deposit Soils in which the residuum left behind after leaching by rainwater is so concentrated in metals that the soil itself becomes an ore deposit.

resonance (seismology) A situation that arises when earthquake waves of a particular frequency cause particularly large-amplitude movements because energy input happens at just the right time.

resurgent dome The new mound, or cone, of igneous rock that grows within a caldera as an eruption begins anew.

retrograde metamorphism Metamorphism that occurs as pressures and temperatures are decreasing; for retrograde metamorphism to occur, water must be added.

return stroke An upward-flowing electric current from the ground that carries positive charges up to a cloud during a lightning flash.

reversed polarity Polarity in which the paleomagnetic dipole points north.

reverse fault A steeply dipping fault on which the hanging-wall block slides up.

rhythmic layering Banding in sediments, shells, trees, corals, or ice that repeats periodically; it may be correlated to annual cycles.

Richter magnitude scale A scale that defines earthquakes on the basis of the amplitude of the largest ground motion recorded on a seismogram.

ridge axis The crest of a mid-ocean ridge; the ridge axis defines the position of a divergent plate boundary.

ridge-push force A process in which gravity causes the elevated lithosphere at a mid-ocean ridge axis to push on the lithosphere that lies farther from the axis, making it move away.

right-lateral strike-slip fault A strike-slip fault in which the block on the opposite fault plane from a fixed spot moves to the right of that spot.

rip current A strong, localized seaward flow of water perpendicular to a beach.

ripple mark Relatively small elongated ridges that form on a sedimentary bed surface at right angles to the direction of current flow.

riprap Loose boulders or concrete piled together along a beach to absorb wave energy before it strikes a cliff face.

roche moutonnée A glacially eroded hill that becomes elongate in the direction of flow and asymmetric; glacial rasping smoothes the upstream part of the hill into a gentle slope, while glacial plucking erodes the downstream edge into a steep slope.

rock A coherent, naturally occurring solid, consisting of an aggregate of minerals or a mass of glass.

rock burst A sudden explosion of rock off the ceiling or wall of an underground mine.

rock cycle The succession of events that results in the transformation of Earth materials from one rock type to another, then another, and so on.

rockfall A mass of rock that separates from a cliff, typically along a joint, and then free-falls downslope.

rock flour Fine-grained sediment produced by glacial abrasion of the substrate over which a glacier flows.

rock glacier A slow-moving mixture of rock fragments and ice.

rockslide A sudden downslope movement of rock.

rocky coast An area of coast where bedrock rises directly from the sea, so beaches are absent.

Rodinia A proposed Precambrian supercontinent that existed around 1 billion years ago.

rogue wave Waves that are two to five times the size of most of the large waves passing a locality in a given time interval.

rotational axis The imaginary line through the center of the Earth around which the Earth spins.

running water Water that flows down the surface of sloping land in response to the pull of gravity.

R-waves (Rayleigh waves) Surface seismic waves that cause the ground to ripple up and down, like water waves in a pond.

sabkah A region of formerly flooded coastal desert in which stranded seawater has left a salt crust over a mire of mud that is rich in organic material.

salinity The degree of concentration of salt in water.

saltation The movement of a sediment in which grains bounce along their substrate, knocking other grains into the water column (or air) in the process.

salt dome A rising bulbous dome of salt that bends up the adjacent layers of sedimentary rock.

salt wedging The process in arid climates by which dissolved salt in groundwater crystallizes and grows in open pore spaces in rocks and pushes apart the surrounding grains.

sand dune A relatively large ridge of sand built up by a current of wind (or water); cross bedding typically occurs within the dune.

sandspit An area where the beach stretches out into open water across the mouth of a bay or estuary.

sandstone Coarse-grained sedimentary rock consisting almost entirely of quartz.

sand volcano (sand blow) A small mound of sand produced when sand layers below the ground surface liquify as a result of seismic shaking, causing the sand to erupt onto the Earth's surface through cracks or holes in overlying clay layers.

saprolite A layer of rotten rock created by chemical weathering in warm, wet climates.

Sargasso Sea The center of North Atlantic Gyre, named for the tropical seaweed sargassum, which accumulates in its relatively noncirculating waters.

saturated solution Water that carries as many dissolved ions as possible under given environmental conditions.

saturated zone The region below the water table where pore space is filled with water.

scattering The dispersal of energy that occurs when light interacts with particles in the atmosphere.

schist A medium-to-coarse-grained metamorphic rock that possesses schistosity.

schistosity Foliation caused by the preferred orientation of large mica flakes.

scientific method A sequence of steps for systematically analyzing scientific problems in a way that leads to verifiable results.

scoria A glassy, mafic, igneous rock containing abundant air-filled holes.

scoria cone An accumulation of lapilli-sized or larger fragments formed from a volcanic eruption that spatters clots of basaltic lava. (Also called cinder cone.)

scouring A process by which running water removes loose fragments of sediment from a streambed.

sea arch An arch of land protruding into the sea and connected to the mainland by a narrow bridge.

sea-floor spreading The gradual widening of an ocean basin as new oceanic crust forms at a mid-ocean ridge axis and then moves away from the axis.

sea ice Ice formed by the freezing of the surface of the sea.

seal A relatively impermeable rock, such as shale, salt, or unfractured limestone, that lies above a reservoir rock and stops the oil from rising further.

seam A sedimentary bed of coal interlayered with other sedimentary rocks.

seamount An isolated submarine mountain.

seasonal floods Floods that appear almost every year during seasons when rainfall is heavy or when winter snows start to melt.

seasonal well A well that provides water only during the rainy season when the water table rises below the base of the well.

sea stack An isolated tower of land just offshore, disconnected from the mainland by the collapse of a sea arch.

seawall A wall of riprap built on the landward side of a backshore zone in order to protect shore cliffs from erosion.

second The basic unit of time measurement, now defined as the time it takes for the magnetic field of a cesium atom to flip polarity 9,192,631,770 times, as measured by an atomic clock.

secondary enrichment The process by which a new ore deposit forms from metals that were dissolved and carried away from preexisting ore minerals.

secondary porosity New pore space in rocks, created some time after a rock first forms.

secondary recovery technique A process used to extract the quantities of oil that will not come out of a reservoir rock with just simple pumping.

sediment An accumulation of loose mineral grains, such as boulders, pebbles, sand, silt, or mud, that are not cemented together.

sediment liquefaction When pressure in the water in the pores push sediment grains apart so that they become surrounded by water and no longer rest against each other, and the sediment becomes able to flow like a liquid.

sedimentary basin A depression, created as a consequence of subsidence, that fills with sediment.

sedimentary rock Rock that forms either by the cementing together of fragments broken off preexisting rock or by the precipitation of mineral crystals out of water solutions at or near the Earth's surface.

sedimentary sequence A grouping of sedimentary units bounded on top and bottom by regional unconformities.

sediment budget The proportion of sand supplied to sand removed from a depositional setting.

sediment load The total volume of sediment carried by a stream.

sediment maturity The degree to which a sediment has evolved from a crushed-up version of the original rock into a sediment that has lost its easily weathered minerals and become well sorted and rounded.

sediment sorting The segregation of sediment by size.

seep A place where oil-filled reservoir rock intersects the ground surface, or where fractures connect a reservoir to the ground surface, so that oil flows out onto the ground on its own.

seiche Rhythmic movement in a body of water caused by ground motion.

seismic belts (seismic zones) The relatively narrow strips of crust on Earth under which most earthquakes occur.

seismicity Earthquake activity.

seismic-moment magnitude scale A scale that defines earthquake size on the basis of calculations involving the amount of slip, length of rupture, depth of rupture, and rock strength.

seismic ray The changing position of an imaginary point on a wave front as the front moves through rock.

seismic-reflection profile A cross-sectional view of the crust made by measuring the reflection of artificial seismic waves off boundaries between different layers of rock in the crust.

seismic retrofitting The strengthening of an already existing structure (building, bridge, etc.) so that it can withstand earthquake vibrations.

seismic tomography Analysis by sophisticated computers of global seismic data in order to create a three-dimensional image of variations in seismic-wave velocities within the Earth.

seismic velocity The speed at which seismic waves travel.

seismic-velocity discontinuity A boundary in the Earth at which seismic velocity changes abruptly.

seismic (earthquake) waves Waves of energy emitted at the focus of an earthquake.

seismogram The record of an earthquake produced by a seismograph.

seismometer (seismograph) An instrument that can record the ground motion from an earthquake.

semipermanent pressure cell A somewhat elliptical zone of high or low atmospheric pressure that lasts much of the year; it forms because high-pressure zones tend to be narrower over land than over sea.

Sevier orogeny A mountain-building event that affected western North America between about 150 Ma and 80 Ma, a result of convergent margin tectonism; a fold-thrust belt formed during this event.

shale Very fine-grained sedimentary rock that breaks into thin sheets.

shale gas Gas that comes directly from a source rock (organic shale).

shatter cones Small, cone-shaped fractures formed by the shock of a meteorite impact.

shear When one part of a material moves sideways, relative to another.

shear strain A change in shape of an object that involves the movement of one part of a rock body sideways past another part so that angular relationships within the body change.

shear stress A stress that moves one part of a material sideways past another part.

shear waves Seismic waves in which particles of material move back and forth perpendicular to the direction in which the wave itself moves.

shear zone A fault in which movement has occurred ductilely.

sheetwash A film of water less than a few mm thick that covers the ground surface during heavy rains.

shell (biology) A relatively hard, protective structure formed of minerals and surrounding the soft part of an invertebrate organism.

shield An older, interior region of a continent.

shield volcano A subaerial volcano with a broad, gentle dome, formed either from low-viscosity basaltic lava or from large pyroclastic sheets.

shocked quartz Grains of quartz that have been subjected to intense pressure such as occurs during a meteorite impact.

shock metamorphism The changes that can occur in a rock due to the passage of a shock wave, generally resulting from a meteorite impact.

shoreline The boundary between the water and land.

shortening The process during which a body of rock or a region of crust becomes shorter.

short-term climate change Climate change that takes place over hundreds to thousands of years.

Sierran arc A large continental volcanic arc along western North America that was initiated at the end of the Jurassic Period and lasted until about 80 million years ago.

silica SiO_2.

silicate rock Rock composed of silicate minerals.

silicates (silicate minerals) Minerals built from silicon-oxygen tetrahedra arranged in chains, sheets, or 3-D networks; they make up most of the Earth's crust and mantle.

siliceous sedimentary rock Sedimentary rock that contains abundant quartz.

silicon-oxygen tetrahedron The SiO_4^{4-} anionic group, in which four oxygen atoms surround a single silicon atom, thereby defining the corners of a tetrahedron.

silicic Rich in silica with relatively little iron and magnesium.

sill A nearly horizontal tabletop-shaped tabular intrusion that occurs between the layers of country rock.

siltstone Fine-grained sedimentary rock generally composed of very small quartz grains.

sinkhole A circular depression in the land that forms when an underground cavern collapses.

slab-pull force The force that downgoing plates (or slabs) apply to oceanic lithosphere at a convergent margin.

slate Fine-grained, low-grade metamorphic rock, formed by the metamorphism of shale.

slaty cleavage The foliation typical of slate, and reflective of the preferred orientation of slate's clay minerals, that allows slate to be split into thin sheets.

slickensides The polished surface of a fault caused by slip on the fault; lineated slickensides also have grooves that indicate the direction of fault movement.

slip face The leeward slope of a dune; sand that builds up at the crest of the dune slides down this face; slip faces are preserved as cross beds within sandstone layers.

slip lineations Linear marks on a fault surface created during movement on the fault; some slip lineations are defined by grooves, some by aligned mineral fibers.

slope failure The downslope movement of material on an unstable slope.

slumping Downslope movement in which a mass of regolith detaches from its substrate along a spoon-shaped, sliding surface and slips downward semicoherently.

smelting The heating of a metal-containing rock to high temperatures in a fire so that the rock will decompose to yield metal plus a nonmetallic residue (slag).

snottite A long gob of bacteria that slowly drips from the ceiling of a cave.

snowball Earth A model proposing that, at times during Earth history, glaciers covered all land, and the entire ocean surface froze.

snow line The boundary above which snow remains all year.

soda straw A hollow stalactite in which calcite precipitates around the outside of a drip.

soil Sediment that has undergone changes at the surface of the Earth, including reaction with rainwater and the addition of organic material.

soil erosion The removal of soil by wind and runoff.

soil horizon Distinct zones within a soil, distinguished from each other by factors such as chemical composition and organic content.

soil moisture Underground water that wets the surface of the mineral grains and organic material making up soil, but lies above the water table.

soil profile A vertical sequence of distinct zones of soil.

Solar System Our Sun and all the materials that orbit it (including planets, moons, asteroids, Kuiper Belt objects, and Oort Cloud objects).

solar wind A stream of particles with enough energy to escape from the Sun's gravity and flow outward into space.

solid-state diffusion The slow movement of atoms or ions through a solid.

solidus The highest temperature at which all the components of a material are solid; at the solidus temperature, the material begins to melt.

solifluction The type of creep characteristic of tundra regions; during the summer, the uppermost layer of permafrost melts, and the soggy, weak layer of ground then flows slowly downslope in overlapping sheets.

solstice A day on which the polar ends of the terminator (the boundary between the day hemisphere and the night hemisphere) lie 23.5° away from the associated geographic poles.

solution A material containing dissolved ions.

Sonoma orogeny A convergent-margin mountain-building event that took place on the western coast of North America in the Late Permian and Early Triassic periods.

sorting (1) The range of clast sizes in a collection of sediment; (2) the degree to which sediment has been separated by flowing currents into different-sized fractions.

source rock A rock (organic-rich shale) containing the raw materials from which hydrocarbons eventually form.

southeast tradewinds Tradewinds in the southern hemisphere, which start flowing northward, deflect to the west, and end up flowing from southeast to northwest.

southern oscillation The movement of atmospheric pressure cells back and forth across the Pacific Ocean, in association with El Niño.

specific gravity A number representing the density of a mineral, as specified by the ratio between the weight of a volume of the mineral and the weight of an equal volume of water.

speleothem A formation that grows in a limestone cave by the accumulation of travertine precipitated from water solutions dripping in a cave or flowing down the wall of a cave.

sphericity The measure of the degree to which a clast approaches the shape of a sphere.

spreading boundary *See* Divergent plate boundary.

spreading rate The rate at which sea floor moves away from a mid-ocean ridge axis, as measured with respect to the sea floor on the opposite side of the axis.

spring A natural outlet from which groundwater flows up onto the ground surface.

spring tide An especially high tide that occurs when the Sun is on the same side of the Earth as the Moon.

stable air Air that does not have a tendency to rise rapidly.

stable slope A slope on which downward sliding is unlikely.

stalactite An icicle-like cone that grows from the ceiling of a cave as dripping water precipitates limestone.

stalagmite An upward-pointing cone of limestone that grows when drips of water hit the floor of a cave.

standing wave A wave whose crest and trough remain in place as water moves through the wave.

star An object in the Universe in which fusion reactions occur pervasively, producing vast amounts of energy; our Sun is a star.

star dune A constantly changing dune formed by frequent shifts in wind direction; it consists of overlapping crescent dunes pointing in many different directions.

steady state condition The condition when proportions of a chemical in different reservoirs remain fairly constant even though there is a constant flux (flow) of the chemical among the reservoirs.

stellar nucleosynthesis The production of new, larger atoms by fusion reactions in stars; the process generates more massive elements that were not produced by the Big Bang.

stellar wind The stream of atoms emitted from a star into space.

stick-slip behavior Stop-start movement along a fault plane caused by friction, which prevents movement until stress builds up sufficiently.

stone rings Ridges of cobbles between adjacent bulges of permafrost ground.

stoping A process by which magma intrudes; blocks of wall rock break off and then sink into the magma.

storm An episode of severe weather in which winds, precipitation, and in some cases lightning become strong enough to be bothersome and even dangerous.

storm-center velocity A storm's (hurricane's) velocity along its track.

storm surge Excess seawater driven landward by wind during a storm; the low atmospheric pressure beneath the storm allows sea level to rise locally, increasing the surge.

strain The change in shape of an object in response to deformation (i.e., as a result of the application of a stress).

strata A succession of several layers or beds together.

stratified drift Glacial sediment that has been redistributed and stratified by flowing water.

stratigraphic column A cross-section diagram of a sequence of strata summarizing information about the sequence.

stratigraphic formation A recognizable layer of a specific sedimentary rock type or set of rock types, deposited during a certain time interval, that can be traced over a broad region.

stratigraphic group Several adjacent stratigraphic formations in a succession.

stratigraphic sequence An interval of strata deposited during periods of relatively high sea level, and bounded above and below by regional unconformities.

stratopause The temperature pause that marks the top of the stratosphere.

stratosphere The stable, stratified layer of atmosphere directly above the troposphere.

stratovolcano A large, cone-shaped subaerial volcano consisting of alternating layers of lava and tephra.

stratus cloud A thin, sheet-like, stable cloud.

streak The color of the powder produced by pulverizing a mineral on an unglazed ceramic plate.

stream A ribbon of water that flows in a channel.

streambed The floor of a stream.

stream capacity The total quantity of sediment a stream carries.

stream capture (stream piracy) The situation in which headward erosion causes one stream to intersect the course of another, previously independent stream, so that the intersected stream starts to flow down the channel of the first stream.

stream competence The maximum particle size that a stream can carry.

stream gradient The slope of a stream's channel in the downstream direction.

stream piracy A process that happens when headward erosion by one stream causes the stream to intersect the course of another stream and capture its flow.

stream rejuvenation The renewed downcutting of a stream into a floodplain or peneplain, caused by a relative drop of the base level.

stream terrace When a stream downcuts through the alluvium of a floodplain so that a new, lower floodplain develops and the original floodplain becomes a step-like platform.

stress The push, pull, or shear that a material feels when subjected to a force; formally, the force applied per unit area over which the force acts.

stretching The process during which a layer of rock or a region of crust becomes longer.

striations Linear scratches in rock.

strike-slip fault A fault in which one block slides horizontally past another (and therefore parallel to the strike line), so there is no relative vertical motion.

strip mining The scraping off of all soil and sedimentary rock above a coal seam in order to gain access to the seam.

stromatolite Layered mounds of sediment formed by cyanobacteria; cyanobacteria secrete a mucous-like substance to which sediment sticks, and as each layer of cyanobacteria gets buried by sediment, it colonizes the surface of the new sediment, building a mound upward.

structural control The condition in which geologic structures, such as faults, affect the distribution and drainage of water or the shape of the land surface.

subaerial Pertaining to land regions above sea level (i.e., under air).

subduction The process by which one oceanic plate bends and sinks down into the asthenosphere beneath another plate.

subduction zone The region along a convergent boundary where one plate sinks beneath another.

sublimation The evaporation of ice directly into vapor without first forming a liquid.

submarine canyon A narrow, steep canyon that dissects a continental shelf and slope.

submarine fan A wedge-shaped accumulation of sediment at the base of a submarine slope; fans usually accumulate at the mouth of a submarine canyon.

submarine slump The underwater downslope movement of a semicoherent block of sediment along a weak mud detachment.

submergent coast A coast at which the land is sinking relative to sea level.

subpolar low The rise of air where the surface flow of a polar cell converges with the surface flow of a Ferrel cell, creating a low-pressure zone in the atmosphere.

subsidence The vertical sinking of the Earth's surface in a region, relative to a reference plane.

substrate A general term for material just below the ground surface.

subtropical high (subtropical divergence zone) A belt of high pressure in the atmosphere at 30° latitude formed where the Hadley cell converges with th Ferrel cell, causing cool, dense air to sink.

subtropics Desert climate regions that lie on either side of the equatorial tropics between the lines of 20° and 30° north or south of the equator.

summit eruption An eruption that occurs in the summit crater of a volcano.

sunspot cycle The cyclic appearance of large numbers of sunspots (black spots thought to be magnetic storms on the Sun's surface) every 9 to 11.5 years.

supercontinent cycle The process of change during which supercontinents develop and later break apart, forming pieces that may merge once again in geologic time to make yet another supercontinent.

supernova A short-lived, very bright object in space that results from the cataclysmic explosion marking the death of a very large star; the explosion ejects large quantities of matter into space to form new nebulae.

superplume A huge mantle plume.

superposed stream A stream whose geometry has been laid down on a rock structure and is not controlled by the structure.

superrotation The faster rotation of the core, relative to the rest of the Earth.

supervolcano A volcano that erupts a vast amount (more than 1,000 cubic km) of volcanic material during a single event; none have erupted during recorded human history.

surface current An ocean current in the top 100 m of water.

surface load (bed load) Sediment that rolls and bounce along the ground (under the air) or along a stream bed (under water).

surface water Liquid or seasonally frozen water that resides at the surface of the Earth in oceans, lakes, streams, and marshes.

surface waves Seismic waves that travel along the Earth's surface.

surface westerlies The prevailing surface winds in North America and Europe, which come out of the west or southwest.

surf zone A region of the shore in which breakers crash onto the shore.

surge (glacial) A pulse of rapid flow in a glacier.

suspended load Tiny solid grains carried along by a stream without settling to the floor of the channel.

sustainable growth The ability of society to prosper without depleting the supply of natural resources, and without destroying the environment.

swamp A wetland dominated by trees.

swash The upward surge of water that flows up a beach slope when breakers crash onto the shore.

S-waves Seismic shear waves that pass through the body of the Earth.

S-wave shadow zone A band between 103° and 180° from the epicenter of an earthquake inside of which S-waves do not arrive at seismograph stations.

swelling clay Clay possessing a mineral structure that allows it to absorb water between its layers and thus swell to several times its original size.

symmetry The condition in which the shape of one part of an object is a mirror image of the other part.

syncline A trough-shaped fold whose limbs dip toward the hinge.

systematic joints Long planar cracks that occur fairly regularly throughout a rock body.

tabular intrusions Sheet intrusions that are planar and of roughly uniform thickness.

Taconic orogeny A convergent mountain-building event that took place around 400 million years ago, in which a volcanic island arc collided with eastern North America.

tailings pile A pile of waste rock from a mine.

talus A sloping apron of fallen rock along the base of a cliff.

tar Hydrocarbons that exist in solid form at room temperature.

tarn A lake that forms at the base of a cirque on a glacially eroded mountain.

tar sand Sandstone reservoir rock in which less viscous oil and gas molecules have either escaped or been eaten by microbes, so that only tar remains.

taxonomy The study and classification of the relationships among different forms of life.

tectonic foliation A planar fabric, such as cleavage, schistosity, or gneissic banding, that develops in rocks; caused by compression or shearing during deformation (e.g., during mountain building).

temperature A measure of the hotness or coldness of a material.

tension A stress that pulls on a material and could lead to stretching.

tephra Unconsolidated accumulations of pyroclastic grains.

terminal moraine The end moraine at the farthest limit of glaciation.

terminator The boundary between the half of the Earth that has daylight and the half experiencing night.

terrace The elevated surface of an older floodplain into which a younger floodplain had cut down.

terrestrial planets Planets that are of comparable size and character to the Earth and consist of a metallic core surrounded by a rock mantle.

thalweg The deepest part of a stream's channel.

theory A scientific idea supported by an abundance of evidence that has passed many tests and failed none.

theory of plate tectonics The theory that the outer layer of the Earth (the lithosphere) consists of separate plates that move with respect to one another.

thermal energy The total kinetic energy in a material due to the vibration and movement of atoms in the material.

thermal metamorphism Metamorphism caused by heat conducted into country rock from an igneous intrusion.

thermocline A boundary between layers of water with differing temperatures.

thermohaline circulation The rising and sinking of water driven by contrasts in water density, which is due in turn to differences in temperature and salinity; this circulation involves both surface and deep-water currents in the ocean.

thermosphere The outermost layer of the atmosphere, containing very little gas.

thin section A 3/100-mm-thick slice of rock that can be examined with a petrographic microscope.

thin-skinned deformation A distinctive style of deformation characterized by displacement on faults that terminate at depth along a subhorizontal detachment fault.

thrust fault A gently dipping reverse fault; the hanging-wall block moves up the slope of the fault.

tidal bore A visible wall of water that moves toward shore with the rising tide in quiet waters.

tidal flat A broad, nearly horizontal plain of mud and silt, exposed or nearly exposed at low tide but totally submerged at high tide.

tidal power Energy produced by the daily rise and fall of the tides; people can utilize this energy, for example, by damming a bay or estuary, so that water passes through turbines when the tide changes.

tidal range The difference in sea level between high tide and low tide at a given point.

tide The daily rising or falling of sea level at a given point on the Earth.

tide-generating force The force, caused in part by the gravitational attraction of the Sun and Moon and in part by the centrifugal force created by the Earth's spin, that generates tides.

tidewater glacier A glacier that has entered the sea along a coast.

till A mixture of unsorted mud, sand, pebbles, and larger rocks deposited by glaciers.

tillite A rock formed from hardened ancient glacial deposits and consisting of larger clasts distributed through a matrix of sandstone and mudstone.

toe (terminus) The leading edge or margin of a glacier.

tombolo A narrow ridge of sand that links a sea stack to the mainland.

topographical map A map that uses contour lines to represent variations in elevation.

topography Variations in elevation.

topsoil The top soil horizons, which are typically dark and nutrient-rich.

tornado A near-vertical, funnel-shaped cloud in which air rotates extremely rapidly around the axis of the funnel.

tornado swarm Dozens of tornadoes produced by the same storm.

tower karst A karst landscape in which steep-sided residual bedrock towers remain between sinkholes.

trace fossil Fossilized imprints or debris that an organism leaves behind while moving on or through sediment; examples include footprints, burrows, and fecal matter.

transform fault A fault marking a transform plate boundary; along mid-ocean ridges, transform faults are the actively slipping segment of a fracture zone between two ridge segments.

transform plate boundary A boundary at which one lithosphere plate slips laterally past another.

transgression The inland migration of shoreline resulting from a rise in sea level.

transition zone The middle portion of the mantle, from 400 to 670 km deep, in which there are several jumps in seismic velocity.

transpiration The release of moisture as a metabolic by-product.

transverse dune A simple, wave-like dune that appears when enough sand accumulates for the ground surface to be completely buried, but only moderate winds blow.

trap A subsurface configuration of seal rocks and structures that keep oil and/or gas underground, so it doesn't seep out at the surface.

travel-time curve A graph that plots the time since an earthquake began on the vertical axis and the distance to the epicenter on the horizontal axis.

travertine A rock composed of crystalline calcium carbonate (CaCO3) formed by chemical precipitation from groundwater that has seeped out at the ground surface.

trellis network A drainage system that develops across a landscape of parallel valleys and ridges so that major tributaries flow down the valleys and join a trunk stream that cuts through the ridge; the resulting map pattern resembles a garden trellis.

trench A deep, elongate trough bordering a volcanic arc; a trench defines the trace of a convergent plate boundary.

triangulation The method for determining the map location of a point from knowing the distance between that point and three other points; this method is used to locate earthquake epicenters.

tributary A smaller stream that flows into a larger stream.

triple junction A point where three lithosphere plate boundaries intersect.

tropical depression A tropical storm with winds reaching up to 61 km per hour; such storms develop from tropical disturbances, and may grow to become hurricanes.

tropical disturbance Cyclonic winds that develop in the tropics.

tropopause The temperature pause marking the top of the troposphere.

troposphere The lowest layer of the atmosphere, where air undergoes convection and where most wind and clouds develop.

truncated spur A spur (elongate ridge between two valleys) whose end was eroded off by a glacier.

trunk stream The single larger stream into which an array of tributaries flow.

tsunami A large wave along the sea surface triggered by an earthquake or large submarine slump.

tuff A pyroclastic igneous rock composed of volcanic ash and fragmented pumice, formed when accumulations of the debris cement together.

tundra A cold, treeless region of land at high latitudes, supporting only species of shrubs, moss, and lichen capable of living on permafrost.

turbidite A graded bed of sediment built up at the base of a submarine slope and deposited by turbidity currents.

turbidity current A submarine avalanche of sediment and water that speeds down a submarine slope.

turbulence The chaotic twisting, swirling motion in flowing fluid.

typhoon The equivalent of a hurricane in the western Pacific Ocean.

ultimate base level Sea level; the level below which a trunk stream cannot cut.

ultramafic A term used to describe igneous rocks or magmas that are rich in iron and magnesium and very poor in silica.

unconfined aquifer An aquifer that intersects the surface of the Earth.

unconformity A boundary between two different rock sequences representing an interval of time during which new strata were not deposited and/or were eroded.

unconsolidated Consisting of unattached grains.

undercutting Excavation at the base of a slope that results in the formation of an overhang.

undersaturated A term used to describe a solution capable of holding more dissolved ions.

Universe All of space and all the matter and energy within it.

unsaturated zone The region of the subsurface above the water table.

unstable air Air that is significantly warmer than air above and has a tendency to rise quickly.

unstable ground Land capable of slumping or slipping downslope in the near future.

unstable slope A slope on which sliding will likely happen.

updraft Upward-moving air.

upper mantle The uppermost section of the mantle, reaching down to a depth of 400 km.

upwelling zone A place where deep water rises in the ocean, or where hot magma rises in the asthenosphere.

U-shaped valley A steep-walled valley shaped by glacial erosion into the form of a U.

vacuum Space that contains very little matter in a given volume (e.g., a region in which air has been removed).

valley A trough with sloping walls, cut into the land by a stream.

valley glacier A river of ice that flows down a mountain valley.

Van Allen radiation belts Belts of solar wind particles and cosmic rays that surround the Earth, trapped by Earth's magnetic field.

van der Waals bonding The relatively weak attachment of two elements or molecules due to their polarity and not due to covalent or ionic bonding.

varve A pair of thin layers of glacial lake-bed sediment, one consisting of silt brought in during the spring floods and the other of clay deposited during the winter when the lake's surface freezes over and the water is still.

vascular plant A plant with woody tissue and seeds and veins for transporting water and food.

vein A seam of minerals that forms when dissolved ions carried by water solutions precipitate in cracks.

vein deposit A hydrothermal deposit in which the ore minerals occur in veins that fill cracks in preexisting rocks.

velocity-versus-depth curve A graph that shows the variation in the velocity of seismic waves with increasing depth in the Earth.

ventifact (faceted rock) A desert rock whose surface has been faceted by the wind.

vesicles Open holes in igneous rock formed by the preservation of bubbles in magma as the magma cools into solid rock.

viscosity The resistance of material to flow.

volatiles (volatile materials) Elements or compounds such as H_2O and CO_2 that evaporate at relatively low temperatures and can exist in gaseous forms at the Earth's surface.

volatility A specification of the ease with which a material evaporates.

volcanic agglomerate An accumulation consisting dominantly of volcanic bombs and other relatively large chunks of igneous material.

volcanic arc A curving chain of active volcanoes formed adjacent to a convergent plate boundary.

volcanic ash Tiny glass shards formed when a fine spray of exploded lava freezes instantly upon contact with the atmosphere.

volcanic bomb A large piece of pyroclastic debris thrown into the atmosphere during a volcanic eruption.

volcanic breccia A pyroclastic igneous rock that consists of fragments of volcanic debris, which either fall through the air and accumulate, or form when solidifying lava breaks up during flow.

volcanic danger-assessment map A map delineating areas that lie in the path of potential lava flows, lahars, debris flows, or pyroclastic flows of an active volcano.

volcanic debris flow A mixture of water and pyroclastic debris that moves downslope like wet concrete.

volcanic gas Elements or compounds that bubble out of magma or lava in gaseous form.

volcanic island arc The volcanic island chain that forms on the edge of the overriding plate where one oceanic plate subducts beneath another oceanic plate.

volcaniclastic deposit An accumulation of large quantities of fragmental igneous material (including both pyroclastic debris, and water-transported debris).

volcaniclastic rock A material composed of cemented-together grains of volcanic material; it includes both pyroclastic rocks and rocks formed from accumulations of water-transported volcanic debris.

volcano (1) A vent from which melt from inside the Earth spews out onto the planet's surface; (2) a mountain formed by the accumulation of extrusive volcanic rock.

V-shaped valley A valley whose cross-sectional shape resembles the shape of a V; the valley probably has a river running down the point of the V.

Wadati-Benioff zone A sloping band of seismicity defined by intermediate- and deep-focus earthquakes that occur in the downgoing slab of a convergent plate boundary.

wadi The name used in the Middle East and North Africa for a dry wash.

warm front A front in which warm air rises slowly over cooler air in the atmosphere.

waste rock Rock dislodged by mining activity yet containing no ore minerals.

waterfall A place where water drops over an escarpment.

water gap An opening in a resistant ridge where a trunk river has cut through the ridge.

watershed The region that collects water that feeds into a given drainage network.

water table The boundary, approximately parallel to the Earth's surface, that separates substrate in which groundwater fills the pores from substrate in which air fills the pores.

wave A disturbance that transmits energy from one point to another in the form of periodic motions.

wave base The depth, approximately equal in distance to half a wavelength in a body of water, beneath which there is no wave movement.

wave-cut bench A platform of rock, cut by wave erosion, at the low-tide line that was left behind a retreating cliff.

wave-cut notch A notch in a coastal cliff cut out by wave erosion.

wave erosion The combined effects of the shattering, wedging, and abrading of a cliff face by waves and the sediment they carry.

wave front The boundary between the region through which a wave has passed and the region through which it has not yet passed.

wavelength The horizontal difference between two adjacent wave troughs or two adjacent crests.

wave refraction (ocean) The bending of waves as they approach a shore so that their crests make no more than a 5° angle with the shoreline.

weather Local-scale conditions as defined by temperature, air pressure, relative humidity, and wind speed.

weathered rock Rock that has reacted with air and/or water at or near the Earth's surface.

weathering The processes that break up and corrode solid rock, eventually transforming it into sediment.

weather system A specific set of weather conditions, reflecting the configuration of air movement in the atmosphere, that affects a region for a period of time.

welded tuff Tuff formed by the welding together of hot volcanic glass shards at the base of pyroclastic flows.

well A hole in the ground dug or drilled in order to obtain water.

Western Interior Seaway A north-south-trending seaway that ran down the middle of North America during the Late Cretaceous Period.

wet-bottom (temperate) glacier A glacier with a thin layer of water at its base, over which the glacier slides.

wetted perimeter The area in which water touches a stream channel's walls.

wind abrasion The grinding away at surfaces in a desert by windblown sand and dust.

wind gap An opening through a high ridge that developed earlier in geologic history by stream erosion, but that is now dry.

xenolith A relict of wall rock surrounded by intrusive rock when the intrusive rock freezes.

yardang A mushroom-like column with a resistant rock perched on an eroding column of softer rock; created by wind abrasion in deserts where a resistant rock overlies softer layers of rock.

yazoo stream A small tributary that runs parallel to the main river in a floodplain because the tributary is blocked from entering the main river by levees.

Younger Dryas An interval of cooler temperatures that took place 4,500 years ago during a general warming/glacier-retreat period.

zeolite facies The metamorphic facies just above diagenetic conditions, under which zeolite minerals form.

zone of ablation The area of a glacier in which ablation (melting, sublimation, calving) subtracts from the glacier.

zone of accumulation (1) The layer of regolith in which new minerals precipitate out of water passing through, thus leaving behind a load of fine clay; (2) the area of a glacier in which snowfall adds to the glacier.

zone of aeration *See* Unsaturated zone.

zone of leaching The layer of regolith in which water dissolves ions and picks up very fine clay; these materials are then carried downward by infiltrating water.

Photo Credits

Title page: Stephen Marshak
Author photo: Kurt Burmeister
Table of Contents (in chronological order): Stephen Marshak; NASA; Stephen Marshak; © Julius T. Csotonyi; Images provided by Google Earth mapping services/NASA, © DigitalGlobe, © Terra Metrics, © GeoEye, © Europa Technologies, Copyright 2014; Mark Schneider/Visuals Unlimited/Corbis; Photos 12 / Alamy; Stephen Marshak (5 photos); Richard Roscoe/Stocktrek Images/Corbis; AP Photo/Natacha Pisarenko; USGS; Stephen Marshak (6 photos); NOAA; Reuters/Newscom; Stephen Marshak (2 photos); Emma Marshak; Stephen Marshak (4 photos).

PRELUDE

Page 1: Stephen Marshak; **p. 2 (both):** Stephen Marshak; **p. 3 (all):** Stephen Marshak; **p. 4 (both):** Stephen Marshak; **p. 5 (a):** AP Photo/Hurriyet; **(b):** Stephen Marshak; **p. 7:** Stephen Marshak.

CHAPTER 1

Page 10: NASA; **p. 12:** NASA; **p. 13:** NASA; **p. 14 (a):** Deyan Georgiev Creative collection / Alamy; **(b):** Gordon Garradd/Science Source; **(c):** Nasa/JPL-Caltech; **p. 15(a):** Peter Apian, *Cosmographia*, Antwerp, 1524; **(b):**Picture Library/Alamy; **p. 16 (all):** Stephen Marshak; **p. 17 (a):** © Miloslav Druckmuller; **(b):** NASA; **(c):** NASA/JPL-Caltech; **p. 20 (left):** 2/Christoph Wilhelm /Ocean/Corbis; **(right):** Stephen Marshak; **p. 23:** Moonrunner Design; **p. 26:** NASA, ESA, and M. Livio and the Hubble 20th Anniversary Team (STScl); **p. 27 (a):** SOHO (ESA & NASA); **(b):** J. Hester and P. Scowen/NASA; **p. 29 (a):** NASA; **(b):** R. Pelisson, © SaharaMet; **(c):** NASA; **p. 32:** Images provided by Google Earth mapping services/NASA, © DigitalGlobe, ©Terra Metrics, © GeoEye, © Europa Technologies, Copyright 2014; **p. 33:** Images provided by Google Earth mapping services/ NASA, © DigitalGlobe, © Terra Metrics, © GeoEye, © Europa Technologies, Copyright 2014; **p. 35 (a):** NASA/ESA; **(b):** © Robert Gendler.

CHAPTER 2

Page 36: Stephen Marshak; **p. 39 (a):** NASA/JPL/Cal Tech; **(b):** Science Source; **(c):** NASA/JPL-Caltech/UMD; **p. 40 (all but Mars):** JPL/NASA; **(Mars):** NASA and The Hubble Heritage Team (STScI/AURA); **p. 41:** NASA/Science Source; **p. 42:** NASA; **p. 44 (both):** Images provided by Google Earth mapping services/NASA, © DigitalGlobe, © Terra Metrics, © GeoEye, © Europa Technologies, Copyright 2014; **p. 45:** Images provided by Google Earth mapping services/NASA, © DigitalGlobe, © Terra Metrics, © GeoEye, © Europa Technologies, Copyright 2014; **p. 47:** Tate Gallery, London/ Art Resource, N.Y; **p. 48:** Stephen Marshak; **p. 50 (a):** Fred Espenak / Science Source; **(b):** ©Tom Bean; Marli Miller/Visuals Unlimited, Inc.; **(d):** AP Photo; **p. 52 (from left to right):** Susan E. Degginger / Alamy (3 photos); The Natural History Museum/Alamy (2 photos); **p. 54:** Stephen Marshak; **p. 60:** NASA/JPL-Caltech.

CHAPTER 3

Page 61: © Julius T. Csotonyi; **p. 63 (a):** Alfred Wegener Institute for Polar and Marine Research; **(b):** Ron Blakey, Colorado Plateau Geosystems; **(bottom):** Images provided by Google Earth mapping services/NASA, © DigitalGlobe, © Terra Metrics, © GeoEye, © Europa Technologies, Copyright 2014; **p. 64:** Images provided by Google Earth mapping services/NASA, © DigitalGlobe, © Terra Metrics, © GeoEye, © Europa Technologies, Copyright 2014; **p. 65 (a):** Pete M. Wilson / Alamy; **(c):** Courtesy of Thomas N. Taylor; **p. 66:** Ron Blakey, Colorado Plateau Geosystems; **p. 73:** NOAA; **p. 74:** Images provided by Google Earth mapping services/NASA, © DigitalGlobe, © Terra Metrics, © GeoEye, © Europa Technologies, Copyright 2014; **p. 78 (all):** Images Courtesy of Gary A.Glatzmaier (University of California, Santa Cruz) And Paul H. Roberts (University of California, Los Angeles), Taken from their computer simulation; **p. 82:** Magnetic Anomaly Map of the World, 2007 Equatorial scale: 1: 50 000 000; © CCGM-CGMW; Authors: J.V. Korhonen,J. Derek Fairhead, M. Hamoudi, K. Hemant, V. Lesur, M. Mandea, S. Maus, M. Purucker, D. Ravat, T. Sazonova & E. Thébault; **p. 85:** Christoph Hormann.

CHAPTER 4

Page 86: Images provided by Google Earth mapping services/NASA, © DigitalGlobe, © Terra Metrics, © GeoEye, © Europa Technologies, Copyright 2014; **p. 89:** Images provided by Google Earth mapping services/NASA, © Digital-Globe, © Terra Metrics, © GeoEye, © Europa Technologies, Copyright 2014; **p. 92:** Images provided by Google Earth mapping services/NASA, © Digital-Globe, © Terra Metrics, © GeoEye, © Europa Technologies, Copyright 2014; **p. 93:** EOS TRANSACTIONS, AMERICAN GEOPHYSICAL UNION, VOL. 78, PAGE 265, JULY 1, 1997, D. Smith et al: Viewing the Morphology of the Mid-Atlantic Ridge from a New perspective; **p. 94 (both):** NOAA / University of Washington; **p. 95:** J.R. Delaney and D.S. Kelley, University of Washington; **p. 97:** Images provided by Google Earth mapping services/NASA, © DigitalGlobe, © Terra Metrics, © GeoEye, © Europa Technologies, Copyright 2014; **p. 99:** Images provided by Google Earth mapping services/NASA, © Digital-Globe, © Terra Metrics, © GeoEye, © Europa Technologies, Copyright 2014; **p. 100:** Kevin Schafer / Alamy; **p. 104 (d):** Johnson Space Center, NASA; **p. 105:** Images provided by Google Earth mapping services/NASA, © DigitalGlobe, © Terra Metrics, © GeoEye, © Europa Technologies, Copyright 2014; **p. 107:** Tomography shear velocity model SAVANI by Ludwig Auer (ETH Zurich) and Lapo Boschi (Université Pierre et Marie Curie Paris).

CHAPTER 5

Page 114: Stephen Marshak; **p. 116:** Mark Schneider/Visuals Unlimited/Corbis; **p. 117:** Yakub88/Dreamstime.com; **p. 118 (left):** Ken Lucas/ Visuals Unlimited; **(right):** Stephen Marshak; **p. 119 (left):** Mark A. Schneider/Science Source; **(center):** incamerastock / Alamy; **(right):** Darryl Brooks/Dreamstime; **p. 123 (a):** Erich Schrempp/Science Source; **(c):** Courtesy of Prof. Huifang Xu, Dept. of Geology and Geophysics, University of Wisconsin, Madison; **p. 125 (a):** Charles O'Rear /Corbis; **(b):** John A. Jaszczak, Michigan Technological University; **(a-d):** Stephen Marshak; **(e):** Wikimedia Commons, pd; **p. 126:** ©1996 Jeff Scovil; **p. 128 (a-b, d, f):** Richard P. Jacobs /JLM Visuals; **(c):** Breck P. Kent/JLM Visuals; **(e top):** Stephen Marshak; **(e bottom):** Andrew Silver/USGS; **(g):** Scientifica/ Visuals Unlimited/Corbis; **p. 130 (a):** Richard P. Jacobs/JLM Visuals; **(b):** 1992 Jeff Scovil; **(c):** Marli Miller/Visuals Unlimited, Inc.; **(d-e):** Richard P. Jacobs/ JLM Visuals; **(h, left):** Arco Images GmbH/Alamy; **(h, right):** Ann Bryant www. Geology.com; **(g):** Stephen Marshak; **p. 131:** Javier Trueba/MSF/Science Source; **p. 132 (a):** Farbled/Dreamstime.com; **(b):** Huguette Roe/Dreamstime.com; **(c):** Dennis Kunkel Microscopy, Inc./Visuals Unlimited/Corbis; **p. 132 (d):** Construction Photography/Alamy; **p. 134 (a-b):** Stephen Marshak; **p. 134:** Images provided by Google Earth mapping services/NASA, © DigitalGlobe, © Terra Metrics, © GeoEye, © Europa Technologies, Copyright 2014; **p. 135 (left):** Michael Langford/Gallo Images/Getty Images; **(right):** Ken Lucas/Visuals Unlimited; **p. 136: (left):** Images provided by Google Earth mapping services/NASA, © Digital-Globe, © Terra Metrics, © GeoEye, © Europa Technologies, Copyright 2014; **(right):** 1996 Smithsonian Institution; **p. 138 (a):** Albert Copley/Visuals Unlimited; **(d):** © 1998 Jeff Scovil; **(inset):** Stephen J. Krasemann/Science Source; **p. 140:** Chip Clark / Smithsonian Institution.

INTERLUDE A

Page 141: Photos 12 / Alamy; **p. 142 (both):** Stephen Marshak; **p. 143 (a, left):** Richard P. Jacobs/JLM Visuals; **(a, center):** Courtesy David W. Houseknecht, USGS; **(b, left):** sciencephotos/Alamy; **(b, center):** Courtesy of Kent Ratajeski, Dept. of Geology and Geophysics, U of Wisconsin, Madison; **p. 145 (all):** Stephen Marshak; **p. 146 (all):** Stephen Marshak; **p. 148 (a):** © Tom Bean; **(b):** Stephen Marshak; **p. 149 (all):** Stephen Marshak; **p. 150 top: (c):** Stephen Marshak; **(d):**

Scenics & Science / Alamy; **(bottom, a):** Product photo courtesy of JEOL, USA; **(bottom, b):** Courtesy of Joseph H. Reibenspies, Texas A & M University.

CHAPTER 6

Page 152: Stephen Marshak; **p. 154 (a):** Liysa/Pacific Stock/Agefotostock; **(b):** J.D. Griggs/ U.S. Geological Survey; **(c):** Stephen Marshak; **p. 155 (top left):** Images provided by Google Earth mapping services/NASA, © DigitalGlobe, © Terra Metrics, © GeoEye, © Europa Technologies, Copyright 2014; **(bottom, b-c):** Stephen Marshak; **p. 158:** USGS; **p. 162:** Stephen Marshak; **p. 163 (a):** USGS; **(b-e):** Stephen Marshak; **p. 165 (both):** Stephen Marshak; **p. 166:** Stephen Marshak; **p. 167:** Stephen Marshak; **p. 168:** Stephen Marshak; **p. 169 (both):** Stephen Marshak; **p. 170 (top, left to right):** Dr. Kent Ratajeski; Omphacite. 2006. Wikimedia; http://en.wikipedia.org/wiki/Public_domain; Dr. Matthew Genge; **(bottom, from left to right):** Stephen Marshak; **(b):** Mark A. Schneider/ Science Source; **(c):** Doug Sokell/Visuals Unlimited; **p. 173: (clockwise from top left):** geoz / Alamy; Stephen Marshak; Siim Sepp /Alamy; Wally Eberhart/Visuals Unlimited/Corbis; Joyce Photographics/Science Source; Mark A. Schneider/ Science Source; **p. 174 (all):** Stephen Marshak; **p. 176:** Images provided by Google Earth mapping services/NASA, © DigitalGlobe, © Terra Metrics, © GeoEye, © Europa Technologies, Copyright 2014; **p. 177 (both):** Stephen Marshak; **p. 179 (both):** Stephen Marshak; **p. 182 (top):** Tony Linck/SuperStock; **(bottom):** Jacques Descloitres, MODIS team, NASA Visible Earth.

INTERLUDE B

Page: 183: Stephen Marshak; **p. 184:** Corbis; **p. 185 (a-b, bottom):** Stephen Marshak; **(c):** Emma Marshak; **p. 186 (all):** Stephen Marshak; **p. 187 (top, both):** Stephen Marshak; **(middle right):** Visuals Unlimited; **(bottom, both):** Stephen Marshak; **p. 188 (all):** Stephen Marshak; **p. 189 (a):** Stephen Marshak; **(b):** Carlo Giovanella, UBC, 1997-2005; **(c):** British Geology Survey; **p. 190:** Stephen Marshak; **p. 192 (both):** Stephen Marshak; **p. 193 (all):** Stephen Marshak; **p. 194 (all):** Stephen Marshak; **p. 195:** Stephen Marshak; **p. 198:** US Dept. of Agriculture, Natural Resources Conservation Services; **p. 199:** Stephen Marshak; **p. 200 (both):** Stephen Marshak; **p. 201:** Jim Richardson/Corbis.

CHAPTER 7

Page 202: Stephen Marshak; **p. 204:** Stephen Marshak; **p. 205 (all):** Stephen Marshak; **p. 208 (all but bottom right):** Stephen Marshak; **(bottom right):** Scottsdale Community College; **p. 209 (top left):** Stephen Marshak; **(top right):** E.R. Degginger/Color-Pic, Inc.; **(bottom left):** Stephen Marshak; **(center both):** Stephen Marshak; **(bottom right):** D. G. F. Long; **p. 211; (a-c, e):** Stephen Marshak; **(d):** Emma Marshak; **p. 212 (both):** Stephen Marshak; **p. 213 (a):** Visuals Unlimited; **(inset):** Stephen Marshak; **(c):** Jason Bye / Alamy; **p. 214 (a):** Marli Miller; **(b):** Photo by Yukinobu Zengame, 2005. http://commons.wikimedia.org/ wiki/File:Limestone_towers_at_Mono_Lake,_California.jpg; http://creativecommons.org/licenses/by/2.0/deed.en; **(c):** M.W. Schmidt; **p. 215 (all):** Stephen Marshak; **p. 216:** Images provided by Google Earth mapping services/NASA, © DigitalGlobe, © Terra Metrics, © GeoEye, © Europa Technologies, Copyright 2014; **p. 217 (both):** Stephen Marshak; **p. 218 (inset):** Stephen Marshak; **(a):** All Canada Photos / Alamy; **(b):** 1980 Grand Canyon Natural History Association; **p. 219 (both):** Stephen Marshak; **p. 220 (a):** Imagina Photography /Alamy; **(b-c):** Stephen Marshak; **p. 221 (top, a, b):** Stephen Marshak; **(c):** Marli Miller/ Visuals Unlimited; **p. 222:** Stephen Marshak; **p. 225 (a):** Emma Marshak; **(b):** Stephen Marshak; **(c):** Marli Miller/Visuals Unlimited; **(d):** Stephen Marshak; **(e):** Stephen Marshak; **(f):** John S. Shelton; **p. 226:** Images provided by Google Earth mapping services/NASA, © DigitalGlobe, © Terra Metrics, © GeoEye, © Europa Technologies, Copyright 2014; **p. 227 (a):** Stephen Marshak; **(b):** Corbis; **p. 228: (a):** Belgian Federal Science Policy Office; **(b):** G.R. "Dick" Roberts © / Natural Sciences Image Library; **p. 232:** Corbis.

CHAPTER 8

Page 233: Stephen Marshak; **p. 235:** Stephen Marshak; **p. 236 (top left):** Stephen Marshak; **(top right):** Visuals Unlimited; **(bottom, all):** Stephen Marshak; **p. 237 (left):** Corbis; **(inset):** Stephen Marshak; **(right):** Kurt Freihauf; **p. 240:** Stephen Marshak; **p. 242 (inset):** Emma Marshak; **(a):** Corbis; **(b):** Stephen Marshak; **p. 243 (both, left):** Stephen Marshak; **(top right):** Dr. Jane A. Gilotti; **(bottom right):** Stephen Marshak; **(bottom, c):** Visuals Unlimited/Corbis; **p. 245:** Stephen Marshak; **p. 246 (all):** Stephen Marshak; **p. 252 (all):** Stephen Marshak; **p. 253:** Stephen Marshak; **p. 256 (top):** Dr. Terry Wright; (bottom, both): Images provided by Google Earth mapping services/NASA, © DigitalGlobe, © Terra Metrics, © GeoEye, © Europa Technologies, Copyright 2014; **p. 258 (both):** Stephen Marshak; **p. 260:** Stephen Marshak.

INTERLUDE C

Page 261 (all): Stephen Marshak; **p. 265:** Stephen Marshak; **p. 269:** Stephen Marshak.

CHAPTER 9

Page 270: Stephen Marshak; **p. 272:** Richard Roscoe/Stocktrek Images/Corbis; **p. 273 (left):** Bridgeman Art Library; **(right):** Source: ZeroOne Animation in association with Museum Victoria; **p. 274 (a):** Images provided by Google Earth mapping services/NASA, © DigitalGlobe, © Terra Metrics, © GeoEye, © Europa Technologies, Copyright 2014; **(b-c, e1):** Stephen Marshak; **(d, e2 3):** Jack Repcheck; **p. 275 (b):** Images provided by Google Earth mapping services/NASA, © DigitalGlobe, © Terra Metrics, © GeoEye, © Europa Technologies, Copyright 2014; **(c):** Marli Miller / Visuals Unlimited; **p. 277 (a):** Robert Francis/Agefotostock; **(b, d-f):** Stephen Marshak; **(c):** USGS; **p. 278 (a):** Thomas Hallstein / Alamy; **(b):** Stephen Marshak; **p. 279 (top, a):** AP Photo; **(top, b):** Stephen Weaver; **(top, c):** Stephen Marshak; **(bottom, a):** AFP/Getty Images; **(bottom, b):** Photo by Suzanne MacLachlan, British Ocean Sediment Core Research Facility, National Oceanography Centre, Southampton; **(bottom, c-d):** Stephen Marshak; **p. 280 (a-b, d, f):** Stephen Marshak; **(c):** USGS; **(e):** Anthony Phelps/Reuters/Corbis; **p. 281:** Images provided by Google Earth mapping services/NASA, © DigitalGlobe, © Terra Metrics, © GeoEye, © Europa Technologies, Copyright 2014; **p. 282 (top, both):** Stephen Marshak; **(right, a):** Sunshine Pics /Alamy; **(right, b):** USGS; **p. 283 (a):** John J. Bangma/Getty Images; **(b):** Marli Miller/ Visuals Unlimited; **p. 284 (a):** Marli Miller/Visuals Unlimited; **(b):** © Tom Bean 1985; **(c):** Robert Harding World Imagery / Alamy; **p. 288 (a):** USGS; **(b):** George Dimijian/Science Source; **(c):** Stephen Marshak; **p. 289 (a):** Westend61 GmbH / Alamy; **(b):** © Tom Pfeiffer / www.volcanodiscovery.com; **(c):** AFP/Getty Images; **(d):** USGS; **p. 291 (top):** Gary Braasch/Corbis; **(bottom):** Stephen Marshak; **p. 292 (b):** USGS; **(c):** AFP/Getty Images; **p. 294 (b):** Images provided by Google Earth mapping services/NASA, © DigitalGlobe, © Terra Metrics, © GeoEye, © Europa Technologies, Copyright 2014; **p. 295:** Images provided by Google Earth mapping services/NASA, © DigitalGlobe, © Terra Metrics, © GeoEye, © Europa Technologies, Copyright 2014; **p. 296:** Arctic Images / Alamy; **p. 297:** Stephen Marshak; **p. 298:** USGS; **p. 300 (a):** USGS; **(b):** Vittoriano Rastelli/Corbis; **(c):** AP Photo; **(d):** Roy Whiddon; **(e):** Alberto Garcia / Corbis; **(f):** Philippe Bourseiller/Getty Images; **(g):** Photo: Magnus T. Gudmundsson, University of Iceland; **(h):** USGS; **p. 301 (b):** Thierry Orban/Sygma/Corbis; **(c):** Peter Turnley/Corbis; **p. 302:** Stephen Marshak; **p. 304:** Jennifer L. Lewicki , USGS; **p. 305 (a):** Vittoriano Rastelli/Corbis; **(b):** Sigurgeir Jonasson/Frank Lane Picture Agency/Corbis; **p. 306:** NASA; **p. 307 (a):** Gail Mooney/Corbis; **(b):** Santorini Caldera. Photo by Steve Jurvetson. 2012. http://creativecommons. org/licenses/by/2.0/deed.en; **(c):** Earth Sciences and Image Analysis Laboratory, NASA; **p. 308 (a):** Julian Baum / Science Source; **(b-e):** NASA/JPL; **p. 311 (left):** Reuters/Landov; **(right):** Arctic Images/Corbis.

CHAPTER 10

Page 312: AP Photo/Natacha Pisarenko; **p. 314 (a):** AFP/Getty Images; **(b):** JIJI Press / AFP / Getty Images; **(c):** AP Photo/ Kyodo News; **p. 317 (both):** Photo Courtesy of Paul ""Kip"" Otis-Diehl, USMC, 29 Palms CA; **p. 319 (a):** Created by Lou Estey with UNAVCO's Jules Verne Voyager, Earth edition, using the "Face of the Earth" dataset from ARC Science Simulations and U.S. Geological Survey earthquake hypocenters (2011). Credit: UNAVCO/NSF; **(left):** Images provided by Google Earth mapping services/NASA, © DigitalGlobe, © Terra Metrics, © GeoEye, © Europa Technologies, Copyright 2014; **p. 321:** Peltzer et al. (1999), Evidence of nonlinear elasticity of the Crust. *Science*, v. 286. Copyright © 1999, AAAS; **p. 322:** Created by Lou Estey with UNAVCO's Jules Verne Voyager, Earth edition, using the "Face of the Earth" dataset from ARC Science Simulations and U.S. Geological Survey earthquake hypocenters (2011). Credit: UNAVCO/NSF; **p. 326:** Inga Spence/Visuals Unlimited; **p. 334 (b):** Corbis; **(c):** George Hall/Corbis; **p. 335:** Patrick Robert/Corbis Sygma; **p. 338:** New Madrid earthquake woodcut from Deven's Our First Century (1877); **p. 340 (from top down):** AP Photo; Pacific Press Service / Alamy; M. Celebi, U.S. Geographical Survey; Reuters; **p. 341 (a):** Barry Lewis / Alamy; **(b):** Bettmann/Corbis; **p. 342 (a):** NOAA / National Geophysical Data Center (NGDC); **(b):** Rob Grange/ Getty Images; **(c):** James Mori, Research Center for Earthquake Prediction, Disaster Prevention Institute Kyoto University; **(e):** Courtesy of the National Information Service for Earthquake Engineering, PEER-NISEE, University of California, Berkeley; **p. 343 (a):** Karl V. Steinbrugge Collection, University of California, Berkeley; **(b):** National Geophysical Data Center (NGDC); **p. 346:** Images provided by Google Earth mapping services/NASA, © DigitalGlobe, © Terra Metrics, © GeoEye, © Europa Technologies, Copyright 2014; **p. 347**

(MBARI) and S.E. Beaulieu (WHOI); **p. 554 (center):** Art Directors & TRIP / Alamy; (bottom): Hemis / Alamy; **p. 555:** Images provided by Google Earth mapping services/NASA, © DigitalGlobe, © Terra Metrics, © GeoEye, © Europa Technologies, Copyright 2014; **p. 556 (a):** Robert W. Gerling/ Visual Unlimited; **(b):** USGS; **(d):** Stephen Marshak; **p. 558:** Rodrigo Arangua / AFP / Getty Images; **p. 559 (a):** Richard P. Jacobs/JLM Visuals; **(b-d):** Stephen Marshak; **p. 560 (all):** Stephen Marshak; **p. 561 (a):** Stephen Marshak; **(b):** Bloomberg/ Getty Images; **p. 566 (a):** Stephen Marshak; **(b):** Doug Sokell/Visuals Unlimited; **(c):** A.J. Copley/Visual Unlimited; **p. 569:** Lucidio Studio, Inc/Getty Images.

INTERLUDE F

Page 570: Stephen Marshak; **p. 572:** NOAA; **p. 573 (all):** Stephen Marshak; **p. 574: (a):** KennethTownsend (artist); http://www.shadedreliefarchive.com/Europe_townsend.html http://www.shadedreliefarchive.com/Kenneth_Townsend.html **p. 574 (b):** JLP/NASA; **p. 576 (both):** Stephen Marshak; **p. 577 (a):** G.R. 'Dick' Roberts © Natural Sciences Image Library; **(b):** Photo by Davie Pierson/ St. Petersburg Times; **p. 578: (all):** Stephen Marshak; **p. 579:** Stephen Marshak; **p. 583 (a):** NASA; **p. 583 (b):** JSC/NASA; **(c):** Dr. David Smith, NASA Goddard Space Flight Center/MOLA Science Team; **(d):** NASA/USGS Flagstaff; **p. 584 (a):** ESA/DLR/FU Berlin (G. Neukum); **(b):** ESA/DLR/FU Berlin; **(c):** AP Photo/NASA; **p. 585 (left):** NASA/JPL/Space Science Institute; **(right):** NASA/JPL/University of Arizona.

CHAPTER 16

Page 586: Reuters/Newscom; **p. 587 (both):** Lloyd Cluff/Corbis; **p. 589 (inset):** Stephen Marshak; **(d):** Marli Miller/Visuals Unlimited, Inc.; **(e):** George Herben Photo/ Visuals Unlimited; **p. 590 (all):** Stephen Marshak; **p. 591 (a):** Shana Reis/ EPA /Landov; **(b):** Cascades Volcano Observatory /USGS; **(c):** Stephen Marshak; **(left):** Images provided by Google Earth mapping services/NASA, © DigitalGlobe, © Terra Metrics, © GeoEye, © Europa Technologies, Copyright 2014; **p. 592 (a):** Bob Schuster, USGS, **(b):** National Geographic Image Collection / Alamy; **(c):** Ron Varela/Ventura County Star; **p. 593 (top, a):** AP Photo/Ted S. Warren; **(bottom, a):** Stephen Marshak; **(bottom, b):** Guido Alberto Rossi/Age Fotostock; **p. 594:** Images provided by Google Earth mapping services/NASA, © DigitalGlobe, © Terra Metrics, © GeoEye, © Europa Technologies, Copyright 2014; **p. 595:** Stephen Marshak; **p. 596: (a):** AFP Photos; **(b):** Stephen Marshak; **(c):** Alaska Stock/Alamy; **p. 597 (a-c):** Stephen Marshak; **(d):** AP Photo; **p. 598 (a):** USGS/Barry W. Eakins; **(b):** USGS, Geologic Investigations Series I-2809 by Barry W. Eakins, Joel E. Robinson, Toshiya Kanamatsu, Jiro Naka, John R. Smith, Eiichi Takahashi, and David A. Clague; **(c):** Jerome Neufeld and Stephen Morris, Nonlinear Physics, University of Toronto; **p. 600:** Stephen Marshak; **p. 604:** Gorm Kallestad/Scan Pix/Sipa USA; **p. 605:** Breck P.Kent/JLM Visuals; **p. 606:** Stephen Marshak; **p. 607 (both):** Images provided by Google Earth mapping services/NASA, © DigitalGlobe, © Terra Metrics, © GeoEye, © Europa Technologies, Copyright 2014; **p. 608:** Stephen Marshak; **p. 609:** AP Photo/The Deseret News, Ravell Call, File; **p. 610:** © 2008, All Rights Reserved, City of Seattle; **p. 613:** Heino Kalis/Reuters/Corbis.

CHAPTER 17

Page 614: Stephen Marshak; **p. 616 (a):** Helen H. Richardson/ The Denver Post/ Getty Images; **(b):** John Gibson/Getty Images; **(c):** Andy Clark/Reuters/Corbis; **p. 617 (b):** Stephen Marshak; **(left):** Images provided by Google Earth mapping services/NASA, © DigitalGlobe, © Terra Metrics, © GeoEye, © Europa Technologies, Copyright 2014; **p. 618 (left):** Images provided by Google Earth mapping services/NASA, © DigitalGlobe, © Terra Metrics, © GeoEye, © Europa Technologies, Copyright 2014; **(right):** Stephen Marshak; **p. 619:** Stephen Marshak; **p. 620:** NASA; **p. 621 (both):** Stephen Marshak; **p. 623:** Stephen Marshak; **p. 624 (a):** Stephen Marshak; **(b):** © Ron Niebrugge; **p. 625 (all):** Stephen Marshak; **p. 628 (a, c):** Stephen Marshak; **(b):** Amar and Isabelle Guillen - Guillen Photography / Alamy; **p. 630:** Stephen Marshak; **p. 631: (all):** Stephen Marshak; **p. 632 (top, b):** Stephen Marshak; **(d):** courtesy of Jenny Jackson, Caltech; **(bottom, a):** Marti Miller/Visuals Unlimited; **(bottom, b):** Stephen Marshak; **p. 633 (a):** 1998 Tom Bean; **(b, e):** Stephen Marshak; **p. 634 (top):** NASA Earth Observatory; **(bottom):** NASA/GSFC/meti/ersdac/jaros, and US/Japan ASTER Science Team; **p. 635:** James Parker; **p. 638:** 1997 Tom Bean; **p. 640 (a):** Stephen Marshak; **(b):** AP Photo/Binsar Bakkara; **(c):** Mike Hollingshead/Corbis; **p. 641 (a, both):** NASA images created by Jesse Allen, Earth Observatory, using data provided courtesy of the Landsat Project Science Office- copyright 2008; **(b):** AP Photo/ United Nations, Evan Schneider; **p. 642 (all):** Stephen Marshak; **p. 643 (both):** Stephen Marshak; **p. 644 (a):** Reuters/Yoray Cohen/Eilat Rescue Unit /Landov; **(b):** USGS; **p. 645 (left):** Courtesy Johnstown Area Heritage Association; **(right):**

Mary Evans Picture Library / The Image Works; **p. 646:** ©Tom Foster; **p. 647 (both):** Stephen Marshak; **p. 648 (b):** AP Photo/The News-Star, Margaret Croft; **(c):** NASA; **p. 649: (a):** Fred Lynch/Southeast Missourian; **(b):** Scott Olson/Getty Images; **p. 652:** Photo courtesy of the Bureau of Reclamation.

CHAPTER 18

Page 655: Stephen Marshak; **p. 656 (a):** Rod Catanach, Woods Hole Oceanographic Institution; **(b):** Topham/ The Image Works; **(c, left):** Woods Hole Oceanographic Institution; **(c, right):** © Harbor Branch Oceanographic Institution; **(e):** Photo by Chris Griner, Woods Hole Oceanographic Institution; **p. 657:** NOAA; **p. 660 (all):** NOAA; **p. 661 (b):** © 2007 MBARI; **(c):** Hannes Grobe / Alfred Wegener Institute for Polar and Marine Research; **(bottom):** Images provided by Google Earth mapping services/NASA, © DigitalGlobe, © Terra Metrics, © GeoEye, © Europa Technologies, Copyright 2014; **p. 662:** U.S. Embassy/ Wikimedia; **p. 664:** NASA; **p. 665:** NASA/ SeaWIFS; **p. 668 (c):** Stephen Marshak; **(e-f):** www.michaelmarten.com; **p. 669 (a):** Wikimedia; http://en.wikipedia.org/wiki/Public_domain; **(b):** Imaginechina/Corbis; **p. 673 (b):** NOAA; **(c):** Stephen Marshak; **p. 674 (both):** Stephen Marshak; **p. 675 (inset):** NOAA; **(a-c):** Stephen Marshak; **(d):** Manfred Gottschalk / Alamy; **p. 677 (a, b, f):** Stephen Marshak; **(e):** NASA; **p. 679:** Stephen Marshak; **p. 680 (a):** G.R. 'Dick' Roberts © Natural Sciences Image Library; **(b, e1):** Stephen Marshak; **(e2):** Emma Marshak; **p. 681:** Cody Duncan / Alamy; **p. 682 (a-b):** Stephen Marshak; **(bottom):** Images provided by Google Earth mapping services/NASA, © DigitalGlobe, © Terra Metrics, © GeoEye, © Europa Technologies, Copyright 2014; **p. 683 (inset):** Steve Bloom Images / Alamy; **(both):** Stephen Marshak; **p. 689 (both, left):** USGS; **(b):** Stephen Marshak; **(top right):** Images provided by Google Earth mapping services/NASA, © DigitalGlobe, © Terra Metrics, © GeoEye, © Europa Technologies, Copyright 2014; **p. 690 (both):** Stephen Marshak.

CHAPTER 19

Page 694: Emma Marshak; **p. 695:** Images provided by Google Earth mapping services/NASA, © DigitalGlobe, © Terra Metrics, © GeoEye, © Europa Technologies, Copyright 2014; **p. 696 (a):** GeoPhoto Publishing Company; **(c):** Images provided by Google Earth mapping services/NASA, © DigitalGlobe, © Terra Metrics, © GeoEye, © Europa Technologies, Copyright 2014; **(d):** Red Huber/ Orlando Sentinel/MCT via Getty Images; **p. 697** **(b):** Photo courtesy of Eric Prokacki and Jim Best, University of Illinois; **(d):** Stephen Marshak; **p. 702 (a):** Stephen Marshak; (b): Larry W. Smith/EPA/Alamy; **p. 707:** Stephen Marshak; **p. 708:** Vince Streano/ Corbis; **p. 709:** Food and Agriculture Organization of the United Nations; **p. 710:** Stephen Marshak; **p. 711 (a):** Allan Tuchman; **(b):** Stephen Marshak; **(c):** Emma Marshak; **(d):** Stephen Marshak; **p. 712 (a-c):** Allan Tuchman; **(d):** Stephen Marshak; **p. 713 (both):** Stephen Marshak; **p. 714:** Döll, P., Fiedler, K. (2008): Global-scale modeling of groundwater recharge. Hydrol. Earth Syst. Sci., 12, 863-885; http://creativecommons.org/licenses/by-sa/3.0/deed.en; **p. 717 (all):** Stephen Marshak; **p. 721 (a, c):** Stephen Marshak; **(b):** Francesco Tomasinelli / Science Source; **p. 722 (top, both):** Stephen Marshak; **(a):** George Steinmetz/Corbis; **(b):** Stephen Marshak; Images provided by Google Earth mapping services/NASA, © DigitalGlobe, © Terra Metrics, © GeoEye, © Europa Technologies, Copyright 2014; **p. 723 (a):** Kjell B. Sandyed /Visuals Unlimited; **(b):** Photo by Jim Pisarowicz/NPS; **p. 724 (left):** Paul F. Hudson, University of Texas; **(right):** ML Sinibaldi / Corbis; **(top):** Lois Kent; **p. 725 (left):** Lois Kent; **(right):** Ashley Cooper/Corbis.

CHAPTER 20

Page 728: Stephen Marshak; **p. 729:** AFP/Getty images; **p. 732:** Reuters NewMedia Inc./Corbis; **p. 733:** Stephen Marshak; **p. 734 (a):** Stephen Marshak; **(b):** Kathryn Marshak; **p. 737:** Stocktrek Images, Inc. / Alamy; **p. 741:** CIMSS; **p. 742 (both):** NASA; **p. 747:** Stephen Marshak; **p. 748:** Stephen Marshak; **p. 749:** Stephen Marshak; **p. 750:** Stephen Marshak; **p. 753 (a):** Eric Nguyen/Corbis; **(b):** NOAA, photo by Brian Bill; **(c):** AFP/Getty images; **(d):** Photo By Miami Herald/Getty Images; **(e):** FEMA Photo by Greg Henshall; **(f):** NOAA; **p. 755 (b):** NASA; **(c):** NOAA; **p. 756 (b):** Photograph by Robert Simmon, ASA Earth Observatory and NASA/NOAA GOES Project Science team; **p. 758 (a):** NOAA; **(b):** AP Photo; **(c):** Images & Stories / Alamy; **(d):** Universal Images Group/Getty Images; **p. 759 (left, both):** Images provided by Google Earth mapping services/NASA, © DigitalGlobe, © Terra Metrics, © GeoEye, © Europa Technologies, Copyright 2014; **(right):** NASA; **p. 760 (a):** NOAA; **(b):** AP Photo/Susan Walsh; **(c):** New York Times Photos & Graphics; **(d):** Vincent Laforet /Pool /Reuters / Corbis; **(e):** Stephen Marshak; **p. 761 (both):** Images provided by Google Earth mapping services/NASA, © DigitalGlobe, © Terra Metrics, © GeoEye, ©

Europa Technologies, Copyright 2014; **p. 764 (a):** FAO-SDRN Agrometeorology Group; **(b):** NASA/ GSFC.

CHAPTER 21

Page 768: Stephen Marshak; **p. 770 (top):** Stephen Marshak; **(bottom):** O. Alamany & E. Vicens / Corbis; **p. 771:** Stephen Marshak; **p. 772:** Professor Andre Danderfer; **p. 773:** Stephen Marshak; **p. 774 (all):** Stephen Marshak; **p. 775 (all):** Stephen Marshak; **p. 776 (a):** Liba Taylor/Corbis; **(b):** Shannon Arledge / USMC / Getty Images; **(c):** Mike Olbinski; **p. 777 (left):** Stephen Marshak; **(a):** Stephen Marshak; **(b):** O. Alamany & E. Vicens / Corbis; **p. 778 (left):** JPL/ NASA; **(right):** Images provided by Google Earth mapping services/NASA, © DigitalGlobe, © Terra Metrics, © GeoEye, © Europa Technologies, Copyright 2014; **p. 779 (all):** Stephen Marshak; **p. 780 (a-b, d, inset):** Stephen Marshak; **(c):** Amar and Isabelle Guillen - Guillen Photography / Alamy; **p. 781:** Stephen Marshak; **p. 782 (b-c):** Stephen Marshak; **(d):** Photo by Flicka. Sept 2007. Wikimedia. http://creativecommons.org/licenses/by-sa/3.0/deed.en; **p. 783 (both):** Stephen Marshak; **p. 786:** Stephen Marshak; **p. 787 (all):** Stephen Marshak; **p. 788 (c-d):** Stephen Marshak; **(bottom right):** Images provided by Google Earth mapping services/NASA, © DigitalGlobe, © Terra Metrics, © GeoEye, © Europa Technologies, Copyright 2014; **p. 790 (top a-c):** Stephen Marshak; **(bottom, a):** Eitan Simanor / Alamy; **(bottom, b):** Stephen Marshak; **p. 791 (a):** Mark Phillips / Alamy; **(c):** Charles & Josette Lenars / Corbis; **p. 792 (top, both):** USGS; **(bottom, left):** BDR / Alamy; **(bottom, right):** Stocktrek Images, Inc. / Alamy; **p. 793 (left):** Library of Congress; **(right):** Image courtesy Jacques Descloitres MODIS Rapid Response Team; **p. 797:** USGS.

CHAPTER 22

Page 795: Stephen Marshak; **p. 796 (a):** Stephen Marshak; **(b):** Tom Bean/Corbis; **p. 798 (a):** Shutterstock; **(b, both):** Stephen Marshak; **(c, center):** Emma Marshak; **(c, bottom):** Ted Spiegel/National Geographic Creative; **p. 800 (b-d):** Stephen Marshak; **(e):** Galen Rowell/Corbis; **(f):** 1996 Galen Rowell; **p. 801:** NASA; **p. 802 (a):** NASA-JPL; **(b):** NASA; **p. 803 (top):** Stephen Marshak; **(bottom):** National Geophysical Data Center/NOAA; **p. 804 (a):** NASA/USGS Landsat 7; **(c):** Emma Marshak; **p. 806:** Antarctic Search for Meteorites Program, Linda Martel; **p. 807 (a):** Stephen Marshak; **(b):** Ted Scambos, National Snow and Ice Data Center, Clevenger/ Corbis; **(d):** Ralph A. Clevenger /Corbis; **(e-f):** Stephen Marshak; **(g, both):** ESA; **p. 809 (a):** Shutterstock; **(b-f):** Stephen Marshak; **p. 810:** Images provided by Google Earth mapping services/NASA, © Digital-Globe, © Terra Metrics, © GeoEye, © Europa Technologies, Copyright 2014; **p. 811 (a-c):** Stephen Marshak; **(d):** 1986 Keith S. Walklet/Quietworks; **p. 812: (a-b):** Stephen Marshak; **(d):** Marli Miller/Visuals Unlimited, Inc.; **p. 812:** Images provided by Google Earth mapping services/NASA, © DigitalGlobe, © Terra Metrics, © GeoEye, © Europa Technologies, Copyright 2014; **p. 813 (a):** Stephen Marshak; **(b):** Wolfgang Meier/zefa/Corbis; **p. 814 (both):** Stephen Marshak; **p. 816 (a-e1):** Stephen Marshak; **(e2):** Kevin Schafer/Alamy; **p. 817:** Stephen Marshak; **p. 818: (b):** Stephen Marshak; **(e):** Glenn Oliver/Visuals Unlimited; **p. 819:** Tom Bean/Corbis; **p. 822:** Michael Beauregard; **p. 823:** © Tom Foster; **p. 824:** Stephen Marshak; **p. 825 (b):** Lynda Dredge /Geological Survey of Canada; **(c):** Shutterstock; **p. 834: (b):** Hendrick Averkamp, Winter Scene with Ice Skaters, ca. 1600. Courtesy of Rijksmusuem, Amsterdam; **(c):** Stephen Marshak; **p. 835 (a):** Ed Stockard; **(b):** Roger J. Braithwaite; **(c):** Michael Melford / Getty Images; **(d):** NASA.

CHAPTER 23

Page 838: Stephen Marshak; **p. 845: (all, b):** Ron Blakey, Colorado Plateau Geosystems, Inc.; **(c):** Bournemouth News and Picture Service; **p. 847 (all):** Stephen Marshak; **p. 850: (top):** ISM/PhotoTakeUSA.com; **(bottom):** Bob Sacha/ Corbis; **(inset):** NOAA; **p. 851 (a):** Nick Cobbing / Alamy; **(c):** Core Repository Lab at Lamont-Doherty Geological Observatory of Columbia University; **(inset):** NOAA; **(bottom, a):** Stephen Marshak; **(bottom, inset):** Provided courtesy of JRTC & Fort Polk. Photography by Bruce Martin, Natural Resources Management Branch, ENRMD; **(bottom, b):** Courtesy of the Climatic Research Unit, University of East Anglia; **p. 853: (all):** Ron Blakey, Colorado Plateau Geosystems, Inc.; **p. 854 (all):** NASA; **p. 856 (b):** Crown Copyright. Reproduced courtesy of Historic Scotland, Edinburgh; **(both, c):** Stephen Marshak; **p. 857 (a):** NASA; **(b-c):** Stephen Marshak; **p. 858:** Mark Garlick / Science Source; **p. 859 (top, left):** Stephen Marshak; **(top, right):** Images provided by Google Earth mapping services/NASA, © DigitalGlobe, © Terra Metrics, © GeoEye, © Europa Technologies, Copyright 2014; **(bottom, left):** Courtesy P&H Mining Equipment; **(center):** Richard Hamilton Smith/Corbis; **(right):** Stephen Marshak; **p. 860 (b):** Fearnside, P. M. 2008. The roles and movements of actors in the deforestation of Brazilian Amazonia. Ecology and Society 13(1): 23, Deforestation data from Brazil's National Institute for Space Research (INPE); **(c):** Nigel Dickinson / Alamy; **(d, both):** Images provided by Google Earth mapping services/NASA, © DigitalGlobe, © Terra Metrics, © GeoEye, © Europa Technologies, Copyright 2014; **p. 862:** Oliver Strewe / Getty Images; **p. 863:** NASA; **p. 865 (a, both):** NASA/Goddard Space Flight Center Scientific Visualization Studio; **p. 866 (a, both):** National Snow and Ice Data Center; **(b, all):** Jacques Descloitres, MODIS Rapid Response Team, NASA/GSFC; (bottom): Images provided by Google Earth mapping services/NASA, © DigitalGlobe, © Terra Metrics, © GeoEye, © Europa Technologies, Copyright 2014, **p. 867 (a, top):** USGS; **(b, bottom):** USGS photograph by Bruce Molnia; **(b):** Stephen Marshak; **(c):** Steven Kaziowski/Alamy; **p. 868:** IPCC; **p. 872 (c):** Stephen Marshak.

APPENDIX

Page A-6 (both): NOAA / NGDC; **A-7** USGS.

Index

Page numbers in *italics* refer to illustrations, tables, and figures. Page numbers in **boldface** refer to key words.

a'a' flow, **276**, *277*
abandoned meander, 634
abbreviations, in discussion of Earth history, 468
ABE (robotic submersible), *656*
ablation, **805**
abrasion, *623*, 679, *812*
  glacial, **808**
  wind, 777, *777*
abrasive water jet, *557*
absolute age, 437, 453, 461
  see also numerical age
absolute plate velocity, **107**, *110*, *111*
absolute zero, 54
abyssal plain, *56*, **72**, *73*, **658**, 660–61, *661*
  and seamounts, 661
  sediment in formation of, *661*
abyssal plains, 44–45, **44**
Acadia National Park, Maine, *812*, 827
Acadian orogeny, 412, 483, *484*
acceleration, *372*, *373*
accreted crust, *478*
accreted terranes, *400*, 401, *489*
accretion, 29, **401**
accretionary coasts, 687
accretionary lapilli, *278*, *279*
accretionary orogens, 477, *478*
accretionary prism, **97**, *98*, *253*, *255*, *255*, *659*
  in Cretaceous North America, 490
  and mass movement, 607
  and ocean, *657*, 684–85
accumulation, 195
acidic water, 189–90
acid mine runoff, **539**, *563*, 566, **566**
acid rain, **539**, *733*, 861, *862*
  from K-T impact, 495
  weathering from, 194
acid runoff, 861
acids, groundwater contamination from, 716
active continental margins, *56*, **89**, **659**, *660*
active faults, 316
active rifts, 105
active volcanoes, 302–3, **302**, *303*
adiabatic cooling, **735**, *736*, 741
adiabatic heating, **735**, 741
Adirondack Mountains, N.Y., earthquake in, 337
adit, 556

advection, **54**, 55
aerobic metabolism, 480
aerosols, **281**, 306, *306*, *732*, **732**, 734
  threat from, 301
Africa:
  during Cretaceous Period, 490
  in Gondwana, *496*
  Karoo region of, 179, 297
  mountain belts in, *66*
  Olduvai Gorge in, *420*
  and Pangaea breakup, 495
  union of, with South America (Mesozoic Era), 636, *638*
African Plate, 89, *111*, 184
African Shield, *258*
aftershocks, **319**, 348
Agassiz, Louis, 796–97, *796*, 826, 827
agate, 214, *215*
age, relative vs. numerical, 437, *437*
  see also relative age
agents of erosion, 577–79, **577**
agents of metamorphism, 236–40
Age of Dinosaurs, 451, 463, 850
  see also dinosaurs
Age of Enlightenment, 144
Age of Mammals, 451, 464, 500
aggregate, 557
Agricola, Georgius, 117
agriculture:
  and disappearance of rain forests, *860*
  erosion from, 579
  erosion from and Dust Bowl, 791–92, *793*
  and "fertile crescent," 792
  in global warming, 862
  groundwater contaminants from, 716
  as landscape modification, 858
  river water for, 651
Agriculture Department, U.S., map of soil types from, *198*
A-horizon, *195*, 196, *198*
air, **729**
air density, 732
air-fall tuff, 292
air mass, 744–45, **744**, *745*
airplanes, volcanic ashes as hazard for, 299
air pollution:
  and fossil fuels, 539
  from ore-processing plants, *563*, 566
air pressure, 42, *42*, 732–33, **732**
  and temperature, 735
  and tides, 670–71
air quality, **733**
air quality index (AQI), 733
Airy, George, 407

Alabama, tornado in, *753*, 755
Alaska, *3*, 657
  avalanche in, *596*
  Denali region of, *625*, *632*
  earthquakes in, 322, 334, 341, *341*, 348
  fjords of, 812
  Glacier Bay National Park in, 827
  glaciers in, 799, *800*, *816*, *817*
    vanished ice of, *834*
    visit to, 827
  Lituya Bay landslide in, 601
  oil in, 825
  in Pleistocene ice ages, 826
  Redoubt Volcano in, 299
  rock glacier in, *589*
  tsunami damage in, *348*
  volcanic activity in, *282*
  volcanoes in, *155*
albedo, *797*, 830, 831, **855**, 860–61
Alberta, flooding in, 615, *616*
albite, 117
Aldrin, Buzz, 37
Aleutian Arc, *294*
Aleutian Islands, 175, *175*, *176*
Aleutian Trench, *74*
Alexandria, Egypt, 18–19
alfisol, 197, *198*
algae, 424
  blue-green (cyanobacteria), 475, 731
  from coastal pollution, 690
  and coral reef destruction, 691
  ethanol from, 532
  oil and gas from, *508*, *509*
  photosynthetic, 731
  in symbiotic relationship with corals, 682
  in upwelling zones, *665*
algae bloom, 690, *691*
Alleghenian orogeny, 413, 484, 485, *485*, *488*
alloy, 46, 547–48, **547**
  see also iron alloy
alluvial fan environments, 224
alluvial fans, *207*, *208*, **224**, *225*, 228, *266*, **630**, 778, *779*, 785
  in Basin and Range Province, *784*
  in death Valley, *632*
alluvium, **626**, *778*, *783*, *785*, 786, 787
alluvium-filled valley, 629, *630*
Alpha Centauri, 19
Alpine Fault, New Zealand, 99, 333, 607
alpine glaciers, 799
  see also mountain glaciers
Alpine-Himalayan chain (orogen), **495**, *496*

Alps Mountains, 105, *382*, 796–97, *809*
  from continental collision, 401
  exposed rocks in, fault in, *397*
  French, *573*, 814
  glacier in, *798*, *799*, 856
  glacier visit to, 826, 827
  mass movements in, 594
  in Pleistocene ice ages, 827
altitude, 762
aluminum, *53*, *124*, 191, 547–49, 553, 555
  in cement, 557
  consumption of, 191
  as metallic mineral resource, 546
  in potassium feldspar, 549
  in shale, 245
  in soils, 198–99, *198*
aluminum oxide, in cement, 557
aluminum silicate, 237, *238*
Alvarez, Luis, 494
Alvarez, Walter, 494
Alvin (research submersible), 656–57, *656*
Amasia, 873
Amazon, *573*
  cleared land in, *860*
Amazon River, 622, 636, *638*
amber, 136, *137*, 422–23, *422*, **422**, 425
American Indians, *see* Native Americans
amethyst, 117, 137
  crystals of, 116
  name origin of, 117
Amethyst (woman), 117
ammonia, 472, 477
  in volcanic gas, 730
ammonites, *429*
amphibians, 483, 485
amphibole, 131, *132*, *146*, 165, *172*, 253
  in gneiss, 241
  in schists, 241
  stability of, *174*
amphibolite, 244–45, 248, *248*
amphibolite-facies rocks, 248
amplitude of ground motion, 330
amplitude of waves, *672*
Amsden Shale, 604, *605*
Amundsen, Roald, 203, *204*
anaerobic metabolism, 480
analytical equipment, for rock study, 149–50
Anatolian Faults, Turkey, 333, 351–52
Anatolian Plate, *111*
Ancestral Rockies, *485*, **485**
ancient (early) civilizations, in river valleys, 650

andalusite, 237, *238*, *251*, 288
Andean orogen, 405, *405*
andesite, *172*, *173*, 176
andesitic eruptions, 276
    pyroclastic debris from, 278
andesitic lava, 276, 288–92
andesitic lava flow, *275*, 276–78
andesitic magma, 288–92
Andes mountains, *36*, 97, *400*, 401
    and Amazon River course, *638*
    as continental arc, 175, *175*
    and convergent-boundary
        tectonism, *496*
    glaciers in, 799
        visit to, 826
    gold in, 554
    ice caps in, 826
    lahar in, 593
    mining in, 558
    rise of, 636
    shortening of, 405, *405*
    and Sierran arc, 490
    and tectonic setting, 686
    Yungay landslide in, 587–88, *587*
andisol, 197, *198*
Andrew, Hurricane, 758, *758*
Andromeda, 19, *35*
angiosperms, 493
angle of repose, **601**
    in sand dune, 788
angularity, 206, *207*
angular unconformities, **443**, *444*,
    445, *449*
anhedral grains, 126, *126*
animal attack, physical weathering
        from, 189
animal behavior, as earthquake
        precursor, 353
Animalia, 425
anion, 121, 123, 124, *124*, 129, 131
annual probability of flooding, 647–
    50, **648**, *649*, 651
anomalies, age of, *82*
anoxic (oxygen-free) environment,
        fossils created in, 424, 425
Antarctica, 203, *659*, *863*
    as below sea level, 819
    British Expedition to, 203
    and creation of Pannotia, 477, 480
    during Cretaceous Period, 490
    glaciers and glacial ice in, 306,
        579, 799, *800*, *801*, 806
        beginning of ice sheet, 832
        icebergs from, 806, *807*, 834
        ice cores from, 849, *864*
        as place to view ice sheet, 826
        in Pleistocene ice ages, 826
    ice sheets in, *796*, 819
    ozone hole in, 861
Antarctic Bottom Water, 667
Antarctic Circle, 742–43
Antarctic Dry Valleys, *204*
Antarctic Plate, *111*
Antarctic Shield, 258
antecedent streams, **639**, 640
anthracite coal, 525
Anthropocene, **498**, 500
anticlines, **394**, *395*, *396*, 397
    and synclines, *396*
anticline trap, 514, *514*
anticyclonic flow, **746**
antimony, 548

Antler orogeny, 483, **484**
apatite, *127*, 564
aphanitic rocks, 168
Appalachian fold-thrust belt, **485**,
    488
Appalachian Mountains, 105, 249,
        401, *405*, *412*
    and Alleghenian orogeny, 484
    from continental collision, 401
    development of, 482, 484–85
    life story of, 412–13, *412*
    and rock cycle changes, 263
Appalachian region, 483
apparent polar-wander paths, 70–71,
        70, **70**
    and continental drift, 70–71, *70*
    and Earth history, *471*
    and "true polar-wander" model, *70*
aquamarine, 136, 137
aquicludes, 699
aquifers, 699–700, *699*, **699**, *700*
    confined vs. unconfined, 699, *699*
    contaminants in, 716–19
    hydraulic head in, 704–5
    for Sahara oasis, 709
    and wells, 708–10
aquitards, 699–700, *699*, **699**
    clay as, 717
    and wells, 708–10
Arabian Desert, *770*, 771
Arabian Peninsula, Empty Quarter
        of, 788
Arabian Plate, *111*
aragonite, 210
Aral Sea, 790–91, *792*
Archaea, 191, 425, *427*
    as earliest life, 474, 477
    fossils of, 428, *430*, 474
    in geothermal waters, 711–12, *711*
Archaeopteryx, 425, *426*
Archean crust, 66
    remnants of, 476
Archean Eon, 6, 7, 449–50, 472–75,
        473, *498*, 854
    and BIF, 480
    changes during, 842–43
    Moon's distance during, *671*
    photosynthesis in, 843
    tillites from, 830
Archean rock:
    atmosphere of, 480
    bacteria in, 474
    in U.S. Midwest, 477
arches, natural, 782, *785*
Arches National Park, Utah, *387*,
        388, 782
Archimedes, 88, 90
Archimedes principle, 374, **374**
Archimedes' principle of buoyancy,
        88, 90, *90*
Arc rock, 473, *474*
arcs, volcanic, **175**
Arctic Circle, 742–43, 832
Arctic Ocean, 497, *659*, 799, 806,
        807, 822
    ice melt in, 865
arête, **810**, *821*
Argentina, Iguazu (Iguaçu) Falls on
        border of, *179*, 631
argillaceous rocks, 204
aridisol, 197, *198*
aridity, 769–70

Arizona, *50*, *186*, 216, *379*, 434, 451,
        453
    canyon wall in, *624*
    desert, 2, *2*
    desert varnish in, *774*
    ground ifssures in, *717*
    mining claims in, 555
    Monument Valley in, *573*, 782,
        782
    national parks of, *452*
    Painted Desert of, *452*, 773, *774*
    Sonoran Desert in, *790*
    Sunset Crater National
        Monument in, *282*
Arkansas, tornado in, *755*
arkose, *207*, 208, 209
Armenia, earthquake in, 340, 354
Armero, Colombia, lahar hits, 299,
        300, 593
Armstrong, Neil, 37
Army Corps of engineers, U.S.,
        646–47
arrival time, of earthquakes, 326, *328*
arroyos, 622, **775**, *784*
arsenic, 548
    in groundwater, 706
Artemis, 117
artesian springs, **708**
artesian well, **709**, *710*
Arthropoda, 427
artificial levees, 646, 647, *647*
asbestos, 132–33, 564
    brown, 132, 133
    and cancer, 132
    as hazardous, 132–33, *132*
    white, 132, 133
asbestosis, 132
ash, **162**, *172*, *173*, **278**
    rhyolitic, 178
ash clouds, 163
ash flow, 162
ash umbrella, 289–92
Asia:
    in collision with Australia, 495
    in collision with India, 105, *402–
        3*, 490, *492*
        warm ocean currents cut off
            by, 833
    ice sheet over, 826
    land bridge to Australia from,
        497, 822, 827
    land bridge to North America
        from, 497, *497*, 822, *823*
    in late Cretaceous Period, 492
    monsoons of, 763–65, *764*
Asian Plate, *403*
asperities, *318*
assay, 555
assimilation, *159*, **159**
asteroid belt, 38
asteroids, 18, **18**, 28, 39, **39**
    and mass extinction, 432, 857
    material from to Earth, 50–51
asthenosphere, 53–55, **53**, *55*, 56, 57,
        88, **88**, 105, *108*, *109*, *111*,
        176, 177, 369, 839
    and forces driving plate motion,
        106, *107*
    in glacial loading and rebound,
        819
    and igneous magma, 178
    and isostacy, 406

    in mid-ocean ridge formation,
        92, *93*
    as mobile, 839
    plasticity of, 839
    and plate tectonics, *92*
    and subduction, 96, *98*
asthenospheric mantle, 108–9
astronauts:
    earth in perspective for, 11, *11*, *13*
    Moon, *13*
    space walk by, *11*
astronomers, *14*
astronomical unit, *see* AU
astronomy, *see* cosmology; solar
        system; universe
Aswan High Dam, 183, 709, 859
Atacama Desert, 771, *771*, 788
Atchafalaya Basin, 647
Atchafalaya River, *636*
Atlantic City, N.J., *572*
Atlantic Ocean, *659*
    creation of, 413, 487, *492*, 495
    in future, 873
    growth of, 495
    and ice age, 833
    opening of, *638*
    spreading rate in, **79**
    water masses of, 665, 667
    see also North Atlantic Ocean;
        South Atlantic Ocean
Atlantis legend, 307
atmosphere, **41**, *42*, 43, 500, 728–66,
        728, **729**, 840–41
    and blueness of sky, 733, *734*
    convection in, *740*, 839
    cooling of, *833*
    evolution of, 844
        formation of, 730–32, *731*
        during Hadean era, 472, *472*
        stages of, *731*
    first, 730
    greenhouse gases in, 848
    layers of, 736–37, *736*
    outer edge of, 733
    oxygen in, 480, *481*, 731, 732, 843
        in Archean Eon, 475
        increase in, 475, 481, 731
        and photosynthesis, 7
    regional circulation in, 738–40
    residence time for, 582
    second, 730
    temperature of, 866
    third, 731–32
    and Venus, 582, 583
    and volcanic gases, 306
    as water reservoir, 581
    see also air pressure; storms;
        weather; winds
atmospheres (atm), 42, 732
atmospheres of other planets, 41
atoll, 682, *683*
atomic clock, 436
atomic mass, 25, **121**
atomic number, 25, **120**, 453
atomic weight (atomic mass), 120–21,
        121, 453
atoms, 24–25, 120, *120*, **120**
    of metals, 547
    nature of, 24
    passing through rock cycle, 263,
        846
    structure of, *120*

AU (astronomical unit), 37, *43*
aureole, metamorphic (contact), *250*, **250**
aurora australis, 737
aurora borealis, 737, *737*
aurorae, *40*, **41**
Australia, *398*
   ancient sandstone found in, 463
   Blue Mountains of, *573*
   and creation of Pannotia, 477, 480
   during Cretaceous Period, 490
   fossil from, *65*
   in future, 873
   Glossopteris fossils found in, *65*
   in Gondwana, *496*
   land bridge to Asia from, 497, 822, 827
   Nullarbor Plain of, 675, *675*
   outcrop in, *385*
   in Pangaea breakup, 495
   sea stacks in, *680*
   separation of from Antarctica, 832
   stromatolites in, *475*
   Uluru (Ayers Rock) in, 783, 786, *786*
   underground coalbed fire in, 528
   zircon found in, *472*
Australian Aborigines, and Uluru, 786
Australian Desert, 771
Australian Shield, *258*
Australopithecus, *420*
Austria, avalanche in, 594, *596*
avalanche chutes, 595
avalanches, **588**, 594–95, *596*, 610, *611*
   glacier formation checked by, 797
   underwater, *267*
avalanche shed, 610, *611*
Avalonia microcontinent, *484*
avulsion, 634, *637*
axial plane, **394**
axis (centerline) of ridge, 92, *95*
Ayles Ice Shelf, 864
azurite, *549*, 552

back-arc basins (marginal sea), 98, *98*
backscattered light, 734
backwash, **673**
   in hurricane, 688
bacteria, *427*
   anaerobic, 521
   as earliest life, 474, 477
   eubacteria, 425
   fossils of, 428, *430*, 474
      microfossils, 424
   in geothermal waters, 711–12, *711*
   in groundwater, 718
      for bioremediation, 718
   hydrocarbon-eating, 512, 520
   weathering process supported by, 191
badlands, 774, *775*
Baffin Island, Canada, *812*
Bahariya oasis, 709
bajada, 778
baked contacts, principle of, 438, *440*
Bakken Shale, 520
balloon travel, 729, *729*
Baltica, 480, *482*, *484*
Baltic Shield, *258*
Banda Aceh, Sumatra, 346–47, *346*, *347*

banded-iron formation (BIF), *479*, *480*, **552–53**, 555
banding, in gneiss, 241, *244*
Bangladesh, cyclone-flood deaths in, 641, 688, 758
banks, of streams, 615
bar, *222*, *625*, **626**
   baymouth, 677, 678
barchan dunes, 784, 787, 792
barrel of oil, **513**
barrier islands, 677, 678
barrier reefs, 682
Barringer meteor crater, *50*
basal sliding, 698, **800**, 801
basalt, **46**, *52*, 69–70, *171*, *172*, *173*, *176*, 177, *178*, *178*, *204*
   beach sand from, 676
   and cooling, 176
   in effusive eruptions, 298
   flood, *179*, **179**, *296*, 298, 309
   in isostasy, 407
   from lava flow, *155*
   and marine anomalies, 78–79
   metamorphism of, 244, *247*, *256*
   from mid-ocean ridge eruptions, 293
   in oceanic crust, 49, 51, 74, 83, 263
   pillow, 94, *94*, 177, 276, *278*, 293, *295*
   seismic wave speed through, *361*
   soil formed on, 196
   weathering of, 263
basalt dike, *165*, 177
   in geologic history illustration, *438*, *441*
basalt flows, in India, 858
basaltic eruptions, 276, *279*
basaltic lava, 177, *275*, 276, *277*, 285–88, 293, *308*, 309
   freezing of, *305*
   threat from, 298
basaltic magma, *157*, 176, 177, 285–88
basalt pillow, 276, *278*
basalt plateau, 296
base level of stream, 627, **627**
basement, 266
   Precambrian, *488*
basement uplifts, *485*, **491**, *492*
base metals, 549
basic (or mafic) metamorphic rocks, 245
   see also mafic rocks
basin, *395*, **395**, 404, *404*, 410–11, *411*
   alluvium-filled, *784*
   on geologic map of eastern U.S., *411*
   between mountains, *784*
   oceanic, 657
   regional, 410–11
   and tides, *669*
   see also sedimentary basins
Basin and Range Province, *104*, 105, *168*, 404, 495–96, **495**, *496*, 497, *699*, *784*, 824, *824*
   rift, *495*
batholiths, *162*, *167*, *168*
   along west coast, *168*
   granitic (Sierra Nevada), 490, 827

bathymetric map, 657, *657*
   of hot-spot tracks in Pacific Ocean, *103*
bathymetric profile, 72, *73*
bathymetry, **44**, 45, **72**, **658**, *661*, *684–85*
   of mid-Atlantic ridge, *93*
   of oceanic plate boundaries, 659–60
Batu Tara Volcano, 272
bauxite, 199, 553, 555
baymouth bar, 677, 678
Bay of Fundy, 667
bays, and tides, 667
beach drift, 678, *678*
beach erosion, 688–90, *689*, **689**, *690*, *691*
beaches, 224, 226, *674*, 676–78, **676**, *677*, *678*, *681*, *684–85*
   oil spill threat to, 540
   protection of, 688–90
   sediment budget of, 678
beach face, 678
beachfront homes, after hurricane, *689*
beach management, 688–90
beach nourishment, 690
beach profile, 677, **678**
Beagle, HMS, 430, 682
Beardmore Glacier, 203
bearing, **389**
Becquerel, Henri, 462
bed, 216, **216**
   coal seam as, 526–27
   and fold, 398
   thrust or reverse faults in, 391
bedded chert, 210
bedding, *147*, *148*, 216–18, **216**, *217*, *222*, *242*, *399*
   disrupted, *352*
   fossils in, *442*
bedding plane, 216, *242*
   fossils in, *420*
   as prone to become failure surfaces, 601, *601*
   vs. fault surfaces, 640
bedforms, 216–18
bed load, 624, **624**
bedrock, *144*, *151*, *185*, *185*, *187*, *198*
bed-surface markings, 219
Beneixama photovoltaic power plant, Spain, *544*
Bergeron, Tor, 748
Bergeron process, 748, *748*
bergy bits, 808
Bering Strait, land bridge across, 497, *497*, 822, *823*
berms, 678, *681*
beryl, 136, 137, *138*
Berzelius, Jöns Jacob, 129
Beston, Henry, 656
beveling topography, 636
Bhartrihari, *839*
B-horizon, 196, 199
BIF (banded-iron formation), *479*, 480, 481, **552–53**, 555
big bang, 23
   aftermath of, 25–26
big bang nucleosynthesis, 23
big bang theory, 13, 23–25, **23**
Big Thompson River flood, 615, 644–45, *644*, 748

Bingham Mine, Utah, 555
biochemical chert, 210
biochemical sedimentary rocks, **203**, 210, *212*
biodegradation, 520–21
biodiesel fuel, 532
biodiversity, 432, **432**, 857
   decrease in, *859*
biofuels, **532**
biogenic minerals, **118**, 121
biogeochemical cycle, 265, 268, 842, **846**
   carbon cycle, 846
   hydrologic cycle, *840*, 846
biography of Earth, *see* Earth history
biomarkers, *424*, 474, *475*
biomass, 507
biomineralization, *125*, 126, *136*
   as chemical weathering, 191
bioremediation, 718
biosphere, 43, *840*
biotite, 119, *172*, *236*, 245
   in gneiss, 241
   and igneous intrusion, *251*
   and metamorphism, 245, *247*
   in Onawa Pluton, *251*
   in schist, 255
   stability of, *190*
bioturbation, 216, **678**
birds:
   appearance of, 451
   and dinosaurs, 490, 500
   in history of Earth, 463
   wind farms as danger to, 534
bird's-foot deltas, 635
Bissel, George, 513
bitumen (heavy oil), 520
bituminous coal, 525, *529*
bivalves, 427, *429*, 483
   fossil record of, 424
black chert, 214, *215*
Black Death, *859*
black hole, 873
black-lung disease, 528
Black Sea, 85, 852
   sediment in, *625*
black smokers, *93*, *94*, 293, 306
   first organisms at, 477
   and life on "snowball Earth," 480
   sulfide ore around, *551*, *551*, *554*, *562*
blasting of unstable slopes, 610
blind faults, 316, 389
blocks (volcanic), *172*, **278**, *279*
blocky lava, 276
blowout preventer, 540
blowouts, 516, **540**, 776
Blue Mountains, Australia, *573*
Blue Ridge, *488*
blueschist, *247*, 248, *248*, 253, 255, *255*
blue shift, **21**
body fossils, **421**
body-wave magnitude (mb), 330
body waves, **323**, 362
bogs, 681
Bolivia, 561, *564*
bolting rock, *611*
bombs (volcanic), *172*, *173*, **278**, *279*, *286*, *287*
bonding in minerals, 124, *125*
Bonneville, Lake, 824, *825*

Bonneville Salt Flats, 212
borax, 778
bornhardts, 783, 786
bottled gas, carbon in, 508
bottomset beds, *226*
boulders, *187, 192,* 205, 206
  in glacial abrasion, 810
  left by glaciers, 796, *796*
Bowen, Norman L., 162, 164–65
Bowen's reaction series, 164–65, *164,*
    164, *190*
BP oil, 540
brachiopods, *211, 385,* 427, *429,* 483,
    560
brackish water, 681
braided stream, **630,** *632,* 815, *816,*
    817
Brazil, *199, 215*
  cliff in, *385*
  coastline of, 675
  coast of, *396*
  deforestation and slumping, *606*
  Iguazu (Iguaçu) Falls on border
    of, *179, 631*
  meandering stream in, *633*
  Paraná Basin and Plateau of, 179,
    179, 297
  point bars in streams in, *625*
  Rio de Janeiro mountains, 675,
    675
  Rio de Janeiro mudflows in, 591,
    591
  rock and sand seascape along, *573*
  sand dunes in, *225*
  sand in, *193*
  valleys in, *409*
Brazilian Shield, *258*
breakers, 673, 679
breakwater, 689
breccia, *207,* **209,** *391,* 698
  collapsed, *724*
  sedimentary, *208*
Breitling Orbiter 3, 729
Bretz, J. Harlan, 646
bricks, 561, *561*
bristlecone pines, 458, 849
Britain, *see* England; Scotland; United
    Kingdom; Wales
British Antarctic Expedition, 203
British Columbia, fjords of, 681
Brittany, 655
brittle deformation, 323, 383–86,
    383, *385*
brittle-ductile transition, 385
brittle-plastic transition, 385, 801
bronze, 545, 548
Bronze Age, volcanic eruptions in, 306
Brunhes polarity chron, 77, 78, *79,* 81
Brunton compass, *389*
Bryant, Edward, 559
Bryce Canyon, Utah, *221, 232,* 451,
    452, 782, *782*
bryozoans, 427, *429,* 560
buckle, *398*
Buckskin Mountains, Arizona, *621*
Buenos Aires, in future of Earth, 873
building codes, for earthquake
    protection, 351
building stone, 546, *563*
Bullard, Edward, *64*
"Bullard fit," *64*
buoyancy, 90

Archimedes' principle of, 88,
    90, *90*
  of crustal roots, *396*
  in hydrocarbon migration, 512,
    512
  and isostacy, 406
buoyancy force, 90
  and magma rise, 159
Burgess Shale, 425, *426*
burial metamorphism, **250**
burning, 528
buttes, **782,** *785*
  at Monument Valley, *573*
Byron, Lord, 305

Cairo, Ill., flood protection for, 647
calcite, *122, 128, 130,* 131, *146, 147,*
    245
  and biochemical limestone, 210
  as cement, 210
  chalk from, 226
  and chemical weathering, 189
  in desert rock, 773
  and groundwater, 720
  in joints, 388
  in limestone, 559
  and marble, 244, 255
  metamorphic rock from, 235
  in soil, 197
  stability of, *190*
  in travertine, 213
calcite mud, *227*
calcite sand, *227*
calcium, 53, 158
calcrete (caliche), 197, *198,* 773
caldera, 283–84, *283,* **283,** 293, 309
  of Olympus Mons, 309
  of Santorini, 307, *307*
  tsunamis from collapse of, 301
  at Yellowstone, 296, *297*
Caledonian orogeny, 483
Calgary, flooding in, 615, *616*
caliche (calcrete), 197, *198,* 773
California:
  alluvial fan in, *225*
  bathymetric map of slump at, *598*
  beach erosion in, *690*
  Death Valley in, *260, 632,* 770
  earthquake expected for, 351
  earthquakes in, *317,* 319–22, *333,*
    334, 340, 341, 343, 354
  in future, 873
  geologic map of, *449*
  Geysers Geothermal Field of, 713
  gold rush in, 141, 546, *546*
  Joshua Tree National Monument
    in, *177*
  landslides in, *185,* 341
  Mojave Desert in, *167,* 773, *773*
  mud flow in, 592, *592*
  Pacific Palisades slumping in, 590
  Racetrack Playa in, *780*
  Salton Sea in, 712
  San Joaquin Valley of, 714, *715,*
    717
  slumping in (Southern
    California), 607–8
  Yosemite National Park in, 595,
    808, *809,* 827
  see also Los Angeles; San
    Francisco region; Sierra
    Nevada Mountains

calving, 806, *807,* 815, 817, *818,* 834
Cambrian explosion, **429,** *430,* **450,**
    451, **482–83**
Cambrian Period, *466,* 482–83, *482,*
    498
  artist's reconstruction of, *426*
  continent distribution in, *482*
  in correlation of strata, *452*
  first animals in, 482–83
camels, and desert conditions, 769,
    770, 789
Cameroon, Nyos Lake disaster in,
    301–2, *301*
Canada, *569,* 823, 864, 901
  ancient gneiss found in, *463*
  Burgess Shale in, 425, *426*
  fjords of, 681
  flooding in, 615, *616*
  glaciers across, 560
  during ice age, *821, 826, 827,* 833
  Laurentide ice sheet of, 826
  Northwest Territories of
    (patterned ground), *825*
  tar sand in, 520, *521*
  tornadoes in, 754
  Waterton Lakes National Park
    in, 827
Canadian Rockies, 391, *406,* 490,
    492, *493,* 597, *795*
  flooding in, 615, *616*
Canadian Shield, *256, 258,* 476, 477,
    808, 822, 827
cancer, 132
Canyon de Chelly, Ariz., *628*
Canyonlands National Park, Utah,
    782
canyons, **629**
  exploring, *1*
  formation of, 628–29, *629*
  shape of, *628*
  slot, 629
capacity (of stream), **626**
Cape Cod, Mass., *219, 573*
  salt marsh on, *682*
  as terminal moraine, 815, *817*
  wave erosion, 689
Cape Girardeau, Mo., concrete
    floodwall for, *647*
Cape Verde Island, *662*
capillary fringe, 701, **701**
Ca-plagioclase, 164–65, *190*
carat, 135
carbon, 135
  in atmosphere, 846
  in coal, 212, 524, *525*
  isotopes of, 474
  in peat, 524
  sequestration of, **539**
  storage of in or near surface, 846
carbonaceous chondrites, 51
carbonate environments, shallow-
    water, 226, *227*
carbonate rock, 210, *211*
carbonate rocks, 204
carbonates (carbonate minerals), **131,**
    190, 213, 706, 730
carbonate sand, 676
carbon capture and sequestration
    (CCS), 870
carbon cycle, 268, **846,** *847*
carbon dating, 456, 457
  see also isotopic dating

carbon dequestration, 539
carbon dioxide (CO$_2$), *863*
  in atmosphere, 732, *732,* 843, 848,
    855–56, 862–64
    early (Archean) atmosphere,
      475
    and forest fires, *732*
    and fossil fuels, 539, 733
    and global warming, 833, 834,
      862–64 *see also* greenhouse
      effect
    and Goldilocks effect, 854
    and ice age, 831, 832
    removal of, 855
    and uplift, 853
  in Cretaceous atmosphere, 493
  as greenhouse gas, 480, 493, 732,
    733, 831, 846, 848–49,
    862–64
  in Hadean atmosphere, 472
  human production vs. volcanic
    production of, 862–64
  Lake Nyos disaster from, 301–2,
    301
  and snowball Earth, 480
  in volcanic gas, 730
carbonic acid, and cave formation,
    719
Carboniferous Period, 450, *499*
  biodiversity in, *858*
  and coal formation, 524, *524*
  coal swamp in, *488*
  evolution of life in, 484, 485
  paleogeography of, *484*
carbonized impression, as fossils,
    422, 423
carbon monoxide:
  in coal gasification, 528
  from fossil fuels, 539
  in iron smelting, 548
Caribbean arc, *175*
Caribbean Plate, *349*
Carina nebula, *26*
Carlsbad Caverns, 720
Carolinas, hurricane, *756*
Cascade arc, 175
Cascade Mountains, 302, 771, *772,*
    826
Cascade volcanic arc, *294*
Caspian Sea, *85*
castle bergs, 806
cast of fossil shell, *422,* **423**
catastrophic change or events, 842,
    855, 857–58
  carbon dioxide mass suffocation
    (Cameroon), 301–2
  and future, 873–74
  Great Missoula Flood, 645–46
  Haiti earthquake, 348–49
  hurricanes classed as, 758
  Lisbon earthquake, 338–39
  mass extinction events, 51, 432–
    33, 486, 855, 857–58, *858*
    see also mass extinction event
  from rising water table, 718–19,
    719
  Sahel desertification, 789
  volcano eruptions, 298–302, *300*
  see also earthquakes; floods; mass
    movement; storms
catastrophism, **431**
catchments, 716, *717*

cations, 121, 124, *124*, 131
CAT scans, 367
Cat's Eye Nebula, *35*
Catskill Deltas, *484*
Catskill Mountains, 483, *484*
Caucasus Mountains, *85*
cave networks, *720*
    formation of, *720*
caverns, 695
caves, 695, 719–26
    and karst landscapes, 720–23,
        723, 724–25
    life in, 723–24, *723*
Cedar Breaks National Monument,
    Utah, 451
celestial objects, **13**, *14*
celestial sphere, 14
cellulose, ethanol from, 532
Celsius scale, 54
cement, **142**, *143*, 203, *206*, 208,
        557–59, **557**, 560
    composition of, 557
cementation, 206, **206**
Cenozoic Era, 7, *7*, 410, 450, 480,
        495–98, *498–99*
    in correlation of strata, *452*
    icehouse conditions in, 855
    life forms in, 451, 497
    long-term cooling in, 832–33
    ore in plutons of, 554
center of mass, 670, *671*
Central America Trench, *74*
Central Arizona Project canal, 652
Centralia, Pa., coalbed fire in, 528, *529*
centrifugal force, 670
    and tides, *671*
centripetal force, 670
cephalopods, 427, *429*
chain reaction, **529**
Chaitén Volcano, Chile, *279*
chalk, *193*, 210, 226
    Huxley's explanation of, 468
chalk beds, along English coast, *228*,
        468, *469*
Challenger, H.M.S., 72, *656*, 657, 663
change of state, 735
channeled scablands, 646, 823
channels, **615**
charge, 25, 120
Charleston, S.C.:
    earthquake in, 337
    Mercalli intensity map for, *329*
Charleston, W.V., water
        contamination, 717
chatter marks, 809
Chelyabinsk, Russia, *50*, 51
chemical bond, *24*, 119, **121**
chemical burning, 529
chemical formula, **121**
    for minerals, 119
chemical fossils, **424**
chemical reactions, **24**, **121**
    energy from, *507*, *508*
chemicals, **121**
    organic, 45, 119
chemical sedimentary rocks, **203**,
        212–15
    dolostone, 213–14
    evaporates, 212–13, *213*
    replacement and precipitated
        chemical sedimentary
        rocks, travertine, 213

chemical weathering, 189–91, *189*,
        191, 833
    and carbon dioxide absorption, 846
    and climate, 196
    and desert, 770, 773
    and physical weathering, 189–91,
        191
    and sea salt, 662
    and surface area, 194
chemistry, basics of, 120–21
Chernobyl nuclear disaster, 531
chert, 210, *214*
    in Archean cratons, 473
    bacteria found in, *475*
    bedded, 210
    biochemical, 210
    black, 215, *215*
    as deep-marine deposit, 226
    nodular, 214
    replacement, 213, *215*
Chesapeake Bay, 681, *681*
Chicago, Ill., dinosaur fossil in (Field
        Museum), 421
Chicxulub crater, 495, 857–58
Chihuahua, Mexico, minerals in cave
        near, *131*
Chile, 564
    Andes in, *400*
    earthquakes in, *312*, 330, 335, 346
    fjords of, 809
    rain-free coastal areas of, 771
    volcano in, *311*
Chilean mine rescue, 558
chilled margin, *440*
chimney cap rock, *785*
chimneys, **782**
China, 47, 545, 641, 669, *859*
    earthquakes in, 354
    Guilin tower karst landscape of,
        722, *723*
    human intervention in, *838*
    ocean exploration by, 657
    in Pangaea, 485
    REE in, 565
    squeezing of, *403*
    underground coalbed fire in, 528
    Yangtze River flood in, 641
    Yellow River Delta in, *634*
Chinese Shield, *258*
chloride, 123, *124*
chlorine, *120*
    and ozone breakdown, 733
chlorite, 248, *249*
chloritic, *247*
chlorofluorocarbons (CFCs), 733, 861
chlorophyll, 507
chondrites, 51
C-horizon, 196
Christchurch, New Zealand, 314
Christchurch earthquake, *342*
chromite, 117
chromium, 565
chrysotile, *128*
ciliate protozoans, 479
cinder cones, 283, *284*, **284**, 285,
        286, 295
cinders (volcanic), 286, 287
circum-Antarctic current, 665, 832
cirque glaciers, 799, *821*
cirques, **799**, 810, *821*
cirrus clouds, 748

cities, sea-level rise threat to, *688*
city water tank system, 709, *710*
Clark, William, 626, 629
class, 425
classification schemes, 144, 145–46
    of fossils, 426–27
    for life forms, 425–27, *428*
clastic deposits, shallow-marine, 226
clastic rocks, **142**, *143*
    origin of, 208–10
clastic (detrital) sedimentary rocks,
        203–10, **203**, *206*, *222*
    classification of, *207*
clasts, **186**, *187*, **203**, *206*, 783
    composition of, 206
    lithic, 206
    and porosity, 697–98
    size of, 204–8, *207*
    see also detritus
clay, 134, 197, *562*
    in A-horizon, 196
    and beaches, 676
    and bricks, 559, 561, *562*
    and chemical weathering, *191*
    in desert, 776
        on playa, *778*
    flakes of, *207*
    from glacier (in varve), 815
    in groundwater, 717
        contamination prevented by,
            717
    from hydration, 190
    at K-T boundary, 494
    in lake environment, 225
    and liquefaction, 342
    in loess, 815
    and marble, 244
    and metamorphosis, 241, *247*
    on ocean floor, *72*
        deep sea floor, 226
    in oil and gas creation, 508, 510
    in pelagic sediment, 660
    potters', 252
    quick clay, 342, *604*, **604**
    in red shale, *236*
    in rock cycle, 263, *269*
    and sedimentary rocks, 206, 209,
        210
    and shale, 246, 398
    and slate, *255*, 382
    and soil erosion, 216
    stability of, *190*
    swelling clays, 606
    from weathering, 190, 191, *191*
clay layers (wet), as prone to become
        failure surfaces, 601, 604,
        604
cleavage, **129**, 134, *146*, 399
cleavage plane, 137, *146*, 399, 901
    in slaty cleavage, 241, *242*
cliff:
    of jointed rock, 387
    stair-step, *781*
cliff retreat, 780–81, **780**, *781*, *782*
climate, 728–66, **761**, **847**
    alternation of, 468
    and California mass movements,
        607
    in Cenozoic Era, 497
    and coastal variability, 687
    in Cretaceous Period, 490, 494,
        497

cycles of (Milankovitch cycles),
        831–32
    in deserts, 769–72
    and eruption of LIP, 179
    and evolution of genus *Homo*, 500
    factors controlling, 762–63, *763*
    and glacier ice, 458
    global changes in, 842, 847–58
        global warming, 539, 848–49,
            862–72; *see also* global
            warming
    in Jurassic and Cretaceous,
        487–89
    and landscape development,
        577–78
    in Late Miocene Epoch, 497
    and mass movements, 607
    paleoclimate, 849
    pollen as indicator of, 424
    and proximity to ocean current,
        762
    and proximity to water, 762
    recognizing of past, 470
    shifts in (Proterozoic Eon), 480
    as soil-forming factor, 195, 196,
        198
    and transport of heat by currents,
        665
    and tree rings, 458, *459*
    types of, 763–65, *763*
    and volcanoes, 305–8
    vs. weather, 847
    see also ice ages
climate belts, 64–66, 762–63, *764*
    and global warming, 870, *871*
climate change:
    challenges of, 869
    greenhouse gases in, 848–49
    methods of study for, 849, 853–54
    models of, **849**
    uncertainty over data on, 866–69
climate change, global, 847–58
    and extinction events, 431, 432,
        855, 857–58, *858*
climatologists, 763
clinker, 559
clinometer, 389
cloning, and extinct species, 425
Cloos, Hans, 273
closure temperature, **457**, 463
clouds, **735**, *736*, 747–48, *747*
    distribution of, *749*
    precipitation, 745–49
    types of, *749*
coal, 5, 135, 211–12, **211**, *212*, 504,
        504, 521, 524–28, *524*,
        524, 536–37
    classification of, 525
    and climate change, 855
    consumption of, *526*
    and continental drift, 64
    finding and mining, 525–28, *527*
    acid runoff from, *539*
    formation of, 229, 484, 485,
        524–25, *524*
    low-sulfur, 537
    sulfur dioxide from, 539
coal-bed fire, 528
coalbed fires, underground, 528, *529*
coalbed methane, **528**
coal gasification, **528**
coal mining, dangers of, 527–28

coal rank, **525**
coal reserves, **525**
distribution of, *527*
coal seams, 524
coal swamps, 485, *488*, 524–28, *524*, **524**
coarse-grained" rocks, 168, 170, *173*
coastal beach sands, 224
coastal deserts, 771, *772*
coastal landforms, 675–83, *675*, *676*, *684–85*
beaches and tidal flats, 676–78, *677*
coral reefs, *683*
fjords, *681*
rocky coast, 679, *680*
coastal plain (U.S.), 490, 686–87, **686**
as threatened by sea level rise, 819, *823*
coastal problems and solutions, 688–91
coastal variability, causes of, 683–87
coastal wetland, **681**, *682*
coasts, 655–93, **657**, 675, *675*, *684–85*
emergent, **686**, *687*
organic, 681
pollution of, 690–91
submergent, *687*, **687**
cobalt, 565
cobbles, *187*, 205, 206, 208
Coconino Sandstone (Grand Canyon), *218*
Cocos Plate, 89, 355
coesite, 235, 237, 253
cold deserts, 769
cold fronts, 745, *745*
Cold War, seismograph stations during, 367
collision, 102, 105, *106*
exhumation from, 256
between India and Asia, 105, *402–3*, *492*, *496*
in late Paleozoic Era, 484
and mountain building, 400, 408–9, *408*
and orogens, 405
collisional mountain range (belt, orogen), 105, *106*, *108*, *250*, 401, *407*, *478*
Himalayas as, *402–3*
collision and coalescence, **747**, *748*
collision zones, earthquakes at, 335–38, *336*, *337*
Colombia:
lahar in, 593
Nevada del Ruiz in, 299, *300*
color:
of mineral, **127**
in rock identification, 167
Colorado:
flooding in, 615, *616*
Gunnison River in, *258*
highway road cut in, *578*
Mesa Verde dwellings in, 204, *205*
mountain stream in, *225*
Rocky Mountains in, *269*, 615
see also Rocky Mountains
stone dam in, *578*
Colorado Plateau, *434*, 451, *452*, *497*, 628, *784*
Colorado River, *790*
as desert river, *784*
Grand Canyon of, 628
see also Grand Canyon
water diverted from, 650

Columbia River, saltwater wedges in, 681
Columbia River Plateau, 179, *179*, 297
Columbia River Valley, and Great Missoula Flood, 646, *646*
columnar jointing, **276**, *278*
comets, 39, **39**, 730
and mass extinction, 432
material from to earth, 50–51
samples from, 39
compaction, *206*, **206**
compasses, 40, *40*, 67–68
competence (of stream), **525**, 625–26
during flood, 640
competition, for minerals, 565
composite volcanoes, *284*
composition, and rock deformation, 385
compositional banding, 255
compositional layering, 241
compound, **24**, 121
Comprehensive Soil Classification System, U.S., 197, *197*
compressibility, 361
compression, 238–39, *239*, *386*, **387**, *399*, 401
and mountain scenery, *394*
compressional waves, **323**, *324*
computer modelling, in earthquake prediction, 352
computers, geology and, *4*
concentration, of solute, **121**
concepts, of geology, 8
conchoidal fractures, **129**
concrete, 557–59, *563*
condensation, 582
and temperature, 735
condensation nuclei, **747**
conduction, 54, *54*
conduits, 282–83
in permeable material, 698–99, 705, 712
Conemaugh River Valley, 645
cone of depression, **708**
confined aquifers, 699
conglomerate, *207*, *209*, 223
and alluvial-fan sediments, 224
and bedding, *217*
flattened-clast, 241, *243*
of stream gravel, *208*
Congo River, 622
Congo volcano disaster, 298, *300*
conodonts, 483
constellations, 14
consuming boundaries, 96
see also convergent plate boundaries
contact, geologic, 445, **445**
contact aureole, 250
contact metamorphism, *247*, *248*, 250, **250**, 255, *255*, *266*, *267*, *404*, 405
contacts, *449*
contaminant plume, 716, **716**, *718*
contamination of water, 860–61
of groundwater, 716–19, **717**, *718*, 860–61
continental arc, *98*
continental collision, *see* collision
continental crust, **51**, *52*, 53, 55, 57, *108*, *360*
in Archean Eon, 472–73, *473*, *474*, 475

in Asia-India collision, *402*
brittle deformation in, 385
earthquakes in, 324
formation of, *497*
igneous rocks in, 153–54
during Proterozoic Eon, 476–77
recognizing growth of, 469
continental divides, *619*, **619**
continental drift, 62, **63**, 87, 839
in apparent polar-wander paths, 70–71
change from, 839, *839*
and climate change (long-term), 853
criticism of, 66–67
evidence for, 63–67, *63*, *64*, 87
in apparent polar-wander paths, 72
in deep-sea drilling, 82–83
in fossil distribution, 61, 65, *65*
and paleomagnetism, 67–71
and sea-floor spreading, 74–75
continental glaciers, **799**, 805, 817
see also ice sheets
continental ice sheets, **796**
continental-interior deserts, 772
continental lithosphere, 53, 55, 87, *88*, *91*, *98*, *104*, *108*, *402–3*, *406*, 658, *658*
continental margin, 74, 266
active, **89**, *659*, *660*, 684
continental platform, 477
continental rift, **102**
Basin and Range Province as, 495–96, *497*
earthquakes at, 335, *336*
continental rifting, 102–5, *103*, *104*, 105, 112
and diamonds, 135
in mountain building, 404
mountains related to, 404, 410–11
and passive margins, 658
and salt precipitation, 212, 213
continental rise, 658–59
continental shelf, *44*, *56*, *64*, 89, *222*, 658–59, **658**
ice-age exposure of, 822
continental slope, 658–59
continental transform fault earthquakes, *334*
continental volcanic arc, 97, 175
and mountain building, 400, 401
continents:
in Archean Eon, 475
basement of, *266*
birth of, 472–75
fit of, 63, 66
formation of, 476–77, 843
history of, 469–70, *471*
continuous reaction series, *164*, 166–67
contour interval, **575**
contour line, **575**
contours (seismology), 329, 330
control rods, 530
convection, **52**, 54, *54*
in asthenosphere, 369
in atmosphere, *740*
in mantle and outer core, 56, 57
plate motion and, 106
convection cells, 367, *369*
and El Niño, 765
inside Earth, 67, 106

convective, plume, 289
convective cells, 55
convective flow, in mantle, *369*
convective lifting, 747, *748*
convergence lifting, 747, *748*
convergence zone, **739**
intertropical, 740
convergent plate boundaries (convergent margins), 90, *92*, 96, *108*, *109*, *110*, 112
in Americas, 495, *496*
and collision, 101, 105
earthquakes at, *332*, 343, *442*
eruptions along, 293
geologic features of, 97–98
on map of relative velocities, *110*
during Mesozoic Era, *489*, *491*
metamorphism at, 251, *255*
mountain belts at, 400–401, *400*
and ocean, 659, *685*
orogens at, *400*
seismicity at, 333–35
and subduction, 96–98
volcano activity at, 174–75, *175*, 176
on volcano map, *172*
on western margin of North America, 489, 495, *496*
Coordinated Universal Time (CUT), 436
Copernicus, Nicolaus, 15
copper, *118*, *127*, 131, 547, *549*, 555, *564*
of Andes, 555
as base metal, 549
in coins, *547*
consumption of, 564
crystal structure of, 547, *548*
as metallic mineral resource, 546
as native metal, 547, *548*
coprolites, 424
coral reefs, 210, *223*, 227, 676, 682–83, *682*, **682**, *683*
and climate, *684*
destruction of, 691
through wind-blown dust, 792
formation of, *683*
in shallow-water environment, 226
corals, 427, *429*
and beaches, 676
Cordillera, glaciers in, *799*, *800*
Cordilleran ice sheet, 826
Cordillera range, *496*
core, *46*, 47–48, *52*–53, *56*, *56*, *57*, *58*, 360–61, *360*
and differentiation, 470, 471
discovering nature of, 365–67, *366*
formation of, 842, *842*
rotation of, 369
temperature in, 52
core-mantle boundary, **365**
discovery of, 365, 367
Coriolis effect, **665**, 739, *740*
in atmosphere, 739, *740*, 741, *741*, 745–46, 755
in oceans, 665, *666*
and upwelling, 666
correlation, 445–48, *448*
corundum, *127*, 136, 137, *138*
cosmic rays, *37*, 40, 856

cosmology, **13**
    ancient views of, 14
    and birth of Earth, 12–35
    Renaissance and modern views of, 14–15
Costa Rica, slumping in, *590*
country rock, **161**
    metamorphism of, *251*
covalent bonds (bonding), 121
Crab Nebula, *27*
cracks, *385*
crag and tail, 812
crater eruptions, *282*
Crater Lake caldera, *283*
Crater Lake volcano, *290*
craters, **283**
    Chicxulub, *494*, *495*, 857–58
    from Midwest meteorite impact, *9*
craton, 409–11, **409**, *410*, *474*, **476**
    North American, *410*, *411*
cratonic platform, 410, **476**
creep, **588**, *589*, 602
Cretaceous Period, 461, 498–99, 700, *845*
    biodiversity in, *858*
    chalk deposits in, 468
    cooling of atmosphere since, *833*, 852
    in correlation of strata, *452*
    and dinosaurs, 451, 493–94
    greenhouse conditions in, 490, 497
    and K-T boundary event, 494–95, 857
    late, 517, *519*
    mass extinction during, 433
    North America in, *491*
    paleogeography of, 490–93, *492*
        climate, 490
    and sea floor, *95*
    and U.S. coastal plain, *477*
    volcanoes in, 853
Cretaceous sandstone, *461*, 699
crevasse, 801, **801**, *803*, *820*, *821*, 827
crinoids, 427, *483*, 560
cross bedding, 216–18
cross beds, *216*, *220*, *784*, *788*, *789*
cross-cutting relations, principle of, 438, *440*, 460–61
crown jewels, *117*
crude oil, 517
crushed stone, 559, *563*
crust, *46*, 48, 49–51, **49**, *52*, 56, *56*, *57*, 58, 360–61, *360*
    composition of, 49–51, 143
    and dikes, *168*
    elements in, *53*
    igneous rocks in, 407
    and isostacy, 406
    and mountain building, *404*
    rock deformation in, 382–87
    shortening and thickening of, 406–7
    and subsidence, *574*
    temperature of, 404, 405
    and uplift, *574*
    see also continental crust; oceanic crust
crustal blocks, in Archean Eon, *474*
crustal fragments, in growth of North America, *489*
crustal rocks, in continental crust formation, *472*, *472*, *473*
crustal root, **406**, *407*

crustal thickening, 401
crust-mantle boundary, discovery of, 362–63, *363*
cryosphere, 43
cryptocrystalline quartz, 210
crystal faces, **122**
crystal habit, *128*, **129**
crystal(line) lattice, **119**
    of metals, 547
    quartz as, 119
crystalline (nonglassy) igneous rocks, **167**, 170, *170*, *172*
    and cooling time, 168
crystalline rocks, **142**, *143*
crystalline solid, 119, **119**
crystallization, of rocks, 457
crystal mush, 93
crystals, **45**, 118, *119*, *122*, **122**, *123*, *125*
    destruction of, 125–26
    formation of, 124–27, *126*
crystal structure, 122–27, **122**, *124–27*, *124*, *125*
    of ice, *797*, *798*, 799
    and weathering, 191
cuesta, *783*, *783*
Cullinan Diamond, 135
cumulonimbus clouds, *748*
cumulus clouds, *748*
currents, oceanic, 663–64, **663**, *664*, *684*
    and climate change, 855
        continental drift as cause, 853
    and coastal deserts, 771
    and El Niño, 765
    and glaciation, 831, 832
    global-warming effect on, 870
    and Isthmus of Panama, 497
cut bank, 631
cutoff, *633*, *634*, *642*
Cuvier, Georges, 419, 425, 431, 432
cyanobacteria, 475, 731
cycle, **265**
    biogeochemical, 265, 268
    carbon, 265
    geochemical, 265, 268
    hydrological, 268
    temporal, 265
cycle of transgression and regressions, *229*
cycles in Earth history:
    biogeochemical cycles, 846
    physical, 844, 846
cyclic change, 842
cyclones, 745–47, *746*
    and nor'easters, 754
cyclonic flow, 745–47, **746**
cyclothems, 844–46, *845*
Cynognathus, 61, 65

Dakota aquifer, 699, *700*
Dalton, John, 24
dams, 650
    and earthquake flooding danger, 355
    environmental issues over, 650, 859
    for hydroelectric power, 532–33, *533*
    ice sheets as, 822–24
        and Black Sea flooding, 852
        and Great Missoula Flood, 645–46, 823

Johnstown Flood from collapse of, 645
    on Nile River (Aswan High Dam), 859
    and sediment carried downriver, 650
    stone dam in Colorado, *578*
    Vaiont Dam disaster, 594
Darcy, Henry, 706
Darcy's law, 705, **706**
"Darkness" (Byron), 305
Darwin, Charles, 87, 430–31, 462, 682
Daryasa gas crater, *513*
daughter isotope, 453, 456–57, *456*, *457*
Davenport, Iowa, flooding of, 644
Dawn spacecraft, 39
Dead Sea, 44
Dead Vlei, Namibia, *791*
dead zone, 690
Death Valley, Calif., *213*, *220*, *225*, *260*, *632*, *768*, *770*, *775*, *779*
    alluvial fan in, *779*
    playa in, *780*
    Racetrack Playa in, 779
    wind erosion in, *777*
debris:
    in Colorado River, *631*
    in desert environment, *266*
    in desert pavements, *785*
    exceeding angle of repose, *605*
    Johnstown dam spillway blocked by, 645
    as Midwest covering, 144
    in Nile River canyon, 183, *184*
    in orbit around Earth, 470
    scattered by tornado, *753*
    unconsolidated, 195
    volcanic (pyroclastic), 154, *155*, 278–81, **278**, *290*
    from Yungay avalanches, 588
    see also pyroclastic debris; regolith
debris avalanche, 595, *603*
debris falls, 595–96
debris flow, 210, *591*, **591**, *603*
    submarine, **596**
debris slide, **594**
Deccan region, India, 179, 297
declamation, 408, *408*
decompression melting, *156*, 177, *178*
deep currents, 664, 665–67
deep earthquakes, *332*, 333, 334
Deep Impact spacecraft, 39
deep-marine deposits, 226
deep-ocean trenches, **45**, *56*, *74*
deep-sea drilling, 82–83
Deep Sea Drilling Project, 184
deep time, 435
    see also geologic time
Deepwater Horizon Disaster, 540–41, *541*
deflation, **776**, *777*
deforestation, 606, *860*
    in tropical forests, 607
deformation, rock, **349**, 382–87, *383*, 398, *405*, *405*
    brittle, 323
    cause of, 386
    of cratonic platforms, 410
    and deposition, 438
    geologic map of, *405*
    plastic, 323, 383–86, 398
    thin-skinned, 485

in western North America, *492*
    see also brittle deformation; ductile deformation
deformation rate, 385
deformation vs. topography, *405*
delamination, *574*
delta plain, 640
delta-plain floods, 640
deltas, 225–26, **225**, **626**, 634–35, *634*, *635*, 640, *642*
    deposition on, *223*
    of Mississippi, *635*, *636*
    of Niger, *634*
    of Nile, 634–35, *634*, *635*, 859
    shape of, *634*
    soil deposition on, 209
    swampy, *676*
    see also Catskill Deltas
deltias, "Gilbert-type," 226
Democritus, 24
Denali National Park region, Alaska, *316*, *625*, *632*
dendritic drainage network, 618, *618*
dendrochronologists, 458, *459*, 849
density, 361
density (atmospheric), 41
density currents, 595
Denver, Colo., earthquake in, 338
deposition, **205**, *206*, 486–87, **576**, *642*
    deltas, 634–35
    from glaciation, 813–18, *817*
    in landscape evolution, 576, *577*
    of sediment, 576
depositional environment, 224–27, **224**
    recognizing changes in, 469
depositional landform, **577**, 815–18, *817*
    and passage of time, 579
depositional processes, 623–26
depositional sequence, 229, *230*
De Re Metallica (Agricola), 117
derricks, 517, *518*
desertification, 790–93, **790**, 860
desert pavement, 777, 783–84, **783**, *785*
desert plateau, 785
deserts and desert regions, *763*, 769–72, **769**, *770*
    of Basin and Range Province, *784*
    cold, 769
    and continental drift, 64
    depositional environments in, *777*, *779*, *783*
    extent of, 771
    geology of, 768–94
    hot, 769
    human inhabitants of, 789–90, *790*
    landscapes of, 779–89, *782*
    life in, 789–93, *790*
    nature and location of, *769–72*
    problems of, 789–93
    and rock cycle, *266*
    Sahara Desert, *see* Sahara Desert
    Sonoran Desert, *790*
    types of, 771–72
    urbanization of, 792
    weathering and erosional processes in, 773–77, *774*, 775–77, *777*
desert varnish, 773–74, **773**, *774*

Index I-7

Des Moines, Iowa, flooding of, 644
despersants, 540
detachment, 485
detachment fault, **391**, *394*
    and fold-thrust belt, 401
detrital (clastic) grains, age of, 458
detritus, 186
    and weathering, 204
    see also Clasts
developing countries, river pollution
    in, 650
development, and southern California
    mass movement, 607
Devils Postpile, Calif., *278*
Devil's Tower, Wyo., 302, *302*
devolatilization, 160
Devonian Period, 483, *484*, *498–99*
    animals in, 483
    Antler orogeny in, 483, *484*
    in correlation of strata, *452*
    late (paleogeographical map of
        North America), *484*
    life forms in, 450
dewpoint temperature, 735
diagenesis, **229**, 248
    and metamorphism, 229
diamicite, *207*, 209, *210*
diamond anvil, 364, *365*
diamonds, 124, *125*, 134–36, 137,
        138, 364, *365*
    Cullinan Diamond, 135
    hardness of, 127, *127*
    Hope Diamond, 134, 136, *136*
    mining, 134–36
    placer deposits of, 553
    regions, 136
    shape of, *125*
    in simulation of mantle, 364
    as ultra-high pressure
        metamorphic rock, 237
diapir, 163
Dickinsonia, *479*
Dietz, Robert, 62, 75
differential stress, **238**, 254
differential weathering, 194, *194*
differentiation, *32*, **32**, 155, *244*, **470**,
        472, **842**
diffraction, **122**
digital elevation maps, *572*, **576**
dike intrusion, *168*
dikes, *93*, **162**, *165*, 166, *166*, *167*,
        168, *287*
    basalt, *165*, 177, 438, *441*
    composition of, 169
    and cross-cutting relations, *440*
    formation of, *168*
    at Shiprock, *166*
    in volcano, *286*
dimension stone, **557**
Dinosaur Ridge, Colo., *219*
dinosaurs, 489, 493–94, 497–500
    and climate, 850–52
    extinction of, 451, *494*
    first appearance of, 451, 489
    fossilization of, *421*, *423*
    during Jurassic Period, *490*, 494
Dionysus, 117
diorite, *172*, *173*
dip, 387, 388–89, **388**
dipole, *40*, **40**, 69, 71, *81*, 375, *375*
    magnetic, **67**
    paleomagnetic, 69, 71, 76
dip-slip faults, *390*, *390*

dip slope, 783
directional drilling, **516**, 520, *523*
"dirty snowballs" (comets), 39
disappearing streams, **721**, *722*
disasters, *see* catastrophic change or
        event
discharge (groundwater), 704
discharge area (groundwater), *704*,
        704
discharge of stream, 621–23, **621**
    and capacity, 625
disconformity, 444, *444*
discontinuous reaction series, *164*,
        166–67
disease, from earthquakes, 350
displacement, 383, *384*
displacement of fault, **316**, *317*, *390*,
        391, *392*
    on San Andreas Fault, *317*
disseminated deposit, 550, *551*
dissolution, 189–90, *189*, **189**, *191*
    of minerals, 126
    of minerals in stream, 624
dissolved load, **624**
distance per second per second (d/
        s2), *394*
distillation column (tower), 517, *518*
distortion, 383, *384*
distributaries, **635**, *643*, 778
divergence zone, **739**
divergent plate boundaries, 90, *92*,
        92, *93*, *108*, *110*
    and sea-floor spreading, 92–95,
        93, *96*
    seismicity at, 333
diversification, 450, 451, 489
diversity, *451*
D" layer, 369
DNA (deoxyribonucleic acid), 425,
        427, 429, *430*, 431
Dobson units, *863*
doldrums, **741**
dolomite, 131, *214*, 255, 552, 706
dolostone, 213–14, 255, 629, *632*
domains, 425
dome, *395*, **395**, *411*
    on geologic map of eastern U.S.,
        411
    regional, 410–11
Donau glaciation, 828
Doppler, C. J., 21
Doppler effect, **21**, *22*, 32
Doppler radar, 754
dormant volcanoes, *302*, *303*
double-chain silicates, 131, *133*
Dover, England, chalk cliffs (White
        Cliffs) of, *228*, 468
downcutting, **617**, 628
downgoing plate, *96*, *97*
downslope force, 600–601
down slope movement, **576**
downstream region, 615
downwelling, **106**
downwelling zones, 369, *369*, 374,
        665–67, *665*, *667*
dowsers, 708
drag folds, 391
drag lines, 525–26
drainage:
    evolution of, 636–39
    and soil characteristics, 196, *197*
drainage basin, 619, *619*, 627
drainage divide, *619*, **619**, 637

drainage network, 615–18, **617**, *618*,
        618
    and rock cycle, *266*
drainage reversal, **636**
    from ice age, 822–24, *823*
Drake, Edwin, 513
Drake Passage, 665
drift, 814–15
drilling, for coal, 527
drilling mud, **516**, *518*
drills, oil, 515–17, *518*
dripstone, 720, *723*
dropstones, *807*, **808**, *821*
droughts, 200, 790–93
drumlins, **817**, *818*, *821*, 827
dry-based glaciers, 801, *802*, 804
dry-snow avalanches, 594
dry wash, 620, *621*, *775*, **775**, *784*, *785*
dry well, 708
ductile deformation, 323, 383, *385*
dunes, 126, **779**
    coastal, *676*
    sand, 220, *784*, 787–89, *787*, *788*
dung, as fuel, *505*
dust, wind-blown, 792
"dust" (cosmic), 28, *29*
"dust bowl" of Oklahoma, 200, *201*,
        791–92, *793*
dust devils, 775
dust storm, 776, *776*, 791
dwarf planets, 18
dynamic metamorphism, **250**, *253*
    along fault zone, 250–51
dynamo, **376**
dynamothermal (regional)
        metamorphism, **252**,
        255, 405

Earth, 11, *11*, *13*
    age of, 6, 434–66, *462*, 470–71,
        497–500, *498–99*
    atmosphere of, 41, *42*
    axis of
        precession of, 831, *832*
        tilt of, *740*, 742–43, 831, *832*,
            855
    basic components of, 115
    biography of, 467–501
    as blue, 657
    changes in from continental drift,
        839, *839*
    circumference of, 19, *28*
    cosmology and birth of, 12–35
    density of, 47–48
    differentiation of, 29–32, *32*
    elemental composition of, 45–46
    energy resources of, 503, 504–44
    evolution of, 819–24
    future of, 870, 873–74
        see also global warming
    heat of (early life), 154–56
    history of, 417
    history of in sedimentary rocks,
        223
    interior of, 36–60, *48*, *52*, 360–78
    layers of, 47–48, *48*, *52*, *52*, 56–57,
        56–57, 87, *360*
    magnetic field of, 40–41, 67–68,
        68, 78, *79*
        reversal of, 77–78, *77*; *see also*
            magnetic reversals
    map of, *44*
    meteorite bombardment of, 472

new discoveries on, 367
orbit of, 831, *832*, 855
other-world explorers' view of,
        37–38
as planet, 6, 15, 18, 32–33, *40*,
        42, 44
power resources from, *536–37*
pressure in, 46
rotation of, 14–15, *666*
seismic tomography image of, *368*
as setting for life, 854
shape of, 14, 32–33
size of, 19
surface of, 43–45, *45*, 53
    see also landforms; landscape
surface veneer of, 182–201
temperature in, 46, *46*
temperature of, 854
    and greenhouse effect, *848*
tilt of axis of, *740*, 742–43
topography of, *44*
Earth history, 467–501, *467*, *839*, *843*
    biogeochemical cycles in, 846
    causes of change in, 839
    end of, 873
    in geologic column, 449–53, *451*
        see also geologic time
    and Grand Canyon, 435
    and human history, 464, 468
    ice ages in, 497, 796, 827–29, *830*,
        850–53
        see also ice ages
    methods for studying, 468–70
    physical cycles in
        rock cycle, *840*, 846
        sea-level cycle, 844–46, *845*
        supercontinent cycle, 844
    in sedimentary rocks, 223
    time periods of
        Archean Eon, 6, *7*, 472–75
        Cenozoic Era, *7*, 7, 494, 500
        Hadean Eon, 6, *7*, 470–72
        Mesozoic Era, *7*, 7, 487–95
        Paleozoic Era, *7*, 7, 482–86
        Proterozoic Eon, 6, *7*, 476–81
        see also specific periods
Earth-Moon system, 669, *670*, *671*
earthquake belts, 90, *92*
earthquake energy (vibration), *318*
earthquake engineering, 354–55, *356*
earthquake magnitude scales, 330–31
earthquakes, 48–49, *49*, **49**, 271,
        312–29, *312*, *313*,
        314, **314**
    adjectives for describing, *331*
    causes of, 315–23
        and location, 332–38
    classes of, 333–34
    in continental crust, 324
    damage from, *331*, 338–50, 354
        in Turkey, *5*, *340*
    deaths from, *315*
    energy released from, 331–32, *331*
    engineering and zoning for,
        354–55
    epicenter maps of, *901*, *909*, *910*
    faulting and, 316–23
    and fracture zones, 98
    household protections from, 355,
        356
    and induced seismicity, 337–38
    in Japan, 313–14, *314*
    magnitude (size) of, 328–32

mass movement triggered by, *586*, 601–2, 608
measuring and locating, 323–27
at mid-ocean ridges, *333*
notable, *315*
number per year of, *331*
and plate boundaries, *91*
precursors of, 352–53
predicting of, 348, 351–54
preventing damage and injury from, *356*
resonance in waves of, 355
and sea-floor spreading, 74–75, *75*
and seismic waves, 323–27
and subducted plates, 96, *98*
in Turkey, *5*, *340*, 351–52
as volcano threat, 301, 303
see also seismic waves
earthquake warning system, 354–56
earthquake waves, 49
see also seismic waves
earthquake zoning, 354
"Earthrise," *13*
EarthScope, *336*, *368*, *370*, **370**
Earth System, 6, **6**, *11*, **43**, 195, 468, 475, 500, **839**, *840–41*, 873–74
anthropogenic changes in, 858–59, 863, *863*
deserts in, 769
effects of global warming on, 870
and global change, 838–75
and life processes, 7
rocks as insight into, 142
sources of energy in, 507–8, *507*
and transfer cycle, 265
East African Rift, *104*, *105*, *175*, 408
and continental rifting, 404
and earthquakes, *336*
and volcanoes, *175*, 300
East Pacific Ridge, *74*
East Pacific Rise, 79–82
eccentricity cycle, 831, *832*
echinoderms, 483
echo sounding (sonar), 72, *73*
ecliptic, 38
eclogite, 248, *248*
ecology, and hydroelectric power, 533
economic minerals, 549, 550
ecosystems, **859**
estuaries as, 681
human modification of, 859–61
eddies, 623, *623*, 665
Ediacaran fauna, *479*, **479**, 480
effusive eruptions, **285**, 288, 298–99
threat from, 298–99
Egypt, 641
Alexandria, 18–19
pyramids of, *578*
Sahara Desert of, *see* Sahara Desert
Syene, 18–19
E horizon, *196*
Einstein, Albert, 87
Ekati Diamond Mine, 136, *136*
Ekman, V. W., 666
Ekman spiral, 666
Ekman transport, *665*, 666
elastic behavior, **317**
elastic-rebound theory, **319**
elastic strain, 317
electonic seismograph, 325, *325*
electromagnet, **375**

electromagnetic radiation, **21**
electron, 120, *120*
electron clouds, 25, 120
electron microprobes, 149, *150*
electrons, 25
element, **24**, 120, **120**
origin of, 26–27
elevation:
and landscape development, 576
world map of, 901
elevation model, of Oahu, Hawaii, *574*
Elk River contamination, 717
Ellesmere Island, 864
El Niño, 763–65, *765*, **765**
elongate (cigar-shaped) grains, 238
El Salvador, *283*
embayments, 679
emerald, 136, 137, *138*
emergent coasts, 686–87, **686**, *687*
Emerson, Ralph Waldo, 313
Emperor seamount chain, 102, 110
emplacement, 408
Enceladus, *308*, 583, *585*
end moraine, *813*, 815
energy, *13*, *504*, **505**
for landscape evolution (internal, external and gravitational), 577
need for, 505–6, *506*
energy density, **508**
energy grid, 542
energy resources, **505**, *506*, 535–42
alternative, 531–34, *532*, *533*, *534*, 539–42
choices over, 535–42
coal, 524–28
see also coal
of Earth, 503, 504–44, *509*
Earth system as sources of, 507–8, *507*
environmental consequences of, 506, *509*, 520
and environmental issues, 535–42
fuel cells, 534
in future, *538*, 539–42
geothermal energy, 531–32
see also geothermal energy
global reserves of, *535*
hydrocarbons (alternative sources), 517–23, *521*
hydrocarbons (oil and gas), 508–9
hydroelectric power, 532–33, *533*
see also hydroelectric energy
nuclear power, 529–31, *530*
see also nuclear energy
and oil crisis of 1970s, 535
and oil crunch, 535–38
and alternative sources, 535–38
problems of, 535–42
renewable vs. nonrenewable, 535–39
and society, *536–37*
solar power, 534
sources of
external (solar), 577
internal, 577
wind power, 532–34
Engelder, Terry, 520
England, *215*
chalk beds along coast of, 468, *469*

chalk in, *193*
gravestones in, *188*
groins to prevent beach erosion in, *689*
wind farm in, *533*
see also United Kingdom
Enlightenment, 436
enrichment of uranium, 530
entisol, 197, *198*
environment, and soil development, 197
environment issues:
atmospheric pollution, 733
geological phenomena affecting, 7
global warming, *see* global warming
groundwater contamination, 716–19, *718*, *719*
human modification of landscapes, 858–59
and hydroelectric power, 533
and metallic mineral resources, 566
and mining, 566–67, 579
nuclear waste, 530–31
pollution, 861
and rivers, 650
and strip mining, 527
Eocene Epoch, 495
and ice ages, 832, 852
and sea floor, 95
warm temperature of, 869
Eolian sand deposit, *785*
Eons, **450**
epeirogeny, **411**
ephemeral streams, **620**, *621*, 774, 775
epicenter of earthquake, 90, *315*, *316*, 324, *334*, *335*, *337*
finding, 326–27, *328*
map of, *332*
maps of, *351*, *353*, 901, *909*, *910*
Epic of Gilgamesh, 852
epicontinental seas, 482
epochs, **450**
EPOXI spacecraft, *39*
Epsilon Eridani, artist's conception of, *60*
equant grains, **146**, 239, *239*
equator, **19**
equatorial climatic belts, and continental drift hypothesis, 64
equatorial low, 740
equilibrium line, *805*, **805**
equinox, 742–43
equipotential surface, *372*, **372**, *373*
equitorial bulge, 372
eras, **450**
Eratosthenes, 18–19, *18*, 32
erg, 780, 787, 788
Rub al Khali as, 788
erosion, *165*, *168*, *193*, 204–5, **204**, *572*, **576**
agents of, 577–79, *577*
from agriculture, 579
in Arches National Park, 388
coastal, 222
and disconformities, 444
of hanging-wall block, *390*
and landscape development, 577
from landscape modification, 859
of mountains, 381, 404, 408, 469, *470*

relief diminished by, *576*
and rock cycle, 265, 268
of sedimentary rock, *232*
in sedimentary rock development, 204–5, *206*
soil, 199–200, *199*, 200
and streams (fluvial), 623, *624*
headward erosion, 617, *617*, 629, 636, *643*, *785*
and meanders, 631–34, *633*
of volcano, 302–3, *302*, *303*
wave, 577, 674, 678, 679, 680, 687
of Nile Delta, 859
by water (desert), 774–75
wind, 775–77, *785*, 790
erosional coasts, 687
erosional landforms, **577**
and passage of time, 575
erratics, *796*, **796**, 815, *816*, *817*, *821*, 826
eruption column, **289**
eruptions, *see* volcanic eruptions
eruptive styles, 282–92, *285*, *287*, *289*
escarpment:
and cliff retreat, 782
from erosion of loess, 815
in mid-floor ridge, 661
Niagara Escarpment, *632*
eskers, 815, 817, **817**, 818, *819*, *821*
estuary(ies), 222, 679–81, **679**, *681*
pollution of, 690–91
ethanol, 507, 532
Etna, Mt., *277*, *289*, *300*, 304–5, *305*
Étretat, France, cliffs of, *680*
euhedral crystal, 126
eukarya (eukaryotes), 425, **425**, 429, 477–79
eukaryotic cells, *430*
Eurasian Plate, *111*
Europe:
during last ice age, 796–97
Pleistocene climatic belts in, *829*
European Plate, 184
eustatic (worldwide) sea-level changes, *487*, 686, **686**, 846
evacuations:
against earthquakes, 351
against floods, 640–41
against mass movements, 608
against volcanic eruption, 304
evaporate, 121, *125*
evaporation, 582
and temperature, 735
evaporites, 212–13, **212**, *213*, 561, 564
as prone to become failure surfaces, 601
evapotranspiration, **582**
Everest, George, 406–7
Everest, Mt., 42, 44, 380, *402*, 406, 407
air pressure at peak of, 42, 732, *733*
height of, 295
Everglades, 714, *715*
evolution (life), **431**, 500, 843–44, *843*
of Cenozoic Era, 497–500
and ecosystems, 859–60
and extinction, 430–33
fossils and, 418–33, 450–51
in geologic column, *451*
of life on Earth, 843–44, *843*

evolution (life), *(continued)*
    in Mesozoic Era
        early and middle, 489–90
        late, 493–94
    in Paleozoic Era
        early, 482–83
        late, 486, *488*
        middle, 483–85
    and paradigm change, 87
    in studying Earth history, 470
    see also life forms
evolution of atmosphere, *731*
exfoliation, 187, *187*
exfoliation joints, as prone to become
    failure surfaces, 601, *601*
exhumation, 175, **186, 256,** *257,* **409**
exoplanets, 18
exosphere, 737
exothermic chemical reactions, 508
exotic terranes, **489**
expanding universe theory, **22,** *23*
exploration of oceans, 656–57
    see also research vessels
*Exploration of the Colorado River and
    Its Canyons* (Powell), 435
explosive eruptions, 285–88, **285**
explosive (pyroclastic) eruptions, 285,
    289
external energy, **577**
external processes, 7
extinction (of species), 419, **432,** 441
    of dinosaurs, 451
    evolution and, 430–33
    see also mass extinction event
extinct volcanoes, *302, 303*
extraordinary fossils, **425,** *426*
extratropical cyclone, *746,* **746,** 754,
    761
extremophiles, 425
extrusive environment, *163, 171*
extrusive igneous rock, **154,** *155*
extrusive realm, *155*
Exxon Corporation, 846
eyes, of hurricanes, *756,* 757
eye walls, of hurricanes, 757
Eyjafjallajökull volcano, 295, 299, *311*

facets (gem), **136**
facies, metamorphic, 245, *248,* **248**
Fahrenheit scale, 54
failure surface, **590,** 601, *601*
faint young Sun paradox, 854
"falling-rock zone" signs, 595
families, 426
Farallon-Pacific Ridge, *496*
far-field tsunami, 346
"father of geology" (James Hutton),
    144
fathoms, 631
fault block mountains, 404
fault breccia, 391, *393*
fault creep, **323**
fault gouge, 391
faulting, 102, *104,* 389, *390*
    breccia from, 698
fault line, 316, 390
faults, 48, *49,* 313–23, **313,** 320, 381,
    382–83, *383,* 389–93, **389,**
    394, 493
    active, 316
    in Alpine outcrop, *383*
    classification of, 390–91, *390*
    in cratonic platform, 410–11

in Cretaceous North America, 491
in crust, 315–16, 320–21
in Desert and Range Province,
    784
earthquakes generated from,
    319–23
formation of, 316–19, *318*
in geologic history illustration,
    438, *441*
inactive, 316
from meteorite impact, 8
and oasis, 709
preexisting, 317, *322*
reactivated, 318
surfaces of, *393*
types of, 315–16
fault scarps, *93, 95,* **316,** 320, **391,** *393*
    and waterfalls, 629
fault systems, 391–93, **391**
fault trace, *316,* **316,** *390*
fault trap, 514, *514*
Federal Emergency Management
    Agency (FEMA), 650
feedback mechanisms, **849**
feldspar, *128,* 134, 165, 167, *172,* 236
    in alluvial fans, 224
    and chemical weathering, *191*
    in gneiss, 241
    in granite, 549, 557
    and hydrolysis, 190, 191
    in quartzo-feldspathic
        metamorphic rocks, 245
    in schists, 241
    in weathering of clastic
        sedimentary rocks, *207*
    see also K-feldspar; orthoclase
felsic igneous rocks, 172
felsic lava flow, *275*
felsic lavas, 159, 160, *173,* 176
felsic (silicic) magma, 158, *158,* *159,*
    176
felsic minerals, *190, 244, 245*
felsic rocks, 46, 51, 159, 160
felsic tuffs, 297
Ferrel, William, 739
Ferrel cells, 739, *740,* 741
fertile crescent, 792, 855
fertilizers, 199, 200
fetch of wind, 672
field force, 16, *16,* **372**
finches, and Darwin's theory, 431
"fine-grained" rocks, 168, 170, *173*
Finger Lakes, *813,* 827
fire:
    discovering of, 827
    earthquake damage from, 342–43,
        343
    in gemstones, 136
    in underground coalbeds, 528
firn, *798,* **799**
first atmosphere, 730
Firth of Forth, 559
fish, 493
    in caves, 723, *723*
    and El Niño, 765
    jawless, 483
fission, nuclear, 24, **24,** 25
fission track, 458–59, **458,** *460*
fissure, **282**
fissure eruptions, 282, *282, 296*
fjords, **681,** 812–13, *812, 813,* 827
    glacial, *676*
    in Iceland, *819*

flank eruptions, 283
flash floods, 644–45, **644,** 774, 775
    in Big Thompson Canyon, 615,
        644–45, *748*
    in desert in Israel, *644*
flattened-clast conglomerate, 241, *243*
flat-topped seamount (guyot), 72, *74*
    see also seamounts
flexural slip, 397, *397*
flood basalt, *179,* **179, 296,** *298*
    on moon, 309
flood-basalt eruptions, 296–97, *298*
flood crest, 640
flood frequency and peak discharge
    graphs, *651*
flood-frequency graph, *651*
flood-hazard maps, *649,* **650**
flooding, calculating threat posed
    by, 651
floodplain, 224, *227,* 626, *633,* **634,**
    637, *643*
    and anticipation of flooding,
        646–50
    development of, 629, *630*
    sediment deposition on, *227*
    soil deposition on, 209
floodplain floods, 640
    in Midwestern U.S. (2011), 641–
        42, *641,* 646–47
floods, **615,** 640–50
    protection measures for, 646–47,
        647, 649
    seasonal, 640–44, **640**
flood stage, 640
floodways, **647**
Florida, *185*
    Everglades in, 714, *715*
    mangrove swamps in, *682*
    sinkhole collapse in, *695, 695, 696*
Florida aquifers, 700
flow folds, *397*
flowstone, 720, *721, 723, 724, 725*
flow velocity, *622,* 804
fluorite, *127, 130,* 131
fluvial deposits, 626
fluvial sediments, **224**
flux melting, 157, **157,** *548*
fly ash, 528
focus of earthquake, 90, **315**
fog, 747
folds, 381, *383,* 386, 393–400, **393,**
    395, 398, 514
    causes of, *398*
    characteristics of, *396*
    in cratonic platform, 410–11
    from flexural slip, *397*
    formation of, 397–98, *404*
    in geological history illustration,
        441
    geometry of, 393–97, *395*
fold-thrust belt, **391,** 401, 413, *491*
    Appalachian, 484, *488*
    Canadian Rockies as, 490
    Sevier, *492*
foliated metamorphic rocks, 241–42,
    243
foliation, *241,* 393–400
    from deformation, *255,* 381
    metamorphic, **147, 235,** 241–42,
        255, *255,* 398
        and failure surfaces, 601, *601*
    tectonic, *398, 399*
food chain, 483

flank eruptions, 283
footwall, 316, 320
footwall block, *390,* 391
force, *16,* 386, **386**
    field, 16, *16*
    mechanical, 16, *16*
    vs. stress, 386–87
forearc basin, *98*
foreland basins, 229
foreland sedimentary basins, 469
foreshocks, **319,** 352
forest fire, atmospheric effect of, *732*
forests:
    percentage of Earth covered by,
        860
    shrinking of, *860*
    see also rainforests
formation, 445–46
    see also stratigraphic formation
"forty-niners," 546, *546*
fossil assemblage, **441**
fossil correlation, 446
fossil fuels, 503, 506, 507, **507**
    and carbon cycle, 846
    and carbon dioxide in atmosphere,
        862
    supplies of, 538–39
fossiliferous limestone, 210, *223,* 226,
    235, *236, 237,* 467, 483,
    484
fossiliferous sediment, 480, 482
fossilization, 420–21, **420,** *421,*
    429–30
fossilized shark tooth, 436
fossils, *61, 65,* 219, 417, *419,* **419, 436**
    and age of sedimentary rocks, *455*
    in bedding surface, *442*
    in Carboniferous coal deposits,
        524, *524*
    in chalk beds, 468
    chemical (molecular)
        (biomarkers), 474, 475
    classification of, 426–27
    and dating of older glaciations,
        827–29
    Dickinsonia, *479*
    different kinds of, 422–24, *422*
    discovery of, 418, *420*
        by da Vinci, 419
    and Earth history, 418–33
    as evidence for continental drift,
        61, 65, 65
    extraordinary, **425,** *426*
    formation of (fossilization),
        420–21, *421*
    and ice-age climate shift, 826, 827
    in identification of early life, 474,
        475
    invertebrate, 427, *429*
    in museums, *419,* 421, *421*
    plankton as, 424, *424,* 828–29
    preservation of, 424–25
    record provided by, 428–30
    rocks containing, 418
    shell of (brachiopod), *385*
    of Silurian and Devonian Period,
        484
    Steno's explanation of, 436
    in study of Earth history, 470
    in volcanic ash, 306
fossil succession, principle of, 438–42,
    442
Foucault, Jean Bernard Léon, 20, *20*
fracking, *see* hydraulic fracturing

fracking fluid, 522
fractional crystallization, 161–62, *161*, **161**, 176
fractured rock:
    geysers at, 712
    and mass movement, 600
    secondary porosity of, 535, *697*
fracture zones, 56, 72, **72**, 98–99, **98**, 659–60
fracturing, *287*
    in setting stage for mass movement, 600
fragmental igneous rocks, **168**, *170*, *172*
fragmentation, as mass movement setting, 600
fragmentation (brecciation), 281
fragmented lava deposits, 281
framework silicates, *133*, 134
France, *439*, 655
    Étretat cliffs in, *680*
    glacial till in, *225*
    glacier visit to, 826
Frankenstein (Shelley), 306
Franklin, Benjamin, 305, 672
freezing, of liquid melt, 153
French Alps, glacier in, *804*
frequency, 21, **21**
frequency content, 355
friction, *317*, *318*, 600, 601
fringing reef, 682
frogs, in desert, 789
front (weather), 744–45, *745*, **745**
frontal lifting, 747
frost wedging, **187**, *188*
fuel, **505**
fuel cells, 508, 534
fuel rods, 529
Fuji, Mt., *175*, *284*, 285, *285*, 293, 303
Fujita, T. T., 754
Fujita scale (Enhanced), *754*, **754**
Fukushima nuclear power plant, 349–50, *350*, 531
fungi:
    and soil formation, *195*
    in wind-blown dust, 792
Fungi (life-form kingdom), 425
fusion (nuclear), *24*, **24**, 26, 507, 542
future of the Earth, *870*, 873
    see also global warming

Ga (giga-annum), 7
gabbro, *46*, *52*, 93, *171*, *172*, *173*, 177, *256*
    metamorphism of, 244
    in oceanic crust, 51, 263
Gagarin, Yuri, 37
gaining stream, 620
Gal, 373
Galápagos finches, and Darwin's theory, 431
Galápagos Islands, Darwin's visit to, 431
galaxies, *15*, *17*, 21
galaxy, planetary systems in, *60*
galena (lead ore), 123, *123*, 129, 549, *549*
Galileo, 15, 373
Galileo spacecraft, 309
Galveston, Tex., Hurricane Ike in, *689*
Ganges Chasma, Mars, *613*

Ganges River, and delta plain, 688
garnet, *122*, *125*, 131, 136, 137, 235, *236*, 241, 246, *249*
    and metamorphism, *247*
gas, *120*, 121
    shale, 517, 519–20, *520*, **520**, 522
    volcanic, *see* volcanic gas
    see also natural gas
gas boom, in U.S., 520
gases, and plate interactions, 11
gas giant planets, *17*, **17**, *18*
gas hydrates, 521–22, **521**
gasification of coal, **528**
gasoline, 508–12, *510*, 513, 535
gas-thrust region, 289, 292
gas trap, *see* trap
Gastropoda (class), 427
gastropods, *429*, 483
Gauss polarity chron, 77, 78, *79*, 81
gelisol, 197, *198*
gems, 118, 134–38, **136**, *138*
    diamonds, 135, *135*, 137
        Hope Diamond, 134, 136, *136*
gemstone, 134–38
    cutting of, 136–37, *138*
general circulation models (GCM), 849
generating force, 671
genes, 425
genetics, 425, 431
genus, 426
Geobiology:
    and the biosphere
        biogeochemical cycles, 839, *840–41*, 846
        ecosystems, 859–61
    and the history of life
        Cambrian explosion, 450, 482–83
        evolution of life on Earth, 474–75, 483–86, 489–90, 493–94, 497–500, *498–99*, 843–44
        extinction events, 432–33, 441, 451, 486, 489, 855, 857–58
        fossil record, 428–30
        fossils, 418–33
        life diversifies, 482
        stomatolites, *475*, **475**
    and microbes
        in biogeochemical cycles, 846
        in desert varnish formation, 773
        in groundwater, 711–12
        in mineral formation, 126, 127
        in soil formation, 196
        in weathering, *189*, 191
geocentric model, **14**, *15*, 32
    and Galileo, 15
geochemical cycle, 265, 268
geochronology, **453**
    see also isotopic dating
geographical poles, 40, *40*, 67, 68
geoid, 372–73, *373*, **373**
geological biography, *see* Earth history
geological province map, *901*, *911*
geological structures, **381**
geologic column, 449–53, **449**, *450*, *451*
    adding numerical ages to, 460–64, *462*
    and numerical ages, 458, *460*

geologic contact, **445**
geologic cross section, **575**
geologic history, 438–42, 443, 444
    geologic principles of, *441*
    reconstruction of, *454–55*
geologic map, **216**, *446*, *449*
geologic structures, orientation of, *388*
geologic time, 6, **6**, *7*, 435–37, **435**, *462*, 463–64, *463*, *843*
    and fossil succession, 441
    picturing of, 463–64, *463*
    and principles of relative age, 438–42, *439*, *442*
    sea-level change in, *845*
geologic time scale, **461**
geologic units, and continental drift, 65
geologists, **2**
    and energy industry, 506
    and fossils, 419
    and landscape, 573–74
    in protecting against mass movements, 608
    in search for ores, 555
geology, **3**
    computers in, *4*
    plate tectonics as paradigm change in, 11, 87
    principal subdisciplines of, **4**
    reasons for studying, 3–5
geomorphologists, tools used by, 575
geophones, 515, *515*
geophysics, **360**
geoscience, 3, *4*
    1960s study of plate tectonics in, 87
geotechnical engineers, 388
geotherm, **156**, *366*, 367
geothermal energy, 507, *507*, 508, 531–32, *532*, 536–37, 540
geothermal gradients, *46*, **46**, 531
    of different crustal regions, 248
    and metamorphic environments, 249, *249*
geothermal regions, *711*, **711**, 712–13, *712*
geothermal resources, 507, *507*
Germany, fossils in, 425, *426*
geysers, 712–13, *712*, **712**
    of Yellowstone, 102
gibbsite, 190
Gilbert, G. K., 226, *226*
Gilbert polarity chron, 77, 78, *79*, 81
Giotto spacecraft, 39
glacial advance, *805*, **805**
glacial drift, **814**
glacial environments, 224
glacial erosion, landscapes formed by, *811*
glacial fjords, 676
glacial ice, *798*
    and evidence of climate change, 849, 851
glacial incorporation, 809
Glacial Lake Agassiz, 824, *824*
Glacial Lake Missoula, 645–46, *823*
glacially polished surfaces, **809**
glacial marine, 815
glacial outwash, **815**, *816*, *821*
glacial outwash plain, **818**
glacial plucking, 809–10
glacial rebound, 819, *822*

glacial retreat, *805*, **805**
glacial striations, *809*, **809**
glacial subsidence, *819*, 822
glacial till, 64, 210, 224, *225*, 479, **815**, *816*
glacial torrents, 646
Glaciated Peaks, *810*
glaciations, 497, **828**, *829*, *830*, 855
    carbon dioxide and, 864
    and continental drift hypothesis, 63–64, *65*
    future prospects for advent of, 833–35
    in North America, 827–29, *829*
    and plate tectonics, 831, 832
    of Pleistocene ice age, 826
    timing of, *830*
Glacier Bay National Park, Alaska, 827
glacier ice, 797
Glacier National Park, Mont., 827
glaciers, 497, 795–837, *795*, **796**, *820–21*
    in Alaska, 799, *800*, 816
        vanished ice of, *834*
        visit to, 826
    calving, 805
    and continental drift hypothesis, 63–64, *65*
    current opportunities to visit, 827
    diminishing of, 856
    in early Oligocene Epoch, 497
    erosion by, *394*, 405
    flow velocity of, *804*
    formation of, 797–99
    and hydrologic cycle, 819–22
    and icebergs, *807*
    on Iceland, *819*
    incorporation in, *810*
    and isostatic compensation, 408
    layering in, 458, *459*
    melting of, 800, 805, 819–22, *823*, *835*, *866*, *867*
    mountain erosion by, 408
    movement of, 801, *802*, *803*, *804*
    nature of, 797–813
    and sea-level changes (eustatic), *686*
    sedimentary deposits of, 814–15, *817*
    sediment carried by, 223, *813*, *814*
    shrinking of, *864*, *866*, *870*
    types of, 799, *800*
        continental (ice sheets), **799**
        dry-based, 801, *802*, *804*
        mountain, 797, *800*
        mountain ice caps, 799, *800*
        piedmont, 799, *800*
        polar, **799**, 801
        temperate, **799**
        valley, 799, *800*
        wet-based, 801, *802*, *804*
        see also specific types
    vanished ice in Alaskan glacier, *834*
    and volcanic activity, 306
    volcanoes under, 295, 299
    as water reservoir, 579, *581*, 819–22, *823*
    see also ice ages
glasses, 45, 46, **119**, *146*, *172*, 561
    and quenching, 252
    sand in, 557

glassy igneous rocks, **168**, 170, *170*, 172
    and cooling, 168
glassy rocks, 142–43, **142**
glaucophane, 248
global change, **839**
    global warming and, 870
global circulation, 739
global climate, 761–65
global climate change, 847–58, **847**, 854
    human intervention in, 846
global cooling, **848**
    from uplift, 853
global positioning satelite (GPS), *336*
    in seismograph calibration, 326
global positioning system (GPS), 110, *405*, **405**
    and Coordinated Universal Time, 436
    and glacial rebound, 819, *822*
    and plate motions, 111, *111*
    for volcano eruption prediction, 303
Global seismic-hazard Map, *352*
global warming, 539, 848–49, **848**, 865
    biological changes as evidence of, 864, 865–66
    climate belts and, 870, *871*
    consequences of, 870
    evidence of, 864–66
    flooding from, 870
    and greenhouse (hot-house) periods, 852, 853
    human impact on, 869
    hurricanes and, 870
    ice sheet melting from, 870
    Intergovernmental Panel on Climate Change (IPCC) and, 867–68, 869
    oceanic currents affected by, 870
    potential solutions for, 870
    recent, 862–72
    sea ice and, 808
    sea level change from, 870
    temperature anomalies in, *868*
    vegetation affected by, *871*
    see also greenhouse effect; greenhouse gases
Glomar Challenger (drilling ship), 82, 184, *184*
Glossopteris, fossils of, *65*, 203
gneiss, *236*, **241**, 244, 246, 252, *254*, 255, *255*
    ancient specimen of, *463*
    in Archean cratons, 473
    foliation of, 398
    in migmatite, 242
    in New York bedrock, 560
    as oldest whole rock, 473
gneissic banding, 241–42, *244*, 255
gneissic layering, 241
Gobi Desert, 772
gold, 131, 546
    of Inca Empire, 553–54
    as metallic mineral resource, 546
    as native metal, 547
    nuggets of, *548*
    panning of, 553, 554, *554*
    placer deposits of, 553
    as precious metal, 547, 549

Golden Gate Bridge, San Francisco, 210
Goldilocks effect, *854*
gold rush of 1849 in Sierra Nevada, 546, *546*
Goma, Congo, disaster in, 298, *300*
Gondwana, *482*, **482**, 484, 495, *496*
Google Earth, 203
"goosenecks," *638*, *638*
Gorda Ridge, *76*
gouge, 391
GPS, *see* global positioning system
graben, 393, *394*, *784*
GRACE satallites, 373, 714, *715*
grade, metamorphic, **245**, *247*, 248, 249, 255
graded bed, 219, *221*
graded stream, **628**
grade of ore, **549**
gradualism, **431**
grain, 45–46, 142, *143*
grain growth, 239
grain size, in classification of rocks, 145–46, *147*
Grand Canyon, 205, 216, 218, 628–29
    in correlation of strata, *452*
    fossiliferous sediment in, 482
    and Milo Limestone, 446
    and Monte Cristo Limestone, 446
    Powell's exploration of, 435, *435*
    rapids in, *631*
    and Redwall Limestone, 446
    sedimentary rock at base of, 451
    spring at, *707*
    and stratigraphic formation, 445–46, *447*, *448*
Grand Coulee Dam, 646
granite, *46*, 52, *52*, 143, *143*, 152, 167, 170, *170*, *171*, *172*, *173*, 245
    of Andes, 555
    in Archean cratons, 473
    architectural definition of, 557
    architectural use of, 167
    and bauxite deposits, 555
    and chemical weathering, 191, 194
    composition of, 549
    in desert, *784*
    hardness of, 167
    headstone of, *194*
    iron-oxide minerals in, *550*
    in Onawa Pluton, 250
    sheared and unsheared, *393*
    of Sierra Nevada, 808
    soil formed on, 196
    weathering of, *186*, 190
granofels, 244
granulite, 248, *248*
granulite-facies rocks, 248
Grapes of Wrath, The (Steinbeck), 791
graphite, 124, *125*, 135
    and oil window, *510*
graptolites, 427
Grasberg Mine, 555
grasses, 497
grasslands, 497
gravel, 206
    and bedding, *217*
    conglomerate from, 209
    and deltas, 226
    in desert, 774, *775*
    glacial sources of, 818

in icebergs, 808
    as nonmetallic mineral resource, 546
    in river, *226*
    from stream, *208*
graveyard:
    salt wedging in, *188*
    weathering in, 194
    see also headstone
gravimeter, **373**
gravitational energy, **577**
gravitational potential energy, **372**
gravitational spreading, *803*, 804
gravity (gravitational pull), **7**, 15, **15**, 16, *16*, 25, *30*, 47–48, *48*, 360, 372–75, *373*, *374–75*, 617, 669, *671*
    energy from, 507
    in formation of stars, 25
    and glacier movement, *803*, 804
    and groundwater depletion, 714
    and potential energy, 704
    and running water, 623
    and tides, *668*, 669, *669*, 670–71
gravity anomalies, *373*, **373**, *379*, **381**
    and isostasy, 374–75
gravity waves, 672
graywacke, 207, *210*, 473
    see also wacke
Great Britain, *see* England; Scotland; United Kingdom; Wales
Great Falls, Mont., flood damage in, *640*
Great Glen fault, Scotland, *321*
Great Lakes, 827
Great Missoula Flood, 645–46, *646*, 823
great oxygenation event, **480**
great oxygen event, 731
Great Plains, *828*
    desertification of, 791–92, *793*
    exposed rocks in, *383*
Great Recession (S-8), oil prices in, 535
Great Salt Lake, Utah, 212, 779, *780*, *824*, 825
Greek mythology, 117
Greek philosophers, cosmological views of, 14
greenhouse effect, 483, 490, 497, **539**, *848*
    and climate change, 856–57
    from uplift, 853
greenhouse gases, **539**, **730**, 846, 848–49, **848**, 854, 862–64
    carbon dioxide as, 480, 493, 730, 831, 846, 862–64
    in climate change, 848–49
    and global warming, 862–64, *868*
    methane as, 732
    and solar energy, 732
greenhouse (hot-house) periods, **850**
Greenland, 62, 473, *866*
    as below sea level, 819
    glacial melting in, *835*
    glaciers and glacial ice in, 306, 799, *801*
        icebergs from, 808
        ice cores from, 849
        visit to, 826
    ice sheet in, 864
    ice sheets in, 819

and Pangaea breakup, 495
    Viking settlements on, 855, *856*
    Wegener's final expedition to, 67
Greenland ice cap, 458
Greenland Shield, *258*
Green Monster (car), 212
"green revolution," 650, 713
greenschist, *247*, 248, *248*
greenstone, 256, 473
Greenwich Mean Time (GMT), 436
Grenville orogeny, 412, **477**
groins, 689, *689*
Gros Ventre slide, 604, *605*
groundmass, 168
ground moraine, 817
ground shaking, as earthquake damage, 339–41, *339*, *341*, *342*
groundwater, 44, **531**, *611*, 616, 617, 694–727, **695**
    for agriculture, 792
    and caves, 719–26
    contamination of, **717**, *718*, 860–61
    depletion of, 713–16, *715*, *717*
    and desert plants, 789
    extraction of, 708–10, *708*, *710*
    flow of, 703–4, *704*, 723
    and geothermal energy, 531–32, *532*
    global usage of, 713
    and hot springs or geysers, *711*, 712–13, *712*
    human intervention in, 716–19
    and hydrologic cycle, *581*, 582
    and hydrothermal fluids, 239–40
    and induced seismicity, 337–38
    in joints, 388
    in lithification, *206*
    lowering level of (protecting against mass movement), *611*
    as magma coolant, 172
    and nuclear waste storage, 531
    and oases, 709
    and permeability, 698–99, *698*
    rates of flow of, 704–5, *706*
    as reservoir, *579*
    residence time for, 582
    and secondary-enrichment deposits, 552, *562*
    sources of, 696–700
    into stream, 616
    in travertine formation, 213
    usage problems with, 713–19, *716*, *718*
    and water in atmosphere, 730
    and water table, 701–3, *701*
groundwater sapping, 511, 617
growlers, *807*, 808
growth bands, **458**
growth rings, *459*
    and climate change record, 458, 849, *851*
grunerite, 132
Guadeloupe, volcano on, 304
Guiana Shield, *258*
Guilin region of China, tower karst landscape in, *723*
Gulf Coast, 688
Gulf of Aden, *104*
Gulf of California, 713

Gulf of Mexico:
offshore drilling in, 540–41, *541*
sediment along coast of, 490
Gulf of Suez, *104*
Gulf Stream, *664, 762, 833, 855*
Gulf War, Kuwaiti oil wells set afire after, *509*
Gunnison River, *258*
Gunz glaciation, 828
gushers, *515*, 516
at Lakeview, Calif., *516*
Gutenberg, Beno, 365
guyot (flat-topped seamount), 72, *74, 103*, **683**, *685*
gymnosperms, 485, 489
gypsum, *127*, 131, *131*, 557, *563*
beneath Mediterranean Sea, 184
as evaporite, 212, *213*
as nonmetallic mineral resource, 546
on playa, *778*
gypsum board, 561, *563*
gyres, **665**

habitable zone, 43, **43**, 854
haboob, 776, *776*
Hadean Eon, 6, *7*, 449–50, *451*, 463, *470–72, 472, 473, 498*, 842
Hadley, George, 739
Hadley cells, 739, *740*, 741
hail, 750–51, **750**
Haiti, earthquakes in, 348–49, *349*
Haiyan, Typhoon, 758, 761
Hale-Bopp comet, *39*
Half Dome, 808, *809*
half-graben, 393
half-life, 453–56, **453**, *456*
halide, 129–31
halite (rock salt), *122*, 123, *123*, 124, 129, *130*, 146
beneath Mediterranean Sea, 184
as evaporite, 213, 561
on playa, 778, *780*
stability of, *190*
weathering of, 190
Halley's comet, 39, *39*
halocline, 662
hamada, 780
hand specimen, **148**, *150*
hanging valley, 629, *810*, *811*
in national parks, 827
in New Zealand, *631*
hanging wall, 316, 320
hanging-wall block, *390*, 391
hardness, of mineral, **127**
hard water, **706**
Hartley 2 comet, *39*
Hawaiian eruptive style, 285, *286*, *288*
Hawaiian Islands, 101, 102, *103*, 153, *175*, 178
basaltic lava flows in, *275*, 276, *277*
bathymetric map of slides of, *598*
coral reefs in, 682
digital elevation map (DEM) of, *574*
elevation model of Oahu, *574*
Kilauea volcano in, 285
landscape of, *574*
lava-flow destruction in, 298, *300*
and Pelé, 307

shield volcanoes in, 293, 295, *295*
slumps along margins of, *597, 598*
tsunami hits, 346
volcanic activity in, 153, *279*, 284, *284*, 285, *288*, *574*
as volcanic island, 72
volcanic lava in, *261*
volcanoes in, 863
Hawaiian seamounts, 102, 110
Hayabusa asteroid, 39
headland, erosion of, *632*
head scarp, 589, **591**, *602, 608, 609*
headstone:
granite, *194*
marble, *194*
see also gravestones
headward erosion, **617**, 629, 636, *643, 785*
headwaters, **615**
heat, 54, 733–35
of early Earth, 154–56
internal, 155, 265, 577
from radiation, 7, 507
vs. temperature, 733–35
heat capacity, 663
heat flow, **74**, 303
and sea-floor spreading, 74
heating, metamorphism due to, 236–37, *238*
heat transfer, 54
heat-transfer melting, 157, **157**
heavy oil (bitumen), 520
heliocentric model, *14*, *15*, 32
heliosphere, **37**, *38*
helium, 730
in aftermath of big bang, 25–26
hematite, 70, *128*, 190, *479*, *553*, 584
herbicides, 200
Herculaneum, 273–74, *273*
Hercynian orogen(y), 485, *485*
Hermit Shale (Grand Canyon), *217*
Herodotus, 419, 635
Hess, Harry, 62, 74–76, *75*, 87
heterosphere, *737*
hiatus, 443
hidden faults, 316
high-altitude westerlies, 742
high-grade rocks, 245
High Plains aquifer (Ogalla aquifer), *700*, 713
high-pressure system, 746
high-pressure zones, in climate, *762*, *763*
high-tech analytical equipment, for rock study, 149–50
Hillary, Sir Edmund, 380
Himalayan Mountains, 799
from continental collision, 105, *402, 403*, 495
and crustal root discovery, 406, 407, *407*
Mt. Everest in, 380
see also Everest, Mt.
hinge, **394**
Hiroshima atomic bomb, 51
energy released from, 331
history:
of Earth vs. humans, 463, 468
see also Earth history
histosol, 197, *198*
hogbacks, 783

Holmes, Arthur, 67
Holocene climatic maximum, 855
Holocene Epoch, **497**, 500, 826, *856*
Homeric age, 14
Homo (genus), 500
Homo erectus, 500
Homo neanderthalis, 500
Homo sapiens, 500, 827
arrival of, 463
successor to, 873
homosphere, 737
Honshu earthquake, Japan, 313–14, *343*, *350*
hoodoos, 782, *782*
Hooke, Robert, 419
Hope, Henry, 134
Hope Diamond, 134, 136, *136*
horizons (soils), 195–99
horizontal-motion seismograph, 325, *325*
horn, **810**, *821*
hornblende, 245, 248
hornfels, *243*, *248*, 250, *251*, *254*, *255*
in Onawa Pluton, 250, *251*
horst, 393, *394*
hospital waste, oceanwide drifting of, 690
hot deserts, 769
hot spots (hot-spot volcanoes), 100–102, **100**, *101, 108, 109*, 112, 175, 178, 263, 293, 295–96
in Archean Eon, 473, *474*
continental, 295–96
in Cretaceous Period, 492
deep-mantle plume model of, 102, *103*, 112, 178, 296
in formation of igneous rocks, 178
non-plume model of, 102
oceanic, 293, 295, *295*, 661, *685*
shallow-mantle plume model of, 102
hot-spot track, *102*, *103*
hot springs, *694*, 710–13, **710**, *711*
and life on "snowball Earth," 480
Hot Springs, Ark., 712
Houchins (Mammoth Cave discoverer), 719
Howard, Luke, 748
Hubbert's Peak, *538*, **538**
Hubble, Edwin, 21–22
Hubble Space Telescope, *12, 17*, 21, *26*
Hudson River, saltwater wedges in, 681
Hudson Valley, N.Y., *722*
human evolution, 500
see also evolution
human history:
and climate changes, 850, *851*
and Earth history, 463
Humbolt Current, *665*, 666, 771
humidity, relative, 737
humus, **195**, *198*, *199*
Hurricane Hugo, 756
Hurricane Ike, *689*
Hurricane Katrina, 635, 758–61, *759–60*
hurricanes, 688, 755–61, *758*
damage from, 757
erosion from, *577*
from global warming, 870
landscape damage from, 688

Saffir-Simpson scale for, 756, *757*
structure and distribution of, *756*
tracks of, *756*
wave amplitudes in, 672
Hurricane Sandy, 688
Hutton, James, 144, 234, 430, 431, 436–38, *442*, 462
Huxley, Thomas Henry, 468
Hwang (Yellow) River, China, 641
early civilization on, 650
flood of, 641
hyaloclastite, **173**, 281, 293
hydration, 190
hydraulic conductivity, 706
hydraulic fracturing (hydrofracturing), **517**, 520, **522**, *523*
hydraulic gradient (HG), *706*, **706**
hydraulic head, 704–5, **704**
hydrocarbon chains:
diversity of, *508*
and temperature, *510*
hydrocarbon generation, 509–10, **510**
hydrocarbon migration, 511–12
hydrocarbon reserve, **510**
hydrocarbons, 508–9, *508*, 539–42
alternative reserves of, 517–23, *521*
hydrocarbon seeps, 513, *513*
hydrocarbon system, *511*, **511**, 512, *512*
hydrochloric acid, 901
hydroelectric energy (power), 507, 532–33, *533*, 539–42
hydroelectric power plant, 532
hydrogen:
in atmosphere, 730
in Hadean atmosphere, 472
in Sun, 854
hydrogen atoms, in aftermath of big bang, 25–26
hydrogen bond, 120
hydrogen fuel cells, 508
hydrogen sulfide, 720
and groundwater, 706
hydrogeologists, 699
hydrogeology, 696
hydrographs, *652*
hydrologic cycle, 268, 573, 579–82, **579**, *580–81*, 584, 696, *840*, 846
glacial reservoir in, 819–22
runoff in, 615, *616*
hydrolysis, 190
hydrosphere, 43, **579**, *840*
hydrothermal deposit, 550–51, **550**, *551, 554, 555, 562*
hydrothermal fluids, 234
and metamorphic environments, 249, 250, 254
in metamorphic reactions, 239–40, 246, 256
hydrothermal metamorphism, **253**, *256*
hydrothermal (hot-water) vents, 293
and archaea development, 428
hydrovolcanic eruptive style, *286*
hypocenter (focus) of earthquake, 315, *316, 337*
hyporheic zone, 620, *621*
hypothesis, **8**
hypsometric curve, 45, **45**

ice, 797, *798*, 812
    carving and carrying by, 808–13
    glacial, 458, 797
    properties of, 797
    slipperiness of, 797
"ice" (cosmic), 18, 28
ice ages, 458, **796**, 850
    causes of, 831–35
    and fjords, 681
    and future, 873
    last (Pleistocene), 797, 813, 814,
        819–24, 826, 827–29
        and glaciers as reservoirs,
            819–22
        ice dam effects in, 822–24, *823*
        and North American
            continental shelf, *686*
    mega-floods of, 645–46, *646*
    and mountain glaciers, *821*
    in Proterozoic Eon, *479*, 480
    and Sahara Desert, 709
        and sand in New York
            concrete, 560
    and sea level, 812, 819–22, *821*,
        823, 827
    See also glaciers
iceberg alley, 808
icebergs, 806–8, **806**, *807*, *821*
    and Archimedes' principle of
        buoyancy, 90, *90*
icebreakers, 808
ice bubbles, and CO2 record, 849
ice-cap regions, *763*
ice caps, 799, 826
ice crystals, as mineral, 124
ice dams, 822
icehouse periods, 850–53, **850**, *852*,
    853, 855
    late Paleozoic, 483
Iceland, 102, *175*, 178, *711*, *712*, 713
    bathymetric map of, *296*
    cool climate from eruption in, 305
    geothermal energy in, 532
    geysers in, *712*
    glaciers and fjords of, *819*
    hot springs in, *711*
    lava-flow spraying in, 305, *305*
    and Surtsey, 288, 295
    volcanic activity on, 295, *296*, *300*
ice loading, 819
ice-margin lake, 818, 823
ice quakes, **805**
ice-rafted sediment, 808
ice sheets (continental glaciers), **799**,
    814–15, *820*, *821*, 822, 832
    of Antarctica, 799, 826
    consequences of melting of, 819
    as dams, 822–24
    equilibrium line on, 805
    erosion from, 812, *828*
    and global warming, 870
    of Greenland, 799, 826
    movement of, 801, 805
    in North America, *828*
    in Permian Period, 483
    Pleistocene, *497*, 826
ice shelves, **806**, *807*, *821*
ice streams, 805
ice tongue, 806, *807*
Ida (asteroid), *39*
Idaho Batholith, *168*
idealized sequence, 486

igneous activity, 174–79, **174**, 404,
    405
igneous intrusion:
    and principle of baked contacts,
        438
    and principle of inclusions, 438
    and rock cycle, 267
    See also plutons
    and thermal metamorphism, 250,
        250
igneous rocks, 45, 115, *146*, **146**,
    153–54, *153*, *261*, 698
    classifying of, 170–74, *172*
    and cooling time, 168
    in crust, 407
    crystalline (nonglassy), **167**, *170*,
        171, *172*
    describing and identifiying of,
        166–74, *170*
    extrusive, *154*, *155*
    formation of, 153, *171*
    and fossils, 420
    in geologic history illustration,
        438
    glassy, *168*, *170*
    and inclusions, *440*
    intrusive, *154*, *155*
    and isometric dating, 457, 461
    and lava flow, 153–54
    in nonconformity, 444
    ore deposits in, 553
    and orogeny, *404*, 407
    porosity of, 698
    in rock cycle, 262–65, *262*
        see also rock cycle
    settings for creation of, 174–79
    texture of, 166–69, *170*, *172*
ignimbrite, *285*, **292**
Iguazu (Iguaçu) Falls, *179*, *631*
Illinoian glaciation, 828, *830*
Illinois, *467*
    canyons in, *1*
    ice sheet effects in, 826
    loess in, *816*
    tornado in, 752, *753*
Illinois Basin, 228, 411
Illinois River, 823
    Valley, 646
illite, 117
inactive faults, 316
Inca Empire, 553–54
inceptisol, 197, *198*
incised meanders, 638, *638*
inclusions, principle of, 438, *440*
incorporation, in glaciers, *810*
independent (isolated) tetrahedra,
    131, *133*
index fossils, **441**
index minerals, **249**
India:
    basalt flows in, 858
    in continental collision, 105,
        402–3, *492*, 495
    and creation of Pannotia, 477, 480
    in Cretaceous Period, 490, *492*
    Deccan region of, 179, 297
    dung as fuel in, *505*
    earthquakes in, 335–36
    folklore of, on earthquakes, 315
    groundwater depletion in, *715*
    mountains in, 407
    tsunami in, 346

Indiana, *504*
    silt in, *192*
    tornado in, 752
Indian Ocean, and monsoons, 763–65
Indian Plate, *402*
Indians, American, *See* Native
    Americans
Indian Shield, *258*
Indonesia, 641
    flood in, 641
    and flooding, *640*
    gold in, 555
    Krakatau (Krakatoa), *175*, 290–91,
        301, 306
    land bridge to Australia from,
        822, 827
    Mt. Tambora in, 305
    volcanic activity in, *272*
induced seismicity, 337–38, **337**
Indus River, Pakistan, flooding of,
    641
industrialization, in global warming,
    862
industrial minerals, 118, **557**
industry, groundwater contamination
    from, 716
Indus Valley, 650
inequant grains, *146*, 238, *239*
inertia, and centrifugal force, 670
infiltration, 696
infrared radiation, 848
ingots, 557
inlet, 678
inner core, 52–53, *56*, *57*, **366**
inner planets, 18
InSAR (Interferometric Synthetic
    Aperture Radar), *322*, 323
inselberg, 783, *783*, *785*
insolation, 739, **739**, 831, *832*
intensity, of earthquake, 328–29, **328**,
    330
interference (waves), 672
interglacials, **828**, 834, 855
interglaicals, 497
Intergovernmental Panel on Climate
    Change (IPCC), 867–68,
    869
interior basin, 778, *779*
intermediate earthquakes, *332*, 333,
    334
intermediate-grade rocks, 245
intermediate magma, 158, *158*, *159*,
    164
intermediate rocks, 46, 159, *160*
    and crystalline rocks, *172*
internal energy, **577**
internal processes, 7
International Commission on
    Stratigraphy, *462*
International Space Station, *11*
intertidal zone, 667
intertropical convergence zone
    (ITCZ), 763
intracontinental basins, 228
intraplate earthquakes, 332, 337, **337**,
    338
intrusive contact, **162**, 166, *167*
intrusive environment, *171*
intrusive igneous rock, *154*, *155*, 162,
    166, *167*, *168*
    and magma chamber, 287
    and metamorphic rock, 234

intrusive realm, *155*
Io (moon of Jupiter), *308*, *309*, *585*
ionic bonds, *120*, 121
ionosphere, **737**
ions, *120*, *121*, **121**, 123, 124, *124*
Iran, Zagros Mountains of, 495
Iraq, dust storm in, 776
Ireland, *423*
    Cenozoic dikes in, *166*
    climate of, *762*
    erosion in, *193*
    glacial till in, *816*
    during last ice age, 826
    limestone pavement in, *724*
    sandstone cliffs in, *187*
    sea cliffs in, *396*
    unconsolidated sediment on coast
        of, *185*
    undeformed rock beds on sea cliff
        in, *383*
    wave disaster off coast of, 672
iridium, 494
iron, 45, *124*, 547–49, 553
    in basic metamorphic rocks, 244,
        245
    in biotite, 119
    consumption of, 564
    in crust, *53*
        continental crust, 549
    in groundwater, 706
    in magma, 158
    as metallic mineral resource, 546
    rusting of, 190
    in soils, 197, 198, *198*
    and weathering, 190, 191
iron alloy:
    in core, 52, 365, *842*
    and magnetic field, 67, *80*
    seismic wave speed through
        (molten), *361*
iron-nickel alloy, 52
iron ore:
    from BIF, 480
    iron-oxide percentage in, *550*
    smelting of, 548, *548*
iron oxide, 158
    in cement, 557
    in desert varnish, 773
    and marble, 244
irrigation:
    problem of source of, 792
    as soil misuse, 200
isobar, **738**
isograd, 249, *249*
isometric dating:
    and ages of younger glaciations,
        828
    and crater site, 495
    and Earth's formation, 470
    and growth of continents, 469
    uncertainty in, 469
isostasy (isostatic equilibrium), 374–
    75, *374*, **374**, *406*, 407,
    407, 408, 660, 661
    and uplift, 406
    see also buoyancy
isostatic equilibrium, **374**
isotherms, *250*, 762, **762**
isotope ratios, 424
isotopes, 121, **453**
    calculating isotopic data for,
        456–57

calculating radiometric data for, 453

half-life of, 453–56, *456*

as measure of past temperatures, 470, 849, *851*

isotopic dating, 76, 78, 435, 453–56, **453**, *455*, 456–58, *456*

carbon-14 dating, 457

and discovery of radioactivity, 462

of meteors and Moon rocks, 463

and Permian-Triassic extinction event, 857

and sedimentary rocks, 458, 460–61

isotopic signatures, 474

Israel, flash flood in desert in, *644*

Italy:

Amalfi coast of, *675*

cave in, *721*

marble from, 244

marble quarry in, *151*

Mt. Vesuvius in, *163*

rocky shore in, *675*

Vaiont Dam disaster in, 594

Ithaca, N.Y., vertical joints near, *387*

Izalco Colcano, 155

Jackson Hole, Wyo., Gros Ventre slide near, 604

jade, 137

Jamaica, Hurricane Sandy and, 761

Japan, *97*

earthquakes in, 49, 313–14, *314*, 334, 335, *335*, *340*, 341, *342*, 343, *343*, 346–50, *358*

folklore of, on earthquakes, 315

nuclear power in, 531

protecting against mass movements in, 610

tsunami in, 346–50, *350*, *358*

Japan Trench, *74*, *97*

Jaramillo normal subchron, 77

jasper, 210, *479*

Java, air liner near-crash from volcanic ash over, 299

Java (Sunda) Trench, *74*

jawless fish, 483

Jefferson, Thomas, 626

jet stream, **742**, *744*

and ash from Mt. Saint Helens, *290*, *291*

jetties, 689, *690*

jewels:

crown, *117*

see also gems

Johnson, Isaac, 559

Johnston, David, 290

Johnstown flood, 645, *645*

JOIDES Resolution (oceanic exploration ship), 657

joints, 186–87, **186**, *187*, 191, *287*, 387–89, *387*, **387**

in cave networks, 720, *721*

dissolved, 724

joint set, 388

jokulhlaupt, 295

Jones, Brian, 729–30, *729*

Joshua Tree National Monument, Calif., *177*

Journey to the Center of the Earth (Verne), 47

Juan de Fuca Plate, 89, *496*

Juan de Fuca Ridge, *76*, *94*

Jupiter, 15

in geocentric image, *15*

moons of, 309, *585*

Jurassic Park (movie), 425

Jurassic Period, *498*, 517

in correlation of strata, *452*

dinosaurs in, *490*, 494

Earth's appearance in, 839

North America in, *489* and Sierran arc, 490

Pangaea breakup in, 487, *489*

and sea floor, *95*

Sierran Arc initiated in, 490

Ka (kila-annum), 7

Kaibab Limestone, *218*, 446, *447*, 453, *453*

Kalahari Desert, 771

kame, 815, 817, *818*

Kansan glaciation, 828

Kansas, tornado in, *753*

Kant, Immanuel, 339

kaolinite, 190, 252

Karakoram Range, *403*

karat, 135

Karoo region, Africa, 179, 297

karst landscapes, 720–23, *722*, *725*

Kashmir earthquake, 335

katabatic wind, 815

Katrina, Hurricane, 758

Kazakstan, Aral Sea in, 790–91, *792*

Keeling, Charles, 863

Keene Valley, N.Y., slumping, 590

Keewatin ice sheet, 826

Kelvin, Lord William, 461–62

Kelvin scale, 54

Kepler, Johannes, 15

Kepler Space Telescope, 18

Kermandec Trench, *74*

kerogen, 212, *509*, **510**, *511*, 521

kerosene, 513

kettle holes, 817, *818*, *821*

K-feldspar, 134, 240, *247*

Kilauea, 275, 285

Kilimanjaro, Mt., *104*, *105*, *175*, *285*, 293

Kimberley Diamond Mine, 134, *134*

kimberlite, *135*

kimberlite pipes, 135

kinetic energy, 154, 733

King, Clarence, 769

kingdoms, 426

Kingston, N.Y., beds of limestone in, *211*

knob-and-kettle topography, 817, *818*

Kobe, Japan earthquake, 335, *335*, *340*

Kodiak, Alaska, tsunami damage of, 348

Koeppen, W., 764

komatiite, *172*

Krakatau (Krakatoa), *175*, *290*, 291, *291*, 301, 306, *331*

Kras Plateau, sinkholes in, *722*

K-T boundary event, 494–95, *494*, 857

Kuhn, Thomas, 87

Kuiper Belt, *37*, 38, *38*, 39

Kuril Trench, *74*

Kuwait, oil fields in, set ablaze after Gulf War, *509*

kyanite, *122*, *128*, 237, *238*, 241

Kyoto Accord (1997), 870

La Brea Tar Pits, Los Angeles, 422, *422*, *513*

laccoliths, **162**, *167*, *171*, *287*

La Conchita, California, *591*, 592, *592*

Laetoli, Tanzania, fossil footprints at, *420*

lag deposit, **776**, *777*, 783

lagoons, 226, *227*, 678

extraordinary fossils in, 425

lahars, 281, 299, *591*, *593*, **593**, *603*

in danger-assessment map for Mt. Rainier, *304*

from Mt. St. Helens, *593*

threat from, 299, *300*

lake environments, 225–26, *226*

lakes and lake beds:

between recessional moraines, 818

extraordinary fossils in, 425

glacial sediments in, 815

after ice age, 822–24

of ice age, 822–24

as local base levels, 627

and numerical age determinants, 458

pluvial, 824–25, *824*, **824**

residence time for, 582

in rift basins, 228

as water reservoir, 581

and water table, 701

Lake Tekapo, New Zealand, *635*

Lakeview, Calif, gusher at, *516*

Lake Vostok, 799

Lamesa, Tex., *516*

Lamotte Sandstone, *466*

land bridge, **497**

land bridges, 833

landfills, 578

landforms, **573**

coastal

of beaches and tidal flats, 676–78, *677*, *684*

coastal wetlands, *682*, 691

coral reefs, 682–83, *682*

estuaries, 679–81, *681*

fjords, 681, *681*

rocky coasts, *680*

depositional, **577**, 815–18, *817*

erosional, **577**

glacial, *817*, *818*

of Mars, 584

mass movements identified through, 608, *609*

of meandering stream, *633*

landmass, and climate change, 853

landscape, **572**

construction and, 592

factors controlling development of, 577–79

human activities, 578–79, 607

human modification of, 858–59

and hydrolic cycle, 571–85

karst, 720–23, *725*

and mass movement, 586, 588–98

of Moon, 582, *583*

of mountain range, 407

of other planets, 466, 582–85

Mars, 582–85, *584*

of sea floor, 657–61

shaping surface of, 574–77

landslide hazard maps, *610*

landslide-potential maps, **609**

landslides, 341, *341*, *586*, **588**, 594, 609

from earthquakes, 341, *341*, 343

from landscape modification, 859

and mountain scenery, *394*

Peru town covered by, 587–88, *587*, *594*

submarine, 559

Taiwan, *586*

as volcano threat, 301

land subsidence, 714–16, *717*

lanthanides, 564

Laozi (Chinese philosopher), 615

lapilli, **162**, 172, 173, *278*, *279*, *287*

as Pelé's tears, 307

pumice, 289, 292

threat from, 299, *300*

lapis, 136

Laramide orogeny, *490*, 491–92, **491**, *492*, 495

Laramide uplifts, *492*, *493*

large igneous provinces (LIPs), 178–79, *178*, **178**, 296–97, **296**

impact of eruption of, 178–79

Larsen B Ice Shelf, 864

Larsen Ice Shelf, 807

lasers, *364*

Las Vegas, Nev.:

erosion near, *775*

irrigation of, *790*

sedimentary rock hills near, *217*

sedimentary rocks near, 446

late heavy bombardment, 842

latent heat of condensation, **735**

lateral continuity, principlc of, 438, *440*

lateral moraines, **813**, *814*, *821*

laterite soil, *199*, **199**

latitude, 762

Laue, Max von, 122

Laurentia, *478*, *482*, *482*, **482**, 483, *484*, *484*

Laurentide ice sheet, 826, *828*, 833

lava, 46, 69, *146*, *152*, 153–54, **153**, *155*

basaltic *see* basaltic lava

cooling of, 154, 168

movement of, 159–60

in tuff, 173

viscosity of, *160*

lava dome, 276

lava flows, **153**, *171*, 275–76, *275*, *277*, *286*, 309, *439*

andesitic, 275

basaltic, 276, *277*

composition of, 170, 173

diverting of, 304–5

fossil tree trunks in, 420

magnetic reversals recorded in, *81*

rhyolitic, 275

structures within, *278*

threat from, 298–99, *300*

lava fountains, 153, *285*, 288, *289*

lava lakes, 153, 285

lava spire, 276

lava tube, **276**, *277*

layering of rocks, *147*, *148*, 383

See also bedding; foliation

leaching, 195

and climate, 196

mineral deposits formed by, *554*

lead, *549*
  as base metal, 549
  in dolomite beds, 552
legends, global climate change in, 852
Lehmann, Inge, 366
Leonardo da Vinci, 405, 419, 662
levees, *648*
  artificial, 646, 647, *647*
  natural, *633*, **634**
Lewis, Meriwether, 626, 629
Lewison Gneiss, *233*
Libya, hottest temperature in, 770
life:
  first appearance of, 474–75
  and Goldilocks effect, *854*
  origin of, 474–75
life forms:
  and carbon cycle, 846
  classification of, 425–27
  in desert, 789, *790*
  diversity of environments in
    support of, 468
  evolution of, 430–33, 843–44, *843*
    See also evolution
  extinction of, 432–33
  during Pleistocene ice ages, 827
life processes or activity:
  and landscape development, 578
  and physical aspects of Earth
    system, 7
light:
  backscattered, 734
  Doppler effect for, 21, *22*
light energy, 21
lightning, 751–52, *751*
lightning bolt, *751*, **751**
lightning rods, 752
light rays, reflection and refraction
    of, *362*
light year, **19**
lignite, 525, *526*
limbs, **394**
lime, in cement, 557, *563*
limestone, 210, *211*, *236*, *260*, 446,
    557, *563*, 719, 720–23
  biochemical, **210**
  calcareous metamorphic rocks
    from, 245
  and caves, 719, 720, 722, *725*
  in cement, 557, *563*
  chalk as, 468
  and coral, 682
  and dolostone, 213–14
  fossiliferous, 210, *223*, 226, 235,
    236, *237*, *467*, 483, 484
  fossils in, *426*
  in geologic history illustrations,
    441
  and groundwater, 706
  at K-T boundary, 494
  limestone columns, 720, *724*, *725*
  and marble, 255
  at Midwest meteorite impact
    site, 8
  and Monte Toc landslide, 594
  natural, 559
  as prolith, *246*
  and replacement chert, 213, *215*
  shatter cones in, *9*
  tilted bed of, *211*
linear structures, *388–89*
Linnaeus, Carolus, 425
lipids, 509, 532

liquefaction, **341**, *342*, **604**
liquid, *120*, 121
liquidus, *156*
Lisbon, Portugal earthquake, 338–39
lithic clasts, 206
lithification, *202*, *206*, **206**, *208*, 209,
    209, 210
lithium, in salt, *561*, 564, 565
lithologic correlation, 446, *448*
lithosphere, 53–55, **53**, *55*, 56, *57*, 75,
    87, **88**, 102, 112, 132, 156,
    175, 177, 178, 374, *840*
  and accretionary prism, *98*
  aging of, *96*
  in Archean Eon, 473
  of Asia and India, *402*
  and collision, *106*
  in continental rifting, *104*
  and earthquakes, 333, 334, 337,
    337
  in glacial loading and rebound,
    819
  and isostacy, 406
  and mantle plume, 296, *298*
  at mid-ocean ridge, *93*
  nature and behavior of, *88*
  plates of, 62, 87–89, *92*
    See also plate tectonics
  and ridge-push force, 106–7
  and sedimentary basins, 228
  and slab-pull force, 107
  and subduction, *96*
  and subsidence, *574*
  thinning and heating of, 408
  and uplift, *574*
  see also continental lithosphere;
    oceanic lithosphere
lithosphere plates, **89**
  see also plates
lithospheric mantle, **53**, 88, *88*, *93*,
    108, *109*
  in Asia-India collision, *402–3*
  and crustal root, 406, *407*
  diamond in, 135
  formation of at mid-ocean ridge,
    95
  and mid-ocean ridges, 95
  removal of, 407–8
little ice age, *834*, **834**, *855*, *856*
Lituya Bay, Alaska, landslide, 601
loading, *574*
loam, **197**
local-scale loads, 408
lodestone, 67
lodgment till, 817
loess, **779**, *815*
Loma Prieta, California, earthquake,
    333, *334*, 354
London:
  in future of Earth, 873
  as threatened by sea-level rise, 835
Longfellow, Henry Wadsworth, 572,
    579
Long Island, N.Y., as terminal
    moraine, 815, *817*
longitudinal dunes, 788, *788*, 789
longitudinal profile, **626**, *627*
longshore current, *674*, **674**, *678*
longshore drift, **674**
long-term climate change, 847,
    850–55
long-term predictions of earthquakes,
    351–52

Los Angeles:
  Colorado River water diverted
    to, 650
  La Brea Tar Pits in, 422, *422*, *513*
losing stream, 620
Louisiana, offshore drilling near, 540
Louisiana Territory, 626
Louis IV (king of France), and Hope
    Diamond, 134
Lowell, Percival, 584
lower mantle, *52*, **52**, **364**
low-grade rocks, 245
low-pressure system, 745
low-pressure zones, in climate, *762*,
    763
low-sulfur coal, 539
low-velocity zone (LVZ), 55, **363**
luster (mineral), **127**
  metallic, 127
  nonmetallic, 127
L-waves (Love waves), 323, *324*, 326,
    339, *339*, 361
Lyell, Charles, 430, 431, 438, 462
Lystrosaurus, 65

Ma (mega-annum), 7
macerals, 212, 524
macrofossils, **424**
Madison Canyon, Mont., landslide
    in, 601
mafic lava, *173*, 276
  and viscosity, 160
mafic magma, 158, *158*, *159*, 161,
    164, 176
mafic minerals, 244, 245, *245*
  stability of, *190*, 242
mafic rocks, 46, 51, 88, 160, *160*
  as crystalline rocks, 170, *171*, 176
Magellan, Ferdinand, 657
magma, 46, 92–94, **153**
  basaltic, *157*, 176, 177, 285–88
  composition of, 158–59
  cooling of, 160–61, *167*, **167**, 170,
    171, 176
  Earth's surface as (Hadean Eon),
    471–72, *472*
  felsic, *164*, 176
  formation of, 154–57
  and hydrothermal fluids, 240
  intermediate, 158, *158*, *159*, 164
  mafic, 158, *158*, *159*, 160, *164*,
    178
  major types of, 158–59, 162
  at mid-ocean ridge axis, 74, *75*
  and migmatite, 242
  movement of, 162, 163
  of Mt. St. Helens, *291*
  rhyolitic, *157*, 295, 296
  ultramafic, 158, *158*, **159**
  viscosity of, 160
  volcanic gas in, 281
magma chamber, 93, **154**, 177, *282*,
    282, *283*, 286, *287*
  and hot springs, 711
  massive-sulfide deposit in, *551*,
    552, *562*
magma mixing, *159*, **159**
magmatic deposit, **550**, 554
magnesium, *53*, *124*
  in basic metamorphic rocks, 245
  in biotite, 119
  in magma, 158
  and weathering, 191

magnesium oxide, 158
magnetic anomalies, 76, *76*, 79, 82,
    376
  and plate movement, 107
magnetic-anomaly map, ore bodies
    shown on, *556*
magnetic declination, **67**, 68, *68*,
    71, *71*
  maps of, 901, *908*
magnetic dipole, **67**
magnetic field, 40–41, *40*, **40**, 360,
    375–77, *376*
  of Earth, 40–41, 67–68, *68*, 79, *79*
  as generated by rock flow, 53
  reversal of, 77–78, *77*
    see also magnetic reversals
magnetic field lines, *40*, **40**, 68, 375
magnetic force, lines of, *57*
magnetic inclination, **68**, 69, *70*,
    71, *71*
magnetic poles, 40, *40*, *57*, **67**, *68*, **375**
magnetic-reversal chronology, 77,
    78, **78**
magnetic reversals, 77–78, *77*, **78**
  lava flows as recording, *81*
  and marine magnetic anomalies,
    80–81
  and sea-floor spreading, 78, 79
magnetism, *16*, **16**
magnetite, 67, 70, 78
magnetization, *376*
magnetometer, 76, *76*
magnetosphere, *40*, **40**
magnetostratigraphy, 458, *460*
magnitude, of earthquakes, *316*,
    330–31, **330**, *331*, 348
  and Richter scale, 330
Mahomet aquifer, 699
Maimi, tornado in, *753*
main bedding, *220*
Maine:
  Acadia National Park in, 812, 827
  fjords of, 681
  Onawa Pluton in, 250, *251*
mainshock (earthquake), 319
malachite, *118*, 136, 549, 552
malleability, 547
mammals, 500
  appearance of, 451
  in desert, 789
  development of, 494
  diversification of, 500
  earliest ancestors of, 490
  extinction of many species of, 860
  forerunners of, 485
  in history of Earth, 463–64
  huge, 500
  during Pleistocene Epoch, 827,
    829
Mammoth Cave, discovery of, 719
Mammoth Hot Springs, Yellowstone,
    213, *214*, 711
Mammoth reverse subchron, 77
mammoths, as fossils, 422, *422*, 432
manganese:
  on ocean floor, 553
  as strategic, 565
  supply of, 565
manganese nodules, **553**
  on ocean floor, 553, *553*
mangrove swamps, 676, 682, *684*, 688
Manicouagen Crater, Quebec, 33
Manitoba, Canada, 823, *825*

mantle, 32, *46*, 48, 51–52, **51**, *52*, 56, *56*, *57*, 58, *108*, 360–61, 367
  convective flow in, 367, *369*
  defining structure of, 363–65, *364*
  diamonds formed in, 135
  early history of, 471
  and rock cycle, 262, 265
mantle plume, **102**, *103*, *175*, 178, 296, *298*
  and D" zone, 369
  in early Earth history, 843
  in Hawaiian Islands, *103*
  and hot spots, 102, *103*, 112, *175*, 178, 296
  and Siberian basalt, 857
  superplumes, **179**, *492*, 857
maps and charts, 901–14
marble, 143, *151*, 194, **244**, *246*, 255, 282, 283
  architectural definition of, 557
Marble Canyon, *453*
Marcellus Gas Play, 520
Marcellus Shale, 520
marginal sea (back-arc basins), 98, *98*
maria, **309**
Mariana arc, 175
Mariana Trench, *74*, 660
maria of the Moon, *308*, 309
marine biology, 657
marine geology, 657
marine life, Paleozoic, *483*
marine magnetic anomalies, 76–82, **77**, *80*
  and Earth history, 469–70, *471*
  and sea-floor spreading, 76–83
marine oil spill, 539, *539*
marine sedimentary environments, 226
  delta deposits, 226, *227*
marker bed, **446**
Mars, 15, 18, *40*, 583, 839
  atmosphere of, *40*
  Earth contrasted with, 7
  in geocentric image, *15*
  landscape of, 582–85, *583*, *584*, *613*
  layers of, *56*
  material from to Earth, 50
  Olympus Mons on, *409*
  polar ice caps on, 799, *802*
  surface of, 777, *778*
  temperature of, *854*
  volcano on, *308*, 309
  and water, *584*
Marshall, James, 546
Marshall Islands, 682
marshes, 681
Mars rovers, 584
Martin, John, *47*
Martinique, lava flow on, *299*
Maskelyne, Nevil, *47*
mass, **13**, *25*
Massachusetts, *3*
  see Cape Cod
mass extinction event, 431, 432–33, *432*, **432**, *857*, *858*
  and K-T boundary event, 494–95, *494*
  during late Paleozoic, 486
  Permian, 486, 489
  and punctuated equilibrium, 432
massive-sulfide deposit, **550**, *551*

mass movement (mass wasting), **588**, 598–606, *602–3*, 609
  factors in classification of, 588
  and plate tectonics, 607
  protecting against, 608–10, *611*
  safety-structures as protection from, 610
  settings of, 600–601, *602–3*, *606*, 609
  submarine, 596–98, *598*
  types of, 588–98
    avalanches, 594–95, *596*
    creep, 588, *589*
    debris falls, 595–96
    debris flows, 591, *591*
    mudflows, **588**
    rockfalls, 595–96, *597*
    rock glaciers, **588**
    slumping, 590–91, *590*
    solifluction, 588
  see also specific types
mass spectrometers, 149, *457*
mass-transfer cycle, **265**
master beds, 220
mastodons, 432
matrix, of rock, 241
matter, **13**
  four states of, *120*
matter, nature of, 24–25
Matterhorn, 810
Matuyama polarity chron, 77, 78, *79*, 81
Mauna Loa caldera, *295*
meander, **630**, *633*, 638, *638*, *642*
  antecedent streams as, *639*, *642–43*
meandering stream, 630–34, **630**
meander neck, 631
mechanical force, 16, *16*
mechanical weathering, 186–89
  see also physical weathering
medial moraine, **813**, *814*, 827
Medieval Warm Period, 855
Mediterranean climates, *763*
Mediterranean Sea, 184, *184*, 212
melt (molten rock), 153–59, *156*
  composition of, 164
  freezing of, 153
  in mantle, 363
  in partial melting, 158
meltdown (nuclear), **530**
melting (of rock), 153
  decompression, *156*, **157**
  heat-transfer, **157**
  partial, 158
melting curve, 367
melting temperature, 126
melts (molten material in general), 46
meltwater lakes, 815
Mendel, Gregor, 431
Mendelév, Dmitri, 24
Merapi, Mt., Indonesia, volcanic eruption of, *292*
Mercalli, Giuseppe, 328
Mercalli Intensity Scale, 328–29, 348, *349*
Mercury, 15, 18, *40*
  atmosphere of, *40*
  in geocentric image, *15*
  layers of, *56*
mercury (metal), 547, 549
mesas, **782**
Mesa Verde, Colo., 204, *205*

Mesosaurus, 65
mesosphere, *42*, 43, **737**
Mesozoic Era, 7, *7*, 63, 65, 450, 480, *498–99*
  Africa-South America highlands during, 636, *638*
  as Age of Dinosaurs, 451
  batholiths in, *168*
  climate of, 853
  in correlation of strata, *452*
  early and middle, 487–90
    life forms in, 489–90
  early and middle, paleography of, 487–89
  granite of from Sierra Nevada, 808
  late, 487–95
  late, evolution in, 493–94
  late, paleography of, 487–89
  mass extinction during, 493–94, *494*
  North America in, 490
  ore in plutons of, 554
  rifting events in, 135
  and Southern California accretionary prism, 607
Mesozoic rocks, 410, *788*
Messel, Germany, fossils near, 425, *426*
metaconglomerate, **241**, *243*
  flattened-clast conglomerate as, 241, *243*
metallic bonds (bonding), 547
metallic luster, 127
metallic mineral resources, 546, 565
  and ore, 549–55, *554*
metallurgy, 547
metals, 46, **547**
  base and precious, 549
  consumption of, *565*
  description of, 547–49, *547*, *548*
  discovery of, 547–49
  native, 131
metamorphic aureole, *250*, 251
metamorphic conditions, 234
metamorphic differentiation, 242, *244*
metamorphic facies, 245, 248, **248**
metamorphic foliation, *147*, *148*, 235, **235**, 241–42, 255, 260, 401
  and failure surfaces, 601, *601*
  in flattened-clast conglomerate, 243
metamorphic grade, **245**, *247*, 248, 255, *255*
metamorphic mineral assemblages, 235, 245
metamorphic minerals, **235**
metamorphic reaction (neocrystallization), 235
metamorphic rocks, 45, 115, **144**, *146*, **234**, 235–36, *236*, *238*, *698*
  chemical composition in classifying of, 245
  classifying of, 245–46, 249
  determining age of, *455*
  exhumation of, *257*
  failure surfaces in, *601*
  formation temperature of, 236–37
  glacier ice as, 797
  and isometric date, 457
  location of, 256–57

  locations of, *258*
  in nonconformity, 444
  and orogen, 382, *404*
  plastic deformation of, 801
  porosity of, 698
  in rock cycle, *261*, 262–65, *262*
  in shield areas, 410
  types of, 241–45
    foliated, 241–42, *243*, 255; *see also* metamorphic foliation
    nonfoliated, 243–45, 254
  in volcano, 287
metamorphic texture, **235**
metamorphic zones, *249*, **249**
metamorphism, 229, 233–60, **234**, *267*, 398, 405, *510*
  causes of, 236–40
  consequences and causes of, 235–40
  contact, 250, 255, *255*, 266, 267, *287*, 401, *404*
  environments of, 249–58, *254–55*
  and fossils, 420
  intensity of, 245–49, *247*, 249
  mountain belt, *247*
  and plate tectonics, 234
  regional (dynamothermal), **252**, 255, 267
  as seen through microscope, *238*
metasandstone, 247
metasomatism, **240**
Meteor Crater, Ariz., 32
meteoric water, 622
meteorite impact, 500
  in American Midwest, 8
meteorites, 8, 29, 32, 39, *50*, **50**, 253
  and age of Earth, 473
  Earth bombarded by, 29, 253, 473, 500
    at Midwest cornfield, 8
  on glacier, in Antarctic ice, 806
  iron from, 547
  and mass extinction, *858*
meteoroid, 50
meteors, 50
  during Hadean Eon, *472*
  isometric dating of, 463
  mass extinction from, *494*, 494
  scars from on Mars, 582
meteor showers, *50*
methane, 472, 477, 521–22, *864*
  in atmosphere, 730, 846, *864*
  coalbed, **528**
  in explosions, 528
metric conversion chart, 901, *912*
Mexico:
  earthquakes in, 354, 355
  underground pool in, *724*
Miami:
  in future of Earth, 873
  as threatened by sea-level rise, 835
mica, *130*, 134, 190, 191, 235
  from clay, *241*, *243*
  in gneiss, 241
  and metamorphism, 247
  in phyllite, 241, *243*, 246
  in schist, 255
mica schist, 241, *243*
Michelangelo, 244, *246*
Michigan:
  Iron Ranges of, *479*
  migmatite outcrop in, *245*
Michigan Basin, 228

micrite, 210, *211*
microbes:
    hydrocarbon-eating, 512, 520
    oil-eating, 540
    in travertine formation, 213
    weathering process supported
      by, 191
    in wind-blown dust, 792
microcontinents, 401
microfossils, *424*, **424**
micrograph, *460*
microplates, 89
microscopes:
    in mineral identification, 118
    petrographic, **148**, *150*
Mid-Atlantic Ridge, 74, 79–82, 92,
    111
    and Iceland, 102, 295, *296*, 713
    and Pangaea breakup, 495
Middle Ages, 14
    Medieval Warm Period in, 855
Middle East, oil reserves in, 517, *519*
Mid-Indian Ocean Ridge, 100
mid-ocean ridges, **45**, **72**, *73*, 92–94,
    93, 100–101, *104*, 105,
    107, *108–9*, 178, 658, 659,
    661, *684*
    bathymetry of, *93*
    as divergent boundary, *93*, 101
    earthquake distribution at, *333*
    and earthquakes, 74–75
    earthquakes at, 333, *333*
    formation of igneous rocks at, 178
    formation of lithospheric mantle
      at, 95
    formation of oceanic crust at,
      92–95
    and fracture zones, 98
    hydrothermal metamorphism at,
      235, 252–53
    magnetic anomalies near, 78, *79*
    and Pangaea breakup, 492
    and plate tectonics, 107
    and ridge-push force, 106–7, *107*
    and rifting, 105
    rise of, 95, *96*
    and rise of heat, 74
    and sea-floor spreading, *75*, *93*,
      94–95
    and sea-level changes, *487*, 686
    volcanic activity at, 176–77, 293,
      295
      hot spots, 100–102, *101*, 102
      plate-boundary volcanoes,
        100–102
Midway Island, 102
Midwestern United States:
    aquifers for, 699–700, *700*
    Archean rocks under, 477
    climate of, *763*
    Dust Bowl of, 791–92, *793*
    floodplain flood in (2011), *640*,
      641–42, 647
    ice-age results in, 819–24
    meteorite impact in, 8
    outcrops rare in, 144
    plutons in, 477
    and tornadoes, 754
migmatite, **242**, *245*, *254*, 255, *255*
migrate, **512**
migration (human), and Pleistocene
    ice age, 497, 822, 827

migration pathway (hydrocarbons),
    512
Milankovic, Milutin, 831
Milankovitch cycles, 831–32, **831**,
    832, 833, 855, 862
Milford Sound, New Zealand,
    waterfall in, *631*
Milky Way, *17*, 19, 33, *35*
Milo Limestone, 446
Milton, John, 47, *47*
Mindel glaciation, 828
mine, 555–57
    deepest in world, 47, 360
mine disasters, rescue of Chilean
    miners in, 558, *558*
mineral assemblage, in metamorphic
    rocks, 248
mineral classes, 129–31, **129**
mineralogist, **117**
mineralogy, **118**
mineral resources, 545–69, **546**
    formation and processing of,
      562–63
    industrial countries consumption
      of, 564
    metallic, 554, 565
    nonmetallic, 546, 557–64, *562*,
      565
    as nonrenewable, 565–66
minerals, 45, 116–51, **118**
    biogenic, **118**
    and Bowen's reaction series,
      164–65
    classification of, 129–37, 145–47
    criteria for, 118–21
    as crystals, 118
    felsic, *190*
    formation and destruction of,
      124–27
    as fossils, 423
    gems, 118, 134–38
    geologic, 118
    in groundwater, 705
    in hot springs, 711
    ice as, 797
    identification flowcharts of, 901,
      903–5
    identification of, 118–19
    industrial, 118
    inorganic, 119
    mafic, *190*
    naming of, 117
    naturally occurring, 118
    ore, 118
    and paleomagnetism, 67
    physical properties of, 127–29, *128*
    on playas, 778
    properties chart of, *906–7*
    relative stability of, *190*
    rock as aggregate of, 142
    rock-forming, 131–34
    silicate, 46, 131, 143
    solid, 118–19
    special properties of, 129
    synthetic, 118
mineral specimen, 142
mining:
    for cement ingredients, 557
    of coal
      strip, 526, *527*
      underground, 527–28, *527*
    dangers of, 557

    and environment, 566, 578
    as landscape modification, 859
    new ways of, 567
    for ore minerals, 555–57, *562*
      open-pit, 556, 560, *562*, 566
      underground, 555–56, *562*, 566
      waste rock from, 566
    see also mine
Minnesota, 822
    ice-age lakes of, 822
    Voyageurs National Park in, 827
Minoan people, 306, *307*
Miocene Epoch:
    and Antarctic ice sheet, 832
    Late (climate), 497
    and sea floor, 95
    and sea level, *845*
mirage, 770, *770*
Mississippian Period, in correlation of
    strata, 452
Mississippi Delta, 635, *636*
Mississippi drainage basin, *619*
Mississippi-Missouri network, 822
Mississippi River, 296
    at Cape Girardeau, *647*
    discharge of, 622
    drainage basin of, *619*
    flooding of, 641–44, *641*, 646–47,
      647
    and Mark Twain, 631, 646
    as meander, 631
    Mississippi River Flood Control
      Act (1927), 646
    peak annual discharge of, 651
    sediment load of, 565, 624
Mississippi Valley, 337
Mississippi Valley-type (MVT) ores,
    552, **552**
Missouri, 411, *466*
Missouri, tornado in, 752–53
Missouri River, flooding of, *640*,
    641–44, 646–47
mixture, 121
Modified Mercalli Scale, 328–29,
    328, *329*
Moenkopi Formation, 453
Moho, **51**, *98*, *157*, *363*, **363**, 367
Mohorovicic, Andrija, 51, 363
Mohs, Friedrich, 127
Mohs hardness scale, 127–28, *127*, **127**
Mojave Desert, *167*, 773, *773*
    weathering in, *194*
mold of fossils, **423**
mold of fossil shell, *422*
molecule, 121, **121**
molecules, 24
mollisol, 197, *198*
mollusks, 483
moment magnitude (Mw) scale, **330**
monoclines, **394**, *395*, *398*, 411
Mono Lake, California, 213, *214*
monsoons, 763–65, *763*, **763**, *764*
    floods from, 641
Montana:
    Glaciated Peaks in, *810*
    Glacier National Park in, 827
    Madison Canyon landslide in, 601
Monte Cristo Limestone, 446, *450*
Monte Toc, rockslide from, 594
Montréal, Canada:
    earthquake in, 337
    during last ice age, 797

Mont St. Michel, France, *669*
Monument Valley, Ariz., *573*, 782, *782*
Moon, 13, *13*, *14*, 18, 37, 839, 842
    changing distance from Earth
      of, *668*
    cratering on, 472
    Earth's distance to, 19
    formation of, 29–32, 471, 854
    forming of, *30–31*
    in geocentric image, *15*
    knowledge of (vs. knowledge of
      ocean), 657
    lack of change in, 839
    landscape of, 582, *583*
    layers of, *56*
    material from to Earth, 50
    shock metamorphism on, 253
    and tides, 507, 669, 670, *671*
    volcanic activity on, *308*
moon rock, isometric dating of, 463
moons, *18*
moraine, **813**, 815
    lithium, 564, *565*
    setting of, 815, *817*, 818
morphology, 426
mortar, 557, *559*
Mosaic Canyon, *260*, *775*
mother lode, 553
Mouna Kea volcano, 863
mountain belts or ranges (orogens),
    381, 405
    accretionary, 401, *478*
    collisional, 105, *106*, *250*, 401,
      402–3, *478*, 482
    collisional, Himalayas as, *402–3*
    crustal roots of, 401
    digital map of, *380–81*
    identifying of, 469
    life story of (Appalachians),
      412–13
    metamorphic rocks in, 256
    topography of, 405–9, *470*
    see also uplift; volcanoes
mountain building, *see* orogeny
mountain (alpine) glaciers, 797, **799**,
    800, 810
    visit to, 827
mountains, *44*
    erosion of, 408
    fascination of, 380–81
    height of, 406, 407–8
    mountain stream environments,
      224
mouth, **615**
Mt. Erebus, *see* Erebus, Mt.
Mt. Everest, 409
Mt. Everest, *see* Everest, Mt.
Mt. St. Helens, *see* St. Helens, Mt.
Mt. Vesuvius, Italy, *163*
mud, *187*, 225
mud cracks, **219**, *221*, 224
    on floodplain, 224
    and uniformitarianism, *439*
mudflow, *286*, *559*, **591**, 593
    at La Conchita beach, *591*, 592,
      592
    at Oso, Wash., 593, *593*, 604
    in Peru landslide, 587, *594*
    as volcano threat, 299
mud pots, *711*
mudslides, 604, *604*
    see also mudflow

mudstone, *207, 208,* **210,** *226*
dinosaur footprints in, *423*
tillites in, 831
Muir, John, 380, 405
Muir Glacier, *867*
muscovite, *146, 165,* 255
and igneous intrusion, *251*
and metamorphism, 240, 245, 246
in Onawa Pluton, *251*
stability of, *190*
museums, fossils in, *419, 421, 421*
Mycenaeans, 307
mylonite, **250,** 251, *253, 254,* 255,
391, *393*

Namazu, 315
Namibia, Dead Vlei of, *791*
NAPL (non-aqueous phase liquids),
716
Na-plagioclase, *190*
Nashville, Tenn., flooding in, 649
Nassar, Lake, 709
national parks, of U.S., *452*
National Weather Service, U.S.,
tornado warnings by, U.S.,
752
Native Americans:
desert-varnished rock used by, *774*
folklore of, on earthquakes, 315
on Mesa Verde, Colo., *204, 205*
Onondaga, 214
stratigraphic sequences named
after, *487*
native metals, 131, **547**
natural bridge, 722, *724, 725*
Natural Bridge, Va., *725*
natural gas, 519
migration of, 512
world supply of, 538–39
natural hazard, 586, **588**
see also earthquakes; floods;
mass movement; storms;
volcanic eruptions
natural levees, *633,* **634,** 640
natural selection, **431,** 462
Neah Bay, Wash., GPS measurements
from, *336*
Neanderthal man, 500
neap tides, *668, 669*
near-field tsunami, 346
Nebraskan glaciation, 828
nebulae, *12,* **25,** *26*
nebular theory, **28,** 154
negative anomalies, 76–77, *76,* 78–79,
79
negative feedback, 849
neocrystallization, 235, 236, 237, 245
Neogene Period, 499
Neptune, 15, 38
Neptunists, 144
Netherlands, little ice age in, 834,
834, 855
Netherlands, sea level rise threat to,
688
neutrons, 25, 121
Nevada, 202
Yucca Mountain in, 531, 717
Nevado del Ruiz, 299, *300*
Nevado Huascarán Mountain,
landslide on, 587, *587*
New England, metamorphic zones
and isograds in, *249*

Newfoundland, submarine slide along
coast of, 559
New Guinea, 495
New Jersey, Hurricane Sandy and, 761
New Jersey, submarine canyon off, *660*
New Madrid, Mo., earthquakes, 337,
338
New Mexico:
boulders in, *192*
Shiprock volcano in, *166, 303*
volcanic evidence in, *163*
weathered sedimentary rock in,
194
New Orleans, La., 635, 758–61,
759–60
Newton, Sir Isaac, 15, 20, 25, 372,
386, 436
New York City:
Central Park in, 826
concrete sidewalks of, *560*
in future of Earth, 873
glacially polished surface in, *809*
Hurricane Sandy and, 761
during last ice age, 797
sea-level rise threat to, 835
New York State, 214, *215,* 446
drumlins in, *818*
redbeds in, *484*
New Zealand:
Alpine Fault in, 99, 333, 607
earthquakes in, 314
fjords of, 681
geothermal energy in, 532, *532*
glacier visit to, 827
Rotorua in, *694, 713*
waterfall in, *631*
Niagara Escarpment, *632*
Niagara Falls, 629, *632*
Niagara Gorge, *632*
Niger Delta, *634*
Nile Delta, 634–35, *634, 635,* 859
Nile River, 641, 650, 709, 792
damming of, 859
Nile River canyon, 183–84, *184*
"nimbus," 749
nitrogen, 42
in atmosphere, 730, 843
in Hadean atmosphere, 472
in thermosphere, 737
in volcanic gas, 730
Noah's Ark, 852
nodules, 214
non-aqueous phase liquids (NAPL),
716
nonconformity, *444,* **444,** 445
nonfoliated metamorphic rocks, 241,
243–45
nonmetallic luster, 127
nonmetallic mineral resources, 546,
557–64, *561, 562*
for homes and farms, 557–64
nonmetallic resources, common, *559*
nonplunging fold, *395*
nonrenewable resources, 535–39,
565–66
mineral resources as, 565–66
nonsystematic joints, 388
nor'easters, **754**
Norgay, Tenzing, 380
normal fault, 316, *317, 320, 320, 333,*
352, 391, **391,** *394, 402,*
404

normal faults, *333*
normal polarity, 77, *79, 80, 81, 82*
normal stress, 238
North America:
asthenosphere beneath, 369
Cenozoic convergent-boundary
activity in, 495
continental divides of, 619
convergent boundary tectonics in
(Mesozoic Era), *489, 491*
in Cretaceous Period, *491*
and Farallon-Pacific Ridge, 496
geographical provinces of, *477, 477*
and ice age, 819–20, 823, 826–29
drainage reversals in, 822–24,
823
glaciations in, 827–29, *828,*
829
in Jurassic Period, *489*
land bridge to Asia from, 497, *497,*
822, *823*
in late Cretaceous Period, *491*
in Mesozoic Era
growth of, 489, *490*
and Sierran arc, 490
paleogeographic maps of, *482,*
489, 506
and Pangaea breakup, 495
Pleistocene climatic belt in, 829
and Pleistocene ice ages, *497,*
826–28
sea-level change in, *845*
stratigraphic sequences, *486*
tapestry of time and terrain map,
901, *911*
see also Canada; Mexico;
Midwestern United States;
United States
North American Cordillera, 400,
401, 410
North American craton, 476, *478*
North American Plate, 89, 100, *100,*
101, 348, 496
and San Andreas Fault, 100, *100*
North Anatolian Fault, 351–52
North Atlantic Current, *664*
North Atlantic Deep Water, 667
North Atlantic gyre, 665
North Atlantic Ocean, 487
iceberg alley in, 808
in Jurassic Period, 839
North Carolina, Outer Banks of, *677*
North Dakota, 823
northern hemisphere, *14*
north magnetic pole, **375**
North Pacific gyre, 665
North Poles, *57,* 799
Northridge, California earthquake,
322, 340
North Sea, 495
North Star, *14*
Northwest Territories, Canada,
patterned ground in, *825*
Norway:
fjords of, 681, *681, 813*
mudslide in, *604*
slump off coast of, 597
and Storegga Slide, *559*
nuclear bomb testing, 326
nuclear energy (power), 508, 529–31,
530, 539–40
nuclear fission, 507, 508, 529

nuclear fusion, 26, 507, 542
nuclear power, challenges of, 530–31
nuclear reactions, *24*
nuclear reactor, **487–88,** *530*
nuclear waste, **531,** 717
"nuclear winter," 308
nucleosynthesis:
big bang, **23,** 26
stellar, **26**
nucleus, 120, *120*
of atom, 25
nuée ardente, 287, 292
Nullarbor Plain, Australia, 675, *675*
numerical age, 437, *450,* 461
and dating of periods, 460–64
and fission tracks, 458–59
in geologic column, 460–64
isotopic dating on, 453, *455*
and magnetostratigraphy, 458
tree rings in determining, 458
nutrients in soil, 199
removal of, 199
Nyirangongo Volcano, Zaire, 298
Nyos, Lake, Cameroon, 301–2, *301*

oases, 709, *709,* 779
in Sahara Desert, *709*
oasis, **708**
oblique-slip fault, *390,* 391
obsidian, *170,* **170,** *171, 172*
occluded front, 745, *745*
ocean:
currents in, and global warming,
762, 870
physical characteristics of, *663*
ocean floor, *see* sea floor
oceanic crust, **51,** *52, 55, 57, 72–74, 360*
age of, *80*
formation of at mid-ocean ridge,
92–95
and magnetic polarity, *80*
and rock cycle, 263, 267
oceanic currents, *see* currents, oceanic
oceanic fracture zone, 99
oceanic islands, 72, 661
oceanic lithospere, 132
oceanic lithosphere, 53, *55,* 87–89, *89,*
91, 98, 102, 108, 109, 178
and lithospheric mantle, 95
and oceans, 657–58, *658*
olivine from, 564
oceanic plateau, *661,* **661**
oceanic plate boundaries, bathymetry
of, 659–60
oceanography, 657
oceans, 44–45, 655–93
cold, 770
currents in, 663–64, *684*
see also currents, oceanic
evolution of, 842, 844, 846
exploration of, 656–57
extraordinary fossils in, 425
formation of, *472*
global circulation in, 663–64
and ice, 799
icebergs in, 806–8
as reservoir, *581*
residence time for, 582
temperature in, 866
tides in, 667–72
water masses in, *665, 667, 667*
of the world, *659*

ocean water, 662–67, *662*
    composition of, 662, *662*
    temperature of, 662–63, *663*
Odyssey satellite, 584
offshore bar, 678
offshore drilling, *515*, 516, 540–41
Oglalla Formation, 699–700, *700*
O-horizon, 196, *198*
oil, 507, 508–12, *510*, *511*, *519*, 535–38, *536–37*
    and climate change, 855
    depletion of supplies of, 535–38
    distribution of resources of, *519*
    drilling for, 250
    drilling of, 515–17
    exploration and production of, *512*, 513–17, *518*
    formation of, *509*, *510*
    modern technology in search for, 513–17, *515*, *520*
    1970s crisis over, 535
    pricing of, *535*
    refining of, 517, *518*
    transportation of, 517, *518*
    vs. natural gas, *519*
Oil Age, 535–38
oil consumption, by U.S., 535
oil industry, birth of, 513
oil reserves, global, 517
oil sand, 520
oil seep, *511*, **512**, 513
oil shale, 212, *510*, *520*, *521*, **521**
oil spills, 539, 690
oil trap, *see* trap
oil wells, in Kuwait, 540
oil window, *510*, **510**
Okeechobee, Lake, 715
Oklahoma, "dust bowl" of, 200, *201*, 791–92, *793*
Old Faithful geyser, *712*
Olduvai normal subchron, 77
Olgas, The, 786
Oligocene Epoch:
    glaciers during, 497, 832
    and sea floor, *95*
    and sea level, *845*
olivine, 117, 131, 132, *134*, 136, 164–65, *164*
    and serpentine, 564
    stability of, *190*
Olympic Peninsula, Wash., 677
Olympus Mons, 308, 309, 409, 582, *583*
On a Piece of Chalk (Huxley), 468
Onawa Pluton, Maine, 250, *251*
Onondaga Indian tribe, 214
Ontario, Canada, 823
Ontong Java Oceanic Plateau, 178
Oort Cloud, 37–38, **37**, *38*, 39
opal, 136, 137
OPEC (Organization of Petroleum-Exporting Countries), 535
open-pit mine, 555–56, **555**, *556*
    environmental damage by, 566
ophiolite, 555
Opportunity (Mars rover), 584
orbital eccentricity, 831, *832*, 855
orbital forcing, 831
orders, 425
ordinary well, 708, *708*
Ordovician Period, 482, *482*, 483, *498*
    land plants in, 483

life forms in, 450
and stratigraphic sequences, *483*
and Taconic orogeny, 482
ore, 549–55, **549**
    stained rock as indicator of, *556*
ore body, *556*, 557
ore deposit, 549–55, **550**, *554*, *562*
    formation of, 549–55, *551*
    location of, 553–55
ore minerals, 118, 549–55, *549*, **549**, *550*
    exploration and production of, 555–57
    in groundwater, 716
ore resources, expected lifetimes of, *565*, **565**
organic chemicals, **45**, 119, 507, 508
organic coasts, **681**, 690–91
organic matter:
    carbon-14 dating for, 457
    in cave development, 719–20
    in soil formation, *195*
organic sedimentary rocks, **203**, 210–12, *212*
organisms:
    chemical weathering by, 191
    in Phanerozoic Eon, 480
    and shallow-water carbonate environments, 226
Organization of Petroleum-Exporting Countries (OPEC), 535
orientation, of geologic structures, *388*
original horizontality, principle of, 438, *439*
Origin of the Continents and Oceans, The (Wegener), 62
Orion Nebula, *12*
orogenic collapse, 409, *410*
orogens, *see* mountain belts or ranges
orogeny (mountain building), 379, 381–82, **381**, 405, 412–13, 468, 469
    causes of, 400–405
    crustal deformation and, 392–415
    and dynamothermal metamorphism, *254*, 255
    exhumation from, 256, *257*
    and greenhouse gases, 853
    measuring of, 405, 495
    and rock cycle, 263
    rock formation during, *404*
    and sedimentary rock deformation, *222*
orographic barrier, **763**
orographic lifting, 747, *748*
orthoclase (K-feldspar), 117, *127*, *134*
    stability of, *190*
    see also feldspar; K-feldspar
Oso, Wash., mudflow in, 593, *593*, *604*
Oswaldo Sandstone, 446
Ouachita Mountains, 484
outcrop, **144**, *145*, 147–48
    jointed rock in, *600*
    observation of, 147–48
    stromatolite deposit in, *475*
Outer Banks, N.C., 677
outer core, 53, 56, 57, **366**
outer planets, 28

outgassing, 472
outwash, glacial, *816*, *817*, 818
overhang, *606*
overriding plate, 96
Owens, Rosa May, 695
oxbow lake, 633, **634**
oxidation reaction, 190
oxides, 129, 158, *190*
    ore minerals as, 549
oxisol, 197, 198, *198*
oxygen, 42, 45, 124
    in atmosphere, 480, 731, *731*, 843
        in Archean Eon, 475
        increase in, 475, 731
        and photosynthesis, 7
        in thermosphere, 737
    in crust, 53, 143
    in hydrogen fuel cell, 534
    and pollutants in rivers, 650
    in quartz, 119
    in silicon-oxygen tetrahedron, 131
    and underground coalbed fires, 528
oxygen-isotope ratios, and climate change, 849, *851*
Ozark Dome, 411
ozone, 480, 731, 732
    breakdown of, 733
    and stratosphere, 737
ozone hole, 733, **861**, *863*

Pacific Ocean:
    floor of, *657*
    in future, 873
    hot-spot tracks in, *103*
Pacific Palisades, slumping of, 590
Pacific Plate, 97, *97*, 110, *111*, 495, *496*
    and Hawaii, 101
    and Japan earthquake, 313
    and San Andreas Fault, 100, *100*, 607
Pacific Rim, earthquakes in region of, *332*, *333*, 334
pahoehoe flows, **276**, *277*
Painted Desert, *452*, 773, *774*
Pakistan:
    earthquakes in, 335–36
    flooding in, *641*
    glacier in, *800*
    Indus River flood in, 641, *641*
Paleocene bed, *461*
Paleocene Epoch, and sea floor, *95*
paleoclimate, **849**
paleoequator, *485*, *488*
Paleogene Period, *499*
paleogeography, 480, 482
    of Cenozoic Era, 495–98
    of early and middle Mesozoic Era, 487–89
    of early Paleozoic Era, 482
    of late Mesozoic Era, 487–89
    of late Paleozoic Era, 483–95
    of middle Paleozoic Era, 483, *485*
paleolatitude, 71
paleolongitude, 71
paleomagnetic dipole, 69, 71, 76
paleomagnetism, 62, 67–71, **67**, 68–70, *69*, 80–81, 469
    development of, 68–70, *69*
paleontological evidence, on climate change, 849
paleontologists, **419**, *420*, *421*

paleontology, **419**
    and theory of evolution, 430–33
paleopole, *70*, **70**, 71
paleoseismology, 351
paleosol, 445, *445*, 827
Paleozoic Era, 7, *7*, 450, 480, 482–86, *485*, 498–99
    Appalachian Mountains from, 401
    and continental drift hypothesis, 64–65, *65*
    global cooling in, 855
    and ice age in, 64
    and Las Vegas vs. Grand Canyon strata, *448*
    mass extinction in, *298*
    and Pangaea, 844
    parts of
        early, 482–83
        late, 483–86
        middle, 483
Paleozoic rocks, 396, 399, 410
Palisades, mineral distribution in, *182*
Palo Verdes area, "Portuguese Bend slide" in, 607–8, *608*
Panama, Isthmus of, 497, 833, *853*
pandemics, *859*
Pangaea, **62**, 63–64, *63*, *65*, *66*, 111, 484, *485*, *487*, *638*
    Appalachian region in, 413
    breakup of, *112*, 495, 540, 853
    continents existing previous to, 105
    formation of, collisions involved in, 485
    glaciation in, 830
    and history of Earth, 463
    and North America, 105
    in Paleozoic era, 844
    and pole-wander paths, *70*
    and sedimentary rock layers, 64–65, *65*
Pannotia, 477, *478*, 482
Panthéon in Paris, Foucault's pendulum in, *16*, *20*
Panum Crater, Calif., *275*
Papua New Guinea, 495, 597
parabolic dunes, 788, *788*
Paracutín volcano, birth of, 276
paradigm, scientific, 87
Paradise Lost (Milton), 47
parallel drainage network, 618
parallelism, 238–39
Paraná Basin, Brazil, 297
Paraná Plateau, Brazil, 179, *179*
parent isotope, 453, 456–57, *456*
Parthenon, 557
partial melting, 158–59, **158**, *159*
passive continental margins, **89**, **658**, *660*, *684*
    oil from, 540
    as oil sources, 517
passive-flow folds, *397*, *398*
passive-margin basins, 89, 228
    in Mesozoic Era, 490
    western North America as (middle Paleozoic), 483
passive-margin basis, 658–59
Patagonian Shield, *258*
patterned ground, 825, **825**
pauses, 737
pearl, 136, 137
peat, 524–25, *524*, **525**

pebbles, *187, 206*
pedestal, *785*
pediments, *785, 786–87*, **786**
peds, 197
pegmatites, **169**
pelagic sediment, **660**
Pelé, 307
Pelée, Mt., *175*, 291, 304
"Pelé's hair," 278
"Pelé's tears," 278, 307
pelitic metamorphic rocks, 245
pendulum experiment of Foucault, 20, *20*
peneplain, 636, *637*
Peninsular Batholith, *168*
Pennsylvania:
    Centralia coalbed fire in, 528, *529*
    Valley and Ridge Province in, 413
Pennsylvanian Period:
    in correlation of strata, *452*
    life forms in, 451
    and sedimentary sequence, 844, *845*
Pennsylvania Valley and Ridge, *488*
perched water table, 702–3, *703*, **703**
peridotite, **46**, 51, 52, *172*, 363
    seismic wave speed through, 361
periglacial environments, 825, *825*
periodic table of elements, 24, 901, *902*
periods, geologic, **450**
permafrost, 422, 588, **825**, *825*, 826, *865, 866, 867,* 870
permanent strain, 383
permanent streams, **620**, *621*
permeability, 511–12, *511*, **511**, 698–99, *698*, **698**
    in Darcy's law, 706
    and pore collapse, 714–16
Permian mass extinction, 486
Permian Period, *499*
    biodiversity in, *858*
    climate of, 483
    in correlation of strata, *452*
    plants during, 483
    tillites from, 829
    volcanoes in, 489
Permian-Triassic extinction event, 857
permineralization, **423**
perovskite structure, 364
Perranporth, Cornwall (England), *668*
Persian Gulf, oil fields around, 517
Peru, *3, 36*
    rain-free coastal areas of, 771
    Yungay landslide in, 587–88, *587, 594*
Peru-Chile trench, *74*
pesticides, 200
petrified wood, 214, *215*, *422*, **423**
petroglyphs, *774*, **774**
petrographic microscope, **148**, *150*
petroleum, 513
    see also oil
phaneritic rocks, 168
Phanerozoic Eon, 6, 7, 450, 480, **480**
    life evolution in, 843
    and sea level, 844
    time periods of
        Cenozoic Era, 495–98
        Mesozoic Era, 487–95
        Paleozoic Era, 482–86

Phanerozoic orogenic belts, *477*
Phanerozoic Period, in correlation of strata, *452*
Phanerozoic sediment, 410
phase change, 235
phase diagram, 237, *238*
phenocrysts, 168
Philippines, Mt. Pinatubo in, 299, *300*, 303, 304
    see also Pinatubo, Mt.
Philippines, Typhoon Haiyan in, 761
Philippine Trench, *74*
Phlegrean Fields, *290*
Phoenicians, as ocean explorers, 657
Phoenix, Ariz., 699
    Colorado River water diverted to, *604*, 650
    dust storm in, 776, *776*
Phoenix Basin aquifer, 699
phosphate, 546, 564
Phosphoria Formation, 564
photochemical smog, 733
photomicrograph, **148**, *237*
photosynthesis, 475, 481, 507, 839, 842
    and carbon-14 ingestion, 457
    and carbon absorption, 846
    energy from, 507
    and K-T extinction, 494
    and oxygen in atmosphere, 7, 731
photovoltaic cells, **534**
    quartz in, 564
phreatic eruptions, 288
phreatic zone, 701
phyla, 425
phyllite, **241**, *243*, 246, *247*
phyllitic luster, 241, *243*
phylogenetic tree, **429**, *430*
phylogeny, 429
physical weathering, 186–89, **186**
    and chemical weathering, 189–91, *191*
Piccard, Bertrand, 729–30, *729*
Piedmont, *488*
piedmont glacier, 799, *800, 821*
pillow basalt, 93, *94, 177, 256,* 276, *278, 293, 295*
pillow lava, **276**
Pinatubo, Mt., *175, 289, 290,* 299, *300,* 303, 304, 306, *306,* 863
pinnacle bergs, 808
pipelines, 517
pitch, 21
pitchblende, 529
placer deposits, **553**, *554, 562*
    locations of, 553
plagioclase, *128,* 134, 164–65, *164,* 245, 901
planar structures, *388–89*
Planetary Geology, 43–44, 50–51, 56, *56,* 155, 463, 582–85, 843, *844, 854*
    asteroids, 39, *39,* 50–51
    comets, 39, *39,* 50–51
    differentiation, 32
    diversity of planets, 43, *56*
    evolution of life, 843–44, *843*
    evolution of the atmosphere and oceans, 843
    formation of Earth, *30–31*
    formation of planets, 27–29, 33, 155

formation of the Solar System, 27–29, *30–31,* 33
formation of the Sun, *31*
meteors, 50–51, *50,* 463, 582
nebular hypothesis, *30–31*
nebular theory of Solar System formation, 29–32, 33, 154
planetary atmospheres, 41–43
planetary interiors, 56
planetary surfaces, 582, 583, 584, *584*
volcanoes on other planets, *308,* 309
    see also specific planets
planetary systems, *60*
planetisimals, **28**, 29, *29,* 32–33, 51
    and igneous rocks, 154
planets, *14,* **15**
    definition of, 15
    discovery of, 15
    dwarf, 18
    formation of, 27–29
    forming of, 30–31, *31*
    inner (terrestrial), **18**
    landscapes of, 582–85
        Mars, 582–85, *584*
    moons of, 18
    outer (Jovian, gas giant), 28
    relative sizes of, *17*
    stars distinguished from (age of Homer), 14
plankton, *228*
    and carbon dioxide, 832
    in chalk, 468
    and climate change record, 849
    and food chain, 765
    in fossil record of glaciations, 828, 832
    in gas hydrate formation, 521
    and K-T boundary event, 494
    in limestone, 210
    as microfossils, 424, *424*
    on ocean (sea) floor, 72, 184, *184,* 226
    oil and gas from, *508, 509–10*
plantae, 425
plants:
    of desert, 789
    in middle Paleozoic, 483
    in Phanerozoic Eon, 480
plasma, *120,* 121
plastic deformation, 235–36, 239, *240,* 323, 383–86, **384**, *398*
plastic materials, 53
plastics, 517
plate boundaries, **89**, 90–92, *92,* 101, *108–9*
    and earthquakes, *91*
    earthquakes at, 332–35, *332, 335*
    formation and death of, 102–5
    igneous activity at, 178
    transform, 98–100
    triple junctions of, 100, *101*
Plate-Boundary Observatory, 370
plate-boundary volcanoes, 100
plate-driving mechanisms, 106
plate graveyards, 369
plate interiors, 90
plate motion:
    forces behind, 106–12, *107*
    manifestations of, 110–11
    velocity of, 6, *6,* 107–11

plates, 6, **6**, **89**, 110–11
    major, 89
    micro-, 89
    shapes and sizes of, *91*
plate tectonics, **62**, 86–113, *86, 87,* 174–79, *264,* 500, 839, *839*
    in Archean Eon, 473
    beginning operation of, 842
    and coastal variability, 683–87
    collisions, 102
    continental rifting, 102, 105
    convergent plate boundaries and subduction, 96–98
    and distribution of deserts, 772
    divergent plate boundaries and sea-floor spreading, 92–95, *96*
    and Earth as dynamic planet, 271, *271*
    and Earth history, *841*
    and Earth's future, 873
    and Eceladus, 583
    forces driving plate motion, 106–12, *110*
    in glaciation, 831, 832
    and hot spots, *see* hot spots
    and lithosphere plates, 87–89, 90
    and Mars, 582
    and metamorphism, 234, 250
    and Moon, 582
    and mountain building, 386–87, 400
    and ore deposits, 553
    plate boundary identification, 90
    and river courses, 636
    and sedimentary basins, 228–29
    and supercontinent cycle, 844
    theory of, *108–9*
    transform plate boundaries, 98–100
    triple junctions, 100
    velocity of plate motion, 107–11
    and violent hazards
        mass movements, 607
        and volcanic eruptions, 293, *294; see also* catastrophic change or events; earthquakes; storm
    and Wegener's vision, 67
    see also specific areas
plate velocities, global map of, *111*
platforms, 476
platinum:
    as precious metal, 549
    as strategic, 565
    supply of, 565
Plato, and Atlantis, 307
platy (pancake-shaped) grains, 238
playa, 778–79, **778**, *780*
playa lake, *784, 785*
Playfair, John, 460
Pleistocene climatic belts, 826–27, *829*
Pleistocene Epoch:
    and human evolution, 500
    and sea floor, 95
Pleistocene Era, stratigraphic record in, 855
Pleistocene ice ages, *497,* **497**, 699, *797,* 826–30, *828, 830*
    explanatory model for, 833
    life and climate in, 826–27
    timing of, 827–29

Plesiosaur, *845*
Plinean eruptions, 288–92, *289*
Plinian eruptive style, *286, 289*
Pliny the Younger, 288
Pliocene Epoch, 826
    and sea floor, *95*
    and sea level, *845*
Plio-Pleistocene ice ages, 826
plowing, in glaciers, *810*
plume, *see* mantle plume
plunge, *389*
plunge pool, 629
plunging fold, 395, *395, 397*
Plutarch, 306
Pluto, 15, 18
    as dwarf planet, 18
Plutonists, 144
plutons, **162**, *163, 166, 167, 171,* 176,
    255, *287*
    of Andes, 554
    and baked contact, *440*
    composition of, *169*
    cooling of, 168, *169*
    in geological history illustration,
        438, *441*
    joints in, *187*
    in Midwest, 477
    and thermal metamorphism, 250,
        251
    uranium in, 529
    see also igneous intrusion
pluvial lakes, 824–25, *824,* **824**
point bars, 625, **626,** 631
polar cells, 739, *740*
polar deserts, 772
polar easterlies, 741
polar front, *739,* 741
polar glaciers, **799,** 801
    flow in, 801
polar ice cap, 65, 229, 852, 853
    and global warming, 870
polarity chrons, *78, 80*
polarity reversal, 458, *460*
polarity subchrons, *77, 78*
Polar Plateau, 203, *204*
polar-wander paths, 70–71, *70*
    apparent, 70–71, *70,* **70**
        and continental drift, 70–71,
        70
        and Earth history, *471*
    true, *70*
poles, 40, *40*
    geographic, 67
    magnetic, *67, 68,* 70
    reversal of, 458, *460*
pollen, and ancient climates, 424
    and climate change record, 849,
        850
pollution, **861**
    of air, *563,* 566, 567, 733
        and fossil fuels, 539; *see also*
            greenhouse gases
    coastal, 690–91
    of groundwater, 714
    of rivers, 650
    see also environment issues
polymorphs, **124**
Polynesians, as ocean explorers, 657
polynyas, 808
polyps, 682
Pompeii, 273–74, *273*
ponds, residence time for, 582
Pontchartrain, Lake, *763*

population (human), increase in, 858
pore collapse, 714–16
pores, 511–12, **511,** *697,* **697**
    and oil or gas reservoirs, 511
porosity, 511–12, *511,* **511,** 696–98,
    697, **697**
    and pore collapse, 714–16, *717*
    secondary, *697,* 698
    vs. permeability, 699
porphyritic rocks, 168
porphyroblasts, 241
porphyry copper deposits, 551
Port-au-Prince, Haiti, 348–49, *349*
Portland cement, **559,** 560
Portugal (Lisbon), earthquake in,
    338–39
"Portuguese Bend slide," 607–8, *608*
positive anomalies, 76–77, *76,* 78–79,
    79
positive feedback, 849
positive feedback mechanisms, 832
potash, 564
potassium, *53, 124,* 158, 190, 191
potassium feldspar, 549
potential energy, *372,* 703
potentiometric surface, 710, *710*
pothole, **623,** *624*
pottery, 561
pottery making, thermal
    metamorphism as
    comparable to, 252, *252*
Powell, John Wesley, 435, *435*
Precambrian basement, *488*
Precambrian crust, *476*
Precambrian Period, 6, 7, **450,** 453,
    498–99
    in correlation of strata, *452*
    rifting events in, 135
    and sea level, 844
Precambrian rocks, 65, 257, *258, 396,*
    410–11, *410*
    and banded-iron formation, 555
    Cretaceous faults in, *491*
    in North America, 476, *478*
    in Scotland, *233*
    in uplift, *492*
Precambrian shields, rocks unchanged
    in, 263
precession of Earth's axis, 20, 831,
    832
precious metals, 549
precious stones, 137
precipitate, **121**
precipitation, **121**
    in speleothem formation, 720
precipitation from a solution, 45, 124
    and pressure solution, 235
    of salt, 212, *213*
    travertine from, 213
precipitation of water vapor, 582,
    747–48
    in polar regions, 772
    see also rainfall
prediction of earthquakes, 351–54
predictions, of tsunamis, 351, *358*
preferred orientation, 238–39, **238,**
    239, 240
pre-Illinoian glaciation, 828, *830*
preservation potential, **425**
pressure, *237, 385,* **387,** 388
    atmospheric (air), *42,* 732–33
        and temperature, 737
        and tides, 670–71

metamorphism due to, 237, *238*
    rock deformation from, 385
pressure gradient, *738*
pressure solution, 235
primary atmosphere, 730
primary porosity, 697–98
primates, 500
    arrival of, 500
principle of baked contacts, 438, *440*
principle of cross-cutting relations,
    438, *440,* 460–61
principle of fossil succession, 441, *442*
principle of inclusions, 438, *440*
principle of lateral continuity, 438,
    440
principle of original horizontality,
    438
principle of superposition, 438, *440*
principle of uniformitarianism, **437,**
    438, 442–45, 646
Principles of Geology (Lyell), 430, 438
products, chemical, 121
prograde metamorphism, 245–46,
    248, *249*
prokarya, **425**
prokaryotic cells, *430,* 477
propane, 508
Proterozoic Eon, 6, 7, 450, 476–81,
    476, *499*
    atmosphere of, 480, *481,* 843
    and growth of continental crust,
        474
    ice ages in, *479,* 480
    life forms in, 450, 843
    mountain belts of, *66*
    tillites from, 830
Protista, 425, *428*
protocontinents, 473, *474*
"proto-life," 428, *430*
protolith, **234,** *236, 237, 238,* 240,
    242, 243, *244,* 254
proton, 121
protons, 25
protoplanetary disk, **28,** *29*
protoplanets, **29**
    and differentiation, 32, *32*
protostar, **26**
proto-Sun, 28
protozoans, ciliate, 479
pterosaurs, 494
Ptolemy, 14, 15
P-T-t (pressure-temperature-time),
    249
puddle, 616
Puerto Rico, sand beach in, *677*
Puerto Rico Trench, *74*
Puget Sound, *816*
Pulido, Dionisio, 276
Pulpit rock, *813*
pumice, *172,* **172,** *174,* 282
pumice lapilli, 278, *279, 289,* 292
pumps, oil, 517
punctuated equilibrium, **431**
    and catastrophic collisions, 513
Puyehue-Cordón Caulle volcano,
    311
P-waves (primary waves), 323–24,
    324, 326, *328,* 330, 339,
    361, *361, 364*
P-wave shadow zone, **365,** *366*
pyramids of Egypt, *578*
pyrite, *128, 129, 189,* 190, 481, 539
pyroclastic bubbles, 288–92

pyroclastic debris, **154,** 278–81, **278,**
    279
    threat from, *300*
    for various eruptions, *290*
    at Yellowstone, 295–96, *297*
pyroclastic deposits, 281
pyroclastic flows, *162, 163, 171,* 287,
    292, 296
    prediction of, 303
    threat from, 298–99
    see also nuée ardente
pyroclastic rocks, 172–74, **172**
pyroxene, 131, *146, 164,* 165, *172*
    in gneiss, 241
    stability of, *190*

Qaidam Basin, *403*
Qilian Mountains, *403*
quarry, 557, *559, 563*
    crushed-stone, 557, *559*
quarry face, of limestone, *211*
quartz, 117, 119, *126, 128,* 134, *134,*
    146, 172
    as amethyst, 137
    of Andes, 555
    and Bowen's reaction series, *164,*
        165, 166–67
    as cement, 208
    in cement, 559
    and coesite, 235, 237, 253
    compression of, *399*
    cryptocrystalline, 210
    crystal of, *119, 122*
    in desert rock, 773
    in gneiss, *236,* 241
    in granite, 549
    hardness of, *127,* 128
    and hydrolysis, 190
    in joints, 388
    and marble, 244, 557
    in metamorphic rocks, 245
    and metamorphism, 240, *247*
    milky white, *387*
    in New York cement, 560
    in Onawa Pluton, *251*
    in photovoltaic cells, 561
    and recrystallization, 235
    in sand, 676
    and sandstone, 204, 208, 255
    in schists, 241
    in sediment transported by
        glaciers, 560
    in shale, *236*
    shape of, *122*
    shocked, 494
    as silicate, 134
    stability of, *190,* 191
    and weathering, 190, 191, *207,* 209
quartzite, **244,** *246,* 255
    quartz grains in, 382
quartzo-feldspathic metamorphic
    rocks, 245
quartz sandstone (quartz arenite),
    207, 208, 209, *246*
quartz veins, gold in, *548*
Quaternary Period, *499*
Queen Charlotte fault system, 495
Queen Elizabeth II, 673
quenching, 164
    in pottery making, 252
questions about universe, 13
quick clay, 342, *604,* **604**
quicksand, 341

Racetrack Playa, 779
radial drainage networks, 618, *618*
radiation, 54, *54*
radiation sickness, 531
radioactive decay, 453, *456*
   heat produced by, 473, 508
radioactive element, **453**
radioactive elements, 32
radioactive isotopes, 453–56
radioactive materials, transfer of to
      human environments, 861
radioactive waste, groundwater
      contamination from, 716
radioactivity, 349
   energy from, 507
radioactivity, and heating of earth,
      156, 462
radiometric dating, 435, 453
radio transmissions, and ionosphere,
      737
railroad building, through Sierra
      Nevada, 141, *141*
rain, 750, *750*
raindrops, formation of, 747, *748*
rainfall, 616, 747, *748*
   atmospheric water extracted by,
      730
   in desert, 769, 773, 774
   and equatorial lows, 740
   in hurricane, 757
   during last ice age, 827
   from low-pressure mass, 745
   and soil formation, 197–98
   and water table, 701, *701*
rainforest destruction, and soil, 200
rainforests:
   temperate, *763*
   tropical, 741, *763*, 860, *860*
Rainier, Mt., *175*, 302
   danger assessment map for, *304*
rain shadow, **771**
   deserts in, 771, *772*
range of fossils, 441
rapids, **629**
   in Grand Canyon, *631*
rare earth elements (REE), **564**,
      565–66
reach, 615
reach, of stream, 628
reactants, 121
reactivated faults, 318
recessional moraines, **815**, *821*
recharge area, *704*, **704**
recrystallization, 235, 236, 241, 244,
      245, 391
   during dynamic metamorphism,
      250
   and fossils, 420
   of mylonite, 391
   of quartzite, 244
rectangular drainage network, 618,
      618
recurrence interval, **331**, *332*, 648–50,
      648, *651*
   in volcanic activity, 303
redbeds, *217*, 222, 224, 481, *484*
Red Cross, 645
red giant, *873*
Redoubt Volcano, Alaska, *292*, 299
Red Sea, *104*
red shift, **21**
   and expanding universe theory, 21

Redwall Limestone, 446, *448*
REE (rare earth elements), **564**,
      565–66
reef bleaching, 691
reefs:
   and continental drift, 64
   coral, 210, *211*, 223, 227, **682**, *683*
   see also coral reefs
reference geoid, 372–73, **372**
refineries, 517
reflection, *362*, **362**
refraction, *362*, **362**
refractory materials, **28**, 29
reg, 780
regional basins, 411
regional domes, 411
regional metamorphism, **252**, 405
   in Orogenic belt, 254
   and rock cycle, *267*
regional unconformities, and
      stratigraphic sequences,
      486
regolith, **185**, 195, 600
   and creep, 588
   in desert, 773
   and mass movement, 600, *602*,
      604, *605*
regrading, 610
regression, **229**, *230*, 486–87, *845*
relative age, *437*, **437**
   and fossil succession, *442*
   physical principles for defining,
      438–42, *441*
relative humidity, **735**, 848
relative motions, 90
relative plate velocity, **107**, *110*, *111*
relief, **576**, 577
   and plate tectonics, 607
renewable resource, 535
   and groundwater, 713
replacement chert, 213, *215*
repose, angle of, see angle of repose
reptiles, 485, 490
   of desert, 789
   see also dinosaurs
re-radiation, 848
research vessels:
   Alvin (submersible), 656–57, *656*
   Glomar Challenger, 82
   H.M.S. *Challenger*, 72, 656, 657,
      663
   JOIDES Resolution, 657
   seismic data collecting, 371
reserves of mineral deposits, **565**
reservoir rocks, 511–12, *511*, **511**, *512*,
      513, *514*
reservoirs, 265, 268
   and global change, 842
   residence time of water in, *579*
   for water, 579–82, *581*, 650
      environmental problems in,
         650
      glaciers as, 819–22, *823*
   and local base levels, 627
residence time, **265**, *579*, **579**
residual mineral deposits, **553**, *554*
resistance force, 600–601
resonance, of earthquake waves, **355**
resources, **505**
   increase in per capita use of, 858
   nonrenewable, 535–39, 565–66
   renewable, 535

see also energy resources; mineral
      resources
resurgent domes, of Krakatau, *291*
retaining wall, *611*
retrograde metamorphism, 246, 248,
      249
revegetation, 610, *611*
reversed polarity, 77, 79, *79*, *81*, *82*
reverse fault, 316, *317*, 320, *321*, **391**
rhyolite, 170, *171*, *172*, *173*, 178
   in rock cycle, 263
   at Yucca Mountain, 531
rhyolite dome, *275*
rhyolitic:
   ash, 178
   tuff, 178
rhyolitic eruptions, 276
   pyroclastic debris from, 278–81
rhyolitic lava, 275, 288
rhyolitic lava flow, *275*, 276–78
rhyolitic (fine-grained) magma, *157*,
      172, 288, 295, 296
rhythmic layering, 458, *459*
Richter, Charles, 330
Richter scale, *330*, **330**
   calculation using, *330*
ridge, 45
   as divergent boundary, *93*
   on map of relative velocities, *110*
   and sea-floor spreading, *75*
   on volcano map, *175*
   see also Mid-Atlantic Ridge; mid-
      ocean ridges
ridge axis, 72–74, **72**, *74*
ridge-push force, 106–7, **106**
rift basins, *228*, 495
   in Pangaea, 487
rifting, **102**, *103*, *104*, 105
rifts, Iceland as, *295*, *296*
rifts or rifting, 177
   active, 105
   in Archean Eon, 474
   in landscape evolution, 577
rigidity, 361
rigid materials, 53
ring dikes, *171*
"Ring of Fire," 293, 307
Rio de Janeiro:
   in future of Earth, 873
   mudflows in, *591*, *591*
   sugar-loaf mountains in, *675*
Rio Grande Rift, *497*
rip current, 674, *675*
ripple marks (ripples), **216**
ripples (ripple marks), on floodplain,
      224
riprap, *610*, *611*, 689, *690*
Riss glaciation, 828
river environments, 224, *225*, 227
rivers:
   in Archean Eon, 474
   deltas of, 226, 628, 634–35, *634*,
      635, *636*, *643*
   see also deltas
   in desert, 770
   environmental issues over, 650
   erosion by, 408, *409*
   in human history, 650–51
   irrigation from, 789, 790, 792, *792*
   relocating of, 610, *611*
   residence time for, 582
   undercutting by, 604

as water reservoir, *581*, 582
   and water table, 701
   see also streams
river sediment, 226, *226*
river systems, 540
roadcuts, *145*
roche moutonnée, *812*, **812**, 827
Rochester, N.Y., drumlins near, *818*
rock:
   fresh (unweathered), 185, **186**
   transforming magma and lava
      into, 160–61
rock assemblages, and continental
      drift, 65
rock avalanche, *603*
rock bolts, *611*
rock bursts, 557
rock composition, **146**
rock cycle, 115, 261–69, *261*, *262*,
      262, *269*, 840, 846
   case study of, 263–65
   causes of, 265
   and plate tectonics, 263, *264*,
      265, *267*
   rates of movement through, 263
   and rock-forming environments,
      266–67
   sand in, *269*
   sandstone in, *269*
rock deformation, *see* deformation,
      rock
Rockefeller, John D., 513
rock exposure, Earth history told
      by, 417
rockfalls, 595–96, *597*, *603*
rock flour, 808
rock formation, during orogeny,
      404–5
rock glaciers, **588**
   in Alaska, *589*
rock groups, 141–51
Rock Island, Ill., flood hazard map of
      region near, *649*
"rock oil," 513
rocks, 115, *142*, **142**
   and carbon-14 dating, 457
   chemical composition of, 143
   classification of, 144–47
   in desert, 770
   determining age of, 78
   in Earth's history, 472–73
      beginning of, 463, *463*
   and Earth's magnetic field, 78–79
   as flowing, 52, 53–55
   as geological record, 142, 263
   geologic setting formation of, *146*
   in glacier, 805, *821*
   and groundwater, 716
   as insulators, 789
   intact vs. fractured, 600
   intermediate, 46, 159, *164*, *172*
   mafic, 46, 160, *160*, *164*, 170,
      171, *176*
   magnetization of, 69–70
   melting of, 153–59
   from Moon, 463
   naming of, 147
   pores of, *511*, 697, **697**
      and porosity, 697–98, *697*
   seismic wave speed through, 361,
      361
   silicate, **46**, 805

rocks, (continued)
  and soil, 196
  study of
    through high-tech equipment, 149–50
    through outcrop observations, 147–48
    through thin-section study, 148–49
  surface occurrences of, 144
  types of
    igneous, 45, **144**, *146*, 153–54
    metamorphic, 45, **144**, *146*
    sedimentary, 45, **144**, *146*, 185; see also specific types
  ultramafic, 46, *172*
  and water table, 701
  see also boulders
rock saws, 148
rock slide, 594, **594**, 610
rocky cliffs, in desert, 780–83
rocky coasts, 679, *680*
Rocky Mountains, *269*
  Ancestral Rockies, 485, *485*
  Big Thompson River flood in, 615, 644–45
  formation of, 491, *492*
  front of, *490*
  glaciers in, 826
Rodinia, **477**, *478*
rogue waves, 672–73
Romans, cement used by, 559
Rome, ancient, 273–74
roof collapse, 723
root wedging, 188, *188*
Rosendale Formation, 560
Rosetti, Cristina, 729
Ross Ice Shelf, 203, *204*
Ross Sea, *807*
rotation:
  of Earth, 372
  in rock deformation, 383, *384*
Rotorua, New Zealand, *694*, 713
Rub al Khali, 788
ruby, 137, *138*
runaway, greenhouse effect, 854
runaway greenhouse effect (Venus), 854
running water, 615
  geology of, 614–54, *614*
  work of, 623–26
runoff, 616, **617**
  in hydrologic cycle, 617
rupturing, *318*
Russia:
  deepest drill hole in, 360
  during ice age, *821*, 826
Rutherford, Ernest, 25
R-waves (Rayleigh waves), 323, *324*, 326, 330, 339, *339*, 361

sabkhas, 771
Saffir-Simpson scale, **756**, *757*
sag pond, 320
Saguaro cactus, *790*
Sahara Desert, *709*, 713, 769, *770*
  and current climate trends, 792
  and Sahel, 790, *791*, 792, *792*
  as subtropical, 771
  and types of landscape, 780
  yardangs in, *777*
Sahel, Africa, 790, *791*, 792, *792*

St. Helens, Mt., *281*, 285, 290–91, *291*, 292, 293, *294*, 299
  and eruptive style, 285
  lahar on, *593*
  mass of pumice from, 305
  mass of pyroclastic debris from, *290*
St. John, U.S. Virgin Islands, *675*
St. Louis, Mo., flooding of, 644
saline intrusion, into groundwater, 714, *716*
salinity, 662, *662*, **662**, 663, 832
salt, 121, *125*, 557
  in desert, 770
  in evaporites, 212, 213
  in groundwater, 705, 706, 714
  as nonmetallic mineral resource, 546
  precipitation of, 212, 213, *213*
  in sea water, 662, *662*
saltation, 624, **624**, **776**
salt crystal, 189
salt deposits, and continental drift, 64
salt-dome trap, 514, *514*
salt flats, 212
Salt Lake City, Utah, mine near, *555*
salt lakes, 778–79
salt marsh, *682*
Salton Sea, 712
salts, *190*
saltwater wedges, 679
salt wedging, 188, *188*
San Andreas fault, 100, *111*, *319*, 323, 592, 607
  as continental transform fault, *334*
  and convergent tectonics, 495
  displacement from, *317*, *392*
  and earthquake projection, *352*
  earthquakes along, 333
    1906 earthquake, *317*, 322, *334*
  and ground surface, 389
  and transform boundary, 100, *100*, *495*, *496*
sand, *187*, *193*, 206, *209*, 557
  and angle of repose, 601
  on beaches, *146*, 676, *677*, *678*, 678
    in hurricane, 689
    loss of, 689
    replenishment of, 689
  calcite, *227*
  coastal beach, 224
  and deltas, 226
  in desert, 770
    abrasion by, *777*
    as dunes, *779*
  in iceberg, 808
  in New York concrete, 560
  as nonmetallic mineral resource, 546
  in plain at end of Firth of Forth, 559
  in river, *209*, *227*
  in rock cycle, *269*
  saltating, 776
sand blows, 341, *342*
sand-dune environments, 224
sand dunes, *220*, *784*, 787–89, **787**, *788*
  archan, *788*
sand layers, as prone to become failure surfaces, 601, *601*
sand pit, *677*

sand spit, **678**
sandstone, *143*, *146*, 202, **204**, *205*, 207, *207*, *209*, 223, 446, *466*, 575
  in Arches National Park, *387*, 388
  and beach environment, 224
  and calculation of Earth's age, 463
  and cement, 559, 560, 561
  and cliff retreat, *782*
  and coal seam, *525*
  and desert realm, *785*
  formation of, 204
  in geologic history illustrations, *441*
  and Gros Ventre slide, 604
  horizontal bedding in, *439*
  maturity of, 208
  metamorphosis of, *247*
  at Midwest meteorite impact site, 8
  protolith of, 246
  and quartz, 204, *209*, 255
  and quartzite, 246, 382
  recrystallization of, 235
  from river sediments, 224
  in rock cycle, 269
  seismic wave speed through, 361, *361*
  and shale, *194*, *209*
  tillites in, 829
  under dike, *165*
  zircon found in, 470
sand volcanoes, 341, *342*, 352
Sandy, Hurricane, 688, 761
San Francisco region:
  active faults in, *334*
  earthquakes in, *317*, 322, 333, *334*, 343, 389
  Golden Gate Bridge in, 210
  and gold rush, 141
  slumps near, *607*
San Joaquin Valley, California, land subsidence in, 714, *715*, *717*
San José mine, 558
San Juan River, Utah, 638, *638*
Santa Ana Volcano, *283*
Santorini volcano, 306, *307*
Santuit Sandstone, 446
sapphire, 136, *138*
Saraha Desert, dust cloud from, *793*
Sargasso Sea, *664*, 665
satellite exploration, of Mars, *777*, *778*
satellite interferometry (InSAR), for volcano eruption prediction, 303, *304*
satellite measurements, of ocean floor, *660*
saturated zone, 701, *701*
Saturn, 15
  in geocentric image, *15*
  moons of, *308*, 583, *585*
savanna, *763*
scale, 706
Scandinavia:
  erosion by glaciers in, 248
  ice sheet over, 826
  see also Norway
scandium, 564
scanning electron microscope, *209*
scattering (light), *734*
Schiaparelli, Giovanni, 584

schist, **241**, *243*, 246, *254*, 255
  foliation of, 398
  from metamorphism, *247*, 252
  in New York bedrock, 560
  in rock cycle, 263
schistosity, 241, 398
Schoharie Formation, 446
science, **8**
science literacy, from study of geology, 8
scientific laws, **9**
scientific method, 8–9, **8**
scientific paradigm, 87
scientific revolution, **87**
scoria, 171, *172*, **172**, 278, 281
scoria cones, 284
Scotia arc, *175*
Scotland, *233*
  Cenozoic dikes in, *166*
  Firth of Forth in, 559
  flow folds exposed in, *397*
  landscape of, 234
  during last ice age, 826
  rock exposures in, 437
  Siccar Point (Hutton's observations at), 442–43, *443*
Scott, Robert Falcon, 203, *204*
Scott, Walter, 2
scouring, **623**, 823
scour marks, 219
sculpture, and marble, 244, *246*
sea arch, 679, *680*
sea (ocean) floor, 72, *73*
  age of, *95*
  bathymetric provinces of, *660*
  cores of sediments in, 68
  dropstones collected on, *807*
  manganese-oxide minerals on, 553, *553*
  maps of, *44*, 72, *73*, *74*
    see also bathymetric map
  as proportion of Earth's area, *44*
  sediment on, 72–73, 82, 226
sea-floor spreading, 62, **75**, 76–83, *93*, *99*, *111*
  and Cretaceous mid-ocean ridges, 492–93
  and divergent plate boundaries, 92–95, *92*, *93*, *94*–95
  evidence for, 79
    in deep-sea drilling, 82–83
    in marine magnetic anomalies and magnetic reversals, 76–83
  and glaciation, 831
  Hess's argument for, 74–76
  and Pangaea breakup, 495
  and plate tectonics, 87
  rate (velocity) of, 79
  and sea-level changes, *487*
sea ice, **808**
sea level:
  changes in, 229, 500, 686–87, *686*, 831
    causes of, 486, *486*, *487*
    contemporary, 688
    in Cretaceous period, 490, 492–93, 495, 497
    and glacier melting, 681
    from global warming, 870, *872*
    during ice age, 812, 819–22, *821*, *823*, 827

in early Paleozoic Era, 482
and growth of large undersea
  basalt plateau, 179
as manifested in stratigraphic
  record, *486*
in Middle Jurassic Period, 489
during Ordovician Period, 483
recognizing past changes in, 469
rise as threat, 688, *688*
and stratigraphic sequences, 486,
  486
as ultimate base level, *627*
sea-level change:
  over geologic time, *845*
  global warming and, 870
sea-level cycle, 844–46, *845*
seal rock (oil and gas), *512*, **512**, 513,
  514
seamount chains, 65
seamount/island chains, 110
seamounts, *72*, *74*, 101, *103*, 661,
  661, *685*
  and abyssal plain, 661
  and hot spot volcanoes, 101, 102
  and plate tectonics, 110
seas:
  and numerical age determinants,
    458
  see also oceans
seasonal floods, 640–44, **640**
  and dam construction, 650
seasons, 739
seasons, and tilt of Earth's axis,
  742–43
sea stacks, 679, *680*
seawalls, 689–90, *691*
secondary-enrichment deposits, 552,
  552
secondary porosity, 697, 698
second atmosphere, 730
sediment, *208*
  in desert, 787
  and soil production, *192–93*
sedimentary basins, 185, 228–29,
  228, *229*
  and burial metamorphism, 250
  and correlation, *448*
  eastern North America as, 411
  foreland, 469
  oil from, 540
  as oil sources, 517
  and plate tectonics theory, 228–29
sedimentary bedding, *see* bedding
sedimentary breccia, *208*
sedimentary cycle chart, *845*
sedimentary deposits of metals, 552,
  553
sedimentary deposits on Mars, 584
sedimentary environment, 220–27
  terrestrial (nonmarine), 224–26,
    225, 226
sedimentary maturity, *207*, 208
sedimentary rock, erosion of, *232*
sedimentary rocks, 45, 115, **144**, *146*,
  185, *202*, **203**, 204, 205,
  208
  biochemical, **203**, 210
    chert, 210
    dolostone, 213–14
    limestone, 210
    replacement and precipitated
      chert, 214, *215*

chemical, **203**, 212–15
  evaporites, 212–13, *213*
  travertine, 213
classes of, 203–15
clastic, 203–10, **203**, *206*, 222
under Colorado Plateau, *784*
and continental drift, 63–64
determining age of, *455*, 460–61
failure surfaces in, *601*
formation of, 204–6, 223
fossils in, 418, 428–30, 437
in Grand Canyon, 451
history of Earth in, *223*
and isometric dating, 458, 460–61
in nonconformity, 444
organic, **203**, 210–12, *212*
  coal, 211–12
primary porosity of, 698
in redbeds, 481
and rock cycle, 261, 262–65, *262*,
  269
  see also rock cycle
and snow, 797
source rocks as, 513
systematic joints in, 388
uranium in, 530
in volcanoes, *287*
see also specific types of rock
sedimentary sequence, **844**, *845*, 846
sedimentary strata, *202*
  and climate, 849
sedimentary structure, 215–20, **215**
  bedding and stratification, 216,
    217
  bed-surface markings, 219
  ripples, dunes and cross bedding,
    216–18, *219*, *220*
  turbidity currents and graded
    beds, 218–19
  value of, 224
sedimentation, 404–5
  in glaciers, *816*
sediment budget, 678, *678*
sediment deposition, *see* deposition
sediment liquefaction, from
  earthquakes, 341–42, **341**,
  342
sediment load, 624, *624*
sediments, 46, 183–201, *183*, **185**
  in accretionary prism, 97, 607
  alluvial-fan, **224**
  along Gulf Coast, 490
  in Archean Eon, *474*
  in Basin and Range Province, 496
  in basins, 404, *404*, 411
  and bedding, 216
  in carbonate environments, 226
  chalk from, *469*
  and climate change record, 849
  and coastal plain, 490
  and coastal pollution, 690
  and coastal variability, 687
  conditions for accumulation of,
    429–30
  on continental crust, 265
  and continental shelf, *267*, 658
  deposition of, 576
  in desert, 770, 773, 774, *785*
  and disconformity, 444, *444*
  evidence of life in, 473
  and fossils, 418, 419–20, 423, 424,
    480, 482, 828–29

in generation of oil and gas, 510
and geologic history, *455*
in glacial lakes, 815
in glacial outwash, 818
and glaciation record, *830*, 831
in glaciers, 808, 809, 813, *814*
  deposition of, 814–15, *821*,
    828
  distinct layers of, 827
  Laurentide ice sheet, *828*
  Proterozoic, *479*
  and tillites, 831
and groundwater, 705
from icebergs, *807*, 808
in lake, 226
marine, 828–29, *830*
at Midwest meteorite impact
  site, 8
moraines from, 815, *821*
from mountain erosion, 381
in Nile River canyon, 184, *184*
and numerical age determinants,
  458
in oceanic crust, 51
on ocean (sea) floor, 72–73, 82, 83,
  656, *657*, *659*, *661*
  and waves, 673
and orogeny, *404*
and oxygen in atmosphere, 481
and Ozark Dome, 411
pores of, 698
  and primary porosity, 698
and principles for defining relative
  age, 437, 438, *439*, 442
in river, 209, 224, *226*, *227*
in rock cycle, 263, *266*, *267*
and sedimentary rock, 209
shallow-marine deposits of, 226
and snow, 797
in streams, 623–26
  and alluvial fan, 630
  in beach erosion, 690
  braided stream from, *632*, *642*
  and damming, 650
  and deltas, 635; *see also* deltas
  and floodplains, 634
  and graded stream, 629
  and river pollutants, 651
  transported by, 615, *624*, *625*
  transported by (Mississippi),
    565
transportation of, 205
  by Canadian glaciers, 560
  and drainage networks, *266*
  and landscape modification,
    860
  by mass movement, 588
  by streams, 615, *624*, *625*
in turbidity currents, 659
and unconformities, 444
on U.S. coastal plain, *477*
and water table, 702
and waves, 676
weathering in formation of,
  185–94
weathering of, physical, 186–89,
  186
in western U.S. (middle Paleozoic),
  483
sediment sorting, 626
seed (crystal), 126
seeps, hydrocarbon, 513, *513*

seiche, 340
seif dunes, 788
seismic belts, **74**, *75*, **90**, **332**
seismic gaps, 352
seismic-hazard assessment, 351
seismic-hazard maps, **351**, *353*
seismicity, **315**
  induced, 337–38, **337**
seismic ray, **361**
seismic record, digital, 326
seismic record, digitalseismic-
  reflection profile, 371, 515,
  515, 657, *657*
seismic retrofitting, 355
seismic tomography, 367–69, *367*, *368*
seismic-velocity discontinuities, **364**
seismic waves, 49, 320, 323–27, **323**,
  361–62, 515
  and defining of structure of
    mantle, 364
  and discovery of core-mantle
    boundary, 365, *366*
  and discovery of crust-mantle
    boundary, 362–63, *363*
  and discovery of nature of core,
    365–67, *366*
  movement of, 361–62
  and new discoveries about Earth,
    367
  period of, 330
  propagation of, *361*
  reflection and refraction of, 362,
    362
  and seismic-reflection profiling,
    371, *371*
  and seismic tomography, 367–69,
    368
  and study of Earth's interior, *359*,
    360–72
seismic wavestypes of, *324*
seismic zones, 332, **351**
seismogram, **326**, *327*
seismograph, 325, **325**
  Cold War deployment of, 367
  and Richter scale, *330*
seismologists, **314**
seismometers, 325–26, *325*, *326*
self-exciting dynamo, **376**
semiprecious stones, 137
septic tanks, 716, *718*
serpentine, 564
serpentinite, 132
Sevier fold-thrust belt, 490–91, *492*
Sevier orogeny, **490**
Shackleton, Earnest, 796
shaded-relief map, 574, *574*
shale, *202*, *207*, *208*, *209*, **210**, *223*,
  255, 387, 575
  black organic (oil and gas from),
    508, 510
  and cement, 559, 560
  and cliff retreat, 781, *781*
  and coal seam, *525*
  in geologic history illustration,
    441
  as impermeable, 702
  lake-bed, 225
  by lithification, 210
  metamorphism of, 235, *236*, 245,
    247
  and Monte Toc landslide, 594
  at Niagara Falls, 629, *632*

shale, *(continued)*
oil, 212, *520, 521*
organic material in, 212
prograde metamorphism of, 246
as prone to become failure
surfaces, 601
from river sediments, 224
in rock cycle, 263, *269*
and sandstone, *209*
and slate, *231*, 382
undeformed, 398
unmetamorphosed, *231*
shale gas, 517, 519–20, *520*, **520**, 522
shallow earthquakes, 332, *332*, 334
shallow-marine clastic deposits, 226
shallow-water carbonate environment,
226
Shanghai, China, *838*
shark tooth, fossilized, 436, *436*
Shasta, Mt., *285*
shatter cones, 8, *9*
shear, *240*, 391
between wind and water, *673*
shearing, 251, 253, *253, 254, 255*,
398, *399*
under metamorphic conditions,
240
shear strain, 383, *385*
shear stress, 238, *239, 385*, 387, 388,
398
shear waves, 323, *324, 361, 362*
shear zone, **391**, *393*
Sheep Mountain, *396*, 604
sheet silicates, 131–34, *133*
sheetwash, *616*, **616**, 617
pediments from, 787
shell, with growth rings, *459*
Shelley, Mary, 306
Shell Oil Company, *371*
shells (orbitals), 120, *120*
shell-secreting organisms, 480
and numerical age determinants,
458
shield, **257**, **410**, *476*, **476**
shield volcanoes, **284**, 285, *285*,
302–3
oceanic hot-spot volcanoes as,
293, *295*
Olympus Mons as, 309
shingles, slate, *242*
Shiprock, N.Mex., *166*, 303
ships for research, *see* research vessels
shock metamorphism, 253, **253**
shortening, 383, *385*
short-term climate change, 847,
855–58
short-term predictions of
earthquakes, 351, 352–53
shows of ore, 555
Siberia, 297, *298*, 901
basalt in, 857
in Cambrian Period, 482
and creation of Pannotia, *478*
mammoth found in, 422, *422*
in Paleozoic Era, 484, 485
in Pangaea, 484
in Pleistocene ice ages, 826
Siberian Shield, *258*
Siberian Traps, 297, *298*
Siccar Point, Scotland, Hutton's
observations at, 442–43,
443

Sicily, Mt. Etna on, *300, 305*
side-scan sonar, 657, *657*
Sierran arc, 489, 490
Sierra Nevada, Calif., *497*
roche moutonnée in, *812*
Sierra Nevada Batholith, *168*
Sierra Nevada Mountains, 490, *784*,
808
glaciers in, 826, 827
gold rush in, 141, 546, *546*
and railroad building through,
141
Yosemite in, 827
silica, 46, 158, *159*, 160
in basic metamorphic rocks, 245
in cement, 560
and chert, 210
in glass, 561
and lava viscosity, 275–76
and magma type, 165
and viscosity, 160
silicate minerals, 46, 131, 143
silicate rocks, **46**, 129
silicates, 131–34, **131**
siliceous rocks, 204
silicic lava, 275–76
silicic melt, *159*
silicic rocks, *172*
silicon, *124*
in crust, *53*, 143
in quartz, 119
silicon-oxide tetrahedron, **131**
silicon-oxygen tetrahedron, 160
Silliman, Benjamin, 117
sillimanite, 117, 237, *238*, 240, *247,
249, 251*
sills, **162**, *165, 168, 182, 286, 287*
cooling of, *172*
in geologic history illustration,
438, *441*
and inclusions, *441*
silt, *187, 192, 196*, 205, 206
and bedding, *217*
and deltas, 226
in desert, 775
in floodplains, 209
lithification of, 209
in river, 227
siltstone, *207*, **209**, *223*, 224, 226
fossils in, 420
Silurian Period, 483, *484*, 498
and coal formation, 524
life forms in, 450
silver:
of Andes, 555
as native metal, 547
as precious metal, 547, 549
Sinai Peninsula, *104*
single-chain silicates, 131, *133*
sinkholes, *342*, *695*, **695**, *696*, 721,
722, 724, *725*
in Florida, 695, *695*, 696
sky, blueness of, 733, *734*
slab-pull force, **107**
slag, 547
slash-and-burn agriculture, 860, *860*
slate, **241**, 246, *251*, *255*, **255**
and shale, 246, *247*
slaty cleavage, 241, *242*, 255, 382,
398, *399*
sleet, 747
slickensides, 391

slip, 94
in faults, 315, *317, 318*, 319–22,
389–93
flexural, *397*
in Turkish earthquakes, 351–52
slip face, 218, **789**
slip lineations, 391
slope failure, **600**
slope failure, factors in causing of,
601–6, 610
slope stability, 600–601, *600*, 604–6,
605, 606
slope steepness, and soil
characteristics, *196*, 197
Sloss, Larry, 486–87, *487*
slot canyons, 629
Slovenia, Kras Plateau in, 674
slumping, *341*, 590–91, *590*, 602,
609, 610
around Hawaiian Islands, *103*
near San Francisco, 607
in southern California, 607–8
slumps (volcanoes), 295
smelting, **547**, *548*, 557
Smith, William, 419, 441, 445, 446
Smithsonian Institution, 136
smog, 733, 861
Snake River Plain, Idaho, 295, 297,
496
snotites, 724
snow, 616, 747, *748*
in formation of glacier, 797–99
snow avalanche, *603*
snow avalanches, **594**
snowball Earth, 479, **480**, *830*
Snowdon, Mt. (Wales), *187*
snowflakes, *798*
soda straw, 720, *724*
sodium, *53*, 120, *123*, *124*, 157, *158*,
191
soil, 185
soil color, 197
soil contamination, 200
soil creep, *589*, 602
soil erosion, 199–200, **199**, *200*
soil formation, in desert, 772–73
soil-forming factors, 196
soil horizons, *195*, **195**
soil moisture, 696–700, *701*
soil profile, **195**
soils, 183–89, *183*, 195–200
ancient (paleosol), 827
classification scheme for, 197–99,
198
desert, 776
formation of, *192–93*, 195
map of types of, *198*
moisture in, 616
as ore deposit, 553, *554*
organisms in, 196
use and misuse of, 199–200
variety of, 197–99
weathering and sediment in
production of, *192–93*
world map of, 901
soil structure, 197
soil texture, 197
solar collector, 534
solar energy, 507, *507*, 544, 738–39
and greenhouse gases, 730
and rock cycle, 265
and solar panel arrays, *544*

as stored in fossil fuels, 507
in wind, 507
solar power, 534, *534*, 540, *544*
solar radiation, in climate change,
856, *857*
Solar System, **15**, 28, 29–32, **29**, 43
age of, 29
distance of planets from Sun in,
43
Earth's place in, 854
formation of, 27–29
forming of, *30–31*
habitable zone of, *43*
nature of, 15–18
solar variability, 831
solar wind, 40, *40*
Solenhofen Limestone, 425, *426*
solid, *120*, 121
solidification of a melt, 124
solid-state, 234
solid-state diffusion, 124
solidus, *156*
solifluction, **588**, *589, 602, 603*
solstice, 742–43
solution, 121
solution cavities, 698
sonar (echo sounding), 72, *73*, 657,
657
Sonoran Desert, *788*
sorting:
of clasts, *207*, 208
of sediment, 208
sound, Doppler effect for, 21, *22*
source rock, **510**, *512*, 513, *514*
South Africa:
Karoo region of, 179, 297
world's deepest mine in, 557
world's deepest mine shaft in,
47, 360
South America:
and Andean orogen, 405, *405*, 495
and Cretaceous Period, 490
separation of from Antarctica, 832
union of, with Africa (Mesozoic
Era), 636, *638*
South American Plate, 97
South Atlantic Ocean:
in Jurassic Period, *839*
at Pangaea breakup, *489*
Southeast Asia, squeezing of, *403*
Southeast Indian Ocean Ridge, *74*
Southern Alps, 44
Southern Alps (New Zealand), 799
southern oscillation, 765
south magnetic pole, **375**
South Poles, *57*, 203, *204*, 799
South Sandwich Trench, *74*
Southwest Indian Ocean Ridge, 100
space, interplanetary, 38
space, interstellar, 25, *37*
space shuttle, views from, *42*
species, 426
specific gravity, 128–29, **128**
speleothems, 213, **214**, 720, **720**, *725*
spelunkers, 720, *725*
sphericity, 206
spheroid, 372
Spirit (Mars rover), 584
Spirit of America (car), 212
Spitzbergen, stone circles of, 826
spodosol, 197, *198*
spreading boundaries, 92, *93*

spreading rate, **79**
springs, *581*, **705**, *707–8*, *707*, *708*
    emerging, *724*, *725*
"spring water," 705
spruce forests, and climate change, *850*
S-P (S minus P) time, 326, *330*
squashing, *255*
Sri Lanka, tsunami in, 346, *347*
stability field, 237, *238*
stable slopes, 600
stair-step cliffs, *782*
stair-step slope, *781*, *782*
stalactite, **720**, *721*, *724*, *725*
stalagmite, **720**, *721*, *724*, *725*
Standard Oil Company, 513
star dune, *785*
star dunes, 788, *788*, *789*
Stardust spacecraft, 39
stars:
    death of, 26–27
    as element factories, 26
    formation of, 25–26
    generations of, 27
    planets distinguished from (age of Homer), 14
staurolite, *122*, *247*, *249*
steady-state condition, **846**
steel, 548
    consumption of, 564
Steinbeck, John, 791
stellar nucleosynthesis, **26**
stellar wind, **26**, *27*
Steno, Nicholas, 117, 122, 419, 436, *436*, 438
steppe regions, *763*
stibnite, *122*
stick-slip behavior, *318*, **318**
stone, 557
    crushed, 557, *559*
stone circles, 826
stony plains, 786–87
stoping, **166**
Storegga Slide, 559, *559*
storms, **729**, 750–61
    global warming and, 870
    hurricanes, 755–61, *758*
    nor'easters, 754
    thunderstorms, 750–52, *750*, **750**, *751*
    tornadoes, 752–54, **752**, *754*
storm surge, **688**, *757*
strain, **383**, *385*
Strait of Gibraltar, 184
strata, **216**
    tilted beds of, *384*
    and unconformities, 442–45, *443*
strategic minerals, **565**
stratification, 216
stratified drift, 815
stratigrapher, 216
stratigraphic column, **445**
stratigraphic formation, **216**, *217*, 445–48, **445**
    correlation of, 445–48, 449–50, *450*, *452*
    and Grand Canyon, 445–46, *447*, *448*
stratigraphic record, on climate change, 849
stratigraphic records, in recording ice ages, 829

stratigraphic sequence, 486–87, **486**, *487*
stratigraphic trap, 514, *514*
stratigraphy, 216
    of sea floor, 808
stratographic correlation, **446**
stratopause, 737
stratosphere, *42*, *43*, **737**
    volcanic materials in, 306
stratovolcanoes, 284–85, **284**, *285*, *287*, *288*, *293*
stratus clouds, 748
streak (mineral), **127**
streambed, 615
stream cut, *145*
stream gradient, **626**
stream piracy, 636–368, **636**, *637*
stream rejuvenation, 637–38
streams, **615**, *616*
    antecedent, 638–39, *639*, **639**
    banks of, 615
    base level of, 627, *627*
    braided, **630**, *632*, 815, 817
    clasts in, *625*
    deltas of, 626, 634–35, *639*
    depositional processes of, *223*, 623–26
    in deserts, *784*
    disappearing, *722*
    discharge of, 621–23, *621*, *622*
    drainage network of, *639*
    ephemeral, **620**, *621*, 789
    and erosion, 624, *624*
        headward erosion, 617, *617*, 629, 636, *643*, 785
        and meanders, 631–34, *633*
        in stream creation, 615–18, *617*
    formation of, 615–18
    gaining, 620
    longitudinal profile of, **626**, *627*
    losing, 620
    meandering, 630–34, *633*, *642*
    permanent, **620**, *621*
    rapids in, 629, *631*
    sediment loads of, 624, *624*, *625*
    seepage into increased by paving, 578
    sources of, 615–18
    superposed, **638**, *639*
    turbulence of, *623*, *625*
    waterfalls in, 629, *631*
    as water reservoir, *581*
    see also rivers
stream terraces, 629, *630*
stress, **316**, 317–19, *318*, 386–87, *386*, **386**
    differential, **238**
    and earthquake building codes, 354–55
    normal, 238
    shear, 238, *385*, *386*, 387
stress-triggering models, 353
stretching, 383, *385*
striations, 64
    glacial, **809**, *821*
strike, 387, 388–89, **388**
strike line, 316
strike-slip faults, 316, *317*, 320, *321*, 323, 333, *333*, *352*, *390*, **391**, *393*
strip mining, 526, *527*

stromatolites, *475*, **475**
Stromboli, Italy, 288
Strombolian eruptions, 288, *289*
Strombolian eruptive style, *286*, 288, *289*
Stromboli island, volcanic activity on, 288
subduction, 62, **75**, **96**, *98*, 105, *106*, 112, 334, *336*, *337*
    along West Coast, *496*
    and carbon in mantle, 135
    and convergent plate boundaries, 96–98
    and dynamothermal metamorphism, 255
    in early Earth history, 843
    and formation of igneous rocks, 175–76
    in mountain building, 400–401
    and volcanic activity, *175*
subduction zone, *92*, 96, *98*
    and future, 873
    metamorphism in, 253
    and rock cycle, 263
    and subaerial zones, 293
sublimate, **799**
sublimation, 805
submarine avalanche, 595
submarine basaltic lava, 276
submarine canyons, *223*, **659**, *660*
submarine debris flows, *596*
submarine fan, 222, *267*, 659
submarine landslides, 559
submarine mass movements, 596–98, *598*
submarine plateaus, during Cretaceous Period, 492–93
submarine slumps, **596**, *598*
submarine volcanoes, 276, 292
submergent coasts, *687*, **687**
submersibles, 656–57, *656*
subsea volcanoes, 292
subsidence, **228**, 574–76, **574**
    causes, *574*
    glacial, **819**, *822*
    thermal, 228
subsoil, **196**
substrate, **576**, **588**, 600
substrate composition, and landscape development, 575, 578
substrate composition and soil characteristics, 196, *196*
subsurface water:
    reducing of, 610
    see also groundwater
"subterranean heat," 144
subtropical deserts, 771, *771*
succulents, 789
Sudbury, Ontario, smoke from smelters in, 566, *567*
Sue (dinosaur fossil), 421
sugarcane, ethanol from, 532
sulfates, 131
sulfides (sulfide minerals), 129, 554–55, 566
    around black smokers, 551, *551*, 554, *562*
    ore minerals as, 549
sulfur crystals, *125*
sulfur dioxide, 472, 539
    in volcanic gas, 730
sulfuric-acid speleogenesis, 720

Sumatra, earthquakes in, *254*, 335, 346
Sumatra, tsunami in, *347*
Sumeria, bronze discovered in, 547
summit eruptions, 282
Sun:
    burning of, 27
    and change in Earth, 839
    color of, 734
    Earth's distance to, 19
    future of, 873–74
    in geocentric image, *15*
    mass of, 15
    temperature of, 854
    and tides, *668*, 669
Sunda Trench, 343
sunlight, *734*
Sunset Crater, *277*, *282*, *285*
sunspot cycle, **856**, *857*
Supai Group, *218*
supercell, **750**
supercontinent, 873
supercontinent cycle, 844, **844**
supercontinents, *478*
    Gondwana, 482, *482*, 484, 490, 495, *496*
    Pangaea, *62*, *63*, *63*, 485
        see also Pangaea
    Pannotia, 477, 482
    Rodinia, 477, *478*, 482
supercritical fluid, 239
supernova, 26
supernova explosion, 26, *27*
    and future of Sun, 873–74
supernova explosions, 45
superplumes, **179**, *492*, 661, 857
superposed streams, **638**, *639*
superposition, principle of, 438, *440*
Superstorm Sandy, 761
supervolcanoes, **292**
surface currents, 663–65, *664*
surface load, **776**
surface tension, 600
surface water, 44–45, **44**
surface-wave magnitude (MS), 330
surface waves, **323**, 362
surface westerlies, 741
surge (glacier), **805**
Surtsey, *175*, 295
Surtseyan eruptions, 288, *289*
Surtseyan eruptive style, 288, *289*
suspended load, 624, **624**, **775**
sustainable growth, **873**
Sutter, John, 546
suture, 105, *106*, **401**
    between Indian and Asian Plates, *402*
swamps, 229, 616, 681, *682*, 715
    coal, 485, *524*
    Everglades as, 714, *715*
    mangrove, *676*, *682*, *682*, *684*
    between recessional moraines, 818
swampy deltas, *676*
swash, **673**
S-waves (secondary waves), 323, *324*, 326, *328*, 339, 361, *361*
S-wave shadow zone, 366, **366**
swelling clays, 606
swells, **672**
Switzerland, *866*
    chatter marks in, *809*
    Earth System in, *7*

Switzerland, *(continued)*
    glacially carved peaks in, *409*
    glacier visit to, *827*
    Matterhorn in, *810, 811*
    stream in, *625*
Syene, Egypt, 18–19
symbiotic relationship, of corals and
    algae, 682
symmetry, 123, *123*
synclines, **394**, *395, 396, 397*
    Uluru as, *786*
synthetic minerals, 118
systematic joints, 388
    in Arches National Park, *387*

tabular bergs, *807, 808*
tabular intrusions, 162, *287*
    cooling of, 160
tachylite, *172*
Taconic orogeny, 412, 482
tailing piles, *555, 566, 578, 604*
tailings, 556
Taiwan, landslide in, *586*
talc, *127*, 128
talus, *187*, **187**, **595**, *597*
talus apron, **778**, *779, 782*
Tambora, Mt., *290*, 305, 855
tankers, 517, *518*
Tanzania, Olduvai Gorge in, *420*
tar, 509, *513*, 517
    and fossilization, 422
Tarim Basin, *403*
tarn, 810
tar sands (oil sands), 520–21, **520**
Tavernier, Jean Baptiste, 134
taxonomy, 425–27, **425**, *428*
technology, REEs used in, 564,
    565–66
tectonic activity:
    and antecedent streams, 639
    in geologic history, *455*
    and global climate change, 853
    and mass extinction, 432
    see also plate tectonics
tectonic foliation, **398**, *399*, 405, *711*
    see also foliation
tectonic processes, and soil
    production, *192–93*
tectonics (subdiscipline), 6
Teddy-bear cholla, *790*
temperate glaciers, **799**
temperature, 54
    and air pressure, 737
    in climate change, *852*
    of Earth's surface, 843, 854
    inside Earth, 52
    of ocean water, 662–63, *663*
    and rock deformation, 384–85
    vs. heat, 733–35
    see also global warming
tempering, 547
temporal cycle, 265
Tennessee, earthquake in, 337
tension, 94, 238, 387, 388
Tensleep Formation, 604, *605*
tephra, *275*, **278**, 285, 287, *287*
terminal moraine, **815**
terminator (day-night boundary),
    742–43
terrace, *676*
    at emergent coasts, 686
terraces, stream, *630*
terra firma, 588

terranes, accreted, *400*, 401, *489*
terrestrial planets, *17*, **18**, *40*, 41
terrestrial sedimentary environments,
    224–26
Tertiary Period:
    in correlation of strata, *452*
    and K-T boundary event, 494–95,
    857
testing of hypothesis, 8–9
Tethys Ocean (Sea), 495
Tethys Sea, 832
tetrahedra, 131, 134
    independent, 131
texture, in rock identification,
    166–69, *170*
texture, of rocks, 147, 234
Thailand, tsunami in, *347*
thalweg, **622, 623**
Tharsis Ridge, 582, *583*
Theia (protoplanet), 471
theories, **9**
theory of evolution, **431**
    see also evolution
theory of plate tectonics, **6**, 87
    see also plate tectonics
Theory of the Earth (Hutton), 437
Thera, 307
thermal energy, 54
    and greenhouse gases, 848
thermal expansion, physical
    weathering from, 188
thermal lance, 557
thermal metamorphism, 250, **250**,
    252
thermal subsidence, 228
thermocline, 663
thermohaline circulation, 667, **667**,
    832, 870
thermosphere, *42*, 43, **737**
thin section, **148**, *150*
thin-skinned deformation, 485
third atmosphere, 731–32
Thoreau, Henry David, 37
Three Gorges Dam, 533, *533*
Three Mile Island nuclear incident,
    531
thrust fault, 316, *317*, **391**, *394, 402*
    and fold-thrust belt, 401
thrust faults, 335
thrust sheets, 391
thunder, **752**
thunderstorm, 750, **750**, *751*
Tibetan Plateau, *403*, 495, 763, 833
tidal bore, 667, *669*
tidal bulge, *668*, 669
tidal flat, 667, *669*, 677, *678*
tidal power, **533**
tidal range, **667**, *668*
tidal waves, 344
    see tsunamis
tide-generating force, **669**
tides, 507, 667–72, **667**, *684*
    causes of, 670–71
    forces causing, 669, 672
    hydroelectric power from, 533,
    533, 536–37
    manifestation of, *668*
tidewater glaciers, **806**
Tien Shan Mountains, *403*
Tigris and Euphrates Rivers, 650,
    792
Tiktaalik, *484*
til, glacial, 64

till, 815, *817*
    glacial, 222, 224, *225*, **815**
    lodgment, 815
tillites, 829–30, **829**
tilt of Earth's axis, 739, 742–43, 831,
    832, 855
    and tides, 669
time, 435, 436
    geologic vs. historical, *see* geologic
    time
Timpanogos Cave, Utah, *214*
tin, 549
Titanic sinking, 662, 806
Titusville, Pa., oil drilling at, 513
Toba volcanic eruption, *290*
Toba volcano, 307–8
    eruption of, 855, 858
toe of glacier, *805*, **805**, 808, 810,
    814, *817*
Tohoku-Oki earthquake, 319, 335,
    343, 531
    and tsunami, 314, *314*, *358*
Tohoku-Oki tsunami, 348, *350*
Tokyo:
    earthquake in, 343
    see also Japan
Tolstoy, Leo, 546
tombolo, 679, *680*
Tonga, subsea volcano near, *289*
Tonga Trench, 74
topaz, *127*, 136, 137
topographic map, 574, **575**
topographic profile, **575**
topography, **44**, 574
    and deformation, *405*
topsoil, **196**
"tornado alley," *755*
tornadoes, 752–54, *752*, **752**, *753*
    Fujita scale for, 754, *754*
    and thunderstorms, 752–54
tornado outbreak, 752
tornado swarm, 752
tornado warning, **754**
tornado watch, 754
Toroweep Formation, *218*
torrents, 823
tourmaline, 136, 137, *140*
tower karst, 722, *722*
trace fossils, **423**
trade winds, **741**
Trans-Alaska Pipeline, *518*, 825
Transantarctica Mountains, 203, *204*
Transantarctic Mountains, 806, *806*
transform, on volcanic map, *175*
transform boundary (transform fault),
    99, **99**, *100*
transform fault, 56, 92, 112
    oceanic, 659, *684*
transform fault (transform boundary),
    99, *99*, *100*
transform plate boundaries,
    earthquake at, 348
transform plate boundary, **90**,
    98–100, 108, *109*, 112
    earthquakes along, *333*
    and western margin of North
    America, *496*
transform plate boundary seismicity,
    333, 348
transforms, 92, 98–100, 659
    on map of relative velocities, *110*
transgression, **229**, *230*, 486–87, *524*
transition zone, *52*, 57, **364**

transport, grain size and, *207*
transportation, 582
transporting agents, and landscape
    development, 577
transverse dunes, 788
trap (oil and gas), **512**, 513–15
travel-time curve, 326, *328*
travel time (seismic wave), 361–62,
    361
travertine (chemical limestone), 213,
    214, 458
    dripstone as, 720
"tree of life," 429, *430*
treeline, highland polar climate
    above, *763*
tree rings, **458**
    and climate change record, 458,
    459, 849, *851*
    in determining numerical age, 458
trees, in creeping soil, 588, *589*
trellis drainage network, 618, *618*
trellis drainage pattern, 255
trenches, 44, 56, **72**, *96*, 659, 685
    as convergent boundary, *92*
    and sea-floor spreading, *75*
triangulation, 327
Triassic Period, **498**
    in correlation of strata, *452*
    dinosaurs in, 489
    life forms in, 451
    Pangaea breakup in, 487, *489*
    sea level during, *845*
tributaries, *618*, 627
    base level for, 627
    yazoo streams, 634
trilobites, 427, *429*, 483
triple junctions, 100, **100**, 112
    East African Rift at, *104*
tropical rainforests, 741, *763*, 860,
    860
Tropic of Cancer, 742–43
Tropic of Capricorn, 742–43
tropopause, 736–37, *741*
troposhere, *42*, 43
troposphere, **736**
    winds in, 742–43
trough, 672
"true polar-wander" model, *70*
truncated spurs, 810
trunk stream, **618**
tsnuami, damage from, 343–50
tsunamis, 307, 314, *314*, 335, *336*,
    343–50, *344*, 531, 559,
    559, 597
    causes of, 346
    earthquake damage from, 343–50
    far-field (distant), 346
    formation of, *344*
    from Indian Ocean earthquake,
    343
    in Japan, *358*
    from K-T impact, 494
    near-field (local), 346
    predicting of, 351, *358*
    from Storegga Slide, *559*
    tsunami buoy, *356*
    as volcano threat, 301
tufa, 213, *214*
tuff, *163*, *170*, **173**, *174*, *278*, 287
    air-fall, *175*, 292
    rhyolitic, 178
    volcanic, *171*
    welded, 281

tundra, *763*, **826**, *829*
turbidite, 219, *227*
turbidity current, 218–19, *221, 222, 223, 226*, **596**, *598, 659*
turbulence, **623**
during flood, 640
Turkey, earthquakes in, *5*, 340, 351–52
Turkmenistan, *513*
Turnagain Heights earthquake, 341–42, *342, 343*
Turner, Joseph M. W., *273*, 305
turquoise, 137
Twain, Mark, 435, 631, 646
Twelve Apostles (Australian sea stacks), *680*
typhoons, 754, 758
Tyrannosaurus Rex, 421, 494

Uinta Mountains, *597*
Ukraine, nuclear power in, 531
ultisol, 197, *198*
ultra-high pressure metamorphic rocks, 237
ultramafic magma, 158, *158, 159*
ultramafic rocks, 46, 88
as crystalline rocks, *172*
Uluru (Ayers Rock), Australia, 783, 786, *786*
Umnak Island, *176*
unconfined aquifers, 699
unconformity, *222, 223*, 442–45, *443*, **443**, *444, 445, 447, 448*
in geologic history, *455*
and Grand Canyon, 446
regional, 486, *486, 487*
and superposed stream, *639*
types of
angular, 443, *444, 445, 449*
disconformity, 442–45, *444*
nonconformity, *444*
unconsolidated debris, 195
unconsolidated sediment, 184, *185*
soil development on, 196
unconventional hydrocarbon reserve, **517**
unconventional reserves, **510**
undeformed rock beds, *383*
undercutting, **604**, *606, 611*, 679
prevention of, 610
underground coalbed fires, 528
underground mine, **556**, *562*, 566
unidirectional change, 842, *842*
uniformitarianism, principle of, **437**, 438, 439, 442–45, 646
Union Pacific railroad, 141
United Kingdom:
climate of, *762*
dikes in, *166*
Cenozoic, *166*
during last ice age, 826
see also England; Scotland; Wales
United States:
and continental platform, 477
during ice age, *821, 826*, 828
Pleistocene deposits in, *829*
see also coastal plain; Midwestern United States; North America; *individual states*
seismic-hazard map of, *353*
stockpiling of resources by, 565
tornadoes in, 752–54

yearly per capita usage of geologic materials in, *565*
"universal ocean," 144
Universe, **13**
formation of, 21–26
image of, 13–19
modern image of, 15–18
size of, 18–19, 21–22
structure of, 13–15
unloading, *574*
unsaturated zone, 701, *701*, 703
unstable slopes, 600
blasting of, 610
unstratified drift, 814
uplift, 381, 382, 405–6, **405**, 407–9, *408*, 574–76, **574**, 607
during Alleghenian orogeny, 485
and antecedent streams, 639, *639*
basement, 485, 491, *492*
causes of, *574*
and chemical weathering reactions, 843
vs. erosion, 409
and global climate change, 853
in Southern California, 607
upper mantle, *52*, **52**, **364**
heat of, 156–57
upstream region, 615
upwelling, **106**, 369, *369*, 374
upwelling zones, 665–67, *665, 667*
Ural Mountains, 485
uranium, 529–30, 539
uranium enrichment, 530
Uranus, 15
urbanization:
ecosystem destruction from, 858
and Everglades, 715
and river-water supply, 652
USArray, **370**
U.S. Geological Survey (USGS), 901
U-shaped valley, **810**, *811, 821*, 827
Ussher, James, 436
Utah, *3*, 261, 271, 434, 451, *591*
Arches National Park in, *387*, 388
Bonneville Salt Flats in, 212
Bryce Canyon in, *452*, 782, *782*
Canyonlands National Park in, 782
cave in, *721*
clouds in, *748*
exposed sedimentary rock in, *217*
floodplain in, *633*
Great Salt Lake in, 212, 779, *780*, 824, *824*
San Juan River in, 638, *638*
slumping in, *590*
Zion National Park in, *452*, 788
Utan:
debris flow in, *591*
Wasatch Mountains in, *625*

vacuum, **37**
vadose zone, 701
Vaiont Dam, rockslide disaster at, 594
Valdez, Alaska, tsunami damage of, 348, *348*
Valles Marineris, 582, *583, 613*
valley, shape of, *629*
Valley and Ridge Province, Pa., *405*, 413
valley glaciers, 799, *800, 803*, 805, 810, *821*, 827

shrinking of, 864
Valley of the Mummies, 709
valleys, **629**
alluvium-filled, 91
drowned, *676, 687*
formation of, 628–29
from rivers, 408, *409*
and glacial erosion, 810, *821*
hanging, 629, *631*, **810**, 827
U-shaped, **810**, *811, 821*, 827
V-shaped, *628*, 629, 810, *811*
Van Allen radiation belts, 40, *40*, 58
Vancouver, air temperature in, 663
varve, **815**
vascular plants, 483
vectors, 670–71
vegetation, and slope stability, 606, *606*
vegetation types:
distribution of (and global warming effect), *871*
and soil formation, 195, 197
vein deposit, 551, *551*
veins, **240**, 387–89, *387*, **388**
velocity-versus-depth curve, **367**
velocity-versus-depth profile, 367, *367*
Venezuela, tar sand in, 520
Venice, flooding of, 716, *717*
vent, **282**
ventifacts, 777, **777**
Venus, 15, 18, *40*, 839
atmosphere of, *40*, 582, 730
Earth contrasted with, 7
in geocentric image, *15*
landscape of, 582–83, *583*
layers of, 56
runaway greenhouse effect on, 854
temperature of, 854
volcanic edifices on, *308, 309*
Vermilion Cliffs, *453*
Vermont, quarry face in, *211*
Verne, Jules, 47
vertical-motion seismograph, 325, *325*
vertisol, 197, *198*
vesicle, **170**, *174*, **281**, 698
Vesuvius, Mt., *175*, 273, *273, 285*, 288, *290*
Victoria, British Columbia, glacial striations in, *809*
Vietnam, cave in, *721*
Vikings, 855, *856*
Virginia, Natural Bridge in, *725*
Virgin Islands, 675
viscosity, **160**, 272–75, *275*, **275**, 508
of lavas or magmas, 160, *160*, 272–75
and temperature, 53–55
volatile materials, **28**
volatiles (volatile materials), 46, 157, 176
in magma, 157
and viscosity, 160
volatility, 508
volcanic activity, 152–82, 271, 272–311
volcanic agglomerate, **173**
volcanic arcs, **72**, 97–98, *97, 98, 175*, **175**, 400, 401, *402*
and growth of North America, 490
and ocean currents, 853

volcanic ash, 278, *287, 292*
and fossils, *420*
threat from, 299, *300*
volcanic blast, as volcano threat, 299
volcanic breccia, **173**, *174*
volcanic danger (volcanic hazard) assessment map, 303–4, *304*
volcanic debris flow, 278–81
volcanic edifice, 282
volcanic emissions, in climate change, 855
volcanic eruptions, 271, 272–311, *289*
along convergent boundaries, 293
along mid-ocean ridges, *175*, 293
carbon dioxide produced by, 862–64
cool climate due to, 305–8
explosive, 278–81, *292*
through fissures vs. circular vents, *282*
geologic settings of, 292–97
as hypothesis for anomalies at Midwest site, 8
lava flows from, 272–75
and long-term climate change, 853
memorable examples of, *290–91*
in mountain building, 407
prediction of, 303
products of, 275–82
protection from, 302–5
pyroclastic debris from, 278–81, *279*
volcanic explosivity index (VEI), **290**
volcanic gas, 281–82
in atmosphere, *304, 306*, 730, 843
secondary atmosphere, 730
and eruptive style, 288–92
threat from, 301–2, *301*
in volcano prediction, 303
volcanic island arc, **97–98**
volcaniclastic deposits, 278
volcaniclastic rock, **173**
volcaniclastic sediment, 281
volcanic neck, *166*
volcanic-sedimentary deposits, 281
volcanic tuff, *171*
volcanoes, 139–67, **153**, 272–311, *273*, 473, 839
in ancient Rome, 273–75
architecture and shape of, 282–85, *284, 285*, 303
and civilization, 305–8
and climate, 305–8
eruptions of, *163*
eruptive styles of, 282–92, *289*
extinction of, 101–2, *103*, 112
hazards from, 298–302
controlling of, 302–5
hot-spot, *see* hot spots
lava flow from, *152*
location map of, *901, 909*
magma and igneous rocks in, 152–81
and magma movement, 162
on other planets, *308, 309*
in Permian and Mesozoic, 489
and Scottish outcrops, 234
subaerial, *175, 176*
subsea, 292
worldwide distribution of, *294*

Voltaire, and Lisbon earthquake, 339
Voyager spacecraft, 37
Voyageurs National Park, Minn., 827
V-shaped valley, *628*, 629, 810, *811*
Vulcan, 273
Vulcanian eruptions, 288
Vulcanian eruptive style, *286*, 288, *290*
Vulcano island, 273

wacke, *207*, 210
Wadati-Benioff zone, **96**, *97*, **334**, *335*
wadi, **775**, *784*
See also dry wash
Wales:
metamorphic rocks in, *240*
Mt. Snowdon in, *187*
rock beds on coast of, *383*
tidal flat along coast of, *677*
Wallace, Alfred Russel, 431
wall rock, **154**, 157, 162
See country rock
warm front, 745, *745*
wars, and Earth's resources, 565
Wasatch Mountains, 408
Wasatch Mountains, Utah, *624*
washes, 775
Washington State, *281*
Cascade Mountains in, 771, *772*
Olympic Peninsula in, *677*
scouring of, 823
waste rock, 556
water:
atmospheric, 695, 730, 735, 843
and latent heat, 735
as greenhoue gas, 848
groundwater proportion of, 714
See also groundwater
in Hadean atmosphere, 472, *472*
and hydrologic cycle, 579–82, *581*
in joints, 388
and Mars, *584*
meteoric, 622
molecular structure of, 121
molecule of, 797
of ocean, 662–67, *662*
overuse of, 650
reservoirs of, *579*, *581*, 582
rise of, 718–19, *719*
running, 615
and slope stability, *605*, 606
subsurface (reducing of), 610
surface, *see* lakes and lake beds;
rivers; streams
in volcanic gas, 730
water erosion, 774–75
waterfall, 629, *642*
from hanging valley, *828*
from Mediterranean into Black
Sea, 852
at Yosemite, 827

water gap, 618, 636, *637*, *639*
water masses, in ocean, 667
watershed, **619**, 622, 627, *642*
See also drainage basin
water table, *611*, 616, 701–3, *701*, **701**, *702*, 722
and cave formation, *725*
human intervention in, *715*
lowering of, 714
topography of, 701–2, *703*
Waterton Lakes National Park,
Canada, 827
water vapor, atmospheric, 855
wave base, **672**
wave-cut bench, *679*, *680*, 687
wave-cut notch, **679**, *680*
wave cyclone, 747
wave erosion, 679, *687*
wave front, 361–62, **361**
wave interference, 672
wavelength, 21, *22*, 672
wave refraction, **673**, *674*
waves, 21
erosion from, *577*, 674, 679, *680*, *684*, *859*
in hurricanes, 757
oceanic, 672–75, *673*, *674*
at beach, *674*, *676*, *685*
tidal, *see* tsunami
reflection and refraction of, 362, *362*
resonance of, 355
and rip current, 674, *675*
rogue, 672–73
seismic (earthquake), 49, *49*, 51
tidal, *see* tsunami
wind-driven, 344
weather, **729**, **847**
extremes of, *744*
storms, 750–61
vs. climate, 847
weathering, 135, *152*, *186*, **190**, 203
chemical, 189–91, *189*, *191*
and carbon dioxide absorption, 846, 853
and climate, 196
in desert, 770, 773, *774*
and mountain building, *853*
and physical, 189–91, *191*
and sea salt, 662
and surface area, *191*
in deserts, 773–77
as detritus formation, 204
differential, 194
grain size and, *207*
as mass movement setting, 600
mineral deposits formed by, *554*
and mineral stability, *190*
physical (mechanical), 186–89, **186**, *188*
and chemical, 189–91, *191*

and rock cycle, 265, 268
in setting stage for mass
movement, 600
and slope stability, 604–6
in soil formation, *192–93*
and soil production, *192–93*
weather layer of atmosphere, 737
wedging:
frost, 187
root, 188, *188*
salt, 188, *188*
Wegener, Alfred, *61*, 62–67, *63*, 71, 82, 87, 90, 106, 111, 203, 485, 830
"weight percent," *159*
welded tuff, 172, *175*, 281
wells, 708–10, *710*
extraction, 718
injection, 716
Werner, Abraham, 144
Western Interior Seaway (North
America), 491
West Virginia, 717
wet-based glaciers, 801, *802*, 804
wetlands:
floodplains transformed into, 647
oil spill threat to, 540
pollution of, 690–91
wetness, and soil characteristics, 197
wet-snow avalanches, 594
Whitby, England, gravestones in, *188*
whitewater, 629, *631*
wildcatters, 513
wildfires, global warming and, 870
Wilson, Edmund, 203, *204*
Wilson, J. Tuzo, 98–99, 101–2, 413
Wilson cycle, 413
wind, in surface currents, 664–65
wind erosion, 775–77, *776*
wind gap, 636
wind power, 532–34, *533*, *536–37*, 542
Wind River Mountains, 152, 256, *493*, *621*
winds, 507, **729**, 738–43
and balloon travel, 729–30
deposition from, 779
in hurricanes, 757
katabatic, 815
prevailing (surface), 741–42, **741**
in troposphere, 742–43
Winnipeg, air temperature in, 663
Winston, Harry, 136
Winter Park, Fla., sinkhole collapse
in, 695, *696*
wireline saw, 557
Wisconsinan glaciation, 828, *830*
wood, as energy source, 507
World Meteorological Organization, 868
world topography (map), *380–81*

world-wide seismic network, 326
Wrangelia, *400*
Würm glaciation, 828
Wyoming, *256*, *493*, *621*
anticline and syncline in, *396*
Devil's Tower in, 302, *302*
erratics in, *816*
Gros Ventre slide in, 604, *605*
moraine in, *817*
sandstone and shale ridges in, *575*
and Sevier orogeny, 490
Wind River Range in, *152*
See also Yellowstone National Park

xenolith, **166**, 438
X-ray diffractometers, 149, *150*
X-rays, in mineral study, 118, 122

Yangtze River:
flooding of, 641
and Three Gorges Dam, 532, *533*
yardangs, 777, *777*
yazoo streams, 634
"year without a summer," 305, 855
Yellow River, 641
Yellow River Delta, China, *634*
Yellowstone Canyon, *297*
Yellowstone National Park, 101, 102, 178, 213, *214*, 295–96, 297, *297*, 497, *711*, *712*, *818*
supervolcano caldera at, 292
volcanic activity at, 290, *290*
volcanic eruptions at, *297*
Yosemite National Park, Calif., 827
Half Dome in, 808, *809*
rock fall in, 595
Younger Dryas, 855
yttrium, 564
Yucatán peninsula, extraterrestrial-
object impact on, 494, *494*, 495, *857–58*
Yucca Mountain, Nev., 531, 717
Yungay, Peru, landslide destroys, 587–88, *587*, *594*

Zagreb, Croatia, 51
Zagros Mountains, 495
zeolite, 248, *248*
zinc, 549, 552
Zion Canyon, *452*, 453
Zion National Park, Utah, 220, 489, *489*, *788*
zircons, 470
and calculation of Earth's age, 470
oxygen isotopes in, 470
zone of ablation, **805**
zone of accumulation (glacier), **805**
zone of accumulation (soil formation), *195*, **195**
zone of leaching, *195*, **195**
zone of saturation, *701*